Advanced Organic Chemistry
FIFTH EDITION

Part B: Reactions and Synthesis

Advanced Organic Chemistry

PART A: Structure and Mechanisms
PART B: Reactions and Synthesis

Advanced Organic Chemistry

Chemistry

FIFTH EDITION

Part B: Reactions and Synthesis

FRANCIS A. CAREY
and RICHARD J. SUNDBERG

University of Virginia
Charlottesville, Virginia

 Springer

Francis A. Carey
Department of Chemistry
University of Virginia
Charlottesville, VA 22904

Richard J. Sundberg
Department of Chemistry
University of Virginia
Charlottesville, VA 22904

Library of Congress Control Number: 2006939782

ISBN-13: 978-0-387-68350-8 (hard cover)   e-ISBN-13: 978-0-387-44899-2
ISBN-13: 978-0-387-68354-6 (soft cover)

Printed on acid-free paper.

9 8 7

springer.com

Preface

The methods of organic synthesis have continued to advance rapidly and we have made an effort to reflect those advances in this Fifth Edition. Among the broad areas that have seen major developments are enantioselective reactions and transition metal catalysis. Computational chemistry is having an expanding impact on synthetic chemistry by evaluating the energy profiles of mechanisms and providing structural representation of unobservable intermediates and transition states.

The organization of Part B is similar to that in the earlier editions, but a few changes have been made. The section on introduction and removal of protecting groups has been moved forward to Chapter 3 to facilitate consideration of protecting groups throughout the remainder of the text. Enolate conjugate addition has been moved from Chapter 1 to Chapter 2, where it follows the discussion of the generalized aldol reaction. Several new sections have been added, including one on hydroalumination, carboalumination, and hydrozirconation in Chapter 4, another on the olefin metathesis reactions in Chapter 8, and an expanded discussion of the carbonyl-ene reaction in Chapter 10.

Chapters 1 and 2 focus on enolates and other carbon nucleophiles in synthesis. Chapter 1 discusses enolate formation and alkylation. Chapter 2 broadens the discussion to other carbon nucleophiles in the context of the generalized aldol reaction, which includes the Wittig, Peterson, and Julia olefination reactions. The chapter considers the stereochemistry of the aldol reaction in some detail, including the use of chiral auxiliaries and enantioselective catalysts.

Chapters 3 to 5 focus on some fundamental functional group modification reactions. Chapter 3 discusses common functional group interconversions, including nucleophilic substitution, ester and amide formation, and protecting group manipulations. Chapter 4 deals with electrophilic additions to double bonds, including the use of hydroboration to introduce functional groups. Chapter 5 considers reductions by hydrogenation, hydride donors, hydrogen atom donors, and metals and metal ions.

Chapter 6 looks at concerted pericyclic reactions, including the Diels-Alder reaction, 1,3-dipolar cycloaddition, [3,3]- and [2,3]-sigmatropic rearrangements, and thermal elimination reactions. The carbon-carbon bond-forming reactions are emphasized and the stereoselectivity of the reactions is discussed in detail.

v

Chapters 7 to 9 deal with organometallic reagents and catalysts. Chapter 7 considers Grignard and organolithium reagents. The discussion of organozinc reagents emphasizes their potential for enantioselective addition to aldehydes. Chapter 8 discusses reactions involving transition metals, with emphasis on copper- and palladium-mediated reactions. Chapter 9 considers the use of boranes, silanes, and stannanes in carbon-carbon bond formation. These three chapters focus on reactions such as nucleophilic addition to carbonyl groups, the Heck reaction, palladium-catalyzed cross-coupling, olefin metathesis, and allyl- boration, silation, and stanny-lation. These organometallic reactions currently are among the more important for construction of complex carbon structures.

Chapter 10 considers the role of reactive intermediates—carbocations, carbenes, and radicals—in synthesis. The carbocation reactions covered include the carbonyl-ene reaction, polyolefin cyclization, and carbocation rearrangements. In the carbene section, addition (cyclopropanation) and insertion reactions are emphasized. Catalysts that provide both selectivity and enantioselectivity are discussed. The section on radicals considers both intermolecular and intramolecular (cyclization) addition reactions of radicals are dealt with. The use of atom transfer steps and tandem sequences in synthesis is also illustrated.

Chapter 11 focuses on aromatic substitution, including electrophilic aromatic substitution, reactions of diazonium ions, and palladium-catalyzed nucleophilic aromatic substitution. Chapter 12 discusses oxidation reactions and is organized on the basis of functional group transformations. Oxidants are subdivided as transition metals, oxygen and peroxides, and other oxidants.

Chapter 13 illustrates applications of synthetic methodology by multistep synthesis and perhaps provides some sense of the evolution of synthetic capabilities. Several syntheses of two relatively simple molecules, juvabione and longifolene, illustrate some classic methods for ring formation and functional group transformations and, in the case of longifolene, also illustrate the potential for identification of relatively simple starting materials by retrosynthetic analysis. The syntheses of Prelog-Djerassi lactone highlight the methods for control of multiple stereocenters, and those of the Taxol precursor Baccatin III show how synthesis of that densely functionalized tricyclic structure has been accomplished. The synthesis of epothilone A illustrates both control of acyclic stereochemistry and macrocyclization methods, including olefin metathesis. The syntheses of (+)-discodermolide have been added, illustrating several methods for acyclic stereoselectivity and demonstrating the virtues of convergency. The chapter ends with a discussion of solid phase synthesis and its application to syntheses of polypeptides and oligonucleotides, as well as in combinatorial synthesis.

There is increased emphasis throughout Part B on the representation of transition structures to clarify stereoselectivity, including representation by computational models. The current practice of organic synthesis requires a thorough knowledge of molecular architecture and an understanding of how the components of a structure can be assembled. Structures of enantioselective reagents and catalysts are provided to help students appreciate the three-dimensional aspects of the interactions that occur in reactions.

A new feature of this edition is a brief section of commentary on the reactions in most of the schemes, which may point out a specific methodology or application. Instructors who want to emphasize the broad aspects of reactions, as opposed to specific examples, may wish to advise students to concentrate on the main flow of the text, reserving the schemes and commentary for future reference. As mentioned in the

Acknowledgment and Personal Statement, the selection of material in the examples and schemes does not reflect priority, importance, or generality. It was beyond our capacity to systematically survey the many examples that exist for most reaction types, and the examples included are those that came to our attention through literature searches and reviews.

Several computational studies have been abstracted and manipulable three-dimensional images of reactants, transition structures, intermediates, and products provided. This material provides the opportunity for detailed consideration of these representations and illustrates how computational chemistry can be applied to the mechanistic and structural interpretation of reactivity. This material is available in the Digital Resource at springer.com/carey-sundberg.

As in previous editions, the problems are drawn from the literature and references are given. In this addition, brief answers to each problem have been provided and are available at the publishers website.

Acknowledgment
and Personal Statement

The revision and updating of *Advanced Organic Chemistry* that appears as the Fifth Edition spanned the period September 2002 through December 2006. Each chapter was reworked and updated and some reorganization was done, as described in the Prefaces to Parts A and B. This period began at the point of conversion of library resources to electronic form. Our university library terminated paper subscriptions to the journals of the American Chemical Society and other journals that are available electronically as of the end of 2002. Shortly thereafter, an excavation mishap at an adjacent construction project led to structural damage and closure of our departmental library. It remained closed through June 2007, but thanks to the efforts of Carol Hunter, Beth Blanton-Kent, Christine Wiedman, Robert Burnett, and Wynne Stuart, I was able to maintain access to a few key print journals including the *Journal of the American Chemical Society*, *Journal of Organic Chemistry*, *Organic Letters*, *Tetrahedron*, and *Tetrahedron Letters*. These circumstances largely completed an evolution in the source for specific examples and data. In the earlier editions, these were primarily the result of direct print encounter or search of printed *Chemical Abstracts* indices. The current edition relies mainly on electronic keyword and structure searches. Neither the former nor the latter method is entirely systematic or comprehensive, so there is a considerable element of circumstance in the inclusion of specific material. There is no intent that specific examples reflect either priority of discovery or relative importance. Rather, they are interesting examples that illustrate the point in question.

Several reviewers provided many helpful corrections and suggestions, collated by Kenneth Howell and the editorial staff of Springer. Several colleagues provided valuable contributions. Carl Trindle offered suggestions and material from his course on computational chemistry. Jim Marshall reviewed and provided helpful comments on several sections. Michal Sabat, director of the Molecular Structure Laboratory, provided a number of the graphic images. My co-author, Francis A. Carey, retired in 2000 to devote his full attention to his text, *Organic Chemistry*, but continued to provide valuable comments and insights during the preparation of this edition. Various users of prior editions have provided error lists, and, hopefully, these corrections have been made. Shirley Fuller and Cindy Knight provided assistance with many aspects of the preparation of the manuscript.

This Fifth Edition is supplemented by the *Digital Resource* that is available at springer.com/carey-sundberg. The *Digital Resource* summarizes the results of several computational studies and presents three-dimensional images, comments, and exercises based on the results. These were developed with financial support from the Teaching Technology Initiative of the University of Virginia. Technical support was provided by Michal Sabat, William Rourk, Jeffrey Hollier, and David Newman. Several students made major contributions to this effort. Sara Higgins Fitzgerald and Victoria Landry created the prototypes of many of the sites. Scott Geyer developed the dynamic representations using IRC computations. Tanmaya Patel created several sites and developed the measurement tool. I also gratefully acknowledge the cooperation of the original authors of these studies in making their output available. *Problem Responses* have been provided and I want to acknowledge the assistance of R. Bruce Martin, David Metcalf, and Daniel McCauley in helping work out some of the specific kinetic problems and in providing the attendant graphs.

It is my hope that the text, problems, and other material will assist new students to develop a knowledge and appreciation of structure, mechanism, reactions, and synthesis in organic chemistry. It is gratifying to know that some 200,000 students have used earlier editions, hopefully to their benefit.

Richard J. Sundberg
Charlottesville, Virginia
June 2007

Introduction

The focus of Part B is on the closely interrelated topics of *reactions* and *synthesis*. In each of the first twelve chapters, we consider a group of related reactions that have been chosen for discussion primarily on the basis of their usefulness in synthesis. For each reaction we present an outline of the mechanism, its regio- and stereochemical characteristics, and information on typical reaction conditions. For the more commonly used reactions, the schemes contain several examples, which may include examples of the reaction in relatively simple molecules and in more complex structures. The goal of these chapters is to develop a fundamental base of knowledge about organic reactions in the context of synthesis. We want to be able to answer questions such as: What transformation does a reaction achieve? What is the mechanism of the reaction? What reagents and reaction conditions are typically used? What substances can catalyze the reaction? How sensitive is the reaction to other functional groups and the steric environment? What factors control the stereoselectivity of the reaction? Under what conditions is the reaction enantioselective?

Synthesis is the application of one or more reactions to the preparation of a particular target compound, and can pertain to a single-step transformation or to a number of sequential steps. The selection of a reaction or series of reactions for a synthesis involves making a judgment about the most effective possibility among the available options. There may be a number of possibilities for the synthesis of a particular compound. For example, in the course of learning about the reactions in Chapter 1 to 12, we will encounter a number of ways of making ketones, as outlined in the scheme that follows.

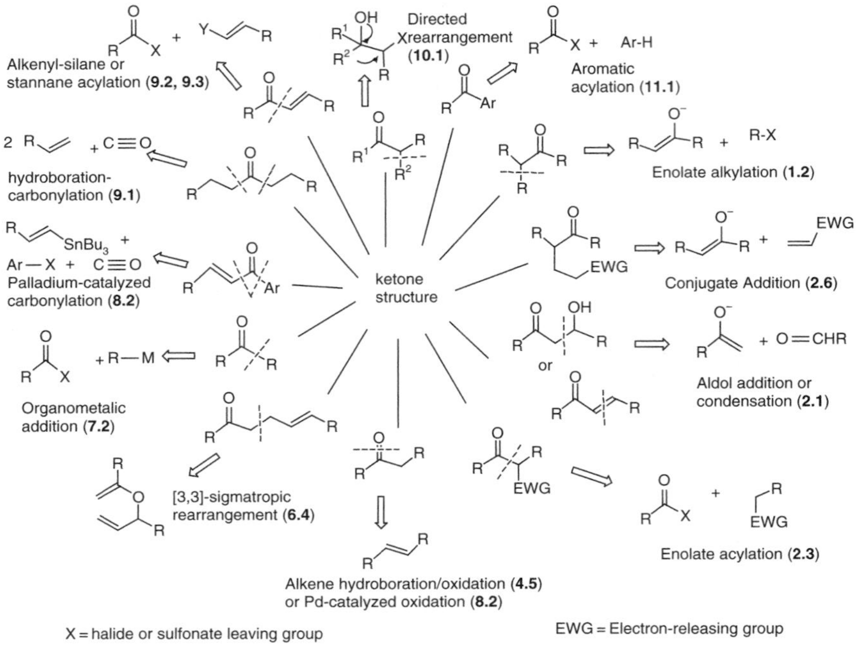

X = halide or sulfonate leaving group

EWG = Electron-releasing group

The focus of Chapters 1 and 2 is enolates and related carbon nucleophiles such as silyl enol ethers, enamines, and imine anions, which can be referred to as *enolate equivalents*.

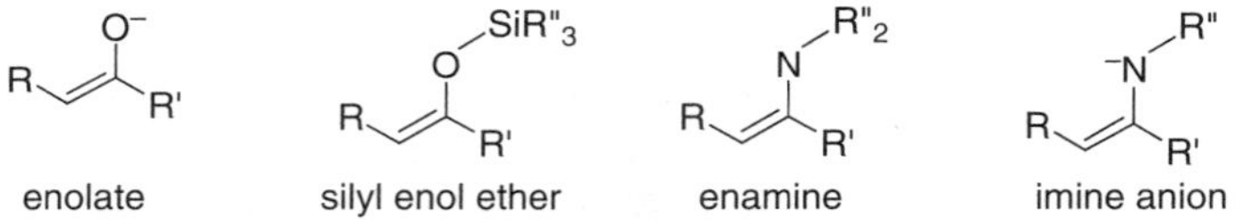

Chapter 1 deals with alkylation of carbon nucleophiles by alkyl halides and tosylates. We discuss the major factors affecting stereoselectivity in both cyclic and acyclic compounds and consider intramolecular alkylation and the use of chiral auxiliaries.

Aldol addition and related reactions of enolates and enolate equivalents are the subject of the first part of Chapter 2. These reactions provide powerful methods for controlling the stereochemistry in reactions that form hydroxyl- and methyl-substituted structures, such as those found in many antibiotics. We will see how the choice of the nucleophile, the other reagents (such as Lewis acids), and adjustment of reaction conditions can be used to control stereochemistry. We discuss the role of open, cyclic, and chelated transition structures in determining stereochemistry, and will also see how chiral auxiliaries and chiral catalysts can control the enantioselectivity of these reactions. Intramolecular aldol reactions, including the Robinson annulation are discussed. Other reactions included in Chapter 2 include Mannich, carbon acylation, and olefination reactions. The reactivity of other carbon nucleophiles including phosphonium ylides, phosphonate carbanions, sulfone anions, sulfonium ylides, and sulfoxonium ylides are also considered.

$$R'_3{}^+P\!-\!C^-HR \qquad (R'O)_2\overset{\displaystyle O}{\overset{\|}{P}}C^-HR \qquad RC^-\overset{\displaystyle O}{\underset{\displaystyle \|}{\overset{\|}{S}}}R' \qquad R'_2{}^+S\!-\!C^-HR \qquad R'_2{}^+\overset{\displaystyle O}{\overset{\|}{S}}\!-\!C^-HR$$

| phosphonium ylide | phosphonate carbanion | sulfone anion | sulfonium ylide | sulfoxonium ylide |

Among the olefination reactions, those of phosphonium ylides, phosphonate anions, silylmethyl anions, and sulfone anions are discussed. This chapter also includes a section on conjugate addition of carbon nucleophiles to α, β-unsaturated carbonyl compounds. The reactions in this chapter are among the most important and general of the carbon-carbon bond-forming reactions.

Chapters 3 to 5 deal mainly with introduction and interconversion of functional groups. In Chapter 3, the conversion of alcohols to halides and sulfonates and their subsequent reactions with nucleophiles are considered. Such reactions can be used to introduce functional groups, invert configuration, or cleave ethers. The main methods of interconversion of carboxylic acid derivatives, including acyl halides, anhydrides, esters, and amides, are reviewed. Chapter 4 discusses electrophilic additions to alkenes, including reactions with protic acids, oxymercuration, halogenation, sulfenylation, and selenylation. In addition to introducing functional groups, these reagents can be used to effect cyclization reactions, such as iodolactonization. The chapter also includes the fundamental hydroboration reactions and their use in the synthesis of alcohols, aldehydes, ketones, carboxylic acids, amines, and halides. Chapter 5 discusses reduction reactions at carbon-carbon multiple bonds, carbonyl groups, and certain other functional groups. The introduction of hydrogen by hydrogenation frequently establishes important stereochemical relationships. Both heterogeneous and homogeneous catalysts are discussed, including examples of enantioselective catalysts. The reduction of carbonyl groups also often has important stereochemical consequences because a new stereocenter is generated. The fundamental hydride transfer reagents $NaBH_4$ and $LiAlH_4$ and their derivatives are considered. Examples of both enantioselective reagents and catalysts are discussed, as well as synthetic applications of several other kinds of reducing agents, including hydrogen atom donors and metals.

In Chapter 6 the focus returns to carbon-carbon bond formation through cycloadditions and sigmatropic rearrangements. The Diels-Alder reaction and 1,3-dipolar cycloaddition are the most important of the former group. The predictable regiochemistry and stereochemistry of these reactions make them very valuable for ring formation. Intramolecular versions of these cycloadditions can create at least two new rings, often with excellent stereochemical control. Although not as broad in scope, $[2+2]$ cycloadditions, such as the reactions of ketenes and photocycloaddition reactions of enones, also have important synthetic applications. The $[3,3]$- and $[2,3]$-sigmatropic rearrangements also proceed through cyclic transition structures and usually provide predictable stereochemical control. Examples of $[3,3]$-sigmatropic rearrangements include the Cope rearrangement of 1,5-dienes, the Claisen rearrangement of allyl vinyl ethers, and the corresponding reactions of ester enolate equivalents.

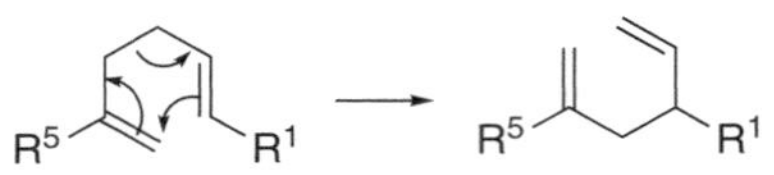

Cope rearrangement

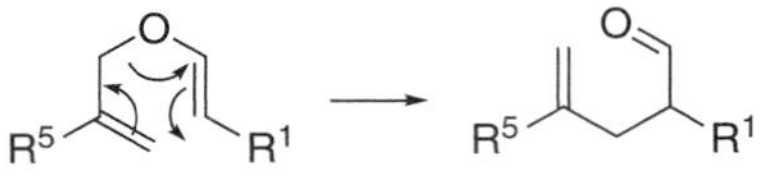

Claisen rearrangement

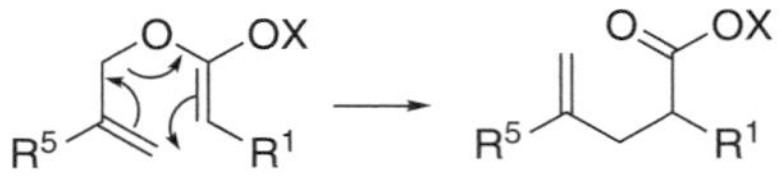

X = (–), R, SiR'$_3$

Claisen-type rearrangements of
ester enolates, ketene acetals,
and silyl ketene acetals

Synthetically valuable [2,3]-sigmatropic rearrangements include those of allyl sulfonium and ammonium ylides and α'-carbanions of allyl vinyl ethers.

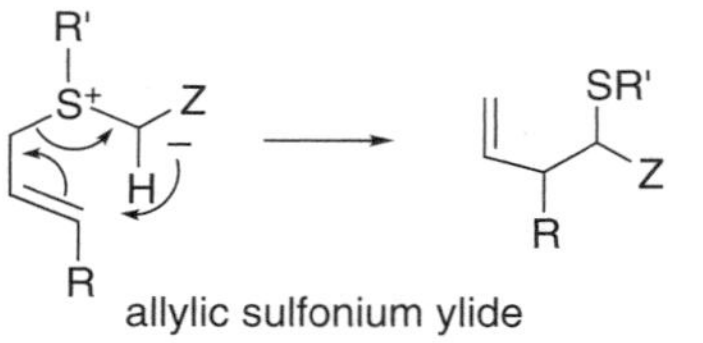

allylic sulfonium ylide

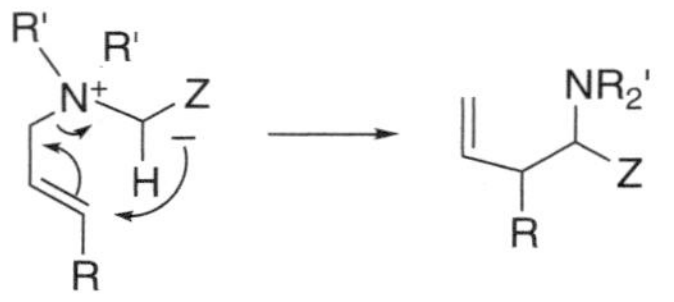

allylic ammonium ylide

allylic ether anion

This chapter also discusses several β-elimination reactions that proceed through cyclic transition structures.

In Chapters 7, 8, and 9, the focus is on organometallic reagents. Chapter 7 considers the Group I and II metals, emphasizing organolithium, -magnesium, and -zinc reagents, which can deliver saturated, unsaturated, and aromatic groups as nucleophiles. Carbonyl compounds are the most common co-reactants, but imines and nitriles are also reactive. Important features of the zinc reagents are their adaptability to enantioselective catalysis and their compatibility with many functional groups. Chapter 8 discusses the role of transition metals in organic synthesis, with the emphasis on copper and palladium. The former provides powerful nucleophiles that can react by displacement, epoxide ring opening, and conjugate addition, while organopalladium compounds are usually involved in catalytic processes. Among the important applications are allylic substitution, coupling of aryl and vinyl halides with alkenes (Heck reaction), and cross coupling with various organometallic reagents including magnesium, zinc, tin, and boron derivatives. Palladium catalysts can also effect addition of organic groups to carbon monoxide (carbonylation) to give ketones, esters, or amides. Olefin metathesis reactions, also discussed in this chapter, involve ruthenium or molybdenum catalysts and both intermolecular and ring-closing metathesis have recently found applications in synthesis.

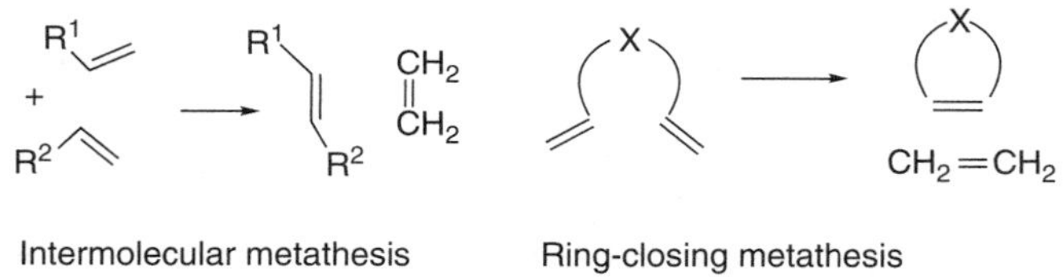

Intermolecular metathesis   Ring-closing metathesis

Chapter 9 discusses carbon-carbon bond-forming reactions of boranes, silanes, and stannanes. The borane reactions usually involve B → C migrations and can be used to synthesize alcohols, aldehydes, ketones, carboxylic acids, and amines. There are also stereoselective alkene syntheses based on organoborane intermediates. Allylic boranes and boronates provide stereospecific and enantioselective addition reactions of allylic groups to aldehydes. These reactions proceed through cyclic transition structures and provide a valuable complement to the aldol reaction for stereochemical control of acyclic systems. The most important reactions of silanes and stannanes involve vinyl and allyl derivatives. These reagents are subject to electrophilic attack, which is usually followed by demetallation, resulting in net substitution by the electrophile, with double-bond transposition in the allylic case. Both these reactions are under the regiochemical control of the β-carbocation–stabilizing ability of the silyl and stannyl groups.

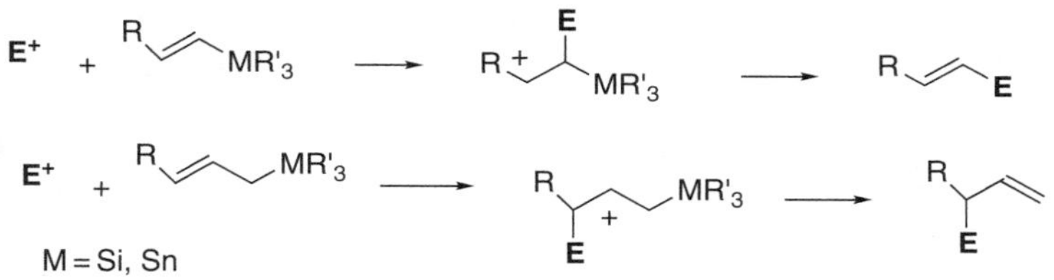

$M = Si, Sn$

In Chapter 10, the emphasis is on synthetic application of carbocations, carbenes, and radicals in synthesis. These intermediates generally have high reactivity and short lifetimes, and successful application in synthesis requires taking this factor into account. Examples of reactions involving carbocations are the carbonyl-ene reaction, polyene cyclization, and directed rearrangements and fragmentations. The unique divalent character of the carbenes and related intermediates called carbenoids can be exploited in synthesis. Both addition (cyclopropanation) and insertion are characteristic reactions. Several zinc-based reagents are excellent for cyclopropanation, and rhodium catalysts have been developed that offer a degree of selectivity between addition and insertion reactions.

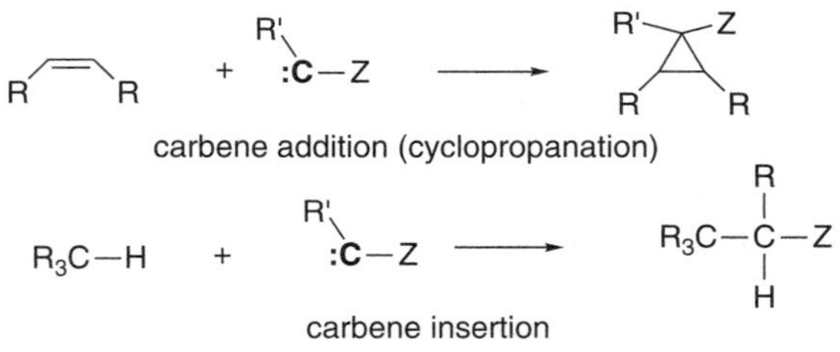

carbene addition (cyclopropanation)

carbene insertion

Radical reactions used in synthesis include additions to double bonds, ring closure, and atom transfer reactions. Several sequences of tandem reactions have been developed that can close a series of rings, followed by introduction of a substituent. Allylic stannanes are prominent in reactions of this type.

Chapter 11 reviews aromatic substitution reactions including electrophilic aromatic substitution, substitution via diazonium ions, and metal-catalyzed nucleophilic substitution. The scope of the latter reactions has been greatly expanded in recent years by the development of various copper and palladium catalysts. Chapter 12 discusses oxidation reactions. For the most part, these reactions are used for functional group transformations. A wide variety of reagents are available and we classify them as based on metals, oxygen and peroxides, and other oxidants. Epoxidation reactions have special significance in synthesis. The introduction of the epoxide ring can set the stage for subsequent nucleophilic ring opening to introduce a new group or extend the carbon chain. The epoxidation of allylic alcohols can be done enantioselectively, so epoxidation followed by ring opening can control the configuration of three contiguous stereocenters.

The methods available for synthesis have advanced dramatically in the past half-century. Improvements have been made in selectivity of conditions, versatility of transformations, stereochemical control, and the efficiency of synthetic processes. The range of available reagents has expanded. Many reactions involve compounds of boron, silicon, sulfur, selenium, phosphorus, and tin. Catalysis, particularly by transition metal complexes, has also become a key part of organic synthesis. The mechanisms of catalytic reactions are characterized by _catalytic cycles_ and require an understanding not only of the ultimate bond-forming and bond-breaking steps, but also of the mechanism for regeneration of the active catalytic species and the effect of products, by-products, and other reaction components in the catalytic cycle.

Over the past decade enantioselectivity has become a key concern in reactivity and synthesis. Use of _chiral auxiliaries_ and/or _enantioselective catalysts_ to control configuration is often a crucial part of synthesis. The analysis and interpretation of enantioselectivity depend on consideration of diastereomeric intermediates and transition structures on the reaction pathway. Often the differences in free energy of competing reaction pathways are on the order of 1 kcal, reflecting small and subtle differences in structure. We provide a number of examples of the structural basis for enantioselectivity, but a good deal of unpredictability remains concerning the degree of enantioselectivity. Small changes in solvent, additives, catalyst structure, etc., can make large differences in the observed enantioselectivity.

Mechanistic insight is a key to both discovery of new reactions and to their successful utilization in specific applications. Use of reactions in a synthetic context often entails optimization of reaction conditions based on mechanistic interpretations. Part A of this text provides fundamental information about the reactions discussed here. Although these mechanistic concepts may be recapitulated briefly in Part B, the details may not be included; where appropriate, reference is made to relevant sections in Part A. In addition to experimental mechanistic studies, many reactions of synthetic interest are now within the range of computational analysis. Intermediates and transition structures on competing or alternative reaction pathways can be modeled and compared on the basis of MO and/or DFT calculations. Such computations can provide intricate structural details and may lead to mechanistic insight. A number of such studies are discussed in the course of the text.

A key skill in the practice of organic synthesis is the ability to recognize important aspects of molecular structure. Recognition of all aspects of stereochemistry, including conformation, ring geometry, and configuration are crucial to understanding reactivity and applying reactions to synthesis. We consider the stereochemical aspects of each reaction. For most reactions, good information is available on the structure of key intermediates and the transition structure. Students should make a particular effort to understand the consequences of intermediates and transition structures for reactivity.

Applying the range of reactions to synthesis involves planning and foreseeing the outcome of a particular sequence of reactions. Planning is best done on the basis of *retrosynthetic analysis*, the identification of key subunits of the target molecule that can be assembled by feasible reactions. The structure of the molecule is studied to identify bonds that are amenable to formation. For example, a molecule containing a carbon-carbon double bond might be disconnected at that bond, since there are numerous ways to form a double bond from two separate components. β-Hydroxy carbonyl units suggest the application of the aldol addition reaction, which assembles this functionality from two separate carbonyl compounds.

$$R^1CH{=}O \quad + \quad R^2CH_2\overset{\overset{\displaystyle O}{\|}}{C}R^3 \quad \xrightarrow[\text{acid}]{\text{base or}} \quad R^1\overset{}{\underset{\displaystyle OH}{\diagup}}\overset{R^2}{\underset{\displaystyle O}{\diagdown}}R^3$$

electrophilic   nucleophilic
reactant   reactant

The construction of the overall molecular skeleton, that is, the carbon-carbon and other bonds that constitute the framework of the molecule, is the primary challenge. Molecules also typically contain a number of functional groups and they must be compatible with the projected reactivity at each step in the synthesis. This means that it may be necessary to modify or protect functional groups at certain points. Generally speaking, the protection and interconversion of functional groups is a less fundamental challenge than construction of the molecular framework because there are numerous methods for functional group interconversion.

As the reactions discussed in Chapters 1 to 12 illustrate, the methodology of organic synthesis is highly developed. There are many possible means for introduction and interconversion of functional groups and for carbon-carbon bond formation, but putting them together in a multistep synthesis requires more than knowledge of the reactions. A plan that orchestrates the sequence of reactions toward the final goal is necessary.

In Chapter 13, we discuss some of the generalizations of multistep synthesis. Retrosynthetic analysis identifies bonds that can be broken and key intermediates. Various methods of stereochemical control, including intramolecular interactions. Chiral auxiliaries, and enantioselective catalysts, can be used. Protective groups can be utilized to prevent functional group interferences. Ingenuity in synthetic planning can lead to efficient construction of molecules. We take a retrospective look at the synthesis of six molecules of differing complexity. Juvabione is an oxidized terpene with one ring and two stereocenters. Successful syntheses date from the late 1960s to the present. Longifolene is a tricyclic sesquiterpene and its synthesis poses the problem of ring construction. The Prelog-Djerassi lactone, the lactone of (2*R*,3*S*,4*R*,6*R*)-3-hydroxy-2,4,6-trimethylheptanedioic acid, is a degradation product isolated from various antibiotics. Its alternating methyl and hydroxy groups are typical of structural features found in many antibiotics and other natural substances biosynthetically derived from polypropionate units. Its synthesis illustrates methods of acyclic stereochemical control.

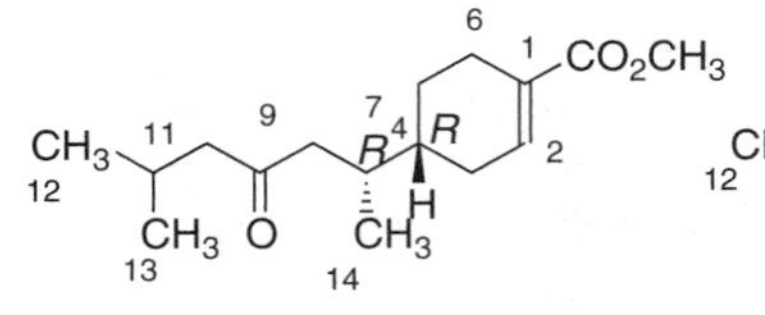

threo-Juvabione

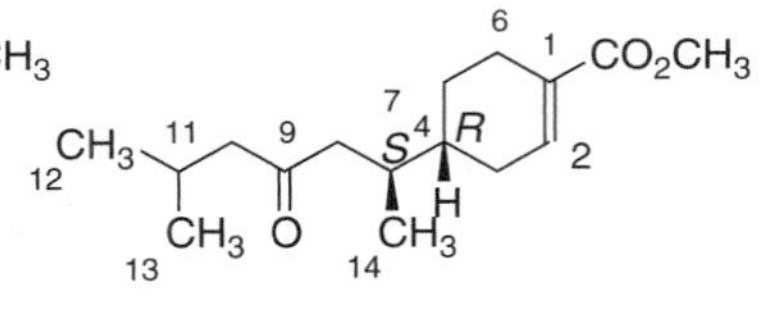

erythro-Juvabione

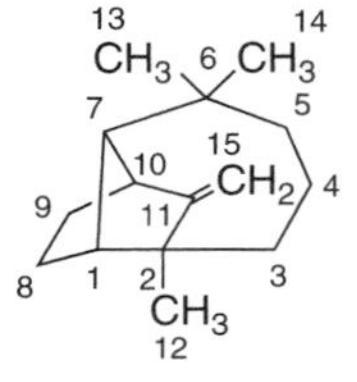

Longifolene

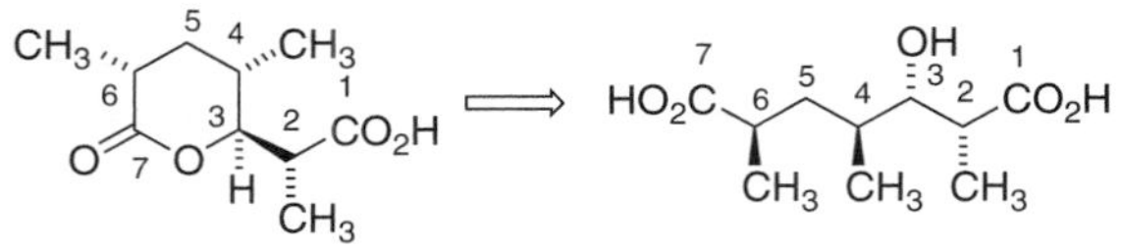

Prelog-Djerassi Lactone

Synthetic methodology is applied to molecules with important biological activity such as the prostaglandins and steroids. Generally speaking, the stereochemistry of these molecules can be controlled by relationships to the ring structure.

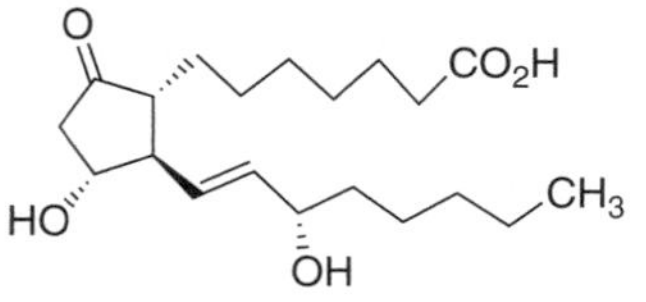

prostaglandin E₁

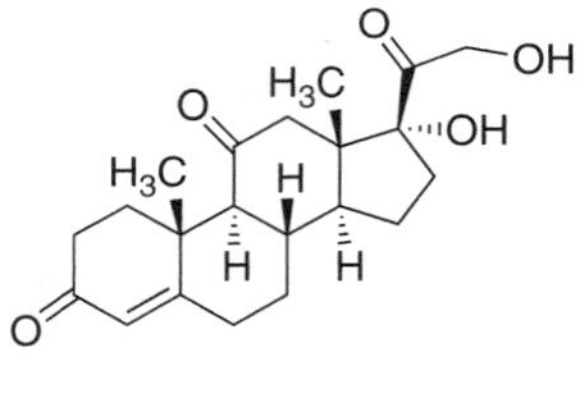

cortisone

A somewhat more complex molecule, both in terms of the nature of the rings and the density of functionality is Baccatin III, a precursor of the antitumor agent Taxol®. We summarize syntheses of Baccatin III that involve sequences of 40–50 reactions. Baccatin III is a highly oxygenated diterpene and these syntheses provide examples of ring construction and functional group manipulations. Despite its complexity, the syntheses of Baccatin III, for the most part, also depend on achieving formation of rings and use of the ring structure to control stereochemistry.

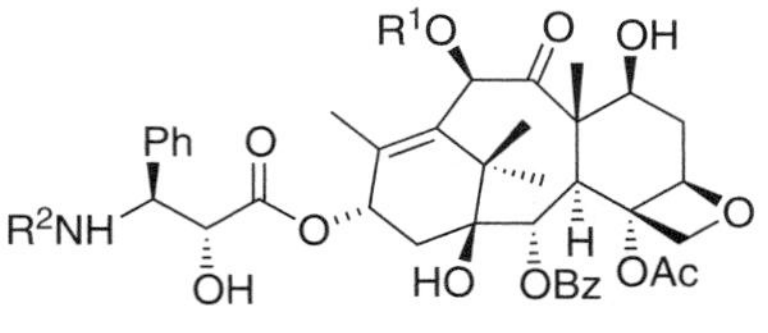

taxol R$_1$ = Ac, R^2 = PhCO

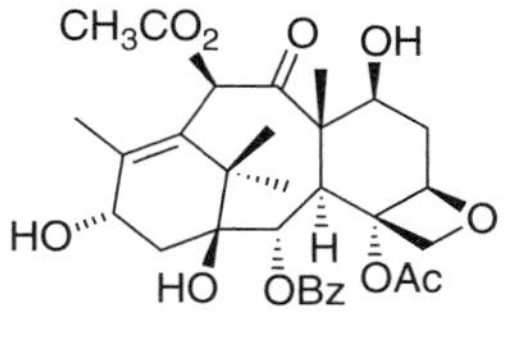

baccatin III

Macrocyclic antibiotics such as the erythronolide present an additional challenge.

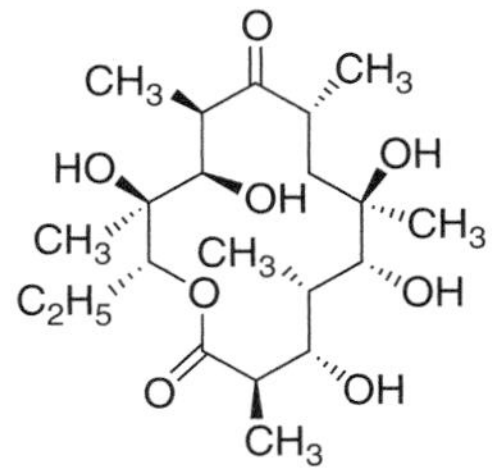

erythronolide

These molecules contain many stereogenic centers and they are generally constructed from acyclic segments, so the ability to control configuration in acyclic systems is necessary. Solutions to this problem developed beginning in the 1960s are based on analysis of transition structures and the concepts of cyclic transition structure and facial selectivity. The effect of nearby stereogenic centers has been studied carefully and resulted in concepts such as the Felkin model for carbonyl addition reactions and Cram's model of chelation control. In Chapter 13, several syntheses of epothilone A, a 16-membered lactone that has antitumor activity, are summarized. The syntheses illustrate methods for both acyclic stereochemical control and macrocyclization, including the application of the olefin metathesis reaction.

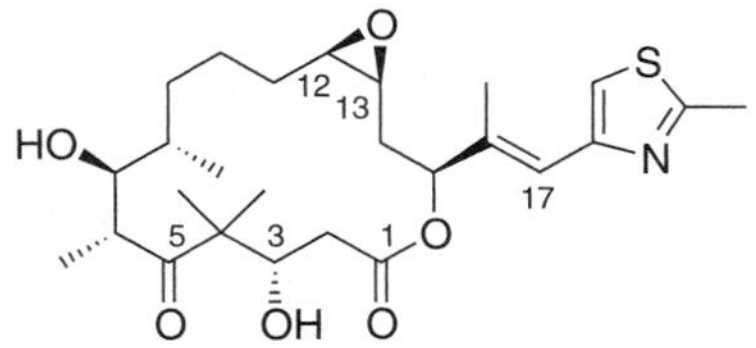

Epothilone A

We also discuss the synthesis of (+)-discodermolide, a potent antitumor agent isolated from a deep-water sponge in the Caribbean Sea. The first synthesis was reported in the mid-1990s, and synthetic activity is ongoing. Discodermolide is a good example of the capability of current synthetic methodology to produce complex molecules. The molecule contains a 24-carbon chain with a single lactone ring connecting C(1) and C(5). There are eight methyl substituents and six oxygen substituents, one of which is carbamoylated. The chain ends with a diene unit. By combining and refining elements of several earlier syntheses, it was possible to carry out a 39-step synthesis. The early stages were done on a kilogram scale and the entire effort provided 60 grams of the final product for preliminary clinical evaluation.

(+)–Discodermolide

There is no synthetic path that is uniquely "correct," but there may be factors that recommend particular pathways. The design of a synthesis involves applying one's knowledge about reactions. Is the reaction applicable to the particular steric and electronic environment under consideration? Is the reaction compatible with other functional groups and structures that are present elsewhere in the molecule? Will the reaction meet the regio- and stereochemical requirements that apply? Chemists rely on mechanistic considerations and the precedent of related reactions to make these judgments. Other considerations may come into play as well, such as availability and/or cost of starting materials, and safety and environmental issues might make one reaction preferable to another. These are critical concerns in synthesis on a production scale.

Certain types of molecules, especially polypeptides and polynucleotides, lend themselves to synthesis on solid supports. In such syntheses, the starting material is attached to a small particle (bead) or a surface and the molecule remains attached during the course of the synthetic sequence. Solid phase synthesis also plays a key role in creation of combinatorial libraries, that is, collections of many molecules synthesized by a sequence of reactions in which the subunits are systematically varied to create a range of structures (*molecular diversity*).

There is a vast amount of knowledge about reactions and how to use them in synthesis. The primary source for this information is the published chemical literature that is available in numerous journals, and additional information can be found in patents, theses and dissertations, and technical reports of industrial and governmental organizations. There are several means of gaining access to information about specific reactions. The series *Organic Syntheses* provides examples of specific transformations with detailed experimental procedures. Another series, *Organic Reactions*, provides fundamental information about the scope and mechanism as well as comprehensive literature references to many examples of a specific reaction type. Various review journals, including *Accounts of Chemical Research* and *Chemical Reviews*, provide overviews of particular reactions. A traditional system of organization is based on *named reactions*. Many important reactions bear well-recognized names of the chemists involved in their discovery or development. Other names such as dehydration, epoxidation, enolate alkylation, etc., are succinct descriptions of the structural changes associated with the reaction. This vocabulary is an important tool for accessing information about organic reactions. There are large computerized databases of organic reactions, most notably those of *Chemical Abstracts* and *Beilstein*. Chemical structures can be uniquely described and these databases can be searched for complete or partial structures. Systematic ways of searching for reactions are also incorporated into the databases. Another database, *Science Citation Index*, allows search for subsequent citations of published work.

A major purpose of organic synthesis at the current time is the discovery, understanding, and application of biological activity. Pharmaceutical laboratories, research foundations, and government and academic institutions throughout the world are engaged in this research. Many new compounds are synthesized to discover useful biological activity, and when activity is discovered, related compounds are synthesized to improve it. Syntheses suitable for production of drug candidate molecules are developed. Other compounds are synthesized to explore the mechanisms of biological processes. The ultimate goal is to apply this knowledge about biological activity for treatment and prevention of disease. Another major application of synthesis is in agriculture for control of insects and weeds. Organic synthesis also plays a part in the development of many consumer products, such as fragrances.

The unique power of synthesis is the ability to create new molecules and materials with valuable properties. This capacity can be used to interact with the natural world, as in the treatment of disease or the production of food, but it can also produce compounds and materials beyond the capacity of living systems. Our present world uses vast amounts of synthetic polymers, mainly derived from petroleum by synthesis. The development of nanotechnology, which envisions the application of properties at the molecular level to catalysis, energy transfer, and information management has focused attention on multimolecular arrays and systems capable of self-assembly. We can expect that in the future synthesis will bring into existence new substances with unique properties that will have impacts as profound as those resulting from syntheses of therapeutics and polymeric materials.

Contents

Chapter 4.   Electrophilic Additions to Carbon-Carbon Multiple Bonds...   289

Alkylation of Enolates and Other Carbon Nucleophiles

Introduction

Carbon-carbon bond formation is the basis for the construction of the molecular framework of organic molecules by synthesis. One of the fundamental processes for carbon-carbon bond formation is a reaction between a nucleophilic and an electrophilic carbon. The focus in this chapter is on *enolates*, *imine anions*, and *enamines*, which are carbon nucleophiles, and their reactions with *alkylating agents*. Mechanistically, these are usually S_N2 reactions in which the carbon nucleophile displaces a halide or other leaving group with inversion of configuration at the alkylating group. Efficient carbon-carbon bond formation requires that the S_N2 alkylation be the dominant reaction. The crucial factors that must be considered include: (1) the conditions for generation of the carbon nucleophile; (2) the effect of the reaction conditions on the structure and reactivity of the nucleophile; and (3) the regio- and stereo-selectivity of the alkylation reaction. The reaction can be applied to various carbonyl compounds, including ketones, esters, and amides.

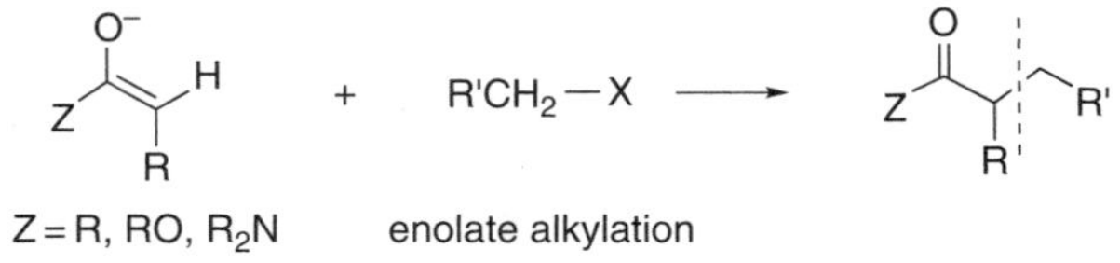

$Z = R, RO, R_2N$ enolate alkylation

These reactions introduce a new substituent α to the carbonyl group and constitute an important method for this transformation. In the retrosynthetic sense, the disconnection is between the α-carbon and a potential alkylating agent.

There are similar reactions involving nitrogen analogs called *imine anions*. The alkylated imines can be hydrolyzed to the corresponding ketone, and this reaction is discussed in Section 1.3.

Either enolate or imine anions can be used to introduce alkyl α-substituents to a carbonyl group. Because the reaction involves a nucleophilic substitution, primary groups are the best alkylating agents, with methyl, allyl, and benzyl compounds being particularly reactive. Secondary groups are less reactive and are likely to give lower yields because of competing elimination. Tertiary and aryl groups cannot be introduced by an S_N2 mechanism.

1.1. Generation and Properties of Enolates and Other Stabilized Carbanions

1.1.1. Generation of Enolates by Deprotonation

The fundamental aspects of the structure and stability of carbanions were discussed in Chapter 6 of Part A. In the present chapter we relate the properties and reactivity of carbanions stabilized by carbonyl and other EWG substituents to their application as nucleophiles in synthesis. As discussed in Section 6.3 of Part A, there is a fundamental relationship between the stabilizing functional group and the acidity of the C−H groups, as illustrated by the pK data summarized in Table 6.7 in Part A. These pK data provide a basis for assessing the stability and reactivity of carbanions. The acidity of the reactant determines which bases can be used for generation of the anion. Another crucial factor is the distinction between *kinetic or thermodynamic control of enolate formation by deprotonation* (Part A, Section 6.3), which determines the enolate composition. Fundamental mechanisms of S_N2 alkylation reactions of carbanions are discussed in Section 6.5 of Part A. A review of this material may prove helpful.

A primary consideration in the generation of an enolate or other stabilized carbanion by deprotonation is the choice of base. In general, reactions can be carried out under conditions in which the enolate is *in equilibrium* with its conjugate acid or under which the reactant is *completely converted* to its conjugate base. The key determinant is the amount and strength of the base. For complete conversion, the base must be derived from a substantially weaker acid than the reactant. Stated another way, the reagent must be a stronger base than the anion of the reactant. Most current procedures for alkylation of enolates and other carbanions involve complete conversion to the anion. Such procedures are generally more amenable to both regiochemical and stereochemical control than those in which there is only a small equilibrium concentration of the enolate. The solvent and other coordinating or chelating additives also have strong effects on the structure and reactivity of carbanions formed by deprotonation. The nature of the solvent determines the degree of ion pairing and aggregation, which in turn affect reactivity.

Table 1.1 gives approximate pK data for various functional groups and some of the commonly used bases. The strongest acids appear at the top of the table and the strongest bases at the bottom. The values listed as pK_{ROH} are referenced to water and are appropriate for hydroxylic solvents. Also included in the table are pK values determined in dimethyl sulfoxide (pK_{DMSO}). The range of acidities that can be measured directly in DMSO is greater than that in protic media, thereby allowing direct comparisons between weakly acidic compounds to be made more confidently. The pK values in DMSO are normally larger than in water because water stabilizes anions more effectively, by hydrogen bonding, than does DMSO. Stated another way, many anions are more strongly basic in DMSO than in water. This relationship is particularly apparent for the oxy anion bases, such as acetate, hydroxide, and the alkoxides, which are much more basic in DMSO than in protic solvents. At the present time, the pK_{DMSO} scale includes the widest variety of structural types of synthetic interest.[1] The pK values collected in Table 1.1 provide an ordering of some important

Table 1.1. Approximate pK Values from Some Compounds with Carbanion Stabilizing Groups and Some Common Bases[a]

Compound	pK_{ROH}	pK_{DMSO}	Base	pK_{ROH}	pK_{DMSO}
$O_2NCH_2NO_2$	3.6		$CH_3CO_2^-$	4.2	11.6
$CH_3COCH_2NO_2$	5.1				
$CH_3CH_2NO_2$	8.6	16.7	HCO_3^-	6.5	
$CH_3COCH_2COCH_3$	9				
$PhCOCH_2COCH_3$	9.6		PhO^-	9.9	16.4
CH_3NO_2	10.2	17.2			
$CH_3COCH_2CO_2C_2H_5$	10.7	14.2	CO_3^{2-}	10.2	
$NCCH_2CN$	11.2	11.0	$(C_2H_5)_3N$	10.7	
$PhCH_2NO_2$		12.3	$(CH_3CH_2)_2NH$	11	
$CH_2(SO_2CH_3)_2$	12.2	14.4			
$CH_2(CO_2C_2H_5)_2$	12.7	16.4			
Cyclopentadiene	15		CH_3O^-	15.5	29.0
$PhSCH_2COCH_3$		18.7	HO^-	15.7	31.4
$CH_3CH_2CH(CO_2C_2H_5)_2$	15		$C_2H_5O^-$	15.9	29.8
$PhSCH_2CN$		20.8	$(CH_3)_2CHO^-$		30.3
$(PhCH_2)_2SO_2$		23.9	$(CH_3)_3CO^-$	19	32.2
$PhCOCH_3$	15.8	24.7			
$PhCH_2COCH_3$	19.9				
CH_3COCH_3	20	26.5			
$CH_3CH_2COCH_2CH_3$		27.1			
Fluorene	20.5	22.6			
$PhSO_2CH_3$		29.0			
$PhCH_2SOCH_3$	29.0		$[(CH_3)_3Si]_2N^-$	30[b]	
CH_3CN	25	31.3			
Ph_2CH_2		32.2			
Ph_3CH	33	30.6	NH_2^-	35	41
			$CH_3SOCH_2^-$	35	35.1
			$(CH_3CH_2)_2N^-$	36	
$PhCH_3$		43			
CH_4		56			

a. From F. G. Bordwell, *Acc. Chem. Res.*, **21**, 456 (1988).
b. In THF; R. R. Fraser and T. S. Mansour, *J. Org. Chem.*, **49**, 3442 (1984).

[1.] F. G. Bordwell, *Acc. Chem. Res.*, **21**, 456 (1988).

substituents with respect to their ability to stabilize carbanions. The order indicated is $NO_2 > COR > CN \sim CO_2R > SO_2R > SOR > Ph \sim SR > H > R$. Familiarity with the relative acidity and approximate pK values is important for an understanding of the reactions discussed in this chapter.

There is something of an historical division in synthetic procedures involving carbanions as nucleophiles in alkylation reactions.[2] As can be seen from Table 1.1, β-diketones, β-ketoesters, malonates, and other compounds with two stabilizing groups have pK values slightly below ethanol and the other common alcohols. As a result, these compounds can be converted completely to enolates by sodium or potassium alkoxides. These compounds were the usual reactants in carbanion alkylation reactions until about 1960. Often, the second EWG is extraneous to the overall purpose of the synthesis and its removal requires an extra step. After 1960, procedures using aprotic solvents, especially THF, and amide bases, such as lithium di-isopropylamide (LDA) were developed. The dialkylamines have a pK around 35. These conditions permit the conversion of monofunctional compounds with p$K > 20$, especially ketones, esters, and amides, completely to their enolates. Other bases that are commonly used are the anions of hexaalkyldisilylamines, especially hexamethyldisilazane.[3] The lithium, sodium, and potassium salts are abbreviated LiHMDS, NaHMDS, and KHMDS. The disilylamines have a pK around 30.[4] The basicity of both dialkylamides and hexaalkyldisilylamides tends to increase with branching in the alkyl groups. The more branched amides also exhibit greater steric discrimination. An example is lithium tetramethylpiperidide, LiTMP, which is sometimes used as a base for deprotonation.[5] Other strong bases, such as amide anion ($^-NH_2$), the conjugate base of DMSO (sometimes referred to as the "dimsyl" anion),[6] and triphenylmethyl anion, are capable of effecting essentially complete conversion of a ketone to its enolate. Sodium hydride and potassium hydride can also be used to prepare enolates from ketones, although the reactivity of the metal hydrides is somewhat dependent on the means of preparation and purification of the hydride.[7]

By comparing the approximate pK values of the bases with those of the carbon acid of interest, it is possible to estimate the position of the acid-base equilibrium for a given reactant-base combination. For a carbon acid C$-$H and a base B$-$H,

$$K_{a_{(C-H)}} = \frac{[C^-][H^+]}{[C-H]} \text{ and } K_{a_{(B-H)}} = \frac{[B^-][H^+]}{[B-H]}$$

at equilibrium

$$\frac{K_{a_{(C-H)}}[C-H]}{[C^-]} = \frac{K_{a_{(B-H)}}[B-H]}{[B^-]}$$

for the reaction

$$C-H + B^- \rightleftharpoons B-H + C^-$$

2. D. Seebach, *Angew. Chem. Int. Ed. Engl.*, **27**, 1624 (1988).
3. E. H. Amonoco-Neizer, R. A. Shaw, D. O. Skovlin, and B. C. Smith, *J. Chem. Soc.*, 2997 (1965); C. R. Kruger and E. G. Rochow, *J. Organomet. Chem.*, **1**, 476 (1964).
4. R. R. Fraser and T. S. Mansour, *J. Org. Chem.*, **49**, 3442 (1984).
5. M. W. Rathke and R. Kow, *J. Am. Chem. Soc.*, **94**, 6854 (1972); R. A. Olofson and C. M. Dougherty, *J. Am. Chem. Soc.*, **95**, 581, 582 (1973).
6. E. J. Corey and M. Chaykovsky, *J. Am. Chem. Soc.*, **87**, 1345 (1965).
7. C. A. Brown, *J. Org. Chem.*, **39**, 1324 (1974); R. Pi, T. Friedl, P. v. R. Schleyer, P. Klusener, and L. Brandsma, *J. Org. Chem.*, **52**, 4299 (1987); T. L. Macdonald, K. J. Natalie, Jr., G. Prasad, and J. S. Sawyer, *J. Org. Chem.*, **51**, 1124 (1986).

$$K = \frac{[B-H][C^-]}{[C-H][B^-]} = \frac{K_{a(C-H)}}{K_{a(B-H)}}$$

If we consider the case of a simple alkyl ketone in a protic solvent, for example, we see that hydroxide ion or primary alkoxide ions will convert only a fraction of a ketone to its anion.

$$\underset{\underset{\text{RCCH}_3}{\overset{\text{O}}{\|}}}{} + \text{RCH}_2\text{O}^- \;\rightleftharpoons\; \underset{\underset{\text{RC}=\text{CH}_2}{\overset{\text{O}^-}{|}}}{} + \text{RCH}_2\text{OH} \quad K < 1$$

The slightly more basic tertiary alkoxides are comparable to the enolates in basicity, and a more favorable equilibrium will be established with such bases.

$$\underset{\underset{\text{RCCH}_3}{\overset{\text{O}}{\|}}}{} + \text{R}_3\text{CO}^- \;\rightleftharpoons\; \underset{\underset{\text{RC}=\text{CH}_2}{\overset{\text{O}^-}{|}}}{} + \text{R}_3\text{COH} \quad K \sim 1$$

Note also that dialkyl ketones such as acetone and 3-pentanone are slightly *more acidic* than the simple alcohols in DMSO. Use of alkoxide bases in DMSO favors enolate formation. For the amide bases, $K_{a(B-H)} << K_{a(C-H)}$, and complete formation of the enolate occurs.

$$\underset{\underset{\text{RCCH}_3}{\overset{\text{O}}{\|}}}{} + \text{R}_2\text{N}^- \;\rightleftharpoons\; \underset{\underset{\text{RC}=\text{CH}_2}{\overset{\text{O}^-}{|}}}{} + \text{R}_2\text{NH} \quad K >> 1$$

It is important to keep the position of the equilibria in mind as we consider reactions of carbanions. The base and solvent used determine the extent of deprotonation. Another important physical characteristic that has to be kept in mind is the degree of aggregation of the carbanion. Both the solvent and the cation influence the state of aggregation. This topic is discussed further in Section 1.1.3.

1.1.2. Regioselectivity and Stereoselectivity in Enolate Formation from Ketones and Esters

Deprotonation of the corresponding carbonyl compound is a fundamental method for the generation of enolates, and we discuss it here for ketones and esters. An unsymmetrical dialkyl ketone can form two *regioisomeric* enolates on deprotonation.

$$\underset{\underset{\text{R}_2\text{CHCCH}_2\text{R}'}{\overset{\text{O}}{\|}}}{} \quad \overset{\text{B}^-}{\longrightarrow} \quad \underset{\underset{\text{R}_2\text{C}=\text{CCH}_2\text{R}'}{\overset{\text{O}^-}{|}}}{} \quad \text{or} \quad \underset{\underset{\text{R}_2\text{CHC}=\text{CHR}'}{\overset{\text{O}^-}{|}}}{}$$

Full exploitation of the synthetic potential of enolates requires control over the regioselectivity of their formation. Although it may not be possible to direct deprotonation so as to form one enolate to the exclusion of the other, experimental conditions can often be chosen to favor one of the regioisomers. The composition of an enolate mixture can be governed by kinetic or thermodynamic factors. The enolate ratio is governed by *kinetic control* when the product composition is determined by the *relative rates of the competing proton abstraction reactions.*

$$R_2CHCCH_2R' \;+\; B^- \quad \overset{k_a}{\underset{k_b}{\rightleftarrows}} \quad \begin{matrix} R_2C=CCH_2R' \quad \mathbf{A} \\[2mm] R_2CHC=CHR' \quad \mathbf{B} \end{matrix} \qquad \frac{[\mathbf{A}]}{[\mathbf{B}]} = \frac{k_a}{k_b}$$

Kinetic control of isomeric enolate composition

By adjusting the conditions of enolate formation, it is possible to establish either kinetic or thermodynamic control. *Conditions for kinetic control of enolate formation are those in which deprotonation is rapid, quantitative, and irreversible.*[8] This requirement is met experimentally by using a very strong base such as LDA or LiHMDS in an aprotic solvent in the absence of excess ketone. Lithium is a better counterion than sodium or potassium for regioselective generation of the kinetic enolate, as it maintains a tighter coordination at oxygen and reduces the rate of proton exchange. Use of an aprotic solvent is essential because protic solvents permit enolate equilibration by reversible protonation-deprotonation, which gives rise to the thermodynamically controlled enolate composition. Excess ketone also catalyzes the equilibration by proton exchange.

Scheme 1.1 shows data for the regioselectivity of enolate formation for several ketones under various reaction conditions. A consistent relationship is found in these and related data. *Conditions of kinetic control usually favor formation of the less-substituted enolate*, especially for methyl ketones. The main reason for this result is that removal of a less hindered hydrogen is faster, for steric reasons, than removal of a more hindered hydrogen. Steric factors in ketone deprotonation are accentuated by using bulky bases. The most widely used bases are LDA, LiHMDS, and NaHMDS. Still more hindered disilylamides such as hexaethyldisilylamide[9] and *bis*-(dimethylphenylsilyl)amide[10] may be useful for specific cases.

The equilibrium ratios of enolates for several ketone-enolate systems are also shown in Scheme 1.1. Equilibrium among the various enolates of a ketone can be established by the presence of an excess of ketone, which permits reversible proton transfer. Equilibration is also favored by the presence of dissociating additives such as HMPA. The composition of the equilibrium enolate mixture is usually more closely balanced than for kinetically controlled conditions. In general, the more highly substituted enolate is the preferred isomer, but if the alkyl groups are sufficiently branched as to interfere with solvation, there can be exceptions. This factor, along with CH_3/CH_3 steric repulsion, presumably accounts for the stability of the less-substituted enolate from 3-methyl-2-butanone (Entry 3).

[8] For reviews, see J. d'Angelo, *Tetrahedron*, **32**, 2979 (1976); C. H. Heathcock, *Modern Synthetic Methods*, **6**, 1 (1992).

[9] S. Masamune, J. W. Ellingboe, and W. Choy, *J. Am. Chem. Soc.*, **104**, 5526 (1982).

[10] S. R. Angle, J. M. Fevig, S. D. Knight, R. W. Marquis, Jr., and L. E. Overman, *J. Am. Chem. Soc.*, **115**, 3966 (1993).

Scheme 1.1. Composition of Enolate Mixtures Formed under Kinetic and Thermodynamic Control[a]

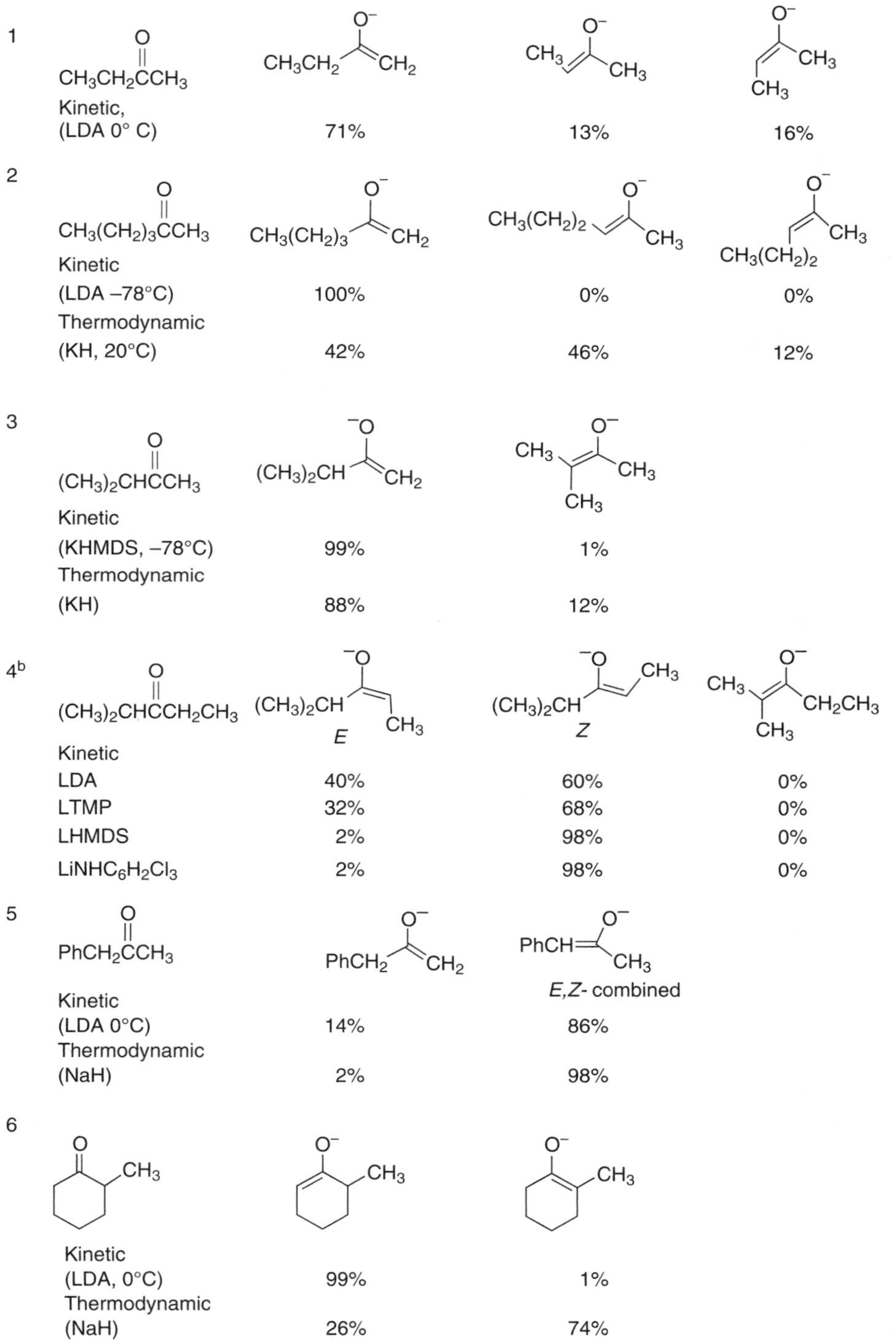

1 $CH_3CH_2CCH_3$ (with =O)

Kinetic, (LDA 0° C)

CH_3CH_2–C(O$^-$)=CH$_2$	CH_3–C(O$^-$)=CH–CH_3	CH_3–C(O$^-$)=C(CH_3)(CH_3)
71%	13%	16%

2 $CH_3(CH_2)_3CCH_3$ (with =O)

	$CH_3(CH_2)_3$–C(O$^-$)=CH$_2$	$CH_3(CH_2)_2$–CH=C(O$^-$)–CH_3	$CH_3(CH_2)_2$–CH=C(O$^-$)–CH_3
Kinetic (LDA −78°C)	100%	0%	0%
Thermodynamic (KH, 20°C)	42%	46%	12%

3 $(CH_3)_2CHCCH_3$ (with =O)

	$(CH_3)_2CH$–C(O$^-$)=CH$_2$	CH_3–C(CH$_3$)=C(O$^-$)–CH_3
Kinetic (KHMDS, −78°C)	99%	1%
Thermodynamic (KH)	88%	12%

4[b] $(CH_3)_2CHCCH_2CH_3$ (with =O)

	$(CH_3)_2CH$–CH=C(O$^-$)–CH_3 *E*	$(CH_3)_2CH$–CH=C(O$^-$)–CH_3 *Z*	CH_3–C(CH$_3$)=C(O$^-$)–CH_2CH_3
Kinetic			
LDA	40%	60%	0%
LTMP	32%	68%	0%
LHMDS	2%	98%	0%
LiNHC$_6$H$_2$Cl$_3$	2%	98%	0%

5 $PhCH_2CCH_3$ (with =O)

	$PhCH_2$–C(O$^-$)=CH$_2$	$PhCH$=C(O$^-$)–CH_3 *E,Z*-combined
Kinetic (LDA 0°C)	14%	86%
Thermodynamic (NaH)	2%	98%

6   2-methylcyclohexanone

Kinetic (LDA, 0°C)	99%	1%
Thermodynamic (NaH)	26%	74%

(Continued)

7

	Kinetic (Ph₃CLi)	Thermodynamic (Ph₃CK)
	100%	0%
	35%	65%

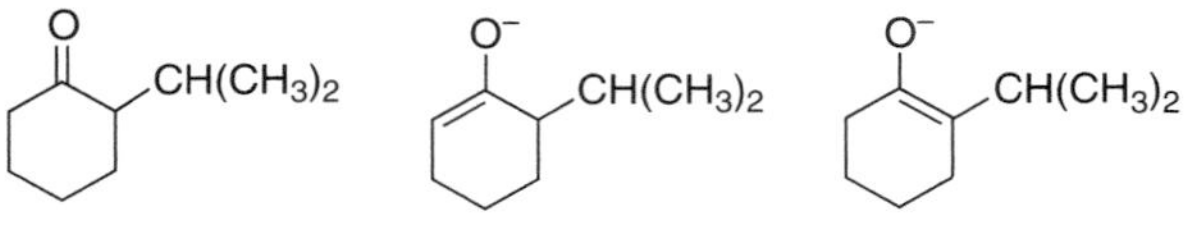

Kinetic
(Ph₃CLi)   100%   0%
Thermodynamic
(Ph₃CK)   35%   65%

8

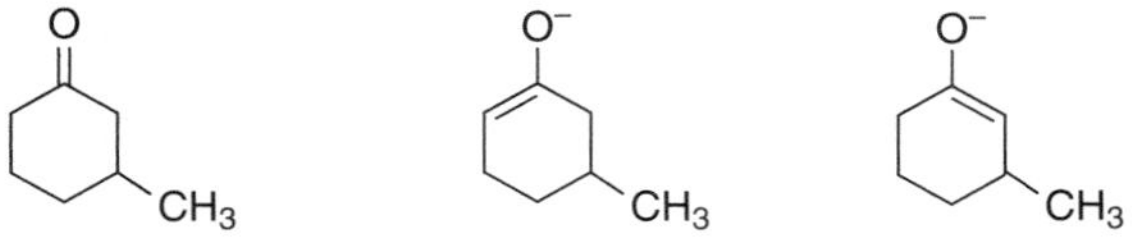

Kinetic
(Ph₃CLi)   82%   18%
Thermodynamic
(Ph₃CK)   52%   48%

9

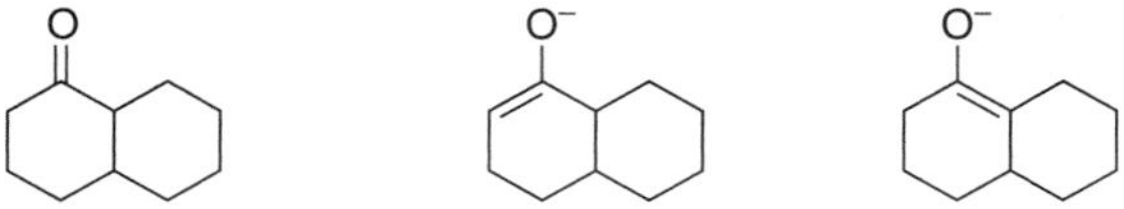

Kinetic
(LDA)   98%   2%
Thermodynamic
(NaH)   50%   50%

a. Selected from a more complete compilation by D. Caine, in *Carbon-Carbon Bond Formation*, R. L. Augustine, ed., Marcel Dekker, New York, 1979.
b. C. H. Heathcock, C. T. Buse, W. A. Kleschick, M. C. Pirrung, J. E. Sohn, and J. Lampe, *J. Org. Chem.*, **45**, 1066 (1980); L. Xie, K. Vanlandeghem, K. M. Isenberger, and C. Bernier, *J. Org. Chem.* **68**, 641 (2003).

88%   12%

The acidifying effect of an adjacent phenyl group outweighs steric effects in the case of 1-phenyl-2-propanone, and as a result the conjugated enolate is favored by both kinetic and thermodynamic conditions (Entry 5).

For cyclic ketones conformational factors also come into play in determining enolate composition. 2-Substituted cyclohexanones are kinetically deprotonated at the C(6) methylene group, whereas the more-substituted C(2) enolate is slightly favored at equilibrium (Entries 6 and 7). A 3-methyl group has a significant effect on the regiochemistry of kinetic deprotonation but very little effect on the thermodynamic stability of the isomeric enolates (Entry 8).

Many enolates can exist as both *E*- and *Z*-isomers.[11] The synthetic importance of LDA and HMDS deprotonation has led to studies of enolate stereochemistry under various conditions. In particular, the stereochemistry of some enolate reactions depends on whether the *E*- or *Z*-isomer is involved. Deprotonation of 2-pentanone was examined with LDA in THF, with and without HMPA. C(1) deprotonation is favored under both conditions, but the *Z:E* ratio for C(3) deprotonation is sensitive to the presence of HMPA.[12] More *Z*-enolate is formed when HMPA is present.

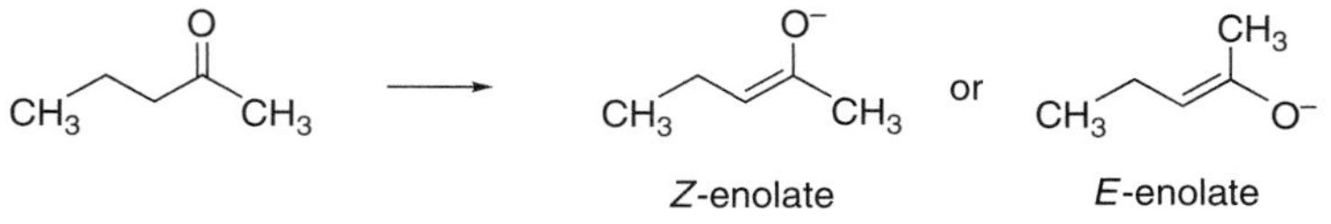

	Ratio C(1):C(3) deprotonation	Ratio *Z:E* for C(3) deprotonation
0° C, THF alone	7.9	0.20
−60° C, THF alone	7.1	0.15
0° C, THF-HMPA	8.0	1.0
−60° C, THF-HMPA	5.6	3.1

These and other related enolate ratios are interpreted in terms of a tight, reactant-like cyclic TS in THF and a looser TS in the presence of HMPA. The cyclic TS favors the *E*-enolate, whereas the open TS favors the *Z*-enolate. The effect of the HMPA is to solvate the Li$^+$ ion, reducing the importance of Li$^+$ coordination with the carbonyl oxygen.[13]

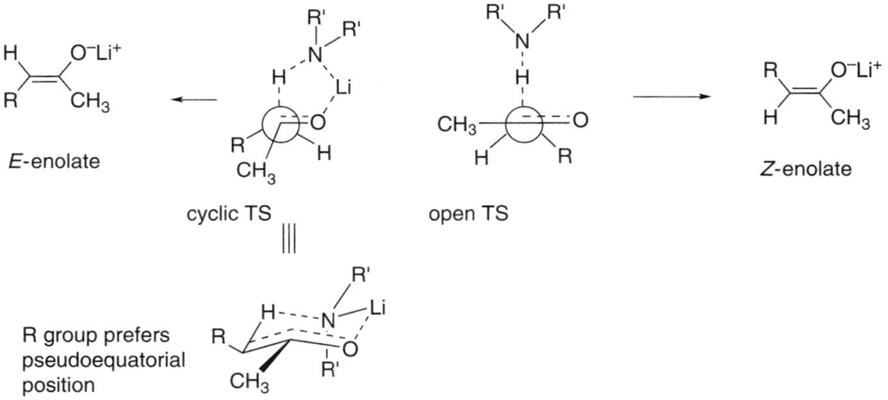

[11.] The enolate oxygen is always taken as a high-priority substituent in assigning the *E*- or *Z*-configuration.

[12.] L. Xie and W. H. Saunders, Jr., *J. Am. Chem. Soc.*, **113**, 3123 (1991).

[13.] R. E. Ireland and A. K. Willard, *Tetrahedron Lett.*, 3975 (1975); R. E. Ireland, R. H. Mueller, and A. K. Willard, *J. Am. Chem. Soc.*, **98**, 2868 (1972); R. E. Ireland, P. Wipf, and J. Armstrong, III, *J. Org. Chem.*, **56**, 650 (1991).

In contrast to LDA, LiHMDS favors the *Z*-enolate.[14] Certain other bases show a preference for formation of the *Z*-enolate. For example, lithium 2,4,6-trichloroanilide, lithium diphenylamide, and lithium trimethylsilylanilide show nearly complete *Z*-selectivity with 2-methyl-3-pentanone.[15]

	Z-enolate	*E*-enolate
LiNH(C$_6$H$_2$Cl$_3$)	98%	2%
LiNPh$_2$	100%	0%
LiN(Ph)Si(CH$_3$)$_3$	95%	5%

The *Z*-selectivity seems to be associated primarily with reduced basicity of the amide anion. It is postulated that the shift to *Z*-stereoselectivity is the result of a looser TS, in which the steric effects of the chair TS are reduced.

Strong effects owing to the presence of lithium halides have been noted. With 3-pentanone, the *E*:*Z* ratio can be improved from 10:1 to 60:1 by addition of one equivalent of LiBr in deprotonation by LiTMP.[16] (Note a similar effect for 2-methyl-3-pentanone in Table 1.2) NMR studies show that the addition of the halides leads to formation of mixed 1:1 aggregates, but precisely how this leads to the change in stereoselectivity has not been unraveled. A crystal structure has been determined for a 2:1:4:1 complex of the enolate of methyl *t*-butyl ketone, with an HMDS anion, four lithium cations, and one bromide.[17] This structure, reproduced in Figure 1.1, shows that the lithium ions are clustered around the single bromide, with the enolate oxygens bridging between two lithium ions. The amide base also bridges between lithium ions.

Very significant acceleration in the rate of deprotonation of 2-methylcyclohexanone was observed when triethylamine was included in enolate-forming reactions in toluene. The rate enhancement is attributed to a TS containing LiHMDS dimer and triethylamine. Steric effects in the amine are crucial in selective stabilization of the TS and the extent of acceleration that is observed.[18]

14. C. H. Heathcock, C. T. Buse, W. A. Kleschick, M. C. Pirrung, J. E. Sohn, and J. Lampe, *J. Org. Chem.*, **45**, 1066 (1980).

15. L. Xie, K. M. Isenberger, G. Held, and L. M. Dahl, *J. Org. Chem.*, **62**, 7516 (1997); L. Xie, K. Vanlandeghem, K. M. Isenberger, and C. Bernier, *J. Org. Chem.*, **68**, 641 (2003).

16. P. L. Hall, J. H. Gilchrist, and D. B. Collum, *J. Am. Chem. Soc.*, **113**, 9571 (1991); P. L. Hall, J. H. Gilchrist, A. T. Harrison, D. J. Fuller, and D. B. Collum, **113**, 9575 (1991).

17. K. W. Henderson, A. E. Dorigo, P. G. W. Williard, and P. R. Bernstein, *Angew. Chem. Int. Ed. Engl.*, **35**, 1322 (1996).

18. P. Zhao and D. B. Collum, *J. Am. Chem. Soc.*, **125**, 4008, 14411 (2003).

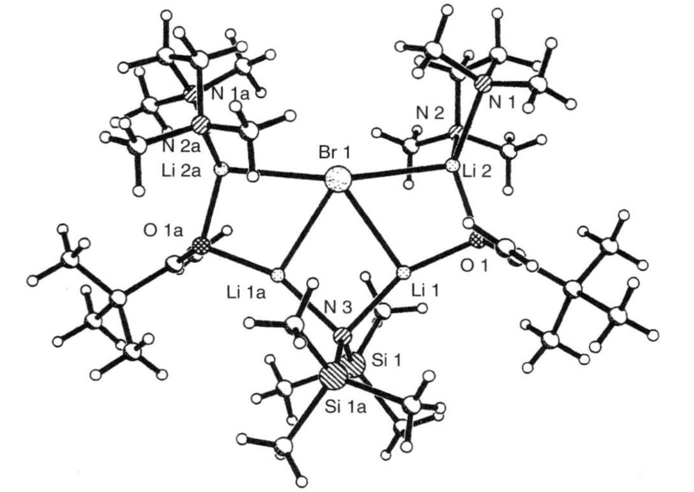

Fig. 1.1. Crystal structure of lithium enolate of methyl *t*-butyl ketone in a structure containing four Li$^+$, two enolates, and one HMDA anions, one bromide ion, and two TMEDA ligands. Reproduced from *Angew. Chem. Int. Ed. Engl.*, **35**, 1322 (1996), by permission of Wiley-VCH.

These effects of LiBr and triethylamine indicate that there is still much to be learned about deprotonation and that there is potential for further improvement in regio- and stereoselectivity.

Some data on the stereoselectivity of enolate formation from both esters and ketones is given in Table 1.2. The switch from *E* to *Z* in the presence of HMPA is particularly prominent for ester enolates. There are several important factors in determining regio- and stereoselectivity in enolate formation, including the strength of the base, the identity of the cation, and the nature of the solvent and additives. In favorable cases such as 2-methyl-3-pentanone and ethyl propanoate, good selectivity is possible for both stereoisomers. In other cases, such as 2,2-dimethyl-3-pentanone, the inherent stability difference between the enolates favors a single enolate, regardless of conditions.

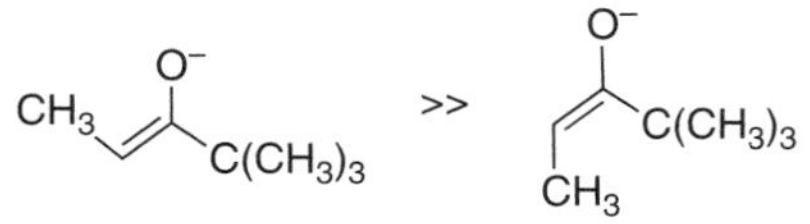

Chelation affects the stereochemistry of enolate formation. For example, the formation of the enolates from α-siloxyesters is *Z* for LiHMDS, but *E* for LiTMP.[19]

[19.] K. Hattori and H. Yamamoto, *J. Org. Chem.*, **58**, 5301 (1993); K. Hattori and H. Yamamoto, *Tetrahedron*, **50**, 3099 (1994).

Table 1.2. Stereoselectivity of Enolate Formation[a]

Reactant	Base	THF (hexane) (Z:E)	THF (23% HMPA) (Z:E)
Ketones			
$CH_3CH_2COCH_2CH_3$[b,c]	LDA	30:70	92:8
$CH_3CH_2COCH_2CH_3$[b]	LiTMP	20:80	
$CH_3CH_2COCH_2CH_3$[b]	LiHMDS	34:66	
$CH_3CH_2COCH(CH_3)_2$[b]	LDA	56:44	
$CH_3CH_2COCH(CH_3)_2$[b]	LiHMDS	> 98:2	
$CH_3CH_2COCH(CH_3)_2$[d]	$LiNPh_2$	100:0	
$CH_3CH_2COCH(CH_3)_2$[e]	LiTMP.LiBr	4:96	
$CH_3CH_2COC(CH_3)_3$[b]	LDA	< 2:98	
CH_3CH_2COPh[b]	LDA	> 97:3	
Esters			
$CH_3CH_2CO_2CH_2CH_3$[f]	LDA	6:94	88:15
$CH_3CO_2C(CH_3)_3$[g]	LDA	5:95	77:23
$CH_3(CH_2)_3CO_2CH_3$[g]	LDA	9:91	84:16
$PhCH_2CO_2CH_3$[h]	LDA	19:81	91:9
Amides			
$CH_3CH_2CON(C_2H_5)_2$[i]	LDA[i]	> 97:3	
$CH_3CH_2CON(CH_2)_4$[i]	LDA	> 97:3	

a. From a more extensive compilation given by C. H. Heathcock, *Modern Synthetic Methods*, **6**, 1 (1992).
b. C. H. Heathcock, C. T. Buse, W. A. Kleschick, M. C. Pirrung, J. E. Sohn, and J. Lampe, *J. Org. Chem.*, **45**, 1066 (1980).
c. Z. A. Fataftah, I. E. Kopka, and M. W. Rathke, *J. Am. Chem. Soc.*, **102**, 3959 (1980).
d. L. Xie, K. Vanlandeghem, K. M. Isenberger, and C. Bernier, *J. Org. Chem.*, **68**, 641 (2003).
e. P. L. Hall, J. H. Gilchrist, and D. B. Collum, *J. Am. Chem. Soc.*, **113**, 9571 (1991).
f. R. E. Ireland, P. Wipf, and J. D. Armstrong, III, *J. Org. Chem.*, **56**, 650 (1991).
g. R. E. Ireland, R. H. Mueller, and A. K. Willard, *J. Am. Chem. Soc.*, **98**, 2868 (1976).
h. F. Tanaka and K. Fuji, *Tetrahedron Lett.*, **33**, 7885 (1992).
i. J. M. Takacs, Ph. D. Thesis, California Institute of Technology, 1981.

It has been suggested that this stereoselectivity might arise from a chelated TS in the case of the less basic LiHMDS.

Kinetically controlled deprotonation of α,β-unsaturated ketones usually occurs preferentially at the α'-carbon adjacent to the carbonyl group. The polar effect of the carbonyl group is probably responsible for the faster deprotonation at this position.

(only enolate)

Ref. 20

20. R. A. Lee, C. McAndrews, K. M. Patel, and W. Reusch, *Tetrahedron Lett.*, 965 (1973).

Under conditions of thermodynamic control, however, it is the enolate corresponding to deprotonation of the γ-carbon that is present in the greater amount.

| major enolate (more stable) | | 2 (less stable) |

1

Ref. 21

These isomeric enolates differ in that **1** is fully conjugated, whereas the π system in **2** is cross-conjugated. In isomer **2**, the delocalization of the negative charge is restricted to the oxygen and the α′-carbon, whereas in the conjugated system of **1** the negative charge is delocalized on oxygen and both the α- and γ-carbon.

It is also possible to achieve *enantioselective enolate formation* by using chiral bases. Enantioselective deprotonation requires discrimination between two enantiotopic hydrogens, such as in *cis*-2,6-dimethylcyclohexanone or 4-(*t*-butyl)cyclohexanone. Among the bases that have been studied are chiral lithium amides such as **A** to **D**.[22]

A[23] **B**[24] **C**[25] **D**[26]

Enantioselective enolate formation can also be achieved by *kinetic resolution* through preferential reaction of one of the enantiomers of a racemic chiral ketone such as 2-(*t*-butyl)cyclohexanone (see Section 2.1.8 of Part A to review the principles of kinetic resolution).

45% yield, 90% e.e.   51% yield, 94% e.e.

Ref. 25a

[21.] G. Buchi and H. Wuest, *J. Am. Chem. Soc.*, **96**, 7573 (1974).

[22.] P. O'Brien, *J. Chem. Soc., Perkin Trans. 1*, 1439 (1998); H. J. Geis, *Methods of Organic Chemistry*, Vol. E21a, Houben-Weyl, G. Thieme Stuttgart, 1996, p. 589.

[23.] P. J. Cox and N. S. Simpkins, *Tetrahedron: Asymmetry*, **2**, 1 (1991); N. S. Simpkin, *Pure Appl. Chem.*, **68**, 691 (1996); B. J. Bunn and N. S. Simpkins, *J. Org. Chem.*, **58**, 533 (1993).

[24.] C. M. Cain, R. P. C. Cousins, G. Coumbarides, and N. S. Simpkins, *Tetrahedron*, **46**, 523 (1990).

[25.] (a) D. Sato, H. Kawasaki, T. Shimada, Y. Arata, K. Okamura, T. Date, and K. Koga, *J. Am. Chem. Soc.*, **114**, 761 (1992); (b) T. Yamashita, D. Sato, T. Kiyoto, A. Kumar, and K. Koga, *Tetrahedron Lett.*, **37**, 8195 (1996); (c) H. Chatani, M. Nakajima, H. Kawasaki, and K. Koga, *Heterocycles*, **46**, 53 (1997); (d) R. Shirai, D. Sato, K. Aoki, M. Tanaka, H. Kawasaki, and K. Koga, *Tetrahedron*, **53**, 5963 (1997).

[26.] M. Asami, *Bull. Chem. Soc. Jpn.*, **63**, 721 (1996).

Such enantioselective deprotonations depend upon kinetic selection between prochiral or enantiomeric hydrogens and the chiral base, resulting from differences in diastereomeric TSs.[27] For example, transition structure **E** has been proposed for deprotonation of 4-substituted cyclohexanones by base **D**.[28] This structure includes a chloride generated from trimethylsilyl chloride.

1.1.3. Other Means of Generating Enolates

Reactions other than deprotonation can be used to generate specific enolates under conditions in which lithium enolates do not equilibrate with regio- and stereoisomers. Several methods are shown in Scheme 1.2. Cleavage of trimethylsilyl enol ethers or enol acetates by methyllithium (Entries 1 and 3), depends on the availability of these materials in high purity. Alkoxides can also be used to cleave silyl enol ethers and enol acetates.[29] When KO-*t*-Bu is used for the cleavage, subsequent alkylation occurs at the more-substituted position, regardless of which regioisomeric silyl enol ether is used.[30] Evidently under these conditions, the potassium enolates equilibrate and the more highly substituted enolate is more reactive.

Trimethylsilyl enol ethers can also be cleaved by tetraalkylammonium fluoride (Entry 2) The driving force for this reaction is the formation of the very strong Si−F bond, which has a bond energy of 142 kcal/mol.[31] These conditions, too, lead to enolate equilibration.

27. A. Corruble, J.-Y. Valnot, J. Maddaluno, Y. Prigent, D. Davoust, and P. Duhamel, *J. Am. Chem. Soc.*, **119**, 10042 (1997); D. Sato, H. Kawasaki, and K. Koga, *Chem. Pharm. Bull.*, **45**, 1399 (1997); K. Sugasawa, M. Shindo, H. Noguchi, and K. Koga, *Tetrahedron Lett.*, **37**, 7377 (1996).
28. M. Toriyama, K. Sugasawa, M. Shindo, N. Tokutake, and K. Koga, *Tetrahedron Lett.*, **38**, 567 (1997).
29. D. Cahard and P. Duhamel, *Eur. J. Org. Chem.*, 1023 (2001).
30. P. Duhamel, D. Cahard, Y. Quesnel, and J.-M. Poirier, *J. Org. Chem.*, **61**, 2232 (1996); Y. Quesnel, L. Bidois-Sery, J.-M. Poirier, and L. Duhamel, *Synlett*, 413 (1998).
31. For reviews of the chemistry of O-silyl enol ethers, see J. K. Rasmussen, *Synthesis*, 91 (1977); P. Brownbridge, *Synthesis*, 1, 85 (1983); I. Kuwajima and E. Nakamura, *Acc. Chem. Res.*, **18**, 181 (1985).

Scheme 1.2. Other Means of Generating Specific Enolates

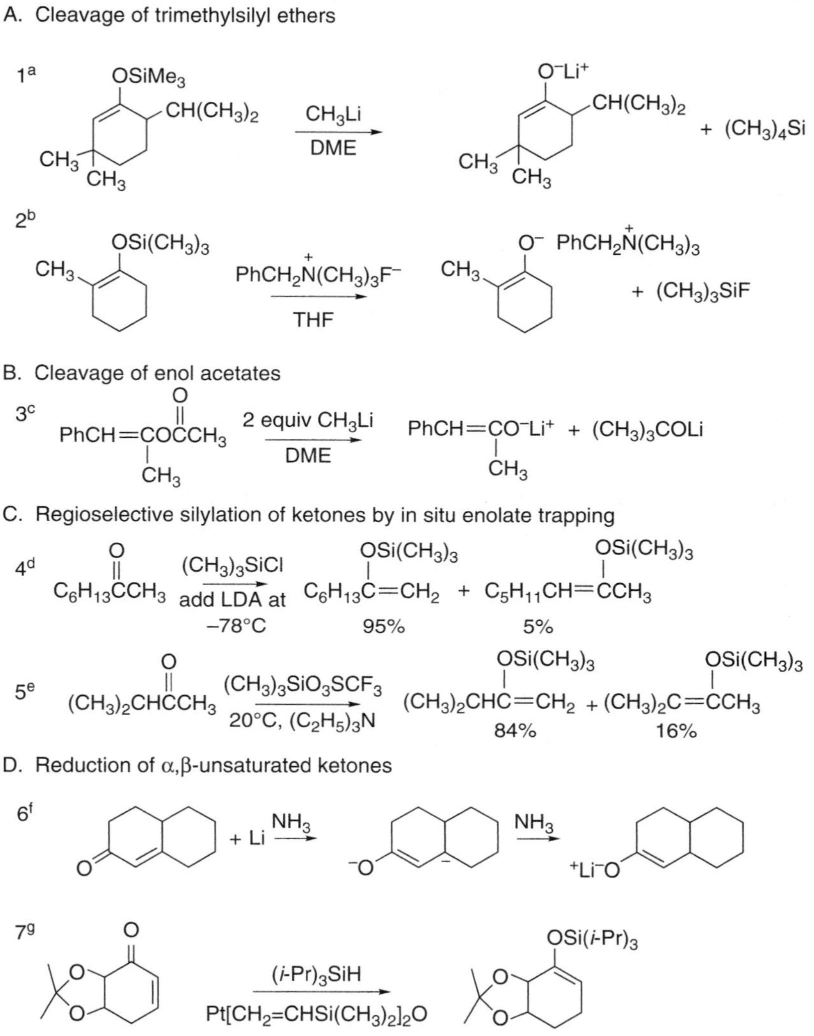

15

A. Cleavage of trimethylsilyl ethers

1[a]

$+ (CH_3)_4Si$

2[b]

$+ (CH_3)_3SiF$

B. Cleavage of enol acetates

3[c]

$$PhCH=C(CH_3)OCCH_3 \xrightarrow[DME]{2 \text{ equiv } CH_3Li} PhCH=CO^-Li^+ + (CH_3)_3COLi$$
with CH_3 substituent

C. Regioselective silylation of ketones by in situ enolate trapping

4[d]

$$C_6H_{13}CCH_3 \xrightarrow[\substack{\text{add LDA at} \\ -78°C}]{(CH_3)_3SiCl} C_6H_{13}C=CH_2 + C_5H_{11}CH=CCH_3$$
OSi(CH_3)_3   95%   OSi(CH_3)_3   5%

5[e]

$$(CH_3)_2CHCCH_3 \xrightarrow[20°C, (C_2H_5)_3N]{(CH_3)_3SiO_3SCF_3} (CH_3)_2CHC=CH_2 + (CH_3)_2C=CCH_3$$
OSi(CH_3)_3   84%   OSi(CH_3)_3   16%

D. Reduction of α,β-unsaturated ketones

6[f]

7[g]

a. G. Stork and P. Hudrlik, *J. Am. Chem. Soc.*, **90**, 4464 (1968); H. O. House, L. J. Czuba, M. Gall, and H. D. Olmstead, *J. Org. Chem.*, **34**, 2324 (1969).
b. I. Kuwajima and E. Nakamura, *J. Am. Chem. Soc.*, **97**, 3258 (1975).
c. G. Stork and S. R. Dowd, *Org. Synth.*, **55**, 46 (1976); see also H. O. House and B. M. Trost, *J. Org. Chem.*, **30**, 2502 (1965).
d. E. J. Corey and A. W. Gross, *Tetrahedron Lett.*, **25**, 495 (1984).
e. E. Emde, A. Goetz, K. Hofmann, and G. Simchen, *Justus Liebigs Ann. Chem.*, 1643 (1981).
f. G. Stork, P. Rosen, N. Goldman, R. V. Coombs, and J. Tsuji, *J. Am. Chem. Soc.*, **87**, 275 (1965).
g. C. R. Johnson and R. K. Raheja, *J. Org. Chem.*, **59**, 2287 (1994).

The composition of the trimethylsilyl enol ethers prepared from an enolate mixture reflects the enolate composition. If the enolate formation can be done with high regio-selection, the corresponding trimethylsilyl enol ether can be obtained in high purity. If not, the silyl enol ether mixture must be separated. Trimethylsilyl enol ethers can be prepared directly from ketones. One procedure involves reaction with trimethylsilyl chloride and a tertiary amine.[32] This procedure gives the regioisomers in a ratio favoring the thermodynamically more stable enol ether. Use of *t*-butyldimethylsilyl chloride with potassium hydride as the base also seems to favor the thermodynamic product.[33] Trimethylsilyl trifluoromethanesulfonate (TMS-OTf), which is more reactive, gives primarily the less-substituted trimethylsilyl enol ether.[34] Higher ratios of the less-substituted enol ether are obtained by treating a mixture of ketone and trimethylsilyl chloride with LDA at −78° C.[35] Under these conditions the kinetically preferred enolate is immediately trapped by reaction with trimethylsilyl chloride. Even greater preferences for the less-substituted silyl enol ether can be obtained by using the more hindered lithium amide from *t*-octyl-*t*-butylamine (LOBA).

Lithium-ammonia reduction of α, β-unsaturated ketones (Entry 6) provides a very useful method for generating specific enolates.[36] The starting enones are often readily available and the position of the double bond in the enone determines the structure of the resulting enolate. For acyclic enones, the TMS-Cl trapping of enolates generated by conjugate reduction gives a silyl enol ether having a composition that reflects the conformation of the enone.[37] (See Section 2.2.1 of Part A to review enone conformation.)

Trimethylsilyl enol ethers can also be prepared by 1,4-reduction of enones using silanes as reductants. Several effective catalysts have been found,[38] of which the most versatile appears to be a Pt complex of divinyltetramethyldisiloxane.[39] This catalyst gives good yields of substituted silyl enol ethers (e.g., Scheme 1.2, Entry 7).

32. H. O. House, L. J. Czuba, M. Gall, and H. D. Olmstead, *J. Org. Chem.*, **34**, 2324 (1969); R. D. Miller and D. R. McKean, *Synthesis*, 730 (1979).
33. J. Orban, J. V. Turner, and B. Twitchin, *Tetrahedron Lett.*, **25**, 5099 (1984).
34. H. Emde, A. Goetz, K. Hofmann, and G. Simchen, *Liebigs Ann. Chem.*, 1643 (1981); see also E. J. Corey, H. Cho, C. Ruecker, and D. Hua, *Tetrahedron Lett.*, 3455 (1981).
35. E. J. Corey and A. W. Gross, *Tetrahedron Lett.*, **25**, 495 (1984).
36. For a review of α, β-enone reduction, see D. Caine, *Org. React.*, **23**, 1 (1976).
37. A. R. Chamberlin and S. H. Reich, *J. Am. Chem. Soc.*, **107**, 1440 (1985).
38. I. Ojima and T. Kogure, *Organometallics*, **1**, 1390 (1982); T. H. Chan and G. Z. Zheng, *Tetrahedron Lett.*, **34**, 3095 (1993); D. E. Cane and M. Tandon, *Tetrahedron Lett.*, **35**, 5351 (1994).
39. C. R. Johnson and R. K Raheja, *J. Org. Chem.*, **59**, 2287 (1994).

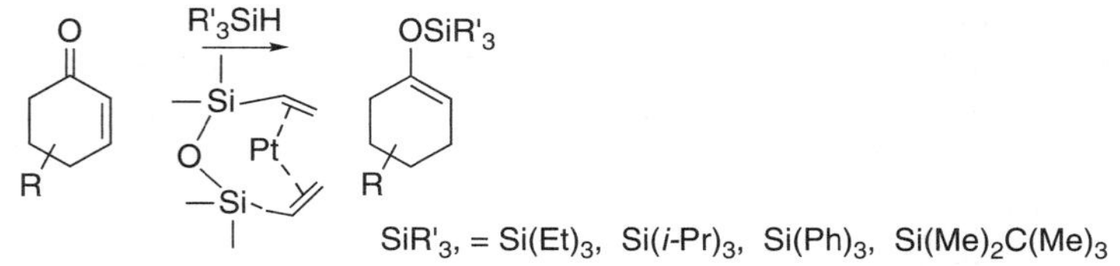

$$SiR'_3, = Si(Et)_3, \ Si(i\text{-}Pr)_3, \ Si(Ph)_3, \ Si(Me)_2C(Me)_3$$

Excellent yields of silyl enol have also been obtained from enones using $B(C_6F_5)_3$ as a catalyst.[40] *t*-Butyldimethylsilyl, triethylsilyl, and other silyl enol ethers can also be made under these conditions.

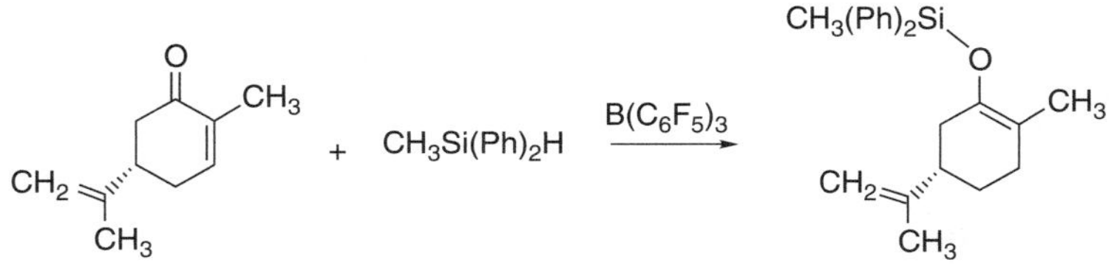

These and other reductive methods for generating enolates from enones are discussed more fully in Chapter 5.

Another very important method for specific enolate generation is the conjugate addition of organometallic reagents to enones. This reaction, which not only generates a specific enolate, but also adds a carbon substituent, is discussed in Section 8.1.2.3.

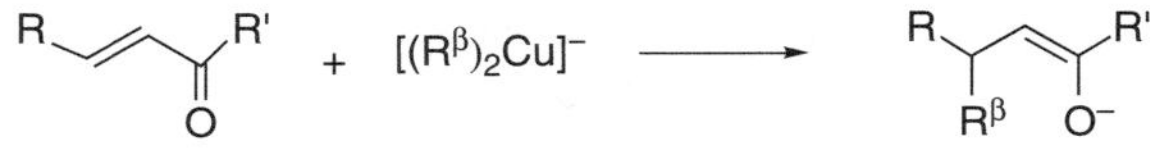

1.1.4. Solvent Effects on Enolate Structure and Reactivity

The rate of alkylation of enolate ions is strongly dependent on the solvent in which the reaction is carried out.[41] The relative rates of reaction of the sodium enolate of diethyl *n*-butylmalonate with *n*-butyl bromide are shown in Table 1.3. Dimethyl sulfoxide (DMSO) and *N,N*-dimethylformamide (DMF) are particularly effective in enhancing the reactivity of enolate ions. Both of these are *polar aprotic solvents*. Other

**Table 1.3. Relative Alkylation Rates of Sodium Diethyl
n-Butylmalonate in Various Solvents[a]**

Solvent	Dielectric constant ϵ	Relative rate
Benzene	2.3	1
Tetrahydrofuran	7.3	14
Dimethoxyethane	6.8	80
N,N-Dimethylformamide	37	970
Dimethyl sulfoxide	47	1420

a. From H. E. Zaugg, *J. Am. Chem. Soc.*, **83**, 837 (1961).

40. J. M. Blackwell, D. J. Morrison, and W. E. Piers, *Tetrahedron*, **58**, 8247 (2002).
41. For reviews, see (a) A. J. Parker, *Chem. Rev.*, **69**, 1 (1969); (b) L. M. Jackmamn and B. C. Lange, *Tetrahedron*, **33**, 2737 (1977).

compounds that are used as cosolvents in reactions between enolates and alkyl halides include *N*-methylpyrrolidone (NMP), hexamethylphosphoric triamide (HMPA) and *N*,*N*′-dimethylpropyleneurea (DMPU).[42] Polar aprotic solvents, as the name indicates, are materials that have high dielectric constants but lack hydroxy or other hydrogen-bonding groups. Polar aprotic solvents possess excellent metal cation coordination ability, so they can solvate and dissociate enolates and other carbanions from ion pairs and clusters.

dimethyl sulfoxide (DMSO) $\varepsilon = 47$; *N*,*N*-dimethylformamide (DMF) $\varepsilon = 37$; *N*-methylpyrrolidone (NMP) $\varepsilon = 32$; hexamethylphosphoric triamide (HMPA) $\varepsilon = 30$; *N*,*N*′-dimethylpropyleneurea (DMPU)

The reactivity of alkali metal (Li^+, Na^+, K^+) enolates is very sensitive to the state of aggregation, which is, in turn, influenced by the reaction medium. The highest level of reactivity, which can be approached but not achieved in solution, is that of the "bare" unsolvated enolate anion. For an enolate-metal ion pair in solution, the maximum reactivity is expected when the cation is strongly solvated and the enolate is very weakly solvated. Polar aprotic solvents are good cation solvators and poor anion solvators. Each one has a negatively polarized oxygen available for coordination to the metal cation. Coordination to the enolate anion is less effective because the positively polarized atoms of these molecules are not nearly as exposed as the oxygen. Thus, these solvents provide a medium in which enolate-metal ion aggregates are dissociated to give a less encumbered, more reactive enolate.

aggregated ions $\longrightarrow$ dissociated ions

Polar protic solvents such as water and alcohols also possess a pronounced ability to separate ion aggregates, but are less favorable as solvents in enolate alkylation reactions because they can coordinate to both the metal cation and the enolate anion. Solvation of the enolate anion occurs through hydrogen bonding. The solvated enolate is relatively less reactive because the hydrogen bonding must be disrupted during alkylation. Enolates generated in polar protic solvents such as water, alcohols, or ammonia are therefore less reactive than the same enolate in a polar aprotic solvent such as DMSO. Of course, hydroxylic solvents also impose limits on the basicity of enolates that are stable.

solvated ions

42. T. Mukhopadhyay and D. Seebach, *Helv. Chim. Acta*, **65**, 385 (1982).

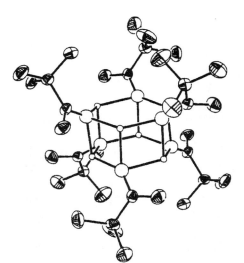

Fig. 1.2. Unsolvated hexameric aggregate of lithium enolate of methyl *t*-butyl ketone; the open circles represent oxygen and the small circles are lithium. Reproduced from *J. Am. Chem. Soc.*, **108**, 462 (1986), by permission of the American Chemical Society.

Tetrahydrofuran (THF) and dimethoxyethane (DME) are slightly polar solvents that are moderately good cation solvators. Coordination to the metal cation involves the oxygen unshared electron pairs. These solvents, because of their lower dielectric constants, are less effective at separating ion pairs and higher aggregates than are the polar aprotic solvents. The structures of the lithium and potassium enolates of methyl *t*-butyl ketone have been determined by X-ray crystallography. The structures are shown in Figures 1.2 and 1.3.[43] Whereas these represent the solid state structures,

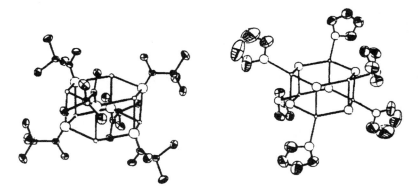

Fig. 1.3. Potassium enolate of methyl *t*-butyl ketone; open circles are oxygen and small circles are potassium. (a) left panel shows only the enolate structures; (b) right panel shows only the solvating THF molecules. The actual structure is the superposition of both panels. Reproduced from *J. Am. Chem. Soc.*, **108**, 462 (1986), by permission of the American Chemical Society.

43. P. G. Williard and G. B. Carpenter, *J. Am. Chem. Soc.*, **108,** 462 (1986).

the hexameric clusters are a good indication of the nature of the enolates in relatively weakly coordinating solvents. In both structures, series of alternating metal cations and enolate oxygens are assembled in two offset hexagons. The cluster is considerably tighter with Li^+ than with K^+. The $M-O$ bonds are about $1.9\,\text{Å}$ for Li^+ and $2.6\,\text{Å}$ for K^+. The enolate $C-O$ bond is longer ($1.34\,\text{Å}$) for Li^+ than for K^+ ($1.31\,\text{Å}$), whereas the $C=C$ bond is shorter for Li^+ ($1.33\,\text{Å}$) than for K^+ ($1.35\,\text{Å}$). Thus, the Li^+ enolate has somewhat more of oxy-anion character and is expected to be a "harder" than the potassium enolate.

Despite the somewhat reduced reactivity of aggregated enolates, THF and DME are the most commonly used solvents for synthetic reactions involving enolate alkylation. They are the most suitable solvents for *kinetic enolate generation* and also have advantages in terms of product workup and purification over the polar aprotic solvents. Enolate reactivity in these solvents can often be enhanced by adding a reagent that can bind alkali metal cations more strongly. Popular choices are HMPA, DMPU, tetramethylethylenediamine (TMEDA), and the crown ethers. TMEDA chelates metal ions through the electron pairs on nitrogen. The crown ethers encapsulate the metal ions through coordination with the ether oxygens. The 18-crown-6 structure is of such a size as to allow sodium or potassium ions to fit in the cavity. The smaller 12-crown-4 binds Li^+ preferentially. The cation complexing agents lower the degree of aggregation of the enolate and metal cations, which results in enhanced reactivity.

The effect of HMPA on the reactivity of cyclopentanone enolate has been examined.[44] This enolate is primarily a dimer, even in the presence of excess HMPA, but the reactivity increases by a factor of 7500 for a tenfold excess of HMPA at $-50°\,C$. The kinetics of the reaction with CH_3I are consistent with the dimer being the active nucleophile. It should be kept in mind that the reactivity of regio- and stereoisomeric enolates may be different and the alkylation product ratio may not reflect the enolate composition. This issue was studied with 2-heptanone.[45] Although kinetic deprotonation in THF favors the 1-enolate, a nearly equal mixture of C(1) and C(3) alkylation was observed. The inclusion of HMPA improved the C(1) selectivity to 11:1 and also markedly accelerated the rate of the reaction. These results are presumably due to increased reactivity and less competition from enolate isomerization in the presence of HMPA.

C(3) alkylation

The effect of chelating polyamines on the rate and yield of benzylation of the lithium enolate of 1-tetralone was compared with HMPA and DMPU. The triamine

44. M. Suzuki, H. Koyama, and R. Noyori, *Bull. Chem. Soc. Jpn.*, **77**, 259 (2004); M. Suzuki, H. Koyama, and R. Noyori, *Tetrahedron*, **60**, 1571 (2004).
45. C. L. Liotta and T. C. Caruso, *Tetrahedron Lett.*, **26**, 1599 (1985).

and tetramine were even more effective than HMPA in promoting reaction.[46] These results, too, are presumably due to disaggregation of the enolate by the polyamines.

Additive (3eq)	Yield (%)
none	6
HMPA	34
DMPU	3
$Me_2NCH_2CH_2NMe_2$	6
$(Me_2NCH_2CH_2)_2NMe$	50
$(Me_2NCH_2CH_2NCH_2)_2$ Me	72
$Me_2N(CH_2CH_2N)_3CH_2CH_2NMe_2$ Me	33

The reactivity of enolates is also affected by the metal counterion. For the most commonly used ions the order of reactivity is $Mg^{2+} < Li^+ < Na^+ < K^+$. The factors that are responsible for this order are closely related to those described for solvents. The smaller, harder Mg^{2+} and Li^+ cations are more tightly associated with the enolate than are the Na^+ and K^+ ions. The tighter coordination decreases the reactivity of the enolate and gives rise to more highly associated species.

1.2. Alkylation of Enolates[47]

1.2.1. Alkylation of Highly Stabilized Enolates

Relatively acidic compounds such as malonate esters and β-ketoesters were the first class of compounds for which reliable conditions for carbanion alkylation were developed. The alkylation of these relatively acidic compounds can be carried out in alcohols as solvents using metal alkoxides as bases. The presence of two electron-withdrawing substituents facilitates formation of the resulting enolate. Alkylation occurs by an S_N2 process, so the alkylating agent must be reactive toward nucleophilic displacement. Primary halides and sulfonates, especially allylic and benzylic ones, are the most reactive alkylating agents. Secondary systems react more slowly and often give only moderate yields because of competing elimination. Tertiary halides give only elimination products. Methylene groups can be dialkylated if sufficient base and alkylating agent are used. Dialkylation can be an undesirable side reaction if the monoalkyl derivative is the desired product. Sequential dialkylation using two different alkyl groups is possible. Use of dihaloalkanes as alkylating reagents leads to ring formation. The relative rates of cyclization for ω-haloalkyl malonate esters

---

[46.] M. Goto, K. Akimoto, K. Aoki, M. Shindo, and K. Koga, *Chem. Pharm. Bull.*, **48**, 1529 (2000).

[47.] For general reviews of enolate alkylation, see D. Caine, in *Carbon-Carbon Bond Formation*, Vol. 1, R. L. Augustine, ed., Marcel Dekker, New York, 1979, Chap. 2; C. H. Heathcock, *Modern Synthetic Methods*, **6**, 1 (1992).

are 650,000:1:6500:5 for formation of three-, four-, five-, and six-membered rings, respectively.[48] (See Section 4.3 of Part A to review the effect of ring size on S_N2 reactions.)

Some examples of alkylation reactions involving relatively acidic carbon acids are shown in Scheme 1.3. Entries 1 to 4 are typical examples using sodium ethoxide as the base. Entry 5 is similar, but employs sodium hydride as the base. The synthesis of diethyl cyclobutanedicarboxylate in Entry 6 illustrates ring formation by *intramolecular alkylation reactions*. Additional examples of intramolecular alkylation are considered in Section 1.2.5. Note also the stereoselectivity in Entry 7, where the existing branched substituent leads to a *trans* orientation of the methyl group.

The 2-substituted β-ketoesters (Entries 1, 4, 5, and 7) and malonic ester (Entries 2 and 6) prepared by the methods illustrated in Scheme 1.3 are useful for the synthesis

Scheme 1.3. Alkylation of Enolates Stabilized by Two Functional Groups

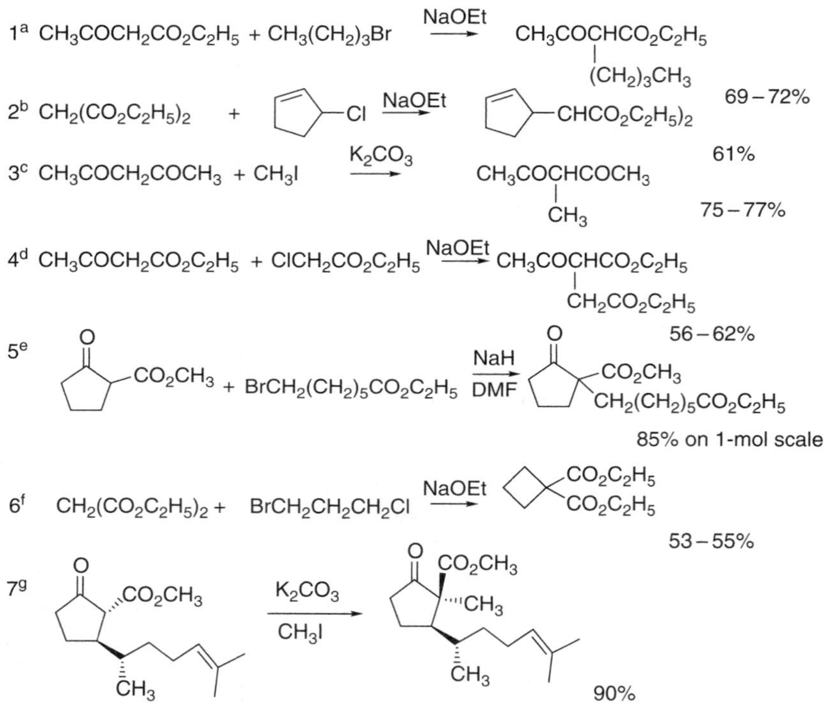

a. C. S. Marvel and F. D. Hager, *Org. Synth.*, **I**, 248 (1941).
b. R. B. Moffett, *Org. Synth.*, **IV**, 291 (1963).
c. A. W. Johnson, E. Markham, and R. Price, *Org. Synth.*, **42**, 75 (1962).
d. H. Adkins, N. Isbell, and B. Wojcik, *Org. Synth.*, **II**, 262 (1943).
e. K. F. Bernardy, J. F. Poletto, J. Nocera, P. Miranda, R. E. Schaub, and M. J. Weiss, *J. Org. Chem.*, **45**, 4702 (1980).
f. R. P. Mariella and R. Raube, *Org. Synth.*, **IV**, 288 (1963).
g. D. F. Taber and S. C. Malcom, *J. Org. Chem.*, **66**, 944 (2001).

48. A. C. Knipe and C. J. Stirling, *J. Chem. Soc. B*, 67 (1968); For a discussion of factors that affect intramolecular alkylation of enolates, see J. Janjatovic and Z. Majerski, *J. Org. Chem.*, **45**, 4892 (1980).

of ketones and carboxylic acids. Both β-keto acids and malonic acids undergo facile decarboxylation.

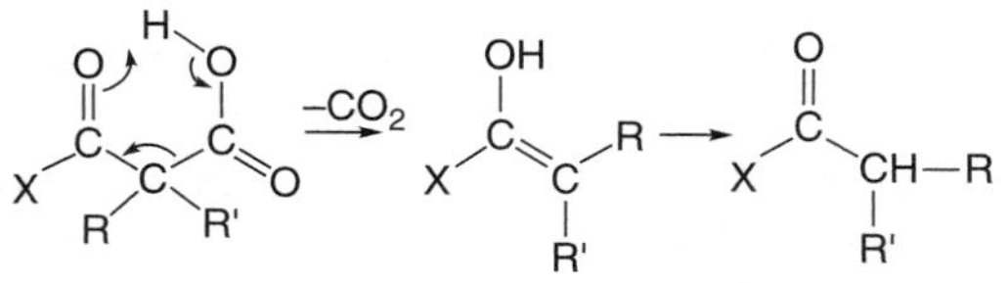

β-keto acid: X = alkyl or aryl = ketone
substituted malonic acid: X = OH = substituted acetic acid

Examples of this approach to the synthesis of ketones and carboxylic acids are presented in Scheme 1.4. In these procedures, an ester group is removed by hydrolysis and decarboxylation after the alkylation step. The malonate and acetoacetate carbanions are the *synthetic equivalents* of the simpler carbanions that lack the additional ester substituent. In the preparation of 2-heptanone (Entry 1), for example, ethyl acetoacetate functions

Scheme 1.4. Synthesis by Decarboxylation of Malonates and other β-Dicarbonyl Compounds

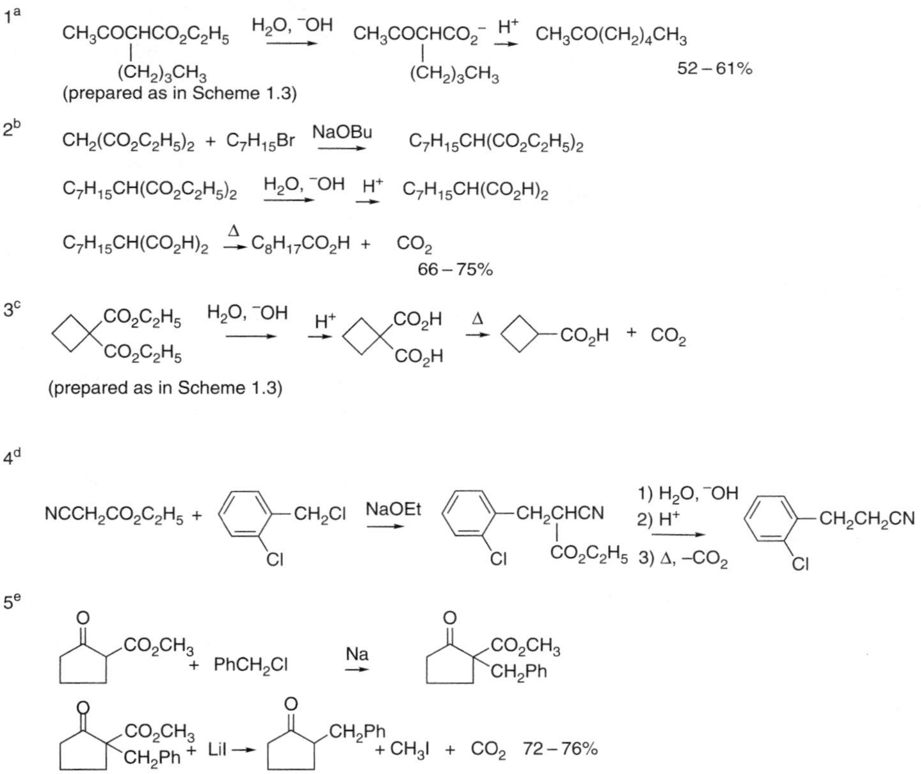

a. J. R. Johnson and F. D. Hager, *Org. Synth.*, **I**, 351 (1941).
b. E. E. Reid and J. R. Ruhoff, *Org. Synth.*, **II**, 474 (1943).
c. G. B. Heisig and F. H. Stodola, *Org. Synth.*, **III**, 213 (1955).
d. J. A. Skorcz and F. E. Kaminski, *Org. Synth.*, **48**, 53 (1968).
e. F. Elsinger, *Org. Synth.*, **V**, 76 (1973).

as the synthetic equivalent of acetone. Entries 2 and 3 show synthesis of carboxylic acids via the malonate ester route. Entry 4 is an example of a nitrile synthesis, starting with ethyl cyanoacetate as the carbon nucleophile. The cyano group also facilitates decarboxylation. Entry 5 illustrates an alternative decarboxylation procedure in which lithium iodide is used to cleave the β-ketoester by nucleophilic demethylation.

It is also possible to use the dilithium derivative of acetoacetic acid as the synthetic equivalent of acetone enolate.[49] In this case, the hydrolysis step is unnecessary and decarboxylation can be done directly on the alkylation product.

$$CH_3\overset{O}{\overset{\|}{C}}CH_2CO_2H \xrightarrow{2n\text{-BuLi}} CH_3\overset{O^-Li^+}{\overset{|}{C}}{=}CHCO_2{}^-Li^+ \xrightarrow[\substack{2)\ H^+ \\ (-CO_2)}]{1)\ R{-}X} CH_3\overset{O}{\overset{\|}{C}}CH_2R$$

Similarly, the dilithium dianion of monoethyl malonate is easily alkylated and the product decarboxylates after acidification.[50]

$$n\text{-}C_4H_9Br \quad + \quad \underset{\underset{CO_2C_2H_5}{|}}{LiCHCO_2Li} \quad \xrightarrow[\substack{2)\ 68°C,\ 18\,h \\ (-CO_2)}]{1)\ 25°C,\ 2\,h} \quad CH_3(CH_2)_4CO_2H \atop 80\%$$

1.2.2. Alkylation of Ketone Enolates

The preparation of ketones and ester from β-dicarbonyl enolates has largely been supplanted by procedures based on selective enolate formation. These procedures permit direct alkylation of ketone and ester enolates and avoid the hydrolysis and decarboxylation of keto ester intermediates. The development of conditions for stoichiometric formation of both kinetically and thermodynamically controlled enolates has permitted the extensive use of enolate alkylation reactions in multistep synthesis of complex molecules. One aspect of the alkylation reaction that is crucial in many cases is the stereoselectivity. The alkylation has a stereoelectronic preference for approach of the electrophile perpendicular to the plane of the enolate, because the π electrons are involved in bond formation. A major factor in determining the stereoselectivity of ketone enolate alkylations is the difference in steric hindrance on the two faces of the enolate. The electrophile approaches from the less hindered of the two faces and the degree of stereoselectivity depends on the steric differentiation. Numerous examples of such effects have been observed.[51] In ketone and ester enolates that are exocyclic to a conformationally biased cyclohexane ring there is a small preference for

[49.] R. A. Kjonaas and D. D. Patel, *Tetrahedron Lett.*, **25**, 5467 (1984).

[50.] J. E. McMurry and J. H. Musser, *J. Org. Chem.*, **40**, 2556 (1975).

[51.] For reviews, see D. A. Evans, in *Asymmetric Synthesis*, Vol. 3, J. D. Morrison, ed., Academic Press, New York, 1984, Chap. 1; D. Caine, in *Carbon-Carbon Bond Formation*, R. L. Augustine, ed., Marcel Dekker, New York, 1979, Chap. 2.

the electrophile to approach from the equatorial direction.[52] If the axial face is further hindered by addition of a substituent, the selectivity is increased.

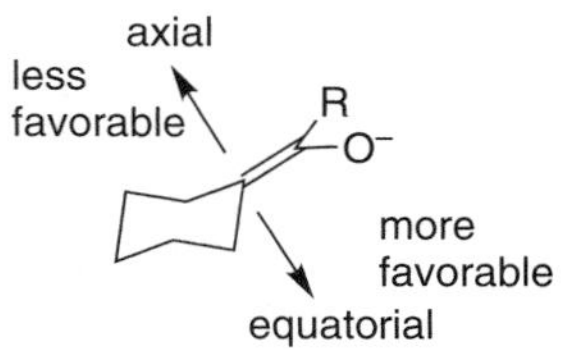

For simple, conformationally biased cyclohexanone enolates such as that from 4-*t*-butylcyclohexanone, there is little steric differentiation. The alkylation product is a nearly 1:1 mixture of the *cis* and *trans* isomers.

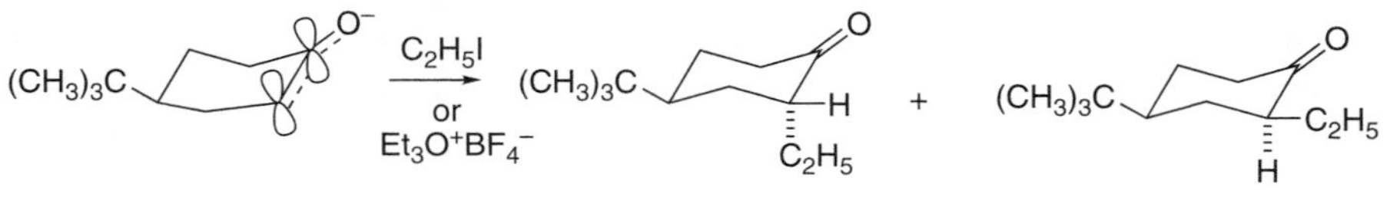

Ref. 53

The *cis* product must be formed through a TS with a twistlike conformation to adhere to the requirements of stereoelectronic control. The fact that this pathway is not disfavored is consistent with other evidence that the TS in enolate alkylations occurs *early* and reflects primarily the structural features of the reactant, not the product. A late TS would disfavor the formation of the *cis* isomer because of the strain associated with the nonchair conformation of the product.

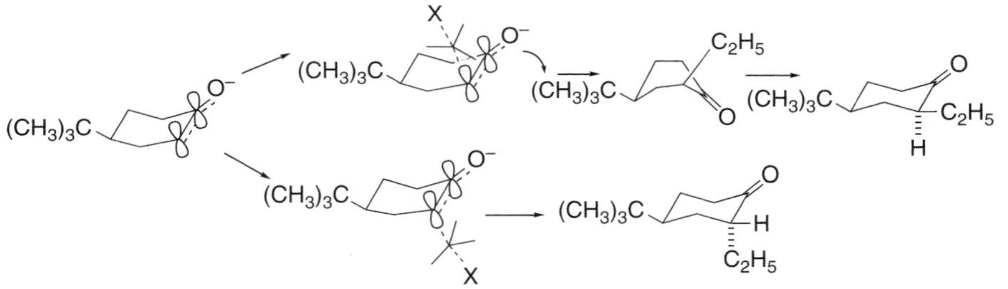

The introduction of an alkyl substituent at the α-carbon in the enolate enhances stereoselectivity somewhat. This is attributed to a steric effect in the enolate. To minimize steric interaction with the solvated oxygen, the alkyl group is distorted somewhat from coplanarity, which biases the enolate toward attack from the axial direction. The alternate approach from the upper face increases the steric interaction by forcing the alkyl group to become eclipsed with the enolate oxygen.[54]

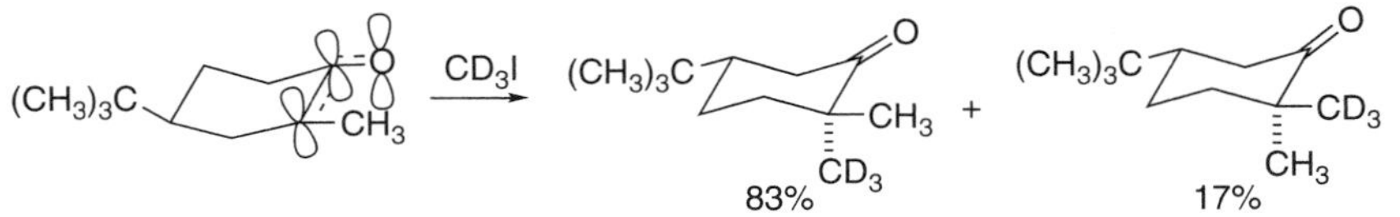

83%   17%

[52.] A. P. Krapcho and E. A. Dundulis, *J. Org. Chem.*, **45**, 3236 (1980); H. O. House and T. M. Bare, *J. Org. Chem.*, **33**, 943 (1968).

[53.] H. O. House, B. A. Terfertiller, and H. D. Olmstead, *J. Org. Chem.*, **33**, 935 (1968).

[54.] H. O. House and M. J. Umen, *J. Org. Chem.*, **38**, 1000 (1973).

When an additional methyl substituent is placed at C(3), there is a strong preference for alkylation *anti* to the 3-methyl group. This is attributed to the conformation of the enolate, which places the C(3) methyl in a pseudoaxial orientation because of allylic strain (see Part A, Section 2.2.1). The axial C(3) methyl then shields the lower face of the enolate.[55]

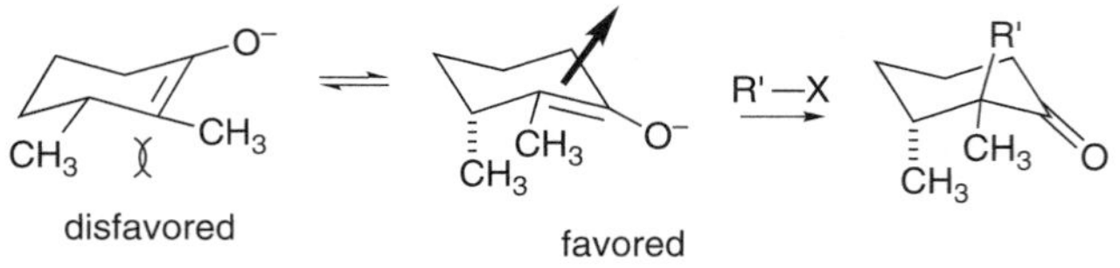

The enolates of 1- and 2-decalone derivatives provide further insight into the factors governing stereoselectivity in enolate alkylations. The 1(9)-enolate of 1-decalone shows a preference for alkylation to give the *cis* ring juncture, and this is believed to be due primarily a steric effect. The upper face of the enolate presents three hydrogens in a 1,3-diaxial relationship to the approaching electrophile. The corresponding hydrogens on the lower face are equatorial.[56]

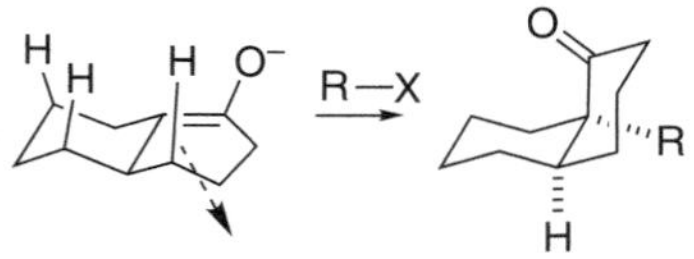

The 2(1)-enolate of *trans*-2-decalone is preferentially alkylated by an axial approach of the electrophile.

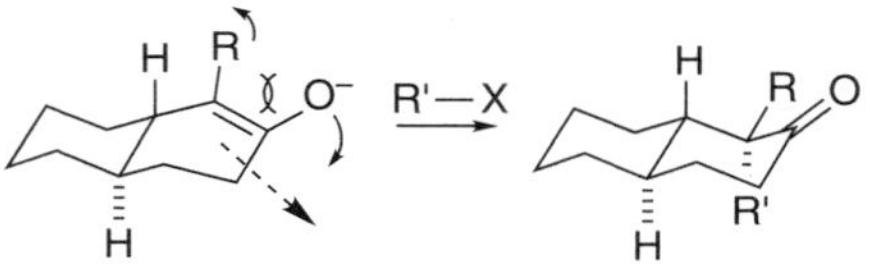

The stereoselectivity is enhanced if there is an alkyl substituent at C(1). The factors operating in this case are similar to those described for 4-*t*-butylcyclohexanone. The *trans*-decalone framework is conformationally rigid. Axial attack from the lower face leads directly to the chair conformation of the product. The 1-alkyl group enhances this stereoselectivity because a steric interaction with the solvated enolate oxygen distorts the enolate to favor the axial attack.[57] The placement of an axial methyl group at C(10) in a 2(1)-decalone enolate introduces a 1,3-diaxial interaction with the approaching electrophile. The preferred alkylation product results from approach on the opposite side of the enolate.

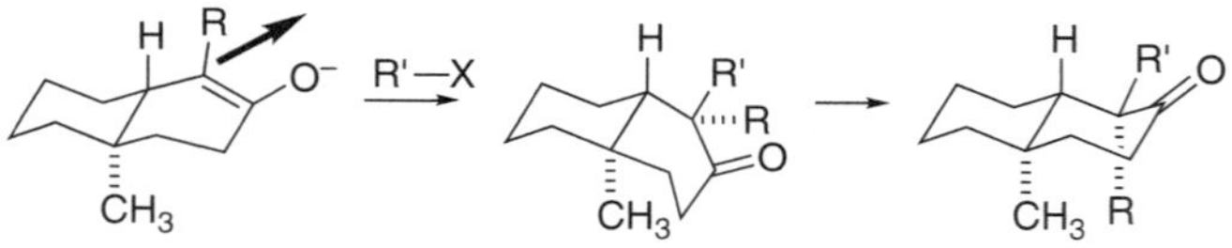

55. R. K. Boeckman, Jr., *J. Org. Chem.*, **38**, 4450 (1973).
56. H. O. House and B. M. Trost, *J. Org. Chem.*, **30**, 2502 (1965).
57. R. S. Mathews, S. S. Grigenti, and E. A. Folkers, *J. Chem. Soc., Chem. Commun.*, 708 (1970); P. Lansbury and G. E. DuBois, *Tetrahedron Lett.*, 3305 (1972).

The prediction and interpretation of alkylation stereochemistry requires consideration of conformational effects in the enolate. The decalone enolate **3** was found to have a strong preference for alkylation to give the *cis* ring junction, with alkylation occurring *cis* to the *t*-butyl substituent.[58]

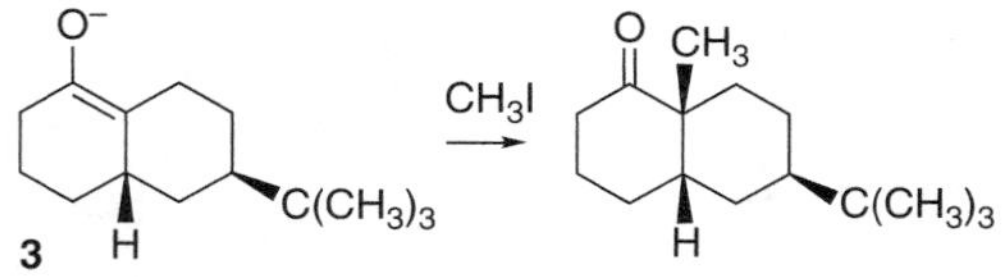

According to molecular mechanics (MM) calculations, the minimum energy conformation of the enolate is a twist-boat (because the chair leads to an axial orientation of the *t*-butyl group). The enolate is convex in shape with the second ring shielding the bottom face of the enolate, so alkylation occurs from the top.

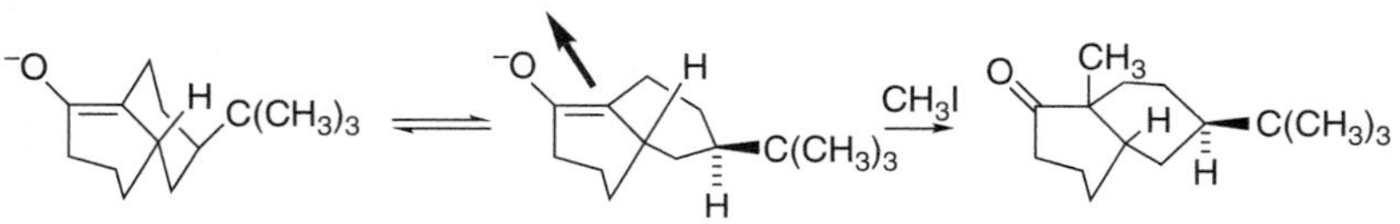

Houk and co-workers examined the role of torsional effects in the stereoselectivity of enolate alkylation in five-membered rings, and their interpretation can explain the preference for C(5) alkylation *syn* to the 2-methyl group in *trans*-2,3-dimethylcyclopentanone.[59]

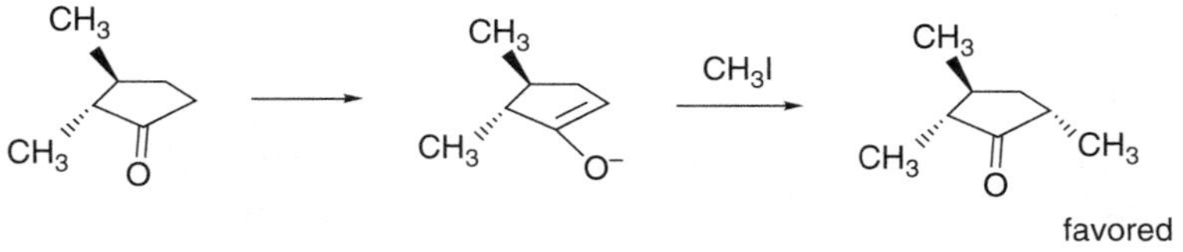

favored

The *syn* TS is favored by about 1 kcal/mol, owing to reduced eclipsing, as illustrated in Figure 1.4. An experimental study using the kinetic enolate of 3-(*t*-butyl)-2-methylcyclopentanone in an alkylation reaction with benzyl iodide gave an 85:15 preference for the predicted *cis*-2,5-dimethyl derivative.

In acyclic systems, the enolate conformation comes into play. β,β-Disubstituted enolates prefer a conformation with the hydrogen eclipsed with the enolate double bond. In unfunctionalized enolates, alkylation usually takes place *anti* to the larger substituent, but with very modest stereoselectivity.

[58] H. O. House, W. V. Phillips, and D. Van Derveer, *J. Org. Chem.*, **44**, 2400 (1979).
[59] K. Ando, N. S. Green, Y. Li, and K. N. Houk, *J. Am. Chem. Soc.*, **121**, 5334 (1999).

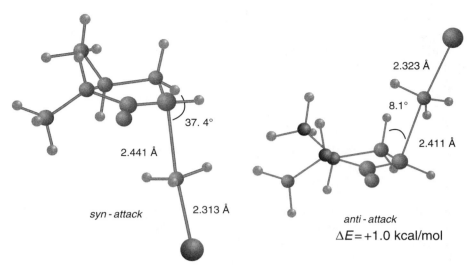

Fig. 1.4. Transition structures for *syn* and *anti* attack on the kinetic enolate of *trans*-2,3-dimethylcyclopentanone showing the staggered versus eclipsed nature of the newly forming bond. Reproduced from *J. Am. Chem. Soc.*, **121**, 5334 (1999), by permission of the American Chemical Society.

Ref. 60

These examples illustrate the issues that must be considered in analyzing the stereoselectivity of enolate alkylation. *The major factors are the conformation of the enolate, the stereoelectronic requirement for an approximately perpendicular trajectory, the steric preference for the least hindered path of approach, and minimization of torsional strain.* In cyclic systems the ring geometry and positioning of substituents are often the dominant factors. For acyclic enolates, the conformation and the degree of steric discrimination govern the stereoselectivity.

For enolates with additional functional groups, chelation may influence stereoselectivity. Chelation-controlled alkylation has been examined in the context of the synthesis of a polyol lactone (-)-discodermolide. The lithium enolate **4** reacts with the allylic iodide **5** in a hexane:THF solvent mixture to give a 6:1 ratio favoring the desired stereoisomer. Use of the sodium enolate gives the opposite stereoselectivity, presumably because of the loss of chelation.[61] The solvent seems to be quite important in promoting chelation control.

60. I. Fleming and J. J. Lewis, *J. Chem. Soc., Perkin Trans. 1*, 3257 (1992).
61. S. S. Harried, G. Yang, M. A. Strawn, and D. C. Myles, *J. Org. Chem.*, **62**, 6098 (1997).

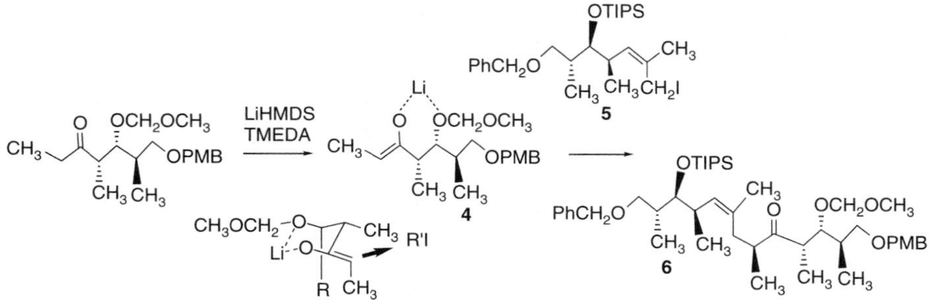

chelated enolate transition structure

6:1 *S:R* in 55:45 hexane-THF

Previous studies with related enolates having different protecting groups also gave products with the opposite C(16)–R configuration.[62]

Scheme 1.5 gives some examples of alkylation of ketone enolates. Entries 1 and 2 involve formation of the enolates by deprotonation with LDA. In Entry 2, equilibration

Scheme 1.5. Alkylation of Ketone Enolates

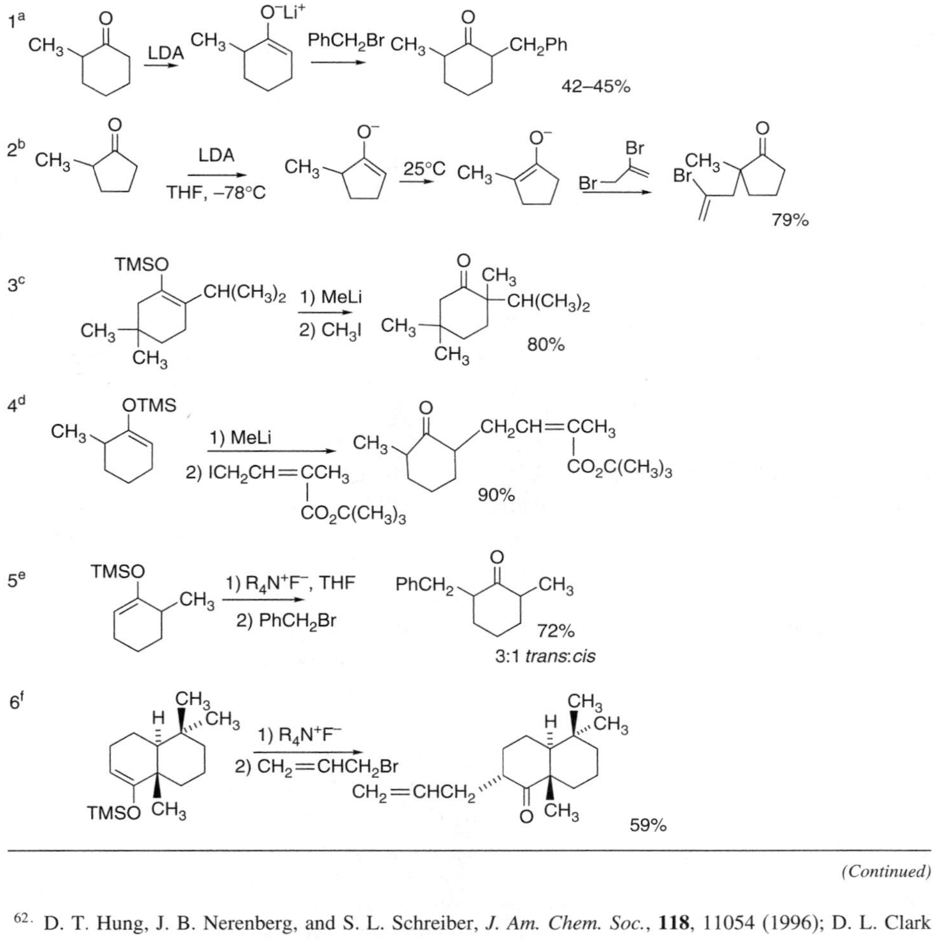

(Continued)

[62] D. T. Hung, J. B. Nerenberg, and S. L. Schreiber, *J. Am. Chem. Soc.*, **118**, 11054 (1996); D. L. Clark and C. H. Heathcock, *J. Org. Chem.*, **58**, 5878 (1993).

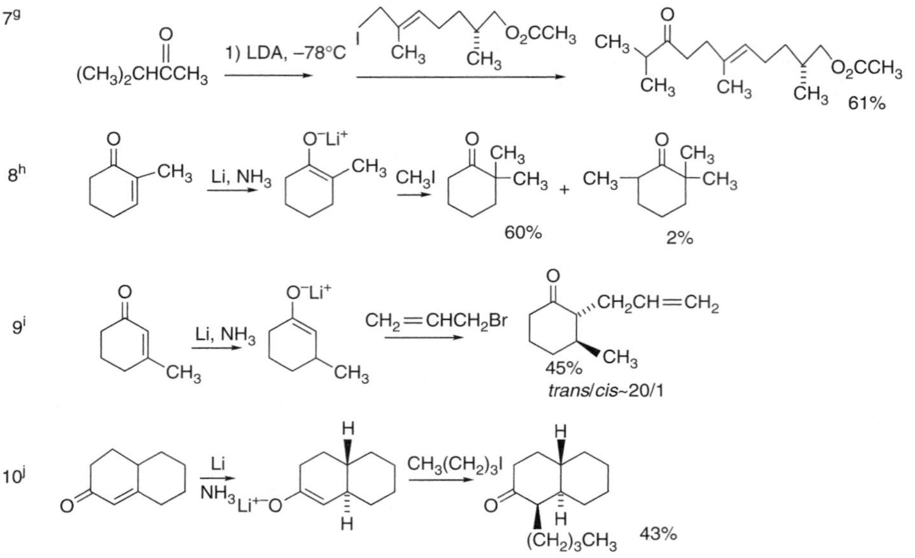

a. M. Gall and H. O. House, *Org. Synth.*, **52**, 39 (1972).
b. S. C. Welch and S. Chayabunjonglerd, *J. Am. Chem. Soc.*, **101**, 6768 (1979).
c. G. Stork and P. F. Hudrlik, *J. Am. Chem. Soc.*, **90**, 4464 (1968).
d. P. L. Stotter and K. A. Hill, *J. Am. Chem. Soc.*, **96**, 6524 (1974).
e. I. Kuwajima, E. Nakamura, and M. Shimizu, *J. Am. Chem. Soc.*, **104**, 1025 (1982).
f. A. B. Smith, III, and R. Mewshaw, *J. Org. Chem.*, **49**, 3685 (1984).
g. Y. L. Li, C. Huang, W. Li, and Y. Li, *Synth. Commun.*, **27**, 4341 (1997).
h. H. A. Smith, B. J. L. Huff, W. J. Powers, III, and D. Caine, *J. Org. Chem.*, **32**, 2851 (1967).
i. D. Caine, S. T. Chao, and H. A. Smith, *Org. Synth.*, **56**, 52 (1977).
j. G. Stork, P. Rosen, N. Goldman, R. V. Coombs, and J. Tsujii, *J. Am. Chem. Soc.*, **87**, 275 (1965).

to the more-substituted enolate precedes alkylation. Entries 3 and 4 show regiospecific generation of enolates by reaction of silyl enol ethers with methyllithium. Alkylation can also be carried out using silyl enol ethers by generating the enolate by fluoride ion.[63] Anhydrous tetraalkylammonium fluoride salts in anhydrous are normally the fluoride ion source.[64] Entries 5 and 6 illustrate this method. Entry 7 shows the kinetic deprotonation of 3-methylbutanone, followed by alkylation with a functionalized allylic iodide. Entries 8, 9, and 10 are examples of alkylation of enolates generated by reduction of enones. Entry 10 illustrates the preference for axial alkylation of the 2-(1)-decalone enolate.

In enolates formed by proton abstraction from α,β-unsaturated ketones, there are three potential sites for attack by electrophiles: the oxygen, the α-carbon, and the γ-carbon. The kinetically preferred site for both protonation and alkylation is the α-carbon.[65]

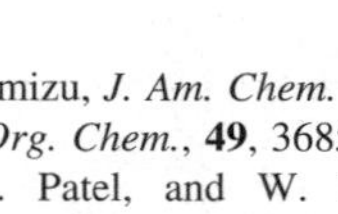

[63.] I. Kuwajima, E. Nakamura, and M. Shimizu, *J. Am. Chem. Soc.*, **104**, 1025 (1982).

[64.] A. B. Smith, III, and R. Mewshaw, *J. Org. Chem.*, **49**, 3685 (1984).

[65.] R. A. Lee, C. McAndrews, K. M. Patel, and W. Reusch, *Tetrahedron Lett.*, 965 (1973); J. A. Katzenellenbogen and A. L. Crumrine, *J. Am. Chem. Soc.*, **96**, 5662 (1974).

The selectivity for electrophilic attack at the α-carbon presumably reflects a greater negative charge, as compared with the γ-carbon.

Protonation of the enolate provides a method for converting α,β-unsaturated ketones and esters to the less stable β,γ-unsaturated isomers.

Ref. 66

1.2.3. Alkylation of Aldehydes, Esters, Carboxylic Acids, Amides, and Nitriles

Among the compounds capable of forming enolates, the alkylation of ketones has been most widely studied and applied synthetically. Similar reactions of esters, amides, and nitriles have also been developed. Alkylation of aldehyde enolates is not very common. One reason is that aldehydes are rapidly converted to aldol addition products by base. (See Chapter 2 for a discussion of this reaction.) Only when the enolate can be rapidly and quantitatively formed is aldol formation avoided. Success has been reported using potassium amide in liquid ammonia[67] and potassium hydride in tetrahydrofuran.[68] Alkylation via enamines or enamine anions provides a more general method for alkylation of aldehydes. These reactions are discussed in Section 1.3.

Ref. 68

Ester enolates are somewhat less stable than ketone enolates because of the potential for elimination of alkoxide. The sodium and potassium enolates are rather unstable, but Rathke and co-workers found that the lithium enolates can be generated at $-78°$ C.[69] Alkylations of simple esters require a strong base because relatively weak bases such as alkoxides promote condensation reactions (see Section 2.3.1). The successful formation of ester enolates typically involves an amide base, usually LDA or LiHDMS, at low temperature.[70] The resulting enolates can be successfully alkylated with alkyl bromides or iodides. HMPA is sometimes added to accelerate the alkylation reaction.

66. H. J. Ringold and S. K. Malhotra, *Tetrahedron Lett.*, 669 (1962); S. K. Malhotra and H. J. Ringold, *J. Am. Chem. Soc.*, **85**, 1538 (1963).

67. S. A. G. De Graaf, P. E. R. Oosterhof, and A. van der Gen, *Tetrahedron Lett.*, 1653 (1974).

68. P. Groenewegen, H. Kallenberg, and A. van der Gen, *Tetrahedron Lett.*, 491 (1978).

69. M. W. Rathke, *J. Am. Chem. Soc.*, **92**, 3222 (1970); M. W. Rathke and D. F. Sullivan, *J. Am. Chem. Soc.*, **95**, 3050 (1973).

70. (a) M. W. Rathke and A. Lindert, *J. Am. Chem. Soc.*, **93**, 2318 (1971); (b) R. J. Cregge, J. L. Herrmann, C. S. Lee, J. E. Richman, and R. H. Schlessinger, *Tetrahedron Lett.*, 2425 (1973); (c) J. L. Herrmann and R. H. Schlessinger, *J. Chem. Soc., Chem. Commun.*, 711 (1973).

In acyclic systems, the stereochemistry of alkylation depends on steric factors. Stereoselectivity is low for small substituents.[71]

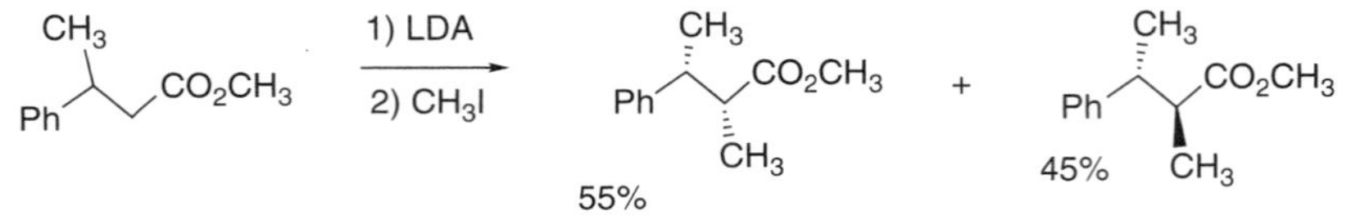

When a larger substituent is present, the reaction becomes much more selective. For example, a β-dimethylphenylsilyl substituent leads to more than 95:5 *anti* alkylation in ester enolates.[72]

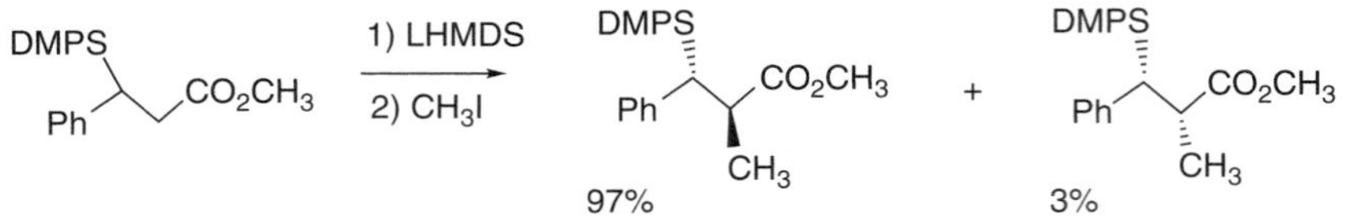

This stereoselectivity is the result of the conformation of the enolate and steric shielding by the silyl substituent.

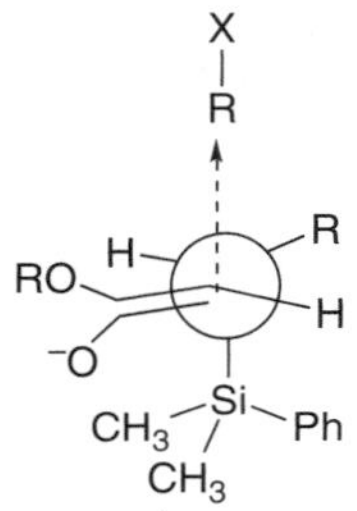

This directive effect has been employed in stereoselective synthesis.

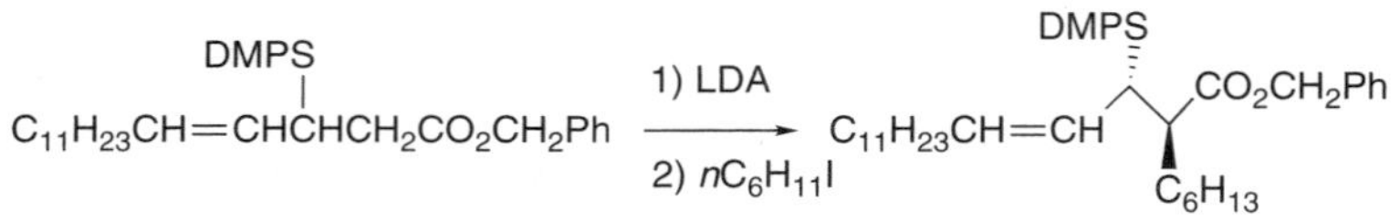

Ref. 72

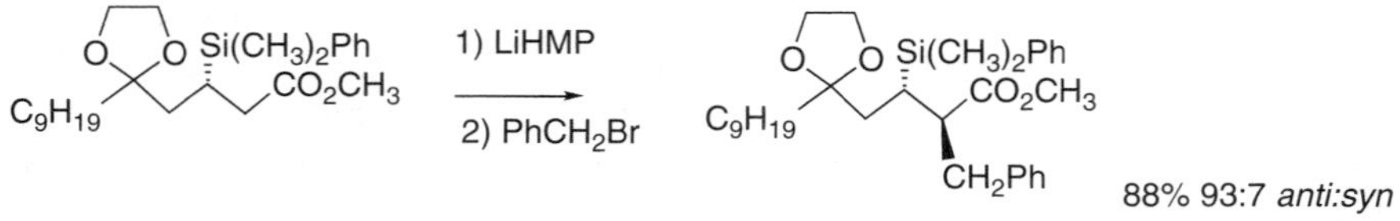

88% 93:7 *anti:syn*

Ref. 73

A careful study of the alkylation of several enolates of dialkyl malate esters has been reported.[74] These esters form dianions resulting from deprotonation of the hydroxy

71. R. A. N. C. Crump, I. Fleming, J. H. M. Hill, D. Parker, N. L. Reddy, and D. Waterson, *J. Chem. Soc., Perkin Trans. 1*, 3277 (1992).

72. I. Fleming and N. J. Lawrence, *J. Chem. Soc., Perkin Trans. 1*, 2679 (1998).

73. R. Verma and S.K. Ghosh, *J. Chem. Soc., Perkin Trans. 2*, 265(1999).

74. M. Sefkow, A. Koch, and E. Kleinpeter, *Helv. Chim. Acta*, **85**, 4216 (2002).

Fig. 1.5. Minimum energy structure of dilithium derivative of di-*iso*-propyl malate. Reproduced from *Helv. Chim. Acta*, **85**, 4216 (2002), by permission of Wiley-VCH.

group as well as the C(3). HF/6-31G* computations indicate that tricoordinate structures are formed, such as that shown for the di-*iso*-propyl ester in Figure 1.5. Curiously, the highest diastereoselectivity (19:1) is seen with the di-*iso*-propyl ester. For the dimethyl, diethyl, and di-*t*-butyl esters, the ratios are about 8:1. The diastereoselectivity is even higher (40:1) with the mixed *t*-butyl-*iso*-propyl ester. This result can be understood by considering the differences in the *si* and *re* faces of the enolates. In the di-*t*-butyl ester, both faces are hindered and selectivity is low. The di-*iso*-propyl ester has more hindrance to the *re* face, and this is accentuated in the mixed ester.

favored by 19:1	favored by 7:1	favored by 4.5:1	favored by 40:1
	increased hindrance at both faces	increased hindrance at *si* face	increased hindrance at *re* face

Alkylations of this type also proved to be sensitive to the cation. Good stereoselectivity (15:1) was observed for the lithium enolate, but the sodium and potassium enolates were much less selective.[75] This probably reflects the weaker coordination of the latter metals.

base	yield	*anti:syn*
LiHMDS	80	15:1
NaHMDS	45	1:2
KHMDS	20	1:1

Carboxylic acids can be directly alkylated by conversion to dianions with two equivalents of LDA. The dianions are alkylated at the α-carbon, as would be expected, because the enolate carbon is more strongly nucleophilic than the carboxylate anion.[76]

75. M. Sefkow, *J. Org. Chem.*, **66**, 2343 (2001).
76. P. L. Creger, *J. Org. Chem.*, **37**, 1907 (1972); P. L. Creger, *J. Am. Chem. Soc.*, **89**, 2500 (1967); P. L. Creger, *Org. Synth.*, **50**, 58 (1970).

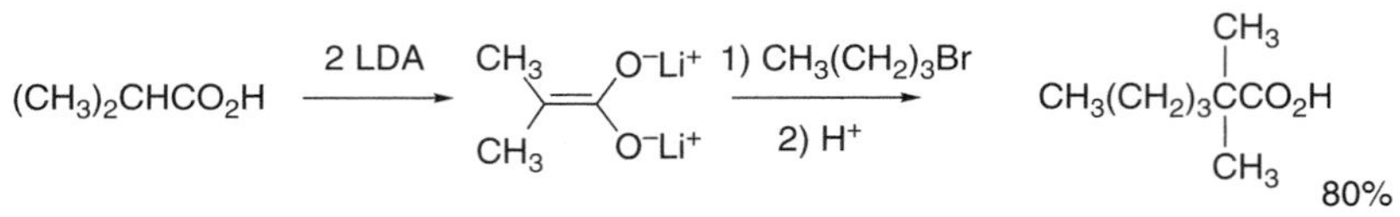

$$(CH_3)_2CHCO_2H \xrightarrow{2\ LDA} \underset{CH_3}{\overset{CH_3}{\mathrel{\;}}}C=C\overset{O^-Li^+}{\underset{O^-Li^+}{\mathrel{\;}}} \xrightarrow[2)\ H^+]{1)\ CH_3(CH_2)_3Br} CH_3(CH_2)_3\overset{CH_3}{\underset{CH_3}{\overset{|}{\underset{|}{C}}}}CO_2H \quad 80\%$$

Nitriles can also be converted to anions and alkylated. Acetonitrile ($pK_{DMSO} = 31.3$) can be deprotonated, provided a strong nonnucleophilic base such as LDA is used.

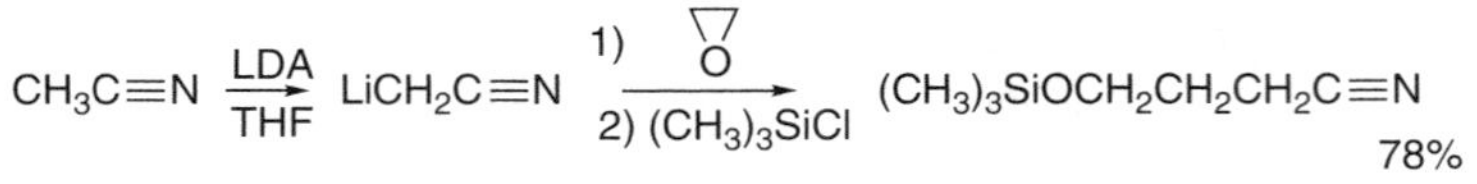

$$CH_3C\equiv N \xrightarrow[THF]{LDA} LiCH_2C\equiv N \xrightarrow[2)\ (CH_3)_3SiCl]{1)\ \triangle O} (CH_3)_3SiOCH_2CH_2CH_2C\equiv N \quad 78\%$$

Ref. 77

Phenylacetonitrile ($pK_{DMSO} = 21.9$) is considerably more acidic than acetonitrile. Dialkylation has been used in the synthesis of meperidine, an analgesic substance.[78]

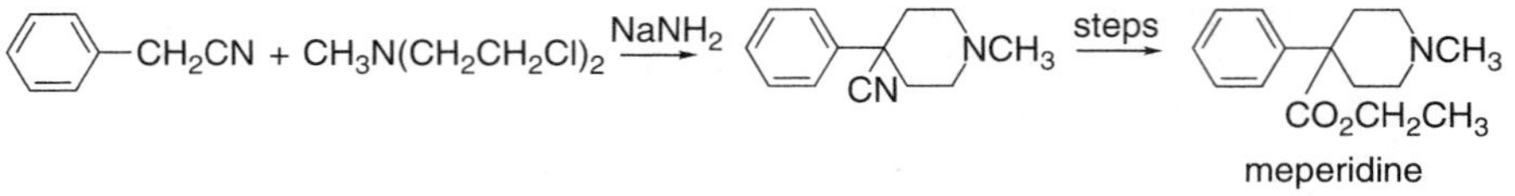

Scheme 1.6 gives some examples of alkylation of esters, amides, and nitriles. Entries 1 and 2 are representative ester alkylations involving low-temperature

We will see in Section 1.2.6 that the enolates of *imides* are very useful in synthesis. Particularly important are the enolates of chiral *N*-acyloxazolidinones.

Scheme 1.6. Alkylation of Esters, Amides, and Nitriles

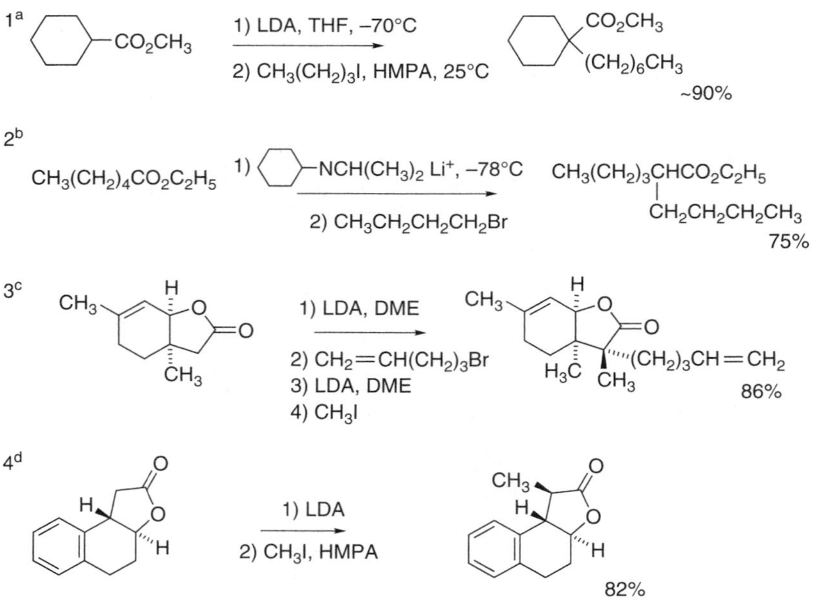

(Continued)

77. S. Murata and I. Matsuda, *Synthesis*, 221 (1978).
78. O. Eisleb, *Ber.*, **74**, 1433 (1941); cited in H. Kagi and K. Miescher, *Helv. Chim. Acta*, **32**, 2489 (1949).

Scheme 1.6. *(Continued)*

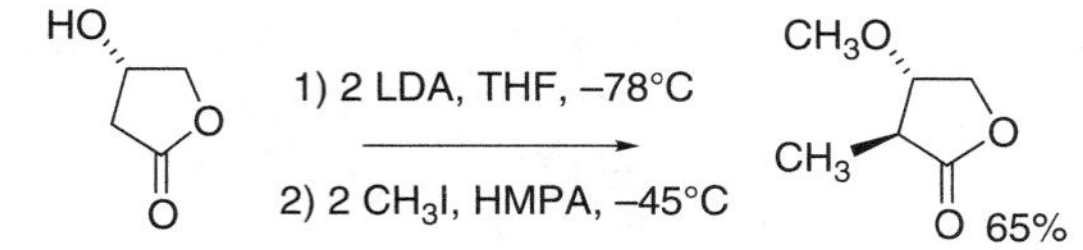

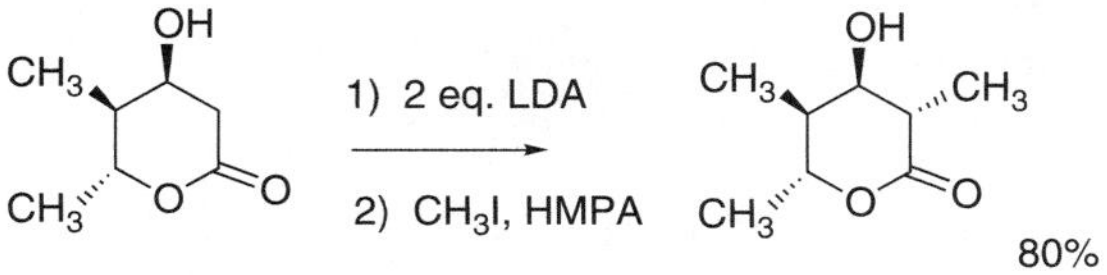

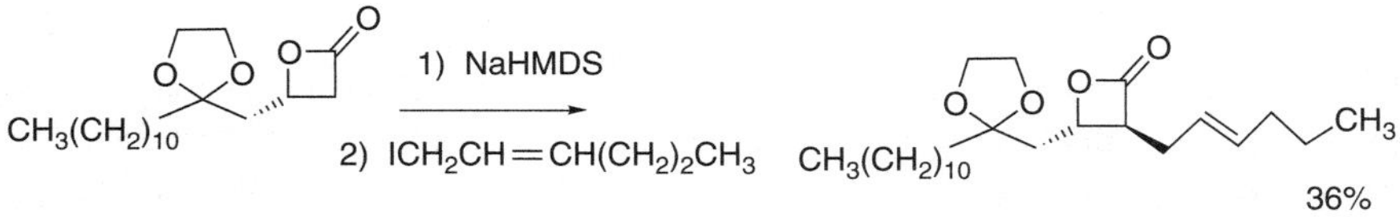

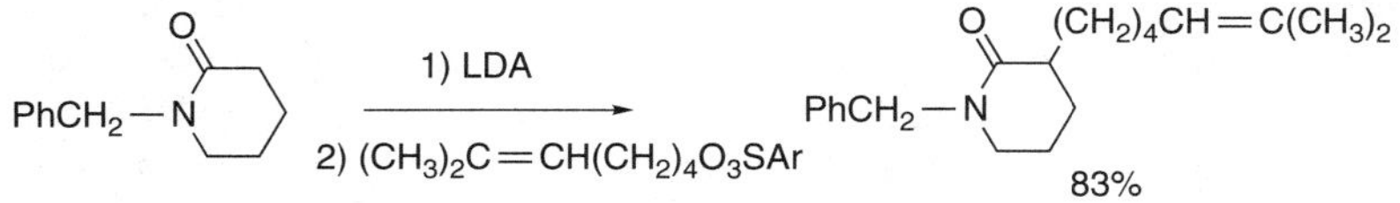

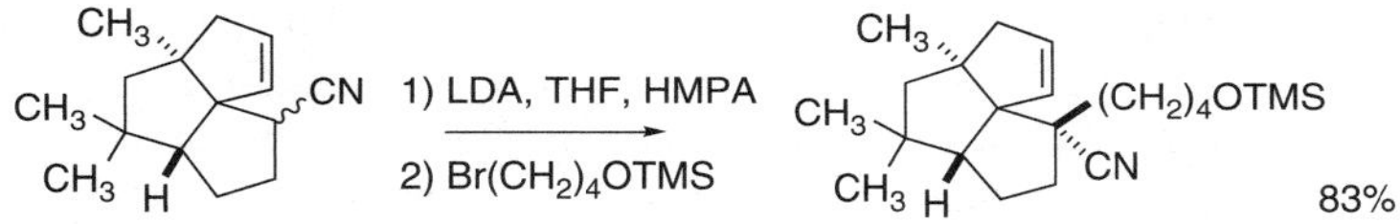

35

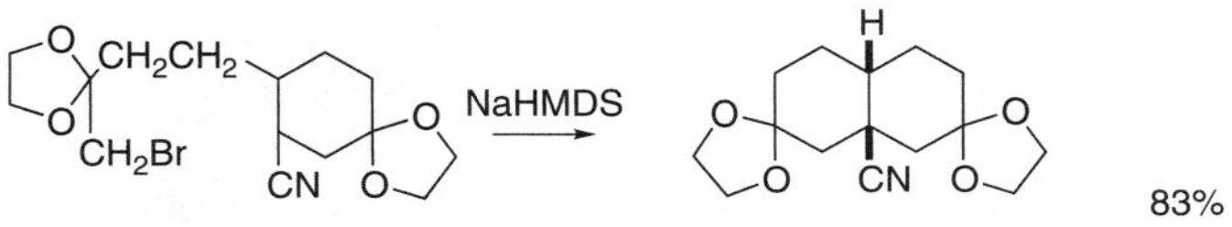

a. T. R. Williams and L. M. Sirvio, *J. Org. Chem.*, **45**, 5082 (1980).
b. M. W. Rathke and A. Lindert, *J. Am. Chem. Soc.*, **93**, 2320 (1971).
c. S. C. Welch, A. S. C. Prakasa Rao, G. G. Gibbs, and R. Y. Wong, *J. Org. Chem.*, **45**, 4077 (1980).
d. W. H. Pirkle and P. E. Adams, *J. Org. Chem.*, **45**, 4111 (1980).
e. H.-M. Shieh and G. D. Prestwich, *J. Org. Chem.*, **46**, 4319 (1981).
f. J. Tholander and E. M. Carriera, *Helv. Chim. Acta*, **84**, 613 (2001).
g. P. J. Parsons and J. K. Cowell, *Synlett*, 107 (2000).
h. D. Kim, H. S. Kim, and J. Y. Yoo, *Tetrahedron Lett.*, **32**, 1577 (1991).
i. L. A. Paquette, M. E. Okazaki, and J.-C. Caille, *J. Org. Chem.*, **53**, 477 (1988).
j. G. Stork, J. O. Gardner, R. K. Boeckman, Jr., and K. A. Parker, *J. Am. Chem. Soc.*, **95**, 2014 (1973).

deprotonation by hindered lithium amides. Entries 3 to 7 are lactone alkylations. Entry 3 involves two successive alkylation steps, with the second group being added from the more open face of the enolate. Entry 4 also illustrates stereoselectivity based on a steric effect. Entry 5 shows alkylation at both the enolate and a hydroxy group. Entry 6 is a step in the synthesis of the C(33)–C(37) fragment of the antibiotic amphotericin B. Note that in this case although the hydroxy group is deprotonated it is not methylated under the reaction conditions being used. Entry 7 is a challenging alkylation of a sensitive β-lactone. Although the corresponding saturated halide was not reactive enough, the allylic iodide gave a workable yield. Entry 8 is an alkylation of a lactam. Entries 9 and 10 are nitrile alkylations, the latter being intramolecular.

1.2.4. Generation and Alkylation of Dianions

In the presence of a very strong base, such as an alkyllithium, sodium or potassium hydride, sodium or potassium amide, or LDA, 1,3-dicarbonyl compounds can be converted to their *dianions* by two sequential deprotonations.[79] For example, reaction of benzoylacetone with sodium amide leads first to the enolate generated by deprotonation at the more acidic methylene group between the two carbonyl groups. A second equivalent of base deprotonates the benzyl methylene group to give a dienediolate.

$$PhCH_2CCH_2CCH_3 \xrightarrow{2\ NaNH_2} PhCH=C-CH=C-CH_3 \xrightarrow[Cl]{PhCHCH_3} PhCHCCH_2CCH_3$$

Ref. 80

Alkylation of dianions occurs at the *more basic carbon*. This technique permits alkylation of 1,3-dicarbonyl compounds to be carried out cleanly at the less acidic position. Since, as discussed earlier, alkylation of the monoanion occurs at the carbon between the two carbonyl groups, the site of monoalkylation can be controlled by choice of the amount and nature of the base. A few examples of the formation and alkylation of dianions are collected in Scheme 1.7. In each case, alkylation occurs at the less stabilized anionic carbon. In Entry 3, the α-formyl substituent, which is removed after the alkylation, serves to direct the alkylation to the methyl-substituted carbon. Entry 6 is a step in the synthesis of artemisinin, an antimalarial component of a Chinese herbal medicine. The sulfoxide serves as an anion-stabilizing group and the dianion is alkylated at the less acidic α-position. Note that this reaction is also stereoselective for the *trans* isomer. The phenylsulfinyl group is removed reductively by aluminum. (See Section 5.6.2 for a discussion of this reaction.)

1.2.5. Intramolecular Alkylation of Enolates

There are many examples of formation of three- through seven-membered rings by intramolecular enolate alkylation. The reactions depend on attainment of a TS having an approximately linear arrangement of the nucleophilic carbon, the electrophilic carbon, and the leaving group. Since the HOMO of the enolate (ψ_2) is involved, the approach must be approximately perpendicular to the enolate.[81] In intramolecular alkylation, these stereoelectronic restrictions on the direction of approach of the electrophile to the enolate become important. Baldwin has summarized the general principles that govern the energetics of intramolecular ring-closure reactions.[82] Analysis of the stereochemistry of intramolecular enolate alkylation requires consideration of both the direction of approach and enolate conformation. The intramolecular alkylation reaction of **7** gives exclusively **8**, having the *cis* ring juncture.[83] The alkylation probably occurs through a TS like **F**. The TS geometry permits the π electrons of the enolate to achieve an approximately colinear alignment with the sulfonate leaving group. The TS **G** for

[79] For reviews, see (a) T. M. Harris and C. M. Harris, *Org. React.*, **17**, 155 (1969); E. M. Kaiser, J. D. Petty, and P. L. A. Knutson, *Synthesis*, 509 (1977); C. M. Thompson and D. L. C. Green, *Tetrahedron*, **47**, 4223 (1991); C. M. Thompson, *Dianion Chemistry in Organic Synthesis*, CRC Press, Boca Raton, FL, 1994.

[80] D. M. von Schriltz, K. G. Hampton, and C. R. Hauser, *J. Org. Chem.*, **34**, 2509 (1969).

[81] J. E. Baldwin and L. I. Kruse, *J. Chem. Soc., Chem. Commun.*, 233 (1977).

[82] J. E. Baldwin, R. C. Thomas, L. I. Kruse, and L. Silberman, *J. Org. Chem.*, **42**, 3846 (1977).

[83] J. M. Conia and F. Rouessac, *Tetrahedron*, **16**, 45 (1961).

Scheme 1.7. Generation and Alkylation of Dianions

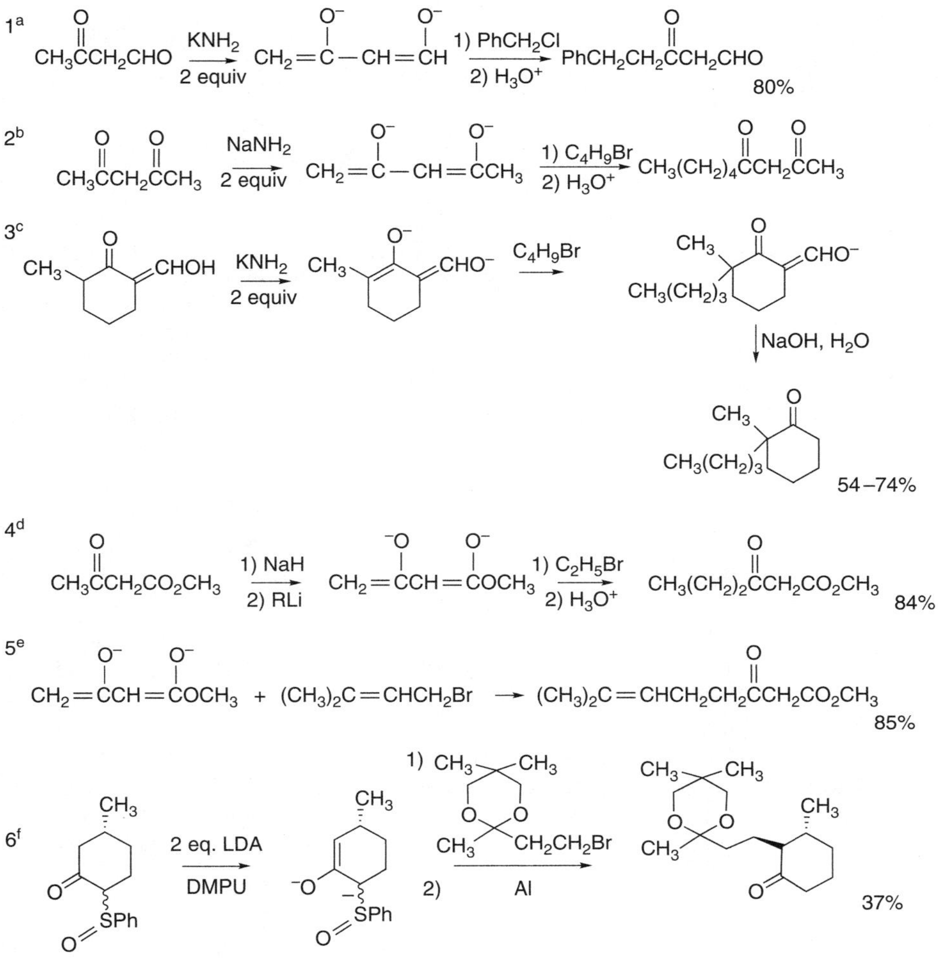
1[a]
CH3CCH2CHO
KNH2
2 equiv
CH2=C—CH=CH
1) PhCH2Cl
2) H3O+
PhCH2CH2CCH2CHO
80%
2[b]
CH3CCH2CCH3
NaNH2
2 equiv
CH2=C—CH=CCH3
1) C4H9Br
2) H3O+
CH3(CH2)4CCH2CCH3
3[c]
CHOH
KNH2
2 equiv
CHO−
C4H9Br
CH3(CH2)3
NaOH, H2O
54–74%
4[d]
CH3CCH2CO2CH3
1) NaH
2) RLi
CH2=CCH=COCH3
1) C2H5Br
2) H3O+
CH3(CH2)2CCH2CO2CH3
84%
5[e]
CH2=CCH=COCH3 + (CH3)2C=CHCH2Br
(CH3)2C=CHCH2CH2CCH2CO2CH3
85%
6[f]
SPh
2 eq. LDA
DMPU
CH2CH2Br
AI
37%

a. T. M. Harris, S. Boatman, and C. R. Hauser, *J. Am. Chem. Soc.*, **85**, 3273 (1963); S. Boatman, T. M. Harris, and C. R. Hauser, *J. Am. Chem. Soc.*, **87**, 82 (1965); K. G. Hampton, T. M. Harris, and C. R. Hauser, *J. Org. Chem.*, **28**, 1946 (1963).
b. K. G. Hampton, T. M. Harris, and C. R. Hauser, *Org. Synth.*, **47**, 92 (1967).
c. S. Boatman, T. M. Harris, and C. R. Hauser, *Org. Synth.*, **48**, 40 (1968).
d. S. N Huckin and L. Weiler, *J. Am. Chem. Soc.*, **96**,1082 (1974).
e. F. W. Sum and L. Weiler, *J. Am. Chem. Soc.*, **101**, 4401 (1979).
f. M. A. Avery, W. K. M. Chong, and C. Jennings-White, *J. Am. Chem. Soc.*, **114**, 974 (1992).

formation of the *trans* ring junction would be more strained because of the necessity to span the distance to the opposite face of the enolate π system.

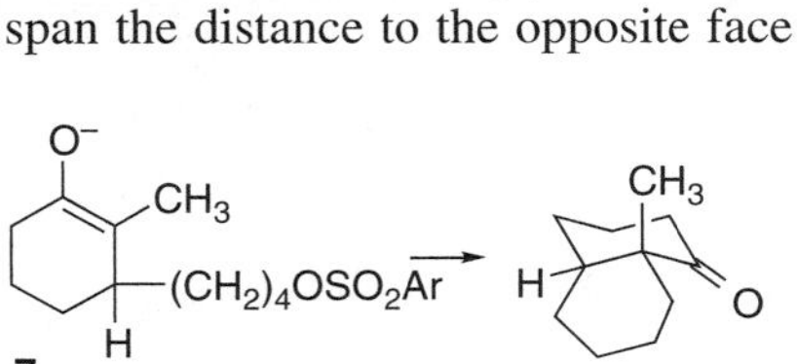
CH3
(CH2)4OSO2Ar
H
CH3
O

7 **8**

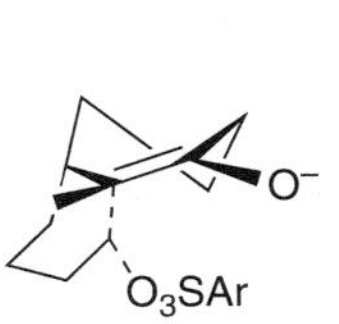
O−
O3SAr

F

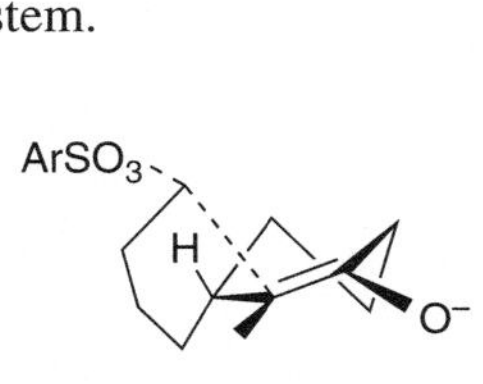
ArSO3
H
O−

G

Geometric factors in the TS are also responsible for differences in the case of cyclization of enolates **9** and **10**.[84]

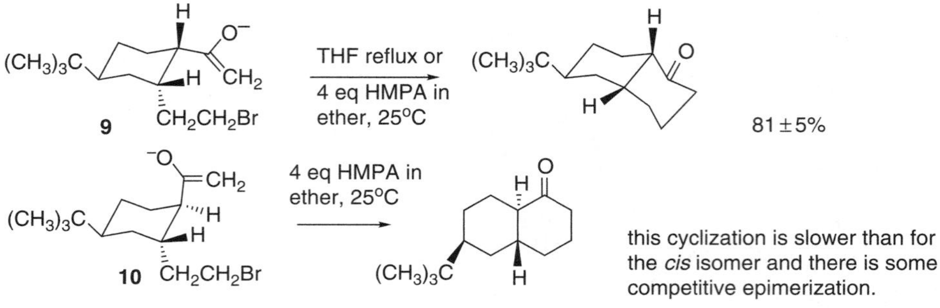

81±5%

this cyclization is slower than for the *cis* isomer and there is some competitive epimerization.

A number of examples of good stereoselectivity based on substituent control of reactant conformation have been identified. For example, **11** gives more than 96% stereoselectivity for the isomer in which the methyl and 2-propenyl groups are *cis*.[85]

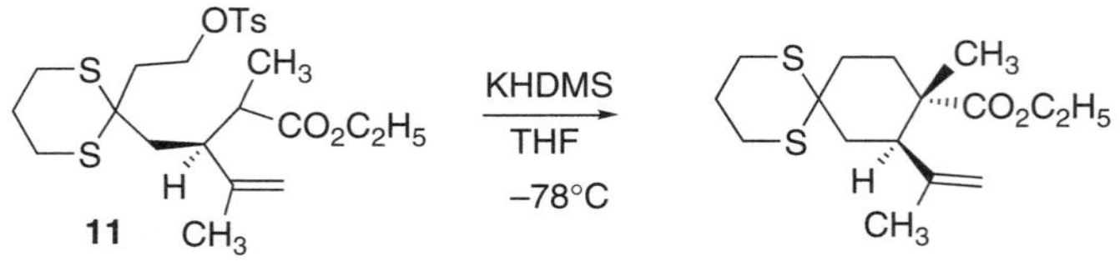

Similar *cis* stereoselectivity was observed in formation of four- and five-membered rings.[86] The origin of this stereoselectivity was probed systematically by a study in which a methyl substituent was placed at the C(3), C(4), C(5), and C(6) positions of ethyl 7-bromoheptanoate. Good (>93%) stereoselectivity was noted for all but the C(5) derivative.[87] These results are consistent with a chairlike TS with the enolate in an equatorial-like position. In each case the additional methyl group can occupy an equatorial position. The reduced selectivity of the 5-methyl isomer may be due to the fact that the methyl group is farther from the reaction site than in the other cases.

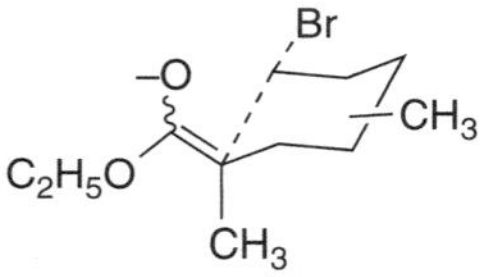

An intramolecular alkylation following this stereochemical pattern was used in the synthesis of (-)-fumagillol, with the alkadienyl substituent exerting the dominant conformational effect.[88]

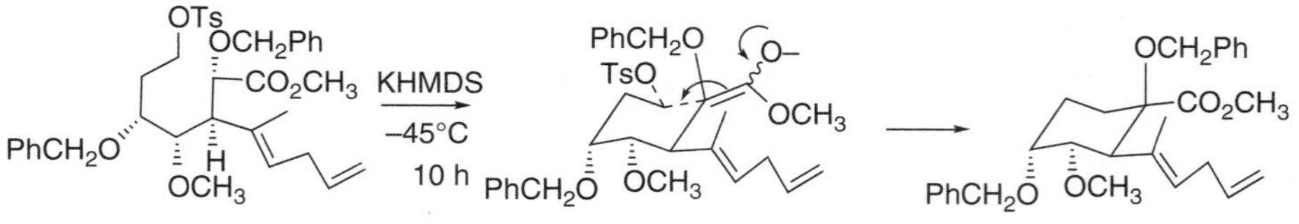

84. H. O. House and W. V. Phillips, *J. Org. Chem.*, **43**, 3851 (1978).
85. D. Kim and H. S. Kim, *J. Org. Chem.*, **52**, 4633 (1987).
86. D. Kim, Y. M. Jang, I. O. Kim, and S. W. Park, *J. Chem. Soc., Chem. Commun.*, 760 (1988).
87. T. Tokoroyama and H. Kusaka, *Can. J. Chem.*, **74**, 2487 (1996).
88. D. Kim, S. K. Ahn, H. Bae, W. J. Choi, and H. S. Kim, *Tetrahedron Lett.*, **38**, 4437 (1997).

Scheme 1.8 shows some intramolecular enolate alkylations. The reactions in Section A involve alkylation of ketone enolates. Entry 1 is a case of α-alkylation of a conjugated dienolate. In this case, the α-alkylation is also favored by ring strain effects because γ-alkylation would lead to a four-membered ring. The intramolecular alkylation in Entry 2 was used in the synthesis of the terpene seychellene.

Scheme 1.8. Intramolecular Enolate Alkylation

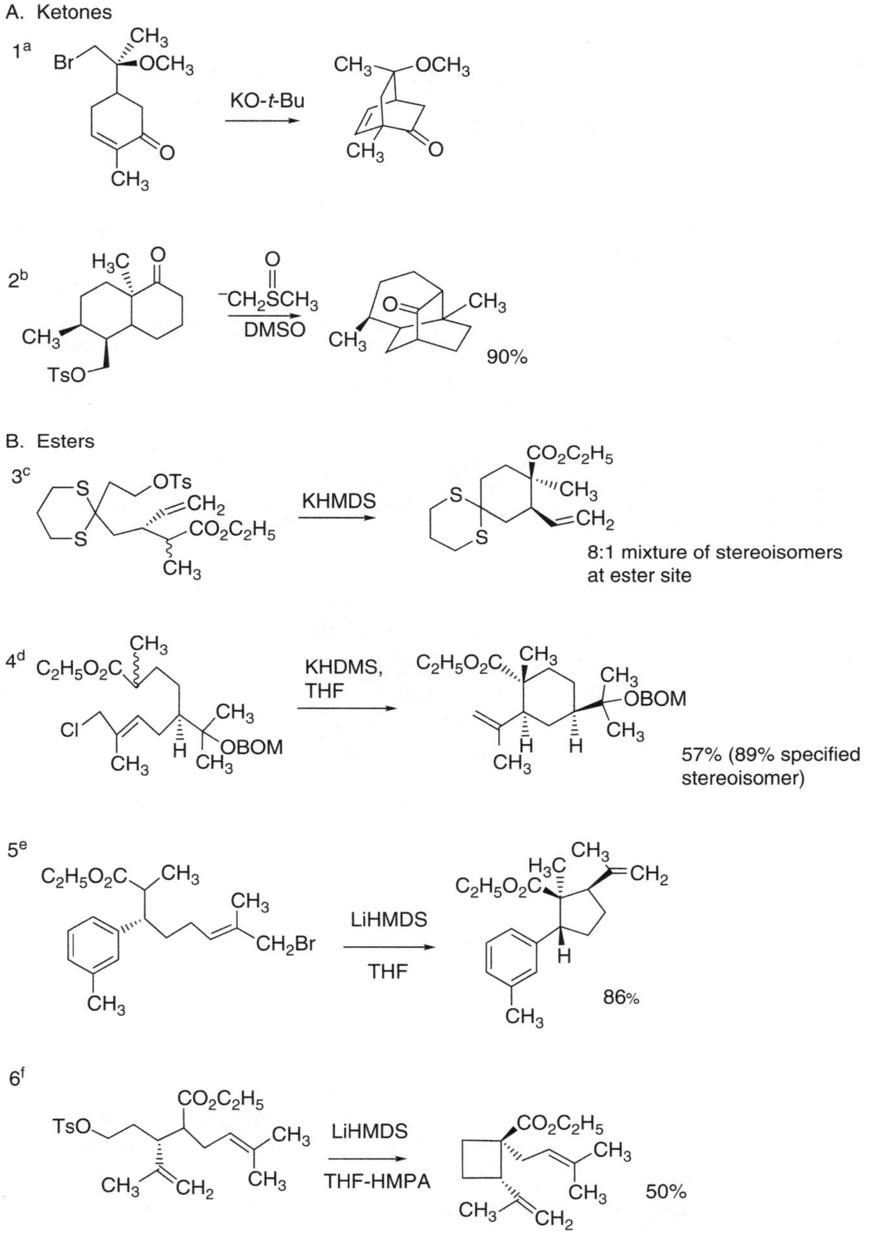

(Continued)

7[g]

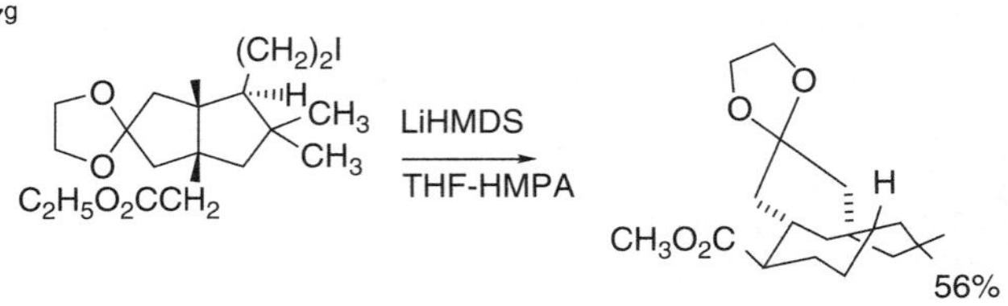

8[h]

70% on
2-kg scale

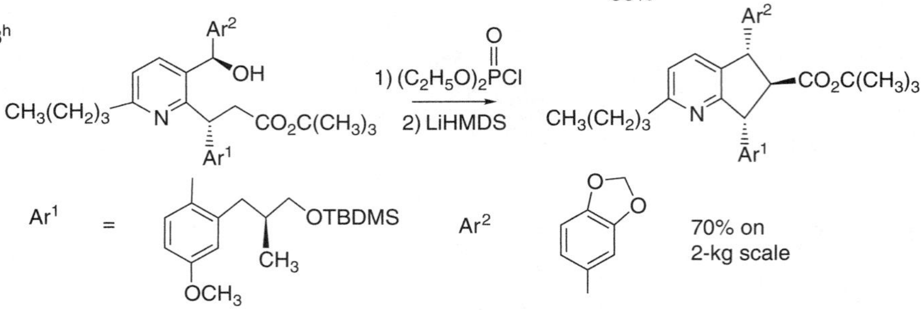

a. A. Srikrishna, G. V. R. Sharma, S. Danieldoss, and P. Hemamalini, *J. Chem. Soc., Perkin Trans. 1*, 1305 (1996).
b. E. Piers, W. de Waal, and R. W. Britton, *J. Am. Chem. Soc.*, **93**, 5113 (1971).
c. D. Kim, S. Kim, J.-J. Lee, and H. S. Kim, *Tetrahedron Lett.*, **31**, 4027 (1990).
d. D. Kim, J. I. Lim, K. J. Shin, and H. S. Kim, *Tetrahedron Lett.*, **34**, 6557 (1993).
e. J. Lee and J. Hong, *J. Org. Chem.*, **69**, 6433 (2004).
f. F.-D. Boyer and P.-H. Ducrot, *Eur. J. Org. Chem.*, 1201 (1999).
g. S. Danishefsky, K. Vaughan, R. C. Gadwood, and K. Tsuzuki, *J. Am. Chem. Soc.*, **102**, 4262 (1980).
h. Z. J. Song, M. Zhao, R. Desmond, P. Devine, D. M. Tschaen, R. Tillyer, L.Frey, R. Heid, F. Xu;, B. Foster, J. Li,
R. Reamer, R. Volante, E. J. Grabowski, U. H. Dolling, P. J. Reider, S. Okada, Y. Kato and E. Mano, *J. Org. Chem.*,
64, 9658 (1999).

Entries 3 to 6 are examples of ester enolate alkylations. These reactions show
stereoselectivity consistent with cyclic TSs in which the hydrogen is eclipsed with the
enolate and the larger substituent is pseudoequatorial. Entries 4 and 5 involve S_N2'
substitutions of allylic halides. The formation of the six- and five-membered rings,
respectively, is the result of ring size preferences with $5 > 7$ and $6 > 8$. In Entry 4,
reaction occurs through a chairlike TS with the tertiary C(5) substituent controlling
the conformation. The cyclic TS results in a *trans* relationship between the ester and
vinylic substituents.

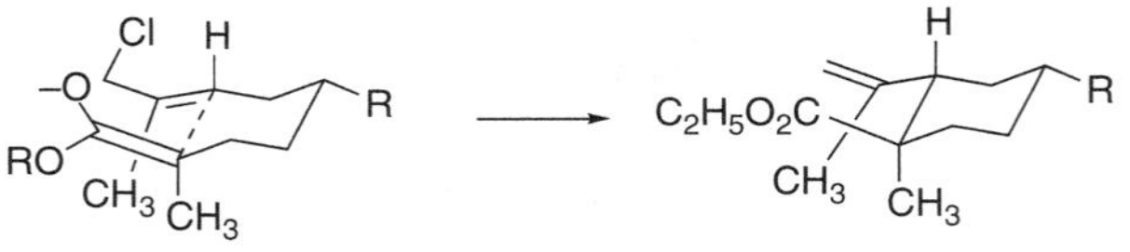

Entry 6 results in the formation of a four-membered ring and shows good stereoselectivity. Entry 7 is a step in the synthesis of a tetracyclic lactone, quadrone, that
is isolated from a microorganism. Entry 8 is a step in a multikilo synthesis of an
endothelin receptor antagonist called cyclopentapyridine I. The phosphate group was
chosen as a leaving group because sulfonates were too reactive at the diaryl carbinol
site. The reaction was shown to go with inversion of configuration.

1.2.6. Control of Enantioselectivity in Alkylation Reactions

The alkylation of an enolate creates a new stereogenic center when the α-substituents are nonidentical. In enantioselective synthesis, it is necessary to control the direction of approach and thus the configuration of the new stereocenter.

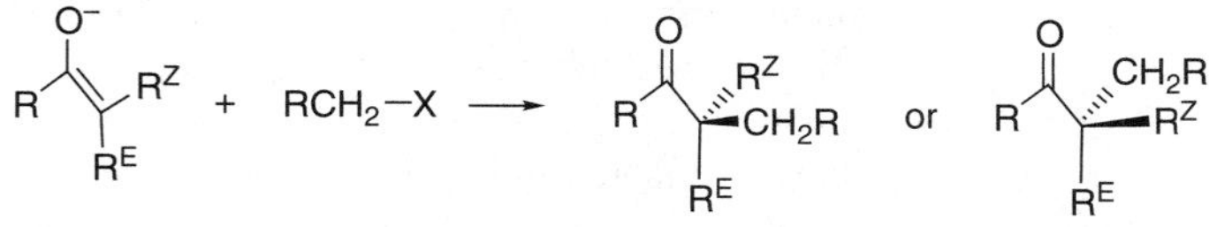

Enantioselective enolate alkylation can be done using chiral auxiliaries. (See Section 2.6 of Part A to review the role of chiral auxiliaries in control of reaction stereochemistry.) The most frequently used are the *N*-acyloxazolidinones.[89] The 4-isopropyl and 4-benzyl derivatives, which can be obtained from valine and phenylalanine, respectively, and the *cis*-4-methyl-5-phenyl derivatives are readily available. Another useful auxiliary is the 4-phenyl derivative.[90]

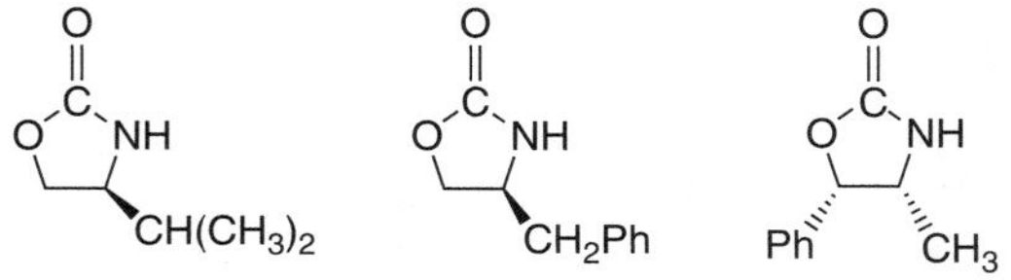

Several other oxazolidinones have been developed for use as chiral auxiliaries. The 4-isopropyl-5,5-dimethyl derivative gives excellent enantioselectivity.[91] 5,5-Diaryl derivatives are also quite promising.[92]

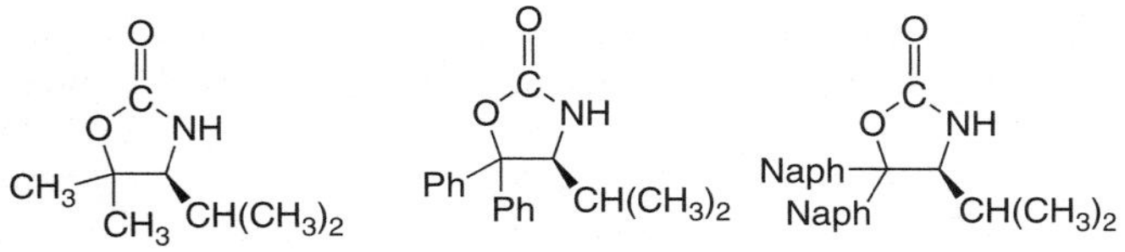

The reactants are usually *N*-acyl derivatives. The lithium enolates form chelate structures with *Z*-stereochemistry at the double bond. The ring substituents then govern the preferred direction of approach.

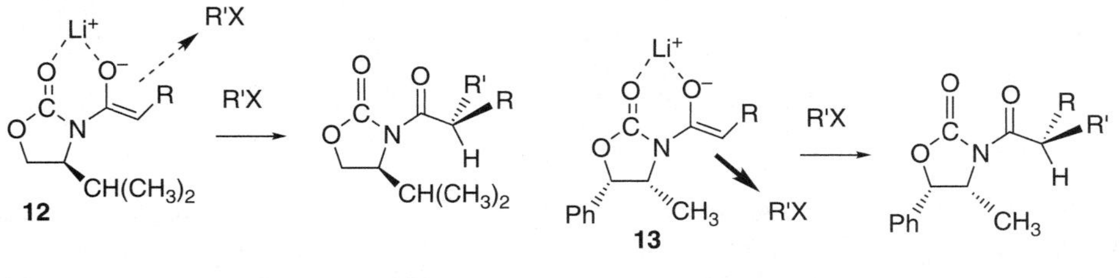

[89]. D. A. Evans, M. D. Ennis, and D. J. Mathre, *J. Am. Chem. Soc.*, **104**, 1737 (1982); D. J. Ager, I. Prakash, and D. R. Schaad, *Chem. Rev.*, **96**, 835 (1996); D. J. Ager, I. Prakash, and D. R. Schaad, *Aldrichimica Acta*, **30**, 3 (1997).

[90]. E. Nicolas, K. C. Russell, and V. J. Hruby, *J. Org. Chem.*, **58**, 766 (1993).

[91]. S. D. Bull, S. G. Davies, S. Jones, and H. J. Sanganee, *J. Chem. Soc., Perkin Trans. 1*, 387 (1999); S. G. Davies and H. J. Sangaee, *Tetrahedron: Asymmetry*, **6**, 671 (1995); S. D. Bull, S. G. Davies, R. L. Nicholson, H. J. Sanganee, and A. D. Smith, *Org. Biomed. Chem.*, **1**, 2886 (2003).

[92]. T. Hintermann and D. Seebach, *Helv. Chim. Acta*, **81**, 2093 (1998); C. L. Gibson, K. Gillon, and S. Cook, *Tetrahedron Lett.*, **39**, 6733 (1998).

In **12** the upper face is shielded by the isopropyl group, whereas in **13** the lower face is shielded by the methyl and phenyl groups. As a result, alkylation of the two derivatives gives products of the opposite configuration. The initial alkylation product ratios are typically 95:5 in favor of the major isomer. Since these products are diastereomeric mixtures, they can be separated and purified. Subsequent hydrolysis or alcoholysis provides acids or esters in enantiomerically enriched form. Alternatively, the acyl imides can be reduced to alcohols or aldehydes. The final products can often be obtained in greater than 99% enantiomeric purity.

A number of other types of chiral auxiliaries have been employed in enolate alkylation. Excellent results are obtained using amides of pseudoephedrine. Alkylation occurs *anti* to the α-oxybenzyl group.[93] The reactions involve the *Z*-enolate and there is likely bridging between the two lithium cations, perhaps by di-(isopropyl)amine.[94]

Both enantiomers of the auxiliary are available, so either enantiomeric product can be obtained. This methodology has been applied to a number of enantioselective syntheses.[95] For example, the glycine derivative **14** can be used to prepare α-amino acid analogs.[96]

Enolates of phenylglycinol amides also exhibit good diastereoselectivity.[97] A chelating interaction with the deprotonated hydroxy group is probably involved here as well.

The *trans*-2-naphthyl cyclohexyl sulfone **15** can be prepared readily in either enantiomeric form. The corresponding ester enolates can be alkylated in good yield and diastereoselectivity.[98] In this case, the steric shielding is provided by the naphthyl

93. A. G. Myers, B. H. Yang, H. Chen, L. McKinstry, D. J. Kopecky, and J. L. Gleason, *J. Am. Chem. Soc.*, **119**, 6496 (1997); A. G. Myers, M. Siu, and F. Ren, *J. Am. Chem. Soc.*, **124**, 4230 (2002).
94. J. L. Vicario, D. Badia, E. Dominguez, and L. Carrillo, *J. Org. Chem.*, **64**, 4610 (1999).
95. S. Karlsson and E. Hedenstrom, *Acta Chem. Scand.*, **53**, 620 (1999).
96. A. G. Myers, P. S. Schnider, S. Kwon, and D. W. Kung, *J. Org. Chem.*, **64**, 3322 (1999).
97. V. Jullian, J.-C. Quirion, and H.-P. Husson, *Synthesis*, 1091 (1997).
98. G. Sarakinos and E. J. Corey, *Org. Lett.*, **1**, 1741 (1999).

group and there is probably also a $\pi - \pi$ interaction between the naphthalene ring and the enolate.

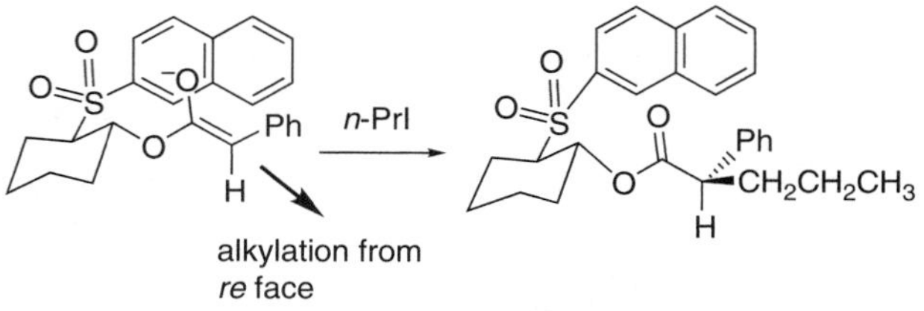

alkylation from
re face

As with the acyl oxazolidinone auxiliaries, each of these systems permits hydrolytic removal and recovery of the chiral auxiliary.

Scheme 1.9 gives some examples of diastereoselective enolate alkylations. Entries 1 to 6 show the use of various *N*-acyloxazolidinones and demonstrate the

Scheme 1.9. Diastereoselective Enolate Alkylation Using Chiral Auxiliaries

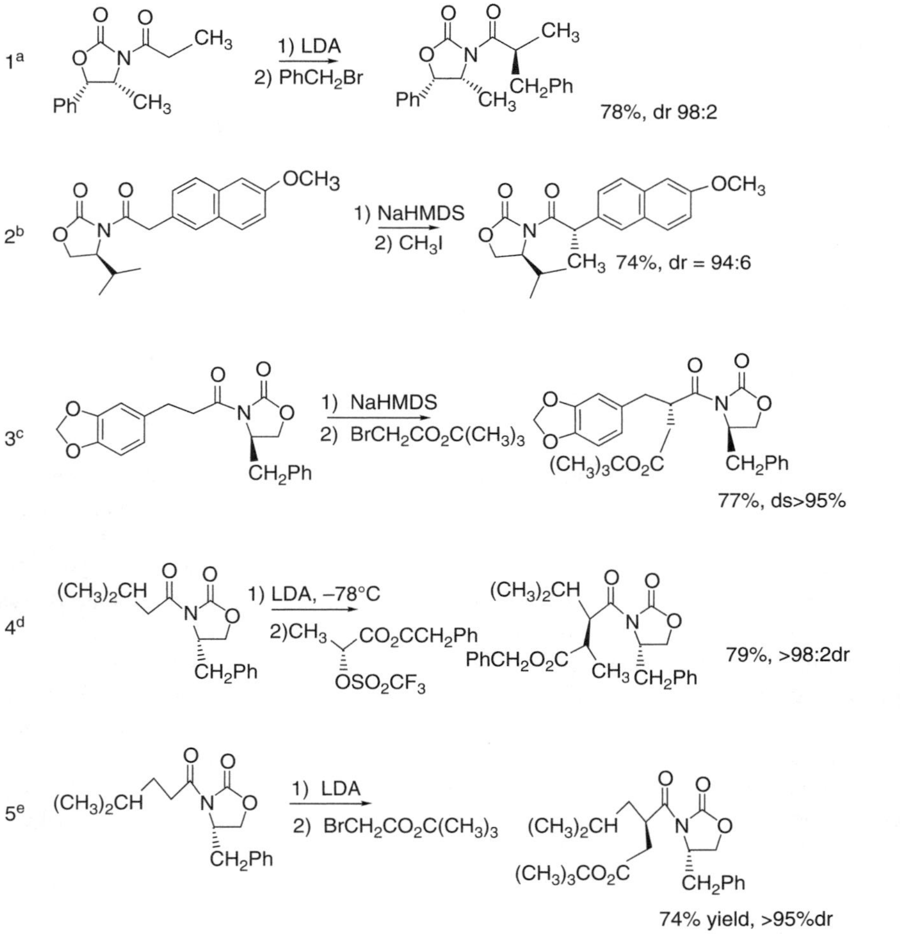

(Continued)

Scheme 1.9. *(Continued)*

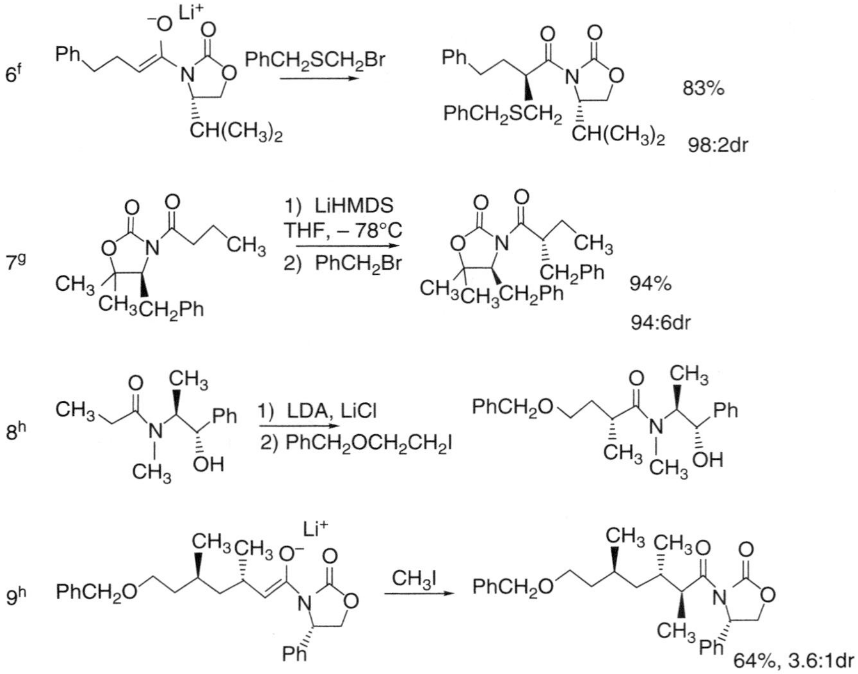

a. D. A. Evans, M. D. Ennis, and D. J. Mathre, *J. Am. Chem. Soc.*, **104**, 1737 (1982).
b. A. Fadel, *Synlett*, 48 (1992).
c. J. L. Charlton and G-L. Chee, *Can. J. Chem.*, **75**, 1076 (1997).
d. C. P. Decicco, D. J. Nelson, B. L. Corbett, and J. C. Dreabit, *J. Org. Chem.*, **60**, 4782 (1995).
e. R. P. Beckett, M. J. Crimmin, M. H. Davis, and Z. Spavold, *Synlett*, 137 (1993).
f. D. A. Evans, D. J. Mathre, and W. L. Scott, *J. Org. Chem.*, **50**, 1830 (1985).
g. S. D. Bull, S. G. Davies, R. L. Nicholson, H. J. Sanganee, and A. D. Smith, *Organic and Biomolec. Chem.*, **1**, 2886 (2003).
h. J. D. White, C.-S. Lee and Q. Xu, *Chem. Commun.* 2012 (2003).

stereochemical control by the auxiliary ring substituent. Entry 2 demonstrated the feasibility of enantioselective synthesis of α-aryl acetic acids such as the structure found in naproxen. Entries 3 to 6 include ester groups in the alkylating agent. In the case of Entry 4, it was shown that inversion occurs in the alkylating reagent. Entry 7 is an example of the use of one of the more highly substituted oxazolidinone derivatives. Entries 8 and 9 are from the synthesis of a neurotoxin isolated from a saltwater bacterium. The pseudoephedrine auxiliary shown in Entry 8 was used early in the synthesis and the 4-phenyloxazolidinone auxiliary was used later, as shown in Entry 9.

The facial selectivity of a number of more specialized enolates has also been explored, sometimes with surprising results. Schultz and co-workers compared the cyclic enolate **H** with **I**.[99] Enolate **H** presents a fairly straightforward picture. Groups such as methyl, allyl, and benzyl all give selective β-alkylation, and this is attributed to steric factors. Enolate **I** can give either α- or β-alkylation, depending on the conditions. The presence of NH_3 or use of LDA favors α-alkylation, whereas the use

[99] A. G. Schultz, M. Macielag, P. Sudararaman, A. G. Taveras, and M. Welch, *J. Am. Chem. Soc.*, **110**, 7828 (1988).

of *n*-butyllithium as the base favors β-alkylation. Other changes in conditions also affect the stereoselectivity. This is believed to be due to alternative aggregated forms of the enolate.

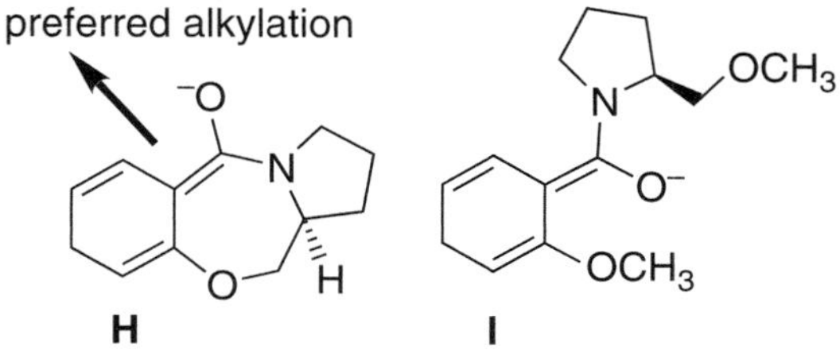

The compact bicyclic lactams **15** and **16** are examples of chiral systems that show high facial selectivity. Interestingly, **15** is alkylated from the convex face. When two successive alkylations are done, both groups are added from the *endo* face, so the configuration of the newly formed quaternary center can be controlled. The closely related **16** shows *exo* stereoselectivity. [100]

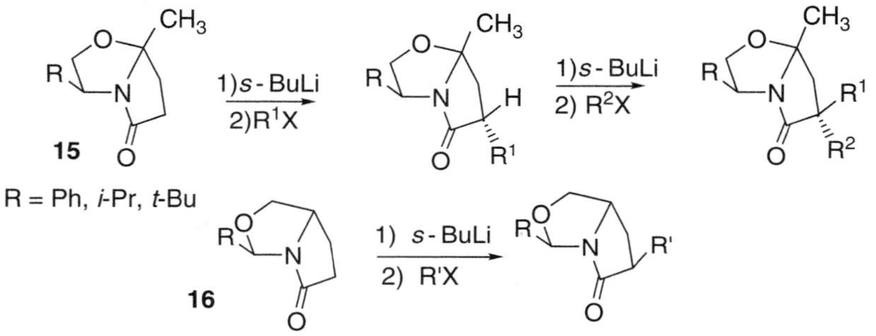

Crystal structure determination and computational studies indicate substantial pyramidalization of both enolates with the higher HOMO density being on the *endo* face for both **15** and **16**. However, the TS energy [MP3/6-31G+(*d*)] correlates with experiment, favoring the *endo* TS for **15** (by 1.3 kcal/mol) and *exo* for **16** (by 0.9 kcal/mol). A B3LYP/6-31G(*d*) computational study has also addressed the stereoselectivity of **16**.[101] As with the ab intitio calculation, the Li$^+$ is found in the *endo* position with an association with the heterocyclic oxygen. The *exo* TS is favored but the energy difference is very sensitive to the solvent model. The differences between the two systems seems to be due to the *endo* C(4) hydrogen that is present in **16** but not in **15**.

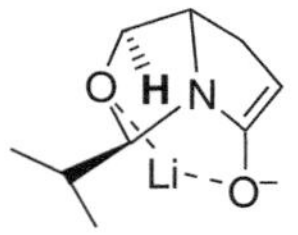

[100] A. I. Meyers, M. A. Seefeld, B. A. Lefker, J. F. Blake, and P. G. Williard, *J. Am. Chem. Soc.*, **120**, 7429 (1998).

[101] Y. Ikuta and S. Tomoda, *Org. Lett.*, **6**, 189 (2004).

1.3. The Nitrogen Analogs of Enols and Enolates: Enamines and Imine Anions

The nitrogen analogs of ketones and aldehydes are called imines, azomethines, or Schiff bases, but *imine* is the preferred name and we use it here. These compounds can be prepared by condensation of primary amines with ketones or aldehydes.[102] The equilibrium constants are unfavorable, so the reaction is usually driven forward by removal of water.

$$
\underset{\substack{\| \\ R-C-R}}{O} + H_2NR' \longrightarrow \underset{\substack{\| \\ R-C-R}}{N-R'} + H_2O
$$

When secondary amines are heated with ketones or aldehydes in the presence of an acidic catalyst, a related reaction occurs, and the product is a substituted vinylamine or *enamine*.

$$
\underset{R'}{\overset{O}{\|}}CH_2R + HNR''_2 \longrightarrow \underset{R'}{\overset{NR'_2}{\|}}CHR + H_2O
$$

There are other methods for preparing enamines from ketones that utilize strong chemical dehydrating reagents. For example, mixing carbonyl compounds and secondary amines followed by addition of titanium tetrachloride rapidly gives enamines. This method is especially applicable to hindered amines.[103] Triethoxysilane can also be used.[104] Another procedure involves converting the secondary amine to its N-trimethylsilyl derivative. Owing to the higher affinity of silicon for oxygen than nitrogen, enamine formation is favored and takes place under mild conditions.[105]

$$
(CH_3)_2CHCH_2CH{=}O + (CH_3)_3SiN(CH_3)_2 \longrightarrow (CH_3)_2CHCH{=}CHN(CH_3)_2
$$
$$
88\%
$$

The β-carbon atom of an enamine is a nucleophilic site because of conjugation with the nitrogen atom. Protonation of enamines takes place at the β-carbon, giving an iminium ion.

$$
\underset{\substack{| \\ R}}{R'_2N-C{=}CR_2} \longleftrightarrow \underset{\substack{| \\ R}}{R'_2\overset{+}{N}{=}C-\overset{-}{C}R_2} \overset{H^+}{\longrightarrow} \underset{\substack{| \\ R}}{R'_2\overset{+}{N}{=}C-CHR_2}
$$

---

102. For general reviews of imines and enamines, see P. Y. Sollenberger and R. B. Martin, in *Chemistry of the Amino Group*, S. Patai, ed., Interscience, New York, 1968, Chap. 7; G. Pitacco and E. Valentin, in *Chemistry of Amino, Nitroso and Nitro Groups and Their Derivatives*, Part 1, S. Patai, ed., Interscience, New York, 1982, Chap. 15; P. W. Hickmott, *Tetrahedron*, **38**, 3363 (1982); A. G. Cook, ed., *Enamines, Synthesis, Structure and Reactions*, Marcel Dekker, New York, 1988.
103. W. A. White and H. Weingarten, *J. Org. Chem.*, **32**, 213 (1967); R. Carlson, R. Phan-Tan-Luu, D. Mathieu, F. S. Ahounde, A. Babadjamian, and J. Metzger, *Acta Chem. Scand.*, **B32**, 335 (1978); R. Carlson, A. Nilsson, and M. Stromqvist, *Acta Chem. Scand.*, **B37**, 7 (1983); R. Carlson and A. Nilsson, *Acta Chem. Scand.*, **B38**, 49 (1984); S. Schubert, P. Renaud, P.-A. Carrupt, and K. Schenk, *Helv. Chim. Acta*, **76**, 2473 (1993).
104. B. E. Love and J. Ren, *J. Org. Chem.*, **58**, 5556 (1993).
105. R. Comi, R. W. Franck, M. Reitano, and S. M. Weinreb, *Tetrahedron Lett.*, 3107 (1973).

The nucleophilicity of the β-carbon atoms permits enamines to be used synthetically for alkylation reactions.

The enamines derived from cyclohexanones are of particular interest. The pyrrolidine enamine is most frequently used for synthetic applications. The enamine mixture formed from pyrrolidine and 2-methylcyclohexanone is predominantly isomer **17**.[106] A steric effect is responsible for this preference. Conjugation between the nitrogen atom and the π orbitals of the double bond favors coplanarity of the bonds that are darkened in the structures. In isomer **17** the methyl group adopts a quasi-axial conformation to avoid steric interaction with the amine substituents.[107] A serious nonbonded repulsion ($A^{1,3}$ strain) in **18** destabilizes this isomer.

Owing to the predominance of the less-substituted enamine, alkylations occur primarily at the less-substituted α-carbon. Synthetic advantage can be taken of this selectivity to prepare 2,6-disubstituted cyclohexanones. The iminium ions resulting from C-alkylation are hydrolyzed in the workup procedure.

Ref. 108

Alkylation of enamines requires relatively reactive alkylating agents for good results. Methyl iodide, allyl and benzyl halides, α-halo esters, α-halo ethers, and α-halo ketones are the most successful alkylating agents. The use of enamines for selective alkylation has largely been supplanted by the methods for kinetic enolate formation described in Section 1.2.

Some enamine alkylation reactions are shown in Scheme 1.10. Entries 1 and 2 are typical alkylations using reactive halides. In Entries 3 and 4, the halides are secondary with α-carbonyl substituents. Entry 5 involves an unactivated primary bromide and the yield is modest. The reaction in Entry 6 involves introduction of two groups. This

106. W. D. Gurowitz and M. A. Joseph, *J. Org. Chem.*, **32**, 3289 (1967).
107. F. Johnson, L. G. Duquette, A. Whitehead, and L. C. Dorman, *Tetrahedron*, **30**, 3241 (1974); K. Muller, F. Previdoli, and H. Desilvestro, *Helv. Chim. Acta*, **64**, 2497 (1981); J. E. Anderson, D. Casarini, and L. Lunazzi, *Tetrahedron Lett.*, **25**, 3141 (1988).
108. P. L. Stotter and K. A. Hill, *J. Am. Chem. Soc.*, **96**, 6524 (1974).

Scheme 1.10. Alkylation of Enamines

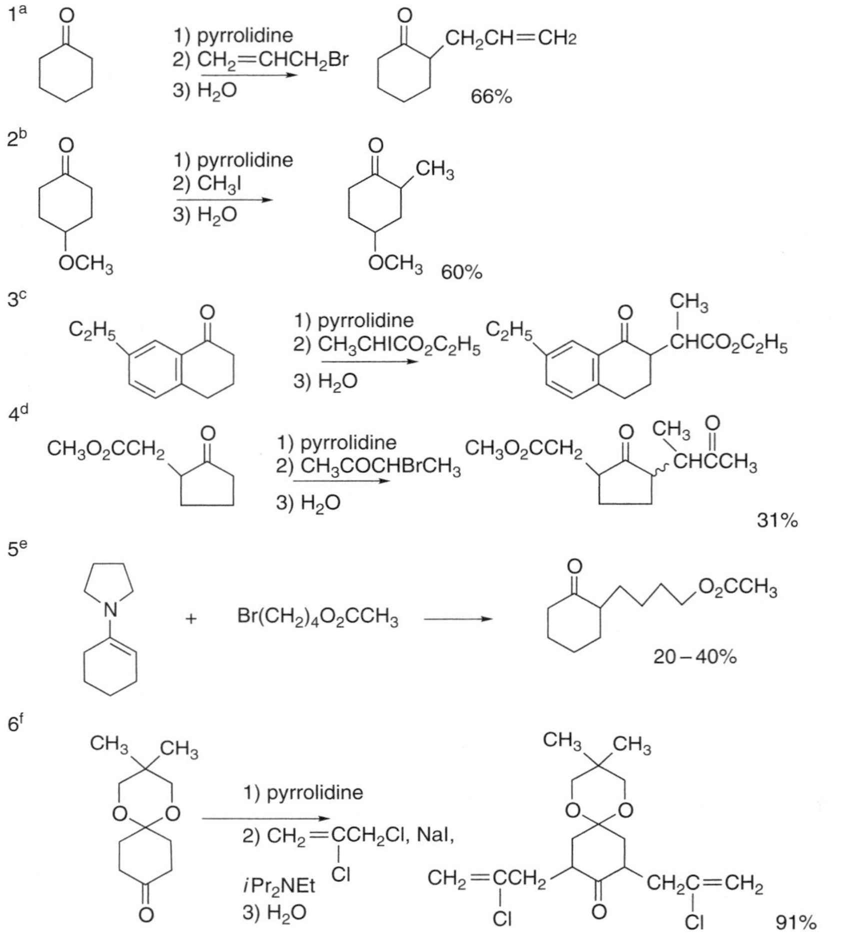

a. G. Stork, A. Brizzolara, H. Landesman, J. Szmuszkovicz, and R. Terrell, *J. Am. Chem. Soc.*, **85**, 207 (1963).
b. G. Stork and S. D. Darling, *J. Am. Chem. Soc.*, **86**, 1761 (1964).
c. D. M. Locke and S. W. Pelletier, *J. Am. Chem. Soc.*, **80**, 2588 (1958).
d. K. Sisido, S. Kurozumi, and K. Utimoto, *J. Org. Chem.*, **34**, 2661 (1969).
e. I. J. Borowitz, G. J. Williams, L. Gross, and R. Rapp, *J. Org. Chem.*, **33**, 2013 (1968).
f. J. A. Marshall and D. A. Flynn, *J. Org. Chem.*, **44**, 1391 (1979).

was done by carrying out the reaction in the presence of an amine, which deprotonates
the iminium ion and permits the second alkylation to occur.

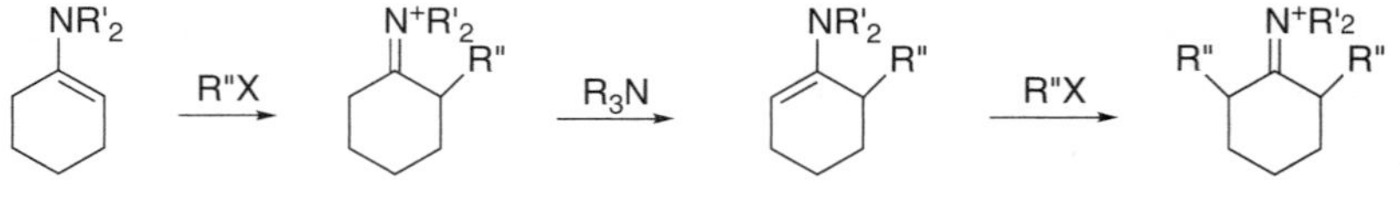

Imines can be deprotonated at the α-carbon by strong bases to give the nitrogen
analogs of enolates. Originally, Grignard reagents were used for deprotonation but
lithium amides are now usually employed. These anions, referred to as *imine anions*

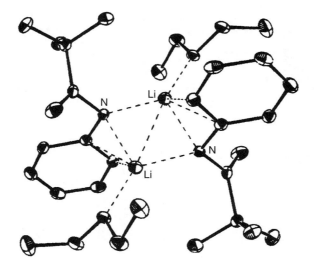

Fig. 1.6. Crystal structure of dimer of lithium salt of *N*-phenylimine of methyl *t*-butyl ketone. Two molecules of diethyl ether are present. Reproduced from *J. Am. Chem. Soc.*, **108**, 2462 (1986), by permission of the American Chemical Society.

or *metalloenamines*,[109] are isoelectronic and structurally analogous to both enolates and allyl anions; they can also be called *azaallyl anions*.

$$\underset{RC-CHR''_2}{\overset{\overset{NR'}{\|}}{}} \xrightarrow{\text{base}} \underset{RC-C^-R''_2}{\overset{\overset{NR'}{\|}}{}} \longleftrightarrow \underset{RC=CR''_2}{\overset{^-NR'}{|}}$$

Spectroscopic investigations of the lithium derivatives of cyclohexanone *N*-phenylimine indicate that it exists as a dimer in toluene and that as a better donor solvent, THF, is added, equilibrium with a monomeric structure is established. The monomer is favored at high THF concentrations.[110] A crystal structure determination was done on the lithiated *N*-phenylimine of methyl *t*-butyl ketone, and it was found to be a dimeric structure with the lithium cation positioned above the nitrogen and closer to the phenyl ring than to the β-carbon of the imine anion.[111] The structure, which indicates substantial ionic character, is shown in Figure 1.6.

Just as enamines are more nucleophilic than enol ethers, imine anions are more nucleophilic than enolates and react efficiently with alkyl halides. One application of imine anions is for the alkylation of aldehydes.

109. For a general review of imine anions, see J. K. Whitesell and M. A. Whitesell, *Synthesis*, 517 (1983).
110. N. Kallman and D. B. Collum, *J. Am. Chem. Soc.*, **109**, 7466 (1987).
111. H. Dietrich, W. Mahdi, and R. Knorr, *J. Am. Chem. Soc.*, **108**, 2462 (1986); P. Knorr, H. Dietrich, and W. Mahdi, *Chem. Ber.*, **124**, 2057 (1991).

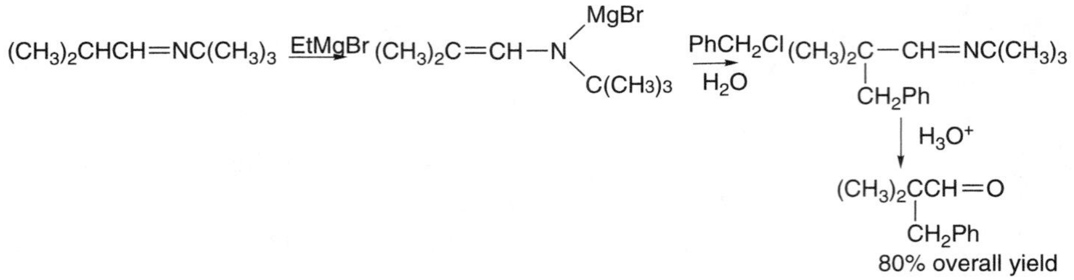

$(CH_3)_2CHCH=NC(CH_3)_3 \xrightarrow{\text{EtMgBr}} (CH_3)_2C=CH-N \overset{MgBr}{\underset{C(CH_3)_3}{|}} \xrightarrow[H_2O]{PhCH_2Cl} (CH_3)_2C-CH=NC(CH_3)_3$
$\underset{CH_2Ph}{|}$

$\downarrow H_3O^+$

$(CH_3)_2CCH=O$
$\underset{CH_2Ph}{|}$

80% overall yield

Ref. 112

$CH_3CH=CH-CH=N-\bigcirc \xrightarrow[\substack{2) \\ ICH_2CH_2 \\ 3) H_2O}]{1) LDA}$

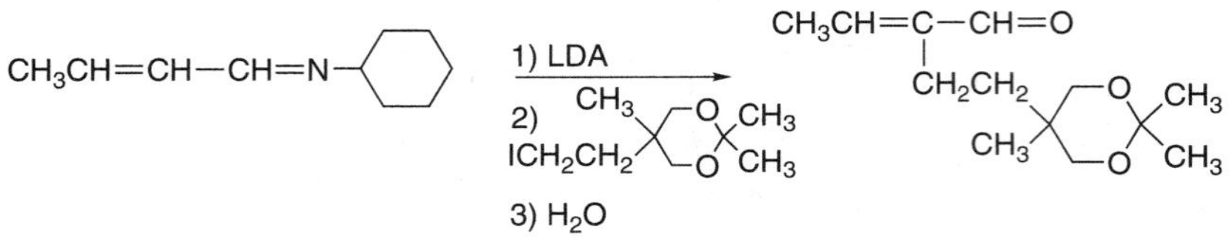

Ref. 113

Ketone imine anions can also be alkylated. The prediction of the regioselectivity of lithioenamine formation is somewhat more complex than for the case of kinetic ketone enolate formation. One of the complicating factors is that there are two imine stereoisomers, each of which can give rise to two regioisomeric imine anions. The isomers in which the nitrogen substituent R' is *syn* to the double bond are the more stable.[114]

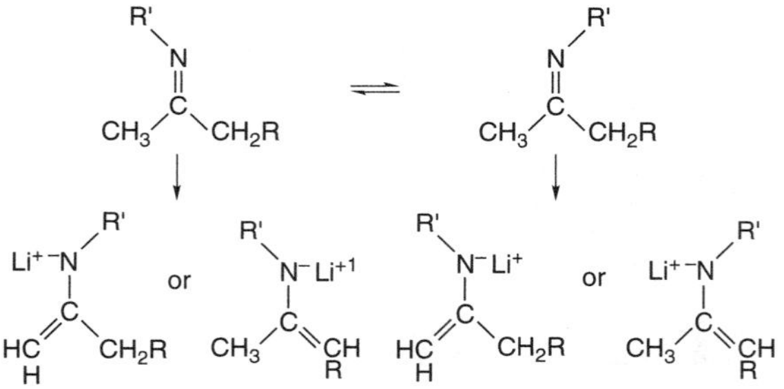

For methyl ketimines good regiochemical control in favor of methyl deprotonation, regardless of imine stereochemistry, is observed using LDA at $-78°$C. With larger *N*-substituents, deprotonation at $25°$C occurs *anti* to the nitrogen substituent.[115]

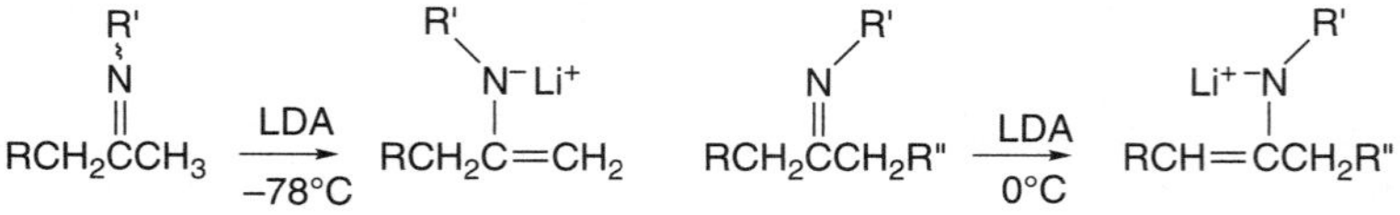

112. G. Stork and S. R. Dowd, *J. Am. Chem. Soc.*, **85**, 2178 (1963).

113. T. Kametani, Y. Suzuki, H. Furuyama, and T. Honda, *J. Org. Chem.*, **48**, 31 (1983).

114. K. N. Houk, R. W. Stozier, N. G. Rondan, R. R. Frazier, and N. Chauqui-Ottermans, *J. Am. Chem. Soc.*, **102**, 1426 (1980).

115. J. K. Smith, M. Newcomb, D. E. Bergbreiter, D. R. Williams, and A. I. Meyer, *Tetrahedron Lett.*, **24**, 3559 (1983); J. K. Smith, D. E. Bergbreiter, and M. Newcomb, *J. Am. Chem. Soc.*, **105**, 4396 (1983); A. Hosomi, Y. Araki, and H. Sakurai, *J. Am. Chem. Soc.*, **104**, 2081 (1982).

The thermodynamic composition is established by allowing the lithiated ketimines to come to room temperature. The most stable structures are those shown below, and each case represents the less-substituted isomer.

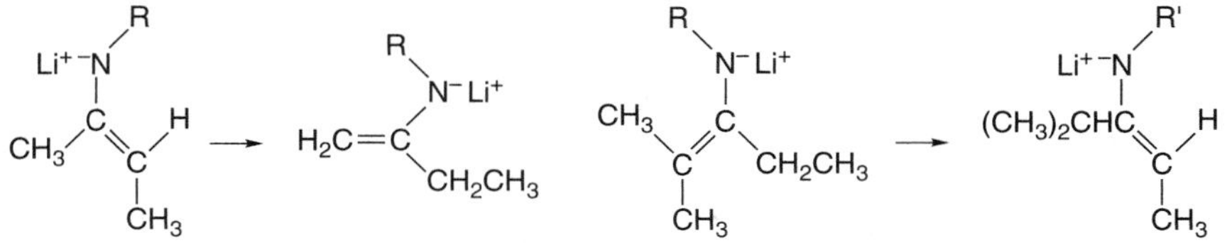

The complete interpretation of regiochemistry and stereochemistry of imine deprotonation also requires consideration of the state of aggregation and solvation of the base.[116]

A thorough study of the factors affecting the rates of formation of lithiated imines from cyclohexanone imines has been carried out.[117] Lithiation occurs preferentially *anti* to the N-substituent and with a preference for abstraction of an axial hydrogen.

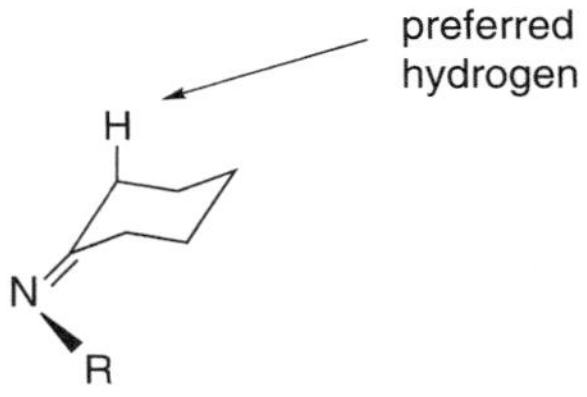

preferred hydrogen

If the amine carries a chelating substituent, as for 2-methoxyethylamine, the rate of deprotonation is accelerated. For any specific imine, ring substituents also influence the imine conformation and rate of deprotonation. These relationships reflect steric, stereoelectronic, and chelation influences, and sorting out each contribution can be challenging.

One of the potentially most useful aspects of the imine anions is that they can be prepared from enantiomerically pure amines. When imines derived from chiral amines are alkylated, the new carbon-carbon bond is formed with a bias for one of the two possible stereochemical configurations. Hydrolysis of the imine then leads to enantiomerically enriched ketone. Table 1.4 lists some examples that have been reported.[118]

The interpretation and prediction of the relationship between the configuration of the newly formed chiral center and the configuration of the amine is usually based on steric differentiation of the two faces of the imine anion. Most imine anions that show high stereoselectivity incorporate a substituent that can engage the metal cation in a

116. M. P. Bernstein and D. B. Collum, *J. Am. Chem. Soc.*, **115**, 8008 (1993).
117. S. Liao and D. B. Collum, *J. Am. Chem. Soc.*, **125**, 15114 (2003).
118. For a review, see D. E. Bergbreiter and M. Newcomb, in *Asymmetric Synthesis*, Vol. 2, J. D. Morrison, ed., Academic Press, New York, 1983, Chap. 9.

Table 1.4. Enantioselective Alkylation of Ketimines

	Amine	Ketone	Alkyl group	Yield%	e.e.
1[a]	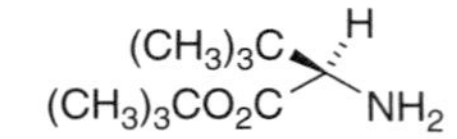 $(CH_3)_3C$ H / $(CH_3)_3CO_2C$ NH_2	Cyclohexanone	$CH_2=CHCH_2Br$	75	84
2[b]	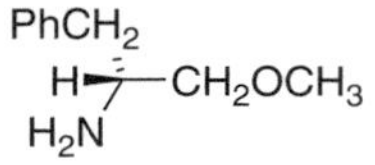 $PhCH_2$ / $H-C-CH_2OCH_3$ / H_2N	Cyclohexanone	$CH_2=CHCH_2Br$	80	>99
3[c]	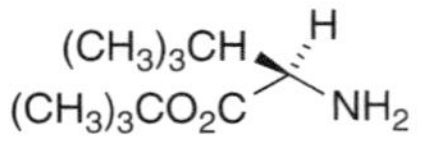 $(CH_3)_3CH$ H / $(CH_3)_3CO_2C$ NH_2	2-Carbomethoxy-cyclohexanone	CH_3I	57	>99
4[d]	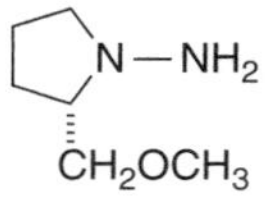 $N-NH_2$ / CH_2OCH_3	3-pentanone	$CH_3CH_2CH_2I$	57	97
5[e]	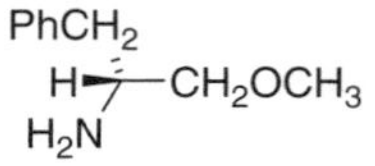 $PhCH_2$ / $H-C-CH_2OCH_3$ / H_2N	5-Nonanone	$CH_2=CHCH_2Br$	80	94

a. S. Hashimoto and K. Koga, *Tetrahedron Lett.*, 573 (1978).
b. A. I. Meyers, D. R. Williams, G. W. Erickson, S. White, and M. Druelinger, *J. Am. Chem. Soc.*, **103**, 3081 (1981).
c. K. Tomioka, K. Ando, Y. Takemasa, and K. Koga, *J. Am. Chem. Soc.*, **106**, 1718 (1984).
d. D. Enders, H. Kipphardt, and P. Fey, *Org. Synth.*, **65**, 183 (1987).
e. A. I. Meyers, D. R. Williams, S. White, and G. W. Erickson, *J. Am. Chem. Soc.*, **103**, 3088 (1981).

compact TS by chelation. In the case of Entry 2 in Table 1.4, for example, the TS **J** rationalizes the observed enantioselectivity.

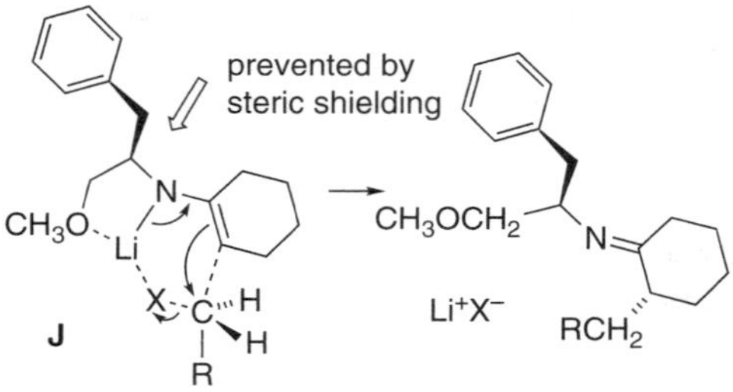

The important features of this transition structure are: (1) the chelation of the methoxy group with the lithium ion, which establishes a rigid structure; (2) the interaction of the lithium ion with the bromide leaving group, and (3) the steric effect of the benzyl group, which makes the underside the preferred direction of approach for the alkylating agent.

Hydrazones can also be deprotonated to give lithium salts that are reactive toward alkylation at the β-carbon. Hydrazones are more stable than alkylimines and therefore have some advantages in synthesis.[119] The *N,N*-dimethylhydrazones of methyl ketones are kinetically deprotonated at the methyl group. This regioselectivity is independent

119. D. Enders, in *Asymmetric Synthesis*, J. D. Morrison, ed., Academic Press, Orlando, FL, 1984.

of the stereochemistry of the hydrazone.[120] Two successive alkylations of the N,N-dimethylhydrazone of acetone provides unsymmetrical ketones.

Ref. 121

The anion of cyclohexanone N,N-dimethylhydrazone shows a strong preference for axial alkylation.[122] 2-Methylcyclohexanone N,N-dimethylhydrazone is alkylated by methyl iodide to give *cis*-2,6-dimethylcyclohexanone. The 2-methyl group in the hydrazone occupies a pseudoaxial orientation. Alkylation apparently occurs *anti* to the lithium cation, which is on the face opposite the 2-methyl substituent.

The N,N-dimethylhydrazones of α,β-unsaturated aldehydes give α-alkylation, similarly to the enolates of enones.[123]

Chiral hydrazones have also been developed for enantioselective alkylation of ketones. The hydrazones are converted to the lithium salt, alkylated, and then hydrolyzed to give alkylated ketone in good chemical yield and with high diastereoselective[124] (see Table 1.4, Entry 4). Several procedures have been developed for conversion of the hydrazones back to ketones.[125] Mild conditions are necessary to maintain the configuration at the enolizable position adjacent to the carbonyl group. The most frequently used hydrazones are those derived from N-amino-2-methoxymethypyrrolidine, known as SAMP. The (R)-enantiomer is called RAMP. The crystal structure of the lithium anion of the SAMP hydrazone from 2-acetylnaphthalene has been determined[126] (Figure 1.7). The lithium cation is chelated by the exocyclic nitrogen and the methoxy group.

120. D. E. Bergbreiter and M. Newcomb, *Tetrahedron Lett.*, 4145 (1979); M. E. Jung, T. J. Shaw, R. R. Fraser, J. Banville, and K. Taymaz, *Tetrahedron Lett.*, 4149 (1979).
121. M. Yamashita, K. Matsumiya, M. Tanabe, and R. Suetmitsu, *Bull. Chem. Soc. Jpn.*, **58**, 407 (1985).
122. D. B. Collum, D. Kahne, S. A. Gut, R. T. DePue, F. Mohamadi, R. A. Wanat, J. Clardy, and G. Van Duyne, *J. Am. Chem. Soc.*, **106**, 4865 (1984); R. A. Wanat and D. B. Collum, *J. Am. Chem. Soc.*, **107**, 2078 (1985).
123. M. Yamashita, K. Matsumiya, and K. Nakano, *Bull. Chem. Soc. Jpn.*, **60**, 1759 (1993).
124. D. Enders, H. Eichenauer, U. Baus, H. Schubert, and K. A. M. Kremer, *Tetrahedron*, **40**, 1345 (1984); D. Enders, H. Kipphardt, and P. Fey, *Org. Synth.*, **65**, 183 (1987); D. Enders and M. Klatt, *Synthesis*, 1403 (1996).
125. D. Enders, L. Wortmann, and R. Peters, *Acc. Chem. Res.*, **33**, 157 (2000).
126. D. Enders, G. Bachstadtler, K. A. M. Kremer, M. Marsch, K. Hans, and G. Boche, *Angew. Chem. Int. Ed. Engl.*, **27**, 1522 (1988).

Scheme 1.11. Alkylation of Imine and Hydrazone Anions

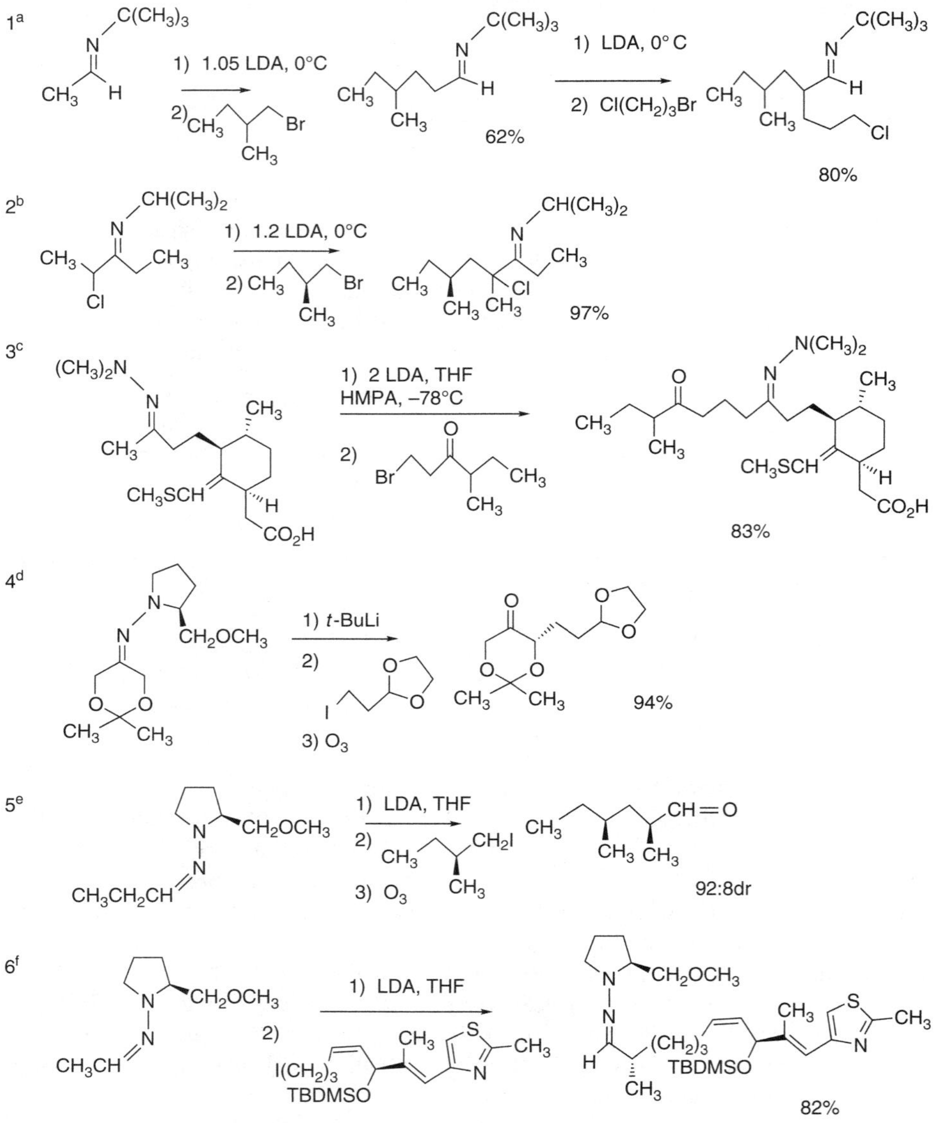

a. C. Stevens and N. De Kimpe, *J. Org. Chem.*, **58**, 132 (1993).
b. N. De Kimpe and W. Aelterman, *Tetrahedron*, **52**, 12815 (1996).
c. M. A. Avery, S. Mehrotra, J. D. Bonk, J. A. Vroman, D. K. Goins, and R. Miller, *J. Med. Chem.*, **39**, 2900 (1996).
d. M. Majewski and P. Nowak, *Tetrahedron Asymmetry*, **9**, 2611 (1998).
e. K. C. Nicolaou, E. W. Yue, S. LaGreca, A. Nadin, Z. Yang, J. E. Leresche, T. Tsuri, Y. Naniwa, and F. De Riccardis, *Chem. Eur. J.*, **1**, 467 (1995).
f. K. C. Nicolaou, F. Sarabia, S. Ninkovic, M. Ray, V. Finlay, and C. N. C. Body, *Angew. Chem. Int. Ed. Engl.*, **37**, 81 (1998).

Scheme 1.11 provides some examples of alkylation of imine and hydrazone anions. Entries 1 and 2 involve alkylation of anions derived from *N*-alkylimines. In Entry 1, two successive alkyl groups are added. In Entry 2, complete regioselectivity

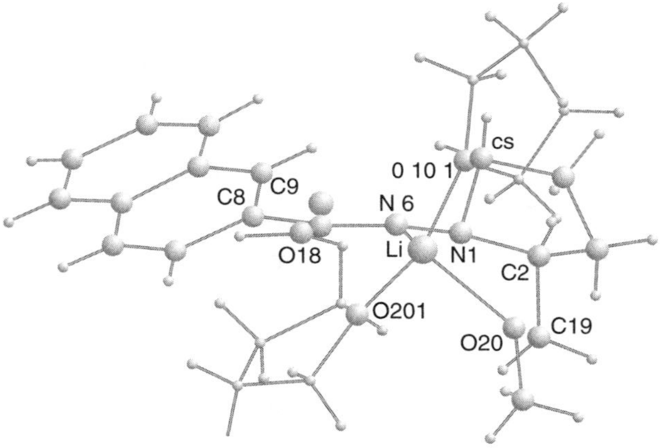

Fig. 1.7. Crystal structure of lithium salt of SAMP hydrazone of 2-acetylnaphthalene. Two molecules of THF are present. Reproduced from *Angew. Chem. Int. Ed. Engl.*, **27**, 1522 (1988), by permission of Wiley-VCH.

for the chloro-substituted group is observed. This reaction was used in the synthesis of an ant alarm pheromone called (*S*)-manicone. Entry 3 is an alkylation of a methyl group in an *N*,*N*-dimethylhydrazone. This reaction was used to synthesize analogs of the antimalarial substance arteminsinin. Entries 4 to 6 take advantage of the SAMP group to achieve enantioselective alkylations in the synthesis of natural products. Note that in Entries 4 and 5 the hydrazone was cleaved by ozonolysis. The reaction in Entry 6 was done in the course of synthesis of epothilone analogs. (See Section 13.2.5. for several epothilone syntheses.) In this case, the hydrazone was first converted to a nitrile by reaction with magnesium monoperoxyphthalate and then reduced to the aldehyde using DiBAlH.[127]

General References

D. E. Bergbreiter and M. Newcomb, in *Asymmetric Synthesis*, J. D. Morrison, ed., Academic Press, New York, 1983, Chap. 9.

D. Caine, in *Carbon-Carbon Bond Formation*, Vol. 1, R. L. Augustine, ed., Marcel Dekker, New York, 1979, Chap. 2.

A. G. Cook, ed., *Enamines: Synthesis, Structure and Reactions*, 2d Edition, Marcel Dekker, New York, 1988

C. H. Heathcock, *Modern Synthetic Methods,* **6**, 1 (1992).

V. Snieckus, ed., *Advances in Carbanion Chemistry*, Vol. 1, JAI Press, Greenwich, CT, 1992.

J. C. Stowell, *Carbanions in Organic Synthesis*, Wiley-Interscience, New York, 1979.

[127.] D. Enders, D. Backhaus, and J. Runsink, *Tetrahedron*, **52**, 1503 (1996).

Problems

(References for these problems will be found on page 1271.)

1.1. Arrange each series of compounds in order of decreasing acidity.

(a) $CH_3CH_2NO_2$, $(CH_3)_2CHCPh$, CH_3CH_2CN, $CH_2(CN)_2$

(b) $[(CH_3)_2CH]_2NH$, $(CH_3)_2CHOH$, $(CH_3)_2CH_2$, $(CH_3)_2CHPh$

(c) $CH_3CCH_2CO_2CH_3$, $CH_3CCH_2CCH_3$, CH_3OCCH_2Ph, CH_3COCH_2Ph

(d) $PhCCH_2Ph$, $(CH_3)_3CCCH_3$, $(CH_3)_3CCCH(CH_3)_2$, $PhCCH_2CH_2CH_3$

1.2. Write the structures of all possible enolates for each ketone. Indicate which you expect to be favored in a kinetically controlled deprotonation. Indicate which you would expect to be the most stable enolate.

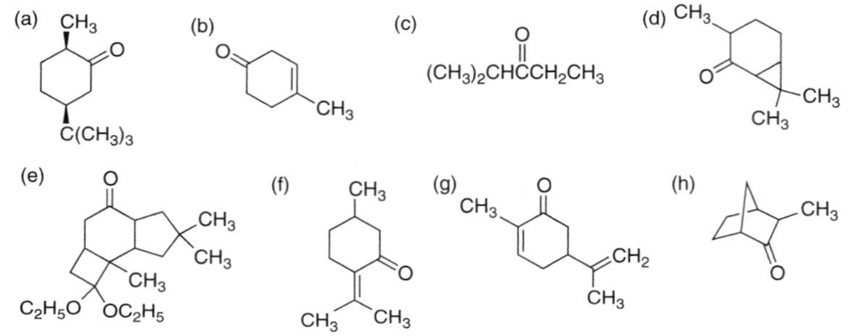

1.3. Suggest reagents and reaction conditions that would be suitable for effecting each of the following conversions.

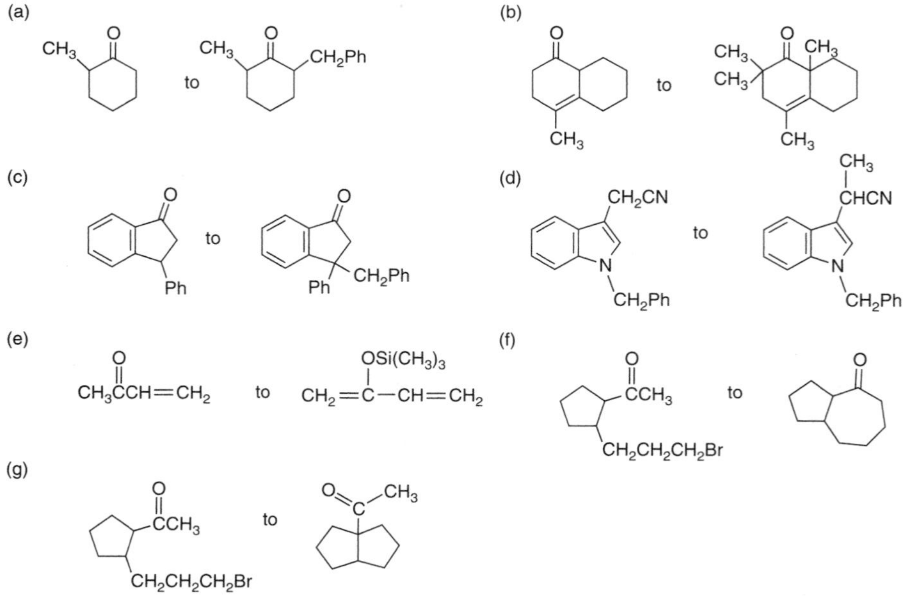

1.4. Intramolecular alkylation of enolates can be used to synthesize bi- and tricyclic compounds. Identify all the bonds in the following compounds that could be formed by intramolecular enolate alkylation. Select the one that you think is most likely to succeed and suggest reasonable reactants and reaction conditions for cyclization.

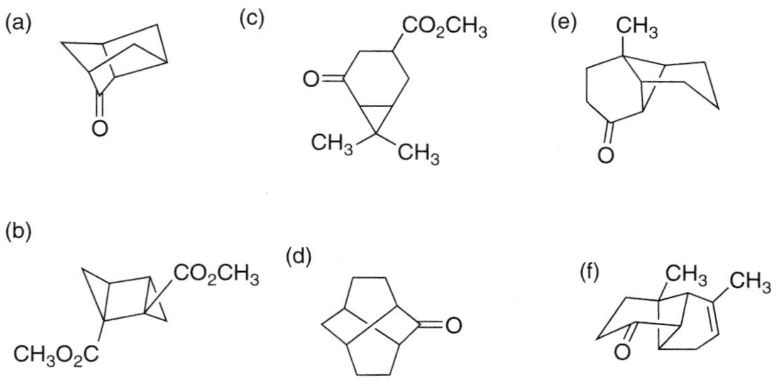

1.5. Predict the major product of each of the following reactions:

(a) PhCHCO₂Et   (1) 1 equiv LiNH₂/NH₃
    | (2) CH₃I
    CH₂CO₂Et

(b) PhCHCO₂Et   (1) 2 equiv LiNH₂/NH₃
    | (2) CH₃I
    CH₂CO₂Et

(c) PhCHCO₂H   (1) 2 equiv LiNH₂/NH₃
    | (2) CH₃I
    CH₂CO₂Et

1.6. Treatment of 2,3,3-triphenylpropanonitrile with one equivalent of KNH₂ in liquid ammonia, followed by addition of benzyl chloride, gives 2-benzyl-2,3,3-triphenylpropanonitrile in 97% yield. Use of two equivalents of KNH₂ gives an 80% yield of 2,3,3,4-tetraphenylbutanonitrile under the same reaction conditions. Explain.

1.7. Suggest readily available starting materials and reaction conditions suitable for obtaining each of the following compounds by a procedure involving alkylation of a carbon nucleophile.

(a) $PhCH_2CH_2CHPh$
                    |
                    CN

(b) $(CH_3)_2C{=}CHCH_2CH_2\overset{\overset{\displaystyle O}{\|}}{C}CH_2CO_2CH_3$

(c)

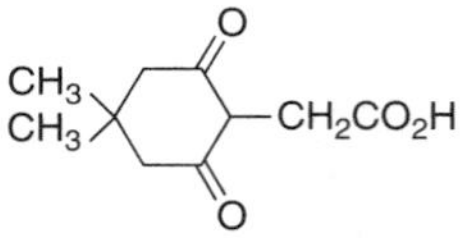

CH_3
CH_3 — ring with two O and $-CH_2CO_2H$

(d) $CH_3CH{=}CHCHCH_2CH_2CO_2H$

(e)   2,2,3-triphenylpropanonitrile

(f)   2,6-diallylcyclohexanone

(g)

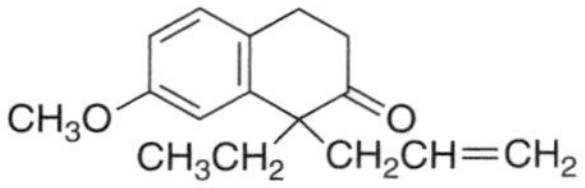

CH_3O
CH_3CH_2 $CH_2CH{=}CH_2$

(h) CN
                          |
      $H_2C{=}CHCH_2CPh$
                          |
                          CNH_2
                          ‖
                          O

(i) $CH_2{=}CHCHCH_2C{\equiv}CH$
                |
                $CO_2CH_2CH_3$

1.8. Perform a retrosynthetic dissection of each of the following compounds to the suggested starting material using reactions that involve alkylation of an enolate or an enolate equivalent. Then suggest a sequence of reactions that you think would succeed in converting the suggested starting material to the desired product.

(a)

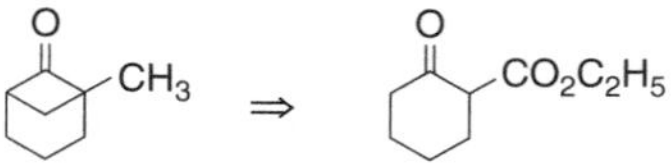

(b)

(c)

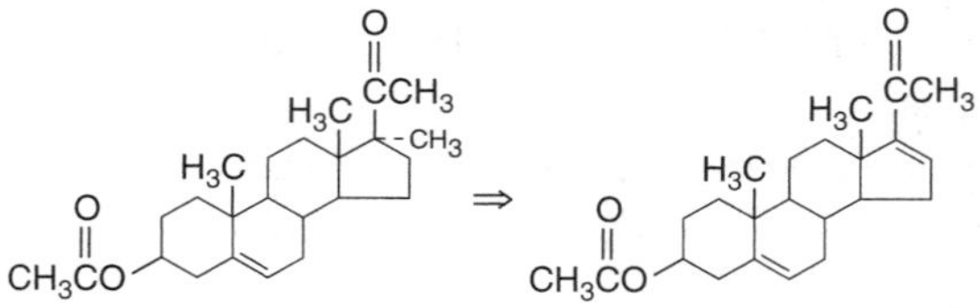

(d) $(CH_3O)_2PCH_2C(CH_2)_4CH_3 \Rightarrow (CH_3O)_2PCH_2CCH_3$

(e) $PhCH_2CH_2CHCO_2C_2H_5 \Rightarrow PhCH_2CO_2C_2H_5$
        |
        Ph

(f)

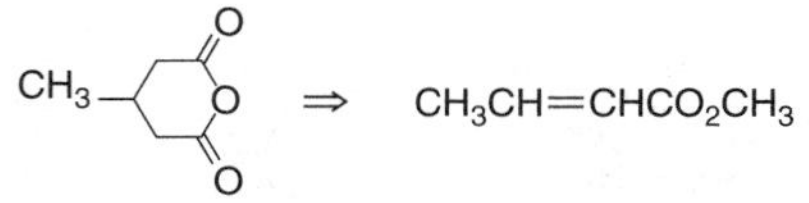

CH_3 ⬡(structure) $\Rightarrow CH_3CH=CHCO_2CH_3$

(g)

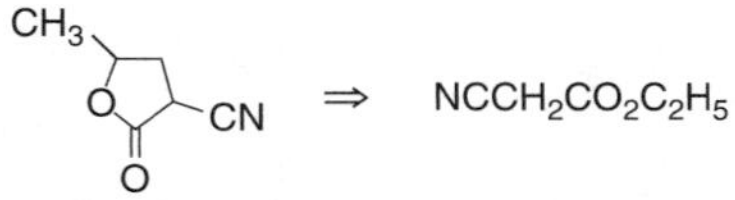

CH_3 (structure) CN $\Rightarrow NCCH_2CO_2C_2H_5$

(h)

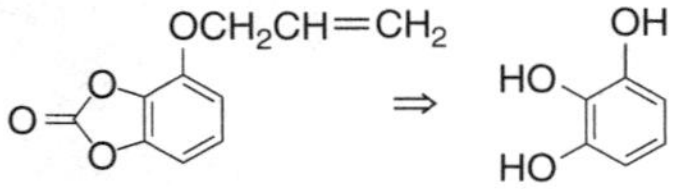

$OCH_2CH=CH_2$ (structure) $\Rightarrow$ (structure with OH, HO, HO)

(i)

CH_3 (structure) $CCH_2CH_2C=CH_2 \Rightarrow$ CH_3 (structure) $CCH_2CO_2CH_2CH_3$
                       ‖ | ‖
                       O CH_3 O

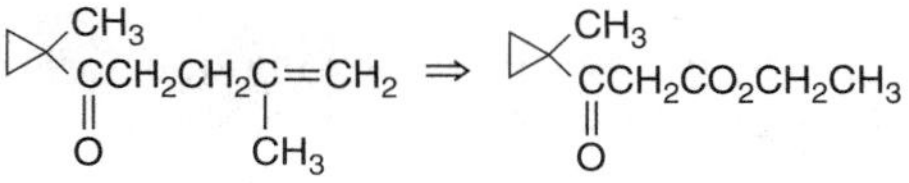

1.9. The carbon skeleton in structure **9-B** is found in certain natural substances, such as **9-C**. Outline a strategy to synthesize **9-B** from **9-A.**

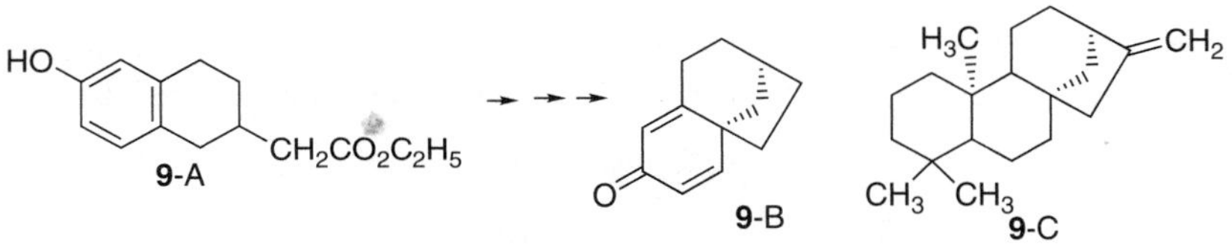

9-A $CH_2CO_2C_2H_5$

9-B

H_3C CH_2 (structure) CH_3 CH_3 **9-C**

1.10. Analyze the factors that you expect to control the stereochemistry of the following reactions:

(a)

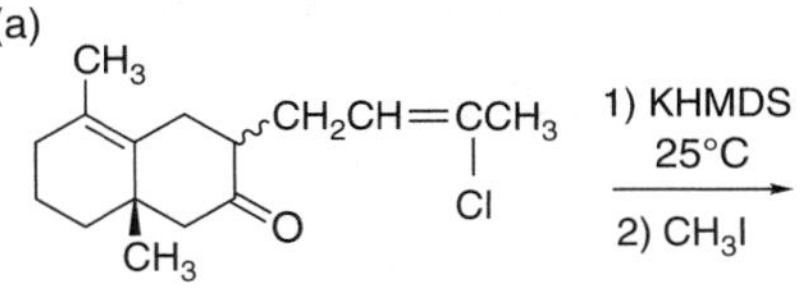

CH_3 (structure) $CH_2CH=CCH_3$
                              |
                              Cl
CH_3

1) KHMDS
   25°C
2) CH_3I

(b)

$CH(CH_3)_2$
CN
$C-CH_2CH(CH_3)_2$
$PhCH_2OCH_2$ RO

1) NaH
2) CH_3I

CH_3
|
R = CH_3CH_2OCH-

(c)

CH_3O_2C N (structure) Ph
H_3C O

1) LDA
2) $BrCH_2CH=CH_2$

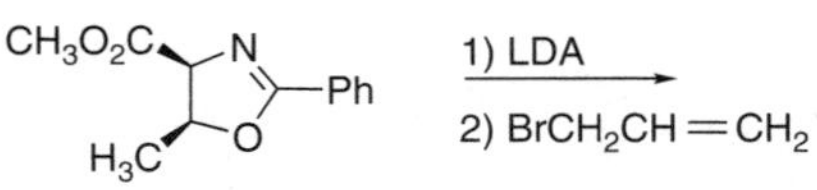

(d)

H_3C CO_2CH_3
(structure) CO_2CH_3
        OH
H

1) NaH
2) $BrCH_2C=CH_2$
              |
              CH_3

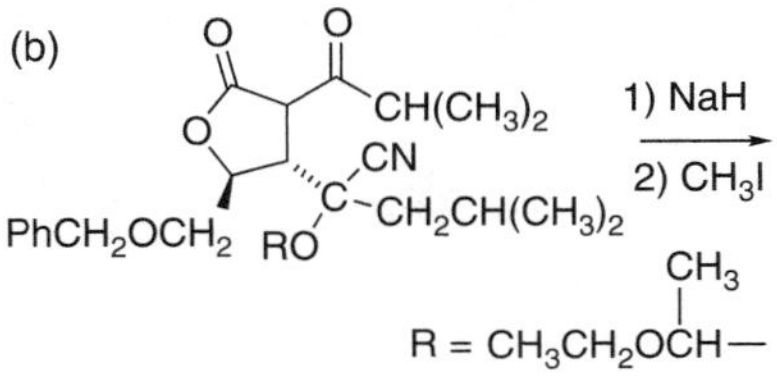

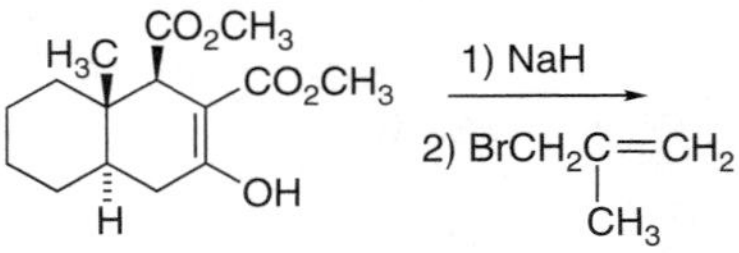

(e)

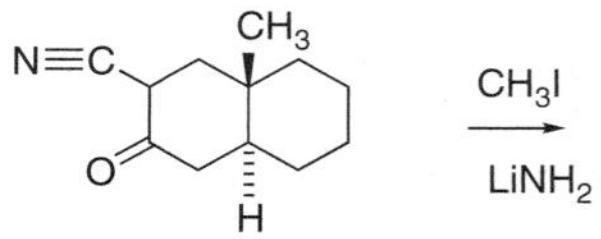

(f)

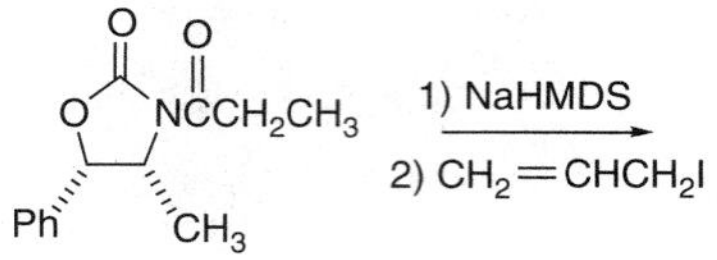

(g)

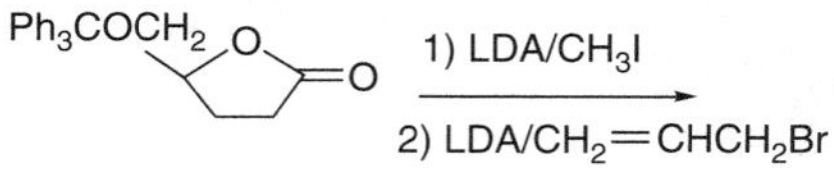

(h)

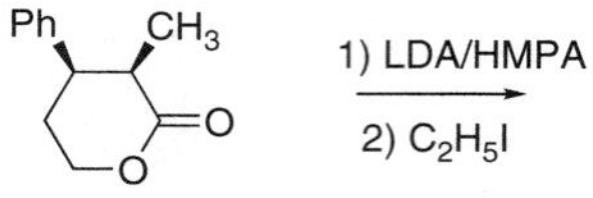

(i)

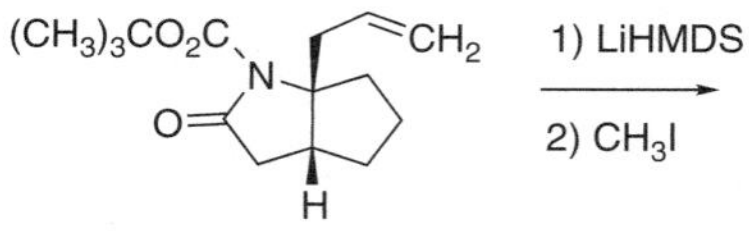

(j)

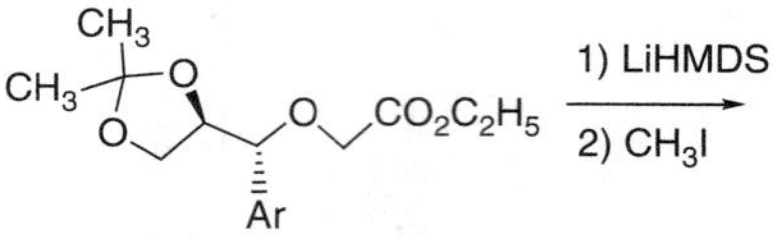

Ar = 4-methoxyphenyl

1.11. Suggest methodology for carrying out the following transformations in a way that high enantioselectivity could be achieved.

a. 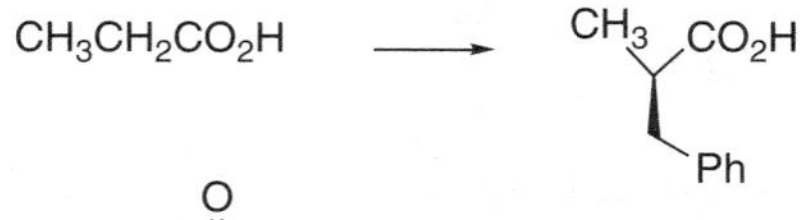 $CH_3CH_2CO_2H$ $\longrightarrow$

b.

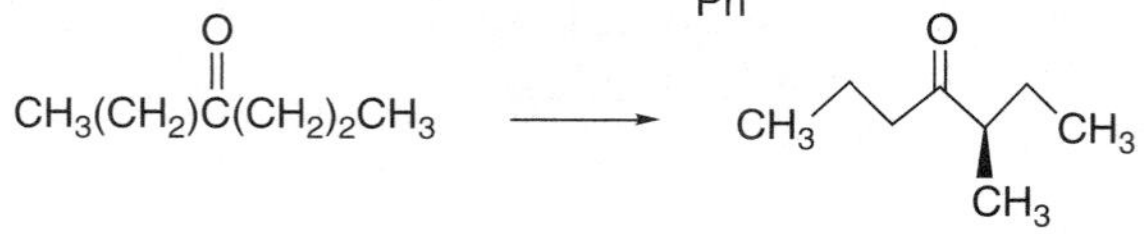

1.12. Indicate reagents and approximate reaction conditions that could be used to effect the following transformations. More than one step may be required.

(a)

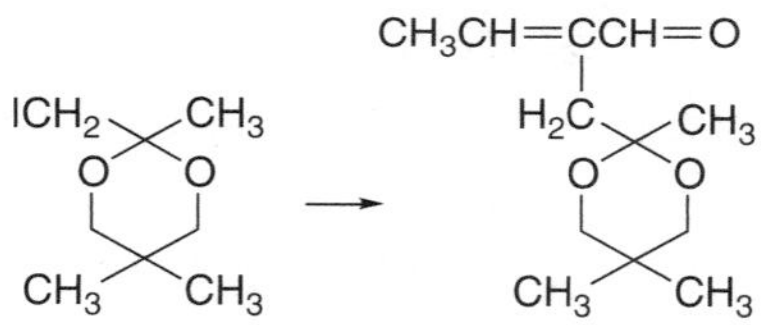

(b)

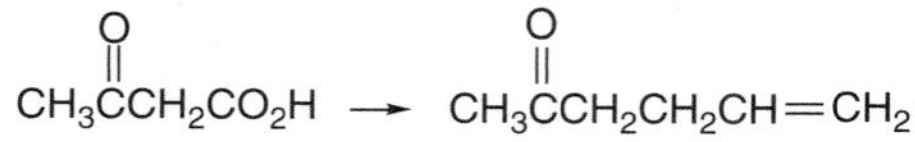

(c)

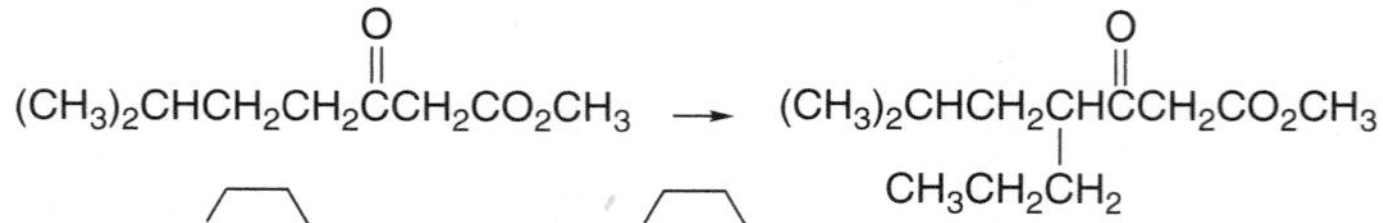

(d)

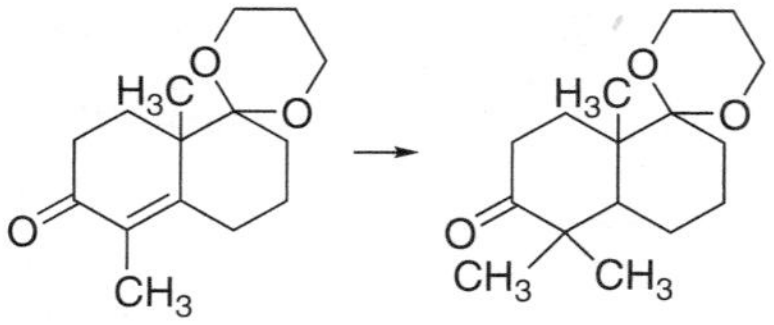

(e)

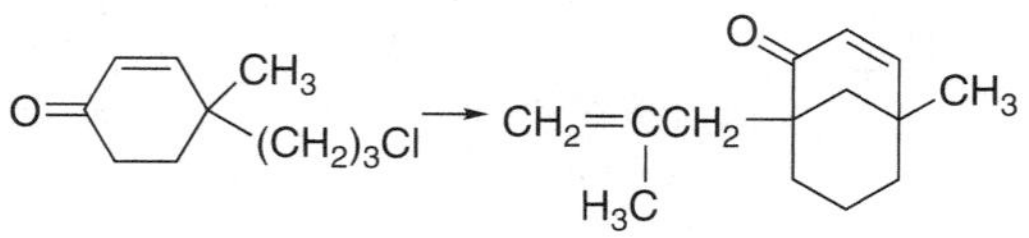

1.13. The observed stereoselectivity of each of the following reactions is somewhat enigmatic. Discuss factors that could contribute to stereoselectivity in these reactions.

(a)

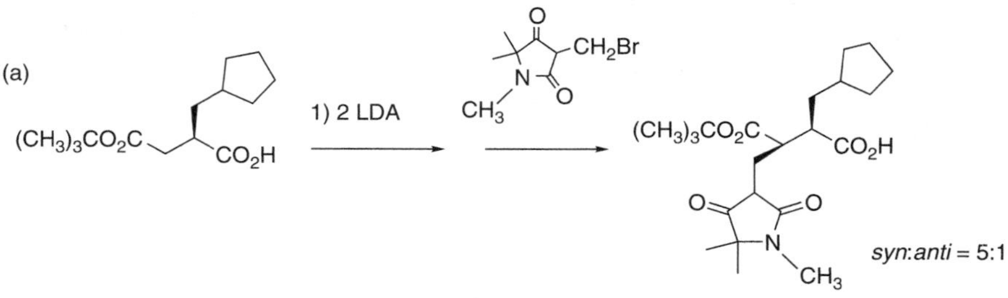

syn:anti = 5:1

(b)

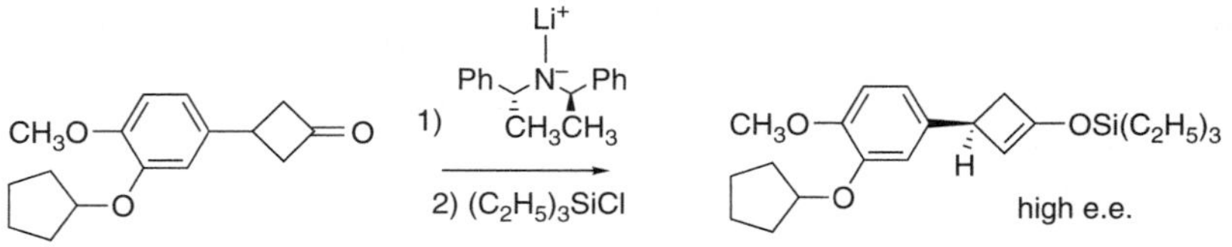

high e.e.

(c)

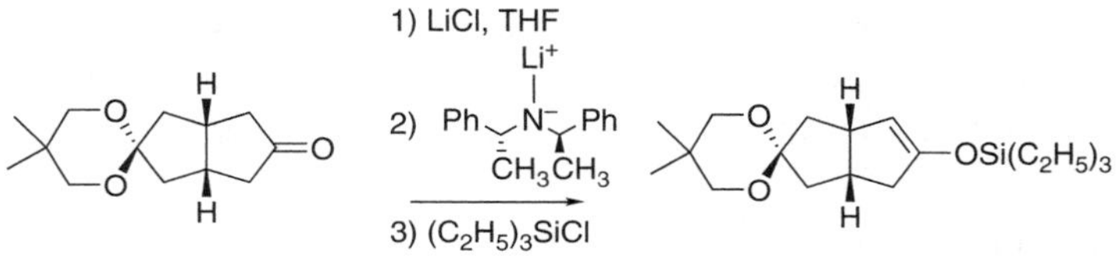

(d)

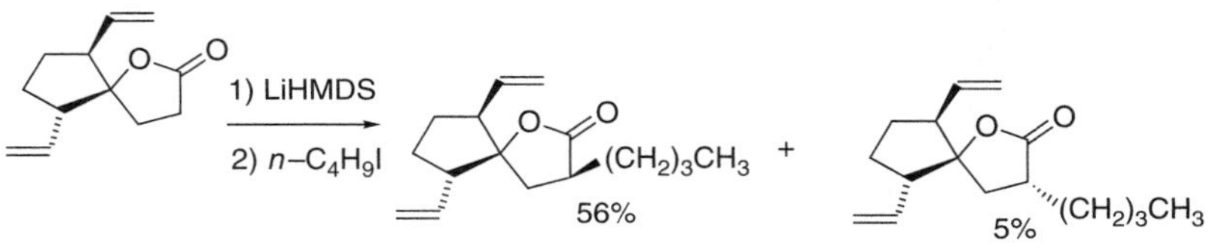

56%   +   5%

(e)

$CH_3CH(CH_2)_3CO_2C(CH_3)_3$
          |
          OH

1) 2 LiNEt$_2$

2) CH$_3$X

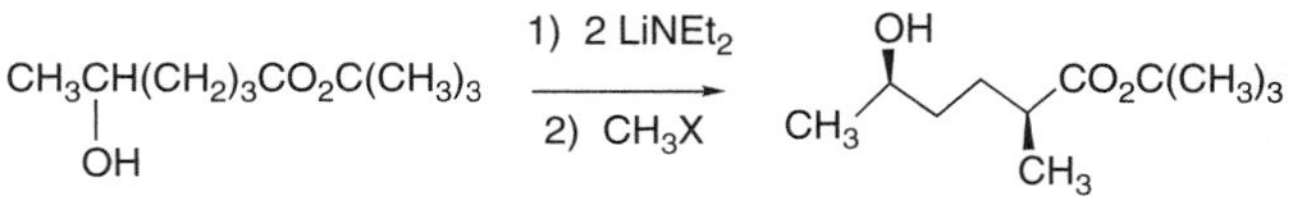

CH$_3$X	HMPA	syn	anti
CH$_3$I	no	56	44
CH$_3$I	yes	87	13
(CH$_3$O)$_2$SO$_2$	yes	90	10

1.14. One of the compounds shown below undergoes intramolecular cyclization to give a tricyclic ketone on being treated with NaHMDS, but the other does not cyclize. Indicate which compound will cyclize more readily and offer and explanation.

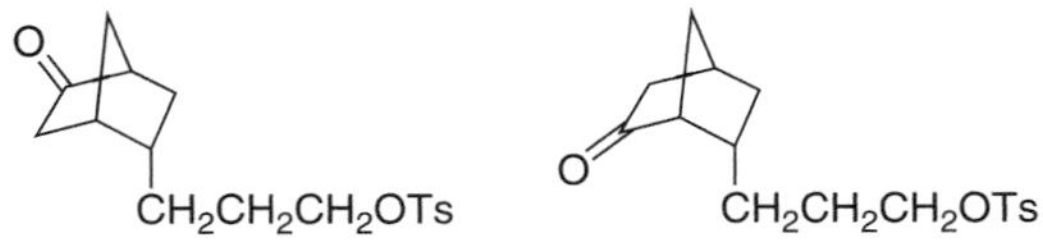

1.15. The alkylation of the enolate of 3-methyl-2-cyclohexenone with several different dibromides led to the products shown below. Discuss the course of each reaction and offer an explanation for the dependence of the product structure on the chain length of the dihalide.

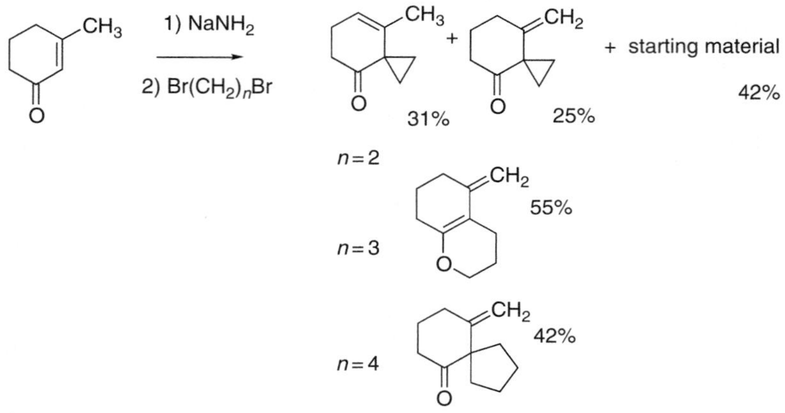

$n = 2$

$n = 3$

$n = 4$

1.16. Treatment of ethyl 2-azidobutanoate with a catalytic amount of lithium ethoxide in THF leads to evolution of nitrogen. Quenching the resulting solution with 3 N HCl gives ethyl 2-oxobutanoate in 86% yield. Suggest a mechanism for this process.

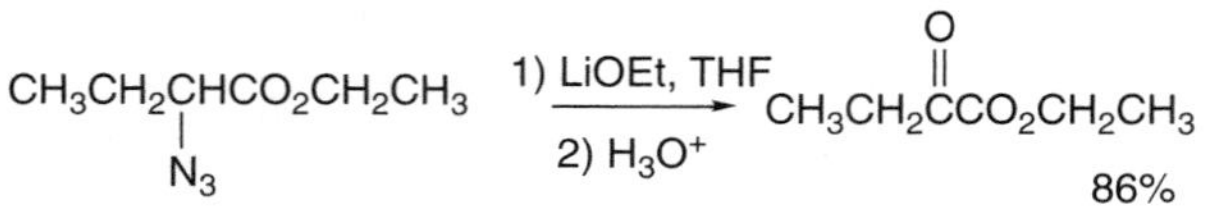

Reactions of Carbon Nucleophiles with Carbonyl Compounds

Introduction

The reactions described in this chapter include some of the most useful methods for carbon-carbon bond formation: *the aldol reaction*, the *Robinson annulation*, the *Claisen condensation* and other *carbon acylation* methods, and the *Wittig reaction* and other *olefination methods*. All of these reactions begin with the addition of a stabilized carbon nucleophile to a carbonyl group. The product that is isolated depends on the nature of the stabilizing substituent (Z) on the carbon nucleophile, the substituents (A and B) at the carbonyl group, and the ways in which A, B, and Z interact to complete the reaction pathway from the addition intermediate to the product. Four fundamental processes are outlined below. Aldol addition and condensation lead to β-hydroxyalkyl or α-alkylidene derivatives of the carbon nucleophile (Pathway **A**). The acylation reactions follow Pathway **B**, in which a group leaves from the carbonyl electrophile. In the Wittig and related olefination reactions, the oxygen in the adduct reacts with the group Z to give an elimination product (Pathway **C**). Finally, if the enolate has an α-substituent that is a leaving group, cyclization can occur, as in Pathway **D**. This is observed, for example, with enolates of α-haloesters. The fundamental mechanistic concepts underlying these reactions were introduced in Chapter 7 of Part A. Here we emphasize the scope, stereochemistry, and synthetic utility of these reactions.

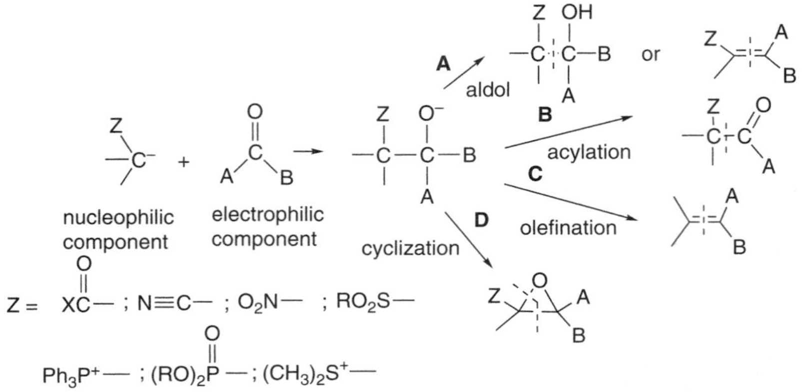

A second important reaction type considered in this chapter is *conjugate addition*, which involves addition of nucleophiles to electrophilic double or triple bonds. A crucial requirement for this reaction is an electron-withdrawing group (EWG) that can stabilize the negative charge on the intermediate. We focus on reactions between enolates and α,β-unsaturated carbonyl compounds and other electrophilic alkenes such as nitroalkenes.

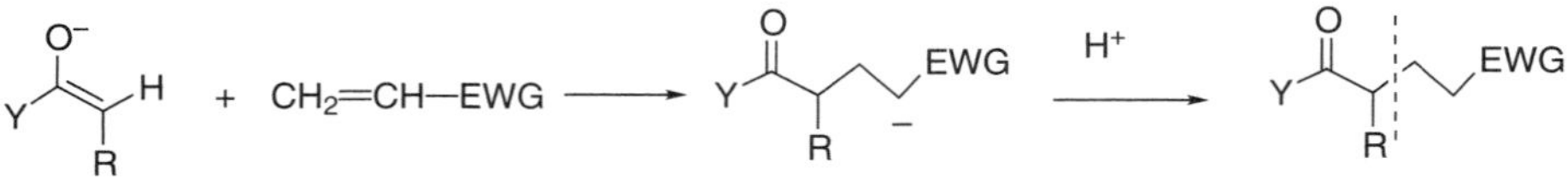

enolate conjugate addition

The retrosynthetic dissection is at a bond that is α to a carbonyl and β to an anion-stabilizing group.

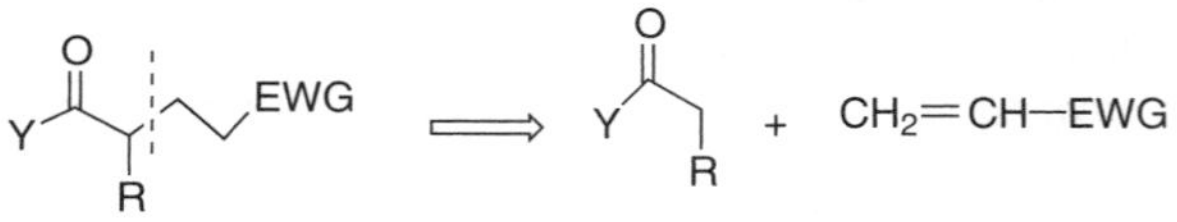

2.1. Aldol Addition and Condensation Reactions

2.1.1. The General Mechanism

The general mechanistic features of the aldol addition and condensation reactions of aldehydes and ketones were discussed in Section 7.7 of Part A, where these general mechanisms can be reviewed. That mechanistic discussion pertains to reactions occurring in *hydroxylic solvents* and under *thermodynamic control*. These conditions are useful for the preparation of aldehyde dimers (aldols) and certain α,β-unsaturated aldehydes and ketones. For example, the mixed condensation of aromatic aldehydes with aliphatic aldehydes and ketones is often done under these conditions. The conjugation in the β-aryl enones provides a driving force for the elimination step.

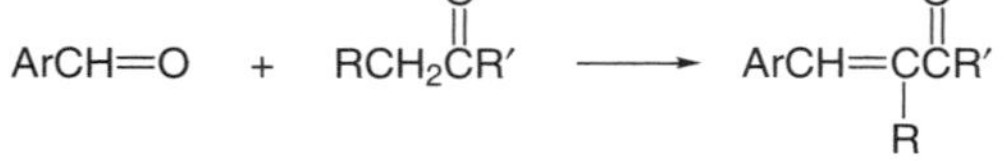

The aldol reaction is also important in the synthesis of more complex molecules and in these cases control of both regiochemistry and stereochemistry is required. In most cases, this is accomplished under conditions of *kinetic control*. In the sections that follow, we discuss how variations of the basic mechanism and selection of specific reagents and reaction conditions can be used to control product structure and stereochemistry.

The addition reaction of enolates and enols with carbonyl compounds is of broad scope and of great synthetic importance. Essentially all of the stabilized carbanions mentioned in Section 1.1 are capable of adding to carbonyl groups, in what is known as the *generalized aldol reaction*. Enolates of aldehydes, ketones, esters, and amides, the carbanions of nitriles and nitro compounds, as well as phosphorus- and sulfur-stabilized carbanions and ylides undergo this reaction. In the next section we emphasize the fundamental regiochemical and stereochemical aspects of the reactions of ketones and aldehydes.

2.1.2. Control of Regio- and Stereoselectivity of Aldol Reactions of Aldehydes and Ketones

The synthetic utility of the aldol reaction depends on both the versatility of the reactants and the control of the regio- and stereochemistry. The term *directed aldol addition* is applied to reactions that are designed to achieve specific regio- and stereochemical outcomes.[1] Control of product structure requires that one reactant act exclusively as the *nucleophile* and the other exclusively as the *electrophile*. This requirement can be met by pre-forming the nucleophilic enolate by deprotonation, as described in Section 1.1. The enolate that is to serve as the nucleophile is generated stoichiometrically, usually with lithium as the counterion in an aprotic solvent at low temperature. Under these conditions, the kinetic enolate does not equilibrate with the other regio- or stereoisomeric enolates that can be formed from the ketone. The enolate gives a specific adduct, provided that the addition step is fast relative to proton exchange between the nucleophilic and electrophilic reactants. The reaction is under *kinetic control*, at both the stage of formation of the enolate and the addition step.

Under other reaction conditions, the product can result from *thermodynamic control*. Aldol reactions can be effected for many compounds using less than a stoichiometric amount of base. In these circumstances, the aldol reaction is reversible and the product ratio is determined by the relative stability of the various possible products. Thermodynamic conditions also permit equilibration among the enolates of the nucleophile. The conditions that lead to equilibration include higher reaction temperatures, protic or polar dissociating solvents, and the use of weakly coordinating cations. Thermodynamic conditions can be used to enrich the composition in the *most stable* of the isomeric products.

Reaction conditions that involve other enolate derivatives as nucleophiles have been developed, including boron enolates and enolates with titanium, tin, or zirconium as the metal. These systems are discussed in detail in the sections that follow, and in Section 2.1.2.5, we discuss reactions that involve *covalent enolate equivalents*, particularly silyl enol ethers. Scheme 2.1 illustrates some of the procedures that have been developed. A variety of carbon nucleophiles are represented in Scheme 2.1, including lithium and boron enolates, as well as titanium and tin derivatives, but in

[1.] T. Mukaiyama, *Org. React.*, **28**, 203 (1982).

Scheme 2.1. Examples of Directed Aldol Reactions

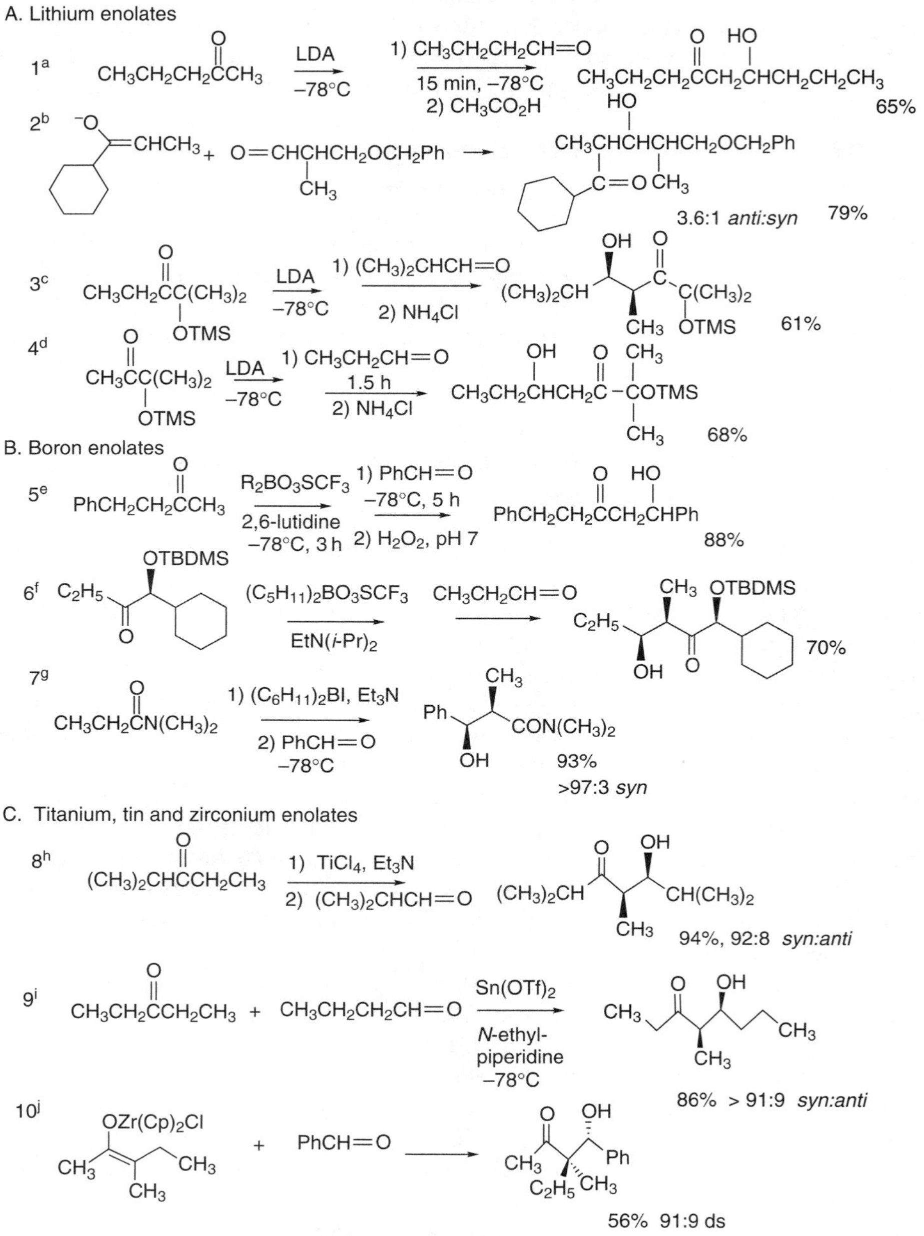

A. Lithium enolates

B. Boron enolates

C. Titanium, tin and zirconium enolates

a. G. Stork, G. A. Kraus, and G. A. Garcia, *J. Org. Chem.*, **39**, 3459 (1974).
b. S. Masamune, J. W. Ellingboe, and W. Choy, *J. Am. Chem. Soc.*, **104**, 5526 (1982).
c. R. Bal, C. T. Buse, K. Smith, and C. Heathcock, *Org. Synth.*, **63**, 89 (1984).
d. P. J. Jerris and A. B. Smith, III, *J. Org. Chem.*, **46**, 577 (1981).
e. T. Inoue, T. Uchimaru, and T. Mukaiyama, *Chem. Lett.*, 153 (1977).
f. S. Masamune, W. Choy, F. A. J. Kerdesky, and B. Imperiali, *J. Am. Chem. Soc.*, **103**, 1566 (1981).
g. K. Ganesan and H. C. Brown, *J. Org. Chem.*, **59**, 7346 (1994).
h. D. A. Evans, D. L. Rieger, M. T. Bilodeau, and F. Urpi, *J. Am. Chem. Soc.*, **113**, 1047 (1991).
i. T. Mukaiyama, N. Iwasawa, R. W. Stevens, and T. Hagu, *Tetrahedron*, **40**, 1381 (1984).
j. S. Yamago, D. Machii, and E. Nakamura, *J. Org. Chem.*, **56**, 2098 (1991).

each case the electrophile is an aldehyde. Pay particular attention to the retrosynthetic relationship between the products and the reactants, which corresponds in each case to Path **A** (p. 64). We see that the aldol addition reaction provides β-hydroxy carbonyl compounds or, more generally, adducts with a hydroxy group β to the stabilizing group Z of the carbon nucleophile.

$$
\begin{array}{c}
\underset{\substack{| \\ R^2}}{\overset{\substack{Z \quad OH \\ | \quad |}}{-C-C-R^1}} \quad\Longrightarrow\quad -\overset{Z}{\underset{|}{\overset{|}{C}}}{}^- \;+\; \underset{R^1 \quad R^2}{\overset{O}{\|}}
\end{array}
$$

Note also the stereochemistry. In some cases, two new stereogenic centers are formed. The hydroxy group and any C(2) substituent on the enolate can be in a *syn* or *anti* relationship. For many aldol addition reactions, the stereochemical outcome of the reaction can be predicted and analyzed on the basis of the detailed mechanism of the reaction. Entry 1 is a mixed ketone-aldehyde aldol addition carried out by kinetic formation of the less-substituted ketone enolate. Entries 2 to 4 are similar reactions but with more highly substituted reactants. Entries 5 and 6 involve boron enolates, which are discussed in Section 2.1.2.2. Entry 7 shows the formation of a boron enolate of an amide; reactions of this type are considered in Section 2.1.3. Entries 8 to 10 show titanium, tin, and zirconium enolates and are discussed in Section 2.1.2.3.

2.1.2.1. Aldol Reactions of Lithium Enolates. Entries 1 to 4 in Scheme 2.1 represent cases in which the nucleophilic component is a lithium enolate formed by kinetically controlled deprotonation, as discussed in Section 1.1. Lithium enolates are usually highly reactive toward aldehydes and addition occurs rapidly when the aldehyde is added, even at low temperature. The low temperature ensures kinetic control and enhances selectivity. When the addition step is complete, the reaction is stopped by neutralization and the product is isolated.

The fundamental mechanistic concept for diastereoselectivity of aldol reactions of lithium enolates is based on a cyclic TS in which both the carbonyl and enolate oxygen are coordinated to the lithium cation.[2] The Lewis acid character of the lithium ion promotes reaction by increasing the carbonyl group electrophilicity and by bringing the reactants together in the TS. Other metal cations and electrophilic atoms can play the role of the Lewis acid, as we will see when we discuss reactions of boron and other metal enolates. The fundamental concept is that the aldol addition normally occurs through a chairlike TS. It is assumed that the structure of the TS is sufficiently similar to a chair cyclohexane that the conformational concepts developed for cyclohexane rings can be applied. In the structures that follow, the reacting aldehyde is shown with R rather than H in the equatorial-like position, which avoids a 1,3-diaxial interaction with the enolate C(1) substituent. A consequence of this mechanism is that the reaction

[2.] (a) H. E. Zimmerman and M. D. Traxler, *J. Am. Chem. Soc.*, **79**, 1920 (1957); (b) P. Fellman and J. E. Dubois, *Tetrahedron*, **34**, 1349 (1978); (c) C. H. Heathcock, C. T. Buse, W. A. Kleschick, M. C. Pirrung, J. E. Sohn, and J. Lampe, *J. Org. Chem.*, **45**, 1066 (1980).

is *stereospecific with respect to the E- or Z-configuration of the enolate*. The *E*-enolate gives the *anti* aldol product, whereas the *Z*-enolate gives the *syn*-aldol.[3]

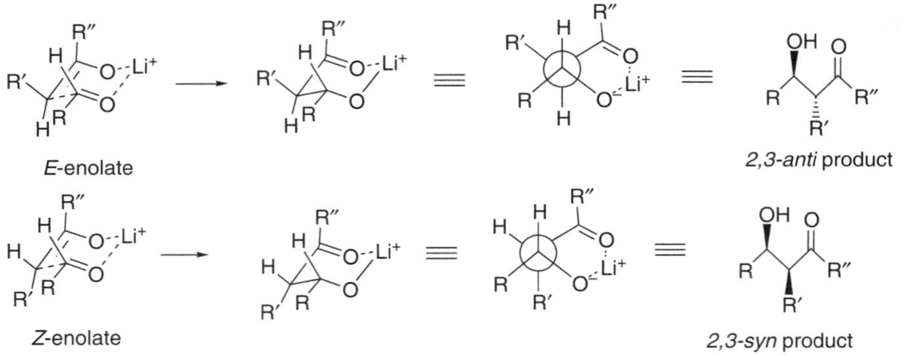

The preference for chairlike TSs has been confirmed by using deuterium-labeled enolates prepared from the corresponding silyl enol ethers. The ratio of the location of the deuterium corresponds closely to the ratio of the stereoisomeric enolates for several aldehydes.[4]

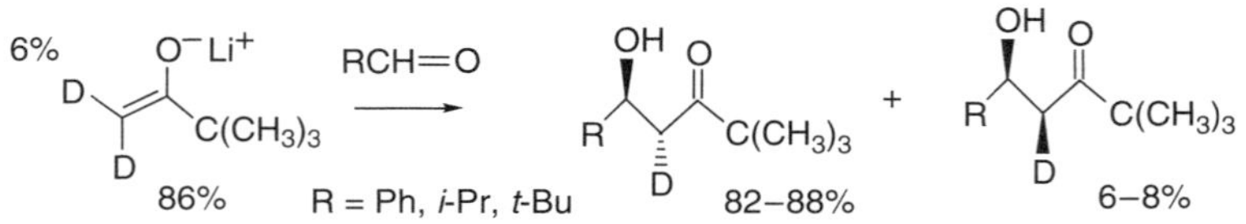

Provided that the reaction occurs through a chairlike TS, the $E \rightarrow anti/Z \rightarrow syn$ relationship will hold. There are three cases that can lead to departure from this relationship. These include a nonchair TS, that can involve either an open TS or a nonchair cyclic TS. Internal chelation of the aldehyde or enolate can also cause a change in TS structure.

The first element of stereocontrol in aldol addition reactions of ketone enolates is the enolate structure. Most enolates can exist as two stereoisomers. In Section 1.1.2, we discussed the factors that influence enolate composition. The enolate formed from 2,2-dimethyl-3-pentanone under kinetically controlled conditions is the *Z*-isomer.[5] When it reacts with benzaldehyde only the *syn* aldol is formed.[4] The product stereochemistry is correctly predicted if the TS has a conformation with the phenyl substituent in an equatorial position.

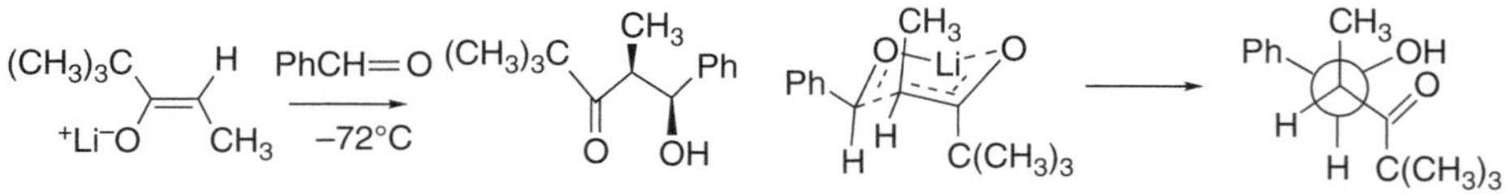

---

3. For consistency in designating the relative configuration the carbonyl group is numbered (1). The newly formed bond is labeled 2,3- and successive carbons are numbered accordingly. The carbons derived from the enolate are numbered 2′,3′, etc., starting with the α′-carbon.

4. C. M. Liu, W. J. Smith, III, D. J. Gustin, and W. R. Roush, *J. Am. Chem. Soc.*, **127**, 5770 (2005).

5. To avoid potential uncertainties in the application of the Cahn-Ingold-Prelog priority rules, by convention the enolate oxygen is assigned the higher priority.

A similar preference for formation of the *syn* aldol is found for other *Z*-enolates derived from ketones in which one of the carbonyl substituents is bulky. Ketone enolates with less bulky substituents show a decreasing stereoselectivity in the order *t*-butyl > *i*-propyl > ethyl.[2c] This trend parallels a decreasing preference for stereoselective formation of the *Z*-enolate.

R =	C_2H_5	70:30	36:64
	$CH(CH_3)_2$	40:60	18:82
	$C(CH_3)_3$	2:98	2:98

The enolates derived from cyclic ketones are necessarily *E*-isomers. The enolate of cyclohexanone reacts with benzaldehyde to give both possible stereoisomeric products. The stereoselectivity is about 5:1 in favor of the *anti* isomer under optimum conditions.[6]

From these and many related examples the following generalizations can be made about kinetic stereoselection in aldol additions of lithium enolates. (1) The chair TS model provides a basis for analyzing the stereoselectivity observed in aldol reactions of ketone enolates having one bulky substituent. The preference is *Z*-enolate → *syn* aldol; *E*-enolate → *anti* aldol. (2) When the enolate has no bulky substituent, stereoselectivity is low. (3) *Z*-Enolates are more stereoselective than *E*-enolates. Table 2.1 gives some illustrative data.

The requirement that an enolate have at least one bulky substituent restricts the types of compounds that give highly stereoselective aldol additions via the lithium enolate method. Furthermore, only the enolate formed by kinetic deprotonation is directly available. Whereas ketones with one tertiary alkyl substituent give mainly the *Z*-enolate, less highly substituted ketones usually give mixtures of *E*- and *Z*-enolates.[7] (Review the data in Scheme 1.1.) Therefore efforts aimed at increasing the stereo-selectivity of aldol additions have been directed at two facets of the problem: (1) better control of enolate stereochemistry, and (2) enhancement of the degree of stereoselectivity in the addition step, which is discussed in Section 2.1.2.2.

The *E:Z* ratio can be modified by the precise conditions for formation of the enolate. For example, the *E:Z* ratio for 3-pentanone and 2-methyl-3-pentanone can be increased by use of a 1:1 lithium tetramethylpiperidide(LiTMP)-LiBr mixture for

6. M. Majewski and D. M. Gleave, *Tetrahedron Lett.*, **30**, 5681 (1989).
7. R. E. Ireland, R. H. Mueller, and A. K. Willard, *J. Am. Chem. Soc.*, **98**, 2868 (1976); W. A. Kleschick, C. T. Buse, and C. H. Heathcock, *J. Am. Chem. Soc.*, **99**, 247 (1977); Z. A. Fataftah, I. E. Kopka, and M. W. Rathke, *J. Am. Chem. Soc.*, **102**, 3959 (1980).

Table 2.1. Diastereoselectivity of Addition of Lithium Enolates to Benzaldehyde

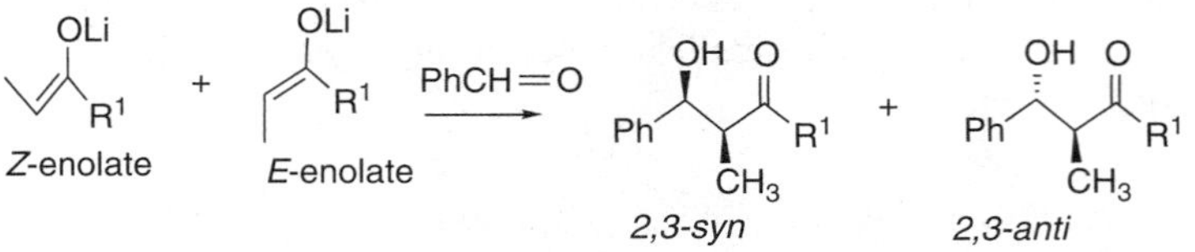

R^1	$Z:E$ ratio	*syn:anti* ratio
H	100:0	50:50
H	0:100	65:35
C_2H_5	30:70	64:36
C_2H_5	66:34	77:23
$(CH_3)_2CH$	>98:2	90:10
$(CH_3)_2CH$	0:100	45:55
$(CH_3)_3C$	>98:2	>98:2
1-Adamantyl	>98:2	>98:2
C_6H_5	>98:2	88:12
Mesityl	8:92	8:92
Mesityl	87:13	88:12

a. From C. H. Heathcock, in *Asymmetric Synthesis*, Vol. 3, J. D. Morrison, ed., Academic Press, New York, 1984, Chap. 2.

kinetic enolization.[8] The precise mechanism of this effect is still a matter of investigation, but it is probably due to an aggregate species containing bromide acting as the base (see Section 1.1.1).[9]

	E:Z Stereoselectivity		
	LDA	LiTMP	LiTMP + LiBr
$CH_3CH_2CCH_2CH_3$ (C=O)	3.3:1	5:1	50:1
$(CH_3)_2CHCCH_2CH_3$ (C=O)	1.7:1	2:1	21:1
$(CH_3)_3CCCH_2CH_3$ (C=O)	1: >50	1: >20	1:>20

Other changes in deprotonation conditions can influence enolate composition. Relatively weakly basic lithium anilides, specifically lithium 2,4,6-trichloroanilide and lithium diphenylamide, give high $Z:E$ ratios.[10] Lithio 1,1,3,3-tetramethyl-1,3-diphenyldisilylamide is also reported to favor the Z-enolate.[11] On the other hand, lithium N-trimethylsilyl-*iso*-propylamide and lithium N-trimethylsilyl-*tert*-butylamide give selectivity for the E-enolate[12] (see Scheme 1.1).

8. P. L. Hall, J. H. Gilchrist, and D. B. Collum, *J. Am. Chem. Soc.*, **113**, 9571 (1991).
9. F. S. Mair, W. Clegg, and P. A. O'Neil, *J. Am. Chem. Soc.*, **115**, 3388 (1993).
10. L. Xie, K. Vanlandeghem, K. M. Isenberger, and C. Bernier, *J. Org. Chem.*, **68**, 641 (2003).
11. S. Masamune, J. W. Ellingboe, and W. Choy, *J. Am. Chem. Soc.*, **104**, 5526 (1982).
12. L. Xie, K. M. Isenberger, G. Held, and L. M. Dahl, *J. Org. Chem.*, **62**, 7516 (1997).

When aldol addition is carried out under thermodynamic conditions, the product stereoselectivity is usually not as high as under kinetic conditions. All the regio- and stereoisomeric enolates can participate as nucleophiles. The adducts can return to reactants, so the difference in *stability* of the stereoisomeric *anti* and *syn* products determines the product composition. In the case of lithium enolates, the adducts can be equilibrated by keeping the reaction mixture at room temperature. This has been done, for example, with the product from the reaction of the enolate of 2,2-dimethyl-3-pentanone and benzaldehyde. The greater stability of the *anti* isomer is attributed to the pseudoequatorial position of the methyl group in the chairlike product chelate. With larger substituent groups, the thermodynamic preference for the *anti* isomer is still greater.[13]

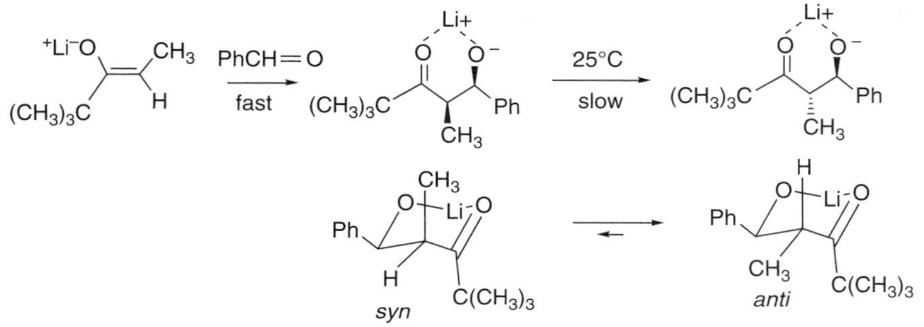

For synthetic efficiency, it is useful to add $MgBr_2$, which accelerates the equilibration.

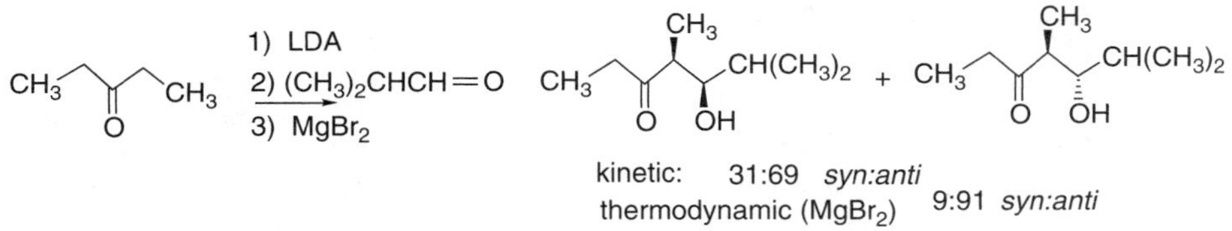

kinetic: 31:69 *syn:anti*
thermodynamic ($MgBr_2$) 9:91 *syn:anti*

Ref. 14

2.1.2.2. Aldol Reactions of Boron Enolates. The matter of increasing stereoselectivity in the addition step can be addressed by using other reactants. One important version of the aldol reaction involves the use of boron enolates.[15] A cyclic TS similar to that for lithium enolates is involved, and the same relationship exists between enolate config- uration and product stereochemistry. In general, the stereoselectivity is higher than for lithium enolates. The O–B bond distances are shorter than for lithium enolates, and this leads to a more compact structure for the TS and magnifies the steric interactions that control stereoselectivity.

13. C. H. Heathcock and J. Lampe, *J. Org. Chem.*, **48**, 4330 (1983).
14. K. A. Swiss, W.-B. Choi, D. C. Liotta, A. F. Abdel-Magid, and C. A. Maryanoff, *J. Org. Chem.*, **56**, 5978 (1991).
15. C. J. Cowden and I. A. Paterson, *Org. React.*, **51**, 1 (1997); E. Tagliavini, C. Trombini, and A. Umani-Ronchi, *Adv. Carbanion Chem.*, **2**, 111 (1996).

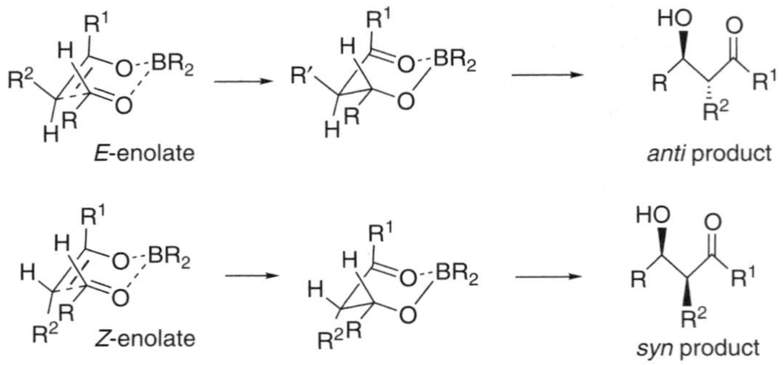

Boron enolates can be prepared by reaction of the ketone with a dialkylboron trifluoromethanesulfonate (triflate) and a tertiary amine.[16] Use of boron triflates and a bulky amine favors the Z-enolate. The resulting aldol products are predominantly the *syn* stereoisomers.

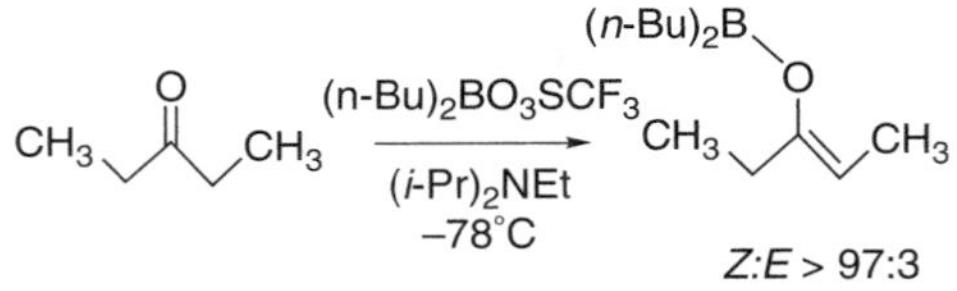

The E-boron enolates of some ketones can be preferentially obtained by using dialkylboron chlorides.[17]

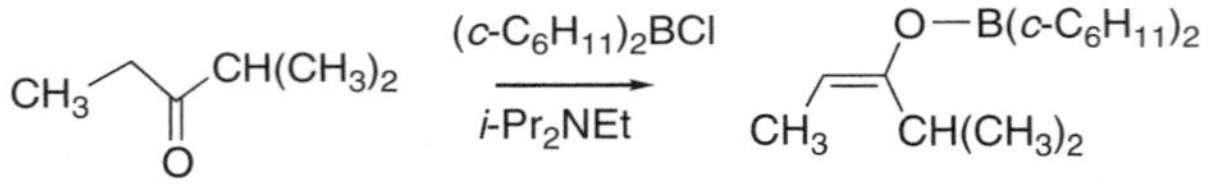

The contrasting stereoselectivity of the boron triflates and chlorides has been discussed in terms of reactant conformation and the stereoelectronic requirement for alignment of the hydrogen being removed with the carbonyl group π orbital.[18] With the triflate reagents, the boron is *anti* to the enolizable group. With the bulkier dicyclohexylboron chloride, the boron favors a conformation *cis* to the enolizable group. A computational study of the reaction also indicates that the size of the boron ligand and the resulting conformational changes are the dominant factors in determining stereoselectivity.[19] There may also be a distinction between the two types of borylation reagents in the extent of dissociation of the leaving group. The triflate is probably an ion pair, whereas with the less reactive chloride, the deprotonation may be a concerted (E2-like) process.[18b] The two proposed TSs are shown below.

[16] D. A. Evans, E. Vogel, and J. V. Nelson, *J. Am. Chem. Soc.*, **101**, 6120 (1979); D. A. Evans, J. V. Nelson, E. Vogel, and T. R. Taber, *J. Am. Chem. Soc.*, **103**, 3099 (1981).

[17] H. C. Brown, R. K. Dhar, R. K. Bakshi, P. K. Pandiarajan, and B. Singaram, *J. Am. Chem. Soc.*, **111**, 3441 (1989); H. C. Brown, R. K. Dhar, K. Ganesan, and B. Singaram, *J. Org. Chem.*, **57**, 499 (1992); H. C. Brown, R. K. Dhar, K. Ganesan, and B. Singaram, *J. Org. Chem.*, **57**, 2716 (1992); H. C. Brown, K. Ganesan, and R. K. Dhar, *J. Org. Chem.*, **58**, 147 (1993); K. Ganesan and H. C. Brown, *J. Org. Chem.*, **58**, 7162 (1993).

[18] (a) J. M. Goodman and I. Paterson, *Tetrahedron Lett.*, **33**, 7223 (1992); (b) E. J. Corey and S. S. Kim, *J. Am. Chem. Soc.*, **112**, 4976 (1990).

[19] J. Murga, E. Falomir, M. Carda, and J. A. Marco, *Tetrahedron*, **57**, 6239 (2001).

Z-Boron enolates can also be obtained from silyl enol ethers by reaction with the bromoborane derived from 9-BBN (9-borabicyclo[3.3.1]nonane). This method is necessary for ketones such as 2,2-dimethyl-3-pentanone, which give *E*-boron enolates by other methods. The *Z*-stereoisomer is formed from either the *Z*- or *E*-silyl enol ether.[20]

The *E*-boron enolate from cyclohexanone shows a preference for the *anti* aldol product. The ratio depends on the boron alkyl groups and is modest (2:1) with di-*n*-butylboron but greater than 20:1 for cyclopentyl-*n*-hexylboron.[16]

The general trend is that boron enolates *parallel* lithium enolates in their stereoselectivity but show *enhanced stereoselectivity*. There also are some advantages in terms of access to both stereoisomeric enol derivatives. Another important characteristic of boron enolates is that they are not subject to internal chelation. The tetracoordinate dialkylboron in the cyclic TS is not able to accept additional ligands, so there is no tendency to form a chelated TS when the aldehyde or enolate carries a donor substituent. Table 2.2 gives some typical data for boron enolates and shows the strong correspondence between enolate configuration and product stereochemistry.

2.1.2.3. Aldol Reactions of Titanium, Tin, and Zirconium Enolates. Metals such as Ti, Sn, and Zr give enolates that are intermediate in character between the ionic Li$^+$ enolates and covalent boron enolates. The Ti, Sn, or Zr enolates can accommodate additional ligands. Tetra-, penta-, and hexacoordinate structures are possible. This permits the formation of chelated TSs when there are nearby donor groups in the enolate or electrophile. If the number of anionic ligands exceeds the oxidation state of the metal, the complex has a formal negative charge on the metal and is called an "ate" complex. Such structures enhance the nucleophilicity of enolate ligands. Depending on the nature of the metal ligands, either a cyclic or an acyclic TS can be involved. As we will see in Section 2.1.3.5, the variability in the degree and nature of coordination provides an additional factor in analysis and control of stereoselectivity.

[20.] J. L. Duffy, T. P. Yoon, and D. A. Evans, *Tetrahedron Lett.*, **36**, 9245 (1993).

Table 2.2. Diastereoselectivity of Boron Enolates toward Aldehydes[a]

R^1	L	X	R^2	Z : E	syn:anti
C$_2$H$_5$[b]	n-C$_4$H$_9$	OTf	Ph	>97:3	>97:3
C$_2$H$_5$[b]	n-C$_4$H$_9$	OTf	Ph	69:31	72:28
C$_2$H$_5$[b]	n-C$_4$H$_9$	OTf	n-C$_3$H$_7$	>97:3	>97:3
C$_2$H$_5$[b]	n-C$_4$H$_9$	OTf	t-C$_4$H$_9$	>97:3	>97:3
C$_2$H$_5$[b]	n-C$_4$H$_9$	OTf	CH$_2$=CHCH$_3$	>97:3	92:8
C$_2$H$_5$[b]	n-C$_4$H$_9$	OTf	E-C$_4$H$_7$	>97:3	93:7
i-C$_3$H$_7$[b]	n-C$_4$H$_9$	OTf	Ph	45:55	44:56
i-C$_4$H$_9$[b]	n-C$_4$H$_9$	OTf	Ph	>99:1	>97:3
t-C$_4$H$_9$[b]	n-C$_4$H$_9$	OTf	Ph	>99:1	>97:3
n-C$_5$H$_{11}$[c]	n-C$_4$H$_9$	OTf	Ph	95:5	94:6
n-C$_9$H$_{19}$[c]	n-C$_4$H$_9$	OTf	Ph	91:9	91:9
c-C$_6$H$_{11}$[c]	n-C$_4$H$_9$	OTf	Ph	95:5	94:6
PhCH$_2$[c]	n-C$_4$H$_9$	OTf	Ph	98:2	>99:1
Ph[b]	n-C$_4$H$_9$	OTf	Ph	96:4	95:5
C$_2$H$_5$[d]	c-C$_6$H$_{11}$	Cl	Ph		21:79
i-C$_3$H$_7$[d]	c-C$_6$H$_{11}$	Cl	Ph		<3:97
c-C$_6$H$_{11}$[d]	c-C$_6$H$_{11}$	Cl	Ph		<1:99
t-C$_4$H$_9$[d]	c-C$_6$H$_{11}$	Cl	Ph		<3:97

a. From a more complete compilation, see C. H. Heathcock, in *Asymmetric Synthesis*, Vol. 3, J. D. Morrison, ed., Academic Press, New York, 1984, Chap. 3.
b. D. A. Evans, J. V. Nelson, E. Vogel, and T. R. Taber, *J. Am. Chem. Soc.*, **103**, 3099 (1981).
c. I. Kuwajima, M. Kato, and A. Mori, *Tetrahedron Lett.*, **21**, 4291 (1980).
d. H. C. Brown, R. K. Dhar, R. K. Bakshi, P. K. Pandiarajan, and P. Singaram, *J. Am. Chem. Soc.*, **111**, 3441 (1989); H. C. Brown, K. Ganesan, and R. K. Dhar, *J. Org. Chem.*, **58**, 147 (1993).

Titanium enolates can be prepared from lithium enolates by reaction with a trialkoxytitanium(IV) chloride, such as *tris*-(isopropoxy)titanium chloride.[21] Titanium enolates are usually prepared directly from ketones by reaction with TiCl$_4$ and a tertiary amine.[22] Under these conditions, the *Z*-enolate is formed and the aldol adducts have *syn* stereochemistry. The addition step proceeds through a cyclic TS assembled around titanium.

Z-enolate, R^E = H, *syn*
E-enolate, R^Z = H, *anti*

Entry 8 in Scheme 2.1 is an example of this method. Titanium enolates are frequently employed in the synthesis of complex molecules and with other carbonyl derivatives,

[21.] C. Siegel and E. R. Thornton, *J. Am. Chem. Soc.*, **111**, 5722 (1989).
[22.] D. A. Evans, D. L. Rieger, M. T. Bilodeau, and F. Urpi, *J. Am. Chem. Soc.*, **113**, 1047 (1991).

such as the *N*-acyloxazolidinones that serve as chiral auxiliaries (see Section 2.1.3.4).

Mixed aldehyde-aldehyde additions have been carried out using TiCl$_4$ and TMEDA. The reaction gives *syn* adducts, presumably through a cyclic TS. Treatment of the *syn* adducts with 1 mol % Ti(O-*i*-Pr)$_4$ leads to equilibration to the more stable *anti* isomer.[23]

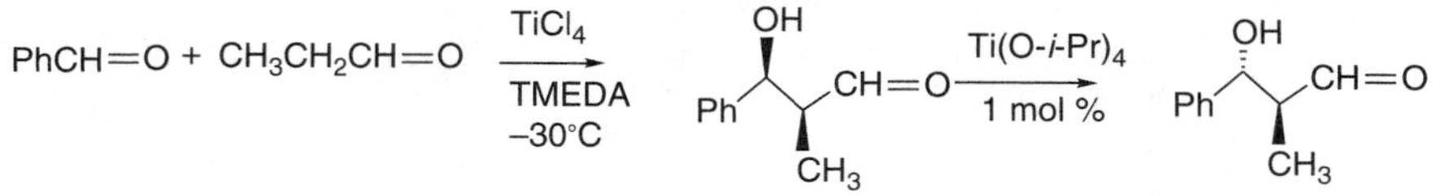

The equilibration in this case is believed to involve oxidation-reduction at the alcohol center, rather than reversal of the addition. (See Section 5.3.2 for a discussion of Ti(O-*i*-Pr)$_4$ as an oxidation-reduction catalyst.)

Ketone-aldehyde additions have been effected using TiCl$_4$ in toluene.[24] These reactions exhibit the same stereoselectivity trends as other titanium-mediated additions. With unsymmetrical ketones, this procedure gives the product from the more-substituted enolate.[25]

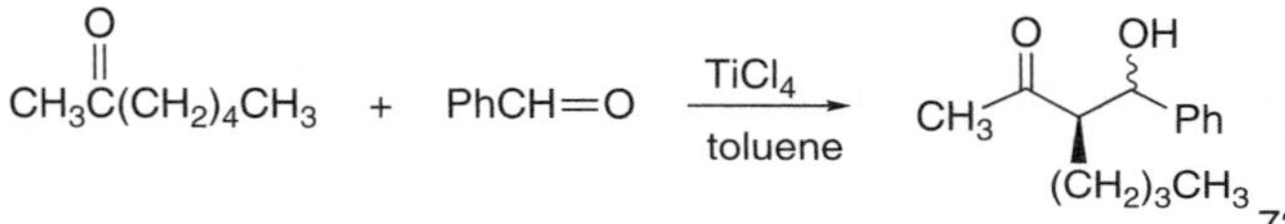

Titanium enolates can also be used under conditions in which the titanium exists as an "ate" species. Crossed aldehyde-aldehyde additions have been accomplished starting with trimethylsilyl enol ethers, which are converted to lithium enolates and then to "ate" species by addition of Ti(O-*n*-Bu)$_4$.[26] These conditions show only modest stereoselectivity.

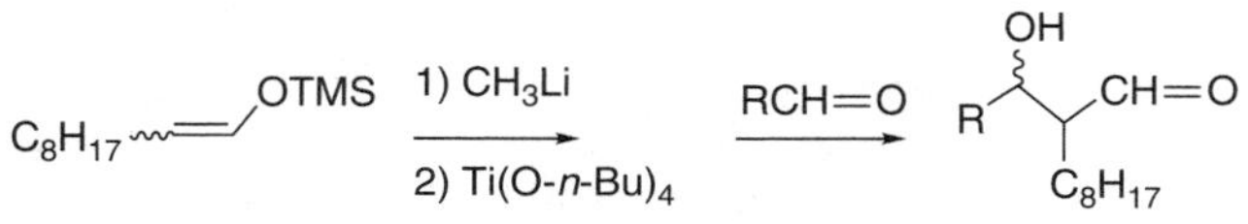

Silyl enol ether	R	*syn:anti*
Z	C$_2$H$_5$	28:72
Z	(CH$_3$)$_2$CH	20:80
Z	(CH$_3$)$_3$C	10:90
Z	Ph	54:46
E	(CH$_3$)$_2$CH	47:53
E	(CH$_3$)$_3$C	28:72

Titanium "ate" species have also been used to add aldehyde enolates to ketones. This reaction is inherently difficult because of the greater reactivity of aldehyde

23. R. Mahrwald, B. Costisella, and B. Gundogan, *Synthesis*, 262 (1998).
24. R. Mahrwald, *Chem. Ber.*, **128**, 919 (1995).
25. R. Mahrwald and B. Gundogan, *J. Am. Chem. Soc.*, **120**, 413 (1998).
26. K. Yachi, H. Shinokubo, and K. Oshima, *J. Am. Chem. Soc.*, **121**, 9465 (1999).

carbonyls over ketone carbonyls. The reaction works best with ketones having EWG substituents such as alkynones and α-haloketones. The reaction is thought to proceed through a cyclic intermediate that is stable until hydrolysis. This cyclic intermediate may be necessary to drive the normally unfavorable equilibrium of the addition step.

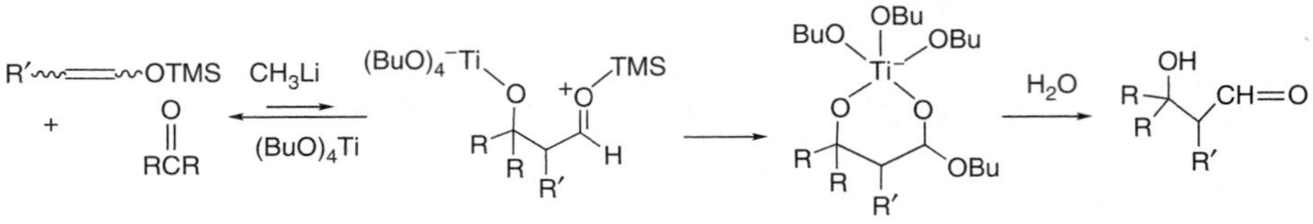

Tin enolates are also used in aldol reactions.[27] Both the Sn(II) and Sn(IV) oxidation states are reactive. Tin(II) enolates can be generated from ketones and $Sn(II)(O_3SCF_3)_2$ in the presence of tertiary amines.[28] The subsequent aldol addition is *syn* selective and independent of enolate configuration.[29] This preference arises from avoidance of *gauche* interaction of the aldehyde group and the enolate β-substituent. The *syn* stereoselectivity indicates that reaction occurs through an open TS.

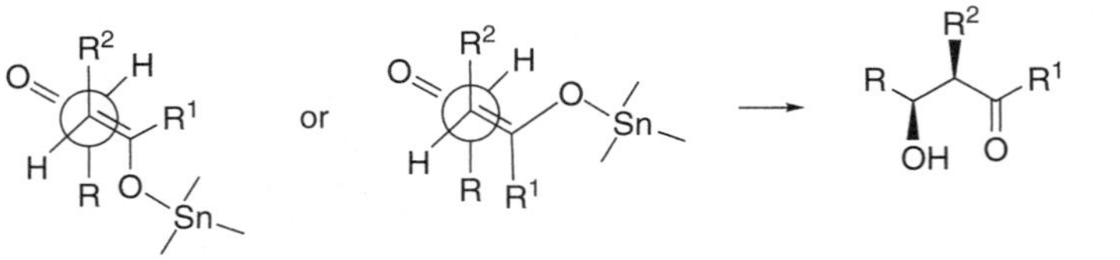

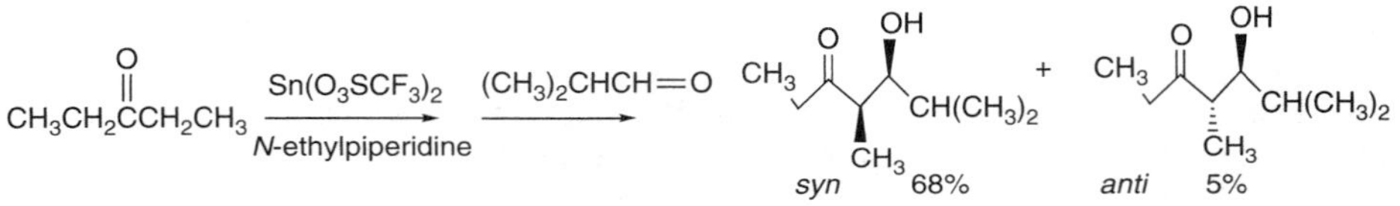

Even cyclohexanone gives the *syn* product.

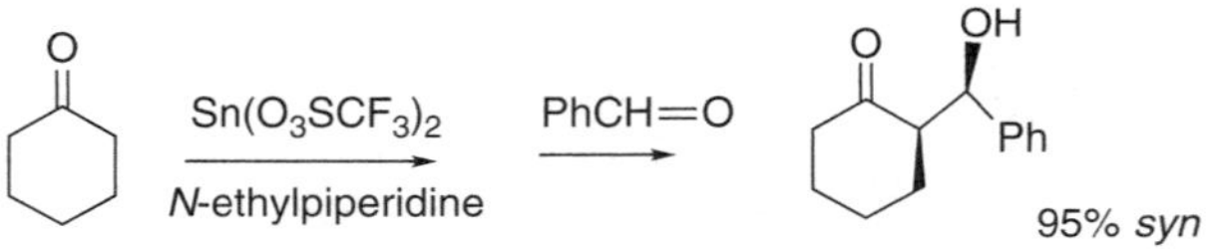

Entry 9 of Scheme 2.1 is an example of application of these conditions. Tin(II) enolates prepared in this way also show good reactivity toward ketones as the electrophilic component.

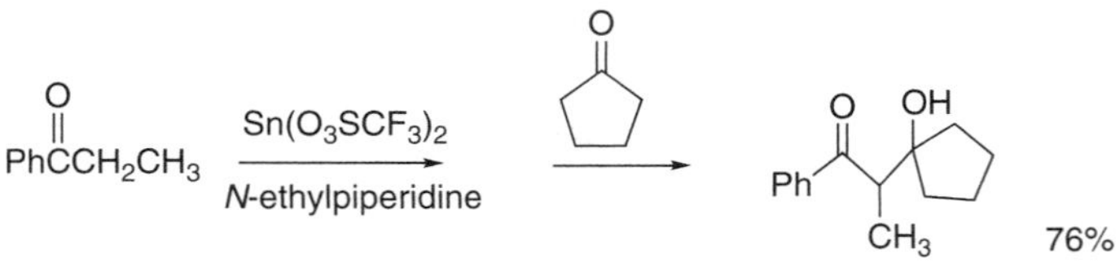

Ref. 30

27. T. Mukaiyama and S. Kobayashi, *Org. React.*, **46**, 1 (1994).
28. T. Mukaiyama, N. Iwasawa, R. W. Stevens, and T. Haga, *Tetrahedron*, **40**, 1381 (1984); I. Shibata and A. Babu, *Org. Prep. Proc. Int.*, **26**, 85 (1994).
29. T. Mukaiyama, R. W. Stevens, and N. Iwasawa, *Chem. Lett.*, 353 (1982).
30. R. W. Stevens, N. Iwasawa, and T. Mukaiyama, *Chem. Lett.*, 1459 (1982).

Trialkylstannyl enolates can be prepared from enol acetates by reaction with trialkyltin alkoxides and are sufficiently reactive to add to aldehydes. Uncatalyzed addition of trialkylstannyl enolates to benzaldehyde shows *anti* stereoselectivity.[31]

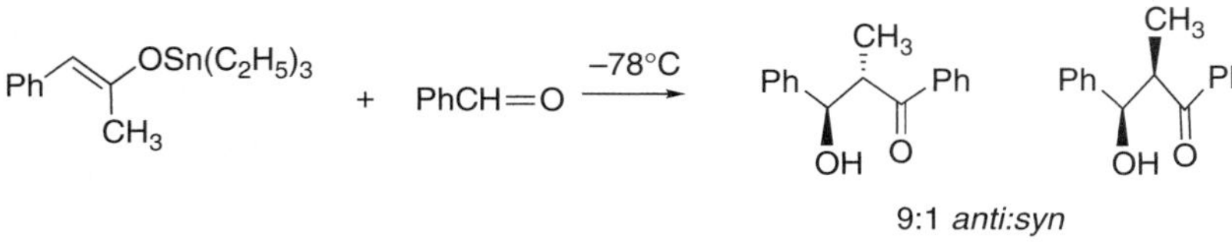

9:1 *anti:syn*

Isolated tri-*n*-butylstannyl enolates react with benzaldehyde under the influence of metal salts including $Pd(O_3SCF_3)_2$, $Zn(O_3SCF_3)_2$, and $Cu(O_3SCF_3)_2$.[32] The tri-*n*-butylstannyl enol derivative of cyclohexanone gives mainly *anti* product. The *anti:syn* ratio depends on the catalyst, with $Pd(O_3SCF_3)_2$ giving the highest ratio.

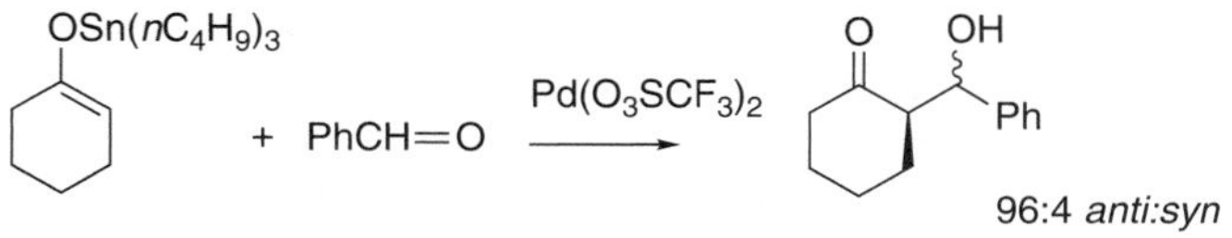

96:4 *anti:syn*

Zirconium *tetra-t*-butoxide is a mildly basic reagent that has occasionally been used to effect aldol addition.[33]

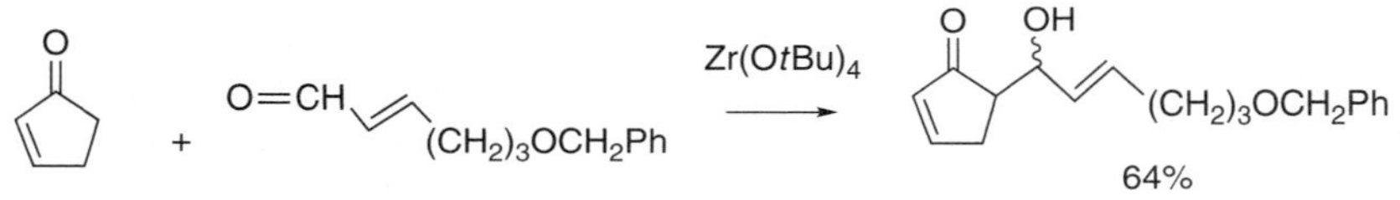

64%

Zirconium enolates can also prepared by reaction of lithium enolates with $(Cp)_2ZrCl_2$, and they act as nucleophiles in aldol addition reactions.[34]

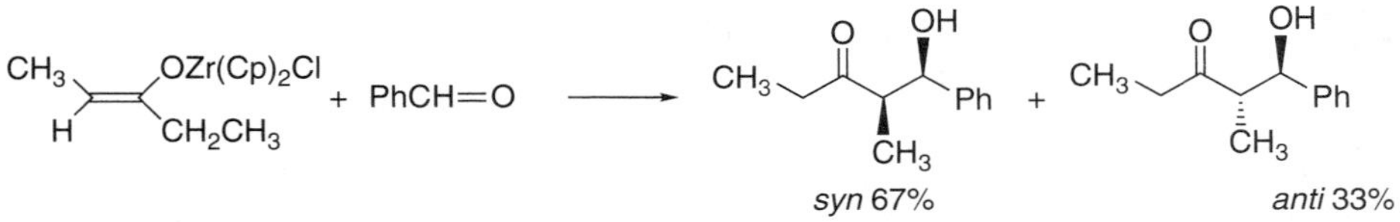

syn 67%   *anti* 33%

Ref. 34d

anti 83%   *syn* 17%

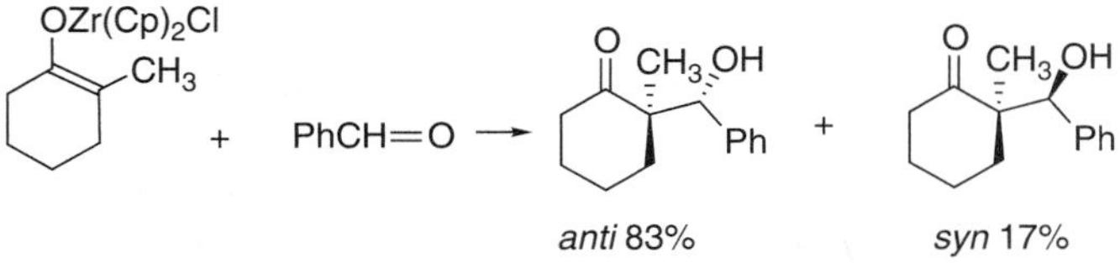

Ref. 34d

31. S. S. Labadie and J. K. Stille, *Tetrahedron*, **40**, 2329 (1984).
32. A. Yanagisawa, K. Kimura, Y. Nakatsuka, and M. Yamamoto, *Synlett*, 958 (1998).
33. H. Sasai, Y. Kirio, and M. Shibasaki, *J. Org. Chem.*, **55**, 5306 (1990).
34. (a) D. A. Evans and L. R. McGee, *Tetrahedron Lett.*, **21**, 3975 (1980); (b) Y. Yamamoto and K. Maruyama, *Tetrahedron Lett.*, **21**, 4607 (1980); (c) M. Braun and H. Sacha, *Angew. Chem. Int. Ed. Engl.*, **30**, 1318 (1991); (d) S. Yamago, D. Machii, and E. Nakamura, *J. Org. Chem.*, **56**, 2098 (1991).

A comparison of the *anti:syn* diastereoselectivity of the lithium, dibutylboron, and $(Cp)_2Zr$ enolates of 3-methyl-2-hexanone with benzaldehyde has been reported.[34d] The order of stereoselectivity is $Bu_2B > (Cp)_2Zr > Li$. These results suggest that the reactions of the zirconium enolates proceed through a cyclic TS.

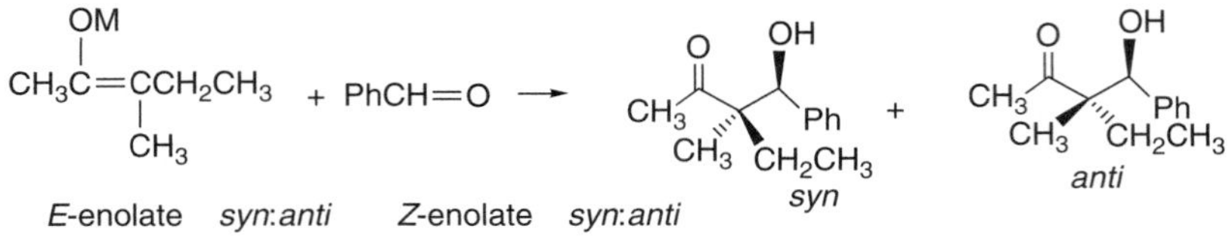

E-enolate	syn:anti	Z-enolate	syn:anti
Li	17:83	Li	45:55
Bu₂B	3:97	Bu₂B	94:6
(Cp)₂ZrCl	9:91	(Cp)₂ZrCl	86:14

2.1.2.4. Summary of the Relationship between Diastereoselectivity and the Transition Structure. In this section we considered *simple diastereoselection* in aldol reactions of ketone enolates. Numerous observations on the reactions of enolates of ketones and related compounds are consistent with the general concept of a chairlike TS.[35] These reactions show a consistent $E \rightarrow anti : Z \rightarrow syn$ relationship. Noncyclic TSs have more variable diastereoselectivity. The prediction or interpretation of the specific ratio of *syn* and *anti* product from any given reaction requires assessment of several variables: (1) What is the stereochemical composition of the enolate? (2) Does the Lewis acid promote tight coordination with both the carbonyl and enolate oxygen atoms and thereby favor a cyclic TS? (3) Does the TS have a chairlike conformation? (4) Are there additional Lewis base coordination sites in either reactant that can lead to reaction through a chelated TS? Another factor comes into play if either the aldehyde or the enolate, or both, are chiral. In that case, facial selectivity becomes an issue and this is considered in Section 2.1.5.

2.1.3. Aldol Addition Reactions of Enolates of Esters and Other Carbonyl Derivatives

The enolates of other carbonyl compounds can be used in mixed aldol reactions. Extensive use has been made of the enolates of esters, thiol esters, amides, and imides, including several that serve as chiral auxiliaries. The methods for formation of these enolates are similar to those for ketones. Lithium, boron, titanium, and tin derivatives have all been widely used. The silyl ethers of ester enolates, which are called *silyl ketene acetals*, show reactivity that is analogous to silyl enol ethers and are covalent equivalents of ester enolates. The silyl thioketene acetal derivatives of thiol esters are also useful. The reactions of these enolate equivalents are discussed in Section 2.1.4.

Because of their usefulness in aldol additions and other synthetic methods (see especially Section 6.4.2.3), there has been a good deal of interest in the factors that

35. C. H. Heathcock, *Modern Synthetic Methods*, **6**, 1 (1992); C. H. Heathcock, in *Asymmetric Syntheses*, Vol. 3, J. D. Morrison, ed., 1984, Chap. 2, Academic Press; C. H. Heathcock, in *Comprehensive Carbanion Chemistry*, Part B, E. Buncel and T. Durst, ed., Elsevier, Amsterdam, 1984, Chap. 4; D. A. Evans, J. V. Nelson, and T. R. Taber, *Top. Stereochem.*, **13**, 1 (1982); A. T. Nielsen and W. J. Houlihan, *Org. React.*, **16**, 1 (1968); R. Mahrwald, ed., *Modern Aldol Reactions*, Wiley-VCH (2004).

control the stereoselectivity of enolate formation from esters. For simple esters such as ethyl propanoate, the *E*-enolate is preferred under kinetic conditions using a strong base such as LDA in THF solution. Inclusion of a strong cation-solvating cosolvent, such as HMPA or DMPU, favors the *Z*-enolate.[36] These enolates can be trapped and analyzed as the corresponding silyl ketene acetals. The relationships are similar to those discussed for formation of ketone enolates in Section 1.1.2.

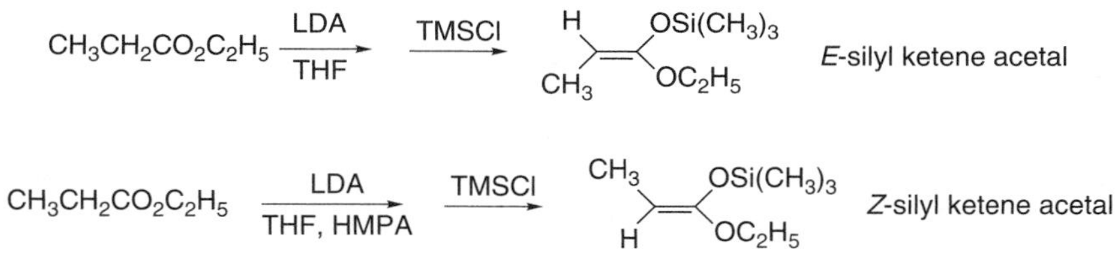

These observations are explained in terms of a chairlike TS for the LDA/THF conditions and a more open TS in the presence of an aprotic dipolar solvent.

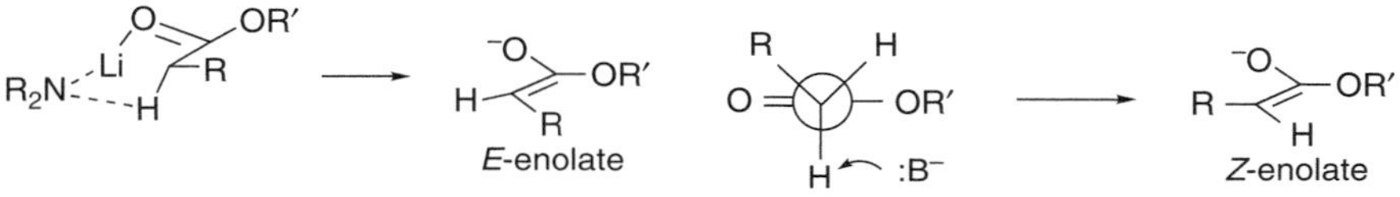

Despite the ability to control ester enolate geometry, the aldol addition reactions of unhindered ester enolate are not very stereoselective.[37]

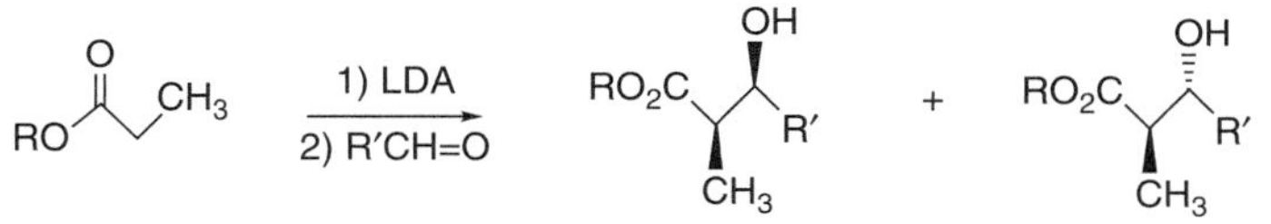

R	R′	syn:anti
CH_3	$(CH_3)_2CH$	45:55
CH_3	Ph	45:55
$(CH_3)_3C$	Ph	49:51

This stereoselectivity can be improved by use of a very bulky group. 2,6-Dimethylphenyl esters give *E*-enolates and *anti* aldol adducts.[38]

36. R. E. Ireland and A. K. Willard, *Tetrahedron Lett.*, 3975 (1975); R. E. Ireland, R. H. Mueller, and A. K. Willard, *J. Am. Chem. Soc.*, **98**, 2868 (1976); R. E. Ireland, P. Wipf, and J. D. Armstrong, III, *J. Org. Chem.*, **56**, 650 (1991).

37. A. I. Meyers and P. J. Reider, *J. Am. Chem. Soc.*, **101**, 2501 (1979); C. H. Heathcock, C. T. Buse, W. A. Kleschick, M. C. Pirrung, J. E. Sohn, and J. Lampe, *J. Org. Chem.*, **45**, 1066 (1980).

38. M. C. Pirrung and C. H. Heathcock, *J. Org. Chem.*, **45**, 1728 (1980).

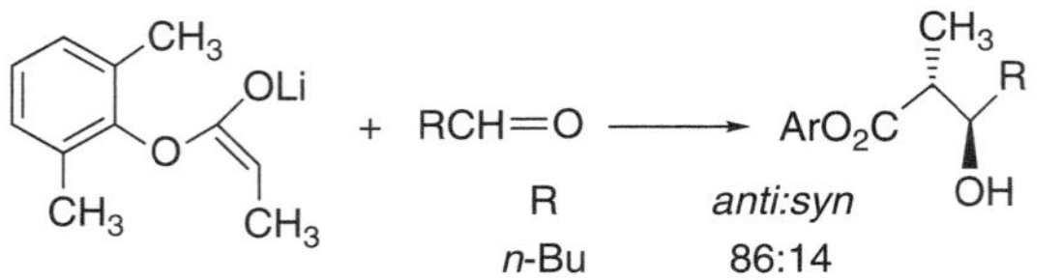

R	anti:syn
n-Bu	86:14
i-Pr	>98:2
t-Bu	>98:2
Ph	88:12

The lithium enolates of α-alkoxy esters exhibit high stereoselectivity, which is consistent with involvement of a chelated enolate.[37a,39] The chelated ester enolate is approached by the aldehyde in such a manner that the aldehyde R group avoids being between the α-alkoxy and methyl groups in the ester enolate. A *syn* product is favored for most ester groups, but this shifts to *anti* with extremely bulky groups.

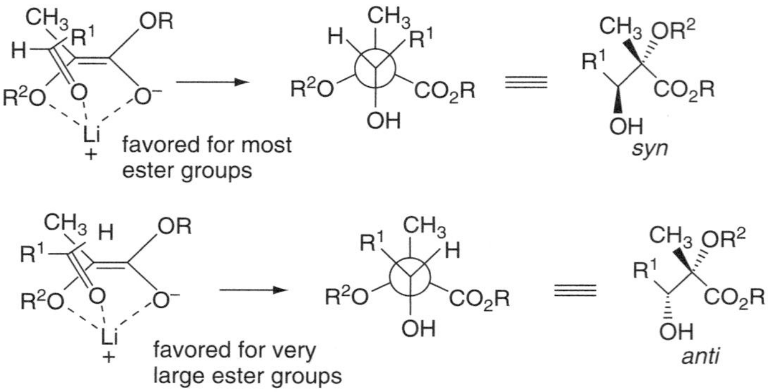

RO	syn:anti
Methyl	70:30
2,6-Dimethylphenyl	83:17
2,6-Di-(i-propyl)phenyl	33:67
2,6-Di-(t-butyl)-4-methylphenyl	< 3:97

Boron enolates can be obtained from esters[40,41] and amides[42] by methods that are similar to those used for ketones. Various combinations of borylating reagents and amines have been used and the *E:Z* ratios are dependent on the reagents and conditions. In most cases esters give *Z*-enolates, which lead to *syn* adducts, but there are exceptions. Use of branched-chain alcohols increases the amount of *anti* enolate, and with *t*-butyl esters the product ratio is higher than 97:3.

39. C. H. Heathcock, M. C. Pirrung, S. D. Young, J. P. Hagen, E. T. Jarvi, U. Badertscher, H.-P. Marki, and S. H. Montgomery, *J. Am. Chem. Soc.*, **106**, 8161 (1984).
40. K. Ganesan and H. C. Brown, *J. Org. Chem.*, **59**, 2336 (1994).
41. A. Abiko, J.-F. Liu, and S. Masamune, *J. Org. Chem.*, **61**, 2590 (1996); T. Inoue, J.-F. Liu, D. C. Buske, and A. Abiko, *J. Org. Chem.*, **67**, 5250 (2002).
42. K. Ganesan and H. C. Brown, *J. Org. Chem.*, **59**, 7346 (1994).

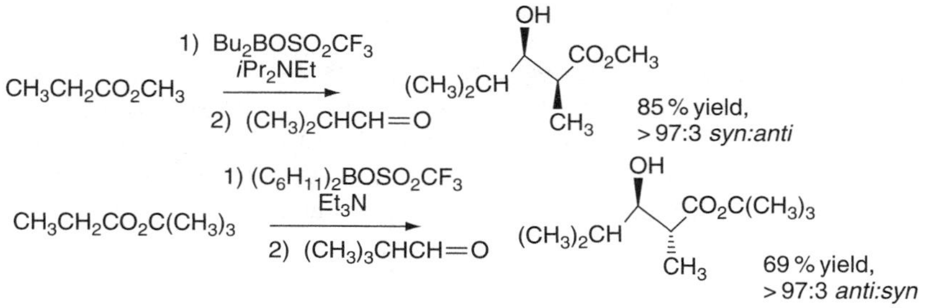

Ref. 41

Branched-chain esters also give mainly *anti* adducts when the enolates are formed using dicyclohexyliodoborane.

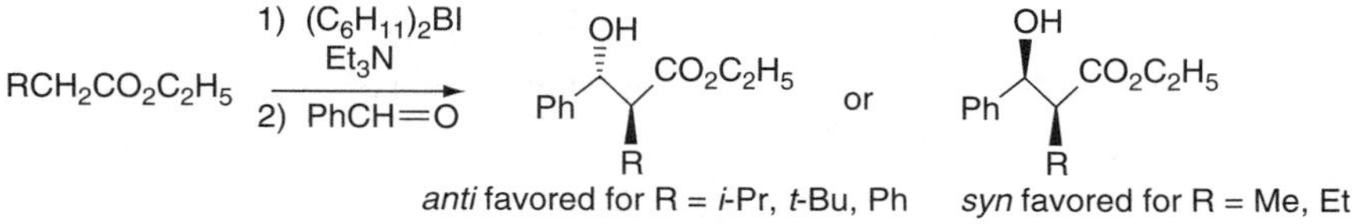

anti favored for R = *i*-Pr, *t*-Bu, Ph   *syn* favored for R = Me, Et

Ref. 40

Phenyl and phenylthio esters have proven to be advantageous in $TiCl_4$-mediated additions, perhaps because they are slightly more acidic than the alkyl analogs. The reactions show *syn* diastereoselectivity, indicating that Z-enolates are formed.[43]

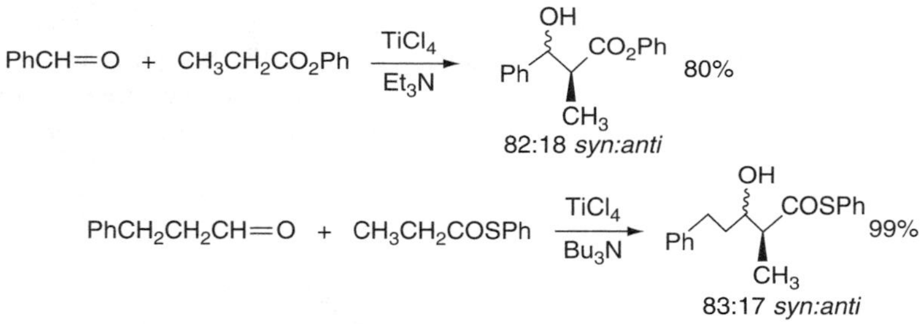

80%

82:18 *syn:anti*

99%

83:17 *syn:anti*

Among the most useful carbonyl derivatives are N-acyloxazolidinones, and as we shall see in Section 2.3.4, they provide facial selectivity in aldol addition reactions. 1,3-Thiazoline-2-thiones constitute another useful type of chiral auxiliary, and they can be used in conjunction with $Bu_2BO_3SCF_3$,[44] $Sn(O_3SCF_3)_2$,[45] or $TiCl_4$[46] for generation of enolates. The stereoselectivity of the reactions is consistent with formation of a Z-enolate and reaction through a cyclic TS.

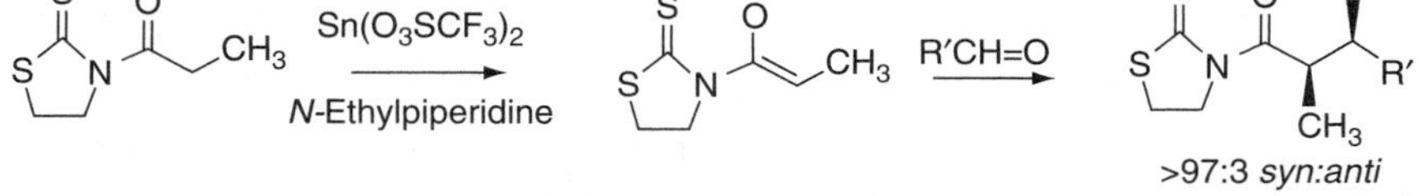

>97:3 *syn:anti*

Ref. 47

43. Y. Tanabe, N. Matsumoto, S. Funakoshi, and N. Manta, *Synlett*, 1959 (2001).

44. C.-N. Hsiao, L. Liu, and M. J. Miller, *J. Org. Chem.*, **52**, 2201 (1987).

45. Y. Nagao, Y. Hagiwara, T. Kumagai, M. Ochiai, T. Inoue, K. Hashimoto, and E. Fujita, *J. Org. Chem.*, **51**, 2391 (1986); Y. Nagao, Y. Nagase, T. Kumagai, H. Matsunaga, T. Abe, O. Shimada, T. Hayashi, and Y. Inoue, *J. Org. Chem.*, **57**, 4243 (1992).

46. D. A. Evans, S. J. Miller, M. D. Ennis, and P. L. Ornstein, *J. Org. Chem.*, **57**, 1067 (1992).

47. T. Mukaiyama and N. Isawa, *Chem. Lett.*, 1903 (1982); N. Isawa, H. Huang, and T. Mukaiyama, *Chem. Lett.*, 1045 (1985).

2.1.4. The Mukaiyama Aldol Reaction

The *Mukaiyama aldol reaction* refers to Lewis acid–catalyzed aldol addition reactions of silyl enol ethers, silyl ketene acetals, and similar *enolate equivalents*.[48] Silyl enol ethers are not sufficiently nucleophilic to react directly with aldehydes or ketones. However, Lewis acids cause reaction to occur by coordination at the carbonyl oxygen, activating the carbonyl group to nucleophilic attack.

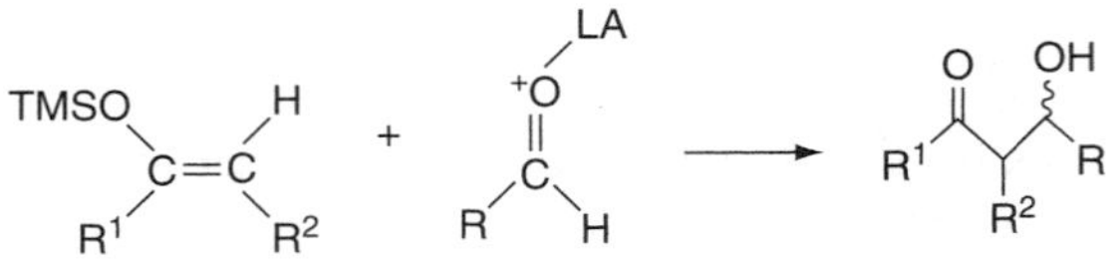

Lewis acids such as $TiCl_4$ and $SnCl_4$ induce addition of both silyl enol ethers and ketene silyl acetals to aldehydes.[49]

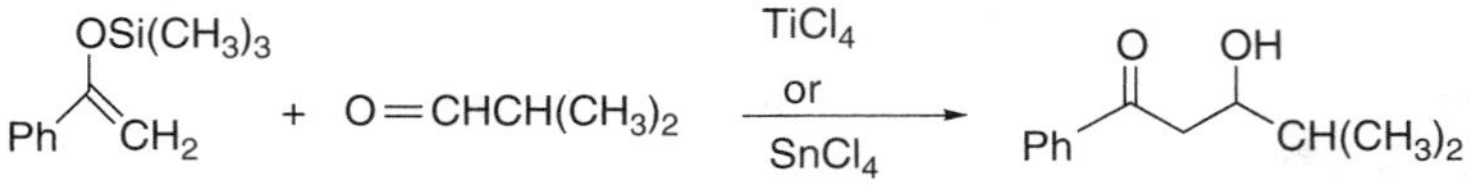

If there is no other interaction, the reaction proceeds through an acyclic TS and steric factors determine the amount of *syn* versus *anti* addition. This is the case with BF_3, where the tetracoordinate boron-aldehyde adduct does not offer any free coordination sites for formation of a cyclic TS. Stereoselectivity increases with the steric bulk of the silyl enol ether substituent R^1.[50]

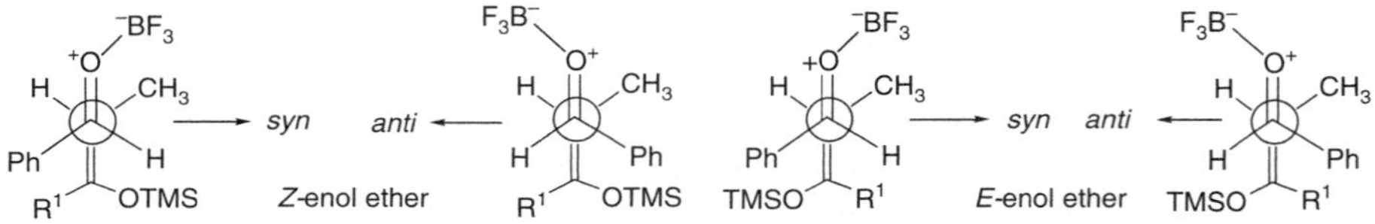

	Z-silyl enol ether	E-silyl enol ether
R^1	*syn:anti*	*syn:anti*
Et	60:40	57:43
i-Pr	56:44	35:65
t-Bu	<5 : 95	–
Ph	47:53	30:70

Quite a number of other Lewis acids can catalyze the Mukaiyama aldol reaction, including $Bu_2Sn(O_3SCF_3)_2$,[51] Bu_3SnClO_4,[52] $Sn(O_3SCF_3)_2$,[53] $Zn(O_3SCF_3)_2$,[54] and

48. R. Mahrwald, *Chem. Rev.*, **99**, 1095 (1999).
49. T. Mukaiyama, K. Banno, and K. Narasaka, *J. Am. Chem. Soc.*, **96**, 7503 (1974).
50. C. H. Heathcock, K. T. Hug, and L. A. Flippin, *Tetrahedron Lett.*, **25**, 5973 (1984).
51. T. Sato, J. Otera, and H. Nozaki, *J. Am. Chem. Soc.*, **112**, 901 (1990).
52. J. Otera and J. Chen, *Synlett*, 321 (1996).
53. T. Oriyama, K. Iwanami, Y. Miyauchi, and G. Koga, *Bull. Chem. Soc. Jpn.*, **63**, 3716 (1990).
54. M. Chini, P. Crotti, C. Gardelli, F. Minutolo, and M. Pineschi, *Gazz. Chim. Ital.*, **123**, 673 (1993).

LiClO$_4$.[55] Cerium, samarium, and other lanthanide halides promote addition of silyl ketene acetals to aldehydes.[56] Triaryl perchlorate salts are also very active catalysts.[57] In general terms, there are at least three possible mechanisms for catalysis. One is through Lewis acid activation of the electrophilic carbonyl component, similar to that discussed for BF$_3$, TiCl$_4$, and SnCl$_4$. Another is by exchange with the enolate equivalent to generate a more nucleophilic species. A third is activation of a catalytic cycle that generates trimethylsilyl cation as the active catalysts.

Aldol additions of silyl enol ethers and silyl ketene acetals can be catalyzed by (Cp)$_2$Zr^{2+} species including [(Cp)$_2$ZrO-t-Bu]$^+$ and (Cp)$_2$Zr(O$_3$SCF$_3$)$_2$.[58]

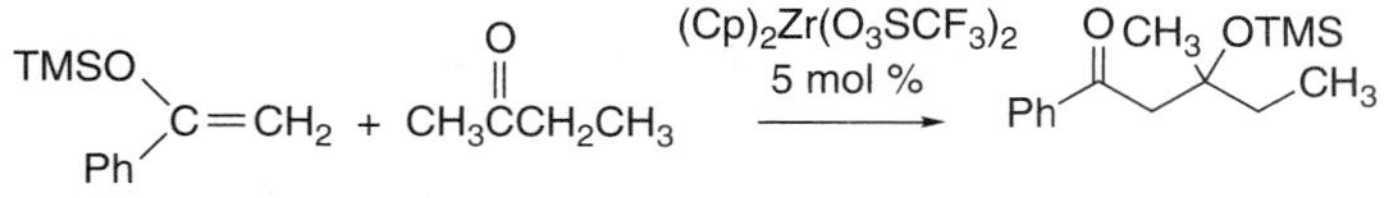

The catalytic cycle involves transfer of the silyl group to the adduct.

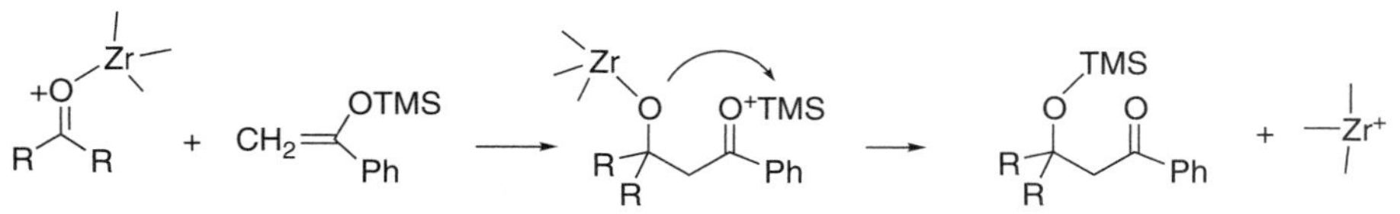

Trialkylsilyl cations may play a key role in other Lewis acid–catalyzed reactions.[59] For example, trimethylsilyl triflate can be formed by intermolecular transfer of the silyl group. When this occurs, the trimethylsilyl triflate can initiate a catalytic cycle that does not directly involve the Lewis acid.

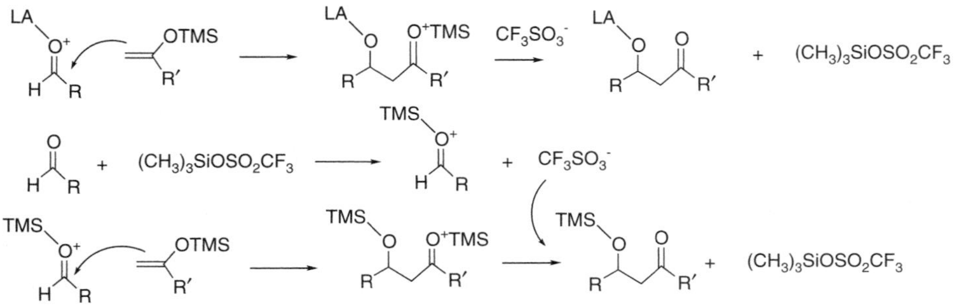

55. M. T. Reetz and D. N. A. Fox, *Tetrahedron Lett.*, **34**, 1119 (1993).

56. P. Van de Weghe and J. Colin, *Tetrahedron Lett.*, **34**, 3881 (1993); A. E. Vougioukas and H. B. Kagan, *Tetrahedron Lett.*, **28**, 5513 (1987).

57. T. Mukaiyama, S. Kobayashi, and M. Murakami, *Chem. Lett.*, 447 (1985); T. Mukaiyama, S. Kobayashi, and M. Murakami, *Chem. Lett.*, 1759 (1984); S. E. Denmark and C.-T. Chen, *Tetrahedron Lett.*, **35**, 4327 (1994).

58. (a) T. K. Hollis, N. P. Robinson, and B. Bosnich, *Tetrahedron Lett.*, **33**, 6423 (1992); (b) Y. Hong, D. J. Norris, and S. Collins, *J. Org. Chem.*, **58**, 3591 (1993).

59. E. M. Carreira and R. A. Singer, *Tetrahedron Lett.*, **35**, 4323 (1994); T. K. Hollis and B. Bosnich, *J. Am. Chem. Soc.*, **117**, 4570 (1995).

Hindered *bis*-phenoxyaluminum derivatives are powerful cocatalysts for reactions mediated by TMS triflate and are believed to act by promoting formation of trimethylsilyl cations by sequestering the triflate anion.[60]

MABR = *bis*-(4-bromo-2,6-di-*tert*-butylphenoxy)methylaluminum

The lanthanide salts are unique among Lewis acids in that they can be effective as catalysts in aqueous solution.[61] Silyl enol ethers react with formaldehyde and benzaldehyde in water-THF mixtures using lanthanide triflates such as $Yb(O_3SCF_3)_3$. The catalysis reflects the strong affinity of lanthanides for carbonyl oxygen, even in aqueous solution.

91% yield, 73:27 *syn:anti*

Ref. 62

Certain other metal ions also exhibit catalysis in aqueous solution. Two important criteria are rate of ligand exchange and the acidity of the metal hydrate. Metal hydrates that are too acidic lead to hydrolysis of the silyl enol ether, whereas slow exchange limits the ability of catalysis to compete with other processes. Indium(III) chloride is a borderline catalysts by these criteria, but nevertheless is effective. The optimum solvent is 95:5 isopropanol-water. Under these conditions, the reaction is *syn* selective, suggesting a cyclic TS.[63]

96:4 *syn:anti*

In addition to aldehydes, acetals can serve as electrophiles in Mukaiyama aldol reactions.[64] Effective catalysts include $TiCl_4$,[65] $SnCl_4$,[66] $(CH_3)_3SiO_3SCF_3$,[67] and

60. M. Oishi, S. Aratake, and H. Yamamoto, *J. Am. Chem. Soc.*, **120**, 8271 (1998).
61. S. Kobayashi and K. Manabe, *Acc. Chem. Res.*, **35**, 209 (2002).
62. S. Kobayashi and I. Hachiya, *J. Org. Chem.*, **59**, 3590 (1994).
63. O. Munoz-Muniz, M. Quintanar-Audelo, and E. Juaristi, *J. Org. Chem.*, **68**, 1622 (2003).
64. Y. Yamamoto, H. Yatagai, Y. Naruta, and K. Maruyama, *J. Am. Chem. Soc.*, **102**, 7107 (1980);
    T. Mukaiyama and M. Murakami, *Synthesis*, 1043 (1987).
65. T. Mukaiyama and M. Hayashi, *Chem. Lett.*, 15 (1974).
66. R. C. Cambie, D. S. Larsen, C. E. F. Rickard, P. S. Rutledge, and P. D. Woodgate, *Austr. J. Chem.*, **39**, 487 (1986).
67. S. Murata, M. Suzuki, and R. Noyori, *Tetrahedron*, **44**, 4259 (1988).

$Bu_2Sn(O_3SCF_3)_2$.[68] The Lewis acids promote ionization of the acetal to an oxonium ion that acts as the electrophile. The products are β-alkoxy ketones.

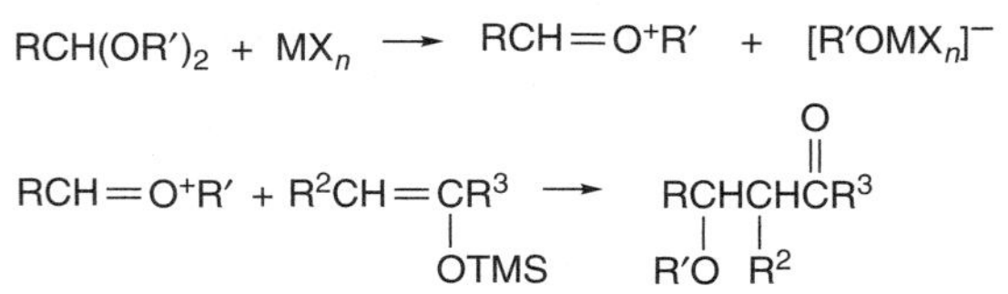

$$RCH(OR')_2 + MX_n \longrightarrow RCH{=}O^+R' + [R'OMX_n]^-$$

$$RCH{=}O^+R' + R^2CH{=}CR^3 \longrightarrow \underset{\underset{R^2}{\overset{\overset{O}{\parallel}}{|}}}{RCHCHCR^3}$$

(with OTMS and R'O substituents shown)

In some cases, the enolate can be formed directly in the presence of the acetal with the Lewis acid also activating the acetal.[69]

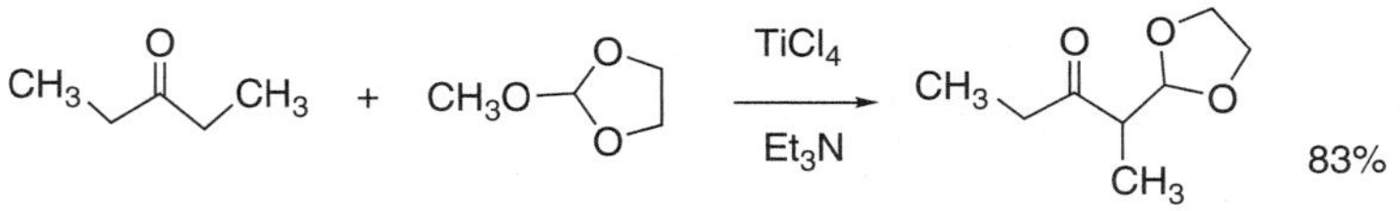

83%

Dibutylboron triflate promotes both enol borinate formation and addition.[70]

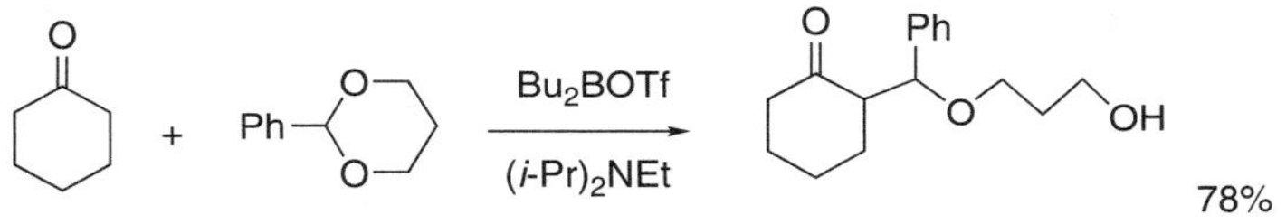

78%

Reactions with acetals can serve to introduce β-alkoxy groups into complex molecules, as in the following reaction.[71]

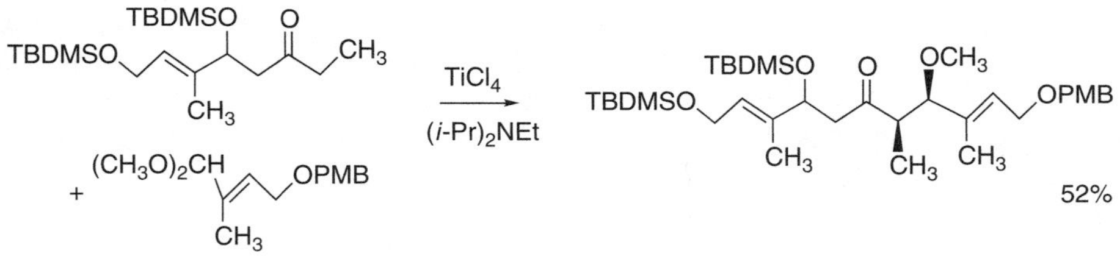

52%

It has been proposed that there may be a single electron transfer mechanism for the Mukaiyama reaction under certain conditions.[72] For example, photolysis of benzaldehyde dimethylacetal and 1-trimethylsilyloxycyclohexene in the presence of a

68. T. Sato, J. Otera, and H. Nozaki, *J. Am. Chem. Soc.*, **112**, 901 (1990).

69. D. A. Evans, F. Urpi, T. C. Somers, J. S. Clark, and M. T. Bilodeau, *J. Am. Chem. Soc.*, **112**, 8215 (1990).

70. L.-S. Li, S. Das, and S. C. Sinha, *Org. Lett.*, **6**, 127 (2004).

71. G. E. Keck, C. A. Wager, T. T. Wager, K. A. Savin, J. A. Covel, M. D. McLaws, D. Krishnamurthy, and V. J. Cee, *Angew. Chem. Int. Ed. Engl.*, **40**, 231 (2001).

72. T. Miura and Y. Masaki, *J. Chem. Soc., Perkin Trans. 1*, 1659 (1994); T. Miura and Y. Masaki, *J. Chem. Soc., Perkin Trans. 1*, 2155 (1995); J. Otera, Y. Fujita, N. Sakuta, M. Fujita, and S. Fukuzumi, *J. Org. Chem.*, **61**, 2951 (1996).

typical photoelectron acceptor, triphenylpyrylium cation, gives an excellent yield of the addition product.

94% yield, 66:34 *syn:anti*

Ref. 73

These reactions may operate by providing a source of trimethylsilyl cations, which serve as the active catalyst by a cycle similar to that for Lewis acids.

The Mukaiyama aldol reaction can provide access to a variety of β-hydroxy carbonyl compounds and use of acetals as reactants can provide β-alkoxy derivatives. The issues of stereoselectivity are the same as those in the aldol addition reaction, but the tendency toward acyclic rather than cyclic TSs reduces the influence of the *E*- or *Z*-configuration of the enolate equivalent on the stereoselectivity.

Scheme 2.2 illustrates several examples of the Mukaiyama aldol reaction. Entries 1 to 3 are cases of addition reactions with silyl enol ethers as the nucleophile and $TiCl_4$ as the Lewis acid. Entry 2 demonstrates steric approach control with respect to the silyl enol ether, but in this case the relative configuration of the hydroxy group was not assigned. Entry 4 shows a fully substituted silyl enol ether. The favored product places the larger C(2) substituent *syn* to the hydroxy group. Entry 5 uses a silyl ketene thioacetal. This reaction proceeds through an open TS and favors the *anti* product.

Entries 6 to 9 involve reactions conducted under catalytic conditions. Entry 6 uses an yttrium catalyst that is active in aqueous solution. Entries 7 and 8 are examples of the use of $(Cp)_2Ti(O_3SCF_3)_2$ as a Lewis acid. Entry 9 illustrates the TMS triflate-MABR catalytic combination.

Entries 10 to 14 show reactions involving acetals. Interestingly, Entry 10 shows much-reduced stereoselectivity compared to the corresponding reaction of the aldehyde (The BF_3-catalyzed reaction of the aldehyde is reported to be 24:1 in favor of the *anti* product; ref. 80, p. 91). There are no stereochemical issues in Entries 11 or 12. Entry 13, involving two cyclic reactants, gave a 2:1 mixture of stereoisomers. Entry 14 is a step in a synthesis directed toward the taxane group of diterpenes. Four stereoisomeric products were produced, including the *Z:E* isomers at the new enone double bond.

2.1.5. Control of Facial Selectivity in Aldol and Mukaiyama Aldol Reactions

In the discussion of the stereochemistry of aldol and Mukaiyama reactions, the most important factors in determining the *syn* or *anti* diastereoselectivity were identified as the nature of the TS (cyclic, open, or chelated) and the configuration (*E* or *Z*) of the enolate. If either the aldehyde or enolate is chiral, an additional factor enters the picture. The aldehyde or enolate then has two nonidentical faces and the stereochemical outcome will depend on *facial selectivity*. In principle, this applies to any stereocenter in the molecule, but the strongest and most studied effects are those of α- and β-substituents. If the aldehyde is chiral, particularly when the stereogenic center is adjacent to the carbonyl group, the competition between the two diastereotopic faces of the carbonyl group determines the stereochemical outcome of the reaction.

[73.] M. Kamata, S. Nagai, M. Kato, and E. Hasegawa, *Tetrahedron Lett.*, **37**, 7779 (1996).

A. Reactions of silyl end ethers with aldehydes and ketones

1[a]

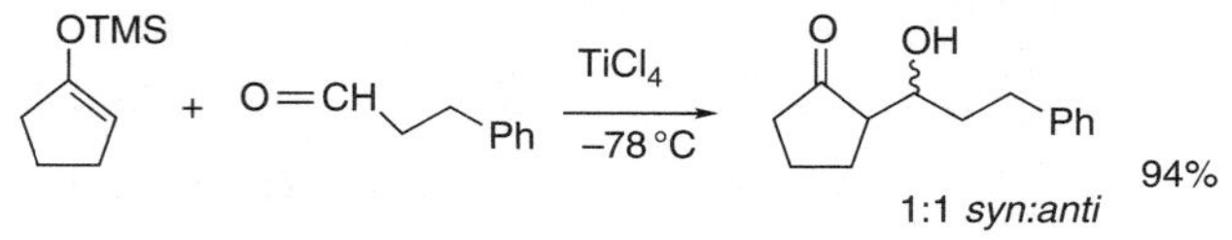

94%

1:1 *syn:anti*

2[b]

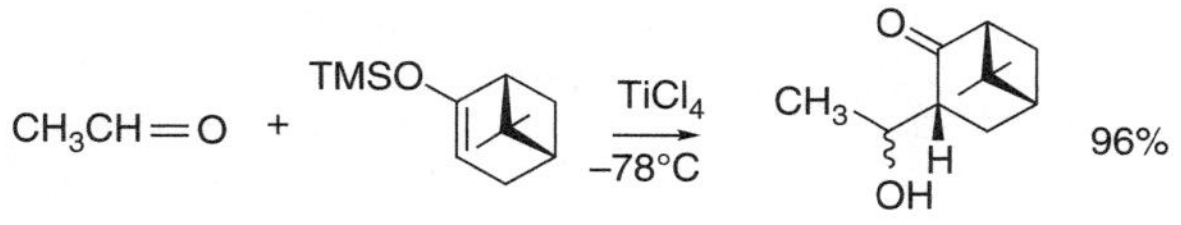

96%

3[c]

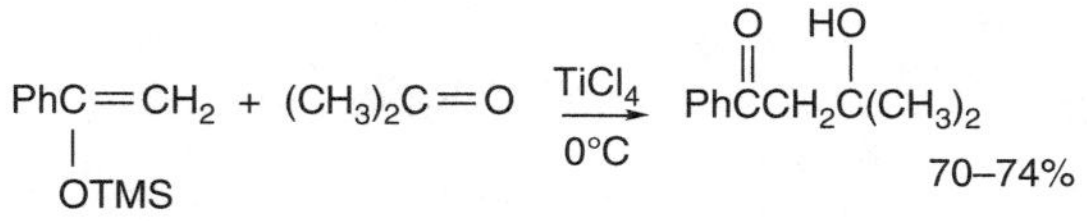

70–74%

4[d]

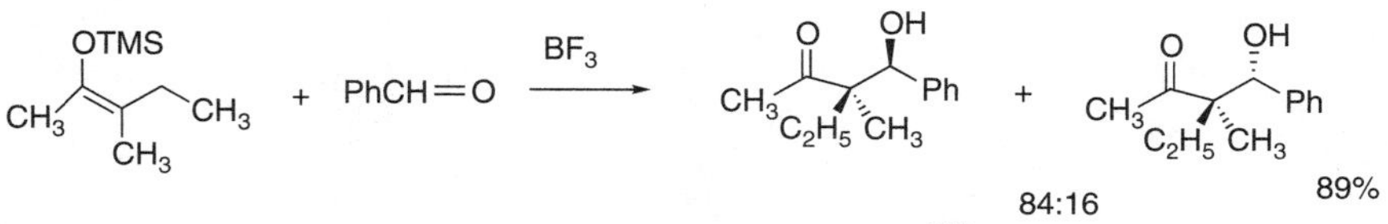

84:16

89%

5[e]

96%

19:1 *anti:syn*

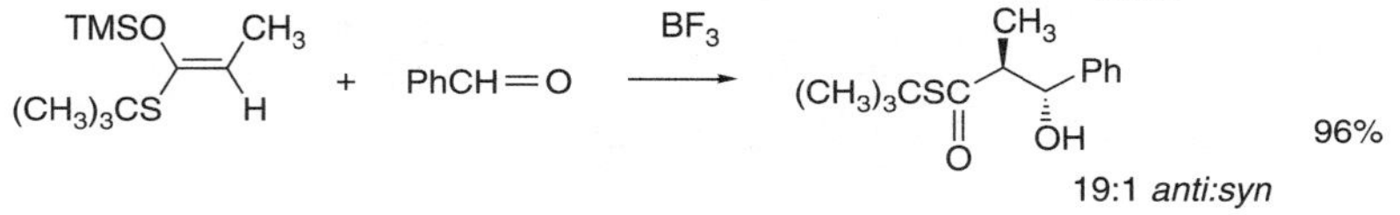

B. Catalytic Mukaiyama Reactions

6[f]

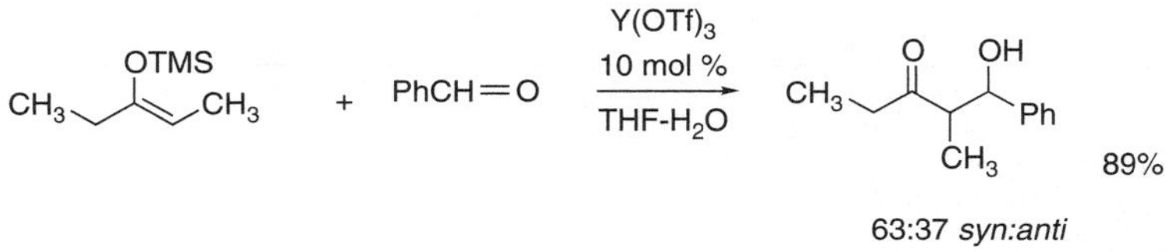

89%

63:37 *syn:anti*

7[g]

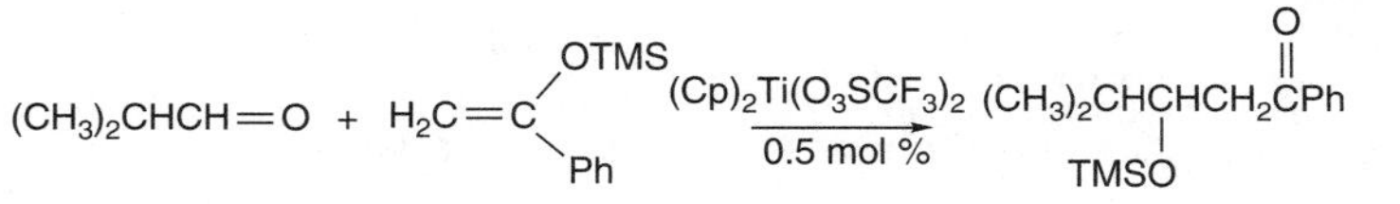

8[h]

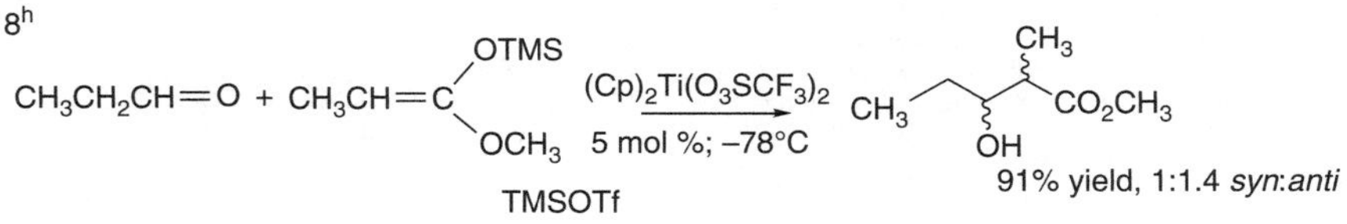

91% yield, 1:1.4 *syn:anti*

9[i]

90%

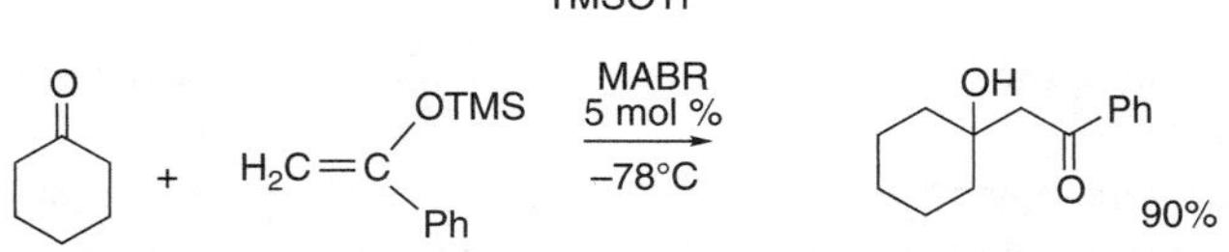

MABR = *bis*(4-bromo-2,6-di-*tert* butylphenoxy) methyl aluminum

(Continued)

Scheme 2.2. (*Continued*)

C. Reactions with acetals

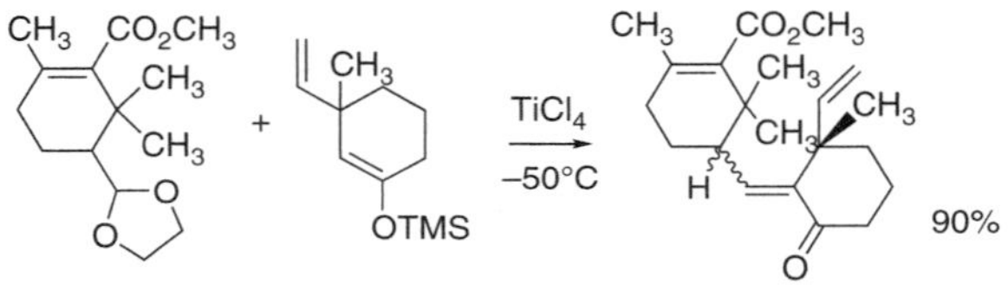

a. T. Mukaiyama, K. Banno, and K. Narasaka, *J. Am. Chem. Soc.*, **96**, 7503 (1974).
b. T. Yanami, M. Miyashita, and A. Yoshikoshi, *J. Org. Chem.*, **45**, 607 (1980).
c. T. Mukaiyama and K. Narasaka, *Org. Synth.*, **65**, 6 (1987).
d. S. Yamago, D. Machii, and E. Nakamura, *J. Org. Chem.*, **56**, 2098 (1991)
e. C. Gennari, A. Bernardi, S. Cardani, and C. Scolastico, *Tetrahedron Lett.*, **26**, 797 (1985).
f. S. Kobayashi and I. Hachiya, *J. Org. Chem.*, **59**, 3590 (1994).
g. T. K. Hollis, N. Robinson, and B. Bosnich, *Tetrahedron Lett.*, **33**, 6423 (1992).
h. Y. Hong, D. J. Norris, and S. Collins, *J. Org. Chem.*, **58**, 3591 (1993).
i. M. Oishi, S. Aratake, and H. Yamamoto, *J. Am. Chem. Soc.*, **120**, 8271 (1998).
j. I. Mori, K. Ishihara, L. A. Flippin, K. Nozaki, H. Yamamoto, P. A. Bartlett, and C. H. Heathcock, *J. Org. Chem.*, **55**, 6107 (1990).
k. S. Murata, M. Suzuki, and R. Noyori, *Tetrahedron*, **44**, 4259 (1998).
l. T. Satay, J. Otera, and H. N. Zaki, *J. Am. Chem. Soc.*, **112**, 901 (1990).
m. T. M. Meulemans, G. A. Stork, B. J. M. Jansen, and A. de Groot, *Tetrahedron Lett.*, **39**, 6565 (1998).
n. A. S. Kende, S. Johnson, P. Sanfilippo, J. C. Hodges, and L. N. Jungheim, *J. Am. Chem. Soc.*, **108**, 3513 (1986).

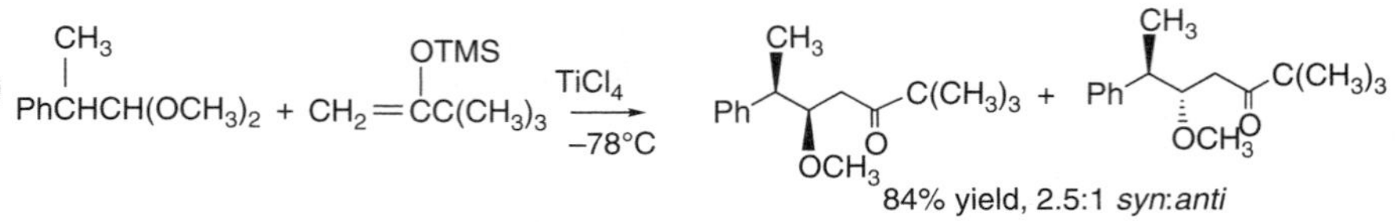

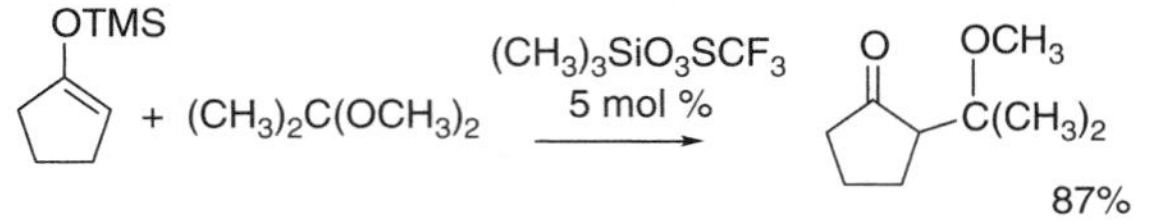

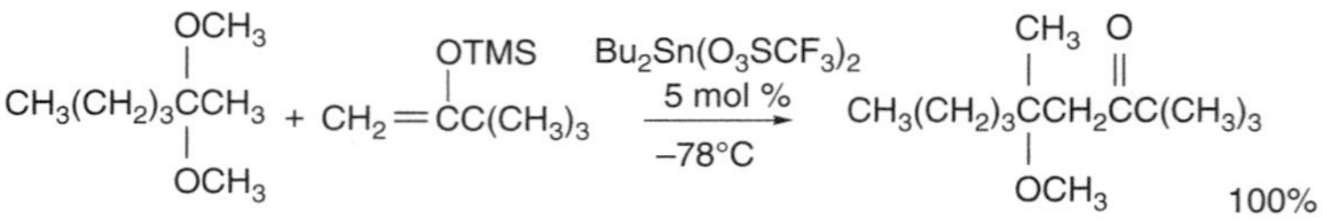

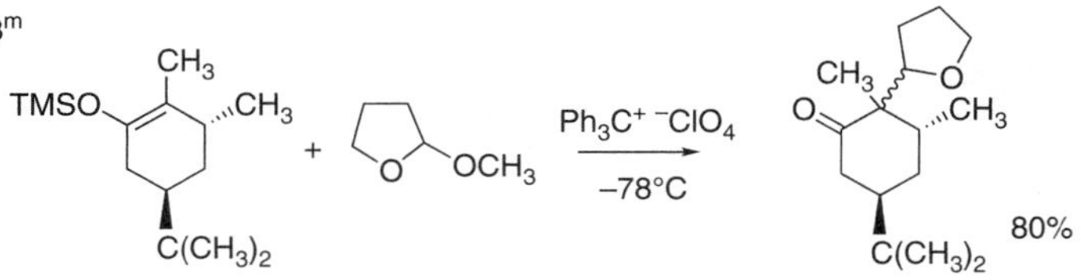

Similarly, there will be a degree of selectivity between the two faces of the enolate if it contains a stereocenter.

The stereogenic centers may be integral parts of the reactants, but chiral auxiliaries can also be used to impart facial diastereoselectivity and permit eventual isolation of enantiomerically enriched product. Alternatively, use of chiral Lewis acids as catalysts can also achieve facial selectivity. Although the general principles of control of the stereochemistry of aldol addition reactions have been well developed for simple molecules, the application of the principles to more complex molecules and the selection of the optimum enolate system requires analyses of the individual cases.[74] Often, one of the available reactant systems proves to be superior.[75] Sometimes a remote structural feature strongly influences the stereoselectivity.[76] The issues that have to be addressed in specific cases include the structure of the reactants, including its configuration and potential sites for chelation; the organization of the TS (cyclic, open, or chelated); and the steric, electronic, and polar factors affecting the facial selectivity.

2.1.5.1. Stereochemical Control by the Aldehyde. A chiral center in an aldehyde can influence the direction of approach by an enolate or other nucleophile. This facial selectivity is in addition to the simple *syn, anti* diastereoselectivity so that if either the aldehyde or enolate contains a stereocenter, four stereoisomers are possible. There are four possible chairlike TSs, of which two lead to *syn* product from the *Z*-enolate and two to *anti* product from the *E*-enolate. The two members of each pair differ in the facial approach to the aldehyde and give products of opposite configuration at *both of the newly formed stereocenters*. If the substituted aldehyde is racemic, the enantiomeric products will be formed, making a total of eight stereoisomers possible.

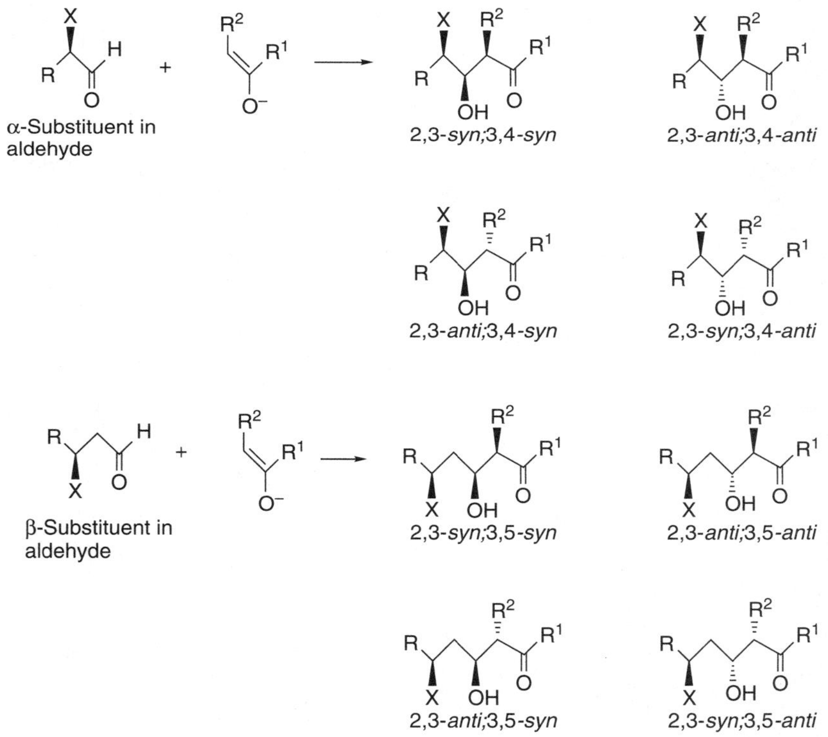

α-Substituent in aldehyde

2,3-*syn*;3,4-*syn*   2,3-*anti*;3,4-*anti*

2,3-*anti*;3,4-*syn*   2,3-*syn*;3,4-*anti*

β-Substituent in aldehyde

2,3-*syn*;3,5-*syn*   2,3-*anti*;3,5-*anti*

2,3-*anti*;3,5-*syn*   2,3-*syn*;3,5-*anti*

[74] (a) W. R. Roush, *J. Org. Chem.*, **56**, 4151 (1991); (b) C. Gennari, S. Vieth, A. Comotti, A. Vulpetti, J. M. Goodman, and I. Paterson, *Tetrahedron*, **48**, 4439 (1992); (c) D. A. Evans, M. J. Dart, J. L. Duffy, and M. G. Yang, *J. Am. Chem. Soc.*, **118**, 4322 (1996); (d) A. S. Franklin and I. Paterson, *Contemp. Org. Synth.*, **1**, 317 (1994).

[75] E. J. Corey, G. A. Reichard, and R. Kania, *Tetrahedron Lett.*, **34**, 6977 (1993).

[76] A. Balog, C. Harris, K. Savin, X.-G. Zhang, T. C. Chou, and S. J. Danishefsky, *Angew. Chem. Int. Ed. Engl.*, **37**, 2675 (1998).

If the substituents are nonpolar, such as an alkyl or aryl group, the control is exerted mainly by steric effects. In particular, for α-substituted aldehydes, the Felkin TS model can be taken as the starting point for analysis, in combination with the cyclic TS. (See Section 2.4.1.3, Part A to review the Felkin model.) The analysis and prediction of the direction of the preferred reaction depends on the same principles as for simple diastereoselectivity and are done by consideration of the attractive and repulsive interactions in the presumed TS. In the Felkin model for nucleophilic addition to carbonyl centers the larger α-substituent is aligned *anti* to the approaching enolate and yields the 3,4-*syn* product. If reaction occurs by an alternative approach, the stereochemistry is reversed, and this is called an anti-Felkin approach.

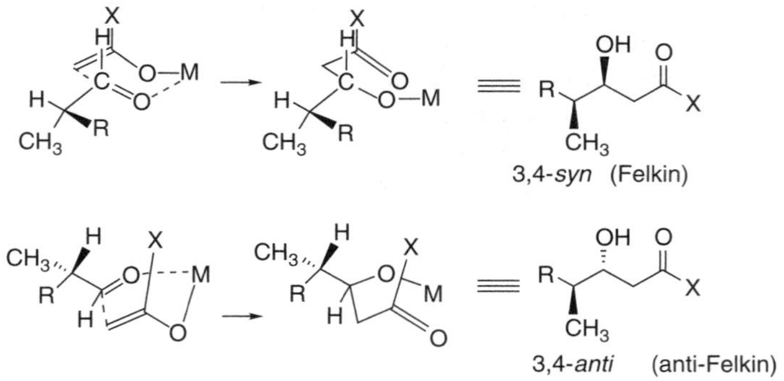

A study of the lithium enolate of pinacolone with several α-phenyl aldehydes gave results generally consistent with the Felkin model. Steric, rather than electronic, effects determine the conformational equilibria.[77] If the alkyl group is branched, it occupies the "large" position. Thus, the *t*-butyl group occupies the "large" position, not the phenyl.

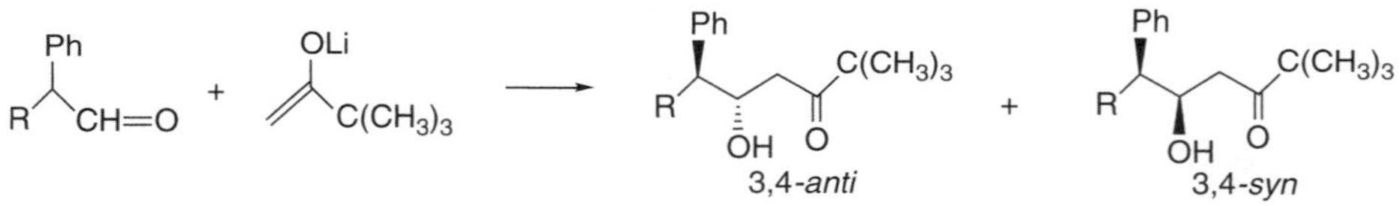

R	3,4-*anti:syn* ratio
CH$_3$	3.64:1
C$_2$H$_5$	6.05:1
(CH$_3$)$_2$CH	2.25:1
(CH$_3$)$_3$C	1:1.7

The situation encounters another factor with enolates having a C(2) substituent. The case of steric control has been examined carefully. The stereoselectivity depends on the orientation of the stereocenter relative to the remainder of the TS. The Felkin TS is **A**. TS **B** represents a non-Felkin conformer, but with the same facial approach as **A**. The preferred TS for the Z-enolate is believed to be structure **C**. This TS is preferred to **A** because of the interaction between the R^M group and the R^2 group of the enolate

[77.] E. P. Lodge and C. H. Heathcock, *J. Am. Chem. Soc.*, **109**, 3353 (1987).

in **A**.[78] This *double-gauche* interaction is analogous to the 1,3-diaxial relationship in chair cyclohexane. TS **C** results in the anti-Felkin approach. The relative energy of TS **B** and TS **C** depends on the size of R^L, with larger R groups favoring TS **C** because of an increased R^2/R^L interaction.

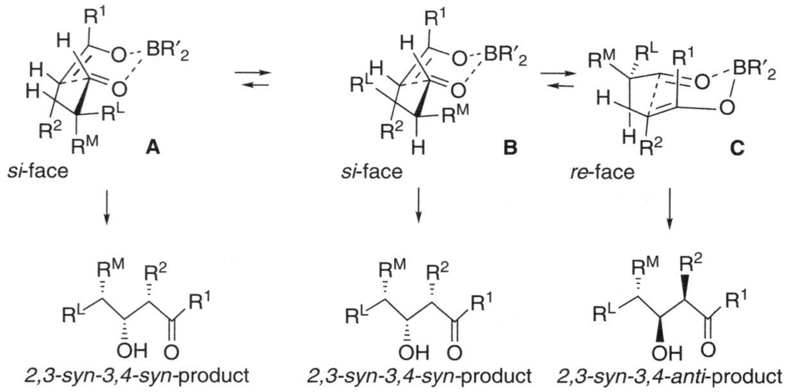

si-face   **A**   *si*-face   **B**   *re*-face   **C**

2,3-syn-3,4-syn-product   *2,3-syn-3,4-syn*-product   *2,3-syn-3,4-anti*-product

For *E*-enolates the Felkin TS is preferred, the enolate approaches opposite the largest aldehyde substituent, and the preferred product is 2,3-*anti*-3,4-*syn*. TS **D** is preferred for *E*-enolates because of the *gauche* interaction between R^2 and R^L in TS **E**.

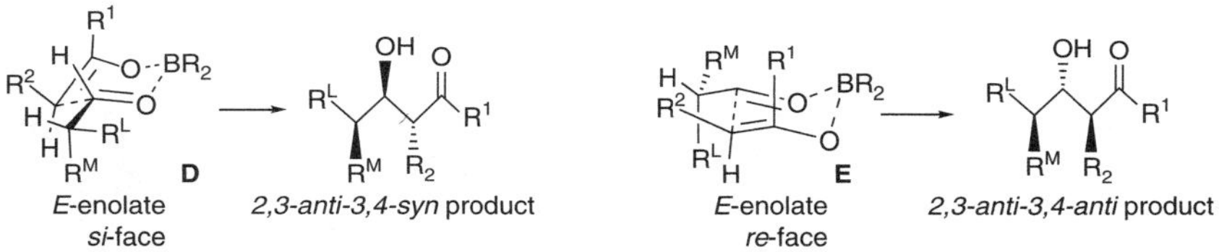

E-enolate   *2,3-anti-3,4-syn* product   *E*-enolate   *2,3-anti-3,4-anti* product
si-face   **D**   *re*-face   **E**

The qualitative application of these models depends on evaluating the magnitude of the steric interactions among the various groups. In this regard, phenyl and vinyl groups seem to be smaller than alkyl groups, perhaps because of their ability to rotate into conformations in which the π dimension minimizes steric repulsions. These concepts have been quantitatively explored using force field models. For nonpolar substituents, steric interactions are the controlling factor in the stereoselectivity, but there is considerable flexibility for adjustment of the TS geometry in response to the specific interactions.[79]

Mukaiyama reactions of α-methyl aldehydes proceed through an open TS and show a preference for the 3,4-*syn* stereoisomer, which is consistent with a Felkin TS.[80]

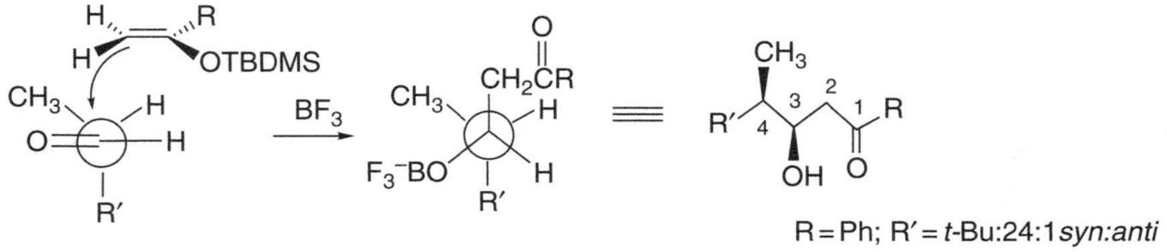

$R = Ph; R' = t\text{-}Bu$; 24:1 *syn:anti*

78. W. R. Roush, *J. Org. Chem.*, **56**, 4151 (1991).
79. C. Gennari, S. Vieth, A. Comotti, A. Vulpetti, J. M. Goodman, and I. Paterson, *Tetrahedron*, **48**, 4439 (1992).
80. C. H. Heathcock and L. A. Flippin, *J. Am. Chem. Soc.*, **105**, 1667 (1983); D. A. Evans and J. R. Gage, *Tetrahedron Lett.*, **31**, 6129 (1990).

The stereoselectivity of aldol addition is also affected by chelation.[81] α- and β-Alkoxy aldehydes can react through chelated structures with Li^+ and other Lewis acids that can accommodate two donor groups.

α–alkoxy aldehyde   β-alkoxy aldehyde

The potential for coordination depends on the oxy substituents.[82] Alkoxy substituents are usually chelated, whereas highly hindered silyloxy groups usually do not chelate. Trimethylsilyloxy groups are intermediate in chelating ability. The extent of chelation also depends on the Lewis acid. Studies with α-alkoxy and β-alkoxy aldehydes with lithium enolates found only modest diastereoselectivity.[83]

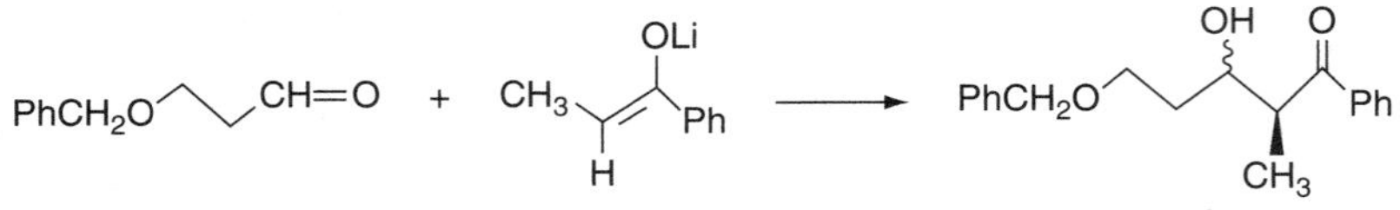

66:34 *anti:syn*

Ref. 84

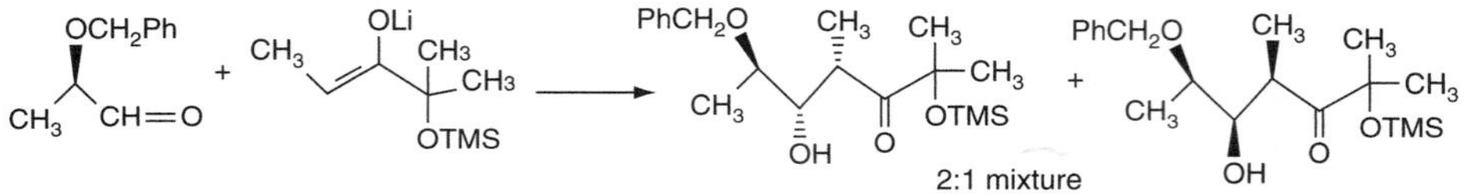

2:1 mixture

Ref. 83b

Several α-methyl-β-alkoxyaldehydes show a preference for 2,3-*syn*-3,4-*anti* products on reaction with Z-enolates. A chelated TS can account for the observed stereochemistry.[85] The chelated aldehyde is most easily approached from the face opposite the methyl and R′ substituents.

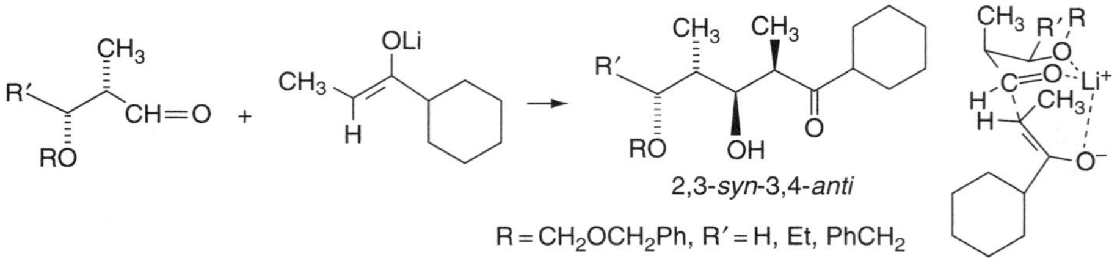

2,3-*syn*-3,4-*anti*

R = CH₂OCH₂Ph, R′ = H, Et, PhCH₂

Dialkylboron enolates cannot accommodate an additional aldehyde ligand group and chelated TSs are not expected. When BF_3 is used as the Lewis acid, chelation is

81. M. T. Reetz, *Angew. Chem. Int. Ed. Engl.*, **23**, 556 (1984); R. Mahrwald, *Chem. Rev.*, **99**, 105 (1999).

82. X. Chen, E. R. Hortelano, E. L. Eliel, and S. V. Frye, *J. Am. Chem. Soc.*, **114**, 1778 (1992).

83. (a) C. H. Heathcock, S. D. Young, J. P. Hagen, M. C. Pirrung, C. T. White, and D. Van Derveer, *J. Org. Chem.*, **45**, 3846 (1980); (b) C. H. Heathcok, M. C. Pirrung, J. Lampe, C. T. Buse, and S. D. Young, *J. Org. Chem.*, **46**, 2290 (1981).

84. M. T. Reetz, K. Kesseler, and A. Jung, *Tetrahedron*, **40**, 4327 (1984).

85. S. Masamune, J. W. Ellingboe, and W. Choy, *J. Am. Chem. Soc.*, **104**, 5526 (1982).

also precluded in Mukaiyama reactions. Chelation control does occur in the Mukaiyama reaction using other Lewis acids. Both α- and β-alkoxy aldehydes give chelation-controlled products with $SnCl_4$ and $TiCl_4$, but not with BF_3.[86] If there is an additional substituent on the aldehyde, the chelate establishes a facial preference for the approach of the nucleophile.[87]

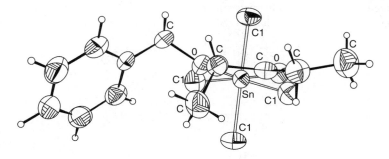

In each instance, the silyl enol ether approaches *anti* to the methyl substituent on the chelate. This results in a 3,4-*syn* relationship between the hydroxy and alkoxy groups for α-alkoxy aldehydes and a 3,5-*anti* relationship for β-alkoxy aldehydes with the main chain in the extended conformation.

97 % 2,3-*syn*-3,4-*syn*
3 % 2,3-*anti*-3,4-*anti*

Ref. 88

92:8 3,5-*anti:syn*

Ref. 84

A crystal structure is available for the $SnCl_4$ complex of 2-benzyloxy-3-pentanone.[89] The steric shielding by the methyl group with respect to the C=O is evident in this structure (Figure 2.1). NMR studies indicate that the reaction involves

Fig. 2.1. Structure of the $SnCl_4$ complex of 2-benzyloxy-3-pentanone. Reproduced from *Acc. Chem. Res.*, **26**, 462 (1993) by permission of the American Chemical Society.

86. C. H. Heathcock, S. K. Davidsen, K. T. Hug, and L. A. Flippin, *J. Org. Chem.*, **51**, 3027 (1986).
87. M. T. Reetz and A. Jung, *J. Am. Chem. Soc.*, **105**, 4833 (1983); C. H. Heathcock, S. Kiyooka, and T. A. Blumenkopf, *J. Org. Chem.*, **51**, 4214 (1984).
88. M. T. Reetz, K. Kesseler, S. Schmidtberger, B. Wenderoth, and P. Steinbach, *Angew. Chem. Int. Ed. Engl.*, **22**, 989 (1983).
89. M. T. Reetz, K. Harms, and W. Reif, *Tetrahedron Lett.*, **29**, 5881 (1988).

formation of trimethylsilyl chloride from the chelated intermediate. This step is followed by conversion to the more stable aldol chelate.[90]

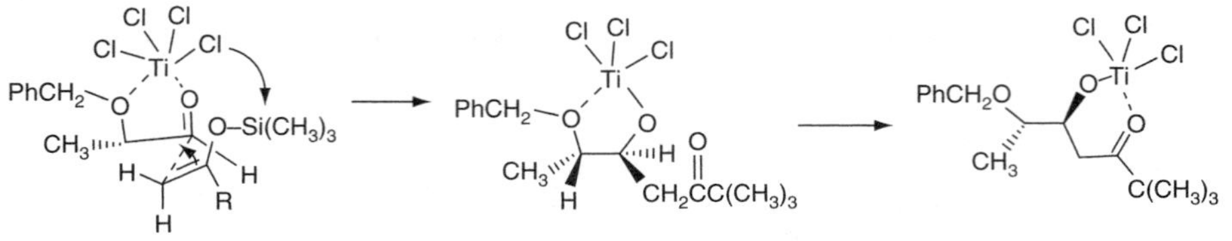

With α- and β-benzyloxyaldehydes, the *t*-butylthio ketene acetals also gave chelation-controlled addition.[91]

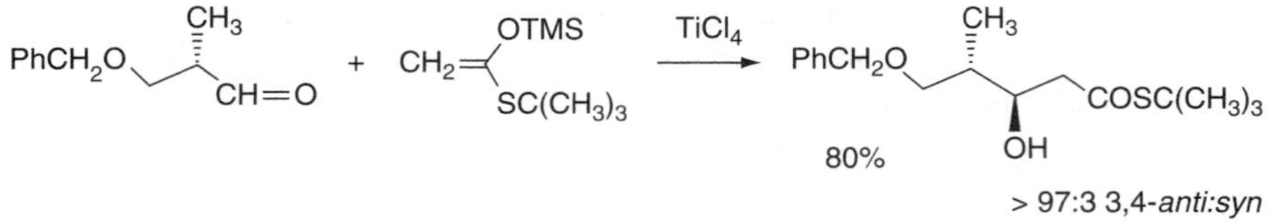

80%

> 97:3 3,4-*anti:syn*

This reaction occurs through a TS in which the aldehyde is chelated, but the silyl thioketene acetal is not coordinated to the Ti (open TS).

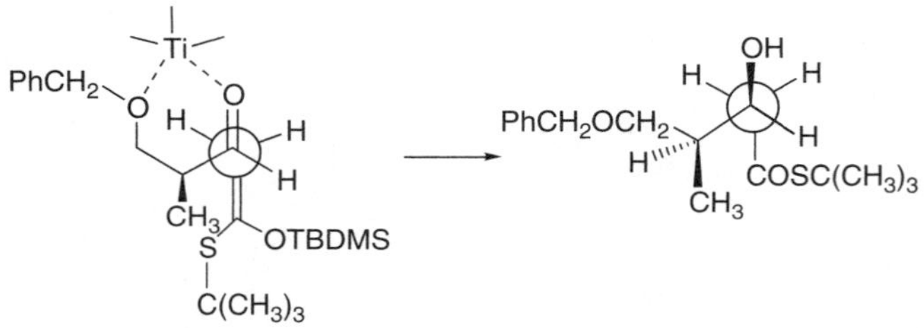

The choice of Lewis acid can determine if a chelated or open TS is involved. For example, all four possible stereoisomers of **1** were obtained by variation of the Lewis acid and the stereochemistry in the reactant.[92] The BF_3-catalyzed reactions occur through an open TS, whereas the $TiCl_4$ reactions are chelation controlled.

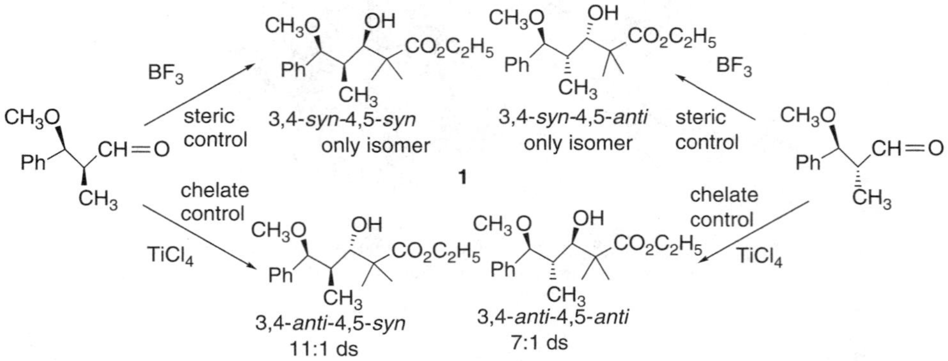

90. M. T. Reetz, B. Raguse, C. F. Marth, H. M. Hügel, T. Bach, and D. N. A. Fox, *Tetrahedron*, **48**, 5731 (1992); M. T. Reetz, *Acc. Chem. Res.*, **26**, 462 (1993).
91. C. Gennari and P. G. Cozzi, *Tetrahedron*, **44**, 5965 (1988).
92. S. Kiyooka, M. Shiinoki, K. Nakata, and F. Goto, *Tetrahedron Lett.*, **43**, 5377 (2002).

In the reaction of α-methylthiobutanal, where the methylthio group has the potential for chelation, BF_3 gave 100% of *anti* product, whereas $TiCl_4$ gave a 5:1 *syn:anti* ratio.[93]

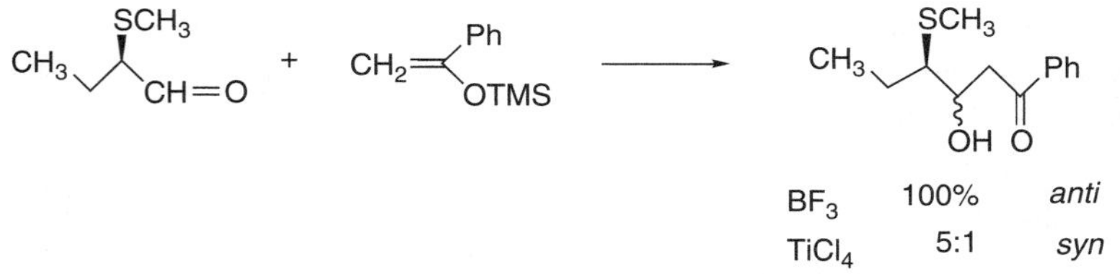

BF_3	100%	*anti*
$TiCl_4$	5:1	*syn*

Chelation-controlled product is formed from reaction of α-benzyloxypropanal and the TBDMS silyl ketene acetal derived from ethyl acetate using 3% $LiClO_4$ as catalyst.[94]

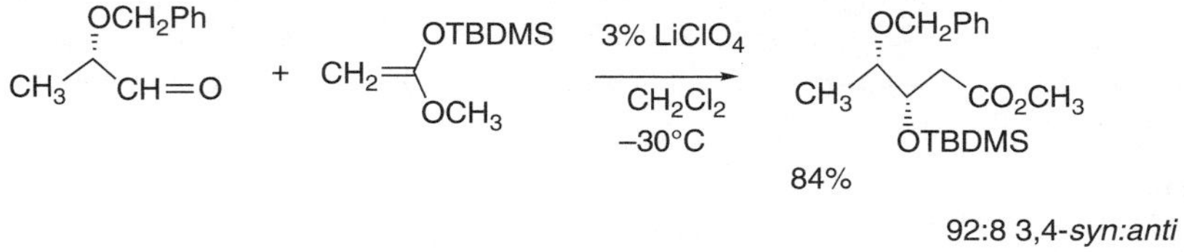

84%

92:8 3,4-*syn:anti*

Recently, $(CH_3)_2AlCl$ and CH_3AlCl_2 have been shown to have excellent chelation capacity. These catalysts effect chelation control with both 3-benzyloxy- and 3-(t-butyldimethylsilyloxy)-2-methylpropanal, whereas BF_3 leads to mainly *syn* product.[95] The reaction is believed to occur through a cationic complex, with the chloride ion associated with a second aluminum as $[(CH_3)_2AlCl_2]^-$. Interestingly, although $TiCl_4$ induced chelation control with the benzyloxy group, it did not do so with the TBDMS group.

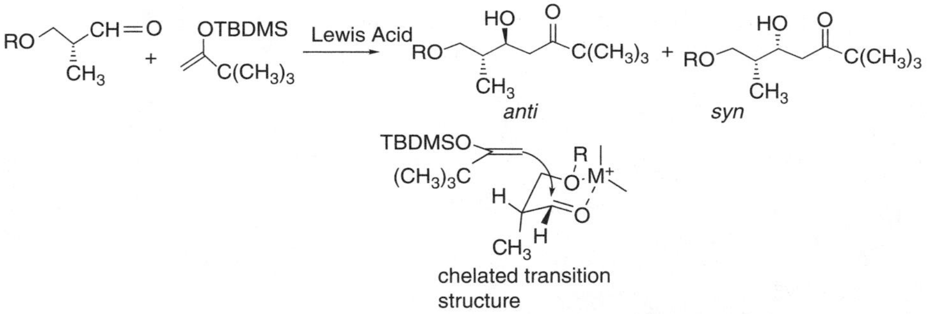

chelated transition
structure

Lewis acid	R = CH₂Ph	R = OTBDMS
	anti:syn	*anti:syn*
BF_3	26:74	9:91
$SnCl_4$	50:50	7:93
$TiCl_4$	97:3	7:93
$(CH_3)_2AlCl$	90:10	97:3
CH_3AlCl_2	78:22	77:23

93. R. Annuziata, M. Cinquini, F. Cozzi, P. G. Cozzi, and E. Consolandi, *J. Org. Chem.*, **57**, 456 (1992).

94. M. T. Reetz and D. N. A. Fox, *Tetrahedron Lett.*, **34**, 1119 (1993).

95. D. A. Evans, B. D. Allison, and M. G. Yang, *Tetrahedron Lett.*, **40**, 4457 (1999); D. A. Evans, B. D. Allison, M. G. Yang, and C. E. Masse, *J. Am. Chem. Soc.*, **123**, 10840 (2001).

Heteroatom substituents also introduce polar effects. In the case of α-alkoxy aldehydes the preferred TS appears to be **F** and **G** for the *E*- and *Z*-enolates, respectively. These differ from the normal Felkin TS for nucleophilic addition. The reactant conformation is believed to be determined by minimization of dipolar repulsion between the alkoxy substituent and the carbonyl group.[96] This model predicts higher 3,4-*anti* ratios for *Z*-enolates, and this is observed.

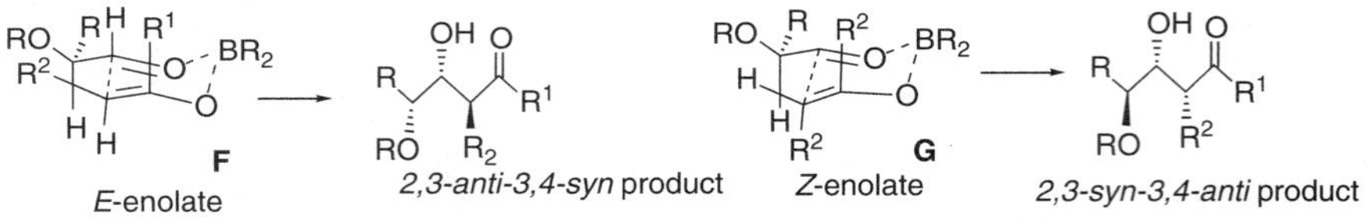

Dipole-dipole interactions may also be important in determining the stereoselectivity of Mukaiyama aldol reactions proceeding through an open TS. A BF$_3$-catalyzed reaction was found to be 3,5-*anti* selective for several β-substituted 5-phenylpentanals. This result can be rationalized by a TS that avoids an unfavorable alignment of the C=O and C–X dipoles.[97]

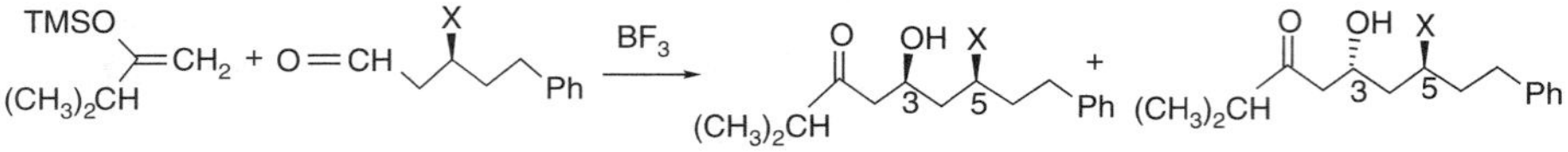

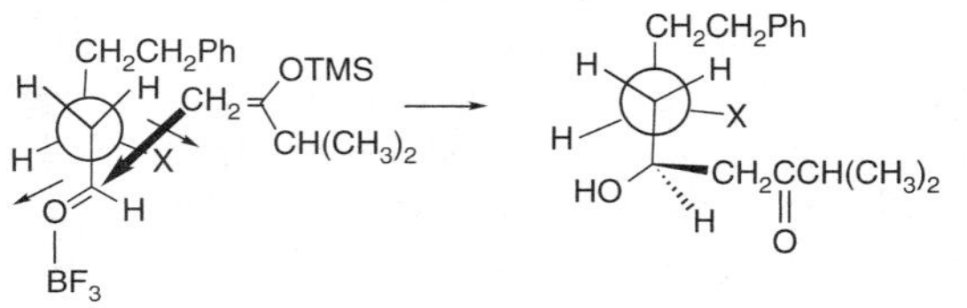

X	3,5-*anti:syn*
PMBO	81:19
OTBDMS	73:27
OAc	43:57
Cl	83:17

The same stereoselectivity was observed with a more complex pair of reactants in which the β-substituent is a cyclic siloxy oxygen.[98]

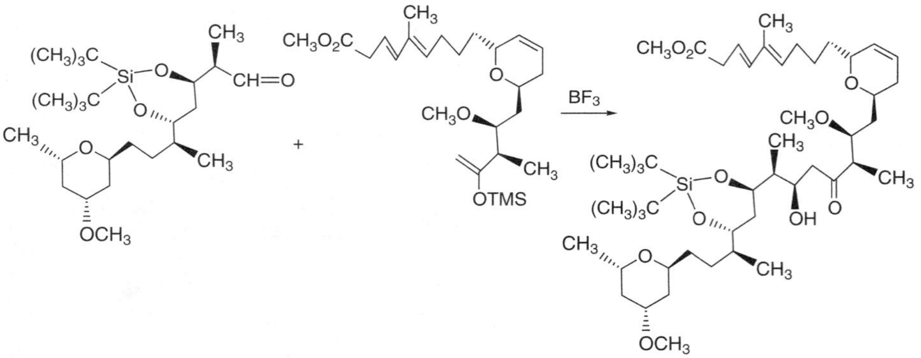

Thus we see that steric effects, chelation, and the polar effects of α- and β-substituents can influence the facial selectivity in aldol additions to aldehydes. These relationships provide a starting point for prediction and analysis of stereoselectivity

96. D. A. Evans, S. J. Siska, and V. J. Cee, *Angew. Chem. Int. Ed. Engl.*, **42**, 1761 (2003).
97. D. A. Evans, M. J. Dart, J. L. Duffy, and M. G. Yang, *J. Am. Chem. Soc.*, **118**, 4322 (1996).
98. I. Paterson, R. A. Ward, J. D. Smith, J. G. Cumming, and K.-S. Yeung, *Tetrahedron*, **51**, 9437 (1995).

Table 2.3. Summary of Stereoselectivity for Aldol Addition Reactions

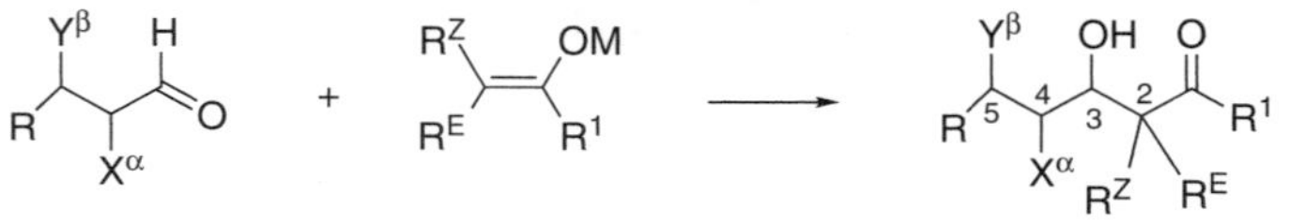

	Aldehyde Steric (Felkin) Control	Aldehyde Chelate TS	Aldehyde Polar Substituent Control
Cyclic TS	3,4-*syn* for X^{α} = medium *E*-enolate 2,3-anti, 3,4-*syn* *Z*-enolate 2,3-syn, 3,4-anti	X^{α} = alkoxy 3,4-*syn* Y^{β} = alkoxy 3,5-*anti*	X^{α} = alkoxy *E*-enolate 2,3-*anti,* 3,4-weak *Z*-enolate 2,3-*syn,*3,4-anti Y^{β} = alkoxy 3,5-*anti*
Open TS	3,4-*syn* for X^{α} = medium		

based on structural effects in the reactant aldehyde. These general principles have been applied to the synthesis of a number of more complex molecules. Table 2.3 summarizes the relationships discussed in this section.

Scheme 2.3 shows reactions of several substituted aldehydes of varying complexity that illustrate aldehyde facial diastereoselectivity in the aldol and Mukaiyama reactions. The stereoselectivity of the new bond formation depends on the effect that reactant substituents have on the detailed structure of the TS. The 3,4-*syn* stereoselectivity of Entry 1 derives from a Felkin-type acyclic TS.

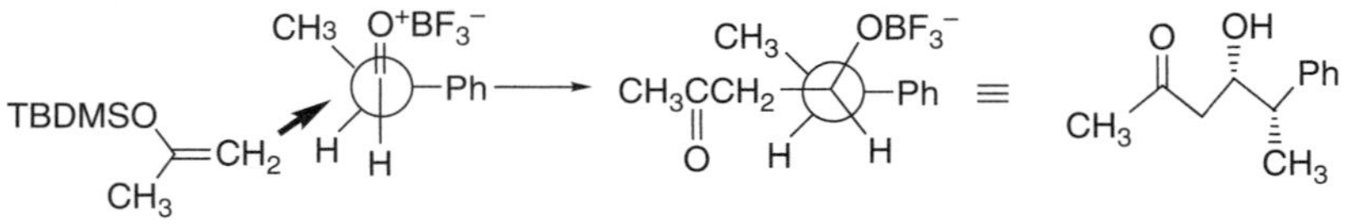

Entry 2 shows an *E*-enolate of a hindered ester reacting with an aldehyde having both an α-methyl and β-methoxy group. The reaction shows a 13:1 preference for the Felkin approach product (3,4-*syn*) and is controlled by the steric effect of the α-methyl substituent. Another example of steric control with an ester enolate is found in a step in the synthesis of (+)-discodermolide.[99] The *E*-enolate of a hindered aryl ester was generated using LiTMP and LiBr. Reaction through a Felkin TS resulted in *syn* diastereoselectivity for the hydroxy and ester groups at the new bond.

99. I. Paterson, G. J. Florence, K. Gerlach, J. P. Scott, and N. Sereinig, *J. Am. Chem. Soc.*, **123**, 9535 (2001).

**Scheme 2.3. Examples of Aldol and Mukaiyama Reactions with Stereoselectivity Based
on Aldehyde Structure**

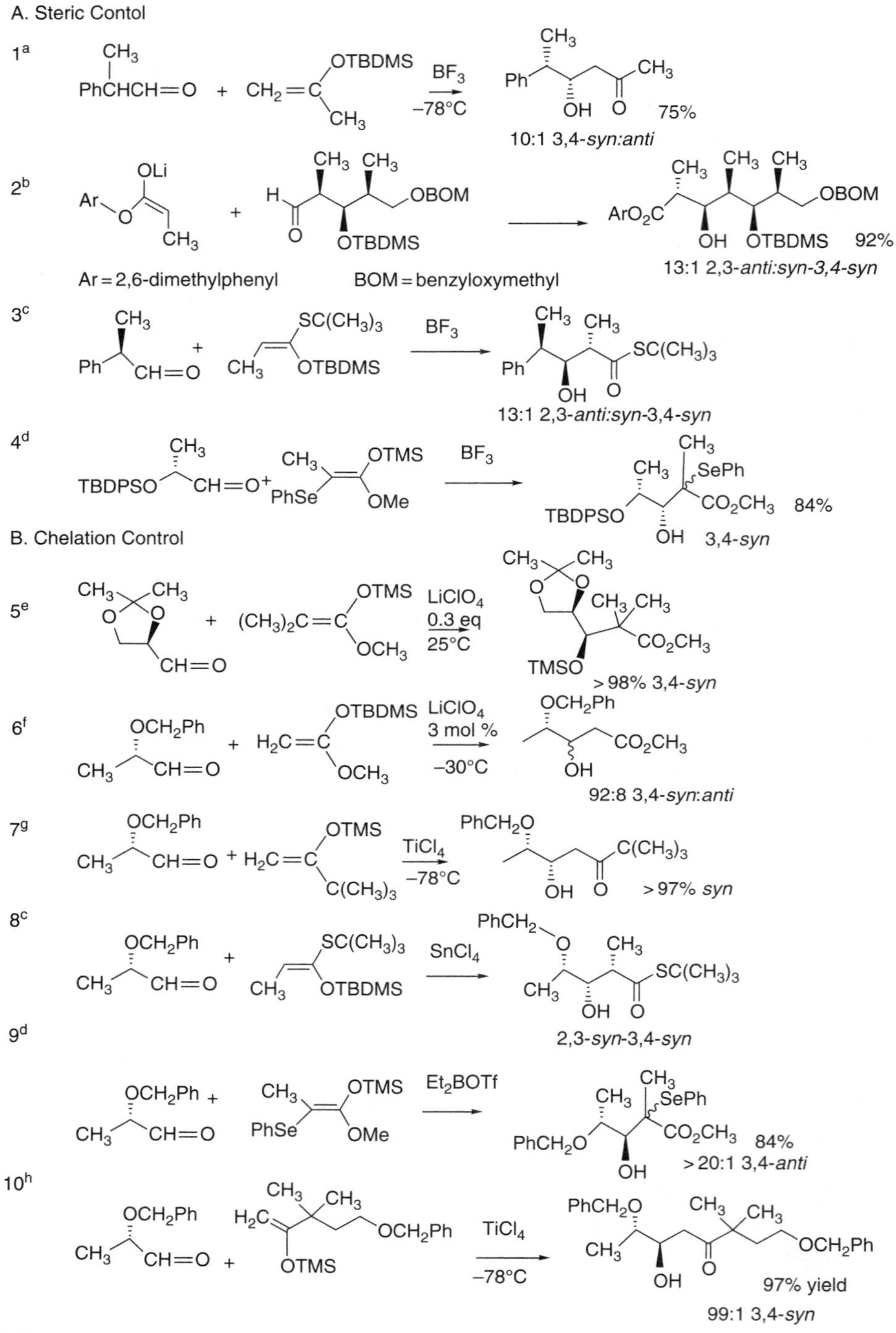

A. Steric Contol

B. Chelation Control

(Continued)

C. Polar Control

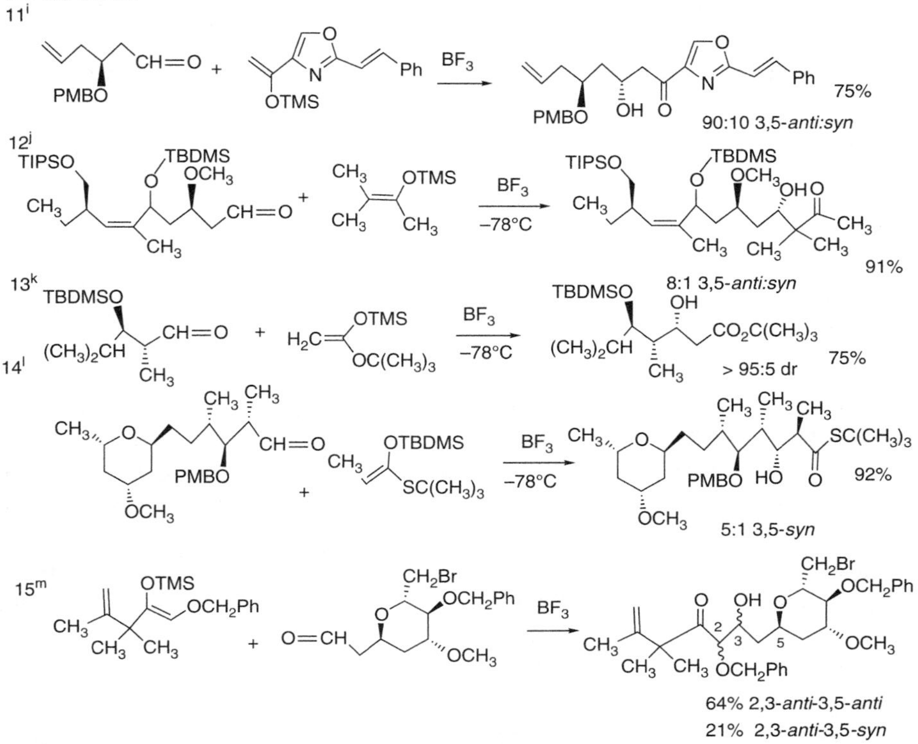

a. C. H. Heathcock and L. A. Flippin, *J. Am. Chem. Soc.*, **105**, 1667 (1983).
b. I. Paterson, *Tetrahedron Lett.*, **24**, 1311 (1983).
c. C. Gennari, M. G. Beretta, A. Bernardi, G. Moro, C. Scolastico, and R. Todeschini, *Tetrahedron*, **42**, 893 (1986).
d. Y. Guindon, M. Prevost, P. Mochirian, and B. Guerin, *Org. Lett.*, **4**, 1019 (2002).
e. J. Ipaktschi and A. Heydari, *Chem. Ber.*, **126**, 1905 (1993).
f. M. T. Reetz and D. N. A. Fox, *Tetrahedron Lett.*, **34**, 1119 (1993).
g. M. T. Reetz, B. Raguse, C. F. Marth, H. M. Hügel, T. Bach, and D. N. A. Fox, *Tetrahedron*, **48**, 5731 (1992).
h. C. Q. Wei, X. R. Jiang, and Y. Ding, *Tetrahedron*, **54**, 12623 (1998).
i. F. Yokokawa, T. Asano, and T. Shioiri, *Tetrahedron*, **57**, 6311 (2001).
j. R. E. Taylor and M. Jin, *Org. Lett.*, **5**, 4959 (2003).
k. L. C. Dias, L. J. Steil and V. de A. Vasconcelos, *Tetrahedron: Asymmetry*, **15**, 147 (2004).
l. G. E. Keck and G. D. Lundquist, *J. Org. Chem.*, **64**, 4482 (1999).
m. D. W. Engers, M. J. Bassindale, and B. L. Pagenkopf, *Org. Lett.*, **6**, 663 (2004).

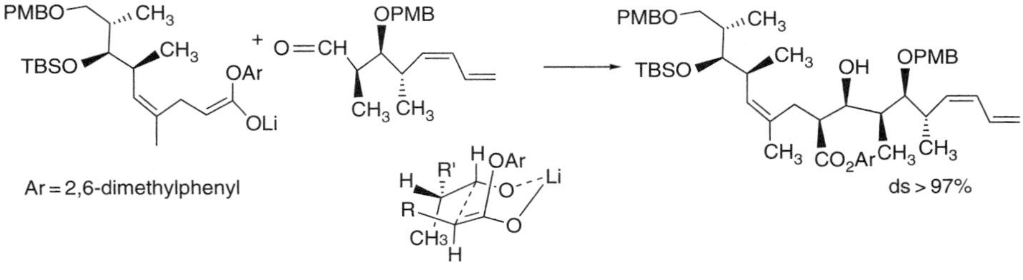

Ar = 2,6-dimethylphenyl

ds > 97%

Entries 3 and 8 show additions of a silyl thioketene acetal to α-substituted aldehydes. Entry 3 is under steric control and gives an 13:1 2,3-*anti-syn* ratio. The reaction proceeds through an open TS with respect to the nucleophile and both the

E- and *Z*-silyl thioketene acetals give the 2,3-*anti* product. The 3,4-*syn* ratio is 50:1, and is consistent with the Felkin model. When this nucleophile reacts with 2-benzyloxypropanal (Entry 8), a chelation product results. The facial selectivity with respect to the methyl group is now reversed. Both isomers of the silyl thioketene acetal give mainly the 2,3-*syn*-3,4-*syn* product. The ratio is higher than 30:1 for the *Z*-enolate but only 3:1 for the *E*-enolate.

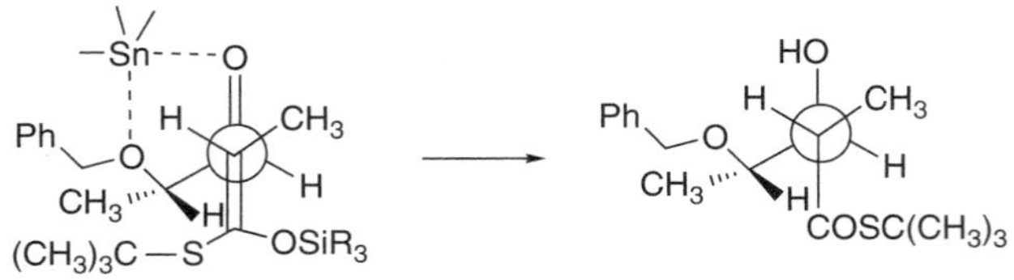

Entries 4 and 9 are closely related structures that illustrate the ability to control stereochemistry by choice of the Lewis acid. In Entry 4, the Lewis acid is BF$_3$ and the β-oxygen is protected as a *t*-butyldiphenylsilyl derivative. This leads to reaction through an open TS, and the reaction is under steric control, resulting in the 3,4-*syn* product. In Entry 9, the enolate is formed using di-*n*-butylboron triflate (1.2 equiv.), which permits the aldehyde to form a chelate. The chelated aldehyde then reacts via an open TS with respect to the silyl ketene acetal, and the 3,4-*anti* isomer dominates by more than 20:1.

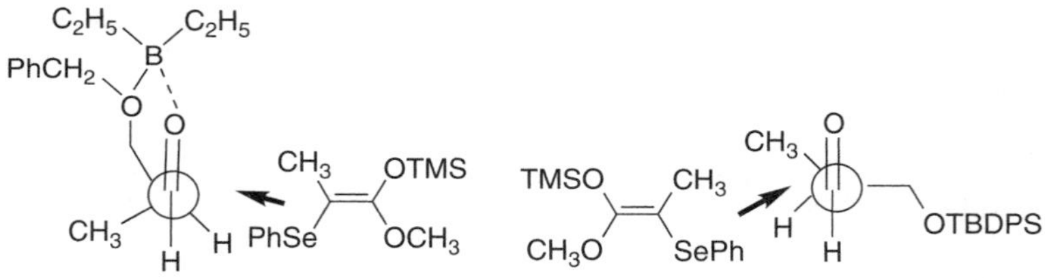

TS for chelate control   TS for steric control

Entry 5 is an example of LiClO$_4$ catalysis and results in very high stereoselectivity, consistent with a chelated structure for the aldehyde.

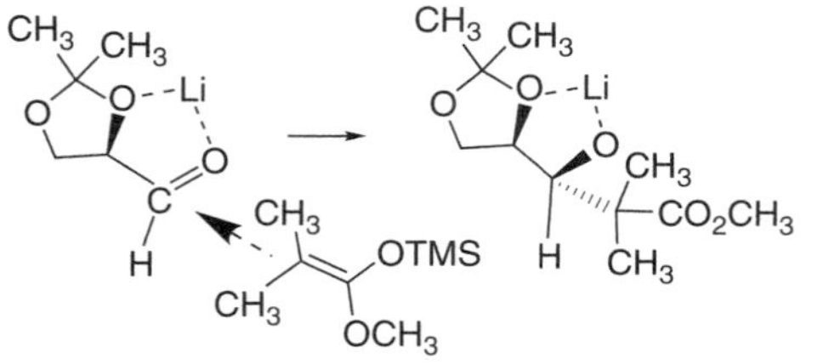

Entries 6 and 7 are examples of reactions of α-benzyloxypropanal. In both cases, the product stereochemistry is consistent with a chelated TS.

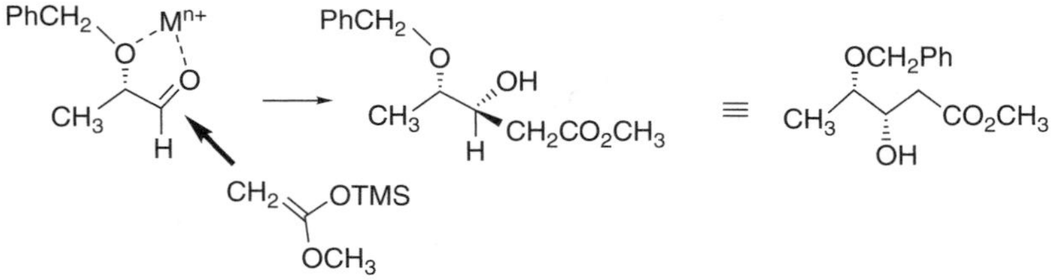

Entry 10 is an example of the application of chelate-controlled stereoselectivity using TiCl$_4$. Entry 11 also involves stereodirection by a β-(*p*-methoxybenzyloxy) substituent. In this case, the BF$_3$-catalyzed reaction should proceed through an open TS and the β-polar effect described on p. 96 prevails, resulting in the *anti*-3,5-isomer.

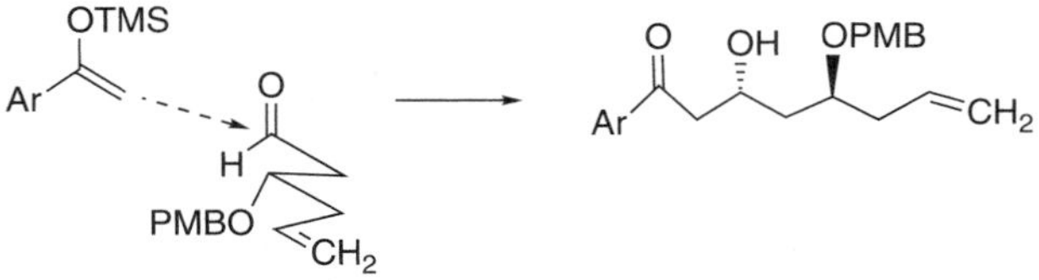

The β-methoxy group in Entry 12 has a similar effect. The aldehydes in Entries 13 and 14 have α-methyl-β-oxy substitution and the reactions in these cases are with a silyl ketene acetal and silyl thioketene acetal, respectively, resulting in a 3,4-*syn* relationship between the newly formed hydroxy and α-methyl substituents.

Entry 15 involves a benzyloxy group at C(2) and is consistent with control by a β-oxy substituent, which in this instance is part of a ring. The *anti* relationship between the C(2) and the C(3) groups results from steric control by the branched substituent in the silyl enol ether. The stereogenic center in the ring has only a modest effect.

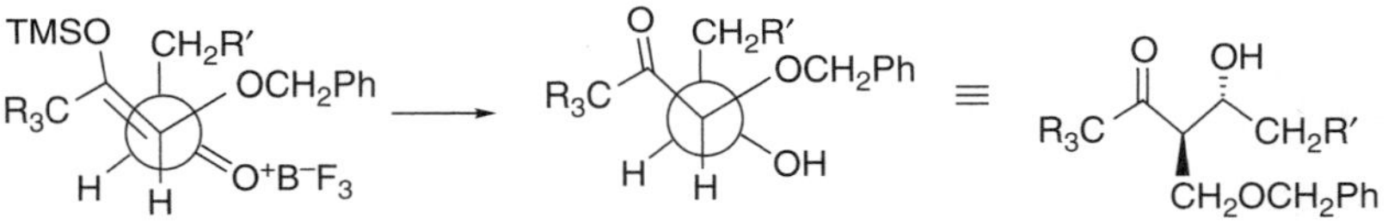

2.1.5.2. Stereochemical Control by the Enolate or Enolate Equivalent. The facial selectivity of aldol addition reactions can also be controlled by stereogenic centers in the nucleophile. A stereocenter can be located at any of the adjacent positions on an enolate or enolate equivalent. The configuration of the substituent can influence the direction of approach of the aldehyde.

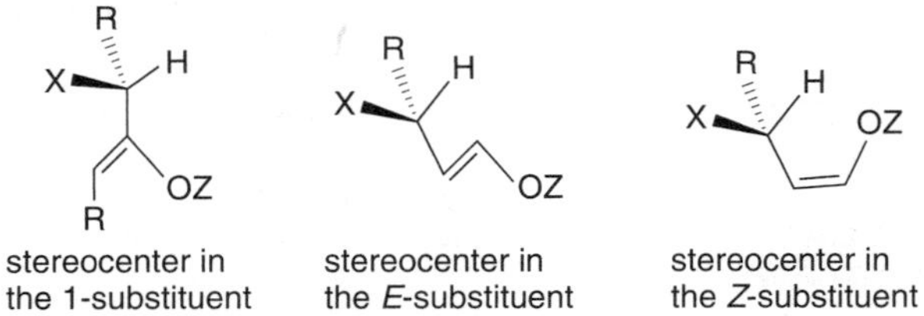

stereocenter in
the 1-substituent

stereocenter in
the *E*-substituent

stereocenter in
the *Z*-substituent

When there is a nonchelating stereocenter at the 1-position of the enolate, the two new stereocenters usually adopt a 2,2'-*syn* relationship to the M substituent. This result is consistent with a cyclic TS having a conformation of the chiral group with the hydrogen pointed toward the boron and the approach to the aldehyde from the smaller of the other two substituents as in TS **H**.[100]

This stereoselectivity, for example, was noted with enolate **2**.[101]

The same effects are operative with titanium enolates.[100a]

Little steric differentiation is observed with either the lithium or boron enolates of 2-methyl-2-pentanone.[102]

M = Li   57:43 2′,3-*anti:syn*
M = BBu₂   64:36 2′,3-*anti:syn*

α-Oxygenated enolates show a strong dependency on the nature of the oxygenated substituent. TBDMS derivatives are highly selective for 2, 2′-*syn*-2,3-*syn* product, but benzyloxy substituents are much less selective. This is attributed to involvement of two competing chelated TSs in the case of benzyloxy, but of a nonchelated TS for the siloxy substituent.[103] The contrast between the oxy substituents is consistent with the tendency for alkoxy groups to be better donors toward Ti(IV) than siloxy groups.

[100] (a) D. A. Evans, D. L. Rieger, M. T. Bilodeau, and F. Urpi, *J. Am. Chem. Soc.*, **113**, 1047 (1991); (b) A. Bernardi, A. M. Capelli, A. Comotti, C. Gennari, M. Gardner, J. M. Goodman, and I. Paterson, *Tetrahedron*, **47**, 3471 (1991).

[101] I. Paterson and A. N. Hulme, *J. Org. Chem.*, **60**, 3288 (1995).

[102] D. Seebach, V. Ehrig, and M. Teschner, *Liebigs Ann. Chem.*, 1357 (1976); D. A. Evans, J. V. Nelson, E. Vogel, and T. R. Taber, *J. Am. Chem. Soc.*, **103**, 3099 (1981).

[103] S. Figueras, R. Martin, P. Romea, F. Urpi, and J. Vilarrasa, *Tetrahedron Lett.*, **38**, 1637 (1997).

Oxy substituent	R	2, 2'-*syn*-2,3-*syn*:2,2'-*anti*-2,3-*syn*
TBDMS	CH_3	30:1
TBDMS	$PhCH_2$	35:1
TBDMS	$(CH_3)_2CH$	> 95:1
$PhCH_2$	CH_3	5:1
$PhCH_2$	$PhCH_2$	4:1
$PhCH_2$	$(CH_3)_2CH$	1:1

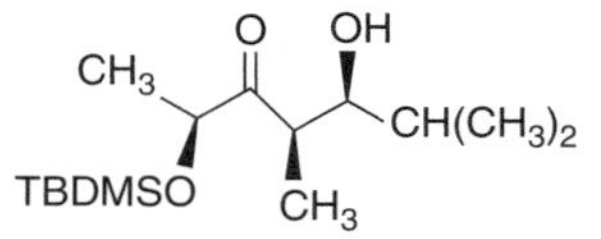

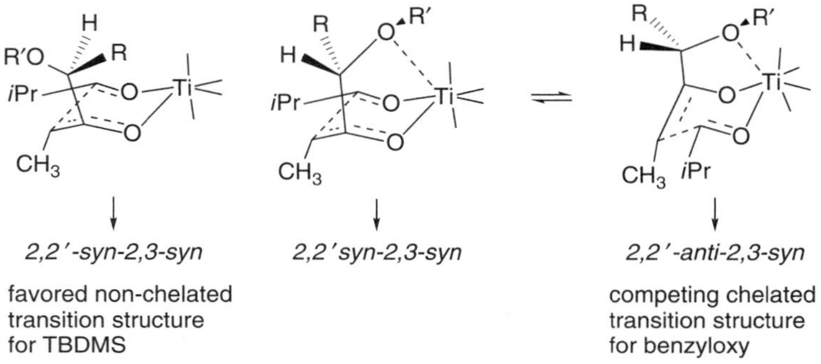

2,2'-*syn*-2,3-*syn*

favored non-chelated
transition structure
for TBDMS

2,2'*syn*-2,3-*syn*

2,2'-*anti*-2,3-*syn*

competing chelated
transition structure
for benzyloxy

The stereoselectivity of this reaction also depends on the titanium reagent used to prepare the enolate.[104] When the substituent is benzyloxy, the 2, 2'-*anti*-2,3-*syn* product is preferred when $(i\text{-PrO})TiCl_3$ is used as the reagent, as would be expected for a chelated TS. However, when $TiCl_4$ is used, the 2, 2'-*syn*-2,3-*syn* product is formed. A detailed explanation for this observation has not been established, but it is expected that the benzyloxy derivative would still react through a chelated TS. The reversal on use of $TiCl_4$ indicates that the identity of the titanium ligands is also an important factor.

High facial selectivity attributable to chelation was observed with the TMS silyl ethers of 3-acyloxy-2-butanone.[105]

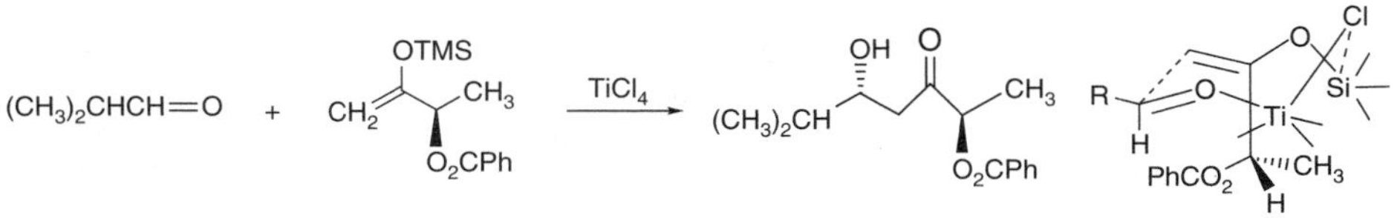

Several enolates of 4,4-dimethyl-3-(trimethylsiloxy)-2-pentanone have been investigated.[106] The lithium enolate reacts through a chelated TS with high 2,2'-*anti* stereoselectivity, based on the steric differentiation by the *t*-butyl group.

[104.] J. G. Solsona, P. Romea, F. Urpi, and J. Villarrasa, *Org. Lett.*, **5**, 519 (2003).

[105.] B. M. Trost and H. Urabe, *J. Org. Chem.*, **55**, 3982 (1990).

[106.] C. H. Heathcock and S. Arseniyadis, *Tetrahedron Lett.*, **26**, 6009 (1985) and Erratum *Tetrahedron Lett.*, **27**, 770 (1986); N. A. Van Draanen, S. Arseniyadis, M. T. Crimmins, and C. H. Heathcock, *J. Org. Chem.*, **56**, 2499 (1991).

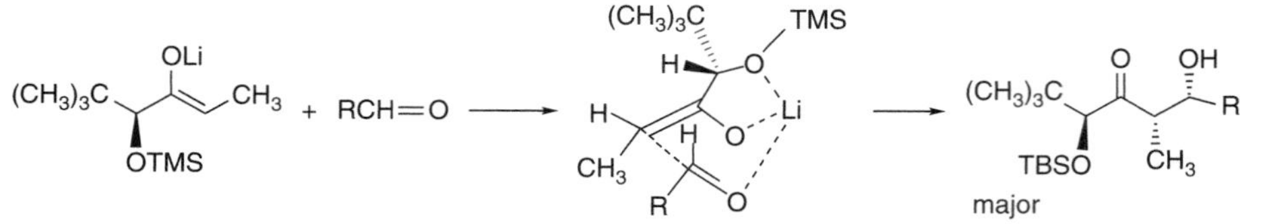

R	2,2'-anti:syn
i-Pr	> 95:5
t-Bu	> 95:5
Ph	> 95:5
PhCH$_2$OCH$_2$	> 95:5

The corresponding di-*n*-butylboron enolate gives the 2,2'-*syn* adduct. The nonchelating boron is thought to react through a TS in which the conformation of the substituent is controlled by a dipolar effect.

The *E*-titanium enolate was prepared by deprotonation with TMP-MgBr, followed by reaction with (*i*-PrO)$_3$TiCl in the presence of HMPA. The TS for addition is also dominated by a polar effect and gives and 2,2'-*anti* product.

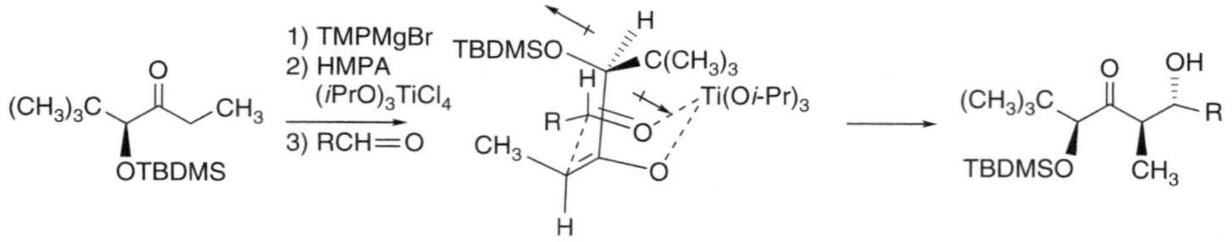

An indication of the relative effectiveness of oxygen substituent in promoting chelation of lithium enolates is found in the enolates **3a–d**. The order of preference for the chelation-controlled product is CH$_3$OCH$_2$O > TMSO > PhCH$_2$O > TBDMSO, with the nonchelation product favored for TBDMSO.[107]

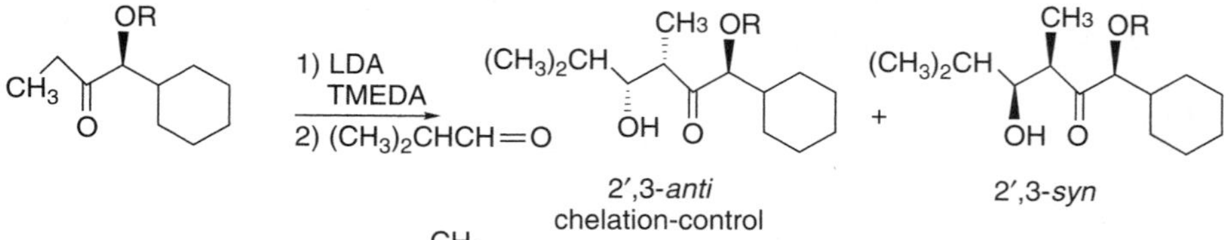

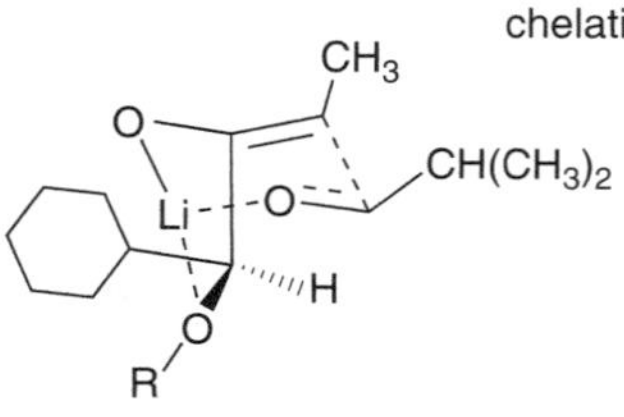

	R	
3a	CH$_3$OCH$_2$	93:7
b	PhCH$_2$	75:25
c	TMS	88:12
d	TBDMS	24:76

107. C. Siegel and E. R. Thornton, *Tetrahedron Lett.*, **27**, 457 (1986); A Choudhury and E. R. Thornton, *Tetrahedron Lett.*, **34**, 2221 (1993).

Tin(II) enolates having 3′-benzyloxy substituents are subject to chelation control. The enolate from 2-(benzyloxymethyl)-3-pentanone gave mainly 2,2′-*syn*-2,3-*syn* product, a result that is consistent with a chelated TS.[108]

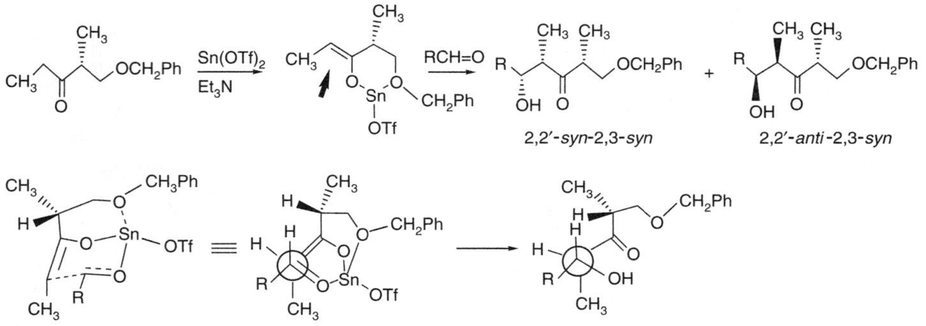

2,2′-*syn*-2,3-*syn*   2,2′-*anti*-2,3-*syn*

Polar effects appear to be important for 3′-alkoxy substituents in enolates. 3-Benzyloxy groups enhance the facial selectivity of *E*-boron enolates, and this is attributed to a TS **I** in which the benzyloxy group faces toward the approaching aldehyde. This structure is thought to be preferable to an alternate conformation **J**, which may be destabilized by electron pair repulsions between the benzyloxy oxygen and the enolate oxygen.[109]

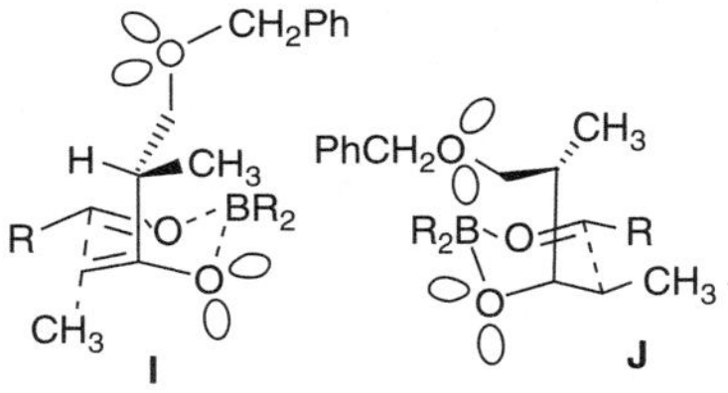

I   **J**

This effect is seen in the case of ketone **4**, where the stereoselectivity of the benzyloxy derivative is much higher than the compound lacking the benzyloxy group.[110]

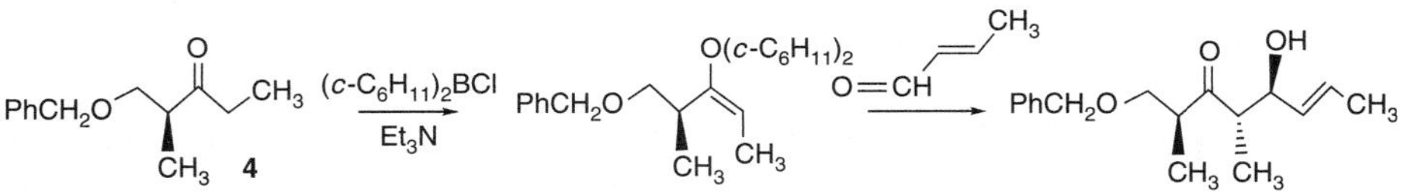

The same β-alkoxy effect appears to be operative in a 2'-methoxy substituted system.[111]

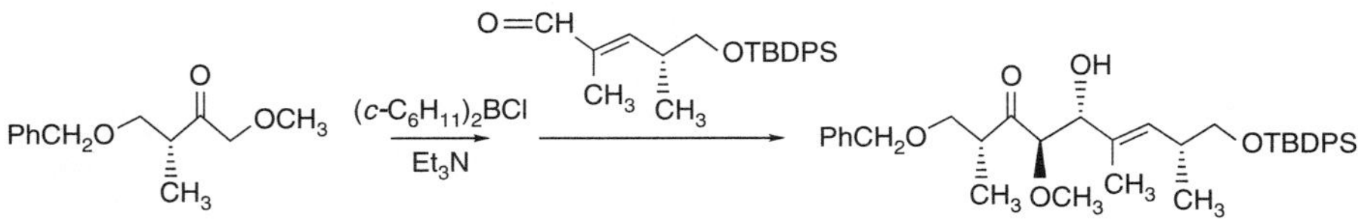

[108.] I. Paterson and R. D. Tillyer, *Tetrahedron Lett.*, **33**, 4233 (1992).

[109.] A. Bernardi, C. Gennari, J. M. Goodman, and I. Paterson, *Tetrahedron: Asymmetry*, **6**, 2613 (1995).

[110.] I. Paterson, J. M. Goodman, and M. Isaka, *Tetrahedron Lett.*, **30**, 7121 (1989).

[111.] I. Paterson and R. D. Tillyer, *J. Org. Chem.*, **58**, 4182 (1993).

A 3′-benzyloxy ketone gives preferential 2,2′-*syn* stereochemistry through a chelated TS for several titanium enolates. The best results were obtained using isopropoxytitanium trichloride.[112] The corresponding *E*-boron enolate gives the 2,2′-*anti*-2,3-*anti* isomer as the main product through a nonchelated TS.[110]

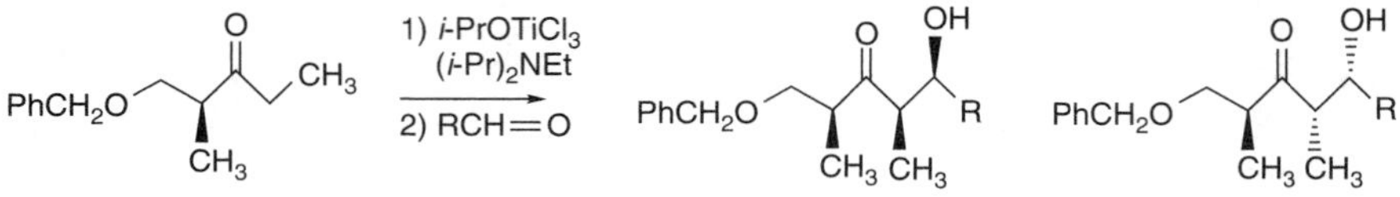

2,2′-*syn*-2,3-*syn*   2,2′-*anti*-2,3-*syn*

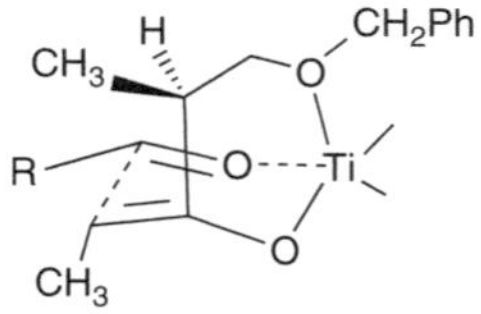

R	ratio
C_2H_5	93:7
$(CH_3)_2CH$	97:3
$(CH_3)_2CHCH_2$	94:6
Ph	94:6

In summary, the same factors that operate in the electrophile, namely steric, chelation, and polar effects, govern facial selectivity for enolates. The choice of the Lewis acid can determine if the enolate reacts via a chelate. The final outcome depends upon the relative importance of these factors within the particular TS.

Scheme 2.4 provides some specific examples of facial selectivity of enolates. Entry 1 is a case of steric control with Felkin-like TS with approach *anti* to the cyclohexyl group.

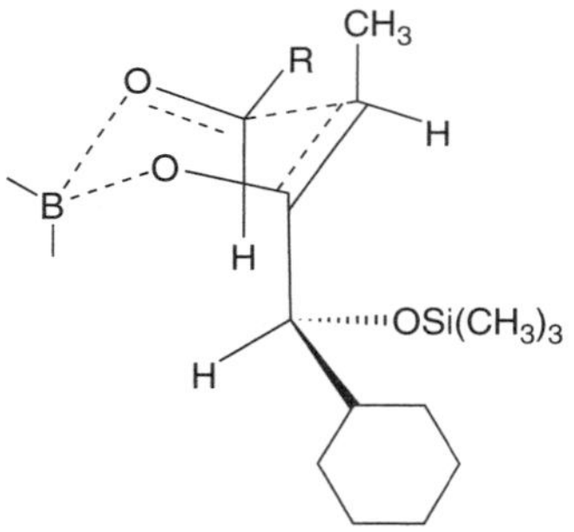

Entry 2 is an example of the polar β-oxy directing effect. Entries 3 and 4 involve formation of *E*-enolates using dicyclohexylboron chloride. The stereoselectivity is consistent with a cyclic TS in which a polar effect orients the benzyloxy group away from the enolate oxygen.

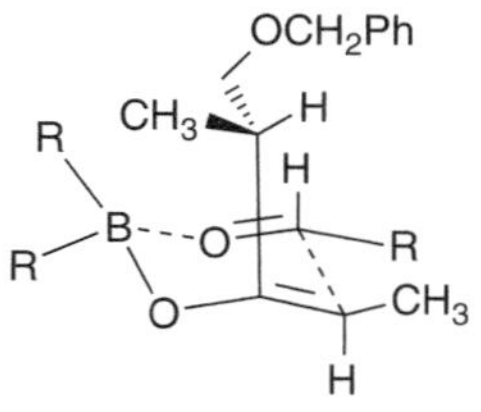

[112.] J. G. Solsona, J. Nebot, P. Romea, and F. Urpi, *J. Org. Chem.*, **70**, 6533 (2005).

Scheme 2.4. Examples of Facial Selectivity in Aldol and Mukaiyama Reactions Based on Enolate Structure

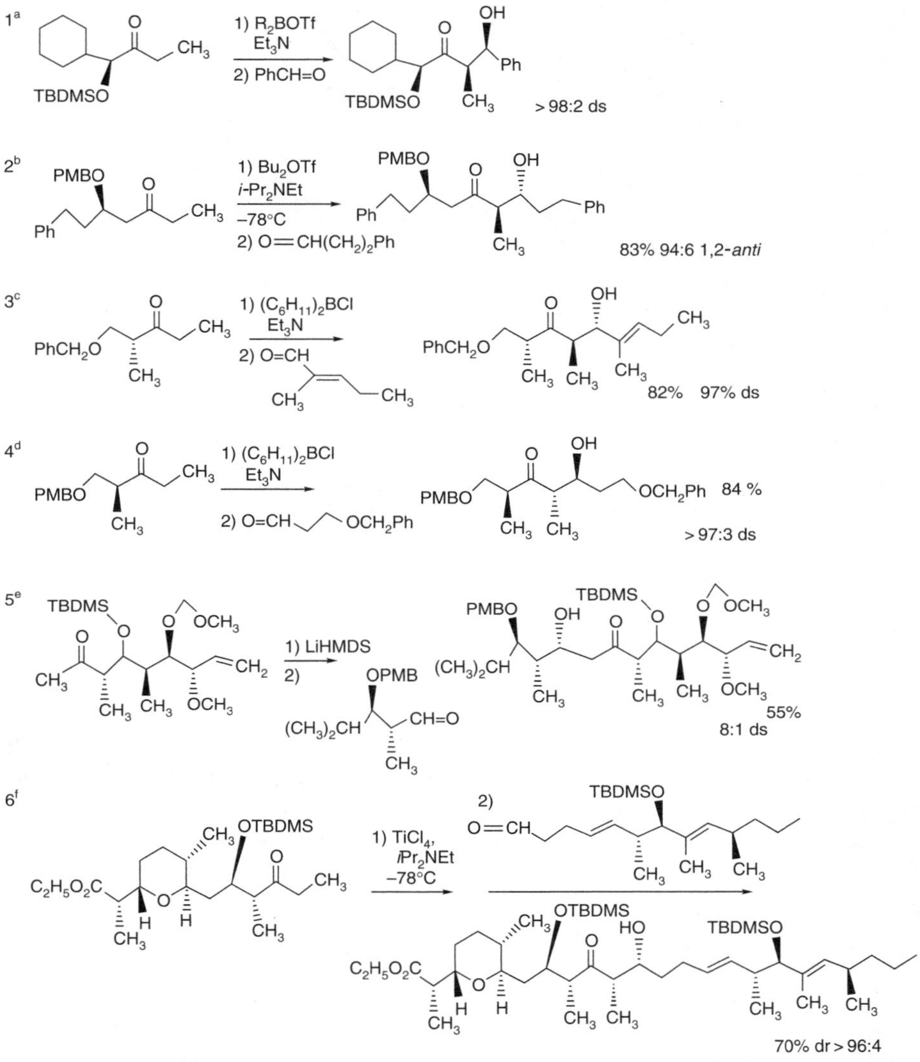

a. D. A. Evans, D. L. Rieger, M. T. Bilodeau, and F. Urpi, *J. Am. Chem. Soc.*, **113**, 1047 (1991).
b. D. A. Evans, P. J. Coleman, and B. Cote, *J. Org. Chem.*, **62**, 788 (1997).
c. I. Paterson and M. V. Perkins, *Tetrahedron*, **52**, 1811 (1996).
d. I. Paterson and I. Lyothier, *J. Org. Chem.*, **70**, 5454 (2005).
e. W. R. Roush, T. D. Bannister, M. D. Wendt, J. A. Jablonsowki, and K. A. Scheidt, *J. Org. Chem.*, **67**, 4275 (2002).
f. M. Defosseux, N. Blanchard, C. Meyer, and J. Cossy, *J. Org. Chem.*, **69**, 4626 (2004).

Entry 5, where the same stereochemical issues are involved was used in the synthesis of (+)-discodermolide. (See Section 13.5.6 for a more detailed discussion of this synthesis.) There is a suggestion that this entry involves a chelated lithium enolate and there are two stereogenic centers in the aldehyde. In the next section, we discuss how the presence of stereogenic centers in both reactants affects stereoselectivity.

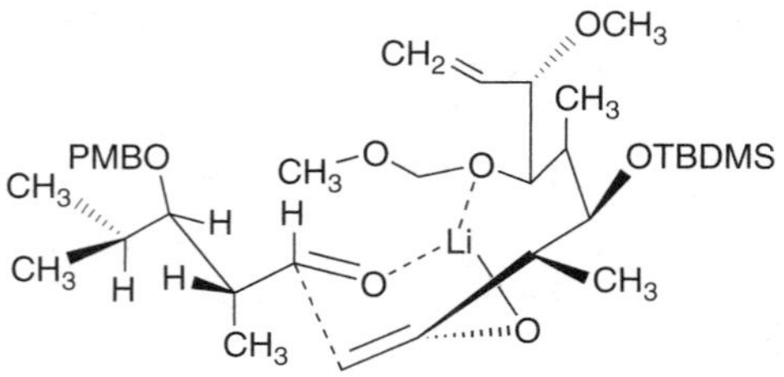

Entry 6 involves a titanium enolate of an ethyl ketone. The aldehyde has no nearby stereocenters. Systems with this substitution pattern have been shown to lead to a 2,2'-*syn* relationship between the methyl groups flanking the ketone, and in this case, the β-siloxy substituent has little effect on the stereoselectivity. The configuration (*Z*) and conformation of the enolate determines the 2,3-*syn* stereochemistry.[113]

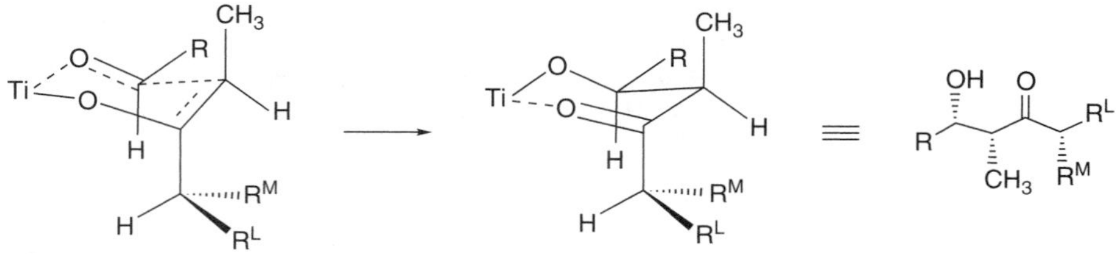

2.1.5.3. Complementary/Competitive Control: Double Stereodifferentiation. If both the aldehyde and the enolate in an aldol addition are chiral, mutual combinations of stereoselectivity come into play. The chirality in the aldehyde and enolate each impose a bias toward one absolute configuration. The structure of the chairlike TS imposes a bias toward the relative configuration (*syn* or *anti*) of the newly formed stereocenters as described in Section 2.1.2. One combination of configurations, e.g., (*R*)-aldehyde/(*S*)-enolate, provides complementary, reinforcing stereoselection, whereas the alternative combination results in opposing preferences and leads to diminished overall stereoselectivity. The combined interaction of stereocenters in both the aldehyde and the enolate component is called *double stereodifferentiation*.[114] The reinforcing combination is called *matched* and the opposing combination is called *mismatched*.

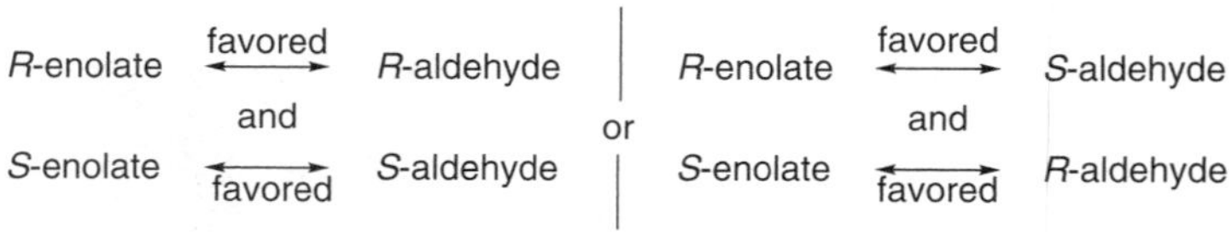

113. D. A. Evans, D. L. Rieger, M. T. Bilodeau, and F. Urpi, *J. Am. Chem. Soc.*, **113**, 1047 (1991).
114. S. Masamune, W. Choy, J. S. Petersen, and L. R. Sita, *Angew. Chem. Int. Ed. Engl.*, **24**, 1 (1985).

For example the aldol addition of (S)-2-cyclohexylpropanal is more stereoselective with the enolate (S)-**5** than with the enantiomer (R)-**5**. The stereoselectivity of these cases derives from relative steric interactions in the matched and mismatched cases.

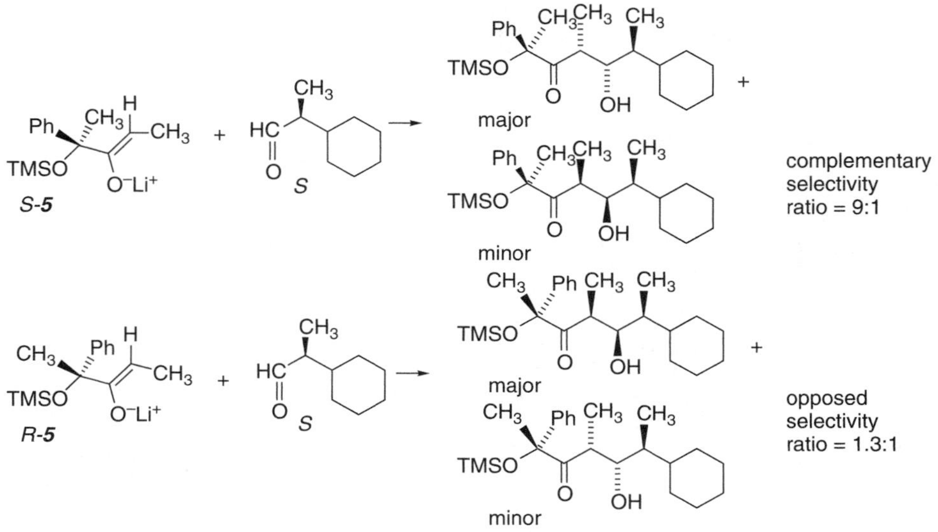

Ref. 115

Chelation can also be involved in double stereodifferentiation. The lithium enolate of the ketone **7** reacts selectively with the chiral aldehyde **6** to give a single stereoisomer.[116] The enolate is thought to be chelated, blocking one face and leading to the observed product.

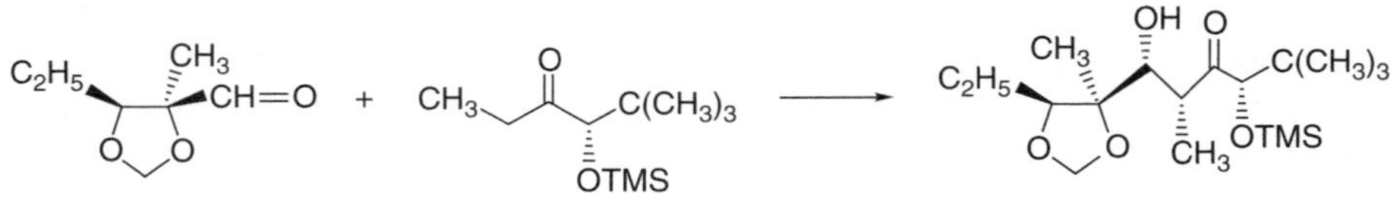

There can be more than two stereocenters, in which case there are additional combinations. For example with three stereocenters, there will be one fully matched set, one fully mismatched set, and two partially matched sets. In the latter two, one of the factors may dominate the others. For example, the ketone **8** and the four stereoisomers of the aldehyde **9** have been examined.[117] Both the *E*-boron and the *Z*-titanium enolates were studied. The results are shown below.

115. S. Masamune, S. A. Ali, D. L. Snitman, and D. S. Garvey, *Angew. Chem. Int. Ed. Engl.*, **19**, 557 (1980).
116. C. H. Heathcock, M. C. Pirrung, C. T. Buse, J. P. Hagen, S. D. Young, and J. E. Sohn, *J. Am. Chem. Soc.*, **101**, 7077 (1979).
117. D. A. Evans, M. J. Dart, J. L. Duffy, and D. L. Rieger, *J. Am. Chem. Soc.*, **117**, 9073 (1995).

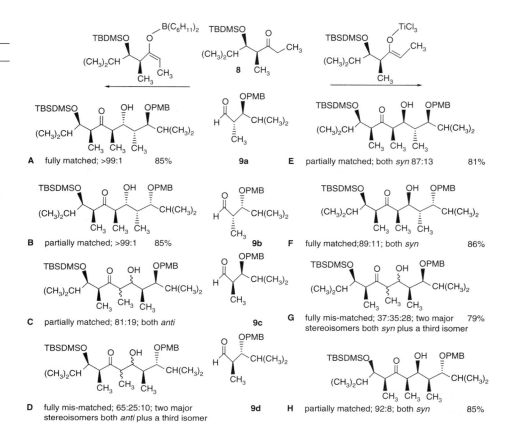

A fully matched; >99:1   85%

9a

E partially matched; both *syn* 87:13   81%

B partially matched; >99:1   85%

9b

F fully matched; 89:11; both *syn*   86%

C partially matched; 81:19; both *anti*

9c

G fully mis-matched; 37:35:28; two major   79%
stereoisomers both *syn* plus a third isomer

D fully mis-matched; 65:25:10; two major
stereoisomers both *anti* plus a third isomer

9d

H partially matched; 92:8; both *syn*   85%

The results for the boron enolates show that when the aldehyde and enolate centers are matched the diastereoselectivity is high (Cases **A** and **B**). In Case **C**, the enolate is matched with respect to the β-alkoxy group but mismatched with the α-methyl group. The result is an 81:19 dominance of the anti-Felkin product. For the titanium enolates, Cases **E** and **F** correspond to a matched relationship with the α-stereocenter. Case **G** is fully mismatched and shows little selectivity. In Case **H**, the matched relationship between the enolate and the β-alkoxy group overrides the α-methyl effect and a 2,3-*syn* (Felkin) product is formed. The corresponding selectivity ratios have also been determined for the lithium enolates.[118] Comparison with the boron enolates shows that although the stereoselectivity of the fully matched system is higher with the boron enolate, in the mismatched cases for the lithium enolate, the aldehyde bias overrides the enolate bias and gives modest selectivity for the alternative *anti* isomer.

In general, BF$_3$-catalyzed Mukaiyama reactions lack a cyclic organization because of the maximum coordination of four for boron. In these circumstances, the reactions show a preference for the Felkin type of approach and exhibit a preference for *syn* stereoselectivity that is independent of silyl enol ether structure.[119]

118. D. A. Evans, M. G. Yang, M. J. Dart, and J. L. Duffy, *Tetrahedron Lett.*, **37**, 1957 (1996).
119. D. A. Evans, M. G. Yang, M. J. Dart, J. L. Duffy, and A. S. Kim, *J. Am. Chem. Soc.*, **117**, 9598 (1995).

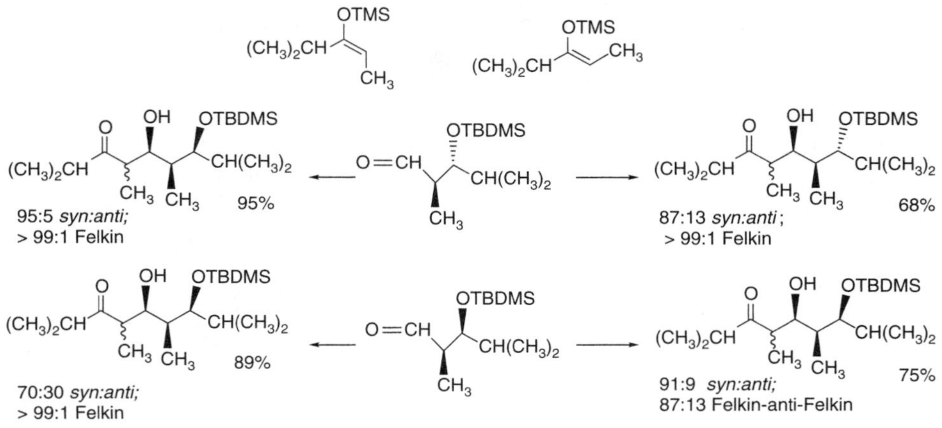

When there is also a stereogenic center in the silyl enol ether, it can enhance or detract from the underlying stereochemical preferences. The two reactions shown below possess reinforcing structures with regard to the aldehyde α-methyl and the enolate TBDMSO groups and lead to high stereoselectivity. The stereochemistry of the β-TBDMSO group in the aldehyde has little effect on the stereoselectivity.

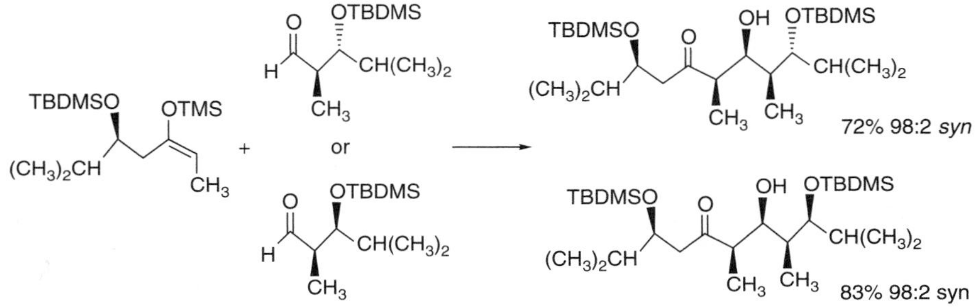

Scheme 2.5 gives some additional examples of double stereodifferentiation. Entry 1 combines the steric (Felkin) facial selectivity of the aldehyde with the facial selectivity of the enolate, which is derived from chelation. In reaction with the racemic aldehyde, the (*R*)-enantiomer is preferred.

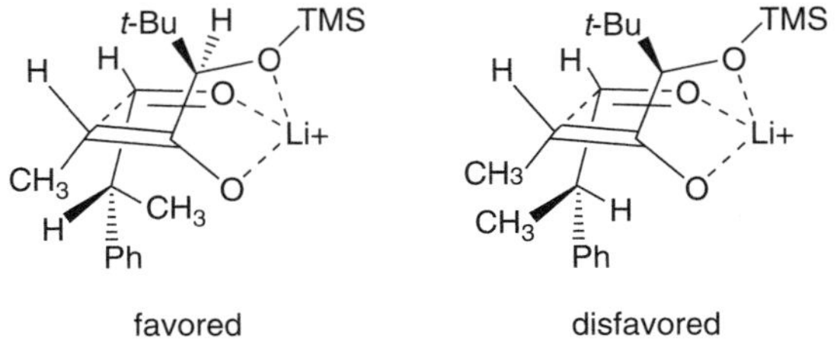

Entry 2 involves the use of a sterically biased enol boronate with an α-substituted aldehyde. The reaction, which gives 40:1 facial selectivity, was used in the synthesis of 6-deoxyerythronolide B and was one of the early demonstrations of the power of double diastereoselection in synthesis. In Entry 3, the *syn* selectivity is the result of a chelated TS, in which the β-*p*-methoxybenzyl substituent interacts with the tin ion.[120]

120. I. Paterson and R. D. Tillyer, *Tetrahedron Lett.*, **33**, 4233 (1992).

Scheme 2.5. Examples of Double Stereodifferentiation in Aldol and Mukaiyama Reactions

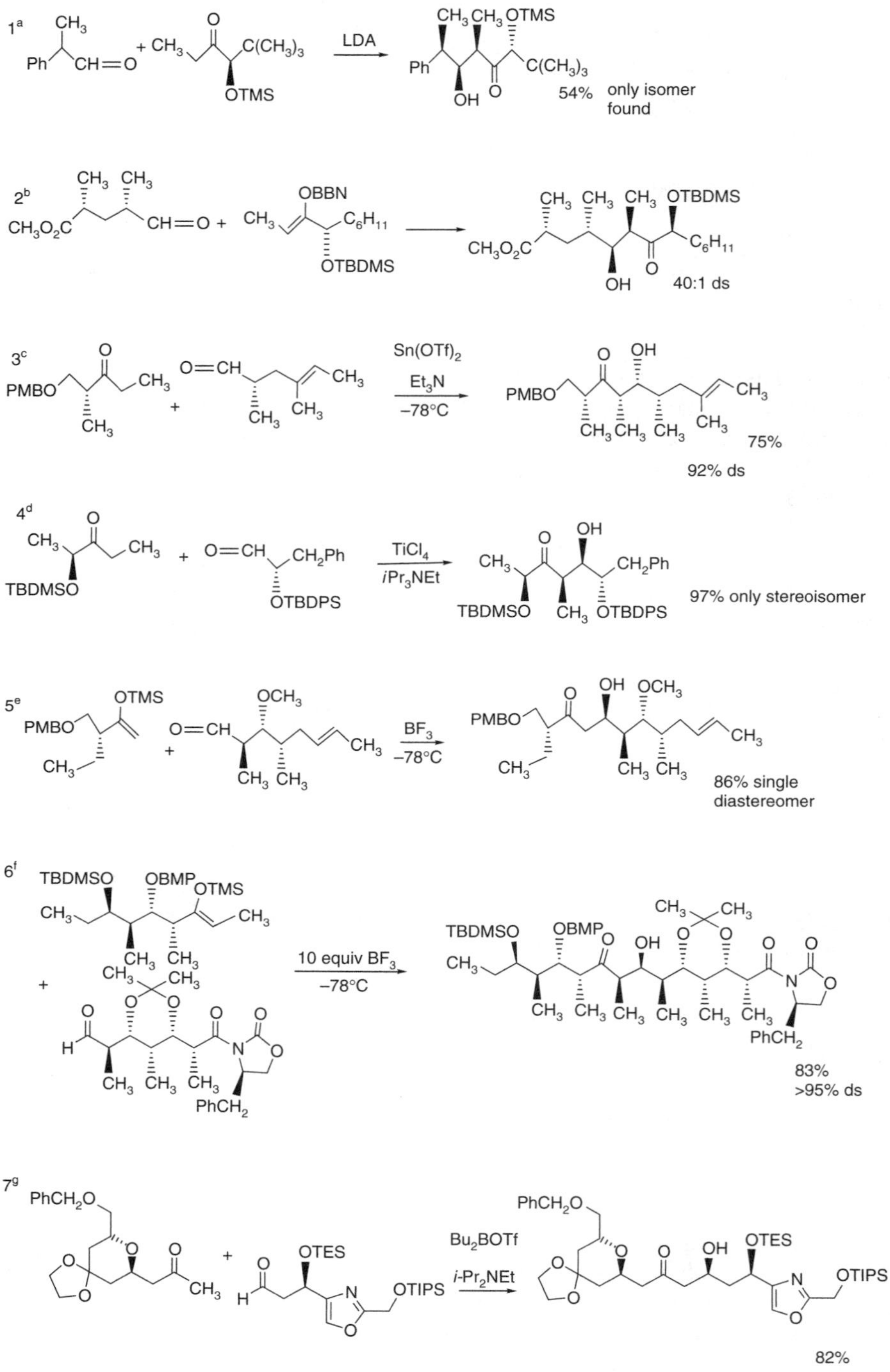

(*Continued*)

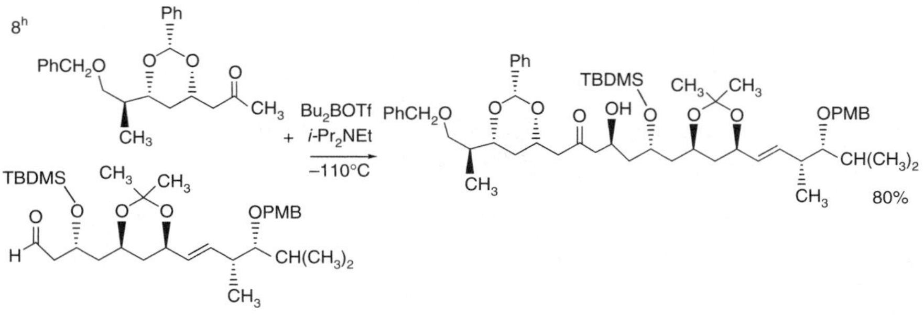

a. C. H. Heathcock, M. C. Pirrung, J. Lampe, C. T. Buse, and S. D. Young, *J. Org. Chem.*, **46**, 2290 (1981).
b. S. Masamune, M. Hirama, S. Mori, S. A. Ali, and D. S. Garvey, *J. Am. Chem. Soc.*, **103**, 1568 (1981).
c. I. R. Correa, Jr., and R. A. Pilli, *Angew. Chem. Int. Ed. Engl.*, **42**, 3017 (2003).
d. C. Esteve, M. Ferrero, P. Romea, F. Urpi, and J. Vilarrasa, *Tetrahedron Lett.*, **40**, 5083 (1999).
e. G. E. Keck, C. E. Knutson, and S. A. Wiles, *Org. Lett.*, **3**, 707 (2001).
f. D. A. Evans, A. S. Kim, R. Metternich, and V. J. Novack, *J. Am. Chem. Soc.*, **120**, 5921 (1998).
g. D. A. Evans, D. M. Fitch, T. E. Smith, and V. J. Cee, *J. Am. Chem. Soc.*, **122**, 10033 (2000).
h. D. A. Evans, B. Cote, P. J. Coleman, and B. T. Connell, *J. Am. Chem. Soc.*, **125**, 10893 (2003).

The aldehyde α-methyl substituent determines the facial selectivity with respect to the aldehyde.

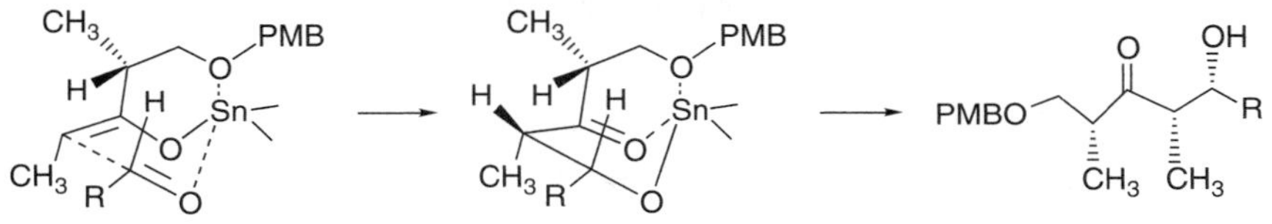

Entry 4 has siloxy substituents in both the (titanium) enolate and the aldehyde. The TBDPSO group in the aldehyde is in the "large" Felkin position, that is, perpendicular to the carbonyl group.[121] The TBDMS group in the enolate is nonchelated but exerts a steric effect that governs facial selectivity.[122] In this particular case, the two effects are matched and a single stereoisomer is observed.

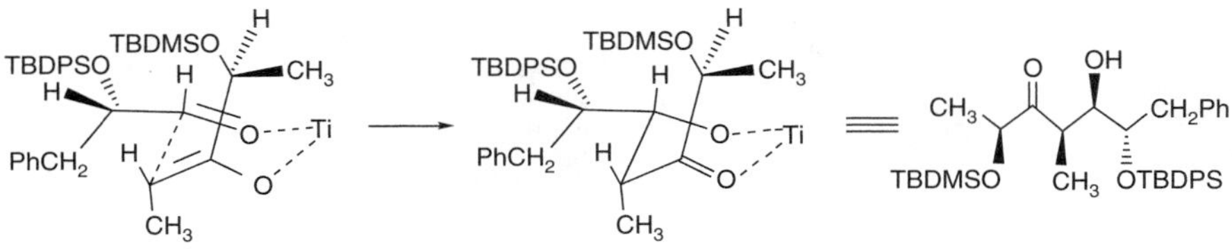

Entry 5 is a case in which the α- and β-substituents reinforce the stereoselectivity, as shown below. The largest substituent is perpendicular to the carbonyl, as in the Felkin model. When this conformation is incorporated into the TS, with the α-methyl

[121.] C. Esteve, M. Ferrero, P. Romea, F. Urpi, and J. Vilarrasa, *Tetrahedron Lett.*, **40**, 5079 (1999).
[122.] S. Figueras, R. Martin, P. Romea, F. Urpi, and J. Vilarrasa, *Tetrahedron Lett.*, **38**, 1637 (1997).

group in the "medium position," the predicted approach leads to the observed 3,4-*syn* stereochemistry.

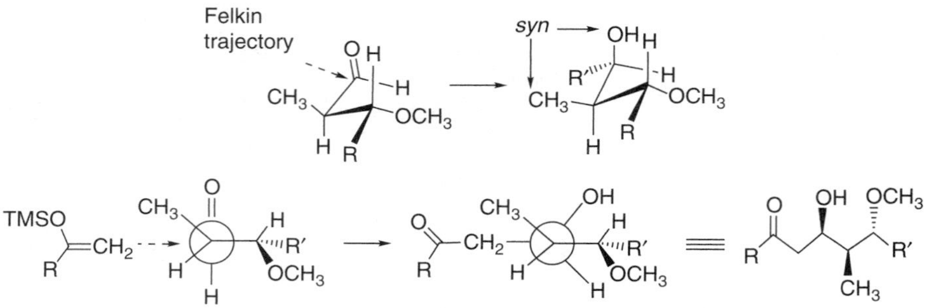

Entry 6 is an example of the methodology incorporated into a synthesis of 6-deoxyerythronolide.[123] Entries 7 and 8 illustrates the operation of the β-alkoxy group in cyclic structures. The reaction in Entry 7 was used in the synthesis of phorboxazole B.

2.1.5.4. Stereochemical Control Through Chiral Auxiliaries. Another approach to control of stereochemistry is installation of a *chiral auxiliary*, which can achieve a high degree of facial selectivity.[124] A very useful method for enantioselective aldol reactions is based on the oxazolidinones **10**, **11**, and **12**. These compounds are available in enantiomerically pure form and can be used to obtain either enantiomer of the desired product.

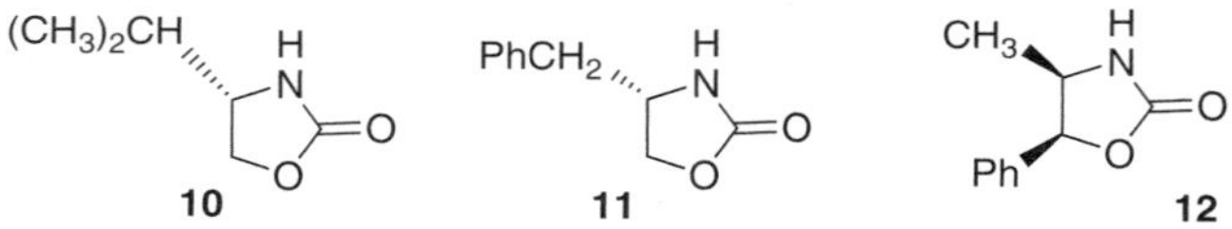

These oxazolidinones can be acylated and converted to the lithium, boron, tin, or titanium enolates by the same methods applicable to ketones and esters. For example, when they are converted to boron enolates using di-*n*-butylboron triflate and triethylamine, the enolates are the Z-stereoisomers.[125]

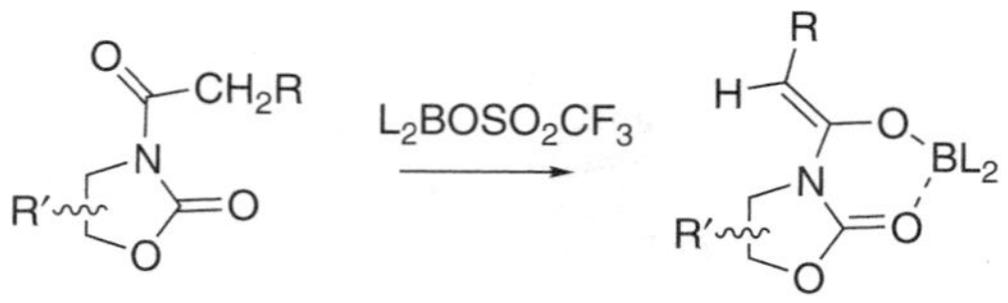

The substituents direct the approach of the aldehyde. The acyl oxazolidinones can be solvolyzed in water or alcohols to give the enantiomeric β-hydroxy acid or ester. Alternatively, they can be reduced to aldehydes or alcohols.

[123.] D. A. Evans, A. S. Kim, R. Metternich, and V. J. Novack, *J. Am. Chem. Soc.*, **120**, 5921 (1998).

[124.] M. Braun and H. Sacha, *J. Prakt. Chem.*, **335**, 653 (1993); S. G. Nelson, *Tetrahedron: Asymmetry*, **9**, 357 (1998); E. Carreira, in *Catalytic Asymmetric Synthesis*, 2nd Edition, I. Ojima, ed., Wiley-VCH, 2000, pp. 513–541; F. Velazquez and H. F. Olivo, *Curr. Org. Chem.*, **6**, 303 (2002).

[125.] D. A. Evans, J. Bartoli, and T. L. Shih, *J. Am. Chem. Soc.*, **103**, 2127 (1981).

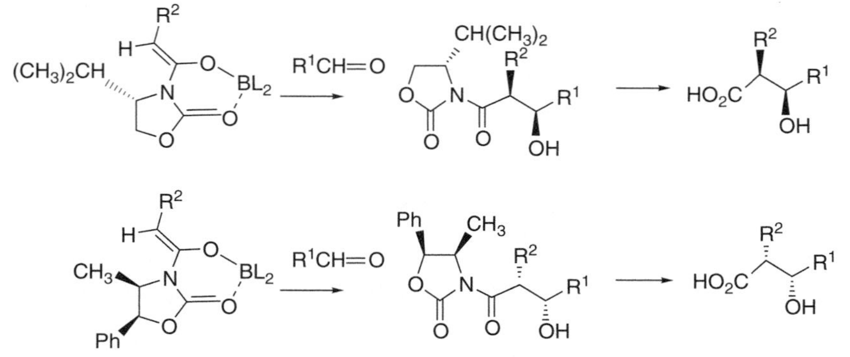

The reacting aldehyde displaces the oxazolidinone oxygen at the tetravalent boron in the reactive TS. The conformation of the addition TS for boron enolates is believed to have the oxazolidinone ring oriented with opposed dipoles of the ring and the aldehyde carbonyl groups.

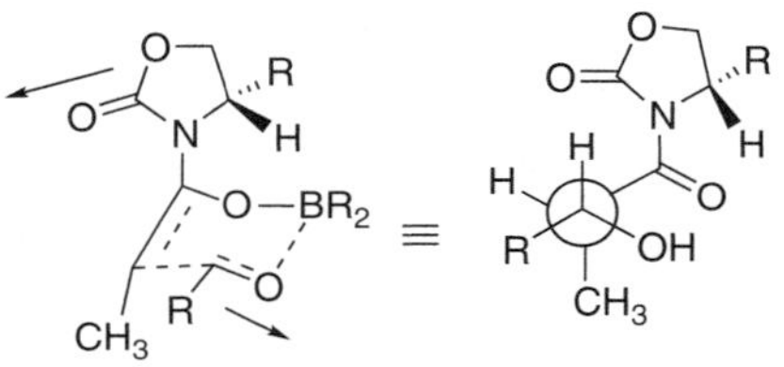

The chiral auxiliary methodology using boron enolates has been successfully applied to many complex structures (see also Scheme 2.6).

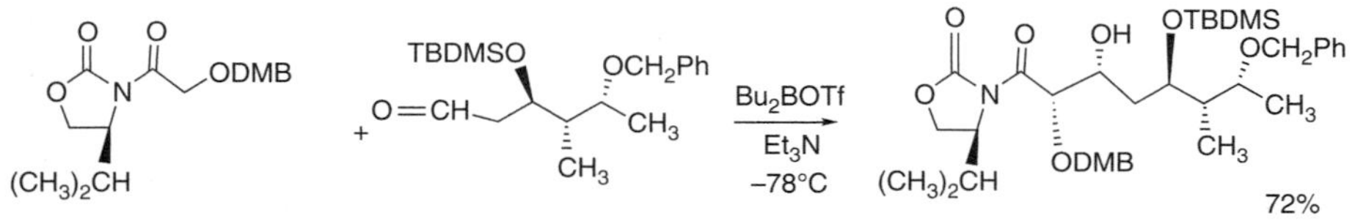

72%

Ref. 126

90%

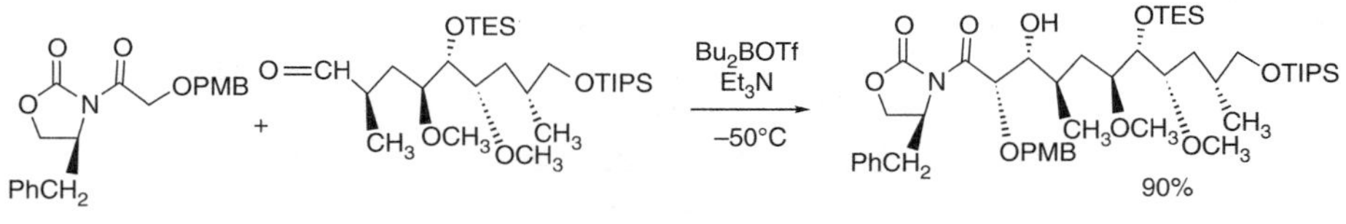

Ref. 127

126. W. R. Roush, T. G. Marron, and L. A. Pfeifer, *J. Org. Chem.*, **62**, 474 (1997).
127. T. K. Jones, R. A. Reamer, R. Desmond, and S. G. Mills, *J. Am. Chem. Soc.*, **112**, 2998 (1990).

Titanium enolates also can be prepared from *N*-acyloxazolidinones. These *Z*-enolates, which are chelated with the oxazolidinone carbonyl oxygen,[128] show *syn* stereoselectivity, and the oxazolidinone substituent exerts facial selectivity.

87% yield, 94:6 *syn:anti*

The *N*-acyloxazolidinones give *anti* products when addition is effected by a catalytic amount of MgCl$_2$ in the presence of a tertiary amine and trimethylsilyl chloride. Under these conditions the adduct is formed as the trimethylsilyl ether.[129]

91% 32:1 dr

Under similar conditions, the corresponding thiazolidinethione derivatives give *anti* product of the *opposite absolute configuration*, at least for cinnamaldehyde.

87% 10:1 dr

The mechanistic basis for the stereoselectivity of these conditions remains to be determined. The choice of reactant and conditions can be used to exert a substantial degree of control of the stereoselectivity.

Recently several other molecules have been developed as chiral auxiliaries. These include derivatives of ephedrine and pseudoephedrine. The *N*-methylephedrine [(1*R*,2*S*)-2-dimethyamino-1-phenyl-1-propanol] chiral auxiliary **13** has been examined with both the (*S*)- and (*R*)-enantiomers of 2-benzyloxy-2-methylpropanal.[130] The two enantiomers reacted quite differently. The (*R*)-enantiomer gave a 60% yield of a pure enantiomer with a *syn* configuration at the new bond. The (*S*)-enantiomer gave a combined 22% yield of two diastereomeric products in a 1.3:1 ratio. The aldehyde is known from NMR studies to form a chelated complex with TiCl$_4$,[131] and presumably reacts through a chelated TS. The TS **J** from the (*R*)-enantiomer has the methyl groups from both the chiral auxiliary and the silyl enol ether in favorable environments (matched pair). The products from the (*S*)-enantiomer arise from TS **K** and

128. D. A. Evans, D. L. Rieger, M. T. Bilodeau, and F. Urpi, *J. Am. Chem. Soc.*, **113**, 1047 (1991).
129. D. A. Evans, J. S. Tedrow, J. T. Shaw, and C. W. Downey, *J. Am. Chem. Soc.*, **124**, 392 (2002).
130. G. Gennari, L. Colombo, G. Bertolini, and G. Schimperna, *J. Org. Chem.*, **52**, 2754 (1987).
131. G. E. Keck and S. Castellino, *J. Am. Chem. Soc.*, **108**, 3847 (1986).

TS **L**, each of which has one of the methyl groups in an unfavorable environment. (mismatched pairs).

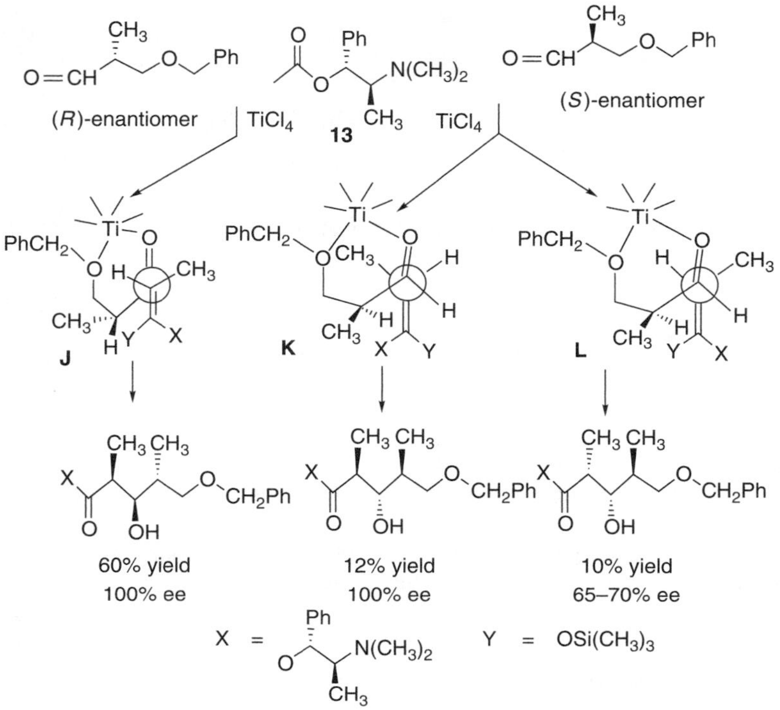

Enantioselectivity can also be induced by use of chiral boron enolates. Both the (+) and (−) enantiomers of diisopinocampheylboron triflate have been used to generate *syn* addition through a cyclic TS.[132] The enantioselectivity was greater than 80% for most cases that were examined. Z-Boron enolates are formed under these conditions and the products are 2,3-*syn*.

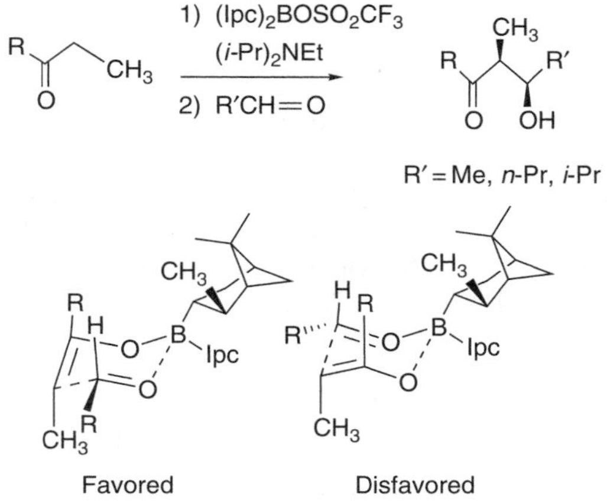

R′ = Me, *n*-Pr, *i*-Pr

Favored   Disfavored

132. I. Paterson, J. M. Goodman, M. A. Lister, R. C. Schumann, C. K. McClure, and R. D. Norcross, *Tetrahedron*, **46**, 4663 (1990).

Another promising boron enolate is derived from $(-)$-menthone.[133] It yields E-boron enolates that give good enantioselectivity in the formation of *anti* products.[134]

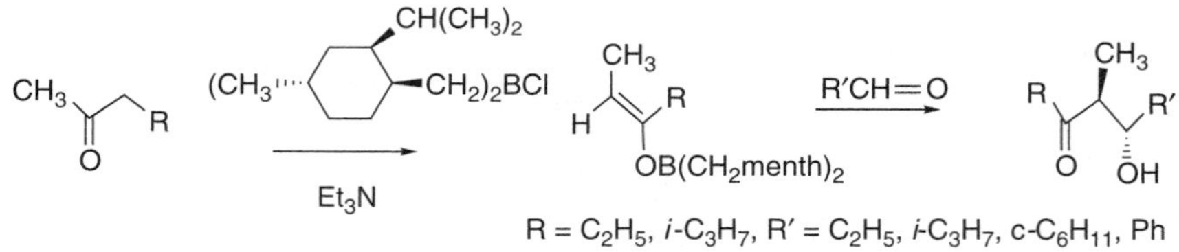

R = C₂H₅, i-C₃H₇, R′ = C₂H₅, i-C₃H₇, c-C₆H₁₁, Ph

The boron enolates of α-substituted thiol esters also give excellent facial selectivity.[135]

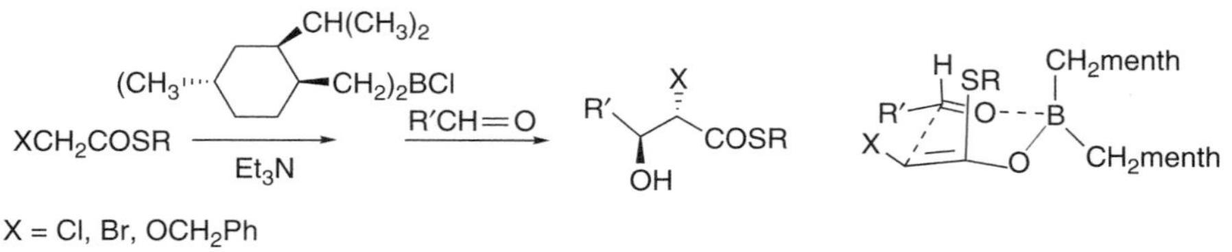

X = Cl, Br, OCH₂Ph

The facial selectivity in these chiral boron enolates has its origin in the steric effects of the boron substituents.

Several chiral heterocyclic borylating agents have been found useful for enantio-selective aldol additions. The diazaborolidine **14** is an example.[136]

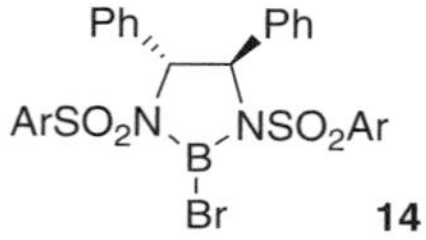

Ar = 3,5-di(trifluoromethyl)phenyl

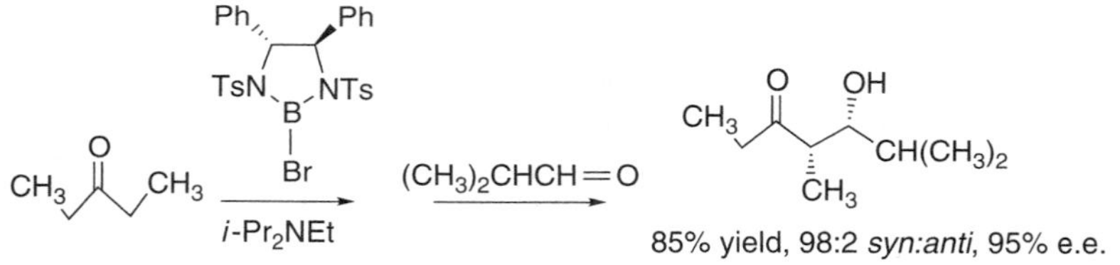

85% yield, 98:2 *syn:anti*, 95% e.e.

Derivatives with various substituted sulfonamides have been developed and used to form enolates from esters and thioesters.[137] An additional feature of this chiral auxiliary is the ability to select for *syn* or *anti* products, depending upon choice of reagents and reaction conditions. The reactions proceed through an acyclic TS, and diastereoselectivity is determined by whether the E- or Z-enolate is formed.[138] t-Butyl esters give E-enolates and *anti* adducts, whereas phenylthiol esters give *syn* adducts.[136]

---

133. C. Gennari, *Pure Appl. Chem.*, **69**, 507 (1997).
134. G. Gennari, C. T. Hewkin, F. Molinari, A. Bernardi, A. Comotti, J. M. Goodman, and I. Paterson, *J. Org. Chem.*, **57**, 5173 (1992).
135. C. Gennari, A. Vulpetti, and G. Pain, *Tetrahedron*, **53**, 5909 (1997).
136. E. J. Corey, R. Imwinkelried, S. Pikul, and Y. B. Xiang, *J. Am. Chem. Soc.*, **111**, 5493 (1989).
137. E. J. Corey and S. S. Kim, *J. Am. Chem. Soc.*, **112**, 4976 (1990).
138. E. J. Corey and D. H. Lee, *Tetrahedron Lett.*, **34**, 1737 (1993).

Ph,,   ,,Ph

ArSO$_2$N   NSO$_2$Ar
        B
        |
        Br

CH$_3$CH$_2$CO$_2$C(CH$_3$)$_3$ ⟶

◯—CH=O ⟶

OH
◯—◡—CO$_2$C(CH$_3$)$_3$
        |
        CH$_3$

Ar = 3,5-di(trifluoromethyl)phenyl

96:4 *syn:anti*, 75% e.e.

(CH$_3$)$_2$CHCH=O +

O
‖
CH$_3$CH$_2$CSPh

14
⟶

OH O
| ‖
(CH$_3$)$_2$CH—◡—SPh
        |
        CH$_3$

72%
97% e.e.

Scheme 2.6 shows some examples of the use of chiral auxiliaries in the aldol and Mukaiyama reactions. The reaction in Entry 1 involves an achiral aldehyde and the chiral auxiliary is the only influence on the reaction diastereoselectivity, which is very high. The *Z*-boron enolate results in *syn* diastereoselectivity. Entry 2 has both an α-methyl and a β-benzyloxy substituent in the aldehyde reactant. The 2,3-*syn* relationship arises from the *Z*-configuration of the enolate, and the 3,4-*anti* stereochemistry is determined by the stereocenters in the aldehyde. The product was isolated as an ester after methanolysis. Entry 3, which is very similar to Entry 2, was done on a 60-kg scale in a process development investigation for the potential antitumor agent (+)-discodermolide (see page 1244).

Entries 4 and 5 are cases in which the oxazolidinone substituent is a β-ketoacyl group. The α-hydrogen (between the carbonyls) does not react as rapidly as the γ-hydrogen, evidently owing to steric restrictions to optimal alignment. The all-*syn* stereochemistry is consistent with a TS in which the exocyclic carbonyl is chelated to titanium.

Oxaz

CH$_3$
H   O
 ◡   Cl
   O—Ti—Cl
  O   Cl
R   ◡
CH$_3$

⟶

O   O   O   OH
‖   ‖   ‖ |
O◡N—◡—◡—◡—R
        | |
      CH$_3$ CH$_3$
    CH$_2$Ph

In Entry 5, the aldehyde is also chiral and double stereodifferentiation comes into play. Entry 6 illustrates the use of an oxazolidinone auxiliary with another highly substituted aldehyde. Entry 7 employs conditions that were found effective for α-alkoxyacyl oxazolidinones. Entries 8 and 9 are examples of the application of the thiazolidine-2-thione auxiliary and provide the 2,3-*syn* isomers with diastereofacial control by the chiral auxiliary.

2.1.5.5. Stereochemical Control Through Reaction Conditions. In the early 1990s it was found that the stereochemistry of reactions of boron enolates of *N*-acyloxazolidinones can be altered by using a Lewis acid complex of the aldehyde or an excess of the Lewis acid. These reactions are considered to take place through an open TS, with the stereoselectivity dependent on the steric demands of the Lewis acid. With various aldehydes, TiCl$_4$ gave a *syn* isomer, whereas the reaction was

Scheme 2.6. Control of Stereochemistry of Aldol and Mukaiyama Aldol Reactions Using Chiral Auxiliaries

1[a]

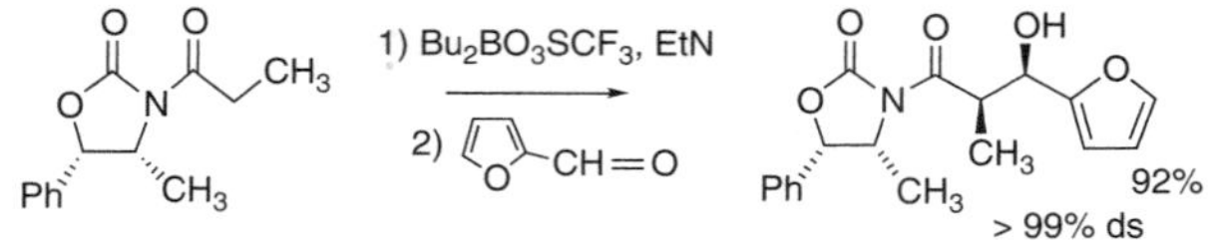

2[b]

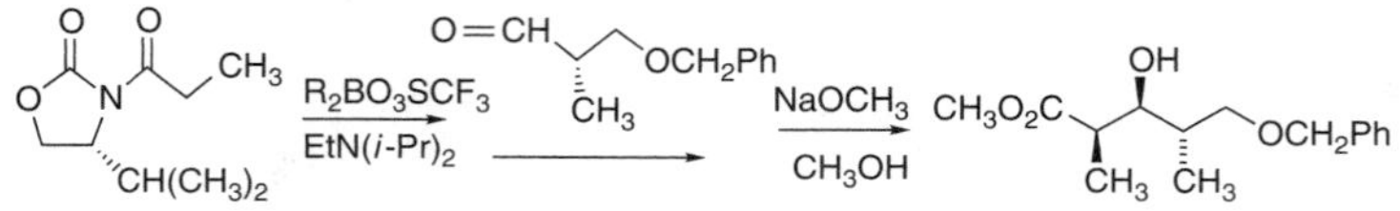

3[c]

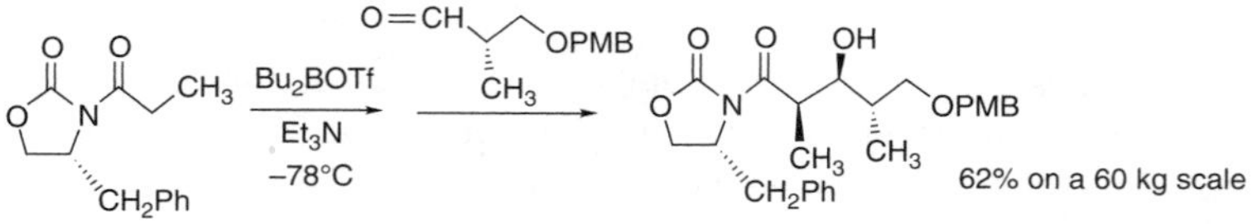

4[d]

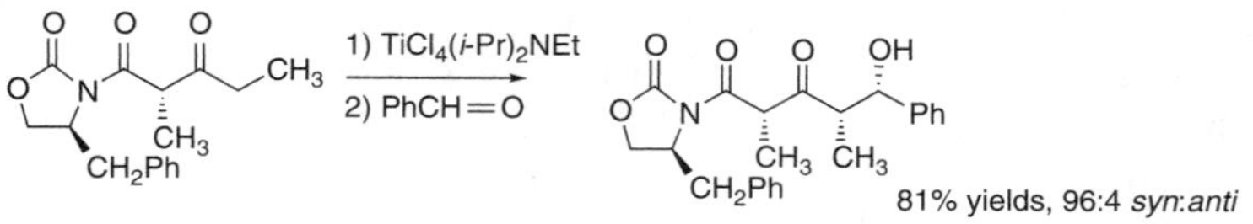

5[e]

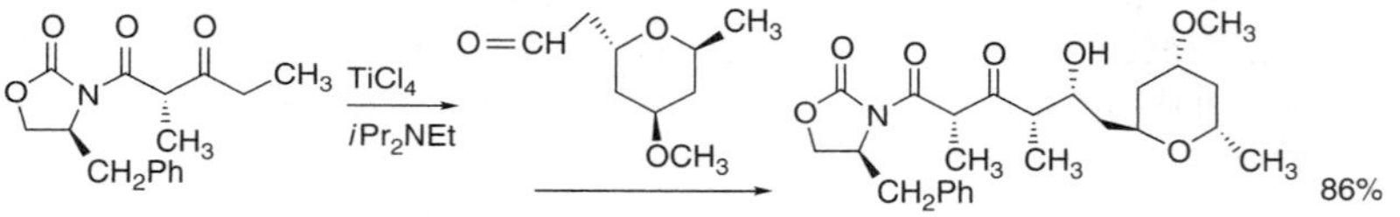

6[f]

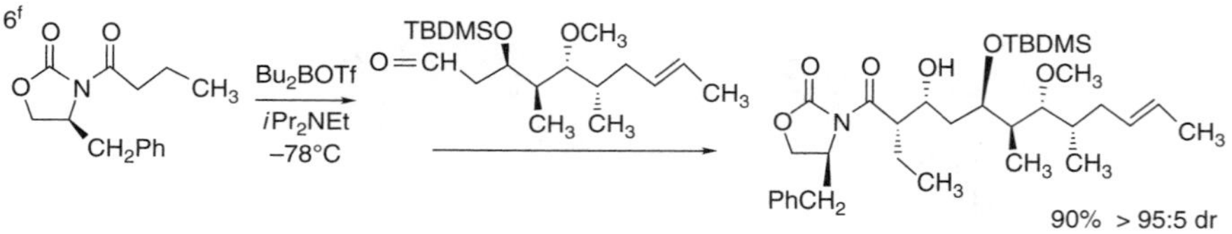

7[g]

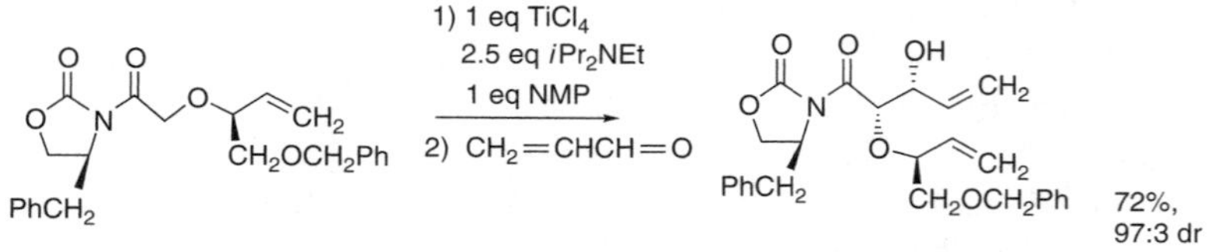

8[h]

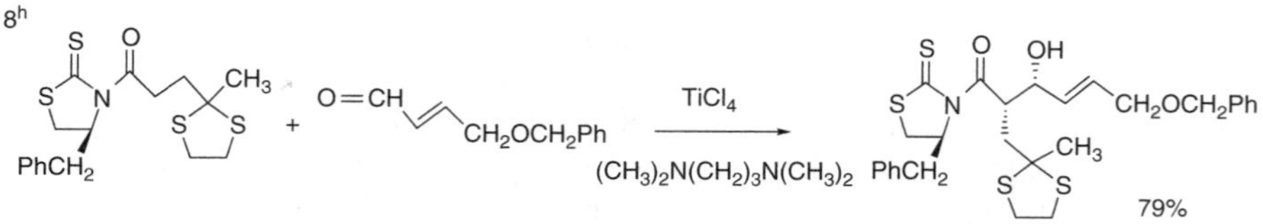

(*Continued*)

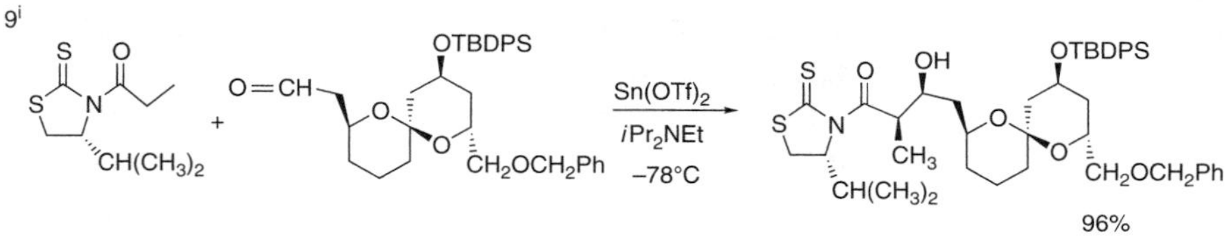

9i
OTBDPS
O=CH
CH(CH3)2
CH2OCH2Ph
Sn(OTf)2
iPr2NEt
–78°C
OH
CH3
96%

a. S. F. Martin and D. E. Guinn, *J. Org. Chem.*, **52**, 5588 (1987).
b. D. Seebach, H.-F. Chow, R. F. W. Jackson, K. Lawson, M. A. Sutter, S. Thaisrivongs, and J. Zimmerman, *J. Am. Chem. Soc.*, **107**, 5292 (1985).
c. S. J. Mickel, G. H. Sedelmeier, D. Niererer, R. Daeffler, A. Osmani, K. Schreiner, M. Seeger-Weibel, B. Berod, K. Schaer, R. Gamboni, S. Chen, W. Chen, C. T. Jagoe, F. Kinder, M. Low, K. Prasad, O. Repic, W. C. Shieh, R. M. Wang, L. Wakole, D. Xu, and S. Xue, *Org. Proc. Res. Dev.*, **8**, 92 (2004).
d. D. A. Evans, J. S. Clark, R. Metternich, V. J. Novack, and G. S. Sheppard, *J. Am. Chem. Soc.*, **112**, 866 (1990).
e. G. E. Keck and G. D. Lundquist, *J. Org. Chem.*, **64**, 4482 (1999).
f. L. C. Dias, L. G. de Oliveira, and M. A. De Sousa, *Org. Lett.*, **5**, 265 (2003).
g. M. T. Crimmins and J. She, *Synlett*, 1371 (2004).
h. J. Wu, X. Shen, Y.-Q. Yang, Q. Hu, and J.-H. Huang, *J. Org. Chem.*, **69**, 3857 (2004).
i. D. Zuev and L. A. Paquette, *Org. Lett.*, **2**, 679 (2000).

anti selective using $(C_2H_5)_2AlCl$.[139] The *anti* selectivity is proposed to arise as a result of the greater size requirement for the complexed aldehyde with $(C_2H_5)_2AlCl$. These reactions both give a different stereoisomer than the reaction done *without the additional Lewis acid*. The chiral auxiliary is the source of facial selectivity.

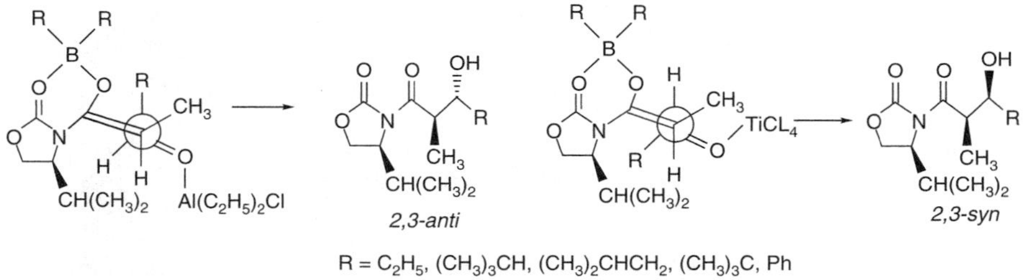

CH3
CH(CH3)2
Al(C2H5)2Cl
2,3-anti
TiCL4
2,3-syn
R = C2H5, (CH3)3CH, (CH3)2CHCH2, (CH3)3C, Ph

With titanium enolates it was found that use of excess (3 equiv.) of the titanium reagent reversed facial selectivity of oxazolidinone enolates.[140] This was attributed to generation of a chelated TS in the presence of the excess Lewis acid. The chelation rotates the oxazolidinone ring and reverses the facial preference, while retaining the *Z*-configuration *syn* diastereoselectivity.

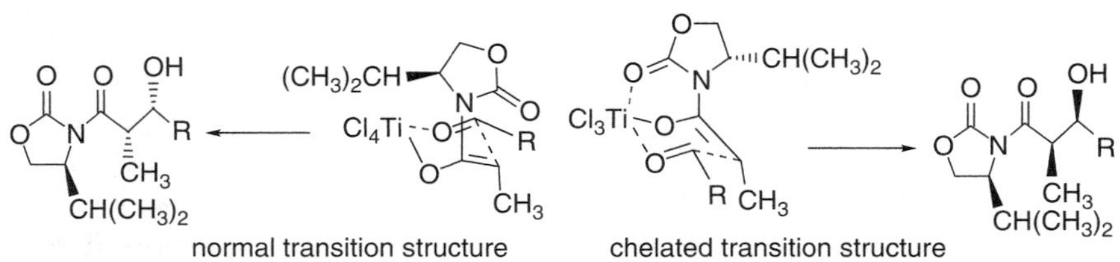

(CH3)2CH
Cl4Ti
Cl3Ti
CH(CH3)2
normal transition structure
chelated transition structure

[139] M. A. Walker and C. H. Heathcock, *J. Org. Chem.*, **56**, 5747 (1991).
[140] M. Nerz-Stormes and E. R. Thornton, *Tetrahedron Lett.*, **27**, 897 (1986); M. Nerz-Stormes and E. R. Thornton, *J. Org. Chem.*, **56**, 2489 (1991).

Crimmins and co-workers have developed *N*-acyloxazolidinethiones as chiral auxiliaries. These reagents show excellent 2,3-*syn* diastereoselectivity and enantioselectivity in additions to aldehydes. The titanium enolates are prepared using $TiCl_4$, with (−)-sparteine being a particularly effective base.[141]

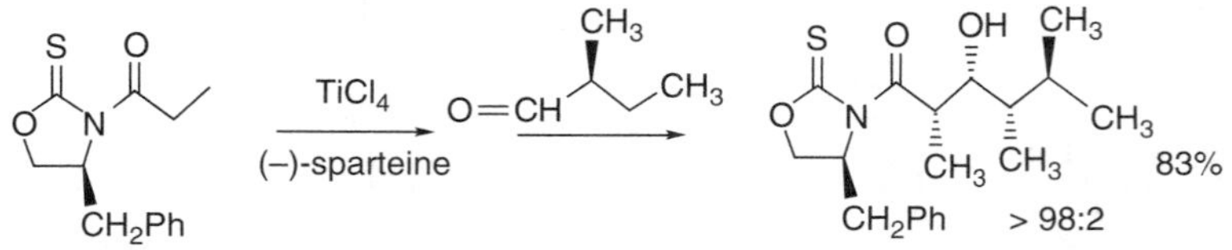

The facial selectivity of these compounds is also dependent on the amount of $TiCl_4$ that is used. With two equivalents, the facial selectivity is reversed. This reversal is also achieved by adding $AgSbF_6$. It was suggested that the excess reagent or the silver salt removes a Cl^- from the titanium coordination sphere and promotes chelation with the thione sulfur.[142] This changes the facial selectivity of the enolate by causing a reorientation of the oxazolidinethione ring. The greater affinity of titanium for sulfur over oxygen makes the oxazolidinethiones particularly effective in these circumstances. The increased tendency for chelation has been observed with other chiral auxiliaries having thione groups.[143]

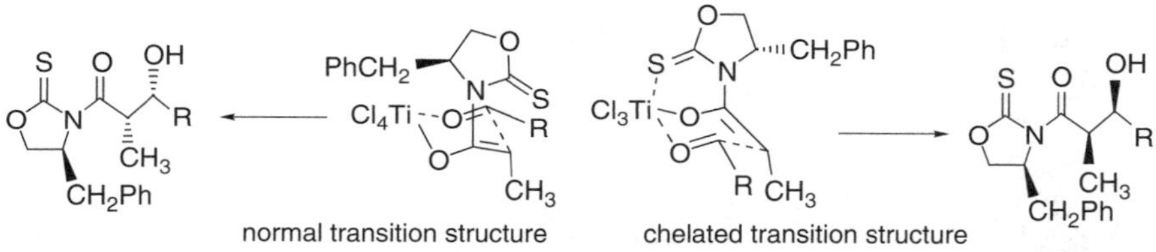

normal transition structure   chelated transition structure

A related effect is noted with α-alkoxyacyl derivatives. These compounds give mainly the *anti* adducts when a second equivalent of $TiCl_4$ is added prior to the aldehyde.[144] The *anti* addition is believe to occur through a TS in which the alkoxy oxygen is chelated. In the absence of excess $TiCl_4$, a nonchelated cyclic TS accounts for the observed *syn* selectivity.

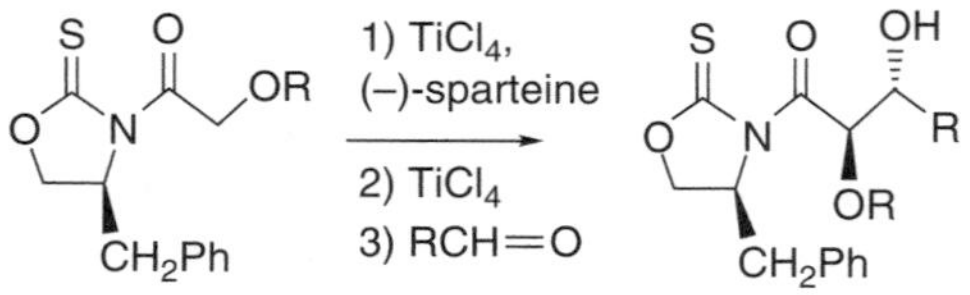

141. M. T. Crimmins and B. W. King, *J. Am. Chem. Soc.*, **120**, 9084 (1998); M. T. Crimmins, B. W. King, E. A. Tabet, and C. Chaudhary, *J. Org. Chem.*, **66**, 894 (2001); M. T. Crimmins and J. She, *Synlett*, 1371 (2004).
142. M. T. Crimmins, B. W. King, and E. A. Tabet, *J. Am. Chem. Soc.*, **119**, 7883 (1997).
143. T. H. Yan, C. W. Tan, H. C. Lee, H. C. Lo, and T. Y. Huang, *J. Am. Chem. Soc.*, **115**, 2613 (1993).
144. M. T. Crimmins and P. J. McDougall, *Org. Lett.*, **5**, 591 (2003).

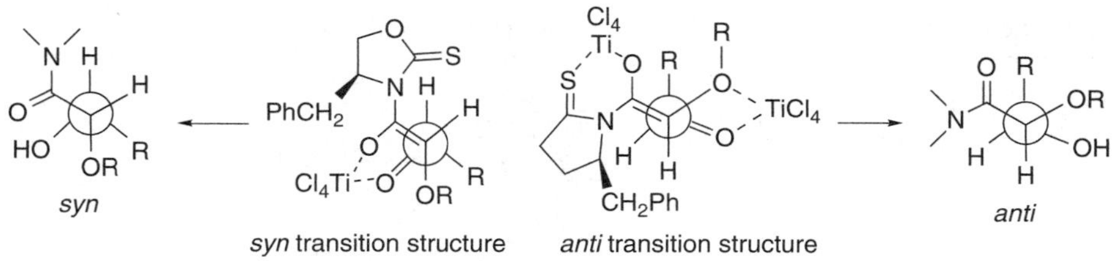

syn transition structure   *anti transition structure*

Camphor-derived sulfonamide can also permit control of enantioselectivity by use of additional Lewis acid. These chiral auxiliaries can be used under conditions in which either cyclic or noncyclic TSs are involved. This frequently allows control of the *syn* or *anti* stereoselectivity.[143] The boron enolates give *syn* products, but inclusion of $SnCl_4$ or $TiCl_4$ gave excellent selectivity for *anti* products and high enantioselectivity for a range of aldehydes.[145]

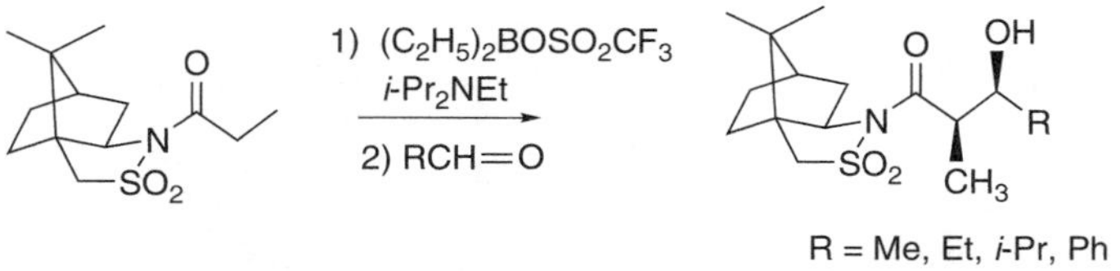

$R = Me, Et, i\text{-}Pr, Ph$ Ref. 146

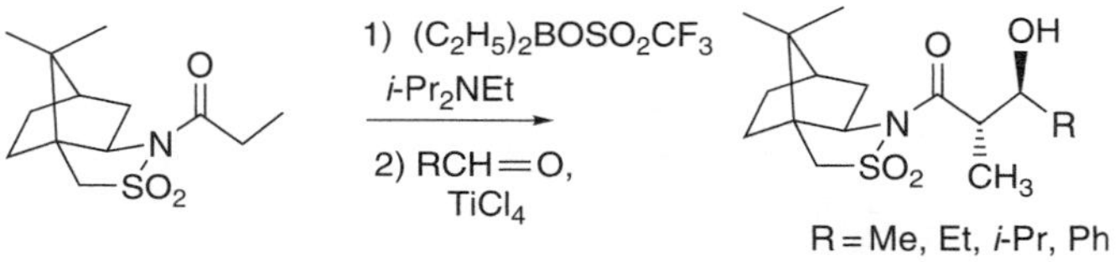

$R = Me, Et, i\text{-}Pr, Ph$ Ref. 147

In the case of boron enolates of the camphor sulfonamides, the $TiCl_4$-mediated reaction is believed to proceed through an open TS, whereas in its absence, the reaction proceeds through a cyclic TS.

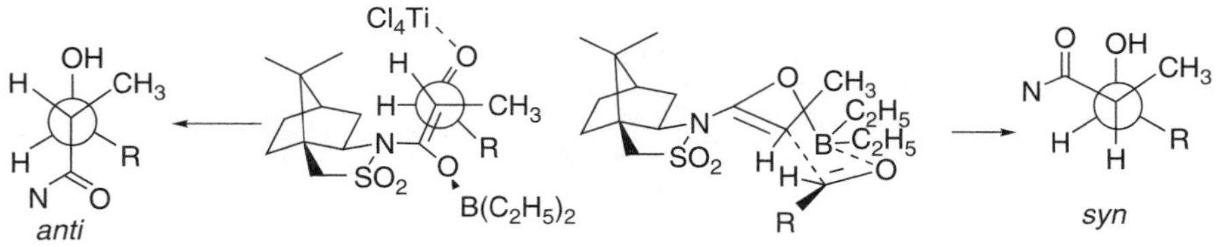

Scheme 2.7 gives some examples of the control of stereoselectivity by use of additional Lewis acid and related methods. Entry 1 shows the effect of the use of excess $TiCl_4$. Entry 2 demonstrates the ability of $(C_2H_5)_2AlCl$ to shift the boron enolate toward formation of the 2,3-*anti* diastereomer. Entries 3 and 4 compare the use of one versus two equivalents of $TiCl_4$ with an oxazoldine-2-thione auxiliary. There is a nearly complete shift of facial selectivity. Entry 5 shows a subsequent application of this methodology. Entries 6 and 7 show the effect of complexation of the aldehyde

[145.] Y.-C. Wang, A.-W. Hung, C.-S. Chang, and T.-H. Yan, *J. Org. Chem.*, **61**, 2038 (1996).
[146.] W. Oppolzer, J. Blagg, I. Rodriguez, and E. Walther, *J. Am. Chem. Soc.*, **112**, 2767 (1990).
[147.] W. Oppolzer and P. Lienhard, *Tetrahedron Lett.*, **34**, 4321 (1993).

Scheme 2.7. Examples of Control of Stereoselectivity by Use of Additional Lewis Acid

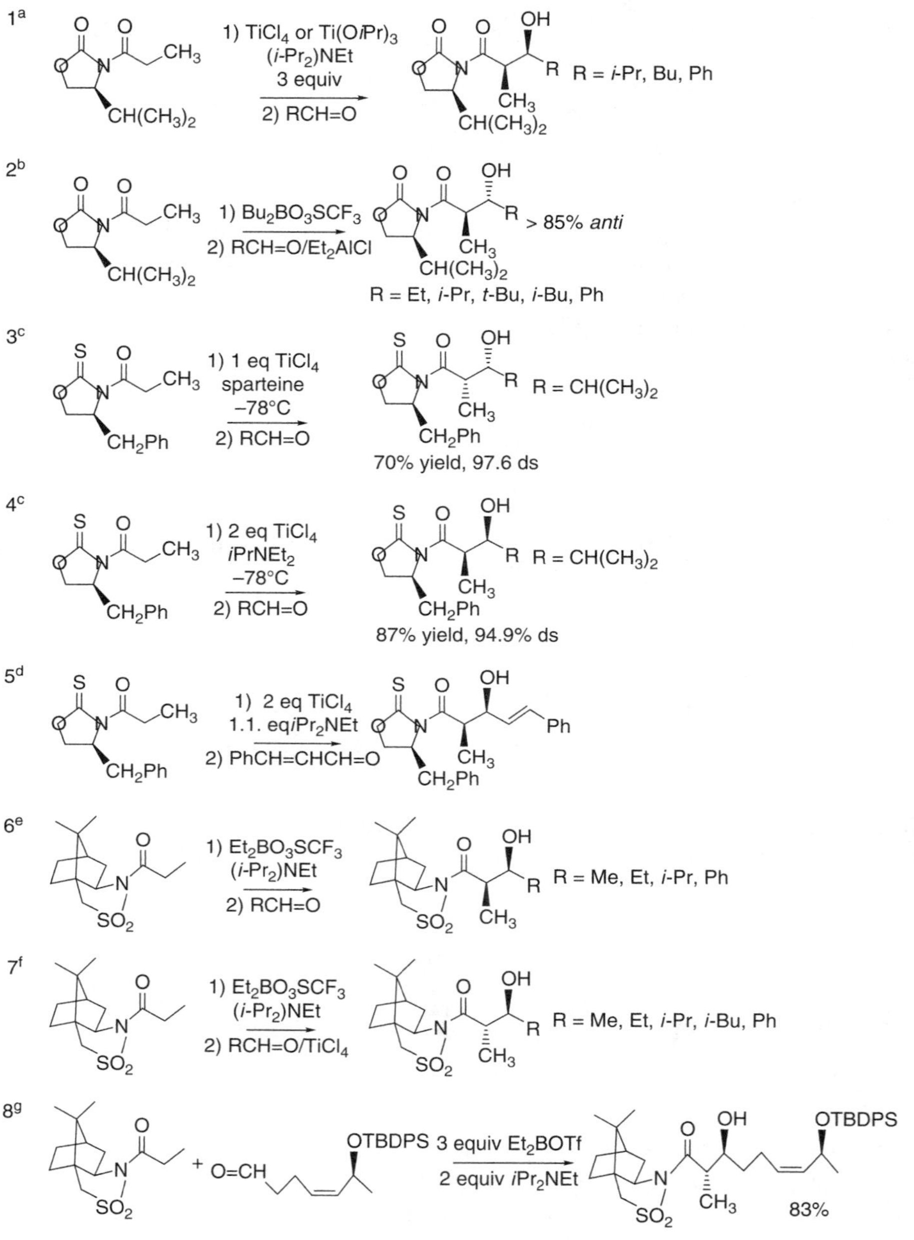

1[a]

1) TiCl$_4$ or Ti(OiPr)$_3$
(i-Pr$_2$)NEt
3 equiv
2) RCH=O

R = i-Pr, Bu, Ph

2[b]

1) Bu$_2$BO$_3$SCF$_3$
2) RCH=O/Et$_2$AlCl

> 85% *anti*

R = Et, i-Pr, t-Bu, i-Bu, Ph

3[c]

1) 1 eq TiCl$_4$
sparteine
−78°C
2) RCH=O

R = CH(CH$_3$)$_2$

70% yield, 97.6 ds

4[c]

1) 2 eq TiCl$_4$
iPrNEt$_2$
−78°C
2) RCH=O

R = CH(CH$_3$)$_2$

87% yield, 94.9% ds

5[d]

1) 2 eq TiCl$_4$
1.1. eq$i$Pr$_2$NEt
2) PhCH=CHCH=O

6[e]

1) Et$_2$BO$_3$SCF$_3$
(i-Pr$_2$)NEt
2) RCH=O

R = Me, Et, i-Pr, Ph

7[f]

1) Et$_2$BO$_3$SCF$_3$
(i-Pr$_2$)NEt
2) RCH=O/TiCl$_4$

R = Me, Et, i-Pr, i-Bu, Ph

8[g]

3 equiv Et$_2$BOTf
2 equiv iPr$_2$NEt

83%

a. M. Nerz-Stormes and E. R. Thornton, *J. Org. Chem.*, **56**, 2489 (1991).
b. M. A. Walker and C. H. Heathcock, *J. Org. Chem.*, **56**, 5747 (1991).
c. M. T. Crimmins, B. W. King, and E. A. Tabet, *J. Am. Chem. Soc.*, **119**, 7883 (1997).
d. T. K. Chakraborty, S. Jayaprakash, and P. Laxman, *Tetrahedron*, **57**, 9461 (2001).
e. W. Oppolzer, J. Blagg, I. Rodriguez, and E. Walther, *J. Am. Chem. Soc.*, **112**, 2767 (1990).
f. W. Oppolzer and P. Lienhard, *Tetrahedron Lett.*, **34**, 4321 (1993).
g. B. Fraser and P. Perlmutter, *J. Chem. Soc.*, *Perkin Trans. 1*, 2896 (2002).

with TiCl$_4$ using the camphor sultam auxiliary. Entry 8 is an example of the use of excess diethylboron triflate to obtain the *anti* stereoisomer in a step in the synthesis of epothilone.

These examples and those in Scheme 2.6 illustrate the key variables that determine the stereochemical outcome of aldol addition reactions using chiral auxiliaries. The first element that has to be taken into account is the configuration of the ring system that is used to establish steric differentiation. Then the nature of the TS, whether it is acyclic, cyclic, or chelated must be considered. Generally for boron enolates, reaction proceeds through a cyclic but nonchelated TS. With boron enolates, excess Lewis acid can favor an acyclic TS by coordination with the carbonyl electrophile. Titanium enolates appear to be somewhat variable but can be shifted to chelated TSs by use of excess reagent and by auxiliaries such as oxazolidine-2-thiones that enhance the tendency to chelation. Ultimately, all of the factors play a role in determining which TS is favored.

2.1.5.6. Enantioselective Catalysis of the Aldol Addition Reaction.   There are also several catalysts that can effect enantioselective aldol addition. The reactions generally involve enolate equivalents, such as silyl enol ethers, that are unreactive toward the carbonyl component alone, but can react when activated by a Lewis acid. The tryptophan-based oxazaborolidinone **15** has proven to be a useful catalyst.[148]

15

This catalyst induces preferential *re* facial attack on simple aldehydes, as indicated in Figure 2.2. The enantioselectivity appears to involve the shielding of the *si* face by the indole ring through a π-stacking interaction.

The *B*-3,5-*bis*-(trifluoromethyl)phenyl derivative was found to be a very effective catalyst.[149]

R = 3,5-di(trifluoromethyl)phenyl

> 99:1 *syn*; > 99% e.e.

148.   E. J. Corey, C. L. Cywin, and T. D. Roper, *Tetrahedron Lett.*, **33**, 6907 (1992); E. J. Corey, T.-P. Loh, T. D. Roper, M. D. Azimioara, and M. C. Noe, *J. Am. Chem. Soc.*, **114**, 8290 (1992); S. G. Nelson, *Tetrahedron: Asymmetry*, **9**, 357 (1998).
149.   K. Ishihara, S. Kondo, and H. Yamamoto, *J. Org. Chem.*, **65**, 9125 (2000).

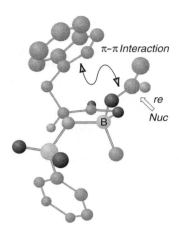

Fig. 2.2. Origin of facial selectivity in indolylmethyloxazaborolidinone structure. Reproduced from *Tetrahedron: Asymmetry*, **9**, 357 (1998), by permission of Elsevier. (See also color insert.)

An oxazaborolidinone derived from valine is also an effective catalyst. In one case, the two enantiomeric catalysts were completely enantioselective for the newly formed center.[150]

Another group of catalysts consist of cyclic borinates derived from tartaric acid. These compounds give good reactivity and enantioselectivity in Mukaiyama aldol reactions. Several structural variations such as **16** and **17** have been explored.[151]

150. S. Kiyooka, K. A. Shahid, F. Goto, M. Okazaki, and Y. Shuto, *J. Org. Chem.*, **68**, 7967 (2003).
151. K. Ishihara, T. Maruyama, M. Mouri, Q. Gao, K. Furuta, and H. Yamamoto, *Bull. Chem. Soc. Jpn.*, **66**, 3483 (1993).

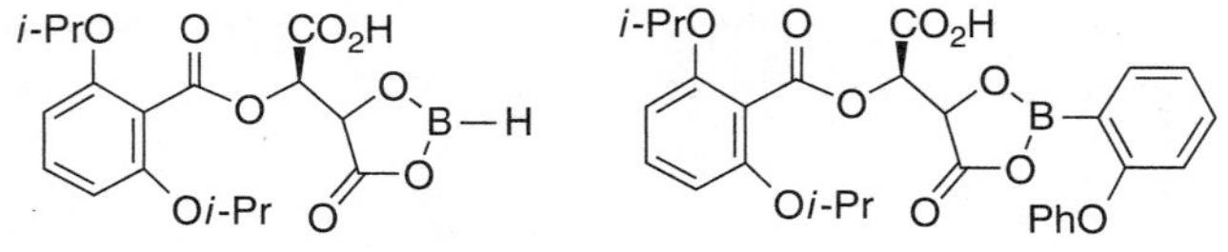

16   **17**

These catalysts are believed to function through an acyclic TS. In addition to the normal steric effects of the open TS, the facial selectivity is probably influenced by π stacking with the aryl ring and possibly hydrogen bonding by the formyl hydrogen.[152]

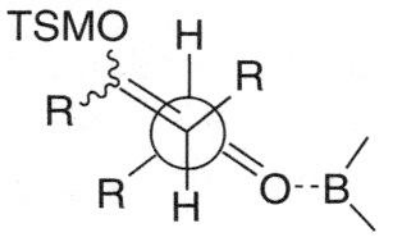

An interesting example of the use of this type of catalysis is a case in which the addition reaction of 3-methylcyclohex-2-enone to 5-methyl-2-hexenal was explored over a range of conditions. The reaction was investigated using both the lithium enolate and the trimethylsilyl enol ether. The yield and stereoselectivity are given for several sets of conditions.[153] Whereas the lithium enolate and achiral Lewis acids TiCl$_4$ and BF$_3$ gave moderate *anti* diastereoselectivity, the catalyst **17** induces good *syn* selectivity, as well as high enantioselectivity.

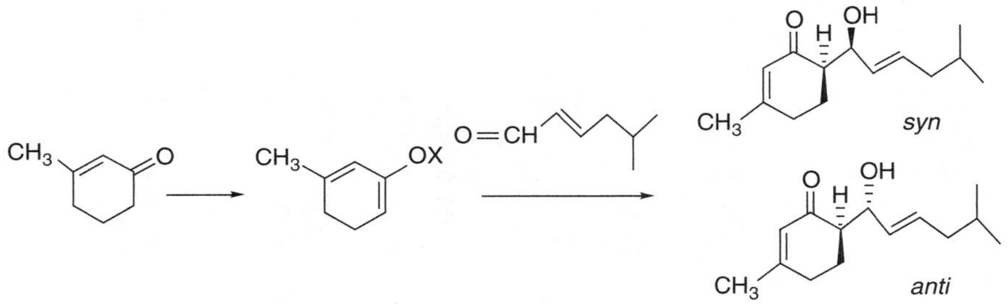

X	Conditions	Yield	*syn*	*anti*	e.e.
Li	(kinetic)	63	18	82	–
Li	(thermo)	66	55	45	–
TMS	TiCl$_4$	53	15	85	–
TMS	BF$_3$	68	25	75	–
TMS	Cat **16**	51	42	58	24(*R*)
TMS	Cat **17**	94	91	9	99(*R*)

The lesson from this case is that reactions that are quite unselective under simple Lewis acid catalysis can become very selective with chiral catalysts. Moreover, as this particular case also shows, they can be very dependent on the specific structure of the catalyst.

[152.] K. Furuta, T. Maruyama, and H. Yamamoto, *J. Am. Chem. Soc.*, **113**, 1041 (1991); K. Ishihara, Q. Gao, and H. Yamamoto, *J. Am. Chem. Soc.*, **115**, 10412 (1993).

[153.] K. Takao, T. Tsujita, M. Hara, and K. Tadano, *J. Org. Chem.*, **67**, 6690 (2002).

Another effective group of catalysts is composed of copper *bis*-oxazolines.[154] The chirality is derived from the 4-substituents on the ring.

This and similar catalysts are effective with silyl ketene acetals and silyl thioketene acetals.[155] One of the examples is the tridentate pyridine-BOX-type catalyst **18**. The reactivity of this catalyst has been explored using α- and β-oxy substituted aldehydes.[154] α-Benzyloxyacetaldehyde was highly enantioselective and the α-trimethylsilyloxy derivative was weakly so (56% e.e.). Nonchelating aldehydes such as benzaldehyde and 3-phenylpropanal gave racemic product. 3-Benzyloxypropanal also gave racemic product, indicating that the β-oxy aldehydes do not chelate with this catalyst.

The Cu-BOX catalysts function as Lewis acids at the carbonyl oxygen. The chiral ligands promote facial selectivity, as shown in Figure 2.3.

Several catalysts based on Ti(IV) and BINOL have shown excellent enantioselectivity in Mukaiyama aldol reactions.[156] A catalyst prepared from a 1:1 mixture of BINOL and Ti(O-*i*-Pr)$_4$ gives good results with silyl thioketene acetals in ether, but is very solvent sensitive.[157]

The structure of the active catalyst and the mechanism of catalysis have not been completely defined. Several solid state complexes of BINOL and Ti(O-*i*-Pr)$_4$ have been characterized by X-ray crystallography.[158] Figure 2.4 shows the structures of complexes having the composition (BINOLate)Ti$_2$(O-*i*-Pr)$_6$ and (BINOLate)Ti$_3$(O-*i*-Pr)$_{10}$.

154. D. A. Evans, J. A. Murry, and M. C. Kozlowski, *J. Am. Chem. Soc.*, **118**, 5814 (1996).

155. D. A. Evans, D. W. C. MacMillan, and K. R. Campos, *J. Am. Chem. Soc.*, **119**, 10859 (1997); D. A. Evans, M. C. Kozlowski, C. S. Burgey, and D. W. C. MacMillan, *J. Am. Chem. Soc.*, **119**, 7893 (1997).

156. S. Matsukawa and K. Mikami, *Tetrahedron: Asymmetry*, **6**, 2571 (1995); H. Matsunaga, Y. Yamada, T. Ide, T. Ishizuka, and T. Kunieda, *Tetrahedron: Asymmetry*, **10**, 3095 (1999).

157. G. E. Keck and D. Krishnamurthy, *J. Am. Chem. Soc.*, **117**, 2363 (1995).

158. T. J. Davis, J. Balsells, P. J. Carroll, and P. J. Walsh, *Org. Lett.*, **3**, 699 (2001).

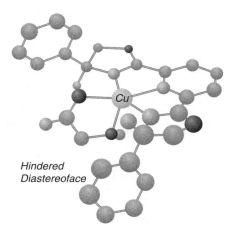

Fig. 2.3. Origin of facial selectivity of *bis*-oxazoline catalyst. Reproduced from *Tetrahedron: Asymmetry*, **9**, 357 (1998), by permission of Elsevier. (See also color insert.)

Halogenated BINOL derivatives of $Zr(O\text{-}t\text{-}Bu)_4$ such as **19** also give good yields and enantioselectivity.[159]

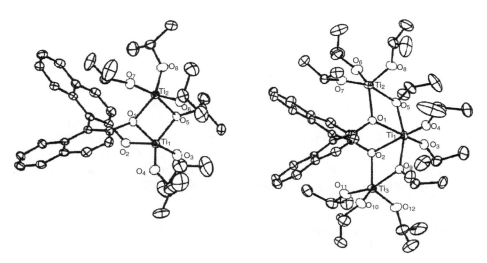

Fig. 2.4. Left: dinuclear complex of composition $(BINOLate)Ti_2(O\text{-}i\text{-}Pr)_6$. Right: trinuclear complex of composition $(BINOLate)Ti_3(O\text{-}i\text{-}Pr)_{10}$. Reproduced from *Org. Lett.*, **3**, 699 (2001), by permission of the American Chemical Society.

[159.] S. Kobayashi, H. Ishitani, Y. Yamashita, M. Ueno, and H. Shimizu, *Tetrahedron*, **57**, 861 (2001).

A titanium catalyst **20** that incorporates binaphthyl chirality along with imine and phenolic (salen) donors is highly active in addition of silyl ketene acetals to aldehydes.[160]

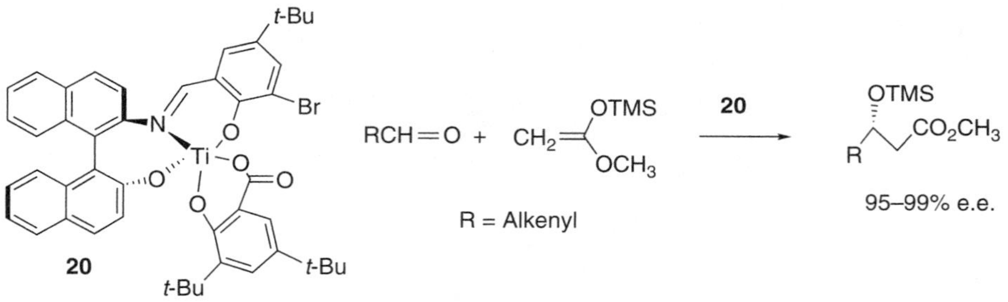

This catalyst is also active toward the simple enol ether 2-methoxypropene.[161]

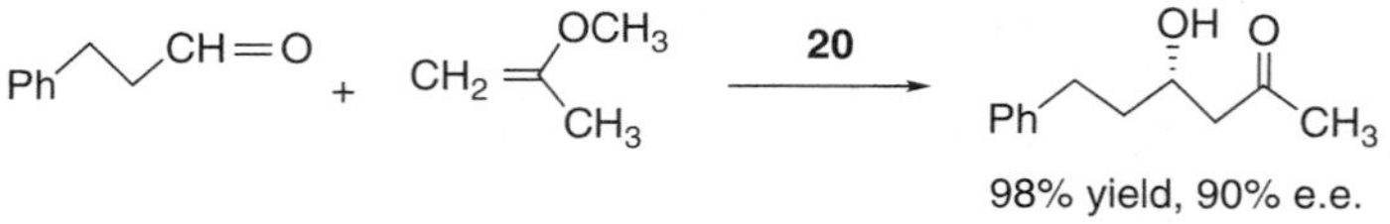

Entry 6 in Scheme 2.9 is an example of the use of this catalyst in a multistep synthesis.

The enantioselectivity of Sn(II) enolate reactions can be controlled by chiral diamine additives. These reagents are particularly effective for silyl thioketene acetals.[162] Several diamines derived from proline have been explored and 1-methyl-2-(1-piperidinomethyl)pyrrolidine **21** is an example. Even higher enantioselectivity can be achieved by attachment of bicyclic amines to the pyrrolidinomethyl group.[163]

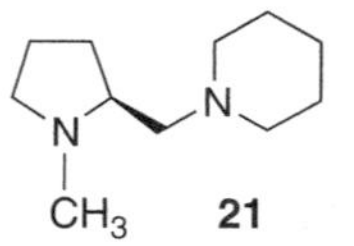

These reactions have been applied to α-benzyloxy and α-(*t*-butyldimethylsiloxy)-thioacetate esters.[164] The benzyloxy derivatives are *anti* selective, whereas the siloxy derivatives are *syn* selective. These differences are attributed to a chelated structure in the case of the benzyloxy derivative and an open TS for the siloxy system.

160. E. M. Carreira, R. A. Singer, and W. Lee, *J. Am. Chem. Soc.*, **116**, 8837 (1994).
161. E. M. Carreira, W. Lee, and R. A. Singer, *J. Am. Chem. Soc.*, **117**, 3649 (1995).
162. S. Kobayashi, H. Uchiro, Y. Fujishita, I. Shiina, and T. Mukaiyama, *J. Am. Chem. Soc.*, **113**, 4247 (1991); S. Kobayashi, H. Uchiro, I. Shiina, and T. Mukaiyama, *Tetrahedron*, **49**, 1761 (1993).
163. S. Kobayashi, M. Horibe, and M. Matsumura, *Synlett*, 675 (1995); S. Kobayashi and M. Horibe, *Chem. Eur. J.*, **3**, 1472 (1997).
164. T. Mukaiyama, I. Shiina, H. Uchiro, and S. Kobayashi, *Bull. Chem. Soc. Jpn.*, **67**, 1708 (1994).

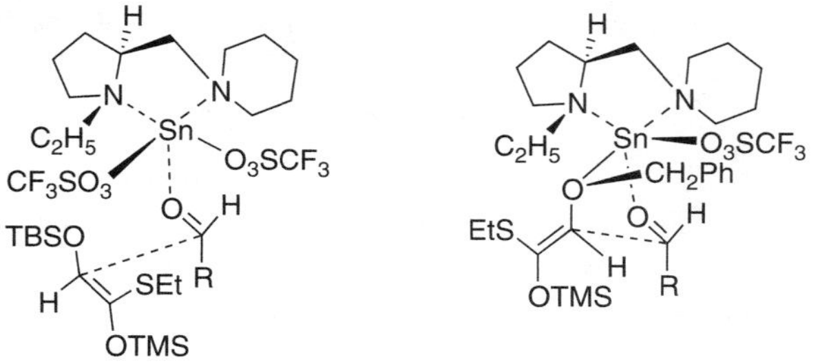

open TS leading to *syn* product   chelated TS leading to *anti* product

White and Deerberg explored this reaction system in connection with the synthesis of a portion of the structure of rapamycin.[165] Better yields were observed from benzyloxy than for a methoxy substituent, and there was a slight enhancement of stereoselectivity with the addition of ERG substituents to the benzyloxy group.

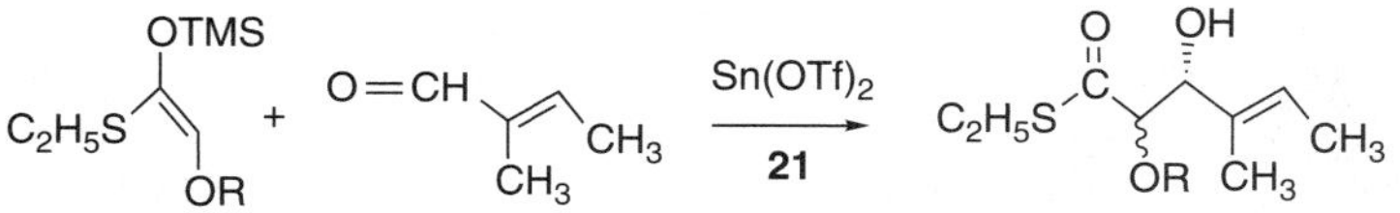

R	syn:anti	e.e.
CH₃O	70:30	87
PhCH₂O	85:15	93
PMB	90:10	96
2,4-DMB	95:5	92

Scheme 2.8 gives some examples of chiral Lewis acids that have been used to catalyze aldol and Mukaiyama reactions.

Scheme 2.9 gives some examples of use of enantioselective catalysts. Entries 1 to 4 are cases of the use of the oxazaborolidinone-type of catalyst with silyl enol ethers and silyl ketene acetals. Entries 5 and 6 are examples of the use of BINOL-titanium catalysts, and Entry 7 illustrates the use of $Sn(OTf)_2$ in conjunction with a chiral amine ligand. The enantioselectivity in each of these cases is determined entirely by the catalyst because there are no stereocenters adjacent to the reaction sites in the reactants.

A different type of catalysis is observed using proline as a catalyst.[166] Proline promotes addition of acetone to aromatic aldehydes with 65–77% enantioselectivity. It has been suggested that the carboxylic acid functions as an intramolecular proton donor and promotes reaction through an enamine intermediate.

165. J. D. White and J. Deerberg, *Chem. Commun.*, 1919 (1997).
166. B. List, R. A. Lerner, and C. F. Barbas, III, *J. Am. Chem. Soc.*, **122**, 2395 (2000); B. List, L. Hoang, and H. J. Martin, *Proc. Natl. Acad. Sci., USA*, **101**, 5839 (2004).

Scheme 2.8. Chiral Catalysts for the Mukaiyama Aldol Reactions

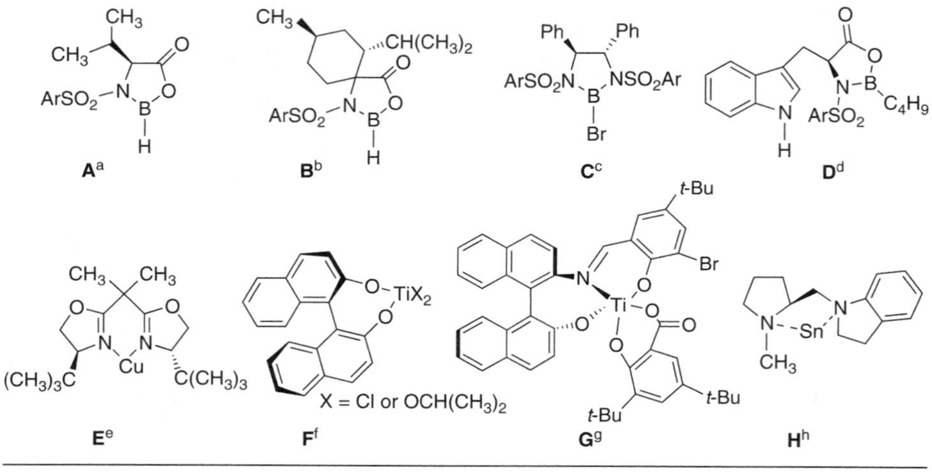

a. S. Kiyooka, Y. Kaneko, M. Komura, H. Matsuo, and M. Nakano, *J. Org. Chem.*, **56**, 2276 (1991).
b. E. R. Parmee, O. Tempkin, S. Masamune, and A. Abiko, *J. Am. Chem. Soc.*, **113**, 9365 (1991).
c. E. J. Corey, R. Imwinkelried, S. Pakul, and Y. B. Xiang, *J. Am. Chem. Soc.*, **111**, 5493 (1989).
d. E. J. Corey, C. L. Cywin, and T. D. Roper, *Tetrahedron Lett.*, **33**, 6907 (1992); E. J. Corey, D. Barnes-Seeman, and T. W. Lee, *Tetrahedron Lett.*, **38**, 1699 (1997).
e. D. A. Evans, J. A. Murry, and M. C. Koslowski, *J. Am. Chem. Soc.*, **118**, 5814 (1996); D. A. Evans, M. C. Koslowski, C. S. Burgey, and D. W. C. MacMillan, *J. Am. Chem. Soc.*, **119**, 7893 (1997); D. A. Evans, D. W. C. MacMillan, and K. R. Campos, *J. Am. Chem. Soc.*, **119**, 10859 (1997).
f. K. Mitami and S. Matsukawa, *J. Am. Chem. Soc.*, **115**, 7039 (1993); K. Mitami and S. Matsukawa, *J. Am. Chem. Soc.*, **116**, 4077 (1994); G. E. Keck and D. Krishnamurthy, *J. Am. Chem. Soc.*, **117**, 2363 (1995); G. E. Keck, X.-Y. Li, and D. Krishnamurthy, *J. Org. Chem.*, **60**, 5998 (1995).
g. E. M. Carreira, R. A. Singer, and W. Lee, *J. Am. Chem. Soc.*, **116**, 8837 (1994).
h. S. Kobayashi and M. Horibe, *Chem. Eur. J.*, **3**, 1472 (1997).

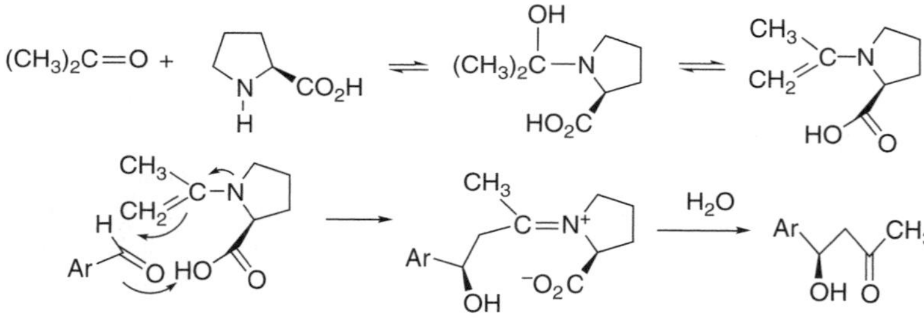

A DFT study found a corresponding TS to be the lowest energy.[167] This study also points to the importance of the solvent, DMSO, in stabilizing the charge buildup that occurs. A further computational study analyzed the stereoselectivity of the proline-catalyzed aldol addition reactions of cyclohexanone with acetaldehyde, isobutyraldehyde, and benzaldehyde on the basis of a similar TS.[168] Another study, which explored the role of proline in intramolecular aldol reactions, is discussed in the next section.[169]

[167] K. N. Rankin, J. W. Gauld, and R. J. Boyd, *J. Phys. Chem. A*, **106**, 5155 (2002).
[168] S. Bahmanyar, K. N. Houk, H. J. Martin, and B. List, *J. Am. Chem. Soc.*, **125**, 2475 (2003).
[169] S. Bahmanyar and K. N. Houk, *J. Am. Chem. Soc.*, **123**, 12911 (2001).

1[a]

$(CH_3)_2C$=C with OTMS, OC_2H_5 + TBSO—CH=O → **cat A** → $C_2H_5O_2C$ ... CH_3 CH_3 ... OTBS, OTMS

88%, > 90% e.e.

2[b]

$PhCH$=O + $(CH_3)_2C$=C with OTMS, OC_2H_5 → **cat A** → Ph—CH(OH)—C(CH_3)(CH_3)—$CO_2C_2H_5$

92% yield, 90% e.e.

3[c]

$(CH_3)_2C$=C with OTMS, OC_2H_5 + CH=O → **cat B** → $C_2H_5O_2C$ ... CH_3 CH_3 ... OTMS

81% yield, >98% e.e.

4[d]

furanyl—CH=O + CH_2=C with OTMS, Ph → **cat D** → furanyl—CH(OH)—CH_2—C(=O)—Ph

100% yield, 92% e.e.

5[e]

$CH_3O_2C(CH_2)_4CH$=O + CH_2=C with OTMS, $SC(CH_3)_3$ → **cat F** → $CH_3O_2C(CH_2)_4$—CH(OH)—CH_2—C(=O)—$SC(CH_3)_3$

65% yield, 96% e.e.

6[f]

TMSO, CH_3O—C=CH_2 + O=CH ... CH_3 ... $CH(CH_3)_2$, OCH_2Ph → **cat G** → CH_3O_2C ... CH_3 ... $CH(CH_3)_2$, OTMS, OCH_2Ph

84%

7[g]

$CH_3(CH_2)_8CH$=O + CH_3—C=C with OTMS, SC_2H_5 → **cat H** → $CH_3(CH_2)_8$—CH(OTMS)—CH(CH_3)—C(=O)—SC_2H_5

75% > 98% e.e.

a. J. Mulzer, A. J. Mantoulidis, and E. Ohler, *Tetrahedron Lett.*, **39**, 8633 (1998).
b. S. Kiyooka, Y. Kaneko, and K. Kume, *Tetrahedron Lett.*, **33**, 4927 (1992).
c. E. J. Corey, C. L. Cywin, and T. D. Roper, *Tetrahedron Lett.*, **33**, 6907 (1992).
d. E. R. Parmee, O. Tempkin, S. Masamune, and A. Abiko, *J. Am. Chem. Soc.*, **113**, 9365 (1991).
e. R. Zimmer, A. Peritz, R. Czerwonka, L. Schefzig, and H.-U. Reissig, *Eur. J. Org. Chem.*, 3419 (2002).
f. S. D. Rychnovsky, U. R. Khire, and G. Yang, *J. Am. Chem. Soc.*, **119**, 2058 (1997).
g. S. Kobayashi, H. Uchiro, I. Shiina, and T. Mukaiyama, *Tetrahedron*, **49**, 1761 (1993).

Visual models, additional information and exercises on Proline-Catalyzed Aldol Reactions can be found in the Digital Resource available at: Springer.com/carey-sundberg.

2.1.5.7. Summary of Facial Stereoselectivity in Aldol and Mukaiyama Reactions. The examples provided in this section show that there are several approaches to controlling the facial selectivity of aldol additions and related reactions. The *E*- or *Z*-configuration of the enolate and the open, cyclic, or chelated nature of the TS are the departure points for prediction and analysis of stereoselectivity. The Lewis acid catalyst and the donor strength of potentially chelating ligands affect the structure of the TS. Whereas dialkyl boron enolates and BF_3 complexes are tetracoordinate, titanium and tin can be hexacoordinate. If the reactants are chiral, facial selectivity must be taken into account. Examples of steric, chelation, and polar effects on TS structure have been described. Chiral auxiliaries can influence facial selectivity not only by their inherent steric effects, but also on the basis of the conformation of their Lewis acid complexes. This can be controlled by the choice of the enolate metal and reaction conditions. Dialkylboron enolates react through a cyclic TS that cannot accommodate additional coordination. Titanium and tin enolates of oxazolidinones are chelated under normal conditions, but the use of excess Lewis acid can modify the TS structure and reverse facial selectivity. Chiral catalysts require that additional stereochemical features be taken into account, and the issue becomes the fit of the reactants within the chiral environment. Although most catalysts rely primarily on steric factors for facial selectivity, hydrogen bonding and π stacking can also come into play.

2.1.6. Intramolecular Aldol Reactions and the Robinson Annulation

The aldol reaction can be applied to dicarbonyl compounds in which the two groups are favorably disposed for intramolecular reaction. Kinetic studies on cyclization of 5-oxohexanal, 2,5-hexanedione, and 2,6-heptanedione indicate that formation of five-membered rings is thermodynamically somewhat more favorable than formation of six-membered rings, but that the latter is several thousand times faster.[170] A catalytic amount of acid or base is frequently satisfactory for formation of five- and six-membered rings, but with more complex structures, the techniques required for directed aldol condensations are used.

Scheme 2.10 illustrates intramolecular aldol condensations. Entries 1 and 2 are cases of formation of five-membered rings, with aldehyde groups serving as the electrophilic center. The regioselectivity in Entry 1 is due to the potential for dehydration of only one of the cyclic aldol adducts.

In Entry 2, the more reactive aldehyde group serves as the electrophilic component in preference to the ketone. Entries 3 to 6 are examples of construction of new rings in preexisting cyclic systems. The structure and stereochemistry of the products of these reactions are dictated by ring geometry and the proximity of reactive groups. Entry 5 is interesting in that it results in the formation of a bridgehead double bond. Entries 7 to 9 are intramolecular Mukaiyama reactions, using acetals as the precursor of the electrophilic center. Entry 9, which is a key step in the synthesis of jatrophones, involves formation of an eleven-membered ring. From a retrosynthetic perspective, bonds between a carbinol (or equivalent) carbon and a carbon that is α to a carbonyl carbon are candidates for formation by intramolecular aldol additions.

A particularly important example of the intramolecular aldol reaction is the *Robinson annulation*, a procedure that constructs a new six-membered ring from a ketone.[171] The reaction sequence starts with conjugate addition of the enolate to methyl

[170] J. P. Guthrie and J. Guo, *J. Am. Chem. Soc.*, **118**, 11472 (1996).

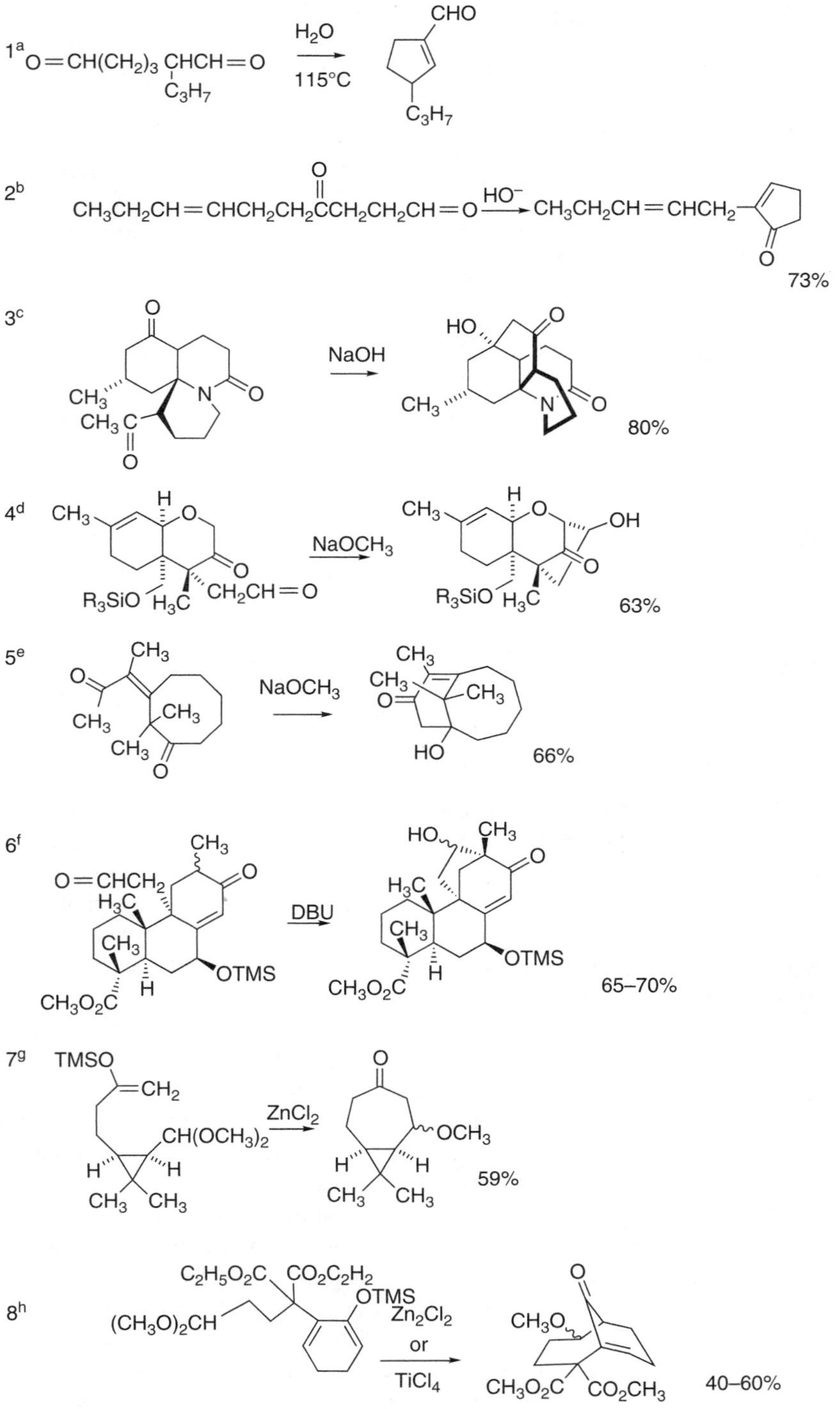

Scheme 2.10. Intramolecular Aldol and Mukaiyama Aldol Reactions

1[a]

2[b]
73%

3[c]
80%

4[d]
63%

5[e]
66%

6[f]
65–70%

7[g]
59%

8[h]
40–60%

(Continued)

Scheme 2.10. (*Continued*)

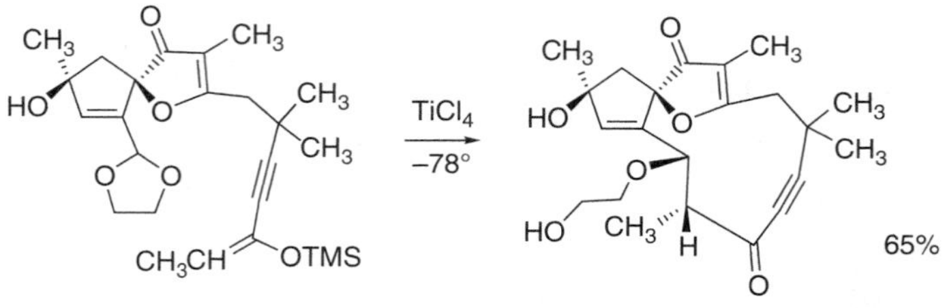

a. J. English and G. W. Barber, *J. Am. Chem. Soc.*, **71**, 3310 (1949).
b. A. I. Meyers and N. Nazarenko, *J. Org. Chem.*, **38**, 175 (1973).
c. K. Wiesner, V. Musil, and K. J. Wiesner, *Tetrahedron Lett.*, 5643 (1968).
d. G. A. Kraus, B. Roth, K. Frazier, and M. Shimagaki, *J. Am. Chem. Soc.*, **104**, 1114 (1982).
e. K. Yamada, H. Iwadare, and T. Mukaiyama, *Chem. Pharm. Bull.*, **45**, 1898 (1997).
f. J. K. Tagat, M. S. Puar, and S. W. McCombie, *Tetrahedron Lett.*, **37**, 8463 (1996).
g. M. D. Taylor, G. Minaskanian, K. N. Winzenberg, P. Santone, and A. B. Smith, III, *J. Org. Chem.*, **47**, 3960 (1962).
h. A. Armstrong, T. J. Critchley, M. E. Gourdel-Martin, R. D. Kelsey, and A. A. Mortlock, *J. Chem. Soc., Perkin Trans. 1*, 1344 (2002).
i. A. B. Smith, III, A. T. Lupo, Jr., M. Ohba, and K. Chen, *J. Am. Chem. Soc.*, **111**, 6648 (1989).

vinyl ketone or a similar enone. This is followed by cyclization by an intramolecular aldol addition. Dehydration usually occurs to give a cyclohexenone derivative.

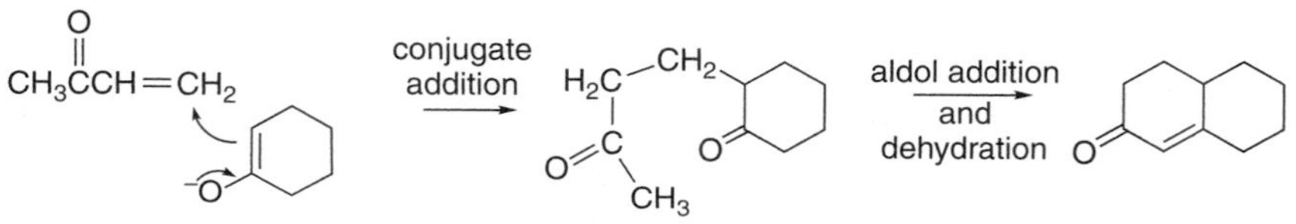

Other α,β-unsaturated enones can be used, but the reaction is somewhat sensitive to substitution at the β-carbon and adjustment of the reaction conditions is necessary.[172]

Scheme 2.11 shows some examples of Robinson annulation reactions. Entries 1 and 2 show annulation reactions of relatively acidic dicarbonyl compounds. Entry 3 is an example of use of 4-(trimethylammonio)-2-butanone as a precursor of methyl vinyl ketone. This compound generates methyl vinyl ketone in situ by β-elimination. The original conditions developed for the Robinson annulation reaction are such that the ketone enolate composition is under thermodynamic control. This usually results in the formation of product from the more stable enolate, as in Entry 3. The C(1) enolate is preferred because of the conjugation with the aromatic ring. For monosubstituted cyclohexanones, the cyclization usually occurs at the more-substituted position in hydroxylic solvents. The alternative regiochemistry can be achieved by using an enamine. Entry 4 is an example. As discussed in Section 1.9, the less-substituted enamine is favored, so addition occurs at the less-substituted position.

Conditions for kinetic control of enolate formation can be applied to the Robinson annulation to control the regiochemistry of the reaction. Entries 5 and 6 of Scheme 2.11 are cases in which the reaction is carried out on a preformed enolate. Kinetic

[171.] E. D. Bergmann, D. Ginsburg, and R. Pappo, *Org. React.*, **10**, 179 (1950); J. W. Cornforth and R. Robinson, *J. Chem. Soc.*, 1855 (1949); R. Gawley, *Synthesis*, 777 (1976); M. E. Jung, *Tetrahedron*, **32**, 3 (1976); B. P. Mundy, *J. Chem. Ed.*, **50**, 110 (1973).
[172.] C. J. V. Scanio and R. M. Starrett, *J. Am. Chem. Soc.*, **93**, 1539 (1971).

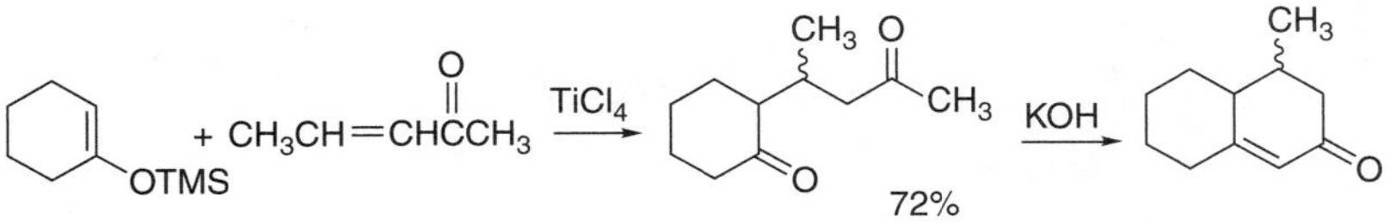

a. F. E. Ziegler, K.-J. Hwang, J. F. Kadow, S. I. Klein, U. K. Pati, and T.-F. Wang, *J. Org. Chem.*, **51**, 4573 (1986).
b. D. L. Snitman, R. J. Himmelsbach, and D. S. Watt, *J. Org. Chem.*, **43**, 4578 (1978).
c. J. W. Cornforth and R. Robinson, *J. Chem. Soc.*, 1855 (1949).
d. G. Stork, A. Brizzolara, H. Landesman, J. Szmuszkovicz, and R. Terrell, *J. Am. Chem. Soc.*, **85**, 207 (1963).
e. G. Stork, J. D. Winkler, and C. S. Shiner, *J. Am. Chem. Soc.*, **104**, 3767 (1982).
f. K. Takaki, M. Okada, M. Yamada, and K. Negoro, *J. Org. Chem.*, **47**, 1200 (1982).
g. J. W. Huffman, S. M. Potnis, and A. V. Smith, *J. Org. Chem.*, **50**, 4266 (1985).

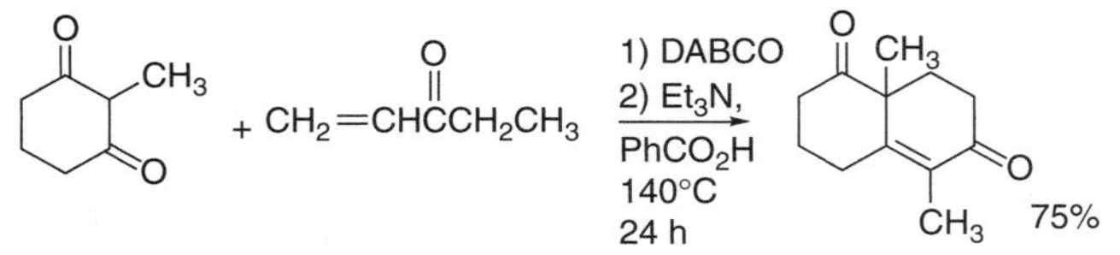

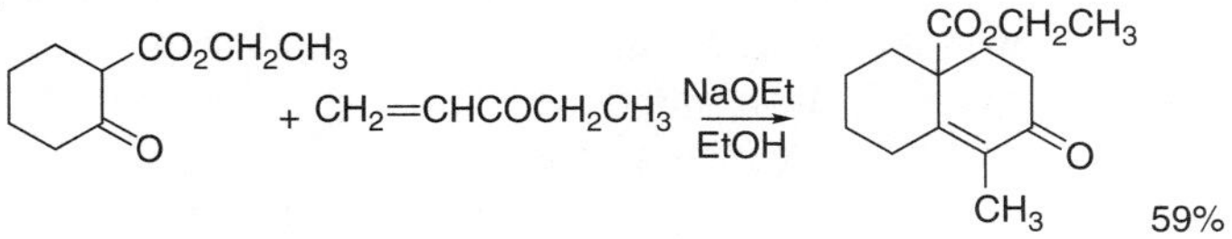

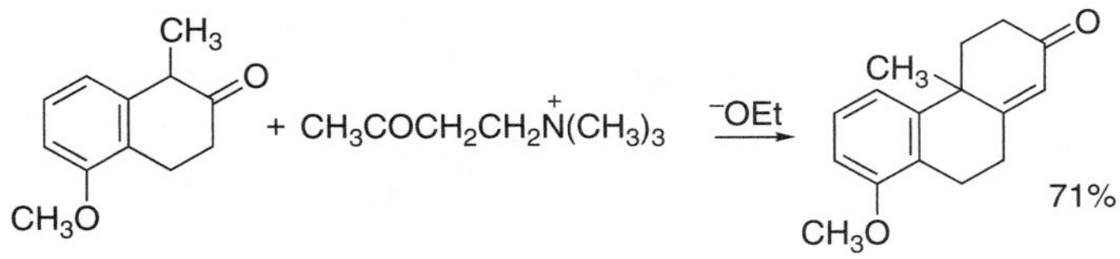

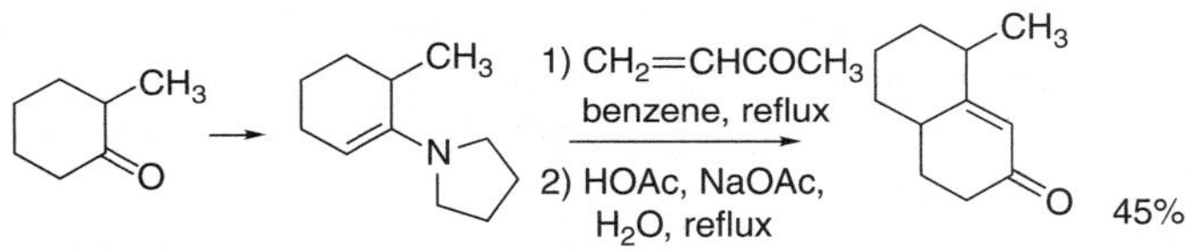

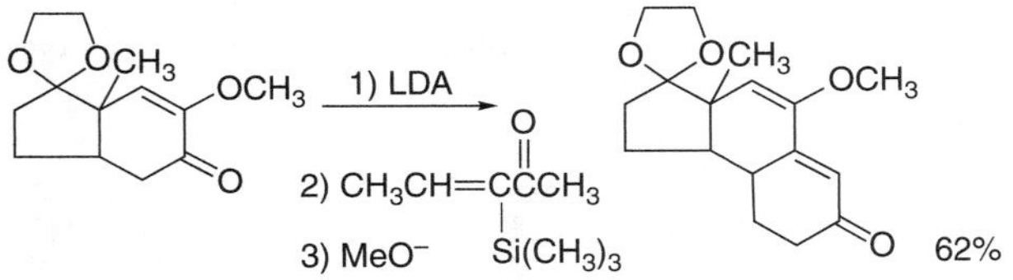

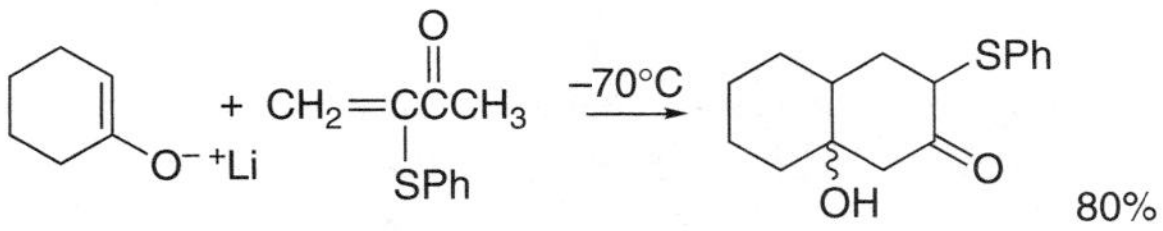

control is facilitated by use of somewhat more activated enones, such as methyl 1-(trimethylsilyl)vinyl ketone.

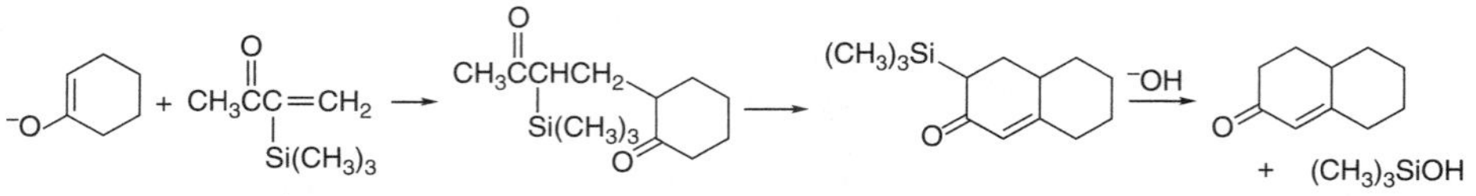

Ref. 173

The role of the trimethylsilyl group is to stabilize the enolate formed in the conjugate addition. The silyl group is then removed during the dehydration step. Methyl 1-trimethylsilylvinyl ketone can be used under aprotic conditions that are compatible with regiospecific methods for enolate generation. The direction of annulation of unsymmetrical ketones can therefore be controlled by the method of enolate formation.

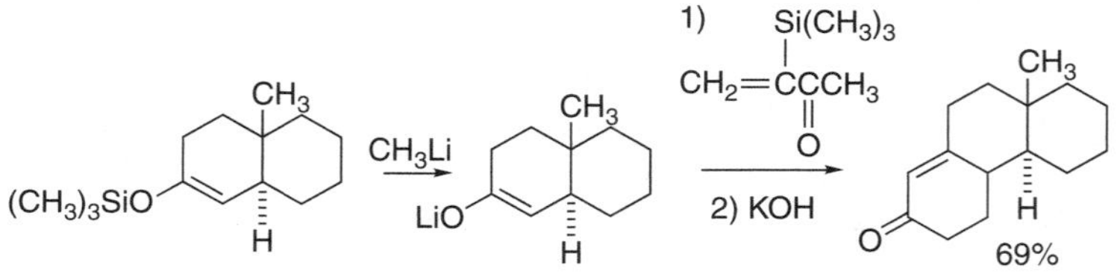

Ref. 174

Methyl 1-phenylthiovinyl ketones can also be used as enones in kinetically controlled Robinson annulation reactions, as illustrated by Entry 6. Entry 7 shows a annulation using silyl enol ether as the enolate equivalent. These reactions are called *Mukaiyama-Michael reactions* (see Section 2.6.3).

The Robinson annulation is a valuable method for preparing bicyclic and tricyclic structures that can serve as starting materials for the preparation of steroids and terpenes.[175] Reaction with 2-methylcyclohexan-1,3-dione gives a compound called the *Wieland-Miescher ketone*.

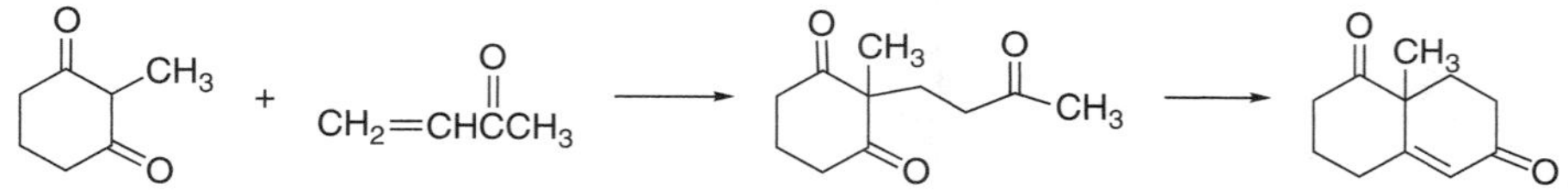

A similar reaction occurs with 2-methylcyclopentane-1,3-dione,[176] and can be done enantioselectively by using the amino acid L-proline to form an enamine intermediate. The (*S*)-enantiomer of the product is obtained in high enantiomeric excess.[177]

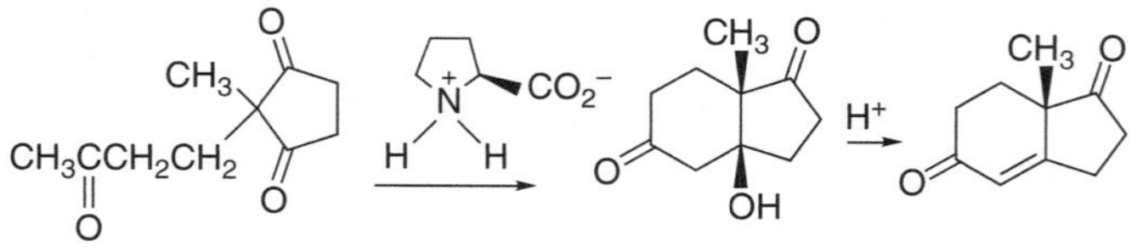

173. G. Stork and B. Ganem, *J. Am. Chem. Soc.*, **95**, 6152 (1973); G. Stork and J. Singh, *J. Am. Chem. Soc.*, **96**, 6181 (1974).

174. R. K. Boeckman, Jr., *J. Am. Chem. Soc.*, **96**, 6179 (1974).

175. N. Cohen, *Acc. Chem. Res.*, **9**, 412 (1976).

176. Z. G. Hajos and D. R. Parrish, *J. Org. Chem.*, **39**, 1615 (1974); U. Eder, G. Sauer, and R. Wiechert, *Angew. Chem. Int. Ed. Engl.*, **10**, 496 (1971); Z. G. Hajos and D. R. Parrish, *Org. Synth.*, **63**, 26 (1985).

177. J. Gutzwiller, P. Buchshacher, and A. Furst, *Synthesis*, 167 (1977); P. Buchshacher and A. Furst, *Org. Synth.*, **63**, 37 (1984); T. Bui and C. F. Barbas, III, *Tetrahedron Lett.*, **41**, 6951 (2000).

The detailed mechanism of this enantioselective transformation remains under investigation.[178] It is known that the acidic carboxylic group is crucial, and the cyclization is believed to occur via the enamine derived from the catalyst and the exocyclic ketone. A computational study suggested that the proton transfer occurs through a TS very similar to that described for the proline-catalyzed aldol reaction (see page 132).[179]

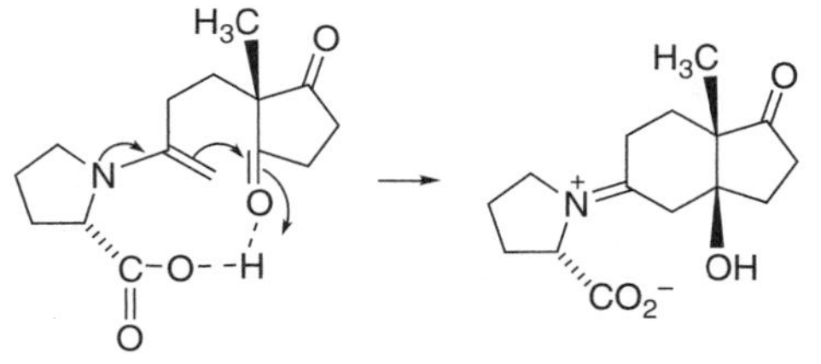

Visual models, additional information and exercises on Proline-Catalyzed Aldol Reactions can be found in the Digital Resource available at: Springer.com/carey-sundberg.

2.2. Addition Reactions of Imines and Iminium Ions

Imines and iminium ions are nitrogen analogs of carbonyl compounds and they undergo nucleophilic additions like those involved in aldol reactions. The reactivity order is $C=NR < C=O < [C=NR_2]^+ < [C=OH]^+$. Because iminium ions are more reactive than imines, the reactions are frequently run under mildly acidic conditions. Under some circumstances, the iminium ion can be the reactive species, even though it is a minor constituent in equilibrium with the amine, carbonyl compound, and unprotonated imine.

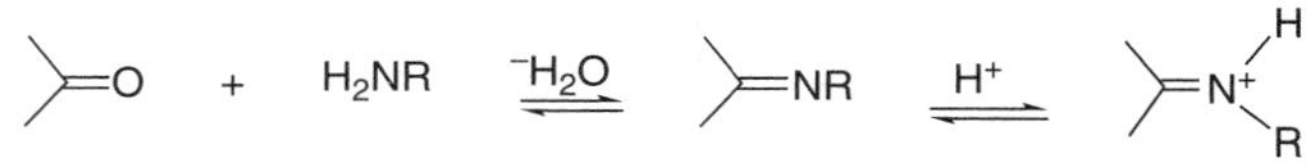

Addition of enols, enolates, or enolate equivalents to imines or iminium ions provides an important route to β-amino ketones.

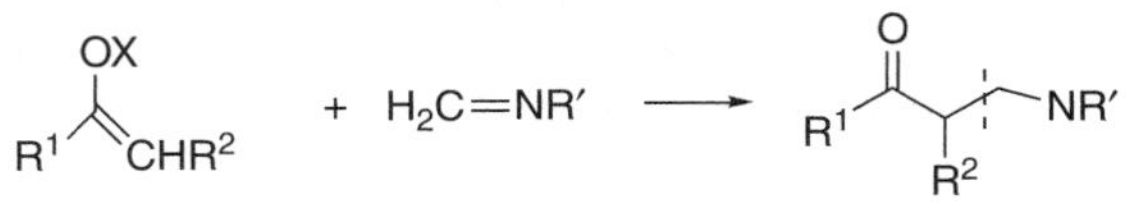

[178.] P. Buchschacher, J.-M. Cassal, A. Furst, and W. Meier, *Helv. Chim. Acta*, **60**, 2747 (1977); K. L. Brown, L. Damm, J. D. Dunitz, A. Eschenmoser, R. Hobi, and C. Kratky, *Helv. Chim. Acta*, **61**, 3108 (1978); C. Agami, F. Meynier, C. Puchot, J. Guilhem, and C. Pascard, *Tetrahedron*, **40**, 1031 (1984); C. Agami, J. Levisalles, and C. Puchot, *J. Chem. Soc., Chem. Commun.*, 441 (1985); C. Agami, *Bull. Soc. Chim. Fr.*, 499 (1988).

[179.] S. Bahmanyar and K. N. Houk, *J. Am. Chem. Soc.*, **123**, 12911 (2001).

2.2.1. The Mannich Reaction

The *Mannich reaction* is the condensation of an enolizable carbonyl compound with an iminium ion.[180] It is usually done using formaldehyde and introduces an α-dialkylaminomethyl substituent.

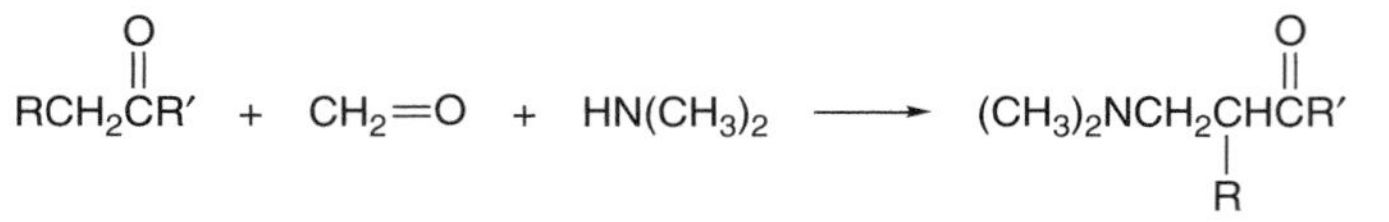

$$RCH_2CR' + CH_2{=}O + HN(CH_3)_2 \longrightarrow (CH_3)_2NCH_2CHCR'$$

The electrophile is often generated in situ from the amine and formaldehyde.

$$CH_2{=}O + HN(CH_3)_2 \rightleftharpoons HOCH_2N(CH_3)_2 \overset{H^+}{\rightleftharpoons} CH_2{=}\overset{+}{N}(CH_3)_2$$

The reaction is normally limited to secondary amines, because dialkylation can occur with primary amines. The dialkylation reaction can be used to advantage in ring closures.

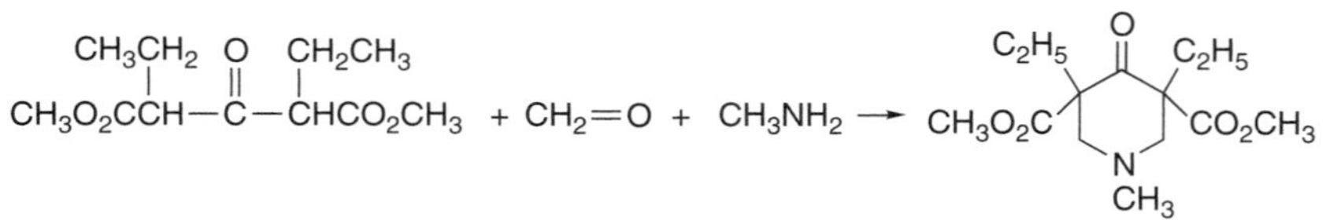

$$CH_3O_2CCH{-}C{-}CHCO_2CH_3 + CH_2{=}O + CH_3NH_2 \rightarrow$$

Ref. 181

Scheme 2.12 shows some representative Mannich reactions. Entries 1 and 2 show the preparation of typical "Mannich bases" from a ketone, formaldehyde, and a dialkylamine following the classical procedure. Alternatively, formaldehyde equivalents may be used, such as *bis*-(dimethylamino)methane in Entry 3. On treatment with trifluoroacetic acid, this aminal generates the iminium trifluoroacetate as a reactive electrophile. *N,N*-(Dimethyl)methylene ammonium iodide is commercially available and is known as *Eschenmoser's salt*.[182] This compound is sufficiently electrophilic to react directly with silyl enol ethers in neutral solution.[183] The reagent can be added to a solution of an enolate or enolate precursor, which permits the reaction to be carried out under nonacidic conditions. Entries 4 and 5 illustrate the preparation of Mannich bases using Eschenmoser's salt in reactions with preformed enolates.

The dialkylaminomethyl ketones formed in the Mannich reaction are useful synthetic intermediates.[184] Thermal elimination of the amines or the derived quaternary salts provides α-methylene carbonyl compounds.

180. F. F. Blicke, *Org. React.*, **1**, 303 (1942); J. H. Brewster and E. L. Eliel, *Org. React.*, **7**, 99 (1953); M. Tramontini and L. Angiolini, *Tetrahedron*, **46**, 1791 (1990); M. Tramontini and L. Angiolini, *Mannich Bases: Chemistry and Uses*, CRC Press, Boca Raton, FL, 1994; M. Ahrend, B. Westerman, and N. Risch, *Angew. Chem. Int. Ed. Engl.*, **37**, 1045 (1998).
181. C. Mannich and P. Schumann, *Chem. Ber.*, **69**, 2299 (1936).
182. J. Schreiber, H. Maag, N. Hashimoto, and A. Eschenmoser, *Angew. Chem. Int. Ed. Engl.*, **10**, 330 (1971).
183. S. Danishefsky, T. Kitahara, R. McKee, and P. F. Schuda, *J. Am. Chem. Soc.*, **98**, 6715 (1976).
184. G. A. Gevorgyan, A. G. Agababyan, and O. L. Mndzhoyan, *Russ. Chem. Rev.* (Engl. Transl.), **54**, 495 (1985).

A. Aminomethylation Using the Mannich Reaction

1^a $PhCOCH_3$ + CH_2O + $(CH_3)_2\overset{+}{N}H_2Cl^-$ ⟶ $PhCOCH_2CH_2\overset{H}{\underset{+}{N}}(CH_3)_2Cl^-$ 70%

2^b CH_3COCH_3 + CH_2O + $(CH_3CH_2)_2\overset{+}{N}H_2Cl^-$ ⟶ $CH_3COCH_2CH_2\overset{H}{\underset{+}{N}}(C_2H_5)_2Cl^-$ 66–75%

3^c $(CH_3)_2CHCOCH_3$ + $[(CH_3)_2N]_2CH_2$ $\xrightarrow{CF_3CO_2H}$ $(CH_3)_2CHCOCH_2CH_2N(CH_3)_2$

4^d

5^e

B. Reactions Involving Secondary Transformations of Aminomethylation Products.

6^f $CH_3CH_2CH_2CH{=}O$ + CH_2O + $(\overset{+}{C}H_3)_2NH_2Cl^-$ $\xrightarrow[\text{2) distill}]{\text{1) 60°C, 6 h}}$ $CH_2{=}CCH{=}O$ with CH_2CH_3 73%

7^g

8^h

9^i $PhCOCH_2CH_2N(CH_3)_2$ + KCN ⟶ $PhCOCH_2CH_2CN$ 67%

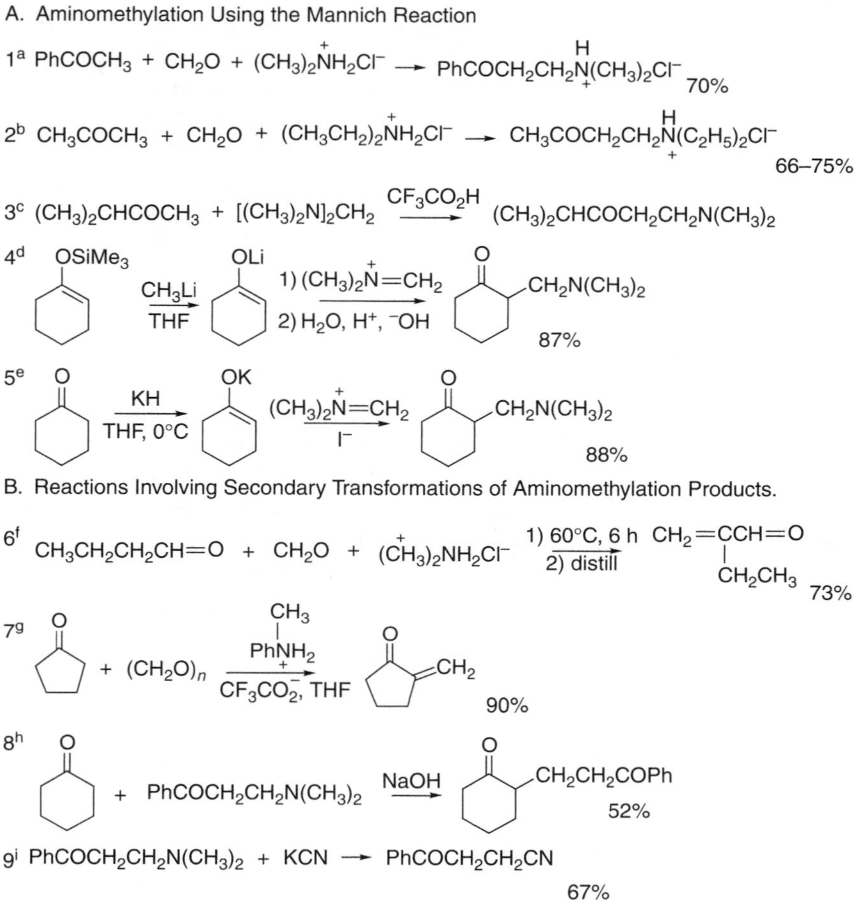

a. C. E. Maxwell, *Org. Synth.*, **III**, 305 (1955).
b. A. L. Wilds, R. M. Novak, and K. E. McCaleb, *Org. Synth.*, **IV**, 281 (1963).
c. M. Gaudry, Y. Jasor, and T. B. Khac, *Org. Synth.*, **59**, 153 (1979).
d. S. Danishefsky, T. Kitahara, R. McKee, and P. F. Schuda, *J. Am. Chem. Soc.*, **98**, 6715 (1976).
e. J. L. Roberts, P. S. Borromeo, and C. D. Poulter, *Tetrahedron Lett.*, 1621 (1977).
f. C. S. Marvel, R. L. Myers, and J. H. Saunders, *J. Am. Chem. Soc.*, **70**, 1694 (1948).
g. J. L. Gras, *Tetrahedron Lett.*, 2111, 2955 (1978).
h. A. C. Cope and E. C. Hermann, *J. Am. Chem. Soc.*, **72**, 3405 (1950).
i. E. B. Knott, *J. Chem. Soc.*, 1190 (1947).

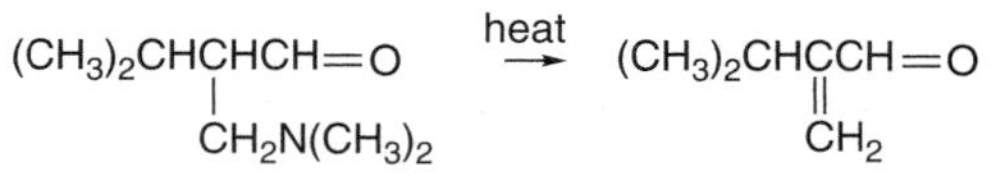

$(CH_3)_2CHCHCH{=}O$ $\xrightarrow{\text{heat}}$ $(CH_3)_2CHCCH{=}O$
with $CH_2N(CH_3)_2$ and CH_2

Ref. 185

These α,β-unsaturated ketones and aldehydes are used as reactants in conjugate additions (Section 2.6), Robinson annulations (Section 2.1.4), and in a number of other reactions that we will encounter later. Entries 8 and 9 in Scheme 2.12 illustrate

185. C. S. Marvel, R. L. Myers, and J. H. Saunders, *J. Am. Chem. Soc.*, **70**, 1694 (1948).

conjugate addition reactions carried out by in situ generation of α,β-unsaturated carbonyl compounds from Mannich bases.

α-Methylenelactones are present in a number of natural products.[186] The reaction of ester enolates with N,N-(dimethyl)methyleneammonium trifluoroacetate,[187] or Eschenmoser's salt,[188] has been used for introduction of the α-methylene group in the synthesis of vernolepin, a compound with antileukemic activity.[189,190]

Mannich reactions, or a mechanistic analog, are important in the biosynthesis of many nitrogen-containing natural products. As a result, the Mannich reaction has played an important role in the synthesis of such compounds, especially in syntheses patterned after the biosynthesis, i.e., *biomimetic synthesis*. The earliest example of the use of the Mannich reaction in this way was Sir Robert Robinson's successful synthesis of tropinone, a derivative of the alkaloid tropine, in 1917.

Ref. 191

As with aldol and Mukaiyama addition reactions, the Mannich reaction is subject to enantioselective catalysis.[192] A catalyst consisting of Ag^+ and the chiral imino aryl phosphine **22** achieves high levels of enantioselectivity with a range of N-(2-methoxyphenyl)imines.[193] The 2-methoxyphenyl group is evidently involved in an interaction with the catalyst and enhances enantioselectivity relative to other N-aryl substituents. The isopropanol serves as a proton source and as the ultimate acceptor of the trimethylsilyl group.

186. S. M. Kupchan, M. A. Eakin, and A. M. Thomas, *J. Med. Chem.*, **14**, 1147 (1971).
187. N. L. Holy and Y. F. Wang, *J. Am. Chem. Soc.*, **99**, 499 (1977).
188. J. L. Roberts, P. S. Borromes, and C. D. Poulter, *Tetrahedron Lett.*, 1621 (1977).
189. S. Danishefsky, P. F. Schuda, T. Kitahara, and S. J. Etheredge, *J. Am. Chem. Soc.*, **99**, 6066 (1977).
190. For reviews of methods for the synthesis of α-methylene lactones, see R. B. Gammill, C. A. Wilson, and T. A. Bryson, *Synth. Comm.*, **5**, 245 (1975); J. C. Sarma and R. P. Sharma, *Heterocycles*, **24**, 441 (1986); N. Petragnani, H. M. C. Ferraz, and G. V. J. Silva, *Synthesis*, 157 (1986).
191. R. Robinson, *J. Chem. Soc.*, 762 (1917).
192. A. Cordova, *Acc. Chem. Res.*, **37**, 102 (2004).
193. N. S. Josephsohn, M. L. Snapper, and A. H. Hoveyda, *J. Am. Chem. Soc.*, **126**, 3734 (2004).

$$RCH{=}N{-}Ar \quad + \quad CH_2{=}C \overset{OTMS}{\underset{R'}{}} \xrightarrow[\text{1 eq. } i\text{-PrOH}]{\overset{\text{cat } 22}{\overset{\text{1-5 mol \%}}{}}} \quad \overset{Ar}{\underset{R}{\overset{NH}{|}}} \overset{O}{\underset{R'}{}}$$

Ar = 2-methoxyphenyl R' = CH₃, Ph 76–96% e.e.
R = alkyl, aryl, alkenyl

cat 22

A zinc catalyst **23** was found effective for aryl hydroxymethyl ketones in reactions with glyoxylic imines. In this case, the 4-methoxy-2-methylphenylimines gave the best results.[194] Interestingly, the 2-methoxyphenyl ketone gave substantially enhanced 2,3-diastereoselectivity (20:1) compared to about 10:1 for most other aryl groups, suggesting that the *o*-methoxy group may introduce an additional interaction with the catalyst. All the compounds gave e.e. > 95%.

$$Ar \overset{O}{\underset{}{}} OH \quad + \quad C_2H_5O_2C{-}CH{=}N{-}Ar' \xrightarrow[\text{4 A MS}]{\text{cat } 23} \quad Ar \overset{O}{\underset{OH}{}} \overset{NHAr'}{\underset{}{}} CO_2C_2H_5$$

Ar' = 4-methoxy-2-methylphenyl

dr = 2:1 to > 20:1
e.e. 95 – > 99%

cat 23

Other types of catalysts that are active in Mannich reactions include the Cu-*bis*-oxazolines.[195] Most of the cases examined to date are for relatively reactive imines, such as those derived from glyoxylate or pyruvate esters.

As already discussed for aldol and Robinson annulation reactions, proline is also a catalyst for enantioselective Mannich reactions. Proline effectively catalyzes the reactions of aldehydes such as 3-methylbutanal and hexanal with *N*-arylimines of ethyl glyoxalate.[196] These reactions show 2,3-*syn* selectivity, although the products with small alkyl groups tend to isomerize to the *anti* isomer.

$$(CH_3)_2CHCH_2CH{=}O \ + \ C_2H_5O_2CCH{=}NAr \xrightarrow[\text{10 mol \%}]{\text{proline}} \ O{=}CH \overset{NHAr}{\underset{CH(CH_3)_2}{}} CO_2C_2H_5$$

Ar = 4 – methoxyphenyl

dr > 10:1   e.e. = 87%

194. B. M. Trost and L. M. Terrell, *J. Am. Chem. Soc.*, **125**, 338 (2003).
195. K. Juhl and K. A. Jorgensen, *J. Am. Chem. Soc.*, **124**, 2420 (2002); M. Marigo, A. Kjaersgaard, K. Juhl, N. Gathergood, and K. A. Jorgensen, *Chem. Eur. J.*, **9**, 2359 (2003).
196. W. Notz, F. Tanaka, S. Watanabe, N. S. Chowdari, J. M. Turner, R. Thayumanavan, and C. F. Barbas, III, *J. Org. Chem.*, **68**, 9624 (2003).

With aromatic aldehydes, d.r. ranged up to more than 10:1 for propanal.

$$CH_3CH_2CH{=}O \quad + \quad ArCH{=}NAr' \quad \xrightarrow[\text{2) NaBH}_4]{\substack{\text{1) proline} \\ \text{30 mol \%}}}$$

Ar′ = 4-methoxyphenyl

The proline-catalyzed reaction has been extend to the reaction of propanal, butanal, and pentanal with a number of aromatic aldehydes and proceeds with high *syn* selectivity.[197] The reaction can also be carried out under conditions in which the imine is formed in situ. Under these conditions, the conjugative stabilization of the aryl imines leads to the preference for the aryl imine to act as the electrophile. A good yield of the expected β-aminoalcohol was obtained with propanal serving as both the nucleophilic and the electrophilic component. The product was isolated as a γ-amino alcohol after reduction with $NaBH_4$.

$$CH_3CH_2CH{=}O \quad + \quad H_2NAr \quad \xrightarrow[\text{2) NaBH}_4]{\substack{\text{1) proline} \\ \text{10 mol \%}}}$$

Ar′ = 4-methoxyphenyl

70% yield
dr > 95:5, 96% e.e.

Ketones such as acetone, hydroxyacetone, and methoxyacetone can be condensed with both aromatic and aliphatic aldehydes.[198]

$$\text{(structure)} + ArCH{=}O + Ar'NH_2 \xrightarrow[\text{proline}]{\text{20–35 mol \%}}$$

Ar′ = 4-methoxyphenyl

The TS proposed for these proline-catalyzed reactions is very similar to that for the proline-catalyzed aldol addition (see p. 132). In the case of imines, however, the aldehyde substituent is directed *toward* the enamine double bond because of the dominant steric effect of the *N*-aryl substituent. This leads to formation of *syn* isomers, whereas the aldol reaction leads to *anti* isomers. This is the TS found to be the most stable by B3LYP/6-31G* computations.[199] The proton transfer is essentially complete at the TS. As with the aldol addition TS, the enamine is oriented *anti* to the proline carboxy group in the most stable TS.

[197.] Y. Hayashi, W. Tsuboi, I. Ashimine, T. Urushima, M. Shoji, and K. Sakai, *Angew. Chem. Int. Ed. Engl.*, **42**, 3677 (2003).

[198.] B. List, P. Pojarliev, W. T. Biller, and H. J. Martin, *J. Am. Chem. Soc.*, **124**, 827 (2002).

[199.] S. Bahmanyar and K. N. Houk, *Org. Lett.*, **5**, 1249 (2003).

Structure **24**, which is a simplification of an earlier catalyst,[200] gives excellent results with *N-t*-butoxycarbonylimines.[201] Catalysts of this type are thought to function through hydrogen-bonding interactions.

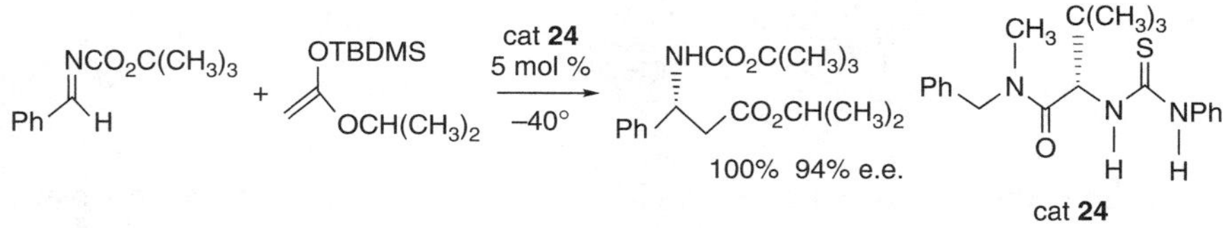

2.2.2. Additions to *N*-Acyl Iminium Ions

Even more reactive C=N bonds are present in *N-acyliminium ions*.[202]

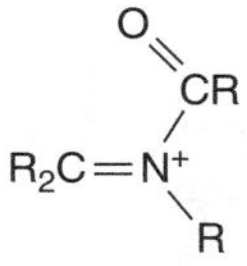

Gas phase reactivity toward allyltrimethylsilane was used to compare the reactivity of several cyclic *N*-acyliminium ions and related iminium ions.[203] Compounds with endocyclic acyl groups were found to be more reactive than compounds with exocyclic acyl substituents. Five-membered ring compounds are somewhat more reactive than six-membered ones. The higher reactivity of the endocyclic acyl derivatives is believed to be due to geometric constraints that maximize the polar effect of the carbonyl group.

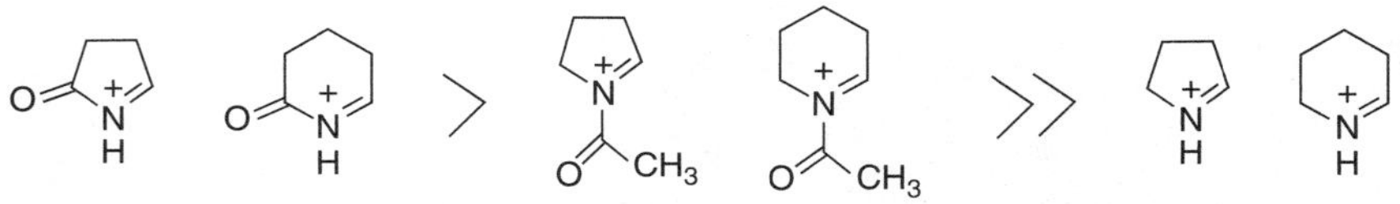

N-Acyliminium ions are usually prepared in situ in the presence of a potential nucleophile. There are several ways of generating acyliminium ions. Cyclic examples can be generated by partial reduction of imides.[204]

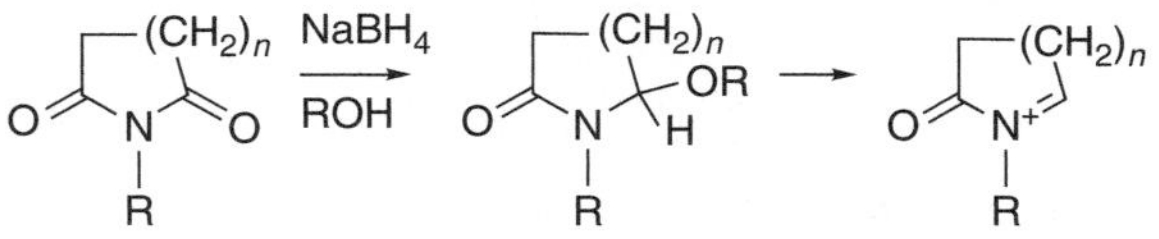

Various oxidations of amides or carbamates can also generate acyliminium ions. An electrochemical oxidation forms α-alkoxy amides and lactams, which then generate

200. P. Vachal and E. N. Jacobsen, *J. Am. Chem. Soc.*, **124**, 10012 (2002).
201. A. G. Wenzel, M. P. Lalonde, and E. N. Jacobsen, *Synlett*, 1919 (2003).
202. H. Hiemstra and W. N. Speckamp, in *Comprehensive Organic Synthesis*, Vol. 2, B. Trost and I. Fleming, eds., 1991, pp. 1047–1082; W. N. Speckamp and M. J. Moolenaar, *Tetrahedron*, **56**, 3817 (2000); B. E. Maryanoff, H.-C. Zhang, J. H. Cohen, I. J. Turchi, and C. A. Maryanoff, *Chem. Rev.*, **104**, 1431 (2004).
203. M. G. M. D'Oca, L. A. B. Moraes, R. A. Pilli, and M. N. Eberlin, *J. Org. Chem.*, **66**, 3854 (2001).
204. J. C. Hubert, J. B. P. A. Wijnberg, and W. Speckamp, *Tetrahedron*, **31**, 1437 (1975); H. Hiemstra, W. J. Klaver, and W. N. Speckamp, *J. Org. Chem.*, **49**, 1149 (1984); P. A. Pilli, L. C. Dias, and A. O. Maldaner, *J. Org. Chem.*, **60**, 717 (1995).

acyliminium ions.[205] *N*-Acyliminium ions can also be obtained by oxidative decarboxylation of *N*-acyl-α-amino acids such as *N*-acyl proline derivatives.[206]

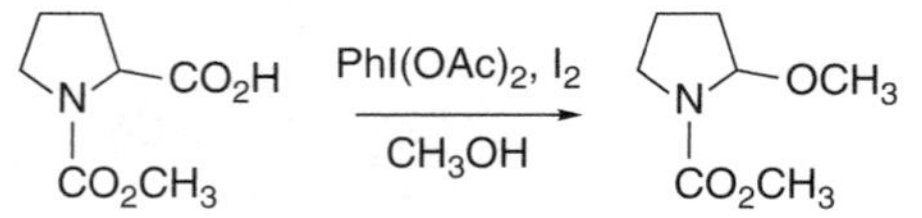

Acyliminium ions are sufficiently electrophilic to react with enolate equivalents such as silyl enol ethers[207] and isopropenyl acetate.[208]

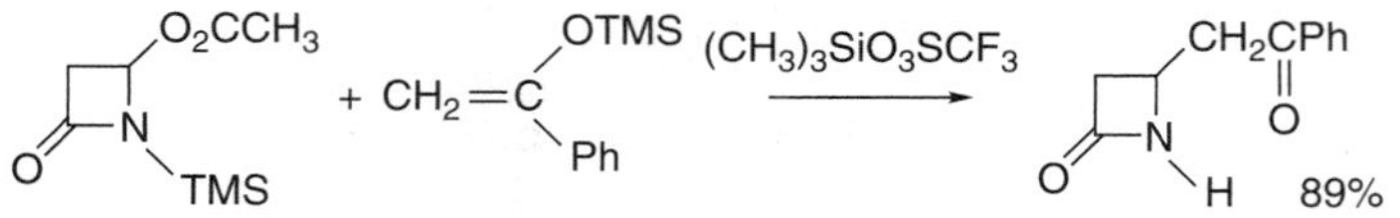

Acyliminium ions can be used in enantioselective additions with enolates having chiral auxiliaries, such as *N*-acyloxazolidinones or *N*-acylthiazolidinethiones.

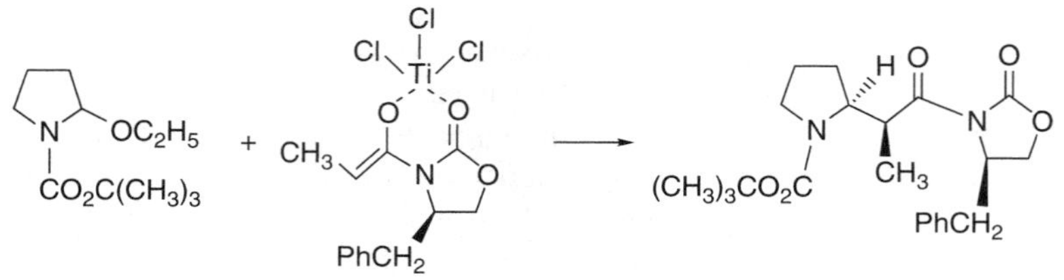

Ref. 209

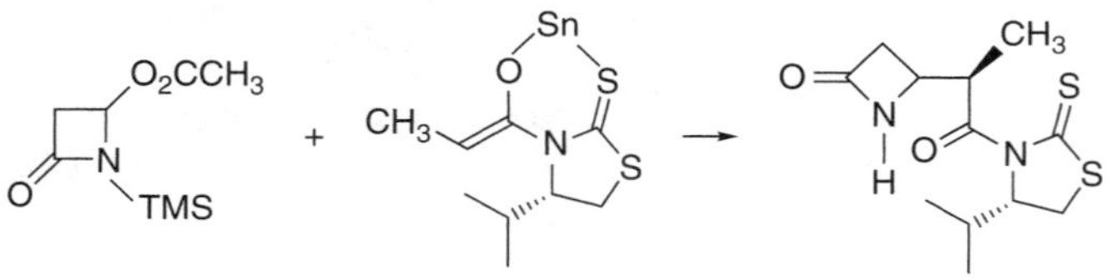

Ref. 210

205. T. Shono, H. Hamaguchi, and Y. Matsumura, *J. Am. Chem. Soc.*, **97**, 4264 (1975); T. Shono, Y. Matsumura, K. Tsubata, Y. Sugihara, S. Yamane, T. Kanazawa, and T. Aoki, *J. Am. Chem. Soc.*, **104**, 6697 (1982); T. Shono, *Tetrahedron*, **40**, 811 (1984).

206. A. Boto, R. Hernandez, and E. Suarez, *J. Org. Chem.*, **65**, 4930 (2000).

207. R. P. Attrill, A. G. M. Barrett, P. Quayle, J. van der Westhuizen, and M. J. Betts, *J. Org. Chem.*, **49**, 1679 (1984); K. T. Wanner, A. Kartner, and E. Wadenstorfer, *Heterocycles*, **27**, 2549 (1988); M. A. Ciufolini, C. W. Hermann, K. H. Whitmire, and N. E. Byrne, *J. Am. Chem. Soc.*, **111**, 3473 (1989); D. S. Brown, M. J. Earle, R. A. Fairhurst, H. Heaney, G. Papageorgiou, R. F. Wilkins, and S. C. Eyley, *Synlett*, 619 (1990).

208. T. Shono, Y. Matsumura, and K. Tsubata, *J. Am. Chem. Soc.*, **103**, 1172 (1981).

209. R. A. Pilli and D. Russowsky, *J. Org. Chem.*, **61**, 3187 (1996); R. A. Pilli, C. de F. Alves, M. A. Boeckelmann, Y. P. Mascarenhas, J. G. Nery, and I. Vencato, *Tetrahedron Lett.*, **40**, 2891 (1999).

210. Y. Nagao, T. Kumagi, S. Tamai, T. Abe, Y. Kuramoto, T. Taga, S. Aoyagi, Y. Nagase, M. Ochiai, Y. Inoue, and E. Fujita, *J. Am. Chem. Soc.*, **108**, 4673 (1986); T. Nagao, W.-M. Dai, M. Ochiai, S. Tsukagoshi, and E. Fujita, *J. Org. Chem.*, **55**, 1148 (1990).

Iminium ions are intermediates in a group of reactions that form α,β-unsaturated compounds having structures corresponding to those formed by mixed aldol addition followed by dehydration. These reactions are catalyzed by amines or buffer systems containing an amine and an acid and are referred to as *Knoevenagel condensations*.[211] The reactive electrophile is probably the protonated form of the imine, since it is a more reactive electrophile than the corresponding carbonyl compound.[212]

$$ArCH{=}NC_4H_9 \xrightarrow{H^+,\ ^-CH_2NO_2} ArCHNHC_4H_9 \longrightarrow ArCH{-}NHC_4H_9 \xrightarrow{H^+} ArCH{=}CHNO_2$$

The carbon nucleophiles in amine-catalyzed reaction conditions are usually rather acidic compounds containing two EWG substituents. Malonate esters, cyanoacetate esters, and cyanoacetamide are examples of compounds that undergo condensation reactions under Knoevenagel conditions.[213] Nitroalkanes are also effective as nucleophilic reactants. The single nitro group activates the α-hydrogens enough to permit deprotonation under the weakly basic conditions. A relatively acidic proton in the nucleophile is important for two reasons. First, it permits weak bases, such as amines, to provide a sufficient concentration of the enolate for reaction. An acidic proton also facilitates the elimination step that drives the reaction to completion. Usually the product that is isolated is the α,β-unsaturated derivative of the original adduct.

$$R_2C{-}C(H)(CO_2R)(CN) \longrightarrow R_2C{=}C(CO_2R)(CN)$$

$$X = OH \text{ or } NR_2$$

Malonic acid or cyanoacetic acid can also be used as the nucleophile. With malonic acid or cyanoacetic acid as reactants, the products usually undergo decarboxylation. This may occur as a concerted fragmentation of the adduct.[214]

$$RCR + CH_2(CO_2H)_2 \longrightarrow R_2C{-}CHCO_2H \longrightarrow R_2C{=}CHCO_2H$$

$$X = OH \text{ or } NR_2$$

Decarboxylative condensations of this type are sometimes carried out in pyridine, which cannot form an imine intermediate, but has been shown to catalyze the decarboxylation of arylidene malonic acids.[215] The decarboxylation occurs by concerted decomposition of the adduct of pyridine to the α,β-unsaturated diacid.

211. G. Jones, *Org. React.*, **15**, 204 (1967); R. L. Reeves, in *The Chemistry of the Carbonyl Group*, S. Patai, ed., Interscience, New York, 1966, pp. 593–599.

212. T. I. Crowell and D. W. Peck, *J. Am. Chem. Soc.*, **75**, 1075 (1953).

213. A. C. Cope, C. M. Hofmann, C. Wyckoff, and E. Hardenbergh, *J. Am. Chem. Soc.*, **63**, 3452 (1941).

214. E. J. Corey, *J. Am. Chem. Soc.*, **74**, 5897 (1952).

215. E. J. Corey and G. Fraenkel, *J. Am. Chem. Soc.*, **75**, 1168 (1953).

Scheme 2.13. Amine-Catalyzed Condensation Reactions of the Knoevenagel Type

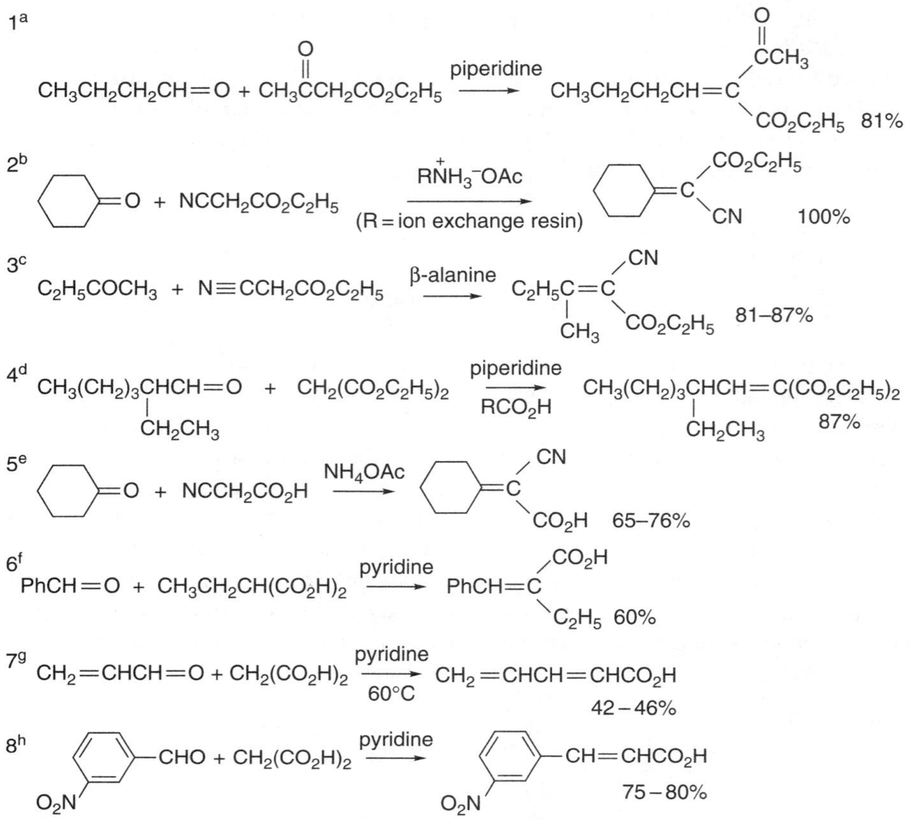

1[a]

$$CH_3CH_2CH_2CH{=}O + CH_3CCH_2CO_2C_2H_5 \xrightarrow{\text{piperidine}} CH_3CH_2CH_2CH{=}C \begin{smallmatrix} CCH_3 \\ \quad \\ CO_2C_2H_5 \end{smallmatrix} \quad 81\%$$

2[b]

$$\bigcirc{=}O + NCCH_2CO_2C_2H_5 \xrightarrow[\text{(R = ion exchange resin)}]{R\overset{+}{N}H_3{}^-OAc} \bigcirc{=}C \begin{smallmatrix} CO_2C_2H_5 \\ \\ CN \end{smallmatrix} \quad 100\%$$

3[c]

$$C_2H_5COCH_3 + N{\equiv}CCH_2CO_2C_2H_5 \xrightarrow{\beta\text{-alanine}} C_2H_5C{=}C \begin{smallmatrix} CN \\ \\ CO_2C_2H_5 \\ | \\ CH_3 \end{smallmatrix} \quad 81\text{–}87\%$$

4[d]

$$CH_3(CH_2)_3CHCH{=}O + CH_2(CO_2C_2H_5)_2 \xrightarrow[RCO_2H]{\text{piperidine}} CH_3(CH_2)_3CHCH{=}C(CO_2C_2H_5)_2 \quad 87\%$$
$$\qquad\; \underset{\displaystyle CH_2CH_3}{|} \qquad\qquad\qquad\qquad\qquad\qquad\qquad\qquad\quad \underset{\displaystyle CH_2CH_3}{|}$$

5[e]

$$\bigcirc{=}O + NCCH_2CO_2H \xrightarrow{NH_4OAc} \bigcirc{=}C \begin{smallmatrix} CN \\ \\ CO_2H \end{smallmatrix} \quad 65\text{–}76\%$$

6[f]

$$PhCH{=}O + CH_3CH_2CH(CO_2H)_2 \xrightarrow{\text{pyridine}} PhCH{=}C \begin{smallmatrix} CO_2H \\ \\ C_2H_5 \end{smallmatrix} \quad 60\%$$

7[g]

$$CH_2{=}CHCH{=}O + CH_2(CO_2H)_2 \xrightarrow[60^\circ C]{\text{pyridine}} CH_2{=}CHCH{=}CHCO_2H \quad 42\text{–}46\%$$

8[h]

$$O_2N{-}\bigcirc{-}CHO + CH_2(CO_2H)_2 \xrightarrow{\text{pyridine}} O_2N{-}\bigcirc{-}CH{=}CHCO_2H \quad 75\text{–}80\%$$

a. A. C. Cope and C. M. Hofmann, *J. Am. Chem. Soc.*, **63**, 3456 (1941).
b. R. W. Hein, M. J. Astle, and J. R. Shelton, *J. Org. Chem.*, **26**, 4874 (1961).
c. F. S Prout, R. J. Harman, E. P.-Y. Huang, C. J. Korpics, and G. R. Tichelaar, *Org. Synth.*, **IV**, 93 (1963).
d. E. F. Pratt and E. Werbie, *J. Am. Chem. Soc.*, **72**, 4638 (1950).
e. A. C. Cope, A. A. D'Addieco, D. E. Whyte, and S. A. Glickman, *Org. Synth.*, **IV**, 234 (1963).
f. W. J. Gensler and E. Berman, *J. Am. Chem. Soc.*, **80**, 4949 (1958).
g. P. J. Jessup, C. B. Petty, J. Roos, and L. E. Overman, *Org. Synth.*, **59**, 1 (1979).
h. R. H. Wiley and N. R. Smith, *Org. Synth.*, **IV**, 731 (1963).

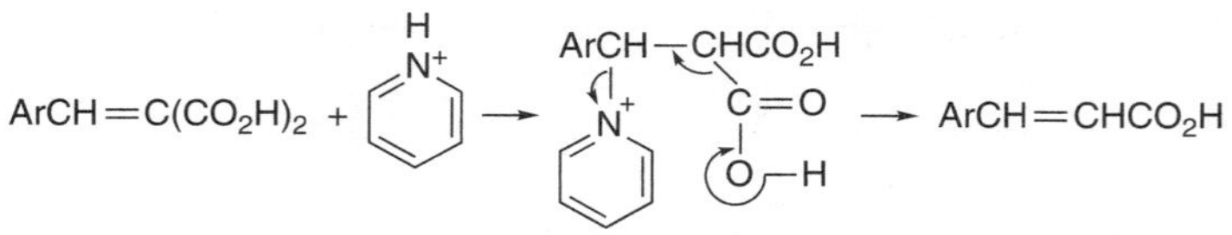

$$ArCH{=}C(CO_2H)_2 + \overset{\displaystyle H}{\underset{\displaystyle \bigcirc}{N^+}} \longrightarrow \cdots \longrightarrow ArCH{=}CHCO_2H$$

Scheme 2.13 gives some examples of Knoevenagel condensation reactions.

2.3. Acylation of Carbon Nucleophiles

The reactions that are discussed in this section involve addition of carbon nucleophiles to carbonyl centers having a potential leaving group. The tetrahedral intermediate formed in the addition step reacts by expulsion of the leaving group. The overall transformation results in the *acylation* of the carbon nucleophile. This transformation corresponds to the general reaction Path **B**, as specified at the beginning of this chapter (p. 64).

$$
\underset{RC-X}{\overset{O}{\overset{\|}{}}} + \quad R'_2\overset{-}{C}-EWG \quad \longrightarrow \quad \underset{RC-CR'_2}{\overset{O\bar{}}{\overset{|}{}}} \quad \longrightarrow \quad \underset{RCCR'_2}{\overset{O}{\overset{\|}{}}}
$$

The reaction pattern can be used for the synthesis of 1,3-dicarbonyl compounds and other systems in which an acyl group is β to an anion-stabilizing group.

$$
\underset{R^1}{\overset{O}{\overset{\|}{}}}\!\!\diagdown_X \quad + \quad R^2CH_2EWG \quad \longrightarrow \quad R^1\overset{O}{\overset{\|}{\diagup}}\!\!\diagdown_{R^2}\!\!\diagup^{EWG}
$$

2.3.1. Claisen and Dieckmann Condensation Reactions

An important group of acylation reactions involves esters, in which case the leaving group is alkoxy or aryloxy. The self-condensation of esters is known as the *Claisen condensation*.[216] Ethyl acetoacetate, for example, is prepared by Claisen condensation of ethyl acetate. All of the steps in the mechanism are reversible, and a full equivalent of base is needed to bring the reaction to completion. Ethyl acetoacetate is more acidic than any of the other species present and is converted to its conjugate base in the final step. The β-ketoester product is obtained after neutralization.

$$
CH_3CO_2CH_2CH_3 + CH_3CH_2O^- \rightleftharpoons {}^-CH_2CO_2CH_2CH_3 + CH_3CH_2OH
$$

$$
\underset{CH_3COCH_2CH_3}{\overset{O}{\overset{\|}{}}} + \quad {}^-CH_2CO_2CH_2CH_3 \quad \rightleftharpoons \quad \underset{\underset{CH_2CO_2CH_2CH_3}{|}}{\overset{O^-}{\overset{|}{CH_3COCH_2CH_3}}}
$$

$$
\underset{\underset{CH_2CO_2CH_2CH_3}{|}}{\overset{O\bar{}}{\overset{|}{CH_3C-OCH_2CH_3}}} \quad \rightleftharpoons \quad \underset{CH_3CCH_2CO_2CH_2CH_3}{\overset{O}{\overset{\|}{}}} + \quad CH_3CH_2O^-
$$

$$
\underset{CH_3CCH_2CO_2CH_2CH_3}{\overset{O}{\overset{\|}{}}} + CH_3CH_2O^- \longrightarrow \underset{CH_3C\overset{}{C}HCO_2CH_2CH_3}{\overset{O}{\overset{\|}{}}} + CH_3CH_2OH
$$

As a practical matter, the alkoxide used as the base must be the same as the alcohol portion of the ester to prevent product mixtures resulting from ester interchange. Sodium hydride with a small amount of alcohol is frequently used as the base for ester condensation. The reactive base is the sodium alkoxide formed by reaction of sodium hydride with the alcohol released in the condensation.

$$
R'OH + NaH \longrightarrow R'ONa + H_2
$$

As the final proton transfer cannot occur when α-substituted esters are used, such compounds do not condense under the normal reaction conditions, but this limitation

216. C. R. Hauser and B. E. Hudson, Jr., *Org. React.*, **1**, 266 (1942).

can be overcome by use of a very strong base that converts the reactant ester completely to its enolate. Entry 2 of Scheme 2.14 illustrates the use of triphenylmethylsodium for this purpose. The sodium alkoxide is also the active catalyst in procedures that use sodium metal, such as in Entry 3 in Scheme 2.14. The alkoxide is formed by reaction of the alcohol that is formed as the reaction proceeds.

The intramolecular version of ester condensation is called the *Dieckmann condensation*.[217] It is an important method for the formation of five- and six-membered rings and has occasionally been used for formation of larger rings. As ester condensation is reversible, product structure is governed by thermodynamic control, and in situations where more than one product can be formed, the product is derived from the most stable enolate. An example of this effect is the cyclization of the diester **25**.[218] Only **27** is formed, because **26** cannot be converted to a stable enolate. If **26**, synthesized by another method, is subjected to the conditions of the cyclization, it is isomerized to **27** by the reversible condensation mechanism.

Entries 3 to 8 in Scheme 2.14 are examples of Dieckmann condensations. Entry 6 is a Dieckmann reaction carried out under conventional conditions, followed by decarboxylation. The product is a starting material for the synthesis of a number of sarpagine-type indole alkaloids and can be carried out on a 100-g scale. The combination of a Lewis acid, such as $MgCl_2$, with an amine can also promote Dieckmann cyclization.[219] Entry 7, which shows an application of these conditions, is a step in the synthesis of a potential drug. These conditions were chosen to avoid the use of $TiCl_4$ in a scale-up synthesis and can be done on a 60-kg scale. The 14-membered ring formation in Entry 8 was carried out under high dilution by slowly adding the reactant to the solution of the NaHMDS base. The product is a mixture of both possible regioisomers (both the 5- and 7-carbomethoxy derivatives are formed) but a single product is obtained after decarboxylation.

Mixed condensations of esters are subject to the same general restrictions as outlined for mixed aldol reactions (Section 2.1.2). One reactant must act preferentially as the acceptor and another as the nucleophile for good yields to be obtained. Combinations that work best involve one ester that cannot form an enolate but is relatively reactive as an electrophile. Esters of aromatic acids, formic acid, and oxalic acid are especially useful. Some examples of mixed ester condensations are shown in Section C of Scheme 2.14. Entries 9 and 10 show diethyl oxalate as the acceptor, and aromatic esters function as acceptors in Entries 11 and 12.

2.3.2. Acylation of Enolates and Other Carbon Nucleophiles

Acylation of carbon nucleophiles can also be carried out with more reactive acylating agents such as acid anhydrides and acyl chlorides. These reactions must

217. J. P. Schaefer and J. J. Bloomfield, *Org. React.*, **15**, 1 (1967).
218. N. S. Vul'fson and V. I. Zaretskii, *J. Gen. Chem. USSR*, **29**, 2704 (1959).
219. S. Tamai, H. Ushitogochi, S. Sano, and Y. Nagao, *Chem. Lett.*, 295 (1995).

Scheme 2.14. Acylation of Nucleophilic Carbon by Esters

A. Intermolecular ester condensations

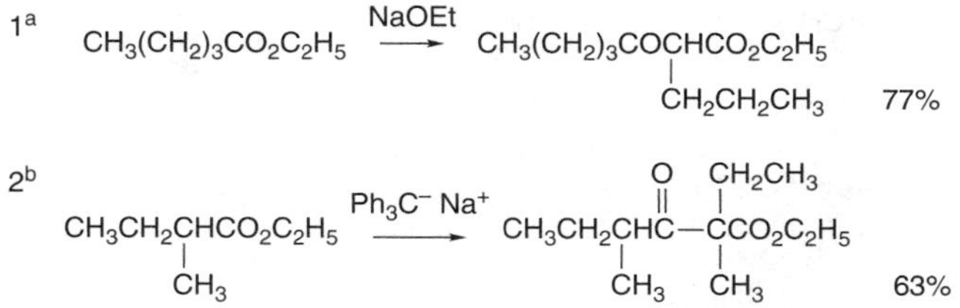
1[a]
NaOEt
77%
2[b]
Ph3C− Na+
63%

B. Cyclization of diesters

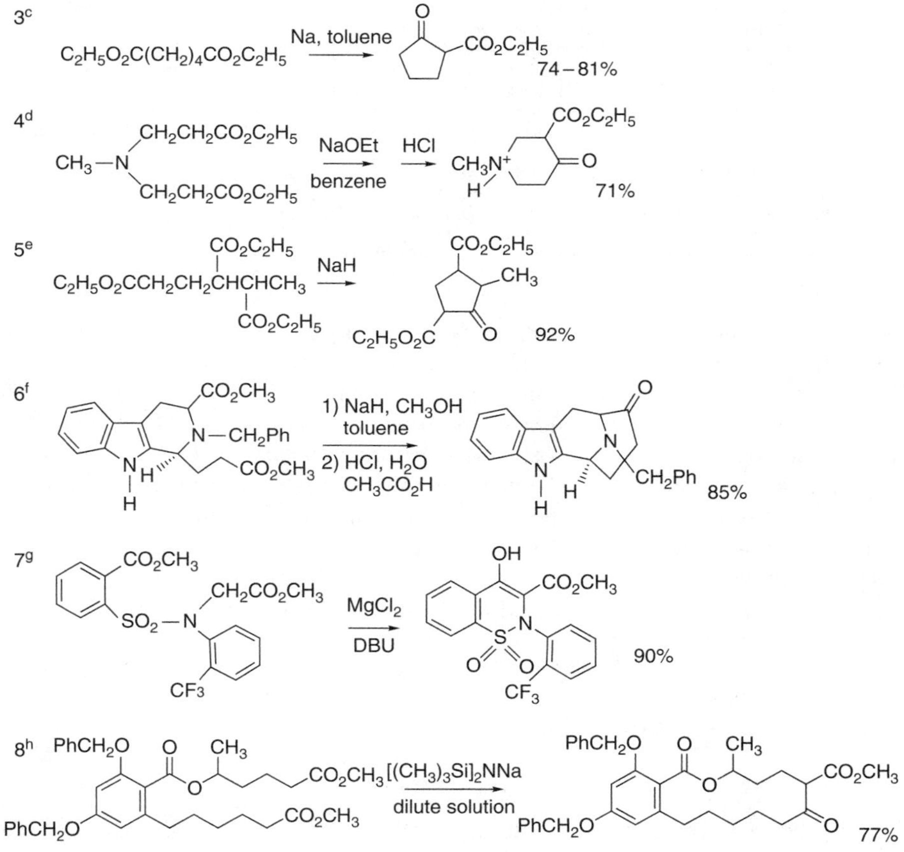
3[c]
Na, toluene
74–81%
4[d]
NaOEt
benzene
HCl
71%
5[e]
NaH
92%
6[f]
1) NaH, CH3OH
toluene
2) HCl, H2O
CH3CO2H
85%
7[g]
MgCl2
DBU
90%
8[h]
[(CH3)3Si]2NNa
dilute solution
77%

C. Mixed ester condensations

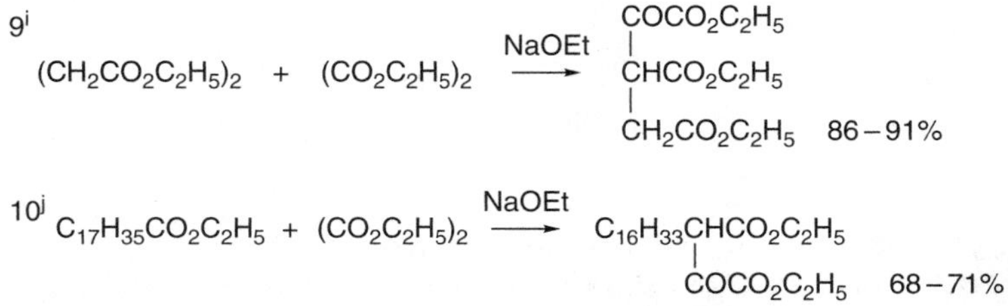
9[i]
NaOEt
86–91%
10[j]
NaOEt
68–71%

(*Continued*)

152

CHAPTER 2

*Reactions of Carbon
Nucleophiles with
Carbonyl Compounds*

Scheme 2.14. (*Continued*)

11^k [structure: pyridine-3-carboxylic acid ethyl ester] —CO$_2$C$_2$H$_5$ + CH$_3$(CH$_2$)$_2$CO$_2$C$_2$H$_5$ $\xrightarrow{\text{NaH}}$ [structure: pyridine] —COCHCO$_2$C$_2$H$_5$ | CH$_2$CH$_3$ 68%

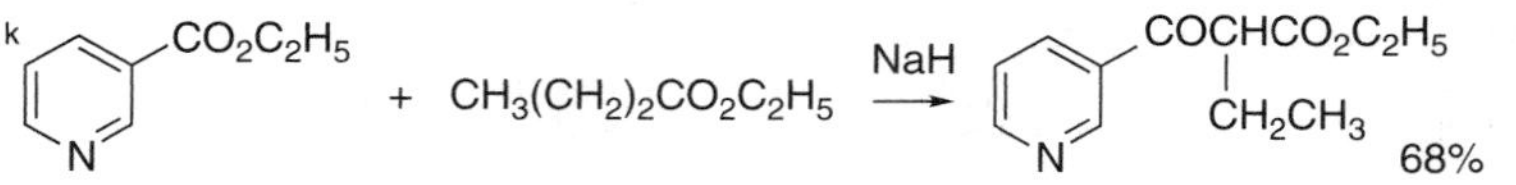

12^l [structure: benzene] —CO$_2$C$_2$H$_5$ + CH$_3$CH$_2$CO$_2$C$_2$H$_5$ $\xrightarrow{(\textit{i}\text{-Pr})_2\text{NMgBr}}$ [structure: benzene] —COCHCO$_2$C$_2$H$_5$ | CH$_3$ 51%

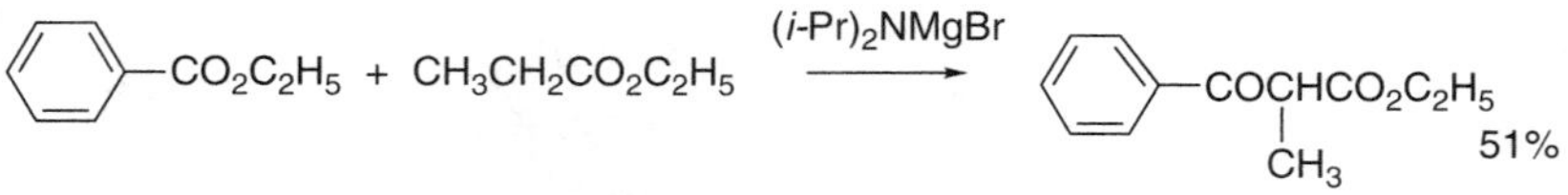

a. R. R. Briese and S. M. McElvain, *J. Am. Chem. Soc.*, **55**, 1697 (1933).
b. B. E. Hudson, Jr., and C. R. Hauser, *J. Am. Chem. Soc.*, **63**, 3156 (1941).
c. P. S. Pinkney, *Org. Synth.*, **II**, 116 (1943).
d. E. A. Prill and S. M. McElvain, *J. Am. Chem. Soc.*, **55**, 1233 (1933).
e. M. S. Newman and J. L. McPherson, *J. Org. Chem.*, **19**, 1717 (1954).
f. J. Yu, T. Wang, X. Liu, J. Deschamps, J. Flippen-Anderson, X. Liao, and J. M. Cook, *J. Org. Chem.*, **68**, 7565 (2003); P. Yu, T. Wang, J. Li, and J. M. Cook, *J. Org. Chem.*, **65**, 3173 (2000).
g. T. E. Jacks, D. T. Belmont, C. A. Briggs, N. M. Horne, G. D. Kanter, G. L. Karrick, J. L. Krikke, R. J. McCabe, J. G. Mustakis, T. N. Nanninga, G. S. Risedorph, R. E. Seamans, R. Skeean, D. D. Winkle, and T. M. Zennie, *Org. Proc. Res. Develop.* **8**, 201 (2004).
h. R. N. Hurd and D. H. Shah, *J. Org. Chem.*, **38**, 390 (1973).
i. E. M. Bottorff and L. L. Moore, *Org. Synth.*, **44**, 67 (1964).
j. F. W. Swamer and C. R. Hauser, *J. Am. Chem. Soc.*, **72**, 1352 (1950).
k. D. E. Floyd and S. E. Miller, *Org. Synth.*, **IV**, 141 (1963).
l. E. E. Royals and D. G. Turpin, *J. Am. Chem. Soc.*, **76**, 5452 (1954).

be done in nonnucleophilic solvents to avoid solvolysis of the acylating agent. The use of these reactive acylating agents can be complicated by competing O-acylation. Magnesium enolates play a prominent role in these C-acylation reactions. The magnesium enolate of diethyl malonate, for example, can be prepared by reaction with magnesium metal in ethanol. It is soluble in ether and undergoes C-acylation by acid anhydrides and acyl chlorides. The preparation of diethyl benzoylmalonate (Entry 1, Scheme 2.15) is an example of the use of an acid anhydride. Entries 2 to 5 illustrate the use of acyl chlorides. Entry 3 is carried out in basic aqueous solution and results in deacylation of the initial product.

Monoalkyl esters of malonic acid react with Grignard reagents to give a chelated enolate of the malonate monoanion.

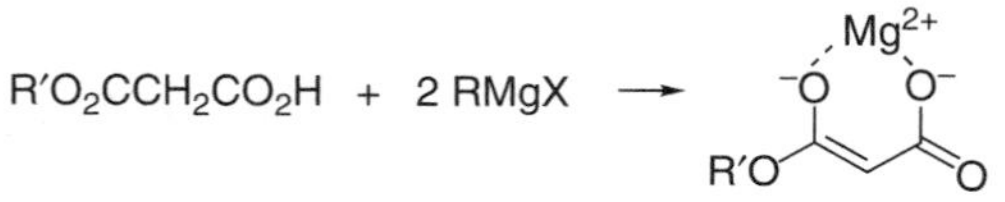

$$\text{R'O}_2\text{CCH}_2\text{CO}_2\text{H} + 2\text{ RMgX} \longrightarrow \text{[chelated Mg}^{2+}\text{ enolate structure]}$$

These carbon nucleophiles react with acyl chlorides[220] or acyl imidazolides.[221] The initial products decarboxylate readily so the isolated products are β-ketoesters.

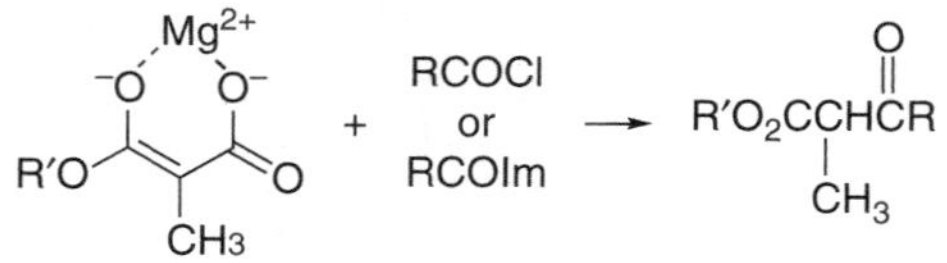

$$\text{[Mg}^{2+}\text{ enolate structure]} + \begin{array}{c}\text{RCOCl}\\ \text{or}\\ \text{RCOIm}\end{array} \longrightarrow \text{R'O}_2\text{CCHCR} \underset{\text{CH}_3}{|} \overset{\text{O}}{||}$$

220. R. E. Ireland and J. A. Marshall, *J. Am. Chem. Soc.*, **81**, 2907 (1959).
221. J. Maibaum and D. H. Rich, *J. Org. Chem.*, **53**, 869 (1988); W. H. Moos, R. D. Gless, and H. Rapoport, *J. Org. Chem.*, **46**, 5064 (1981); D. W. Brooks, L. D.-L. Lu, and S. Masamune, *Angew. Chem. Int. Ed. Engl.*, **18**, 72 (1979).

Scheme 2.15. Acylation of Ester Enolates with Acyl Halides, Anhydrides, and Imidazolides

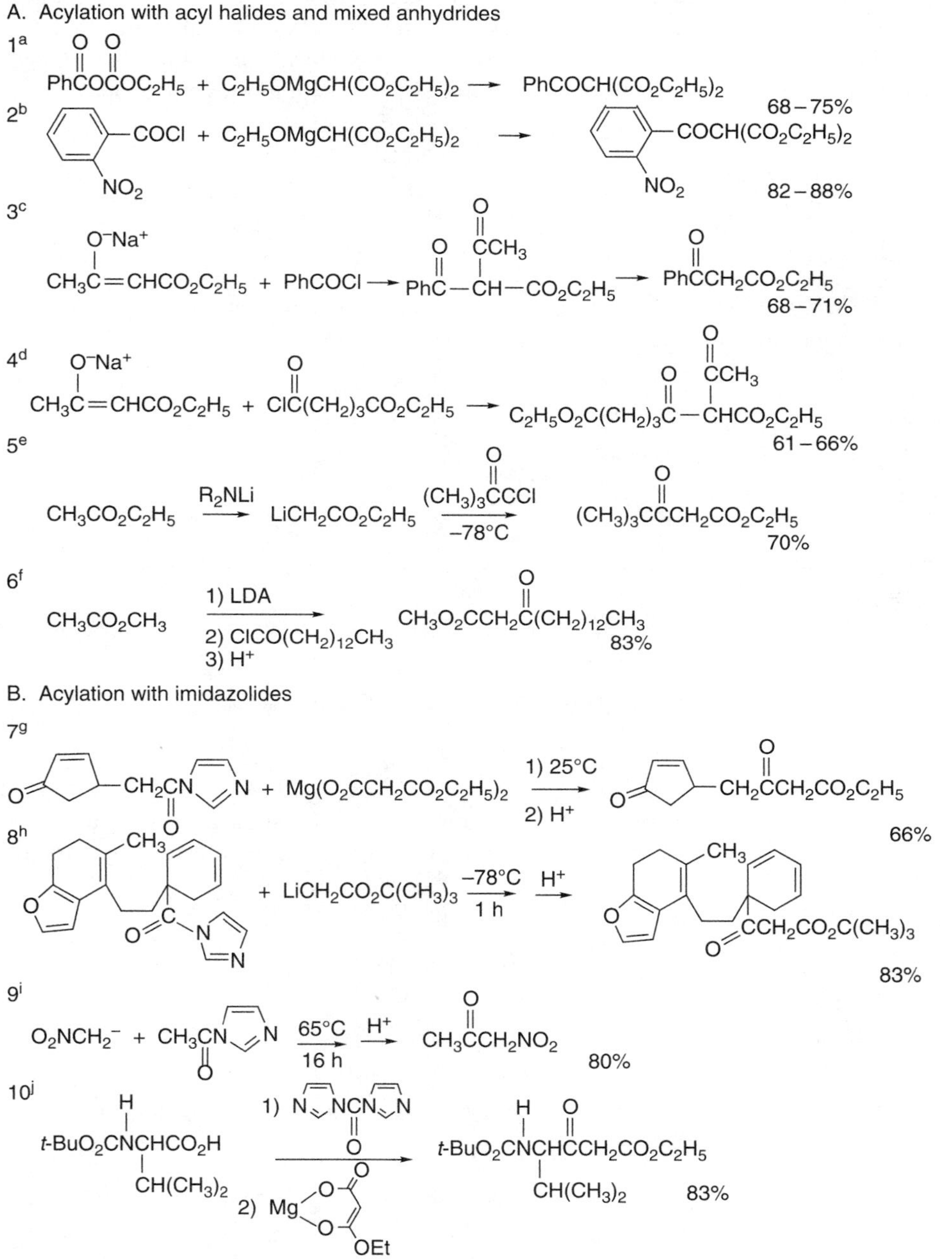

A. Acylation with acyl halides and mixed anhydrides

B. Acylation with imidazolides

a. J. A. Price and D. S. Tarbell, *Org. Synth.*, **IV**, 285 (1963).
b. G. A. Reynolds and C. R. Hauser, *Org. Synth.*, **IV**, 708 (1963).
c. J. M. Straley and A. C. Adams, *Org. Synth.*, **IV**, 415 (1963).
d. M. Guha and D. Nasipuri, *Org. Synth.*, **V**, 384 (1973).
e. M. W. Rathke and J. Deitch, *Tetrahedron Lett.*, 2953 (1971).
f. D. F. Taber, P. B. Deker, H. M. Fales, T. H. Jones, and H. A. Lloyd, *J. Org. Chem.*, **53**, 2968 (1988).
g. A. Barco, S. Bennetti, G. P. Pollini, P. G. Baraldi, and C. Gandolfi, *J. Org. Chem.*, **45**, 4776 (1980).
h. E. J. Corey, G. Wess, Y. B. Xiang, and A. K. Singh, *J. Am. Chem. Soc.*, **109**, 4717 (1987).
i. M. E. Jung, D. D. Grove, and S. I. Khan, *J. Org. Chem.*, **52**, 4570 (1987).
j. J. Maibaum and D. H. Rich, *J. Org. Chem.*, **53**, 869 (1988).

Acyl imidazolides are more reactive than esters but not as reactive as acyl halides. Entry 7 is an example of formation of a β-ketoesters by reaction of magnesium enolate monoalkyl malonate ester by an imidazolide. Acyl imidazolides also are used for acylation of ester enolates and nitromethane anion, as illustrated by Entries 8, 9, and 10. *N*-Methoxy-*N*-methylamides are also useful for acylation of ester enolates.

82%

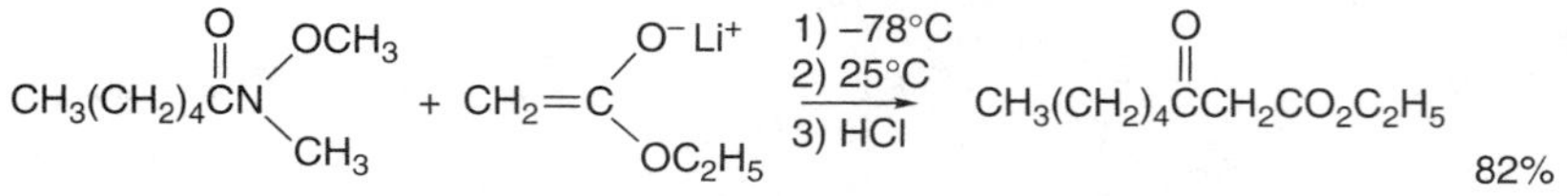

Ref. 222

Both diethyl malonate and ethyl acetoacetate can be acylated by acyl chlorides using magnesium chloride and pyridine or triethylamine.[223]

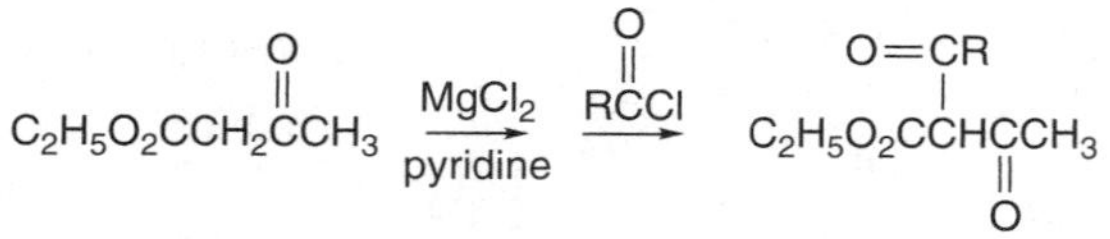

Rather similar conditions can be used to convert ketones to β-keto acids by carboxylation.[224]

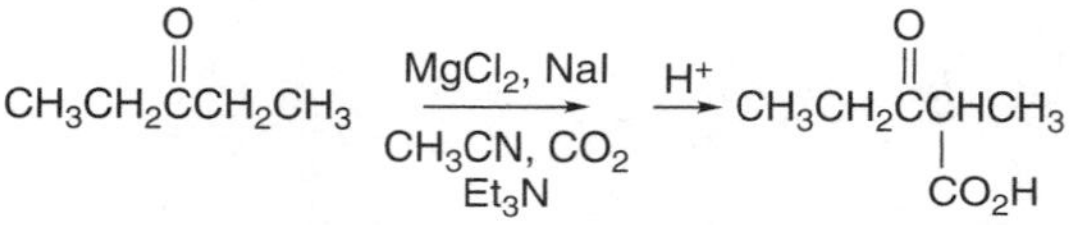

These reactions presumably involve formation of a magnesium chelate of the keto acid. The β-ketoacid is liberated when the reaction mixture is acidified during workup.

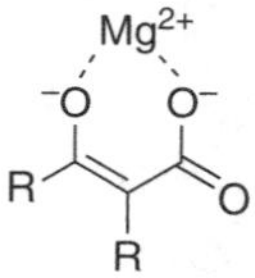

Carboxylation of ketones and esters can also be achieved by using the magnesium salt of monomethyl carbonate.

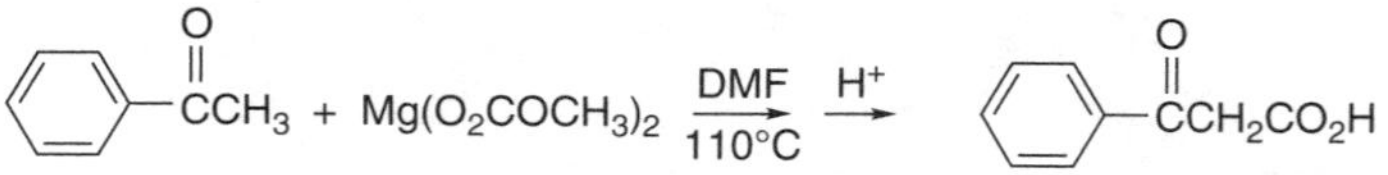

Ref. 225

222. J. A. Turner and W. S. Jacks, *J. Org. Chem.*, **54**, 4229 (1989).
223. M. W. Rathke and P. J. Cowan, *J. Org. Chem.*, **50**, 2622 (1985).
224. R. E. Tirpak, R. S. Olsen, and M. W. Rathke, *J. Org. Chem.*, **50**, 4877 (1985).
225. M. Stiles, *J. Am. Chem. Soc.*, **81**, 2598 (1959).

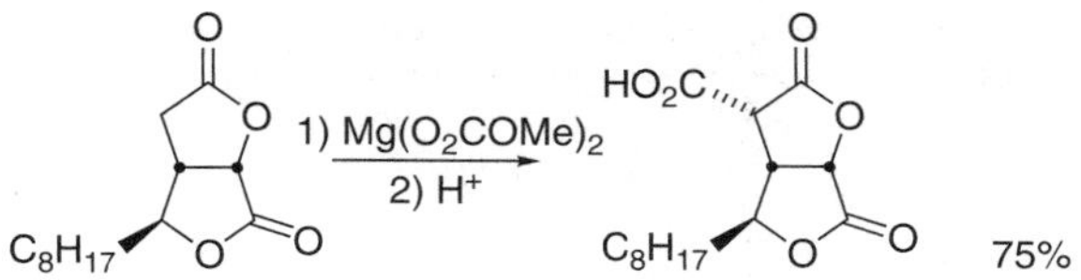

Ref. 226

The enolates of ketones can be acylated by esters and other acylating agents. The products of these reactions are β-dicarbonyl compounds, which are rather acidic and can be alkylated by the procedures described in Section 1.2. Reaction of ketone enolates with formate esters gives a β-ketoaldehyde. As these compounds exist in the enol form, they are referred to as *hydroxymethylene derivatives*. Entries 1 and 2 in Scheme 2.16 are examples. Product formation is under thermodynamic control so the structure of the product can be predicted on the basis of the stability of the various possible product anions.

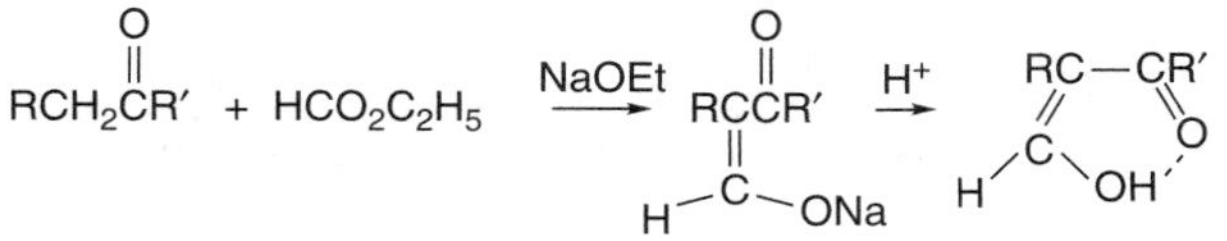

Ketones are converted to β-ketoesters by acylation with diethyl carbonate or diethyl oxalate, as illustrated by Entries 4 and 5 in Scheme 2.16. Alkyl cyanoformate can be used as the acylating reagent under conditions where a ketone enolate has been formed under kinetic control.[227]

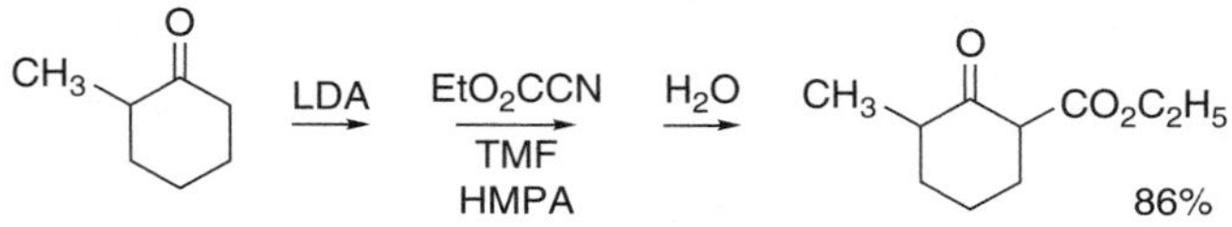

When this type of reaction is quenched with trimethylsilyl chloride, rather than by neutralization, a trimethylsilyl ether of the adduct is isolated. This result shows that the tetrahedral adduct is stable until the reaction mixture is hydrolyzed.

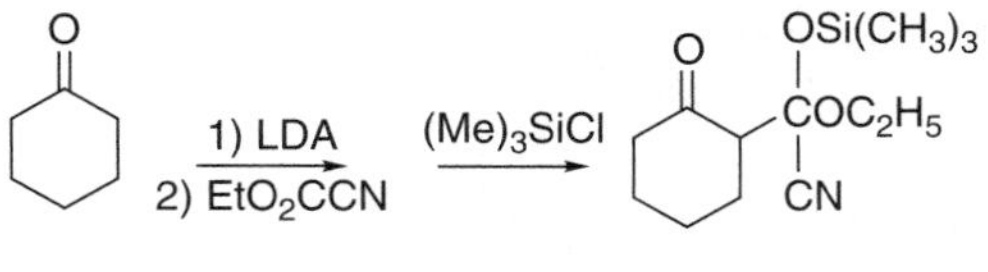

Ref. 228

β-Keto sulfoxides can be prepared by acylation of dimethyl sulfoxide anion with esters.[229]

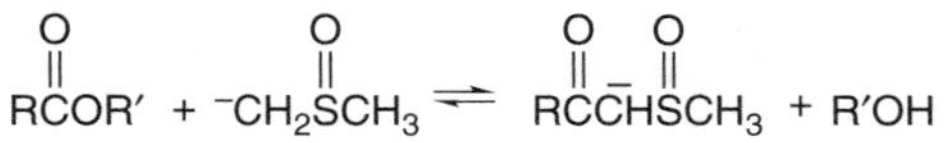

226. W. L. Parker and F. Johnson, *J. Org. Chem.*, **38**, 2489 (1973).
227. L. N. Mander and S. P. Sethi, *Tetrahedron Lett.*, **24**, 5425 (1983).
228. F. E. Ziegler and T.-F. Wang, *Tetrahedron Lett.*, **26**, 2291 (1985).
229. E. J. Corey and M. Chaykovsky, *J. Am. Chem. Soc.*, **87**, 1345 (1965); H. D. Becker, G. J. Mikol, and G. A. Russell, *J. Am. Chem. Soc.*, **85**, 3410 (1963).

Scheme 2.16. Acylation of Ketones by Esters

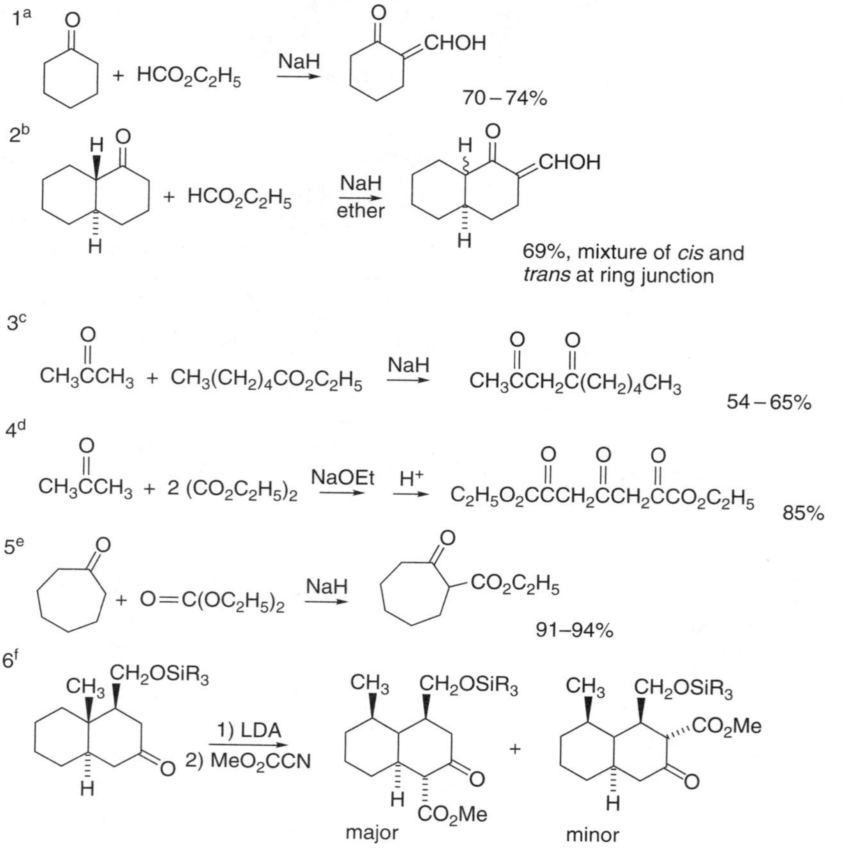
1a
+ HCO2C2H5
NaH
CHOH
70–74%
2b
+ HCO2C2H5
NaH
ether
CHOH
69%, mixture of cis and trans at ring junction
3c
CH3CCH3 + CH3(CH2)4CO2C2H5
NaH
CH3CCH2C(CH2)4CH3
54–65%
4d
CH3CCH3 + 2 (CO2C2H5)2
NaOEt
H+
C2H5O2CCCH2CCH2CCO2C2H5
85%
5e
+ O=C(OC2H5)2
NaH
CO2C2H5
91–94%
6f
CH3
CH2OSiR3
1) LDA
2) MeO2CCN
CO2Me
major
minor

a. C. Ainsworth, *Org. Synth.*, **IV**, 536 (1963).
b. P. H. Lewis, S. Middleton, M. J. Rosser, and L. E. Stock, *Aust. J. Chem.*, **32**, 1123 (1979).
c. N. Green and F. B. La Forge, *J. Am. Chem. Soc.*, **70**, 2287 (1948); F. W. Swamer and C. R. Hauser, *J. Am. Chem. Soc.*, **72**, 1352 (1950).
d. E. R. Riegel and F. Zwilgmeyer, *Org. Synth.*, **II**, 126 (1943).
e. A. P. Krapcho, J. Diamanti, C. Cayen, and R. Bingham, *Org. Synth.*, **47**, 20 (1967).
f. F. E. Ziegler, S. I. Klein, U. K. Pati, and T.-F. Wang, *J. Am. Chem. Soc.*, **107**, 2730 (1985).

Mechanistically, this reaction is similar to ketone acylation. The β-keto sulfoxides have several synthetic applications. The sulfoxide substituent can be removed reductively, which leads to methyl ketones.

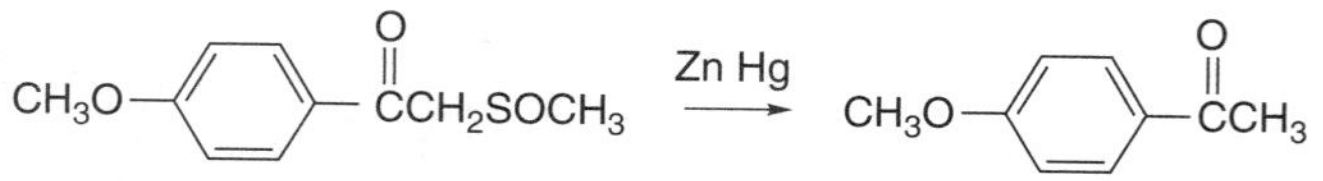
CH3O
CCH2SOCH3
Zn Hg
CH3O
CCH3

Ref. 230

The β-keto sulfoxides can be alkylated via their anions. Inclusion of an alkylation step prior to the reduction provides a route to ketones with longer chains.

[230] G. A. Russell and G. J. Mikol, *J. Am. Chem. Soc.*, **88**, 5498 (1966).

$$PhCOCH_2SOCH_3 \xrightarrow[\text{2) CH}_3\text{I}]{\text{1) NaH}} PhCOCHSOCH_3 \xrightarrow{Zn\ Hg} PhCOCH_2CH_3$$
$$\underset{CH_3}{|}$$

Ref. 231

These reactions accomplish the same overall synthetic transformation as the acylation of ester enolates, but use desulfurization rather than decarboxylation to remove the anion-stabilizing group. Dimethyl sulfone can be subjected to similar reaction sequences.[232]

2.4. Olefination Reactions of Stabilized Carbon Nucleophiles

This section deals with reactions that correspond to Pathway **C,** defined earlier (p. 64), that lead to formation of alkenes. The reactions discussed include those of phosphorus-stabilized nucleophiles (Wittig and related reactions), α-silyl (Peterson reaction) and α-sulfonyl carbanions (Julia olefination) with aldehydes and ketones. These important rections can be used to convert a carbonyl group to an alkene by reaction with a carbon nucleophile. In each case, the addition step is followed by an elimination.

A crucial issue for these reactions is the stereoselectivity for formation of *E*- or *Z*-alkene. This is determined by the mechanisms of the reactions and, as we will see, can be controlled in some cases by the choice of particular reagents and reaction conditions.

2.4.1. The Wittig and Related Reactions of Phosphorus-Stabilized Carbon Nucleophiles

The *Wittig reaction* involves *phosphonium ylides* as the nucleophilic carbon species.[233] An ylide is a molecule that has a contributing resonance structure with opposite charges on adjacent atoms, each of which has an octet of electrons. Although this definition includes other classes of compounds, the discussion here is limited to ylides having the negative charge on the carbon. Phosphonium ylides are stable, but quite reactive, compounds. They can be represented by two limiting resonance structures, which are referred to as the ylide and ylene forms.

$$(CH_3)_3\overset{+}{P}-CH_2^- \longleftrightarrow (CH_3)_3P=CH_2$$

ylide   ylene

231. P. G. Gassman and G. D. Richmond, *J. Org. Chem.*, **31,** 2355 (1966).

232. H. O. House and J. K. Larson, *J. Org. Chem.*, **33,** 61 (1968).

233. For general reviews of the Wittig reaction, see A. Maercker, *Org. React.*, **14,** 270 (1965); I. Gosney and A. G. Rowley, in *Organophosphorus Reagents in Organic Synthesis*, J. I. G. Cadogan, ed., Academic Press, London, 1979, pp. 17–153; B. A. Maryanoff and A. B. Reitz, *Chem. Rev.*, **89,** 863 (1989); A. W. Johnson, *Ylides and Imines of Phosphorus*, John Wiley, New York, 1993; N. J. Lawrence, in *Preparation of Alkenes*, Oxford University Press, Oxford, 1996, pp. 19–58; K. C. Nicolaou, M. W. Harter, J. L. Gunzer, and A. Nadin, *Liebigs Ann. Chem.*, 1283 (1997).

NMR spectroscopic studies (^{1}H, ^{13}C, and ^{31}P) are consistent with the dipolar ylide structure and suggest only a minor contribution from the ylene structure.[234] Theoretical calculations support this view.[235] The phosphonium ylides react with carbonyl compounds to give olefins and the phosphine oxide.

$$R_3\overset{+}{P}-\overset{-}{C}R_2 \ + \ R_2'C{=}O \longrightarrow R_2C{=}CR_2' \ + \ R_3P{=}O$$

There are related reactions involving phosphonate esters or phosphines oxides. These reactions differ from the Wittig reaction in that they involve *carbanions* formed by deprotonation. In the case of the phosphonate esters, a second EWG substituent is usually present.

$$\underset{(R'O)_2PCH_2\text{-}EWG}{\overset{\overset{\displaystyle O}{\parallel}}{}} \xrightarrow{\text{base}} \underset{(R'O)_2P\underset{\sim}{C}H\text{-}EWG}{\overset{\overset{\displaystyle O}{\parallel}}{}} \xrightarrow{R_2C{=}O} R_2C{=}CH\text{-}EWG$$

2.4.1.1. Olefination Reactions Involving Phosphonium Ylides. The synthetic potential of phosphonium ylides was developed initially by G. Wittig and his associates at the University of Heidelberg. The reaction of a phosphonium ylide with an aldehyde or ketone introduces a carbon-carbon double bond in place of the carbonyl bond. The mechanism originally proposed involves an addition of the nucleophilic ylide carbon to the carbonyl group to form a dipolar intermediate (a *betaine*), followed by elimination of a phosphine oxide. The elimination is presumed to occur after formation of a four-membered *oxaphosphetane* intermediate. An alternative mechanism proposes direct formation of the oxaphosphetane by a cycloaddition reaction.[236] There have been several computational studies that find the oxaphosphetane structure to be an intermediate.[237] Oxaphosphetane intermediates have been observed by NMR studies at low temperature.[238] Betaine intermediates have been observed only under special conditions that retard the cyclization and elimination steps.[239]

234. H. Schmidbaur, W. Bucher, and D. Schentzow, *Chem. Ber.*, **106**, 1251 (1973).
235. A. Streitwieser, Jr., A. Rajca, R. S. McDowell, and R. Glaser, *J. Am. Chem. Soc.*, **109**, 4184 (1987); S. M. Bachrach, *J. Org. Chem.*, **57**, 4367 (1992); D. G. Gilheany, *Chem. Rev.*, **94**, 1339 (1994).
236. E. Vedejs and K. A. J. Snoble, *J. Am. Chem. Soc.*, **95**, 5778 (1973); E. Vedejs and C. F. Marth, *J. Am. Chem. Soc.*, **112**, 3905 (1990).
237. R. Holler and H. Lischka, *J. Am. Chem. Soc.*, **102**, 4632 (1980); F. Volatron and O. Eisenstein, *J. Am. Chem. Soc.*, **106**, 6117 (1984); F. Mari, P. M. Lahti, and W. E. McEwen, *J. Am. Chem. Soc.*, **114**, 813 (1992); A. A. Restrepocossio, C. A. Gonzalez, and F. Mari, *J. Phys. Chem. A*, **102**, 6993 (1998); H. Yamataka and S. Nagase, *J. Am. Chem. Soc.*, **120**, 7530 (1998).
238. E. Vedejs, G. P. Meier, and K. A. J. Snoble, *J. Am. Chem. Soc.*, **103**, 2823 (1981); B. E. Maryanoff, A. B. Reitz, M. S. Mutter, R. R. Inners, H. R. Almond, Jr., R. R. Whittle, and R. A. Olofson, *J. Am. Chem. Soc.*, **108**, 7684 (1986).
239. R. A. Neumann and S. Berger, *Eur. J. Org. Chem.*, 1085 (1998).

Phosphonium ylides are usually prepared by deprotonation of phosphonium salts. The phosphonium salts that are used most often are alkyltriphenylphosphonium halides, which can be prepared by the reaction of triphenylphosphine and an alkyl halide. The alkyl halide must be reactive toward S_N2 displacement.

$$Ph_3P \ + \ RCH_2X \ \longrightarrow \ Ph_3\overset{+}{P}-CH_2R \ X^-$$

$$X = I, \ Br, \ or \ Cl$$

$$Ph_3\overset{+}{P}\overset{-}{C}H_2R \ \xrightarrow{\text{base}} \ Ph_3P{=}CHR$$

Alkyltriphenylphosphonium halides are only weakly acidic, and a strong base must be used for deprotonation. Possibilities include organolithium reagents, the anion of dimethyl sulfoxide, and amide ion or substituted amide anions, such as LDA or NaHMDS. The ylides are not normally isolated, so the reaction is carried out either with the carbonyl compound present or with it added immediately after ylide formation. Ylides with nonpolar substituents, e.g., R = H, alkyl, aryl, are quite reactive toward both ketones and aldehydes. Ylides having an α-EWG substituent, such as alkoxycarbonyl or acyl, are less reactive and are called *stabilized ylides*.

The stereoselectivity of the Wittig reaction is believed to be the result of steric effects that develop as the ylide and carbonyl compound approach one another. The three phenyl substituents on phosphorus impose large steric demands that govern the formation of the diastereomeric adducts.[240] Reactions of unstabilized phosphoranes are believed to proceed through an early TS, and steric factors usually make these reactions selective for the *cis*-alkene.[241] Ultimately, however, the precise stereoselectivity is dependent on a number of variables, including reactant structure, the base used for ylide formation, the presence of other ions, solvent, and temperature.[242]

Scheme 2.17 gives some examples of Wittig reactions. Entries 1 to 5 are typical examples of using ylides without any functional group stabilization. The stereoselectivity depends strongly on both the structure of the ylide and the reaction conditions. Use of sodium amide or NaHMDS as bases gives higher selectivity for Z-alkenes than do ylides prepared with alkyllithium reagents as base (see Entries 3 to 6). Benzylidenetriphenylphosphorane (Entry 6) gives a mixture of both *cis*- and *trans*-stilbene on reaction with benzaldehyde. The diminished stereoselectivity is attributed to complexes involving the lithium halide salt that are present when alkyllithium reagents are used as bases.

β-Ketophosphonium salts are considerably more acidic than alkylphosphonium salts and can be converted to ylides by relatively weak bases. The resulting ylides, which are stabilized by the carbonyl group, are substantially less reactive than unfunctionalized ylides. More vigorous conditions are required to bring about reactions with ketones. Stabilized ylides such as (carboethoxymethylidene)triphenylphosphorane (Entries 8 and 9) react with aldehydes to give exclusively *trans* double bonds.

[240] M. Schlosser, *Top. Stereochem.*, **5**, 1 (1970); M. Schlosser and B. Schaub, *J. Am. Chem. Soc.*, **104**, 5821 (1982); H. J. Bestmann and O. Vostrowsky, *Top. Curr. Chem.*, **109**, 85 (1983); E. Vedejs, T. Fleck, and S. Hara, *J. Org. Chem.*, **52**, 4637 (1987).

[241] E. Vedejs, C. F. Marth, and P. Ruggeri, *J. Am. Chem. Soc.*, **110**, 3940 (1988); E. Vedejs and C. F. Marth, *J. Am. Chem. Soc.*, **110**, 3948 (1988); E. Vedejs and C. F. Marth, *J. Am. Chem. Soc.*, **112**, 3905 (1990).

[242] A. B. Reitz, S. O. Nortey, A. D. Jordan, Jr., M. S. Mutter, and B. E. Maryanoff, *J. Org. Chem.*, **51**, 3302 (1986); B. E. Maryanoff and A. B. Reitz, *Chem. Rev.*, **89**, 863 (1989); E. Vedejs and M. J. Peterson, *Adv. Carbanion Chem.*, **2**, 1 (1996); E. Vedejs and M. J. Peterson, *Top. Stereochem.*, **21**, 1 (1994).

Scheme 2.17. The Wittig Reaction

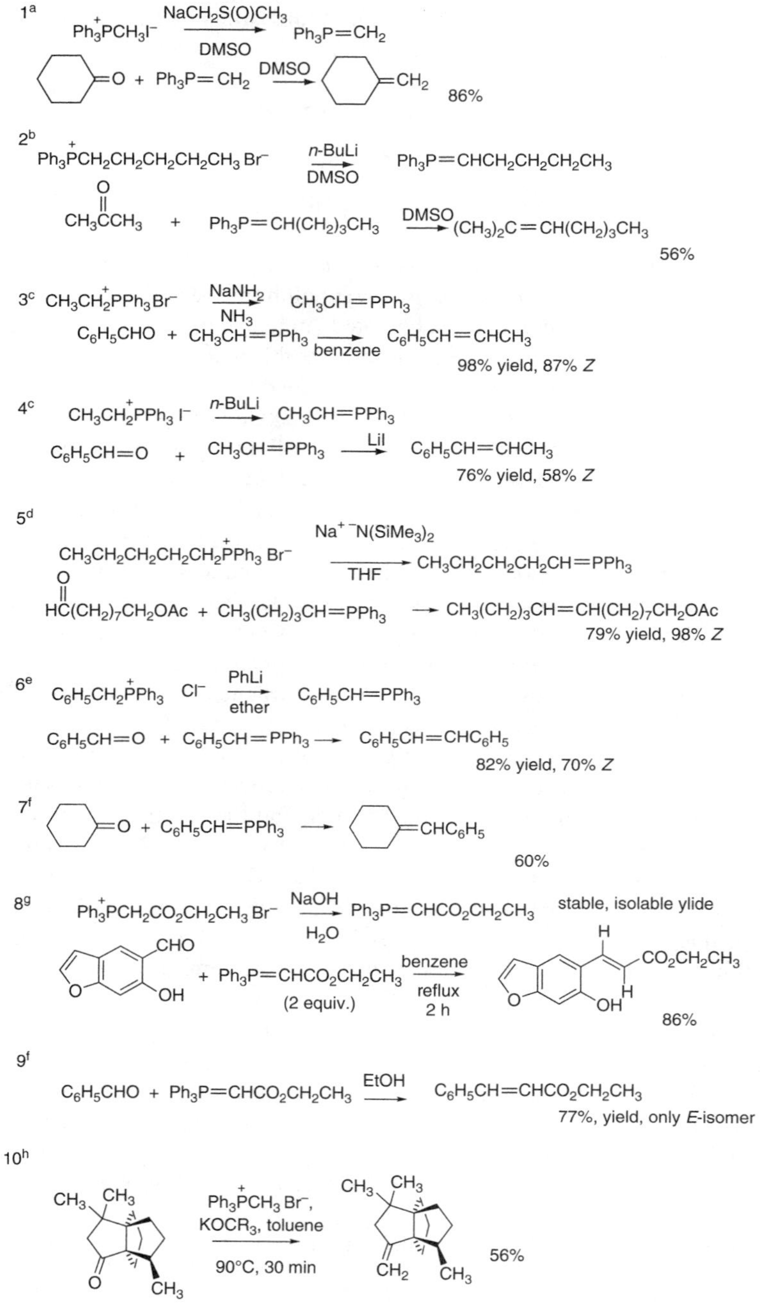
1a
Ph3P+CH3I−
NaCH2S(O)CH3
DMSO
Ph3P=CH2
=O + Ph3P=CH2
DMSO
=CH2
86%
2b
Ph3P+CH2CH2CH2CH2CH3 Br−
n-BuLi
DMSO
Ph3P=CHCH2CH2CH2CH3
CH3CCH3 + Ph3P=CH(CH2)3CH3
DMSO
(CH3)2C=CH(CH2)3CH3
56%
3c CH3CH2P+Ph3Br−
NaNH2
NH3
CH3CH=PPh3
C6H5CHO + CH3CH=PPh3
benzene
C6H5CH=CHCH3
98% yield, 87% Z
4c CH3CH2P+Ph3 I−
n-BuLi
CH3CH=PPh3
C6H5CH=O + CH3CH=PPh3
LiI
C6H5CH=CHCH3
76% yield, 58% Z
5d
CH3CH2CH2CH2CH2P+Ph3 Br−
Na+ −N(SiMe3)2
THF
CH3CH2CH2CH2CH=PPh3
HC(CH2)7CH2OAc + CH3(CH2)3CH=PPh3
CH3(CH2)3CH=CH(CH2)7CH2OAc
79% yield, 98% Z
6e C6H5CH2P+Ph3 Cl−
PhLi
ether
C6H5CH=PPh3
C6H5CH=O + C6H5CH=PPh3
C6H5CH=CHC6H5
82% yield, 70% Z
7f
=O + C6H5CH=PPh3
=CHC6H5
60%
8g Ph3P+CH2CO2CH2CH3 Br−
NaOH
H2O
Ph3P=CHCO2CH2CH3
stable, isolable ylide
CHO
OH
+ Ph3P=CHCO2CH2CH3
(2 equiv.)
benzene
reflux
2 h
CO2CH2CH3
86%
9f
C6H5CHO + Ph3P=CHCO2CH2CH3
EtOH
C6H5CH=CHCO2CH2CH3
77%, yield, only E-isomer
10h
Ph3P+CH3 Br−,
KOCR3, toluene
90°C, 30 min
CH2
56%

(*Continued*)

Scheme 2.17. (*Continued*)

161

SECTION 2.4

*Olefination Reactions of
Stabilized Carbon
Nucleophiles*

11[i]

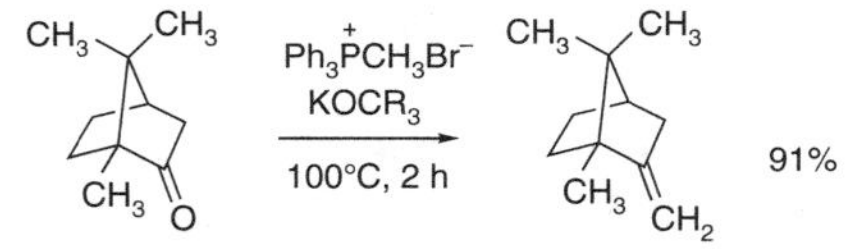

91%

12[b]

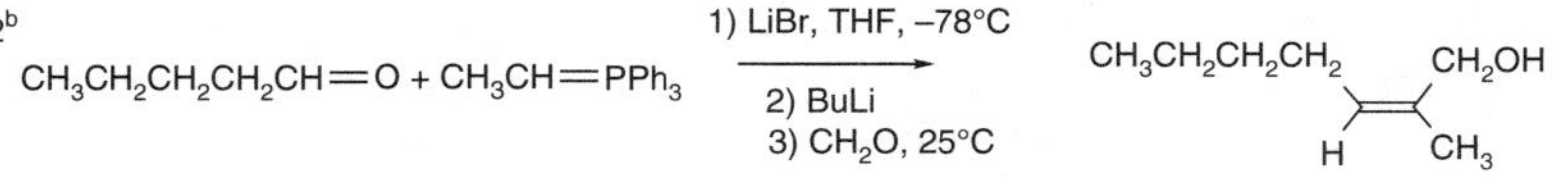

CH$_3$CH$_2$CH$_2$CH$_2$CH=O + CH$_3$CH=PPh$_3$

1) LiBr, THF, –78°C

2) BuLi

3) CH$_2$O, 25°C

13[j]

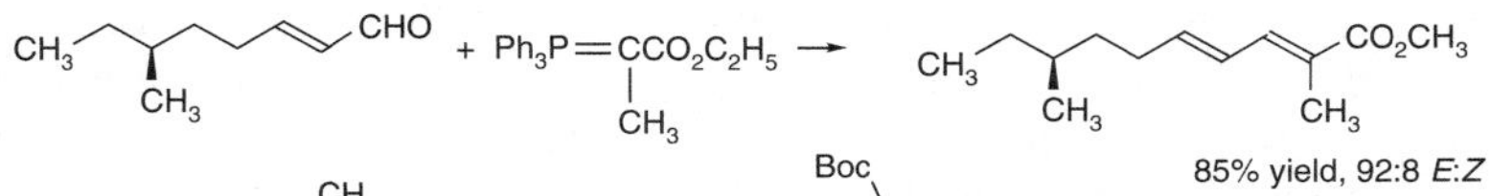

+ Ph$_3$P=CCO$_2$C$_2$H$_5$ →

85% yield, 92:8 *E:Z*

14[k]

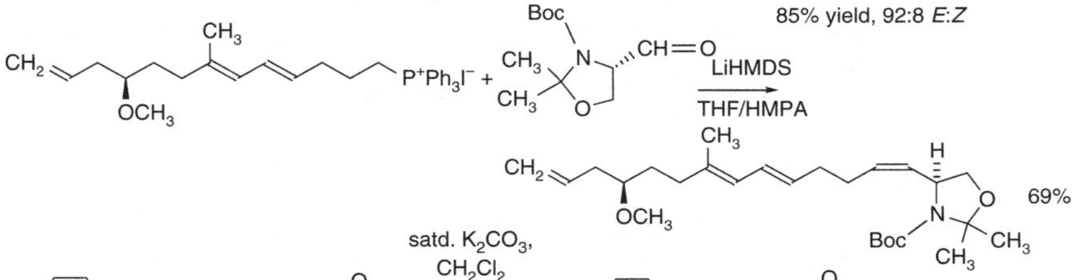

+ LiHMDS

THF/HMPA

69%

15[l]

NC—⟨ ⟩—CHO + Ph$_3$P$^+$CH$_2$

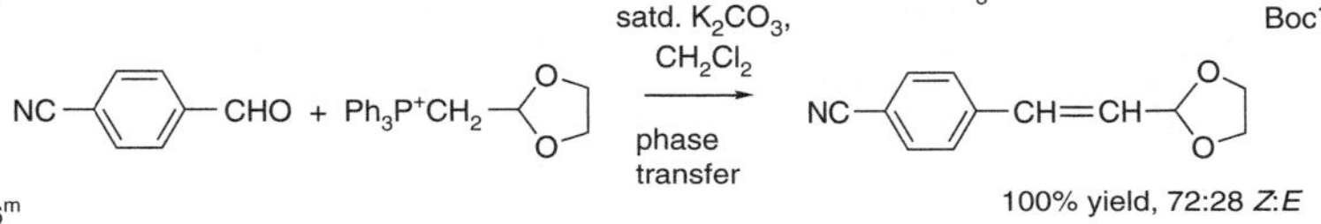

satd. K$_2$CO$_3$,
CH$_2$Cl$_2$

phase
transfer

NC—⟨ ⟩—CH=CH

100% yield, 72:28 *Z:E*

16[m]

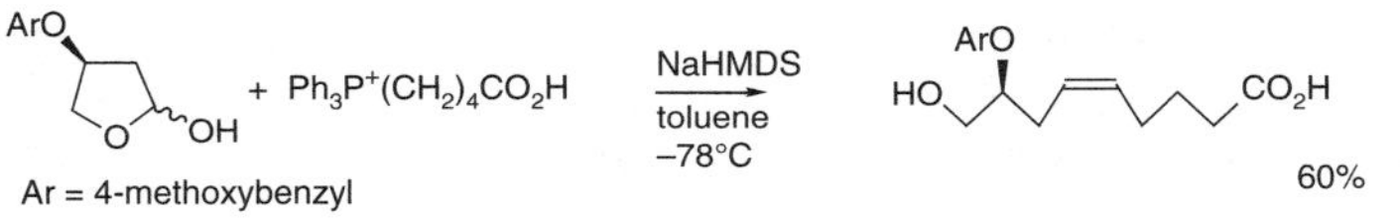

+ Ph$_3$P$^+$(CH$_2$)$_4$CO$_2$H

Ar = 4-methoxybenzyl

NaHMDS
toluene
–78°C

60%

17[n]

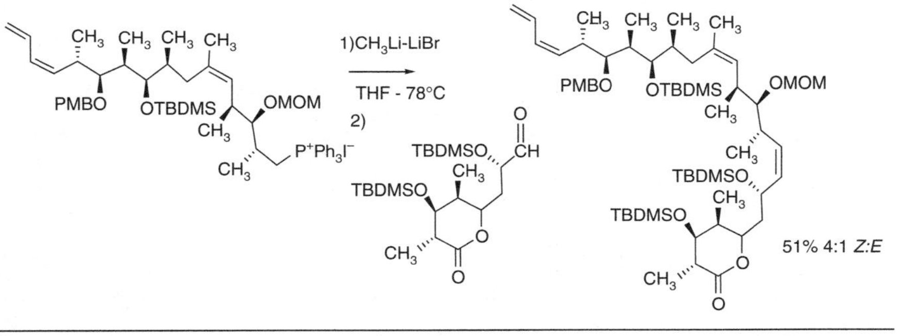

1)CH$_3$Li-LiBr

THF - 78°C

2)

51% 4:1 *Z:E*

a. R. Greenwald, M. Chaykovsky, and E. J. Corey, *J. Org. Chem.*, **28**, 1128 (1963).
b. U. T. Bhalerao and H. Rapoport, *J. Am. Chem. Soc.*, **93**, 4835 (1971).
c. M. Schlosser and K. F. Christmann, *Liebigs Ann. Chem.*, **708**, 1 (1967).
d. H. J. Bestmann, K. H. Koschatzky, and O. Vostrowsky, *Chem. Ber.*, **112**, 1923 (1979).
e. G. Wittig and U. Schollkopf, *Chem. Ber.*, **87**, 1318 (1954).
f. G. Wittig and W. Haag, *Chem. Ber.*, **88**, 1654 (1955).
g. Y. Y. Liu, E. Thom, and A. A. Liebman, *J. Heterocycl. Chem.*, **16**, 799 (1979).
h. A. B. Smith, III, and P. J. Jerris, *J. Org. Chem.*, **47**, 1845 (1982).
i. L. Fitjer and U. Quabeck, *Synth. Commun.*, **15**, 855 (1985).
j. J. D. White, T. S. Kim, and M. Nambu, *J. Am. Chem. Soc.*, **119**, 103 (1997).
k. N. Daubresse, C. Francesch, and G. Rolando, *Tetrahedron*, **54**, 10761 (1998).
l. A. G. M. Barrett, M. Pena, and J. A. Willardsen, *J. Org. Chem.*, **61**, 1082 (1996).
m. D. Critcher, S. Connoll, and M. Wills, *J. Org. Chem.*, **62**, 6638 (1997).
n. A. B. Smith, III, B. S. Freeze, I. Brouard, and T. Hirose, *Org. Lett.*, **5**, 4405 (2003).

When a hindered ketone is to be converted to a methylene derivative, the best results are obtained if potassium *t*-alkoxide is used as the base in a hydrocarbon solvent. Under these conditions the reaction can be carried out at elevated temperatures.[243] Entries 10 and 11 illustrate this procedure.

The reaction of nonstabilized ylides with aldehydes can be induced to yield *E*-alkenes with high stereoselectivity by a procedure known as the *Schlosser modification* of the Wittig reaction.[244] In this procedure, the ylide is generated as a lithium halide complex and allowed to react with an aldehyde at low temperature, presumably forming a mixture of diastereomeric betaine-lithium halide complexes. At the temperature at which the addition is carried out, there is no fragmentation to an alkene and triphenylphosphine oxide. This complex is then treated with an equivalent of strong base such as phenyllithium to form a β-*oxido ylide*. Addition of one equivalent of *t*-butyl alcohol protonates the β-oxido ylide stereoselectivity to give the *syn*-betaine as a lithium halide complex. Warming the solution causes the *syn*-betaine-lithium halide complex to give *trans*-alkene by a *syn* elimination.

An extension of this method can be used to prepare allylic alcohols. Instead of being protonated, the β-oxido ylide is allowed to react with formaldehyde. The β-oxido ylide and formaldehyde react to give, on warming, an allylic alcohol. Entry 12 is an example of this reaction. The reaction is valuable for the stereoselective synthesis of *Z*-allylic alcohols from aldehydes.[245]

The Wittig reaction can be applied to various functionalized ylides.[246] Methoxymethylene and phenoxymethylene ylides lead to vinyl ethers, which can be hydrolyzed to aldehydes.[247]

243. J. M. Conia and J. C. Limasset, *Bull. Soc. Chim. France*, 1936 (1967); J. Provin, F. Leyendecker, and J. M. Conia, *Tetrahedron Lett.*, 4053 (1975); S. R. Schow and T. C. Morris, *J. Org. Chem.*, **44**, 3760 (1979).

244. M. Schlosser and K.-F. Christmann, *Liebigs Ann. Chem.*, **708**, 1 (1967); M. Schlosser, K.-F. Christmann, and A. Piskala, *Chem. Ber.*, **103**, 2814 (1970).

245. E. J. Corey and H. Yamamoto, *J. Am. Chem. Soc.*, **92**, 226 (1970); E. J. Corey, H. Yamamoto, D. K. Herron, and K. Achiwa, *J. Am. Chem. Soc.*, **92**, 6635 (1970); E. J. Corey and H. Yamamoto, *J. Am. Chem. Soc.*, **92**, 6636 (1970); E. J. Corey and H. Yamamoto, *J. Am. Chem. Soc.*, **92**, 6637 (1970); E. J. Corey, J. I. Shulman, and H. Yamamoto, *Tetrahedron Lett.*, 447 (1970).

246. S. Warren, *Chem. Ind. (London)*, 824 (1980).

247. S. G. Levine, *J. Am. Chem. Soc.*, **80**, 6150 (1958); G. Wittig, W. Boll, and K. H. Kruck, *Chem. Ber.*, **95**, 2514 (1962).

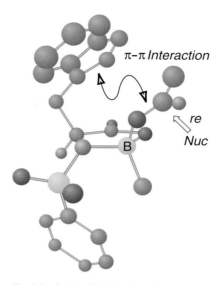

Fig. 2.2. Origin of facial selectivity in indolyl-methyloxaborazolidinone structure. Reproduced from *Tetrahedron: Asymmetry*, **9**, 357 (1998), by permission of Elsevier.

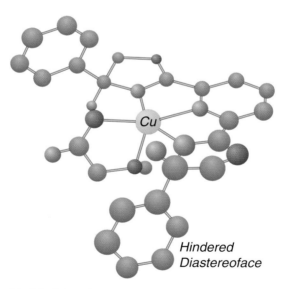

Fig. 2.3. Origin of facial selectivity of *bis*-oxazoline catalyst. Reproduced from *Tetrahedron: Asymmetry*, **9**, 357 (1998), by permission of Elsevier.

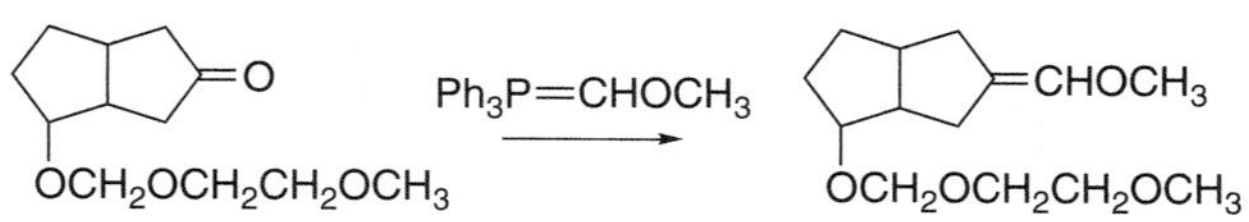

Ref. 248

2-(1,3-Dioxolanyl)methyl ylides can be used for the introduction of α, β-unsaturated aldehydes (see Entry 15, Scheme 2.17). Methyl ketones can be prepared by a reaction using the α-methoxyethylidene phosphorane.

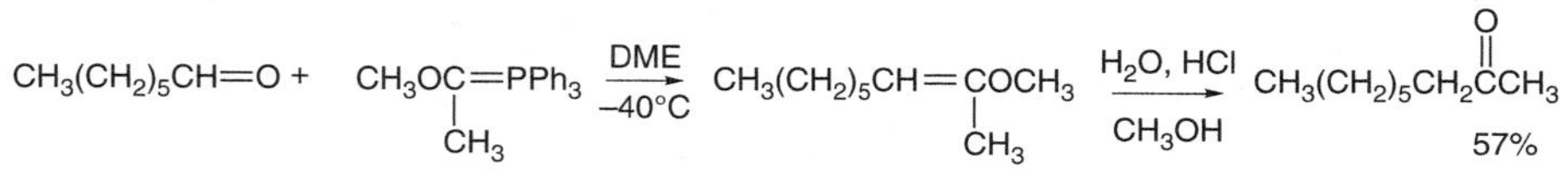

Ref. 249

There have been many applications of the Wittig reaction in multistep syntheses. The reaction can be used to prepare extended conjugated systems, such as crocetin dimethyl ester, which has seven conjugated double bonds. In this case, two cycles of Wittig reactions using stabilized ylides provided the seven double bonds. Note the use of a conjugated stabilized ylide in the second step.[250]

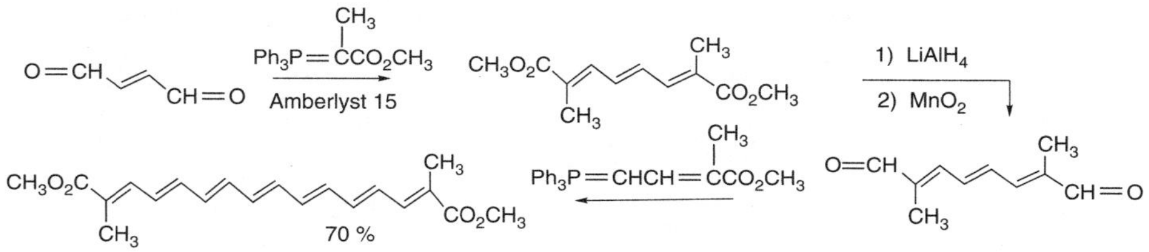

In several cases of syntheses of highly functionalized molecules, use of CH_3Li-LiBr for ylide formation has been found to be advantageous. For example, in the synthesis of milbemycin D, Crimmins and co-workers obtained an 84% yield with 10:1 Z:E selectivity.[251] In this case, the more stable E-isomer was required and it was obtained by I_2-catalyzed isomerization.

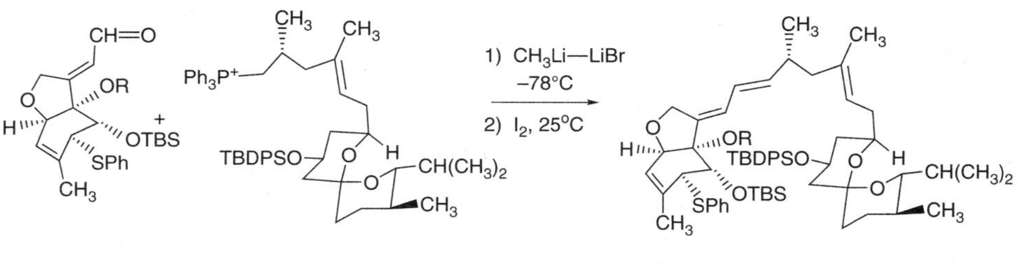

248. M. Yamazaki, M. Shibasaki, and S. Ikegami, *J. Org. Chem.*, **48**, 4402 (1983).

249. D. R. Coulsen, *Tetrahedron Lett.*, 3323 (1964).

250. D. Frederico, P. M. Donate, M. G. Constantino, E. S. Bronze, and M. I. Sairre, *J. Org. Chem.*, **68**, 9126 (2003).

251. M. T. Crimmins, R. S. Al-awar, I. M. Vallin, W. G. Hollis, Jr., R. O'Mahony, J. G. Lever, and D. M. Bankaitis-Davis, *J. Am. Chem. Soc.*, **118**, 7513 (1996).

This methodology was also used in the connecting of two major fragments in the synthesis of spongistatins.[252]

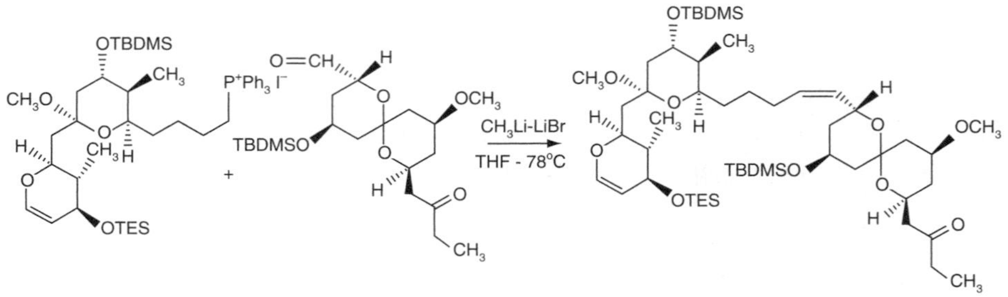

These conditions were also employed for a late stage of the synthesis of (+)-discodermolide (see Entry 17, Scheme 2.17).

2.4.1.2. Olefination Reactions Involving Phosphonate Anions. An important complement to the Wittig reaction involves the reaction of phosphonate carbanions with carbonyl compounds.[253] The alkylphosphonic acid esters are made by the reaction of an alkyl halide, preferably primary, with a phosphite ester. Phosphonate carbanions are generated by treating alkylphosphonate esters with a base such as sodium hydride, *n*-butyllithium, or sodium ethoxide. Alumina coated with KF or KOH has also found use as the base.[254]

$$RCH_2X + P(OC_2H_5)_3 \longrightarrow RCH_2\overset{O}{\overset{\|}{P}}(OC_2H_5)_2 + C_2H_5X$$

$$RCH_2\overset{O}{\overset{\|}{P}}(OC_2H_5)_2 \xrightarrow{\text{base}} R\bar{C}H\overset{O}{\overset{\|}{P}}(OC_2H_5)_2$$

$$R\bar{C}H\overset{O}{\overset{\|}{P}}(OC_2H_5)_2 + R'_2C{=}O \longrightarrow \underset{R'_2C-CHR}{{}^{-}O\quad \overset{O}{\overset{\|}{P}}(OC_2H_5)_2} \longrightarrow R'_2C{=}CHR + (C_2H_5O)_2\overset{O}{\overset{\|}{P}}{-}O^-$$

Reactions with phosphonoacetate esters are used frequently to prepare α,β-unsaturated esters. This reaction is known as the *Wadsworth-Emmons reaction* and usually leads to the *E*-isomer.

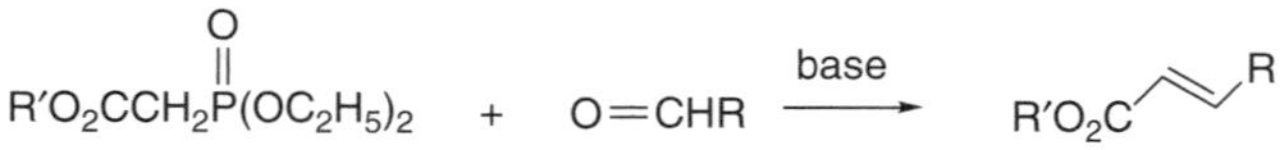

The conditions can be modified to favor the *Z*-isomer. Use of KHMDS with 18-crown-6 favors the *Z*-product.[255] This method was used, for example, to control the

252. M. T. Crimmins, J. D. Katz, D. G. Washburn, S. P. Allwein, and L. F. McAtee, *J. Am. Chem. Soc.*, **124**, 5661 (2002); see also C. H. Heathcock, M. McLaughlin, J. Medina, J. L. Hubbs, G. A. Wallace, R. Scott, M. M. Claffey, C. J. Hayes, and G. R. Ott, *J. Am. Chem. Soc.*, **125**, 12844 (2003).
253. For reviews of reactions of phosphonate carbanions with carbonyl compounds, see J. Boutagy and R. Thomas, *Chem. Rev.*, **74**, 87 (1974); W. S. Wadsworth, Jr., *Org. React.*, **25**, 73 (1977); H. Gross and I. Keitels, *Z. Chem.*, **22**, 117 (1982).
254. F. Texier-Boullet, D. Villemin, M. Ricard, H. Moison, and A. Foucaud, *Tetrahedron*, **41**, 1259 (1985); M. Mikolajczyk and R. Zurawinski, *J. Org. Chem.*, **63**, 8894 (1998).
255. W. C. Still and C. Gennari, *Tetrahedron Lett.*, **24**, 4405 (1983).

stereochemistry in the synthesis of the *Z*- and *E*-isomers of β-santalol, a fragrance that is a component of sandalwood oil.

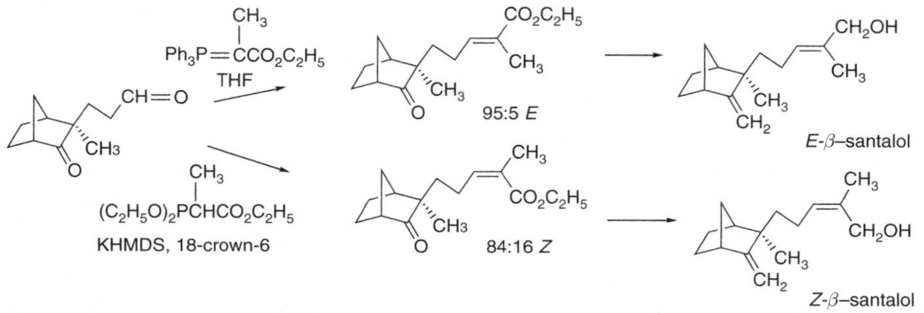

Ref. 256

Several modified phosphonoacetate esters show selectivity for the *Z*-enoate product. Trifluoroethyl,[256] phenyl,[257] 2-methylphenyl,[258] and 2,6-difluorophenyl[259] esters give good *Z*-stereoselectivity with aldehydes. The trifluoroethyl esters also give *Z*-selectivity with ketones.[260]

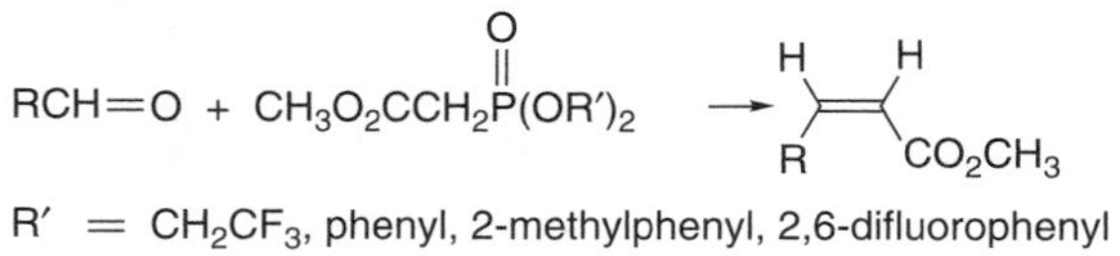

R' = CH_2CF_3, phenyl, 2-methylphenyl, 2,6-difluorophenyl

Several other methodologies have been developed for control of the stereoselectivity of Wadsworth-Emmons reactions. For example, K_2CO_3 in chlorobenzene with a catalytic amount of 18-crown-6 is reported to give excellent *Z*-selectivity.[261] Another group found that use of excess Na^+, added as NaI, improved *Z*-selectivity for 2-methylphenyl esters.

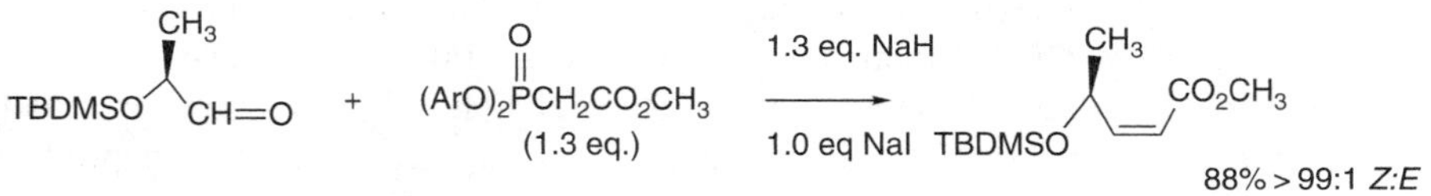

88% > 99:1 *Z:E*

An alternative procedure for effecting the condensation of phosphonoacetates is to carry out the reaction in the presence of lithium chloride and an amine such as diisopropylethylamine. The lithium chelate of the substituted phosphonate is sufficiently acidic to be deprotonated by the amine.[262]

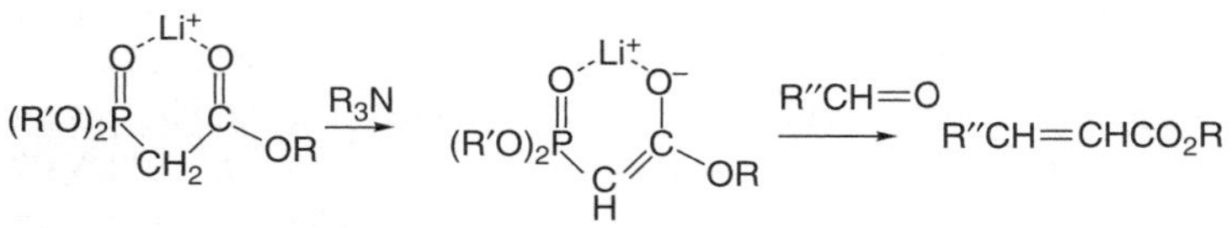

256. A. Krotz and G. Helmchen, *Liebigs Ann. Chem.*, 601 (1994).
257. K. Ando, *Tetrahedron Lett.*, **36**, 4105 (1995); K. Ando, *J. Org. Chem.*, **63**, 8411 (1998).
258. K. Ando, *J. Org. Chem.*, **62**, 1934 (1997); K. Ando, T. Oishi, M. Hirama, H. Ohno, and T. Ibuka, *J. Org. Chem.*, **65**, 4745 (2000).
259. K. Kokin, J. Motoyoshiya, S. Hayashi, and H. Aoyama, *Synth. Commun.*, **27**, 2387 (1997).
260. S. Sano, K. Yokoyama, M. Shiro, and Y. Nagao, *Chem. Pharm. Bull.*, **50**, 706 (2002).
261. F. P. Touchard, *Tetrahedron Lett.*, **45**, 5519 (2004).
262. M. A. Blanchette, W. Choy, J. T. Davis, A. P. Essenfeld, S. Masamune, W. R. Roush, and T. Sakai, *Tetrahedron Lett.*, **25**, 2183 (1984).

This version of the Wadsworth-Emmons reaction has been used in the scaled-up syntheses of drugs and drug-candidate molecules. For example, it is used to prepare a cinnamate ester that is a starting material for pilot plant synthesis of a potential integrin antagonist.[263]

Entries 10 and 11 of Scheme 2.18 also illustrate this procedure.

Scheme 2.18 gives some representative olefination reactions of phosphonate anions. Entry 1 represents a typical preparative procedure. Entry 2 involves formation of a 2,4-dienoate ester using an α,β-unsaturated aldehyde. Diethyl benzylphosphonate can be used in the Wadsworth-Emmons reaction, as illustrated by Entry 3. Entries 4 to 6 show other anion-stabilizing groups. Intramolecular reactions can be used to prepare cycloalkenes.[264]

Ref. 265

Intramolecular condensation of phosphonate carbanions with carbonyl groups carried out under conditions of high dilution have been utilized in macrocycle syntheses. Entries 7 and 8 show macrocyclizations involving the Wadsworth-Emmons reaction. Entries 9 to 11 illustrate the construction of new double bonds in the course of a multistage synthesis. The LiCl/amine conditions are used in Entries 9 and 10.

The stereoselectivity of the reactions of stabilized phosphonate anions is usually considered to be the result of reversible adduct formation, followed by rate/product-controlling elimination that favors the *E*-isomer. This matter has been investigated by computation. The Wadsworth-Emmons reaction between lithio methyl dimethylphosphonoacetate and acetaldehyde has been modeled at the HF/6-31G* level. Energies were also calculated at the B3LYP/6-31G* level.[266] The energy profile for the intermediates and TSs are shown in Figure 2.5. In agreement with the prevailing experimental interpretation, the highest barrier is for formation of the oxaphosphetane and the addition step is reversible. The stereochemistry, then, is determined by the relative ease of formation of the stereoisomeric oxaphosphetanes. The oxaphosphetane species is of marginal stability and proceeds rapidly to product. At the B3LYP/6-31 + G* level, TS2$_{trans}$ is 2.2 kcal/mol more stable than TS2$_{cis}$. The path to the *cis* product encounters two additional small barriers associated with slightly stable stereoisomeric

[263.] J. D. Clark, G. A. Weisenburger, D. K. Anderson, P.-J. Colson, A. D. Edney, D. J. Gallagher, H. P. Kleine, C. M. Knable, M. K. Lantz, C. M. V. Moore, J. B. Murphy, T. E. Rogers, P. G. Ruminski, A. S. Shah, N. Storer, and B. E. Wise, *Org. Process Res. Devel.*, **8**, 51 (2004).
[264.] K. B. Becker, *Tetrahedron*, **36**, 1717 (1980).
[265.] P. A. Grieco and C. S. Pogonowski, *Synthesis*, 425 (1973).
[266.] K. Ando, *J. Org. Chem.*, **64**, 6815 (1999).

Scheme 2.18. Carbonyl Olefination Using Phosphonate Carbanions

1[a]

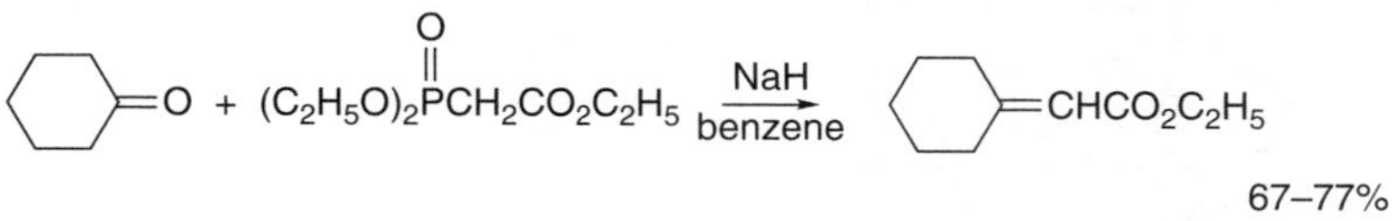

67–77%

2[b]

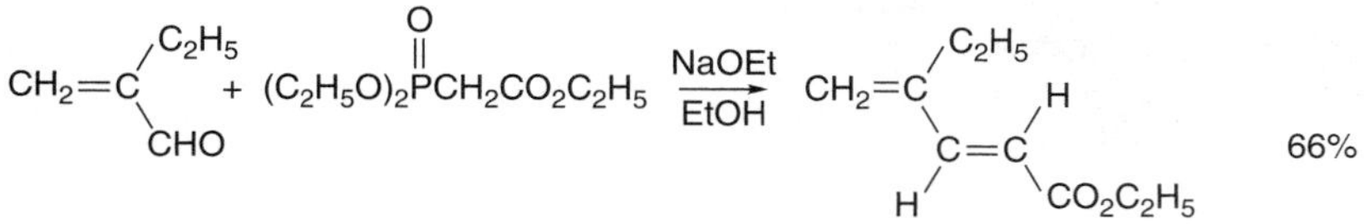

66%

3[c]

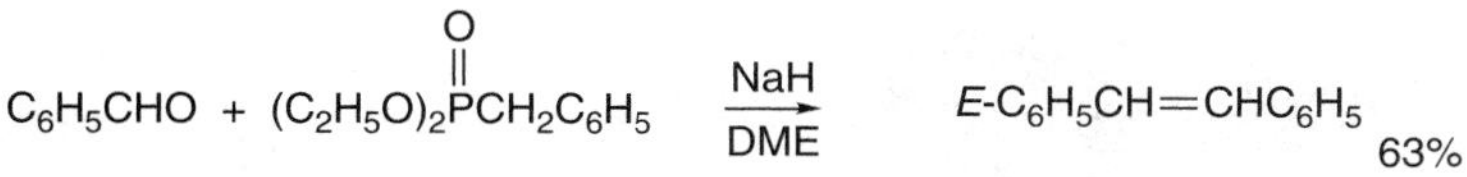

$$C_6H_5CHO + (C_2H_5O)_2\overset{O}{\overset{\|}{P}}CH_2C_6H_5 \xrightarrow[DME]{NaH} E\text{-}C_6H_5CH\text{=}CHC_6H_5$$

63%

4[d]

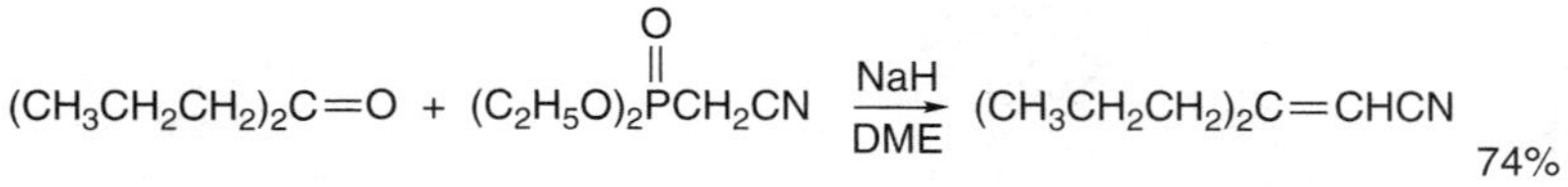

$$(CH_3CH_2CH_2)_2C\text{=}O + (C_2H_5O)_2\overset{O}{\overset{\|}{P}}CH_2CN \xrightarrow[DME]{NaH} (CH_3CH_2CH_2)_2C\text{=}CHCN$$

74%

5[e]

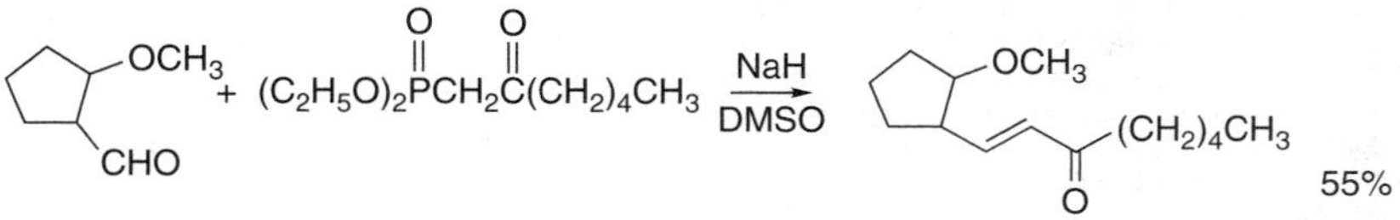

55%

6[f]

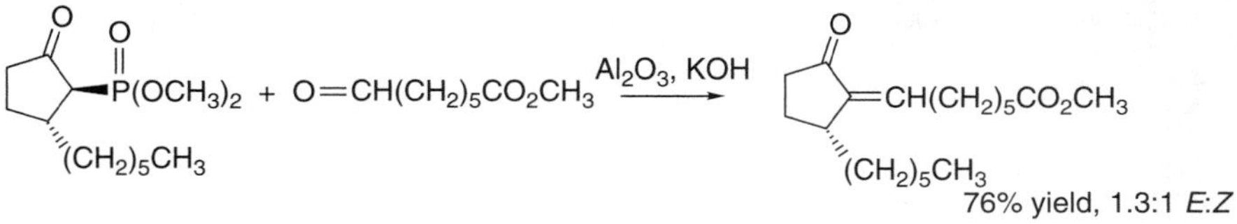

76% yield, 1.3:1 *E:Z*

7[g]

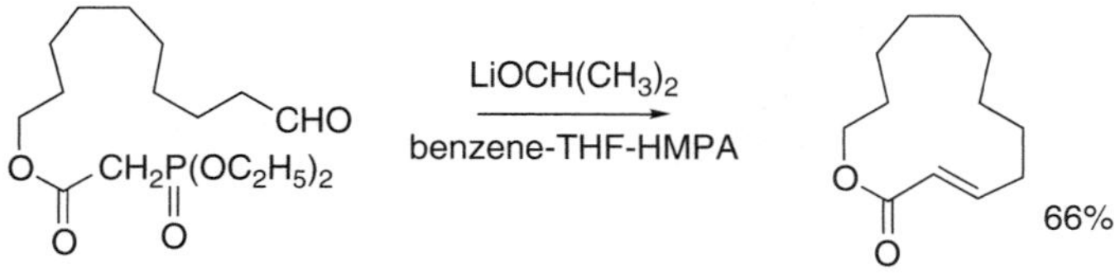

66%

8[h]

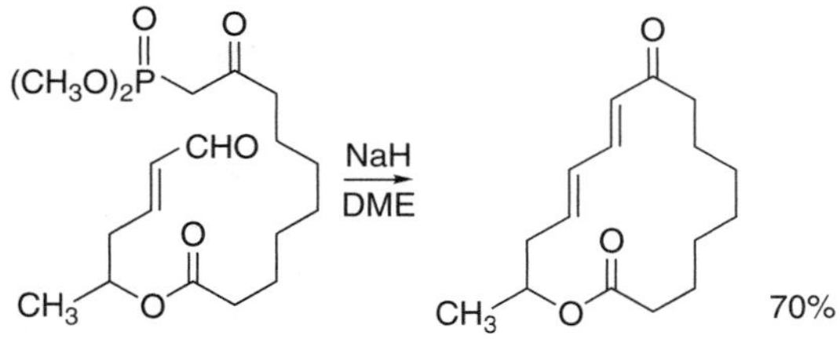

70%

9[i]

70%

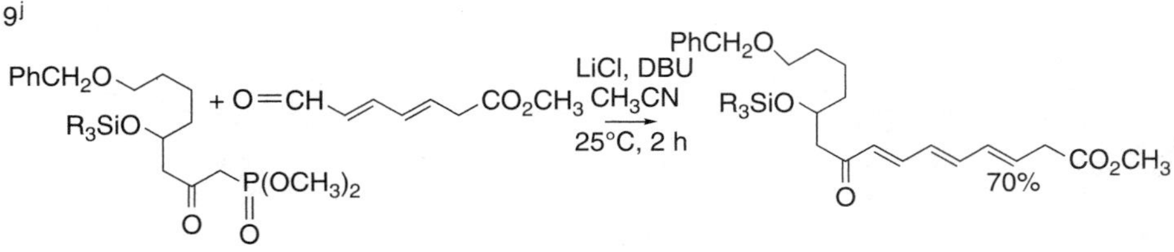

(*Continued*)

Scheme 2.18. (*Continued*)

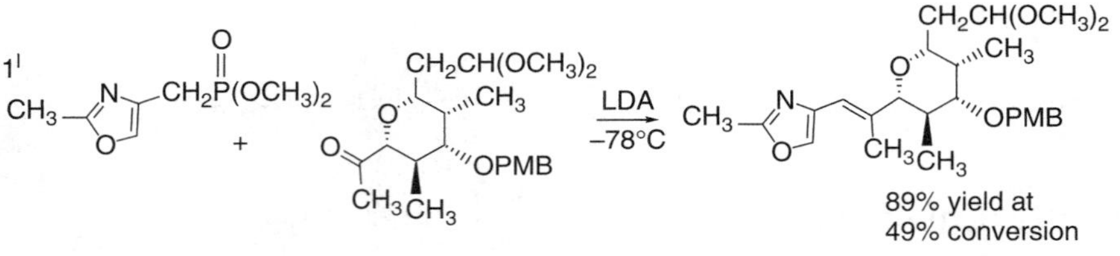

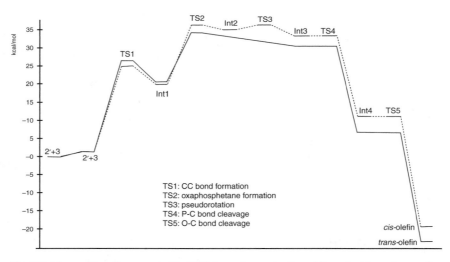

a. W. S. Wadsworth, Jr., and W. D. Emmons, *Org. Synth.*, **45**, 44 (1965).
b. R. J. Sundberg, P. A. Bukowick, and F. O. Holcombe, *J. Org. Chem.*, **32**, 2938 (1967).
c. W. S. Wadsworth, Jr., and W. D. Emmons, *J. Am. Chem. Soc.*, **83**, 1733 (1961).
d. J. A. Marshall, C. P. Hagan, and G. A. Flynn, *J. Org. Chem.*, **40**, 1162 (1975).
e. N. Finch, J. J. Fitt, and I. H. S. Hsu, *J. Org. Chem.*, **40**, 206 (1975).
f. M Mikolajczyk and R. Zurawski, *J. Org. Chem.*, **63**, 8894 (1998).
g. G. M. Stork and E. Nakamura, *J. Org. Chem.*, **44**, 4010 (1979).
h. K. C. Nicolaou, S. P. Seitz, M. R. Pavia, and N. A. Petasis, *J. Org. Chem.*, **44**, 4010 (1979).
i. M. A. Blanchette, W. Choy, J. T. Davis, A. P. Essenfeld, S. Masamune, W. R. Roush, and T. Sakai, *Tetrahedron Lett.*, **25**, 2183 (1984).
j. G. E. Keck and J. A. Murry, *J. Org. Chem.*, **56**, 6606 (1991).
k. G. Pattenden, M. A. Gonzalez, P. B. Little, D. S. Millan, A. T. Plowright, J. A. Tornos, and T. Ye, *Org. Biomolec. Chem.*, **1**, 4173 (2003).

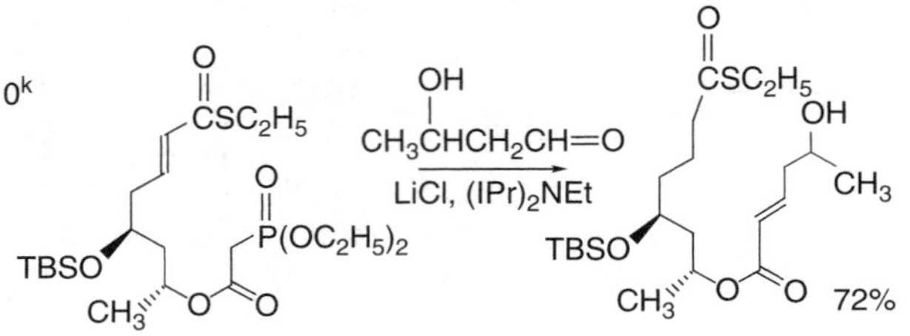

Fig. 2.5. Comparison of energy profile (ΔG) for pathways to *E*- and *Z*-product from the reaction of lithio methyl dimethylphosphonoacetate and acetaldehyde. One molecule of dimethyl ether is coordinated to the lithium ion. Reproduced from *J. Org. Chem.*, **64**, 6815 (1999), by permission of the American Chemical Society.

oxaphosphetane intermediates. The oxaphosphatane is not a stable intermediate on the path to *trans* product.

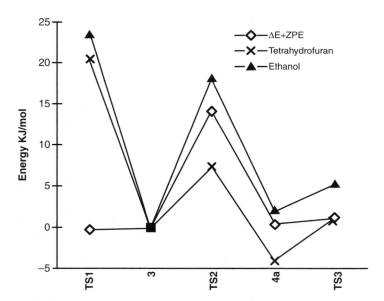

Visual models, additional information and exercises on the Wadsworth-Emmons Reaction can be found in the Digital Resource available at: Springer.com/carey-sundberg.

Fig. 2.6. Free-energy profile (B3LYP/6-31 + G* with ZPE correction) for intermediates and transition structures for Wadsworth-Emmons reactions between the lithium enolate of trimethyl phosphonoacetate anion and formaldehyde in the gas phase and in tetrahydrofuran or ethanol. Adapted from *J. Org. Chem.*, **63**, 1280 (1998), by permission of the American Chemical Society.

Another computational study included a solvation model.[267] Solvation strongly stabilized the oxyanion adduct, suggesting that its formation may be rate and product determining under certain circumstances. When this is true, analysis of stereoselectivity must focus on the addition TS. Figure 2.6 shows the computed energy profile for the TSs and intermediates. TS1 is the structure leading to the oxyanion intermediate. According to the energy profile, its formation is irreversible in solution and therefore determines the product stereochemistry. The structure shows a rather small (30°–35°) dihedral angle and suggests that steric compression would arise with a Z-substituent.

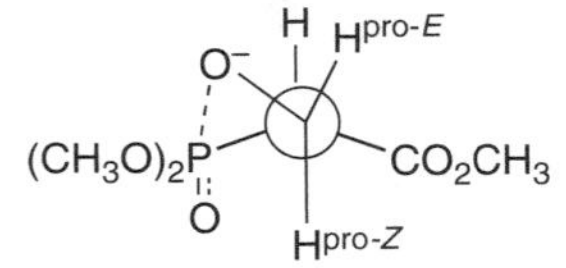

Structure **3** is the intermediate oxyanion adduct. TS2 is the structure leading to cyclization of the oxyanion to the oxaphosphetane. Structure **4a** is the oxaphosphetane, and the computation shows only a small barrier for its conversion to product.

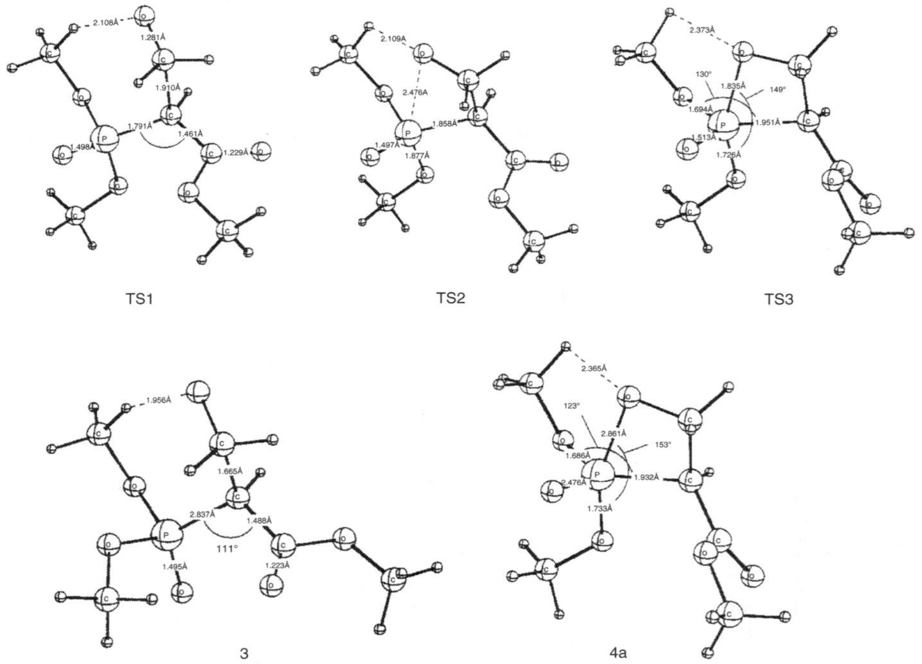

TS1   TS2   TS3

3   4a

Carbanions derived from phosphine oxides also add to carbonyl compounds. The adducts are stable but undergo elimination to form alkene on heating with a base such as sodium hydride. This reaction is known as the *Horner-Wittig* reaction.[268]

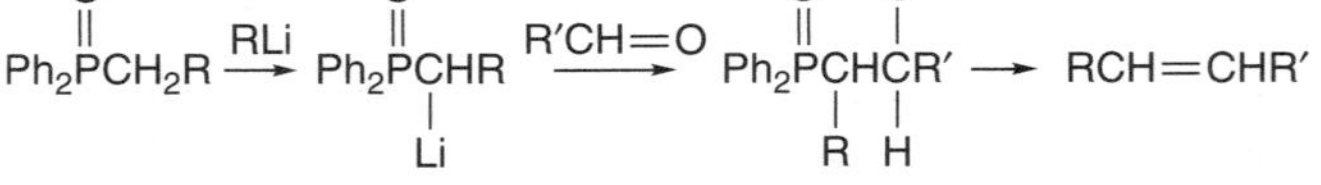

267. P. Brandt, P.-O. Norrby, I. Martin, and T. Rein, *J. Org. Chem.*, **63**, 1280 (1998).
268. For a review, see J. Clayden and S. Warren, *Angew. Chem. Int. Ed. Engl.*, **35**, 241 (1996).

The unique feature of the Horner-Wittig reaction is that the addition intermediate can be isolated and purified, which provides a means for control of the reaction's stereochemistry. It is possible to separate the two diastereomeric adducts in order to prepare the pure alkenes. The elimination process is *syn*, so the stereochemistry of the alkene that is formed depends on the stereochemistry of the adduct. Usually the *anti* adduct is the major product, so it is the *Z*-alkene that is favored. The *syn* adduct is most easily obtained by reduction of β-ketophosphine oxides.[269]

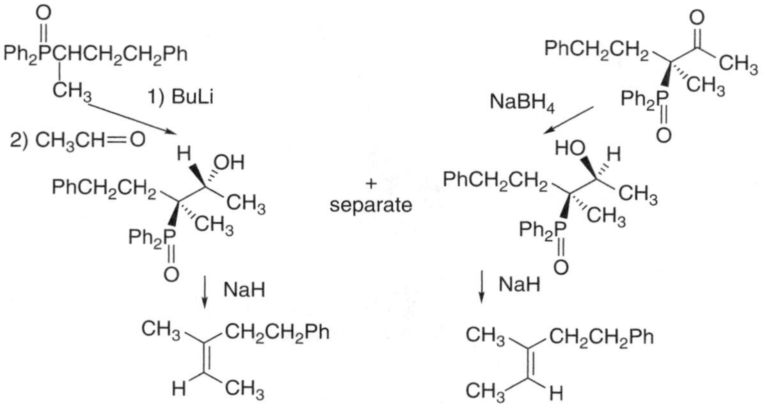

2.4.2. Reactions of α-Trimethylsilylcarbanions with Carbonyl Compounds

Trialkylsilyl groups have a modest stabilizing effect on adjacent carbanions (see Part A, Section 3.4.2). Reaction of the carbanions with carbonyl compounds gives β-hydroxyalkylsilanes. β-Hydroxyalkylsilanes are converted to alkenes by either acid or base.[270] These eliminations provide the basis for a synthesis of alkenes. The reaction is sometimes called the *Peterson reaction*.[271] For example, the Grignard reagent derived from chloromethyltrimethylsilane adds to an aldehyde or ketone and the intermediate can be converted to a terminal alkene by acid or base.[272]

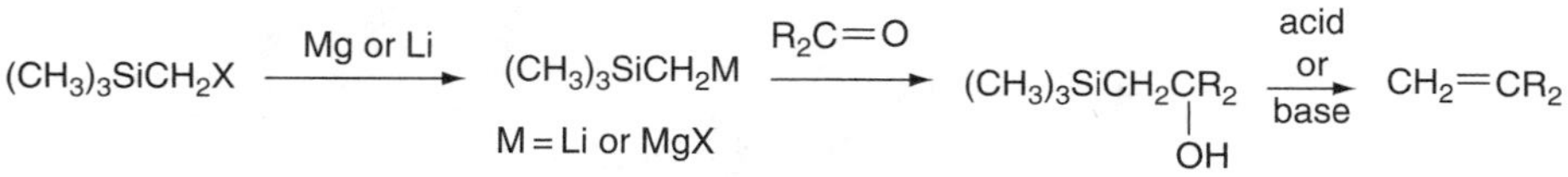

Alternatively, organolithium reagents of the type $(CH_3)_3SiCH(Li)Z$, where Z is a carbanion-stabilizing substituent, can be prepared by deprotonation of $(CH_3)_3SiCH_2Z$ with *n*-butyllithium.

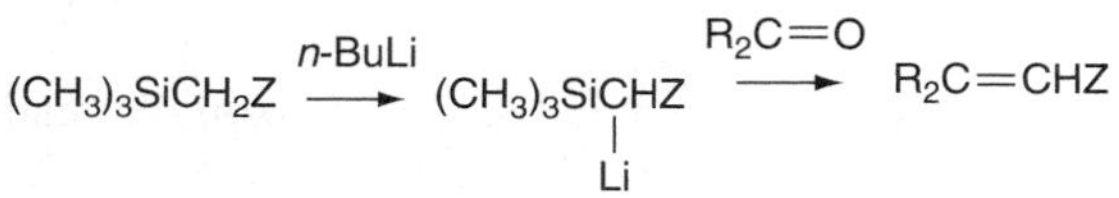

269. A. D. Buss and S. Warren, *J. Chem. Soc., Perkin Trans. 1*, 2307 (1985).
270. P. F. Hudrlik and D. Peterson, *J. Am. Chem. Soc.*, **97**, 1464 (1975).
271. For reviews, see D. J. Ager, *Org. React.*, **38**, 1 (1990); D. J. Ager, *Synthesis*, 384 (1984); A. G. M. Barrett, J. M. Hill, E. M. Wallace, and J. A. Flygare, *Synlett*, 764 (1991).
272. D. J. Peterson, *J. Org. Chem.*, **33**, 780 (1968).

These reagents usually react with aldehydes and ketones to give substituted alkenes directly. No separate elimination step is necessary because fragmentation of the intermediate occurs spontaneously under the reaction conditions.

In general, the elimination reactions are *anti* under acidic conditions and *syn* under basic conditions. This stereoselectivity is the result of a cyclic mechanism under basic conditions, whereas under acidic conditions an acyclic β-elimination occurs.

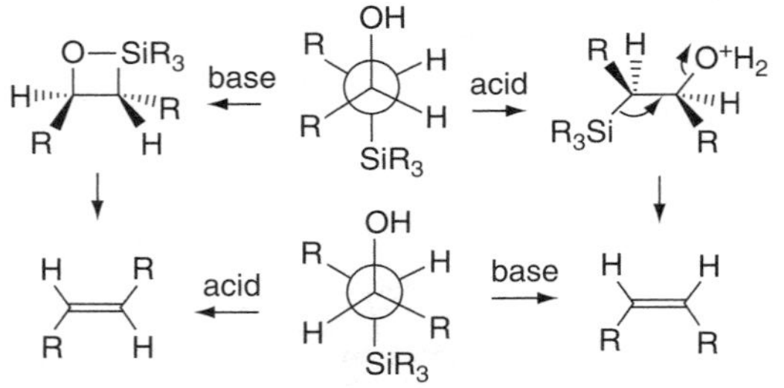

The *anti* elimination can also be achieved by converting the β-silyl alcohols to trifluoroacetate esters.[273] The stereoselectivity of the Peterson olefination depends on the generation of pure *syn* or *anti* β-silylalcohols, so several strategies have been developed for their stereoselective preparation.[274]

There can be significant differences in the rates of elimination of the stereoisomeric β-hydroxysilanes. Van Vranken and co-workers took advantage of such a situation to achieve a highly stereoselective synthesis of a styryl terpene. (The lithiated reactant is prepared by reductive lithiation; see p. 625). The *syn* adduct decomposes rapidly at $-78°$ C but because of steric effects, the *anti* isomer remains unreacted. Acidification then promotes *anti* elimination to the desired *E*-isomer.[275]

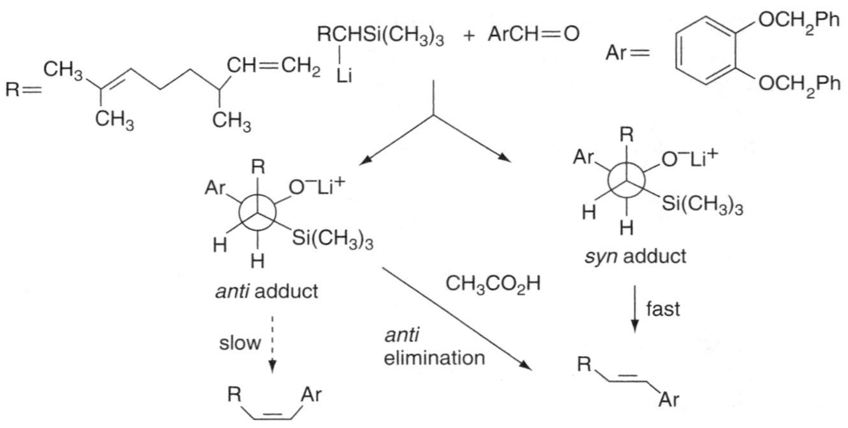

Scheme 2.19 provides some examples of the Peterson olefination. The Peterson olefination has not been used as widely in synthesis as the Wittig and Wadsworth-Emmons reactions, but it has been used advantageously in the preparation of relatively

273. M. F. Connil, B. Jousseaume, N. Noiret, and A. Saux, *J. Org. Chem.*, **59**, 1925 (1994).
274. A. G. M. Barrett and J. A. Flygare, *J. Org. Chem.*, **56**, 638 (1991); L. Duhamel, J. Gralak, and A. Bouyanzer, *J. Chem. Soc., Chem. Commun.*, 1763 (1993).
275. J. B. Perales, N. F. Makino, and D. L. Van Vranken, *J. Org. Chem.*, **67**, 6711 (2002).

Scheme 2.19. Carbonyl Olefination Using Trimethylsilyl-Substituted Organolithium Reagents

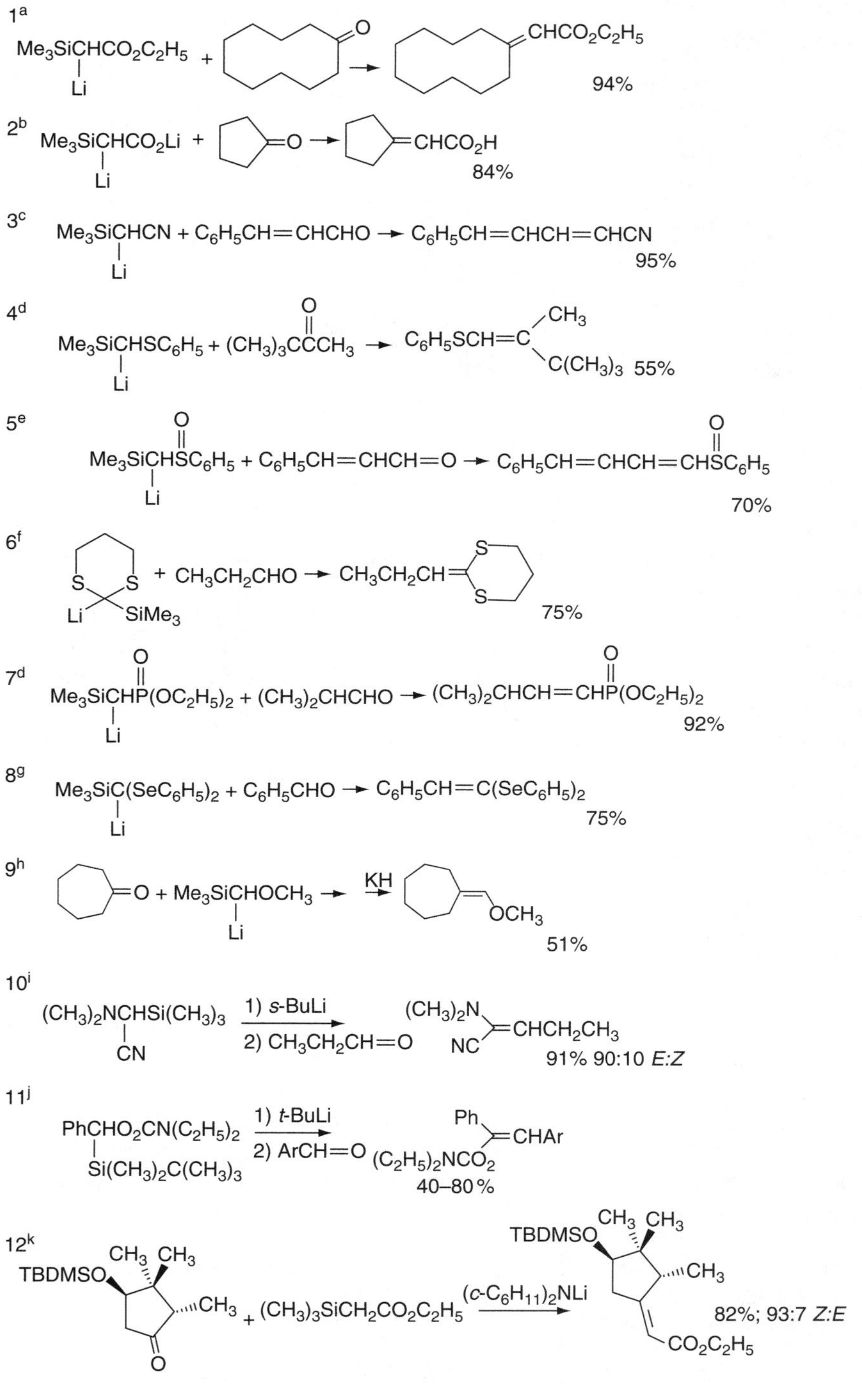

1[a]

$$Me_3SiCHCO_2C_2H_5 \ (Li) + \text{(cyclodecanone)} \longrightarrow \text{(=CHCO}_2C_2H_5) \quad 94\%$$

2[b]

$$Me_3SiCHCO_2Li \ (Li) + \text{(cyclopentanone =O)} \longrightarrow \text{(=CHCO}_2H) \quad 84\%$$

3[c]

$$Me_3SiCHCN \ (Li) + C_6H_5CH=CHCHO \longrightarrow C_6H_5CH=CHCH=CHCN \quad 95\%$$

4[d]

$$Me_3SiCHSC_6H_5 \ (Li) + (CH_3)_3CCCH_3 \ (O) \longrightarrow C_6H_5SCH=C \begin{smallmatrix} CH_3 \\ C(CH_3)_3 \end{smallmatrix} \quad 55\%$$

5[e]

$$Me_3SiCHSC_6H_5 \ (Li) + C_6H_5CH=CHCH=O \ (O) \longrightarrow C_6H_5CH=CHCH=CHSC_6H_5 \ (O) \quad 70\%$$

6[f]

$$\text{(1,3-dithiane Li, SiMe}_3) + CH_3CH_2CHO \longrightarrow CH_3CH_2CH= \text{(1,3-dithiane)} \quad 75\%$$

7[d]

$$Me_3SiCHP(OC_2H_5)_2 \ (O)(Li) + (CH_3)_2CHCHO \longrightarrow (CH_3)_2CHCH=CHP(OC_2H_5)_2 \ (O) \quad 92\%$$

8[g]

$$Me_3SiC(SeC_6H_5)_2 \ (Li) + C_6H_5CHO \longrightarrow C_6H_5CH=C(SeC_6H_5)_2 \quad 75\%$$

9[h]

$$\text{(cycloheptanone =O)} + Me_3SiCHOCH_3 \ (Li) \xrightarrow{\text{KH}} \text{(=CH}OCH_3) \quad 51\%$$

10[i]

$$(CH_3)_2NCHSi(CH_3)_3 \ (CN) \xrightarrow[\text{2) } CH_3CH_2CH=O]{\text{1) } s\text{-BuLi}} (CH_3)_2N-\text{(NC)}C=CHCH_2CH_3 \quad 91\% \ 90{:}10 \ E{:}Z$$

11[j]

$$PhCHO_2CN(C_2H_5)_2 \ (Si(CH_3)_2C(CH_3)_3) \xrightarrow[\text{2) ArCH=O}]{\text{1) } t\text{-BuLi}} \begin{smallmatrix} Ph \\ (C_2H_5)_2NCO_2 \end{smallmatrix} C=CHAr \quad 40\text{--}80\%$$

12[k]

$$\text{(TBDMSO-cyclopentanone)} + (CH_3)_3SiCH_2CO_2C_2H_5 \xrightarrow{(c\text{-}C_6H_{11})_2NLi} \text{(olefination product)} \quad 82\%; \ 93{:}7 \ Z{:}E$$

(Continued)

174

CHAPTER 2

*Reactions of Carbon
Nucleophiles with
Carbonyl Compounds*

Scheme 2.19. (*Continued*)

13[l]

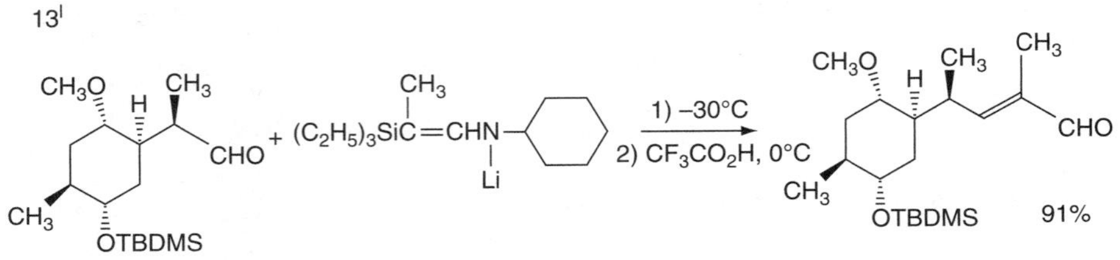

a. K. Shimoji, H. Taguchi, H. Yamamoto, K. Oshima, and H. Hozaki, *J. Am. Chem. Soc.*, **96**, 1620 (1974).
b. P. A. Grieco, C. L. J. Wang, and S. D. Burke, *J. Chem. Soc. Chem. Commun.*, 537 (1975).
c. I. Matsuda, S. Murata, and Y. Ishii, *J. Chem. Soc., Perkin Trans. 1*, 26 (1979).
d. F. A. Carey and A. S. Court, *J. Org. Chem.*, **37**, 939 (1972).
e. F. A. Carey and O. Hernandez, *J. Org. Chem.*, **38**, 2670 (1973).
f. D. Seebach, M. Kolb, and B.-T. Grobel, *Chem. Ber.*, **106**, 2277 (1973).
g. B. T. Grobel and D. Seebach, *Chem. Ber.*, **110**, 852 (1977).
h. P. Magnus and G. Roy, *Organometallics*, **1**, 553 (1982).
i. W. Adam and C. M. Ortega-Schulte, *Synlett*, 414 (2003).
j. L. F. van Staden, B. Bartels-Rahm, J. S. Field, and N. D. Emslie, *Tetrahedron*, **54**, 3255 (1998).
k. J.-M. Galano, G. Audran, and H. Monti, *Tetrahedron Lett.*, **42**, 6125 (2001).
l. S. F. Martin, J. A. Dodge, L. E. Burgess, and M. Hartmann, *J. Org. Chem.*, **57**, 1070 (1992).

unstable olefins. Entries 1 to 8 show the use of lithio silanes having a range of anion-stabilizing groups. The anions are prepared using alkyllithium reagents or lithium amides. Entries 9 to 11 illustrate the utility of the reaction to prepare relatively unstable substituted alkenes. The silyl anions are typically more reactive than stabilized Wittig ylides, and in the case of Entry 12 good results were obtained while the triphenylphosphonium ylide was unreactive. Entry 13 shows the use of Peterson olefination for chain extension with an α-methyl-α, β-unsaturated aldehyde. The preferred reagent for this transformation is a lithio β-trialkylsilylenamine.[276]

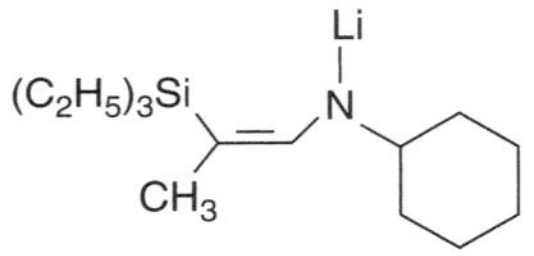

2.4.3. The Julia Olefination Reaction

The *Julia olefination* involves the addition of a sulfonyl-stabilized carbanion to a carbonyl compound, followed by elimination to form an alkene.[277] In the initial versions of the reaction, the elimination was done under *reductive conditions*. More recently, a modified version that avoids this step was developed. The former version is sometimes referred to as the *Julia-Lythgoe olefination*, whereas the latter is called the *Julia-Kocienski olefination*. In the reductive variant, the adduct is usually acylated and then treated with a reducing agent, such as sodium amalgam or samarium diiodide.[278]

276. R. Desmond, S. G. Mills, R. P. Volante, and I. Shinkai, *Tetrahedron Lett.*, **29**, 3895 (1988).
277. P. R. Blakemore, *J. Chem. Soc., Perkin Trans. 1*, 2563 (2002).
278. A. S. Kende and J. Mendoza, *Tetrahedron Lett.*, **31**, 7105 (1990); G. E. Keck, K. A. Savin, and M. A. Weglarz, *J. Org. Chem.*, **60**, 3194 (1995); K. Fukumoto, M. Ihara, S. Suzuki, T. Taniguchi, and Y. Yokunaga, *Synlett*, 895 (1994); I. E. Marko, F. Murphy, and S. Dolan, *Tetrahedron Lett.*, **37**, 2089 (1996); I. E. Marko, F. Murphy, L. Kumps, A. Ates, R. Touillaux, D. Craig, S. Carballares, and S. Dolan, *Tetrahedron*, **57**, 2609 (2001).

The mechanistic details of reductive elimination reactions of this type are considered in Section 5.8.

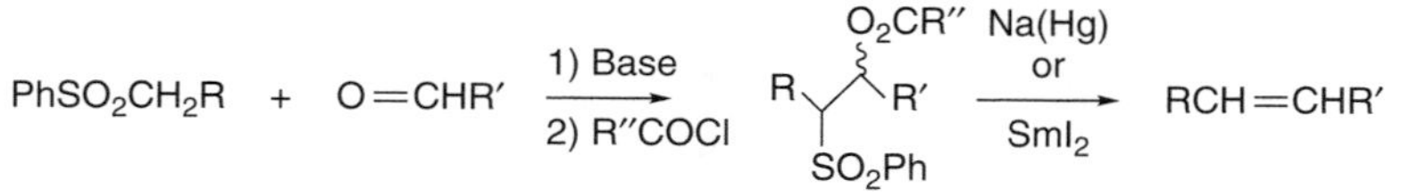

In the modified procedure one of several heteroaromatic sulfones is used. The crucial role of the heterocyclic ring is to provide a nonreductive mechanism for the elimination step, which occurs by an addition-elimination mechanism that results in fragmentation to the alkene. The original example used a benzothiazole ring,[279] but more recently tetrazoles have been developed for this purpose.[280]

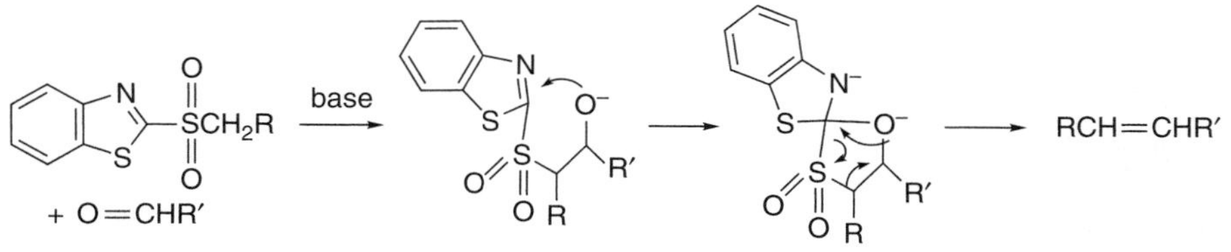

Other aryl sulfones that can accommodate the nucleophilic addition step also react in the same way. For example, excellent results have been obtained using 3,5-*bis*-(trifluoromethyl)phenyl sulfones.[281]

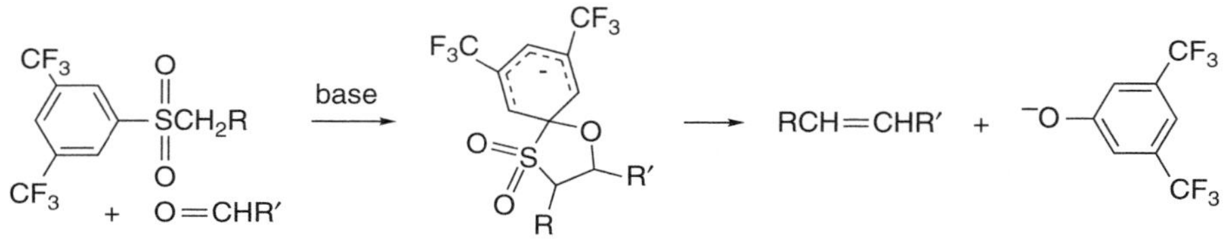

As is the case with the Wittig and Peterson olefinations, there is more than one point at which the stereoselectivity of the reaction can be determined, depending on the details of the mechanism. Adduct formation can be product determining or reversible. Furthermore, in the reductive mechanism, there is the potential for stereorandomization if radical intermediates are involved. As a result, there is a degree of variability in the stereoselectivity. Fortunately, the modified version using tetrazolyl sulfones usually gives a predominance of the *E*-isomer.

Scheme 2.20 gives some examples of the application of the Julia olefination in synthesis. Entry 1 demonstrates the reductive elimination conditions. This reaction gave a good *E*:*Z* ratio under the conditions shown. Entry 2 is an example of the use of the modified reaction that gave a good *E*:*Z* ratio in the synthesis of vinyl chlorides. Entry 3 uses the tetrazole version of the reaction in the synthesis of a long-chain ester. Entries 4 to 7 illustrate the use of modified conditions for the synthesis of polyfunctional molecules.

[279.] J. B. Baudin, G. Hareau, S. A. Julia, and O. Ruel, *Tetrahedron Lett.*, **32**, 1175 (1991).
[280.] P. R. Blakemore, W. J. Cole, P. J. Kocienski, and A. Morley, *Synlett*, 26 (1998); P. J. Kocienski, A. Bell, and P. R. Blakemore, *Synlett*, 365 (2000).
[281.] D.A. Alonso, M. Fuensanta, C. Najera, and M. Varea, *J. Org. Chem.*, **70**, 6404 (2005).

Scheme 2.20. Julia Olefination Reactions

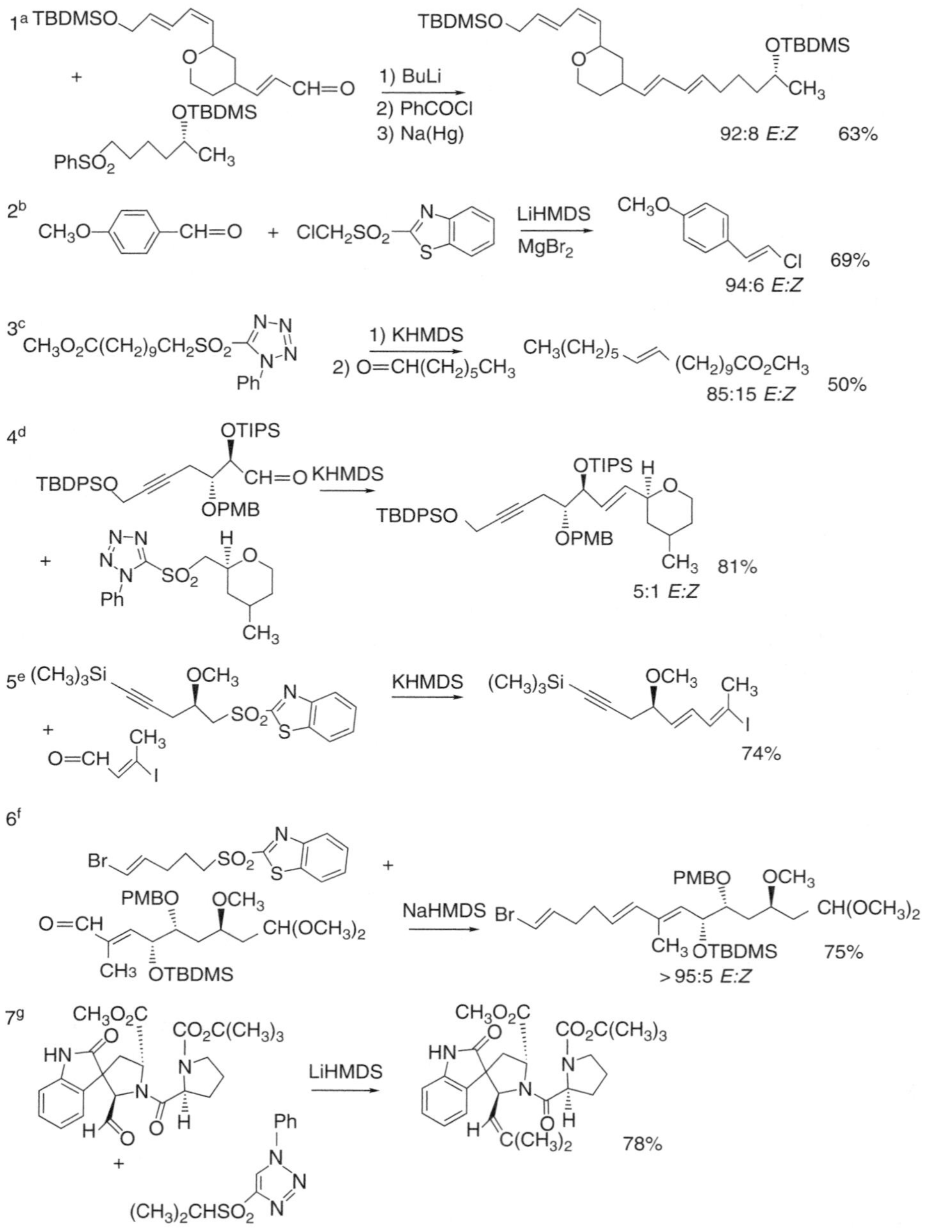

a. J. P. Marino, M. S. McClure, D. P. Holub, J. V. Comasseto, and F. C. Tucci, *J. Am. Chem. Soc.*, **124**, 1664 (2002).
b. M.-E. Lebrun, P. Le Marquand, and C. Berthelette, *J. Org. Chem.*, **71**, 2009 (2006).
c. P. E. Duffy, S. M. Quinn, H. M. Roche, and P. Evans, *Tetrahedron*, **62**, 4838 (2006).
d. A. Sivaramakrishnan, G. T. Nadolski, I. A. McAlexander, and B. S. Davidson, *Tetrahedron Lett.*, **43**, 2132 (2002).
e. G. Pattenden, A. T. Plowright, J. A. Tornos, and T. Ye, *Tetrahedron Lett.*, **39**, 6099 (1998).
f. D. A. Evans, V. J. Cee, T. E. Smith, D. M. Fitch, and P. S. Cho, *Angew. Chem. Int. Ed. Engl.*, **39**, 2533 (2000).
g. C. Marti and E. M. Carreira, *J. Am. Chem. Soc.*, **127**, 11505 (2005).

The reactions in this section correspond to the general Pathway **D** discussed earlier (p. 64), in which the carbon nucleophile contains a potential leaving group. This group can be the same or a different group from the anion-stabilizing group. One group of reagents that reacts according to this pattern are the sulfonium ylides, which react with carbonyl compounds to give epoxides.

$$R_2{}^+S-CH_2{}^- \quad + \quad O{=}CR'_2 \quad \longrightarrow \quad \cdots \longrightarrow \cdots$$

There are related reactions in which the sulfur is at the sulfoxide or sulfilimine oxidation level. Another example of the addition-cyclization route involves α-haloesters, which react to form epoxides by displacement of the halide ion.

$$C_2H_5O_2C\underset{X}{\overset{}{\underset{|}{C}}}{-}R \quad + \quad O{=}CR'_2 \quad \longrightarrow \quad \cdots \longrightarrow \cdots$$

2.5.1. Sulfur Ylides and Related Nucleophiles

Sulfur ylides have several applications as reagents in synthesis.[282] Dimethylsulfonium methylide and dimethylsulfoxonium methylide are particularly useful.[283] These sulfur ylides are prepared by deprotonation of the corresponding sulfonium salts, both of which are commercially available.

$$(CH_3)_2\overset{+}{S}CH_3\ I^- \quad \xrightarrow[\text{DMSO}]{\overset{O}{\overset{\|}{NaCH_2SCH_3}}} \quad (CH_3)_2\overset{+}{S}-CH_2{}^-$$
dimethylsulfonium methylide

$$(CH_3)_2\overset{O}{\underset{+}{\overset{\|}{S}}}CH_3\ I^- \quad \xrightarrow[\text{DMSO}]{\text{NaH}} \quad (CH_3)_2\overset{O}{\underset{+}{\overset{\|}{S}}}-CH_2{}^-$$
dimethylsulfoxonium methylide

Whereas phosphonium ylides normally react with carbonyl compounds to give alkenes, dimethylsulfonium methylide and dimethylsulfoxonium methylide yield epoxides. Instead of a four-center elimination, the adducts from the sulfur ylides undergo intramolecular displacement of the sulfur substituent by oxygen. In this reaction, the sulfur substituent serves both to promote anion formation and as the leaving group.

$$R_2C{=}O + (CH_3)_2\overset{+}{S}{-}\overset{-}{C}H_2 \longrightarrow R_2\overset{O^-}{\underset{}{C}}{-}CH_2{-}\overset{+}{S}(CH_3)_2 \longrightarrow R_2\overset{O}{\overset{/\backslash}{C}}{-}CH_2 + (CH_3)_2S$$

$$R_2C{=}O + (CH_3)_2\overset{O}{\underset{+}{\overset{\|}{S}}}{-}\overset{-}{C}H_2 \longrightarrow R_2\overset{O^-}{\underset{}{C}}{-}CH_2\overset{O}{\underset{+}{\overset{\|}{S}}}(CH_3)_2 \longrightarrow R_2\overset{O}{\overset{/\backslash}{C}}{-}CH_2 + (CH_3)_2S{=}O$$

[282] B. M. Trost and L. S. Melvin, Jr., *Sulfur Ylides*, Academic Press, New York, 1975; E. Block, *Reactions of Organosulfur Compounds*, Academic Press, New York, 1978.

[283] E. J. Corey and M. Chaykovsky, *J. Am. Chem. Soc.*, **87**, 1353 (1965).

Dimethylsulfonium methylide is both more reactive and less stable than dimethylsulfoxonium methylide, so it is generated and used at a lower temperature. A sharp distinction between the two ylides emerges in their reactions with α, β-unsaturated carbonyl compounds. Dimethylsulfonium methylide yields epoxides, whereas dimethylsulfoxonium methylide reacts by conjugate addition and gives cyclopropanes (compare Entries 5 and 6 in Scheme 2.21). It appears that the reason for the difference lies in the relative rates of the two reactions available to the betaine intermediate: (a) reversal to starting materials, or (b) intramolecular nucleophilic displacement.[284] Presumably both reagents react most rapidly at the carbonyl group. In the case of dimethylsulfonium methylide the intramolecular displacement step is faster than the reverse of the addition, and epoxide formation takes place.

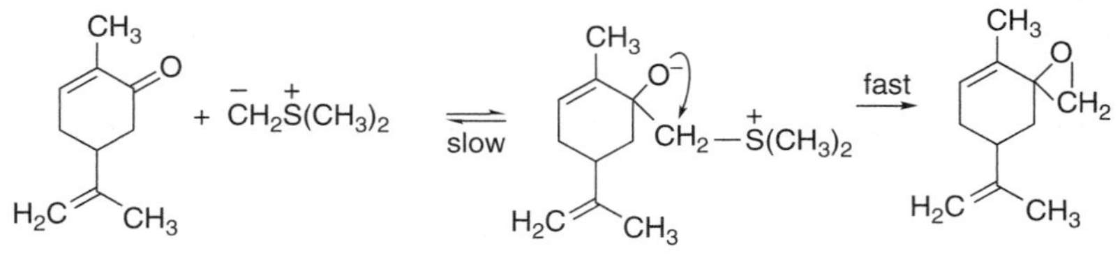

With the more stable dimethylsulfoxonium methylide, the reversal is relatively more rapid and product formation takes place only after conjugate addition.

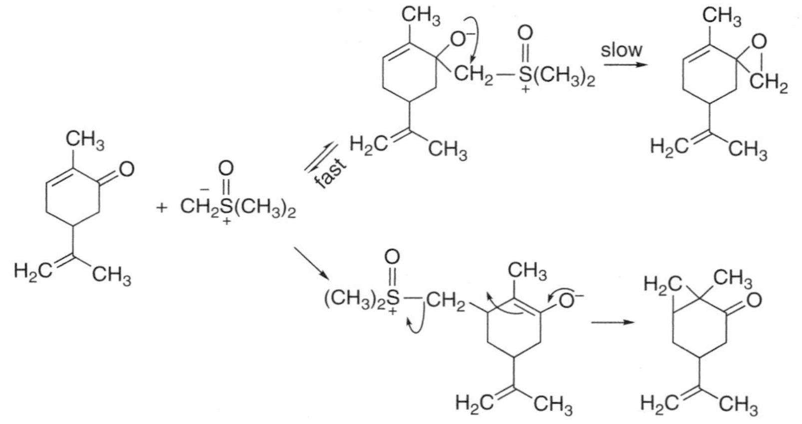

Another difference between dimethylsulfonium methylide and dimethylsulfoxonium methylide concerns the stereoselectivity in formation of epoxides from cyclohexanones. Dimethylsulfonium methylide usually adds from the axial direction whereas dimethylsulfoxonium methylide favors the equatorial direction. This result may also be due to reversibility of addition in the case of the sulfoxonium methylide.[92] The product from the sulfonium ylide is the result the kinetic preference for axial addition by small nucleophiles (see Part A, Section 2.4.1.2). In the case of reversible addition of the sulfoxonium ylide, product structure is determined by the rate of displacement and this may be faster for the more stable epoxide.

[284.] C. R. Johnson, C. W. Schroeck, and J. R. Shanklin, *J. Am. Chem. Soc.*, **95**, 7424 (1973).

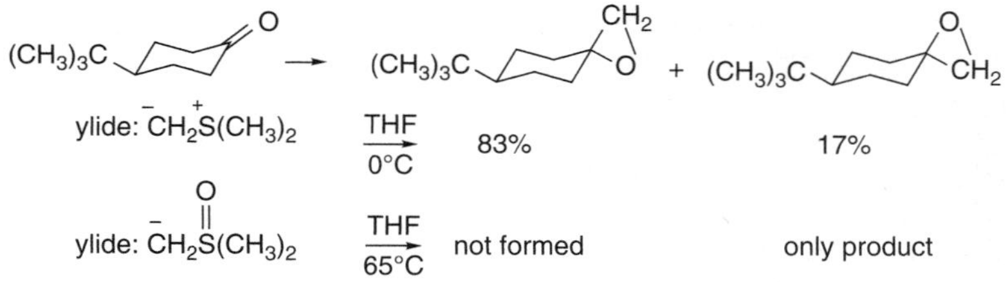

Examples of the use of dimethylsulfonium methylide and dimethylsulfoxonium methylide are listed in Scheme 2.21. Entries 1 to 5 are conversions of carbonyl compounds to epoxides. Entry 6 is an example of cyclopropanation with dimethyl sulfoxonium methylide. Entry 7 compares the stereochemistry of addition of dimethylsulfonium methylide to dimethylsulfoxonium methylide for nornborn-5-en-2-one. The product in Entry 8 was used in a synthesis of α-tocopherol (vitamin E).

Sulfur ylides can also transfer substituted methylene units, such as isopropylidene (Entries 10 and 11) or cyclopropylidene (Entries 12 and 13). The oxaspiropentanes formed by reaction of aldehydes and ketones with diphenylsulfonium cyclopropylide are useful intermediates in a number of transformations such as acid-catalyzed rearrangement to cyclobutanones.[285]

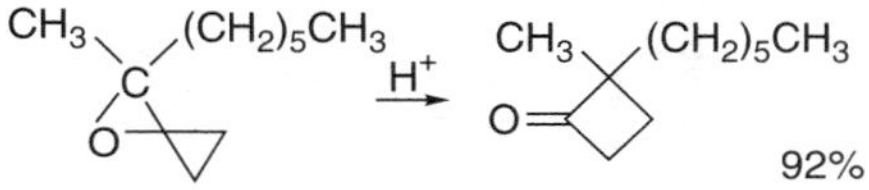

Aside from the methylide and cyclopropylide reagents, the sulfonium ylides are not very stable. A related group of reagents derived from sulfoximines offers greater versatility in alkylidene transfer reactions.[286] The preparation and use of this class of ylides is illustrated below.

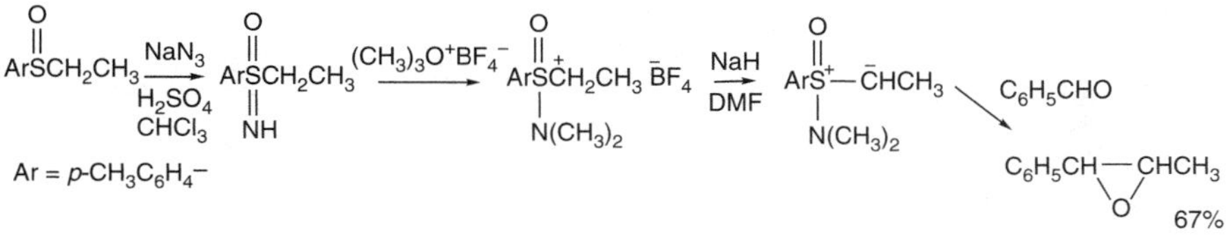

Ar = p-CH$_3$C$_6$H$_4$–

A similar pattern of reactivity has been demonstrated for the anions formed by deprotonation of S-alkyl-N-p-toluenesulfoximines (see Entry 14 in Scheme 2.21).[287]

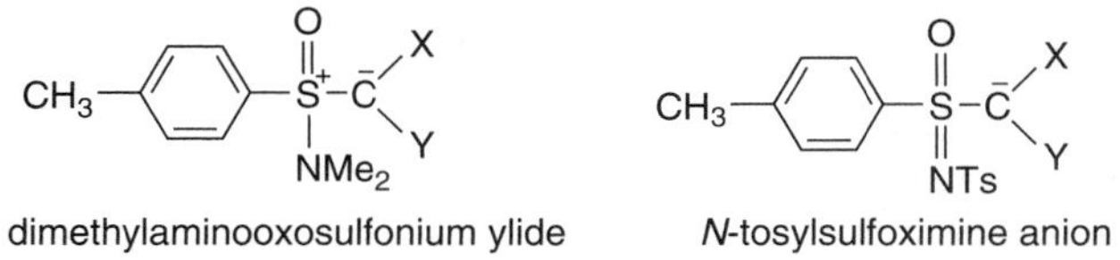

dimethylaminooxosulfonium ylide N-tosylsulfoximine anion

The sulfoximine group provides anion-stabilizing capacity in a chiral environment and a number of synthetic applications have been developed based on these properties.[288]

285. B. M. Trost and M. J. Bogdanowicz, *J. Am. Chem. Soc.*, **95**, 5321 (1973).
286. C. R. Johnson, *Acc. Chem. Res.*, **6**, 341 (1973); C. R.Johnson, *Aldrichimica Acta*, **18**, 3 (1985).
287. C. R. Johnson, R. A. Kirchoff, R. J. Reischer, and G. F. Katekar, *J. Am. Chem. Soc.*, **95**, 4287 (1973).
288. M. Reggelin and C. Zur, *Synthesis*, 1 (2000).

Scheme 2.21. Reactions of Sulfur Ylides

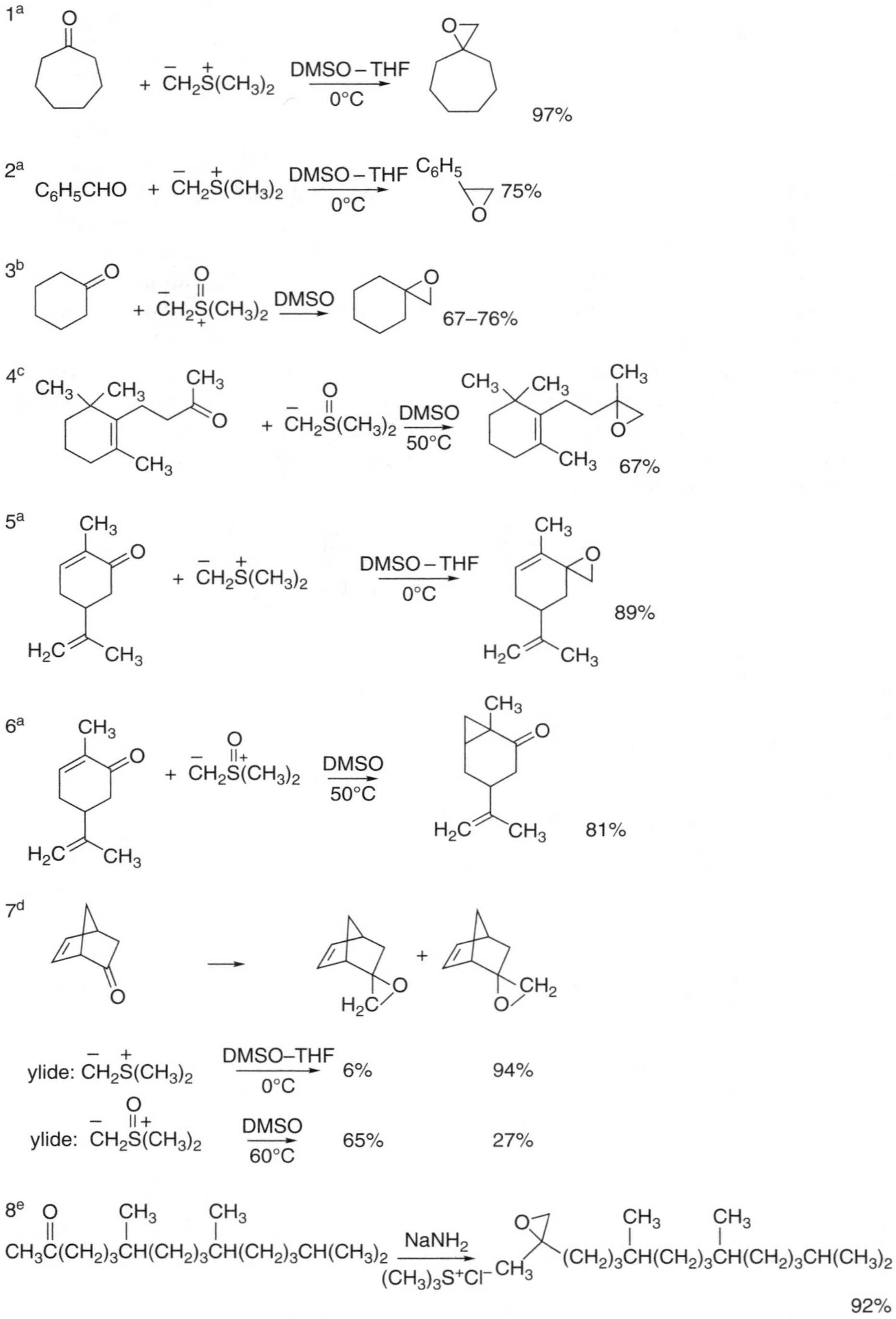

1[a]

$+ \ \bar{C}H_2\overset{+}{S}(CH_3)_2$ $\xrightarrow[0°C]{DMSO-THF}$ 97%

2[a]

C_6H_5CHO $+ \ \bar{C}H_2\overset{+}{S}(CH_3)_2$ $\xrightarrow[0°C]{DMSO-THF}$ 75%

3[b]

$+ \ \bar{C}H_2\overset{+}{\underset{\parallel}{S}}(CH_3)_2$ $\xrightarrow{DMSO}$ 67–76%

4[c]

$+ \ \bar{C}H_2\overset{+}{\underset{\parallel}{S}}(CH_3)_2$ $\xrightarrow[50°C]{DMSO}$ 67%

5[a]

$+ \ \bar{C}H_2\overset{+}{S}(CH_3)_2$ $\xrightarrow[0°C]{DMSO-THF}$ 89%

6[a]

$+ \ \bar{C}H_2\overset{+}{\underset{\parallel}{S}}(CH_3)_2$ $\xrightarrow[50°C]{DMSO}$ 81%

7[d]

ylide: $\bar{C}H_2\overset{+}{S}(CH_3)_2$ $\xrightarrow[0°C]{DMSO-THF}$ 6% 94%

ylide: $\bar{C}H_2\overset{+}{\underset{\parallel}{S}}(CH_3)_2$ $\xrightarrow[60°C]{DMSO}$ 65% 27%

8[e]

$CH_3\overset{O}{\overset{\parallel}{C}}(CH_2)_3\overset{CH_3}{\overset{|}{C}H}(CH_2)_3\overset{CH_3}{\overset{|}{C}H}(CH_2)_3CH(CH_3)_2$ $\xrightarrow[(CH_3)_3S^+Cl^-]{NaNH_2}$ $CH_3\overset{O}{\overset{\diagdown}{C}}(CH_2)_3\overset{CH_3}{\overset{|}{C}H}(CH_2)_3\overset{CH_3}{\overset{|}{C}H}(CH_2)_3CH(CH_3)_2$

92%

(Continued)

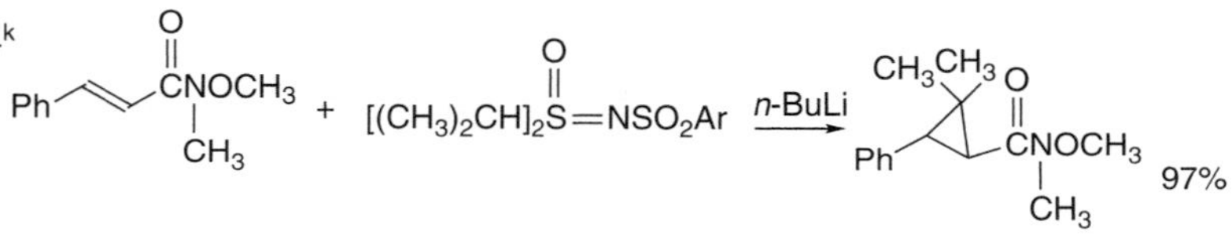

a. E. J. Corey and M. Chaykovsky, *J. Am. Chem. Soc.*, **87**, 1353 (1965).
b. E. J. Corey and M. Chaykovsky, *Org. Synth.*, **49**, 78 (1969).
c. M. G. Fracheboud, O. Shimomura, R. K. Hill, and F. H. Johnson, *Tetrahedron Lett.*, 3951 (1969).
d. R. S. Bly, C. M. DuBose, Jr., and G. B. Konizer, *J. Org. Chem.*, **33**, 2188 (1968).
e. G. L. Olson, H.-C. Cheung, K. Morgan, and G. Saucy, *J. Org. Chem.*, **45**, 803 (1980).
f. M. Rosenberger, W. Jackson, and G. Saucy, *Helv. Chim. Acta*, **63**, 1665 (1980).
g. E. J. Corey, M. Jautelat, and W. Oppolzer, *Tetrahedron Lett.*, 2325 (1967).
h. E. J. Corey and M. Jautelat, *J. Am. Chem. Soc.*, **89**, 3112 (1967).
i. B. M. Trost and M. J. Bogdanowicz, *J. Am. Chem. Soc.*, **95**, 5307 (1973).
j. B. M. Trost and M. J. Bogdanowicz, *J. Am. Chem. Soc.*, **95**, 5311 (1973).
k. K. E. Rodriques, *Tetrahedron Lett.*, **32**, 1275 (1991).

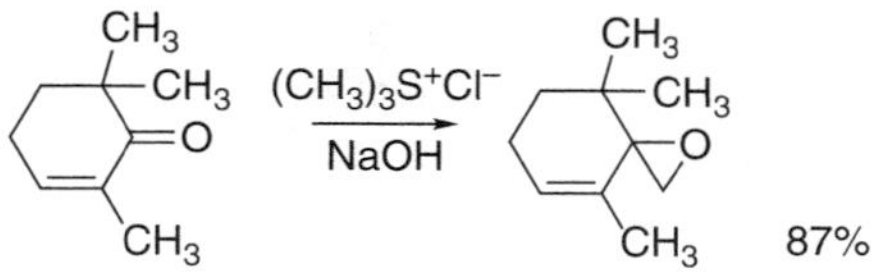

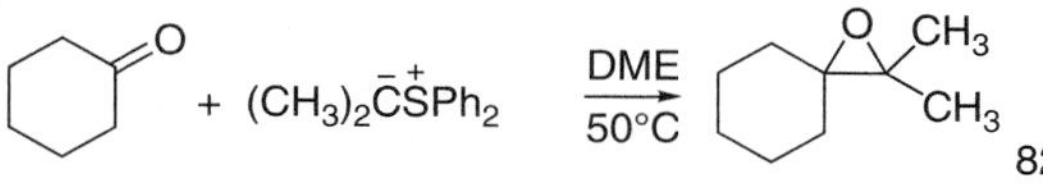

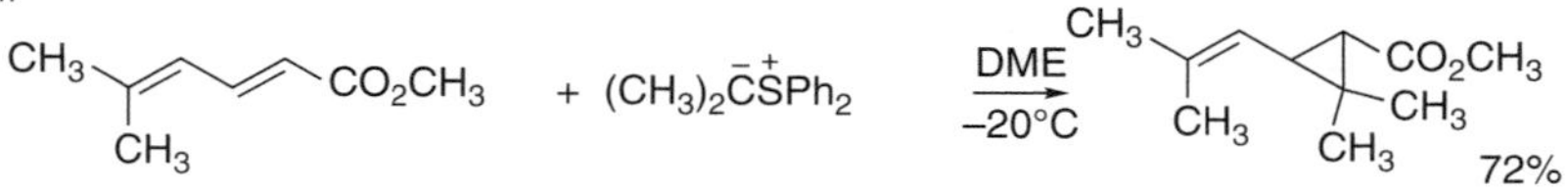

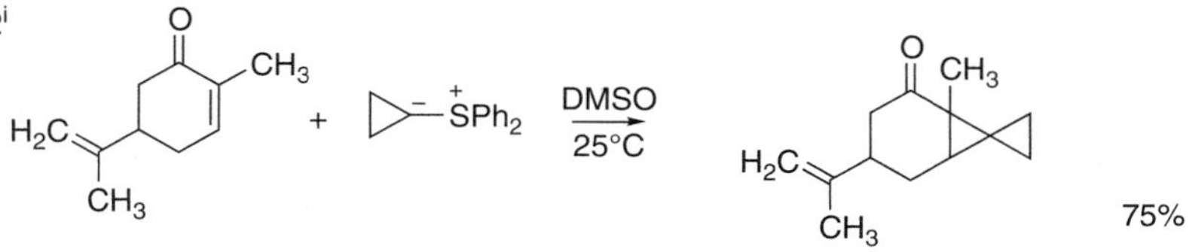

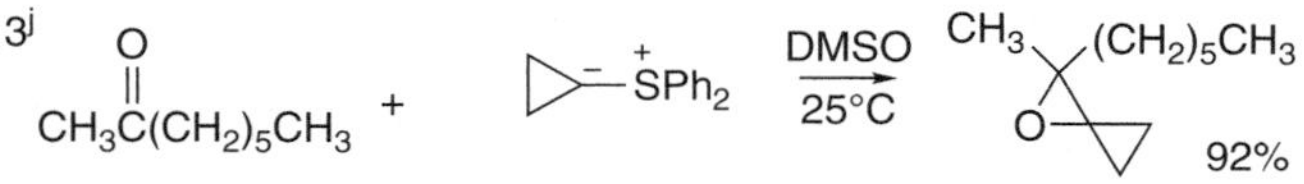

Dimethylsulfonium methylide reacts with reactive alkylating reagents such as allylic and benzylic bromides to give terminal alkenes. A similar reaction occurs with primary alkyl bromides in the presence of LiI. The reaction probably involves alkylation of the ylide, followed by elimination.[289]

$$RCH_2\!-\!X \; + \; CH_2\!=\!S^+(CH_3)_2 \; \longrightarrow \; RCH_2CH_2S^+(CH_3)_2 \; \longrightarrow \; RCH\!=\!CH_2$$

[289.] L. Alcaraz, J. J. Harnett, C. Mioskowski, J. P. Martel, T. LeGall, D.-S. Shin, and J. R. Falck, *Tetrahedron Lett.*, **35**, 5453 (1994).

2.5.2. Nucleophilic Addition-Cyclization of α-Haloesters

The pattern of nucleophilic addition at a carbonyl group followed by intramolecular nucleophilic displacement of a leaving group present in the nucleophile can also be recognized in a much older synthetic technique, the *Darzens reaction*.[290] The first step in this reaction is addition of the enolate of the α-haloester to the carbonyl compound. The alkoxide oxygen formed in the addition then effects nucleophilic attack, displacing the halide and forming an α,β-epoxy ester (also called a glycidic ester).

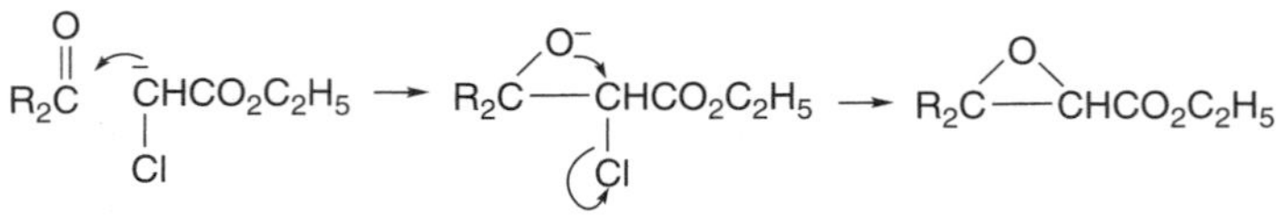

Scheme 2.22 shows some examples of the Darzens reaction.

Trimethylsilylepoxides can be prepared by an addition-cyclization process. Reaction of chloromethyltrimethylsilane with *sec*-butyllithium at very low temperature gives an α-chloro lithium reagent that leads to an epoxide on reaction with an aldehyde or ketone.[291]

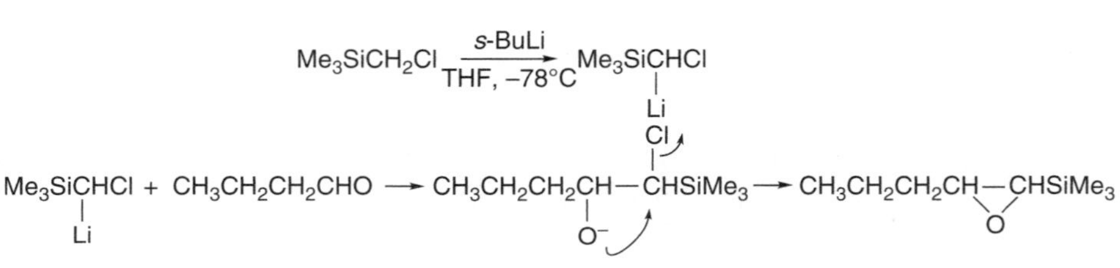

Scheme 2.22. Darzens Condensation Reaction

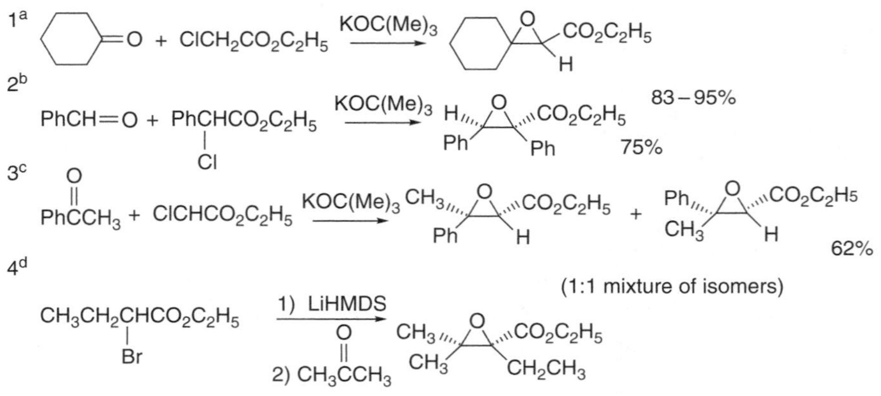

a. R. H. Hunt, L. J. Chinn, and W. S. Johnson, *Org. Synth.*, **IV**, 459 (1963).
b. H. E. Zimmerman and L. Ahramjian, *J. Am. Chem. Soc.*, **82**, 5459 (1960).
c. F. W. Bachelor and R. K. Bansal, *J. Org. Chem.*, **34**, 3600 (1969).
d. R. F. Borch, *Tetrahedron Lett.*, 3761 (1972).

290. M. S. Newman and B. J. Magerlein, *Org. React.*, **5**, 413 (1951).
291. C. Burford, F. Cooke, E. Ehlinger, and P. D. Magnus, *J. Am. Chem. Soc.*, **99**, 4536 (1977).

2.6. Conjugate Addition by Carbon Nucleophiles

The previous sections dealt with reactions in which the new carbon-carbon bond is formed by addition of the nucleophile to a carbonyl group. Another important method for alkylation of carbon nucleophiles involves addition to an electrophilic multiple bond. The electrophilic reaction partner is typically an α,β-unsaturated ketone, aldehyde, or ester, but other electron-withdrawing substituents such as nitro, cyano, or sulfonyl also activate carbon-carbon double and triple bonds to nucleophilic attack. The reaction is called *conjugate addition* or the *Michael reaction.*

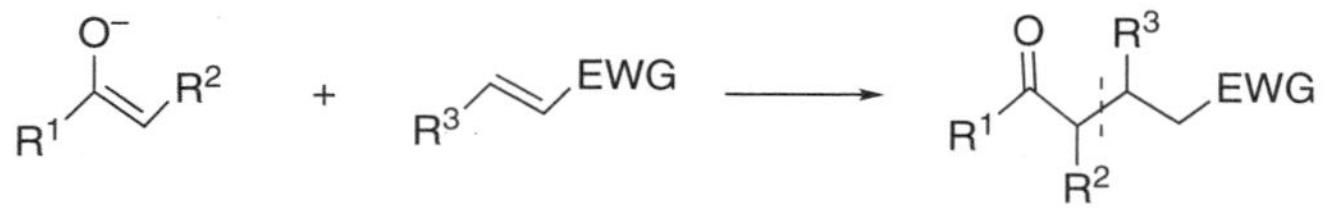

More generally, many combinations of EWG substituents can serve as the anion-stabilizing and alkene-activating groups. Conjugate addition has the potential to form a bond α to one group and β to the other to form a α,γ-disubstituted system.

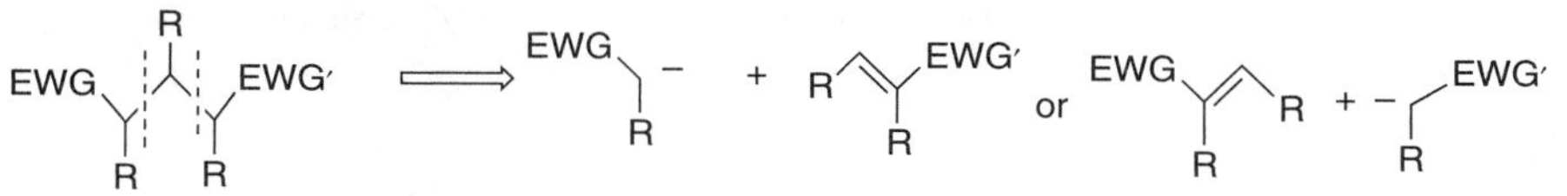

The scope of the conjugate addition reaction can be further expanded by use of Lewis acids in conjunction with enolate equivalents, especially silyl enol ethers and silyl ketene acetals. The adduct is stabilized by a new bond to the Lewis acid and products are formed from the adduct.

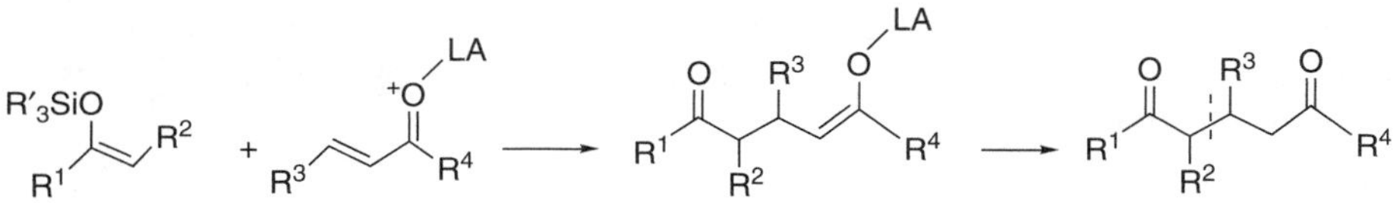

Other kinds of nucleophiles such as amines, alkoxides, and sulfide anions also react with electrophilic alkenes, but we focus on the carbon-carbon bond forming reactions.

2.6.1. Conjugate Addition of Enolates

Conjugate addition of enolates under some circumstances can be carried out with a catalytic amount of base. All the steps are reversible.

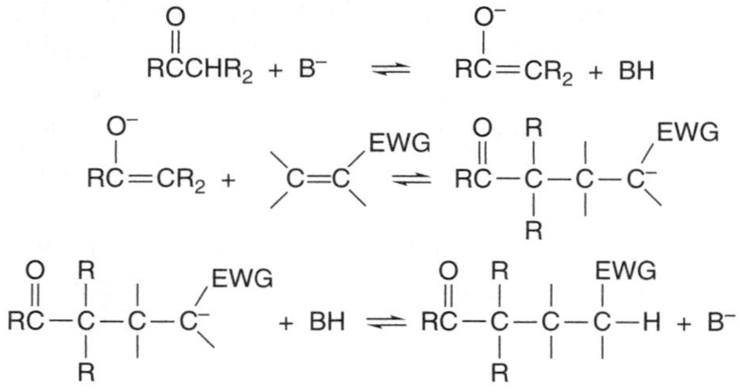

When the EWG is a carbonyl group, there can be competition with 1,2-addition, which is especially likely for aldehydes but can also occur with ketones. With successively less reactive carbonyl groups, 1,4-addition becomes more favorable. Highly reactive, hard nucleophiles tend to favor 1,2-addition and the reaction is irreversible if the nucleophile is a poor leaving group. For example with organometallic reagents, 1,2-addition is usually observed and it is irreversible because there is no tendency to expel an alkyl anion. Section 2.6.5 considers some exceptions in which organometallic reagents are added in the 1,4-manner. With less basic nucleophiles, the 1,2-addition is more easily reversible and the 1,4-addition product is usually more stable.

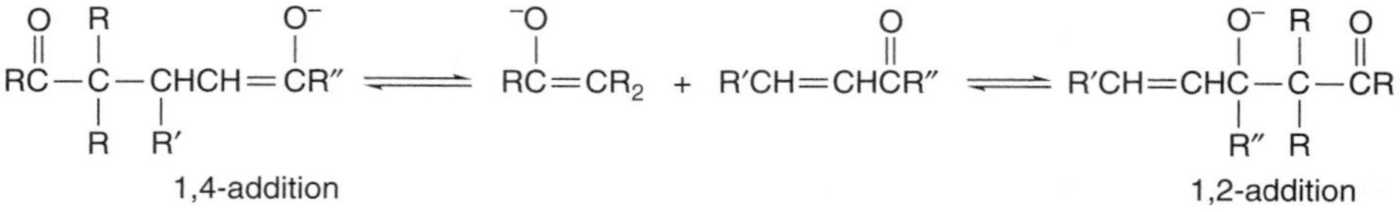

Retrosynthetically, there are inherently two possible approaches to the products of conjugate addition as represented below, where Y and Z represent two different anion-stabilizing groups.

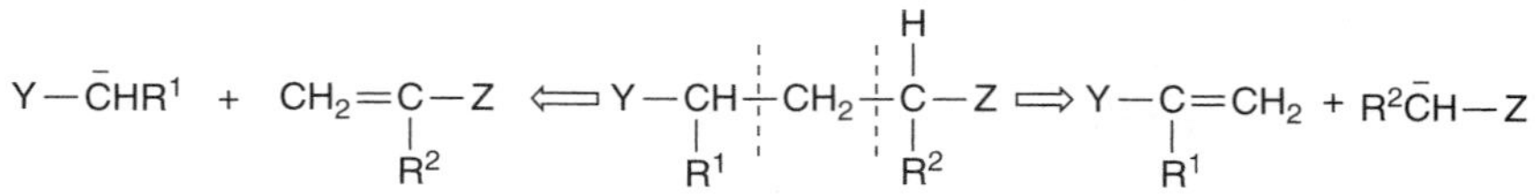

When a catalytic amount of base is used, the most effective nucleophiles are enolates derived from relatively acidic compounds such as β-ketoesters or malonate esters. The adduct anions are more basic than the nucleophile and are protonated under the reaction conditions.

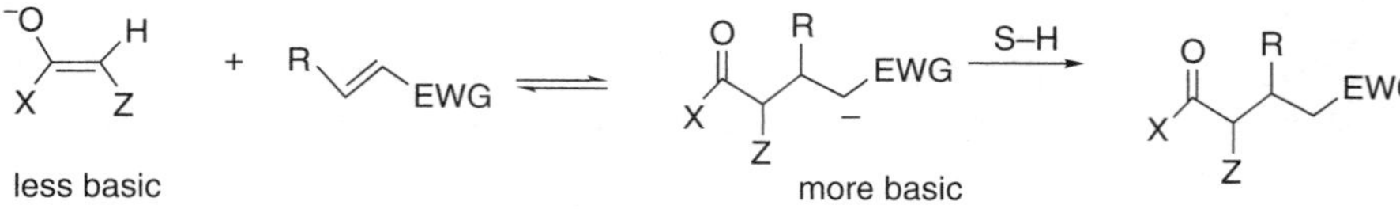

Scheme 2.23 provides some examples of conjugate addition reactions. Entry 1 illustrates the tendency for reaction to proceed through the more stable enolate. Entries 2 to 5 are typical examples of addition of doubly stabilized enolates to electrophilic alkenes. Entries 6 to 8 are cases of addition of nitroalkanes. Nitroalkanes are comparable in acidity to β-ketoesters (see Table 1.1) and are often excellent nucleophiles for conjugate addition. Note that in Entry 8 fluoride ion is used as the base. Entry 9 is a case of adding a zinc enolate (Reformatsky reagent) to a nitroalkene. Entry 10 shows an enamine as the carbon nucleophile. All of these reactions were done under equilibrating conditions.

The fluoride ion is an effective catalyst for conjugate additions involving relatively acidic carbon nucleophiles.[292] The reactions can be done in the presence of excess

292. J. H. Clark, *Chem. Rev.*, **80**, 429 (1980).

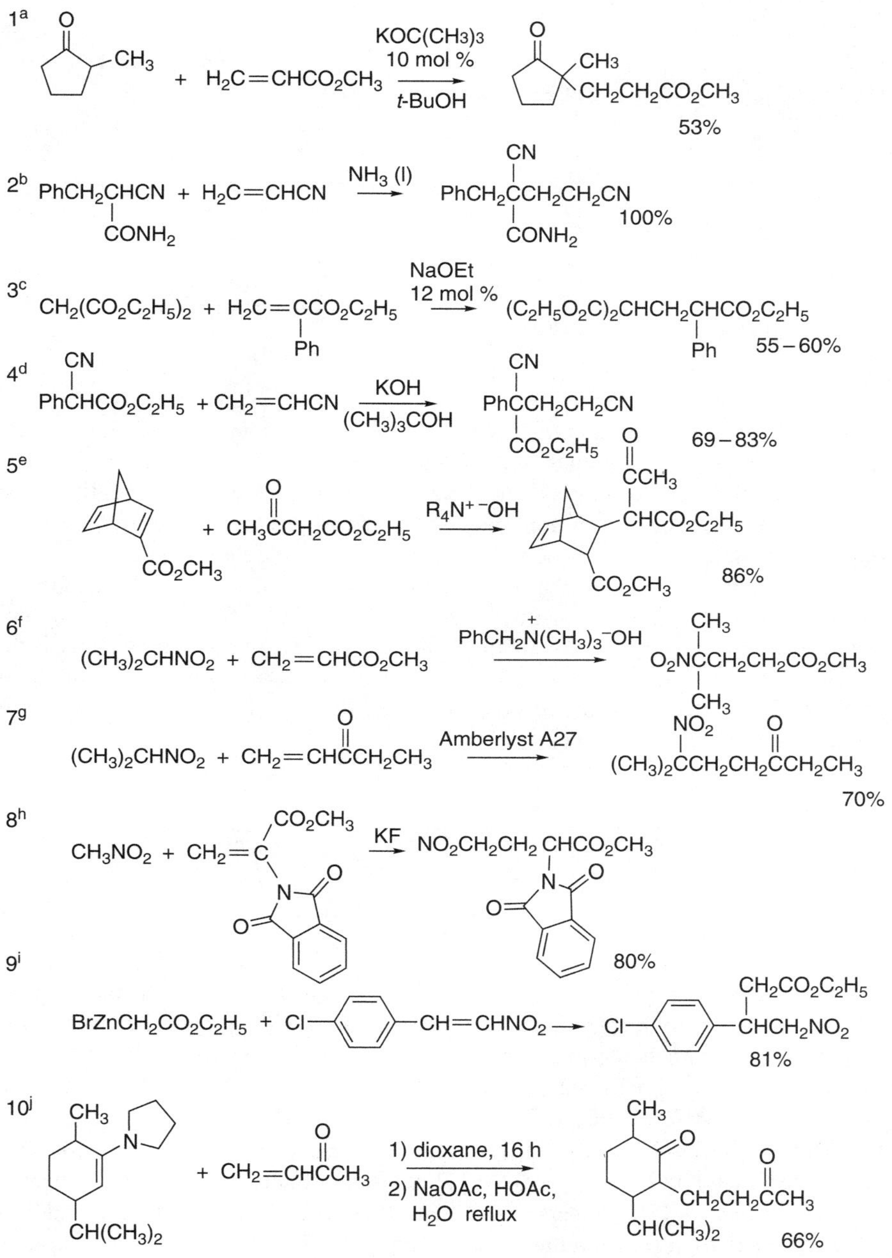

a. H. O. House, W. L. Roelofs, and B. M. Trost, *J. Org. Chem.*, **31**, 646 (1966).
b. S. Wakamatsu, *J. Org. Chem.*, **27**, 1285 (1962).
c. E. M. Kaiser, C. L. Mao, C. F. Hauser, and C. R. Hauser, *J. Org. Chem.*, **35**, 410 (1970).
d. E. C. Horning and A. F. Finelli, *Org. Synth.*, **IV**, 776 (1963).
e. K. Alder, H. Wirtz, and H. Koppelberg, *Liebigs Ann. Chem.*, **601**, 138 (1956).
f. R. B. Moffett, *Org. Synth.*, **IV**, 652 (1963).
g. R. Ballini, P. Marziali, and A. Mozziacafreddo, *J. Org. Chem.*, **61**, 3209 (1996).
h. M. J. Crossley, Y. M. Fung, J. J. Potter, and A. W. Stamford, *J. Chem. Soc., Perkin Trans. 2*, 1113 (1998).
i. R. Menicagli and S. Samaritani, *Tetrahedron*, **52**, 1425 (1996).
j. K. D. Croft, E. L. Ghisalberti, P. R. Jefferies, and A. D. Stuart, *Aust. J. Chem.*, **32**, 2079 (1979).

fluoride, where the formation of the $[F-H-F^-]$ ion occurs, or by use of a tetralkyl-ammonium fluoride in an aprotic solvent.

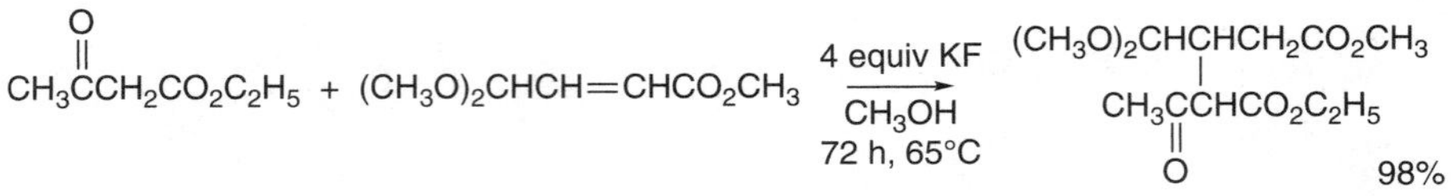

$$CH_3CCH_2CO_2C_2H_5 + (CH_3O)_2CHCH=CHCO_2CH_3 \xrightarrow[\substack{CH_3OH \\ 72\ h,\ 65°C}]{4\ equiv\ KF} \begin{array}{c} (CH_3O)_2CHCHCH_2CO_2CH_3 \\ | \\ CH_3CCHCO_2C_2H_5 \\ || \\ O \end{array}$$

98%

Ref. 293

$$(CH_3)_2CHNO_2 + CH_2=CHCOCH_3 \xrightarrow[\substack{R_4N^+F^- \\ 2\ h,\ 25°C}]{0.5\ equiv} \begin{array}{c} CH_3 \quad\ O \\ | \qquad || \\ O_2NCCH_2CH_2CCH_3 \\ | \\ CH_3 \end{array}$$

95%

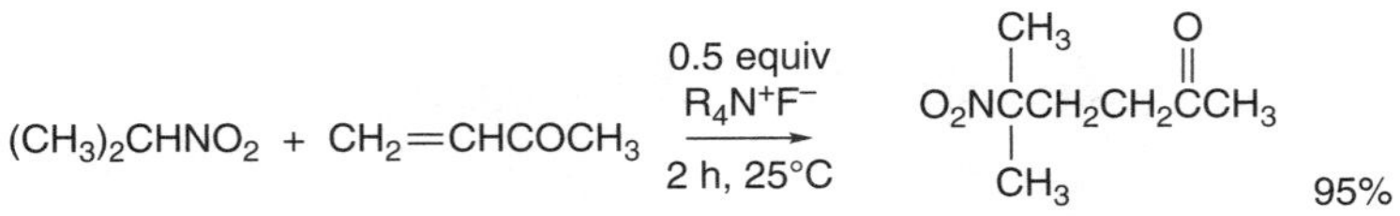

Ref. 294

As in the case of aldol addition, the scope of conjugate addition reactions can be extended by the use of techniques for regio- and stereospecific preparation of enolates and enolate equivalents. If the reaction is carried out with a stoichiometrically formed enolate in the absence of a proton source, the initial product is the enolate of the adduct. The replacement of a π bond by a σ bond ensures a favorable ΔH.

Among Michael acceptors that have been shown to react with ketone and ester enolates under kinetic conditions are methyl α-trimethylsilylvinyl ketone,[295] methyl α-methylthioacrylate,[296] methyl methylthiovinyl sulfoxide,[297] and ethyl α-cyanoacrylate.[298] Each of these acceptors benefits from a second anion-stabilizing substituent. The latter class of acceptors has been found to be capable of generating contiguous quaternary carbon centers.

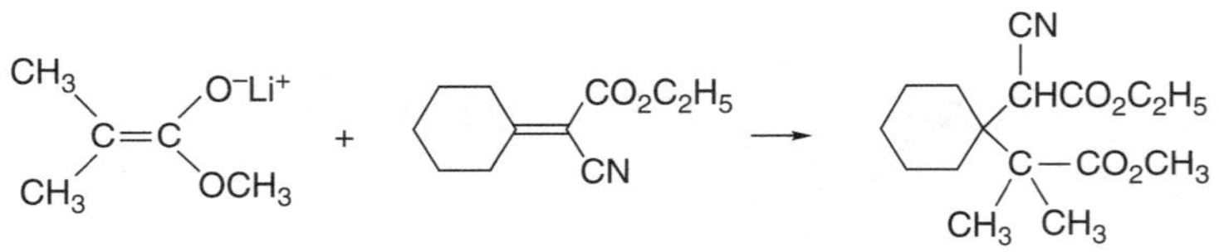

Ref. 298

Several examples of conjugate addition of carbanions carried out under aprotic conditions are given in Scheme 2.24. The reactions are typically quenched by addition of a proton source to neutralize the enolate. It is also possible to trap the adduct by silylation or, as we will see in Section 2.6.2, to carry out a tandem alkylation. Lithium enolates preformed by reaction with LDA in THF react with enones to give 1,4-diketones (Entries 1 and 2). Entries 3 and 4 involve addition of ester enolates to enones. The reaction in Entry 3 gives the 1,2-addition product at −78°C but isomerizes to the 1,4-product at 25°C. Esters of 1,5-dicarboxylic acids are obtained by addition of ester enolates to α,β-unsaturated esters (Entry 5). Entries 6 to 8 show cases of

293. S. Tori, H. Tanaka, and Y. Kobayashi, *J. Org. Chem.*, **42**, 3473 (1977).

294. J. H. Clark, J. M. Miller, and K.-H. So, *J. Chem. Soc., Perkin Trans. I*, 941 (1978).

295. G. Stork and B. Ganem, *J. Am. Chem. Soc.*, **95**, 6152 (1973).

296. R. J. Cregge, J. L. Herrmann, and R. H. Schlessinger, *Tetrahedron Lett.*, 2603 (1973).

297. J. L. Herrmann, G. R. Kieczykowski, R. F. Romanet, P. J. Wepple, and R. H. Schlessinger, *Tetrahedron Lett.*, 4711 (1973).

298. R. A. Holton, A. D. Williams, and R. M. Kennedy, *J. Org. Chem.*, **51**, 5480 (1986).

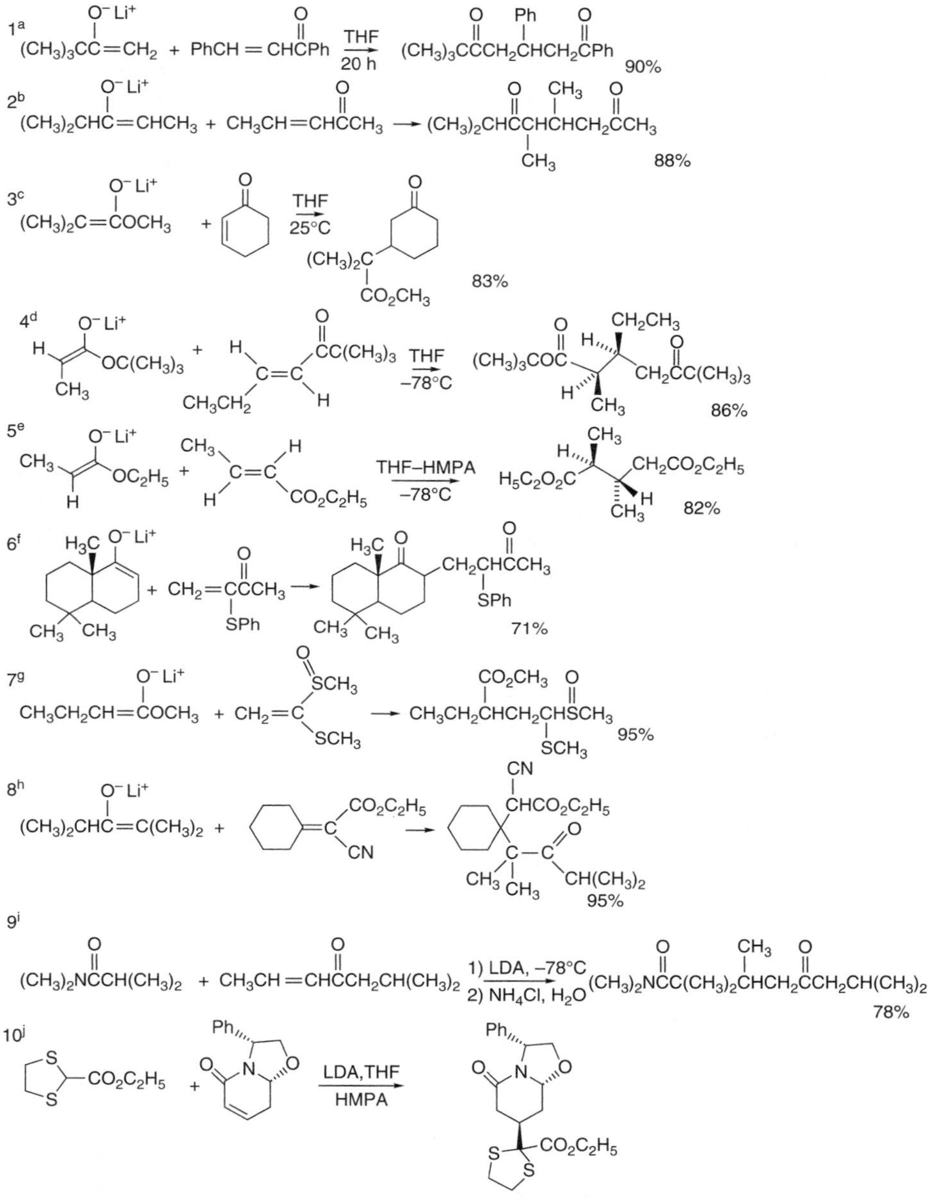

a. J. Bertrand, L. Gorrichon, and P. Maroni, *Tetrahedron*, **40**, 4127 (1984).
b. D. A. Oare and C. H. Heathcock, *Tetrahedron Lett.*, **27**, 6169 (1986).
c. A. G. Schultz and Y. K. Yee, *J. Org. Chem.*, **41**, 4044 (1976).
d. C. H. Heathcock and D. A. Oare, *J. Org. Chem.*, **50**, 3022 (1985).
e. M. Yamaguchi, M. Tsukamoto, S. Tanaka, and I. Hirao, *Tetrahedron Lett.*, **25**, 5661 (1984).
f. K. Takaki, M. Ohsugi, M. Okada, M. Yasumura, and K. Negoro, *J. Chem. Soc., Perkin Trans. 1*, 741 (1984).
g. J. L. Herrmann, G. R. Kieczykowski, R. F. Romanet, P. J. Wepplo, and R. H. Schlessinger, *Tetrahedron Lett.*, 4711 (1973).
h. R. A. Holton, A. D. Williams, and R. M. Kennedy, *J. Org. Chem.*, **51**, 5480 (1986).
i. D. A. Oare, M. A. Henderson, M. A. Sanner, and C. H. Heathcock, *J. Org. Chem.*, **55**, 132 (1990).
j. M. Amat, M. Perez, N. Llor, and J. Bosch, *Org. Lett.*, **4**, 2787 (2002).

enolate addition to acceptors with two anion-stabilizing groups. Entry 8 is noteworthy in that it creates two contiguous quaternary carbons. Entry 9 shows an addition of an amide anion. Entry 10 is a case of an enolate stabilized by both the dithiane ring and ester substituent. The acceptor, an α,β-unsaturated lactam, is relatively unreactive but the addition is driven forward by formation of a new σ bond. The chiral moiety incorporated into the five-membered ring promotes enantioselective formation of the new stereocenter.

There have been several studies of the stereochemistry of conjugate addition reactions. If there are substituents on both the nucleophilic enolate and the acceptor, either *syn* or *anti* adducts can be formed.

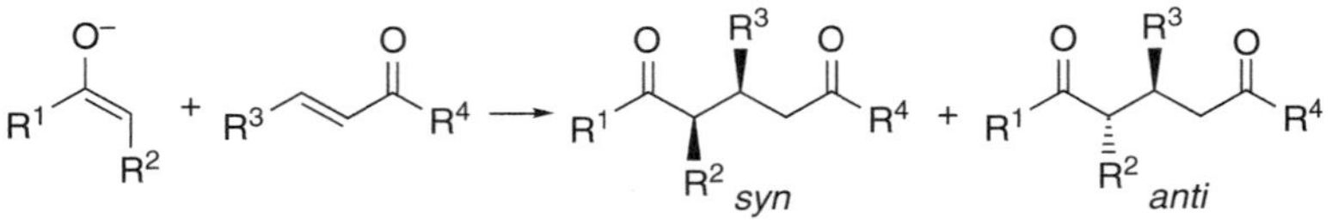

The reaction shows a dependence on the *E*- or *Z*-stereochemistry of the enolate. *Z*-enolates favor *anti* adducts and *E*-enolates favor *syn* adducts. These tendencies can be understood in terms of an eight-membered chelated TS.[299] The enone in this TS is in an *s-cis* conformation. The stereochemistry is influenced by the *s-cis/s-trans* equilibria. Bulky R^4 groups favor the *s-cis* conformer and enhance the stereoselectivity of the reaction. A computational study on the reaction also suggested an eight-membered TS.[300]

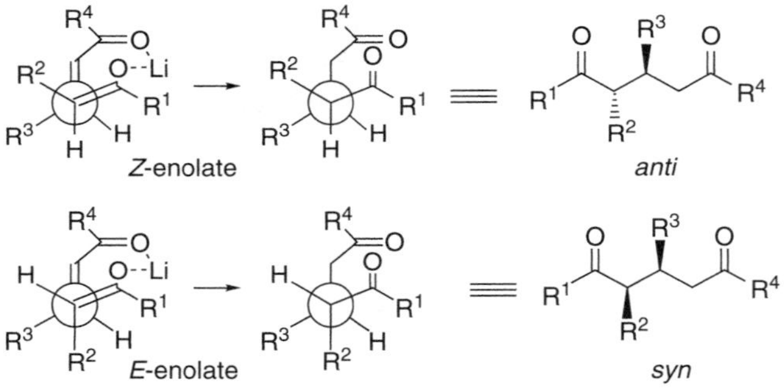

The carbonyl functional groups are the most common both as activating EWG substituents in the acceptor and as the anion-stabilizing group in the enolate, but several other EWGs also undergo conjugate addition reactions. Nitroalkenes are excellent acceptors. The nitro group is a strong EWG and there is usually no competition from nucleophilic attack on the nitro group.

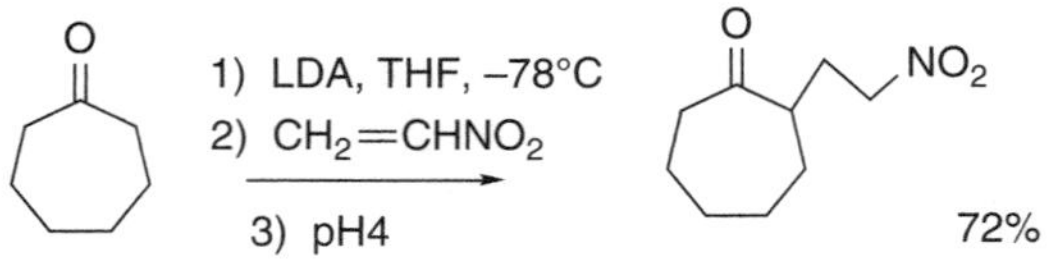

Ref. 301

299. D. Oare and C. H. Heathcock, *J. Org. Chem.*, **55**, 157 (1990); D. A. Oare and C. H. Heathcock, *Top. Stereochem.*, **19**, 227 (1989); A. Bernardi, *Gazz. Chim. Ital.*, **125**, 539 (1995).

300. A. Bernardi, A. M. Capelli, A. Cassinari, A. Comotti, C. Gennari, and C. Scolastico, *J. Org. Chem.*, **57**, 7029 (1992).

301. R. J. Flintoft, J. C. Buzby, and J. A. Tucker, *Tetrahedron Lett.*, **40**, 4485 (1999).

The nitro group can be converted to a ketone by hydrolysis of the nitronate anion, permitting the synthesis of 1,4-dicarbonyl compounds.

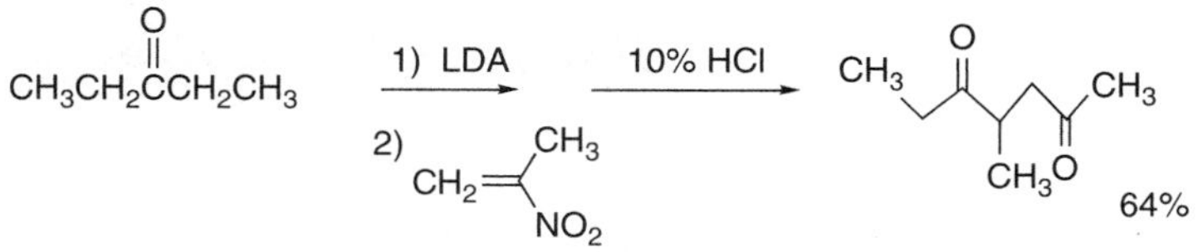

Ref. 302

Anions derived from nitriles can act as nucleophiles in conjugate addition reactions. A range of substituted phenylacetonitriles undergoes conjugate addition to 4-phenylbut-3-en-2-one.

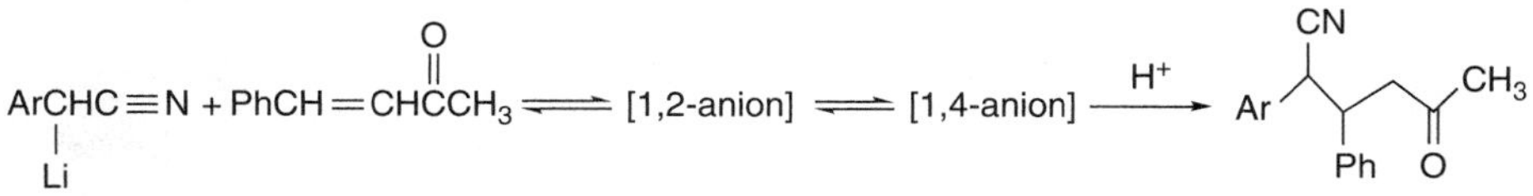

The reaction occurs via the 1,2-adduct, which isomerizes to the 1,4-adduct,[303] and there is an energy difference of about 5 kcal/mol in favor of the 1,4-adduct. With the parent compound in THF, the isomerization reaction has been followed kinetically and appears to occur in two phases. The first part of the reaction occurs with a half-life of a few minutes, and the second with a half-life of about an hour. A possible explanation is the involvement of dimeric species, with the homodimer being more reactive than the heterodimer.

$$[1,2\text{-anion}]_2 \xrightarrow{\;k = 30 \times 10^{-4}s^{-1}\;} [1,2\text{-anion}][1,4\text{-anion}] \xrightarrow{\;k = 2 \times 10^{-4}s^{-1}\;} [1,4\text{-anion}]_2$$

A very important extension of the conjugate addition reaction is discussed in Chapter 8. Organocopper reagents have a strong preference for conjugate addition. Organocopper nucleophiles do not require anion-stabilizing substituents, and they allow conjugate addition of alkyl, alkenyl, and aryl groups to electrophilic alkenes.

2.6.2. Conjugate Addition with Tandem Alkylation

When conjugate addition is carried out under aprotic conditions with stoichiometric formation of the enolate, the adduct is present as an enolate until the reaction mixture is quenched with a proton source. It is therefore possible to effect a second reaction of the enolate by addition of an alkyl halide or sulfonate to the solution of the adduct enolate, which results in an alkylation. This reaction sequence permits the formation of two new C—C bonds.

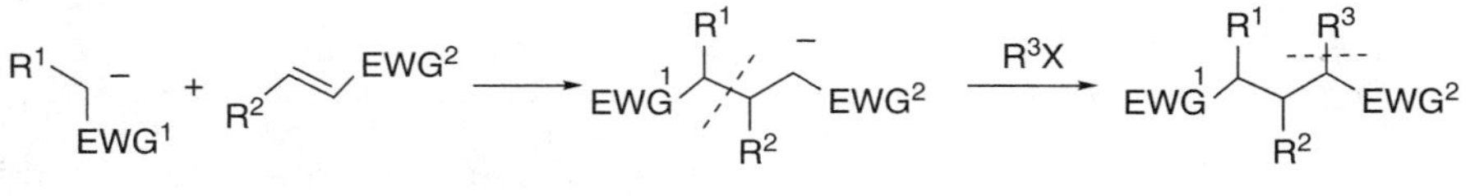

[302.] M. Miyashita, B. Z. Awen, and A. Yoshikoshi, *Synthesis*, 563 (1990).
[303.] H. J. Reich, M. M. Biddle, and R. J. Edmonston, *J. Org. Chem.*, **70**, 3375 (2005).

Several examples of tandem conjugate addition-alkylation follow.

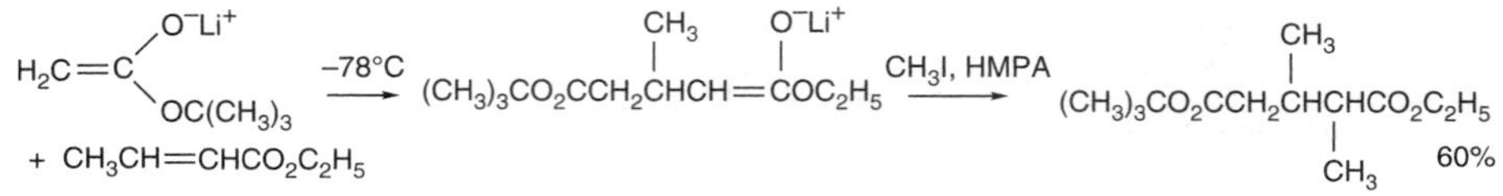

Ref. 304

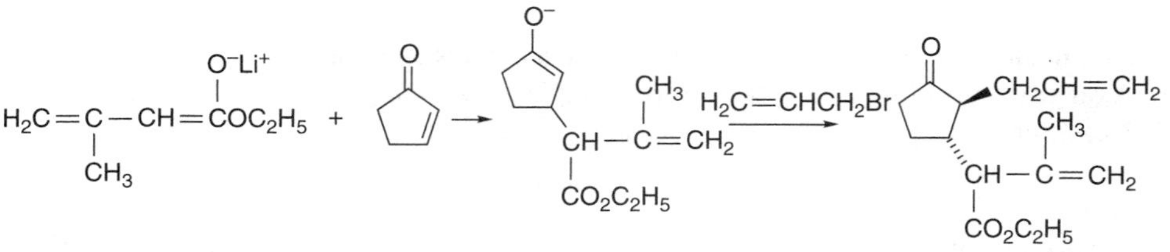

Ref. 305

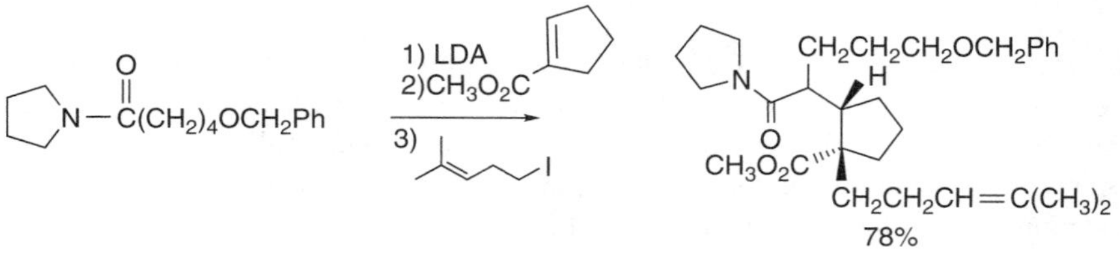

Ref. 306

Tandem conjugate addition-alkylation has proven to be an efficient means of introducing groups at both α- and β-positions at enones.[307] As with simple conjugate addition, organocopper reagents are particularly important in this application, and they are discussed further in Section 8.1.2.3.

2.6.3. Conjugate Addition by Enolate Equivalents

Conditions for effecting conjugate addition of neutral enolate equivalents such as silyl enol ethers in the presence of Lewis acids have been developed and are called *Mukaiyama-Michael reactions*. Trimethylsilyl enol ethers can be caused to react with electrophilic alkenes by use of $TiCl_4$. These reactions proceed rapidly even at $-78°$ C.[308]

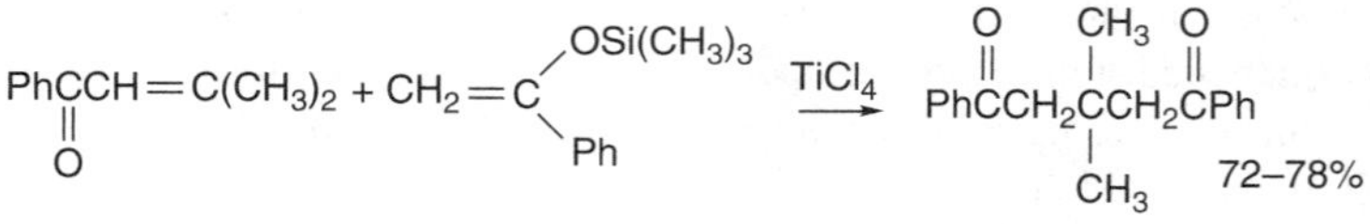

Ref. 309

304. M. Yamaguchi, M. Tsukamoto, and I. Hirao, *Tetrahedron Lett.*, **26**, 1723 (1985).
305. W. Oppolzer, R. P. Heloud, G. Bernardinelli, and K. Baettig, *Tetrahedron Lett.*, **24**, 4975 (1983).
306. C. H. Heathcock, M. M. Hansen, R. B. Ruggeri, and J. C. Kath, *J. Org. Chem.*, **57**, 2544 (1992).
307. For additional examples, see M. C. Chapdelaine and M. Hulce, *Org. React.*, **38**, 225 (1990); E. V. Gorobets, M. S. Miftakhov, and F. A. Valeev, *Russ. Chem. Rev.*, **69**, 1001 (2000).
308. K. Narasaka, K. Soai, Y. Aikawa, and T. Mukaiyama, *Bull. Chem. Soc. Jpn.*, **49**, 779 (1976).
309. K. Narasaka, *Org. Synth.*, **65**, 12 (1987).

Silyl ketene acetals also undergo conjugate addition. For example, $Mg(ClO_4)_2$ and $LiClO_4$ catalyze addition of silyl ketene acetals to enones.

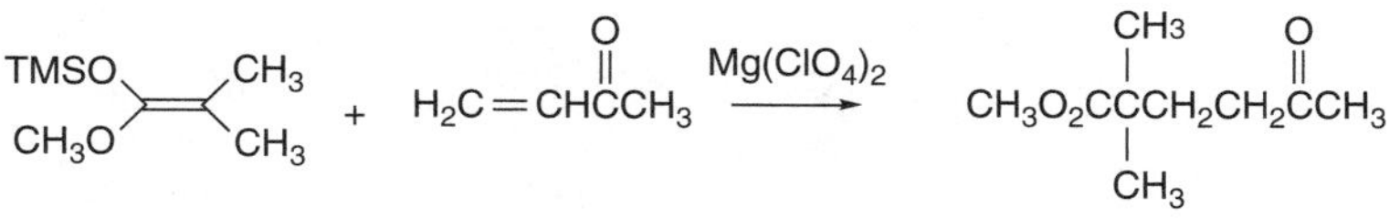

Ref. 310

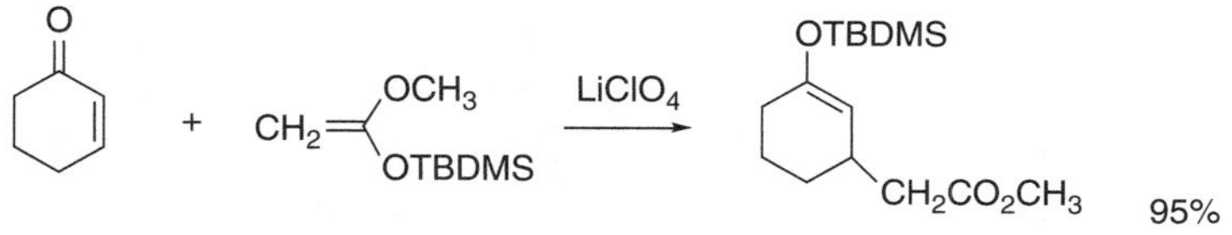

95%

Ref. 311

Initial stereochemical studies suggested that the Mukaiyama-Michael reaction proceeds through an open TS, since there was a tendency to favor *anti* diastereoselectivity, regardless of the silyl enol ether configuration.[312]

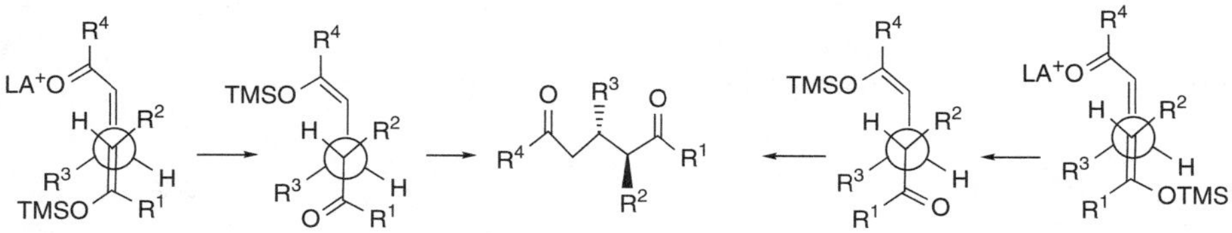

The stereoselectivity can be enhanced by addition of $Ti(O\text{-}i\text{-}Pr)_4$. The active nucleophile under these conditions is expected to be an "ate" complex in which a much larger $Ti(O\text{-}i\text{-}Pr)_4$ group replaces Li^+ as the Lewis acid.[313] Under these conditions, the *syn:anti* ratio is dependent on the stereochemistry of the enolate.

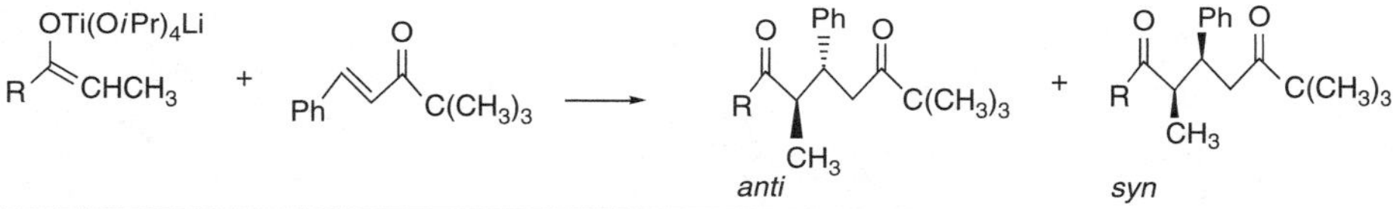

R	Configuration	*anti:syn*	Yield(%)
Et	Z	95:5	69
Ph	Z	> 92:8	85
i-Pr	Z	> 97:3	65
i-Pr	E	17:83	91

Silyl acetals of thiol esters have also been studied. With $TiCl_4$ as the Lewis acid, there is correspondence between the configuration of the silyl thioketene acetal and the adduct stereochemistry.[314] *E*-Isomers show high *anti* selectivity, whereas *Z*-isomers are less selective.

310. S. Fukuzumi, T. Okamoto, K. Yasui, T. Suenobu, S. Itoh, and J. Otera, *Chem. Lett.*, 667 (1997).

311. P. A. Grieco, R. J. Cooke, K. J. Henry, and J. M. Vander Roest, *Tetrahedron Lett.*, **32**, 4665 (1991).

312. C. H. Heathcock, M. H. Norman, and D. E. Uehling, *J. Am. Chem. Soc.*, **107**, 2797 (1985).

313. A. Bernardi, P. Dotti, G. Poli, and C. Scolastico, *Tetrahedron*, **48**, 5597 (1992); A. Bernardi, M. Cavicchioi, and C. Scolastico, *Tetrahedron*, **49**, 10913 (1993).

314. Y. Fujita, J. Otera, and S. Fukuzumi, *Tetrahedron*, **52**, 9419 (1996).

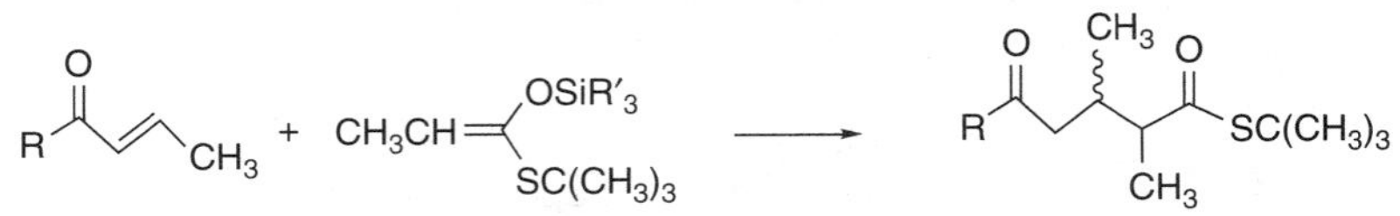

R	SiR′$_3$	configuration	*syn:anti*
t-Bu	TBDMS	E	5:95
t-Bu	TBDMS	Z	91:9
Ph	TBDMS	E	7:93
Ph	TBDMS	Z	54:46
CH$_3$	TBDMS	E	8:92
CH$_3$	TBDMS	Z	40:60

Stannyl enolates give good addition yields in the presence of a catalytic amount of $n\text{-}(C_4H_9)_4N^+Br^-$.[315] The bromide ion plays an active role in this reaction by forming a more reactive species via coordination at the tin atom.

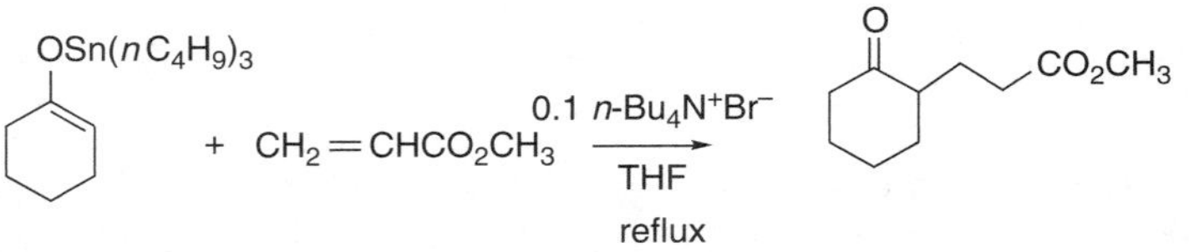

It is believed that this reaction involves the formation of the α-stannyl ester. Metals such as lithium that form ionic enolates would be more likely to reverse the addition step.

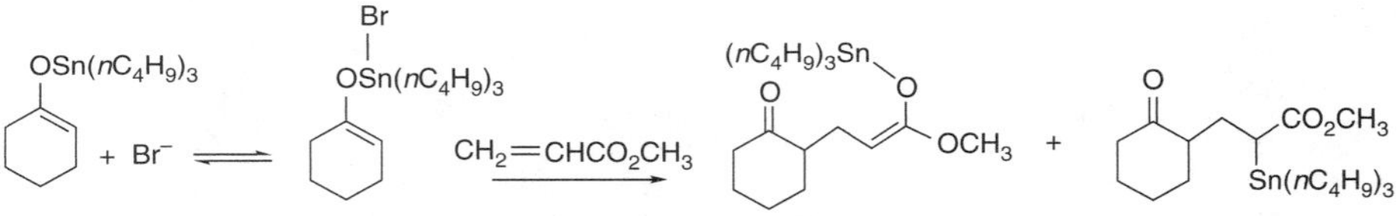

Nitroalkenes are also reactive Michael acceptors under Lewis acid–catalyzed conditions. Titanium tetrachloride or stannic tetrachloride can induce addition of silyl enol ethers. The initial adduct is trapped in a cyclic form by trimethylsilylation.[316] Hydrolysis of this intermediate regenerates the carbonyl group and also converts the *aci*-nitro group to a carbonyl.[317]

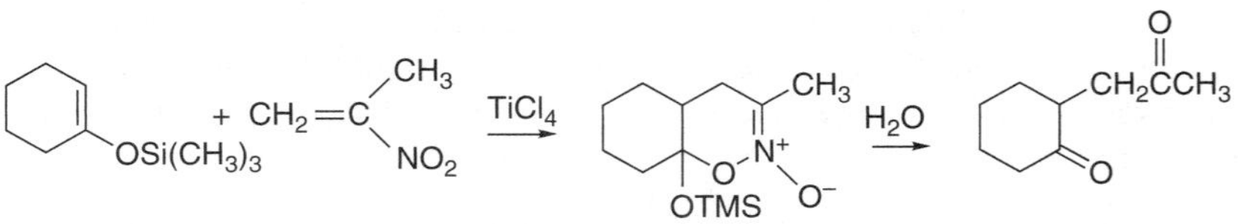

315. M. Yasuda, N. Ohigashi, I. Shibata, and A. Baba, *J. Org. Chem.*, **64**, 2180 (1999).
316. A. F. Mateos and J. A. de la Fuento Blanco, *J. Org. Chem.*, **55**, 1349 (1990).
317. M. Miyashita, T. Yanami, T. Kumazawa, and A. Yoshikoshi, *J. Am. Chem. Soc.*, **106**, 2149 (1984).

Fluoride ion can also induce reaction of silyl ketene acetals with electrophilic alkenes. The fluoride source in these reactions is *tris*-(dimethylamino)sulfonium difluorotrimethylsilicate (TASF).

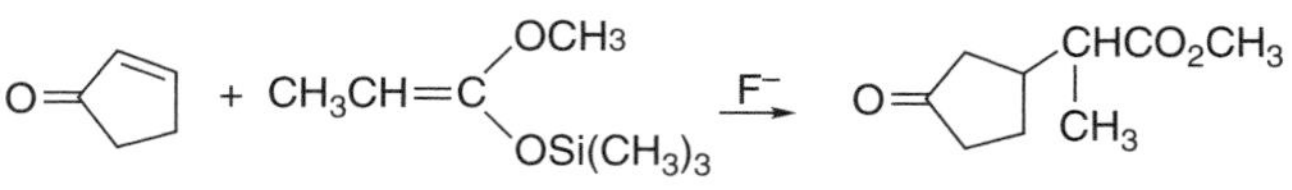

Ref. 318

Enamines also react with electrophilic alkenes to give conjugate addition products. The addition reactions of enamines of cyclohexanones show a strong preference for attack from the axial direction.[319] This is anticipated on stereoelectronic grounds because the π orbital of the enamine is the site of nucleophilicity.

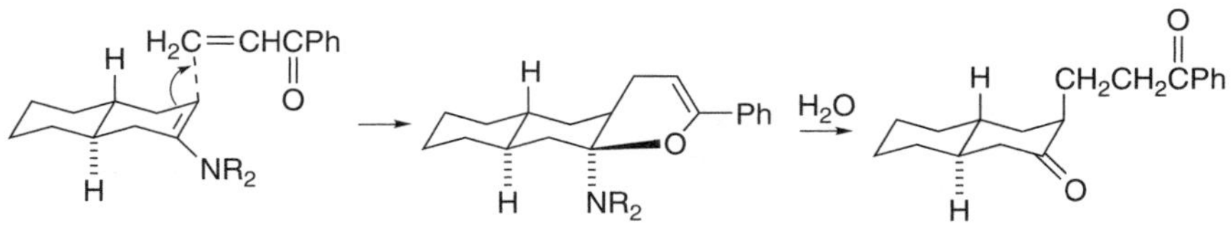

Scheme 2.25 shows some examples of additions of enolate equivalents. A range of Lewis acid catalysts has been used in addition to $TiCl_4$ and $SnCl_4$. Entry 1 shows uses of a lanthanide catalyst. Entry 2 employs $LiClO_4$ as the catalyst. The reaction in Entry 3 includes a chiral auxiliary that controls the stereoselectivity; the chiral auxiliary is released by a cyclization using *N*-methylhydroxylamine. Entries 4 and 5 use the triphenylmethyl cation as a catalyst and Entries 6 and 7 use trimethylsilyl triflate and an enantioselective catalyst, respectively.

2.6.4. Control of Facial Selectivity in Conjugate Addition Reactions

As is the case for aldol addition, chiral auxiliaries and catalysts can be used to control stereoselectivity in conjugate addition reactions. Oxazolidinone chiral auxiliaries have been used in both the nucleophilic and electrophilic components under Lewis acid–catalyzed conditions. *N*-Acyloxazolidinones can be converted to nucleophilic titanium enolates with $TiCl_3(O\text{-}i\text{-Pr})$.[320]

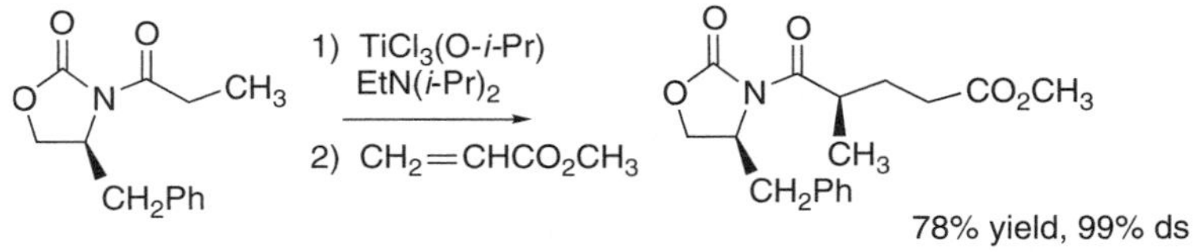

78% yield, 99% ds

318. T. V. Rajan Babu, *J. Org. Chem.*, **49**, 2083 (1984).
319. E. Valentin, G. Pitacco, F. P. Colonna, and A. Risalti, *Tetrahedron*, **30**, 2741 (1974); M. Forchiassin, A. Risalti, C. Russo, M. Calligaris, and G. Pitacco, *J. Chem. Soc.*, 660 (1974).
320. D. A. Evans, M. T. Bilodeau, T. C. Somers, J. Clardy, D. Cherry, and Y. Kato, *J. Org. Chem.*, **56**, 5750 (1991).

Scheme 2.25. Conjugate Addition of Enolate Equivalents

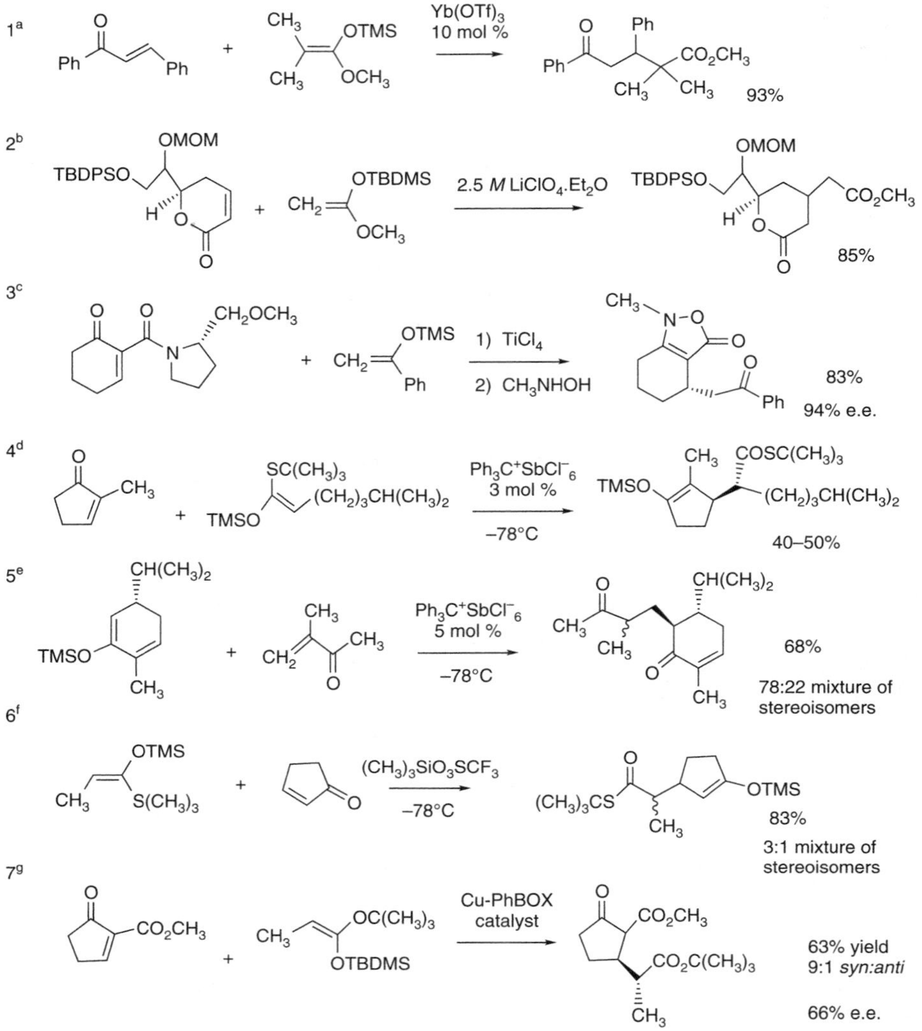

a. S. Kobayahi, I. Hachiya, T. Takahori, M. Araki, and H. Ishitani, *Tetrahedron Lett.*, **33**, 6815 (1992).
b. P. A. Grieco, R. J. Cooke, K. J. Henry, and J. M. Vander Roest, *Tetrahedron Lett.*, **32**, 4665 (1991).
c. A. G. Schultz and H. Lee, *Tetrahedron Lett.*, **33**, 4397 (1992).
d. P. Grzywacz, S. Marczak, and J. Wicha, *J. Org. Chem.* **62**, 5293 (1997).
e. A. V. Baranovsky, B. J. M. Jansen, T. M Meulemans, and A. de Groot, *Tetrahedron*, **54**, 5623 (1998).
f. K. Michalak and J. Wicha, *Polish J. Chem.*, **78**, 205 (2004).
g. A. Bernardi, G. Colombo and C. Scolastico, *Tetrahedron Lett.*, **37**, 8921 (1996).

Unsaturated acyl derivatives of oxazolidinones can be used as acceptors, and these reactions are enantioselective in the presence of chiral *bis*-oxazoline catalysts.[321] Silyl ketene acetals of thiol esters are good reactants and the stereochemistry depends on the ketene acetal configuration. The Z-isomer gives higher diastereoselectivity than the E-isomer.

[321.] D. A. Evans, K. A. Scheidt, J. N. Johnston, and M. C. Willis, *J. Am. Chem. Soc.*, **123**, 4480 (2001).

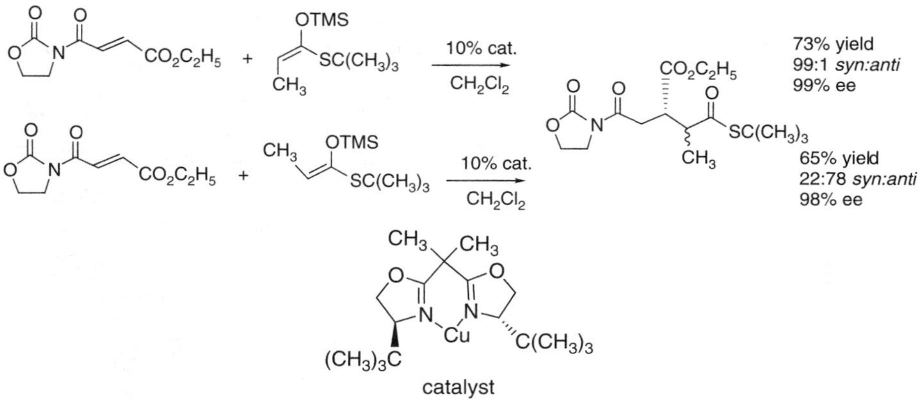

The above examples contain an ester group that acts as a second activating group. The reactions are also accelerated by including one equivalent of $(CF_3)_2CHOH$. This alcohol functions by promoting solvolysis of a dihydropyran intermediate that otherwise inhibits the catalyst.

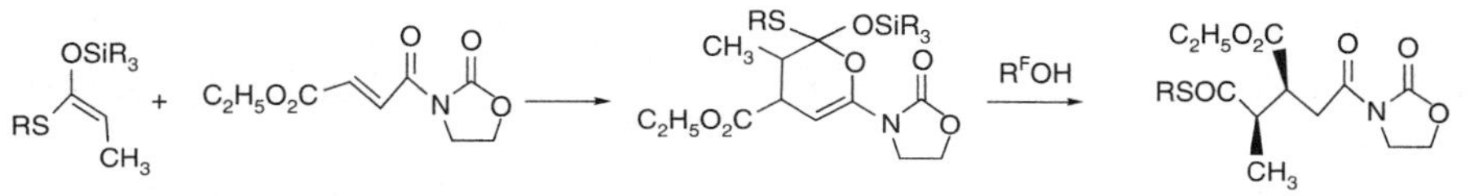

Alkylidenemalonate esters are also good acceptors in reactions with silyl ketene acetals of thiol esters under very similar conditions.[322]

A number of other chiral catalysts can promote enantioselective conjugate additions of silyl enol ethers, silyl ketene acetals, and related compounds. For example, an oxazaborolidinone derived from allothreonine achieves high enantioselectivity in additions of silyl thioketene acetals.[323] The optimal conditions for this reaction also include a hindered phenol and an ether additive.

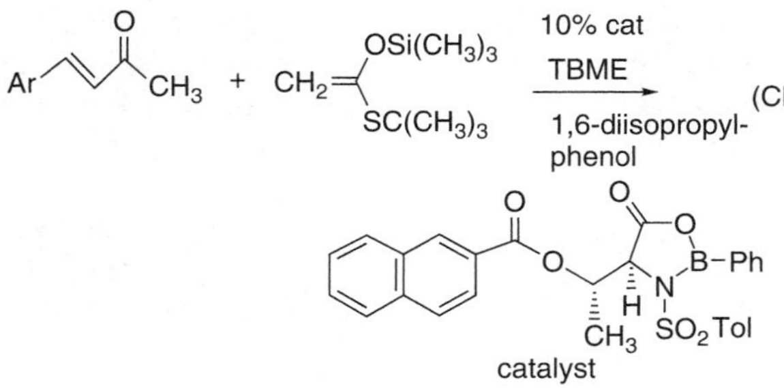

Enantioselectivity has been observed for acyclic ketones, using proline as a catalyst. Under optimum conditions, ds > 80% and e.e. > 70% were observed.[324] These

[322.] D. A. Evans, T. Rovis, M. C. Kozlowski, C. W. Downey, and J. S. Tedrow, *J. Am. Chem. Soc.*, **122**, 9134 (2000).

[323.] X. Wang, S. Adachi, H. Iwai, H. Takatsuki, K. Fujita, M. Kubo, A. Oku, and T. Harada, *J. Org. Chem.*, **68**, 10046 (2003).

[324.] D. Enders and A. Seki, *Synlett*, 26 (2002).

reactions presumably involve the proline-derived enamine. (See Section 2.1.5.6 for a discussion of enantioselective reactions of proline enamines.)

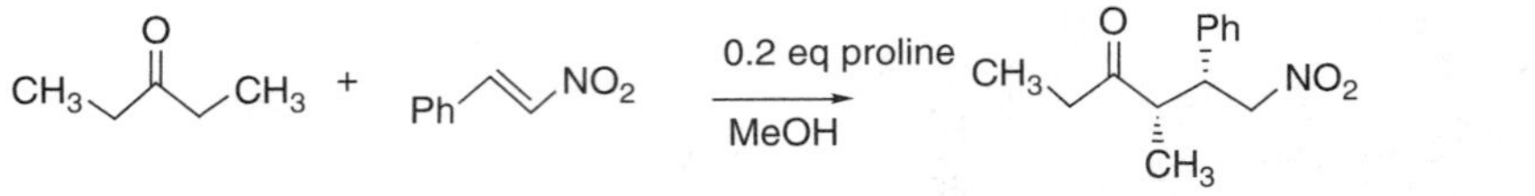

74% yield, 88% ds, 76% ee

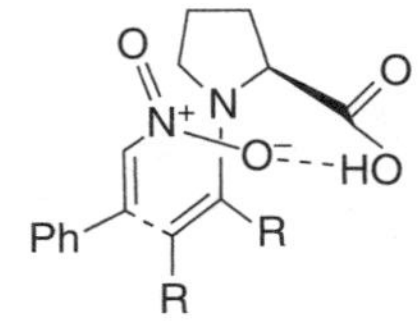

Enantioselective additions of β-dicarbonyl compounds to β-nitrostyrenes have been achieved using *bis*-oxazolidine catalysts. This method was used in an enantio-selective synthesis of the antidepressant drug rolipram.[325]

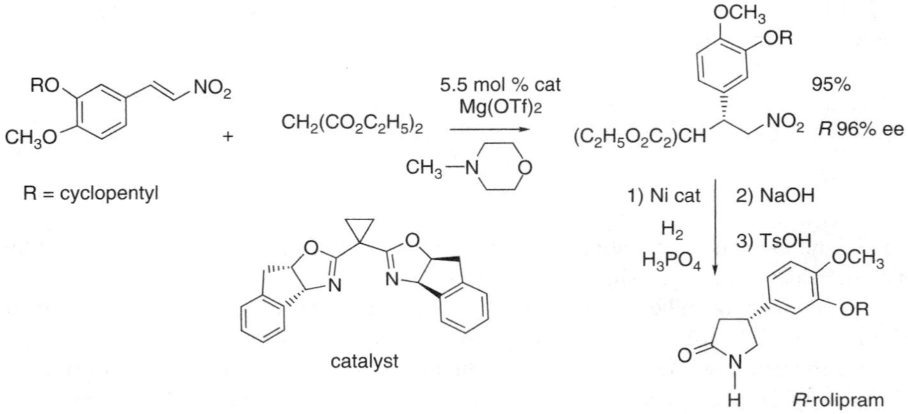

R = cyclopentyl

catalyst

R-rolipram

Enantioselectivity can also be based on structural features present in the reactants. A silyl substituent has been used to control stereochemistry in both cyclic and acyclic systems. The silyl substituent can then be removed by TBAF.[326] As with enolate alkylation (see p. 32), the steric effect of the silyl substituent directs the approach of the acceptor to the opposite face.

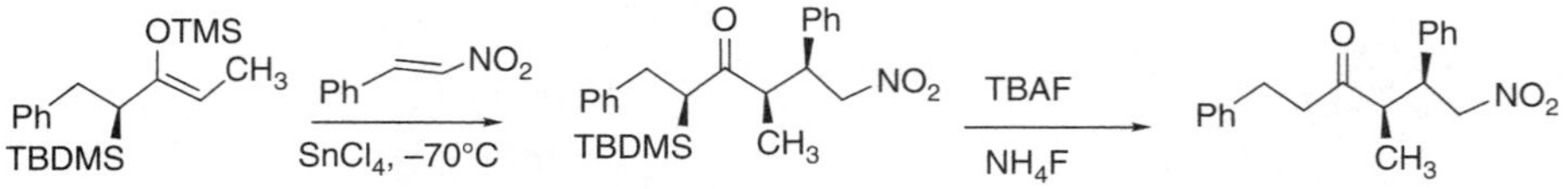

dr > 96%, ee > 96%

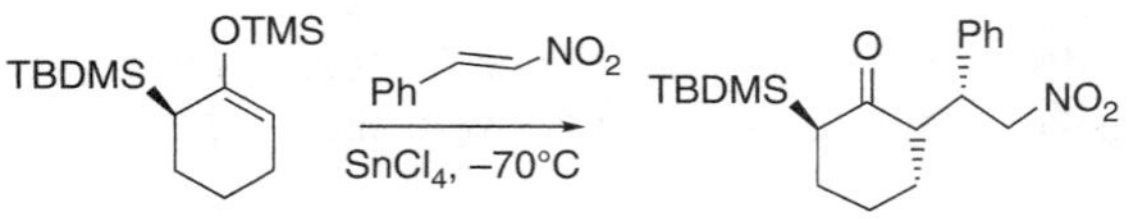

74%, > 91% ee

325. D. M. Barnes, J. Ji, M. G. Fickes, M. A. Fitzgerald, S. A. King, H. E. Morton, F. A. Plagge, M. Preskill, S. H. Wagaw, S. J. Wittenberger, and J. Zhang, *J. Am. Chem. Soc.*, **124**, 13097 (2002).
326. D. Enders and T. Otten, *Synlett*, 747 (1999).

High stereoselectivity is also observed in the addition of an enamine using 2-methoxymethylpyrrolidine as the amine.[327]

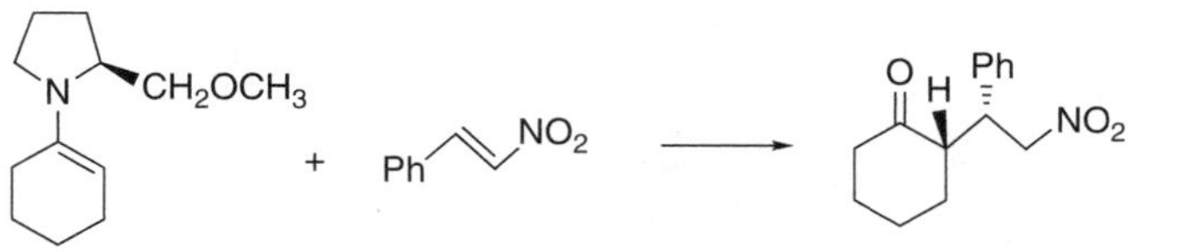

2.6.5. Conjugate Addition of Organometallic Reagents

There are relatively few examples of organolithium compounds acting as nucleophiles in conjugate addition. Usually, organolithium compounds react at the carbonyl group, to give 1,2-addition products. Here, we consider a few cases of organometallic reagents that give conjugate addition products. There are a very large number of copper-mediated conjugate additions, and we discuss these reactions in Section 8.1.2.3.

Alkyl and aryllithium compounds have been found to undergo 1,4-addition with the salts of α,β-unsaturated acids.[328] This result reflects the much reduced reactivity of the carboxylate carbonyl group as an electrophile.

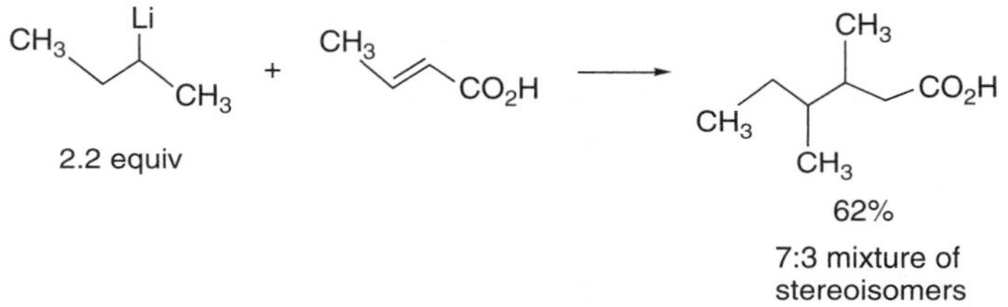

2.2 equiv

62%

7:3 mixture of
stereoisomers

α,β-Unsaturated amides have been found to be good reactants toward organometallic reagents. These reactions involve the deprotonated amide ion, which is less susceptible to 1,2-addition than ketones and esters.

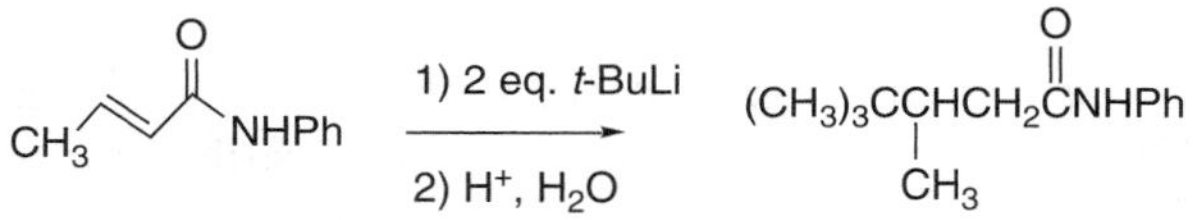

Ref. 329

Similar reactions have also been observed with tertiary amides and the adducts can be alkylated by tandem S_N2 reactions.

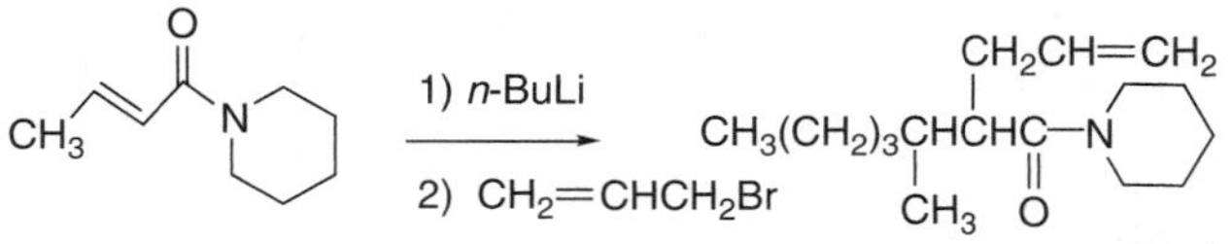

90%

Ref. 330

327. S. J. Blarer, W. B. Schweizer, and D. Seebach, *Helv. Chim. Acta*, **65**, 1637 (1982); S. J. Blarer and D. Seebach, *Chem. Ber.*, **116**, 2250 (1983).
328. B. Plunian, M. Vaultier, and J. Mortier, *Chem. Commun.*, 81 (1998).
329. J. E. Baldwin and W. A. Dupont, *Tetrahedron Lett.*, **21**, 1881 (1980).
330. G. B. Mpango, K. K. Mahalanabis, S. Mahdavi-Damghani, and V. Snieckus, *Tetrahedron Lett.*, **21**, 4823 (1980).

Lithiated *N*-allylcarbamates add to nitroalkenes. In the presence of (−)-sparteine, this reaction is both diastereoselective (*anti*) and enantioselective.[331]

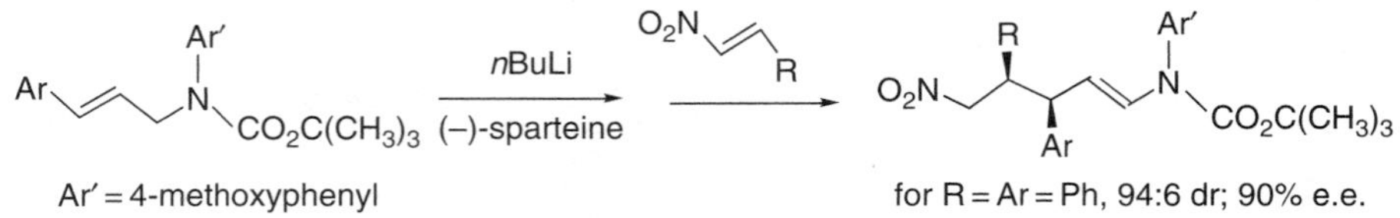

Ar′ = 4-methoxyphenyl

for R = Ar = Ph, 94:6 dr; 90% e.e.

The enantioselectivity is due to the retention of the chiral sparteine in the lithiated reagent. The adducts have been used to synthesize a number of pyrrolidine and piperidine derivatives.

Several mixed organozinc reagents having a trimethylsilylmethyl group as the nonreacting substituent add to enones under the influence of TMS-Br.[332] The types of groups that can be added include alkyl, aryl, heteroaryl, and certain functionalized alkyl groups, including 5-pivaloyloxypentyl and 3-ethoxycarbonylpropyl.

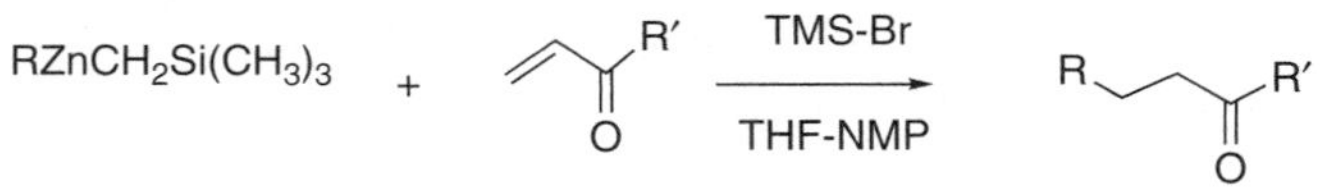

α,β-Unsaturated aldehydes and esters, as well as nitroalkenes, can also function as acceptors under these conditions. Dialkylzinc reagents add to β-nitrostyrene in the presence of TADDOL-TiCl$_2$.[333]

$(C_8H_{17})_2Zn$ + $PhCH=CHNO_2$ $\xrightarrow{\text{TADDOL-TiCl}_2}$

87%, 76% ee

2.6.6. Conjugate Addition of Cyanide Ion

Cyanide ion acts as a carbon nucleophile in the conjugate addition reaction. The pK of HCN is 9.3, so addition in hydroxylic solvents is feasible. An alcoholic solution of potassium or sodium cyanide is suitable for simple compounds.

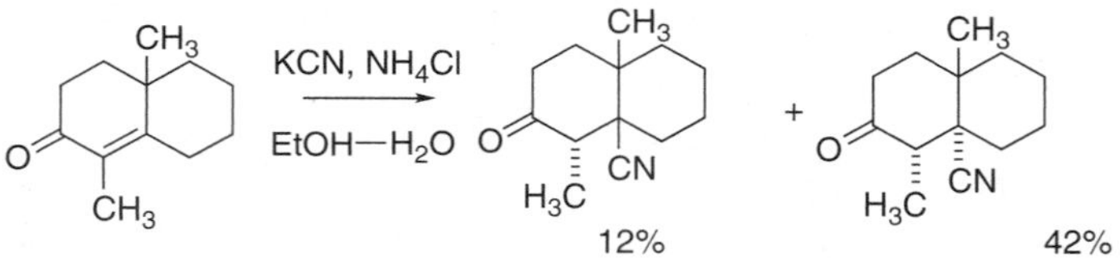

Ref. 334

Cyanide addition has also been done under Lewis acid catalysis. Triethylaluminum-hydrogen cyanide and diethylaluminum cyanide are useful reagents for conjugate

331. T. A. Johnson, D. O. Jang, B. W. Slafer, M. D. Curtis, and P. Beak, *J. Am. Chem. Soc.*, **124**, 11689 (2002).

332. P. Jones, C. K. Reddy, and P. Knochel, *Tetrahedron*, **54**, 1471 (1998).

333. H. Schaefer and D. Seebach, *Tetrahedron*, **51**, 2305 (1995).

334. O. R. Rodig and N. J. Johnston, *J. Org. Chem.*, **34**, 1942 (1969).

addition of cyanide. The latter is the more reactive of the two reagents. These reactions presumably involve the coordination of the aluminum reagent at the carbonyl oxygen.

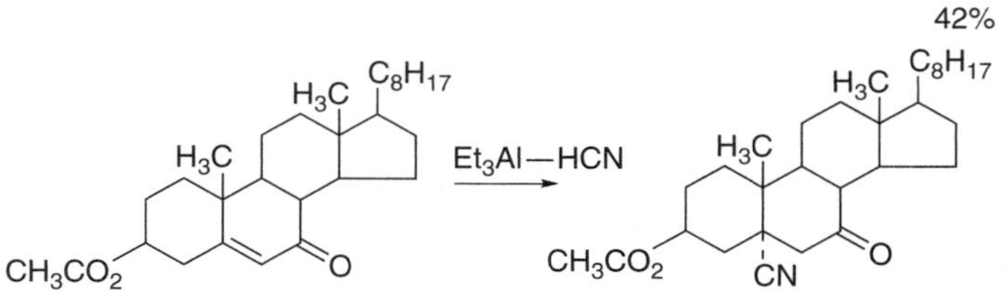

Ref. 335

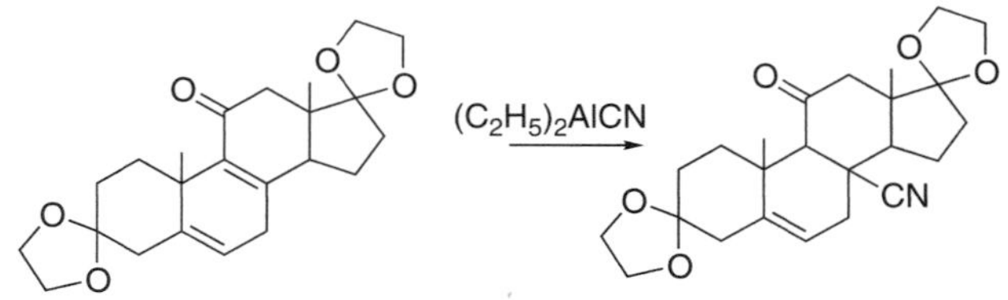

Ref. 336

Diethylaluminum cyanide mediates conjugate addition of cyanide to α, β-unsaturated oxazolines. With a chiral oxazoline, 30–50% diastereomeric excess can be achieved. Hydrolysis gives partially resolved α-substituted succinic acids. The rather low enantioselectivity presumably reflects the small size of the cyanide ion.

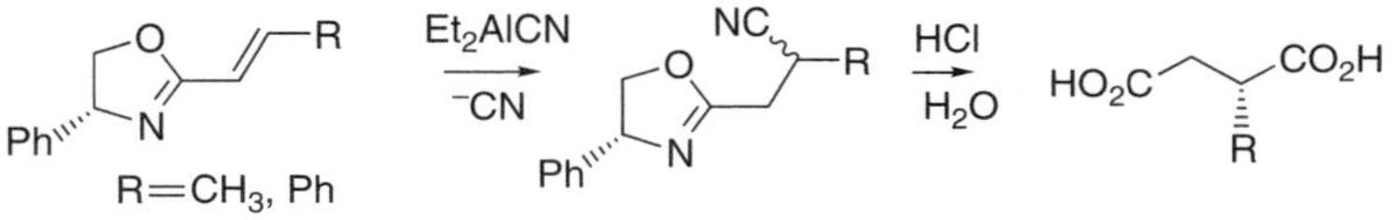

R = CH$_3$, d.e. = 50–56%; e.e. = 45–50%
R = Ph,   d.e. = 45–52%; e.e. = 57%

Ref. 337

A chiral aluminum-salen catalyst gives good enantioselectivity in the addition of cyanide (from TMS-CN) to unsaturated acyl imides.[338]

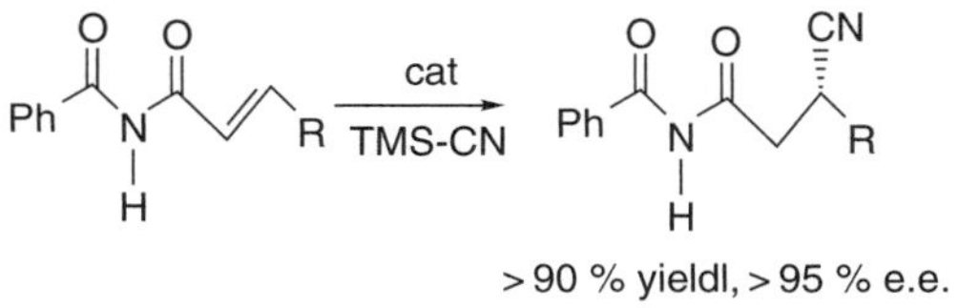

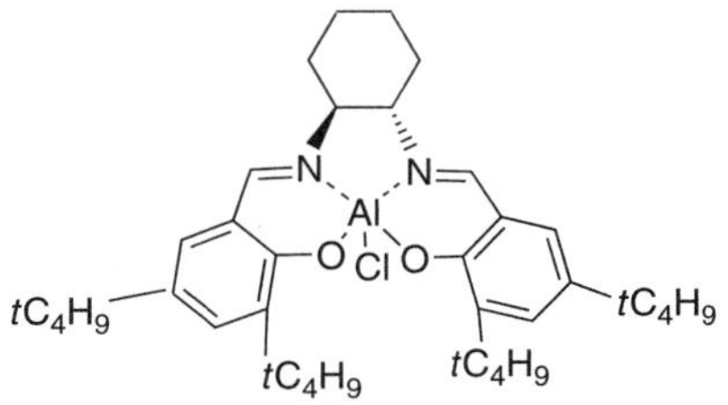

> 90 % yieldl, > 95 % e.e.

catalyst

335. W. Nagata and M. Yoshioka, *Org. Synth.*, **52**, 100 (1972).
336. W. Nagata, M. Yoshioka, and S. Hirai, *J. Am. Chem. Soc.*, **94**, 4635 (1972).
337. M. Dahuron and N. Langlois, *Synlett*, 51 (1996).
338. G. M. Sammis and E. N. Jacobsen, *J. Am. Chem. Soc.*, **125**, 4442 (2003).

General References

Aldol Additions and Condensations

M. Braun in *Advances in Carbanion Chemistry*, Vol. 1. V. Snieckus, ed., JAI Press, Greenwich, CT, 1992.

D. A. Evans, J. V. Nelson, and T. R. Taber, *Top. Stereochem.*, **13**, 1 (1982).

A. S. Franklin and I. Paterson, *Contemp. Org. Synth.*, **1**, 317 (1994).

C. H. Heathcock, in *Comprehensive Carbanion Chemistry*, E. Buncel and T. Durst, ed., Elsevier, Amsterdam, 1984.

C. H. Heathcock, in *Asymmetric Synthesis*, Vol. 3, J. D. Morrison, ed., Academic Press, New York, 1984.

R. Mahrwald, ed. *Modern Aldol Reactions*, Wiley-VCH, 2004.

S. Masamune, W. Choy, J. S. Petersen, and L. R. Sita, *Angew. Chem. Int. Ed. Engl.*, **24**, 1 (1985).

T. Mukaiyama, *Org. React.*, **28**, 203 (1982).

A. T. Nielsen and W. T. Houlihan, *Org. React.*, **16**, 1 (1968).

Annulation Reactions

R. E. Gawley, *Synthesis*, 777 (1976).

M. E. Jung, *Tetrahedron*, **32**, 3 (1976).

Mannich Reactions

F. F. Blicke, *Org. React.*, **1**, 303 (1942).

H. Bohme and M. Heake, in *Iminium Salts in Organic Chemistry*, H. Bohmne and H. G. Viehe, ed., Wiley-Interscience, New York, 1976, pp. 107–223.

M. Tramontini and L. Angiolini, *Mannich Bases: Chemistry and Uses*, CRC Press, Boca Raton, FL, 1994.

Phosphorus-Stabilized Ylides and Carbanions

I. Gosney and A. G. Rowley in *Organophosphorus Reagents in Organic Synthesis*, J. I. G. Cadogan, ed., Academic Press, London, 1979, pp. 17–153.

A. W. Johnson, *Ylides and Imines of Phosphorus*, John Wiley, New York, 1993.

A. Maercker, *Org. React.*, **14**, 270 (1965).

W. S. Wadsworth, Jr., *Org. React.*, **25**, 73 (1977).

Conjugate Addition Reactions

P. Perlmatter, *Conjugate Addition Reactions in Organic Synthesis*, Permagon Press, New York, 1992.

Problems

(*References for these problems will be found on page 1272.*)

2.1 Predict the product formed in each of the following reactions:

(a) γ-butyrolactone + ethyl oxalate $\xrightarrow[\text{2) H}^+]{\text{1) NaOCH}_2\text{CH}_3}$ $C_8H_{10}O_5$

(b) 4-bromobenzaldehyde + ethyl cyanoacetate $\xrightarrow[\text{piperidine}]{\text{ethanol}}$ $C_{12}H_{10}BrNO_2$

(c) $CH_3CH_2CH_2\overset{\overset{\text{O}}{\|}}{C}CH_3$ $\xrightarrow[\begin{array}{c}\text{2) CH}_3\text{CH}_2\text{CHO, 15 min}\\\text{3)H}_2\text{O}\end{array}]{\text{1) LiN(}i\text{-Pr)}_2,\ -78°\text{C}}$ $C_8H_{16}O_2$

(d) furan-CHO + PhCH$_2\overset{\overset{\text{O}}{\|}}{C}CH_3$ $\xrightarrow{\text{NaOH, H}_2\text{O}}$ $C_{13}H_{10}O_2$

(e)

$C_6H_5CH=$ (OAc)(CH₃) $\xrightarrow[\text{2) ZnCl}_2]{\text{1) CH}_3\text{Li, 2 equiv.}}$ $C_{13}H_{17}O_2$
3) n-C₃H₇CHO

(f)

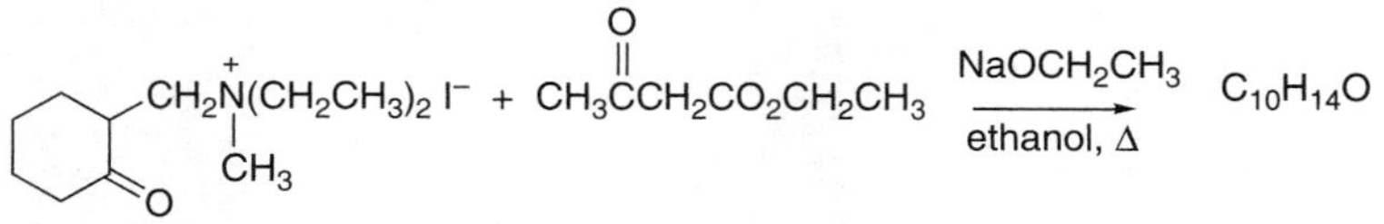

$\text{CH}_2\overset{+}{\text{N}}(\text{CH}_2\text{CH}_3)_2\ I^- + CH_3\overset{O}{\overset{\|}{C}}CH_2CO_2CH_2CH_3$ $\xrightarrow[\text{ethanol, }\Delta]{\text{NaOCH}_2\text{CH}_3}$ $C_{10}H_{14}O$

(g)

$\overset{O}{\underset{}{}}$ –CH₃ + HCO₂CH₂CH₃ $\xrightarrow[\text{ether}]{\text{Na}}$ $C_7H_9O_2Na$

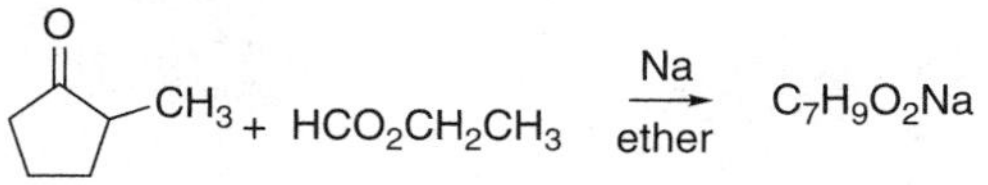

(h)

$\overset{O}{\overset{\|}{C}}CH_3 + (CH_3CH_2O)_2C=O$ $\xrightarrow[\text{toluene}]{\text{NaNH}_2}$ $C_{11}H_{18}O_3$

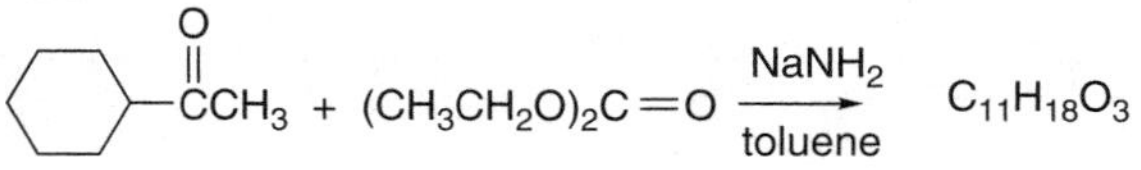

(i)

$C_6H_5\overset{O}{\overset{\|}{C}}CH_3 + (CH_3CH_2O)_2\overset{O}{\overset{\|}{P}}CH_2CN$ $\xrightarrow[\text{THF}]{\text{NaH}}$ $C_{10}H_9N$

(j)

$CH_3CH_2\overset{O}{\overset{\|}{C}}CH_2CH_2CO_2CH_2CH_3$ $\xrightarrow[\text{xylene}]{\text{NaOCH}_3}$ $C_6H_8O_2$

(k)

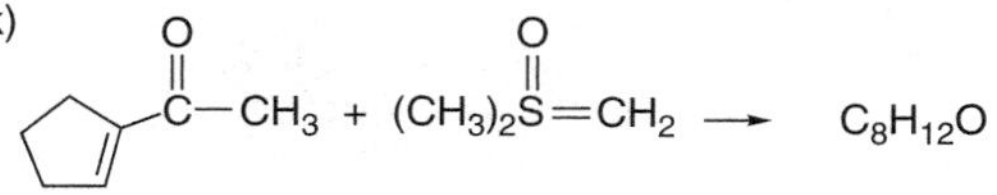

$\overset{O}{\overset{\|}{C}}-CH_3 + (CH_3)_2\overset{O}{\overset{\|}{S}}=CH_2 \longrightarrow C_8H_{12}O$

(l)

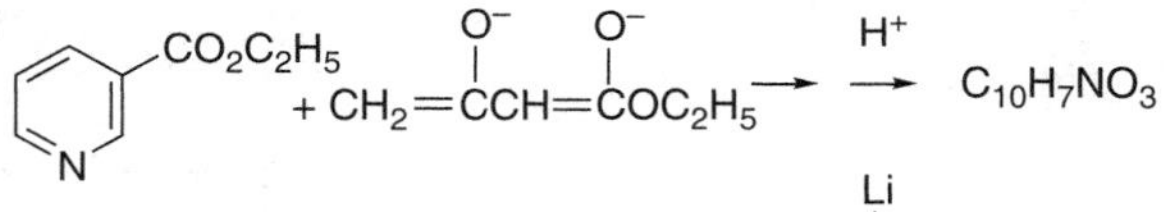

$\overset{CO_2C_2H_5}{\underset{N}{}} + CH_2=\overset{O^-}{C}CH=\overset{O^-}{C}OC_2H_5$ $\xrightarrow{H^+} C_{10}H_7NO_3$

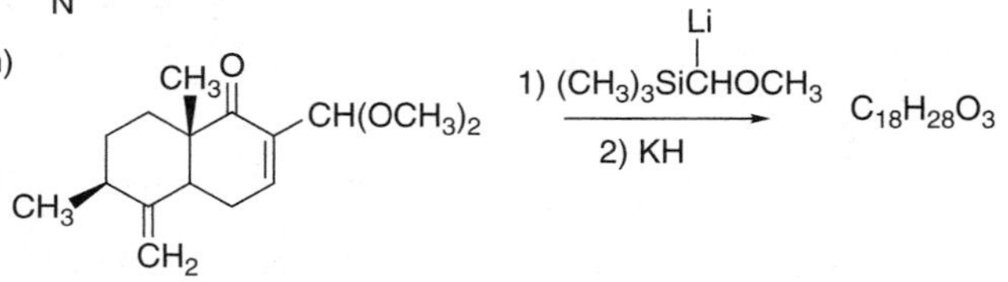

(m)

1) (CH₃)₃SiĊHOCH₃ (Li)
2) KH $\longrightarrow C_{18}H_{28}O_3$

(n)

$\overset{CH_3CO_2}{\underset{O}{}} + CH_3\overset{O}{\overset{\|}{C}}CH=CH_2$ $\xrightarrow{\text{NaOH}} C_{12}H_{16}O_5$

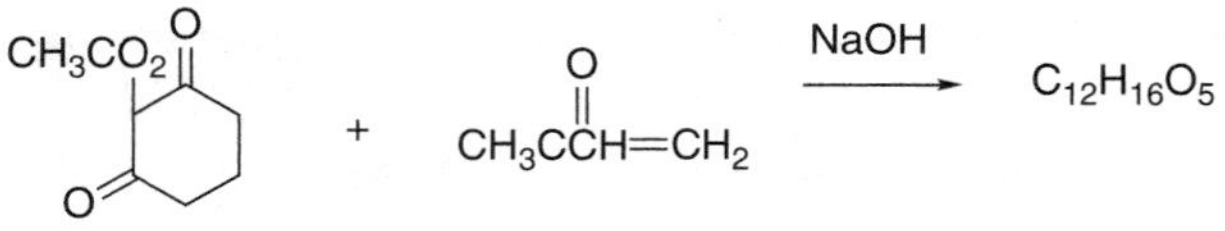

2.2. Indicate reaction conditions or a series of reactions that could effect each of the following synthetic conversions:

(a) $CH_3CO_2C(CH_3)_3 \longrightarrow (CH_3)_2\overset{OH}{\underset{}{C}}CH_2CO_2C(CH_3)_3$

(b) $THPO(CH_2)_3CH=O \longrightarrow THPO(CH_2)_3$ (CH₂OH)(CH₃)

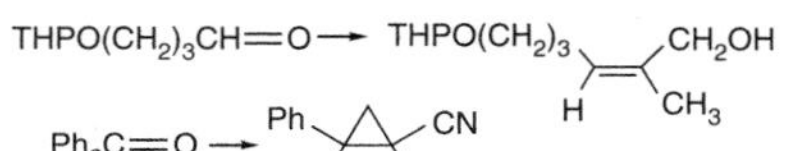

(c) → $\overset{O}{\underset{}{}}$CHOH

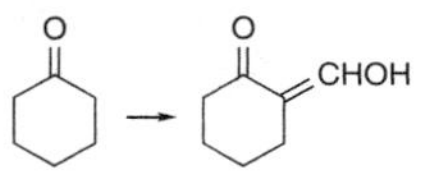

(d) $Ph_2C=O \longrightarrow$ Ph, Ph (CN)(CO₂C₂H₅)

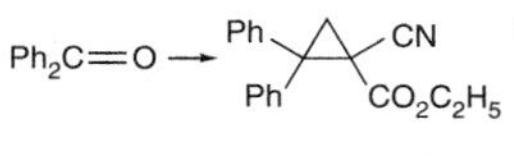

(e) → $\overset{O}{\underset{CO_2C_2H_5}{}}$

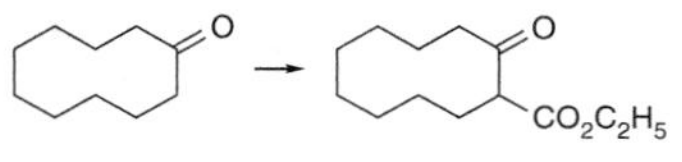

(f) → (CH₂OH)

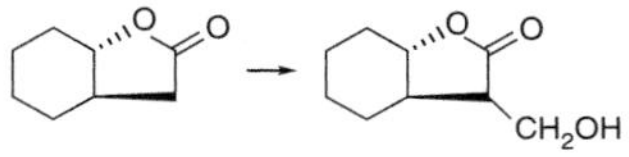

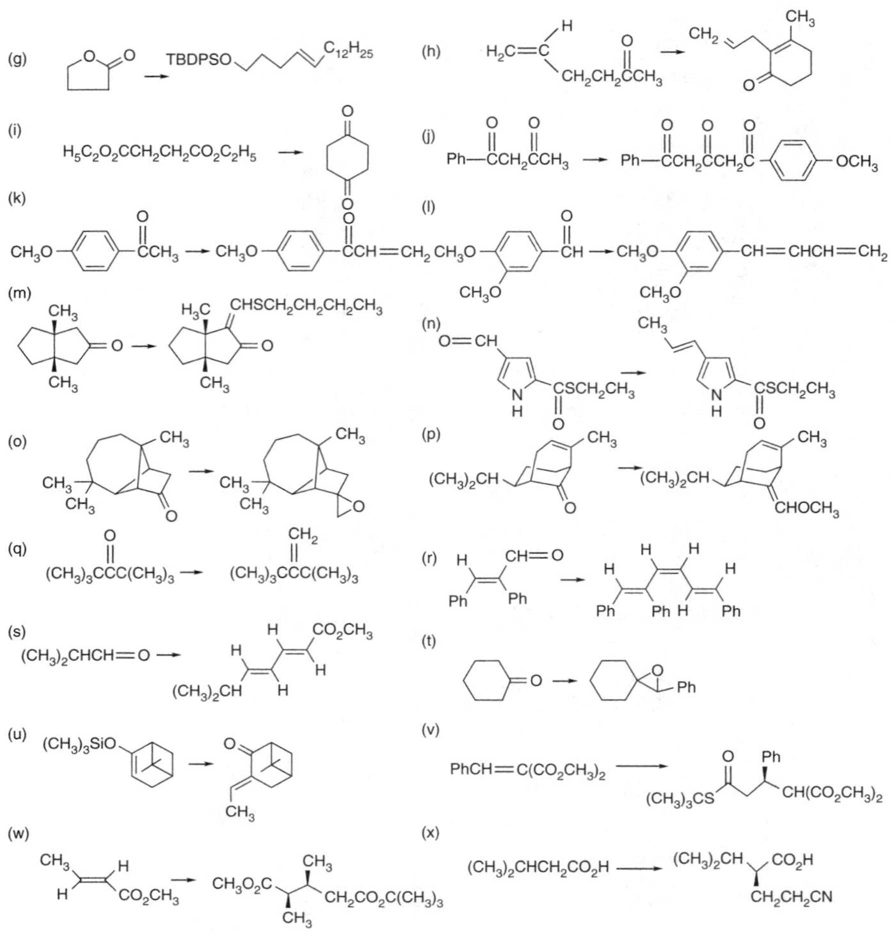

(g)

(h)

(i)

(j)

(k)

(l)

(m)

(n)

(o)

(p)

(q)

(r)

(s)

(t)

(u)

(v)

(w)

(x)

2.3. Step-by-step retrosynthetic analysis of each of the target molecules reveals that they can be efficiently prepared in a few steps from the starting material shown on the right. Do a retrosynthetic analysis and suggest reagents and reaction conditions for carrying out the desired synthesis.

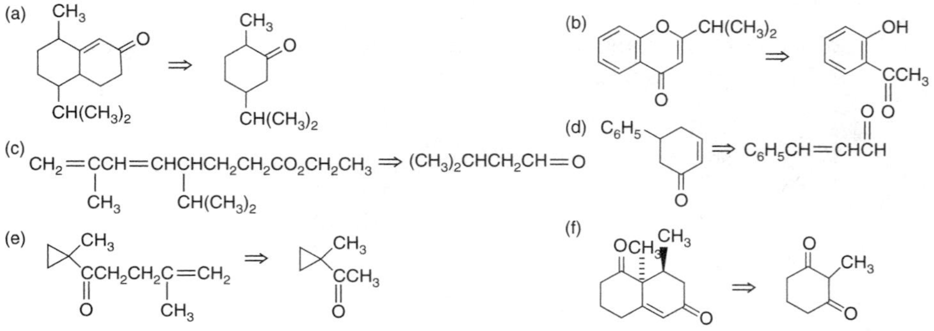

(a)

(b)

(c)

(d)

(e)

(f)

(g) ⟹

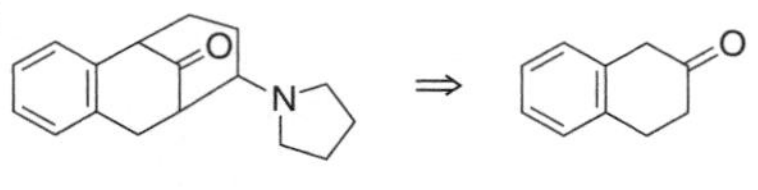

(h)

$$PhC{=}CHCH{=}O \Rightarrow PhCCH_3$$

with CH_3 substituent on the central carbon

(i)

$$CH_3CH_2CCH{=}CHCO_2C_2H_5 \Rightarrow CH_3CH_2CH_2CH{=}O,$$
$$\overset{\|}{CH_2} \qquad\qquad ClCH_2CO_2C_2H_5$$

(j)

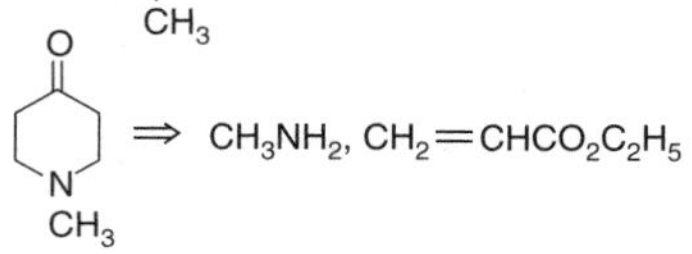

⟹ CH_3NH_2, $CH_2{=}CHCO_2C_2H_5$

(k)

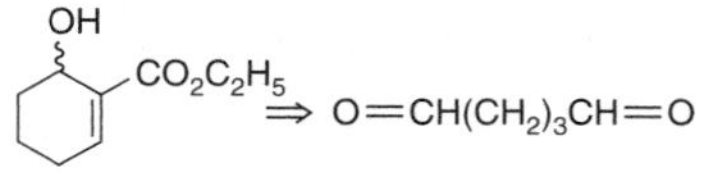

⟹ $O{=}CH(CH_2)_3CH{=}O$

(l)

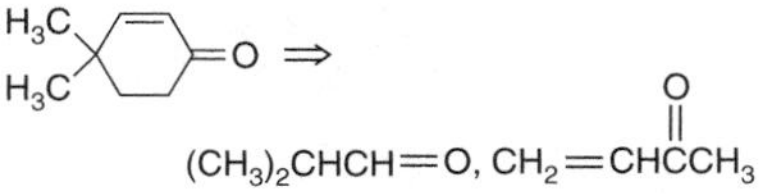

⟹

$(CH_3)_2CHCH{=}O$, $CH_2{=}CHCCH_3$

(m)

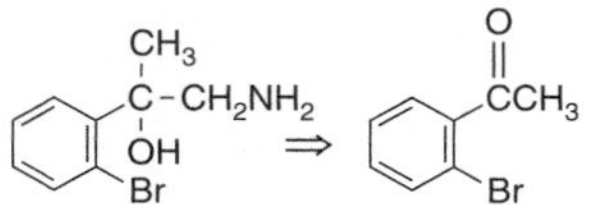

⟹

(n)

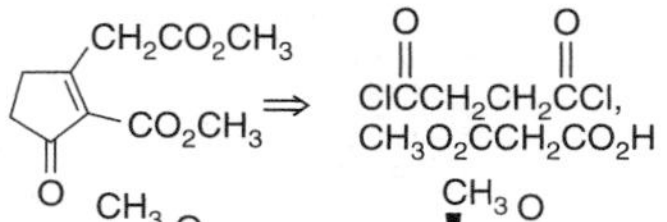

⟹ $ClCCH_2CH_2CCl$, $CH_3O_2CCH_2CO_2H$

(o)

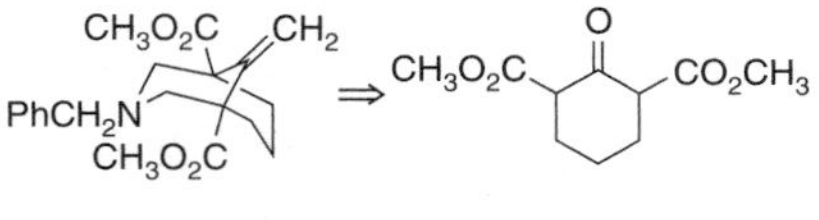

⟹ CH_3O_2C CO_2CH_3

(p)

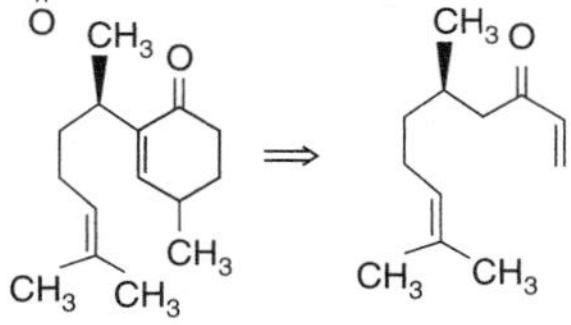

⟹

(q)

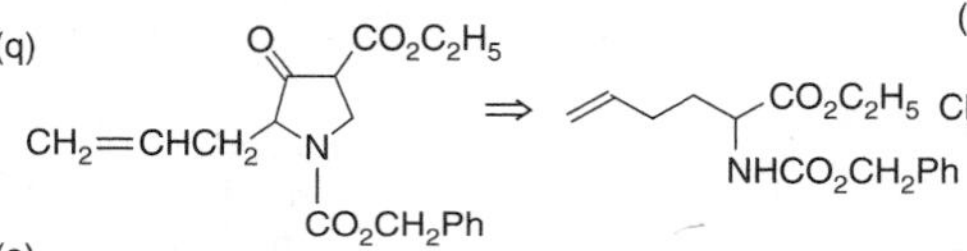

⟹

(r)

CH_3O_2C $CH_2CO_2C(CH_3)_3$ ⟹

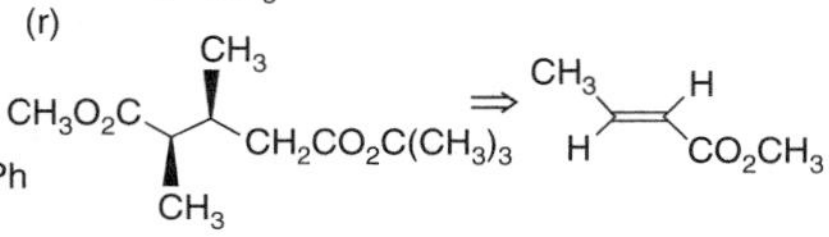

(s)

⟹

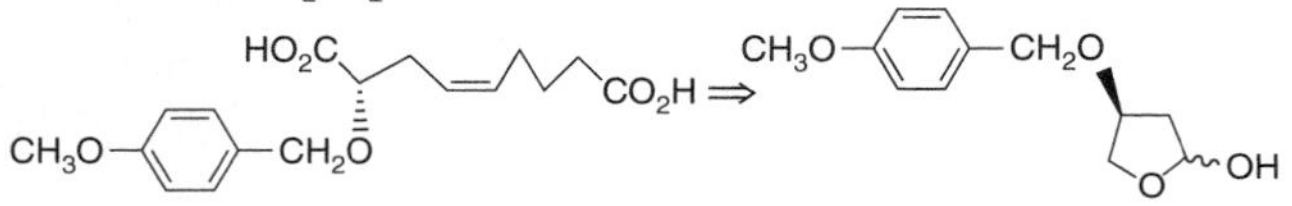

2.4. Offer a mechanism for each of the following reactions:

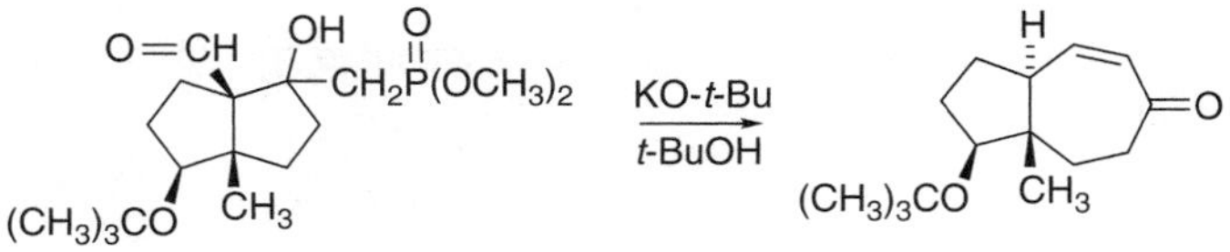

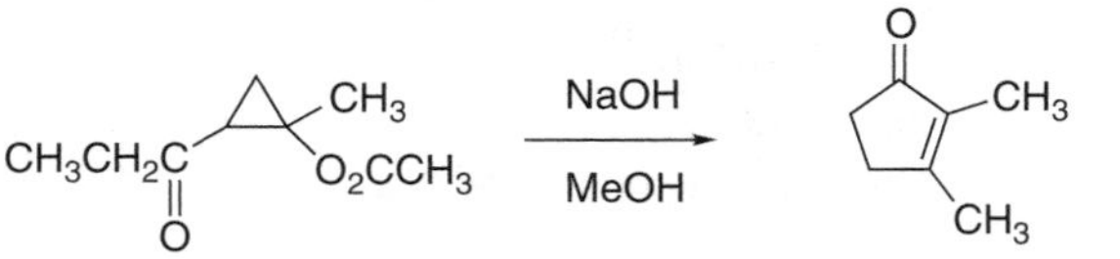

(a)

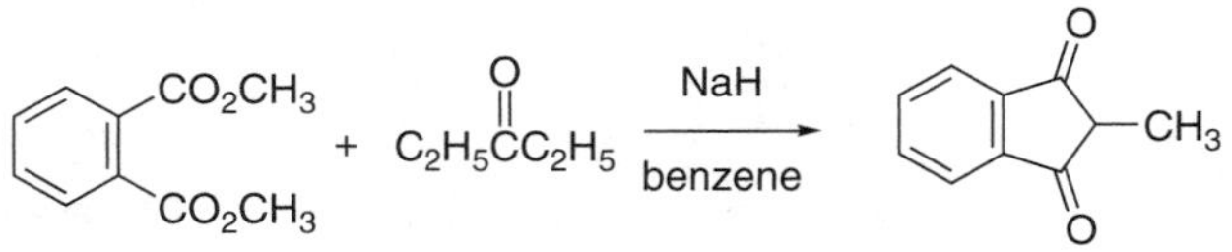

$+$ $C_2H_5CC_2H_5$ $\xrightarrow[\text{benzene}]{\text{NaH}}$

(b)

$\xrightarrow[\text{$t$-BuOH}]{\text{KO-$t$-Bu}}$

(c)

$\xrightarrow[\text{MeOH}]{\text{NaOH}}$

(d)

$\xrightarrow[\text{dioxane, 150°C}]{\text{KOH, H}_2\text{O}}$ $CH_3CH_2CHCH_3$ $+$ $PhCCH_3$

(e)

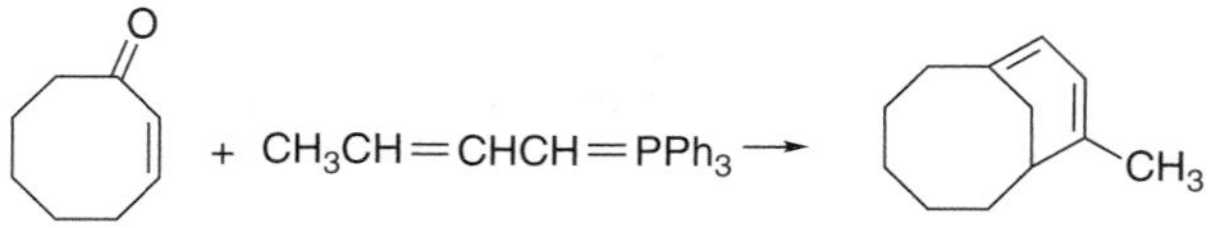

(f)

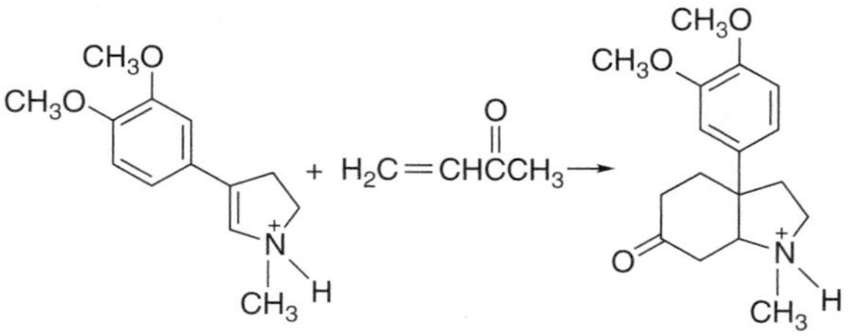

(g)

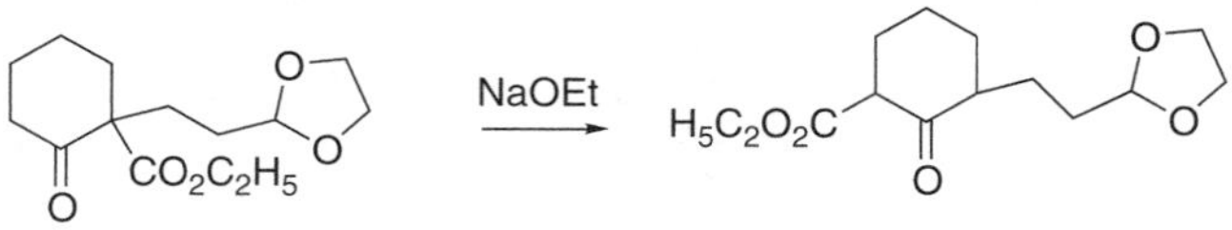

(h)

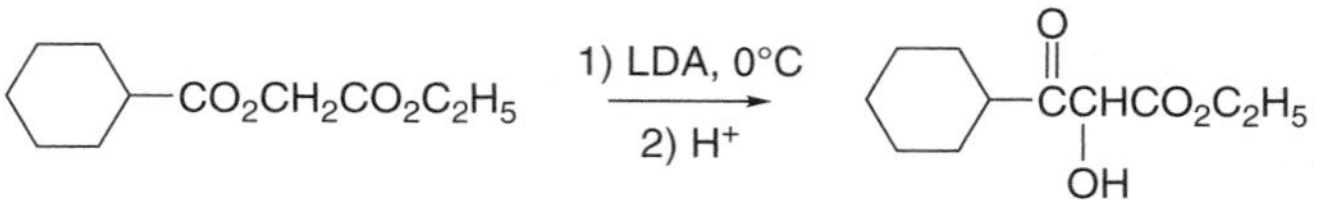

(i)

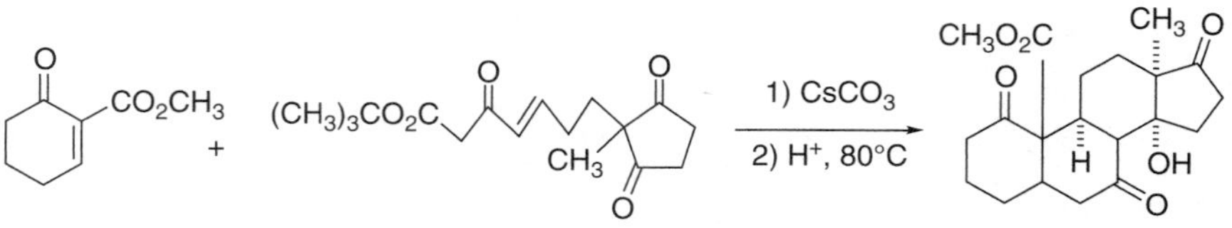

(j)

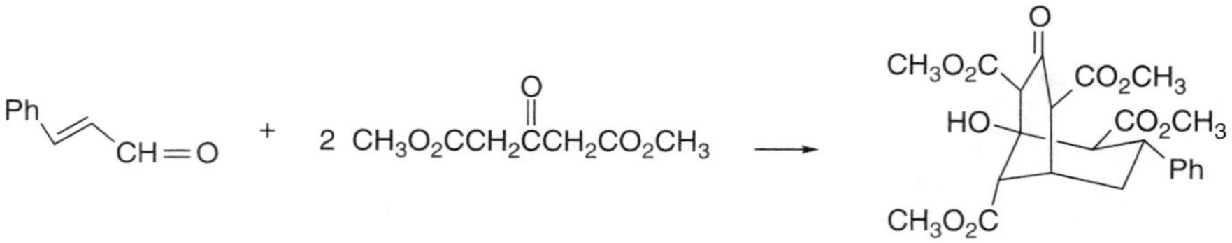

2.5. Tetraacetic acid (or a biological equivalent) is suggested as an intermediate in the biosynthesis of phenolic natural products. In the laboratory, it can be readily converted to orsellinic acid. Suggest a mechanism for this reaction under the conditions specified.

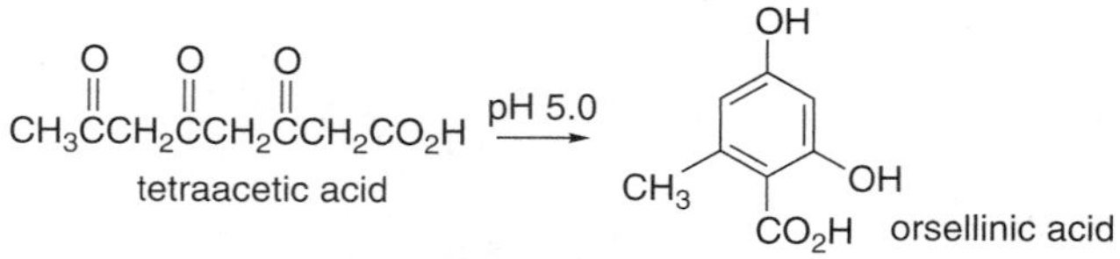

2.6. a. A stereospecific method for deoxygenating epoxides to alkenes involves reaction of the epoxide with the diphenylphosphide ion, followed by methyl iodide. The method results in overall inversion of alkene stereochemistry. Thus, *cis*-cyclooctene epoxide gives *trans*-cyclooctene. Propose a mechanism for this reaction and discuss its relationship to the Wittig reaction.

b. Reaction of the epoxide of *E*-4-octene (*trans*-2,3-dipropyloxirane) with potassium trimethylsilanide gives *Z*-4-octene as the only alkene product in 93% yield. Suggest a reasonable mechanism for this reaction.

2.7. a. A fairly general method for ring closure has been developed that involves vinyltriphenylphosphonium haldides as reactants. Indicate the mechanism of this reaction, as applied to the two examples shown below. Suggest two other types of rings that could be synthesized using vinyltriphenylphosphonium salts.

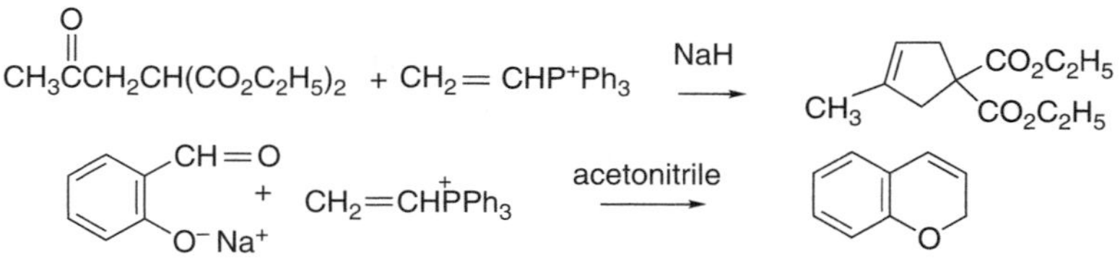

b. Allylphosphonium salts were used as a synthon in the synthesis of cyclo-hexadienes. Suggest an appropriate co-reactant and other reagents that would be expected to lead to cyclohexadienes.

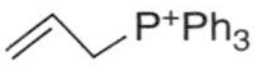

c. The product shown below is formed by the reaction of vinyltriphenylphos-phonium bromide, the lithium enolate of cyclohexanone, and 1,3-diphenyl-2-propen-1-one. Formulate a mechanism.

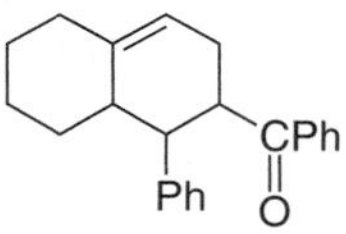

d. The dimethoxy phosphonylmethylcylcopentenone shown below has been used as a starting material for the synthesis of prostaglandin analogs such as **7A**. The reaction involves formation of the anion, reaction with an alkyl halide, and a Wadsworth-Emmons reaction. What reactivity of the anion makes this approach feasible?

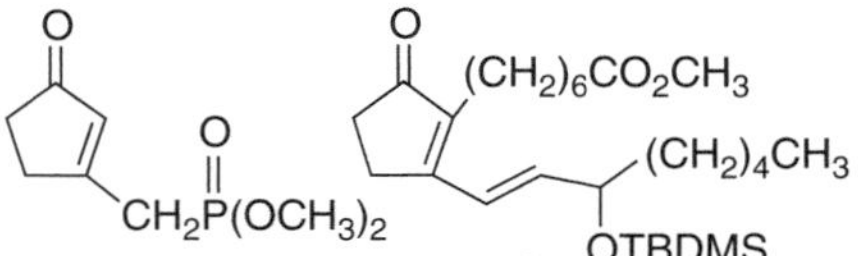

e. The reagent **7B** has found use in the expeditious construction of more complex molecules from simple starting materials. For example, the enolate of 3-pentanone when treated first with **7B** and then with benzaldehyde gives **7C** as a 2:1 mixture of stereoisomers. Explain the mechanism by which this reaction occurs.

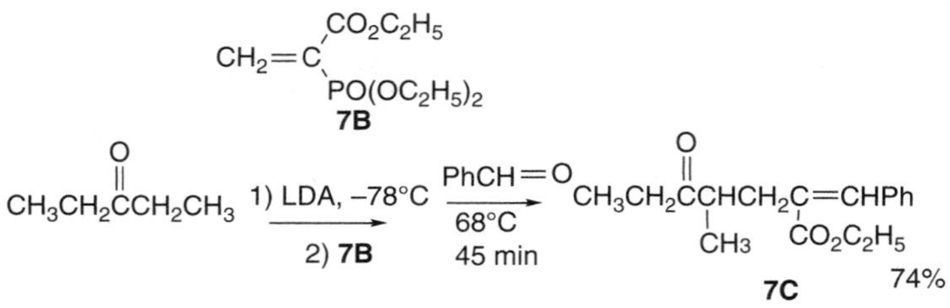

f. The reagent **7D** converts enolates of aldehydes into cyclohexadienyl phosphonates **7E**. Write a mechanism for this reaction. What alternative products might have been observed?

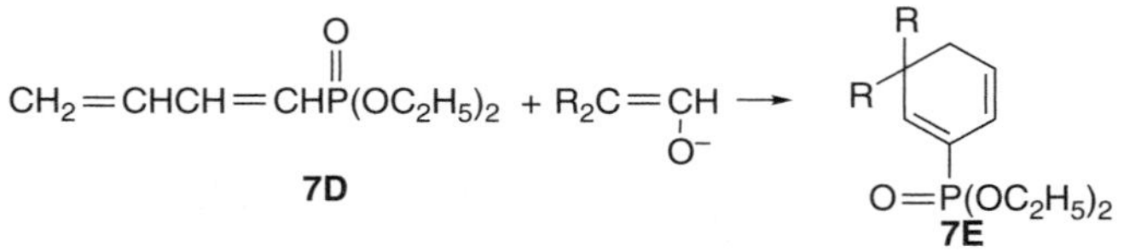

2.8. Compounds **8A** and **8B** were key intermediates in an early total synthesis of cholesterol. Rationalize their formation by the routes shown.

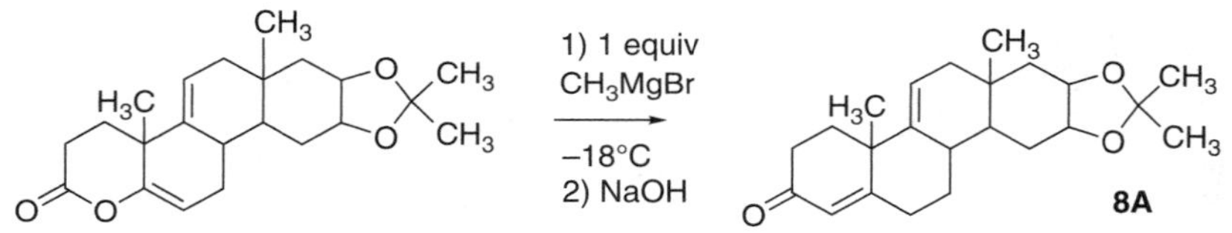

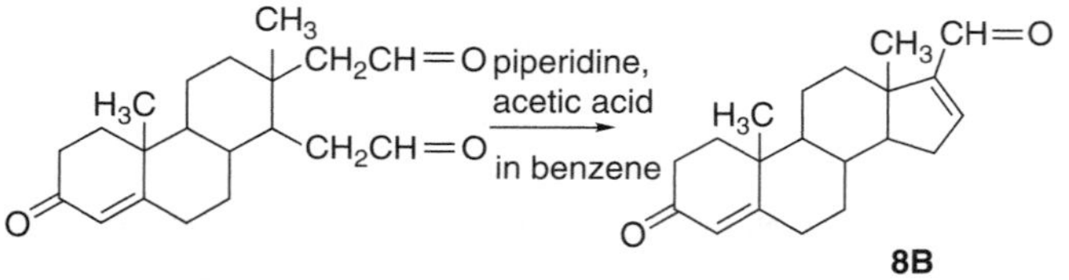

2.9. The first few steps in a synthesis of the alkaloid conessine produce **9B**, starting from **9A**. Suggest a sequence of reactions for effecting this conversion.

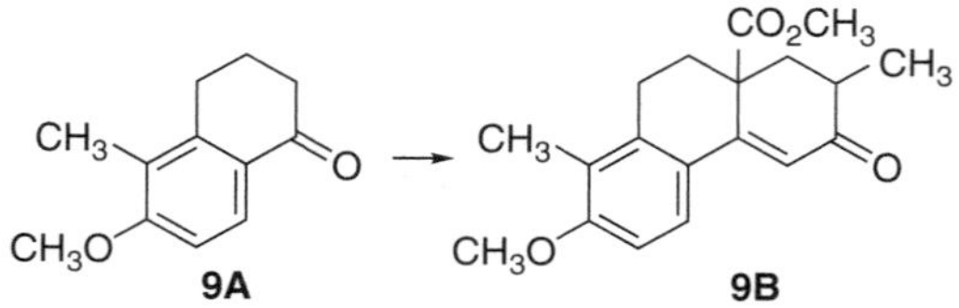

2.10. A substance known as elastase is involved in various inflammatory diseases such as arthritis, pulmonary emphysema, and pancreatitis. Elastase activity can be inhibited by a compound known as elasnin, obtained from a microorganism.

A synthesis of elasnin has been reported that utilizes compound **10A** as a key intermediate. Suggest a synthesis of **10A** from methyl hexanoate and hexanal.

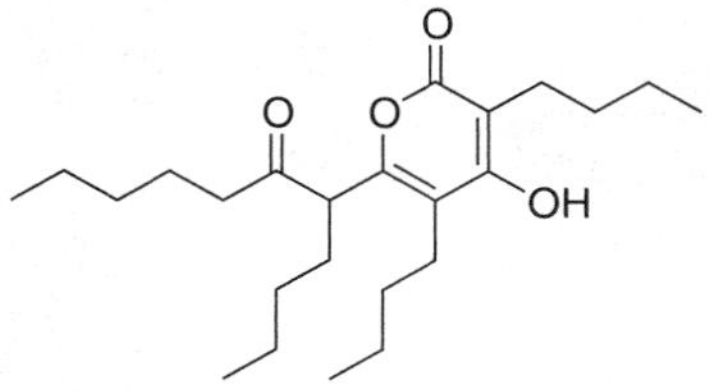

Elasnin 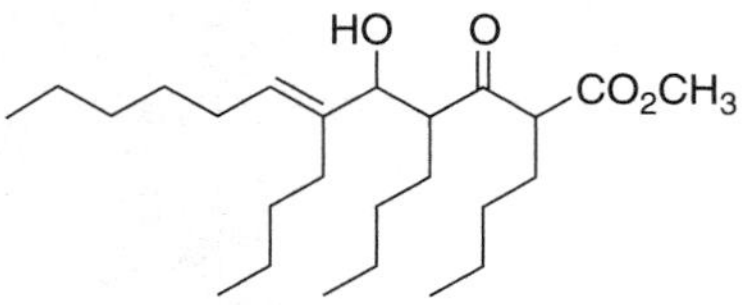   **10A**

2.11. Treatment of compound **11A** with LDA followed by cyclohexanone can give either **11B** or **11C.** Compound **11B** is formed when the aldehyde is added at −78° C, whereas **11C** is formed if the aldehyde is added at 0° C. Treatment of **11B** with LDA at 0° C gives **11C**. Explain these results.

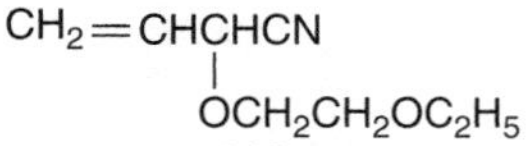

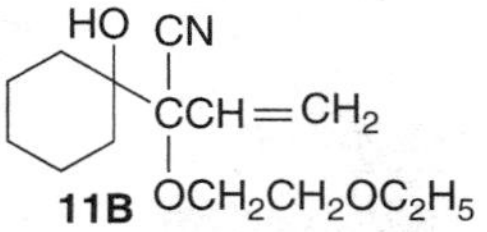

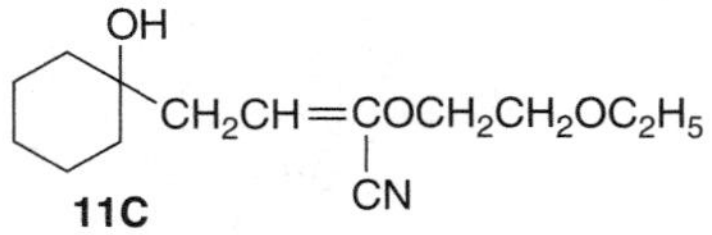

2.12. Dissect the following molecules into potential precursors by locating all bonds that could be made by intramolecular aldol or conjugate addition reactions. Suggest possible starting materials and conditions for performing the desired reactions.

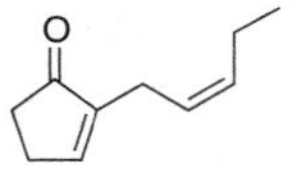

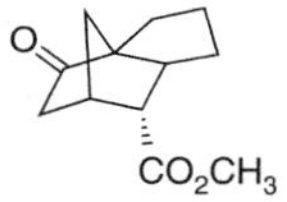

2.13. Mannich condensations permit one-step reactions to form the following substances from substantially less complex starting materials. Identify a potential starting material that would give rise to the product shown in a single step under Mannich reaction conditions.

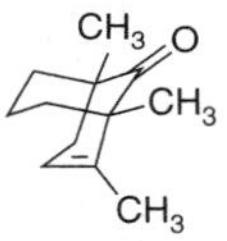

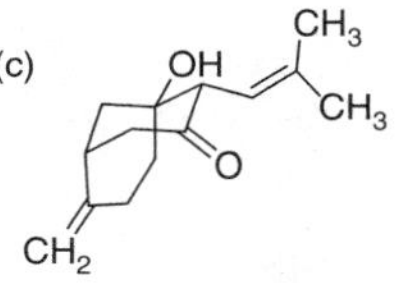

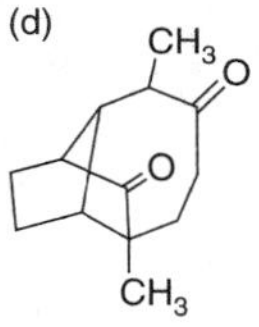

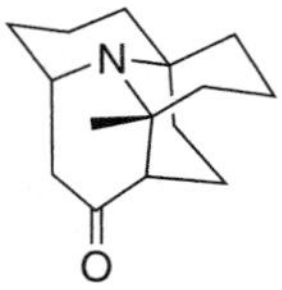

2.14. Indicate whether or not the aldol reactions shown below would be expected to exhibit high stereoselectivity. Show the stereochemistry of the expected product(s).

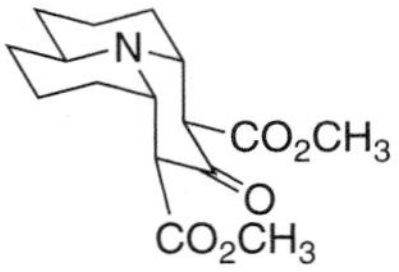

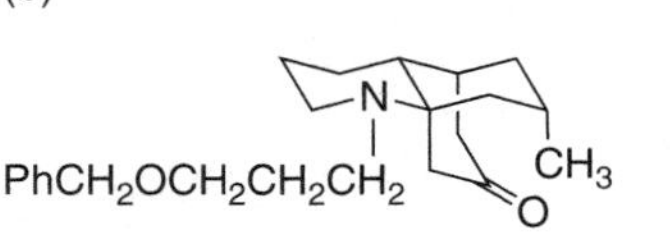

(a)

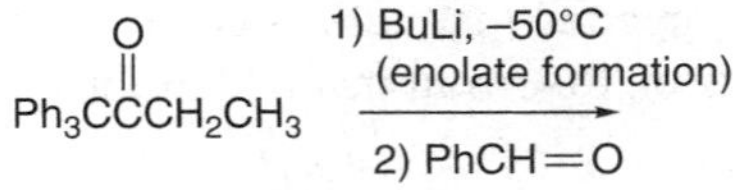

$$Ph_3CCCH_2CH_3$$ (with O double bond)

1) BuLi, −50°C
(enolate formation)

2) PhCH=O

(b)

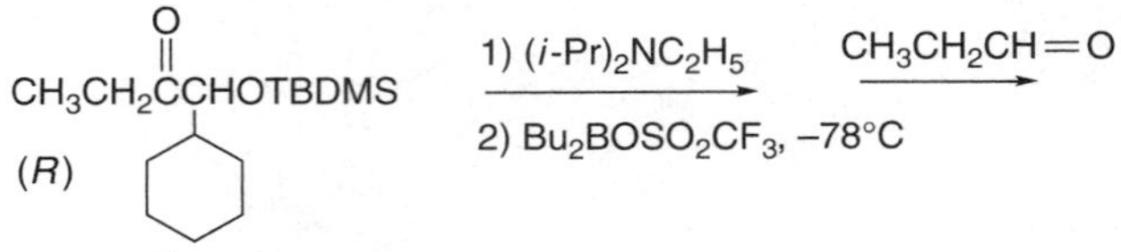

$$CH_3CH_2CCHOTBDMS$$ (with O double bond, cyclohexyl substituent)

(*R*)

1) (*i*-Pr)$_2$NC$_2$H$_5$

2) Bu$_2$BOSO$_2$CF$_3$, −78°C

CH$_3$CH$_2$CH=O

(c)

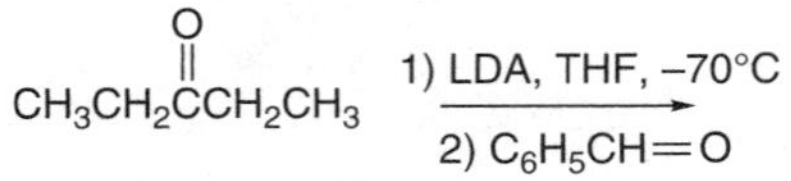

$$CH_3CH_2CCH_2CH_3$$ (with O double bond)

1) LDA, THF, −70°C

2) C$_6$H$_5$CH=O

(d)

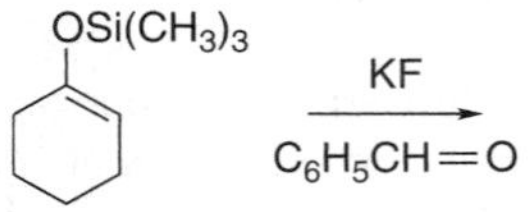

OSi(CH$_3$)$_3$ (on cyclohexene)

KF

C$_6$H$_5$CH=O

(e)

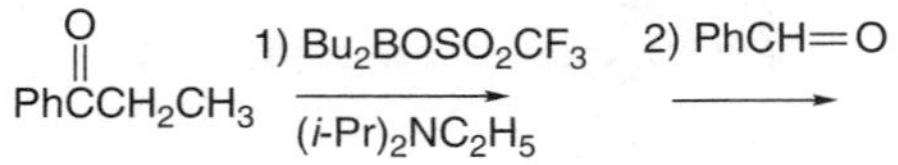

$$PhCCH_2CH_3$$ (with O double bond)

1) Bu$_2$BOSO$_2$CF$_3$ 2) PhCH=O

(*i*-Pr)$_2$NC$_2$H$_5$

(f)

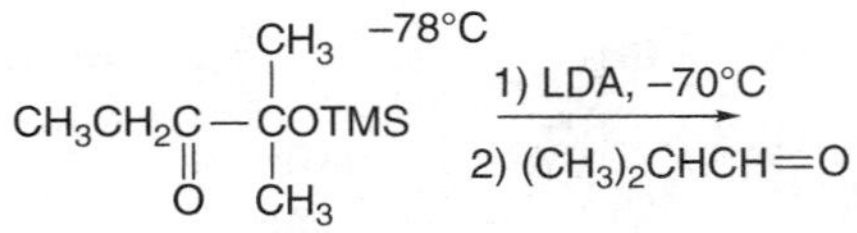

$$CH_3CH_2C-COTMS$$ (with CH$_3$ substituents and O double bond)

−78°C

1) LDA, −70°C

2) (CH$_3$)$_2$CHCH=O

(g)

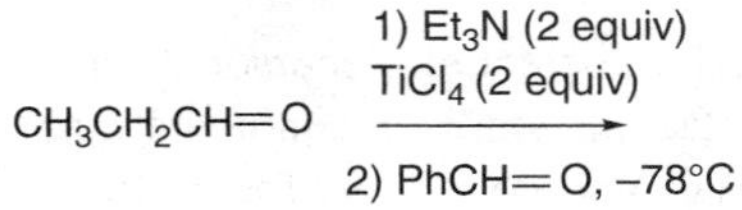

CH$_3$CH$_2$CH=O

1) Et$_3$N (2 equiv)
TiCl$_4$ (2 equiv)

2) PhCH=O, −78°C

(h)

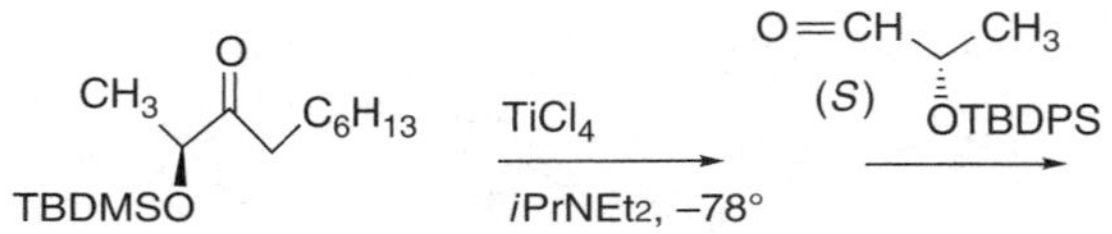

CH$_3$ (with O double bond, C$_6$H$_{13}$, TBDMSO)

TiCl$_4$

*i*PrNEt2, −78°

O=CH $\diagup$ CH$_3$

(*S*)   OTBDPS

2.15. Suggest transition structures that would account for the observed stereoselectivity of the following reactions.

(a)

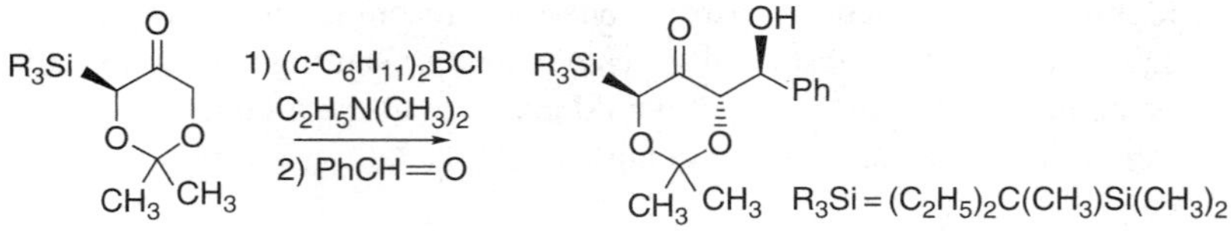

R$_3$Si (on dioxane ring with O double bond, CH$_3$ CH$_3$)

1) (*c*-C$_6$H$_{11}$)$_2$BCl
   C$_2$H$_5$N(CH$_3$)$_2$

2) PhCH=O

R$_3$Si (product with OH, Ph, CH$_3$ CH$_3$)

R$_3$Si = (C$_2$H$_5$)$_2$C(CH$_3$)Si(CH$_3$)$_2$

(b)

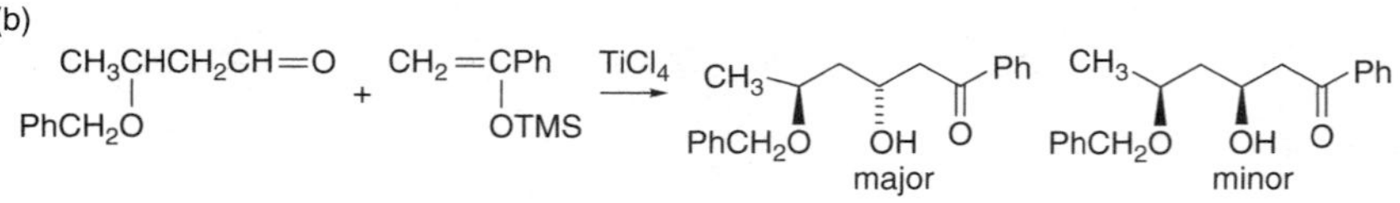

CH$_3$CHCH$_2$CH=O   +   CH$_2$=CPh

PhCH$_2$O   OTMS

TiCl$_4$

CH$_3$ (product) Ph

PhCH$_2$O   OH   O

major

CH$_3$ (product) Ph

PhCH$_2$O   OH   O

minor

2.16. Suggest starting materials and reaction conditions suitable for obtaining each
of the following compounds by a procedure involving conjugate addition.

(a) 4,4-dimethyl-5-nitropentan-2-one
(b) diethyl 2,3-diphenylglutarate
(c) ethyl 2-benzoyl-4-(2-pyridyl)butanoate
(d) 2-phenyl-3-oxocyclohexaneacetic acid

(e)

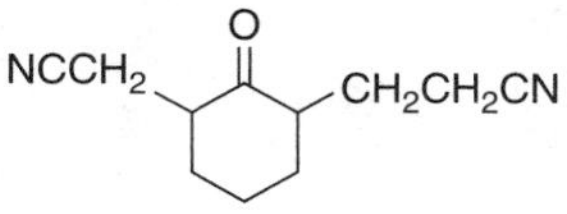

(f)

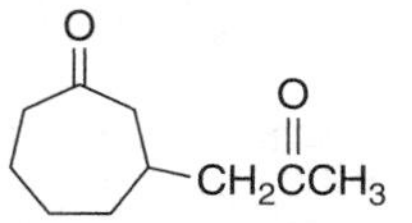

(g) $CH_3CH_2CHCH_2CH_2CCH_3$ with NO_2 substituent and C=O

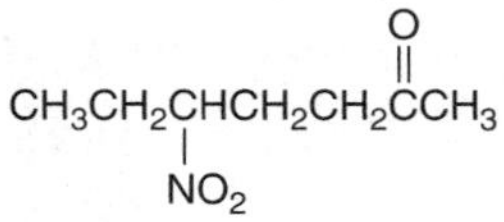

(h) $(CH_3)_2CHCHCH_2CH_2CO_2CH_2CH_3$ with $CH=O$ substituent

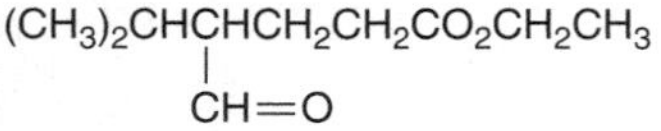

(i)

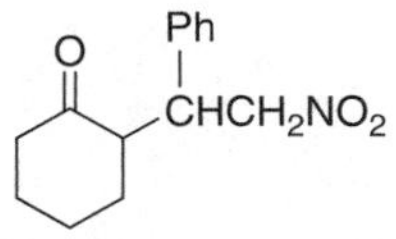

(j) $PhCHCHCH_2CCH_3$ with Ph, CN, and O substituents

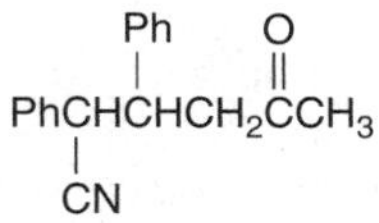

(k)

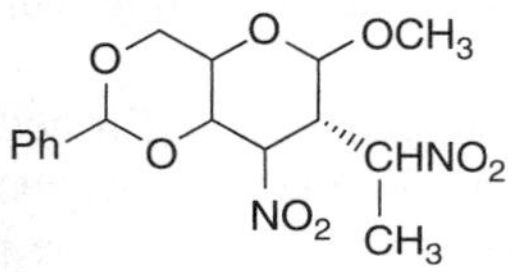

(l)

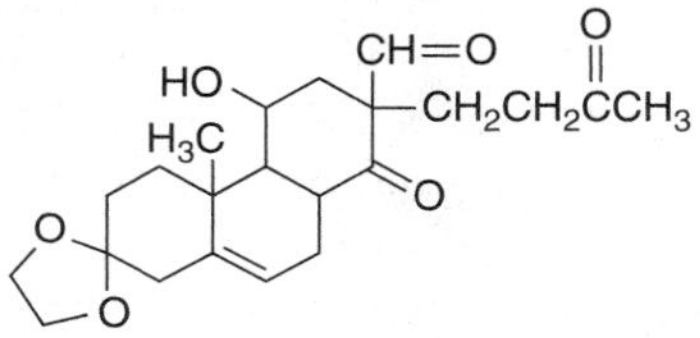

2.17. In the synthesis of a macrolide **17A**, known as latrunculin A, the intermediate **17B** was assembled from components **17C**, **17D**, and **17E** in a "one-pot" tandem process. By a retrosynthetic analysis, show how the synthesis could occur and identify a sequence of reactions and corresponding reagents.

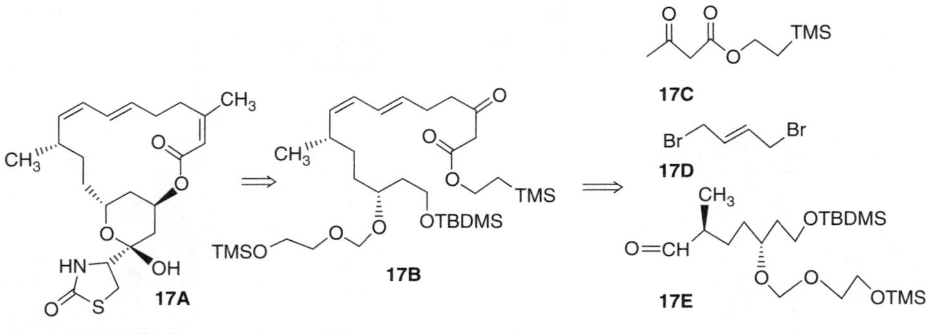

latrunculin A

2.18. The tricyclic substance **18A** and **18B** are both potential synthetic intermediates for synthesis of the biologically active diterpene forskolin. These intermediates can be prepared from the monocyclic precursors shown. Indicate the nature of the reactions involved in these transformations.

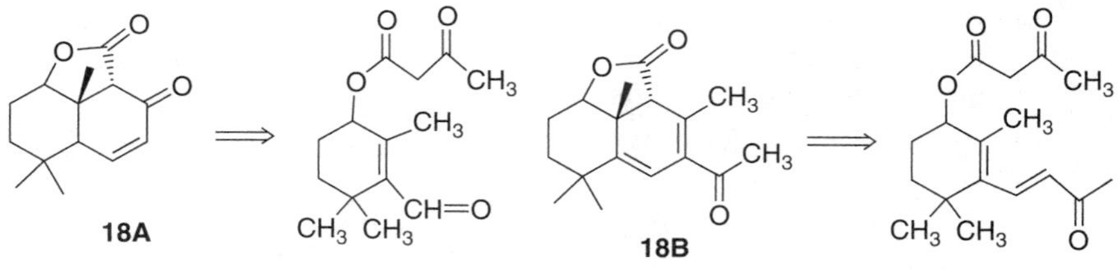

18A   **18B**

2.19. Account for the course of the following reactions:

a. Substituted acetophenones react with ethyl phenylpropynoate under basic conditions to give pyrones. Formulate a mechanism for this reaction.

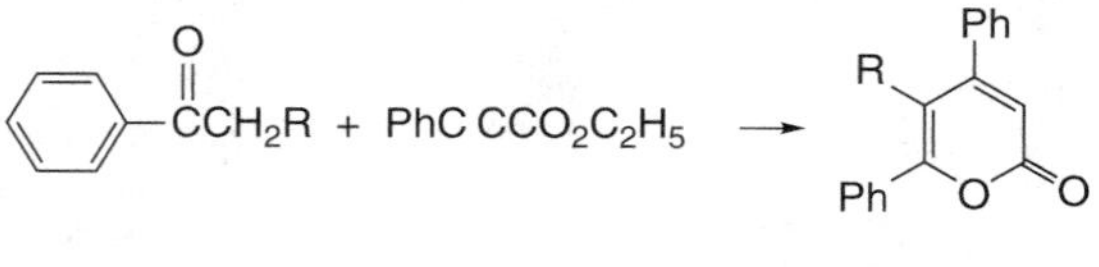

b. The reaction of simple ketones such as 2-butanone or 1-phenyl-2-propanone with α,β-unsaturated ketones gives cyclohexanone on heating with methanol containing potassium methoxide. Indicate how the cyclohexanones could be formed. Can more than one isomeric cyclohexanone be formed? Can you suggest a means for distinguishing between possible cyclohexanones?

c. α-Benzolyloxyphenylacetonitrile reacts with acrylonitrile in the presence of NaH to give 2-cyano-1,4-diphenylbutane-2,4-dione.

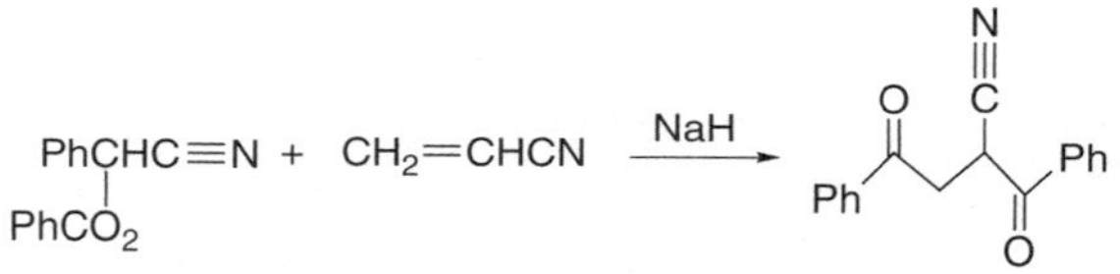

d. Reaction of the lithium anion of 3-methoxy-2-methylcyclopentanone with methyl acrylate gives the two products shown as an 82:18 mixture. Alkaline hydrolysis of the mixture gives a single pure product. How is the minor product formed and how is it converted to the hydrolysis product?

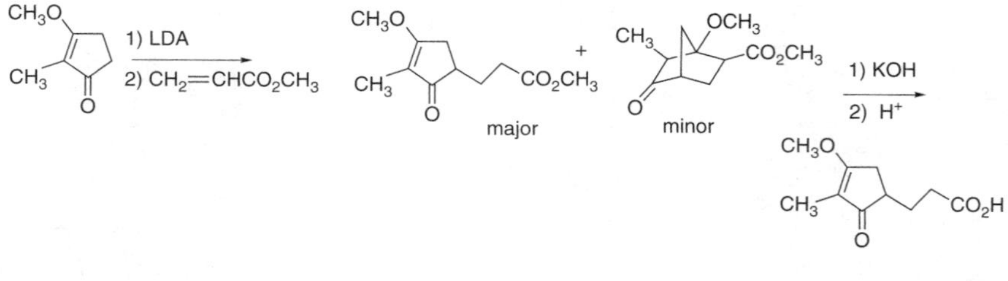

2.20. Explain the stereochemical outcome of the following reactions.

a.

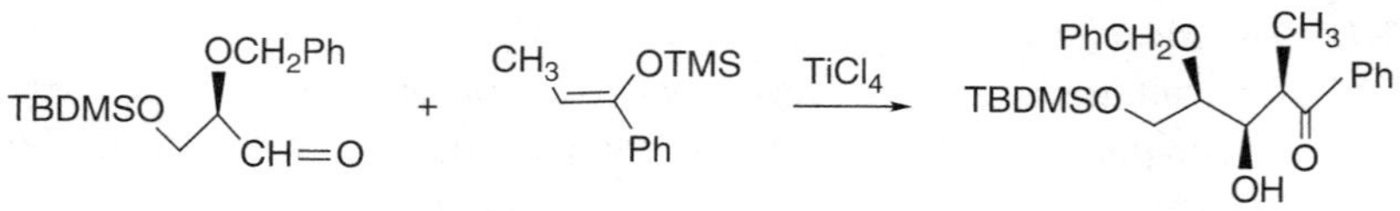

b.

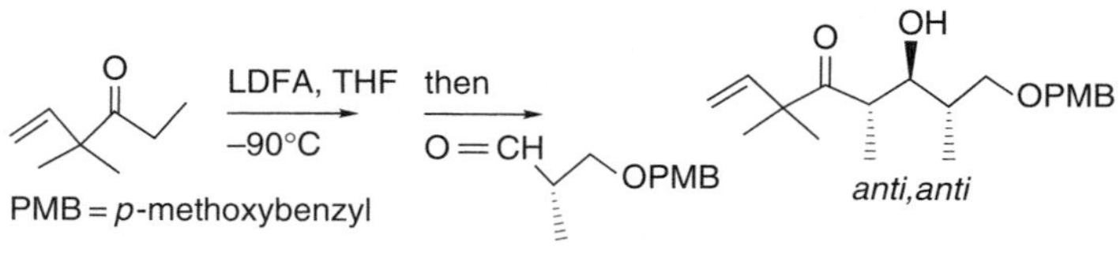

c. The facial selectivity of 2-benzyloxy-3-pentanone toward typical alkyl, alkenyl, and aryl aldehydes is reversed by a change of catalyst from $TiCl_4$ to $[(CH_3)_2CHO]TiCl_3$.

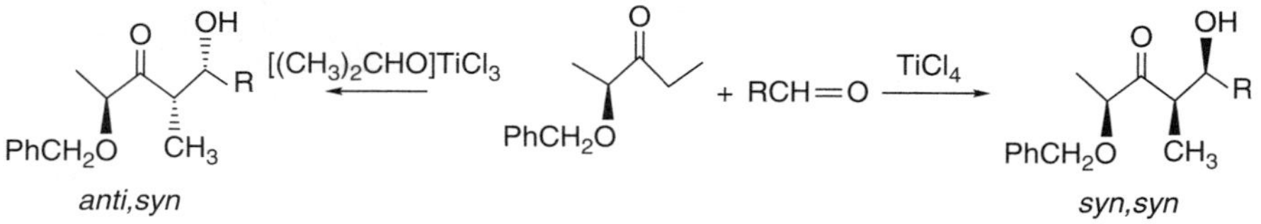

d. The boron enolates generated from ketones **20A** and **20B** give more than 95% selectivity for the *anti,anti* diastereomer.

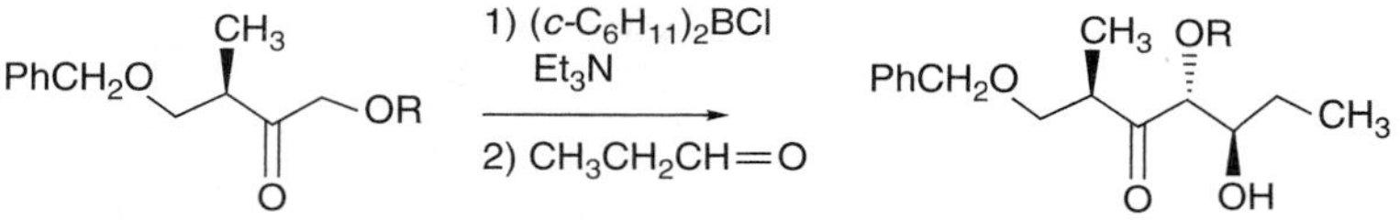

20A R = CH$_3$

20B R = CH$_2$Ph

e.

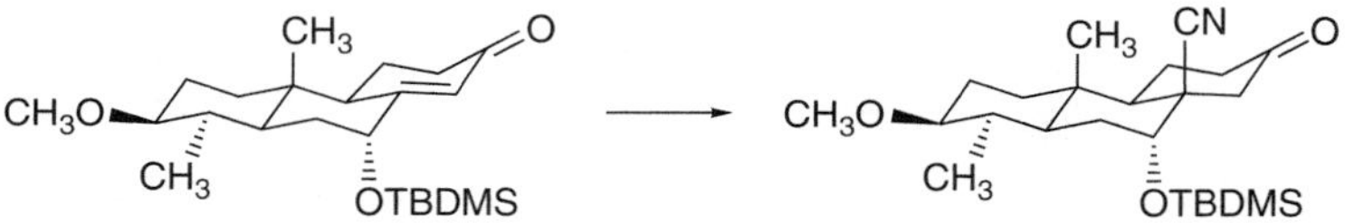

f.

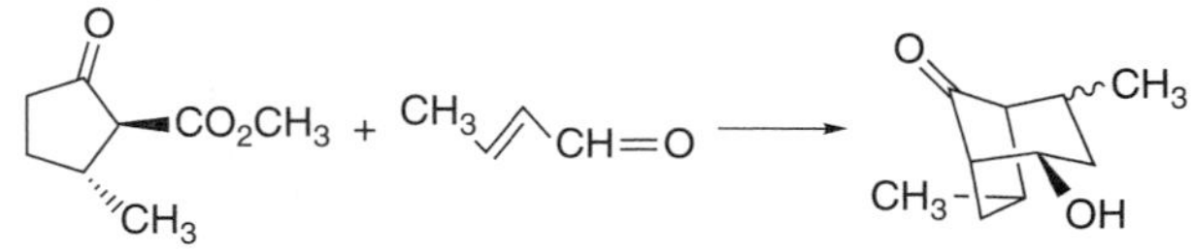

2.21. The camphor sultam derivative **21A** was used in a synthesis of epothilone. The stereoselectivity of the aldol addition was examined with several different aldehydes. Discuss the factors that lead to the variable stereoselectivity in the three cases shown.

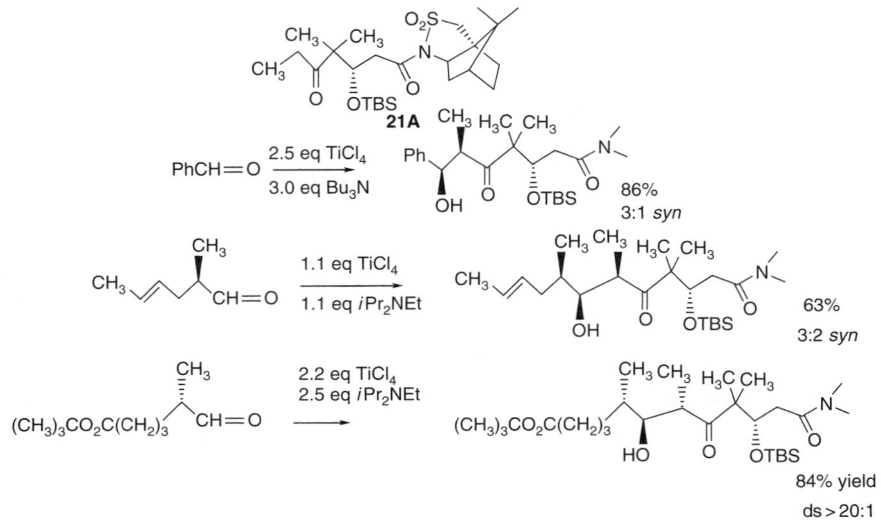

2.22. The facial selectivity of the aldehydes **22A** and **22B** is dependent on both the configuration at the β-center and the nature of the enolate as indicated by the data below. Consider possible transition structures for these reactions and offer a rationale for the observed facial selectivity.

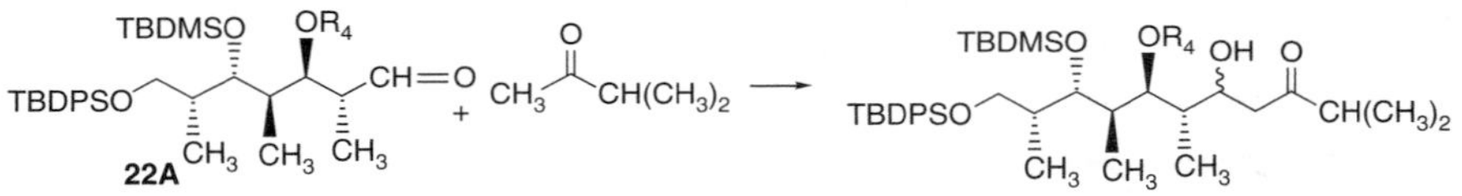

enolate	R⁴	3,4-*syn:anti* ratio	enolate	R⁴	3,4-*syn:anti* ratio
Li	MOM	97:3	Li	TES	66:34
TiCl₄	MOM	84:16	TiCl₄	TES	60:40
Bu₂B	MOM	63:38	Bu₂B	TES	<5:95
(C₅H₁₁)₂B	MOM	52:48	(C₅H₁₁)₂B	TES	14:86

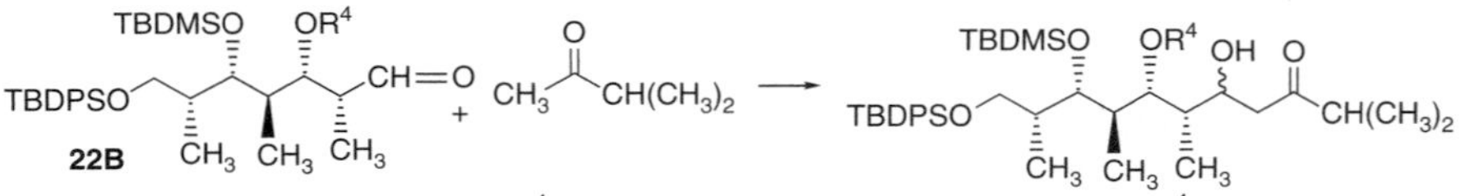

enolate	R⁴	3,4-*syn:anti* ratio	enolate	R⁴	3,4-*syn:anti* ratio
Li	MOM	15:85	Li	TES	7:93
TiCl₄	MOM	62:38	TiCl₄	TES	27:73
Bu₂B	MOM	81:39	Bu₂B	TES	45:55
(C₅H₁₁)₂B	MOM	50:50	(C₅H₁₁)₂B	TES	52:48

2.23. Predict the stereochemical outcome of the following aldol addition reactions involving chiral auxiliaries.

(a)

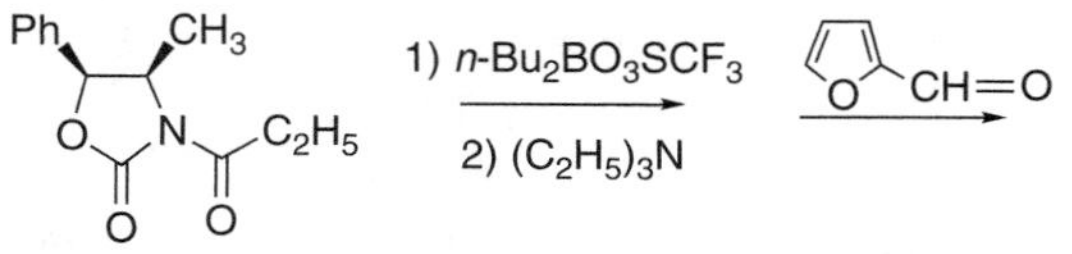

1) *n*-Bu$_2$BO$_3$SCF$_3$

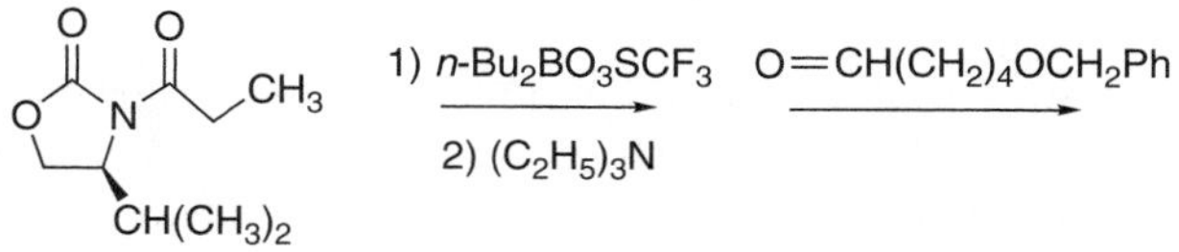

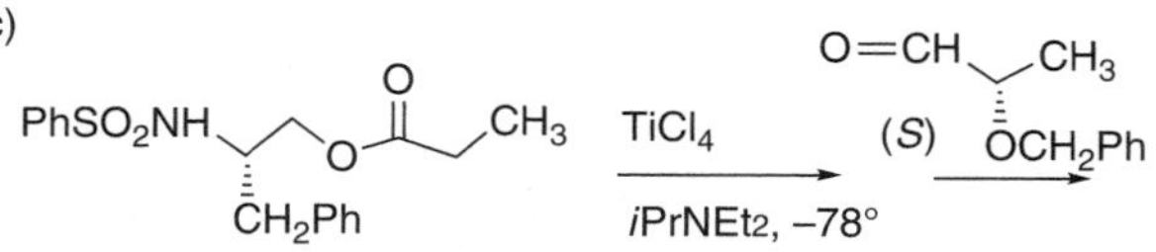

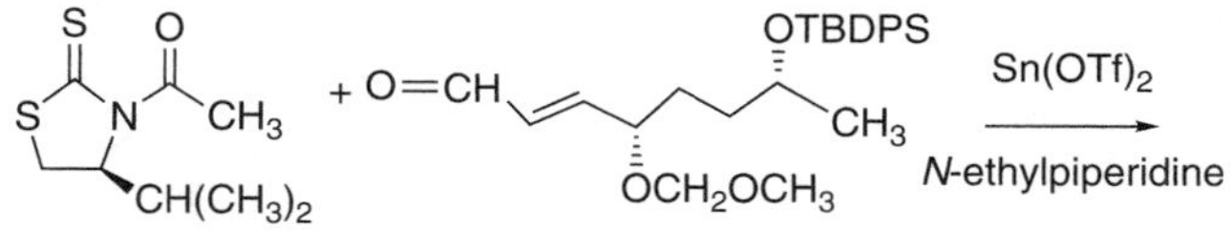

2) (C$_2$H$_5$)$_3$N

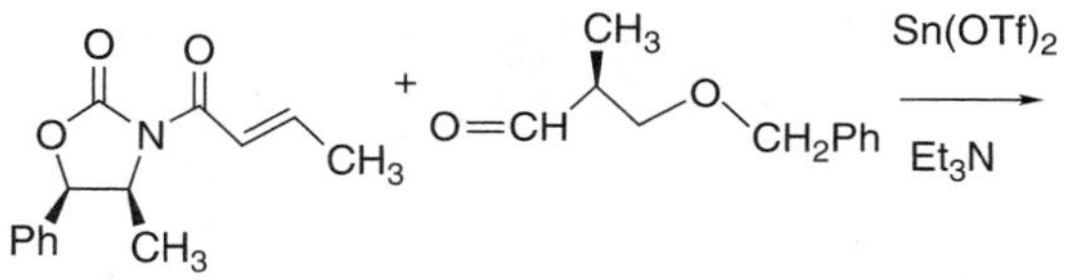

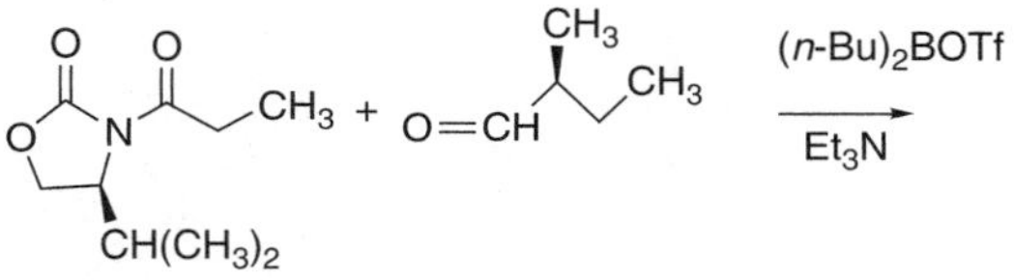

(b)

1) *n*-Bu$_2$BO$_3$SCF$_3$ O=CH(CH$_2$)$_4$OCH$_2$Ph

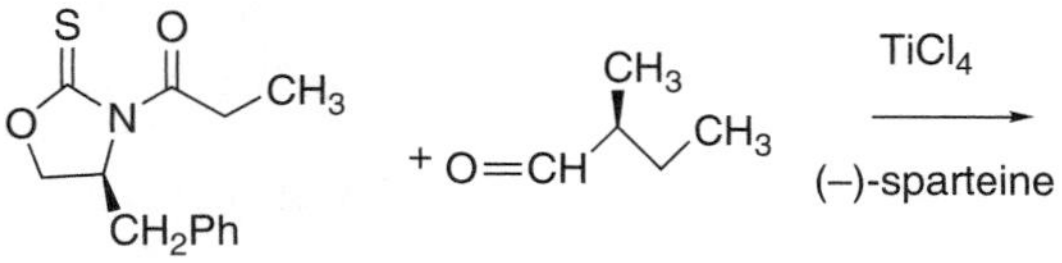

2) (C$_2$H$_5$)$_3$N

(c)

PhSO$_2$NH

CH$_2$Ph

TiCl$_4$

*i*PrNEt2, –78°

O=CH   CH$_3$

(*S*)   OCH$_2$Ph

(d)

+ O=CH

OTBDPS

CH$_3$

OCH$_2$OCH$_3$

Sn(OTf)$_2$

N-ethylpiperidine

(e)

+ O=CH

CH$_3$

O   CH$_2$Ph

Sn(OTf)$_2$

Et$_3$N

(f)

+ O=CH

CH$_3$

CH$_3$

(*n*-Bu)$_2$BOTf

Et$_3$N

(g)

+ O=CH

CH$_3$

CH$_3$

TiCl$_4$

(–)-sparteine

(h)

+ O=CH

CH$_3$

CH$_3$

TiCl$_4$

i-Pr2NEt

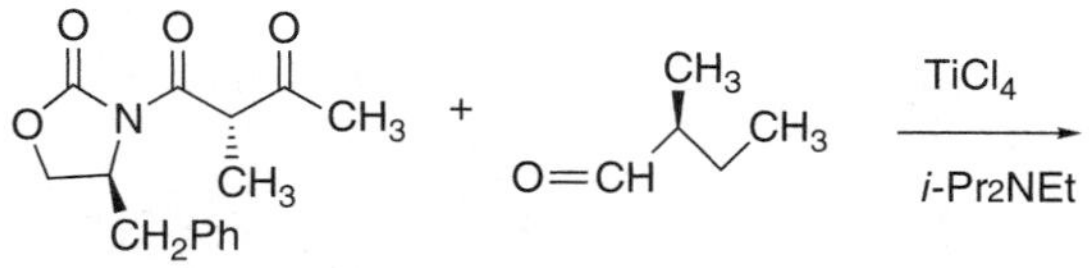

2.24. Suggest an enantioselective synthetic route to the antidepression drug rolipram from the suggested reactant.

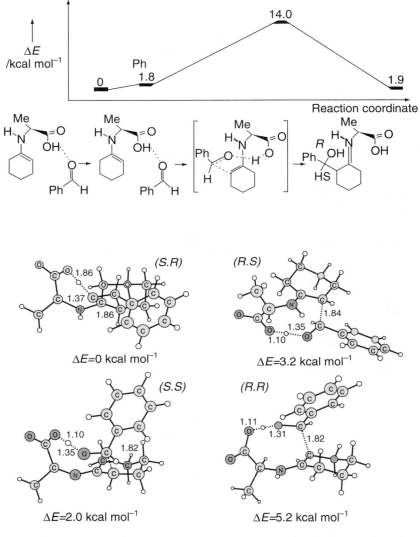

2.25. Figure 2.P25 shows the calculated [B3LYP/6-31G(d,p)] reaction energy profile
for the aldol addition of benzaldehyde and cyclohexanone catalyzed by alanine.
The best TSs leading to (S,R); (R,S); (S,S); and (R,R) products are given. What
factors favor the observed (R,S) product?

Fig. 2.P25. Top: Reaction energy profile for alanine-catalyzed aldol reaction of
benzaldehyde and cyclohexanone. Bottom: Diastereomeric transition structures. Reproduced from *Angew. Chem. Int. Ed. Engl.*, **44**, 7028 (2005), by permission of Wiley-VCH

Functional Group Interconversion by Substitution, Including Protection and Deprotection

Introduction

Chapters 1 and 2 dealt with formation of new carbon-carbon bonds by reactions in which one carbon acts as the nucleophile and another as the electrophile. In this chapter we turn our attention to noncarbon nucleophiles. Nucleophilic substitution is used in a variety of interconversions of functional groups. We discuss substitution at both sp^3 carbon and carbonyl groups. Substitution at saturated carbon usually involves the S_N2 mechanism, whereas substitution at carbonyl groups usually occurs by addition-elimination.

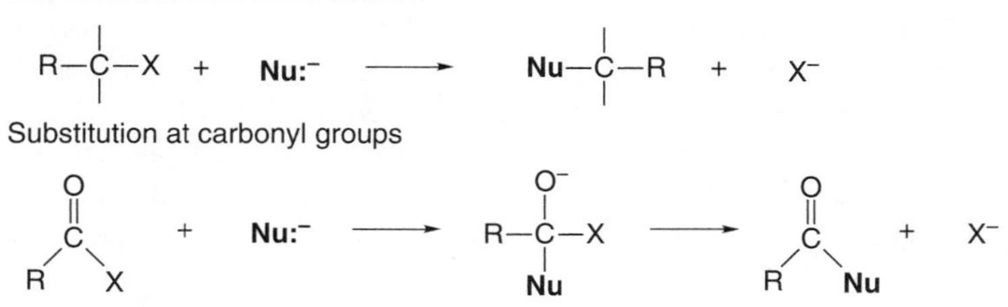

Substitution at saturated carbons

$$R-\overset{|}{\underset{|}{C}}-X \ + \ \textbf{Nu:}^- \ \longrightarrow \ Nu-\overset{|}{\underset{|}{C}}-R \ + \ X^-$$

Substitution at carbonyl groups

$$\underset{R \quad X}{\overset{O}{\underset{\|}{C}}} \ + \ \textbf{Nu:}^- \ \longrightarrow \ R-\overset{O^-}{\underset{\underset{\textbf{Nu}}{|}}{\overset{|}{C}}}-X \ \longrightarrow \ \underset{R \quad \textbf{Nu}}{\overset{O}{\underset{\|}{C}}} \ + \ X^-$$

The mechanistic aspects of nucleophilic substitutions at saturated carbon and carbonyl centers were considered in Part A, Chapters 4 and 7, respectively. In this chapter we discuss some of the important synthetic transformations that involve these types of

216

CHAPTER 3

*Functional Group
Interconversion
by Substitution,
Including Protection and
Deprotection*

reactions. Section 3.1 considers conversion of alcohols to reactive alkylating agents and Section 3.2 discusses the use of S_N2 reactions for various functional group transformations. Substitution reactions can also be used to *break* bonds for synthetic purposes, and Section 3.3 deals with cleavage of C–O bonds in ethers and esters by S_N2 and S_N1 reactions. The carbonyl substitution reactions that interconvert the acyl halides, acid anhydrides, esters, and carboxamides are discussed in Section 3.4. Often, manipulation of protecting groups also involves nucleophilic substitution and carbonyl exchange reactions. We discuss protection and deprotection of the most common functional groups in Section 3.5.

3.1. Conversion of Alcohols to Alkylating Agents

3.1.1. Sulfonate Esters

Alcohols are a very important compounds for synthesis. However, because the hydroxide ion is a very poor leaving group, alcohols are not reactive as alkylating agents. They can be activated to substitution by O-protonation, but the acidity that is required is incompatible with most nucleophiles except those, such as the halides, that are anions of strong acids. The preparation of sulfonate esters from alcohols is an effective way of installing a reactive leaving group on an alkyl chain. The reaction is very general and complications arise only if the resulting sulfonate ester is sufficiently reactive to require special precautions. *p*-Toluenesulfonate (*tosylate*) and methanesulfonate (*mesylate*) esters are used most frequently for preparative work, but the very reactive trifluoromethanesulfonates (*triflates*) are useful when an especially good leaving group is required. The usual method for introducing tosyl or mesyl groups is to allow the alcohol to react with the sulfonyl chloride in pyridine at 0°–25° C.[1] An alternative method is to convert the alcohol to a lithium salt, which is then allowed to react with the sulfonyl chloride.[2]

Trifluoromethanesulfonates of alkyl and allylic alcohols can be prepared by reaction with trifluoromethanesulfonic anhydride in halogenated solvents in the presence of pyridine.[3] Since the preparation of sulfonate esters does not disturb the C–O bond, problems of rearrangement or racemization do not arise in the ester formation step. However, sensitive sulfonate esters, such as allylic systems, may be subject to reversible ionization reactions, so appropriate precautions must be taken to ensure structural and stereochemical integrity. Tertiary alkyl sulfonates are neither as easily prepared nor as stable as those from primary and secondary alcohols. Under the standard preparative conditions, tertiary alcohols are likely to be converted to the corresponding alkene.

[1] R. S. Tipson, *J. Org. Chem.*, **9**, 235 (1944); G. W. Kabalka, M. Varma, R. S. Varma, P. C. Srivastava, and F. F. Knapp, Jr., *J. Org. Chem.*, **51**, 2386 (1986).
[2] H. C. Brown, R. Bernheimer, C. J. Kim, and S. E. Scheppele, *J. Am. Chem. Soc.*, **89**, 370 (1967).
[3] C. D. Beard, K. Baum, and V. Grakauskas, *J. Org. Chem.*, **38**, 3673 (1973).

3.1.2. Halides

The prominent role of alkyl halides in the formation of carbon-carbon bonds by enolate alkylation was evident in Chapter 1. The most common precursors for alkyl halides are the corresponding alcohols and a variety of procedures have been developed for this transformation. The choice of an appropriate reagent is usually dictated by the sensitivity of the alcohol and any other functional groups present in the molecule. In some cases, the hydrogen halides can be used. Unsubstituted primary alcohols can be converted to bromides with hot concentrated hydrobromic acid.[4] Alkyl chlorides can be prepared by reaction of primary alcohols with hydrochloric acid–zinc chloride.[5] Owing to the harsh conditions, these procedures are only applicable to very acid-stable molecules. These reactions proceed by the S_N2 mechanism and elimination, and rearrangements are not a problem for primary alcohols. Reactions of hydrogen halides with tertiary alcohols proceed by the S_N1 mechanism, so these reactions are preparatively useful only when the carbocation intermediate is unlikely to give rise to rearranged product.[6] In general, these methods are suitable only for simple, unfunctionalized alcohols.

Another general method for converting alcohols to halides involves reactions with halides of certain nonmetallic elements. Thionyl chloride, phosphorus trichloride, and phosphorus tribromide are the most common examples of this group of reagents. These reagents are suitable for alcohols that are neither acid sensitive nor prone to structural rearrangement. The reaction of alcohols with thionyl chloride initially results in the formation of a chlorosulfite ester. There are two mechanisms by which the chlorosulfite can be converted to a chloride. In aprotic nucleophilic solvents, such as dioxane, solvent participation can lead to overall retention of configuration.[7]

In the absence of solvent participation, chloride attack on the chlorosulfite ester leads to product with inversion of configuration.

Primary and secondary alcohols are rapidly converted to chlorides by a 1:1 mixture of $SOCl_2$ and benzotriazole in an inert solvent such as CH_2Cl_2.[8]

[4] E. E. Reid, J. R. Ruhoff, and R. E. Burnett, *Org. Synth.*, **II**, 246 (1943).

[5] J. E. Copenhaver and A. M. Wharley, *Org. Synth.*, **I**, 142 (1941).

[6] J. F. Norris and A. W. Olmsted, *Org. Synth.*, **I**, 144 (1941); H. C. Brown and M. H. Rei, *J. Org. Chem.*, **31**, 1090 (1966).

[7] E. S. Lewis and C. E. Boozer, *J. Am. Chem. Soc.*, **74**, 308 (1952).

[8] S. S. Chaudhari and K. G. Akamanchi, *Synlett*, 1763 (1999).

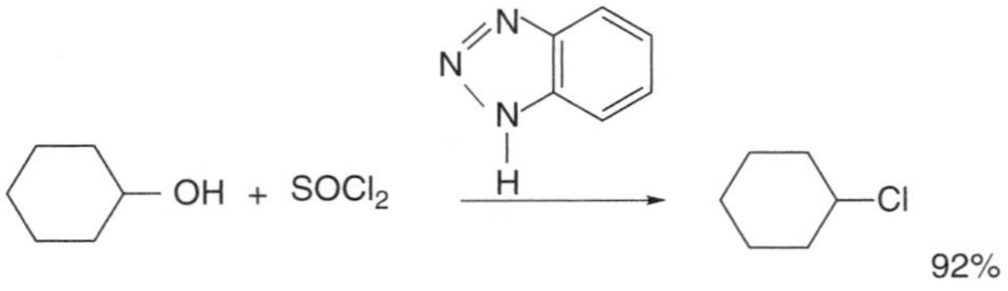

218

CHAPTER 3

*Functional Group
Interconversion
by Substitution,
Including Protection and
Deprotection*

92%

This reagent combination also converts carboxylic acids to acyl chlorides (see Section 3.4.1). The mechanistic basis for the special effectiveness of benzotriazole has not yet been determined, but it seems likely that nucleophilic catalysis is involved. Sulfinyl ester intermediates may be involved, because Z-2-butene-1,4-diol gives a cyclic sulfite ester with one equivalent of reagent but the dichloride with two equivalents.

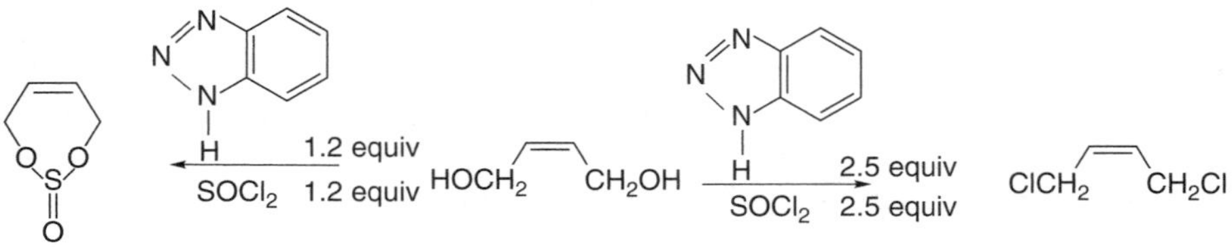

Reaction with the hindered secondary alcohol menthol stops at the dialkyl sulfite ester. The examples reported do not establish the stereochemistry of the reaction.

The mechanism for the reactions of alcohols with phosphorus halides can be illustrated using phosphorus tribromide. Initial reaction between the alcohol and phosphorus tribromide leads to a trialkyl phosphite ester by successive displacements of bromide. The reaction stops at this stage if it is run in the presence of an amine, which neutralizes the hydrogen bromide that is formed.[9] If the hydrogen bromide is not neutralized, the phosphite ester is protonated and each alkyl group is converted to the halide by nucleophilic substitution by bromide ion. The driving force for cleavage of the C–O bond is the formation of a strong phosphoryl double bond.

As C–Br bond formation occurs by back-side attack, inversion of the configuration at carbon is anticipated. However, both racemization and rearrangement are observed as competing processes.[10] For example, conversion of 2-butanol to 2-butyl bromide with PBr$_3$ is accompanied by 10–13% racemization and a small

9. A. H. Ford-Moore and B. J. Perry, *Org. Synth.*, **IV**, 955 (1963).
10. H. R. Hudson, *Synthesis*, 112 (1969).

amount of *t*-butyl bromide is also formed.[11] The extent of rearrangement increases with increasing chain length and branching.

$$CH_3CH_2CHCH_2CH_3 \xrightarrow[\text{ether}]{PBr_3} CH_3CH_2CHCH_2CH_3 + CH_3CH_2CH_2CHCH_3$$

$$\underset{OH}{|} \qquad\qquad \underset{Br}{|} \qquad\qquad \underset{Br}{|}$$

85–90%   10–15%

Ref. 12

$$(CH_3)_3CCH_2OH \xrightarrow[\text{quinoline}]{PBr_3} (CH_3)_3CCH_2Br + (CH_3)_2CCH_2CH_3 + CH_3CHCH(CH_3)_2$$

$$\qquad\qquad\qquad\qquad\qquad\qquad \underset{Br}{|} \qquad\qquad \underset{Br}{|}$$

63%   26%   11%

Ref. 13

Owing to the acidic conditions, these methods are limited to acid-stable molecules. Milder reagents are necessary for many functionally substituted alcohols. A very general and important method for activating alcohols toward nucleophiles is by converting them to *alkoxyphosphonium ions*.[14] The trivalent phosphorus reagents are activated by reaction with a halogen or related electrophile, and the alkoxyphosphonium ions are very reactive toward nucleophilic attack, with the driving force for substitution being formation of the strong phosphoryl bond.

$$R'_3P + E-Y \longrightarrow R'_3P\overset{E}{\underset{Y}{\big<}} \rightleftharpoons R'_3P^+-E + Y^-$$

$$R'_3P^+-E + ROH \longrightarrow R'_3P^+-OR + HE$$

$$R'_3P^+-OR + Nu^- \longrightarrow R'_3P=O + R-Nu$$

A variety of reagents can function as the electrophile E^+ in the general mechanism. The most useful synthetic procedures for preparation of halides are based on the halogens, positive halogens sources, and diethyl azodicarboxylate. A 1:1 adduct formed from triphenylphosphine and bromine converts alcohols to bromides.[15] The alcohol displaces bromide ion from the pentavalent adduct, giving an alkoxyphosphonium intermediate. The phosphonium ion intermediate then undergoes nucleophilic attack by bromide ion, forming triphenylphosphine oxide.

$$PPh_3 + Br_2 \longrightarrow Br_2PPh_3$$

$$Br_2PPh_3 + ROH \longrightarrow ROP^+Ph_3Br^- + HBr$$

$$Br^- + ROP^+Ph_3 \longrightarrow RBr + Ph_3P=O$$

The alkoxy phosphonium intermediate is formed by a reaction that does not break the C−O bond and the second step proceeds by back-side displacement on carbon, so the stereochemistry of the overall process is inversion.

[11.] D. G. Goodwin and H. R. Hudson, *J. Chem. Soc. B*, 1333 (1968); E. J. Coulson, W. Gerrard, and H. R. Hudson, *J. Chem. Soc.*, 2364 (1965).
[12.] J. Cason and J. S. Correia, *J. Org. Chem.*, **26**, 3645 (1961).
[13.] H. R. Hudson, *J. Chem. Soc.*, 664 (1968).
[14.] B. P. Castro, *Org. React.*, **29**, 1 (1983).
[15.] G. A. Wiley, R. L. Hershkowitz, B. M. Rein, and B. C. Chung, *J. Am. Chem. Soc.*, **86**, 964 (1964).

Ref. 16

2,4,4,6-Tetrabromocyclohexa-2,5-dienone is also a useful bromine source.

Ref. 17

Triphenylphosphine dichloride exhibits similar reactivity and can be used to prepare chlorides.[18] The most convenient methods for converting alcohols to chlorides are based on in situ generation of chlorophosphonium ions[19] by reaction of triphenylphosphine with various chlorine compounds such as carbon tetrachloride[20] or hexachloroacetone.[21] These reactions involve formation of chlorophosphonium ions.

$$Ph_3P + CCl_4 \longrightarrow Ph_3P^+-Cl + {}^-CCl_3$$

$$Ph_3P + Cl_3C\overset{O}{\overset{\|}{C}}CCl_3 \longrightarrow Ph_3P^+-Cl + {}^-CCl_2\overset{O}{\overset{\|}{C}}CCl_3$$

The chlorophosphonium ion then reacts with the alcohol to give an alkoxyphosphonium ion that is converted to the chloride.

$$Ph_3P^+-Cl + ROH \longrightarrow Ph_3P^+-OR + HCl$$

$$Ph_3P^+-OR + Cl^- \longrightarrow Ph_3P{=}O + R-Cl$$

Several modifications of procedures based on halophosphonium ion have been developed. Triphenylphosphine and imidazole in combination with iodine or bromine gives good conversion of alcohols to iodides or bromides.[22] An even more reactive system consists of chlorodiphenylphosphine, imidazole, and the halogen,[23] and has the further advantage that the resulting phosphorus by-product diphenylphosphinic acid, can be extracted with base during product workup.

$$Ph_2PCl + H-N\overset{\frown}{\underset{}{}}N + I_2 + ROH \longrightarrow RI + Ph_2\overset{O}{\overset{\|}{P}}H$$

A very mild procedure for converting alcohols to iodides uses triphenylphosphine, diethyl azodicarboxylate (DEAD), and methyl iodide.[24] This reaction occurs

16. D. Levy and R. Stevenson, *J. Org. Chem.*, **30**, 2635 (1965).
17. A. Tanaka and T. Oritani, *Tetrahedron Lett.*, **38**, 1955 (1997).
18. L. Horner, H. Oediger, and H. Hoffmann, *Justus Liebigs Ann. Chem.*, **626**, 26 (1959).
19. R. Appel, *Angew. Chem. Int. Ed. Engl.*, **14**, 801 (1975).
20. J. B. Lee and T. J. Nolan, *Can. J. Chem.*, **44**, 1331 (1966).
21. R. M. Magid, O. S. Fruchey, W. L. Johnson, and T. G. Allen, *J. Org. Chem.*, **44**, 359 (1979).
22. P. J. Garegg, R. Johansson, C. Ortega, and B. Samuelsson, *J. Chem. Soc., Perkin Trans.*, **1**, 681 (1982).
23. B. Classon, Z. Liu, and B. Samuelsson, *J. Org. Chem.*, **53**, 6126 (1988).
24. O. Mitsunobu, *Synthesis*, 1 (1981).

with clean inversion of stereochemistry.[25] The key intermediate is again an alkoxyphosphonium ion.

$$Ph_3P + ROH + C_2H_5O_2CN{=}NCO_2C_2H_5 \longrightarrow Ph_3\overset{+}{P}OR + C_2H_5O_2C\bar{N}NHCO_2C_2H_5$$

$$C_2H_5O_2C\bar{N}NHCO_2C_2H_5 + CH_3I \longrightarrow C_2H_5O_2C\underset{\underset{CH_3}{|}}{N}NHCO_2C_2H_5 + I^-$$

$$Ph_3P^+OR + I^- \longrightarrow RI + Ph_3P{=}O$$

The role of the DEAD is to activate the triphenylphosphine toward nucleophilic attack by the alcohol. In the course of the reaction the N=N double bond is reduced. As is discussed later, this method is applicable for activation of alcohols to substitution by other nucleophiles in addition to halide ions. The activation of alcohols to nucleophilic attack by the triphenylphosphine-DEAD combination is called the *Mitsunobu reaction.*[26]

A very mild method that is useful for compounds that are prone to allylic rearrangement involves prior conversion of the alcohol to a sulfonate, followed by nucleophilic displacement with halide ion.

$$
\begin{array}{c}
CH_3CH_2CH_2 \\
\diagdown \\
C{=}CHCH_2OH \\
\diagup \\
CH_3CH_2CH_2
\end{array}
\xrightarrow[\text{2) LiCl, DMF}]{\text{1) CH}_3\text{SO}_2\text{Cl}}
\begin{array}{c}
CH_3CH_2CH_2 \\
\diagdown \\
C{=}CHCH_2Cl \\
\diagup \\
CH_3CH_2CH_2
\end{array}
$$

83%   Ref. 27

Another very mild procedure involves reaction of the alcohol with the heterocyclic 2-chloro-3-ethylbenzoxazolium cation.[28] The alcohol adds to the electrophilic heterocyclic ring, displacing chloride. The alkoxy group is thereby activated toward a nucleophilic substitution that forms a stable product, 3-ethylbenzoxazolinone.

The reaction can be used for making either chlorides or bromides by using the appropriate tetraalkylammonium salt as a halide source.

Scheme 3.1 gives some examples of the various alcohol to halide conversions that have been discussed. Entries 1 and 2 are examples of synthesis of primary bromides using PBr_3. Entry 3 is an example of synthesis of a chloride using $Ph_3P{-}Cl_2$. The reactant, neopentyl alcohol, is often resistant to nucleophilic substitution and prone to rearrangement, but reacts well under these conditions. Entries 4 and 5 illustrate the use of halogenated solvents as chlorine sources in Ph_3P-mediated reactions. The reactions in Entries 6 and 7 involve synthesis of bromides by nucleophilic substitution on tosylates. The reactant in Entry 7 is prone to rearrangement via ring expansion,

25. H. Loibner and E. Zbiral, *Helv. Chim. Acta,* **59**, 2100 (1976).
26. D. L. Hughes, *Org. React.,* **42**, 335 (1992).
27. E. W. Collington and A. I. Meyers, *J. Org. Chem.,* **36**, 3044 (1971).
28. T. Mukaiyama, S. Shoda, and Y. Watanabe, *Chem. Lett.,* 383 (1977); T. Mukaiyama, *Angew. Chem. Int. Ed. Engl.,* **18**, 707 (1979).

CHAPTER 3

*Functional Group
Interconversion
by Substitution,
Including Protection and
Deprotection*

Scheme 3.1. Preparation of Alkyl Halides

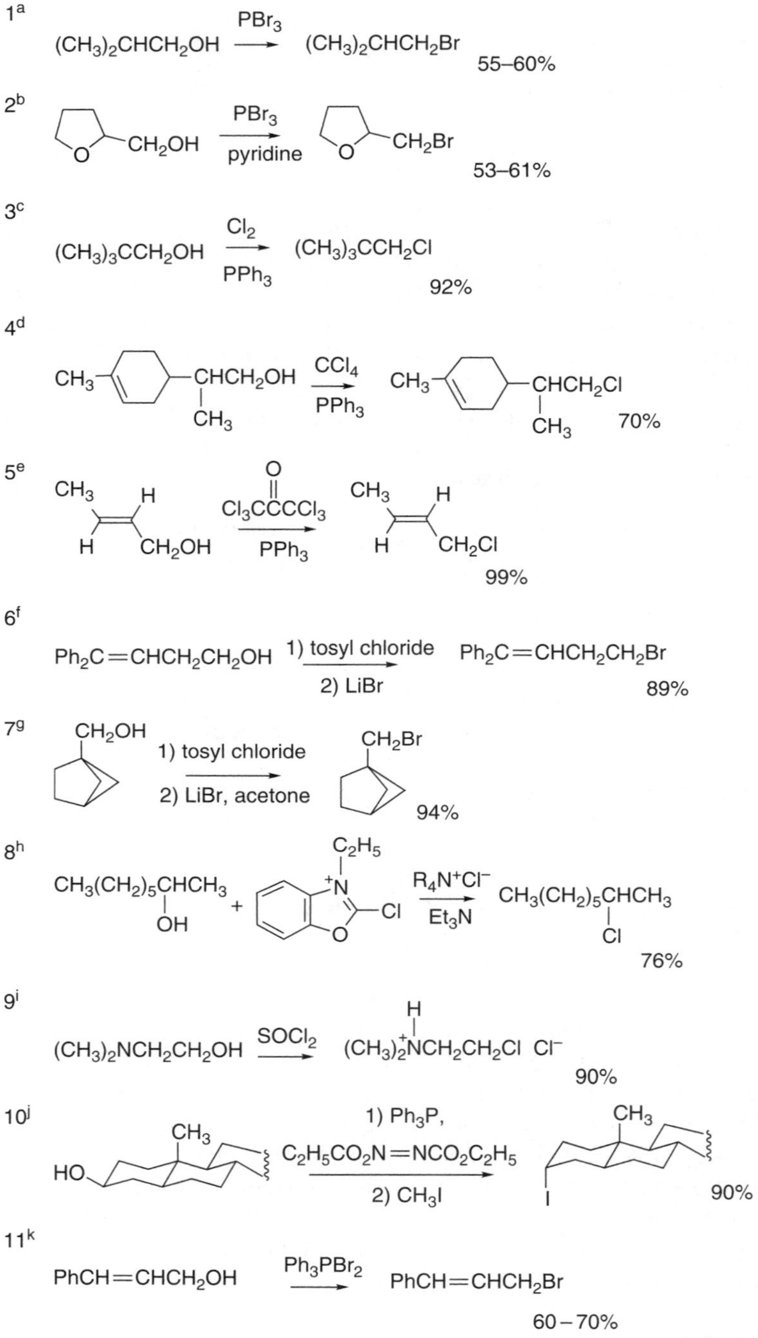

1[a]

$(CH_3)_2CHCH_2OH \xrightarrow{PBr_3} (CH_3)_2CHCH_2Br$ 55–60%

2[b]

$\xrightarrow[\text{pyridine}]{PBr_3}$ 53–61%

3[c]

$(CH_3)_3CCH_2OH \xrightarrow[PPh_3]{Cl_2} (CH_3)_3CCH_2Cl$ 92%

4[d]

$\xrightarrow[PPh_3]{CCl_4}$ 70%

5[e]

$\xrightarrow[PPh_3]{Cl_3CCCl_3}$ 99%

6[f]

$Ph_2C=CHCH_2CH_2OH \xrightarrow[\text{2) LiBr}]{\text{1) tosyl chloride}} Ph_2C=CHCH_2CH_2Br$ 89%

7[g]

$\xrightarrow[\text{2) LiBr, acetone}]{\text{1) tosyl chloride}}$ 94%

8[h]

$CH_3(CH_2)_5\underset{OH}{CHCH_3} +$ $\xrightarrow[Et_3N]{R_4N^+Cl^-} CH_3(CH_2)_5\underset{Cl}{CHCH_3}$ 76%

9[i]

$(CH_3)_2NCH_2CH_2OH \xrightarrow{SOCl_2} (CH_3)_2\overset{H}{\underset{}{N}}{}^+CH_2CH_2Cl\ Cl^-$ 90%

10[j]

$\xrightarrow[\text{2) CH}_3\text{I}]{\text{1) Ph}_3\text{P,}\ C_2H_5CO_2N=NCO_2C_2H_5}$ 90%

11[k]

$PhCH=CHCH_2OH \xrightarrow{Ph_3PBr_2} PhCH=CHCH_2Br$ 60–70%

(Continued)

Scheme 3.1. (*Continued*)

223

SECTION 3.2

*Introduction of
Functional Groups by
Nucleophilic Substitution
at Saturated Carbon*

12[l]

75%

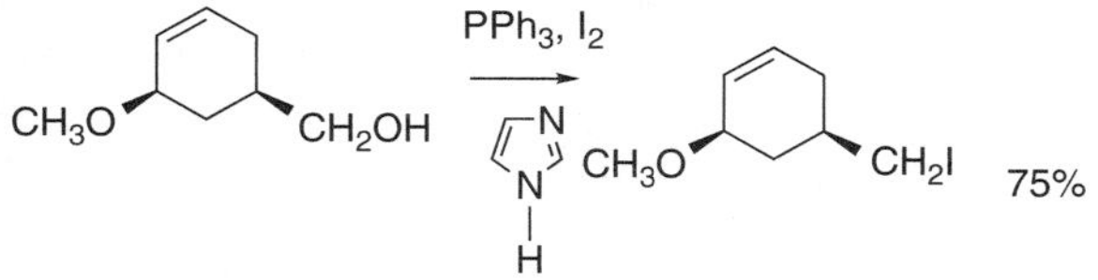

a. C. R. Noller and R. Dinsomore, *Org. Synth.*, **II**, 358 (1943).
b. L. H. Smith, *Org. Synth.*, **III**, 793 (1955).
c. G. A. Wiley, R. L. Hershkowitz, B. M. Rein, and B. C. Chung, *J. Am. Chem. Soc.*, **86**, 964 (1964).
d. D. B. MacKenzie, M. M. Angelo, and J. Wolinsky, *J. Org. Chem.*, **44**, 4042 (1979).
e. R. M. Magid, O. S. Fruchy, W. L. Johnson, and T. G. Allen, *J. Org. Chem.*, **44**, 359 (1979).
f. M. E. H. Howden, A. Maerker, J. Burdon, and J. D. Roberts, *J. Am. Chem. Soc.*, **88**, 1732 (1966).
g. K. B. Wiberg and B. R. Lowry, *J. Am. Chem. Soc.*, **85**, 3188 (1963).
h. T. Mukaiyama, S. Shoda, and Y. Watanabe, *Chem. Lett.*, 383 (1977).
i. L. A. R. Hall, V. C. Stephens, and J. H. Burkhalter, *Org. Synth.*, **IV**, 333 (1963).
j. H. Loibner and E. Zviral, *Helv. Chim. Acta*, **59**, 2100 (1976).
k. J. P. Schaefer, J. G. Higgins, and P. K. Shenoy, *Org. Synth.*, **V**, 249 (1973).
l. R. G. Linde II, M. Egbertson, R. S. Coleman, A. B. Jones, and S. J. Danishefsky, *J. Org. Chem.*, **55**, 2771 (1990).

but no rearrangement was observed under these conditions. Entry 8 illustrates the use of a chlorobenzoxazolium cation for conversion of a secondary alcohol to a chloride. This reaction was shown to proceed with inversion of configuration. Entry 9 involves conversion of a primary alcohol to a chloride using $SOCl_2$. In this particular example, the tertiary amino group captures the HCl that is formed by the reaction of the alcohol with $SOCl_2$. There is also some suggestion from the procedure that much of the reaction proceeds through a chlorosulfite intermediate. After the reactants are mixed (exothermic reaction), the material is heated in ethanol, during which time gas evolution occurs. This suggests that much of the chlorosulfite ester survives until the heating stage.

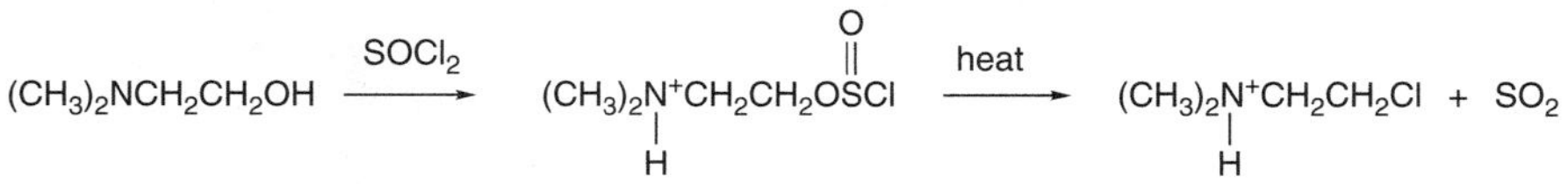

Entry 10 illustrates the application of the Mitsunobu reaction to synthesis of a steroidal iodide and demonstrates that inversion occurs. Entry 11 shows the use of the isolated $Ph_3P–Br_2$ complex. The reaction in Entry 12 involves the preparation of a primary iodide using the $Ph_3P–I_2$-imidazole reagent combination.

3.2. Introduction of Functional Groups by Nucleophilic Substitution at Saturated Carbon

The mechanistic aspects of nucleophilic substitution reactions were treated in detail in Chapter 4 of Part A. That mechanistic understanding has contributed to the development of nucleophilic substitution reactions as important synthetic processes. Owing to its stereospecificity and avoidance of carbocation intermediates, the S_N2 mechanism is advantageous from a synthetic point of view. In this section we discuss

224

CHAPTER 3

*Functional Group
Interconversion
by Substitution,
Including Protection and
Deprotection*

the role of S_N2 reactions in the preparation of several classes of compounds. First, however, it is desirable to review the important role that solvent plays in S_N2 reactions. The knowledgeable manipulation of solvent and related medium effects has led to significant improvement of many synthetic procedures that proceed by the S_N2 mechanism.

3.2.1. General Solvent Effects

The objective in selecting the reaction conditions for a preparative nucleophilic substitution is to enhance the mutual reactivity of the leaving group and nucleophile so that the desired substitution occurs at a convenient rate and with minimal competition from other possible reactions. The generalized order of leaving-group reactivity $RSO_3^- \sim I^- > Br^- > Cl^-$ pertains for most S_N2 processes. (See Section 4.2.3 of Part A for more complete data.) Mesylates, tosylates, iodides, and bromides are all widely used in synthesis. Chlorides usually react rather slowly, except in especially reactive systems, such as allyl and benzyl.

The overall synthetic objective normally governs the choice of the nucleophile. Optimization of reactivity therefore must be achieved by selection of the reaction conditions, particularly the solvent. Several generalizations about solvents can be made. Hydrocarbons, halogenated hydrocarbons, and ethers are usually unsuitable solvents for reactions involving ionic metal salts. Acetone and acetonitrile are somewhat more polar, but the solubility of most ionic compounds in these solvents is low. Solubility can be considerably improved by use of salts of cations having substantial hydrophobic character, such as those containing tetraalkylammonium ions. Alcohols are reasonably good solvents for salts, but the nucleophilicity of hard anions is relatively low in alcohols because of extensive solvation. The polar aprotic solvents, particularly dimethylformamide (DMF) and dimethylsulfoxide (DMSO), are good solvents for salts and, by virtue of selective cation solvation, anions usually show enhanced nucleophilicity in these solvents. Hexamethylphosphoric triamide (HMPA), N,N-dimethylacetamide, and N-methylpyrrolidinone are other examples of polar aprotic solvents.[29] The high water solubility of these solvents and their high boiling points can sometimes cause problems in product separation and purification. Furthermore, HMPA is toxic. In addition to enhancing reactivity, polar aprotic solvents also affect the order of reactivity of nucleophilic anions. In DMF the halides are all of comparable nucleophilicity,[30] whereas in hydroxylic solvents the order is $I^- > Br^- > Cl^-$ and the differences in reactivity are much greater.[31]

There are two other approaches to enhancing reactivity in nucleophilic substitutions by exploiting solvation effects on reactivity: the use of *crown ethers* as catalysts and the utilization of *phase transfer conditions*. The crown ethers are a family of cyclic polyethers, three examples of which are shown below.

29. A. F. Sowinski and G. M. Whitesides, *J. Org. Chem.*, **44**, 2369 (1979).
30. W. M. Weaver and J. D. Hutchinson, *J. Am. Chem. Soc.*, **86**, 261 (1964).
31. R. G. Pearson and J. Songstad, *J. Org. Chem.*, **32**, 2899 (1967).

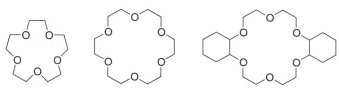

15-crown-5   18-crown-6   dicyclohexano-18-crown-6

225

SECTION 3.2

*Introduction of
Functional Groups by
Nucleophilic Substitution
at Saturated Carbon*

The first number designates the ring size and the second the number of oxygen atoms in the ring. By complexing the cation in the cavity of the crown ether, these compounds can solubilize salts in nonpolar solvents. In solution, the anions are more reactive as nucleophiles because they are weakly solvated. Tight ion pairing is also precluded by the complexation of the cation by the nonpolar crown ether. As a result, nucleophilicity approaches or exceeds that observed in aprotic polar solvents,[32] but the crown ethers do present some hazards. They are toxic and also have the potential to transport toxic anions, such as cyanide, through the skin.

Another method of accelerating nucleophilic substitution is to use phase transfer catalysts,[33] which are ionic substances, usually quaternary ammonium or phosphonium salts, in which the hydrocarbon groups in the cation are large enough to convey good solubility in nonpolar solvents. In other words, the cations are highly *lipophilic*. Phase transfer catalysis usually is done in a two-phase system. The reagent is dissolved in a water-insoluble solvent such as a hydrocarbon or halogenated hydrocarbon. The salt of the nucleophile is dissolved in water. Even with vigorous mixing, such systems show little tendency to react, because the nucleophile and reactant remain separated in the water and organic phases, respectively. When a phase transfer catalyst is added, the lipophilic cations are transferred to the nonpolar phase and anions are attracted from the water to the organic phase to maintain electrical neutrality. The anions are weakly solvated in the organic phase and therefore exhibit enhanced nucleophilicity. As a result, the substitution reactions proceed under relatively mild conditions. The salts of the nucleophile are often used in high concentration in the aqueous solution and in some procedures the solid salts are used.

3.2.2. Nitriles

The replacement of a halide or sulfonate by cyanide ion, extending the carbon chain by one atom and providing an entry to carboxylic acid derivatives, has been a reaction of synthetic importance since the early days of organic chemistry. The classical conditions for preparing nitriles involve heating a halide with a cyanide salt in aqueous alcohol solution.

[32.] M. Hiraoka, *Crown Compounds: Their Characteristics and Application*, Elsevier, Amsterdam, 1982.

[33.] E. V. Dehmlow and S. S. Dehmlow, *Phase Transfer Catalysis*, 3rd Edition, Verlag Chemie, Weinheim 1992; W. P. Weber and G. W. Gokel, *Phase Transfer Catalysis in Organic Synthesis*, Springer Verlag, New York, 1977; C. M. Stark, C. Liotta, and M. Halpern, *Phase Transfer Catalysis: Fundamentals, Applications and Industrial Perspective*, Chapman and Hall, New York, 1994.

226

CHAPTER 3

*Functional Group
Interconversion
by Substitution,
Including Protection and
Deprotection*

$$\text{C}_6\text{H}_5\text{—CH}_2\text{Cl} + \text{NaCN} \xrightarrow[\text{reflux 4h}]{\text{H}_2\text{O, C}_2\text{H}_5\text{OH}} \text{C}_6\text{H}_5\text{—CH}_2\text{CN}$$

80–90%

Ref. 34

$$\text{ClCH}_2\text{CH}_2\text{CH}_2\text{Br} + \text{KCN} \xrightarrow[\text{reflux 1.5h}]{\text{H}_2\text{O, C}_2\text{H}_5\text{OH}} \text{ClCH}_2\text{CH}_2\text{CH}_2\text{CN}$$

40–50%

Ref. 35

These reactions proceed more rapidly in polar aprotic solvents. In DMSO, for example, primary alkyl chlorides are converted to nitriles in 1 h or less at temperatures of 120°–140° C.[36] Phase transfer catalysis by hexadecyltributylphosphonium bromide permits conversion of 1-chlorooctane to octyl cyanide in 95% yield in 2 h at 105° C.[37]

$$\text{CH}_3\text{CH}_2\text{CH}_2\text{CH}_2\text{Cl} \xrightarrow[\text{90–160°C}]{\substack{\text{NaCN} \\ \text{DMSO}}} \text{CH}_3\text{CH}_2\text{CH}_2\text{CH}_2\text{CN} \quad 93\%$$

$$\text{CH}_3(\text{CH}_2)_6\text{CH}_2\text{Cl} \xrightarrow[\substack{\text{C}_{16}\text{H}_{33}\text{P}^+(\text{C}_4\text{H}_9)_3 \\ \text{105°C, 2h}}]{\substack{\text{NaCN} \\ \text{H}_2\text{O, decane}}} \text{CH}_3(\text{CH}_2)_6\text{CH}_2\text{CN} \quad 95\%$$

Catalysis by 18-crown-6 of the reaction of solid potassium cyanide with a variety of chlorides and bromides has been demonstrated.[38] With primary bromides, yields are high and reaction times are 15–30 h at reflux in acetonitrile (83° C). Interestingly, the chlorides are more reactive and require reaction times of only about 2 h. Secondary halides react more slowly and yields drop because of competing elimination. Tertiary halides do not react satisfactorily because elimination dominates.

3.2.3. Oxygen Nucleophiles

The oxygen nucleophiles that are of primary interest in synthesis are the hydroxide ion (or water), alkoxide ions, and carboxylate anions, which lead, respectively, to alcohols, ethers, and esters. Since each of these nucleophiles can also act as a base, reaction conditions are selected to favor substitution over elimination. Usually, a given alcohol is more easily obtained than the corresponding halide so the halide-to-alcohol transformation is not used extensively for synthesis. The hydrolysis of benzyl halides to the corresponding alcohols proceeds in good yield. This can be a useful synthetic transformation because benzyl halides are available either by side chain halogenation or by the chloromethylation reaction (Section 11.1.3).

34. R. Adams and A. F. Thal, *Org. Synth.*, **I**, 101 (1932).
35. C. F. H. Allen, *Org. Synth.*, **I**, 150 (1932).
36. L. Friedman and H. Shechter, *J. Org. Chem.*, **25**, 877 (1960); R. A. Smiley and C. Arnold, *J. Org. Chem.*, **25**, 257 (1960).
37. C. M. Starks, *J. Am. Chem. Soc.*, **93**, 195 (1971); C. M. Starks and R. M. Owens, *J. Am. Chem. Soc.*, **95**, 3613 (1973).
38. F. L. Cook, C. W. Bowers, and C. L. Liotta, *J. Org. Chem.*, **39**, 3416 (1974).

227

SECTION 3.2

*Introduction of
Functional Groups by
Nucleophilic Substitution
at Saturated Carbon*

$$NC-\langle\rangle-CH_2Cl \xrightarrow[\text{H}_2\text{O},100°\text{C}]{K_2CO_3} NC-\langle\rangle-CH_2OH \qquad 85\% \qquad \text{Ref. 39}$$

Ether formation from alkoxides and alkylating reagents is a reaction of wide synthetic importance. The conversion of phenols to methoxyaromatics, for example, is a very common reaction. Methyl iodide, methyl tosylate, or dimethyl sulfate can be used as the alkylating agents. The reaction proceeds in the presence of a weak base, such as Na_2CO_3 or K_2CO_3, which deprotonates the phenol. The conjugate bases of alcohols are considerably more basic than phenoxides, so β-elimination can be a problem. Phase transfer conditions can be used in troublesome cases.[40] Fortunately, the most useful and commonly encountered ethers are methyl and benzyl ethers, where elimination is not a problem and the corresponding halides are especially reactive toward substitution.

Two methods for converting carboxylic acids to esters fall into the mechanistic group under discussion: the reaction of carboxylic acids with diazo compounds, especially diazomethane and alkylation of carboxylate anions by halides or sulfonates. The esterification of carboxylic acids with diazomethane is a very fast and clean reaction.[41] The alkylating agent is the extremely reactive methyldiazonium ion, which is generated by proton transfer from the carboxylic acid to diazomethane. The collapse of the resulting ion pair with loss of nitrogen is extremely rapid.

$$RCO_2H + CH_2N_2 \longrightarrow [RCO_2^- + CH_3\overset{+}{N}_2] \longrightarrow RCO_2CH_3 + N_2$$

The main drawback to this reaction is the toxicity of diazomethane and some of its precursors. Diazomethane is also potentially explosive. Trimethylsilyldiazomethane is an alternative reagent,[42] which is safer and frequently used in preparation of methyl esters from carboxylic acids.[43] Trimethylsilyldiazomethane also O-methylates alcohols.[44] The latter reactions occur in the presence of fluoroboric acid in dichloromethane.

Especially for large-scale work, esters may be more safely and efficiently prepared by reaction of carboxylate salts with alkyl halides or tosylates. Carboxylate anions are not very reactive nucleophiles so the best results are obtained in polar aprotic solvents[45] or with crown ether catalysts.[46] The reactivity order for carboxylate salts is $Na^+ < K^+ < Rb^+ < Cs^+$. Cesium carboxylates are especially useful in polar aprotic solvents. The enhanced reactivity of the cesium salts is due to both high solubility and minimal ion pairing with the anion.[47] Acetone is a good solvent for reaction of carboxylate anions with alkyl iodides.[48] Cesium fluoride in DMF is another useful

[39.] J. N. Ashley, H. J. Barber, A. J. Ewins, G. Newbery, and A. D. Self, *J. Chem. Soc.*, 103 (1942).

[40.] F. Lopez-Calahorra, B. Ballart, F. Hombrados, and J. Marti, *Synth. Commun.*, **28**, 795 (1998).

[41.] T. H. Black, *Aldrichimia Acta*, **16**, 3 (1983).

[42.] N. Hashimoto, T. Aoyama, and T. Shiori, *Chem. Pharm. Bull.*, **29**, 1475 (1981).

[43.] T. Shioiri and T. Aoyama, *Adv. Use Synthons Org. Chem.*, **1**, 51 (1993); A. Presser and A. Huefner, *Monatsh. Chem.*, **135**, 1015 (2004).

[44.] T. Aoyama and T. Shiori, *Tetrahedron Lett.*, **31**, 5507 (1990).

[45.] P. E. Pfeffer, T. A. Foglia, P. A. Barr, I. Schmeltz, and L. S. Silbert, *Tetrahedron Lett.*, 4063 (1972); J. E. Shaw, D. C. Kunerth, and J. J. Sherry, *Tetrahedron Lett.*, 689 (1973); J. Grundy, B. G. James, and G . Pattenden, *Tetrahedron Lett.*, 757 (1972).

[46.] C. L. Liotta, H. P. Harris, M. McDermott, T. Gonzalez, and K. Smith, *Tetrahedron Lett.*, 2417 (1974).

[47.] G. Dijkstra, W. H. Kruizinga, and R. M. Kellog, *J. Org. Chem.*, **52**, 4230 (1987).

[48.] G. G. Moore, T. A. Foglia, and T. J. McGahan, *J. Org. Chem.*, **44**, 2425 (1979).

combination.[49] Carboxylate alkylation procedures are particularly advantageous for preparation of hindered esters, which can be relatively difficult to prepare by the acid-catalyzed esterification method (Fisher esterification), which we discuss in Section 3.4.

During the course of synthesis, it is sometimes necessary to invert the configuration at an oxygen-substituted center. One of the best ways of doing this is to activate the hydroxy group to substitution by a carboxylate anion. The activation is frequently done using the Mitsunobu reaction.[50] Hydrolysis of the resulting ester give the alcohol of inverted configuration.

Ref. 51

Ref. 52

Carboxylate anions derived from somewhat stronger acids, such as *p*-nitrobenzoic acid and chloroacetic acid, seem to be particularly useful in this Mitsunobu inversion reaction.[53] Inversion can also be carried out on sulfonate esters using cesium carboxylates and DMAP as a catalyst in toluene.[54] The effect of the DMAP seems to involve complexation and solubilization of the cesium salts.

Sulfonate esters also can be prepared under Mitsunobu conditions. Use of zinc tosylate in place of the carboxylic acid gives a tosylate of inverted configuration.

Ref. 55

The Mitsunobu conditions also can be used to effect a variety of other important and useful nucleophilic substitution reactions, such as conversion of alcohols to mixed phosphite esters.[56] The active phosphitylating agent is believed to be a mixed phosphoramidite.

49. T. Sato, J. Otera, and H. Nozaki, *J. Org. Chem.*, **57**, 2166 (1992).
50. D. L. Hughes, *Org. React.*, **42**, 335 (1992); D. L. Hughes, *Org. Prep. Proc. Intl.*, **28**, 127 (1996).
51. M. J. Arco, M. H. Trammel, and J. D. White, *J. Org. Chem.*, **41**, 2075 (1976).
52. C.-T. Hsu, N.-Y. Wang, L. H. Latimer, and C. J. Sih, *J. Am. Chem. Soc.*, **105**, 593 (1983).
53. J. A. Dodge, J. I. Tujillo, and M. Presnell, *J. Org. Chem.*, **59**, 234 (1994); M. Saiah, M.Bessodes, and K. Antonakis, *Tetrahedron Lett.*, **33**, 4317 (1992); S. F. Martin and J. A. Dodge, *Tetrahedron Lett.*, **32**, 3017 (1991); P. J. Harvey, M. von Itzstein, and I. D. Jenkins, *Tetrahedron*, **53**, 3933 (1997).
54. N. A. Hawryluk and B. B. Snider, *J. Org. Chem.*, **65**, 8379 (2000).
55. I. Galynker and W. C. Still, *Tetrahedron Lett.*, 4461 (1982).
56. I. D. Grice, P. J. Harvey, I. D. Jenkins, M. J. Gallagher, and M. G. Ranasinghe, *Tetrahedron Lett.*, **37**, 1087 (1996).

229

SECTION 3.2

Introduction of
Functional Groups by
Nucleophilic Substitution
at Saturated Carbon

$$(CH_3O)_2PH + i\text{-}PrO_2CN=NCO_2\text{-}i\text{-}Pr + Ph_3P \longrightarrow (CH_3O)_2PNNHCO_2\text{-}i\text{-}Pr + Ph_3P=O$$
$$\underset{\overset{|}{CO_2\text{-}i\text{-}Pr}}{}$$

$$(CH_3O)_2PNNHCO_2\text{-}i\text{-}Pr + ROH \longrightarrow ROP(OCH_3)_2$$
$$\underset{\overset{|}{CO_2\text{-}i\text{-}Pr}}{}$$

Mixed phosphonate acid esters can also be prepared from alkylphosphonate monoesters, although here the activation is believed occur at the alcohol.[57]

$$ROP^+Ph_3 + R'PO_2^- \longrightarrow R'POR + Ph_3P=O$$
$$\underset{\overset{|}{OCH_3}}{} \qquad \underset{\overset{|}{OCH_3}}{}$$

3.2.4. Nitrogen Nucleophiles

The alkylation of neutral amines by halides is complicated from a synthetic point of view by the possibility of multiple alkylation that can proceed to the quaternary ammonium salt in the presence of excess alkyl halide.

$$RNH_2 + R'—X \longrightarrow R\overset{H+}{\underset{H}{N}}R' + X^-$$

$$R\overset{H+}{\underset{H}{N}}R' + RNH_2 \rightleftharpoons R\underset{H}{N}R' + R\overset{+}{N}H_3$$

$$R\underset{H}{N}R' + R'—X \longrightarrow R\overset{+}{\underset{H}{N}}R'_2 + X^-$$

$$R\overset{+}{\underset{H}{N}}R'_2 + RNH_2 \rightleftharpoons RNR'_2 + R\overset{+}{N}H_3$$

$$RNR'_2 + R'—X \longrightarrow R\overset{+}{N}R'_3 + X^-$$

Even with a limited amount of the alkylating agent, the equilibria between protonated product and the neutral starting amine are sufficiently fast that a mixture of products may be obtained. For this reason, when monoalkylation of an amine is desired, the reaction is usually best carried out by *reductive amination*, a reaction that is discussed in Chapter 5. If complete alkylation to the quaternary salt is desired, use of excess alkylating agent and a base to neutralize the liberated acid normally results in complete reaction.

Amides are weakly nucleophilic and react only slowly with alkyl halides. The anions of amides are substantially more reactive. The classical Gabriel procedure for synthesis of amines from phthalimide is illustrative.[58]

$$\text{(phthalimide)}N^-K^+ + BrCH_2CH_2Br \longrightarrow \text{(phthalimide)}NCH_2CH_2Br$$

70–80%   Ref. 59

[57.] D. A. Campbell, *J. Org. Chem.*, **57**, 6331 (1992); D. A. Campbell and J. C. Bermak, *J. Org. Chem.*, **59**, 658 (1994).

[58.] M. S. Gibson and R. N. Bradshaw, *Angew. Chem. Int. Ed. Engl.*, **7**, 919 (1968).

[59.] P. L. Salzberg and J. V. Supniewski, *Org. Synth.*, **I**, 119 (1932).

230

CHAPTER 3

*Functional Group
Interconversion
by Substitution,
Including Protection and
Deprotection*

The enhanced acidity of the NH group in phthalimide permits formation of the anion, which is readily alkylated by alkyl halides or tosylates. The amine can then be liberated by reaction of the substituted phthalimide with hydrazine.

$$CH_3O_2CCHCH_2CHCO_2CH_3 \longrightarrow CH_3O_2CCHCH_2CHCO_2CH_3 \xrightarrow[CH_3OH]{NH_2NH_2} \xrightarrow[H_2O]{HCl} HO_2CCHCH_2CHCO_2H$$

phthal = phthalimido

Ref. 60

It has been found that the deprotection phase of the Gabriel synthesis is accelerated by inclusion of NaOH.[61]

Secondary amides can be alkylated on nitrogen by using sodium hydride for deprotonation, followed by reaction with an alkyl halide.[62]

Neutral tertiary and secondary amides react with very reactive alkylating agents, such as triethyloxonium tetrafluoroborate, to give O-alkylation.[63] The same reaction occurs, but more slowly, with tosylates and dimethyl sulfate. Neutralization of the resulting salt provides iminoethers.

Sulfonamides are relatively acidic and their anions can serve as nitrogen nucleophiles.[64] Sulfonamido groups can be introduced at benzylic positions with a high level of inversion under Mitsunobu conditions.[65]

60. J. C. Sheehan and W. A. Bolhofer, *J. Am. Chem. Soc.*, **72**, 2786 (1950).

61. A. Ariffin, M. N. Khan, L. C. Lan, F. Y. May, and C. S. Yun, *Synth. Commun.*, **34**, 4439 (2004); M. N. Khan, *J. Org. Chem.*, **61**, 8063 (1996).

62. W. S. Fones, *J. Org. Chem.*, **14**, 1099 (1949); R. M. Moriarty, *J. Org. Chem.*, **29**, 2748 (1964).

63. L. Weintraub, S. R. Oles, and N. Kalish, *J. Org. Chem.*, **33**, 1679 (1968); H. Meerwein, E. Battenberg, H. Gold, E. Pfeil, and G. Willfang, *J. Prakt. Chem.*, **154**, 83 (1939).

64. D. Papaioannou, C. Athanassopoulos, V. Magafa, N. Karamanos, G. Stavropoulos, A Napoli, G. Sindona, D. W. Aksnes, and G. W. Francis, *Acta Chem. Scand.*, **48**, 324 (1994).

65. T. S. Kaufman, *Tetrahedron Lett.*, **37**, 5329 (1996).

The Mitsunobu conditions can be used for alkylation of 2-pyridones, as in the course of synthesis of analogs of the antitumor agent camptothecin.

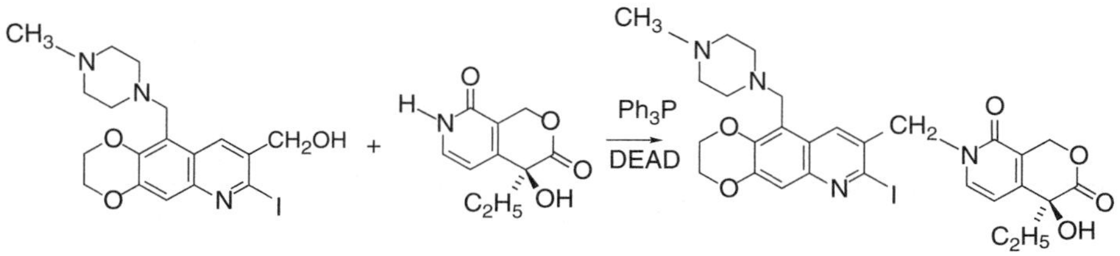

Ref. 66

Proline analogs can be obtained by cyclization of δ-hydroxyalkylamino acid carbamates.

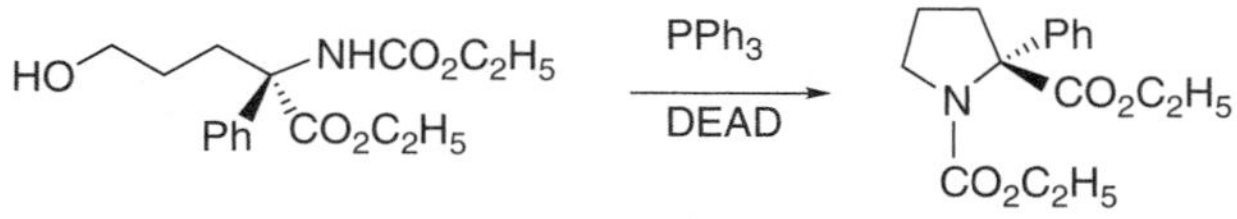

Ref. 67

Mitsunobu conditions are effective for glycosylation of weak nitrogen nucleophiles, such as indoles. This reaction has been used in the synthesis of antitumor compounds.

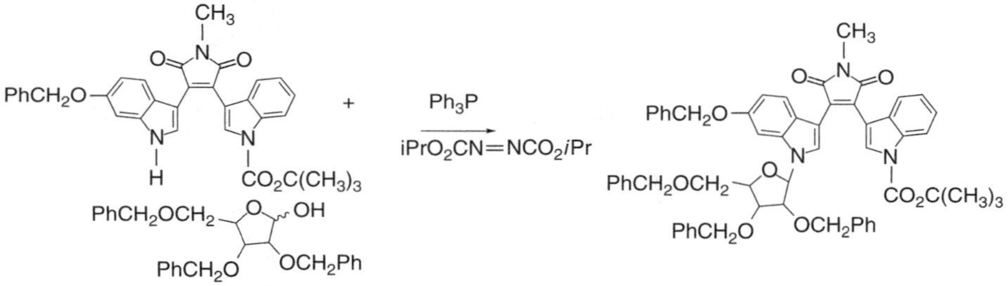

Ref. 68

Azides are useful intermediates for synthesis of various nitrogen-containing compounds. They can also be easily reduced to primary amines and undergo cycloaddition reactions, as is discussed in Section 6.2. Azido groups are usually introduced into aliphatic compounds by nucleophilic substitution.[69] The most reliable procedures involve heating an appropriate halide with sodium azide in DMSO[70] or DMF.[71] Alkyl azides can also be prepared by reaction in high-boiling alcohols.[72]

$$CH_3(CH_2)_3CH_2I + NaN_3 \xrightarrow[H_2O]{CH_3CH_2(OCH_2CH_2)_2OH} CH_3(CH_2)_3CH_2N_3 \quad 84\%$$

66. F. G. Fang, D. D. Bankston, E. M. Huie, M. R. Johnson, M.-C. Kang, C. S. LeHoullier, G. C. Lewis, T. C. Lovelace, M. W. Lowery, D. L. McDougald, C. A. Meerholz, J. J. Partridge, M. J. Sharp, and S. Xie, *Tetrahedron*, **53**, 10953 (1997).

67. J. van Betsbrugge, D. Tourwe, B. Kaptein, H. Kierkals, and R. Broxterman, *Tetrahedron*, **53**, 9233 (1997).

68. M. Ohkubo, T. Nishimura, H. Jona, T. Honma, S. Ito, and H. Morishima, *Tetrahedron*, **53**, 5937 (1997).

69. M. E. C. Biffin, J. Miller, and D. B. Paul, in *The Chemistry of the Azido Group*, S. Patai, ed., Interscience, New York, 1971, Chap. 2.

70. R. Goutarel, A. Cave, L. Tan, and M. Leboeuf, *Bull. Soc. Chim. France*, 646 (1962).

71. E. J. Reist, R. R. Spencer, B. R. Baker, and L. Goodman, *Chem. Ind. (London)*, 1794 (1962).

72. E. Lieber, T. S. Chao, and C. N. R. Rao, *J. Org. Chem.*, **22**, 238 (1957); H. Lehmkuhl, F. Rabet, and K. Hauschild, *Synthesis*, 184 (1977).

232

CHAPTER 3

*Functional Group
Interconversion
by Substitution,
Including Protection and
Deprotection*

Phase transfer conditions are used as well for the preparation of azides.[73]

$$CH_2=CH \overset{CH_3}{\underset{}{\cdots}} \overset{Br}{\underset{}{\cdots}} CO_2CH_3 \quad \xrightarrow[\substack{R_4P^+ \ ^-Br \\ 4\ h,\ 25°C}]{NaN_3} \quad CH_2=CH \overset{CH_3}{\underset{}{\cdots}} \overset{N_3}{\underset{}{\cdots}} CO_2CH_3$$

Tetramethylguanidinium azide, an azide salt that is readily soluble in halogenated solvents, is a useful source of azide ions in the preparation of azides from reactive halides such as α-haloketones, α-haloamides, and glycosyl halides.[74]

There are also useful procedures for preparation of azides directly from alcohols. Reaction of alcohols with 2-fluoro-1-methylpyridinium iodide followed by reaction with lithium azide gives good yields of alkyl azides.[75]

Diphenylphosphoryl azide reacts with alcohols in the presence of triphenylphosphine and DEAD.[76] Hydrazoic acid, HN_3, can also serve as the azide ion source under these conditions.[77] These reactions are examples of the Mitsunobu reaction.

$$ROH + Ph_3P + C_2H_5O_2CN=NCO_2C_2H_5 \longrightarrow RO\overset{+}{P}Ph_3 + C_2H_5O_2C\overset{-}{N}NHCO_2C_2H_5$$

$$RO\overset{+}{P}Ph_3 + N_3^- \longrightarrow RN_3 + Ph_3P=O$$

Diphenylphosphoryl azide also gives good conversion of primary alkyl and secondary benzylic alcohols to azides in the presence of the strong organic base diazabicycloundecane (DBU). These reactions proceed by O-phosphorylation followed by S_N2 displacement.[78]

This reaction can be extended to secondary alcohols with the more reactive *bis*-(4-nitrophenyl)phosphorazidate.[79]

73. W. P. Reeves and M. L. Bahr, *Synthesis*, 823 (1976); B. B. Snider and J. V. Duncia, *J. Org. Chem.*, **46**, 3223 (1981).
74. Y. Pan, R. L. Merriman, L. R. Tanzer, and P. L. Fuchs, *Biomed. Chem. Lett.*, **2**, 967 (1992); C. Li, T.-L. Shih, J. U. Jeong, A. Arasappan, and P. L. Fuchs, *Tetrahedron Lett.*, **35**, 2645 (1994); C. Li, A. Arasappan, and P. L. Fuchs, *Tetrahedron Lett.*, **34**, 3535 (1993); D. A. Evans, T. C. Britton, J. A. Ellman, and R. L. Dorow, *J. Am. Chem. Soc.*, **112**, 4011 (1990).
75. K. Hojo, S. Kobayashi, K. Soai, S. Ikeda, and T. Mukaiyama, *Chem. Lett.*, 635 (1977).
76. B. Lal, B. N. Pramanik, M. S. Manhas, and A. K. Bose, *Tetrahedron Lett.*, 1977 (1977).
77. J. Schweng and E. Zbiral, *Justus Liebigs Ann. Chem.*, 1089 (1978); M. S. Hadley, F. D. King, B. McRitchie, D. H. Turner, and E. A. Watts, *J. Med. Chem.*, **28**, 1843 (1985).
78. A. S. Thompson, G. R. Humphrey, A. M. DeMarco, D. J. Mathre, and E. J. J. Grabowski, *J. Org. Chem.*, **58**, 5886 (1993).
79. M. Mizuno and T. Shioiri, *J. Chem. Soc., Chem. Commun.*, **22**, 2165 (1997).

Anions derived from thiols are strong nucleophiles and are easily alkylated by halides.

$$CH_3S^-Na^+ + ClCH_2CH_2OH \xrightarrow{C_2H_5OH} CH_3SCH_2CH_2OH$$

75–80%   Ref. 80

Neutral sulfur compounds are also good nucleophiles, Sulfides and thioamides readily form salts with methyl iodide, for example.

$$(CH_3)_2S + CH_3I \xrightarrow[12-16\ h]{25°C} (CH_3)_3S^+I^-$$

Ref. 81

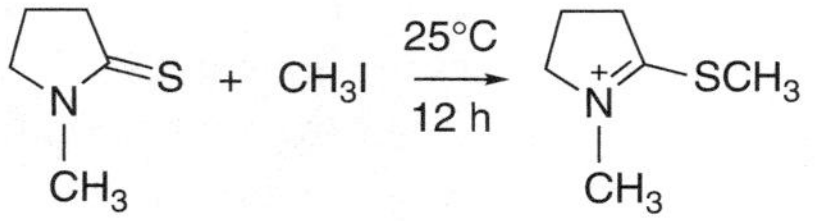

Ref. 82

Even sulfoxides, in which nucleophilicity is decreased by the additional oxygen, can be alkylated by methyl iodide. These sulfoxonium salts have useful synthetic applications as discussed in Section 2.5.1.

$$(CH_3)_2S{=}O + CH_3I \xrightarrow[72\ h]{25°C} (CH_3)_2\overset{+}{S}{=}O\ I^-$$

Ref. 83

3.2.6. Phosphorus Nucleophiles

Both neutral and anionic phosphorus compounds are good nucleophiles toward alkyl halides. We encountered examples of these reactions in Chapter 2 in connection with the preparation of the valuable phosphorane and phosphonate intermediates used for Wittig reactions.

$$Ph_3P + CH_3Br \xrightarrow[2\ days]{room\ temp} Ph_3\overset{+}{P}CH_3\ Br^-$$

Ref. 84

$$[(CH_3)_2CHO]_3P + CH_3I \longrightarrow [(CH_3)_2CHO]_2\overset{O}{\overset{\|}{P}}CH_3 + (CH_3)_2CHI$$

Ref. 85

The reaction with phosphite esters is known as the *Michaelis-Arbuzov reaction* and proceeds through an unstable trialkoxyphopsphonium intermediate. The second stage is another example of the great tendency of alkoxyphosphonium ions to react with nucleophiles to break the O−C bond, resulting in formation of a phosphoryl P−O bond.

$$(R'O)_3P + XCH_2R \longrightarrow (R'O)_3\overset{+}{P}CH_2R \atop X^- \longrightarrow (R'O)_2\overset{O}{\overset{\|}{P}}CH_2R + R'X$$

80. W. Windus and P. R. Shildneck, *Org. Synth.*, **II**, 345 (1943).
81. E. J. Corey and M. Chaykovsky, *J. Am. Chem. Soc.*, **87**, 1353 (1965).
82. R. Gompper and W. Elser, *Org. Synth.*, **V**, 780 (1973).
83. R. Kuhn and H. Trischmann, *Justus Liebigs Ann. Chem.*, **611,** 117 (1958).
84. G. Wittig and U. Schoellkopf, *Org. Synth.*, **V**, 751 (1973).
85. A. H. Ford-Moore and B. J. Perry, *Org. Synth.*, **IV**, 325 (1963).

CHAPTER 3

*Functional Group
Interconversion
by Substitution,
Including Protection and
Deprotection*

3.2.7. Summary of Nucleophilic Substitution at Saturated Carbon

Some of the nucleophilic substitution reactions at sp^3 carbon that are most valuable for synthesis were outlined in the preceding sections, and they all fit into the general mechanistic patterns that were discussed in Chapter 4 of Part A. The order of reactivity of alkylating groups is benzyl $\sim$ allyl > methyl > primary > secondary. Tertiary halides and sulfonates are generally not satisfactory because of the preference for elimination over S_N2 substitution. Owing to their high reactivity toward nucleophilic substitution, α-haloesters, α-haloketones, and α-halonitriles are usually favorable reactants for substitution reactions. The reactivity of leaving groups is sulfonate $\sim$ iodide > bromide > chloride. Steric hindrance decreases the rate of nucleophilic substitution. Thus projected synthetic steps involving nucleophilic substitution must be evaluated for potential steric problems.

Scheme 3.2 gives some representative examples of nucleophilic substitution processes drawn from *Organic Syntheses* and from other synthetic efforts. Entries 1 to 3 involve introduction of cyano groups via tosylates and were all conducted in polar aprotic solvents. Entries 4 to 8 are examples of introduction of the azido functional group by substitution. The reaction in Entry 4 was done under phase transfer conditions. A concentrated aqueous solution of NaN_3 was heated with the alkyl bromide and 5 mol % methyltrioctylammonium chloride. Entries 5 to 7 involve introduction of the azido group at secondary carbons with inversion of configuration in each case. The reactions in Entries 7 and 8 involve formation of phosphoryl esters as intermediates. These conditions were found preferable to the Mitsunobu conditions for the reaction in Entry 7. The electron-rich benzylic reactant gave both racemization and elimination via a carbocation intermediate under the Mitsunobu conditions. Entries 9 and 10 are cases of controlled alkylation of amines. In the reaction in Entry 9, the pyrrolidine was used in twofold excess. The ester EWGs have a rate-retarding effect that slows further alkylation to the quaternary salt. In the reaction in Entry 10, the monohydrochloride of piperazine is used as the reactant. The reaction was conducted in ethanol, and the dihydrochloride salt of the product precipitates as reaction proceeds, which helps minimize quaternization or N,N'-dialkylation. The yield of the dihydrochloride is 97–99%, and that of the amine is 65–75% after neutralization of the salt and distillation. The reaction in Entry 11 is the O-alkylation of an amide. The reaction was done in refluxing benzene, and the product was obtained by distillation after the neutralization.

Sections D through H of Scheme 3.2 involve oxygen nucleophiles. The hydrolysis reactions in Entries 12 and 13 both involve benzylic positions. The reaction site in Entry 13 is further activated by the ERG substituents on the ring. Entries 14 to 17 are examples of base-catalyzed ether formation. The selectivity of the reaction in Entry 17 for the *meta*-hydroxy group is an example of a fairly common observation in aromatic systems. The *ortho*-hydroxy group is more acidic and probably also stabilized by chelation, making it less reactive.

Dialkylation occurs if a stronger base (NaOH) and dimethyl sulfate is used. Entry 18 is a typical diazomethane methylation of a carboxylic acid. The toxicity of diazomethane

A. Nitriles

1[a]

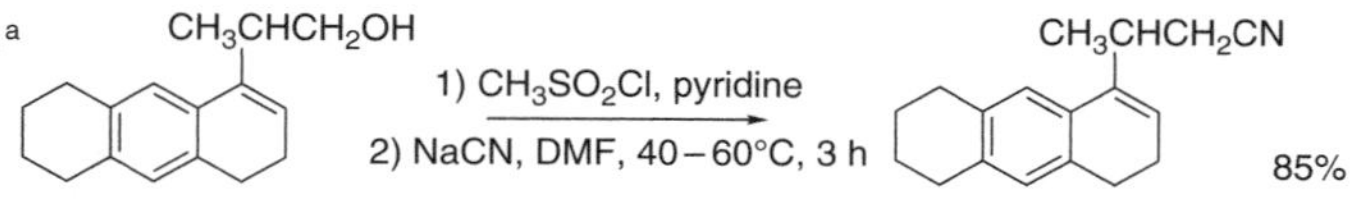

CH$_3$CHCH$_2$OH

1) CH$_3$SO$_2$Cl, pyridine
2) NaCN, DMF, 40–60°C, 3 h

CH$_3$CHCH$_2$CN

85%

2[b]

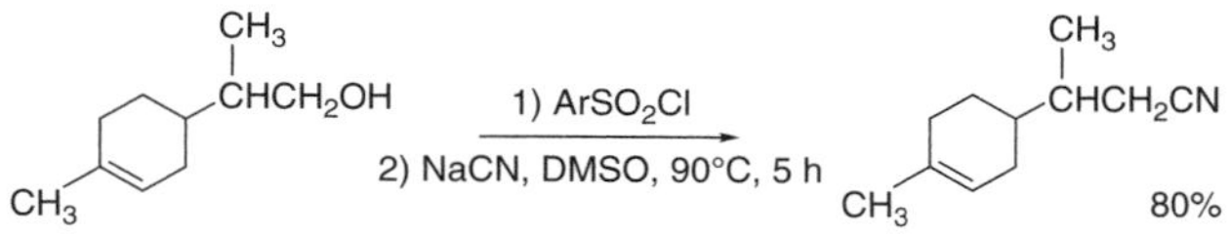

CHCH$_2$OH

1) ArSO$_2$Cl
2) NaCN, DMSO, 90°C, 5 h

CHCH$_2$CN

80%

3[c]

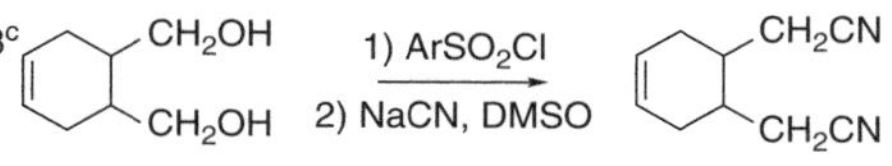

CH$_2$OH
CH$_2$OH

1) ArSO$_2$Cl
2) NaCN, DMSO

CH$_2$CN
CH$_2$CN

B. Azides

4[d]

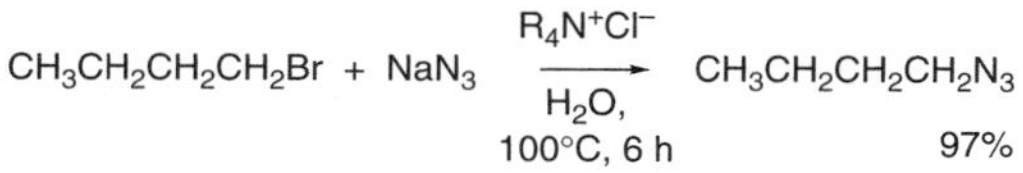

CH$_3$CH$_2$CH$_2$CH$_2$Br + NaN$_3$

R$_4$N$^+$Cl$^-$
———————
H$_2$O,
100°C, 6 h

CH$_3$CH$_2$CH$_2$CH$_2$N$_3$

97%

5[e]

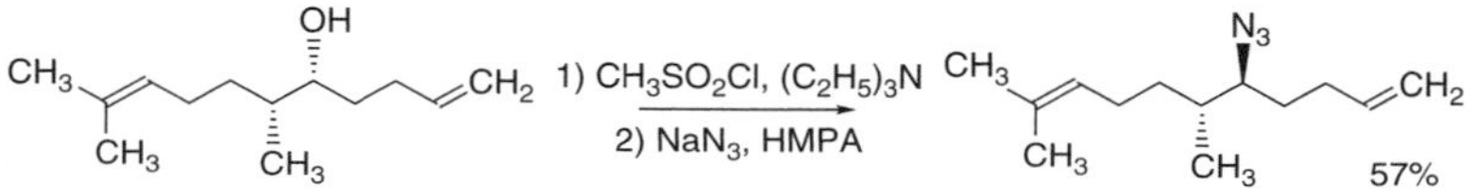

1) CH$_3$SO$_2$Cl, (C$_2$H$_5$)$_3$N
2) NaN$_3$, HMPA

57%

6[f]

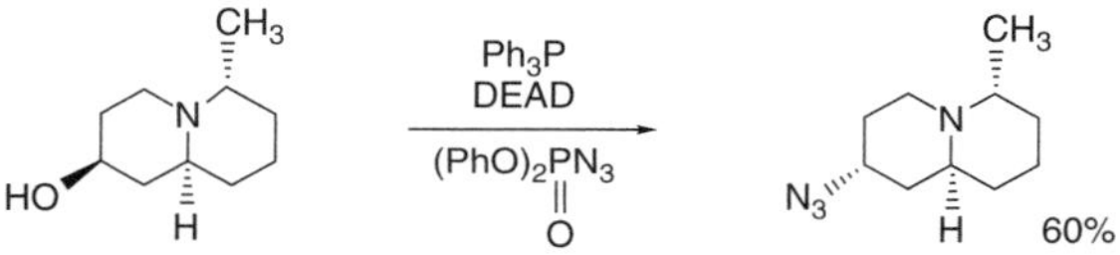

Ph$_3$P
DEAD
———————
(PhO)$_2$PN$_3$

60%

7[g]

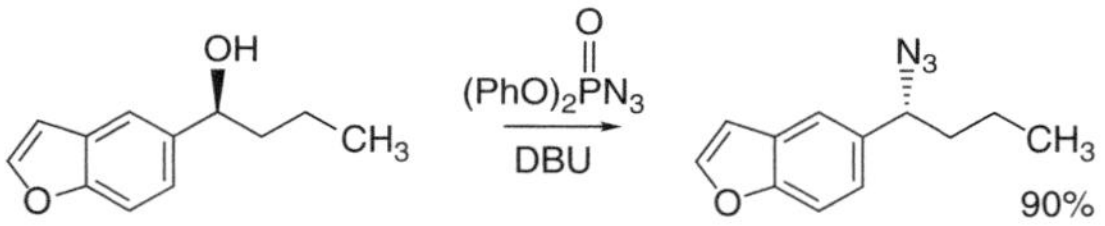

O
‖
(PhO)$_2$PN$_3$
———————
DBU

90%

8[h]

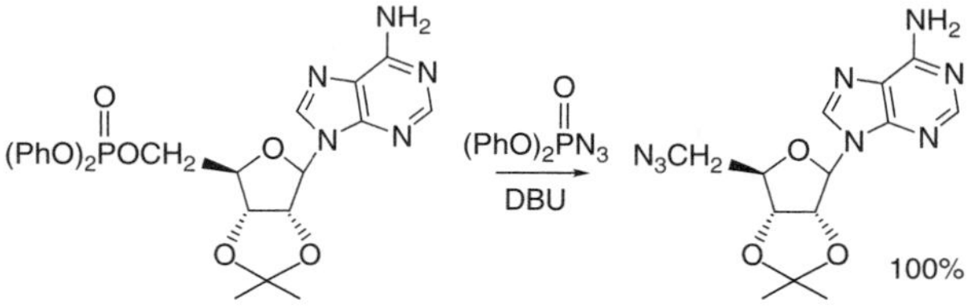

O
‖
(PhO)$_2$PN$_3$
———————
DBU

100%

C. Amines and amides

9[i]

NH + CH$_3$CHCO$_2$C$_2$H$_5$
            |
           Br

NCHCO$_2$C$_2$H$_5$
  |
 CH$_3$ 80–90%

10[j]

HN   NH$_2$
    +

PhCH$_2$Cl $^-$OH

PhCH$_2$N   NH

65–75%

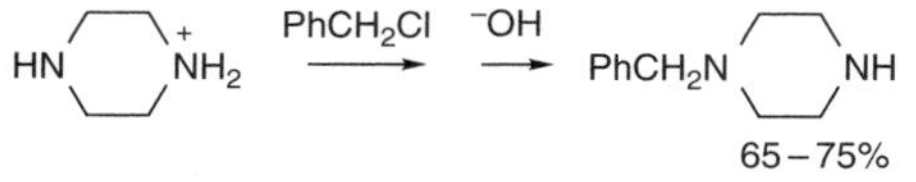

(Continued)

Scheme 3.2. (*Continued*)

CHAPTER 3

*Functional Group
Interconversion
by Substitution,
Including Protection and
Deprotection*

11[k]

60–70%

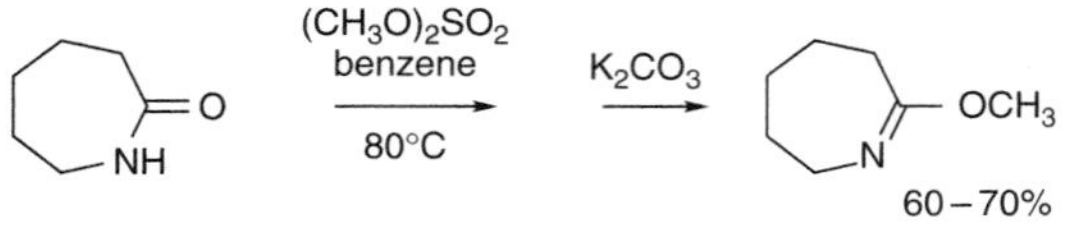

D. Hydrolysis by alkyl halides

12[l]

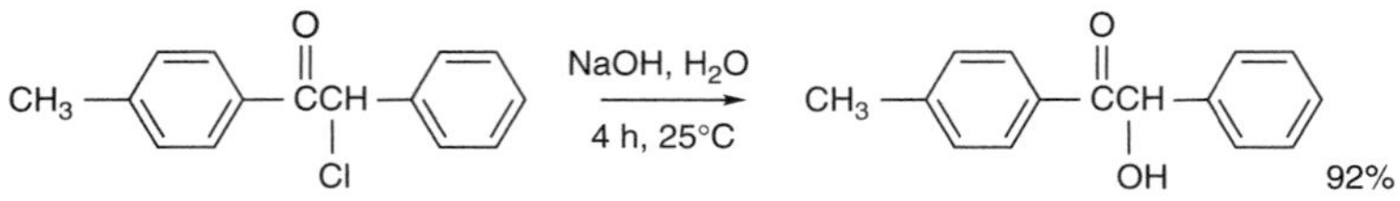

92%

13[m]

92%

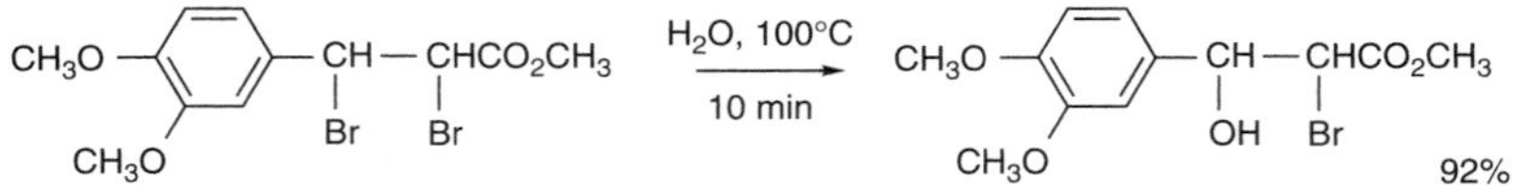

E. Ethers by base – catalyzed alkylation

14[n]

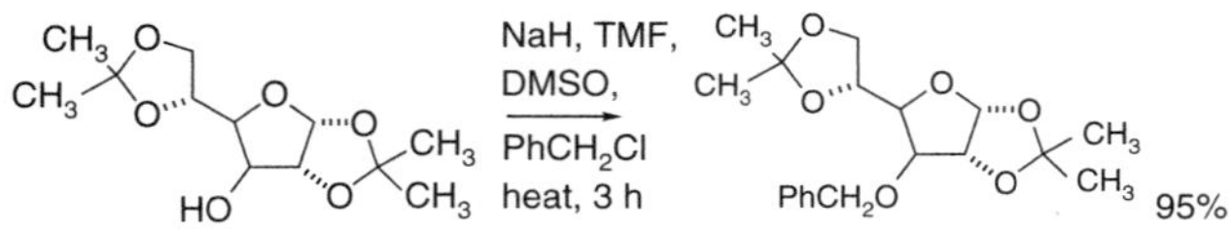

95%

15[o]

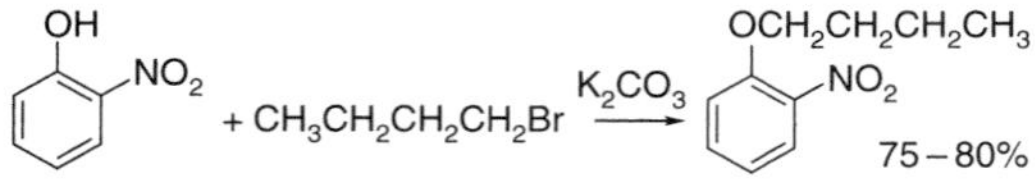

75–80%

16[p]

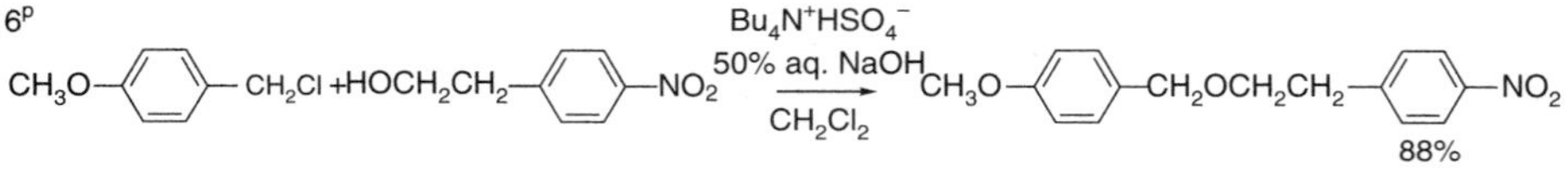

88%

17[q]

55–65%

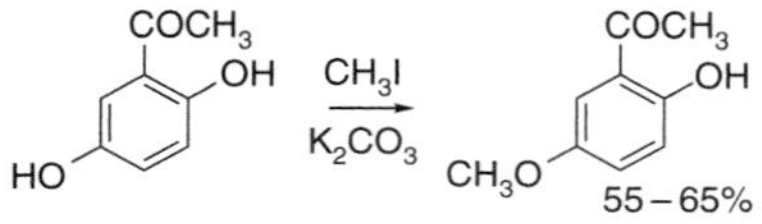

F. Esterification by diazoalkanes

18[r]

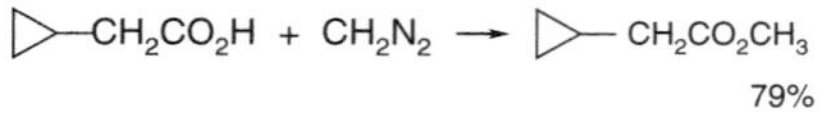

79%

G. Esterification by nucleophilic substitution with carboxylate salts

19[s]

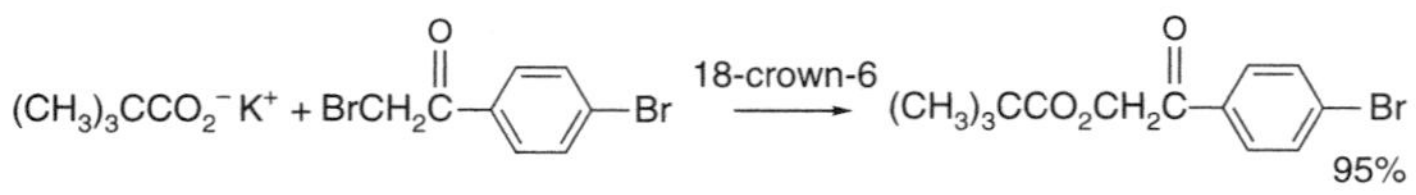

95%

20[t]

100%

(*Continued*)

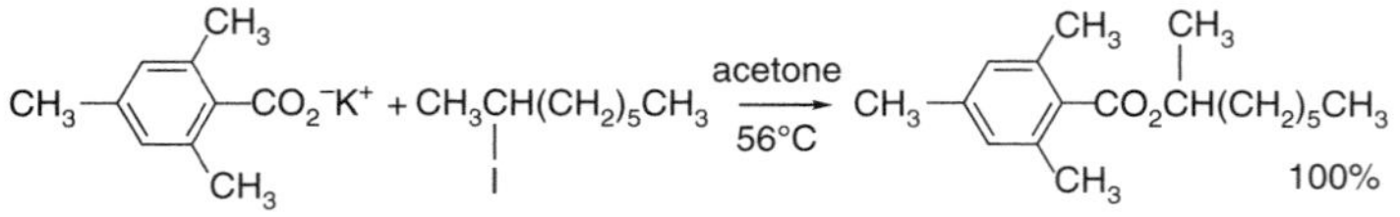

Scheme 3.2. (*Continued*)

237

SECTION 3.2

*Introduction of
Functional Groups by
Nucleophilic Substitution
at Saturated Carbon*

21[u]

CH$_3$I, KF,
DMF, 25°C
―――――
18h

84%

H. Sulfonate esters

22[v]

PPh$_3$, *i*-Pr-O$_2$CN $=$ NCO$_2$-*i*-Pr
――――――――→
p-toluenesulfonic acid,
(C$_2$H$_5$)$_3$N

Ar = *p*-CH$_3$C$_6$H$_5$

I. Phosphorus nucleophiles

23[w]

Ph$_3$P + BrCH$_2$CH$_2$OPh ⟶ Ph$_3$$\overset{+}{\text{P}}CH_2CH_2$OPh Br$^-$

24[x]

[(CH$_3$)$_2$CHO]$_3$P + CH$_3$I ⟶ [(CH$_3$)$_2$CHO]$_2$$\overset{\overset{\text{O}}{\|}}{\text{P}}CH_3$ + (CH$_3$)$_2$CHI

85–90%

J. Sulfur nucleophiles

25[y]

CH$_3$(CH$_2$)$_{10}$CH$_2$Br + S $=$ C(NH$_2$)$_2$ $\xrightarrow{\text{NaOH} \atop \text{H}_2\text{O}}$ CH$_3$(CH$_2$)$_{10}$CH$_2$SH

80%

26[z]

Na$^+$ $^-$SCH$_2$CH$_2$S$^-$ Na$^+$ + BrCH$_2$CH$_2$Br ⟶

55–60%

27[aa]

1) CH$_2$I
2) (CH$_3$)$_3$CO$^-$ K$^+$

62%

a. M. S. Newman and S. Otsuka, *J. Org. Chem.*, **23**, 797 (1958).
b. B. A. Pawson, H.-C. Cheung, S. Gurbaxani, and G. Saucy, *J. Am. Chem. Soc.*, **92**, 336 (1970).
c. J. J. Bloomfield and P. V. Fennessey, *Tetrahedron Lett.*, 2273 (1964).
d. W. P. Reeves and M. L. Bahr, *Synthesis*, 823 (1976).
e. D. F. Taber, M. Rahimizadeh, and K. K. You, *J. Org. Chem.*, **60**, 529 (1995).
f. M. S. Hadley, F. D. King, B. McRitchie, D. H. Turner, and E. A. Watts, *J. Med. Chem.*, **28**, 1843 (1985).
g. A. S. Thompson, G. G. Humphrey, A. M. De Marco, D. J. Mathre, and E. J. J. Grabowski, *J. Org. Chem.*, **58**, 5886 (1993).
h. P. Liu and D. J. Austin, *Tetrahedron Lett.*, **42**, 3153 (2001).
i. R. B. Moffett, *Org. Synth.*, **IV**, 466 (1963).
j. J. C. Craig and R. J. Young, *Org. Synth.*, **V**, 88 (1973).
k. R. E. Benson and T. L. Cairns, *Org. Synth.*, **IV**, 588 (1963).
l. R. N. McDonald and P. A. Schwab, *J. Am. Chem. Soc.*, **85**, 4004 (1963).
m. C. H. Heathcock, C. T. White, J. J. Morrison, and D. Van Derveer, *J. Org. Chem.*, **46**, 1296 (1981).
n. E. Adler and K. J. Bjorkquist, *Acta Chem. Scand.*, **5**, 241 (1951).
o. E. S. West and R. F. Holden, *Org. Synth.*, **III**, 800 (1955).
p. F. Lopez-Calahorra, B. Ballart, F. Hombrados, and J. Marti, *Synth. Commun.*, **28**, 795 (1998).
q. G. N. Vyas and M. N. Shah, *Org. Synth.*, **IV**, 836 (1963).
r. L. I. Smity and S. McKenzie, Jr., *J. Org. Chem.*, **15**, 74 (1950); A. I. Vogel, *Practical Organic Chemistry*, 3rd Edition, Wiley, 1956, p. 973.
s. H. D. Durst, *Tetrahedron Lett.*, 2421 (1974).
t. G. G. Moore, T. A. Foglia, and T. J. McGahan, *J. Org. Chem.*, **44**, 2425 (1979).
u. C. H. Heathcock, C.-T. White, J. Morrison, and D. VanDerveer, *J. Org. Chem.*, **46**, 1296 (1981).
v. N. G. Anderson, D. A. Lust, K. A. Colapret, J. H. Simpson, M. F. Malley, and J. Z. Gougoutas, *J. Org. Chem.*, **61**, 7955 (1996).
w. E. E. Schweizer and R. D. Bach, *Org. Synth.*, **V**, 1145 (1973).
x. A. H. Ford-Moore and B. J. Perry, *Org. Synth.*, **IV**, 325 (1963).
y. G. G. Urquhart, J. W. Gates, Jr., and P. Conor, *Org. Synth.*, **III**, 363 (1965).
z. R. G. Gillis and A. B. Lacey, *Org. Synth.*, **IV**, 396 (1963).
aa. R. Gompper and W. Elser, *Org. Synth.*, **V**, 780 (1973).

238

CHAPTER 3

*Functional Group
Interconversion
by Substitution,
Including Protection and
Deprotection*

and its precursors, as well as the explosion hazard of diazomethane, requires that all recommended safety precautions be taken. Entries 19 to 21 involve formation of esters by alkylation of carboxylate salts. The reaction in Entry 19 was done in the presence of 5 mol % 18-crown-6. A number of carboxylic acids, including pivalic acid as shown in the example, were alkylated in high yield under these conditions. Entry 20 shows the alkylation of the rather hindered mesitoic acid by a secondary iodide. These conditions also gave high yields for unhindered acids and iodides. Entry 21 involves formation of a methyl ester using CH_3I and KF as the base in DMF. Entry 22 involves formation of a sulfonate ester under Mitsunobu conditions with clean inversion of configuration. The conditions reported represent the optimization of the reaction as part of the synthesis of an antihypertensive drug, fosinopril.

Sections I and J of Scheme 3.2 show reactions with sulfur and phosphorus nucleophiles. The reaction in Entry 25 is a useful method for introducing thiol groups. The solid thiourea is a convenient source of sulfur. A thiouronium ion is formed and this avoids competition from formation of a dialkyl sulfide. The intermediate is readily hydrolyzed by base.

$$RCH_2Br \ + \ S{=}C(NH_2)_2 \ \longrightarrow \ RCH_2S{-}\underset{NH_2}{\overset{N^+H_2}{\diagup}} \ \xrightarrow[H_2O]{NaOH} \ RCH_2SH$$

3.3. Cleavage of Carbon-Oxygen Bonds in Ethers and Esters

The cleavage of carbon-oxygen bonds in ethers or esters by nucleophilic substitution is frequently a useful synthetic transformation.

$$R{-}O{-}CH_3 + Nu^- \ \longrightarrow \ RO^- + CH_3{-}Nu$$
$$\overset{O}{\overset{\|}{R}C}{-}O{-}CH_3 + Nu^- \ \longrightarrow \ RCO_2^- + CH_3{-}Nu$$

The alkoxide group is a poor leaving group and carboxy is only slightly better. As a result, these reactions usually require assistance from a protic or Lewis acid. The classical ether cleavage conditions involving concentrated hydrogen halides are much too strenuous for most polyfunctional molecules, so several milder reagents have been developed,[86] including boron tribromide,[87] dimethylboron bromide,[88] trimethylsilyl iodide,[89] and boron trifluoride in the presence of thiols.[90] The mechanism for ether cleavage with boron tribromide involves attack of bromide ion on an adduct formed

86. M. V. Bhatt and S. U. Kulkarni, *Synthesis*, 249 (1983).
87. J. F. W. McOmie, M. L. Watts, and D. E. West, *Tetrahedron*, **24**, 2289 (1968).
88. Y. Guindon, M. Therien, Y. Girard, and C. Yoakim, *J. Org. Chem.*, **52**, 1680 (1987).
89. M. E. Jung and M. A. Lyster, *J. Org. Chem.*, **42**, 3761 (1977).
90. (a) M. Node, H. Hori, and E. Fujita, *J. Chem. Soc., Perkin Trans. 1*, 2237 (1976); (b) K. Fuji, K. Ichikawa, M. Node, and E. Fujita, *J. Org. Chem.*, **44**, 1661 (1979).

from the ether and the electrophilic boron reagent. The cleavage step can occur by either an S_N2 or an S_N1 process, depending on the structure of the alkyl group.

$$R-O-R + BBr_3 \longrightarrow R-\overset{+}{\underset{\underset{-BBr_3}{|}}{O}}-R \rightleftharpoons R-\overset{+}{\underset{\underset{BBr_2}{|}}{O}}-R + Br^-$$

$$R-\overset{+}{\underset{\underset{BBr_2}{|}}{O}}-R \;\overset{\frown}{}\; Br \longrightarrow R-O-BBr_2 + RBr$$

$$R-O-BBr_2 + 3\,H_2O \longrightarrow ROH + B(OH)_3 + 2\,HBr$$

Good yields are generally observed, especially for methyl ethers. The combination of boron tribromide with dimethyl sulfide has been found to be particularly effective for cleaving aryl methyl ethers.[91]

The boron trifluoride–alkyl thiol reagent combination also operates on the basis of nucleophilic attack on an oxonium ion generated by reaction of the ether with boron trifluoride.[90]

$$R-O-R + BF_3 \longrightarrow R-\overset{+}{\underset{\underset{-BF_3}{|}}{O}}-R$$

$$R-\overset{+}{\underset{\underset{-BF_3}{|}}{O}}-R + R'SH \longrightarrow R\overset{-}{O}BF_3 + RSR' + H^+$$

Trimethylsilyl iodide (TMSI) cleaves methyl ethers in a period of a few hours at room temperature.[89] Benzyl and *t*-butyl systems are cleaved very rapidly, whereas secondary systems require longer times. The reaction presumably proceeds via an initially formed silyl oxonium ion.

$$R-O-R' + (CH_3)_3SiI \longrightarrow R-\overset{+}{\underset{\underset{Si(CH_3)_3}{|}}{O}}-R' + I^- \longrightarrow R-O-Si(CH_3)_3 + R'I$$

The direction of cleavage in unsymmetrical ethers is determined by the relative ease of O–R bond breaking by either S_N2 (methyl, benzyl) or S_N1 (*t*-butyl) processes. As trimethylsilyl iodide is rather expensive, alternative procedures that generate the reagent in situ have been devised.

$$(CH_3)_3SiCl + NaI \xrightarrow{CH_3CN} (CH_3)_3SiI + NaCl$$

Ref. 92

$$PhSi(CH_3)_3 + I_2 \longrightarrow (CH_3)_3SiI + PhI$$

Ref. 93

91. P. G. Williard and C. R. Fryhle, *Tetrahedron Lett.*, **21**, 3731 (1980).
92. T. Morita, Y. Okamoto, and H. Sakurai, *J. Chem. Soc., Chem. Commun.*, 874 (1978); G. A. Olah, S. C. Narang, B. G. B. Gupta, and R. Malhotra, *Synthesis*, 61 (1979).
93. T. L. Ho and G. A. Olah, *Synthesis*, 417 (1977); A. Benkeser, E. C. Mozdzen, and C. L. Muth, *J. Org. Chem.*, **44**, 2185 (1979).

CHAPTER 3

*Functional Group
Interconversion
by Substitution,
Including Protection and
Deprotection*

Allylic ethers are cleaved in a matter of a few minutes by TMSI under in situ conditions.

Ref. 94

Diiodosilane, SiH_2I_2, is an especially effective reagent for cleaving secondary alkyl ethers.[95]

TMSI also effects rapid cleavage of esters. The cleavage step involves iodide attack on the O-silylated ester. The first products formed are trimethylsilyl esters, but these are hydrolyzed rapidly on exposure to water.[96]

Benzyl, methyl, and *t*-butyl esters are rapidly cleaved, but secondary esters react more slowly. In the case of *t*-butyl esters, the initial silylation is followed by a rapid ionization to the *t*-butyl cation.

Ether cleavage can also be effected by reaction with acetic anhydride and Lewis acids such as BF_3, $FeCl_3$, and $MgBr_2$.[97] Mechanistic investigations point to acylium ions generated from the anhydride and Lewis acid as the reactive electrophile.

Scheme 3.3 gives some specific examples of ether and ester cleavage reactions. Entries 1 and 2 illustrate the use of boron tribromide for ether cleavage. The reactions are conducted at dry ice-acetone temperature and the exposure to water on workup hydrolyzes residual O−B bonds. In the case of Entry 2, the primary hydroxy group that is deprotected lactonizes spontaneously. The reaction in Entry 3 uses HBr in acetic acid to cleave a methyl aryl ether. This reaction was part of a scale-up of the synthesis of a drug candidate molecule. Entries 4 to 6 are examples of the cleavage of ethers and esters using TMSI. The selectivity exhibited in Entry 6 for

94. A. Kamal, E. Laxman, and N. V. Rao, *Tetrahedron Lett.*, **40**, 371 (1999).
95. E. Keinan and D. Perez, *J. Org. Chem.*, **52**, 4846 (1987).
96. T. L. Ho and G. A. Olah, *Angew. Chem. Int. Ed. Engl.*, **15**, 774 (1976); M. E. Jung and M. A. Lyster, *J. Am. Chem. Soc.*, **99**, 968 (1977).
97. C. R. Narayanan and K. N. Iyer, *J. Org. Chem.*, **30**, 1734 (1965); B. Ganem and V. R. Small, Jr., *J. Org. Chem.*, **39**, 3728 (1974); D. J. Goldsmith, E. Kennedy, and R. G. Campbell, *J. Org. Chem.*, **40**, 3571 (1975).

Scheme 3.3. Cleavage of Ethers and Esters

1[a]

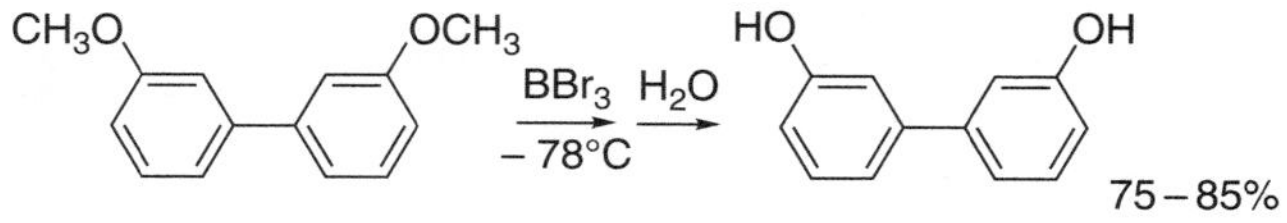

$$\xrightarrow[-78°C]{BBr_3 \quad H_2O}$$

75–85%

2[b]

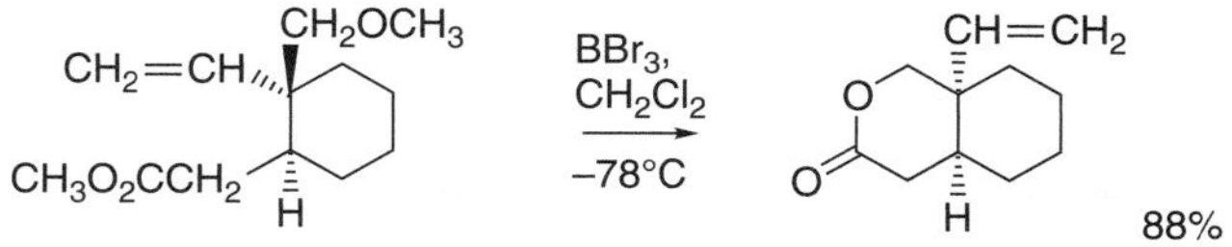

$$\xrightarrow[-78°C]{BBr_3, \quad CH_2Cl_2}$$

88%

3[c]

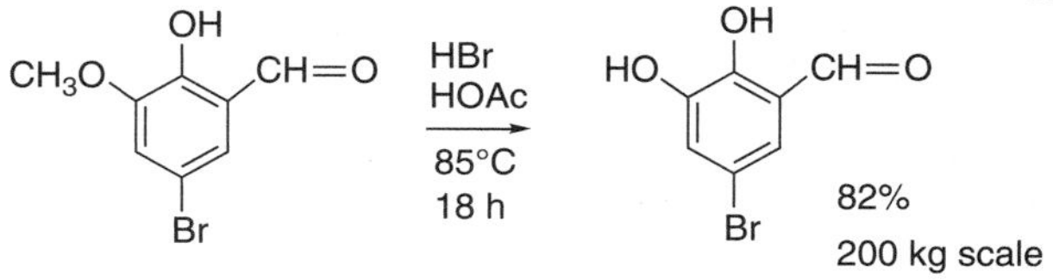

$$\xrightarrow[\substack{85°C \\ 18\,h}]{\substack{HBr \\ HOAc}}$$

82%

200 kg scale

4[d]

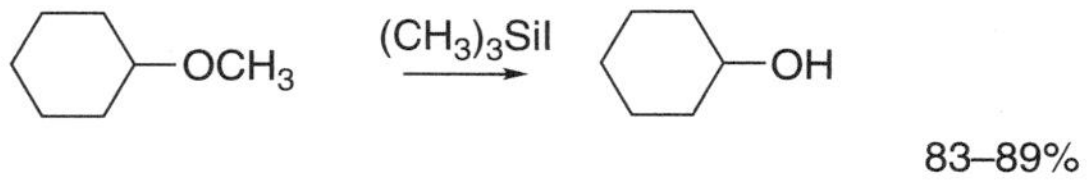

$$\xrightarrow{(CH_3)_3SiI}$$

83–89%

5[e]

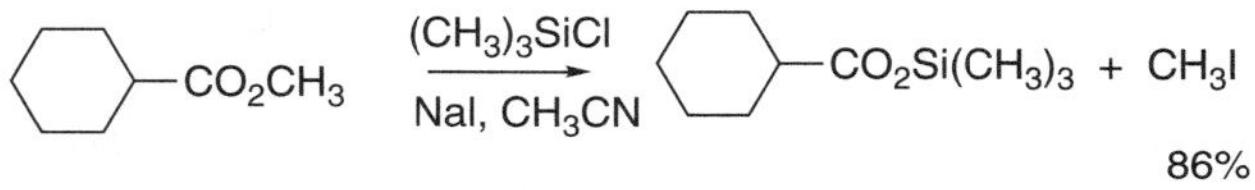

$$\xrightarrow[NaI, \; CH_3CN]{(CH_3)_3SiCl}$$ cyclohexyl–$CO_2Si(CH_3)_3$ + CH_3I

86%

6[f]

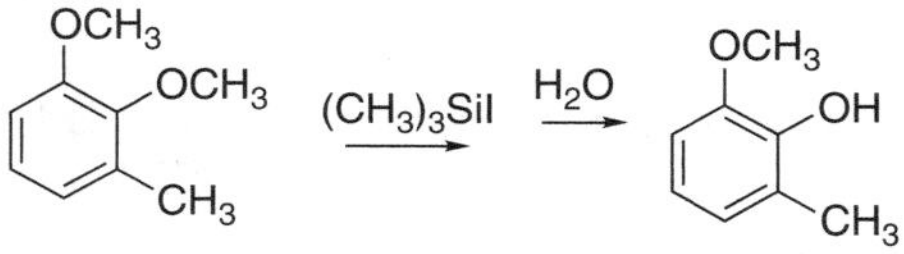

$$\xrightarrow{(CH_3)_3SiI \quad H_2O}$$

7[g]

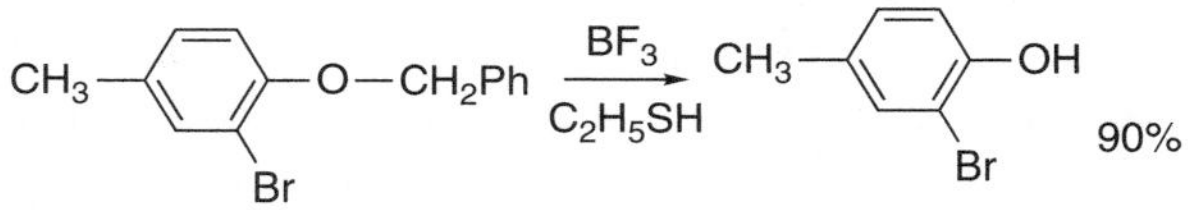

$$CH_3\text{—}\underset{Br}{\bigcirc}\text{—}O\text{—}CH_2Ph \xrightarrow[C_2H_5SH]{BF_3} CH_3\text{—}\underset{Br}{\bigcirc}\text{—}OH$$

90%

8[h]

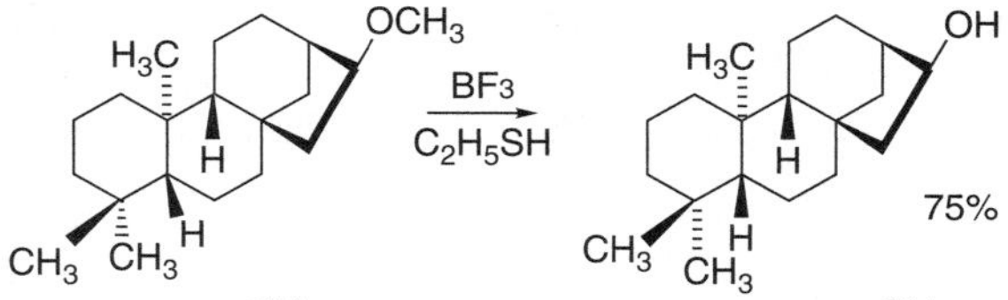

$$\xrightarrow[C_2H_5SH]{BF_3}$$

75%

9[i]

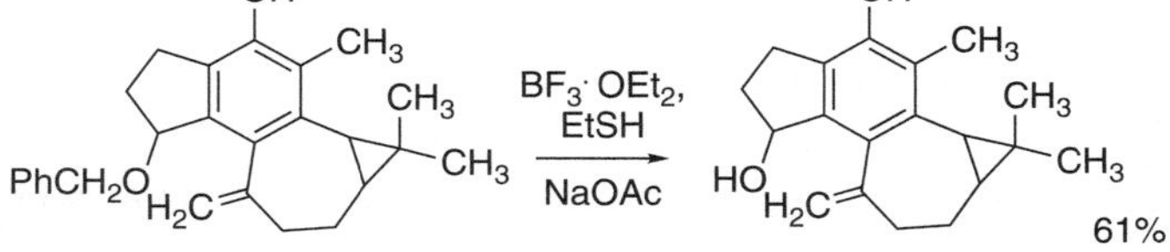

$$\xrightarrow[NaOAc]{\substack{BF_3 \cdot OEt_2, \\ EtSH}}$$

61%

10[j]

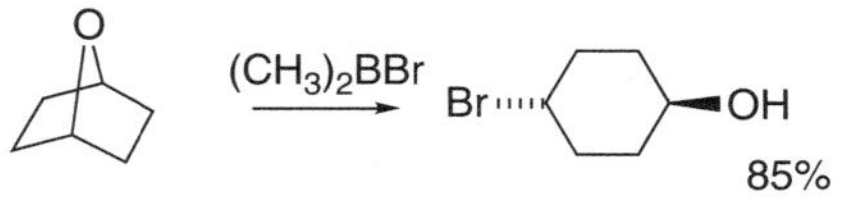

$$\xrightarrow{(CH_3)_2BBr}$$ Br⸱⸱⸱—OH

85%

11[k]

$$(CH_3)_2CHOCH(CH_3)_2 \xrightarrow[(CH_3CO)_2O]{FeCl_3} (CH_3)_2CHO_2CCH_3$$

83%

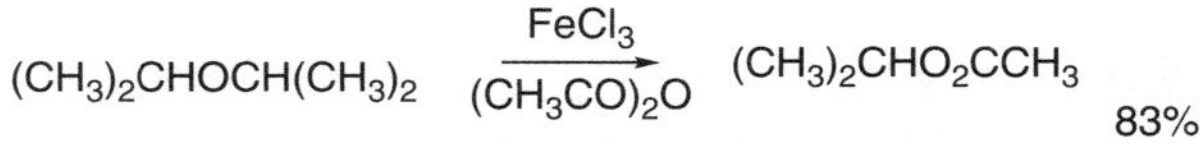

(Continued)

CHAPTER 3

*Functional Group
Interconversion
by Substitution,
Including Protection and
Deprotection*

Scheme 3.3. (*Continued*)

a. J. F. W. McOmie and D. E. West, *Org. Synth.*, **V**, 412 (1973).
b. P. A. Grieco, K. Hiroi, J. J. Reap, and J. A. Noguez, *J. Org. Chem.*, **40**, 1450 (1975).
c. T. E. Jacks, D. T. Belmont, C. A. Briggs, N. M. Horne, G. D. Kanter, G. L. Karrick, J. J. Krikke, R. J. McCabe, J. G. Mustakis, T. N. Nanninga, G. S. Risendorph, R. E. Seamans, R. Skeean, D. D. Winkle, and T. M. Zennie, *Org. Proc. Res. Dev.*, **8**, 201 (2004).
d. M. E. Jung and M. A. Lyster, *Org. Synth.*, **59**, 35 (1980).
e. T. Morita, Y. Okamoto, and H. Sakurai, *J. Chem. Soc., Chem. Commun.*, 874 (1978).
f. E. H. Vickery, L. F. Pahler, and E. J. Eisenbraun, *J. Org. Chem.*, **44**, 4444 (1979).
g. K. Fuji, K. Ichikawa, M. Node, and E. Fujita, *J. Org. Chem.*, **44**, 1661 (1979).
h. M. Nobe, H. Hori, and E. Fujita, *J. Chem. Soc. Perkin Trans.*, 1, 2237 (1976).
i. A. B. Smith, III, N. J. Liverton, N. J. Hrib, H. Sivaramakrishnan, and K. Winzenberg, *J. Am. Chem. Soc.*, **108**, 3040 (1986).
j. Y. Guidon, M. Therien, Y. Girard, and C. Yoakim, *J. Org. Chem.*, **52**, 1680 (1987).
k. B. Ganem and V. R. Small, Jr., *J. Org. Chem.*, **39**, 3728 (1974).

cleavage of the more hindered of the two ether groups may reflect a steric acceleration of the nucleophilic displacement step.

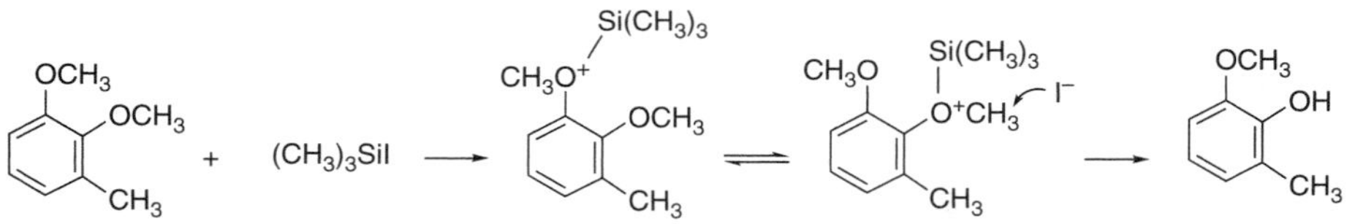

Entries 7 to 9 illustrate the use of the BF_3-EtSH reagent combination. The reaction in Entry 9 was described as "troublesome in the extreme." The problem is that the ether is both a primary benzylic ether and a secondary one, the latter associated with a ring having several ERG substituents. Electrophilic conditions lead to preferential cleavage of the secondary benzylic bond and formation of elimination products. The reaction was done successfully in the presence of excess NaOAc, which presumably allows the nucleophilic S_N2 cleavage of the primary benzyl bond to dominate by reducing the reactivity of the electrophilic species that are present. The cleavage of the cyclic ether shown in Entry 10 occurs with inversion of configuration at the reaction site, as demonstrated by the *trans* stereochemistry of the product. When applied to 2-substituted tetrahydrofurans, the reaction gives mainly cleavage of the C(5)−O bond, indicating that steric access of the nucleophilic component of the reaction is dominant in determining regioselectivity.

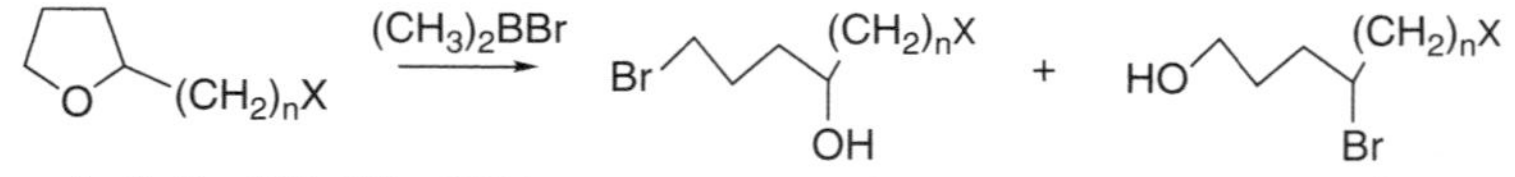

Entry 11 illustrates a cleavage reaction using an acylating agent in conjunction with a Lewis acid.

3.4. Interconversion of Carboxylic Acid Derivatives

The classes of compounds that are conveniently considered together as derivatives of carboxylic acids include the acyl chlorides, carboxylic acid anhydrides, esters, and amides. In the case of simple aliphatic and aromatic acids, synthetic transformations among these derivatives are usually straightforward, involving such fundamental reactions as ester saponification, formation of acyl chlorides, and the reactions of amines with acid anhydrides or acyl chlorides to form amides. The mechanisms of these reactions are discussed in Section 7.4 of Part A.

$$RCO_2CH_3 \xrightarrow[H_2O]{^-OH} RCO_2^- + CH_3OH$$

$$RCO_2H + SOCl_2 \longrightarrow RCOCl + HCl + SO_2$$

$$RCOCl + R'_2NH \longrightarrow RCONR'_2 + HCl$$

When a multistep synthesis is being undertaken with other sensitive functional groups present in the molecule, milder reagents and reaction conditions may be necessary. As a result, many alternative methods for effecting interconversion of the carboxylic acid derivatives have been developed and some of the most useful reactions are considered in the succeeding sections.

3.4.1. Acylation of Alcohols

The traditional method for transforming carboxylic acids into reactive acylating agents capable of converting alcohols to esters or amines to amides is by formation of the acyl chloride. Molecules devoid of acid-sensitive functional groups can be converted to acyl chlorides with thionyl chloride or phosphorus pentachloride. When milder conditions are necessary, the reaction of the acid or its sodium salt with oxalyl chloride provides the acyl chloride. When a salt is used, the reaction solution remains essentially neutral.

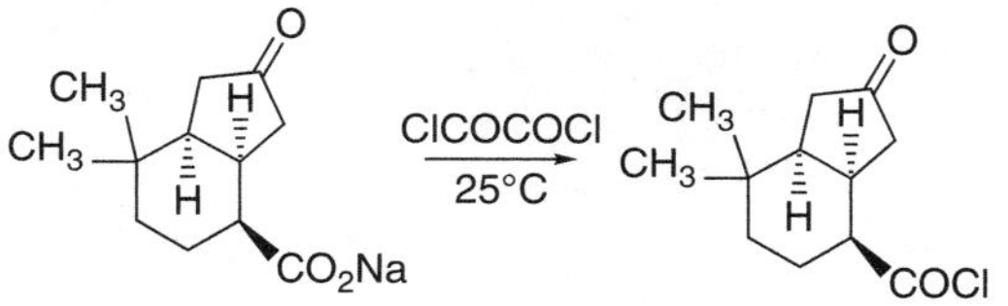

Ref. 98

This reaction involves formation of a mixed anhydride-chloride of oxalic acid, which then decomposes, generating both CO_2 and CO.

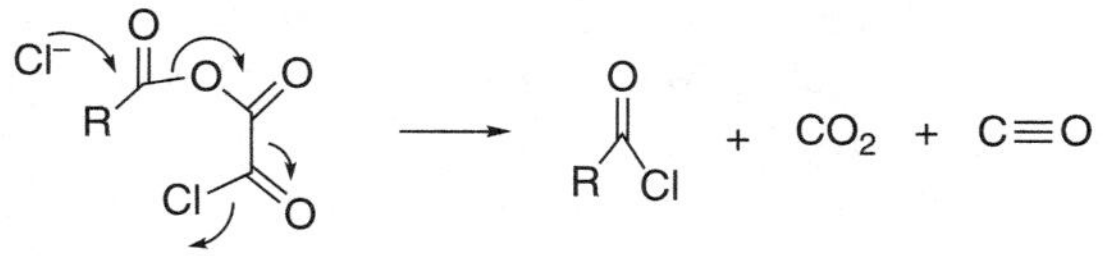

Treatment of carboxylic acids with half an equivalent of oxalyl chloride can generate anhydrides.[99]

$$2\,RCO_2H + \underset{1-1.2\ \text{equiv}}{ClCCCl} \longrightarrow RCOCR + CO_2 + CO + 2\,HCl$$

98. M. Miyano and C. R. Dorn, *J. Org. Chem.*, **37**, 268 (1972).
99. R. Adams and L. H. Urich, *J. Am. Chem. Soc.*, **42**, 599 (1920).

244

CHAPTER 3

*Functional Group
Interconversion
by Substitution,
Including Protection and
Deprotection*

Carboxylic acids can be converted to acyl chlorides and bromides by a combination of triphenylphosphine and a halogen source. Triphenylphosphine and carbon tetrachloride convert acids to the corresponding acyl chloride.[100] Similarly, carboxylic acids react with the triphenyl phosphine-bromine adduct to give acyl bromides.[101] Triphenylphosphine–N-bromosuccinimide also generates acyl bromide in situ.[102] All these reactions involve acyloxyphosphonium ions and are mechanistically analogous to the alcohol-to-halide conversions that are discussed in Section 3.1.2.

$$RCO_2H + Ph_3\overset{+}{P}Br \longrightarrow RC\overset{O}{\overset{\|}{-}}O-\overset{+}{P}Ph_3 + HBr$$

$$Br^- + RC\overset{O}{\overset{\|}{-}}O-\overset{+}{P}Ph_3 \longrightarrow RCBr + Ph_3P=O$$

Acyl chlorides are highly reactive acylating agents and react very rapidly with alcohols and other nucleophiles. Preparative procedures often call for use of pyridine as a catalyst. Pyridine catalysis involves initial formation of an acyl pyridinium ion, which then reacts with the alcohol. Pyridine is a better nucleophile than the neutral alcohol, but the acyl pyridinium ion reacts more rapidly with the alcohol than the acyl chloride.[103]

An even stronger catalytic effect is obtained when 4-dimethylaminopyridine (DMAP) is used.[104] The dimethylamino group acts as an electron donor, increasing both the nucleophilicity and basicity of the pyridine nitrogen.

The inclusion of DMAP to the extent of 5–20 mol % in acylations by acid anhydrides and acyl chlorides increases acylation rates by up to four orders of magnitude and permits successful acylation of tertiary and other hindered alcohols. The reagent combination of an acid anhydride with $MgBr_2$ and a hindered tertiary amine, e.g., $(i\text{-}Pr)_2NC_2H_5$ or 1,2,2,6,6,-pentamethylpiperidine, gives an even more reactive acylation system, which is useful for hindered and sensitive alcohols.[105]

100. J. B. Lee, *J. Am. Chem. Soc.*, **88**, 3440 (1966).

101. H. J. Bestmann and L. Mott, *Justus Liebigs Ann. Chem.*, **693**, 132 (1966).

102. K. Sucheta, G. S. R. Reddy, D. Ravi, and N. Rama Rao, *Tetrahedron Lett.*, **35**, 4415 (1994).

103. A. R. Fersht and W. P. Jencks, *J. Am. Chem. Soc.*, **92**, 5432, 5442 (1970).

104. G. Hoefle, W. Steglich, and H. Vorbruggen, *Angew. Chem. Int. Ed. Engl.*, **17**, 569 (1978); E. F. V. Scriven, *Chem. Soc. Rev.*, **12**, 129 (1983); R. Murugan and E. F. V. Scriven, *Aldrichimica Acta*, **36**, 21 (2003).

105. E. Vedejs and O. Daugulis, *J. Org. Chem.*, **61**, 5702 (1996).

Another efficient catalyst for acylation is $Sc(O_3SCF_3)_3$, which can be used in combination with anhydrides[106] and other reactive acylating agents[107] and is a mild reagent for acylation of tertiary alcohols. Mechanistic investigation of $Sc(O_3SCF_3)_3$-catalyzed acylation indicates that triflic acid is involved. Acylation is stopped by the presence of a sterically hindered base such as 2,6-di-(t-butyl)-4-methylpyridine. The active acylating agent appears to be the acyl triflate. Two catalytic cycles operate. Cycle 2 requires only triflic acid, whereas Cycle 1 involves both the scandium salt and triflic acid.[108]

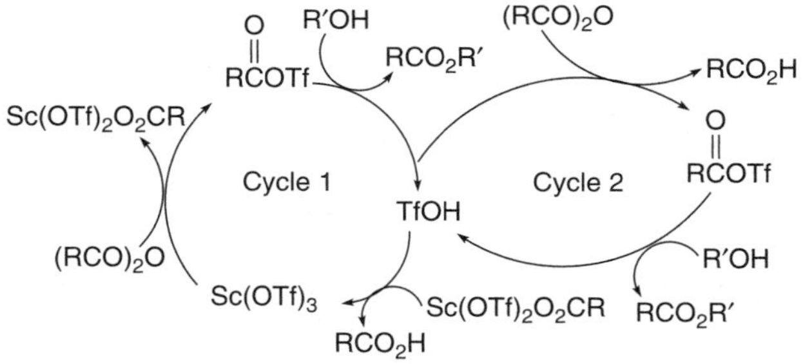

The acylation of tertiary alcohols can be effected by use of $Sc(O_3SCF_3)_3$ with diisopropylcarbodiimide (D-i-PCI) and DMAP.[109]

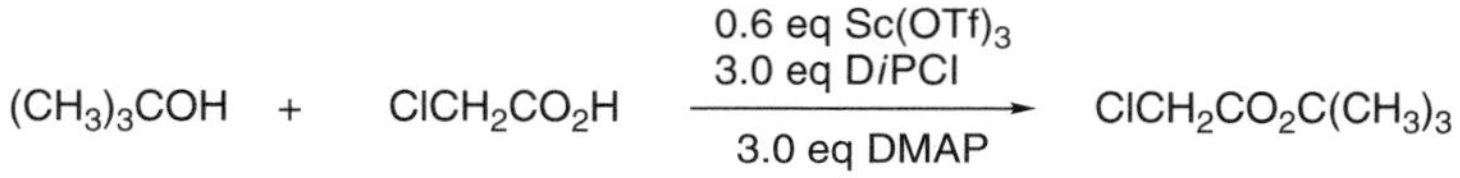

This method was effective for acylation of a hindered tertiary alcohol in the anticancer agent camptothecin by protected amino acids.

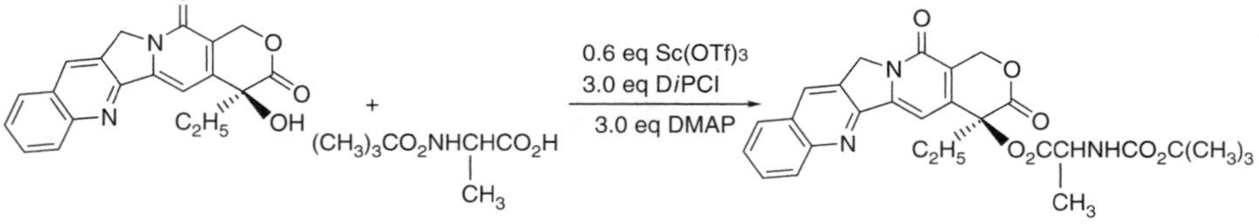

Ref. 110

Lanthanide triflates have similar catalytic effects. $Yb(O_3SCF_3)_3$ and $Lu(O_3SCF_3)_3$, for example, were used in selective acylation of 10-deacetylbaccatin III, an important intermediate for preparation of the antitumor agent paclitaxel.[111]

106. K. Ishihara, M. Kubota, H. Kurihara, and H. Yamamoto, *J. Org. Chem.*, **61**, 4560 (1996); A. G. M. Barrett and D. C. Braddock, *J. Chem. Soc., Chem. Commun.*, 351 (1997).
107. H. Zhao, A. Pendri, and R. B. Greenwald, *J. Org. Chem.*, **63**, 7559 (1998).
108. R. Dummeunier and I. E. Marko, *Tetrahedron Lett.*, **45**, 825 (2004).
109. H. Zhao, A. Pendri, and R. B. Greenwald, *J. Org. Chem.*, **63**, 7559 (1998).
110. R. R. Greenwald, A. Pendri, and H. Zhao, *Tetrahedron: Asymmetry*, **9**, 915 (1998).
111. E. W. P. Damen, L. Braamer, and H. W. Scheeren, *Tetrahedron Lett.*, **39**. 6081 (1998).

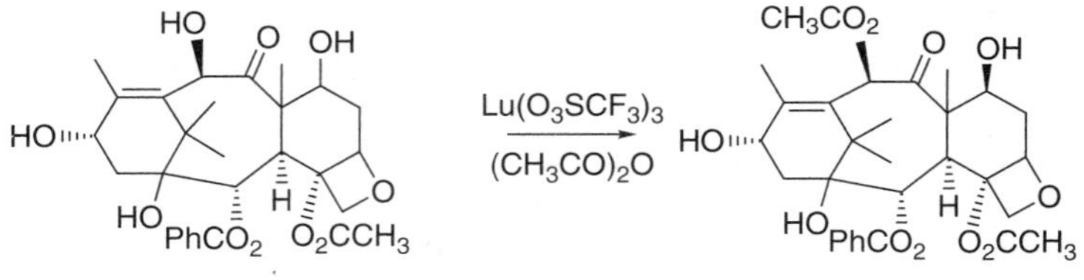

246

CHAPTER 3

*Functional Group
Interconversion
by Substitution,
Including Protection and
Deprotection*

Scandium triflimidate, $Sc[N(SO_2CF_3)_2]_3$, is also a very active acylation catalyst.

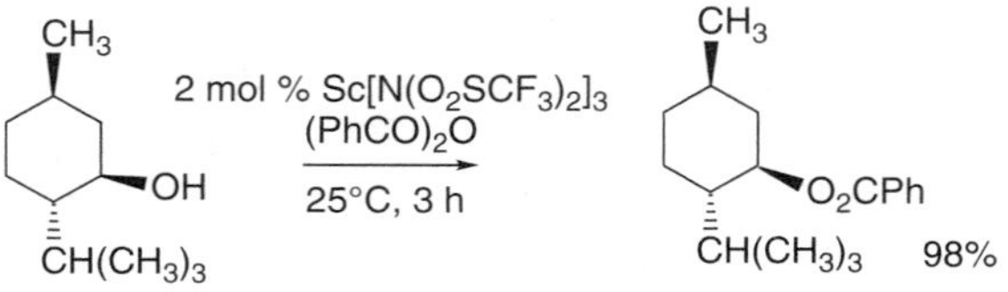

Ref. 112

Bismuth(III) triflate is also a powerful acylation catalyst that catalyzes reactions with acetic anhydride and other less reactive anhydrides such as benzoic and pivalic anhydrides.[113] Good results are achieved with tertiary and hindered secondary alcohols, as well as with alcohols containing acid- and base-sensitive functional groups.

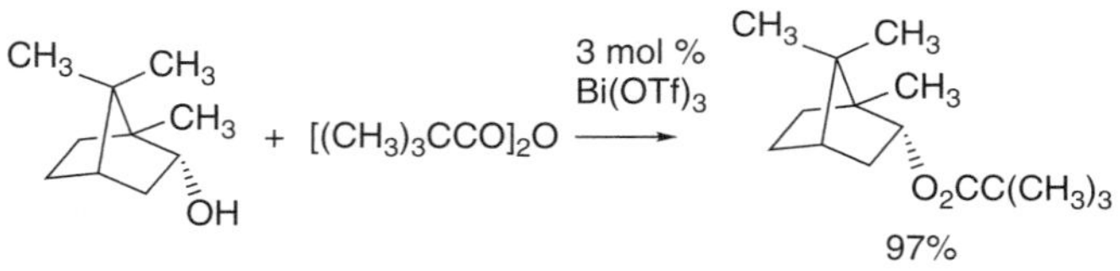

Trimethylsilyl triflate is also a powerful catalyst for acylation by anhydrides. Reactions of alcohols with a modest excess (1.5 equival) of anhydride proceed in inert solvents at 0° C. Even tertiary alcohols react rapidly.[114] The active acylation reagent is presumably generated by O-silylation of the anhydride.

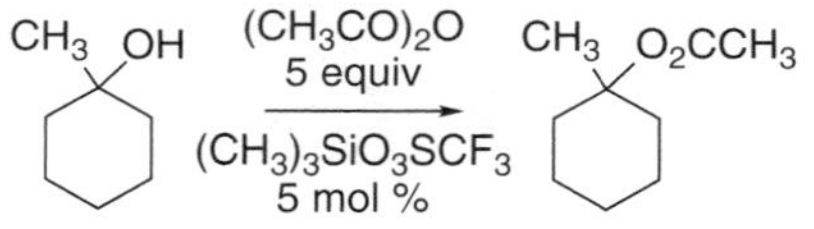

In addition to acyl halides and acid anhydrides, there are a number of milder and more selective acylating agents that can be readily prepared from carboxylic acids. Imidazolides, the *N*-acyl derivatives of imidazole, are examples.[115] Imidazolides are isolable substances and can be prepared directly from the carboxylic acid by reaction with carbonyldiimidazole.

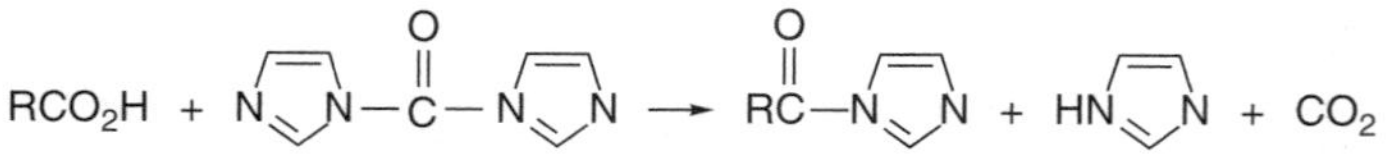

112. K. Ishihara, M. Kubota, and H. Yamamoto, *Synlett*, 265 (1996).

113. A. Orita, C. Tanahashi, A. Kakuda, and J. Otera, *J. Org. Chem.*, **66**, 8926 (2001).

114. P. A. Procopiou, S. P. D. Baugh, S. S. Flack, and G. G. A. Inglis, *J. Org. Chem.*, **63**, 2342 (1998).

115. H. A. Staab and W. Rohr, *Newer Methods Prep. Org. Chem.*, **5**, 61 (1968).

Two factors are responsible for the reactivity of the imidazolides as acylating reagents. One is the relative weakness of the "amide" bond. Owing to the aromatic character of imidazole nitrogens, there is little of the $N \rightarrow C=O$ delocalization that stabilizes normal amides. The reactivity of the imidazolides is also enhanced by protonation of the other imidazole nitrogen, which makes the imidazole ring a better leaving group.

Imidazolides can also be activated by N-alkylation with methyl triflate.[116] Imidazolides react with alcohols on heating to give esters and react at room temperature with amines to give amides. Imidazolides are particularly appropriate for acylation of acid-sensitive materials.

Dicyclohexylcarbodiimide (DCCI) is an example of a reagent that converts carboxylic acids to reactive acylating agents. This compound has been widely applied in the acylation step in the synthesis of polypeptides from amino acids[117] (see also Section 13.3.1). The reactive species is an *O*-acyl isourea. The acyl group is highly reactive because the nitrogen is susceptible to protonation and the cleavage of the acyl-oxygen bond converts the carbon-nitrogen double bond of the isourea to a more stable carbonyl group.[118]

The combination of carboxyl activation by DCCI and catalysis by DMAP provides a useful method for in situ activation of carboxylic acids for reaction with alcohols. The reaction proceeds at room temperature.[119]

2-Chloropyridinium[120] and 3-chloroisoxazolium[121] cations also activate carboxy groups toward nucleophilic attack. In each instance the halide is displaced from the heterocycle by the carboxylate via an addition-elimination mechanism. Nucleophilic attack on the activated carbonyl group results in elimination of the heterocyclic ring, with the departing oxygen being converted to an amidelike structure. The positive

116. G. Ulibarri, N. Choret, and D. C. H. Bigg, *Synthesis*, 1286 (1996).
117. F. Kurzer and K. Douraghi-Zadeh, *Chem. Rev.*, **67**, 107 (1967).
118. D. F. DeTar and R. Silverstein, *J. Am. Chem. Soc.*, **88**, 1013, 1020 (1966); D. F. DeTar, R. Silverstein, and F. F. Rogers, Jr., *J. Am. Chem. Soc.*, **88**, 1024 (1966).
119. A. Hassner and V. Alexanian, *Tetrahedron Lett.*, 4475 (1978); B. Neises and W. Steglich, *Angew. Chem. Int. Ed. Engl.*, **17**, 522 (1978).
120. T. Mukaiyama, M. Usui, E. Shimada, and K. Saigo, *Chem. Lett.*, 1045 (1975).
121. K. Tomita, S. Sugai, T. Kobayashi, and T. Murakami, *Chem. Pharm. Bull.*, **27**, 2398 (1979).

charge on the heterocyclic ring accelerates both the initial addition step and the subsequent elimination of the heterocycle.

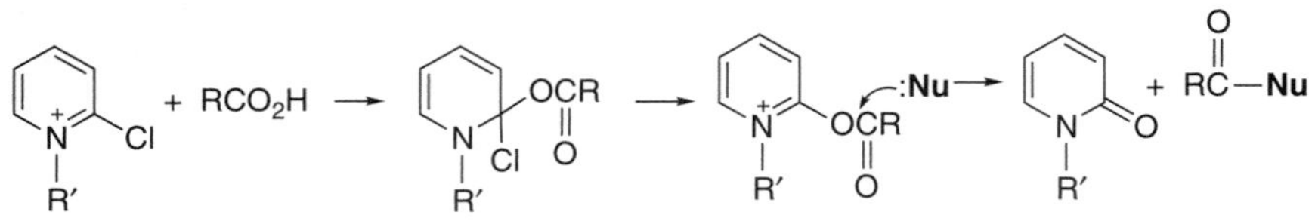

Carboxylic acid esters of thiols are considerably more reactive as acylating reagents than the esters of alcohols. Particularly reactive are esters of pyridine-2-thiol because there is an additional driving force in the formation of the more stable pyridine-2-thione tautomer.

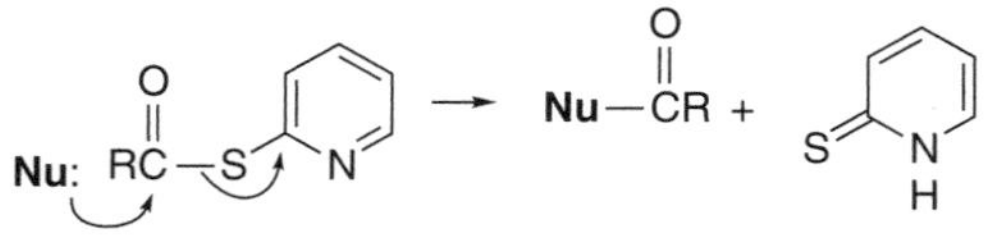

Additional acceleration of acylation can be obtained by inclusion of cupric salts, which coordinate at the pyridine nitrogen. This modification is useful for the preparation of highly hindered esters.[122] Pyridine-2-thiol esters can be prepared by reaction of the carboxylic acid with 2,2'-dipyridyl disulfide and triphenylphosphine[123] or directly from the acid and 2-pyridyl thiochloroformate.[124]

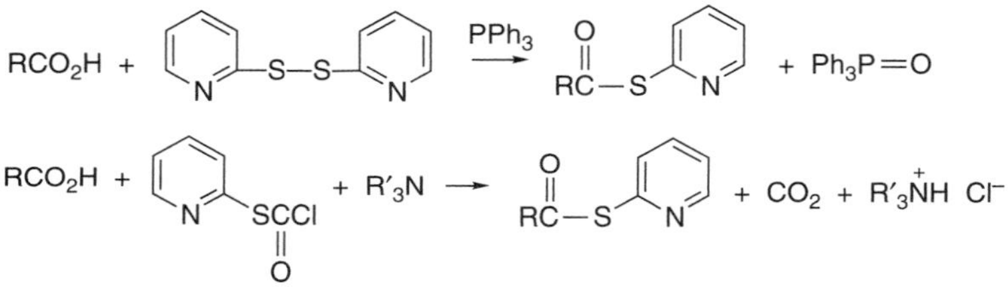

The 2-pyridyl and related 2-imidazolyl disulfides have found special use in the closure of large lactone rings.[125] Structures of this type are encountered in a number of antibiotics and other natural products and require mild conditions for cyclization because numerous other sensitive functional groups are present. It has been suggested that the pyridyl and imidazoyl thioesters function by a mechanism in which the heterocyclic nitrogen acts as a base, deprotonating the alcohol group. This proton transfer provides a cyclic TS in which hydrogen bonding can enhance the reactivity of the carbonyl group.[126]

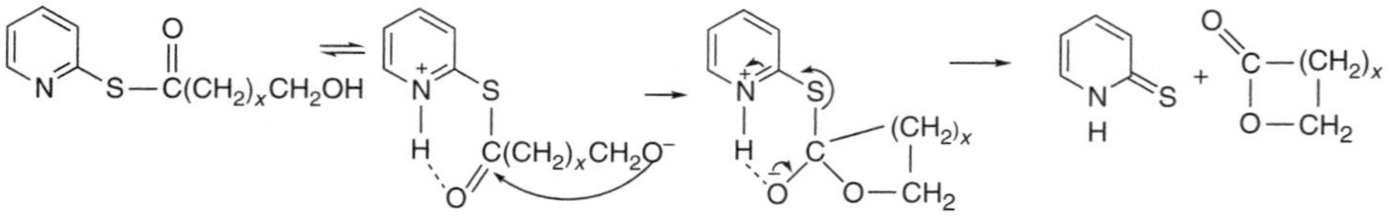

122. S. Kim and J. I. Lee, *J. Org. Chem.*, **49**, 1712 (1984).

123. T. Mukaiyama, R. Matsueda, and M. Suzuki, *Tetrahedron Lett.*, 1901 (1970).

124. E. J. Corey and D. A. Clark, *Tetrahedron Lett.*, 2875 (1979).

125. E. J. Corey and K. C. Nicolaou, *J. Am. Chem. Soc.*, **96**, 5614 (1974); K. C. Nicolaou, *Tetrahedron*, **33**, 683 (1977).

126. E. J. Corey, K. C. Nicolaou, and L. S. Melvin, Jr., *J. Am. Chem. Soc.*, **97**, 654 (1975); E. J. Corey, D. J. Brunelle, and P. J. Stork, *Tetrahedron Lett.*, 3405 (1976).

Good yields of large ring lactones are achieved by this method.

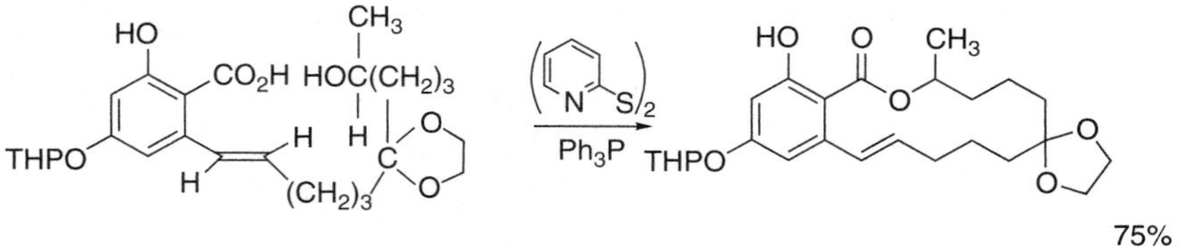

75%

Ref. 96

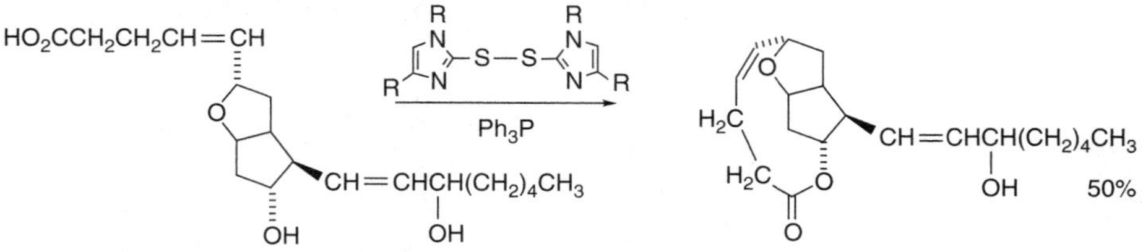

Ref. 127

Use of 2,4,6-trichlorobenzoyl chloride, Et_3N, and DMAP, known as the *Yamaguchi method*,[128] is frequently used to effect macrolactonization. The reaction is believed to involve formation of the mixed anhydride with the aroyl chloride, which then forms an acyl pyridinium ion on reaction with DMAP.[129]

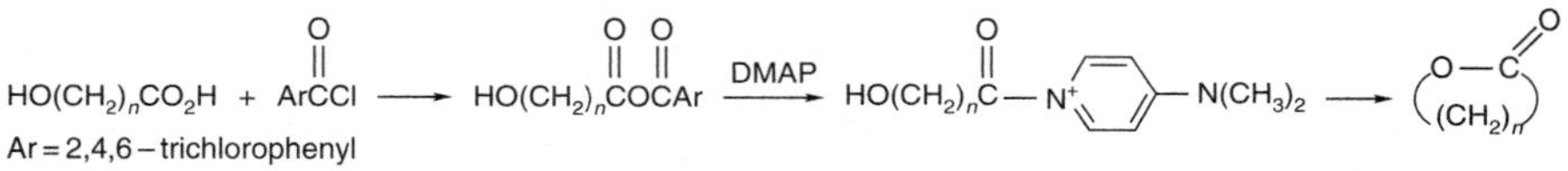

Intramolecular lactonization can also be carried out with DCCI and DMAP. As with most other macrolactonizations, the reactions must be carried out in rather dilute solution to promote the intramolecular cyclization in competition with inter-molecular reaction, which leads to dimers or higher oligomers. A study with 15-hydroxypentadecanoic acid demonstrated that a proton source is beneficial under these conditions and found the hydrochloride of DMAP to be convenient.[130]

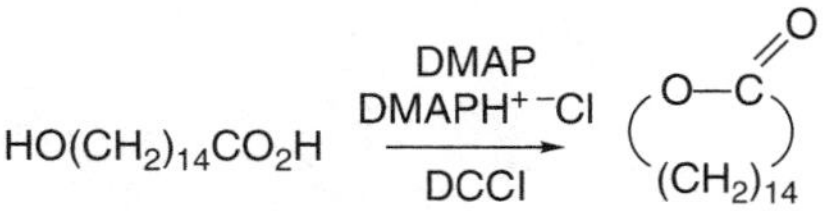

Scheme 3.4 gives some typical examples of the preparation and use of active acylating agents from carboxylic acids. Entries 1 and 2 show generation of acyl chlorides by reaction of carboxylic acids or salts with oxalyl chloride. Entry 3 shows a convenient preparation of 2-pyridylthio esters, which are themselves potential acylating agents (see p. 248). Entries 4 to 6 employ various coupling agents to form esters. Entries 7 and 8 illustrate acylations catalyzed by DMAP. Entries 9 to 13 are

127. E. J. Corey, H. L. Pearce, I. Szekely, and M. Ishiguro, *Tetrahedron Lett.*, 1023 (1978).
128. H. Saiki, T. Katsuki, and M. Yamaguchi, *Bull. Chem. Soc. Jpn.*, **52**, 1989 (1979).
129. M. Hikota, H. Tone, K. Horita, and O. Yonemitsu, *J. Org. Chem.*, **55**, 7 (1990).
130. E. P. Boden and G. E. Keck, *J. Org. Chem.*, **50**, 2394 (1985).

250

CHAPTER 3

Functional Group
Interconversion
by Substitution,
Including Protection and
Deprotection

Scheme 3.4. Preparation and Reactions of Active Acylating Agents

A. Generation of acylation reagents

1[a]

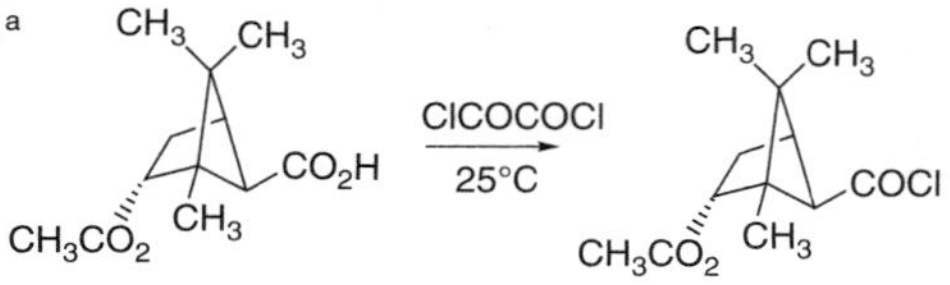

CH₃CO₂H + ClCOCOCl → at 25°C

2[b]

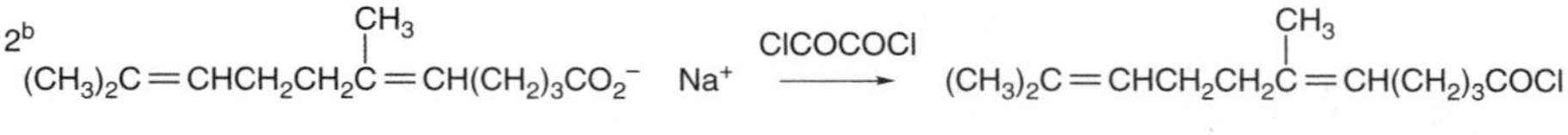

$(CH_3)_2C=CHCH_2CH_2C(CH_3)=CH(CH_2)_3CO_2^-$ Na⁺ + ClCOCOCl → $(CH_3)_2C=CHCH_2CH_2C(CH_3)=CH(CH_2)_3COCl$

3[c]

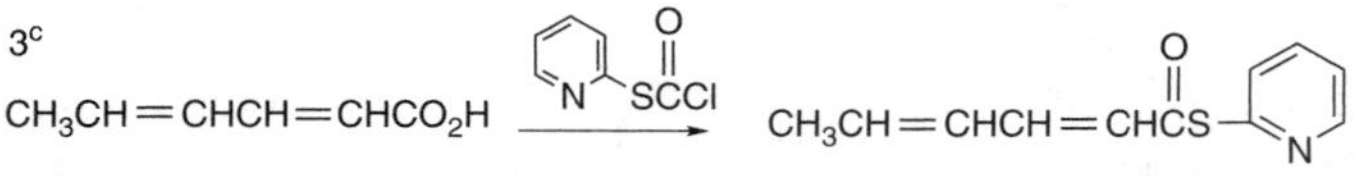

$CH_3CH=CHCH=CHCO_2H$ →

B. Esterification.

4[d]

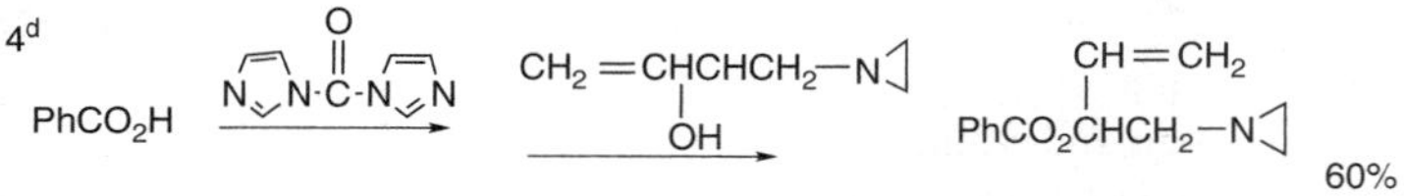

$PhCO_2H$ → 60%

5[e]

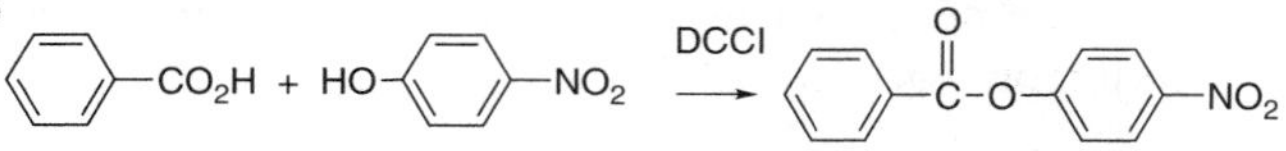

—CO_2H + $HO-$—NO_2 DCCI →

6[f]

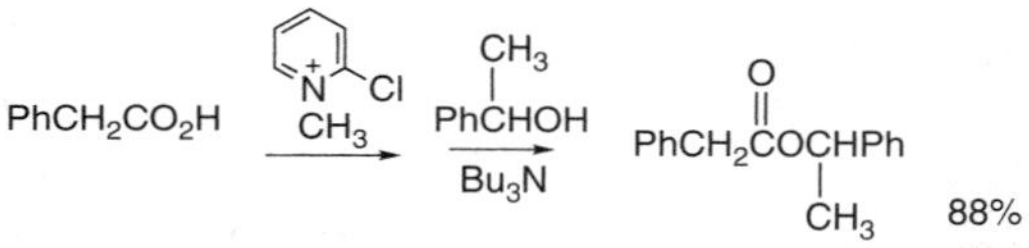

$PhCH_2CO_2H$ → 88%

7[g]

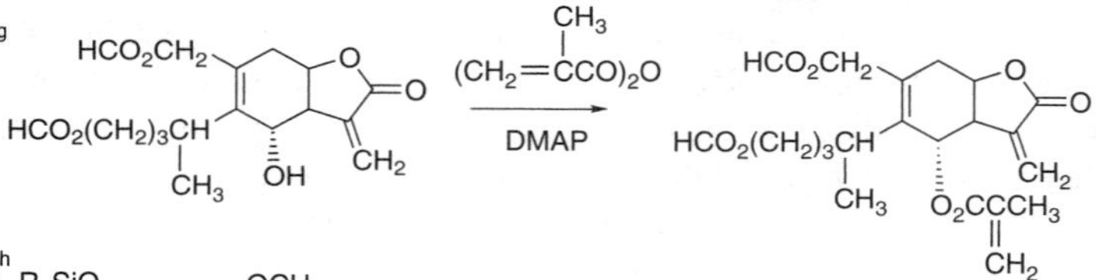

 DMAP →

8[h]

 DCCI, DMAP → 97%

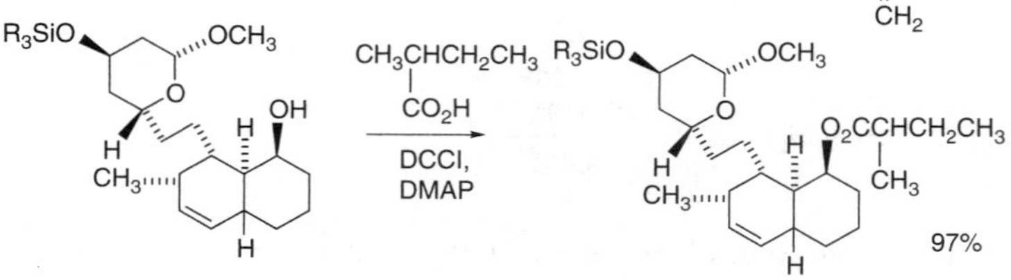

C. Macrolactonization

9[i]

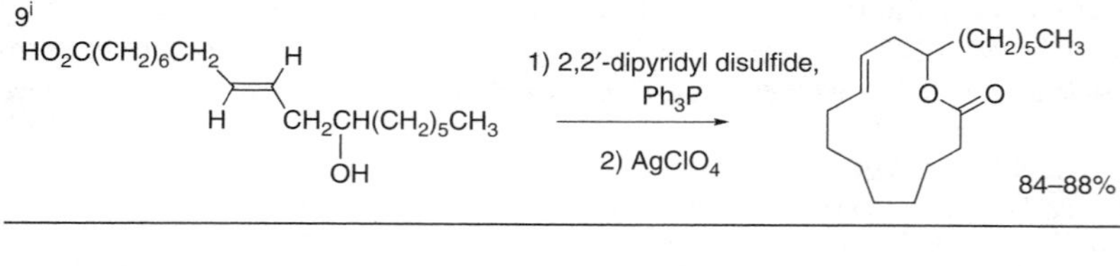

$HO_2C(CH_2)_6CH_2$ 1) 2,2'-dipyridyl disulfide, Ph₃P 2) AgClO₄ → 84–88%

10[j]

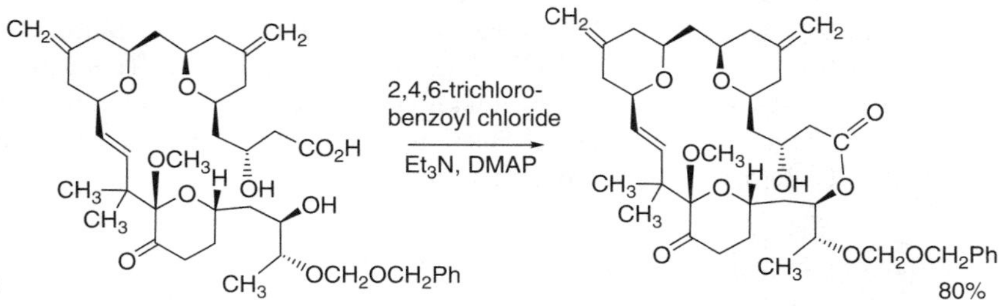

2,4,6-trichloro-benzoyl chloride

Et₃N, DMAP

80%

11[k]

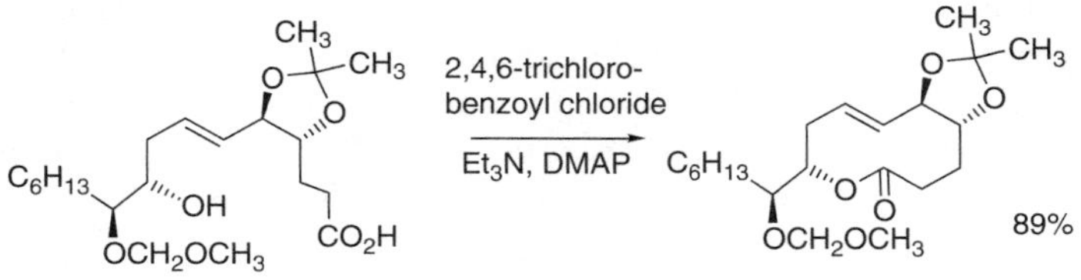

2,4,6-trichloro-benzoyl chloride

Et₃N, DMAP

89%

12[l]

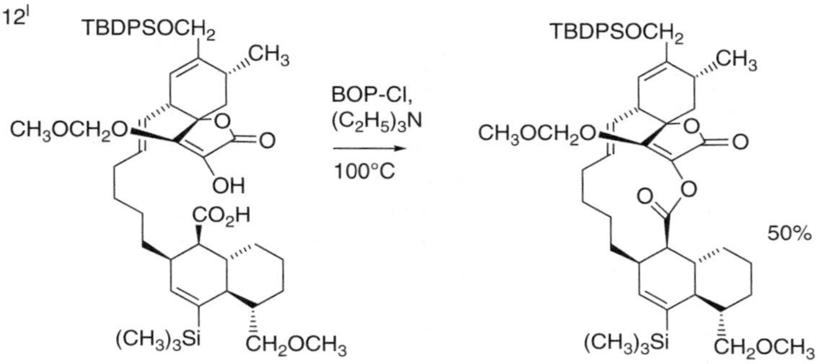

BOP-Cl, (C₂H₅)₃N

100°C

50%

13[m]

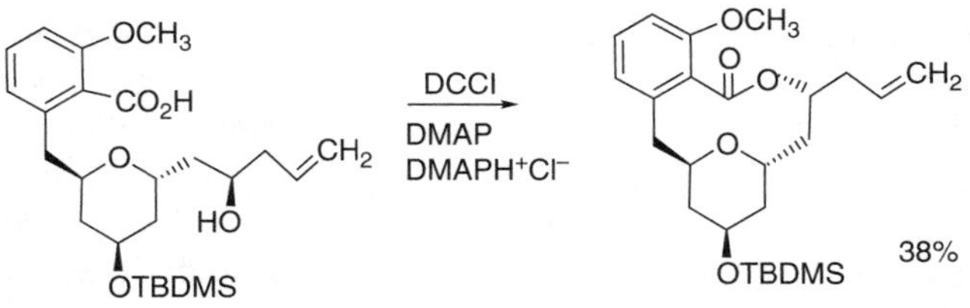

DCCI
DMAP
DMAPH⁺Cl⁻

38%

a. J. Meinwald, J. C. Shelton, G. L. Buchanan, and A. Courtain, *J. Org. Chem.*, **33**, 99 (1968).
b. U. T. Bhalerao, J. J. Plattner, and H. Rapoport, *J. Am. Chem. Soc.*, **92**, 3429 (1970).
c. E. J. Corey and D. A. Clark, *Tetrahedron Lett.*, 2875 (1979).
d. H. A. Staab and Rohr, *Chem. Ber.*, **95**, 1298 (1962).
e. S. Neeklakantan, R. Padmasani, and T. R. Seshadri, *Tetrahedron*, **21**, 3531 (1965).
f. T. Mukaiyama, M. Usui, E. Shimada, and K. Saigo, *Chem. Lett.*, 1045 (1970).
g. P. A. Grieco, T. Oguri, S. Gilman, and G. DeTitta, *J. Am. Chem. Soc.*, **100**, 1616 (1978).
h. Y.-L. Yang, S. Manna, and J. R. Falck, *J. Am. Chem. Soc.*, **106**, 3811 (1984).
i. A. Thalman, K. Oertle, and H. Gerlach, *Org. Synth.*, **63**, 192 (1984).
j. G. E. Keck and A. P. Troung, *Org. Lett.*, **7**, 2153 (2005).
k. P. Kumar and S. V. Naidu, *J. Org. Chem.*, **70**, 4207 (2005).
l. W. R. Roush and and R. J. Sciotti, *J. Am. Chem. Soc.*, **120**, 7411 (1998).
m. A. Lewis, I. Stefanuti, S. A. Swain, S. A. Smith, and R. J. K. Taylor, *Org. Biomol. Chem.*, **1**, 81 (2003)

252

CHAPTER 3

*Functional Group
Interconversion
by Substitution,
Including Protection and
Deprotection*

examples of macrocyclizations. Entry 9 uses the di-2-pyridyl disulfide-Ph$_3$P method. The cyclization was done in approximately 0.02 M acetonitrile by dropwise addition of the disulfide. Entries 10 and 11 are examples of application of the Yamaguchi macrolactonization procedure via the mixed anhydride with 2,4,6-trichlorobenzoyl chloride. The reaction in Entry 12 uses BOP-Cl as the coupling reagent. This particular reagent gave the best results among the several alternatives that were explored. Further discussion of this reagent can be found in Section 13.3.1. Entry 13 is an example of the use of the DCCI-DMAP reagent combination.

3.4.2. Fischer Esterification

As noted in the preceding section, one of the most general methods of synthesis of esters is by reaction of alcohols with an acyl chloride or other activated carboxylic acid derivative. Section 3.2.5 dealt with two other important methods, namely, reactions with diazoalkanes and reactions of carboxylate salts with alkyl halides or sulfonate esters. There is also the acid-catalyzed reaction of carboxylic acids with alcohols, which is called the *Fischer esterification*.

$$RCO_2H + R'OH \xrightarrow{H^+} RCO_2R' + H_2O$$

This is an equilibrium process and two techniques are used to drive the reaction to completion. One is to use a large excess of the alcohol, which is feasible for simple and inexpensive alcohols. The second method is to drive the reaction forward by irreversible removal of water, and azeotropic distillation is one way to accomplish this. Entries 1 to 4 in Scheme 3.5 are examples of acid-catalyzed esterifications. Entry 5 is the preparation of a diester starting with an anhydride. The initial opening of the anhydride ring is followed by an acid-catalyzed esterification.

3.4.3. Preparation of Amides

The most common method for preparation of amides is the reaction of ammonia or a primary or secondary amine with one of the reactive acylating reagents described in Section 3.4.1. Acid anhydrides give rapid acylation of most amines and are convenient if available. However, only one of the two acyl groups is converted to an amide. When acyl halides are used, some provision for neutralizing the hydrogen halide that is formed is necessary because it will react with the amine to form the corresponding salt. The *Schotten-Baumann conditions*, which involve shaking an amine with excess anhydride or acyl chloride and an alkaline aqueous solution, provide a very satisfactory method for preparation of simple amides.

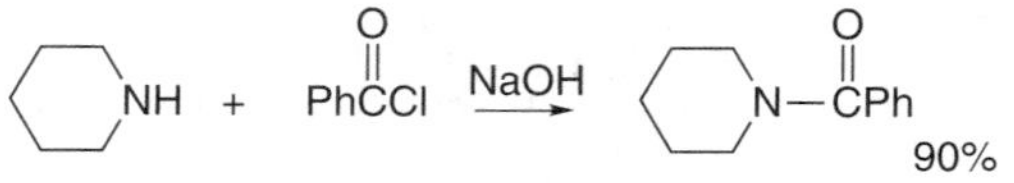

90%

Ref. 131

A great deal of work has been done on the in situ activation of carboxylic acids toward nucleophilic substitution by amines. This type of reaction is fundamental for synthesis of polypeptides (see also Section 13.3.1). Dicyclohexylcarbodiimide

131. C. S. Marvel and W. A. Lazier, *Org. Synth.*, **I**, 99 (1941).

1[a]

$$CH_3CO_2H \ + \ HOCH_2CH_2CH_2Cl \ \xrightarrow[\substack{\text{benzene,} \\ \text{removal of water}}]{ArSO_3H} \ CH_3CO_2CH_2CH_2CH_2Cl \qquad 93\text{–}95\%$$

2[b]

$$HO_2CC \equiv CCO_2H \ + \ \underset{\text{(20 equiv)}}{CH_3OH} \ \xrightarrow[25°C, \ 4 \ days]{H_2SO_4} \ CH_3O_2CC \equiv CCO_2CH_3 \qquad 72\text{–}88\%$$

3[c]

$$CH_3CH{=}CHCO_2H \ + \ \underset{\underset{OH}{|}}{CH_3CHCH_2CH_3} \ \xrightarrow[\substack{\text{benzene,} \\ \text{removal of water}}]{H_2SO_4} \ CH_3CH{=}CHCO_2\underset{\underset{CH_3}{|}}{CHCH_2CH_3} \quad 85\text{–}90\%$$

4[d]

$$\underset{\underset{OH}{|}}{PhCHCO_2H} \ + \ \underset{\text{(excess)}}{C_2H_5OH} \ \xrightarrow[78°C, \ 5 \ h]{HCl} \ \underset{\underset{OH}{|}}{PhCHCO_2C_2H_5} \qquad 82\text{–}86\%$$

5[e]

$$80\text{–}90\%$$

a. C. F. H. Allen and F. W. Spangler, *Org. Synth.*, **III**, 203 (1955).
b. E. H. Huntress, T. E. Lesslie, and J. Bornstein, *Org. Synth.*, **IV**, 329 (1963).
c. J. Munch-Petersen, *Org. Synth.*, **V**, 762 (1973).
d. E. L. Eliel, M. T. Fisk, and T. Prosser, *Org. Synth.*, **IV**, 169 (1963).
e. H. B. Stevenson, H. N. Cripps, and J. K. Williams, *Org. Synth.*, **V**, 459 (1973).

(DCCI) is often used for coupling carboxylic acids and amines to give amides. Since amines are better nucleophiles than alcohols, the leaving group in the acylation reagent need not be as reactive as is necessary for alcohols. The *p*-nitrophenyl[132] and 2,4,5-trichlorophenyl[133] esters of amino acids are sufficiently reactive toward amines to be useful in amide synthesis. Acyl derivatives of *N*-hydroxysuccinimide are also useful for synthesis of peptides and other types of amides.[134, 135] Like the *p*-nitrophenyl esters, the acylated *N*-hydroxysuccinimides can be isolated and purified, but react rapidly with free amino groups.

The *N*-hydroxysuccinimide that is liberated is easily removed because of its solubility in dilute base. The relative stability of the anion of *N*-hydroxysuccinimide is also responsible for the acyl derivative being reactive toward nucleophilic attack by an

132. M. Bodanszky and V. DuVigneaud, *J. Am. Chem. Soc.*, **81**, 5688 (1959).
133. J. Pless and R. A. Boissonnas, *Helv. Chim. Acta*, **46**, 1609 (1963).
134. G. W. Anderson, J. E. Zimmerman, and F. M. Callahan, *J. Am. Chem. Soc.*, **86**, 1839 (1964).
135. E. Wunsch and F. Drees, *Chem. Ber.*, **99**, 110 (1966); E. Wunsch, A. Zwick, and G. Wendlberger, *Chem. Ber.*, **100**, 173 (1967).

254

CHAPTER 3

*Functional Group
Interconversion
by Substitution,
Including Protection and
Deprotection*

amino group. Esters of *N*-hydroxysuccinimide are also used to carry out chemical modification of peptides, proteins, and other biological molecules by acylation of nucleophilic groups in these molecules. For example, detection of estradiol antibodies can be accomplished using an estradiol analog to which a fluorescent label has been attached.

fluorescein

Ref. 136

Similarly, photolabels, such as 4-azidobenzoylglycine can be attached to peptides and used to detect binding sites in proteins.[137]

1-Hydroxybenzotriazole is also useful in conjunction with DCCI.[138] For example, Boc-protected leucine and the methyl ester of phenylalanine can be coupled in 88% yield with these reagents.

Ref. 139

Carboxylic acids can also be activated by the formation of mixed anhydrides with various phosphoric acid derivatives. Diphenyl phosphoryl azide, for example, is an effective reagent for conversion of amines to amides.[140] The proposed mechanism involves formation of the acyl azide as a reactive intermediate.

136. M. Adamczyk, Y.-Y. Chen, J. A. Moore, and P. G. Mattingly, *Biorg. Med. Chem. Lett.*, **8**, 1281 (1998); M. Adamczyk, J. R. Fishpaugh, and K. J. Heuser, *Bioconjugate Chem.*, **8**, 253 (1997).
137. G. C. Kundu, I. Ji, D. J. McCormick, and T. H. Ji, *J. Biol. Chem.*, **271**, 11063 (1996).
138. W. Konig and R. Geiger, *Chem. Ber.*, **103**, 788 (1970).
139. M. Bodanszky and A. Bodanszky, *The Practice of Peptide Synthesis*, 2nd Edition, Springer-Verlag, Berlin, 1994, pp. 119–120.
140. T. Shioiri and S. Yamada, *Chem. Pharm. Bull.*, **22**, 849 (1974); T. Shioiri and S. Yamada, *Chem. Pharm. Bull.*, **22**, 855 (1974); T. Shioiri and S. Yamada, *Chem. Pharm. Bull.*, **22**, 859 (1974).

$$RCO_2^- + (PhO)_2PN_3 \longrightarrow RC\text{-}O\text{-}P(OPh)_2 + N_3^-$$

$$RC\text{-}O\text{-}P(OPh)_2 + N_3^- \longrightarrow RCN_3 + {}^-O_2P(OPh)_2$$

$$RCN_3 + R'NH_2 \longrightarrow RCNHR' + HN_3$$

Another useful reagent for amide formation is compound **1**, known as BOP-Cl,[141] which also proceeds by formation of a mixed carboxylic phosphoric anhydride.

Another method for converting esters to amides involves aluminum amides, which can be prepared from trimethylaluminum and the amine. These reagents convert esters directly to amides at room temperature.[142]

78%

The driving force for this reaction is the strength of the aluminum-oxygen bond relative to the aluminum-nitrogen bond. This reaction provides a good way of making synthetically useful amides of *N*-methoxy-*N*-methylamine.[143] Trialkylaminotin and *bis*-(hexamethyldisilylamido)tin amides, as well as *tetrakis*-(dimethylamino)titanium, show similar reactivity.[144] These reagents can also catalyze exchange reactions between amines and amides under moderate conditions.[145] For example, whereas exchange of benzylamine into *N*-phenylheptanamide occurs very slowly at 90° C in the absence of catalyst (> months), the conversion is effected in 16 h by $Ti[N(CH_3)_2]_4$.

$$CH_3(CH_2)_5CNHPh + PhCH_2NH_2 \xrightarrow[90°C, 16 h]{5 \text{ mol \% Ti(NMe}_2)_4} CH_3(CH_2)_5CNHCH_2Ph$$
99%

141. J. Diago-Mesequer, A. L. Palomo-Coll, J. R. Fernandez-Lizarbe, and A. Zugaza-Bilbao, *Synthesis*, 547 (1980); R. D. Tung, M. K. Dhaon, and D. H. Rich, *J. Org. Chem.*, **51**, 3350 (1986); W. J. Collucci, R. D. Tung, J. A. Petri, and D. H. Rich, *J. Org. Chem.*, **55**, 2895 (1990); J. Jiang, W. R. Li, R. M. Przeslawski, and M. M. Joullie, *Tetrahedron Lett.*, **34**, 6705 (1993).
142. A. Basha, M. Lipton, and S. M. Weinreb, *Tetrahedron Lett.*, 4171 (1977); A. Solladie-Cavallo and M. Bencheqroun, *J. Org. Chem.*, **57**, 5831 (1992).
143. J. I. Levin, E. Turos, and S. M. Weinreb, *Synth. Commun.*, **12**, 989 (1982); T. Shimizu, K. Osako, and T. Nakata, *Tetrahedron Lett.*, **38**, 2685 (1997).
144. G. Chandra, T. A. George, and M. F. Lappert, *J. Chem. Soc. C*, 2565 (1969); W.-B. Wang and E. J. Roskamp, *J. Org. Chem.*, **57**, 6101 (1992); W.-B. Wang, J. A. Restituyo, and E. J. Roskamp, *Tetrahedron Lett.*, **34**, 7217 (1993).
145. S. E. Eldred, D. A. Stone, S. M. Gellman, and S. S. Stahl, *J. Am. Chem. Soc.*, **125**, 3423 (2003).

256

CHAPTER 3

*Functional Group
Interconversion
by Substitution,
Including Protection and
Deprotection*

Tris-(dimethylamino)aluminum also promotes similar exchange reactions. The catalysis by titanium and aluminum amides may involve bifunctional catalysis in which the metal center acts as a Lewis acid while also delivering the nucleophilic amide.

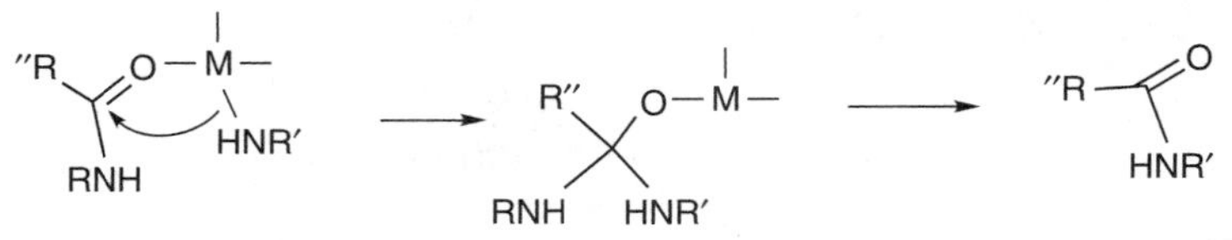

Interestingly, $Sc(O_3SCF_3)_3$ is also an active catalyst for these exchange reactions.

The cyano group is at the carboxylic acid oxidation level, so nitriles are potential precursors of primary amides. Partial hydrolysis is sometimes possible.[146]

A milder procedure involves the reaction of a nitrile with an alkaline solution of hydrogen peroxide.[147] The strongly nucleophilic hydrogen peroxide adds to the nitrile and the resulting adduct gives the amide. There are several possible mechanisms for the subsequent decomposition of the peroxycarboximidic adduct.[148]

In all the mechanisms, the hydrogen peroxide is converted to oxygen and water, leaving the organic substrate hydrolyzed, but at the same oxidation level.

Scheme 3.6 illustrates some of the means of preparation of amides. Entries 1 and 2 are cases of preparation of simple amides by conversion of the carboxylic acid to an acyl chloride using $SOCl_2$. Entry 3 is the acetylation of glycine by acetic anhydride. The reaction is done in concentrated aqueous solution ($\sim 3\,M$) using a twofold excess of the anhydride. The reaction is exothermic and the product crystallizes from the reaction mixture when it is cooled. Entries 4 and 5 are ester aminolysis reactions. The cyano group is an activating group for the ester in Entry 4, and this reaction occurs at room temperature in concentrated ammonia solution. The reaction in Entry 5 involves a less nucleophilic and more hindered amine, but involves a relatively reactive aryl ester. A much higher temperature is required for this reaction. Entries 6 to 8 illustrate the use of several of the coupling reagents for preparation of amides. Entries 9 and 10 show preparation of primary amides by hydrolysis of nitriles. The first reaction involves partial hydrolysis, whereas the second is an example of peroxide-accelerated hydrolysis.

146. W. Wenner, *Org. Synth.*, **IV**, 760 (1963).
147. C. R. Noller, *Org. Synth.*, **II**, 586 (1943); J. S. Buck and W. S. Ide, *Org. Synth.*, **II**, 44 (1943).
148. K. B. Wiberg, *J. Am. Chem. Soc.*, **75**, 3961 (1953); *J. Am. Chem. Soc.*, **77**, 2519 (1955); J. E. McIsaac, Jr., R. E. Ball, and E. J. Behrman, *J. Org. Chem.*, **36**, 3048 (1971).

A. From acyl chlorides and anhydrides

1[a]

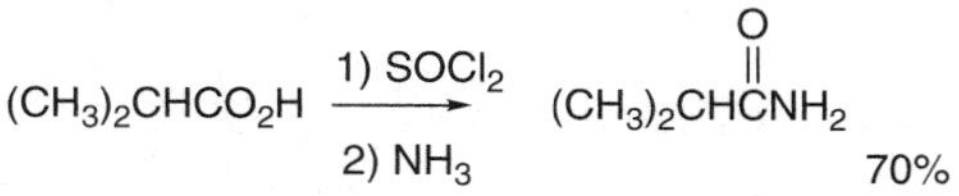

$(CH_3)_2CHCO_2H$ → 1) $SOCl_2$ / 2) NH_3 → $(CH_3)_2CHCNH_2$ 70%

2[b]

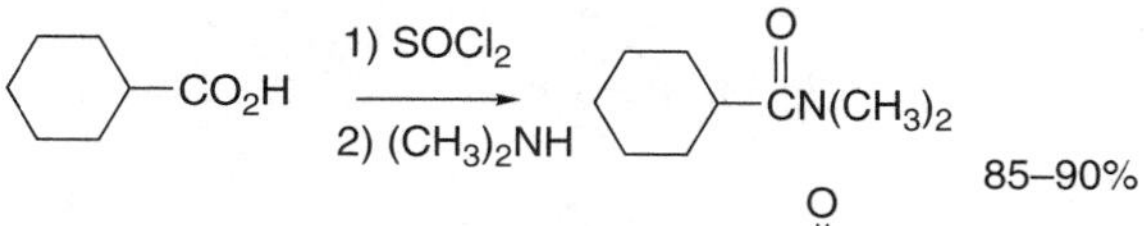

(cyclohexane)—CO_2H → 1) $SOCl_2$ / 2) $(CH_3)_2NH$ → (cyclohexane)—$CN(CH_3)_2$ 85–90%

3[c]

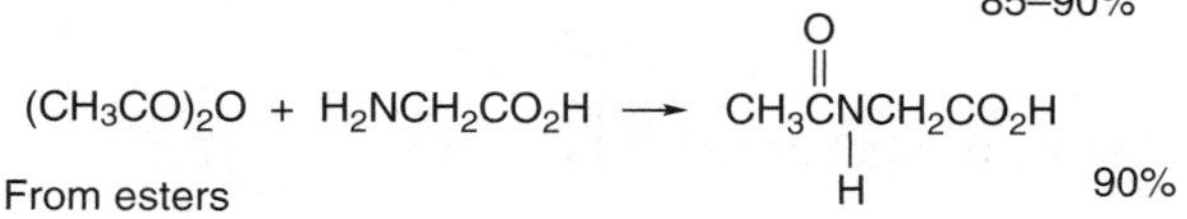

$(CH_3CO)_2O$ + $H_2NCH_2CO_2H$ → $CH_3CNCH_2CO_2H$ 90%

B. From esters

4[d]

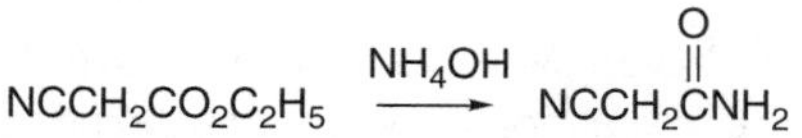

$NCCH_2CO_2C_2H_5$ → NH_4OH → $NCCH_2CNH_2$

5[e]

(benzene-OH-CO_2Ph) + (benzene-CH_3-H_2N) → trichlorobenzene / 185–200°C → (benzene-OH-C(=O)—NH-benzene-CH_3)   75%

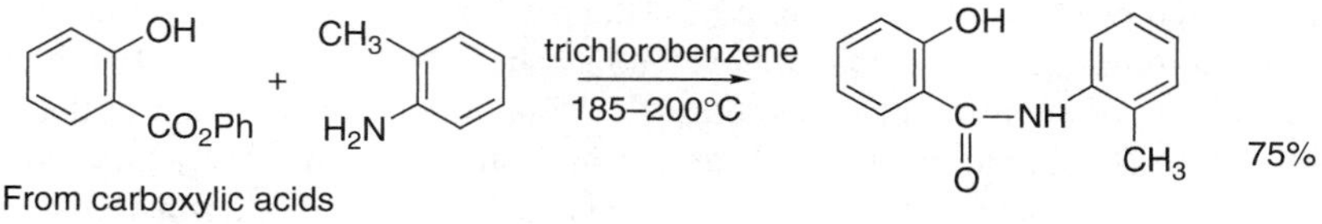

C. From carboxylic acids

6[f]

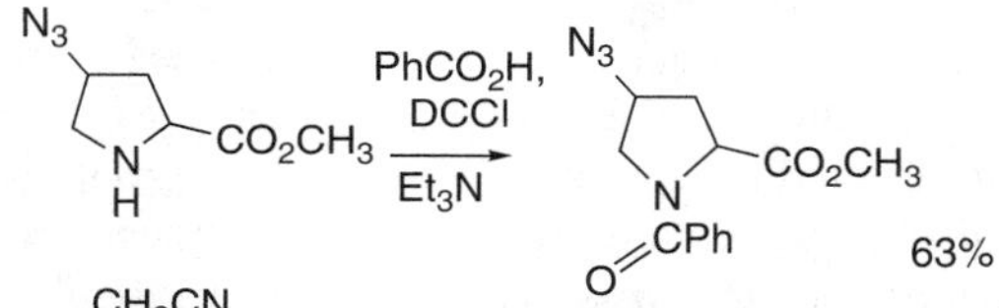

(pyrrolidine with N_3 and CO_2CH_3, NH) → $PhCO_2H$, DCCI / Et_3N → (pyrrolidine with N_3, CO_2CH_3, N—C(=O)Ph)   63%

7[g]

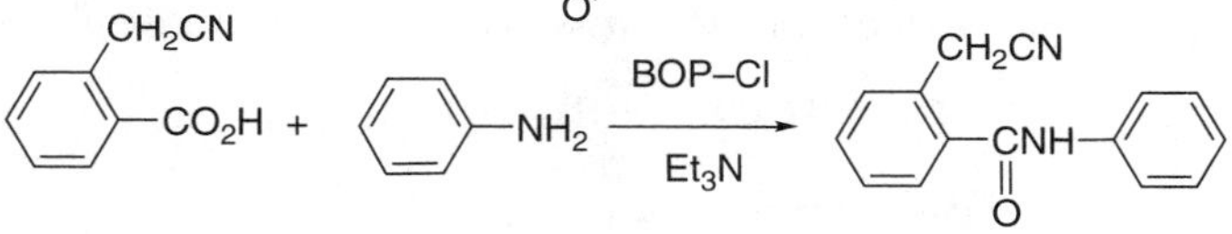

(benzene-CH_2CN-CO_2H) + (benzene-NH_2) → BOP–Cl / Et_3N → (benzene-CH_2CN-CNH-benzene, C=O)

8[h]

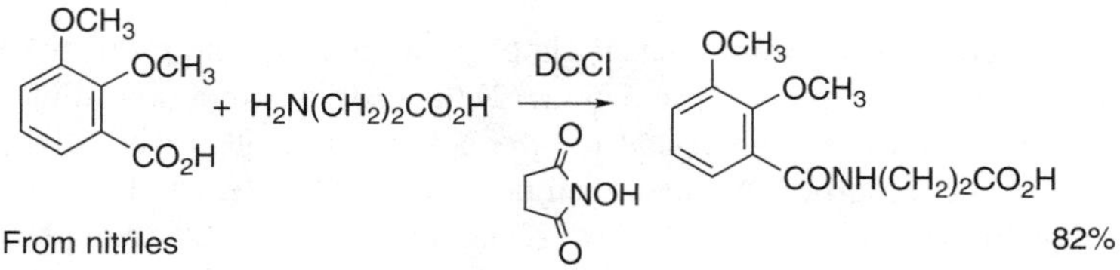

(benzene with OCH_3, OCH_3, CO_2H) + $H_2N(CH_2)_2CO_2H$ → DCCI / NOH (succinimide) → (benzene with OCH_3, OCH_3, $CONH(CH_2)_2CO_2H$)   82%

D. From nitriles

9[i]

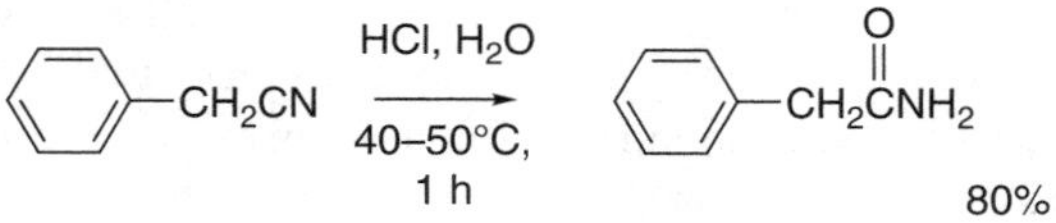

(benzene-CH_2CN) → HCl, H_2O / 40–50°C, 1 h → (benzene-CH_2CNH_2, C=O)   80%

10[j]

(benzene with CH_3, CN) → 30% H_2O_2, NaOH / 40–50°C, 4 h → (benzene with CH_3, CNH_2, C=O)   90%

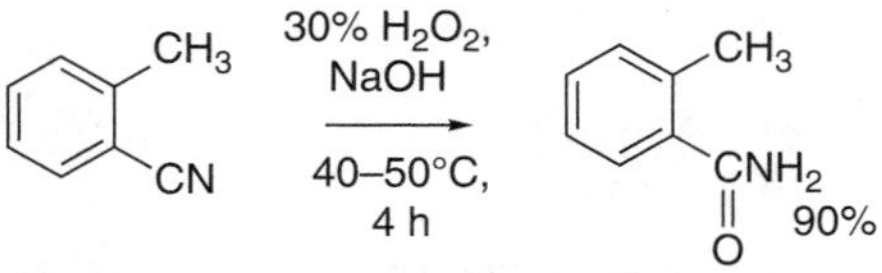

(*Continued*)

CHAPTER 3

*Functional Group
Interconversion
by Substitution,
Including Protection and
Deprotection*

a. R. E. Kent and S. M. McElvain, *Org. Synth.*, **III**, 490 (1955).
b. A. C. Cope and E. Ciganek, *Org. Synth.*, **IV**, 339 (1963).
c. R. M. Herbst and D. Shemin, *Org. Synth.*, **II**, 11 (1943).
d. B. B. Corson, R. W. Scott, and C. E. Vose, *Org. Synth.*, **I**, 179 (1941).
e. C. F. H. Allen and J. Van Allen, *Org. Synth.*, **III**, 765 (1955).
f. D. J. Abraham, M. Mokotoff, L. Sheh, and J. E. Simmons, *J. Med. Chem.*, **26**, 549 (1983).
g. J. Diago-Mesenguer, A. L. Palamo-Coll, J. R. Fernandez-Lizarbe, and A. Zugaza-Bilbao, *Synthesis*, 547 (1980).
h. R. J. Bergeron, S. J. Kline, N. J. Stolowich, K. A. McGovern, and P. S. Burton, *J. Org. Chem.*, **46**, 4524 (1981).
i. W. Wenner, *Org. Synth.*, **IV**, 760 (1963).
j. C. R. Noller, *Org. Synth.*, **II**, 586 (1943).

3.5. Installation and Removal of Protective Groups

Protective groups play a key role in multistep synthesis. When the synthetic target is a relatively complex molecule, a sequence of reactions that would be expected to lead to the desired product must be devised. At the present time, syntheses requiring 15–20 steps are common and many that are even longer have been completed. In the planning and execution of such multistep syntheses, an important consideration is the compatibility of the functional groups that are already present with the reaction conditions required for subsequent steps. It is frequently necessary to modify a functional group in order to prevent interference with some reaction in the synthetic sequence. A protective group can be put in place and then subsequently removed in order to prevent an undesired reaction or other adverse influence. For example, alcohols are often protected as trisubstituted silyl ethers and carbonyl groups as acetals. The silyl group masks both the acidity and nucleophilicity of the hydroxy group. An acetal group can prevent both unwanted nucleophilic additions or enolate formation at a carbonyl group.

$$R-OH + R'_3SiX \longrightarrow R-O-SiR'_3$$

$$R_2C{=}O + R'OH \longrightarrow R_2C(OR')_2$$

Three considerations are important in choosing an appropriate protective group: (1) the nature of the group requiring protection; (2) the reaction conditions under which the protective group must be stable; and (3) the conditions that can be tolerated for removal of the protecting group. No universal protective groups exist. The state of the art has been developed to a high level, however, and the many mutually complementary protective groups provide a great degree of flexibility in the design of syntheses of complex molecules.[149] Protective groups play a passive role in synthesis, but each operation of introduction and removal of a protective group adds steps to the synthetic sequence. It is thus desirable to minimize the number of such operations. Fortunately, the methods for protective group installation and removal have been highly developed and the yields are usually excellent.

3.5.1. Hydroxy-Protecting Groups

3.5.1.1. Acetals as Protective Groups. A common requirement in synthesis is that a hydroxy group be masked as a derivative lacking the proton. Examples of this requirement are reactions involving Grignard or other strongly basic organometallic

---

[149.] T. W. Green and P. G. Wuts, *Protective Groups in Organic Synthesis*, 3rd Edition, Wiley, New York, 1999; P. J. Kocienski, *Protective Groups*, Thieme, New York, 2000.

reagents. The acidic proton of a hydroxy group will destroy one equivalent of a strongly basic organometallic reagent and possibly adversely affect the reaction in other ways. In some cases, protection of the hydroxy group also improves the solubility of alcohols in nonpolar solvents. The choice of the most appropriate group is largely dictated by the conditions that can be tolerated in subsequent removal of the protecting group. The tetrahydropyranyl ether (THP) is applicable when mildly acidic hydrolysis is an appropriate method for deprotection.[150] The THP group, like other acetals and ketals, is inert to basic and nucleophilic reagents and is unchanged under such conditions as hydride reduction, organometallic reactions, or base-catalyzed reactions in aqueous solution. It also protects the hydroxy group against oxidation. The THP group is introduced by an acid-catalyzed addition of the alcohol to the vinyl ether moiety in dihydropyran. *p*-Toluenesulfonic acid or its pyridinium salt are frequently used as the catalyst,[151] although other catalysts are advantageous in special cases.

The THP group can be removed by dilute aqueous acid. The chemistry involved in both the introduction and deprotection stages is the reversible acid-catalyzed formation and hydrolysis of an acetal (see Part A, Section 7.1).

Various Lewis acids also promote hydrolysis of THP groups. Treatment with five equivalents of LiCl and ten equivalents of H_2O in DMSO removes THP groups in high yield.[152] $PdCl_2(CH_3CN)_2$ smoothly removes THP groups from primary alcohols.[153] $CuCl_2$ is also reported to catalyze hydrolysis of the THP group.[154] These procedures may involve generation of protons by interaction of water with the metal cations.

A disadvantage of the THP group is the fact that a new stereogenic center is produced at C(2) of the tetrahydropyran ring. This presents no difficulties if the alcohol is achiral, since a racemic mixture results. However, if the alcohol is chiral, the reaction gives a mixture of diastereomers, which may complicate purification and/or characterization. One way of avoiding this problem is to use methyl 2-propenyl ether in place of dihydropyran (abbreviated MOP, for methoxypropyl). No new chiral center

[150.] W. E. Parham and E. L. Anderson, *J. Am. Chem. Soc.*, **70**, 4187 (1948).
[151.] J. H. van Boom, J. D. M. Herscheid, and C. B. Reese, *Synthesis*, 169 (1973); M. Miyashita, A. Yoshikoshi, and P. A. Grieco, *J. Org. Chem.*, **42**, 3772 (1977).
[152.] G. Maiti and S. C. Roy, *J. Org. Chem.*, **61**, 6038 (1996).
[153.] Y.-G. Wang, X.-X. Wu, and S.-Y. Jiang, *Tetrahedron Lett.*, **45**, 2973 (2004).
[154.] J. K. Davis, U. T. Bhalerao, and B. V. Rao, *Ind. J. Chem. B*, **39B**, 860 (2000); J. Wang, C. Zhang, Z. Qu, Y. Hou, B. Chen, and P. Wu, *J. Chem. Res. Syn.*, 294 (1999).

260

CHAPTER 3

*Functional Group
Interconversion
by Substitution,
Including Protection and
Deprotection*

is introduced, and this acetal offers the further advantage of being hydrolyzed under somewhat milder conditions than those required for THP ethers.[155]

$$ROH + CH_2=C-OCH_3 \xrightarrow{H^+} ROC(CH_3)_2OCH_3$$
$$|$$
$$CH_3$$

Ethyl vinyl ether is also useful for hydroxy group protection. The resulting derivative (1-ethoxyethyl ether) is abbreviated as the EE group.[156] As with the THP group, the EE group introduces an additional stereogenic center.

The methoxymethyl (MOM) and β-methoxyethoxymethyl (MEM) groups are used to protect alcohols and phenols as formaldehyde acetals. These groups are normally introduced by reaction of an alkali metal salt of the alcohol with methoxymethyl chloride or β-methoxyethoxymethyl chloride.[157]

$$CH_3OCH_2Cl \longrightarrow ROCH_2OCH_3$$

$$RO^-M^+$$

$$CH_3OCH_2CH_2OCH_2Cl \longrightarrow ROCH_2OCH_2CH_2OCH_3$$

The MOM and MEM groups can be cleaved by pyridinium tosylate in moist organic solvents.[158] An attractive feature of the MEM group is the ease with which it can be removed under nonaqueous conditions. Reagents such as zinc bromide, magnesium bromide, titanium tetrachloride, dimethylboron bromide, or trimethylsilyl iodide permit its removal.[159] The MEM group is cleaved in preference to the MOM or THP groups under these conditions. Conversely, the MEM group is more stable to acidic aqueous hydrolysis than the THP group. These relative reactivity relationships allow the THP and MEM groups to be used in a complementary fashion when two hydroxy groups must be deprotected at different points in a synthetic sequence.

Ref. 160

The methylthiomethyl (MTM) group is a related alcohol-protecting group. There are several methods for introducing the MTM group. Alkylation of an alcoholate by

155. A. F. Kluge, K. G. Untch, and J. H. Fried, *J. Am. Chem. Soc.*, **94**, 7827 (1972).

156. H. J. Sims, H. B. Parseghian, and P. L. DeBenneville, *J. Org. Chem.*, **23**, 724 (1958).

157. G. Stork and T. Takahashi, *J. Am. Chem. Soc.*, **99**, 1275 (1977); R. J. Linderman, M. Jaber, and B. D. Griedel, *J. Org. Chem.*, **59**, 6499 (1994); P. Kumar, S. V. N. Raju, R. S. Reddy, and B. Pandey, *Tetrahedron Lett.*, **35**, 1289 (1994).

158. H. Monti, G. Leandri, M. Klos-Ringuet, and C. Corriol, *Synth. Commun.*, **13**, 1021 (1983); M. A. Tius and A. M. Fauq, *J. Am. Chem. Soc.*, **108**, 1035 (1986).

159. E. J. Corey, J.-L. Gras, and P. Ulrich, *Tetrahedron Lett.*, 809 (1976); Y. Quindon, H. E. Morton, and C. Yoakim, *Tetrahedron Lett.*, **24**, 3969 (1983); J. H. Rigby and J. Z. Wilson, *Tetrahedron Lett.*, **25**, 1429 (1984); S. Kim, Y. H. Park, and I. S. Kee, *Tetrahedron Lett.*, **32**, 3099 (1991).

160. E. J. Corey, R. L. Danheiser, S. Chandrasekaran, P. Siret, G. E. Keck, and J.-L. Gras, *J. Am. Chem. Soc.*, **100**, 8031 (1978).

methylthiomethyl chloride is efficient if catalyzed by iodide ion.[161] Alcohols are also converted to MTM ethers by reaction with dimethyl sulfoxide in the presence of acetic acid and acetic anhydride,[162] or with benzoyl peroxide and dimethyl sulfide.[163] The latter two methods involve the generation of the methylthiomethylium ion by ionization of an acyloxysulfonium ion (Pummerer reaction).

$$RO^-M^+ \ + \ CH_3SCH_2Cl \ \xrightarrow{\ I^-\ } \ ROCH_2SCH_3$$

$$ROH + CH_3SOCH_3 \ \xrightarrow[(CH_3CO)_2O]{CH_3CO_2H} \ ROCH_2SCH_3$$

$$ROH + (CH_3)_2S + (PhCO_2)_2 \ \longrightarrow \ ROCH_2SCH_3$$

The MTM group is selectively removed under nonacidic conditions in aqueous solutions containing Ag^+ or Hg^{2+} salts. The THP and MOM groups are stable under these conditions.[161] The MTM group can also be removed by reaction with methyl iodide, followed by hydrolysis of the resulting sulfonium salt in moist acetone.[162]

Two substituted alkoxymethoxy groups are designed for cleavage involving β-elimination. The 2,2,2-trichloroethoxymethyl groups can be cleaved by reducing agents, including zinc, samarium diiodide, and sodium amalgam.[164] The β-elimination results in the formation of a formaldehyde hemiacetal, which decomposes easily.

$$Cl_3CCH_2OCH_2OR \ \xrightarrow{\ 2e^-\ } \ Cl^- + Cl_2C{=}CH_2 \ + \ CH_2{=}O \ + \ {}^-OR$$

The 2-(trimethylsilyl)ethoxymethyl group (SEM) can be removed by various fluoride sources, including TBAF, pyridinium fluoride, and HF.[165] This deprotection involves nucleophilic attack at silicon, which triggers β-elimination.

$$F^- + (CH_3)_3SiCH_2{-}CH_2OCH_2OR \ \longrightarrow \ (CH_3)_3SiF + \ CH_2{=}CH_2 + \ CH_2{=}O \ + \ {}^-OR$$

The SEM group can also be cleaved by $MgBr_2$. A noteworthy aspect of this method is that trisubstituted silyl ethers (see below) can survive.

Ref. 166

161. E. J. Corey and M. G. Bock, *Tetrahedron Lett.*, 3269 (1975).
162. P. M. Pojer and S. J. Angyal, *Tetrahedron Lett.*, 3067 (1976).
163. J. C. Modina, M. Salomon, and K. S. Kyler, *Tetrahedron Lett.*, **29**, 3773 (1988).
164. R. M. Jacobson and J. W. Clader, *Synth. Commun.*, **9**, 57 (1979); D. A. Evans, S. W. Kaldor, T. K. Jones, J. Clardy, and T. J. Stout, *J. Am. Chem. Soc.*, **112**, 7001 (1990).
165. B. H. Lipshutz and J. J. Pegram, *Tetrahedron Lett.*, **21**, 3343 (1980); B. H. Lipshutz and T. A. Miller, *Tetrahedron Lett.*, **30**, 7149 (1989); T. Kan, M. Hashimoto, M. Yanagiya, and H. Shirahama, *Tetrahedron Lett.*, **29**, 5417 (1988); J. D. White and M. Kawasaki, *J. Am. Chem. Soc.*, **112**, 4991 (1990); K. Sugita, K. Shigeno, C. F. Neville, H. Sasai, and M. Shibasaki, *Synlett*, 325 (1994).
166. A. Vakalopoulos and H. M. R. Hoffmann, *Org. Lett.*, **2**, 1447 (2000).

262

CHAPTER 3

Functional Group
Interconversion
by Substitution,
Including Protection and
Deprotection

$MgBr_2$ removal of SEM groups is also useful for deprotection of carboxy groups in N-protected amino acids.

$$(CH_3)_3CO_2CNCHCOCH_2O(CH_2)_2Si(CH_3)_3 \xrightarrow[CH_2Cl]{MgCl_2} (CH_3)_3CO_2HNCHCO_2H$$

Ref. 167

3.5.1.2. Ethers as Protective Groups. The simple alkyl groups are generally not very useful for protection of alcohols as ethers. Although they can be introduced readily by alkylation, subsequent cleavage requires strongly electrophilic reagents such as boron tribromide (see Section 3.3). The *t*-butyl group is an exception and has found some use as a hydroxy-protecting group. Owing to the stability of the *t*-butyl cation, *t*-butyl ethers can be cleaved under moderately acidic conditions. Trifluoroacetic acid in an inert solvent is frequently used.[168] *t*-Butyl ethers can also be cleaved by acetic anhydride–$FeCl_3$ in ether.[169] The *t*-butyl group is normally introduced by reaction of the alcohol with isobutylene in the presence of an acid catalyst.[170] Acidic ion exchange resins are effective catalysts.[171]

$$ROH + CH_2=C(CH_3)_2 \xrightarrow{H^+} ROC(CH_3)_3$$

The triphenylmethyl (trityl, abbreviated Tr) group is removed under even milder conditions than the *t*-butyl group and is an important hydroxy-protecting group, especially in carbohydrate chemistry.[172] This group is introduced by reaction of the alcohol with triphenylmethyl chloride via an S_N1 substitution. Owing to their steric bulk, triarylmethyl groups are usually introduced only at primary hydroxy groups. Reactions at secondary hydroxy groups can be achieved using stronger organic bases such as DBU.[173] Hot aqueous acetic acid suffices to remove the trityl group. The ease of removal can be increased by addition of ERG substituents. The *p*-methoxy (PMTr) and *p,p'*-dimethoxy (DMTr) derivatives are used in this way.[174] Trityl groups can also be removed oxidatively using $Ce(NH_3)_6(NO_3)_3$ (CAN) on silica.[175] This method involves a single-electron oxidation and, as expected, the rate of reaction is DMTr > PMTr > Tr. The DMTr group is especially important in the protection of primary hydroxy groups in nucleotide synthesis (see Section 13.3.2).

The benzyl group can serve as a hydroxy-protecting group if acidic conditions for ether cleavage cannot be tolerated. The benzyl C–O bond is cleaved by catalytic hydrogenolysis,[176] or by electron-transfer reduction using sodium in liquid ammonia or

167. W.-C. Chen, M. D. Vera, and M. M. Joullie, *Tetrahedron Lett.*, **38**, 4025 (1997).
168. H. C. Beyerman and G. J. Heiszwolf, *J. Chem. Soc.*, 755 (1963).
169. B. Ganem and V. R. Small, Jr., *J. Org. Chem.*, **39**, 3728 (1974).
170. J. L. Holcombe and T. Livinghouse, *J. Org. Chem.*, **51**, 111 (1986).
171. A. Alexakis, M. Gardette, and S. Colin, *Tetrahedron Lett.*, **29**, 2951 (1988).
172. O. Hernandez, S. K. Chaudhary, R. H. Cox, and J. Porter, *Tetrahedron Lett.*, **22**, 1491 (1981); S. K. Chaudhary and O. Hernandez, *Tetrahedron Lett.*, **20**, 95 (1979).
173. S. Colin-Messager, J.-P. Girard, and J.-C. Rossi, *Tetrahedron Lett.*, **33**, 2689 (1992).
174. M. Smith, D. H. Rammler, I. H. Goldberg, and H. G. Khorana, *J. Am. Chem. Soc.*, **84**, 430 (1962).
175. J. R. Hwu, M. L. Jain, F.-Y. Tsai, S.-C. Tsay, A. Balakumar, and G. H. Hakimelahi, *J. Org. Chem.*, **65**, 5077 (2000).
176. W. H. Hartung and R. Simonoff, *Org. React.*, **7**, 263 (1953).

aromatic radical anions.[177] Benzyl ethers can also be cleaved using formic acid, cyclohexene, or cyclohexadiene as hydrogen sources in transfer hydrogenolysis catalyzed by platinum or palladium.[178] Several nonreductive methods for cleavage of benzyl ether groups have also been developed. Treatment with *s*-butyllithium, followed by reaction with trimethyl borate and then hydrogen peroxide liberates the alcohol.[179] The lithiated ether forms an alkyl boronate, which is oxidized as discussed in Section 4.5.2.

Lewis acids such as $FeCl_3$ and $SnCl_4$ also cleave benzyl ethers.[180]

Benzyl groups having 4-methoxy (PMB) or 3,5-dimethoxy (DMB) substituents can be removed oxidatively by dichlorodicyanoquinone (DDQ).[181] These reactions presumably proceed through a benzylic cation and the methoxy substituent is necessary to facilitate the oxidation.

These reaction conditions do not affect most of the other common hydroxy-protecting groups and the methoxybenzyl group is therefore useful in synthetic sequences that require selective deprotection of different hydroxy groups. 4-Methoxybenzyl ethers can also be selectively cleaved by dimethylboron bromide.[182]

Benzyl groups are usually introduced by the Williamson reaction (Section 3.2.3). They can also be prepared under nonbasic conditions if necessary. Benzyl alcohols are converted to trichloroacetimidates by reaction with trichloroacetonitrile. These then react with an alcohol to transfer the benzyl group.[183]

Phenyldiazomethane can also be used to introduce benzyl groups.[184]

177. E. J. Reist, V. J. Bartuska, and L. Goodman, *J. Org. Chem.*, **29**, 3725 (1964); R. E. Ireland, D. W. Norbeck, G. S. Mandel, and N. S. Mandel, *J. Am. Chem. Soc.*, **107**, 3285 (1985); R. E. Ireland and M. G. Smith, *J. Am. Chem. Soc.*, **110**, 854 (1988); H.-J. Liu, J. Yip, and K.-S. Shia, *Tetrahedron Lett.*, **38**, 2253 (1997).
178. B. El Amin, G. M. Anatharamaiah, G. P. Royer, and G. E. Means, *J. Org. Chem.*, **44**, 3442 (1979); A. M. Felix, E. P. Heimer, T. J. Lambros, C. Tzougraki, and J. Meienhofer, *J. Org. Chem.*, **43**, 4194 (1978); A. E. Jackson and R. A. W. Johnstone, *Synthesis*, 685 (1976); G. M. Anatharamaiah and K. M. Sivandaiah, *J. Chem. Soc., Perkin Trans.*, **I**, 490 (1977).
179. D. A. Evans, C. E. Sacks, W. A. Kleschick, and T. R. Taber, *J. Am. Chem. Soc.*, **101**, 6789 (1979).
180. M. H. Park, R. Takeda, and K. Nakanishi, *Tetrahedron Lett.*, **28**, 3823 (1987).
181. Y. Oikawa, T. Yoshioka, and O. Yonemitsu, *Tetrahedron Lett.*, **23**, 885 (1982); Y. Oikawa, T. Tanaka, K. Horita, T. Yoshioka, and O. Yonemitsu, *Tetrahedron Lett.*, **25**, 5393 (1984); N. Nakajima, T. Hamada, T. Tanaka, Y. Oikawa, and O. Yonemitsu, *J. Am. Chem. Soc.*, **108**, 4645 (1986).
182. N. Hebert, A. Beck, R. B. Lennox, and G. Just, *J. Org. Chem.*, **57**, 1777 (1992).
183. H.-P. Wessel, T. Iverson, and D. R. Bundle, *J. Chem. Soc., Perkin Trans.*, **I**, 2247 (1985); N. Nakajima, K. Horita, R. Abe, and O. Yonemitsu, *Tetrahedron Lett.*, **29**, 4139 (1988); S. J. Danishefsky, S. DeNinno, and P. Lartey, *J. Am. Chem. Soc.*, **109**, 2082 (1987).
184. L. J. Liotta and B. Ganem, *Tetrahedron Lett.*, **30**, 4759 (1989).

264

CHAPTER 3

*Functional Group
Interconversion
by Substitution,
Including Protection and
Deprotection*

4-Methoxyphenyl (PMP) ethers find occasional use as hydroxy protecting groups. Unlike benzylic groups, they cannot be made directly from the alcohol. Instead, the phenoxy group must be introduced by a nucleophilic substitution.[185] Mitsunobu conditions are frequently used.[186] The PMP group can be cleaved by oxidation with CAN.

Allyl ethers can be removed by conversion to propenyl ethers, followed by acidic hydrolysis of the resulting enol ether.

$$ROCH_2CH{=}CH_2 \longrightarrow ROCH{=}CHCH_3 \xrightarrow{H_3O^+} ROH \ + \ CH_3CH_2CH{=}O$$

The isomerization of an allyl ether to a propenyl ether can be achieved either by treatment with potassium *t*-butoxide in dimethyl sulfoxide[187] or by catalysts such as $Rh(PPh_3)_3Cl$[188] or $RhH(PPh_3)_4$.[189] Heating allyl ethers with Pd-C in acidic methanol can also effect cleavage of allyl ethers.[190] This reaction, too, is believed to involve isomerization to the 1-propenyl ether. Other very mild conditions for allyl group cleavage include Wacker oxidation conditions[191] (see Section 8.2.1) and DiBAlH with catalytic $NiCl_2(dppp)$.[192]

3.5.1.3. Silyl Ethers as Protective Groups. Silyl ethers play a very important role as hydroxy-protecting groups.[193] Alcohols can be easily converted to trimethylsilyl (TMS) ethers by reaction with trimethylsilyl chloride in the presence of an amine or by heating with hexamethyldisilazane. Trimethylsilyl groups are easily removed by hydrolysis or by exposure to fluoride ions. *t*-Butyldimethylsilyl (TBDMS) ethers are also very useful. The increased steric bulk of the TBDMS group improves the stability of the group toward such reactions as hydride reduction and Cr(VI) oxidation. The TBDMS group is normally introduced using a tertiary amine as a catalyst in the reaction of the alcohol with *t*-butyldimethylsilyl chloride or triflate. Cleavage of the TBDMS group is slow under hydrolytic conditions, but anhydrous tetra-*n*-butylammonium fluoride (TBAF),[194] methanolic NH_4F,[195] aqueous HF,[196] BF_3,[197] or SiF_4[198] can be used for its removal. Other highly substituted silyl groups, such as dimethyl(1,2,2-trimethylpropyl)silyl[199] and *tris*-isopropylsilyl,[200] (TIPS) are even more

185. Y. Masaki, K. Yoshizawa, and A. Itoh, *Tetrahedron Lett.*, **37**, 9321 (1996); S. Takano, M. Moriya, M. Suzuki, Y. Iwabuchi, T. Sugihara, and K. Ogaswawara, *Heterocycles*, **31**,1555 (1990).
186. T. Fukuyama, A. A. Laird, and L. M. Hotchkiss, *Tetrahedron Lett.*, **26**, 6291 (1985); M. Petitou, P. Duchaussoy, and J. Choay, *Tetrahedron Lett.*, **29**, 1389 (1988).
187. R. Griggs and C. D. Warren, *J. Chem. Soc. C*, 1903 (1968).
188. E. J. Corey and J. W. Suggs, *J. Org. Chem.*, **38**, 3224 (1973).
189. F. E. Ziegler, E. G. Brown, and S. B. Sobolov, *J. Org. Chem.*, **55**, 3691 (1990).
190. R. Boss and R Scheffold, *Angew. Chem. Int. Ed. Engl.*, **15**, 558 (1976).
191. H. B. Mereyala and S. Guntha, *Tetrahedron Lett.*, **34**, 6929 (1993).
192. T. Taniguchi and K. Ogasawara, *Angew. Chem. Int. Ed. Engl.*, **37**, 1136 (1998).
193. J. F. Klebe, in *Advances in Organic Chemistry: Methods and Results*, Vol. 8, E. C. Taylor, ed., Wiley-Interscience, New York, 1972, pp. 97–178; A. E. Pierce, *Silylation of Organic Compounds*, Pierce Chemical Company, Rockford, IL, 1968.
194. E. J. Corey and A. Venkataswarlu, *J. Am. Chem. Soc.*, **94**, 6190 (1972).
195. W. Zhang and M. J. Robins, *Tetrahedron Lett.*, **33**, 1177 (1992).
196. R. F. Newton, D. P. Reynolds, M. A. W. Finch, D. R. Kelly, and S. M. Roberts, *Tetrahedron Lett.*, 3981 (1979).
197. D. R. Kelly, S. M. Roberts, and R. F. Newton, *Synth. Commun.*, **9**, 295 (1979).
198. E. J. Corey and K. Y. Yi, *Tetrahedron Lett.*, **32**, 2289 (1992).
199. H. Wetter and K. Oertle, *Tetrahedron Lett.*, **26**, 5515 (1985).
200. R. F. Cunico and L. Bedell, *J. Org. Chem.*, **45**, 4797 (1980).

sterically hindered than the TBDMS group and can be used when added stability is required. The triphenylsilyl (TPS) and *t*-butyldiphenylsilyl (TBDPS) groups are also used.[201] The hydrolytic stability of the various silyl protecting groups is in the order TMS < TBDMS < TIPS < TBDPS.[202] All the groups are also susceptible to TBAF cleavage, but the TPS and TBDPS groups are cleaved more slowly than the trialkylsilyl groups.[203] Bromine in methanol readily cleaves TBDMS and TBDPS groups.[204]

3.5.1.4. Esters as Protective Groups.   Protection of an alcohol function by esterification sometimes offers advantages over use of acetal or ether groups. Generally, esters are stable under acidic conditions, and they are especially useful in protection during oxidations. Acetates, benzoates, and pivalates, which are the most commonly used derivatives, can be conveniently prepared by reaction of unhindered alcohols with acetic anhydride, benzoyl chloride, or pivaloyl chloride, respectively, in the presence of pyridine or other tertiary amines. 4-Dimethylaminopyridine (DMAP) is often used as a catalyst. The use of *N*-acylimidazolides (see Section 3.4.1) allows the acylation reaction to be carried out in the absence of added base.[205] Imidazolides are less reactive than the corresponding acyl chloride and can exhibit a higher degree of selectivity in reactions with a molecule possessing several hydroxy groups.

78%   Ref. 206

Hindered hydroxy groups may require special acylation procedures. One approach is to increase the reactivity of the hydroxy group by converting it to an alkoxide ion with strong base (e.g., *n*-BuLi or KH). When this conversion is not feasible, a more reactive acylating reagent is used. Highly reactive acylating agents are generated in situ when carboxylic acids are mixed with trifluoroacetic anhydride. The mixed anhydride exhibits increased reactivity because of the high reactivity of the trifluoroacetate ion as a leaving group.[207] Dicyclohexylcarbodiimide is another reagent that serves to activate carboxy groups.

Ester groups can be removed readily by base-catalyzed hydrolysis. When basic hydrolysis is inappropriate, special acyl groups are required. Trichloroethyl carbonate esters, for example, can be reductively removed with zinc.[208]

$$ROCOCH_2CCl_3 \xrightarrow{\text{Zn}} ROH + H_2C{=}CCl_2 + CO_2$$

201. S. Hanessian and P. Lavallee, *Can. J. Chem.*, **53**, 2975 (1975); S. A. Hardinger and N. Wijaya, *Tetrahedron Lett.*, **34**, 3821 (1993).

202. J. S. Davies, C. L. Higginbotham, E. J. Tremeer, C. Brown, and R. S. Treadgold, *J. Chem. Soc., Perkin Trans.*, **1**, 3043 (1992).

203. J. W. Gillard, R. Fortin, H. E. Morton, C. Yoakim, C. A. Quesnelle, S. Daignault, and Y. Guindon, *J. Org. Chem.*, **53**, 2602 (1988).

204. M. T. Barros, C. D. Maycock, and C. Thomassigny, *Synlett*, 1146 (2001).

205. H. A. Staab, *Angew. Chem.*, **74**, 407 (1962).

206. F. A. Carey and K. O. Hodgson, *Carbohydr. Res.*, **12**, 463 (1970).

207. R. C. Parish and L. M. Stock, *J. Org. Chem.*, **30**, 927 (1965); J. M. Tedder, *Chem. Rev.*, **55**, 787 (1955).

208. T. B. Windholz and D. B. R. Johnston, *Tetrahedron Lett.*, 2555 (1967).

266

CHAPTER 3

*Functional Group
Interconversion
by Substitution,
Including Protection and
Deprotection*

Allyl carbonate esters are also useful hydroxy-protecting groups and are introduced using allyl chloroformate. A number of Pd-based catalysts for allylic deprotection have been developed.[209] They are based on a catalytic cycle in which Pd^0 reacts by oxidative addition and activates the allylic bond to nucleophilic substitution. Various nucleophiles are effective, including dimedone,[210] pentane-2,4-dione,[211] and amines.[212]

Table 3.1 gives the structure and common abbreviation of some of the most frequently used hydroxy-protecting groups.

3.5.1.5. Protective Groups for Diols. Diols represent a special case in terms of applicable protecting groups. 1,2- and 1,3-diols easily form cyclic acetals with aldehydes and ketones, unless cyclization is precluded by molecular geometry. The isopropylidene derivatives (also called acetonides) formed by reaction with acetone are a common example.

The isopropylidene group can also be introduced by acid-catalyzed exchange with 2,2-dimethoxypropane.[213]

This acetal protective group is resistant to basic and nucleophilic reagents, but is readily removed by aqueous acid. Formaldehyde, acetaldehyde, and benzaldehyde are also used as the carbonyl component in the formation of cyclic acetals, and they function in the same manner as acetone. A disadvantage in the case of acetaldehyde and benzaldehyde is the possibility of forming a mixture of diastereomers, because of the new stereogenic center at the acetal carbon. Owing to the multiple hydroxy groups present in carbohydrates, the use of cyclic acetal protecting groups is common.

209. F. Guibe, *Tetrahedron*, **53**, 13509 (1997).
210. H. Kunz and H. Waldmann, *Angew. Chem. Int. Ed. Engl.*, **23**, 71 (1984).
211. A. De Mesmaeker, P. Hoffmann, and B. Ernst, *Tetrahedron Lett.*, **30**, 3773 (1989).
212. H. Kunz, H. Waldmann, and H. Klinkhammer, *Helv. Chim. Acta*, **71**, 1868 (1988); S. Friedrich-Bochnitschek, H. Waldman, and H. Kunz, *J. Org. Chem.*, **54**, 751 (1989); J. P. Genet, E. Blart, M. Savignac, S. Lemeune, and J.-M. Paris, *Tetrahedron Lett.*, **34**, 4189 (1993).
213. M. Tanabe and B. Bigley, *J. Am. Chem. Soc.*, **83**, 756 (1961).

Table 3.1. Common Hydroxy-Protecting Groups

Structure	Name	Abbreviation
A. Ethers		
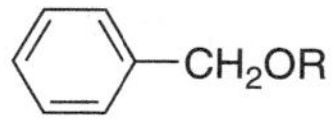 Ph–CH₂OR	Benzyl	Bn
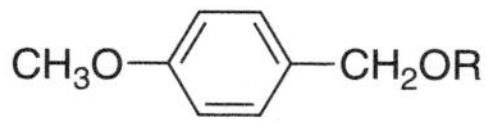 CH₃O–C₆H₄–CH₂OR	*p*-Methoxybenzyl	PMB
CH₂=CHCH₂OR	Allyl	
Ph₃COR	Triphenylmethyl (trityl)	Tr
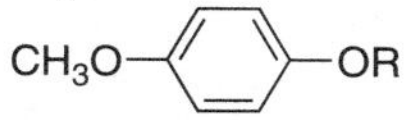 CH₃O–C₆H₄–OR	*p*-Methoxyphenyl	PMP
B. Acetals		
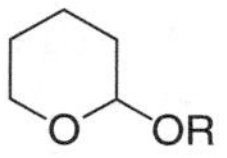 (tetrahydropyran)–OR	Tetrahydropyranyl	THP
CH₃OCH₂OR	Methoxymethyl	MOM
CH₃CH₂OCHOR / CH₃	1-Ethoxyethyl	EE
(CH₃)₂COR / OCH₃	2-Methoxy-2-propyl	MOP
Cl₃CCH₂OCH₂OR	2,2,2-Trichloroethoxymethyl	
CH₃OCH₂CH₂OCH₂OR	2-Methoxyethoxymethyl	MEM
(CH₃)₃SiCH₂CH₂OCH₂OR	2-Trimethylsilylethoxymethyl	SEM
CH₃SCH₂OR	Methylthiomethyl	MTM
C. Silyl ethers		
(CH₃)₃SiOR	Trimethylsilyl	TMS
(C₂H₅)₃SiOR	Triethylsilyl	TES
[(CH₃)₂CH]₃OR	Tri-*i*-propylsilyl	TIPS
Ph₃SiOR	Triphenylsilyl	TPS
(CH₃)₃CSi(CH₃)₂SiOR	*t*-Butyldimethylsilyl	TBDMS
(CH₃)₃CSi(Ph)₂SiOR	*t*-Butyldiphenylsilyl	TBDPS
D. Esters		
CH₃CO₂R	Acetate	Ac
PhCO₂R	Benzoate	Bz
(CH₃)₃CO₂R	Pivalate	Piv
CH₂=CHCH₂O₂COR	Allyl carbonate	
Cl₃CCH₂O₂COR	2,2,2-Trichloroethyl carbonate	Troc
(CH₃)₃SiCH₂CH₂O₂COR	2-Trimethylsilylethyl carbonate	

Cyclic carbonate esters are easily prepared from 1,2- and 1,3-diols. These are commonly prepared by reaction with N,N'-carbonyldiimidazole[214] or by transesterification with diethyl carbonate.

3.5.2. Amino-Protecting Groups

Amines are nucleophilic and easily oxidized. Primary and secondary amino groups are also sufficiently acidic that they are deprotonated by many organometallic reagents. If these types of reactivity are problematic, the amino group must be protected. The

---

[214] J. P. Kutney and A. H. Ratcliffe, *Synth. Commun.*, 547 (1975).

268

CHAPTER 3

*Functional Group
Interconversion
by Substitution,
Including Protection and
Deprotection*

most general way of masking nucleophilicity is by acylations, and carbamates are particularly useful. A most effective group for this purpose is the carbobenzyloxy (Cbz) group,[215] which is introduced by acylation of the amino group using benzyl chloroformate. The amine can be regenerated from a Cbz derivative by hydrogenolysis of the benzyl C–O bond, which is accompanied by spontaneous decarboxylation of the resulting carbamic acid.

$$\text{Ph—CH}_2\text{OCNR}_2 \xrightarrow[\text{cat}]{\text{H}_2} \left[\text{HOCNR}_2 \right] \longrightarrow \text{CO}_2 \ + \ \text{HNR}_2$$
$$+ \text{ toluene}$$

In addition to standard catalytic hydrogenolysis, methods for transfer hydrogenolysis using hydrogen donors such as ammonium formate or formic acid with Pd-C catalyst are available.[216] The Cbz group also can be removed by a combination of a Lewis acid and a nucleophile: for example, boron trifluoride in conjunction with dimethyl sulfide or ethyl sulfide.[217]

The *t*-butoxycarbonyl (*t*Boc) group is another valuable amino-protecting group. The removal in this case is done with an acid such as trifluoroacetic acid or *p*-toluenesulfonic acid.[218] *t*-Butoxycarbonyl groups are introduced by reaction of amines with *t*-butylpyrocarbonate or a mixed carbonate-imidate ester known as "BOC-ON."[219]

$$(\text{CH}_3)_3\text{COCOCOC(CH}_3)_3$$

t-butyl pyrocarbonate

$$(\text{CH}_3)_3\text{COCON}=\text{CPh}$$

"BOC – ON"

2-(*t*-butoxycarbonyloxyimino)-
2-phenylacetonitrile

Another carbamate protecting group is 2,2,2-trichloroethyloxycarbonyl, known as Troc. 2,2,2-Trichloroethylcarbamates can be reductively cleaved by zinc.[220]

Allyl carbamates also can serve as amino-protecting groups. The allyloxy group is removed by Pd-catalyzed reduction or nucleophilic substitution. These reactions involve formation of the carbamic acid by oxidative addition to the palladium. The allyl-palladium species is reductively cleaved by stannanes,[221] phenylsilane,[222] formic acid,[223] and NaBH$_4$,[224] which convert the allyl group to propene. Reagents

215. W. H. Hartung and R. Simonoff, *Org. React.*, **7**, 263 (1953).

216. S. Ram and L. D. Spicer, *Tetrahedron Lett.*, **28**, 515 (1987); B. El Amin, G. Anantharamaiah, G. Royer, and G. Means, *J. Org. Chem.*, **44**, 3442 (1979).

217. I. M. Sanchez, F. J. Lopez, J. J. Soria, M. I. Larraza, and H. J. Flores, *J. Am. Chem. Soc.*, **105**, 7640 (1983); D. S. Bose and D. E. Thurston, *Tetrahedron Lett.*, **31**, 6903 (1990).

218. E. Wunsch, *Methoden der Organischen Chemie*, Vol. 15, 4th Edition, Thieme, Stuttgart, 1975.

219. O. Keller, W. Keller, G. van Look, and G. Wersin, *Org. Synth.*, **63**, 160 (1984); W. J. Paleveda, F. W. Holly, and D. F. Weber, *Org. Synth.*, **63**, 171 (1984).

220. G. Just and K. Grozinger, *Synthesis*, 457 (1976).

221. O. Dangles, F. Guibe, G. Balavoine, S. Lavielle, and A. Marquet, *J. Org. Chem.*, **52**, 4984 (1987).

222. M. Dessolin, M.-G. Guillerez, N. T. Thieriet, F. Guibe, and A. Loffet, *Tetrahedron Lett.*, **36**, 5741 (1995).

223. I. Minami, Y. Ohashi, I. Shimizu, and J. Tsuji, *Tetrahedron Lett.*, **26**, 2449 (1985); Y. Hayakawa, S. Wakabashi, H. Kato, and R. Noyori, *J. Am. Chem. Soc.*, **112**, 1691 (1990).

224. R. Beugelmans, L. Neville, M. Bois-Choussy, J. Chastanet, and J. Zhu, *Tetrahedron Lett.*, **36**, 3129 (1995).

used for nucleophilic cleavage include N,N'-dimethylbarbituric acid,[225] and silylating agents, including TMS-N$_3$/NH$_4$F,[226] TMSN(Me)$_2$,[227] and TMSN(CH$_3$)COCF$_3$.[219] The silylated nucleophiles trap the deallylated product prior to hydrolytic workup.

Allyl groups attached directly to amine or amide nitrogen can be removed by isomerization and hydrolysis. [228] These reactions are analogous to those used to cleave allylic ethers (see p. 266). Catalysts that have been found to be effective include Wilkinson's catalyst,[229] other rhodium catalysts,[230] and iron pentacarbonyl.[45] Treatment of N-allyl amines with Pd(PPh$_3$)$_4$ and N,N'-dimethylbarbituric acid also cleaves the allyl group.[231]

Sometimes it is useful to be able to remove a protecting group by photolysis. 2-Nitrobenzyl carbamates meet this requirement. The photoexcited nitro group abstracts a hydrogen from the benzylic position, which is then converted to a α-hydroxybenzyl carbamate that readily hydrolyzes.[232]

N-Benzyl groups can be removed from tertiary amines by reaction with chloroformates. This can be a useful method for protective group manipulation if the resulting carbamate is also easily cleaved. A particularly effective reagent is α-chloroethyl chloroformate, which can be removed by subsequent solvolysis,[233] and it has been used to remove methyl and ethyl groups. These reactions are related to ether cleavage by acylation reagents (see Section 3.3).

Simple amides are satisfactory protecting groups only if the rest of the molecule can resist the vigorous acidic or alkaline hydrolysis necessary for removal. For this

225. P. Braun, H. Waldmann, W. Vogt, and H. Kunz, *Synlett*, 105 (1990).
226. G. Shapiro and D. Buechler, *Tetrahedron Lett.*, **35**, 5421 (1994).
227. A. Merzouk, F. Guibe, and A. Loffet, *Tetrahedron Lett.*, **33**, 477 (1992).
228. I. Minami, M. Yuhara, and J. Tsuji, *Tetrahedron Lett.*, **28**, 2737 (1987); M. Sakaitani, N. Kurokawa, and Y. Ohfune, *Tetrahedron Lett.*, **27**, 3753 (1986).
229. B. C. Laguzza and B. Ganem, *Tetrahedron Lett.*, **22**, 1483 (1981).
230. J. K. Stille and Y. Becker, *J. Org. Chem.*, **45**, 2139 (1980); R. J. Sundberg, G. S. Hamilton, and J. P. Laurino, *J. Org. Chem.*, **53**, 976 (1988).
231. F. Garro-Helion, A. Merzouk, and F. Guibe, *J. Org. Chem.*, **52**, 6109 (1993).
232. J. F. Cameron and J. M. J. Frechet, *J. Am. Chem. Soc.*, **113**, 4303 (1991).
233. R. A. Olofson, J. T. Martz, J.-P. Senet, M. Piteau, and T. Malfroot, *J. Org. Chem.*, **49**, 2081 (1984).

270

CHAPTER 3

*Functional Group
Interconversion
by Substitution,
Including Protection and
Deprotection*

reason, only amides that can be removed under mild conditions are useful as amino-protecting groups. Phthalimides, which are used to protect primary amino groups, can be cleaved by treatment with hydrazine, as in the Gabriel synthesis of amines (see Section 3.2.4). This reaction proceeds by initial nucleophilic addition at an imide carbonyl, followed by an intramolecular acyl transfer.

A similar sequence that takes place under milder conditions uses 4-nitrophthalimides as the protecting group and N-methylhydrazine for deprotection.[234] Reduction by $NaBH_4$ in aqueous ethanol is an alternative method for deprotection of phthalimides. This reaction involves formation of an o-(hydroxymethyl)benzamide in the reduction step. Intramolecular displacement of the amino group follows.[235]

Owing to the strong EWG effect of the trifluoromethyl group, trifluoroacetamides are subject to hydrolysis under mild conditions. This has permitted trifluoroacetyl groups to be used as amino-protecting groups in some situations. For example, the amino group was protected by trifluoroacetylation during BBr_3 demethylation of **2**.

Ref. 236

Amides can also be deacylated by partial reduction. If the reduction proceeds only to the carbinolamine stage, hydrolysis can liberate the deprotected amine. Trichloroacetamides are readily cleaved by sodium borohydride in alcohols by this mechanism.[237] Benzamides, and probably other simple amides, can be removed by careful partial reduction with diisobutylaluminum hydride (see Section 5.3.1.1).[238]

234. H. Tsubouchi, K. Tsuji, and H. Ishikawa, *Synlett*, 63 (1994).
235. J. O. Osborn, M. G. Martin, and B. Ganem, *Tetrahedron Lett.*, **25**, 2093 (1984).
236. Y.-P. Pang and A. P. Kozikowski, *J. Org. Chem.*, **56**, 4499 (1991).
237. F. Weygand and E. Frauendorfer, *Chem. Ber.*, **103**, 2437 (1970).
238. J. Gutzwiller and M. Uskokovic, *J. Am. Chem. Soc.*, **92**, 204 (1970); K. Psotta and A. Wiechers, *Tetrahedron*, **35**, 255 (1979).

The 4-pentenoyl group is easily removed from amides by I_2 and can be used as a protecting group. The mechanism of cleavage involves iodocyclization and hydrolysis of the resulting iminolactone (see Section 4.2.1).[239]

$$RNCCH_2CH_2CH{=}CH_2 \xrightarrow{I_2} RN{=}\!\!\langle\!\!\langle O \rangle\!\!\rangle\!\!-CH_2I \xrightarrow{H_2O} RNH_2$$

Sulfonamides are very difficult to hydrolyze. However, a photoactivated reductive method for desulfonylation has been developed.[240] Sodium borohydride is used in conjunction with 1,2- or 1,4-dimethoxybenzene or 1,5-dimethoxynaphthalene. The photoexcited aromatic serves as an electron donor toward the sulfonyl group, which then fragments to give the deprotected amine. The $NaBH_4$ reduces the radical cation and the sulfonyl radical.

$$R_2NSO_2Ar + CH_3O{-}\!\!\langle \rangle\!\!-OCH_3 \xrightarrow{h\nu} R_2N^- + CH_3O{-}\!\!\langle \overset{+\cdot}{~} \rangle\!\!-OCH_3 + ArSO_2{\cdot}$$

Table 3.2 summarizes the common amine-protecting groups. Reagents that permit protection of primary amino groups as cyclic *bis*-silyl derivatives have been developed. Anilines, for example, can be converted to disilazolidines.[241] These groups are stable to a number of reaction conditions, including generation and reaction of organometallic reagents.[242] They are readily removed by hydrolysis.

$$ArNH_2 + (CH_3)_2SiCH_2CH_2Si(CH_3)_2 \xrightarrow[100°C]{CsF, HMPA} Ar{-}N\big(Si(CH_3)_2CH_2CH_2Si(CH_3)_2\big)$$

$$ArNH_2 + \begin{array}{c}(CH_3)_2SiH\\(CH_3)_2SiH\end{array} \xrightarrow{(PPh_3)_3RhCl} Ar{-}N\big(Si(CH_3)_2{-}C_6H_4{-}Si(CH_3)_2\big)$$

Amide nitrogens can be protected by 4-methoxy or 2,4-dimethoxyphenyl groups. The protecting group can be removed by oxidation with ceric ammonium nitrate.[243] 2,4-Dimethoxybenzyl groups can be removed using anhydrous trifluoroacetic acid.[244]

239. R. Madsen, C. Roberts, and B. Fraser-Reid, *J. Org. Chem.*, **60**, 7920 (1995).
240. T. Hamada, A. Nishida, and O. Yonemitsu, *Heterocycles*, **12**, 647 (1979); T. Hamada, A. Nishida, Y. Matsumoto, and O. Yonemitsu, *J. Am. Chem. Soc.*, **102**, 3978 (1980).
241. R. P. Bonar-Law, A. P. Davis, and B. J. Dorgan, *Tetrahedron Lett.*, **31**, 6721 (1990); R. P. Bonar-Law, A. P. Davis, B. J. Dorgan, M. T. Reetz, and A. Wehrsig, *Tetrahedron Lett.*, **31**, 6725 (1990); S. Djuric, J. Venit, and P. Magnus, *Tetrahedron Lett.*, **22**, 1787 (1981); T. L. Guggenheim, *Tetrahedron Lett.*, **25**, 1253 (1984); A. P. Davis and P. J. Gallagher, *Tetrahedron Lett.*, **36**, 3269 (1995).
242. R. P. Bonar-Law, A. P. Davis, and J. P. Dorgan, *Tetrahedron*, **49**, 9855 (1993); K. C. Grega, M. R. Barbachyn, S. J. Brickner, and S. A. Mizsak, *J. Org. Chem.*, **60**, 5255 (1995).
243. M. Yamaura, T. Suzuki, H. Hashimoto, J. Yoshimura, T. Okamoto, and C. Shin, *Bull. Chem. Soc. Jpn.*, **58**, 1413 (1985); R. M. Williams, R. W. Armstrong, and J.-S. Dung, *J. Med. Chem.*, **28**, 733 (1985).
244. R. H. Schlessinger, G. R. Bebernitz, P. Lin, and A. J. Pos, *J. Am. Chem. Soc.*, **107**, 1777 (1985); P. DeShong, S. Ramesh, V. Elango, and J. J. Perez, *J. Am. Chem. Soc.*, **107**, 5219 (1985).

Table 3.2. Common Amine-Protecting Groups

Structure	Name	Abbreviation
A. Carbamates		
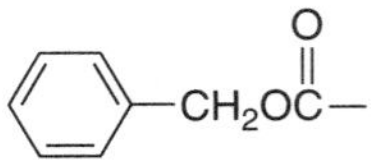 (benzyl)–CH$_2$OC(=O)–	Carbobenzyloxy (Benzyloxycarbonyl)	Cbz
(CH$_3$)$_3$COC(=O)–	*t*-Butoxycarbonyl	*t*-Boc
CH$_2$=CHCH$_2$OC(=O)–	Allyloxycarbonyl	
Cl$_3$CCH$_2$OC(=O)–	Trichloroethoxycarbonyl	Troc
B. *N*-Substituents		
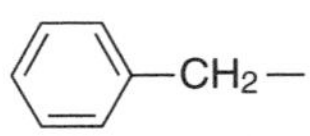 (phenyl)–CH$_2$–	Benzyl	Bn
CH$_2$=CHCH$_2$–	Allyl	
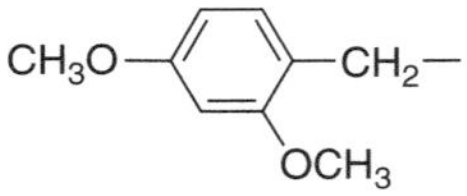 CH$_3$O–(aryl)–CH$_2$– with OCH$_3$	2,4-Dimethoxybenzyl	DMB
C. Amides and Imides		
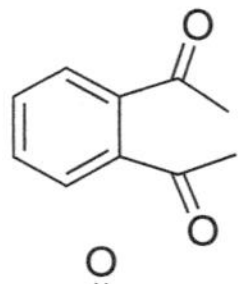 (phthaloyl structure)	Phthaloyl	Phthal
CF$_3$C(=O)–	Trifluoroacetyl	
CH$_2$=CHCH$_2$CH$_2$C(=O)–	4-Pentenoyl	

3.5.3. Carbonyl-Protecting Groups

Conversion to acetals is a very general method for protecting aldehydes and ketones against nucleophilic addition or reduction.[245] Ethylene glycol, which gives a cyclic dioxolane derivative, is frequently employed for this purpose. The dioxolanes are usually prepared by heating a carbonyl compound with ethylene glycol in the presence of an acid catalyst, with provision for azeotropic removal of water.

[245.] A. R. Hajipour, S. Khoee, and A. E. Ruoho, *Org. Prep. Proced. Int.*, **35**, 527 (2003).

$$\underset{\underset{RCR'}{\overset{O}{\parallel}}}{} + HOCH_2CH_2OH \xrightarrow{H^+} \underset{R'}{\overset{R}{>}}\!C\!\underset{O-CH_2}{\overset{O-CH_2}{<}}\!\!\Big| + H_2O$$

Scandium triflate is also an effective catalyst for dioxolane formation.[246]

Dimethyl or diethyl acetals can be prepared by acid-catalyzed exchange with an acetal such as 2,2-dimethoxypropane or an orthoester.[247]

$$\underset{\underset{RCR'}{\overset{O}{\parallel}}}{} + (CH_3O)_2C(CH_3)_2 \xrightarrow{H^+} R\!-\!\underset{\underset{OCH_3}{\overset{OCH_3}{|}}}{C}\!-\!R' + (CH_3)_2C\!=\!O$$

$$\underset{\underset{RCR'}{\overset{O}{\parallel}}}{} + HC(OCH_3)_3 \xrightarrow{H^+} R\!-\!\underset{\underset{OCH_3}{\overset{OCH_3}{|}}}{C}\!-\!R' + HCO_2CH_3$$

Acetals can be prepared under very mild conditions by reaction of the carbonyl compound with a trimethylsilyl ether, using trimethylsilyl trifluoromethylsulfonate as the catalyst.[248]

$$R_2C\!=\!O + 2\,R'OSi(CH_3)_3 \xrightarrow{Me_3SiO_3SCF_3} R_2C(OR')_2 + (CH_3)_3SiOSi(CH_3)_3$$

The carbonyl group can be deprotected by acid-catalyzed hydrolysis by the general mechanism for acetal hydrolysis (see Part A, Section 7.1). A number of Lewis acids have also been used to remove acetal protective groups. Hydrolysis is promoted by $LiBF_4$ in acetonitrile.[249] Bismuth triflate promotes hydrolysis of dimethoxy, diethoxy, and dioxolane acetals.[250] The dimethyl and diethyl acetals are cleaved by 0.1–1.0 mol % of catalyst in aqueous THF at room temperature, whereas dioxolanes require reflux. Bismuth nitrate also catalyzes acetal hydrolysis.[251]

If the carbonyl group must be regenerated under nonhydrolytic conditions, β-halo alcohols such as 3-bromopropane-1,2-diol or 2,2,2-trichloroethanol can be used for acetal formation. These groups can be removed by reduction with zinc, which leads to β-elimination.

$$BrCH_2\!-\!\underset{O}{\overset{O}{\diagdown\!\diagup}}\!\!\underset{R}{\overset{R}{\diagup}} \xrightarrow{Zn} R_2C\!=\!O + HOCH_2CH\!=\!CH_2$$

Ref. 252

[246.] K. Ishihara, Y. Karumi, M. Kubota, and H. Yamamoto, *Synlett*, 839 (1996).

[247.] C. A. MacKenzie and J. H. Stocker, *J. Org. Chem.*, **20**, 1695 (1955); E. C. Taylor and C. S. Chiang, *Synthesis*, 467 (1977).

[248.] T. Tsunoda, M. Suzuki, and R. Noyori, *Tetrahedron Lett.*, **21**, 1357 (1980).

[249.] B. H. Lipshutz and D. F. Harvey, *Synth. Commun.*, **12**, 267 (1982).

[250.] M. D. Carrigan, D. Sarapa, R. C. Smith, L. C. Wieland, and R. S. Mohan, *J. Org. Chem.*, **67**, 1027 (2002).

[251.] N. Srivasta, S. K. Dasgupta, and B. K. Banik, *Tetrahedron Lett.*, **44**, 1191 (2003).

[252.] E. J. Corey and R. A. Ruden, *J. Org. Chem.*, **38**, 834 (1973).

274

CHAPTER 3

*Functional Group
Interconversion
by Substitution,
Including Protection and
Deprotection*

$$RC(OCH_2CCl_3)_2 \quad \xrightarrow[\text{THF}]{\text{Zn}} \quad \overset{\overset{\displaystyle O}{\|}}{RCR'} + CH_2=CCl_2$$
$$\overset{|}{R'}$$

Ref. 253

Another carbonyl-protecting group is the 1,3-oxathiolane derivative, which can be prepared by reaction with mercaptoethanol in the presence of a number of Lewis acids including BF_3[254] and $In(OTf)_3$[255] or by heating with an acid catalyst with azeotropic removal of water.[256] The 1,3-oxathiolanes are particularly useful when nonacidic conditions are required for deprotection. The 1,3-oxathiolane group can be removed by treatment with Raney nickel in alcohol, even under slightly alkaline conditions.[257] Deprotection can also be accomplished by treating with a mild halogenating agent, such as NBS,[258] tetrabutylammonium tribromide,[259] or chloramine-T.[260] These reagents oxidize the sulfur to a halosulfonium salt and activate the ring to hydrolytic cleavage.

X = Br or Cl

Dithioketals, especially the cyclic dithiolanes and dithianes, are also useful carbonyl-protecting groups.[261] These can be formed from the corresponding dithiols by Lewis acid–catalyzed reactions. The catalysts that are used include BF_3, $Mg(O_3SCF_3)_2$, $Zn(O_3SCF_3)_2$, and $LaCl_3$.[262] S-Trimethylsilyl ethers of thiols and dithiols also react with ketones to form dithioketals.[263] *Bis*-trimethylsilyl sulfate in the presence of silica also promotes formation of dithiolanes.[264] Di-*n*-butylstannyldithiolates also serve as sources of dithiolanes and dithianes. These reactions are catalyzed by di-*n*-butylstannyl ditriflate.[265]

$$R_2C=O + (n\text{-Bu})_2Sn\overset{S}{\underset{S}{\diagdown}}(CH_2)_n \quad \xrightarrow{(n\text{-Bu})_2Sn(O_3SCF_3)_2} \quad R_2C\overset{S}{\underset{S}{\diagup}}(CH_2)_n$$

The regeneration of carbonyl compounds from dithioacetals and dithiolanes is often done with reagents that oxidize or otherwise activate the sulfur as a leaving

253. J. L. Isidor and R. M. Carlson, *J. Org. Chem.*, **38**, 544 (1973).

254. G. E. Wilson, Jr., M. G. Huang, and W. W. Scholman, Jr., *J. Org. Chem.*, **33**, 2133 (1968).

255. K. Kazahaya, N. Hamada, S. Ito, and T. Sato, *Synlett*, 1535 (2002).

256. C. Djerassi and M. Gorman, *J. Am. Chem. Soc.*, **75**, 3704 (1953).

257. C. Djerassi, E. Batres, J. Romo, and G. Rosenkranz, *J. Am. Chem. Soc.*, **74**, 3634 (1952).

258. B. Karimi, H. Seradj, and M. H. Tabaei, *Synlett*, 1798 (2000).

259. E. Mondal, P. R. Sahu, G. Bose, and A. T. Khan, *Tetrahedron Lett.*, **43**, 2843 (2002).

260. D. W. Emerson and H. Wynberg, *Tetrahedron Lett.*, 3445 (1971).

261. A. K. Banerjee and M. S. Laya, *Russ. Chem. Rev.*, **69**, 947 (2000).

262. L. F. Fieser, *J. Am. Chem. Soc.*, **76**, 1945 (1954); E. J. Corey and K. Shimoji, *Tetrahedron Lett.*, **24**, 169 (1983); L. Garlaschelli and G. Vidari, *Tetrahedron Lett.*, **31**, 5815 (1990); A. T. Khan, E. Mondal, P. R. Satu, and S. Islam, *Tetrahedron Lett.*, **44**, 919 (2003).

263. D. A. Evans, L. K. Truesdale, K. G. Grimm, and S. L. Nesbitt, *J. Am. Chem. Soc.*, **99**, 5009 (1977).

264. H. K. Patney, *Tetrahedron Lett.*, **34**, 7127 (1993).

265. T. Sato, J. Otero, and H. Nozaki, *J. Org. Chem.*, **58**, 4971 (1993).

group and facilitate hydrolysis. Among the reagents that have been found effective are nitrous acid, *t*-butyl hypochlorite, $NaClO_2$, $PhI(O_2CCF_3)_2$, DDQ, $SbCl_5$, and cupric salts.[266]

3.5.4. Carboxylic Acid–Protecting Groups

If only the O–H, as opposed to the carbonyl, of a carboxyl group has to be masked, it can be readily accomplished by esterification. Alkaline hydrolysis is the usual way for regenerating the acid. *t*-Butyl esters, which are readily cleaved by acid, can be used if alkaline conditions must be avoided. 2,2,2-Trichloroethyl esters, which can be reductively cleaved with zinc, are another possibility.[267] Some esters can be cleaved by treatment with anhydrous TBAF. These reactions proceed best for esters of relatively acidic alcohols, such as 4-nitrobenzyl, 2,2,2-trichloroethyl, and cyanoethyl.[268]

The more difficult problem of protecting the carbonyl group can be accomplished by conversion to a oxazoline derivative. One example is the 4,4-dimethyl derivative, which can be prepared from the acid by reaction with 2-amino-2-methylpropanol or with 2,2-dimethylaziridine.[269]

The heterocyclic derivative successfully protects the acid from attack by Grignard or hydride-transfer reagents. The carboxylic acid group can be regenerated by acidic hydrolysis or converted to an ester by acid-catalyzed reaction with the appropriate alcohol.

Carboxylic acids can also be protected as orthoesters. Orthoesters derived from simple alcohols are very easily hydrolyzed, and the 4-methyl-2,6,7-trioxabicyclo[2.2.2]octane structure is a more useful orthoester protecting group. These

266. M. T. M. El-Wassimy, K. A. Jorgensen, and S. O. Lawesson, *J. Chem. Soc., Perkin Trans. 1*, 2201 (1983); J. Lucchetti and A. Krief, *Synth. Commun.*, **13**, 1153 (1983); G. Stork and K. Zhao, *Tetrahedron Lett.*, **30**, 287 (1989); L. Mathew and S. Sankararaman, *J. Org. Chem.*, **58**, 7576 (1993); J. M. G. Fernandez, C. O. Mellet, A. M. Marin, and J. Fuentes, *Carbohydrate Res.*, **274**, 263 (1995); K. Tanemura, H. Dohya, M. Imamura, T. Suzuki, and T. Horaguchi, *J. Chem. Soc., Perkin Trans. 1*, 453 (1996); M. Kamata, H. Otogawa, and E. Hasegawa, *Tetrahedron Lett.*, **32**, 7421 (1991); T. Ichige, A. Miyake, N. Kanoh, and M. Nakata, *Synlett*, 1686 (2004).
267. R. B. Woodward, K. Heusler, J. Gostelli, P. Naegeli, W. Oppolzer, R. Ramage, S. Ranganathan, and H. Vorbruggen, *J. Am. Chem. Soc.*, **88**, 852 (1966).
268. M. Namikoshi, B. Kundu, and K. L. Rinehart, *J. Org. Chem.*, **56**, 5464 (1991); Y. Kita, H. Maeda, F. Takahashi, S. Fukui, and T. Ogawa, *Chem. Pharm. Bull.*, **42**, 147 (1994).
269. A. I. Meyers, D. L. Temple, D. Haidukewych, and E. Mihelich, *J. Org. Chem.*, **39**, 2787 (1974).

276

CHAPTER 3

*Functional Group
Interconversion
by Substitution,
Including Protection and
Deprotection*

derivatives can be prepared by exchange with other orthoesters,[270] by reaction with iminoethers,[271] or by rearrangement of the ester derived from 3-hydroxymethyl-3-methyloxetane.[272]

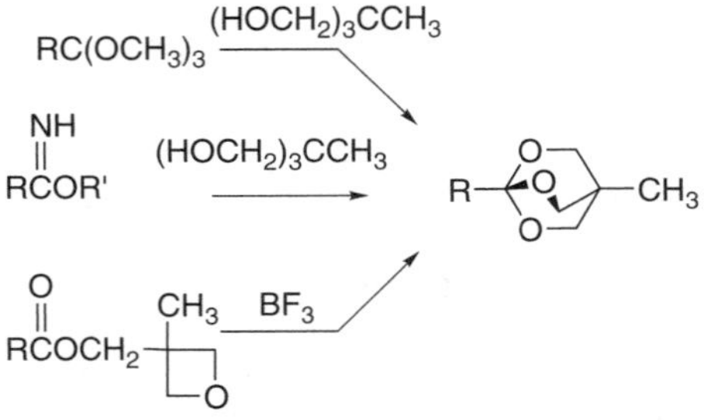

The latter method is improved by use of the 2,2-dimethyl derivative.[273] The rearrangement is faster and the stability of the orthoester to hydrolysis is better. Isotopic labeling showed that the rearrangement occurs by ionization at the tertiary position.

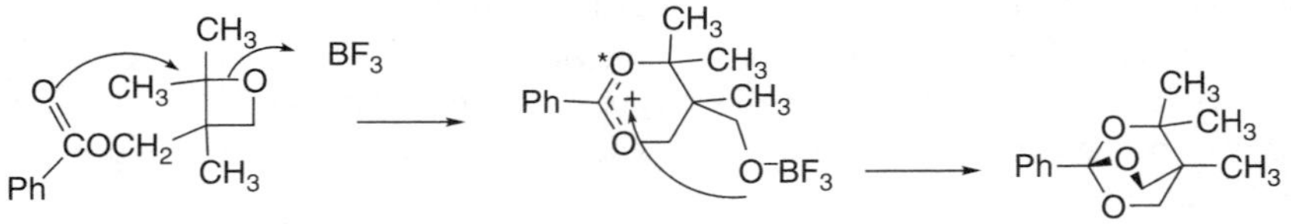

Lactones can be protected as dithiolane derivatives using a method that is analogous to ketone protection. The required reagent is readily prepared from trimethyl-aluminum and ethanedithiol.

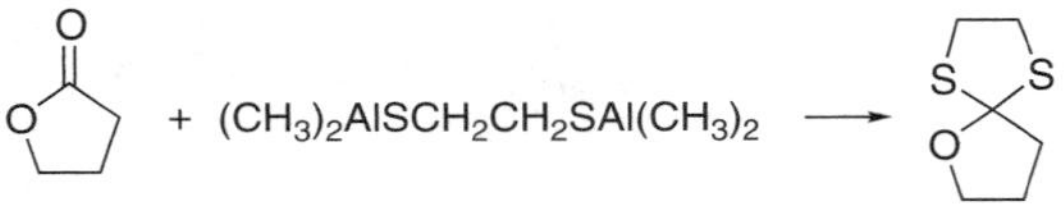

Ref. 274

Acyclic esters react with this reagent to give ketene dithio acetals.

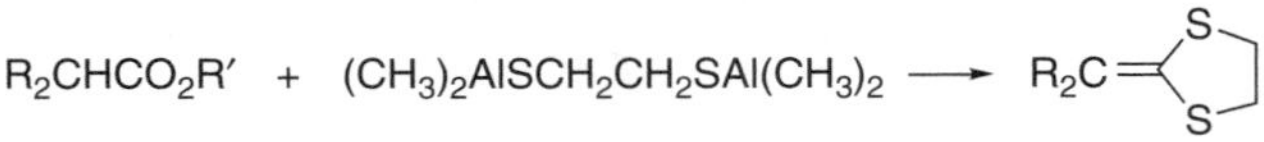

In general, the methods for protection and deprotection of carboxylic acids and esters are not as convenient as for alcohols, aldehydes, and ketones. It is therefore common to carry potential carboxylic acids through synthetic schemes in the form of protected primary alcohols or aldehydes. The carboxylic acid can then be formed at a late stage in the synthesis by an appropriate oxidation. This strategy allows one to utilize the wider variety of alcohol and aldehyde protective groups indirectly for carboxylic acid protection.

270. M. P. Atkins, B. T. Golding, D. A. Howe, and P. J. Sellers, *J. Chem. Soc., Chem. Commun.*, 207 (1980).

271. E. J. Corey and K. Shimoji, *J. Am. Chem. Soc.*, **105**, 1662 (1983).

272. E. J. Corey and N. Raju, *Tetrahedron Lett.*, **24**, 5571 (1983).

273. J.-L. Griner, *Org. Lett.*, **7**, 499 (2005).

274. E. J. Corey and D. J. Beames, *J. Am. Chem. Soc.*, **95**, 5829 (1973).

Problems

(References for these problems will be found on page 1275.)

3.1. Give the products that would be expected to be formed under the specified reaction conditions. Be sure to specify all aspects of the stereochemistry.

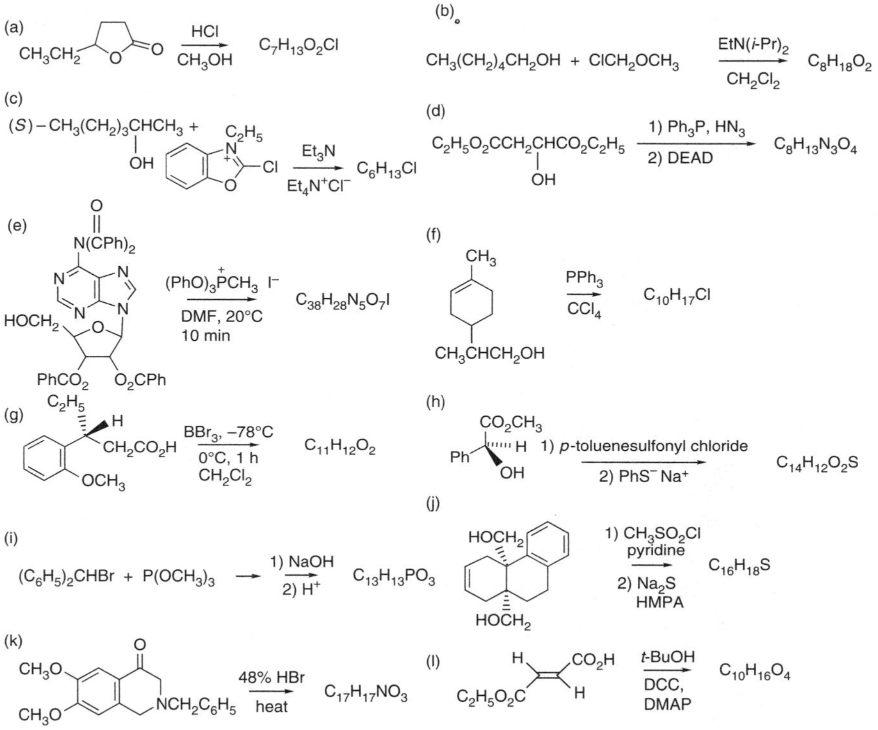

3.2. When (R)-$(-)$-5-hexen-2-ol was treated with Ph_3P in refluxing, CCl_4, $(+)$-5-chloro-1-hexene was obtained. Conversion of (R)-$(-)$-5-hexen-2-ol to its 4-bromobenzenesulfonate ester and subsequent reaction with LiCl gave $(+)$-5-chloro-1-hexene. Reaction of (S)-$(+)$-5-hexen-2-ol with PCl_5 in ether gave $(-)$-5-chloro-1-hexene.

    a. Write chemical equations for each of these reactions and specify whether each occurs with net retention or inversion of configuration.

    b. What is the sign of rotation of (R)-5-chloro-1-hexene?

3.3. A careful investigation of the extent of isomeric products formed by reaction of several alcohols with thionyl chloride has been reported. The product compositions for several of the alcohols are given below. Identify the structural features that promote isomerization and show how each of the rearranged products is formed.

278

CHAPTER 3

*Functional Group
Interconversion
by Substitution,
Including Protection and
Deprotection*

$$ROH \xrightarrow[100°C]{SOCl_2} RCl$$

R	Percent unrearranged RCl	Structure and amount of rearranged RCl
$CH_3CH_2CH_2CH_2$—	100	
$(CH_3)_2CHCH_2$—	99.7	$(CH_3)_2CHCH_3$
$(CH_3)_2CHCH_2CH_2$—	100	$\overset{|}{Cl}$ 0.3%

(The structure and amounts of rearranged RCl continue below):

For $CH_3CH_2\overset{|}{\underset{CH_3}{C}}HCH_2$— (78): $CH_3\overset{|}{\underset{Cl}{C}}HCH_2CH_2CH_3$ 1%, $CH_3CH_2\overset{|}{\underset{Cl}{C}}HCH_2CH_3$ 11%, $CH_3CH_2\overset{|}{\underset{Cl}{C}}(CH_3)_2$ 10%

For $(CH_3)_3CCH_2$— (2): $CH_3CH_2\overset{|}{\underset{Cl}{C}}(CH_3)_2$ 98%

For $CH_3CH_2CH_2\overset{|}{\underset{CH_3}{C}}H$ (98): $CH_3CH_2\overset{|}{\underset{Cl}{C}}HCH_2CH_3$ 2%

For $CH_3CH_2\overset{|}{\underset{CH_3}{C}}HCH_2CH_3$ (90): $CH_3CH_2CH_2\overset{|}{\underset{Cl}{C}}HCH_3$ 10%

For $(CH_3)_2\overset{|}{\underset{CH_3}{C}}HCHCH_3$ (5): $CH_3CH_2\overset{|}{\underset{Cl}{C}}(CH_3)_2$ 95%

3.4. Give a reaction mechanism that would explain the following observations and reactions.

a. Kinetic measurements reveal that solvolytic displacement of sulfonate is about 5×10^5 faster for **4B** than for **4A**.

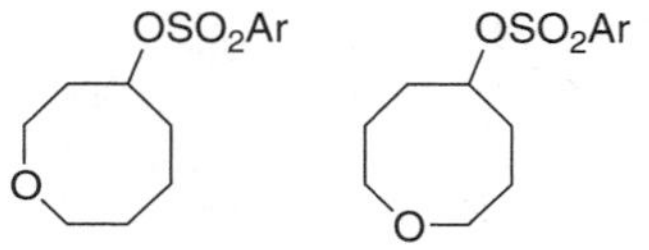

4A   **4B**

b.

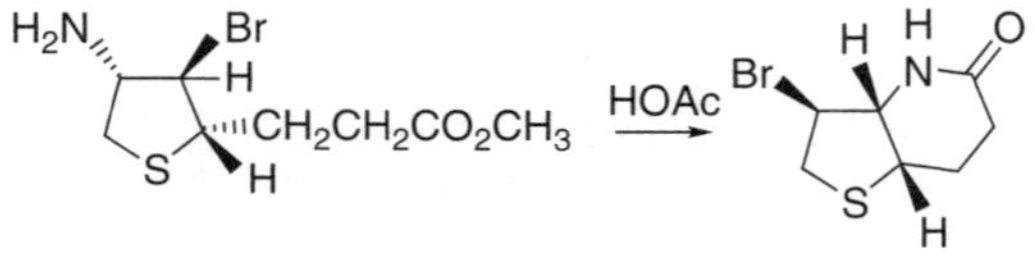

c.

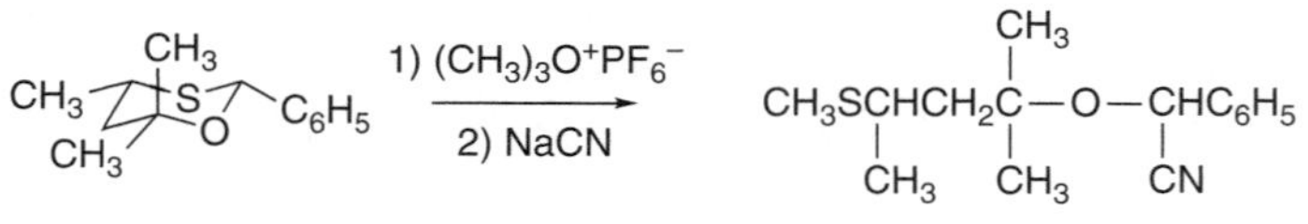

d.

e.

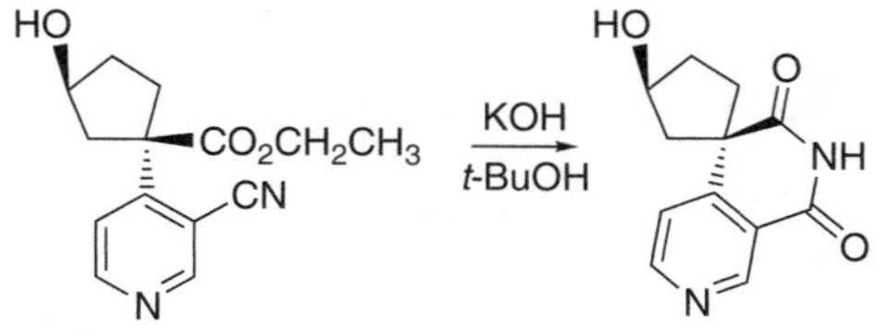

HO—⟨cyclopentane⟩ $CO_2CH_2CH_3$ / CN (pyridine with N) → KOH / *t*-BuOH → HO—⟨spiro bicyclic⟩ O / NH / O (pyridine with N)

f.

$CH_3(CH_2)_6CO_2H$ + $PhCH_2NH_2$ → [o-nitrophenyl isothiocyanate / Bu_3P, 25°C] → $CH_3(CH_2)_6\overset{\text{O}}{\overset{\|}{C}}NHCH_2Ph$ 99%

g.

⟨cyclohexane⟩ ···OH / ''CH_2NO_2 → [$EtO_2CN=NCO_2Et$ / PPh_3] → ⟨bicyclic⟩ '''NO_2   92%

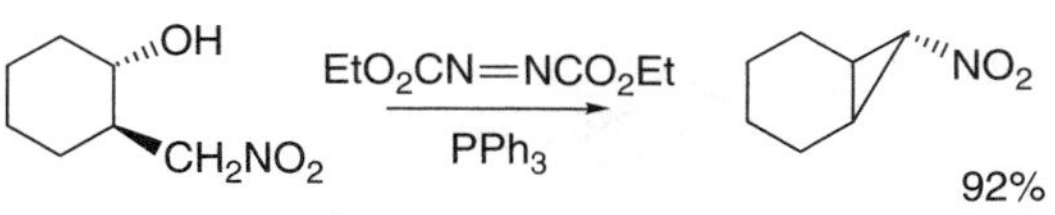

h. Both **4C** and **4D** gave the same product when subjected to Mitsunobu conditions with phenol as the nucleophile.

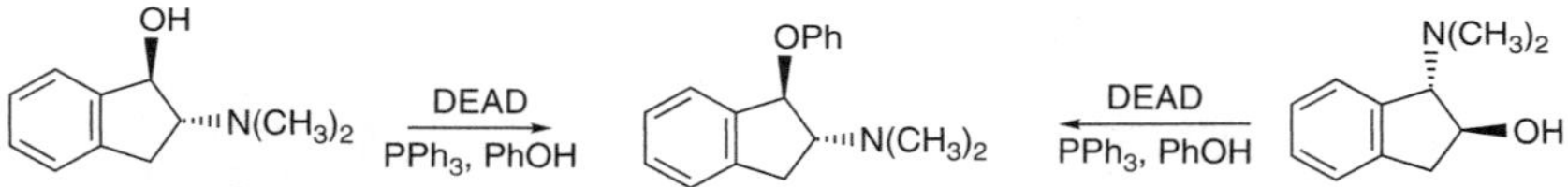

⟨indane⟩ OH / '''N(CH_3)_2   **4C** → [DEAD / PPh_3, PhOH] → ⟨indane⟩ OPh / '''N(CH_3)_2 ← [DEAD / PPh_3, PhOH] ← ⟨indane⟩ N(CH_3)_2 / OH   **4D**

3.5. Substances such as carbohydrates and amino acids as well as other small molecules available from natural sources are valuable starting materials in enantiospecific syntheses. Suggest reagents that could effect the following transformations, taking particular care to ensure that the product will be enantiomerically pure.

(a) $(CH_3)_2N\overset{\text{O}}{\overset{\|}{C}}$—⟨$\overset{H}{\underset{H}{C}}$⟩—⟨$\overset{CH_3O}{\underset{OCH_3}{C}}$⟩—$\overset{\text{O}}{\overset{\|}{C}}N(CH_3)_2$ from CH_3O_2C—⟨$\overset{H}{\underset{}{C}}OH$⟩—⟨$\overset{}{\underset{H}{C}}OH$⟩—$CO_2CH_3$

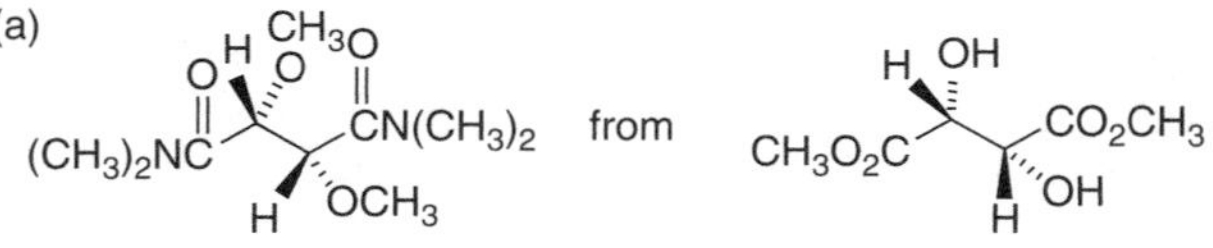

(b) CH_3O···⟨pyrrolidine⟩···OCH_3 / N / CH_3 from $HOCH_2$—⟨$\overset{H}{\underset{}{C}}OCH_3$⟩—⟨$\overset{}{\underset{H}{C}}OCH_3$⟩—$CH_2OH$

(c) CH_3—N⟨$=O$⟩—⟨CH_3O···C···OCH_3, CH_3⟩—N⟨$=O$⟩—CH_3 with O—CH_2CH_2—O bridge   from CH_3NHCH_2—⟨$\overset{H}{\underset{}{C}}OCH_3$⟩—⟨$\overset{}{\underset{H}{C}}OCH_3$⟩—$CH_2NHCH_3$

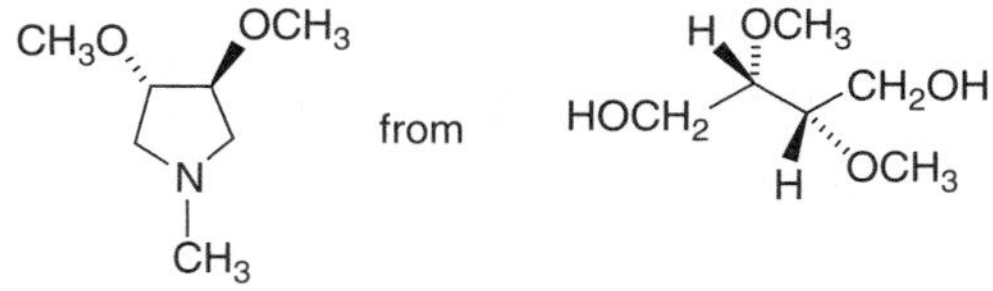

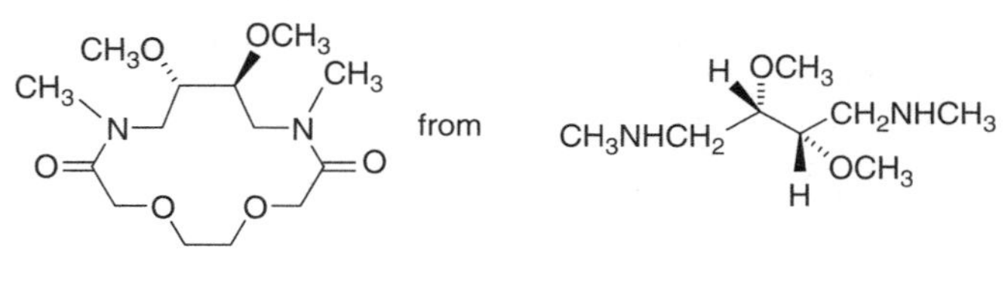

280

CHAPTER 3

*Functional Group
Interconversion
by Substitution,
Including Protection and
Deprotection*

(d)

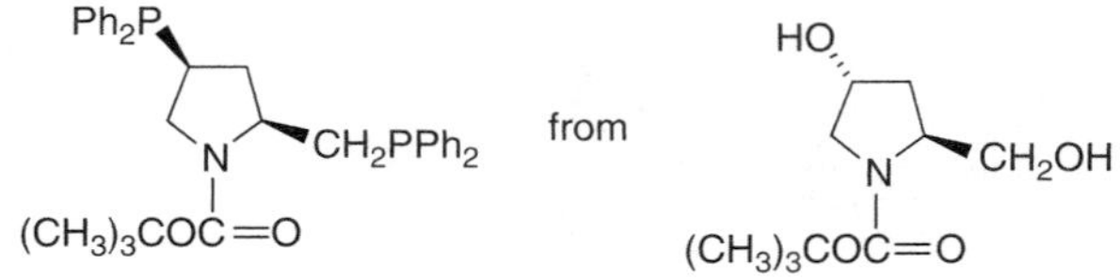

(e)

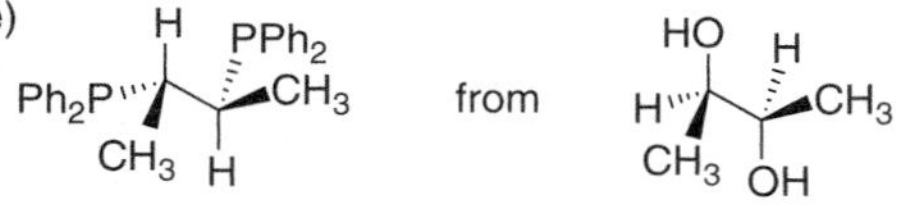

(f)

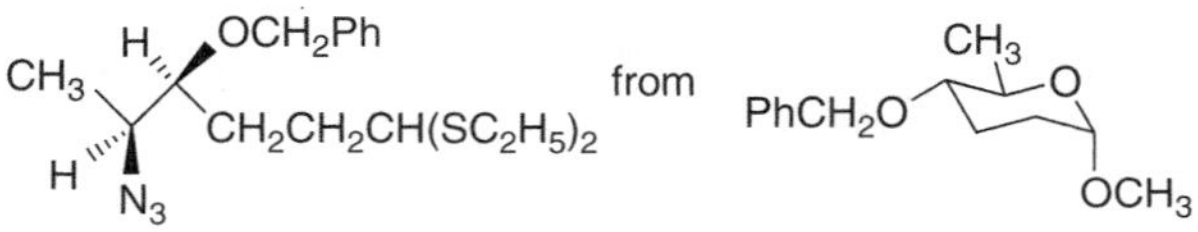

(g)

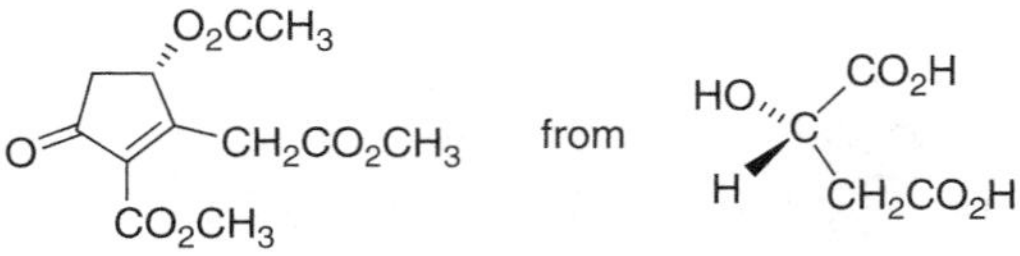

(h)

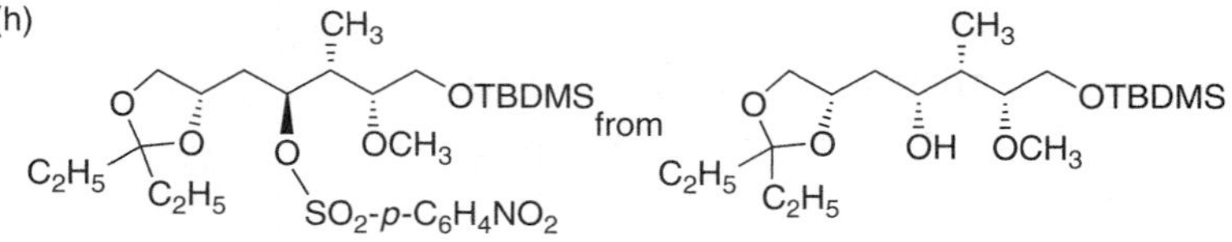

(i)

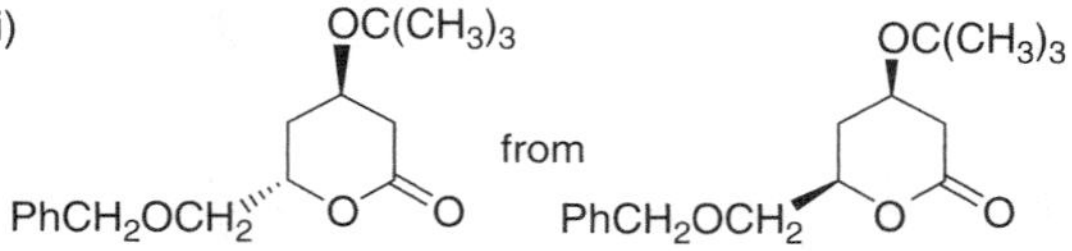

3.6. Indicate conditions that would be appropriate for the following transformations involving introduction or removal of protective groups:

(a)

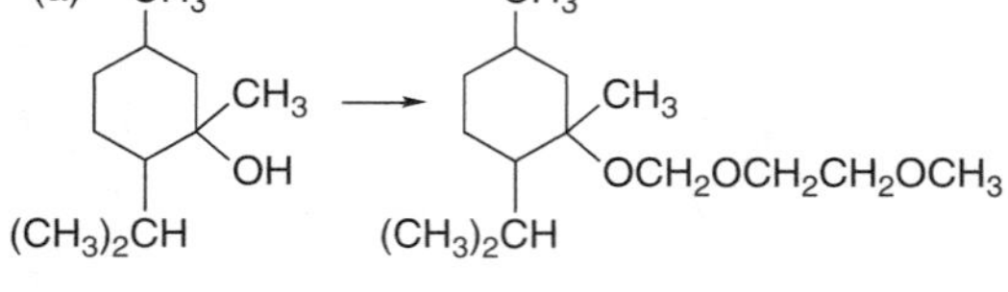

(b)

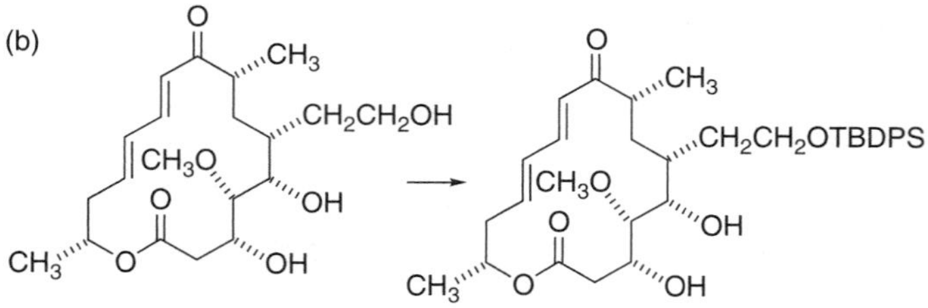

(c)

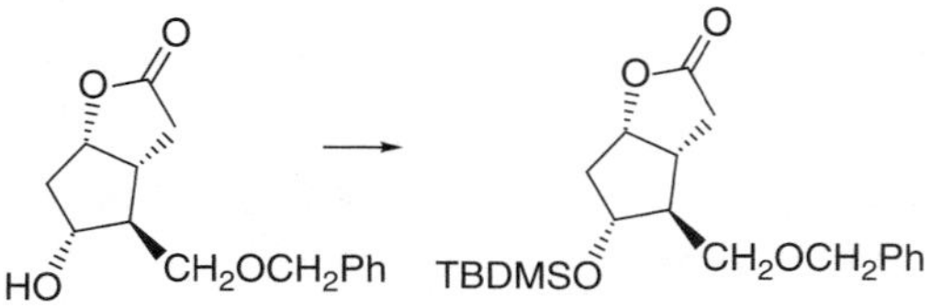

(d)

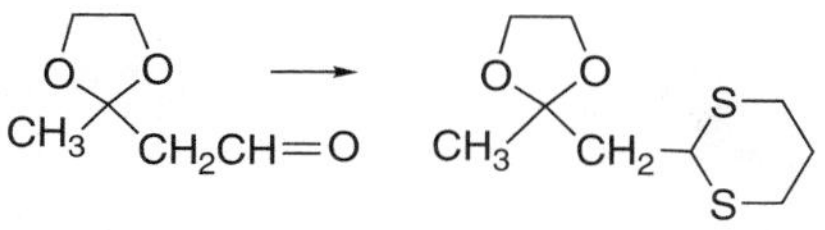

(e)

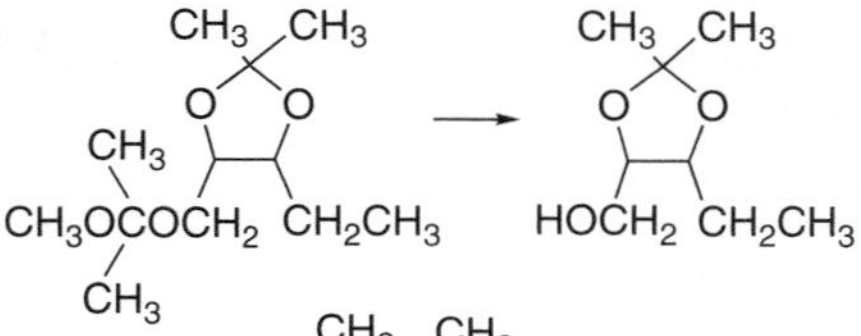

(f)

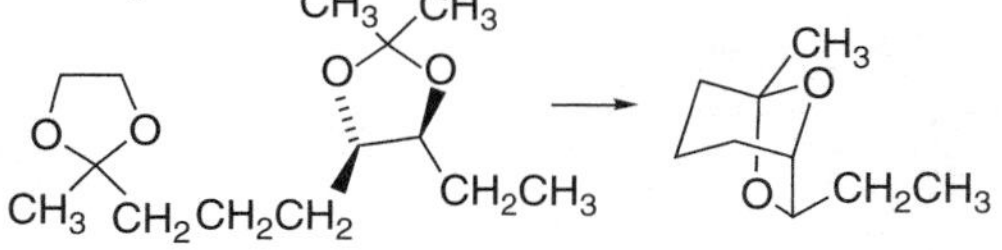

3.7. Suggest reagents and approximate reaction conditions that would effect the following conversions. Note any special features of the reactant that should be taken into account in choosing a reagent system.

(a)

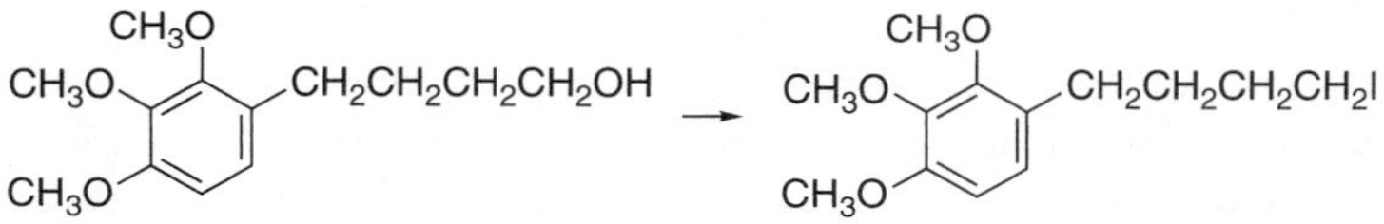

(b)

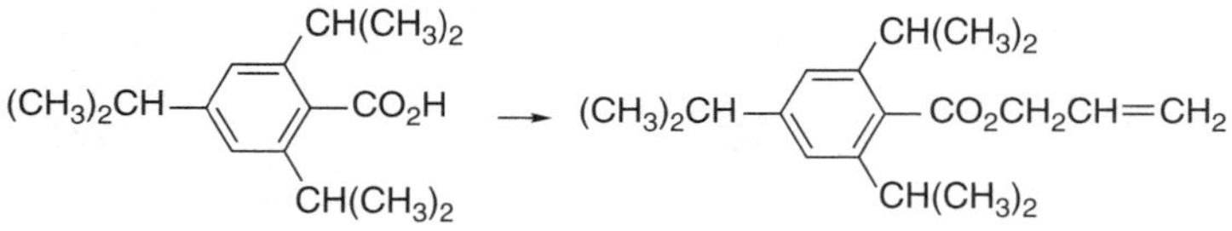

(c)

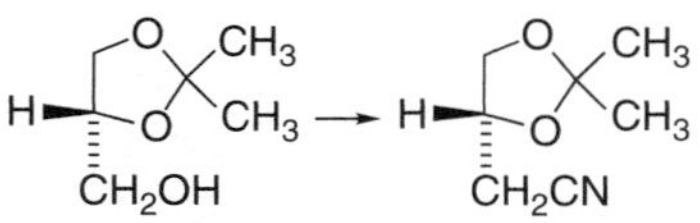

(d)

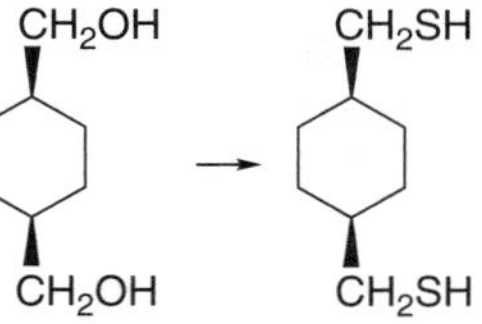

(e)

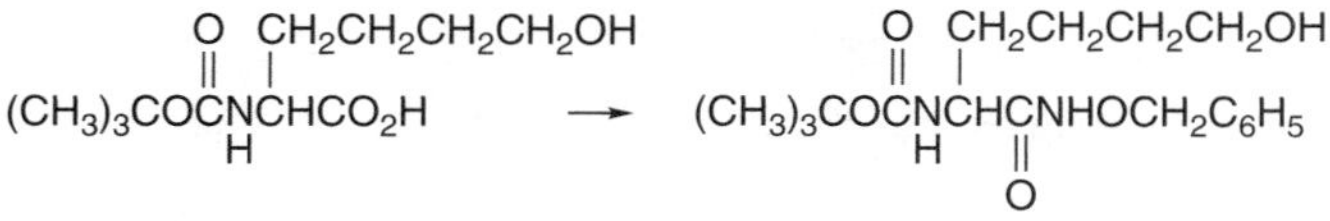

282

CHAPTER 3

*Functional Group
Interconversion
by Substitution,
Including Protection and
Deprotection*

(f)

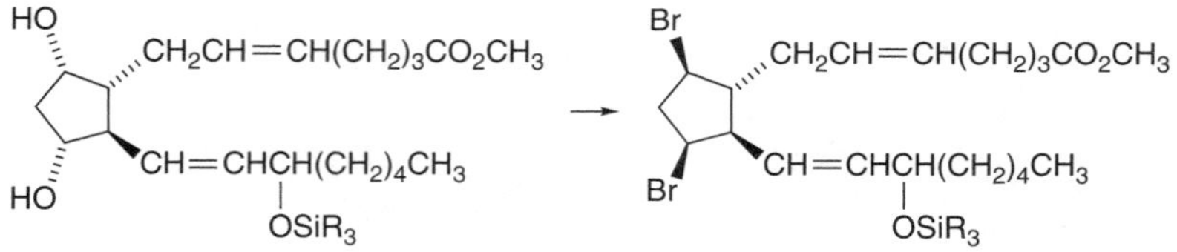

(g)  $(CH_3)_2CCH_2CHCH_3$ ⟶ $(CH_3)_2CCH_2CHCH_3$
      OH OH   Br OH

3.8. Provide a mechanistic interpretation of the following reactions and observations.

a. Show the mechanism for inversion of a hydroxyl site under the Mitsunobu conditions, as illustrated by the reaction of cholesterol.

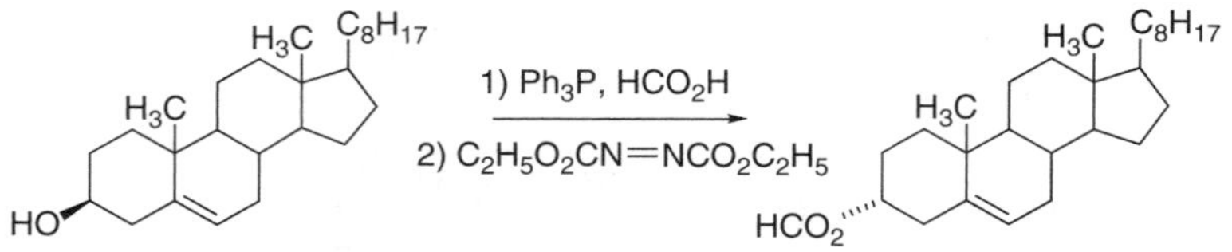

1) Ph_3P, HCO_2H

2) $C_2H_5O_2CN=NCO_2C_2H_5$

b. Triphenylphosphine oxide reacts with trifluoromethylsulfonic anhydride to give an ionic substance having the composition of a 1:1 adduct. When this substance is added to a solution containing a carboxylic acid, followed by addition of an amine, amides are formed in good yield. Similarly, esters are formed on reaction with alcohols. What is the structure of the adduct and how does it activate the carboxylic acids to nucleophilic substitution?

c. Sulfonate esters having quaternary nitrogen substituents, such as **8A** and **8B**, show high reactivity toward nucleophilic substitution. Sulfonates **8A** are comparable in reactivity to 2,2,2-trifluoroethylsulfonate in homogeneous solution and are even more reactive in two-phase solvent mixtures.

$ROSO_2CH_2CH_2\overset{+}{N}(CH_3)_3$

8A

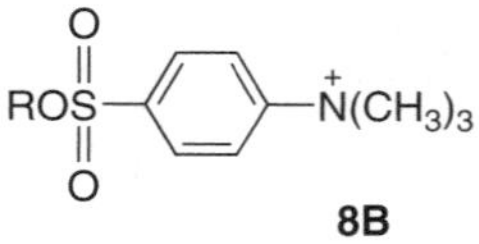

8B

d. Alcohols react with hexachloroacetone in the presence of DMF to give alkyl trichloroacetates in good yield. Primary alcohols react faster than secondary alcohols, but tertiary alcohols are unreactive under these conditions.

e. The β-hydroxy-α-amino acids serine and threonine can be converted to their respective *bis-O-t*-butyl derivatives on reaction with isobutene and H_2SO_4. Subsequent treatment with one equivalent of trimethylsilyl triflate and then water cleaves the ester group, but not the ether group. What is the basis for this selectivity?

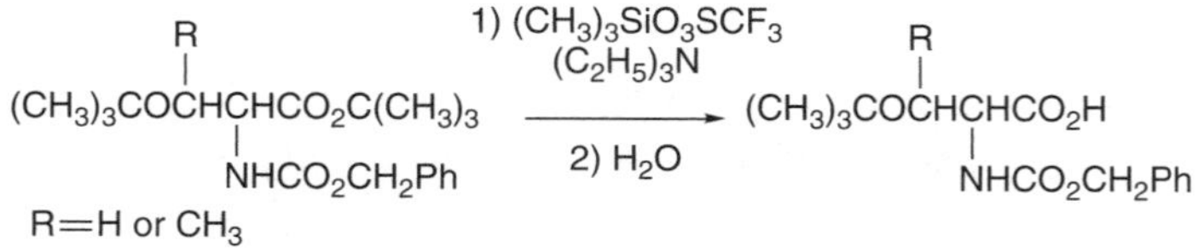

1) $(CH_3)_3SiO_3SCF_3$
   $(C_2H_5)_3N$

2) H_2O

$R=H$ or CH_3

f. 2′-Deoxyadenosine can be cleanly converted to its 5′-chloro analog by reaction with 1.5 equivalent of $SOCl_2$ in HMPA. The reaction proceeds through an intermediate of composition $C_{20}H_{22}N_{10}Cl_2O_5S$, which is converted to the product on exposure to aqueous ammonia. With larger amounts of $SOCl_2$, the 3′5′-dichloro derivative is formed.

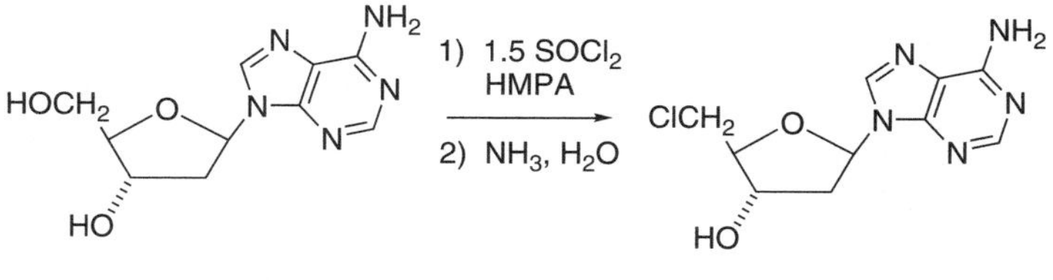

3.9. Short synthetic sequences have been used to obtain the material on the left from the starting material on the right. Suggest an appropriate method. No more than three steps should be required.

(a)

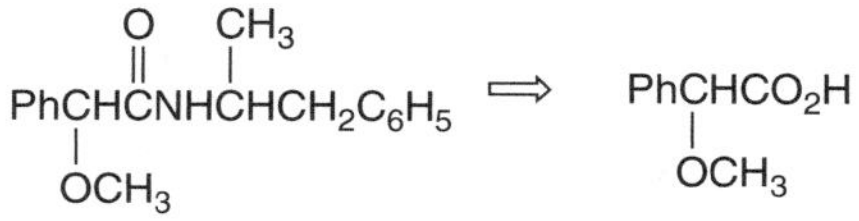

(b)

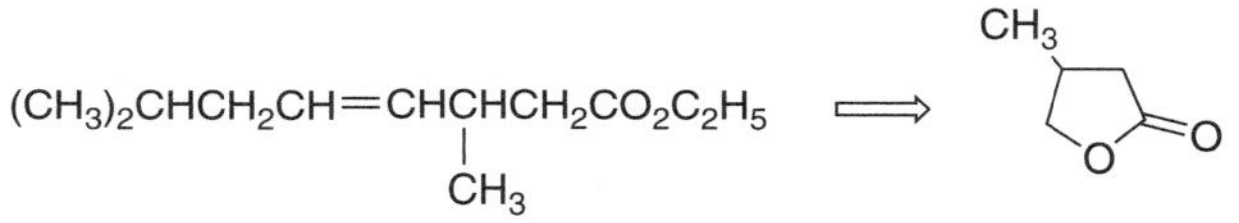

(c)

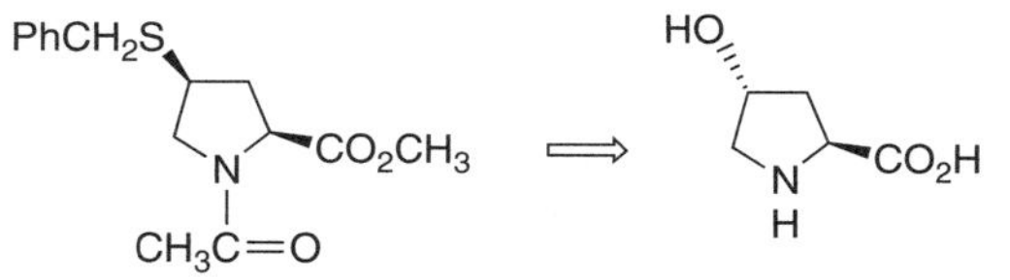

(d)

(e)

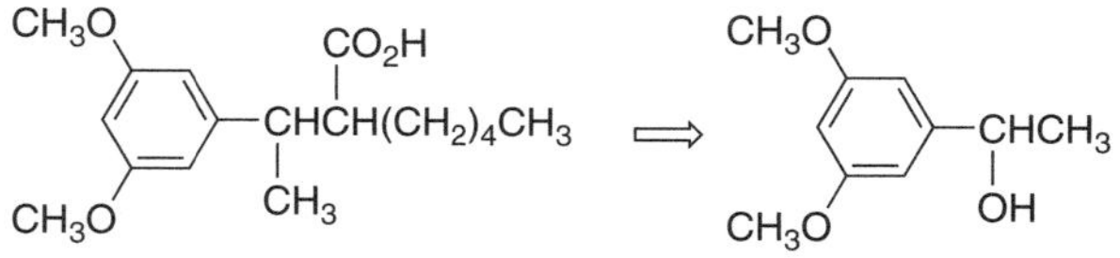

3.10. Amino acids can be converted to epoxides of high enantiomeric purity by the reaction sequence below. Analyze the stereochemistry of each step of the reaction sequence.

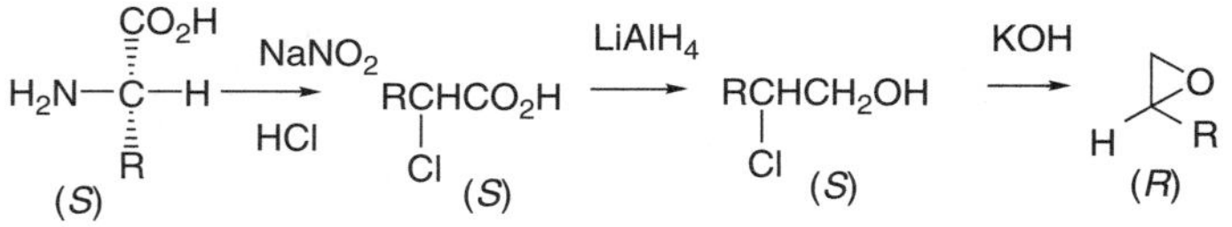

284

CHAPTER 3

*Functional Group
Interconversion
by Substitution,
Including Protection and
Deprotection*

3.11. Indicate the product to be expected under the following reaction conditions:

(a)

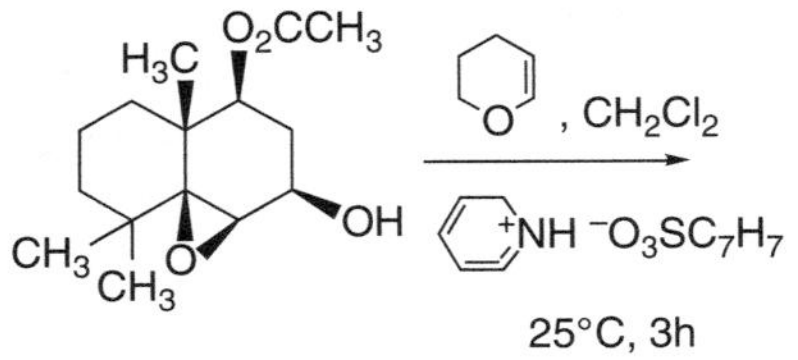

25°C, 3h

(b)

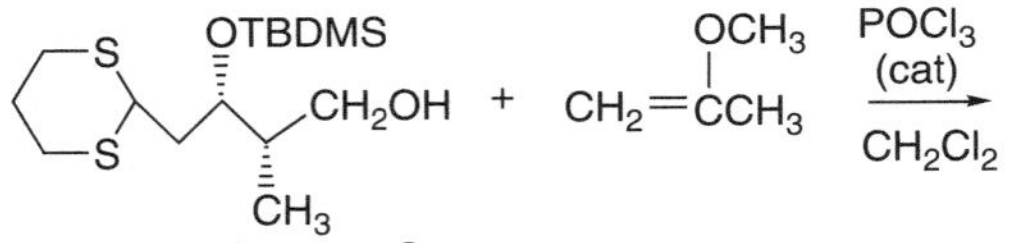

(c)

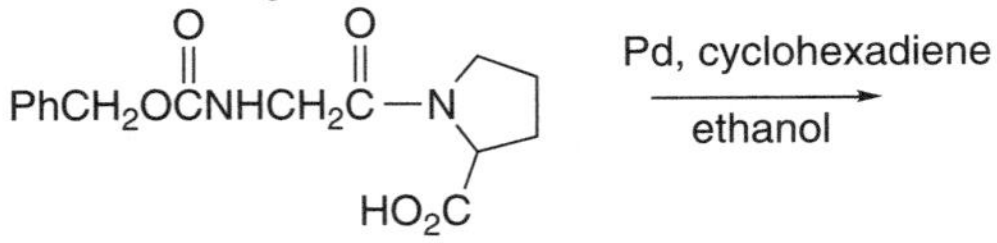

(d)

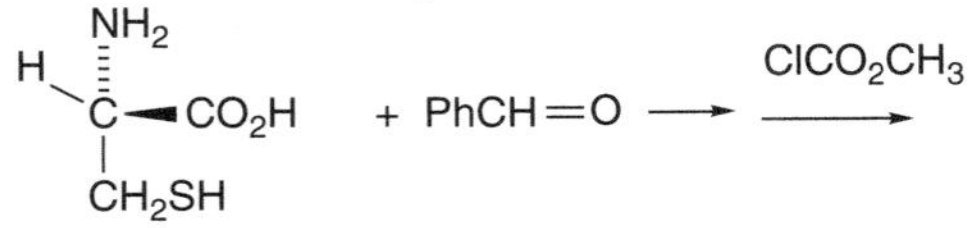

(e)

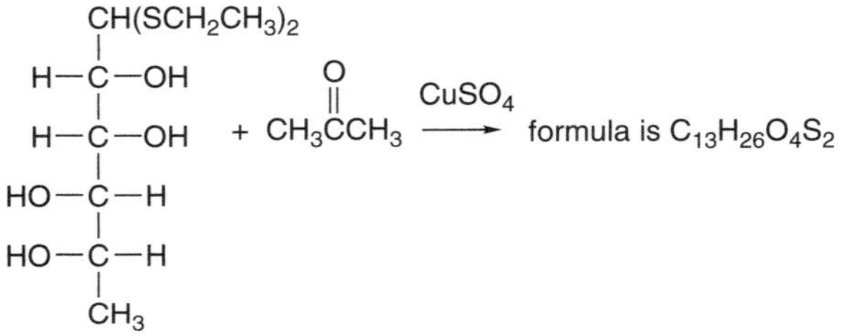

3.12. A reagent that can introduce benzyloxycarbonyl protecting groups on amino groups in nucleosides is prepared by allowing benzyl chloroformate to react first with imidazole and then with trimethyloxonium tetrafluoroborate. What is the structure of the resulting reagent (a salt) and why is it an especially reactive acylating agent?

3.13. Triphenylphosphine reacts with peroxides to give intermediates that are related to those formed in the Mitsunobu reaction. The following reactions are examples:

What properties of the intermediates in the Mitsunobu reaction are suggested by these reactions?

3.14. The scope of the reaction of $Ph_3P-Cl_3CCOCCl_3$ with allylic alcohols has been studied. Primary and some secondary alcohols, such as **14A** and **14B**, give good

yields of unrearranged allylic chlorides. The reaction also exhibits retention of
E,Z-configuration at the allylic double bonds (**14C** and **14D**). Certain other
alcohols, such as **14E** and **14F**, give more complex mixtures. What structural
features determine how cleanly the alcohol is converted to chloride? How are
these structural features related to the mechanism of the reaction?

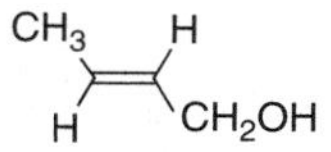

14A 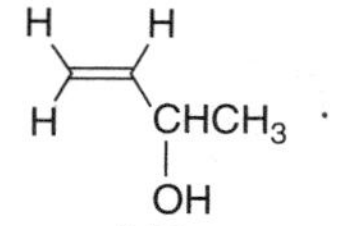   **14B** 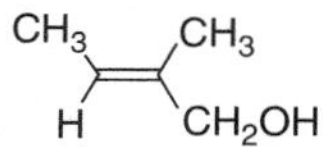   **14C** 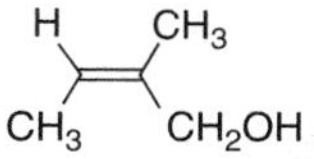   **14D**

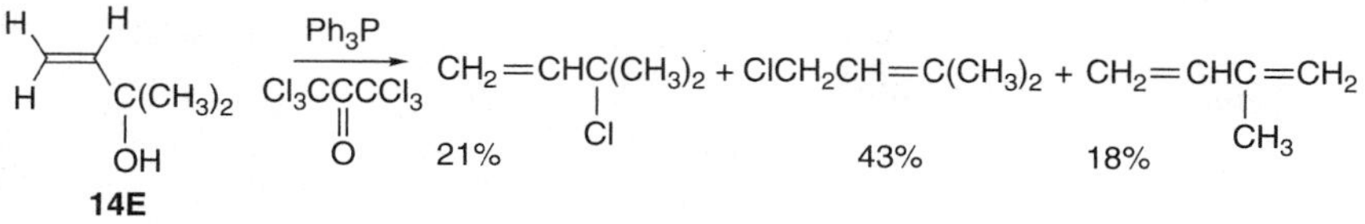

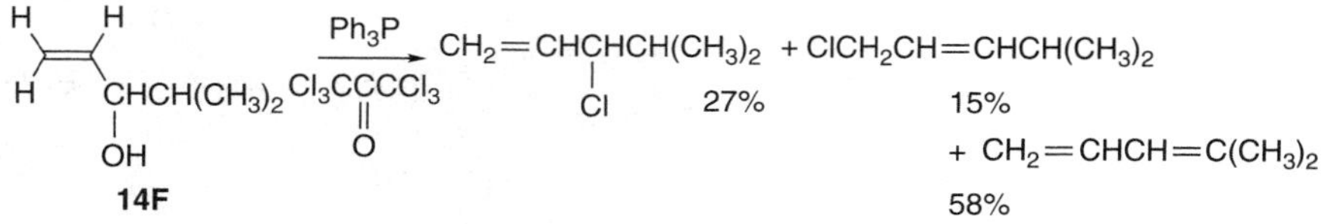

14E

14F

3.15. In each of the synthetic transformations shown, the reagents are appropriate for
the desired transformation but the reaction would not succeed as written. Suggest
a protective group strategy that would permit each transformation to be carried
out to give the desired product.

(a)

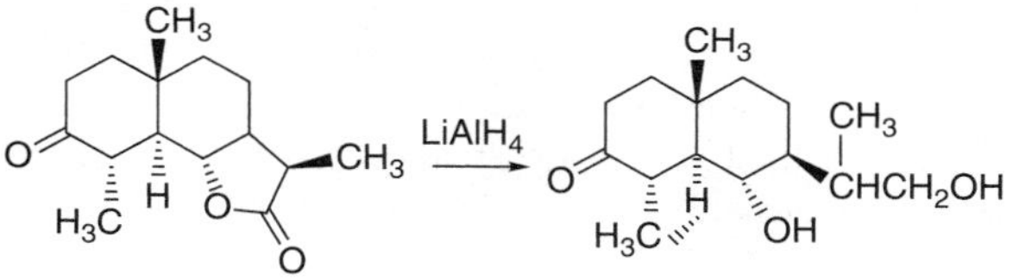

(b)

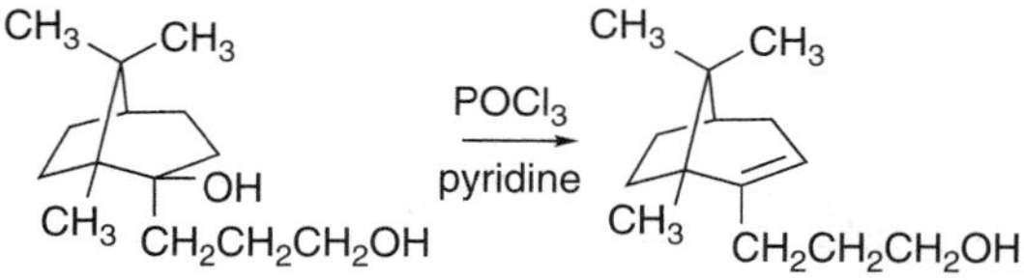

(c)

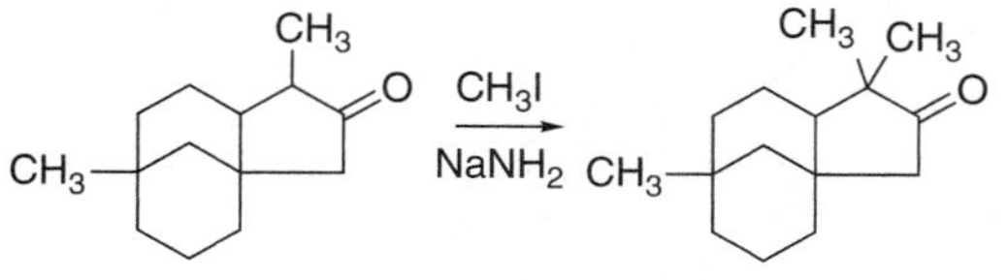

(d)

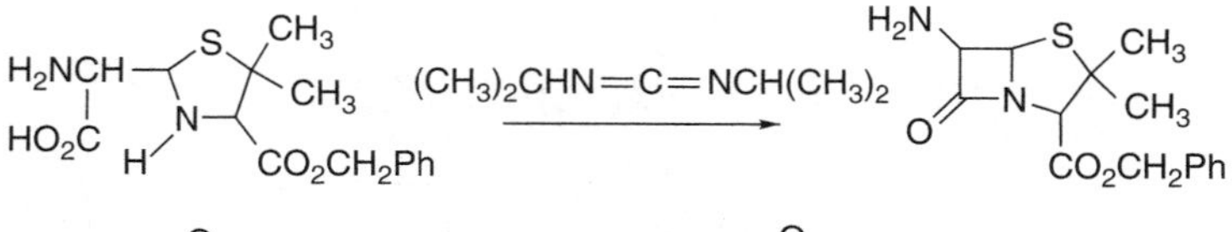

(e)

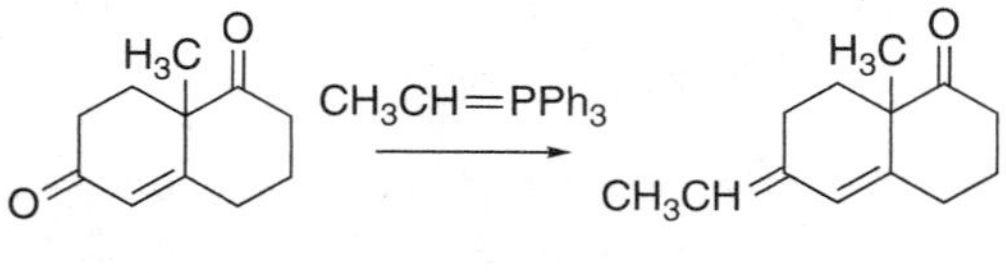

286

CHAPTER 3

*Functional Group
Interconversion
by Substitution,
Including Protection and
Deprotection*

3.16. Two heterocyclic ring systems that have found some use in the formation of amides under mild conditions are *N*-alkyl-5-arylisoxazolium salts (**16A**) and *N*-acyloxy-2-alkoxydihydroquinolines (**16B**).

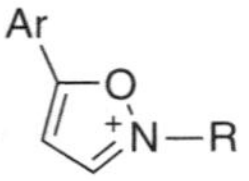

16A 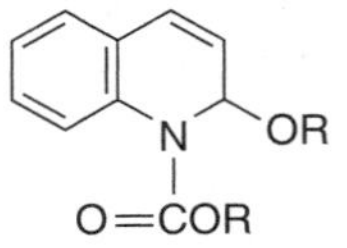   **16B**

Typical reaction conditions for these reagents are shown below. Propose mechanisms by which these heterocyclic molecules can function to activate carboxy groups under these conditions.

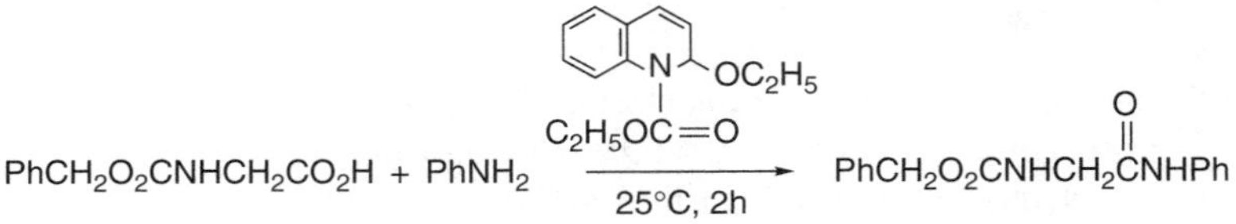

3.17. Either because of potential interference with other functional groups present in the molecule or because of special structural features, the following reactions require careful selection of reagents and reaction conditions. Identify the special requirements in each reactant and suggest appropriate reagents and reaction conditions for each transformation.

(a)

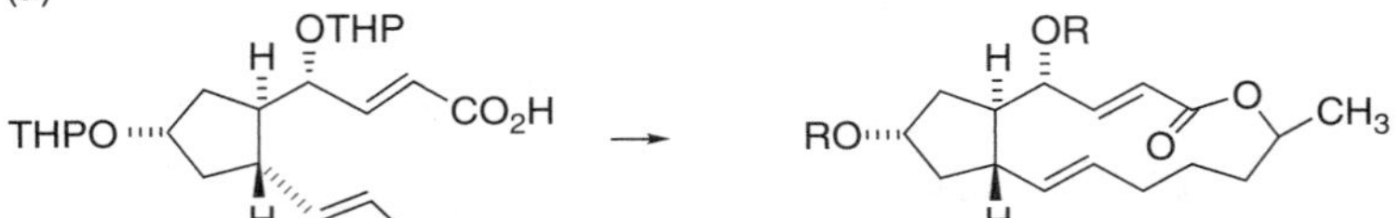

(b)

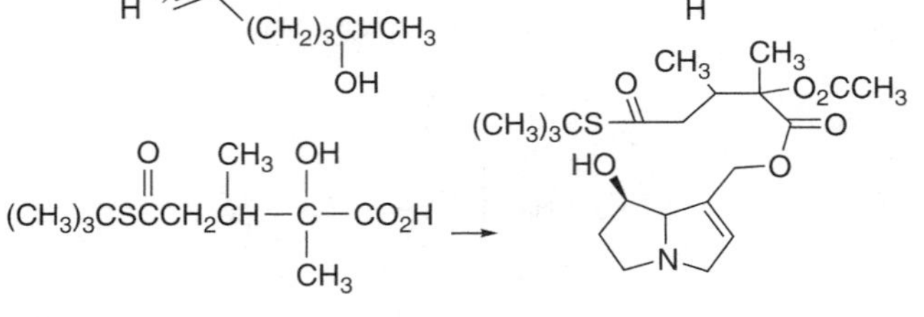

(c)

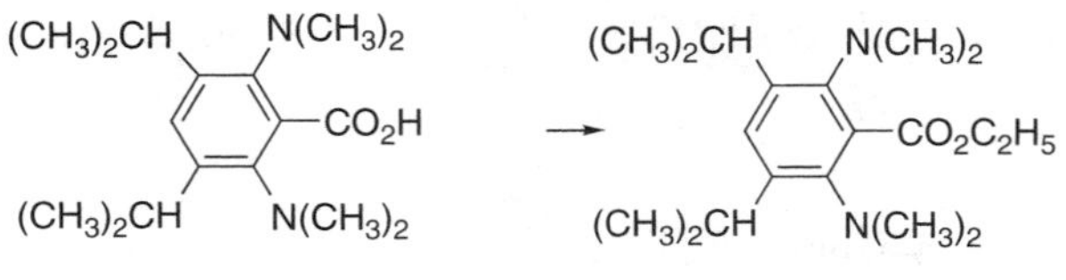

3.18. The preparation of nucleosides by reaction between carbohydrates and heterocyclic bases is fundamental to the study of the important biological activity of these substances. Several methods exist for forming the nucleoside bonds. Application of 2-chloro-3-ethylbenzoxazolium chloride to this reaction was investigated using 2,3,4,6-tetra-O-acetyl-β-D-glucopyranose. Good yields were observed and the reaction was stereospecific for the β-nucleoside. Suggest a mechanism to explain the retention of configuration.

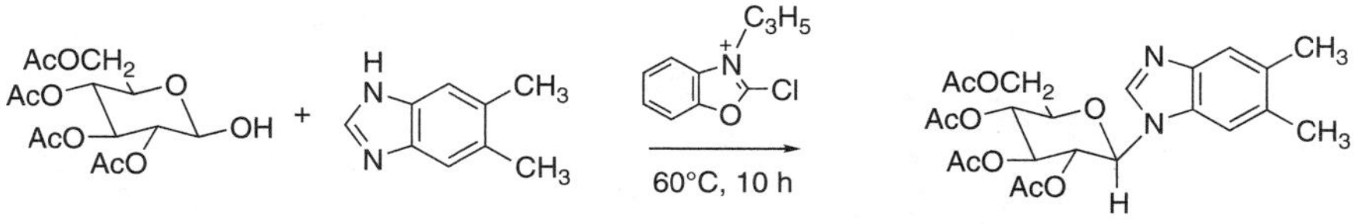

3.19. A route to α-glycosides involves treatment of a 2,3,4,6-tetra-O-benzyl-α-D-glucopyranosyl bromide with an alcohol, tetraethylammonium bromide, and diisopropylethylamine in CH_2Cl_2. Explain the stereoselectivity of this reaction.

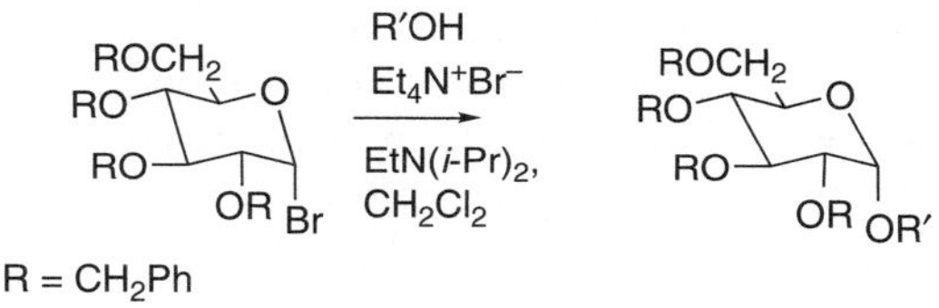

R = CH₂Ph

3.20. Write mechanisms for formation of 2-pyridylthio esters by the following methods:

(a)

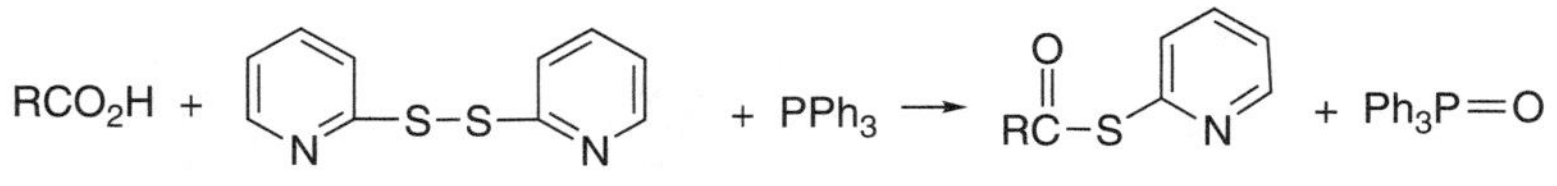

(b)

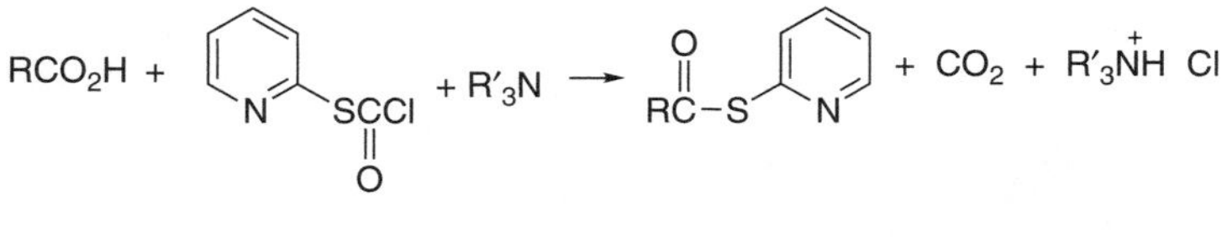

3.21. The ionophoric antibiotic nonactin is a 32-membered macrocycle that contains two units of (−)-nonactic acid and two units of (+)-nonactic acid in an alternating sequence.

a. Assuming that you have access to both (+)- and (−)-nonactic acid, devise a strategy and protecting group sequence that could provide the natural macro-molecule in high stereochemical purity.

288

CHAPTER 3

*Functional Group
Interconversion
by Substitution,
Including Protection and
Deprotection*

b. Suppose you had access to (+)-nonactic acid and the C(8) epimer of (−)-nonactic acid, how could you obtain nonactin?

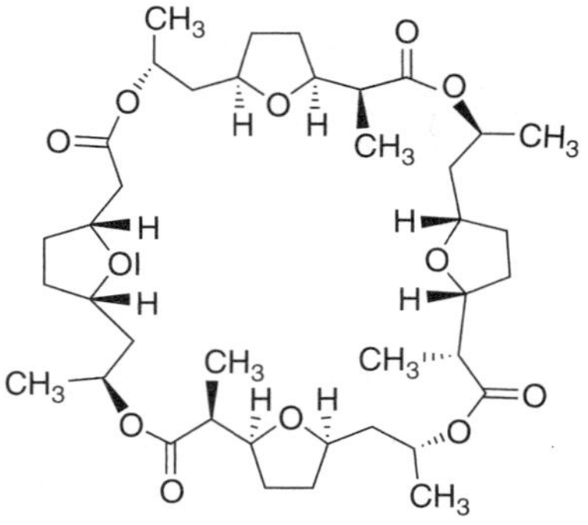

Nonactin

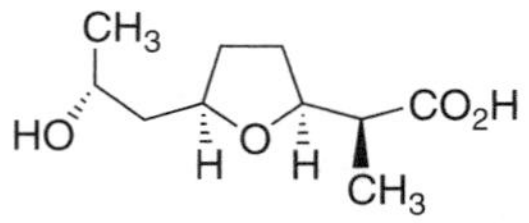

(+)-nonactic acid

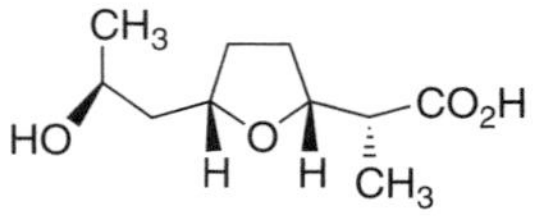

(−)-nonactic acid

3.22. Because they are readily available from natural sources in enantiomerically pure form, carbohydrates are very useful starting materials for the synthesis of other enantiomerically pure substances. However, the high number of similar functional groups present in the carbohydrates requires versatile techniques for protection and deprotection. Show how appropriate manipulation of protecting groups and other selective reactions could be employed to effect the following transformations.

(a)

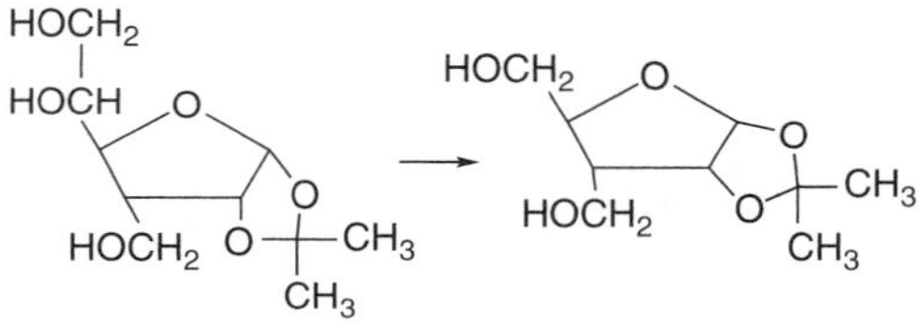

(b)

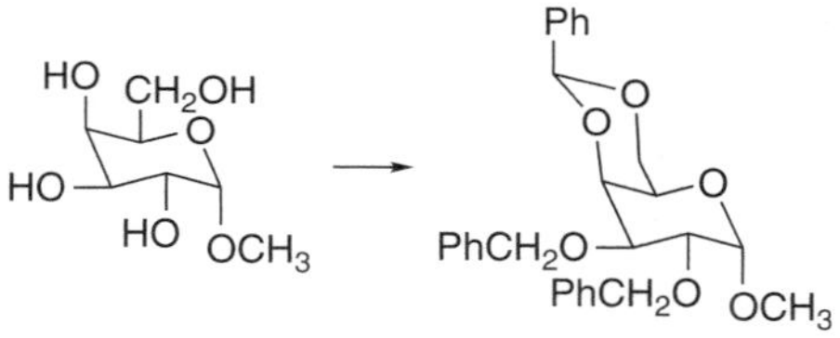

(c)

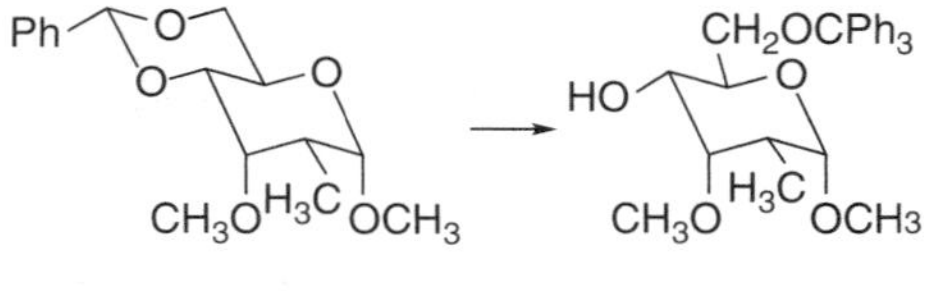

(d)

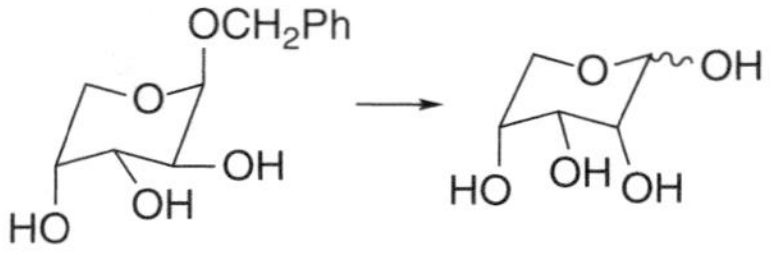

4

Electrophilic Additions to Carbon-Carbon Multiple Bonds

Introduction

Addition of electrophilic reagents is one of the most general and useful reactions of alkenes and alkynes. This chapter focuses on reactions that proceed through polar intermediates or transition structures. We discuss the fundamental mechanistic characteristics of this class of reactions in Chapter 5 of Part A, including proton-catalyzed additions of water and alcohols and the addition of hydrogen halides. Other electrophilic reagents that we consider there are the halogens and positive halogen compounds, electrophilic sulfur and selenium reagents, and mercuric salts. Hydroboration is another important type of electrophilic addition to alkenes. In the present chapter, we emphasize synthetic application of these reactions. For the most part, electrophilic additions are used to introduce functionality at double and triple bonds. When the nucleophile addition step is intramolecular, a new heterocyclic ring is formed, and this is a very useful synthetic method.

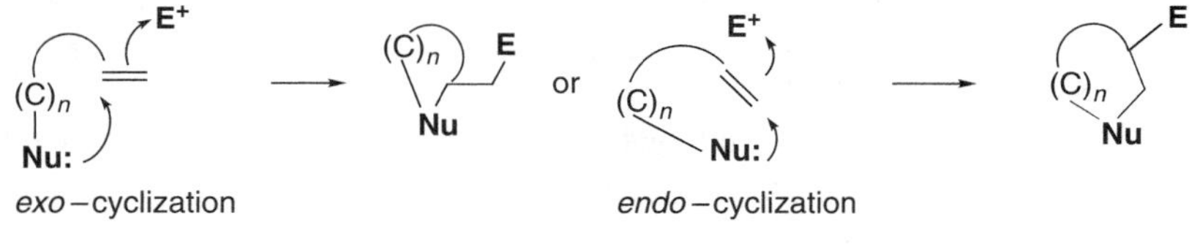

exo-cyclization *endo*-cyclization

Carbonyl compounds can react with electrophiles via their enol isomers or equivalents, and these reactions result in α-substitution.

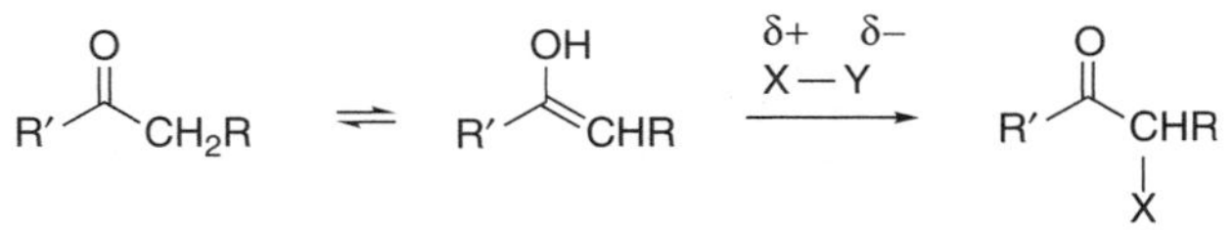

Several other types of addition reactions of alkenes are also of importance and these are discussed elsewhere. Nucleophilic additions to electrophilic alkenes are covered in Section 2.6 and cycloadditions involving concerted mechanisms are encountered in Sections 6.1 to 6.3. Free radical addition reaction are considered in Chapter 11.

4.1. Electrophilic Addition to Alkenes

4.1.1. Addition of Hydrogen Halides

Hydrogen chloride and hydrogen bromide react with alkenes to give addition products. In early work, it was observed that addition usually takes place to give the product with the halogen atom attached to the more-substituted carbon of the double bond. This behavior is sufficiently general that the name *Markovnikov's rule* was given to the statement describing this mode of addition. The term *regioselective* is used to describe addition reactions that proceed selectively in one direction with unsymmetrical alkenes.[1] A rudimentary picture of the reaction mechanism indicates the basis of Markovnikov's rule. The addition involves either protonation or a partial transfer of a proton to the double bond. The relative stability of the two possible carbocations from an unsymmetrical alkene favors formation of the more-substituted intermediate. Addition is completed when the carbocation reacts with a halide anion.

$$R_2C{=}CH_2 + HX \longrightarrow \overset{R}{\underset{R}{\overset{|}{C}}}{}^+{-}CH_3 + X^- \longrightarrow R_2\underset{\underset{X}{|}}{C}CH_3$$

Markovnikov's rule describes a specific case of regioselectivity that is based on the stabilizing effect of alkyl and aryl substituents on carbocations.

$$^+CH_2CH_2R \quad CH_3\overset{+}{C}HR \quad CH_3\overset{+}{C}HAr \quad CH_3\overset{+}{C}R_2 \quad CH_3\overset{+}{C}(Ar)_2$$
$$\longrightarrow \text{increasing stability} \longrightarrow$$

A more complete discussion of the mechanism of addition of hydrogen halides to alkenes is given in Chapter 6 of Part A. In particular, the question of whether or not discrete carbocations are involved is considered there. Even when a carbocation is not involved, the regioselectivity of electrophilic addition is the result of attack of the electrophile at the more *electron-rich* carbon of the double bond. Alkyl substituents increase the electron density of the terminal carbon by hyperconjugation (see Part A, Section 1.1.8).

Terminal and disubstituted internal alkenes react rather slowly with HCl in nonpolar solvents. The rate is greatly accelerated in the presence of silica or alumina in noncoordinating solvents such as dichloromethane or chloroform. Preparatively convenient conditions have been developed in which HCl is generated in situ from $SOCl_2$ or $(ClCO)_2$.[2] These heterogeneous reaction systems also give a Markovnikov orientation.

[1] A. Hassner, *J. Org. Chem.*, **33**, 2684 (1968).
[2] P. J. Kropp, K. A. Daus, M. W. Tubergen, K. D. Kepler, V. P. Wilson, S. L. Craig, M. M. Baillargeon, and G. W. Breton, *J. Am. Chem. Soc.*, **115**, 3071 (1993).

The mechanism is thought to involve an interaction of the silica or alumina surface with HCl that facilitates proton transfer.

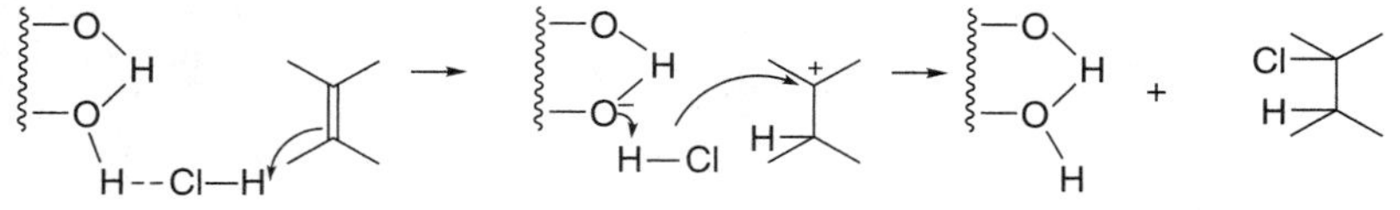

Another convenient procedure for hydrochlorination involves adding trimethylsilyl chloride to a mixture of an alkene and water. Good yields of HCl addition products (Markovnikov orientation) are formed.[3] These conditions presumably involve generation of HCl by hydrolysis of the silyl chloride, but it is uncertain if the silicon plays any further role in the reaction.

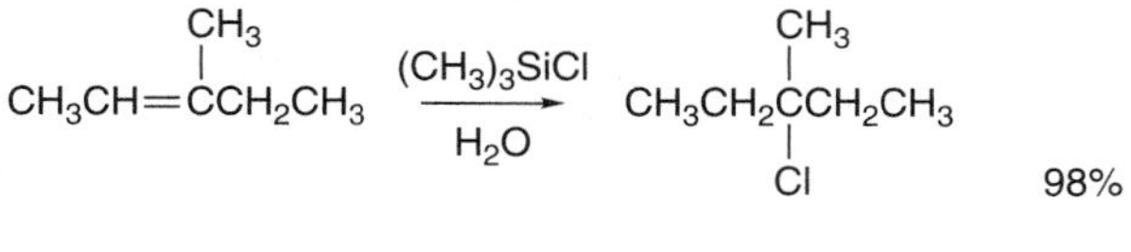

In nucleophilic solvents, products that arise from reaction of the solvent with the cationic intermediate may be formed. For example, reaction of cyclohexene with hydrogen bromide in acetic acid gives cyclohexyl acetate as well as cyclohexyl bromide. This occurs because acetic acid acts as a nucleophile in competition with the bromide ion.

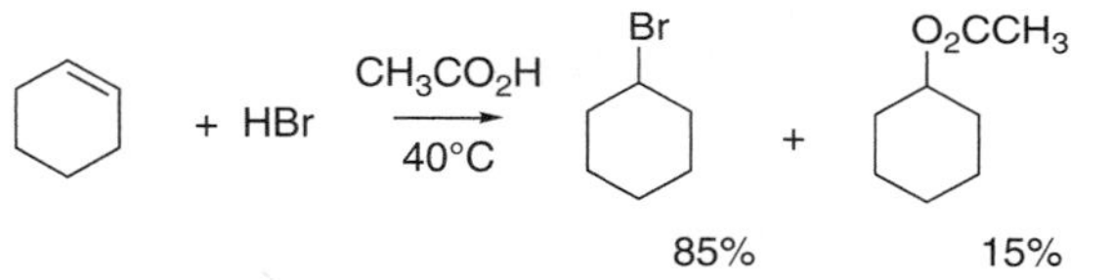

When carbocations are involved as intermediates, carbon skeleton rearrangement can occur during electrophilic addition reactions. Reaction of t-butylethylene with hydrogen chloride in acetic acid gives both rearranged and unrearranged chloride.[5]

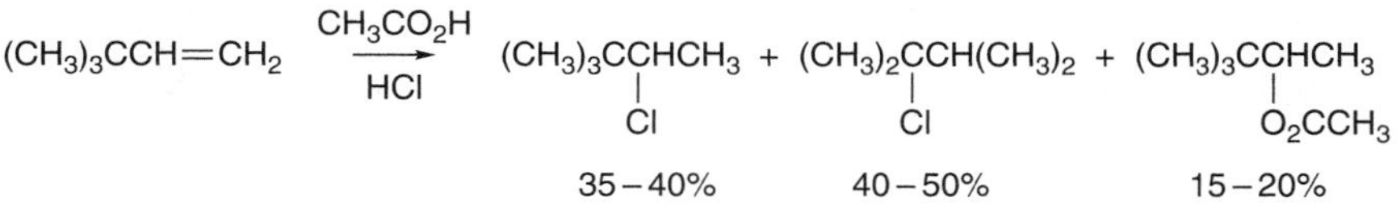

The stereochemistry of addition of hydrogen halides to alkenes depends on the structure of the alkene and also on the reaction conditions. Addition of hydrogen bromide to cyclohexene and to E- and Z-2-butene is *anti*.[6] The addition of hydrogen chloride to 1-methylcyclopentene is entirely *anti* when carried out at 25° C in nitromethane.[7]

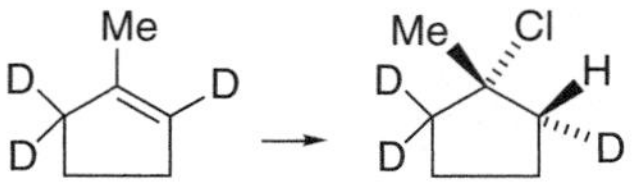

3. P. Boudjouk, B.-K. Kim, and B.-H. Han, *Synth. Commun.*, **26**, 3479 (1996); P. Boudjouk, B.-K. Kim, and B.-H. Han, *J. Chem. Ed.*, **74**, 1223 (1997).
4. R. C. Fahey and R. A. Smith, *J. Am. Chem. Soc.*, **86**, 5035 (1964).
5. R. C. Fahey and C. A. McPherson, *J. Am. Chem. Soc.*, **91**, 3865 (1969).
6. D. J. Pasto, G. R. Meyer, and S. Kang, *J. Am. Chem. Soc.*, **91**, 4205 (1969).
7. Y. Pocker and K. D. Stevens, *J. Am. Chem. Soc.*, **91**, 4205 (1969).

1,2-Dimethylcyclohexene is an example of an alkene for which the stereochemistry of hydrogen chloride addition is dependent on the solvent and temperature. At $-78°$ C in dichloromethane, 88% of the product is the result of *syn* addition, whereas at $0°$ C in ether, 95% of the product results from *anti* addition.[8] *Syn* addition is particularly common with alkenes having an aryl substituent. Table 4.1 lists several alkenes for which the stereochemistry of addition of hydrogen chloride or hydrogen bromide has been studied.

The stereochemistry of addition depends on the details of the mechanism. The addition can proceed through an ion pair intermediate formed by an initial protonation step. Most alkenes, however, react via a complex that involves the alkene, hydrogen halide, and a third species that delivers the nucleophilic halide. This termolecular mechanism is generally pictured as a nucleophilic attack on an alkene-hydrogen halide complex. This mechanism bypasses a discrete carbocation and exhibits a preference for *anti* addition.

The major factor in determining which mechanism is followed is the stability of the carbocation intermediate. Alkenes that can give rise to a particularly stable carbocation

Table 4.1. Stereochemistry of Addition of Hydrogen Halides to Alkenes

Alkene	Hydrogen halide	Stereochemistry
1,2-Dimethylcyclohexene[a]	HBr	*anti*
1,2-Dimethylcyclohexene[a]	HCl	Solvent and temperature dependent
Cyclohexene[b]	HBr	*anti*
Z-2-Butene[c]	DBr	*anti*
E-2-Butene[c]	DBr	*anti*
1-Methylcyclopentene[d]	HCl	*anti*
1,2-Dimethylcyclopentene[e]	HBr	*anti*
Norbornene[f]	HBr	*syn* and rearrangement
Norbornene[g]	HCl	*syn* and rearrangement
E-1-Phenylpropene[h]	HBr	*syn* (9:1)
Z-1-Phenylpropene[h]	HBr	*syn* (8:1)
Bicyclo[3.1.0]hex-2-ene[i]	DCl	*syn*
1-Phenyl-4-(*t*-butyl)cyclohexene[j]	DCl	*syn*

a. G. S. Hammond and T. D. Nevitt, *J. Am. Chem. Soc.*, **76**, 4121 (1954); R. C. Fahey and C. A. McPherson, *J. Am. Chem. Soc.*, **93**, 2445 (1971); K. B. Becker and C. A. Grob, *Synthesis*, 789 (1973).
b. R. C. Fahey and R. A. Smith, *J. Am. Chem. Soc.*, **86**, 5035 (1964).
c. D. J. Pasto, G. R. Meyer, and B. Lepeska, *J. Am. Chem. Soc.*, **96**, 1858 (1974).
d. Y. Pocker and K. D. Stevens, *J. Am. Chem. Soc.*, **91**, 4205 (1969).
e. G. S. Hammond and C. H. Collins, *J. Am. Chem. Soc.*, **82**, 4323 (1960).
f. H. Kwart and J. L. Nyce, *J. Am. Chem. Soc.*, **86**, 2601 (1964).
g. J. K. Stille, F. M. Sonnenberg, and T. H. Kinstle, *J. Am. Chem. Soc.*, **88**, 4922 (1966).
h. M. J. S. Dewar and R. C. Fahey, *J. Am. Chem. Soc.*, **85**, 3645 (1963).
i. P. K. Freeman, F. A. Raymond, and M. F. Grostic, *J. Org. Chem.*, **32**, 24 (1967).
j. K. D. Berlin, R. O. Lyerla, D. E. Gibbs, and J. P. Devlin, *J. Chem. Soc., Chem. Commun.*, 1246 (1970).

8. K. B. Becker and C. A. Grob, *Synthesis*, 789 (1973).

are likely to react via the ion pair mechanism, which is not necessarily stereospecific, as the carbocation intermediate permits loss of stereochemistry relative to the reactant alkene. It might be expected that the ion pair mechanism would lead to a preference for *syn* addition, since at the instant of formation of the ion pair, the halide is on the same side of the alkene as the proton being added. Rapid collapse of the ion pair intermediate would lead to *syn* addition. If the lifetime of the ion pair is longer and the ion pair dissociates, a mixture of *syn* and *anti* addition products can be formed. The termolecular mechanism is expected to give *anti* addition because the nucleophilic attack occurs on the opposite side of the double bond from proton addition. Further discussion of the structural features that affect the competition between the two possible mechanisms can be found in Section 6.1 of Part A.

4.1.2. Hydration and Other Acid-Catalyzed Additions of Oxygen Nucleophiles

Oxygen nucleophiles can be added to double bonds under strongly acidic conditions. A fundamental example is the hydration of alkenes in acidic aqueous solution.

$$R_2C{=}CH_2 + H^+ \longrightarrow R_2\overset{+}{C}CH_3 \xrightarrow{H_2O} R_2\underset{\overset{|}{{}^+OH_2}}{C}CH_3 \xrightarrow{-H^+} R_2\underset{\overset{|}{OH}}{C}CH_3$$

Addition of a proton occurs to give the more-substituted carbocation, so addition is regioselective and in accord with Markovnikov's rule. A more detailed discussion of the reaction mechanism is given in Section 6.2 of Part A. Owing to the strongly acidic and rather vigorous conditions required to effect hydration of most alkenes, these conditions are applicable only to molecules that have no acid-sensitive functional groups. The reaction is occasionally applied to the synthesis of tertiary alcohols.

$$(CH_3)_2C{=}CHCH_2CH_2\overset{\overset{\displaystyle O}{\|}}{C}CH_3 \xrightarrow[H_2O]{H_2SO_4} (CH_3)_2\underset{\overset{|}{OH}}{C}CH_2CH_2CH_2\overset{\overset{\displaystyle O}{\|}}{C}CH_3$$

Ref. 9

Moreover, because of the involvement of cationic intermediates, rearrangements can occur in systems in which a more stable cation can result by aryl, alkyl, or hydrogen migration. *Oxymercuration-reduction*, a much milder and more general procedure for alkene hydration, is discussed in the next section.

Addition of nucleophilic solvents such as alcohols and carboxylic acids can be effected by using strong acids as catalysts.[10]

$$(CH_3)_2C{=}CH_2 + CH_3OH \xrightarrow{HBF_4} (CH_3)_3COCH_3$$

$$CH_3CH{=}CH_2 + CH_3CO_2H \xrightarrow{HBF_4} (CH_3)_2CHO_2CCH_3$$

9. J. Meinwald, *J. Am. Chem. Soc.*, **77**, 1617 (1955).
10. R. D. Morin and A. E. Bearse, *Ind. Eng. Chem.*, **43**, 1596 (1951); D. T. Dalgleish, D. C. Nonhebel, and P. L. Pauson, *J. Chem. Soc. C*, 1174 (1971).

Trifluoroacetic acid (TFA) is strong enough to react with alkenes under relatively mild conditions.[11] The addition is regioselective in the direction predicted by Markovnikov's rule.

$$Cl(CH_2)_3CH{=}CH_2 \xrightarrow[\Delta]{CF_3CO_2H} Cl(CH_2)_3\underset{\underset{O_2CCF_3}{|}}{C}HCH_3$$

Ring strain enhances alkene reactivity. Norbornene, for example, undergoes rapid addition of TFA at 0° C.[12]

4.1.3. Oxymercuration-Reduction

The addition reactions discussed in Sections 4.1.1 and 4.1.2 are initiated by the interaction of a proton with the alkene. Electron density is drawn toward the proton and this causes nucleophilic attack on the double bond. The role of the electrophile can also be played by metal cations, and the mercuric ion is the electrophile in several synthetically valuable procedures.[13] The most commonly used reagent is mercuric acetate, but the trifluoroacetate, trifluoromethanesulfonate, or nitrate salts are more reactive and preferable in some applications. A general mechanism depicts a *mercurinium ion* as an intermediate.[14] Such species can be detected by physical measurements when alkenes react with mercuric ions in nonnucleophilic solvents.[15] The cation may be predominantly bridged or open, depending on the structure of the particular alkene. The addition is completed by attack of a nucleophile at the more-substituted carbon. The nucleophilic capture is usually the rate- and product-controlling step.[13,16]

$$RCH{=}CH_2 + Hg(II) \rightleftharpoons \overset{Hg^{2+}}{RCH{=}CH_2} \text{ or } \underset{+}{\overset{Hg^+}{RCH{-}CH_2}} \xrightarrow{Nu^-} [RCHCH_2{-}Hg]^+ \underset{Nu}{|}$$

The nucleophiles that are used for synthetic purposes include water, alcohols, carboxylate ions, hydroperoxides, amines, and nitriles. After the addition step is complete, the mercury is usually reductively removed by sodium borohydride, the net result being the addition of hydrogen and the nucleophile to the alkene. The regioselectivity is excellent and is in the same sense as is observed for proton-initiated additions.[17]

11. P. E. Peterson, R. J. Bopp, D. M. Chevli, E. L. Curran, D. E. Dillard, and R. J. Kamat, *J. Am. Chem. Soc.*, **89**, 5902 (1967).

12. H. C. Brown, J. H. Kawakami, and K.-T. Liu, *J. Am. Chem. Soc.*, **92**, 5536 (1970).

13. (a) R. C. Larock, *Angew. Chem. Int. Ed. Engl.*, **17**, 27 (1978); (b) W. Kitching, *Organomet. Chem. Rev.*, **3**, 61 (1968).

14. S. J. Cristol, J. S. Perry, Jr., and R. S. Beckley, *J. Org. Chem.*, **41**, 1912 (1976); D. J. Pasto and J. A. Gontarz, *J. Am. Chem. Soc.*, **93**, 6902 (1971).

15. G. A. Olah and P. R. Clifford, *J. Am. Chem. Soc.*, **95**, 6067 (1973); G. A. Olah and S. H. Yu, *J. Org. Chem.*, **40**, 3638 (1975).

16. W. L. Waters, W. S. Linn, and M. C. Caserio, *J. Am. Chem. Soc.*, **90**, 6741 (1968).

17. H. C. Brown and P. J. Geoghegan, Jr., *J. Org. Chem.*, **35**, 1844 (1970); H. C. Brown, J. T. Kurek, M.-H. Rei, and K. L. Thompson, *J. Org. Chem.*, **49**, 2511 (1984); H. C. Brown, J. T. Kurek, M.-H. Rei, and K. L. Thompson, *J. Org. Chem.*, **50**, 1171 (1985).

The reductive replacement of mercury using sodium borohydride is a free radical chain reaction involving a mercuric hydride intermediate.[18]

$$RHg^{II}X + NaBH_4 \longrightarrow RHg^{II}H$$

$$In\cdot + RHg^{II}H \longrightarrow In\text{-}H + RHg^{I}$$

$$RHg^{I} \longrightarrow R\cdot + Hg^{0}$$

$$R\cdot + RHg^{II}H \longrightarrow RH + RHg^{I}$$

The evidence for the free radical mechanism includes the fact that the course of the reaction can be diverted by oxygen, an efficient radical scavenger. In the presence of oxygen, the mercury is replaced by a hydroxy group. Also consistent with a free radical intermediate is the formation of cyclic products when 5-hexenylmercury compounds are reduced with sodium borohydride.[19] This cyclization reaction is highly characteristic of reactions involving 5-hexenyl radicals (see Part A, Section 11.2.3.3). In the presence of oxygen, no cyclic product is formed, indicating that O_2 traps the radical faster than cyclization occurs.

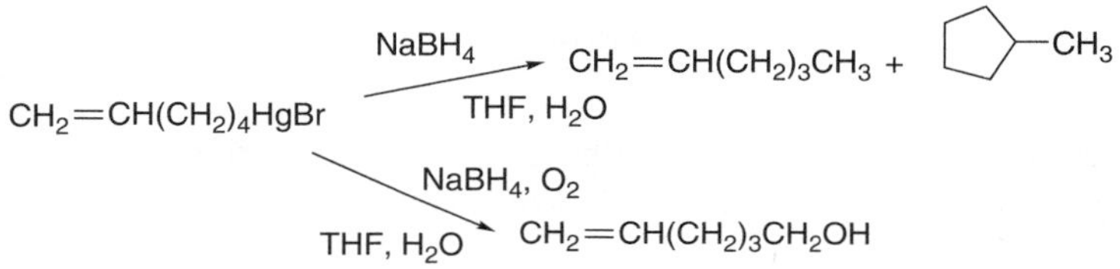

Tri-*n*-butyltin hydride can also be used for reductive demercuration.[20] An alternative reagent for demercuration is sodium amalgam in a protic solvent. Here the evidence is that free radicals are not involved and the mercury is replaced with retention of configuration.[21]

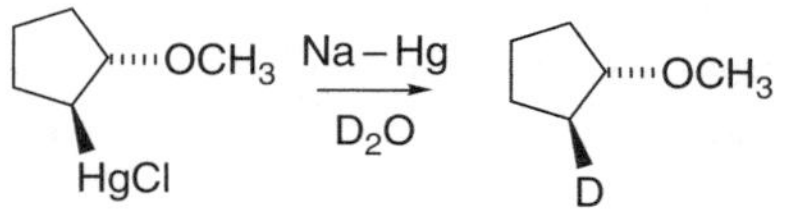

The stereochemistry of oxymercuration has been examined in a number of systems. Conformationally biased cyclic alkenes such as 4-*t*-butylcyclohexene and 4-*t*-butyl-1-methylcyclohexene give exclusively the product of *anti* addition, which is consistent with a mercurinium ion intermediate.[17,22]

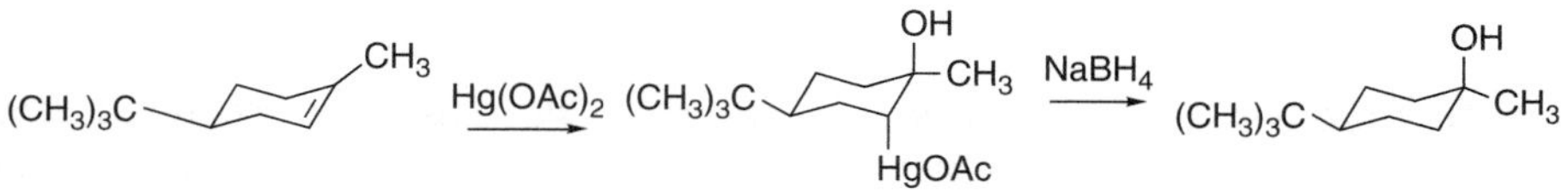

18. C. L. Hill and G. M. Whitesides, *J. Am. Chem. Soc.*, **96**, 870 (1974).
19. R. P. Quirk and R. E. Lea, *J. Am. Chem. Soc.*, **98**, 5973 (1976).
20. G. M. Whiteside and J. San Fillipo, Jr., *J. Am. Chem. Soc.*, **92**, 6611 (1970).
21. F. R. Jensen, J. J. Miller, S. J. Cristol, and R. S. Beckley, *J. Org. Chem.*, **37**, 434 (1972); R. P. Quirk, *J. Org. Chem.*, **37**, 3554 (1972); W. Kitching, A. R. Atkins, G. Wickham, and V. Alberts, *J. Org. Chem.*, **46**, 563 (1981).
22. H. C. Brown, G. J. Lynch, W. J. Hammar, and L. C. Liu, *J. Org. Chem.*, **44**, 1910 (1979).

Norbornene, in contrast reacts by *syn* addition.[23] This is believed to occur by internal transfer of the nucleophile.

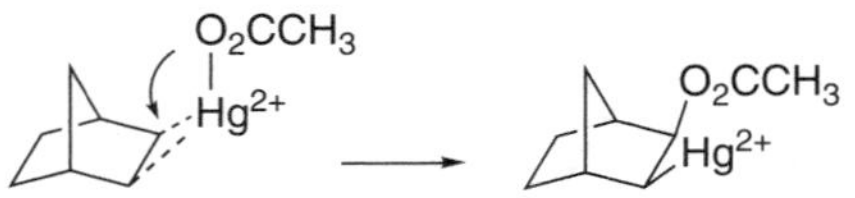

The reactivity of different alkenes toward mercuration spans a considerable range and is governed by a combination of steric and electronic factors.[24] Terminal double bonds are more reactive than internal ones. Disubstituted terminal alkenes, however, are more reactive than monosubstituted cases, as would be expected for electrophilic attack. (See Part A, Table 5.6 for comparative rate data.) The differences in relative reactivities are large enough that selectivity can be achieved with certain dienes.

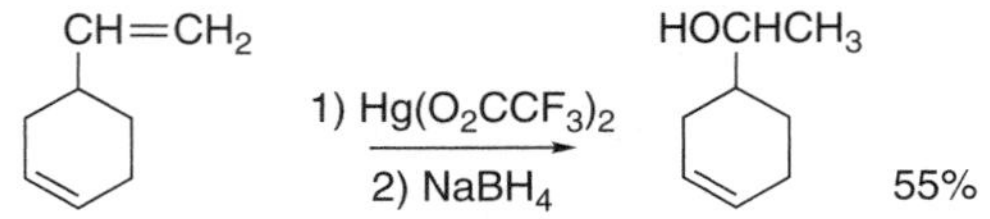

Ref. 24

Diastereoselectivity has been observed in oxymercuration of alkenes having nearby oxygen substituents. Terminal allylic alcohols show a preference for formation of the *anti* 2,3-diols.

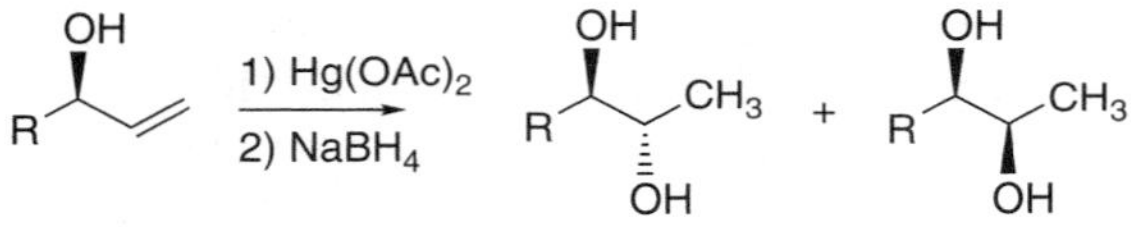

R	anti	syn
Et	76	24
i-Pr	80	20
t-Bu	98	2
Ph	88	12

This result can be explained in terms of a steric preference for conformation **A** over **B**. The approach of the mercuric ion is directed by the hydroxy group. The selectivity increases with the size of the substituent R.[25]

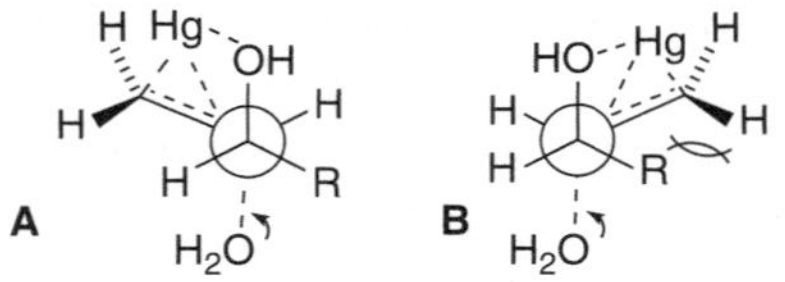

The directive effect of allylic silyoxy groups has also been examined. The reactions are completely regioselective for 1,3-oxygen substitution. The reaction of

23. T. G. Traylor and A. W. Baker, *J. Am. Chem. Soc.*, **85**, 2746 (1963); H. C. Brown and J. H. Kawakami, *J. Am. Chem. Soc.*, **95**, 8665 (1973).
24. H. C. Brown and P. J. Geoghegan, Jr., *J. Org. Chem.*, **37**, 1937 (1972); H. C. Brown, P. J. Geoghegan, Jr., G. J. Lynch, and J. T. Kurek, *J. Org. Chem.*, **37**, 1941 (1972); H. C. Brown, P. J. Geoghegan, Jr., and J. T. Kurek, *J. Org. Chem.*, **46**, 3810 (1981).
25. B. Giese and D. Bartmann, *Tetrahedron Lett.*, **26**, 1197 (1985).

Z-isomers of 2-pentenyloxy ethers show modest stereoselectivity, but the *E*-ethers show no stereoselectivity.[26] Trisubstituted allylic TBDPS ethers show good stereoselectivity.[27]

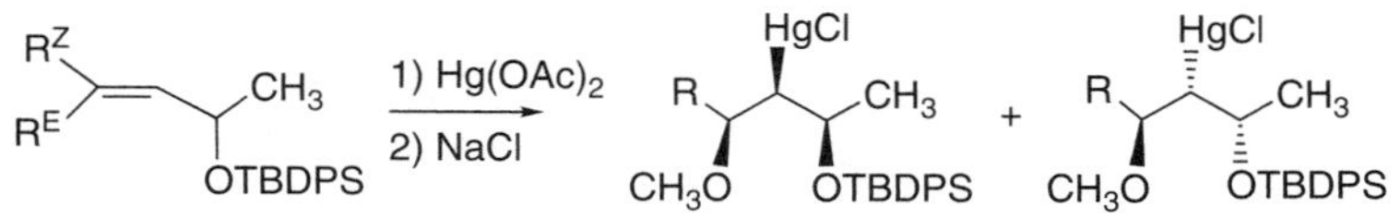

R^E	R^Z	syn	anti
CH$_3$	H	1	1
H	CH$_3$	5	1
CH$_3$	CH$_3$	20	1

These results are consistent with a directive effect by the silyloxy substituent through the sterically favored conformation of the reactant.

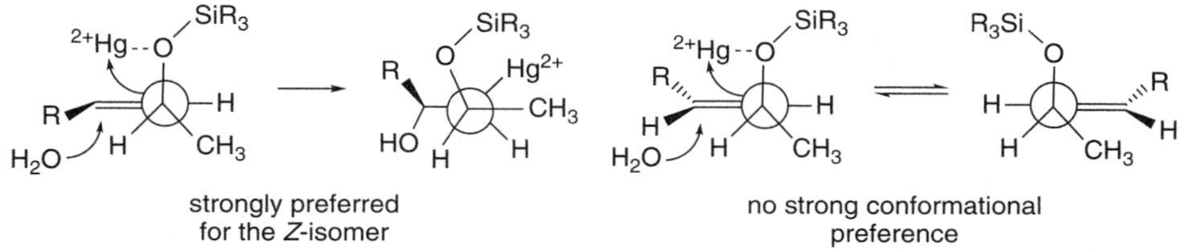

strongly preferred
for the *Z*-isomer

no strong conformational
preference

With acetoxy derivatives, the 2,3-*syn* isomer is preferred as a result of direct nucleophilic participation by the carbonyl oxygen.

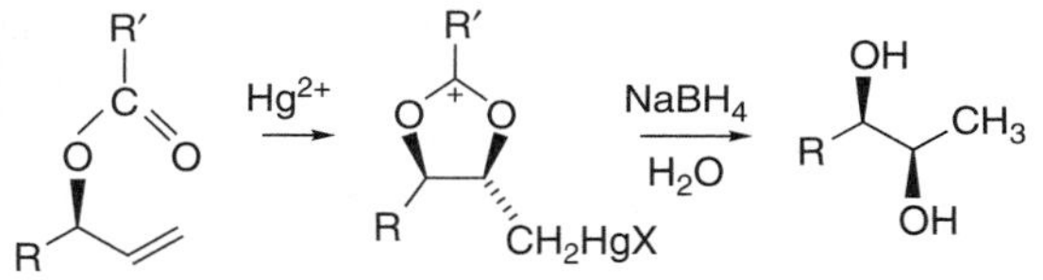

Polar substituents can exert a directing effect. Cyclohexenol, for example, gives high regioselectivity but low stereoselectivity.[28] This indicates that some factor other than hydroxy coordination is involved.

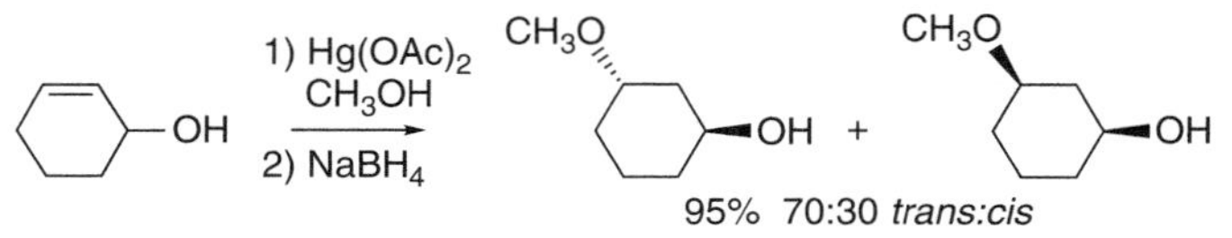

95% 70:30 *trans:cis*

A computational study of remote directing effects was undertaken in substituted norbornenes.[29] It was concluded that polar effects of EWGs favors mercuration at the

26. R. Cormick, J. Loefstedt, P. Perlmutter, and G. Westman, *Tetrahedron Lett.*, **38**, 2737 (1997).
27. R. Cormick, P. Perlmutter, W. Selajarern, and H. Zhang, *Tetrahedron Lett.*, **41**, 3713 (2000).
28. Y. Senda, S. Takayanagi, T. Sudo, and H. Itoh, *J. Chem. Soc., Perkin Trans. 1*, 270 (2001).
29. P. Mayo, G. Orlova, J. D. Goddard, and W. Tam, *J. Org. Chem.*, **66**, 5182 (2001).

carbon that is closer to the substituent, which is attributed to a favorable polar effect that stabilizes the negative charge on the mercurated carbon.

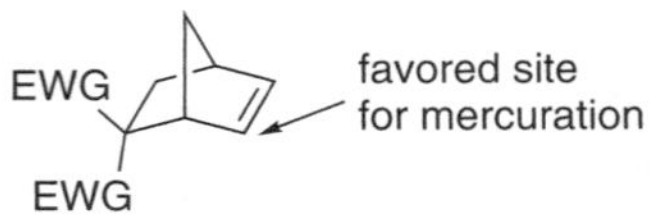

Visual models, additional information and exercises on Oxymercuration can be found in the Digital Resource available at: Springer.com/carey-sundberg.

Scheme 4.1 includes examples of oxymercuration reactions. Entries 1 and 2 illustrate the Markovnikov orientation under typical reaction conditions. The high *exo* selectivity in Entry 3 is consistent with steric approach control on a weakly bridged (or open) mercurinium ion. There is no rearrangement, indicating that the intermediate is a localized cation.

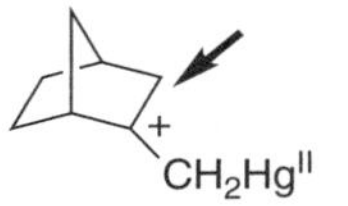

Entries 4 and 5 involve formation of ethers using alcohols as solvents, whereas the reaction in Entry 6 forms an amide in acetonitrile. Entries 7 and 8 show use of other nucleophiles to capture the mercurinium ion.

4.1.4. Addition of Halogens to Alkenes

The addition of chlorine or bromine to an alkene is a very general reaction. Section 6.3 of Part A provides a discussion of the reaction mechanism. Bromination of simple alkenes is extremely fast. Some specific rate data are tabulated and discussed in Section 6.3 of Part A. As halogenation involves electrophilic attack, substituents on the double bond that increase electron density increase the rate of reaction, whereas EWG substituents have the opposite effect. Considerable insight into the mechanism of halogen addition has come from studies of the stereochemistry of the reaction. Most simple alkenes add bromine in a stereospecific manner, giving the product of *anti* addition. Among the alkenes that give *anti* addition products are Z-2-butene, E-2-butene, maleic and fumaric acid, and a number of cycloalkenes.[30] Cyclic, positively charged bromonium ion intermediates provide an explanation for the observed *anti* stereospecificity.

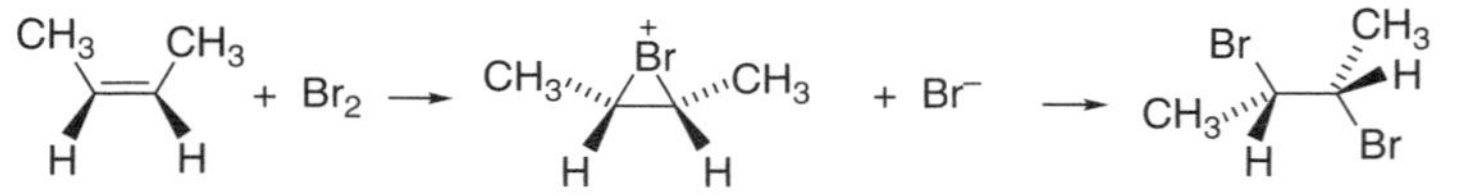

[30.] J. H. Rolston and K. Yates, *J. Am. Chem. Soc.*, **91**, 1469, 1477 (1969).

A. Alcohols

1[a] $(CH_3)_3CCH=CH_2$ $\xrightarrow{\text{1) Hg(OAc)}_2}$ $(CH_3)_3CCHCH_3$ + $(CH_3)_3CCH_2CH_2OH$

$\xrightarrow{\text{2) NaBH}_4}$ $\underset{OH}{|}$ 97%   3%

2[b] $(CH_2)_8CH=CH_2$ (phthalide ring) $\xrightarrow[\text{2) NaBH}_4]{\text{1) Hg(OAc)}_2}$ $(CH_2)_8CHCH_3$, OH (phthalide ring) 80%

3[c] (norbornane with =CH$_2$) $\xrightarrow[\text{2) NaBH}_4]{\text{1) Hg(OAc)}_2}$ (norbornane with OH and CH$_3$) 99.5%

B. Ethers

4[d] (cyclohexene) $\xrightarrow[\text{2) NaBH}_4]{\text{1) Hg(O}_2\text{CCF}_3)_2, (CH_3)_2CHOH}$ (cyclohexyl)$-OCH(CH_3)_2$ 98%

5[e] $CH_3(CH_2)_3CH=CH_2$ $\xrightarrow[\text{EtOH}]{\text{Hg(O}_2\text{CCF}_3)_2}$ $CH_3(CH_2)_3CHCH_3$, OC_2H_5 97%

C. Amides

6[f] $CH_3(CH_2)_3CH=CH_2$ $\xrightarrow[\text{2) NaBH}_4, H_2O]{\text{1) Hg(NO}_3)_2, CH_3CN}$ $CH_3CH_2CH_2CH_2CHCH_3$, $HNCOCH_3$ 92%

D. Peroxides

7[g] $CH_3(CH_2)_4CH=CHCH_3$ $\xrightarrow[\text{2) NaBH}_4]{\text{1) Hg(OAc)}_2, t\text{-BuOOH}}$ $CH_3(CH_2)_4CHCH_2CH_3$, $OOC(CH_3)_3$ 40%

E. Amines

8[h] CH_3O—(phenyl)—$CH_2CH=CH_2$ + $PhCH_2NH_2$ $\xrightarrow[\text{2) NaBH}_4]{\text{1) Hg(ClO}_4)_2}$ CH_3O—(phenyl)—CH_2CHCH_3, $HNCH_2Ph$ 70%

a. H. C. Brown and P. J. Geoghegan, Jr., *J. Org. Chem.*, **35**, 1844 (1970).
b. H. L. Wehrmeister and D. E. Robertson, *J. Org. Chem.*, **33**, 4173 (1968).
c. H. C. Brown and W. J. Hammar, *J. Am. Chem. Soc.*, **89**, 1524 (1967).
d. H. C. Brown and M.-H. Rei, *J. Am. Chem. Soc.*, **91**, 5646 (1969).
e. H. C. Brown, J. T. Kurek, M.-H. Rei, and K. L. Thompson, *J. Org. Chem.*, **50**, 1171 (1985).
f. H. C. Brown and J. T. Kurek, *J. Am. Chem. Soc.*, **91**, 5647 (1969).
g. D. H. Ballard and A. J. Bloodworth, *J. Chem. Soc. C*, 945 (1971).
h. R. C. Griffith, R. J. Gentile, T. A. Davidson, and F. L. Scott, *J. Org. Chem.*, **44**, 3580 (1979).

The bridging by bromine prevents rotation about the remaining bond and back-side nucleophilic opening of the bromonium ion by bromide ion leads to the observed *anti* addition. Direct evidence for the existence of bromonium ions has been obtained from NMR measurements.[31] A bromonium ion salt (with Br_3^- as the counterion) has been isolated from the reaction of bromine with the very hindered alkene adamantylideneadamantane.[32]

[31.] G. A. Olah, J. M. Bollinger, and J. Brinich, *J. Am. Chem. Soc.*, **90**, 2587 (1968); G. A. Olah, P. Schilling, P. W. Westerman, and H. C. Lin, *J. Am. Chem. Soc.*, **96**, 3581 (1974).
[32.] J. Strating, J. H. Wierenga, and H. Wynberg, *J. Chem. Soc., Chem. Commun.*, 907 (1969).

A substantial amount of *syn* addition is observed for *Z*-1-phenylpropene (27–80% *syn* addition), *E*-1-phenylpropene (17–29% *syn* addition), and *cis*-stilbene (up to 90% *syn* addition in polar solvents).

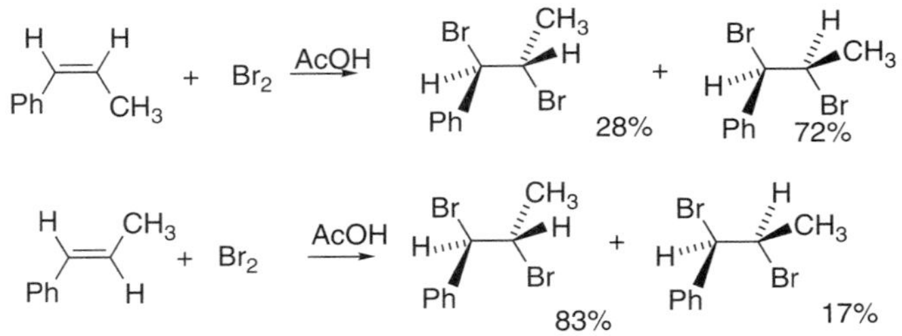

Ref. 30

A common feature of the compounds that give extensive *syn* addition is the presence of a phenyl substituent on the double bond. The presence of a phenyl substituent diminishes the strength of bromine bridging by stabilizing the cationic center. A weakly bridged structure in equilibrium with an open benzylic cation can account for the loss in stereospecificity.

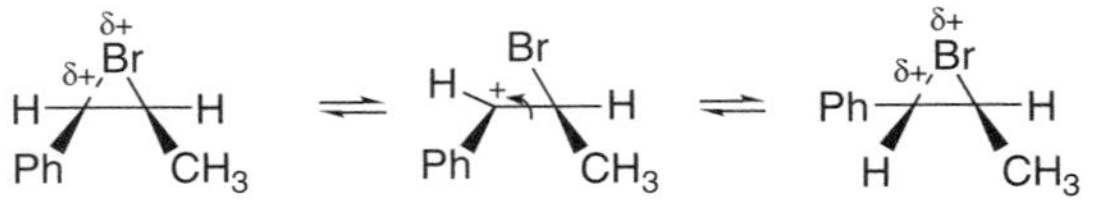

The diminished stereospecificity is similar to that noted for hydrogen halide addition to phenyl-substituted alkenes.

Although chlorination of aliphatic alkenes usually gives *anti* addition, *syn* addition is often dominant for phenyl-substituted alkenes.[33]

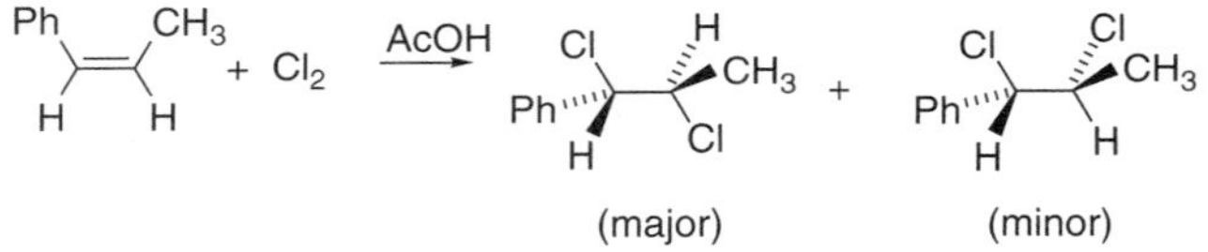

These results, too, reflect a difference in the extent of bridging in the intermediates. With unconjugated alkenes, there is strong bridging and high *anti* stereospecificity. Phenyl substitution leads to cationic character at the benzylic site, and there is more *syn* addition. Because of its smaller size and lesser polarizability, chlorine is not as effective as bromine in bridging for any particular alkene. Bromination therefore generally gives a higher degree of *anti* addition than chlorination, all other factors being the same.[34]

33. M. L. Poutsma, *J. Am. Chem. Soc.*, **87**, 2161, 2172 (1965); R. C. Fahey, *J. Am. Chem. Soc.*, **88**, 4681 (1966); R. C. Fahey and C. Shubert, *J. Am. Chem. Soc.*, **87**, 5172 (1965).
34. R. J. Abraham and J. R. Monasterios, *J. Chem. Soc., Perkin Trans. 1*, 1446 (1973).

Chlorination can be accompanied by other reactions that are indicative of carbocation intermediates. Branched alkenes can give products that are the result of elimination of a proton from a cationic intermediate.[35]

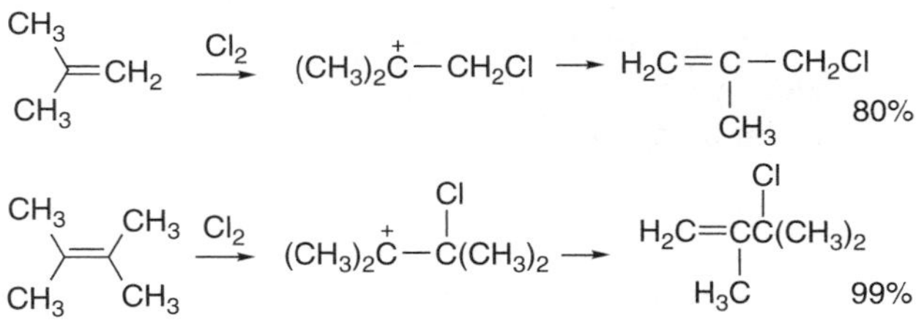

Skeletal rearrangements are observed in systems that are prone toward migration.

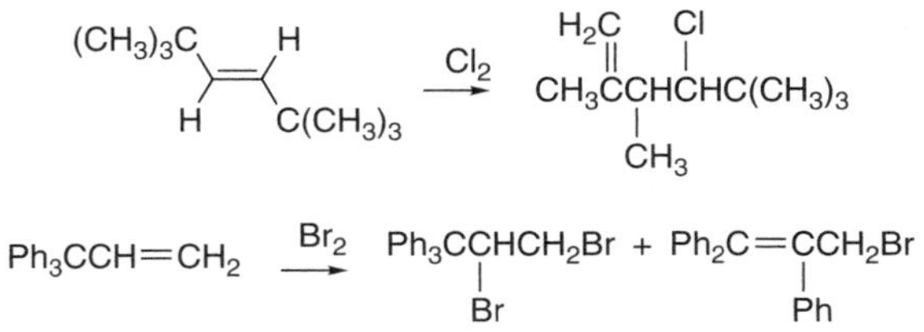

Ref. 35

Ref. 36

Nucleophilic solvents can compete with halide ion for the cationic intermediate. For example, the bromination of styrene in acetic acid leads to significant amounts of the acetoxybromo derivative.

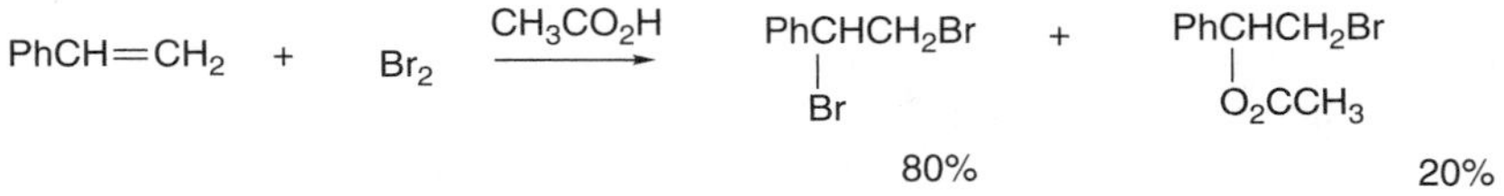

Ref. 30

The acetoxy group is introduced exclusively at the benzylic carbon. This is in accord with the intermediate being a weakly bridged species or a benzylic cation. The addition of bromide salts to the reaction mixture diminishes the amount of acetoxy compound formed by shifting the competition for the electrophile in favor of the bromide ion. Chlorination in nucleophilic solvents can also lead to solvent incorporation, as, for example, in the chlorination of 1-phenylpropene in methanol.[37]

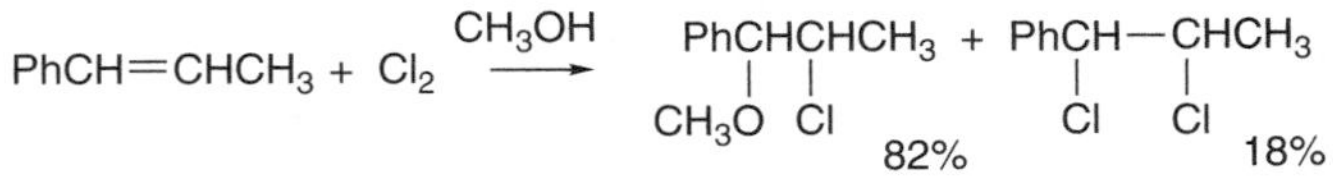

From a synthetic point of view, the participation of water in brominations, leading to bromohydrins, is the most important example of nucleophilic capture of the intermediate by solvent. To favor introduction of water, it is desirable to keep the concentration

35. M. L. Poutsma, *J. Am. Chem. Soc.*, **87**, 4285 (1965).
36. R. O. C. Norman and C. B. Thomas, *J. Chem. Soc. B*, 598 (1967).
37. M. L. Poutsma and J. L. Kartch, *J. Am. Chem. Soc.*, **89**, 6595 (1967).

of the bromide ion as low as possible. One method for accomplishing this is to use *N*-bromosuccinimide (NBS) as the brominating reagent.[38,39] High yields of bromohydrins are obtained by using NBS in aqueous DMSO. The reaction is a stereospecific *anti* addition. As in bromination, a bromonium ion intermediate can explain the *anti* stereospecificity. It has been shown that the reactions in DMSO involve nucleophilic attack by the sulfoxide oxygen. The resulting alkoxysulfonium ion intermediate reacts with water to give the bromohydrin.

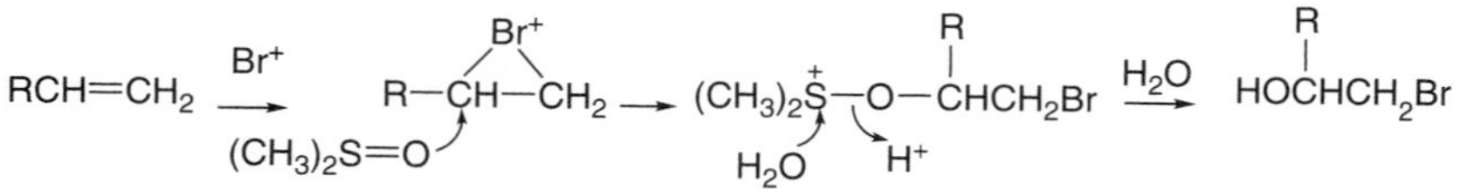

In accord with the Markovnikov rule, the hydroxy group is introduced at the carbon best able to support positive charge.

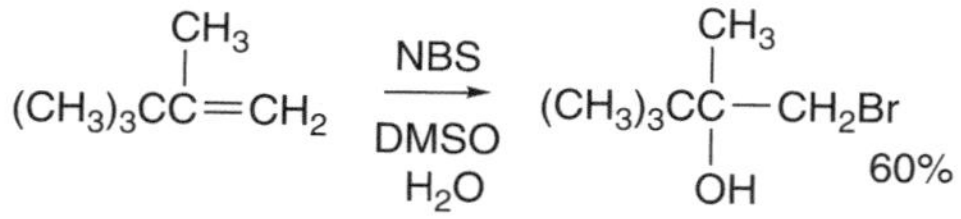

Ref. 40

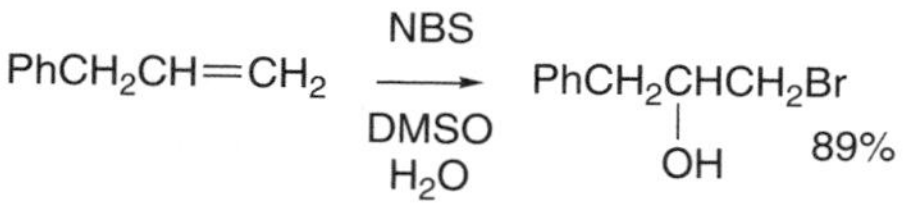

Ref. 41

The participation of sulfoxy groups can be used to control the stereochemistry in acyclic systems. In the reaction shown below, the internal sulfoxide captures the bromonium ion and then undergoes inversion at sulfur in the hydrolytic step.

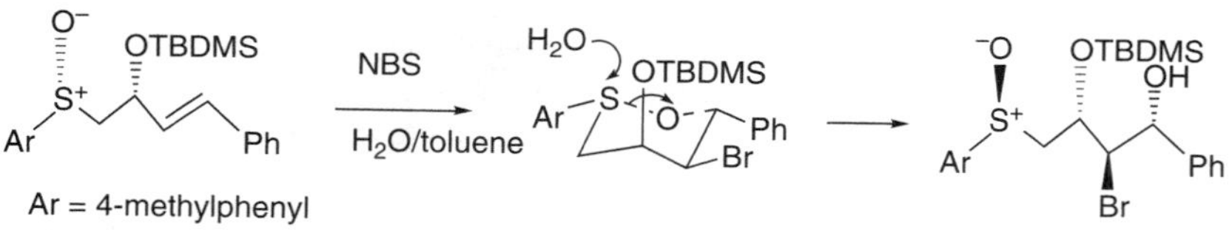

Ar = 4-methylphenyl

Ref. 42

A procedure that is useful for the preparation of both bromohydrins and iodohydrins involves in situ generation of the hypohalous acid from $NaBrO_3$ or $NaIO_4$ by reduction with bisulfite.[43]

38. A. J. Sisti and M. Meyers, *J. Org. Chem.*, **38**, 4431 (1973).
39. C. O. Guss and R. Rosenthal, *J. Am. Chem. Soc.*, **77**, 2549 (1965).
40. D. R. Dalton, V. P. Dutta, and D. C. Jones, *J. Am. Chem. Soc.*, **90**, 5498 (1968).
41. A. W. Langman and D. R. Dalton, *Org. Synth.*, **59**, 16 (1979).
42. S. Raghavan and M. A. Rasheed, *Tetrahedron*, **59**, 10307 (2003).
43. H. Masuda, K. Takase, M. Nishio, A. Hasegawa, Y. Nishiyama, and Y. Ishii, *J. Org. Chem.*, **59**, 5550 (1994).

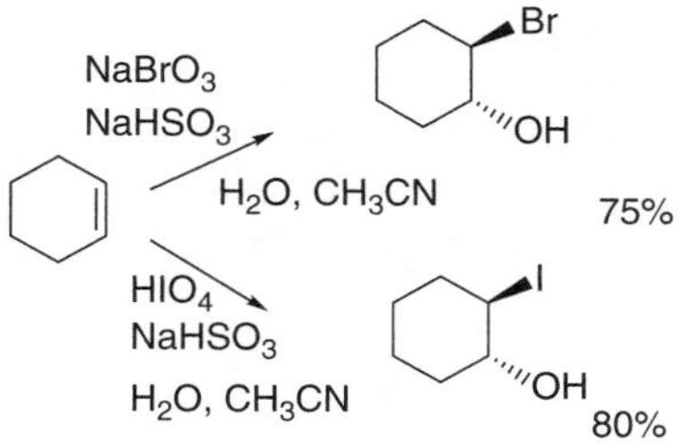

These reactions show the same regioselectivity and stereoselectivity as other reactions that proceed through halonium ion intermediates.

Because of its high reactivity, special precautions must be taken with reactions of fluorine and its use is somewhat specialized.[44] Nevertheless, there is some basis for comparison with the less reactive halogens. Addition of fluorine to *Z*- and *E*-1-propenylbenzene is not stereospecific, but *syn* addition is somewhat favored.[45] This result is consistent with formation of a cationic intermediate.

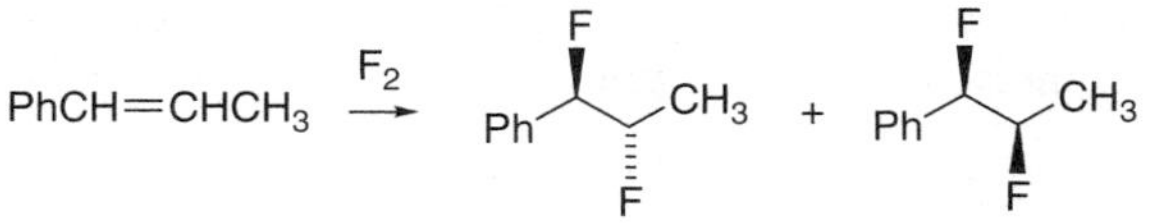

In methanol, the solvent incorporation product is formed, as would be expected for a cationic intermediate.

$$PhCH{=}CHCH_3 \xrightarrow[CH_3OH]{F_2} PhCH(OCH_3)CH(F)CH_3$$

These results are consistent with the expectation that fluorine would not be an effective bridging atom.

There are other reagents, such as CF_3OF and CH_3CO_2F, that transfer an electrophilic fluorine to double bonds. These reactions probably involve an ion pair that collapses to an addition product.

$$PhCH{=}CHPh + CF_3OF \longrightarrow PhCH(OCF_3)CH(F)Ph$$ Ref. 46

$$CH_3(CH_2)_9CH{=}CH_2 + CH_3CO_2F \longrightarrow CH_3(CH_2)_9CH(O_2CCH_3)CH_2F \quad 30\%$$ Ref. 47

The stability of hypofluorites is improved in derivatives having electron-withdrawing substituents, such as 2,2-dichloropropanoyl hypofluorite.[48] Various other fluorinating agents have been developed and used, including *N*-fluoropyridinium salts such as the

44. H. Vypel, *Chimia*, **39**, 305 (1985).
45. R. F. Merritt, *J. Am. Chem. Soc.*, **89**, 609 (1967).
46. D. H. R. Barton, R. H. Hesse, G. P. Jackman, L. Ogunkoya, and M. M. Pechet, *J. Chem. Soc., Perkin Trans. 1*, 739 (1974).
47. S. Rozen, O. Lerman, M. Kol, and D. Hebel, *J. Org. Chem.*, **50**, 4753 (1985).
48. S. Rozen and D. Hebel, *J. Org. Chem.*, **55**, 2621 (1990).

triflate[49] and heptafluorodiborate.[50] The reactivity of these reagents can be "tuned" by varying the pyridine ring substituents. In contrast to the hypofluorites, these reagents are storable.[51] In nucleophilic solvents such as acetic acid or alcohols, the reagents give addition products, whereas in nonnucleophilic solvents, alkenes give substitution products resulting from deprotonation of a carbocation intermediate.

Addition of iodine to alkenes can be accomplished by a photochemically initiated reaction. Elimination of iodine is catalyzed by excess iodine, but the diiodo compounds can be obtained if unreacted iodine is removed.[52]

$$RCH{=}CHR + I_2 \rightleftharpoons RCH{-}CHR$$

The diiodo compounds are very sensitive to light and are seldom used in syntheses.

The elemental halogens are not the only sources of electrophilic halogen, and for some synthetic purposes other "positive halogen" compounds may be preferable as electrophiles. The utility of *N*-bromosuccinimide in formation of bromohydrins was mentioned earlier. Both *N*-chlorosuccinimide and *N*-bromosuccinimide transfer electrophilic halogen with the succinimide anion acting as the leaving group. As this anion is subsequently protonated to give the weak nucleophile succinimide, these reagents favor nucleophilic additions by solvent and cyclization reactions because there is no competition from a halide anion. Other compounds that are useful for specific purposes are indicated in Table 4.2. Pyridinium hydrotribromide (pyridinium hydrobromide perbromide), benzyltrimethyl ammonium tribromide, and dioxane-bromine are examples of complexes of bromine in which its reactivity is somewhat attenuated, resulting in increased selectivity. In 2,4,6-tetrabromocyclohexadienone is a very mild and selective source of electrophilic bromine; the leaving group is 2,4,6-tribromophenoxide ion.

49. T. Umemoto, S. Fukami, G. Tomizawa, K. Harasawa, K. Kawada, and K. Tomita, *J. Am. Chem. Soc.*, **112**, 8563 (1990).
50. A. J. Poss. M. Van Der Puy, D. Nalewajek, G. A. Shia, W. J. Wagner, and R. L. Frenette, *J. Org. Chem.*, **56**, 5962 (1991).
51. T. Umemoto, K. Tomita, and K. Kawada, *Org. Synth.*, **69**, 129 (1990).
52. P. S. Skell and R. R. Pavlis, *J. Am. Chem. Soc.*, **86**, 2956 (1964); R. L. Ayres, C. J. Michejda, and E. P. Rack, *J. Am. Chem. Soc.*, **93**, 1389 (1971).

Table 4.2. Other Sources of Electrophilic Halogen

Reagents	Synthetic applications[a]
A. Chlorinating agents	
Sodium hypochlorite solution	Formation of chlorohydrins from alkenes
N-Chlorosuccinimide	Chlorination with solvent participation and cyclization
Chloramine-T[b]	Formation of chlorohydrins in acidic aqueous solution.
B. Brominating agents	
Pyridinium hydrotribromide (pyidinium hydrobromide perbromide)	Mild and selective substitute for bromine
Dioxane bromine complex	Same as for pyridinium hydrotribromide
N-Bromosuccinimide	Used in place of bromine when low bromide concentration is required.
2,4,4,6-Tetrabromocyclohexadienone[c]	Selective bromination of alkenes and carbonyl compounds
Quaternary ammonium tribromides[d]	Similar to pyridinium hydrotribromide
C. Iodinating agents	
bis-(Pyridinium)iodonium[e] tetrafluoroborate	Selective iodination and iodocyclization.

a. For specific examples, consult M. Fieser and L. F. Fieser, *Reagents for Organic Synthesis*, John Wiley & Sons, New York.
b. B. Damin, J. Garapon, and B. Sillion, *Synthesis*, 362 (1981).
c. F. Calo, F. Ciminale, L. Lopez, and P. E. Todesco, *J. Chem. Soc., C*, 3652 (1971) ;Y. Kitahara, T. Kato, and I. Ichinose, *Chem. Lett.*, 283 (1976)
d. S. Kaigaeshi and T. Kakinami, *Ind. Chem. Libr.*, **7**, 29 (1985); G. Bellucci, C. Chiappe, and F. Marioni, *J. Am. Chem. Soc.*, **109**, 515 (1987).
e. J. Barluenga, J. M. Gonzalez, M. A. Garcia-Martin, P. J. Campos, and G. Asensio, *J. Org. Chem.*, **58**, 2058 (1993).

Electrophilic iodine reagents are extensively employed in iodocyclization (see Section 4.2.1). Several salts of pyridine complexes with I$^+$ such as *bis*-(pyridinium)iodonium tetrafluoroborate and *bis*-(collidine)iodonium hexafluorophosphate have proven especially effective.[53]

4.1.5. Addition of Other Electrophilic Reagents

Many other halogen-containing compounds react with alkenes to give addition products by mechanisms similar to halogenation. A complex is generated and the halogen is transferred to the alkene to generate a bridged cationic intermediate. This may be a symmetrical halonium ion or an unsymmetrically bridged species, depending on the ability of the reacting carbon atoms to accommodate positive charge. The direction of opening of the bridged intermediate is usually governed by electronic factors. That is, the addition is completed by attack of the nucleophile at the more positive carbon atom of the bridged intermediate. The regiochemistry of addition therefore follows Markovnikov's rule. The stereochemistry of addition is usually *anti*, because of the involvement of a bridged halonium intermediate.[54] Several reagents of this type are listed in Entries 1 to 6 of Scheme 4.2. The nucleophilic anions include isocyanate, azide, thiocyanate, and nitrate.

Entries 7 to 9 involve other reagents that react by similar mechanisms. In the case of thiocyanogen chloride and thiocyanogen, the formal electrophile is [NCS]$^+$. The presumed intermediate is a cyanothiairanium ion. The thiocyanate anion is an

53. Y. Brunel and G. Rousseau, *J. Org. Chem.*, **61**, 5793 (1996).
54. A. Hassner and C. Heathcock, *J. Org. Chem.*, **30**, 1748 (1965).

Scheme 4.2. Addition Reactions of Other Electrophilic Reagents

	Reagent	Preparation	Poduct
1[a]	I—N=C=O	AgCNO, I_2	RCH—CHR 　|　　| 　I　NCO
2[b]	Br—N=$\overset{+}{N}$=N⁻	HN_3, Br_2	RCH—CHR 　|　　| 　Br　N_3
3[c]	I—N=$\overset{+}{N}$=N⁻	NaN_3, ICl	RCH—CHR 　|　　| 　I　N_3
4[d]	I—S=C≡N	$(NCS)_2$, I_2	RCH—CHR 　|　　| 　I　S—C≡N
5[e]	I—ONO_2	$AgNO_3$, ICl	RCH—CHR 　|　　| 　I　ONO_2
6[f]	Cl—SCN	$Pb(SCN)_2$, Cl_2	RCH—CHR 　|　　| 　Cl　SCN
7[g]	N≡CS—SC≡N	$Pb(SCN)_2$, Br_2	RCH—CHR 　|　　| N≡CS　SC≡N　and　RCH—CHR 　　　　　　　　　|　　| 　　　　　　　　N≡CS　N=C=S
8[h]	O=N—Cl		RC—CHR 　||　　| 　HON　Cl
9[i]	O=N—O_2CH	C_5H_{11}ONO HCO_2H	RC—CHR 　||　　| 　HON　O_2CH

a. A. Hassner, R. P. Hoblitt, C. Heathcock, J. E. Kropp, and M. Lorber, *J. Am. Chem. Soc.*, **92**, 1326 (1970); A. Hassner, M. E. Lorber, and C. Heathcock, *J. Org. Chem.*, **32**, 540 (1967).
b. A. Hassner, F. P. Boerwinkle, and A. B. Levy, *J. Am. Chem. Soc.*, **92**, 4879 (1970).
c. F. W. Fowler, A. Hassner, and L. A. Levy, *J. Am. Chem. Soc.*, **89**, 2077 (1967).
d. R. J. Maxwell and L. S. Silbert, *Tetrahedron Lett.*, 4991 (1978).
e. J. W. Lown and A. V. Joshua, *J. Chem. Soc., Perkin Trans. 1*, 2680 (1973).
f. R. G. Guy and I. Pearson, *J. Chem. Soc., Perkin Trans. 1*, 281 (1973); *J. Chem. Soc., Perkin Trans. 2*, 1359 (1973).
g. R. Bonnett, R. G. Guy, and D. Lanigan, *Tetrahedron*, **32**, 2439 (1976); R. J. Maxwell, L. S. Silbert, and J. R. Russell, *J. Org. Chem.*, **42**, 1510 (1977).
h. J. Meinwald, Y. C. Meinwald, and T. N. Baker, III, *J. Am. Chem. Soc.*, **86**, 4074 (1964).
i. H. C. Hamann and D. Swern, *J. Am. Chem. Soc.*, **90**, 6481 (1968).

ambient nucleophile and both carbon-sulfur and carbon-nitrogen bond formation can be observed, depending upon the reaction conditions (see Entry 7 in Scheme 4.2).

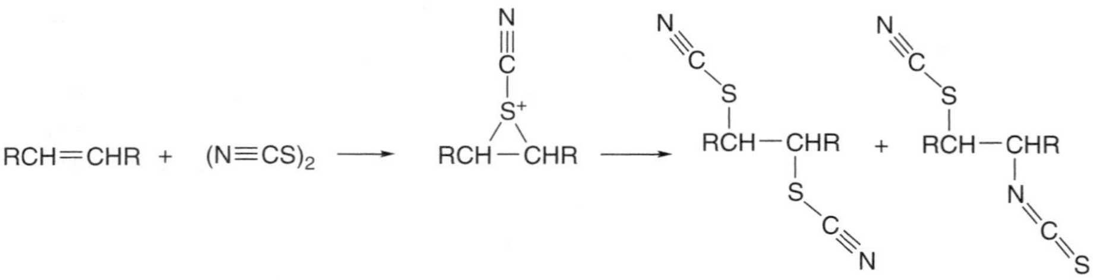

For nitrosyl chloride (Entry 8) and nitrosyl formate (Entry 9), the electrophile is the nitrosonium ion NO^+. The initially formed nitroso compounds can dimerize or isomerize to the more stable oximes.

$$RCH{=}CHR \longrightarrow \underset{\underset{O}{\overset{}{N}}}{RCH}{-}\underset{X}{CHR} \longrightarrow \underset{\underset{HO}{\overset{}{N}}}{RC}{=}\underset{X}{CHR}$$

4.1.6. Addition Reactions with Electrophilic Sulfur and Selenium Reagents

Compounds having divalent sulfur and selenium atoms bound to more electronegative elements react with alkenes to give addition products. The mechanism is similar to that in halogenation and involves of bridged cationic intermediates.

$$R'S{-}Cl + RCH{=}CHR \longrightarrow \underset{\underset{}{\overset{R'}{\underset{S^+}{\diagup\!\diagdown}}}}{RCH}{-}CHR \xrightarrow{Cl^-} \underset{\underset{Cl}{}}{\overset{SR'}{RCHCHR}}$$

$$R'Se{-}Cl + RCH{=}CHR \longrightarrow \underset{\underset{}{\overset{R'}{\underset{^+Se}{\diagup\!\diagdown}}}}{RCH}{-}CHR \xrightarrow{Cl^-} R{-}\underset{}{\overset{SeR'}{CH}}{-}\underset{Cl}{CHR}$$

In many synthetic applications, the sulfur or selenium substituent is subsequently removed by elimination, as is discussed in Chapter 6.

A variety of electrophilic reagents have been employed and several examples are given in Scheme 4.3. The sulfenylation reagents are listed in Section A. Both aryl and alkyl sulfenyl chlorides are reactive (Entries 1 and 2). Dimethyl(methylthio)sulfonium fluoroborate (Entry 3) uses dimethyl sulfide as a leaving group and can be utilized to effect capture of hydroxylic solvents and anionic nucleophiles, such as acetate and cyanide. Entries 4 and 5 are examples of *sulfenamides*, which normally require a Lewis acid catalyst to react with alkenes. Entry 6 represents application of the *Pummerer rearrangement* for in situ generation of a sulfenylation reagent. Sulfoxides react with acid anhydrides to generate sulfonium salts. When a *t*-alkyl group is present, fragmentation occurs and a sulfenylium ion is generated.[55] TFAA is the preferred anhydride in this application.

$$\underset{\overset{\|}{O}}{R{-}S}{-}C(CH_3)_3 \xrightarrow{(CF_3CO)_2O} \underset{\underset{+}{\overset{O_2CCF_3}{|}}}{R{-}S}{-}C(CH_3)_3 \longrightarrow RS^+ + (CH_3)_3CO_2CCF_3$$

The selenylation reagents include the arylselenenyl chlorides and bromides (Entries 7 and 8), selenylium salts with nonnucleophilic counterions (Entry 9), and selenenyl trifluoroacetates, sulfates, and sulfonates (Entries 10 to 13). Diphenyldiselenide reacts with several oxidation reagents to transfer electrophilic phenylselenenylium ions (Entries 14 to 16). *N*-Phenylselenenylphthalimide is a useful synthetic reagent that has the advantage of the nonnucleophilicity of the phthalimido leaving group (Entry 18). The hindered selenenyl bromide in Entry 19 is useful for selenylcyclizations (see Section 4.2.2).

Selenylation can also be done under conditions in which another nucleophilic component of the reaction captures the selenium-bridged ion. For

[55] M.-H. Brichard, M. Musick, Z. Janousek, and H. G. Viehe, *Synth. Commun.*, **20**, 2379 (1990).

Scheme 4.3. Sulfur and Selenium Reagents for Electrophilic Addition Reactions

A. Sulfenylation reagents

1[a]   2[a]   3[b]   4[c]   5[d]

ArSCl   RSCl $(CH_3)_2S^+SCH_3\ BF_4^-$ PhS—N◯O   ArSNHPh, BF_3

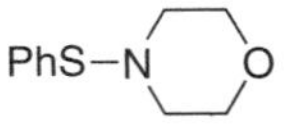

6[e] $\overset{O}{\underset{||}{RSC(CH_3)_3}}$

$(CF_3CO)_2O$

B. Selenenylation reagents

7[f]   8[g]   9[h]   10[i]   11[j]

PhSeCl   PhSeBr $PhSe^+PF_6^-$ $PhSeO_2CCF_3$ $PhSeOSO_2Ar$

12[k]   13[l]   14[m]   15[n]   16[o]

$PhSeOSO_3CF_3$ $PhSeOSO_3^-$ $(PhSe)_2$ $(PhSe)_2$ $(PhSe)_2$

                                       $(NH_4)_2S_2O_8$ DDQ $PhI(OAc)_2$

17[p]   18[q]   19[r]

$PhSeO_2H, H_3PO_2$

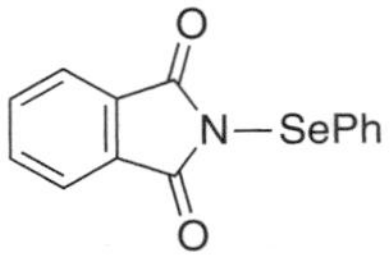

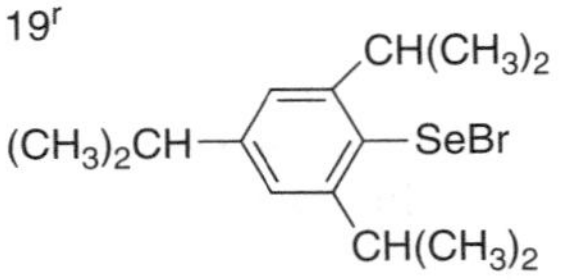

a. G. Capozzi, G. Modena, and L. Pasquato in *The Chemistry of Sulphenic Acids and Their Derivatives*, S. Patai, ed., Wiley, Chichester, 1990, Chap. 10.

b. B. M. Trost, T. Shibata, and S. J. Martin, *J. Am. Chem. Soc.*, **104**, 3228 (1982).

c. P. Brownbridge, *Tetrahedron Lett.*, **25**, 3759 (1984); P. Brownbridge, *J. Chem. Soc. Chem. Commun.*, 1280 (1987); N. S. Zefirov, N. V. Zyk, A. G. Kutateldze, and S. I. Kolbasenko, *Zh. Org. Khim.*, **23**, 227 (1987).

d. L. Benati, P. C. Montevecchi, and P. Spagnolo, *J. Chem. Soc., Perkin Trans. 1*, 1691 (1990).

e. M.-H. Brichard, M. Musick, Z. Janousek, and H. G. Viehe, *Synth. Commun.*, **20**, 2378 (1990).

f. K. B. Sharpless and R. F. Lauer, *J. Org. Chem.*, **39**, 429 (1974).

g. T. G. Back, *The Chemistry of Organic Selenium and Tellurium Compounds*, S. Patai, ed., Wiley, 1987, pp. 91–312.

h. W. P. Jackson, S. V. Ley, and A. J. Whittle, *J. Chem. Soc.* 1173 (1980).

i. H. J. Reich, *J. Org. Chem.*, **39**, 428 (1974).

j. T. G. Back and K. R. Muralidharan, *J. Org. Chem.* **56**, 2781 (1991).

k. S. Murata and T. Suzuki, *Tetrahedron Lett.*, **28**, 4297, 4415 (1987).

l. M. Tiecco, L. Testaferri, M. Tingoli, L. Bagnoli, and F. Marini, *J. Chem. Soc., Perkin Trans. 1*, 1989 (1993).

m. M. Tiecco, L. Testaferri, M. Tingoli, D. Chianelli, and D. Bartoli, *Tetrahedron Lett.*, **30**, 1417 (1989).

n. M. Tiecco, L. Testaferri, A. Temperinik, L. Bagnoli, F. Marini, and C. Santi, *Synlett*, 1767 (2001).

o. M. Tingoli, M. Tiecco, L. Testaferri, and Temperini, *Synth. Commun.*, **28**, 1769 (1998).

p. D. Labar, A. Krief, and L. Hevesi, *Tetrahedron Lett.*, 3967 (1978).

q. K. C. Nicolaou, N. A. Petasis, and D. A. Claremon, *Tetrahedron*, **41**, 4835 (1985).

r. B. H. Lipshutz and T. Gross, *J. Org. Chem.*, **60**, 3572 (1995).

example, the combination phenylselenylphthalimide and trimethylsilyl azide generates β-azido selenides and phenylselenyl chloride used with $AgBF_4$ and ethyl carbamate give β-carbamido selenides.

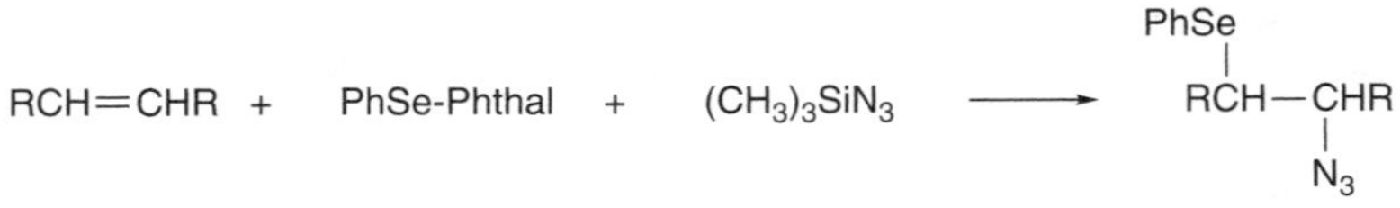

$$RCH=CHR + PhSe\text{-}Phthal + (CH_3)_3SiN_3 \longrightarrow \overset{PhSe}{\underset{N_3}{RCH-CHR}}$$ Ref. 56

56. A. Hassner and A. S. Amarasekara, *Tetrahedron Lett.*, **28**, 5185 (1987); R. M. Giuliano and F. Duarte, *Synlett*, 419 (1992).

$$RCH=CHR + PhSeCl + AgBF_4 + H_2NCO_2C_2H_5 \longrightarrow \underset{\underset{NHCO_2C_2H_5}{|}}{\overset{\overset{PhSe}{|}}{RCH-CHR}}$$

Ref. 57

In the absence of better nucleophiles, solvent can be captured, as in selenenylamidation, which occurs in acetonitrile.

$$CH_3(CH_2)_5CH=CH_2 \xrightarrow[CH_3CN]{PhSeCl} \underset{\underset{NHCOCH_3}{|}}{CH_3(CH_2)_5CHCH_2SePh} + \underset{\underset{SePh}{|}}{CH_3(CH_2)_5CHCH_2NHCOCH_3}$$

85:15

Ref. 58

When reactions with phenylselenenyl chloride are carried out in aqueous acetonitrile solution, β-hydroxyselenides are formed as the result of solvolysis of the chloride.[59]

$$(CH_3)_2C=CH_2 \xrightarrow[CH_3CN-H_2O]{PhSeCl} \underset{\underset{OH}{|}}{(CH_3)_2CCH_2SePh} \quad 87\%$$

Mechanistic studies have been most thorough with the sulfenyl halides.[60] The reactions show moderate sensitivity to alkene structure, with ERGs on the alkene accelerating the reaction. The addition can occur in either the Markovnikov or anti-Markovnikov sense.[61] The variation in regioselectivity can be understood by focusing attention on the sulfur-bridged intermediate, which may range from being a sulfonium ion to a less electrophilic chlorosulfurane.

Compared to a bromonium ion, the C—S bonds are stronger and the TS for nucleophilic addition is reached later. This is especially true for the sulfurane structures. Steric interactions that influence access by the nucleophile are a more important factor in determining the direction of addition. For reactions involving phenylsulfenyl chloride or methylsulfenyl chloride, the intermediate is a fairly stable species and ease of approach by the nucleophile is the major factor in determining the direction of ring opening. In these cases, the product has the anti-Markovnikov orientation.[62]

[57] C. G. Francisco, E. I. Leon, J. A. Salazar, and E. Suarez, *Tetrahedron Lett.*, **27**, 2513 (1986).

[58] A. Toshimitsu, T. Aoai, H. Owada, S. Uemura, and M. Okano, *J. Org. Chem.*, **46**, 4727 (1981).

[59] A. Toshimitsu, T. Aoai, H. Owada, S. Uemura, and M. Okano, *Tetrahedron*, **41**, 5301 (1985).

[60] W. A. Smit, N. S. Zefirov, I. V. Bodrikov, and M. Z. Krimer, *Acc. Chem. Res.*, **12**, 282 (1979); G. H. Schmid and D. G. Garratt, *The Chemistry of Double-Bonded Functional Groups*, S. Patai, ed., Wiley-Interscience, New York, 1977, Chap. 9; G. A. Jones, C. J. M. Stirling, and N. G. Bromby, *J. Chem. Soc., Perkin Trans.*, **2**, 385 (1983).

[61] W. H. Mueller and P. E. Butler, *J. Am. Chem. Soc.*, **90**, 2075 (1968); G. H. Schmid and D. I. Macdonald, *Tetrahedron Lett.*, **25**, 157 (1984).

[62] G. H. Schmid, M. Strukelj, S. Dalipi, and M. D. Ryan, *J. Org. Chem.*, **52**, 2403 (1987).

$$CH_2=CHCH(CH_3)_2 \xrightarrow{CH_3SCl} ClCH_2CHCH(CH_3)_2 + CH_3SCH_2CHCH(CH_3)_2$$

$$\underset{SCH_3}{|} \quad 94\% \qquad \underset{Cl}{|} \quad 6\%$$

Ref. 61a

$$CH_3CH_2CH=CH_2 \xrightarrow{p\text{-}ClPhSCl} ClCH_2CHCH_2CH_3 + ArSCH_2CHCH_2CH_3$$

$$\underset{SAr}{|} \quad 77\% \qquad \underset{Cl}{|} \quad 23\%$$

Ref. 63

Terminal alkenes react with selenenyl halides with Markovnikov regioselectivity.[64] However, the β-selenyl halide addition products readily rearrange to the isomeric products.[65]

$$R_2C=CH_2 + ArSeX \longrightarrow \underset{X}{\overset{}{R_2CCH_2SeAr}} \rightleftharpoons \overset{ArSe}{\underset{}{R_2CCH_2X}}$$

4.2. Electrophilic Cyclization

When unsaturated reactants contain substituents that can participate as nucleophiles, electrophilic reagents frequently bring about cyclizations. Groups that can act as internal nucleophiles include carboxy and carboxylate, hydroxy, amino and amido, as well as carbonyl oxygen. There have been numerous examples of synthetic application of these electrophilic cyclizations.[66] The ring-size preference is usually $5 > 6 > 3 > 4$, but there are exceptions. Both the ring-size preference and the stereoselectivity reactions can usually be traced to structural and conformational features of the cyclization TS. Baldwin called attention to the role of stereoelectronic factors in cyclization reactions.[67] He classified cyclization reactions as *exo* and *endo* and as *tet*, *trig*, and *dig*, according to the hybridization at the cyclization center. The cyclizations are also designated by the size of the ring being formed. For any given separation ($n = 1, 2, 3$, etc.) of the electrophilic and nucleophilic centers, either an *exo* or *endo* mode of cyclization is usually preferred. The preferences for cyclization at trigonal centers are 5-*endo* >> 4-*exo* for $n = 2$; 5-*exo* > 6-*endo* for $n = 3$; and 6-*exo* >> 7-*endo* for $n = 4$. These relationships are determined by the preferred trajectory of the nucleophile to the electrophilic center. Substituents can affect the TS structure by establishing a preferred conformation and by electronic or steric effects.

[63.] G. H. Schmid, C. L. Dean, and D. G. Garratt, *Can. J. Chem.*, **54**, 1253 (1976).
[64.] D. Liotta and G. Zima, *Tetrahedron Lett.*, 4977 (1978); P. T. Ho and R. J. Holt, *Can. J. Chem.*, **60**, 663 (1982).
[65.] S. Raucher, *J. Org. Chem.*, **42**, 2950 (1977).
[66.] M. Frederickson and R. Grigg, *Org. Prep. Proced. Int.*, **29**, 63 (1997).
[67.] J. E. Baldwin, *J. Chem. Soc., Chem. Commun.*, 734, 738 (1976).

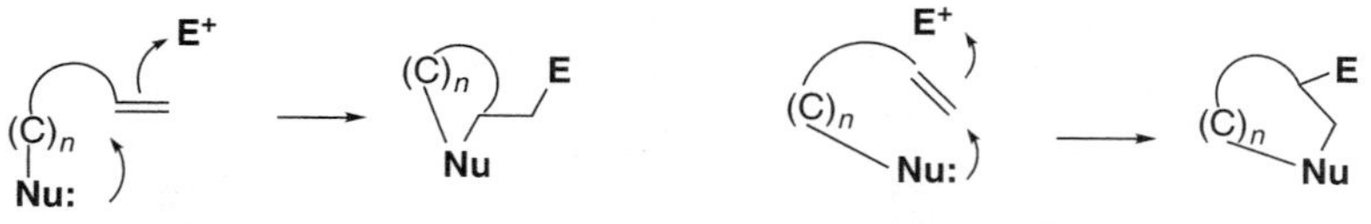

exo-trig cyclization   *endo-trig* cyclization

Electrophilic cyclizations are useful for closure of a variety of oxygen-, nitrogen-, and sulfur-containing rings. The product structure depends on the ring size and the *exo-endo* selectivity. The most common cases are formation of five- and six-membered rings.

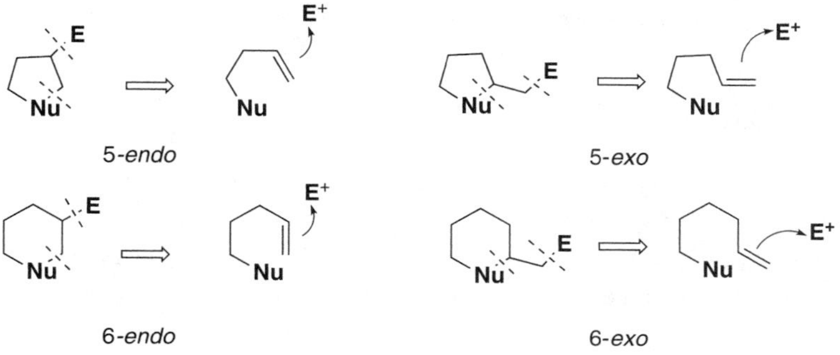

5-*endo*   5-*exo*

6-*endo*   6-*exo*

$Nu = CO_2^-$, OH, C=O, NHR, SH E^+ = Br^+, I^+, RS^+, RSe^+, Hg^{2+}

4.2.1. Halocyclization

Brominating and iodinating reagents effect cyclization of alkenes that have a nucleophilic group situated to permit formation of five-, six-, and, in some cases, seven-membered rings. Hydroxy and carboxylate groups are the most common nucleophiles, but the reaction is feasible for any nucleophilic group that is compatible with the electrophilic halogen source. Amides and carbamates can react at either oxygen or nitrogen, depending on the relative proximity. Sulfonamides are also potential nitrogen nucleophiles. Carbonyl oxygens can act as nucleophiles and give stable products by α-deprotonation.

Intramolecular reactions usually dominate intermolecular addition for favorable ring sizes. Semiempirical (AM1) calculations found the intramolecular TS favorable to a comparable intermolecular reaction.[68] (See Figure 4.1) The intramolecular TS, which is nearly 4 kcal/mol more stable, is quite productlike with a C—O bond distance of 1.6 Å, and a bond order of 0.62. The bromonium ion bridging is unsymmetrical and fairly weak. The bond parameters for the intra- and intermolecular TSs are quite similar.

In general, cyclization can be expected in compounds having the potential for formation of five- or six-membered rings. In addition to the more typical bromination reagents, such as those listed in Table 4.2, the combination of trimethylsilyl bromide, a tertiary amine, and DMSO can effect bromolactonization.

[68.] J. Sperka and D. C. Liotta, *Heterocycles*, **35**, 701 (1993).

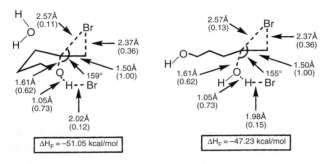

Fig. 4.1. Comparison of intramolecular and intermolecular transition
structures for reaction of Br^+, H_2O and 4-penten-1-ol. The numbers in
parentheses are bond orders. From *Heterocycles*, **35**, 701 (1993)

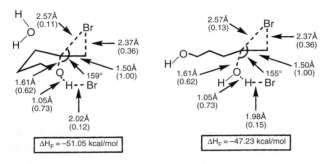

60%   Ref. 69

3-Phenylprop-2-enyl sulfates are cyclized stereospecifically and with Markovnikov
regiochemical control. These are 6-*endo* cyclizations.

Ref. 70

    Iodine is a very good electrophile for effecting intramolecular nucleophilic
addition to alkenes, as exemplified by the *iodolactonization reaction*.[71] Reaction of
iodine with carboxylic acids having carbon-carbon double bonds placed to permit
intramolecular reaction results in formation of iodolactones. The reaction shows a
preference for formation of five- over six-membered[72] rings and is a stereospecific
anti addition when carried out under basic conditions.

Ref. 73

[69] R. Iwata, A. Tanaka, H. Mizuno, and K. Miyashita, *Hetereocycles*, **31**, 987 (1990).

[70] J. G. Steinmann, J. H. Phillips, W. J. Sanders, and L. L. Kiessling, *Org. Lett.*, **3**, 3557 (2001).

[71] M. D. Dowle and D. I. Davies, *Chem. Soc. Rev.*, **8**, 171 (1979); G. Cardillo and M. Orena, *Tetrahedron*, **46**, 3321 (1990); S. Robin and G. Rousseau, *Tetrahedron*, **54**, 13681 (1998); S. Ranganathan, K. M. Muraleedharan, N. K. Vaish, and N. Jayaraman, *Tetrahedron*, **60**, 5273 (2004).

[72] S. Ranganathan, D. Ranganathan, and A. K. Mehrota, *Tetrahedron*, **33**, 807 (1977); C. V. Ramana, K. R. Reddy, and M. Nagarajan, *Ind. J. Chem. B*, **35**, 534 (1996).

[73] L. A. Paquette, G. D. Crouse, and A. K. Sharma, *J. Am. Chem. Soc.*, **102**, 3972 (1980).

The *anti* addition is a kinetically controlled process that results from irreversible back-side opening of an iodonium ion intermediate by the carboxylate nucleophile. Bartlett and co-workers showed that the more stable *trans* product was obtained under acidic conditions in which there is acid-catalyzed equilibration (thermodynamic control).[74]

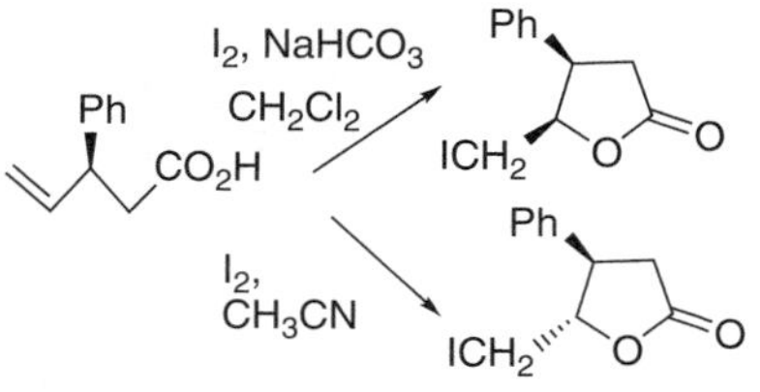

Ref. 75

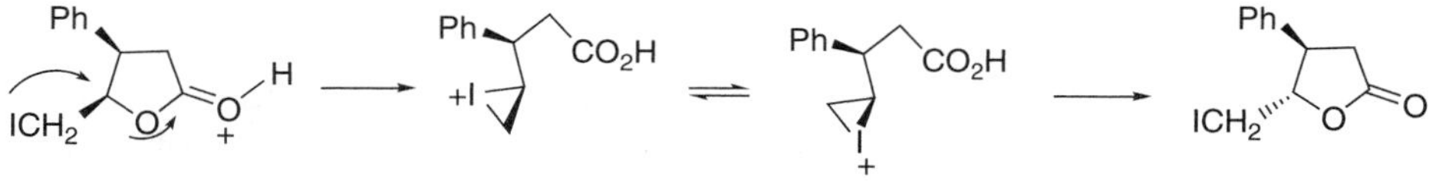

Under kinetic conditions, iodolactonization reflects reactant conformation. Several cases illustrate how the stereoselectivity of iodolactonization can be related to reactant conformation. For example, the high stereoselectivity of **1** corresponds to proximity of the carboxylate group to one of the two double bonds in the preferred reactant conformation.[76]

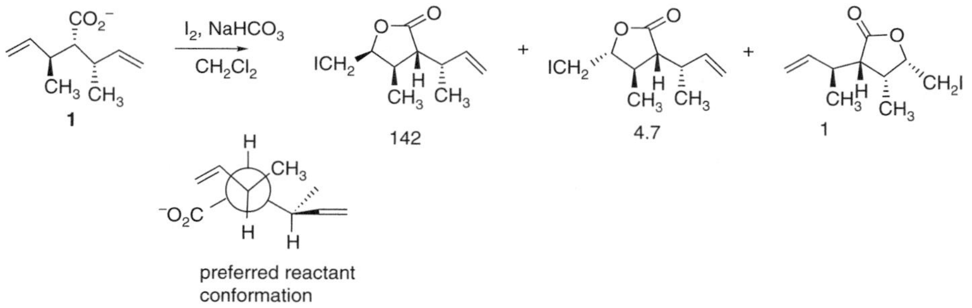

preferred reactant
conformation

Similarly, with reactants **2** and **3** conformational preference dominates in the selectivity between CO_2^- and CH_2OH as the internal nucleophile. This conformational preference even extends to CO_2CH_3, which can cyclize in preference to CH_2OH when it is in the conformationally preferred position.[77]

74. P. A. Bartlett and J. Myerson, *J. Am. Chem. Soc.*, **100**, 3950 (1978).
75. F. R. Gonzalez and P. A. Bartlett, *Org. Synth.*, **64**, 175 (1984).
76. M. J. Kurth and E. G. Brown, *J. Am. Chem. Soc.*, **109**, 6844 (1987).
77. M. J. Kurth, R. L. Beard, M. Olmstead, and J. G. Macmillan, *J. Am. Chem. Soc.*, **111**, 3712 (1989).

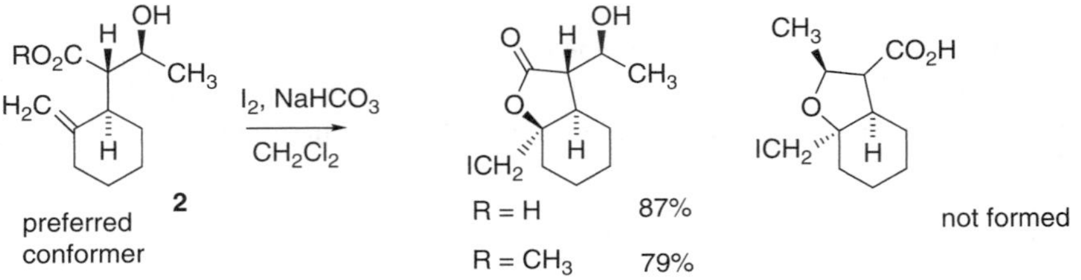

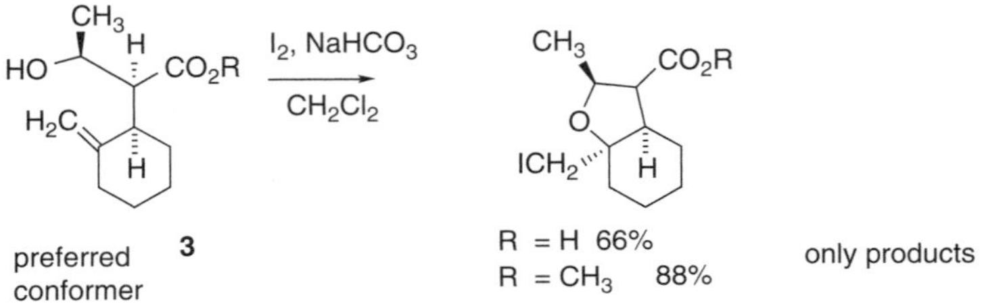

On the other hand, when the competition is between a monosubstituted and a disubstituted double bond, the inherent reactivity difference between the two double bonds overcomes reactant conformational preferences.[78]

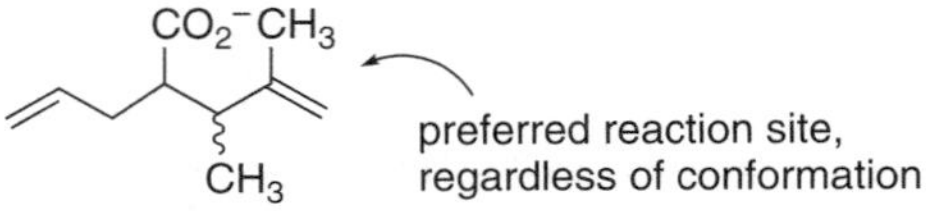

Several other nucleophilic functional groups can be induced to participate in iodocyclization reactions. *t*-Butyl carbonate esters cyclize to diol carbonates.[79]

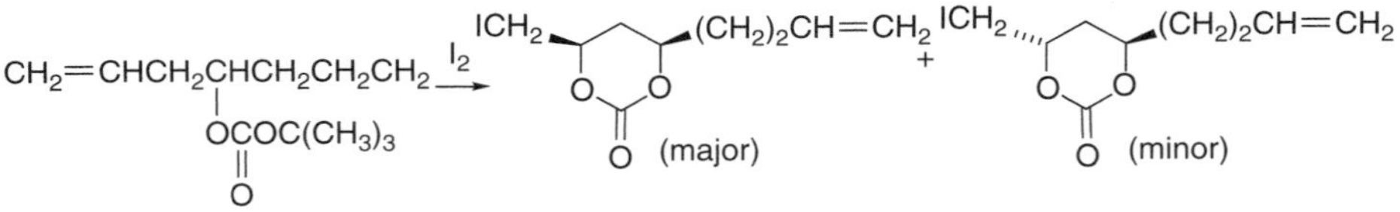

Lithium salts of carbonate monoesters can also be cyclized.[80]

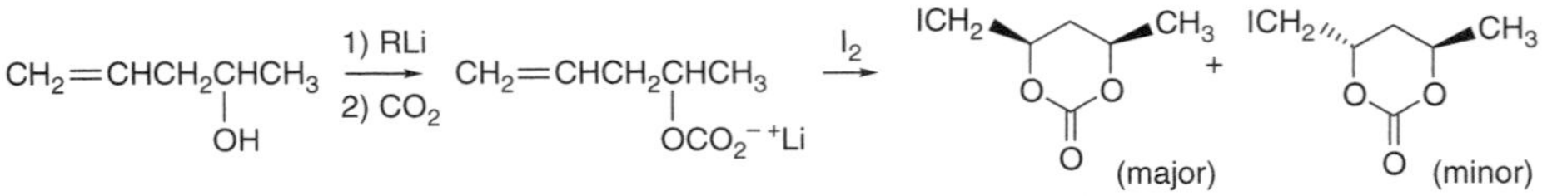

Enhanced stereoselectivity has been found using IBr, which reacts at a lower temperature.[81] (Compare Entries 6 and 7 in Scheme 4.4.) Other reagent systems that generate electrophilic iodine, such as $KI + KHSO_5$,[82] can be used for iodocyclization.

78. M. J. Kurth, E. G. Brown, E. J. Lewis, and J. C. McKew, *Tetrahedron Lett.*, **29**, 1517 (1988).
79. P. A. Bartlett, J. D. Meadows, E. G. Brown, A. Morimoto, and K. K. Jernstedt, *J. Org. Chem.*, **47**, 4013 (1982).
80. A. Bogini, G. Cardillo, M. Orena, G. Ponzi, and S. Sandri, *J. Org. Chem.*, **47**, 4626 (1982).
81. J. J.-W. Duan and A. B. Smith, III, *J. Org. Chem.*, **58**, 3703 (1993).
82. M. Curini, F. Epifano, M. C. Marcotullio, and F. Montanari, *Synlett*, 368 (2004).

Analogous cyclization reactions are induced by brominating reagents but they tend to be less selective than the iodocyclizations.[83] The bromonium ion intermediates are much more reactive and less selective.

The iodocyclization products have a potentially nucleophilic oxygen substituent β to the iodide, which makes them useful in stereospecific syntheses of epoxides and diols.

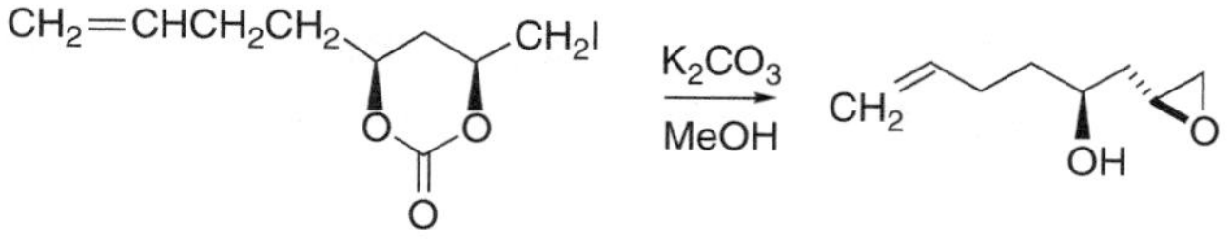

Ref. 79

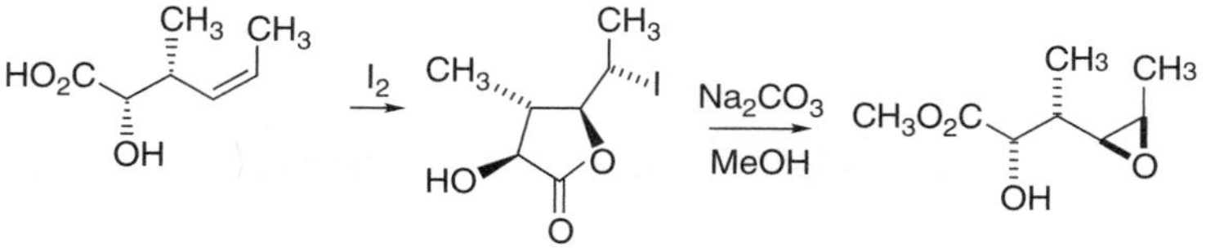

Ref. 84

Positive halogen reagents can cyclize γ- and δ-hydroxyalkenes to tetrahydrofuran and tetrahydropyran derivatives, respectively.[85] Iodocyclization of homoallylic alcohols generates 3-iodotetrahydrofurans when conducted in anhydrous acetonitrile.[86] The reactions are stereospecific, with the *E*-alcohols generating the *trans* and the *Z*-isomer the *cis* product. These are 5-*endo* cyclizations, which are preferred to 4-*exo* reactions.

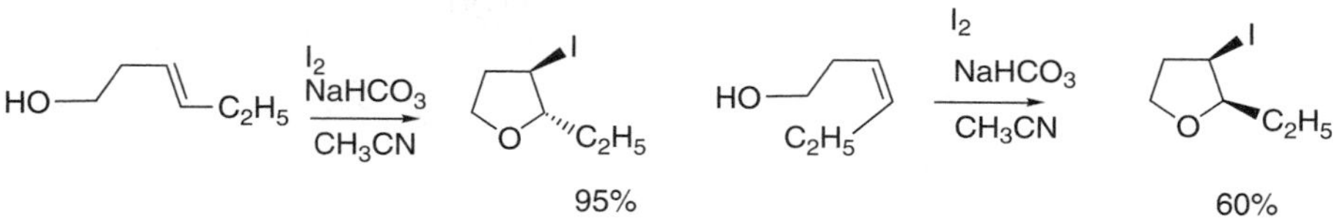

With the corresponding secondary alcohols, the preferred cyclization is via a conformation with a pseudoequatorial conformation.

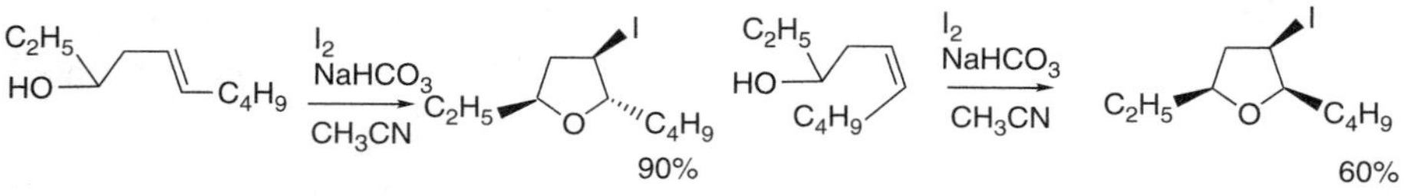

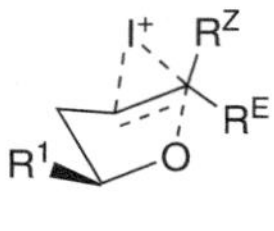

83. B. B. Snider and M. I. Johnston, *Tetrahedron Lett.*, **26**, 5497 (1985).
84. C. Neukome, D. P. Richardson, J. H. Myerson, and P. A. Bartlett, *J. Am. Chem. Soc.*, **108**, 5559 (1986).
85. A. B. Reitz, S. O. Nortey, B. E. Maryanoff, D. Liotta, and R. Monahan, III, *J. Org. Chem.*, **52**, 4191 (1981).
86. J. M. Banks, D. W. Knight, C. J. Seaman, and G. G. Weingarten, *Tetrahedron Lett.*, **35**, 7259 (1994); S. B. Bedford, K. E. Bell, F. Bennett, C. J. Hayes, D. W. Knight, and D. E. Shaw, *J. Chem. Soc., Perkin Trans. 1*, 2143 (1999).

Related *O*-TBS and *O*-benzyl ethers cyclize with loss of the ether substituent.

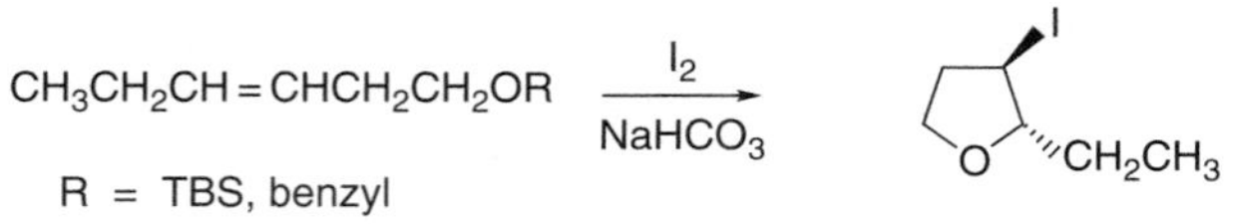
CH3CH2CH = CHCH2CH2OR
R = TBS, benzyl
I2
NaHCO3
I
O
CH2CH3

Ref. 87

Other nucleophilic functional groups can participate in iodocyclization. Amides usually react at oxygen, generating imino lactones that are hydrolyzed to lactones.[88]

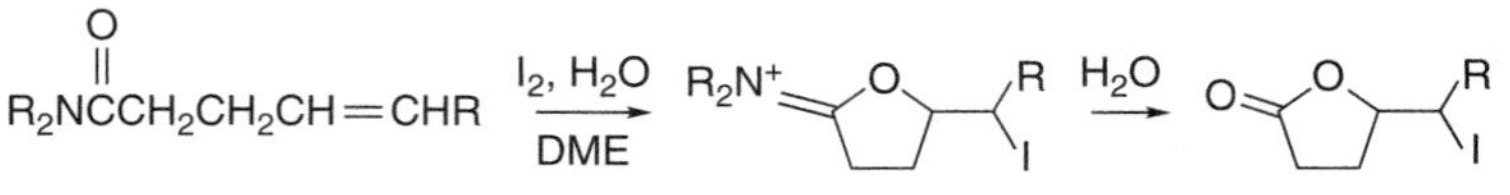
O
R2NCCH2CH2CH=CHR
I2, H2O
DME
R2N+
O
R
I
H2O
O
O
R
I

Ref. 89

Use of a chiral amide can promote enantioselective cyclization.[90]

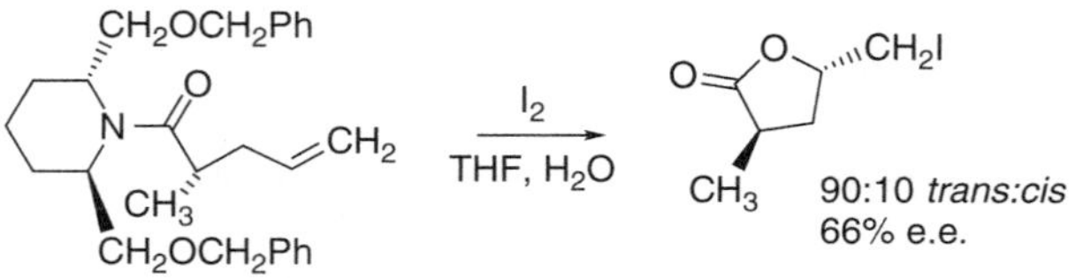
CH2OCH2Ph
O
N
CH2
CH3
CH2OCH2Ph
I2
THF, H2O
O
O
CH2I
CH3
90:10 trans:cis
66% e.e.

The TS preference is influenced by avoidance of $A^{1,3}$ strain between the α-methyl group and the piperidine ring.

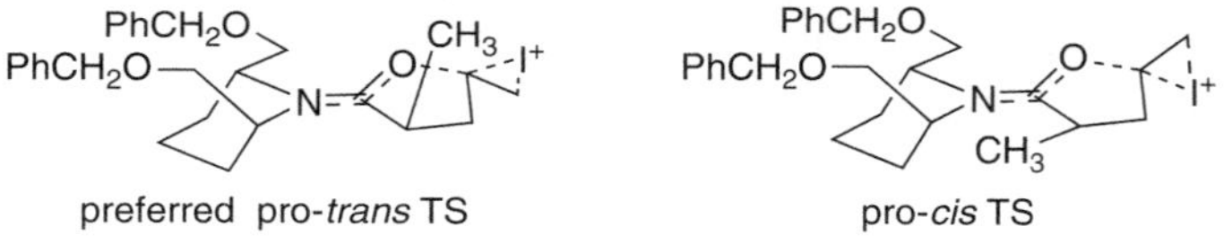
PhCH2O
PhCH2O
CH3
O
I+
N
preferred pro-trans TS
PhCH2O
PhCH2O
O
I+
N
CH3
pro-cis TS

Lactams can be obtained by cyclization of *O,N*-trimethylsilyl imidates.[91]

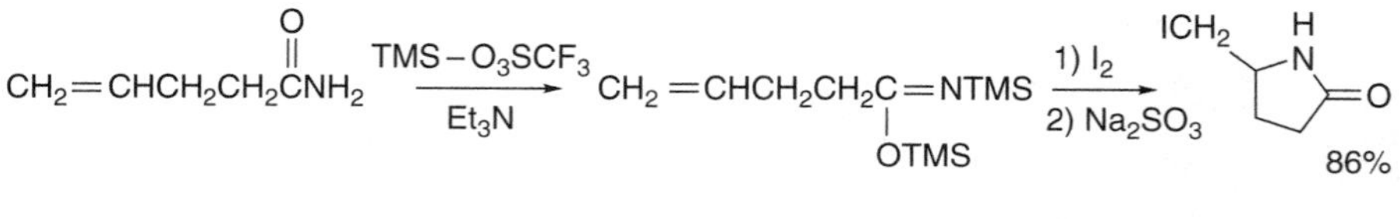
O
CH2=CHCH2CH2CNH2
TMS−O3SCF3
Et3N
CH2 =CHCH2CH2C=NTMS
OTMS
1) I2
2) Na2SO3
ICH2
H
N
O
86%

As compared with amides, where oxygen is the most nucleophilic atom, the silyl imidates are more nucleophilic at nitrogen.

Examples of halolactonization and related halocyclizations can be found in Scheme 4.4. The first entry, which involves NBS as the electrophile, demonstrates the *anti* stereospecificity of the reaction, as well as the preference for five-membered rings.

87. S. P. Bew, J. M. Barks, D. W. Knight, and R. J. Middleton, *Tetrahedron Lett.*, **41**, 4447 (2000).
88. S. Robin and G. Rousseau, *Tetrahedron*, **54**, 13681 (1998).
89. Y. Tamaru, M. Mizutani, Y. Furukawa, S. Kawamura, Z. Yoshida, K. Yanagi, and M. Minobe, *J. Am. Chem. Soc.*, **106**, 1079 (1984).
90. S. Najdi, D. Reichlin, and M. J. Kurth, *J. Org. Chem.*, **55**, 6241 (1990).
91. S. Knapp, K. E. Rodriquez, A. T. Levorse, and R. M. Ornat, *Tetrahedron Lett.*, **26**, 1803 (1985).

Scheme 4.4. Iodolactonizations and Other Halocyclizations

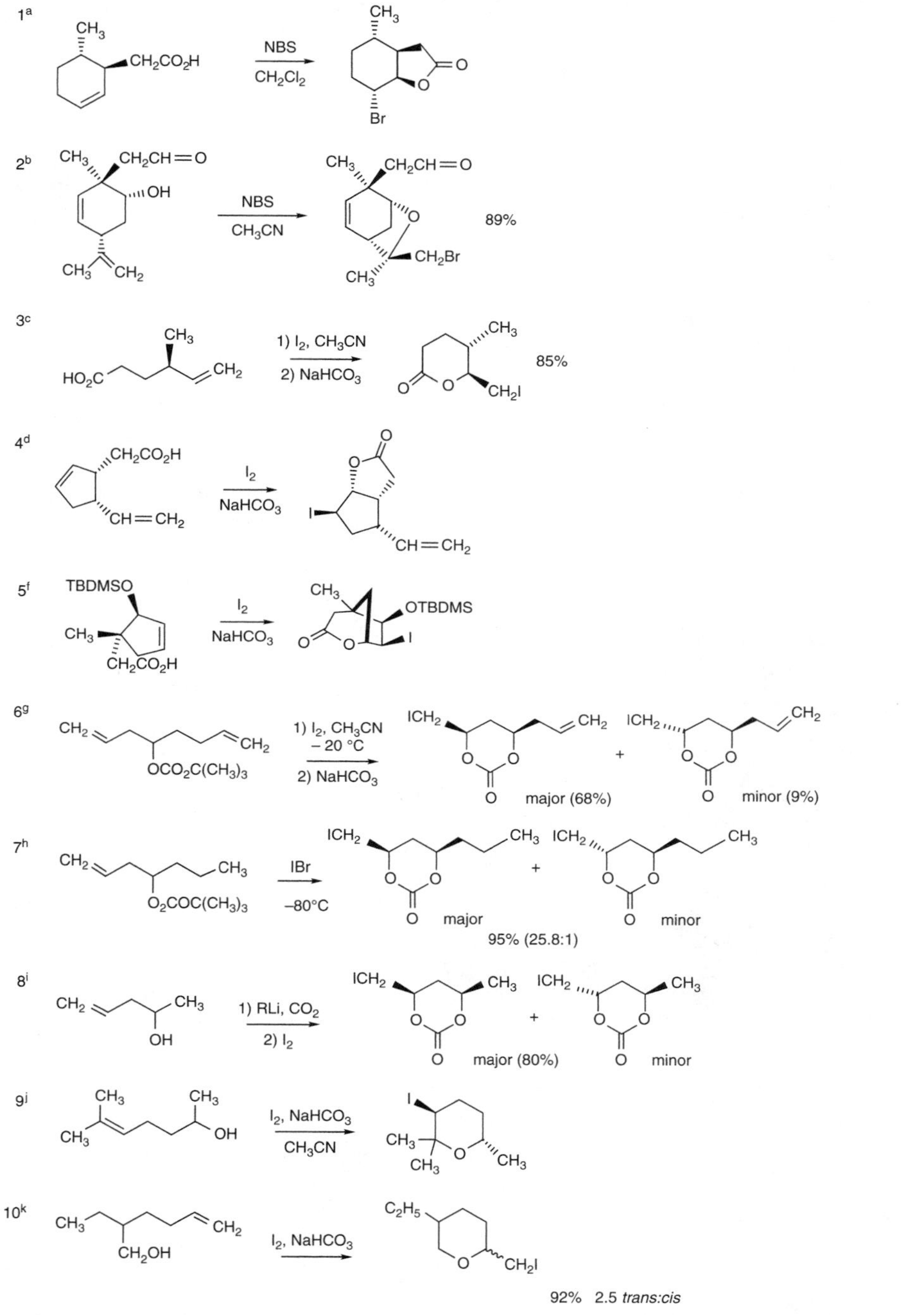

1[a]

2[b]   89%

3[c]   85%

4[d]

5[f]

6[g]   major (68%)   +   minor (9%)

7[h]   major   +   minor   95% (25.8:1)

8[i]   major (80%)   +   minor

9[j]

10[k]   92% 2.5 *trans:cis*

(Continued)

Scheme 4.4. (*Continued*)

11[k]

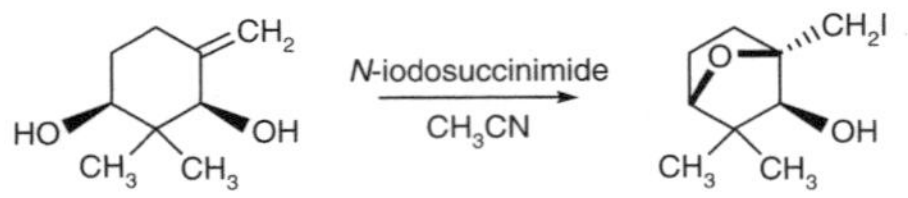

N-iodosuccinimide
CH₃CN

12[l]

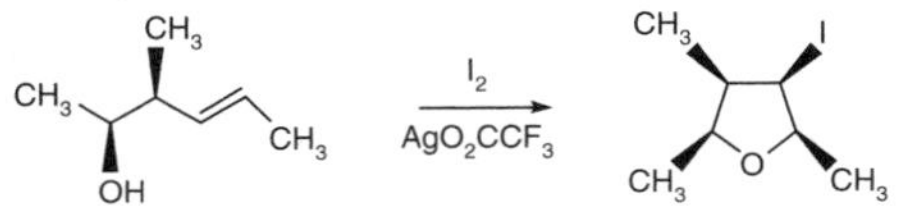

I₂
AgO₂CCF₃

13[m]

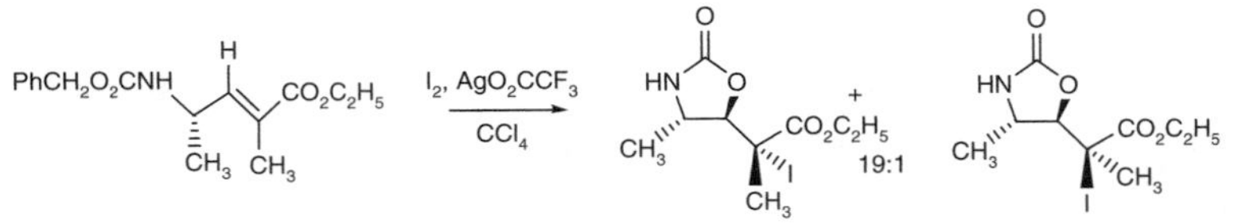

I₂, AgO₂CCF₃
CCl₄

19:1

14[n]

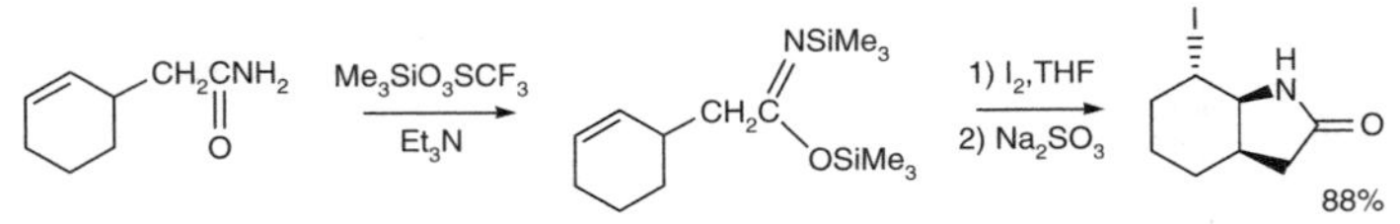

Me₃SiO₃SCF₃
Et₃N

1) I₂,THF
2) Na₂SO₃

88%

15[o]

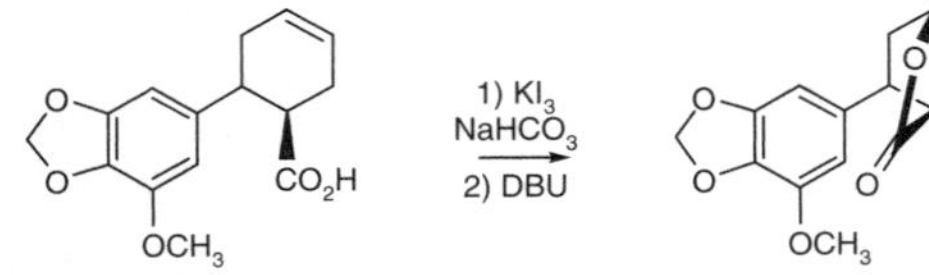

1) KI₃
NaHCO₃
2) DBU

16[p]

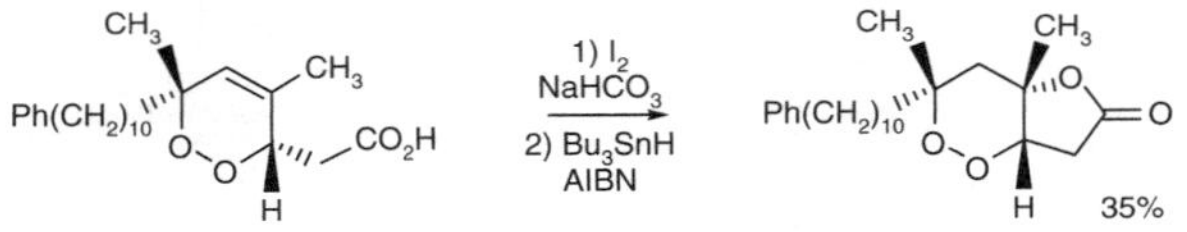

1) I₂
NaHCO₃
2) Bu₃SnH
AIBN

35%

17[q]

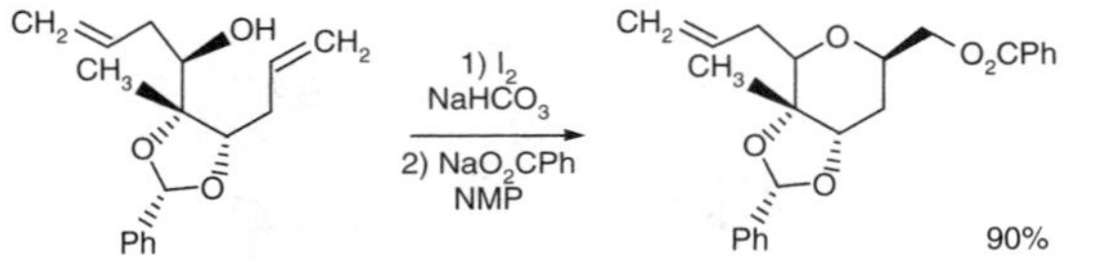

1) I₂
NaHCO₃
2) NaO₂CPh
NMP

90%

18[r]

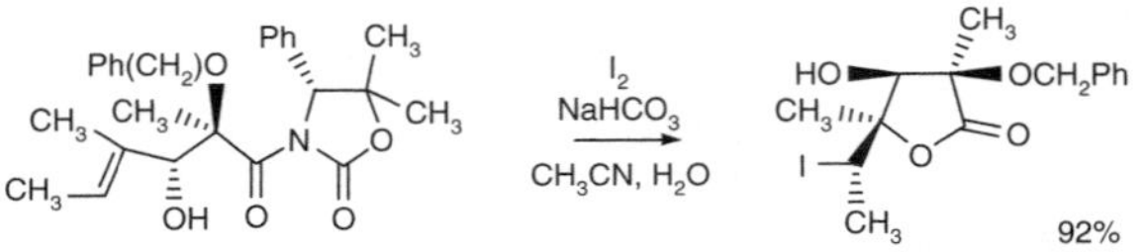

I₂
NaHCO₃
CH₃CN, H₂O

92%

19[s]

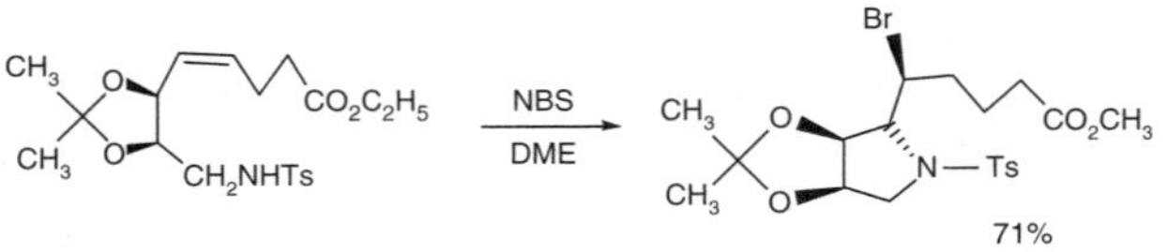

NBS
DME

71%

(*Continued*)

Scheme 4.4. (*Continued*)

319

SECTION 4.2

Electrophilic Cyclization

a. M. F. Semmelhack, W. R. Epa, A. W. H. Cheung, Y. Gu, C. Kim, N. Zhang, and W. Lew, *J. Am. Chem. Soc.*, **116**, 7455 (1994).
b. M. Miyashita, T. Suzuki, and A. Yoshikoshi, *J. Am. Chem. Soc.*, **111**, 3728 (1989).
c. A. G. M. Barrett, R. A. E. Carr, S. V. Atwood, G. Richardson, and N. D. A. Walshe, *J. Org. Chem.*, **51**, 4840 (1986).
d. L. A. Paquette, G. D. Crouse, and A. K. Sharma, *J. Am. Chem. Soc.*, **102**, 3972 (1980).
e. A. J. Pearson and S.-Y. Hsu, *J. Org. Chem.*, **51**, 2505 (1986).
f. P. A. Bartlett, J. D. Meadows, E. G. Brown, A. Morimoto, and K. K. Jernstedt, *J. Org. Chem.*, **47**, 4013 (1982).
g. J. J.-W. Duan and A. B. Smith, III, *J. Org. Chem.*, **58**, 3703 (1993).
h. L. F. Tietze and C. Schneider, *J. Org. Chem.*, **56**, 2476 (1991).
i. G. L. Edwards and K. A. Walker, *Tetrahedron Lett.*, **33**, 1779 (1992).
j. A. Bongini, G. Cardillo, M. Orena, G. Porzi, and S. Sandri, *J. Org. Chem.*, **47**, 4626 (1982).
k. A. Murai, N. Tanimoto, N. Sakamoto, and T. Masamune, *J. Am. Chem. Soc.*, **110**, 1985 (1988).
l. B. F. Lipshutz and J. C. Barton, *J. Am. Chem. Soc.*, **114**, 1084 (1992).
m. Y. Guindon, A. Slassi, E. Ghiro, G. Bantle, and G. Jung, *Tetrahedron Lett.*, **33**, 4257 (1992).
n. S. Knapp and A. T. Levorse, *J. Org. Chem.*, **53**, 4006 (1988).
o. S. Kim, H. Ko, E. Kim, and D. Kim, *Org. Lett.*, **4**, 1343 (2002).
p. M. Jung, J. Han, and J. Song, *Org. Lett.*, **4**, 2763 (2002).
q. S. H. Kang, S. Y. Kang, H. Choi, C. M. Kim, H.-S. Jun, and J.-H. Youn, *Synthesis*, 1102 (2004).
r. Y. Murata, T. Kamino, T. Aoki, S. Hosokawa, and S. Kobayashi, *Angew. Chem. Int. Ed. Engl.*, **43**, 3175 (2004).
s. Y. G. Kim and J. K. Cha, *Tetrahedron Lett.*, **30**, 5721 (1989).

Entry 2 is a 5-*exo* bromocyclization. The reaction in Entry 3 involves formation of a δ-lactone in an acyclic system. This reaction was carried out under conditions that lead to the thermodynamically favored *trans* isomer. Entry 4 shows typical iodolactonization conditions and illustrates both the *anti* stereoselectivity and preference for formation of five-membered rings. In Entry 5, a six-membered lactone is formed, again with *anti* stereospecificity. Entry 6 is a cyclization of a *t*-butyl carbonate ester. The selectivity between the two double bonds is the result of the relative proximity of the nucleophilic group. Entry 7 is a closely related reaction, but carried out at a much lower temperature by the use of IBr. The *cis:trans* ratio was improved to nearly 26:1. The ratio was also solvent dependent, with toluene being the best solvent. Entry 8 is a variation using a lithium carbonate as the nucleophile. Entries 9 and 10 involve hydroxy groups as nucleophiles. Entry 9 is a 6-*endo* iodocyclization. In Entry 10, a primary hydroxy group serves as the nucleophile. Entry 11 is another cyclization involving a hydroxy group, in this case forming a 7-oxabicyclo[2.2.1]heptane structure. Entry 12 is a rather unusual 5-*endo* cyclization.

Entry 13 shows cyclization with concomitant loss of the benzyloxycarbonyl group. The TS for this reaction is 5-*exo* with conformation determined by the pseudoequatorial position of the methyl group.

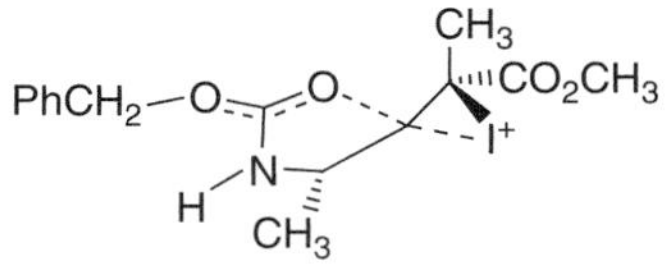

Entry 14 involves formation of a lactam by cyclization of a *bis*-trimethylsilylimidate. The stereoselectivity parallels that of iodolactonization.

Entries 15 to 18 are examples of use of iodocyclization in multistep syntheses. In Entry 15, iodolactonization was followed by elimination of HI from the bicyclic lactone. In Entry 16, a cyclic peroxide group remained unaffected by the standard iodolactonization and subsequent Bu$_3$SnH reductive deiodination. (See Section 5.5 for a discussion of this reaction.) In Entry 17, the primary iodo substituent was replaced by a benzoate group. In Entry 18, the reactant was prepared with high *anti* selectivity by an auxiliary-directed aldol reaction. The acyloxazolidinone auxiliary then participated in the iodocyclization and was cleaved in the process.

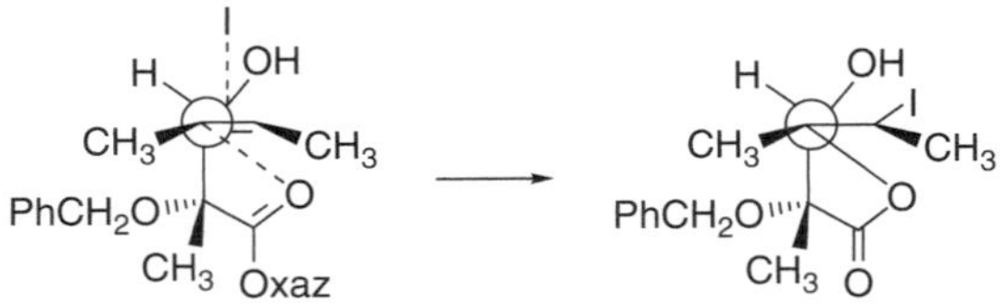

The reaction in Entry 19 was effected using NBS.

4.2.2. Sulfenylcyclization and Selenenylcyclization

Reactants with internal nucleophiles are also subject to cyclization by electrophilic sulfur reagents, a reaction known as *sulfenylcyclization*.[92] As for iodolactonization, unsaturated carboxylic acids give products that result from *anti* addition.[93]

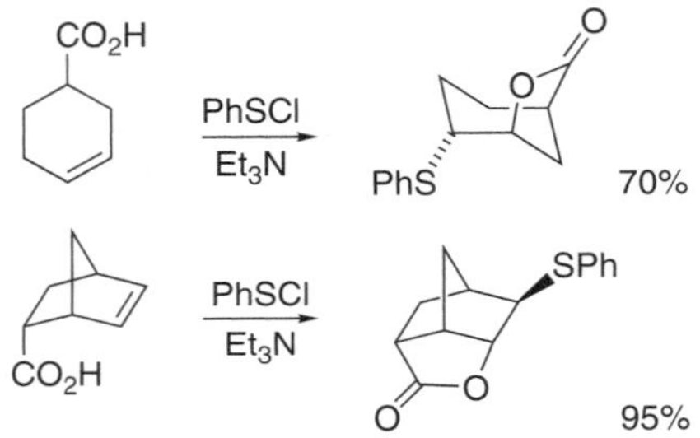

Similarly, alcohols undergo cyclization to ethers.

The corresponding reactions using selenium electrophiles are called selenenylcyclization.[94,95] Carboxylate (selenylactonization), hydroxy (selenyletherification), and nitrogen (selenylamidation) groups can all be captured in appropriate cases.

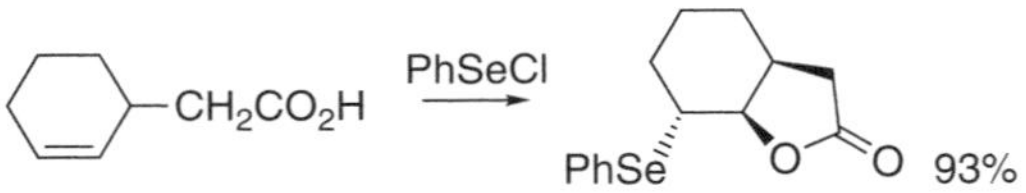

Internal nucleophilic capture of seleniranium ion is governed by general principles similar to those of other electrophilic cyclizations.[96] The stereochemistry of cyclization can usually be predicted on the basis of a cyclic TS with favored pseudoequatorial orientation of the substituents.

92. G. Capozzi, G. Modena, and L. Pasquato, in *The Chemistry of Sulphenic Acids and Their Derivatives*, S. Patai, ed., Wiley, Chichester, 1990, pp. 446–460.
93. K. C. Nicolaou, S. P. Seitz, W. T. Sipio, and J. F. Blount, *J. Am. Chem. Soc.*, **101**, 3884 (1979).
94. K. Fujita, *Rev. Heteroatom. Chem.*, **16**, 101 (1997).
95. K. C. Nicolaou, S. P. Seitz, W. J. Sipio, and J. F. Blount, *J. Am. Chem. Soc.*, **101**, 3884 (1979); M. Tiecco, *Topics Curr. Chem.*, **208**, 7 (2000); S. Raganathan, K. M. Muraleedharan, N. K. Vaish, and N. Jayaraman, *Tetrahedron*, **60**, 5273 (2004).
96. N. Petragnani, H. A. Stefani, and C. J. Valduga, *Tetrahedron*, **57**, 1411 (2001).

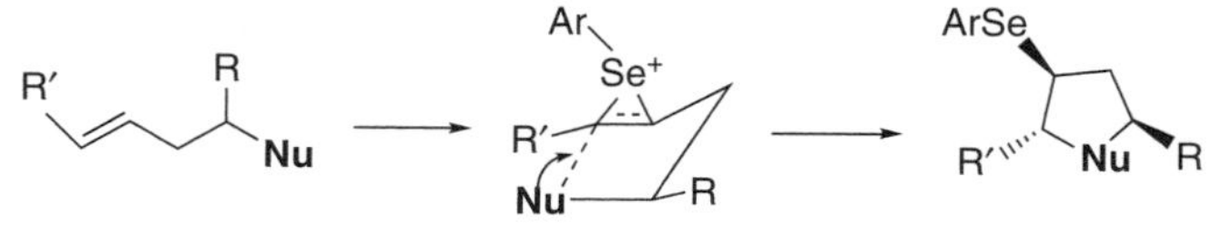

Although *exo* cyclization is usually preferred, there is no strong prohibition of *endo* cyclization and aryl-controlled regioselectivity can override the *exo* preference.

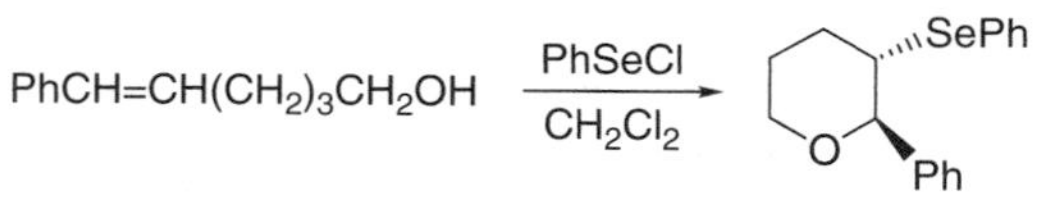

PhCH=CH(CH₂)₃CH₂OH $\xrightarrow[\text{CH}_2\text{Cl}_2]{\text{PhSeCl}}$

Ref. 97

PhSeBr / 2 equiv

93:7

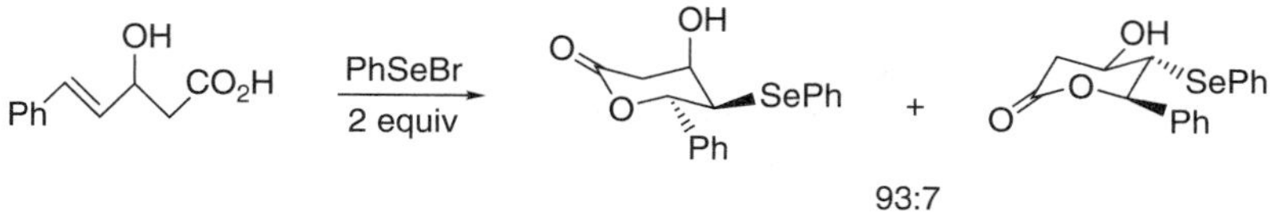

Ref. 98

Various electrophilic selenium reagents such as those described in Scheme 4.3 can be used. *N*-Phenylselenylphthalimide is an excellent reagent for this process and permits the formation of large ring lactones.[99] The advantage of the reagent in this particular application is the low nucleophilicity of phthalimide, which does not compete with the remote internal nucleophile. The reaction of phenylselenenyl chloride or *N*-phenylselenenylphthalimide with unsaturated alcohols leads to formation of β-phenylselenenyl ethers.

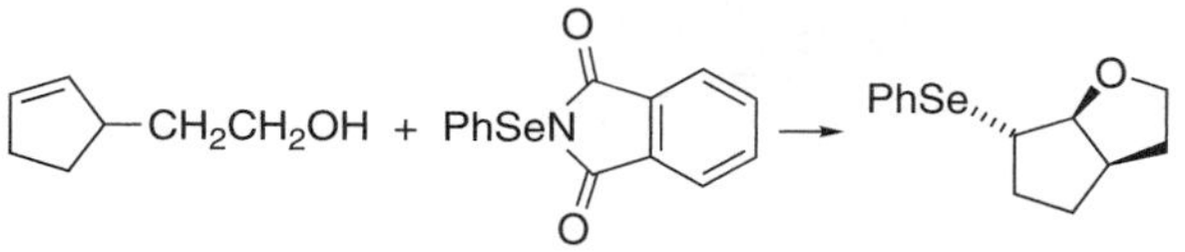

—CH₂CH₂OH + PhSeN

Ref. 100

Another useful reagent for selenenylcyclization is phenylselenenyl triflate. This reagent is capable of cyclizing unsaturated acids[101] and alcohols.[102] Phenylselenenyl sulfate can be prepared in situ by oxidation of diphenyl diselenide with ammonium peroxydisulfate.[103]

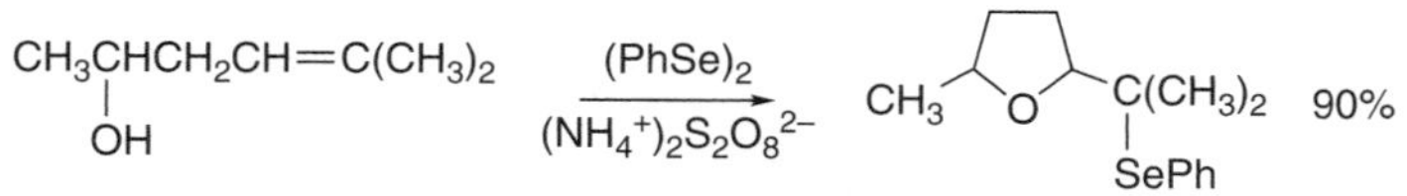

CH₃CHCH₂CH=C(CH₃)₂ $\xrightarrow[(\text{NH}_4^+)_2\text{S}_2\text{O}_8{}^{2-}]{(\text{PhSe})_2}$ 90%

Several examples of sulfenylcyclization are given in Section A of Scheme 4.5. Entry 1 is a *6-exo* sulfenoetherification induced by phenylsulfenyl chloride. Entry 2

[97.] M. A. Brimble, G. S. Pavia, and R. J. Stevenson, *Tetrahedron Lett.*, **43**, 1735 (2002).

[98.] M. Gruttadauria, C. Aprile, and R. Noto, *Tetrahedron Lett.*, **43**, 1669 (2002).

[99.] K. C. Nicolaou, D. A. Claremon, W. E. Barnette, and S. P. Seitz, *J. Am. Chem. Soc.*, **101**, 3704 (1979).

[100.] K. C. Nicolaou, R. L. Magolda, W. J. Sipio, W. E. Barnette, Z. Lysenko, and M. M. Joullie, *J. Am. Chem. Soc.*, **102**, 3784 (1980).

[101.] S. Murata and T. Suzuki, *Chem. Lett.*, 849 (1987).

[102.] A. G. Kutateladze, J. L. Kice, T. G. Kutateladze, N. S. Zefirov, and N. V. Zyk, *Tetrahedron Lett.*, **33**, 1949 (1992).

[103.] M. Tiecco, L. Testaferri, M. Tingoli, D. Bartoli, and R. Balducci, *J. Org. Chem.*, **55**, 429 (1990).

Scheme 4.5. Sulfenyl- and Selenenylcyclization Reactions

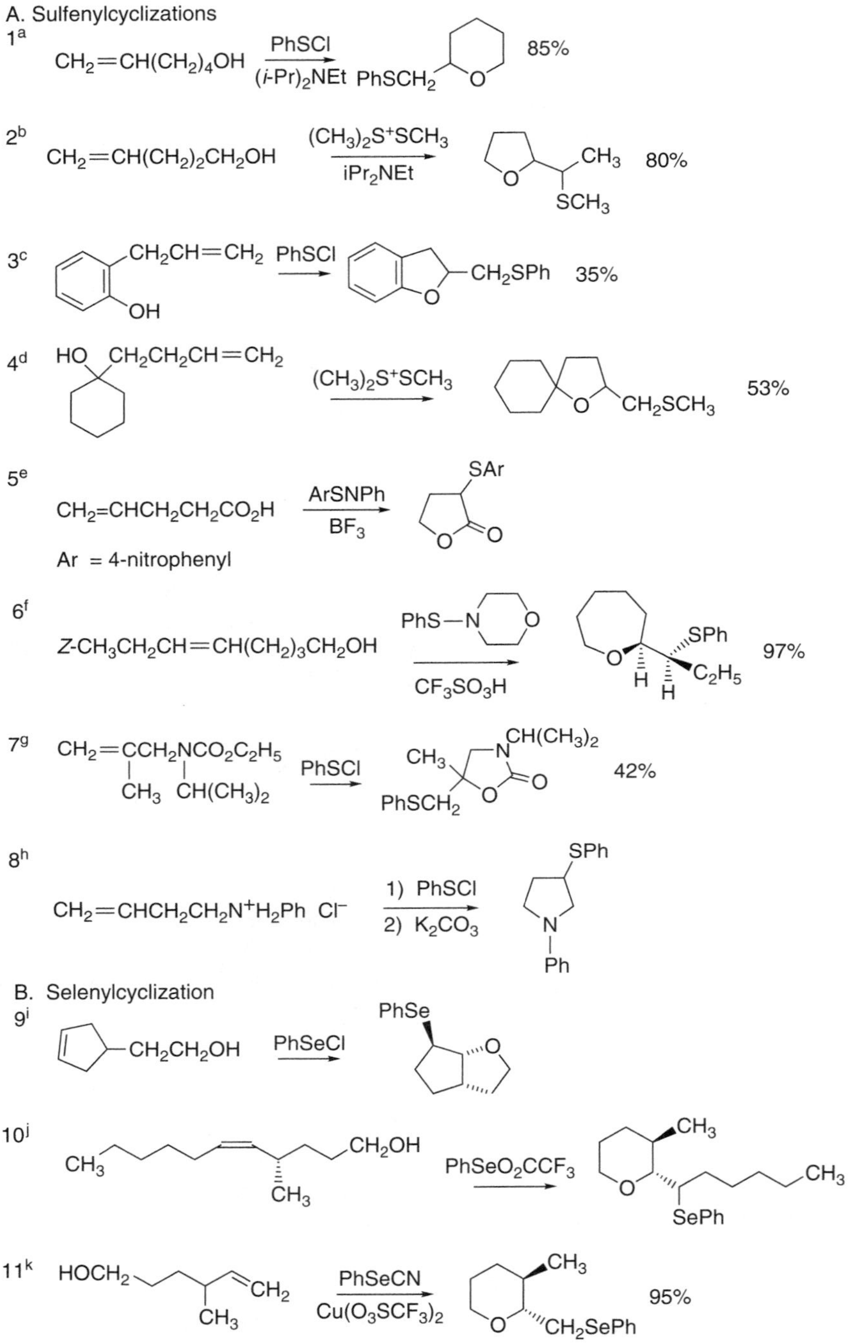
A. Sulfenylcyclizations
1a
CH2=CH(CH2)4OH
PhSCl
(i-Pr)2NEt
PhSCH2
85%
2b
CH2=CH(CH2)2CH2OH
(CH3)2S+SCH3
iPr2NEt
CH3
SCH3
80%
3c
CH2CH=CH2
OH
PhSCl
CH2SPh
35%
4d
HO
CH2CH2CH=CH2
(CH3)2S+SCH3
CH2SCH3
53%
5e
CH2=CHCH2CH2CO2H
ArSNPh
BF3
SAr
Ar = 4-nitrophenyl
6f
Z-CH3CH2CH=CH(CH2)3CH2OH
PhS—N
CF3SO3H
SPh
C2H5
97%
7g
CH2=CCH2NCO2C2H5
CH3 CH(CH3)2
PhSCl
CH(CH3)2
CH3
PhSCH2
42%
8h
CH2=CHCH2CH2N+H2Ph Cl−
1) PhSCl
2) K2CO3
SPh
Ph
B. Selenylcyclization
9i
CH2CH2OH
PhSeCl
PhSe
10j
CH3
CH2OH
PhSeO2CCF3
SePh
11k
HOCH2
CH2
CH3
PhSeCN
Cu(O3SCF3)2
CH2SePh
95%

(Continued)

Scheme 4.5. (*Continued*)

323

SECTION 4.2

Electrophilic Cyclization

12[l]

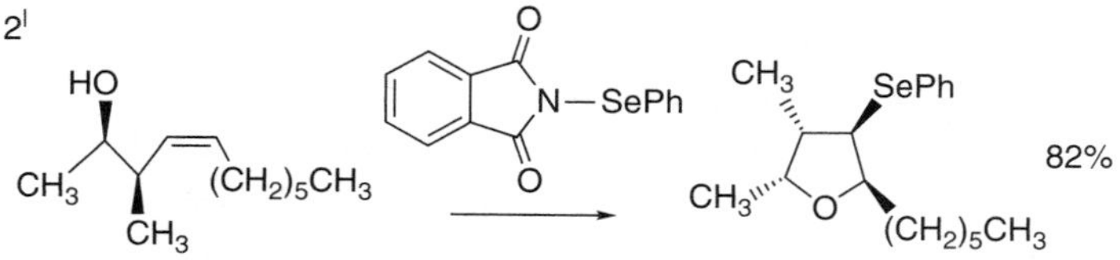

82%

13[m]

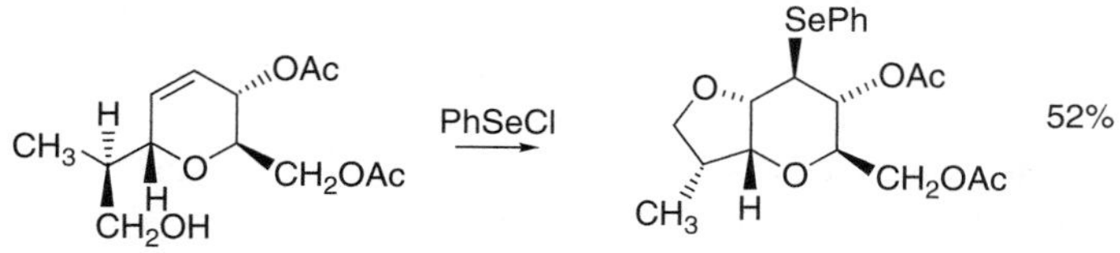

PhSeCl

52%

14[n]

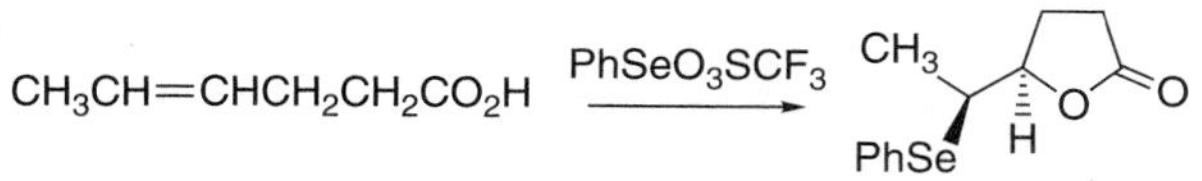

$CH_3CH{=}CHCH_2CH_2CO_2H$

PhSeO$_3$SCF$_3$

15[o]

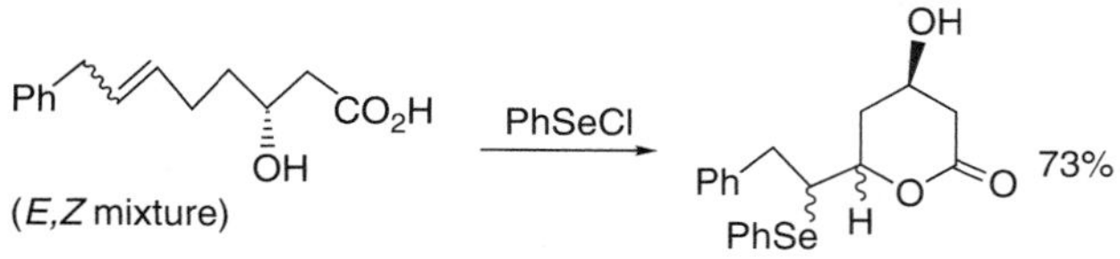

(*E,Z* mixture)

PhSeCl

73%

16[p]

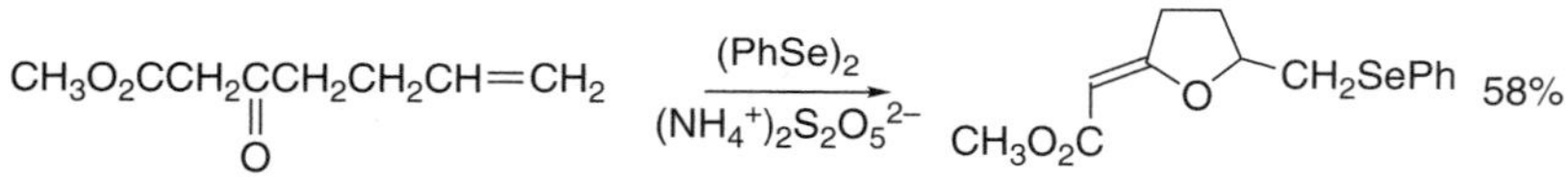

$CH_3O_2CCH_2CCH_2CH_2CH{=}CH_2$

$\overset{O}{\|}$

(PhSe)$_2$

(NH$_4{}^+$)$_2$S$_2$O$_5{}^{2-}$

58%

17[q]

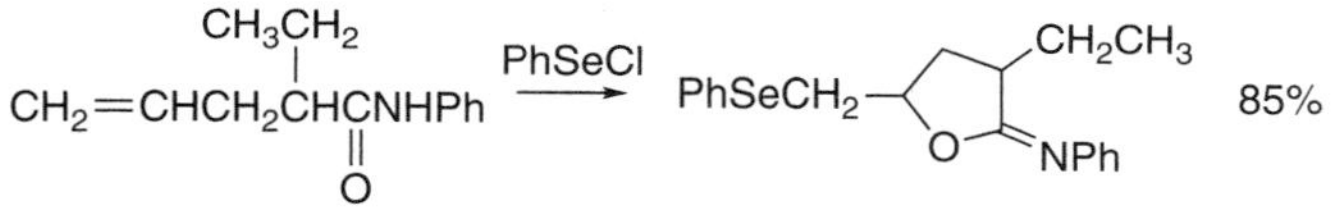

$CH_2{=}CHCH_2CHCNHPh$

with CH_3CH_2 and $\overset{O}{\|}$

PhSeCl

85%

18[r]

$CH_2CH_2CH{=}CH_2$

PhSeBr

68%

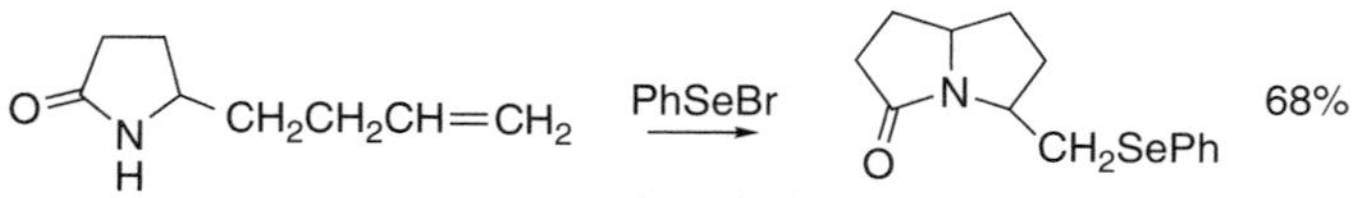

a. S. M. Tuladhar and A. G. Fallis, *Can. J. Chem.*, **65**, 1833 (1987).
b. G. J. O'Malley and M. P. Cava, *Tetrahedron Lett.*, **26**, 6159 (1985).
c. M. Muehlstaedt, C. Shubert, and E. Kleinpeter, *J. Prakt. Chem.*, **327**, 270 (1985).
d. G. Capozzi, S. Menichetti, M. Nicastro, and M. Taddei, *Heterocycles*, **29**, 1703 (1987).
e. L. Benati, L. Capella, P. C. Montevecchi, and P. Spagnolo, *Tetrahedron*, **50**, 12395 (1994).
f. P. Brownbridge, *J. Chem. Soc., Chem. Commun.*, 1280 (1980).
g. M. Muehlstaedt, R. Widera, and B. Olk, *J. Prakt. Chem.*, **324**, 362 (1982).
h. T. Ohsawa, M. Ihara, K. Fukumoto, and T. Kametani, *J. Org. Chem.*, **48**, 3644 (1983).
i. D. L. J. Clive, G. Chittattu, and C. K. Wong, *Can. J. Chem.*, **55**, 3894 (1987).
j. G. Li and W. C. Still, *J. Org. Chem.*, **56**, 6964 (1991).
k. H. Inoue and S. Murata, *Heterocycles*, **45**, 847 (1997).
l. E. D. Mihelich and G. A. Hite, *J. Am. Chem. Soc.*, **114**, 7318 (1992).
m. S. J. Danishefsky, S. DeNinno, and P. Lartey, *J. Am. Chem. Soc.*, **109**, 2082 (1987).
n. S. Murata and T. Suzuki, *Chem. Lett.*, 849 (1987).
o. F. Bennett, D. W. Knight, and G. Fenton, *J. Chem. Soc., Perkin Trans. 1*, 519 (1991).
p. M. Tiecco, L. Testaferri, M. Tingoli, D. Bartoli, and R. Balducci, *J. Org. Chem.*, **55**, 429 (1990).
q. A. Toshimitsu, K. Terao, and S. Uemura, *J. Org. Chem.*, **52**, 2018 (1987).
r. A. Toshimitsu, K. Terao, and S. Uemura, *J. Org. Chem.*, **51**, 1724 (1986).

is mediated by dimethyl(methylthio)sulfonium tetrafluoroborate. Entries 3 and 4 are other examples of 5-*exo* cyclizations. Entries 5 and 6 involve use of sulfenamides as the electrophiles. Entry 7 shows the cyclization of a carbamate involving the carbonyl oxygen. Entry 8 is an 5-*endo* aminocyclization.

Part B of Scheme 4.5 gives some examples of cyclizations induced by selenium electrophiles. Entries 9 to 13 are various selenyletherifications. All exhibit *anti* stereochemistry. Entries 14 and 15 are selenyllactonizations. Entries 17 and 18 involve amido groups as the internal nucleophile. Entry 17 is an 5-*exo* cyclization in which the amido oxygen is the more reactive nucleophilic site, leading to an iminolactone. Geometric factors favor N-cyclization in the latter case.

Chiral selenenylating reagents have been developed and shown to be capable of effecting enantioselective additions and cyclizations. The reagent **4**, for example, achieves more than 90% enantioselectivity in typical reactions.[104]

4.2.3. Cyclization by Mercuric Ion

Electrophilic attack by mercuric ion can effect cyclization by intramolecular capture of a nucleophilic functional group. A variety of oxygen and nitrogen nucleophiles can participate in cyclization reactions, and there have been numerous synthetic applications of the reaction. Mechanistic studies have been carried out on several alkenol systems. The ring-size preference for cyclization of 4-hexenol depends on the mercury reagent that is used. The more reactive mercuric salts favor 6-*endo* addition. It is proposed that reversal of formation of the kinetic *exo* product is responsible.[105] Equilibration to favor the thermodynamic addition products occurs using $Hg(O_3SCF_3)_2$ and $Hg(NO_3)_2$. The equilibration does not seem to be dependent on acid catalysis, since the thermodynamically favored product is also formed in the presence of the acid-scavenger TMU.

104. K. Fujita, K. Murata, M. Iwaoka, and S. Tomoda, *Tetrahedron*, **53**, 2029 (1997); K. Fujita, *Rev. Heteroatom Chem.*, **16**, 101 (1997); T. Wirth, *Tetrahedron*, **55**, 1 (1999).
105. M. Nishizawa, T. Kashima, M. Sakakibara, A. Wakabayashi, K. Takahasi, H. Takao, H. Imagawa, and T. Sugihara, *Heterocycles*, **54**, 629 (2001).

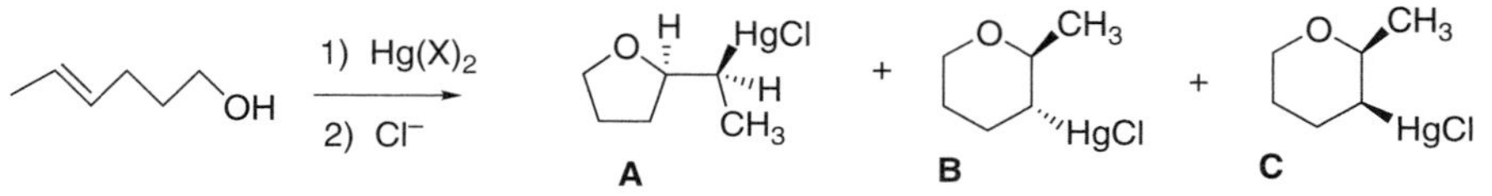

X	A	B	C
O_2CCH_3	92	8	0
O_2CCF_3	13	87	0
O_3SCF_3	0	88	12
O_3SCF_3 (TMU)	0	100 .	0
NO_3	0	100	0

In 5-aryl-4-hexenols with ERG substituents, electronic factors outweigh the *exo* preference.[106] The ERG substituents increase the cationic character at C(5).

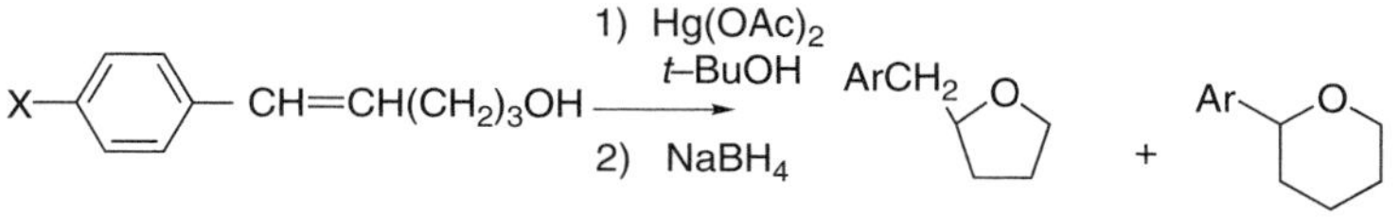

X	
H	81:19
CH_3	47:53
CH_3O	0:100

Cyclization of δ, ε-enols is controlled by a conformation-dependent strain in the *exo* TS.[107] The C(5)–C(6) bond is rotated to minimize $A^{1,3}$ strain.

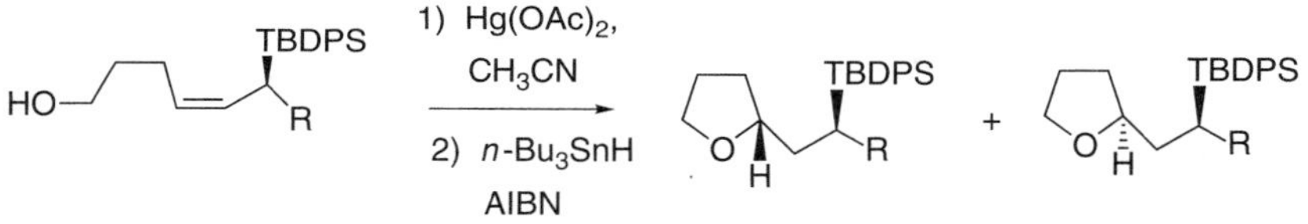

R	ratio
CH_3	19:1
$CH_3CH_2CH_2$	10:1
Ph	10:1

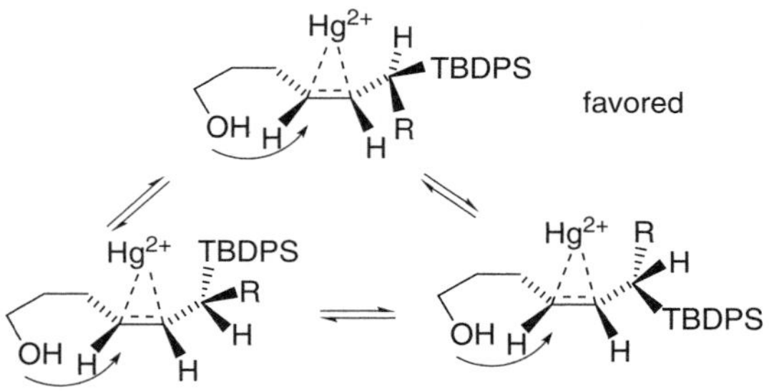

favored

106. Y. Senda, H. Kanto, and H. Itoh, *J. Chem. Soc., Perkin Trans. 2*, 1143 (1997).
107. K. Bratt, A. Garavelas, P. Perlmutter, and G. Westman, *J. Org. Chem.*, **61**, 2109 (1996).

In the corresponding *E*-alkene, where this factor is not present, the cyclization is much less stereoselective. A stabilizing interaction between the siloxy oxygen and the Hg^{2+} center has also been suggested.[108]

Reaction of $Hg(O_2CCF_3)_2$ or $Hg(O_3SCF_3)_2$ with a series of dibenzylcarbinols gave *exo* cyclization for formation of five-, six-, and seven-, but not eight-membered rings.[109]

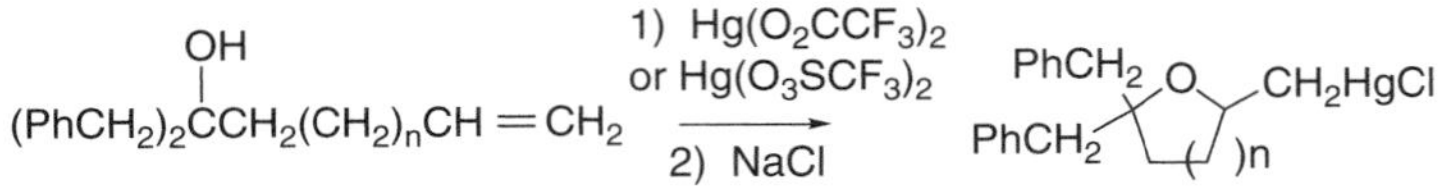

n	ring size	exo: endo
1	5	> 99:1
2	6	>99:1
3	7	>99:1
4	8	-

Benzyl carbamates have been used to form both five- and six-membered nitrogen-containing rings. The selectivity for N over O nucleophilicity in these cases is the result of the nitrogen being able to form a better ring size (5 or 6 versus 7 or 8) than the carbonyl oxygen.

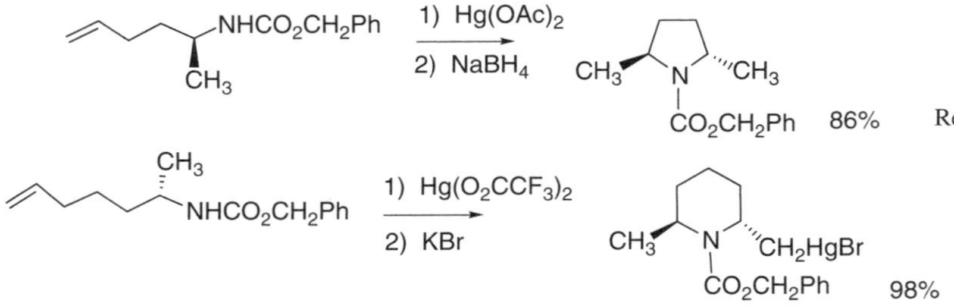

Ref. 110

Ref. 111

The trapping of the radical intermediate in demercuration by oxygen can be exploited as a method for introduction of a hydroxy substituent (see p. 295). The example below and Entries 3 and 4 in Scheme 4.6 illustrate this reaction.

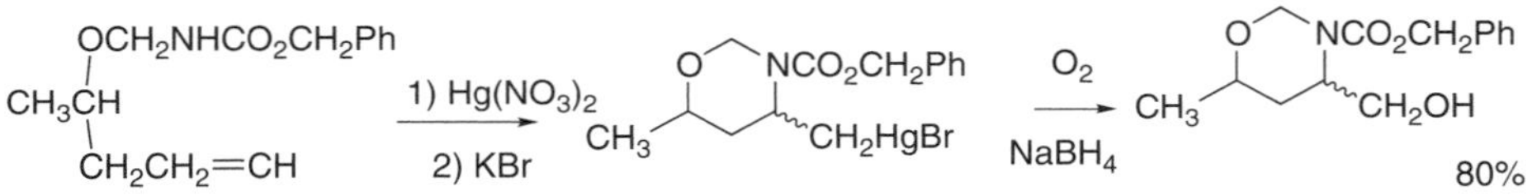

Ref. 112

Cyclization induced by mercuric ion is often used in multistep syntheses to form five- and six-membered hetereocyclic rings, as illustrated in Scheme 4.6. The reactions in Entries 1 to 3 involve acyclic reactants that cyclize to give 5-*exo* products. Entry 4 is an 6-*exo* cyclization. In Entries 1 and 2, the mercury is removed reductively, but in Entries 3 and 4 a hydroxy group is introduced in the presence of oxygen. Inclusion of triethylboron in the reduction has been found to improve yields (Entry 1).[113]

108. A. Garavelas, I. Mavropoulos, P. Permutter, and G. Westman, *Tetrahedron Lett.*, **36**, 463 (1995).
109. H. Imagawa, T. Shigaraki, T. Suzuki, H. Takao, H. Yamada, T. Sugihara, and M. Nishizawa, *Chem. Pharm. Bull.*, **46**, 1341 (1998).
110. T. Yamakazi, R. Gimi, and J. T. Welch, *Synlett*, 573 (1991).
111. H. Takahata, H. Bandoh, and T. Momose, *Tetrahedron*, **49**, 11205 (1993).
112. K. E. Harding, T. H. Marman, and D.-H. Nam, *Tetrahedron Lett.*, **29**, 1627 (1988).
113. S. H. Kang, J. H. Lee, and S. B. Lee, *Tetrahedron Lett.*, **39**, 59 (1998).

1[a]

CH$_2$=CHCH$_2$CHCO$_2$H
(Ph substituent)

1) Hg(O$_2$CCF$_3$)$_2$, K$_2$CO$_3$
2) (C$_2$H$_5$)$_3$B
3) NaBH$_4$

93%

2[b]

PhCH$_2$OCH$_2$CH=CH–CH$_2$
PhCH$_2$O$_2$CNHCH$_2$O

1) Hg(NO$_3$)$_2$
2) NaBH$_4$

PhCH$_2$OCH$_2$CH$_2$
PhCH$_2$O$_2$C–N (oxazolidine)

81%

3[c]

CH$_3$
CbzNH

1) Hg(OAc)$_2$
2) NaBr, NaHCO$_3$
3) O$_2$, NaBH$_4$

CH$_3$ (pyrrolidine) CH$_2$OH
Cbz

67%

4[d]

PhCH$_2$O, PhCH$_2$O, OCH$_2$Ph
H–N–CH$_2$Ph (vinyl)

1) Hg(O$_2$CCF$_3$)$_2$
2) NaBH$_4$, O$_2$

HOCH$_2$
PhCH$_2$O–N–CH$_2$Ph
PhCH$_2$O, OCH$_2$Ph

60:40 mixture

5[e]

CH$_3$ CH$_3$
CH$_3$ O OH

1) Hg[O$_2$CC(CH$_3$)$_3$]$_2$
2) NaBH$_4$

CH$_3$
CH$_3$ O
CH$_3$ O

42%

6[f]

(CH$_3$)$_2$CH, CH$_3$
H, H
CH$_3$, H, H
HO, HO O
CH$_2$

1) Hg(OAc)$_2$
2) NaBH$_4$

(CH$_3$)$_2$CH, CH$_3$
H, H
H, H
CH$_3$ O CH$_3$
OH

47%

7[g]

O
O, CO$_2$H
TBDMSO
(CH$_2$)$_2$CH$_3$

1) Hg(O$_2$CCF$_3$)$_2$
2) NaCl

O
O, HgCl (CH$_2$)$_2$CH$_3$
TBDMSO, O

99%

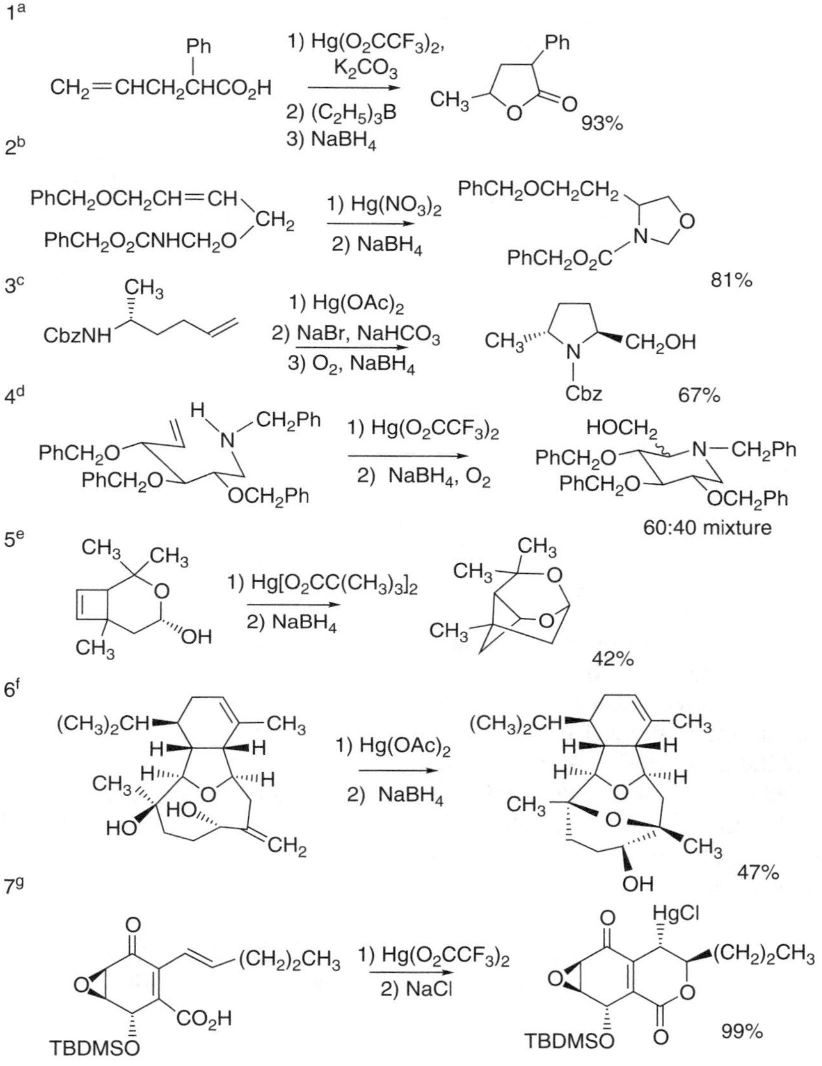

a. S. H. Kang, J. H. Lee, and S. B. Lee, *Tetrahedron Lett.*, **39**, 59 (1998).
b. K. E. Harding and D. R. Hollingsworth, *Tetrahedron Lett.*, **29**, 3789 (1988).
c. H. Takahata, H. Bandoh, and T. Momose, *J. Org. Chem.*, **57**, 4401 (1992).
d. R. C. Bernotas and B. Ganem, *Tetrahedron Lett.*, **26**, 1123 (1985).
e. J. D. White, M. A. Avery and J. P. Carter, *J. Am. Chem. Soc.*, **104**, 5486 (1986).
f. D. W. C. MacMillan, L. E. Overman, and L. D. Pennington, *J. Am. Chem. Soc.*, **123**, 9033 (2001).
g. M. Shoji, T. Uno, and Y. Hayashi, *Org. Lett.*, **6**, 4535 (2004).

The reaction in Entry 5 was used in the syntheses of linetin, which is an aggregation pheromone of the ambrosia beetle. In Entry 6, a transannular 5-*exo* cyclization occurs. Entry 7 is an example of formation of a lactone by carboxylate capture. In this case, the product was isolated as the mercurochloride.

Some progress has been made toward achieving enantioselectivity in mercuration-induced cyclization. Several *bis*-oxazoline (BOX) ligands have been investigated. The diphenyl BOX ligand, in conjunction with $Hg(O_2CCF_3)_2$, results in formation of tetrahydrofuran rings with 80% e.e. Other *bis*-oxazoline ligands derived from tartaric acid were screened and the best results were obtained with a 2-naphthyl ligand, which gave more than 90% e.e. in several cases.

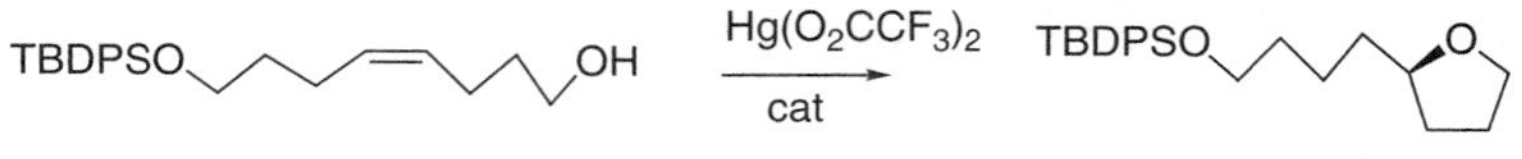

68% yield, 86% e.e.

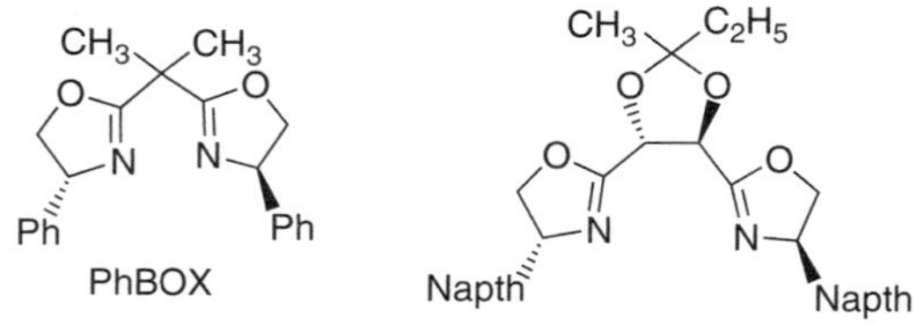

PhBOX

Napth

Ref. 114

4.3. Electrophilic Substitution α to Carbonyl Groups

4.3.1. Halogenation α to Carbonyl Groups

Although the reaction of ketones and other carbonyl compounds with electrophiles such as bromine leads to substitution rather than addition, the mechanism of the reaction is closely related to electrophilic additions to alkenes. An enol, enolate, or enolate equivalent derived from the carbonyl compound is the nucleophile, and the electrophilic attack by the halogen is analogous to that on alkenes. The reaction is completed by restoration of the carbonyl bond, rather than by addition of a nucleophile. The acid- and base-catalyzed halogenation of ketones, which is discussed briefly in Section 6.4 of Part A, provide the most-studied examples of the reaction from a mechanistic perspective.

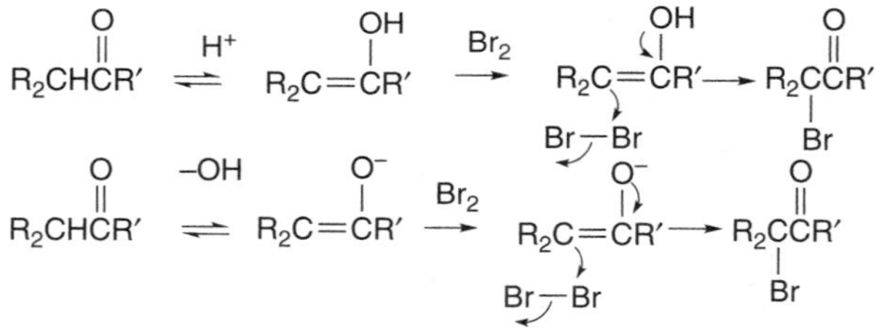

The reactions involving bromine or chlorine generate hydrogen halide and are autocatalytic. Reactions with *N*-bromosuccinimide or tetrabromocyclohexadienone do not form any hydrogen bromide and may therefore be preferable reagents in the case of acid-sensitive compounds. Under some conditions halogenation is faster than enolization. When this is true, the position of substitution in unsymmetrical ketones is governed by the relative rates of formation of the isomeric enols. In general, mixtures are formed with unsymmetrical ketones. The presence of a halogen substituent

[114.] S. H. Kang and M. Kim, *J. Am. Chem. Soc.*, **125**, 4684 (2003).

decreases the rate of acid-catalyzed enolization and thus retards the introduction of a second halogen at the same site, so monohalogenation can usually be carried out satisfactorily. In contrast, in basic solution halogenation tends to proceed to polyhalogenated products because the polar effect of a halogen accelerates base-catalyzed enolization. With methyl ketones, base-catalyzed reaction with iodine or bromine leads ultimately to cleavage to a carboxylic acid.[115] These reactions proceed to the trihalomethyl ketones, which are susceptible to base-induced cleavage.

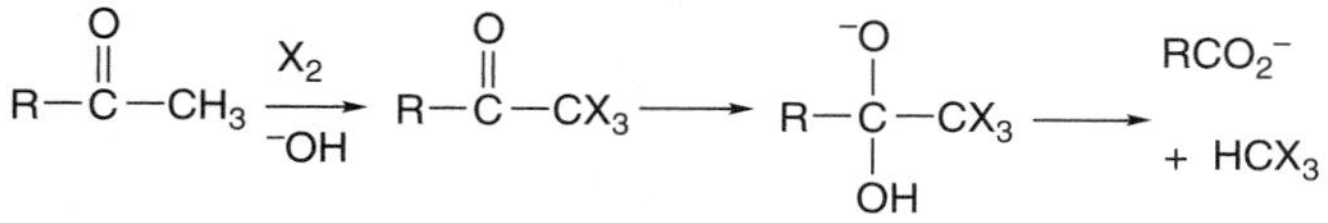

$$
R-\overset{\overset{\text{O}}{\|}}{C}-CH_3 \xrightarrow[^-OH]{X_2} R-\overset{\overset{\text{O}}{\|}}{C}-CX_3 \longrightarrow R-\overset{\overset{\overset{-}{\text{O}}}{|}}{\underset{\text{OH}}{C}}-CX_3 \longrightarrow \begin{array}{l} RCO_2^- \\ + HCX_3 \end{array}
$$

The reaction can also be effected with hypochlorite ion, and this constitutes a useful method for converting methyl ketones to carboxylic acids.

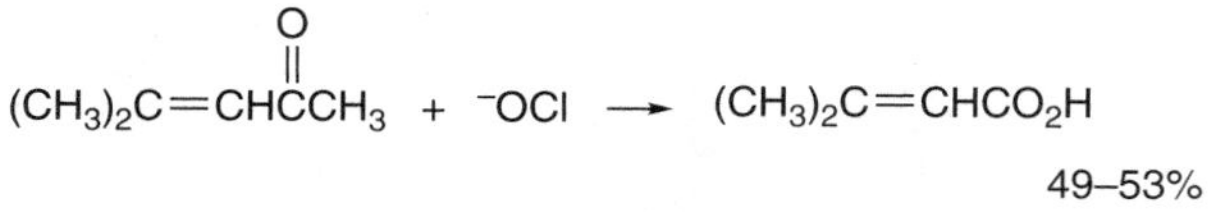

$$
(CH_3)_2C=CH\overset{\overset{\text{O}}{\|}}{C}CH_3 + {}^-OCl \longrightarrow (CH_3)_2C=CHCO_2H
$$

49–53%

Ref. 116

The most common preparative procedures involve use of the halogen, usually bromine, in acetic acid. Other suitable halogenating agents include *N*-bromosuccinimide, tetrabromocyclohexadienone, and sulfuryl chloride.

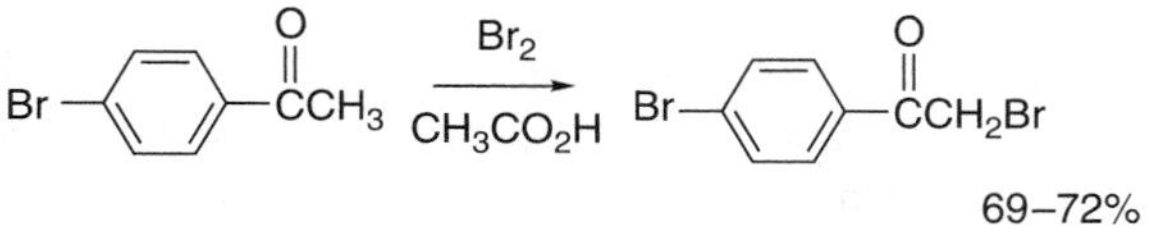

69–72%

Ref. 117

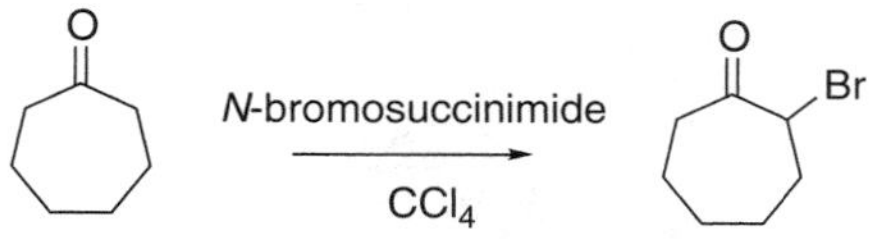

Ref. 118

83–85%

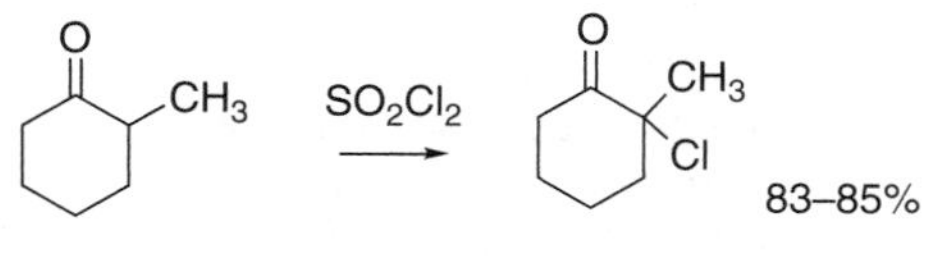

Ref. 119

115. S. J. Chakabartty, in *Oxidations in Organic Chemistry*, Part C, W. Trahanovsky, ed., Academic Press, New York, 1978, Chap. V.
116. L. J. Smith, W. W. Prichard, and L. J. Spillane, *Org. Synth.*, **III**, 302 (1955).
117. W. D. Langley, *Org. Synth.*, **1**, 122 (1932).
118. E. J. Corey, *J. Am. Chem. Soc.*, **75**, 2301 (1954).
119. E. W. Warnhoff, D. G. Martin, and W. S. Johnson, *Org. Synth.*, **IV**, 162 (1963).

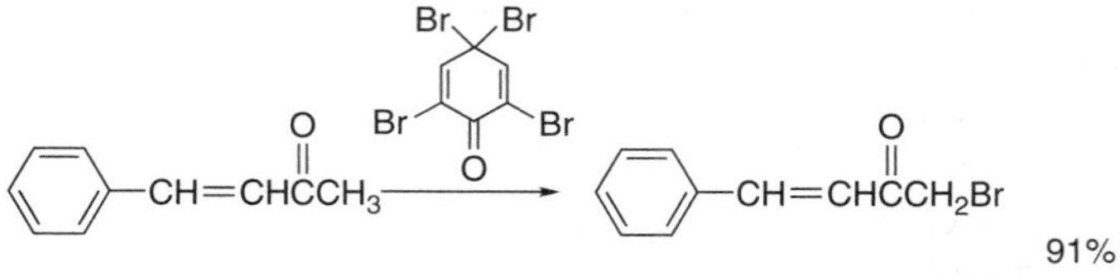

91%   Ref. 120

Another preparatively useful procedure for monohalogenation of ketones involves reaction with cupric chloride or cupric bromide.[121]

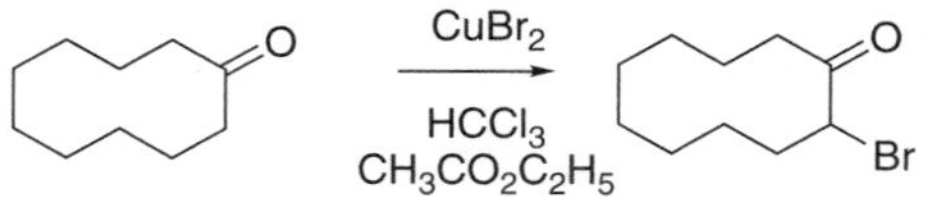

Ref. 122

Instead of direct halogenation of ketones, reactions with more reactive derivatives such as silyl enol ethers and enamines have advantages in certain cases.

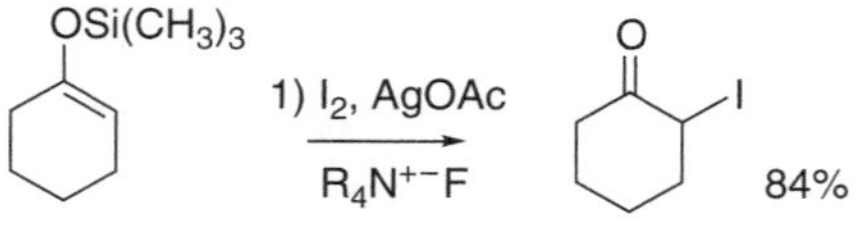

84%   Ref. 123

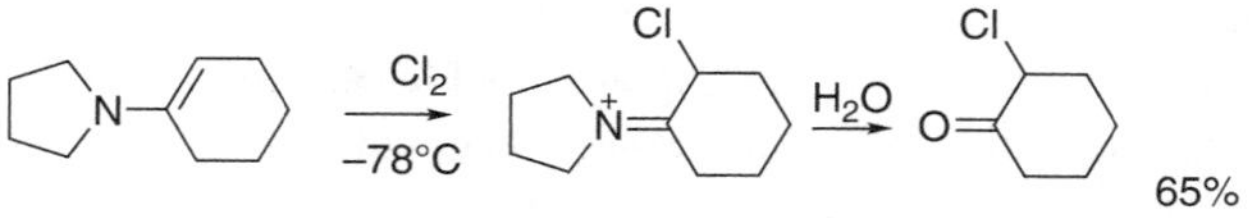

65%   Ref. 124

There are also procedures in which the enolate is generated quantitatively and allowed to react with a halogenating agent. Regioselectivity can then be controlled by the direction of enolate formation. Among the sources of halogen that have been used under these conditions are bromine,[125] *N*-chlorosuccinimide,[126] trifluoromethanesulfonyl chloride,[127] and hexachloroethane.[128]

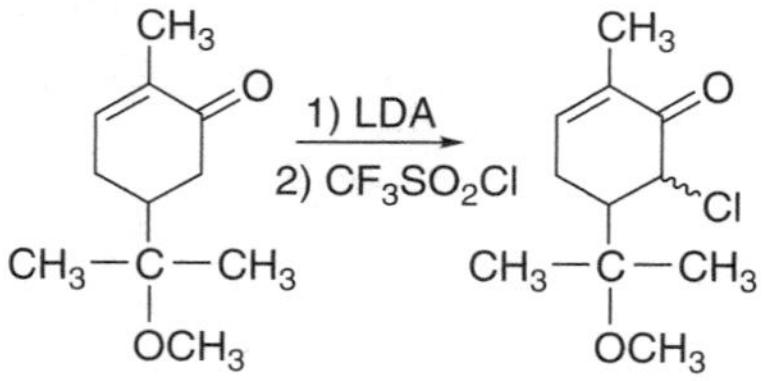

[120] V. Calo, L. Lopez, G. Pesce, and P. E. Todesco, *Tetrahedron*, **29**, 1625 (1973).

[121] E. M. Kosower, W. J. Cole, G.-S. Wu, D. E. Cardy, and G. Meisters, *J. Org. Chem.*, **28**, 630 (1963); E. M. Kosower and G.-S. Wu, *J. Org. Chem.*, **28**, 633 (1963).

[122] D. P. Bauer and R. S. Macomber, *J. Org. Chem.*, **40**, 1990 (1975).

[123] G. M. Rubottom and R. C. Mott, *J. Org. Chem.*, **44**, 1731 (1979); G. A. Olah, L. Ohannesian, M. Arvanaghi, and G. K. S. Prakash, *J. Org. Chem.*, **49**, 2032 (1984).

[124] W. Seufert and F. Effenberger, *Chem. Ber.*, **112**, 1670 (1979).

[125] T. Woolf, A. Trevor, T. Baille, and N. Castagnoli, Jr., *J. Org. Chem.*, **49**, 3305 (1984).

[126] A. D. N. Vaz and G. Schoellmann, *J. Org. Chem.*, **49**, 1286 (1984).

[127] P. A. Wender and D. A. Holt, *J. Am. Chem. Soc.*, **107**, 7771 (1985).

[128] M. B. Glinski, J. C. Freed, and T. Durst, *J. Org. Chem.*, **52**, 2749 (1987).

α-Fluoroketones are made primarily by reactions of enol acetates or silyl enol ethers with fluorinating agents such as CF_3OF[129], XeF_2,[130] or dilute F_2.[131] Other fluorinating reagents that can be used include N-fluoropyridinium salts,[132] 1-fluoro-4-hydroxy-1,4-diazabicyclo[2.2.2]octane,[133] and 1,4-difluoro-1,4-diazabicyclo[2.2.2]octane.[134] These reagents fluorinate readily enolizable carbonyl compounds and silyl enol ethers.

$$PhCCH_2CH_3 + F-N^+N^+OH \longrightarrow PhCCHCH_3 \quad 88\%$$

Ref. 135

The α-halogenation of acid chlorides also has synthetic utility. The mechanism is presumed to be similar to ketone halogenation and to proceed through an enol. The reaction can be effected in thionyl chloride as solvent to give α-chloro, α-bromo, or α-iodo acyl chlorides, using, respectively, N-chlorosuccinimide, N-bromosuccinimide, or molecular iodine as the halogenating agent.[136] Since thionyl chloride rapidly converts carboxylic acids to acyl chlorides, the acid can be used as the starting material.

$$CH_3(CH_2)_3CH_2CO_2H \xrightarrow[\text{SOCl}_2]{\textit{N}\text{-chlorosuccinimide}} CH_3(CH_2)_3CHCOCl \quad 87\%$$
(Cl)

$$PhCH_2CH_2CO_2H \xrightarrow[\text{SOCl}_2]{I_2} PhCH_2CHCOCl \quad 95\%$$
(I)

Direct chlorination can be carried out in the presence of $ClSO_3H$, which acts as a strong acid catalyst. These procedures use various compounds including 1,3-dinitrobenzene, chloranil, and TCNQ to inhibit competing radical chain halogenation.[137]

$$(CH_3)_2CHCH_2CO_2H \xrightarrow[\text{chloranil}]{\substack{Cl_2, ClSO_3H \\ 140°C}} (CH_3)_2CHCHCO_2H$$
(Cl)

4.3.2. Sulfenylation and Selenenylation α to Carbonyl Groups

The α-sulfenylation[138] and α-selenenylation[139] of carbonyl compounds are synthetically important reactions, particularly in connection with the introduction of

129. W. J. Middleton and E. M. Bingham, *J. Am. Chem. Soc.*, **102**, 4845 (1980).
130. B. Zajac and M. Zupan, *J. Chem. Soc., Chem. Commun.*, 759 (1980).
131. S. Rozen and Y. Menahem, *Tetrahedron Lett.*, 725 (1979).
132. T. Umemoto, M. Nagayoshi, K. Adachi, and G. Tomizawa, *J. Org. Chem.*, **63**, 3379 (1998).
133. S. Stavber, M. Zupan, A. J. Poss, and G. A. Shia, *Tetrahedron Lett.*, **36**, 6769 (1995).
134. T. Umemoto and M. Nagayoshi, *Bull. Chem. Soc. Jpn.*, **69**, 2287 (1996).
135. S. Stavber and M. Zupan, *Tetrahedron Lett.*, **37**, 3591 (1996).
136. D. N. Harpp, L. Q. Bao, C. J. Black, J. G. Gleason, and R. A. Smith, *J. Org. Chem.*, **40**, 3420 (1975); Y. Ogata, K. Adachi, and F.-C. Chen, *J. Org. Chem.*, **48**, 4147 (1983).
137. Y. Ogata, T. Harada, K. Matsuyama, and T. Ikejiri, *J. Org. Chem.*, **40**, 2960 (1975); R. J. Crawford, *J. Org. Chem.*, **48**, 1364 (1983).
138. B. M. Trost, *Chem. Rev.*, **78**, 363 (1978).
139. H. J. Reich, *Acc. Chem. Res.*, **12**, 22 (1979); H. J. Reich, J. M. Renga, and I. L. Reich, *J. Am. Chem. Soc.*, **97**, 5434 (1975).

Scheme 4.7. α-Sulfenylation and α-Selenenylation of Carbonyl Compounds

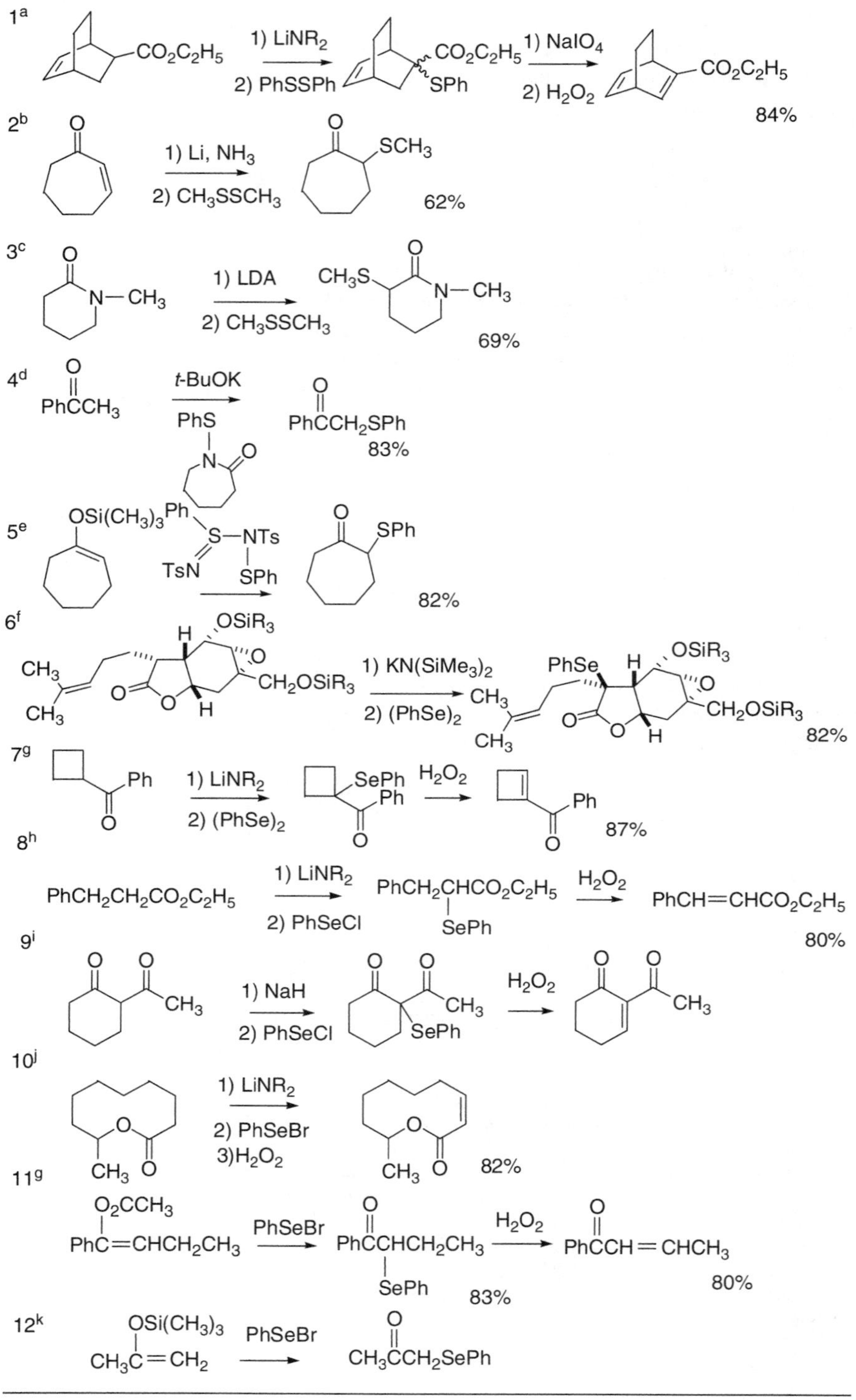

1[a]

2[b]

3[c]

4[d]

5[e]

6[f]

7[g]

8[h]

$$PhCH_2CH_2CO_2C_2H_5 \xrightarrow[\text{2) PhSeCl}]{\text{1) LiNR}_2} PhCH_2\underset{\underset{SePh}{|}}{CH}CO_2C_2H_5 \xrightarrow{H_2O_2} PhCH=CHCO_2C_2H_5 \quad 80\%$$

9[i]

10[j]

11[g]

12[k]

(Continued)

Scheme 4.7. (*Continued*)

333

SECTION 4.4

Additions to Allenes and Alkynes

a. B. M. Trost, T. N. Salzmann, and K. Hiroi, *J. Am. Chem. Soc.*, **98**, 4887 (1976).
b. P. G. Gassman, D. P. Gilbert, and S. M. Cole, *J. Org. Chem.*, **42**, 3233 (1977).
c. P. G. Gassman and R. J. Balchunis, *J. Org. Chem.*, **42**, 3236 (1977).
d. G. Foray, A. Penenory, and A. Rossi, *Tetrahedron Lett.*, **38**, 2035 (1997).
e. P. Magnus and P. Rigollier, *Tetrahedron Lett.*, **33**, 6111 (1992).
f. A. B. Smith, III, and R. E. Richmond, *J. Am. Chem. Soc.*, **105**, 575 (1983).
g. H. J. Reich, J. M. Renga, and I. L. Reich, *J. Am. Chem. Soc.*, **97**, 5434 (1975).
h. J. M. Renga and H. J. Reich, *Org. Synth.*, **59**, 58 (1979).
i. T. Wakamatsu, K. Akasaka, and Y. Ban, *J. Org. Chem.*, **44**, 2008 (1979).
j. H. J. Reich, I. L. Reich, and J. M. Renga, *J. Am. Chem. Soc.*, **95**, 5813 (1973).
k. I. Ryu, S. Murai, I. Niwa, and N. Sonoda, *Synthesis*, 874 (1977).

unsaturation. The products can subsequently be oxidized to sulfoxides and selenoxides that readily undergo elimination (see Section 6.8.3), generating the corresponding α,β-unsaturated carbonyl compound. Sulfenylations and selenenylations are usually carried out under conditions in which the enolate of the carbonyl compound is the reactive species. If a regiospecific enolate is generated by one of the methods described in Chapter 1, the position of sulfenylation or selenenylation can be controlled.[140] Disulfides are the most common sulfenylation reagents, whereas diselenides or selenenyl halides are used for selenenylation.

Scheme 4.7 gives some specific examples of these types of reactions. Entry 1 shows the use of sulfenylation followed by oxidation to introduce a conjugated double bond. Entries 2 and 3 are α-sulfenylations of a ketone and lactam, respectively, using dimethyl disulfide as the sulfenylating reagent. Entries 4 and 5 illustrate the use of alternative sulfenylating reagents. Entry 4 uses *N*-phenylsulfenylcaprolactam, which is commercially available. The reagent in Entry 5 is generated by reaction of diphenyldisulfide with chloramine-T. Entries 6 to 10 are examples of reactions of preformed enolates with diphenyl diselenide or phenylselenenyl chloride. As Entries 11 and 12 indicate, the selenenylation of ketones can also be effected by reactions of enol acetates or enol silyl ethers.

4.4. Additions to Allenes and Alkynes

Both allenes[141] and alkynes[142] require special consideration with regard to mechanisms of electrophilic addition. The attack by a proton on allene might conceivably lead to the allyl cation or the 2-propenyl cation.

$$^{+}CH_2-CH=CH_2 \xleftarrow{\ H^+\ } CH_2=C=CH_2 \xrightarrow{\ H^+\ } CH_3-\overset{+}{C}=CH_2$$

An immediate presumption that the more stable allyl ion will be formed overlooks the stereoelectronic facets of the reaction. Protonation at the center carbon without rotation of one of the terminal methylene groups leads to a primary carbocation

[140] P. G. Gassman, D. P. Gilbert, and S. M. Cole, *J. Org. Chem.*, **42**, 3233 (1977).

[141] H. F. Schuster and G. M. Coppola, *Allenes in Organic Synthesis*, Wiley, New York, 1984 ; W. Smadja, *Chem. Rev.*, **83**, 263 (1983); S. Ma, in *Modern Allene Chemistry*, N. Krause and A. S. K. Hashmi, eds., Wiley-VCH, Weinheim, 2004, pp. 595–699.

[142] W. Drenth, in *The Chemistry of Triple Bonded Functional Groups*, Supplement C2, Vol. 2, S. Patai, ed., John Wiley & Sons, New York, 1994, pp. 873–915.

that is not stabilized by resonance, because the adjacent π bond is orthogonal to the empty p orbital.

As a result, protonation both in solution[143] and gas phase[144] occurs at a terminal carbon to give the 2-propenyl cation, not the allylic cation.

The addition of HCl, HBr, and HI to allene has been studied in some detail.[145] In each case a 2-halopropene is formed, corresponding to protonation at a terminal carbon. The initial product can undergo a second addition, giving rise to 2,2-dihalopropanes. The regiochemistry reflects the donor effect of the halogen. Dimers are also formed, but we have not considered them.

$$CH_2{=}C{=}CH_2 \ + \ HX \ \longrightarrow \ CH_3\overset{\overset{\textstyle X}{|}}{C}{=}CH_2 \ + \ CH_3\overset{\overset{\textstyle X}{|}}{\underset{\underset{\textstyle X}{|}}{C}}CH_3$$

The presence of a phenyl group results in the formation of products from protonation at the center carbon.[146]

$$PhCH{=}C{=}CH_2 \ \xrightarrow[\text{HOAc}]{\text{HCl}} \ PhCH{=}CHCH_2Cl$$

Two alkyl substituents, as in 1,1-dimethylallene, also lead to protonation at the center carbon.[147]

$$(CH_3)_2C{=}C{=}CH_2 \ \longrightarrow \ (CH_3)_2C{=}CHCH_2Cl$$

These substituent effects are due to the stabilization of the carbocations that result from protonation at the center carbon. Even if allylic conjugation is not important, the aryl and alkyl substituents make the terminal carbocation more stable than the alternative, a secondary vinyl cation.

Acid-catalyzed additions to terminal alkynes follow the Markovnikov rule.

$$CH_3(CH_2)_6C{\equiv}CH \ \xrightarrow{Et_4N^+HBr_2} \ CH_3(CH_2)_6\overset{\overset{\textstyle }{}}{C}{=}CH_2 \qquad \text{Ref. 148}$$
$$\underset{\textstyle Br} {\underset{\textstyle |}{}} \quad 77\%$$

The rate and selectivity of the reaction can be considerably enhanced by using an added quaternary bromide salt in 1:1 $TFA:CH_2Cl_2$. Note that the reactions are quite

143. P. Cramer and T. T. Tidwell, *J. Org. Chem.*, **46**, 2683 (1981).
144. M. T. Bowers, L. Shuying, P. Kemper, R. Stradling, H. Webb, D. H. Aue, J. R. Gilbert, and K. R. Jennings, *J. Am. Chem. Soc.*, **102**, 4830 (1980); S. Fornarini, M. Speranza, M. Attina, F. Cacace, and P. Giacomello, *J. Am. Chem. Soc.*, **106**, 2498 (1984).
145. K. Griesbaum, W. Naegele, and G. G. Wanless, *J. Am. Chem. Soc.*, **87**, 3151 (1965).
146. T. Okuyama, K. Izawa, and T. Fueno, *J. Am. Chem. Soc.*, **95**, 6749 (1973).
147. T. L. Jacobs and R. N. Johnson, *J. Am. Chem. Soc.*, **82**, 6397 (1960).
148. J. Cousseau, *Synthesis*, 805 (1980).

slow, even under these favorable conditions, but there is clean formation of the *anti* addition product.[149]

$$CH_3CH_2CH_2C{\equiv}CCH_2CH_2CH_3 \xrightarrow[\substack{1:4\ TFA:CH_2Cl_2 \\ 144\ h}]{1.0\ M\ Bu_4N^+Br^-} \underset{H}{\overset{Br}{CH_3CH_2CH_2}}{=}{CH_2CH_2CH_3} \quad 100\%$$

$$HC{\equiv}C(CH_2)_5CH_3 \xrightarrow[\substack{1:4\ TFA:CH_2Cl_2 \\ 336\ h}]{1.0\ M\ Bu_4N^+Br^-} \underset{Br}{CH_2{=}C(CH_2)_5CH_3} \quad 98\%$$

Surface-mediated addition of HCl or HBr can be carried out in the presence of silica or alumina.[150] The hydrogen halides can be generated from thionyl chloride, oxalyl chloride, oxalyl bromide, phosphorus tribromide, or acetyl bromide. The kinetic products from HCl and 1-phenylpropyne result from *syn* addition, but isomerization to the more stable *Z*-isomer occurs upon continued exposure to the acidic conditions.

$$PhC{\equiv}CCH_3 \xrightarrow[\substack{SiO_2 \\ 0.3\ h}]{SOCl_2} \underset{Ph}{\overset{Cl}{\diagdown}}{=}\underset{CH_3}{\overset{H}{\diagup}} \xrightarrow{3\ h} \underset{Ph}{\overset{Cl}{\diagdown}}{=}\underset{H}{\overset{CH_3}{\diagup}}$$

The initial addition products to alkynes are not always stable. Addition of acetic acid, for example, results in the formation of enol acetates, which are converted to the corresponding ketone under the reaction conditions.[151]

$$C_2H_5C{\equiv}CC_2H_5 \xrightarrow[CH_3CO_2H]{H^+} \underset{O_2CCH_3}{C_2H_5C}{=}CHCH_2CH_3 \longrightarrow \underset{O}{C_2H_5CCH_2CH_2CH_3}$$

The most synthetically valuable method for converting alkynes to ketones is by mercuric ion–catalyzed hydration. Terminal alkynes give methyl ketones, in accordance with the Markovnikov rule. Internal alkynes give mixtures of ketones unless some structural feature promotes regioselectivity. Reactions with $Hg(OAc)_2$ in other nucleophilic solvents such as acetic acid or methanol proceed to β-acetoxy- or β-methoxyalkenylmercury intermediates,[152] which can be reduced or solvolyzed to ketones. The regiochemistry is indicative of a mercurinium ion intermediate that is opened by nucleophilic attack at the more positive carbon, that is, the additions follow the Markovnikov rule. Scheme 4.8 gives some examples of alkyne hydration reactions.

Addition of chlorine to 1-butyne is slow in the absence of light. When addition is initiated by light, the major product is *E*-1,2-dichlorobutene if butyne is present in large excess.[153]

$$CH_3CH_2C{\equiv}CH + Cl_2 \longrightarrow \underset{Cl}{\overset{CH_3CH_2}{\diagdown}}{=}\underset{H}{\overset{Cl}{\diagup}}$$

[149]. H. M. Weiss and K. M. Touchette, *J. Chem. Soc., Perkin Trans. 2*, 1523 (1998).

[150]. P. J. Kropp and S. D. Crawford, *J. Org. Chem.*, **59**, 3102 (1994).

[151]. R. C. Fahey and D.-J. Lee, *J. Am. Chem. Soc.*, **90**, 2124 (1968).

[152]. M. Uemura, H. Miyoshi, and M. Okano, *J. Chem. Soc., Perkin Trans. 1*, 1098 (1980); R. D. Bach, R. A. Woodward, T. J. Anderson, and M. D. Glick, *J. Org. Chem.*, **47**, 3707 (1982); M. Bassetti, B. Floris, and G. Spadafora, *J. Org. Chem.*, **54**, 5934 (1989).

[153]. M. L. Poutsma and J. L. Kartch, *Tetrahedron*, **22**, 2167 (1966).

Scheme 4.8. Ketones by Hydration of Alkynes

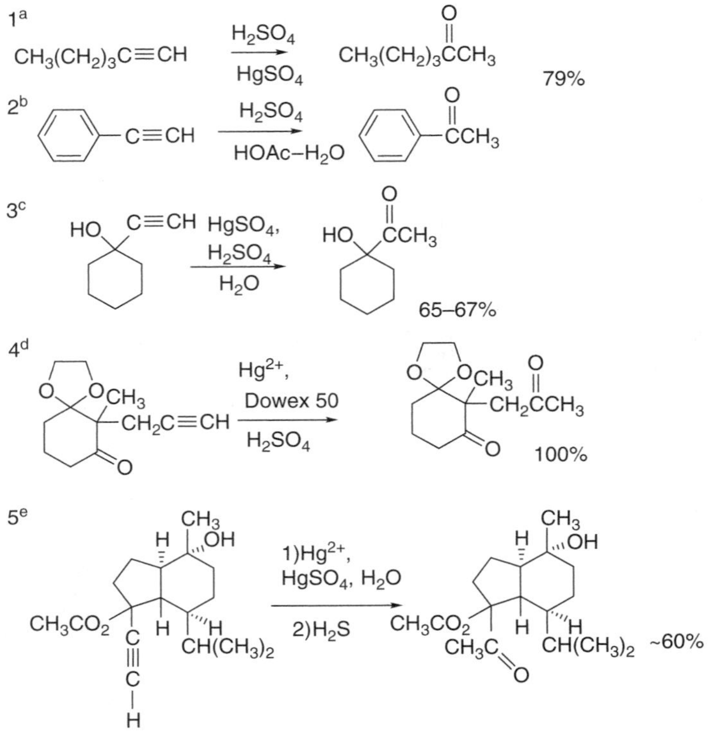

a. R. J. Thomas, K. N. Campbell, and G. F. Hennion, *J. Am. Chem. Soc.*, **60**, 718 (1938).
b. R. W. Bott, C. Eaborn, and D. R. M. Walton, *J. Chem. Soc.*, 384 (1965).
c. G. N. Stacy and R. A. Mikulec, *Org. Synth.*, **IV**, 13 (1963).
d. W. G. Dauben and D. J. Hart, *J. Org. Chem.*, **42**, 3787 (1977).
e. D. Caine and F. N. Tuller, *J. Org. Chem.*, **38**, 3663 (1973).

In acetic acid, both 1-pentyne and 1-hexyne give the *syn* addition product. With 2-butyne and 3-hexyne, the major products are β-chlorovinyl acetates of *E*-configuration.[154] Some of the dichloro compounds are also formed, with more of the *E*- than the *Z*-isomer being observed.

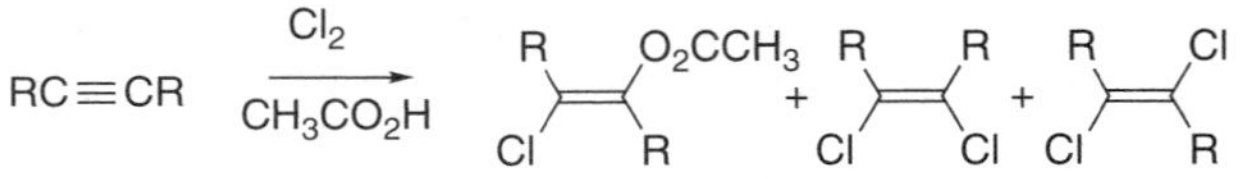

The reactions of the internal alkynes are considered to involve a cyclic halonium ion intermediate, whereas the terminal alkynes seem to react by a rapid collapse of a vinyl cation.

Alkynes react with electrophilic selenium reagents such as phenylselenenyl tosylate.[155] The reaction occurs with *anti* stereoselectivity. Aryl-substituted alkynes are regioselective, but alkyl-substituted alkynes are not.

154. K. Yates and T. A. Go, *J. Org. Chem.*, **45**, 2385 (1980).
155. T. G. Back and K. R. Muralidharan, *J. Org. Chem.*, **56**, 2781 (1991).

$$CH_3(CH_2)_7C\equiv CH \xrightarrow{\text{PhSeOSO}_2\text{Ar}}$$

$$\underset{CH_3(CH_2)_7}{\overset{ArSO_2O}{\diagdown}}C=C\underset{SePh}{\overset{H}{\diagup}} + \underset{CH_3(CH_2)_7}{\overset{PhSe}{\diagdown}}C=C\underset{OSO_2Ar}{\overset{H}{\diagup}}$$

85% 55:45

$$PhC\equiv CCH_3 \xrightarrow{\text{PhSeOSO}_2\text{Ar}}$$

$$\underset{Ph}{\overset{ArSO_2O}{\diagdown}}C=C\underset{SePh}{\overset{CH_3}{\diagup}}$$ 75%

Some of the most synthetically useful addition reactions of alkynes are with organometallic reagents, and these reactions, which can lead to carbon-carbon bond formation, are discussed in Chapter 8.

4.5. Addition at Double Bonds via Organoborane Intermediates

4.5.1. Hydroboration

Borane, BH_3, having only six valence electrons on boron, is an avid electron pair acceptor. Pure borane exists as a dimer in which two hydrogens bridge the borons.

In aprotic solvents that can act as electron pair donors such as ethers, tertiary amines, and sulfides, borane forms Lewis acid-base adducts.

$$R_2\overset{+}{O}-\overset{-}{B}H_3 \qquad R_3\overset{+}{N}-\overset{-}{B}H_3 \qquad R_2\overset{+}{S}-\overset{-}{B}H_3$$

Borane dissolved in THF or dimethyl sulfide undergoes addition reactions rapidly with most alkenes. This reaction, which is known as *hydroboration*, has been extensively studied and a variety of useful synthetic processes have been developed, largely through the work of H. C. Brown and his associates.

Hydroboration is highly *regioselective* and *stereospecific*. The boron becomes bonded primarily to the *less-substituted* carbon atom of the alkene. A combination of steric and electronic effects works to favor this orientation. Borane is an electrophilic reagent. The reaction with substituted styrenes exhibits a weakly negative ρ value (-0.5).[156] Compared with bromination $(\rho^+ = -4.3)$,[157] this is a small substituent effect, but it does favor addition of the electrophilic boron at the less-substituted end of the double bond. In contrast to the case of addition of protic acids to alkenes, it is the *boron, not the hydrogen, that is the more electrophilic atom.* This electronic effect is reinforced by steric factors. Hydroboration is usually done under conditions in which the borane eventually reacts with three alkene molecules to give a trialkylborane. The

156. L. C. Vishwakarma and A. Fry, *J. Org. Chem.*, **45**, 5306 (1980).
157. J. A. Pincock and K. Yates, *Can. J. Chem.*, **48**, 2944 (1970).

second and third alkyl groups would increase steric repulsion if the boron were added at the internal carbon.

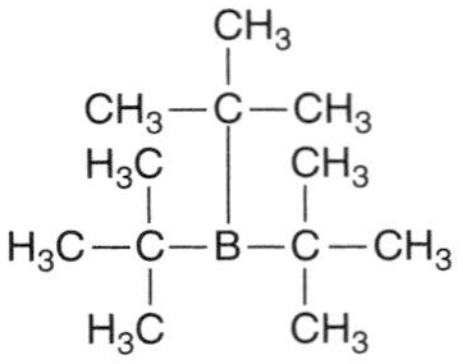

nonbonded
repulsions

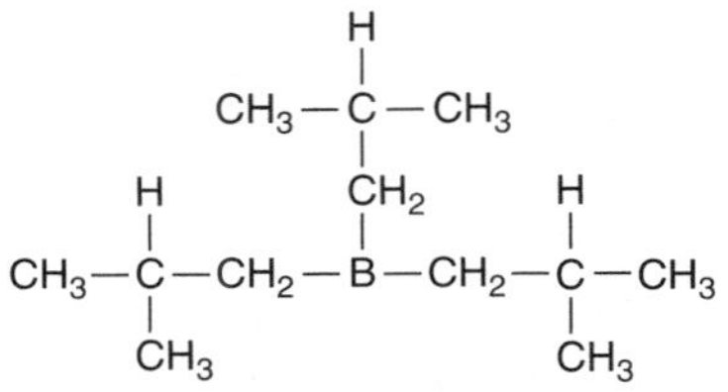

nonbonded repulsions
reduced

Table 4.3 provides some data on the regioselectivity of addition of diborane and several of its derivatives to representative alkenes. Table 4.3 includes data for some mono- and dialkylboranes that show even higher regioselectivity than diborane itself. These derivatives are widely used in synthesis and are frequently referred to by the shortened names shown with the structures.

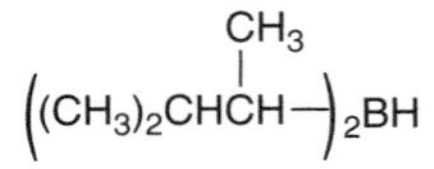

disiamylborane
bis(1,2-dimethylpropyl)
borane

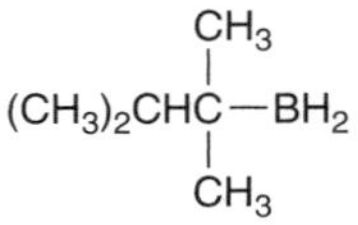

thexylborane
1,1,2-trimethylpropylborane

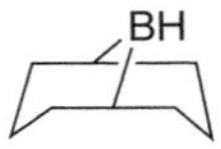

9-BBN
9-borabicyclo[3.3.1]nonane

Table 4.3. Regioselectivity of Diborane and Alkylboranes toward Some Alkenes

	Percent boron at less substituted carbon			
Hydroborating agent	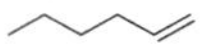 1-Hexene	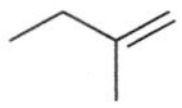 2-Methyl-1-butene	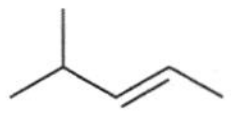 4-Methyl-2-pentene	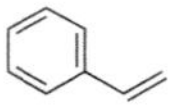 Styrene
Diborane[a]	94	99	57	80
Chloroborane-dimethyl sulfide[b]	99	99.5	–	98
Disiamylborane[a]	99	–	97	98
Thexylborane-dimethyl sulfide[c]	94	–	66	95
Thexylchloroborane-dimethyl sulfide	99	99	97	99
9-Borabicyclo[3.3.1]borane	99.9	99.8[f]	99.3	98.5

a. G. Zweifel and H. C. Brown, *Org. React.*, **13**, 1 (1963).
b. H. C. Brown, N. Ravindran, and S. U. Kulkarni, *J. Org. Chem.*, **44**, 2417 (1969); H. C. Brown and U. S. Racherla, *J. Org. Chem.*, **51**, 895 (1986).
c. H. C. Brown and G. Zweifel, *J. Am. Chem. Soc.*, **82**, 4708 (1960).
d. H. C. Brown, J. A. Sikorski, S. U. Kulkarni, and H. D. Lee, *J. Org. Chem.*, **45**, 4540 (1980).
e. H. C. Brown, E. F. Knight, and C. G. Scouten, *J. Am. Chem. Soc.*, **96**, 7765 (1974).
f. Data for 2-methyl-1-pentene.

These reagents are prepared by hydroboration of the appropriate alkene, using control of stoichiometry to terminate the hydroboration at the desired degree of alkylation.

Hydroboration is a stereospecific *syn* addition that occurs through a four-center TS with simultaneous bonding to boron and hydrogen. The new C−B and C−H bonds are thus both formed from the same face of the double bond. In molecular orbital terms, the addition is viewed as taking place by interaction of the filled alkene π orbital with the empty p orbital on boron, accompanied by concerted C−H bond formation.[158]

As is true for most reagents, there is a preference for approach of the borane from the less hindered face of the alkene. Because diborane itself is a relatively small molecule, the stereoselectivity is not high for unhindered alkenes. Table 4.4 gives some data comparing the direction of approach for three cyclic alkenes. The products in all cases result from *syn* addition, but the mixtures result from both the low regioselectivity and from addition to both faces of the double bond. Even 7,7-dimethylnorbornene shows only modest preference for *endo* addition with diborane. The selectivity is enhanced with the bulkier reagent 9-BBN.

Table 4.4. Stereoselectivity of Hydroboration of Cyclic Alkenes[a]

| | Product composition[b] | | | | | | | | |
| | 3-Methyl cyclopentene | | | 4-Methyl cyclohexene | | | | 7,7-Dimethylbi-cyclo[2.2.1]heptene | |
	trans-2	*cis*-3	*trans*-3	*cis*-2	*trans*-2	*cis*-3	*trans*-3	*exo*	*endo*
Diborane	45	55		16	34	18	32	22	78[c]
Disiamylborane	40	60		18	30	27	25	–	–
9-BBN	25	50	25	0	20	40	40	3	97

a. Data from H. C. Brown, R. Liotta, and L. Brener, *J. Am. Chem. Soc.*, **99**, 3427 (1977), except where otherwise noted.
b. Product composition refers to methylcycloalkanols formed by oxidation.
c. H. C. Brown, J. H. Kawakami, and K.-T. Liu, *J. Am. Chem. Soc.*, **95**, 2209 (1973).

[158.] D. J. Pasto, B. Lepeska, and T.-C. Cheng, *J. Am. Chem. Soc.*, **94**, 6083 (1972); P. R. Jones, *J. Org. Chem.*, **37**, 1886 (1972); S. Nagase, K. N. Ray, and K. Morokuma, *J. Am. Chem. Soc.*, **102**, 4536 (1980); X. Wang, Y. Li, Y.-D. Wu, M. N. Paddon-Row, N. G. Rondan, and K. N. Houk, *J. Org. Chem.*, **55**, 2601 (1990); N. J. R. van Eikema Hommes and P. v. R. Schleyer, *J. Org. Chem.*, **56**, 4074 (1991).

The haloboranes BH_2Cl, BH_2Br, $BHCl_2$, and $BHBr_2$ are also useful hydroborating reagents.[159] These compounds are somewhat more regioselective than borane itself, but otherwise show similar reactivity. A useful aspect of the chemistry of the haloboranes is the potential for sequential introduction of substituents at boron. The halogens can be replaced by alkoxide or by hydride. When halogen is replaced by hydride, a second hydroboration step can be carried out.

$$R_2BX + NaOR' \longrightarrow R_2BOR'$$

$$R_2BX + LiAlH_4 \longrightarrow R_2BH$$

$$RBX_2 + LiAlH_4 \longrightarrow RBH_2$$

$$X = Cl, Br$$

Examples of these transformations are discussed in Chapter 9, where carbon-carbon bond-forming reactions of organoboranes are covered.

Amine-borane complexes are not very reactive toward hydroboration, but the pyridine complex of borane can be activated by reaction with iodine.[160] The active reagent is thought to be the pyridine complex of iodoborane.

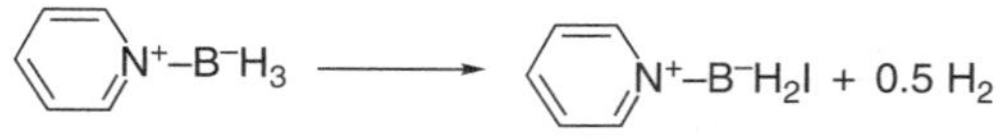

The resulting boranes can be subjected to oxidation or isolated as potassium trifluoroborates.

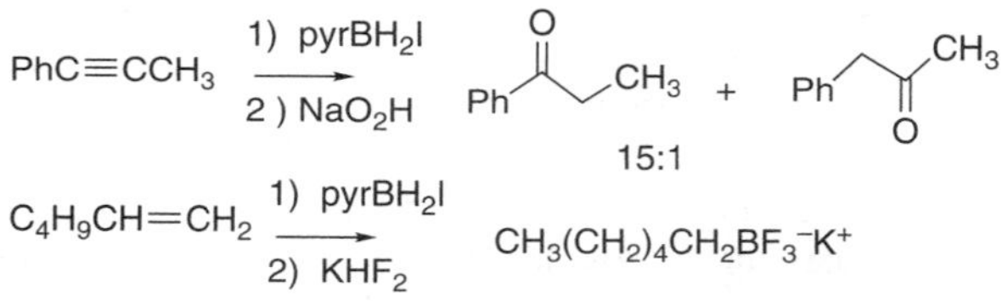

Catecholborane and pinacolborane, in which the boron has two oxygen substituents, are much less reactive hydroborating reagents than alkyl or haloboranes because the boron electron deficiency is attenuated by the oxygen atoms. Nevertheless, they are useful reagents for certain applications.[161] The reactivity of catecholborane has been found to be substantially enhanced by addition of 10–20% of N,N-dimethylacetamide to CH_2Cl_2.[162]

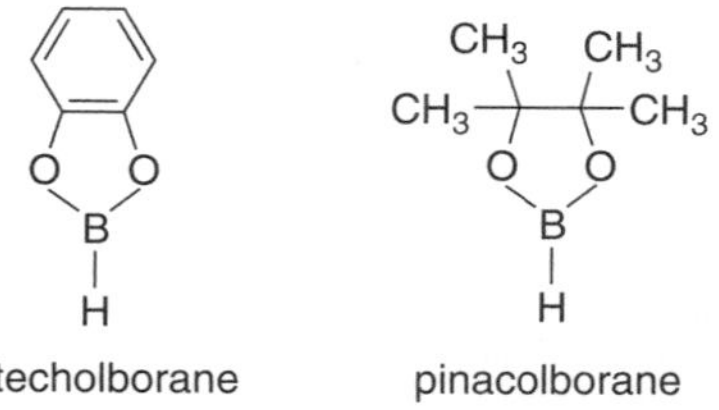

catecholborane   pinacolborane

159. H. C. Brown and S. U. Kulkarni, *J. Organomet. Chem.*, **239**, 23 (1982).
160. J. M. Clay and E. Vedejs, *J. Am. Chem. Soc.*, **127**, 5766 (2005).
161. C. E. Tucker, J. Davidson, and P. Knockel, *J. Org. Chem.*, **57**, 3482 (1992).
162. C. E. Garrett and G. C. Fu, *J. Org. Chem.*, **61**, 3224 (1996).

Catecholborane and pinacolborane are especially useful in hydroborations catalyzed by transition metals.[163] Wilkinson's catalyst $Rh(PPh_3)_3Cl$ is among those used frequently.[164] The general mechanism for catalysis is believed to be similar to that for homogeneous hydrogenation and involves oxidative addition of the borane to the metal, generating a metal hydride.[165]

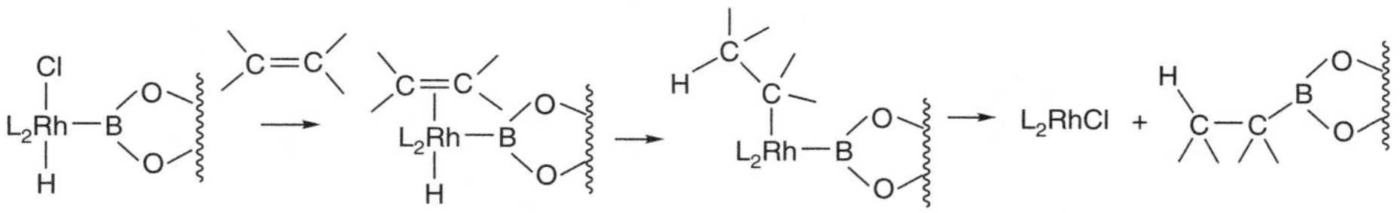

Variation in catalyst and ligand can lead to changes in both regio- and enantioselectivity. For example, the hydroboration of vinyl arenes such as styrene and 6-methoxy-2-vinylnaphthalene can be directed to the internal secondary borane by use of $Rh(COD)_2BF_4$ as a catalyst.[166] These reactions are enantioselective in the presence of a chiral phosphorus ligand.

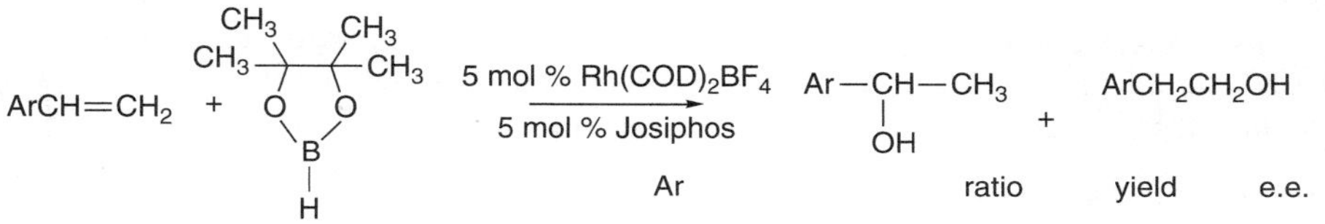

Ar	ratio	yield	e.e.
Phenyl	83:17	87%	84%
6-Methoxynaphthyl	95:5	83%	88%

On the other hand, iridium catalysts give very high selectivity for formation of the primary borane.[167] Several other catalysts have been described, including, for example, dimethyltitanocene.[168]

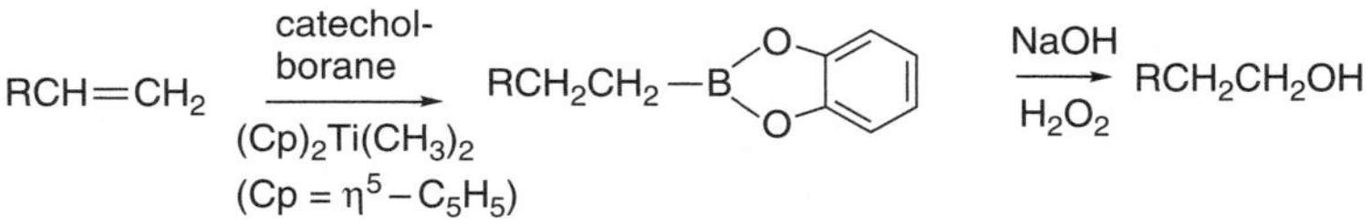

Catalyzed hydroboration has proven to be valuable in controlling the stereoselectivity of hydroboration of functionalized alkenes.[169] For example, allylic alcohols

163. I. Beletskaya and A. Pelter, *Tetrahedron*, **53**, 4957 (1997); H. Wadepohl, *Angew. Chem. Int. Ed. Engl.*, **36**, 2441 (1997); K. Burgess and M. J. Ohlmeyer, *Chem. Rev.*, **91**, 1179 (1991); C. M. Crudden and D. Edwards, *Eur. J. Org. Chem.*, 4695 (2003).
164. D. A. Evans, G. C. Fu, and A. H. Hoveyda, *J. Am. Chem. Soc.*, **110**, 6917 (1988); D. Maenning and H. Noeth, *Angew. Chem. Int. Ed. Engl.*, **24**, 878 (1985).
165. D. A. Evans, G. C. Fu, and B. A. Anderson, *J. Am. Chem. Soc.*, **114**, 6679 (1992).
166. C. M. Crudden, Y. B. Hleba, and A. C. Chen, *J. Am. Chem. Soc.*, **126**, 9200 (2004).
167. Y. Yamamoto, R. Fujikawa, T. Unemoto, and N. Miyaura, *Tetrahedron*, **60**, 10695 (2004).
168. X. He and J. F. Hartwig, *J. Am. Chem. Soc.*, **118**, 1696 (1996).
169. D. A. Evans, G. C. Fu, and A. H. Hoveyda, *J. Am. Chem. Soc.*, **114**, 6671 (1992).

and ethers give mainly *syn* product when catalyzed by Rh(PPh$_3$)$_3$Cl, whereas direct hydroboration with 9-BBN gives mainly *anti* product.

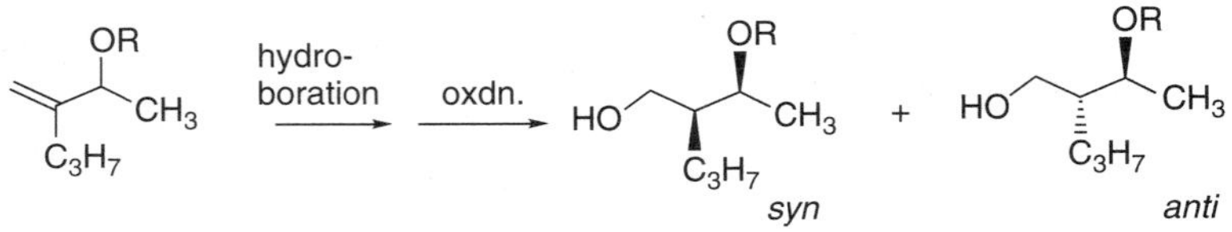

	9-BBN		catecholborane 3 mol % Rh(PPh)$_3$Cl	
R	yield	syn:anti	yield	syn:anti
H	91	17:83	79	81:19
PhCH$_2$	82	25:75	63	80:20
TBDMS	85	13:87	79	93:7

The stereoselectivity of the catalyzed reaction appears to be associated with the complexation step, which is product determining. The preferred orientation of approach of the complex is *anti* to the oxygen substituent, which acts as an electron acceptor and more electronegative groups enhance reactivity. The preferred conformation of the alkene has the hydrogen oriented toward the double bond and this leads to a *syn* relationship between the alkyl and oxygen substituents.[170]

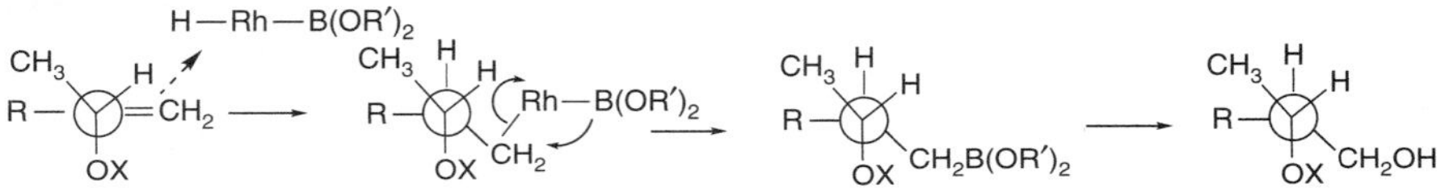

The use of chiral ligands in catalysts can lead to enantioselective hydroboration. Rh-BINAP[171] **C** and the related structure **D**[172] have shown good stereoselectivity in the hydroboration of styrene and related compounds (see also Section 4.5.3).

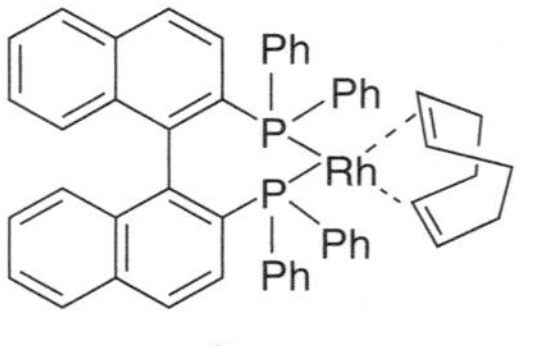

	styrene	indene
C	96% e.e.	13% e.e.
D	67% e.e.	84% e.e.

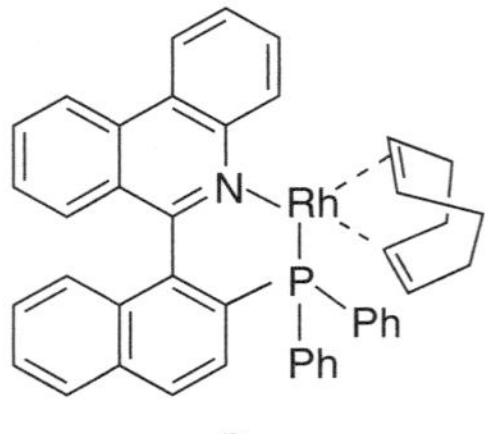

Hydroboration is thermally reversible. B—H moieties are eliminated from alkylboranes at 160° C and above, but the equilibrium still favors of the addition products.

170. K. Burgess, W. A. van der Donk, M. B. Jarstfer, and M. J. Ohlmeyer, *J. Am. Chem. Soc.*, **113**, 6139 (1991).
171. T. Hayashi and Y. Matsumoto, *Tetrahedron: Asymmetry*, **2**, 601 (1991).
172. J. M. Valk, G. A. Whitlock, T. P. Layzell, and J. M. Brown, *Tetrahedron: Asymmetry*, **6**, 2593 (1995).

This provides a mechanism for migration of the boron group along the carbon chain by a series of eliminations and additions.

$$R\!-\!\underset{\underset{B}{H}}{\overset{R}{C}}\!-\!CH\!-\!CH_3 \;\rightleftharpoons\; R\!-\!\overset{R}{C}\!=\!CH\!-\!CH_3 + \underset{H\!-\!B}{H} \; R\!-\!\underset{H}{\overset{R}{C}}\!-\!\overset{H}{C}\!=\!CH_2 \;\rightleftharpoons\; R\!-\!\underset{H}{\overset{R}{C}}\!-\!CH_2\!-\!CH_2\!-\!B$$

Migration cannot occur past a quaternary carbon, however, since the required elimination is blocked. At equilibrium the major trialkyl borane is the least-substituted terminal isomer that is accessible, since this isomer minimizes unfavorable steric interactions.

$$\left[\; \underset{}{\overset{H_3C\;\;H}{\text{(cage)}}} B \;\right]_3 \xrightarrow{160^\circ C} \left[\; \underset{}{\overset{H\;\;CH_2}{\text{(cage)}}} B \;\right]_3$$

Ref. 173

$$CH_3(CH_2)_{13}CH\!=\!CH(CH_2)_{13}CH_3 \xrightarrow[\text{2) }80^\circ C,\,14\,h]{\text{1) }B_2H_6} [CH_3(CH_2)_{29}]_3B$$

Ref. 174

Migrations are more facile for *tetra*-substituted alkenes and occur at 50°–60° C.[175] Bulky substituents on boron facilitate the migration. *bis*-Bicyclo[2.2.2]octanylboranes, in which there are no complications from migrations in the bicyclic substituent, were found to be particularly useful.

$$\text{(bis-bicyclooctyl)B}\!-\!H + CH_3CH\!=\!C(CH_3)_2 \longrightarrow \xrightarrow{\Delta} \text{(bis-bicyclooctyl)B}CH_2CH_2CH(CH_3)_2$$

Ref. 176

There is evidence that boron migration occurs intramolecularly.[177] A TS involving an electron-deficient π complex about 20–25 kcal above the trialkylborane that describes the migration has been located computationally.[178]

$$\underset{\underset{H}{\overset{|}{C}}\!-\!\overset{|}{C}\!-\!H}{\overset{-B\;\;H}{|\;\;|}} \;\rightleftharpoons\; \underset{\underset{H}{\overset{|}{C}}\!-\!\overset{|}{C}\!-\!H}{\overset{-B\!-\!H}{\cdots}} \;\longrightarrow\; \underset{\underset{H}{\overset{|}{C}}\!-\!\overset{|}{C}\!-\!H}{\overset{H\;\;B-}{|\;\;|}}$$

173. G. Zweifel and H. C. Brown, *J. Am. Chem. Soc.*, **86**, 393 (1964).
174. K. Maruyama, K. Terada, and Y. Yamamoto, *J. Org. Chem.*, **45**, 737 (1980).
175. L. O. Bromm, H. Laaziri, F. Lhermitte, K. Harms, and P. Knochel, *J. Am. Chem. Soc.*, **122**, 10218 (2000).
176. H. C. Brown and U. S. Racherla, *J. Am. Chem. Soc.*, **105**, 6506 (1983).
177. S. E. Wood and B. Rickborn, *J. Org. Chem.*, **48**, 555 (1983).
178. N. J. R. van Eikema Hommes and P. v. R. Schleyer, *J. Org. Chem.*, **56**, 4074 (1991).

Migration of boron to terminal positions is observed under much milder conditions in the presence of transition metal catalysts. For example, hydroboration of 2-methyl-3-hexene by pinacolborane in the presence of $Rh(PPh_3)_3Cl$ leads to the terminal boronate ester.

$$(CH_3)_2CHCH=CHCH_2CH_3 \quad \xrightarrow[\text{Rh(PPh}_3)_3\text{Cl}]{\substack{\text{pinacol-}\\\text{borane}}} \quad (CH_3)_2CH(CH_2)_4-B\underset{O}{\overset{O}{\Big\langle}}\substack{CH_3 \\ CH_3 \\ CH_3 \\ CH_3}$$

Ref. 179

4.5.2. Reactions of Organoboranes

The organoboranes have proven to be very useful intermediates in organic synthesis. In this section we discuss methods by which the boron atom can be replaced by hydroxy, carbonyl, amino, or halogen groups. There are also important processes that use alkylboranes in the formation of new carbon-carbon bonds. These reactions are discussed in Section 9.1.

The most widely used reaction of organoboranes is the oxidation to alcohols, and alkaline hydrogen peroxide is the reagent usually employed to effect the oxidation. The mechanism, which is outlined below, involves a series of B to O migrations of the alkyl groups. The R−O−B bonds are hydrolyzed in the alkaline aqueous solution, generating the alcohol.

$$R_3B + HOO^- \longrightarrow R-\underset{\underset{R}{|}}{\overset{\overset{R}{|}}{B}}-O-OH \longrightarrow R-\underset{\underset{R}{|}}{\overset{\overset{R}{|}}{B}}-OR + {}^-OH$$

$$R_2BOR + HOO^- \longrightarrow R-\underset{\underset{R}{|}}{\overset{\overset{R-O}{|}}{B}}-O-O-H \longrightarrow R-\underset{\underset{RO}{|}}{\overset{\overset{RO}{|}}{B}} + {}^-OH$$

$$(RO)_2BR + HOO^- \longrightarrow (RO)_2\underset{\underset{R}{|}}{\overset{-}{B}}-O-O-H \longrightarrow (RO)_3B + {}^-OH$$

$$(RO)_3B + 3H_2O \longrightarrow 3ROH + B(OH)_3$$

The stereochemical outcome is replacement of the C−B bond by a C−O bond with *retention of configuration*. In combination with stereospecific *syn* hydroboration, this allows the structure and stereochemistry of the alcohols to be predicted with confidence. The preference for hydroboration at the least-substituted carbon of a double bond results in the alcohol being formed with regiochemistry that is complementary to that observed by direct hydration or oxymercuration, that is, anti-Markovnikov.

Several other oxidants can be used to effect the borane to alcohol conversion. Oxone® $(2K_2SO_5 \cdot KHSO_4 \cdot K_2SO_4)$ has been recommended for oxidations done on a

179. S. Pereira and M. Srebnik, *J. Am. Chem. Soc.*, **118**, 909 (1996); S. Pereira and M. Srebnik, *Tetrahedron Lett.*, **37**, 3283 (1996).

large scale.[180] Conditions that permit oxidation of organoboranes to alcohols using molecular oxygen,[181] sodium peroxycarbonate[182] or amine oxides[183] as oxidants have also been developed. The reaction with molecular oxygen is particularly effective in perfluoroalkane solvents.[184]

$$
\text{(norbornene)} \quad \xrightarrow[\text{2) } O_2, \text{ Br}(CF_2)_7CF_3]{\text{1) } HB(C_2H_5)_2} \quad \text{(norbornanol) OH} \quad 82\%
$$

More vigorous oxidants such as Cr(VI) reagents effect replacement of boron and oxidation to the carbonyl level.[185]

$$
\text{(1-phenylcyclohexene) Ph} \quad \xrightarrow[\text{2) } K_2Cr_2O_7]{\text{1) } B_2H_6} \quad \text{(2-phenylcyclohexanone) Ph, O}
$$

An alternative procedure for oxidation to ketones involves treatment of the alkylborane with a quaternary ammonium perruthenate salt and an amine oxide[186] (see Entry 6 in Scheme 4.9). Use of dibromoborane-dimethyl sulfide for hydroboration of terminal alkenes, followed by hydrolysis and Cr(VI) oxidation gives carboxylic acids.[187]

$$
RCH=CH_2 \quad \xrightarrow[\text{2) } H_2O]{\text{1) } BHBr_2S(CH_3)_2} \quad RCH_2CH_2B(OH)_2 \quad \xrightarrow[\text{HOAc, } H_2O]{Cr(VI)} \quad RCH_2CO_2H
$$

The boron atom can also be replaced by an amino group.[188] The reagents that effect this conversion are chloramine or hydroxylamine-*O*-sulfonic acid, and the mechanism of these reactions is very similar to that of the hydrogen peroxide oxidation of organoboranes. The nitrogen-containing reagent initially reacts as a nucleophile by adding at boron and a B to N rearrangement with expulsion of chloride or sulfate ion follows. Usually only two of the three alkyl groups migrate. As in the oxidation, the migration step occurs with retention of configuration. The amine is freed by hydrolysis.

$$
R_3B \quad \xrightarrow{NH_2X} \quad R_2\bar{B}-NH-X \quad \longrightarrow R_2B-NH \quad \xrightarrow{NH_2X} \quad RB(NHR)_2 \quad \xrightarrow{H_2O} \quad 2\ RNH_2
$$
$$
\qquad\qquad \underset{R}{|} \qquad\qquad\qquad \underset{R}{|}
$$
$$
X = Cl \text{ or } OSO_3
$$

180. D. H. B. Ripin, W. Cai, and S. T. Brenek, *Tetrahedron Lett.*, **41**, 5817 (2000).

181. H. C. Brown, M. M. Midland, and G. W. Kabalka, *J. Am. Chem. Soc.*, **93**, 1024 (1971).

182. G. W. Kabalka, P. P. Wadgaonkar, and T. M. Shoup, *Tetrahedron Lett.*, **30**, 5103 (1989).

183. G. W. Kabalka and H. C. Hedgecock, Jr., *J. Org. Chem.*, **40**, 1776 (1975); R. Koster and Y. Monta, *Liebigs Ann. Chem.*, **704**, 70 (1967).

184. I. Klement and P. Knochel, *Synlett*, 1004 (1996).

185. H. C. Brown and C. P. Garg, *J. Am. Chem. Soc.*, **83**, 2951 (1961); H. C. Brown, C. Rao, and S. Kulkarni, *J. Organomet. Chem.*, **172**, C20 (1979).

186. M. H. Yates, *Tetrahedron Lett.*, **38**, 2813 (1997).

187. H. C. Brown, S. V. Kulkarni, V. V. Khanna, V. D. Patil, and U. S. Racherla, *J. Org. Chem.*, **57**, 6173 (1992).

188. M. W. Rathke, N. Inoue, K. R. Varma, and H. C. Brown, *J. Am. Chem. Soc.*, **88**, 2870 (1966); G. W. Kabalka, K. A. R. Sastry, G. W. McCollum, and H. Yoshioka, *J. Org. Chem.*, **46**, 4296 (1981).

The alkene can be used more efficiently if the hydroboration is done with dimethyl-borane.[189]

$$RCH{=}CH_2 \xrightarrow{(CH_3)_3BH} RCH_2CH_2B(CH_3)_2 \xrightarrow{NH_2X} (CH_3)_2BNCH_2CH_2R \xrightarrow{H_2O} RCH_2CH_2NH_2$$

Secondary amines are formed by reaction of trisubstituted boranes with alkyl or aryl azides. The most efficient borane intermediates are monoalkyldichloroboranes, which are generated by reaction of an alkene with $BHCl_2 \cdot Et_2O$.[190] The entire sequence of steps and the mechanism of the final stages are summarized by the equation below.

$$BHCl_2 + RCH{=}CH_2 \longrightarrow RCH_2CH_2BCl_2$$

$$RCH_2CH_2BCl_2 + R'{-}N_3 \longrightarrow Cl_2B^-\overset{R'}{\underset{RCH_2CH_2}{\overset{|}{-}N}}\overset{+}{-}N{\equiv}N \longrightarrow Cl_2B\overset{R'}{\overset{|}{N}}CH_2CH_2R \xrightarrow{H_2O} R'NHCH_2CH_2R$$

This reaction has been used to prepare α-*N*-methylamino acids using $(CH_3)_2BBr$.[191]

$$\underset{\overset{\vdots}{N_3}}{PhCHCO_2H} \xrightarrow{(CH_3)_2BBr} \underset{\overset{\vdots}{NHCH_3}}{PhCHCO_2H}$$

Secondary amines can also be made using the *N*-chloro derivatives of primary amines.[192]

$$(CH_3CH_2)_3B + \underset{\overset{|}{Cl}}{HN}(CH_2)_7CH_3 \longrightarrow CH_3CH_2\underset{\overset{|}{H}}{N}(CH_2)_7CH_3$$
$$90\%$$

Organoborane intermediates can also be used to synthesize alkyl halides. Replacement of boron by iodine is rapid in the presence of base.[193] The best yields are obtained using sodium methoxide in methanol.[194] If less basic conditions are desirable, the use of iodine monochloride and sodium acetate gives good yields.[195] As is the case in hydroboration-oxidation, the regioselectivity of hydroboration-halogenation is opposite to that observed by direct ionic addition of hydrogen halides to alkenes. Terminal alkenes give primary halides.

$$RCH{=}CH_2 \xrightarrow[\text{2) Br}_2,\ \text{NaOH}]{\text{1) B}_2\text{H}_6} RCH_2CH_2Br$$

189. H. C. Brown, K.-W. Kim, M. Srebnik, and B. Singaram, *Tetrahedron*, **43**, 4071 (1987).
190. H. C. Brown, M. M. Midland, and A. B. Levy, *J. Am. Chem. Soc.*, **95**, 2394 (1973).
191. R. L. Dorow and D. E. Gingrich, *J. Org. Chem.*, **60**, 4986 (1995).
192. G. W. Kabalka, G. W. McCollum, and S. A. Kunda, *J. Org. Chem.*, **49**, 1656 (1984).
193. H. C. Brown, M. W. Rathke, and M. M. Rogic, *J. Am. Chem. Soc.*, **90**, 5038 (1968).
194. N. R. De Lue and H. C. Brown, *Synthesis*, 114 (1976).
195. G. W. Kabalka and E. E. Gooch, III, *J. Org. Chem.*, **45**, 3578 (1980).

Scheme 4.9 gives some examples of the use of boranes in syntheses of alcohols, aldehydes, ketones, amines, and halides. Entry 1 demonstrates both the regioselectivity and stereospecificity of hydroboration, resulting in the formation of *trans*-2-methylcyclohexanol. Entry 2 illustrates the facial selectivity, with the borane adding *anti* to the *endo* methyl group.

Entry 3 illustrates all aspects of the regio- and stereoselectivity, with *syn* addition occurring *anti* to the dimethyl bridge in the pinene structure. The stereoselectivity in Entry 4 is the result of the preferred conformation of the alkene and approach *syn* to the smaller methyl group, rather than the 2-furyl group.

Entries 5 to 7 are examples of oxidation of boranes to the carbonyl level. In Entry 5, chromic acid was used to obtain a ketone. Entry 6 shows 5 mol % tetrapropylammonium perruthenate with *N*-methylmorpholine-*N*-oxide as the stoichiometric oxidant converting the borane directly to a ketone. Aldehydes were obtained from terminal alkenes using this reagent combination. Pyridinium chlorochromate (Entry 7) can also be used to obtain aldehydes. Entries 8 and 9 illustrate methods for amination of alkenes via boranes. Entries 10 and 11 illustrate the preparation of halides.

4.5.3. Enantioselective Hydroboration

Several alkylboranes are available in enantiomerically enriched or pure form and can be used to prepare enantiomerically enriched alcohols and other compounds available via organoborane intermediates.[196] One route to enantiopure boranes is by hydroboration of readily available terpenes that occur naturally in enantiomerically enriched or pure form. The most thoroughly investigated of these is *bis*-(isopinocampheyl)borane; (Ipc)$_2$BH), which can be prepared in 100% enantiomeric purity from the readily available terpene α-pinene.[197] Both enantiomers are available.

[196] H. C. Brown and B. Singaram, *Acc. Chem. Res.*, **21**, 287 (1988); D. S. Matteson, *Acc. Chem. Res.*, **21**, 294 (1988).

[197] H. C. Brown, P. K. Jadhav, and A. K. Mandal, *Tetrahedron*, **37**, 3547 (1981); H. C. Brown and P. K. Jadhav, in *Asymmetric Synthesis*, Vol. 2, J. D. Morrison, ed., Academic Press, New York, 1983, Chap. 1.

**Scheme 4.9. Synthesis of Alcohols, Aldehydes, Ketones, and Amines from
Organoboranes**

A. Alcohols

1[a]

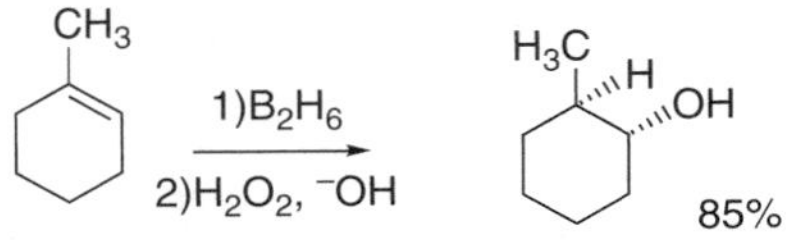

1)B$_2$H$_6$

2)H$_2$O$_2$, $^-$OH

85%

2[b]

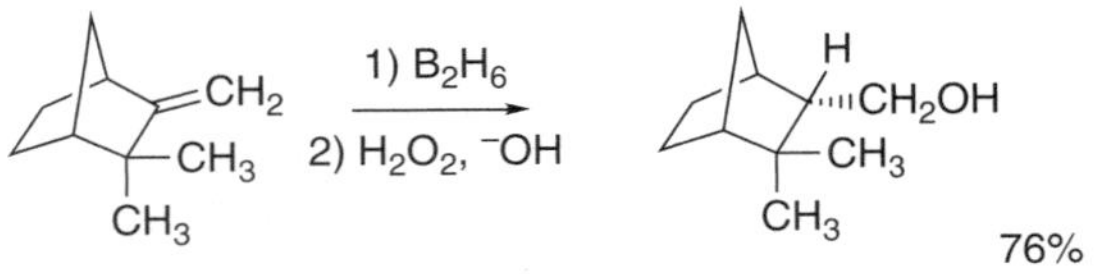

1) B$_2$H$_6$

2) H$_2$O$_2$, $^-$OH

76%

3[c]

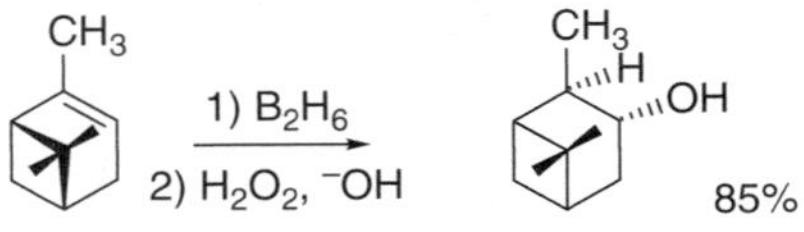

1) B$_2$H$_6$

2) H$_2$O$_2$, $^-$OH

85%

4[d]

1) B$_2$H$_6$, THF

2) H$_2$O$_2$, $^-$OH

85%

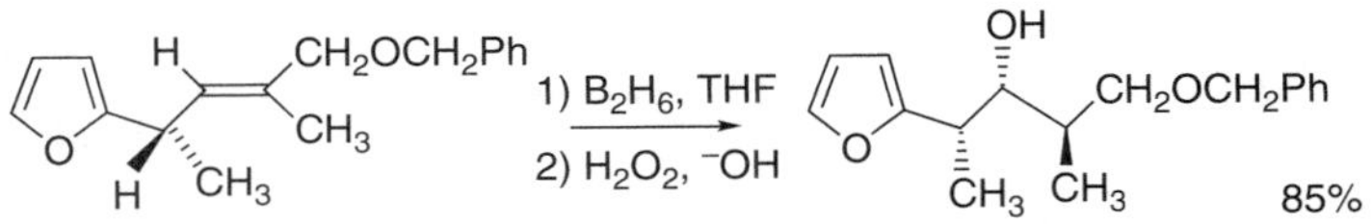

B. Ketones and aldehydes

5[e]

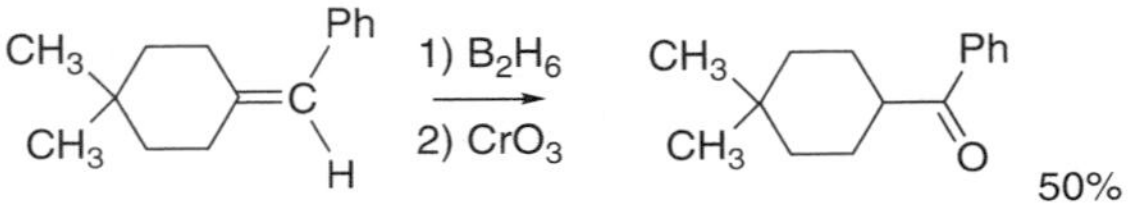

1) B$_2$H$_6$

2) CrO$_3$

50%

6[f]

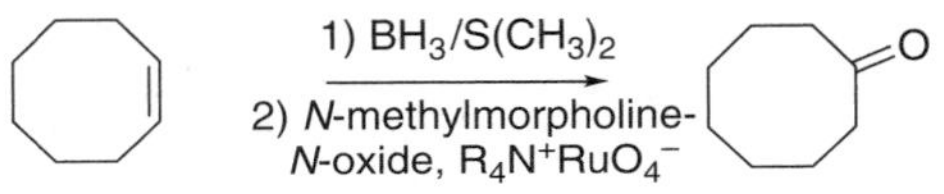

1) BH$_3$/S(CH$_3$)$_2$

2) *N*-methylmorpholine-
N-oxide, R$_4$N$^+$RuO$_4$$^-$

7[g]

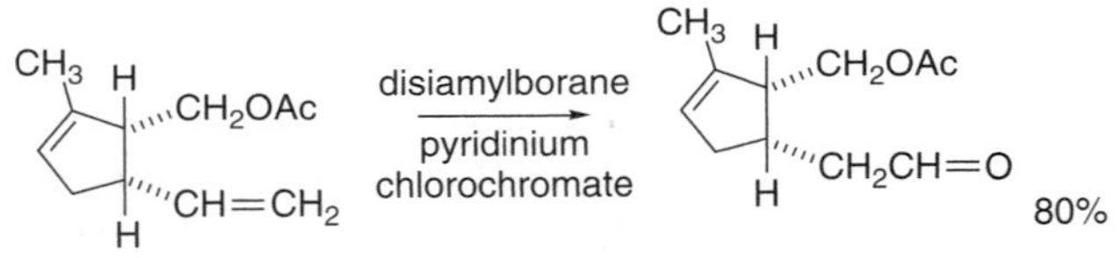

disiamylborane

pyridinium
chlorochromate

80%

C. Amines

8[h]

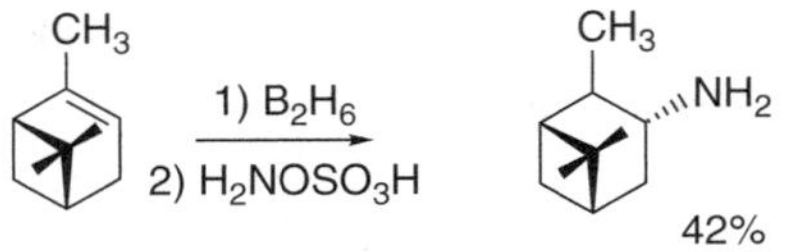

1) B$_2$H$_6$

2) H$_2$NOSO$_3$H

42%

9[i]

1) BHCl$_2$

2) PhN$_3$, H$_2$O

84%

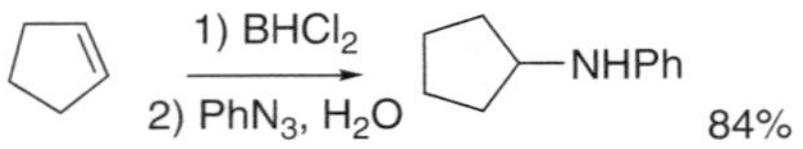

(Continued)

Scheme 4.9. (*Continued*)

349

SECTION 4.5

*Addition at Double
Bonds via Organoborane
Intermediates*

D. Halides

10^j (CH$_3$)$_3$CCH$_2$C=CH$_2$ 1) B$_2$H$_6$, THF   (CH$_3$)$_3$CCH$_2$CHCH$_2$I
                    | ————————→ |
                    CH$_3$ 2) I$_2$ CH$_3$ 92%
                                   3) CH$_3$OH, $^-$OH

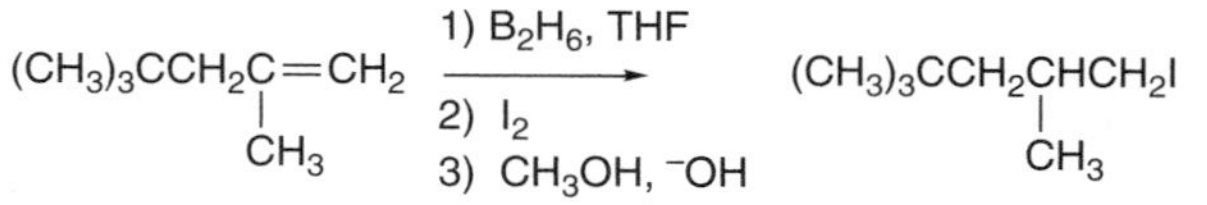

11^k CH$_3$ 1) B$_2$H$_6$, THF   CH$_3$
        TBSO⟍⟋⟍⟋   ————————→   TBSO⟍⟋⟍⟋⟍I
                    2) CH$_3$OH, NaOAc
                    3) ICl   60%

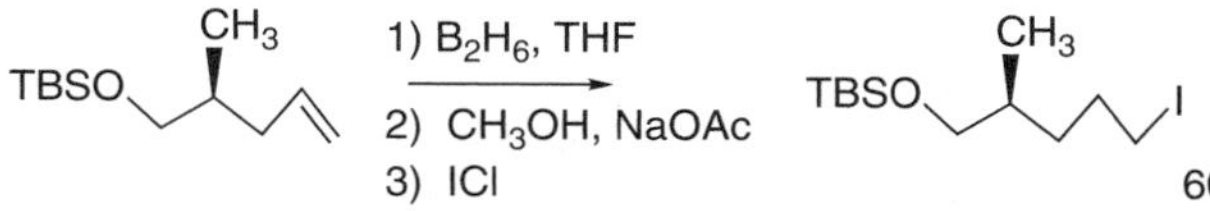

a. H. C. Brown and G. Zweifel, *J. Am. Chem. Soc.*, **83**, 2544 (1961).
b. R. Dulou, Y. Chretien-Bessiere, *Bull. Soc. Chim. Fr.*, 1362 (1959).
c. G. Zweifel and H. C. Brown, *Org. Synth.*, **52**, 59 (1972).
d. G. Schmid, T. Fukuyama, K. Akasaka, and Y. Kishi, *J. Am. Chem. Soc.*, **101**, 259 (1979).
e. W. B. Farnham, *J. Am. Chem. Soc.*, **94**, 6857 (1972).
f. M. H. Yates, *Tetrahedron Lett.*, **38**, 2813 (1997).
g. H. C. Brown, S. U. Kulkarni, and C. G. Rao, *Synthesis*, 151 (1980); T. H. Jones and
   M. S. Blum, *Tetrahedron Lett.*, **22**, 4373 (1981).
h. M. W. Rathke and A. A. Millard, *Org. Synth.*, **58**, 32 (1978).
i. H. C. Brown, M. M. Midland, and A. B. Levy, *J. Am. Chem. Soc.*, **95**, 2394 (1973).
j. H. C. Brown, M. W. Rathke, M. M. Rogic, and N. R. DeLue, *Tetrahedron*, **44**, 2751 (1988).
k. D. Schinzer, A. Bauer, and J. Schreiber, *Chem. Eur. J.*, **5**, 2492 (1999).

Other examples of chiral organoboranes derived from terpenes are **E**, **F**, and **G**, which
are derived from longifolene,[198] 2-carene,[199] and limonene,[200] respectively.

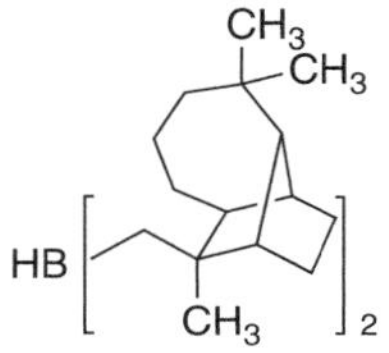

E 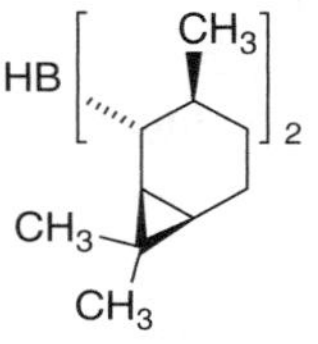   **F** 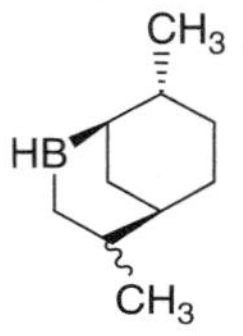   **G**

(Ipc)$_2$BH adopts a conformation that minimizes steric interactions. This conformation can be represented schematically as in **H** and **I**, where the S, M, and L
substituents are, respectively, the 3-H, 4-CH$_2$, and 2-CHCH$_3$ groups of the carbocyclic
structure. The steric environment at boron in this conformation is such that Z-alkenes
encounter less steric encumbrance in TS **I** than in **H**.

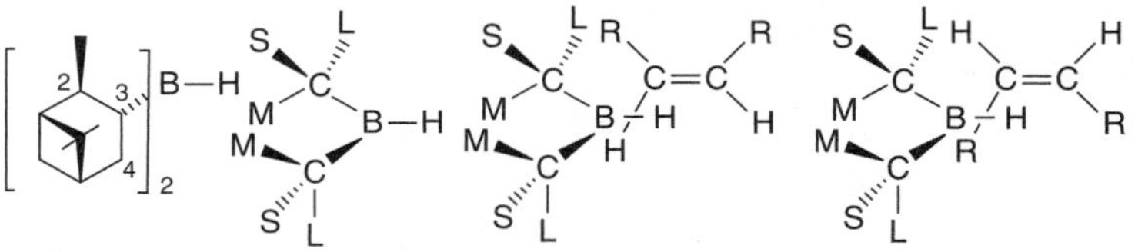

H   **I**

The degree of enantioselectivity of (Ipc)$_2$BH is not high for all simple alkenes.
Z-Disubstituted alkenes give good enantioselectivity (75–90%) but *E*-alkenes and

198. P. K. Jadhav and H. C. Brown, *J. Org. Chem.*, **46**, 2988 (1981).
199. H. C. Brown, J. V. N. Vara Prasad, and M. Zaidlewicz, *J. Org. Chem.*, **53**, 2911 (1988).
200. P. K. Jadhav and S. U. Kulkarni, *Heterocylces*, **18**, 169 (1982).

simple cycloalkenes give low enantioselectivity (5–30%). Interestingly, vinyl ethers exhibit good enantioselectivity for both the *E*- and *Z*-isomers.[201]

72% yield
> 97% e.e.

77% yield
90% e.e.

Monoisocampheylborane (IpcBH$_2$) can be prepared in enantiomerically pure form by separation of a TMEDA adduct.[202] When this monoalkylborane reacts with a prochiral alkene, one of the diastereomeric products is normally formed in excess and can be obtained in high enantiomeric purity by an appropriate separation.[203] Oxidation of the borane then provides the corresponding alcohol having the enantiomeric purity achieved for the borane.

As oxidation also converts the original chiral terpene-derived group to an alcohol, it is not directly reusable as a chiral auxiliary. Although this is not a problem with inexpensive materials, the overall efficiency of generation of enantiomerically pure product is improved by procedures that can regenerate the original terpene. This can be done by heating the dialkylborane intermediate with acetaldehyde. The α-pinene is released and a diethoxyborane is produced.[204]

The usual oxidation conditions then convert this boronate ester to an alcohol.[205]

The corresponding haloboranes are also useful for enantioselective hydroboration. Isopinocampheylchloroborane can achieve 45–80% e.e. with representative alkenes.[206] The corresponding bromoborane achieves 65–85% enantioselectivity with simple alkenes when used at −78° C.[207]

201. D. Murali, B. Singaram, and H. C. Brown, *Tetrahedron: Asymmetry*, **11**, 4831 (2000).

202. H. C. Brown, J. R. Schwier, and B. Singaram, *J. Org. Chem.*, **43**, 4395 (1978); H. C. Brown, A. K. Mandal, N. M. Yoon, B. Singaram, J. R. Schwier, and P. K. Jadhav, *J. Org. Chem.*, **47**, 5069 (1982).

203. H. C. Brown and B. Singaram, *J. Am. Chem. Soc.*, **106**, 1797 (1984); H. C. Brown, P. K. Jadhav, and A. K. Mandal, *J. Org. Chem.*, **47**, 5074 (1982).

204. H. C. Brown, B. Singaram, and T. E. Cole, *J. Am. Chem. Soc.*, **107**, 460 (1985); H. C. Brown, T. Imai, M. C. Desai, and B. Singaram, *J. Am. Chem. Soc.*, **107**, 4980 (1985).

205. D. S. Matteson and K. M. Sadhu, *J. Am. Chem. Soc.*, **105**, 2077 (1983).

206. U. P. Dhokte, S. V. Kulkarni, and H. C. Brown, *J. Org. Chem.*, **61**, 5140 (1996).

207. U. P. Dhokte and H. C. Brown, *Tetrahedron Lett.*, **37**, 9021 (1996).

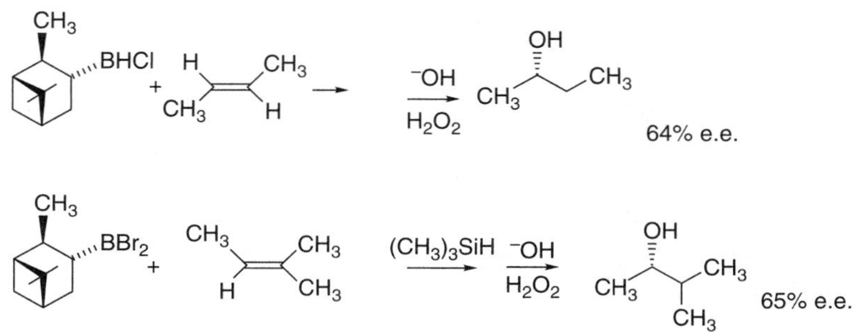

Procedures for synthesis of chiral amines[208] and halides[209] based on chiral alkylboranes involve applying the methods discussed earlier to the enantiomerically enriched organoborane intermediates. For example, enantiomerically pure terpenes can be converted to trialkylboranes and then aminated with hydroxylaminesulfonic acid.

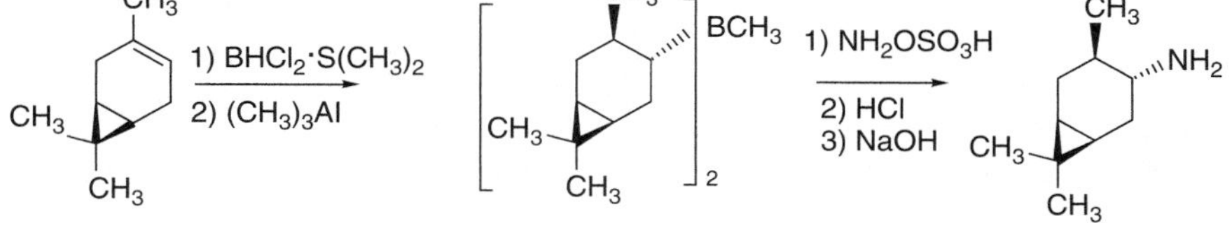

Ref. 210

Combining catalytic enantioselective hydroboration (see p. 342) with amination has provided certain amines with good enantioselectivity. In this procedure the catechol group is replaced by methyl prior to the amination step.

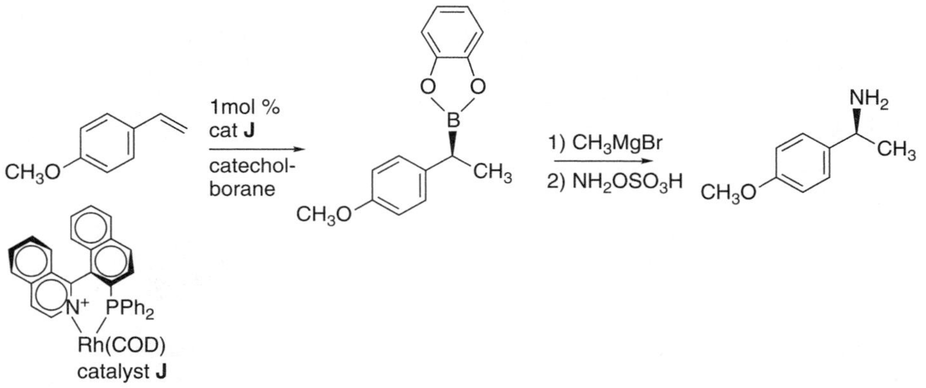

Ref. 211

208. L. Verbit and P. J. Heffron, *J. Org. Chem.*, **32**, 3199 (1967); H. C. Brown, K.-W. Kim, T. E. Cole, and B. Singaram, *J. Am. Chem. Soc.*, **108**, 6761 (1986); H. C. Brown, A. M. Sahinke, and B. Singaram, *J. Org. Chem.*, **56**, 1170 (1991).
209. H. C. Brown, N. R. De Lue, G. W. Kabalka, and H. C. Hedgecock, Jr., *J. Am. Chem. Soc.*, **98**, 1290 (1976).
210. H. C. Brown, S. V. Malhotra, and P. V. Ramachandran, *Tetrahedron: Asymmetry*, **7**, 3527 (1996).
211. E. Fernandez, M. W. Hooper, F. I. Knight, and J. M. Brown, *J. Chem. Soc., Chem. Commun.*, 173 (1997).

4.5.4. Hydroboration of Alkynes

Alkynes are reactive toward hydroboration reagents. The most useful procedures involve addition of a disubstituted borane to the alkyne, which avoids complications that occur with borane and lead to polymeric structures. Catechol borane is a particularly useful reagent for hydroboration of alkynes.[212] Protonolysis of the adduct with acetic acid results in reduction of the alkyne to the corresponding *cis*-alkene. Oxidative workup with hydrogen peroxide gives ketones via enol intermediates.

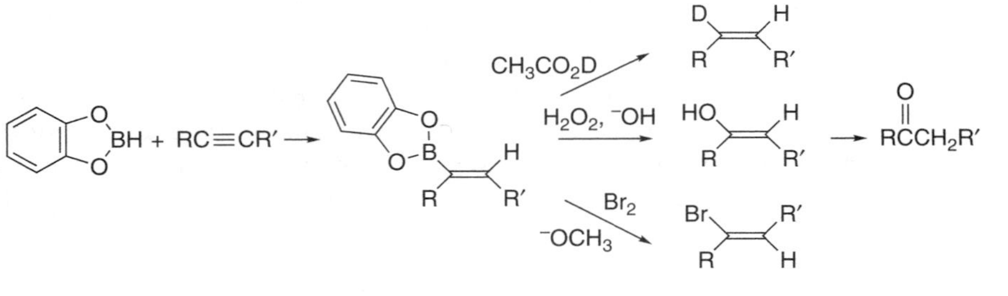

Treatment of the vinylborane with bromine and base leads to vinyl bromides. The reaction occurs with net *anti* addition, and the stereoselectivity is explained on the basis of *anti* addition of bromine followed by a second *anti* elimination of bromide and boron.

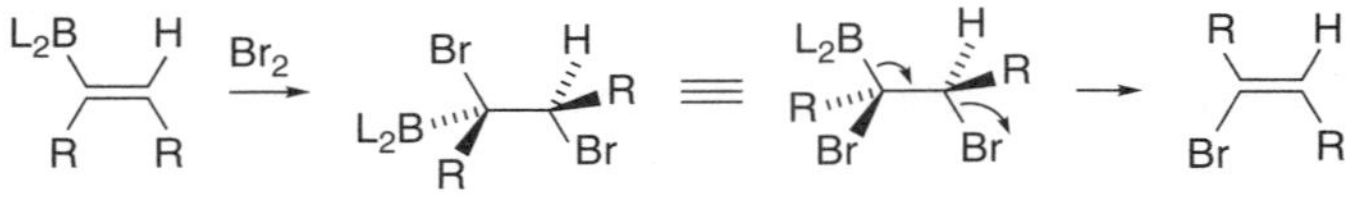

Exceptions to this stereoselectivity have been noted.[213]

The adducts derived from catechol borane are hydrolyzed by water to vinylboronic acids. These materials are useful intermediates for the preparation of terminal vinyl iodides. Since the hydroboration is a *syn* addition and the iodinolysis occurs with retention of the alkene geometry, the iodides have the *E*-configuration.[214]

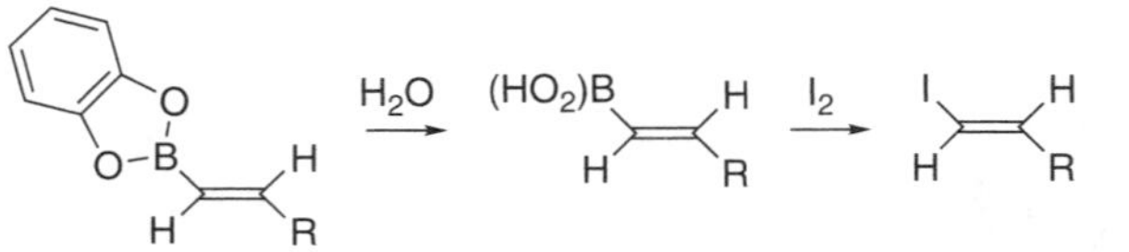

The dimethyl sulfide complex of dibromoborane[215] and pinacolborane[216] are also useful for synthesis of *E*-vinyl iodides from terminal alkynes.

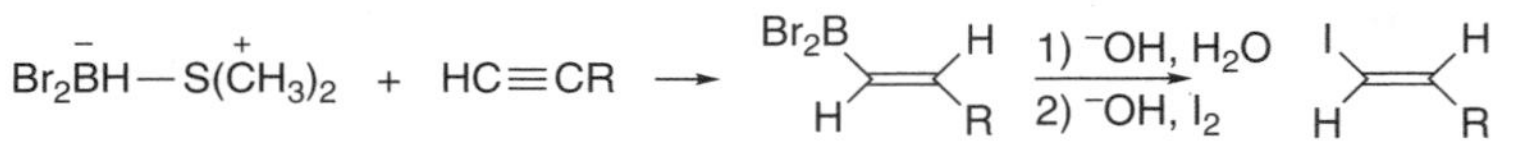

212. H. C. Brown, T. Hamaoka, and N. Ravindran, *J. Am. Chem. Soc.*, **95**, 6456 (1973); C. F. Lane and G. W. Kabalka, *Tetrahedron*, **32**, 981 (1976).
213. J. R. Wiersig, N. Waespe-Sarcevic, and C. Djerassi, *J. Org. Chem.*, **44**, 3374 (1979).
214. H. C. Brown, T. Hamaoka, and N. Ravindran, *J. Am. Chem. Soc.*, **95**, 5786 (1973).
215. H. C. Brown and J. B. Campbell, Jr., *J. Org. Chem.*, **45**, 389 (1980); H. C. Brown, T. Hamaoka, N. Ravindran, C. Subrahmanyam, V. Somayaji, and N. G. Bhat, *J. Org. Chem.*, **54**, 6075 (1989).
216. C. E. Tucker, J. Davidson, and P. Knochel, *J. Org. Chem.*, **57**, 3482 (1992).

Other disubstituted boranes have also been used for selective hydroboration of alkynes. 9-BBN can be used to hydroborate internal alkynes. Protonolysis can be carried out with methanol and this provides a convenient method for formation of a disubstituted Z-alkene.[217]

A large number of procedures that involve carbon-carbon bond formation have been developed based on organoboranes. These reactions are considered in Chapter 9.

4.6. Hydroalumination, Carboalumination, Hydrozirconation, and Related Reactions

Aluminum is the immediate congener of boron, and dialkyl and trialkyl aluminum compounds, which are commercially available, have important industrial applications. They also have some similarities with organoboranes that can be exploited for synthetic purposes. Aluminum is considerably less electronegative than boron and as a result the reagents also share characteristics with the common organometallic reagents such as organomagnesium and organolithium compounds. The addition reactions of alkenes and dialkylaluminum reagents occur much less easily than hydroboration. Only terminal or strained alkenes react readily at room temperature.[218] With internal and branched alkenes, the addition does not go to completion. Addition of dialkylalanes to alkynes occurs more readily, and the regiochemistry and stereochemistry are analogous to hydroboration. The resulting vinylalanes react with halogens with *retention of configuration* at the double bond.[219]

With trialkylaluminum compounds, the addition reaction is called *carboalumination*. As discussed below, this reaction requires a catalyst to proceed.

Computational studies of both hydroalumination and carboalumination have indicated a four-center TS for the addition.[220] The aluminum reagents, however, have more nucleophilic character than do boranes. Whereas the TS for hydroboration is primarily electrophilic and resembles that for attack of CH_3^+ on a double bond, the

217. H. C. Brown and G. A. Molander, *J. Org. Chem.*, **51**, 4512 (1986); H. C. Brown and K. K. Wang, *J. Org. Chem.*, **51**, 4514 (1986).
218. F. Ansinger, B. Fell, and F. Thiessen, *Chem. Ber.*, **100**, 937 (1967); R. Schimpf and P. Heimbach, *Chem. Ber.*, **103**, 2122 (1970).
219. G. Zweifel and C. C. Whitney, *J. Am. Chem. Soc.*, **89**, 2753 (1967).
220. J. W. Bunders and M. M. Francl, *Organometallics*, **12**, 1608 (1993); J. W. Bunders, J. Yudenfreund, and M. M. Francl, *Organometallics*, **18**, 3913 (1999).

reaction with CH_3AlH_2 has a closer resemblance to reaction of CH_3^- with ethene and the strongest interaction is with the ethene LUMO. This interpretation is consistent with relative reactivity trends in which the reactivity of alkenes decreases with increasing alkyl substitution and alkynes are more reactive than alkenes.

Effective catalysts have recently been developed for the addition of trialkyl-aluminum reagents to alkenes (carboalumination). *bis*-(Pentamethylcyclopentadienyl) zirconium dimethylide activated by *tris*-(pentafluorophenyl)boron promotes the addition of trimethylaluminum to terminal alkenes.[221]

$$CH_3(CH_2)_3CH{=}CH_2 \ + \ (CH_3)_3Al \ \xrightarrow[(C_6F_5)_3B]{(Cp^*)_2Zr(CH_3)_2} \ \xrightarrow{O_2} \ CH_3(CH_2)_3 \overset{CH_3}{\underset{}{\diagup}} OH$$

Cp* = 1,2,3,4,5 – pentamethyl-cyclopentadienide

71%

A chiral indene derivative, structure **K**, has been most commonly used.[222] The catalyst interacts with the trialkylaluminum to generate a bimetallic species that is the active catalyst.

$$\begin{array}{c} CH_3 \\ | \\ (-Al-O-)n \\ \text{methylalumoxane} \\ \text{MAO} \end{array} \qquad \begin{array}{c} CH_2CH(CH_3)_2 \\ | \\ (-Al-O-)n \\ \text{isobutylalumoxane} \\ \text{IBAO} \end{array}$$

The detailed mechanism of the catalysis is not known, but it is believed that the Lewis acid character of the zirconium is critical.[223] The reaction is further accelerated by inclusion of partially hydrolyzed trialkylaluminum reagents known as alumoxanes.[224]

$$CH_3(CH_2)_6Al(iBu)_2 \ + \ CH_2{\diagup}{\diagup}{\diagdown}{\diagup}OTBDMS \ \xrightarrow[\text{IBAO}]{5\ mol\%\ cat\ \mathbf{K}\ \ H^+} \ CH_3(CH_2)_6 \overset{CH_3}{\underset{}{\diagup}}{\diagdown}{\diagup}OH$$

IBAO = isobutylaluminoxane

77% yield, 91% e.e.

The adducts can be protonolyzed or converted to halides or alcohols.

$$RCH{=}CH_2 \ + \ R'_3Al \ \xrightarrow[\ ]{\substack{\text{cat} \\ \mathbf{K}}} \ \underset{\underset{R'}{|}}{RCHCH_2AlR'_2} \ \xrightarrow[\substack{\text{or } O_2, \\ X_2}]{H^+} \ \underset{\underset{R'}{|}}{RCHCH_2Z}$$

Z = H, OH, X

221. K. H. Shaugnessy and R. M. Waymouth, *J. Am. Chem. Soc.*, **117**, 5873 (1995).
222. D. Y. Kondakov and E. Negishi, *J. Am. Chem. Soc.*, **118**, 1577 (1996); K. H. Shaugnessy and R. M. Waymouth, *Organometallics*, **17**, 5738 (1998).
223. E. Negishi, D. Y. Kondakov, D. Choueiry, K. Kasai, and T. Takahashi, *J. Am. Chem. Soc.*, **118**, 9577 (1996); E. Negishi, *Chem. Eur. J.*, **5**, 411 (1999).
224. S. Huo, J. Shi, and E. Negishi, *Angew. Chem. Int. Ed. Engl.*, **41**, 2141 (2002).

355

SECTION 4.6

*Hydroalumination,
Carboalumination,
Hydrozirconation,
and Related Reactions*

This methodology has been used to create chiral centers in saturated hydrocarbon chains such as those found in vitamin E, vitamin K, and phytol.[225]

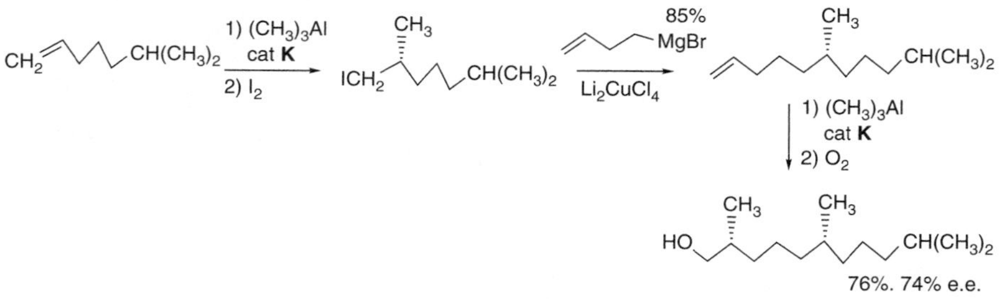

By converting the primary alcohol group to an alkene by oxidation and a Wittig reaction, the reaction can be carried out in iterative fashion to introduce several methyl groups.[226]

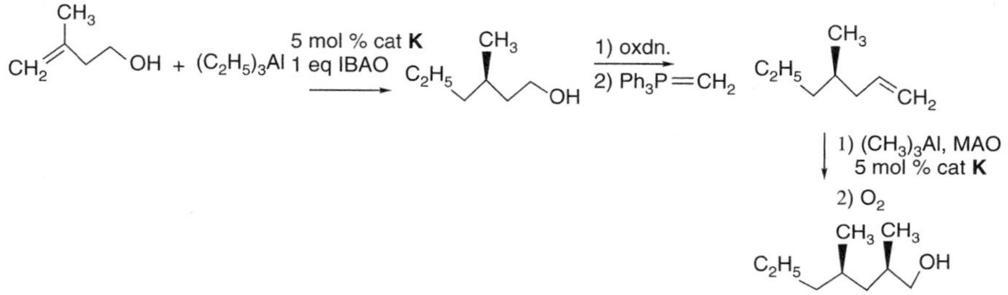

At this point in time carboalumination of alkynes has been more widely applied in synthesis. The most frequently used catalyst is $(Cp)_2ZrCl_2$. It is believed that a bimetallic species is formed.[227]

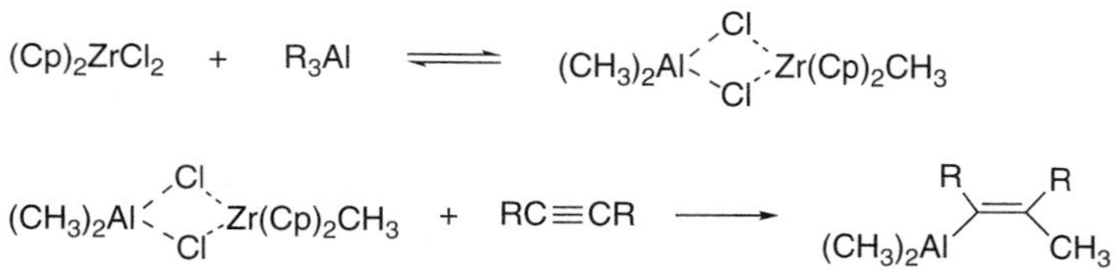

$$(Cp)_2ZrCl_2 + R_3Al \rightleftharpoons (CH_3)_2Al\langle^{Cl}_{Cl}\rangle Zr(Cp)_2CH_3$$

$$(CH_3)_2Al\langle^{Cl}_{Cl}\rangle Zr(Cp)_2CH_3 + RC\equiv CR \longrightarrow$$

Small amounts of water accelerate carboalumination of alkynes.[228] This acceleration may be the result of formation of aluminoxanes.

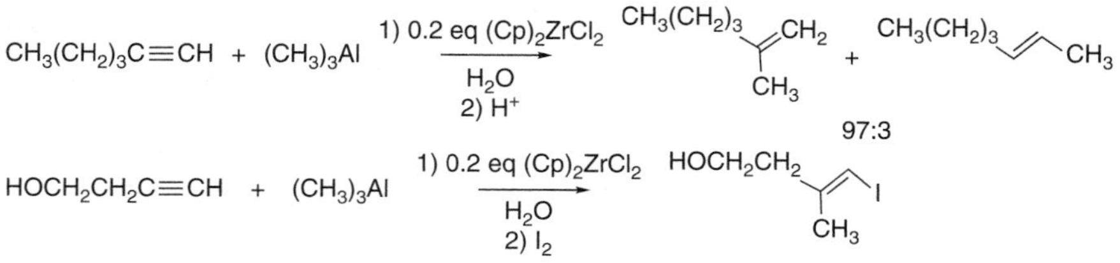

97:3

[225] S. Huo and E. Negishi, *Org. Lett.*, **3**, 3253 (2001).
[226] E. Negishi, Z. Tan, B. Liang, and T. Novak, *Proc. Natl. Acad. Sci. USA*, **101**, 5782 (2004); M. Magnin-Lachaux, Z. Tan, B. Liang, and E. Negishi, *Org. Lett.*, **6**, 1425 (2004).
[227] E. Negishi and D. Y. Kondakov, *Chem. Soc. Rev.*, **25**, 417 (1996).
[228] P. Wipf and S. Lim, *Angew. Chem. Int. Ed. Engl.*, **32**, 1068 (1993).

Scheme 4.10. Carbomethylations of Alkynes

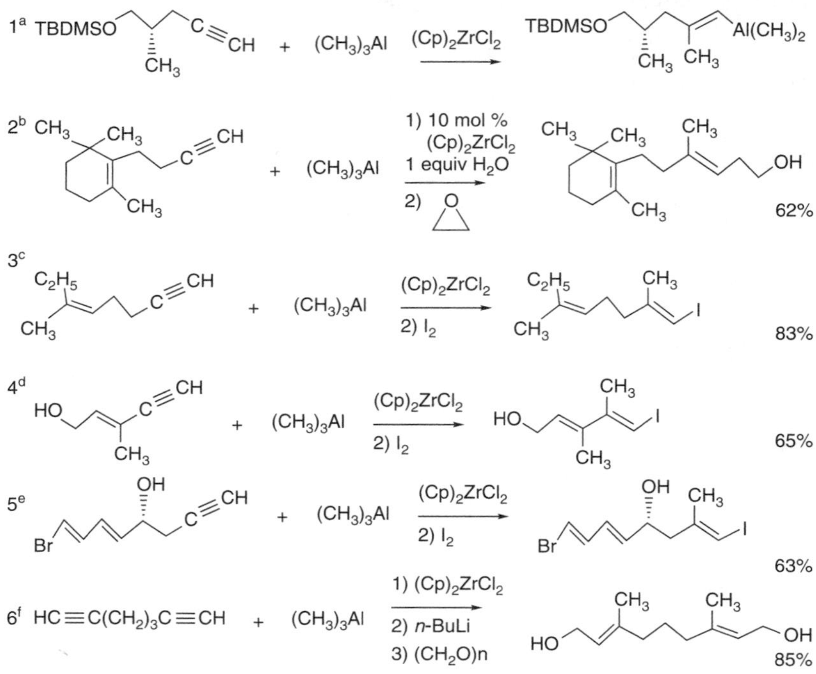

a. R. E. Ireland, L. Liu, and T. D. Roper, *Tetrahedron*, **53**, 13221 (1997).
b. A. Pommier, V. Stephanenko, K. Jarowicki, and P. J. Kocienski, *J. Org. Chem.*, **68**, 4008 (2003).
c. K. Mori and N. Murata, *Liebigs Ann. Chem.*, 2089 (1995).
d. T. K. Chakraborty and D. Thippeswamy, *Synlett*, 150 (1999).
e. M. Romero-Ortega, D. A. Colby, and H. F. Olivo, *Tetrahedron Lett.*, **47**, 6439 (2002).
f. G. Hidalgo-Del Vecchio and A. C. Oehlschlager, *J. Org. Chem.*, **59**, 4853 (1994).

As indicated by the mechanism, carboalumination is a *syn* addition. The resulting vinylalanes react with electrophiles with net retention of configuration. The electrophiles that have been used successfully include iodine, epoxides, formaldehyde, and ethyl chloroformate.[229] We will also see in Chapter 8 that the vinylalanes can undergo exchange reactions with transition metals, opening routes for formation of carbon-carbon bonds.

Scheme 4.10 gives some examples of application of alkyne carboalumination in synthesis. The reaction in Entry 1 was carried out as part of a synthesis of the immunosuppressant drug FK-506. The vinyl alane was subsequently transmetallated to a cuprate reagent (see Chapter 8). In Entry 2, the vinyl alane was used as a nucleophile for opening an epoxide ring and extending the carbon chain by two atoms. In Entries 3 to 5, the vinyl alane adducts were converted to vinyl iodides. In Entry 6, the vinyl alane was converted to an "ate" reagent prior to reaction with formaldehyde.

Derivatives of zirconium with a Zr−H bond also can add to alkenes and alkynes. This reaction is known as *hydrozirconation*.[230] The reagent that is used most frequently

229. N. Okukado and E. Negishi, *Tetrahedron Lett.*, 2357 (1978); M. Kobayashi, L. F. Valente, E. Negishi, W. Patterson, and A. Silveira, Jr., *Synthesis*, 1034 (1980); C. L. Rand, D. E. Van Horn, M. W. Moore, and E. Negishi, *J. Org. Chem.*, **46**, 4093 (1981).
230. P. Wipf and H. Jahn, *Tetrahedron*, **52**, 1283 (1996); P. Wipf and C. Kendall, *Topics Organmetallic Chem.*, **8**, 1 (2004).

in synthesis is *bis*-(cyclopentadienido)hydridozirconium(IV) chloride. Reduction of $(Cp)_2ZrCl_2$ generates a reactive species that can add to alkenes and alkynes.[231] Various reductants such as $LiAlH_4$[232] and $LiEt_3BH$[233] can be used. Alkynes readily undergo hydrozirconation. With internal alkynes, this reagent initially gives a regioisomeric mixture but isomerization occurs to give the less sterically hindered isomer.

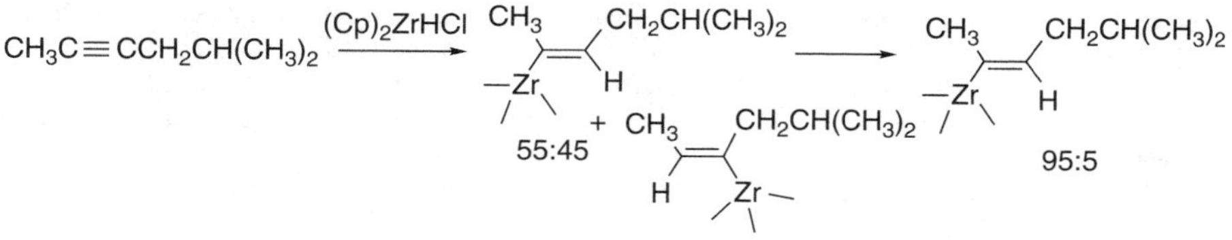

The adducts react with electrophiles such as NCS, NBS, and I_2 to give vinyl halides.

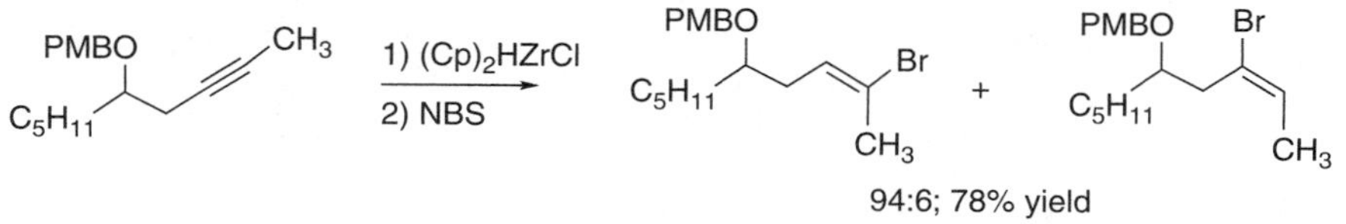

Ref. 234

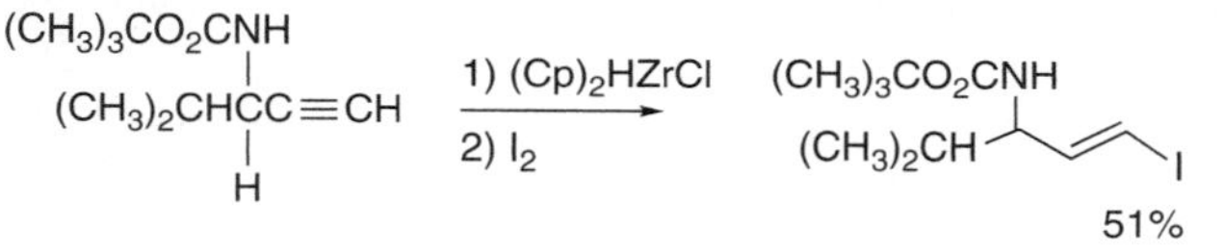

Ref. 235

Alkenes are less reactive and reactivity decreases with increasing substitution. The adducts from internal alkenes undergo isomerization to terminal derivatives.[236]

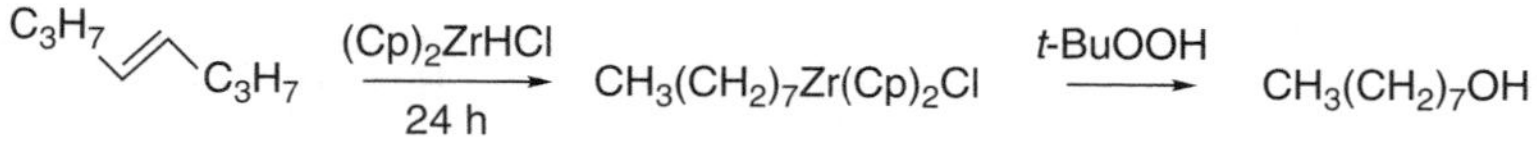

Carbon-carbon bond formation from alkyl and alkenyl zirconium reagents usually involves transmetallation reactions and are discussed in Chapter 8.

231. D. W. Hart, T. F. Blackburn, and J. Schwartz, *J. Am. Chem. Soc.*, **97**, 679 (1975); J. Schwartz and J. A. Labinger, *Angew. Chem. Int. Ed. Engl.*, **15**, 333 (1976).
232. S. L. Buchwald, S. J. La Maire, R. B. Nielsen, B. T. Watson, and S. M. King, *Tetrahedron Lett.*, **28**, 3895 (1987).
233. B. H. Lipshutz, R. Kell, and E. L. Ellsworth, *Tetrahedron Lett.*, **31**, 7257 (1990).
234. A. B. Smith, III, S. S.-Y. Chen, F. C. Nelson, J. M. Reichert, and B. A. Salvatore, *J. Am. Chem. Soc.*, **119**, 10935 (1997).
235. J. R. Hauske, P. Dorff, S. Julin, J. Di Brino, R. Spencer, and R. Williams, *J. Med. Chem.*, **35**, 4284 (1992).
236. D. W. Hart and J. Schwartz, *J. Am. Chem. Soc.*, **96**, 8115 (1974); T. Gibson, *Tetrahedron Lett.*, **23**, 157 (1982).

General References

P. B. de la Mare and R. Bolton, *Electrophilic Additions to Unsaturated Systems*, 2nd ed., Elsevier, New York, 1982.

N. Krause and A. S. K. Hashmi, eds., *Modern Allene Chemistry*, Wiley-VCH, Weinheim, 2004.

C. Paulmier, *Selenium Reagents and Intermediates in Organic Synthesis*, Pergamon, Oxford, 1986.

S. Patai, ed., *The Chemistry of Double-Bonded Functional Groups, Supplement A*, Vol 2, John Wiley & Sons, New York, 1988.

S. Patai, ed., *The Chemistry of Sulphenic Acids and Their Derivatives*, Wiley, Chichester, 1990.

S. Patai, editors, *The Chemistry of Trible-Bonded Functional Groups*, Supplement C2, John Wiley & Sons, New York, 1994.

S. Patai, and Z. Rappoport, eds., *The Chemistry of Organic Selenium and Tellurium Compounds*, John Wiley & Sons, New York, 1986.

A. Pelter, A. Smith, and H. C. Brown, *Borane Reagents*, Academic Press, 1988.

P. V. Ramachandran and H. C. Brown, *Organoboranes for Synthesis*, American Chemical Society, Washington, 2001.

H. F. Schuster and G. M. Coppola, *Allenes in Organic Synthesis*, Wiley, New York, 1984.

P. J. Stang and F. Diederich, eds., *Modern Acetylene Chemistry*, VCH Publishers, Weinheim, 1995.

Problems

(References for these problems will be found on page 1277.)

4.1. Predict the products, including regio- and stereochemistry, for the following reactions:

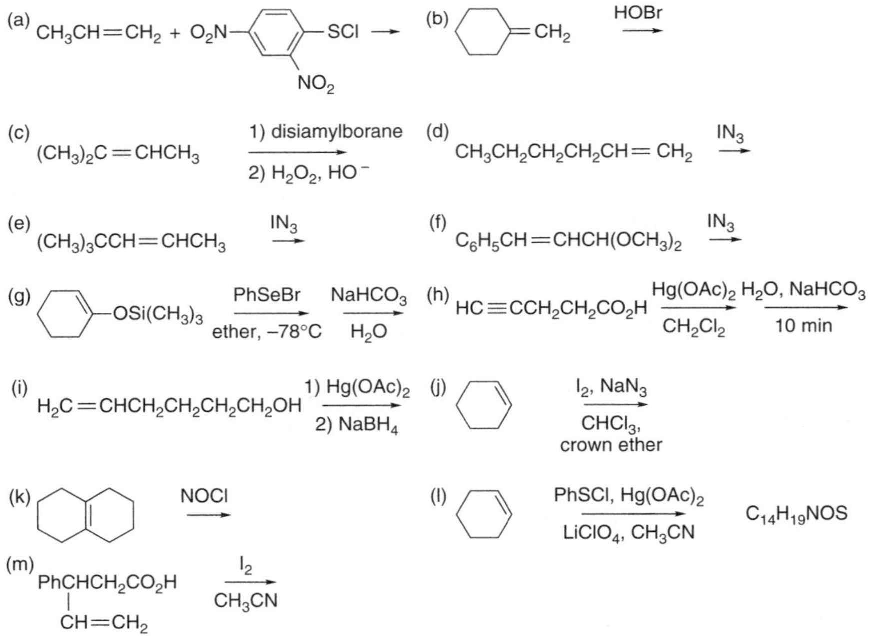

4.2. Bromination of 4-*t*-butylcyclohexene in methanol gives a 45:55 mixture of two compounds, each of composition $C_{11}H_{21}BrO$. Predict the structure and stereochemistry of these two products. How would you confirm your prediction?

4.3. Oxymercuration of 4-*t*-butylcyclohexene, followed by NaBH$_4$ reduction, gives *cis*-4-*t*-butylcyclohexanol and *trans*-3-*t*-butylcyclohexanol in approximately equal amounts. 1-Methyl-4-*t*-butylcyclohexanol under similar conditions gives only *cis*-4-*t*-butyl-1-methylcyclohexanol. Formulate an explanation for these observations.

4.4. Treatment of compound **C** with *N*-bromosuccinimide in acetic acid containing sodium acetate gives a product C$_{13}$H$_{19}$BrO$_3$. Propose a structure, including stereochemistry, and explain the basis for your proposal.

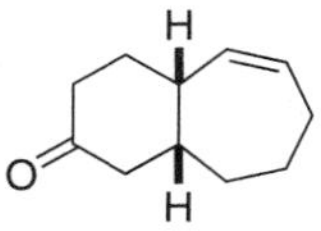

C

4.5. The hydration of 5-undecyn-2-one with HgSO$_4$ and H$_2$SO$_4$ in methanol is regioselective, giving 2,5-undecadione in 85% yield. Suggest an explanation for the high regioselectivity of this internal alkyne.

4.6. A procedure for the preparation of allylic alcohols uses the equivalent of phenylselenenic acid and an alkene. The reaction product is then treated with *t*-butylhydroperoxide. Suggest a mechanistic rationale for this process.

$$CH_3CH_2CH_2CH=CHCH_2CH_2CH_3 \xrightarrow[\text{2) }t\text{-BuOOH}]{\text{1) ``C}_6\text{H}_5\text{SeOH''}} CH_3CH_2CH_2CHCH=CHCH_2CH_3$$

with OH substituent, 88%

4.7. Suggest reaction conditions or short synthetic sequences that could provide the desired compound from the suggested starting material.

(a)

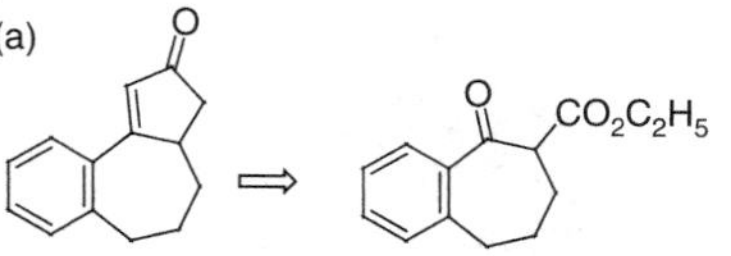

(b)

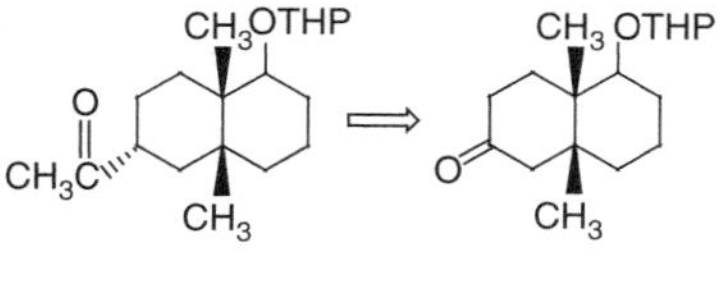

(c)

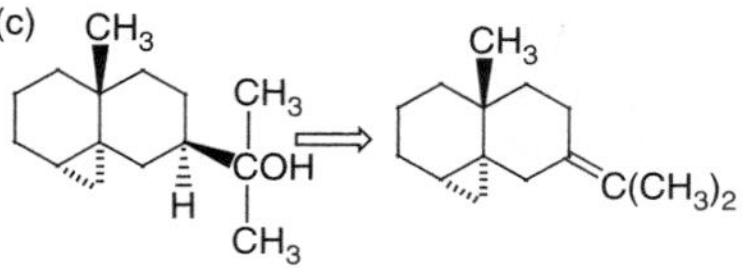

(d)

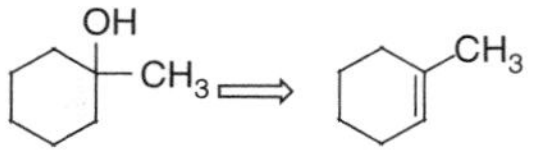

(e)

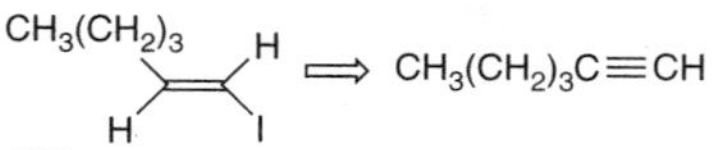

(f)

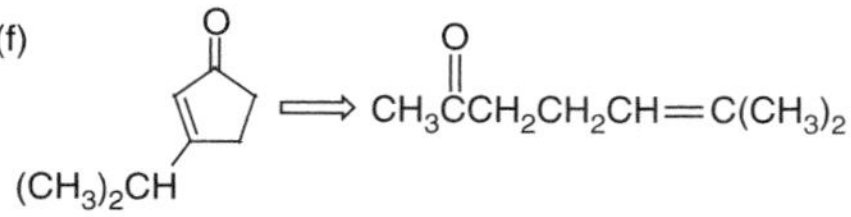

(g)

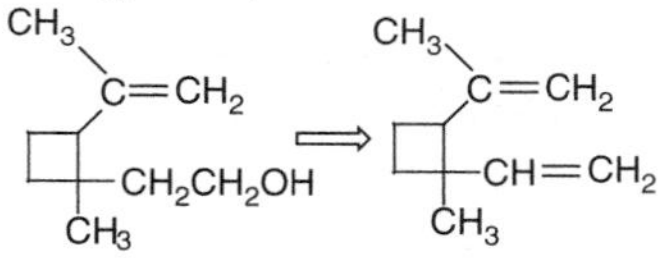

(h)

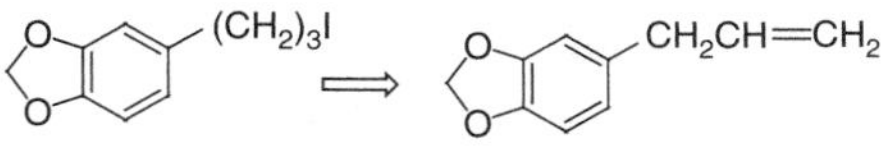

(i)

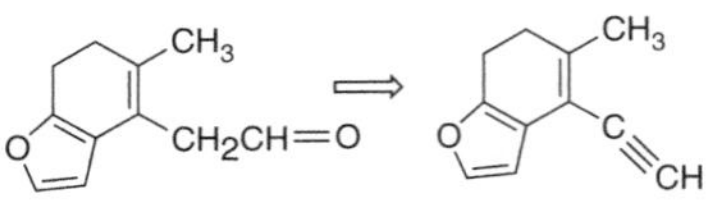

(j)

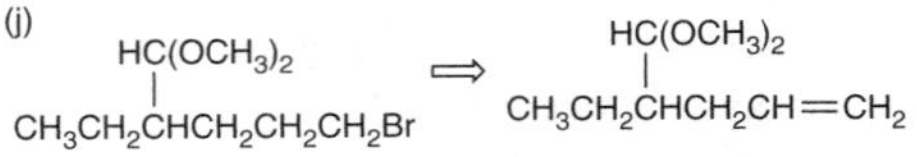

(k)

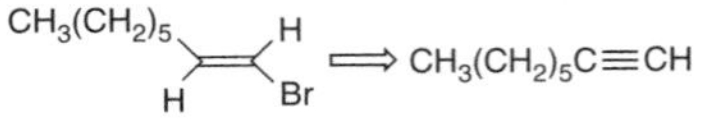

(l)

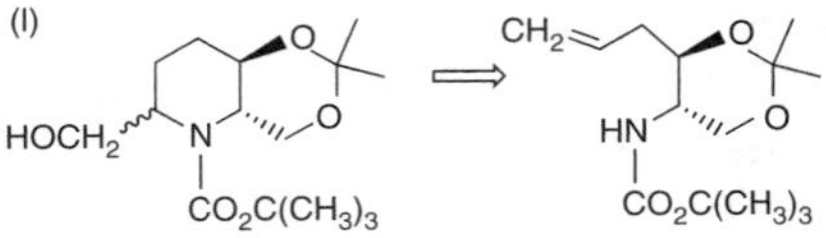

(m)

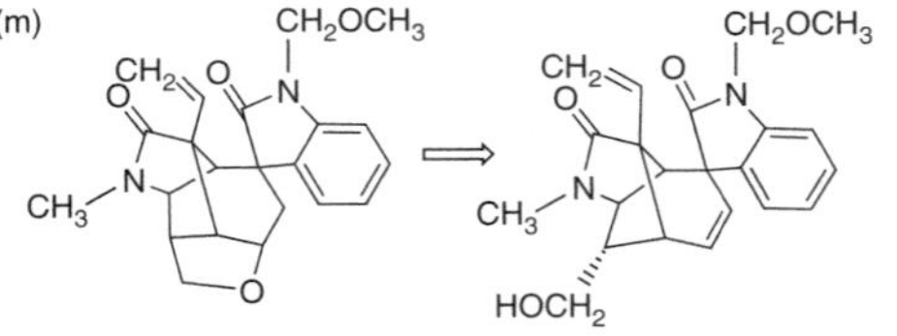

4.8. Three methods for the preparation of nitroalkenes are outlined below. Describe the mechanism by which each of these transformations occurs.

(a)

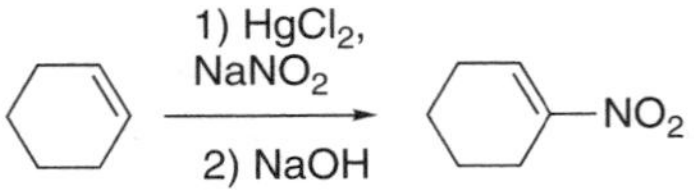

(b)

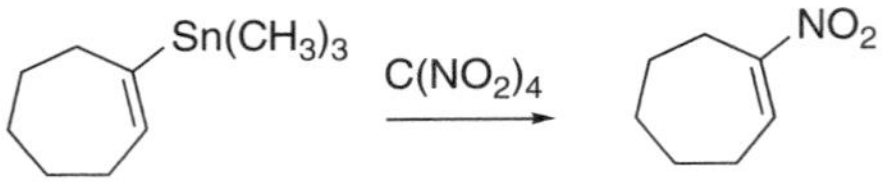

(c)

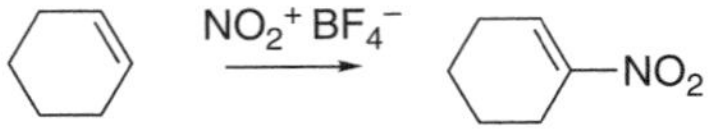

4.9. Hydroboration-oxidation of 1,4-di-*t*-butylcyclohexene gave three alcohols: **9-A** (77%), **9-B** (20%), and **9-C** (3%). Oxidation of **9-A** gave a ketone **9-D** that was readily converted by either acid or base to an isomeric ketone **9-E**. Ketone **9-E** was the only oxidation product of alcohols **9-B** and **9-C**. What are the structures of compounds **9A–9E**?

4.10. Show how by using regioselective enolate chemistry and organoselenium reagents, you could convert 2-phenylcyclohexanone to either 2-phenyl-2-cyclohexen-1-one or 6-phenyl-2-cyclohexen-1-one.

4.11. On the basis of the mechanistic pattern for oxymercuration-demercuration, predict the structure and stereochemistry of the alcohol(s) to be expected by application of the reaction to each of the following substituted cyclohexenes.

(a)

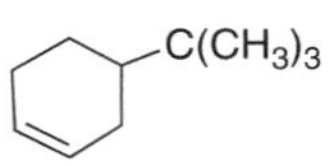

(b)

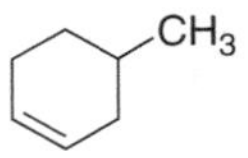

(c)

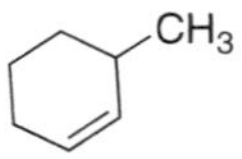

4.12. Give the structure, including stereochemistry, of the expected products of the following reactions. Identify the critical factors that determine the regio- and stereochemistry of the reaction.

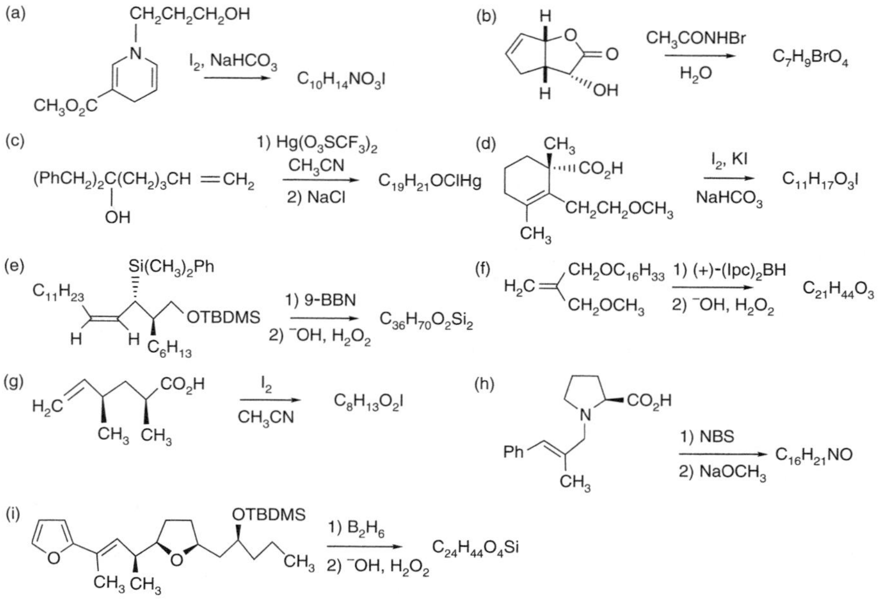

(a)

CH$_2$CH$_2$CH$_2$OH

$\xrightarrow{I_2, \text{NaHCO}_3}$ C$_{10}$H$_{14}$NO$_3$I

CH$_3$O$_2$C

(b)

$\xrightarrow[\text{H}_2\text{O}]{\text{CH}_3\text{CONHBr}}$ C$_7$H$_9$BrO$_4$

(c)

(PhCH$_2$)$_2$C(CH$_2$)$_3$CH =CH$_2$

OH

$\xrightarrow[\text{2) NaCl}]{\substack{\text{1) Hg(O}_3\text{SCF}_3)_2 \\ \text{CH}_3\text{CN}}}$ C$_{19}$H$_{21}$OClHg

(d)

CH$_3$

CO$_2$H

CH$_2$CH$_2$OCH$_3$

CH$_3$

$\xrightarrow[\text{NaHCO}_3]{I_2, \text{KI}}$ C$_{11}$H$_{17}$O$_3$I

(e)

C$_{11}$H$_{23}$

Si(CH$_3$)$_2$Ph

OTBDMS

H H C$_6$H$_{13}$

$\xrightarrow[\text{2) $^-$OH, H}_2\text{O}_2]{\text{1) 9-BBN}}$ C$_{36}$H$_{70}$O$_2$Si$_2$

(f)

H$_2$C

CH$_2$OC$_{16}$H$_{33}$

CH$_2$OCH$_3$

$\xrightarrow[\text{2) $^-$OH, H}_2\text{O}_2]{\text{1) (+)-(Ipc)}_2\text{BH}}$ C$_{21}$H$_{44}$O$_3$

(g)

H$_2$C

CO$_2$H

CH$_3$ CH$_3$

$\xrightarrow{\substack{I_2 \\ \text{CH}_3\text{CN}}}$ C$_8$H$_{13}$O$_2$I

(h)

Ph

N

CO$_2$H

CH$_3$

$\xrightarrow[\text{2) NaOCH}_3]{\text{1) NBS}}$ C$_{16}$H$_{21}$NO

(i)

O

CH$_3$ CH$_3$

OTBDMS

O

CH$_3$

$\xrightarrow[\text{2) $^-$OH, H}_2\text{O}_2]{\text{1) B}_2\text{H}_6}$ C$_{24}$H$_{44}$O$_4$Si

4.13. Some synthetic transformations are shown in the retrosynthetic format. Propose a short series of reactions (no more than three steps should be necessary) that could effect each conversion.

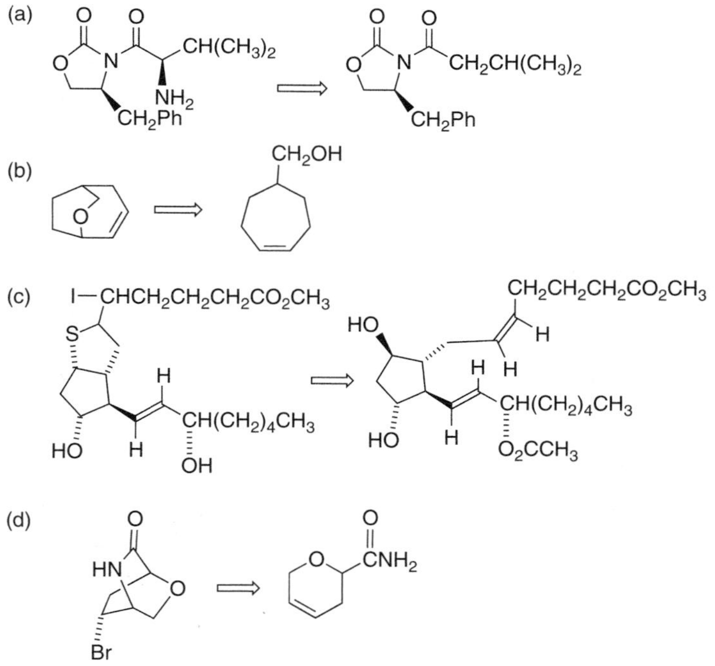

(a)

O O

O N CH(CH$_3$)$_2$

NH$_2$
CH$_2$Ph

$\Longrightarrow$

O O

O N CH$_2$CH(CH$_3$)$_2$

CH$_2$Ph

(b)

$\Longrightarrow$

CH$_2$OH

(c)

I—CHCH$_2$CH$_2$CH$_2$CO$_2$CH$_3$

S

H

HO H

CH(CH$_2$)$_4$CH$_3$

OH

$\Longrightarrow$

HO

H H

CH$_2$CH$_2$CH$_2$CO$_2$CH$_3$

H

HO H

CH(CH$_2$)$_4$CH$_3$

O$_2$CCH$_3$

(d)

O

HN

O

Br

$\Longrightarrow$

O

O CNH$_2$

(e)

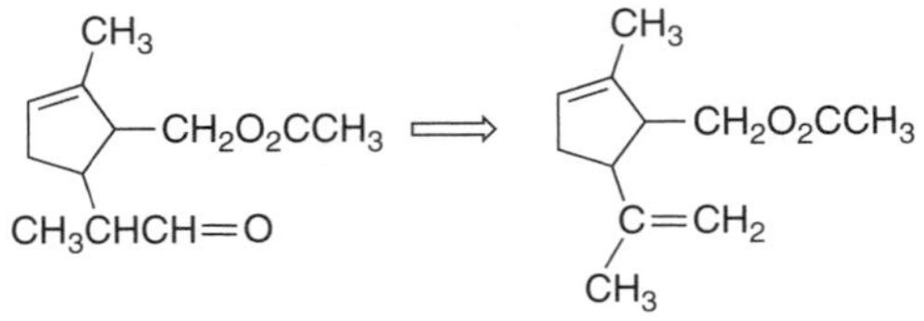

4.14. Write mechanisms for the following reactions:

(a)

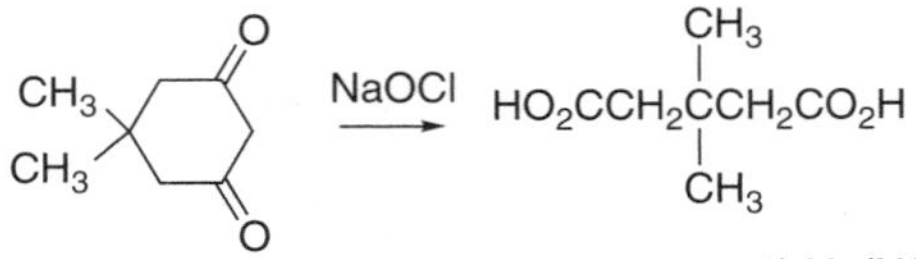

(b) CH₂=CHCH₂CHOCH₂NHCO₂CH₂Ph $\xrightarrow[\text{2) KBr}]{\text{1) Hg(NO}_3)_2}$ $\xrightarrow[\text{O}_2]{\text{NaBH}_4}$

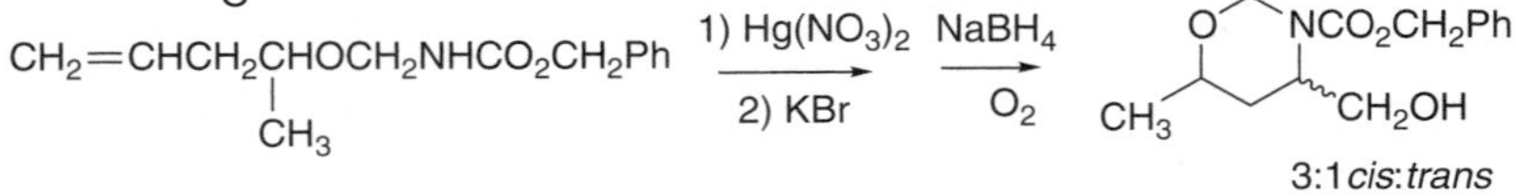

4.15. 4-Pentenyl amides such as **15A** cyclize to lactams **15B** on reaction with phenyl selenenyl bromide. The 3-butenyl compound **15C**, on the other hand, cyclizes to an imino ether **15D**. What is the basis for the differing reactions?

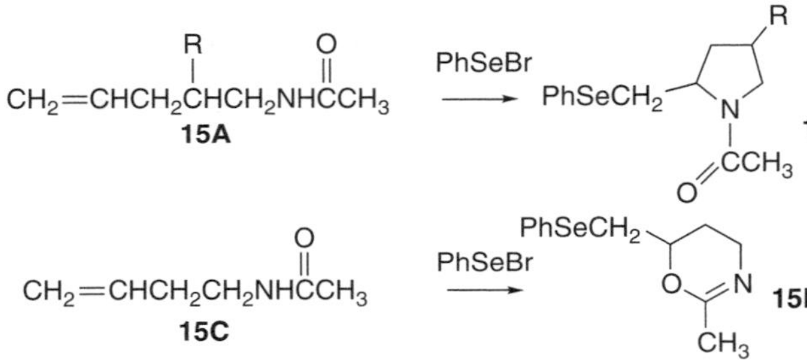

4.16. Procedures for enantioselective preparation of α-bromo acids based on reaction of NBS with enol derivatives **16A** and **16B** have been developed. Predict the absolute configuration of the halogenated compounds produced from both **16A** and **16B**. Explain the basis of your prediction.

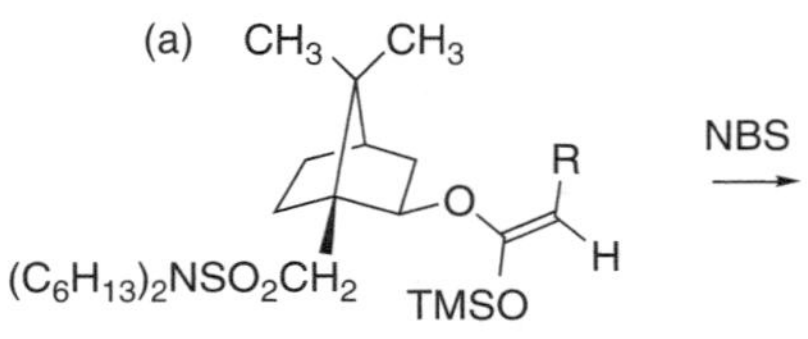

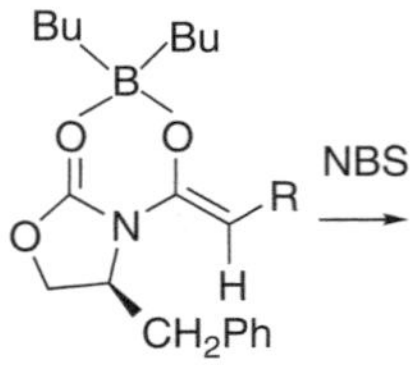

4.17. The stereochemical outcome of the hydroboration-oxidation of 1, 1′-bicyclohexenyl depends on the amount of diborane used. When 1.1 equivalent is used, the product is a 3:1 mixture of **17A** and **17B**. When 2.1 equivalent is used, **17A** is formed nearly exclusively. Offer an explanation of these results.

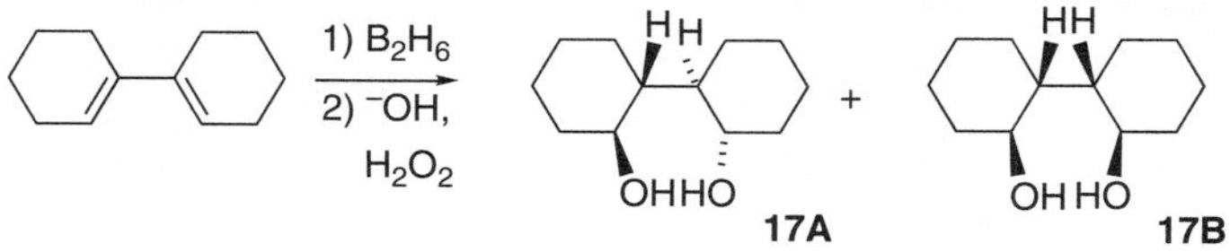

4.18. Predict the absolute configuration of the products obtained from the following enantioselective hydroborations.

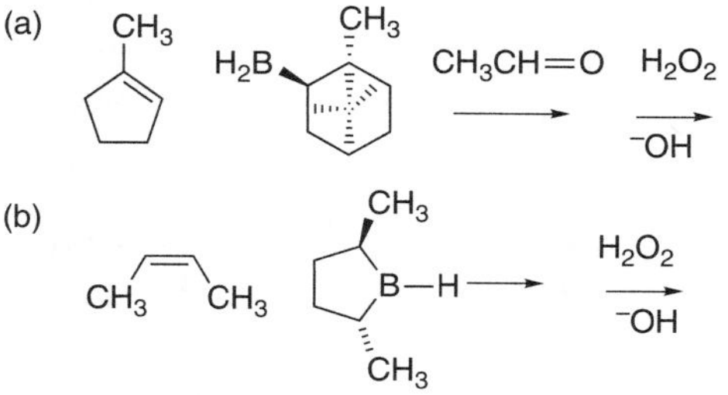

(a)

(b)

4.19. The regioselectivity and stereoselectivity of electrophilic additions to 2-benzyl-3-azabicyclo[2.2.1]hept-5-en-3-one are quite dependent on the specific electrophile. Discuss the factors that could influence the differing selectivity patterns that are observed.

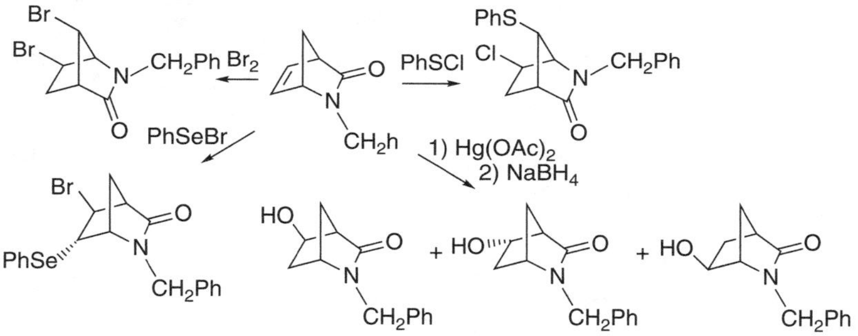

4.20. Offer mechanistic explanations of the following observations:

a. In the cyclization reactions shown below, **20A** is the preferred product for R = H, but **20B** is the preferred product for R = methyl or phenyl.

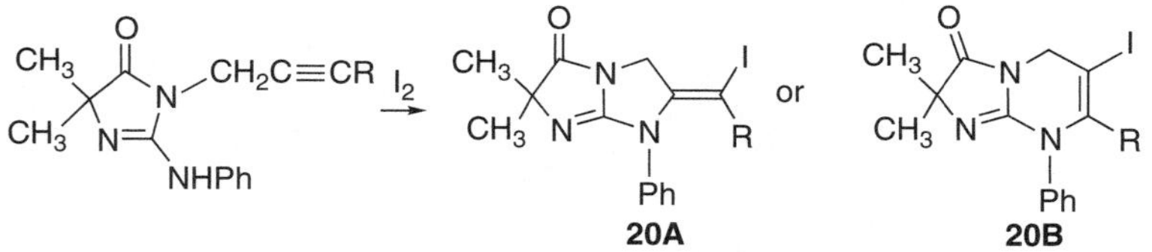

b. The pent-4-enoyl group has been developed as a protecting group for primary and secondary amines. The conditions for cleavage involve treatment with iodine and a aqueous solution with either THF or acetonitrile as the cosolvent. Account for the mild deprotection under these conditions.

4.21. Analyze the data below concerning the effect of allylic and homoallylic benzyloxy substituents on the regio- and stereoselectivity of hydroboration-oxidation. Propose a TS that is consistent with the results.

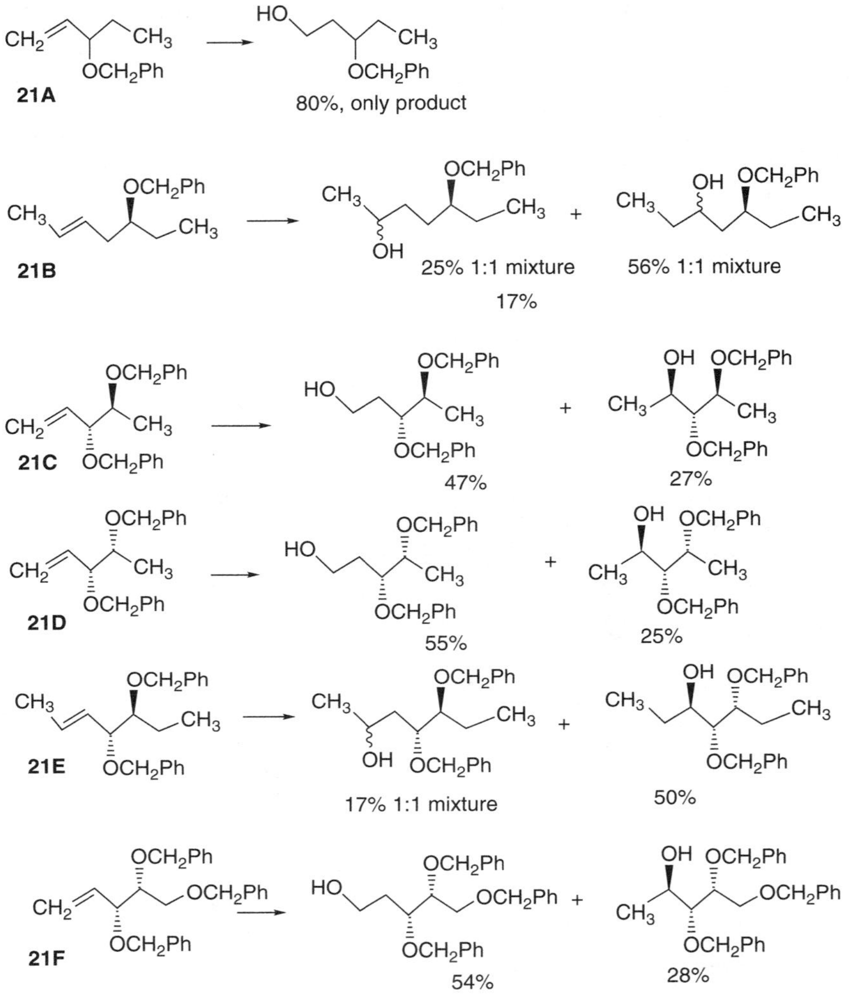

21A

80%, only product

21B

25% 1:1 mixture   56% 1:1 mixture

17%

21C

47%   27%

21D

55%   25%

21E

17% 1:1 mixture   50%

21F

54%   28%

4.22. Propose an enantioselective synthesis of (+) methyl nonactate from the aldehyde shown.

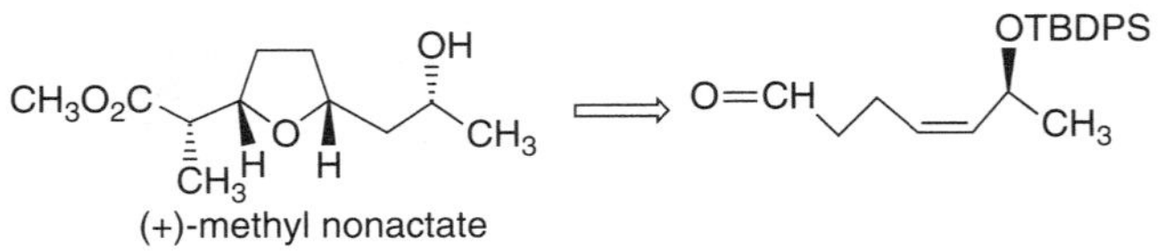

(+)-methyl nonactate

4.23. On page 313, the effect of methyl substitution on the stereoselectivity of α,α-diallylcarboxylic acids under iodolactonization conditions was discussed. Consider the two compounds shown and construct a reaction energy profile for each compound that illustrates the role of conformational equilibrium, facial selectivity, and substituent effects on $\Delta G^{\ddagger}$ on the stereochemical outcome.

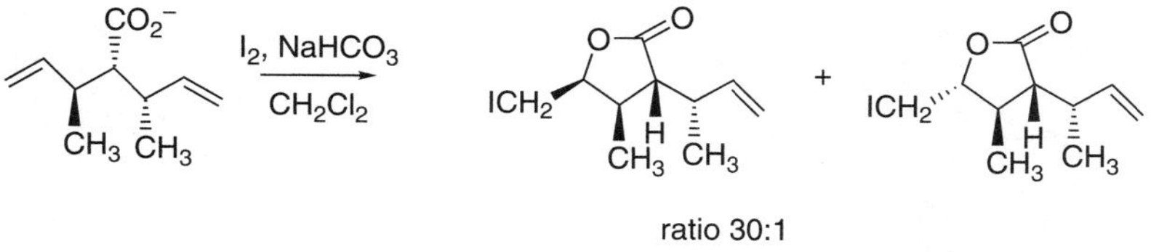

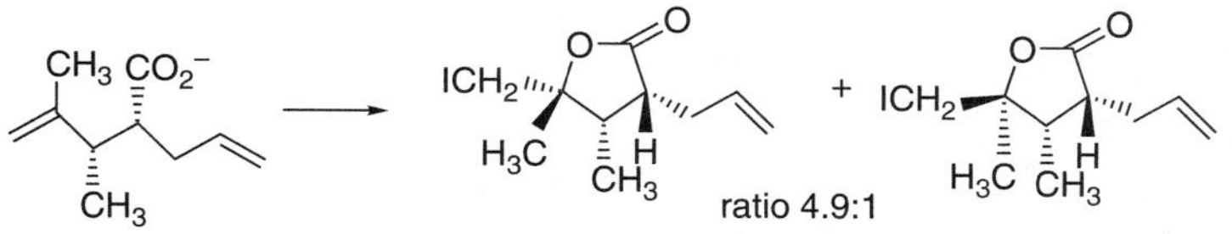

ratio 30:1

ratio 4.9:1

4.24. It has been found that when δ,ε-enolates bearing β-siloxy substituents are subject to iodolactonization, the substituent directs the stereochemistry of cyclization in a manner opposite to an alkyl substituent. Suggest a TS structure that would account for this difference.

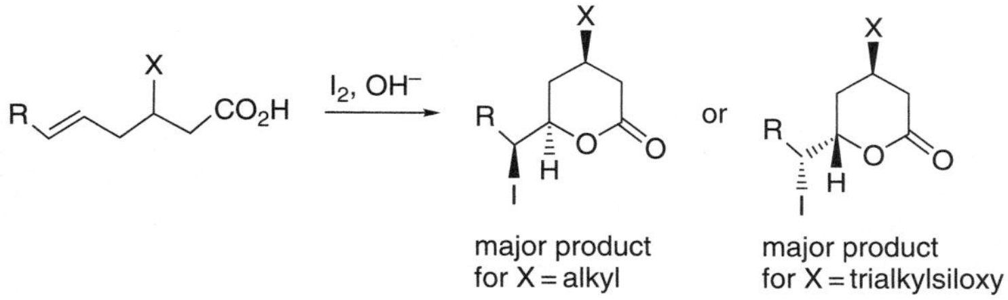

major product
for X = alkyl

or

major product
for X = trialkylsiloxy

5

Reduction of Carbon-Carbon Multiple Bonds, Carbonyl Groups, and Other Functional Groups

Introduction

The subject of this chapter is reduction reactions that are especially important in synthesis. Reduction can be accomplished by several broad methods including addition of hydrogen and/or electrons to a molecule or by removal of oxygen or other electronegative substituents. The most widely used reducing agents from a synthetic point of view are molecular hydrogen and hydride derivatives of boron and aluminum, and these reactions are discussed in Sections 5.1 through 5.3. A smaller group of reactions transfers hydride from silicon or carbon, and these are the topic of Section 5.4. Certain reductions involving a free radical mechanism use silanes or stannanes as hydrogen atom donors, and these reactions are considered in Section 5.5. Other important procedures use metals such as lithium, sodium, or zinc as electron donors. Reduction by metals can be applied to carbonyl compounds and aromatic rings and can also remove certain functional groups.

Addition of Hydrogen
___

$$R_2C{=}X \xrightarrow{\text{H}_2} R_2CH{-}XH$$

catalytic hydrogenation

$X = CR'_2,\ O,\ NR'$

$$R_2C{=}X \xrightarrow{[\text{MH}_4{}^-]} R_2CH{-}XH$$

hydride reduction

$X = O,\ NR'$

$$R_2C{=}X \xrightarrow[\text{2 H}^+]{\text{2 M}\cdot} R_2CH{-}XH$$

reduction by metals

$X = O,\ NR'$

CHAPTER 5

*Reduction of
Carbon-Carbon Multiple
Bonds, Carbonyl
Groups, and Other
Functional Groups*

$$R_3C-Y \xrightarrow[2\,H^+]{2\,M\cdot} R_3C-H + H-Y$$

dissolving metals

Y = halogen, oxygen substituents,
α–to carbonyl groups

$$R_3C-Y \xrightarrow{R'_3ZH} R_3C-H + R'_3Z-Y$$

hydrogen atom donors

Y = halogen, thio ester

Z = Sn, Si

There are also procedures that form carbon-carbon bonds. Most of these reactions begin with an electron transfer that generates a radical intermediate, which then undergoes a coupling or addition reaction. These reactions are discussed in Section 5.6.

$$R_2C=X + M\cdot \longrightarrow R_2\overset{X}{\overset{|}{C}}{}^{\underline{\cdot}} \longrightarrow R_2\overset{^-X}{\overset{|}{C}}-\overset{X^-}{\overset{|}{C}}R_2 \quad \text{or} \quad R_2C=CR_2$$

X=O   M· = Na⁰, Ti^{II}, Sm^{II}

reductive coupling

Reductive removal of oxygen from functional groups such as ketones and aldehydes, alcohols, α-oxy ketones, and diols are also important in synthesis. These reactions, which provide important methods for interconversion of functional groups, are considered in Section 5.7

$$\overset{O}{\underset{R}{\overset{||}{C}}}{}_R \longrightarrow R-CH_2-R \qquad \overset{O}{\underset{R}{\overset{||}{C}}}{}_R \longrightarrow RCH=CHR \qquad \overset{HO\ \ OH}{R_2\overset{|}{C}-\overset{|}{C}R_2} \longrightarrow R_2C=CR_2$$

carbonyl ⟶ methylene   carbonyl ⟶ alkene   diol ⟶ alkene

reductive deoxygenation

5.1. Addition of Hydrogen at Carbon-Carbon Multiple Bonds

The most widely used method for adding the elements of hydrogen to carbon-carbon double bonds is catalytic hydrogenation. Except for very sterically hindered alkenes, this reaction usually proceeds rapidly and cleanly. The most common catalysts are various forms of transition metals, particularly platinum, palladium, rhodium, ruthenium, and nickel. Both the metals as finely dispersed solids or adsorbed on inert supports such as carbon or alumina (*heterogeneous catalysts*) and certain soluble complexes of these metals (*homogeneous catalysts*) exhibit catalytic activity. Depending upon conditions and catalyst, other functional groups are also subject to reduction under these conditions.

$$RCH=CHR + H_2 \xrightarrow{catalyst} RCH_2CH_2R$$

5.1.1. Hydrogenation Using Heterogeneous Catalysts

The mechanistic description of catalytic hydrogenation of alkene is somewhat imprecise, partly because the reactive sites on the metal surface are not as well described as small-molecule reagents in solution. As understanding of the chemistry of soluble hydrogenation catalysts developed, it became possible to extrapolate the mechanistic concepts to heterogeneous catalysts. It is known that hydrogen is adsorbed onto the metal surface, forming metal hydrogen bonds similar to those in transition metal hydrides. Alkenes are also adsorbed on the catalyst surface and at least three types of intermediates have been implicated in hydrogenation. The initially formed intermediate is pictured as attached at both carbon atoms of the double bond by π-type bonding, as shown in **A**. The bonding involves an interaction between the alkene π and π^* orbitals with corresponding acceptor and donor orbitals of the metal. A hydride can be added to the adsorbed group, leading to **B**, which involves a σ-type carbon-metal bond. This species can react with another hydrogen to give the alkane, which is desorbed from the surface. A third intermediate species, shown as **C**, accounts for double-bond isomerization and the exchange of hydrogen that sometimes accompanies hydrogenation. This intermediate is equivalent to an allyl group bound to the metal surface by π bonds. It can be formed from absorbed alkene by abstraction of an allylic hydrogen atom by the metal. The reactions of transition metals with organic compounds are discussed in Chapter 8. There are well-characterized examples of structures corresponding to each of the intermediates **A**, **B**, and **C** that are involved in hydrogenation. However, one issue that is left unresolved by this mechanism is whether there is cooperation between adjacent metal atoms, or if the reactions occur at a single metal center, which is usually the case with soluble catalysts.

A π-complex   **B** σ-bond   **C** π-allyl complex

Catalytic hydrogenations are usually very clean reactions with little by-product formation, unless reduction of other groups is competitive, but careful study reveals that sometimes double-bond migration takes place in competition with reduction. For example, hydrogenation of 1-pentene over Raney nickel is accompanied by some isomerization to both *E*- and *Z*-2-pentene.[1] The isomerized products are converted to pentane, but at a slower rate than 1-pentene. Exchange of hydrogen atoms between the reactant and adsorbed hydrogen can be detected by isotopic exchange. Allylic positions undergo such exchange particularly rapidly.[2] Both the isomerization and allylic hydrogen exchange can be explained by the intervention of the π-allyl intermediate **C** in the general mechanism for hydrogenation. If hydrogen is added at the alternative end of the allyl system, an isomeric alkene is formed. Hydrogen exchange occurs if a hydrogen from the metal surface, rather than the original hydrogen, is transferred prior to desorption.

In most cases, both hydrogen atoms are added to the same face of the double bond (*syn* addition). If hydrogenation occurs by addition of hydrogen in two steps, as

1. H. C. Brown and C. A. Brown, *J. Am. Chem. Soc.*, **85**, 1005 (1963).
2. G. V. Smith and J. R. Swoap, *J. Org. Chem.*, **31**, 3904 (1966).

370

CHAPTER 5

*Reduction of
Carbon-Carbon Multiple
Bonds, Carbonyl
Groups, and Other
Functional Groups*

implied by the above mechanism, the intermediate must remain bonded to the metal surface in such a way that the stereochemical relationship is maintained. Adsorption to the catalyst surface normally involves the less sterically congested side of the double bond, and as a result hydrogen is added from the less hindered face of the double bond. There are many hydrogenations in which hydrogen addition is not entirely *syn*, and independent corroboration of the stereochemistry is normally necessary.

Scheme 5.1 illustrates some hydrogenations in which the *syn* addition from the less hindered side is observed. Some exceptions are also included. Entry 1 shows the hydrogenation of an exocyclic methylene group. This reaction was studied at various H_2 pressures and over both Pt and Pd catalysts. 4-Methyl- and 4-*t*-butylmethylene cyclohexane also give mainly the *cis* product.[3] These results are consistent with a favored (2.3:1) equatorial delivery of hydrogen.

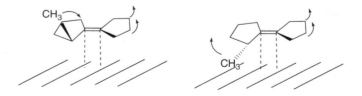

The Entry 2 reactant, 1,2-dimethylcyclohexene, was also studied by several groups and a 2:1–4:1 preference for *syn* addition was noted, depending on the catalyst and conditions. In the reference cited, the catalyst was prepared by reduction of a Pt salt with $NaBH_4$. A higher ratio of the *cis* product was noted at $0°C$ (5.2:1) than at $25°C$ (2.5:1). In Entry 3, the 2,6-dimethycyclohexene gives mainly *cis* product with a Pt catalyst but *trans* product dominates with a Pd catalyst. These three cases indicate that stereoselectivity for unhindered alkenes is modest and dependent on reaction conditions. Entries 4 and 5 involve more rigid and sterically demanding alkenes. In both cases, *syn* addition of hydrogen occurs from the less hindered face of the molecule. Entries 6 to 8 are cases in which hydrogen is added from the more-substituted face of the double bond. The compound in Entry 6 gives mainly *trans* product at high H_2 pressure, where the effects of alkene isomerization are minimized. This result indicates that the primary adsorption must be from the methyl-substituted face of the molecule. This may result from structural changes that occur on bonding to the catalyst surface. In the *cis* approach, the methyl substituent moves away from the cyclopentane ring as rehybridization of the double bond occurs. In the *trans* approach, the methyl group must move closer to the adjacent cyclopentane ring.

The preference for addition from the more hindered of the substituents in Entries 7 and 8 can be attributed to functional group interactions with the catalyst. Polar

[3.] J.-F. Sauvage, R. H. Baker, and A. S. Hussey, *J. Am. Chem. Soc.*, **82**, 6090 (1960).

Scheme 5.1. Stereochemistry of Hydrogenation of Some Alkenes

A. Examples of preferential *syn* addition from less hindered side
1[a]
Pt
H2
70%
30%
2[b]
70–85%
15–30%
3[a]
70%
30%
4[c]
5[b]
Pt(BH4−)—C
95%
B. Exceptions
6[d]
Pt, H2
acetic acid
80%
20%
7[e]
8[f]
Ni, H2
95%
5%

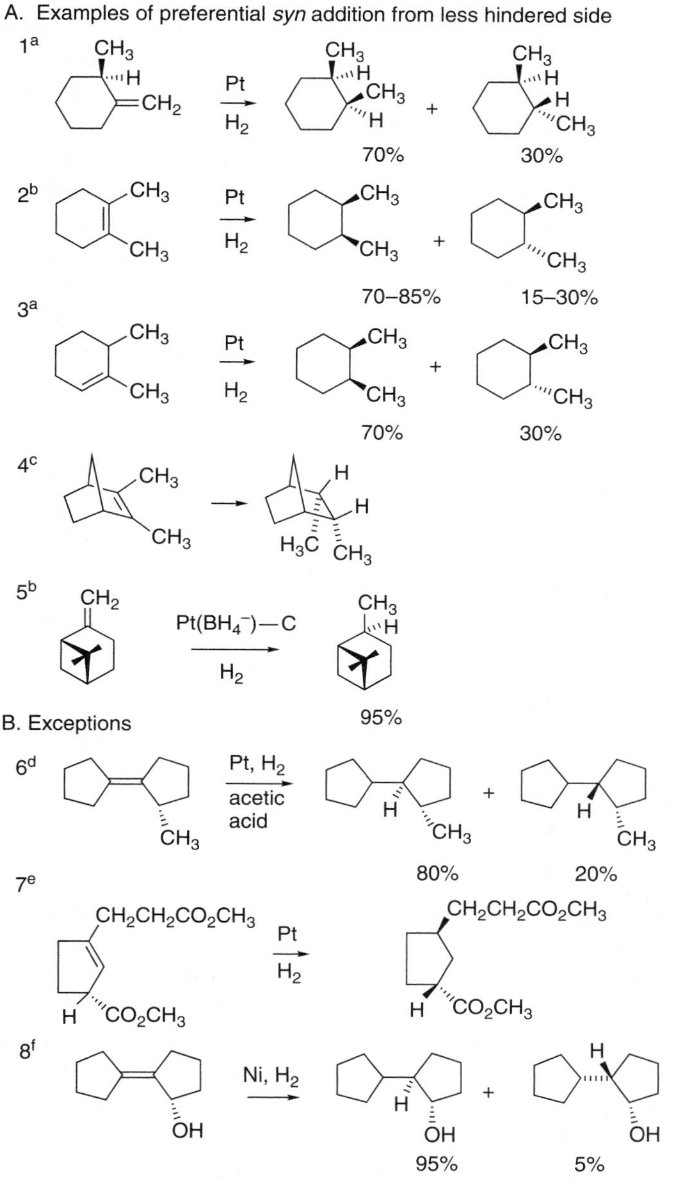

a. S. Siegel and G. V. Smith, *J. Am. Chem. Soc.*, **82**, 6082, 6087 (1960).
b. C. A. Brown, *J. Am. Chem. Soc.*, **91**, 5901 (1969).
c. K. Alder and W. Roth, *Chem. Ber.*, **87**, 161 (1954).
d. S. Siegel and J. R. Cozort, *J. Org. Chem.*, **40**, 3594 (1975).
e. J. P. Ferris and N. C. Miller, *J. Am. Chem. Soc.*, **88**, 3522 (1966).
f. S. Mitsui, Y. Senda, and H. Saito, *Bull. Chem. Soc. Jpn.*, **39**, 694 (1966).

372

CHAPTER 5

Reduction of
Carbon-Carbon Multiple
Bonds, Carbonyl
Groups, and Other
Functional Groups

groups sometimes favor *cis* addition of hydrogen, relative to the substituent. This is a very common observation for hydroxy groups, but less so for esters (*vide infra*).

The facial stereoselectivity of hydrogenation is affected by the presence of polar functional groups that can govern the mode of adsorption to the catalyst surface. For instance, there are many of examples of hydrogen being introduced from the face of the molecule occupied by the hydroxy group, which indicates that the hydroxy group interacts with the catalyst surface. This behavior can be illustrated with the alcohol **1a** and the ester **1b**.[4] Although the overall shapes of the two molecules are similar, the alcohol gives mainly the product with a *cis* ring juncture (**2a**), whereas the ester gives a product with *trans* stereochemistry (**3b**). The stereoselectivity of hydroxy-directed hydrogenation is a function of solvent and catalyst. The *cis*-directing effect is strongest in nonpolar solvents such as hexane. This is illustrated by the results from compound **4**. In ethanol, the competing interaction of the solvent molecules evidently swamps out the effect of the hydroxymethyl group.

| 1a X = CH₂OH | 2a 94% | 3a 6% |
| 1b X = CO₂CH₃ | 2b 15% | 3b 85% |

Solvent	% *cis*	% *trans*
Hexane	61	39
DME	20	80
EtOH	6	94

Thompson and co-workers have explored the range of substituents that can exert directive effects using polycyclic systems. For ring system **1**, hydroxymethyl and formyl showed strong directive effects; cyano, oximino, and carboxylate were moderate; and carboxy, ester, amide, and acetyl groups were not directive (see Table 5.1).[4,5] As with **4**, the directive effects were shown to be solvent dependent. Strong donor solvents, such as ethanol and DMF, minimized the substituent-directing effect. Similar studies were carried out with ring system **5**.[6] The results are given in Table 5.1. It would be expected that the overall shape of the reactant molecule would influence the effectiveness of the directive effect. The trends in ring systems **1** and **5** are similar, although ring system **5** appears to be somewhat less susceptible to directive effects. These hydrogenations were carried out in hydroxylic solvents and it would be expected that the directive effects would be enhanced in less polar solvents.

4. (a) H. W. Thompson, *J. Org. Chem.*, **36**, 2577 (1971); (b) H. W. Thompson, E. McPherson, and B. L. Lences, *J. Org. Chem.*, **41**, 2903 (1976).
5. H. W. Thompson and R. E. Naipawer, *J. Am. Chem. Soc.*, **95**, 6379 (1973).
6. H. W. Thompson and S. Y. Rashid, *J. Org. Chem.*, **67**, 2813 (2002).

Table 5.1. Substituent Directive Effects for Ring Systems 1 and 5

Substituent X	Ring system 1[a]		Ring system 5[b]	
	% cis (Directive)	% trans (Nondirective)	% cis (Directive)	% trans (Nondirective)
CH_2NH_2			87	13
$CH_2N(CH_3)_2$			62	38
CH_2OH	95	5	48	52
$CH{=}O$	93	7	42	58
CN	75	25	20	80
$CH{=}NOH$	65	35	45	55
CH_2OCH_3			44	56
$CH_2NHCOCH_3$			33	67
CO_2Na (or K)	55	45	30	70
CO_2H	18	82	17	83
CO_2CH_3	15	85	16	84
$CONH_2$	10	90	33	67
$COCH_3$	14	86	22	78

a. In methoxyethanol.
b. In ethanol.

The general ordering of aminomethyl > hydroxymethyl > CH=O > ester suggests that Lewis basicity is the dominant factor in the directive effect. Problem 5.2 involves considering the ordering of the various acyl substituents in more detail.

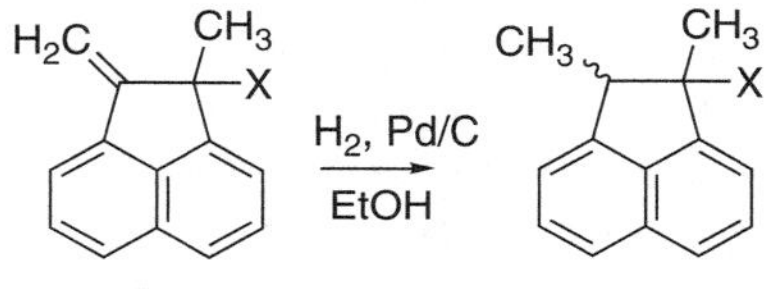

Substituted indenes provide other examples of substituent directive effects. Over Pd-alumina, the indenols **6a-c** show both *cis* stereoselectivity and a *syn* directive effect. The directive effect is reinforced by steric effects as the alkyl group becomes larger.[7]

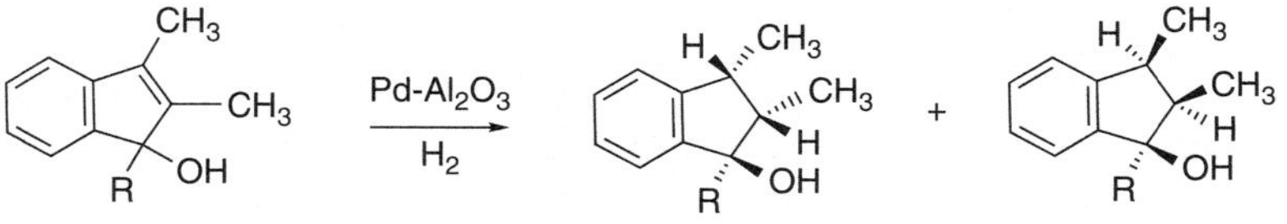

R	trans,trans-**7**	cis,cis-**7**
CH_3	88	12
C_2H_5	97	3
$(CH_3)_2CH$	100	0

Several indanes (**8**) were reduced to hexahydroindanes over Rh-Al_2O_3. The stereochemistry of the ring junction is established at the stage of the reduction of the tetrasubstituted double bonds. Only the amino group shows a strong directive effect.[8]

7. K. Borszeky, T. Mallat, and A. Baiker, *J. Catalysis*, **188**, 413 (1999).
8. V. S. Ranade, G. Consiglio, and R. Prins, *J. Org. Chem.*, **65**, 1132 (2000); V. S. Ranade, G. Consiglio, and R. Prins, *J. Org. Chem.*, **64**, 8862 (1999).

374

CHAPTER 5

*Reduction of
Carbon-Carbon Multiple
Bonds, Carbonyl
Groups, and Other
Functional Groups*

X	cis,cis-**9**	cis,trans-**9**
OH	67	33
CH$_2$OH	52	48
NH$_2$	1.5	98.5
CH$_3$	64	36
OCH$_3$	88	12
CO$_2$CH$_3$	85	15
CONH$_2$	81	19

5.1.2. Hydrogenation Using Homogeneous Catalysts

In addition to solid transition metals, numerous soluble transition metal complexes are active hydrogenation catalysts.[9] One of the first to be used was *tris*-(triphenylphosphine)rhodium chloride, known as *Wilkinson's catalyst*.[10] Hydrogenation by homogeneous catalysts is believed to take place by initial formation of a π complex. The addition of hydrogen to the metal occurs by *oxidative addition* and increases the formal oxidation state of the metal by two. This is followed by transfer of hydrogen from rhodium to carbon to form an alkylrhodium intermediate. The final step is a second migration of hydrogen to carbon, leading to elimination of the saturated product (reductive elimination) and regeneration of active catalyst.

In some cases an alternative sequence involving addition of hydrogen at rhodium prior to complexation of the alkene may operate.[11] The phosphine ligands serve both to provide a stable soluble complex and to adjust the reactivity at the metal center. The σ-bonded intermediates have been observed for Wilkinson's catalyst[12] and for several other related catalysts.[13] For example, a partially hydrogenated structure has been isolated from methyl α-acetamidocinnamate.[14]

9. A. J. Birch and D. H. Williamson, *Org. React.*, **24**, 1 (1976); B. R. Jones, *Homogeneous Hydrogenation*, Wiley, New York, 1973.
10. J. A. Osborn, F. H. Jardine, J. F. Young, and G. Wilkinson, *J. Chem. Soc. A*, 1711 (1966).
11. I. D. Gridnev and T. Imamoto, *Acc. Chem. Res.*, **37**, 633 (2004).
12. D. Evans, J. A. Osborn and G. Wilkinson, *J. Chem. Soc. A*, 3133 (1968); V. S. Petrosyan, A. B. Permin, V. I. Bogdaskina, and D. P. Krutko, *J. Orgmet. Chem.*, **292**, 303 (1985).
13. H. Heinrich, R. Giernoth, J. Bargon, and J. M. Brown, *Chem. Commun.*, 1296 (2001); I. D. Gridnev, N. Higashi and T. Imamoto, *Organometallics*, **20**, 4542, (2001).
14. J. A. Ramsden, T. D. Claridge and J. M. Brown, *J. Chem. Soc., Chem. Commun.*, 2469 (1995).

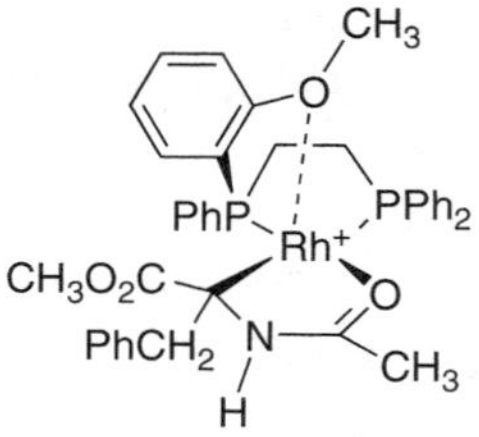

The regioselectivity of the hydride addition step has been probed by searching for deuterium exchange into isomerized alkenes that have undergone *partial* reduction.[15] The results suggest that Rh is electrophilic in the addition step and that the hydride transfer is nucleophilic.

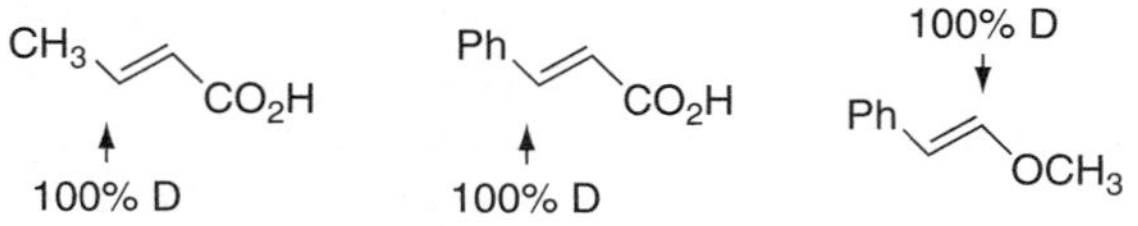

The stereochemistry of reduction by homogeneous catalysts is often controlled by functional groups in the reactant. Delivery of hydrogen occurs *cis* to a polar functional group. This behavior has been found to be particularly characteristic of an iridium-based catalyst that contains cyclooctadiene, pyridine, and tricyclohexylphosphine as ligands, known as the *Crabtree catalyst*.[16] Homogeneous iridium catalysts have been found to be influenced not only by hydroxy groups, but also by amide, ester, and ether substituents.[17]

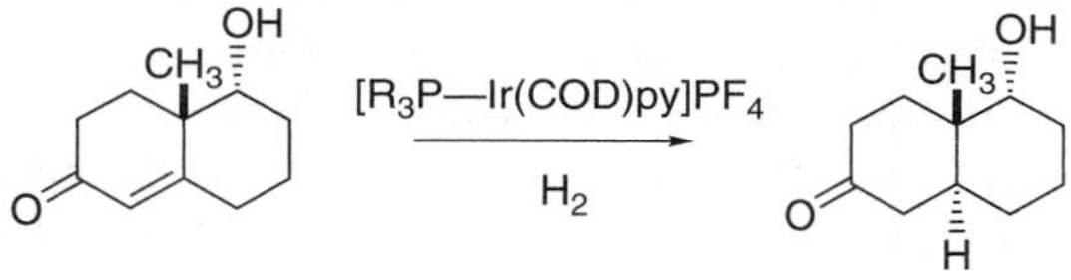

Ref. 18

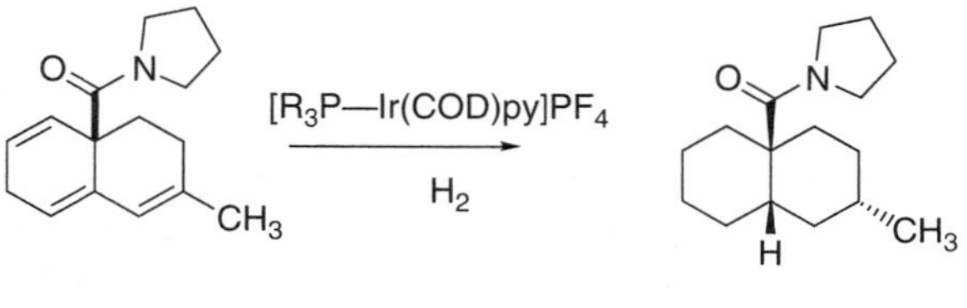

Ref. 19

[15] J. Yu and J. B. Spencer, *J. Am. Chem. Soc.*, **119**, 5257 (1997); J. Yu and J. B. Spencer, *Tetrahedron*, **54**, 15821 (1998).

[16] R. Crabtree, *Acc. Chem. Res.*, **12**, 331 (1979).

[17] R. H. Crabtree and M. W. Davis, *J. Org. Chem.*, **51**, 2655 (1986); P. J. McCloskey and A. G. Schultz, *J. Org. Chem.*, **53**, 1380 (1988).

[18] G. Stork and D. E. Kahne, *J. Am. Chem. Soc.*, **105**, 1072 (1983).

[19] A. G. Schultz and P. J. McCloskey, *J. Org. Chem.*, **50**, 5905 (1985).

376

CHAPTER 5

*Reduction of
Carbon-Carbon Multiple
Bonds, Carbonyl
Groups, and Other
Functional Groups*

The Crabtree catalyst also exhibited superior stereoselectivity in comparison with other catalysts in reduction of an exocyclic methylene group.[20]

$[Ir(cod)(pyr)PR_3] PF_6$	> 99:1
$Rh(nbd)(dppb) BF_4$	90:10
$Rh(Ph_3P)_3Cl$	6:94
Pd/C	5:95

Presumably, the stereoselectivity in these cases is the result of coordination of iridium by the functional group. The crucial property required for a catalyst to be stereodirective is that it be able to coordinate with both the directive group and the double bond and still accommodate the metal hydride bonds necessary for hydrogenation. In the iridium catalyst illustrated above, the cyclooctadiene ligand (COD) in the catalysts is released by hydrogenation, permitting coordination of the reactant and reaction with hydrogen.

Scheme 5.2 gives some examples of hydrogenations carried out with homogeneous catalysts. Entry 1 is an addition of deuterium that demonstrates net *syn* addition with the Wilkinson catalyst. The reaction in Entry 2 proceeds with high stereoselectivity and is directed by steric approach control, rather than a substituent-directing effect. One potential advantage of homogeneous catalysts is the ability to achieve a high degree of selectivity among different functional groups. Entries 3 and 4 are examples that show selective reduction of the unconjugated double bond. Similarly in Entry 5, reduction of the double bond occurs without reduction of the nitro group, which is usually rapidly reduced by heterogeneous hydrogenation. Entries 6 and 7 are cases of substituent-directed hydrogenation using the iridium (Crabtree) catalyst. The catalyst used in Entry 8 is related to the Wilkinson catalyst, but on hydrogenation of norbornadiene (NBD) has two open coordination positions. This catalyst exhibits a strong hydroxy-directing effect. The Crabtree catalyst gave excellent results in the hydrogenation of 3-methylpentadeca-4-enone to *R*-muscone. (Entry 9) A number of heterogeneous catalysts led to 5–15% racemization (by allylic exchange).

5.1.3. Enantioselective Hydrogenation

The fundamental concepts of enantioselective hydrogenation were introduced in Section 2.5.1 of Part A, and examples of reactions of acrylic acids and the important case of α-acetamido acrylate esters were discussed. The chirality of enantioselective hydrogenation catalysts is usually derived from phosphine ligands. A number of chiral phosphines have been explored in the development of enantioselective hydrogenation catalysts,[21] and it has been found that some of the most successful catalysts are derived from chiral 1, 1′-binaphthyldiphosphines, such as BINAP.[22]

20. J. M. Bueno, J. M. Coteron, J. L. Chiara, A. Fernandez-Mayoralas, J. M. Fiandor, and N. Valle, *Tetrahedron Lett.*, **41**, 4379 (2000).

21. B. Bosnich and M. D. Fryzuk, *Top. Stereochem.*, **12**, 119 (1981); W. S. Knowles, W. S. Chrisopfel, K. E. Koenig, and C. F. Hobbs, *Adv. Chem. Ser.*, **196**, 325 (1982); W. S. Knowles, *Acc. Chem. Res.*, **16**, 106 (1983).

22. R. Noyori and H. Takaya, *Acc. Chem. Res.*, **23**, 345 (1990).

Scheme 5.2. Homogeneous Catalytic Hydrogenation

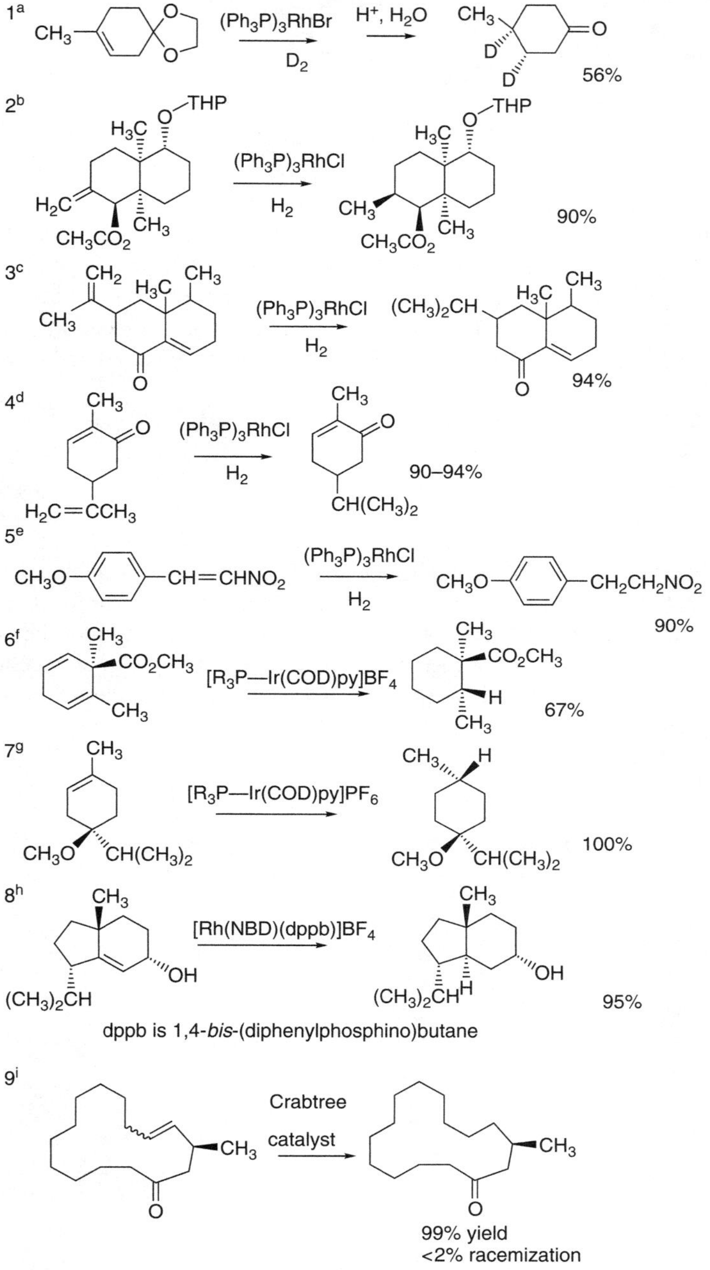

1[a]

56%

2[b]

90%

3[c]

94%

4[d]

90–94%

5[e]

CH_3O—⟨⟩—$CH=CHNO_2$ $\xrightarrow[H_2]{(Ph_3P)_3RhCl}$ CH_3O—⟨⟩—$CH_2CH_2NO_2$

90%

6[f]

67%

7[g]

100%

8[h]

95%

dppb is 1,4-*bis*-(diphenylphosphino)butane

9[i]

99% yield
<2% racemization

a. W. C. Agosta and W. L. Shreiber, *J. Am. Chem. Soc.*, **93**, 3947 (1971).
b. E. Piers, W. de Waal, and R. W. Britton, *J. Am. Chem. Soc.*, **93**, 5113 (1971).
c. M. Brown and L. W. Piszkiewicz, *J. Org. Chem.*, **32**, 2013 (1967).
d. R. E. Ireland and P. Bey, *Org. Synth*, **53**, 63 (1973).
e. R. E. Harmon, J. L. Parsons, D. W. Cooke, S. K. Gupta, and J. Schoolenberg, *J. Org. Chem.*, **34**, 3684 (1969).
f. A. G. Schultz and P. J. McCloskey, *J. Org. Chem.*, **50**, 5905 (1985).
g. R. H. Crabtree and M. W. Davies, *J. Org. Chem.*, **51**, 2655 (1986).
h. D. A. Evans and M. M. Morrissey, *J. Am. Chem. Soc.*, **106**, 3866 (1984).
i. C. Fehr, J. Galindo, I. Farris, and A. Cuenca, *Helv. Chim. Acta*, **87**, 1737 (2004).

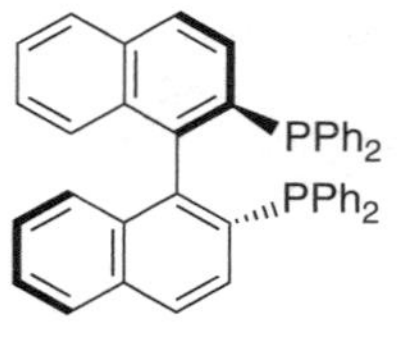

378

CHAPTER 5

*Reduction of
Carbon-Carbon Multiple
Bonds, Carbonyl
Groups, and Other
Functional Groups*

BINAP

Ruthenium complexes containing this ligand are able to reduce a variety of double bonds with e.e. above 95%. In order to achieve high enantioselectivity, the reactant must show a strong preference for a specific orientation when complexed with the catalyst. This ordinarily requires the presence of a functional group that can coordinate with the metal. The ruthenium-BINAP catalyst has been used successfully with unsaturated amides,[23] allylic and homoallylic alcohols,[24] and unsaturated carboxylic acids.[25]

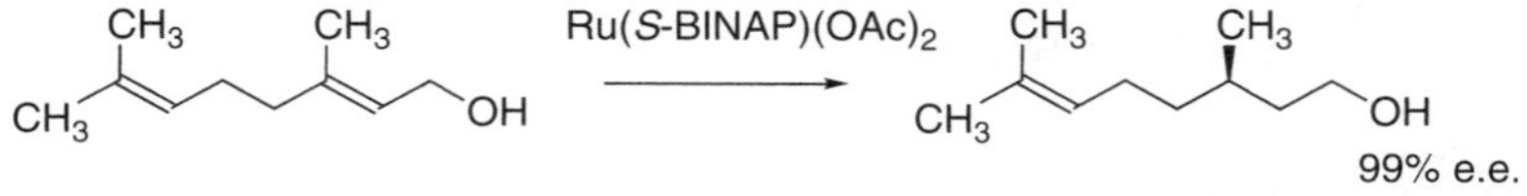

99% e.e. Ref. 12

The mechanism of such reactions using unsaturated carboxylic acids and $Ru(BINAP)(O_2CCH_3)_2$ is consistent with the idea that coordination of the carboxy group establishes the geometry at the metal ion.[26] The configuration of the new stereocenter is then established by the hydride transfer. In this particular mechanism, the second hydrogen is introduced by protonolysis, but in other cases a second hydride transfer step occurs.

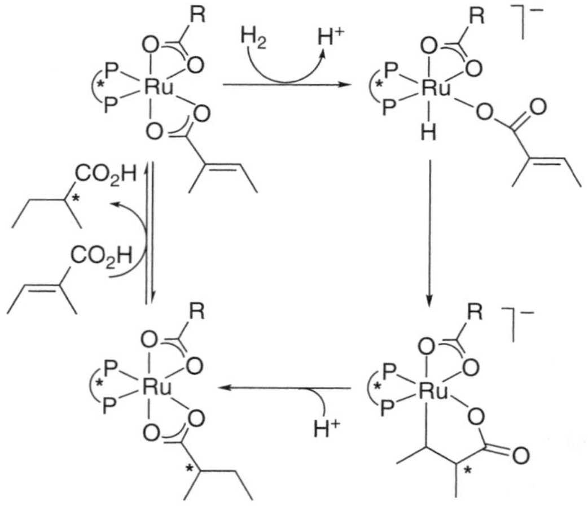

23. R. Noyori, M. Ohta, Y. Hsiao, M. Kitamura, T. Ohta, and H. Takaya, *J. Am. Chem. Soc.*, **108**, 7117 (1986).

24. H. Takaya, T. Ohta, N. Sayo, H. Kumobayashi, S. Akutagawa, S. Inoue, I. Kasahara, and R. Noyori, *J. Am. Chem. Soc.*, **109**, 1596 (1987).

25. T. Ohta, H. Takaya, M. Kitamura, K. Nagai, and R. Noyori, *J. Org. Chem.*, **52**, 3174 (1987).

26. M. T. Ashby and J. T. Halpern, *J. Am. Chem. Soc.*, **113**, 589 (1991).

This reaction has been used in the large-scale preparation of an intermediate in the synthesis of a cholesterol acyl-transferase inhibitor.[27]

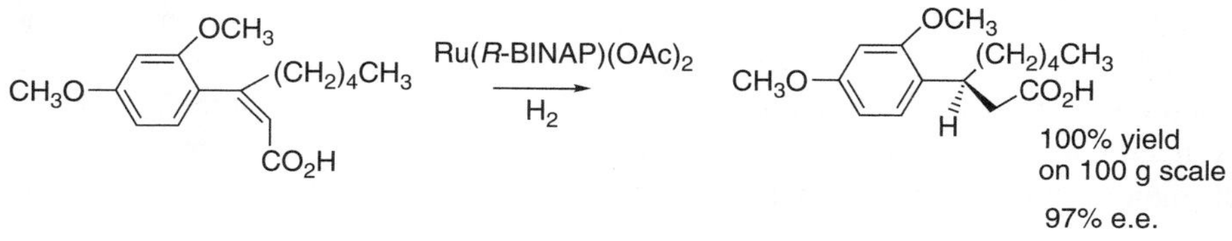

100% yield
on 100 g scale

97% e.e.

An enantioselective hydrogenation of this type is also of interest in the production of α-tocopherol (vitamin E). Totally synthetic α-tocopherol can be made in racemic form from 2,3,5-trimethylhydroquinone and racemic isophytol. The product made in this way is a mixture of all eight possible stereoisomers.

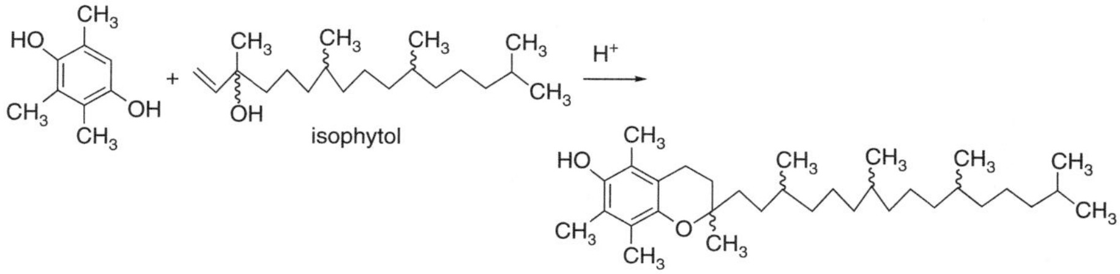

Tocopherol can be produced as the pure $2R, 4'R, 8'R$ stereoisomer from natural vegetable oils. This is the most biologically active of the stereoisomers. The correct side-chain stereochemistry can be obtained using a process that involves two successive enantioselective hydrogenations.[28] The optimum catalyst contains a $6, 6'$-dimethoxybiphenyl phosphine ligand. This reaction has not yet been applied to the enantioselective synthesis of α-tocopherol because the cyclization step with the phenol is not enantiospecific.

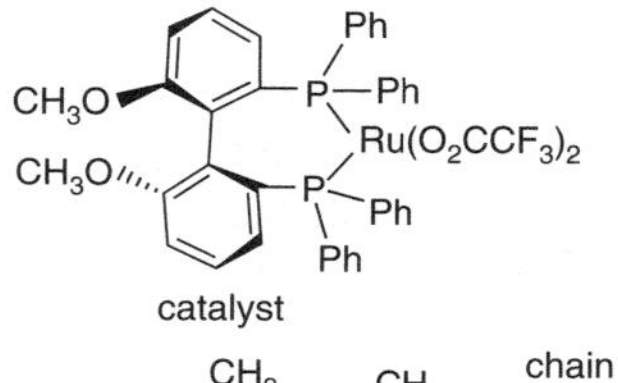

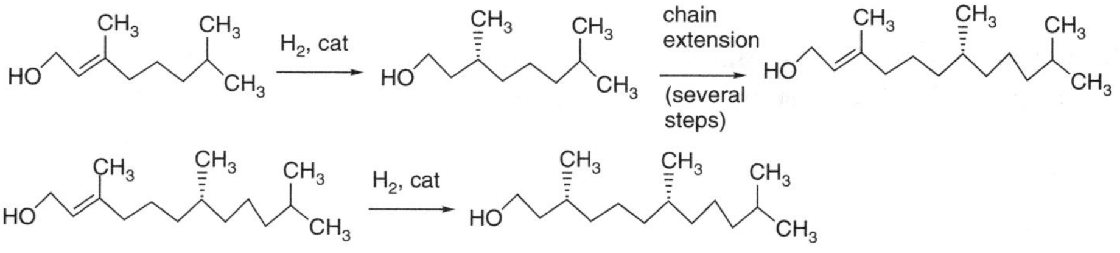

catalyst

27. M. Murakami, K. Kobayashi, and K. Hirai, *Chem. Pharm. Bull.*, **48**, 1567 (2000).
28. T. Netscher, M. Scalione, and R. Schmid, in *Asymmetric Catalysis on an Industrial Scale: Challenges, Approaches and Solutions*, H. U. Blaser and E. Schmidt, eds., Wiley-VCH, Weinhem, 2004, pp. 71–89.

380

CHAPTER 5

*Reduction of
Carbon-Carbon Multiple
Bonds, Carbonyl
Groups, and Other
Functional Groups*

An especially important case is the enantioselective hydrogenation of α-amidoacrylic acids, which leads to α-aminoacids.[29] A particularly detailed study has been carried out on the mechanism of reduction of methyl Z-α-acetamidocinnamate by a rhodium catalyst with a chiral diphosphine ligand DIPAMP.[30] It has been concluded that the reactant can bind reversibly to the catalyst to give either of two complexes. Addition of hydrogen at rhodium then leads to a reactive rhodium hydride and eventually to product. Interestingly, the addition of hydrogen occurs most rapidly in the minor isomeric complex, and the enantioselectivity is due to this kinetic preference.

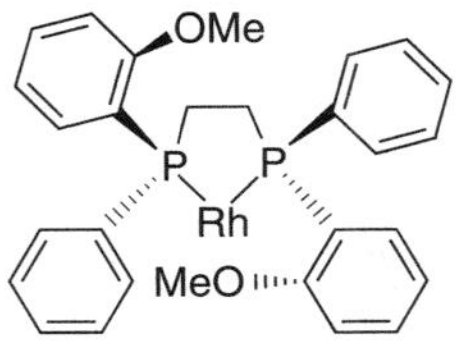

DIPAMP

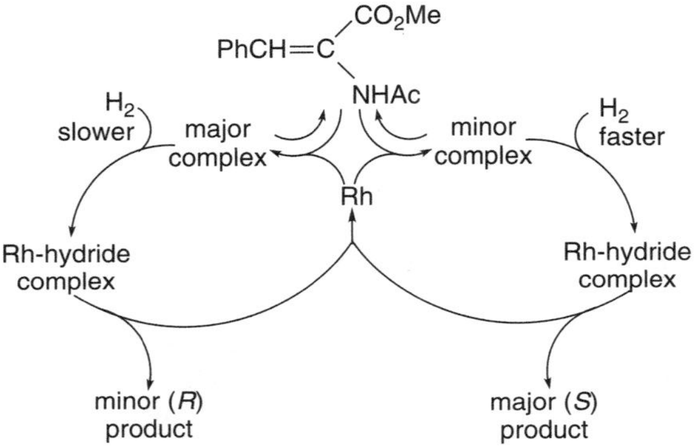

A thorough computational study of this process has been carried out using B3LYP/ONIOM calculations.[31] The rate-determining step is found to be the formation of the rhodium hydride intermediate. The barrier for this step is smaller for the minor complex than for the major one. Additional details on this study can be found at:

Visual models and additional information on Asymmetric Hydrogenation can be found in the Digital Resource available at: Springer.com/carey-sundberg.

29. J. Halpern, in *Asymmetric Synthesis*, Vol. 5, J. D. Morrison, ed., Academic Press, Orlando, FL, 1985; A. Pfaltz and J. M. Brown, in *Stereoselective Synthesis*, G. Helmchen, R. W. Hoffmann, J. Mulzer, and E. Schauman, eds., Thieme, New York, 1996, Part D, Sect. 2.5.1.2; U. Nagel and J. Albrecht, *Catalysis Lett.*, **5**, 3 (1998).
30. C. R. Landis and J. Halpern, *J. Am. Chem. Soc.*, **109**, 1746 (1987).
31. S. Feldgus and C. R. Landis, *J. Am. Chem. Soc.*, **122**, 12714 (2000).

Another mechanistic study, carried out using *S*-BINAP-ruthenium(II) diacetate catalyst, concluded that the mechanism shown in Figure 5.1 was operating.[32] The rate-determining step is the hydrogenolysis of intermediate **13**, which has an E_a of about 19 kcal/mol. This step also determines the enantioselectivity and proceeds with retention of configuration. The prior steps are reversible and the relative stability of $13_R > 13_S$ determines the preference for the *S*-enantiomer. The energy relationships are summarized in Figure 5.2. The major difference between the major and minor pathways is in the precursors 12_{re} (favored) and 12_{si} (disfavored). There is a greater steric repulsion between the carboxylate substituent and the BINAP ligand in 12_{si} than in 12_{re} (Figure 5.3.).

A related study with a similar ruthenium catalyst led to the structural and NMR characterization of an intermediate that has the crucial Ru−C bond in place and also shares other features with the BINAP-ruthenium diacetate mechanism.[33] This mechanism, as summarized in Figure 5.4, shows the formation of a metal hydride prior to the complexation of the reactant. In contrast to the mechanism for acrylic acids shown on p. 378, the creation of the new stereocenter occurs at the stage of the addition of the second hydrogen.

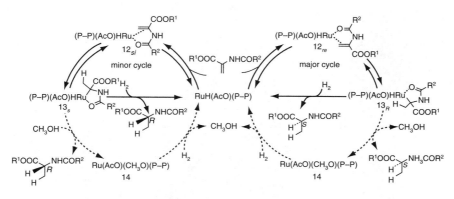

Fig. 5.1. Mechanism of ruthenium catalyzed enantioselective hydrogenation of α-acetamidoacrylate esters. Reproduced from *J. Am. Chem. Soc.*, **124**, 6649 (2002), by permission of the American Chemical Society.

[32] M. Kitamura, M. Tsukamoto, Y. Bessho, M. Yoshimura, U. Kobs, M. Widhalm, and R. Noyori, *J. Am. Chem. Soc.*, **124**, 6649 (2002).

[33] J. A. Wiles and S. H. Bergens, *Organometallics*, **17**, 2228 (1998); J. A. Wiles and S. H. Bergens, *Organometallics*, **18**, 3709 (1999).

CHAPTER 5

*Reduction of
Carbon-Carbon Multiple
Bonds, Carbonyl
Groups, and Other
Functional Groups*

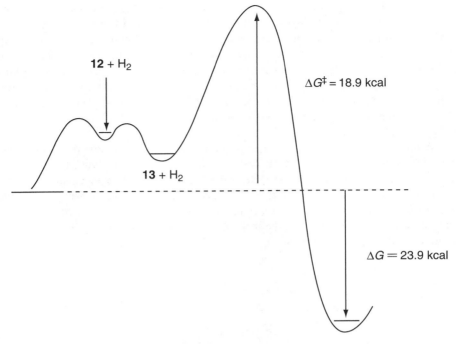

12 + H$_2$

13 + H$_2$

$\Delta G^{\ddagger} = 18.9$ kcal

$\Delta G = 23.9$ kcal

Fig. 5.2. Summary energy diagram for enantioselective ruthenium-catalyzed hydrogenation of α-acetamidoacrylate esters. Reproduced from *J. Am. Chem. Soc.*, **124**, 6649 (2002), by permission of the American Chemical Society.

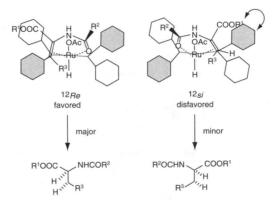

12*Re*
favored

12*si*
disfavored

major

minor

Fig. 5.3. (a) View of (*S*)-BINAP-ruthenium complex showing the chiral environment. (b) Relationship of reactant to chiral environment showing preferred orientation. The binaphthyl rings are omitted for clarity. Adapted from *J. Am. Chem. Soc.*, **124**, 6649 (2002), by permission of the American Chemical Society.

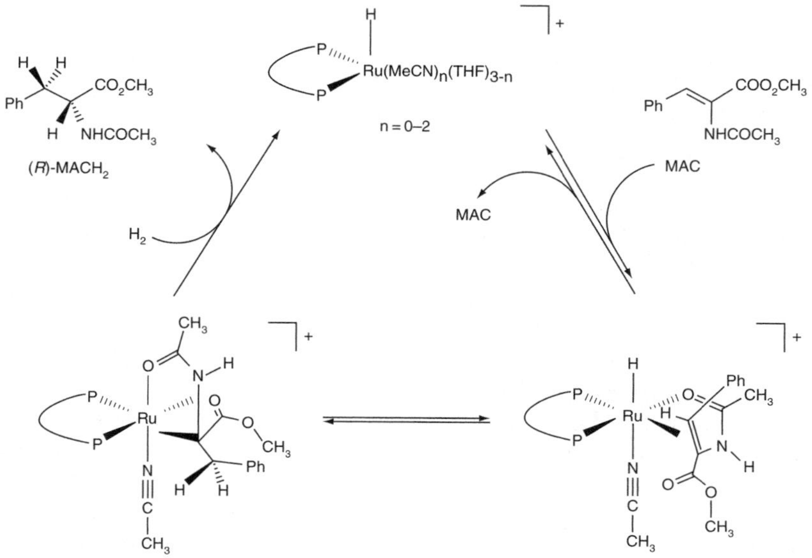

Fig. 5.4. Schematic mechanism for enantioselective hydrogenation of methyl acetamidocinnamate (MAC) over a cationic ruthenium catalyst. Reproduced from *Organometallics*, **18**, 3709 (1999), by permission of the American Chemical Society.

Catalyst reactivity and enantioselectivity can be affected by substituents on ligands. In the Rh-catalyzed hydrogenation of methyl Z-α-acetamidocinnamate, for example, BINOL phosphites with ERGs give much higher enantioselectivity than those with EWGs. The ligand substituents modify the electron density at the metal center and change the energy balance between the competing pathways. This example demonstrates the potential for fine-tuning of the catalysts by changes that are relatively remote from the catalytic site.[34]

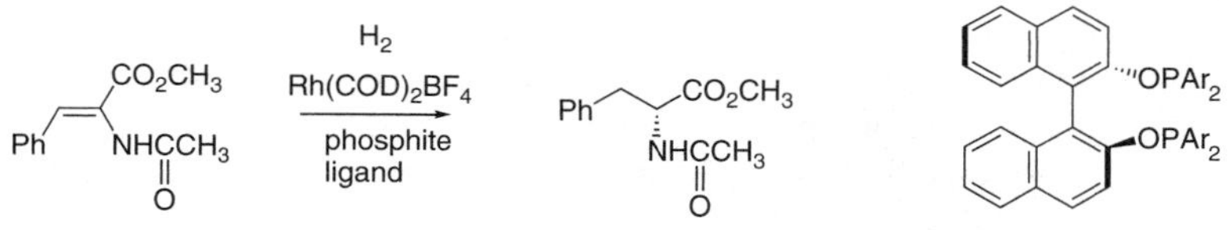

Ar substituent	% e.e.
3,5-di-CF$_3$	31
4-CF$_3$	49
4-CH$_3$	93
3,5-di-CH$_3$	94
4-CH$_3$O	99

Many other catalysts and ligands have been examined for the enantioselective reduction of α-acetamidoacrylates and related substrates. Phosphoramidites derived from BINOL and the cyclic amines piperidine and morpholine give excellent results.[35]

[34] I. Gergely, C. Hegedus, A. Szollosy, A. Monsees, T. Riermeier, and J. Bakos, *Tetrahedron Lett.*, **44**, 9025 (2003).

[35] H. Bernsmann, M. van der Berg, R. Hoen, A. J. Minnaard, G. Mehler, M. T. Reetz, J. G. De Vries, and B. L. Feringa, *J. Org. Chem.*, **70**, 943 (2005).

384

CHAPTER 5

Reduction of
Carbon-Carbon Multiple
Bonds, Carbonyl
Groups, and Other
Functional Groups

These ligands also give excellent results with dimethyl itaconate and α-arylenamides.

Scheme 5.3 shows the enantioselectivity of some hydrogenations of unsaturated acids and amides. Entries 1 to 5 are examples of hydrogenations of α-acetamidoacrylate and α-acetamidocinnamate esters. The catalyst in Entries 1 and 2 uses chiraphos as the chiral phosphine ligand and norbornadiene as the removable ligand. The catalyst in Entry 3 uses DIPAMP as the chiral ligand. BINAP is the ligand in Entry 4. The ligand in Entry 5, known as EtDuPHOS, gave highly selective reduction of the α,β-double bond in the conjugated system. Entries 6 and 7 show reduction of acrylate esters having other types of substituents that give good results with the DIPAMP catalyst. Entries 8 to 10 show examples of several alkylidene succinate half-esters.

There can be significant differences in the detailed structure and mechanism of these catalysts. For example, the geometry of the phosphine ligands may affect the reactivity at the metal ion, but the basic elements of the mechanism of enantioselection are similar. The phosphine ligands establish a chiral environment and provide an appropriate balance of reactivity and stability for the metal center. The reactants bind to the metal through the double bond and at least one other functional group, and mutual interaction with the chiral environment is the basis for enantioselectivity. The new stereocenters are established under the influence of the chiral environment.

The enantioselective hydrogenation of unfunctionalized alkenes presents special challenges. Functionalized reactants such as acrylate esters can coordinate with the metal in the catalyst and this point of contact can serve to favor a specific orientation and promote enantioselectivity. Unfunctionalized alkenes do not have such coordination sites and enantioselectivity is based on steric factors. A number of iridium-based catalysts have been developed. One successful type of catalyst incorporates phosphine or phosphite groups and a chiral oxazoline ring as donors.[36] The catalysts also incorporate cyclooctadiene as a removable ligand. These catalysts are extremely sensitive to even weakly coordinating anions and the preferred anion for alkene hydrogenation is *tetrakis*-[(3,5-trifluoromethyl)phenyl]borate. Most of the examples to date have been with aryl-substituted double bonds.

[36]. G. Helmchen and A. Pfaltz, *Acc. Chem. Res.*, **33**, 336 (2000).

[37]. F. Menges, M. Neuburger, and A. Pfaltz, *Org. Lett.*, **4**, 4713 (2002).

[38]. S. P. Smidt, F. Menges, and A. Pfaltz, *Org. Lett.*, **6**, 2023 (2004).

[39]. D. R. Hou, J. Reibenspies, T. J. Colacot, and K. Burgess, *Chem. Eur. J.*, **7**, 5391 (2001).

Scheme 5.3. Enantioselectivity for Catalytic Hydrogenation of Substituted Acrylic Acids

Reactant	Catalyst	Product	Configuration	% e.e.
1[a] 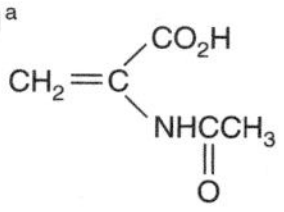	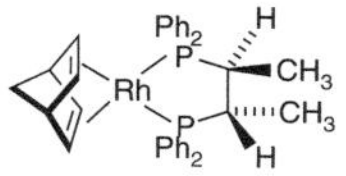	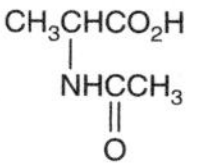 CH$_3$CHCO$_2$H / NHCCH$_3$ / O	R	90
2[a] 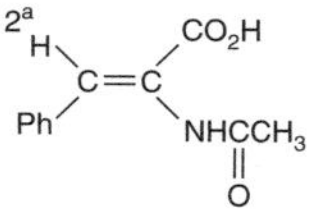	Same as above	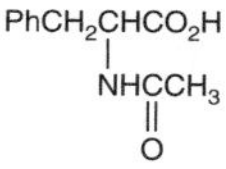 PhCH$_2$CHCO$_2$H / NHCCH$_3$ / O	S	95
3[b] 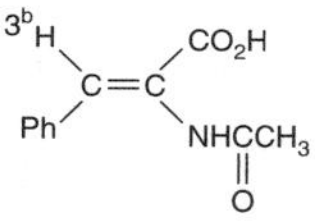	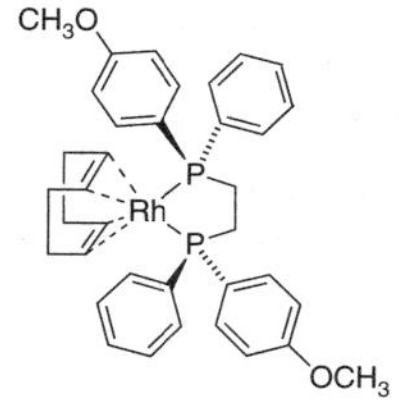	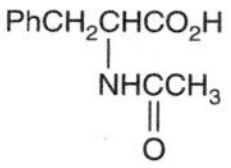 PhCH$_2$CHCO$_2$H / NHCCH$_3$ / O	R	94
4[c] 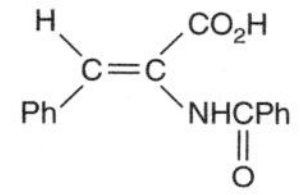	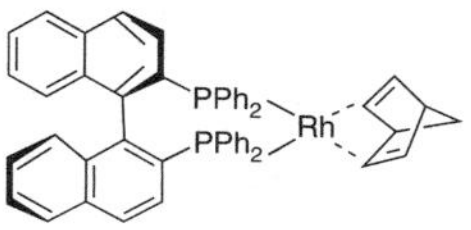	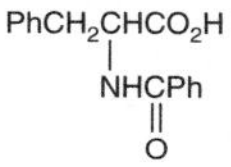 PhCH$_2$CHCO$_2$H / NHCPh / O	S	100
5[d] 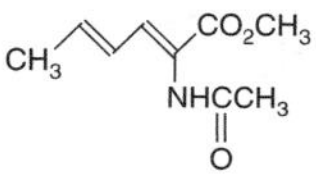	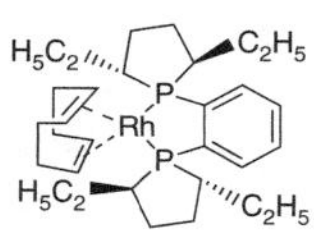	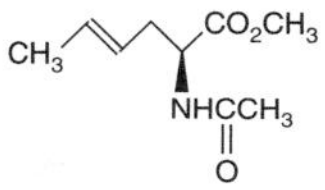 CH$_3$ ... CO$_2$CH$_3$ / NHCCH$_3$ / O	R	99.2
6[b] 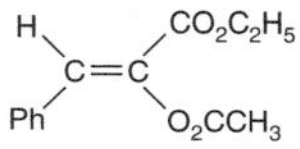	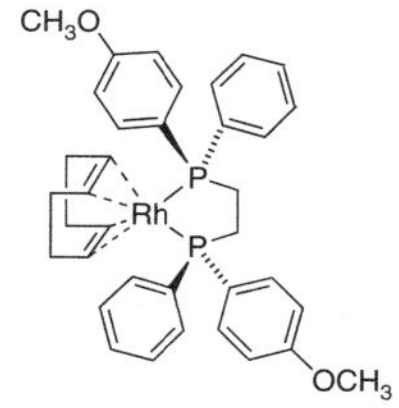	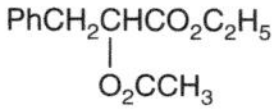 PhCH$_2$CHCO$_2$C$_2$H$_5$ / O$_2$CCH$_3$	S	90
7[e] 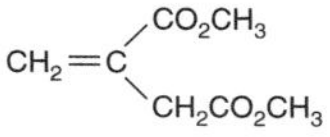	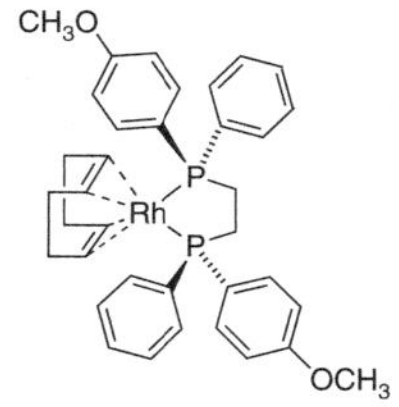	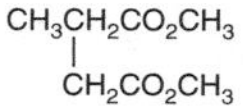 CH$_3$CH$_2$CO$_2$CH$_3$ / CH$_2$CO$_2$CH$_3$	R	88
8[f] 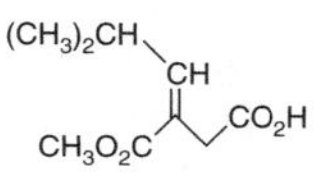	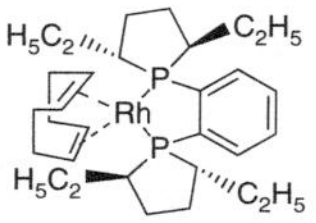	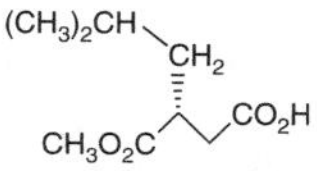 (CH$_3$)$_2$CH / CH$_2$ / CH$_3$O$_2$C / CO$_2$H	R	99

(*Continued*)

CHAPTER 5

*Reduction of
Carbon-Carbon Multiple
Bonds, Carbonyl
Groups, and Other
Functional Groups*

Scheme 5.3. (*Continued*)

Reactant	Catalyst	Product	Configuration	% e.e.
9[g]			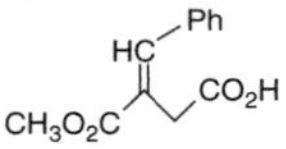 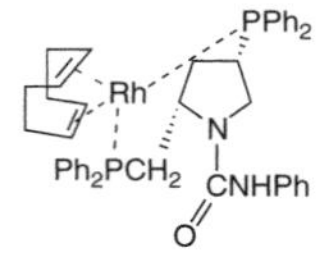 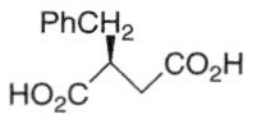 *S*	>95
10[h]			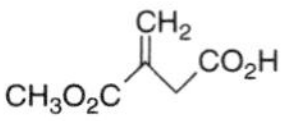 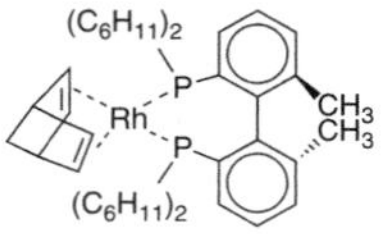 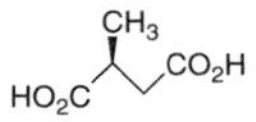 *S*	96

a. M. D. Fryzuk and B. Bosnich, *J. Am. Chem. Soc.* **99**, 6262 (1977).
b. B. D. Vineyard, W. S. Knowles, M. J. Sabacky, G. L. Bachman, and D. J. Weinkauff, *J. Am. Chem. Soc.*, **99**, 5946 (1977).
c. A. Miyashita, H. Takaya, T. Souchi, and R. Noyori, *Tetrahedron*, **40**, 1245 (1984).
d. M. J. Burk, J. G. Allen, and W. F. Kiesman, *J. Am. Chem. Soc.*, **120**, 657 (1998).
e. W. C. Christopfel and B. D. Vineyard, *J. Am. Chem. Soc.*, **101**, 4406 (1979).
f. M. J. Burk, F. Bienewald, M. Harris, and A. Zanotti-Gerosa, *Angew. Chem. Int. Ed. Engl.* **37**, 1931 (1998).
g. H. Jendralla, *Tetrahedron Lett.*, **32**, 3671 (1991).
h. T. Chiba, A. Miyashita, H. Nohira, and H. Takaya, *Tetrahedron Lett.*, **32**, 4745 (1991).

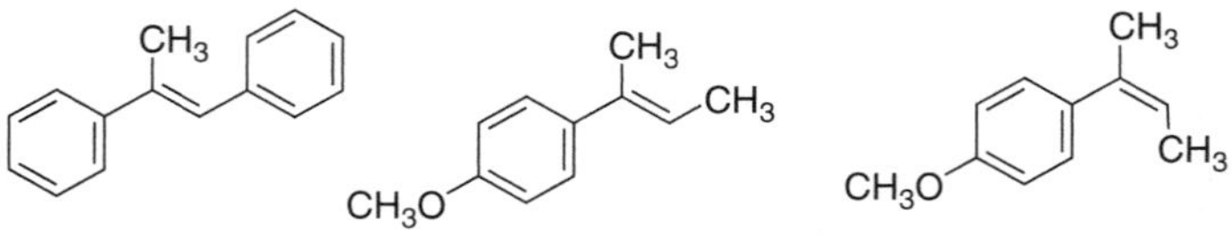

Catalyst	Percent e.e.		
A	98	81	63
B	98	91	66
C	89	86	75

These catalysts also provide excellent results with acrylate esters and allylic alcohols.

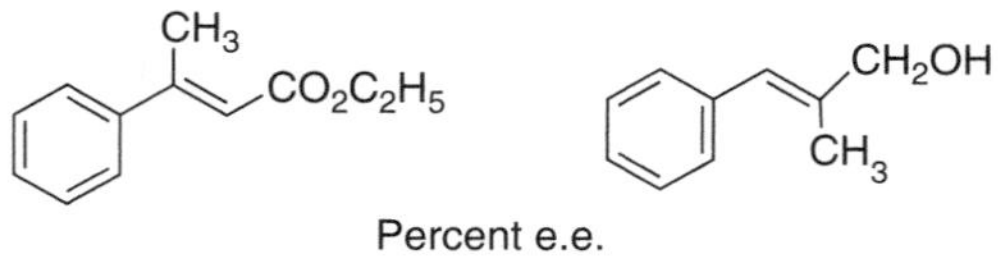

Catalyst	Percent e.e.	
A	84	96
B	94	97

These catalysts are activated by hydrogenation of the cyclooctadiene ligand, which releases cyclooctane and opens two coordination sites at iridium. The mechanism has been probed by computational studies.[40] It is suggested that the catalytic cycle involves

40. P. Brandt, C. Hedberg, and P. G. Andersson, *Chem. Eur. J.*, **9**, 339 (2003).

the addition of two hydrogens to the alkene-catalyst complex, followed by formation of an alkyliridium intermediate and reductive elimination.

The enantioselectivity is thought to result from both steric blocking by the *t*-butyl substituent on the oxazoline ring and an attractive van der Waals interaction of an aryl ring and the oxazoline ring, as shown in Figure 5.5.

5.1.4. Partial Reduction of Alkynes

Partial reduction of alkynes to *Z*-alkenes is an important synthetic application of selective hydrogenation catalysts. The transformation can be carried out under heterogeneous or homogeneous conditions. Among heterogeneous catalysts, the one that

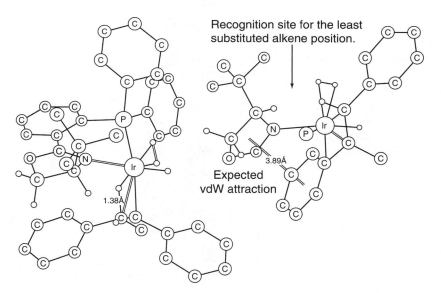

Fig. 5.5. Suggested basis of enantioselectivity in hydrogenation of α-methylstilbene by a phosphinoaryl oxazoline–iridium catalyst. Reproduced from *Chem. Eur. J.*, **9**, 339 (2003), by permission of Wiley-VCH.

388

CHAPTER 5

*Reduction of
Carbon-Carbon Multiple
Bonds, Carbonyl
Groups, and Other
Functional Groups*

is most successful is *Lindlar's catalyst*, a lead-modified palladium-CaCO$_3$ catalyst.[41] A nickel-boride catalyst prepared by reduction of nickel salts with sodium hydride is also useful.[42] Rhodium catalysts have also been reported to show good selectivity.[43]

5.1.5. Hydrogen Transfer from Diimide

Catalytic hydrogenation transfers the elements of molecular hydrogen through a series of complexes and intermediates. Diimide, HN=NH, an unstable hydrogen donor that can be generated in situ, finds specialized application in the reduction of carbon-carbon double bonds. Simple alkenes are reduced efficiently by diimide, but other easily reduced functional groups, such as nitro and cyano are unaffected. The mechanism of the reaction is pictured as a concerted transfer of hydrogen via a nonpolar cyclic TS.

In agreement with this mechanism is the fact that the stereochemistry of addition is *syn*.[44] The rate of reaction with diimide is influenced by torsional and angle strain in the alkene. More strained double bonds react at accelerated rates.[45] For example, the more strained *trans* double bond is selectively reduced in *Z,E*-1,5-cyclodecadiene.

Ref. 46

Diimide selectively reduces terminal over internal double bonds in polyunsaturated systems.[47]

Reduction by diimide can be advantageous when compounds contain functional groups that would be reduced by other methods or when they are unstable to hydrogenation catalysts. There are several methods for generation of diimide and they are illustrated in Scheme 5.4. The method in Entry 1 is probably the one used most frequently in synthetic work and involves the generation and spontaneous decarboxylation of azodicarboxylic acid. Entry 2, which illustrates another convenient method, thermal decomposition of *p*-toluenesulfonylhydrazide, is interesting in that it

[41] H. Lindlar and R. Dubuis, *Org. Synth.*, **V**, 880 (1973).

[42] H. C. Brown and C. A. Brown, *J. Am. Chem. Soc.*, **85**, 1005 (1963); E. J. Corey, K. Achiwa, and J. A. Katzenellenbogen, *J. Am. Chem. Soc.*, **91**, 4318 (1969).

[43] R. R. Schrock and J. A. Osborn, *J. Am. Chem. Soc.*, **98**, 2143 (1976); J. M. Tour, S. L. Pendalwar, C. M. Kafka, and J. P. Cooper, *J. Org. Chem.*, **57**, 4786 (1992).

[44] E. J. Corey, D. J. Pasto, and W. L. Mock, *J. Am. Chem. Soc.*, **83**, 2957 (1961).

[45] E. W. Garbisch, Jr., S. M. Schildcrout, D. B. Patterson, and C. M. Sprecher, *J. Am. Chem. Soc.*, **87**, 2932 (1965).

[46] J. G. Traynham, G. R. Franzen, G. A. Kresel, and D. J. Northington, Jr., *J. Org. Chem.*, **32**, 3285 (1967).

[47] E. J. Corey, H. Yamamoto, D. K. Herron, and K. Achiwa, *J. Am. Chem. Soc.*, **92**, 6635 (1970); E. J. Corey and H. Yamamoto, *J. Am. Chem. Soc.*, **92**, 6636, 6637 (1970).

Scheme 5.4. Reductions with Diimide

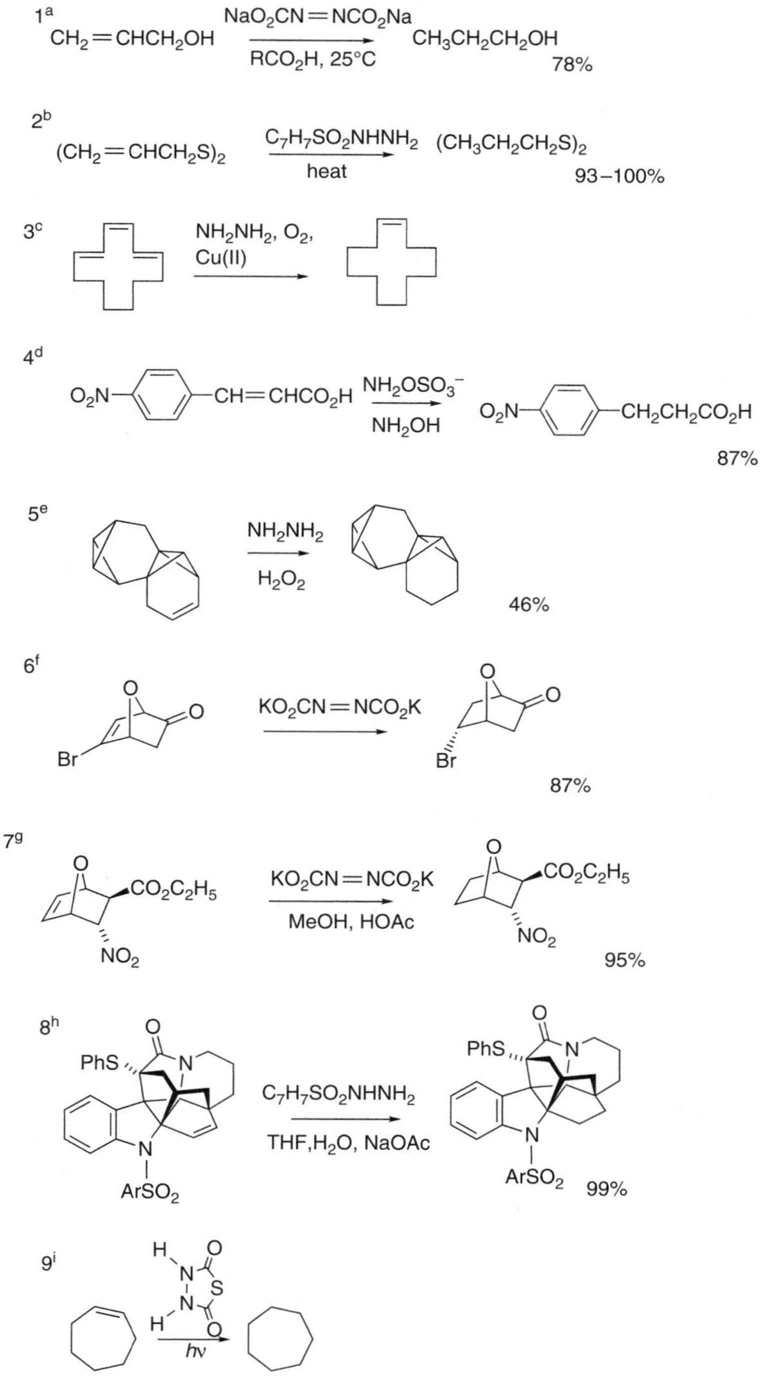
1[a]
$CH_2{=}CHCH_2OH$
$NaO_2CN{=}NCO_2Na$
RCO_2H, 25°C
$CH_3CH_2CH_2OH$
78%
2[b]
$(CH_2{=}CHCH_2S)_2$
$C_7H_7SO_2NHNH_2$
heat
$(CH_3CH_2CH_2S)_2$
93–100%
3[c]
NH_2NH_2, O_2,
Cu(II)
4[d]
O_2N
$CH{=}CHCO_2H$
$NH_2OSO_3^-$
NH_2OH
O_2N
$CH_2CH_2CO_2H$
87%
5[e]
NH_2NH_2
H_2O_2
46%
6[f]
$KO_2CN{=}NCO_2K$
Br
87%
7[g]
$CO_2C_2H_5$
NO_2
$KO_2CN{=}NCO_2K$
MeOH, HOAc
95%
8[h]
PhS
N
$ArSO_2$
$C_7H_7SO_2NHNH_2$
THF,H_2O, NaOAc
99%
9[i]
$h\nu$

(*Continued*)

CHAPTER 5

*Reduction of
Carbon-Carbon Multiple
Bonds, Carbonyl
Groups, and Other
Functional Groups*

Scheme 5.4. (*Continued*)

10[j]

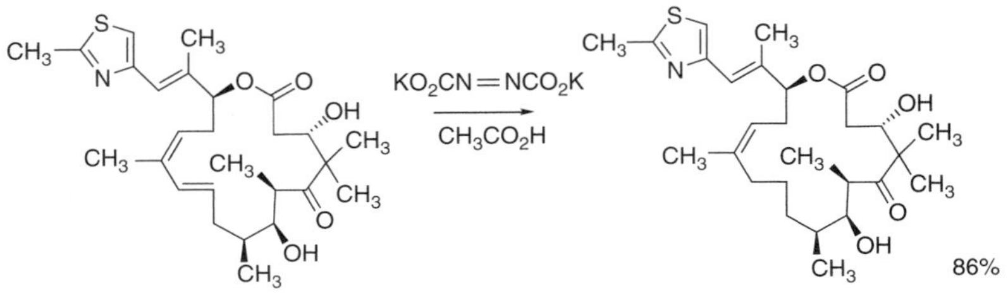

86%

a. E. E. van Tamelen, R. S. Dewey, and R. J. Timmons, *J. Am. Chem. Soc.*, **83**, 3725 (1961).
b. E. E. van Tamelen, R. S. Dewey, M. F. Lease, and W. H. Pirkle, *J. Am. Chem. Soc.*, **83**, 4302 (1961).
c. M Ohno, and M. Okamoto, *Org. Synth.*, **49**, 30 (1969).
d. W. Durckheimer, *Liebigs Ann. Chem.*, **712**, 240 (1969).
e. L. A. Paquette, A. R. Browne, E. Chamot, and J. F. Blount, *J. Am. Chem. Soc.*, **102**, 643 (1980).
f. J.-M. Durgnat and P. Vogel, *Helv. Chim. Acta*, **76**, 222 (1993).
g. P. A. Grieco, R. Lis, R. E. Zelle, and J. Finn, *J. Am. Chem. Soc.*, **108**, 5908 (1986).
h. P. Magnus, T. Gallagher, P. Brown, and J. C. Huffman, *J. Am. Chem. Soc.*, **106**, 2105 (1984).
i. M. Squillacote, J. DeFelippis, and Y. L. Lai, *Tetrahedron Lett.*, **34**, 4137 (1993).
j. K. Biswas, H. Lin, J. T. Njgardson, M. D. Chappell, T.-C. Chou, Y. Guan, W. P. Tong, L. He, S. B. Horwitz, and S. J. Danishefsky, *J. Am. Chem. Soc.*, **124**, 9825 (2002).

demonstrates that the very easily reduced disulfide bond is unaffected by diimide. Entry 3 involves generation of diimide by oxidation of hydrazine and also illustrates the selective reduction of *trans* double bonds in a medium-sized ring. Entry 4 shows that nitro groups are unaffected by diimide. Entries 5 to 7 involve sensitive molecules in which double bonds are reduced successfully. Entry 8, part of a synthesis of the kopsane group of alkaloids, successfully retains a sulfur substituent. Entry 9 illustrates a more recently developed diimide source, photolysis of 1,3,4-thiadiazolin-2,5-dione. Entry 10 is a selective reduction of a *trans* double bond in a macrocyclic lactone and was used in the synthesis of epothilone analogs.[48]

5.2. Catalytic Hydrogenation of Carbonyl and Other Functional Groups

Many other functional groups are also reactive under conditions of catalytic hydrogenation. Ketones, aldehydes, and esters can all be reduced to alcohols, but in most cases these reactions are slower than alkene reductions. For most synthetic applications, the hydride transfer reagents, discussed in Section 5.3, are used for reduction of carbonyl groups. The reduction of nitro compounds to amines, usually proceeds very rapidly. Amides, imines, and also nitriles can be reduced to amines. Hydrogenation of amides requires extreme conditions and is seldom used in synthesis, but reductions of imines and nitriles are quite useful. Table 5.2 gives a summary of the approximate conditions for catalytic hydrogenation of some common functional groups.

[48.] For another example, see J. D. White, R. G. Carter, and K. F. Sundermann, *J. Org. Chem.*, **64**, 684 (1999).

Table 5.2. Conditions for Catalytic Reduction of Various Functional Groups[a]

Reactant	Product	Catalyst	Conditions
$C=C$	$-\overset{\mid}{\underset{H}{C}}-\overset{\mid}{\underset{H}{C}}-$	Pd, Pt, Ni, Ru, Rh	Rapid at room temperature (R.T.) and 1 atm except for highly substituted or hindered cases
$-C\equiv C-$	$\underset{H}{C}=\underset{H}{C}$	Lindlar	R. T. and low pressure, quinoline or lead added to deactivate catalyst
(phenyl)	(cyclohexyl)	Rh, Pt	Moderate pressure (5–10 atm), 50–100°C
(phenyl)	(cyclohexyl)	Ni, Pd	High pressure (100–200 atm), 100–200°C
$RC(O)R$	$RCHR$, OH	Pt, Ru	Moderate rate at R. T. and 1–4 atm. acid-catalyzed
$RC(O)R$	$RCHR$, OH	Cu–Cr, Ni	High pressure, 50–100°C
(aryl)–C(O)R or (aryl)–CHR with OR	(aryl)–CH_2R	Pd	R. T., 1–4 atm. acid-catalyzed
(aryl)–CHR with NR_2	(aryl)–CH_2R	Pd, Ni	50–100°C, 1–4 atm
$RCCl(O)$	$RCH(O)$	Pd	R. T., 1 atm. quinoline or other catalyst moderator used
$RCOH(O)$	RCH_2OH	Pd, Ni, Ru	Very strenuous conditions required
$RCOR(O)$	RCH_2OH	Cu–Cr, Ni	200°C, high pressure
$RC\equiv N$	RCH_2NH_2	Ni, Rh	50–100°C, usually high pressure, NH_3 added to increase yield of primary amine
$RCNH_2(O)$	RCH_2NH_2	Cu–Cr	Very strenuous conditions required
RNO_2	RNH_2	Pd, Ni, Pt	R. T., 1–4 atm
$RCR(NR)$	R_2CHNHR	Pd, Pt	R. T., 4–100 atm
$R-Cl$, $R-Br$, $R-I$	$R-H$	Pd	Order of reactivity: I > Br > Cl > F, bases promote reactions for R = alkyl
epoxide $-C(O)C-$	$-\overset{H}{C}-\overset{OH}{C}-$	Pt, Pd	Proceeds slowly at R. T., 1–4 atm, acid-catalyzed

a. General References: M. Freifelder, *Catalytic Hydrogenation in Organic Synthesis: Procedures and Commentary*, John Wiley & Sons, New York, 1978; P. N. Rylander, *Hydrogenation Methods*, Academic Press, Orlando FL, 1985.

Many enantioselective catalysts have been developed for reduction of functional groups, particularly ketones. BINAP complexes of Ru(II)Cl$_2$ or Ru(II)Br$_2$ give good enantioselectivity in reduction of β-ketoesters.[49] This catalyst system has been shown to be subject to acid catalysis.[50] Thus in the presence of 0.1 mol % HCl, reduction proceeds smoothly at 40 psi of H$_2$ at 40° C.

49. R. Noyori, T. Ohkuma, M. Kitamura, H. Takaya, N. Sayo, H. Kumobayashi, and S. Akutagawa, *J. Am. Chem. Soc.*, **109**, 5856 (1987).
50. S. A. King, A. S. Thompson, A. O. King, and T. R. Verhoeven, *J. Org. Chem.*, **57**, 6689 (1992).

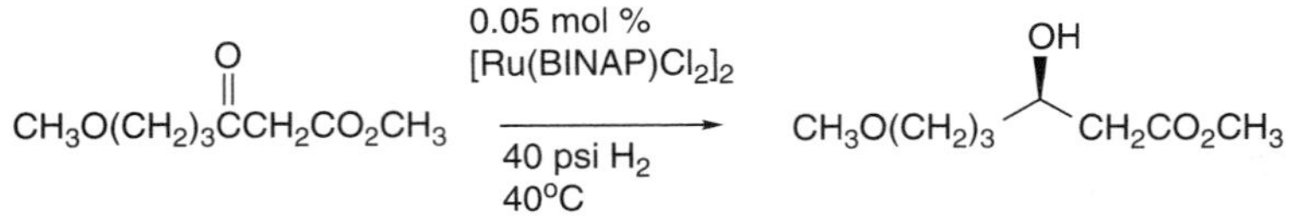

392

CHAPTER 5

*Reduction of
Carbon-Carbon Multiple
Bonds, Carbonyl
Groups, and Other
Functional Groups*

For reduction of monofunctional ketones, the most effective catalysts include diamine ligands. The diamine catalysts exhibit strong selectivity for carbonyl groups over carbon-carbon double and triple bonds. These catalysts have a preference for equatorial approach in the reduction of cyclohexanones and for steric approach control in the reduction of acyclic ketones.[51]

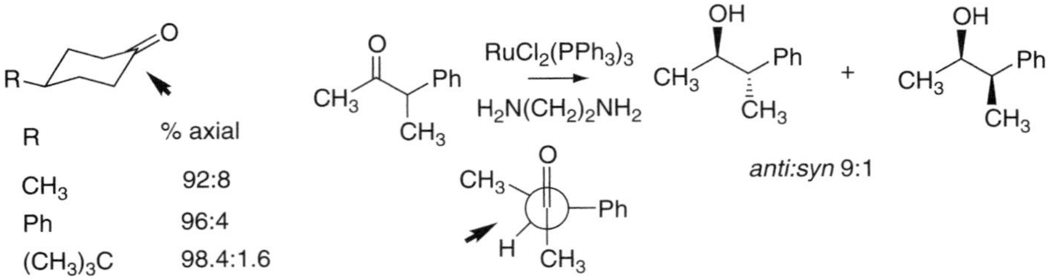

R	% axial
CH_3	92:8
Ph	96:4
$(CH_3)_3C$	98.4:1.6

anti:syn 9:1

Related catalysts include both a chiral BINAP-type phosphine and a chiral diamine ligand. A wide range of aryl ketones gave more than 95% enantioselectivity when substituted-1,1′-binaphthyl and ethylene diamines were used.[52]

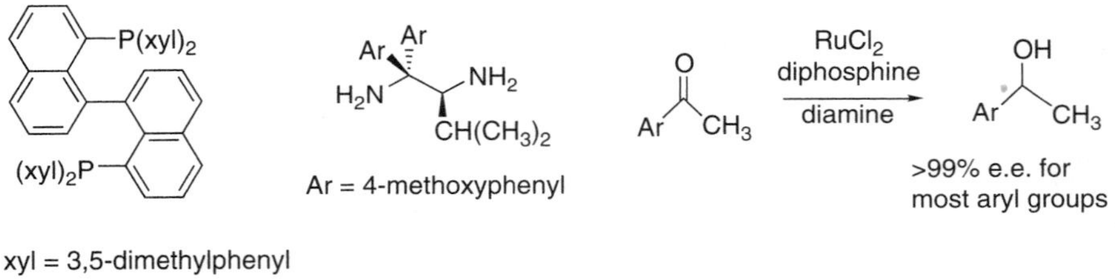

Ar = 4-methoxyphenyl

>99% e.e. for
most aryl groups

xyl = 3,5-dimethylphenyl

Cyclic and α,β-unsaturated ketones also gave high e.e. but straight-chain alkyl ketones did not.

The suggested catalytic cycle for the diamine catalysts indicates that the NH group of the diamine plays a direct role in the hydride transfer through a six-membered TS.[53] A feature of this mechanism is the absence of direct contact between the ketone and the metal. Rather, the reaction is pictured as a nucleophilic delivery of hydride from ruthenium, concerted with a proton transfer from nitrogen.

[51.] T. Ohkuma, H. Ooka, M. Yamakawa, T. Ikariya, and R. Noyori, *J. Org. Chem.*, **61**, 4872 (1996).

[52.] T. Ohkuma, M. Koizuma, H. Doucet, T. Pham, M. Kozawa, K. Murata, E. Katayama, T. Yokozawa, T. Ikariya, and R. Noyori, *J. Am. Chem. Soc.*, **120**, 13529 (1998).

[53.] C. A. Sandoval, T. Ohkuma, Z. Muniz, and R. Noyori, *J. Am. Chem. Soc.*, **125**, 13490 (2003).

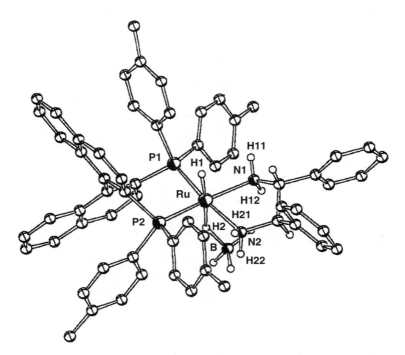

The catalyst used for these mechanistic studies has been characterized by X-ray crystallography, as shown in Figure 5.6. It is obtained as a hydrido ruthenium(II) species that is also coordinated by a $[BH_4]^-$ anion. The catalyst is prepared by exposing the DINAP-diamine $RuCl_2$ complex to excess $NaBH_4$.[54]

Fig. 5.6. Crystal structure of *tetrakis-P,P,P′P′-*(4-methylphenyl)-1,1′-binaphthyldiphosphine-1,2-diphenyl-1,2-ethanediamine ruthenium borohydride catalyst. Reproduced from *J. Am. Chem. Soc.*, **124**, 6508 (2002), by permission of the American Chemical Society.

[54] T. Ohkuma, M. Koizumi, K. Muniz, G. Hilt, C. Kabuto, and R. Noyori, *J. Am. Chem. Soc.*, **124**, 6508 (2002).

394

CHAPTER 5

*Reduction of
Carbon-Carbon Multiple
Bonds, Carbonyl
Groups, and Other
Functional Groups*

Several other versions of these catalysts have been developed. Arene complexes of monotosyl-1,2-diphenylethylenediamine ruthenium chloride give good results with α,β-ynones.[55] The active catalysts are generated by KOH. These catalysts also function by hydrogen transfer, with isopropanol serving as the hydrogen source. Entries 6 to 8 in Scheme 5.3 are examples.

Catalyst **D**: Arene = mesitylene
Catalyst **E**: Arene = *p*-cymene

Scheme 5.5 gives some examples of the application of these Ru(II)-diphosphine and diamine catalysts. Entries 1 and 2 are examples of the hydrogenation of β-dicarbonyl compounds with Ru(BINAP)Cl$_2$. Excellent enantioselectivity is observed, although elevated hydrogen pressure is required. Entry 3 proceeds in fair yield and enantioselectivity, and without reduction of the conjugated carbon-carbon double bond. Entry 4 uses the cymene complex catalyst **E** under hydrogen transfer conditions. Entry 5 involves tandem 1,4- and 1,2-reduction and was done under hydrogen transfer conditions, using formic acid as the hydride donor. Entries 6 to 8 show good yields and enantioselectivity for several alkynyl ketones of increasing structural complexity. In the latter two cases, only a single stereoisomer was observed.

Certain functional groups can be entirely removed and replaced by hydrogen, a reaction known as *hydrogenolysis*. For example, aromatic halogen substituents are frequently removed by hydrogenation over transition metal catalysts. Aliphatic halogens are somewhat less reactive but hydrogenolysis is promoted by base.[56] The most useful type of hydrogenolysis reaction involves removal of oxygen functional groups at benzylic and allylic positions.[57]

Hydrogenolysis of halides and benzylic groups presumably involves intermediates formed by *oxidative addition* to the active metal catalyst to generate intermediates similar to those involved in hydrogenation. The hydrogenolysis is completed by reductive elimination.[58] Many other examples of this pattern of reactivity are discussed in Chapter 8.

---

55. K. Matsumura, S. Hashiguchi, T. Ikariya, and R. Noyori, *J. Am. Chem. Soc.*, **119**, 8738 (1997).
56. A. R. Pinder, *Synthesis*, 425 (1980).
57. W. H. Hartung and R. Simonoff, *Org. React.*, **7**, 263 (1953); P. N. Rylander, *Catalytic Hydrogenation over Platinum Metals*, Academic Press, New York, 1967, Chap. 25; P. N. Rylander, *Catalytic Hydrogenation in Organic Synthesis*, Academic Press, New York, 1979, Chap. 15; P. N. Rylander, *Hydrogenation Methods*, Academic Press, Orlando, FL, 1985, Chap. 13.
58. The mechanism of benzylic hydrogenolysis has not been definitively established. For other possibilities, see R. B. Grossman, *The Art of Writing Reasonable Organic Mechanisms*, 2nd Edition, Springer, New York, 2003, pp. 309–310.

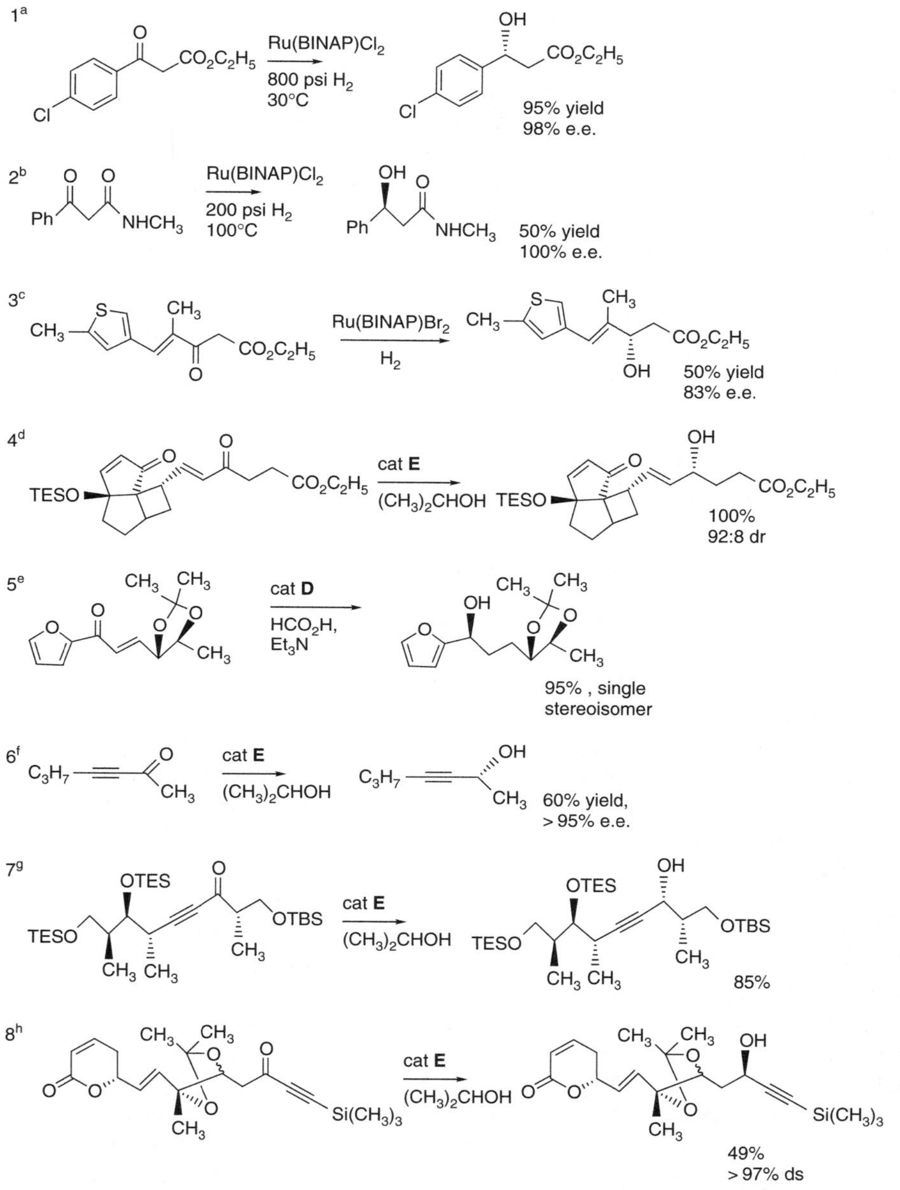

Scheme 5.5. Enantioselective Hydrogenation with Ruthenium Complex Catalysts

1[a]

$$Ru(BINAP)Cl_2$$, 800 psi H_2, 30°C → 95% yield, 98% e.e.

2[b]

$$Ru(BINAP)Cl_2$$, 200 psi H_2, 100°C → 50% yield, 100% e.e.

3[c]

$$Ru(BINAP)Br_2$$, H_2 → 50% yield, 83% e.e.

4[d]

cat **E**, $(CH_3)_2CHOH$ → 100%, 92:8 dr

5[e]

cat **D**, HCO_2H, Et_3N → 95% , single stereoisomer

6[f]

cat **E**, $(CH_3)_2CHOH$ → 60% yield, > 95% e.e.

7[g]

cat **E**, $(CH_3)_2CHOH$ → 85%

8[h]

cat **E**, $(CH_3)_2CHOH$ → 49%, > 97% ds

a. V. V. Thakur, M. D. Nikalje, and A. Sudalai, *Tetrahedron: Asymmetry*, **14**, 581 (2003).

b. H.-L. Huang, L. T. Liu, S.-F. Chen, and H. Ku, *Tetrahedron:Asymmetry*, **9**, 1637 (1998).

c. E. A. Reiff, S. K. Nair, B. S. N. Reddy, J. Inagaki, J. T. Henri, J. F. Greiner, and G. I. Georg, *Tetrahedron Lett.*, **45**, 5845 (2004).

d. H. Ito, M. Hasegawa, Y. Takenaka, T. Kobayashi, and K. Iguchi, *J. Am. Chem. Soc.*, **126**, 4520 (2004).

e. M. Li and G. O'Doherty, *Tetrahedron Lett.*, **45**, 6407 (2004).

f. N. Petry, A. Parenty, and J.-M. Campagne, *Tetrahedron: Asymmetry*, **15**, 1199 (2004).

g. J. A. Marshall and M. P. Bourbeau, *Org. Lett.*, **5**, 3197 (2003).

h. K. Fujii, K. Maki, M. Kanai, and M. Shibasaki, *Org. Lett.*, **5**, 733 (2003).

396

CHAPTER 5

*Reduction of
Carbon-Carbon Multiple
Bonds, Carbonyl
Groups, and Other
Functional Groups*

$$Pd^0 \;+\; H_2 \;\longrightarrow\; Pd^{II}H_2 \;\rightleftharpoons\; [Pd^{II}H]^- \;+\; H^+$$

$$PhCH_2OR \;+\; [Pd^{II}H]^- \;\longrightarrow\; [PhCH_2Pd^{II}H] \;+\; {}^-OR$$

$$[PhCH_2Pd^{II}H]^- \;\longrightarrow\; PhCH_3 \;+\; Pd^0$$

The facile cleavage of the benzyl-oxygen bond has made the benzyl group a useful protecting group in multistep syntheses. A particularly important example is the use of the carbobenzyloxy group in peptide synthesis. The protecting group is removed by hydrogenolysis. The substituted carbamic acid generated by the hydrogenolysis decarboxylates spontaneously to provide the amine (see Section 3.5.2).

$$\underset{\textstyle PhCH_2OCNHR}{\overset{\textstyle O}{\overset{\textstyle \|}{}}} \;\longrightarrow\; PhCH_3 \;+\; \underset{\textstyle HOCNHR}{\overset{\textstyle O}{\overset{\textstyle \|}{}}} \;\longrightarrow\; CO_2 \;+\; H_2NR$$

5.3. Group III Hydride-Donor Reagents

5.3.1. Comparative Reactivity of Common Hydride Donor Reagents

Most reductions of carbonyl compounds are done with reagents that transfer a hydride from boron or aluminum. The various reagents of this type that are available provide a considerable degree of chemo- and stereoselectivity. Sodium borohydride and lithium aluminum hydride are the most widely used of these reagents. Sodium borohydride is a mild reducing agent that reacts rapidly with aldehydes and ketones but only slowly with esters. It is moderately stable in hydroxylic solvents and can be used in water or alcoholic solutions. Lithium aluminum hydride is a much more powerful hydride donor, and it rapidly reduces esters, acids, nitriles, and amides, as well as aldehydes and ketones. Lithium aluminum hydride is strongly basic and reacts very rapidly (*violently*) with water or alcohols to release hydrogen. It must be used in anhydrous solvents, usually ether or tetrahydrofuran. The difference in the reactivity of these two compounds is due to properties of both the cations and the anions. Lithium is a stronger Lewis acid than sodium and AlH_4^- is a more reactive hydride donor than BH_4^-. Neither sodium borohydride nor lithium aluminum hydride reacts with isolated carbon-carbon double bonds. The reactivity of these reagents and some related reducing reagents is summarized in Table 5.3.

The mechanism by which the Group III hydrides effect reduction involves activation of the carbonyl group by coordination with a metal cation and nucleophilic transfer of hydride to the carbonyl group. Hydroxylic solvents also participate in the reaction,[59] and as reduction proceeds and hydride is transferred, the Lewis acid character of boron and aluminum becomes a factor.

59. D. C. Wigfield and R. W. Gowland, *J. Org. Chem.*, **42**, 1108 (1977).

Table 5.3. Reactivity of Hydride-Donor Reducing Agents

	Reactant					
	Iminium ion	Acyl chloride	Aldehyde or ketone	Ester	Amide	Carboxylate salt
	Most reactive ——————————————→ Least reactive					
Hydride donor			Product[a]			
LiAlH$_4$[b]	Amine	Alcohol	Alcohol	Alcohol	Amine	Alcohol
Red-Al[c]		Alcohol	Alcohol	Alcohol	Amine	Alcohol
LiAlH(OtBu)$_3$[d]		Aldehyde[e]	Alcohol	Alcohol	Aldehyde[f]	
NaBH$_4$[b]	Amine		Alcohol	Alcohol[f]		
NaBH$_3$CN[g]	Amine					
B$_2$H$_6$[h]			Alcohol		Amine	Alcohol[i]
AlH$_3$[j]		Alcohol	Alcohol	Alcohol	Amine	Alcohol
Disiamylborane[k]			Alcohol		Aldehyde[e]	
DIBAlH			Alcohol	Aldehyde[e]	Aldehyde[e]	Alcohol

a. Products shown are the usual products of synthetic operations. Where no entry is given, the combination has not been studied or is not of major synthetic utility.
b. J. Seyden-Penne, *Reductions by the Alumino- and Borohydrides in Organic Synthesis*, VCH Publishers, New York, 1991.
c. J. Malek, *Org. React.*, **34**, 1 (1985); **36**, 249 (1989).
d. H. C. Brown and R. F. McFarlin, *J. Am. Chem. Soc.*, **78**, 752 (1956); **80**, 5372 (1958); H. C. Brown and B. C. Subba Rao, *J. Am. Chem. Soc.*, **80**, 5377 (1958); H. C. Brown and A. Tsukamoto, *J. Am. Chem. Soc.*, **86**, 1089 (1964).
e. Reaction must be controlled by use of a stoichiometric amount of reagent and low temperature.
f. Reaction occurs slowly.
g. C. F. Lane, *Synthesis*, 135 (1975).
h. H. C. Brown, P. Heim, and N. M. Yoon, *J. Am. Chem. Soc.*, **92**, 1637 (1970); N. M Yoon, C. S. Park, H. C. Brown, S. Krishnamurthy, and T. P. Stocky, *J. Org. Chem.*, **38**, 2786 (1973); H. C. Brown and P. Heim, *J. Org. Chem.*, **38**, 912 (1973).
i. Reaction occurs through an acyloxyborane.
j. H. C. Brown and N. M. Yoon, *J. Am. Chem. Soc.*, **88**, 1464 (1966).
k. H. C. Brown, D. B. Bigley, S. K. Arora, and N. M. Yoon, *J. Am. Chem. Soc.*, **92**, 7161 (1970); H. C. Brown and V. Varma, *J. Org. Chem.*, **39**, 1631 (1974).
l. E. Winterfeldt, *Synthesis*, 617 (1975); H. Reinheckel, K. Haage, and D. Jahnke, *Organomet. Chem. Res.*, **4**, 47 (1969); N. M. Yoon and Y. S. Gyoung, *J. Org. Chem.*, **50**, 2443 (1985).

As all four of the hydrides can eventually be transferred, there are actually several distinct reducing agents functioning during the course of the reaction.[60] Although this somewhat complicates interpretation of rates and stereoselectivity, it does not detract from the synthetic utility of these reagents. Reduction with NaBH$_4$ is usually done in aqueous or alcoholic solution and the alkoxyboranes formed as intermediates are rapidly solvolyzed.

$$BH_4^- \quad + \quad R_2CO \longrightarrow R_2CHO\bar{B}H_3$$

$$R_2CHO\bar{B}H_3 \quad + \quad R_2CO \longrightarrow [R_2CHO]_2\bar{B}H_2$$

$$[R_2CHO]_2\bar{B}H_2 \quad + \quad R_2CO \longrightarrow [R_2CHO]_3\bar{B}H$$

$$[R_2CHO]_3\bar{B}H \quad + \quad R_2CO \longrightarrow [R_2CHO]_4\bar{B}$$

$$[R_2CHO]_4\bar{B} \quad + \quad 4\,SOH \longrightarrow 4\,R_2CHOH + \bar{B}(OS)_4$$

The mechanism for reduction by LiAlH$_4$ is very similar. However, since LiAlH$_4$ reacts very rapidly with protic solvents to form molecular hydrogen, reductions with this reagent must be carried out in aprotic solvents, usually ether or tetrahydrofuran.

60. B. Rickborn and M. T. Wuesthoff, *J. Am. Chem. Soc.*, **92**, 6894 (1970).

398

CHAPTER 5

*Reduction of
Carbon-Carbon Multiple
Bonds, Carbonyl
Groups, and Other
Functional Groups*

The products are liberated by hydrolysis of the aluminum alkoxide at the end of the reaction. Lithium aluminum hydride reduction of esters to alcohols involves an elimination step in addition to hydride transfers.

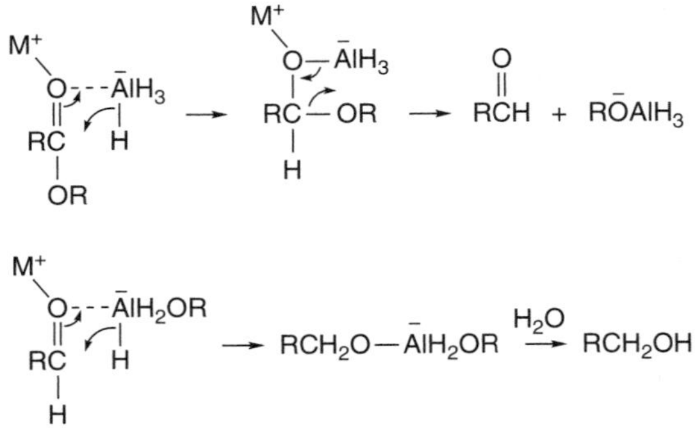

Amides are reduced to amines because the nitrogen is a poorer leaving group than oxygen at the intermediate stage of the reduction. Primary and secondary amides are rapidly deprotonated by the strongly basic LiAlH₄, so the addition step involves the conjugate base.

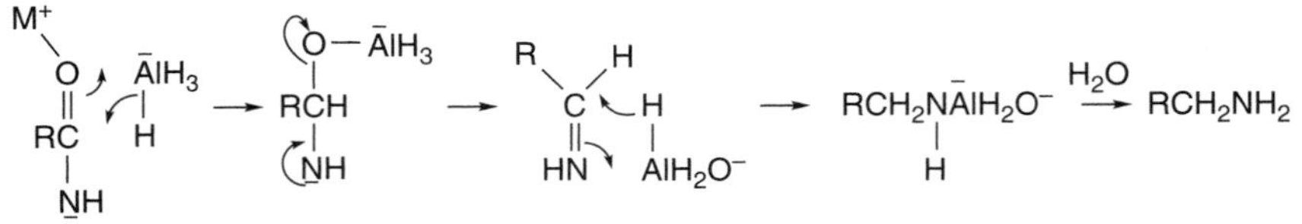

Reduction of amides by LiAlH₄ is an important method for the synthesis of amines.

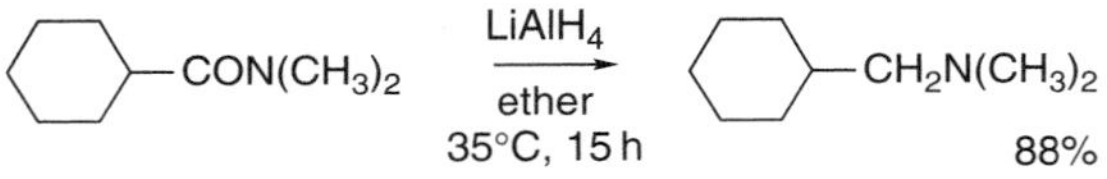

Ref. 61

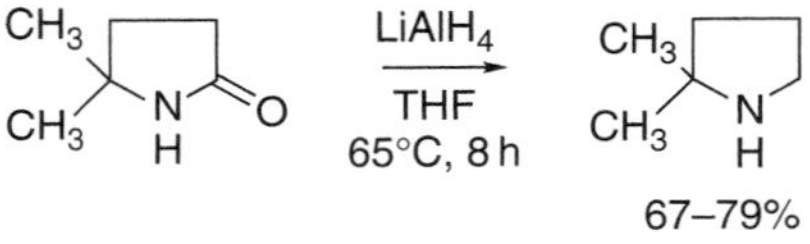

Ref. 62

Several factors affect the reactivity of the boron and aluminum hydrides, including the metal cation present and the ligands, in addition to hydride, in the complex hydride. Some of these effects can be illustrated by considering the reactivity of ketones and aldehydes toward various hydride transfer reagents. Comparison of LiAlH₄ and NaAlH₄ has shown the former to be more reactive,[63] which is attributed to the greater

61. A. C. Cope and E. Ciganek, *Org. Synth.*, **IV**, 339 (1963).
62. R. B. Moffett, *Org. Synth.*, **IV**, 354 (1963).
63. E. C. Ashby and J. R. Boone, *J. Am. Chem. Soc.*, **98**, 5524 (1976); J. S. Cha and H. C. Brown, *J. Org. Chem.*, **58**, 4727 (1993).

Lewis acid strength and hardness of the lithium cation. Both $LiBH_4$ and $Ca(BH_4)_2$ are more reactive than sodium borohydride. This enhanced reactivity is due to the greater Lewis acid strength of Li^+ and Ca^{2+}, compared with Na^+. Both of these reagents can reduce esters and lactones efficiently.

70%

Ref. 64

45%

Ref. 65

Zinc borohydride, which is also a useful reagent,[66] is prepared by reaction of $ZnCl_2$ with $NaBH_4$ in THF. Owing to the stronger Lewis acid character of Zn^{2+}, $Zn(BH_4)_2$ is more reactive than $NaBH_4$ toward esters and amides and reduces them to alcohols and amines, respectively.[67] $Zn(BH_4)_2$ reduces carboxylic acids to primary alcohols.[68] The reagent also smoothly reduces α-aminoacids to β-aminoalcohols.[69]

87%

Sodium borohydride is sometimes used in conjunction with $CeCl_3$ (*Luche's reagent*).[70] The active reductants under these conditions are thought to be alkoxyborohydrides. Sodium cyanoborohydride is a useful derivative of sodium borohydride.[71] The electron-attracting cyano substituent reduces reactivity and only iminium groups are rapidly reduced by this reagent.

Alkylborohydrides are also used as reducing agents. These compounds have greater steric demands than the borohydride ion and therefore are more stereoselective in situations in which steric factors come into play.[72] These compounds are prepared by reaction of trialkylboranes with lithium, sodium, or potassium hydride.[73] Several of the compounds are available commercially under the trade name Selectrides®.[74]

[64] H. C. Brown, S. Narasimhan, and Y. M. Choi, *J. Org. Chem.*

[65] K. Soai and S. Ookawa, *J. Org. Chem.*, **51**, 4000 (1986).

[66] S. Narasimhan and R. Balakumar, *Aldrichimica Acta*, **31**, 19 (1998).

[67] S. Narasimhan, S. Madhavan, R. Balakumar, and S. Swamalakshmi, *Synth. Commun.*, **27**, 391 (1997).

[68] S. Narasimhan, S. Madhavan, and K. G. Prasad, *J. Org. Chem.*, **60**, 5314 (1995); B. C. Ranue and A. R. Das, *J. Chem. Soc., Perkin Trans. 1*, 1561 (1992).

[69] S. Narasimhan, S. Madhavan, and K. G. Prasad, *Synth. Commun.*, **26**, 703 (1996).

[70] A. C. Gemal and J.-L. Luche, *J. Am. Chem. Soc.*, **103**, 5454 (1981).

[71] C. F. Lane, *Synthesis*, 135 (1975).

[72] H. C. Brown and S. Krishnamurthy, *J. Am. Chem. Soc.*, **94**, 7159 (1972); S. Krishnamurthy and H. C. Brown, *J. Am. Chem. Soc.*, **98**, 3383 (1976).

[73] H. C. Brown, S. Krishnamurthy, and J. L. Hubbard, *J. Am. Chem. Soc.*, **100**, 3343 (1978).

[74] Selectride is a trade name of the Aldrich Chemical Company.

400

CHAPTER 5

*Reduction of
Carbon-Carbon Multiple
Bonds, Carbonyl
Groups, and Other
Functional Groups*

$$\text{Li}^+\text{H}\bar{\text{B}}(-\underset{\underset{\text{CH}_3}{|}}{\text{CHCH}_2\text{CH}_3})_3 \quad \text{Li}^+\text{H}\bar{\text{B}}[-\underset{\underset{\text{CH}_3}{|}}{\text{CHCH}(\text{CH}_3)_2}]_3 \quad \text{Na}^+\text{H}\bar{\text{B}}(-\underset{\underset{\text{CH}_3}{|}}{\text{CHCH}_2\text{CH}_3})_3 \quad \text{K}^+\text{H}\bar{\text{B}}(-\underset{\underset{\text{CH}_3}{|}}{\text{CHCH}_2\text{CH}_3})_3$$

L-selectride   LS-selectride   N-selectride   K-selectride

Derivatives of aluminum hydrides in which one or more of the hydrides is replaced by an alkoxide ion can be prepared by addition of the calculated amount of the appropriate alcohol.

$$\text{LiAlH}_4 + 2\,\text{ROH} \longrightarrow \text{LiAlH}_2(\text{OR})_2 + 2\,\text{H}_2$$

$$\text{LiAlH}_4 + 3\,\text{ROH} \longrightarrow \text{LiAlH}(\text{OR})_3 + 3\,\text{H}_2$$

These reagents generally show increased solubility in organic solvents, particularly at low temperatures, and are useful in certain selective reductions.[75] Lithium tri-*t*-butoxyaluminum hydride and sodium *bis*-(2-methoxyethoxy)aluminum hydride (Red-Al)[76] are examples of these types of reagents that have synthetic use. Their reactivity toward carbonyl groups is summarized in Table 5.3.

Closely related to, but distinct from, the anionic boron and aluminum hydrides are the neutral boron (borane, BH_3) and aluminum (alane, AlH_3) hydrides. These molecules also contain hydrogen that can be transferred as hydride. Borane and alane differ from the anionic hydrides in being electrophilic species by virtue of the vacant *p* orbital and are Lewis acids. Reduction by these molecules occurs by an intramolecular hydride transfer in a Lewis acid-base complex of the reactant and reductant.

$$R_2MH + \underset{R}{\overset{O}{\underset{|}{\overset{\|}{\text{C}}}}}\overset{R}{} \longrightarrow \underset{R}{\overset{\overset{+}{\text{O}}\cdots\bar{\text{M}}R_2}{\underset{|}{\overset{\|}{\text{C}}}}}\overset{H}{R} \longrightarrow R-\underset{\underset{R}{|}}{\overset{O-MR_2}{\underset{|}{\text{C}}}}-H$$

Alkyl derivatives of boron and alane can function as reducing reagents in a similar fashion. Two reagents of this type, disiamylborane and diisobutylaluminum hydride (DiBAlH) are included in Table 5.3. The latter is an especially useful reagent.

Diborane also has a useful pattern of selectivity. It reduces carboxylic acids to primary alcohols under mild conditions that leave esters unchanged.[77] Nitro and cyano groups are relatively unreactive toward diborane. The rapid reaction between carboxylic acids and diborane is the result of formation of a triacyloxyborane intermediate by protonolysis of the B—H bonds. The resulting compound is essentially a mixed anhydride of the carboxylic acid and boric acid in which the carbonyl groups have enhanced reactivity toward borane or acetoxyborane.

$$3\,\text{RCO}_2\text{H} + \text{BH}_3 \longrightarrow (\text{RCO}_2)_3\text{B} + 3\,\text{H}_2$$

$$\underset{}{\overset{O}{\underset{}{\overset{\|}{\text{RC}}}}}-\text{O}-\text{B}(\text{O}_2\text{CR})_2 \longleftrightarrow \underset{}{\overset{O}{\underset{}{\overset{\|}{\text{RC}}}}}-\overset{+}{\text{O}}=\bar{\text{B}}(\text{O}_2\text{CR})_2$$

Diborane also reduces amides to amines (see Section 5.3.1.2).

[75.] J. Malek and M. Cerny, *Synthesis*, 217 (1972); J. Malek, *Org. React.*, **34**, 1 (1985).

[76.] Red-Al is a trademark of the Aldrich Chemical Company.

[77.] N. M. Yoon, C. S. Pak, H. C. Brown, S. Krishnamurthy, and T. P. Stocky, *J. Org. Chem.*, **38**, 2786 (1973).

In synthesis, the principal factors that affect the choice of a reducing agent are selectivity among functional groups (chemoselectivity) and stereoselectivity. Chemoselectivity can involve two issues. One may wish to effect a *partial reduction* of a particular functional group or it may be necessary to *reduce one group in preference to another*.[78] In the sections that follow, we consider some synthetically useful partial and selective reductions.

5.3.1.1. Partial Reduction of Carboxylic Acid Derivatives. One of the more difficult partial reductions is the conversion of a carboxylic acid derivative to an aldehyde without overreduction to the alcohol. Aldehydes are inherently more reactive than acids or esters, so the challenge is to stop the reduction at the aldehyde stage. Several approaches have been used to achieve this objective. One is to replace some of the hydrogens in the hydride with more bulky groups, thus modifying reactivity by steric factors. Lithium tri-*t*-butoxyaluminum hydride is an example of this approach.[79] Sodium tri-*t*-butoxyaluminum hydride can be used to reduce acid chlorides to aldehydes without overreduction to the alcohol.[80] The excellent solubility of sodium *bis*-(2-methoxyethoxy)aluminum hydride (Red-Al) makes it a useful reagent for selective reductions. The reagent is soluble in toluene even at −70° C, and selectivity is enhanced by the low temperature. It is possible to reduce esters to aldehydes and lactones to lactols with this reagent.

Ref. 81

Ref. 82

The most widely used reagent for partial reduction of esters and lactones at the present time is diisobutylaluminum hydride (DiBAlH).[83] By use of a controlled amount of the reagent at low temperature, partial reduction can be reliably achieved. The selectivity results from the relative stability of the hemiacetal intermediate that is formed. The aldehyde is not liberated until the hydrolytic workup and is therefore not

78. For more complete discussion of functional group selectivity of hydride reducing agents, see E. R. H. Walter, *Chem. Soc. Rev.*, **5**, 23 (1976).
79. H. C. Brown and B. C. Subba Rao, *J. Am. Chem. Soc.*, **80**, 5377 (1958).
80. J. S. Cha and H. C. Brown, *J. Org. Chem.*, **58**, 4732 (1993).
81. R. Kanazawa and T. Tokoroyama, *Synthesis*, 526 (1976).
82. H. Disselnkoetter, F. Lieb, H. Oediger, and D. Wendisch, *Liebigs Ann. Chem.*, 150 (1982).
83. F. Winterfeldt, *Synthesis*, 617 (1975); N. M. Yoon and Y. G. Gyoung, *J. Org. Chem.*, **50**, 2443 (1985).

402

CHAPTER 5

*Reduction of
Carbon-Carbon Multiple
Bonds, Carbonyl
Groups, and Other
Functional Groups*

subject to overreduction. At higher temperatures, where the intermediate undergoes elimination, diisobutylaluminum hydride reduces esters to primary alcohols.

$$\text{(i-Bu)}_2\text{AlH, toluene} \xrightarrow{-60°C} \quad 83\%$$

Ref. 84

$$\xrightarrow[-90°C]{\text{(i-Bu)}_2\text{AlH}}$$

Ref. 85

$$\xrightarrow[\substack{1)\text{(i-Bu)}_2\text{AlH,}\\ \text{hexane}\\ -78°C \\ 2)\ \text{H}_2\text{O, tartaric acid}}]{} \quad 80\%$$

Ref. 86

Selective reduction to aldehydes can also be achieved using N-methoxy-N-methylamides.[87] $LiAlH_4$ and DiBAlH have both been used as the hydride donor. The partial reduction is again the result of the stability of the initial reduction product. The N-methoxy substituent leads to a chelated structure that is stable until acid hydrolysis occurs during workup.

$$\text{RCNOCH}_3 + \text{M—H} \longrightarrow \xrightarrow[\text{H}_2\text{O}]{\text{H}^+} \text{RCH}=\text{O}$$

Another useful approach to aldehydes is by partial reduction of nitriles to imines. The reduction stops at the imine stage because of the low electrophilicity of the deprotonated imine intermediate. The imines are then hydrolyzed to the aldehyde. Diisobutylaluminum hydride seems to be the best reagent for this purpose.[88,89]

$$\text{CH}_3\text{CH}=\text{CHCH}_2\text{CH}_2\text{CH}_2\text{C}\equiv\text{N} \xrightarrow[2)\text{H}^+,\ \text{H}_2\text{O}]{1)\ \text{(i-Bu)}_2\text{AlH}} \text{CH}_3\text{CH}=\text{CHCH}_2\text{CH}_2\text{CH}_2\text{CH}=\text{O}$$

$$64\%$$

84. C. Szantay, L. Toke, and P. Kolonits, *J. Org. Chem.*, **31**, 1447 (1966).
85. G. E. Keck, E. P. Boden, and M. R. Wiley, *J. Org. Chem.*, **54**, 896 (1989).
86. P. Baeckstrom, L. Li, M. Wickramaratne, and T. Norin, *Synth. Commun.*, **20**, 423 (1990).
87. S. Nahm and S. M. Weinreb, *Tetrahedron Lett.*, **22**, 3815 (1981).
88. N. A. LeBel, M. E. Post, and J. J. Wang, *J. Am. Chem. Soc.*, **86**, 3759 (1964).
89. R. V. Stevens and J. T. Lai, *J. Org. Chem.*, **37**, 2138 (1972); S. Trofimenko, *J. Org. Chem.*, **29**, 3046 (1964).

This method can be used in conjunction with addition of cyanide to prepare α-hydroxy aldehydes from ketones.[90]

$$CH_3(CH_2)_4\overset{\overset{\displaystyle O}{\|}}{C}(CH_2)_4CH_3 \quad \xrightarrow[\text{2) (\textit{i}-Bu)}_2\text{AlH}]{\overset{\text{1) TMS-CN}}{\text{ZnI}_2}} \quad \xrightarrow[\text{2) HCl, H}_2\text{O}]{\text{1) NH}_4\text{Cl}} \quad \underset{\underset{79\ \%}{CH_3(CH_2)_4} \diagdown \overset{}{\diagup} (CH_2)_4CH_3}{\overset{HO}{\diagdown}\overset{}{\diagup}\overset{CH=O}{}}$$

5.3.1.2. Reduction of Imines and Amides to Amines. A second type of chemoselectivity arises in the context of the need to reduce one functional group in the presence of another. If the group to be reduced is more reactive than the one to be left unchanged, it is simply a matter of choosing a reducing reagent with the appropriate level of reactivity. Sodium borohydride, for example, is very useful in this respect since it reduces ketones and aldehydes much more rapidly than esters. Sodium cyanoborohydride is used to reduce imines to amines, but this reagent is only reactive toward iminium ions. At pH 6–7, $NaBH_3CN$ is essentially unreactive toward carbonyl groups. When an amine and ketone are mixed together, equilibrium is established with the imine. At mildly acidic pH only the protonated imine is reactive toward $NaBH_3CN$.[91] This process is called *reductive amination*.

$$R_2C{=}O + R'NH_2 + H^+ \rightleftharpoons R_2C{=}\overset{H}{\underset{+}{N}}R'$$

$$R_2C{=}\overset{H}{\underset{+}{N}}R' + BH_3CN^- \longrightarrow R_2CHNHR'$$

Reductive amination by $NaBH_3CN$ can also be carried out in the presence of $Ti(O\text{-}i\text{-}Pr)_4$. These conditions are especially useful for situations in which it is not practical to use the amine in excess (as is typically done under the acid-catalyzed conditions) or for acid-sensitive compounds. The $Ti(O\text{-}i\text{-}Pr)_4$ may act as a Lewis acid in generation of a tetrahedral adduct, which then may be reduced directly or via a transient iminium intermediate.[92]

$$R_2C{=}O + HNR'_2 \xrightarrow{Ti(O\text{-}i\text{-}Pr)_4} R_2\overset{\overset{\displaystyle OTi(O\text{-}i\text{-}Pr)_3}{|}}{C}{-}NR'_2 \rightleftharpoons R_2C{=}N^+R'_2 \xrightarrow{NaBH_3CN} R_2CHNR'_2$$

Sodium triacetoxyborohydride is an alternative to $NaBH_3CN$ for reductive amination. This reagent can be used with a wide variety of aldehydes or ketones with primary and secondary amines, including aniline derivatives.[93] This reagent has been used successfully to alkylate amino acid esters.[94]

90. M. Hayashi, T. Yoshiga, and N. Oguni, *Synlett*, 479 (1991).
91. R. F. Borch, M. D. Bernstein, and H. D. Durst, *J. Am. Chem. Soc.*, **93**, 2897 (1971).
92. R. J. Mattson, K. M. Pham, D. J. Leuck, and K. A. Cowen, *J. Org. Chem.*, **55**, 2552 (1990).
93. A. F. Abdel-Magid, K. G. Carson, B. H. Harris, C. A. Maryanoff, and R. D. Shah, *J. Org. Chem.*, **61**, 3849 (1996).
94. J. M. Ramanjulu and M. M. Joullie, *Synth. Commun.*, **26**, 1379 (1996).

404

CHAPTER 5

Reduction of
Carbon-Carbon Multiple
Bonds, Carbonyl
Groups, and Other
Functional Groups

$$R_2C{=}O + HNR'_2 \xrightarrow{\text{NaBH(OAc)}_3} R_2CHNR'_2$$

$$PhCH_2CHCO_2CH_3 + CH_3(CH_2)_4CH{=}O \xrightarrow{\text{NaBH(OAc)}_3} PhCH_2CHCO_2CH_3$$

with $NH_3^+Cl^-$ and $NH(CH_2)_5CH_3$ 79%

This method was used in a large-scale synthesis of 1-benzyl-3-methylamino-4-methylpiperidine.[95]

1) CH_3NH_2
2) $NaBH_4$
CH_3CO_2H

92% yield
on 35 kg scale
86:14 *cis:trans*

Zinc borohydride has been found to effect very efficient reductive amination in the presence of silica. The amine and carbonyl compound are mixed with silica and the powder is then treated with a solution of $Zn(BH_4)_2$. Excellent yields are also obtained for unsaturated aldehydes and ketones.[96]

1) SiO_2
2) $Zn(BH_4)_2$

80%

Aromatic aldehydes can be reductively aminated with the combination $Zn(BH_4)_2$-$ZnCl_2$,[97] and the $ZnCl_2$ assists in imine formation.

1) $ZnCl_2$
2) $Zn(BH_4)_2$

77%

Amides are usually reduced to amines using $LiAlH_4$. Amides require vigorous reaction conditions for reduction by $LiAlH_4$, so that little selectivity can be achieved with this reagent. Diborane is also a useful reagent for reducing amides. Tertiary and secondary amides are easily reduced, but primary amides react only slowly.[98] The electrophilicity of borane is involved in the reduction of amides. The boron complexes at the carbonyl oxygen, enhancing the reactivity of the carbonyl center.

[95]. D. H. B. Ripin, S. Abele, W. Cai, T. Blumenkopf, J. M. Casavant, J. L. Doty, M. Flanagan, C. Koecher, K. W. Laue, K. McCarthy, C. Meltz, M. Munchoff, M. Pouwer, B. Shah, J. Sun, J. Teixera, T. Vries, D. A. Whipple, and G. Wilcox, *Org. Proc. Res. Dev.*, **7**, 115 (2003).

[96]. B. C. Ranu, A. Majee, and A. Sarkar, *J. Org. Chem.*, **63**, 370 (1998).

[97]. S. Bhattacharyya, A. Chatterjee, and J. S. Williamson, *Synth. Commun.*, **27**, 4265 (1997).

[98]. H. C. Brown and P. Heim, *J. Org. Chem.*, **38**, 912 (1973).

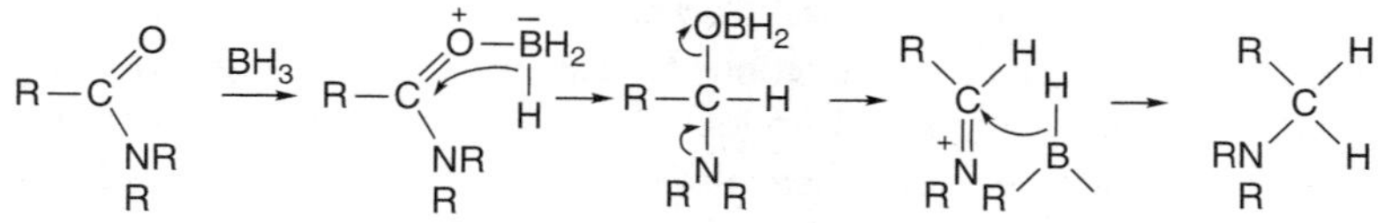

Diborane permits the selective reduction of amides in the presence of ester and nitro groups.

Alane is also a useful group for reducing amides and it, too, can be used to reduce amides to amines in the presence of ester groups.

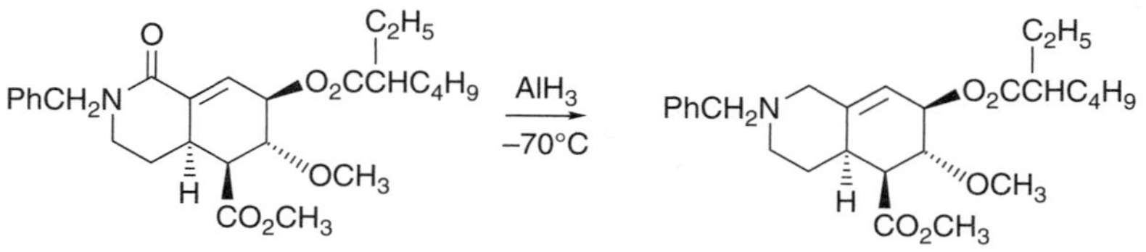

Ref. 99

The electrophilicity of alane is the basis for its selective reaction with the amide group. Alane is also useful for reducing azetidinones to azetidines. Most nucleophilic hydride reducing agents lead to ring-opened products. DiBAlH, AlH_2Cl, and $AlHCl_2$ can also reduce azetinones to azetidines.[100]

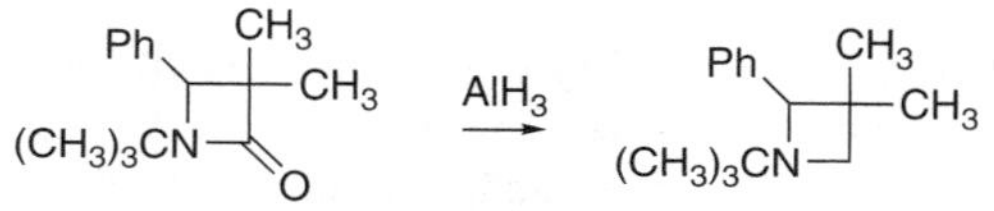

Ref. 101

Another approach to reduction of an amide group in the presence of other groups that are more easily reduced is to convert the amide to a more reactive species. One such method is conversion of the amide to an O-alkyl derivative with a positive charge on nitrogen.[102] This method has proven successful for tertiary and secondary, but not primary, amides.

$$RC(=O)NR_2 + Et_3O^+ \longrightarrow RC(OEt)=\overset{+}{N}R_2$$

$$RC(OEt)=\overset{+}{N}R_2 + NaBH_4 \longrightarrow RCH_2NR_2$$

Other compounds that can be readily derived from amides that are more reactive toward hydride reducing agents are α-alkylthioimmonium ions[103] and α-chloroimmonium ions.[104]

99. S. F. Martin, H. Rueger, S. A. Williamson, and S. Grzejszczak, *J. Am. Chem. Soc.*, **109**, 6124 (1987).
100. I. Ojima, M. Zhao, T. Yamamoto, K. Nakanishi, M. Yamashita, and R. Abe, *J. Org. Chem.*, **56**, 5263 (1991).
101. M. B. Jackson, L. N. Mander, and T. M. Spotswood, *Aust. J. Chem.*, **36**, 779 (1983).
102. R. F. Borch, *Tetrahedron Lett.*, 61 (1968).
103. S. Raucher and P. Klein, *Tetrahedron Lett.*, 4061 (1980); R. J. Sundberg, C. P. Walters, and J. D. Bloom, *J. Org. Chem.*, **46**, 3730 (1981).
104. M. E. Kuehne and P. J. Shannon, *J. Org. Chem.*, **42**, 2082 (1972).

406

CHAPTER 5

*Reduction of
Carbon-Carbon Multiple
Bonds, Carbonyl
Groups, and Other
Functional Groups*

5.3.1.3. Reduction of α,β-Unsaturated Carbonyl Compounds. An important case of chemoselectivity arises in the reduction of α,β-unsaturated carbonyl compounds. Reaction can occur at the carbonyl group, giving an allylic alcohol or at the double bond giving a saturated ketone. These alternative reaction modes are called 1,2- and 1,4-reduction, respectively. If hydride is added at the carbonyl group, the allylic alcohol is usually not susceptible to further reduction. If a hydride is added at the β-position, the initial product is an enolate. In protic solvents this leads to the ketone, which can be reduced to the saturated alcohol. Both $NaBH_4$ and $LiAlH_4$ have been observed to give both types of product, although the extent of reduction to saturated alcohol is usually greater with $NaBH_4$.[105]

1,2-reduction

$$R_2C=CHCR' + [H^-] \longrightarrow R_2C=CHCR' \xrightarrow{H^+} R_2C=CHCHR'$$

1,4-reduction leading to saturated alcohol

$$R_2C=CHCR' + [H^-] \longrightarrow R_2CH-CH=CR' \xrightarrow{H^+} R_2CHCH_2CR'$$

$$R_2CHCH_2CR' + [H^-] \longrightarrow R_2CHCH_2CR' \xrightarrow{H^+} R_2CHCH_2CHR'$$

Several reagents have been developed that lead to exclusive 1,2- or 1,4-reduction. Use of $NaBH_4$ in combination with cerium chloride (*Luche reagent*) results in clean 1,2-reduction.[106] DiBAlH[107] and the dialkylborane 9-BBN[108] also give exclusive carbonyl reduction. In each case the reactivity of the carbonyl group is enhanced by a Lewis acid complexation at oxygen.

Selective reduction of the carbon-carbon double bond can usually be achieved by catalytic hydrogenation. A series of reagents prepared from a hydride reducing agent and copper salts also gives primarily the saturated ketone.[109] Similar reagents have been shown to reduce α,β-unsaturated esters[110] and nitriles[111] to the corresponding saturated compounds. The mechanistic details are not known with certainty, but it is likely that "copper hydrides" are the active reducing agents and that they form an organocopper intermediate by conjugate addition.

105. M. R. Johnson and B. Rickborn, *J. Org. Chem.*, **35**, 1041 (1970); W. R. Jackson and A. Zurqiyah, *J. Chem. Soc.*, 5280 (1965).
106. J.-L. Luche, *J. Am. Chem. Soc.*, **100**, 2226 (1978); J.-L. Luche, L. Rodriguez-Hahn, and P. Crabbe, *J. Chem. Soc., Chem. Commun.*, 601 (1978).
107. K. E. Wilson, R. T. Seidner, and S. Masamune, *J. Chem. Soc., Chem. Commun.*, 213 (1970).
108. K. Krishnamurthy and H. C. Brown, *J. Org. Chem.*, **42**, 1197 (1977).
109. S. Masamune, G. S. Bates, and P. E. Georghiou, *J. Am. Chem. Soc.*, **96**, 3686 (1974); E. C. Ashby, J.-J. Lin, and R. Kovar, *J. Org. Chem.*, **41**, 1939 (1976); E. C. Ashby, J.-J. Lin, and A. B. Goel, *J. Org. Chem.*, **43**, 183 (1978); W. S. Mahoney, D. M. Brestensky, and J. M. Stryker, *J. Am. Chem. Soc.*, **110**, 291 (1988); D. M. Brestensky, D. E. Huseland, C. McGettigan, and J. M. Stryker, *Tetrahedron Lett.*, **29**, 3749 (1988); T. M. Koenig, J. F. Daeuble, D. M. Brestensky, and J. M. Stryker, *Tetrahedron Lett.*, **31**, 3237 (1990).
110. M. F. Semmelhack, R. D. Stauffer, and A. Yamashita, *J. Org. Chem.*, **42**, 3180 (1977).
111. M. E. Osborn, J. F. Pegues, and L. A. Paquette, *J. Org. Chem.*, **45**, 167 (1980).

$$\text{"H—Cu—H"} + RCH{=}CHCR \longrightarrow \overset{\underset{|}{\ \ }}{-}\overset{H}{\underset{|}{C}}u{-}\overset{R}{\underset{|}{C}}H{-}CH_2CR \longrightarrow RCH_2CH_2CR$$

Combined use of Co(acac)$_2$ and DiBAlH also gives selective reduction for α,β-unsaturated ketones, esters, and amides.[112] Another reagent combination that selectively reduces the carbon-carbon double bond is Wilkinson's catalyst and triethylsilane. The initial product is the enol silyl ether.[113]

$$(CH_3)_2C{=}CH(CH_2)_2\overset{CH_3}{\underset{|}{C}}{=}CHCH{=}O \ \xrightarrow[\text{(Ph}_3\text{P)}_3\text{RhCl}]{\text{Et}_3\text{SiH}} \ (CH_3)_2C{=}CH(CH_2)_2\overset{CH_3}{\underset{|}{C}}HCH{=}CHOSiEt_3$$

$$\downarrow H_2O$$

$$(CH_3)_2C{=}CH(CH_2)_2\overset{CH_3}{\underset{|}{C}}HCH_2CH{=}O$$

Unconjugated double bonds are unaffected by this reducing system.[114]

The enol ethers of β-dicarbonyl compounds are reduced to α,β-unsaturated ketones by LiAlH$_4$, followed by hydrolysis.[115] Reduction stops at the allylic alcohol, but subsequent acid hydrolysis of the enol ether and dehydration leads to the isolated product. This reaction is a useful method for synthesis of substituted cyclohexenones.

5.3.2. Stereoselectivity of Hydride Reduction

5.3.2.1. Cyclic Ketones. Stereoselectivity is a very important aspect of reductions by hydride transfer reagents. The stereoselectivity of the reduction of carbonyl groups is affected by the same combination of steric and stereoelectronic factors that control the addition of other nucleophiles, such as enolates and organometallic reagents to carbonyl groups. A general discussion of these factors is given in Section 2.4.1 of Part A. The stereochemistry of hydride reduction has been thoroughly studied with conformationally biased cyclohexanones. Some reagents give predominantly axial cyclohexanols, whereas others give the equatorial isomer. Axial alcohols are most likely to be formed when the reducing agent is a sterically hindered hydride donor because the equatorial direction of approach is more open and is preferred by bulky reagents. This is called *steric approach control*.[116]

112. T. Ikeno, T. Kimura, Y. Ohtsuka, and T. Yamada, *Synlett*, 96 (1999).
113. I. Ojima, T. Kogure, and Y. Nagai, *Tetrahedron Lett.*, 5035 (1972); I. Ojima, M. Nihonyanagi, T. Kogure, M. Kumagai, S. Horiuchi, K. Nakatsugawa, and Y. Nogai, *J. Organomet. Chem.*, **94**, 449 (1973).
114. H.-J. Liu and E. N. C. Browne, *Can. J. Chem.*, **59**, 601 (1981); T. Rosen and C. H. Heathcock, *J. Am. Chem. Soc.*, **107**, 3731 (1985).
115. H. E. Zimmerman and D. I. Schuster, *J. Am. Chem. Soc.*, **84**, 4527 (1962); W. F. Gannon and H. O. House, *Org. Synth.*, **40**, 14 (1960).
116. W. G. Dauben, G. J. Fonken, and D. S. Noyce, *J. Am. Chem. Soc.*, **78**, 2579 (1956).

408

CHAPTER 5

*Reduction of
Carbon-Carbon Multiple
Bonds, Carbonyl
Groups, and Other
Functional Groups*

Steric Approach Control

With less hindered hydride donors, particularly $NaBH_4$ and $LiAlH_4$, conformationally biased cyclohexanones give predominantly the equatorial alcohol, which is normally the more stable of the two isomers. However, hydride reductions are exothermic reactions with low activation energies. The TS should resemble starting ketone, so product stability should not control the stereoselectivity. A major factor in the preference for the equatorial isomer is the torsional strain that develops in the formation of the axial alcohol.[117]

Torsional strain increases as oxygen
passes through an eclipsed conformation

Oxygen moves away from equatorial
hydrogens; no torsional strain

An alternative interpretation is that the carbonyl group π-antibonding orbital, which acts as the LUMO in the reaction, has a greater density on the axial face.[118] At the present time the importance of such orbital effects is not entirely clear. Most of the stereoselectivities that have been reported can be reconciled with torsional and steric effects being dominant.[119]

A large amount of data has been accumulated on the stereoselectivity of reduction of cyclic ketones.[120] Table 5.4 compares the stereoselectivity of reduction of several ketones by hydride donors of increasing steric bulk. The trends in the table illustrate

117. M. Cherest, H. Felkin, and N. Prudent, *Tetrahedron Lett.*, 2205 (1968); M. Cherest and H. Felkin, *Tetrahedron Lett.*, 383 (1971).

118. J. Klein, *Tetrahedron Lett.*, 4307 (1973); N. T. Ahn, O. Eisenstein, J.-M. Lefour, and M. E. Tran Huu Dau, *J. Am. Chem. Soc.*, **95**, 6146 (1973).

119. W. T. Wipke and P. Gund, *J. Am. Chem. Soc.*, **98**, 8107 (1976); J.-C. Perlburger and P. Mueller, *J. Am. Chem. Soc.*, **99**, 6316 (1977); D. Mukherjee, Y.-D. Wu, F. R. Fronczek, and K. N. Houk, *J. Am. Chem. Soc.*, **110**, 3328 (1988).

120. D. C. Wigfield, *Tetrahedron*, **35**, 449 (1979); D. C. Wigfield and D. J. Phelps, *J. Org. Chem.*, **41**, 2396 (1976).

the increasing importance of steric approach control as both the hydride reagent and the ketone become more highly substituted. The alkyl borohydrides have especially high selectivity for the least hindered direction of approach.

When a ketone is relatively hindered, as, for example, in the bicyclo[2.2.1]heptan-2-one system, steric approach control governs stereoselectivity even for small hydride donors.

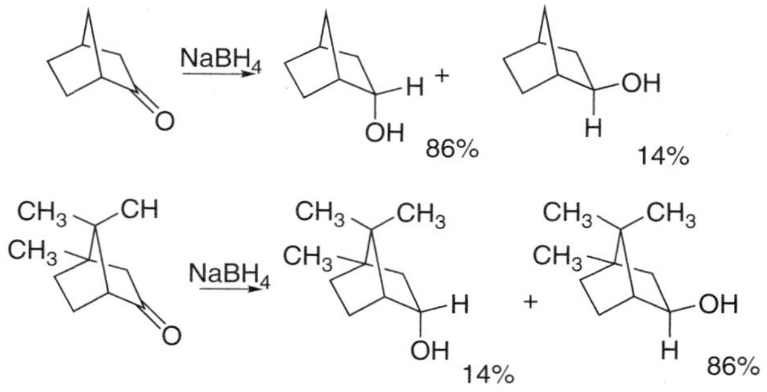

The NaBH$_4$-CeCl$_3$ reagent has been observed to give hydride delivery from the more hindered face of certain bicyclic ketones.[121]

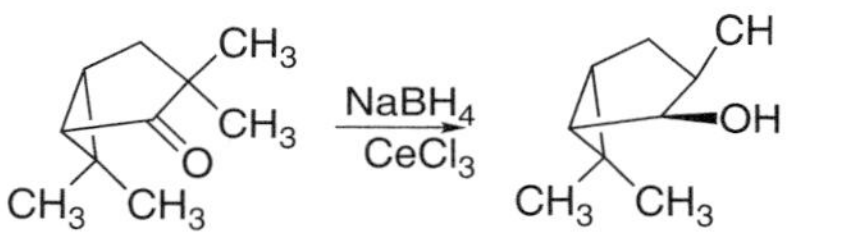

91% 99:1 *exo*

Table 5.4. Stereoselectivity of Hydride Reducing Agent

Reducing agent	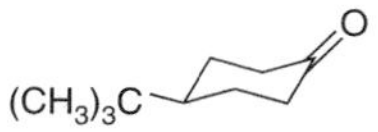 (CH$_3$)$_3$C ketone	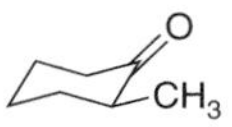 CH$_3$ ketone	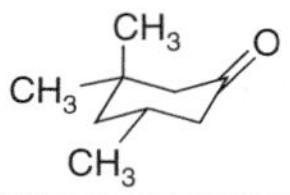 (CH$_3$)$_3$ trisubstituted ketone	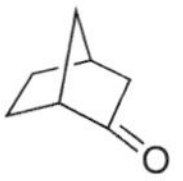 bicyclic ketone	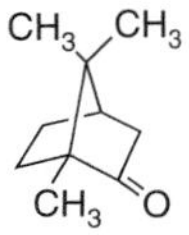 dimethyl bicyclic ketone
	% axial	% axial	% axial	%*endo*	% *exo*
NaBH$_4$	20[b]	25[c]	58[c]	86[d]	86[d]
LiAlH$_4$	8	24	83	89	92
LiAl(OMe)$_3$H	9	69	95	98	99
LiAl(O*t*Bu)$_3$H	9	35[f]		94[f]	94[f]
L-Selectride	93[g]	98[g]	99.8[g]	99.6[g]	99.6[g]
LS-Selectride	>99[h]	>99[h]		>99[h]	NR[h]

a. Except where noted otherwise, data are from H. C. Brown and W. D. Dickason, *J. Am. Chem. Soc.*, **92**, 709 (1970). Data for many other cyclic ketones and other reducing agents are given by A. V. Kamernitzky and A. A. Akhrem, *Tetrahedron*, **18**, 705 (1962) and W. T. Wipke and P. Gund, *J. Am. Chem. Soc.*, **98**, 8107 (1976).
b. P. T. Lansbury, and R. E. MacLeay, *J. Org. Chem.*, **28**, 1940 (1963).
c. B. Rickborn and W. T. Wuesthoff, *J. Am. Chem. Soc.*, **92**, 6894 (1970).
d. H. C. Brown and J. Muzzio, *J. Am. Chem. Soc.*, **88**, 2811 (1966).
e. J. Klein, E. Dunkelblum, E. L. Eliel, and Y. Senda, *Tetrahedron Lett.*, 6127 (1968).
f. E. C. Ashby, J. P. Sevenair, and F. R. Dobbs, *J. Org. Chem.*, **36**, 197 (1971).
g. H. C. Brown and S. Krishnamurthy, *J. Am. Chem. Soc.*, **94**, 7159 (1972).
h. S. Krishnamurthy and H. C. Brown, *J. Am. Chem. Soc.*, **98**, 3383 (1976).

121. A. Krief and D. Surleraux, *Synlett*, 273 (1991).

410

CHAPTER 5

*Reduction of
Carbon-Carbon Multiple
Bonds, Carbonyl
Groups, and Other
Functional Groups*

Similarly, NaBH$_4$-CeCl$_3$ reverses the stereochemistry relative to NaBH$_4$ in the bicyclic ketone **12**.[122]

	α:β
NaBH$_4$	20:80
NaBH$_4$, CeCl$_3$	95:5

Thus, NaBH$_4$-CeCl$_3$ tends to give the *more stable* alcohol, but the origin of this stereoselectivity does not seem to have been established. It is thought that these reductions proceed through alkoxyborohydrides.[123] It is likely that equilibration occurs by reversible hydride transfer.

5.3.2.2. Acyclic Ketones. The stereochemistry of the reduction of acyclic aldehydes and ketones is a function of the substitution on the adjacent carbon atom and can be predicted on the basis of the Felkin conformational model of the TS,[63] which is based on a combination of steric and stereoelectronic effects.

S, M, L = relative size of substituents

From a purely steric standpoint, minimal interaction with the groups L and M by approaching from the direction of the smallest substituent is favorable. The stereo-electronic effect involves the interaction between the approaching hydride ion and the LUMO of the carbonyl group. This orbital, which accepts the electrons of the incoming nucleophile, is stabilized when the group L is perpendicular to the plane of the carbonyl group.[124] This conformation permits a favorable interaction between the LUMO and the antibonding σ* orbital associated with the C−L bond.

In the case of α-substituted phenyl ketones, the order of stereoselectivity is C≡CH > CH=CH$_2$ > CH$_2$CH$_3$.[125] These results indicate a stereoelectronic as well as a steric

122. M. Leclaire and P. Jean, *Bull. Soc. Chim. Fr.*, **133**, 801 (1996).
123. A. C. Gemal and J.-L. Luche, *J. Am. Chem. Soc.*, **103**, 5454 (1981).
124. N. T. Ahn, *Top. Current Chem.*, **88**, 145 (1980).
125. M. Fujita, S. Akimoto, and K. Ogura, *Tetrahedron Lett.*, **34**, 5139 (1993).

component because the stereoselectivity corresponds to placing the unsaturated groups in the perpendicular position.

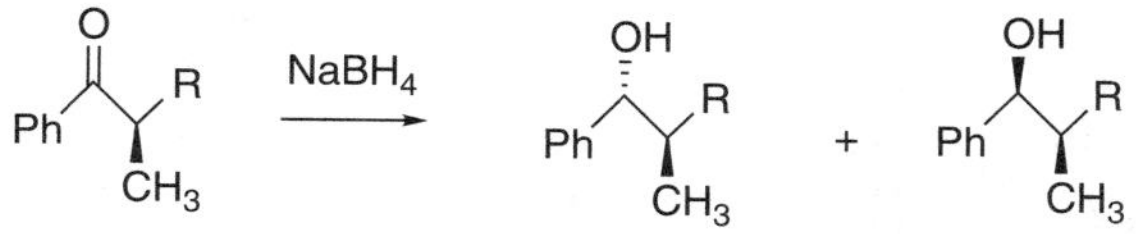

R	anti:syn
C_2H_5	57:43
$CH_2=CH$	70:30
$HC\equiv C$	89:11

Steric factors arising from groups that are more remote from the center undergoing reduction can also influence the stereochemical course of reduction. Such steric factors are magnified by use of bulky reducing agents. For example, a 4.5:1 preference for stereoisomer **14** over **15** is achieved by using the trialkylborohydride **13** as the reducing agent in the reduction of a prostaglandin intermediate.[126]

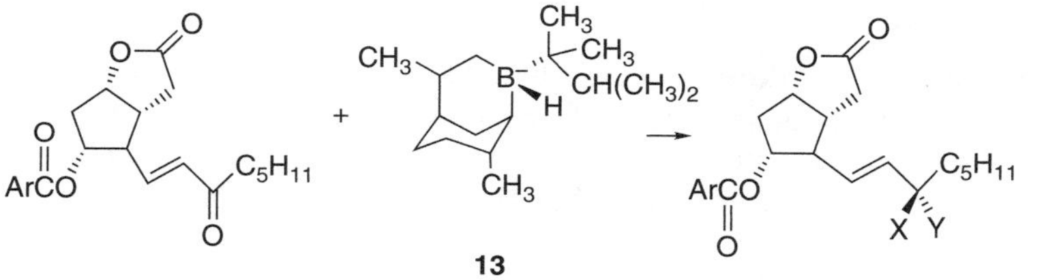

13

14 X = H, Y = OH 82%
15 X = OH, Y = H 18%

5.3.2.3. Chelation Control. The stereoselectivity of reduction of carbonyl groups can be controlled by chelation when there is a nearby donor substituent. In the presence of such a group, specific complexation among the substituent, the carbonyl oxygen, and the Lewis acid can establish a preferred conformation for the reactant. Usually hydride is then delivered from the less sterically hindered face of the chelate so the hydroxy group is *anti* to the chelating substituent.

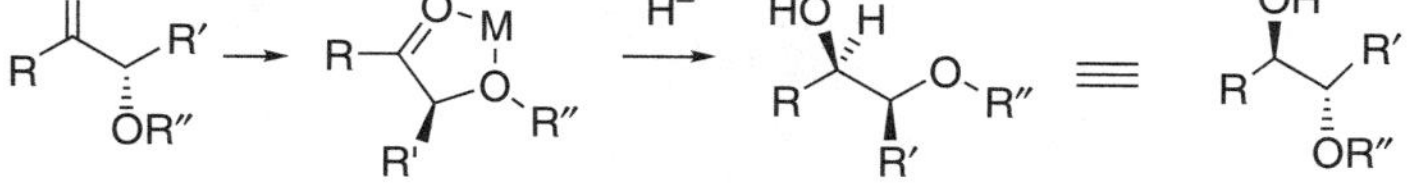

α-Hydroxy[127] and α-alkoxyketones[128] are reduced to *anti* 1,2-diols by $Zn(BH_4)_2$ through a chelated TS. This stereoselectivity is consistent with the preference for TS **F**

126. E. J. Corey, S. M. Albonico, U. Koelliker, T. K. Schaaf, and R. K. Varma, *J. Am. Chem. Soc.*, **93**, 1491 (1971).
127. T. Nakata, T. Tanaka, and T. Oishi, *Tetrahedron Lett.*, **24**, 2653 (1983).
128. G. J. McGarvey and M. Kimura, *J. Org. Chem.*, **47**, 5420 (1982).

412

CHAPTER 5

*Reduction of
Carbon-Carbon Multiple
Bonds, Carbonyl
Groups, and Other
Functional Groups*

over **G**. The stereoselectivity increases with the bulk of substituent R^2. $LiAlH_4$ shows the same trend, but is not as stereoselective.

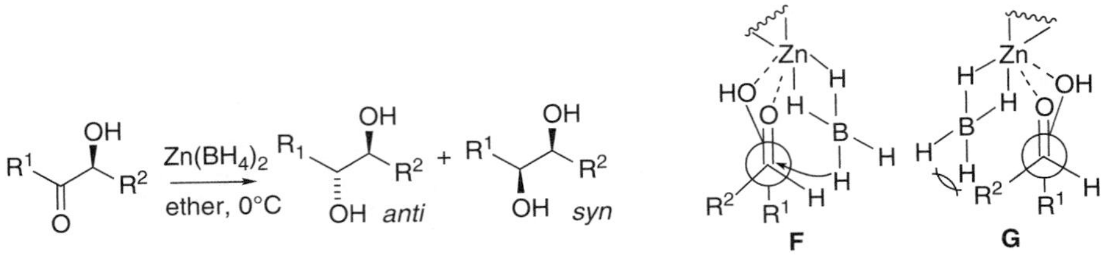

R^1	R^2	$Zn(BH_4)_2$ anti:syn	$LiAlH_4$ anti:syn
$n\text{-}C_5H_{11}$	CH_3	77:23	64:36
CH_3	$n\text{-}C_5H_{11}$	85:15	70:30
$i\text{-}C_3H_7$	CH_3	85:15	58:42
CH_3	$i\text{-}C_3H_7$	96:4	73:27
Ph	CH_3	98:2	87:13
CH_3	Ph	90:10	80:20

Reduction of β-hydroxy ketones through chelated TSs favors *syn*-1,3-diols. Boron chelates have been exploited to achieve this stereoselectivity.[129] One procedure involves in situ generation of diethylmethoxyboron, which then forms a chelate with the β-hydroxyketone. Reduction with $NaBH_4$ leads to the *syn*-diol.[130]

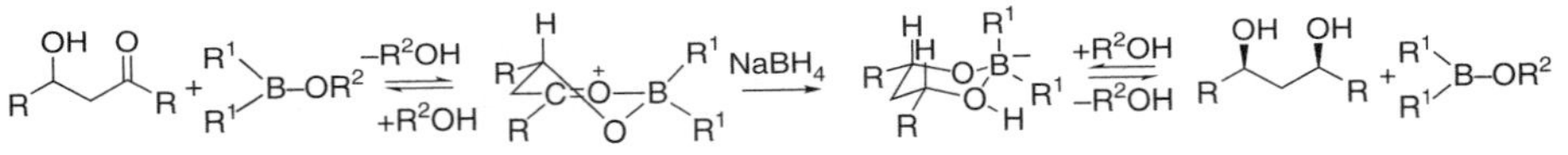

This procedure was used in the synthesis of the cholesterol-reducing drug lescol.[131] The diethylmethoxyboron can be prepared in situ from triethylboron and one equivalent of methanol.

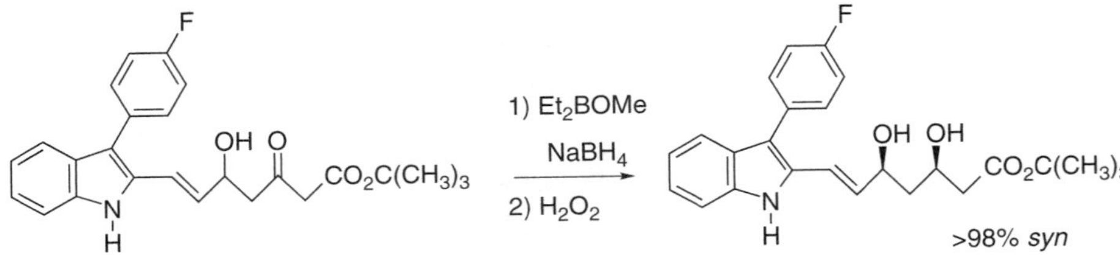

Syn-1,3-diols can be obtained from β-hydroxyketones using LiI-$LiAlH_4$ at low temperatures.[132] β-Hydroxyketones also give primarily *syn*-1,3-diols when

129. K. Narasaka and F.-C. Pai, *Tetrahedron*, **40**, 2233 (1984); K.-M. Chen, G. E. Hardtmann, K. Prasad, O. Repic, and M. J. Shapiro, *Tetrahedron Lett.*, **28**, 155 (1987).

130. K.-M. Chen, K. G. Gunderson, G. E. Hardtmann, K. Prasad, O. Repic, and M. J. Shapiro, *Chem. Lett.*, 1923 (1987).

131. O. Repic, K. Prasad, and G. T. Lee, *Org. Proc. Res. Dev.*, **5**, 519 (2001).

132. Y. Mori, A. Takeuchi, H. Kageyama, and M. Suzuki, *Tetrahedron Lett.*, **29**, 5423 (1988).

chelates prepared with BCl_3 are reduced with quaternary ammonium salts of BH_4^- or BH_3CN^-.[133]

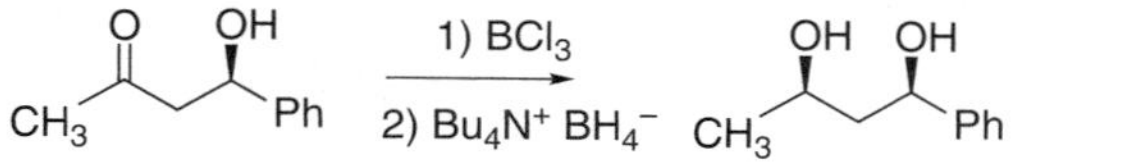

78%, 90:10 *syn:anti*

Similar results are obtained with β-methoxyketones using $TiCl_4$ as the chelating reagent.[134]

The effect of the steric bulk of the hydride reducing agent has been examined in the case of 3-benzyloxy-2-butanone.[135] The ratio of chelation-controlled product increased with the steric bulk of the reductant. This is presumably due to amplification of the steric effect of the methyl group in the chelated TS as the reductant becomes more sterically demanding. In these reactions, the degree of chelation control was also enhanced by use of CH_2Cl_2 as a cosolvent.

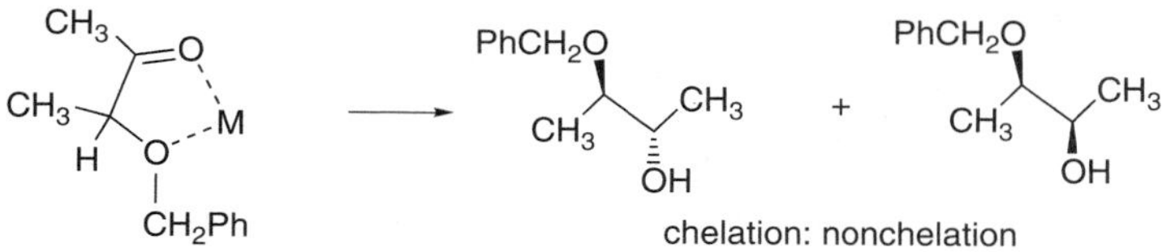

chelation: nonchelation

$Zn(BH_4)_2$	ether/CH_2Cl_2	6:1
$LiBEt_3H$	THF/CH_2Cl_2	28:1
$LiB(n\text{-}Bu)_3H$	ether/CH_2Cl_2	99:1

A survey of several of alkylborohydrides found that $LiBu_3BH$ in ether-pentane gave the best ratio of chelation-controlled reduction products from α- and β-alkoxy ketones.[134] In this case, the Li^+ cation acts as the Lewis acid. The alkylborohydrides provide an added increment of steric discrimination.

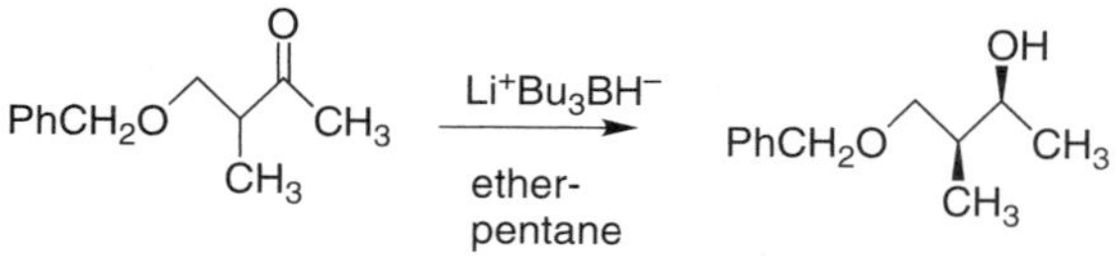

Tetramethylammonium triacetoxyborohydride gives *anti*-1,3-diols from β-hydroxy ketones.[136] These reactions are thought to occur by a rapid exchange that introduces the hydroxy group as a boron ligand.

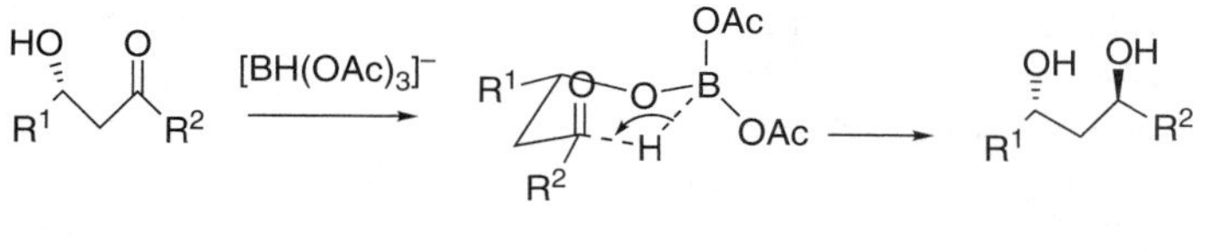

133. C. R. Sarko, S. E. Collibee, A. L. Knorr, and M. DiMare, *J. Org. Chem.*, **61**, 868 (1996).

134. C. R. Sarko, I. C. Guch, and M. DiMare, *J. Org. Chem.*, **59**, 705 (1994); G. Bartoli, M. C. Bellucci, M. Bosco, R. Dalpozzo, E. Marcantoni, and L. Sambri, *Tetrahedron Lett.*, **40**, 2845 (1999).

135. A.-M. Faucher, C. Brochu, S. R. Landry, I. Duchesne, S. Hantos, A. Roy, A. Myles, and C. Legualt, *Tetrahedron Lett.*, **39**, 8425 (1998).

136. D. A. Evans, K. T. Chapman, and E. M. Carreira, *J. Am. Chem. Soc.*, **110**, 3560 (1988).

414

CHAPTER 5

*Reduction of
Carbon-Carbon Multiple
Bonds, Carbonyl
Groups, and Other
Functional Groups*

Similarly, cyclic ketones **16** and **17** both give the *trans*-diol, as anticipated for intramolecular delivery of hydride. In the case of the equatorial alcohol, the reaction must occur through a nonchair conformer.

In 2-hydroxy-2,4-dimethylcyclohexanone there is a strong preference for equatorial attack by LiAlH$_4$, NaBH$_4$, and Zn(BH$_4$)$_2$.[137] In the case of the less conformationally biased 2-hydroxy-2-methylcyclohexanone, stereoselectivity is much weaker for these reductants, but is high for NaB(OAc)$_3$H. These results are attributed to prior complexation of the hydride at the hydroxy group with intramolecular delivery of hydride, leading to *anti*-diol. A 3-hydroxy substituent had a much weaker effect, except with NaB(OAc)$_3$H. This reagent presumably reacts more rapidly with hydroxy groups because of the greater lability of the acetoxy substituents, and in this case the reagent becomes a better hydride donor by replacing acetoxy with an alkoxide.

	% *anti*-diol		% *anti*-diol
NaBH$_4$	100	NaBH$_4$	57
LiAlH$_4$	100	LiAlH$_4$	74
Zn(BH$_4$)$_2$	100	Zn(BH$_4$)$_2$	75
NaB(OAc)$_3$H	100	NaB(OAc)$_3$H	97

Similar studies were carried out with methoxycyclohexanones.[138] 3-Methoxy groups showed no evidence of chelation effects with these reagents and the 2-methoxy group showed an effect only with Zn(BH$_4$)$_2$. This supports the suggestion that the effect of the hydroxy groups operates through deprotonated alkoxide complexes.

Chelation effects also come into play in the reduction of α,β-epoxyketones. Both CaCl$_2$ and LaCl$_3$ lead to enhanced *anti* stereoselectivity.[139] The same stereoselectivity is observed with CeCl$_3$ and with Zn(BH$_4$)$_2$.[140]

137. Y. Senda, N. Kikuchi, A. Inui, and H. Itoh, *Bull. Chem. Soc. Jpn.*, **73**, 237 (2000).
138. Y. Senda, H. Sakurai, S. Nakano, and H. Itoh, *Bull. Chem. Soc. Jpn.*, **69**, 3297 (1996).
139. M. Taniguchi, H. Fujii, K. Oshima, and K. Utimoto, *Tetrahedron*, **51**, 679 (1995).
140. K. Li, L. G. Hamann, and M. Koreeda, *Tetrahedron Lett.*, **33**, 6569 (1992).

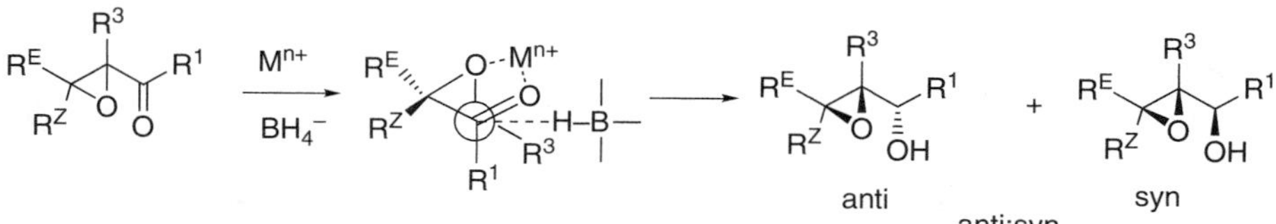

	anti:syn
NaBH$_4$	42:58
n-Bu$_4$NBH$_4$	48:52
NaBH$_4$–CaCl$_2$	92:8
NaBH$_4$–LaCl$_3$	92:8
NaBH$_4$–CeCl$_3$	>99:1
Zn(BH$_4$)$_2$	>99:1

β-Ketosulfoxides are subject to chelation control when reduced by DiBAlH in the presence of ZnCl$_2$.[141] This allows the use of chirality of the sulfoxide group to control the stereochemistry at the ketone carbonyl.

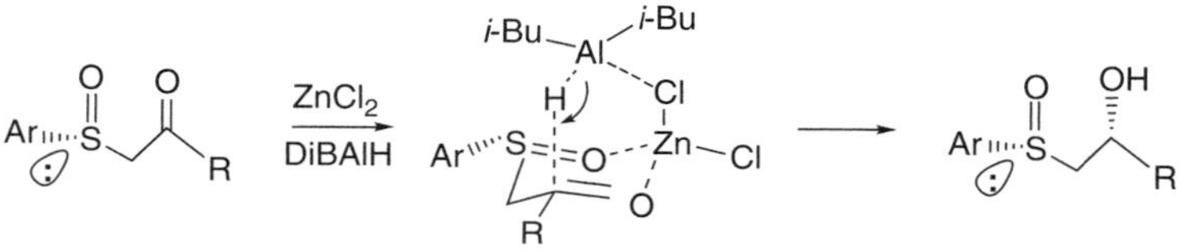

5.3.3. Enantioselective Reduction of Carbonyl Compounds

5.3.3.1. Reduction with Chiral Boranes. The reduction of an unsymmetrical ketone creates a new stereogenic center. Owing to the importance of hydroxy groups both in synthesis and in the properties of molecules, including biological activity, there has been a great deal of effort directed toward enantioselective reduction of ketones. One approach is to use chiral borohydride reagents.[142] Boranes derived from chiral alkenes can be converted to alkylborohydrides, and several such reagents are commercially available.[143]

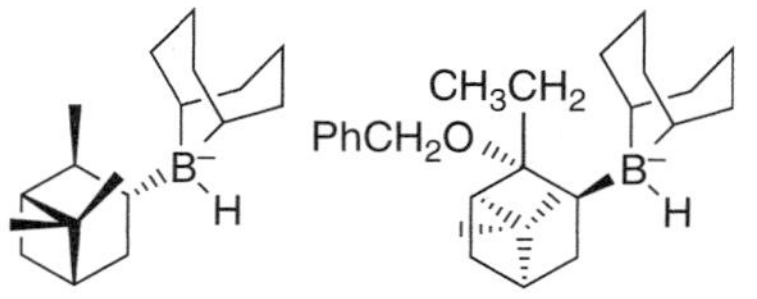

Alpine-Hydride*　　　　NB-Enantride*

Chloroboranes have also been found useful for enantioselective reduction. Di-(isopinocampheyl)chloroborane,[144] (Ipc)$_2$BCl, and *t*-butyl(isopinocampheyl)

---

141. A. Solladie-Cavallo, J. Suffert, A. Adib, and G. Solladie, *Tetrahedron Lett.*, **31**, 6649 (1990).
142. M. M. Midland, *Chem. Rev.*, **89**, 1553 (1989).
143. Alpine-Hydride and NB-Enantride are trademarks of the Sigma-Aldrich Corporation.
144. H. C. Brown, J. Chandrasekharan, and P. V. Ramachandran, *J. Am. Chem. Soc.*, **110**, 1539 (1988); M. Zhao, A. O. King, R. D. Larsen, T. R. Verhoeven, and P. J. Reider, *Tetrahedron Lett.*, **38**, 2641 (1997); N. N. Joshi, C. Pyun, V. K. Mahindroo, B. Singaram, and H. C. Brown, *J. Org. Chem.*, **57**, 504 (1992).

416

CHAPTER 5

*Reduction of
Carbon-Carbon Multiple
Bonds, Carbonyl
Groups, and Other
Functional Groups*

chloroborane[145] achieve high enantioselectivity for aryl and branched dialkyl ketones. Di-(iso-2-ethylapopinocampheyl)chloroborane,[146] (Eap)$_2$BCl, shows good enantio-selectivity for a wider range of alcohols.

(Ipc)$_2$BCl   *t*-BulpcBCl   (Eap)$_2$BCl

For example, (Ipc)$_2$BCl was found to be an advantageous in the enantioselective reduction in the large-scale preparation of L-699,392, a specific leukotriene antagonist of interest in the treatment of asthma.[147]

87% yield
99.5% e.e. on
2.75 kg scale

These reagents react through cyclic TSs and regenerate an alkene.

Table 5.5 gives some typical results for enantioselective reduction of ketones by alkylborohydrides and chloroboranes.

5.3.3.2. Catalytic Enantioselective Reduction of Ketones. An even more efficient approach to enantioselective reduction is to use a chiral catalyst. One of the most developed is the oxazaborolidine **18**, which is derived from the amino acid proline.[148] The enantiomer is also available. These catalysts are called the *CBS-oxazaborolidines*.

18

A catalytic amount (5–20 mol %) of the reagent, along with BH$_3$ as the reductant, can reduce ketones such as acetophenone and pinacolone in more than 95% e.e. An adduct of borane and **18** is the active reductant. This adduct can be prepared, stored,

[145.] H. C. Brown, M. Srebnik, and P. V. Ramachandran, *J. Org. Chem.*, **54**, 1577 (1989).

[146.] H. C. Brown, P. V. Ramachandran, A. V. Teodorovic, and S. Swaminathan, *Tetrahedron Lett.*, **32**, 6691 (1991).

[147.] A. O. King, E. G. Corley, R. K. Anderson, R. D. Larsen, T. R. Verhoeven, P. J. Reider, Y. B. Xiang, M. Belley, Y. Leblanc, M. Labelle, P. Prasit, and R. J. Zamboni, *J. Org. Chem.*, **58**, 3731 (1993).

[148.] E. J. Corey, R. K. Bakhi, S. Shibata, C. P. Chen, and V. K. Singh, *J. Am. Chem. Soc.*, **109**, 7925 (1987); E. J. Corey and C. J. Helal, *Angew. Chem. Int. Ed. Engl.*, **37**, 1987 (1998); V. A. Glushkov and A. G. Tolstikov, *Russ. Chem. Rev.*, **73**, 581 (2004).

Table 5.5. Enantioselective Reduction of Ketones by Borohydrides and Chloroboranes

Reagent	Ketone	% e.e.	Configuration
Alpine-Hydride[a,b]	3-methyl-2-butanone	62	S
NB-Enantride[a,c]	2-octanone	79	S
(Ipc)$_2$BCl[d]	2-acetylnaphthalene	94	S
(tBu)(Ipc)BCl[e]	acetophenone	96	R
(Ipc)$_2$BCl[f]	2,2-dimethylcyclohexanone	91	S
(Eap)$_2$BCl[g]	3-methyl-2-butanone	95	R

a. Trademark of Sigma-Aldrich Corporation.
b. H. C. Brown and G. G. Pai, *J. Org. Chem.*, **50**, 1384 (1985).
c. M. M. Midland and A. Kozubski, *J. Org. Chem.*, **47**, 2495 (1982).
d. M. Zhao, A. O. King, R. D. Larsen, T. R. Verhoeven, and A. J. Reider, *Tetrahedron Lett.*, **38**, 2641 (1997).
e. H. C. Brown, M. Srebnik, and P. V. Ramachandran, *J. Org. Chem.*, **54**, 1577 (1989).
f. H. C. Brown, J. Chandrasekharan, and P. V. Ramachandran, *J. Am. Chem. Soc.*, **110**, 1539 (1988).
g. H. C. Brown, P. V. Ramachandran, A. V. Teodorovic, and S. Swaminathan, *Tetrahedron Lett.*, **32**, 6691 (1991).

and used as a stoichiometric reagent if so desired.[149] The catalytic cycle depends on dissociation of the reduced product.

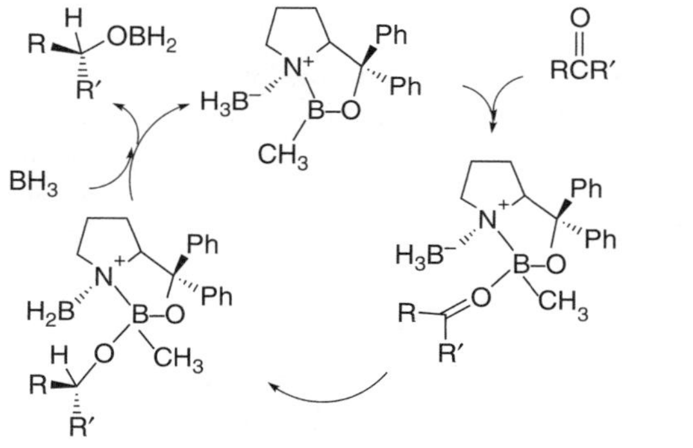

The corresponding *N*-butyloxazaborolidine is also frequently used as a catalyst. The enantioselectivity and reactivity of these catalysts can be modified by changes in substituent groups to optimize selectivity toward a particular ketone.[150] Catecholborane can also be used as the reductant.[151]

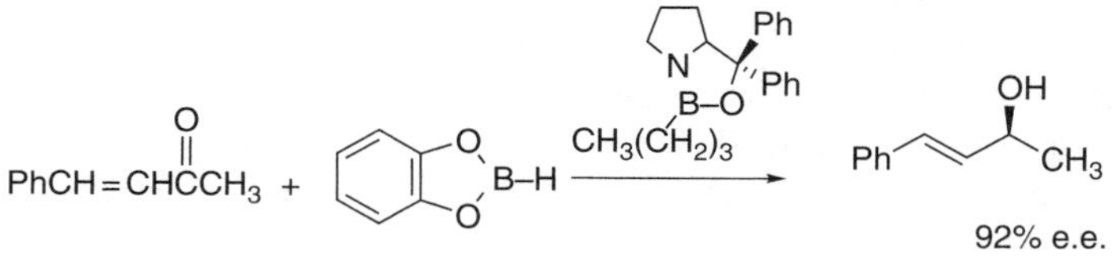

Both mechanistic and computational studies have been used to explore the catalytic process. A crystal structure of the catalysts is available (Figure 5.7).[152] The

149. D. J. Mahre, A. S. Thompson, A. W. Douglas, K. Hoogsteen, J. D. Carroll, E. G. Corley, and E. J. J. Grabowski, *J. Org. Chem.*, **58**, 2880 (1993).
150. A. W. Douglas, D. M. Tschaen, R. A. Reamer, and Y.-J. Shi, *Tetrahedron: Asymmetry*, **7**, 1303 (1996).
151. E. J. Corey and R. K. Bakshi, *Tetrahedron Lett.*, **31**, 611 (1990).
152. E. J. Corey, M. Azimiaora, and S. Sarshar, *Tetrahedron Lett.*, **33**, 3429 (1992).

418

CHAPTER 5

Reduction of
Carbon-Carbon Multiple
Bonds, Carbonyl
Groups, and Other
Functional Groups

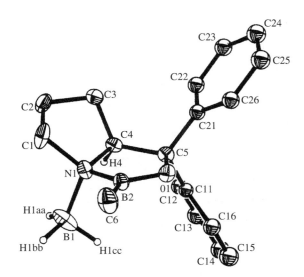

Fig. 5.7. Crystal structure of borane complex of α,α-diphenylprolinol oxazaborolidine catalysts. Reproduced from *Tetrahedron Lett.*, **33**, 3429 (1992), by permission of Elsevier.

orientation of the ketone is dictated by the phenyl groups and the relatively rigid geometry of the ring system. The enantioselectivity in these reductions is proposed to arise from a chairlike TS in which the governing steric interaction is with the alkyl substituent on boron.[153,154] There are experimental data indicating that the steric demand of the boron substituent influences enantioselectivity.[154]

There have been ab initio studies of the transition structure using several model catalysts and calculations at the HF/3-21G, HF/6-31G(d), and MP2/6-31G(d) levels.[155] The enantioselectivity is attributed to the preference for an *exo* rather than an *endo* approach of the ketone, as shown in Figure 5.8.

According to B3LYP/6-31G* computations of the intermediates and TSs, there are no large barriers to the reaction and it is strongly exothermic.[156] Measured E_a values are around 10 kcal/mol.[157] The complexation of borane to the catalyst shifts electron density from nitrogen to boron and enhances the nucleophilicity of the hydride. The

153. D. K. Jones, D. C. Liotta, I. Shikai, and D. J. Mathre, *J. Org. Chem.*, **58**, 799 (1993).
154. T. K. Jones, J. J. Mohan, L. C. Xavier, T. J. Blacklock, D. J. Mathre, P. Sohar, E. T. T. Jones, R. A. Beaner, F. E. Roberts, and E. J. J. Grabowski, *J. Org. Chem.*, **56**, 763 (1991).
155. G. J. Quallich, J. F. Blake, and T. M. Woodall, *J. Am. Chem. Soc.*, **116**, 8516 (1994).
156. G. Alagona, C. Ghio, M. Persico, and S. Tomas, *J. Am. Chem. Soc.*, **125**, 10027 (2003).
157. H. Jockel, R. Schmidt, H. Jope, and H. G. Schmalz, *J. Chem. Soc., Perkin Trans. 2*, 69 (2000).

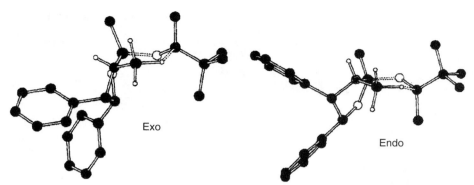

Fig. 5.8. Optimized (HF/3-21G) structures of the *exo* and *endo* transition states for reduction of *t*-butyl methyl ketone by model catalyst. The *exo* structure is favored by 2.1 kcal, in accord with an experimental e.e of 88%. Reproduced from *J. Am. Chem. Soc.*, **116**, 8516 (1994), by permission of the American Chemical Society.

complexation also diminishes the N–B delocalization present in the oxazaborolidine ring, with the bond length increasing from 1.410 to 1.498 Å, according to the computations. The computed structural parameters are close to those found by crystallography.

Scheme 5.6 shows some examples of enantioselective reduction of ketones using CBS-oxazaborolidine catalysts. The reaction in Entry 1 was carried out in the course of synthesis of a potential drug candidate. Entry 2 employs the catalyst to achieve stereoselective reduction at the C(15) center in a prostaglandin precursor. Entries 3 and 4 report high enantioselectivity in the reduction of cyclic ketones. Entries 5 and 6 are cases of acyclic ketones with adjacent functionality and are reduced with high enantioselectivity. Entries 7 and 8 are applications of the reaction to aromatic ketones done on a relatively large scale in the course of drug development. Entry 7 used an indane-derived aminoalcohol as the oxazaborolidine precursor, whereas the procedure in Entry 8 involves in situ generation of the CBS catalyst. Entries 9 to 14 show other examples of the reaction that were carried out in the course of multistage syntheses of complex molecules.

Enantioselective 1,4-reduction of enones can be done using a copper-BINAP catalyst in conjunction with silicon hydride donors.[158] Polymethylhydrosilane (PMHS) is one reductants that is used.

The reduction can also be effected with diphenylsilane and the intermediate silyl enol ethers can be alkylated in a tandem process.[159]

[158.] Y. Moritani, D. H. Appella, V. Jurkauskas, and S. L. Buchwald, *J. Am. Chem. Soc.*, **122**, 6797 (2000).
[159.] J. Yun and S. L. Buchwald, *Org. Lett.*, **3**, 1129 (2001).

Scheme 5.6. Enantioselective Reduction of Ketones Using CBS-Oxazaborolidine Catalysts

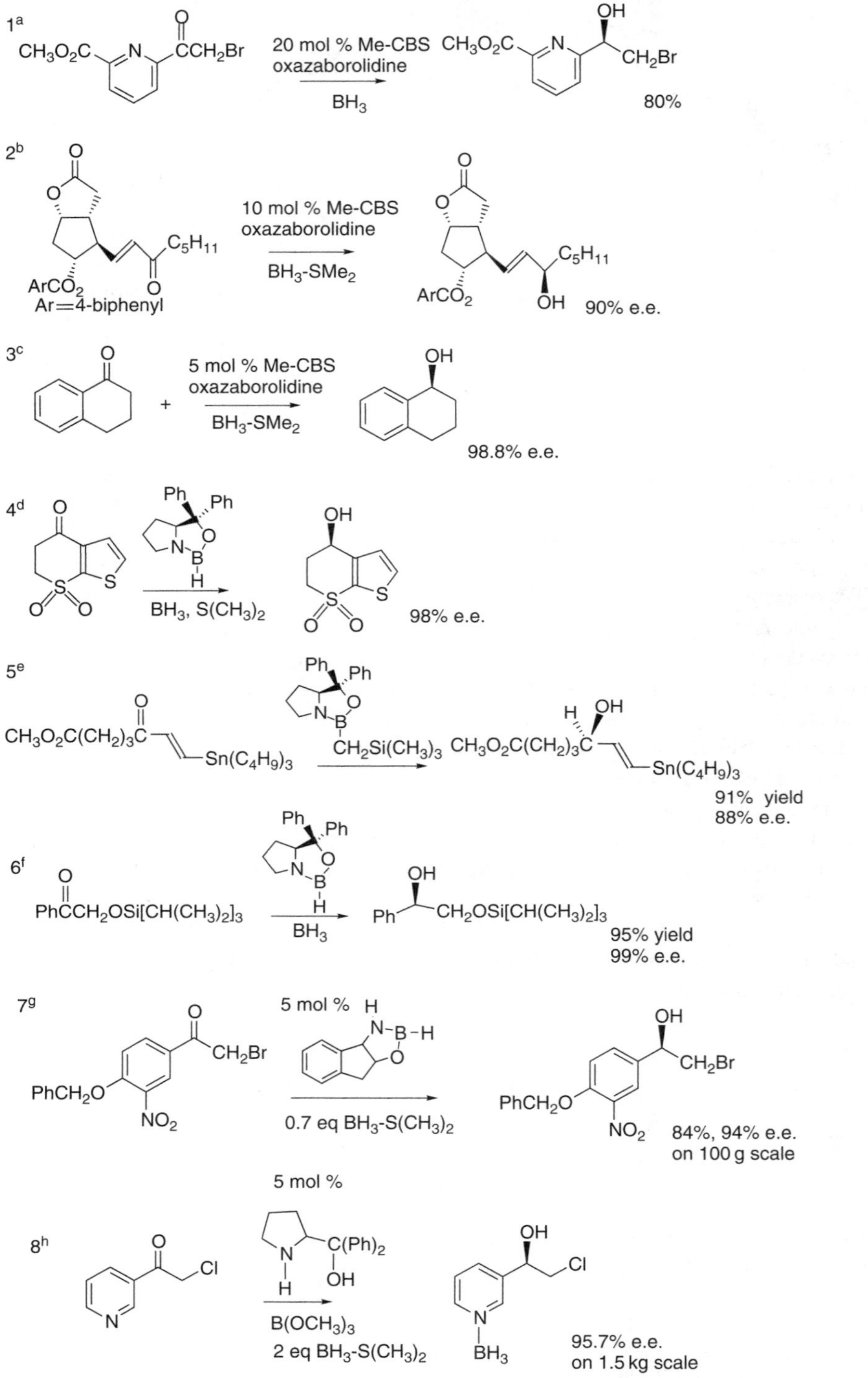

1[a]

20 mol % Me-CBS oxazaborolidine

BH_3

80%

2[b]

10 mol % Me-CBS oxazaborolidine

BH_3-SMe_2

Ar=4-biphenyl

90% e.e.

3[c]

5 mol % Me-CBS oxazaborolidine

BH_3-SMe_2

98.8% e.e.

4[d]

BH_3, $S(CH_3)_2$

98% e.e.

5[e]

$CH_3O_2C(CH_2)_3C$—$Sn(C_4H_9)_3$

91% yield
88% e.e.

6[f]

$PhCCH_2OSi[CH(CH_3)_2]_3$

BH_3

95% yield
99% e.e.

7[g]

5 mol %

0.7 eq BH_3-$S(CH_3)_2$

84%, 94% e.e.
on 100 g scale

8[h]

5 mol %

$B(OCH_3)_3$
2 eq BH_3-$S(CH_3)_2$

95.7% e.e.
on 1.5 kg scale

(Continued)

Scheme 5.6. (*Continued*)

421

SECTION 5.3

Group III
Hydride-Donor Reagents

9^i

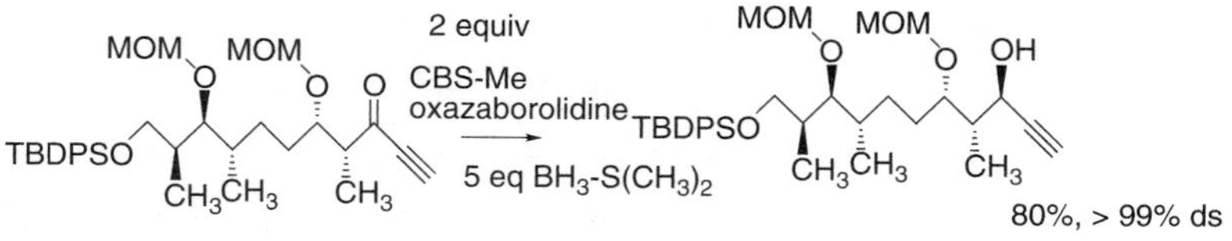

2 equiv
CBS-Me
oxazaborolidine

5 eq BH_3-$S(CH_3)_2$

80%, > 99% ds

10^j

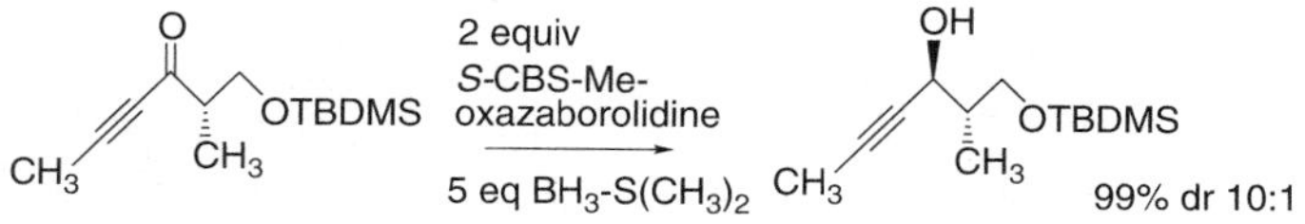

2 equiv
S-CBS-Me-
oxazaborolidine

5 eq BH_3-$S(CH_3)_2$

99% dr 10:1

11^k

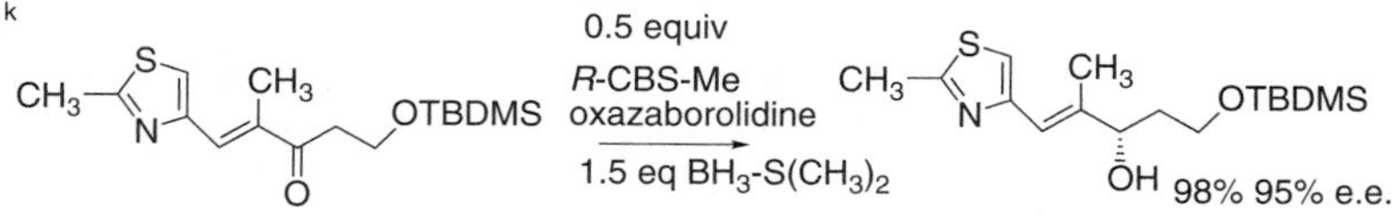

0.5 equiv
R-CBS-Me
oxazaborolidine

1.5 eq BH_3-$S(CH_3)_2$

98% 95% e.e.

12^l

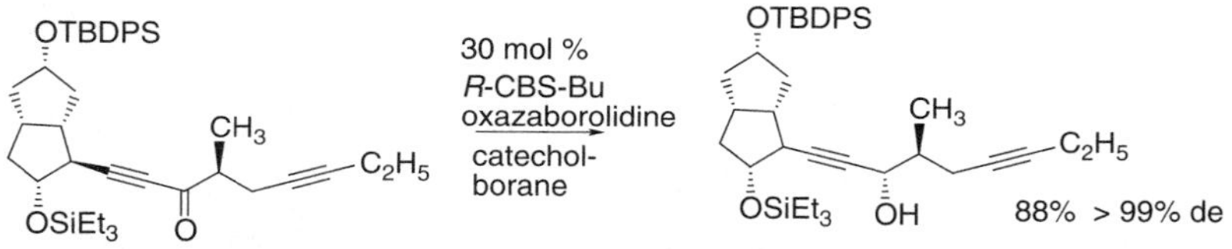

30 mol %
R-CBS-Bu
oxazaborolidine
catechol-
borane

88% > 99% de

13^m

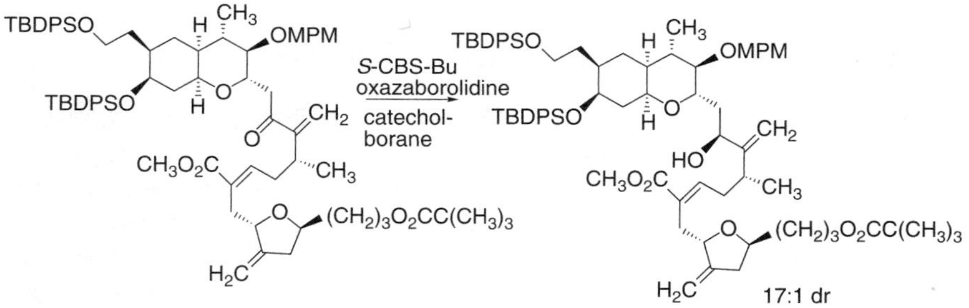

S-CBS-Bu
oxazaborolidine
catechol-
borane

17:1 dr

14^n

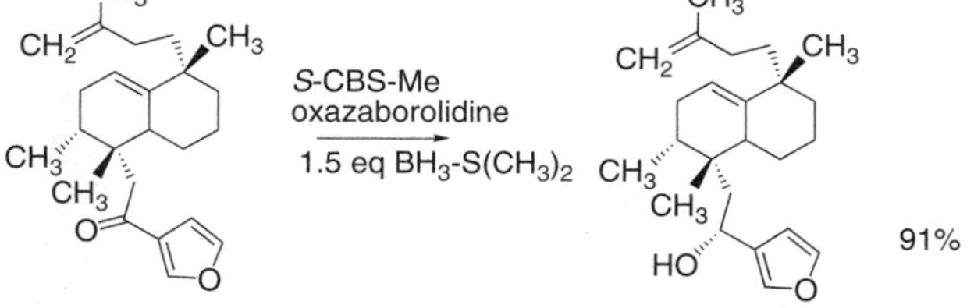

S-CBS-Me
oxazaborolidine

1.5 eq BH_3-$S(CH_3)_2$

91%

CHAPTER 5

Reduction of
Carbon-Carbon Multiple
Bonds, Carbonyl
Groups, and Other
Functional Groups

a. K. G. Hull, M. Visnick, W. Tautz, and A. Sheffron, *Tetrahedron*, **53**, 12405 (1997).
b. E. J. Corey, R. K. Bakshi, S. Shibata, C.-P. Chen, and V. K. Singh, *J. Am. Chem. Soc.*, **109**, 7925 (1987).
c. D. J. Mathre, A. S. Thompson, A. W. Douglas, K. Hoogsteen, J. D. Carroll, E. G. Corley, and E. J. J. Grabowski, *J. Org. Chem.*, **58**, 2880 (1993).
d. T. K. Jones, J. J. Mohan, L. C. Xavier, T. J. Blacklock, D. J. Mathre, P. Sohar, E. T. T. Jones, R. A. Reamer, F. E. Roberts, and E. J. J. Grabowski, *J. Org. Chem.*, **56**, 763 (1991).
e. E. J. Corey, A. Guzman-Perez, and S. E. Lazerwith, *J. Am. Chem. Soc.*, **119**, 11769 (1997).
f. B. T. Cho and Y. S. Chun, *J. Org. Chem.*, **63**, 5280 (1998).
g. R. Hett, Q. K. Fang, Y. Gao, S. A. Wald, and C. H. Senanayake, *Org. Proc. Res. Dev.*, **2**, 96 (1998).
h. J. Duquette, M Zhang, L. Zhu, and R. S. Reeves, *Org. Proc. Res. Dev.*, **7**, 285 (2003).
i. L. Bialy and H. Waldmann, *Chem. Eur. J.*, **10**, 2759 (2004).
j. B. M. Trost, J. L. Guzner, O. Dirat, and Y. H. Rhee, *J. Am. Chem. Soc.*, **124**, 10396 (2002).
k. E. A. Reiff, S. K. Nair, B. S. N. Reddy, J. Inagaki, J. T. Henri, J. F. Greiner, and G. I. Georg, *Tetrahedron Lett.*, **45**, 5845 (2004).
l. M. Lerm, H.-J. Gais, K. Cheng, and C. Vermeeren, *J. Am. Chem. Soc.*, **125**, 9653 (2003).
m. D. P. Stamos, S. S. Chen, and Y. Kishi, *J. Org. Chem.*, **62**, 7552 (1997).
n. E. J. Corey and B. E. Roberts, *J. Am. Chem. Soc.*, **119**, 12425 (1997).

e.e. > 90%
dr > 15:1

When necessary, the *trans:cis* ratio can be improved by base-catalyzed equilibration.

5.3.4. Reduction of Other Functional Groups by Hydride Donors

Although reductions of the common carbonyl and carboxylic acid derivatives are the most prevalent uses of hydride donors, these reagents can reduce a number of other groups in ways that are of synthetic utility. Halogen and sulfonate leaving groups can undergo replacement by hydride. Both aluminum and boron hydrides exhibit this reactivity, and lithium trialkylborohydrides are especially reactive.[160] The reduction is particularly rapid and efficient in polar aprotic solvents such as DMSO, DMF, and HMPA. Table 5.6 gives some indication of the reaction conditions. The normal factors in susceptibility to nucleophilic attack govern reactivity with I > Br > Cl being the order in terms of the leaving group and benzyl ∼ allyl > primary > secondary > tertiary in terms of the substitution site.[161] For primary alkyl groups, it is likely that the reaction proceeds by an S_N2 mechanism. However, the range of halides that can be reduced includes aryl halides and bridgehead halides, which cannot react by the S_N2 mechanism.[162] The loss of stereochemical integrity in the reduction of vinyl halides suggests the involvement of radical intermediates.[163] Formation and subsequent

[160.] S. Krishnamurthy and H. C. Brown, *J. Org. Chem.*, **45**, 849 (1980).
[161.] S. Krishnamurthy and H. C. Brown, *J. Org. Chem.*, **47**, 276 (1982).
[162.] C. W. Jefford, D. Kirkpatrick, and F. Delay, *J. Am. Chem. Soc.*, **94**, 8905 (1972).
[163.] S. K. Chung, *J. Org. Chem.*, **45**, 3513 (1980).

Table 5.6. Reaction Conditions for Reductive Replacement of Halogen and Sulfonate Groups by Hydride Donors

	Approximate conditions for complete reduction	
Hydride donor	Halides	Sulfonates
NaBH$_3$CN[a]	C$_{12}$H$_{23}$I, HMPA, 25°C, 4 h	C$_{12}$H$_{23}$O$_3$SC$_7$H$_7$, HMPA, 70°C, 8 h
NaBH$_4$[b]	C$_{12}$H$_{23}$Br, DMSO, 85°C, 1.5 h	C$_{12}$H$_{23}$O$_3$SC$_7$H$_7$, DMSO, 85°C, 2 h
LiAlH$_4$[c,d]	C$_8$H$_{17}$Br, THF, 25°C, 1 h	C$_8$H$_{17}$O$_3$SC$_7$H$_7$, DME, 25°C, 6 h
LiB(C$_2$H$_5$)$_3$H[c]	C$_8$H$_{17}$Br, THF, 25°C, 3 h	

a. R. O. Hutchins, D. Kandasamy, C. A. Maryanoff, D. Masilamani, and B. E. Maryanoff, *J. Org. Chem.*, **42**, 82 (1977).
b. R. O. Hutchins, D. Kandasamy, F. Dux, III, C. A. Maryanoff, D. Rotstein, B. Goldsmith, W. Burgoyne, F. Cistone, J. Dalessandro, and J. Puglis, *J. Org. Chem.*, **43**, 2259 (1978).
c. S. Krishnamurthy and H. C. Brown, *J. Org. Chem.*, **45**, 849 (1980).
d. S. Krishnamurthy, *J. Org. Chem.*, **45**, 2550 (1980).

dissociation of a radical anion by one-electron transfer is a likely mechanism for reductive dehalogenation of compounds that cannot react by an S$_N$2 mechanism.

$$R-X + e^- \longrightarrow R-X^{\overset{\cdot}{-}}$$
$$R-X^{\overset{\cdot}{-}} \longrightarrow R\cdot + X^-$$
$$R\cdot + H^- \longrightarrow R-H + e^-$$

One experimental test for the involvement of radical intermediates is to study 5-hexenyl systems and look for the characteristic cyclization to cyclopentane derivatives (see Part A, Section 11.2.3). When 5-hexenyl bromide or iodide reacts with LiAlH$_4$, no cyclization products are observed. However, the more hindered 2,2-dimethyl-5-hexenyl iodide gives mainly cyclic product.[164]

CH$_2$=CH(CH$_2$)$_3$CH$_2$I + LiAlH$_4$ $\xrightarrow[1\,h]{24°C}$ CH$_2$=CH(CH$_2$)$_3$CH$_3$ 94%

Some cyclization also occurs with the bromide, but not with the chloride or the tosylate. The secondary iodide, 6-iodo-1-heptene, gives a mixture of cyclic and acyclic product in THF.[165]

CH$_2$=CH(CH$_2$)$_3$CHCH$_3$ $\xrightarrow[THF]{LiAlH_4}$ CH$_2$=CH(CH$_2$)$_3$CH$_2$CH$_3$ + ... 21% / 72%, 3.7:1 *cis:trans*

164. E. C. Ashby, R. N. DePriest, A. B. Goel, B. Wenderoth, and T. N. Pham, *J. Org. Chem.*, **49**, 3545 (1984).
165. E. C. Ashby, T. N. Pham, and A. Amrollah-Madjadabadi, *J. Org. Chem.*, **56**, 1596 (1991).

424

CHAPTER 5

*Reduction of
Carbon-Carbon Multiple
Bonds, Carbonyl
Groups, and Other
Functional Groups*

The occurrence of a radical intermediate is also indicated in the reduction of 2-octyl iodide by $LiAlD_4$ since, in contrast to the chloride or bromide, extensive racemization accompanies reduction.

The presence of transition metal ions has a catalytic effect on reduction of halides and tosylates by $LiAlH_4$.[166] Various "copper hydride" reducing agents are effective for removal of halide and tosylate groups.[167] The primary synthetic value of these reductions is for the removal of a hydroxy function after conversion to a halide or tosylate.

Epoxides are converted to alcohols by $LiAlH_4$ in a reaction that occurs by nucleophilic attack, and hydride addition at the less hindered carbon of the epoxide is usually observed.

Cyclohexene epoxides are preferentially reduced by an axial approach by the nucleophile.[168]

Lithium triethylborohydride is a superior reagent for the reduction of epoxides that are relatively unreactive or prone to rearrangement.[169]

Alkynes are reduced to *E*-alkenes by $LiAlH_4$.[170] This stereochemistry is complementary to that of partial hydrogenation, which gives *Z*-isomers. Alkyne reduction by $LiAlH_4$ is greatly accelerated by a nearby hydroxy group. Typically, propargylic alcohols react in ether or tetrahydrofuran over a period of several hours,[171] whereas forcing conditions are required for isolated triple bonds.[172] This is presumably the result of coordination of the hydroxy group at aluminum and formation of a cyclic intermediate. The involvement of intramolecular Al–H addition has been demonstrated by use of $LiAlD_4$ as the reductant. When reduction by $LiAlD_4$ is followed by quenching with normal water, propargylic alcohol gives *Z*-3-^{2}H-prop-2-enol. Quenching with D_2O gives 2-^{2}H-3-^{2}H-prop-2-enol.[173]

166. E. C. Ashby and J. J. Lin, *J. Org. Chem.*, **43**, 1263 (1978).
167. S. Masamune, G. S. Bates, and P. E. Georghiou, *J. Am. Chem. Soc.*, **96**, 3686 (1974); E. C. Ashby, J. J. Lin, and A. B. Goel, *J. Org. Chem.*, **43**, 183 (1978).
168. B. Rickborn and J. Quartucci, *J. Org. Chem.*, **29**, 3185 (1964); B. Rickborn and W. E. Lamke, II, *J. Org. Chem.*, **32**, 537 (1967); D. K. Murphy, R. L. Alumbaugh, and B. Rickborn, *J. Am. Chem. Soc.*, **91**, 2649 (1969).
169. H. C. Brown, S. C. Kim, and S. Krishnamurthy, *J. Org. Chem.*, **45**, 1 (1980); H. C. Brown, S. Narasimhan, and V. Somayaji, *J. Org. Chem.*, **48**, 3091 (1983).
170. E. F. Magoon and L. H. Slaugh, *Tetrahedron*, **23**, 4509 (1967).
171. N. A. Porter, C. B. Ziegler, Jr., F. F. Khouri, and D. H. Roberts, *J. Org. Chem.*, **50**, 2252 (1985).
172. H. C. Huang, J. K. Rehmann, and G. R. Gray, *J. Org. Chem.*, **47**, 4018 (1982).
173. J. E. Baldwin and K. A. Black, *J. Org. Chem.*, **48**, 2778 (1983).

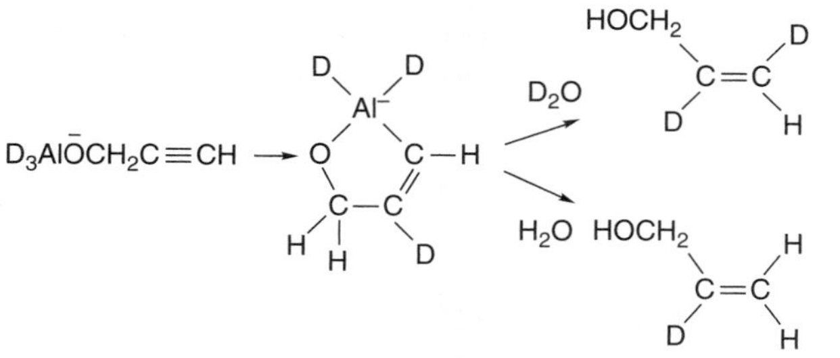

The efficiency and stereospecificity of reduction is improved by using a 1:2 mixture of LiAlH$_4$-NaOCH$_3$ as the reducing agent.[174] The mechanistic basis of this effect has not been explored in detail.

Scheme 5.7 illustrates these and other applications of the hydride donors. Entries 1 and 2 are examples of reduction of alkyl halides, whereas Entry 3 shows removal of an aromatic halogen. Entries 4 to 6 are sulfonate displacements, with the last example using a copper hydride reagent. Entry 7 is an epoxide ring opening. Entries 8 and 9 illustrate the difference in ease of reduction of alkynes with and without hydroxy participation.

5.4. Group IV Hydride Donors

5.4.1. Reactions Involving Silicon Hydrides

Both Si$-$H and C$-$H compounds can function as hydride donors under certain circumstances. The silicon-hydrogen bond is capable of transferring a hydride to carbocations. Alcohols that can be ionized in trifluoroacetic acid are reduced to hydrocarbons in the presence of a silane.

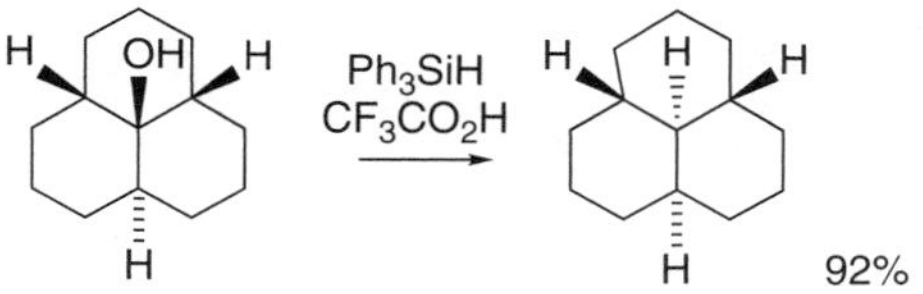

Ref. 175

Aromatic aldehydes and ketones are reduced to alkylaromatics under similar conditions through reactions involving benzylic cations.[176]

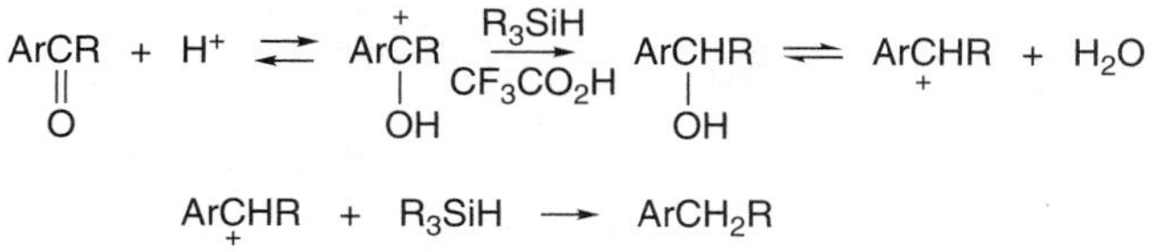

174. E. J. Corey, J. A. Katzenellenbogen, and G. H. Posner, *J. Am. Chem. Soc.*, **89**, 4245 (1967); B. B. Molloy and K. L. Hauser, *J. Chem. Soc., Chem. Commun.*, 1017 (1968).

175. F. A. Carey and H. S. Tremper, *J. Org. Chem.*, **36**, 758 (1971).

176. C. T. West, S. J. Donnelly, D. A. Kooistra, and M. P. Doyle, *J. Org. Chem.*, **38**, 2675 (1973); M. P. Doyle, D. J. DeBruyn, and D. A. Kooistra, *J. Am. Chem. Soc.*, **94**, 3659 (1972); M. P. Doyle and C. T. West, *J. Org. Chem.*, **40**, 3821 (1975).

Scheme 5.7. Reduction of Other Functional Groups by Hydride Donors

Halides

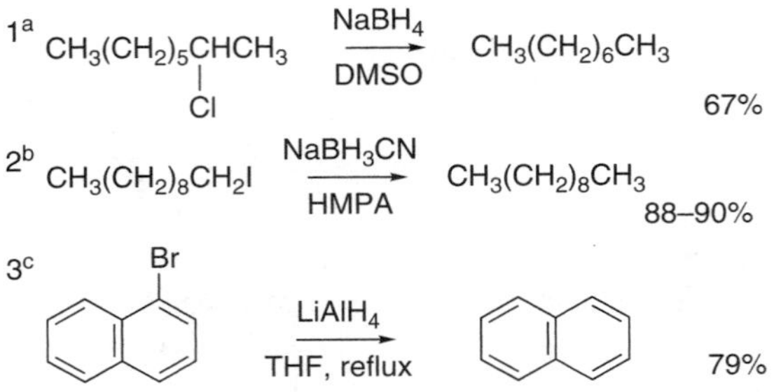
1[a]
CH3(CH2)5CHCH3
Cl
NaBH4
DMSO
CH3(CH2)6CH3
67%
2[b]
CH3(CH2)8CH2I
NaBH3CN
HMPA
CH3(CH2)8CH3
88–90%
3[c]
Br
LiAlH4
THF, reflux
79%

Sulfonates

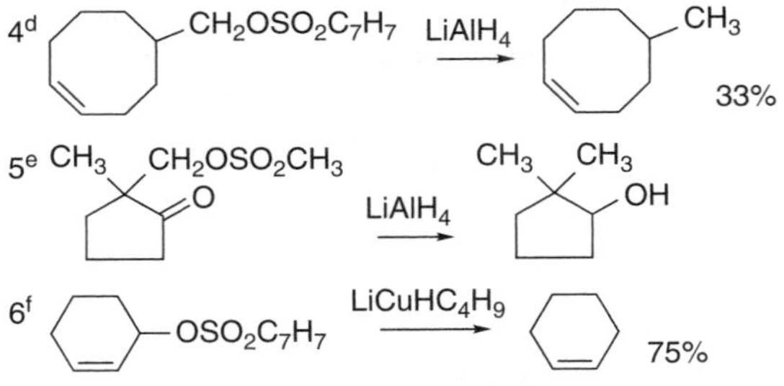
4[d]
CH2OSO2C7H7
LiAlH4
CH3
33%
5[e]
CH3
CH2OSO2CH3
O
LiAlH4
CH3
CH3
OH
6[f]
OSO2C7H7
LiCuHC4H9
75%

Epoxides

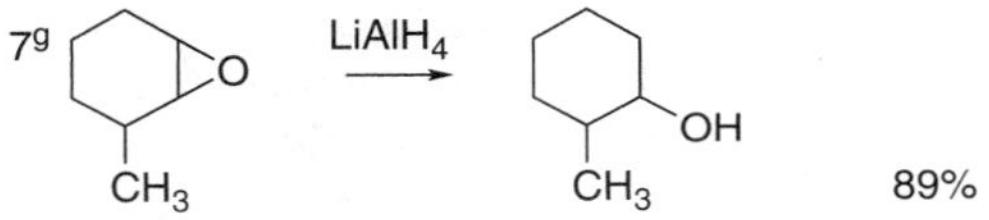
7[g]
O
CH3
LiAlH4
OH
CH3
89%

Acetylenes

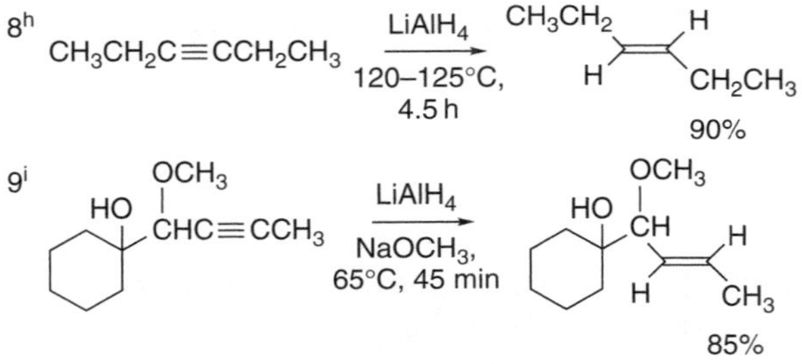
8[h]
CH3CH2C≡CCH2CH3
LiAlH4
120–125°C,
4.5 h
CH3CH2
H
H
CH2CH3
90%
9[i]
OCH3
HO
CHC≡CCH3
LiAlH4
NaOCH3,
65°C, 45 min
OCH3
HO
CH
H
H
CH3
85%

a. R. O. Hutchins, D. Hoke, J. Keogh, and D. Koharski, *Tetrahedron Lett.*, 3495 (1969); H. M. Bell, C. W. Vanderslice, and A. Spehar, *J. Org. Chem.*, **34**, 3923 (1969).
b. R. O. Hutchins, C. A. Milewski, and B. E. Maryanoff, *Org. Synth.*, **53**, 107 (1973).
c. H. C. Brown and S. Krishnamurthy, *J. Org. Chem.*, **34**, 3918 (1969).
d. A. C. Cope and G. L. Woo, *J. Am. Chem. Soc.*, **85**, 3601 (1963).
e. A. Eshenmoser and A. Frey, *Helv. Chim. Acta*, **35**, 1660 (1952).
f. S. Masamune, G. S. Bates, and P. E. Geoghiou, *J. Am. Chem. Soc.*, **96**, 3686 (1974).
g. B. Rickborn and W. E. Lamke, II, *J. Org. Chem.*, **32**, 537 (1967).
h. E. F. Magoon and L. H. Slaugh, *Tetrahedron*, **23**, 4509 (1967).
i. D. A. Evans and J. V. Nelson, *J. Am. Chem. Soc.*, **102**, 774 (1980).

Aryl ketones are also reduced with triethylsilane and $TiCl_4$. This method can be used to prepare γ-arylaminoacids.[177]

$$\underset{\underset{CO_2H}{|}}{ArCCH_2CHNHCO_2CH_3} \xrightarrow[\substack{2)\ (C_2H_5)_3SiH, \\ TiCl_4}]{\substack{1)\ TMSCl \\ Et_3N}} \underset{\underset{CO_2H}{|}}{ArCH_2CH_2CHNHCO_2CH_3}$$

Aliphatic ketones can be reduced to hydrocarbons by triethylsilane and gaseous BF_3.[178] The BF_3 is a sufficiently strong Lewis acid to promote formation of a carbocation from the intermediate alcohol.

$$\underset{RCR}{\overset{BF_3}{\underset{\|}{\overset{/}{\underset{O}{\overset{+}{}}}}}} \xrightarrow{Et_3SiH} \underset{H}{\overset{OBF_3}{\underset{|}{\overset{|}{R-C-R}}}} \longrightarrow \underset{H}{\overset{R}{\underset{|}{\overset{+}{C}}}}\overset{R}{\diagdown} \xrightarrow{Et_3SiH} RCH_2R$$

A combination of Friedel-Crafts alkylation and reduction can be achieved using $InCl_3$ and chlorodimethylsilane. The Lewis acid presumably promotes both the Friedel-Craft reaction and the subsequent reduction.[179]

$$Br-\langle C_6H_4 \rangle \quad + \quad PhCH=O \xrightarrow[(CH_3)_2SiHCl]{5\ mol\%\ InCl_3} Br-\langle C_6H_4 \rangle CH_2Ph \quad 93\%$$

$$34{:}6{:}60\ o{:}m{:}p$$

There are several procedures for reductive condensation of silyl ethers with carbonyl compounds to form ethers. One method uses TMSOTf as the catalyst.[180]

$$\underset{PhCCH_3}{\overset{O}{\overset{\|}{}}} \quad + \quad PhCH_2OTMS \xrightarrow[Et_3SiH]{TMSOTf} \underset{\underset{CH_3}{|}}{PhCHOCH_2Ph} \quad 100\%$$

A number of related procedures have been developed. For example, TMSI can be used.[181]

$$\langle \rangle{=}O \quad + \quad TMSO{-}\langle \rangle \xrightarrow[Et_3SiH]{TMSI} \langle \rangle{-}O{-}\langle \rangle \quad 75\%$$

The trimethylsilyl group can be replaced by a dialkylsilyloxy group, in which case the silyl ether serves as the hydride donor.

$$PhCH{=}CHCH{=}O \quad + \quad (CH_3)_2HSiO(CH_2)_3CH_3 \xrightarrow{TMSI} PhCH{=}CHCH_2O(CH_2)_3CH_3 \quad 88\%$$

Ref. 182

177. M. Yato, K. Homma, and A. Ishida, *Heterocycles*, **49**, 233 (1998).
178. J. L. Frey, M. Orfanopoulos, M. G. Adlington, W. R. Dittman, Jr., and S. B. Silverman, *J. Org. Chem.*, **43**, 374 (1978).
179. T. Miyai, Y. Onishi, and A. Baba, *Tetrahedron Lett.*, **39**, 6291 (1998).
180. S. Hatakeyama, H. Mori, K. Kitano, H. Yamada, and M. Nishizawa, *Tetrahedron Lett.*, **35**, 4367 (1994).
181. M. B. Sassaman, K. D. Kotian, G. K. S. Prakash, and G. Olah, *J. Org. Chem.*, **52**, 4314 (1987).
182. K. Miura, K. Ootsuka, S. Suda, H. Nishikori, and A. Hosomi, *Synlett*, 313 (2002).

428

CHAPTER 5

*Reduction of
Carbon-Carbon Multiple
Bonds, Carbonyl
Groups, and Other
Functional Groups*

$$Ph(CH_2)_2OSiH[CH(CH_3)_2]_2 \quad + \quad PhCH{=}CHCH{=}O \quad \longrightarrow \quad Ph(CH_2)_2OCH_2CH{=}CHPh$$

89%

Ref. 183

These reactions presumably proceed by catalytic cycles in which the carbonyl component is silylated. The silyl ether can then act as a nucleophile, and an oxonium ion is generated by elimination of a disilyl ether. The reduction of the oxonium ion regenerates the silyl cation, which can continue the catalytic cycle.

$$RCH{=}O \quad + \quad {}^+SiR''_3 \quad \rightleftharpoons \quad RCH{=}O^+SiR''_3$$

$$RCH{=}O^+SiR''_3 + R'OSiR''_3 \longrightarrow \underset{\underset{R'}{\overset{\|}{O^+}}{\overset{\displaystyle RCH-OSiR''_3}{|}}}{\overset{}{}} \longrightarrow \cdots \longrightarrow RCH{=}O^+R'$$

$$RCH{=}O^+R' \quad + \quad H{-}SiR''_3 \longrightarrow RCH_2OR' \quad + \quad R''_3Si^+$$

Various other kinds of Lewis acids can also promote the reaction. For example, $Cu(OTf)_2$ and Et_3SiH have been used to prepare a number of benzyl and alkyl ethers.[184]

72%

The reductive condensation can also be carried out using $BiBr_3$ and Et_3SiH. The active catalyst under these conditions is Et_3SiBr, which is generated in situ.[185]

Reduction of ketones to triphenylsilyl ethers is effected by the unique Lewis acid perfluorotriphenylborane. Mechanistic and kinetic studies have provided considerable insight into the mechanism of this reaction.[186] The salient conclusion is that the hydride is delivered from a borohydride ion, not directly from the silane. Although the borane forms a Lewis acid-base complex with the ketone, its key function is in delivery of the hydride.

183. X. Jiang, J. S. Bajwa, J. Slade, K. Prasad, O. Repic, and T. J. Blacklock, *Tetrahedron Lett.*, **43**, 9225 (2002).
184. W.-C. Yang, X.-A. Lu, S. S. Kulkarni, and S.-C. Huang, *Tetrahedron Lett.*, **44**, 7837 (2003).
185. N. Komatsu, J. Ishida, and H. Suzuki, *Tetrahedron Lett.*, **38**, 7219 (1997).
186. D. J. Parks, J. M. Blackwell, and W. E. Piers, *J. Org. Chem.*, **65**, 3090 (2000).

Copper-catalyzed systems have been developed that reduce ketones directly to silyl ethers. The reactions involve chiral biphenyl diphosphine type ligands and silane or siloxane hydride donors.[187]

H Ar = 3,5-dimethylphenyl

I Ar = 3,5-*bis*-(*t*-butyl)phenyl

The reactions proceed with an e.e. of about 80% when the enantiopure ligand is used. Similar conditions using poly[oxy(methylsilylene)] (PMHS) as the hydride donor lead to reduction of aryl ketones with up to 98% e.e.[188]

98% yield
98% e.e.

5.4.2. Hydride Transfer from Carbon

There are also reactions in which hydride is transferred from carbon. The carbon-hydrogen bond has little intrinsic tendency to act as a hydride donor, so especially favorable circumstances are required to promote this reactivity. Frequently these reactions proceed through a cyclic TS in which a new C−H bond is formed simultaneously with the C–H cleavage. Hydride transfer is facilitated by high electron density at the carbon atom. Aluminum alkoxides catalyze transfer of hydride from an alcohol to a ketone. This is generally an equilibrium process and the reaction can be driven to completion if the ketone is removed from the system, by, e.g., distillation, in a process known as the *Meerwein-Pondorff-Verley reduction*.[189] The reverse reaction in which the ketone is used in excess is called the *Oppenauer oxidation*.

$$3\ R_2C{=}O\ +\ Al[OCH(CH_3)_2]_3\ \rightleftharpoons\ [R_2CHO]_3Al\ +\ 3\ CH_3\overset{\underset{\|}{O}}{C}CH_3$$

The reaction proceeds via a cyclic TS involving coordination of both the alcohol and ketone oxygens to the aluminum. Computational (DFT) and isotope effect studies are consistent with the cyclic mechanism.[190] Hydride donation usually takes place from

[187] B. H. Lipshutz, C. C. Caires, P. Kuipers, and W. Chrisman, *Org. Lett.*, **5**, 3085 (2003).

[188] B. H. Lipshutz, K. Noson, W. Chrisman, and A. Lower, *J. Am. Chem. Soc.*, **125**, 8779 (2003).

[189] A. L. Wilds, *Org. React.*, **2**, 178 (1944); C. F. de Graauw, J. A. Peters, H. van Bekkum, and J. Huskens, *Synthesis*, 1007 (1994).

[190] R. Cohen, C. R. Graves, S. T. Nguyen, J. M. L. Martin, and M. A. Ratner, *J. Am. Chem. Soc.*, **126**, 14796 (2004).

430

CHAPTER 5

*Reduction of
Carbon-Carbon Multiple
Bonds, Carbonyl
Groups, and Other
Functional Groups*

the less hindered face of the carbonyl group.[191] However, these conditions frequently promote equilibration of the alcohol stereoisomers.

Recently, enantioselective procedures involving chiral catalysts have been developed. The combination of BINOL and $Al(CH_3)_3$ can achieve 80% e.e. in the reduction of acetophenone.[192] Compound **J** is also an effective catalyst.[193]

J

Certain lanthanide alkoxides, such as $t\text{-BuOSmI}_2$, have also been found to catalyze hydride exchange between alcohols and ketones.[194] Isopropanol can serve as the reducing agent for aldehydes and ketones that are thermodynamically better hydride acceptors than acetone.

$$O_2N \text{—} \langle \rangle \text{—CH=O} \xrightarrow[t\text{-BuOSmI}_2]{\overset{\overset{\displaystyle CH_3CHCH_3}{|}}{\underset{\displaystyle OH}{}}} O_2N \text{—} \langle \rangle \text{—CH}_2OH$$

94%

Samarium metal in isopropanol also achieves reduction.[195] Like the Meerwein-Pondorff-Verley procedure, these conditions are believed to be under thermodynamic control and the more stable stereoisomer is the main product.[196]

Another reduction process, catalyzed by iridium chloride, is characterized by very high axial:equatorial product ratios for cyclohexanones and apparently involves hydride transfer from isopropanol.[197]

$$(CH_3)_3C \xrightarrow[\substack{(CH_3O)_3P, \ H_2O \\ (CH_3)_2CHOH}]{IrCl_4, \ HCl} (CH_3)_3C \text{—} \langle \rangle \text{—OH}$$

Formic acid can also act as a donor of hydrogen, and the driving force in this case is the formation of carbon dioxide. A useful application is the Clark-Eschweiler

[191.] F. Nerdel, D. Frank, and G. Barth, *Chem. Ber.*, **102**, 395 (1969).
[192.] E. J. Campbell, H. Zhou, and S. T. Nguyen, *Angew. Chem. Int. Ed. Engl.*, **41**, 1020 (2002).
[193.] T. Ooi, H. Ichikawa, and K. Maruoka, *Angew. Chem. Int. Ed. Engl.*, **40**, 3610 (2001).
[194.] J. L. Namy, J. Souppe, J. Collin, and H. B. Kagan, *J. Org. Chem.*, **49**, 2045 (1984).
[195.] S. Fukuzawa, N. Nakano, and T. Saitoh, *Eur. J. Org. Chem.*, 2863 (2004).
[196.] D.A. Evans, S. W. Kaldor, T. K. Jones, J. Clardy, and T. J. Stout, *J. Am. Chem. Soc.*, **112**, 7001 (1990).
[197.] E. L. Eliel, T. W. Doyle, R. O. Hutchins, and E. C. Gilbert, *Org. Synth.*, **50**, 13 (1970).

reductive methylation of amines, in which heating a primary or secondary amine with formaldehyde and formic acid results in complete methylation to the tertiary amine.[198]

$$RNH_2 + CH_2{=}O + HCO_2H \longrightarrow RN(CH_3)_2 + CO_2$$

The hydride acceptor is the iminium ion that results from condensation of the amine with formaldehyde.

5.5. Reduction Reactions Involving Hydrogen Atom Donors

Reduction by hydrogen atom donors involves free radical intermediates and usually proceeds by chain mechanisms. Tri-*n*-butylstannane is the most prominent example of this type of reducing agent. Other synthetically useful hydrogen atom donors include hypophosphorous acid, dialkyl phosphites, and *tris*-(trimethylsilyl)silane. The processes that have found most synthetic application are reductive replacement of halogen and various types of thiono esters.

Tri-*n*-butylstannane is able to reductively replace halogen by hydrogen. Mechanistic studies indicate a free radical chain mechanism.[199] The order of reactivity for the halides is RI > RBr > RCl > RF, which reflects the relative ease of the halogen atom abstraction.[200]

$$In\cdot + Bu_3SnH \longrightarrow In{-}H + Bu_3Sn\cdot \quad (In\cdot = \text{initiator})$$
$$Bu_3Sn\cdot + R{-}X \longrightarrow R\cdot + Bu_3SnX$$
$$R\cdot + Bu_3SnH \longrightarrow RH + Bu_3Sn\cdot$$

Scheme 5.8 gives several examples of dehalogenation using tri-*n*-butylstannane. Entries 1 and 2 are examples from the early studies of this method. Entries 3 and 4 illustrate selective dehalogenation of polyhalogenated compounds. The stabilizing effect of the remaining halogen on the radical intermediate facilitates partial dehalogenation. These reactions also demonstrate stereoselectivity. In Entry 3, the stereochemical preference is for hydrogen abstraction from the more accessible face of the radical intermediate. Entry 4 shows retention of configuration at the fluorocyclopropyl carbon. (The stereoisomeric compound also reacts with retention of configuration.) This result indicates that hydrogen abstraction is faster than inversion for these cyclopropyl radicals (see Part A, Section 11.1.5).

A procedure that is catalytic in Bu$_3$SnH and uses NaBH$_4$ as the stoichiometric reagent has been developed.[201] This method has advantages in the isolation and purification of product. Entry 5 is an example of this procedure. The reaction was carried

[198.] M. L. Moore, *Org. React.*, **5**, 301 (1949); S. H. Pine and B. L. Sanchez, *J. Org. Chem.*, **36**, 829 (1971).
[199.] L. W. Menapace and H. G. Kuivila, *J. Am. Chem. Soc.*, **86**, 3047 (1964).
[200.] H. G. Kuivila and L. W. Menapace, *J. Org. Chem.*, **28**, 2165 (1963).
[201.] E. J. Corey and J. W. Suggs, *J. Org. Chem.*, **40**, 2554 (1975).

CHAPTER 5

*Reduction of
Carbon-Carbon Multiple
Bonds, Carbonyl
Groups, and Other
Functional Groups*

Scheme 5.8. Dehalogenation with Stannanes

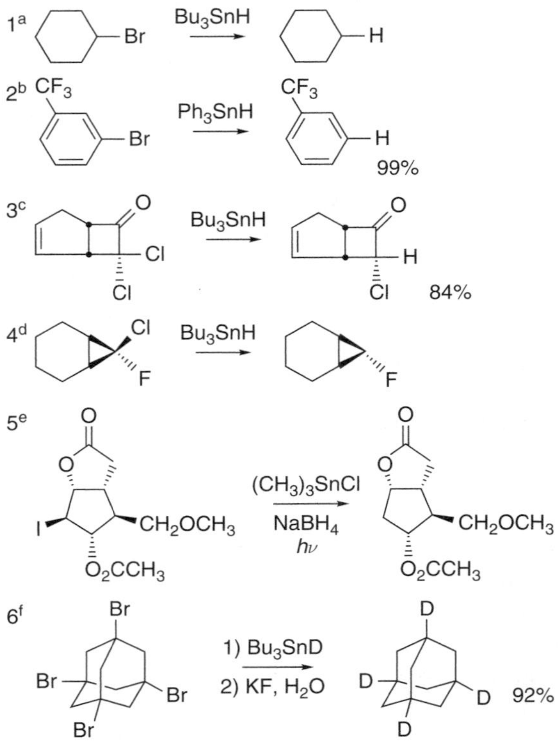

a. H. G. Kuivila, L. W. Menapace, and C. R. Warner, *J. Am. Chem. Soc.*,
   84, 3584 (1962).
b. D. H. Lorenz, P. Shapiro, A. Stern, and E. J. Becker, *J. Org. Chem.*
   28, 2332 (1963).
c. W. T. Brady and E. F. Hoff, Jr., *J. Org. Chem.*, **35**, 3733 (1970).
d. T. Ando, F. Namigata, H. Yamanaka, and W. Funasaka, *J. Am. Chem.
   Soc.*, **89**, 5719 (1967).
e. E. J. Corey and J. W. Suggs, *J. Org. Chem.*, **40**, 2554 (1975).
f. J. E. Leibner and J. Jacobson, *J. Org. Chem.*, **44**, 449 (1979).

out under illumination to provide for chain initiation, and the reactant was prepared
by an iodolactonization reaction. The sequence iodolactonization-dehalogenation is
frequently used in the synthesis of five-membered lactones. Entry 6 illustrates the
use of dehalogenation with deuterium incorporation. The addition of the fluoride salt
facilitates workup by precipitation of tin by-products.

Hypophosphorous acid has been used as a hydrogen atom donor in the dehalogenation of nucleosides.[202]

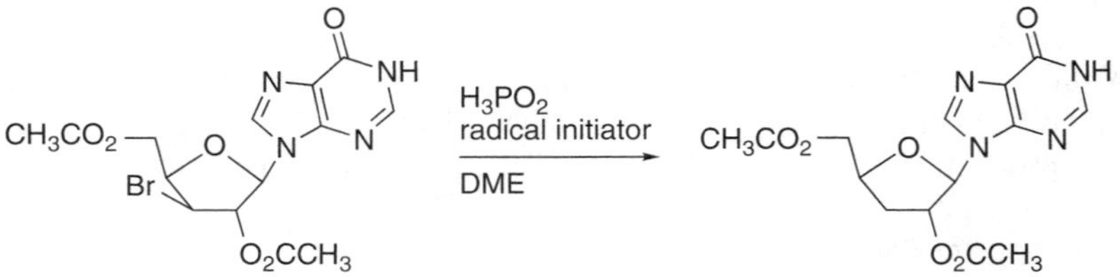

[202.] S. Takamatsu, S. Katayama, N. Hirose, M. Naito, and K. Izawa, *Tetrahedron Lett.*, **42**, 7605 (2001).

Tri-*n*-butyltin hydride also serves as a hydrogen atom donor in radical-mediated methods for reductive deoxygenation of alcohols via thiono esters.[203] The alcohol is converted to a thiocarbonyl derivative. These thiono esters undergo a radical reaction with tri-*n*-butyltin hydride. The resulting radicals fragment to give the alkyl radical, and the chain is propagated by hydrogen atom abstraction.

$$R-OCX + Bu_3Sn\cdot \longrightarrow R\overset{S-SnBu_3}{\underset{\cdot}{OCX}} \longrightarrow R\cdot + XCS-SnBu_3$$

$$R\cdot + Bu_3SnH \longrightarrow R-H + Bu_3Sn\cdot$$

This procedure gives good yields from secondary alcohols and by appropriate adjustment of conditions can also be adapted to primary alcohols.[204]

Owing to the expense, toxicity, and purification problems associated with use of stoichiometric amounts of tin hydrides, there has been interest in finding other hydrogen atom donors.[205] The trialkylboron-oxygen system for radical generation (see Part A, Section 11.1.4) has been used with *tris-*(trimethylsilyl)silane or diphenylsilane as a hydrogen donor.[206]

$$c\text{-}C_{12}H_{23}OCO-\!\!\!\bigotimes\!\!\!-F \xrightarrow[\text{(Ph)}_2SiH_2]{Et_3B, O_2} c\text{-}C_{12}H_{24}$$

96%

Chain reaction mechanism

$$(C_2H_5)_3B + O_2 \longrightarrow C_2H_5\cdot$$

$$C_2H_5\cdot + R_3SiH \longrightarrow C_2H_6 + R_3Si\cdot$$

$$R_3Si\cdot + R'O\overset{\parallel}{\underset{S}{C}}OR' \longrightarrow R'O\overset{\cdot}{\underset{SSiR_3}{C}}OR'$$

$$R'O\overset{\cdot}{\underset{SSiR_3}{C}}OR' \longrightarrow R'\cdot + R'O_2CSSiR_3$$

$$R'\cdot + R_3SiH \longrightarrow R'-H + R_3Si\cdot$$

The alcohol derivatives that have been successfully deoxygenated include thionocarbonates and xanthates.[207] Peroxides can be used as initiators.[208]

Scheme 5.9 illustrates some of the conditions that have been developed for the reductive deoxygenation of alcohols. Entries 1 to 4 illustrate the most commonly used methods for generation of thiono esters and their reduction by tri-*n*-butylstannane. These include formation of thiono carbonates (Entry 1), xanthates (Entry 2), and thiono imidazolides (Entries 3 and 4). Entry 5 is an example of use of dimethyl phosphite as the hydrogen donor. Entry 6 uses *tris-*(trimethylsilyl)silane as the hydrogen atom donor.

203. D. H. R. Barton and S. W. McCombie, *J. Chem. Soc., Perkin Trans. 1*, 1574 (1975).For reviews of this method, see W. Hartwig, *Tetrahedron*, **39**, 2609 (1983); D. Crich and L. Quintero, *Chem. Rev.*, **89**, 1413 (1989).
204. D. H. R. Barton, W. B. Motherwell, and A. Stange, *Synthesis*, 743 (1981).
205. A. Studer and S. Amrein, *Synthesis*, 835 (2002).
206. D. H. R. Barton, D. O. Jang, and J. C. Jaszberenyi, *Tetrahedron Lett.*, **31**, 4681 (1990).
207. J. N. Kirwan, B. P. Roberts, and C. R. Willis, *Tetrahedron Lett.*, **31**, 5093 (1990).
208. D. H. Barton, D. O. Jang, and J. C. Jaszberenyi, *Tetrahedron Lett.*, **33**, 7187 (1991).

434

CHAPTER 5

*Reduction of
Carbon-Carbon Multiple
Bonds, Carbonyl
Groups, and Other
Functional Groups*

Scheme 5.9. Deoxygenation of Alcohols via Thiono Esters and Related Derivatives

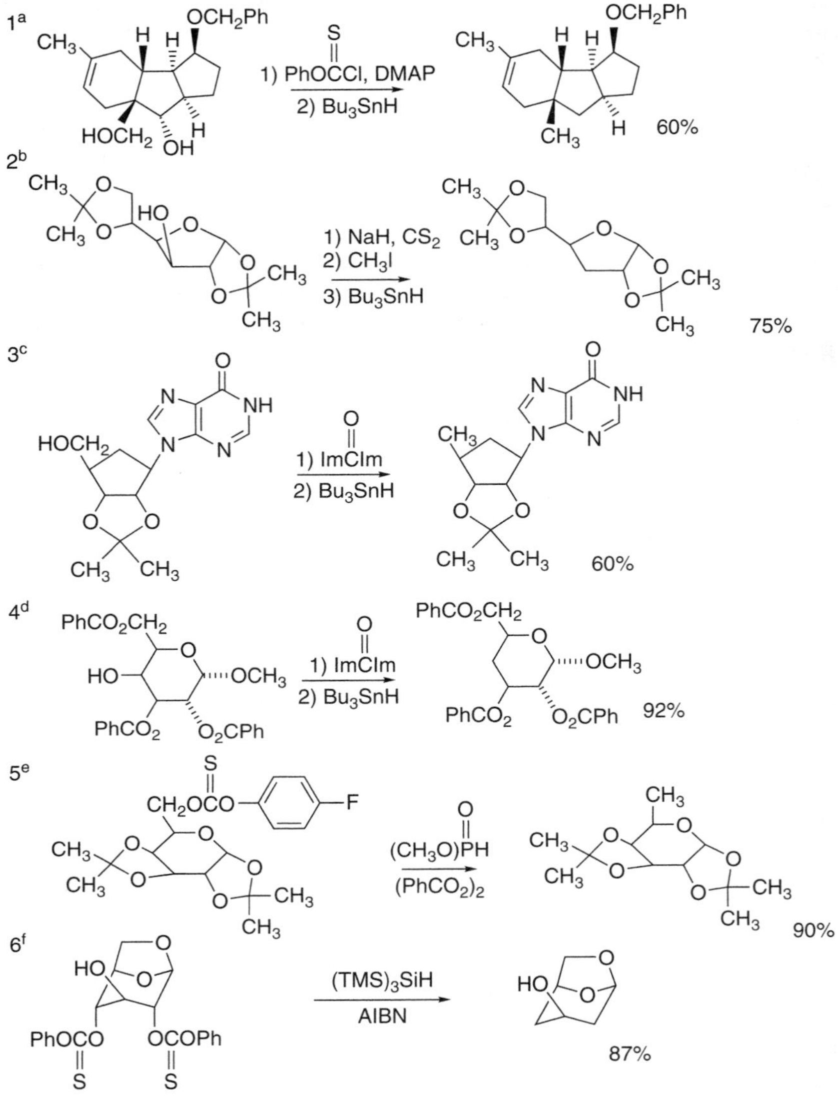

a. H. J. Liu and M. G. Kulkarni, *Tetrahedron Lett.*, **26**, 4847 (1985).
b. S. Iacono and J. R. Rasmussen, *Org. Synth.*, **64**, 57 (1985).
c. O. Miyashita, F. Kasahara, T. Kusaka, and R. Marumoto, *J. Antibiot.*, **38**, 98 (1985).
d. J. R. Rasmussen, C. J. Slinger, R. J. Kordish, and D. D. Newman-Evans, *J. Org. Chem.*, **46**, 4843 (1981).
e. D. H. R. Barton, D. O. Jang, and J. C. Jaszberenyi, *Tetrahedron Lett.*, **33**, 2311 (1992).
f. D. H. R. Barton, D. O. Jang, and J. C. Jaszberenyi, *Tetrahedron Lett.*, **33**, 6629 (1992).

5.6. Dissolving-Metal Reductions

Another group of synthetically useful reductions employs a metal as the reducing agent. The organic reactant under these conditions accepts one or more electrons from the metal. The subsequent course of the reaction depends on the structure of the reactant and reaction conditions. Three broad types of reactions can be recognized and these are discussed separately. They include reactions in which the overall change involves: (a) net addition of hydrogen, (b) reductive removal of a functional group, and (c) formation of carbon-carbon bonds.

5.6.1. Addition of Hydrogen

5.6.1.1. Reduction of Ketones and Enones. Although the method has been supplanted for synthetic purposes by hydride donors, the reduction of ketones to alcohols in ammonia or alcohols provides mechanistic insight into dissolving-metal reductions. The outcome of the reaction of ketones with metal reductants is determined by the fate of the initial ketyl radical formed by a single-electron transfer. The radical intermediate, depending on its structure and the reaction medium, may be protonated, disproportionate, or dimerize.[209] In hydroxylic solvents such as liquid ammonia or in the presence of an alcohol, the protonation process dominates over dimerization. Net reduction can also occur by a disproportionation process. As is discussed in Section 5.6.3, dimerization can become the dominant process under conditions in which protonation does not occur rapidly.

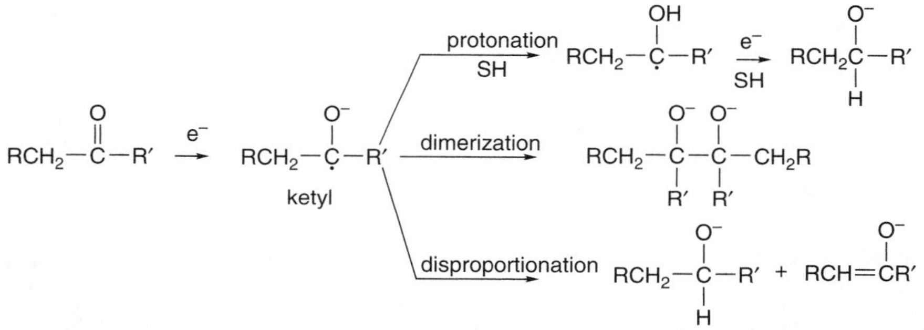

α, β-Unsaturated carbonyl compounds are cleanly reduced to the enolate of the corresponding saturated ketone on reduction with lithium in ammonia.[210] Usually an alcohol is added to the reduction solution to serve as the proton source.

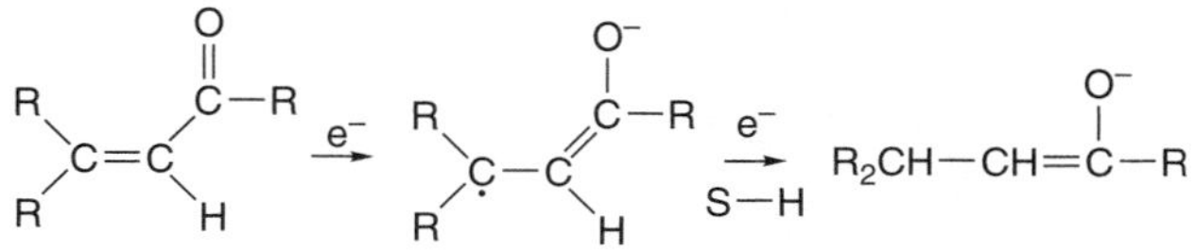

As noted in Chapter 1, this is one of the best methods for generating a specific enolate of a ketone. The enolate generated by conjugate reduction can undergo the characteristic alkylation and addition reactions that are discussed in Chapters 1 and 2. When this is the objective of the reduction, it is important to use only one equivalent of the proton donor. Ammonia, being a weaker acid than an aliphatic ketone, does

209. V. Rautenstrauch and M. Geoffroy, *J. Am. Chem. Soc.*, **99**, 6280 (1977); J. W. Huffman and W. W. McWhorter, *J. Org. Chem.*, **44**, 594 (1979); J. W. Huffman, P. C. Desai, and J. E. LaPrade, *J. Org. Chem.*, **48**, 1474 (1983).

210. D. Caine, *Org. React.*, **23**, 1 (1976).

436

CHAPTER 5

*Reduction of
Carbon-Carbon Multiple
Bonds, Carbonyl
Groups, and Other
Functional Groups*

not act as a proton donor toward an enolate, and the enolate remains available for subsequent reaction, as in the tandem alkylations shown below. If the saturated ketone is the desired product, the enolate is protonated either by use of excess proton donor during the reduction or on workup.

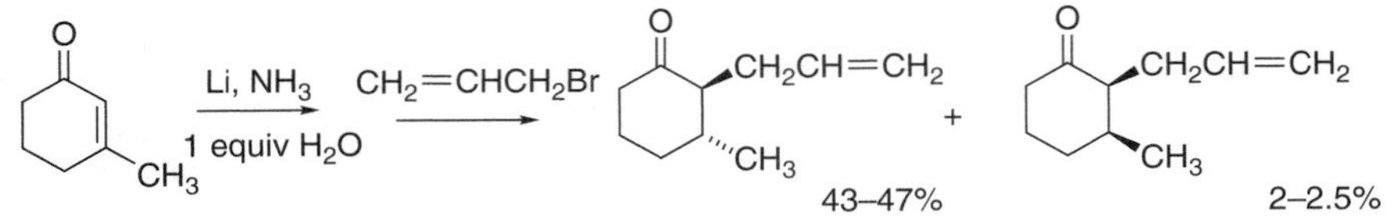

43–47%   2–2.5%

Ref. 211

47%

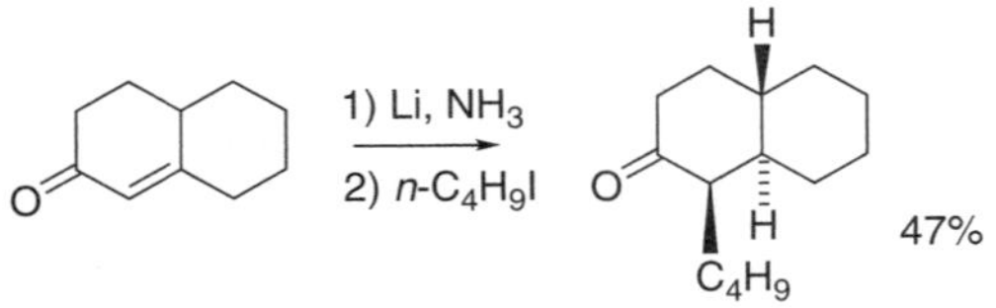

Ref. 212

The stereochemistry of conjugate reduction is established by the proton transfer to the β-carbon. In the well-studied case of $\Delta^{1,9}$-2-octalones, the ring junction is usually *trans*.[213]

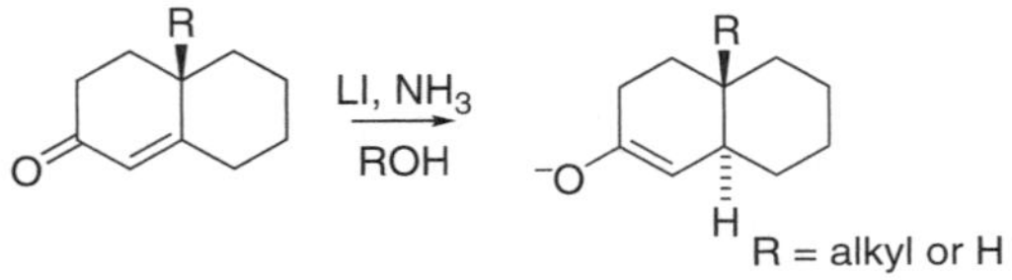

R = alkyl or H

The stereochemistry is controlled by a stereoelectronic preference for protonation perpendicular to the enolate system and, given that this requirement is met, the stereochemistry normally corresponds to protonation of the most stable conformation of the dianion intermediate from its least hindered side.

5.6.1.2. Dissolving-Metal Reduction of Aromatic Compounds and Alkynes. Dissolving-metal systems constitute the most general method for partial reduction of aromatic rings. The reaction is called the *Birch reduction*,[214] and the usual reducing medium is lithium or sodium in liquid ammonia. An alcohol is usually added to serve as a proton source. The reaction occurs by two successive electron transfer/protonation steps.

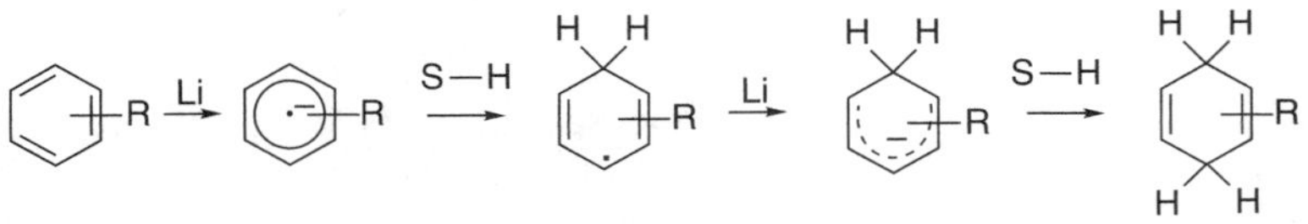

211. D. Caine, S. T. Chao, and H. A. Smith, *Org. Synth.*, **56**, 52 (1977).
212. G. Stork, P. Rosen, and N. L. Goldman, *J. Am. Chem. Soc.*, **83**, 2965 (1961).
213. G. Stork, P. Rosen, N. Goldman, R. V. Coombs, and J. Tsuji, *J. Am. Chem. Soc.*, **87**, 275 (1965); M. J. T. Robinson, *Tetrahedron*, **21**, 2475 (1965).
214. A. J. Birch and G. Subba Rao, *Adv. Org. Chem.*, **8**, 1 (1972); R. G. Harvey, *Synthesis*, 161 (1980); J. M. Hook and L. N. Mander, *Nat. Prod. Rep.*, **3**, 35 (1986); P. W. Rabideau, *Tetrahedron*, **45**, 1599 (1989); A. J. Birch, *Pure Appl. Chem.*, **68**, 553 (1996).

The isolated double bonds in the dihydro product are much less easily reduced than the conjugated ring, so the reduction stops at the dihydro stage. Alkyl and alkoxy aromatics, phenols, and benzoate anions are the most useful reactants for Birch reduction. In aromatic ketones and nitro compounds, the substituents are reduced in preference to the aromatic ring. Substituents also govern the position of protonation. Alkyl and alkoxy aromatics normally give the 2,5-dihydro derivative. Benzoate anions give 1,4-dihydro derivatives.

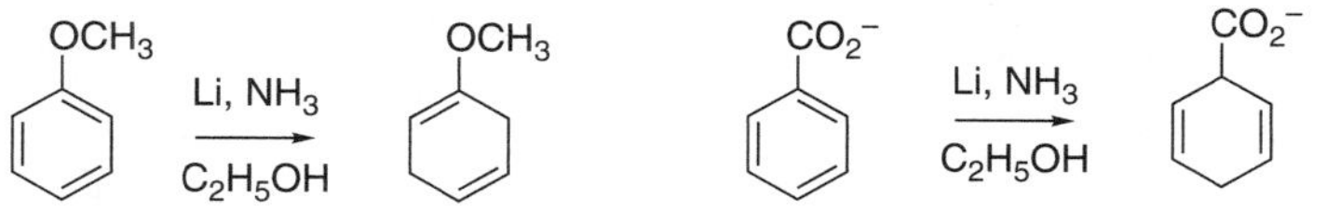

The structure of the products is determined by the site of protonation of the radical anion intermediate formed after the first electron transfer step. In general, ERG substituents favor protonation at the *ortho* position, whereas EWGs favor protonation at the *para* position.[215] Addition of a second electron gives a pentadienyl anion, which is protonated at the center carbon. As a result, 2,5-dihydro products are formed with alkyl or alkoxy substituents and 1,4-products are formed from EWG substituents. The preference for protonation of the central carbon of the pentadienyl anion is believed to be the result of the greater 1,2 and 4,5 bond order and a higher concentration of negative charge at C(3).[216] The reduction of methoxybenzenes is of importance in the synthesis of cyclohexenones via hydrolysis of the intermediate enol ethers.

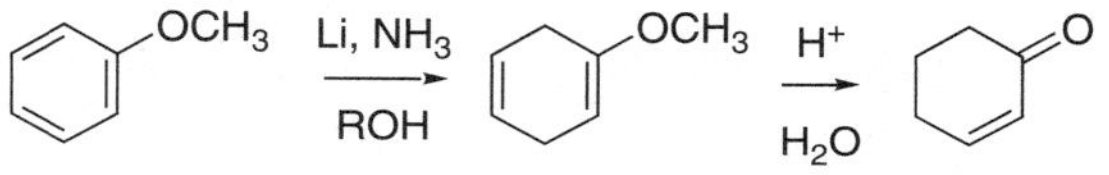

The anionic intermediates formed in Birch reductions can be used in tandem alkylation reactions.

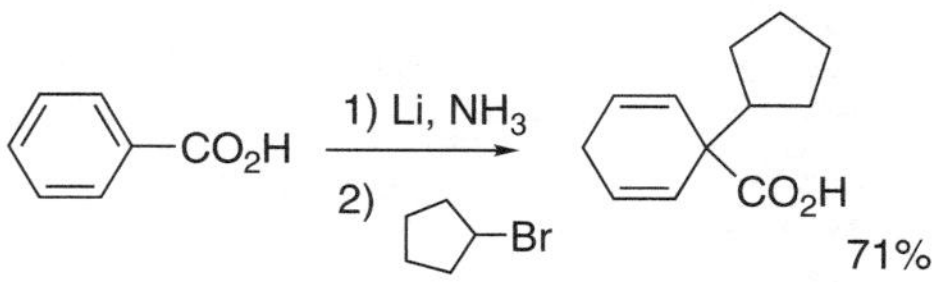

Ref. 217

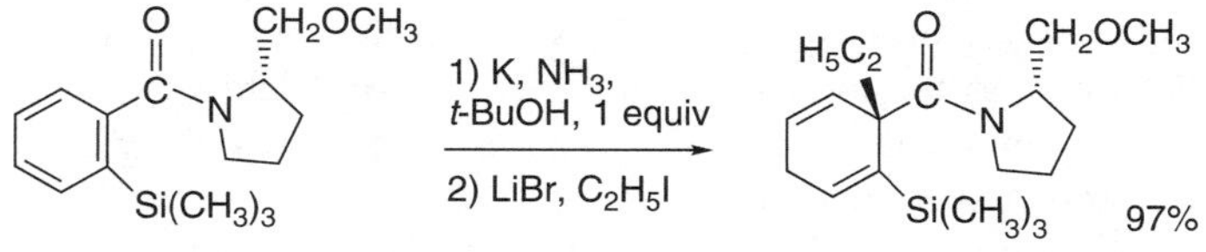

Ref. 218

215. A. J. Birch, A. L. Hinde, and L. Radom, *J. Am. Chem. Soc.*, **102**, 2370 (1980); H. E. Zimmerman and P. A. Wang, *J. Am. Chem. Soc.*, **112**, 1280 (1990).
216. P. W. Rabideau and D. L. Huser, *J. Org. Chem.*, **48**, 4266 (1983); H. E. Zimmerman and P. A. Wang, *J. Am. Chem. Soc.*, **115**, 2205 (1993).
217. P. A. Baguley and J. C. Walton, *J. Chem. Soc., Perkin Trans. 1*, 2073 (1998).
218. A. G. Schultz and L. Pettus, *J. Org. Chem.*, **62**, 6855 (1997).

CHAPTER 5

*Reduction of
Carbon-Carbon Multiple
Bonds, Carbonyl
Groups, and Other
Functional Groups*

Scheme 5.10. Birch Reduction of Aromatic Rings

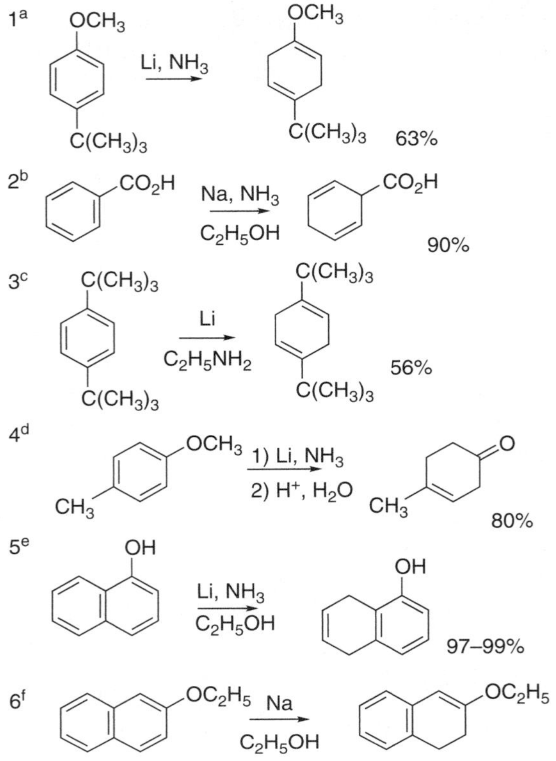

a. D. A. Bolon, *J. Org. Chem.* **35**, 715 (1970).
b. M. E. Kuehne and B. F. Lambert, *Org. Synth.*, **V**, 400 (1973).
c. H. Kwart and R. A. Conley, *J. Org. Chem.*, **38**, 2011 (1973).
d. E. A. Braude, A. A. Webb, and M. U. S. Sultanbawa, *J. Chem. Soc.*, 3328 (1958); W. C. Agosta and W. L. Schreiber, *J. Am. Chem. Soc.*, **93**, 3947 (1971).
e. C. D. Gutsche and H. H. Peter, *Org. Synth.*, **IV**, 887 (1963).
f. M. D. Soffer, M. P. Bellis, H. E. Gellerson, and R. A. Stewart, *Org. Synth.*, **IV**, 903 (1963).

Scheme 5.10 lists some examples of the use of the Birch reduction. Entries 1 and 2 illustrate the usual regioselectivity for alkoxy aromatics and for benzoic acid. Entry 3 uses an alkylamine as the solvent. In the case cited, the yield was much better than that obtained using ammonia. Entry 4 illustrates the preparation of a cyclohex-3-enone via the Birch reduction route. Entries 5 and 6 show an interesting contrast in the regioselectivity of naphthalene derivatives. The selective reduction of the unsubstituted ring may reflect the more difficult reduction of the ring having a deprotonated oxy substituent. On the other hand, empirical evidence indicates that ERG substituents in the 2-position direct reduction to the substituted ring.[219] The basis of this directive effect does not seem to have been developed in modern electronic terms.

219. M. D. Soffer, R. A. Stewart, J. C. Cavagnol, H. E. Gellerson, and E. A. Bowler, *J. Am. Chem. Soc.*, **72**, 3704 (1950).

Reduction of acetylenes can be done with sodium in ammonia,[220] lithium in low molecular weight amines,[221] or sodium in HMPA containing *t*-butanol as a proton source,[222] all of which lead to the *E*-alkene. The reaction is assumed to involve successive electron transfer and protonation steps.

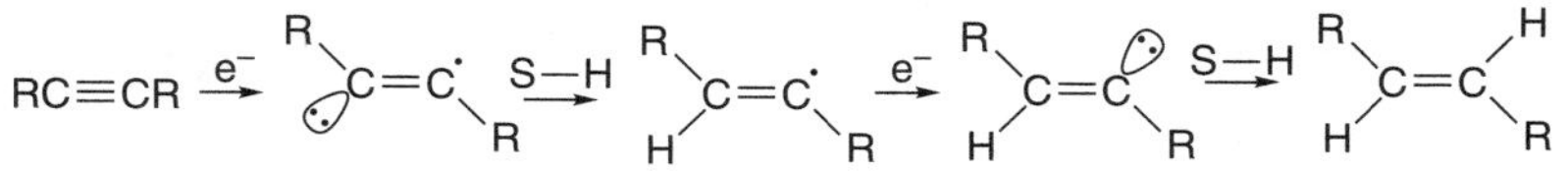

5.6.2. Reductive Removal of Functional Groups

The reductive removal of halogen can be accomplished with lithium or sodium. Tetrahydrofuran containing *t*-butanol is a useful reaction medium. Good results have also been achieved with polyhalogenated compounds by using sodium in ethanol.

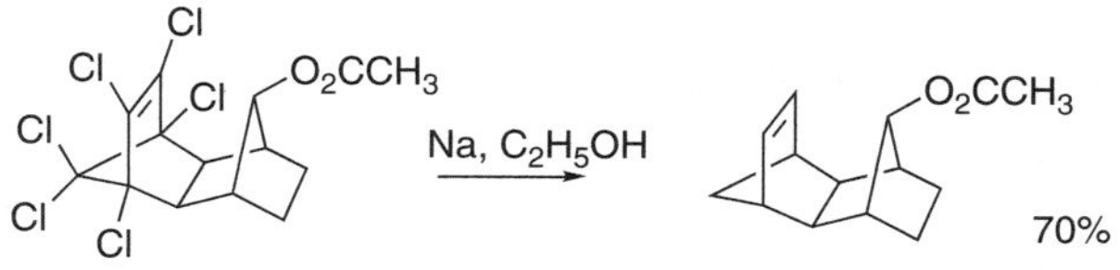

Ref. 223

An important synthetic application of this reaction is in dehalogenation of dichloro- and dibromocyclopropanes. The dihalocyclopropanes are accessible via carbene addition reactions (see Section 10.2.3). Reductive dehalogenation can also be used to introduce deuterium at a specific site. The mechanism of the reaction involves electron transfer to form a radical anion, which then fragments with loss of a halide ion. The resulting radical is reduced to a carbanion by a second electron transfer and subsequently protonated.

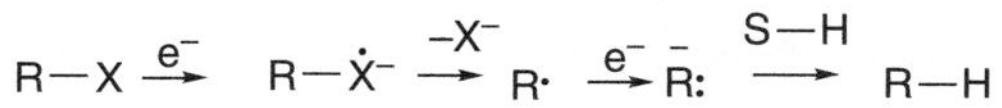

Phosphate groups can also be removed by dissolving-metal reduction. Reductive removal of vinyl phosphate groups is one method for conversion of a carbonyl compound to an alkene.[224] (See Section 5.7.2 for other methods.) The required vinyl phosphate esters are obtained by phosphorylation of the enolate with diethyl phosphorochloridate or *N,N,N′,N′*-tetramethyldiamidophosphorochloridate.[225]

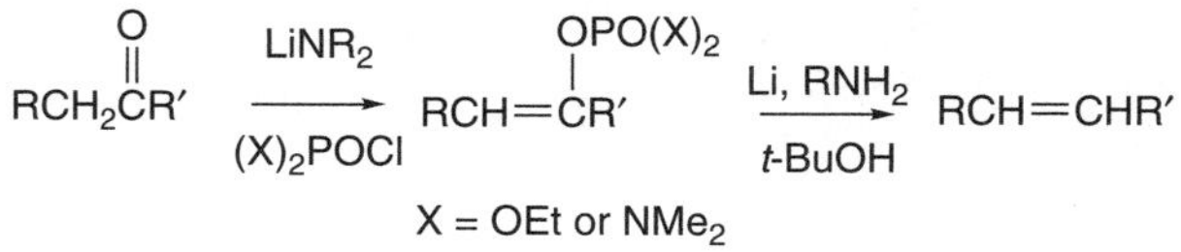

220. K. N. Campbell and T. L. Eby, *J. Am. Chem. Soc.*, **63**, 216, 2683 (1941); A. L. Henne and K. W. Greenlee, *J. Am. Chem. Soc.*, **65**, 2020 (1943).
221. R. A. Benkeser, G. Schroll, and D. M. Sauve, *J. Am. Chem. Soc.*, **77**, 3378 (1955).
222. H. O. House and E. F. Kinloch, *J. Org. Chem.*, **39**, 747 (1974).
223. B. V. Lap and M. N. Paddon-Row, *J. Org. Chem.*, **44**, 4979 (1979).
224. R. E. Ireland and G. Pfister, *Tetrahedron Lett.*, 2145 (1969).
225. R. E. Ireland, D. C. Muchmore, and U. Hengartner, *J. Am. Chem. Soc.*, **94**, 5098 (1972).

440

CHAPTER 5

*Reduction of
Carbon-Carbon Multiple
Bonds, Carbonyl
Groups, and Other
Functional Groups*

Ketones can also be reduced to alkenes via enol triflates. The use of $Pd(OAc)_2$ and triphenylphosphine as the catalyst and tertiary amines as the hydrogen donors is effective.[226]

Ref. 227

Reductive removal of oxygen from aromatic rings can also be achieved by reductive cleavage of aryl diethyl phosphate esters.

Ref. 228

There are also examples in which phosphate esters of saturated alcohols are reductively deoxygenated.[229] Mechanistic studies of the cleavage of aryl dialkyl phosphates have indicated that the crucial C−O bond cleavage occurs after transfer of two electrons.[230]

For preparative purposes, titanium metal can be used in place of sodium or lithium in liquid ammonia for both the vinyl phosphate[231] and aryl phosphate[232] cleavages. The titanium metal is generated in situ from $TiCl_3$ by reduction with potassium metal in tetrahydrofuran.

Scheme 5.11 shows some examples of these reductive reactions. Entry 1 is an example of conditions that have been applied to both alkyl and aryl halides. The reaction presumably proceeds through formation of a Grignard reagent, which then undergoes protonolysis. Entries 2 and 3 are cases of the dehalogenation of polyhalogenated compounds by sodium in *t*-butanol. Entry 4 illustrates conditions that were found useful for monodehalogenation of dibromo- and dichlorocyclopropanes. This method is not very stereoselective. In the example given, the ratio of *cis:trans* product was 1.2:1. Entries 5 to 7 are cases of dissolving-metal reduction of vinyl and aryl phosphates.

226. W. J. Scott and J. K. Stille, *J. Am. Chem. Soc.*, **108**, 3033 (1986); L. A. Paquette, P. G. Meister, D. Friedrich, and D. R. Sauer, *J. Am. Chem. Soc.*, **115**, 49 (1993).
227. K. I. Keverline, P. Abraham, A. H. Lewin, and F. I. Carroll, *Tetrahedron Lett.* **36**, 3099 (1995).
228. R. A. Rossi and J. F. Bunnett, *J. Org. Chem.*, **38**, 2314 (1973).
229. R. R. Muccino and C. Djerassi, *J. Am. Chem. Soc.*, **96**, 556 (1974).
230. S. J. Shafer, W. D. Closson, J. M. F. van Dijk, O. Piepers, and H. M. Buck, *J. Am. Chem. Soc.*, **99**, 5118 (1977).
231. S. C. Welch and M. E. Walters, *J. Org. Chem.*, **43**, 2715 (1978).
232. S. C. Welch and M. E. Walters, *J. Org. Chem.*, **43**, 4797 (1978).

Scheme 5.11. Reductive Dehalogenation and Deoxygenation by Dissolving Metals

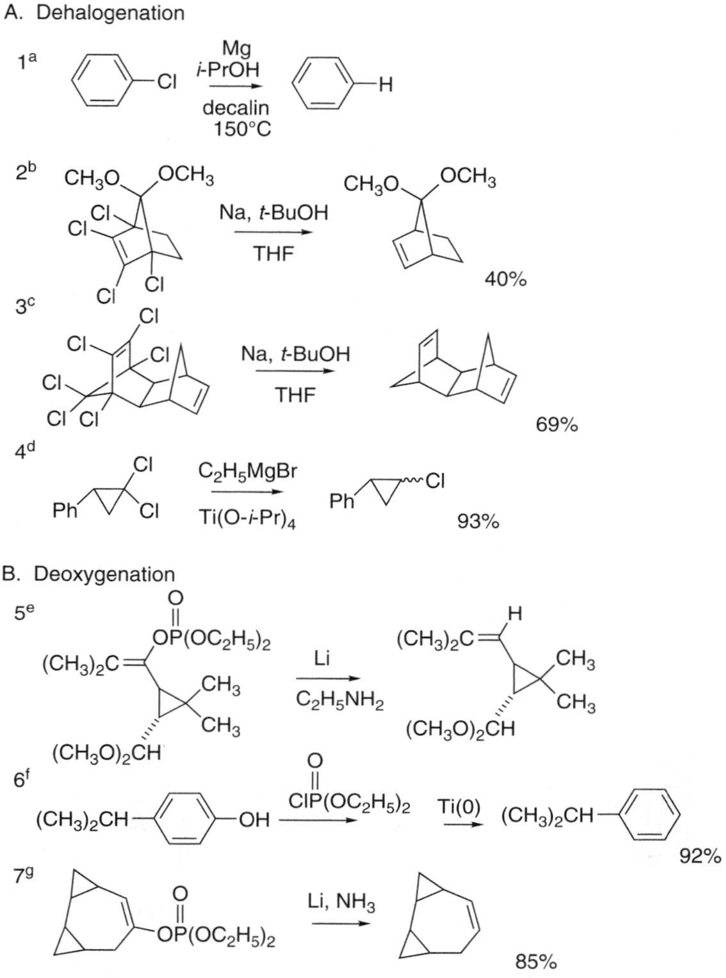

A. Dehalogenation

B. Deoxygenation

a. D. Bryce-Smith and B. J. Wakefield, *Org. Synth.*, **47**, 103 (1967).
b. P. G. Gassman and J. L. Marshall, *Org. Synth.*, **48**, 68 (1968).
c. B. V. Lap and M. N. Paddon-Row, *J. Org. Chem.*, **44**, 4979 (1979).
d. J. R. Al Duyayymi, M. S. Baird, I. G. Bolesov, V. Tversovsky, and M. Rubin, *Tetrahedron Lett.*, **37**, 8933 (1996).
e. S. C. Welch and T. A. Valdes, *J. Org. Chem.*, **42**, 2108 (1977).
f. S. C. Welch and M. E. Walter, *J. Org. Chem.*, **43**, 4797 (1978).
g. M. R. Detty and L. A. Paquette, **99**, 821 (1977).

Both metallic zinc and aluminum amalgam are milder reducing agents than the alkali metals. These reductants selectively remove oxygen and sulfur functional groups α to carbonyl groups. The mechanistic picture that seems most generally applicable is a net two-electron reduction with expulsion of the oxygen or sulfur substituent as an anion. The reaction must be a concerted process, because the isolated functional groups are not reduced under these conditions.

442

CHAPTER 5

*Reduction of
Carbon-Carbon Multiple
Bonds, Carbonyl
Groups, and Other
Functional Groups*

Another useful reagent for reduction of α-acetoxyketones and similar compounds is samarium diiodide.[233] SmI_2 is a strong one-electron reducing agent, and it is believed that the reductive elimination occurs after a net two-electron reduction of the carbonyl group.

These conditions were used, for example, in the preparation of the anticancer compound 10-deacetoxytaxol.

Ref. 234

Scheme 5.12 gives some examples of the reductive removal of functional groups adjacent to carbonyl groups. Entry 1 is an application of this reaction as it was used in an early steroid synthesis. The reaction in Entry 2 utilizes calcium in ammonia for the reduction. The reaction in Entry 3 converts the acyloin derived from dimethyl decanedicarboxylate into cyclodecanone. In the reaction in Entry 4, a sulfonate group is removed. In Entry 5 an epoxide is opened using aluminum amalgam, and in Entry 6 a lactone ring is opened. The latter reaction was part of a synthetic sequence in which the lactone intermediate was used to establish the stereochemistry of the acyclic product. The reaction in Entry 7 removes a sulfinyl group. Keto sulfoxides can be obtained by acylation of the anion of dimethylsulfoxide, so this reaction constitutes a general route to ketones (see Section 2.3.2). The reaction in Entry 8 is a *vinylogous* version of the reduction. The reductant in Entries 9 and 10 is SmI_2. In Entry 9, the 2-phenylcyclohexyloxy group that is removed was used earlier in the synthesis as a chiral auxiliary. Samarium diiodide is useful for deacetoxylation or dehydroxylation of α-oxygenated lactones derived from carbohydrates (Entry 10).[235] The reaction is also applicable to protected hydroxy groups, such as in acetonides. The reactions in Scheme 5.12 include quite a broad range of reductable groups, including some (e.g., ether) that are modest leaving groups.

233. G. A. Molander and G. Hahn, *J. Org. Chem.*, **51**, 1135 (1986).
234. R. A. Holton, C. Somoza, and K.-B. Chai, *Tetrahedron Lett.*, **35**, 1665 (1994).
235. S. Hanessian, C. Girard, and J. L. Chiara, *Tetrahedron Lett.*, **33**, 573 (1992).

Scheme 5.12. Reductive Removal of Functional Groups from α-Substituted Carbonyl Compounds

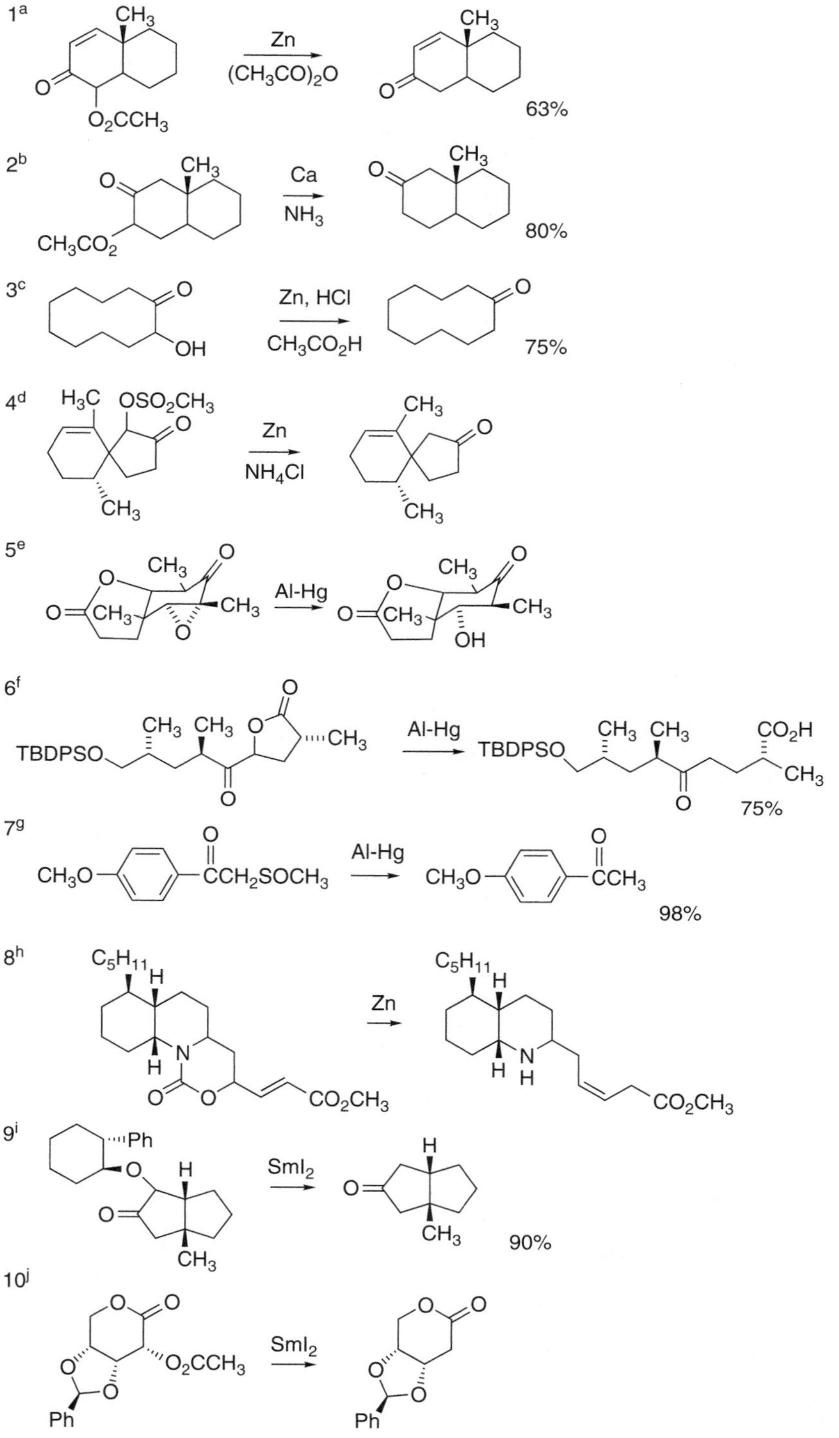

1[a]

$$\xrightarrow[(CH_3CO)_2O]{Zn}$$

63%

2[b]

$$\xrightarrow[NH_3]{Ca}$$

80%

3[c]

$$\xrightarrow[CH_3CO_2H]{Zn, HCl}$$

75%

4[d]

$$\xrightarrow[NH_4Cl]{Zn}$$

5[e]

$$\xrightarrow{Al\text{-}Hg}$$

6[f]

$$\xrightarrow{Al\text{-}Hg}$$

75%

7[g]

$$\xrightarrow{Al\text{-}Hg}$$

98%

8[h]

$$\xrightarrow{Zn}$$

9[i]

$$\xrightarrow{SmI_2}$$

90%

10[j]

$$\xrightarrow{SmI_2}$$

(*Continued*)

Scheme 5.12. (*Continued*)

CHAPTER 5

*Reduction of
Carbon-Carbon Multiple
Bonds, Carbonyl
Groups, and Other
Functional Groups*

a. R. B. Woodward, F. Sondheimer, D. Taub, K. Heusler, and M. W. McLamore, *J. Am. Chem. Soc.*, **74**, 4223 (1952).
b. J. A. Marshall and H. Roebke, *J. Org. Chem.*, **34**, 4188 (1969).
c. A. C. Cope, J. W. Barthel, and R. D. Smith, *Org. Synth.*, **IV**, 218 (1963).
d. T. Ibuka, K. Hayashi, H. Minakata, and Y. Inubushi, *Tetrahedron Lett.*, 159 (1979).
e. E. J. Corey, E. J. Trybulski, L. S. Melvin, Jr., K. C. Nicolaou, J. A. Secrist, R. Lett, P. W. Sheldrake, J. R. Falck,
   D. J. Brunelle, M. F. Haslanger, S. Kim, and S. Yoo, *J. Am. Chem. Soc.*, **100**, 4618 (1978).
f. P. A. Grieco, E. Williams, H. Tanaka, and S. Gilman, *J. Org. Chem.*, **45**, 3537 (1980).
g. E. J. Corey and M. Chaykovsky, *J. Am. Chem. Soc.*, **86**, 1639 (1964).
h. L. E. Overman and C. Fukaya, *J. Am. Chem. Soc.*, **102**, 1454 (1980).
i. J. Castro, H. Sorensen, A. Riera, C. Morin, A. Moyano, M. A. Pericas, and A. E. Greene, *J. Am. Chem. Soc.*, **112**, 9388
   (1990).
j. S. Hanessian, C. Girard, and J. L. Chiara, *Tetrahedron Lett.*, **33**, 573 (1992).

5.6.3. Reductive Coupling of Carbonyl Compounds

As reductions by metals often occur by one-electron transfers, radicals are involved as intermediates. When the reaction conditions are adjusted so that coupling competes favorably with other processes, the formation of a carbon-carbon bond can occur. The reductive coupling of acetone to 2,3-dimethylbutane-2,3-diol (pinacol) is an example of such a reaction.

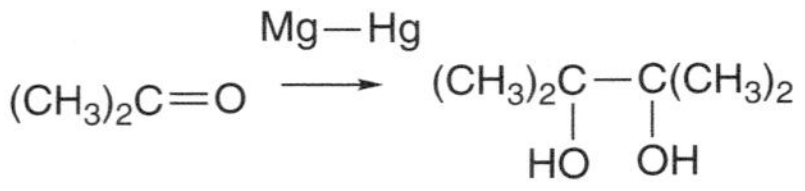

Ref. 236

Reduced forms of titanium are currently the most versatile and dependable reagents for reductive coupling of carbonyl compounds. These reagents are collectively referred to as *low-valent titanium*. Either diols or alkenes can be formed, depending on the conditions.[237] Several different procedures have evolved for titanium-mediated coupling. One procedure involves prereduction of $TiCl_3$ with strong reducing agents such as $LiAlH_4$,[238] potassium on graphite (C_8K),[239] or Na-naphthalenide.[240b] The reductant prepared in this way is quite effective at coupling reactants with several oxygen substituents.

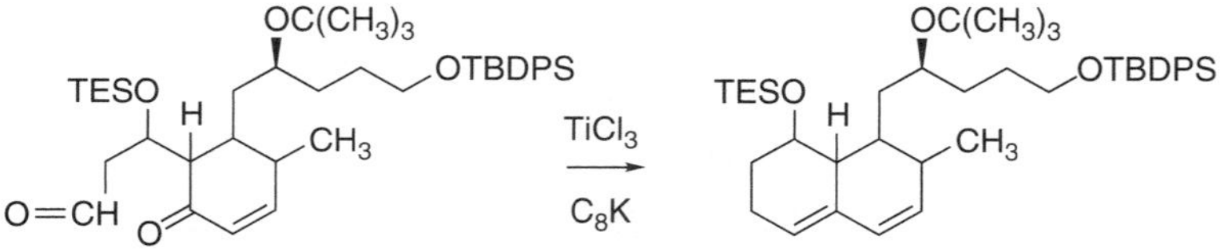

Ref. 240

236. R. Adams and E. W. Adams, *Org. Synth.*, **I**, 448 (1932).
237. J. E. McMurry, *Chem. Rev.*, **89**, 1513 (1989).
238. J. E. McMurry and M. P. Fleming, *J. Org. Chem.*, **41**, 896 (1976); J. E. McMurry and L. R. Krepski,
     J. Org. Chem., **41**, 3929 (1976); J. E. McMurry, M. P. Fleming, K. L. Kees, and L. R. Krepski, *J. Org.
     Chem.*, **43**, 3255 (1978); J. E. McMurry, *Acc. Chem. Res.*, **16**, 405 (1983).
239. (a) A. Furstner and H. Weidmann, *Synthesis*, 1071 (1987); (b) D. L. J. Clive, C. Zhang, K. S. K. Murthy,
     W. D. Hayward, and S. Daigneault, *J. Org. Chem.*, **56**, 6447 (1991).
240. D. L. J. Clive, K. S. K. Murthy, A. G. H. Wee, J. S. Prasad, G. V. J. Da Silva, M. Majewski,
     P. C. Anderson, C. F. Evans, R. D. Haugen, L. D. Heerze, and J. R. Barrie, *J. Am. Chem. Soc.*, **112**,
     3018 (1990).

Another particularly reactive form of titanium is generated by including 0.25 equivalent of I_2. This reagent permits low-temperature reductive deoxygenation to alkenes.[241]

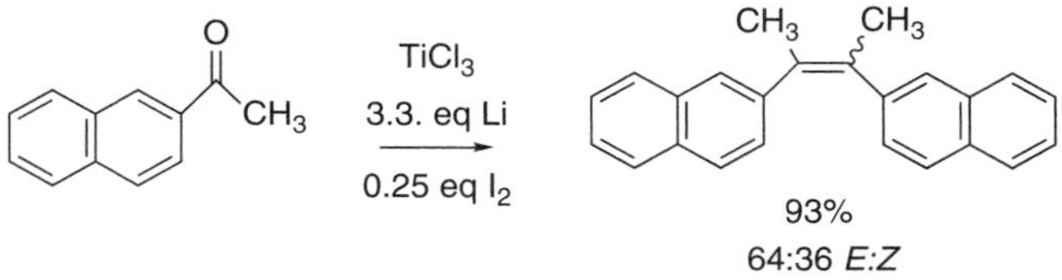

93%
64:36 E:Z

Titanium metal is also activated by TMS-Cl.[242] These conditions were used in a number of dimerizations and cyclizations, including the formation of a 36-membered ring.

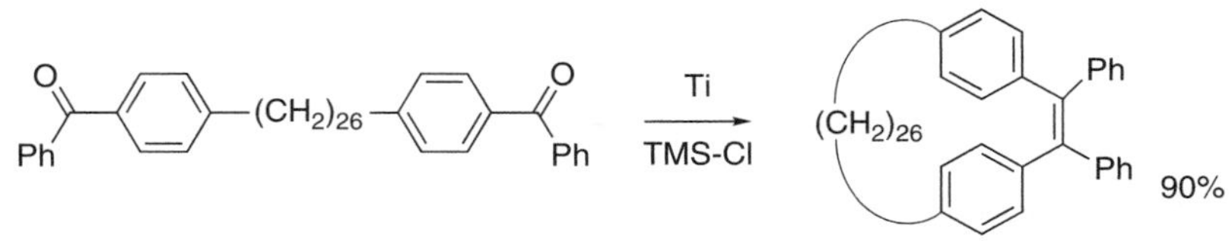

90%

Another process that is widely used involves reduction by Zn-Cu couple. This reagent is especially reliable when prepared from $TiCl_3$ purified as a DME complex,[243] and is capable of forming normal, medium, and large rings with comparable efficiency.

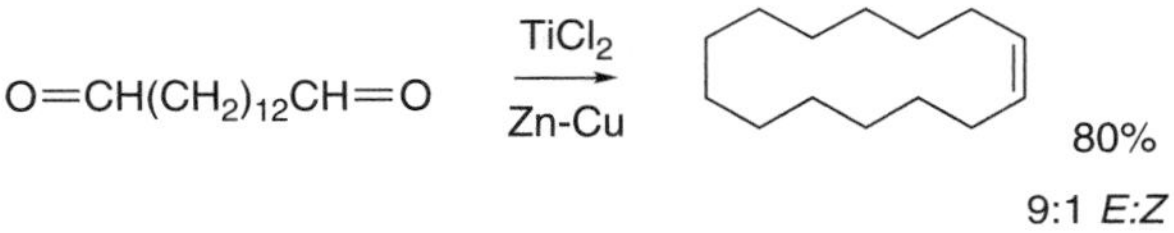

$$O=CH(CH_2)_{12}CH=O \xrightarrow[\text{Zn-Cu}]{TiCl_2}$$

80%

9:1 E:Z

Ref. 244

The macrocyclization has proven useful in the formation of a number of natural products.[245] These conditions have been used to prepare 36- and 72-membered rings.

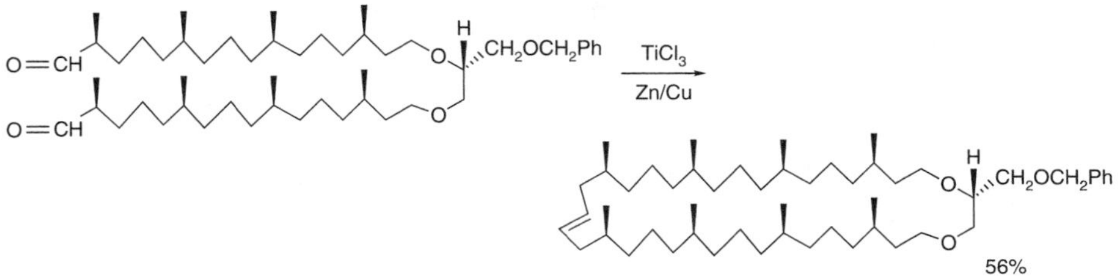

56%

Ref. 246

241. S. Talukadar, S. K. Nayak, and A. Banerji, *J. Org. Chem.*, **63**, 4925 (1998).
242. A. Furstner and A. Hupperts, *J. Am. Chem. Soc.*, **117**, 4468 (1995).
243. J. E. McMurry, T. Lectka, and J. G. Rico, *J. Org. Chem.*, **54**, 3748 (1989).
244. J. E. McMurry, J. R. Matz, K. L. Kees, and P. A. Bock, *Tetrahedron Lett.*, **23**, 1777 (1982).
245. J. E. McMurry, J. G. Rico, and Y. Shih, *Tetrahedron Lett.*, **30**, 1173 (1989); J. E. McMurry and R. G. Dushin, *J. Am. Chem. Soc.*, **112**, 6942 (1990).
246. T. Eguchi, K. Arakawa, T. Terachi, and K. Kakinuma, *J. Org. Chem.*, **62**, 1924 (1997).

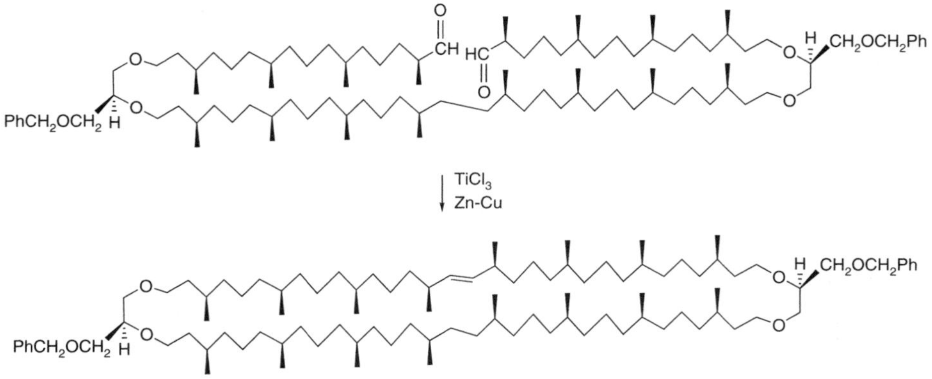

446

CHAPTER 5

Reduction of
Carbon-Carbon Multiple
Bonds, Carbonyl
Groups, and Other
Functional Groups

Ref. 247

The double bonds were reduced to the give the saturated compounds, so the double-bond configuration was not an immediate issue. It appears, however, that the *E*-double bonds are formed. The debenzylated derivatives of propan-1,2,3-triol occur as lipid components in various prokaryotes (archaebacteria) that grow under extreme thermal conditions.

Under other conditions, reduction leads to diols. Reductive coupling to diols can be done using magnesium amalgam[248] or zinc dust.[249]

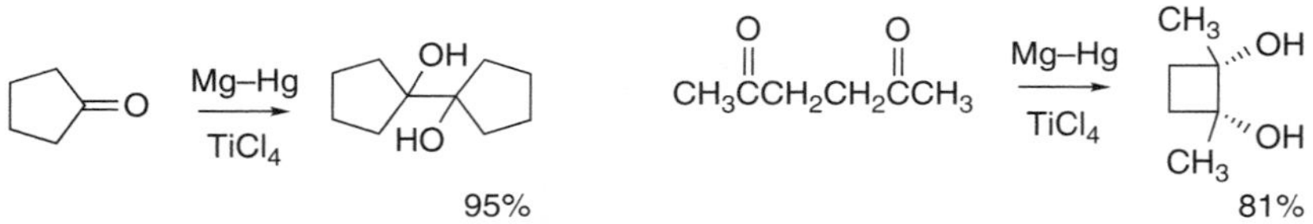

The most general procedures are based on low-valent titanium. Good yields of diols are obtained from aromatic aldehydes and ketones by adding catechol to the $TiCl_3$-Mg reagent prior to coupling.[250]

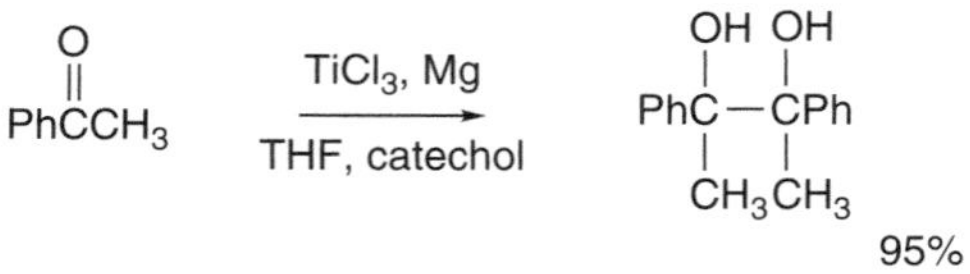

Both unsymmetrical alkenes and diols can be prepared by applying these methods to mixtures of two different carbonyl compounds. An excess of one component can be used to achieve a high conversion of the more valuable reactant. A mixed reductive

247. T. Eguchi, K. Ibaragi, and K. Kakinuma, *J. Org. Chem.*, **63**, 2689 (1998).
248. E. J. Corey, R. L. Danheiser, and S. Chandrasekaran, *J. Org. Chem.*, **41**, 260 (1976).
249. A. Furstner, A. Hupperts, A. Ptock, and E. Janssen, *J. Org. Chem.*, **59**, 5215 (1994).
250. N. Balu, S. K. Nayak, and A. Banerji, *J. Am. Chem. Soc.*, **118**, 5932 (1996).

deoxygenation with TiCl$_4$-Zn was used to prepare 4-hydroxytamoxifen, the active antiestrogenic metabolite of tamoxifen.

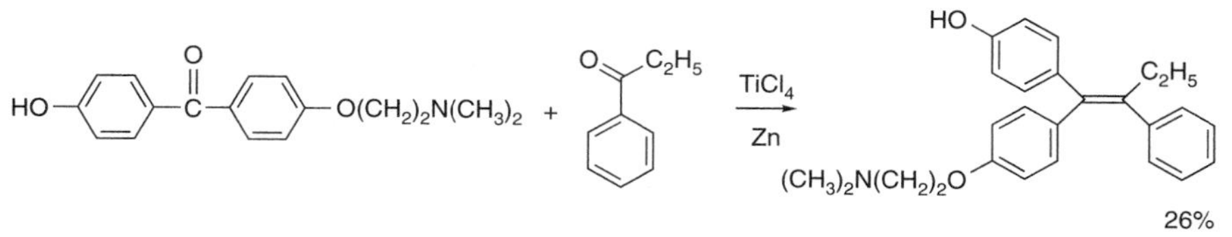

26%

Ref. 251

Stereoselectivity has been observed in some coupling reactions of this type. For example, coupling with 4-hydroxy-3'-pivaloyoxybenzophenone was stereoselective for the *E*-isomer.

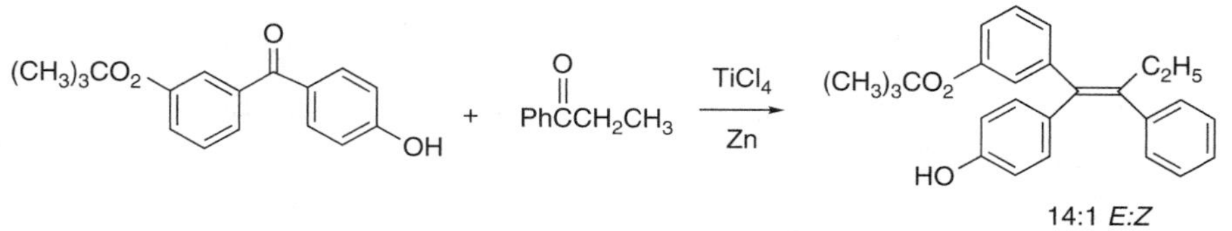

14:1 *E:Z*

Ref. 252

It is not clear at this time what factors determine stereoselectivity.

Titanium-mediated reductive couplings are normally heterogeneous, and it was originally thought that the reactions take place at the metal surface.[253] However, mechanistic study has suggested that Ti(II) may be the active species. Hydride reducing agents generate a solid having the composition $(HTi^{II}Cl)_n$ that effects reductive couplings. This species is believed to react with carbonyl compounds with elimination of hydrogen to generate a complexed form of the carbonyl compound. The ketone in this complex is considered to be analogous to a "ketone dianion"[254] and is strongly nucleophilic. This mechanism accounts for the characteristic "template effect" of the titanium reagents in promoting ring formation because it involves cooperating titanium ions.

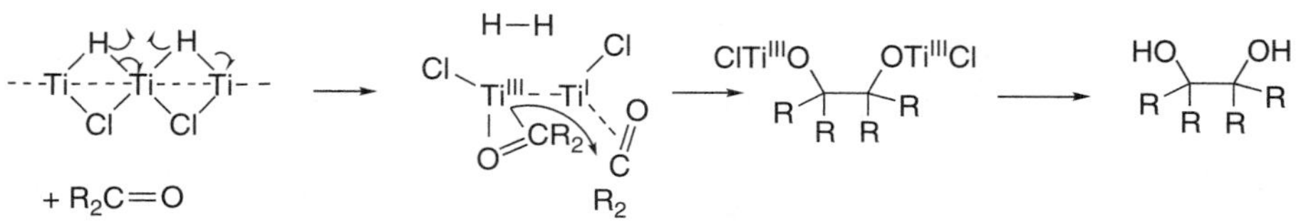

It has been suggested that a similar mechanism operates under some conditions in which the reductant is generated in situ by a Zn-Cu couple.[255] The key intermediate in this mechanism is a complex of the carbonyl compound with TiCl$_2$. The formation

251. S. Gauthier, J. Mailhot, and F. Labrie, *J. Org. Chem.*, **61**, 3890 (1996).
252. S. Gauthier, J.-Y. Sanceau, J. Mailhot, B. Caron, and J. Cloutier, *Tetrahedron*, **56**, 703 (2000).
253. R. Dams, M. Malinowski, I. Westdrop, and H. Y. Geise, *J. Org. Chem.*, **47**, 248 (1982).
254. B. Bogdanovic, C. Kruger, and B. Wermeckes, *Angew. Chem. Int. Ed. Engl.*, **19**, 817 (1980).
255. A. Furstner and B. Bogdanovic, *Angew. Chem. Int. Ed. Engl.*, **35**, 2442 (1996).

448

CHAPTER 5

*Reduction of
Carbon-Carbon Multiple
Bonds, Carbonyl
Groups, and Other
Functional Groups*

of alkene involves a second reduction step, which can occur at elevated temperature in the presence of excess reactant.

According to a DFT computational study, this mechanism is plausible.[256]

Samarium diiodide is another powerful one-electron reducing agent that can effect carbon-carbon bond formation under appropriate conditions.[257] Aromatic aldehydes and aliphatic aldehydes and ketones undergo pinacol-type coupling with SmI_2 or $SmBr_2$.

δ-Ketoaldehydes and 1,5-diketones are reduced to *cis*-cyclopentanediols.[258] 1,6-Diketo compounds can be cyclized to cyclohexanediols, again with a preference for *cis*-diols.[259] These reactions are believed to occur through successive one-electron transfer, radical cyclization, and a second electron transfer with Sm^{2+} serving as a tether and Lewis acid, as well as being the reductant.

Many of the compounds used have additional functional groups, including ester, amide, ether, and acetal. These groups may be involved in coordination to samarium and thereby influence the stereoselectivity of the reaction.

The ketyl intermediates in SmI_2 reductions can be trapped by carbon-carbon double bonds, leading, for example, to cyclization of δ,ε-enones to cyclopentanols.

Ref. 260

256. M. Stahl, U. Pidun, and G. Frenking, *Angew. Chem. Int. Ed. Engl.*, **36**, 2234 (1997).

257. G.A. Molander, *Org. React.*, **46**, 211 (1994); J. L. Namy, J. Souppe, and H. B. Kagan, *Tetrahedron Lett.*, **24**, 765 (1983); A. Lebrun, J.-L. Namy, and H. B. Kagan, *Tetrahedron Lett.*, **34**, 2311 (1993); H. Akane, T. Hatano, H. Kusui, Y. Nishiyama, and Y. Ishii, *J. Org. Chem.*, **59**, 7902 (1994).

258. G. A. Molander and C. Kemp, *J. Am. Chem. Soc.*, **111**, 8236 (1989); J. Uenishi, S. Masuda, and S. Wakabashi, *Tetrahedron Lett.*, **32**, 5097 (1991).

259. J. L. Chiara, W. Cabri, and S. Hanessian, *Tetrahedron Lett.*, **32**, 1125 (1991); J. P. Guidot, T. Le Gall, and C. Mioskowski, *Tetrahedron Lett.*, **35**, 6671 (1994).

260. G. Molander and C. Kenny, *J. Am. Chem. Soc.*, **111**, 8236 (1989).

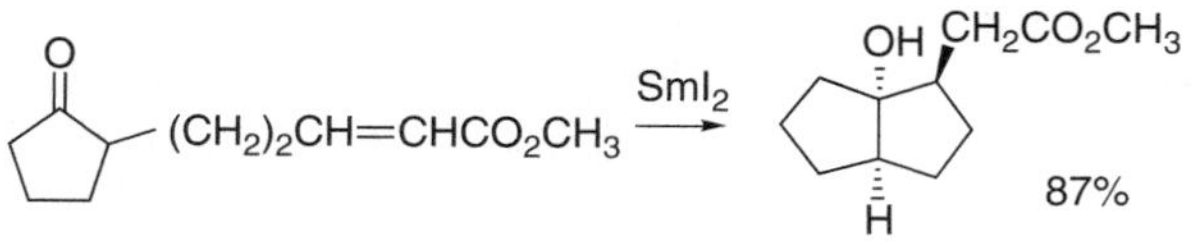

Ref. 261

SmI$_2$ has also been used to form cyclooctanols by cyclization of 7,8-enones.[262] These alkene addition reactions presumably proceed by addition of the ketyl radical to the double bond, followed by a second electron transfer.

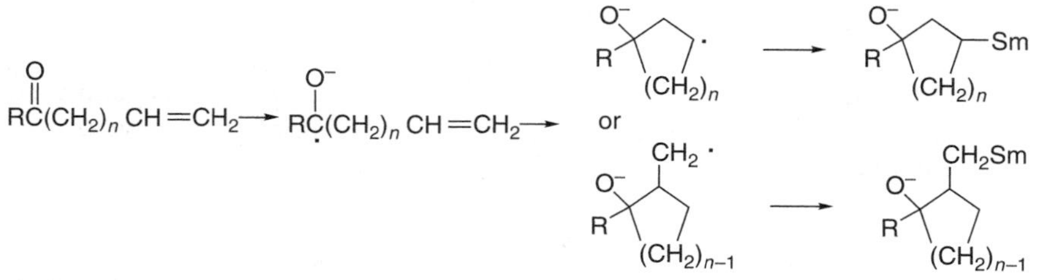

The initial products of such additions under aprotic conditions are organosamarium reagents and further (tandem) transformations are possible, including addition to ketones, anhydrides, or carbon dioxide.

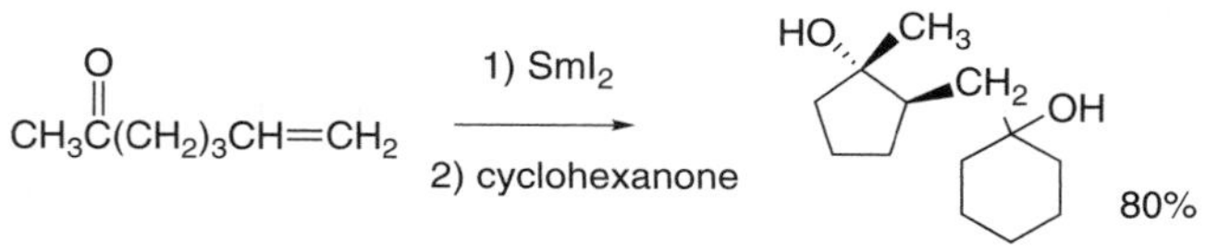

Ref. 263

Another reagent that has found use in pinacolic coupling is prepared from VCl$_3$ and zinc dust.[264] This reagent is selective for aldehydes that can form chelated intermediates, such as β-formylamides, α-amidoaldehydes, α-phosphinoylaldehydes,[265] and δ-ketoaldehydes.[266] The vanadium reagent can be used for both homodimerization and heterodimerization. In the latter case, the reactive aldehyde is added to an excess of the second aldehyde. Under these conditions, the ketyl intermediate formed from the chelated aldehyde reacts with the second aldehyde.

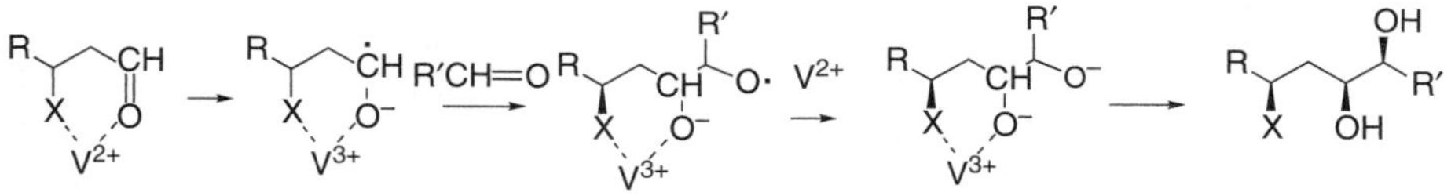

The VCl$_3$-Zn reagent has also been used in cyclization reactions, as in Entries 4 and 5 in Scheme 5.13.

261. E. J. Enholm and A. Trivellas, *Tetrahedron Lett.*, **30**, 1063 (1989).
262. G. A. Molander and J. A. McKie, *J. Org. Chem.*, **59**, 3186 (1994).
263. G. A. Molander and J. A. McKie, *J. Org. Chem.*, **57**, 3132 (1992).
264. J. H. Freudenberg, A. W. Konradi, and S. F. Pedersen, *J. Am. Chem. Soc.*, **111**, 8014 (1989).
265. J. Park and S. F. Pedersen, *J. Org. Chem.*, **55**, 5924 (1990).
266. A. S. Raw and S. F. Pedersen, *J. Org. Chem.*, **56**, 830 (1991).

450

CHAPTER 5

*Reduction of
Carbon-Carbon Multiple
Bonds, Carbonyl
Groups, and Other
Functional Groups*

Another important reductive coupling is the conversion of esters to α-hydroxyketones (*acyloin condensation*).[267] This reaction is usually carried out with sodium metal in an inert solvent. Good results have also been obtained for sodium metal dispersed on solid supports.[268] Diesters undergo intramolecular reactions and this is also an important method for the preparation of medium and large carbocyclic rings.

$$CH_3O_2C(CH_2)_8CO_2CH_3 \xrightarrow[\text{2) } CH_3CO_2H]{\text{1) Na}}$$

Ref. 269

There has been considerable discussion of the mechanism of the acyloin condensation. One formulation of the reaction envisages coupling of radicals generated by one-electron transfer.

An alternative mechanism bypasses the postulated α-diketone intermediate because its involvement is doubtful.[270]

Regardless of the details of the mechanism, the product prior to neutralization is the dianion of an α-hydroxy ketone, namely an enediolate. It has been found that the overall yields are greatly improved if trimethylsilyl chloride is present during the reduction to trap these dianions as trimethylsilyl ethers.[271] The silylated derivatives are much more stable to the reaction conditions than the enediolates. Hydrolysis during workup gives the acyloin product. This modified version of the reaction has been applied to cyclizations leading to small, medium, and large rings, as well as to intermolecular couplings.

Scheme 5.13 provides several examples of reductive carbon-carbon bond formation, including formation of diols, alkenes, and acyloins. Entry 1 uses magnesium amalgam in the presence of dichlorodimethylsilane. The role of the silane may be to

[267] J. J. Bloomfield, D. C. Owsley, and J. M. Nelke, *Org. React.*, **23**, 259 (1976).

[268] M. Makosza and K. Grela, *Synlett*, 267 (1997); M. Makosza, P. Nieczypor, and K. Grela, *Tetrahedron*, **54**, 10827 (1998).

[269] N. Allinger, *Org. Synth.*, **IV**, 840 (1963).

[270] J. J. Bloomfield, D. C. Owsley, C. Ainsworth, and R. E. Robertson, *J. Org. Chem.*, **40**, 393 (1975).

[271] K. Ruhlmann, *Synthesis*, 236 (1971).

Scheme 5.13. Reductive Coupling of Carbonyl Compound

A. Diol formation

1[a]

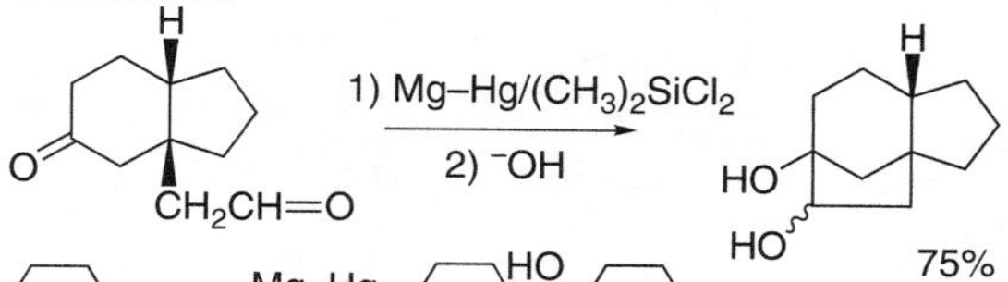

2[b]

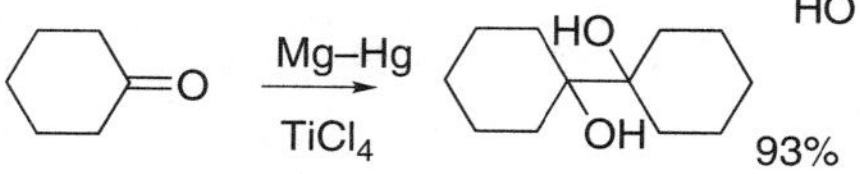

3[c]

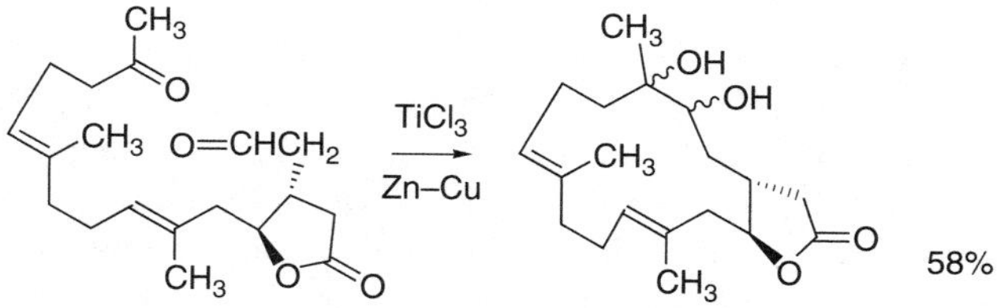

4[d]

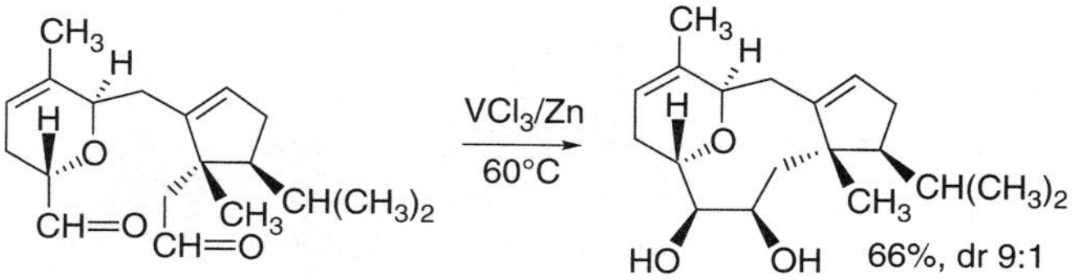

5[e]

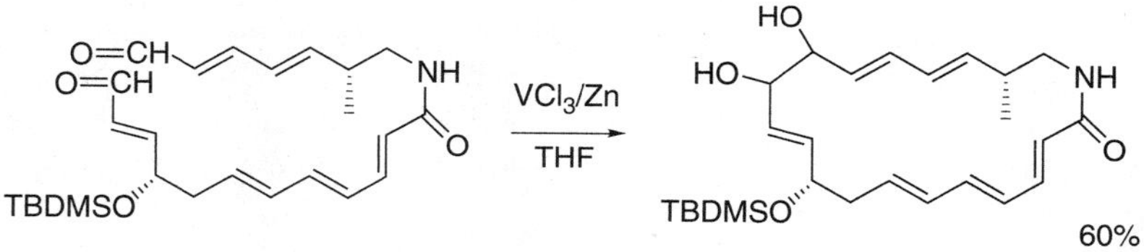

B. Alkene formation

6[f]

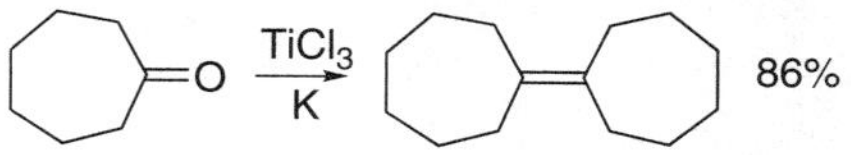

7[g]

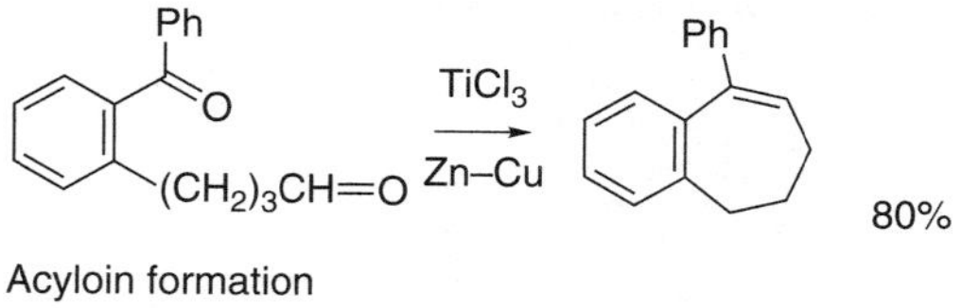

C. Acyloin formation

8[h] $CH_3O_2C(CH_2)_8CO_2CH_3$

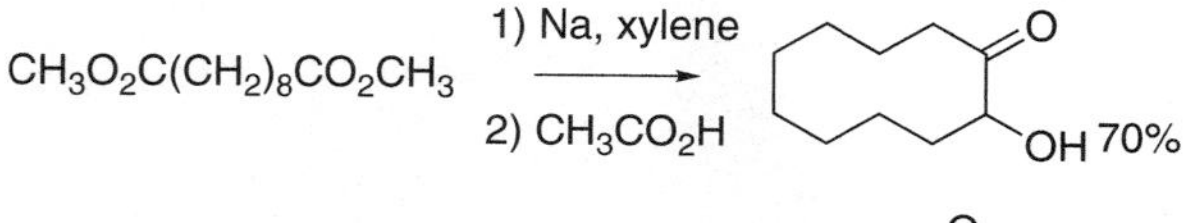

1) Na, xylene

2) CH_3CO_2H 70%

9[i] $C_2H_5O_2CH_2CH_2CO_2C_2H_5$

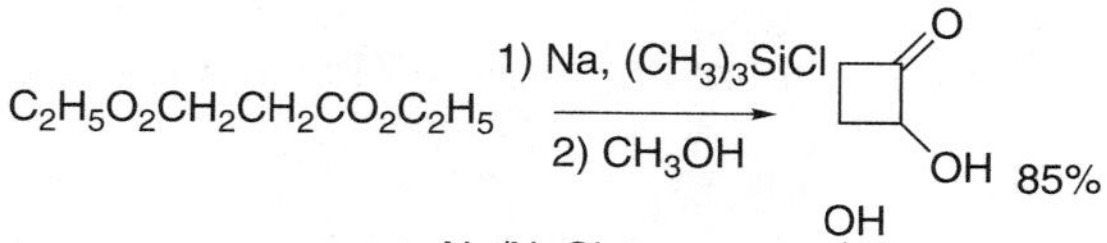

1) Na, $(CH_3)_3SiCl$

2) CH_3OH 85%

10[j] $CH_3(CH_2)_6CO_2C_2H_5$

Na/NaCl

benzene

$CH_3(CH_2)_6\overset{OH}{\underset{O}{C}}HC(CH_2)_6CH_3$ 78%

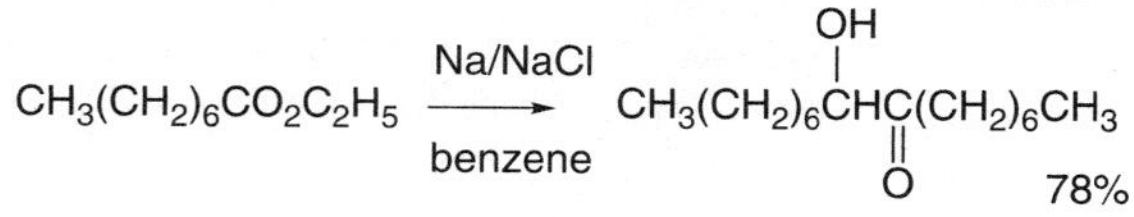

(Continued)

Scheme 5.13. (*Continued*)

CHAPTER 5

*Reduction of
Carbon-Carbon Multiple
Bonds, Carbonyl
Groups, and Other
Functional Groups*

a. E. J. Corey and R. L. Carney, *J. Am. Chem. Soc.*, **93**, 7318 (1971).
b. E. J. Corey, R. L. Danheiser, and S. Chandrasekaran, *J. Org. Chem.*, **41**, 260 (1976).
c. J. E. McMurry and R. G. Dushin, *J. Am. Chem. Soc.*, **112**, 6942 (1990).
d. D. R. Williams and R. W. Heidebrecht, Jr., *J. Am. Chem. Soc.*, **125**, 1843 (2003).
e. M. Nazare and H. Waldmann, *Chem. Eur. J.*, **7**, 3363 (2001).
f. J. E. McMurry, M. P. Fleming, K. L. Kees, and L. R. Krepski, *J. Org. Chem.*, **43**, 3255 (1978).
g. C. B. Jackson and G. Pattenden, *Tetrahedron Lett.*, **26**, 3393 (1985).
h. N. L. Allinger, *Org. Synth.*, **IV**, 340 (1963).
i. J. J. Bloomfield and J. M. Nelke, *Org. Synth.*, **57**, 1 (1977).
j. M. Makosza and K. Grela, *Synlett*, 267 (1997).

trap the pinacol as a cyclic siloxane. The reaction in Entry 2 is thought to involve Ti(II) as the active reductant and to proceed by a mechanism of the type described on p. 447. These conditions were also successful for the reaction shown in Entry 1. Entry 3 involves formation of a 14-membered ring using a low-valent titanium reagent. The product is a mixture of all four possible diastereomeric diols in yields ranging from 7 to 21%. Entry 4 is an example of a pinacol reduction using a vanadium reagent prepared in situ from VCl_3 and Zn, which tends to give a high proportion of *cis*-diol as a result of chelation with vanadium. Entry 5 shows the synthesis of a sensitive polyunsaturated lactam. The *cis*-diol was formed in 60% yield. In this particular case, various low-valent titanium reagents were unsuccessful. Entries 6 and 7 describe conditions that lead to alkene formation. Entries 8 to 10 are acyloin condensations. The reaction in Entry 8 illustrates the classical conditions. Entry 9 is an example of the reaction conducted in the presence of TMS-Cl to trap the enediolate intermediate and make the reaction applicable to formation of a four-membered ring. The example in Entry 10 uses sodium in the form of a solid deposit on an inert material. This is an alternative to the procedures that require dispersion of molten sodium in the reaction vessel (Entries 8 and 9).

5.7. Reductive Deoxygenation of Carbonyl Groups

Several methods are available for reductive removal of carbonyl groups from organic compounds. Reduction to methylene groups or conversion to alkenes can be achieved.

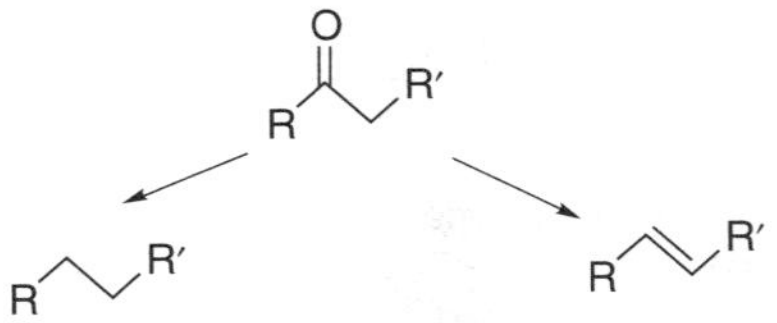

5.7.1. Reductive Deoxygenation of Carbonyl Groups to Methylene

Zinc and hydrochloric acid form a classical reagent combination for conversion of carbonyl groups to methylene groups, a reaction known as the *Clemmensen reduction*.[272] The corresponding alcohols are not reduced under the conditions of the

[272] E. Vedejs, *Org. React.*, **22**, 401 (1975).

reaction, so they are evidently not intermediates. The Clemmensen reaction works best for aryl ketones and is less reliable with unconjugated ketones. The mechanism is not known in detail but may involve formation of carbon-zinc bonds at the metal surface.[273] The reaction is commonly carried out in hot concentrated hydrochloric acid with ethanol as a cosolvent. These conditions preclude the presence of acid-sensitive or hydrolyzable functional groups. A modification in which the reaction is run in ether saturated with dry hydrogen chloride gave good results in the reduction of steroidal ketones.[274]

The *Wolff-Kishner reaction*[275] is the reduction of carbonyl groups to methylene groups by base-catalyzed decomposition of the hydrazone of the carbonyl compound. It is thought that alkyldiimides are formed and then collapse with loss of nitrogen.[276]

The reduction of tosylhydrazones by $LiAlH_4$ or $NaBH_4$ also converts carbonyl groups to methylene.[277] It is believed that a diimide is involved, as in the Wolff-Kishner reaction.

Excellent yields can also be obtained using $NaBH_3CN$ as the reducing agent.[278] The $NaBH_3CN$ can be added to a mixture of the carbonyl compound and *p*-toluenesulfonylhydrazide. Hydrazone formation is faster than reduction of the carbonyl group by $NaBH_3CN$ and the tosylhydrazone is reduced as it is formed. Another reagent that can reduce tosylhydrazones to give methylene groups is $CuBH_4(PPh_3)_2$.[279]

Reduction of tosylhydrazones of α, β-unsaturated ketones by $NaBH_3CN$ gives alkenes with the double bond located between the former carbonyl carbon and the α-carbon.[280] This reaction is believed to proceed by an initial conjugate reduction, followed by decomposition of the resulting vinylhydrazine to a vinyldiimide.

273. M. L. Di Vona and V. Rosnatti, *J. Org. Chem.*, **56**, 4269 (1991).
274. M. Toda, M. Hayashi, Y. Hirata, and S. Yamamura, *Bull. Chem. Soc. Jpn.*, **45**, 264 (1972).
275. D. Todd, *Org. React.*, **4**, 378 (1948); Huang-Minlon, *J. Am. Chem. Soc.*, **68**, 2487 (1946).
276. T. Tsuji and E. M. Kosower, *J. Am. Chem. Soc.*, **93**, 1992 (1971). Alkyldiimides are also converted to hydrocarbons by a free radical mechanism; A. G. Myers, M. Movassaghi and B. Zheng, *Tetrahedron Lett.*, **38**, 6569 (1997).
277. L. Caglioti, *Tetrahedron*, **22**, 487 (1966).
278. R. O. Hutchins, C. A. Milewski, and B. E. Maryanoff, *J. Am. Chem. Soc.*, **95**, 3662 (1973).
279. B. Milenkov and M. Hesse, *Helv. Chim. Acta*, **69**, 1323 (1986).
280. R. O. Hutchins, M. Kacher, and L. Rua, *J. Org. Chem.*, **40**, 923 (1975).

454

CHAPTER 5

*Reduction of
Carbon-Carbon Multiple
Bonds, Carbonyl
Groups, and Other
Functional Groups*

Catecholborane or sodium borohydride in acetic acid can also be used as a reducing reagent in this reaction.[281]

Carbonyl groups can be converted to methylene groups by desulfurization of thioketals. The cyclic thioketal from ethanedithiol is commonly used. Reaction with excess Raney nickel causes hydrogenolysis of both C–S bonds.

Ref. 282

Other reactive forms of nickel including nickel boride[283] and nickel alkoxide complexes[284] can also be used for desulfurization. Tri-*n*-butyltin hydride is an alternative reagent for desulfurization.[285]

Scheme 5.14 illustrates some representative carbonyl deoxygenations. Entries 1 and 2 are Clemmensen reductions of acyl phenols. Entry 3 is an example of the Wolff-Kishner reaction. Entry 4 describes modified conditions for the Wolff-Kishner reaction that take advantage of the strong basicity of the KO*t*Bu-DMSO combination. Entries 5 to 7 are examples of conversion of sulfonylhydrazones to methylene groups (Caglioti reaction). In addition to LiAlH$_4$, which was used in the original procedure, NaBH$_3$CN (Entry 6) and catecholborane (Entry 7) can be used as reducing agents. Entries 8 and 9 are thioketal desulfurizations.

5.7.2. Reduction of Carbonyl Compounds to Alkenes

Ketone *p*-toluenesulfonylhydrazones are converted to alkenes on treatment with strong bases such as an alkyllithium or lithium dialkylamide.[286] Known as the *Shapiro reaction*,[287] this proceeds through the anion of a vinyldiimide, which decomposes to a vinyllithium reagent. Treatment of this intermediate with a proton source gives the alkene.

The Shapiro reaction has been particularly useful for cyclic ketones, but its scope includes acyclic systems as well. In the case of unsymmetrical acyclic ketones,

281. G. W. Kabalka, D. T. C. Yang, and J. D. Baker, Jr., *J. Org. Chem.*, **41**, 574 (1976); R. O. Hutchins and N. R. Natale, *J. Org. Chem.*, **43**, 2299 (1978).

282. F. Sondheimer and S. Wolfe, *Can. J. Chem.*, **37**, 1870 (1959).

283. W. E. Truce and F. M. Perry, *J. Org. Chem.*, **30**, 1316 (1965).

284. S. Becker, Y. Fort, and P. Caubere, *J. Org. Chem.*, **55**, 6194 (1990).

285. C. G. Gutierrez, R. A. Stringham, T. Nitasaka, and K. G. Glasscock, *J. Org. Chem.*, **45**, 3393 (1980).

286. R. H. Shapiro and M. J. Heath, *J. Am. Chem. Soc.*, **89**, 5734 (1967).

287. R. H. Shapiro, *Org. React.*, **23**, 405 (1976); R. M. Adington and A. G. M. Barrett, *Acc. Chem. Res.*, **16**, 53 (1983); A. R. Chamberlin and S. H. Bloom, *Org. React.*, **39**, 1 (1990).

Scheme 5.14. Carbonyl to Methylene Reductions

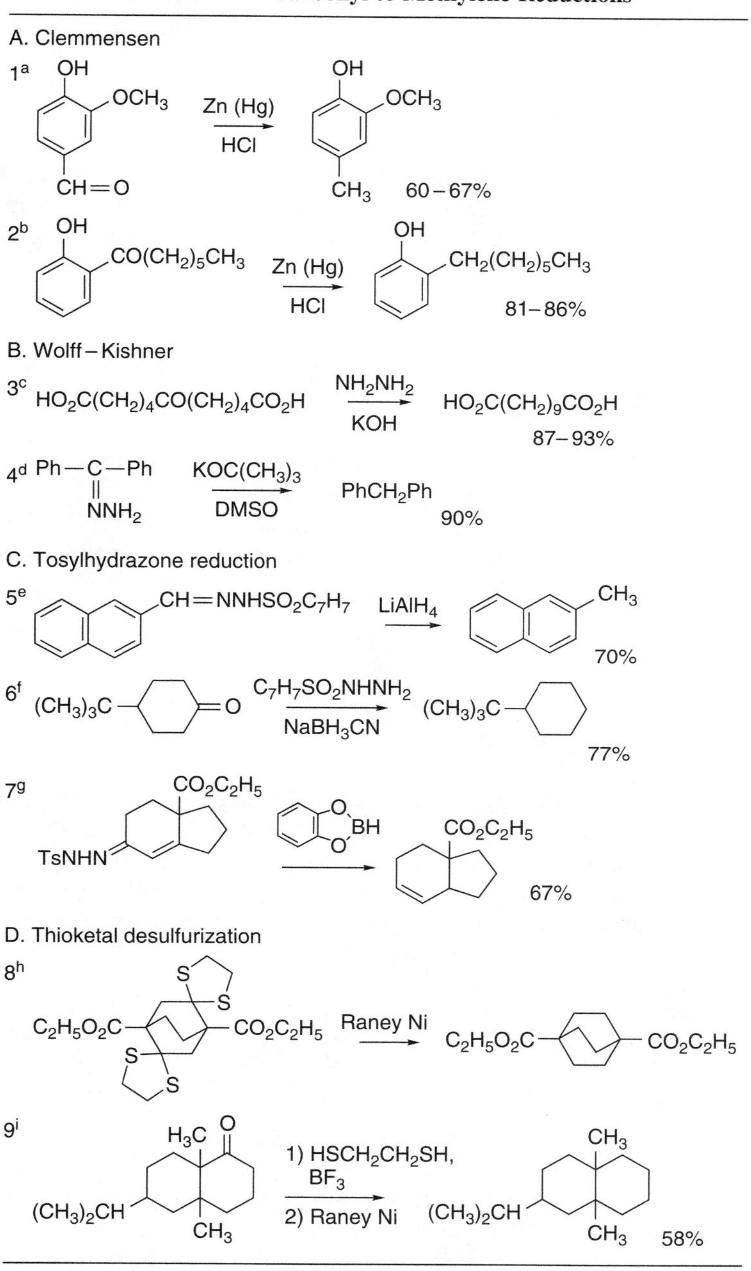
A. Clemmensen
1a
OH
OCH3
CH=O
Zn (Hg)
HCl
CH3
60–67%
2b
CO(CH2)5CH3
CH2(CH2)5CH3
81–86%
B. Wolff–Kishner
3c
HO2C(CH2)4CO(CH2)4CO2H
NH2NH2
KOH
HO2C(CH2)9CO2H
87–93%
4d
Ph—C—Ph
NNH2
KOC(CH3)3
DMSO
PhCH2Ph
90%
C. Tosylhydrazone reduction
5e
CH=NNHSO2C7H7
LiAlH4
70%
6f
(CH3)3C
C7H7SO2NHNH2
NaBH3CN
77%
7g
CO2C2H5
TsNHN
BH
67%
D. Thioketal desulfurization
8h
C2H5O2C
CO2C2H5
Raney Ni
9i
H3C
O
(CH3)2CH
1) HSCH2CH2SH,
BF3
2) Raney Ni
58%

a. R. Schwarz and H. Hering, *Org. Synth.*, **IV**, 203 (1963).
b. R. R. Read and J. Wood, Jr., *Org. Synth.*, **III**, 444 (1955).
c. L. J. Durham, D. J. McLeod, and J. Cason, *Org. Synth.*, **IV**, 510 (1963).
d. D. J. Cram, M. R. V. Sahyun, and G. R. Knox, *J. Am. Chem. Soc.*, **84**, 1734 (1962).
e. L. Caglioti and M. Magi, *Tetrahedron*, **19**, 1127 (1963).
f. R. O. Hutchins, B. E. Maryanoff, and C. A. Milewski, *J. Am. Chem. Soc.*, **93**, 1793 (1971).
g. M. N. Greco and B. E. Maryanoff, *Tetrahedron Lett.*, **33**, 5009 (1992).
h. J. D. Roberts and W. T. Moreland, Jr., *J. Am. Chem. Soc.*, **75**, 2167 (1953).
i. P. N. Rao, *J. Org. Chem.*, **36**, 2426 (1971).

456

CHAPTER 5

*Reduction of
Carbon-Carbon Multiple
Bonds, Carbonyl
Groups, and Other
Functional Groups*

questions of both regiochemistry and stereochemistry arise. 1-Octene is the exclusive product from 2-octanone.[288]

This regiospecificity has been shown to depend on the stereochemistry of the C=N bond in the starting hydrazone. There is evidently a strong preference for abstracting the proton *syn* to the arenesulfonyl group, probably because this permits chelation with the lithium ion.

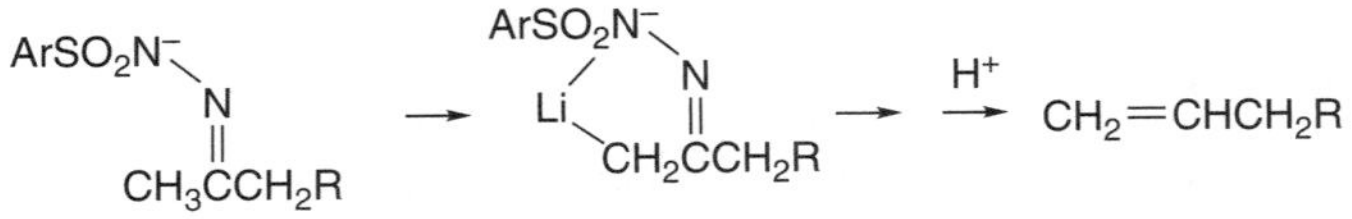

The Shapiro reaction converts the *p*-toluenesulfonylhydrazones of α,β-unsaturated ketones to dienes (see Entries 3 to 5 in Scheme 5.15).[289]

The vinyl lithium reagents generated in the Shapiro reaction can be used in tandem reactions. In the reaction shown below, a hydroxymethyl group was added by formylation followed by reduction.

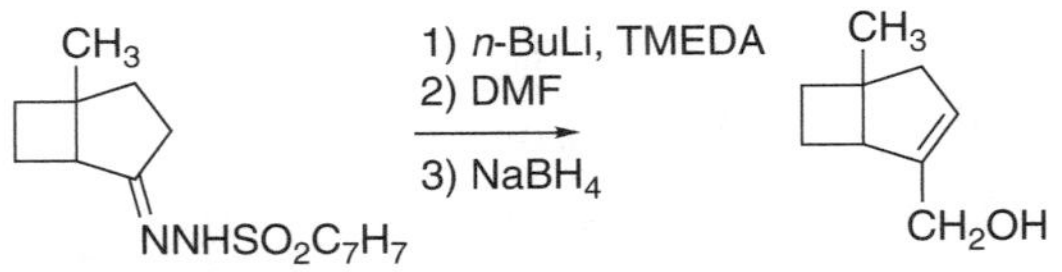

In another example, a sequence of methylation-elimination-hydroxymethylation was used to install the functionality pattern found in the A-ring of taxol. The hydrazone dianion was generated and methylated at low temperature. The hydrazone was then deprotonated again using excess *n*-butyllithium and allowed to warm to room temperature, at which point formation of the vinyllithium occurred. Reaction with paraformaldehyde generated the desired product.[290]

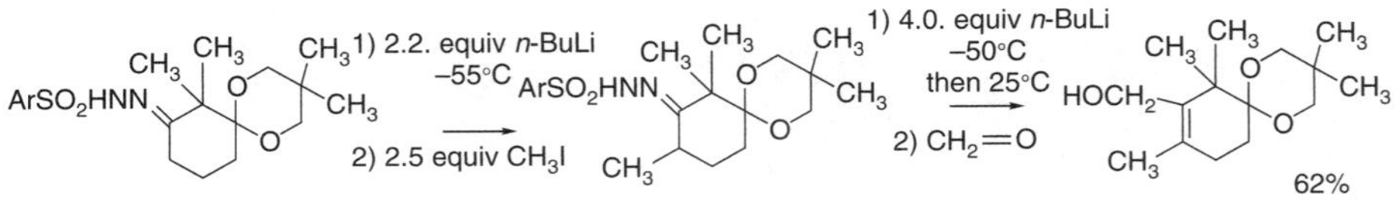

Ar = 2,4,6-trimethylphenyl

Scheme 5.15 shows some examples of the Shapiro reaction. Entry 1 is an example of the standard procedure, as documented in *Organic Syntheses*. Entry 2 illustrates the preference for the formation of the less-substituted double bond. Entries 3, 4, and 5 involve tosylhydrazone of α,β-unsaturated ketones. The reactions proceed by α'-deprotonation. Entry 6 illustrates the applicability of the reaction to a highly strained system.

288. K. J. Kolonko and R. H. Shapiro, *J. Org. Chem.*, **43**, 1404 (1978).
289. W. G. Dauben, G. T. Rivers, and W. T. Zimmerman, *J. Am. Chem. Soc.*, **99**, 3414 (1977).
290. O. P. Tormakangas, R. J. Toivola, E. K. Karvinen, and A. M. P. Koskinen, *Tetrahedron*, **58**, 2175 (2002).

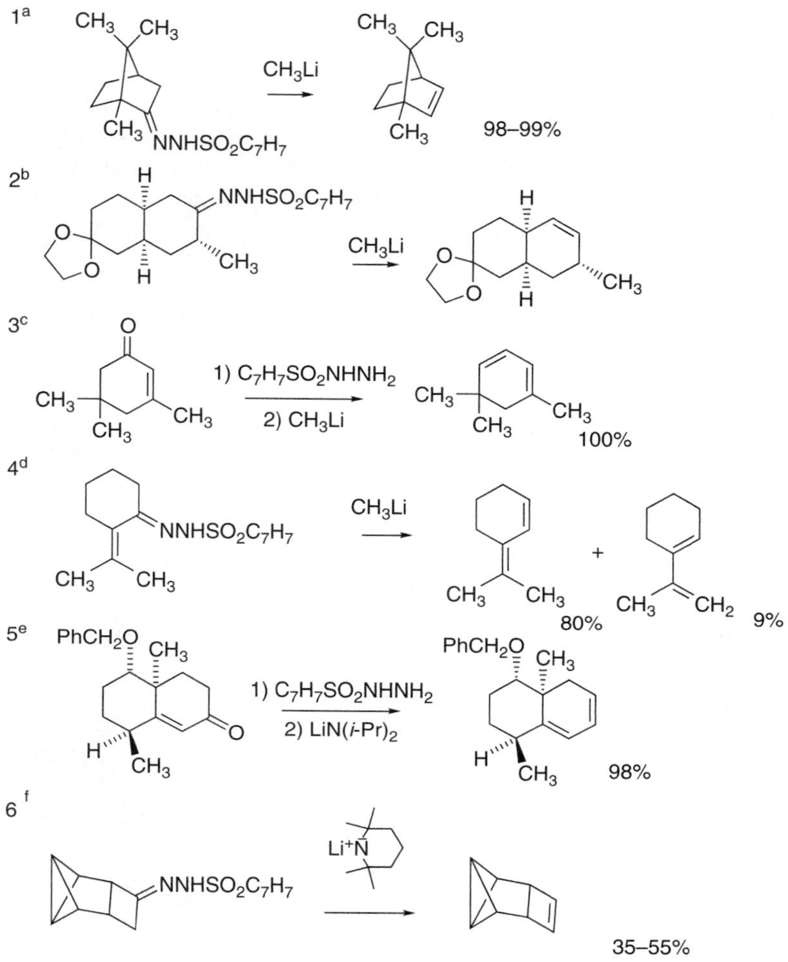

a. R. H. Shapiro and J. H. Duncan, *Org. Synth.*, **51**, 66 (1971).
b. W. L. Scott and D. A. Evans, *J. Am. Chem. Soc.*, **94**, 4779 (1972).
c. W. G. Dauben, M. E. Lorber, N. D. Vietmeyer, R. H. Shapiro, J. H. Duncan, and K. Tomer, *J. Am. Chem. Soc.*, **90**, 4762 (1968).
d. W. G. Dauben, G. T. Rivers, and W. T. Zimmerman, *J. Am. Chem. Soc.*, **99**, 3414 (1977).
e. P. A. Grieco, T. Oguri, C.-L. J. Wang, and E. Williams, *J. Org. Chem.*, **42**, 4113 (1977).
f. L. R. Smith, G. R. Gream, and J. Meinwald, *J. Org. Chem.*, **42**, 927 (1977).

5.8. Reductive Elimination and Fragmentation

The presence of a potential leaving group β to the site of carbanionic character usually leads to β-elimination. In some useful synthetic procedures, the carbanionic character is generated by a reductive process.

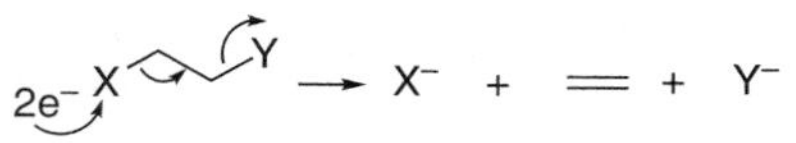

458

CHAPTER 5

*Reduction of
Carbon-Carbon Multiple
Bonds, Carbonyl
Groups, and Other
Functional Groups*

Similarly, carbanionic character δ to a leaving group can lead to β, γ-fragmentation.

A classical example of the β-elimination reaction is the reductive debromination of vicinal dibromides. Zinc metal is the traditional reducing agent.[291] A multitude of other reducing agents have been found to give this and similar reductive eliminations. Some examples are given in Table 5.7. Some of the reagents exhibit *anti* stereospecificity, whereas others do not. A stringent test for *anti* stereospecificity is the extent of Z-alkene formed from a *syn* precursor.

Anti stereospecificity is associated with a concerted reductive elimination, whereas single-electron transfer fragmentation leads to loss of stereospecificity and formation of the more stable E-stereoisomer.

As vicinal dibromides are usually made by bromination of alkenes, their utility for synthesis is limited, except for temporary masking of a double bond. Much more frequently it is desirable to convert a diol to an alkene, and several useful procedures have been developed. The reductive deoxygenation of diols via thiono carbonates was

**Table 5.7. Reagents for Reductive Dehalo-
genation**

Reagent	*Anti* stereoselectivity
Zn, cat TiCl$_4$[a]	Yes
Zn, H$_2$NSNH$_2$[b]	?
SnCl$_2$, DiBAlH[c]	?
Sm, CH$_3$OH[d]	No
Fe, graphite[e]	Yes
C$_2$H$_5$MgBr, cat Ni(dppe)Cl$_2$[f]	No

a. F. Sato, T. Akiyama, K. Ida, and M. Sato, *Synthesis*, 1025 (1982).
b. R. N. Majumdar and H. J. Harwood, *Synth. Commun.*, **11**, 901 (1981).
c. T. Oriyama and T. Mukaiyama, *Chem. Lett.*, 2069 (1984).
d. R. Yanada, N. Negoro, K. Yanada, and T. Fujita, *Tetrahedron Lett.*, **37**, 9313 (1996).
e. D. Savoia, E. Tagliavini, C. Trombini, and A. Umani-Ronchi, *J. Org. Chem.*, **47**, 876 (1982).
f. C. Malanga, L. A. Aronica, and L. Lardicci, *Tetrahedron Lett.*, **36**, 9189 (1995).

291. J. C. Sauer, *Org. Synth.*, **IV**, 268 (1965).

developed by Corey and co-workers.[292] Triethyl phosphite is useful for many cases, but the more reactive 1,3-dimethyl-2-phenyl-1,3,2-diazaphospholidine can be used when milder conditions are required.[293] The reaction presumably occurs by initial P−S bonding followed by a concerted elimination of carbon dioxide and the thiophosphoryl compound.

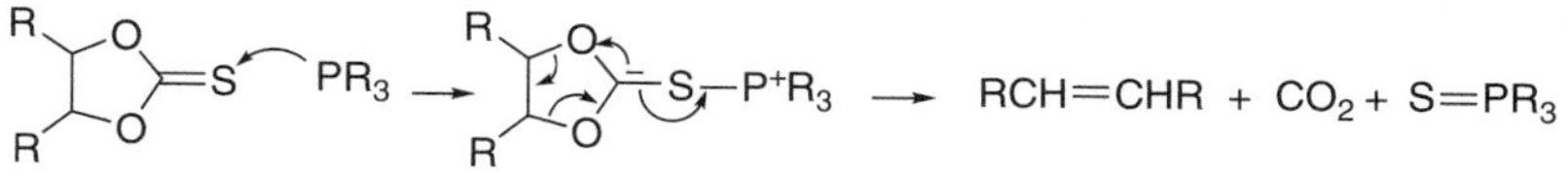

$$RCH{=}CHR + CO_2 + S{=}PR_3$$

Diols can also be deoxygenated via *bis*-sulfonate esters using sodium naphthalenide.[294] Cyclic sulfate esters are also cleanly reduced by lithium naphthalenide.[295]

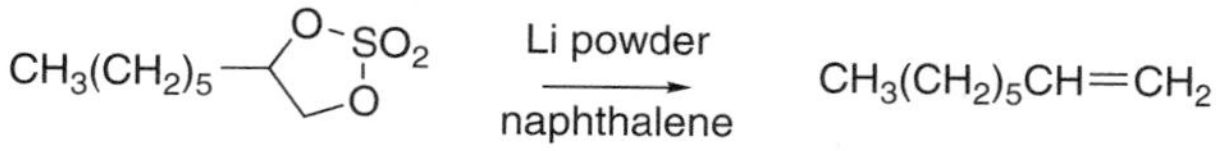

This reaction, using sodium naphthalenide, has been used to prepare unsaturated nucleosides.

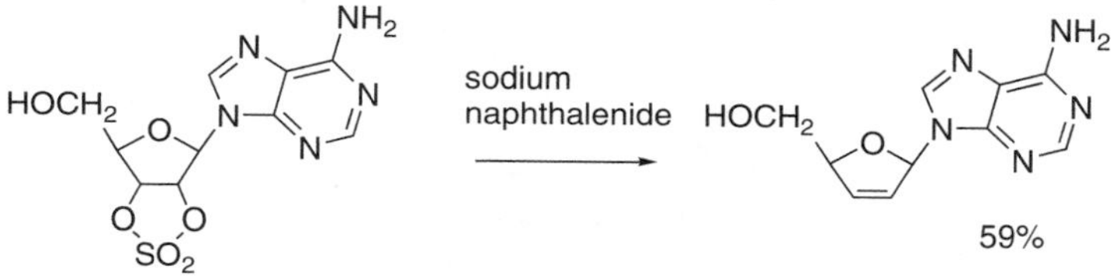

59%

Ref. 296

It is not entirely clear whether these reactions involve a redox reaction at sulfur or if they proceed by organometallic intermediates.

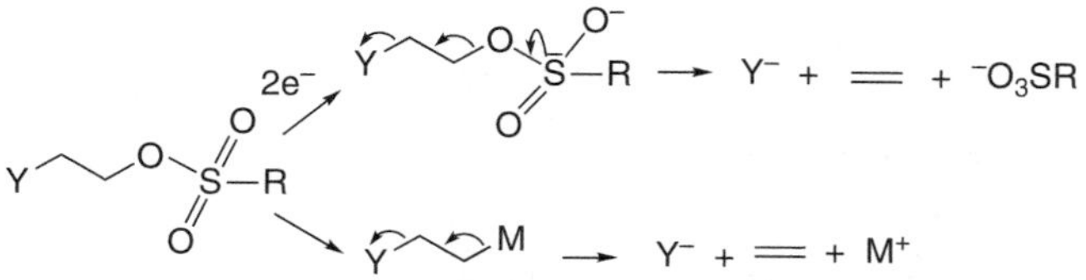

Iodination reagents combined with aryl phosphines and imidazole can also effect reductive conversion of diols to alkenes. One such combination is 2,4,5-triiodoimidazole, imidazole, and triphenylphosphine.[297] These reagent combinations

292. E. J. Corey and R. A. E. Winter, *J. Am. Chem. Soc.*, **85**, 2677 (1963); E. J. Corey, F. A. Carey, and R. A. E. Winter, *J. Am. Chem. Soc.*, **87**, 934 (1965).
293. E. J. Corey and P. B. Hopkins, *Tetrahedron Lett.*, **23**, 1979 (1982).
294. J. C. Carnahan, Jr., and W. D. Closson, *Tetrahedron Lett.*, 3447 (1972); R. J. Sundberg and R. J. Cherney, *J. Org. Chem.*, **55**, 6028 (1990).
295. D. Guijarro, B. Mancheno, and M. Yus, *Tetrahedron Lett.*, **33**, 5597 (1992).
296. M. J. Robins, E. Lewandowska, and S. F. Wnuk, *J. Org. Chem.*, **63**, 7375 (1998).
297. P. J. Garegg and B. Samuelsson, *Synthesis*, 813 (1979); Y. Watanabe, M. Mitani, and S. Ozaki, *Chem. Lett.*, 123 (1987).

460

CHAPTER 5

*Reduction of
Carbon-Carbon Multiple
Bonds, Carbonyl
Groups, and Other
Functional Groups*

are believed to give oxyphosphonium intermediates, which then can serve as leaving groups, forming triphenylphosphine oxide as in the Mitsunobu reaction (see Section 3.2.3). The iodide serves as both a nucleophile and reductant.

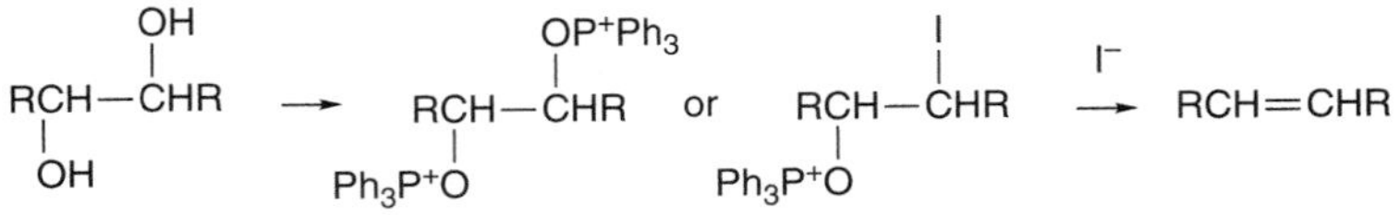

In a related procedure, chlorodiphenylphosphine, imidazole, iodine, and zinc cause reductive elimination of diols.[298] β-Iodophosphinate esters can be shown to be intermediates in some cases.

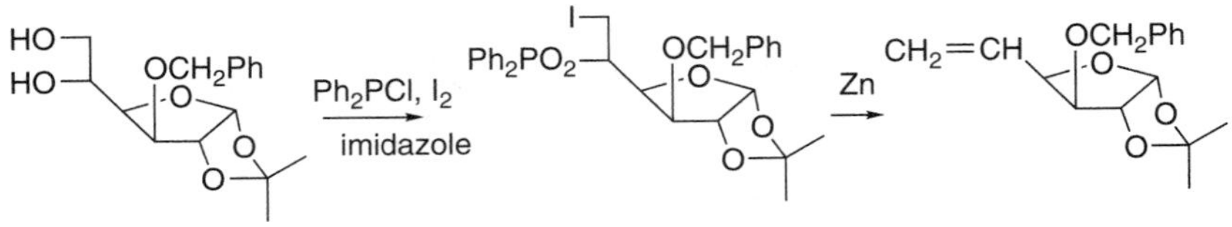

Another alternative for conversion of diols to alkenes is the use of the Barton radical fragmentation conditions (see Section 5.5) with a silane hydrogen atom donor.[299]

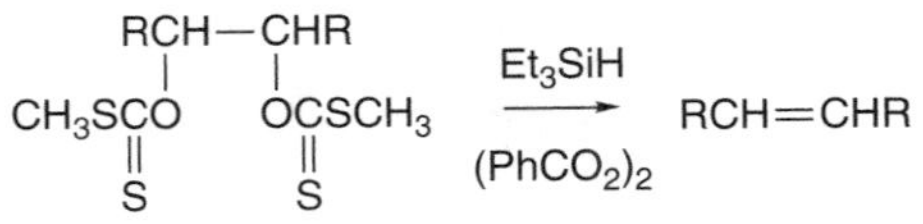

N-Ethylpiperidinium hypophosphite has been used as a reductant in deoxygenation of nucleoside diol xanthates in aqueous solution.[300]

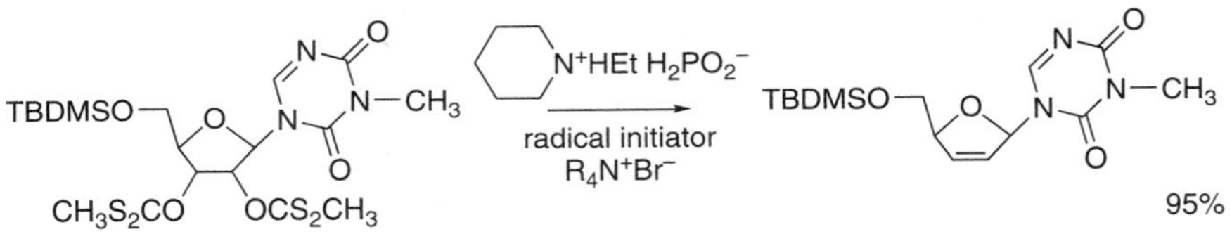

The reductive elimination of β-hydroxysulfones is the final step in the *Julia-Lythgoe alkene synthesis* (see Section 2.4.3).[301] The β-hydroxysulfones are normally obtained by an aldol addition.

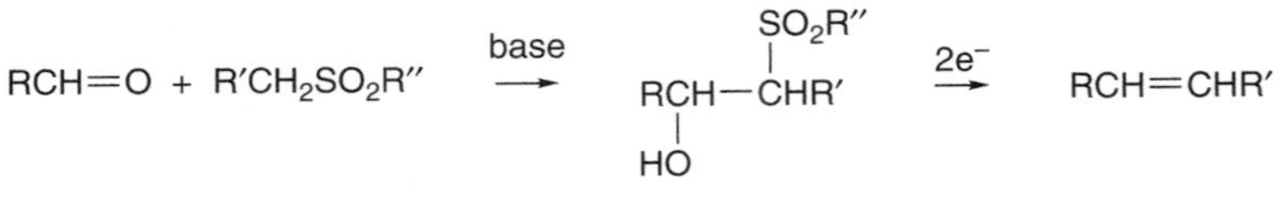

298. Z. Liu, B. Classon, and B. Samuelsson, *J. Org. Chem.*, **55**, 4273 (1990).
299. D. H. R. Barton, D. O. Jang, and J. C. Jaszberenyi, *Tetrahedron Lett.*, **32**, 2569 (1991); D. H. R. Barton, D. O. Jang, and J. C. Jaszberenyi, *Tetrahedron Lett.*, **32**, 7187 (1991).
300. D. O. Jang and D. H. Cho, *Tetrahedron Lett.*, **43**, 5921 (2002).
301. P. Kocienski, *Phosphorus and Sulfur*, **24**, 97 (1985).

Several reducing agents have been used for the elimination, including sodium amalgam[302] and samarium diiodide.[303] The elimination can also be done by converting the hydroxy group to a xanthate or thiocarbonate and using radical fragmentation.[304]

Reductive elimination from 2-en-1,4-diol derivatives has been used to generate 1,3-dienes. Low-valent titanium generated from TiCl$_3$-LiAlH$_4$ can be used directly with the diols. This reaction has been used successfully to create extended polyene conjugation.[305]

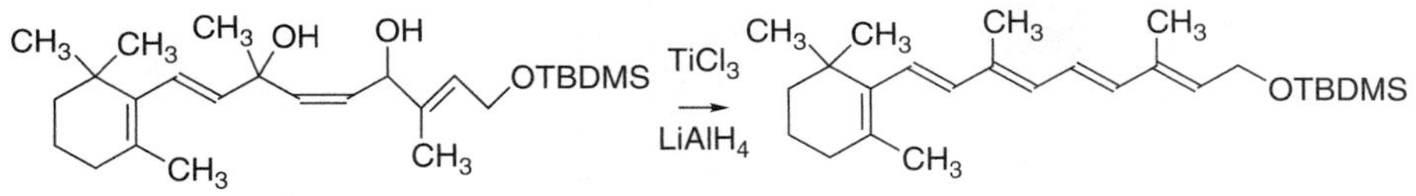

Benzoate esters of 2-en-1,4-diols undergo reductive elimination with sodium amalgam.[306]

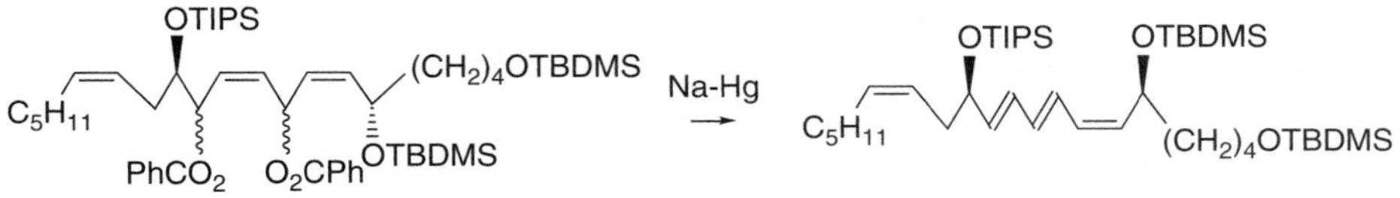

The β,γ-fragmentation is known as Grob fragmentation. Its synthetic application is usually in the construction of medium-sized rings by fragmentation of fused-ring systems. The reaction below results in both a reductive fragmentation and deoxygenation via a cyclic sulfate.

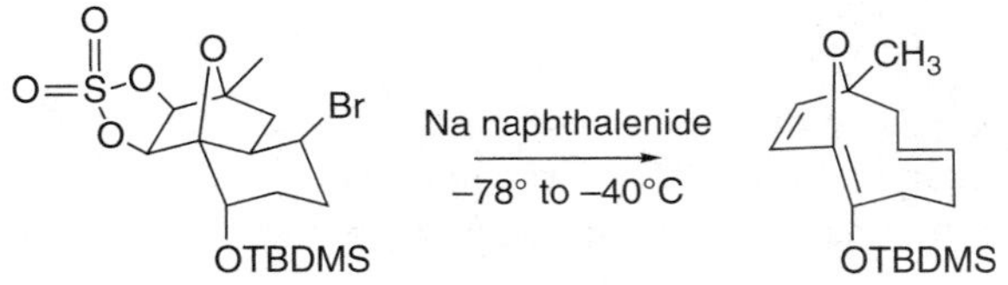

Ref. 307

302. P. J. Kocienski, B. Lythgoe, and I. Waterhouse, *J. Chem. Soc., Perkin Trans. 1*, 1045 (1980); A. Armstrong, S. V. Ley, A. Madin, and S. Mukherjee, *Synlett*, 328 (1990); M. Kagayama, T. Tamura, M. H. Nantz, J. C. Roberts, P. Somfai, D. C. Whritenour, and S. Masamune, *J. Am. Chem. Soc.*, **112**, 7407 (1990).

303. A. S. Kende and J. S. Mendoza, *Tetrahedron Lett.*, **31**, 7105 (1990); I. E. Marko, F. Murphy, and S. Dolan, *Tetrahedron Lett.*, **37**, 2089 (1996); G. E. Keck, K. A. Savin, and M. A. Weglarz, *J. Org. Chem.*, **60**, 3194 (1995).

304. D. H. R. Barton, J. C. Jaszberenyi, and C. Tachdjian, *Tetrahedron Lett.*, **32**, 2703 (1991).

305. G. Solladie, A. Givardin, and G. Lang, *J. Org. Chem.*, **54**, 2620 (1989); G. Solladie and V. Berl, *Tetrahedron Lett.*, **33**, 3477 (1992).

306. G. Solladie, A. Urbano, and G. B. Stone, *Tetrahedron Lett.*, **34**, 6489 (1993).

307. W. B. Wang and E. J. Roskamp, *Tetrahedron Lett.*, **33**, 7631 (1992).

Problems

CHAPTER 5

*Reduction of
Carbon-Carbon Multiple
Bonds, Carbonyl
Groups, and Other
Functional Groups*

(*References for these problems will be found on page 1278.*)

5.1. Give the product(s) to be expected from the following reactions. Be sure to specify all facets of stereochemistry.

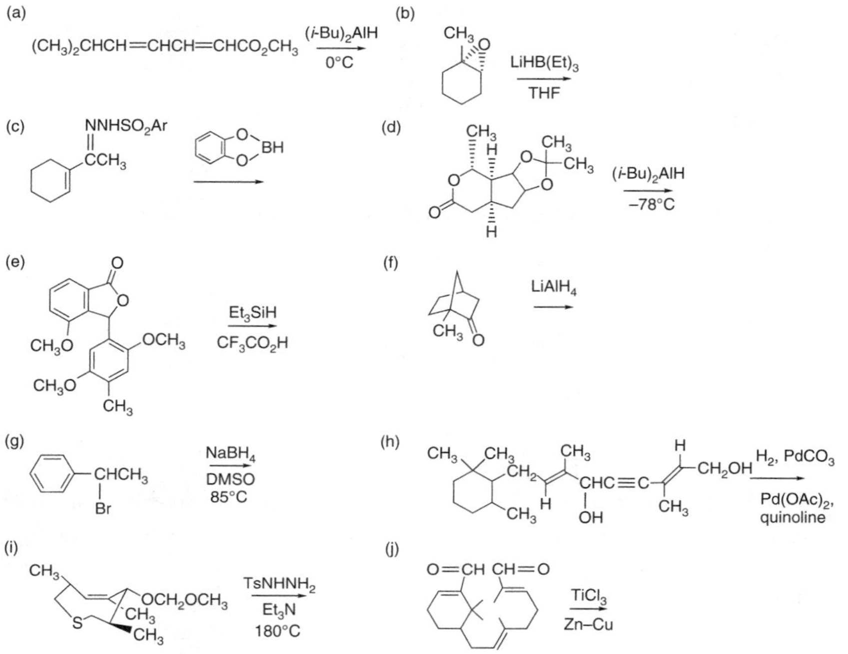

(a)

$(CH_3)_2CHCH{=}CHCH{=}CHCO_2CH_3$ $\xrightarrow[0°C]{(i\text{-}Bu)_2AlH}$

(b)

$\xrightarrow[THF]{LiHB(Et)_3}$

(c)

(d)

$\xrightarrow[-78°C]{(i\text{-}Bu)_2AlH}$

(e)

$\xrightarrow[CF_3CO_2H]{Et_3SiH}$

(f)

$\xrightarrow{LiAlH_4}$

(g)

$\xrightarrow[\substack{DMSO \\ 85°C}]{NaBH_4}$

(h)

$\xrightarrow[\substack{Pd(OAc)_2, \\ quinoline}]{H_2,\ PdCO_3}$

(i)

$\xrightarrow[\substack{Et_3N \\ 180°C}]{TsNHNH_2}$

(j)

$\xrightarrow[Zn{-}Cu]{TiCl_3}$

5.2. The data below give the ratio of equatorial:axial alcohol by $NaBH_4$ reduction of each cyclohexanone derivative under conditions in which 4-*t*-butylcyclohexanone gives an approximately 85:15 ratio. Analyze the effect of the substituents in each case.

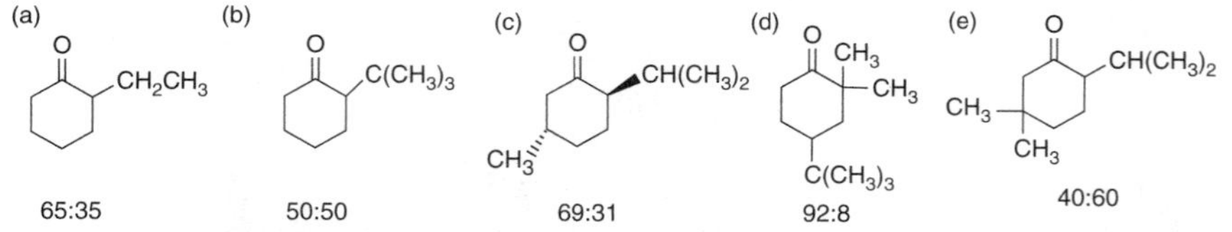

(a)

65:35

(b)

50:50

(c)

69:31

(d)

92:8

(e)

40:60

5.3. Indicate reaction conditions that would accomplish each of the following transformations in a single step:

(a)

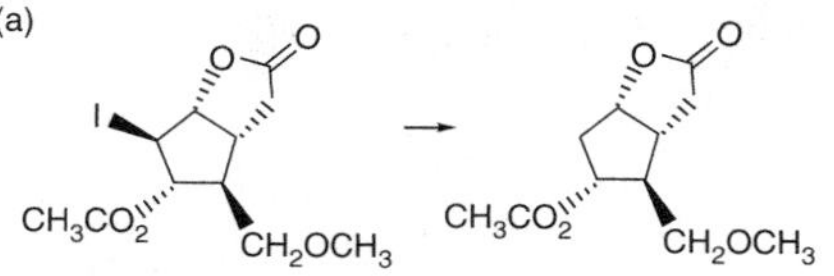

(b)

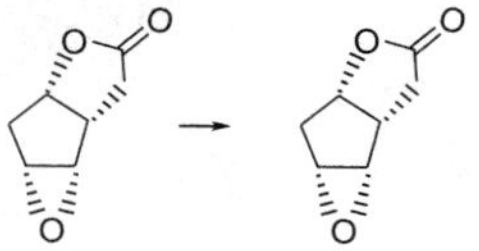

(c)

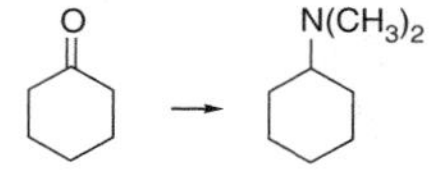

(d)

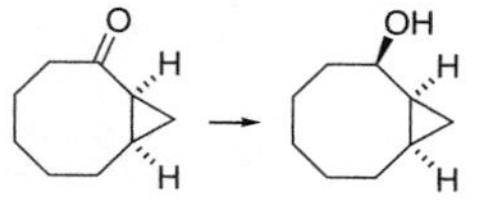

(e)

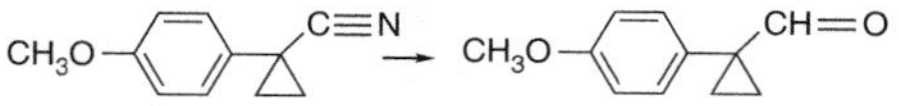

(f)

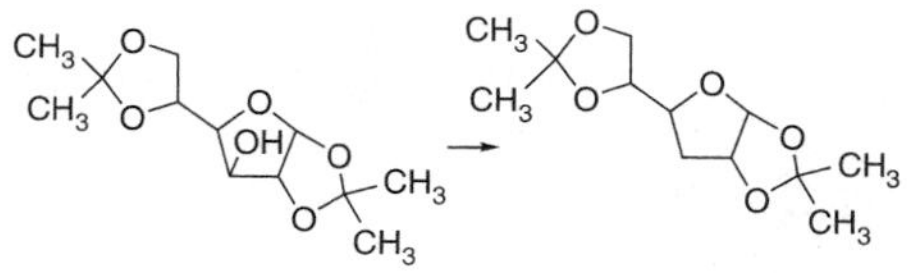

(g)

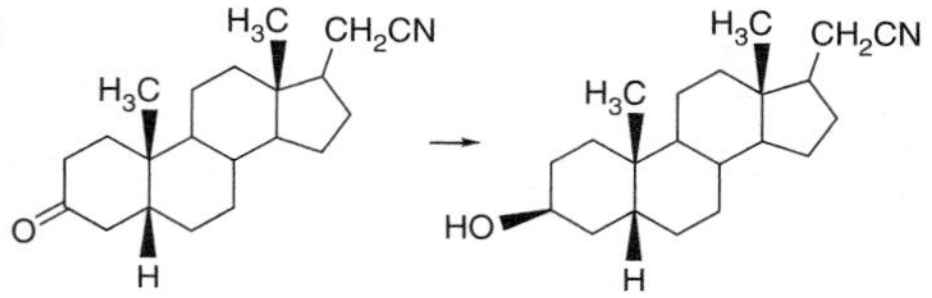

(h)

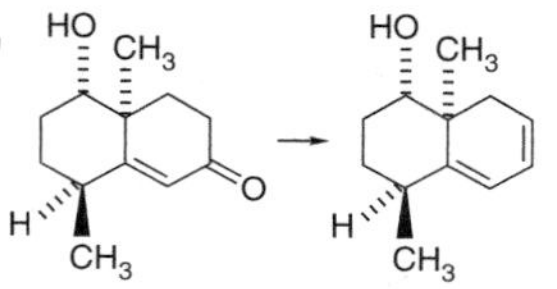

(i)

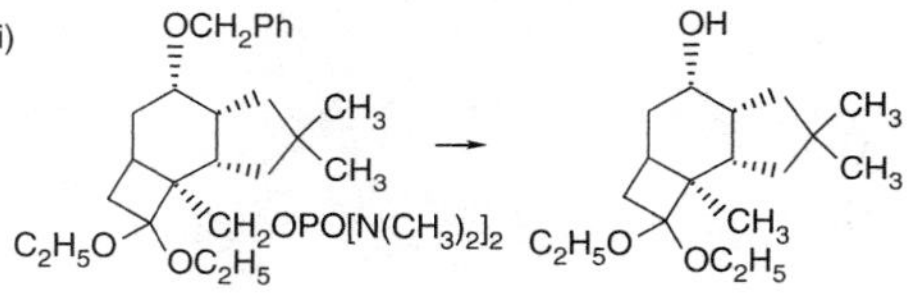

(j)

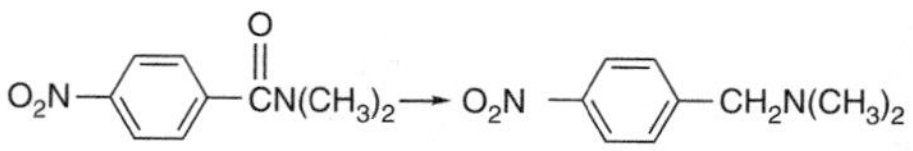

(k)

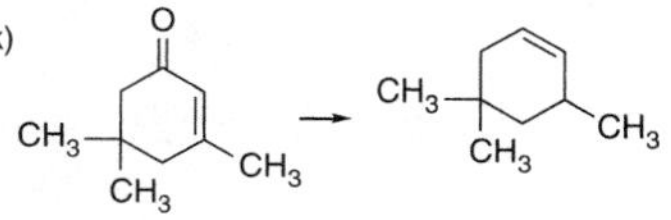

(l)

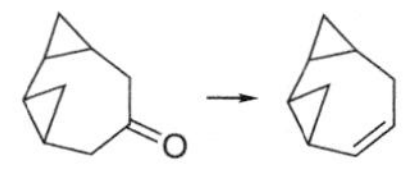

(m)

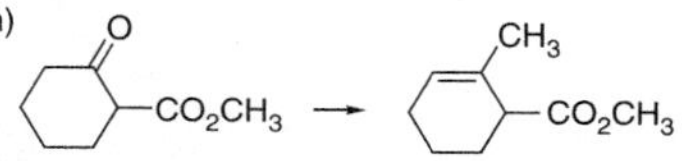

(n)

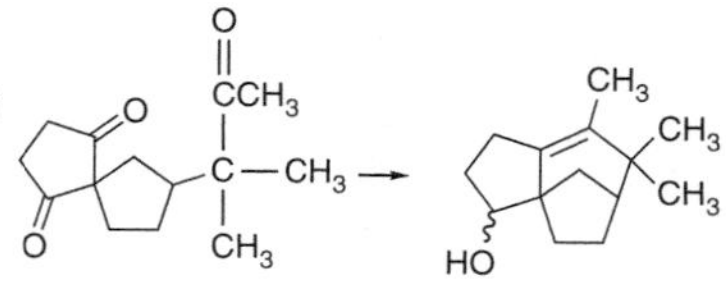

(o)

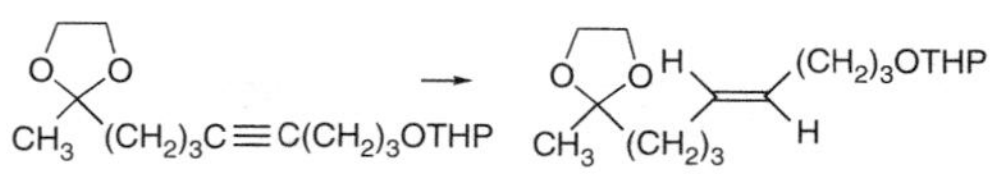

464

CHAPTER 5

*Reduction of
Carbon-Carbon Multiple
Bonds, Carbonyl
Groups, and Other
Functional Groups*

5.4. Predict the stereochemistry of the products from the following reactions and justify your prediction.

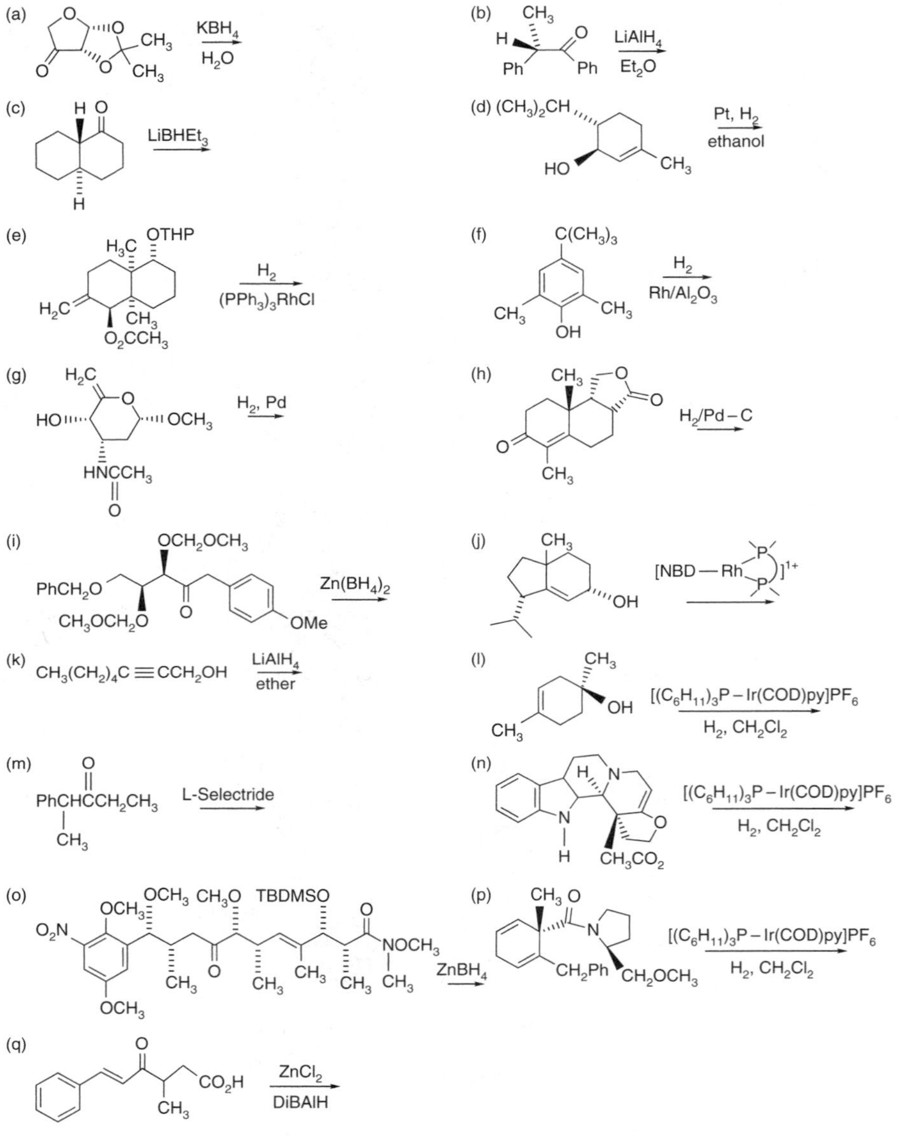

(a)

KBH₄
H₂O

(b)

LiAlH₄
Et₂O

(c)

LiBHEt₃

(d)

Pt, H₂
ethanol

(e)

H₂
(PPh₃)₃RhCl

(f)

H₂
Rh/Al₂O₃

(g)

H₂, Pd

(h)

H₂/Pd–C

(i)

Zn(BH₄)₂

(j)

[NBD—Rh(P)₂]¹⁺

(k)

CH₃(CH₂)₄C≡CCH₂OH

LiAlH₄
ether

(l)

[(C₆H₁₁)₃P–Ir(COD)py]PF₆
H₂, CH₂Cl₂

(m)

L-Selectride

(n)

[(C₆H₁₁)₃P–Ir(COD)py]PF₆
H₂, CH₂Cl₂

(o)

ZnBH₄

(p)

[(C₆H₁₁)₃P–Ir(COD)py]PF₆
H₂, CH₂Cl₂

(q)

ZnCl₂
DiBAlH

5.5. Suggest a convenient method for carrying out the following syntheses. The compound on the left is to be made from the one on the right (retrosynthetic notation). No more than three steps should be necessary.

(a)

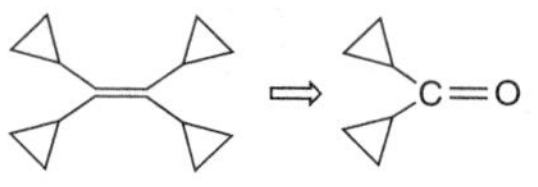

(b)

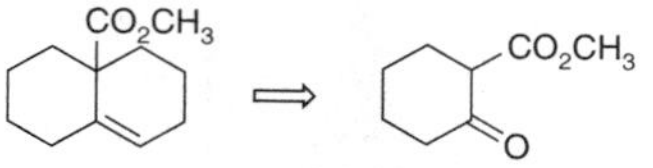

(c)

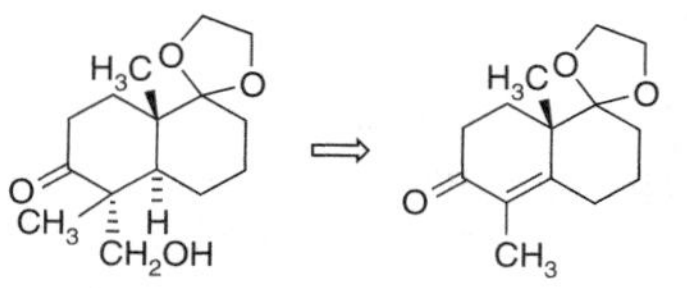

(d)

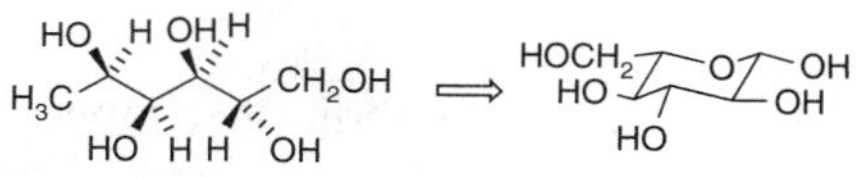

(e)

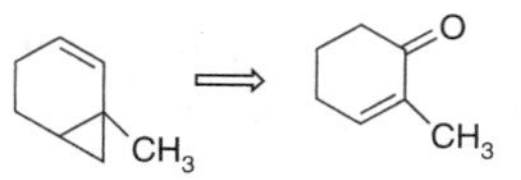

(f)

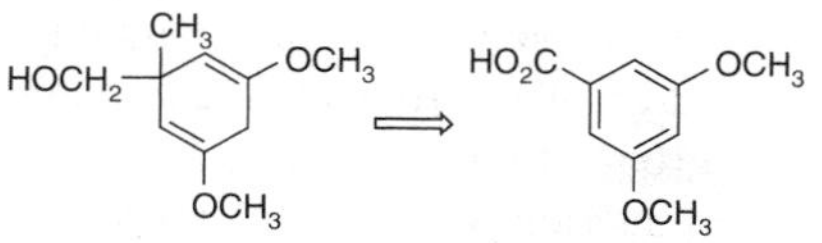

(g)

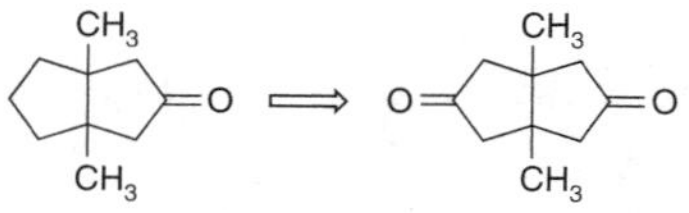

(h) $meso\text{-}(CH_3)_2CHCHCHCH(CH_3)_2$ with HO, OH $\Rightarrow$ $(CH_3)_2CHCO_2CH_3$

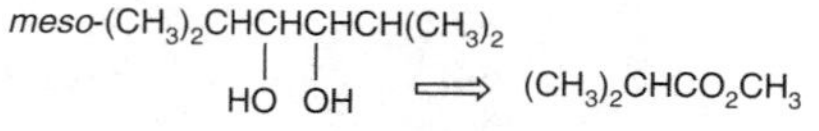

(i) CH_3O ... $CH_2CH(CH_2OH)_2$ ... OCH_3 $\Rightarrow$ CH_3O ... CH_2Cl ... OCH_3

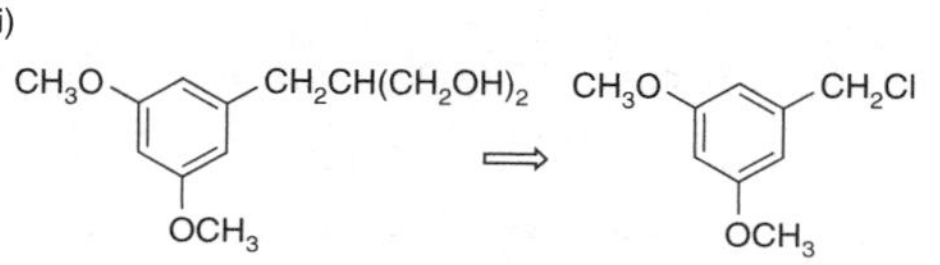

(j) $C_6H_5CHCH_2CHCH_3$ with SC_6H_5, OH $\Rightarrow$ $C_6H_5CH=CHCCH_3$ (with carbonyl O)

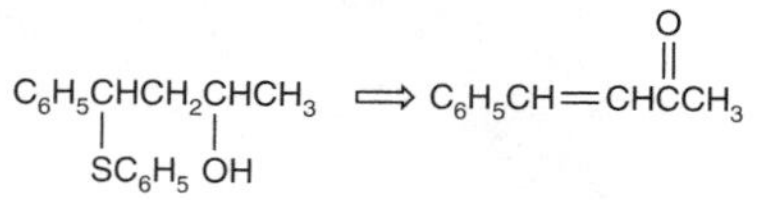

5.6. Offer an explanation to account for the observed differences in the rate of the following reactions:

   a. LiAlH$_4$ reduces camphor about 30 times faster than does NaAlH$_4$.
   b. The rate of reduction of camphor by LiAlH$_4$ is decreased by a factor of about 4 when a crown ether is added to the reaction mixture.
   c. For reduction of cyclohexanones by LiAlH(t-OBu)$_3$, the addition of one methyl group at C(3) has little effect, but a second group on the same carbon has a large effect. The addition of a third methyl group at C(5) has no effect and the addition of a second methyl at C(5) has only a small effect.

Ketone	Rel. rate
Cyclohexanone	439
3-Methylcyclohexanone	280
3,3-Dimethylcyclohexanone	17.5
3,3,5-Trimethylcyclohexanone	17.4
3,3,5,5-Tetramethylcyclohexanone	8.9

5.7. Suggest reaction conditions appropriate for stereoselective conversion of the octalone shown to each of the diastereomeric decalones.

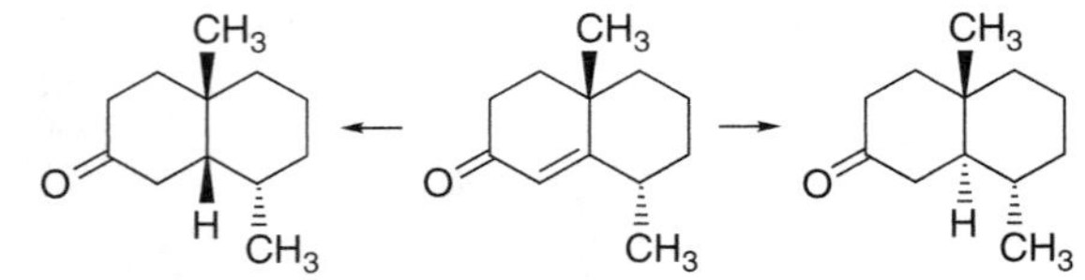

466

CHAPTER 5

*Reduction of
Carbon-Carbon Multiple
Bonds, Carbonyl
Groups, and Other
Functional Groups*

5.8. The fruit of a shrub that grows in Sierra Leone is very toxic and has been used as a rat poison. The toxic principal has been identified as Z-18-fluoro-9-octadecenoic acid. Suggest a synthesis from 8-fluorooctanol, 1-chloro-7-iodoheptane, acetylene, and any other necessary organic or inorganic reagents.

5.9. Each of the following compounds contains more than one potentially reducible group. Indicate a reducing agent that will be suitable for effecting the desired reduction. Explain the basis for the expected selectivity.

(a)

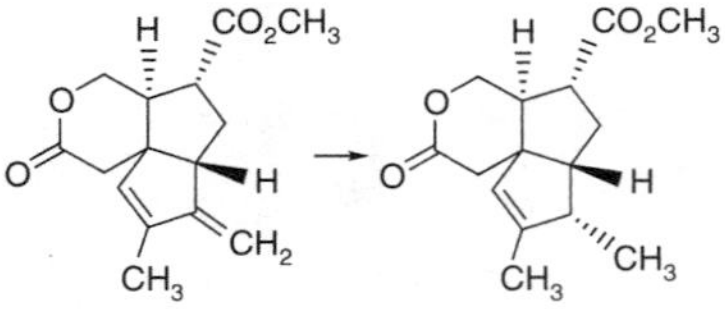

(b)

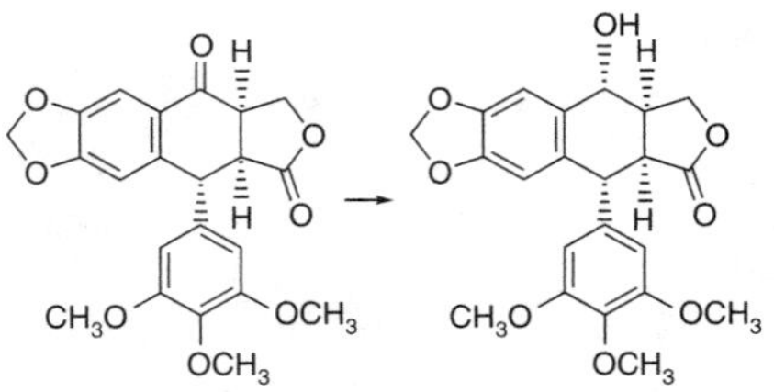

(c)

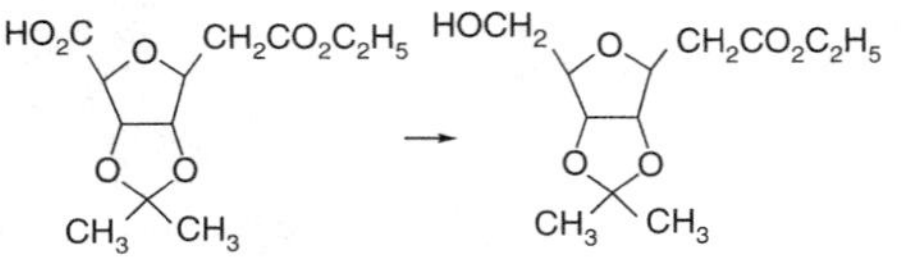

(d)

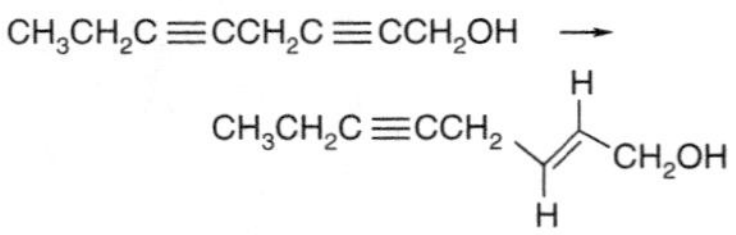

(e)

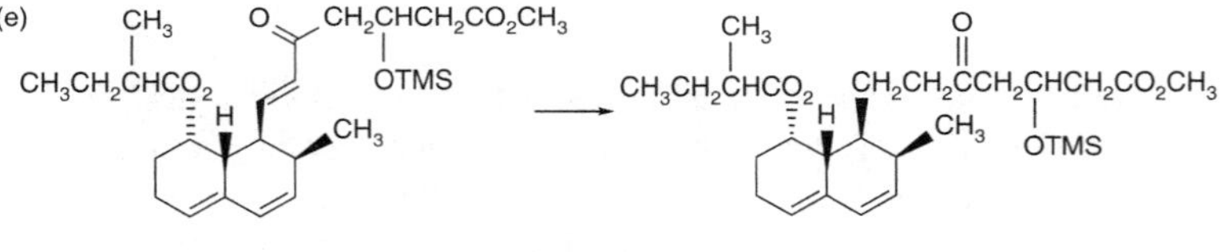

(f)

(g)

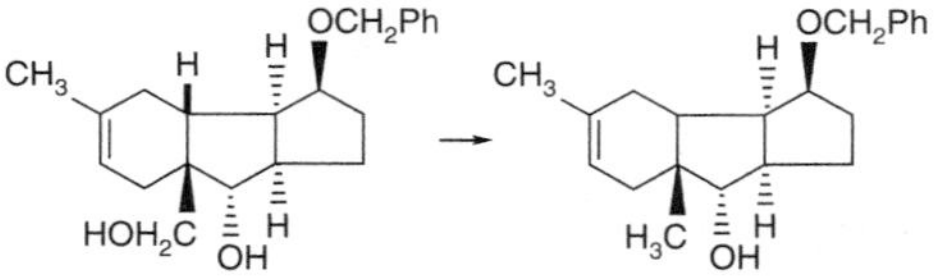

(h)

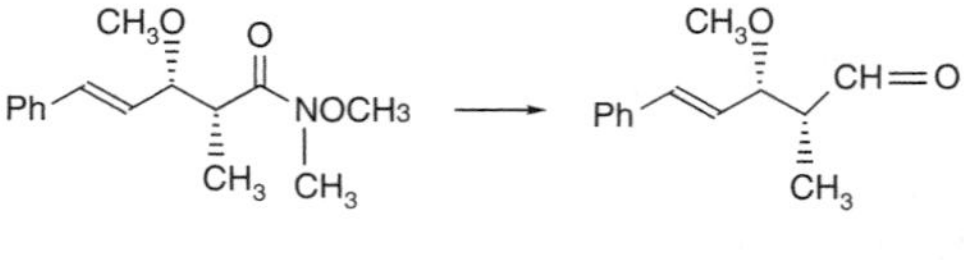

(i)

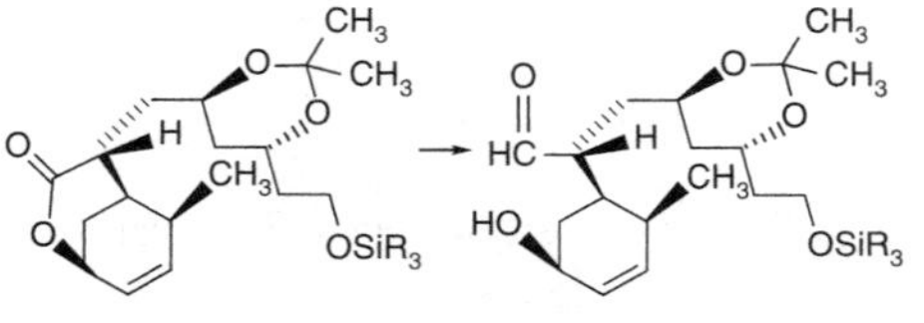

5.10. Explain the basis of the observed stereoselectivity for the following reactions:

(a)

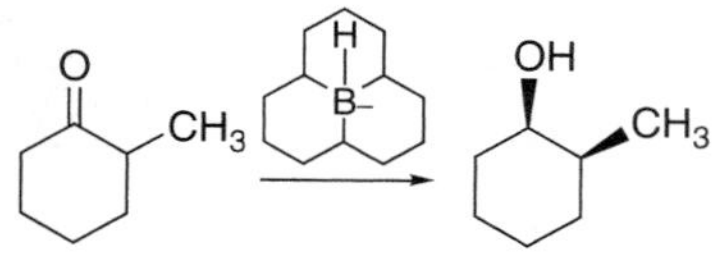

(b)

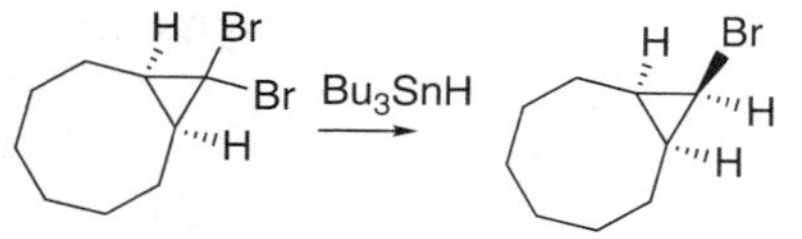

(c)

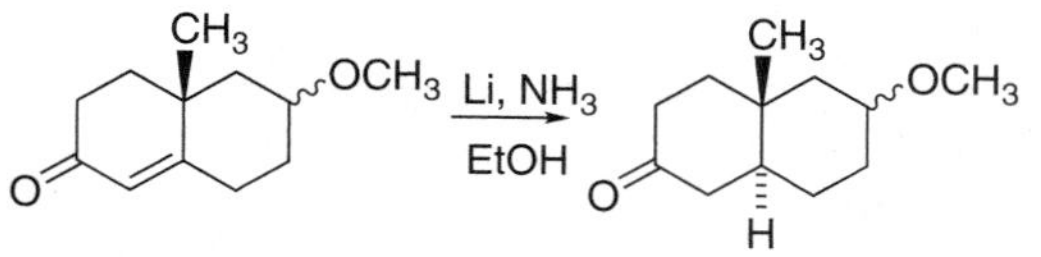

5.11. A valuable application of sodium cyanoborohydride is in the synthesis of amines by reductive amination. What combination of carbonyl and amine components would give the following amines by this method?

(a)

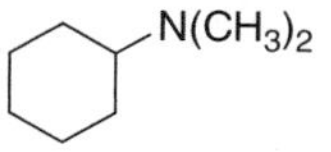

(b)

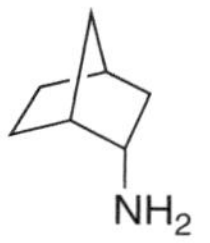

5.12. The reduction of allyl o-bromphenyl ether by $LiAlH_4$ has been studied in several solvents. In ether, two products **12-A** and **12-B** are formed. The ratio **12-A:12-B** increases with increasing $LiAlH_4$ concentration. When $LiAlD_4$ is used as the reductant, about half of product **12-B** is monodeuterated. Provide a mechanistic rationale for these results. What is the predicted location of the deuterium in the **12-B**? Why is the product not completely deuterated?

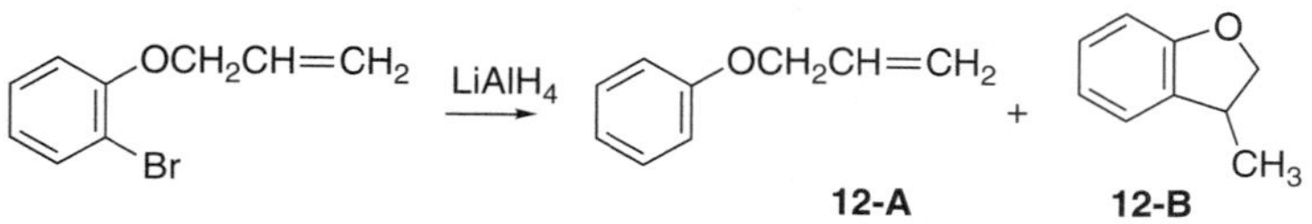

12-A   **12-B**

5.13. Each of the following parts describes a synthetic sequence in which Birch reduction is employed to convert aromatic rings to partially saturated products.

a. A simple synthesis of 2-substituted cyclohexenones from 2-methoxybenzoic acid has been developed. The reaction sequence entails Birch reduction, tandem alkylation, and acid hydrolysis. Although the yields are only 25–30%, it can be carried out as a one-pot process using the sequence of reactions shown below. Explain the mechanistic basis of this synthesis and identify the intermediate present after each stage of the sequence.

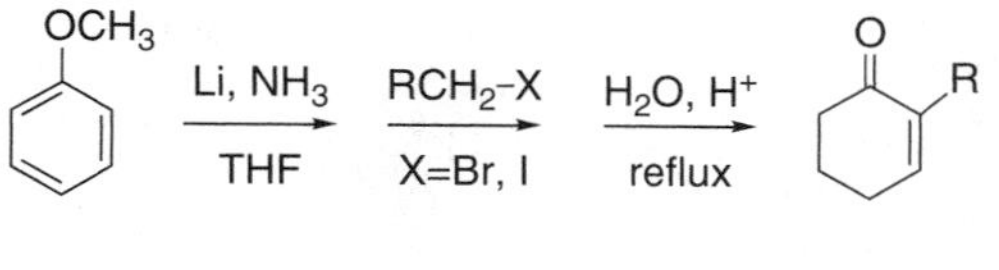

468

CHAPTER 5

Reduction of
Carbon-Carbon Multiple
Bonds, Carbonyl
Groups, and Other
Functional Groups

b. Birch reduction of 3,4,5-trimethoxybenzoic acid gives a dihydrobenzoic acid in 94% yield, but it has only *two* methoxy substituents. Suggest a plausible structure for this product based on the mechanism of the Birch reduction.

c. The cyclohexenone **13-C** has been prepared in a one-pot process starting with 4-methylpent-3-en-2-one. The reagents that are added in succession are 4-methoxyphenyllithium, Li, and NH_3, followed by acidic workup. Show the intermediates that are involved in the process.

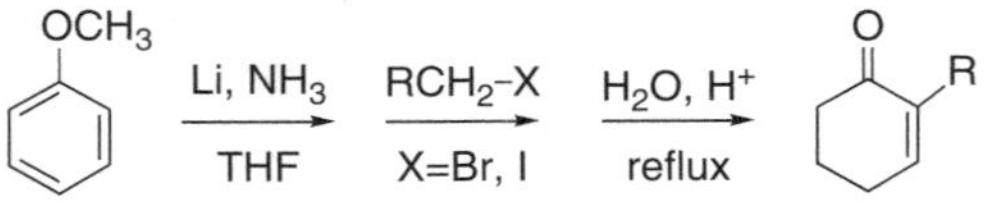

5.14. Ketones can be converted to nitriles by the following sequence of reagents. Indicate the intermediate stages of the reaction.

5.15. In the synthesis of fluorinated analogs of the acetylcholinesterase inhibitor, huperzine A, it was necessary to accomplish reductive elimination of the diol **15-D** to **15-E**. Of the methods for diol reduction, which seems most compatible with the other functional groups in this compound?

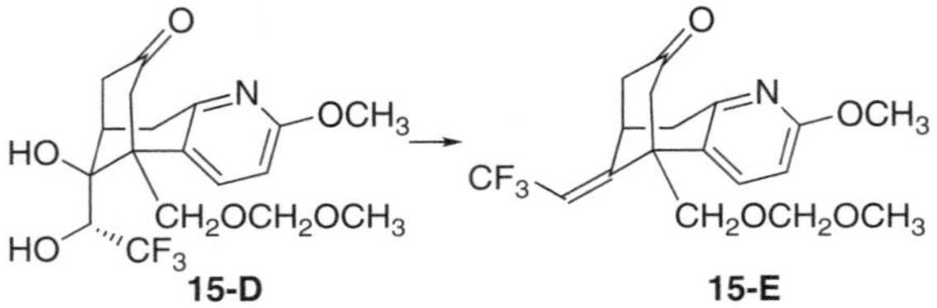

5.16. Wolff-Kishner reduction of ketones bearing other functional groups sometimes gives products other than the expected methylene reduction product. Several examples are given below. Indicate a mechanism for each reaction.

(a)

$(CH_3)_3CCCH_2OPh \longrightarrow (CH_3)_3CCH{=}CH_2$

(b)

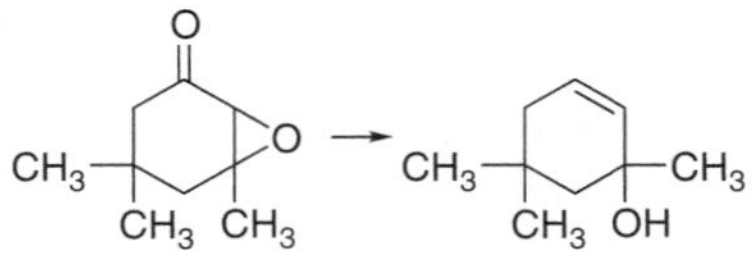

(c)

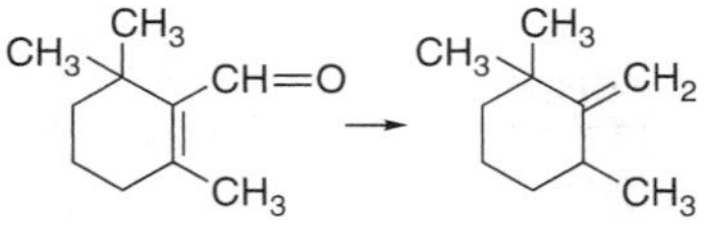

(d)

$PhCH{=}CHCH{=}O \longrightarrow Ph{-}\triangle$

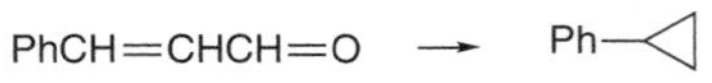

5.17. Suggest reagents and reaction conditions that would be suitable for each of the following selective or partial reductions:

(a)

$$HO_2C(CH_2)_4CO_2C_2H_5 \longrightarrow HOCH_2(CH_2)_4CO_2C_2H_5$$

(b)

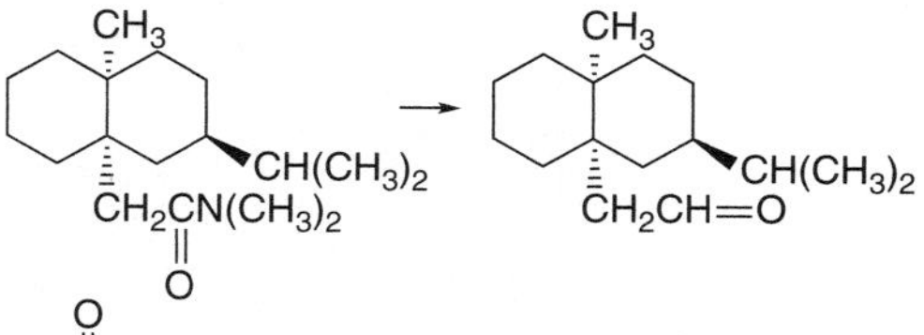

(c)

$$CH_3\overset{O}{\overset{\|}{C}}(CH_2)_2CO_2C_8H_{17} \longrightarrow CH_3(CH_2)_3CO_2C_8H_{17}$$

(d)

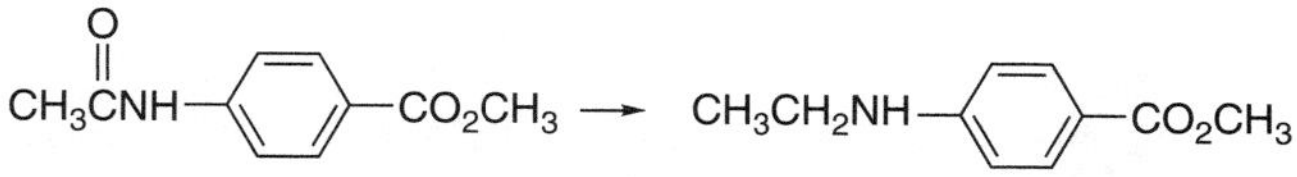

(e)

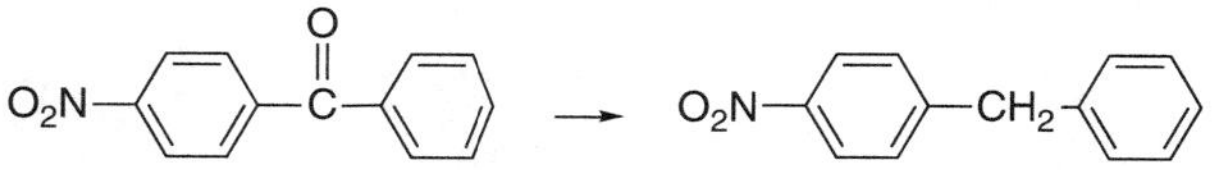

(f)

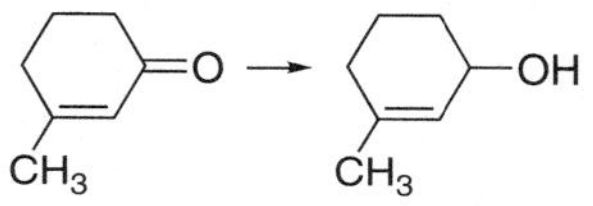

(g)

5.18. The reduction of the ketone **18-F** gives product **18-G** in preference to **18-H** with *increasing* stereoselectivity in the order $NaBH_4 < LiAlH_2(OCH_2CH_2OCH_3)_2 < Zn(BH_4)_2$. With L-Selectride, however, **18-H** is favored. Account for the dependence of the stereoselectivity on the various reducing agents.

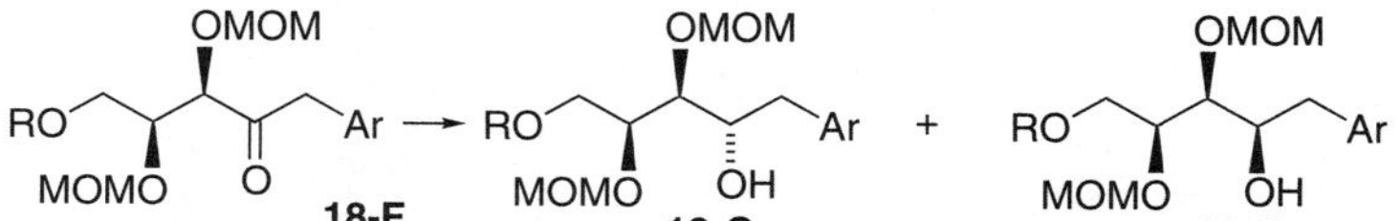

Ar = 4-methoxyphenyl
R = benzyl
MOM = methoxymethyl

5.19. The following reducing agents effect enantioselective reduction of ketones. Propose a transition structure that is in accord with the observed enantioselectivity.

470

CHAPTER 5

*Reduction of
Carbon-Carbon Multiple
Bonds, Carbonyl
Groups, and Other
Functional Groups*

(a)

$\longrightarrow$ *R*-α-hydroxyester in 90% e.e.

(b)

+ BH₃ +

$\longrightarrow$ *R*-alcohol in 97% e.e.

(0.6 equiv)   (0.1 equiv)

(c)

$\longrightarrow$ *S*-alcohol in 97% e.e.

5.20. By retrosynthetic analysis, devise a sequence of reactions that would accomplish the following transformations:

(a)

from

(b)

from

(c)

from

5.21. A group of topologically unique molecules called "betweenanenes" has been synthesized. Successful synthesis of such molecules depends on effective means of closing large rings. Suggest an overall strategy (details not required) to synthesize such molecules. Suggest types of reactions that might be considered for formation of the large rings.

5.22. Give the products expected from the following reactions with Sm(II) reagents.

(a)

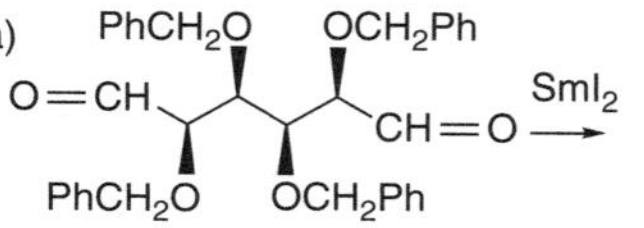

(b)

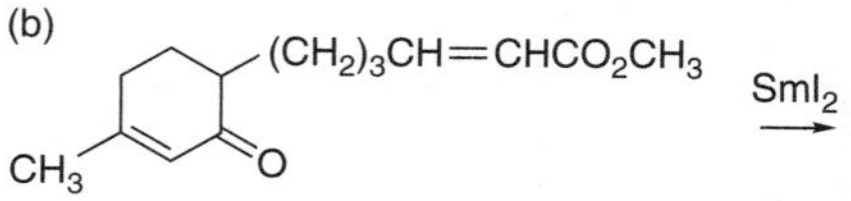

(c)

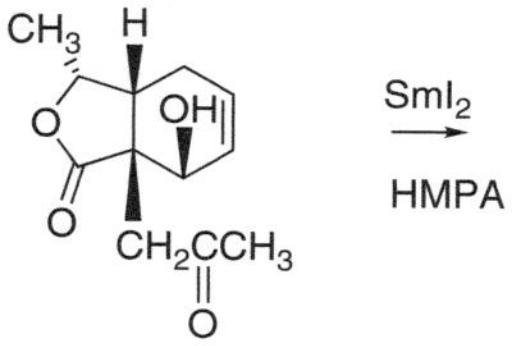

(d)

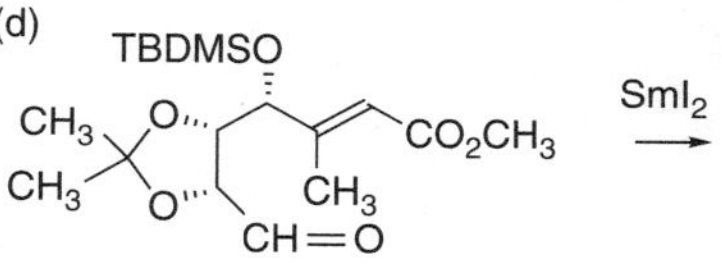

(e)

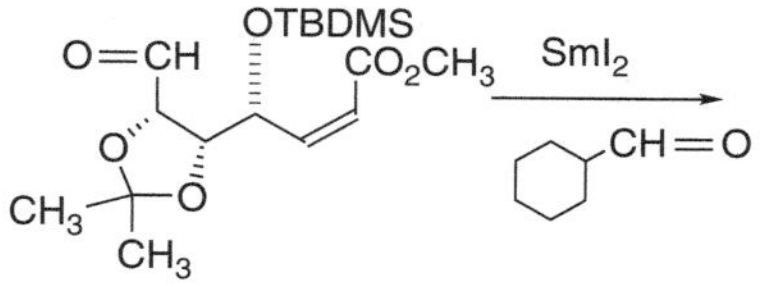

(f)

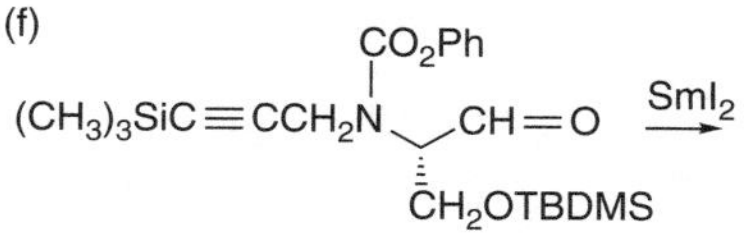

5.23. Provide an explanation based on a transition structure for the trends in stereoselectivity revealed by the following data.

(a)

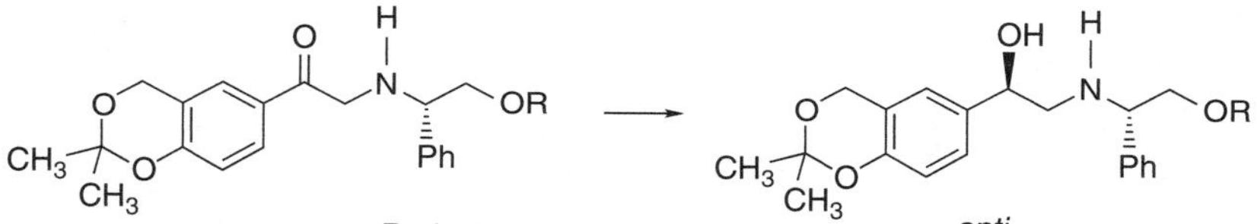

anti

Reducing agent	R	*anti:syn*
$NaBH_4$	H	2:1
$NaBH_4/CaCl_2$	H	10:1
$NaBH_4/CaCl_2$	CH_3	4.5:1

(b)

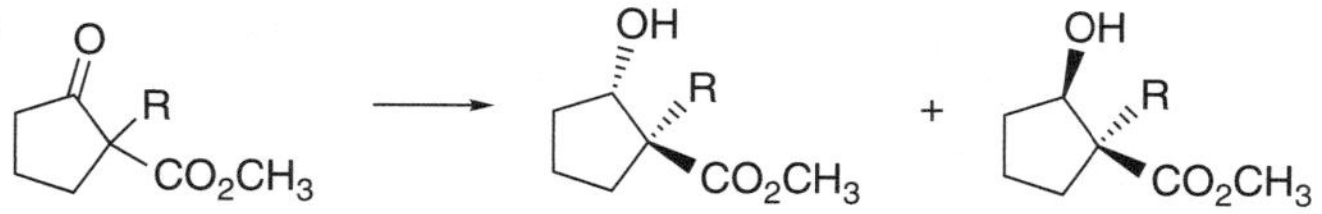

trans *cis*

R	$NaBH_4$	$NaBH_4/CaCl_2$
	trans:cis	*trans:cis*
$CH_3CH_2CH_2$	1:1.9	1:99
$PhCH_2$	1:2.0	1:12
$CH_3CH_2CH{=}CH$	1:2.3	1:7

6

Concerted Cycloadditions, Unimolecular Rearrangements, and Thermal Eliminations

Introduction

Most of the reactions described in the preceding chapters involve polar or polarizable reactants and proceed through polar intermediates and/or transition structures. One reactant can be identified as nucleophilic and the other as electrophilic. Carbanion alkylations, nucleophilic additions to carbonyl groups, and electrophilic additions to alkenes are examples of such reactions. The reactions to be examined in this chapter, on the other hand, occur via a reorganization of electrons through transition structures that may not be much more polar than the reactants. These reactions proceed through cyclic transition structures. The activation energy can be provided by thermal or photochemical excitation of the reactant(s) and often no other reagents are involved. Most of the transformations fall into the category of *concerted pericyclic reactions*, in which there are no intermediates and the transition structures are stabilized by favorable orbital interactions, as discussed in Chapter 10 of Part A. These reactions can be classified into three broad types: cycloadditions, unimolecular rearrangements, and eliminations. We also discuss some reactions that effect closely related transformations, but which on mechanistic scrutiny are found to proceed through discrete intermediates.

474

CHAPTER 6

Concerted
Cycloadditions,
Unimolecular
Rearrangements, and
Thermal Eliminations

6.1. Diels-Alder Reactions

6.1.1. The Diels-Alder Reaction: General Features

Cycloaddition reactions result in the formation of a new ring from two reactants. A concerted mechanism requires that a single transition state, and therefore no intermediate, lie on the reaction path between reactants and adduct. The most important example of cycloaddition is the *Diels-Alder (D-A) reaction*. The cycloaddition of alkenes and dienes is a very useful method for forming substituted cyclohexenes.[1]

A clear understanding of concerted cycloaddition reactions developed as a result of the formulation of the mechanisms within the framework of molecular orbital theory. Consideration of the MOs of reactants and products reveals that in many cases a smooth transformation of the orbitals of the reactants to those of products is possible. In other cases, reactions that might appear feasible if no consideration is given to the symmetry and spatial orientation of the orbitals are found to require high-energy TSs when the orbitals are considered in detail. (Review Section 10.1 of Part A for a discussion of the orbital symmetry analysis of cycloaddition reactions.) The relationships between reactants and TS orbitals permit description of potential cycloaddition reactions as "allowed" or "forbidden" and indicate whether specific reactions are likely to be energetically favorable. The same orbital symmetry relationships that are informative as to the feasibility of a reaction are often predictive of the regiochemistry and stereochemistry. This predictability is an important feature for synthetic purposes. Another attractive aspect of the D-A reaction is the fact that *two new carbon-carbon bonds* are formed in a single reaction.

In the terminology of orbital symmetry classification, the Diels-Alder reaction is a $[4\pi_s + 2\pi_s]$ cycloaddition, an allowed process. There have been a large number of computational studies of the D-A reaction, and as it is a fundamental example of a concerted reaction, it has frequently been the subject of advanced calculations.[2] These studies support a concerted mechanism, which is also supported by good agreement between experimental and calculated (B3LYP/6-31G*) kinetic isotope effects.[3] The TS for a concerted reaction requires that the diene adopt the *s-cis* conformation. The diene and substituted alkene (called the *dienophile*) approach each other in approximately parallel planes. The symmetry properties of the π orbitals permit stabilizing interactions between C(1) and C(4) of the diene and the dienophile. Usually, the strongest bonding

1. L. W. Butz and A. W. Rytina, *Org. React.*, **5**, 136 (1949); M. C. Kloetzel, *Org. React.*, **4**, 1 (1948); A. Wasserman, *Diels-Alder Reactions*, Elsevier, New York (1965); F. Fringuelli and A. Tatacchi, *Diels-Alder Reactions: Selected Practical Methods*, Wiley, New York, 2001.
2. P. D. Karadakov, D. L. Cooper, and J. Gerratt, *J. Am. Chem. Soc.*, **120**, 3975 (1998); H. Lischka, E. Ventura, and M. Dallows, *Chem. Phys. Phys. Chem.*, **5**, 1365 (2004); E. Kraka, A. Wu, and D. Cremer, *J. Phys. Chem. A*, **107**, 9008 (2003); S. Berski, J. Andres, B. Silvi, and L. R. Domingo, *J. Phys. Chem. A*, **107**, 6014 (2003); H. I. Sobe, Y. Takano, Y. Kitagawa, T. Kawakami, S. Yamanaka, K. Yamagushi, and K. N. Houk, *J. Phys. Chem. A*, **107**, 682 (2003).
3. E. Goldstein, B. Beno, and K. N. Houk, *J. Am. Chem. Soc.*, **118**, 6036 (1996); D. R. Singleton, S. R. Merrigan, B. R. Beno, and K. N. Houk, *Tetrahedron Lett.*, **40**, 5817 (1999).

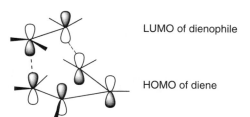

Fig. 6.1. Interaction between LUMO of dienophile and HOMO of diene in the Diels-Alder reaction.

interaction is between the HOMO of the diene and the LUMO of the dienophile. The interaction between the frontier orbitals is depicted in Figure 6.1.

6.1.2. Substituent Effects on the Diels-Alder Reaction

There is a strong electronic substituent effect on the D-A reaction. The most reactive dienophiles for simple dienes are those having electron-attracting groups. Thus, quinones, maleic anhydride, and nitroalkenes are among the most reactive dienophiles. α,β-Unsaturated aldehydes, esters, ketones, and nitriles are also effective dienophiles. It is significant that if an electron-poor diene is utilized, the preference is reversed and electron-rich alkenes, such as vinyl ethers, are the best dienophiles. Such reactions are called *inverse electron demand Diels-Alder reactions*, and the relationships involved are readily understood in terms of frontier orbital theory. Electron-rich dienes have high-energy HOMOs and interact strongly with the LUMOs of electron-poor dienophiles. When the substituent pattern is reversed and the diene is electron-poor, the strongest interaction is between the dienophile HOMO and the diene LUMO.

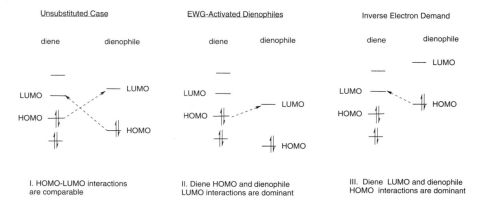

Frontier orbital theory can also explain the regioselectivity observed when both the diene and alkene are unsymmetrically substituted.[4] Generally, there is a preference

---

[4.] K. N. Houk, *Acc. Chem. Res.*, **8**, 361 (1975); I. Fleming, *Frontier Orbitals and Organic Chemical Reactions*, Wiley-Interscience, New York, 1976; O. Eisenstein, J. M. LeFour, N. T. Anh, and R. F. Hudson, *Tetrahedron*, **33**, 523 (1977).

476

CHAPTER 6

*Concerted
Cycloadditions,
Unimolecular
Rearrangements, and
Thermal Eliminations*

for the "*ortho*" product when the diene has a donor (ERG) substituent at C(1) and for "*para*" product when the diene has an ERG at C(2), as in the examples shown.[5]

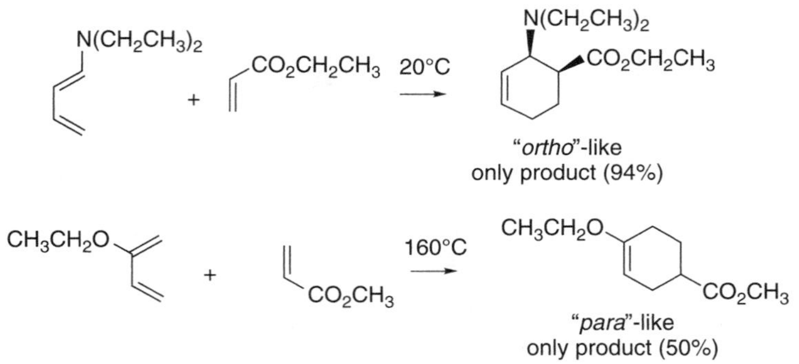

"*ortho*"-like
only product (94%)

"*para*"-like
only product (50%)

When the dienophile bears an EWG substituent and the diene an ERG, the strongest interaction is between the HOMO of the diene and the LUMO of the dienophile. The reactants are oriented so that the carbons having the highest coefficients of these two frontier orbitals can begin the bonding process, and this leads to the observed regiochemical preference as summarized in Figure 6.2.

Diels-Alder reactions are *stereospecific* with respect to the *E*- and *Z*-relationships in both the dienophile and the diene. For example, addition of dimethyl fumarate and dimethyl maleate with cyclopentadiene is completely stereospecific with respect to the *cis* or *trans* orientation of the ester substituents.

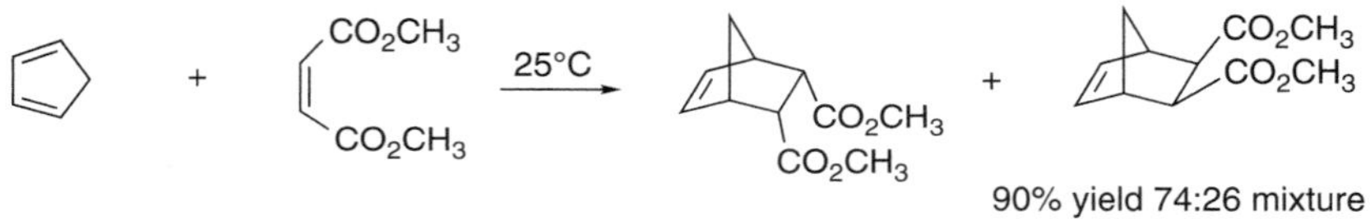

90% yield 74:26 mixture

Ref. 6

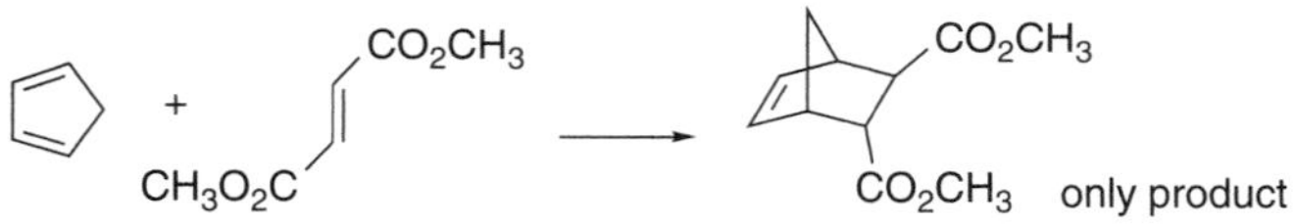

only product

Ref. 7

Similarly, *E,E*-2,4-hexadiene gives a product that is stereospecific with respect to the diene methyl groups.

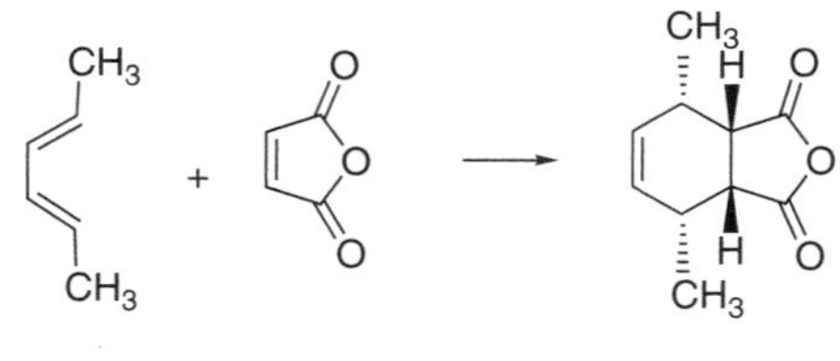

Ref. 8

5. J. Sauer, *Angew. Chem. Int. Ed. Engl.*, **6**, 16 (1967).
6. W. Kirmse, U. Mrotzeck, and R. Siegfried, *Chem. Ber.*, **124**, 238 (1991).
7. C. Girard and R. Bloch, *Tetrahedron Lett.*, **23**, 3683 (1982).
8. G. Berube and P. Deslongchamps, *Bull. Soc. Chim. Fr.*, 103 (1987).

a) Coefficient at C(2) is higher than at C(1) in the LUMO of a dienophile bearing and electron-withdrawing substituent.

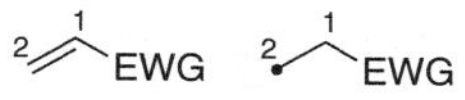

(b) Coefficient at C(4) is higher than at C(1) in HOMO of a diene bearing an electron-releasing substituent at C(1).

(c) Coefficient at C(1) is higher than at C(4) in HOMO of a diene bearing an electron-releasing substituent at C(1).

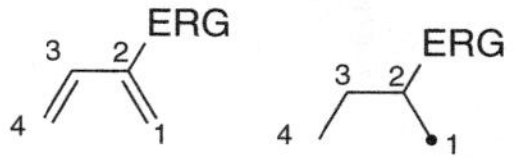

(d) The regioselectivity of the Diels-Alder reaction corresponds to matching the carbon atoms having the largest coefficients of the frontier orbitals.

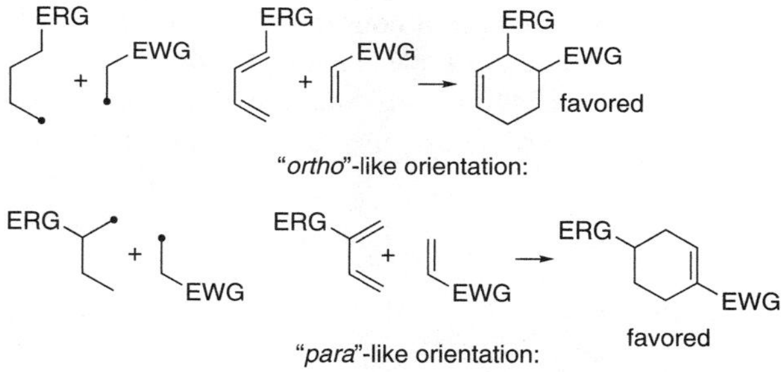

"*ortho*"-like orientation:

"*para*"-like orientation:

Fig. 6.2. HOMO-LUMO interactions rationalize regioselectivity of Diels-Alder reactions.

Stereospecificity also is exhibited for dienes having stronger electron-releasing groups, such as trimethylsiloxy.

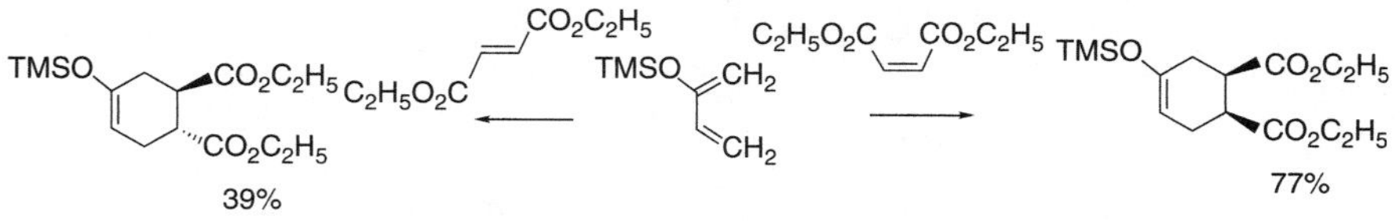

39%

77%

Ref. 9

9. M. E. Jung and C. A. McCombs, *Org. Synth.*, **58**, 163 (1978); M. E. Jung and C. A. McCombs, *Tetrahedron Lett.*, 2935 (1976).

478

CHAPTER 6

*Concerted
Cycloadditions,
Unimolecular
Rearrangements, and
Thermal Eliminations*

For an unsymmetrical dienophile there are two possible stereochemical orientations with respect to the diene, *endo* and *exo*, as illustrated in Figure 6.3. In the *endo* TS the reference substituent on the dienophile is oriented toward the π orbitals of the diene. In the *exo* TS the substituent is oriented away from the π system. For many substituted butadiene derivatives, the TSs lead to two different stereoisomeric products. The *endo* mode of addition is usually preferred when an electron-attracting substituent such as a carbonyl group is present on the dienophile. The empirical statement that describes this preference is called the *Alder rule*. Frequently a mixture of both stereoisomers is formed and sometimes the *exo* product predominates, but the Alder rule is a useful initial guide to prediction of the stereochemistry of a D-A reaction. The *endo* product is often the more sterically congested. The preference for the *endo* TS is strongest for relatively rigid dienophiles such as maleic anhydride and benzoquinone. For methyl acrylate, methyl methacrylate, and methyl crotonate the selectivity ratios are not high.[10] The preference for the *endo* TS increases somewhat with increasing solvent polarity.[11] This has been attributed to a higher polarity of the *endo* TS, resulting from alignment of the dipoles.

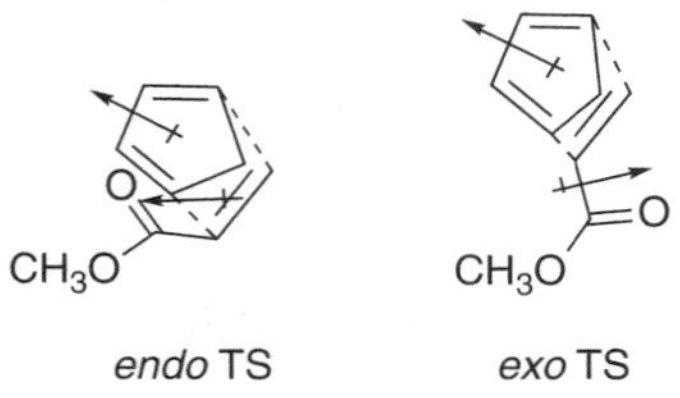

endo TS   *exo* TS

The preference for the *endo* TS is considered to be the result of interaction between the dienophile substituent and the π electrons of the diene. These are called *secondary orbital interactions*. Dipolar attractions and van der Waals attractions may also be involved.[12] Some *exo-endo* ratios for thermal D-A reactions of cyclopentadiene are

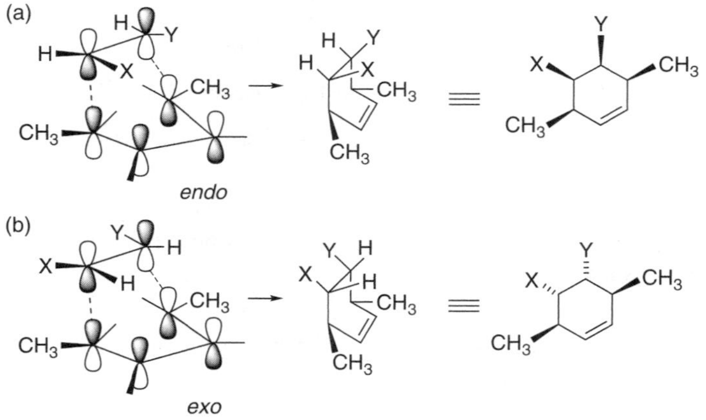

Fig. 6.3. *Endo* (a) and *exo* (b) stereochemistry in Diels-Alder reactions.

10. K. N. Houk and L. J. Lusku, *J. Am. Chem. Soc.*, **93**, 4606 (1971).
11. J. A. Berson, Z. Hamlet, and W. A. Mueller, *J. Am. Chem. Soc.*, **84**, 297 (1962).
12. Y. Kobuke, T. Sugimoto, J. Furukawa, and T. Funeo, *J. Am. Chem. Soc.*, **94**, 3633 (1972); K. L. Williamson and Y.-F. L. Hsu, *J. Am. Chem. Soc.*, **92**, 7385 (1970).

Table 6.1. *Endo:Exo* Stereoselectivity toward Cyclopentadiene

Dienophile	*Endo:exo* ratio
$CH_2{=}CHCH{=}O^a$	80:20
$CH_2{=}CHCOCH_3{}^a$	82:18
$CH_2{=}CHCO_2CH_3{}^b$	73:27
$CH_2{=}C(CH_3)CO_2CH_3{}^b$	30:70
$CH_3CH{=}CHCO_2CH_3{}^b$	52:48
$CH_2{=}CHSO_2CH_3{}^c$	75:25
$CH_2{=}CHPO(OCH_3)_2{}^d$	55:45
$CH_2{=}CHCN^e$	58:42
$CH_2{=}C(CH_3)CN^e$	12:88
$CH_3CH{=}CHCN^e$	34:66

a. O. F. Guner, R. M Ottenbrite and D. D. Shillady, *J. Org. Chem.*, **53**, 5348 (1988).
b. K. N. Houk and L. J. Lusku, *J. Am. Chem. Soc.*, **93**, 4606 (1971).
c. J. C. Philips and M. Oku, *J. Org. Chem.*, **37**, 4479 (1972).
d. H. J. Callot and C. Berezra, *J. Chem. Soc., Chem. Commun.*, 485 (1970).
e. A. I. Konovalov and G. I. Kamasheva, *Russ. J. Org Chem (Engl. Trans.)*, **8**, 1879 (1972)

given in Table 6.1. Most of the data pertain to dienophiles with carbonyl substituents. Note that tetrahedral noncarbonyl EWGs such as sulfonyl and phosphonyl also exhibit a small preference for the *endo* TS. The cyano group shows little *endo:exo* preference. Both α- and β-methyl groups result in more *exo* product, as seen for the methyl-substituted esters and nitriles. As we will see shortly, the use of Lewis acid catalysts usually increases the preference for the *endo* TS.

Steric effects play a dominant role with more highly substituted dienes. Hexachloro-cyclopentadiene, for example, shows a higher *endo* preference than cyclopentadiene because the 5-chlorine causes steric interference with *exo* substituents.[13]

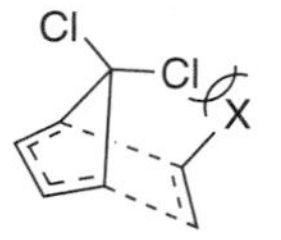

Cyclic α-methylene ketones and lactones, in which the *syn* conformation is enforced, give predominantly *exo* adducts.[14]

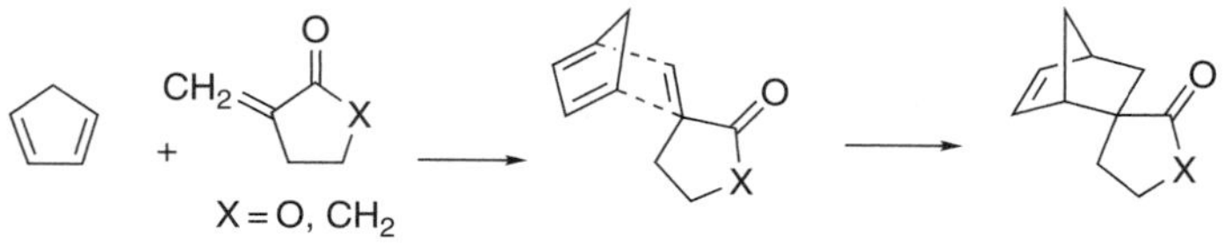

[13.] K. L. Williamson, Y.-F. L. Hsu, R. Lacko, and C. H. Youn, *J. Am. Chem. Soc.*, **91**, 6129 (1969).
[14.] F. Fotiadu, F. Michel, and G. Buono, *Tetrahedron Lett.*, **31**, 4863 (1990); J. Mattay, J. Mertes, and G. Maas, *Chem. Ber.*, **122**, 327 (1989).

480

CHAPTER 6

*Concerted
Cycloadditions,
Unimolecular
Rearrangements, and
Thermal Eliminations*

It has been suggested that this is due to a more favorable alignment of dipoles in the *exo* TS.[15]

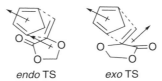

endo TS   exo TS

Computational studies predict a preference for the *endo* TS.[16] There have been several computational efforts to dissect the various factors that contribute to the differences between the *exo* and *endo* TS.[17] These generally are in agreement with the experimental preference for the *endo* TS, but there is no consensus on the dominant factors in this preference.[18]

Diels-Alder cycloadditions are sensitive to steric effects of two major types in the diene. Bulky substituents on the termini of the diene hinder approach of the two components to each other and decrease the rate of reaction. This effect can be seen in the relative reactivity of 1-substituted butadienes toward maleic anhydride.[19]

R	k_{rel} (25°C)
–H	1
–CH$_3$	4.2
–C(CH$_3$)$_3$	<0.05

Substitution of hydrogen by methyl results in a slight rate *increase* as a result of the electron-releasing effect of the methyl group. A *t*-butyl substituent produces a large rate *decrease* because the steric effect is dominant.

Another type of steric effect results from interactions between diene substituents. Adoption of the *s-cis* conformation of the diene in the TS brings the *cis*-oriented 1- and 4-substituents on a diene close together. *E*-1,3-Pentadiene is 10^3 times more reactive than 4-methyl-1,3-pentadiene toward the very reactive dienophile tetracyanoethylene. This is because the unfavorable interaction between the additional methyl substituent and the C(1) hydrogen in the *s-cis* conformation raises the energy of the TS.[20]

R	k_{rel}
–H	1
–CH$_3$	10^{-3}

Relatively small substituents at C(2) and C(3) of the diene exert little steric influence on the rate of D-A addition. 2,3-Dimethylbutadiene reacts with maleic anhydride about ten times faster than butadiene owing to the electronic effect of the methyl

15. W. R. Roush and B. B. Brown, *J. Org. Chem.*, **57**, 3380 (1992).
16. (a) R. J. Loncharich, T. R. Schwartz, and K. N. Houk, *J. Am. Chem. Soc.*, **109**, 14 (1987); (b) R. J. Loncharich, T. R. Schwartz, and K. N. Houk, *J. Org. Chem.*, **54**, 1129 (1989); (c) D. M. Birney and K. N. Houk, *J. Am. Chem. Soc.*, **112**, 4127 (1990); (d) J. I. Garcia, V. Martinez-Merino, J. A. Mayoral, and L. Salvatella, *J. Am. Chem. Soc.*, **120**, 2415 (1998).
17. W. L. Jorgensen, D. Lim, and J. F. Blake, *J. Am. Chem. Soc.*, **115**, 2936 (1993); A. Arrieta, F. P. Cossio, and B. Lecea, *J. Org. Chem.*, **66**, 6178 (2001); J. I. Garcia, J. A. Mayoral, and L. Salvatella, *Eur. J. Org. Chem.*, 85, (2004).
18. J. I. Garcia, J. A. Mayoral, and L. Salvatella, *Acc. Chem. Res.*, **33**, 658 (2000).
19. D. Craig, J. J. Shipman, and R. B. Fowler, *J. Am. Chem. Soc.*, **83**, 2885 (1961).
20. C. A. Stewart, Jr., *J. Org. Chem.*, **28**, 3320 (1963).

groups. 2-*t*-Butyl-1,3-butadiene is 27 times more reactive than butadiene. The *t*-butyl substituent favors the *s-cis* conformation because of steric repulsions in the *s-trans* conformation.

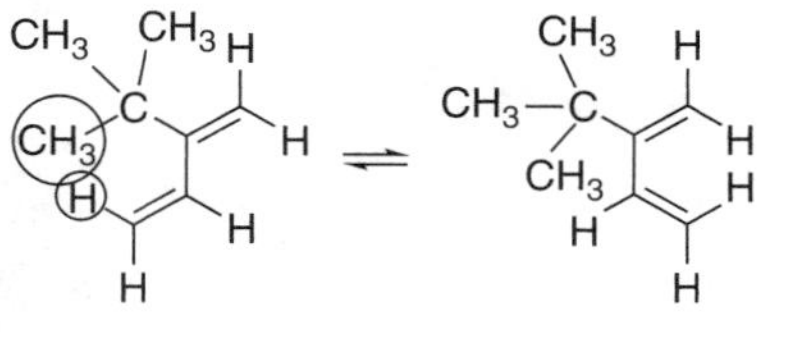

The presence of a *t*-butyl substituent on *both* C(2) and C(3), however, prevents attainment of the *s-cis* conformation, and D-A reactions of 2,3-di-(*t*-butyl)-1,3-butadiene have not been observed.[21]

6.1.3. Lewis Acid Catalysis of the Diels-Alder Reaction

Lewis acids such as zinc chloride, boron trifluoride, tin tetrachloride, aluminum chloride, methylaluminum dichloride, and diethylaluminum chloride catalyze Diels-Alder reactions.[22] The catalytic effect is the result of coordination of the Lewis acid with the dienophile. The complexed dienophile is more electrophilic and more reactive toward electron-rich dienes. The mechanism of the addition is believed to be concerted and enhanced regio- and stereoselectivity is often observed.[23]

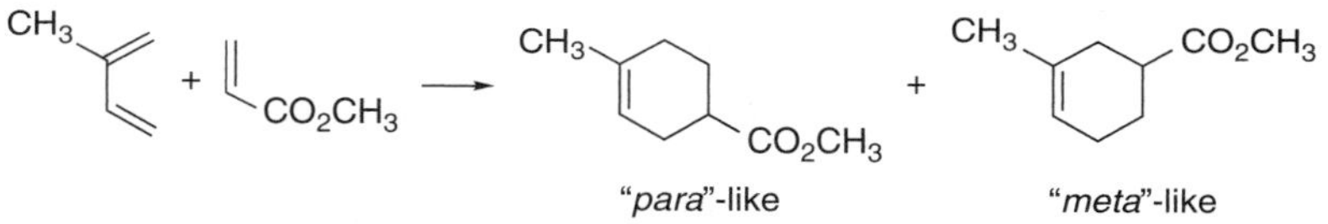

"*para*"-like   "*meta*"-like

	Product ratio	
Uncatalyzed reaction: 120°C, 6 h	70%	30%
Aluminum chloride catalyzed: 20°C, 3 h	95%	5%

Ref. 24

Among the catalysts currently in use, CH_3AlCl_2 was the most effective when employed with *Z*-dienes, which often exhibit low reactivity.

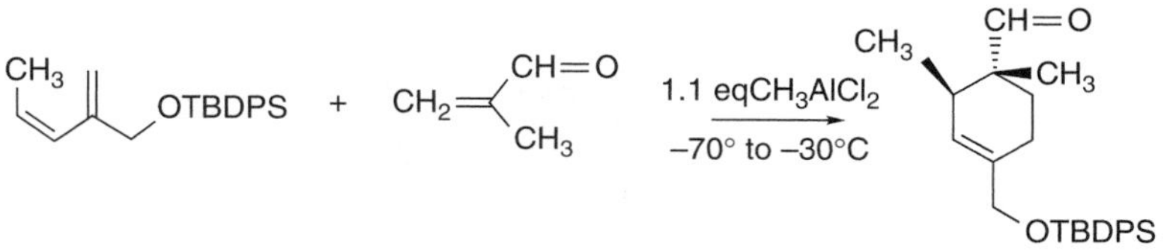

Ref. 22g

21. H. J. Backer, *Rec. Trav. Chim. Pays-Bas*, **58**, 643 (1939).
22. (a) P. Yates and P. Eaton, *J. Am. Chem. Soc.*, **82**, 4436 (1960); (b) T. Inukai and M. Kasai, *J. Org. Chem.*, **30**, 3567 (1965); (c) T. Inukai and T. Kojima, *J. Org. Chem.*, **31**, 2032 (1966); (d) T. Inukai and T. Kojima, *J. Org. Chem.*, **32**, 869, 872 (1967); (e) F. Fringuelli, F. Pizzo, A. Taticchi, and E. Wenkert, *J. Org. Chem.*, **48**, 2802 (1983); (f) F. K. Brown, K. N. Houk, D. J. Burnell, and Z. Valenta, *J. Org. Chem.*, **52**, 3050 (1987); (g) W. R. Roush and D. A. Barda, *J. Am. Chem. Soc.*, **119**, 7402 (1997).
23. K. N. Houk, *J. Am. Chem. Soc.*, **95**, 4094 (1973).
24. T. Inukai and Kojima, *J. Org. Chem.*, **31**, 1121 (1966).

482

CHAPTER 6

*Concerted
Cycloadditions,
Unimolecular
Rearrangements, and
Thermal Eliminations*

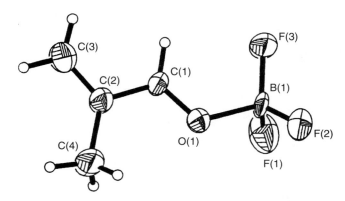

Fig. 6.4. Structure of the BF_3–2-methylpropenal complex. Reproduced from *Tetrahedron Lett.*, **33**, 6945 (1992), by permission of Elsevier.

The stereoselectivity of any particular reaction depends on the details of the structure of the TS. The structures of several enone–Lewis acid complexes have been determined by X-ray crystallography.[25] The site of complexation is the carbonyl oxygen, which maintains a trigonal geometry, but with somewhat expanded angles (130°–140°). The Lewis acid is normally *anti* to the larger carbonyl substituent. Boron trifluoride complexes are tetrahedral, but Sn(IV) and Ti(IV) complexes can be tetrahedral, bipyramidal or octahedral. The structure of the 2-methylpropenal–BF_3 complex in Figure 6.4 is illustrative.[26] Chelation can favor a particular structure. For example, *O*-acryloyl lactates adopt a chelated hexacoordinate structure with $TiCl_4$, as shown in Figure 6.5.[27]

Computational studies have explored the differences between thermal and Lewis acid–catalyzed D-A reactions. Ab initio calculations (HF/6-31G*) have been used to compare the energy of four possible TSs for the D-A reaction of the BF_3 complex of propenal with 1,3-butadiene.[16d] The TSs are designated *endo* and *exo* and *s-cis* and *s-trans*. The latter designations refer to the dienophile conformation. The results are summarized in Figure 6.6. In the thermal reaction, the *endo-cis* and *exo-cis* TSs are nearly equal in total and activation energies. In the BF_3-catalyzed reaction, the

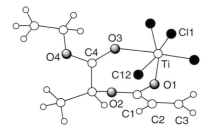

Fig. 6.5. Structure of the $TiCl_4$ complex of *O*-acryloyl ethyl lactate. Reproduced from *Angew. Chem. Int. Ed. Engl.*, **24**, 112 (1985), by permission of Wiley-VCH.

25. S. Shambayati, W. E. Crowe, and S. L. Schreiber, *Angew. Chem. Int. Ed. Engl.*, **29**, 256 (1990).
26. E. J. Corey, T.-P. Loh, S. Sarshar, and M. Azimioara, *Tetrahedron Lett.*, **33**, 6945 (1992).
27. T. Poll, J. O. Metter, and G. Helmchen, *Angew. Chem. Int. Ed. Engl.*, **24**, 112 (1985).

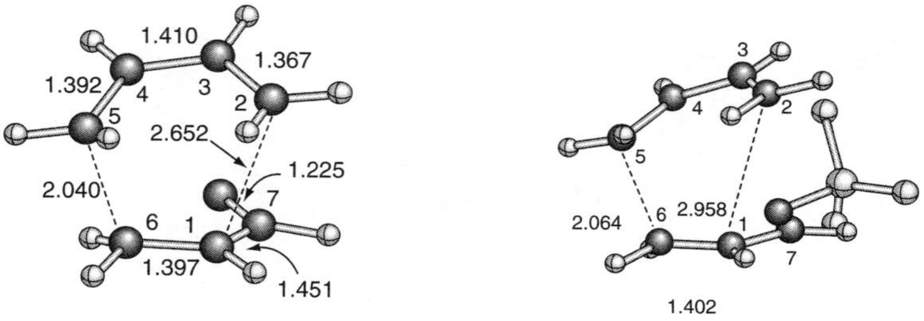

Relative Energies and Activation Energies

Thermal	$\Delta\Delta E_{298}$	ΔG^*_{298}	BF$_3$-catalyzed	$\Delta\Delta E_{298}$	ΔG^*_{298}
Endo-cis	0.00	32.7	*Endo-cis*	0.00	23.2
Endo-trans	1.24	33.9	*Endo-trans*	2.25	25.7
Exo-cis	0.06	32.7	*Exo-cis*	1.72	24.3
Exo-trans	1.93	34.5	*Exo-trans*	5.61	28.3

Fig. 6.6. Relative energies of four possible transition structures for Diels-Alder reaction of 1,3-butadiene and propenal, with and without BF$_3$ catalyst. Geometric parameters of the most stable transition structures (*endo-cis*) are shown. Adapted from *J. Am. Chem. Soc.*, **120**, 2415 (1998), by permission of the American Chemical Society.

endo-cis TS is favored by 1.7 kcal/mol. The calculated ΔG^* is reduced by nearly 10 kcal/mol for the catalyzed reaction, relative to the thermal reaction. The catalyzed reaction shows significantly greater asynchronicity than the thermal reaction. In the BF$_3$-catalyzed reaction, the forming bond distances are 2.06 and 2.96 Å, whereas in the thermal reaction they are 2.04 and 2.65 Å. (See Topic 10.1 of Part A for discussion of asynchronicity.)

A similar study was done with methyl acrylate as the dienophile.[28] The uncatalyzed and catalyzed TSs are shown in Figure 6.7. As with propenal, the catalyzed reaction is quite asynchronous with C(2)−C(3) bonding running ahead of C(1)−C(6) bonding. In this system, there is a shift from favoring the *exo-s-cis* TS in the thermal reaction to the *endo-s-trans* TS in the catalyzed reaction. A large component in this difference is the relative stability of the free and complexed dienophile. The free dienophile favors the *s-cis* conformation, whereas the BF$_3$ complex favors the *s-trans* conformation.

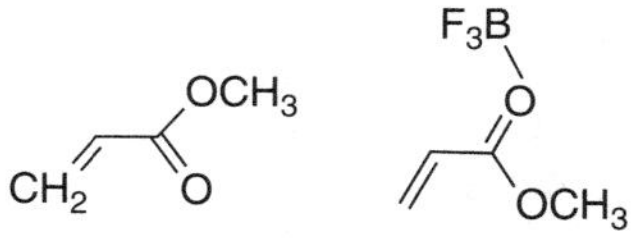

Visual models, additional information and exercises on the Diels-Alder Reaction can be found in the Digital Resource available at: Springer.com/carey-sundberg.

In terms of both the effect of substituents and Lewis acid catalysis, the rates of D-A reactions *increase as the donor-acceptor character of the reactive*

28. J. I. Garcia, J. A. Mayoral, and L. Salvatella, *Tetrahedron*, **53**, 6057 (1997).

484

CHAPTER 6

*Concerted
Cycloadditions,
Unimolecular
Rearrangements, and
Thermal Eliminations*

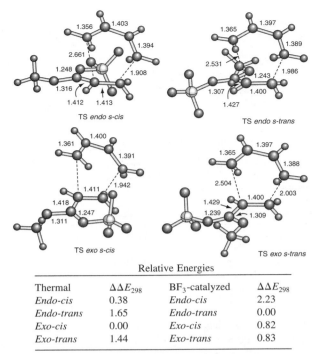

Relative Energies			
Thermal	$\Delta\Delta E_{298}$	BF$_3$-catalyzed	$\Delta\Delta E_{298}$
Endo-cis	0.38	*Endo-cis*	2.23
Endo-trans	1.65	*Endo-trans*	0.00
Exo-cis	0.00	*Exo-cis*	0.82
Exo-trans	1.44	*Exo-trans*	0.83

Fig. 6.7. Transition structures for the reaction between 1,3-butadiene and the methyl acrylate–BF$_3$ complex calculated at the ab initio HF/6-31G* level. Relative energies are in kcal/mol. Adapted from *Tetrahedron*, **53**, 6057 (1997), by permission of Elsevier.

complex increases. That is, the better the donor substituents in the diene and the stronger the acceptor substituents in the dienophile, the faster the reaction. Similarly, the more electrophilic the Lewis acid, the faster the reaction. In extreme cases, cycloaddition may become stepwise.

Such a stepwise reaction would not be expected to change the regiochemistry of cycloaddition, but it could lead to loss of stereospecificity if the zwitterionic intermediate has a long enough lifetime. In most reactions where only carbon-carbon bonds are being formed, the D-A reaction remains stereospecific.

In one study, the mechanisms of the reaction of methyl cinnamate and cyclopentadiene with BF$_3$, AlCl$_3$, and catecholborane bromide as catalysts were compared.[29] According to these computations (B3LYP/6-31G*), the uncatalyzed and BF$_3$- and AlCl$_3$-catalyzed reactions proceed by asynchronous concerted mechanisms, but a

29. C. N. Alves, F. F. Camilo, J. Gruber, and A. B. F. da Silva, *Chem. Phys.*, **306**, 35 (2004).

stepwise mechanism is found with catecholborane bromide. Experimentally, this is the only catalyst that is effective for this reaction.[30]

Metal cations can catalyze reactions of certain dienophiles. For example, Cu^{2+} strongly catalyzes addition reactions of 2-pyridyl styryl ketones, presumably through a chelate involving the carbonyl oxygen and pyridine nitrogen.[31]

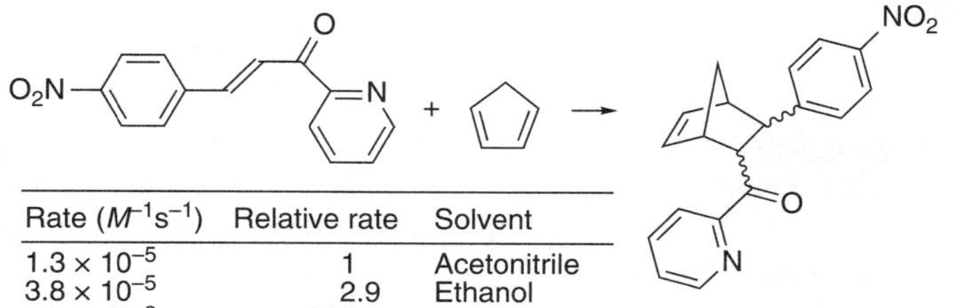

Rate ($M^{-1}s^{-1}$)	Relative rate	Solvent
1.3×10^{-5}	1	Acetonitrile
3.8×10^{-5}	2.9	Ethanol
4.0×10^{-3}	310	Water
3.25	250,000	Water + 0.01 M Cu(NO$_3$)$_2$

This reaction has been studied computationally with Zn^{2+} as the metal cation.[32] The calculations indicate that a stepwise reaction occurs, beginning with electrophilic attack of the complexed dienophile on the diene.

Some D-A reactions are catalyzed by high concentrations of LiClO$_4$ in ether,[33] a catalysis that involves Lewis acid complexation of Li$^+$ with the dienophile.[34]

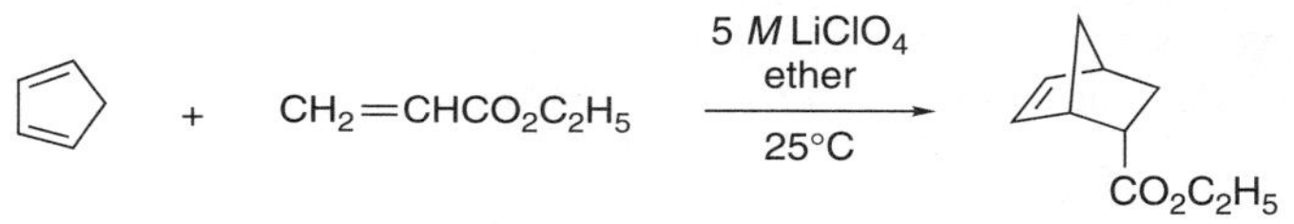

The LiClO$_4$-diethyl ether system shows a considerable dependency on concentration, with the maximal effect around 5 M, which may be due to the detailed structure of LiClO$_4$ in ether. The optimum reactivity may be associated with a monosolvate. Dilute solutions have more of the dietherate, whereas in more concentrated solution LiClO$_4$ may form less reactive aggregates.[35] LiN(SO$_2$SCF$_3$)$_2$ has been recommended as an alternative to avoid the use of a perchlorate salt.[36]

Lithium *tetrakis*-(3,5-ditrifluoromethyl)borate, which provides an unsolvated lithium cation in noncoordinating solvents, exhibits a several thousandfold catalysis of the reaction of cyclopentadiene and methyl vinyl ketone.[37] Lithium tetrafluoroborate is also an effective catalyst and in some instances has worked when LiClO$_4$ has failed, such as in the intramolecular reaction shown below.[38]

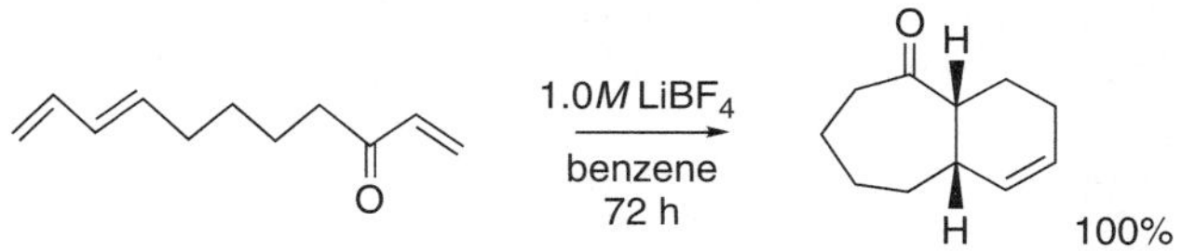

30. F. Camilo and J. Gruber, *Quim. Nova*, **22**, 382 (1999).
31. S. Otto and J. B. F. N. Engberts, *Tetrahedron Lett.*, **36**, 2645 (1995).
32. L. R. Domingo, J. Andres, and C. N. Alves, *Eur. J. Org. Chem.*, **15**, 2557 (2002).
33. P. A. Grieco, J. J. Nunes, and M. D. Gaul, *J. Am. Chem. Soc.*, **112**, 4595 (1990).
34. M. A. Forman and W. P. Dailey, *J. Am. Chem. Soc.*, **113**, 2761 (1991).
35. A. Kumar and S. S. Pawar, *J. Org. Chem.*, **66**, 7646 (2001).
36. S. T. Handy, P. A. Grieco, C. Mineur, and L. Ghosez, *Synlett*, 565 (1995).
37. K. Fujiki, S.-Y. Ikeda, H. Kobayashi, M. Hiroshi, A. Nagira, J. Nie, T. Sonoda, and Y. Yagupolskii, *Chem. Lett.*, 62 (2000).
38. D. A. Smith and K. N. Houk, *Tetrahedron Lett.*, **32**, 1549 (1991).

486

CHAPTER 6

*Concerted
Cycloadditions,
Unimolecular
Rearrangements, and
Thermal Eliminations*

Scandium triflate has been found to catalyze D-A reactions.[39] For example, with 10 mol % $Sc(O_3SCF_3)_3$ present, isoprene and methyl vinyl ketone react to give the expected adduct in 91% yield after 13 h at 0°C.

Among the unique features of $Sc(O_3SCF_3)_3$ is its ability to function as a catalyst in hydroxylic solvents. Other dienophiles, including *N*-acryloyloxazolidinones, also are subject to catalysis by $Sc(O_3SCF_3)_3$. Indium trichloride is another Lewis acid that can act as a catalyst in aqueous solution.[40]

Reversible O-silylation also enhances the electrophilicity of carbonyl dienophiles. For example, 10 mol % *N*-trimethylsilyl triflimide catalyzes the reaction of pent-3-en-2-one with cyclopentadiene. A hindered base, such as 2,6-*bis-t*-butyl-4-methylpyridine improves the yield in cases in which the catalyst causes the occurrence of reactant degradation.

Ref. 41

The solvent also has an important effect on the rate of D-A reactions. The traditional solvents were nonpolar organic solvents such as aromatic hydrocarbons. However, water and other highly polar solvents, such as ethylene glycol and formamide, accelerate a number of D-A reactions.[42] The accelerating effect of water is attributed to "enforced hydrophobic interactions." That is, the strong hydrogen-bonding network in water tends to exclude nonpolar solutes and force them together, resulting in higher effective concentrations and relative stabilization of the developing TS.[43] More specific hydrogen bonding with the TS also contributes to the rate acceleration.[44]

39. S. Kobayashi, I. Hachiya, M. Araki, and H. Ishitani, *Tetrahedron Lett.*, **34**, 3755 (1993); S. Kobayashi, H. Ishitani, M. Araki, and I. Hachiya, *Tetrahedron Lett.*, **35**, 6325, (1994); S. Kobayahsi, *Eur. J. Org. Chem.*, 15 (1999).
40. T.-P. Loh, J. Pei, and M. Lin, *Chem. Commun.*, 2315 (1995); 505 (1996).
41. B. Mathieu and L. Ghosez, *Tetrahedron*, **58**, 8219 (2002).
42. D. Rideout and R. Breslow, *J. Am. Chem. Soc.*, **102**, 7816 (1980); R. Breslow and T. Guo, *J. Am. Chem. Soc.*, **110**, 5613 (1988); T. Dunams, W. Hoekstra, M. Pentaleri, and D. Liotta, *Tetrahedron Lett.*, **29**, 3745 (1988).
43. R. Breslow and C. J. Rizzo, *J. Am. Chem. Soc.*, **113**, 4340 (1991).
44. W. Blokzijl, M. J. Blandamer, and J. B. F. N. Engberts, *J. Am. Chem. Soc.*, **113**, 4241 (1991); W. Blokzijl and J. B. F. N. Engberts, *J. Am. Chem. Soc.*, **114**, 5440 (1992); S. Otto, W. Blokzijl, and J. B. F. N. Engberts, *J. Org. Chem.*, **59**, 5372 (1994); A. Meijer, S. Otto, and J. B. F. N. Engberts, *J. Org. Chem.*, **65**, 8989 (1998).

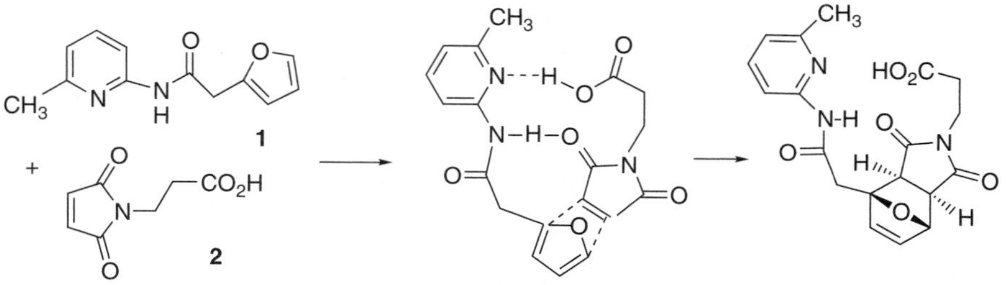

Fig. 6.8. Proposed hydrogen bonding in TS for addition of **1** and **2**. Reproduced from *Tetrahedron Lett.*, **45**, 4777 (2004), by permission of Elsevier.

Hydrogen-bonding interactions can be designed into reaction systems. For example, the reactants **1** and **2** were found to react much more rapidly than the corresponding ester and to give exclusively the *exo* product.[45] Molecular mechanics and spectroscopic studies indicate that the hydrogen-bonding pattern shown in Figure 6.8 is responsible.

To summarize the key points, D-A reactions are usually concerted processes. The regio- and stereoselectivity can be predicted by applying FMO analysis. The reaction between electron donor dienes and electron acceptor dienophiles is facilitated by Lewis acids, polar solvents, and favorable hydrogen-bonding interactions. The D-A reaction is quite sensitive to steric factors, which can retard the reaction and also influence the stereoselectivity with respect to *exo* or *endo* approach.

6.1.4. The Scope and Synthetic Applications of the Diels-Alder Reaction

Schemes 10.1 and 10.4 of Part A, respectively, give the structure of a number of typical dienophiles and show representative D-A reactions involving relatively simple reactants. The D-A reaction is frequently used in synthesis and can either be utilized early in a process to construct basic ring structures or to bring together two subunits in a convergent synthesis. The intramolecular version, which will be discussed in section 6.1.7, can be used to construct two new rings.

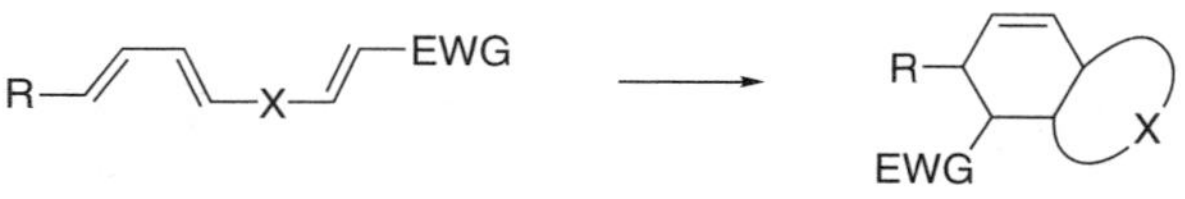

The virtues of the D-A reaction include its ability to create a cyclohexene ring by formation of *two new bonds with predictable regiochemistry*. The reaction can also create as many as four contiguous stereogenic centers. The stereoselectivity is also often predictable on the basis of the *supra-supra* stereospecificity and considerations of the preference for the *endo* or *exo* TS.

6.1.4.1. Examples of Dienes and Dienophiles. The synthetic value of D-A reactions can be enhanced in various ways. In addition to hydrocarbon dienes, substituted dienes can be used to introduce functional groups into the products. One example that illustrates the versatility of such reagents is 1-methoxy-3-trimethylsiloxy-1,3-butadiene

[45.] R. J. Pearson, E. Kassianidis, and D. Philip, *Tetrahedron Lett.*, **45**, 4777 (2004).

488

CHAPTER 6

*Concerted
Cycloadditions,
Unimolecular
Rearrangements, and
Thermal Eliminations*

(*Danishefsky's diene*).[46] The two donor substituents provide strong regiochemical control. The D-A adducts are trimethylsilyl enol ethers that can be readily hydrolyzed to ketones. The β-methoxy group is often eliminated during hydrolysis, resulting in formation of cyclohexenones.

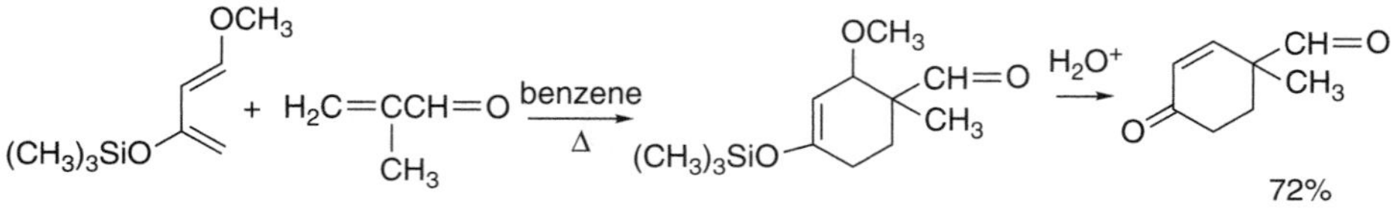

A milder protocol for the conversion to enones involves use of a catalytic amount of TMSOTf and a pyridine base.[47]

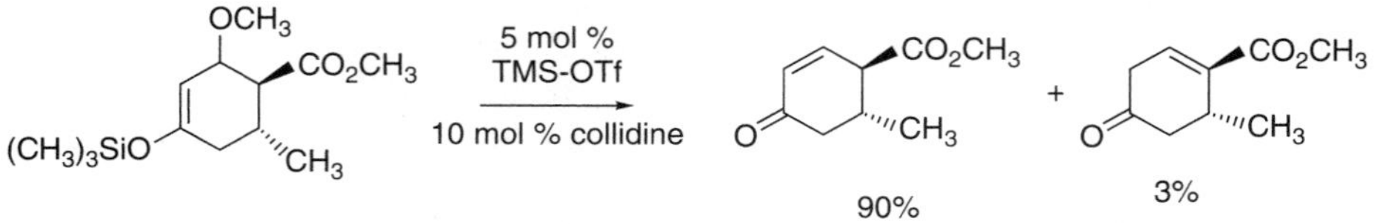

The desilylation is also promoted by various Lewis acids, $Yb(OTf)_3$ being among the most effective. This catalyst can be used in a one-pot sequence in which it promotes both the cycloaddition and subsequent elimination.[48]

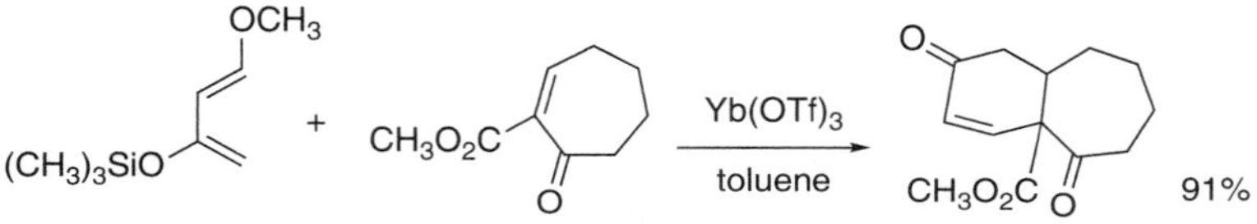

An analogous silyoxydienamine shows a similar reactivity pattern.[49]

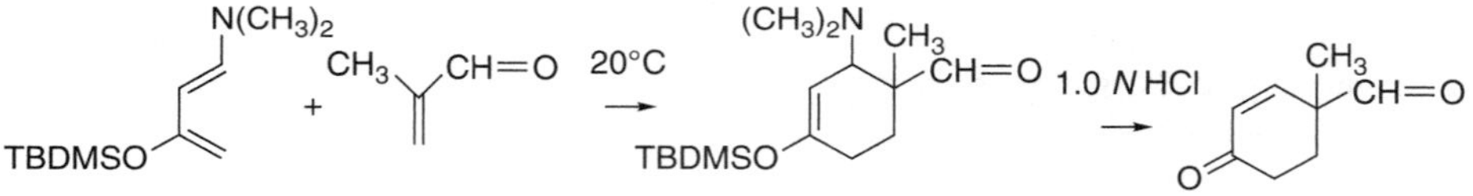

2-(Diethoxyphosphoryloxy)-1,3-butadiene and 2-(diethoxyphosphoryloxy)-1,3-pentadiene are good dienes and are compatible with Lewis acid catalysts.[50] They exhibit the regioselectivity expected for a donor substituent and show a preference for *endo* addition with enones.

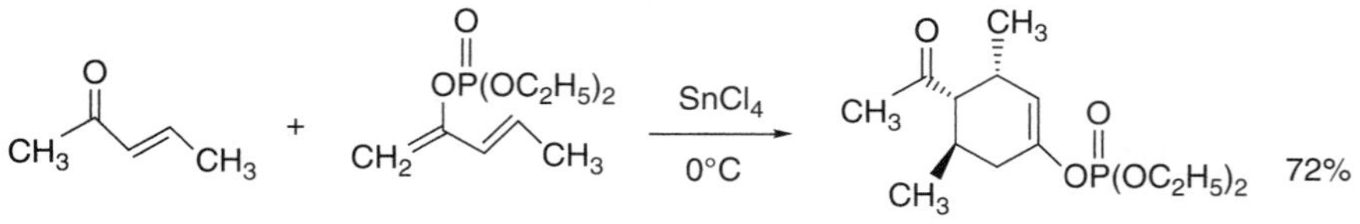

46. S. Danishefsky and T. Kitahara, *J. Am. Chem. Soc.*, **96**, 7807 (1974).
47. P. E. Vorndam, *J. Org. Chem.*, **55**, 3693 (1990).
48. T. Inokuchi, M. Okano, T. Miyamoto, H. B. Madon, and M. Takagi, *Synlett*, 1549 (2000); T. Inokuchi, M. Okano, and T. Miyamoto, *J. Org. Chem.*, **66**, 8059 (2001).
49. S. A. Kozmin and V. H. Rawal, *J. Org. Chem.*, **62**, 5252 (1997).
50. H.-J. Liu, W. M. Feng, J. B. Kim, and E. N. C. Browne, *Can. J. Chem.*, **72**, 2163 (1994).

Unstable dienes can be generated in situ in the presence of a dienophile. Among the most useful examples are the *ortho*-quinodimethanes. These compounds are exceedingly reactive as dienes because the cycloaddition reestablishes a benzenoid ring and results in aromatic stabilization.[51]

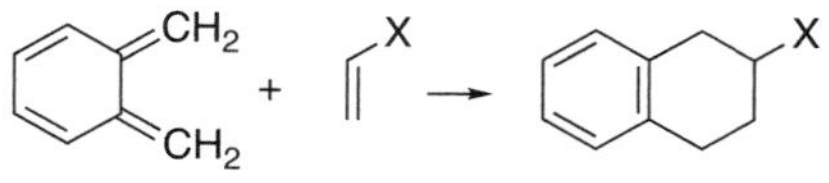

quinodimethane

There are several general routes to quinodimethanes. One is pyrolysis of benzocyclobutenes.[52]

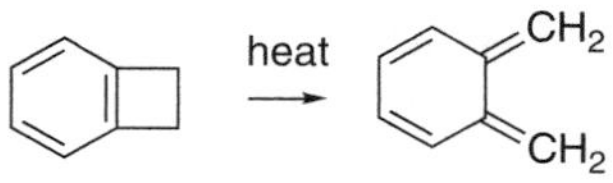

This reaction can be applied to substituted benzocyclobutenes. For example, the reaction has been used to form an array of five linear rings containing most of the functionality for the antibiotic tetracycline.

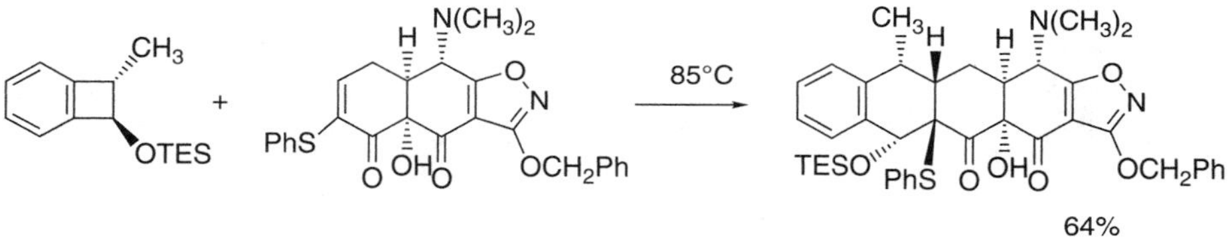

64%

Ref. 53

1,4-Eliminations from α,α'-*ortho*-disubstituted benzenes can be carried out with various potential leaving groups. Benzylic silyl substituents can serve as the carbanion precursors.

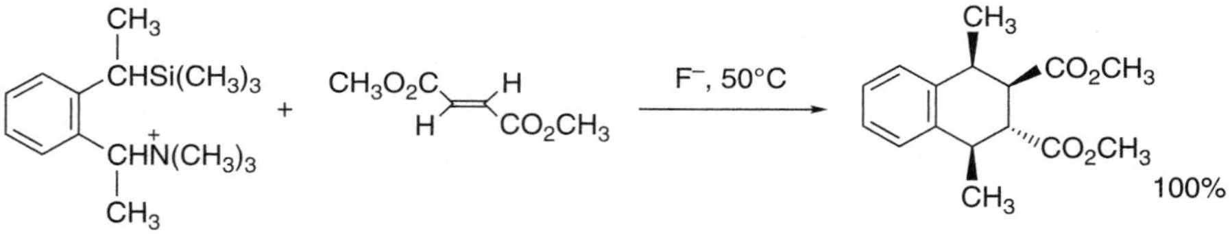

100%

Ref. 54

51. W. Oppolzer, *Angew. Chem. Int. Ed. Engl.*, **16**, 10 (1977); T. Kametani and K. Fukumoto, *Heterocycles*, **3**, 29 (1975); J. J. McCullogh, *Acc. Chem. Res.*, **13**, 270 (1980); W. Oppolzer, *Synthesis*, 793 (1978); J. L. Charlton and M. M. Alauddin, *Tetrahedron*, **43**, 2873 (1987); H. N. C. Wong, K.-L. Lau, and K. F. Tam, *Top. Curr. Chem.*, **133**, 85 (1986); P. Y. Michellys, H. Pellissier, and M. Santelli, *Org. Prep. Proced. Int.*, **28**, 545 (1996).

52. M. P. Cava and M. J. Mitchell, *Cyclobutadiene and Related Compounds*, Academic Press, New York, 1967, Chap. 6; I. L. Klundt, *Chem. Rev.*, **70**, 471 (1970); R. P. Thummel, *Acc. Chem. Res.*, **13**, 70 (1980).

53. M. G. Charest, D. R. Siegel, and A. G. Myers, *J. Am. Chem. Soc.*, **127**, 8292 (2005).

54. Y. Ito, M. Nakatsuka, and T. Saegusa, *J. Am. Chem. Soc.*, **104**, 7609 (1982).

490

CHAPTER 6

*Concerted
Cycloadditions,
Unimolecular
Rearrangements, and
Thermal Eliminations*

Several procedures have been developed for obtaining quinodimethane intermediates from *o*-substituted benzylstannanes. The reactions occur by generating an electrophilic center at the adjacent benzylic position, which triggers a 1,4-elimination.

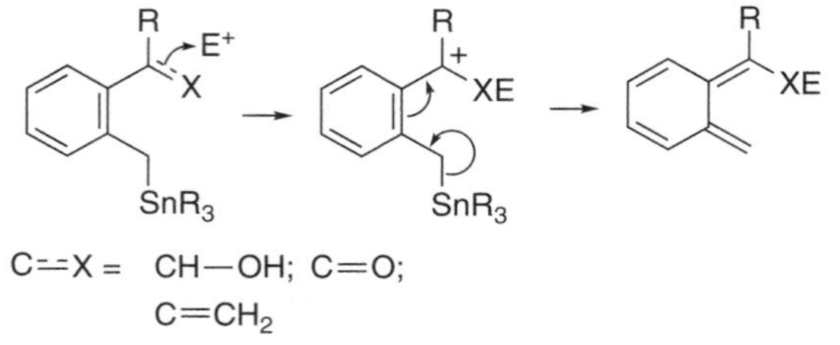

$$C\dot{=}X = \quad CH-OH; \quad C=O;$$
$$C=CH_2$$

Specific examples include treatment of *o*-stannyl benzyl alcohols with TFA,[55] reactions of ketones and aldehydes with Lewis acids,[56] and electrophilic selenation of styrenes.[57]

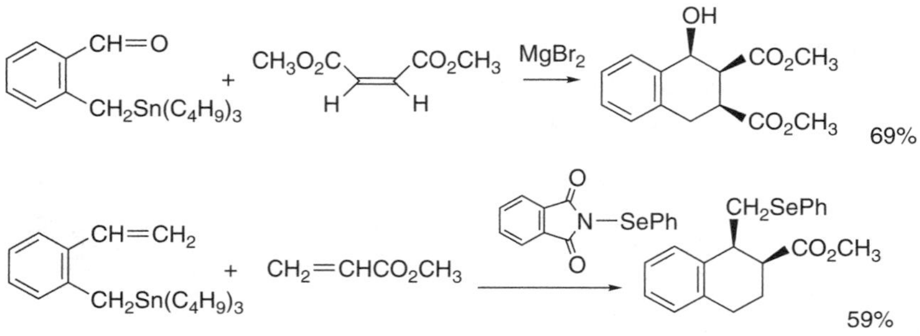

o-bis-(Bromomethyl)benzenes can be converted to quinodimethanes with reductants such as zinc, nickel, chromous ion, and tri-*n*-butylstannide.[58]

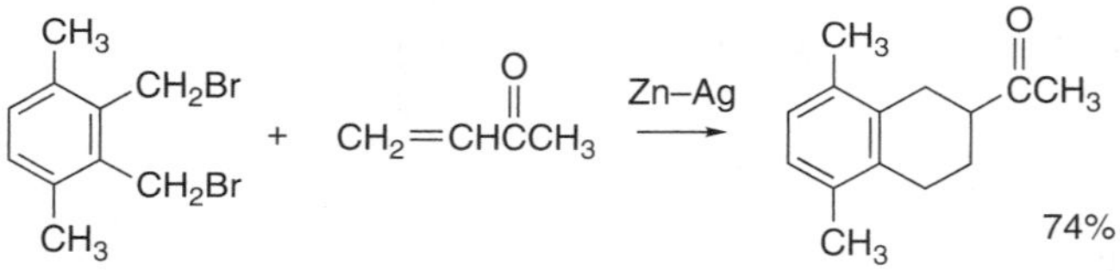

Quinodimethanes have been especially useful in intramolecular D-A reactions, as is illustrated in Section 6.1.7.

Pyrones are useful dienes, although they are not particularly reactive. The adducts have the potential for elimination of carbon dioxide, resulting in the formation of an aromatic ring. Pyrones react best with electron-rich dienophiles. Vinyl ethers are frequently used as dienophiles with pyrones. The regiochemical preference places the dienophile donor *ortho* to the pyrone carbonyl.

[55.] H. Sans, H. Ohtsuka, and T. Migita, *J. Am. Chem. Soc.*, **110**, 2014 (1988).

[56.] S. H. Woo, *Tetrahedron Lett.*, **35**, 3975 (1994).

[57.] S. H. Woo, *Tetrahedron Lett.*, **34**, 7587 (1993).

[58.] G. M. Rubottom and J. E. Wey, *Synth. Commun.*, **14**, 507 (1984); S. Inaba, R. M. Wehmeyer, M. W. Forkner, and R. D. Rieke, *J. Org. Chem.*, **53**, 339 (1988); D. Stephan, A. Gorques, and A. LeCoq, *Tetrahedron Lett.*, **25**, 5649 (1984); H. Sato, N. Isono, K. Okamura, T. Date, and M. Mori, *Tetrahedron Lett.*, **35**, 2035 (1994).

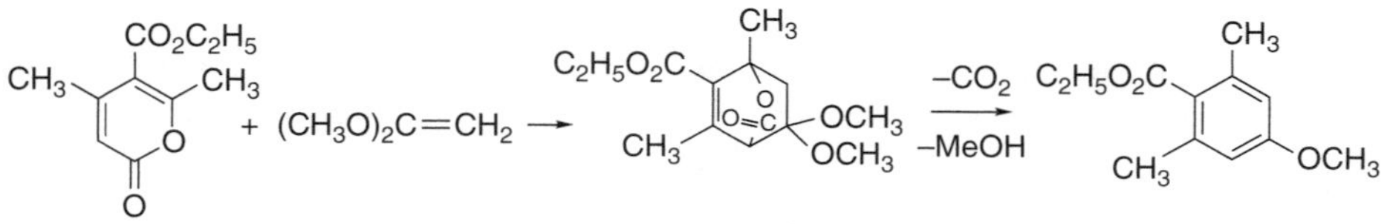

Ref. 59

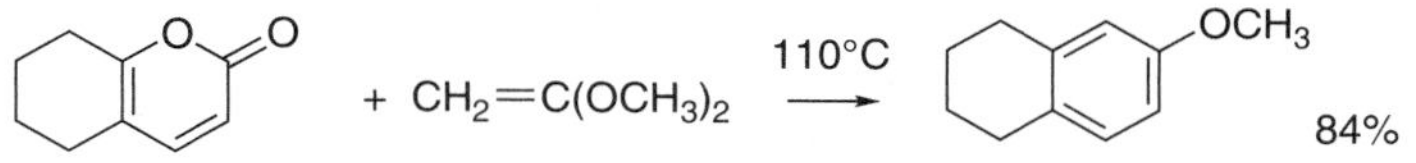

Ref. 60

These reactions can be catalyzed by Lewis acids such as *bis*-alkoxytitanium dichlorides[61] and lanthanide salts.[62]

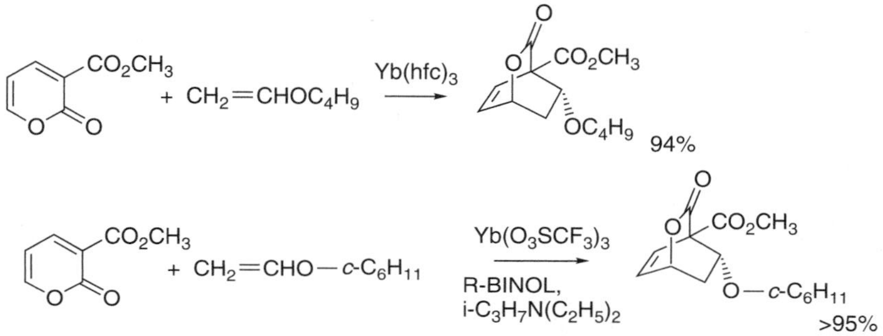

Yb(hfc)$_3$ ... 94%

Yb(O$_3$SCF$_3$)$_3$, R-BINOL, i-C$_3$H$_7$N(C$_2$H$_5$)$_2$ ... >95%

Another type of special diene, the polyaza benzene heterocyclics, such as triazines and tetrazines, is discussed in Section 6.6.2.

The synthetic utility of the D-A reaction can be expanded by the use of dienophiles that contain *masked functionality* and are the *synthetic equivalents* of unreactive or inaccessible compounds. (See Section 13.1.2 for a more complete discussion of the concept of synthetic equivalents.) For example, α-chloroacrylonitrile shows satisfactory reactivity as a dienophile. The α-chloronitrile functionality in the adduct can be hydrolyzed to a carbonyl group. Thus, α-chloroacrylonitrile can function as the equivalent of ketene, CH$_2$=C=O,[63] which is not a suitable dienophile because it has a tendency to react with dienes by [2+2] cycloaddition, rather than the desired [4+2] fashion.

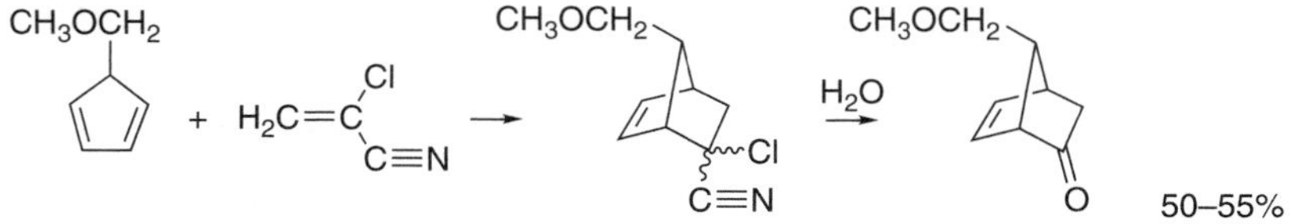

50–55%

Ref. 64

59. M. E. Jung and J. A. Hagenah, *J. Org. Chem.*, **52**, 1889 (1987).
60. D. L. Boger and M. D. Mullican, *Org. Synth.*, **65**, 98 (1987).
61. G. H. Posner, J.-C. Carry, J. K. Lee, D. S. Bull, and H. Dai, *Tetrahedron Lett.*, **35**, 1321 (1994); G. H. Posner, H. Dai, D. S. Bull, J.-K. Lee, F. Eydoux, Y. Ishihara, W. Welsh, N. Pryor, and S. Petr, Jr., *J. Org. Chem.*, **61**, 671 (1996).
62. G. H. Posner, J.-C. Carry, T. E. N. Anjeh, and A. N. French, *J. Org. Chem.*, **57**, 7012 (1992).
63. V. K. Aggarwal, A. Ali, and M. P. Coogan, *Tetrahedron*, **55**, 293 (1999).
64. E. J. Corey, N. M. Weinshenker, T. K. Schaff, and W. Huber, *J. Am. Chem. Soc.*, **91**, 5675 (1969).

492

CHAPTER 6

*Concerted
Cycloadditions,
Unimolecular
Rearrangements, and
Thermal Eliminations*

Nitroalkenes are good dienophiles and the variety of transformations available for nitro groups makes them versatile intermediates.[65] Nitro groups can be converted to carbonyl groups by reductive hydrolysis, so nitroethylene can be used as a ketene equivalent.[66]

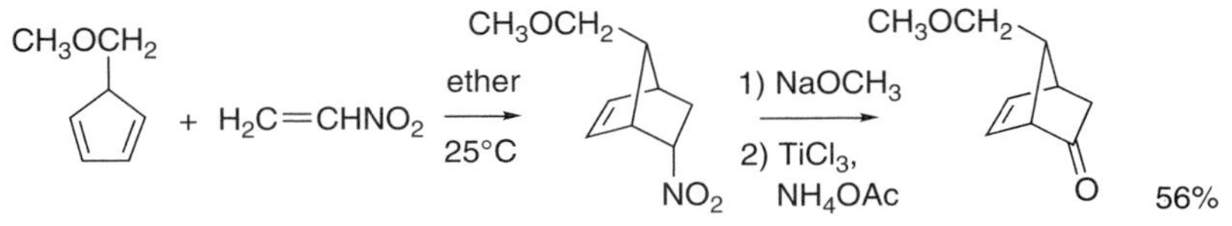

Ref. 67

Vinyl sulfones are reactive as dienophiles. The sulfonyl group can be removed reductively with sodium amalgam (see Section 5.6.2). In this two-step reaction sequence, the vinyl sulfone functions as an ethylene equivalent. The sulfonyl group also permits alkylation of the adduct, via the carbanion. This three-step sequence permits the vinyl sulfone to serve as the synthetic equivalent of a terminal alkene.[68]

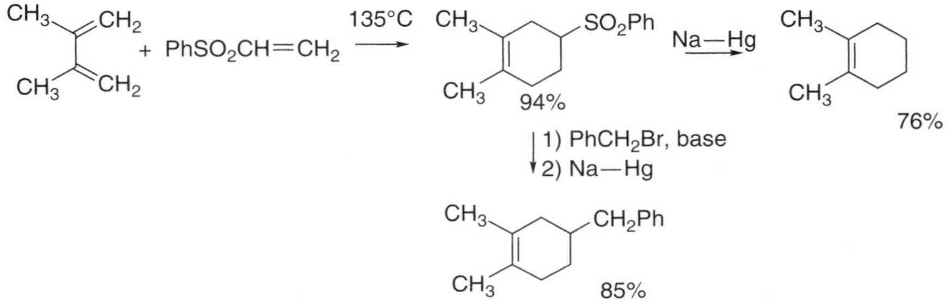

Phenyl vinyl sulfoxide can serve as an acetylene equivalent. Its D-A adducts can undergo thermal elimination of benzenesulfenic acid.

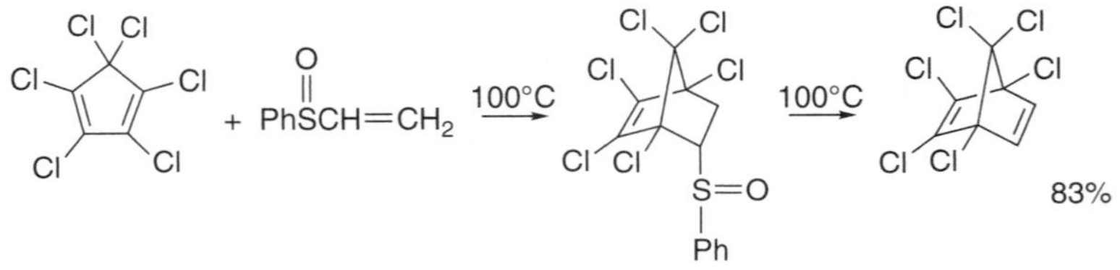

Ref. 69

65. D. Ranganathan, C. B. Rao, S. Ranganathan, A. K. Mehrotra, and R. Iyengar, *J. Org. Chem.*, **45**, 1185 (1980).
66. For a review of ketene equivalents, see S. Ranganathan, D. Ranganathan, and A. K. Mehrotra, *Synthesis*, 289 (1977).
67. S. Ranganathan, D. Ranganathan, and A. K. Mehrotra, *J. Am. Chem. Soc.*, **96**, 5261 (1974).
68. R. V. C. Carr and L. A. Paquette, *J. Am. Chem. Soc.*, **102**, 853 (1980); R. V. C. Carr, R. V. Williams, and L. A. Paquette, *J. Org. Chem.*, **48**, 4976 (1983); W. A. Kinney, G. O. Crouse, and L. A. Paquette, *J. Org. Chem.*, **48**, 4986 (1983).
69. L. A. Paquette, R. E. Moerck, B. Harirchian, and P. D. Magnus, *J. Am. Chem. Soc.*, **100**, 1597 (1978).

Cis- and *trans-bis*-benzenesulfonylethene are also acetylene equivalents. The two sulfonyl groups undergo reductive elimination on reaction with sodium amalgam.

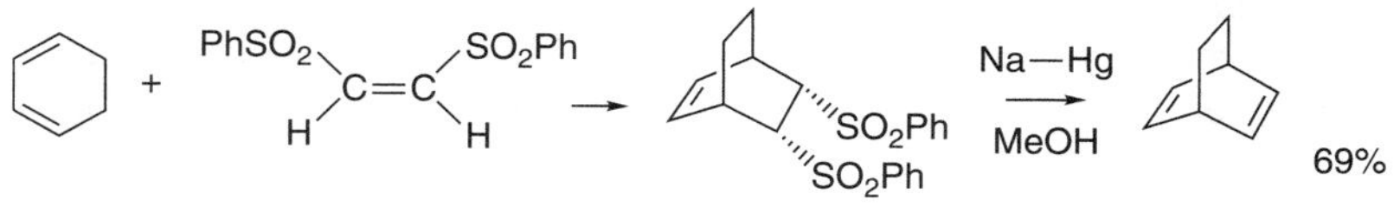

Ref. 70

Vinylphosphonium salts are reactive as dienophiles as a result of the EWG character of the phosphonium substituent. The D-A adducts can be deprotonated to give ylides that undergo the Wittig reaction to introduce an exocyclic double bond. This sequence of reactions corresponds to a D-A reaction employing allene as the dienophile.[71]

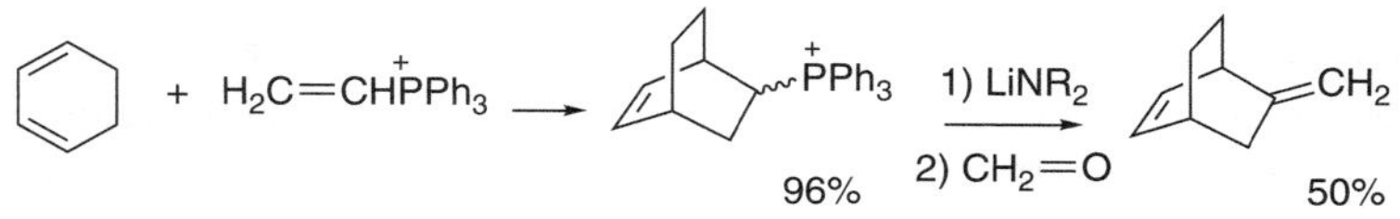

The use of 2-vinyldioxolane, the ethylene glycol acetal of acrolein, as a dienophile illustrates application of the masked functionality concept in a different way. The acetal itself would not be expected to be a reactive dienophile, but in the presence of a catalytic amount of acid the acetal is in equilibrium with the electrophilic oxonium ion.

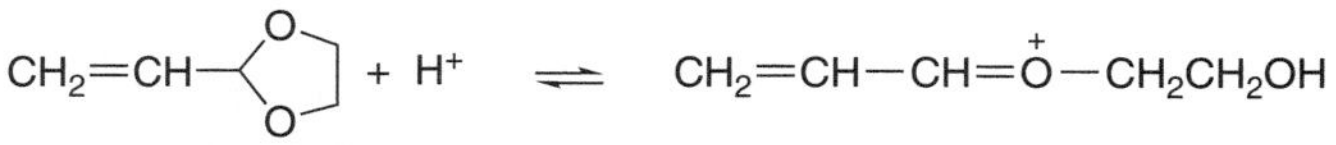

Diels-Alder addition occurs through this cationic intermediate at room temperature.[72] Similar reactions occur with substituted alkenyldioxolanes.

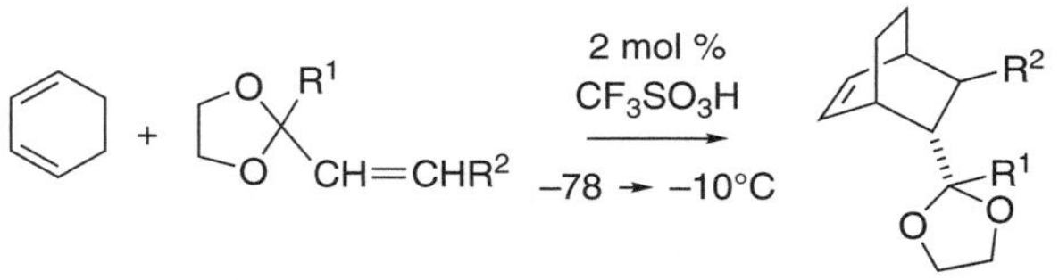

This reaction has been used to construct the carbon skeleton found in dysidiolide, a cell cycle inhibitor isolated from a marine sponge.[73] In this case, the reactive oxonium ion intermediate was generated by O-silylation.

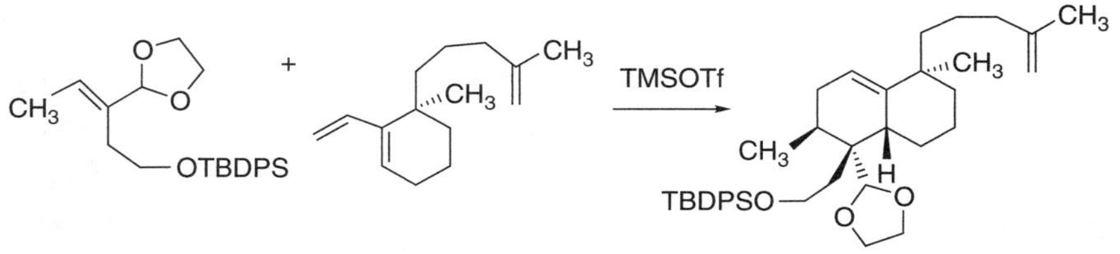

70. O. DeLucchi, V. Lucchini, L. Pasquato, and G. Modena, *J. Org. Chem.*, **49**, 596 (1984).
71. R. Bonjouklian and R. A. Ruden, *J. Org. Chem.*, **42**, 4095 (1977).
72. P. G. Gassman, D. A. Singleton, J. J. Wilwerding, and S. P. Chavan, *J. Am. Chem. Soc.*, **109**, 2182 (1987).
73. S. R. Magnuson, L. Sepp-Lorenzino, N. Rosen, and S. J. Danishefsky, *J. Am. Chem. Soc.*, **120**, 1615 (1998).

494

CHAPTER 6

*Concerted
Cycloadditions,
Unimolecular
Rearrangements, and
Thermal Eliminations*

6.1.4.2. Synthetic Applications of the Diels-Alder Reaction. Diels-Alder reactions have long played an important role in synthetic organic chemistry.[74] The reaction of a substituted benzoquinone and 1,3-butadiene, for example, was the first step in one of the early syntheses of steroids. The angular methyl group was introduced by the methyl group on the quinone and the other functional groups were used for further elaboration.

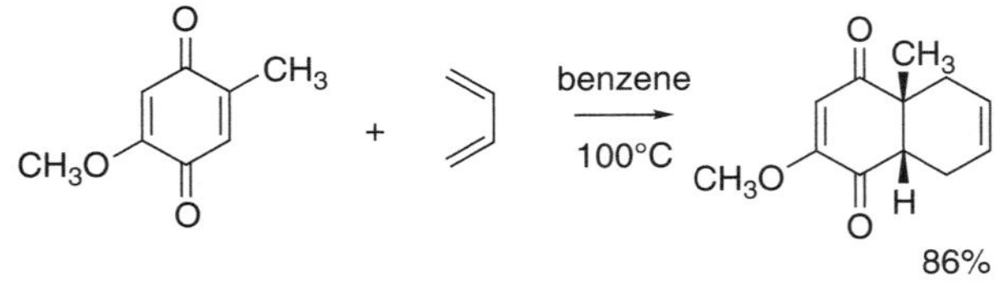

Ref. 75

In a synthesis of gibberellic acid, a diene and quinone, both with oxygen-substituted side chains, gave the initial intermediate. Later in the synthesis, an intramolecular D-A reaction was used to construct the A-ring.

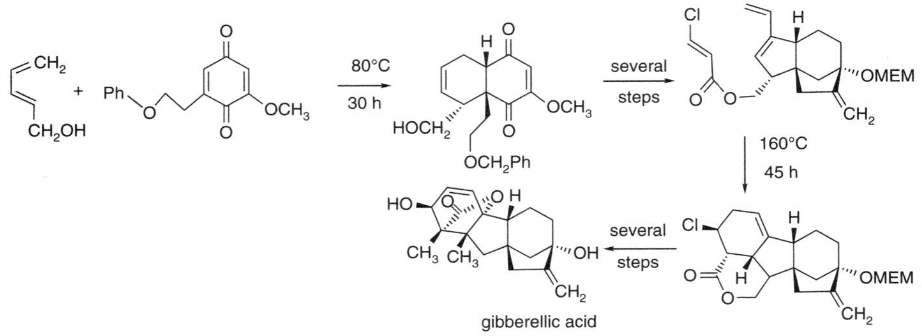

Ref. 76

Functionality can be built into either the diene or dienophile for purposes of subsequent transformations. For example, in the synthesis of prephenic acid, the diene has the capacity to generate an enone. The dienophile contains a sulfoxide substituent that is subsequently used to introduce a second double bond by elimination.

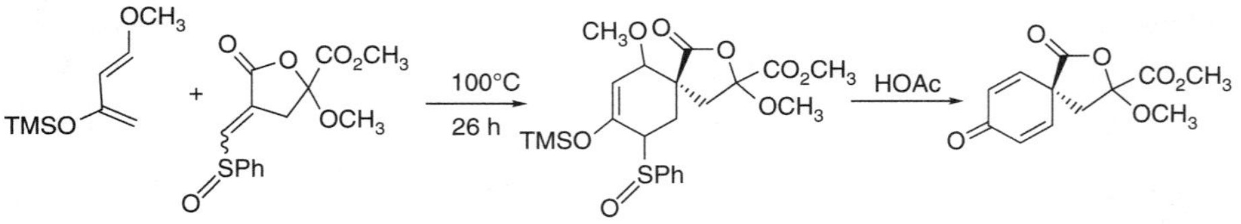

Ref. 77

74. K. C. Nicolaou, S. A. Snyder, T. Montagnon, and G. Vassilikogiannakis, *Angew. Chem. Int. Ed. Engl.*, **41**, 1668 (2002).

75. R. B. Woodward, F. Sondheimer, D. Taub, K. Heusler, and W. M. McLamore, *J. Am. Chem. Soc.*, **74**, 4223 (1952).

76. E. J. Corey, R. L. Danheiser, S. Chandrasekaran, P. Siret, G. E. Keck, and J.-L. Gras, *J. Am. Chem. Soc.*, **100**, 8031 (1978); E. J. Corey, R. L. Danheiser, S. Chandrakeskaran, G. E. Keck, B. Gopalan, S. D. Larsen, P. Siret, and J.-L. Gras, *J. Am. Chem. Soc.*, **100**, 8034 (1978).

77. S. J. Danishefsky, M. Hirama, N. Fitsch, and J. Clardy, *J. Am. Chem. Soc.*, **101**, 7013 (1979).

Scheme 6.1. Examples of Thermal Diels-Alder Reactions

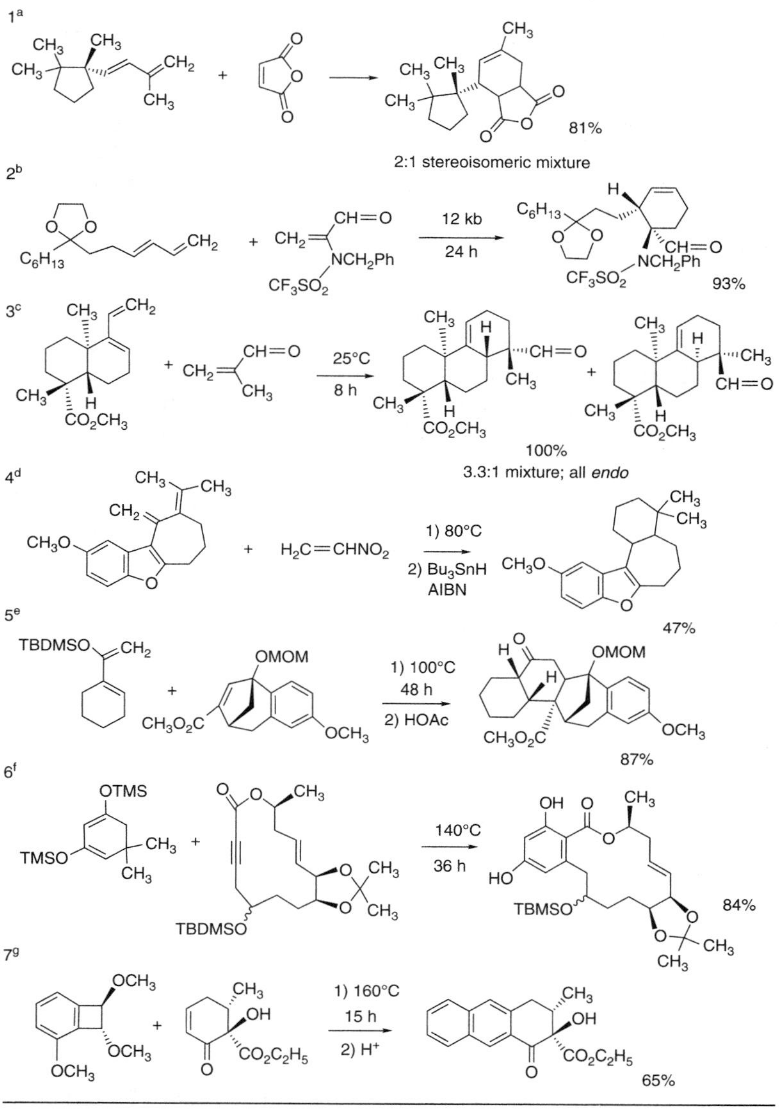

a. A. Nayek and S. Ghosh, *Tetrahedron Lett.*, **43**, 1313 (2002).
b. J.-H. Maeng and R. L. Funk, *Org. Lett.*, **4**, 331 (2002).
c. T. Ling, B. A. Kramer, M. A. Palladino, and E. A. Theodorakis, *Org. Lett.*, **2**, 2073 (2000).
d. M. Inoue, M. W. Carson, A. J. Frontier, and S. J. Danishefsky, *J. Am. Chem. Soc.*, **123**, 1878 (2001).
e. P. D. O'Connor, L. N. Mander, and M. W. McLachlan, *Org. Lett.*, **6**, 703 (2004).
f. X. Geng and S. J. Danishefsky, *Org. Lett.*, **6**, 413 (2004).
g. K. Yamamoto, M. F. Hentemann, J. G. Allen, and S. J. Danishefsky, *Chem. Eur. J.*, **9**, 3242 (2003).

Scheme 6.1 gives some additional examples of application of thermal D-A reactions in syntheses. The reaction in Entry 1 was eventually used to construct an aromatic ring by decarboxylation and aromatization. The reaction did not exhibit much facial selectivity, but this was irrelevant for the particular application. Entry

496

CHAPTER 6

*Concerted
Cycloadditions,
Unimolecular
Rearrangements, and
Thermal Eliminations*

2 illustrates the use of high pressure to accelerate reaction. This reaction gives only an *endo* product, since both the electronic effect of the formyl group and the steric effect of the sulfonamido group favor this orientation. The reaction in Entry 3 involves a typical diene and dienophiles. The reaction is completely regiospecific in the direction expected [donor alkyl groups at C(1) and C(3) of the diene unit] and is also completely *endo* selective. The facial selectivity with respect to the diene, however, is only 3.3:1. Entry 4 is an example of the use of nitroethene as an ethene equivalent. The nitro group was removed by reduction with Bu_3SnH. The reaction in Entry 5 involves a diene unit activated by a 2-siloxy substituent. On exposure to acid, this provides the product as a ketone. The reaction is evidently completely regio- and stereoselective. Entry 6 involves a doubly activated diene. The aromatic ring is formed by extrusion of isobutylene from a bicyclic intermediate. Entry 7 involves the ring opening of a benzocyclobutene to a quinodimethane. In this case, aromatization occurs as the result of the loss of two methoxy groups.

Owing to their advantages in terms of the lower temperature required and the higher regio- and stereoselectivity, Lewis acid–catalyzed D-A reactions are often preferable to the corresponding thermal version. Scheme 6.2 gives some examples of D-A reactions catalyzed by Lewis acids. Entries 1 and 2 are cases with substituent groups on the reacting bonds. Systems of this type are often relatively unreactive in thermal D-A reactions. The reaction in Entry 2 is an inverse electron demand case, and the catalyst activates the *diene* rather than the dienophile. Entry 3 involves a relatively highly substituted diene. The reaction was used to create a structure corresponding to the A-ring of the antitumor substance taxol. Entries 4, 5, and 6 involve dienes that have donor substituents that impart regioselectivity. The products of each of the reactions result from *endo* addition. The reaction in Entry 4 involves a cyclohexenone dienophile. 5-Substituted cyclohexenones have a strong preference for *anti* approach relative to the substituent.[78] The isopropenyl substituent establishes a conformational preference and the diene approaches from the *anti* direction.

Entries 5 and 6 exhibit the "ortho" regioselectivity expected for a 1-ERG on the diene. These dienes also present the possibility for competing Lewis acid coordination sites in the diene that would be expected to be *deactivating*. In Entry 6, the phenyl substituent on the oxazolidinone ring establishes a facial preference. The dienophiles in Entries 7 and 8 have both ERG and EWG substituents (sometimes called capto-dative dienophiles). The regiochemistry is consistent with the acceptor substituent having the dominant influence.[79] Entry 9 illustrates the excellent regio- and stereoselectivity often seen for Lewis acid–catalyzed reactions. Only a single product was found.

78. F. Fringuelli, L. Minuti, F. Pizzo, and A. Taticchi, *Acta Chem. Scand.*, **47**, 255 (1993).
79. R. Herrera, H. A. Jiminez-Vazquez, A. Modelli, D. Jones, B. C. Soderberg, and J. Tamariz, *Eur. J. Org. Chem.*, 4657 (2001).

Scheme 6.2. Diels-Alder Reactions Catalyzed by Lewis-Acids

1[a]

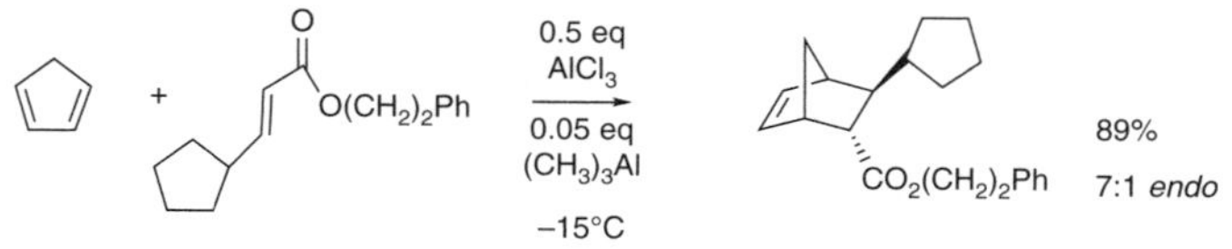

0.5 eq AlCl₃

0.05 eq (CH₃)₃Al

−15°C

89%
7:1 *endo*

2[b]

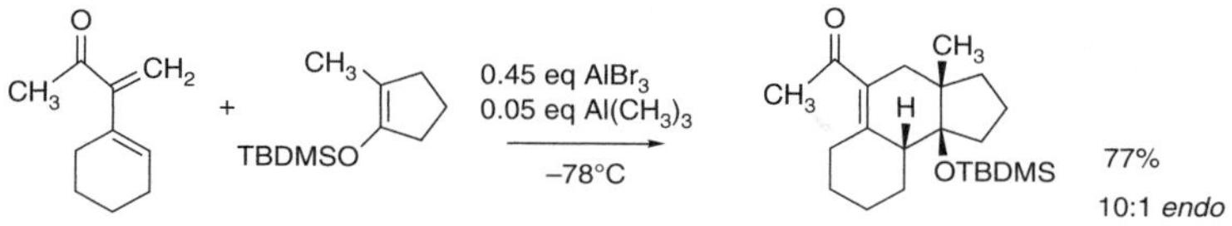

0.45 eq AlBr₃
0.05 eq Al(CH₃)₃

−78°C

77%
10:1 *endo*

3[c]

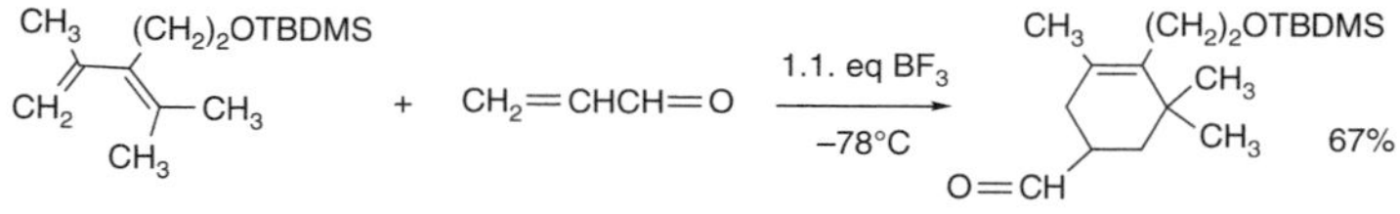

1.1. eq BF₃

−78°C

67%

4[d]

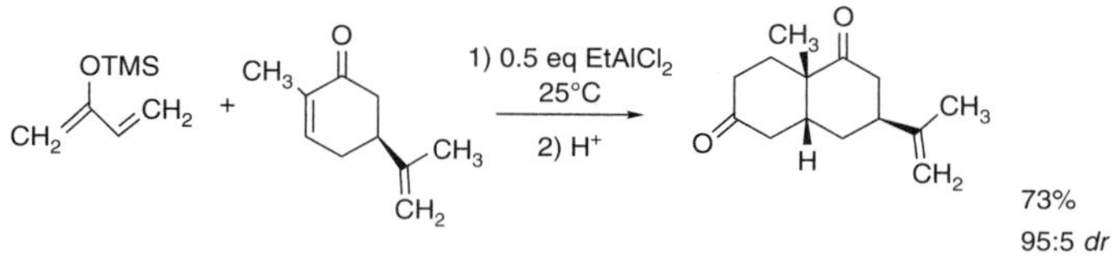

1) 0.5 eq EtAlCl₂
25°C

2) H⁺

73%
95:5 *dr*

5[e]

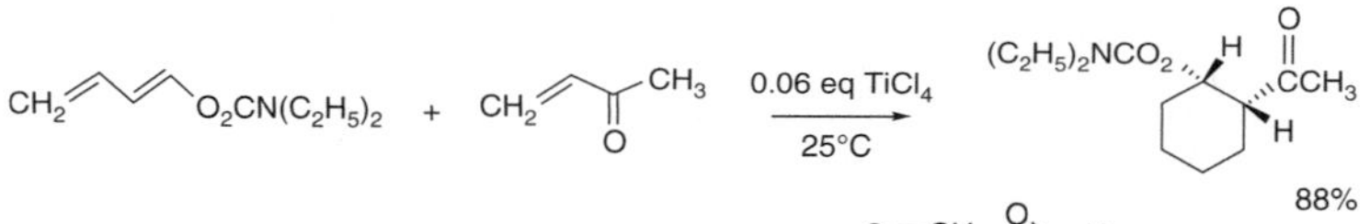

0.06 eq TiCl₄

25°C

88%

6[f]

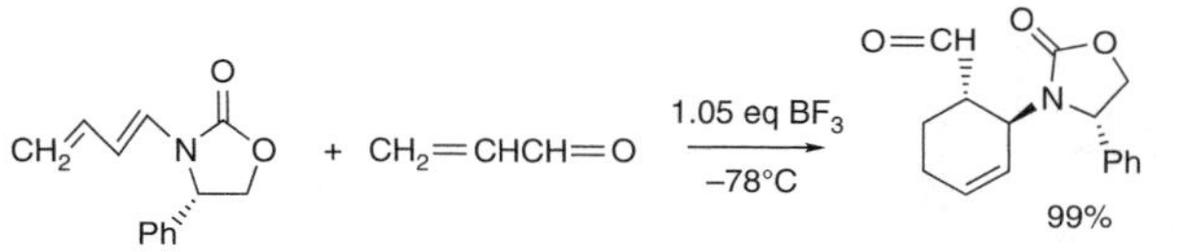

1.05 eq BF₃

−78°C

99%

7[g]

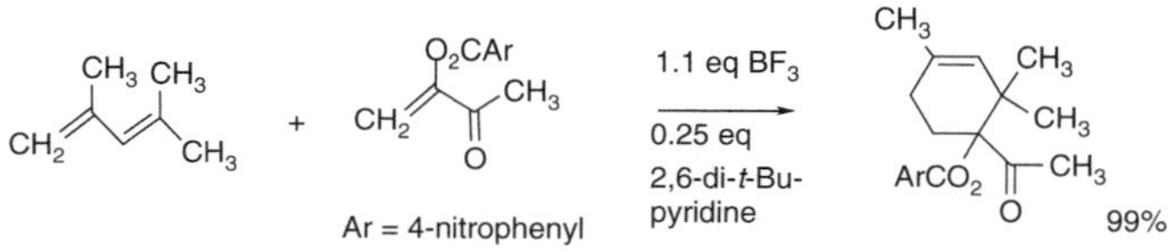

Ar = 4-nitrophenyl

1.1 eq BF₃

0.25 eq
2,6-di-*t*-Bu-pyridine

99%

8[h]

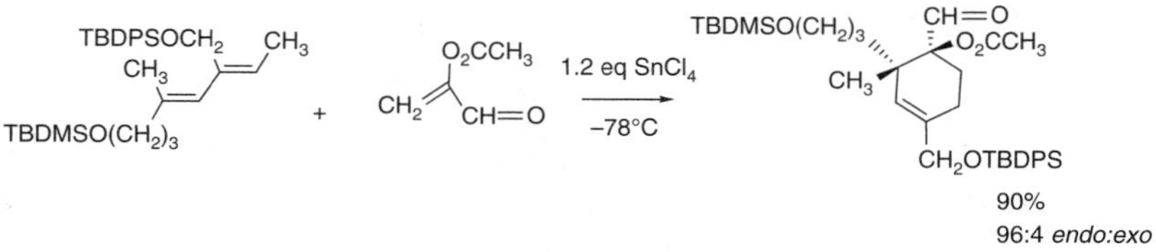

1.2 eq SnCl₄

−78°C

90%
96:4 *endo:exo*

9[i]

AlCl₃

−50°C → −30°C

95%

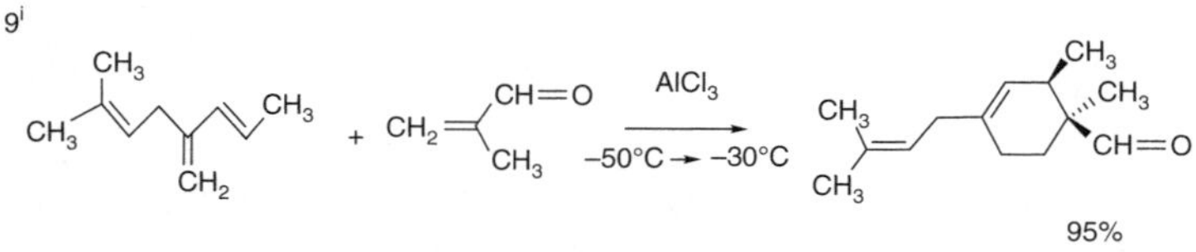

(*Continued*)

Scheme 6.2. *(Continued)*

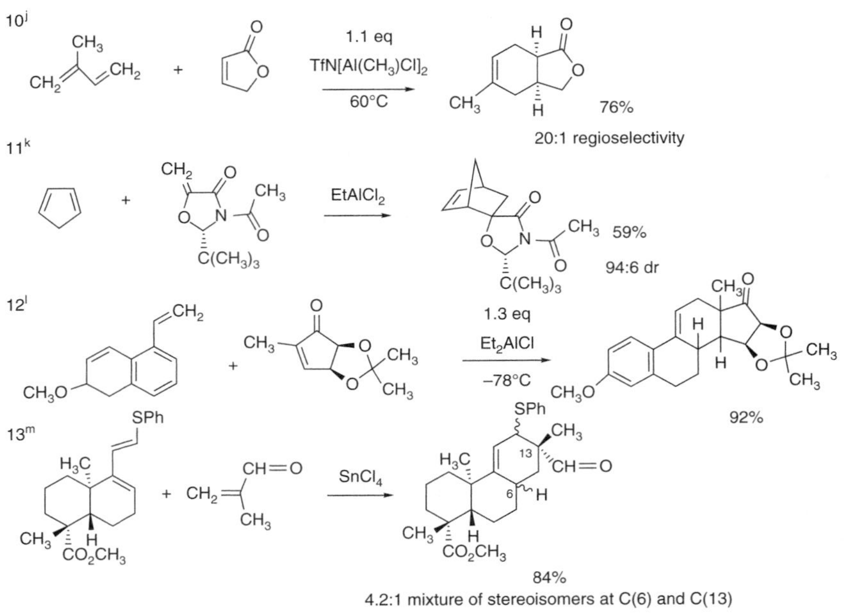

CHAPTER 6

*Concerted
Cycloadditions,
Unimolecular
Rearrangements, and
Thermal Eliminations*

a. R. D. Hubbard and B. L. Miller, *J. Org. Chem.*, **63**, 4143 (1998).
b. M. E. Jung and P. Davidov, *Angew. Chem. Int. Ed. Engl.*, **41**, 4125 (2002).
c. M. W. Tjepkema, P. D. Wilson, H. Audrain, and A. G. Fallis, *Can. J. Chem.*, **75**, 1215 (1997).
d. A. A. Haaksma, B. J. M. Jansen, and A. de Groot, *Tetrahedron*, **48**, 3121 (1992).
e. P. F. De Cusati and R. A. Olofson, *Tetrahedron Lett.*, **31**, 1409 (1990).
f. D. A. Vosburg, S. Weiler, and E. J. Sorensen, *Chirality*, **15**, 156 (2003).
g. J. D. Dudones and P. Sampson, *J. Org. Chem.*, **62**, 7508 (1997).
h. W. R. Roush and D. A. Barda, *J. Am. Chem. Soc.*, **119**, 7402 (1997).
i. G. Frater, U. Mueller, and F. Schroeder, *Tetrahedron: Asymmetry*, **15**, 3967 (2004).
j. A. Saito, H. Yanai, and T. Taguchi, *Tetrahedron Lett.*, **45**, 9439 (2004).
k. W. R. Roush, A. P. Essenfeld, J. S. Warmus, and B. B. Brown, *Tetrahedron Lett.*, **30**, 7305 (1989).
l. K. Tanaka, H. Nakashima, T. Taniguchi, and K. Ogasawara, *Org. Lett.*, **2**, 1915 (2000).
m. T. Ling, B. A. Kramer, M. A. Palladino, and E. A. Theodorakis, *Org. Lett.*, **2**, 2073 (2000).

Entries 10 and 11 involve lactones and lactams, respectively. The catalyst used in Entry 10 is thought to be capable of interaction with both the carbonyl and ether oxygens.

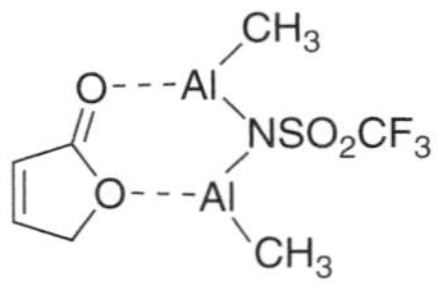

In Entry 11 the dienophile is an α-methylene lactam. As noted for this class of dienophiles, the stereoselectivity results from preferred *exo* addition (see p. 471). The reaction in Entry 12 was used in an enantiospecific synthesis of estrone. The dienophile was used in enantiomerically pure form and the dioxolane ring imparts a high facial selectivity to the dienophile. The reaction occurs through an *endo* TS.

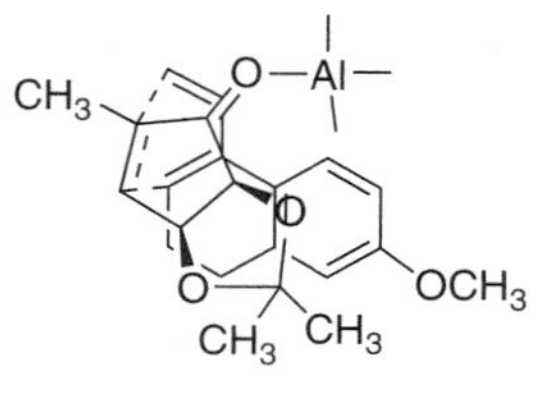

The reaction in Entry 13 is completely regioselective and both stereoisomers are formed through an *endo* TS. The two stereoisomers result from competing facial approaches to the diene.

6.1.5. Diastereoselective Diels-Alder Reactions Using Chiral Auxiliaries

The highly ordered cyclic TS of the D-A reaction permits design of diastereo- or enantioselective reactions. (See Section 2.4 of Part A to review the principles of diastereoselectivity and enantioselectivity.) One way to achieve this is to install a chiral auxiliary.[80] The cycloaddition proceeds to give two diastereomeric products that can be separated and purified. Because of the lower temperature required and the greater stereoselectivity observed in Lewis acid–catalyzed reactions, the best diastereoselectivity is observed in catalyzed reactions. Several chiral auxiliaries that are capable of high levels of diastereoselectivity have been developed. Chiral esters and amides of acrylic acid are particularly useful because the auxiliary can be recovered by hydrolysis of the purified adduct to give the enantiomerically pure carboxylic acid. Early examples involved acryloyl esters of chiral alcohols, including lactates and mandelates. Esters of the lactone of 2,4-dihydroxy-3,3-dimethylbutanoic acid (pantolactone) have also proven useful.

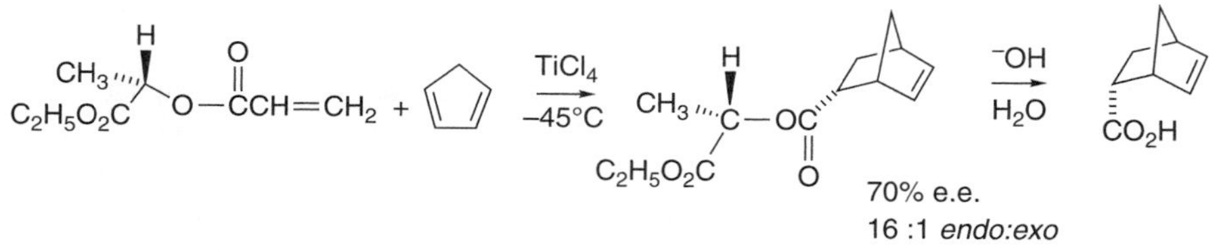

Ref. 81

Prediction and analysis of diastereoselectivity are based on steric, stereoelectronic, and complexing interactions in the TS.[82] In the case of the lactic acid auxiliary, a chelated structure promotes facial selectivity. In the $TiCl_4$ complex of *O*-acryloyl ethyl lactate,

[80] W. Oppolzer, *Angew. Chem. Int. Ed. Engl.*, **23**, 876 (1984); M. J. Tascher, in *Organic Synthesis: Theory and Applications*, Vol. 1, T. Hudlicky, ed., JAI Press, Greenwich, CT, 1989, pp. 1–101; H. B. Kagan and O. Riant, *Chem. Rev.*, **92**, 1007 (1992); K. Narasaka, *Synthesis*, 16 (1991).

[81] T. Poll, G. Helmchen, and B. Bauer, *Tetrahedron Lett.*, **25**, 2191 (1984).

[82] For example, see T. Poll, A. Sobczak, H. Hartmann, and G. Helmchen, *Tetrahedron Lett.*, **26**, 3095 (1985).

500

CHAPTER 6

*Concerted
Cycloadditions,
Unimolecular
Rearrangements, and
Thermal Eliminations*

one of the chlorines attached to titanium shields one face of the double bond (see also Figure 6.5).

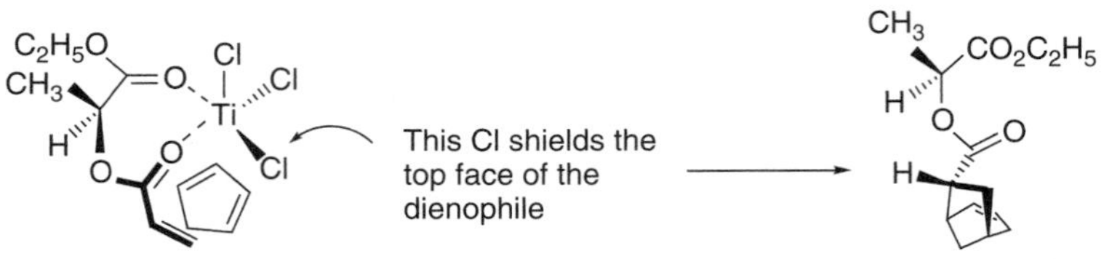

This Cl shields the top face of the dienophile

An 8-phenylmenthol ester was employed as the chiral auxiliary to achieve enantioselectivity in the synthesis of prostaglandin precursors.[83] The crucial features of the TS are the *anti* disposition of the Lewis acid relative to the alcohol moiety and a π stacking with the phenyl ring that provides both stabilization and steric shielding of the α-face.

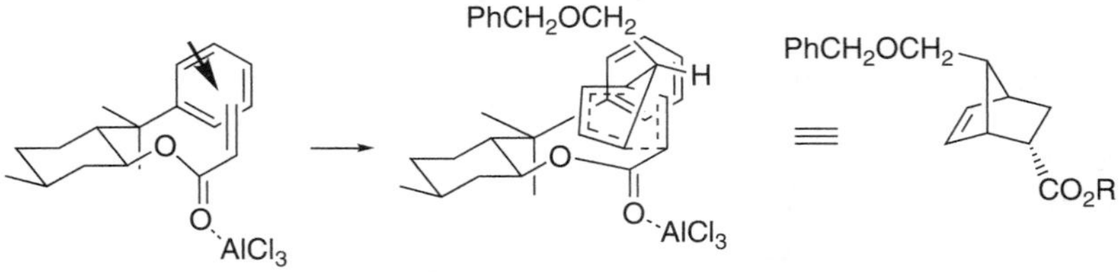

The cyclic α-hydroxylactone, pantolactone, has been used extensively as a chiral auxiliary in D-A reactions.[84] Reactions involving $TiCl_4$ and $SnCl_4$ occur through chelated TSs.[85]

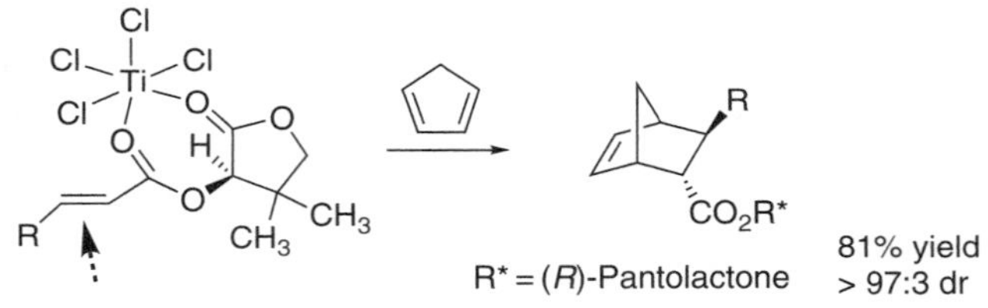

R* = (*R*)-Pantolactone

81% yield
> 97:3 dr

Several other Lewis acids including BF_3, Et_2AlCl, and $EtAlCl_2$ gave somewhat reduced levels of diastereoselectivity, but still favored the chelation-controlled product.[86] However, use of two equivalents of a highly hindered monodentate Lewis acid of the MAD type favored the other diastereoisomer. These reactions are thought to proceed through an open 2:1 complex exhibiting the opposite facial selectivity.

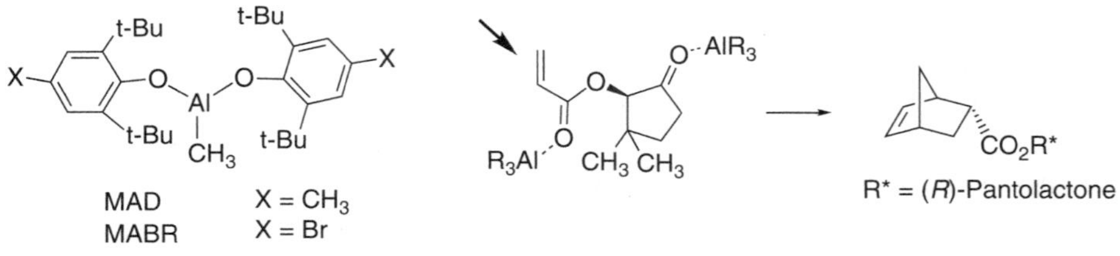

| MAD | X = CH_3 |
| MABR | X = Br |

R* = (*R*)-Pantolactone

83. E. J. Corey, T. K. Schaaf, W. Huber, H. Koelliker, and N. M. Weinshenker, *J. Am. Chem. Soc.*, **92**, 397 (1970).

84. P. Campos and D. Munoz-Torreno, *Curr. Org. Chem.*, **8**, 1339 (2004).

85. T. Poll, A. F. Abdel Hady, R. Karge, G. Linz, J. Weetman, and G. Helmchen, *Tetrahedron Lett.*, **30**, 5595 (1989).

86. R. Maruoka, M. Oishi, and H. Yamamoto, *Synlett*, 683 (1993).

For the diester of fumaric acid, EtAlCl$_2$ was the most effective catalyst and the reaction proceeded with more than 90% diastereoselectivity.[87]

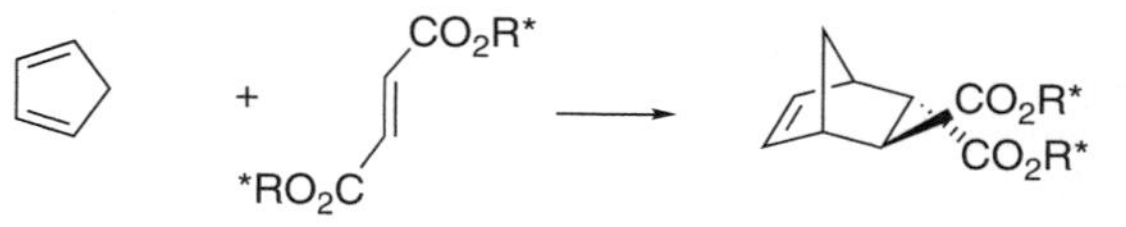

R* = (R)-Pantolactone

Mandelate and lactate esters have been found to generate diastereoselectivity in reactions of hydroxy-substituted quinodimethanes generated by thermolysis of benzo-cyclobutenols.[88] The reactions are thought to proceed by an *exo* TS with a crucial hydrogen bond between the hydroxy group and a dienophile carbonyl. The phenyl (or methyl in the case of lactate) group promotes facial selectivity.

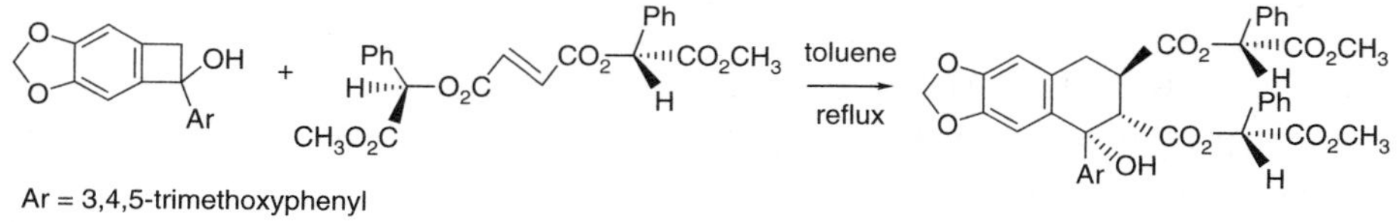

Ar = 3,4,5-trimethoxyphenyl

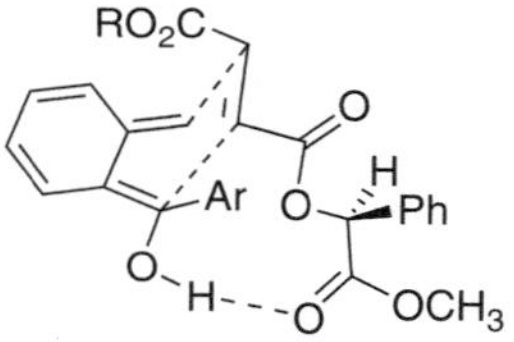

Several aspects of this reaction are intriguing. Despite the relatively high temperature (105° C), the nine-membered ring seems to have a strong influence on the stereoselectivity. The tendency for planarity at the ester bond may also contribute to the stability of the TS.

α, β-Unsaturated derivatives of chiral oxazolidinones have proven to be especially useful chiral auxiliaries for D-A additions. Reaction occurs at low temperatures in the presence of Lewis acids. The most effective catalyst for this system is (C$_2$H$_5$)$_2$AlCl.[89]

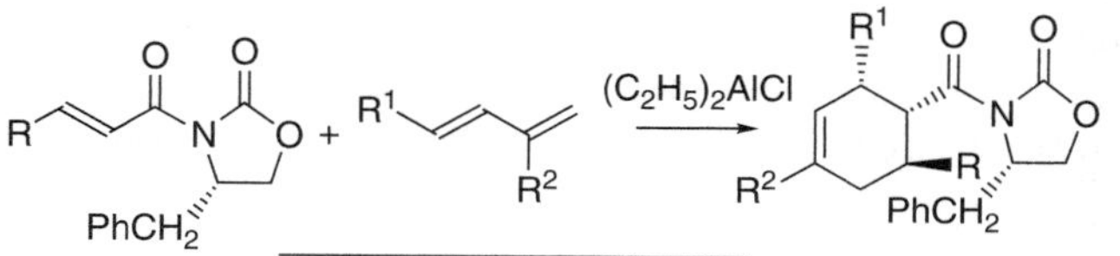

R	R^1	R^2	Yield	dr
H	H	CH$_3$	85%	95:5
H	CH$_3$	H	84%	>100:1
CH$_3$	H	CH$_3$	83%	94:6
CH$_3$	CH$_3$	H	77%	95:5

[87.] G. Helmchen, A. F. A. Hady, H. Hartmann, R. Karge, A. Krotz, K. Sartor, and M. Urmann, *Pure Appl. Chem.*, **61**, 409 (1989).

[88.] D. E. Bogucki and J. L. Charlton, *J. Org. Chem.*, **60**, 588 (1995); J. L. Charlton and S. Maddaford, *Can. J. Chem.*, **71**, 827 (1993).

[89.] D. A. Evans, K. T. Chapman, and J. Bisaha, *J. Am. Chem. Soc.*, **110**, 1238 (1988).

502

CHAPTER 6

*Concerted
Cycloadditions,
Unimolecular
Rearrangements, and
Thermal Eliminations*

The highest level of enantioselectivity is obtained using 1.5–2.0 equivalents of $(C_2H_5)_2AlCl$. Under these conditions the reactions are thought to proceed through a chelated TS having the vinyl substituent in the *s-cis*-conformation. For oxazolidinones having *S*-configuration at C(4) of the ring, this structure exposes the *si* face at the α-carbon of the dienophile.

This complex is formed with more than 1.0 equivalents of $(C_2H_5)_2AlCl$ with concomitant formation of $[Et_2AlCl_2]^-$. The open and chelated structures have been characterized by NMR.[90] The chelated structure is substantially *more reactive* than the open complex, which accounts for the increase in enantioselectivity with more than 1.0 equivalents of catalyst.

Chelation alone, however, is not sufficient to induce high enantioselectivity since other Lewis acids capable of chelation, such as $SnCl_4$ and $TiCl_4$, give lower enantioselectivity.

Scheme 6.3 gives some other examples of use of chiral auxiliaries in D-A reactions.[91] Entries 1 and 2 show two chiral auxiliaries developed from terpene precursors. The acrylate shown in Entry 1 gave excellent enantioselectivity with cyclopentadiene and 1,3-butadiene, but introduction of a methyl substituent on the dienophile (crotonyl derivative) resulted in a very slow reaction owing to steric problems. The sulfonamide auxiliary shown in Entry 2 has been exploited in other contexts (see, e.g., p. 123). The acyl derivatives give very good facial selectivity and are thought to react through a chelated TS. The carbocyclic ring establishes facial selectivity.

90. S. Castellino and W. J. Dwight, *J. Am. Chem. Soc.*, **115**, 2986 (1993).
91. For additional examples, see W. Oppolzer, *Tetrahedron*, **43**, 1969, 4057 (1987).

Scheme 6.3. Diels-Alder Reactions with Chiral Auxiliaries

Entry	Dienophile	Diene	Catalyst, temperature	Yield (%)	dr
1[a]	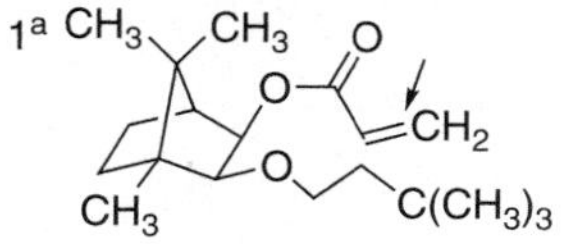		TiCl$_2$(*i*-OPr)$_2$, −20°C 1.5 equiv	90	>99:1
2[b]	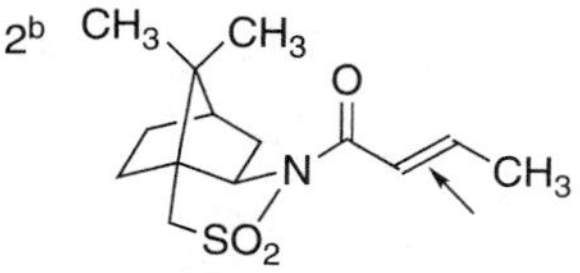		TiCl$_4$, −78°C 0.5 equiv	88	99:1
3[c]	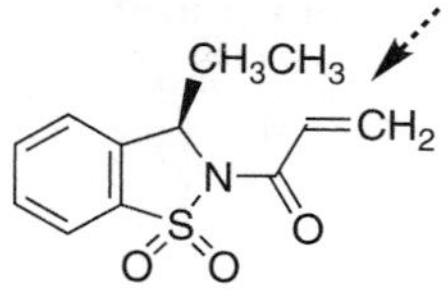		(C$_2$H$_5$)$_2$AlCl, −40°C	94	98:2
4[d]	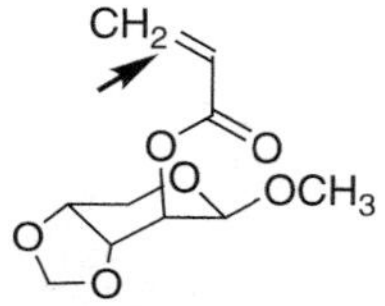		SnCl$_4$, −78°C 2 equiv	93	96:4
5[e]	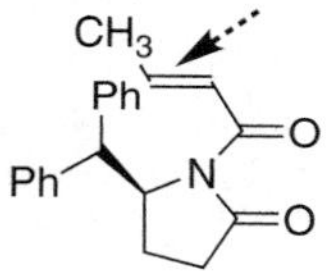		ZrCl$_4$, −78°C	86	>99:1
6[f]	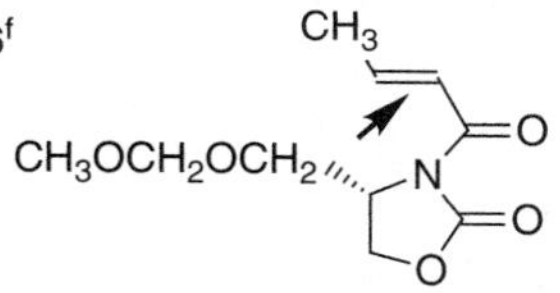		(C$_2$H$_5$)$_2$AlCl, 78°C 1.1 equiv	62	97:3
7[g]	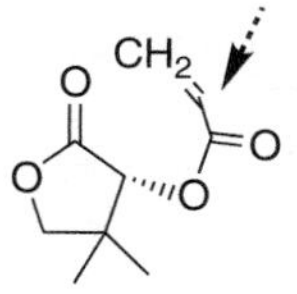	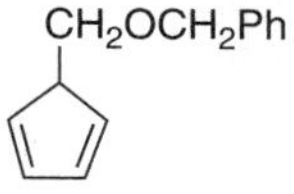	TiCl$_4$, −55° to −20°C	79	96:2

a. W. Oppolzer, C. Chapuis, D. Dupuis, and M. Guo, *Helv. Chim. Acta*, **68**, 2100 (1985).
b. W. Oppolzer, C. Chapuis, and G. Bernardinelli, *Helv. Chim. Acta*, **67**, 1397 (1984); M. Vanderwalle, J. Van der Eycken, W. Oppolzer, and C. Vullioud, *Tetrahedron*, **42**, 4035 (1986).
c. W. Oppolzer, B. M. Seletsky, and G. Bernardinelli, *Tetrahedron Lett.*, **35**, 3509 (1994).
d. R. Nougier, J.-L. Gras, B. Giraud, and A. Virgilli, *Tetrahedron Lett.*, **32**, 5529 (1991).
e. M. P. Sibi, P. K. Deshpande, and J. Ji, *Tetrahedron Lett.*, **36**, 8965 (1995).
f. M. Ikota, *Chem. Pharm. Bull.*, **37**, 2219 (1989).
g. K. Miyaji, Y. Ohara, Y. Takahashi, T. Tsuruda, and K. Arai, *Tetrahedron Lett.*, **32**, 4557 (1991).

504

CHAPTER 6

*Concerted
Cycloadditions,
Unimolecular
Rearrangements, and
Thermal Eliminations*

Entry 3 involves another sultam auxiliary. The chirality of the product is consistent with approach of the diene from the *re* face of a conformation in which the carbonyl oxygen is *syn* to the sulfonyl group.

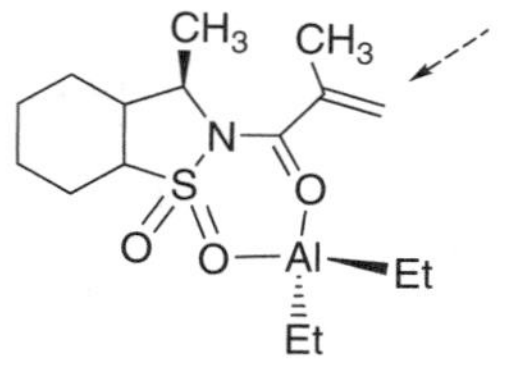

Entry 4 shows a carbohydrate-derived auxiliary with $SnCl_4$ as the Lewis acid. This dienophile also gives good enantioselectivity using $TiCl_4$ as the Lewis acid. Entry 5 is a proline-derived oxazolidinone auxiliary used in conjunction with $ZrCl_4$. The observed diastereoselectivity is consistent with a chelated TS having an *s-cis* conformation at the carbonyl group.

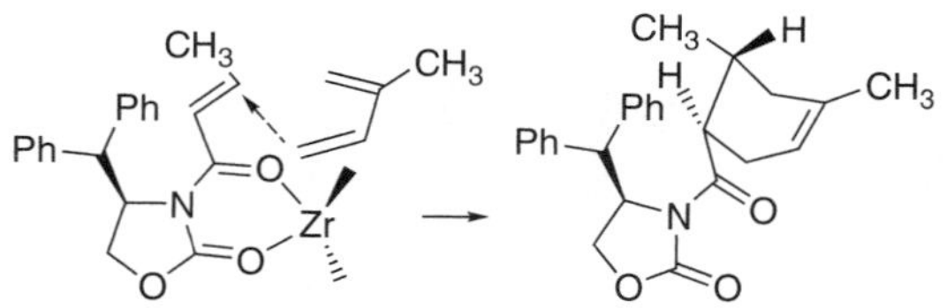

Entry 6 uses a chiral auxiliary derived from pyroglutamic acid. Entry 7 is an example of the use of pantolactone as a chiral auxiliary to form a prostaglandin precursor.

The alkenyl oxonium ion dienophiles generated from dioxolanes can be made diastereoselective by use of chiral diols. For example, acetals derived from *anti*-pentane-2,4-diol react under the influence of $TiCl_4/Ti(i\text{-}OPr)_4$ with stereoselectivity ranging from 3:1 to 15:1.

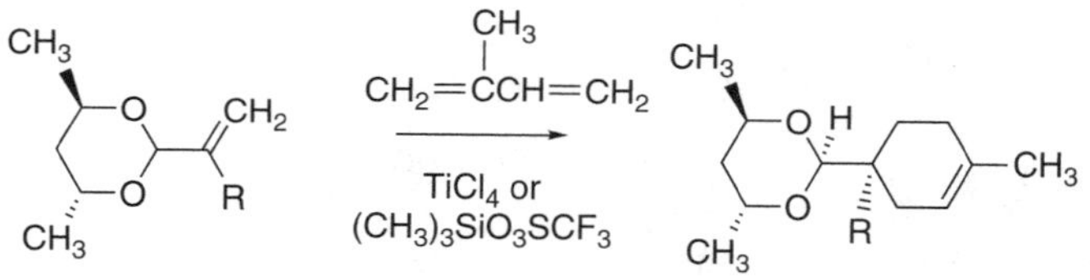

Ref. 92

Dioxolanes derived from *syn*-1,2-diphenylethane-1,2-diol react with dienes such as cyclopentadiene and isoprene, but in most cases the diastereoselectivity is low.

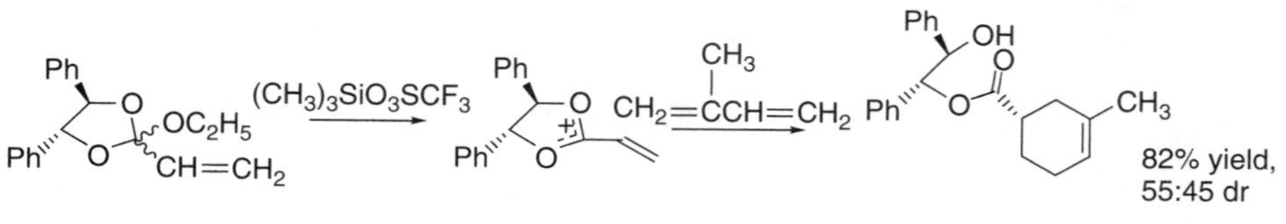

82% yield,
55:45 dr

Ref. 93

92. T. Sammakia and M. A. Berliner, *J. Org. Chem.*, **59**, 6890 (1994).
93. A. Haudrechy, W. Picoul, and Y. Langlois, *Tetrahedron: Asymmetry*, **8**, 129 (1997).

6.1.6. Enantioselective Catalysts for Diels-Alder Reactions

Enantioselectivity can also be achieved with chiral catalysts. The chiral oxazaboro-lidinones introduced in Section 2.1.5.6 as enantioselective aldol addition catalysts have been found to be useful in D-A reactions. The tryptophan-derived catalyst **A** can achieve 99% enantioselectivity in the cycloaddition between 5-benzyloxymethyl-1,3-cyclopentadiene and 2-bromopropenal. The indole ring provides π stacking and steric shielding. There is also believed to be a *formyl hydrogen bond* to the ring oxygen. A significant feature of this reaction is that the product is *exo* with respect to the formyl group. The adduct can be converted to an important intermediate for the synthesis of prostaglandins.[94]

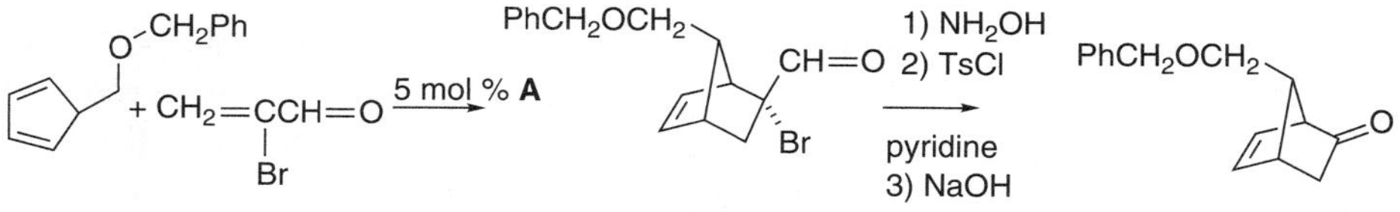

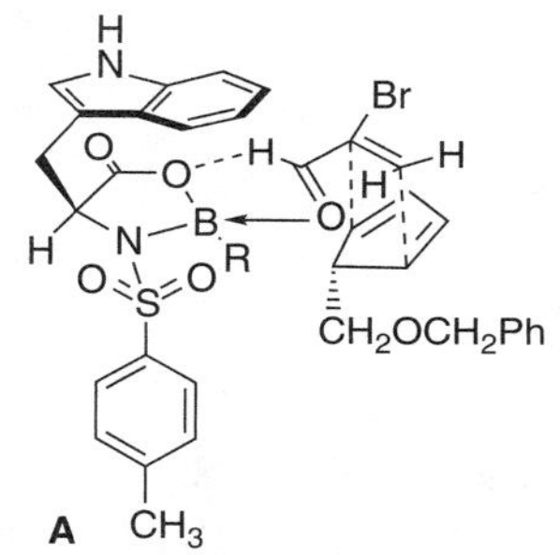

The oxazaborolidines **B** and **C** derived from proline are also effective catalysts. The protonated forms of these catalysts, generated using triflic acid or triflimide, are very active catalysts,[95] and the triflimide version is more stable above 0° C. Another protonated catalyst **D** is derived from 2-cyclopentenylacetic acid.

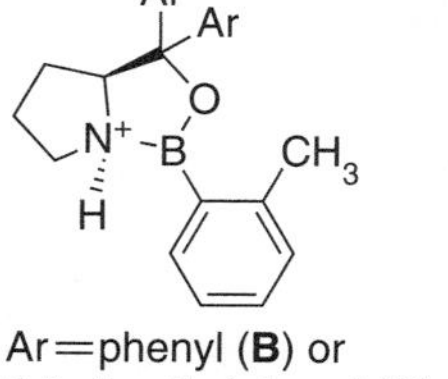

Ar=phenyl (**B**) or
3,5-dimethylphenyl (**C**)

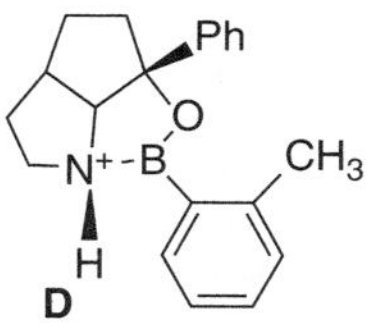

[94] E. J. Corey and T. P. Loh, *J. Am. Chem. Soc.*, **113**, 8966 (1991).

[95] E. J. Corey, T. Shibata, and T. W. Lee, *J. Am. Chem. Soc.*, **124**, 3808 (2002); D. H. Ryu and E. J. Corey, *J. Am. Chem. Soc.*, **125**, 6388 (2003); E. J. Corey, *Angew. Chem. Int. Ed.*, **41**, 1650 (2002).

506

CHAPTER 6

*Concerted
Cycloadditions,
Unimolecular
Rearrangements, and
Thermal Eliminations*

α, β-Unsaturated aldehydes react via TS **E**, whereas α,β-unsaturated ketones and esters react via TS **F**.

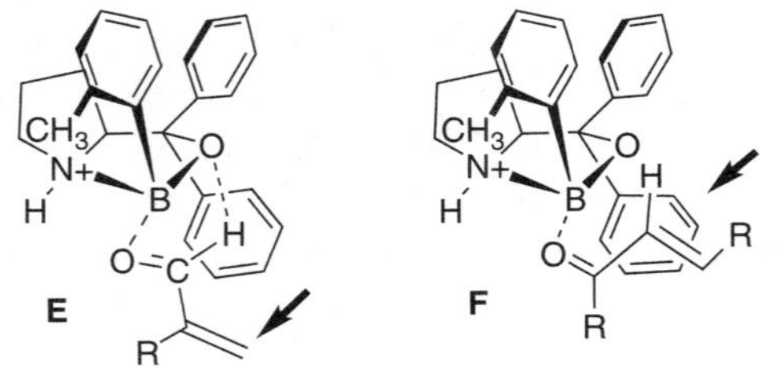

With trisubstituted benzoquinones and use of the cationic oxazaborolidinium catalyst **B**, 2-[*tris*-(isopropyl)silyloxy]-1,3-butadiene reacts at the monosubstituted quinone double bond. The reactions exhibit high regioselectivity and more than 95% e.e. With 2-mono- and 2,3-disubstituted quinones, reaction occurs at the unsubstituted double bond. The regiochemistry is directed by coordination to the catalyst at the more basic carbonyl oxygen.

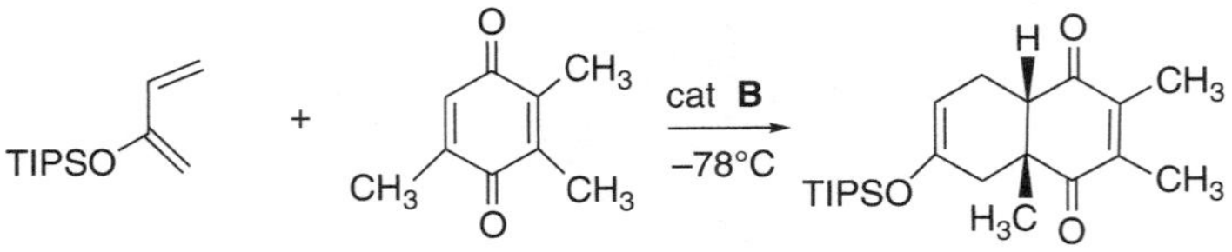

The enantioselectivity is consistent with a TS in which the less-substituted double bond of the quinone is oriented toward the catalyst, as in TS **G**.

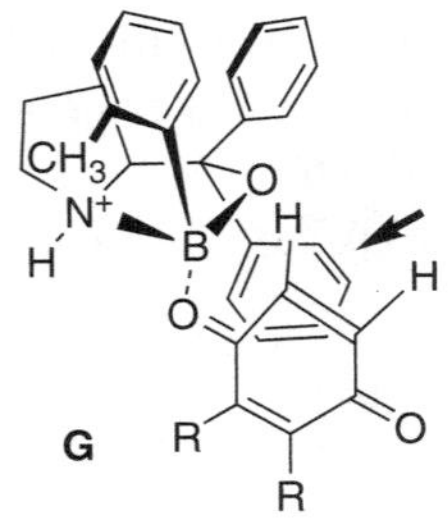

These catalysts have been applied to D-A reactions that are parts of several important synthetic routes, thereby making them enantioselective.[96] For example, key intermediates in the synthesis of cortisone and coriolin were prepared in enantiomerically pure form using catalyst **B**.

96. Q.-Y. Hu, G. Zhou, and E. J. Corey, *J. Am. Chem. Soc.*, **126**, 13708 (2004).

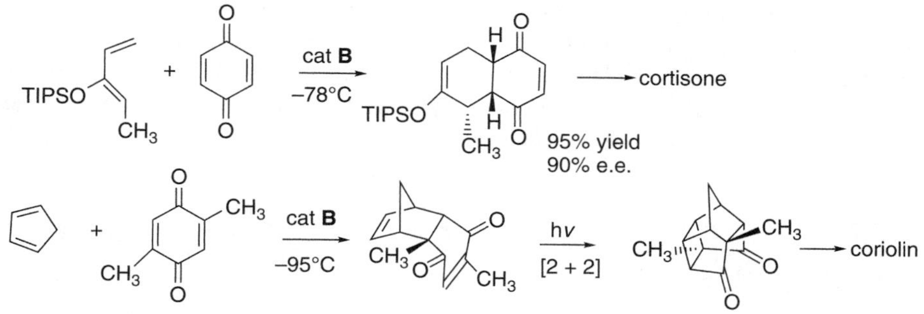

Similarly, an enantioselective synthesis of estrone is based on catalyst **D**.[97]

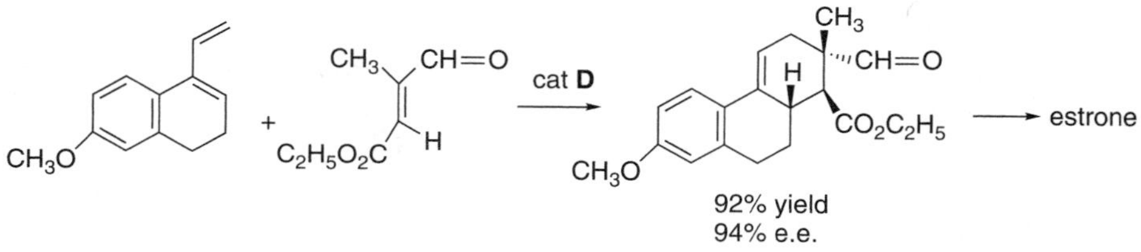

A valine-derived oxazaborolidine derivative has been found to be subject to activation by Lewis acids, with $SnCl_4$ being particularly effective.[98] This catalyst combination also has reduced sensitivity to water and other Lewis bases.

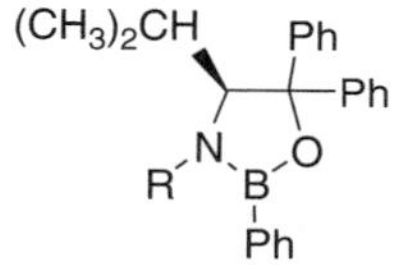

H R = *n*-octyl

I R = 1-naphthylmethyl

Catalyst **H** and the corresponding *N*-(1-naphthylmethyl) derivative **I** give high e.e. and good *endo* stereoselectivity for several typical dienophiles with cyclopentadiene.

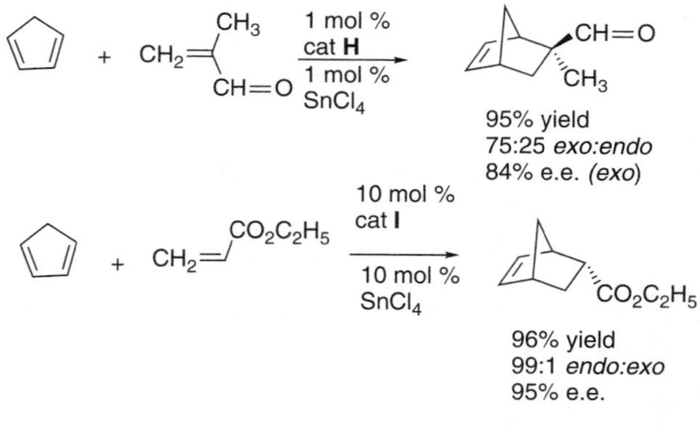

[97.] Q.-Y. Hu, P. D. Rege, and E. J. Corey, *J. Am. Chem. Soc.*, **126**, 5984 (2004).
[98.] K. Futatsugi and H. Yamamoto, *Angew. Chem. Int. Ed. Engl.*, **44**, 1484 (2005).

508

CHAPTER 6

*Concerted
Cycloadditions,
Unimolecular
Rearrangements, and
Thermal Eliminations*

Cationic oxazaborolidines derived from α,α-diphenylpyrrolidine-2-methanol have been examined and shown to considerably extend the range of dienophiles that are responsive to the catalysts.[99] The best proton source for activation of these catalysts is triflimide, $(CF_3SO_2)_2NH$.[100] For example, cyclohexenone and cyclopentadiene react with 93% enantioselectivity using catalyst **J**.

97% yield
91:9 *endo:exo*
93% *e.e.* (endo)

Another cyclic boron catalyst **K**, derived from *trans*-2-aminocyclohexane-methanol, can be prepared with a quaternary nitrogen that enhances activity.[101] This particular catalyst is not very stable, but it is highly active.

Ar=3,5-dimethylphenyl

K

99% yield
96% e.e.

Another useful group of catalysts for D-A reactions is made up of Cu^{2+} chelates of *bis*-oxazolines.[102] The copper salts are the most effective of the first transition metal series because they offer both strong Lewis acid activation and fast ligand exchange. The anion is also important and must be noncoordinating. The triflates can be used, but the hexafluoroantimonates are even more active.[103] These catalysts have been applied to dienophiles with two donor sites, in particular *N*-acyloxazolidinones. The chelated structures provide strong facial differentiation, as shown in Figure 6.9.[104] Installing chirality into the oxazolidinone results in matched and mismatched combinations. In addition to the *t*-butyl derivative, the 4-isopropyl-5,5-phenyl derivatives have also been explored.[105] The *bis*-oxazolines derived from *cis*-2-aminoindanol have also proven to be effective catalysts.[106] Various solid-supported forms of these BOX catalysts have been developed.[107]

99. E. J. Corey, T. Shibata, and T. W. Lee, *J. Am. Chem. Soc.*, **124**, 3808 (2002); D. H. Ryu, T. W. Lee, and E. J. Corey, *J. Am. Chem. Soc.*, **124**, 9992 (2002).
100. D. H. Ryu and E. J. Corey, *J. Am. Chem. Soc.*, **125**, 6388 (2003).
101. Y. Hayashi, J. J. Rohde, and E. J. Corey, *J. Am. Chem. Soc.*, **118**, 5502 (1996).
102. J. J. Johnson and D. A. Evans, *Acc. Chem. Res.*, **33**, 325 (2000).
103. D. A. Evans, D. M. Barnes, J. S. Johnson, T. Lectka, P. von Matt, S. J. Miller, J. A. Murry, R. D. Norcross, E. A. Shaughnessy, and K. R. Campos, *J. Am. Chem. Soc.*, **121**, 7582 (1999).
104. D. A. Evans, S. J. Miller, T. Lectka, and P. von Matt, *J. Am. Chem. Soc.*, **121**, 7559 (1999).
105. T. Hintermann and D. Seebach, *Helv. Chim. Acta*, **81**, 2093 (1998).
106. A. K. Ghosh, S. Fidanze, and C. H. Senanayake, *Synthesis*, 937 (1998); C. H. Senanayake, *Aldrichimica Acta*, **31**, 3 (1998).
107. D. Rechavi and M. Lemaine, *Chem. Rev.*, **102**, 3467 (2002).

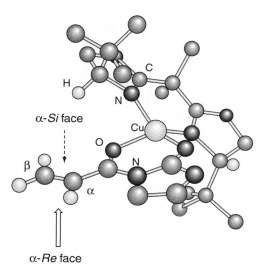

α-*Si* face

β

α

α-*Re* face

Fig. 6.9. Model of Cu(*S,S-t*-BuBOX) catalyst with *N*-acryloyloxazolidinone showing facial stereodifferentiation. Reproduced from *J. Am. Chem. Soc.*, **121**, 7559 (1999), by permission of the American Chemical Society.

Ref. 108

The related PyBOX ligands incorporate a pyridine ring that provides an additional coordination site and are tridentate. The Sc^{3+} and lanthanide ions with the PyBOX ligand can accommodate seven to nine donors. In these complexes, the enantioselectivity is influenced by the number and identity of the coordinating species.[109] Figure 6.10 shows examples of a monohydrated Sc^{3+} triflate[110] having seven contacts and a tetrahydrated lanthanide cation with a total of nine contacts, including two triflate anions.[111]

The basis of the enantioselectivity of the BOX catalysts has been probed using B3LYP/6-31G* calculations.[112] It has been proposed that in the case of the *t*-butyl

108. D. A. Evans and D. M. Barnes, *Tetrahedron Lett.*, **38**, 57 (1997).

109. G. Desimoni, G. Faita, M. Guala, and C. Pratelli, *J. Org. Chem.*, **68**, 7862 (2003).

110. D. A. Evans, Z. K. Sweeney, T. Rovis, and J. S. Tedrow, *J. Am. Chem. Soc.*, **123**, 12095 (2001).

111. G. Desimoni, G. Faita, S. Filippone, M. Mella, M. G. Zampori, and M. Zema, *Tetrahedron*, **57**, 10203 (2001).

112. J. DeChancie, O. Acevedo, and J. D. Evanseck, *J. Am. Chem. Soc.*, **126**, 6043 (2004).

510

CHAPTER 6

*Concerted
Cycloadditions,
Unimolecular
Rearrangements, and
Thermal Eliminations*

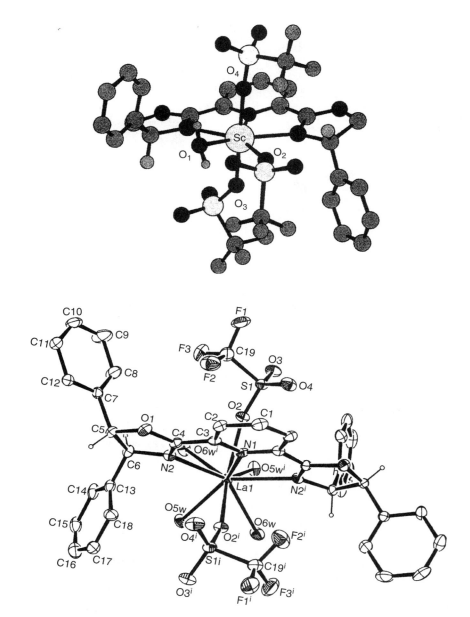

Fig. 6.10. (top) Scandium[S,S-phenylPyBOX(H$_2$O)(CF$_3$SO$_3^-$)$_3$. Reproduced from *J. Am. Chem. Soc.*, **123**, 12095 (2001), by permission of the American Chemical Society. (bottom) Lanthanum[R,R-phenylPyBOX(H$_2$O)$_4$(CF$_3$SO$_3^-$)$_2$ cation. Reproduced from *Tetrahedron*, **57**, 10203 (2001), by permission of Elsevier.

derivatives, catalyst activity and enantioselectivity are governed by the degree to which solvent or anions can approach the copper ion. The most active catalysts are those in which nucleophilic coordination is restricted by a *t*-butyl group.

Several catalysts for enantioselective D-A reactions are based on BINOL. For example, additions of *N*-acryloyloxazolidinones can be made enantioselective using

$Sc(O_3SCF_3)_3$ in the presence of a BINOL ligand.[113] Optimized conditions involved use of 5–20 mol % of the catalyst along with a hindered amine such as *cis*-1,2,6-trimethylpiperidine. A hexacoordinate TS in which the amine is hydrogen bonded to the BINOL has been proposed.

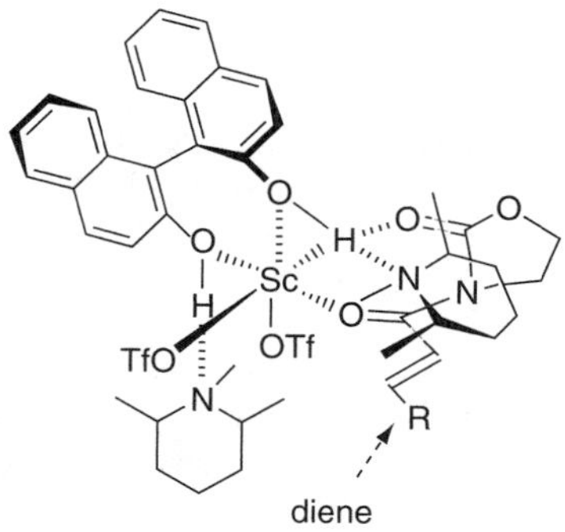

diene

Enantioselective D-A reactions of acrolein are also catalyzed by 3-(2-hydroxyphenyl) derivatives of BINOL in the presence of an aromatic boronic acid. The optimum boronic acid is 3,5-di-(trifluoromethyl)benzeneboronic acid, with which more than 95% e.e. can be achieved. The TS is believed to involve Lewis acid complexation of the boronic acid at the carbonyl oxygen and hydrogen bonding with the hydroxy substituent. In this TS π-π interactions between the dienophile and the hydroxybiphenyl substituent can also help to align the dienophile.[114]

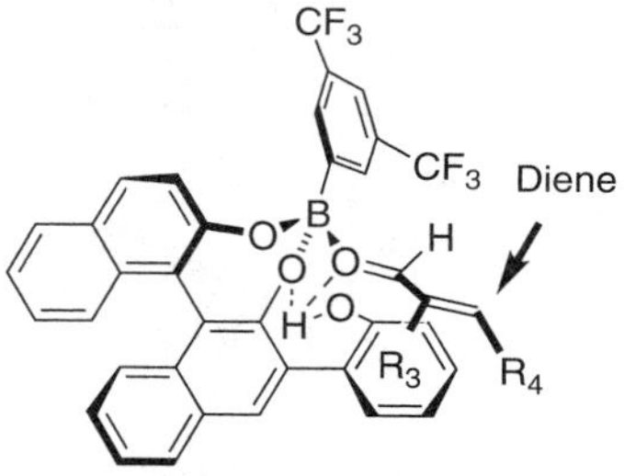

Dienophile	Yield (%)	*exo:endo*	e.e. (%)
$CH_2{=}CHCH{=}O$	84	3:97	95
$CH_2{=}CCH{=}O$ $\vert$ Br	99	90:10	>99
E-$CH_3CH{=}CHCH{=}O$	94	10:90	95
E-$PhCH{=}CHCH{=}O$	94	26:74	80

BINOL has also been used in conjunction with Ti(IV). (*S*)-BINOL-TiCl$_2$ provided an enantiomerically enriched starting material in the synthesis of (–)colombiasin A.[115]

113. S. Kobayashi, M. Araki, and I. Hachiya, *J. Org. Chem.*, **59**, 3758 (1994).
114. K. Ishihara, H. Kurihara, M. Matsumoto, and H. Yamamoto, *J. Am. Chem. Soc.*, **120**, 6920 (1995).
115. K. C. Nicolaou, G. Vassilikogiannakis, W. Magerlein, and R. Kranich, *Angew. Chem. Int. Ed. Engl.*, **40**, 2482 (2001); K. C. Nicolaou, G. Vassilikogiannakis, W. Magerlein, and R. Kranich, *Chem. Eur. J.*, **7**, 5359 (2001).

512

CHAPTER 6

*Concerted
Cycloadditions,
Unimolecular
Rearrangements, and
Thermal Eliminations*

BINOL in conjunction with $TiCl_2(O\text{-}i\text{-}Pr)_2$ gives good enantioselectivity in a D-A reaction with a pyrone as the diene.[116] This is a case of an inverse electron demand reaction and the catalysts would be complexed to the diene.

The $\alpha,\alpha,\alpha,\alpha$-tetraaryl-1,3-dioxolane-4,5-dimethanol (TADDOL) chiral ligands have also been the basis of enantioselective catalysis of the D-A reaction. In a study using 2-methoxy-6-methylquinone as the dienophile, evidence was found that the chloride-ligated form of the catalysts was more active than the dimeric oxy-bridged form.[117]

active form of catalyst

A computational study [B3LYP/3-21G(*d*)] examined a related aspect of the mechanism of TADDOL-$TiCl_2$ catalysis of reactions with *N*-acryloyloxazolidinone.[118] The TS model does not address the steric shielding provided by the ligand substituents but rather the role of the coordination geometry at Ti. The results of this study suggest that the reaction may proceed through a *nonminimum energy complex*. Three different TSs corresponding to different coordination geometries of the ligands were characterized, as shown in Figure 6.11. Although complex MA is lowest in energy, MB has the lowest LUMO. This structure places the exocyclic carbonyl *trans* to a chloride. The authors suggest that it may therefore be the *most reactive* complex. This issue

116. G. H. Posner, H. Dai, D. S. Bull, J.-K. Lee, F. Eydoux, Y. Ishihara, W. Welsh, N. Pryor, and S. Peter, Jr., *J. Org. Chem.*, **61**, 671 (1996).
117. S. M. Moharrram, G. Hirai, K. Koyama, H. Oguri, and M. Hirama, *Tetrahedron Lett.*, **41**, 6669 (2000).
118. J. I. Garcia, V. Martinez-Merino, and J. A. Mayoral, *J. Org. Chem.*, **63**, 2321 (1998).

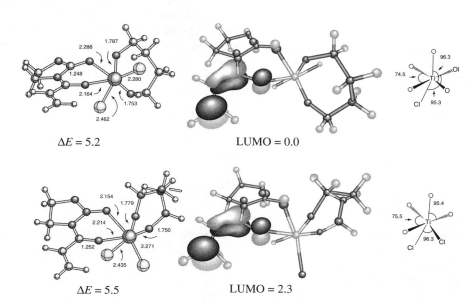

Fig. 6.11. Representation of transition structure and the LUMO orbitals for three stereoisomeric complexes of *N*-acryloyloxazolidinone with a TADDOL model, Ti[O(CH$_2$)$_4$O]Cl$_2$. The LUMO energies (B3LYP/6-3111+G(d)) in kcal/mol. Reproduced from *J. Org. Chem.*, **63**, 2321 (1998), by permission of the American Chemical Society.

has not been resolved, but there is some experimental evidence that the reaction may proceed through a minor complex.[119]

Visual models and additional information on Asymmetric Diels-Alder Reactions can be found in the Digital Resource available at: Springer.com/carey-sundberg.

These examples serve to illustrate several general points about use of chiral catalysts for D-A reactions. A cationic metal center is present in nearly all of the catalysts developed to date and has several functions. It is the anchor for the chiral ligands and also serves as a Lewis acid with respect to the dienophile. The chiral ligands establish the facial selectivity of the complexed dienophile. There are several indications of the importance of the anions to catalytic activity. Anions, in general,

[119.] D. Seebach, R. Dahinden, R. E. Marti, A. K. Beck, D. A. Plattner, and F. N. M. Kuhnle, *J. Org. Chem.*, **60**, 1788 (1995); D. Seebach, R. E. Marti, and T. Hinterman, *Helv. Chim. Acta*, **79**, 710 (1996); C. Haase, C. R. Sarko, and M. Di Mare, *J. Org. Chem.*, **60**, 1777 (1995).

514

CHAPTER 6

Concerted
Cycloadditions,
Unimolecular
Rearrangements, and
Thermal Eliminations

can compete for the ligand binding sites on the metal so that catalytic activity is improved with weakly coordinating anions. Finally, there are some indications in the TADDOL-type catalysts that the anions may exert electronic effects and serve to distinguish between reactivity of dienophiles in *cis* or *trans* positions in the octahedral coordination complex.

Several examples of catalytic enantioselective D-A reactions are given in Scheme 6.4. Entries 1 to 6 involve *N*-acyloxazolidinones and *N*-acylthiazolidinones as dienophiles. Note that there are no stereogenic centers in the reactants, so racemic mixtures would result from reaction in the absence of a chiral catalyst. The metal ions used in these reactions can accommodate two additional ligands in addition to those present in the catalyst. The reactions are believed to involve a chelated TS similar to those involved when chiral oxazolidinone are used (see p. 509). The catalyst in Entry 1 has a BOX-type ligand. The phenyl substituents and the tetrahedral coordination geometry at magnesium give rise to a well-defined geometry. Note that the catalyst has c_2 symmetry. The phenyl substituents cause differential facial shielding.

The enantioselectivity of this catalyst, which is prepared as the iodide salt, is somewhat dependent on the anion that is present. If $AgSbF_6$ is used as a cocatalyst, the iodide is removed by precipitation and the e.e. increases from 81 to 91%. These results indicate that the absence of a coordinating anion improved enantioselectivity. Entry 2 shows the extensively investigated *t*-BuBOX ligand with an *N*-acryloylthiazolidinone dienophile. With Cu^{2+} as the metal, the coordination geometry is square planar. The complex exposes the *re* face of the dienophile.

Entry 3 involves a catalyst derived from (*R,R*)-*trans*-cyclohexane-1,2-diamine. The square planar Cu^{2+} complex exposes the *re* face of the dienophile. As with the BOX catalysts, this catalyst has c_2 symmetry.

Scheme 6.4. Catalytic Enantioselective Diels-Alder Reactions

Entry	Dienophile	Diene	Catalyst	Amount	Product	Yield (%)	e.e.
1[a]			Ph Mg Ph	10 mol %		82	95
2[b]			*t*-Bu Cu *t*-Bu	10 mol %	CH₃	79	94
3[c]			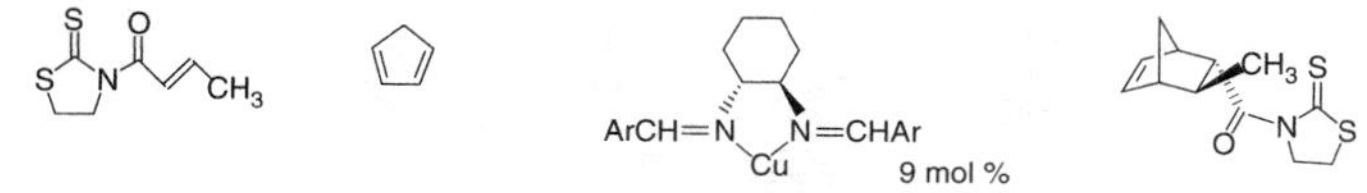 ArCH=N N=CHAr Cu; Ar =2,6-dichlorophenyl	9 mol %	CH₃	86	91
4[d]			Cu	10 mol %	CH₃	88	84
5[e]			TiCl₂	2 equiv	CH₃	93	92
6[f]			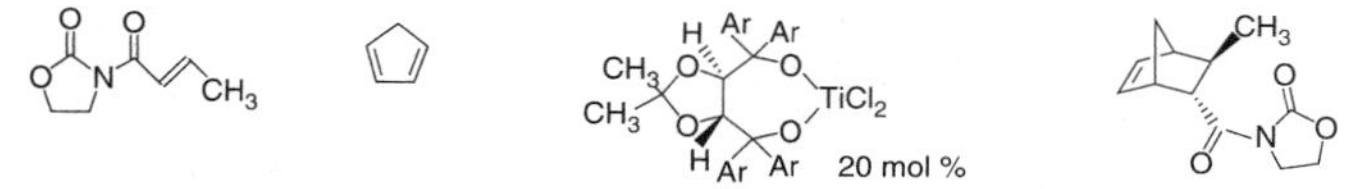 TiCl₂; Ar =2,6-dimethylphenyl	20 mol %	CH₃	92	93
7[g]			Ti(IV)			94	80
8[h]			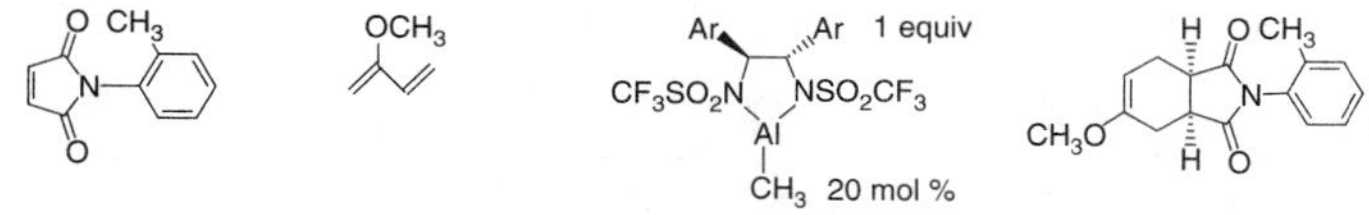 CF₃SO₂N NSO₂CF₃ Al CH₃; Ar =3,5-dimethylphenyl	1 equiv; 20 mol %		98	93
9[i]	CH₂=CCH=O, Br		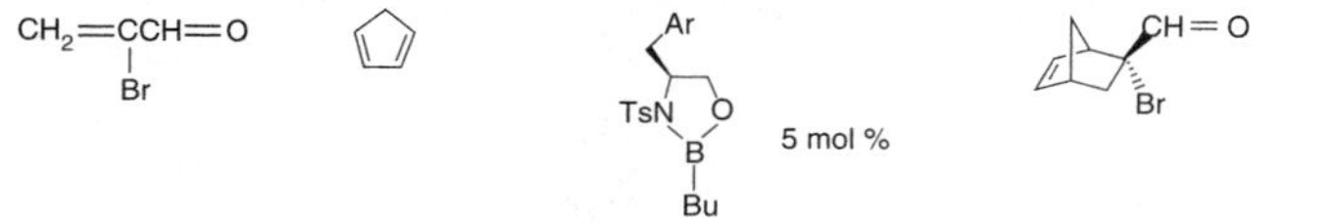 TsN B Bu; Ar = 3-indolyl	5 mol %	CH=O, Br		>99.5
10[j]			⁻N(SO₂CF₃)₂	20 mol %		97%	91%

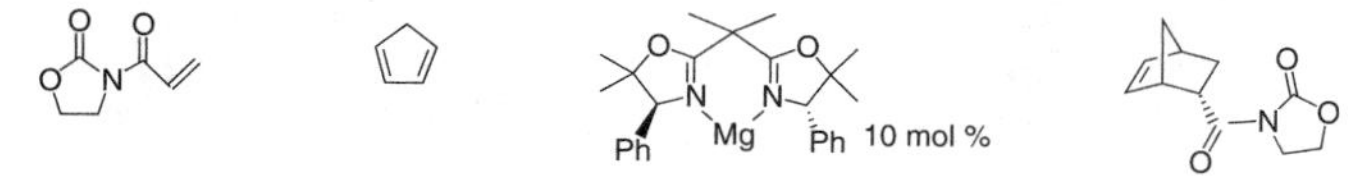

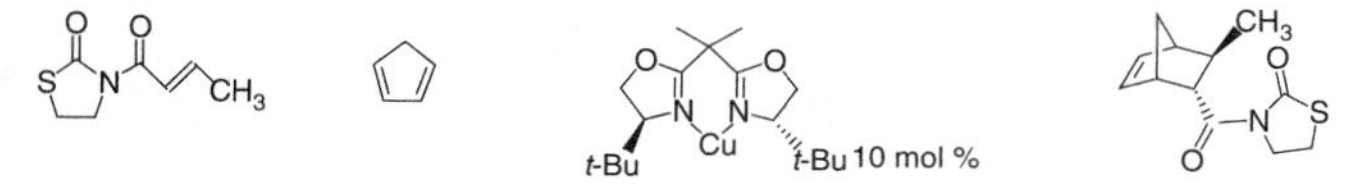

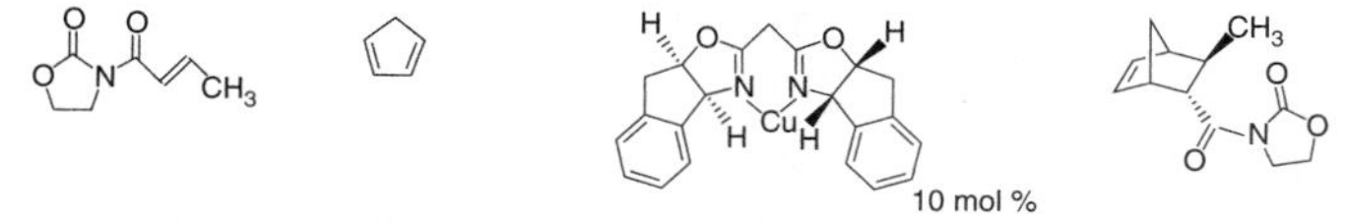

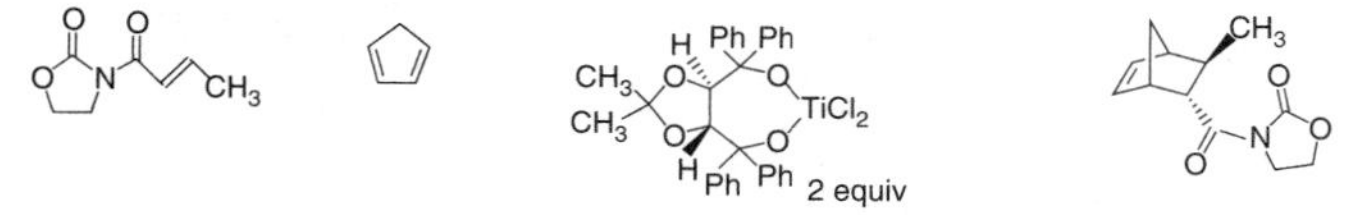

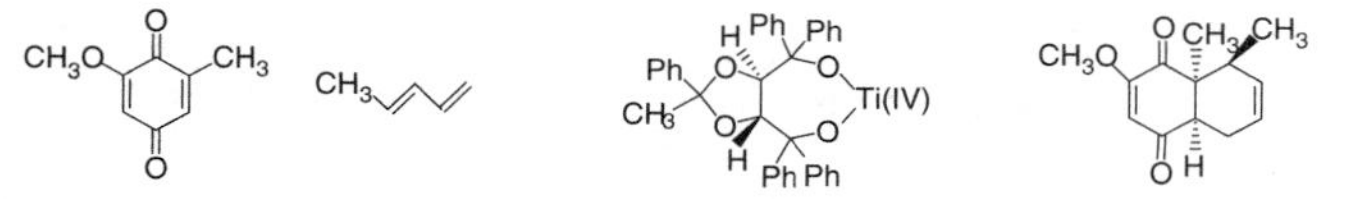

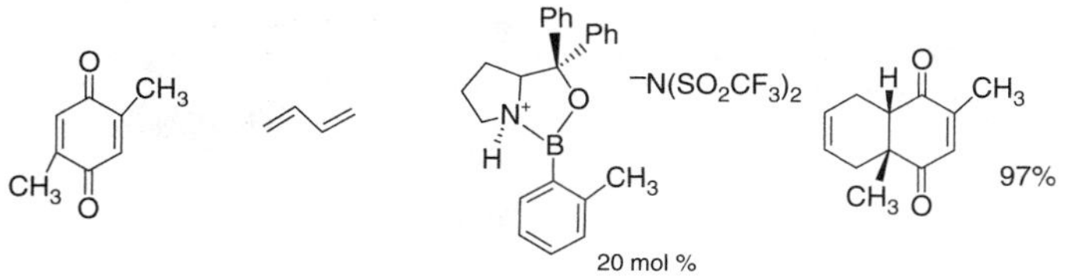

(Continued)

516

CHAPTER 6

*Concerted
Cycloadditions,
Unimolecular
Rearrangements, and
Thermal Eliminations*

Scheme 6.4. (*Continued*)

Entry	Dienophile	Diene	Catalyst	Amount	Product	Yield (%)	e.e.

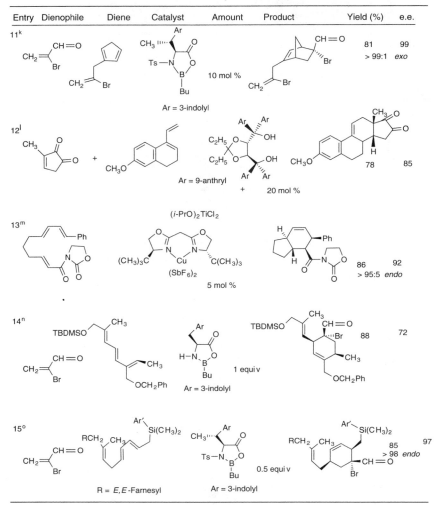

Entry 11[k] — Ar = 3-indolyl — 10 mol % — Yield 81, e.e. 99, > 99:1 *exo*

Entry 12[l] — Ar = 9-anthryl — + (i-PrO)$_2$TiCl$_2$ — 20 mol % — Yield 78, e.e. 85

Entry 13[m] — (CH$_3$)$_3$C ... Cu ... C(CH$_3$)$_3$ (SbF$_6$)$_2$ — 5 mol % — Yield 86, e.e. 92, > 95:5 *endo*

Entry 14[n] — Ar = 3-indolyl — 1 equiv — Yield 88, e.e. 72

Entry 15[o] — R = *E,E*-Farnesyl — Ar = 3-indolyl — 0.5 equiv — Yield 85, e.e. 97, > 98 *endo*

a. E. J. Corey and K. Ishihara, *Tetrahedron Lett.*, **33**, 6807 (1992).

b. D. A. Evans, S. J. Miller, and T. Lectka, *J. Am. Chem. Soc.*, **115**, 6460 (1993).

c. D. A. Evans, T. Lectka, and S. J. Miller, *Tetrahedron Lett.*, **34**, 7027 (1993).

d. A. K. Ghosh, H. Cho, and J. Cappiello, *Tetrahedron: Asymmetry*, **9**, 3687 (1998).

e. K. Narasaka, N. Iwasawa, M. Inoue, T. Yamada, M. Nakashima, and J. Sugimori, *J. Am. Chem. Soc.*, **111**, 5340 (1989).

f. E. J. Corey and Y. Matsumura, *Tetrahedron Lett.*, **32**, 6289 (1991).

g. T. A. Engler, M. A. Letavic, K. O. Lynch, Jr., and F. Takusagawa, *J. Org. Chem.*, **59**, 1179 (1994).

h. E. J. Corey, S. Sarshar, and D.-H. Lee, *J. Am. Chem. Soc.*, **116**, 12089 (1994).

i. E. J. Corey, T.-P. Loh, T. D. Roper, M. D. Azimioara, and M. C. Noe, *J. Am. Chem. Soc.*, **114**, 8290 (1992).

j. D. H. Ryu and E. J. Corey, *J. Am. Chem. Soc.*, **125**, 6388 (2003).

k. E. J. Corey, A. Guzman-Perez, and T.-P. Loh, *J. Am. Chem. Soc.*, **116**, 3611 (1994).

l. G. Quinkert, A. Del Grosso, A. Doering, and W. Doering, R. I. Schenkel, M. Bauch, G. T. Dambacher, J. W. Bats, G. Zimmerman, and G. Durrer, *Helv. Chim. Acta*, **78**, 1345 (1995).

m. D. A. Evans, D. M. Barnes, J. S. Johnson, T. Lectka, P. von Matt, S. J. Miller, J. A. Murry, R. D. Norcross, E. A. Shaugnessy and K. R. Campos, *J. Am. Chem. Soc.*, **121**, 7582 (1999).

n. J. A. Marshall and S. Xie, *J. Org. Chem.*, **57**, 2987 (1992).

o. T. W. Lee and E. J. Corey, *J. Am. Chem. Soc.*, **123**, 1872 (2001).

Entry 4 is a BOX-type catalyst derived from *cis*-1-aminoindan-2-ol. This is a somewhat more rigid ligand than the monocyclic BOX ligands. The chiral ligands in Entries 5 to 7 are TADDOLS (see p. 512) derived from tartaric acid. In Entry 5 the catalyst is prepared from $TiCl_2(O\text{-}i\text{-}Pr)_2$ and 4A molecular sieves. About 0.10 equivalent of the catalyst is used. In Entry 6, the catalyst was prepared using $Ti(O\text{-}i\text{-}Pr)_4$ and $SiCl_4$. In this catalyst, the aryl groups are carry 3,5-dimethyl groups. The 3,5-di-CF_3 and 3,5-di-Cl derivatives, which were also studied, gave high *exo:endo* ratios, but much reduced enantioselectivity. This is thought to be due to the reduced π donor character of the rings with EWG substituents. As mentioned on p. 513, the presence of chlorides at the Ti center is also probably an important factor in the reactivity of the catalyst.

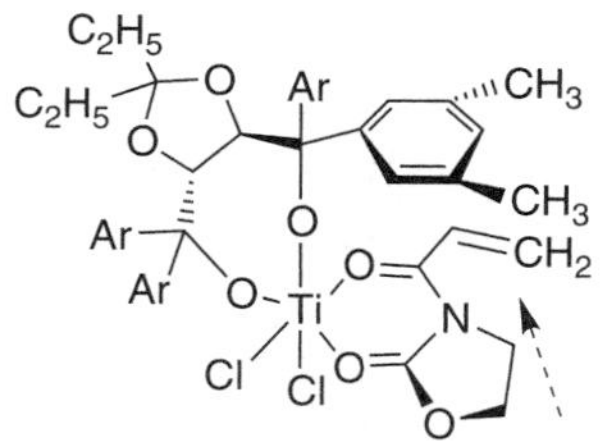

Ar = 3,5-dimethylphenyl

Entry 7 features a quinone dienophile. The reaction exhibits the expected selectivity for the more electrophilic quinone double bond (see p. 506). The reaction is also regioselective with respect to the diene, with the methyl group acting as a donor substituent. The enantioselectivity is 80%.

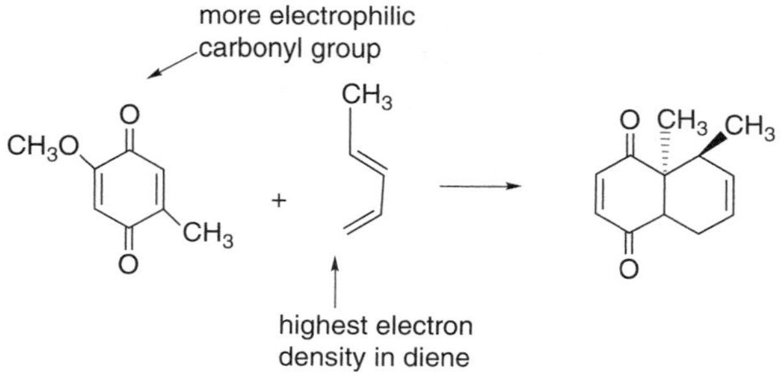

In this case, the catalyst was formed by premixing $Ti(O\text{-}i\text{-}Pr)_4$ and $TiCl_4$ and adding the TADDOL ligand. These conditions also gave good regioselectivity with isoprene, although the e.e. was not as high.

Entry 8 uses a *bis*-trifluoromethanesulfonamido chelate of methylaluminum as the catalyst. As in Entry 6, the use of a 3,5-dimethylphenyl group in place of phenyl improved enantioselectivity. The *ortho*-methylphenyl substituent on the maleimide dienophile restricts the potential coordination sites at the metal center. NMR characterization of the reactant-catalyst complex suggests that reaction occurs through the TS shown below.

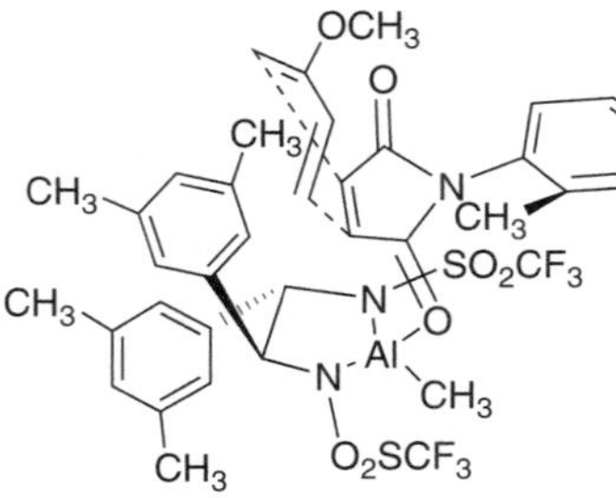

CHAPTER 6

*Concerted
Cycloadditions,
Unimolecular
Rearrangements, and
Thermal Eliminations*

Entry 9 uses the oxaborazolidine catalysts discussed on p. 505 with 2-bromopropenal as the dienophile. The aldehyde adopts the *exo* position in each case, which is consistent with the proposed TS model. Entry 10 illustrates the use of a cationic oxaborazolidine catalyst. The chirality is derived from *trans*-1,2-diaminocyclohexane. Entry 12 shows the use of a TADDOL catalyst in the construction of the steroid skeleton. Entry 13 is an intramolecular D-A reaction catalyzed by a Cu-*bis*-oxazoline. Entries 14 and 15 show the use of the oxazaborolidinone catalyst with more complex dienes.

6.1.7. Intramolecular Diels-Alder Reactions

Intramolecular Diels-Alder (IMDA) reactions are very useful in the synthesis of polycyclic compounds.[120] The stereoselectivity of a number of IMDA reactions has been analyzed and conformational factors in the TS often play the dominant role in determining product structure.[121] It has also been noted in certain systems that the stereoselectivity is influenced by the activating substituent on the dienophile double bond, both for thermal and Lewis acid–catalyzed reactions.[122] The general trends in regioselectivity are in agreement with frontier orbital concepts, with conformational effects being the main factors in determining stereoselectivity. Since the conformational interactions depend on the substituent pattern in the specific case, no general rules for stereoselectivity can be put forward. Molecular modeling can frequently identify the controlling structural features.[123]

It is possible to introduce substituents that can influence the conformational equilibria to favor a particular product. In the reactions shown below, the addition of the trimethylsilyl substituent leads to a single stereoisomer in 85% yield, whereas in the unsubstituted system two stereoisomers are formed in ratios from 4:1 to 8:1.[124]

120. W. Oppolzer, *Angew. Chem. Int. Ed. Engl.*, **16**, 10 (1977); G. Brieger and J. N. Bennett, *Chem. Rev.*, **80**, 63 (1980); E. Ciganek, *Org. React.*, **32**, 1 (1984); D. F. Taber, *Intramolecular Diels-Alder and Alder Ene Reactions*, Springer-Verlag, Berlin, 1984.
121. W. R. Roush, A. I. Ko, and H. R. Gillis, *J. Org. Chem.*, **45**, 4264 (1980); R. K. Boeckman, Jr., and S. K. Ko, *J. Am. Chem. Soc.*, **102**, 7146 (1980); W. R. Roush and S. E. Hall, *J. Am. Chem. Soc.*, **103**, 5200 (1981); K. A. Parker and T. Iqbal, *J. Org. Chem.*, **52**, 4369 (1987).
122. J. A. Marshall, J. E. Audia, and J. Grote, *J. Org. Chem.*, **49**, 5277 (1984); W. R. Roush, A. P. Essenfeld, and J. S. Warmus, *Tetrahedron Lett.*, **28**, 2447 (1987); T.-C. Wu and K. N. Houk, *Tetrahedron Lett.*, **26**, 2293 (1985).
123. K. J. Shea, L. D. Burke, and W. P. England, *J. Am. Chem. Soc.*, **110**, 860 (1988); L. Raimondi, F. K. Brown, J. Gonzalez, and K. N. Houk, *J. Am. Chem. Soc.*, **114**, 4796 (1992); D. P. Dolata and L. M. Harwood, *J. Am. Chem. Soc.*, **114**, 10738 (1992); F. K. Brown, U. C. Singh, P. A. Kollman, L. Raimondi, K. N. Houk, and C. W. Bock, *J. Org. Chem.*, **57**, 4862 (1992); J. D. Winkler, H. S. Kim, S. Kim, K. Ando, and K. N. Houk, *J. Org. Chem.*, **62**, 2957 (1997).
124. R. K. Boeckman, Jr., and T. E. Barta, *J. Org. Chem.*, **50**, 3421 (1985).

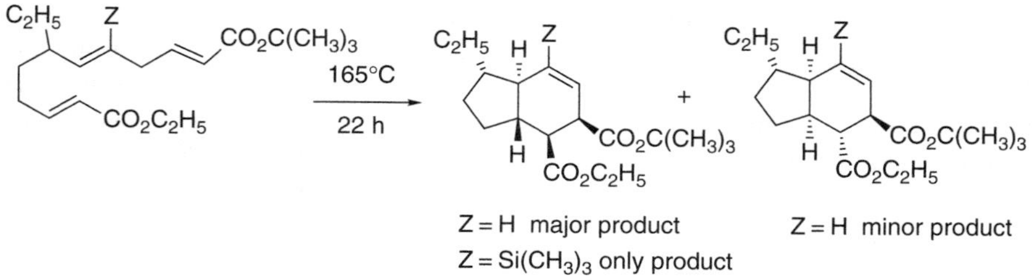

Z = H major product   Z = H minor product

Z = Si(CH$_3$)$_3$ only product

Similarly, the 2,8,10-triene **3a** gives a mixture of four isomers, but introduction of a TMS group as in **3b** gives a single stereoisomer in 89% yield. The reason for the improved stereoselectivity is that the steric effect introduced by the TMS substituent favors a single conformer.

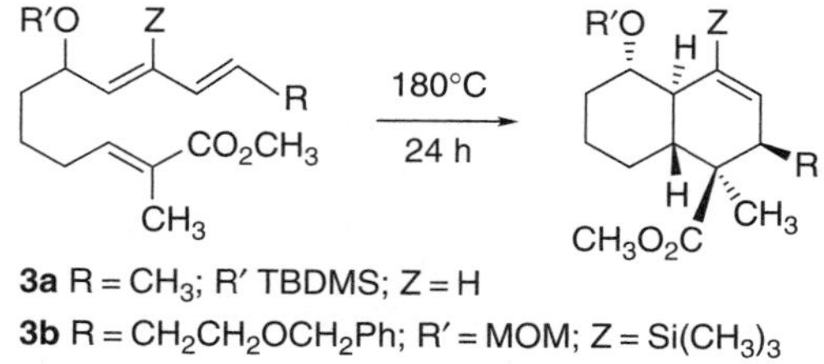

3a R = CH$_3$; R′ TBDMS; Z = H

3b R = CH$_2$CH$_2$OCH$_2$Ph; R′ = MOM; Z = Si(CH$_3$)$_3$

Lewis acid catalysis usually substantially improves the stereoselectivity of IMDA reactions, just as it does in intermolecular cases. For example, the thermal cyclization of **4** at 160°C gives a 50:50 mixture of two stereoisomers, but the use of (C$_2$H$_5$)$_2$AlCl as a catalyst permits the reaction to proceed at room temperature and *endo* addition is favored by 7:1.[125]

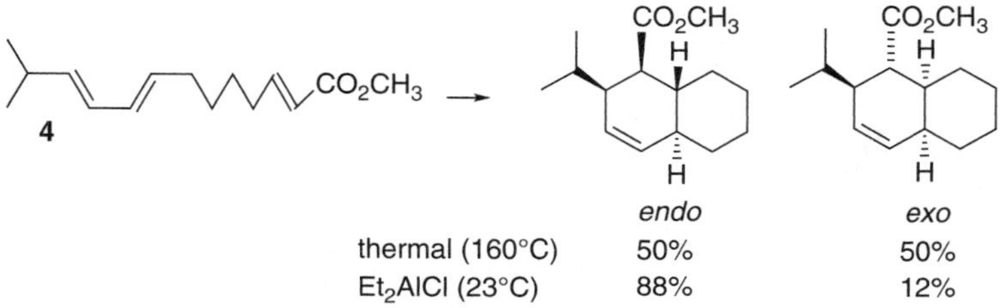

	endo	*exo*
thermal (160°C)	50%	50%
Et$_2$AlCl (23°C)	88%	12%

There has been quite thorough study of 3,5-hexadienyl acrylates, where the ester functions both as part of the link and an activating substituent. The reaction tends to be quite slow, even though at first glance it would appear to encounter little strain. The *cis* ring juncture is favored by 9:1.

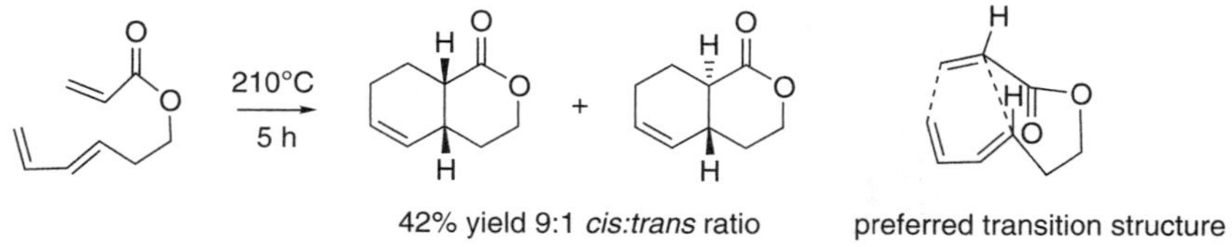

42% yield 9:1 *cis:trans* ratio   preferred transition structure

Ref. 126

125. W. R. Roush and H. R. Gillis, *J. Org. Chem.*, **47**, 4825 (1982).
126. S. F. Martin, S. A.Williamson, R. P. Gist, and K. M. Smith, *J. Org. Chem.*, **48**, 5170 (1983).

520

CHAPTER 6

*Concerted
Cycloadditions,
Unimolecular
Rearrangements, and
Thermal Eliminations*

One factor that is believed to contribute to the sluggishness of the reaction is that a chairlike arrangement of the linking group causes a twist in the ester group from the preferred planarity. The TS also requires that the ester alkyl group be in an *anti* relationship to the carbonyl group, rather than the preferred *syn* conformation. Several substituted systems have been studied and they react primarily through a boatlike *endo* TS.[127] The size of the α-substituent R controls the degree of preference for the TS.

R = H	R = CH_3	R = C(CH_3)_3
52% yield; 6:4 *cis:trans* ring junction	76% yield; 5:1 *cis:trans* ring junction	55% yield, only *cis* ring junction

This system has been studied computationally at the B3LYP/6-31 + G* level.[128] In agreement with the experimental results, the *endo* boat TS was found to be the most stable. The *endo* chair and *exo* boat were about 1.3 kcal/mol higher in energy, and the *exo* chair still higher. This study confirmed that the boatlike TS allows the ester group to stay closer to planarity. Eclipsing interactions also contribute to the higher energy of the chairlike TS. In accordance with the idea that a bidentate Lewis acid might both effect Lewis acid catalysis and promote a planar geometry at the ester group, it was found that the reaction could be effectively catalyzed by a bidentate Lewis acid.[129] Use of one equivalent of the catalyst gave 95% yield after 2 h at 0° C. The catalyst is believed to be coordinated with both the carbonyl and the ester oxygens.

Some examples of IMDA reactions are given in Scheme 6.5. In Entry 1 the dienophilic portion bears a carbonyl substituent and cycloaddition occurs easily. Two stereoisomeric products are formed, but both have the *cis* ring fusion, which is the stereochemistry expected for an *endo* TS, with the major diastereomer being formed from the TS with an equatorial isopropyl group.

127. M. E. Jung, A. Huang, and T. W. Johnson, *Org. Lett.*, **2**, 1835 (2000); P. Kim, M. H. Nantz, M. J. Kurth, and M. M. Olmstead, *Org. Lett.*, **2**, 1831 (2000).
128. D. J. Tantillo, K. N. Houk, and M. E. Jung, *J. Org. Chem.*, **66**, 1938 (2001).
129. A. Saito, H. Ito, and T. Taguchi, *Org. Lett.*, **4**, 4619 (2002).

Scheme 6.5. Intramolecular Diels-Alder Reactions

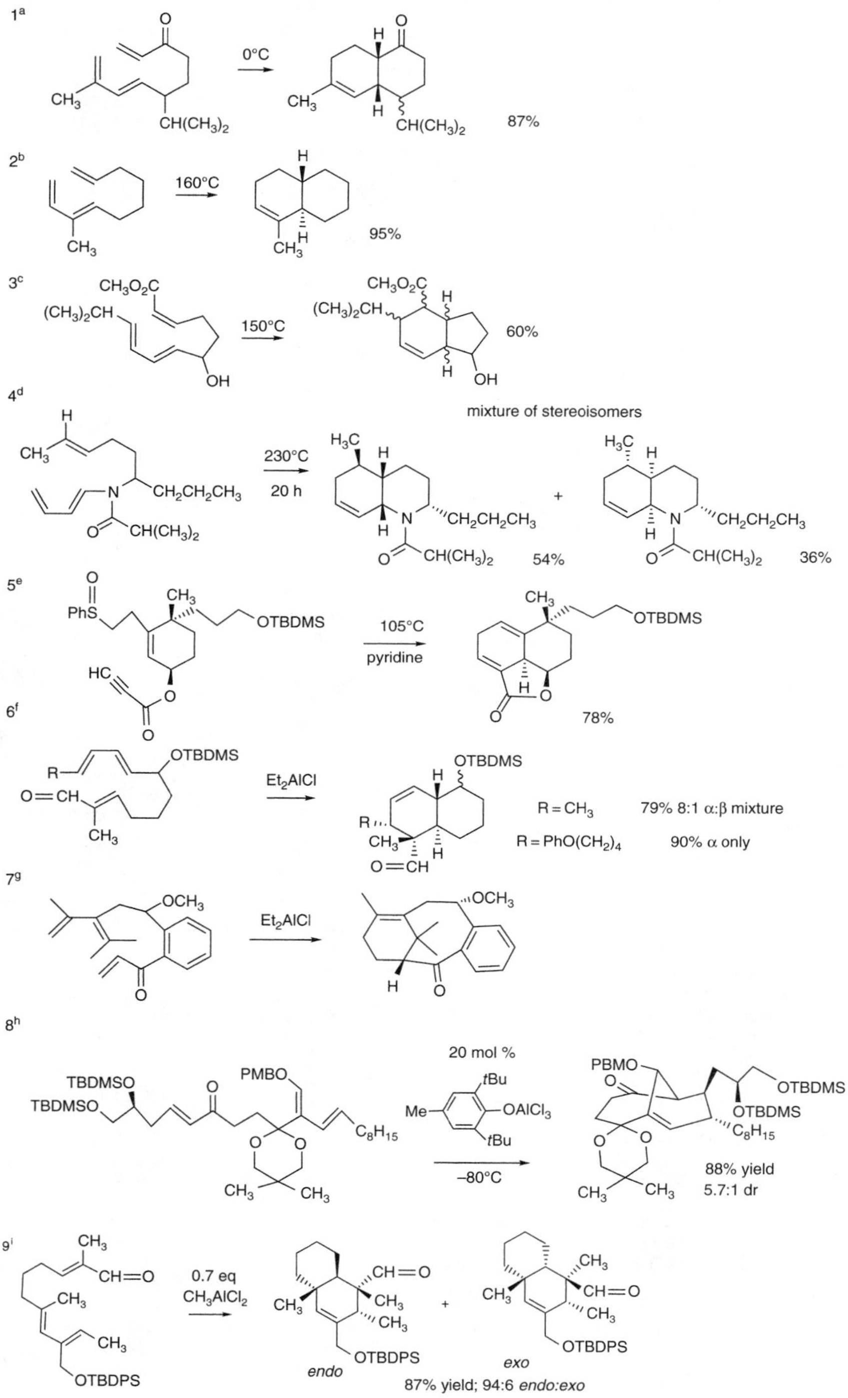

1[a]

0°C

87%

2[b]

160°C

95%

3[c]

150°C

60%

mixture of stereoisomers

4[d]

230°C
20 h

54%   +   36%

5[e]

105°C
pyridine

78%

6[f]

Et₂AlCl

R = CH₃   79% 8:1 α:β mixture
R = PhO(CH₂)₄   90% α only

7[g]

Et₂AlCl

8[h]

20 mol %

−80°C

88% yield
5.7:1 dr

9[i]

0.7 eq
CH₃AlCl₂

endo   +   exo

87% yield; 94:6 *endo:exo*

(*Continued*)

522

CHAPTER 6

*Concerted
Cycloadditions,
Unimolecular
Rearrangements, and
Thermal Eliminations*

Scheme 6.5. (*continued*)

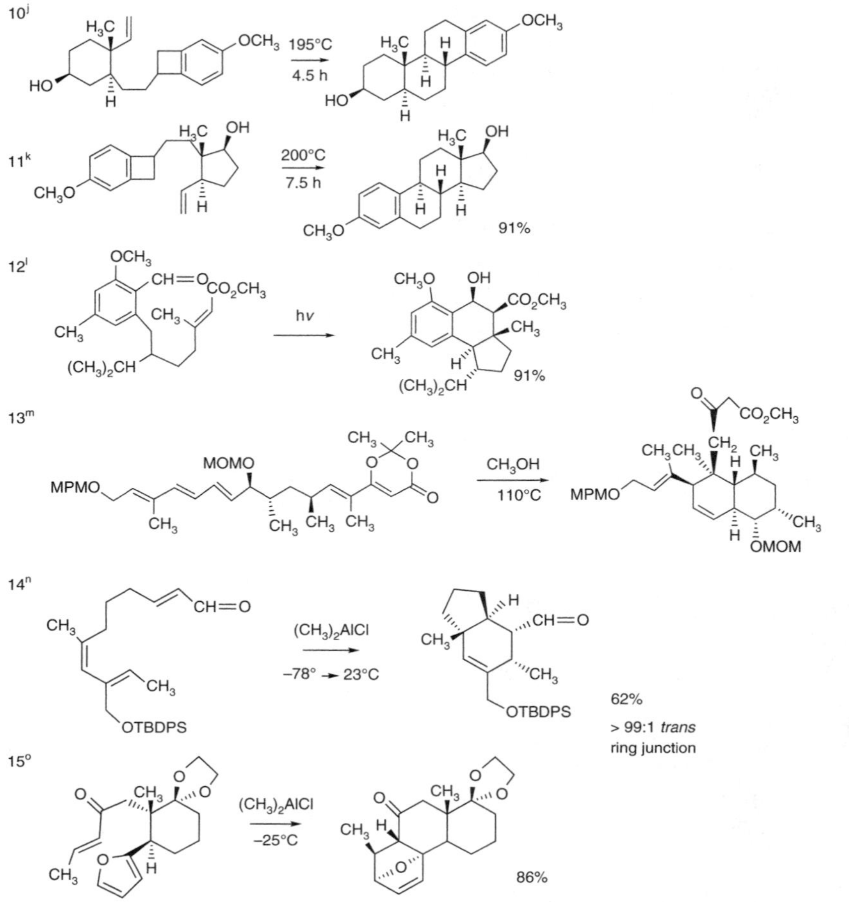

a. D. F. Taber and B. P. Gunn, *J. Am. Chem. Soc.*, **101**, 3992 (1979).
b. S. R. Wilson and D. T. Mao, *J. Am. Chem. Soc.*, **100**, 6289 (1978).
c. W. R. Roush, *J. Am. Chem. Soc.*, **102**, 1390 (1980).
d. W. Oppolzer and E. Flaskamp, *Helv. Chim. Acta*, **60**, 204 (1977); W. Oppolzer, E. Flaskamp, and L. W. Bieber, *Helv. Chim. Acta*, **84**, 141 (2001).
e. H. Miyaoka, Y. Kajiwara, and Y. Yamada, *Tetrahedron Lett.*, **41**, 911 (2000).
f. J. A. Marshall, J. E. Audia, and J. Grote, *J. Org. Chem.*, **49**, 5277 (1984).
g. D. V. Smil, A. Laurent, N. S. Spassova, and A. G. Fallis, *Tetrahedron Lett.*, **44**, 5129 (2003).
h. K. C. Nicolaou, J. Jung, W. H. Yoon, K. C. Fong, H.-S. Choi, Y. He, Y.-L. Zhong, and P. S. Baran, *J. Am. Chem. Soc.*, **124**, 2183 (2002).
i. N. A. Yakelis and W. R. Roush, *Org. Lett.*, **3**, 957 (2001).
j. T. Kametani, K. Suzuki, and H. Nemoto, *J. Org. Chem.*, **45**, 2204 (1980); *J. Am. Chem. Soc.*, **103**, 2890 (1981).
k. P. A. Grieco, T. Takigawa, and W. J. Schillinger, *J. Org. Chem.*, **45**, 2247 (1980).
l. K. C. Nicolaou, D. Gray, and J. Tae, *Angew. Chem. Int. Ed. Engl.*, **40**, 3679 (2001); K. C. Nicolaou, D. L. F. Gray, and J. Tae, *J. Am. Chem. Soc.*, **126**, 613 (2004).
m. R. K. Boeckman, Jr., T. E. Barta, and S. G. Nelson, *Tetrahedron Lett.*, **32**, 4091 (1991).
n. N. A. Yakelis and W. R. Roush, *Org. Lett.*, **3**, 957 (2001).
o. S. Claeys, D. Van Haver, P. J. De Clerc, M. Milanesio, and D. Viterbo, *Eur. J. Org. Chem.*, 1051 (2002).

In Entry 2 a similar triene that lacks the activating carbonyl group undergoes reaction but a much higher temperature is required. In this case the ring junction is *trans*, which corresponds to an *exo* TS and may reflect the absence of secondary orbital interaction between the diene and dienophile.

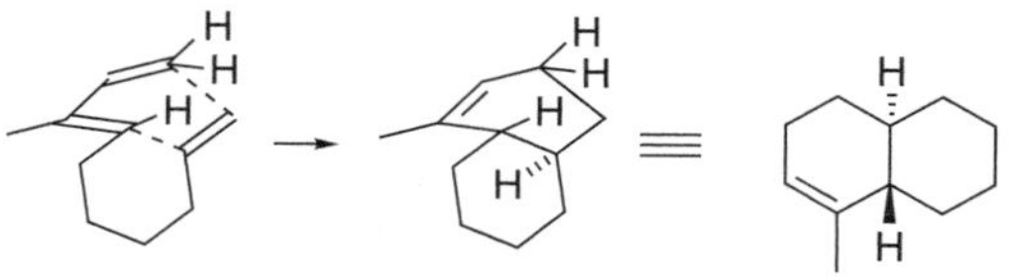

In Entry 3 the dienophilic double bond bears an EWG substituent, but a higher temperature is required than for Entry 1 because the connecting chain contains one less methylene group, which leads to a more strained TS. A mixture of stereoisomers is formed, reflecting a conflict between the Alder rule, which favors *endo* addition, and conformational factors, which favor the *exo* TS. The reaction in Entry 4 was carried out as a key step in the synthesis of the frog neurotoxin, pumiliotoxin C. The isolated double bond has no activating substituents and the reaction requires forcing conditions. Nevertheless, the yield is excellent and both products are formed with a *cis* ring juncture, but there is minimal facial selectivity. In Entry 5, the diene system is generated in situ by thermal elimination of the sulfoxide group and then reacts with the acetylenic dienophile.

Entry 6 shows a stereoselective formation of a highly substituted *trans*-decalin system. The reaction in Entry 7 establishes a taxanelike structure. The stereochemistry is consistent with a TS in which both the carbonyl oxygen and the methoxy group are coordinated to aluminum.

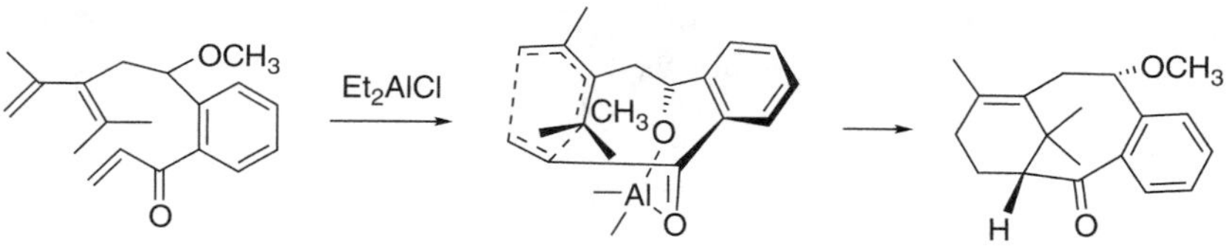

The reaction in Entry 8 was used in the synthesis of members of the phomoidrides. The cyclohexene ring that is constructed creates a bicyclo[4.3.1]skeleton containing seven- and nine-membered rings.

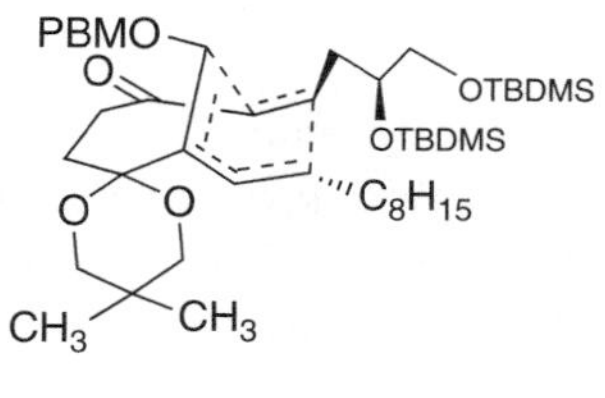

Entry 9 is a Lewis acid–catalyzed example, and the major stereoisomer is formed through a TS having an *endo* orientation of the complexed formyl group. Interestingly, the thermal version of this reaction favors the *exo* stereoisomer.

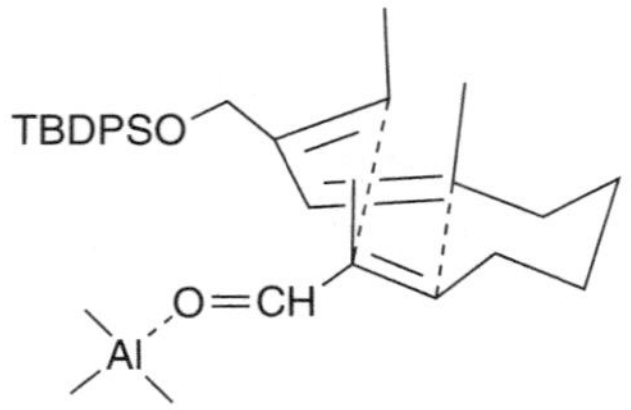

524

CHAPTER 6

*Concerted
Cycloadditions,
Unimolecular
Rearrangements, and
Thermal Eliminations*

Entries 10 and 11 are examples of reactions involving thermal generation of quinodimethanes. In Entry 12 a quinodimethane is generated by photoenolization and used in conjunction with an IMDA reaction to create the carbon skeleton found in the hamigerans, which are marine natural products having antiviral activity.

In Entry 13, the dioxinone ring undergoes thermal decomposition to an acyl ketene that is trapped by the solvent methanol. The resulting β-keto-γ,δ-enoate ester then undergoes stereoselective cyclization. The stereoselectivity is controlled by the preference for pseudoequatorial conformations of the C(6) and C(9) substituents.

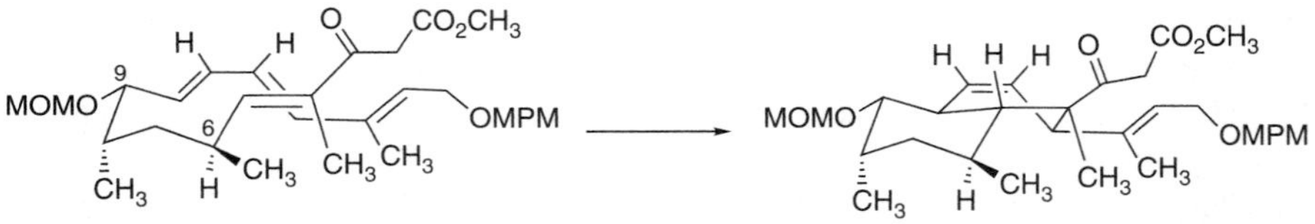

Entry 14 forms a *trans* ring juncture with greater than 99:1 selectivity. In contrast, the thermal reaction in this case shows a 2:1 preference for the *cis* ring juncture. Evidently the Lewis acid changes the structure of the TS sufficiently that the steric effects that control the thermal reaction are diminished.

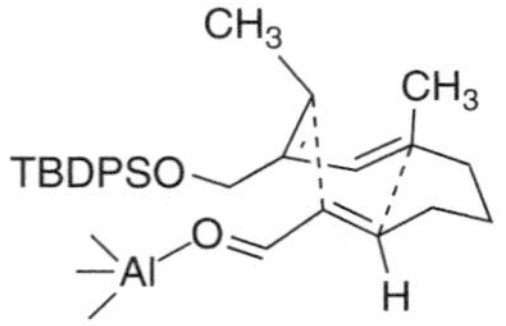

Entry 15 creates a portion of the steroid skeleton and also illustrates the use of a furan ring as a diene.

As in intermolecular reactions, enantioselectivity can be achieved in IMDA additions by use of chiral components. For example, the dioxolane ring in **5** and **6** results in TS structures that lead to enantioselective reactions.[130] The chirality in the dioxolane ring is reflected in the respective TSs, both of which have an *endo* orientation of the carbonyl group.

[130] T. Wong, P. D. Wilson, S. Woo, and A. G. Fallis, *Tetrahedron Lett.*, **40**, 7045 (1997).

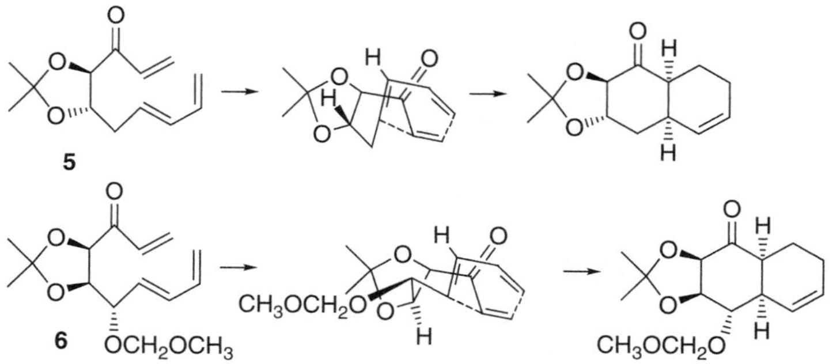

5

6

Chiral catalysts (see Section 6.1.6) can also achieve enantioselectivity in IMDA reactions.

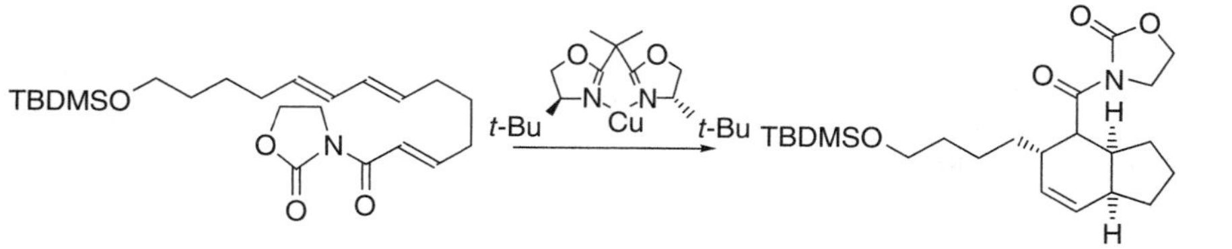

96% e.e.

Ref. 131

The kinetic advantages of IMDA additions can be exploited by installing temporary links (tethers) between the diene and dienophile components.[132] After the addition reaction, the tether can be broken. Siloxy derivatives have been used in this way, since silicon-oxygen bonds can be readily cleaved by solvolysis or by fluoride ion.[133] The silyl group can also be used to introduce a hydroxy function by oxidation.

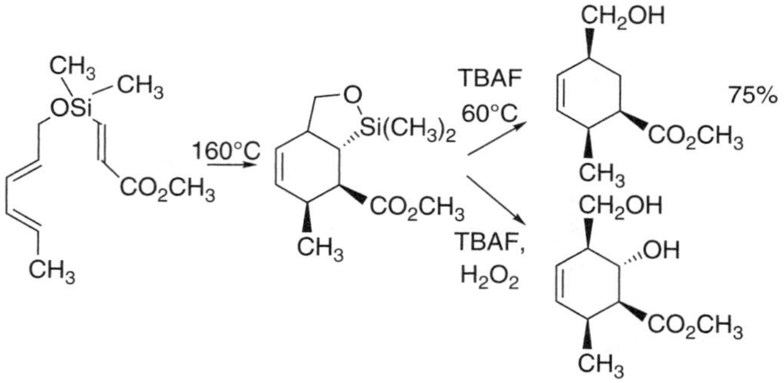

Ref. 133a

131. D. A. Evans and J. S. Johnson, *J. Org. Chem.*, **62**, 786 (1997).

132. L. Fensterbank, M. Malacria, and S. McN. Sieburth, *Synthesis*, 813 (1997); M. Bols and T. Skrydstrup, *Chem. Rev.*, **95**, 1253 (1995).

133. (a) G. Stork, T. Y. Chan, and G. A. Breault, *J. Am. Chem. Soc.*, **114**, 7578 (1992); (b) S. McN. Sieburth and L. Fensterbank, *J. Org. Chem.*, **57**, 5279 (1992); (c) J. W. Gillard, R. Fortin, E. L. Grimm, M. Maillard, M. Tjepkema, M. A. Bernstein, and R. Glaser, *Tetrahedron Lett.*, **32**, 1145 (1991); (d) D. Craig and J. C. Reader, *Tetrahedron Lett.*, **33**, 4073 (1992).

526

CHAPTER 6

*Concerted
Cycloadditions,
Unimolecular
Rearrangements, and
Thermal Eliminations*

Ref. 133d

Acetals have also been used as removable tethers.

2.7:1

Ref. 134

The activating capacity of boronate groups can be combined with the ability for facile transesterification at boron to permit intramolecular reactions between vinyl-boronates and 2,4-dienols.

Ref. 135

6.2. 1,3-Dipolar Cycloaddition Reactions

In Chapter 10 of Part A, the mechanistic classification of *1,3-dipolar cycloadditions* as concerted cycloadditions was developed. Dipolar cycloaddition reactions are useful both for syntheses of heterocyclic compounds and for carbon-carbon bond formation. Table 6.2 lists some of the types of molecules that are capable of dipolar cycloaddition. These molecules, which are called *1,3-dipoles*, have π electron systems that are isoelectronic with allyl or propargyl anions, consisting of two filled and one empty orbital. Each molecule has at least one charge-separated resonance structure with opposite charges in a 1,3-relationship, and it is this structural feature that leads to the name 1,3-dipolar cycloadditions for this class of reactions.[136]

134. P. J. Ainsworth, D. Craig, A. J. P. White, and D. J. Williams, *Tetrahedron*, **52**, 8937 (1996).

135. R. A. Batey, A. N. Thadani, and A. J. Lough, *J. Am. Chem. Soc.*, **121**, 450 (1999).

136. For comprehensive reviews of 1,3-dipolar cycloaddition reactions, see R. Huisgen, R. Grashey and J. Sauer in *The Chemistry of Alkenes*, S. Patai, ed., Interscience London, 1965, pp. 806–878; G. Bianchi, C. DeMicheli, and R. Gandolfi, in *The Chemistry of Double Bonded Functional Groups*, Part I, Supplement A, S. Patai, ed., Wiley-Interscience, New York, 1977, pp. 369–532; A. Padwa, ed., *1,3-Dipolar Cycloaddition Chemistry*, Wiley, New York, 1984.

Table 6.2. 1,3-Dipolar Compounds

:N̈=N̈—C̈R₂ ↔ :N≡N̈—C̈R₂		Diazoalkane
:N̈=N̈—N̈R ↔ :N≡N̈—N̈R		Azide
RC̈=N—C̈R₂ ↔ RC≡N̈—C̈R₂		Nitrile ylide
RC̈=N̈—N̈R ↔ RC≡N̈—N̈R		Nitrile imine
RC̈=N—Ö: ↔ RC≡N̈—Ö:		Nitrile oxide
R₂C̈—N̈—C̈R₂ (R) ↔ R₂C=N(R)—C̈R₂		Azomethine ylide
R₂C̈—N̈(R)—Ö: ↔ R₂C=N(R)—Ö:		Nitrone
R₂C̈—Ö—Ö: ↔ R₂C=Ö—Ö:		Carbonyl oxide

The other reactant in a dipolar cycloaddition, usually an alkene or alkyne, is referred to as the *dipolarophile*. Other multiply bonded functional groups such as imine, azo, and nitroso can also act as dipolarophiles. The 1,3-dipolar cycloadditions involve four π electrons from the 1,3-dipole and two from the dipolarophile. As in the D-A reaction, the reactants approach one another in parallel planes to permit interaction between the π and π* orbitals.

Mechanistic studies have shown that the TSs for 1,3-dipolar cycloadditions (1,3-DCA) are not very polar, the rate of reaction is not strongly sensitive to solvent polarity, and in most cases the reaction is a concerted $[2\pi_s + 4\pi_s]$ cycloaddition.[137] The destruction of charge separation that is implied is more apparent than real because

---

[137.] P. K. Kadaba, *Tetrahedron*, **25**, 3053 (1969); R. Huisgen, G. Szeimes, and L. Mobius, *Chem. Ber.*, **100**, 2494 (1967); P. Scheiner, J. H. Schomaker, S. Deming, W. J. Libbey, and G. P. Nowack, *J. Am. Chem. Soc.*, **87**, 306 (1965).

528

CHAPTER 6

Concerted
Cycloadditions,
Unimolecular
Rearrangements, and
Thermal Eliminations

most 1,3-dipolar compounds are not highly polar. The polarity implied by any single structure is balanced by other contributing structures.

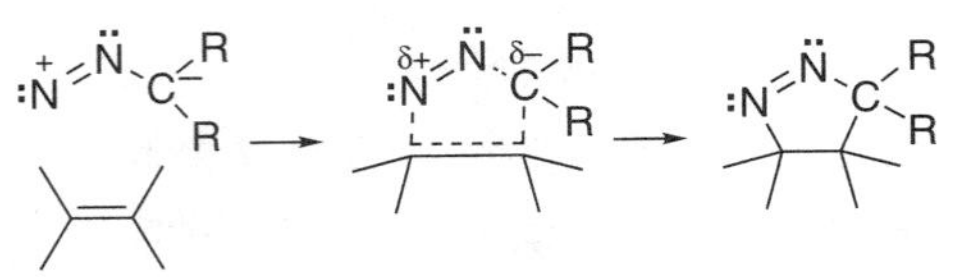

6.2.1. Regioselectivity and Stereochemistry

Two issues are essential for predicting the structure of 1,3-DCA products: (1) What is the regiochemistry? and (2) What is the stereochemistry? Many specific examples demonstrate that 1,3-dipolar cycloaddition is a stereospecific *syn* addition with respect to the dipolarophile, as expected for a concerted process.

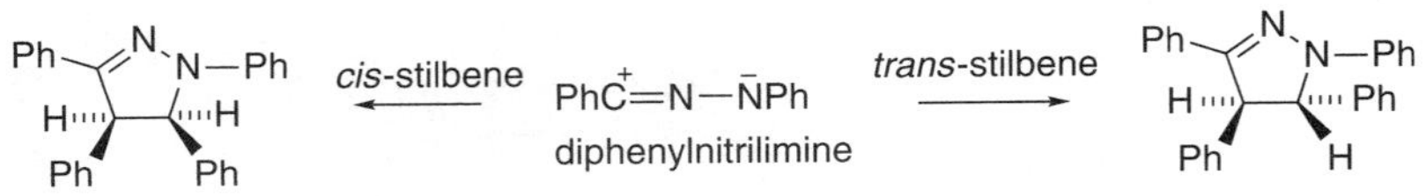

Ref. 138

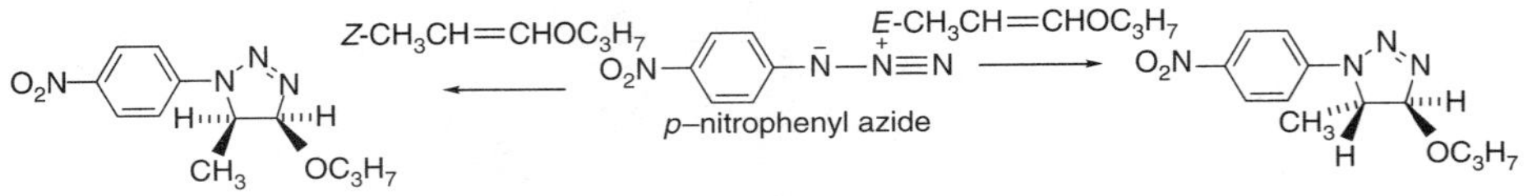

Ref. 139

With some 1,3-dipoles, two possible stereoisomers can be formed by *syn* addition. These result from two differing orientations of the reacting molecules that are analogous to the *endo* and *exo* TS in D-A reactions. Phenyldiazomethane, for example, can add to unsymmetrical dipolarophiles to give two diastereomers.

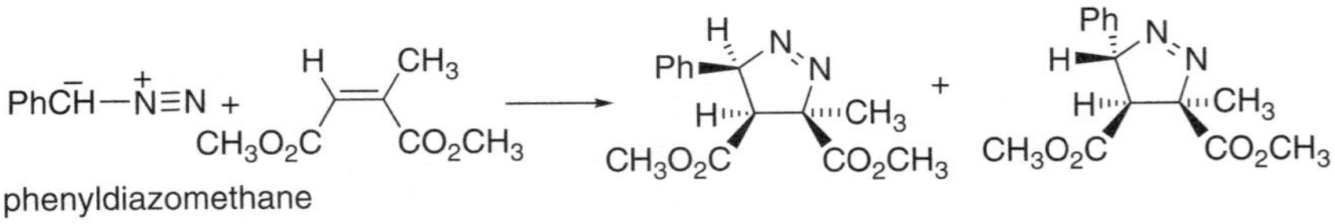

Ref. 140

Each 1,3-dipole exhibits a characteristic regioselectivity toward different types of dipolarophiles. The dipolarophiles can be grouped, as were dienophiles, depending upon whether they have ERG or EWG substituents. The regioselectivity can be

[138.] R. Huisgen, M. Seidel, G. Wallibillich, and H. Knupfer, *Tetrahedron*, **17**, 3 (1965).
[139.] R. Huisgen and G. Szeimies, *Chem. Ber.*, **98**, 1153 (1965).
[140.] R. Huisgen and P. Eberhard, *Tetrahedron Lett.*, 4343 (1971).

interpreted in terms of frontier orbital theory. Depending on the relative orbital energies in the 1,3-dipole and dipolarophile, the strongest interaction may be between the HOMO of the dipole and the LUMO of the dipolarophile or vice versa. Usually for dipolarophiles with EWGs the dipole-HOMO/dipolarophile-LUMO interaction is dominant. The reverse is true for dipolarophiles with ERG substituents. In some circumstances the magnitudes of the two interactions may be comparable.[141] When HOMO-LUMO interactions control regioselectivity, the reaction is said to be under *electronic control*. If steric effects are dominant, the reaction is under *steric control*.

The prediction of regiochemistry requires estimation or calculation of the energies of the orbitals that are involved, which permits identification of the frontier orbitals. The energies and orbital coefficients for the most common dipoles and dipolarophiles have been summarized.[141] Figure 10.15 of Part A gives the orbital coefficients of some representative 1,3-dipoles. Regioselectivity is determined by the preference for the orientation that results in bond formation between the atoms having the largest coefficients in the two frontier orbitals. This analysis is illustrated in Figure 6.12.

Apart from the role of substituents in determining regioselectivity, several other structural features affect the reactivity of dipolarophiles. Strain increases reactivity; norbornene, for example, is consistently more reactive than cyclohexene in 1,3-DCA reactions. Conjugated functional groups usually increase reactivity. This increased reactivity has most often been demonstrated with electron-attracting substituents, but for some 1,3-dipoles, enol ethers, enamines, and other alkenes with donor substituents are also quite reactive. Some reactivity data for a series of alkenes with several 1,3-dipoles are given in Table 10.6 of Part A. Additional discussion of these reactivity trends can be found in Section 10.3.1 of Part A.

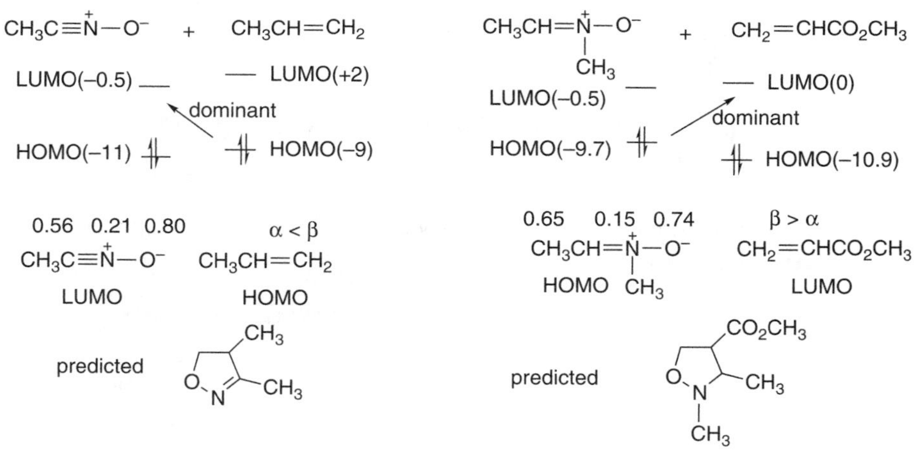

Fig. 6.12. Prediction of regioselectivity of 1,3-dipolar cycloaddition on the basis of FMO theory. The energies of the HOMO and LUMO of the reactants (in eV) are indicated in parentheses.

141. K. N. Houk, J. Sims, B. E. Duke, Jr., R. W. Strozier, and J. K. George, *J. Am. Chem. Soc.*, **95**, 7287 (1973); I. Fleming, *Frontier Orbitals and Organic Chemical Reactions*, Wiley, New York, 1977; K. N. Houk, in *Pericyclic Reactions*, Vol. II, A. P. Marchand and R. E. Lehr, eds., Academic Press, New York, 1977, pp. 181–271.

530

CHAPTER 6

*Concerted
Cycloadditions,
Unimolecular
Rearrangements, and
Thermal Eliminations*

1,3-Dipoles can be embedded in heterocyclic structures, just as diene units are present in pyrones and other ring structures (see p. 491). N-Substituted pyridinium-3-ols can be deprotonated to give 3-oxidopyridinium betaines that have 1,3-dipolar character.[142]

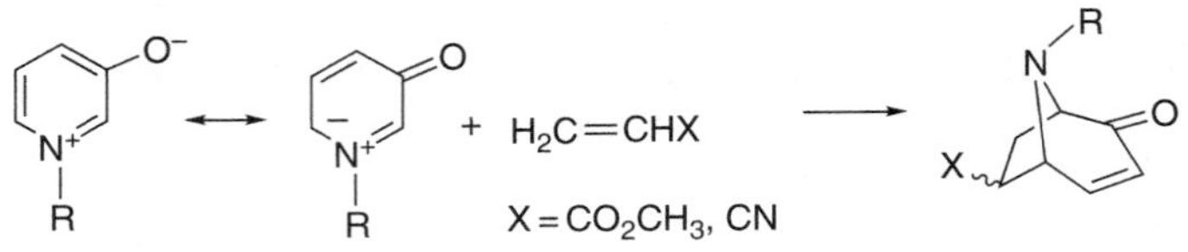

A reaction of this type was used to prepare an intermediate in the synthesis of a natural compound with antiglaucoma activity.[143]

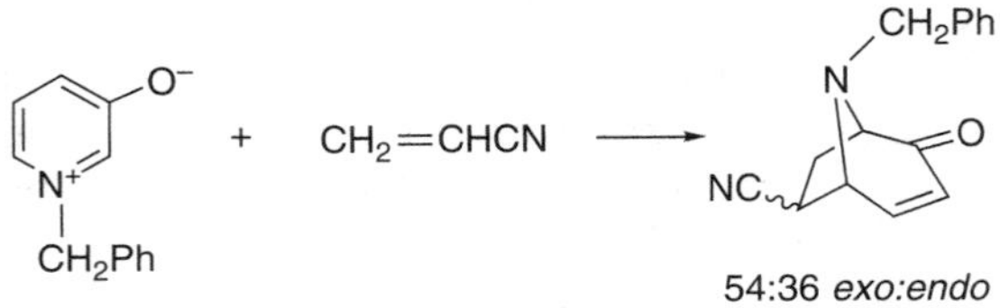

54:36 *exo:endo*

Oxazolium oxides, which can be generated by cyclization of α-amido acids, give pyrroles on reaction with acetylenic dipolarophiles.[144] These reactions proceed by formation of oxazolium oxide intermediates. The bicyclic adduct can then undergo a concerted (retro 4 + 2) decarboxylation.

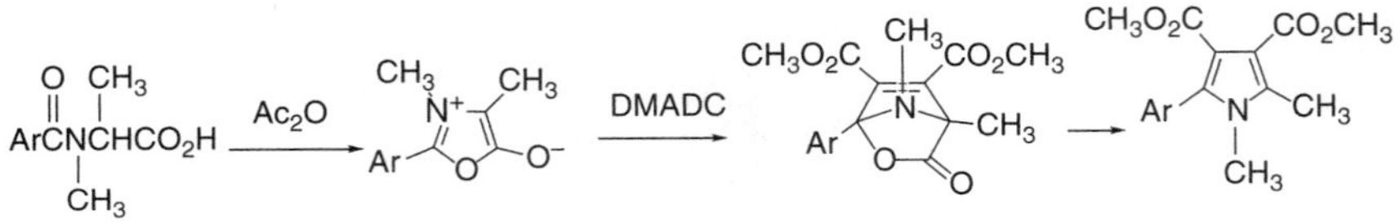

Oxazolium oxides can also be made by *N*-alkylation of oxazolinones.[145]

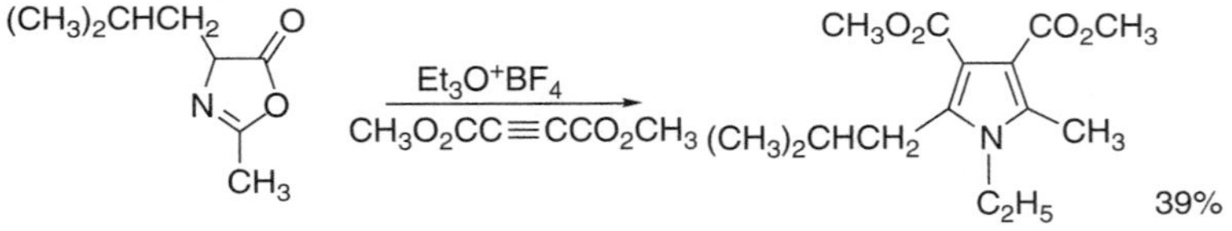

39%

Pyrroles are also formed from dipolarophiles such as α-acetoxy esters and α-chloroacrylonitrile that have potential leaving groups.

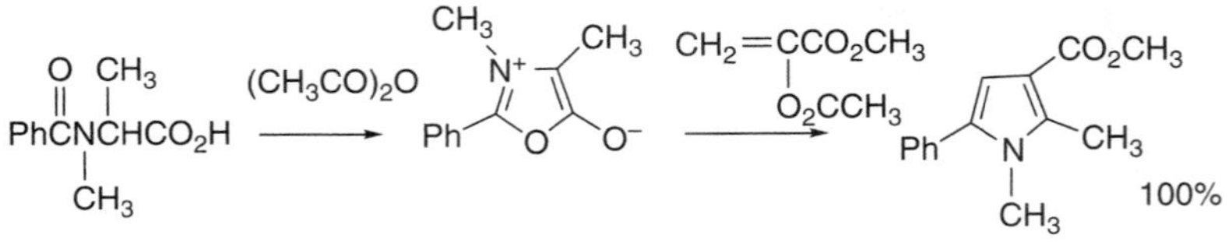

100%

Ref. 146

142. N. Dennis, A. R. Katritzky, and Y. Takeuchi, *Angew. Chem. Int. Ed. Engl.*, **15**, 1 (1976).

143. M. E. Jung, Z. Longmei, P. Tangsheng, Z. Huiyan, L. Yan, and S. Jingyu, *J. Org. Chem.*, **57**, 3528 (1992).

144. H. Gotthardt, R. Huisgen, and H. O. Bayer, *J. Am. Chem. Soc.*, **92**, 4340 (1970).

145. F. M. Hershenson and M. R. Pavia, *Synthesis*, 999 (1988).

146. G. Grassi, F. Foti, F. Risitano, and D. Zona, *Tetrahedron Lett.*, **46**, 1061 (2005).

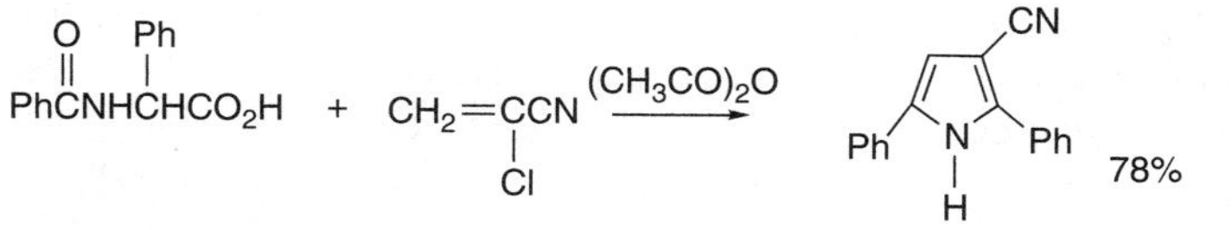

Ref. 147

Another interesting variation of the 1,3-dipolar cycloaddition involves generation of 1,3-dipoles from three-membered rings. As an example, aziridines **7** and **8** give adducts derived from apparent formation of 1,3-dipoles **9** and **10**, respectively.[148]

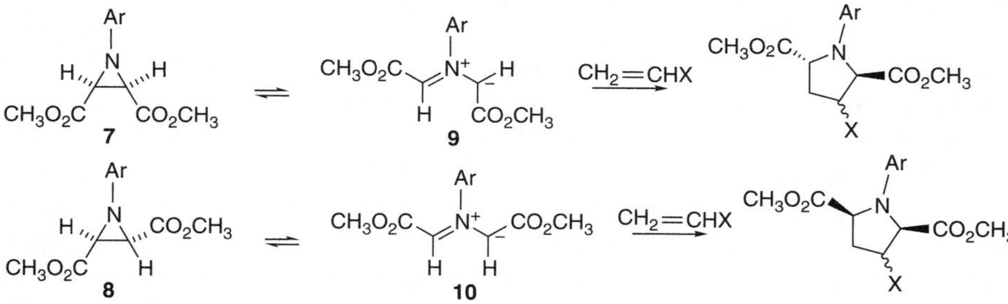

The evidence for the involvement of 1,3-dipoles as discrete intermediates includes the observation that the reaction rates are independent of dipolarophile concentration. This fact indicates that the ring opening is the rate-determining step in the reaction. Ring opening is most facile for aziridines that have an electron-attracting substituent to stabilize the carbanion center in the dipole.

6.2.2. Synthetic Applications of Dipolar Cycloadditions

1,3-DCA reactions are an important means of synthesis of a wide variety of heterocyclic molecules, some of which are useful intermediates in multistage syntheses. Pyrazolines, which are formed from alkenes and diazo compounds, for example, can be pyrolyzed or photolyzed to give cyclopropanes.

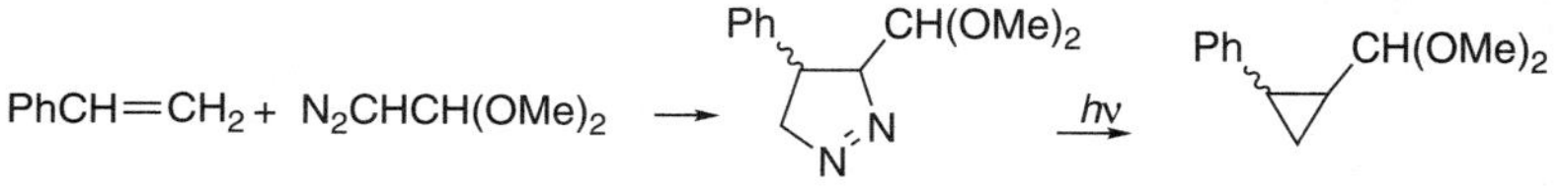

Ref. 149

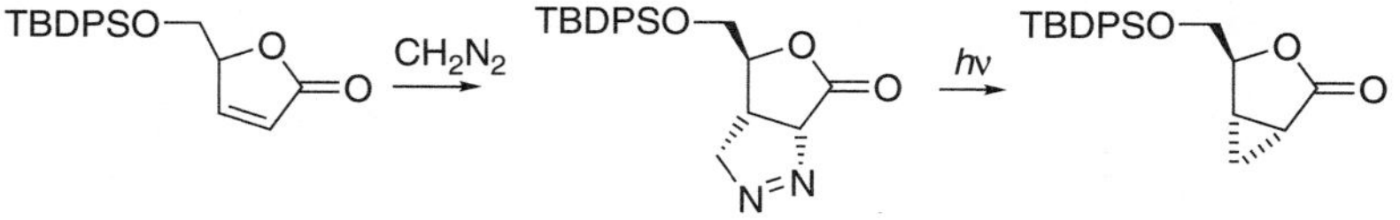

Ref. 150

147. I. A. Benages and S. M. Albonico, *J. Org. Chem.*, **43**, 4273 (1978).

148. R. Huisgen and H. Mader, *J. Am. Chem. Soc.*, **93**, 1777 (1971).

149. P. Carrie, *Heterocycles*, **14**, 1529 (1980).

150. M. Martin-Villa, N. Hanafi, J. M. Jiminez, A. Alvarez-Larena, J. F. Piniella, V. Branchadell, A. Oliva, and R. M. Ortuno, *J. Org. Chem.*, **63**, 3581 (1998).

532

CHAPTER 6

*Concerted
Cycloadditions,
Unimolecular
Rearrangements, and
Thermal Eliminations*

Scheme 6.6 gives some examples of 1,3-DCA reactions. Entry 1 is an addition of an aryl azide to norbornene. The EWG nitro group is rate enhancing and the reaction occurs with a rate constant of $6.3 \times 10^{-3} \, M^{-1} \, s^{-1}$ at 25° C. Owing to steric approach control, the product is the *exo* stereoisomer. Entry 2 involves an acetylenic dipolarophile and gives an aromatic triazole as the product. Entry 3 is an addition of diazomethane to the dioxolane derivative of acrolein. The reaction is carried out in a closed vessel at room temperature. Entry 4 involves a nitrone as the 1,3-dipole. Nitrone cycloadditions are particularly useful in synthesis because a new carbon-carbon bond is formed and the adducts can be reduced to β-amino alcohols. Nitrile oxides, which are formed by dehydration of nitroalkanes or by oxidation of oximes with hypochlorite,[151] are also useful 1,3-dipoles. They are highly reactive, must be generated in situ,[152] and react with both alkenes and alkynes. The product in Entry 5 is an example in an isoxazole that was eventually converted to a prostaglandin derivative.

Intramolecular 1,3-dipolar cycloadditions have proven to be especially useful in synthesis.[153] The products of nitrone-alkene cycloadditions are isoxazolines and the oxygen-nitrogen bond can be cleaved by reduction, leaving both an amino and hydroxy function in place. A number of imaginative syntheses have employed this strategy. Entry 6 shows the formation of a new six-membered carbocyclic ring. The nitrone **11** is generated by condensation of the aldehyde group with *N*-methylhydroxylamine and then goes on to product by intramolecular cycloaddition.

11

These reactions are highly stereoselective, provided a substituent is present at C(3). The stereochemistry is consistent with a chairlike TS having the 3-subsituent in an equatorial position.

151. G. A. Lee, *Synthesis*, 508 (1982).

152. K. Torssell, *Nitrile Oxides, Nitrones and Nitronates in Organic Synthesis*, VCH Publishers, New York, 1988.

153. For reviews of nitrone cycloadditions, see D. St. C. Black, R. F. Crozier, and V. C. Davis, *Synthesis*, 205 (1975); J. J. Tufariello, *Acc. Chem. Res.*, **12**, 396 (1979); P. N. Confalone and E. M. Huie, *Org. React.*, **36**, 1 (1988); K. V. Gothelf and K. A. Jorgensen, *Chem. Rev.*, **98**, 863 (1998).

Scheme 6.6. Typical 1,3-Dipolar Cycloaddition Reactions

A. Intermolecular cycloaddition

1[a]

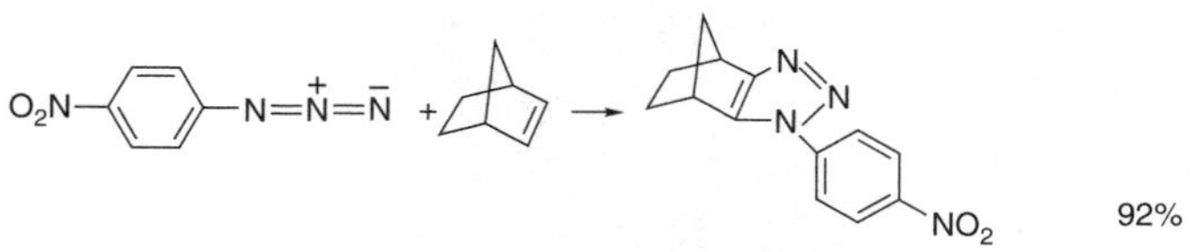

92%

2[b]

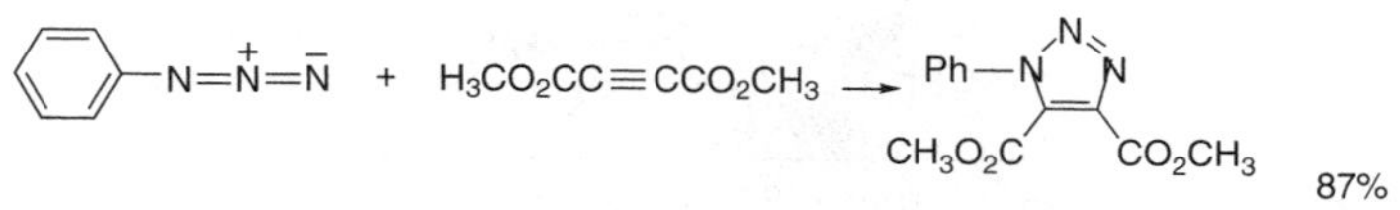

87%

3[c]

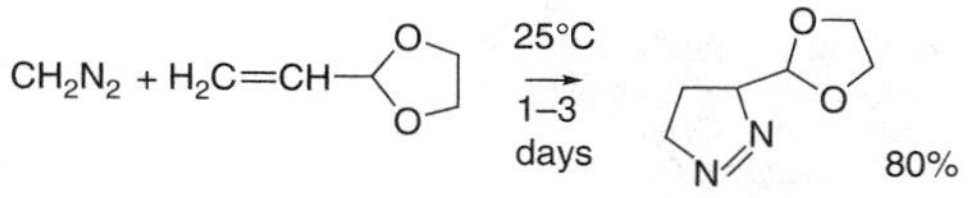

80%

4[d]

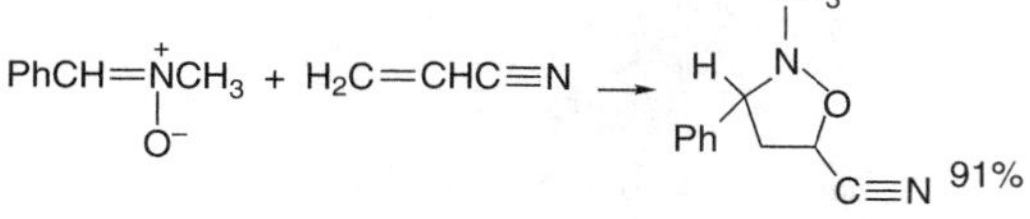

91%

5[e]

$R = -(CH_2)_6CO_2(CH_2)_3CH_3$

60%

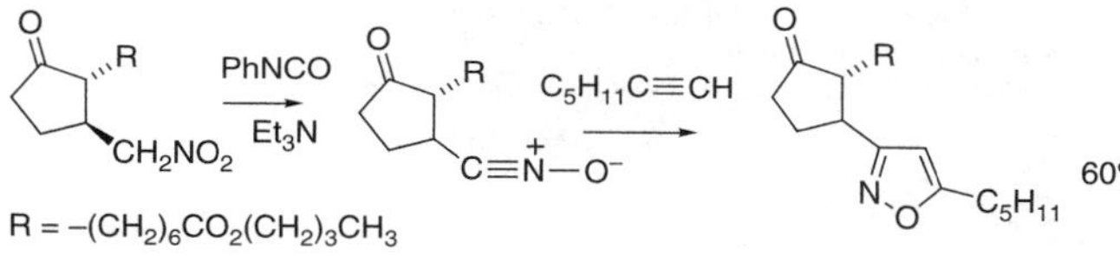

B. Intramolecular cycloaddition

6[f]

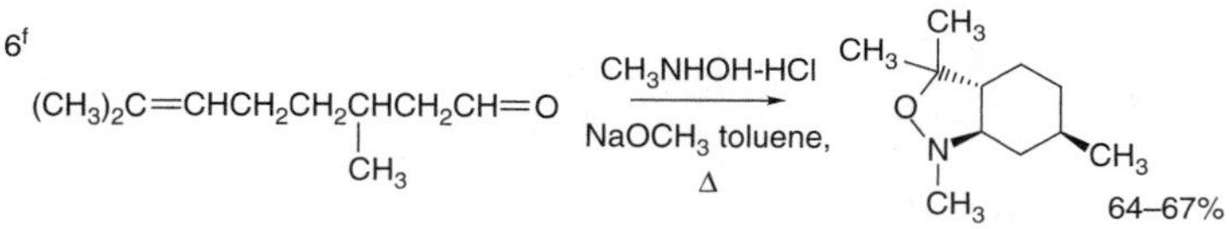

64–67%

7[g]

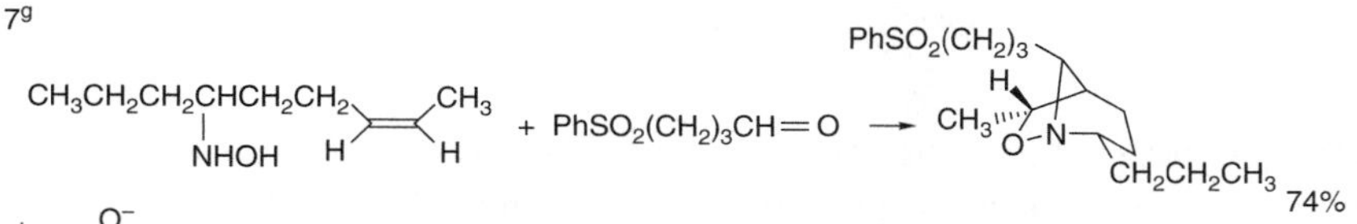

74%

8[h]

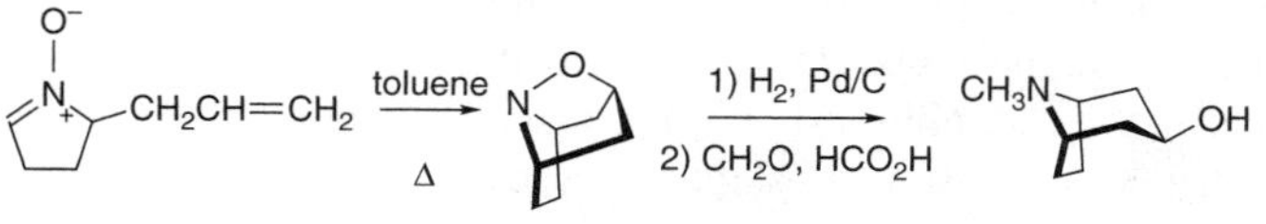

9[i]

66%   96%

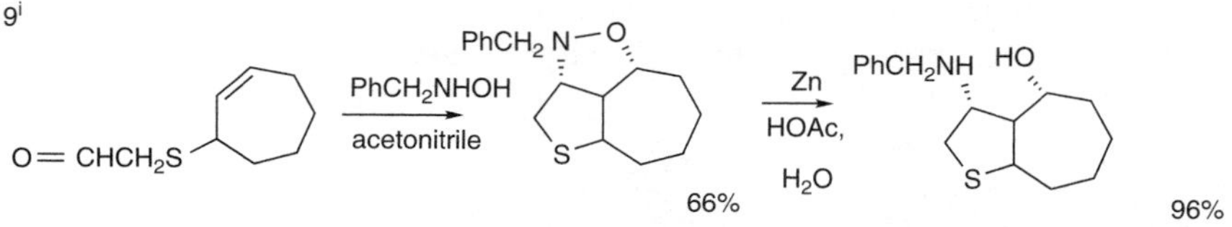

(*Continued*)

534

CHAPTER 6

*Concerted
Cycloadditions,
Unimolecular
Rearrangements, and
Thermal Eliminations*

Scheme 6.6. (*Continued*)

10[j]

11[k]

96%

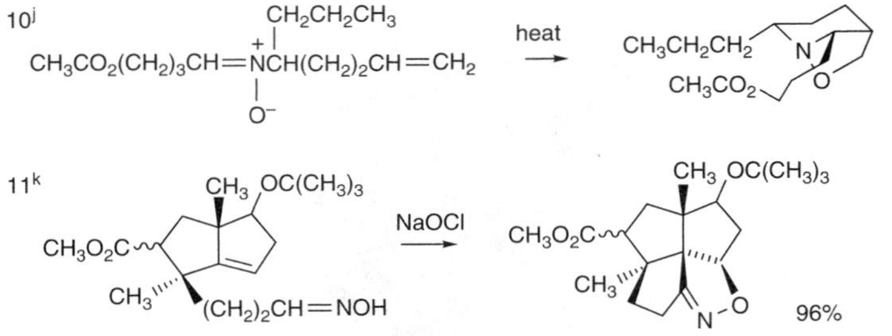

a. P. Scheiner, J. H. Schomaker, S. Deming, W. J. Libbey, and G. P. Nowack, *J. Am. Chem. Soc.*, **87**, 306 (1965).
b. R. Huisgen, R. Knorr, L. Mobius, and G. Szeimies, *Chem. Ber.*, **98**, 4014 (1965).
c. J. M. Stewart, C. Carlisle, K. Kem, and G. Lee, *J. Org. Chem.*, **35**, 2040 (1970).
d. R. Huisgen, H. Hauck, R. Grashey, and H. Seidl, *Chem. Ber.*, **101**, 2568 (1968).
e. A. Barco, S. Benetti, G. P. Pollini, P. G. Baraldi, M. Guarneri, D. Simoni, and C. Gandolfi, *J. Org. Chem.*, **46**, 4518 (1981).
f. N. A. LeBel and D. Hwang, *Org. Synth.*, **58**, 106 (1978); N. A. LeBel, M. E. Post, and J. J. Whang, *J. Am. Chem. Soc.*, **86**, 3759 (1964).
g. N. A. LeBel and N. Balasubramanian, *J. Am. Chem. Soc.*, **111**, 3363 (1989).
h. J. J. Tufariello, G. B. Mullen, J. J. Tegeler, E. J. Trybulski, S. C. Wong, and S. A. Ali, *J. Am. Chem. Soc.*, **101**, 2435 (1979).
i. P. N. Confalone, G. Pizzolato, D. I. Confalone, and M. R. Uskokovic, *J. Am. Chem. Soc.*, **102**, 1954 (1980).
j. A. L. Smith, S. F. Williams, A. B. Holmes, L. R. Hughes, Z. Lidert, and C. Swithenbank, *J. Am. Chem. Soc.*, **110**, 8696 (1988).
k. M. Ihara, Y. Tokunaga, N. Taniguchi, K. Fukumoto, and C. Kabuto, *J. Org. Chem.*, **56**, 5281 (1991).

Entry 7 is another intramolecular nitrone cycloaddition, but in this case the hydroxylamine function is present in the alkene.

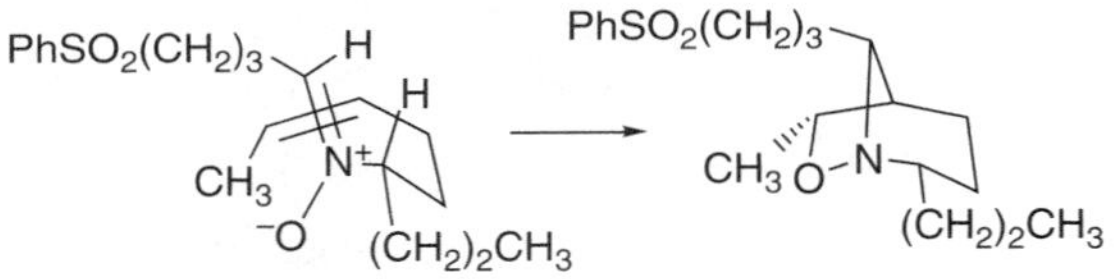

The product of the reaction in Entry 8 was used in the synthesis of the alkaloid pseudotropine. The proper stereochemical orientation of the hydroxy group is determined by the structure of the oxazoline ring formed in the cycloaddition. Entry 9 portrays the early stages of synthesis of the biologically important molecule biotin. The reaction in Entry 10 was used to establish the carbocyclic skeleton and stereochemistry of a group of toxic indolizidine alkaloids found in dart poisons from frogs. Entry 11 involves generation of a nitrile oxide. Three other stereoisomers are possible. The observed isomer corresponds to approach from the less hindered convex face of the molecule.

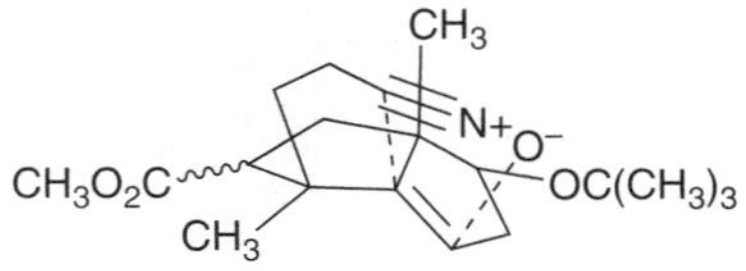

6.2.3. Catalysis of 1,3-Dipolar Cycloaddition Reactions

The role of Lewis acid catalysts in 1,3-DCA reactions is similar to that in D-A reactions. The catalysis results from a lowering of the LUMO energy of the dipolarophile, which is analogous to the Lewis acid catalysis of D-A reactions. The more organized TS, incorporating the metal ion and associated ligands, then enforces a preferred orientation of the reagents. In contrast to the D-A reaction involving hydrocarbon dienes, 1,3-DCA reactions may encounter competing complexation at the 1,3-dipole. Lewis acid interaction with the 1,3-dipole is likely to be detrimental if the dipole is the more nucleophilic component of the reaction. For example, with nitrones and enones, formation of a Lewis acid adduct with the nitrone in competition with the enone is detrimental. One approach to the need for selectivity is to use highly substituted catalysts that are selective for the less-substituted reactant. Bulky aryloxyaluminum compounds are excellent catalysts for nitrone cycloaddition and also enhance regioselectivity. The reaction of diphenylnitrone with enones is usually subject to steric regiochemical control. With the catalyst **L** high *electronic regiochemical control* is achieved and reactivity is greatly enhanced. The catalyst does not, however, strongly affect the *exo*:*endo* selectivity, which is 23:77 for propenal.

electronically-controlled product

sterically-controlled product

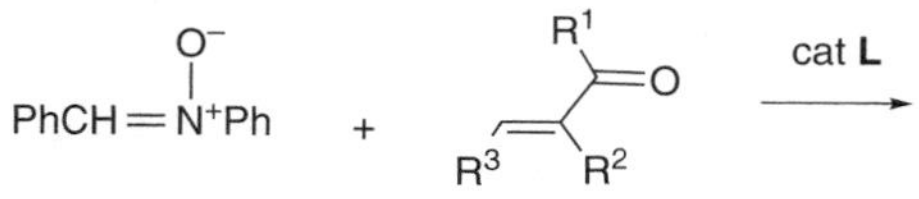

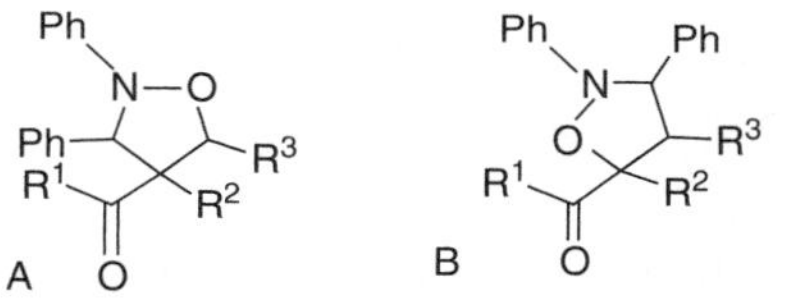

R1	R2	R3	catalyst	Yield (%)	A:B ratio
H	H	H	no	5	20:80
			yes	100	>99:1
CH_3	H	H	no	7	8:92
			yes	82	100:0
H	CH_3	H	no	5	0:100
			yes	100	91:9
H	H	CH_3	no	2	100:0
			yes	100	100:0

catalyst **L**

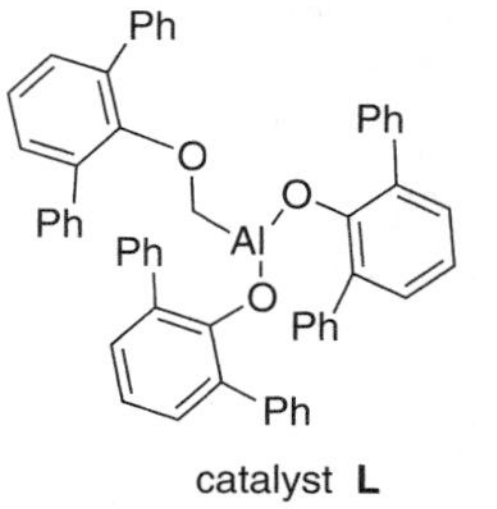

Lithium perchlorate and lithium triflate in acetonitrile catalyze intramolecular cycloaddition reactions of nitrones of allyloxybenzaldehydes and unsaturated aldehydes.[154]

154. J. S. Yadav, B. V. S. Reddy, D. Narsimhaswamy, K. Narsimulu, and A. C. Kumar, *Tetrahedron Lett.*, **44**, 3697 (2003).

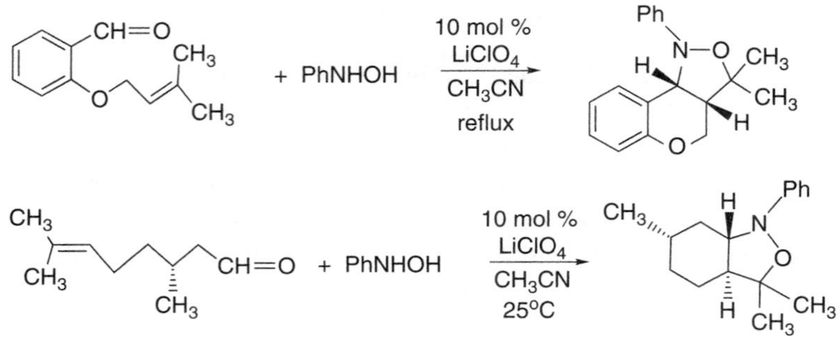

536

CHAPTER 6

*Concerted
Cycloadditions,
Unimolecular
Rearrangements, and
Thermal Eliminations*

A series of similar reactions was examined in the course of synthesis of substituted chromanes.[155] The reactions are thought to proceed through TS **M** in preference to **N** because of steric interactions with the phenyl ring on the chiral hydroxylamine.

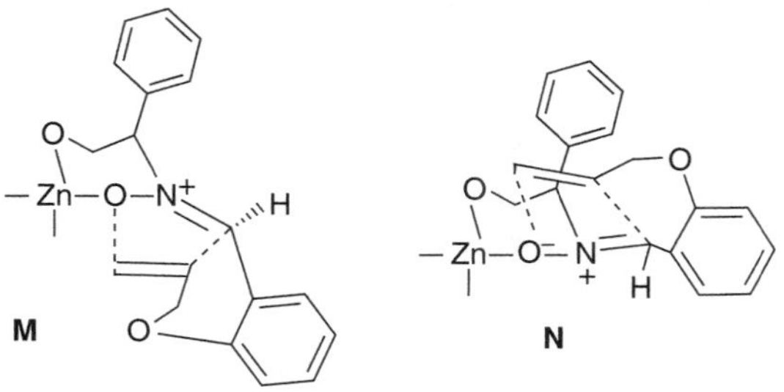

The best Lewis acid found was $Zn(OTf)_2$, which improved stereoselectivity from 6:1 to 22:1.

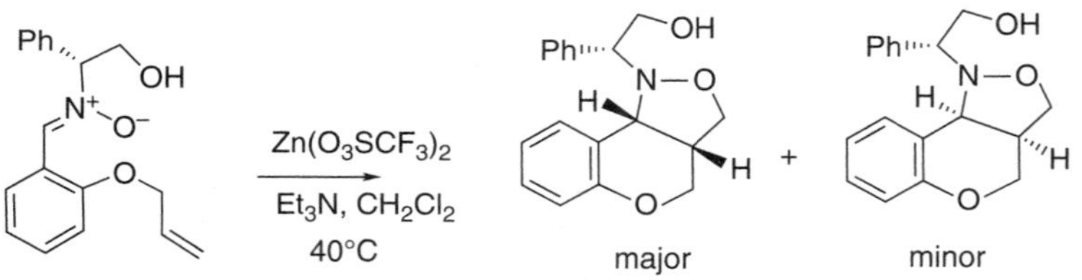

Interestingly, the reactions were modestly *slower* in the presence of the Lewis acid. It is suggested that the catalyst inverts the HOMO-LUMO relationships, making the complexed nitrone the electrophilic reactant. In agreement with this interpretation, the reaction is favored by EWGs on the aromatic ring.

As with D-A reactions, it is possible to achieve enantioselective cycloaddition in the presence of chiral catalysts.[156] Many of the catalysts are similar to those used in enantioselective D-A reactions. The catalysis usually results from a lowering of the LUMO energy of the dipolarophile, which is analogous to the Lewis acid catalysis of D-A reactions. The more organized TS, incorporating a metal ion and associated

155. Q. Zhao, F. Han, and D. L. Romero, *J. Org. Chem.*, **67**, 3317 (2002).
156. K. V. Gothelf and K. A. Jorgensen, *Chem. Rev.*, **98**, 863 (1998); M. Frederickson, *Tetrahedron*, **53**, 503 (1997).

ligands, then enforces a preferred orientation of the reagents. For example, the bulky aryl groups in the catalysts **O** and **P** favor one direction of approach of the nitrone reactant.[157]

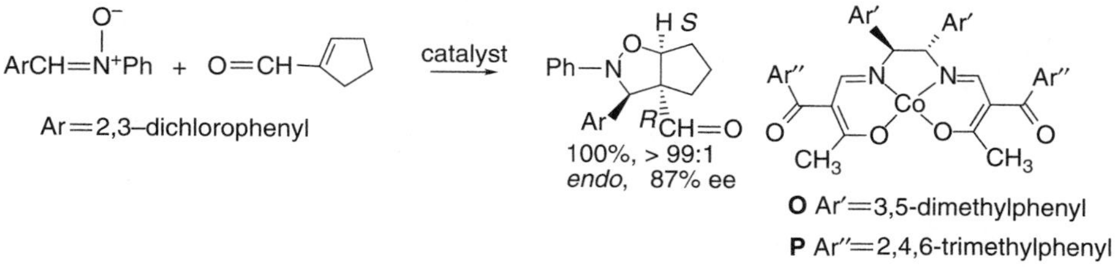

ArCH=N⁺Ph + O=CH—⟨cyclopentene⟩

Ar = 2,3–dichlorophenyl

100%, > 99:1
endo, 87% ee

O Ar′=3,5-dimethylphenyl
P Ar″=2,4,6-trimethylphenyl

The Ti(IV) TADDOL catalyst **Q** leads to moderate enantioselectivity in nitrone-alkene cycloaddition.[158]

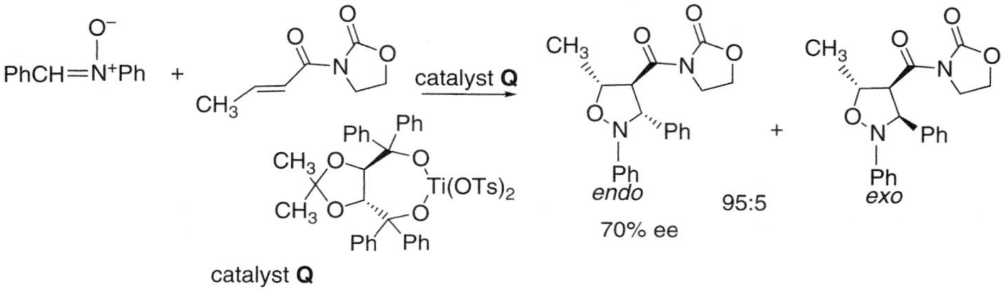

PhCH=N⁺Ph +

catalyst **Q**

endo   95:5   *exo*

70% ee

Favorable results have also been achieved using PyBOX type catalysts. Acryloyl and crotonoyloxazolidinones gave 80–95% yields, 90–98% e.e., and more than 9:1 *endo*-diastereoselectivity in reactions with *N*-phenylbenzylidene nitrones.[159]

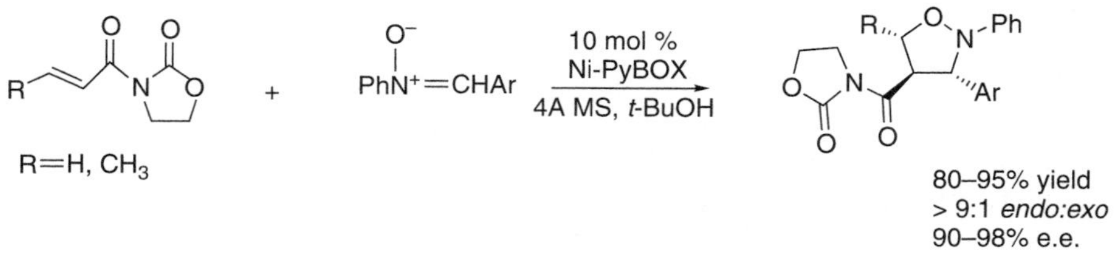

R=H, CH₃

10 mol %
Ni-PyBOX
4A MS, *t*-BuOH

80–95% yield
> 9:1 *endo:exo*
90–98% e.e.

Other effective enantioselective catalysts include Yb(OTf)₃ with BINOL,[160] Mg²⁺-*bis*-oxazolines,[161] and oxazaborolidinones.[162]

[157.] T. Mita, N. Ohtsuki, T. Ikeno, and T. Yamada, *Org. Lett.*, **4**, 2457 (2002).

[158.] K. V. Gothelf and K. A. Jorgensen, *Acta Chem. Scand.*, **50**, 652 (1996); K. B. Jensen, K. V. Gothelf, R. G. Hazell, and K. A. Jorgensen, *J. Org. Chem.*, **62**, 2471 (1997); K. B. Jensen, K. V. Gothelf, and K. A. Jorgensen, *Helv. Chim. Acta*, **80**, 2039 (1997).

[159.] S. Iwasa, H. Maeda, K. Nishiyama, S. Tsushima, Y. Tsukamoto, and H. Nishiyama, *Tetrahedron*, **58**, 8281 (2002).

[160.] M. Kawamura and S. Kobayashi, *Tetrahedron Lett.*, **40**, 3213 (1999).

[161.] G. Desimoni, G. Faita, A. Mortoni, and P. Righetti, *Tetrahedron Lett.*, **40**, 2001 (1999); K. V. Gothelf, R. G. Hazell, and K. A. Jorgensen, *J. Org. Chem.*, **63**, 5483 (1998).

[162.] J. P. G. Seerden, M. M. M. Boeren, and H. W. Scheeren, *Tetrahedron*, **53**, 11843 (1997).

Scheme 6.7. Catalytic Enantioselective 1,3-Dipolar Cycloaddition Reactions

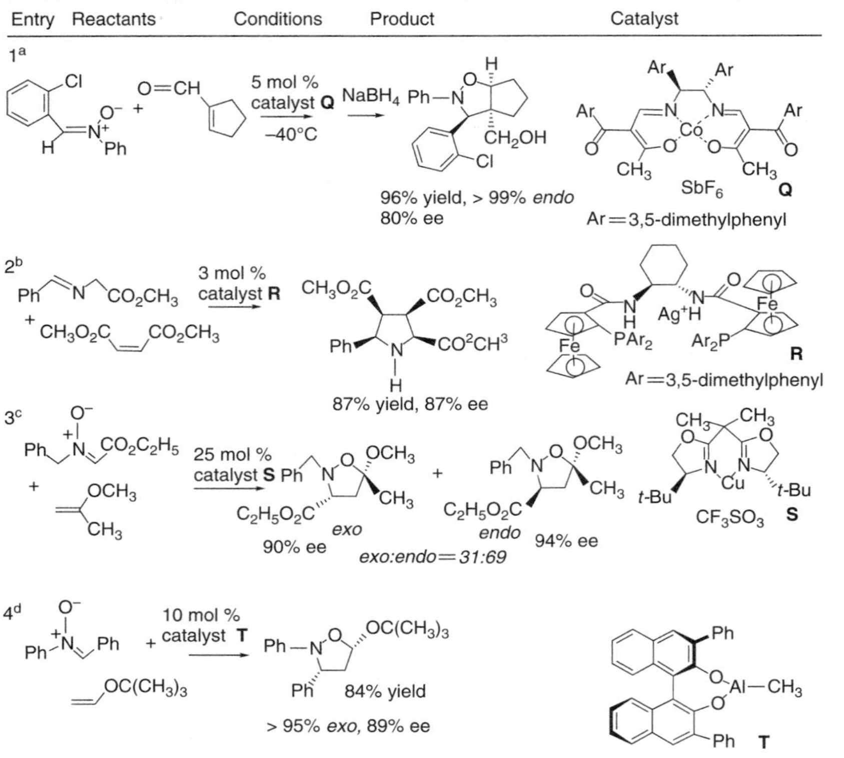

CHAPTER 6

*Concerted
Cycloadditions,
Unimolecular
Rearrangements, and
Thermal Eliminations*

Entry	Reactants	Conditions	Product	Catalyst

1[a] 5 mol % catalyst Q, −40°C, NaBH₄
96% yield, > 99% endo, 80% ee
Ar = 3,5-dimethylphenyl   Q

2[b] 3 mol % catalyst R
87% yield, 87% ee
Ar = 3,5-dimethylphenyl   R

3[c] 25 mol % catalyst S
exo 90% ee + endo 94% ee
exo:endo = 31:69   S

4[d] 10 mol % catalyst T
84% yield
> 95% exo, 89% ee   T

a. T. Mitra, N. Ohtsuki, T. Ikeno, and T. Yamada. *Org. Lett.*, **4**, 2457 (2002).
b. J. M. Longmire, B. Wang, and X. M. Zhang, *J. Am. Chem. Soc.*, **124**, 13400 (2002).
c. K. B. Jensen, R. G. Hazell, and K. A. Jorgensen, *J. Org. Chem.*, **64**, 2353 (1999).
d. K. B. Simonsen, B. Bayon, R. G. Hazell, D. V. Gothelf, and K. A. Jorgensen, *J. Am. Chem. Soc.*, **121**, 3845 (1999).

Scheme 6.7 shows some other examples of enantioselective catalysts. Entry 1 illustrates the use of a Co(III) complex, with the chirality derived from the diamine ligand. Entry 2 is a silver-catalyzed cycloaddition involving generation of an azomethine ylide. The ferrocenylphosphine groups provide a chiral environment by coordination of the catalytic Ag^+ ion. Entries 3 and 4 show typical Lewis acid catalysts in reactions in which nitrones are the electrophilic component.

6.3. [2 + 2] Cycloadditions and Related Reactions Leading to Cyclobutanes

As discussed in Section 10.4 of Part A, concerted suprafacial $[2\pi + 2\pi]$ cycloadditions are forbidden by orbital symmetry rules. Two types of $[2+2]$ cycloadditions are of synthetic value: addition reactions of ketenes and photochemical additions. The latter group includes reactions of alkenes, dienes, enones, and carbonyl compounds, and these additions are discussed in the sections that follow.

6.3.1. Cycloaddition Reactions of Ketenes and Alkenes

[2 + 2] Cycloadditions of ketenes and alkenes have synthetic utility for the preparation of cyclobutanones.[163] The stereoselectivity of ketene-alkene cycloaddition can be analyzed in terms of the Woodward-Hoffmann rules.[164] To be an allowed process, the $[2\pi + 2\pi]$ cycloaddition must be suprafacial in one component and antarafacial in the other. An alternative description of the TS is a $2\pi_s + (2\pi_s + 2\pi_s)$ addition.[165] Figure 6.13 illustrates these combinations. Note that both representations predict formation of the *cis*-substituted cyclobutanone.

Ketenes are especially reactive in [2 + 2] cycloadditions and an important reason is that they offer a low degree of steric interaction in the TS. Another reason is the electrophilic character of the ketene LUMO. As discussed in Section 10.4 of Part A, there is a large net charge transfer from the alkene to the ketene, with bond formation at the ketene *sp* carbon running ahead of that at the sp^2 carbon. The stereoselectivity of ketene cycloadditions is the result of steric effects in the TS. Minimization of interaction between the substituents R and R′ leads to a cyclobutanone in which these substituents are *cis*, which is the stereochemistry usually observed in these reactions.

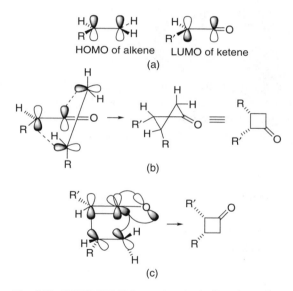

Fig. 6.13. HOMO-LUMO interactions in the [2 + 2] cycloadditions of an alkene and a ketene: (a) frontier orbitals of the alkene and ketene; (b) $[2\pi_s + 2\pi_a]$ representation of suprafacial addition to the alkene and antarafacial addition to the ketene; (c) $[2\pi_s + (2\pi_s + 2\pi_s)]$ alignment of orbitals.

163. For reviews, see W. T. Brady, in *The Chemistry of Ketenes, Allenes, and Related Compounds*, S. Patai, ed., Wiley-Interscience, New York, 1980, Chap. 8; W. T. Brady, *Tetrahedron*, **37**, 2949 (1981); J. Hyatt and R. W. Reynolds, *Org. React.*, **45**, 159 (1994); T. T. Tidwell, *Ketenes*, Wiley, New York, 1995.
164. R. B. Woodward and R. Hoffman, *Angew. Chem. Int. Ed. Engl.*, **8**, 781 (1969).
165. D. J. Pasto, *J. Am. Chem. Soc.*, **101** 37 (1979); E. Valenti, M. A. Pericas, and A. Moyano, *J. Org. Chem.*, **55**, 3582 (1990).

540

CHAPTER 6

*Concerted
Cycloadditions,
Unimolecular
Rearrangements, and
Thermal Eliminations*

Ref. 166

The best yields are obtained when the ketene has an electronegative substituent, such as halogen. Simple ketenes are not very stable and must usually be generated in situ. The most common method for generating ketenes for synthesis is by dehydrohalogenation of acyl chlorides. This is usually done with an amine such as triethylamine.[167] Other activated carboxylic acid derivatives, such as acyloxypyridinium ions, have also been used as ketene precursors.[168] Ketene itself and certain alkyl derivatives can be generated by pyrolysis of carboxylic anhydrides.[169]

Intramolecular ketene cycloadditions are possible if the ketene and alkene functionalities can achieve an appropriate orientation.[170]

43%

Ref. 171

Some trends in relative reactivity for intramolecular ketene cycloadditions have been examined by internal competitions.[172] For example, **12** gives exclusively **13**, pointing to a preference for five-membered rings over six-membered ones.

When two different aryl substituents are compared, the double bond with an ERG substituent is more reactive, as would be expected if the alkene acts primarily as an electron donor.

166. M. Rey, S. M. Roberts, A. S. Dreiding, A. Roussel, H. Vanlierde, S. Toppert, and L. Ghosez, *Helv. Chim. Acta*, **65**, 703 (1982).
167. K. Shishido, T. Azuma, and M. Shibuya, *Tetrahedron Lett.*, **31**, 219 (1990).
168. R. L. Funk, P. M. Novak, and M. M. Abelman, *Tetrahedron Lett.*, **29**, 1493 (1988).
169. G. J. Fisher, A. F. MacLean, and A. W. Schnizer, *J. Org. Chem.*, **18**, 1055 (1953).
170. B. B. Snider, *Chem. Rev.*, **88**, 793 (1988).
171. E. J. Corey and M. C. Desai, *Tetrahedron Lett.*, **26**, 3535 (1985).
172. G. Belanger, F. Levesque, J. Paquet, and G. Barbe, *J. Org. Chem.*, **70**, 291 (2005).

Comparison of *E*- and *Z*-double bonds indicates that the former are about 30 times more reactive.[173]

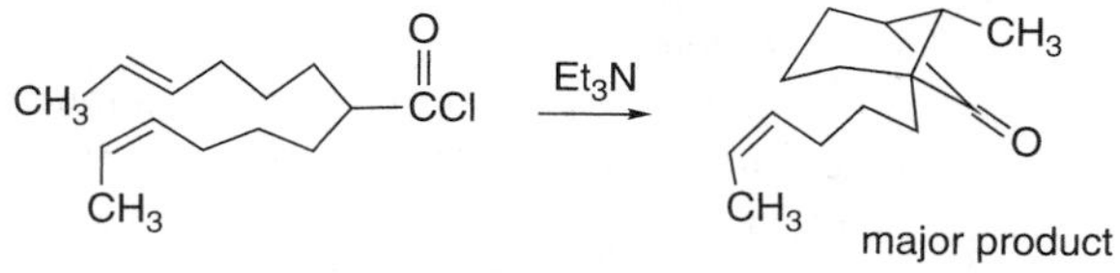

This relative reactivity results from larger steric interactions in the TS for the *Z*-double bond.

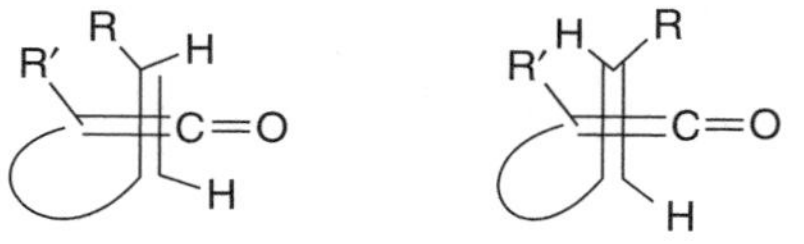

The competition between formation of bicyclo[3.2.0] and bicyclo[3.1.1] products is determined by substitution on the alkene.

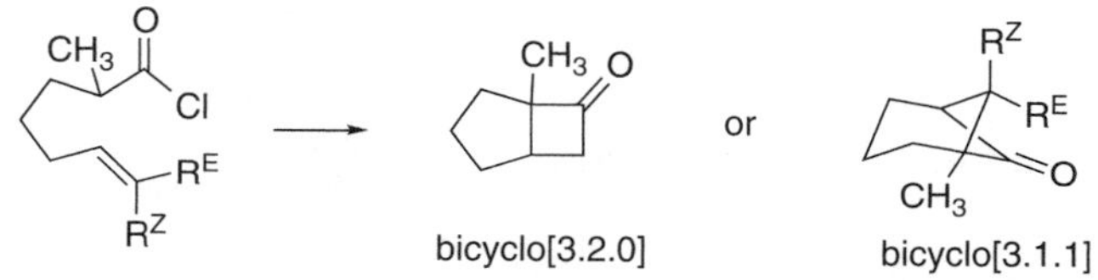

R^E	R^Z	bicyclo[3.2.0]	bicyclo[3.1.1]
H	H	45% (only product)	
CH₃	H		23% (only product)
CH₃	CH₃		45% (only product)

Initial bond formation occurs between the ketene carbonyl and the more nucleophilic end of the alkene double bond. This is related to the charge separation in the TS and results in the second bond being formed between the terminal ketene carbon and the carbon that is best able to support positive character.[174]

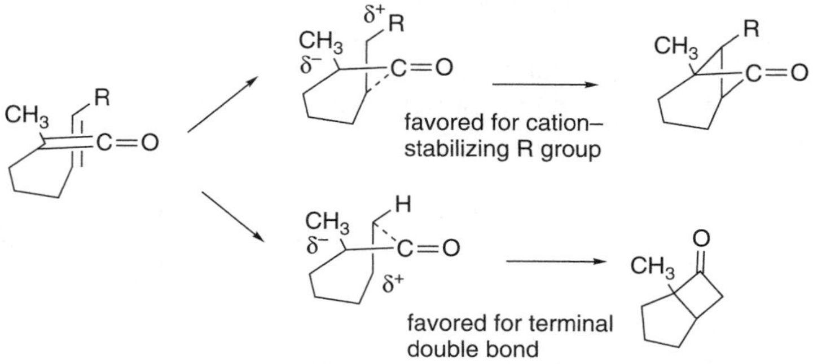

favored for cation–stabilizing R group

favored for terminal double bond

173. B. B. Snider, A. J. Allentoff, and M. B. Walner, *Tetrahedron*, **46**, 8031 (1990).
174. B. B. Snider, R. A. H. F. Hui, and Y. S. Kulkarni, *J. Am. Chem. Soc.*, **107**, 2194 (1985).

542

CHAPTER 6

*Concerted
Cycloadditions,
Unimolecular
Rearrangements, and
Thermal Eliminations*

Scheme 6.8 gives some examples of ketene-alkene cycloadditions. In Entry 1, dimethylketene was generated by pyrolysis of the dimer, 2,2,4,4-tetramethylcyclobutane-1,3-dione and passed into a solution of the alkene maintained at 70°C. Entries 2 and 3 involve generation of chloromethylketene by dehydrohalogenation of α-chloropropanoyl chloride. Entry 4 involves formation of dichloroketene. Entry 5 is an intramolecular addition, with the ketene being generated from a 2-pyridyl ester. Entries 6, 7, and 8 are other examples of intramolecular ketene additions.

Cyclobutanes can also be formed by nonconcerted processes involving zwitterionic intermediates. The combination of an electron-rich alkene (enamine, enol ether) and an electrophilic one (nitro- or polycyanoalkene) is required for such processes.

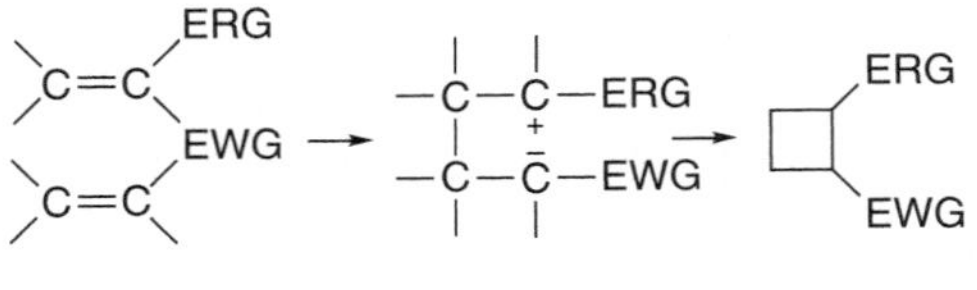

ERG = electron releasing group (−OR, −NR$_2$)
EWG = electron withdrawing group (−NO$_2$, −C≡N)

Two examples of this reaction type are shown below.

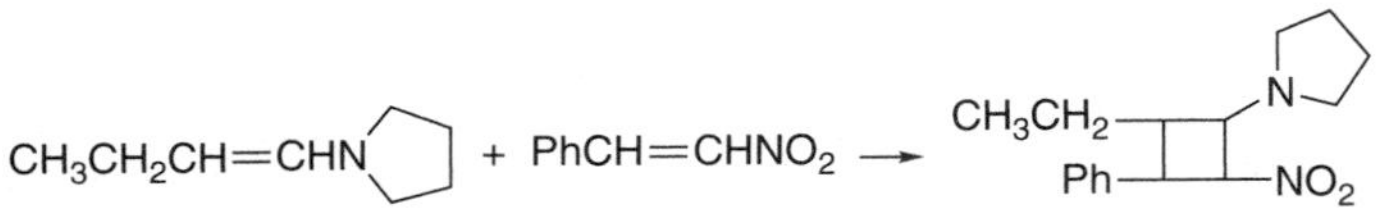

$$CH_3CH_2CH=CHN\underset{}{\bigcirc} + PhCH=CHNO_2 \longrightarrow$$

100%

Ref. 175

$$H_2C=CHOCH_3 + (NC)_2C=C(CN)_2 \longrightarrow$$

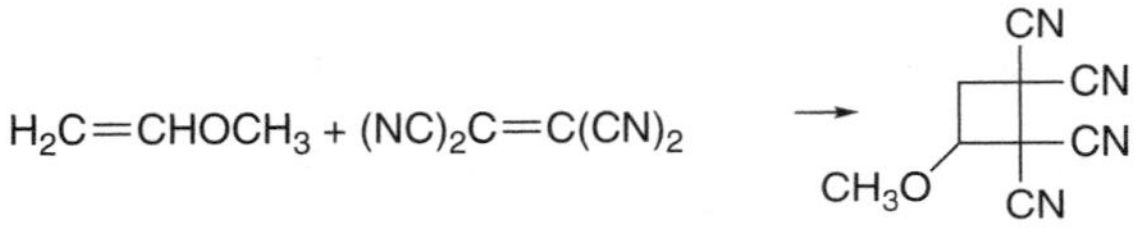

90%

Ref. 176

The stereochemistry of these reactions depends on the lifetime of the dipolar intermediate, which, in turn, is influenced by the polarity of the solvent. In the reactions of enol ethers with tetracyanoethylene, the stereochemistry of the enol ether is retained in nonpolar solvents. In polar solvents, cycloaddition is nonstereospecific, as a result of a longer lifetime for the zwitterionic intermediate.[177]

Lewis acid catalysis has been used to promote stepwise [2 + 2] cycloaddition of silyl enol ethers and unsaturated esters.[178] The best catalyst is (C$_2$H$_5$)$_2$AlCl and polyfluoroalkyl esters give the highest stereoselectivity. The reactions give the more stable *trans* products.

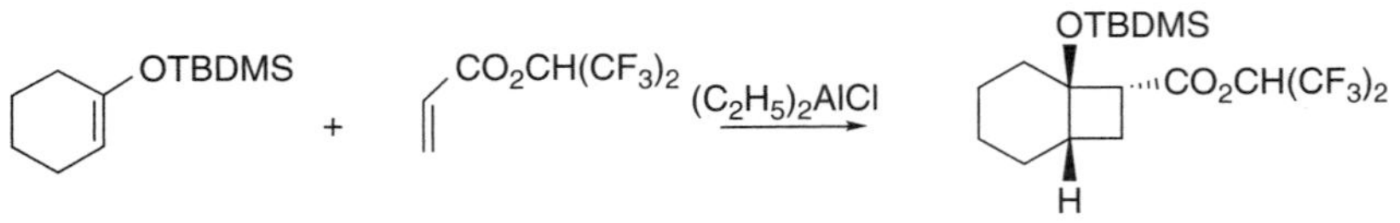

175. M. E. Kuehne and L. Foley, *J. Org. Chem.*, **30**, 4280 (1965).
176. J. K. Williams, D. W. Wiley, and B. C. McKusick, *J. Am. Chem. Soc.*, **84**, 2210 (1962).
177. R. Huisgen, *Acc. Chem. Res.*, **10**, 117, 199 (1977).
178. K. Takasu, M. Ueno, K. Inanaga, and M. Ihara, *J. Org. Chem.*, **69**, 517 (2004).

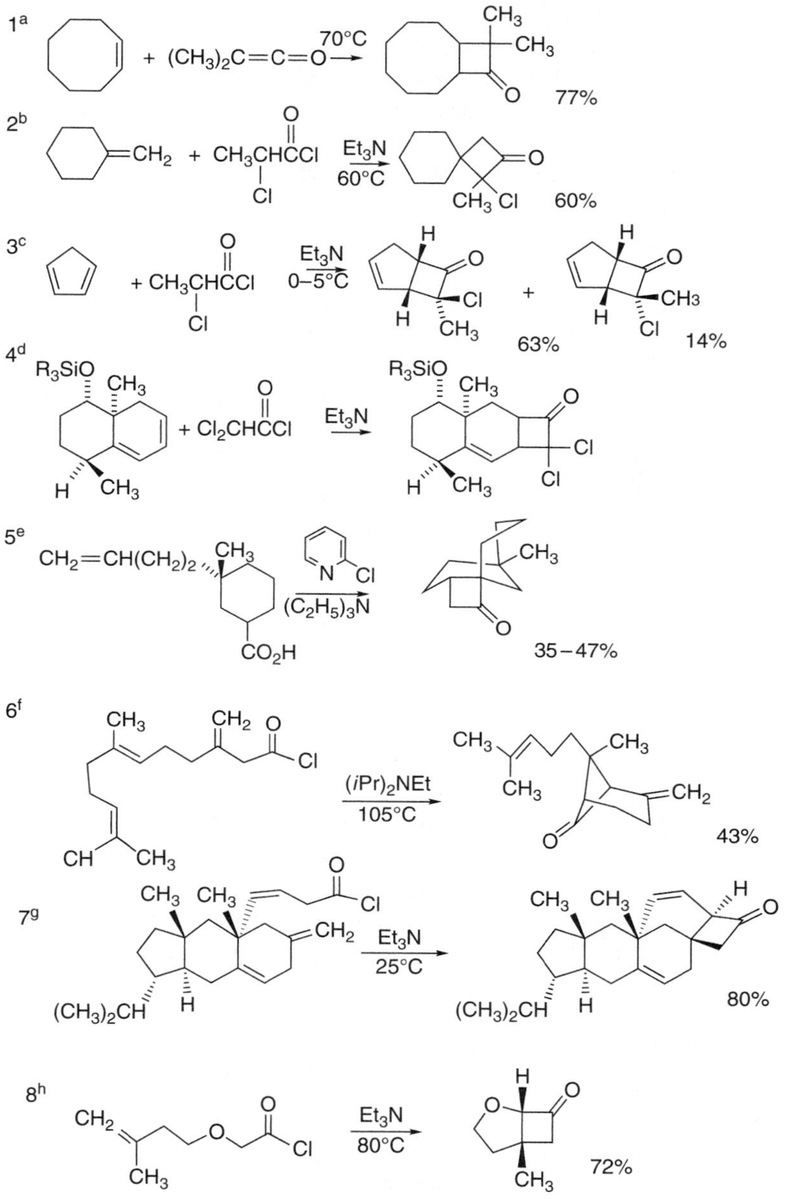

Scheme 6.8. [2 + 2] Cycloadditions of Ketenes

1[a]

2[b]

3[c]

4[d]

5[e]

6[f]

7[g]

8[h]

a. A. P. Krapcho and J. H. Lesser, *J. Org. Chem.*, **31**, 2030 (1966).
b. W. T. Brady and A. D. Patel, *J. Org. Chem.*, **38**, 4106 (1973).
c. W. T. Brady and R. Roe, *J. Am. Chem. Soc.*, **93**, 1662 (1971).
d. P. A. Grieco, T. Oguri, and S. Gilman, *J. Am. Chem. Soc.*, **102**, 5886 (1980).
e. R. L. Funk, P. M. Novak, and M. M. Abraham, *Tetrahedron Lett.*, **29**, 1493 (1988).
f. E. J. Corey and M. C. Desai, *Tetrahedron Lett.*, **26**, 3535 (1985).
g. E. J. Corey, M. C. Desai, and T. A. Engler, *J. Am. Chem. Soc.*, **107**, 4339 (1985).
h. B. B. Snider, R. A. H. F. Hui, and Y. S. Kulkarni, *J. Am. Chem. Soc.*, **107**, 2194 (1985).

CHAPTER 6

Concerted
Cycloadditions,
Unimolecular
Rearrangements, and
Thermal Eliminations

6.3.2. Photochemical Cycloaddition Reactions

6.3.2.1. Photocycloaddition of Alkenes and Dienes. Photochemical cycloadditions provide a method that is often complementary to thermal cycloadditions with regard to the types of compounds that can be prepared. The theoretical basis for this complementary relationship between thermal and photochemical modes of reaction lies in orbital symmetry relationships, as discussed in Chapter 10 of Part A. The reaction types permitted by photochemical excitation that are particularly useful for synthesis are [2 + 2] additions between two carbon-carbon double bonds and [2 + 2] additions of alkenes and carbonyl groups to form oxetanes. Photochemical cycloadditions are often not concerted processes because in many cases the reactive excited state is a triplet. The initial adduct is a triplet 1,4-diradical that must undergo spin inversion before product formation is complete. Stereospecificity is lost if the intermediate 1,4-diradical undergoes bond rotation faster than ring closure.

Intermolecular photocycloadditions of alkenes can be carried out by photosensitization with mercury or directly with short-wavelength light.[179] Relatively little preparative use has been made of this reaction for simple alkenes. Dienes can be photosensitized using benzophenone, butane-2,3-dione, and acetophenone.[180] The photodimerization of derivatives of cinnamic acid was among the earliest photochemical reactions to be studied.[181] Good yields of dimers are obtained when irradiation is carried out in the crystalline state. In solution, *cis-trans* isomerization is the dominant reaction.

The presence of Cu(I) salts promotes intermolecular photocycloaddition of simple alkenes. Copper(I) triflate is especially effective.[182] It is believed that the photoreactive species is a 2:1 alkene:Cu(I) complex in which the two alkene molecules are brought together prior to photoexcitation.[183]

179. H. Yamazaki and R. J. Cvetanovic, *J. Am. Chem. Soc.*, **91**, 520 (1969).
180. G. S. Hammond, N. J. Turro, and R. S. H. Liu, *J. Org. Chem.*, **28**, 3297 (1963).
181. A. Mustafa, *Chem. Rev.*, **51**, 1 (1962).
182. R. G. Salomon, *Tetrahedron*, **39**, 485 (1983); R. G. Salomon and S. Ghosh, *Org. Synth.*, **62**, 125 (1984).
183. R. G. Salomon, K. Folking, W. E. Streib, and J. K. Kochi, *J. Am. Chem. Soc.*, **96**, 1145 (1974).

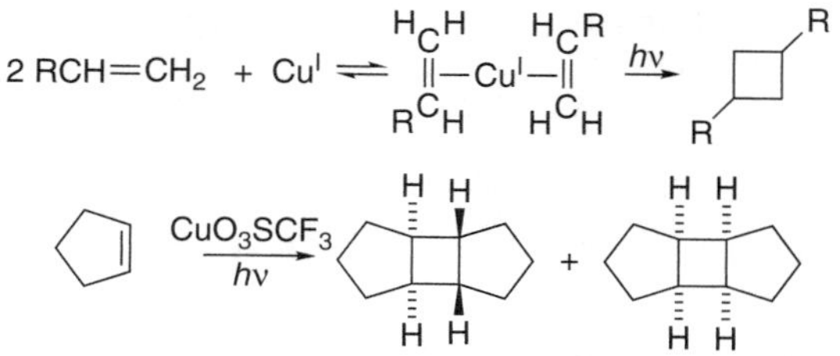

Intramolecular $[2+2]$ photocycloadditions of alkenes is an important method of formation of compounds containing four-membered rings.[184] Direct irradiation of simple nonconjugated dienes leads to cyclobutanes.[185] Strain makes the reaction unfavorable for 1,4-dienes but when the alkene units are separated by at least two carbon atoms cycloaddition becomes possible.

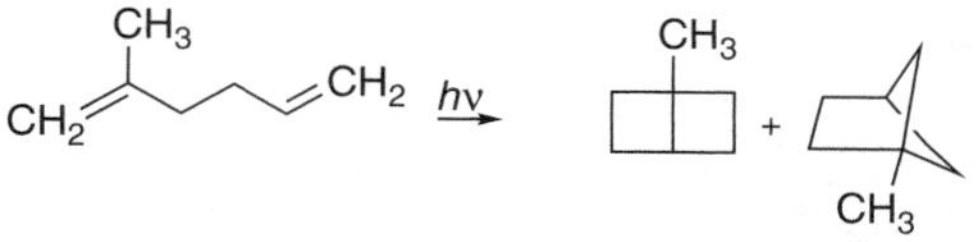

Ref. 186

Copper(I) triflate can facilitate these intramolecular additions, as is the case for intermolecular reactions.

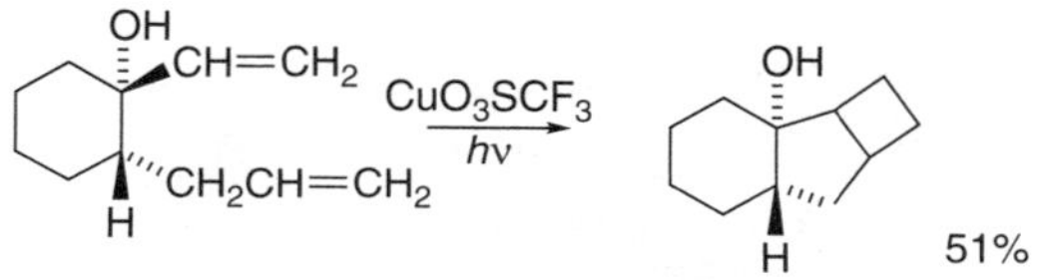

51%

Ref. 187

The most widely exploited photochemical cycloadditions involve irradiation of dienes in which the two double bonds are fairly close and result in formation of polycyclic cage compounds. Some examples of alkene photocyclizations are given in Scheme 6.9. Entry 1 is a transannular cyclization. The preference for the observed product over tricyclo[4.2.0.0^{2,5}]octane does not seem to have been analyzed in detail. Entries 2, 3, and 4 involve photolysis in the presence of CuO_3SCF_3. Entries 5 and 6 are cases in which the double bonds are in close proximity and can cyclize to caged structures.

6.3.2.2. Photocycloaddition Reactions of Enones.

Cyclic α,β-unsaturated ketones are another class of molecules that undergo photochemical cycloadditions.[188] The reactive

184. P. de Mayo, *Acc. Chem. Res.*, **4**, 41 (1971).
185. R. Srinivasan, *J. Am. Chem. Soc.*, **84**, 4141 (1962); *J. Am. Chem. Soc.*, **90**, 4498 (1968).
186. J. Meinwald and G. W. Smith, *J. Am. Chem. Soc.*, **89**, 4923 (1967); R. Srinivasan and K. H. Carlough, *J. Am. Chem. Soc.*, **89**, 4932 (1967).
187. K. Avasthi and R. G. Salomon, *J. Org. Chem.*, **51**, 2556 (1986).
188. A. C. Weedon, in *Synthetic Organic Photochemistry*, W. M. Horspool, ed., Plenum Press, New York, 1984, Chap. 2; D. I. Schuster, G. Lem, and N. A. Kaprinidis, *Chem. Rev.*, **93**, 3 (1993); M. T. Crimmins and T. L. Reinhold, *Org. React.*, **44**, 297 (1993); D. I. Schuster, in *CRC Handbook of Organic Photochemistry and Photobiology*, W. Horspool and F. Lanci, eds., CRC Press, Boca Raton, FL, 2002, pp. 72-1–72-24.

CHAPTER 6

*Concerted
Cycloadditions,
Unimolecular
Rearrangements, and
Thermal Eliminations*

Scheme 6.9. Intramolecular [2 + 2] Photocycloadditions of Dienes

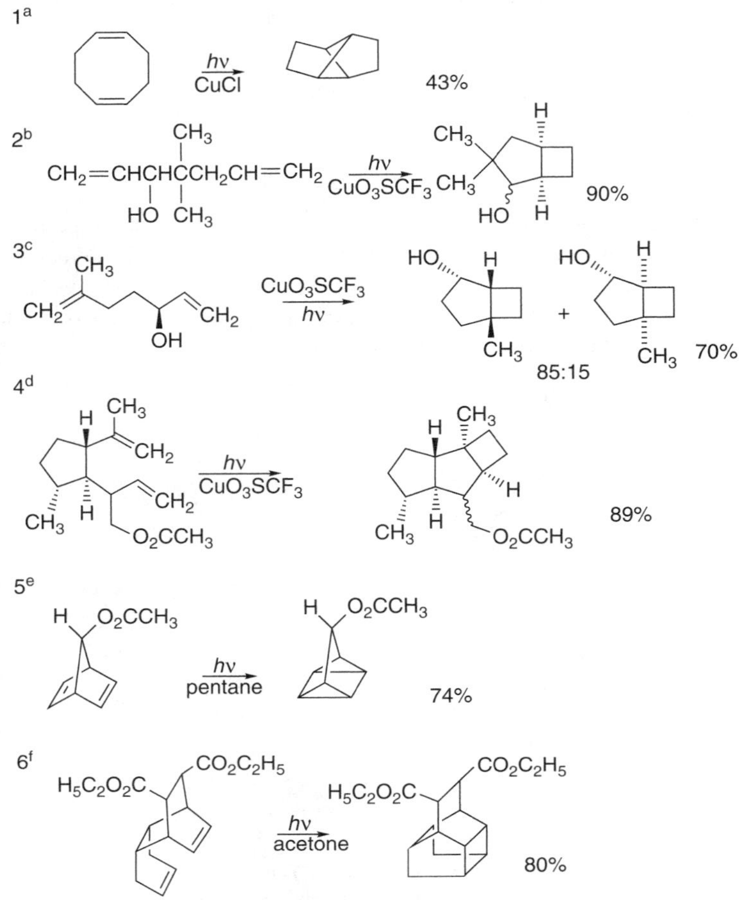

1[a]

$\xrightarrow[\text{CuCl}]{h\nu}$ 43%

2[b]

$CH_2=CHCHCH_2CH=CH_2$ $\xrightarrow[\text{CuO}_3\text{SCF}_3]{h\nu}$ 90%

3[c]

$\xrightarrow[h\nu]{\text{CuO}_3\text{SCF}_3}$ 85:15 70%

4[d]

$\xrightarrow[\text{CuO}_3\text{SCF}_3]{h\nu}$ 89%

5[e]

$\xrightarrow[\text{pentane}]{h\nu}$ 74%

6[f]

$\xrightarrow[\text{acetone}]{h\nu}$ 80%

a. P. Srinivasan, *J. Am. Chem. Soc.*, **86**, 3318 (1964); *Org. Photochem. Synth.*, **1**, 101 (1971).
b. R. G. Salomon and S. Ghosh, *Org. Synth.*, **62**, 125 (1984).
c. K. Lange and J. Mattay, *J. Org. Chem.*, **60**, 7256 (1995).
d. T. Bach and A. Spiegel, *Synlett*, 1305 (2002).
e. P. G. Gassman and D. S. Patton, *J. Am. Chem. Soc.*, **90**, 7276 (1968).
f. B. M. Jacobson, *J. Am. Chem. Soc.*, **95**, 2579 (1973).

excited state is a π-π^* triplet of the enone. The reaction is most successful with cyclopentenones and cyclohexenones. The excited states of acyclic enones and larger ring compounds are rapidly deactivated by *cis-trans* isomerization and do not readily add to alkenes. Photoexcited enones can also add to alkynes.[189] Unsymmetrical alkenes can undergo two regioisomeric modes of addition. It is generally observed that alkenes with donor groups are oriented such that the substituted carbon becomes bound to the β-carbon, whereas with acceptor substituents the other orientation is preferred.[190]

189. R. L. Cargill, T. Y. King, A. B. Sears, and M. R. Willcott, *J. Org. Chem.*, **36**, 1423 (1971); W. C. Agosta and W. W. Lowrance, *J. Org. Chem.*, **35**, 3851 (1970).
190. E. J. Corey, J. D. Bass, R. Le Mahieu, and R. B. Mitra, *J. Am. Chem. Soc.*, **86**, 5570 (1984); T. Suishu, T. Shimo, and K. Somekawa, *Tetrahedron*, **53**, 3545 (1997).

| X = CN | 76:24 |
| X = OC₂H₅ | 19:28 |

$X = CN \quad 76:24$

$X = OC_2H_5 \quad 19:28$

The photoadditions proceed through 1,4-diradical intermediates. Trapping experiments with hydrogen atom donors indicate that the initial bond formation can take place at either the α- or β-carbon of the enone. The excited enone has its highest nucleophilic character at the β-carbon. The initial bond formation occurs at the β-carbon for electron-poor alkenes but at the α-carbon for electron-rich alkenes.[191] Selectivity is low for alkenes without strong donor or acceptor substituents.[192] The final product ratio also reflects the rate and efficiency of ring closure relative to fragmentation of the biradical.[193]

Other structural factors can influence regioselectivity. Comparison of 2-propenol, 3-butenol, and 4-pentenol in various solvents suggests that hydrogen bonding can orient the reactants.[194] The reversal of regioselectivity between hexane and methanol suggests that the hydrogen bonding effects are swamped in the hydroxylic solvent methanol.

n	yield	ratio hexane	methanol
1	84	71:26	33:67
2	86	65:35	34:66
3	79	60:40	32:68

Intramolecular enone-alkene cycloadditions are also possible. In the case of β-(5-pentenyl) substituents, there is a general preference for *exo*-type cyclization to form a five-membered ring.[195] This is consistent with the general pattern for radical cyclizations and implies initial bonding at the β-carbon of the enone.

191. J. L. Broecker, J. E. Eksterowicz, A. J. Belk, and K. N. Houk, *J. Am. Chem. Soc.*, **117**, 1847 (1995).

192. J. D. White and D. N. Gupta, *J. Am. Chem. Soc.*, **88**, 5364 (1966); P. E. Eaton, *Acc. Chem. Res.*, **1**, 50 (1968).

193. D. I. Schuster, G. E. Heibel, P. B. Brown, N. J. Turro, and C. V. Kumar, *J. Am. Chem. Soc.*, **110**, 8261 (1988); N. A. Kaprinidis, G. Lem, S. H. Courtney, and D. I. Schuster, *J. Am. Chem. Soc.*, **115**, 3324 (1993); D. Andrew, D. J. Hastings, and A. C. Weedon, *J. Am. Chem. Soc.*, **116**, 10870 (1994).

194. L. K. Syudnes, K. I. Hansen, D. L. Oldroyd, A. C. Weedon, and E. Jorgensen, *Acta Chem. Scand.*, **47**, 916 (1993).

195. (a) W. C. Agosta and S. Wolff, *J. Org. Chem.*, **45**, 3139 (1980); (b) M. C. Pirrung, *J. Am. Chem. Soc.*, **103**, 82 (1981); (c) P. J. Connolly and C. H. Heathcock, *J. Org. Chem.*, **50**, 4135 (1985).

548

CHAPTER 6

*Concerted
Cycloadditions,
Unimolecular
Rearrangements, and
Thermal Eliminations*

Ref. 195c

Scheme 6.10 gives some examples of enone cycloaddition reactions. The reaction in Entry 1 was done by direct irradiation ($\lambda > 290$ nm) in benzene. No regiochemical issues arise and the cyano group does not change the course of the reaction. The reaction in Entry 2 was used to construct [4.2.2]propellane, and was done at low temperature. The reaction in Entry 3 presumably occurs by initial bonding at the β-carbon. The preference for the *syn* orientation of the cyclohexane ring appears to be due to a steric interaction with the isopropyl group. The closure of the cyclobutane ring shows little stereoselectivity, resulting in a 2:1 mixture of stereoisomers.

The stereochemistry of the adduct formed in Entry 4 is evidently *cis* at the cyclopentane ring but it is not clear if the cyclobutane ring is *syn* or *anti*. The reaction in Entry 6 gave a mixture of stereoisomers that was subjected to reductive elimination of the vicinal dichloride. The reaction in Entry 7 exhibited complete facial stereoselectivity based on the convex shape of the ring and the presence of the methyl group on the concave face. Entries 8 to 13 are intramolecular additions that generate polycyclic rings. The reaction in Entry 8 was used in the synthesis of longifolene, a tricyclic terpene. Entry 9 gave a single stereoisomer that was used in the synthesis of a sesquiterpene, isocomene. Entry 10 was part of a synthetic route to [5.5.5.4]fenestrane. The fenestranes are tetracyclic compounds that share a central carbon. The reaction in Entry 11 was used in the synthesis of a nitrogenous terpene, incarvilline. In Entry 12, a furan ring is involved in the photocyclization. The stereochemistry seems to be determined by the reactant conformation. Other conformations of the reactant have more destabilizing steric interactions.

6.3.2.3. Photocycloaddition Reactions of Carbonyl Compounds and Alkenes.

Photocycloaddition of ketones and aldehydes with alkenes can result in formation of four-membered cyclic ethers (oxetanes), a process often referred to as the *Paterno-Buchi reaction*.[196]

[196.] D. R. Arnold, *Adv. Photochem.*, **6**, 301 (1968); H. A. J. Carless, in *Synthetic Organic Photochemistry*, W. M. Horspool, ed., Plenum Press, New York, 1984, Chap. 8; T. Bach, *Synthesis*, 683 (1998).

Scheme 6.10. Photocycloadditions of Enones with Alkenes and Alkynes

A. Intrermolecular additions

1[a]

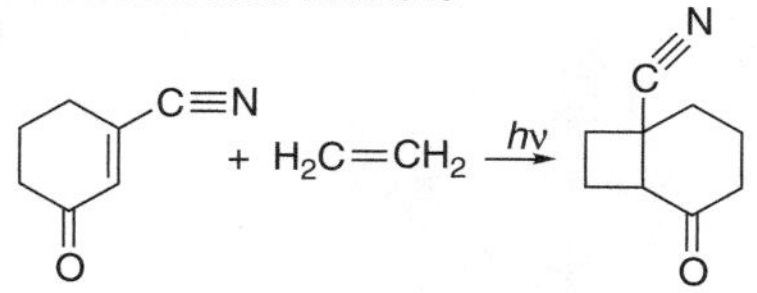

62%

2[b]

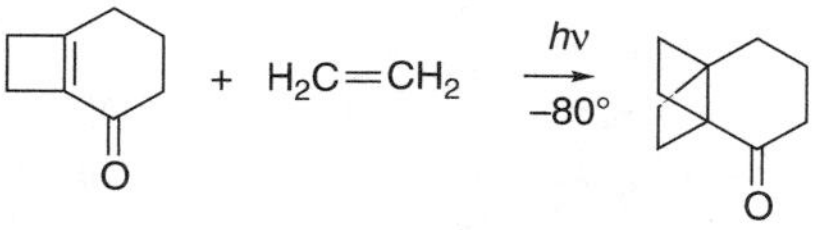

50%

3[c]

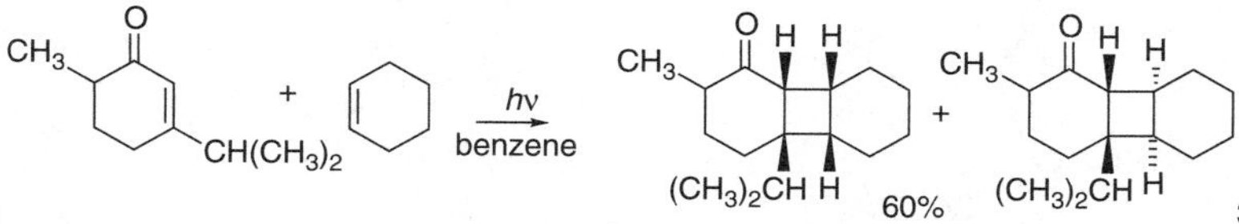

60%   30%

4[d]

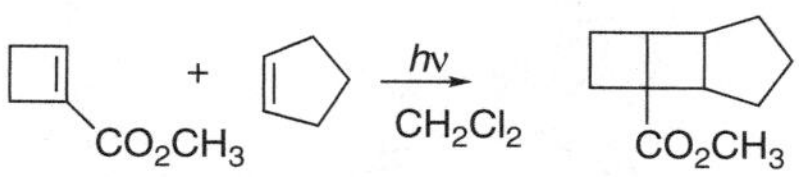

67%

5[e]

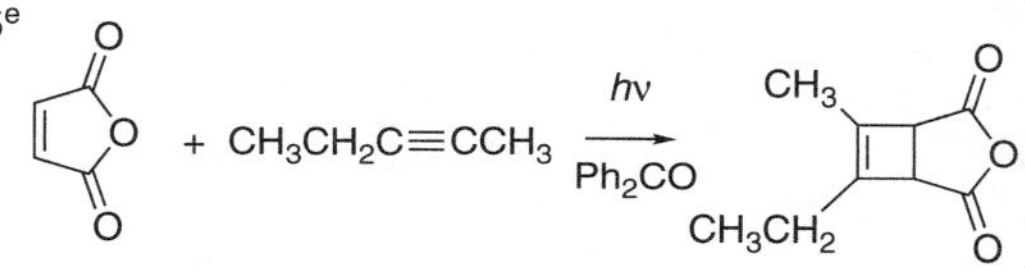

79%

6[f]

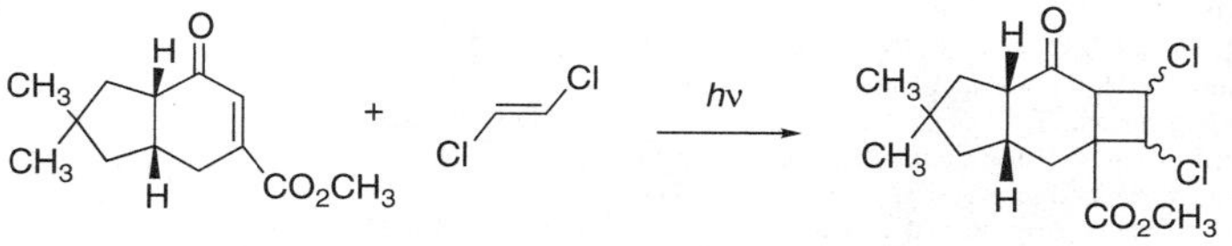

95%

7[g]

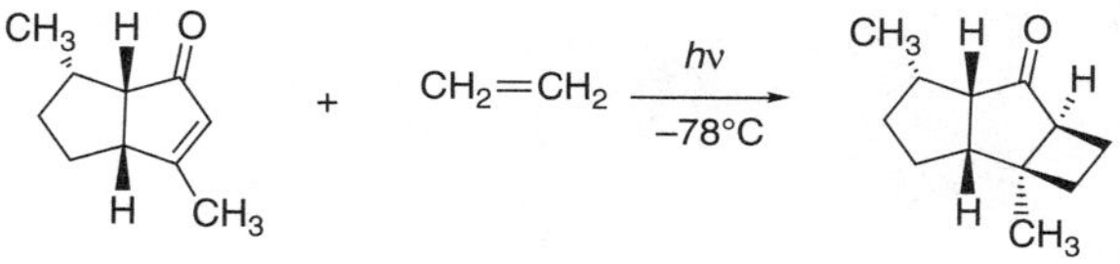

B. Intramolecular Additions

8[h]

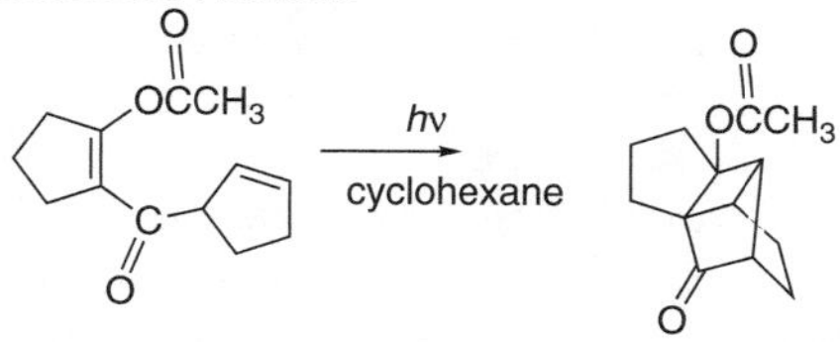

78%

9[i]

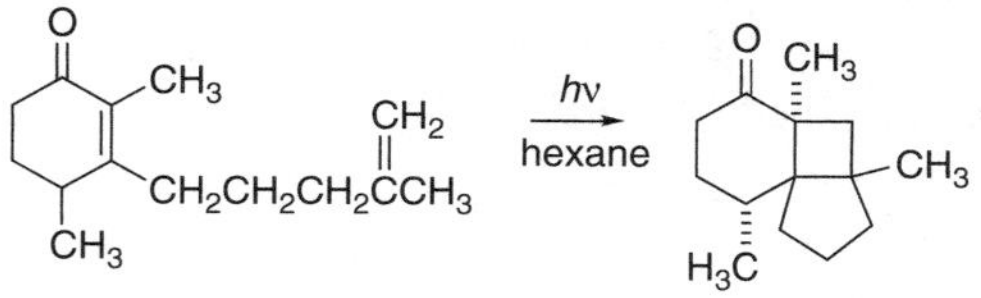

77%

(Continued)

550

CHAPTER 6

*Concerted
Cycloadditions,
Unimolecular
Rearrangements, and
Thermal Eliminations*

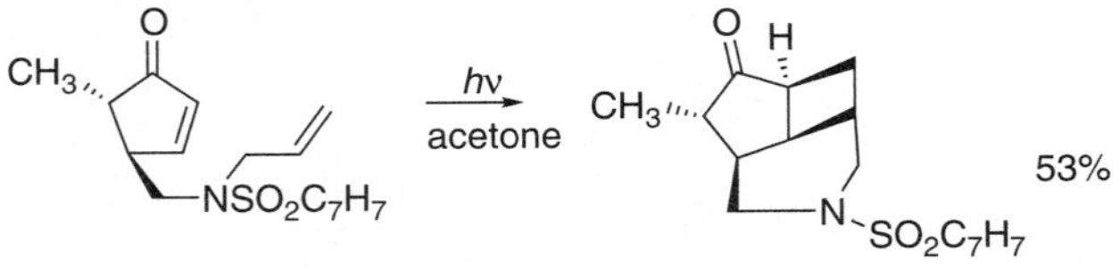

Scheme 6.10. (*Continued*)

10ʲ

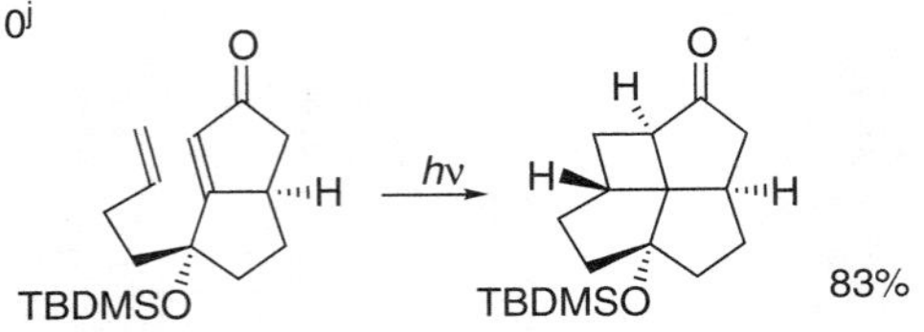

83%

11ᵏ

53%

12ˡ

85%

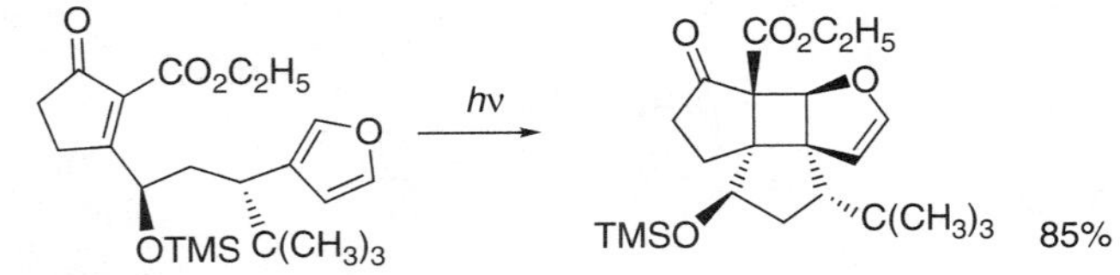

a. W. C. Agosta and W. W. Lowrance, Jr., *J. Org. Chem.*, **35**, 3851 (1970).
b. P. E. Eaton and K. Nyi, *J. Am. Chem. Soc.*, **93**, 2786 (1971).
c. P. Singh, *J. Org. Chem.*, **36**, 3334 (1971).
d. P. A. Wender and J. C. Lechleiter, *J. Am. Chem. Soc.*, **99**, 267 (1977).
e. R. M. Scarborough, Jr., B. H. Toder, and A. B. Smith, III, *J. Am. Chem. Soc.*, **102**, 3904 (1980).
f. G. Mehta and K. Sreenivas, *Tetrahedron Lett.*, **43**, 703 (2002).
g. E. Piers and A. Orellana, *Synthesis*, 2138 (2001).
h. W. Oppolzer and T. Godel, *J. Am. Chem. Soc.*, **100**, 2583 (1978).
i. M. C. Pirrung, *J. Am. Chem. Soc.*, **103**, 82 (1981).
j. M. Thommen and R. Keese, *Synlett*, 231 (1997).
k. M. Ichikawa, S. Aoyagi, and C. Kibayashi, *Tetrahedron Lett.*, **46**, 2327 (2005).
l. M. T. Crimmins, J. M. Pace, P. G. Naternet, A. S. Kim-Meade, J. B. Thomas, S. H. Watterson, and A. S. Wagman, *J. Am. Chem. Soc.*, **122**, 8453 (2000).

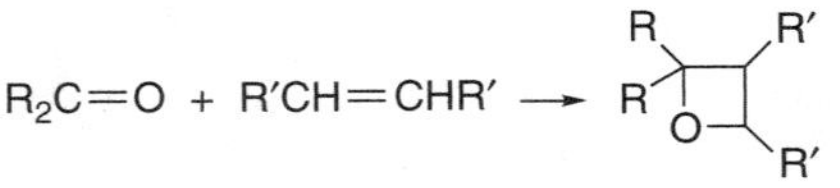

$$R_2C{=}O \ + \ R'CH{=}CHR' \ \longrightarrow \ \begin{array}{c} R \quad R' \\ R \overset{\displaystyle}{\underset{O}{\square}} R' \end{array}$$

The reaction is stereospecific for at least some aliphatic ketones but not for aromatic carbonyls.[197] This result suggests that the reactive excited state is a singlet for aliphatics and a triplets for aromatics. With aromatic ketones, the regioselectivity of addition can usually be predicted on the basis of formation of the more stable of the two possible diradical intermediates obtained by bond formation between oxygen and the alkene.[198]

197. N. C. Yang and W. Eisenhardt, *J. Am. Chem. Soc.*, **93**, 1277 (1971); D. R. Arnold, R. L. Hinman, and A. H. Glick, *Tetrahedron Lett.*, 1425 (1964); N. J. Turro and P. A. Wriede, *J. Am. Chem. Soc.*, **90**, 6863 (1968); J. A. Barltrop and H. A. J. Carless, *J. Am. Chem. Soc.*, **94**, 8761 (1972).
198. A. Griesbach, S. Buhr, M. Fiegel, J. Lex, and H. Schmickler, *J. Org. Chem.*, **63**, 3847 (1998).

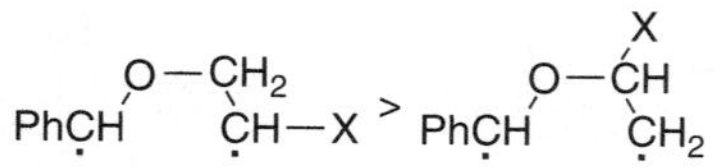

Stereochemistry can be interpreted in terms of conformation effects in the 1,4-biradical intermediates.[199] Vinyl enol ethers and enamides add to aromatic ketones to give 3-substituted oxetanes, usually with the *cis* isomer preferred.[200]

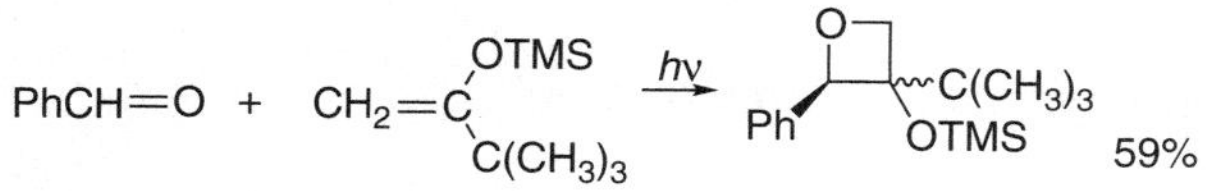

Ref. 200a

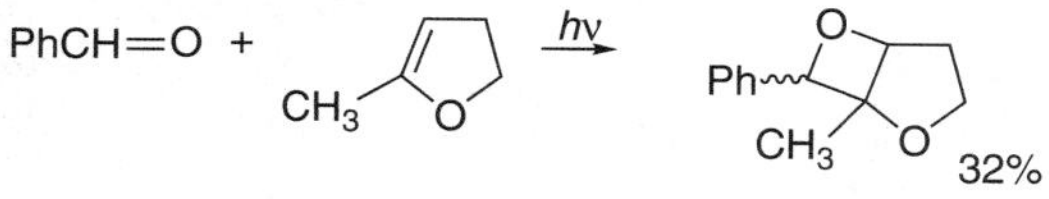

Ref. 199

Scheme 6.11. Photocycloaddition Reactions of Carbonyl Compounds and Alkenes

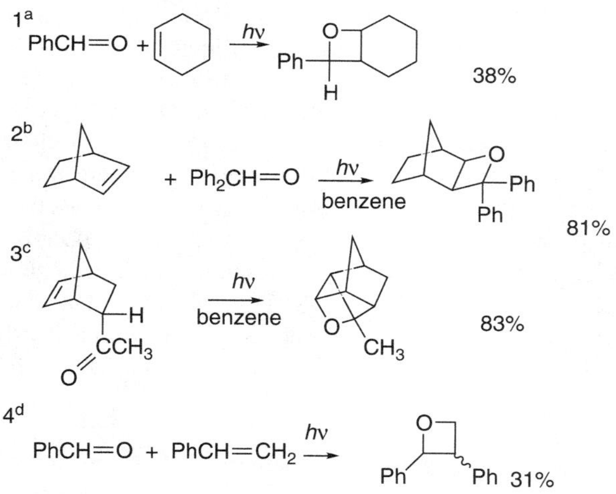

a. J. S. Bradshaw, *J. Org. Chem.*, **31**, 237 (1966).
b. D. R. Arnold, A. H. Glick, and V. Y. Abraitys, *Org. Photochem. Synth.*, **1**, 51 (1971).
c. R. R. Sauers, W. Schinksi, and B. Sickles, *Org. Photochem. Synth.*, **1**, 76 (1971).
d. H. A. J. Carless, A. K. Maitra, and H. S. Trivedi *J. Chem. Soc., Chem. Commun.*, 984 (1979).

199. A. G. Griesbach and S. Stadtmuller, *J. Am. Chem. Soc.*, **113**, 6923 (1991).
200. (a) T. Bach, *Tetrahedron Lett.*, **32**, 7037 (1991); (b) A. G. Griesbeck and S. Stadtmuller, *J. Am. Chem. Soc.*, **113**, 6923 (1991); (c) T. Bach, *Liebigs Ann. Chem.*, 1627 (1997); T. Bach, *Synthesis*, 683 (1998).

552

CHAPTER 6

*Concerted
Cycloadditions,
Unimolecular
Rearrangements, and
Thermal Eliminations*

$$PhCH{=}O \ + \ \text{(pyrroline ring)} \ \xrightarrow{h\nu} \ \text{(bicyclic product)}$$

Ref. 200c

Some other examples of Paterno-Buchi reactions are given in Scheme 6.11.

6.4. [3,3]-Sigmatropic Rearrangements

The mechanistic basis of sigmatropic rearrangements was introduced in Chapter 10 of Part A. The sigmatropic process that is most widely applied in synthesis is the [3,3]-sigmatropic rearrangement. The principles of orbital symmetry establish that concerted [3,3]-sigmatropic rearrangements are allowed processes. Stereochemical predictions and analyses are based on the cyclic transition structure for a concerted reaction mechanism. Some of the various [3,3]-sigmatropic rearrangements that are used in synthesis are presented in outline form in Scheme 6.12.[201] We discuss these reactions in succeeding sections.

6.4.1. Cope Rearrangements

The Cope rearrangement is the conversion of a 1,5-hexadiene derivative to an isomeric 1,5-hexadiene by the [3,3]-sigmatropic mechanism. For unstrained compounds, the reaction occurs in the range of 150°–250°C. The reaction is both stereospecific and stereoselective. It is stereospecific in that a *Z*- or *E*-configurational relationship at either double bond is maintained in the TS and governs the relative configuration at the newly formed single bond in the product.[202] However, the relationship depends on the conformation of the TS. When a chair TS is favored the *E,E*- and *Z,Z*-dienes lead to *anti*-3,4-diastereomers, whereas the *E,Z*- and *Z,E*-isomers give the 3,4-*syn* product. TS conformation also determines the stereochemistry of the new double bond. If both *E*- and *Z*-stereoisomers are possible for the product, the product ratio reflects product (and TS) stability. The *E*-arrangement is normally favored for the newly formed double bonds. The stereochemical aspects of the Cope rearrangements for simple acyclic reactants are consistent with a chairlike TS in which the larger substituent at C(3) [or C(4)] adopts an equatorial-like conformation.

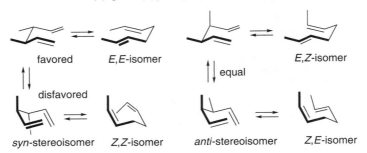

favored   *E,E*-isomer   *E,Z*-isomer

disfavored   equal

syn-stereoisomer   *Z,Z*-isomer   *anti*-stereoisomer   *Z,E*-isomer

201. For reviews of synthetic application of [3,3]sigmatropic rearrangements, see G. B. Bennett, *Synthesis*, 589 (1977); F. E. Ziegler, *Acc. Chem. Res.*, **10**, 227 (1977).
202. W. v. E. Doering and W. R. Roth, *Tetrahedron*, **18**, 67 (1962).

1[a] Cope rearrangement

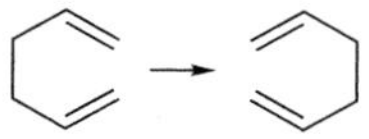

2[b] Oxy-Cope rearrangement

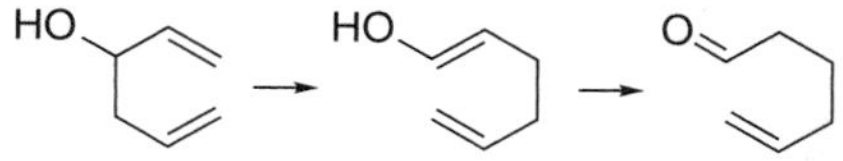

3[c] Anionic oxy-Cope rearrangement

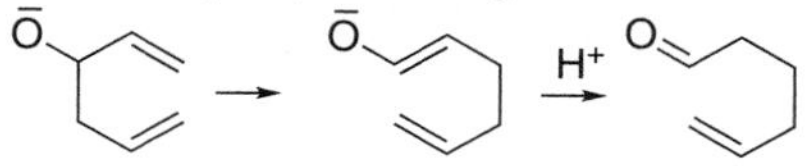

4[d] Claisen rearrangement of allyl vinyl ethers

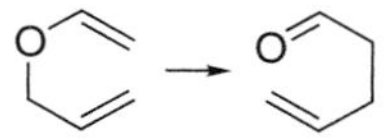

5[d] Claisen rearrangement of allyl phenyl ethers

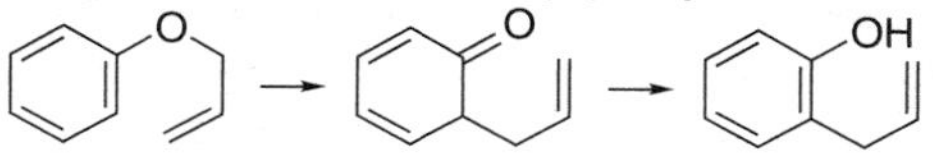

6[e] Ortho ester Claisen rearrangement

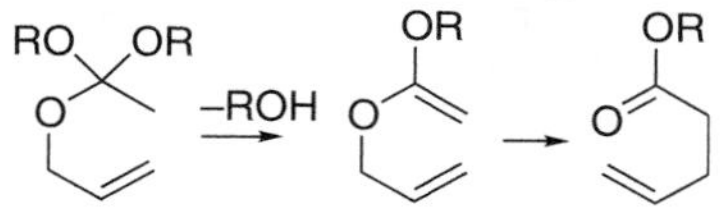

7[f] Ireland-Claisen rearrangement of *O*-allyl-*O'*-trimethylsilyl ketene acetals

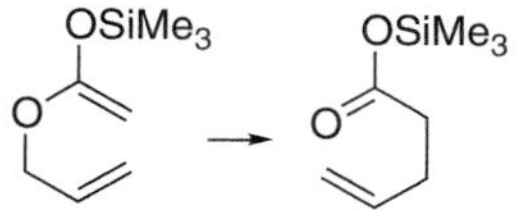

8[g] Ester enolate Claisen rearrangement

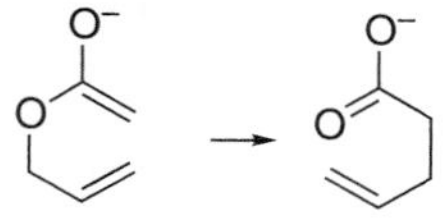

9[h] Claisen rearrangement of *O*-allyl-*N,N*-dialkyl ketene aminals

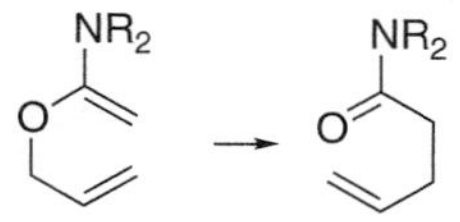

(*Continued*)

CHAPTER 6

*Concerted
Cycloadditions,
Unimolecular
Rearrangements, and
Thermal Eliminations*

Scheme 6.12. (*Continued*)

10^i Aza-Claisen rearrangement of *O*-allyl imidates

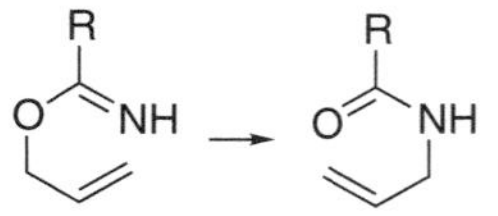

a. S. J. Rhoads and N. R. Raulins, *Org. React.*, **22**, 1 (1975).
b. J. A. Berson and M. Jones, Jr., *J. Am. Chem. Soc.*, **86**, 5019 (1964).
c. D. A. Evans and A. M. Golob, *J. Am. Chem. Soc.*, **97**, 4765 (1975).
d. D. S. Tarbell, *Org. React.*, **2**, 1 (1944).
e. W. S. Johnson, L. Werthemann, W. R. Bartlett, T. J. Brocksom, T. Li, D. J. Faulkner, and M. R. Petersen, *J. Am. Chem. Soc.*, **92**, 741 (1970).
f. R. E. Ireland and R. H. Mueller, *J. Am. Chem. Soc.*, **94**, 5898 (1972).
g. R. E. Ireland, R. H. Mueller, and A. K. Willard, *J. Am. Chem. Soc.*, **98**, 2868 (1976).
h. D. Felix, K. Gschwend-Steen, A. E. Wick, and A. Eschenmoser, *J. Am. Chem. Soc.*, **98**, 2868 (1976).
i. L. E. Overman, *Acc. Chem. Res.*, **13**, 218 (1980).

Owing to the concerted mechanism, chirality at C(3) [or C(4)] leads to enantiospecific formation of new stereogenic centers formed at C(1) [or C(6)].[203] These relationships are illustrated in the example below. Both the configuration of the new stereocenter and the new double bond are those expected on the basis of a chairlike TS. Since there are two stereogenic centers, the double bond and the asymmetric carbon, there are four possible stereoisomers of the product. Only two are formed. The *E*-double bond isomer has the *S*-configuration at C(4) and the *Z*-isomer has the *R*-configuration. These are the products expected for a chair TS. The stereochemistry of the new double bond is determined by the relative stability of the two chair TSs. TS **B** is less favorable than **A** because of the axial placement of the larger phenyl substituent.

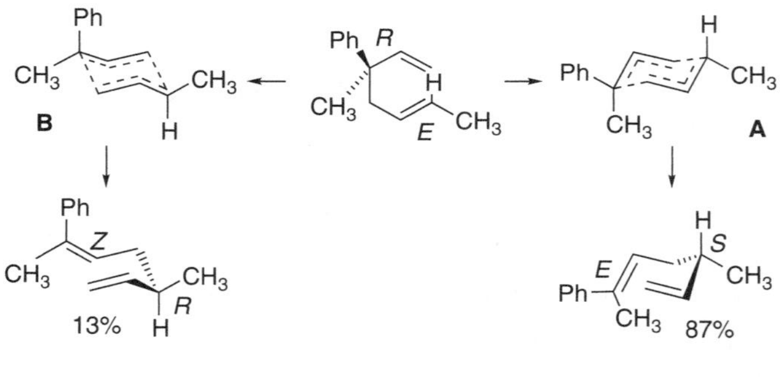

The products corresponding to boatlike TSs are usually not observed for acyclic dienes. However, this TS is allowed and if steric factors make a boat TS preferable to a chair, reaction can proceed through a boat. Thermochemical[204] and computational[205] studies indicate that the boat TS is intrinsically 6–10 kcal/mol higher in energy. Reactions that proceed through a boat TS have the reverse stereochemical relationships between the configuration at the stereogenic center and the double bond.

203. R. K. Hill and N. W. Gilman, *Chem. Commun.*, 619 (1967); R. K. Hill, in *Asymmetric Synthesis*, Vol. 4, J. D. Morrison, ed., Academic Press, New York, 1984, pp. 503–572.
204. M. Goldstein and M. S. Benzon, *J. Am. Chem. Soc.*, **94**, 7147 (1972).
205. O. Wiest, K. A. Black, and K. N. Houk, *J. Am. Chem. Soc.*, **116**, 10336 (1995).

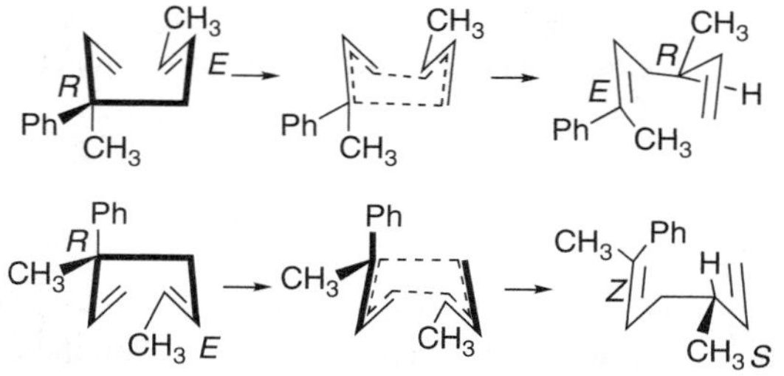

Cope rearrangements are reversible reactions and, as there is no change in the number or types of bonds as a result of the reaction, to a first approximation the total bond energy is unchanged. The position of the final equilibrium is governed by the relative stability of the starting material and product. In the example cited above, the equilibrium is favorable because the product is stabilized by conjugation of the alkene with the phenyl ring.

When ring strain is relieved, Cope rearrangements can occur at much lower temperatures and with complete conversion to ring-opened products. A striking example is the conversion of *cis*-divinylcyclopropane to 1,4-cycloheptadiene, a reaction that occurs readily below −40° C.[206]

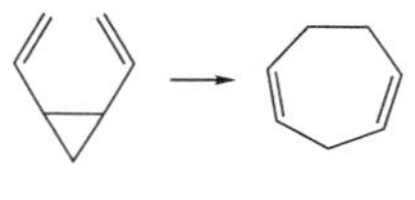

Several transition metal ions and complexes, especially Pd(II) salts, have been found to catalyze Cope rearrangements.[207] The catalyst that has been adopted for synthetic purposes is $PdCl_2(CH_3CN)_2$, and with it the rearrangement of **14** to **15** and **16** occurs at room temperature, as contrasted to 240° C in its absence.[208] The catalyzed reaction shows enhanced stereoselectivity and is consistent with a chairlike TS.

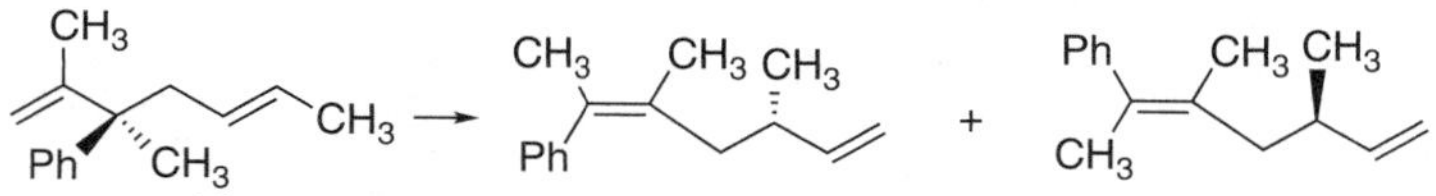

14 → **15** thermal 1:1
catalyzed 7:3

+ **16** >90% enantioselectivity
under both conditions

The mechanism for catalysis is formulated as a stepwise process in which the electrophilic character of Pd(II) facilitates the bond formation.[209]

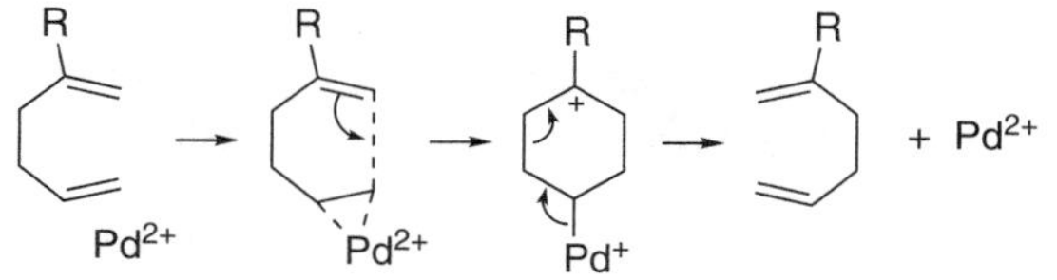

When there is a hydroxy substituent at C(3) of the diene system, the Cope rearrangement product is an enol that is subsequently converted to the corresponding

206. W. v. E. Doering and W. R. Roth, *Tetrahedron*, **19**, 715 (1963).
207. R. P. Lutz, *Chem. Rev.*, **84**, 205 (1984).
208. L. E. Overman and F. M. Knoll, *J. Am. Chem. Soc.*, **102**, 865 (1980).
209. L. E. Overman and A. F. Renaldo, *J. Am. Chem. Soc.*, **112**, 3945 (1990).

556

CHAPTER 6

Concerted
Cycloadditions,
Unimolecular
Rearrangements, and
Thermal Eliminations

carbonyl compound. This is called the *oxy-Cope rearrangement*.[210] The formation of the carbonyl compound provides a net driving force for the reaction.[211]

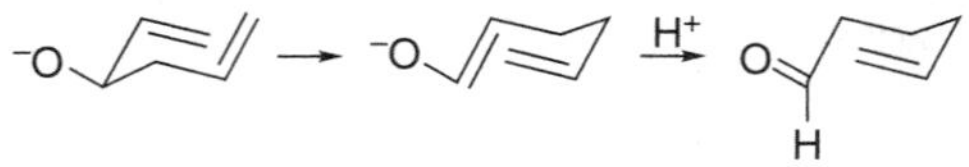

An important improvement in the oxy-Cope reaction was made when it was found that the reaction is strongly catalyzed by base.[212] When the C(3) hydroxy group is converted to its alkoxide, the reaction is accelerated by a factor of 10^{10}–10^{17}. These base-catalyzed reactions are called *anionic oxy-Cope rearrangements*, and their rates depend on the degree of cation coordination at the oxy anion. The reactivity trend is $K^+ > Na^+ > Li^+$. Catalytic amounts of tetra-*n*-butylammonium salts lead to accelerated rates in some cases. This presumably results from the dissociation of less reactive ion pair species promoted by the tetra-*n*-butylammonium ion.[213]

The stereochemistry of acyclic anionic oxy-Cope rearrangements is consistent with a chair TS having a conformation that favors equatorial placement of both alkyl and oxy substituents and minimizes the number of 1,3-diaxial interactions.[214] For the reactions shown below, the double-bond configuration is correctly predicted on the basis of the most stable TS available in the first three reactions. In the fourth reaction, the TSs are of comparable energy and a 2:1 mixture of *E*- and *Z*-isomers is formed.

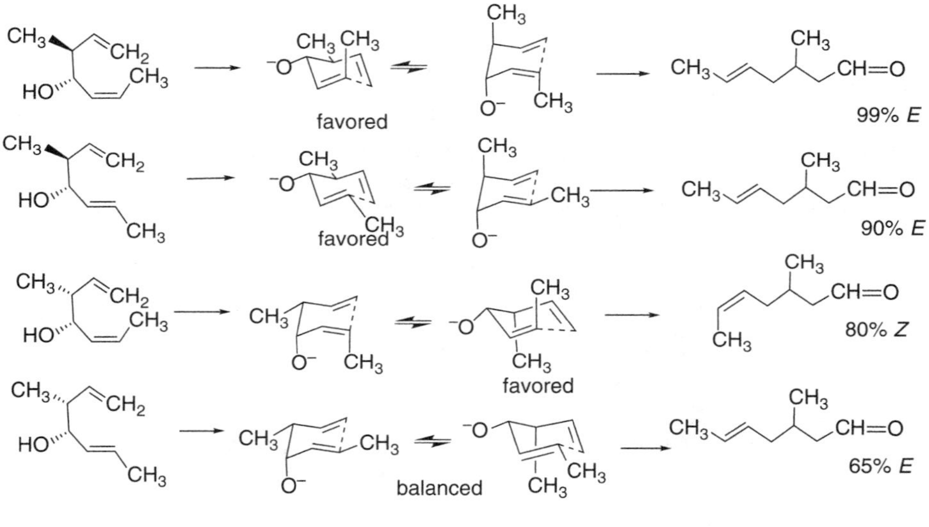

Silyl ethers of vinyl allyl alcohols can also be used in oxy-Cope rearrangements.[215] Known as the *siloxy-Cope rearrangement*, this methodology has been used in

210. S. R. Wilson, *Org. React.*, **43**, 93 (1993); L. A. Paquette, *Angew. Chem. Int. Ed. Engl.*, **29**, 609 (1990); L. A. Paquette, *Tetrahedron*, **53**, 13971 (1997).

211. A. Viola, E. J. Iorio, K. K. Chen, G. M. Glover, U. Nayak, and P. J. Kocienski, *J. Am. Chem. Soc.*, **89**, 3462 (1967).

212. D. A. Evans and A. M. Golob, *J. Am. Chem. Soc.*, **97**, 4765 (1975); D. A. Evans, D. J. Balillargeon, and J. V. Nelson, *J. Am. Chem. Soc.*, **100**, 2242 (1978).

213. M. George, T.-F. Tam, and B. Fraser-Reid, *J. Org. Chem.*, **50**, 5747 (1985).

214. K. Tomooka, S.-Y. Wei, and T. Nakai, *Chem. Lett.*, 43 (1991).

215. R. W. Thies, M. T. Wills, A. W. Chin, L. E. Schick, and E. S. Walton, *J. Am. Chem. Soc.*, **95**, 5281 (1973).

connection with *syn*-selective aldol additions in stereoselective synthesis.[216] The use of the silyloxy group prevents reversal of the aldol addition, which would otherwise occur under anionic conditions. The reactions proceed at convenient rates at 140°–180° C.

Ref. 217

Ref. 218

>95%

Scheme 6.13 gives some examples of Cope and oxy-Cope rearrangements. Entry 1 shows a reaction that was done to compare the energy of chair and boat TSs. The chiral diastereomer shown can react through a chair TS and has a ΔG^* about 8 kcal/mol lower than the *meso* isomer, which must react through a boat TS. The equilibrium is biased toward product by the fact that the double bonds in the product are more highly substituted, and therefore more stable, than those in the reactant.

$\Delta G^* = 34$ kcal/mol

$\Delta G^* = 42$ kcal/mol

Entry 2 illustrates the reversibility of the Cope rearrangement. In this case, the equilibrium is closely balanced with the reactant benefiting from a more-substituted double bond, whereas the product is stabilized by conjugation. The reaction in Entry 3 involves a *cis*-divinylcyclopropane and proceeds at much lower temperature than the previous examples. The reaction was used in the preparation of an intermediate for the synthesis of pseudoguiane-type natural products.

Entries 4 and 5 illustrate the use of the oxy-Cope rearrangement in formation of medium-size rings. The *trans*-double bond in the product for Entry 4 arises from a chair TS.

216. C. Schneider and M. Rehfeuter, *Synlett*, 212 (1996); C. Schneider and M. Rehfeuter, *Tetrahedron*, **53**, 133 (1997); W. C. Black, A. Giroux, and G. Greidanus, *Tetrahedron Lett.*, **37**, 4471 (1996).
217. C. Schneider, *Eur. J. Org. Chem.*, 1661 (1998).
218. M. M. Bio and J. L. Leighton, *J. Am. Chem. Soc.*, **121**, 890 (1999).

558

CHAPTER 6

*Concerted
Cycloadditions,
Unimolecular
Rearrangements, and
Thermal Eliminations*

Scheme 6.13. Cope and Oxy-Cope Rearrangements of 1,5-Dienes

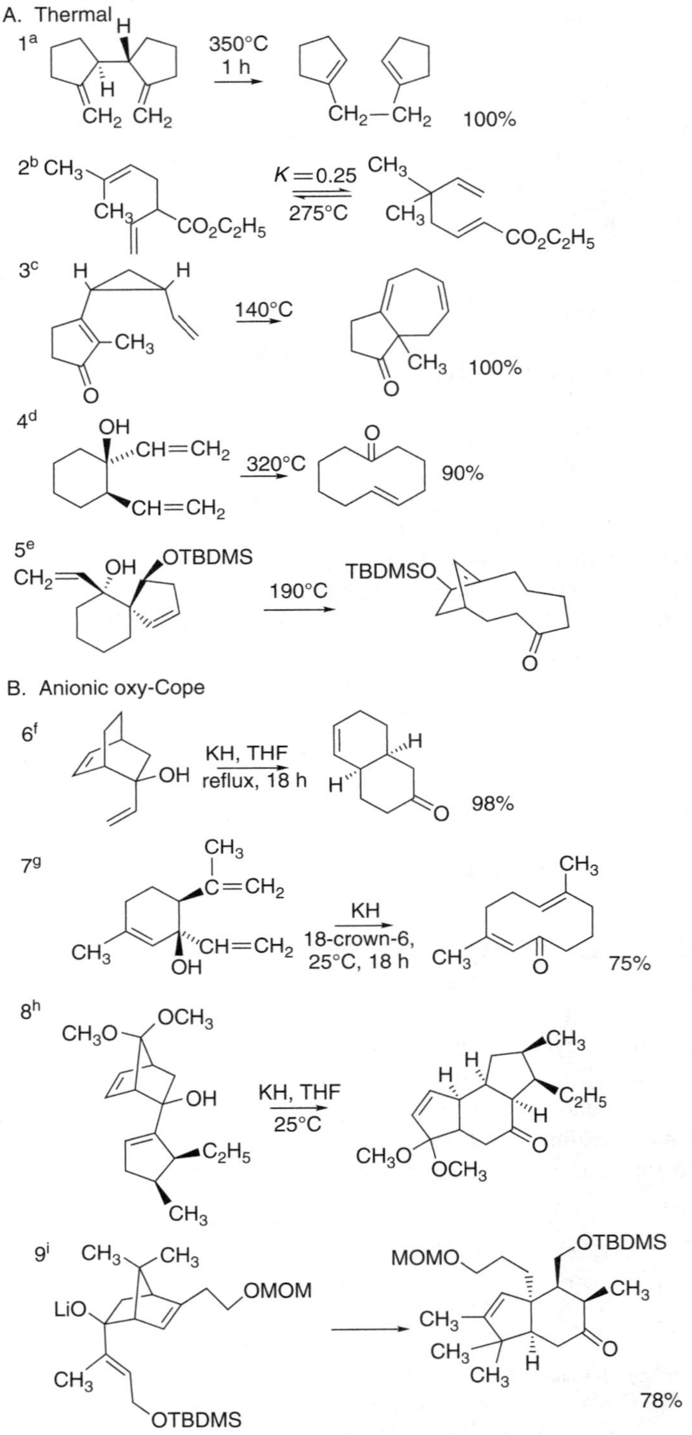

A. Thermal

B. Anionic oxy-Cope

(*Continued*)

Scheme 6.13. (*Continued*)

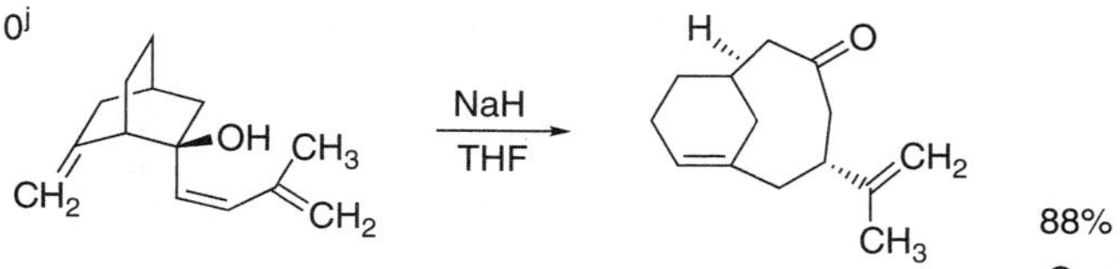

559

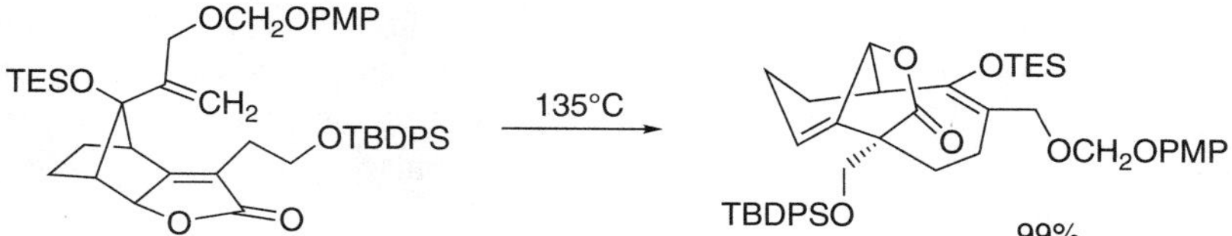

SECTION 6.4

*[3,3]-Sigmatropic
Rearrangements*

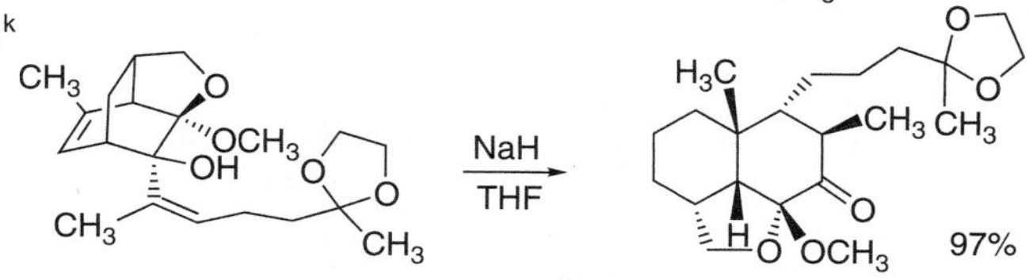

C. Siloxy-Cope

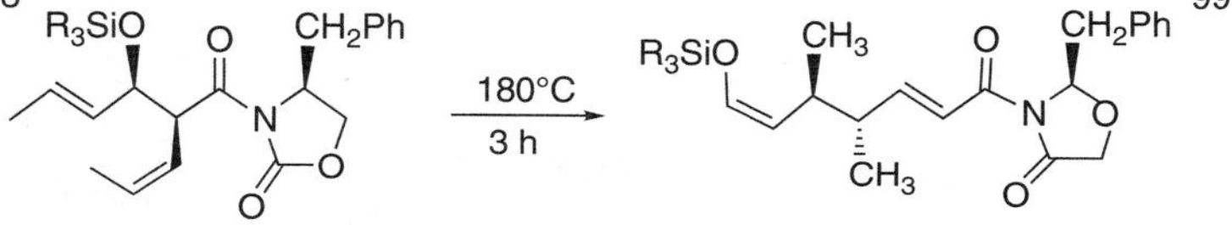

a. K. J. Shea and R. B. Phillips, *J. Am. Chem. Soc.*, **102**, 3156 (1980).
b. F. E. Ziegler and J. J. Piwinski, *J. Am. Chem. Soc.*, **101**, 1612 (1979).
c. P. A. Wender, M. A. Eissenstat, and M. P. Filosa, *J. Am. Chem. Soc.*, **101**, 2196 (1979).
d. E. N. Marvell and W. Whalley, *Tetrahedron Lett.*, 509 (1970).
e. G. Ladouceur and L.A. Paquette, *Synthesis*, 185 (1992)
f. D. A. Evans, A. M. Golob, N. S. Mandel, and G. S. Mandel, *J. Am. Chem. Soc.*, **100**, 8170 (1978).
g. W. C. Still, *J. Am. Chem. Soc.*, **99**, 4186 (1977).
h. L. A. Paquette, K. S. Learn, J. L. Romine, and H.-S. Lin, *J. Am. Chem. Soc.*, **110**, 879 (1988); L. A. Paquette,
   J. L. Romine, H.-S. Lin, and J. Wright, *J. Am. Chem. Soc.*, **112**, 9284 (1990).
i. L. A. Paquette and F.-T. Hong, *J. Org. Chem.*, **68**, 6905 (2003).
j. L. Gentric, I. Hanna, A. Huboux, and R. Zaghdoudi, *Org. Lett.*, **5**, 3631 (2003).
k. D. S. Hsu and C-C. Liao, *Org. Lett.*, **5**, 3631 (2003).
l. D. L. J. Clive, S. Sun, V. Gagliardini, and M. K. Sano, *Tetrahedron Lett.*, **41**, 6259 (2000).
m. C. Schneider, *Eur. J. Org. Chem.*, 1661 (1998).

The reaction in Entry 5 is a case in which the thermal conditions were preferable to the basic conditions because of the base sensitivity of the product. Entries 6 to 10 show anionic oxy-Cope reactions. Entries 6 and 7 are early examples of the application of the reaction in synthesis. Entries 8 and 9 involve rearrangements of bicyclo[2.2.1]hept-2-en-2-ol derivatives to give *cis*-fused bicyclo[4.3.0]non-7-en-3-ones.

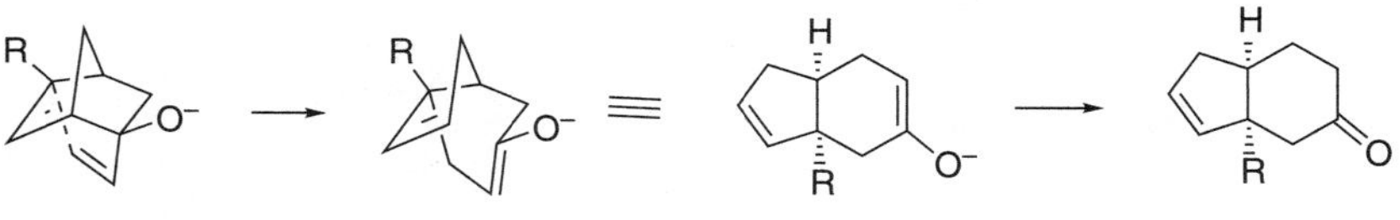

The rearrangement in Entry 9 occurs spontaneously on warming of the reaction mixture from addition of an organolithium reagent to form the vinyl carbinol unit. This is a very general means of constructing reactants for oxy-Cope rearrangements that leads

560

CHAPTER 6

*Concerted
Cycloadditions,
Unimolecular
Rearrangements, and
Thermal Eliminations*

to carbon-carbon bond formation between C(2) of the vinyllithium reagent and C(4) of the β,γ-enone.

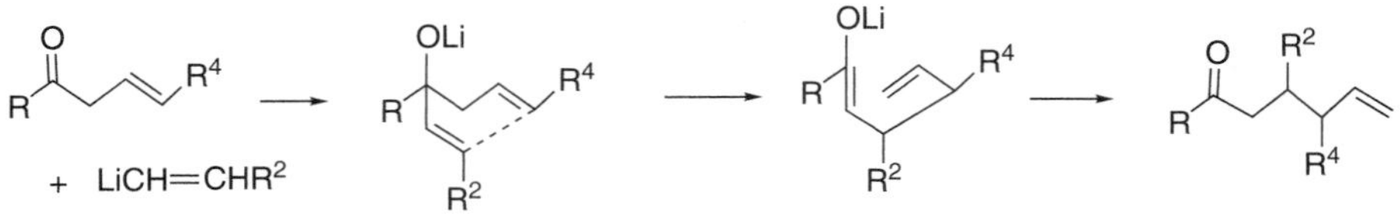

The reaction in Entry 10 demonstrated that a vinyl substituent in conjugation with the vinyl carbinol accelerates rearrangement. The reaction was considerably more facile than the corresponding reaction with a saturated isopropyl group. The reaction in Entry 11 was used in the synthesis of terpene derivatives. Entries 12 and 13 are examples of the siloxy-Cope version of the reaction. These entries illustrate the utility of the oxy-Cope reaction in the synthesis of ring systems. Some of these transformations may be difficult to recognize, at least at first glance. The retrosynthetic transformation can be recognized by identifying the δ,ε-enone and locating the bond that is formed in the rearrangement. For example, the retrosynthetic formulation of the reaction in Entry 9 identifies the precursor.

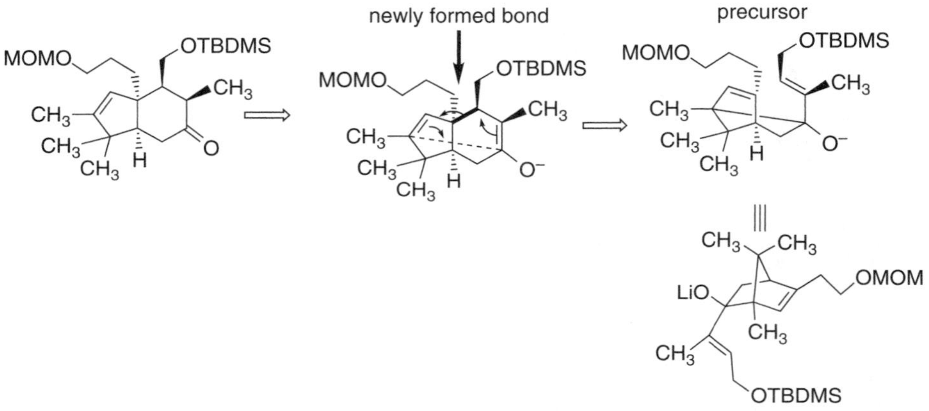

6.4.2. Claisen and Modified Claisen Rearrangements

The basic pattern of the Claisen rearrangement is the conversion of a vinyl allyl ether to a γ,δ-enone. The reaction is also observed for allyl phenyl ethers, in which case the products are *o*-allylphenols.

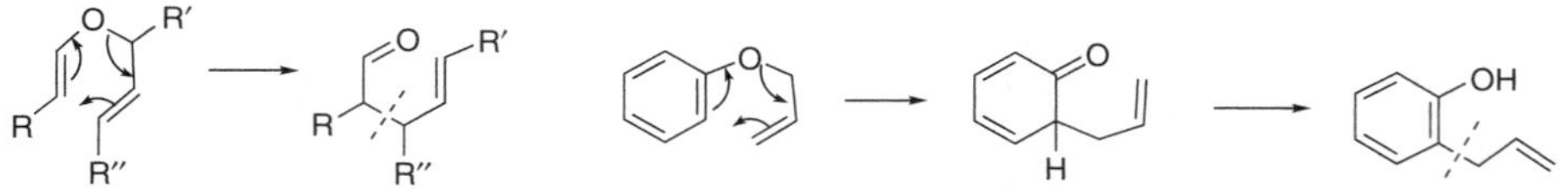

There are several synthetically important adaptations of the reaction. It can be applied to orthoesters (Section 6.4.2.2) or silyl ketene acetals (Section 6.4.2.3), in which case the products are γ,δ-unsaturated acids or esters. An analogous reaction using amide acetals gives γ,δ-unsaturated amides (Section 6.4.2.4). In all cases, the reactions occur with 1,3-transposition of the allylic group.

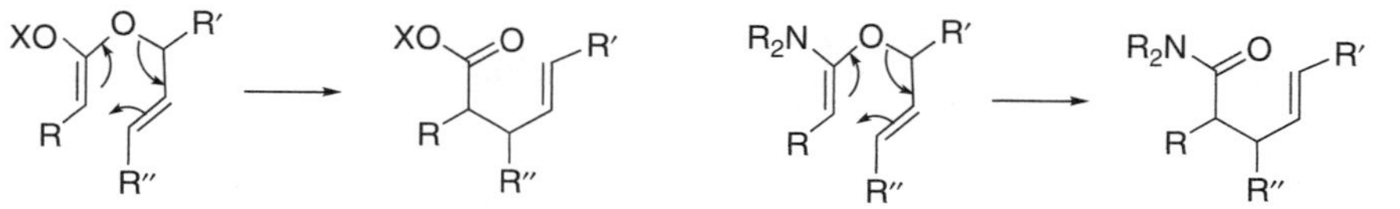

X = alkyl, silyl

6.4.2.1. Claisen Rearrangements of Allyl Vinyl Ethers. The [3,3]-sigmatropic rearrangement of allyl vinyl ethers leads to γ,δ-enones and is known as the *Claisen rearrangement.*[219] The reaction is mechanistically analogous to the Cope rearrangement and occurs at temperatures above 150° C. As the product is a carbonyl compound, the equilibrium is usually favorable. The reaction introduces an α-acyl alkyl group at the γ-carbon of the allylic alcohol, with 1,3-transposition of the allylic double bond.

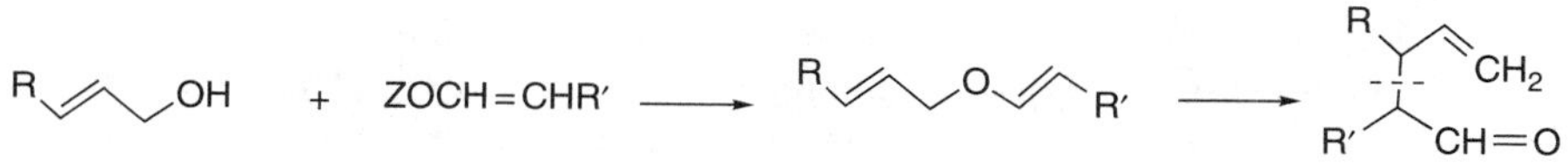

The reactants can be made from allylic alcohols by mercuric ion-catalyzed exchange with ethyl vinyl ether.[220] The allyl vinyl ether need not be isolated and is often prepared under conditions that lead to its rearrangement. The simplest of all Claisen rearrangements, the conversion of allyl vinyl ether to 4-pentenal, typifies this process.

$$CH_2{=}CHCH_2OH + CH_2{=}CHOCH_2CH_3 \xrightarrow[\Delta]{Hg(OAc)_2} [CH_2{=}CHCH_2OCH{=}CH_2] \rightarrow CH_2{=}CHCH_2CH_2\overset{O}{\overset{\|}{C}}H \quad 96\%$$

Ref. 221

Acid-catalyzed exchange can also be used to prepare the vinyl ethers.

$$RCH{=}CHCH_2OH + CH_3CH_2OCH{=}CH_2 \xrightarrow{H^+} RCH{=}CHCH_2OCH{=}CH_2$$

Ref. 222

Vinyl ethers can also be generated by thermal elimination reactions. For example, base-catalyzed conjugate addition of allyl alcohols to phenyl vinyl sulfone generates 2-(phenylsulfinyl)ethyl ethers that can undergo elimination at 200° C.[223] The sigmatropic

219. F. E. Ziegler, *Chem. Rev.*, **88**, 1423 (1988); A. M. M. Castro, *Chem. Rev.*, **104**, 2939 (2004).
220. W. H. Watanabe and L. E. Conlon, *J. Am. Chem. Soc.*, **79**, 2828 (1957); D. B. Tulshian, R. Tsang, and B. Fraser-Reid, *J. Org. Chem.*, **49**, 2347 (1984).
221. S. E. Wilson, *Tetrahedron Lett.*, 4651 (1975).
222. G. Saucy and R. Marbet, *Helv. Chim. Acta*, **50**, 2091 (1967); R. Marbet and G. Saucy, *Helv. Chim. Acta*, **50**, 2095 (1967).
223. T. Mandai, S. Matsumoto, M. Kohama, M. Kawada, J. Tsuji, S. Saito, and T. Moriwake, *J. Org. Chem.*, **55**, 5671 (1990); T. Mandai, M. Ueda, S. Hagesawa, M. Kawada, J. Tsuji, and S. Saito, *Tetrahedron Lett.*, **31**, 4041 (1990).

562

CHAPTER 6

*Concerted
Cycloadditions,
Unimolecular
Rearrangements, and
Thermal Eliminations*

rearrangement proceeds under these conditions. Allyl vinyl ethers can also be prepared by Wittig reactions using ylides generated from allyloxymethylphosphonium salts.[224]

$$RCH=CHCH_2OH + CH_2=CHSPh \xrightarrow{NaH} RCH=CHCH_2OCH_2CH_2SPh \xrightarrow{200°C} RCH=CHCH_2OCH=CH_2$$

$$R_2C=O + Ph_3P^+CH_2OCH_2CH=CH_2 \xrightarrow{K^+ {}^-O\text{-}t\text{-}Bu} R_2C=CHOCH_2CH=CH_2$$

As with the Cope rearrangement, $PdCl_2$ can catalyze the Claisen rearrangement.

65%

Ref. 225

However, it can also catalyze competing reactions and works best for relatively highly substituted systems.[226] Catalysis of Claisen rearrangements has been achieved using highly hindered *bis*-(phenoxy)methylaluminum as Lewis acids.[227] These reagents also have the ability to control the E:Z ratio of the products. Very bulky catalysts tend to favor the Z-isomer by forcing the α-substituent of the allyl group into an axial conformation.

tris-Aryloxyaluminum compounds are also effective catalysts for the Claisen rearrangement.[228] When used in a 1.2 molar ratio, the rearrangement occurs at −78° C.

Some representative Claisen rearrangements are shown in Scheme 6.14. Entry 1 illustrates the application of the Claisen rearrangement in the introduction of a substituent at the junction of two six-membered rings. Introduction of a substituent at this type of position is frequently necessary in the synthesis of steroids and terpenes. In Entry 2, formation and rearrangement of a 2-propenyl ether leads to formation of a methyl ketone. Entry 3 illustrates the use of 3-methoxyisoprene to form the allylic ether. The rearrangement of this type of ether leads to introduction of isoprene structural units into the reaction product. Entry 4 involves an allylic ether prepared by O-alkylation of a β-keto enolate. Entry 5 was used in the course of synthesis of a diterpene lactone. Entry 6 is a case in which $PdCl_2$ catalyzes both the formation and rearrangement of the reactant.

224. M. G. Kulkarni, D. S. Pendharkar, and R. M. Rasne, *Tetrahedron Lett.*, **38**, 1459 (1997).
225. J. L. van der Baan and F. Bickelhaupt, *Tetrahedron Lett.*, **27**, 6267 (1986).
226. M. Hiersemann and L. Abraham, *Eur. J. Org. Chem.*, 1461 (2002).
227. K. Nonoshita, H. Banno, K. Maruoka, and H. Yamamoto, *J. Am. Chem. Soc.*, **112**, 316 (1990).
228. S. Saito, K. Shimada, and H. Yamamoto, *Synlett*, 720 (1996).

Scheme 6.14. Claisen Rearrangements of Allyl Vinyl Ethers and Related Compounds

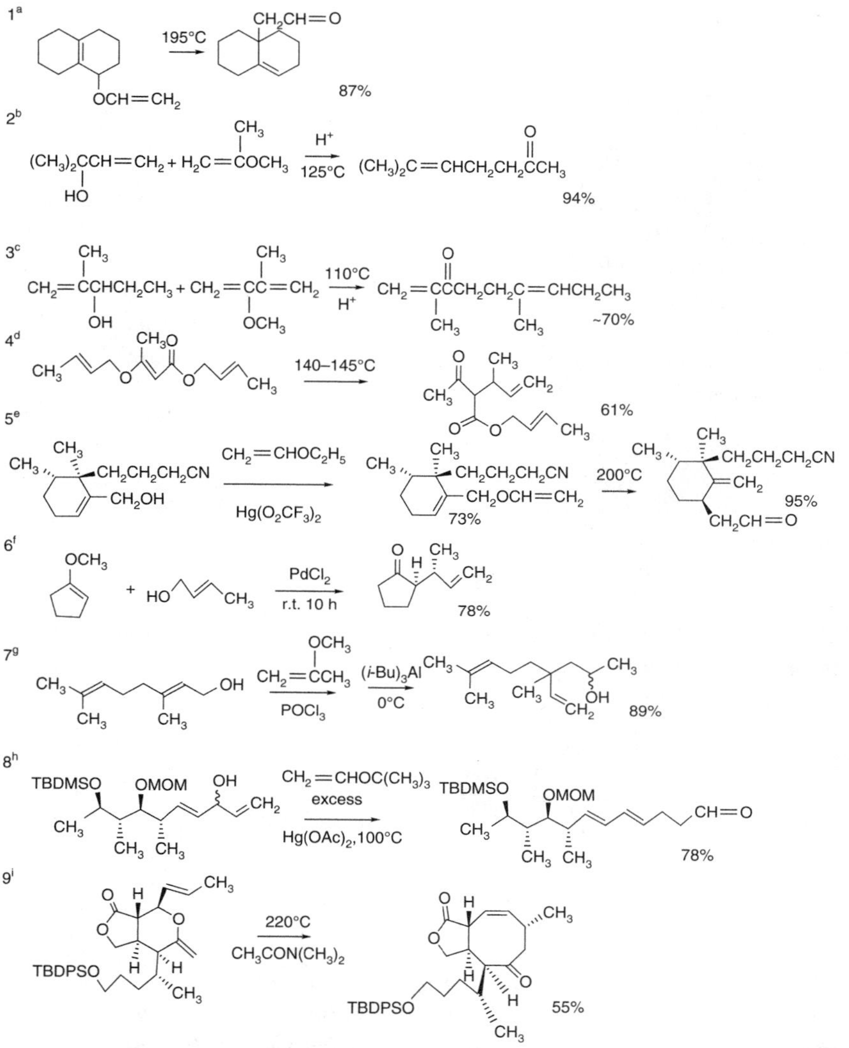

a. A. W. Burgstahler and I. C. Nordin, *J. Am. Chem. Soc.*, **83**, 198 (1961).
b. G. Saucy and R. Marbet, *Helv. Chim. Acta*, **50**, 2091 (1967).
c. D. J. Faulkner and M. R. Petersen, *J. Am. Chem. Soc.*, **95**, 553 (1973).
d. J. W. Ralls, R. E. Lundin, and G. F. Bailey, *J. Org. Chem.*, **28**, 3521 (1963).
e. L. A. Paquette, T.-Z. Wang, S. Nang and C. M. G. Philippo, *Tetrahedron Lett.*, **34**, 3523 (1993).
f. K. Mitami, K. Takahashi, and T. Nakai, *Tetrahedron Lett.*, **28**, 5879 (1987).
g. S. D. Rychnovsky and J. L. Lee, *J. Org. Chem.*, **60**, 4318 (1995).
h. T. Berkenbusch and R. Brueckner, *Chem. Eur. J.*, **10**, 1545 (2004).
i. T.-Z. Wang, E. Pinard, and L. A. Paquette, *J. Am. Chem. Soc.*, **118**, 1309 (1996).

Entry 7 illustrates reaction conditions that were applicable to formation and rearrangement of an isopropenyl allylic ether. The tri-isopropylaluminum is thought to both catalyze the sigmatropic rearrangement and reduce the product ketone.

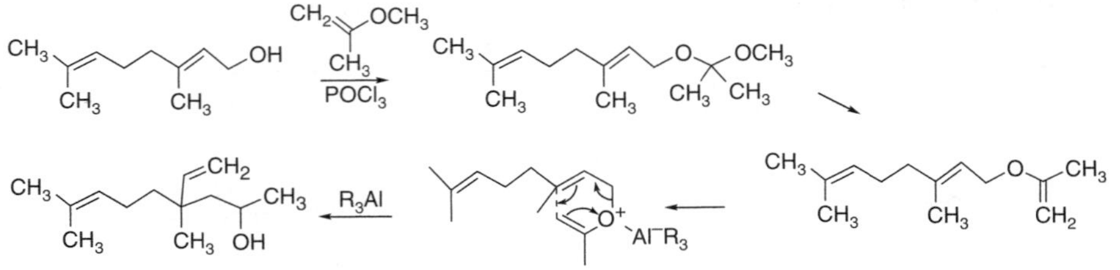

564

CHAPTER 6

*Concerted
Cycloadditions,
Unimolecular
Rearrangements, and
Thermal Eliminations*

The reaction in Entry 8 was conducted in excess refluxing vinyl *t*-butyl ether, using 1.1 equivalent of $Hg(OAc)_2$ to catalyze the exchange reaction. In Entry 9 a thermal reaction leads to formation of an eight-membered ring.

Aryl allyl ethers can also undergo [3,3]-sigmatropic rearrangement. In fact, Claisen rearrangements of allyl phenyl ethers to *ortho*-allyl phenols were the first [3,3]-sigmatropic rearrangements to be thoroughly studied.[229] The reaction proceeds through a cyclohexadienone that enolizes to the stable phenol.

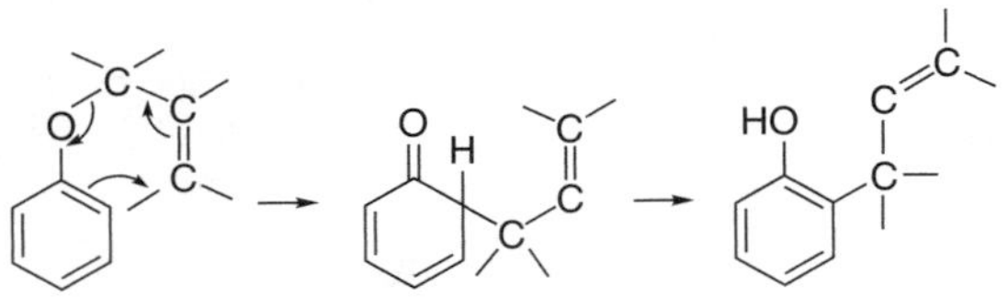

If both *ortho*-positions are substituted, the allyl group undergoes a second migration, giving the *para*-substituted phenol:

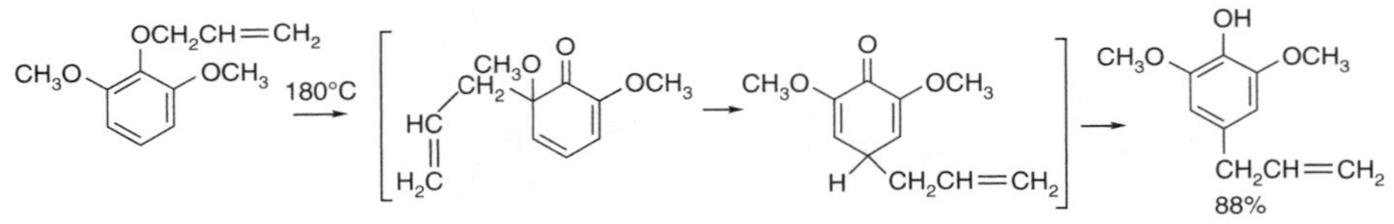

Ref. 230

6.4.2.2. Orthoester Claisen Rearrangements. There are several variations of the Claisen rearrangement that make it a powerful tool for the synthesis of γ,δ-unsaturated carboxylic acids. The *orthoester modification of the Claisen rearrangement* allows carboalkoxymethyl groups to be introduced at the γ-position of allylic alcohols.[231] A mixed orthoester is formed as an intermediate and undergoes sequential elimination and sigmatropic rearrangement.

229. S. J. Rhoads, in *Molecular Rearrangements*, Vol. 1, P. de Mayo, ed., Interscience, New York, 1963, pp. 655–684.

230. I. A. Pearl, *J. Am. Chem. Soc.*, **70**, 1746 (1948).

231. W. S. Johnson, L. Werthemann, W. R. Bartlett, T. J. Brocksom, T. Li, D. J. Faulkner, and M. R. Petersen, *J. Am. Chem. Soc.*, **92**, 741 (1970).

Both the exchange and elimination are catalyzed by the addition of a small amount of a weak acid, such as propanoic acid. These reactions are usually conducted at the reflux temperature of the orthoester, which is about 110° C for the trimethyl ester and 140° C for the triethyl ester. Microwave heating has been used and is reported to greatly accelerate orthoester-Claisen rearrangements.[232]

The mechanism and stereochemistry of the orthoester Claisen rearrangement is analogous to the Cope rearrangement. The reaction is stereospecific with respect to the double bond present in the initial allylic alcohol. In acyclic molecules, the stereochemistry of the product can usually be predicted on the basis of a chairlike TS.[233] When steric effects or ring geometry preclude a chairlike structure, the reaction can proceed through a boatlike TS.[234]

High levels of enantiospecificity have been observed in the rearrangement of chiral reactants. This method can be used to establish the configuration of the newly formed carbon-carbon bond on the basis of the configuration of the C−O bond in the starting allylic alcohol. Treatment of $(2R, 3E)$-3-penten-2-ol with ethyl orthoacetate gives the ethyl ester of $(3R, 4E)$-3-methyl-4-hexenoic acid in 90% enantiomeric purity.[235] The configuration of the new stereocenter is that predicted by a chairlike TS with the methyl group occupying a pseudoequatorial position.

Scheme 6.15 gives some representative examples of the orthoester Claisen rearrangement. Entry 1 is an example of the standard conditions for the orthoester Claisen rearrangement using triethyl orthoacetate as the reactant. The allylic alcohol is heated in an excess of the orthoester (5.75 equivalents) with 5 mol % of propanoic acid. Ethanol is distilled from the reaction mixture. The E-double bond arises from the chair TS.

The reaction in Entry 2, involving trimethyl orthoacetate, was effected in the course of synthesis of an insect juvenile hormone. The reaction is highly stereoselective ($> 98\%$) for the E-isomer at the new double bond. The reactions in Entries 3 and 4 were used to introduce ester substituents on the nitrogen-containing rings. Note that in Entry 4 an orthobutanoate ester is used, demonstrating that longer-chain orthoesters

232. A. Srikrishna, S. Nagaraju, and P. Kondaiah, *Tetrahedron*, **51**, 1809 (1995).
233. G. W. Daub, J. P. Edwards, C. R. Okada, J. W. Allen, C. T. Makey, M. S. Wells, A. S. Goldstien, M. J. Dibley, C. J. Wang, D. P. Ostercamp, S. Chung, P. S. Lunningham, and M. A. Berliner, *J. Org. Chem.*, **62**, 1976 (1997).
234. R. J. Cave, B. Lythgoe, D. A. Metcalf, and I. Waterhouse, *J. Chem. Soc., Perkin Trans. 1*, 1218 (1977); G. Buchi and J. E. Powell, Jr., *J. Am. Chem. Soc.*, **92**, 3126 (1970); J. J. Gajewski and J. L. Jiminez, *J. Am. Chem. Soc.*, **108**, 468 (1986).
235. R. K. Hill, R. Soman, and S. Sawada, *J. Org. Chem.*, **37**, 3737 (1972); **38**, 4218 (1973).

Scheme 6.15. Orthoester-Claisen Rearrangements

CHAPTER 6

*Concerted
Cycloadditions,
Unimolecular
Rearrangements, and
Thermal Eliminations*

1[a]

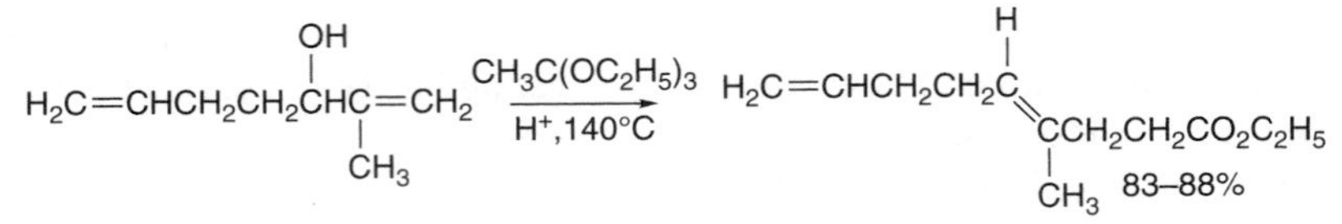

$H_2C=CHCH_2CH_2CHC=CH_2$ with OH and CH_3 → $CH_3C(OC_2H_5)_3$, H^+, 140°C → $H_2C=CHCH_2CH_2C$ ... $CCH_2CH_2CO_2C_2H_5$, CH_3 83–88%

2[b]

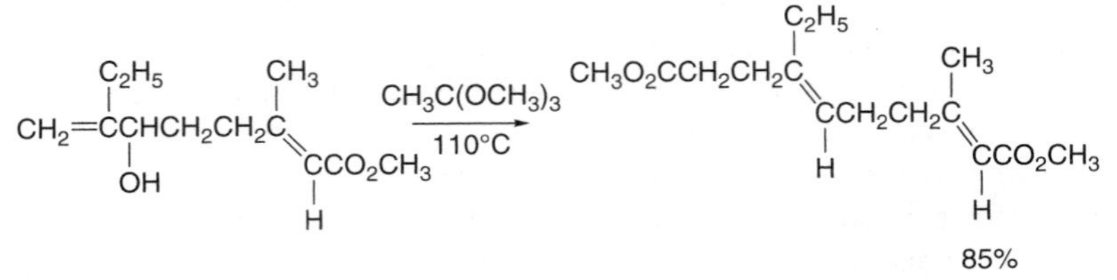

85%

3[c]

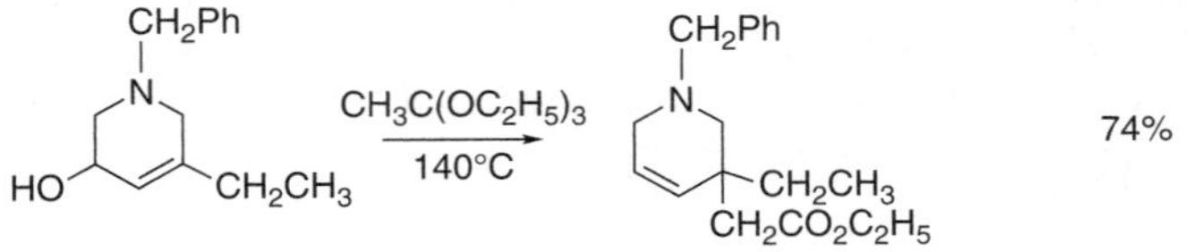

74%

4[d]

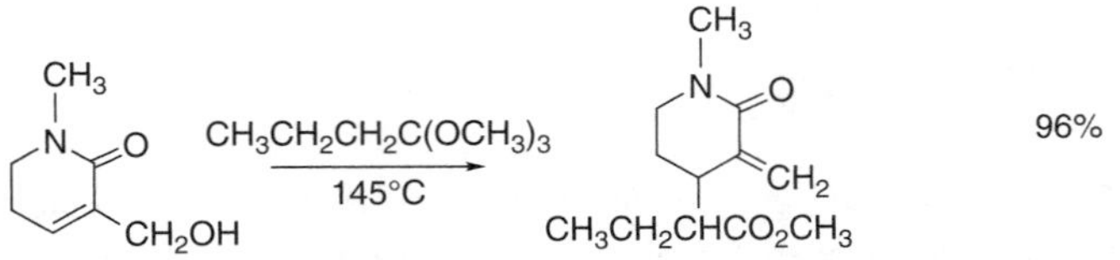

96%

5[e]

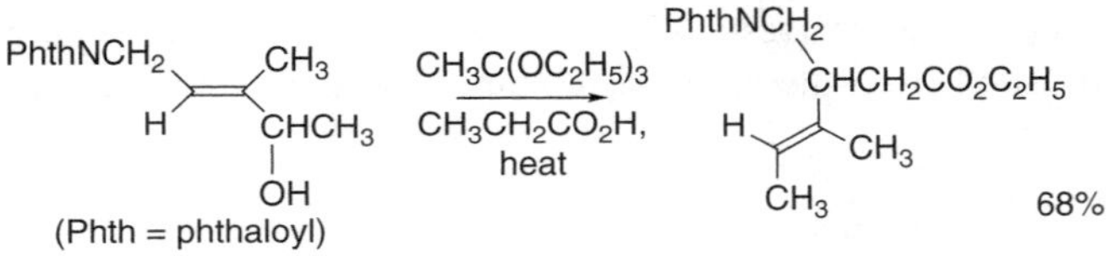

(Phth = phthaloyl)   68%

6[f]

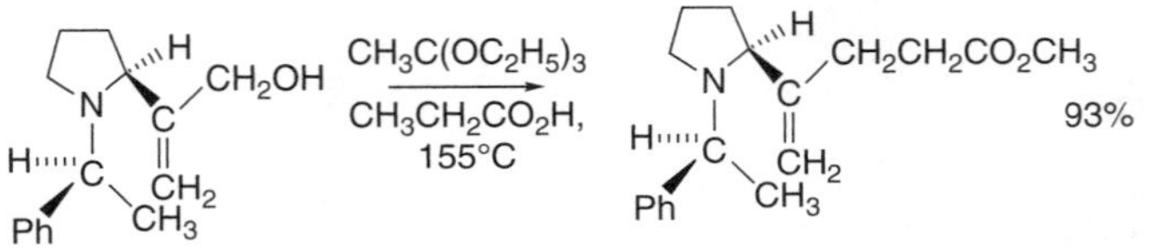

93%

7[g]

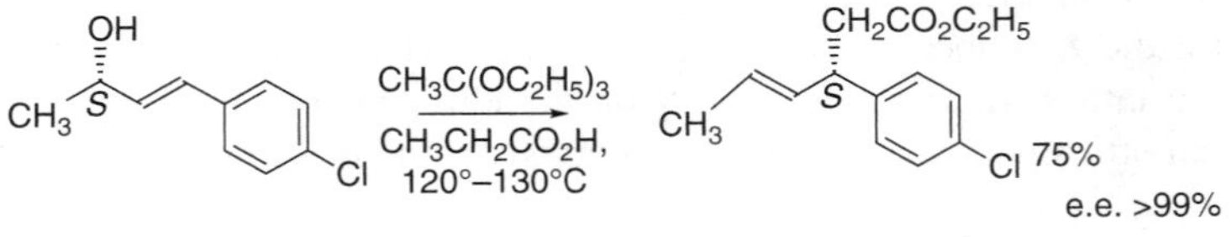

75%
e.e. >99%

8[h]

$CH_3(CH_2)_3$—≡— with OH and $=CH_2$ → $CH_3C(OC_2H_5)_3$, $CH_3CH_2CO_2H$, 110°C → $CH_3(CH_2)_3$—≡—$(CH_2)_2CO_2C_2H_5$

67%

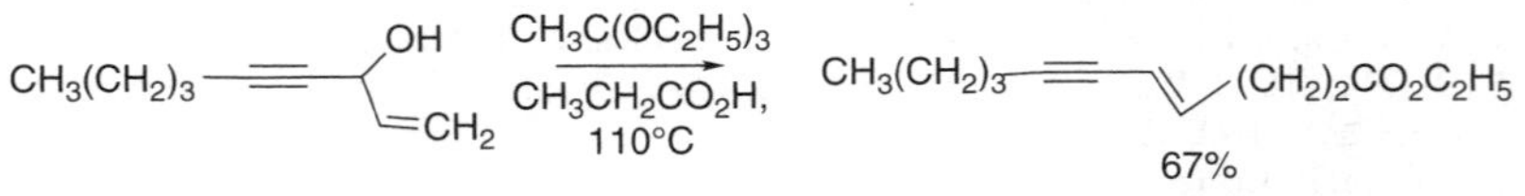

(Continued)

Scheme 6.15. *(Continued)*

567

SECTION 6.4

[3,3]-Sigmatropic
Rearrangements

9[i]

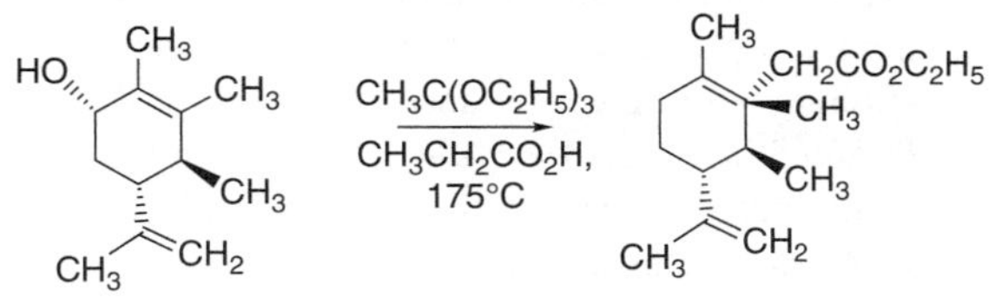

CH₃C(OC₂H₅)₃ / CH₃CH₂CO₂H, 175°C → 65%

a. R. I. Trust and R. E. Ireland, *Org. Synth.*, **53**, 116 (1973).
b. C. A. Hendrick, R. Schaub, and J. B. Siddall, *J. Am. Chem. Soc.*, **94**, 5374 (1972).
c. F. E. Ziegler and G. B. Bennett, *J. Am. Chem. Soc.*, **95**, 7458 (1973).
d. J. J. Plattner, R. D. Glass, and H. Rapoport, *J. Am. Chem. Soc.*, **94**, 8614 (1972).
e. L. Serfass and P. J. Casara, *Bioorg. Med. Chem. Lett.*, **8**, 2599 (1998).
f. D. N. A. Fox, D. Lathbury, M. F. Mahon, K. C. Molloy, and T. Gallagher, *J. Am. Chem. Soc.*, **113**, 2652 (1991).
g. E. Brenna, N. Caraccia, C. Fuganti, and P. Graselli, *Tetrahedron: Asymmetry*, **8**, 3801 (1997).
h. L. C. Passaro and F. X. Webster, *Synthesis*, 1187 (2003).
i. A. Srikrishna and D. Vijaykumar, *J. Chem. Soc., Perkin Trans. 1*, 2583 (2000).

are suitable for the reaction and permit the synthesis of α, α-disubstituted esters. The reaction in Entry 5 was used in the synthesis of protected analogs of γ-amino acids. The reaction gave the expected *E*-double bond. The reaction in Entry 6 was used in an enantiospecific synthesis of a pumiliotoxin alkaloid. Entry 7 presents a case of chirality transfer. The *S*-allylic alcohol generates the *S*-configuration at the new C–C bond with an e.e. of more than 99%. The reaction in Entry 8 was used in the synthesis of an insect pheromone, and the triple bond was eventually reduced to a *Z*-double bond. The reaction in Entry 9 was part of enantiospecific synthesis of more complex terpenoids from *R*-carvone. Note that in this case, the cyclic TS results in introduction of the ester substituent *syn* to the hydroxy group on the ring, which is a general result for cyclic reactants.

6.4.2.3. Rearrangements of Silyl Ketene Acetals and Ester Enolates. Esters of allylic alcohols can be rearranged to γ, δ-unsaturated carboxylic acids via the *O*-trimethylsilyl ethers of the ester enolate.[236] These intermediates are called *silyl ketene acetals*. This version of the reaction, known as the *Ireland-Claisen rearrangement*,[237] takes place under much milder conditions than the orthoester method. The reaction occurs at room temperature or slightly above. The stereochemistry of the silyl ketene acetal Claisen rearrangement is controlled not only by the configuration of the double bond in the allylic alcohol but also by the stereochemistry of the silyl ketene acetal. The chair TS predicts that the relative configuration at the newly formed C–C bond will be determined by the *E*- or *Z*-stereochemistry of the silyl ketene acetal.

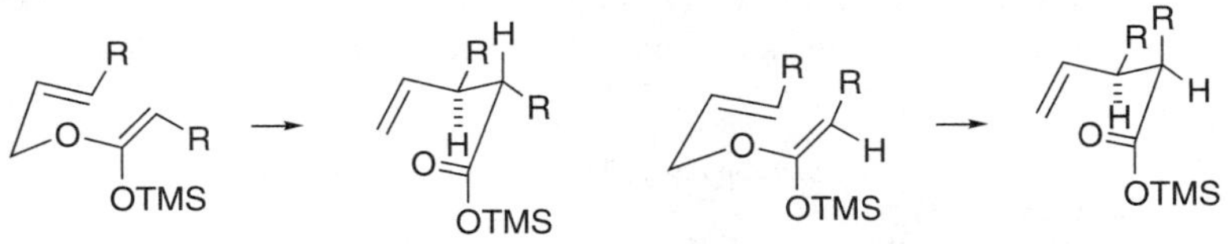

Z-silyl ketene acetal   *syn* isomer   *E*-silyl ketene acetal   *anti* isomer

[236.] R. E. Ireland, R. H. Mueller, and A. K. Willard, *J. Am. Chem. Soc.*, **98**, 2868 (1976).
[237.] For reviews, see S. Pereira and M. Srebnik, *Aldrichimica Acta*, **26**, 17 (1993); Y. Chai, S. Hong, H. A. Lindsay, C. McFarland, and M. C. McIntosh, *Tetrahedron*, **58**, 2905 (2002).

568

CHAPTER 6

*Concerted
Cycloadditions,
Unimolecular
Rearrangements, and
Thermal Eliminations*

The stereochemistry of the silyl ketene acetal can be controlled by the conditions of preparation. The base that is usually used for enolate formation is lithium diisopropylamide (LDA). If the enolate is prepared in pure THF, the *E*-enolate is generated and this stereochemistry is maintained in the silyl derivative. The preferential formation of the *E*-enolate can be explained in terms of a cyclic TS in which the proton is abstracted from the stereoelectronically preferred orientation perpendicular to the carbonyl plane. The carboxy substituent is oriented away from the alkyl groups on the amide base.

transition structure   transition structure
for *E*-enolate   for *Z*-enolate

If HMPA is included in the solvent, the *Z*-enolate predominates.[236,238] DMPU also favors the *Z*-enolate. The switch to the *Z*-enolate with HMPA or DMPU is attributed to a looser, perhaps acyclic TS being favored as the result of strong solvation of the lithium ion. The steric factors favoring the *E*-TS are therefore diminished.[239] These general principles of solvent control of enolate stereochemistry are applicable to other systems.[240] For example, by changing the conditions for silyl ketene acetal formation, the diastereomeric compounds **17a** and **17b** can be converted to the same product with high diastereoselectivity.[241]

A number of steric effects on the rate of rearrangement have been observed and can be accommodated by the chairlike TS model.[242] The *E*-silyl ketene acetals rearrange

238. R. E. Ireland and A. K. Willard, *Tetrahedron Lett.*, 3975 (1975); R. E. Ireland, P. Wipf, and J. Armstrong, III, *J. Org. Chem.*, **56**, 650 (1991).

239. C. H. Heathcock, C. T. Buse, W. A. Kleschick, M. C. Pirrung, J. E. Sohn, and J. Lamp, *J. Org. Chem.*, **45**, 1066 (1980).

240. J. Corset, F. Froment, M.-F. Lautie, N. Ratovelomanana, J. Seyden-Penne, T. Strzalko, and M. C. Roux-Schmitt, *J. Am. Chem. Soc.*, **115**, 1684 (1993).

241. S. D. Hiscock, P. B. Hitchcock, and P. J. Parsons, *Tetrahedron*, **54**, 11567 (1998).

242. C. S. Wilcox and R. E. Babston, *J. Am. Chem. Soc.*, **108**, 6636 (1986).

somewhat more slowly than the corresponding Z-isomer. This is interpreted as resulting from the pseudoaxial placement of the methyl group in the E-transition structure.

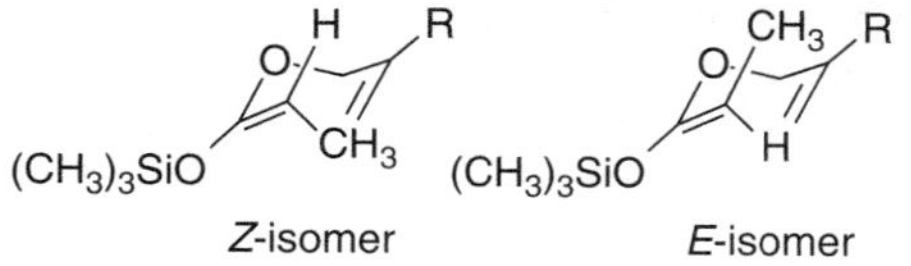

Z-isomer   E-isomer

The size of the substituent R also influences the rate, with the rate increasing somewhat for both isomers as R becomes larger. It is believed that steric interactions with R are relieved as the C−O bond stretches. The rate acceleration reflects the higher ground state energy resulting from these steric interactions.

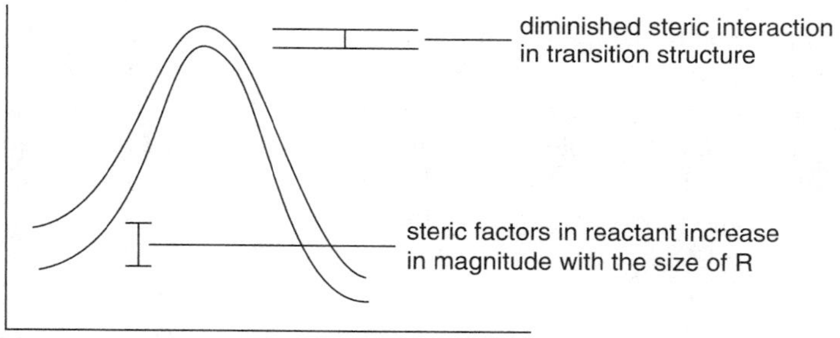

diminished steric interaction in transition structure

steric factors in reactant increase in magnitude with the size of R

The silyl ketene acetal rearrangement can also be carried out by reaction of the ester with a silyl triflate and tertiary amine, without formation of the ester enolate. Optimum results are obtained with bulky silyl triflates and amines, e.g., *t*-butyldimethylsilyl triflate and *N*-methyl-*N*, *N*-dicyclohexylamine. Under these conditions the reaction is stereoselective for the Z-silyl ketene acetal and the stereochemistry of the allylic double bond determines the *syn* or *anti* configuration of the product.[243]

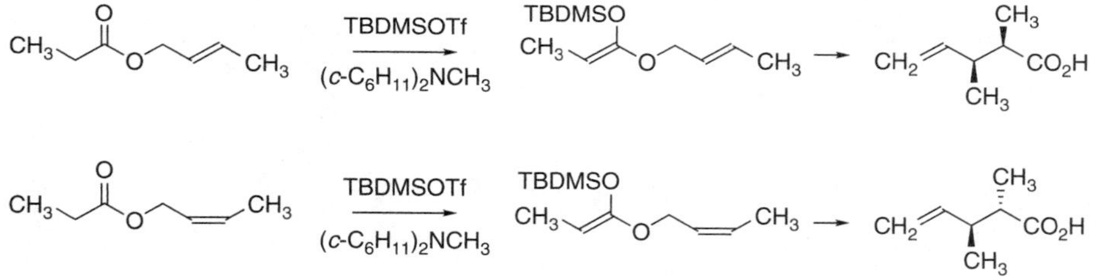

The stereochemistry of Ireland-Claisen rearrangements of cyclic compounds is sometimes indicative of reaction through a boat TS. For example, the major product from 2-cyclohexenyl propanoate is formed through a boat TS.[244]

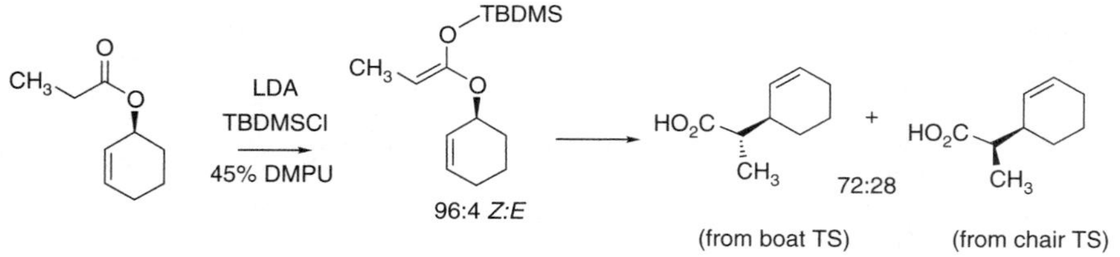

96:4 Z:E   72:28

(from boat TS)   (from chair TS)

243. M. Kobayashi, K. Matsumoto, E. Nakai, and T. Nakai, *Tetrahedron Lett.*, **37**, 3005 (1996).
244. (a) R. E. Ireland and P. Maienfisch, *J. Org. Chem.*, **53**, 640 (1988); (b) R. E. Ireland, P. Wipf, and J.D. Armstrong, *J. Org. Chem.*, **56**, 650 (1991); (c) R. E. Ireland, P. Wipf, and J.-N. Xiang, *J. Org. Chem.*, **56**, 3572 (1991).

570

CHAPTER 6

Concerted
Cycloadditions,
Unimolecular
Rearrangements, and
Thermal Eliminations

The reason for the trend toward boat TSs in cyclic systems is the introduction of additional steric factors. For example, addition of methyl and isopropenyl substituents leads to a TS in which the cyclohexene ring adopts a boat conformation, whereas the TS is chairlike.

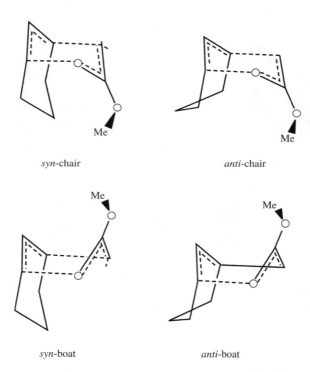

Heteroatoms, particularly oxygen, introduce electronic factors that favor boat TSs.

Computational modeling (B3LYP/6-31G*) of rearrangement of cyclohexenol identified the four potential TS geometries shown in Figure 6.14.[245] Using the *O*-methyl enol ether as a model, a 2-cyclohexenyl ester prefers a *syn*-boat TS, in agreement with the experimental results. As in the experimental work, the placement of additional substituents alters the relative energies of these TSs.

syn-chair   *anti*-chair

syn-boat   *anti*-boat

Fig. 6.14. Possible transition structures for [3,3]-sigmatropic rearrangement of 2-cyclohexenyl ester enol ethers. Adapted from *J. Org. Chem.*, **68**, 572 (2003), by permission of the American Chemical Society.

245. M. M. Khaledy, M. Y. S. Kalani, K. S. Khuong, K. N. Houk, V. Aviyente, R. Neier, N. Soldermann, and J. Velker, *J. Org. Chem.*, **68**, 572 (2003).

The stereoselectivity of silyl ketene acetal Claisen rearrangements can also be controlled by specific intramolecular interactions.[246] The enolates of α-alkoxy esters adopt the Z-configuration because of chelation by the alkoxy substituent.

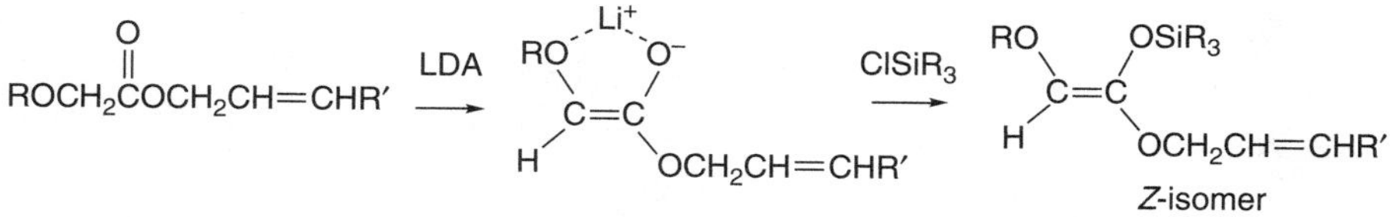

The configuration at the newly formed C−C bond is then controlled by the stereochemistry of the double bond in the allylic alcohol. The *E*-isomer gives a *syn* orientation, whereas the *Z*-isomer gives rise to *anti* stereochemistry.[247]

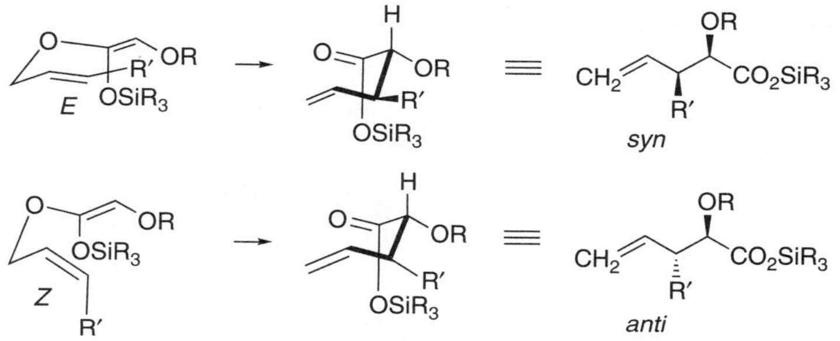

Similar chelation effects are present in α-alkoxymethyl derivatives. Magnesium enolates give predominantly the Z-enolate as a result of this chelation. The corresponding trimethylsilyl ketene acetals give *E,Z* mixtures.[248]

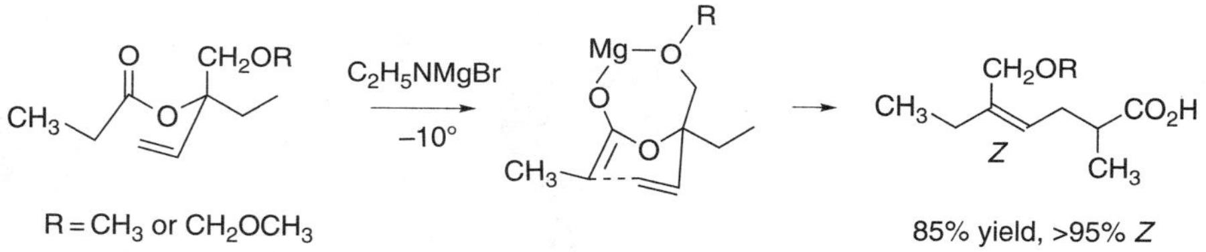

R = CH₃ or CH₂OCH₃

85% yield, >95% Z

Enolates of allyl esters of α-amino acids are also subject to chelation-controlled Claisen rearrangement.[249]

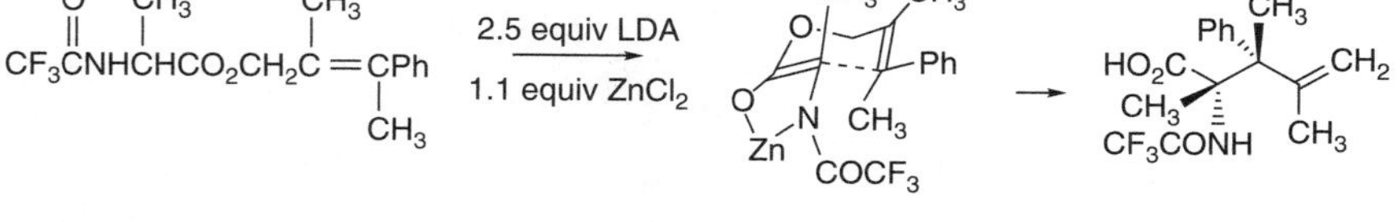

246. H. Frauenrath, in *Stereoselective Synthesis*, G. Helmchen, R. W. Hoffmann, J. Mulzer, and E. Schaumann, eds., Georg Thieme Verlag, Stuttgart, 1996.

247. T. J. Gould, M. Balestra, M. D. Wittman, J. A. Gary, L. T. Rossano, and J. Kallmerten, *J. Org. Chem.*, **52**, 3889 (1987); S. D. Burke, W. F. Fobare, and G. J. Pacofsky, *J. Org. Chem.*, **48**, 5221 (1983); P. A. Bartlett, D. J. Tanzella, and J. F. Barstow, *J. Org. Chem.*, **47**, 3941 (1982).

248. M. E. Krafft, O. A. Dasse, S. Jarrett, and A. Fierve, *J. Org. Chem.*, **60**, 5093 (1995).

249. U. Kazmaier, *Liebigs Ann. Chem.*, 285 (1997); U. Kazmaier, *J. Org. Chem.*, **61**, 3694 (1996); U. Kazmaier and S. Maier, *Tetrahedron*, **52**, 941 (1996).

572

CHAPTER 6

*Concerted
Cycloadditions,
Unimolecular
Rearrangements, and
Thermal Eliminations*

Various salts can achieve chelation but $ZnCl_2$ and $MgCl_2$ are suitable for most cases. The rearrangement is a useful reaction for preparing amino acid analogs and has also been applied to synthesis of modified dipeptides.[250]

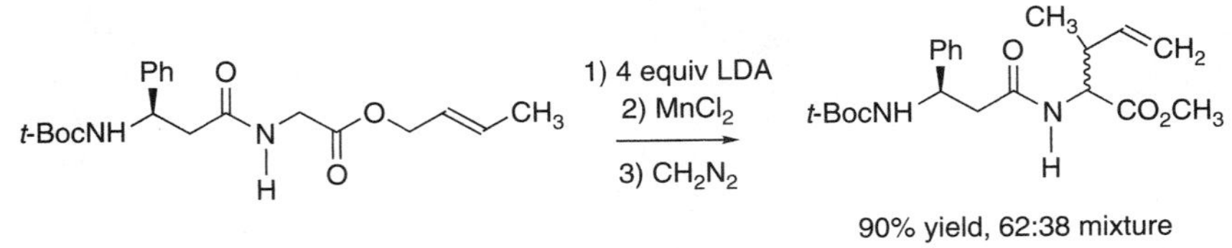

90% yield, 62:38 mixture

Lewis acid catalysis of Ireland-Claisen rearrangements by $TiCl_4$ has been observed.[251] This methodology was employed in the synthesis of a novel type of anti-inflammatory drug candidate.[252]

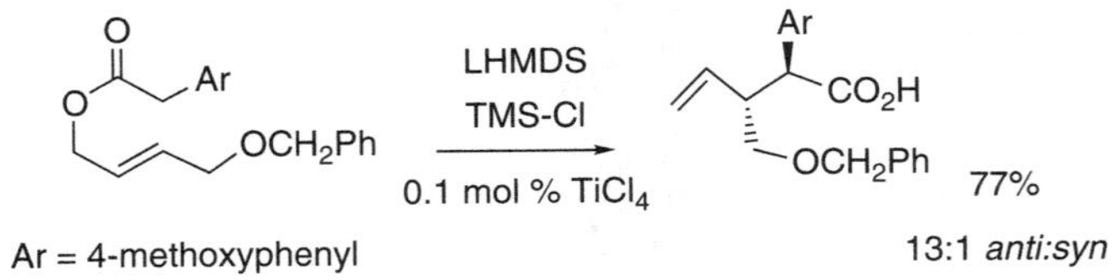

Ar = 4-methoxyphenyl

77%

13:1 *anti:syn*

The possibility of using chiral auxiliaries or chiral catalysts to achieve enantio-selective Claisen rearrangements has been explored.[253] One approach is to use chiral boron enolates. For example, enolates prepared with the chiral diazaborolidine bromide **O** lead to rearranged products of more than 95% enantiomeric excess.[254]

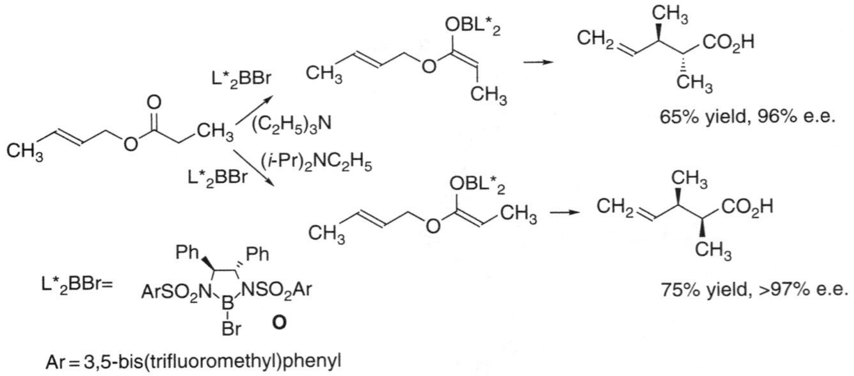

65% yield, 96% e.e.

75% yield, >97% e.e.

Ar = 3,5-bis(trifluoromethyl)phenyl

The enantioselectivity is consistent with a chairlike TS in which the stereocenters control the rotational preference for the sulfonyl groups that provide stereodifferentiation at the boron center.

[250.] U. Kazmaier and S. Maier, *J. Chem. Soc., Chem. Commun.*, 2535 (1998).

[251.] G. Koch, P. Janser, G. Kottirsch, and E. Romero-Giron, *Tetrahedron Lett.*, **43**, 4837 (2002).

[252.] G. Koch, G. Kottirsch, B. Wiefeld, and E. Kuesters, *Org. Proc. Res. Dev.*, **6**, 652 (2002).

[253.] D. Enders, M. Knopp, and R. Schiffers, *Tetrahedron: Asymmetry*, **7**, 1847 (1996).

[254.] E. J. Corey and D.-H. Lee, *J. Am. Chem. Soc.*, **113**, 4026 (1991).

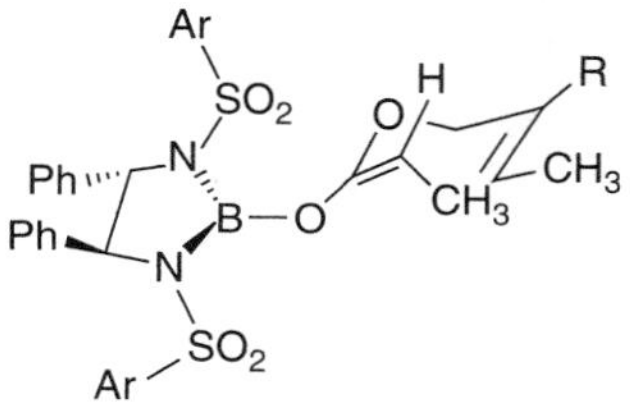

This methodology has been applied to both acyclic esters and macrocyclic lactones.

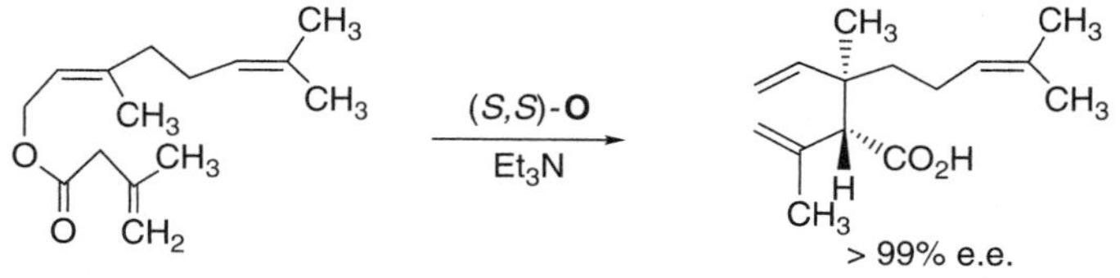

Ref. 255

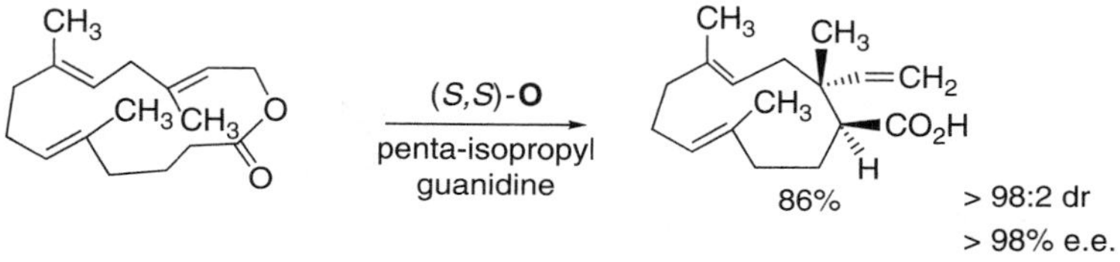

Ref. 256

Scheme 6.16 gives some examples of Ireland-Claisen rearrangements of silyl ketene acetals and related intermediates. Entry 1 is an example from an early investigation of this version of the rearrangement. Entry 2 involves direct rearrangement of the enolate without silylation. The reaction in Entry 3 was used for stereoselective synthesis of the γ, δ-unsaturated acid, which was used in the synthesis of a butterfly pheromone. The TBDMS derivative gave a somewhat higher yield than the TMS derivative in this case. The reaction in Entry 4 was used in the conversion of carbohydrate-derived starting materials to structures found in ionophore antibiotics. The reaction conditions, which involved use of *premixed* LDA and TMS-Cl, were designed to avoid a competing β-elimination of the enolate by rapid silylation of the enolate.

In Entry 5, the chirality at an alkylated succinate ester is maintained and a 9:1 dr favoring the *anti* product is achieved, based on a preferred orientation relative to the branched substituent.

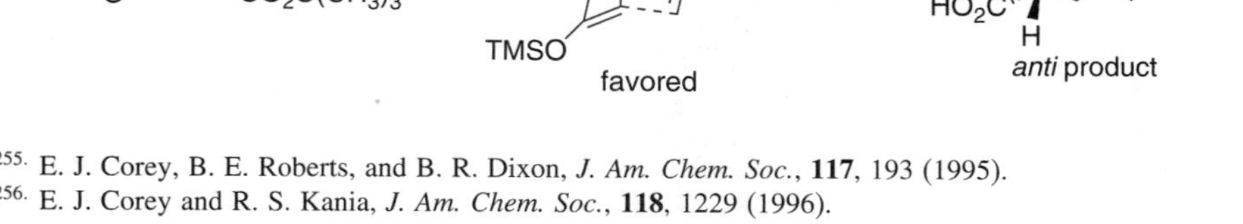

255. E. J. Corey, B. E. Roberts, and B. R. Dixon, *J. Am. Chem. Soc.*, **117**, 193 (1995).
256. E. J. Corey and R. S. Kania, *J. Am. Chem. Soc.*, **118**, 1229 (1996).

Scheme 6.16. Rearrangement of Silyl Ketene Acetals and Ester Enolates

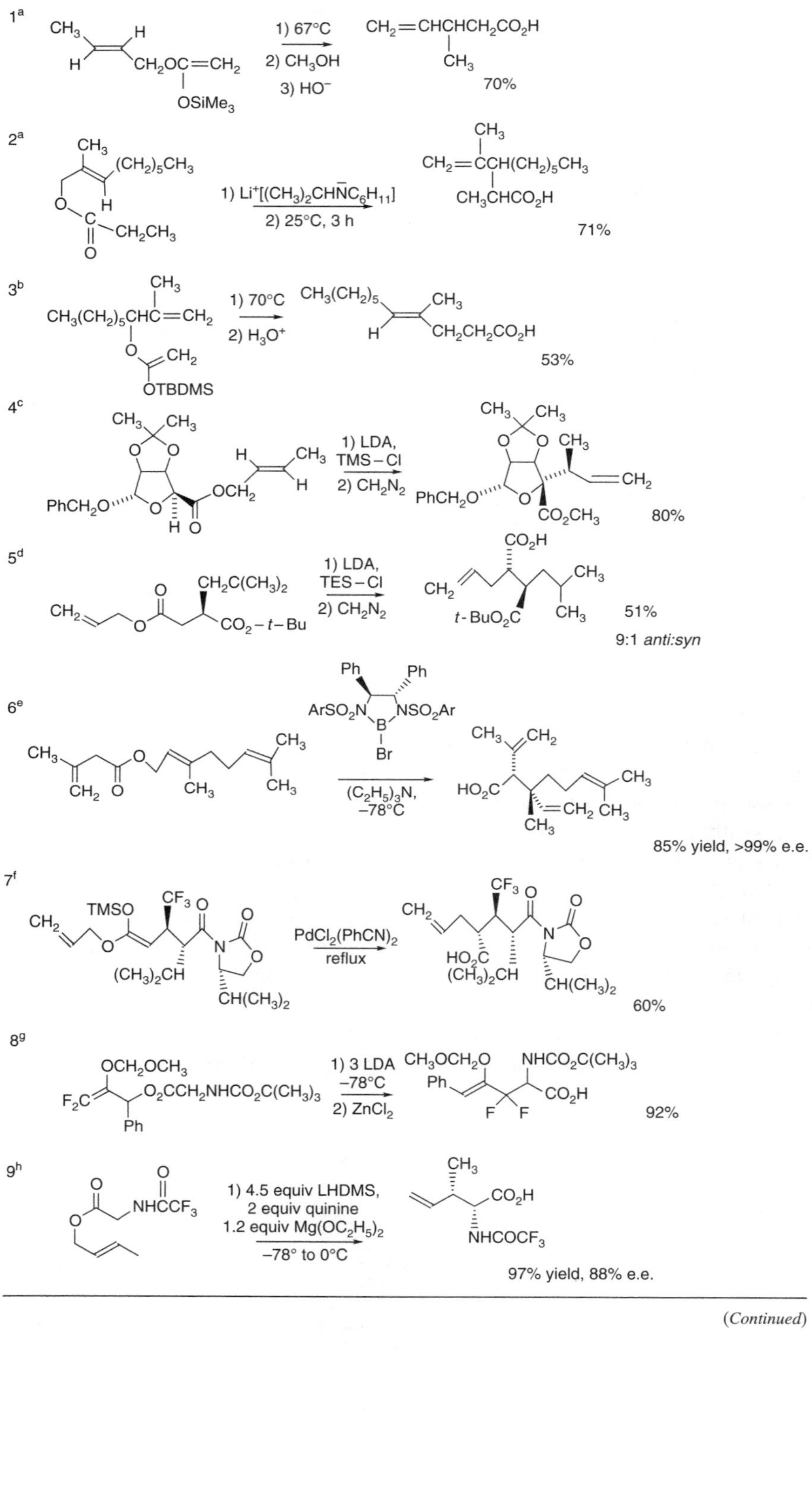

CHAPTER 6

*Concerted
Cycloadditions,
Unimolecular
Rearrangements, and
Thermal Eliminations*

1[a]

1) 67°C
2) CH$_3$OH
3) HO$^-$

70%

2[a]

1) Li$^+$[(CH$_3$)$_2$CHN̄C$_6$H$_{11}$]
2) 25°C, 3 h

71%

3[b]

1) 70°C
2) H$_3$O$^+$

53%

4[c]

1) LDA, TMS–Cl
2) CH$_2$N$_2$

80%

5[d]

1) LDA, TES–Cl
2) CH$_2$N$_2$

51%
9:1 *anti:syn*

6[e]

(C$_2$H$_5$)$_3$N,
–78°C

85% yield, >99% e.e.

7[f]

PdCl$_2$(PhCN)$_2$
reflux

60%

8[g]

1) 3 LDA
–78°C
2) ZnCl$_2$

92%

9[h]

1) 4.5 equiv LHDMS,
2 equiv quinine
1.2 equiv Mg(OC$_2$H$_5$)$_2$
–78° to 0°C

97% yield, 88% e.e.

(Continued)

Scheme 6.16. (*Continued*)

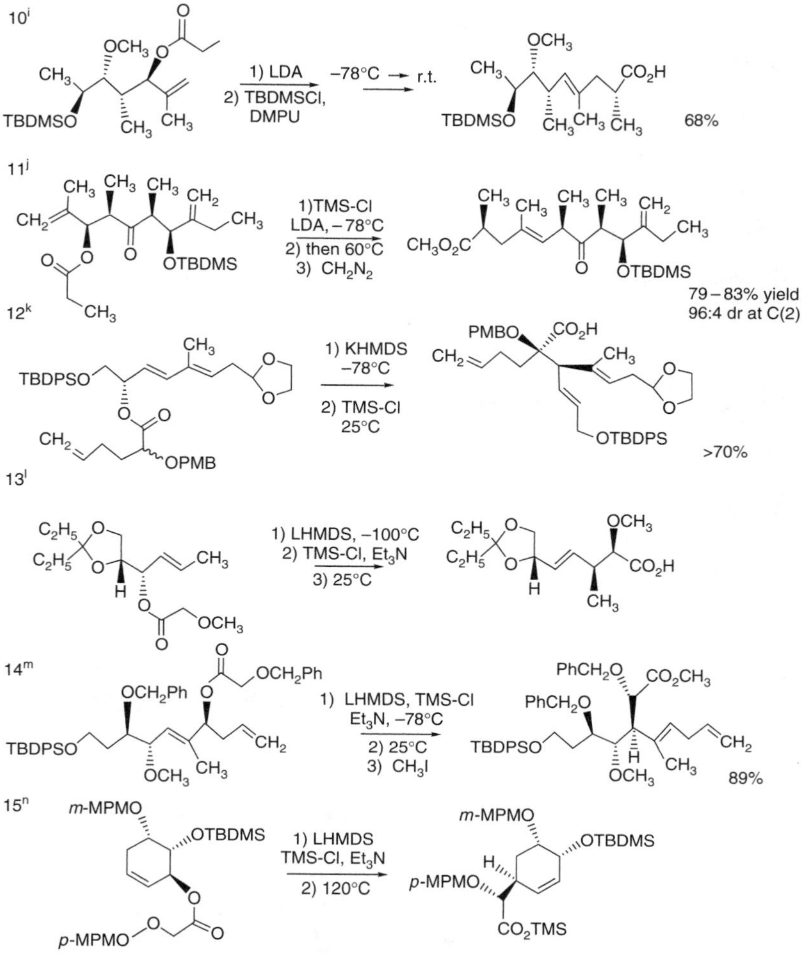

575

SECTION 6.4

[3,3]-Sigmatropic Rearrangements

a. R. E. Ireland, R. H. Mueller, and A. K. Willard, *J. Am. Chem. Soc.*, **98**, 2868 (1976).
b. J. A. Katzenellenbogen and K. J. Cristy, *J. Org. Chem.*, **39**, 3315 (1974).
c. R. E. Ireland and D. W. Norbeck, *J. Am. Chem. Soc.*, **107**, 3279 (1985).
d. L. M. Pratt, S. A. Bowler, S. F. Courney, C. Hidden, C. N. Lewis, F. M. Martin, and R. S. Todd, *Synlett*, 531 (1998).
e. E. J. Corey, B. E. Roberts, and B. R. Dixon, *J. Am. Chem. Soc.*, **117**, 193 (1995).
f. T. Yamazaki, N. Shinohara, T. Katzume, and S. Sato, *J. Org. Chem.*, **60**, 8140 (1995).
g. J. M. Percy, M.E. Prime, and M J. Broadhurst, *J. Org. Chem.*, **63**, 8049 (1998).
h. A. Kazmaier and A. Krebs, *Tetrahedron Lett.*, **40**, 479 (1999).
i. P. R. Blakemore, P. J. Kocienski, A. Morley, and K. Muir, *J. Chem. Soc., Perkin Trans. 1*, 955 (1999).
j. I. Paterson and A. N. Hulme, *J. Org. Chem.*, **60**, 3288 (1995).
k. O. Bedell, A. Haudrecky, and Y. Langlois, *Eur. J. Org. Chem.*, 3813 (2004).
l. S. D. Burke, J. Hong, J. R. Lennox, and A. P. Mongin, *J. Org. Chem.*, **63**, 6952 (1998).
m. D. Kim, S. K. Ahn, H. Bae, W. J. Choi, and H. S. Kim, *Tetrahedron Lett.*, **38**, 4437 (1997).
n. S. D. Burke, J. J. Letourneau, and M. Matulenko, *Tetrahedron Lett.*, **40**, 9 (1999).

Entry 6 is an example of application of the chiral diazaborolidine enolate method (see p. 572). Entry 7 involves generation of the silyl ketene acetal by silylation after conjugate addition of the enolate of 3-methylbutanoyloxazolidinone to allyl 3,3,3-trifluoroprop-2-enoate. A palladium catalyst improved the yield in the rearrangement

576

CHAPTER 6

*Concerted
Cycloadditions,
Unimolecular
Rearrangements, and
Thermal Eliminations*

step. Entry 8 involves another fluorinated reactant. The reaction is an adaptation of the rearrangement of α-amido ester enolates, as discussed on p. 572, and involves a chelated enolate. Entry 9 is another example of this type of reaction. Use of quinine or quinidine with the chelating metal leads to enantioselectivity.

Entries 10 to 15 involve use of the Ireland-Claisen rearrangement in multistep syntheses. An interesting feature of Entry 11 is the presence of an unprotected ketone. The reaction was done by adding LDA to the ester, which was premixed with TMS-Cl and Et₃N. The reaction generates the *E*-silyl ketene acetal, which rearranges through a chair TS.

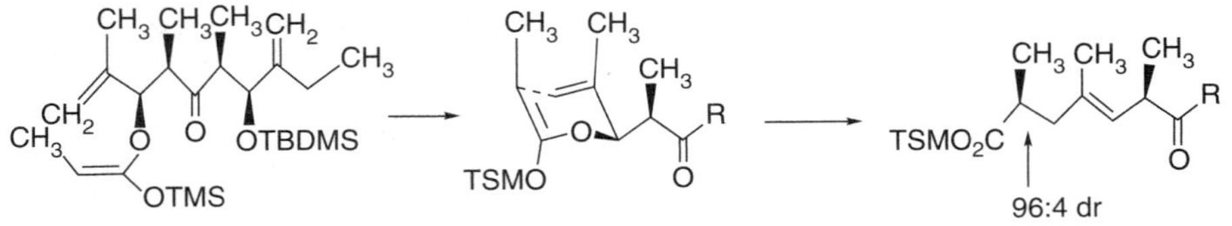

Entries 12 to 15 are examples of α-alkoxy (protected glycolate) esters. These reactions proceed through chelated TSs. (See the discussion on p. 571.) The TS for Entries 13 and 14 are shown below.

Entry 15 also demonstrates the suprafacial specificity with a cyclic allylic alcohol.

6.4.2.4. Claisen Rerrangements of Ketene Aminals and Imidates. A reaction that is related to the orthoester Claisen rearrangement utilizes an amide acetal, such as dimethylacetamide dimethyl acetal, in the exchange reaction with allylic alcohols.[257] The products are γ, δ-unsaturated amides. The stereochemistry of the reaction is analogous to the other variants of the Claisen rearrangement.[258]

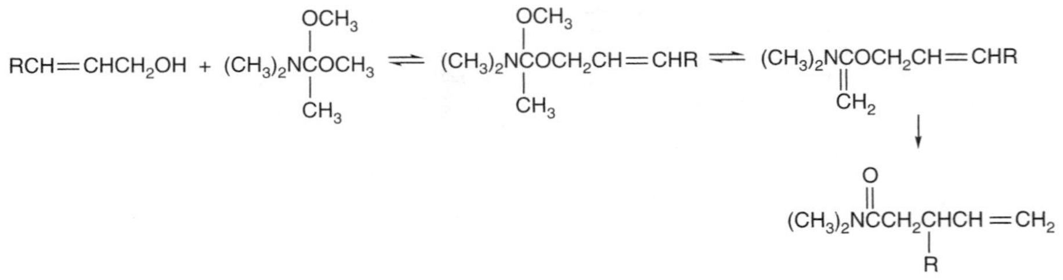

257. A. E. Wick, D. Felix, K. Steen, and A. Eschenmoser, *Helv. Chim. Acta*, **47**, 2425 (1964); D. Felix, K. Gschwend-Steen, A. E. Wick, and A. Eschenmoser, *Helv. Chim. Acta*, **52**, 1030 (1969).
258. W. Sucrow, M. Slopianka, and P. P. Calderia, *Chem. Ber.*, **108**, 1101 (1975).

The rearrangement can be applied to other secondary amines by prior equilibration, which is driven forward by removal of the more volatile dimethylamine.[259]

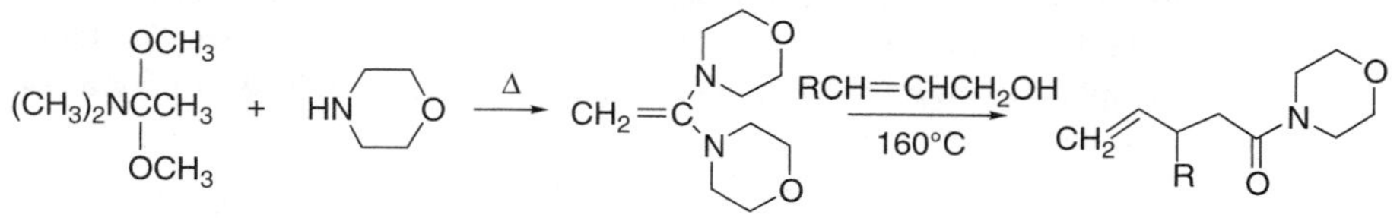

O-Allyl imidate esters undergo [3,3]-sigmatropic rearrangements to *N*-allyl amides. Trichloromethyl imidates can be made easily from allylic alcohols by reaction with trichloroacetonitrile. The rearrangement then provides trichloroacetamides of *N*-allylamines.[260]

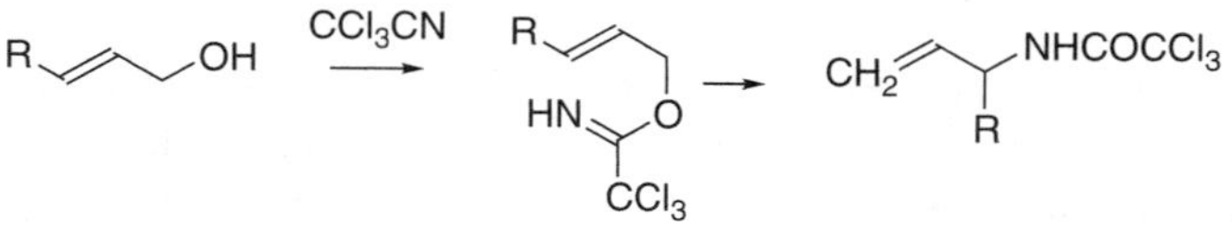

Yields in the reaction are sometimes improved by inclusion of K_2CO_3 in the reaction mixture.[261]

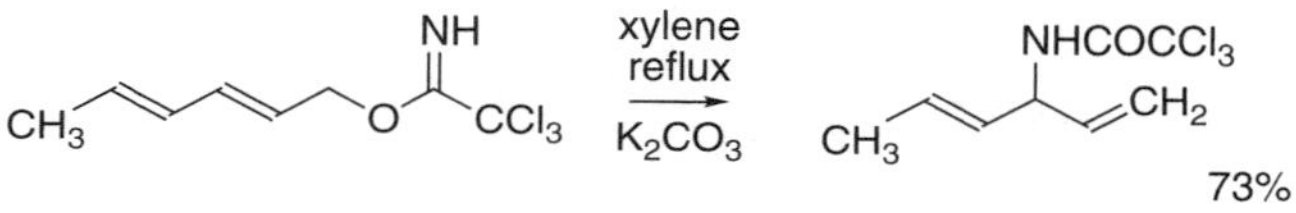

Trifluoromethyl imidates show similar reactivity.[262] Imidate rearrangements are catalyzed by palladium salts.[263] The mechanism is presumably similar to that for the Cope rearrangement (see p. 555).

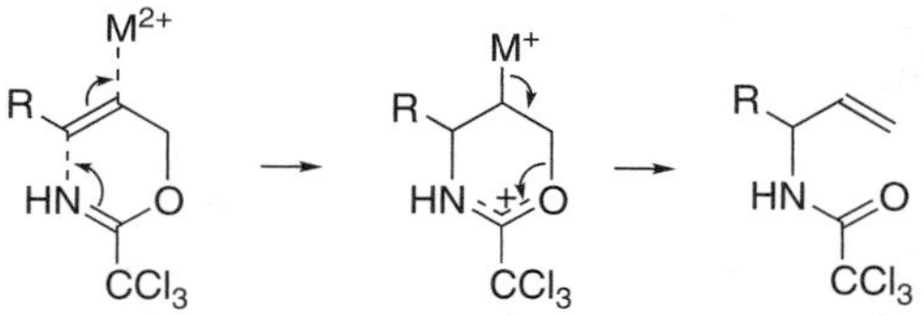

Chiral Pd catalysts can achieve enantioselectivity. The best catalysts developed to date are dimeric ferrocenyl derivatives.[264]

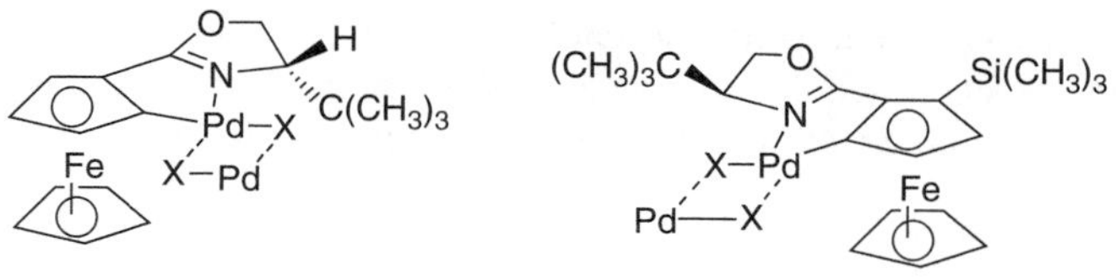

259. S. N. Gradl, J. J. Kennedy-Smith, J. Kim, and D. Trauner, *Synlett*, 411 (2002).
260. L. E. Overman, *J. Am. Chem. Soc.*, **98**, 2901 (1976); L. E. Overman, *Acc. Chem. Res.*, **13**, 218 (1980).
261. T. Nishikawa, M. Asai, N. Ohyabu, and M. Isobe, *J. Org. Chem.*, **63**, 188 (1998).
262. A. Chen, J. Savage, E. D. Thomas, and P. D. Wilson, *Tetrahedron Lett.*, **34**, 6769 (1993).
263. L. E. Overman, *Angew. Chem. Int. Ed. Engl.*, **23**, 579 (1984); T. G. Schenck and B. Bosnich, *J. Am. Chem. Soc.*, **107**, 2058 (1985); P. Metz, C. Mues, and A. Schoop, *Tetrahedron*, **48**, 1071 (1992).
264. Y. Donde and L. E. Overman, *J. Am. Chem. Soc.*, **121**, 2933 (1999).

578

CHAPTER 6

Concerted
Cycloadditions,
Unimolecular
Rearrangements, and
Thermal Eliminations

Imidate esters can also be generated by reaction of imidoyl chlorides and allylic alcohols. The lithium anions of these imidates, prepared using lithium diethylamide, rearrange at around 0° C. When a chiral amine is used, this reaction can give rise to enantioselective formation of γ, δ-unsaturated amides. Good results were obtained with a chiral binaphthylamine.[265] The methoxy substituent is believed to play a role as a Li$^+$ ligand in the reactive enolate.

Enolates of *N*-allyl amides undergo [3,3]-sigmatropic rearrangement. This reaction is analogous to the ester enolate Claisen rearrangement, but the conditions required are more vigorous.[266] An attractive feature of this reaction is that it permits introduction of a chiral group at nitrogen, which then has the potential to effect enantioselective formation of a new C−C bond. For example, α-arylethyl substituents induced enantioselectivity ranging from 3:1 to 11:1.

Analogous rearrangement occurs under much milder conditions when the reactant is a zwitterion generated by deprotonation of an acylammonium ion. Substituted pyrrolidines were used as the chiral auxiliary, with the highest enantioselectivity being achieved with a 2-TBDMS derivative.[267]

91% yield, > 95% de

The preferred TS is a chair with the enolate oriented *syn* to the bulky pyrrolidine substituent. It was suggested that the *syn* acylation occurs through an envelope conformation of the pyrrolidine ring with the nitrogen electron pair oriented axially.

265. P. Metz and B. Hungerhoff, *J. Org. Chem.*, **62**, 4442 (1997).
266. T. Tsunoda, M. Sakai, O. Sasaki, Y. Sato, Y. Hondo, and S. Ito, *Tetrahedron Lett.*, **33**, 1651 (1992).
267. S. Laabs, W. Munch, J.-W. Bats, and U. Nubbemeyer, *Tetrahedron*, **58**, 1317 (2002).

Another promising variant involves thioamides, which provide *Z*-thioenolates on deprotonation.[268] Use of *trans*-2,4-diphenylpyrrolidine as the chiral auxiliary leads to good enantioselectivity.[269] Allyl groups with *E*-configuration give mainly *anti* products with somewhat reduced diastereoselectivity. These results indicate that a steric interaction between the pyrrolidine substituent and the *Z*-allyl group is a controlling factor in diastereoselectivity.

favored TS

The 2-azonia analog of the Cope rearrangement is estimated to be accelerated by 10^6, relative to the unsubstituted system.[270] The product of the rearrangement is an isomeric iminium ion, which is a mild electrophile. In synthetic applications, the reaction is often designed to generate this electrophilic site in a position that can lead to a cyclization by reaction with a nucleophilic site. For example, the presence of a 4-hydroxy substituent generates an enol that can react with the iminiun ion intermediate to form a five-membered ring.[271]

Scheme 6.17 gives some examples of the orthoamide and imidate versions of the Claisen rearrangement. Entry 1 applied the reaction in the synthesis of a portion of the alkaloid tabersonine. The reaction in Entry 2 was used in an enantiospecific synthesis of pravastatin, one of a family of drugs used to lower cholesterol levels. The product from the reaction in Entry 3 was used in a synthesis of a portion of the antibiotic rampamycin. Entries 4 and 5 were used in the synthesis of polycyclic natural products. Note that the reaction in Entry 4 also leads to isomerization of the double bond into conjugation with the ester group. Entries 1 to 5 all involve cyclic reactants, and the concerted TS ensures that the substituent is introduced *syn* to the original hydroxy substituent.

Entry 6 is analogous to a silyl ketene acetal rearrangement. The reactant in this case is an imide. Entry 7 is an example of PdCl$_2$-catalyzed imidate rearrangement. Entry 8 is an example of an azonia-Cope rearrangement, with the monocylic intermediate then undergoing an intramolecular Mannich condensation. (See Section 2.2.1 for a discussion of the Mannich reaction). Entry 9 shows a thioimidate rearrangement.

268. Y. Tamaru, Y. Furukawa, M. Mizutani, O. Kitao, and Z. Yoshida, *J. Org. Chem.*, **48**, 3631 (1983).
269. S. He, S. A. Kozmin, and V. H. Rawal, *J. Am. Chem. Soc.*, **122**, 190 (2000).
270. L. A. Overman, *Acc. Chem. Res.*, **25**, 353 (1992).
271. L. E. Overman and M. Kakimoto, *J. Am. Chem. Soc.*, **101**, 1310 (1979); L. E. Overman, M. Kakimoto, M. Okazaki, and G. P. Meier , *J. Am. Chem. Soc.*, **105**, 6622 (1983).

580

CHAPTER 6

*Concerted
Cycloadditions,
Unimolecular
Rearrangements, and
Thermal Eliminations*

Scheme 6.17. Rearrangements of Orthoamides and Imidates

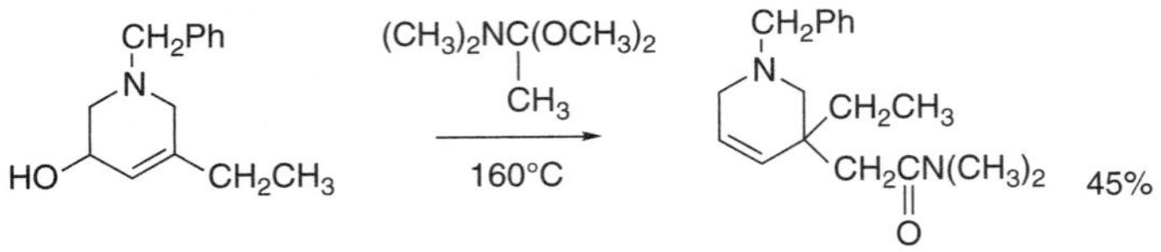

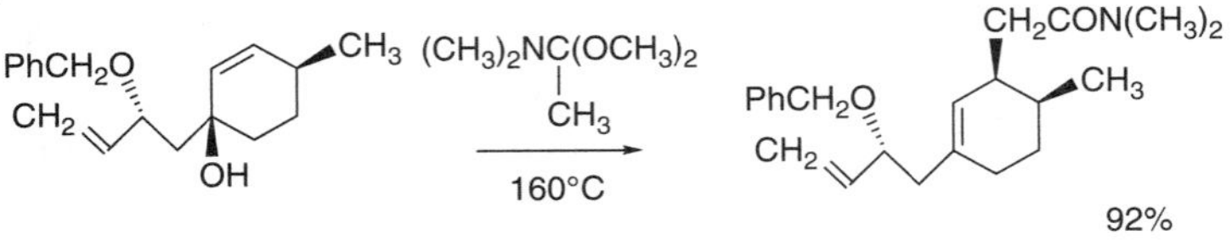

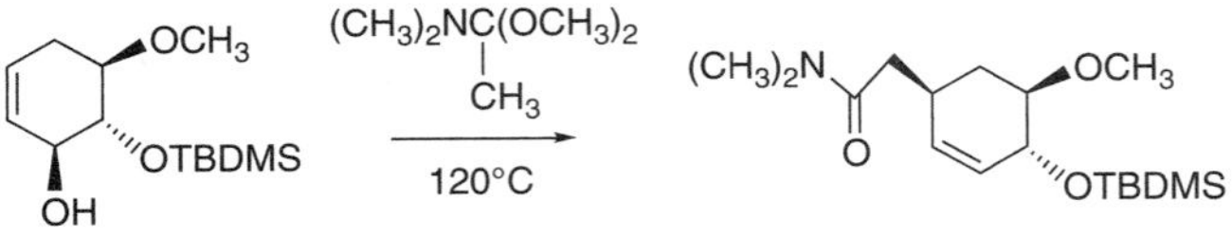

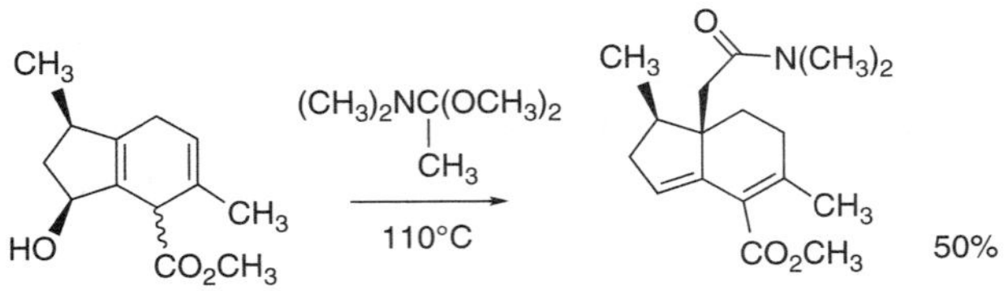

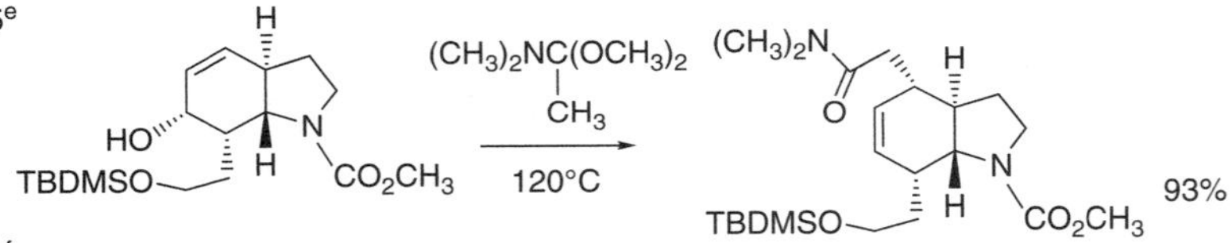

Ar = 2,3-methylenedioxyphenyl

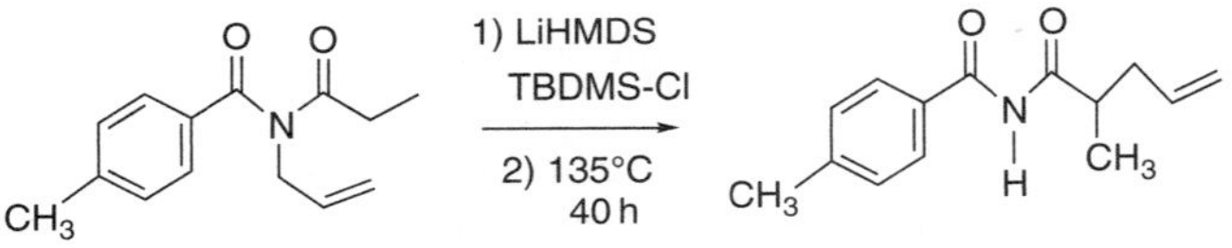

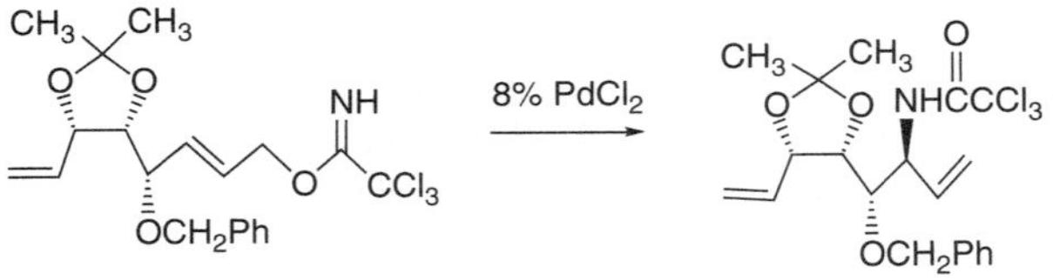

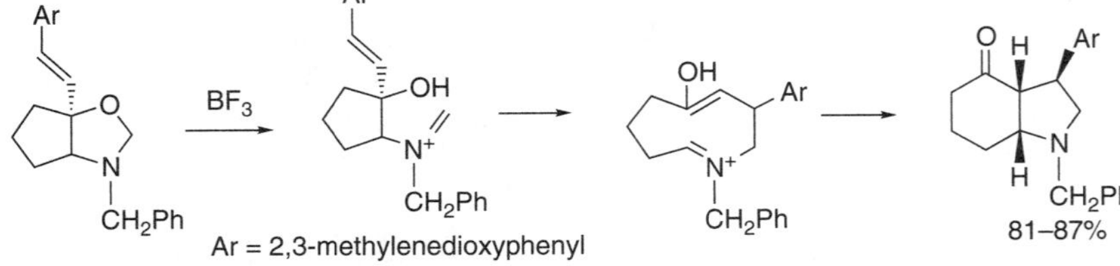

(Continued)

Scheme 6.17. (Continued)

70%; 85:15 syn:anti

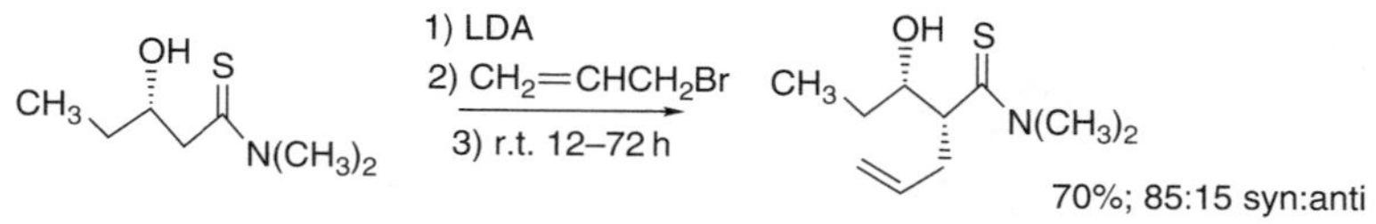

a. F. E. Ziegler and G. B. Bennett, *J. Am. Chem. Soc.*, **95**, 7458 (1973).
b. A. R. Daniewski, P. M. Wovkulich, and M. R. Uskokovic, *J. Org. Chem.*, **57**, 7133 (1992).
c. K. C. Nicolaou, P. Bertinato, A. D. Piscopio, T. K. Chakraborty, and N. Minowa, *J. Chem. Soc., Chem. Commun.*, 619 (1993).
d. T.-P. Loh and Q.-Y. Hu, *Org. Lett.*, **3**, 279 (2001).
e. C.-Y. Chen and D. J. Hart, *J. Org. Chem.*, **58**, 3840 (1993).
f. K. Neuschutz, J.-M. Simone, T. Thyrann, and R. Neier, *Helv. Chim. Acta*, **83**, 2712 (2000).
g. H. Ovaa, J. D. C. Codee, B. Lastdrager, H. Overkleeft, G. A.van der Marel, and J. H. van Boom, *Tetrahedron Lett.*, **40**, 5063 (1999).
h. L. E. Overman and J. Shim, *J. Org. Chem.*, **58**, 4662 (1993).
i. P. Beslin and B. Lelong, *Tetrahedron*, **53**, 17253 (1997).

6.5. [2,3]-Sigmatropic Rearrangements

The [2,3]-sigmatropic class of rearrangements is represented by two generic charge types, neutral and anionic.

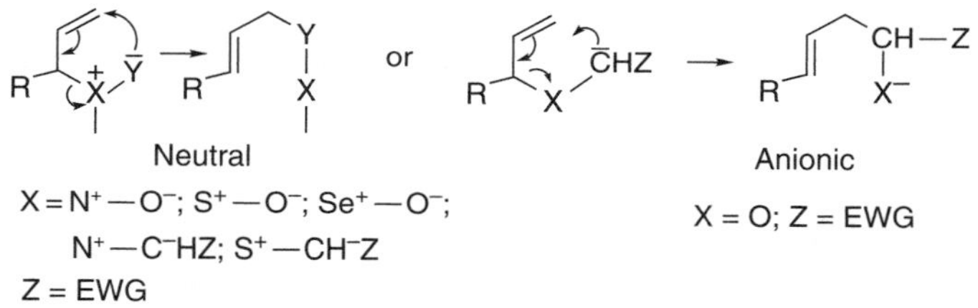

The rearrangements of allylic sulfoxides, selenoxides, and amine oxides are an example of the first type. Allylic sulfonium ylides and ammonium ylides also undergo [2,3]-sigmatropic rearrangements. Rearrangements of carbanions of allylic ethers are the major example of the anionic type. These reactions are considered in the following sections.

6.5.1. Rearrangement of Allylic Sulfoxides, Selenoxides, and Amine Oxides

The rearrangement of allylic sulfoxides to allylic sulfenates was first studied in connection with the mechanism of racemization of allyl aryl sulfoxides.[272] Although the allyl sulfoxide structure is strongly favored at equilibrium, rearrangement through the achiral allyl sulfenate provides a low-energy pathway for racemization.

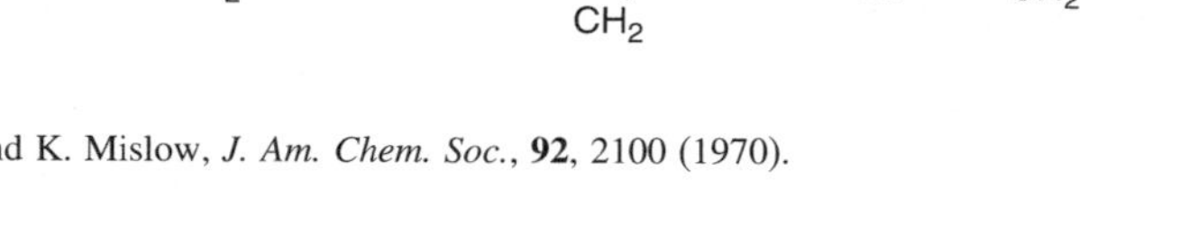

[272.] R. Tang and K. Mislow, *J. Am. Chem. Soc.*, **92**, 2100 (1970).

582

CHAPTER 6

Concerted
Cycloadditions,
Unimolecular
Rearrangements, and
Thermal Eliminations

The reactions occur preferentially though an *endo* TS in which the sulfur substituent is oriented toward the allylic group.[273] Computational studies (MP2/6-31G*) found the *endo* TS to be favored over the *exo* by 1.5–2.2. kcal/mol.[274]

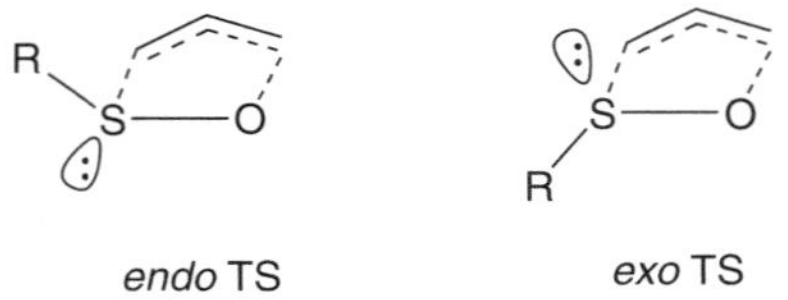

endo TS   exo TS

The allyl sulfoxide–allyl sulfenate rearrangement can be used to prepare allylic alcohols.[275] The reaction is carried out in the presence of a reagent, such as phenylthiolate or trimethyl phosphite, that reacts with the sulfenate to cleave the S–O bond.

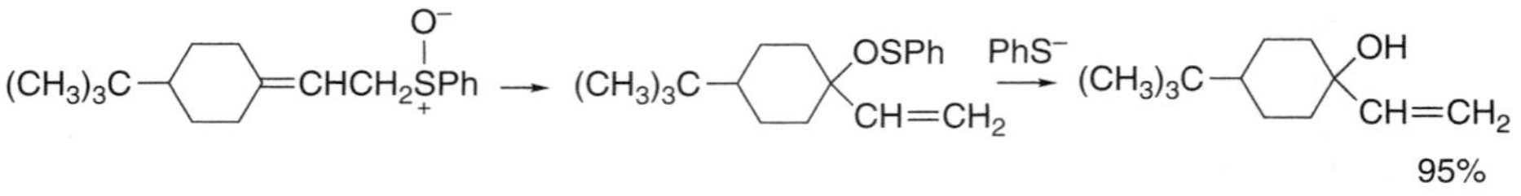

95%

Ref. 276

An analogous reaction occurs when allylic selenoxides are generated in situ by oxidation of allylic selenyl ethers.[277]

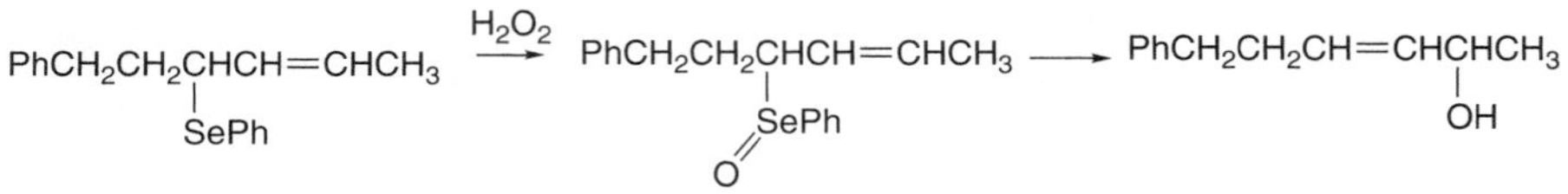

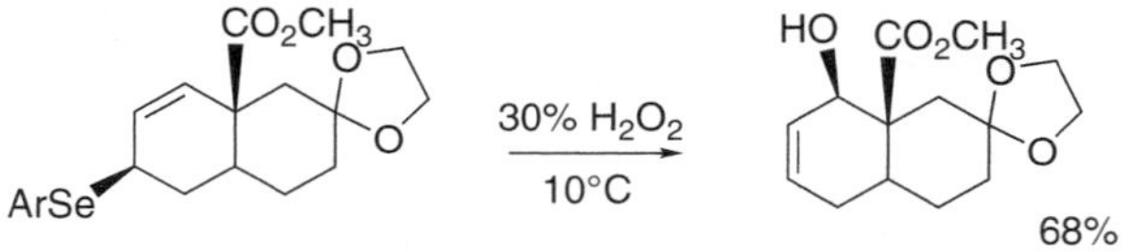

68%

Ref. 278

N-Allylamine oxides represent the general pattern for [2,3]-sigmatropic rearrangement where X = N and Y = O⁻ The rearrangement provides *O*-allyl hydroxylamine derivatives.

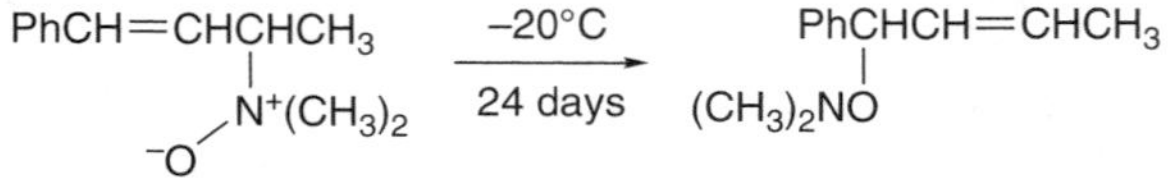

Ref. 279

273. P. Bickart, F. W. Carson, J. Jacobus, E. G. Miller, and K. Mislow, *J. Am. Chem. Soc.*, **90**, 4869 (1968).
274. D. K. Jones-Hertzog and W. L. Jorgensen, *J. Am. Chem. Soc.*, **117**, 9077 (1995).
275. D. A. Evans and G. C. Andrews, *Acc. Chem. Res.*, **7**, 147 (1974).
276. D. A. Evans, G. C. Andrews, and C. L. Sims, *J. Am. Chem. Soc.*, **93**, 4956 (1971).
277. H. J. Reich, *J. Org. Chem.*, **40**, 2570 (1975); D. L. J. Clive, G. Chittatu, N. J. Curtis, and S. M. Menchen, *Chem. Commun.*, 770 (1978).
278. P. A. Zoretic, R. J. Chambers, G. D. Marbury, and A. A. Riebiro, *J. Org. Chem.*, **50**, 2981 (1985).
279. Y. Yamamoto, J. Oda, and Y. Inouye, *J. Org. Chem.*, **41**, 303 (1976).

6.5.2. Rearrangement of Allylic Sulfonium and Ammonium Ylides

Allylic sulfonium ylides readily undergo [2,3]-sigmatropic rearrangement.[280] The ylides are usually formed by deprotonation of the *S*-allyl sulfonium salts.

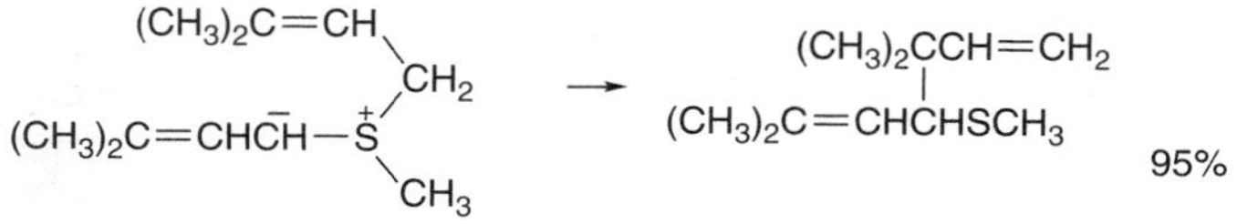

95%

The reaction proceeds best when the ylide has a carbanion-stabilizing substituent. This reaction results in carbon-carbon bond formation and has found synthetic application in ring-expansion sequences for generation of medium-sized rings.

Sulfonium ylides can also be generated by *in situ* alkylation with diazo compounds. The alkylation can be carried out by reaction of a diazo compound with HBF_4 and DBU.[281] The reagents are added alternately in small portions and the reaction presumably proceeds by trapping of the carbocation generated by dediazonization and deprotonation.

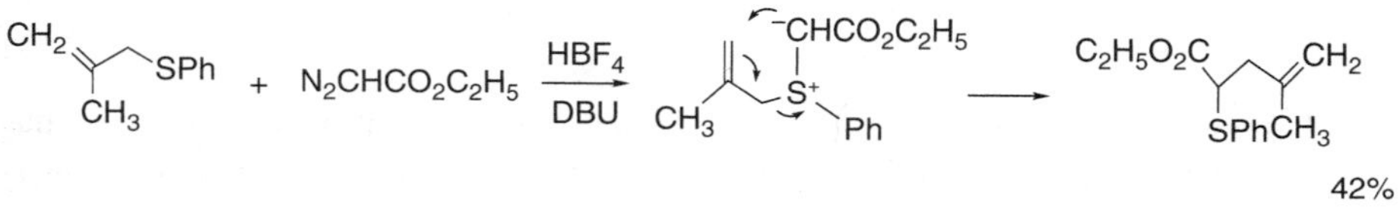

42%

Reactions of this type lead to preferential formation of the *anti* stereochemistry at the new C−C bond.

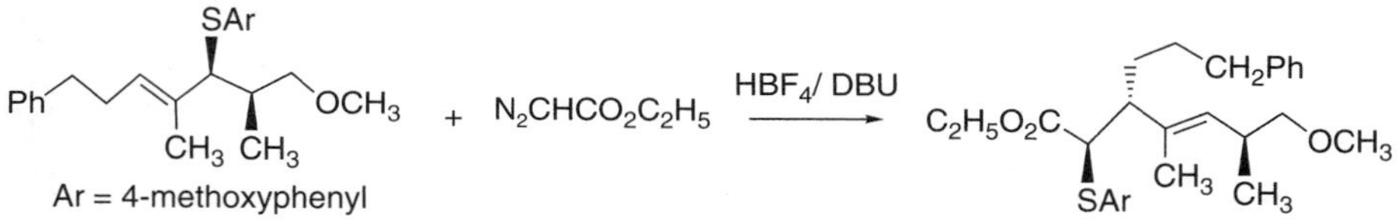

Ar = 4-methoxyphenyl

Sulfonium ylides can also be generated from diazo compounds under carbenoid conditions by using metal catalysts. (See Section 10.2.3.2 for discussion of this means of carbene generation.) The reaction results in transposition of the ester fragment and the sulfide group to the γ-carbon of the allylic group. This reaction has been investigated using chiral catalysts such as Cu(*t*-BuBOX)PF$_6$. Modest enantioselectivity has been achieved using ethyl diazoacetate[282] and methyl phenyldiazoacetate[283] as the carbene precursors.

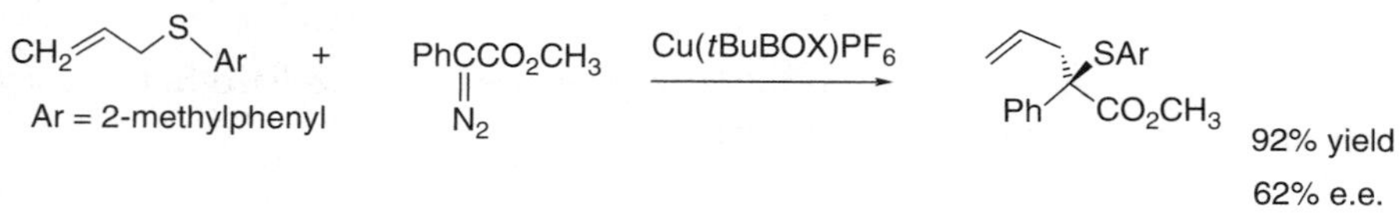

Ar = 2-methylphenyl

92% yield

62% e.e.

280. J. E. Baldwin, R. E. Hackler, and D. P. Kelly, *Chem. Commun.*, 537 (1968).
281. M. J. Kurth, S. H. Tahir, and M. M. Olmstead, *J. Org. Chem.*, **55**, 2286 (1990); R. C. Hartley, S. Warren, and I. C. Richards, *J. Chem. Soc., Perkin Trans. 1*, 507 (1994).
282. D. W. McMillen, N. Varga, B. A. Reed, and C. King, *J. Org. Chem.*, **65**, 2532 (2000).
283. X. Zhang, Z. Qu, Z. Ma, W. Shi, X. Jin, and J. Wang, *J. Org. Chem.*, **67**, 5621 (2002).

CHAPTER 6

*Concerted
Cycloadditions,
Unimolecular
Rearrangements, and
Thermal Eliminations*

Rhodium catalysis have been used for formation of ylides by intramolecular reactions.

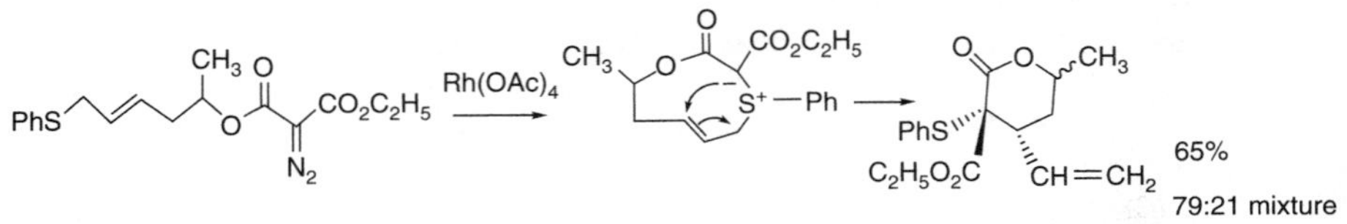

65%
79:21 mixture

Ref. 284

64 %

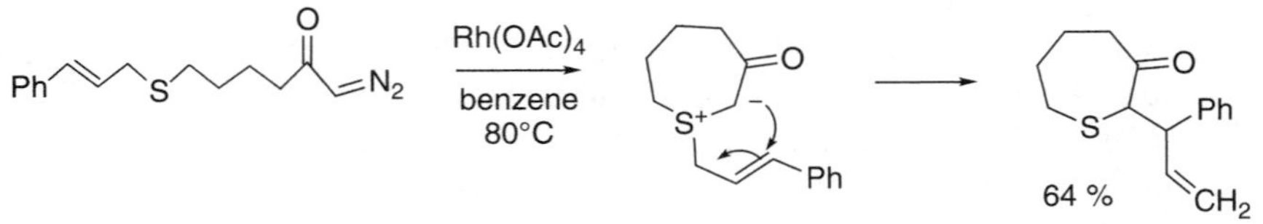

Ref. 285

Ammonium ylides can also be generated when one of the nitrogen substituents has an anion stabilizing group on the α-carbon. For example, quaternary salts of *N*-allyl α-aminoesters readily rearrange to γ,δ-unsaturated α-aminoesters.[286]

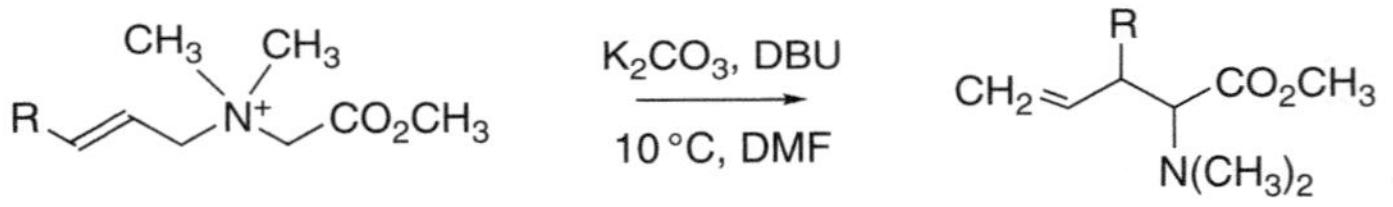

Ammonium ylides can also be generated by the carbenoid route.

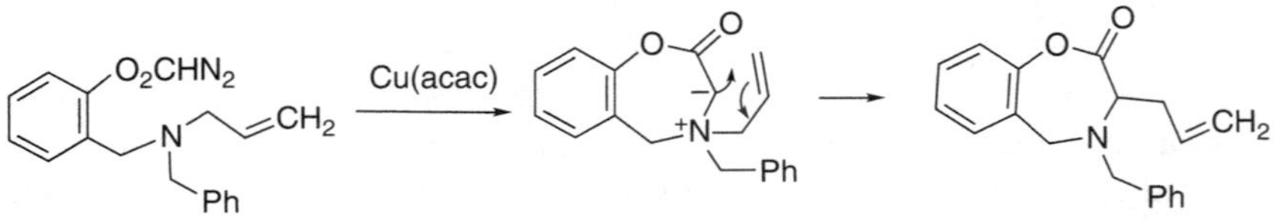

Ref. 287

Copper-catalyzed reactions are particularly effective with α-diazo-β-dicarbonyl compounds such as diethyl diazomalonate.

[284.] F. Kido, S. C. Sinha, T. Abiko, M. Watanabe, and A. Yoshikoshi, *Tetrahedron*, **46**, 4887 (1990).

[285.] C. J. Moody and R. J. Taylor, *Tetrahedron*, **46**, 6501 (1990).

[286.] I. Coldham, M. L. Middleton, and P. L. Taylor, *J. Chem. Soc., Perkin Trans. 1*, 2951 (1997); I. Coldham, M. L. Midleton, and P. L. Taylor, *J. Chem. Soc., Perkin Trans. 1*, 2817 (1998).

[287.] J. S. Clark and M. L. Middleton, *Org. Lett.*, **4**, 765 (2002).

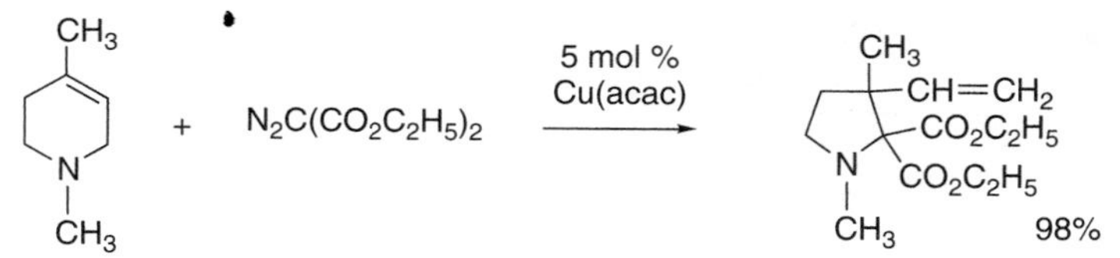

Ref. 288

Scheme 6.18 illustrates typical reaction conditions for [2,3]-sigmatropic rearrangements of sulfonium and ammonium ylides. The reactant sulfonium salt used in Entry 1 is generated by alkylation of ethyl methylthioacetate and rearrangement occurs in the presence of potassium carbonate. Entries 2 and 3 show ring-expansion reactions. The reactant in Entry 2 has no activating group and the reaction presumably proceeds through a small equilibrium concentration of the methylide.

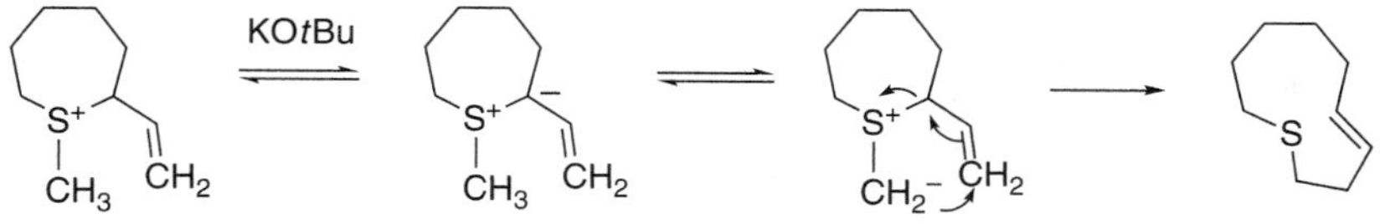

Entries 5 to 8 involve ammonium ylides. These reactions effect an N to C transfer of the substituent with 1,3-allylic transposition. In the case of Entry 7, the anionic stabilization is provided by a vinylogous ester group. The reaction in Entry 8 begins with N-allylation, which takes place *syn* to the ester group because of the *trans* orientation of the ester and benzyl groups, and the chirality is thereby induced at the nitrogen atom. The [2,3]-rearrangement then transfers chirality to C(2) of the pyrrolidine ring.

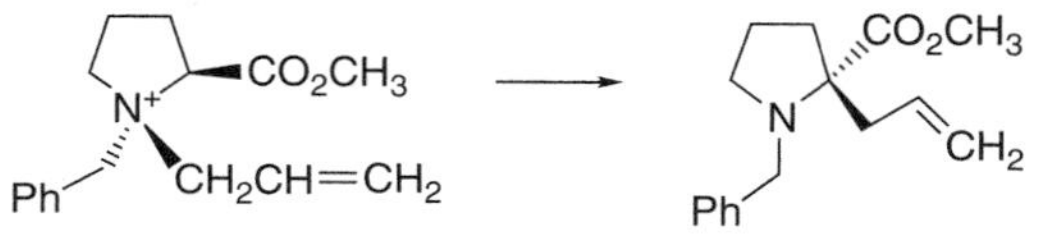

A useful method for *ortho*-alkylation of aromatic amines is based on [2,3]-sigmatropic rearrangement of *S*-anilinosulfonium ylides. These ylides are generated from anilinosulfonium ions, which can be prepared from *N*-chloroanilines and sulfides.[289]

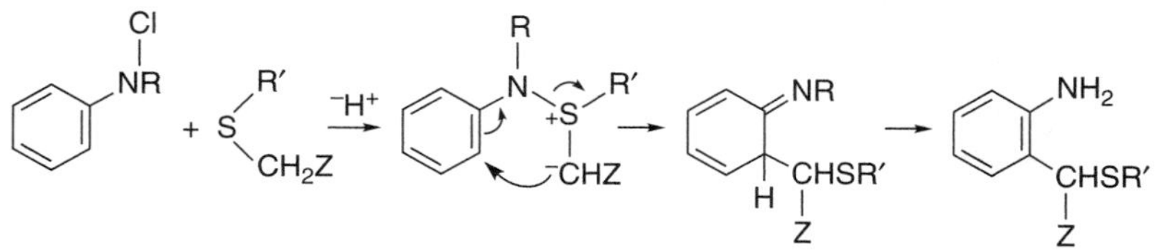

This method is the basis for synthesis of nitrogen-containing heterocyclic compounds when Z is a carbonyl-containing substituent.[290]

[288.] E. Roberts, J. P. Sancon, J. B. Sweeney, and J. A. Workman, *Org. Lett.*, **5**, 4775 (2003).

[289.] P. G. Gassman and G. D. Gruetzmacher, *J. Am. Chem. Soc.*, **96**, 5487 (1974); P. G. Gassman and H. R. Drewes, *J. Am. Chem. Soc.*, **100**, 7600 (1978).

[290.] P. G. Gassman, T. J. van Bergen, D. P. Gilbert, and B. W. Cue, Jr., *J. Am. Chem. Soc.*, **96**, 5495 (1974); P. G. Gassman and T. J. van Bergen, *J. Am. Chem. Soc.*, **96**, 5508 (1974); P. G. Gassman, G. Gruetzmacher, and T. J. van Bergen, *J. Am. Chem. Soc.*, **96**, 5512 (1974).

586

CHAPTER 6

Concerted
Cycloadditions,
Unimolecular
Rearrangements, and
Thermal Eliminations

Scheme 6.18. Carbon-Carbon Bond Formation via [2,3]-Sigmatropic Rearrangements of Sulfonium and Ammonium Ylides

A. Sulfonium ylides

1[a]

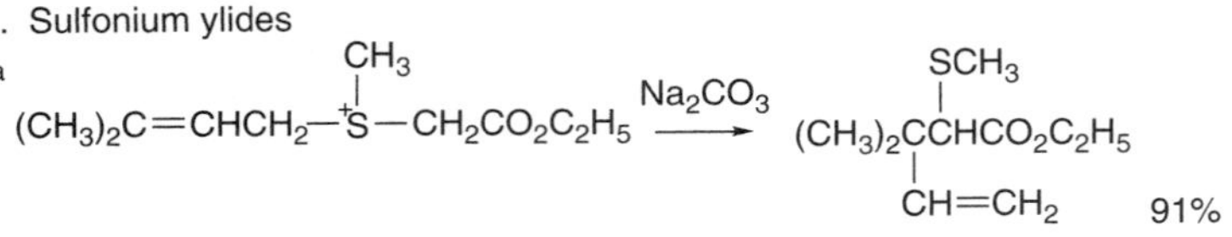

2[b]

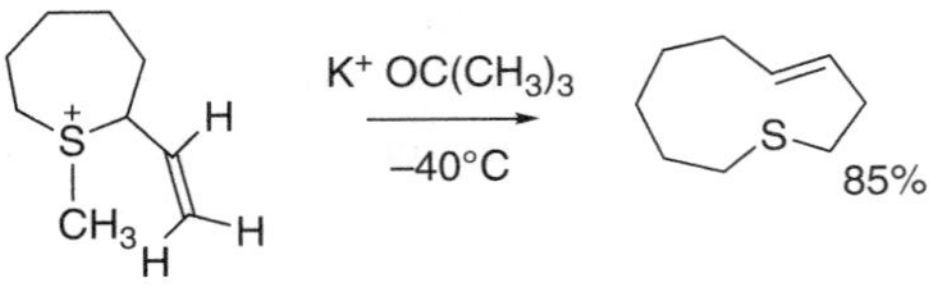

3[c]

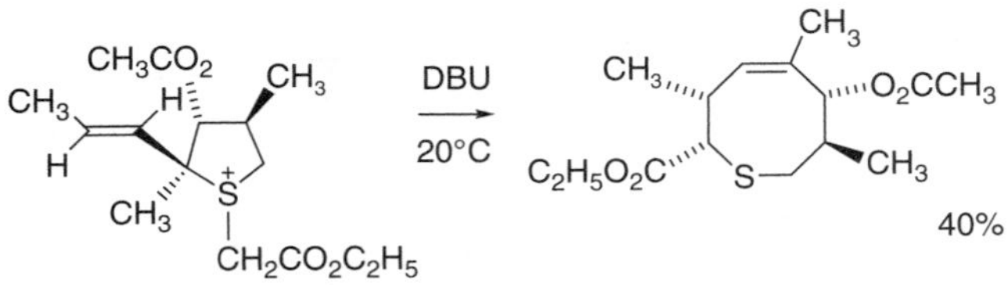

4[d]

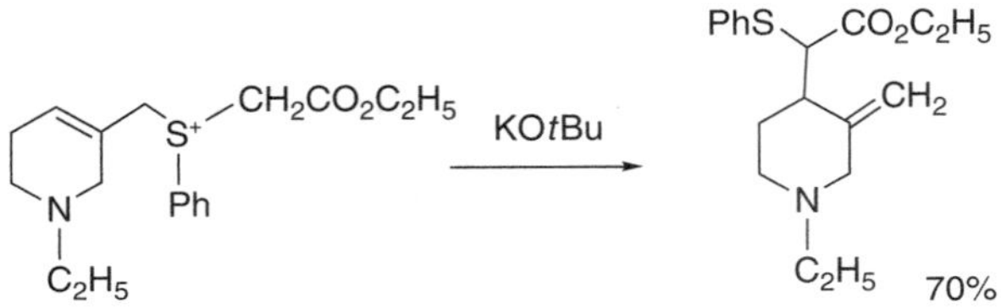

B. Ammonium ylides

5[e]

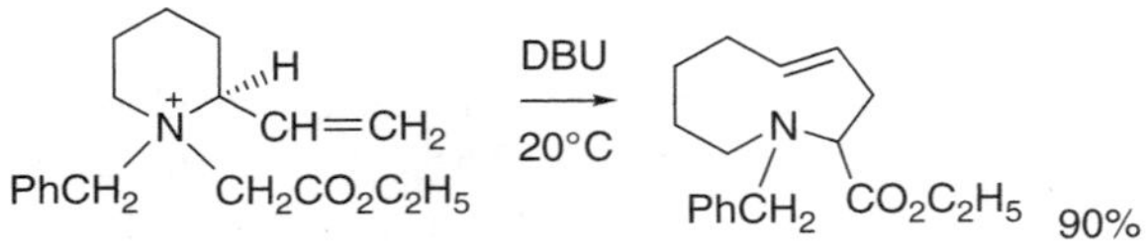

6[f]

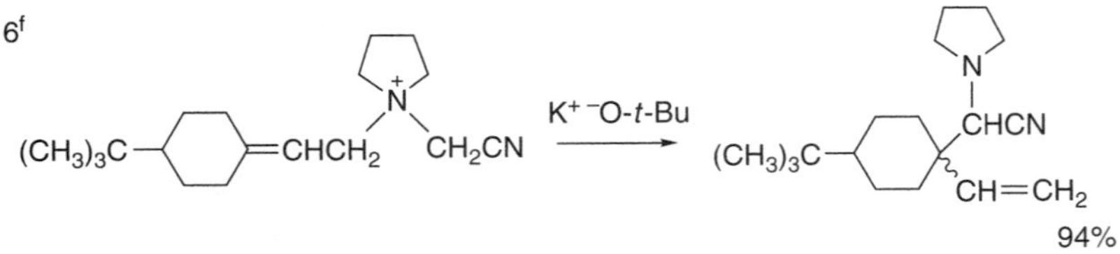

7[g]

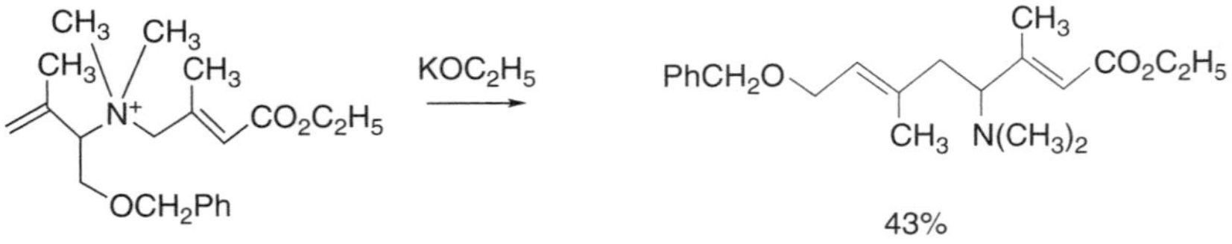

(Continued)

Scheme 6.18. (*Continued*)

587

SECTION 6.5

[2,3]-Sigmatropic Rearrangements

8^h

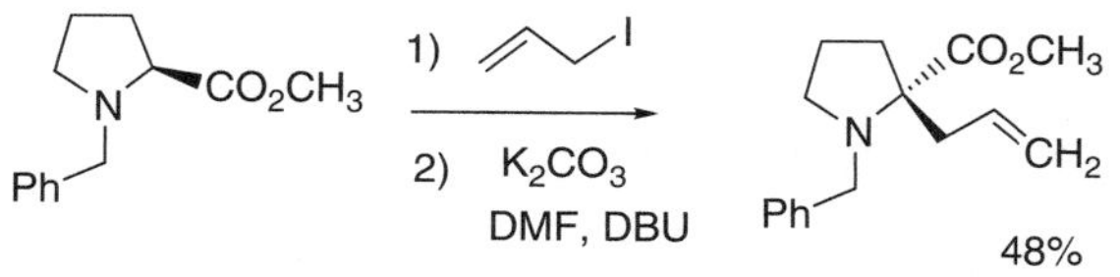

a. K. Ogura, S. Furukawa, and G. Tsuchihashi, *J. Am. Chem. Soc.*, **102**, 2125 (1980).
b. V. Cere, C. Paolucci, S. Pollicino, E. Sandri, and A. Fava, *J. Org. Chem.*, **43**, 4826 (1978).
c. E. Vedejs and M. J. Mullins, *J. Org. Chem.*, **44**, 2947 (1979).
d. R. C. Hartley, S. Warren, and I. C. Richards, *J. Chem. Soc., Perkin Trans. 2*, 507 (1994).
e. E. Vedejs, M. J. Arco, D. W. Powell, J. M. Renga, and S. P. Singer, *J. Org. Chem.*, **43**, 4831 (1978).
f. L. N. Mander and J. V. Turnerk, *Aust. J. Chem.*, **33**, 1559 (1980).
g. K. Honda, I. Yoshii, and S. Inoue, *Chem. Lett.*, 671 (1996).
h. A. P. A. Arbore, D. J. Cane-Honeysett, I. Coldham, and M. L. Middleton, *Synlett*, 236 (2000).

6.5.3. Anionic Wittig and Aza-Wittig Rearrangements

The [2,3]-sigmatropic rearrangement pattern is also observed with anionic species. The most important case for synthetic purposes is the *Wittig rearrangement*, in which a strong base converts allylic ethers to α-allylalkoxides.[291] Since the deprotonation at the α′-carbon must compete with deprotonation of the α-carbon in the allyl group, most examples involve a conjugated or EWG substituent Z.[292]

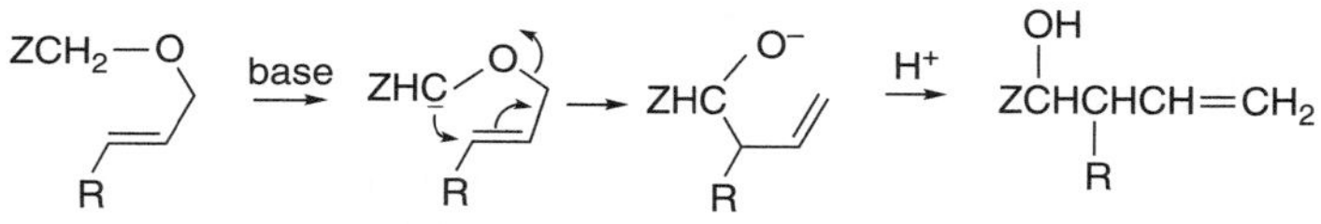

The stereochemistry of the Wittig rearrangement can be predicted in terms of a cyclic five-membered TS in which the α-substituent prefers an equatorial orientation.[293]

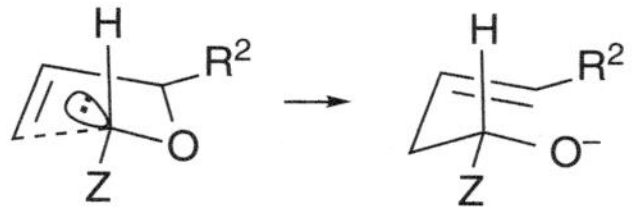

A consistent feature of the stereoselectivity is a preference for *E*-configuration at the newly formed double bond. The reaction can also show stereoselectivity at the newly formed single bond. This stereoselectivity has been carefully studied for the case in

[291.] J. Kallmarten, in *Stereoselective Synthesis: Houben Weyl Methods in Organic Chemistry*, Vol E21d, R. W. Hoffmann, J. Mulzer, and E. Schaumann, eds., G. Thieme Verlag, Stuttgart, 1996, pp. 3810 ff.
[292.] For a review of [2,3]-sigmatropic rearrangement of allyl ethers, see T. Nakai and K. Mikami, *Chem. Rev.*, **86**, 885 (1986).
[293.] R. W. Hoffmann, *Angew. Chem. Int. Ed. Engl.*, **18**, 563 (1979); K. Mikami, Y. Kimura, N. Kishi, and T. Nakai, *J. Org. Chem.*, **48**, 279 (1983); K. Mikami, K. Azuma, and T. Nakai, *Tetrahedron*, **40**, 2303 (1984); Y.-D. Wu, K. N. Houk, and J. A. Marshall, *J. Org. Chem.*, **55**, 1421 (1990).

588

CHAPTER 6

*Concerted
Cycloadditions,
Unimolecular
Rearrangements, and
Thermal Eliminations*

which the substituent Z is an alkynyl group. The *E*-isomer leads to *anti* product and the *Z*-isomer to the *syn* product.

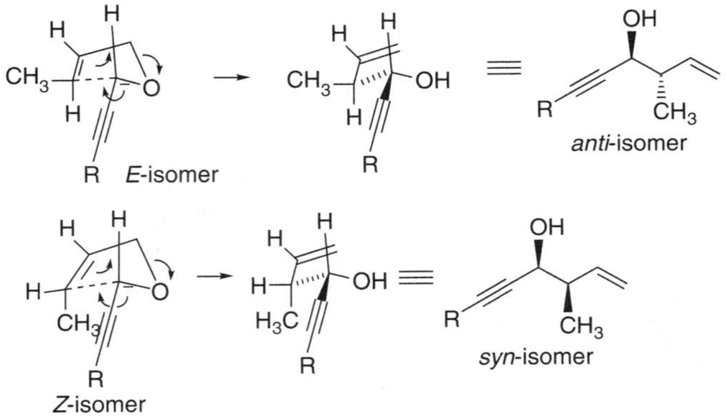

The preferred TS minimizes interaction between the Z and allylic substituents. This stereoselectivity is illustrated in the rearrangement of **18** to **19**.

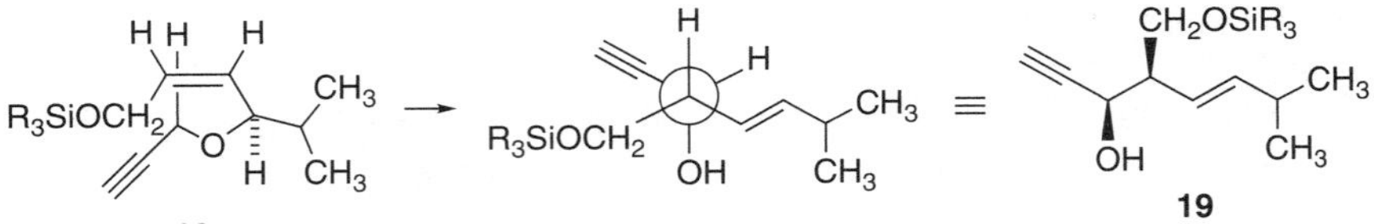

Ref. 294

There are other means of generating the anions of allyl ethers. One of the most useful for synthetic purposes involves a lithium-tin exchange on stannylmethyl ethers (see Section 7.1.2.4).[295]

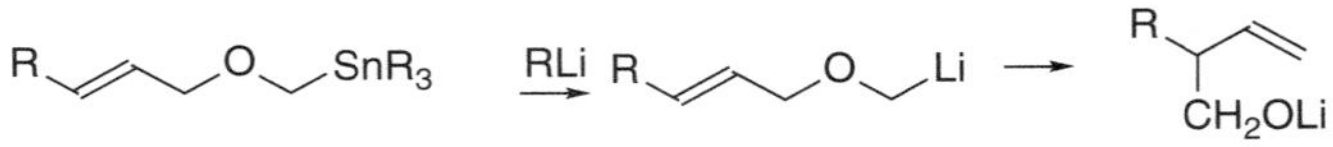

Another means involves reduction of allylic acetals of aromatic aldehydes by SmI_2.[296]

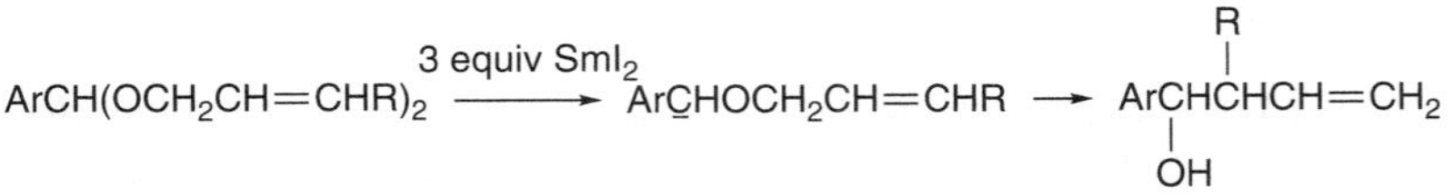

[2,3]-Sigmatropic rearrangements of anions of *N*-allyl amines have also been observed and are known as *aza-Wittig rearrangements*.[297] The reaction requires anion stabilizing substituents and is favored by *N*-benzyl and by silyl or sulfenyl substituents

294. M. M. Midland and J. Gabriel, *J. Org. Chem.*, **50**, 1143 (1985).
295. W. C. Still and A. Mitra, *J. Am. Chem. Soc.*, **100**, 1927 (1978).
296. H. Hioki, K. Kono, S. Tani, and M. Kunishima, *Tetrahedron Lett.*, **39**, 5229 (1998).
297. C. Vogel, *Synlett*, 497 (1997).

on the allyl group.[298] The trimethylsilyl substituents can also influence the stereoselectivity of the reaction. The steric interactions between the benzyl group and allyl
substituent govern the stereoselectivity and it is markedly improved in the trimethylsilyl
derivatives.[299]

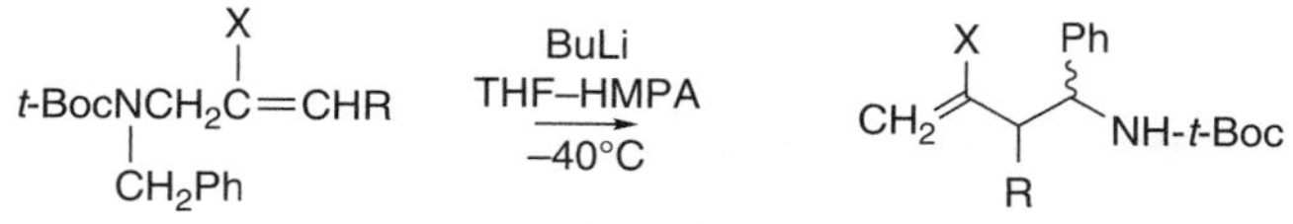

R	X	*anti:syn*
CH₃	H	3:2
C₂H₅	H	1:1
(CH₃)₂CH	H	4:3
CH₃	Si(CH₃)₃	<1:20
C₂H₅	Si(CH₃)₃	1:18
(CH₃)₂CH	Si(CH₃)₃	1:11

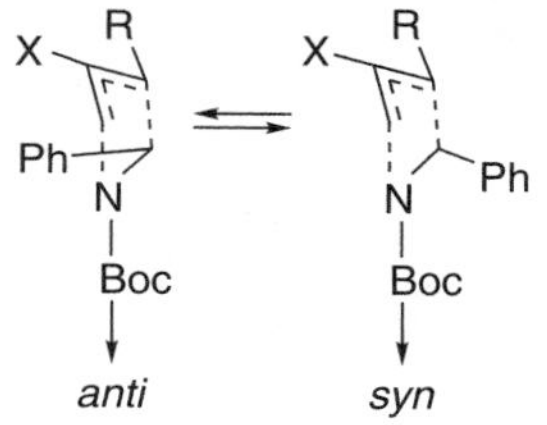

Some examples of synthetic application of the anionic Wittig rearrangement are
given in Scheme 6.19. The reaction in Entry 1 provided a 93:7 ratio favoring the *syn*
isomer, as expected for the preferred *endo* TS. Entry 2 is an example that employs the
lithium-stannane exchange to generate the anion. The reaction in Entry 3 accomplishes
a ring contraction. Under normal conditions, it is selective for the *trans* stereoisomer,
as would be expected from steric factors in the TS. In the presence of HMPA, the *cis*
isomer dominates, but the reason for the change is not known.

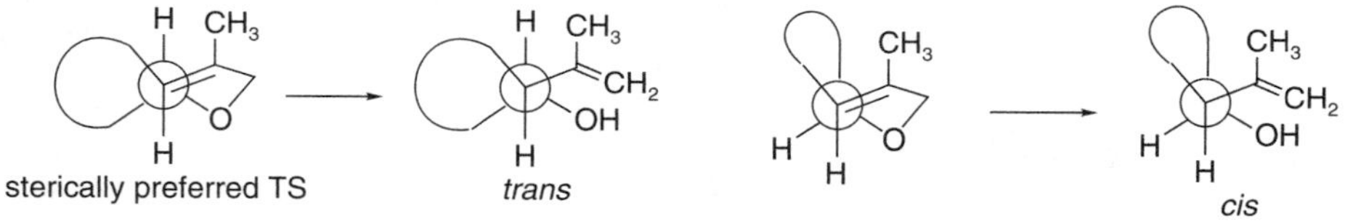

sterically preferred TS   *trans*   *cis*

In Entry 4 the silyl group appears to introduce a controlling steric factor, leading to
the observed stereoisomer. The unsubstituted terminal alkyne, which reacts through
the dianion, gives the alternate isomer.

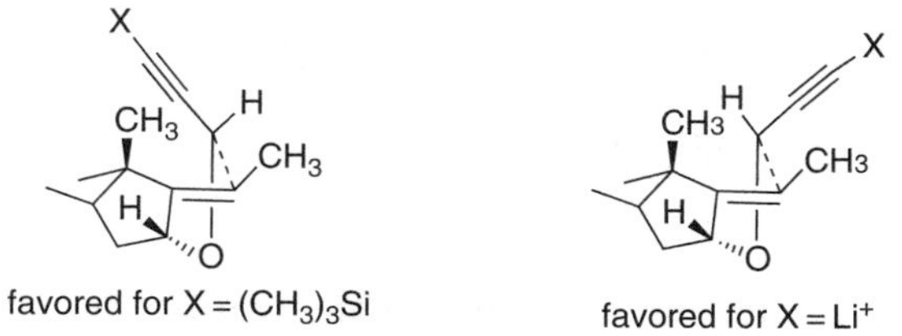

favored for X = (CH₃)₃Si   favored for X = Li⁺

[298.] J. C. Anderson, S. C. Smith, and M. E. Swarbrick, *J. Chem. Soc., Perkin Trans. 1*, 1517 (1997).
[299.] J. C. Anderson, D. C. Siddons, S. C. Smith, and M. E. Swarbrick, *J. Org. Chem.*, **61**, 4820 (1996).

Scheme 6.19. [2,3]-Anionic Wittig Rearrangements

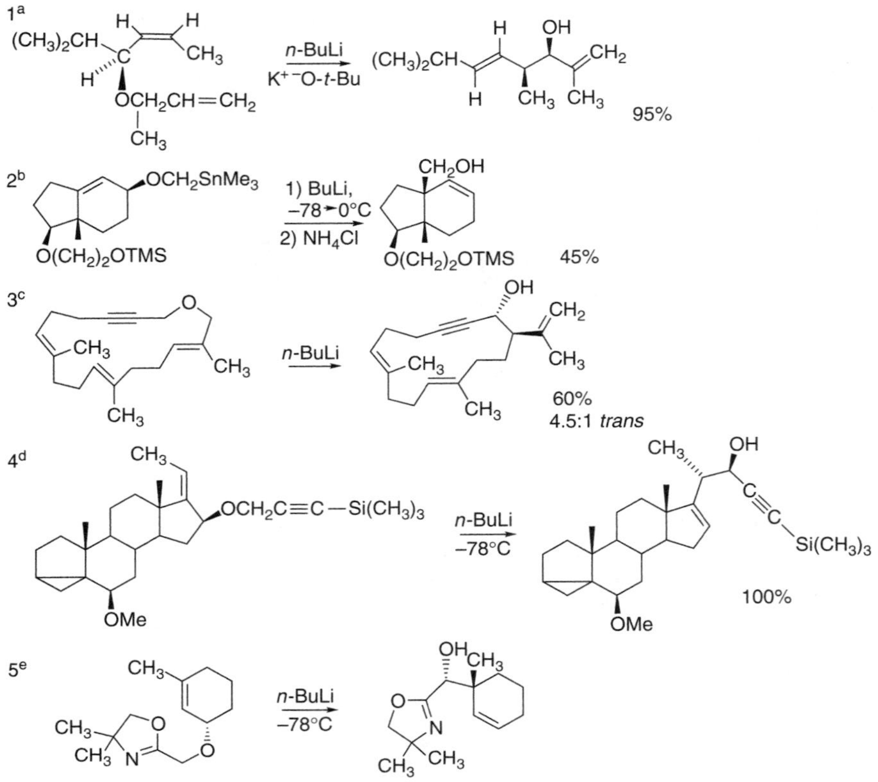

CHAPTER 6

*Concerted
Cycloadditions,
Unimolecular
Rearrangements, and
Thermal Eliminations*

a. D. J.-S. Tsai and M. M. Midland, *J. Am. Chem. Soc.*, **107**, 3915 (1985).
b. T. Sugimura and L. A. Paquette, *J. Am. Chem. Soc.*, **109**, 3017 (1987).
c. J. A. Marshall, T. M. Jenson, and D. S. De Hoff, *J. Org. Chem.*, **51**, 4316 (1986).
d. K. Mikami, K. Kawamoto, and T. Nakai, *Tetrahedron Lett.*, **26**, 5799 (1985).
e. M. H. Kress, B. F. Kaller, and Y. Kishi, *Tetrahedron Lett.*, **34**, 8047 (1993).

The stereoselectivity of the reaction in Entry 5 is also determined by steric factors. Note also that in this case the oxazoline ring serves to stabilize the anion.

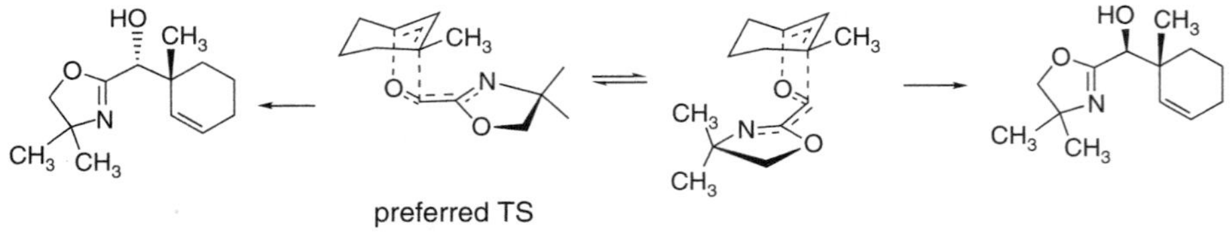

preferred TS

6.6. Unimolecular Thermal Elimination Reactions

This section describes reactions in which elimination to form a double bond or a new ring occurs as a result of thermal activation. There are several such thermal elimination reactions that are used syntheses, some of which are concerted processes. The activation energy requirements and stereochemistry of concerted elimination processes can be analyzed in terms of orbital symmetry considerations. Cheletropic eliminations are discussed in Section 6.6.1 and elimination of nitrogen from azo compounds in Section 6.6.2. We consider an important group of unimolecular β-elimination reactions in Section 6.6.3.

6.6.1. Cheletropic Elimination

Cheletropic processes are defined as reactions in which two bonds are broken at a single atom. Concerted cheletropic reactions are subject to orbital symmetry analysis in the same way as cycloadditions and sigmatropic processes. In the elimination processes of interest here, the atom X is normally bound to other atoms in such a way that elimination gives rise to a stable molecule. In particular, elimination of SO_2, N_2, or CO from five-membered 3,4-unsaturated rings can be a facile process.

$$X=Y=C=O, N=N, SO_2$$

A good example of a concerted cheletropic elimination is the reaction of 3-pyrroline with *N*-nitrohydroxylamine, which gives rise the the diazene **21**, which then undergoes elimination of nitrogen.

$$CH_2=CHCH=CH_2 + N_2$$

Use of substituted systems has shown that the reaction is stereospecific.[300] The groups on C(2) and C(5) of the pyrroline ring rotate in the disrotatory mode on going to product. This stereochemistry is consistent with conservation of orbital symmetry.

The most synthetically useful cheletropic elimination involves 2,5-dihydrothiophene-1,1-dioxides (sulfolene dioxides). At moderate temperatures they fragment to give dienes and sulfur dioxide.[301] The reaction is stereospecific. For example, the dimethyl derivatives **22** and **23** give the *E,E*- and *Z,E*-isomers of 2,4-hexadiene, respectively, at temperatures of 100°–150° C.[302] This stereospecificity corresponds to disrotatory elimination.

22   **23**

[300] D. M. Lemal and S. D. McGregor, *J. Am. Chem. Soc.*, **88**, 1335 (1966).

[301] W. L. Mock, in *Pericyclic Reactions*, Vol. II, A. P. Marchand and R. E. Lehr, eds., Academic Press, New York, 1977, Chap. 3.

[302] W. L. Mock, *J. Am. Chem. Soc.*, **88**, 2857 (1966); S. D. McGregor and D. M. Lemal, *J. Am. Chem. Soc.*, **88**, 2858 (1966).

592

CHAPTER 6

Concerted
Cycloadditions,
Unimolecular
Rearrangements, and
Thermal Eliminations

Elimination of sulfur dioxide has proven to be a useful method for generating dienes that can undergo subsequent D-A addition.

Ref. 303

Ref. 304

Sulfolene dioxide is subject to α-lithiation and alkylation, and this reaction has been used to introduce the ring into more complex molecules.

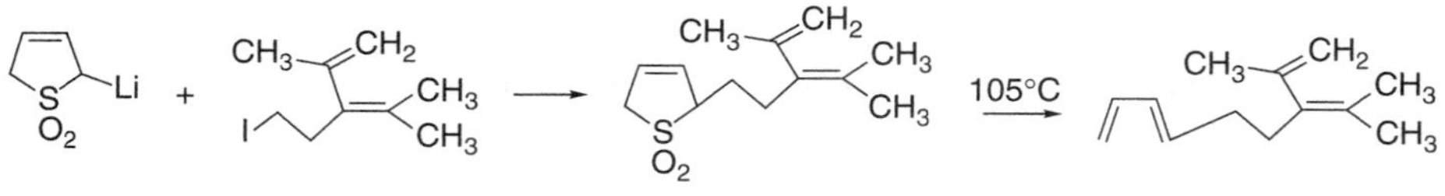

Ref. 305

Sulfolene dioxide thermolysis has also been applied to formation of *o*-quinodimethanes.

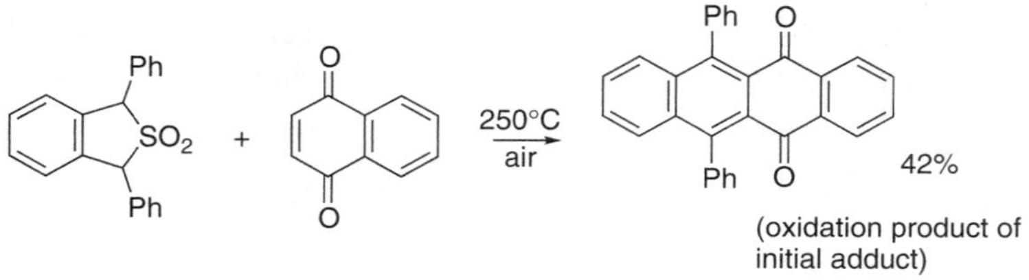

(oxidation product of initial adduct)

Ref. 306

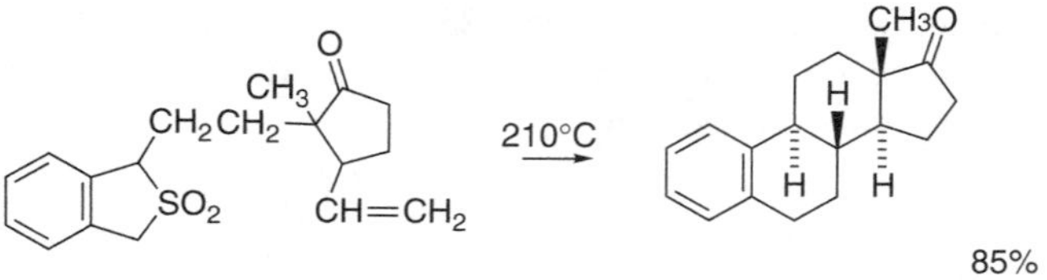

Ref. 307

303. J. M. McIntosh and R. A. Sieler, *J. Org. Chem.*, **43**, 4431 (1978).
304. A. M. Gomez, J. C. Lopez, and B. Fraser-Reid, *Synthesis*, 943 (1993).
305. J. D. Winkler, H. S. Kim, S. Kim, K. Ando, and K. N. Houk, *J. Org. Chem.*, **62**, 2957 (1997).
306. M. P. Cava, M. J. Mitchell, and A. A. Deana, *J. Org. Chem.*, **25**, 1481 (1960).
307. K. C. Nicolaou, W. E. Barnette, and P. Ma, *J. Org. Chem.*, **45**, 1463 (1980).

The elimination of carbon monoxide can occur by a concerted process in some cyclic ketones. The elimination of carbon monoxide from bicyclo[2.2.1]heptadien-7-ones is very facile. In fact, generation of bicyclo[2.2.1]heptadien-7-ones is usually accompanied by spontaneous decarbonylation.

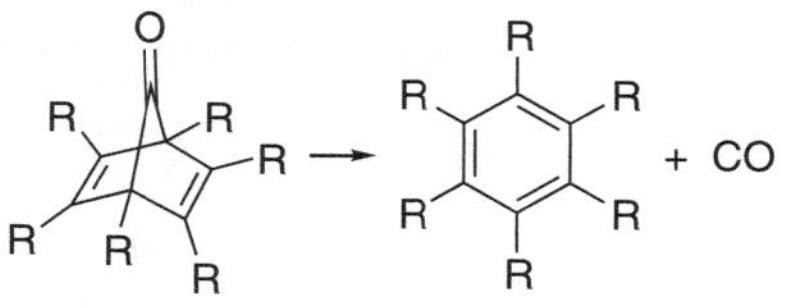

The ring system can be generated by D-A addition of a substituted cyclopentadienone and an alkyne. A reaction sequence involving addition followed by CO elimination can be used for the synthesis of highly substituted benzene rings.[308]

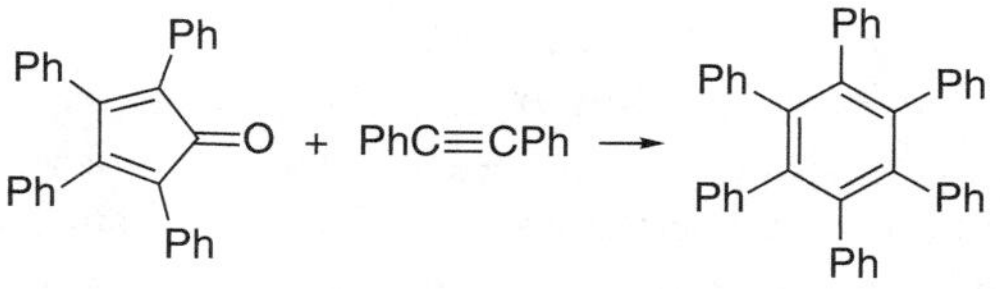

Ref. 309

The synthetic utility of cyclopentadienones is limited, however, because they are quite unstable. Exceptionally facile elimination of CO also takes place from **22**, in which homoaromaticity can facilitate elimination.

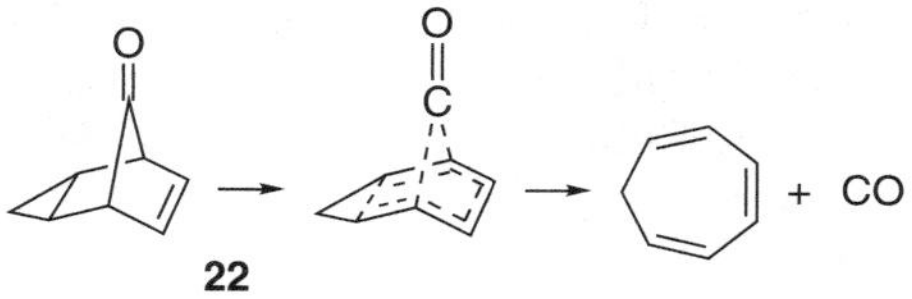

Ref. 310

6.6.2. Decomposition of Cyclic Azo Compounds

Another significant group of elimination reactions involves processes in which a small molecule is eliminated from a ring system and the two reactive sites that remain re-form a ring.

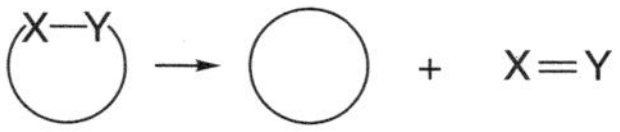

The most common example is decomposition of azo compounds, where $-X-Y-$ is $-N=N-$.[311] The elimination of nitrogen from cyclic azo compounds can be carried

308. M. A. Ogliaruso, M. G. Romanelli, and E. I. Becker, *Chem. Rev.*, **65**, 261 (1965).
309. L. F. Fieser, *Org. Synth.*, **V**, 604 (1973).
310. B. A. Halton, M. A. Battiste, R. Rehberg, C. L. Deyrup, and M. E. Brennan, *J. Am. Chem. Soc.*, **89**, 5964 (1967).
311. P. S. Engel, *Chem. Rev.*, **80**, 99 (1980).

594

CHAPTER 6

*Concerted
Cycloadditions,
Unimolecular
Rearrangements, and
Thermal Eliminations*

out either photochemically or thermally. Although the reaction usually does not proceed by a concerted mechanism, there are some special cases in which concerted elimination is possible. We consider these cases first and then move on to the more general case. An interesting illustration of the importance of orbital symmetry effects is the contrasting stability of azo compounds **23** and **24**. Compound **23** decomposes to norbornene and nitrogen only above $100°$ C. In contrast **24** eliminates nitrogen immediately on preparation, even at $-78°$ C.[312]

The reason for this difference is that if **23** were to undergo a concerted elimination it would have to follow the forbidden (high-energy) $[2\pi_s + 2\pi_s]$ pathway. For **24**, the elimination can take place by the allowed $[2\pi_s + 4\pi_s]$ pathway. Thus, these reactions are the reverse, respectively, of the $[2\pi_s + 2\pi_s]$ and $[2\pi_s + 4\pi_s]$ cycloadditions, and only the latter is an allowed concerted process. The temperature at which **23** decomposes is fairly typical for strained azo compounds and it presumably proceeds by a nonconcerted diradical mechanism. Since a C—N bond must be broken without concomitant compensation by carbon-carbon bond formation, the activation energy is higher than for a concerted process.

Although the concerted mechanism described in the preceding paragraph is available only to those azo compounds with appropriate orbital arrangements, the nonconcerted mechanism occurs at low enough temperatures to be synthetically useful. The elimination can also be carried out photochemically. These reactions presumably occur by stepwise elimination of nitrogen, and the ease of decomposition depends on the stability of the radical R·.

The stereochemistry of the nonconcerted reaction has been a topic of considerable study. Frequently, there is partial stereorandomization, indicating a short-lived diradical intermediate. The details vary from case to case, and both preferential inversion and retention of relative stereochemistry have been observed.

66:33 from *cis*
25:72 from *trans*

predominant inversion

Ref. 313

312. N. Rieber, J. Alberts, J. A. Lipsky, and D. M. Lemal, *J. Am. Chem. Soc.*, **91**, 5668 (1969).
313. R. J. Crawford and A. Mishra, *J. Am. Chem. Soc.*, **88**, 3963 (1966).

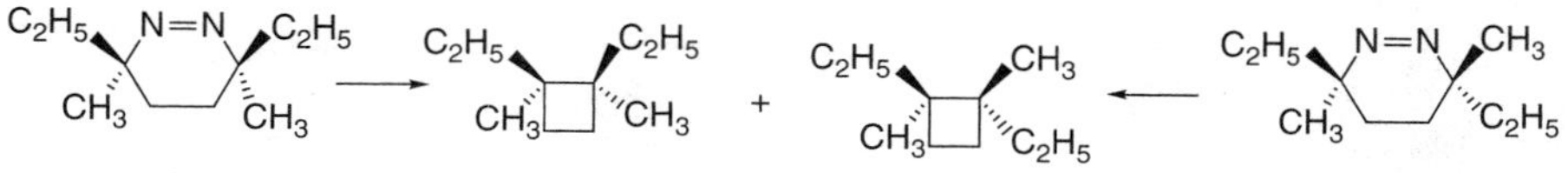

43:2.5 from *cis*
3.5:42 from *trans*

predominant retention

Ref. 314

These results can be interpreted in terms of competition between recombination of the diradical intermediate and conformational equilibration, which would destroy the stereochemical relationships present in the azo compound. The main synthetic application of azo compound decomposition is in the synthesis of cyclopropanes and other strained-ring systems. Some of the required azo compounds can be made by 1,3-dipolar cycloadditions of diazo compounds (see Section 6.2).

Elimination of nitrogen from D-A adducts of certain heteroaromatic rings has been useful in syntheses of substituted aromatic compounds.[315] Pyrazines, triazines, and tetrazines react with electron-rich dienophiles in inverse electron demand cycloadditions. The adducts then aromatize with loss of nitrogen and a dienophile substituent.[316]

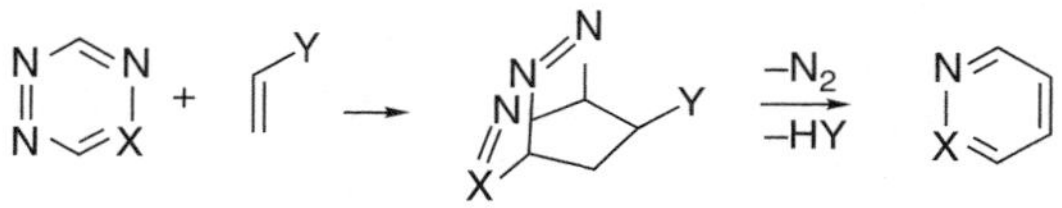

Pyridazine-3,6-dicarboxylate esters react with electron-rich alkenes to give adducts that undergo subsequent elimination to give terephthalate derivatives.[317]

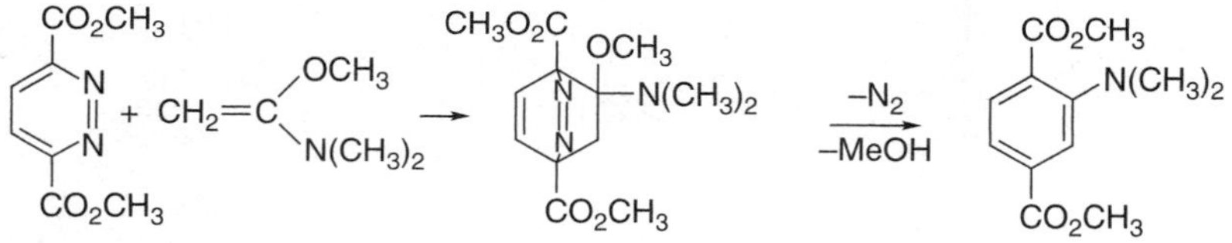

Similar reactions have been developed for 1,2,4-triazines and 1,2,4,5-tetrazines.

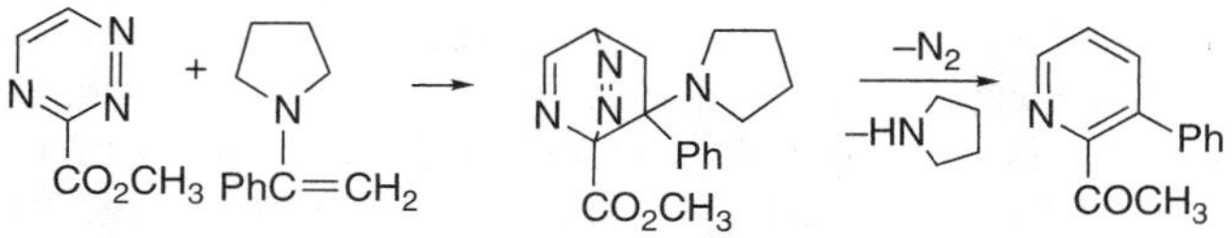

Ref. 318

314. P. D. Bartlett and N. A. Porter, *J. Am. Chem. Soc.*, **90**, 5317 (1968).
315. D. L. Boger, *Chem. Rev.*, **86**, 781 (1986).
316. D. L. Boger, *J. Heterocycl. Chem.*, **33**, 1519 (1996).
317. H. Neunhoeffer and G. Werner, *Liebigs Ann. Chem.*, 437, 1955 (1973).
318. D. L. Boger and J. S. Panek, *J. Am. Chem. Soc.*, **107**, 5745 (1985).

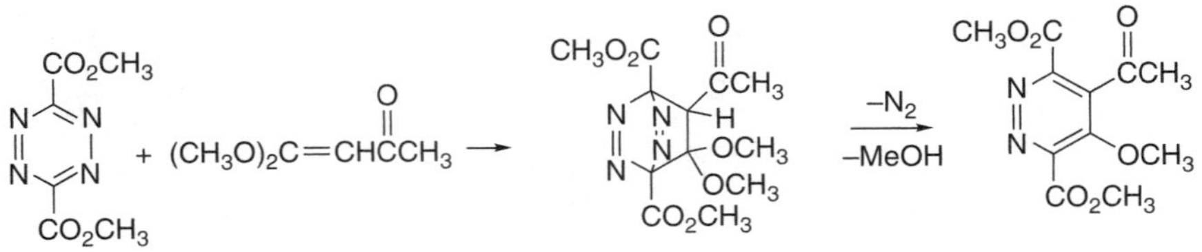

596

CHAPTER 6

*Concerted
Cycloadditions,
Unimolecular
Rearrangements, and
Thermal Eliminations*

Ref. 319

The heterocycles frequently carry substituents such as chloro, methylthio, or alkoxy-carbonyl.

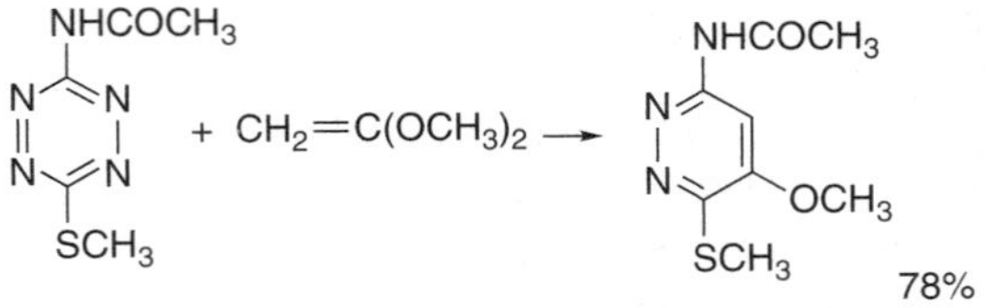

78%

Ref. 320

66%

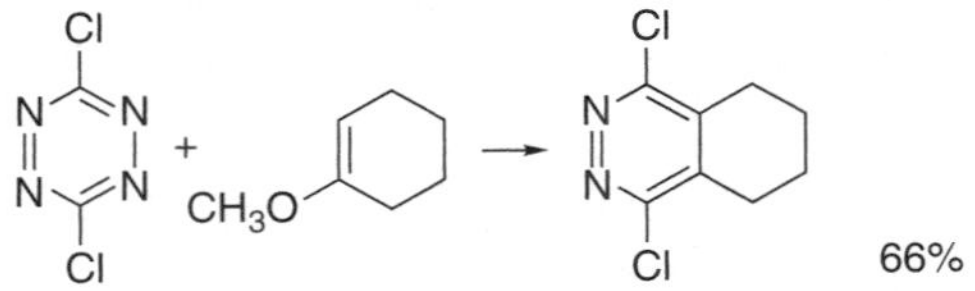

Ref. 321

Acetylenic dienophiles lead directly to aromatic adducts on loss of nitrogen.

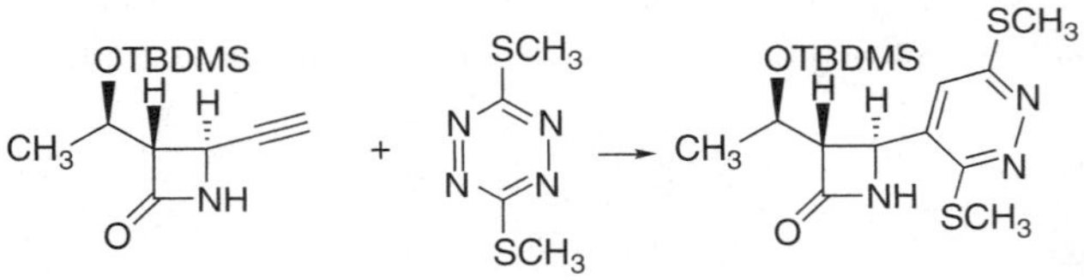

Ref. 322

6.6.3. β-Eliminations Involving Cyclic Transition Structures

Another important family of elimination reactions has as its common mechanistic feature cyclic TSs in which an intramolecular hydrogen transfer accompanies elimination to form a new carbon-carbon double bond. Scheme 6.20 depicts examples of these reaction types. These are thermally activated unimolecular reactions that normally do not involve acidic or basic catalysts. There is, however, a wide variation in the temperature at which elimination proceeds at a convenient rate. The cyclic TS dictates that elimination occurs with *syn* stereochemistry. At least in a formal sense, all the reactions can proceed by a concerted mechanism. The reactions, as a group, are often referred to as *thermal syn eliminations*.

319. D. L. Boger and R. S. Coleman, *J. Am. Chem. Soc.*, **109**, 2717 (1987).
320. D. L. Boger, R. P. Schaum, and R. M. Garbaccio, *J. Org. Chem.*, **63**, 6329 (1998).
321. T. J. Sparey and T. Harrison, *Tetrahedron Lett.*, **39**, 5893 (1998).
322. S. M. Sakya, T. W. Strohmeyer, S. A. Lang, and Y.-I. Lin, *Tetrahedron Lett.*, **38**, 5913 (1997).

1[a]

$$\underset{R-CH-CHR}{\overset{\overset{H}{|}}{\overset{\bar{O}-\overset{+}{N}(CH_3)_2}{|}}} \longrightarrow \underset{\underset{H}{\overset{|}{RC}}-\overset{|}{CHR}}{\overset{\overset{H}{\frown}\overset{\delta-}{O}\overset{\delta+}{N}(CH_3)_2}{}} \longrightarrow RCH{=}CHR + \quad 100°{-}150°C$$
$$HON(CH_3)_2$$

2[b]

$$\underset{R-CH-CHR}{\overset{\overset{H}{|}}{\overset{\bar{O}-\overset{+}{Se}R'}{|}}} \longrightarrow \underset{\underset{H}{\overset{|}{RC}}-\overset{|}{CHR}}{\overset{\overset{H}{\frown}\overset{\delta-}{O}\overset{\delta+}{Se}R'}{}} \longrightarrow RCH{=}CHR + \quad 0°{-}100°C$$
$$HOSeR'$$

3[c]

$$\underset{R-CH-CHR}{\overset{\overset{CH_3}{|}}{\overset{C}{\underset{H}{\overset{O}{\diagdown}}\overset{}{\diagup}O}}} \longrightarrow \underset{\underset{R}{\overset{}{\diagup}}\overset{C}{\underset{H}{\diagdown}}H}{\overset{\overset{CH_3}{|}}{\overset{C}{O\diagup\,\diagdown O}}\atop{H\frown CHR}} \longrightarrow RCH{=}CHR + \quad 400°{-}600°C$$
$$CH_3CO_2H$$

4[d]

$$\underset{R-CH-CHR}{\overset{\overset{SCH_3}{|}}{\overset{C}{\underset{H}{\overset{S}{\diagdown}}\overset{}{\diagup}O}}} \longrightarrow \underset{\underset{R}{\overset{}{\diagup}}\overset{C}{\underset{H}{\diagdown}}H}{\overset{\overset{SCH_3}{|}}{\overset{C}{S\diagup\,\diagdown O}}\atop{H\frown CHR}} \longrightarrow RCH{=}CHR + \quad 150°{-}250°C$$
$$CH_3SH + SCO$$

a. A. C. Cope and E. R. Trumbull, *Org. React.*, **11**, 317 (1960).
b. D. L. J. Clive, *Tetrahedron*, **34**, 1049 (1978).
c. C. H. De Puy and R. W. King, *Chem. Rev.*, **60**, 431 (1960).
d. H. R. Nace, *Org. React.*, **12**, 57 (1962).

Amine oxide pyrolysis occurs at temperatures of 100°–150° C. The reaction can proceed at room temperature in DMSO.[323] If more than one type of β-hydrogen can attain the eclipsed conformation of the cyclic TS, a mixture of alkenes is formed. The product ratio parallels the relative stability of the competing TSs. Usually more of the *E*-alkene is formed because of the larger steric interactions present in the TS leading to the *Z*-alkene, but the selectivity is generally not high.

more favorable   less favorable

In cyclic systems, conformational effects and the requirement for a cyclic TS determine the product composition. This effect can be seen in the product ratios from pyrolysis of *N*,*N*-dimethyl-2-phenylcyclohexylamine-*N*-oxide.

323. D. J. Cram, M. R. V. Sahyun, and G. R. Knox, *J. Am. Chem. Soc.*, **84**, 1734 (1962).

598

CHAPTER 6

*Concerted
Cycloadditions,
Unimolecular
Rearrangements, and
Thermal Eliminations*

from *trans*   85:15
from *cis*   2:98

In the *trans* isomer, elimination to give a double bond conjugated with an aromatic ring is especially favorable. This presumably reflects both the increased acidity of the proton α to the phenyl ring and the stabilizing effect of the developing conjugation in the TS. In the *cis* isomer there is no *syn* hydrogen at the phenyl-substituted carbon and the nonconjugated regioisomer is formed. Amine oxides can be readily prepared from amines by oxidation with hydrogen peroxide or a peroxycarboxylic acid. Some typical examples of amine oxide elimination are given in Section A of Scheme 6.21.

Sulfoxides also undergo thermal elimination reactions. The elimination tends to give β, γ-unsaturation from β-hydroxysulfoxides and can be used to prepare allylic alcohols.

Ref. 324

Sulfoxide elimination in conjunction with [2,3]-sigmatropic rearrangement has been used to convert allylic alcohols to dienes.

54% yield; 60:40 mixture

Ref. 325

EWG substituents promote the removal of hydrogen, and sulfoxide eliminations are particularly favorable for β-keto and similar sulfoxides.

Selenoxides are even more reactive than sulfoxides toward β-elimination. In fact, many selenoxides react spontaneously when generated at room temperature. Synthetic procedures based on selenoxide eliminations usually involve synthesis of the corresponding selenide followed by oxidation and in situ elimination. We have already discussed examples of these procedures in Section 4.3.2, where the conversion of ketones and esters to their α, β-unsaturated derivatives is considered. Selenides can

324. J. Nokami, K. Ueta, and R. Okawara, *Tetrahedron Lett.*, 4903 (1978).
325. H. J. Reich and S. Wollowitz, *J. Am. Chem. Soc.*, **104**, 7051 (1982).

also be prepared by electrophilic addition of selenenyl halides and related compounds to alkenes (see Section 4.1.6). Selenide anions are powerful nucleophiles and can displace halides or tosylates and open epoxides.[326] Selenide substituents stabilize an adjacent carbanion, so α-selenenyl carbanions can be prepared. One procedure involves conversion of a ketone to a *bis*-selenoketal, which can then be cleaved by *n*-butyllithium.[327] The carbanions in turn add to ketones to give β-hydroxyselenides.[328] Elimination gives an allylic alcohol.

Alcohols can be converted to *o*-nitrophenylselenides by reaction with *o*-nitrophenyl selenocyanate and tri(*n*-butyl)phosphine.[329]

The selenides prepared by any of these methods can be converted to selenoxides by such oxidants as hydrogen peroxide, sodium metaperiodate, peroxycarboxylic acids, *t*-butyl hydroperoxide, or ozone.

Like amine oxide elimination, selenoxide eliminations normally favor formation of the *E*-isomer in acyclic structures. In cyclic systems the stereochemical requirements of the cyclic TS govern the product composition. Section B of Scheme 6.21 gives some examples of selenoxide eliminations.

Amine oxide and sulfoxide elimination TS structures have been compared by computations at the MP2/6-31G(*d*) level.[330] The calculated E_a values are 26 and 33 kcal/mol, respectively. Kinetic isotope effects have also been calculated[331] and are in good agreement with experimental values. The experimental E_a values for sulfoxide eliminations are typically near 30 kcal/mol.[332] For aryl sulfoxides, the E_a is somewhat lower, around 25–28 kcal/mol. Several sulfoxide elimination reactions have been examined computationally.[333] MP2/6-311+G(3*df*,2*p*) calculations gave generally good agreement with experimental values for ΔH, $\Delta H^{\ddagger}$, and kinetic isotope effects.

[326] D. L. J. Clive, *Tetrahedron*, **34**, 1049 (1978).

[327] W. Dumont, P. Bayet, and A. Krief, *Angew. Chem. Int. Ed. Engl.*, **13**, 804 (1974).

[328] D. Van Ende, W. Dumont, and A. Krief, *Angew. Chem. Int. Ed. Engl.*, **14**, 700 (1975); W. Dumont and A. Krief, *Angew. Chem. Int. Ed. Engl.*, **14**, 350 (1975).

[329] P. A. Grieco, S. Gilman, and M. Nishizawa, *J. Org. Chem.*, **41**, 1485 (1976); A. Krief and A.-M. Laval, *Bull. Soc. Chim. Fr.*, **134**, 869 (1997).

[330] B. S. Jursic, *Theochem*, **389**, 257 (1997).

[331] R. D. Bach, C. Gonzalez, J. L. Andres, and H. B. Schlegel, *J. Org. Chem*, **60**, 4653 (1995).

[332] D. W. Emerson, A. P. Craig, and I. W. Potts, Jr., *J. Org. Chem.*, **32**, 102, 3725 (1967); C. Walling and L. Bollyky, *J. Org. Chem.*, **29**, 2699 (1964).

[333] J. W. Cubbage, Y. Guo, R. D. McCulla, and W. S. Jenks, *J. Org. Chem.*, **66**, 8722 (2001).

600

CHAPTER 6

*Concerted
Cycloadditions,
Unimolecular
Rearrangements, and
Thermal Eliminations*

The minimum-energy TSs are planar and the O—H and C—H bond orders were usually less than 0.4 and less than 0.5, respectively, and the S—C bond order was less than 0.5. The C—C bond order was around 1.3. The reaction can be described as a concerted intramolecular proton transfer, with the sulfoxide oxygen acting as a base and the sulfur as a leaving group.

The TS for selenoxide elimination has also been examined computationally.[334] The C—H bond cleavage runs ahead of the C—Se cleavage.

A third category of *syn* eliminations involves pyrolytic decomposition of esters with elimination of a carboxylic acid. The pyrolysis of acetate esters normally requires temperatures above 400° C and is usually a vapor phase reaction. In the laboratory this is done by using a glass tube in the heating zone of a small furnace. The vapors of the reactant are swept through the hot chamber by an inert gas and into a cold trap. Similar reactions occur with esters derived from long-chain acids. If the boiling point of the ester is above the decomposition temperature, the reaction can be carried out in the liquid phase, with distillation of the pyrolysis product.

Ester pyrolysis has been shown to be a *syn* elimination in the case of formation of stilbene by the use of deuterium labels.[335]

Although recognizing the existence of the concerted cyclic mechanism, it has been proposed that most preparative pyrolyses proceed as surface-catalyzed reactions.[336]

Mixtures of alkenes are formed when more than one type of β-hydrogen is present. In acyclic compounds the product composition often approaches that expected on a statistical basis from the number of each type of hydrogen. The *E*-alkene usually predominates over the *Z*-alkene for a given isomeric pair. In cyclic structures, elimination is in the direction that the cyclic mechanism can operate most favorably.

334. N. Kondo, H. Fueno, H. Fujimoto, M. Makino, H. Nakaoka, I. Aoki, and S. Uemura, *J. Org. Chem.*, **59**, 5254 (1994).

335. D. Y. Curtin and D. B. Kellom, *J. Am. Chem. Soc.*, **75**, 6011 (1953).

336. D. H. Wertz and N. L. Allinger, *J. Org. Chem.*, **42**, 698 (1977).

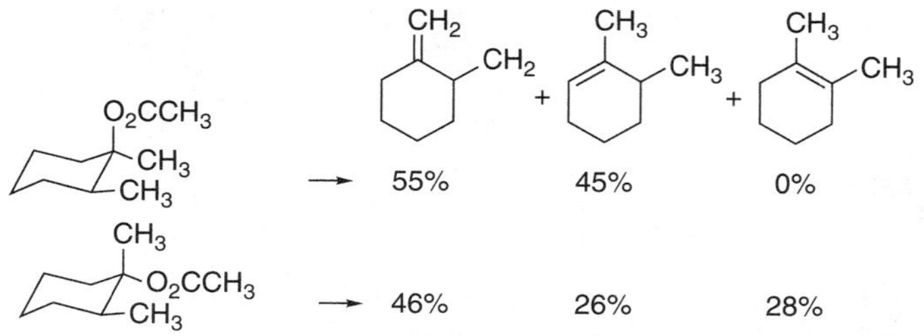

Ref. 336

Alcohols can be dehydrated via xanthate esters at temperatures that are much lower than those required for acetate pyrolysis. The preparation of xanthate esters involves reaction of the alkoxide with carbon disulfide. The resulting salt is alkylated with methyl iodide.

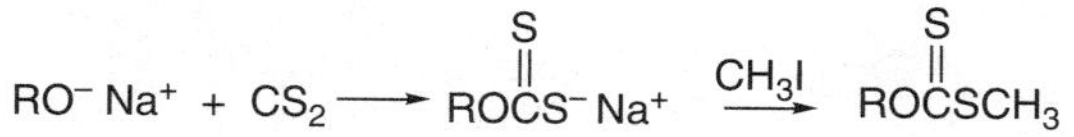

The elimination is often effected simply by distillation.

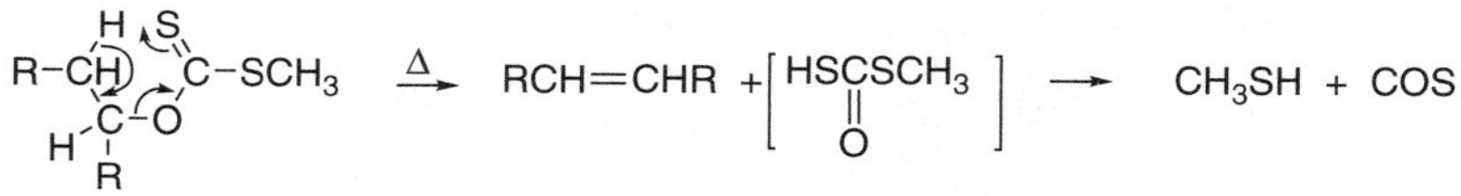

Product mixtures are observed when more than one type of β-hydrogen can participate in the reaction. As with the other *syn* thermal eliminations, there are no intermediates that are prone to skeletal rearrangement.

Scheme 6.21 gives some examples of thermal elimination reactions. Entries 1 to 3 show amine-oxide decompositions. The reaction in Entry 1 shows a preference for the conjugated product. This reaction was also conducted in dry DMSO, where it was found to proceed at 25° C.[338] Entry 2 illustrates the use of the reaction to prepare methylenecyclohexane. The method is particularly useful in this case because there is no tendency for competing elimination or rearrangement to the more stable 1-methylcyclohexene. Entries 4 and 5 are sulfoxide eliminations. Entry 4 is favored by the conjugation of the phenyl group and occurs under very mild conditions. The conditions for elimination in Entry 5 are more typical. Entries 6 to 9 are selenoxide eliminations. In Entries 6 and 7, the selenide group is introduced by nucleophilic substitution. In Entry 8, electrophilic selenolactonization was used to synthesize the reactant. Although the yield of the product, oxete, in Entry 9 is quite low, this was one of the first preparations of this compound. Entries 10 to 12 are high-temperature acetate pyrolyses. Entries 13 to 17 are xanthate pyrolyses. In Entry 15, the use of DMSO as the solvent for the preparation of the dialcoholate was found to be advantageous.

336. D. H. Froemsdorf, C. H. Collins, G. S. Hammond, and C. H. DePuy, *J. Am. Chem. Soc.*, **81**, 643 (1959).
338. D. J. Cram, M. R. V. Sahyun, and G. R. Knox, *J. Am. Chem. Soc.*, **84**, 1734 (1962).

602

CHAPTER 6

*Concerted
Cycloadditions,
Unimolecular
Rearrangements, and
Thermal Eliminations*

Scheme 6.21. Thermal Eliminations Via Cyclic Transition Structures

A. Amine oxide pyrolyses

1[a]

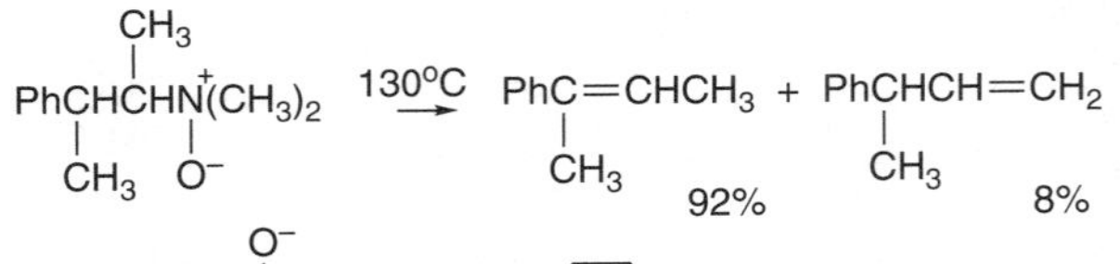

PhCHCHCH_3 + PhCHCH=CH_2
    92%   8%

2[b]

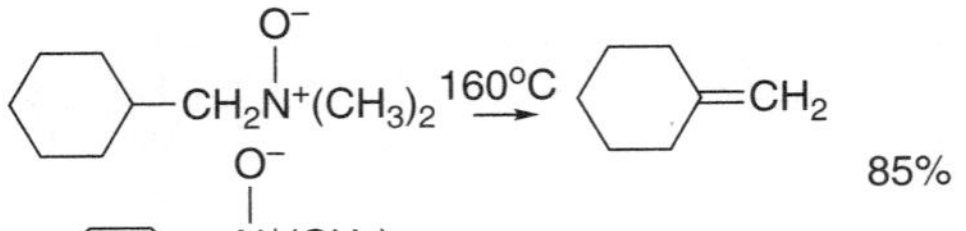

85%

3[c]

$\xrightarrow{165°C}$ 50%

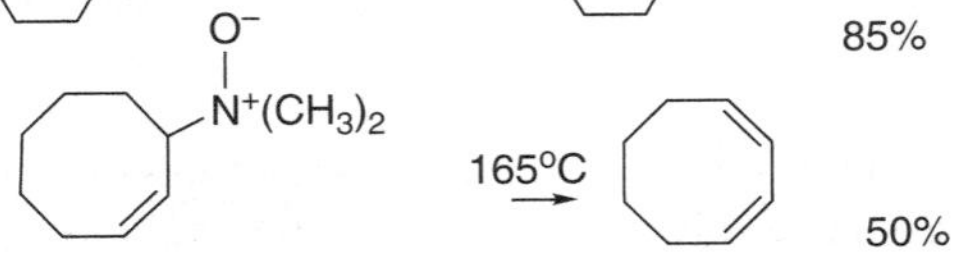

B. Sulfoxide elimination

4[d]

$\xrightarrow{m\text{-CPBA}}$ 94%

5[e]

1) m–CPBA
2) 105°C
   12.5 h

32–66%

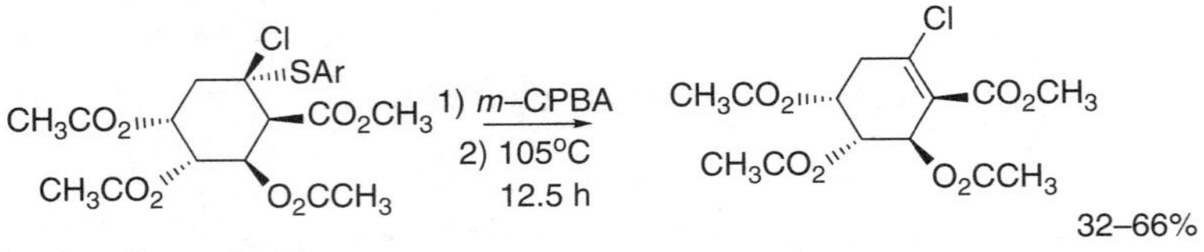

Ar = 4-methoxyphenyl

C. Selenoxide elimination

6[f]

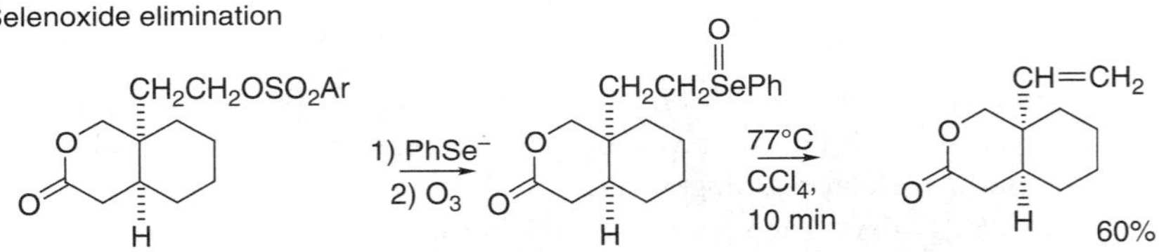

1) PhSe⁻
2) O_3

$\xrightarrow[\text{10 min}]{\substack{77°C \\ CCl_4,}}$ 60%

7[g]

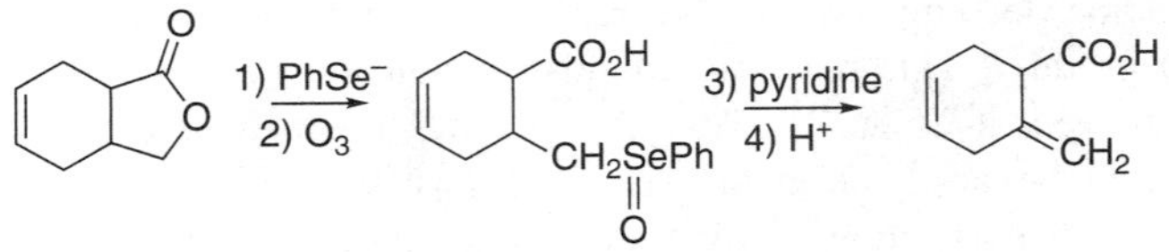

1) PhSe⁻
2) O_3

3) pyridine
4) H⁺

8[h]

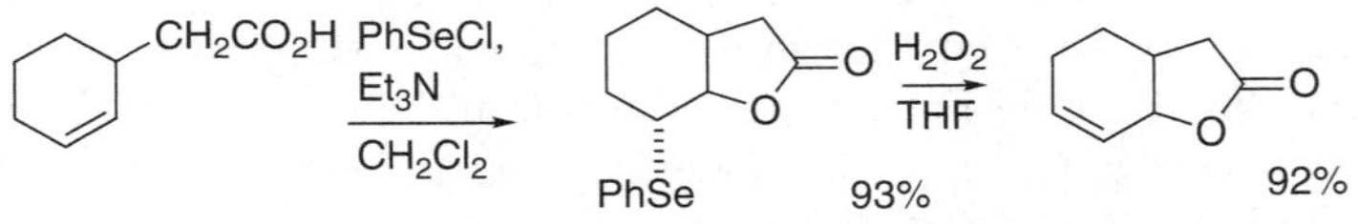

PhSeCl,
Et_3N
$\overline{\text{CH}_2\text{Cl}_2}$

$\xrightarrow{\substack{H_2O_2 \\ THF}}$

93%   92%

9[i]

1) 2-nitrophenyl SeCN, Ph_3P
2) O_3

$\xrightarrow{\text{DBU}}$ 5%

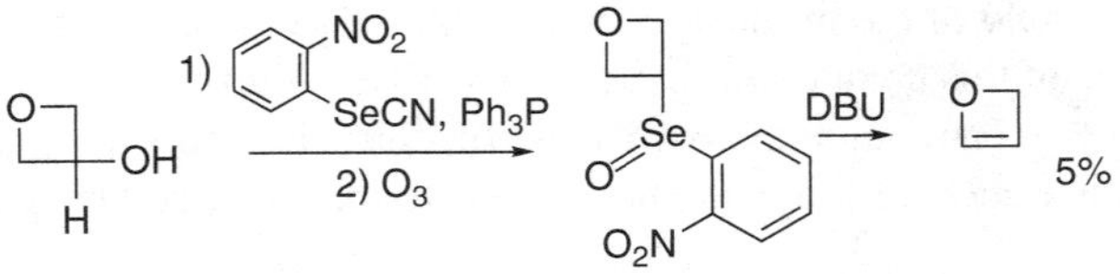

(Continued)

D. Acetate pyrolyses

10[j]

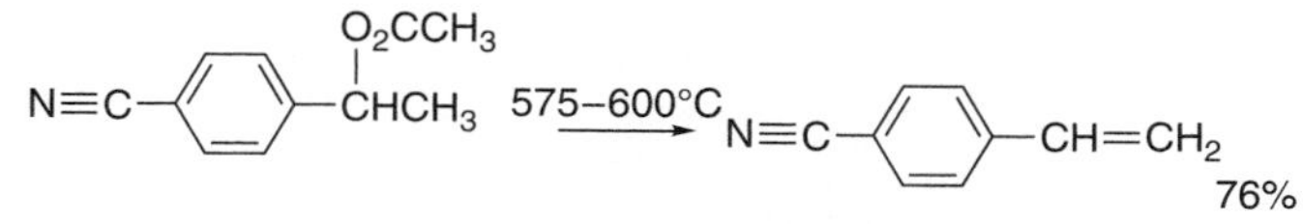
O2CCH3
N≡C–
–CHCH3
575–600°C
N≡C–
–CH=CH2
76%

11[k]

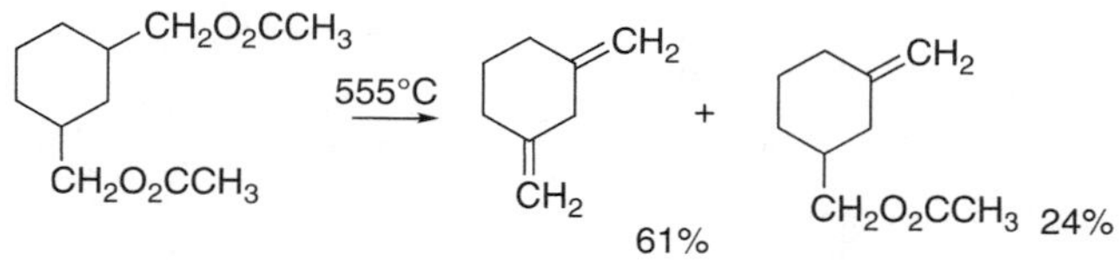
CH2O2CCH3
CH2O2CCH3
555°C
CH2
CH2
61%
+
CH2
CH2O2CCH3 24%

12[l]

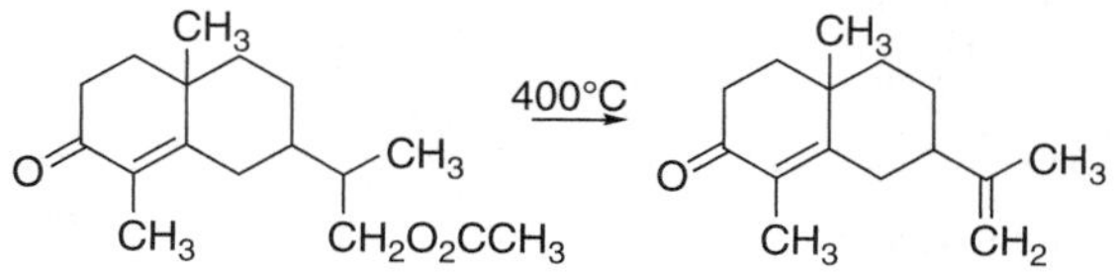
CH3
O
CH3
CH3
CH2O2CCH3
400°C
CH3
O
CH3
CH3
CH2

E. Xanthate ester pyrolyses

13[m]

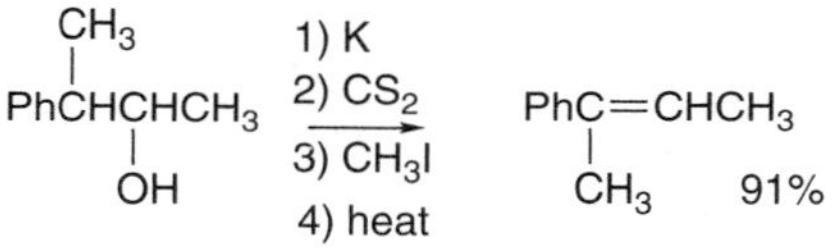
CH3
PhCHCHCH3
OH
1) K
2) CS2
3) CH3I
4) heat
PhC=CHCH3
CH3
91%

14[n]

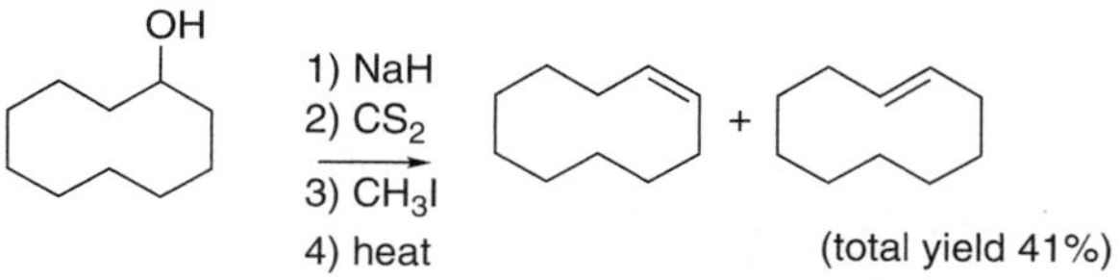
OH
1) NaH
2) CS2
3) CH3I
4) heat
+
(total yield 41%)

15[o]

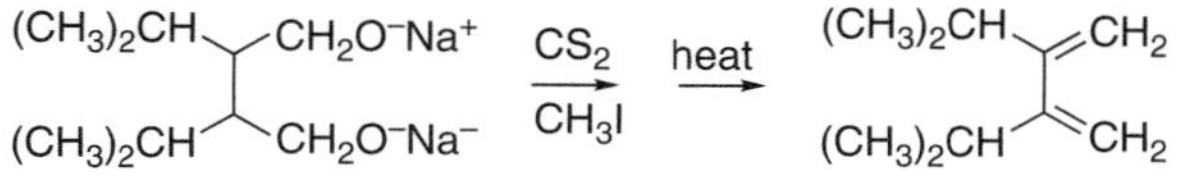
(CH3)2CH
CH2O⁻Na⁺
(CH3)2CH
CH2O⁻Na⁻
CS2
CH3I
heat
(CH3)2CH
CH2
(CH3)2CH
CH2

16[p]

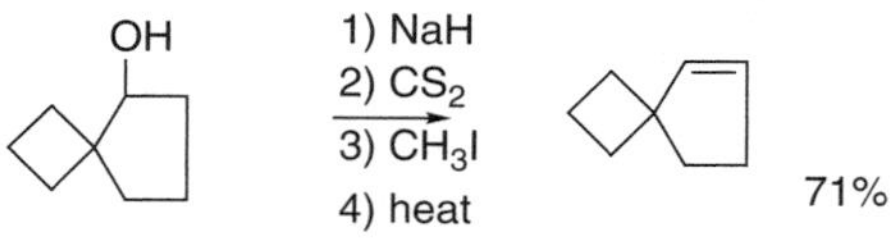
OH
1) NaH
2) CS2
3) CH3I
4) heat
71%

(*Continued*)

CHAPTER 6

Concerted
Cycloadditions,
Unimolecular
Rearrangements, and
Thermal Eliminations

Scheme 6.21. (*Continued*)

17[q]

1) NaH, CS$_2$,
2) CH$_3$I
3) heat

60%

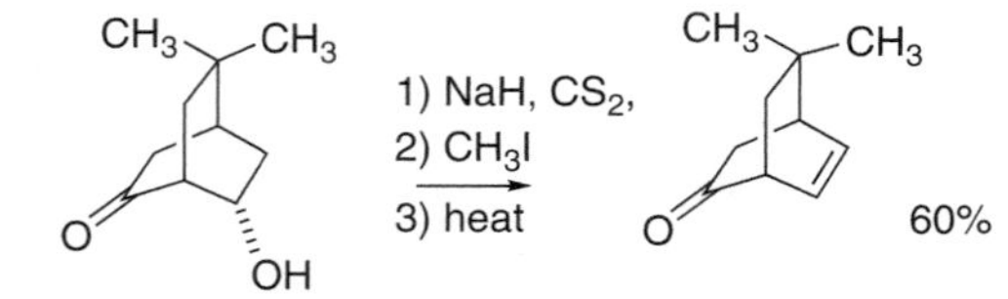

a. D. J. Cram and J. E. McCarty, *J. Am. Chem. Soc.*, **76**, 5740 (1954).
b. A. C. Cope, E. Ciganek, and N. A. LeBel, *J. Am. Chem. Soc.*, **81**, 2799 (1959); A. C. Cope and E. Ciganek, *Org. Synth.*, **IV**, 612 (1963).
c. A. C. Cope and C. L. Bumgardner, *J. Am. Chem. Soc.*, **78**, 2812 (1956).
d. J.-X. Gu and H. L. Holland, *Synth. Commun.*, **28**, 3305 (1998).
e. R. H. Rich, B. M. Lawrence, and P. A. Bartlett, *J. Org. Chem.*, **59**, 693 (1994).
f. R. D. Clark and C. H. Heathcock, *J. Org. Chem.*, **41**, 1396 (1976).
g. D. Liotta and H. Santiesteban, *Tetrahedron Lett.*, 4369 (1977); R. M. Scarborough, Jr., and A. B. Smith, III, *Tetrahedron Lett.*, 4361 (1977).
h. K. C. Nicolaou and Z. Lysenko, *J. Am. Chem. Soc.*, **99**, 3185 (1977).
i. L. E. Friedrich and P. Y. S. Lam, *J. Org. Chem.*, **46**, 306 (1981).
j. C. G. Overberger and R. E. Allen, *J. Am. Chem. Soc.*, **68**, 722 (1946).
k. W. J. Bailey and J. Economy, *J. Org. Chem.*, **23**, 1002 (1958).
l. E. Piers and K. F. Cheng, *Can. J. Chem.*, **46**, 377 (1968).
m. D. J. Cram, *J. Am. Chem. Soc.*, **71**, 3883 (1949).
n. A. T. Blomquist and A. Goldstein, *J. Am. Chem. Soc.*, **77**, 1001 (1955).
o. A. de Groot, B. Evenhuis, and H. Wynberg, *J. Org. Chem.*, **33**, 2214 (1968).
p. C. F. Wilcox, Jr., and C. G. Whitney, *J. Org. Chem.*, **32**, 2933 (1967).
q. L. A. Paquette and H.-C. Tsai, *J. Org. Chem.*, **61**, 142 (1996).

Problems

(References for these problems will be found on page 1280.)

6.1. Predict the products of the following reactions on the basis of the reaction mechanism and anticipated transition structure. Be sure to consider all elements of stereochemistry. Unless otherwise specified, the reactants and reagents are racemic.

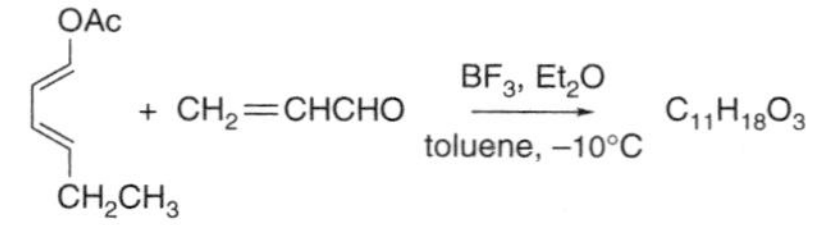

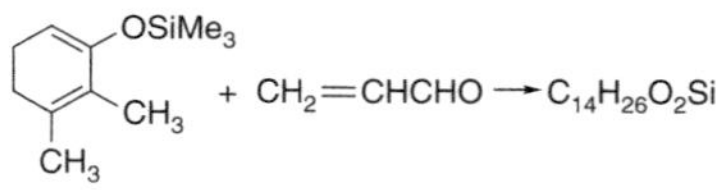

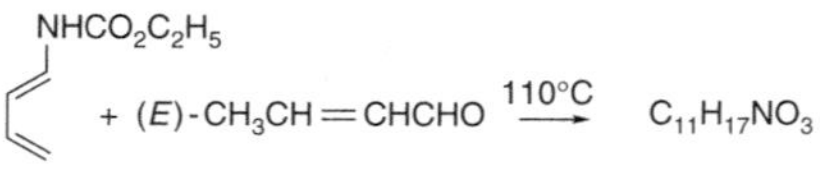

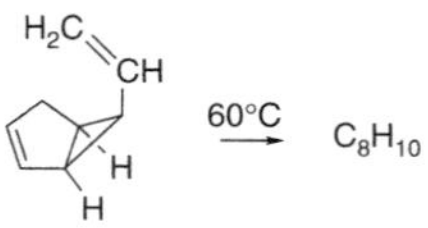

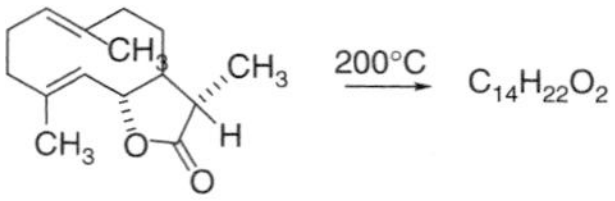

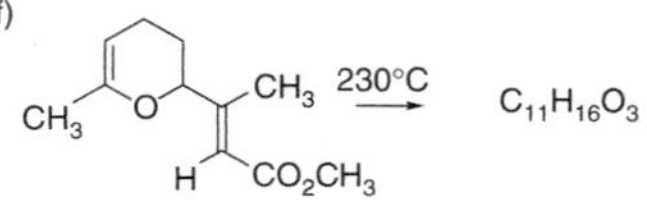

(g)

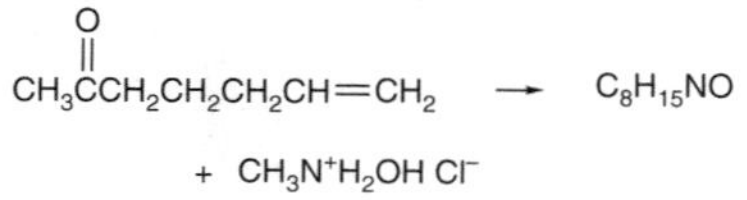

$$CH_3CCH_2CH_2CH_2CH=CH_2 \longrightarrow C_8H_{15}NO$$

$$+ \ CH_3N^+H_2OH \ Cl^-$$

(h)

1) $C_2H_5OCH=CH_2$, Hg^{2+}
2) 210°C $\longrightarrow$ $C_{10}H_{16}O$

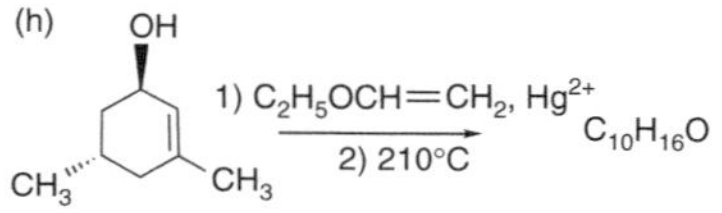

(i)

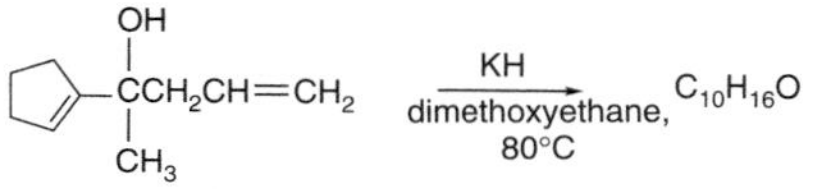

$$\xrightarrow[\substack{\text{dimethoxyethane,} \\ 80°C}]{\text{KH}} C_{10}H_{16}O$$

(j)

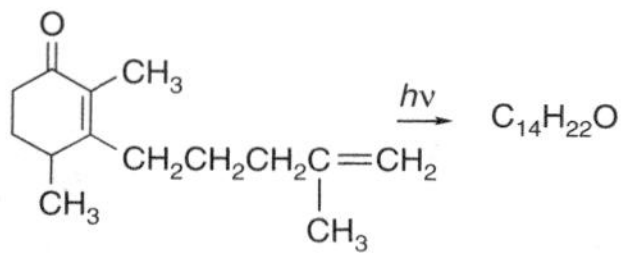

$$\xrightarrow{h\nu} C_{14}H_{22}O$$

(k)

$$C_6H_5CH(SeCH_3)_2 \xrightarrow[\substack{\text{2) 1,2-epoxyhexane} \\ \text{3) } H_2O_2}]{\text{1) } n\text{-BuLi}} C_{13}H_{18}O$$

(l)

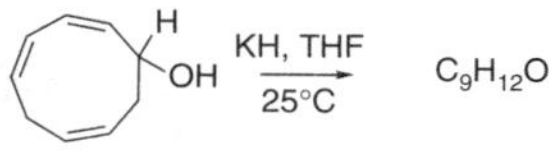

$$\xrightarrow[25°C]{\text{KH, THF}} C_9H_{12}O$$

(m)

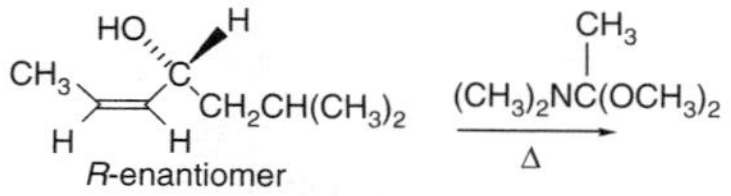

R-enantiomer

$$\xrightarrow[\Delta]{(CH_3)_2NC(OCH_3)_2} C_{11}H_{21}NO$$

$$\xrightarrow{100°C} C_{10}H_{16}O$$

(n)

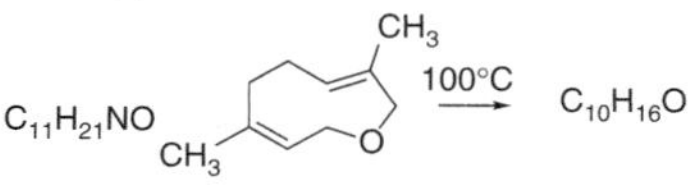

(o)

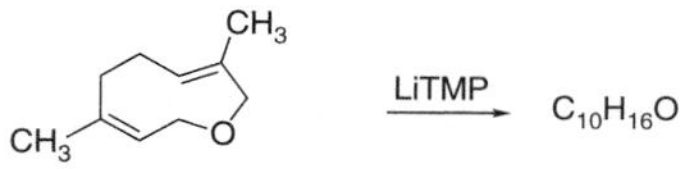

$$\xrightarrow{\text{LiTMP}} C_{10}H_{16}O$$

(p)

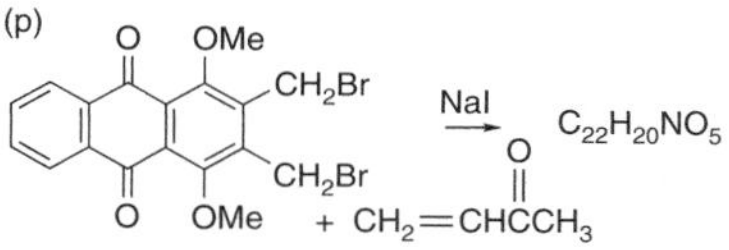

$$\xrightarrow{\text{NaI}} C_{22}H_{20}NO_5$$

$+ \ CH_2=CHCCH_3$

(q)

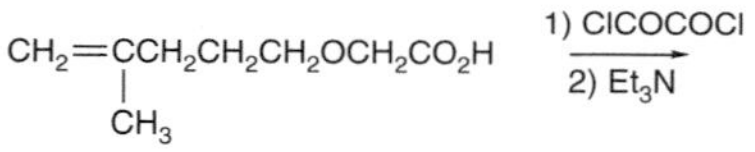

$$CH_2=CCH_2CH_2CH_2OCH_2CO_2H \xrightarrow[\text{2) Et}_3\text{N}]{\text{1) ClCOCOCl}} C_8H_{12}O_2$$

CH_3

(r)

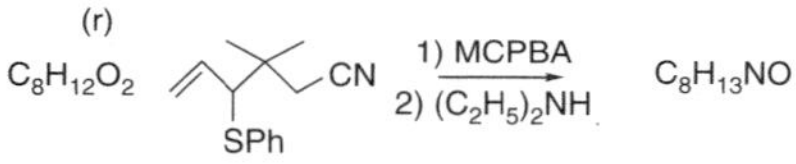

$$\xrightarrow[\text{2) } (C_2H_5)_2NH]{\text{1) MCPBA}} C_8H_{13}NO$$

(s)

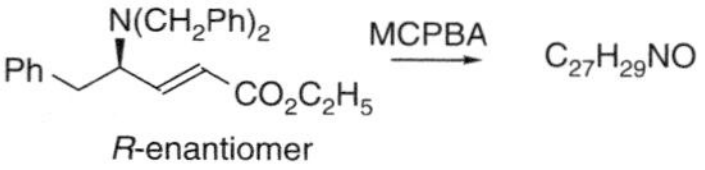

R-enantiomer

$$\xrightarrow{\text{MCPBA}} C_{27}H_{29}NO$$

(t)

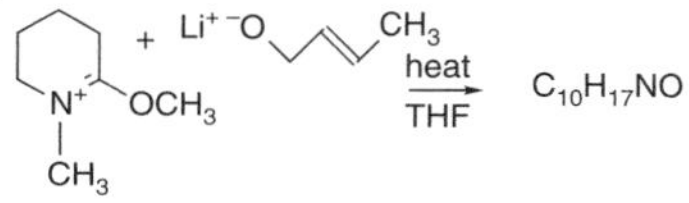

$$+ \ Li^+ \ O \ CH_3 \xrightarrow[\text{THF}]{\text{heat}} C_{10}H_{17}NO$$

(u)

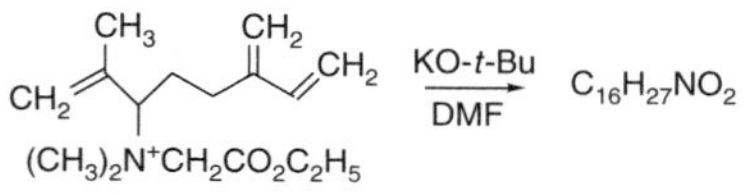

$$(CH_3)_2N^+CH_2CO_2C_2H_5 \xrightarrow[\text{DMF}]{\text{KO-}t\text{-Bu}} C_{16}H_{27}NO_2$$

(v)

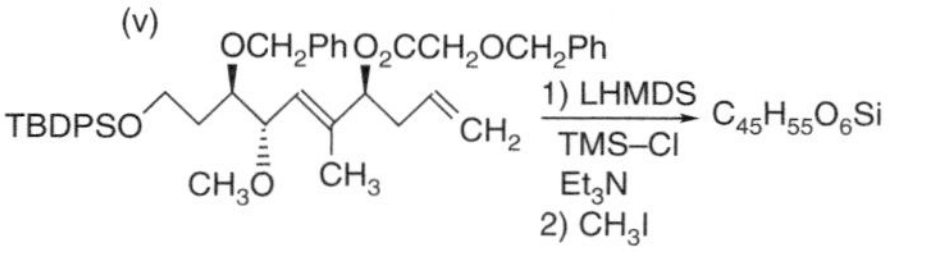

$$\xrightarrow[\substack{\text{TMS–Cl} \\ \text{Et}_3\text{N} \\ \text{2) CH}_3\text{I}}]{\text{1) LHMDS}} C_{45}H_{55}O_6Si$$

(y)

$$\longrightarrow C_{14}H_{17}NO_3$$

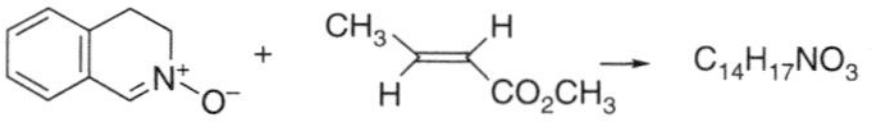

6.2. Intramolecular cycloaddition reactions occur under the conditions specified for each of the following reactions. Show the structures of the products of each reaction, including all aspects of stereochemistry and indicate the structure of the product-determining TS and any key intermediates.

(a)

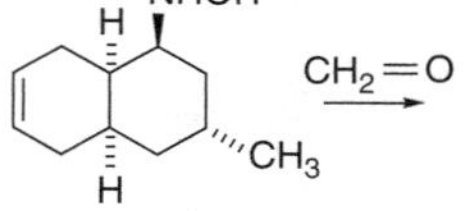

$$CH_2=O$$

(b)

$$\xrightarrow{h\nu}$$

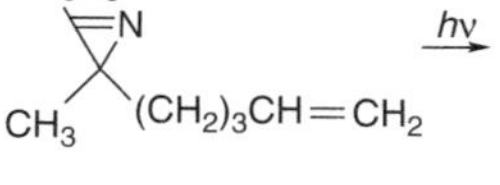

606

CHAPTER 6

*Concerted
Cycloadditions,
Unimolecular
Rearrangements, and
Thermal Eliminations*

(c)

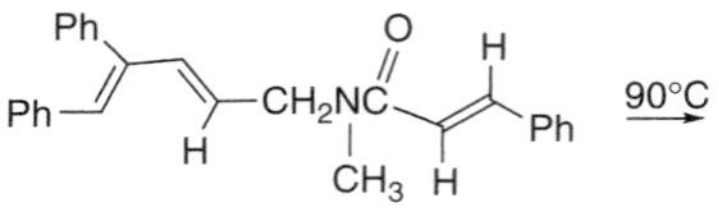

(d)

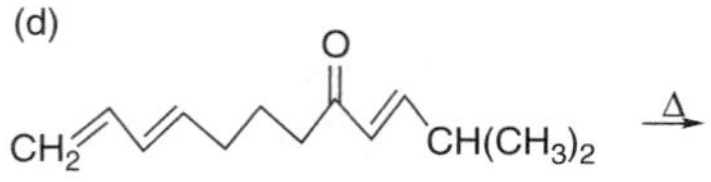

(e)

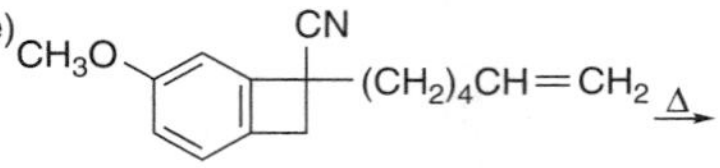

6.3. Indicate the mechanistic type to which each of these reactions belongs and write out a mechanism showing any intermediates.

(a)

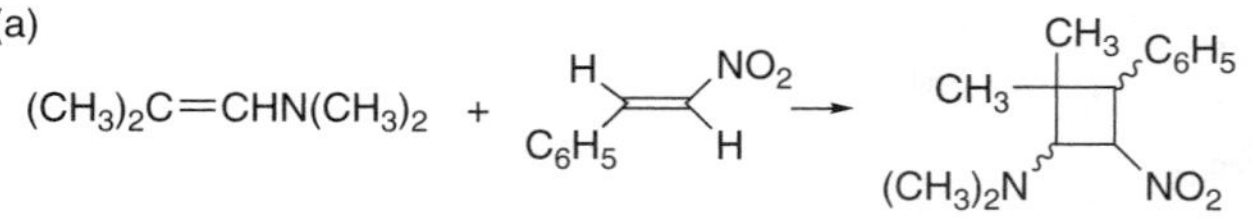

(b)

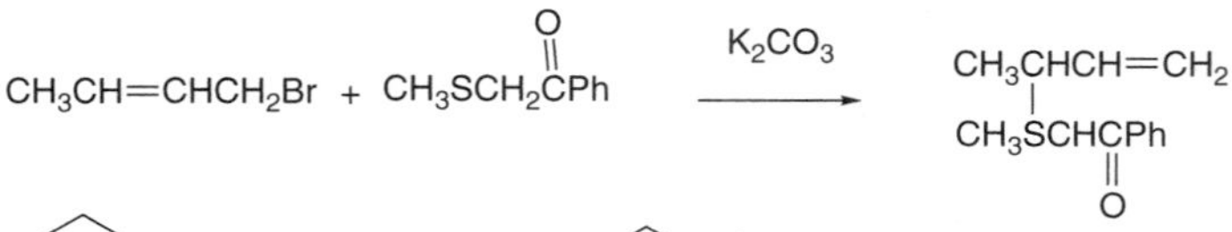

(c)

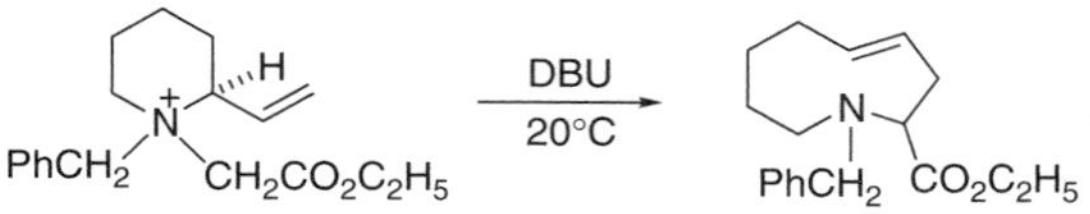

(d)

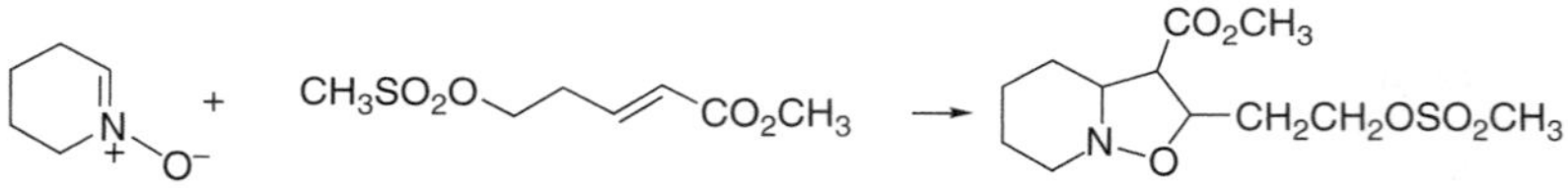

(e)

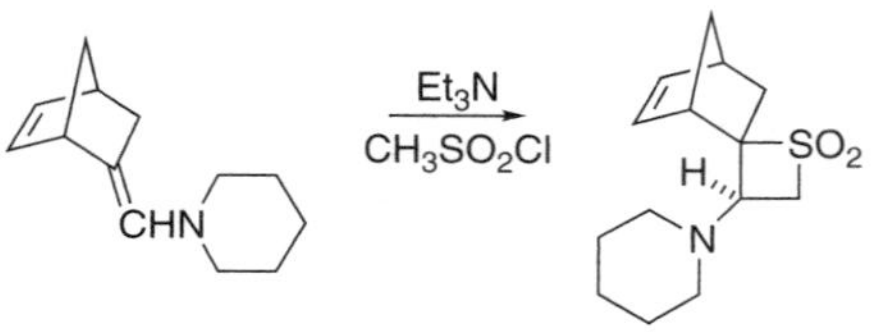

(f)

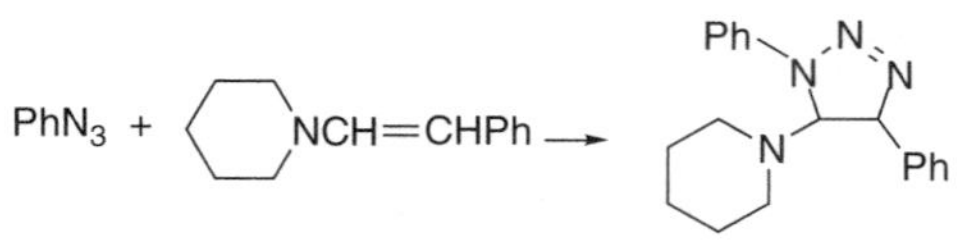

(g)

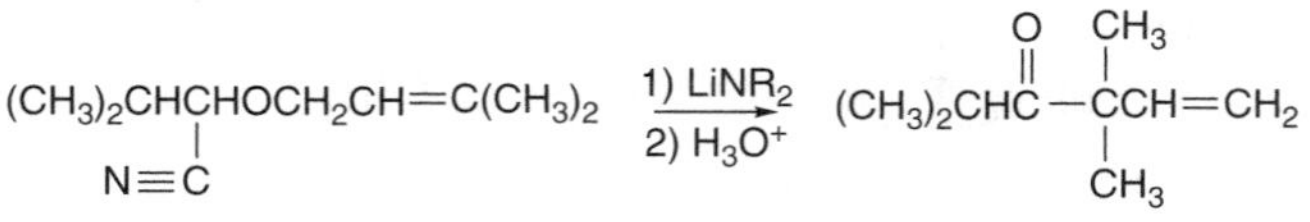

(h)

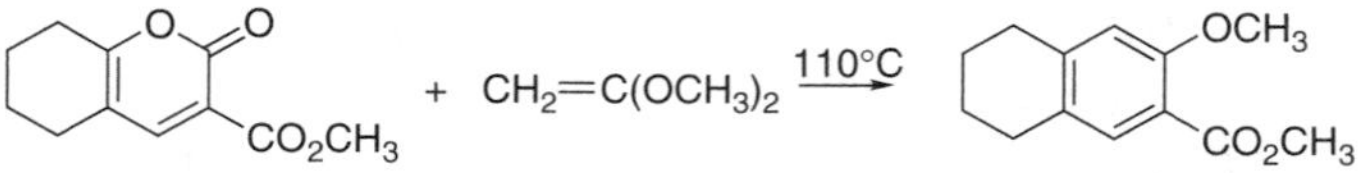

6.4. Apply retrosynthetic analysis to the following transformation and show how each of the target molecules could be prepared from the starting materials given. No more than three separate steps are needed in any of the syntheses.

(a)

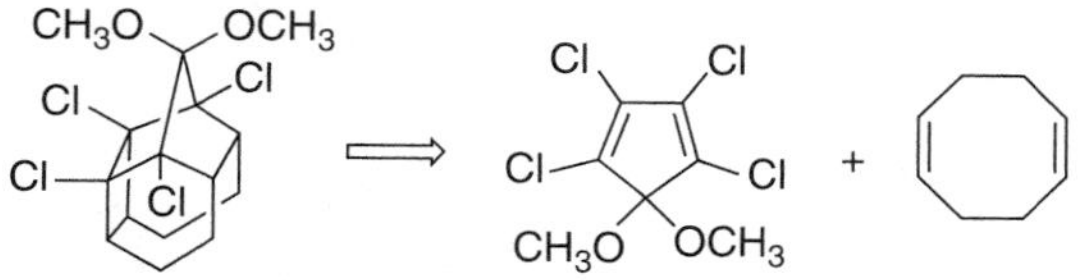

(b)

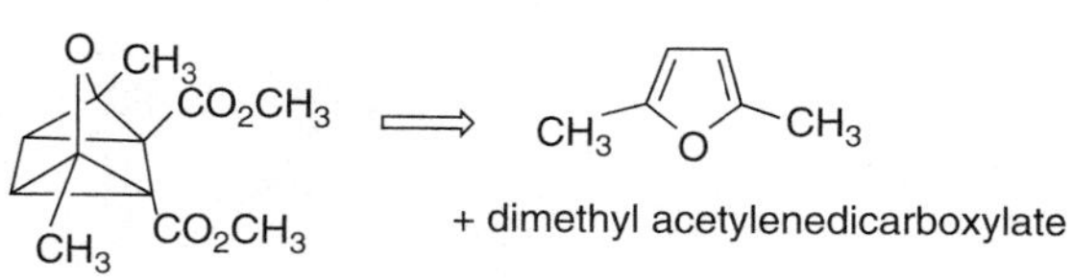

+ dimethyl acetylenedicarboxylate

(c)

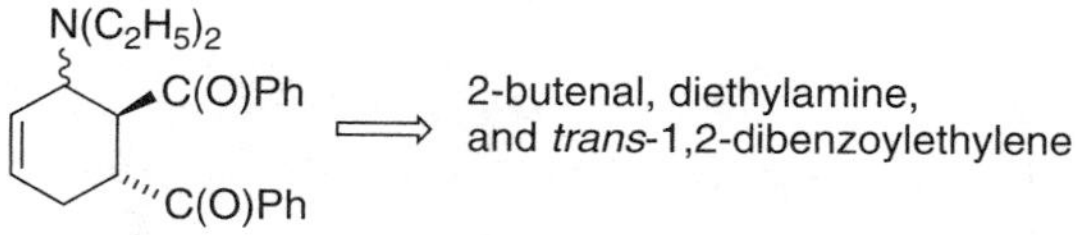

2-butenal, diethylamine, and *trans*-1,2-dibenzoylethylene

(d)

$$CH_3CH_2C{\equiv}CCH{=}CHCH_2CH_2CO_2CH_3 \implies$$ propenal, 1-butyne and triethyl orthoacetate

(e)

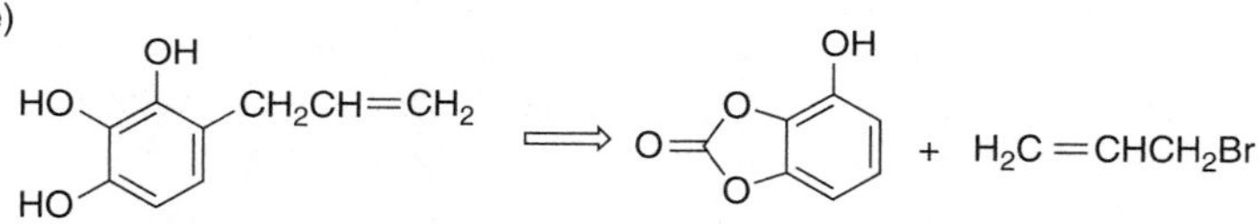

+ $H_2C{=}CHCH_2Br$

(f)

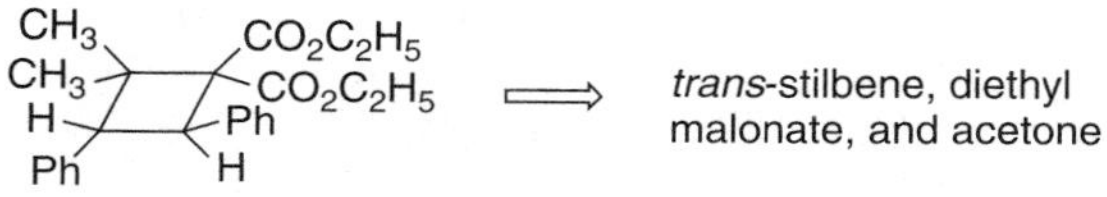

trans-stilbene, diethyl malonate, and acetone

(g)

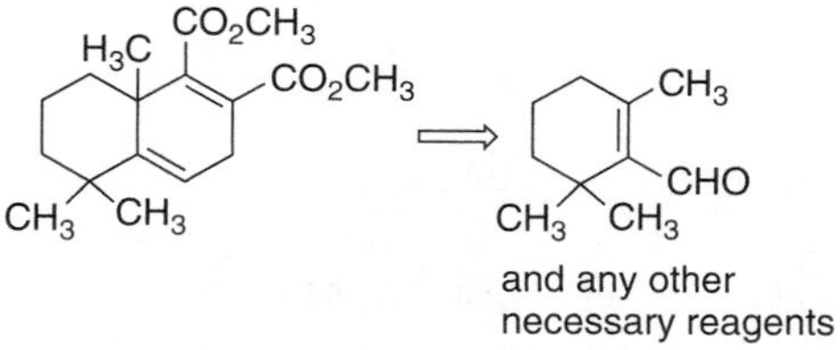

and any other necessary reagents

(h)

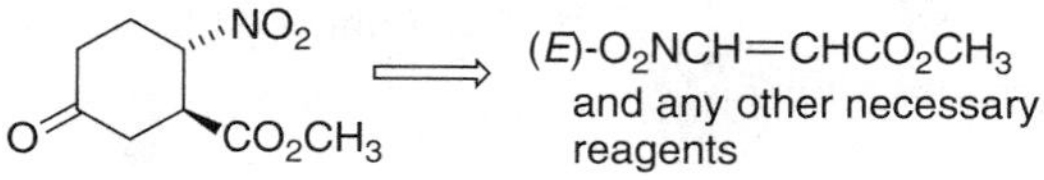

$(E){-}O_2NCH{=}CHCO_2CH_3$
and any other necessary reagents

(i)

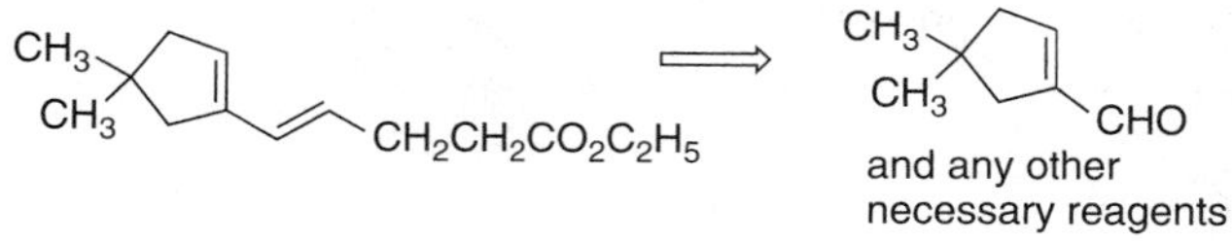

and any other necessary reagents

(j)

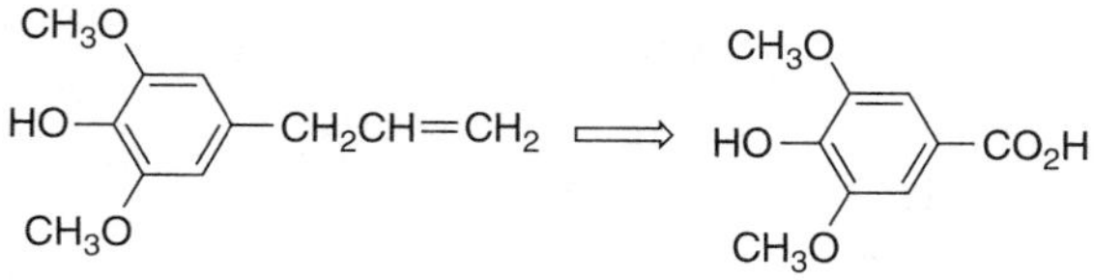

608

CHAPTER 6

Concerted
Cycloadditions,
Unimolecular
Rearrangements, and
Thermal Eliminations

(k)

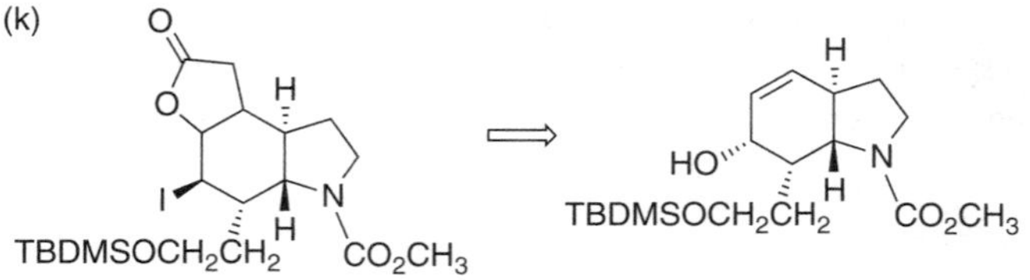

(l)

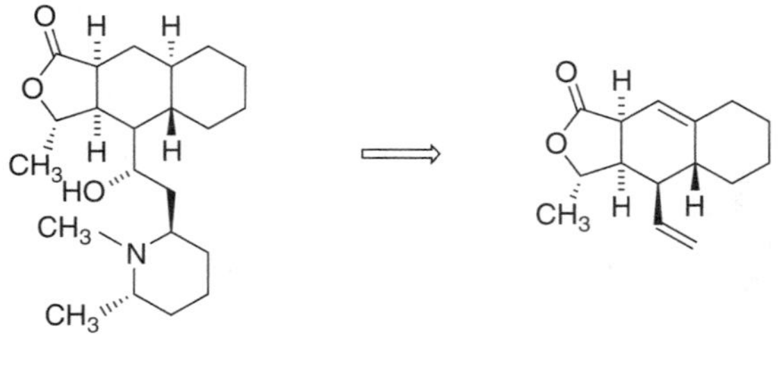

6.5. Suggest mechanisms by which the following transformations occur.

    a. The addition reaction of tetracyanoethylene and ethyl vinyl ether in acetone gives 94% of a [2+2] adduct and 6% of an adduct having the composition tetracyanoethylene + ethyl vinyl ether + acetone. If the [2+2] adduct is kept in contact with acetone for several days, it is completely converted to the minor product. What is a likely structure for the minor product? How is it formed in the original reaction and on standing in acetone?

    b. When vinylcylopropane is irradiated with benzophenone or benzaldehyde both oxetane and oxepene products are obtained. How are the oxepenes formed?

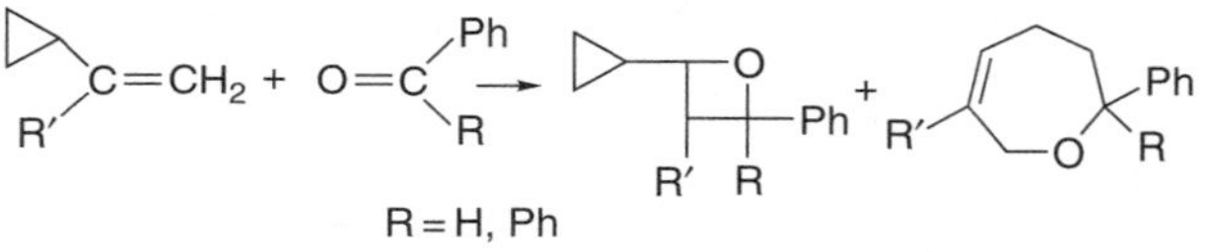

    c. A convenient preparation of 2-allylcyclohexanone involves simply heating the diallyl acetal of cyclohexanone in toluene containing a trace of *p*-toluenesulfonic acid and collecting a distillate of toluene and allyl alcohol.

    d. A solution of 2-butenal, 2-acetoxypropene, and dimethyl acetylenedicarboxylate refluxed in the presence of a small amount of an acid catalyst gives an 80% yield of dimethyl phthalate.

6.6. The following syntheses were carried by short tandem reaction sequences starting with the Diels-Alder reaction shown. Show the reagents and approximate reaction conditions required to complete the transformation.

    (a)

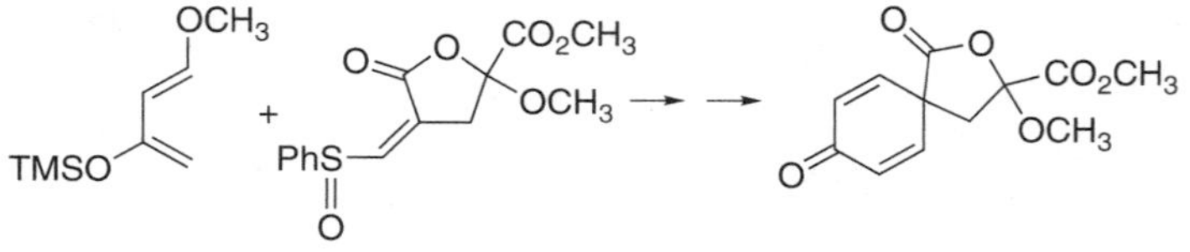

(b)

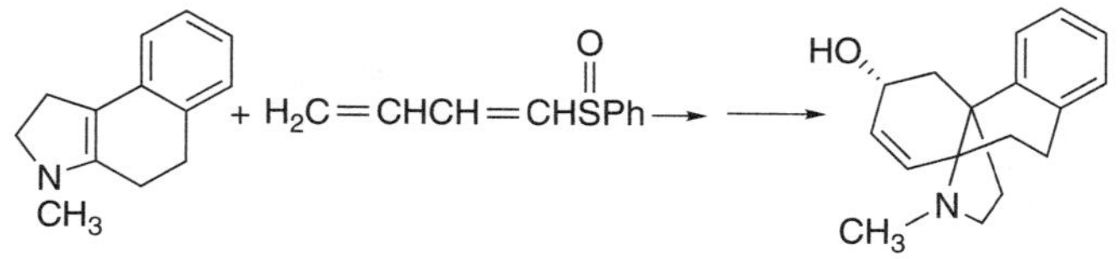

6.7. The ester **7-1** gives alternative stereoisomers when subjected to Claisen rearrangement as the lithium enolate or as the silyl ketene acetal. Analyze the respective transition structures and develop a rationale to explain these results.

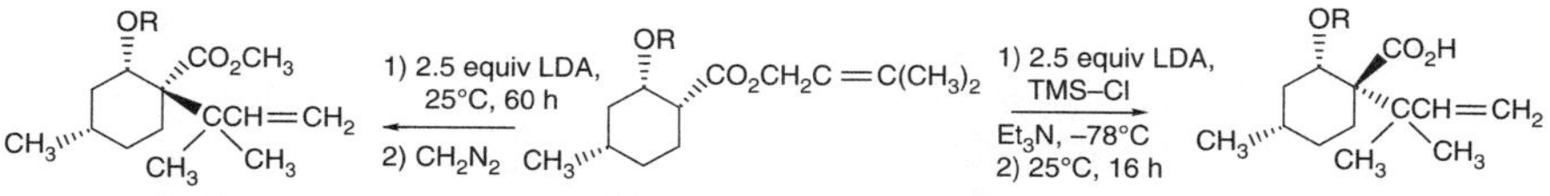

R = H　　　　**7-1**　　　　R = CH_3

6.8. Photolysis of **8-1** gives an isomeric compound **8-2** in 83% yield. Alkaline hydrolysis of **8-2** affords a hydroxy carboxylic acid, **8-3**, $C_{25}H_{32}O_4$. Treatment of **8-2** with silica gel in hexane yields **8-4**, $C_{24}H_{28}O_2$. **8-4** is converted by $NaIO_4$-$KMnO_4$ to a mixture of **8-5** and **8-6**. What are the structures of **8-2**, **8-3**, and **8-4**?

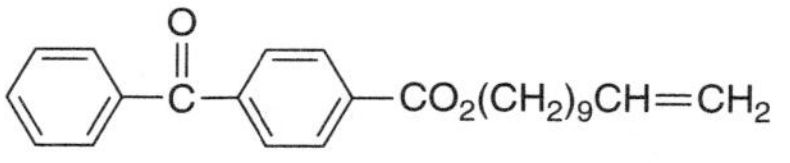

8-1

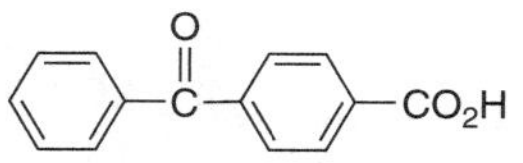

8-5

$HO(CH_2)_9CO_2H$

8-6

6.9. Suggest mechanisms for the following reactions that involve loss of N_2.

a. 1,2,4,5-Tetrazines react with alkenes to give dihydropyridazines, as in the example below.

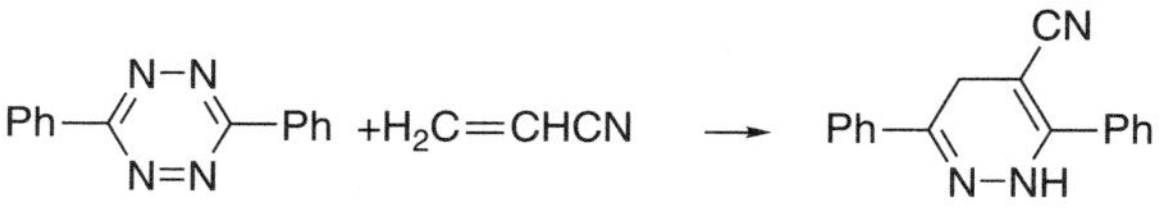

b. Compounds **9-1** and **9-2** are both unstable toward loss of nitrogen at room temperature and both give **9-3** as the product.

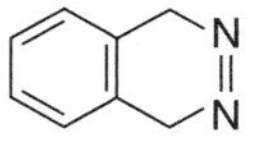

9-1

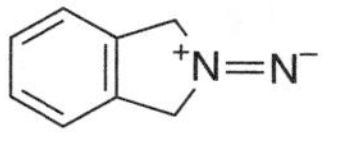

9-2

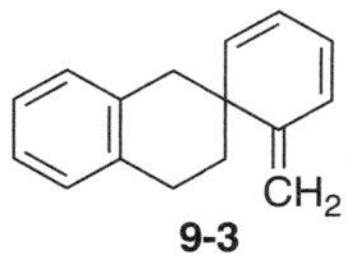

9-3

6.10. For each of the following reactions propose a transition structure that would account for the observed stereoselectivity. Identify important conformational and other features of the proposed transition structure.

610

CHAPTER 6

*Concerted
Cycloadditions,
Unimolecular
Rearrangements, and
Thermal Eliminations*

(a)

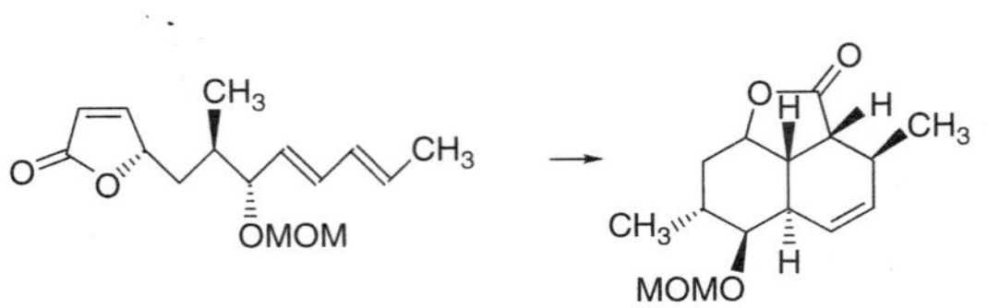

(b)

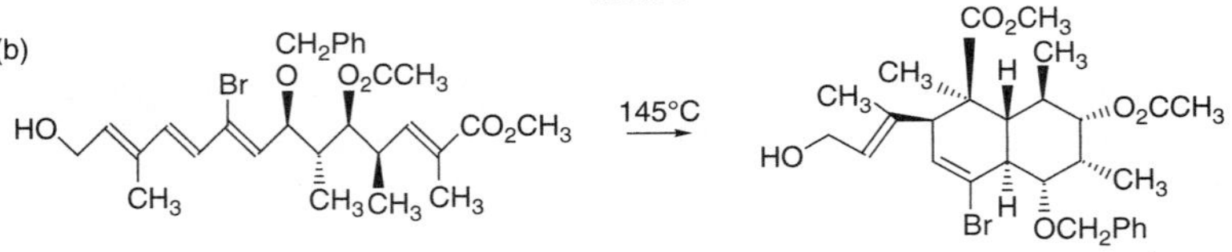

(c)

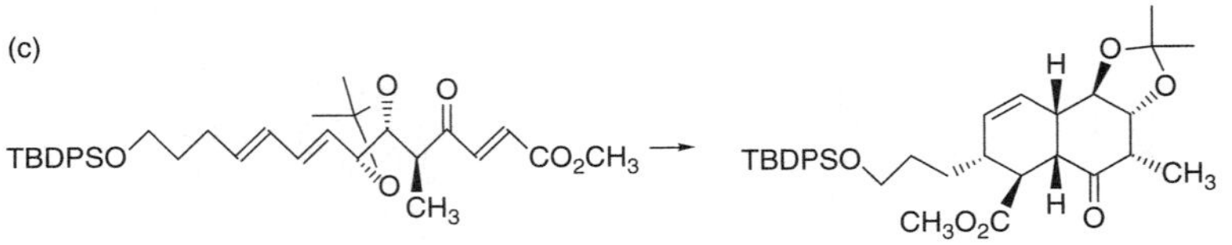

6.11. Provide an outline of the mechanisms of the following transformations.

(a)

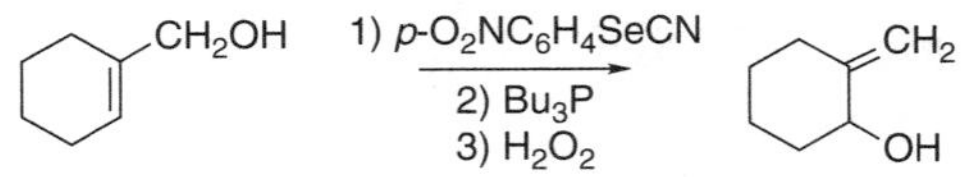

(b)

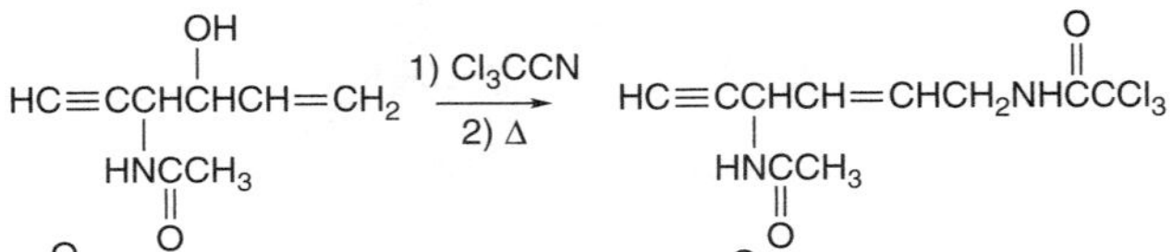

(c)

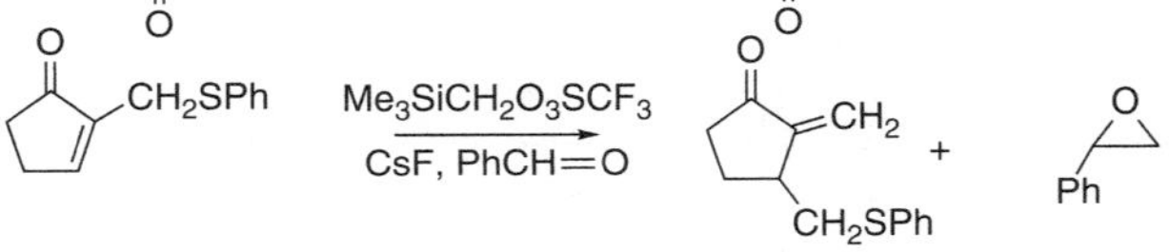

(d)

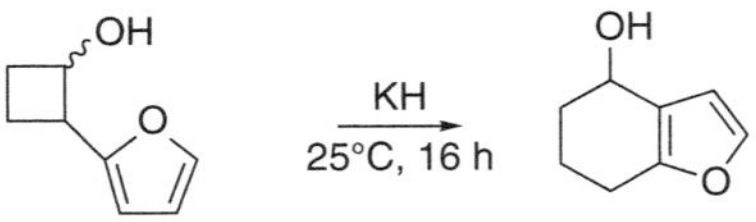

(e)

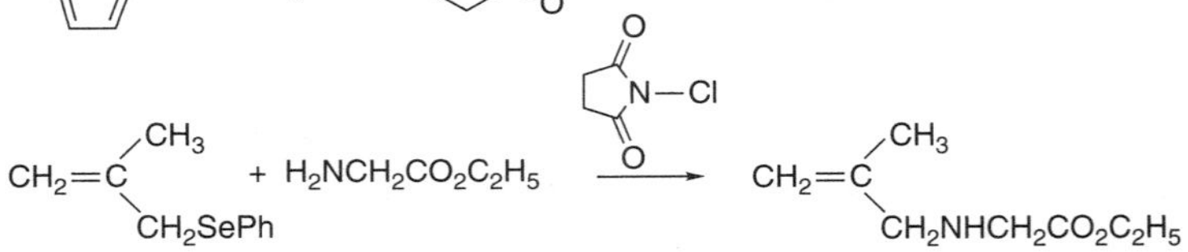

(f)

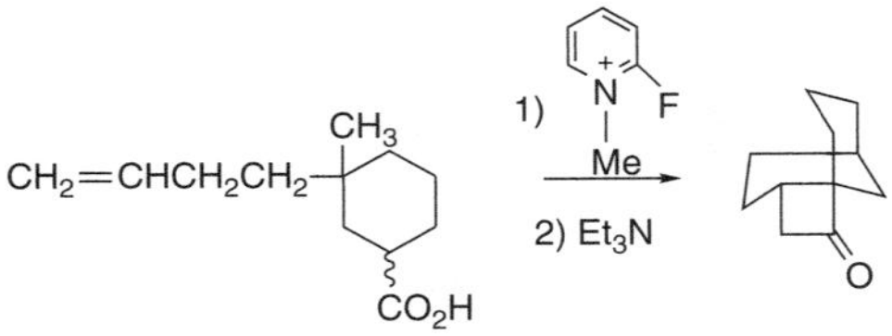

(g)

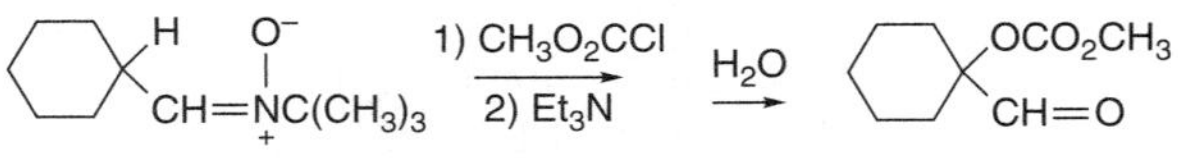

(h)

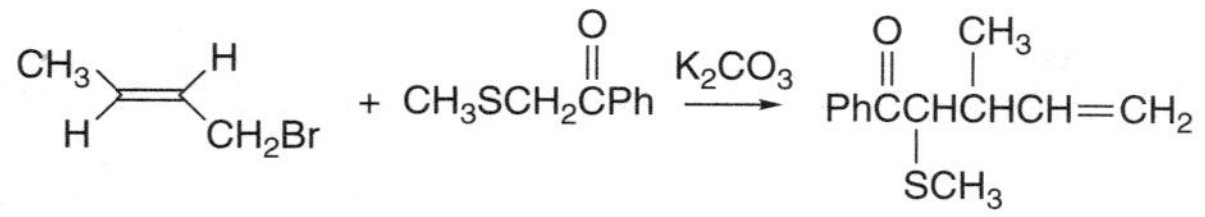

(i)

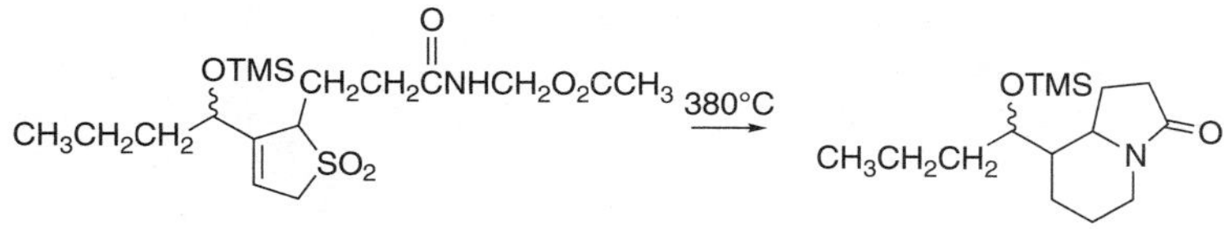

(j)

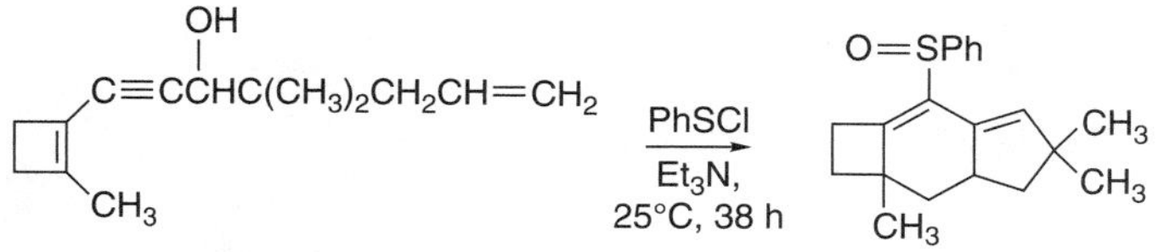

(k)

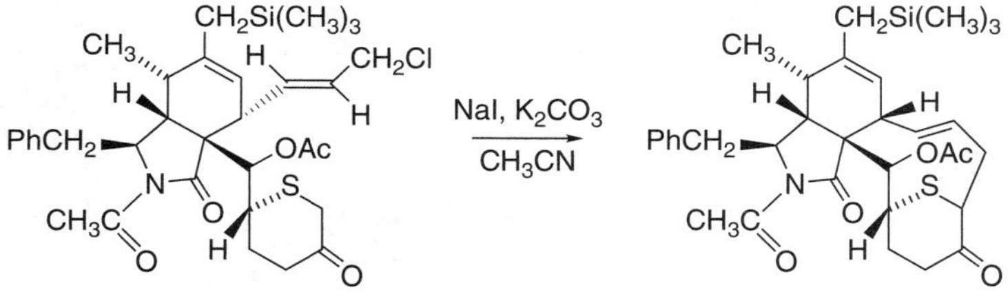

(l)

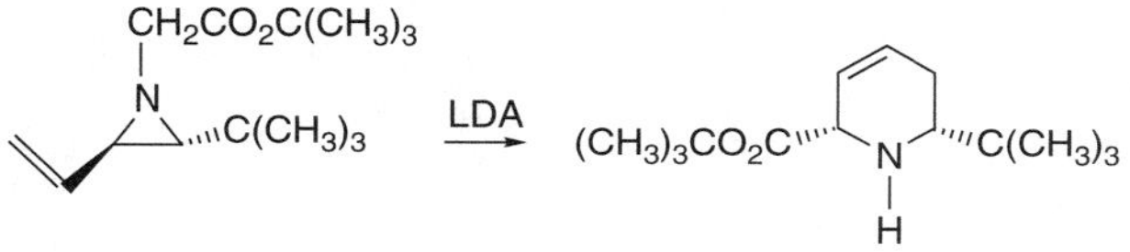

(m)

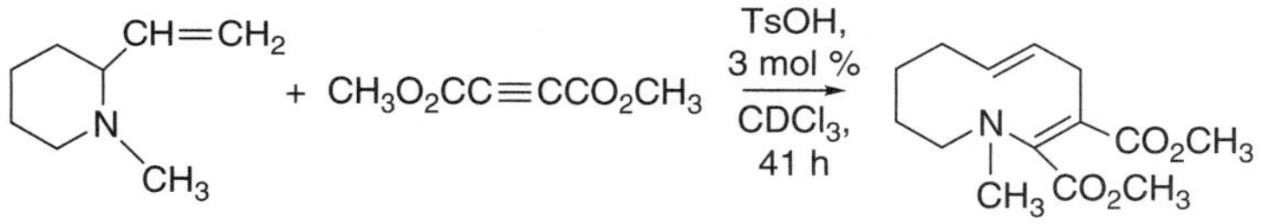

6.12. In each part, the molecule shown was employed as a synthetic equivalent in a cycloaddition reaction. Show a sequence of reactions by which the adduct can be converted to the desired product.

(a)

RC≡CHSO₂Ph
|
NO₂   as an alkyne equivalent in reaction with 1,3-pentadiene.

(b)

PhSO₂CH=CHSi(CH₃)₃ as an acetylene equivalent in reaction with anthracene.

(c)

CH₂=CCN
|
O₂CCH₃   as a ketene equivalent in reaction with 5-(isopropylidene)-1,3-cyclopentadiene (dimethylfulvene).

612

CHAPTER 6

*Concerted
Cycloadditions,
Unimolecular
Rearrangements, and
Thermal Eliminations*

(d) $CH_2{=}CHNO_2$ as a ketene equivalent in reaction with 5-methoxymethyl-1,3-cyclopentadiene.

6.13. Suggest reaction sequences for accomplishing each of the following synthetic transformations.

(a) Squalene from succinaldehyde, 2-bromopropene, and 3-methoxy-2-methyl-1,3-butadiene.

(b)

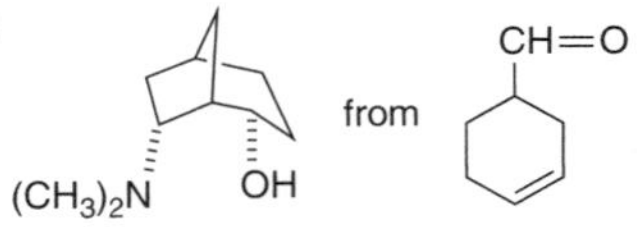

(c)

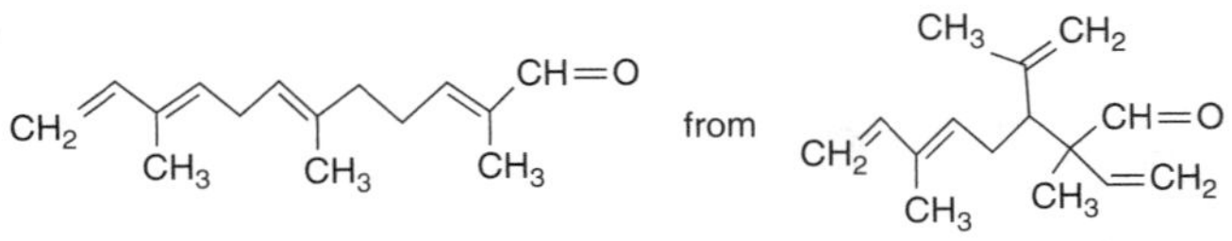

(d)

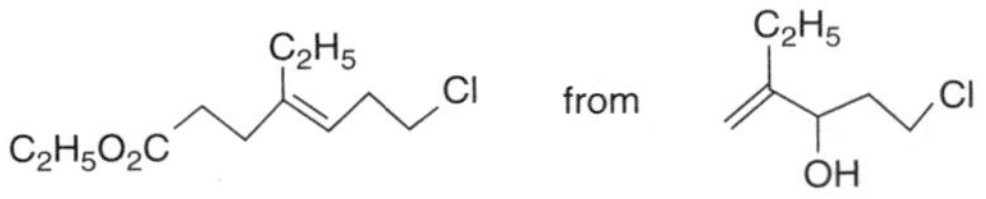

(e)

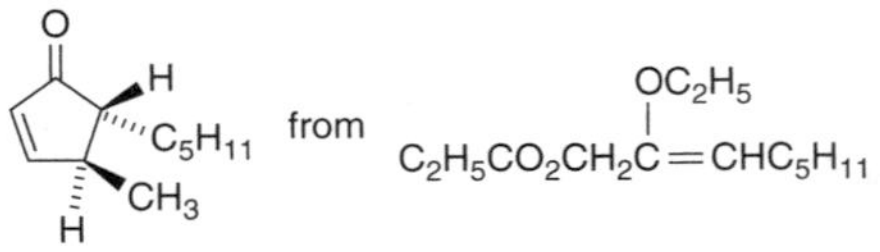

(f)

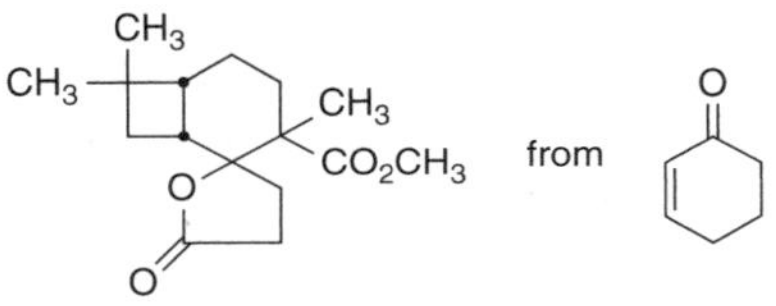

(g)

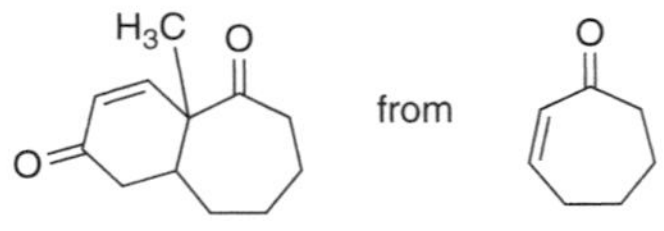

(h)

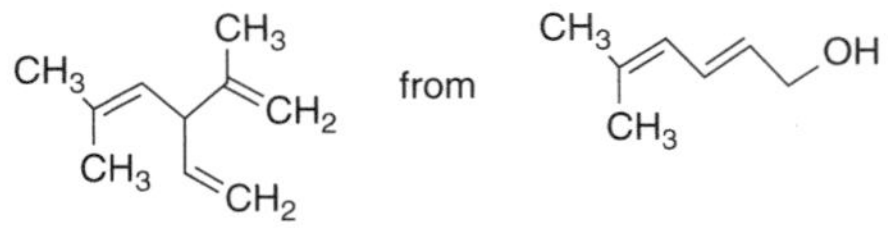

(i)

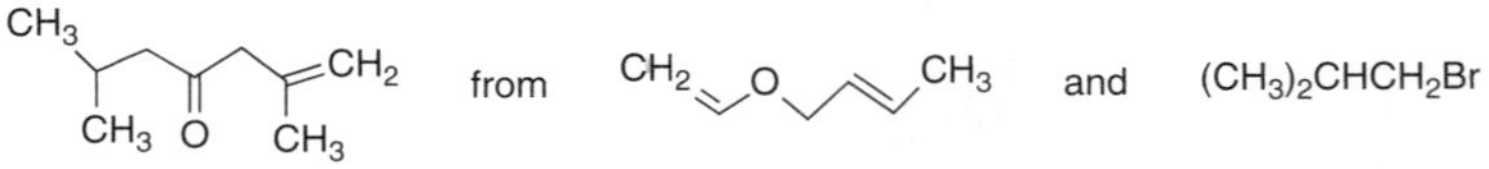

(j)

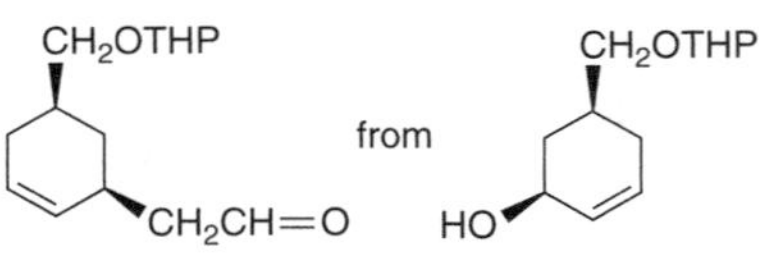

(l)

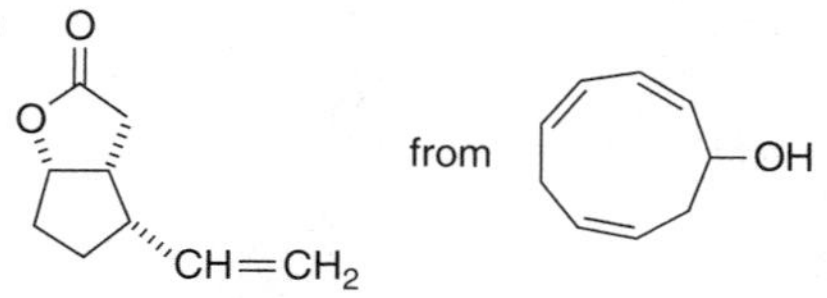

from —OH

(m)

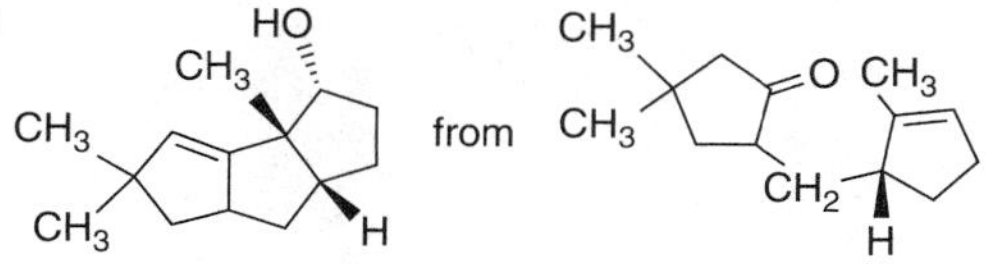

HO
CH$_3$
CH$_3$
CH$_3$
CH$_3$
H   from   CH$_3$ CH$_3$
                CH$_2$
                     H

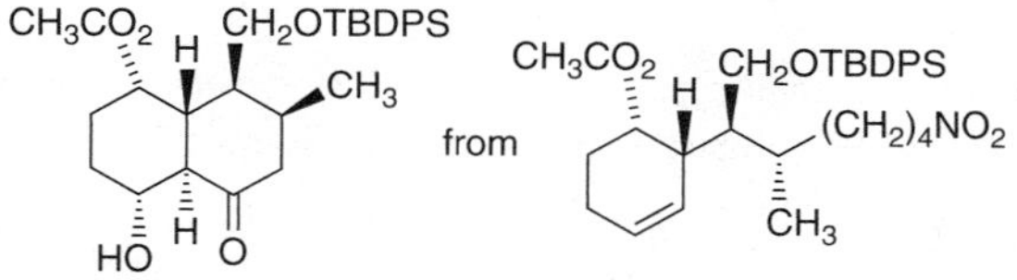

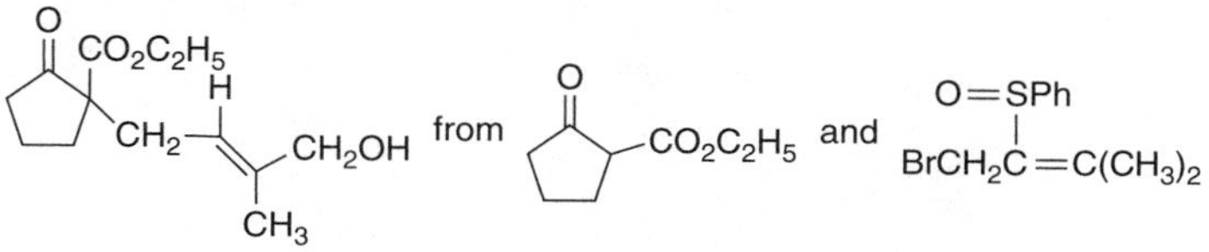

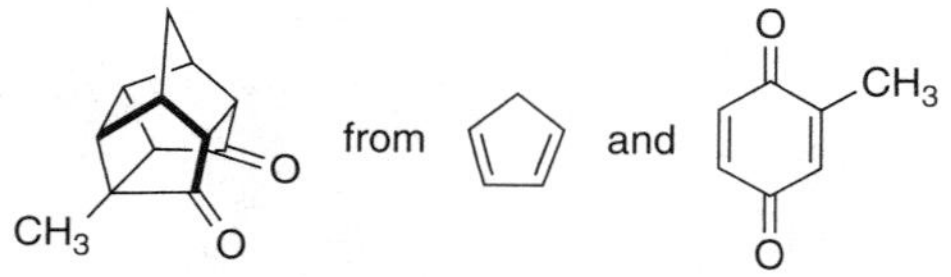

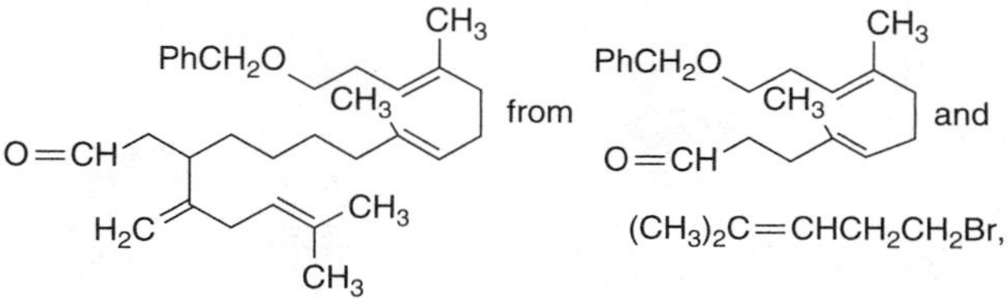

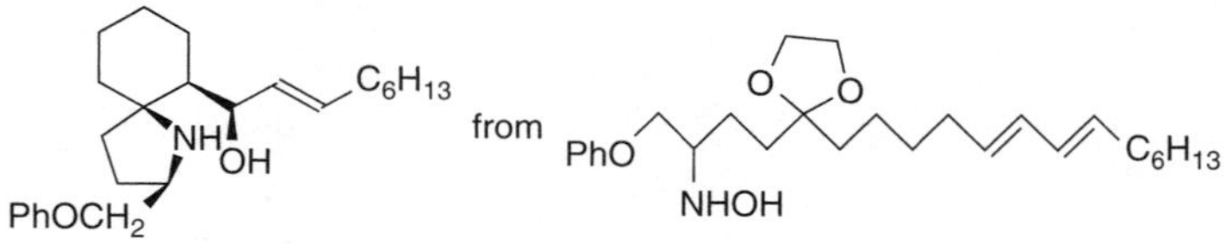

(n) CH$_3$CO$_2$ CH$_2$OTBDPS
H
CH$_3$

HO   H   O

CH$_3$CO$_2$ CH$_2$OTBDPS
H
(CH$_2$)$_4$NO$_2$
from
CH$_3$

(o)

O   CO$_2$C$_2$H$_5$
H
CH$_2$ CH$_2$OH   from
CH$_3$

O
CO$_2$C$_2$H$_5$ and

O=SPh
|
BrCH$_2$C=C(CH$_3$)$_2$

(p)

CH$_3$

O

CH$_3$
O

from   and

O
CH$_3$

O

(q)

PhCH$_2$O   CH$_3$
CH$_3$

O=CH

H$_2$C
CH$_3$
CH$_3$

from

PhCH$_2$O   CH$_3$
CH$_3$

O=CH   and

(CH$_3$)$_2$C=CHCH$_2$CH$_2$Br,

(r)

C$_6$H$_13$
NH   OH
PhOCH$_2$

from

PhO
O   O
C$_6$H$_13$
NHOH

(s)

CO$_2$C$_2$H$_5$

O
O   N

O

from

CO$_2$C$_2$H$_5$
O

O
O   N

O

O

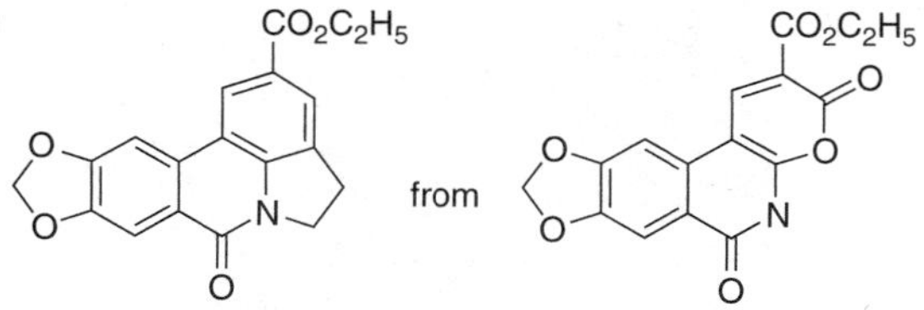

(t)

CH$_3$CH$_3$ CH$_3$

CH$_3$O$_2$C
OTBDMS   from
SC(CH$_3$)$_3$

PhCH$_2$
N   O
CH$_3$
O   O

and

CH$_3$
(CH$_3$)$_3$CS
CH=O

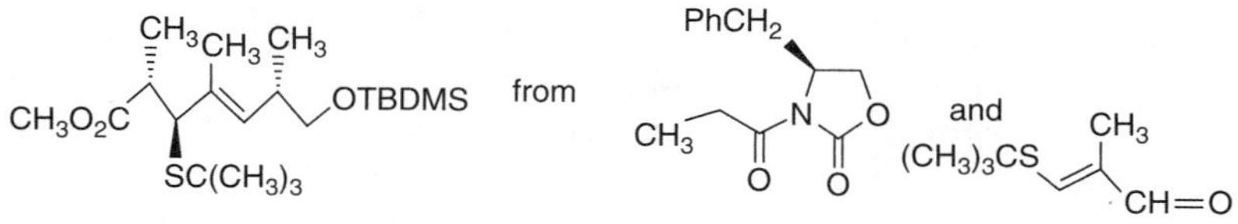

6.14. By retrosynthetic analysis, identify a precursor that could provide the desired product by a single pericyclic reaction. Indicate appropriate reaction conditions for the transformation you identify.

614

CHAPTER 6

*Concerted
Cycloadditions,
Unimolecular
Rearrangements, and
Thermal Eliminations*

(a)

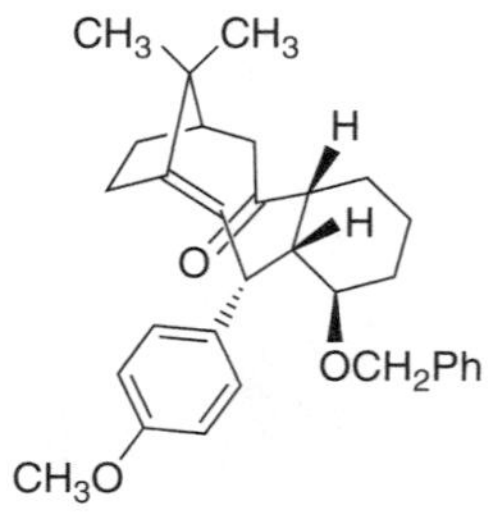

(b)

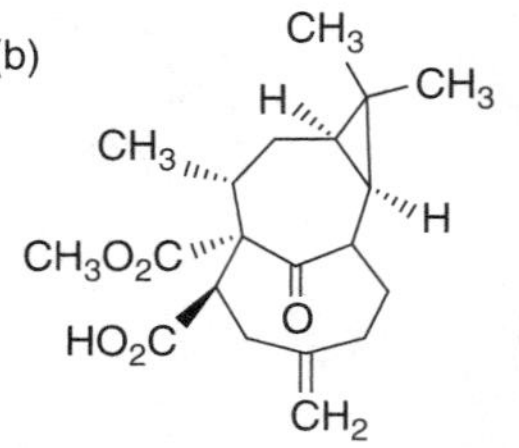

6.15. Predict the structure of the major product, including stereochemistry, of the following reactions. Draw the transition structures and identify the features that control the stereochemistry of the reaction.

(a)

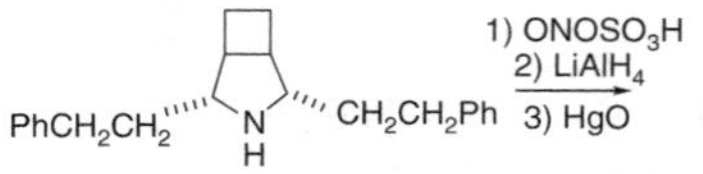

1) ONOSO₃H
2) LiAlH₄
3) HgO

(b)

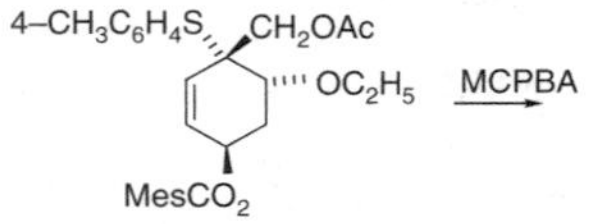

MCPBA

(c)

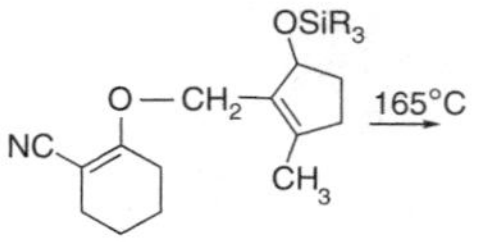

165°C

(d)

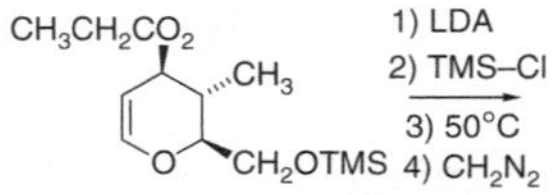

1) LDA
2) TMS–Cl
3) 50°C
4) CH₂N₂

(e)

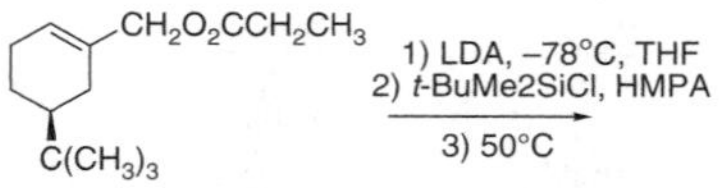

1) LDA, –78°C, THF
2) t-BuMe2SiCl, HMPA
3) 50°C

(f)

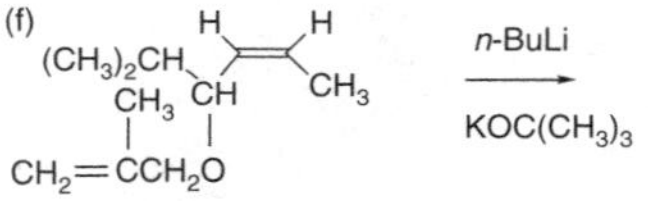

n-BuLi
KOC(CH₃)₃

(g)

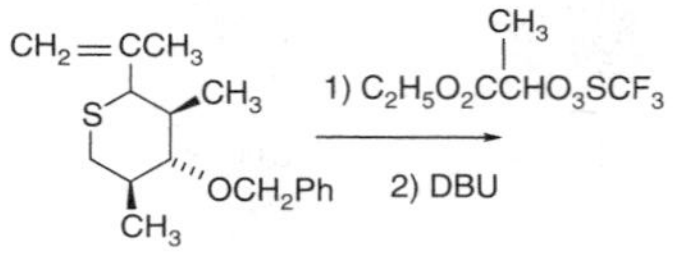

1) C₂H₅O₂CCHO₃SCF₃
2) DBU

(h)

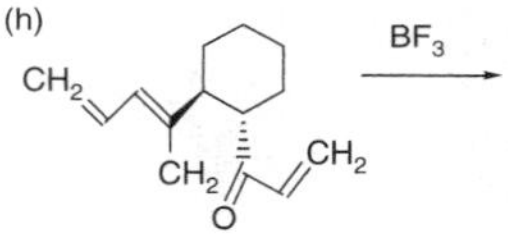

BF₃

(i)

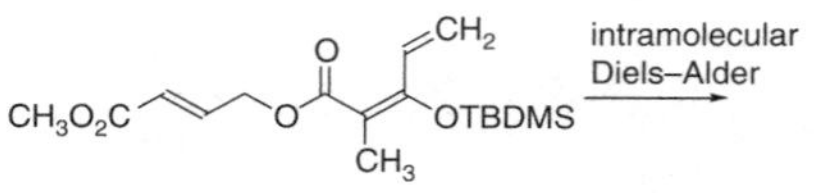

intramolecular
Diels–Alder

(j)

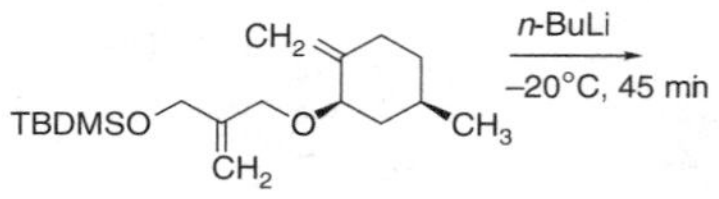

n-BuLi
–20°C, 45 min

(k)

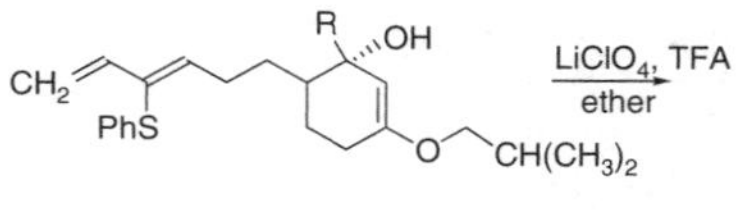

LiClO₄; TFA
ether

R= CH₃, CH₂CO₂-t-Bu, H₂C—C=CH₂
                                         |
                                        CH₃

(l)

PhCH₂NOH
K₂CO₃
60°–70°C

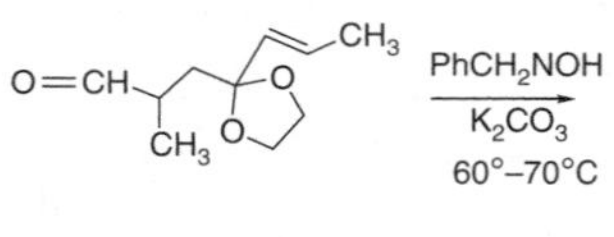

6.16. Oxepin is in equilibrium with benzene oxide by a [3,3]-sigmatropic shift. Advantage has been taken of this equilibrium to develop a short synthesis of barrelene. Outline a way that this could be done.

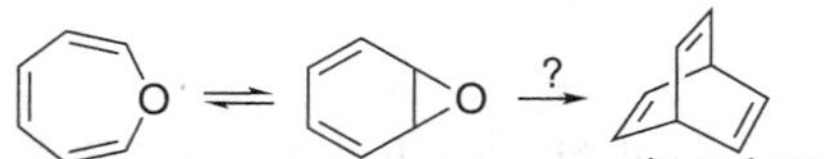

barrelene

6.17. The following transformations involve generation of anionic intermediates that then undergo cycloaddition reactions. Identify the anion intermediate and outline the mechanism for each transformation.

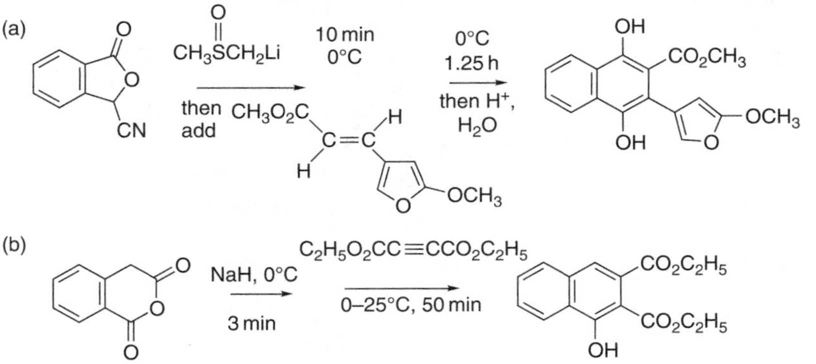

(a)

(b)

6.18. When the lactone silyl ketene acetal **18-1** is heated to 135° C a mixture of four stereoisomers is obtained. Although the major one is the expected [3,3]-sigmatropic rearrangement product, lesser amounts of other possible C(4a) and C(5) epimers are also formed. When the reaction mixture is heated to 100° C, partial conversion to the same mixture of stereoisomers is observed, but most of the product at this temperature is an acyclic triene ester. Suggest a structure for the triene ester and show how it can be formed. Discuss the significance of the observation of the triene ester for the lack of complete stereospecificity in the rearrangement.

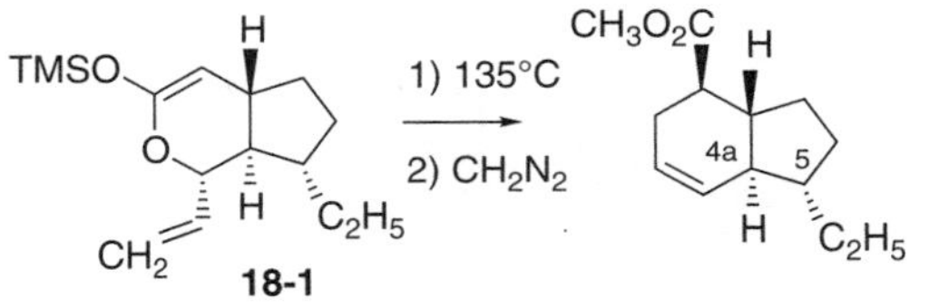

6.19. The following cycloaddition reactions involve chiral auxiliaries and proceed with a good degree of diastereoselectivity. Provide a rationalization of the formation of the preferred product on the basis of a TS.

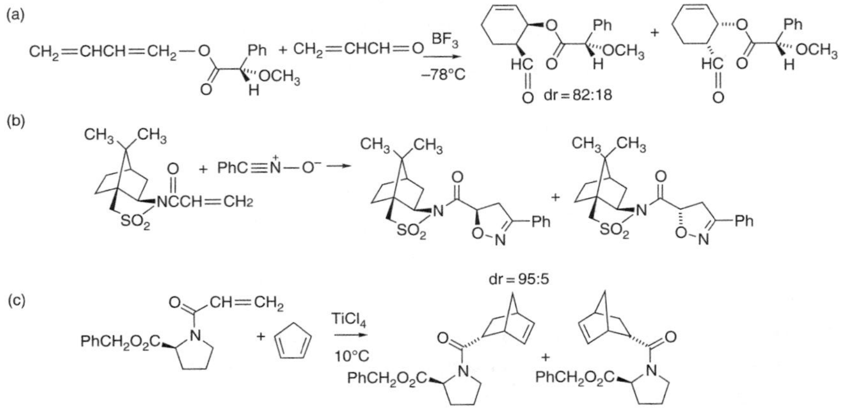

(a)

(b)

(c)

616

CHAPTER 6

*Concerted
Cycloadditions,
Unimolecular
Rearrangements, and
Thermal Eliminations*

6.20. The following transformations involve two or more pericyclic reactions occurring in tandem during the process. Suggest a plausible sequence of reactions that can lead to the observed product.

(a)

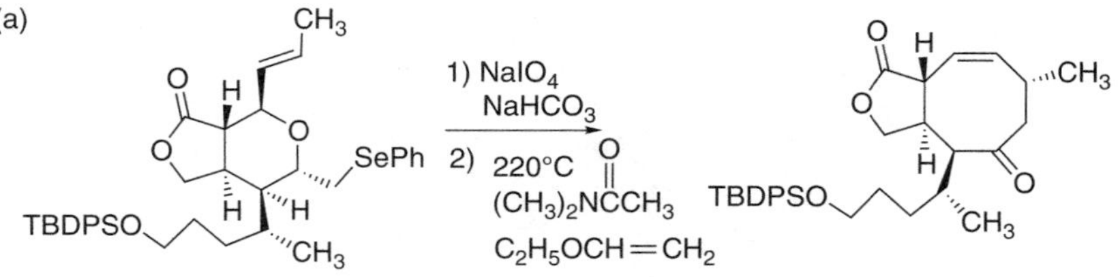

1) NaIO₄
   NaHCO₃

2) 220°C
   (CH₃)₂NCCH₃
   C₂H₅OCH=CH₂

(b)

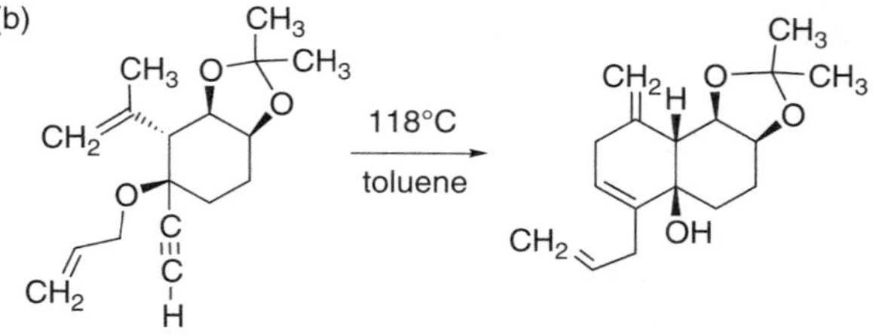

118°C

toluene

(c)

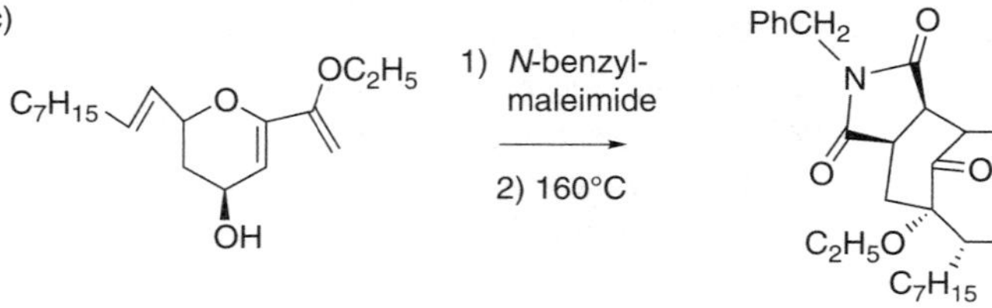

1) *N*-benzyl-
   maleimide

2) 160°C

(d)

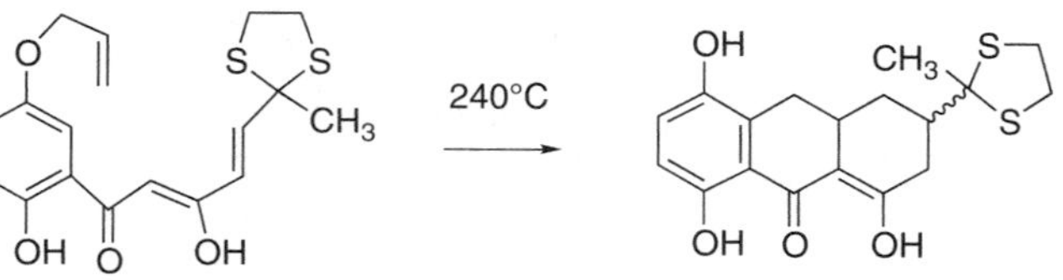

240°C

6.21. The Diels-Alder reaction of *N*-acryloyloxazolidinone catalyzed by Cu(*t*-Bu)BOX shows a reversal of stereoselectivity between 1-acetoxybutadiene and 1-acetoxy-3-methylbutadiene. The former gives a 85:15 *endo:exo* ratio, whereas the latter is 27:73 *endo:exo*. Explain this reversal in terms of the transition structure model given on p. 509.

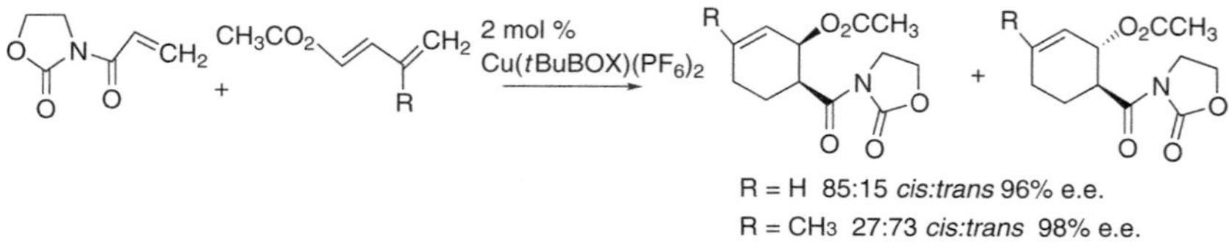

R = H 85:15 *cis:trans* 96% e.e.
R = CH₃ 27:73 *cis:trans* 98% e.e.

6.22. The alkenyl cyclopentenone **22a-c** have been subjected to photolysis with the results shown below. Analyze these results in terms of the mechanistic interpretation give on p. 547.

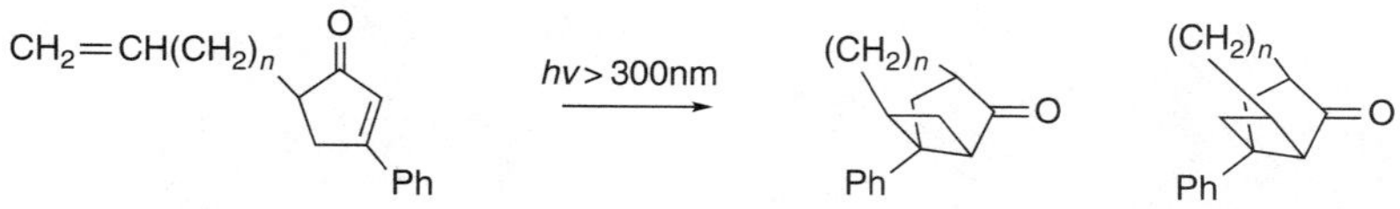

	n	parallel	crossed
22a	1	only product	not found
22b	2	minor product	major product
22c	3	only product	not found

6.23. The intramolecular Diels-Alder reaction of **23-1** carried out under LiClO$_4$ catalysis is rather nonselective. Use a molecular mechanics program to assess the energies of the competing TSs and products. Are the results in agreement with the experimental outcome?

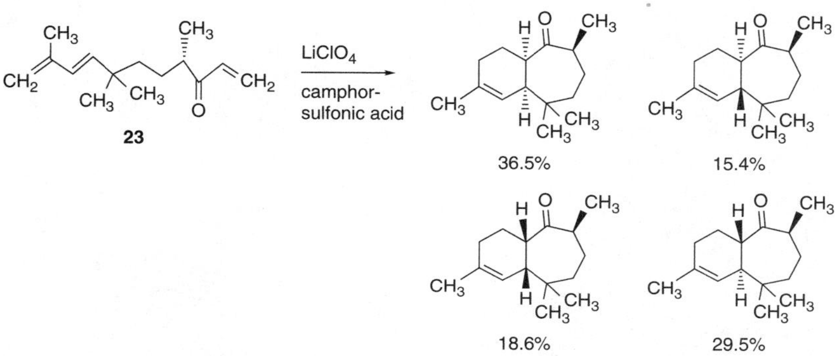

Organometallic Compounds of Group I and II Metals

Introduction

The use of organometallic reagents in organic synthesis had its beginning around 1900 when Victor Grignard discovered that alkyl and aryl halides react with magnesium metal to give homogeneous solutions containing organomagnesium compounds. The "Grignard reagents" proved to be highly reactive carbon nucleophiles and are still very useful synthetic reagents. Organolithium reagents came into synthetic use somewhat later, but are also very important for synthesis. The present chapter focuses on Grignard reagents and organolithium compounds. We also consider zinc, cadmium, mercury, indium, and lanthanide organometallics, which have more specialized places in synthetic methodology. Certain of the transition metals, such as copper, palladium, and nickel, which are also important in synthetic methodology, are discussed in Chapter 8.

The composition of the organolithium compounds is RLi or more accurately $(RLi)_n$. The organomagnesium compounds are usually formulated as RMgX, with X being a halide. The organometallic derivatives of Group I and II metals provide reactive carbon nucleophiles. Reactivity increases in the order Li < Na < K and MgX < CaX, but the lithium and magnesium reactions are by far the most commonly used. Organolithium and magnesium reagents react with polar multiple bonds, especially carbonyl groups, and provide synthetic routes to a variety of alcohols. Other electrophiles, such as acyl halides, nitriles, and CO_2 provide routes to ketones and carboxylic acids.

$$CH_2{=}O$$

$$RCH_2{-}OH \Longrightarrow$$

$$R{-}M \qquad \overset{CO_2}{\Longleftarrow} \qquad RCO_2H$$

$$M = Li,\ MgX$$

$$\overset{R'COY}{\Longleftarrow}$$

$$R'CH{=}O$$

$$or\ R'CN \qquad \overset{O}{\underset{RCR'}{\parallel}}$$

$$RCH{-}OH \atop \overset{|}{R'}$$

$$R'_2C{=}O$$

$$R'CO_2R''$$

$$R'_2C{-}OH \atop \overset{|}{R}$$

$$R_2C{-}OH \atop \overset{|}{R'}$$

The Group IIB organometallics derived from zinc, cadmium, and mercury are considerably less reactive. The carbon-metal bonds in these compounds have more covalent character than for lithium or magnesium reagents. Zinc, cadmium, and mercury are distinct from other transition metals in having a d^{10} shell in the $+2$ oxidation state and their reactions usually do not involve changes in oxidation state. Although organozinc and cadmium reagents react with acyl chloride, reactions with other carbonyl compounds require either Lewis acids or chelates as catalysts. These catalyzed reactions make organozinc reagents particularly useful in additions to aldehydes. The lanthanides and indium organometallics are usually in the $+3$ oxidation state, which are also filled valence shells, and have a number of specialized applications that depend on their strong oxyphilic character.

7.1. Preparation and Properties of Organomagnesium and Organolithium Reagents

The compounds of lithium and magnesium are the most important of Group IA and IIA organometallics. The metals in these two groups are the most electropositive of the elements, and the polarity of the metal-carbon bond increases the electron density on carbon. This electronic distribution is responsible for the strong nucleophilicity and basicity of these compounds. There is a high ionic character in the carbon-metal bonds, but the compounds tend to exist as aggregates and have good solubility in some nonpolar solvents.

7.1.1. Preparation and Properties of Organomagnesium Reagents

The reaction of magnesium metal with an alkyl or aryl halide in diethyl ether is the standard method for synthesis of Grignard reagents. The order of reactivity of the halides is $RI > RBr > RCl$. The formation of Grignard reagents takes place at the metal surface. Reaction commences with an electron transfer to the halide and decomposition of the radical ion, followed by rapid combination of the organic group with a magnesium ion.[1] It

[1] H. R. Rogers, C. L. Hill, Y. Fujuwara, R. J. Rogers, H. L. Mitchell, and G. M. Whitesides, *J. Am. Chem. Soc.*, **102**, 217 (1980); J. F. Garst, J. E. Deutch, and G. M. Whitesides, *J. Am. Chem. Soc.*, **108**, 2490 (1986); E. C. Ashby and J. Oswald, *J. Org. Chem.*, **53**, 6068 (1988); H. M. Walborsky,

has been suggested that the reactions may involve reduction of the halide by clusters of magnesium atoms.[2]

$$R-Br + Mg \longrightarrow R-Br^{\cdot -} + Mg(I)$$
$$R-Br^{\cdot -} \longrightarrow R\cdot + Br^-$$
$$R\cdot + Mg(I) + Br^- \longrightarrow R-Mg-Br$$

Solutions of several Grignard reagents such as methylmagnesium bromide, ethylmagnesium bromide, and phenylmagnesium bromide are available commercially. Some Grignard reagents are formed more rapidly in tetrahydrofuran than in ether. This is true of vinylmagnesium bromide, for example.[3] Other ether solvents such as dimethoxyethane can be used. For industrial purposes, where less volatile solvents are needed for reasons of safety, *bis*-2-butoxyethyl ether (butyl diglyme), bp 256° C, can be used. The solubility of Grignard reagents in ethers is the result of Lewis acid-base complex formation between the magnesium ion and the ether oxygens.

Under normal laboratory conditions magnesium metal is coated with an unreactive layer of $Mg(OH)_2$, and the reactions do not start until the organic halide diffuses through it. The reaction appears to begin at discrete sites,[4] and accelerates as the surface coating breaks up, exposing more active surface. The ether solvents are probably involved and may assist dissociation of the metal ions from the surface. Various techniques for initiating the reactions, such as addition of small amounts of I_2 or $BrCH_2CH_2Br$, appear to involve the generation of Mg^{2+} salts, which serve to facilitate the reaction. Sonication or mechanical pretreatment can also be used to activate magnesium.[5] Organic halides that are unreactive toward magnesium shavings can often be induced to react by using an extremely reactive form of magnesium that is obtained by reducing magnesium salts with sodium or potassium metal.[6] Even alkyl fluorides, which are normally unreactive, form Grignard reagents under these conditions.

One of the fundamental questions about the mechanism is whether the radical is really "free" in the sense of diffusing from the metal surface.[7] For alkyl halides, there is considerable evidence that the radicals behave similarly to alkyl free radicals.[8] One test for the involvement of radical intermediates is to determine whether cyclization occurs in the 6-hexenyl system, where radical cyclization is rapid (see Part A, Section 12.2.2).

Acc. Chem. Res., **23**, 286 (1990); H. M. Walborsky and C. Zimmerman, *J. Am. Chem. Soc.*, **114**, 4996 (1992); C. Hamdouchi, M. Topolski, V. Goedken, and H. M. Walborsky, *J. Org. Chem.*, **58**, 3148 (1993); C. Hamdouchi and H. M. Walborsky, *Handbook of Grignard Reagents*, G. S. Silverman and P. E. Rakita, eds., Marcel Dekker, New York, 1996, pp. 145–218.

2. E. Paralez, J.-C. Negrel, A. Goursot, and M. Chanon, *Main Group Metal Chem.*, **21**, 69 (1998); E. Peralez, J.-C. Negrel, A. Goussot, and M. Chanon, *Main Group Metal Chem.*, **22**, 185 (1999).

3. D. Seyferth and F. G. A. Stone, *J. Am. Chem. Soc.*, **79**, 515 (1957); H. Normant, *Adv. Org. Chem.*, **2**, 1 (1960).

4. C. E. Teerlinck and W. J. Bowyer, *J. Org. Chem.*, **61**, 1059 (1996).

5. K. V. Baker, J. M. Brown, N. Hughes, A. J. Skarnulis, and A. Sexton, *J. Org. Chem.*, **56**, 698 (1991); J.-L. Luche and J.-C. Damaino, *J. Am. Chem. Soc.*, **102**, 7926 (1980).

6. R. D. Rieke and S. E. Bales, *J. Am. Chem. Soc.*, **96**, 1775 (1974); R. D. Rieke, *Acc. Chem. Res.*, **10**, 301 (1977).

7. C. Walling, *Acc. Chem. Res.*, **24**, 255 (1991); J. F. Garst, F. Ungvary, R. Batlaw, and K. E. Lawrence, *J. Am. Chem. Soc.*, **113**, 5392 (1991).

8. J. F. Garst and M. P. Soriaga, *Coord. Chem. Rev.*, **248**, 623 (2004); J. F. Garst and U. Ferenc, in *Grignard Reagents: New Developments*, H. G. Richey, Jr., ed., Wiley, Chichester, 2000, pp. 185–275.

Small amounts of cyclized products are obtained after the preparation of Grignard reagents from 5-hexenyl bromide.[9] This indicates that cyclization of the intermediate radical competes to a small extent with combination of the radical with the metal. Quantitative kinetic models that compare competing processes are consistent with diffusion of the radicals from the surface.[10] Alkyl radicals can be trapped with high efficiency by the nitroxide radical TMPO.[11] Nevertheless, there remains disagreement about the extent to which the radicals diffuse away from the metal surface.[12]

It seems likely that aryl, vinyl, and cyclopropyl halides react by an alternative mechanism, since the corresponding radicals are less stable than alkyl radicals. It has been suggested that these halides may react through a dianion.[13]

$$ArX \xrightarrow{\text{Mg}^0} [ArX]^{2-} \xrightarrow{\text{Mg}^{2+}} ArMgX$$

The radical cyclization test has been applied and although 2-(3-butenyl)phenyl halides give little if any cyclization, substituents that are expected to increase the rate of cyclization to around $10^9\,\text{s}^{-1}$ do give some cyclic product.[14]

The stereochemistry of Grignard reagents having sterogenic centers is another means of probing the structure and lifetime of intermediates. The preparation of Grignard reagents from alkyl halides normally occurs with stereochemical randomization at the reaction site. Stereoisomeric halides give rise to organomagnesium compounds of identical composition.[15] The main exceptions to this generalization are cyclopropyl and alkenyl systems, which react with partial retention of configuration.[16] Once formed, secondary alkylmagnesium compounds undergo stereochemical inversion only slowly. *Endo-* and *exo*-norbornylmagnesium bromide, for example, require 1 day at room temperature to reach equilibrium.[17] NMR studies have demonstrated that inversion of configuration is quite slow, on the NMR time scale, even

9. R. C. Lamb, P. W. Ayers, and M. K. Toney, *J. Am. Chem. Soc.*, **85**, 3483 (1963); R. C. Lamb and P. W. Ayers, *J. Org. Chem.*, **27**, 1441 (1962); C. Walling and A. Cioffari, *J. Am. Chem. Soc.*, **92**, 6609 (1970); H. W. H. J. Bodewitz, C. Blomberg, and F. Bickelhaupt, *Tetrahedron*, **31**, 1053 (1975); J. F. Garst and B. L. Swift, *J. Am. Chem. Soc.*, **111**, 241 (1989).

10. J. F. Garst, B. L. Swift, and D. W. Smith, *J. Am. Chem. Soc.*, **111**, 234 (1989); J. F. Garst, *Acc. Chem. Res.*, **24**, 95 (1991).

11. K. S. Root, C. L. Hill, L. M. Lawrence, and G. M. Whitesides, *J. Am. Chem. Soc.*, **111**, 5405 (1989); L. M. Lawrence and G. M. Whitesides, *J. Am. Chem. Soc.*, **102**, 2493 (1980).

12. C. Hamdouchi and H. M. Walborsky, in *Handbook of Grignard Reagents*, G. S. Silverman and P. E. Rakita, eds., Marcel Dekker, New York, 1996, pp. 145–218; H. M. Walborsky, *Acc. Chem. Res.*, 286 (1990).

13. J. F. Garst, J. R. Boone, L. Webb, K. E. Lawrence, J. T. Baxter, and F. Ungavary, *Inorg. Chim. Acta*, **296**, 52 (1999).

14. N. Bodineau, J.-M. Mattalia, V. Thimokhin, K. Handoo, J.-C. Negrel, and M. Chanon, *Org. Lett.*, **2**, 2303 (2000).

15. N. G. Krieghoff and D. O. Cowan, *J. Am. Chem. Soc.*, **88**, 1322 (1966).

16. T. Yoshino and Y. Manabe, *J. Am. Chem. Soc.*, **85**, 2860 (1963); H. M. Walborsky and A. E. Young, *J. Am. Chem. Soc.*, **86**, 3288 (1964); H. M. Walborsky and B. R. Banks, *Bull. Soc. Chim. Belg.*, **89**, 849 (1980); H. M. Walborsky and J. Rachon, *J. Am. Chem. Soc.*, **111**, 1896 (1989); J. Rachon and H. M. Walborsky, *Tetrahedron Lett.*, **30**, 7345 (1988).

17. F. R. Jensen and K. L. Nakamaye, *J. Am. Chem. Soc.*, **88**, 3437 (1966); N. G. Krieghoff and D. O. Cowan, *J. Am. Chem. Soc.*, **88**, 1322 (1966).

up to 170° C.[18] In contrast, the inversion of configuration of primary alkylmagnesium halides is very fast.[19] This difference in the primary and secondary systems may be the result of a mechanism for inversion that involves exchange of alkyl groups between magnesium atoms.

If bridged intermediates are involved, the larger steric bulk of secondary systems would retard the reaction. Steric restrictions may be further enhanced by the fact that organomagnesium reagents are often present as clusters (see below).

The usual designation of Grignard reagents as RMgX is a useful but incomplete representation of the composition of the compounds in ether solution. An equilibrium exists with magnesium bromide and the dialkylmagnesium. [20]

$$2\ RMgX \rightleftharpoons R_2Mg + MgX_2$$

The position of the equilibrium depends upon the solvent and the identity of the specific organic group, but in ether lies well to the left for simple aryl-, alkyl-, and alkenylmagnesium halides.[21] Solutions of organomagnesium compounds in diethyl ether contain aggregated species.[22] Dimers predominate in ether solutions of alkyl-magnesium chlorides.

The corresponding bromides and iodides show concentration-dependent behavior and in very dilute solutions they exist as monomers. In tetrahydrofuran, there is less tendency to aggregate, and several alkyl and aryl Grignard reagents have been found to be monomeric in this solvent.

A number of Grignard reagents have been subjected to X-ray structure determination.[23] Ethylmagnesium bromide has been observed in both monomeric and dimeric forms in crystal structures.[24] Figures 7.1a and b show, respectively, the crystal structure

[18.] E. Pechold, D. G. Adams, and G. Fraenkel, *J. Org. Chem.*, **36**, 1368 (1971).
[19.] G. M. Whitesides, M. Witanowski, and J. D. Roberts, *J. Am. Chem. Soc.*, **87**, 2854 (1965); G. M. Whitesides and J. D. Roberts, *J. Am. Chem. Soc.*, **87**, 4878 (1965); G. Fraenkel and D. T. Dix, *J. Am. Chem. Soc.*, **88**, 979 (1966).
[20.] K. C. Cannon and G. R. Krow, in *Handbook of Grignard Reagents*, G. S. Silverman and P. E. Rakita, eds., Marcel Dekker, New York, 1996, pp. 271–289.
[21.] G. E. Parris and E. C. Ashby, *J. Am. Chem. Soc.*, **93**, 1206 (1971); P. E. M. Allen, S. Hagias, S. F. Lincoln, C. Mair, and E. H. Williams, *Ber. Bunsenges. Phys. Chem.*, **86**, 515 (1982).
[22.] E. C. Ashby and M. B. Smith, *J. Am. Chem. Soc.*, **86**, 4363 (1964); F. W. Walker and E. C. Ashby, *J. Am. Chem. Soc.*, **91**, 3845 (1969).
[23.] C. E. Holloway and M. Melinik, *Coord. Chem. Rev.*, **135**, 287 (1994); H. L. Uhm, in *Handbook of Grignard Reagents*, G. S. Silverman and P. E. Rakita, eds., Marcel Dekker, New York, 1996, pp. 117–144; F. Bickelhaupt, in *Grignard Reagents: New Developments*, H. G. Richey, Jr., ed., Wiley, New York, 2000, pp. 175–181.
[24.] L. J. Guggenberger and R. E. Rundle, *J. Am. Chem. Soc.*, **90**, 5375 (1968); A. L. Spek, P. Voorbergen, G. Schat, C. Blomberg, and F. Bickelhaupt, *J. Organomet. Chem.*, **77**, 147 (1974).

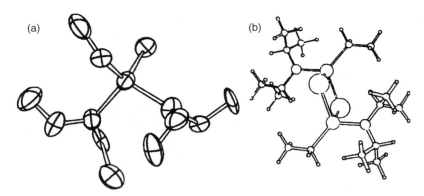

Fig. 7.1. Crystal structures of ethylmagnesium bromide: (a) Monomeric $C_2H_5MgBr[O(C_2H_5)_2]_2$. Reproduced from *J. Am. Chem. Soc.*, **90**, 5375 (1968), by permission of the American Chemical Society. (b) Dimeric $\{C_2H_5MgBr\ [O-(i-C_3H_7)_2]\}_2$. Reproduced from *J. Organomet. Chem.*, 77, 147 (1974), by permission of Elsevier.

of the monomer with two diethyl ether molecules coordinated to magnesium and a dimeric structure with one diisopropyl ether molecule per magnesium.

7.1.2. Preparation and Properties of Organolithium Compounds

7.1.2.1. Preparation Using Metallic Lithium. Most simple organolithium reagents can be prepared by reaction of an appropriate halide with lithium metal. The method is applicable to alkyl, aryl, and alkenyl lithium reagents.

$$R-X + 2\,Li \longrightarrow RLi + LiX$$

As with organomagnesium reagents, there is usually loss of stereochemical integrity at the site of reaction during the preparation of alkyllithium compounds.[25] Alkenyllithium reagents can usually be prepared with retention of configuration of the double bond.[26,27]

For some halides, it is advantageous to use finely powdered lithium and a catalytic amount of an aromatic hydrocarbon, usually naphthalene or 4, 4'-di-*t*-butylbiphenyl (DTBB).[28] These reaction conditions involve either radical anions or dianions generated by reduction of the aromatic ring (see Section 5.6.1.2), which then convert the halide to a radical anion. Several useful functionalized lithium reagents have been prepared by this method. In the third example below, the reagent is trapped in situ by reaction with benzaldehyde.

$$ClCH{=}C(OC_2H_5)_2 \quad \xrightarrow[\text{DTBB}]{\text{Li, 5 equiv}} \quad LiCH{=}C(OC_2H_5)_2$$

Ref. 29

25. W. H. Glaze and C. M. Selman, *J. Org. Chem.*, **33**, 1987 (1968).
26. M. Yus, R. P. Herrera, and A. Guijarro, *Chem. Eur. J.*, **8**, 2574 (2002).
27. J. Millon, R. Lorne, and G. Linstrumelle, *Synthesis*, 434 (1975).
28. M. Yus, *Chem. Soc. Rev.*, 155 (1996); D. J. Ramon and M. Yus, *Tetrahedron*, **52**, 13739 (1996).
29. M. Si-Fofil, H. Ferrrerira, J. Galak, and L. Duhamel, *Tetrahedron Lett.*, **39**, 8975 (1998).

625

SECTION 7.1

*Preparation and
Properties of
Organomagnesium
and Organolithium
Reagents*

Ref. 30

Ref. 31

Alkyllithium reagents can also be generated by reduction of sulfides.[32] Alkenyl-lithium and substituted alkyllithium reagents can be prepared from sulfides,[33] and sulfides can be converted to lithium reagents by the catalytic electron transfer process described for halides.[34]

$$PhCH_2CH_2SPh \xrightarrow[Li^+Naph^-]{Li \text{ or }} PhCH_2CH_2Li$$

Ref. 35

This technique is especially useful for the preparation of α-lithio ethers, sulfides, and silanes.[36] The lithium radical anions of naphthalene, 4, 4′-di-*t*-butyldiphenyl (DTBB) or dimethylaminonaphthalene (LDMAN) are used as the reducing agent.

The simple alkyllithium reagents exist mainly as hexamers in hydrocarbon solvents.[37] In ethers, tetrameric structures are usually dominant.[38] The tetramers,

30. A. Bachki, F. Foubelo, and M. Yus, *Tetrahedron*, **53**, 4921 (1997).

31. A. Guijarro, B. Mancheno, J. Ortiz, and M. Yus, *Tetrahedron*, **52**, 1643 (1993).

32. T. Cohen and M. Bhupathy, *Acc. Chem. Res.*, **22**, 152 (1989).

33. T. Cohen and M. D. Doubleday, *J. Org. Chem.*, **55**, 4784 (1990); D. J. Rawson and A. I. Meyers, *Tetrahedron Lett.*, **32**, 2095 (1991); H. Liu and T. Cohen, *J. Org. Chem.*, **60**, 2022 (1995).

34. F. Foubelo, A. Gutierrez, and M. Yus, *Synthesis*, 503 (1999).

35. C. G. Screttas and M. Micha-Screttas, *J. Org. Chem.*, **43**, 1064 (1978); C. G. Screttas and M. Micha-Screttas, *J. Org. Chem.*, **44**, 113 (1979).

36. T. Cohen and J. R. Matz, *J. Am. Chem. Soc.*, **102**, 6900 (1980); T. Cohen, J. P. Sherbine, J. R. Matz, R. R. Hutchins, B. M. McHenry, and P. R. Wiley, *J. Am. Chem. Soc.*, **106**, 3245 (1984); S. D. Rychnovsky, K. Plzak, and D. Pickering, *Tetrahedron Lett.*, **35**, 6799 (1994); S. D. Rychnovsky and D. J. Skalitzky, *J. Org. Chem.*, **57**, 4336 (1992).

37. G. Fraenkel, W. E. Beckenbaugh, and P. P. Yang, *J. Am. Chem. Soc.*, **98**, 6878 (1976); G. Fraenkel, M. Henrichs, J. M. Hewitt, B. M. Su, and M. J. Geckle, *J. Am. Chem. Soc.*, **102**, 3345 (1980).

38. H. L. Lewis and T. L. Brown, *J. Am. Chem. Soc.*, **92**, 4664 (1970); P. West and R. Waack, *J. Am. Chem. Soc.*, **89**, 4395 (1967); J. F. McGarrity and C. A. Ogle, *J. Am. Chem. Soc.*, **107**, 1085 (1985); D. Seebach, R. Hassig, and J. Gabriel, *Helv. Chim. Acta*, **66**, 308 (1983); T. L. Brown, *Adv. Organomet. Chem.*, **3**, 365 (1965); W. N. Setzer and P. v. R. Schleyer, *Adv. Organomet. Chem.*, **24**, 354 (1985); W. Bauer, T. Clark, and P. v. R. Schleyer, *J. Am. Chem. Soc.*, **109**, 970 (1987).

in turn, are solvated with ether molecules.[39] Phenyllithium is tetrameric in cyclohexane and a mixture of monomer and dimer in THF.[40] Chelating ligands such as tetramethylethylenediamine (TMEDA) reduce the degree of aggregation.[41] Strong donor molecules such as hexamethylphosphorotriamide (HMPA) and *N,N*-dimethylpropyleneurea (DMPU) also lead to more dissociated and more reactive organolithium reagents.[42] NMR studies on phenyllithium show that TMEDA, other polyamine ligands, HMPA, and DMPU favor monomeric solvated species.[43]

$$O=P[N(CH_3)_2]_3$$

HMPA

DMPU

The crystal structures of many organolithium compounds have been determined.[44] Phenyllithium has been crystallized as an ether solvate. The structure is tetrameric with lithium and carbon atoms at alternating corners of a highly distorted cube. The lithium atoms form a tetrahedron and the carbons are associated with the faces of the tetrahedron. Each carbon is 2.33 Å from the three neighboring lithium atoms and an ether molecule is coordinated to each lithium atom. Figures 7.2a and b show, respectively, the Li–C cluster and the complete array of atoms, except for hydrogen.[45] Section 6.2 of Part A provides additional information on the structure of organolithium compounds.

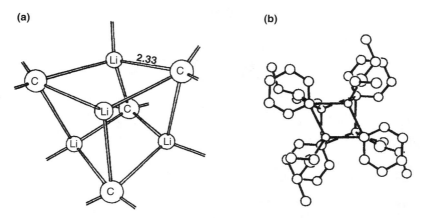

Fig. 7.2. Crystal structure of tetrameric phenyllithium diethyl etherate: (a) tetrameric C_4Li_4 cluster; (b) complete structure except for hydrogens. Reproduced from *J. Am. Chem. Soc.*, **105**, 5320 (1983), by permission of the American Chemical Society.

39. P. D. Bartlett, C. V. Goebel, and W. P. Weber, *J. Am. Chem. Soc.*, **91**, 7425 (1969).
40. L. M. Jackman and L. M. Scarmoutzos, *J. Am. Chem. Soc.*, **106**, 4627 (1984); O. Eppers and H. Gunther, *Helv. Chim. Acta*, **75**, 2553 (1992).
41. W. Bauer and C. Griesinger, *J. Am. Chem. Soc.*, **115**, 10871 (1993); D. Hofffmann and D. B. Collum, *J. Am. Chem. Soc.*, **120**, 5810 (1998).
42. H. J. Reich and D. P. Green, *J. Am. Chem. Soc.*, **111**, 8729 (1989).
43. H. J. Reich, D. P. Green, M. A. Medina, W. S. Goldenberg, B. O. Gudmundsson, R. R. Dykstra, and N. H. Phillips, *J. Am. Chem. Soc.*, **120**, 7201 (1998).
44. E. Weiss, *Angew. Chem. Int. Ed. Engl.*, **32**, 1501 (1993).
45. H. Hope and P. P. Power, *J. Am. Chem. Soc.*, **105**, 5320 (1983).

7.1.2.2. Preparation by Lithiation. There are three other general methods that are very useful for preparing organolithium reagents. The first of these is *hydrogen-metal exchange* or *metallation*, which for the specific case of lithium is known as *lithiation*. This reaction is the usual method for preparing alkynylmagnesium and alkynyllithium reagents. The reactions proceed readily because of the relative acidity of the hydrogen bound to *sp* carbon.

$$H-C\equiv C-R + R'MgBr \longrightarrow BrMgC\equiv C-R + R'H$$

$$H-C\equiv C-R + R'Li \longrightarrow LiC\equiv C-R + R'H$$

Although of limited utility for other types of Grignard reagents, metallation is an important means of preparing a variety of organolithium compounds. The position of lithiation is determined by the relative acidity of the available hydrogens and the directing effect of substituent groups. Benzylic and allylic hydrogens are relatively reactive toward lithiation because of the resonance stabilization of the resulting anions.[46] Substituents that can coordinate to the lithium atom, such as alkoxy, amido, sulfoxide, and sulfonyl, have a powerful influence on the position and rate of lithiation of aromatic compounds.[47] Some substituents, such as *t*-butoxycarbonylamido and carboxy, undergo deprotonation during the lithiation process.[48] The methoxymethoxy substituent is particularly useful among the alkoxy directing groups. It can provide selective lithiation and, being an acetal, is readily removed by hydrolysis.[49] In heteroaromatic compounds the preferred site for lithiation is usually adjacent to the heteroatom.

The features that characterize the activating groups include an electron pair that can coordinate lithium and polarity that can stabilize the anionic character.[50] Geometric factors are also important. For amido groups, for example, it has been deduced by comparison of various cyclic systems that the preferred geometry is for the activating amide group to be coplanar with the position of lithiation.[51] If competing nucleophilic attack is a possibility, as in tertiary amides, steric bulk is also an important factor. Consistent with the importance of polar and electrostatic effects in lithiation, a fluoro substituent is a good directing substituent. Amide bases such as LDA and LTMP give better results than alkyllithium reagents. With these bases, fluorine was found to promote *ortho* lithiation selectively over such directing groups as methoxy and diethylaminocarbonyloxy.[52]

[46] R. D. Clark and A. Jahangir, *Org. React.*, **47**, 1 (1995).

[47] D. W. Slocum and C. A. Jennings, *J. Org. Chem.*, **41**, 3653 (1976); J. M. Mallan and R. C. Rebb, *Chem. Rev.*, **69**, 693 (1969); H. W. Gschwend and H. R. Rodriguez, *Org. React.*, **26**, 1 (1979); V. Snieckus, *Chem. Rev.*, **90**, 879 (1990); C. Quesnelle, T. Iihama, T. Aubert, H. Perrier, and V. Snieckus, *Tetrahedron Lett.*, **33**, 2625 (1992); M. Iwao, T. Iihama, K. K. Mahalandabis, H. Perrier, and V. Snieckus, *J. Org. Chem.*, **54**, 24 (1989); L. A. Spangler, *Tetrahedron Lett.*, **37**, 3639 (1996).

[48] J. M. Muchowski and M. C. Venuti, *J. Org. Chem.*, **45**, 4798 (1980); P. Stanetty, H. Koller, and M. Mihovilovic, *J. Org. Chem.*, **57**, 6833 (1992); J. Mortier, J. Moyroud, B. Benneteau, and P.A. Cain, *J. Org. Chem.*, **59**, 4042 (1994).

[49] C. A. Townsend and L. M. Bloom, *Tetrahedron Lett.*, **22**, 3923 (1981); R. C. Ronald and M. R. Winkle, *Tetrahedron*, **39**, 2031 (1983); M. R. Winkle and R. C. Ronald, *J. Org. Chem.*, **47**, 2101 (1982).

[50] (a) N. J. R. van Eikema Hommes and P. v. R. Schleyer, *Angew. Chem. Int. Ed. Engl.*, **31**, 755 (1992); (b) N. J. R. van Eikema Hommes and P. v. R. Schleyer, *Tetrahedron*, **50**, 5903 (1994).

[51] P. Beak, S. T. Kerrick, and D. J. Gallagher, *J. Am. Chem. Soc.*, **115**, 10628 (1993).

[52] A. J. Bridges, A. Lee, E. C. Maduakor, and C. E. Schwartz, *Tetrahedron Lett.*, **33**, 7495 (1992); D. C. Furlano, S. N. Calderon, G. Chen, and K. L. Kirk, *J. Org. Chem.*, **53**, 3145 (1988).

Scheme 7.1 gives some examples of the preparation of organolithium compounds by lithiation. A variety of directing groups is represented, including methoxy (Entry 1), diethylaminocarbonyl (Entry 2), *N,N*-dimethylimidazolinyl (Entry 3), *t*-butoxycarbonylamido (Entry 4), carboxy (Entry 5), and neopentoxycarbonyl (Entry 6). In the latter case, LDA is used as the base to avoid nucleophilic addition to the carbonyl group. The tri-*i*-propyl borate serves to trap the lithiation product as it is formed and prevent further reactions with the ester carbonyl. Entry 7 is a typical lithiation of a heteroaromatic molecule, and Entry 8 shows the lithiation of methyl vinyl ether. The latter reaction is dependent on the coordination and polar effect of the methoxy group and the relative acidity of the sp^2 C—H bond. Entry 9 is an allylic lithiation, promoted by the trimethylsiloxy group. Entry 10 is an interesting lithiation of an epoxide. The silyl substituent also has a modest stabilizing effect (see Part A, Section 3.4.2).

Reaction conditions can be modified to accelerate the rate of lithiation when necessary. Addition of tertiary amines, especially TMEDA, facilitates lithiation[53] by coordination at the lithium and promoting dissociation of aggregated structures. Kinetic and spectroscopic evidence indicates that in the presence of TMEDA lithiation of methoxybenzene involves the solvated dimeric species $(BuLi)_2(TMEDA)_2$.[54] The reaction shows an isotope effect for the *o*-hydrogen, establishing that proton abstraction is rate determining.[55] It is likely that there is a precomplexation between the methoxybenzene and organometallic dimer.

The lithiation process has been modeled by MP2/6-31+G* calculations. The TSs for lithiation of fluorobenzene and methoxybenzene have lithium nearly in the aromatic plane and coordinated to the directing group as shown in Figure 7.3.[56] Although these structures represent lithiations as occurring through a monomeric species, similar effects are present in dimers or aggregates.[50b] There is a considerable electrostatic component to the stabilization of the TS.[50a] It has also been pointed out that the coordination of the Lewis acid Li^+ at the methoxy or fluorine group decreases the π-donor capacity of the groups and accentuates their σ-EWG capacity. The combination of these interactions is responsible for the activating effects of these groups.

π-donor capacity is reduced

Lithiation of alkyl groups is also possible and again a combination of donor chelation and polar stabilization of anionic character is required. Amides and carbamates can be lithiated α to the nitrogen.

53. G. G. Eberhardt and W. A. Butte, *J. Org. Chem.*, **29**, 2928 (1964); R. West and P. C. Jones, *J. Am. Chem. Soc.*, **90**, 2656 (1968); S. Akiyama and J. Hooz, *Tetrahedron Lett.*, 4115 (1973); D. W. Slocum, R. Moon, J. Thompson, D. S. Coffey, J. D. Li, M. G. Slocum, A. Siegel, and R. Gayton-Garcia, *Tetrahedron Lett.*, **35**, 385 (1994); M. Khaldi, F. Chretien, and Y. Chapleur, *Tetrahedron Lett.*, **35**, 401 (1994); D. B. Collum, *Acc. Chem. Res.*, **25**, 448 (1992).

54. R. A. Rennels, A. J. Maliakal, and D. B. Collum, *J. Am. Chem. Soc.*, 120, 421 (1998).

55. M. Stratakis, *J. Org. Chem.*, **62**, 3024 (1997).

56. J. M. Saa, *Helv. Chim. Acta*, **85**, 814 (2002).

629

SECTION 7.1

*Preparation and
Properties of
Organomagnesium
and Organolithium
Reagents*

Scheme 7.1. Preparation of Organolithium Compounds by Metallation

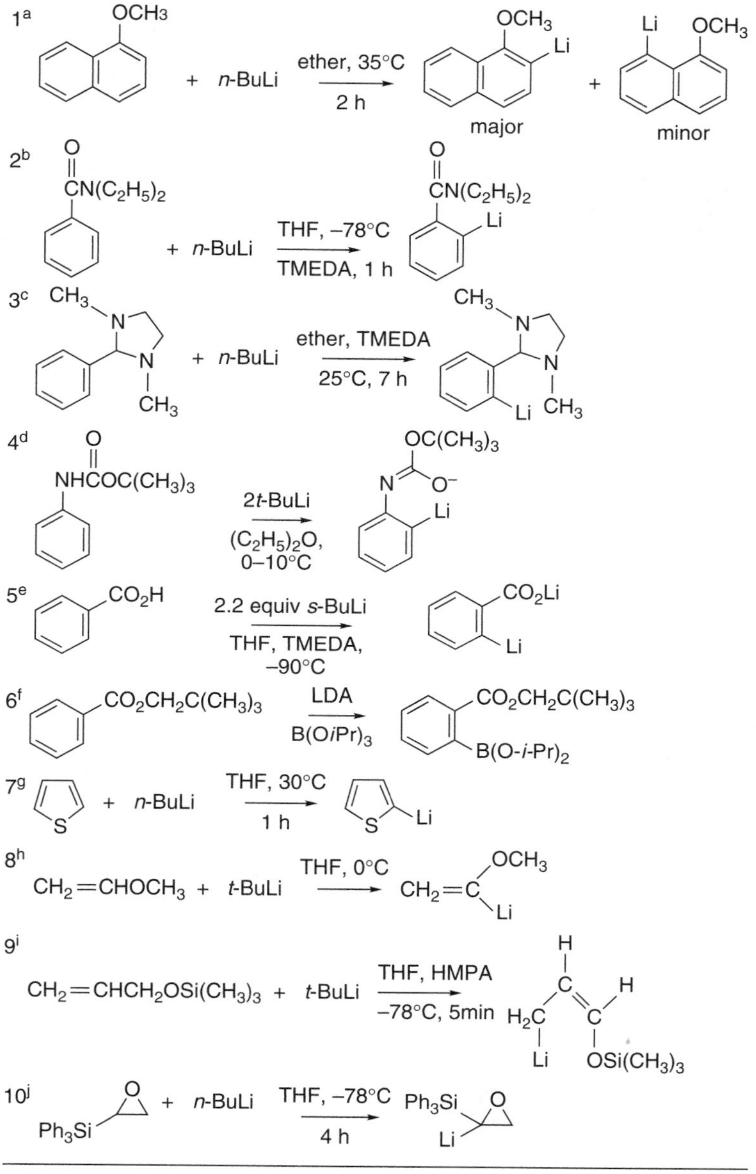

a. B. M. Graybill and D. A. Shirley, *J. Org. Chem.*, **31**, 1221 (1966).
b. P. A. Beak and R. A. Brown, *J. Org. Chem.*, **42**, 1823 (1977); *J. Org. Chem.*, **44**, 4463 (1979).
c. T. D. Harris and G. P. Roth, *J. Org. Chem.*, **44**, 2004 (1979).
d. P. Stanetty, H. Koller, and M. Mihovilovic, *J. Org. Chem.*, **57**, 6833 (1992).
e. B. Bennetau, J. Mortier, J. Moyroud, and J.-L. Guesnet, *J. Chem. Soc., Perkin Trans. 1*, 1265 (1995).
f. S. Caron and J. M. Hawkins, *J. Org. Chem.*, **63**, 2054 (1998).
g. E. Jones and I. M. Moodie, *Org. Synth.*, **50**, 104 (1970).
h. J. E. Baldwin, G. A. Hofle, and O. W. Lever, Jr., *J. Am. Chem. Soc.*, **96**, 7125 (1974).
i. W. C. Still and T. L. Macdonald, *J. Org. Chem.*, **41**, 3620 (1976).
j. J. J. Eisch and J. E. Galle, *J. Am. Chem. Soc.*, **98**, 4646 (1976).

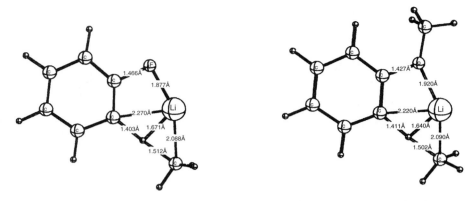

Fig. 7.3. Transition structures for lithiation of fluorobenzene (left) and methoxybenzene (right). Reproduced from *Tetrahedron*, **50**, 5903 (1994), by permission of Elsevier.

CH$_3$CH$_2$NCO$_2$C(CH$_3$)$_3$ 1) *s*-BuLi, TMEDA   CH$_3$CH$_2$NCO$_2$C(CH$_3$)$_3$
|
CH$_3$ 2) (CH$_3$)$_3$SiCl   CH$_2$Si(CH$_3$)$_3$

Ref. 57

1) *s*-BuLi, TMEDA
2) E$^+$

Ref. 58

Studies with bicyclic carbamates of general structure **1** indicated that proximity and alignment of the carbonyl oxygen to the lithiation site is a major factor in determining the rate of lithiation.[59]

1

n = 1,2,3

Bicyclic structures of this type are more reactive than monocyclic or acyclic carbamates, indicating that a relatively rigid orientation of the carbonyl group is favorable to lithiation. Substituted formamidines can also be lithiated.[60]

t-BuLi

57. V. Snieckus, M. Rogers-Evans, P. Beak, W. K. Lee, E. K. Yum, and J. Freskos, *Tetrahedron Lett.*, **35**, 4067 (1994).
58. P. Beak and W. K. Lee, *J. Org. Chem.*, **58**, 1109 (1993).
59. K. M. B. Gross and P. Beak, *J. Am. Chem. Soc.*, **123**, 315 (2001).
60. A. I. Meyers and G. Milot, *J. Am. Chem. Soc.*, **115**, 6652 (1993).

Tertiary amides with carbanion stabilization at the β-carbon give β-lithiation.[61]

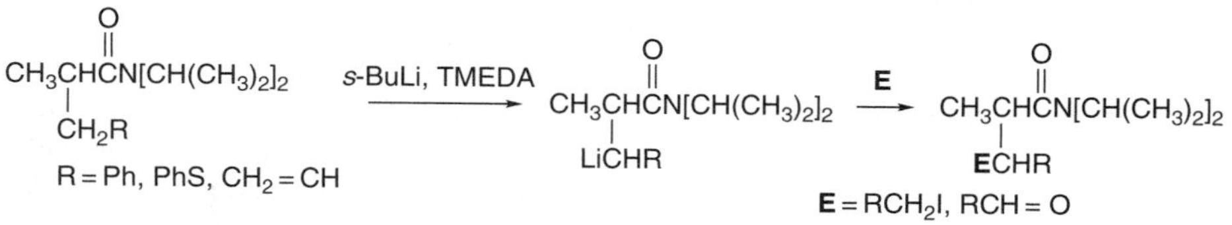

β-Lithiation has also been observed for deprotonated secondary amides of 3-phenylpropanoic acid.

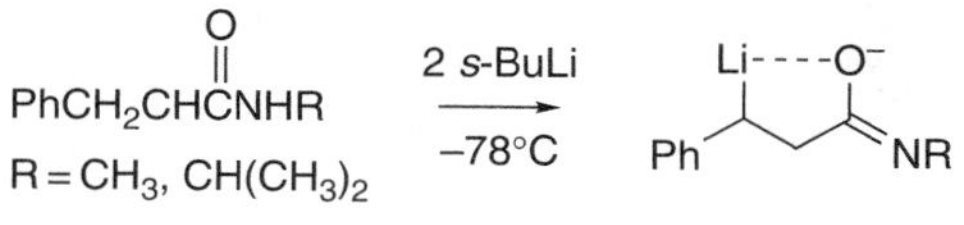

Ref. 62

As with aromatic lithiation, the mechanism of directed lithiation in these systems appears to involve an association between the activating substituent and the lithiating agent.[63]

Alkenyllithium compounds are intermediates in the *Shapiro reaction*, which is discussed in Section 5.7.2. The reaction can be run in such a way that the organolithium compound is generated in high yield and subsequently allowed to react with a variety of electrophiles.[64] This method provides a route to vinyllithium compounds starting from a ketone.

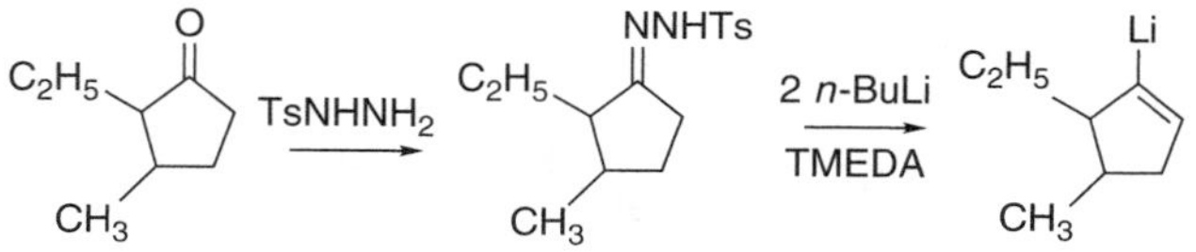

Ref. 65

Hydrocarbons lacking directing substituents are not very reactive toward metallation, but it has been found that a mixture of *n*-butyllithium and potassium *t*-butoxide[66] is sufficiently reactive to give allyl anions from alkenes such as isobutene.[67]

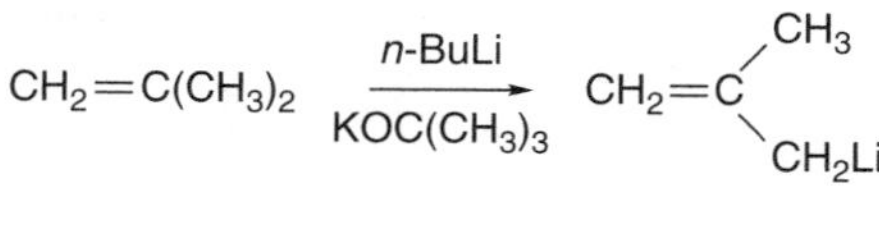

[61] P. Beak, J. E. Hunter, Y. M. Jun, and A. P. Wallin, *J. Am. Chem. Soc.*, **109**, 5403 (1987); G. P. Lutz, A. P. Wallin, S. T. Kerrick, and P. Beak, *J. Org. Chem.*, **56**, 4938 (1991).

[62] G. P. Lutz, H. Du, D. J. Gallagher, and P. Beak, *J. Org. Chem.*, **61**, 4542 (1996).

[63] W. Bauer and P. v. R. Schleyer, *J. Am. Chem. Soc.*, **111**, 7191 (1989); P. Beak, S. T. Kerrick, and D. J. Gallagher, *J. Am. Chem. Soc.*, **115**, 10628 (1993).

[64] F. T. Bond and R. A. DiPietro, *J. Org. Chem.*, **46**, 1315 (1981); T. H. Chan, A. Baldassarre, and D. Massuda, *Synthesis*, 801 (1976); B. M. Trost and T. N. Nanninga, *J. Am. Chem. Soc.*, **107**, 1293 (1985).

[65] W. Barth and L. A. Paquette, *J. Org. Chem.*, **50**, 2438 (1985).

[66] L. Lochmann, J. Pospisil, and D. Lim, *Tetrahedron Lett.*, 257 (1966).

[67] M. Schlosser and J. Hartmann, *Angew. Chem. Int. Ed. Engl.*, **12**, 508 (1973); J. J. Bahl, R. B. Bates, and B. Gordon, III, *J. Org. Chem.*, **44**, 2290 (1979); M. Schlosser and G. Rauchshwalbe, *J. Am. Chem. Soc.*, **100**, 3258 (1978).

7.1.2.3. Preparation by Halogen-Metal Exchange. Halogen-metal exchange is another important method for preparation of organolithium reagents. The reaction proceeds in the direction of forming the more stable organolithium reagent, that is, the one derived from the more acidic organic compound. Thus, by use of the very basic organolithium compounds *n*-butyl- or *t*-butyllithium, halogen substituents at more acidic sp^2 carbons are readily exchanged to give the corresponding lithium compound. Halogen-metal exchange is particularly useful for converting aryl and alkenyl halides to the corresponding lithium compounds.

Ref. 68

Ref. 69

Halogen-metal exchange is a very fast reaction and is usually carried out at -60 to $-120°$ C. This makes it possible to prepare aryllithium compounds containing functional groups, such as cyano and nitro, that react under the conditions required for preparation from lithium metal. Halogen-metal exchange is restricted for alkyl halides by competing reactions, but primary alkyllithium reagents can be prepared from iodides under carefully controlled conditions.[70]

Retention of configuration is sometimes observed when organolithium compounds are prepared by halogen-metal exchange. The degree of retention is low for exchange of most alkyl systems,[71] but it is normally high for cyclopropyl and vinyl halides.[72] Once formed, both cyclopropyl and vinyllithium reagents retain their configuration at room temperature.

Scheme 7.2 gives some examples of preparation of organolithium compounds by halogen-metal exchange. Entries 1, 2, and 3 are representative low-temperature preparations of alkenyllithium reagents. Entry 4 involves a cyclopropyl bromide. Both the *cis* and *trans* isomers react with retention of configuration. In Entries 1, 3, and 4, two equivalents of *t*-butyllithium are required because the *t*-butyl halide formed by exchange consumes one equivalent. Entry 5 is an example of retention of configuration at a double bond. Entries 6 and 7 show aryl bromides with functional groups that

68. N. Neumann and D. Seebach, *Tetrahedron Lett.*, 4839 (1976).
69. T. R. Hoye, S. J. Martin, and D. R. Peck, *J. Org. Chem.*, **47**, 331 (1982).
70. W. F. Bailey and E. R. Punzalan, *J. Org. Chem.*, **55**, 5404 (1990); E. Negishi, D. R. Swanson, and C. J. Rousset, *J. Org. Chem.*, **55**, 5406 (1990).
71. R. L. Letsinger, *J. Am. Chem. Soc.*, **72**, 4842 (1950); D. Y. Curtin and W. J. Koehl, Jr., *J. Am. Chem. Soc.*, **84**, 1967 (1962).
72. H. M. Walborsky, F. J. Impastato, and A. E. Young, *J. Am. Chem. Soc.*, **86**, 3283 (1964); D. Seyferth and L. G. Vaughan, *J. Am. Chem. Soc.*, **86**, 883 (1964); M. J. S. Dewar and J. M. Harris, *J. Am. Chem. Soc.*, **91**, 3652 (1969); E. J. Corey and P. Ulrich, *Tetrahedron Lett.*, 3685 (1975); N. Neumann and D. Seebach, *Tetrahedron Lett.*, 4839 (1976); R. B. Miller and G. McGarvey, *J. Org. Chem.*, **44**, 4623 (1979).

Scheme 7.2. Preparation of Organolithium Reagents by Halogen-Metal Exchange

1[a]

2[b]

3[c]

4[d]

5[e]

6[f]

7[g]

8[h]

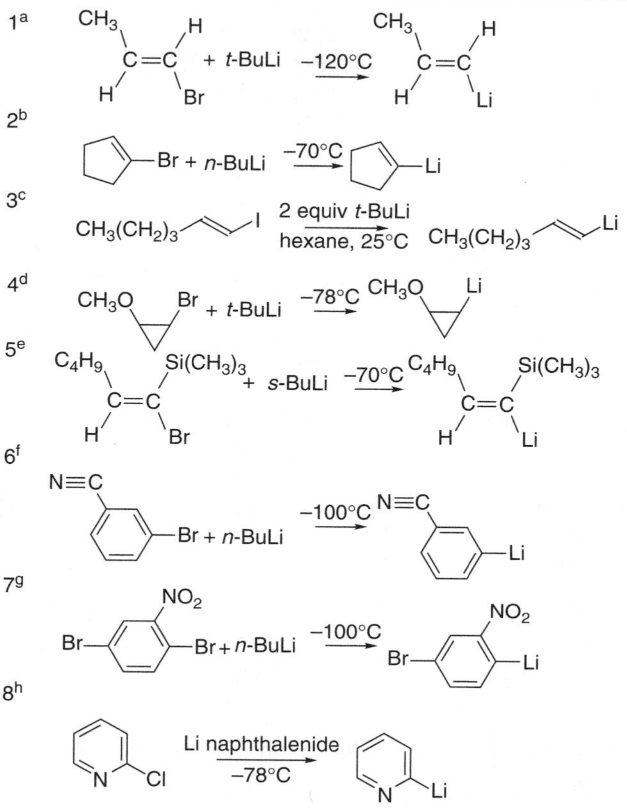

a. H. Neuman and D. Seebach, *Tetrahedron Lett.*, 4839 (1976).
b. J. Milton, R. Lorne, and G. Linsturmelle, *Synthesis*, 434 (1975).
c. M. A. Peterson and R. Polt, *Synth. Commun.* **22**, 477 (1992).
d. E. J. Corey and P. Ulrich, *Tetrahedron Lett.*, 3685 (1975).
e. R. B. Miller and G. McGarvey, *J. Org. Chem.*, **44**, 4623 (1979).
f. W. E. Parham and L. D. Jones, *J. Org. Chem.*, **41**, 1187 (1976).
g. W. E. Parham and R. M. Piccirilli, *J. Org. Chem.*, **42**, 257 (1977).
h. Y. Kondo, N. Murata, and T. Sakamoto, *Heterocycles*, **37**, 1467 (1994).

are reactive toward organometallic compounds at higher temperature, but which can undergo the halogen-metal reaction successfully at low temperature. Entry 8 is an example of the use of lithium naphthalenide for halogen-metal exchange.

7.1.2.4. Preparation by Metal-Metal Exchange. A third useful method of preparing organolithium reagents involves *metal-metal exchange* or *transmetallation*. The reaction between two organometallic compounds proceeds in the direction of placing the more electropositive metal at the more acidic carbon position. Exchanges between organotin reagents and alkyllithium reagents are particularly significant from a synthetic point of view. Terminal alkenyllithium compounds can be made from vinylstannanes, which are available by addition of stannanes to terminal alkynes (see Section 9.3.1).

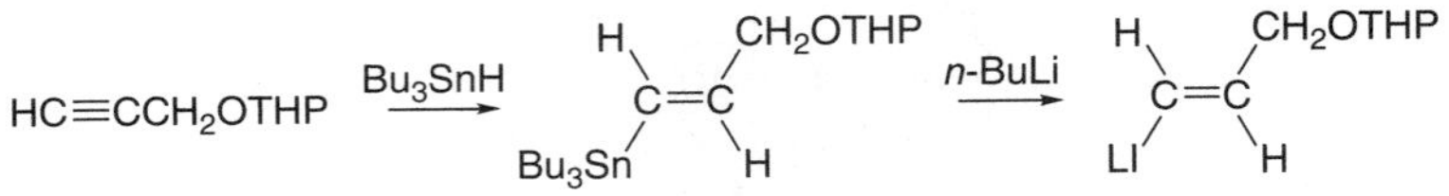

$$HC{\equiv}CCH_2OTHP \xrightarrow{Bu_3SnH} \underset{Bu_3Sn}{\overset{H}{\underset{\diagdown}{}}}C{=}C\underset{H}{\overset{CH_2OTHP}{\diagup}} \xrightarrow{n\text{-BuLi}} \underset{Li}{\overset{H}{\underset{\diagdown}{}}}C{=}C\underset{H}{\overset{CH_2OTHP}{\diagup}}$$

Ref. 73

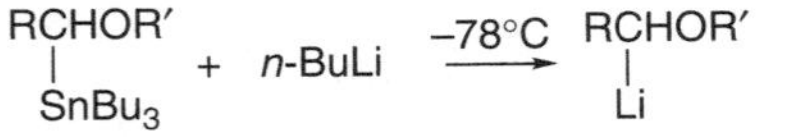

$$\underset{SnBu_3}{\overset{RCHOR'}{\underset{|}{|}}} + n\text{-BuLi} \xrightarrow{-78°C} \underset{Li}{\overset{RCHOR'}{\underset{|}{|}}}$$

Ref. 74

$$R_2NCH_2SnBu_3 + n\text{-BuLi} \xrightarrow{0°C} R_2NCH_2Li$$

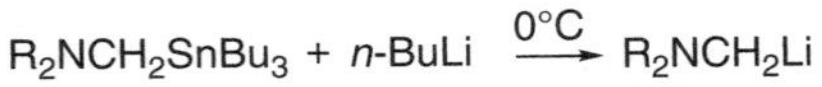

Ref. 75

The α-tri-*n*-butylstannyl derivatives needed for the latter two examples are readily available.

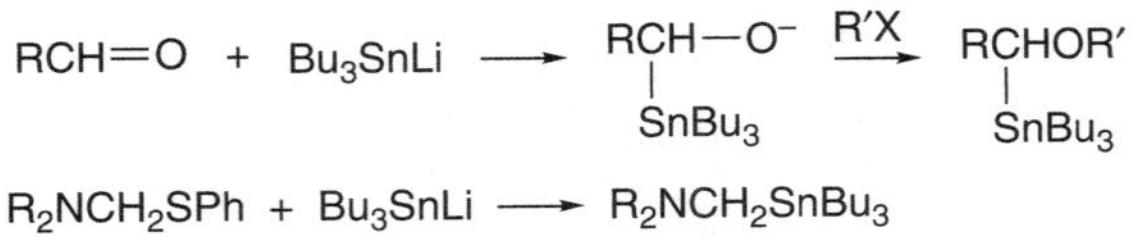

$$RCH{=}O + Bu_3SnLi \longrightarrow \underset{SnBu_3}{\overset{RCH-O^-}{\underset{|}{|}}} \xrightarrow{R'X} \underset{SnBu_3}{\overset{RCHOR'}{\underset{|}{|}}}$$

$$R_2NCH_2SPh + Bu_3SnLi \longrightarrow R_2NCH_2SnBu_3$$

The exchange reactions of α-alkoxystannanes occur with retention of configuration at the carbon-metal bond.[76]

$$RCH_2\underset{SnBu_3}{\overset{H}{\diagup}}\text{OCH}_2OR' \xrightarrow{RLi} RCH_2\underset{Li}{\overset{H}{\diagup}}\text{OCH}_2OR'$$

7.2. Reactions of Organomagnesium and Organolithium Compounds

7.2.1. Reactions with Alkylating Agents

Organomagnesium and organolithium compounds are strongly basic and nucleophilic. Despite their potential to react as nucleophiles in S_N2 substitution reactions, this reaction is of limited utility in synthesis. One limitation on alkylation reactions is competition from electron transfer processes, which can lead to radical reactions. Methyl and other primary iodides usually give the best results in alkylation reactions.

73. E. J. Corey and R. H. Wollenberg, *J. Org. Chem.*, **40**, 2265 (1975).

74. W. C. Still, *J. Am. Chem. Soc.*, **100** 1481 (1978).

75. D. J. Peterson, *J. Am. Chem. Soc.*, **93**, 4027 (1971).

76. W. C. Still and C. Sreekumar, *J. Am. Chem. Soc.*, **102**, 1201 (1980); J. S. Sawyer, A. Kucerovy, T. L. Macdonald, and G. J. McGarvey, *J. Am. Chem. Soc.*, **110**, 842 (1988).

635

SECTION 7.2

*Reactions of
Organomagnesium and
Organolithium
Compounds*

HMPA can accelerate the reaction and improve yields when electron transfer is a complication.[77]

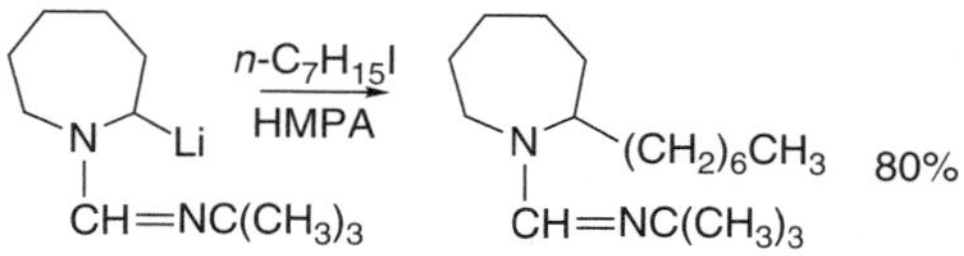

Organolithium reagents in which the carbanion is delocalized are more useful than alkyllithium reagents in alkylation reactions. Allyllithium and benzyllithium reagents can be alkylated and with secondary alkyl bromides and a high degree of inversion of configuration is observed.[78]

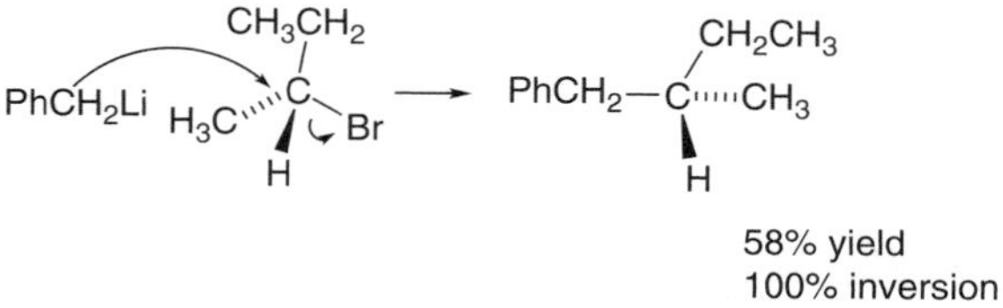

58% yield
100% inversion

Alkenyllithium reagents can be alkylated in good yields by alkyl iodides and bromides.[79]

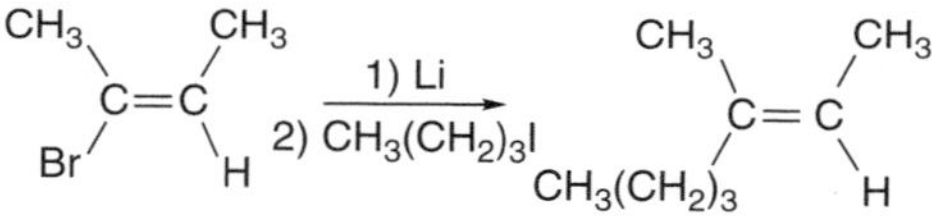

The reactions of aryllithium reagents are accelerated by inclusion of potassium alkoxides.[80]

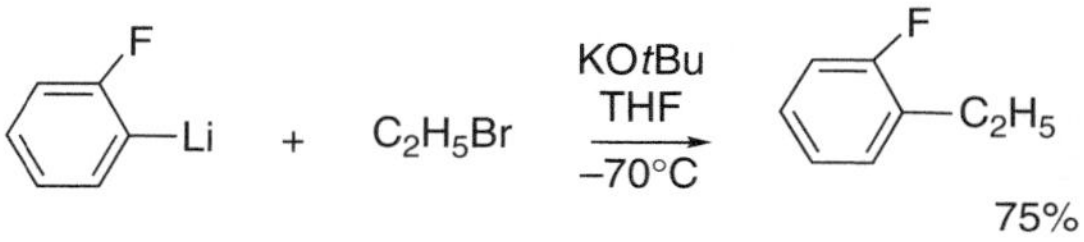

75%

Alkylation by allylic halides is usually a satisfactory reaction, and in this case the reaction may proceed through a cyclic mechanism.[81] For example, when 1-^{14}C-allyl chloride reacts with phenyllithium, about three-fourths of the product has the labeled carbon at the terminal methylene group.

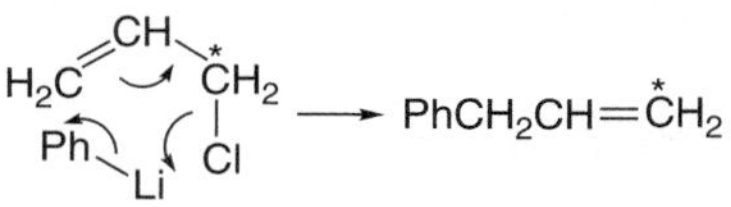

$PhCH_2CH=\overset{*}{C}H_2$

77. A. I. Meyers, P. D. Edwards, W. F. Rieker, and T. R. Bailey, *J. Am. Chem. Soc.*, **106**, 3270 (1984); A. I. Meyers and G. Milot, *J. Am. Chem. Soc.*, **115**, 6652 (1993).
78. L. H. Sommer and W. D. Korte, *J. Org. Chem.*, **35**, 22 (1970).
79. J. Millon, R. Lorne, and G. Linstrumelle, *Synthesis*, 434 (1975).
80. L. Brandsma, A. G. Mal'kina, L. Lochmann, and P. v. R. Schleyer, *Rec. Trav. Chim. Pays-Bas*, **113**, 529 (1994); L. Lochmann and J. Trekoval, *Coll. Czech. Chem. Commun.*, **51**, 1439 (1986).
81. R. M. Magid and J. G. Welch, *J. Am. Chem. Soc.*, **90**, 5211 (1968); R. M. Magid, E. C. Nieh, and R. D. Gandour, *J. Org. Chem.*, **36**, 2099 (1971); R. M. Magid and E. C. Nieh, *J. Org. Chem.*, **36**, 2105 (1971).

Coupling of certain lithiated reagents with aryl and vinyl halides is also possible.[82] These reactions probably proceed by a fast halogen-lithium exchange, generating the alkyl halide, which then undergoes substitution. This reaction has been applied to β-lithiobenzamides.[83]

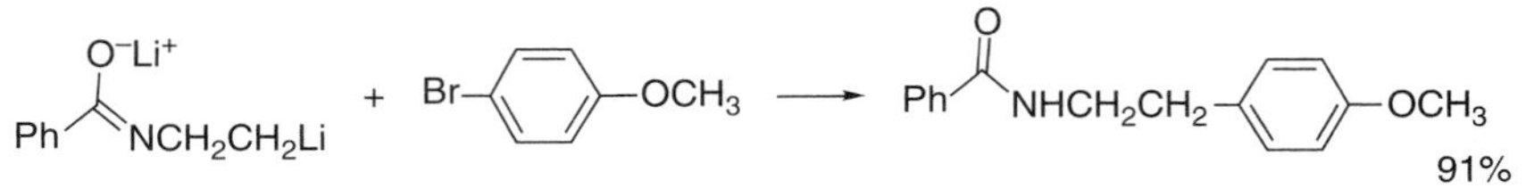

91%

Intramolecular reactions are useful for forming small rings. The reaction of 1,3-, 1,4- , and 1,5-diiodides with *t*-butyllithium is an effective means of ring closure, but 1,6-diiodides give very little cyclization.[84]

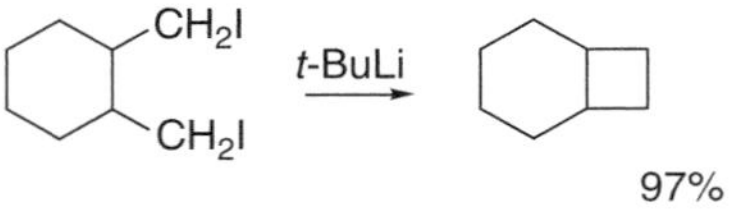

97%

Functionalized organolithium reagents can be prepared and alkylated. The configuration of the dioxanyl reagent **2** proved to be subject to control.[85] The kinetically favored *trans* lithio derivative is converted to the more stable *cis* isomer at 20° C. Both isomers were methylated with *retention* of configuration at saturated carbon.

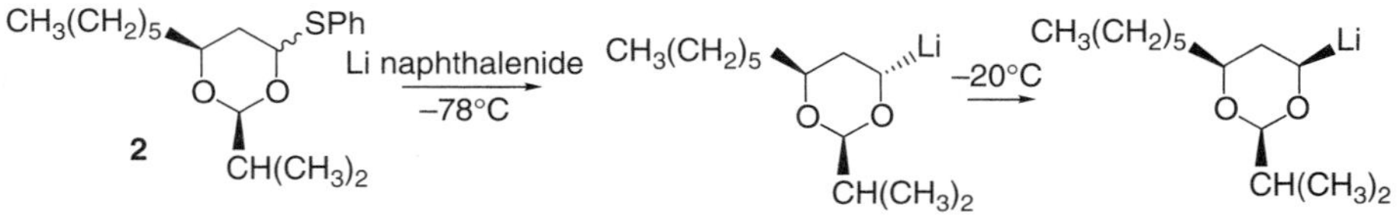

Both trialkylsilyl and trialkylstannyl halides usually give high yields of substitution products with organolithium reagents, and this is an important route to silanes and stannanes (see Section 9.2.1 and 9.3.1).

Grignard reagents are somewhat less reactive toward alkylation but can be of synthetic value, especially when methyl, allyl, or benzyl halides are involved.

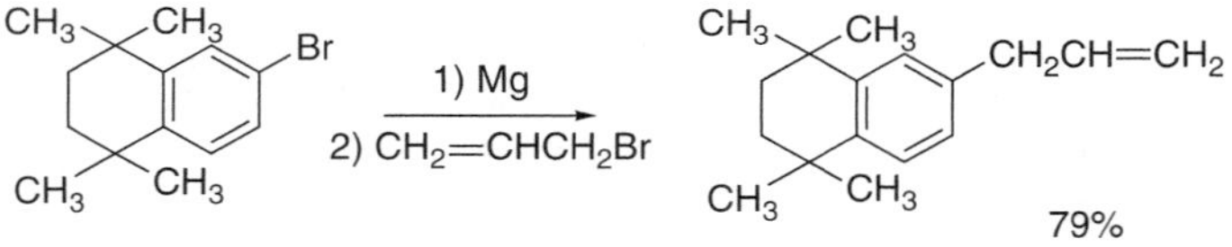

79%   Ref. 86

Synthetically useful alkylation of Grignard reagents can also be carried out with alkyl sulfonates and sulfates.

$$PhCH_2MgCl + CH_3CH_2CH_2CH_2OSO_2C_7H_7 \longrightarrow PhCH_2CH_2CH_2CH_2CH_3$$

50–59%

Ref. 87

82. R. E. Merrill and E. Negishi, *J. Org. Chem.*, **39**, 3452 (1974).
83. J. Barluenga, J. M. Montserrat, and J. Florez, *J. Org. Chem.*, **58**, 5976 (1993).
84. W. F. Bailey, R. P. Gagnier, and J. J. Patricia, *J. Org. Chem.*, **49**, 2098 (1984).
85. S. D. Rychnovsky and D. J. Skalitzky, *J. Org. Chem.*, **57**, 4336 (1992).
86. J. Eustache, J.-M. Barnardon, and B. Shroot, *Tetrahedron Lett.*, **28**, 4681 (1987).
87. H. Gilman and J. Robinson, *Org. Synth.*, **II**, 47 (1943).

637

SECTION 7.2

*Reactions of
Organomagnesium and
Organolithium
Compounds*

CH₃—[ring with CH₃ groups]—MgBr + (CH₃O)₂SO₂ ⟶ CH₃—[ring with CH₃ groups]—CH₃ 52–60%   Ref. 88

7.2.2. Reactions with Carbonyl Compounds

7.2.2.1. Reactions of Grignard Reagents. The most important reactions of Grignard reagents for synthesis involve addition to carbonyl groups. The TS for addition of Grignard reagents is often represented as a cyclic array containing the carbonyl group and two molecules of the Grignard reagent. There is considerable evidence favoring this mechanism involving a termolecular complex.[89]

When the carbonyl carbon is substituted with a potential leaving group, the tetrahedral adduct can break down to regenerate a C=O bond and a second addition step can occur. Esters, for example are usually converted to tertiary alcohols, rather than ketones, in reactions with Grignard reagents.

$$RMgX + R'\overset{O}{\overset{\|}{C}}OR'' \longrightarrow R-\overset{OMgX}{\underset{R'}{\overset{|}{C}}}-OR'' \longrightarrow R\overset{O}{\overset{\|}{C}}R' + R''OMgX$$

$$R\overset{O}{\overset{\|}{C}}R' + RMgX \xrightarrow{fast} R_2\overset{OMgX}{\overset{|}{C}}R' \xrightarrow{H_2O} R_2\overset{OH}{\overset{|}{C}}R'$$

Grignard reagents add to nitriles and, after hydrolysis of the reaction mixture, a ketone is obtained, with hydrocarbons being the preferred solvent for this reaction.[90]

$$RMgX + R'C\equiv N \longrightarrow R\overset{NMgX}{\overset{\|}{C}}R' \xrightarrow{H_2O} R\overset{O}{\overset{\|}{C}}R'$$

Ketones can also be prepared from acyl chlorides by reaction at low temperature using an excess of acyl chloride. Tetrahydrofuran is the preferred solvent.[91] The reaction conditions must be carefully controlled to prevent formation of tertiary alcohol by addition of a Grignard reagent to the ketone as it is formed.

88. L. I. Smith, *Org. Synth.*, **II**, 360 (1943).
89. E. C. Ashby, R. B. Duke, and H. M. Neuman, *J. Am. Chem. Soc.*, **89**, 1964 (1967); E. C. Ashby, *Pure Appl. Chem.*, **52**, 545 (1980).
90. P. Canonne, G. B. Foscolos, and G. Lemay, *Tetrahedron Lett.*, **21**, 155 (1980).
91. F. Sato, M. Inoue, K. Oguro, and M. Sato, *Tetrahedron Lett.*, 4303 (1979).

$$CH_3(CH_2)_5MgBr + CH_3CH_2CH_2\overset{\overset{\displaystyle O}{\|}}{C}Cl \xrightarrow{-30°C} CH_3(CH_2)_5\overset{\overset{\displaystyle O}{\|}}{C}(CH_2)_2CH_3$$

$$92\%$$

2-Pyridinethiolate esters, which are easily prepared from acyl chlorides, also react with Grignard reagents to give ketones (see Entry 6 in Scheme 7.3).[92] *N*-Methoxy-*N*-methylamides are also converted to ketones by Grignard reagents (see Entries 17 and 18).

Aldehydes can be obtained by reaction of Grignard reagents with triethyl orthoformate. The addition step is preceded by elimination of one of the alkoxy groups to generate an electrophilic oxonium ion. The elimination is promoted by the magnesium ion acting as a Lewis acid.[93] The acetals formed by the addition are stable to the reaction conditions, but are hydrolyzed to aldehydes by aqueous acid.

$$\begin{array}{c} C_2H_5O \\ | \\ H-C-OC_2H_5 \\ | \\ C_2H_5O \end{array} \xrightarrow{\overset{R}{\underset{}{Mg-X}}} \begin{array}{c} OC_2H_5 \\ HC{+} \\ OC_2H_5 \end{array} \xrightarrow{RMgX} \begin{array}{c} OC_2H_5 \\ | \\ RCH \\ | \\ OC_2H_5 \end{array} \xrightarrow[H_2O]{H^+} RCH=O$$

$$+ \ C_2H_5OMgR + X^-$$

Aldehydes can also be obtained from Grignard reagents by reaction with formamides, such as *N*-formylpiperidine. In this case, the initial adducts are stable and the aldehyde is not formed until hydrolysis during workup.

$$PhCH_2CH_2MgCl + H\overset{\overset{\displaystyle O}{\|}}{C}-N\bigcirc \longrightarrow \xrightarrow{H_2O} PhCH_2CH_2CH=O$$

$$66-76\%$$

Ref. 94

The addition of Grignard reagents to aldehydes, ketones, and esters is the basis for the synthesis of a wide variety of alcohols, and several examples are given in Scheme 7.3. Primary alcohols can be made from formaldehyde (Entry 1) or, with addition of two carbons, from ethylene oxide (Entry 2). Secondary alcohols are obtained from aldehydes (Entries 3 to 6) or formate esters (Entry 7). Tertiary alcohols can be made from esters (Entries 8 and 9) or ketones (Entry 10). Lactones give diols (Entry 11). Aldehydes can be prepared from trialkyl orthoformate esters (Entries 12 and 13). Ketones can be made from nitriles (Entries 14 and 15), pyridine-2-thiol esters (Entry 16), *N*-methoxy-*N*-methyl carboxamides (Entries 17 and 18), or anhydrides (Entry 19). Carboxylic acids are available by reaction with CO_2 (Entries 20 to 22). Amines can be prepared from imines (Entry 23). Two-step procedures that involve formation and dehydration of alcohols provide routes to certain alkenes (Entries 24 and 25).

92. T. Mukaiyama, M. Araki, and H. Takei, *J. Am. Chem. Soc.*, **95**, 4763 (1973); M. Araki, S. Sakata, H. Takai, and T. Mukaiyama, *Bull. Chem. Soc. Jpn.*, **47**, 1777 (1974).

93. E. L. Eliel and F. W. Nader, *J. Am. Chem. Soc.*, **92**, 584 (1970).

94. G. A. Olah and M. Arvanaghi, *Org. Synth.*, **64**, 114 (1985); G. A. Olah, G. K. S. Prakash, and M. Arvanaghi, *Synthesis*, 228 (1984).

Scheme 7.3. Synthetic Procedures Involving Grignard Reagents.

A. Primary alcohols from formaldehyde

1[a]

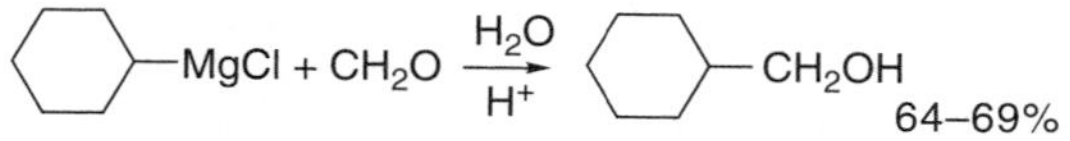

64–69%

B. Primary alcohols from ethylene oxide

2[b] $CH_3(CH_2)_3MgBr + H_2C\!-\!CH_2 \longrightarrow \xrightarrow[H^+]{H_2O} CH_3(CH_2)_5OH$

60–62%

C. Secondary alcohols from aldehydes

3[c] $PhCH\!=\!CHCH\!=\!O + HC\!\equiv\!CMgBr \longrightarrow \xrightarrow[H^+]{H_2O} HC\!\equiv\!CCHCH\!=\!CHPh$

58–69%

4[d]

82–85%

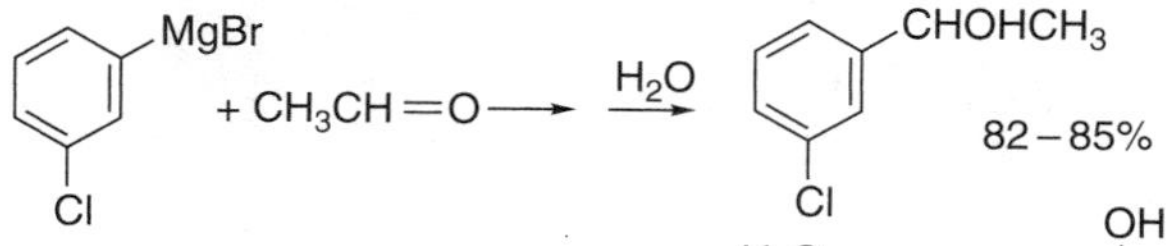

5[e] $CH_3CH\!=\!CHCH\!=\!O + CH_3MgCl \longrightarrow \xrightarrow{H_2O} CH_3CH\!=\!CHCHCH_3$

81–86%

6[f] $(CH_3)_2CHMgBr + CH_3CH\!=\!O \longrightarrow (CH_3)_2CHCHCH_3$

53–54%

D. Secondary alcohols from formate esters

7[g] $2\ CH_3(CH_2)_3MgBr + HCO_2C_2H_5 \longrightarrow \xrightarrow[H^+]{H_2O} (CH_3CH_2CH_2CH_2)_2CHOH$

83–85%

E. Tertiary alcohols from ketones, esters, and lactones

8[h] $3\ C_2H_5MgBr + (C_2H_5O)_2CO \longrightarrow \xrightarrow[NH_4Cl]{H_2O} (C_2H_5)_3COH$

82–88%

9[i] $2\ PhMgBr + PhCO_2C_2H_5 \longrightarrow \xrightarrow{H_2O} Ph_3COH$

89–93%

10[j]

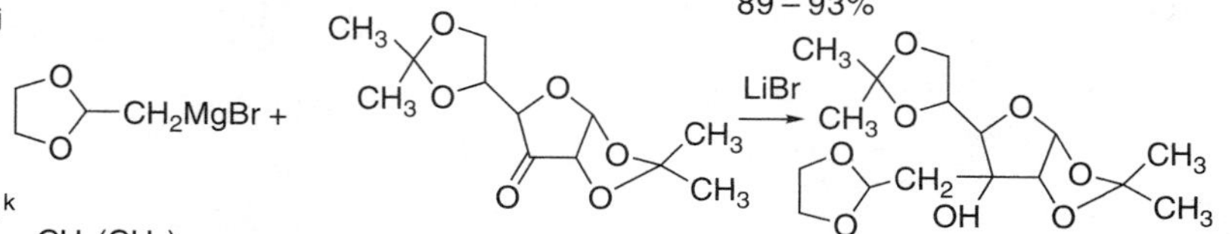

89%

11[k]

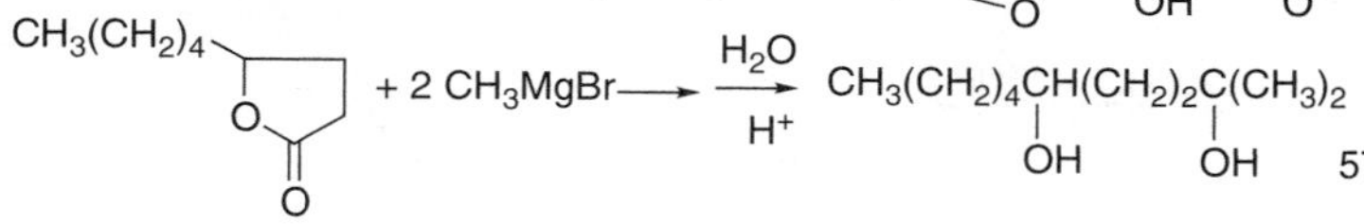

$CH_3(CH_2)_4CH(CH_2)_2C(CH_3)_2$

57%

F. Aldehydes from triethyl orthoformate

12[l]

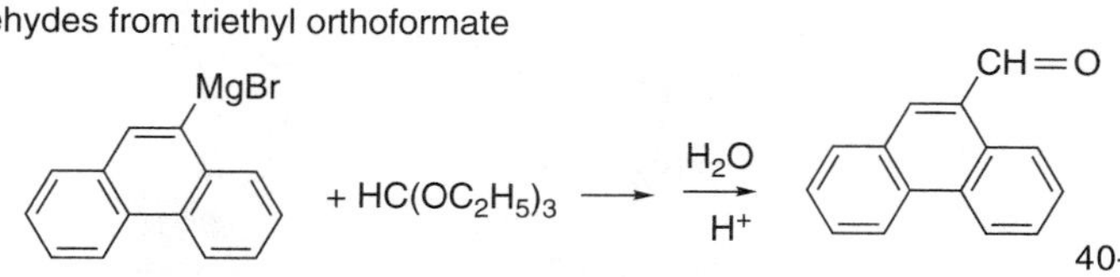

40–42%

13[m] $CH_3(CH_2)_4MgBr + HC(OC_2H_5)_3 \longrightarrow \xrightarrow[H^+]{H_2O} CH_3(CH_2)_4CH\!=\!O$

45–50%

(Continued)

640

CHAPTER 7

*Organometallic
Compounds of Group I
and II Metals*

Scheme 7.3. (*Continued*)

G. Ketones from nitriles, thioesters, amides, and anhydrides

14[n]

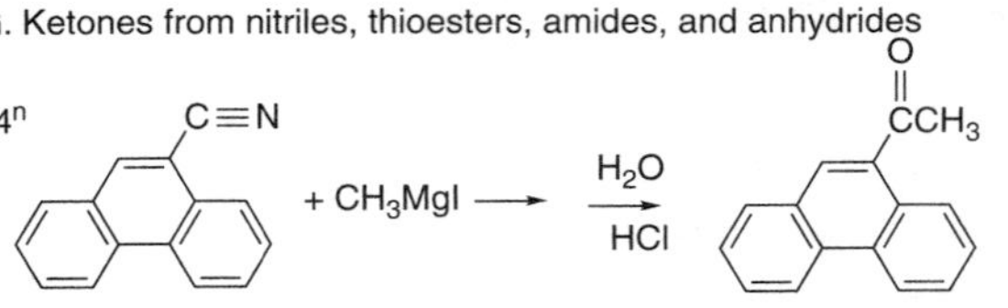

+ CH$_3$MgI $\longrightarrow$ $\xrightarrow[\text{HCl}]{\text{H}_2\text{O}}$ ... 52–59%

15[o]

CH$_3$OCH$_2$C$\equiv$N + PhMgBr $\longrightarrow$ $\xrightarrow[\text{HCl}]{\text{H}_2\text{O}}$ PhCCH$_2$OCH$_3$ 71–78%

16[p]

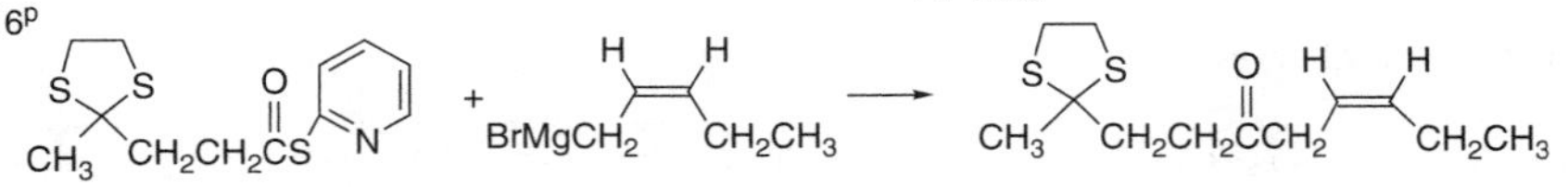

93%

17[q]

PhMgBr + ClCH$_2$CNCH$_3$ $\longrightarrow$ PhCCH$_2$Cl 92%
                     |
                     OCH$_3$

18[r]

HC$\equiv$CCH$_2$CH$_2$CNCH$_3$ + CH$_2$=CHMgBr $\longrightarrow$ HC$\equiv$CCH$_2$CH$_2$CCH=CH$_2$
                         |
                         OCH$_3$

19[s]

PhC$\equiv$CMgBr + (CH$_3$CO)$_2$O $\longrightarrow$ PhC$\equiv$CCCH$_3$ 80%

H. Carboxylic acids by carbonation

20[t]

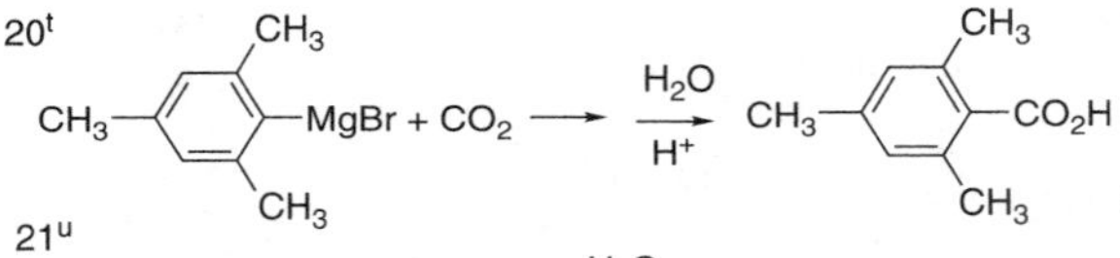

86–87%

21[u]

CH$_3$CH$_2$CHCH$_3$ + CO$_2$ $\longrightarrow$ $\xrightarrow{\text{H}_2\text{O}}{\text{H}^+}$ CH$_3$CH$_2$CHCH$_3$ 76–86%
       | |
       MgBr   CO$_2$H

22[v]

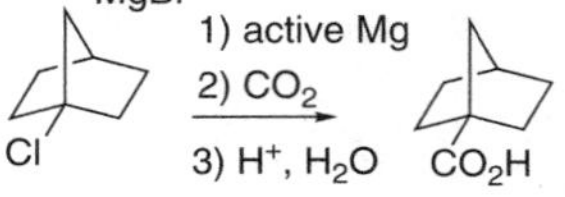

1) active Mg
2) CO$_2$
3) H$^+$, H$_2$O   60–70%

I. Amines from imines

23[w]

PhCH=NCH$_3$ + PhCH$_2$MgCl $\longrightarrow$ $\xrightarrow{\text{H}_2\text{O}}$ PhCHCH$_2$Ph 96%
                                              |
                                              CH$_3$NH

J. Alkenes after dehydration of intermediate alcohols

24[x]

PhCH=CHCH=O + CH$_3$MgBr $\longrightarrow$ $\xrightarrow{\text{H}_2\text{SO}_4}$ PhCH=CHCH=CH$_2$ 75%

25[y]

2 PhMgBr + CH$_3$CO$_2$C$_2$H$_5$ $\longrightarrow$ $\xrightarrow[\text{H}_2\text{O}]{\text{H}^+}$ Ph$_2$C=CH$_2$ 67–70%

(*Continued*)

Scheme 7.3. (*Continued*)

641

SECTION 7.2

*Reactions of
Organomagnesium and
Organolithium
Compounds*

a. H. Gilman and W. E. Catlin, *Org. Synth.*, **I**, 182 (1932).
b. E. E. Dreger, *Org. Synth.*, **I**, 299 (1932).
c. L. Skattebol, E. R. H. Jones, and M. C. Whiting, *Org. Synth.*, **IV**, 792 (1963).
d. C. G. Overberger, J. H. Saunders, R. E. Allen, and R. Gander, *Org. Synth.*, **III**, 200 (1955).
e. E. R. Coburn, *Org. Synth.*, **III**, 696 (1955).
f. N. L. Drake and G. B. Cooke, *Org. Synth.*, **II**, 406 (1943).
g. G. H. Coleman and D. Craig, *Org. Synth.*, **II**, 179 (1943).
h. W. W. Moyer and C. S. Marvel, *Org. Synth.*, **II**, 602 (1943).
i. W. E. Bachman and H. P. Hetzner, *Org. Synth.*, **III**, 839 (1955).
j. M. Schmeichel and H. Redlich, *Synthesis*, 1002 (1996).
k. J. Colonge and R. Marey, *Org. Synth.*, **IV**, 601 (1963).
l. C. A. Dornfeld and G. H. Coleman, *Org. Synth.*, **III**, 701 (1955).
m. G. B. Bachman, *Org. Synth.*, **II**, 323 (1943).
n. J. E. Callen, C. A. Dornfield, and G. H. Coleman, *Org. Synth.*, **III**, 26 (1955).
o. R. B. Moffett and R. L. Shriner, *Org. Synth.*, **III**, 562 (1955).
p. T. Mukaiyama, M. Araki, and H. Takei, *J. Am. Chem. Soc.*, **95**, 4763 (1973); M. Araki, S. Sakata, H. Takei, and T. Mukaiyama, *Bull. Chem. Soc. Jpn.*, **47**, 1777 (1974).
q. R. Tillyer, L. F. Frey, D. M. Tschaen, and U.-H. Dolling, *Synlett*, 225 (1996).
r. B. M. Trost and Y. Sih, *J. Am. Chem. Soc.*, **115**, 942 (1993).
s. A. Zanka, *Org. Proc. Res. Dev.*, **2**, 60 (1998).
t. D. M. Bowen, *Org. Synth.*, **III**, 553 (1955).
u. H. Gilman and R. H. Kirby, *Org. Synth.*, **I**, 353 (1932).
v. R. D. Rieke, S. E. Bales, P. M. Hudnall, and G. S. Poindexter, *Org. Synth.*, **59**, 85 (1977).
w. R. B. Moffett, *Org. Synth.*, **IV**, 605 (1963).
x. O. Grummitt and E. I. Beckner, *Org. Synth.*, **IV**, 771 (1963).
y. C. F. H. Allen and S. Converse, *Org. Synth.*, **I**, 221 (1932).

Several Grignard reactions are used on an industrial scale in drug synthesis.[95] The syntheses of both tamoxifen and droloxifene, which are estrogen antagonists used in treatment of breast cancer and osteoporosis, respectively, involve Grignard addition reactions.[96]

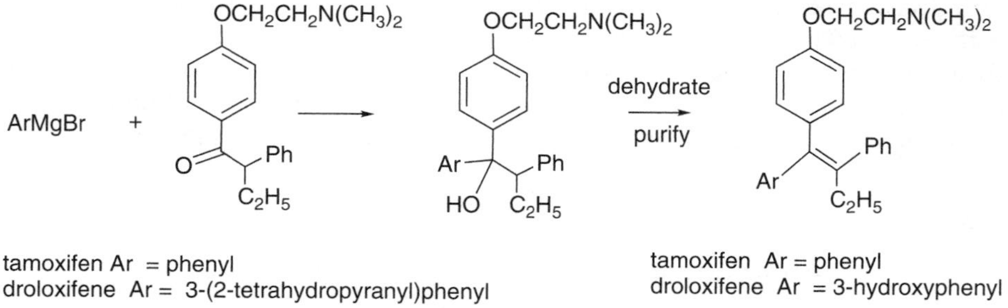

tamoxifen Ar = phenyl
droloxifene Ar = 3-(2-tetrahydropyranyl)phenyl

tamoxifen Ar = phenyl
droloxifene Ar = 3-hydroxyphenyl

Grignard reagents are quite restricted in the types of functional groups that can be present in either the organometallic or the carbonyl compound. Alkene, ether, and acetal functionality usually causes no difficulty but unprotected OH, NH, SH, or carbonyl groups cannot be present and CN and NO_2 groups cause problems in many cases.

Grignard additions are sensitive to steric effects and with hindered ketones a competing process leading to reduction of the carbonyl group can occur. A cyclic TS is involved.

[95.] F. R. Busch and D. M. DeAntonis, in *Grignard Reagents: New Developments*, H. G. Richey, Jr., ed., Wiley, New York, 2000, pp. 175–181.
[96.] R. McCaque, *J. Chem. Soc., Perkin Trans. 1*, 1011 (1987); M. Schickaneder, R. Loser, and M. Grill, US Patent, 5,047,431 (1991).

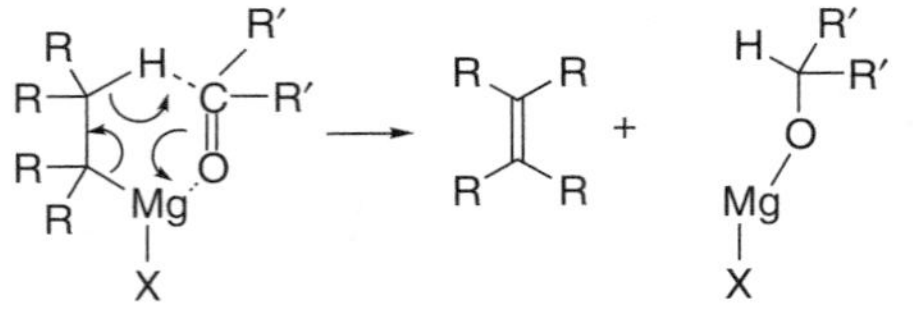

The extent of this reaction increases with the steric bulk of the ketone and Grignard reagent. For example, no addition occurs between diisopropyl ketone and isopropylmagnesium bromide, and the reduction product diisopropylcarbinol is formed in 70% yield.[97] Competing reduction can be minimized in troublesome cases by using benzene or toluene as the solvent.[98] Alkyllithium compounds are much less prone to reduction and are preferred for the synthesis of highly substituted alcohols. This is illustrated by the comparison of the reaction of ethyllithium and ethylmagnesium bromide with adamantone. A 97% yield of the tertiary alcohol is obtained with ethyllithium, whereas the Grignard reagent gives mainly the reduction product.[99]

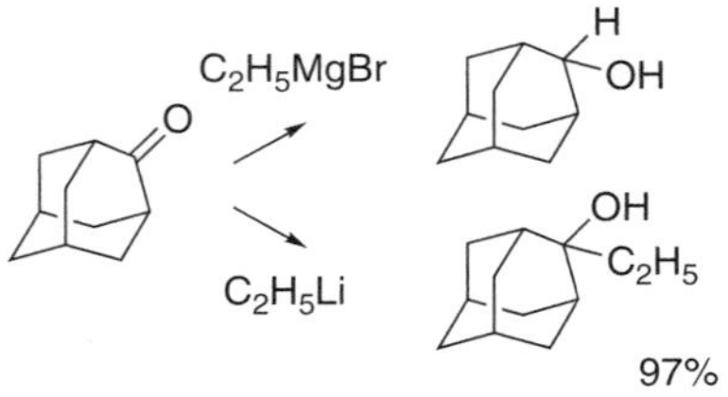

Enolization of the ketone is also sometimes a competing reaction. Since the enolate is unreactive toward nucleophilic addition, the ketone is recovered unchanged after hydrolysis. Enolization has been shown to be especially important when a considerable portion of the Grignard reagent is present as an alkoxide.[100] Alkoxides are formed as the addition reaction proceeds but can also be present as the result of oxidation of some of the Grignard reagent by oxygen during preparation or storage. As with reduction, enolization is most seriously competitive in cases where addition is retarded by steric factors.

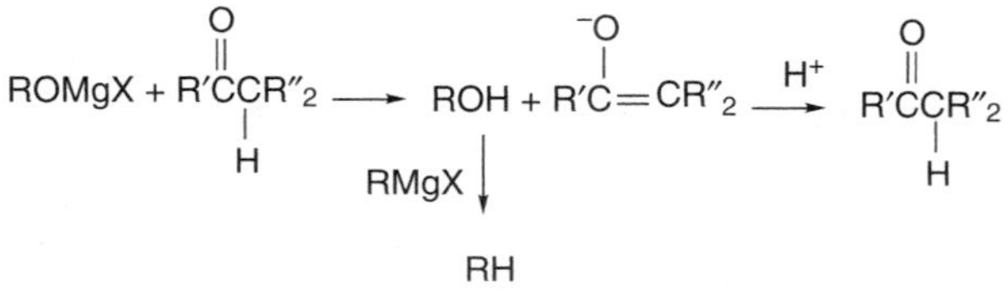

Structural rearrangements are not encountered with saturated Grignard reagents, but allylic and homoallylic systems can give products resulting from isomerization. NMR studies indicate that allylmagnesium bromide exists as a σ-bonded structure in which there is rapid equilibration of the two terminal carbons.[101] Similarly,

97. D. O. Cowan and H. S. Mosher, *J. Org. Chem.*, **27**, 1 (1962).
98. P. Caronne, G. B. Foscolos, and G. Lemay, *Tetrahedron Lett.*, 4383 (1979).
99. S. Landa, J. Vias, and J. Burkhard, *Coll. Czech. Chem. Commun.*, **72**, 570 (1967).
100. H. O. House and D. D. Traficante, *J. Org. Chem.*, **28**, 355 (1963).
101. M. Schlosser and N. Stahle, *Angew. Chem. Int. Ed. Engl.*, **19**, 487 (1980); M. Stahle and M. Schlosser, *J. Organomet. Chem.*, **220**, 277 (1981).

643

SECTION 7.2

*Reactions of
Organomagnesium and
Organolithium
Compounds*

2-butenylmagnesium bromide and 1-methyl-2-propenylmagnesium bromide are in equilibrium in solution.

$$CH_3CH=CHCH_2MgBr \rightleftharpoons CH_3CHCH=CH_2$$
$$| \\ MgBr$$

Addition products are often derived from the latter compound, although it is the minor component at equilibrium.[102] Addition is believed to occur through a cyclic process that leads to an allylic shift.

3-Butenylmagnesium bromide is in equilibrium with a small amount of cyclopropylmethylmagnesium bromide. The existence of the mobile equilibrium has been established by deuterium-labeling techniques.[103] Cyclopropylmethylmagnesium bromide[104] (and cyclopropylmethyllithium[105]) can be prepared by working at low temperature. At room temperature, the ring-opened 3-butenyl reagents are formed.

When the double bond is further removed, as in 5-hexenylmagnesium bromide, there is no evidence of a similar equilibrium.[106]

The corresponding lithium reagent remains uncyclized at $-78°$ C, but cyclizes on warming.[107] γ-, δ-, and ε-Alkynyl lithium reagents undergo *exo* cyclization to α-cycloalkylidene isomers.[108] Anion-stabilizing substituents are required for the strained three- and four-membered rings, but not for the 5-*exo* cyclization. The driving

102. R. A. Benkeser, W. G. Young, W. E. Broxterman, D. A. Jones, Jr., and S. J. Piaseczynski, *J. Am. Chem. Soc.*, **91**, 132 (1969).
103. M. E. H. Howden, A. Maercker, J. Burdon, and J. D. Roberts, *J. Am. Chem. Soc.*, **88**, 1732 (1966).
104. D. J. Patel, C. L. Hamilton, and J. D. Roberts, *J. Am. Chem. Soc.*, **87**, 5144 (1965).
105. P. T. Lansbury, V. A. Pattison, W. A. Clement, and J. D. Sidler, *J. Am. Chem. Soc.*, **86**, 2247 (1964).
106. R. C. Lamb, P. W. Ayers, M. K. Toney, and J. F. Garst, *J. Am. Chem. Soc.*, **88**, 4261 (1966).
107. W. F. Bailey, J. J. Patricia, V. C. Del Gobbo, R. M. Jarrett, and P. J. Okarma, *J. Org. Chem.*, **50**, 1999 (1985); W. F. Bailey, T. T. Nurmi, J. J. Patricia, and W. Wang, *J. Am. Chem. Soc.*, **109**, 2442 (1987); W. F. Bailey, A. D. Khanolkar, K. Gavaskar, T. V. Ovaska, K. Rossi, Y. Thiel, and K. B. Wiberg, *J. Am. Chem. Soc.*, **113**, 5720 (1991).
108. W. F. Bailey and T. V. Ovaska, *J. Am. Chem. Soc.*, **115**, 3080 (1993).

force for cyclization is the formation of an additional C–C σ-bond and the formation of a more stable (sp^2 versus sp^3) carbanion.

$$Li(CH_2)_nC \equiv C-X \longrightarrow (CH_2)_n \quad C=C \overset{Li}{\underset{X}{}}$$

$$X = Ph, TMS; n = 2,3$$

$$Li(CH_2)_4C \equiv C(CH_2)_3CH_3 \longrightarrow \quad =C \overset{(CH_2)_3CH_3}{\underset{Li}{}}$$

An alternative to preparation of organometallic reagents followed by reaction with a carbonyl compound is to generate the organometallic intermediate in situ in the presence of the carbonyl compound. The organometallic compound then reacts immediately with the carbonyl compound. This procedure is referred to as the *Barbier reaction*.[109] This technique has no advantage over the conventional one for most cases for magnesium or lithium reagents. However, when the organometallic reagent is very unstable, it can be a useful method. Allylic halides, which can be difficult to convert to Grignard reagents in good yield, frequently give better results in the Barbier procedure. Since solid metals are used, one of the factors affecting the rate of the reaction is the physical state of the metal. Ultrasonic irradiation has been found to have a favorable effect on the Barbier reaction, presumably by accelerating the generation of reactive sites on the metal surface.[110]

$$(CH_3)_2CHCH_2CH=O + CH_2=\overset{CH_3}{\underset{|}{C}}CH_2Cl \xrightarrow[\text{ether}]{Mg} (CH_3)_2CHCH_2\overset{OH}{\underset{|}{C}}HCH_2\overset{CH_3}{\underset{|}{C}}=CH_2$$
$$92\%$$

7.2.2.2. Reactions of Organolithium Compounds. The reactivity of organolithium reagents toward carbonyl compounds is generally similar to that of Grignard reagents. The lithium reagents are less likely to undergo the competing reduction reaction with ketones, however.

Organolithium compounds can add to α, β-unsaturated ketones by either 1,2- or 1,4-addition. The most synthetically important version of the 1,4-addition involves organocopper intermediates, and is discussed in Chap 8. However, 1,4-addition is observed under some conditions even in the absence of copper catalysts. Highly reactive organolithium reagents usually react by 1,2-addition, but the addition of small amounts of HMPA has been found to favor 1,4-addition. This is attributed to solvation of the lithium ion, which attenuates its Lewis acid character toward the carbonyl oxygen.[111]

One reaction that is quite efficient for lithium reagents but poor for Grignard reagents is the synthesis of ketones from carboxylic acids.[112] The success of the

109. C. Blomberg and F. A. Hartog, *Synthesis*, 18 (1977).
110. J.-L. Luche and J.-C. Damiano, *J. Am. Chem. Soc.*, **102**, 7926 (1980).
111. H. J. Reich and W. H. Sikorski, *J. Org. Chem.*, **64**, 14 (1999).
112. M. J. Jorgenson, *Org. React.*, **18**, 1 (1971).

reaction depends on the stability of the dilithio adduct that is formed. This intermediate does not break down until hydrolysis, at which point the ketone is liberated. Some examples of this reaction are shown in Section B of Scheme 7.4.

$$RLi + R'\overset{O}{\underset{}{C}}O^-Li^+ \longrightarrow R'\overset{O^-Li^+}{\underset{R}{C}}O^-Li^+ \xrightarrow[H_2O]{H^+} R'\overset{OH}{\underset{R}{C}}OH \longrightarrow R\overset{O}{\underset{}{C}}R'$$

A study aimed at optimizing yields in this reaction found that carbinol formation was a major competing process if the reaction was not carried out in such a way that all of the lithium compound was consumed prior to hydrolysis.[113] Any excess lithium reagent that is present reacts extremely rapidly with the ketone as it is formed by hydrolysis. Another way to avoid the problem of carbinol formation is to quench the reaction mixture with trimethylsilyl chloride.[114] This procedure generates the disilyl acetal, which is stable until hydrolysis.

$$\text{CO}_2\text{H} \xrightarrow[\substack{2)\ \text{TMS-Cl} \\ 3)\ \text{H}_2\text{O},\ \text{H}^+}]{1)\ 4\ \text{equiv MeLi}} \overset{O}{\overset{\|}{C}}\text{CH}_3 \quad 92\%$$

The synthesis of unsymmetrical ketones can be carried out in a tandem one-pot process by successive addition of two different alkyllithium reagents.[115]

$$RLi + CO_2 \longrightarrow RCO_2^-Li^+ \xrightarrow[]{R'Li} \xrightarrow[H^+]{H_2O} R\overset{O}{\overset{\|}{C}}R'$$

N-Methyl-*N*-methoxyamides are also useful starting materials for preparation of ketones Again, the reaction depends upon the stability of the tetrahedral intermediate against elimination and a second addition step. In this case chelation with the *N*-methoxy substituent is responsible.

$$R\overset{O}{\overset{\|}{C}}\underset{OCH_3}{\overset{|}{N}}CH_3 + R'Li \longrightarrow \underset{R\ \ N\ \ }{\overset{^-O--Li^+}{R'\diagdown\diagup OCH_3}} \underset{CH_3}{\overset{|}{N}} \xrightarrow{H^+,\ H_2O} R\overset{O}{\overset{\|}{C}}R'$$

Scheme 7.4 illustrates some of the important synthetic reactions in which organolithium reagents act as nucleophiles. The range of reactions includes S_N2-type alkylation (Entries 1 to 3), epoxide ring opening (Entry 4), and formation of alcohols by additions to aldehydes and ketones (Entries 5 to 10). Note that in Entry 2, alkylation takes place mainly at the γ-carbon of the allylic system. The ratio favoring γ-alkylation

[113.] R. Levine, M. J. Karten, and W. M. Kadunce, *J. Org. Chem.*, **40**, 1770 (1975).
[114.] G. M. Rubottom and C. Kim, *J. Org. Chem.*, **48**, 1550 (1983).
[115.] G. Zadel and E. Breitmaier, *Angew. Chem. Int. Ed. Engl.*, **31**, 1035 (1992).

Scheme 7.4. Synthetic Procedures Involving Organolithium Reagents

A. Alkylation

1[a]

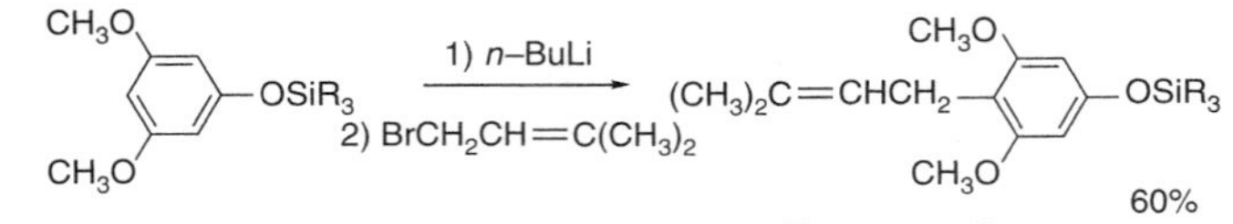

60%

2[b]

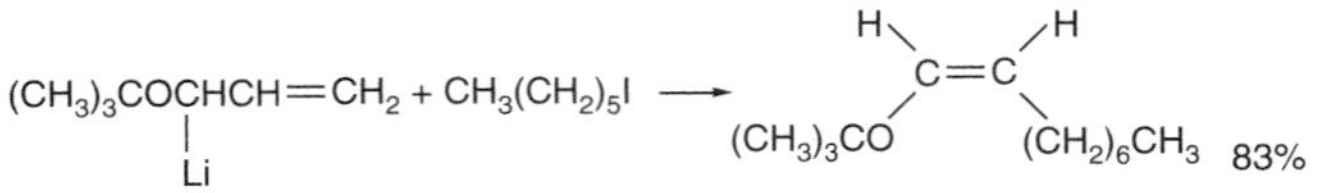

$(CH_3)_3COCHCH=CH_2 + CH_3(CH_2)_5I \longrightarrow$

83%

3[c]

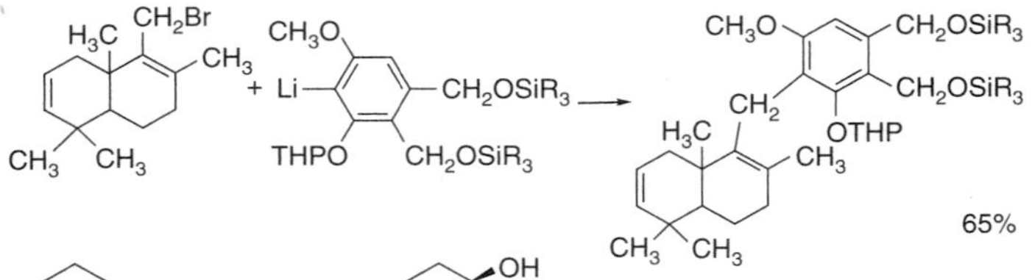

65%

4[d]

97%

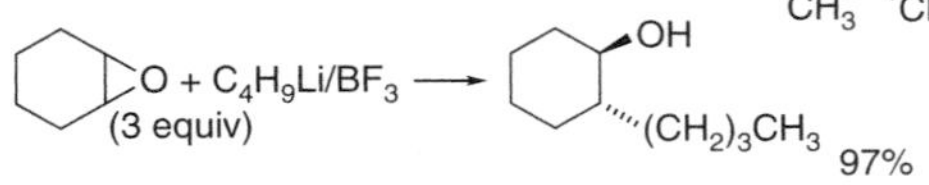

B. Reactions with aldehydes and ketones to give alcohols

5[e]

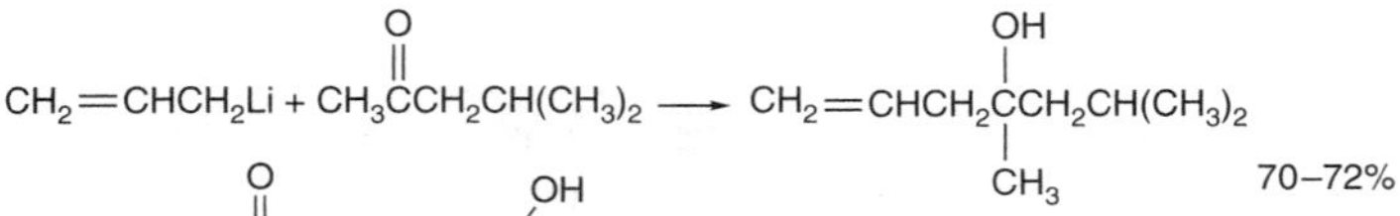

$CH_2=CHCH_2Li + CH_3CCH_2CH(CH_3)_2 \longrightarrow CH_2=CHCH_2CCH_2CH(CH_3)_2$

70–72%

6[f]

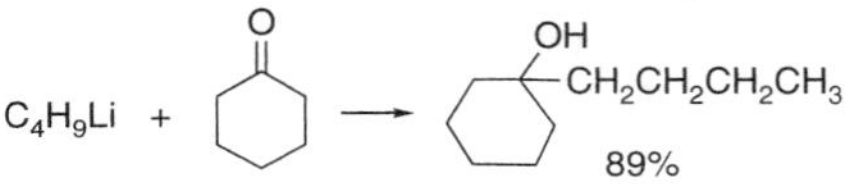

$C_4H_9Li +$ 89%

7[g]

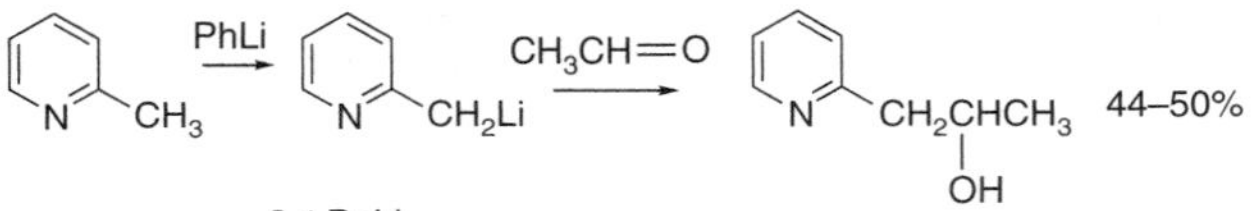

44–50%

8[h]

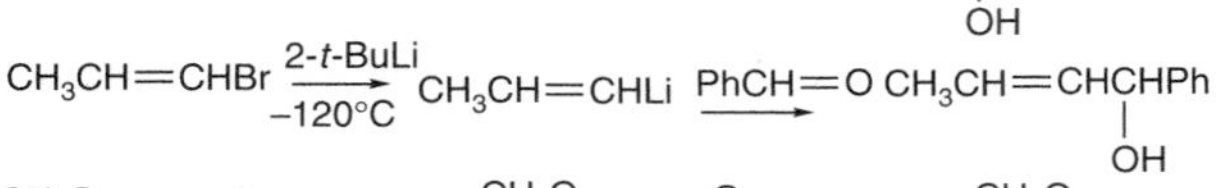

$CH_3CH=CHBr \xrightarrow[-120°C]{2\text{-}t\text{-}BuLi} CH_3CH=CHLi \xrightarrow{PhCH=O} CH_3CH=CHCHPh$

9[i]

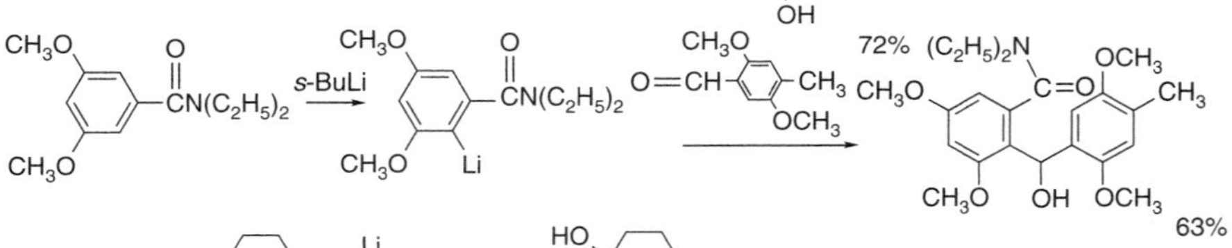

72% 63%

10[j]

$CH_3OCH_2Cl + O=$

90%

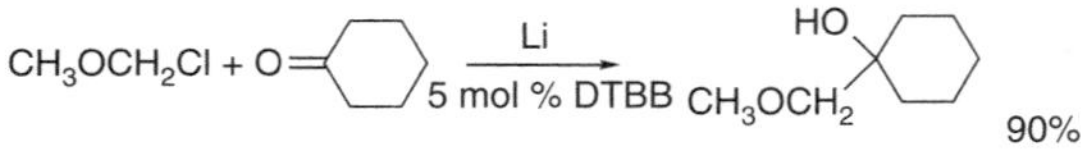

C. Reactions with carboxylic acids, acyl chlorides, acid anhydrides, and *N*-methoxyamides to give ketones

11[k]

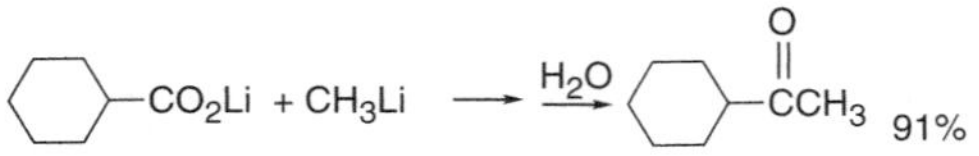

91%

12[l]

$(CH_3)_3CCO_2H + 2 PhLi \xrightarrow{H_2O}$ 65%

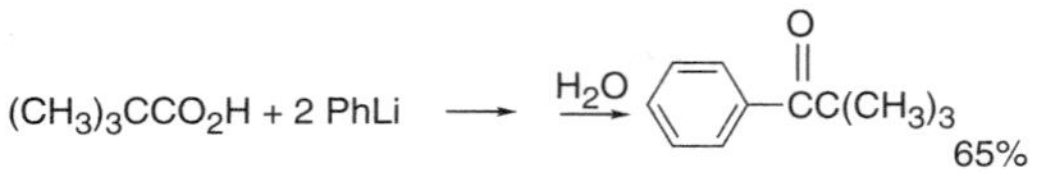

(*Continued*)

Scheme 7.4. (*Continued*)

647

SECTION 7.2

*Reactions of
Organomagnesium and
Organolithium
Compounds*

13[m]

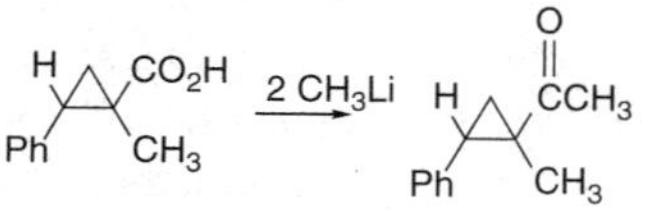

14[n]

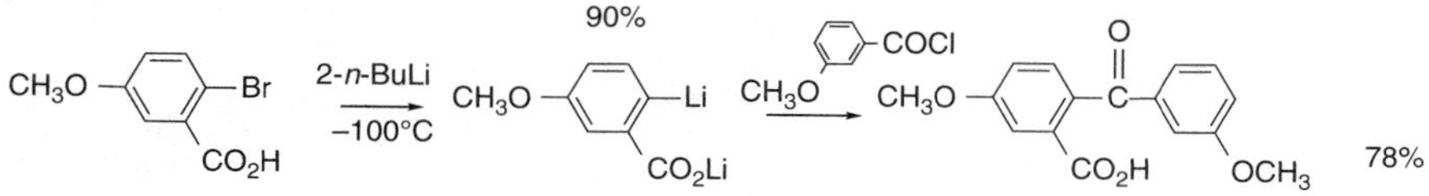

15[o]

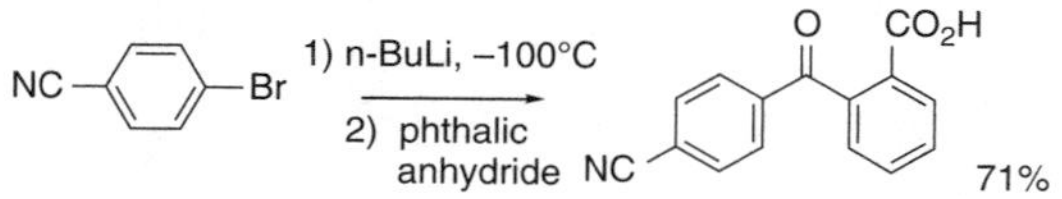

16[p]

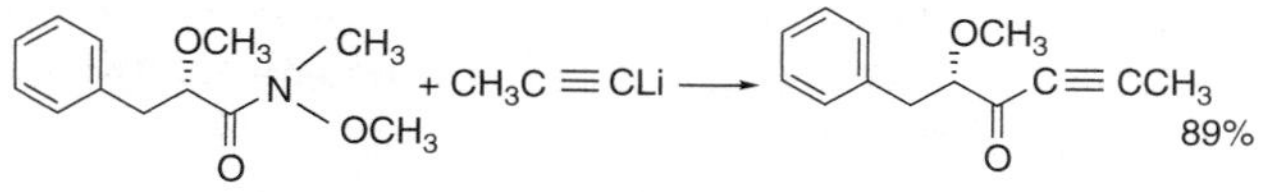

D. Reactions with carbon dioxide to give carboxylic acids

17[q]

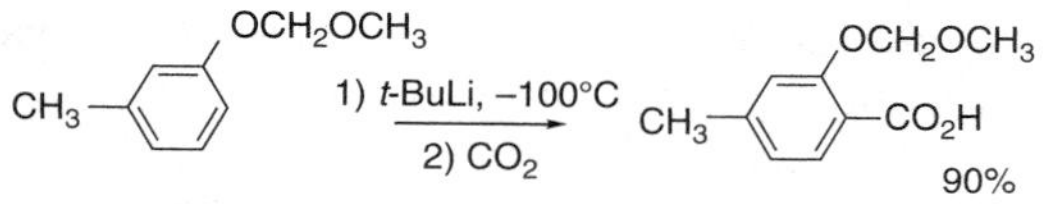

18[r]

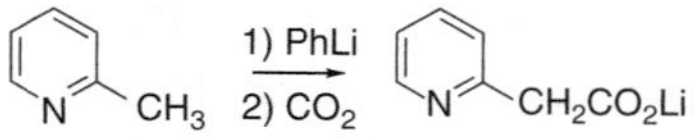

E. Other reactions

19[s]

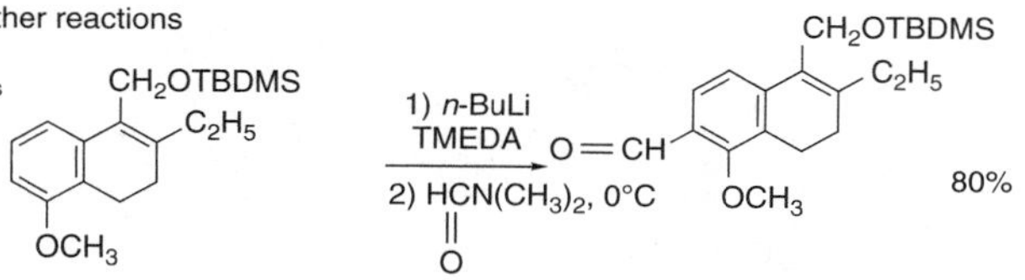

20[t]

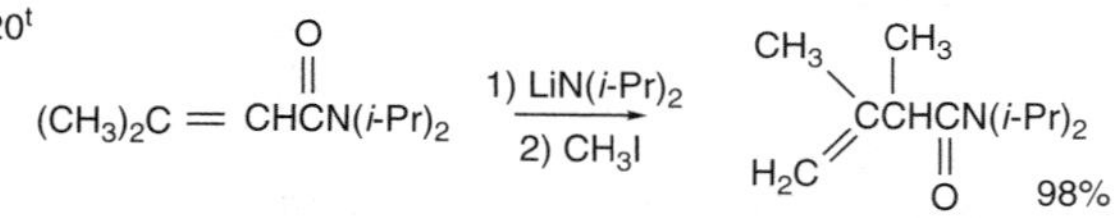

a. T. L. Shih, M. J. Wyvratt, and H. Mrozik, *J. Org. Chem.*, **52**, 2029 (1987).
b. D. A. Evans, G. C. Andrews, and B. Buckwalter, *J. Am. Chem. Soc.*, **96**, 5560 (1974).
c. J. E. McMurry and M. D. Erion, *J. Am. Chem. Soc.*, **107**, 2712 (1985).
d. M. J. Eis, J. E. Wrobel, and B. Ganem, *J. Am. Chem. Soc.*, **106**, 3693 (1984).
e. D. Seyferth and M. A. Weiner, *Org. Synth.*, **V**, 452 (1973).
f. J. D. Buhler, *J. Org. Chem.*, **38**, 904 (1973).
g. L. A. Walker, *Org. Synth.*, **III**, 757 (1955).
h. H. Neumann and D. Seebach, *Tetrahedron Lett.*, 4839 (1976).
i. S. O. diSilva, M. Watanabe, and V. Snieckus, *J. Org. Chem.*, **44**, 4802 (1979).
j. A. Guijarro, B. Mandeno, J. Ortiz, and M. Yus, *Tetrahedron*, **52**, 1643 (1993).
k. T. M. Bare and H. O. House, *Org. Synth.*, **49**, 81 (1969).
l. R. Levine and M. J. Karten, *J. Org. Chem.*, **41**, 1176 (1976).
m. C. H. DePuy, F. W. Breitbeil, and K. R. DeBruin, *J. Am. Chem. Soc.*, **88**, 3347 (1966).
n. W. E. Parham, C. K. Bradsher, and K. J. Edgar, *J. Org. Chem.*, **46**, 1057 (1981).
o. W. E. Parham and R. M. Piccirilli, *J. Org. Chem.*, **41**, 1268 (1976).
p. F. D'Aniello, A. Mann, and M. Taddei, *J. Org. Chem.*, **61**, 4870 (1996).
q. R. C. Ronald, *Tetrahedron Lett.*, 3973 (1975).
r. R. B. Woodward and E. C. Kornfeld, *Org. Synth.*, **III**, 413 (1955).
s. A. S. Kende and J. R. Rizzi, *J. Am. Chem. Soc.*, **103**, 4247 (1981).
t. M. Majewski, G. B. Mpango, M. T. Thomas, A. Wu, and V. Snieckus, *J. Org. Chem.*, **46**, 2029 (1981).

is higher for the *t*-butoxy ether than for ethers with smaller groups. There are several means of preparing ketones using organolithium reagents. Apart from addition to carboxylate salts (Entries 11 to 13), acylation with acyl chlorides (Entry 14), anhydrides (Entry 15), or *N*-methoxy-*N*-methylcarboxyamides (Entry 16) can be used. Carboxylic acids can be made by carbonation with CO_2 (Entries 17 and 18). Aldehydes can be prepared by reactions with DMF (Entry 19). Entry 20 is the alkylation of a stabilized allylic lithium reagent.

In addition to applications as nucleophiles, the lithium reagents have enormous importance in synthesis as bases and as lithiating reagents. The commercially available methyl, *n*-butyl, *s*-butyl, and *t*-butyl reagents are used frequently in this context.

7.2.2.3. Stereoselectivity of Addition to Ketones. The stereochemistry of the addition of both organomagnesium and organolithium compounds to cyclohexanones is similar.[116] With unhindered ketones, the stereoselectivity is not high but there is generally a preference for attack from the equatorial direction to give the axial alcohol. This preference for the equatorial approach increases with the size of the alkyl group. With alkyllithium reagents, added salts improve the stereoselectivity. For example, one equivalent of $LiClO_4$, enhances the proportion of the axial alcohol in the addition of methyllithium to 4-*t*-butylcyclohexanone.[117]

no $LiClO_4$	65%	35%
1 equiv $LiClO_4$	92%	8%

Bicyclic ketones react with organometallic reagents to give the products of addition from the less hindered face of the carbonyl group.

The stereochemistry of addition of organometallic reagents to chiral carbonyl compounds parallels the behavior of the hydride reducing agents, as discussed in Section 5.3.2. Organometallic compounds were included in the early studies that established the preference for addition according to Cram's rule.[118]

S, M, L = relative size of substituents

116. E. C. Ashby and J. T. Laemmle, *Chem. Rev.*, **75**, 521 (1975).
117. E. C. Ashby and S. A. Noding, *J. Org. Chem.*, **44**, 4371 (1979).
118. D. J. Cram and F. A. A. Elhafez, *J. Am. Chem. Soc.*, **74**, 5828 (1952).

The interpretation of the basis for this stereoselectivity can be made in terms of the steric, torsional, and stereoelectronic effects discussed in connection with reduction by hydrides. It has been found that crown ethers enhance stereoselectivity in the reaction of both Grignard reagents and alkyllithium compounds.[119] This effect was attributed to decreased electrophilicity of the metal cations in the presence of the crown ether. The attenuated reactivity leads to greater selectivity.

For ketones and aldehydes in which adjacent substituents permit the possibility of chelation with a metal ion, the stereochemistry can often be interpreted in terms of the steric requirements of the chelated TS. In the case of α-alkoxyketones, for example, an assumption that both the alkoxy and carbonyl oxygens are coordinated with the metal ion and that addition occurs from the less hindered face of this chelate correctly predicts the stereochemistry of addition. The predicted product dominates by as much as 100:1 for several Grignard reagents.[120] Further supporting the importance of chelation is the correlation between rate and stereoselectivity. Groups that facilitate chelation cause an increase in both rate and stereoselectivity.[121] This indicates that chelation not only favors a specific TS geometry, but also lowers the reaction barrier by favoring metal ion complexation.

$R=C_7H_{15}$ $R'=CH_2OCH_3$
$CH_2OCH_2CH_2OCH_3$
CH_2Ph
CH_2OCH_2Ph

The addition of a Grignard reagent to an unsymmetrical ketone generates a new stereogenic center and is potentially enantioselective in the presence of an element of chirality. Perhaps because the reactions are ordinarily very fast, there are relatively few cases in which such reactions are highly enantioselective. The magnesium salt of TADDOL promotes enantioselective additions to acetophenone.[122] These particular reactions occur under heterogeneous conditions and are quite slow at $-100°$ C. Although the details of the mechanism are unclear, the ligand must establish a chiral environment that controls the facial selectivity of the additions.

R	% yield	e.e.(%)
C_2H_5	62	98
n-C_3H_7	84	> 98
n-C_4H_9	75	> 98
n-C_8H_{17}	58	> 98

[119.] Y. Yamamoto and K. Maruyama, *J. Am. Chem. Soc.*, **107**, 6411 (1985).
[120.] W. C. Still and J. H. McDonald, III, *Tetrahedron Lett.*, 1031 (1980).
[121.] X. Chen, E. R. Hortelano, E. L. Eliel, and S. V. Frye, *J. Am. Chem. Soc.*, **112**, 6130 (1990).
[122.] B. Weber and D. Seebach, *Tetrahedron*, **50**, 6117 (1994).

7.3. Organometallic Compounds of Group IIB and IIIB Metals

In this section we discuss organometallic derivatives of zinc, cadmium, mercury, and indium. These Group IIB and IIIB metals have the d^{10} electronic configuration in the $+2$ and $+3$ oxidation states, respectively. Because of the filled d level, the $+2$ or $+3$ oxidation states are quite stable and reactions of these organometallics do not usually involve changes in oxidation level. This property makes the reactivity patterns of Group IIB and IIIB organometallics more similar to derivatives of Group IA and IIA metals than to transition metals having vacancies in the d levels. The IIB metals, however, are less electropositive than the IA and IIA metals and the nucleophilicity of the organometallics is less than for organolithium or organomagnesium compounds. Many of the synthetic applications of these organometallics are based on this attenuated reactivity and involve the use of a specific catalyst to promote reaction.

7.3.1. Organozinc Compounds

Organozinc reagents have become the most useful of the Group IIB organometallics in terms of synthesis.[123] Although they are much less reactive than organolithium or organomagnesium reagents, their addition to aldehydes can be catalyzed by various Lewis acids or by coordinating ligands. They have proven particularly adaptable to enantioselective additions. There are also important reactions of organozinc reagents that involve catalysis by transition metals, and these reactions are discussed in Chapter 8.

7.3.1.1. Preparation of Organozinc Compounds. Organozinc compounds can be prepared by reaction of Grignard or organolithium reagents with zinc salts. When Grignard reagents are treated with $ZnCl_2$ and dioxane, a dioxane complex of the magnesium halide precipitates, leaving a solution of the alkylzinc reagent. A one-pot process in which the organic halide, magnesium metal, and zinc chloride are sonicated is another method for their preparation.[124] Organozinc compounds can also be prepared from organic halides by reaction with highly reactive zinc metal.[125] Simple alkylzinc compounds, which are distillable liquids, can also be prepared from alkyl halides and a Zn-Cu couple.[126] Dimethyl-, diethyl-, di-n-propyl-, and diphenylzinc are commercially available.

Arylzinc reagents can be made from aryl halides with activated zinc[127] or from Grignard reagents by metal-metal exchange with zinc salts.[128]

$$C_2H_5O_2C\text{—}\langle\text{—}\rangle\text{—}I + Zn \longrightarrow C_2H_5O_2C\text{—}\langle\text{—}\rangle\text{—}ZnI$$

$$2\ PhMgBr\ +\ ZnCl_2 \longrightarrow Ph_2Zn\ +\ 2\ MgBrCl$$

123. E. Erdik, *Organozinc Reagents in Organic Synthesis*, CRC Publishing, Boca Raton, FL, 1996.

124. J. Boersma, *Comprehensive Organometallic Chemistry*, G. Wilkinson, ed., Vol. 2, Pergamon Press, Oxford, 1982, Chap. 16; G. E. Coates and K. Wade, *Organometallic Compounds*, Vol. 1, 3rd Edition, Methuen, London, 1967, pp. 121–128.

125. R. D. Rieke, P. T.-J. Li, T. P. Burns, and S. T. Uhm, *J. Org. Chem.*, **46**, 4323 (1981).

126. C. R. Noller, *Org. Synth.*, **II**, 184 (1943).

127. L. Zhu, R. M. Wehmeyer, and R. D. Rieke, *J. Org. Chem.*, **56**, 1445 (1991); T. Sakamoto, Y. Kondo, N. Murata, and H. Yamanaka, *Tetrahedron Lett.*, **33**, 5373 (1992).

128. K. Park, K. Yuan, and W. J. Scott, *J. Org. Chem.*, **58**, 4866 (1993).

Allylic zinc reagents can be prepared in situ in aqueous solution in the presence of aldehydes.[129] These reactions show a strong preference for formation of the more branched product. This suggests that the reactions occur by coordination of the zinc reagent at the carbonyl oxygen and that addition proceeds by a cyclic mechanism, similar to that for allylic Grignard reagents. The kinetic isotope of the reaction measured under these conditions is consistent with a cyclic mechanism.[130]

An attractive feature of organozinc reagents is that many functional groups that would interfere with organomagnesium or organolithium reagents can be present in organozinc reagents.[131,132] Functionalized reagents can be prepared by halogen-metal exchange reactions with diethylzinc.[133] The reaction equilibrium is driven to completion by use of excess diethylzinc and removal of the ethyl iodide by distillation. The pure organozinc reagent can be obtained by removal of the excess diethylzinc under vacuum.

$$2\ X(CH_2)_nI\ +\ (C_2H_5)_2Zn\ \longrightarrow\ X(CH_2)_nZnC_2H_5\ \rightleftharpoons\ [X(CH_2)_n]_2Zn\ +\ 2\ C_2H_5I$$

$$n = 2\text{--}5 \quad X = CH_3CO_2,\ (CH_3)_3CCO_2,\quad N\equiv C,\ Cl$$

These reactions are subject to catalysis by certain transition metal ions and with small amounts of $MnBr_2$ or CuCl the reaction proceeds satisfactorily with alkyl bromides.[134]

$$X(CH_2)_nBr + (C_2H_5)_2Zn \xrightarrow[3\%\ CuCl]{5\%\ MnBr_2} X(CH_2)_nZnBr$$

$$n = 3,\ 4;\quad X = C_2H_5O_2C,\ N\equiv C,\ Cl$$

Another effective catalyst is $Ni(acac)_2$.[135]

129. C. Petrier and J.-L. Luche, *J. Org. Chem.*, **50**, 910 (1985).
130. J. J. Gajewski, W. Bocain, N. L. Brichford, and J. L. Henderson, *J. Org. Chem.*, **67**, 4236 (2002).
131. P. Knochel, J. J. A. Perea, and P. Jones, *Tetrahedron*, **54**, 8275 (1998).
132. P. Knochel and R. D. Singer, *Chem. Rev.*, **93**, 2117 (1993); A. Boudier, L. O. Bromm, M. Lotz, and P. Knochel, *Angew. Chem. Int. Ed. Engl.*, **39**, 4415 (2000); P. Knochel, N. Millot, A. L. Rodriguez, and C. E. Tucker, *Org. React.*, **58**, 417 (2001).
133. M. J. Rozema, A. R. Sidduri, and P. Knochel, *J. Org. Chem.*, **57**, 1956 (1992).
134. I. Klemment, P. Knochel, K. Chau, and G. Cahiez, *Tetrahedron Lett.*, **35**, 1177 (1994).
135. S. Vettel, A. Vaupel, and P. Knochel, *J. Org. Chem.*, **61**, 7473 (1996).

Organozinc reagents can also be prepared from trialkylboranes by exchange with dimethylzinc.[136]

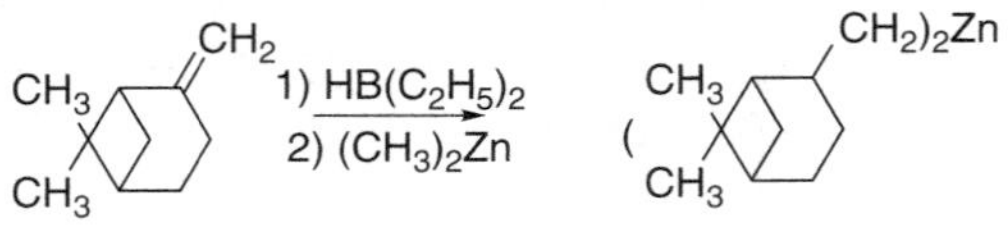

This route can be used to prepare enantiomerically enriched organozinc reagents by asymmetric hydroboration (see Section 4.5.3), followed by exchange with diisopropylzinc. Trisubstituted cycloalkenes such as 2-methyl or 2-phenylcyclohexene give an enantiomeric purity greater than 95%. The exchange reaction takes place with retention of configuration.[137]

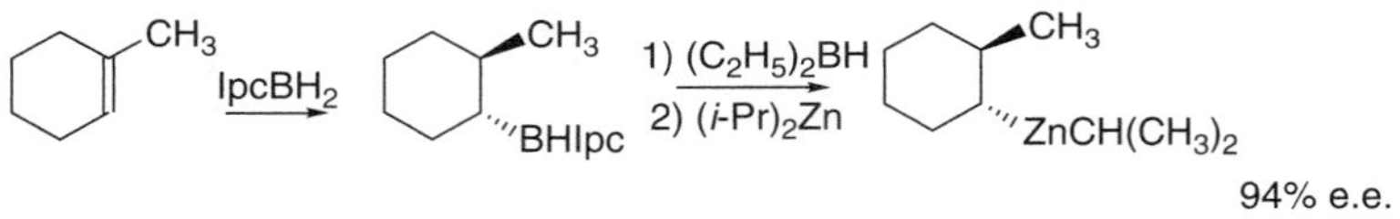

94% e.e.

Exchange with boranes can also be used to prepare alkenylzinc reagents.[138]

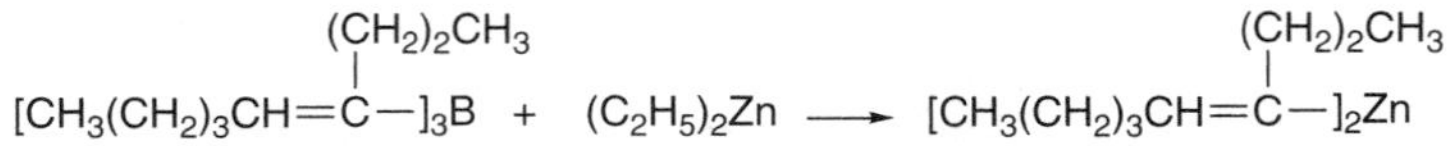

Alkenylzinc reagents can also be made from alkynes by $(Cp)_2TiCl_2$-catalyzed hydrozincation (see Section 4.6).[139] The reaction proceeds with high *syn* stereoselectivity, and the regioselectivity corresponds to relative carbanion stability.

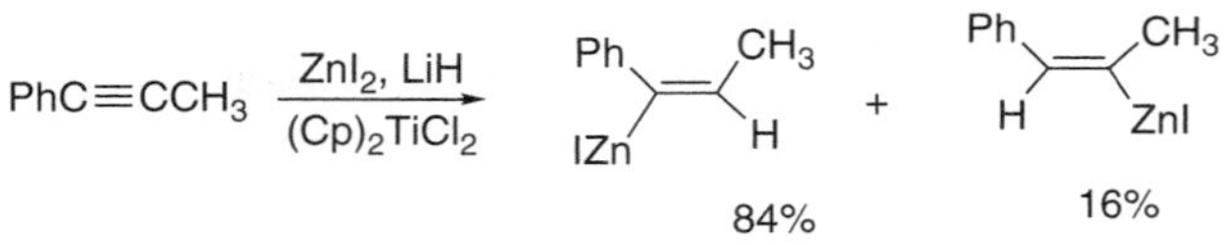

84%   16%

7.3.1.2. Reactions of Organozinc Compounds. Pure organozinc compounds are relatively unreactive toward addition to carbonyl groups, but the reactions are catalyzed by both Lewis acids and chelating ligands. When prepared in situ from $ZnCl_2$ and Grignard reagents, organozinc reagents add to carbonyl compounds to give carbinols.[140]

136. F. Langer, J. Waas, and P. Knochel, *Tetrahedron Lett.*, **34**, 5261 (1993); L. Schwink and P. Knochel, *Tetrahedron Lett.*, **35**, 9007 (1994); F. Langer, A. Devasagayari, P.-Y. Chavant, and P. Knochel, *Synlett*, 410 (1994); F. Langer, L. Schwink, A. Devasagayari, P.-Y. Chavant, and P. Knochel, *J. Org. Chem.*, **61**, 8229 (1996).

137. A. Boudier, F. Flachsmann, and P. Knochel, *Synlett*, 1438 (1998).

138. M. Srebnik, *Tetrahedron Lett.*, **32**, 2449 (1991); K. A. Agrios and M. Srebnik, *J. Org. Chem.*, **59**, 5468 (1994).

139. Y. Gao, K. Harada, T. Hata, H. Urabe, and F. Sato, *J. Org. Chem.*, **60**, 290 (1995).

140. P. R. Jones, W. J. Kauffman, and E. J. Goller, *J. Org. Chem.*, **36**, 186 (1971); P. R. Jones, E. J. Goller, and W. J. Kaufmann, *J. Org. Chem.*, **36**, 3311 (1971).

This must reflect activation of the carbonyl group by magnesium ion, since ketones are less reactive to pure dialkylzinc reagents and tend to react by reduction rather than addition.[141] The addition of alkylzinc reagents is also promoted by trimethylsilyl chloride, which leads to isolation of silyl ethers of the alcohol products.[142]

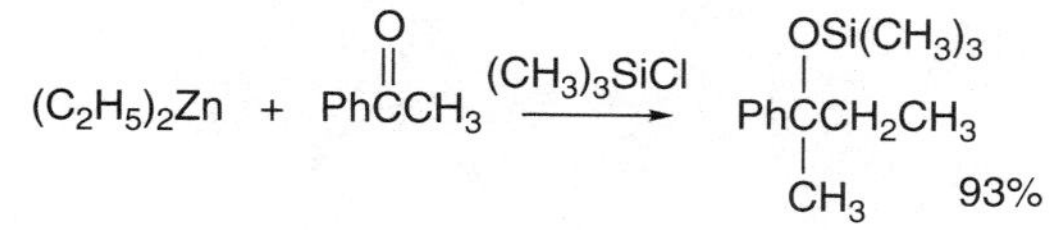

$$(C_2H_5)_2Zn \; + \; PhCCH_3 \; \xrightarrow{(CH_3)_3SiCl} \; PhCCH_2CH_3 \qquad 93\%$$

High degrees of enantioselectivity have been obtained when alkylzinc reagents react with aldehydes in the presence of chiral ligands.[143] Among several compounds that have been used as ligands are *exo*-(dimethylamino)norborneol (**A**),[144] its morpholine analog (**B**),[145] diphenyl(1-methylpyrrolin-2-yl)methanol (**C**),[146] as well as ephedrine derivatives **D**[147] and **E**.[148]

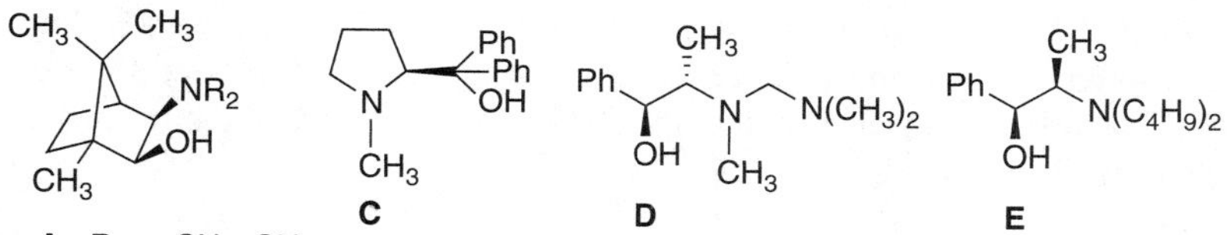

A $R_2 = CH_3, CH_3$

B $R_2 = -(CH_2CH_2)_2O$

The enantioselectivity is the result of chelation of the chiral ligand to the zinc. The TS of the addition is believed to involve two zinc atoms. One zinc functions as a Lewis acid by coordination at the carbonyl oxygen and the other is the source of the nucleophilic carbon. The proposed TS for aminoalcohol **A**, for example, is shown below.[149]

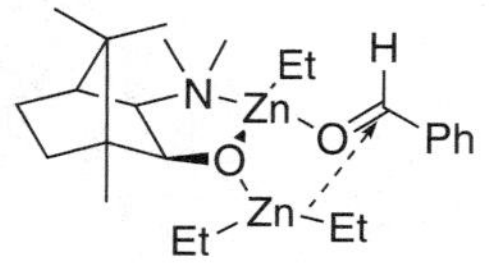

141. G. Giacomelli, L. Lardicci, and R. Santi, *J. Org. Chem.*, **39**, 2736 (1974).

142. S. Alvisi, S. Casolari, A. L. Costa, M. Ritiani, and E. Tagliavini, *J. Org. Chem.*, **63**, 1330 (1998).

143. K. Soai, A. Ookawa, T. Kaba, and K. Ogawa, *J. Am. Chem. Soc.*, **109**, 7111 (1987); M. Kitamura, S. Suga, K. Kawai, and R. Noyori, *J. Am. Chem. Soc.*, **108**, 6071 (1986); W. Oppolzer and R. N. Rodinov, *Tetrahedron Lett.*, **29**, 5645 (1988); K. Soai and S. Niwa, *Chem. Rev.*, **92**, 833 (1992).

144. M. Kitamura, S. Suga, K. Kawai, and R. Noyori, *J. Am. Chem. Soc.*, **108**, 6071 (1986); M. Kitamura, H. Oka, and R. Noyori, *Tetrahedron*, **55**, 3605 (1999).

145. W. A. Nugent, *Chem. Commun.*, 1369 (1999).

146. K. Soai, A. Ookawa, T. Kaba, and E. Ogawa, *J. Am. Chem. Soc.*, **109**, 7111 (1987).

147. E. J. Corey and F. J. Hannon, *Tetrahedron Lett.*, **28**, 5233 (1987).

148. K. Soai, S. Yokoyama, and T. Hayasaka, *J. Org. Chem.*, **56**, 4264 (1991).

149. D. A. Evans, *Science*, **240**, 420 (1988); E. J. Corey, P.-W. Yuen, F. J. Hannon, and D. A. Wierda, *J. Org. Chem.*, **55**, 784 (1990); B. Goldfuss and K. N. Houk, *J. Org. Chem.*, **63**, 8998 (1998).

The catalytic cycle for these reactions is believed to involve dinuclear complexes formed among the zinc chelate, the aldehyde, and the zinc atom that releases the nucleophile.

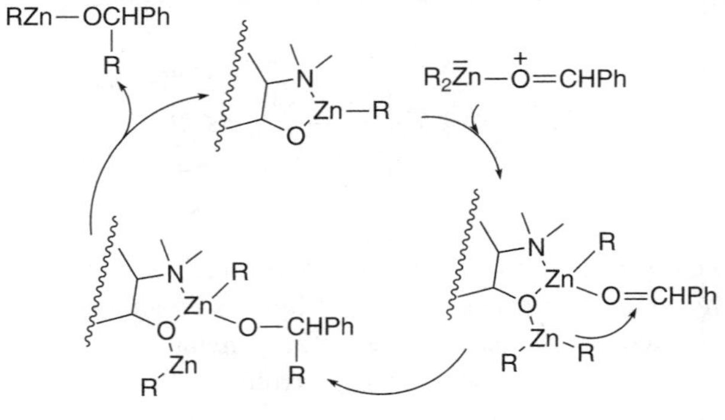

The structures of the TSs have been explored computationally using combined B3LYP-MM methods.[150] There are four stereochemically distinct TSs, as shown in Figure 7.4. For the aminoalcohol ligands, the *anti-trans* arrangement is preferred. Steric factors destabilize the other TSs. The substituents on the ligand determine the facial selectivity of the aldehydes.

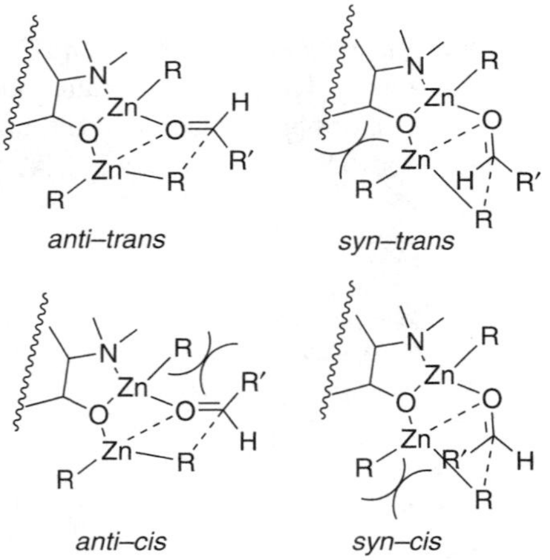

Fig. 7.4. Tricyclic transition structures for aminoalcohol catalysts: *syn* and *anti* refer to the relationship between the transferring group and the bidentate ligand; *cis* and *trans* refer to the relationship between the aldehyde substituent and the coordinating zinc. Reproduced from *J. Am. Chem. Soc.*, **125**, 5130 (2003), by permission of the American Chemical Society.

150. T. Rasmussen and P.-O. Norrby, *J. Am. Chem. Soc.*, **125**, 5130 (2003).

Visual models and additional information on Dialkylzinc Addition can be found in the Digital Resource available at: Springer.com/carey-sundberg.

655

SECTION 7.3

Organometallic Compounds of Group IIB and IIIB Metals

Aryl zinc reagents are considerably more reactive than alkylzinc reagents in these catalyzed additions to aldehydes.[151] Within the same computational framework, phenyl transfer is found to have about a 10 kcal/mol advantage over ethyl transfer.[152] This is attributed to participation of the π orbital of the phenyl ring and to the greater electronegativity of the phenyl ring, which enhances the Lewis acid character of the catalytic zinc.

Aspects of the scale-up of aminoalcohol-catalyzed organozinc reactions with aldehydes have been investigated using N,N-diethylnorephedrine as a catalyst.[153] In addition to examples with aromatic aldehydes, 3-hexanol was prepared in 80% e.e.

$$CH_3(CH_2)_2CH{=}O \quad + \quad (C_2H_5)_2Zn \quad \longrightarrow \quad CH_3(CH_2)_2CHCH_2CH_3$$

80% e.e.

Additions to aldehydes are also catalyzed by Lewis acids, especially $Ti(i\text{-}OPr)_4$ and trimethylsilyl chloride.[154] Reactions of β-, γ-, δ-, and ε-iodozinc esters with benzaldehyde are catalyzed by $Ti(i\text{-}OPr)_3Cl$.[155]

$$PhCH{=}O \quad + \quad IZn(CH_2)_nCO_2C_2H_5 \quad \xrightarrow{Ti(i\text{-}OPr)_3Cl} \quad PhCH(CH_2)_nCO_2C_2H_5$$

n	% yield
2	90*
3	88*
4	80
5	95

* product is lactone

[151.] C. Bohm, N. Kesselgruber, N. Hermanns, J. P. Hildebrand, and G. Raabe, *Angew. Chem. Int. Ed. Engl.*, **40**, 1488 (2001); C. Bohm, J. P. Hildebrand, K. Muniz, and N. Hermanns, *Angew. Chem. Int. Ed. Engl.*, **40**, 3284 (2001).

[152.] J. Rudolph, T. Rasmussen, C. Bohm, and P.-O. Norrby, *Angew. Chem. Int. Ed. Engl.*, **42**, 3002 (2003).

[153.] J. Blacker, *Scale-Up of Chemical Processes,* Conference Proc., 1998; *Chem. Abstr.*, **133**, 296455 (2000).

[154.] D. J. Ramon and M. Yus, *Recent Res. Devel. Org. Chem.*, **2**, 489 (1998).

[155.] H. Ochiai, T. Nishihara, Y. Tamaru, and Z. Yoshida, *J. Org. Chem.*, **53**, 1343 (1988).

Lewis acid–catalyzed additions can be carried out in the presence of other chiral ligands that induce enantioselectivity.[156] Titanium TADDOL induces enantioselectivity in alkylzinc additions to aldehydes. A variety of aromatic, alkyl, and α, β-unsaturated aldehydes give good results with primary alkylzinc reagents.[157]

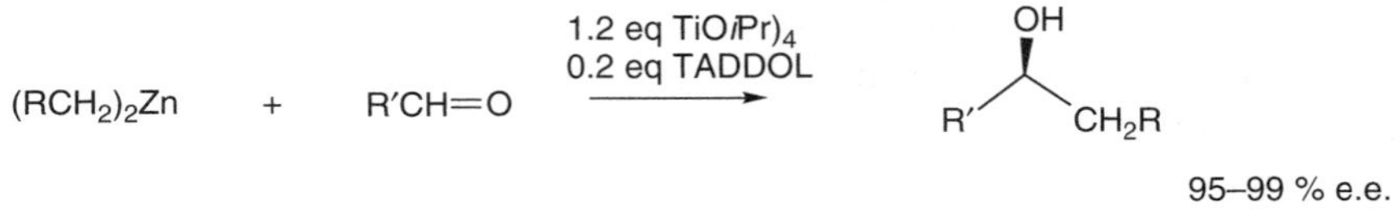

1.2 eq TiO*i*Pr)4
0.2 eq TADDOL
(RCH2)2Zn
+
R′CH=O
OH
R′
CH2R
95–99 % e.e.

The *bis*-trifluoromethanesulfonamide of *trans*-cyclohexane-1,2-diamine also leads to enantioselective additions in 80% or greater e.e.[158]

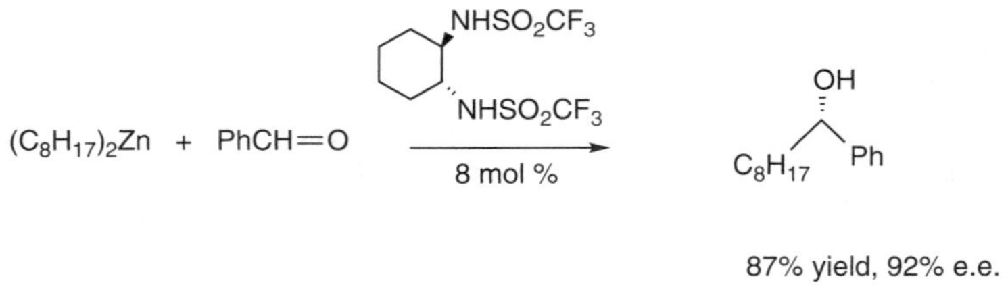

NHSO2CF3
NHSO2CF3
(C8H17)2Zn + PhCH=O
8 mol %
OH
C8H17
Ph
87% yield, 92% e.e.

Ketones are less reactive than aldehydes toward organozinc reagents, and they are inherently less stereoselective because the differentiation is between two carbon substituents, rather than between a carbon substituent and hydrogen. Recently, a diol incorporating both *trans*-cyclohexanediamine and camphorsulfonic acid has proven effective in conjunction with titanium tetraisopropoxide.[159]

O2S
OH
NH
NH
OH
O2S

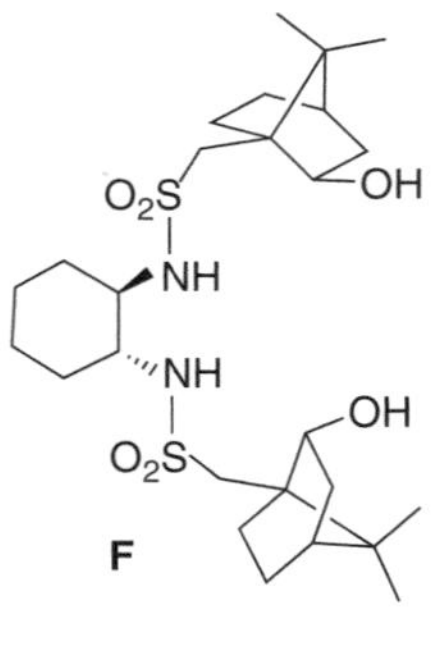

F

The active catalyst is probably a dinuclear species in which the chiral ligand replaces isopropoxide.

[156] D. Seebach, D. A. Plattner, A. K. Beck, Y. M. Wand, D. Hunziker, and W. Petter, *Helv. Chim. Acta*, **75**, 2171 (1992).
[157] D. Seebach, A. K. Beck, B. Schmidt, and Y. M. Wang, *Tetrahedron*, **50**, 4363 (1994); B. Weber and D. Seebach, *Tetrahedron*, **50**, 7473 (1994).
[158] F. Langer, L. Schwink, A. Devasagayaraj, P.-Y. Chavant, and P. Knochel, *J. Org. Chem.*, **61**, 8229 (1996); C. Lutz and P. Knochel, *J. Org. Chem.*, **62**, 7895 (1997).
[159] D. J. Ramon and M. Yus, *Angew. Chem. Int. Ed. Engl.*, **43**, 284 (2004); M. Yus, D. J. Ramon, and O. Prieto, *Tetrahedron: Asymmetry*, **13**, 2291 (2002); C. Garcia, L. K. La Rochelle, and P. J. Walsh, *J. Am. Chem. Soc.*, **124**, 10970 (2002); S.-J. Jeon and P. J. Walsh, *J. Am. Chem. Soc.*, **125**, 9544 (2003).

$$\text{PhCCH}_3 \quad + \quad (\text{C}_2\text{H}_5)_2\text{Zn} \quad \xrightarrow[\text{Ti(O}i\text{Pr})_4 \ 120\%]{10\% \ \mathbf{F}} \quad \overset{\text{HO}\diagdown \ \diagup\text{CH}_3}{\underset{\text{Ph}\diagup \ \diagdown\text{C}_2\text{H}_5}{}}$$

80% yield, 98% e.e.

Lewis acids catalyze the reaction of alkylzinc reagents with acyl chlorides.[160] The reaction is also catalyzed by transition metals, as is discussed in Chapter 8.

$$\text{C}_7\text{H}_{15}\overset{\text{O}}{\overset{\|}{\text{C}}}\text{Cl} \quad \xrightarrow[\substack{2)\ (\text{C}_2\text{H}_5)_2\text{Zn} \\ -30°\text{C then} \\ 25°\text{C}}]{1)\ \text{AlCl}_3} \quad \text{C}_7\text{H}_{15}\overset{\text{O}}{\overset{\|}{\text{C}}}\text{C}_2\text{H}_5$$

94%

Immonium salts are sufficiently reactive to add organozinc halides in the absence of a catalyst.[161] Diallylamines were used because of the ease of subsequent deallylation (see Section 3.5.2).

$$\text{RZnCl} \quad + \quad \text{CH}_2=\text{N}^+(\text{CH}_2\text{CH}=\text{CH}_2)_2 \quad \longrightarrow \quad \text{RCH}_2\text{N}(\text{CH}_2\text{CH}=\text{CH}_2)_2$$

70–90%

The *Reformatsky reaction* is a classical reaction in which metallic zinc, an α-haloester, and a carbonyl compound react to give a β-hydroxyester.[162] The zinc and α-haloester react to form an organozinc reagent. Because the carboxylate group can stabilize the carbanionic center, the product is essentially the zinc enolate of the dehalogenated ester.[163] The enolate effects nucleophilic attack on the carbonyl group.

$$\text{C}_2\text{H}_5\text{O}_2\text{CCH}_2\text{Br} \ + \ \text{Zn} \ \longrightarrow \ \overset{\text{O}^-\text{Zn}^{2+}}{\underset{}{\text{C}_2\text{H}_5\text{O}\overset{|}{\text{C}}=\text{CH}_2}} \ + \ \text{Br}^- \ \longrightarrow$$

With 2-alkylcyclohexanones, the reaction shows a modest preference for equatorial attack.[164]

R = CH₃, C₂H₅, C₃H₇

4:1 preference for equatorial attack

160. M. Arisawa, Y. Torisawa, M. Kawahara, M. Yamanaka, A. Nishida, and M. Nagakawa, *J. Org. Chem.*, **62**, 4327 (1997).
161. N. Millot, C. Piazza, S. Avolio, and P. Knochel, *Synthesis*, 941 (2000).
162. R. L. Shriner, *Org. React.*, **1**, 1 (1942); M. W. Rathke, *Org. React.*, **22**, 423 (1975); A. Furstner, *Synthesis*, 371 (1989); A. Furstner, in *Organozinc Reagents*, P. Knochel and P. Jones, eds., Oxford University Press, New York, 1999, pp. 287–305.
163. W. R. Vaughan and H. P. Knoess, *J. Org. Chem.*, **35**, 2394 (1970).
164. T. Matsumoto and K. Fukui, *Bull. Chem. Soc. Jpn.*, **44**, 1090 (1971).

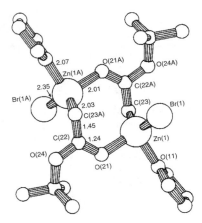

Fig. 7.5. Crystal structure of Reformatsky reagent of *t*-butyl bromoacetate crystallized from THF. Reproduced from *J. Chem. Soc., Chem. Commun.*, 553 (1983), by permission of the Royal Society of Chemistry.

The Reformatsky reaction is related to both organometallic and aldol addition reactions and probably involves a cyclic TS. The Reformatsky reagent from *t*-butyl bromoacetate crystallizes as a dimer having both O—Zn (enolate-like) and C—Zn (organometallic-like) bonds (see Figure 7.5).[165]

It is believed that the reaction occurs through the monomer.[166] Semiempirical MO (PM3) calculations suggest a boat TS.[167] There do not seem to be any definitive experimental studies that define the mechanism precisely.

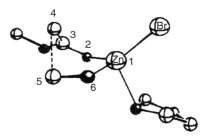

Several techniques have been used to "activate" the zinc metal and improve yields. For example, pretreatment of zinc dust with a solution of copper acetate gives a more reactive zinc-copper couple.[168] Exposure to trimethylsilyl chloride also activates the zinc.[169] Wilkinson's catalyst, RhCl(PPh₃)₃ catalyzes formation of Reformatsky reagents from diethylzinc, and reaction occurs under very mild conditions.[170]

165. J. Dekker, J. Boersma, and G. J. M. van der Kerk, *J. Chem. Soc., Chem. Commun.*, 553 (1983).
166. M. J. S. Dewar and K. M. Merz, Jr., *J. Am. Chem. Soc.*, **109**, 6553 (1987).
167. J. Maiz, A. Arrieta, X. Lopez, J. M. Ugalde, F. P. Cossio, and B. Lecea, *Tetrahedron Lett.*, **34**, 6111 (1993).
168. E. Le Goff, *J. Org. Chem.*, **29**, 2048 (1964); L. R. Krepski, L. E. Lynch, S. M. Heilmann, and J. K. Rasmussen, *Tetrahedron Lett.*, **26**, 981 (1985).
169. G. Picotin and P. Miginiac, *J. Org. Chem.*, **52**, 4796 (1987).
170. K. Kanai, H. Wakabayshi, and T. Honda, *Org. Lett.*, **2**, 2549 (2000).

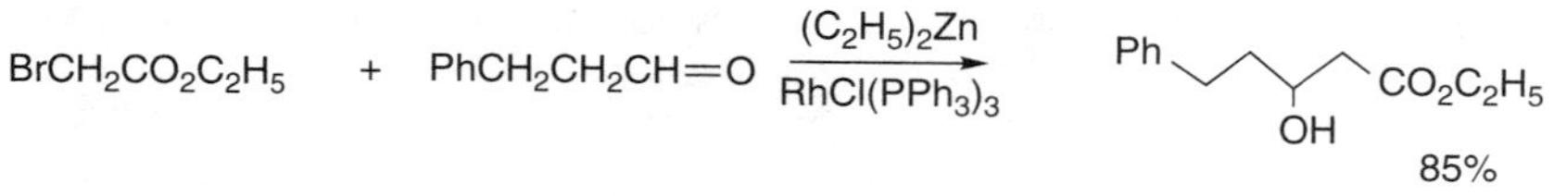

$$BrCH_2CO_2C_2H_5 \quad + \quad PhCH_2CH_2CH{=}O \xrightarrow[RhCl(PPh_3)_3]{(C_2H_5)_2Zn}$$

85%

These conditions also provide good yields in intramolecular reactions. There is a preference for formation of the *cis* product for five- and six-membered rings.

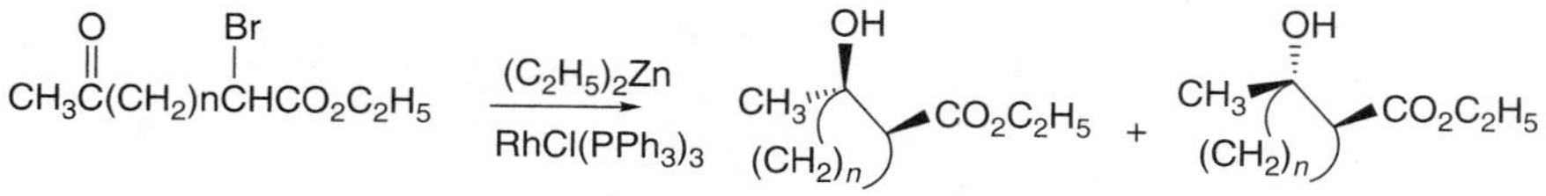

n	cis	trans
3	59%	5%
4	65%	26%

Scheme 7.5 gives some examples of the Reformatsky reaction. Zinc enolates prepared from α-haloketones can be used as nucleophiles in mixed aldol condensations (see Section 2.1.3). Entry 7 is an example. This type of reaction can be conducted in the presence of the Lewis acid diethylaluminum chloride, in which case addition occurs at $-20°$ C.[171]

7.3.1.3. Related Reactions Involving Organozinc Compounds. Organozinc reagents can be converted to anionic "zincate" species by reaction with organolithium compounds.[172] These reagents react directly with aldehydes and ketones to give addition products.[173]

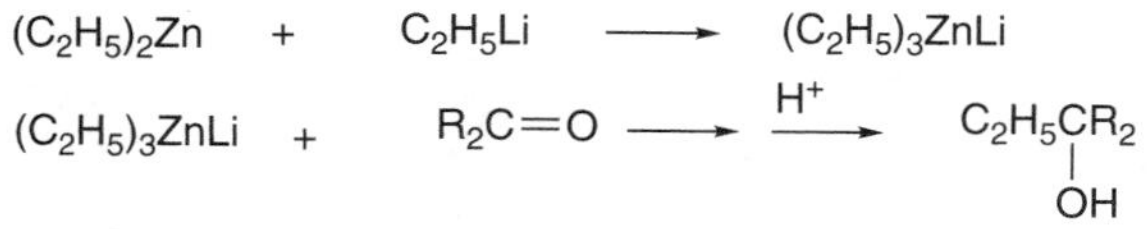

$$(C_2H_5)_2Zn \quad + \quad C_2H_5Li \longrightarrow (C_2H_5)_3ZnLi$$

The 1:1 zincate reagent is believed to be dimeric. At higher ratios of organolithium compounds, 2:1 and 3:1 species can be formed.[174]

Zincate reagents can add to imines with or without Lewis acid catalysis. Alkylimines require BF_3 but imines of pyridine-2-carboxaldehyde react directly. If the imines are derived from chiral amines, diastereoselectivity is observed. Both α-phenylethyl amine and ethyl valinate have been tried. Higher enantioselectivity was observed with mixed magnesium reagents.[175]

[171] K. Maruoka, S. Hashimoto, Y. Kitagawa, H. Yamamoto, and H. Nozaki, *J. Am. Chem. Soc.*, **99**, 7705 (1977).

[172] D. J. Linton, P. Shooler, and A. E. H. Wheatley, *Coord. Chem. Rev.*, **223**, 53 (2001).

[173] C. A. Musser and H. G. Richey, Jr., *J. Org. Chem.*, **65**, 7750 (2000).

[174] M. Uchiyama, M. Kameda, O. Mishima, N. Yokoyama, M. Koike, Y. Kondo, and T. Sakamoto, *J. Am. Chem. Soc.*, **120**, 4934 (1998).

[175] G. Alvaro, P. Pacioni, and D. Savoia, *Chem. Eur. J.*, **3**, 726 (1997).

Scheme 7.5. Addition of Zinc Enolates to Carbonyl Compounds: the Reformatsky Reaction

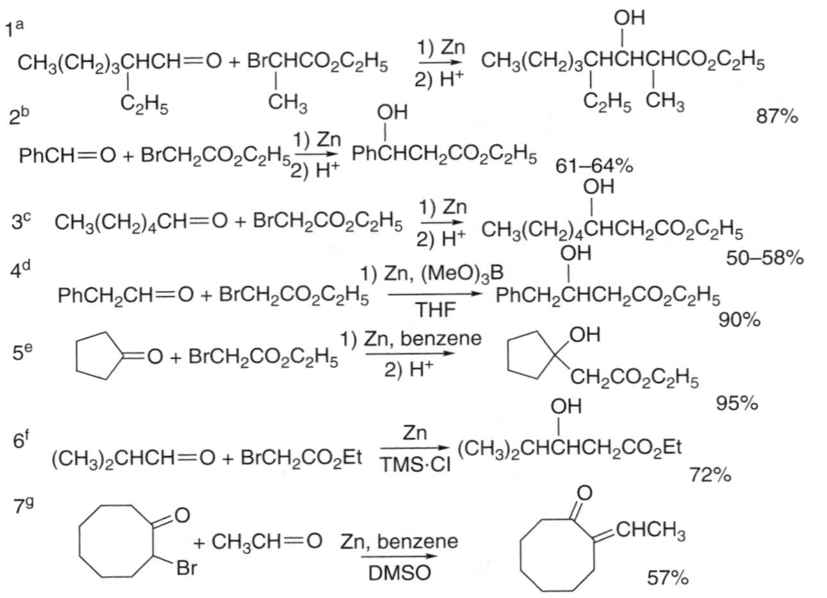

1[a] $CH_3(CH_2)_3CHCH=O + BrCHCO_2C_2H_5 \xrightarrow[\text{2) H}^+]{\text{1) Zn}}$ ... $CH_3(CH_2)_3CHCHCHCO_2C_2H_5$
with C_2H_5 and CH_3 substituents, OH; 87%

2[b] $PhCH=O + BrCH_2CO_2C_2H_5 \xrightarrow[\text{2) H}^+]{\text{1) Zn}} PhCHCH_2CO_2C_2H_5$ (OH) 61–64%

3[c] $CH_3(CH_2)_4CH=O + BrCH_2CO_2C_2H_5 \xrightarrow[\text{2) H}^+]{\text{1) Zn}} CH_3(CH_2)_4CHCH_2CO_2C_2H_5$ (OH) 50–58%

4[d] $PhCH_2CH=O + BrCH_2CO_2C_2H_5 \xrightarrow[\text{THF}]{\text{1) Zn, (MeO)}_3B} PhCH_2CHCH_2CO_2C_2H_5$ (OH) 90%

5[e] cyclopentanone $=O + BrCH_2CO_2C_2H_5 \xrightarrow[\text{2) H}^+]{\text{1) Zn, benzene}}$ cyclopentane with OH and $CH_2CO_2C_2H_5$ 95%

6[f] $(CH_3)_2CHCH=O + BrCH_2CO_2Et \xrightarrow[\text{TMS·Cl}]{\text{Zn}} (CH_3)_2CHCHCH_2CO_2Et$ (OH) 72%

7[g] cyclooctanone (O) with Br $+ CH_3CH=O \xrightarrow[\text{DMSO}]{\text{Zn, benzene}}$ cyclooctanone with =CHCH_3 57%

a. K. L. Rinehart, Jr., and E. G. Perkins, *Org. Synth.*, **IV**, 444 (1963).
b. C. R. Hauser and D. S. Breslow, *Org. Synth.*, **III**, 408 (1955).
c. J. W. Frankenfeld and J. J. Werner, *J. Org. Chem.*, **34**, 3689 (1969).
d. M. W. Rathke and A. Lindert, *J. Org. Chem.*, **35**, 3966 (1971).
e. J. F. Ruppert and J. D. White, *J. Org. Chem.*, **39**, 269 (1974).
f. G. Picotin and P. Migniac, *J. Org. Chem.*, **52**, 4796 (1987).
g. T. A. Spencer, R. W. Britton, and D. S. Watt, *J. Am. Chem. Soc.*, **89**, 5727 (1967).

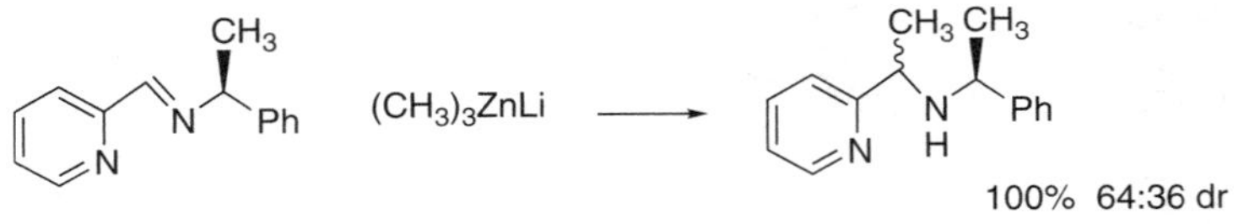

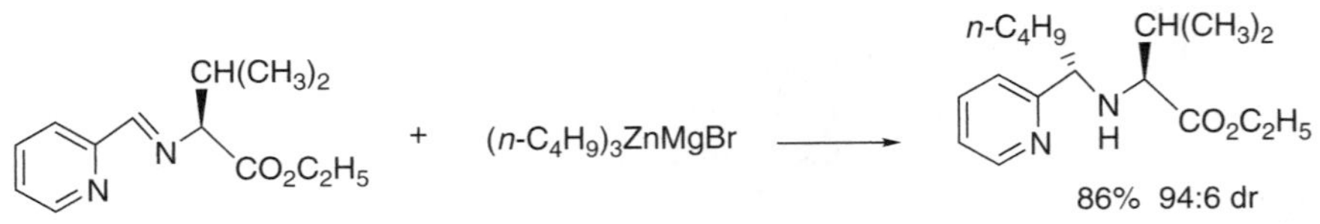

pyridyl-CH=N-CH(CH_3)Ph $+ (CH_3)_3ZnLi \longrightarrow$ pyridyl-CH(CH_3)-NH-CH(CH_3)Ph   100% 64:36 dr

pyridyl-CH=N-CH(CH(CH_3)_2)CO_2C_2H_5 $+ (n\text{-}C_4H_9)_3ZnMgBr \longrightarrow$ pyridyl-CH(n-C_4H_9)-NH-CH(CH(CH_3)_2)CO_2C_2H_5   86% 94:6 dr

Organozinc reagents have been used in conjunction with α-bromovinylboranes in a tandem route to Z-trisubstituted allylic alcohols. After preparation of the vinylborane, reaction with diethylzinc effects migration of a boron substituent with inversion of configuration and exchange of zinc for boron.[176] Addition of an aldehyde then gives the allylic alcohol. The reaction is applicable to formaldehyde; alkyl and aryl aldehydes; and to methyl, primary, and secondary boranes.

176. Y. K. Chen and P. J. Walsh, *J. Am. Chem. Soc.*, **126**, 3702 (2004).

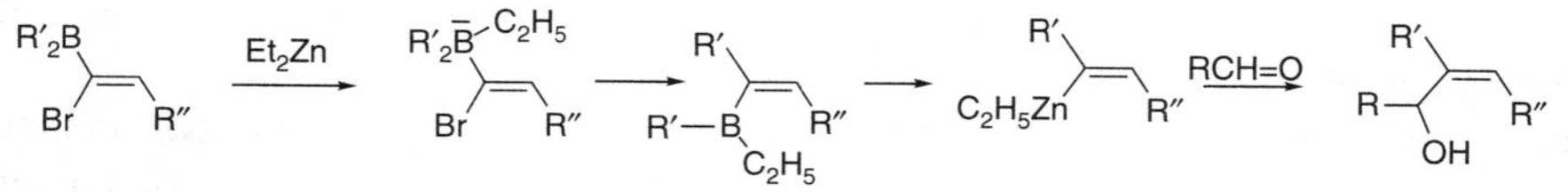

The reagent combination Zn-CH_2Br_2-$TiCl_4$ gives rise to an organometallic reagent known as *Lombardo's reagent*, which converts ketones to methylene groups.[177] The active reagent is presumed to be a dimetallated species that adds to the ketone under the influence of the Lewis acidity of titanium. β-Elimination then generates the methylene group.

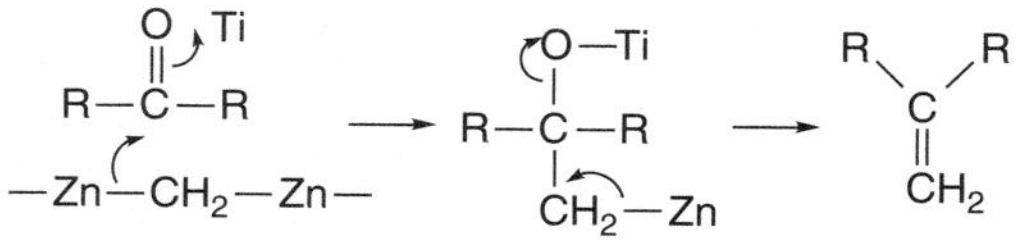

Use of esters and 1,1-dibromoalkanes as reactants gives enol ethers.[178]

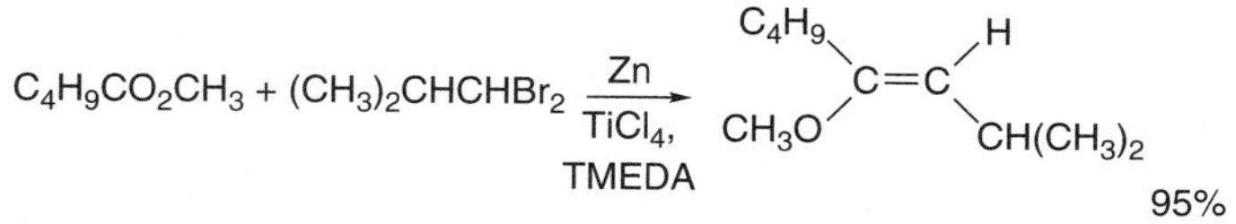

$$C_4H_9CO_2CH_3 + (CH_3)_2CHCHBr_2 \xrightarrow[\text{TMEDA}]{\text{Zn} \atop \text{TiCl}_4,}$$

95%

A similar procedure starting with trimethylsilyl esters generates trimethylsilyl enol ethers.[179]

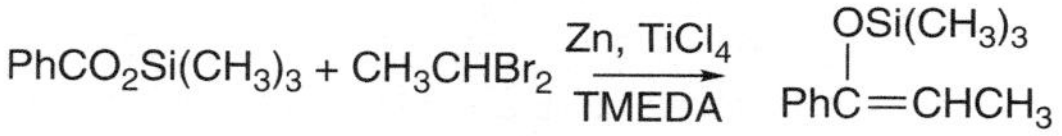

$$PhCO_2Si(CH_3)_3 + CH_3CHBr_2 \xrightarrow[\text{TMEDA}]{\text{Zn, TiCl}_4}$$

Organozinc reagents are also used extensively in conjunction with palladium in a number of carbon-carbon bond-forming processes that are discussed in Section 8.2.

7.3.2. Organocadmium Compounds

Organocadmium compounds can be prepared from Grignard reagents or organolithium compounds by reaction with Cd(II) salts.[180] They can also be prepared directly from alkyl, benzyl, and aryl halides by reaction with highly reactive cadmium metal generated by reduction of Cd(II) salts.[181]

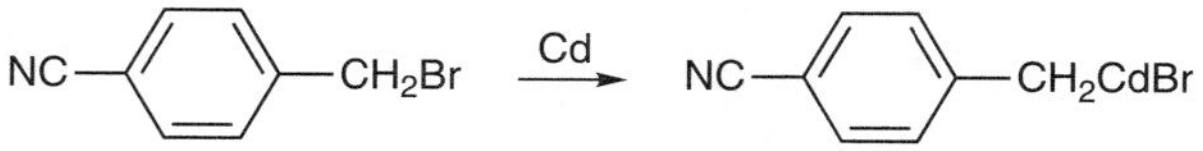

$$NC{-}\langle\text{ }\rangle{-}CH_2Br \xrightarrow{\text{Cd}} NC{-}\langle\text{ }\rangle{-}CH_2CdBr$$

The reactivity of these reagents is similar to the corresponding organozinc compounds.

[177.] K. Oshima, K. Takai, Y. Hotta, and H. Nozaki, *Tetrahedron Lett.*, 2417 (1978); L. Lombardo, *Tetrahedron Lett.*, **23**, 4293 (1982); L. Lombardo, *Org. Synth.*, **65**, 81 (1987).
[178.] T. Okazoe, K. Takai, K. Oshima, and K. Utimoto, *J. Org. Chem.*, **52**, 4410 (1987).
[179.] K. Takai, Y. Kataoka, T. Okazoe, and K. Utimoto, *Tetrahedron Lett.*, **29**, 1065 (1988).
[180.] P. R. Jones and P. J. Desio, *Chem. Rev.*, **78**, 491 (1978).
[181.] E. R. Burkhardt and R. D. Rieke, *J. Org. Chem.*, **50**, 416 (1985).

The most common application of organocadmium compounds has been in the preparation of ketones by reaction with acyl chlorides. A major disadvantage of the use of organocadmium reagents is the toxicity and environmental problems associated with use of cadmium, and this has limited the recent use of organocadmium reagents.

$$[(CH_3)_2CHCH_2CH_2]_2Cd + ClCCH_2CH_2CO_2CH_3 \longrightarrow (CH_3)_2CHCH_2CH_2COCH_2CH_2CO_2CH_3$$

73–75%

Ref. 182

60%

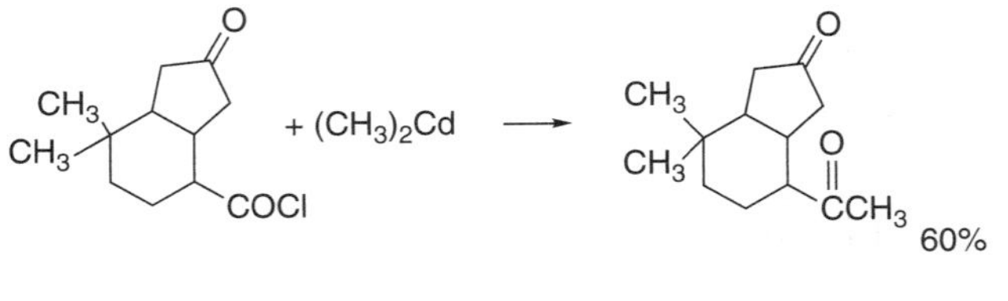

Ref. 183

7.3.3. Organomercury Compounds

There are several useful methods for preparation of organomercury compounds. The general metal-metal exchange reaction between mercury(II) salts and organolithium or magnesium compounds is applicable. The oxymercuration reaction discussed in Section 4.1.3 provides a means of acquiring certain functionalized organomercury reagents. Organomercury compounds can also be obtained by reaction of mercuric salts with trialkylboranes, although only primary alkyl groups react readily.[184] Other organoboron compounds, such as boronic acids and boronate esters also react with mercuric salts.

$$R_3B + 3\ Hg(O_2CCH_3)_2 \longrightarrow 3\ RHgO_2CCH_3$$

$$RB(OH)_2 + Hg(O_2CCH_3)_2 \longrightarrow RHgO_2CCH_3$$

$$RB(OR')_2 + Hg(O_2CCH_3)_2 \longrightarrow RHgO_2CCH_3$$

Alkenylmercury compounds can be prepared by hydroboration of an alkyne with catecholborane, followed by reaction with mercuric acetate.[185]

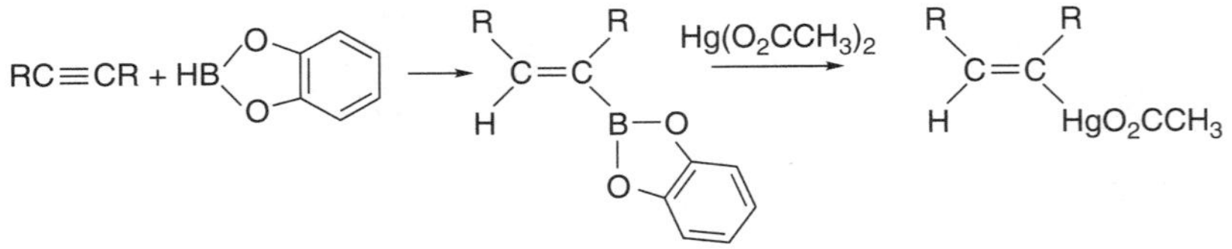

182. J. Cason and F. S. Prout, *Org. Synth.*, **III**, 601 (1955).
183. M. Miyano and B. R. Dorn, *J. Org. Chem.*, **37**, 268 (1972).
184. R. C. Larock and H. C. Brown, *J. Am. Chem. Soc.*, **92**, 2467 (1970); J. J. Tufariello and M. M. Hovey, *J. Am. Chem. Soc.*, **92**, 3221 (1970).
185. R. C. Larock, S. K. Gupta, and H. C. Brown, *J. Am. Chem. Soc.*, **94**, 4371 (1972).

The organomercury compounds can be used in situ or isolated as organomercuric halides.

Organomercury compounds are weak nucleophiles and react only with very reactive electrophiles. They readily undergo electrophilic substitution by halogens.

$$CH_3(CH_2)_6CH=CH_2 \xrightarrow[\substack{2)\ Hg(O_2CCH_3)_2 \\ 3)\ Br_2}]{1)\ B_2H_6} CH_3(CH_2)_8Br$$

69%

Ref. 184

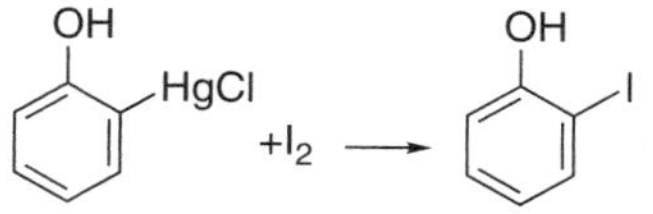

Ref. 186

Organomercury reagents do not react with ketones or aldehydes but Lewis acids cause reaction with acyl chlorides.[187] With alkenyl mercury compounds, the reaction probably proceeds by electrophilic attack on the double bond with the regiochemistry being directed by the stabilization of the β-carbocation by the mercury.[188]

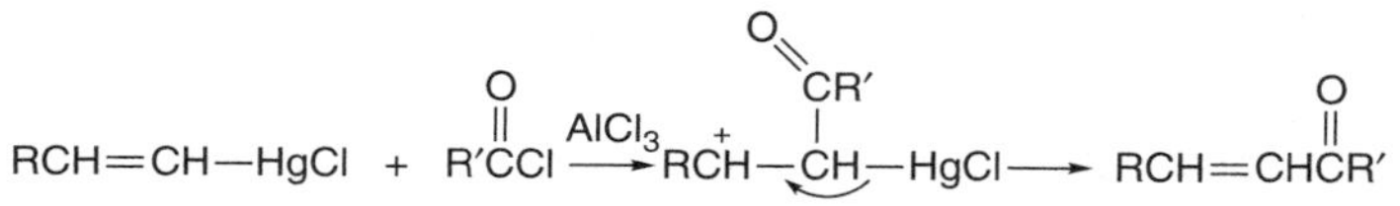

Most of the synthetic applications of organomercury compounds are in transition metal–catalyzed processes in which the organic substituent is transferred from mercury to the transition metal in the course of the reaction. Examples of this type of reaction are considered in Chapter 8.

7.3.4. Organoindium Reagents

Indium is a Group IIIB metal and is a congener of aluminum. Considerable interest has developed recently in the synthetic application of organoindium reagents.[189] One of the properties that makes indium useful is that its first oxidation potential is less than that of zinc and even less than that of magnesium, making it quite reactive as an electron donor to halides. Indium metal reacts with allylic halides in the presence of aldehydes to give the corresponding carbinols.

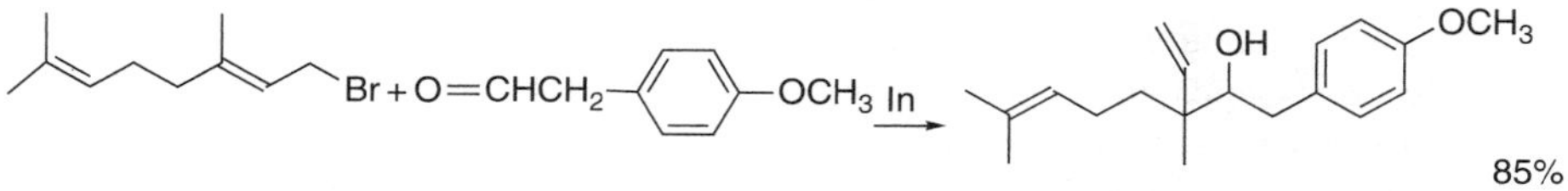

85%

Ref. 190

186. F. C. Whitmore and E. R. Hanson, *Org. Synth.*, **I**, 326 (1941).
187. A. L. Kurts, I. P. Beletskaya, I. A. Savchenko, and O. A. Reutov, *J. Organomet. Chem.*, **17**, 21 (1969).
188. R. C. Larock and J. C. Bernhardt, *J. Org. Chem.*, **43**, 710 (1978).
189. P. Cintas, *Synlett*, 1087 (1995).
190. S. Araki and Y. Butsugan, *J. Chem. Soc., Perkin Trans. 1*, 2395 (1991).

It is believed that the reaction proceeds through a cyclic TS and that the reagent is an In(I) species.[191]

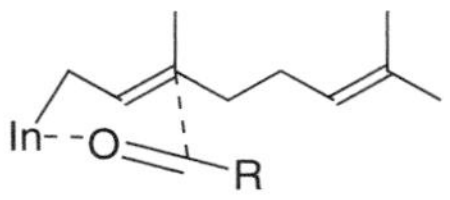

A striking feature of the reactions of indium and allylic halides is that they can be carried out in aqueous solution.[192] The aldehyde traps the organometallic intermediate as it is formed.

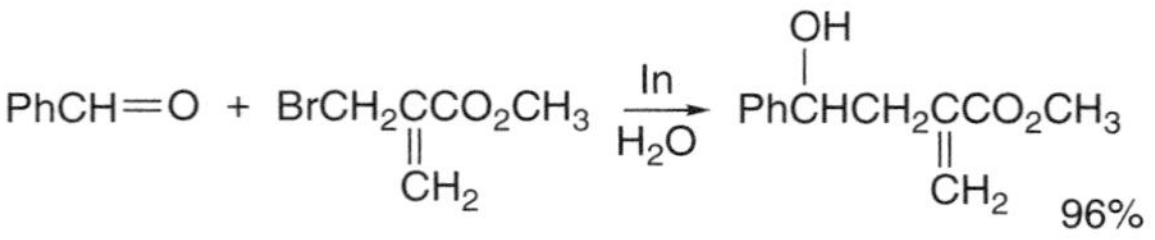

The reaction has been found to be applicable to functionalized allylic halides and aldehydes.

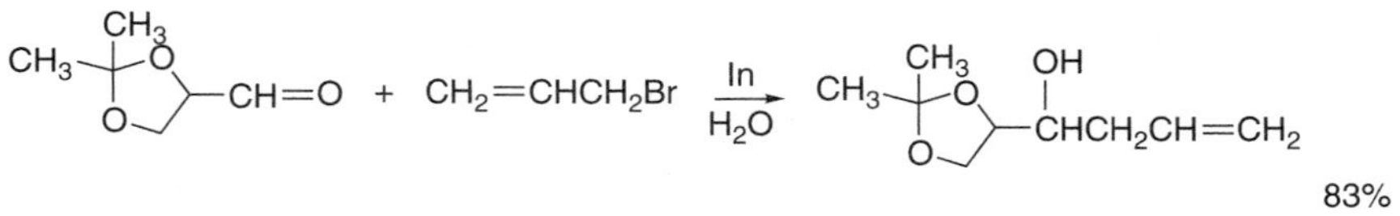

Ref. 193

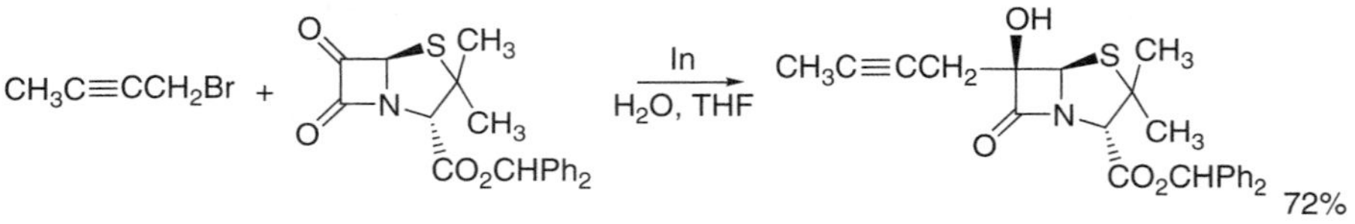

Ref. 194

7.4. Organolanthanide Reagents

The lanthanides are congeners of the Group IIIA metals scandium and yttrium, with the +3 oxidation state usually being the most stable. These ions are strong oxyphilic Lewis acids and catalyze carbonyl addition reactions by a number of nucleophiles. Recent years have seen the development of synthetic procedures involving lanthanide metals, especially cerium.[195] In the synthetic context, organocerium

191. T. H. Chan and Y. Yang, *J. Am. Chem. Soc.*, **121**, 3228 (1999).
192. C.-J. Li and T. H. Chan, *Tetrahedron Lett.*, **32**, 7017 (1991); C.-J. Li, *Tetrahedron*, **52**, 5643 (1996).
193. L. A. Paquette and T. M. Mitzel, *J. Am. Chem. Soc.*, **118**, 1931 (1996); L. A. Paquette and R. R. Rothhaar, *J. Org. Chem.*, **64**, 217 (1999).
194. Y. S. Cho, J. E. Lee, A. N. Pae, K. I. Choi, and H. Y. Yok, *Tetrahedron Lett.*, **40**, 1725 (1999).
195. H. J. Liu, K.-S. Shia, X. Shange, and B.-Y. Zhu, *Tetrahedron*, **55**, 3803 (1999); R. Dalpozzo, A. De Nino, G. Bartoli, L. Sambri, and E. Marcantonio, *Recent Res. Devel. Org. Chem.*, **5**, 181 (2001).

compounds are usually prepared by reaction of organolithium compounds with $CeCl_3$.[196] The precise details of preparation of the $CeCl_3$ and its reaction with the organolithium compound can be important to the success of individual reactions.[197] The organocerium compounds are useful for addition to carbonyl compounds that are prone to enolization or are sterically hindered.[198] The organocerium reagents retain strong nucleophilicity but show a much reduced tendency to effect deprotonation. For example, in addition of trimethylsilylmethyllithium to relatively acidic ketones such as 2-indanone, the yield was greatly increased by use of the organocerium intermediate.[199]

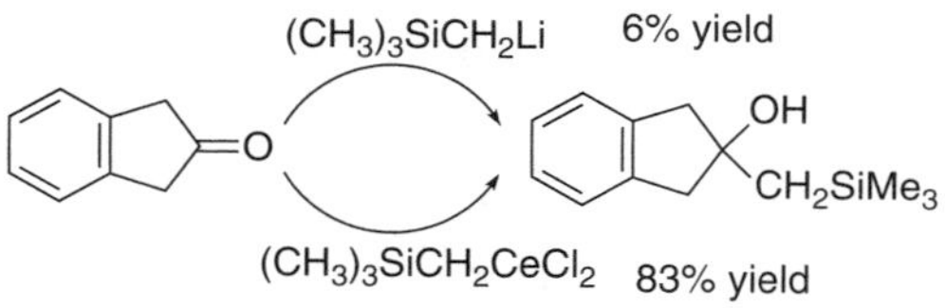

Organocerium reagents have been found to improve yields in additions to bicyclo[3.3.1]nonan-3-ones.[200]

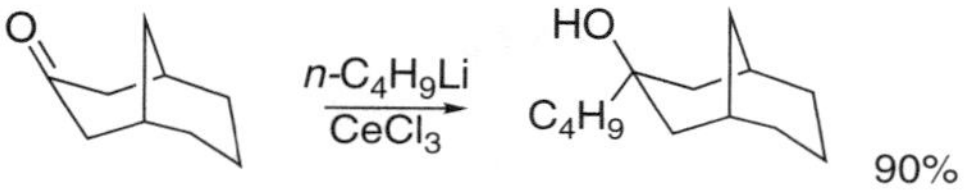

An organocerium reagent gave better yields than either the lithium or Grignard reagents in addition to carbonyl at the 17-position on steroids.[201] Additions of both Grignard and organolithium reagents can be catalyzed by 5–10 mol % of $CeCl_3$.

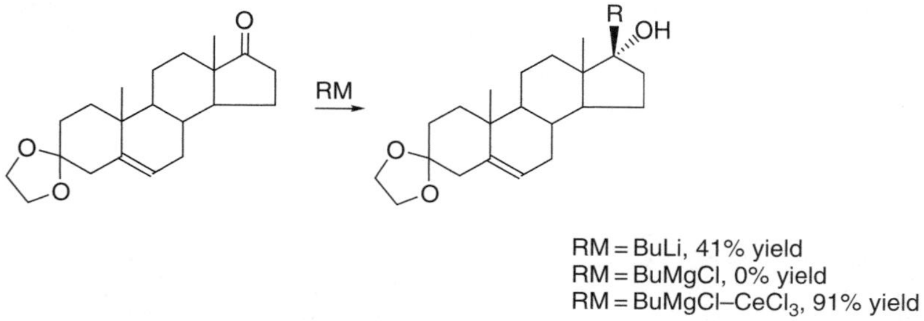

RM = BuLi, 41% yield
RM = BuMgCl, 0% yield
RM = BuMgCl–$CeCl_3$, 91% yield

[196]. T. Imamoto, T. Kusumoto, Y. Tawarayama, Y. Sugiura, T. Mita, Y. Hatanaka, and M. Yokoyama, *J. Org. Chem.*, **49**, 3904 (1984).

[197]. D. J. Clive, Y. Bu, Y. Tao, S. Daigneault, Y.-J. Wu, and G. Meignan, *J. Am. Chem. Soc.*, **120**, 10332 (1998); W. J. Evans, J. D. Feldman, and T. W. Ziller, *J. Am. Chem. Soc.*, **118**, 4581 (1996); V. Dimitrov, K. Kostova, and M. Genov, *Tetrahedron Lett.*, **37**, 6787 (1996).

[198]. T. Inamoto, N. Takiyama, K. Nakamura, T. Hatajma, and Y. Kamiya, *J. Am. Chem. Soc.*, **111**, 4392 (1989).

[199]. C. R. Johnson and B. D. Tait, *J. Org. Chem.*, **52**, 281 (1987).

[200]. T. Momose, S. Takazawa, and M. Kirihara, *Synth. Commun.*, **27**, 3313 (1997).

[201]. V. Dimitrov, S. Bratovanov, S. Simova, and K. Kostova, *Tetrahedron Lett.*, **36**, 6713 (1994); X. Li, S. M. Singh, and F. Labrie, *Tetrahedron Lett.*, **35**, 1157 (1994).

Cerium reagents have also been found to give improved yields in the reaction of organolithium reagents with carboxylate salts to give ketones.

$$CH_3(CH_2)_2Li + CH_3(CH_2)_4CO_2Li \xrightarrow{\text{2 equiv CeCl}_3} CH_3(CH_2)_2\overset{\overset{\displaystyle O}{\|}}{C}(CH_2)_4CH_3$$

83%

Ref. 202

Amides, especially of piperidine and morpholine, give good yields of ketones on reaction with organocerium reagents.[203] It has been suggested that the morpholine oxygen may interact with the oxyphilic cerium to stabilize the addition intermediate.

This procedure has been used with good results to prepare certain long-chain ketones that are precursors of pheromones.[204]

$$CH_3(CH_2)_7CH=CHCH_2\overset{\overset{\displaystyle O}{\|}}{C}N\bigcirc O + CH_3(CH_2)_2MgBr \xrightarrow{CeCl_3} CH_3(CH_2)_7CH=CHCH_2\overset{\overset{\displaystyle O}{\|}}{C}(CH_2)_2CH_3$$

90%

Organocerium reagents also show excellent reactivity toward nitriles and imines,[205] and organocerium compounds were found to be the preferred organometallic reagent for addition to hydrazones in an enantioselective synthesis of amines.[206]

$$RLi \xrightarrow{CeCl_3} RCeCl_2 \longrightarrow \xrightarrow{ClCO_2CH_3} R'CH_2CH-\underset{\underset{R}{|}}{N}-\underset{\underset{CO_2CH_3}{|}}{N} \xrightarrow{H_2, \text{ Raney Ni}} R'CH_2\underset{\underset{R}{|}}{CH}NH_2$$

General References

E. Erdik, *Organozinc Reagents in Organic Synthesis*, CRC Press, Boca Raton, Fl, 1996.

P. Knochel and P. Jones, Editors, *Organozinc Reagents*, Oxford University Press, Oxford, 1999.

R. C. Larock, *Organomercury Compounds in Organic Synthesis*, Springer-Verlag, Berlin, 1985.

H. G. Richey, Jr., ed., *Grignard Reagents; New Developments*, Wiley, New York, 2000.

M. Schlosser, ed., *Organometallic in Synthesis; A Manual*, Wiley, New York, 1994.

G. S. Silverman and P. E. Rakita, eds., *Handbook of Grignard Reagents*, Marcel Dekker, New York, 1996.

B. J. Wakefield, *The Chemistry of Organolithium Compounds*, Pergamon Press, Oxford, 1974.

B. J. Wakefield, *Organolithium Methods*, Academic Press, Orlando, FL, 1988.

B. J. Wakefield, *Organomagnesium Methods in Organic Synthesis*, Academic Press, London, 1995.

202. Y. Ahn and T. Cohen, *Tetrahedron Lett.*, **35**, 203 (1994).

203. M. Kurosu and Y. Kishi, *Tetrahedron Lett.*, **39**, 4793 (1998).

204. M. Badioli, R. Ballini, M. Bartolacci, G. Bosica, E. Torregiani, and E. Marcantonio, *J. Org. Chem.*, **67**, 8938 (2002).

205. E. Ciganek, *J. Org. Chem.*, **57**, 4521 (1992).

206. S. E. Denmark, T. Weber, and D. W. Piotrowski, *J. Am. Chem. Soc.*, **109**, 2224 (1987).

(References for these problems will be found on page 1283.)

7.1. Predict the product of each of the following reactions. Be sure to consider and specify all aspects of stereochemistry involved in the reaction.

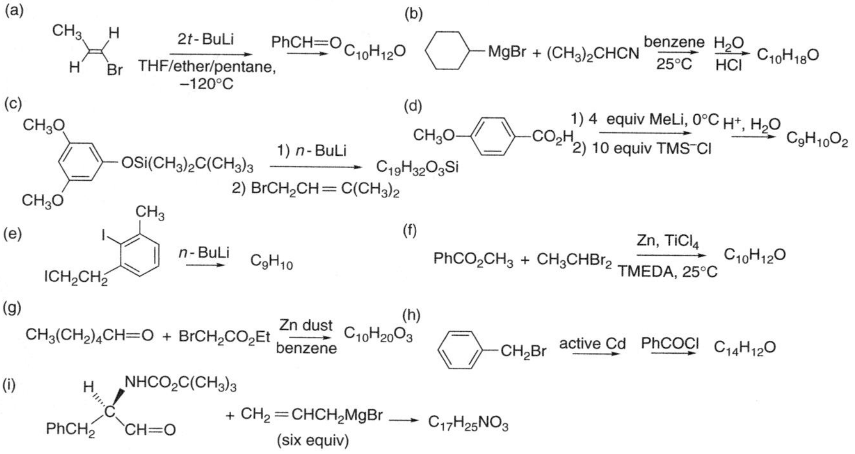

(a)

$$CH_3CH=CHBr \xrightarrow[\text{THF/ether/pentane,} \\ -120°C]{2 t\text{-BuLi}} PhCH=O \quad C_{10}H_{12}O$$

(b)

$$\text{cyclohexyl-MgBr} + (CH_3)_2CHCN \xrightarrow[25°C]{\text{benzene}} \xrightarrow[\text{HCl}]{H_2O} C_{10}H_{18}O$$

(c)

$$\text{(3,5-dimethoxyphenyl)-}OSi(CH_3)_2C(CH_3)_3 \xrightarrow[\text{2) BrCH}_2CH=C(CH_3)_2]{\text{1) } n\text{-BuLi}} C_{19}H_{32}O_3Si$$

(d)

$$CH_3O-\text{C}_6H_4-CO_2H \xrightarrow[\text{2) 10 equiv TMS}^-Cl]{\text{1) 4 equiv MeLi, 0°C}} \xrightarrow[\text{H}_2O]{H^+,} C_9H_{10}O_2$$

(e)

$$\text{(2-iodo-3-(2-iodoethyl)toluene)} \xrightarrow{n\text{-BuLi}} C_9H_{10}$$

(f)

$$PhCO_2CH_3 + CH_3CHBr_2 \xrightarrow[\text{TMEDA, 25°C}]{Zn, TiCl_4} C_{10}H_{12}O$$

(g)

$$CH_3(CH_2)_4CH=O + BrCH_2CO_2Et \xrightarrow[\text{benzene}]{Zn \text{ dust}} C_{10}H_{20}O_3$$

(h)

$$\text{PhCH}_2Br \xrightarrow{\text{active Cd}} \xrightarrow{\text{PhCOCl}} C_{14}H_{12}O$$

(i)

$$PhCH_2-\overset{H\cdots NHCO_2C(CH_3)_3}{\underset{}{C}}-CH=O + CH_2=CHCH_2MgBr \underset{\text{(six equiv)}}{\longrightarrow} C_{17}H_{25}NO_3$$

7.2. Reactions of the epoxide of 1-butene with CH₃Li gives a 90% yield of 3-pentanol. In contrast, reaction with CH₃MgBr under similar conditions gives an array of products, as indicated below. What is the basis for the difference in reactivity of these two organometallic compounds toward this epoxide?

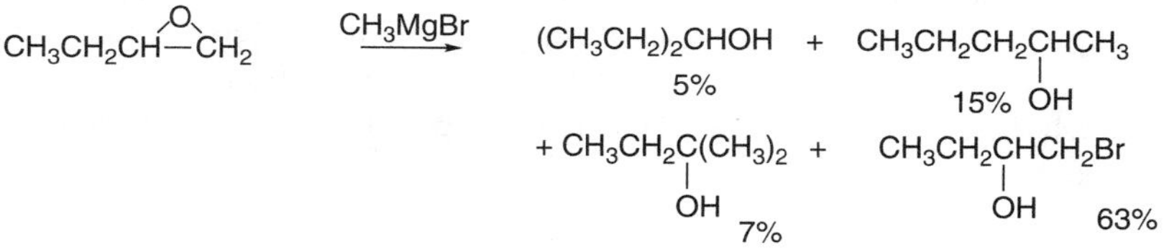

$$CH_3CH_2CH\overset{O}{-}CH_2 \xrightarrow{CH_3MgBr} (CH_3CH_2)_2CHOH + CH_3CH_2CH_2CHCH_3$$
$$5\% \qquad\qquad 15\% \; OH$$

$$+ \; CH_3CH_2C(CH_3)_2 + CH_3CH_2CHCH_2Br$$
$$OH \qquad\qquad OH$$
$$7\% \qquad\qquad\qquad 63\%$$

7.3. Devise an efficient synthesis for the following organometallic compound from the specified starting material.

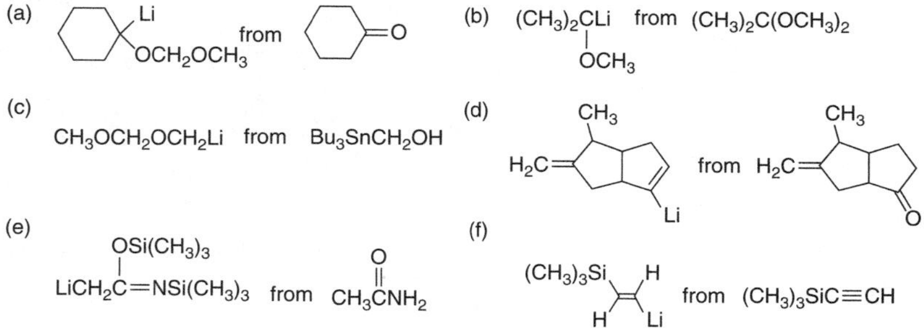

(a)

$$\text{(1-lithio-1-(methoxymethoxy)cyclohexane)} \quad \text{from} \quad \text{cyclohexanone}$$

(b)

$$(CH_3)_2CLi \quad \text{from} \quad (CH_3)_2C(OCH_3)_2$$
$$\quad OCH_3$$

(c)

$$CH_3OCH_2OCH_2Li \quad \text{from} \quad Bu_3SnCH_2OH$$

(d)

$$\text{(bicyclic with CH}_3, H_2C=, Li) \quad \text{from} \quad \text{(bicyclic with CH}_3, H_2C=, O)$$

(e)

$$\underset{}{OSi(CH_3)_3} \atop LiCH_2C=NSi(CH_3)_3 \quad \text{from} \quad CH_3\overset{O}{\overset{\|}{C}}NH_2$$

(f)

$$\underset{H \quad Li}{\overset{(CH_3)_3Si \quad H}{C=C}} \quad \text{from} \quad (CH_3)_3SiC\equiv CH$$

7.4. Each of the following compounds gives a product in which one or more lithium atoms has been introduced under the conditions specified. Predict the structure of the lithiated product on the basis of structural features known to promote lithiation and/or stabilization of lithiated species. The number of lithium atoms introduced is equal to the number of moles of lithium reagent used in each case.

(a)

$$H_2C=CCNHC(CH_3)_3$$ with carbonyl O, and CH$_3$ substituent $\xrightarrow[\text{THF, }-20°C]{\text{2 } n\text{-BuLi, TMEDA,}}$

(b) $(CH_3)_2C=CH_2$ $\xrightarrow[\text{hexane}]{n\text{-BuLi, TMEDA,}}$

(c) $CH_2N(CH_3)_2$ on benzene ring $\xrightarrow[\text{ether, 25°C, 24 h}]{n\text{-BuLi}}$

(d) naphthalene with OCH_3 $\xrightarrow[\text{20 h}]{n\text{-BuLi, ether, 38°C}}$

(e) $HC\equiv CCO_2CH_3$ $\xrightarrow[\text{THF/pentane/ether}]{n\text{-BuLi, }-120°C}$

(f) benzene ring–$NHCC(CH_3)_3$ with carbonyl O $\xrightarrow[\text{0°C, 2 h}]{\text{2 } n\text{-BuLi, THF,}}$

(g) $(CH_3)_2CH$–benzene ring–OCH_3 $\xrightarrow[\text{TMEDA, ether}]{n\text{-BuLi}}$

(h) $\underset{H}{\overset{Ph}{\diagdown}}C=C\underset{CN}{\overset{H}{\diagup}}$ $\xrightarrow[-113°C]{\text{LDA}}$

(i) $CH_2=CCH_2OH$ with CH_3 substituent $\xrightarrow[\text{2 } n\text{-BuLi, 0°C}]{\text{2 K}^+{}^-\text{O-}t\text{-Bu}}$

(j) indole with $PhSO_2$ on N $\xrightarrow[-5°C]{\text{2 } t\text{-BuLi}}$

(k) $Ph_2NCH_2CH=CH_2$ $\xrightarrow{n\text{-BuLi}}$

(l) CH_3O–benzene ring–OCH_3 $\xrightarrow[\text{TMEDA}]{\text{2 } n\text{-BuLi}}$

7.5. Each of the following compounds can be prepared by reactions of organometallic reagents and readily available starting materials. By retrosynthetic analysis, identify an appropriate organometallic reagent in each case and show how it can be prepared. Show how the desired product can be obtained from the organometallic reagent.

(a) $H_2C=CHCH_2CH_2CH_2OH$

(b) $H_2C=CC(CH_2CH_2CH_2CH_3)_2$ with OH and CH_3 substituents

(c) $PhC(CH_2OCH_3)_2$ with OH

(d) benzene ring with $N(CH_3)_2$, CPh_2 with OH, and CH_3 substituents

(e) $(CH_3)_3CCCH_2CH_2CH_3$ with carbonyl O

(f) $H_2C=CHCH=CHCH=CH_2$

7.6. Identify an organometallic reagent that would permit formation of the product on the left of each equation from the specified starting material in a one-pot process.

(a)

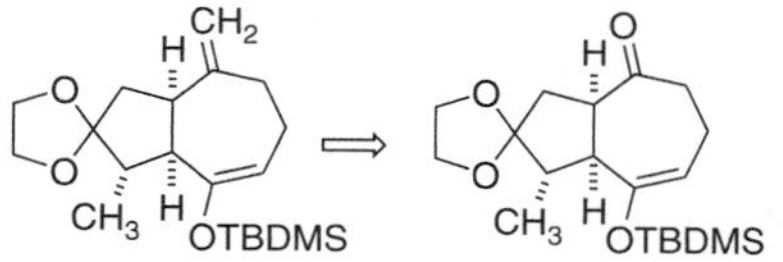

(b)

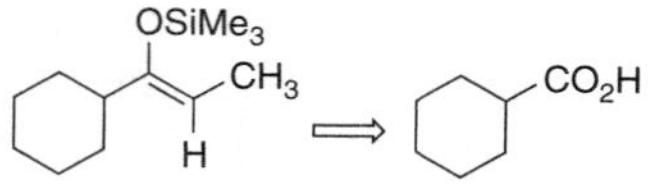

(c)

$$\underset{\|}{\overset{O}{PhCCH_2CH_2CO_2C_2H_5}} \implies PhCOCl$$

(d)

$$(CH_3)_2CH(CH_2)_2\overset{O}{\underset{\|}{C}}(CH_2)_6CO_2C_2H_5 \implies Cl\overset{O}{\underset{\|}{C}}(CH_2)_6CO_2C_2H_5$$

7.7. The solvomercuration reaction (Section 4.1.3) provides a convenient source of organomercury compounds such as **7-1** and **7-2**. How can these be converted to functionalized lithium compounds such as **7-3** and **7-4**?

$$\underset{\underset{R}{|}}{HOCHCH_2HgBr} \quad \underset{\underset{\underset{R}{|}}{|}}{\overset{H}{PhNCH_2HgBr}} \quad \underset{\underset{R}{|}}{LiOCHCH_2Li} \quad \underset{\underset{R}{|}}{\overset{Li}{PhNCH_2CH_2Li}}$$

7-1 **7-2** **7-3** **7-4**

Would the procedure you have suggested also work for the following transformation? Explain your reasoning.

$$\underset{\underset{R}{|}}{CH_3OCHCH_2HgBr} \longrightarrow \underset{\underset{R}{|}}{CH_3OCHCH_2Li}$$

7.8. Predict the stereochemical outcome of the following reactions and indicate the basis for your prediction.

(a)

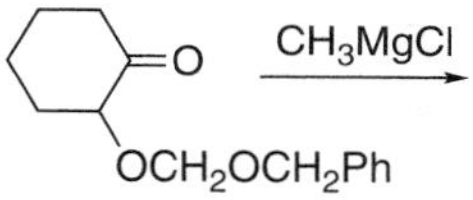

(b)

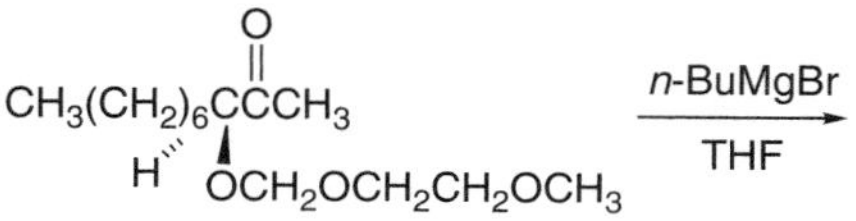

$$CH_3(CH_2)_6\overset{O}{\underset{\|}{C}}CCH_3 \quad \xrightarrow[\text{THF}]{n\text{-BuMgBr}}$$

$$\underset{H^{\prime\prime}}{\quad} \underset{OCH_2OCH_2CH_2OCH_3}{\quad}$$

(c)

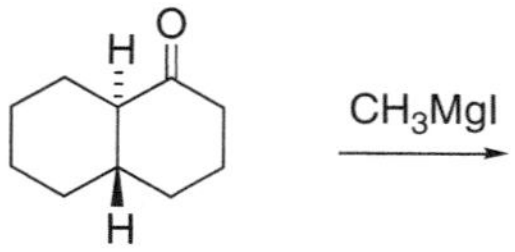

7.9. Tertiary amides **9-1**, **9-2**, and **9-3** are lithiated at the β-carbon, rather than the α-carbon by s-butyllithium-TMEDA. It is estimated that the intrinsic acidity of the α-position exceeds that of the β-position by about 9 pK units. What causes the β-deprotonation to be kinetically preferred?

$$\underset{O=CN(i\text{-}Pr)_2}{\overset{}{CH_3CHCH_2R}} \longrightarrow \underset{O=CN(i\text{-}Pr)_2}{\overset{R}{\underset{|}{CH_3CHCHLi}}}$$

9-1 R = Ph
9-2 R = CH=CH$_2$
9-3 R = SPh

7.10. The following reaction sequence converts esters to bromomethyl ketones. Show the intermediates that are involved in each step of the sequence.

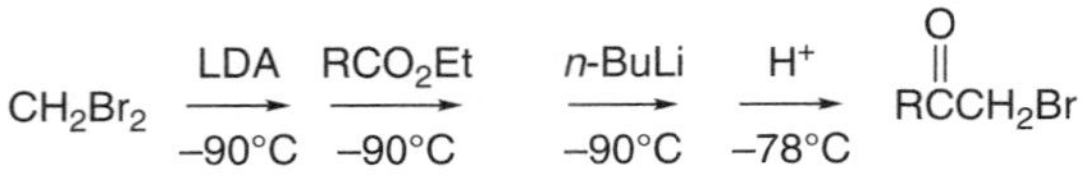

$$CH_2Br_2 \xrightarrow[-90°C]{LDA} \xrightarrow[-90°C]{RCO_2Et} \xrightarrow[-90°C]{n\text{-BuLi}} \xrightarrow[-78°C]{H^+} RCCH_2Br$$

7.11. Normally, the reaction of an ester with one equivalent of a Grignard reagent leads to a mixture of tertiary alcohol, ketone, and unreacted ester. However, when allylic Grignard reagents are used in the presence of one equivalent of LDA, good yields of ketones are obtained. What is the role of the LDA in this process?

7.12. Several examples of intramolecular additions to carbonyl groups by organo-lithium reagents generated by halogen-metal exchange have been reported, such as the two examples shown below. What relative reactivity relationships must hold in order for such procedures to succeed?

(a)

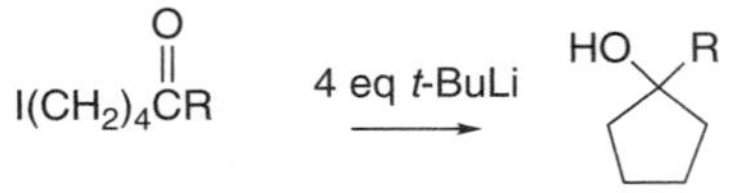

$$I(CH_2)_4CR \xrightarrow{4 \text{ eq } t\text{-BuLi}}$$

R	% yield
CH_3	26
CH_3CH_2CH_2	49
(CH_3)_2CH	78
Ph (2.2.eq t-BuLi)	66

(b)

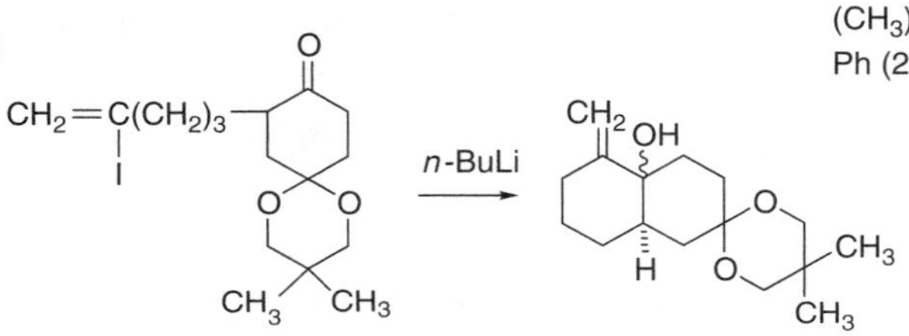

7.13. Short synthetic sequences (three steps or less) involving functionally substituted organometallic reagents can effect the following transformations. Suggest reaction sequences that would be effective for each case. Show how the required organometallic reagent can be prepared.

(a)

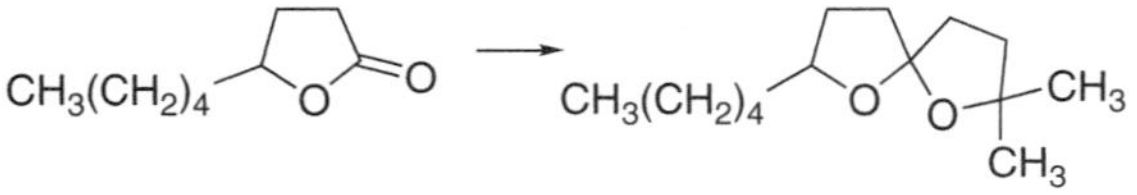

(b)

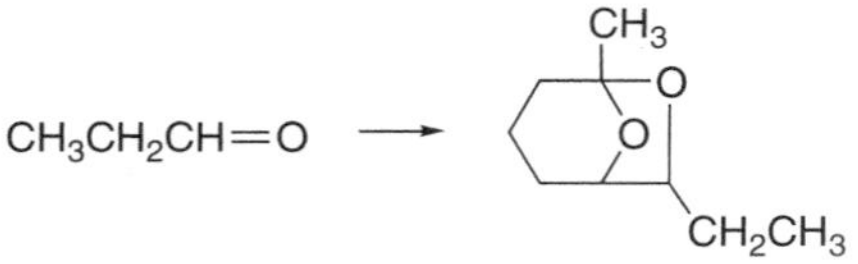

(c)

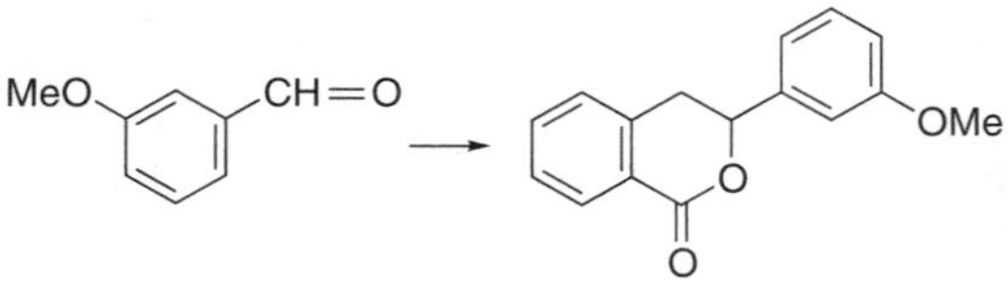

(d)

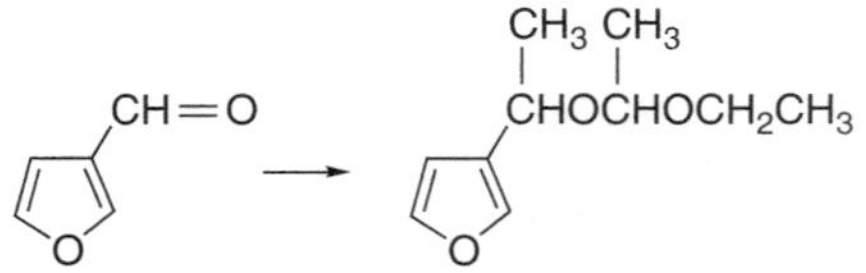

(e)

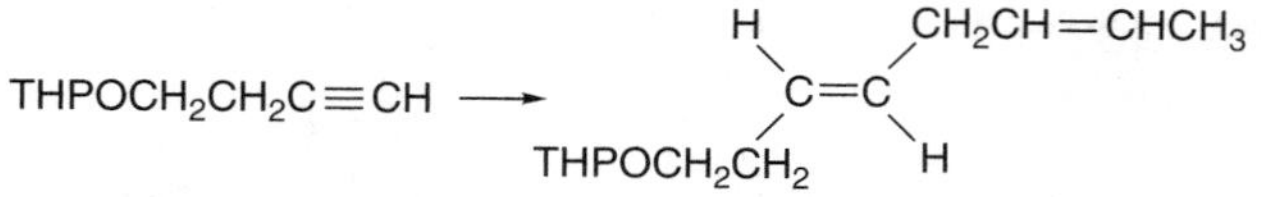

THPOCH$_2$CH$_2$C≡CH ⟶ [structure: C=C with H, CH$_2$CH=CHCH$_3$, THPOCH$_2$CH$_2$, H substituents]

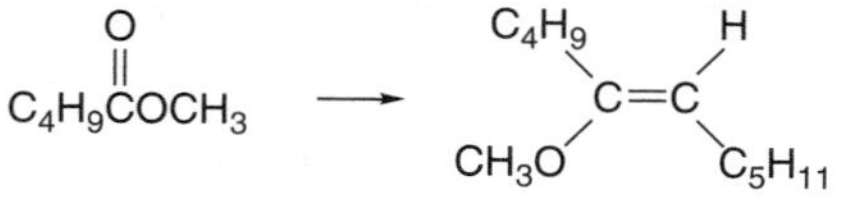

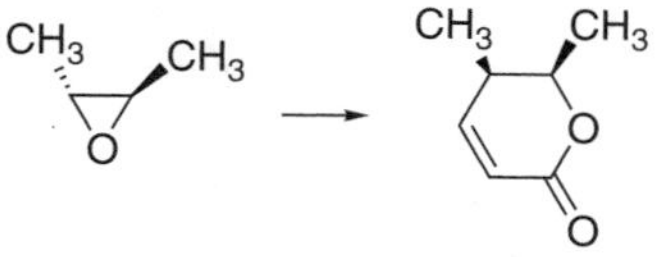

(f)

C$_4$H$_9$C(=O)OCH$_3$ ⟶ [structure: C=C with C$_4$H$_9$, H, CH$_3$O, C$_5$H$_{11}$ substituents]

(g) [epoxide structure with CH$_3$, CH$_3$] ⟶ [pyranone ring with CH$_3$, CH$_3$]

(h) [aromatic aldehyde structure with CH=O, OCH$_3$, CH$_3$O, CH$_3$] ⟶ [isobenzofuranone structure with CH$_3$O, O, CH$_3$O, H, OCH$_3$, CH$_3$O, CH$_3$]

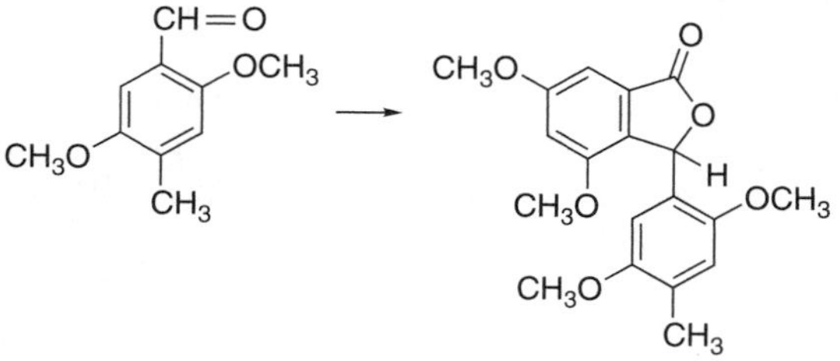

7.14. Catalytic amounts of chiral amino alcohols both catalyze the reactions of alkylzinc reagents with aldehydes and induce a high degree of enantioselectivity. Two examples are given below. Formulate a mechanism for this catalysis. Suggest transition structures consistent with the observed enantioselectivity.

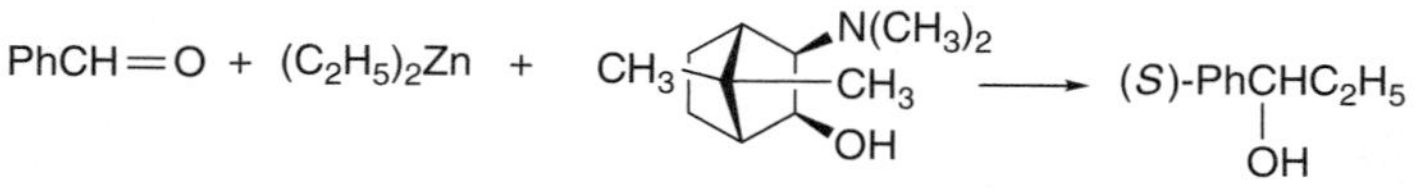

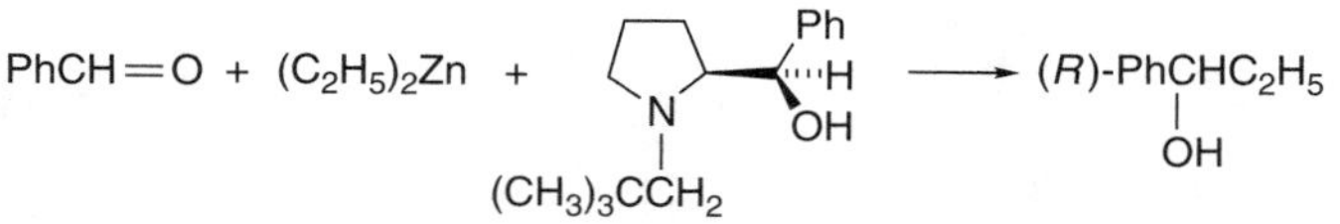

PhCH=O + (C$_2$H$_5$)$_2$Zn + [bicyclic structure with N(CH$_3$)$_2$, CH$_3$, CH$_3$, OH] ⟶ (S)-PhCHC$_2$H$_5$ with OH

PhCH=O + (C$_2$H$_5$)$_2$Zn + [pyrrolidine structure with Ph, H, OH, N, (CH$_3$)$_3$CCH$_2$] ⟶ (R)-PhCHC$_2$H$_5$ with OH

7.15. When 4-substituted 2,2-dimethyl-1,3-dioxolanes react with Grignard reagents, the bond that is broken is the one at the oxygen attached to the less-substituted α-carbon. What factor(s) are likely the cause for this regioselectivity?

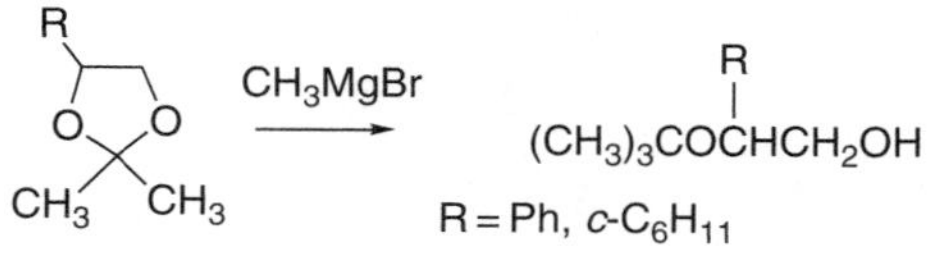

[dioxolane structure with R, O, O, CH$_3$, CH$_3$] + CH$_3$MgBr ⟶ (CH$_3$)$_3$COCHCH$_2$OH with R

R = Ph, c-C$_6$H$_{11}$

However, with **15-A** and **15-B**, the regioselectivity is reversed.

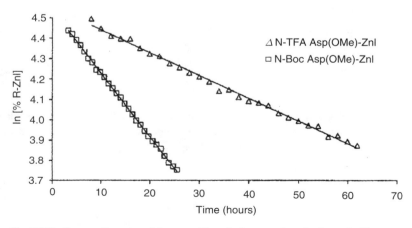

What factors might lead to the reversal in regioselectivity?

7.16. List several features of organocerium reagents that make them applicable to specific synthetic transformations. Give a specific example illustrating each feature.

7.17. Normally, organometallic reagents with potential leaving groups in the β-position decompose readily by elimination. Two examples of reagents with greater stability are described below. Indicate what structural feature(s) may be contributing to the relative stability of these reagents.

 a. Organozinc reagents with β-t-butoxycarbonylamino groups exhibit marginal stability. Replacement of the *t*-butoxycarbonyl by trifluoroacetamido groups *improves* the stability, as illustrated by the rate of decomposition shown in the Figure 7.P17.

$$CH_3O_2C\diagdown\diagup^{NHCY}_{\diagdown}ZnI$$

$$Y = OC(CH_3)_3 \text{ or } CF_3$$

Fig. 7.P17. Comparative rates of decomposition of *t*-butoxycarbonylamino and trifluoroacetamido groups.

b. Certain β-lithio derivatives of cyclic amines are stable.

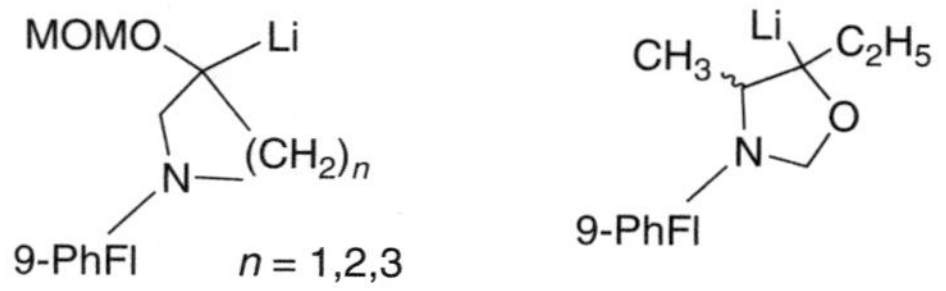

9-PhFl = 9-Phenyl-9-fluorenyl

Reactions Involving Transition Metals

Introduction

In this chapter we discuss important synthetic reactions that involve transition metal compounds and intermediates. Reactions involving copper and palladium, the transition metals that have the widest applications in synthesis, are discussed in the first two sections. In the third section, we consider several other transition metals, including nickel, rhodium, and cobalt. In contrast to lithium, magnesium, and zinc, where the organometallic reagents are used in stoichiometric quantities, many of the transition metal reactions are catalytic processes. The mechanisms are described in terms of *catalytic cycles* that show the role of the catalytic species in the reaction and its regeneration. Another distinguishing feature of transition metal reactions is that they frequently involve changes in oxidation state at the metal atom. In the final two sections we deal with transition metal–catalyzed alkene exchange (*metathesis*) reactions and organometallic compounds that feature π bonding of the organic component.

8.1. Organocopper Intermediates

8.1.1. Preparation and Structure of Organocopper Reagents

The synthetic application of organocopper compounds received a major impetus from the study of the catalytic effect of copper salts on reactions of Grignard reagents with α, β-unsaturated ketones.[1] Although Grignard reagents normally add to such compounds to give the 1,2-addition product, the presence of catalytic amounts of Cu(I) results in conjugate addition. Mechanistic study pointed to a very fast reaction by an organocopper intermediate.

[1.] H. O. House, W. L. Respess, and G. M. Whitesides, *J. Org. Chem.*, **31**, 3128 (1966).

$$CH_3CH=CHCCH_3 \xrightarrow[]{CH_3MgBr} \begin{array}{c} \xrightarrow[H^+]{H_2O} CH_3CH=CHC(CH_3)_2 \\ \underset{OH}{|} \end{array}$$

$$\xrightarrow[CH_3MgBr]{CuI,} \xrightarrow[H^+]{H_2O} (CH_3)_2CHCH_2CCH_3 \underset{O}{\overset{\|}{}}$$

Subsequently, much of the development of organocopper chemistry focused on stoichiometric reagents prepared from organolithium compounds. Several types of organometallic compounds can result from reactions of organolithium reagents with copper(I) salts.[2] Metal-metal exchange reactions using a 1:1 ratio of lithium reagent and a copper(I) salt give alkylcopper compounds that tend to be polymeric and are less useful in synthesis than the 2:1 or 3:1 "ate" compounds.

$$RLi + Cu(I) \longrightarrow [RCu]_n + Li^+$$
$$2\,RLi + Cu(I) \longrightarrow [R_2CuLi] + Li^+$$
$$3\,RLi + Cu(I) \longrightarrow [R_3CuLi_2] + Li^+$$

The 2:1 species are known as *cuprates* and are the most common synthetic reagents. Disubstituted Cu(I) species have the $3d^{10}$ electronic configuration and would be expected to have linear geometry. The Cu is a center of high electron density and nucleophilicity, and in solution, lithium dimethylcuprate exists as a dimer $[LiCu(CH_3)_2]_2$.[3] The compound is often represented as four methyl groups attached to a tetrahedral cluster of lithium and copper atoms. However, in the presence of LiI, the compound seems to be a monomer of composition $(CH_3)_2CuLi$.[4]

Discrete diarylcuprate anions have been observed in crystals in which the lithium cation is complexed by crown ethers.[5] Both tetrahedral Ph_4Cu_4 and linear $[Ph_2Cu]^-$ units have been observed in complex cuprates containing $(CH_3)_2S$ as a ligand. $[(Ph)_3Cu]^{2-}$ units have also been observed as parts of larger aggregates.[6] Larger clusters of composition $[(Ph_6Cu_4)Li]^-$ and $[Ph_6Cu_4Mg\cdot OEt_2]$ have been characterized by crystallography,[7] as shown in Figure 8.1.

Cuprates with two different copper ligands have been developed. These compounds have important advantages in cases in which one of the substituents

2. E. C. Ashby and J. J. Lin, *J. Org. Chem.*, **42**, 2805 (1977); E. C. Ashby and J. J. Watkins, *J. Am. Chem. Soc.*, **99**, 5312 (1977).
3. R. G. Pearson and C. D. Gregory, *J. Am. Chem. Soc.*, **98**, 4098 (1976); B. H. Lipshutz, J. A. Kozlowski, and C. M. Breneman, *J. Am. Chem. Soc.*, **107**, 3197 (1985).
4. A. Gerold, J. T. B. H. Jastrezebski, C. M. P. Kronenburg, N. Krause, and G. Van Koten, *Angew. Chem. Int. Ed. Engl.*, **36**, 755 (1997).
5. H. Hope, M. M. Olmstead, P. P. Power, J. Sandell, and X. Xu, *J. Am. Chem. Soc.*, **107**, 4337 (1985).
6. M. M. Olmstead and P. P. Power, *J. Am. Chem. Soc.*, **112**, 8008 (1990).
7. S. I. Khan, P. G. Edwards, H. S. H. Yuan, and R. Bau, *J. Am. Chem. Soc.*, **107**, 1682 (1985).

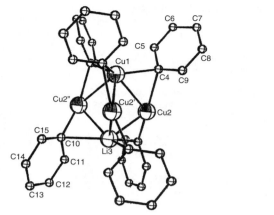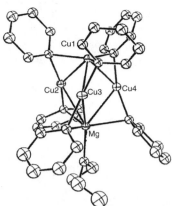

Fig. 8.1. Crystal structures of [(Ph$_6$Cu$_4$)Li]$^-$ (left) and [Ph$_6$Cu$_4$Mg·OEt$_2$]) (right). Reproduced from *J. Am. Chem. Soc.*, **107**, 1682 (1985), by permission of the American Chemical Society.

is derived from a valuable synthetic intermediate. The group R, representing alkyl, alkenyl, or aryl, is normally transferred in preference to the other copper ligand. Table 8.1 presents some of these mixed cuprate reagents and summarizes their reactivity. The group listed first is the nonreactive copper ligand and R is the organic group that is delivered as a nucleophile.

There has been a great deal of study concerning the effect of solvents and other reaction conditions on the stability and reactivity of organocuprate species.[8] These studies have found, for example, that (CH$_3$)$_2$S-CuBr, a readily prepared and purified complex of CuBr, is an especially reliable source of Cu(I) for cuprate preparation.[9] Copper(I) cyanide and iodide are also generally effective and, in some cases, preferable.[10]

An important type of mixed cuprate is prepared from a 2:1 ratio of an alkyllithium and CuCN.[11] Called *higher-order cyanocuprates*, their composition is R$_2$CuCNLi$_2$ in THF solution, but it is thought that most of the molecules are probably present as dimers. The cyanide does not seem to be bound directly to the copper, but rather to the lithium cations.[12] The dimers most likely adopt an eight-membered ring motif.[13]

8. R. H. Schwartz and J. San Filippo, Jr., *J. Org. Chem.*, **44**, 2705 (1979).

9. H. O. House, C.-Y. Chu, J. M. Wilkins, and M. J. Umen, *J. Org. Chem.*, **40**, 1460 (1975).

10. B. H. Lipshutz, R. S. Wilhelm, and D. M. Floyd, *J. Am. Chem. Soc.*, **103**, 7672 (1981); S. H. Bertz, C. P. Gibson, and G. Dabbagh, *Tetrahedron Lett.*, **28**, 4251 (1987); B. H. Lipshutz, S. Whitney, J. A. Kozlowski, and C. M. Breneman, *Tetrahedron Lett.*, **27**, 4273 (1986).

11. B. H. Lipshutz, R. S. Wilhelm, and J. Kozlowski, *Tetrahedron*, **40**, 5005 (1984); B. H. Lipshutz, *Synthesis*, 325 (1987).

12. T. M. Barnhart, H. Huang, and J. E. Penner-Hahn, *J. Org. Chem.*, **60**, 4310 (1995); J. P. Snyder and S. H. Bertz, *J. Org. Chem.*, **60**, 4312 (1995); T. L. Semmler, T. M. Barnhart, J. E. Penner-Hahn, C. E. Tucker, P. Knochel, M. Bohme, and G. Frenking, *J. Am. Chem. Soc.*, **117**, 12489 (1995); S. H. Bertz, G. B. Miao, and M. Eriksson, *J. Chem. Soc., Chem. Commun.*, 815 (1996).

13. E. Nakamura and S. Mori, *Angew. Chem. Int. Ed. Engl.*, **39**, 3750 (2000).

Table 8.1. Mixed-Ligand Organocopper Reagents

Mixed ligand reagent	Reactivity and properties	Reference
$[R'C{\equiv}C{-}Cu{-}R]Li$	Conjugate addition to α,β-unsaturated ketones and certain esters	a
$[ArS{-}Cu{-}R]Li$	Nucleophilic substitution and conjugate addition to unsaturated ketones; ketones from acyl chlorides	b,c
$[(CH_3)_3CO{-}Cu{-}R]Li$	Nucleophilic substitution and conjugate addition to α,β-unsaturated ketones	b
$[(c\text{-}C_6H_{11})_2N{-}Cu{-}R]Li$	Normal range of nucleophilic reactivity; improved thermal stability	d
$[Ph_2P{-}Cu{-}R]Li$	Normal range of nucleophilic reactivity; improved thermal stability	d
$[CH_3\overset{\overset{\displaystyle O}{\|}}{S}CH_2{-}Cu{-}R]Li$	Normal range of nucleophilic reactivity; improved thermal stability	e
$[N{\equiv}C{-}Cu{-}R]Li$	Efficient opening of epoxides	f
$[N{\equiv}C{-}CuR_2]Li_2$	Nucleophilic substitution, conjugate addition	g
S—Cu^-—R	Nucleophilic substitution, conjugate addition and epoxide ring-opening	h
$[(CH_3)_3CCH_2{-}Cu{-}R]Li$	Conjugate addition	i
$[(CH_3)_3SiCH_2{-}Cu{-}R]Li$	High reactivity, thermal stability	j
$\{[(CH_3)_3Si]_2N{-}Cu{-}R]\}Li$	High reactivity, thermal stability	j
$BF_3{-}Cu{-}R$	Conjugate addition, including acrylate esters and acrylonitrile; S_N2' substitution of allylic halides	k

a. H. O. House and M. J. Umen, *J. Org. Chem.*, **38**, 3893 (1973); E. J. Corey, D. Floyd, and B. H. Lipshutz, *J. Org. Chem.* **43**, 3418 (1978).
b. G. H. Posner, C. E. Whitten, and J. J. Sterling, *J. Am. Chem. Soc.*, **95**, 7788 (1973).
c. G. H. Posner and C. E. Whitten, *Org. Synth.*, **58**, 122 (1975).
d. S. H. Bertz, G. Dabbagh, and G. M. Villacorta, *J. Am. Chem. Soc.*, **104**, 5824 (1982).
e. C. R. Johnson and D. S. Dhanoa, *J. Org. Chem.*, **52**, 1885 (1987).
f. R. D. Acker, *Tetrahedron Lett.*, 3407 (1977); J. P. Marino and N. Hatanaka, *J. Org. Chem.*, **44**, 4667 (1979).
g. B. H. Lipshutz and S. Sengupta, *Org. React.*, **41**, 135 (1992).
h. H. Malmberg, M. Nilsson, and C. Ullenius, *Tetrahedron Lett.*, **23**, 3823 (1982); B. H. Lipshutz, M. Koernen, and D. A. Parker, *Tetrahedron Lett.*, **28**. 945 (1987).
i. C. Lutz, P. Jones, and P. Knochel, *Synthesis*, 312 (1999).
j. S. H. Bertz, M. Eriksson, G. Miao, and J. P. Snyder, *J. Am. Chem. Soc.*, **118**, 10906 (1996).
k. K. Maruyama and Y. Yamamoto, *J. Am. Chem. Soc.*, **99**, 8068 (1977); Y. Yamamoto and K. Maruyama, *J. Am. Chem. Soc.*, **100**, 3240 (1978).

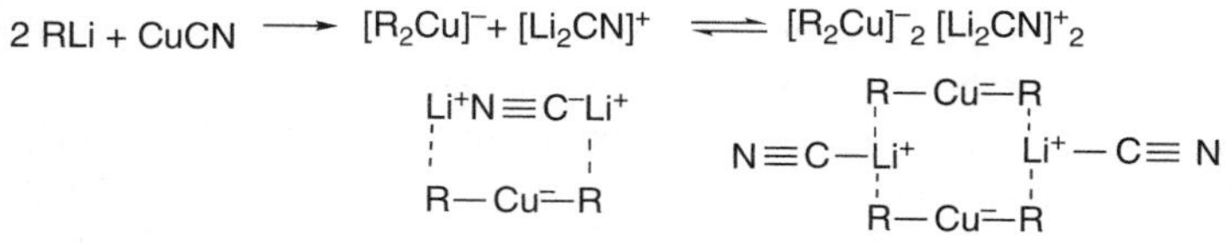

$$2\ RLi + CuCN \longrightarrow [R_2Cu]^- + [Li_2CN]^+ \rightleftharpoons [R_2Cu]^-_2\ [Li_2CN]^+_2$$

These reagents are qualitatively similar in reactivity to other cuprates but they are more stable than the dialkylcuprates. As cyanocuprate reagents usually transfer only one of the two organic groups, it is useful to incorporate a group that does not transfer, and the 2-thienyl group has been used for this purpose.[14] Usually, these reagents are prepared from an organolithium reagent, 2-thienyllithium, and CuCN. These reagents can also be prepared by reaction of an alkyl halide with 2-thienylcopper. The latter method is compatible with functionalized alkyl groups.[15]

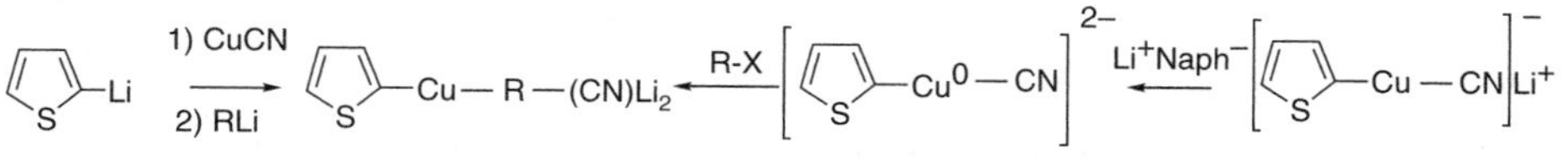

In a mixed alkyl-thienyl cyanocuprate, only the alkyl substituent is normally transferred as a nucleophile.

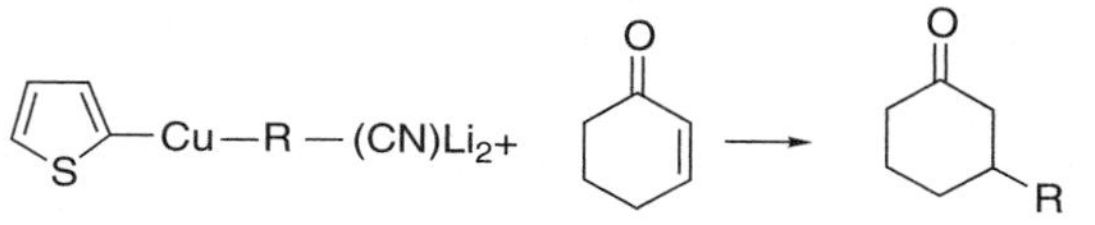

Another type of mixed cyanocuprate has both methyl and alkenyl groups attached to copper. Interestingly, these reagents selectively transfer the alkenyl group in conjugate addition reactions.[16] These reagents can be prepared from alkynes via hydrozirconation, followed by metal-metal exchange.[17]

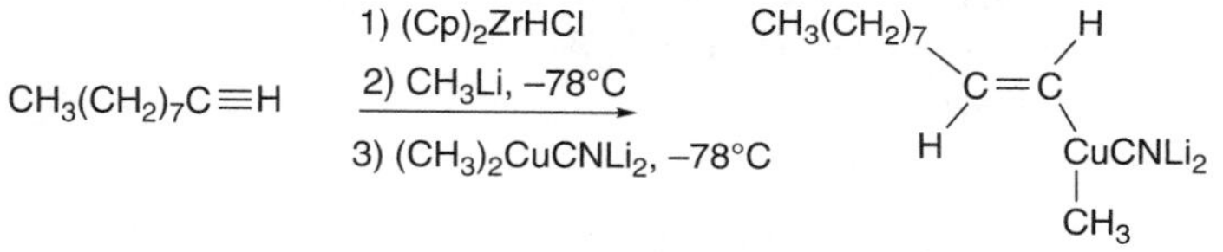

Alkenylcyanocuprates can also be made by metal-metal exchange from alkenylstannanes.[18]

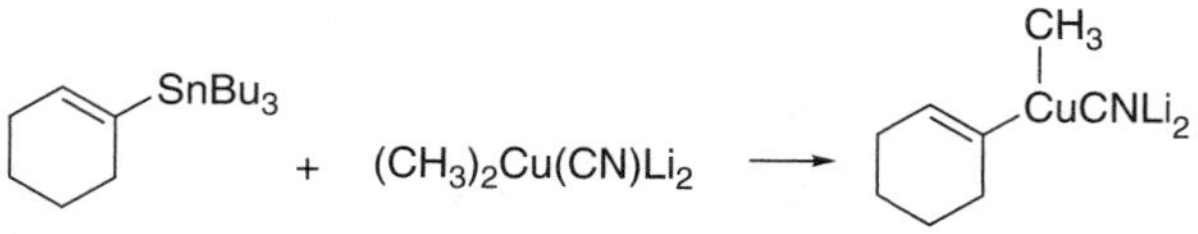

14. B. H. Lipshutz, J. A. Kozlowski, D. A. Parker, S. L. Nguyen, and K. E. McCarthy, *J. Organomet. Chem.*, **285**, 437 (1985); B. H. Lipshutz, M. Koerner, and D. A. Parker, *Tetrahedron Lett.*, **28**, 945 (1987).
15. R. D. Rieke, W. R. Klein, and T.-S. Wu, *J. Org. Chem.*, **58**, 2492 (1993).
16. B. H. Lipshutz, R. S. Wilhelm, and J. A. Kozlowski, *J. Org. Chem.*, **49**, 3938 (1984).
17. B. H. Lipshutz and E. L. Ellsworth, *J. Am. Chem. Soc.*, **112**, 7440 (1990).
18. J. R. Behling, K. A. Babiak, J. S. Ng, A. L. Campbell, R. Moretti, M. Koerner, and B. H. Lipshutz, *J. Am. Chem. Soc.*, **110**, 2641 (1988).

The 1:1 organocopper reagents can be prepared directly from the halide and highly reactive copper metal prepared by reducing Cu(I) salts with lithium naphthalenide.[19] This method of preparation is advantageous for organocuprates containing substituents that are incompatible with organolithium compounds. For example, nitrophenyl and cyanophenyl copper reagents can be prepared in this way, as can alkylcopper reagents having ester and cyano substituents.[20] Allylic chlorides and acetates can also be converted to cyanocuprates by reaction with lithium naphthalenide in the presence of CuCN and LiCl.[21]

$$(CH_3)_2C{=}CHCH_2Cl \xrightarrow[\text{CuCN, LiCl}]{\text{Li naphthalenide}} [(CH_3)_2C{=}CHCH_2]_2CuCNLi_2$$

Organocopper reagents can also be prepared from Grignard reagents, which are generated and used in situ by adding a Cu(I) salt, typically the bromide, iodide, or cyanide.

8.1.2. Reactions Involving Organocopper Reagents and Intermediates

The most characteristic feature of the organocuprate reagents is that they are excellent *soft nucleophiles*, showing greater reactivity in S_N2, S_N2', and conjugate addition reactions than toward direct addition at carbonyl groups. The most important reactions of organocuprate reagents are nucleophilic displacements on halides and sulfonates, epoxide ring opening, conjugate additions to α,β-unsaturated carbonyl compounds, and additions to alkynes.[22] These reactions are discussed in more detail in the following sections.

8.1.2.1. S_N2 and S_N2' Reactions with Halides and Sulfonates. Corey and Posner discovered that lithium dimethylcuprate can replace iodine or bromine by methyl in a wide variety of compounds, including aryl, alkenyl, and alkyl derivatives. This halogen displacement reaction is more general and gives higher yields than displacements with Grignard or lithium reagents.[23]

PhCH=CHBr + (CH₃)₂CuLi ⟶ PhCH=CHCH₃
81%

19. G. W. Ebert and R. D. Rieke, *J. Org. Chem.*, **49**, 5280 (1984); *J. Org. Chem.*, **53**, 4482 (1988); G. W. Ebert, J. W. Cheasty, S. S. Tehrani, and E. Aouad, *Organometallics*, **11**, 1560 (1992); G. W. Ebert, D. R. Pfennig, S. D. Suchan, and T. J. Donovan, Jr., *Tetrahedron Lett.*, **34**, 2279 (1993).
20. R. M. Wehmeyer and R. D. Rieke, *J. Org. Chem.*, **52**, 5056 (1987); T.-C. Wu, R. M. Wehmeyer, and R. D. Rieke, *J. Org. Chem.*, **52**, 5059 (1987); R. M. Wehmeyer and R. D. Rieke, *Tetrahedron Lett.*, **29**, 4513 (1988).
21. D. E. Stack, B. T. Dawson, and R. D. Rieke, *J. Am. Chem. Soc.*, **114**, 5110 (1992).
22. For reviews of the reactions of organocopper reagents, see G. H. Posner, *Org. React.*, **19**, 1 (1972); G. H. Posner, *Org. React.*, **22**, 253 (1975); G. H. Posner, *An Introduction to Synthesis Using Organocopper Reagents*, Wiley, New York, 1980; N. Krause and A. Gerold, *Angew. Chem. Int. Ed. Engl.*, **36**, 187 (1997).
23. E. J. Corey and G. H. Posner, *J. Am. Chem. Soc.*, **89**, 3911 (1967).

Secondary bromides and tosylates react with inversion of stereochemistry, as in the classical S_N2 substitution reaction.[24] Alkyl iodides, however, lead to racemized product. Aryl and alkenyl halides are reactive, even though the direct displacement mechanism is not feasible. For these halides, the overall mechanism probably consists of two steps: an *oxidative addition* to the metal, after which the oxidation state of the copper is +3, followed by combination of two of the groups from the copper. This process, which is very common for transition metal intermediates, is called *reductive elimination*. The $[R'_2Cu]^-$ species is linear and the oxidative addition takes place perpendicular to this moiety, generating a T-shaped structure. The reductive elimination occurs between adjacent R and R' groups, accounting for the absence of R' − R' coupling product.

$$R-X + \quad R'-Cu^I-R' \quad \longrightarrow \quad R'-\underset{\underset{X}{|}}{\overset{\overset{R}{|}}{Cu^{III}}}-R' \quad \longrightarrow \quad R-R' + R'Cu^IX$$

Allylic halides usually give both S_N2 and S_N2' products, although the mixed organocopper reagent RCu-BF$_3$ is reported to give mainly the S_N2' product.[25] Other leaving groups can also be used, including acetate and phosphate esters. Allylic acetates undergo displacement with an allylic shift (S_N2' mechanism).[26] The allylic substitution process may involve initial coordination with the double bond.[27]

$$[R_2Cu]^- + \ CH_2{=}CHCH_2X \ \rightleftharpoons \ \begin{array}{c} R{-}Cu{=}R \\ CH_2{=}CH{\sim}CH_2{-}X \end{array} \longrightarrow \begin{array}{c} R-Cu-R \\ | \\ CH_2CH{=}CH_2 \\ | \\ RCH_2CH{=}CH_2 + RCu \end{array}$$

For substituted allylic systems, both α- and γ-substitution can occur. Reaction conditions can influence the α- versus γ-selectivity. For example, the reaction of geranyl acetate with several butylcopper reagents was explored. Essentially complete α- or γ-selectivity could be achieved by modification of conditions.[28] In ether both CuCN and CuI led to preferential γ-substitution, whereas α-substitution was favored for all anions in THF.

X	solvent	Ratio α:γ	solvent	Ratio α:γ
CN	ether	<1:99	THF	>99:1
Cl	ether	>99:1	THF	>99:1
Br	ether	>99:1	THF	>99:1
I	ether	6 :96	THF	96:4

24. C. R. Johnson and G. A. Dutra, *J. Am. Chem. Soc.*, **95**, 7783 (1973); B. H. Lipshutz and R. S. Wilhelm, *J. Am. Chem. Soc.*, **104**, 4696 (1982); E. Hebert, *Tetrahedron Lett.*, **23**, 415 (1982).
25. K. Maruyama and Y. Yamamoto, *J. Am. Chem. Soc.*, **99**, 8068 (1977).
26. R. J. Anderson, C. A. Henrick, and J. B. Siddall, *J. Am. Chem. Soc.*, **92**, 735 (1970); E. E. van Tamelen and J. P. McCormick, *J. Am. Chem. Soc.*, **92**, 737 (1970).
27. H. L. Goering and S. S. Kantner, *J. Org. Chem.*, **49**, 422 (1984).
28. E. S. M. Persson and J. E. Backvall, *Acta Chem. Scand.*, **49**, 899 (1995).

3-Acetoxy-2-methyl-1-alkenes react primarily at C(1), owing to steric factors.[29]

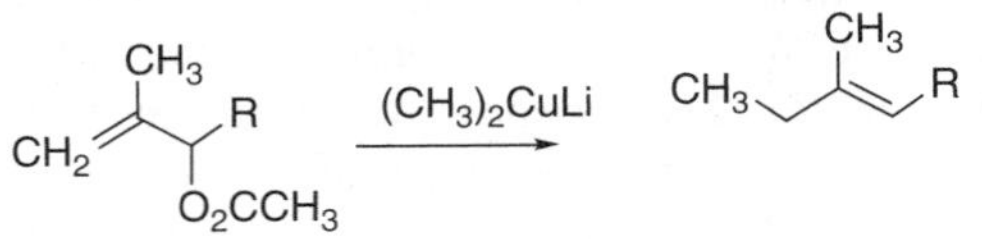

5-Acetoxy-1,3-alkadienes give mainly ϵ-alkylation with dialkylcopper-magnesium reagents.[30]

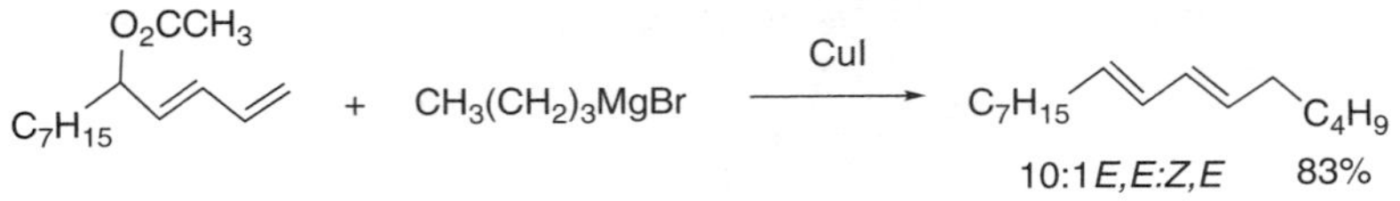

$$10:1\,E,E:Z,E \quad 83\%$$

High γ-selectivity has been observed for allylic diphenyl phosphate esters.[29]

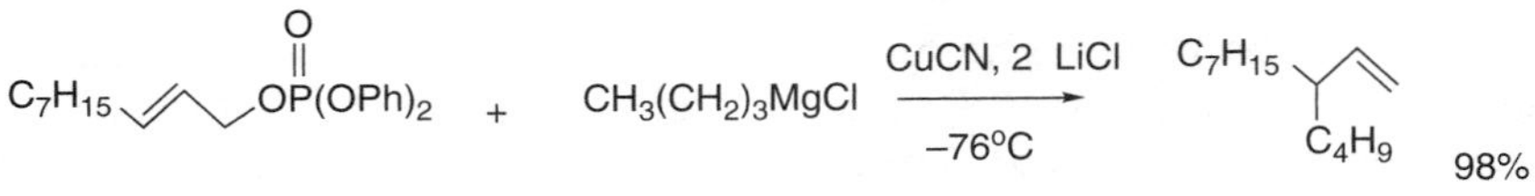

$$98\%$$

The reaction of cyclic allylic acetates shows a preference for *anti* stereochemistry.[31]

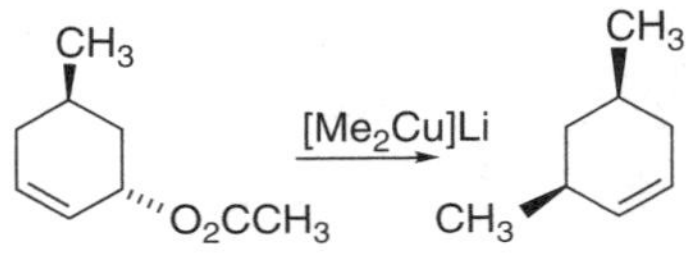

The preferred stereoelectronic arrangement is perpendicular alignment of the acetate with respect to the double bond. For example, the *cis* and *trans* isomers of 1-vinyl-2-methylcyclohexyl acetate show divergent stereochemical results. Only the exocyclic *E*-isomer is formed from the *cis* compound, whereas the *trans* compound gives a 1:1 mixture of the *E*- and *Z*-isomers. This is the result of a strongly preferred conformation for the *cis* isomer, as opposed to a mixture of conformations for the *trans* isomer.[32]

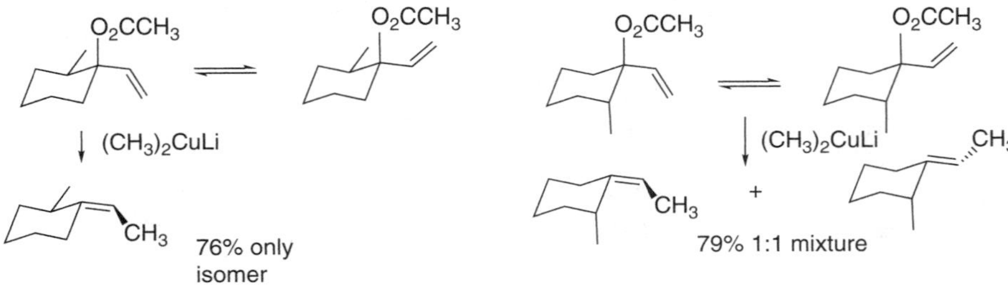

29. R. J. Anderson, C. A. Hendrick, and J. B. Siddall, *J. Am. Chem. Soc.*, **92**, 735 (1970).

30. N. Nakanishi, S. Matsubara, K. Utimoto, S. Kozima, and R. Yamaguchi, *J. Org. Chem.*, **56**, 3278 (1991).

31. H. L. Goering and V. D. Singleton, Jr., *J. Am. Chem. Soc.*, **98**, 7854 (1976); H. L. Goering and C. C. Tseng, *J. Org. Chem.*, **48**, 3986 (1983).

32. P. Crabbe, J. M. Dollat, J. Gallina, J. L. Luche, E. Velarde, M. L. Maddox, and L. Tokes, *J. Chem. Soc., Perkin Trans. 1*, 730 (1978).

Excellent diastereoselectivity is observed for δ-oxy allylic acetates. The stereo-selectivity is attributed to a Felkin-type TS with addition *anti* to the oxy substituent.

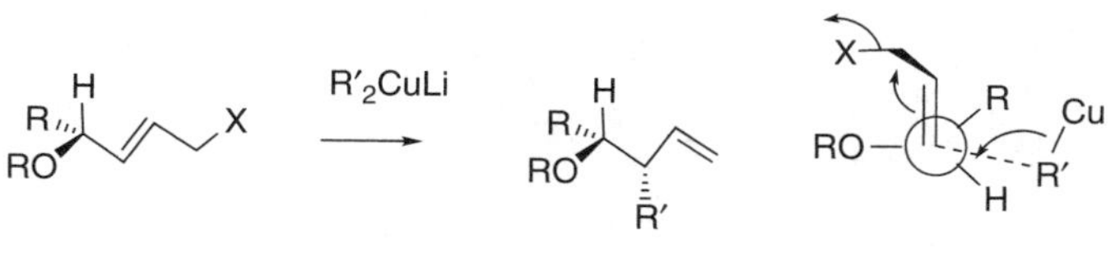

Similar results were obtained using *n*-BuMgBr-CuCN and tertiary allylic acetates, although under these conditions there is competition from S_N2' substitution with primary acetates.[33] The stereoselectivity is reversed with a hydroxy group, indicating a switch to a chelated TS.

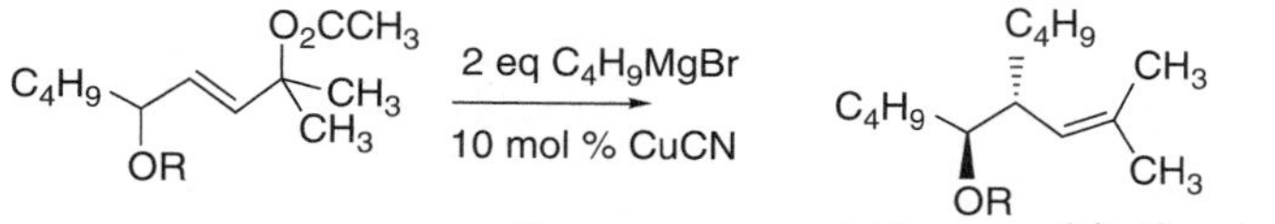

R	yield	dr(anti:syn)
PhCH$_2$	72	86:4
CH$_3$OCH$_2$	80	>98:2
TBDMS	89	90:10
H	84	7:93

Propargylic acetates, halides, and sulfonates usually react with a double-bond shift to give allenes.[34] Some direct substitution product can be formed as well. A high ratio of allenic product is usually found with CH$_3$Cu-LiBr-MgBrI, which is prepared by addition of methylmagnesium bromide to a 1:1 LiBr-CuI mixture.[35]

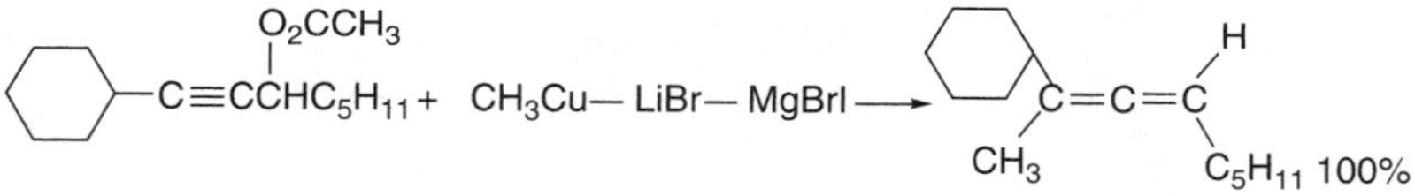

Halogens α to carbonyl groups can be successfully coupled using organocopper reagents. For example, 3,9-dibromocamphor is selectively arylated α to the carbonyl.

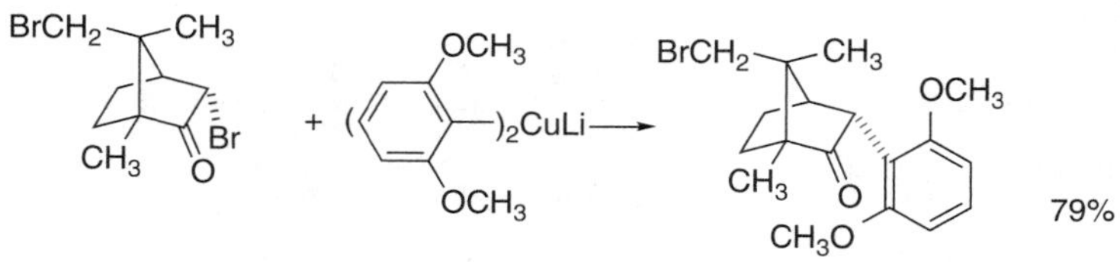

Ref. 36

Scheme 8.1 gives several examples of the use of coupling reactions of organocuprate reagents with halides and acetates. Entries 1 to 3 are examples of the

33. J. L. Belelie and J. M. Chong, *J. Org. Chem.*, **67**, 3000 (2002).
34. P. Rona and P. Crabbe, *J. Am. Chem. Soc.*, **90**, 4733 (1968); R. A. Amos and J. A. Katzenellenbogen, *J. Org. Chem.*, **43**, 555 (1978); D. J. Pasto, S.-K. Chou, E. Fritzen, R. H. Shults, A. Waterhouse, and G. F. Hennion, *J. Org. Chem.*, **43**, 1389 (1978).
35. T. L. Macdonald, D. R. Reagan, and R. S. Brinkmeyer, *J. Org. Chem.*, **45**, 4740 (1980).
36. V. Vaillancourt and F. F. Albizatti, *J. Org. Chem.*, **51**, 3627 (1992).

Scheme 8.1. Nucleophilic Substitution Reactions of Organocopper Reagents

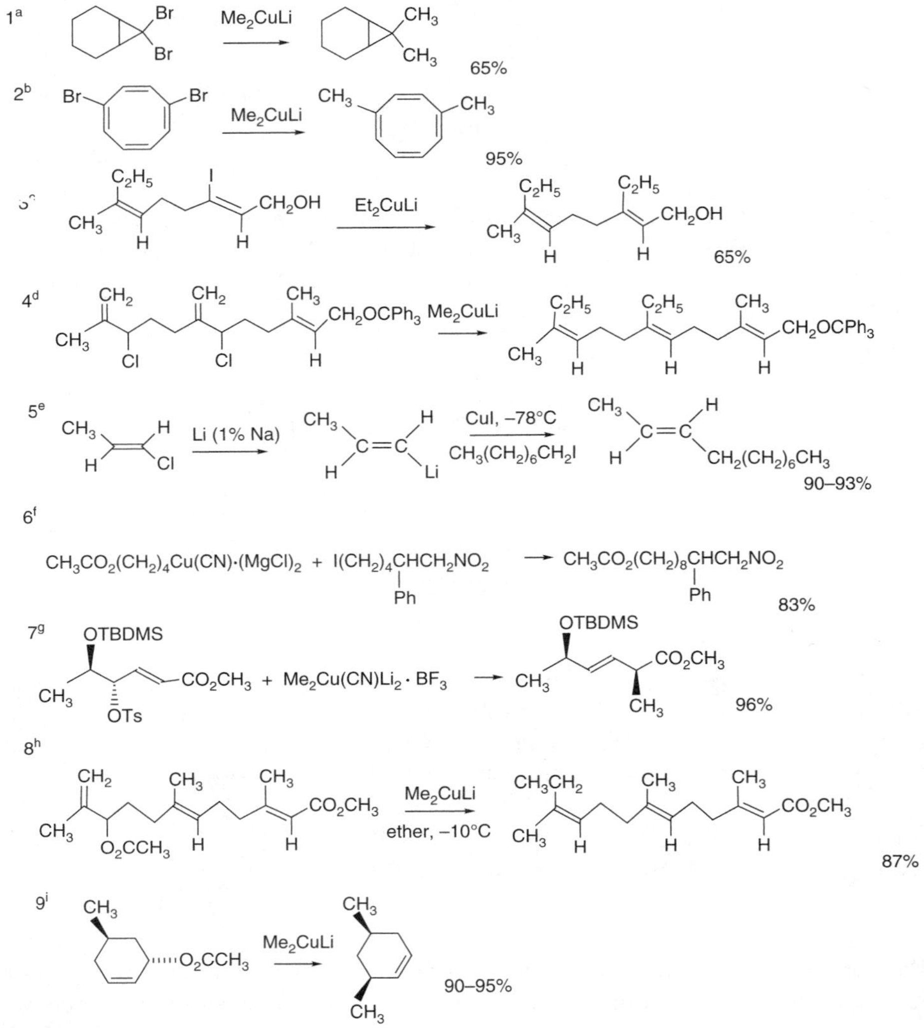

a. E. J. Corey and G. H. Posner, *J. Am. Chem. Soc.*, **89**, 3911 (1967).
b. W. E. Konz, W. Hechtl, and R. Huisgen, *J. Am. Chem. Soc.*, **92**, 4104 (1970).
c. E. J. Corey, J. A. Katzenellenbogen, N. W. Gilman, S. A. Roman, and B. W. Erickson, *J. Am. Chem. Soc.*, **90**, 5618 (1968).
d. E. E. van Tamelen and J. P. McCormick, *J. Am. Chem. Soc.*, **92**, 737 (1970).
e. G. Linstrumelle, J. K. Krieger, and G. M. Whitesides, *Org. Synth.*, **55**, 103 (1976).
f. C. E. Tucker and P. Knochel, *J. Org. Chem.*, **58**, 4781 (1993).
g. T. Ibuka, T. Nakao, S. Nishii, and Y. Yamamoto, *J. Am. Chem. Soc.*, **108**, 7420 (1986).
h. R. L. Anderson, C. A. Henrick, J. B. Siddall, and R. Zurfluh, *J. Am. Chem. Soc.*, **94**, 5379 (1972).
i. H. L. Goering and V. D. Singleton, Jr., *J. Am. Chem. Soc.*, **98**, 7854 (1976).

use of dialkylcuprates. In each case the halide is not susceptible to S_N2 substitution, but the oxidative addition mechanism is feasible. Entry 4 is an example of S_N2' substitution. This reaction, carried out simultaneously at two allylic chloride moieties, was used in the synthesis of the "juvenile hormone" of the moth *Cecropia*. Entry

5 illustrates the alkylation of a vinyl halide with retention of configuration at each stage of the reaction. Entry 6 is an example of a functionalized mixed magnesium-cyanocuprate reagent, which was prepared from an organozinc reagent by treatment with $(CH_3)_2CuCNMg_2Cl_2$. Entry 7 is an S_N2' displacement on a tosylate that occurs stereospecifically. Entries 8 and 9 are S_N2' displacements of allylic acetates.

8.1.2.2. Opening of Epoxides.

Organocopper reagents are excellent nucleophiles for opening epoxide rings. Saturated epoxides are opened in good yield by lithium dimethylcuprate.[37] The methyl group is introduced at the less hindered carbon of the epoxide ring.

$$CH_3CH_2\text{—epoxide} + (CH_3)_2CuLi \longrightarrow CH_3CH_2CHCH_2CH_3 \quad 88\%$$
$$\overset{|}{OH}$$

Even mixed reagents with Lewis acids attack at the less-substituted position, indicating dominance of the nucleophilic bond making over the electrophilic component of ring opening.[38]

$$R\text{—epoxide} + R'_2CuLi\text{-}BF_3 \longrightarrow R\overset{OH}{\diagup}R'$$

The predictable regio- and stereochemistry make these reactions valuable in establishing stereochemistry in both acyclic and cyclic systems.

$$PhCH_2O\text{—}CH(CH_3)\text{—epoxide—}CH_2OH \xrightarrow{Me_2Cu(CN)Li_2} PhCH_2O\text{—}CH(CH_3)CH(OH)CH(CH_3)CH_2OH \quad \text{Ref. 39}$$

With cyclohexene epoxides, the ring opening is *trans*-diaxial.

$$\text{(epoxide) } CH_3O\cdots\text{—O—}CH_2OTBDMS \xrightarrow[\text{0°C}]{(CH_3)_2CuCNLi_2 \text{ ether}} HO\text{—}CH_3 \cdots CH_3O\cdots\text{—O—}CH_2OTBDMS \quad \text{Ref. 40}$$

Epoxides with alkenyl substituents undergo alkylation at the double bond with a double-bond shift accompanying ring opening, leading to formation of allylic alcohols.

$$(CH_3)_2CuLi + H_2C=C(CH_3)\text{—epoxide—}CH_3 \longrightarrow CH_3CH_2C(CH_3)=CHCHCH_3 \quad \text{Ref. 41}$$
$$\overset{|}{OH}$$

37. C. R. Johnson, R. W. Herr, and D. M. Wieland, *J. Org. Chem.*, **38**, 4263 (1973).
38. A. Alexis, D. Jachiet, and J. F. Normant, *Tetrahedron*, **42**, 5607 (1986).
39. A. B. Smith, III, B. A. Salvatore, K. G. Hull, and J. J.-W. Duan, *Tetrahedron Lett.*, **32**, 4859 (1991).
40. R. G. Linde, M. Egbertson, R. S. Coleman, A. B. Jones, and S. J. Danishefsky, *J. Org. Chem.*, **55**, 2771 (1990).
41. R. J. Anderson, *J. Am. Chem. Soc.*, **92**, 4978 (1970); R. W. Herr and C. R. Johnson, *J. Am. Chem. Soc.*, **92**, 4979 (1970); J. A. Marshall, *Chem. Rev.*, **89**, 1503 (1989).

Ref. 42

8.1.2.3. Conjugate Addition Reactions. All of the types of mixed cuprate reagents described in Scheme 8.1 react by conjugate addition with enones. A number of improvements in methodology for carrying out the conjugate addition reactions have been introduced. The addition is accelerated by trimethylsilyl chloride alone or in combination with HMPA.[43] Under these conditions the initial product is a silyl enol ether. The mechanism of the catalysis remains uncertain, but it appears that the silylating reagent intercepts an intermediate and promotes carbon-carbon bond formation, as well as trapping the product by O-silylation.[44]

This technique also greatly improves yields of conjugate addition of cuprates to α, β-unsaturated esters and amides.[45] Trimethylsilyl cyanide also accelerates conjugate addition.[46] Another useful reagent is prepared from a 1:1:1 ratio of organo-lithium reagent, CuCN, and $BF_3\text{-}O(C_2H_5)_2$.[47] The BF_3 appears to interact with the cyanocuprate reagent, giving a more reactive species.[48] The efficiency of the conjugate addition reaction is also improved by the inclusion of trialkylphosphines.[49] Even organocopper reagents prepared from a 1:1 ratio of organolithium compounds are reactive in the presence of phosphines.[50]

84%

42. J. A. Marshall, T. D. Crute, III, and J. D. Hsi, *J. Org. Chem.*, **57**, 115 (1992).
43. E. J. Corey and N. W. Boaz, *Tetrahedron Lett.*, **26**, 6019 (1985); E. Nakamura, S. Matsuzawa, Y. Horiguchi, and I. Kuwajima, *Tetrahedron Lett.*, **27**, 4029 (1986); S. Matsuzawa, Y. Horiguchi, E. Nakamura, and I. Kuwajima, *Tetrahedron*, **45**, 449 (1989); C. R. Johnson and T. J. Marren, *Tetrahedron Lett.*, **28**, 27 (1987); S. H. Bertz and G. Dabbagh, *Tetrahedron*, **45**, 425 (1989); S. H. Bertz and R. A. Smith, *Tetrahedron*, **46**, 4091 (1990); K. Yamamoto, H. Ogura, J. Jukuta, H. Inoue, K. Hamada, Y. Sugiyama, and S. Yamada, *J. Org. Chem.*, **63**, 4449 (1998); M. Kanai, Y. Nakagawa, and K. Tomioka, *Tetrahedron*, **55**, 3831 (1999).
44. M. Eriksson, A. Johansson, M. Nilsson, and T. Olsson, *J. Am. Chem. Soc.*, **118**, 10904 (1996).
45. A. Alexakis, J. Berlan, and Y. Besace, *Tetrahedron Lett.*, **27**, 1047 (1986).
46. B. H. Lipshutz and B. James, *Tetrahedron Lett.*, **34**, 6689 (1993).
47. T. Ibuka, N. Akimoto, M. Tanaka, S. Nishii, and Y. Yamamoto, *J. Org. Chem.*, **54**, 4055 (1989).
48. B. H. Lipshutz, E. L. Ellsworth, and T. J. Siahaan, *J. Am. Chem. Soc.*, **111**, 1351 (1989); B. H. Lipshutz, E. L. Ellsworth, and S. H. Dimock, *J. Am. Chem. Soc.*, **112**, 5869 (1990).
49. M. Suzuki, T. Suzuki, T. Kawagishi, and R. Noyori, *Tetrahedron Lett.*, 1247 (1980).
50. T. Kawabata, P. A. Grieco, H.-L. Sham, H. Kim, J. Y. Jaw, and S. Tu, *J. Org. Chem.*, **52**, 3346 (1987).

The mechanism of conjugate addition reactions probably involves an initial complex between the cuprate and enone.[51] The key intermediate for formation of the new carbon-carbon bond is an adduct formed between the enone and the organocopper reagent. The adduct is formulated as a Cu(III) species, which then undergoes reductive elimination. The lithium ion also plays a key role, presumably by Lewis acid coordination at the carbonyl oxygen.[52] Solvent molecules also affect the reactivity of the complex.[53] The mechanism can be outlined as occurring in three steps.

Isotope effects indicate that the collapse of the adduct by reductive elimination is the rate-determining step.[54] Theoretical treatments of the mechanism suggest similar intermediates. (See Section 8.1.2.7 for further discussion of the computational results.)[55]

There is a correlation between the reduction potential of the carbonyl compounds and the ease of reaction with cuprate reagents.[56] The more easily it is reduced, the more reactive the compound toward cuprate reagents. Compounds such as α, β-unsaturated esters and nitriles, which are not as easily reduced as the corresponding ketones, do not react as readily with dialkylcuprates, even though they are good acceptors in classical Michael reactions with carbanions. α, β-Unsaturated esters are marginal in terms of reactivity toward standard dialkylcuprate reagents, and β-substitution retards reactivity. The RCu-BF_3 reagent combination is more reactive toward conjugated esters and nitriles,[57] and additions to hindered α, β-unsaturated ketones are accelerated by BF_3.[58]

There have been many applications of conjugate additions in synthesis. Some representative reactions are shown in Scheme 8.2. Entries 1 and 2 are examples of addition of lithium dimethylcuprate to cyclic enones. The stereoselectivity exhibited in Entry 2 is the result of both steric and stereoelectronic effects that favor the approach *syn* to the methyl substituent. In particular, the axial hydrogen at C(6) hinders the α approach.

51. S. R. Krauss and S. G. Smith, *J. Am. Chem. Soc.*, **103**, 141 (1981); E. J. Corey and N. W. Boaz, *Tetrahedron Lett.*, **26**, 6015 (1985); E. J. Corey and F. J. Hannon, *Tetrahedron Lett.*, **31**, 1393 (1990).
52. H. O. House, *Acc. Chem. Res.*, **9**, 59 (1976); H. O. House and P. D. Weeks, *J. Am. Chem. Soc.*, **97**, 2770, 2778 (1975); H. O. House and K. A. J. Snoble, *J. Org. Chem.*, **41**, 3076 (1976); S. H. Bertz, G. Dabbagh, J. M. Cook, and V. Honkan, *J. Org. Chem.*, **49**, 1739 (1984).
53. C. J. Kingsbury and R. A. J. Smith, *J. Org. Chem.*, **62**, 4629, 7637 (1997).
54. D. E. Frantz, D. A. Singleton, and J. P. Snyder, *J. Am. Chem. Soc.*, **119**, 3383 (1997).
55. E. Nakamura, S. Mori, and K. Morukuma, *J. Am. Chem. Soc.*, **119**, 4900 (1997); S. Mori and E. Nakamura, *Chem. Eur. J.*, **5**, 1534 (1999).
56. H. O. House and M. J. Umen, *J. Org. Chem.*, **38**, 3893 (1973); B. H. Lipshutz, R. S. Wilhelm, S. T. Nugent, R. D. Little, and M. M. Baizer, *J. Org. Chem.*, **48**, 3306 (1983).
57. Y. Yamamoto and K. Maruyama, *J. Am. Chem. Soc.*, **100**, 3240 (1978); Y. Yamamoto, *Angew. Chem. Int. Ed. Engl.*, **25**, 947 (1986).
58. A. B. Smith, III, and P. J. Jerris, *J. Am. Chem. Soc.*, **103**, 194 (1981).

Scheme 8.2. Conjugate Addition Reactions of Organocopper Reagents

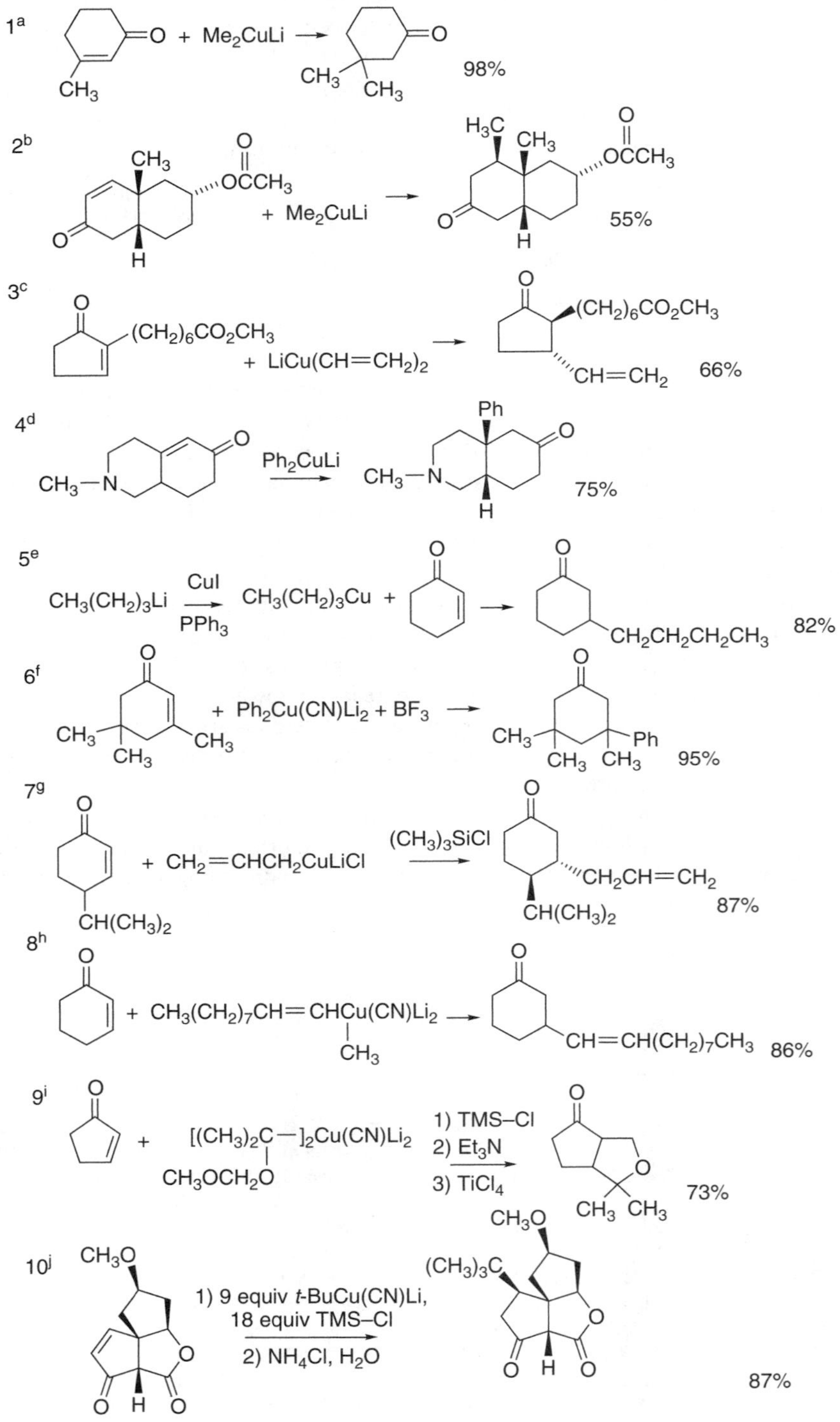

1[a]

98%

2[b]

55%

3[c]

66%

4[d]

75%

5[e]

82%

6[f]

95%

7[g]

87%

8[h]

86%

9[i]

73%

10[j]

87%

(Continued)

Scheme 8.2. (*Continued*)

689

SECTION 8.1

Organocopper Intermediates

11[k]

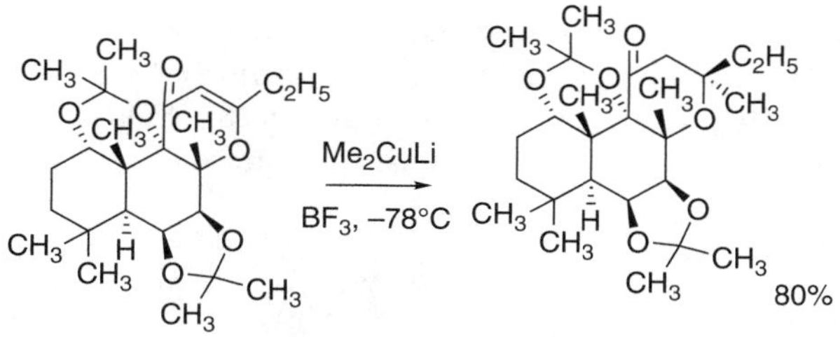

Me$_2$CuLi

BF$_3$, –78°C

80%

12[l]

CH$_2$=CHCu, LiI,
Bu$_3$P

–78°C

95%

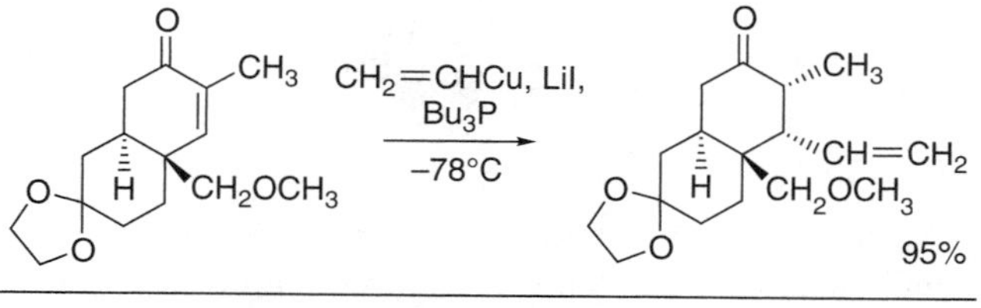

a. H. O. House, W. L. Respess, and G. M. Whitesides, *J. Org. Chem.*, **31**, 3128 (1966).
b. J. A. Marshall and G. M. Cohen, *J. Org. Chem.*, **36**, 877 (1971).
c. F. S. Alvarez, D. Wren, and A. Prince, *J. Am. Chem. Soc.*, **94**, 7823 (1972).
d. N. Finch, L. Blanchard, R. T. Puckett, and L. H. Werner, *J. Org. Chem.*, **39**, 1118 (1974).
e. M. Suzuki, T. Suzuki, T. Kawagishi, and R. Noyori, *Tetrahedron Lett.*, **21**, 1247 (1980).
f. B. H. Lipshutz, D. A. Parker, J. A. Kozlowski, and S. L. Nguyen, *Tetrahedron Lett.* **25**, 5959 (1984).
g. B. H. Lipshutz, E. L. Ellsworth, S. H. Dimock, and R. A. J. Smith, *J. Am. Chem. Soc.*, **112**, 4404 (1990).
h. B. H. Lipshutz and E. L. Ellsworth, *J. Am. Chem. Soc.*, **112**, 7440 (1990).
i. R. J. Linderman and A. Godfrey, *J. Am. Chem. Soc.*, **110**, 6249 (1988).
j. E. J. Corey and K. Kamiyama, *Tetrahedron Lett.*, **31**, 3995 (1990).
k. B. Delpech and R. Lett, *Tetrahedron Lett.*, **28**, 4061 (1987).
l. T. Kawabata, P. Grieco, H. L. Sham, H. Kim, J. Y. Jaw, and S. Tu, *J. Org. Chem.*, **52**, 3346 (1987).

In Entry 3, the *trans* stereochemistry arises at the stage of the protonation of the enolate. Entry 4 gives rise to a *cis* ring juncture, as does the corresponding carbocyclic compound.[59] Models suggest that this is the result of a steric differentiation arising from the axial hydrogens on the α-face of the molecule.

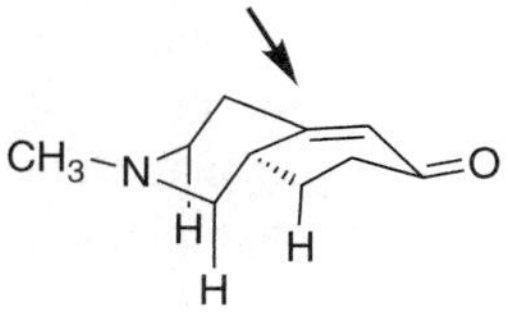

Entries 5 to 9 illustrate some of the modified reagents and catalytic procedures. Entry 5 uses a phosphine-stabilized reagent, whereas Entry 6 includes BF$_3$. Entry 7 involves use of TMS-Cl. Entries 8 and 9 involve cyanocuprates. In Entry 9, the furan ring is closed by a Mukaiyama-aldol reaction subsequent to the conjugate addition (Section 2.1.4).

[59.] S. M. McElvain and D. C. Remy, *J. Am. Chem. Soc.*, **82**, 3960 (1960).

Entries 10 to 12 illustrate the use of organocopper conjugate addition in the synthesis of relatively complex molecules. The installation of a *t*-butyl group adjacent to a quaternary carbon in Entry 10 requires somewhat forcing conditions, but proceeds in good yield. In Entry 11, the addition is to a vinylogous ester, illustrating the ability of the BF_3-modified reagents to react with less electrophilic systems. Steric shielding by the axial methoxymethyl substituent accounts for the stereoselectivity observed in Entry 12.

Prior to protonolysis, the products of conjugate addition are enolates and, therefore, potential nucleophiles. A useful extension of the conjugate addition method is to combine it with an alkylation step that adds a substituent at the α-position.[60] Several examples of this *tandem conjugate addition-alkylation* method are given in Scheme 8.3. In Entry 1 the characteristic β-attack on the *cis* decalone ring is observed (see Scheme 8.2, Entry 2). The alkylation gives a $\beta : \alpha$ ratio of 60:40. In Entry 2, the methylation occurs *anti* to the 4-substituent, presumably because of steric factors. These reactions are part of the synthesis of the cholesterol-lowering drug compactin. Entry 3 illustrates a pattern that has been extensively developed for the synthesis of prostaglandins. In this case, the dioxolane ring controls the stereoselectivity of the conjugate addition step and steric factors lead to *anti* alkylation and formation the *trans* product. Entry 4 is a part of a steroid synthesis. This reaction shows a 4:1 preference for methylation from the β-face (*syn* to the substituent). In Entry 5, the conjugate addition is followed by a Robinson annulation. The product provides a C,D-ring segment of the steroid skeleton.

8.1.2.4. Copper-Catalyzed Reactions. The cuprate reagents that were discussed in the preceding sections are normally prepared by reaction of an organolithium reagent with a copper(I) salt, using a 2:1 ratio of lithium reagent to copper(I). There are also valuable synthetic procedures that involve organocopper intermediates that are generated in the reaction system by use of only a catalytic amount of a copper salt.[61] Coupling of Grignard reagents and primary halides and tosylates can be catalyzed by Li_2CuCl_4.[62] This method was used, for example, to synthesize long-chain carboxylic acids in more than 90% yield.[63]

60. For a review of such reactions, see R. J. K. Taylor, *Synthesis*, 364 (1985).
61. For a review, see E. Erdik, *Tetrahedron*, **40**, 641 (1984).
62. M. Tamura and J. Kochi, *Synthesis*, 303 (1971); T. A. Baer and R. L. Carney, *Tetrahedron Lett.*, 4697 (1976).
63. S. B. Mirviss, *J. Org. Chem.*, **54**, 1948 (1989); see also M. R. Kling, C. J. Eaton, and A. Poulos, *J. Chem. Soc., Perkin Trans. 1*, 1183 (1993).

Scheme 8.3. Tandem Conjugate Addition-Alkylation Using Organocopper Reagents

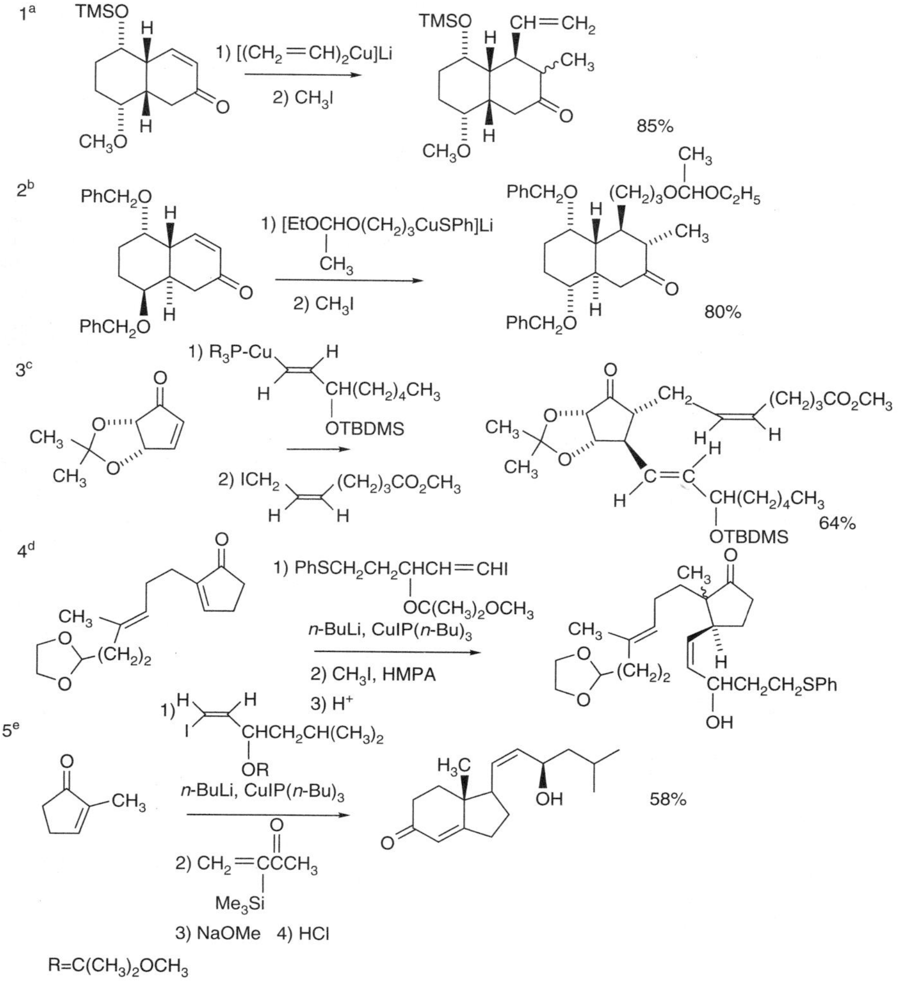

a. N. N. Girotra, R. A. Reamer, and N. L. Wendler, *Tetrahedron Lett.*, **25**, 5371 (1984).
b. N.-Y. Wang, C.-T. Hsu, and C. J. Sih, *J. Am. Chem. Soc.*, **103**, 6538 (1981).
c. C. R. Johnson and T. D. Penning, *J. Am. Chem. Soc.*, **110**, 4726 (1988).
d. T. Takahashi, K. Shimizu, T. Doi, and J. Tsuji, *J. Am. Chem. Soc.*, **110**, 2674 (1988).
e. T. Takahashi, H. Okumoto, J. Tsuji, and N. Harada, *J. Org. Chem.*, **49**, 948 (1984).

$$CH_2=CH(CH_2)_9MgCl + Br(CH_2)_{11}CO_2MgBr \xrightarrow[\text{2) } H^+]{\text{1) } Li_2CuCl_4} CH_2=CH(CH_2)_{20}CO_2H$$

Another excellent catalyst for coupling is a mixture of $CuBr\text{-}S(CH_3)_2$, LiBr, and LiSPh. This catalyst can effect coupling of a wide variety of Grignard reagents with tosylates and mesylates and is superior to Li_2CuCl_4 in coupling with secondary sulfonates.[64]

[64] D. H. Burns, J. D. Miller, H.-K. Chan, and M. O. Delaney, *J. Am. Chem. Soc.*, **119**, 2125 (1997).

Scheme 8.4. Copper-Catalyzed Reactions of Grignard Reagents

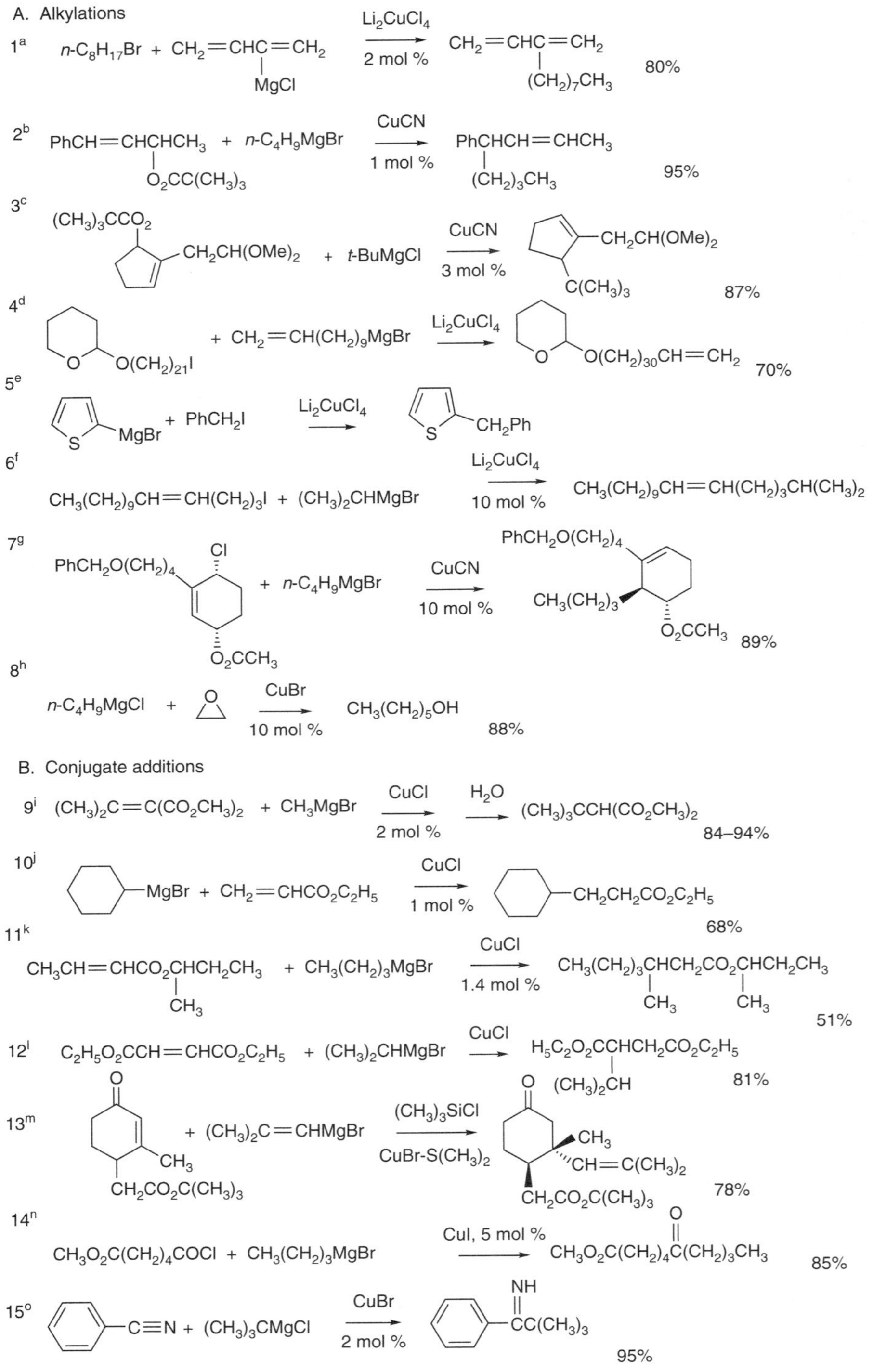

A. Alkylations

B. Conjugate additions

(Continued)

Scheme 8.4. (*Continued*)

693

SECTION 8.1

*Organocopper
Intermediates*

a. S. Nunomoto, Y. Kawakami, and Y. Yamashita, *J. Org. Chem.*, **48**, 1912 (1983).
b. C. C. Tseng, S. D. Paisley, and H. L. Goering, *J. Org. Chem.*, **51**, 2884 (1986).
c. E. J. Corey and A. V. Gavai, *Tetrahedron Lett.*, **29**, 3201 (1988).
d. U. F. Heiser and B. Dobner, *J. Chem. Soc, Perkin Trans. 1*, 809 (1997).
e. Y.-T. Ku, R. R. Patel, and D. P. Sawick, *Tetrahedron Lett.*, **37**, 1949 (1996).
f. E. Keinan, S. C. Sinha, A. Sinha-Bagchi, Z.-M. Wang, X.-L. Zhang, and K. B. Sharpless, *Tetrahedron Lett.*, **33**, 6411 (1992).
g. D. Tanner, M. Sellen, and J. Backvall, *J. Org. Chem.*, **54**, 3374 (1989).
h. G. Huynh, F. Derguini-Boumechal, and G. Linstrumelle, *Tetrahedron Lett.*, 1503 (1979).
i. E. L. Eliel, R. O. Hutchins, and M. Knoeber, *Org. Synth.*, **50**, 38 (1971).
j. S.-H. Liu, *J. Org. Chem.*, **42**, 3209 (1977).
k. T. Kindt-Larsen, V. Bitsch, I. G. K. Andersen, A. Jart, and J. Munch-Petersen, *Acta Chem. Scand.*, **17**, 1426 (1963).
l. V. K. Andersen and J. Munch-Petersen, *Acta Chem. Scand.*, **16**, 947 (1962).
m. Y. Horiguchi, E. Nakamura, and I. Kuwajima, *J. Am. Chem. Soc.*, **111**, 6257 (1989).
n. T. Fujisawa and T. Sato, *Org. Synth.*, **66**, 116 (1988).
o. F. J. Weiberth and S. S. Hall, *J. Org. Chem.*, **52**, 3901 (1987).

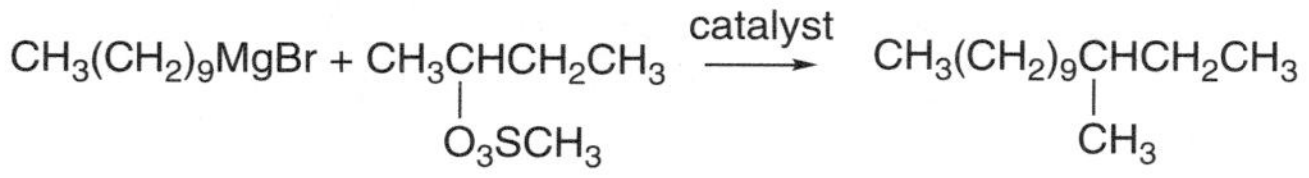

$$CH_3(CH_2)_9MgBr + CH_3CHCH_2CH_3 \xrightarrow{\text{catalyst}} CH_3(CH_2)_9CHCH_2CH_3$$

$$O_3SCH_3 \qquad\qquad CH_3$$

Catalyst	Yield
Li$_2$CuCl$_4$	17%
CuBr/HMPA	30%
CuBr–S(CH$_3$)$_2$, LiBr, LiSPH	62%

These reactions presumably involve fast metal-metal exchange (see Section 7.1.2.4) generating a more nucleophilic organocopper intermediate. The reductive elimination regenerates an active Cu(I) species.

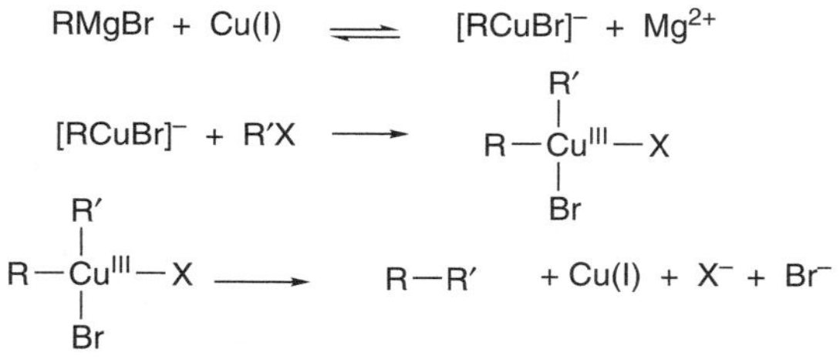

$$RMgBr + Cu(I) \rightleftharpoons [RCuBr]^- + Mg^{2+}$$

$$[RCuBr]^- + R'X \longrightarrow R-\underset{\underset{Br}{|}}{\overset{\overset{R'}{|}}{Cu^{III}}}-X$$

$$R-\underset{\underset{Br}{|}}{\overset{\overset{R'}{|}}{Cu^{III}}}-X \longrightarrow R-R' + Cu(I) + X^- + Br^-$$

Other examples of catalytic substitutions can be found in Section A of Scheme 8.4.

Conjugate addition to α, β-unsaturated esters can often be effected by copper-catalyzed reaction with a Grignard reagent. Other reactions, such as epoxide ring opening, can also be carried out under catalytic conditions. Some examples of catalyzed additions and alkylations are given in Scheme 8.4. These reactions are similar to those carried out with the stoichiometric reagents and presumably involve catalytic cycles that regenerate the active organocopper species. A remarkable aspect of these reactions is that the organocopper cycle must be fast compared to normal organomagnesium reactions, since in many cases there is a potential for competing reactions. The alkylations include several substitutions on allylic systems (Entries 2, 3, and 7). Entry 8 shows that the catalytic process is also applicable to epoxide ring opening. The latter example is a case in which an allylic chloride is displaced in preference to an acetate. The conditions have been observed in related systems to be highly regio-(S_N2') and stereo- (*anti*) specific.[65] The conjugate additions in Entries 9 to 12 show

---

65. J.-E. Backvall, *Bull. Soc. Chim. Fr.*, 665 (1987).

that esters and enones (Entry 13) are reactive to the catalytic processes involving Grignard reagents. Entries 14 and 15 illustrate ketone syntheses from acyl chlorides and nitriles, respectively.

8.1.2.5. Mixed Organocopper-Zinc Reagents. The preparation of organozinc reagents is discussed in Section 7.3.1. Many of these reagents can be converted to mixed copper-zinc organometallics that have useful synthetic applications.[66] A virtue of these reagents is that they can contain a number of functional groups that are not compatible with the organolithium route to cuprate reagents. The mixed copper-zinc reagents are not very basic and can be prepared and allowed to react in the presence of weakly acidic functional groups that would protonate more basic organometallic reagents; for example, reagents containing secondary amide or indole groups can be prepared.[67] They are good nucleophiles and are especially useful in conjugate addition. Mixed zinc reagents can also be prepared by addition of CuCN to organozinc iodides.[68] They are analogous to the cyanocuprates prepared from alkyllithium and CuCN, but with Zn^{2+} in place of Li^+, and react with enones, nitroalkenes, and allylic halides.[69]

In addition to the use of stoichiometric amounts of cuprate or cyanocuprate reagents for conjugate addition, there are also procedures that require only a catalytic amount of copper and use organozinc reagents as the stoichiometric reagent.[70] Simple organozinc reagents, such as diethylzinc, undergo conjugate addition with 0.5 mol % CuO_3SCF_3 in the presence of a phoshine or phosphite.

Ref. 71

In the presence of LiI, TMS-Cl, and a catalytic amount of $(CH_3)_2CuCNLi_2$, conjugate addition of functionalized organozinc reagents occurs in good yield.

Ref. 72

Either CuI or CuCN (10 mol %) in conjunction with BF_3 and TMS-Cl catalyze addition of alkylzinc bromides to enones.

66. P. Knochel and R. D. Singer, *Chem. Rev.*, **93**, 2117 (1993); P. Knochel, *Synlett*, 393 (1995).
67. H. P. Knoess, M. T. Furlong, M. J. Rozema, and P. Knochel, *J. Org. Chem.*, **56**, 5974 (1991).
68. P. Knochel, J. J. Almena Perea, and P. Jones, *Tetrahedron*, **54**, 8275 (1998).
69. P. Knochel, M. C. P. Yeh, S. C. Berk, and J. Talbert, *J. Org. Chem.*, **53**, 2390 (1988); M. C. P. Yeh and P. Knochel, *Tetrahedron Lett.*, **29**, 2395 (1988); S. C. Berk, P. Knochel, and M. C. P. Yeh, *J. Org. Chem.*, **53**, 5789 (1988); H. G. Chou and P. Knochel, *J. Org. Chem.*, **55**, 4791 (1990).
70. B. H. Lipshutz, *Acc. Chem. Res.*, **30**, 277 (1997).
71. A. Alexakis, J. Vastra, and P. Mageney, *Tetrahedron Lett.*, **38**, 7745 (1997).
72. B. H. Lipshutz, M. R. Wood, and R. J. Tirado, *J. Am. Chem. Soc.*, **117**, 6126 (1995).

Ref. 73

Several examples of mixed organocopper-zinc reagents in synthesis are given in Scheme 8.5. Entries 1 and 2 show the use of functionalized reagents prepared from the corresponding iodides by reaction with zinc, followed by CuCN-LiCl. Entry 3 uses a similar reagent to prepare a prostaglandin precursor. Note the slightly different pattern from Entry 3 in Scheme 8.4; in the present case the addition is to an exocyclic methylene group rather than to an endocyclic cyclopentenone. Entry 4 involves generation of a mixed reagent directly from an iodide, followed by conjugate addition to methyl acrylate. Entries 5 and 6 are substitutions on allylic systems. The arylzinc reagent used in Entry 5 was prepared from 2-nitrophenyllithium, which was prepared by halogen-metal exchange, as discussed on p. 632. Entry 7 is a stereospecific S_N2' displacement on an allylic methanesulfonate. Entry 8 is a substitution on a β-sulfonyloxy enone. The zinc reagent is mixed dialkyl zinc. This reaction may proceed by conjugate addition to give the enolate, followed by elimination of the triflate group. Entry 9 shows the use of a tertiary mixed zinc reagent in the preparation of a ketone.

8.1.2.6. Carbometallation with Mixed Organocopper Compounds. Mixed copper-magnesium reagents analogous to the lithium cuprates can be prepared.[74] The precise structural nature of these compounds, often called *Normant reagents*, has not been determined. Individual species with differing Mg:Cu ratios may be in equilibrium.[75] These reagents undergo addition to terminal acetylenes to generate alkenylcopper reagents. The addition is stereospecifically *syn*.

The alkenylcopper adducts can be worked up by protonolysis, or they can be subjected to further elaboration by alkylation or electrophilic substitution.

Mixed copper-zinc reagents also react with alkynes to give alkenylcopper species that can undergo subsequent electrophilic substitution.

73. R. D. Rieke, M. V. Hanson, J. D. Brown, and Q. J. Niu, *J. Org. Chem.*, **61**, 2726 (1996).

74. J. F. Normant and M. Bourgain, *Tetrahedron Lett.*, 2583 (1971); J. F. Normant, G. Cahiez, M. Bourgain, C. Chuit, and J. Villieras, *Bull. Soc. Chim. Fr.*, 1656 (1974); H. Westmijze, J. Meier, H. J. T. Bos, and P. Vermeer, *Recl. Trav. Chim. Pays Bas*, **95**, 299, 304 (1976).

75. E. C. Ashby, R. S. Smith, and A. B. Goel, *J. Org. Chem.*, **46**, 5133 (1981); E. C. Ashby and A. B. Goel, *J. Org. Chem.*, **48**, 2125 (1983).

Scheme 8.5. Conjugate Addition and Substitution Reactions of Mixed Organocopper-Zinc Reagents

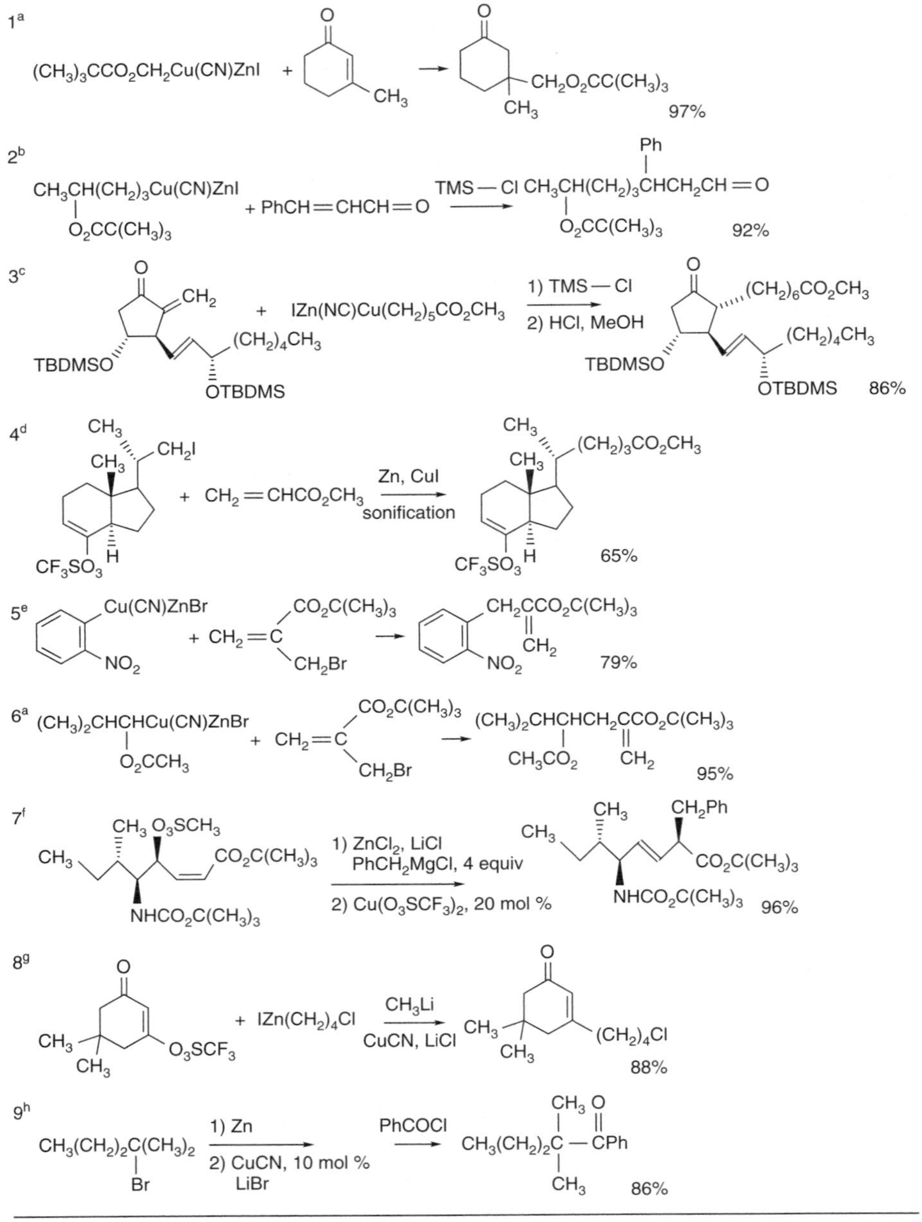

1[a]

$(CH_3)_3CCO_2CH_2Cu(CN)ZnI$ + [cyclohexenone with CH₃] → [product] $CH_2O_2CC(CH_3)_3$ CH₃   97%

2[b]

$CH_3CH(CH_2)_3Cu(CN)ZnI$ with $O_2CC(CH_3)_3$ + $PhCH=CHCH=O$ $\xrightarrow{TMS-Cl}$ $CH_3CH(CH_2)_3CHCH_2CH=O$ with Ph and $O_2CC(CH_3)_3$ 92%

3[c]

[cyclopentanone structure] + $IZn(NC)Cu(CH_2)_5CO_2CH_3$ $\xrightarrow[\text{2) HCl, MeOH}]{\text{1) TMS-Cl}}$ [product $(CH_2)_6CO_2CH_3$, $(CH_2)_4CH_3$] TBDMSO OTBDMS   86%

4[d]

[bicyclic structure with CH₃, CH₂I, CF₃SO₃] + $CH_2=CHCO_2CH_3$ $\xrightarrow[\text{sonification}]{\text{Zn, CuI}}$ [product $(CH_2)_3CO_2CH_3$]   65%

5[e]

[aryl Cu(CN)ZnBr with NO₂] + $CH_2=C$ with $CO_2C(CH_3)_3$ and CH_2Br → [product $CH_2CCO_2C(CH_3)_3$, CH_2, NO_2]   79%

6[a]

$(CH_3)_2CHCHCu(CN)ZnBr$ with O_2CCH_3 + $CH_2=C$ with $CO_2C(CH_3)_3$ and CH_2Br → $(CH_3)_2CHCHCH_2CCO_2C(CH_3)_3$ with CH_3CO_2 and CH_2 95%

7[f]

[structure with CH₃ O₃SCH₃, CO₂C(CH₃)₃, NHCO₂C(CH₃)₃] $\xrightarrow[\text{2) Cu(O_3SCF_3)_2, 20 mol \%}]{\text{1) ZnCl_2, LiCl, PhCH_2MgCl, 4 equiv}}$ [product with CH₃, CH₂Ph, CO₂C(CH₃)₃, NHCO₂C(CH₃)₃]   96%

8[g]

[cyclohexenone with CH₃, CH₃, O₃SCF₃] + $IZn(CH_2)_4Cl$ $\xrightarrow[\text{CuCN, LiCl}]{CH_3Li}$ [product with CH₃, CH₃, $(CH_2)_4Cl$]   88%

9[h]

$CH_3(CH_2)_2C(CH_3)_2$ with Br $\xrightarrow[\text{2) CuCN, 10 mol \% LiBr}]{\text{1) Zn}}$ $\xrightarrow{PhCOCl}$ $CH_3(CH_2)_2C-CPh$ with CH₃ O and CH₃   86%

a. P. Knochel, T. S. Chou, C. Jubert, and D. Rajagopal, *J. Org. Chem.*, **58**, 588 (1993).
b. M. C. P. Yeh, P. Knochel, and L. E. Santa, *Tetrahedron*, **29**, 3887 (1988).
c. H. Tsujiyama, N. Ono, T. Yoshino, S. Okamoto, and F. Sato, *Tetrahedron Lett.*, **31**, 4481 (1990).
d. J. P. Sestalo, J. L. Mascarenas, L. Castedo, and A. Mourina, *J. Org. Chem.*, **58**, 118 (1993).
e. C. Tucker, T. N. Majid, and P. Knochel, *J. Am. Chem. Soc.*, **114**, 3983 (1992).
f. N. Fujii, K. Nakai, H. Habashita, H. Yoshizawa, T. Ibuka, F. Garrido, A. Mann, Y. Chounann, and Y. Yamamoto, *Tetrahedron Lett.*, **34**, 4227 (1993).
g. B. H. Lipshutz and R. W. Vivian, *Tetrahedron Lett.*, **40**, 2871 (1999).
h. R. D. Rieke, M. V. Hanson, and Q. J. Niu, *J. Org. Chem.*, **61**, 2726 (1996).

$$Z(CH_2)_nCu(CN)ZnI \xrightarrow{(CH_3)_2Cu(CN)Li} Z(CH_2)_nCu(CN)Li \cdot Zn(CH_3)_2 \xrightarrow{PhC \equiv CH}$$

Z(CH₂)ₙ / Cu(CN)·Zn(CH₃)₂, C=C, Ph, H $\xrightarrow{CH_2=CHCH_2Br}$ Z(CH₂)ₙ / CH₂CH=CH₂, C=C, Ph, H

Ref. 76

The mechanism of carbometallation has been explored computationally.[77] The reaction consists of an oxidative addition to the triple bond forming a cyclic Cu(III) intermediate. The rate-determining step is reductive elimination to form a vinyl magnesium (or zinc) reagent, which then undergoes transmetallation to the alkenyl-copper product.

R—Cu—R, HC≡CR′ ⟶ (cyclic Mg, R, R, Cu, HC=CR′) ⟶ Mg, Cu, R, H, R′ ⟶ R, Mg, Cu, R, H, R′

Some additional examples are given in Scheme 8.6. The electrophiles that have been used successfully include iodine (Entries 2 and 3) and cyanogen chloride (Entry 4). The adducts can undergo conjugate addition (Entry 5), alkylation (Entry 6), or epoxide ring opening (Entries 7 and 8). The latter reaction is an early step of a synthesis of epothilone B.

The lithium cuprate reagents are not as reactive toward terminal alkynes as mixed magnesium or zinc reagents. The stronger Lewis acid character of Mg^{2+}, as compared to Li^+, is believed to be the reason for the enhanced reactivity of the magnesium reagents. However, lithium dialkylcuprates do react with conjugated acetylenic esters, with *syn* addition being kinetically preferred.[78]

$$(C_4H_9)_2CuLi + CH_3C \equiv CCO_2CH_3 \xrightarrow{H^+}$$

CH₃ / CO₂CH₃, C=C, C₄H₉ / H   86%

The intermediate adduct can be substituted at the α-position by a variety of electrophiles, including acyl chlorides, epoxides, aldehydes, and ketones.[79]

8.1.2.7. Mechanistic Interpretation of the Reactivity of Organocopper Compounds.
The coupling with halides and tosylates, epoxide ring openings, and conjugate additions discussed in the preceding sections illustrate the nucleophilicity of the organocopper reagents. The nucleophilicity is associated with relatively high-energy filled d orbitals that are present in Cu(I), which has a $3d^{10}$ electronic configuration. The role of

76. S. A. Rao and P. Knochel, *J. Am. Chem. Soc.*, **113**, 5735 (1991).
77. S. Mori, A. Hirai, M. Nakamura, and E. Nakamura, *Tetrahedron*, **56**, 2805 (2000).
78. R. J. Anderson, V. L. Corbin, G. Cotterrell, G. R. Cox, C. A. Henrick, F. Schaub, and J. B. Siddall, *J. Am. Chem. Soc.*, **97**, 1197 (1975).
79. J. P. Marino and R. G. Linderman, *J. Org. Chem.*, **48**, 4621 (1983).

Scheme 8.6. Generation and Reactions of Alkenylcopper Reagents from Alkynes

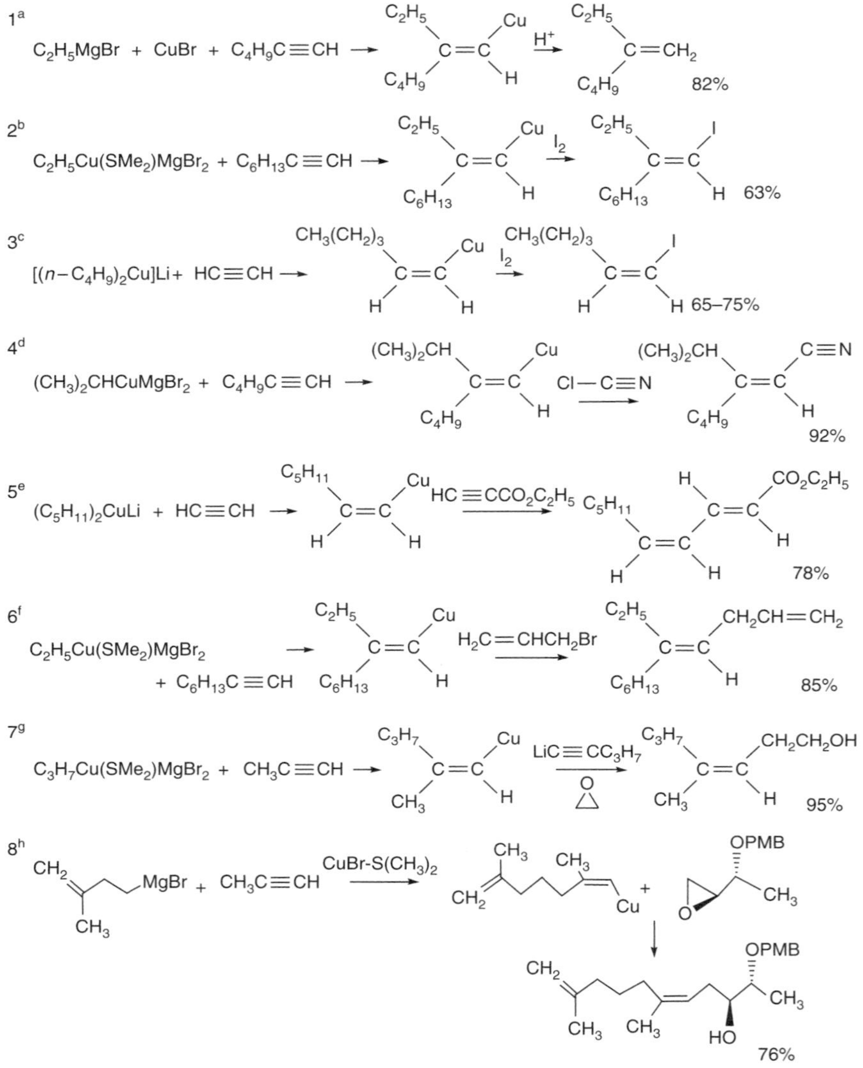

1[a] $C_2H_5MgBr + CuBr + C_4H_9C{\equiv}CH \rightarrow$ ... $\xrightarrow{H^+}$ ... 82%

2[b] $C_2H_5Cu(SMe_2)MgBr_2 + C_6H_{13}C{\equiv}CH \rightarrow$ ... $\xrightarrow{I_2}$ ... 63%

3[c] $[(n{-}C_4H_9)_2Cu]Li + HC{\equiv}CH \rightarrow$ ... $\xrightarrow{I_2}$ ... 65–75%

4[d] $(CH_3)_2CHCuMgBr_2 + C_4H_9C{\equiv}CH \rightarrow$ ... $Cl{-}C{\equiv}N \rightarrow$ ... 92%

5[e] $(C_5H_{11})_2CuLi + HC{\equiv}CH \rightarrow$ ... $HC{\equiv}CCO_2C_2H_5$ ... 78%

6[f] $C_2H_5Cu(SMe_2)MgBr_2 + C_6H_{13}C{\equiv}CH \rightarrow$ ... $H_2C{=}CHCH_2Br \rightarrow$ ... 85%

7[g] $C_3H_7Cu(SMe_2)MgBr_2 + CH_3C{\equiv}CH \rightarrow$ ... $\xrightarrow{LiC{\equiv}CC_3H_7}$ ... 95%

8[h] ... $\xrightarrow{CuBr{-}S(CH_3)_2}$ ... 76%

a. J. F. Normant, G. Cahiez, M. Bourgain, C. Chuit, and J. Villerias, *Bull. Chim. Soc. Fr.*, 1656 (1974).
b. N. J. LaLima, Jr., and A. B. Levy, *J. Org. Chem.*, **43**, 1279 (1978).
c. A. Alexakis, G. Cahiez, and J. F. Normant, *Org. Synth.*, **62**, 1 (1984).
d. H. Westmijze and P. Vermeer, *Synthesis*, 784 (1977).
e. A. Alexakis, J. Normant, and J. Villeras, *Tetrahedron Lett.*, 3461 (1976).
f. R. S. Iyer and P. Helquist, *Org. Synth.*, **64**, 1 (1985).
g. P. R. McGuirk, A. Marfat, and P. Helquist, *Tetrahedron Lett.*, 2465 (1978).
h. M. Valluri, R. M. Hindupur, P. Bijou, G. Labadie, J.-C. Jung, and M. A. Avery, *Org. Lett.*, **3**, 3607 (2001).

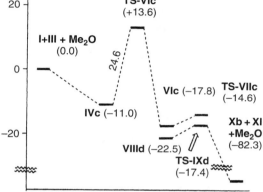

Fig. 8.2. Computational energy profile (B3LYP/631A) for reaction of $(CH_3)_2CuLi$-LiCl with CH_3Br including one solvent (CH_3OCH_3) molecule. Adapted from *J. Am. Chem. Soc.*, **122**, 7294 (2000), by permission of the American Chemical Society.

the copper-lithium clusters has been explored computationally (B3LYP/631A) for reactions with methyl bromide,[80] ethylene oxide,[80] acrolein,[81] and cyclohexenone.[82] In the case of methyl bromide, the reaction was studied both with and without a solvation model. The results in the case of inclusion of one molecule of solvent (CH_3OCH_3) are shown in Figure 8.2. The rate-determining step is the conversion of a complex of the reactant cluster, $[(CH_3)_2CuLi\text{-}LiCl)\text{-}CH_3Br]$, to a tetracoordinate Cu(III) species. The calculated barrier is 13.6 kcal/mol. The reductive elimination step has a very low barrier (~ 5 kcal/mol).

The ring opening of ethylene oxide was studied with CH_3SCH_3 as the solvent molecule and is summarized in Figure 8.3. The crucial TS again involves formation of the C–Cu bond and occurs with assistance from Li^+. As with methyl bromide, the reductive elimination has a low barrier. Incorporation of BF_3 leads to a structure **TS-XXXi** (insert in Figure 8.3) in which BF_3 assists the epoxide ring opening. The

[80.] S. Mori, E. Nakamura, and K. Morokuma, *J. Am. Chem. Soc.*, **122**, 7294 (2000).
[81.] E. Nakamura, S. Mori, and K. Morokuma, *J. Am. Chem. Soc.*, **119**, 4900 (1997).
[82.] S. Mori and E. Nakamura, *Chem. Eur. J.*, **5**, 1534 (1999).

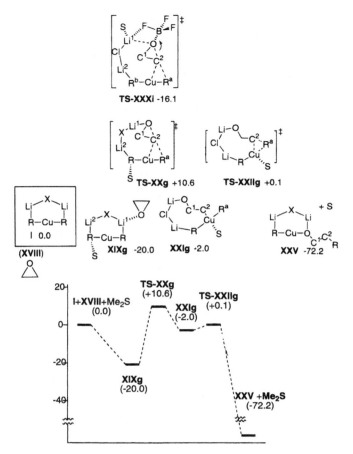

Fig. 8.3. Computational energy profile (B3LYP/631A) for reaction of $(CH_3)_2Cu$-LiCl-$(CH_3)_2S$ with ethylene oxide. The insert (**TS-XXXi**) is a TS that incorporates BF_3, but not $(CH_3)_2S$. Adapted from *J. Am. Chem. Soc.*, **122**, 7294 (2000), by permission of the American Chemical Society.

stabilization of the TS leads to a reduction of almost 37 kcal/mol in the computed E_a relative to **TSXXg**.

The nucleophilicity of the organocuprate cluster derives mainly from the filled copper $3d_z^2$ orbital, in combination with the carbon orbital associated with bonding to copper. These orbitals for the TS for reaction with methyl bromide and ethylene oxide are shown in Figure 8.4.

The conjugate addition reaction has also been studied computationally. B3LYP/631A calculations of the reaction of $[(CH_3)_2CuLi]_2$ with acrolein gives the TS and intermediates depicted in Figure 8.5.[81] Three intermediates and three TSs are represented. The first structure is a complex of the reactants (**CP1i**), which involves coordination of the acrolein oxygen to a lithium cation in the reactant. The second intermediate (**CPcl**) is a π complex in which the cluster is opened. A key feature of the mechanism is the third intermediate **CPop**, which involves interaction of *both* lithium ions with the carbonyl oxygen. Moreover, in contrast to the reactions with halides and epoxides, it is the *reductive elimination step that is rate determining*. The calculated barrier for this step is 10.4 kcal/mol.

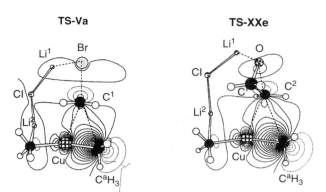

Fig. 8.4. Representation of the orbital involved in C–Cu bond formation in the reaction of $(CH_3)_2CuLi\text{-}LiCl$ with methyl bromide (left) and ethylene oxide (right). Reproduced from *J. Am. Chem. Soc.*, **122**, 7294 (2000), by permission of the American Chemical Society.

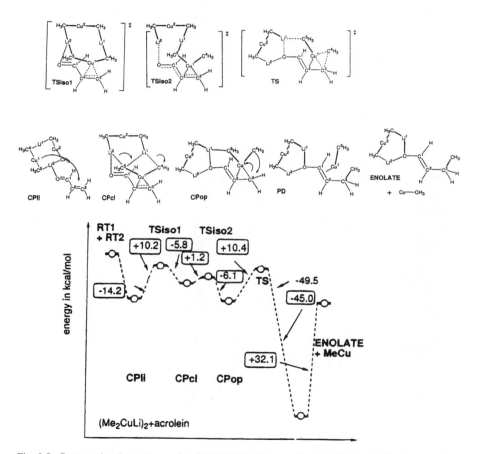

Fig. 8.5. Computational reaction profile (B3LYP/631A) for reaction of $[(CH_3)_2CuLi]_2$ with acrolein. Adapted from *J. Am. Chem. Soc.*, **119**, 4900 (1997), by permission of the American Chemical Society.

The role of BF$_3$ catalysis in the conjugate addition was also explored.[83] Inclusion of BF$_3$ results in a considerable stabilization of the reaction complex, but there is also a lowered barrier for the rate-determining reductive elimination. This suggests that BF$_3$ functions primarily at the Cu(III) stage by facilitating the decomposition of the Cu(III) intermediate.

A similar sequence of intermediates and TSs was found for the reaction of cyclohexenone.[82] In this case, both axial and equatorial approaches were examined. At the crucial rate- and product-determining TS for C—C bond formation, the axial pathway is favored by 1.7 kcal/mol, in agreement with experimental results from conformationally biased cyclohexenones. Nearly all of the difference is due to factors in the cyclohexenone ring and transferring methyl group. This result suggests that analysis of stereoselectivity of cuprate conjugate additions should focus on the relative energies of the competing TS for the C—C bond-forming step. These computational studies comport well with a variety of product, kinetic, and spectroscopic studies that have been applied to determining the mechanism of organocuprates and related reagents.[84]

Visual models and additional information on Organocuprate Intermediates can be found in the Digital Resource available at: Springer.com/carey-sundberg.

8.1.2.8. Enantioselective Reactions of Organocopper Reagents. Several methods have been developed for achieving enantioselectivity with organocopper reagents. Chiral auxiliaries can be used; for example, oxazolidinone auxiliaries have been utilized in conjugate additions. The outcome of these reactions can be predicted on the basis of steric control of reactant approach, as for other applications of the oxazolidinone auxiliaries.

Ref. 85

Conjugate addition reactions involving organocopper intermediates can be made enantioselective by using chiral ligands.[86] Several mixed cuprate reagents containing

83. E. Nakamura, M. Yamanaka, and S. Mori, *J. Am. Chem. Soc.*, **122**, 1826 (2000).

84. E. Nakamura and S. Mori, *Angew. Chem. Int. Ed. Engl.*, **39**, 3750 (2000).

85. M. P. Sibi, M. D. Johnson, and T. Punniyamurthy, *Can. J. Chem.*, **79**, 1546 (2001).

86. N. Krause and A. Gerold, *Angew. Chem. Int. Ed. Engl.*, **36**, 186 (1997); N. Krause, *Angew. Chem. Int. Ed. Engl.*, **37**, 283 (1998).

chiral ligands have been investigated to determine the degree of enantioselectivity that can be achieved. The combination of diethylzinc and cyclohexenone has been studied extensively, and several amide and phosphine ligands have been explored. Enantioselectivity can also be observed using Grignard reagents with catalytic amounts of copper. Scheme 8.7 shows some examples of these reactions using various chiral ligands.

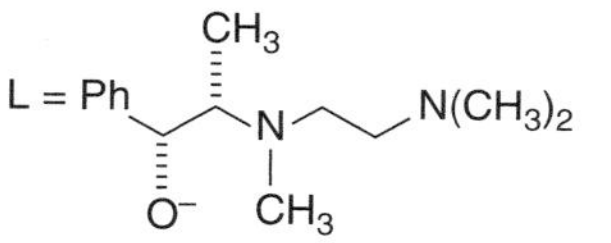

Ref. 87

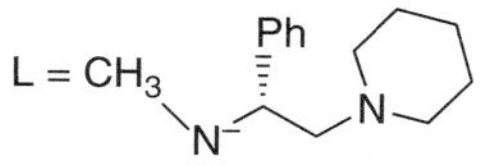

Ref. 88

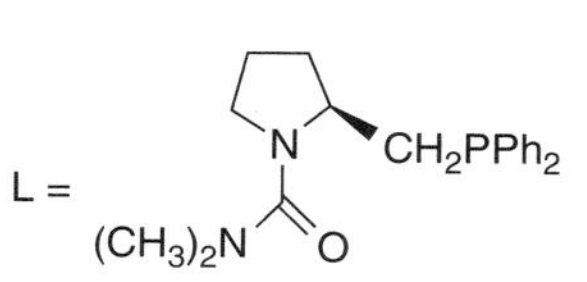

Ref. 89

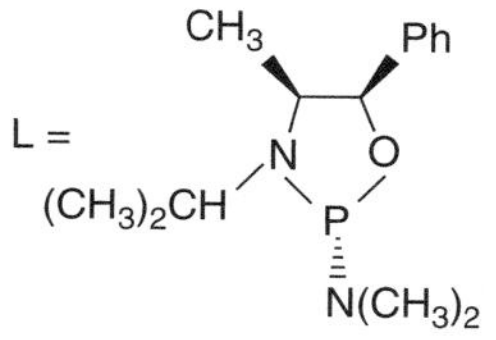

Ref. 90

Enantioselective catalysis of S_N2' alkylation has been achieved.[91] A BINOL-phosphoramidite catalyst (*o*-methoxyphenyl analog) similar to that in Entry 3 in Scheme 8.7 gave good results.

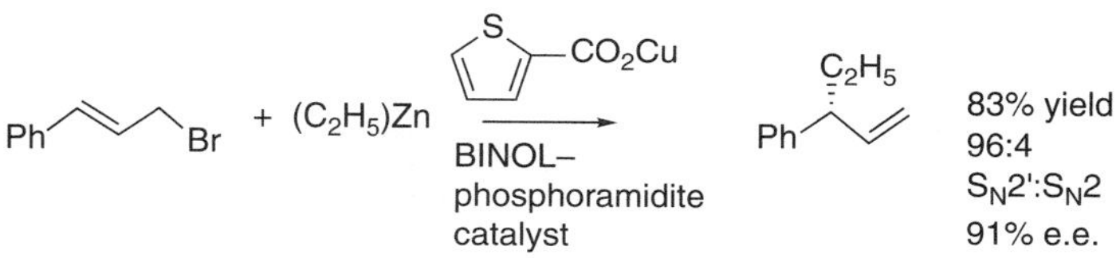

8.1.2.9. Aryl-Aryl Coupling Using Organocopper Reagents.

Organocopper intermediates are also involved in several procedures for coupling of two aromatic reactants to form a new carbon-carbon bond. A classic example of this type of reaction is the *Ullman coupling* of aryl halides, which is done by heating an aryl halide with a copper-bronze alloy.[92] Good yields by this method are limited to halides with EWG substituents.[93] Mechanistic studies have established the involvement of arylcopper

[87.] E. J. Corey, R. Naef, and F. J. Hannon, *J. Am. Chem. Soc.*, **108**, 7144 (1986).
[88.] N. M. Swingle, K. V. Reddy, and B. L. Rossiter, *Tetrahedron*, **50**, 4455 (1994); G. Miao and B. E. Rossiter, *J. Org. Chem.*, **60**, 8424 (1995).
[89.] M. Kanai and K. Tomioka, *Tetrahedron Lett.*, **35**, 895 (1994); **36**, 4273, 4275 (1995).
[90.] A. Alexakis, J. Frutos, and P. Mageney, *Tetrahedron: Asymmetry*, **4**, 2427 (1993).
[91.] K. Tissot-Croset, D. Polet, and A. Alexakis, *Angew. Chem. Int. Ed. Engl.*, **43**, 2426 (2004).
[92.] P. E. Fanta, *Chem. Rev.*, **64**, 613 (1964); P. E. Fanta, *Synthesis*, 9 (1974).
[93.] R. C. Fuson and E. A. Cleveland, *Org. Synth.*, **III**, 339 (1955).

Scheme 8.7. Catalytic Enantioselective Conjugate Addition to Cyclohexenone

Entry	Reactant	Catalyst	Ligand	Yield	e.e.
1[a]	n-C_4H_9MgCl	CuI 10 mol %	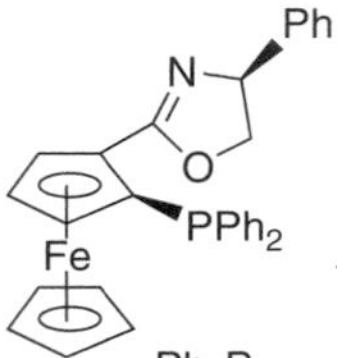 12 mol %	97	83
2[b]	C_2H_5MgBr	CuCl 3 mol %	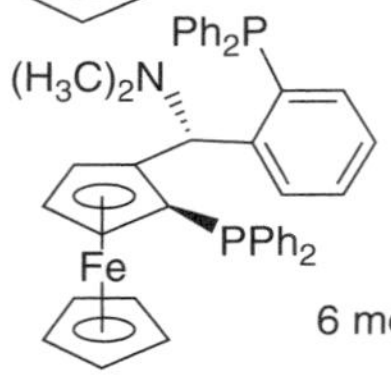 6 mol %	69	96
3[c]	$(C_2H_5)_2Zn$	$Cu(O_3SCF_3)_2$ 2 mol %	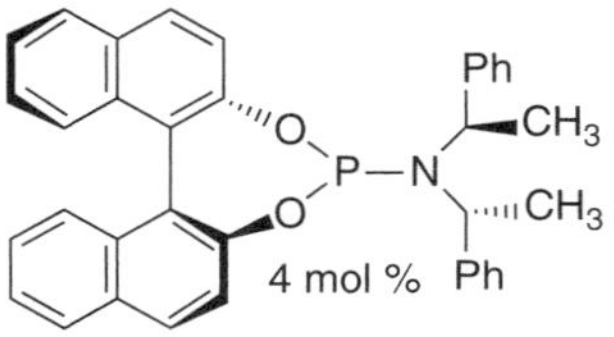 4 mol %	94	>98
4[d]	$(C_2H_5)_2Zn$	$Cu(O_3SCF_3)_2$ 2 mol %	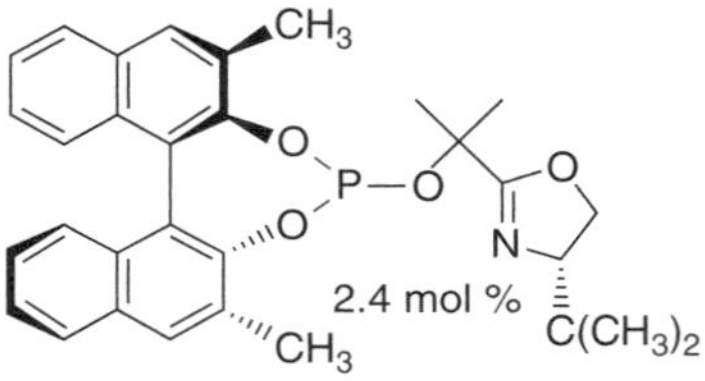 2.4 mol %	96	90
5[e]	$(C_2H_5)_2Zn$	$Cu(O_3SCF_3)_2$ 1.2 mol %	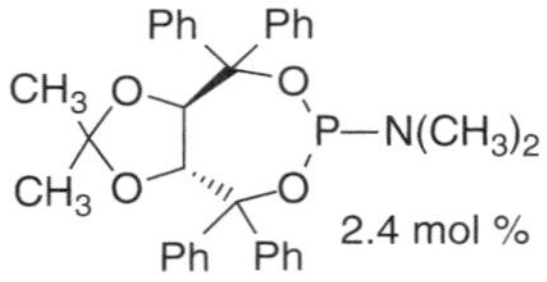 2.4 mol %	90	71
6[f]	n-C_4H_9MgCl	CuI 8 mol %	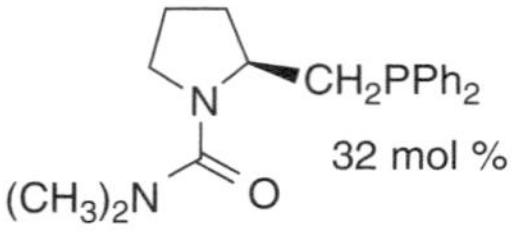 32 mol %	92	90

a. E. L. Stangeland and T. Sammakia, *Tetrahedron*, **53**, 16503 (1997).
b. B. L. Feringa, R. Badorrey, D. Pena, S. R. Harutyunyan, and A. J. Minnaard, *Proc. Natl. Acad. Sci. USA*, **101**, 5834 (2004).
c. B. L. Feringa, M. Pineschi, L. A. Arnold, R. Imbos, and A. H. M. de Vries, *Angew. Chem. Int. Ed. Engl.*, **36**, 2620 (1997).
d. A. K. H. Knobel, I. H. Escher, and A. Pfaltz, *Synlett*, 1429 (1997); I. H. Escher and A. Pfaltz, *Tetrahedron*, **56**, 2879 (2000).
e. E. Keller, J. Maurer, R. Naasz, T. Schader, A Meetsma, and B. L. Feringa, *Tetrahedron: Asymmetry*, **9**, 2409 (1998).
f. M. Kanai, Y. Nakagawa, and K. Tomioka, *Tetrahedron*, **55**, 3843 (1999).

intermediates. Soluble Cu(I) salts, particularly the triflate, effect coupling of aryl halides at much lower temperatures and under homogeneous conditions.[94]

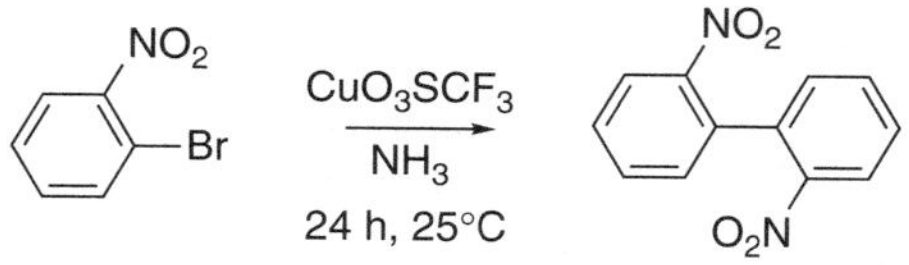

Arylcopper intermediates can be generated from organolithium compounds, as in the preparation of cuprates.[95] These compounds react with a second aryl halide to provide unsymmetrical biaryls in a reaction that is essentially a variant of the cuprate alkylation process discussed on p. 680. An alternative procedure involves generation of a mixed diarylcyanocuprate by sequential addition of two different aryllithium reagents to CuCN, which then undergo decomposition to biaryls on exposure to oxygen.[96] The second addition must be carried out at very low temperature to prevent equilibration with the symmetrical diarylcyanocuprates.

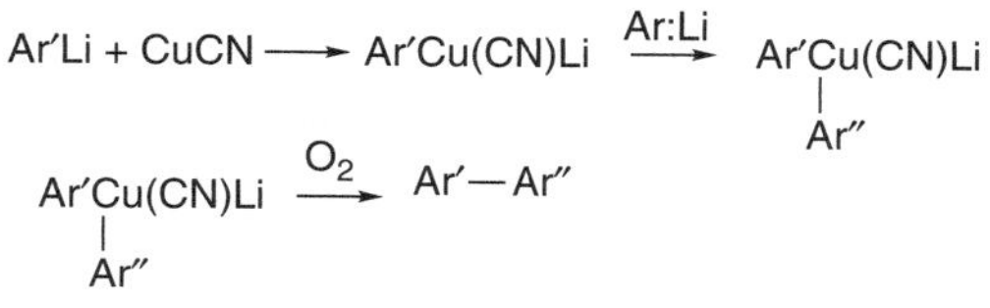

Intramolecular variations of this reaction have been achieved.

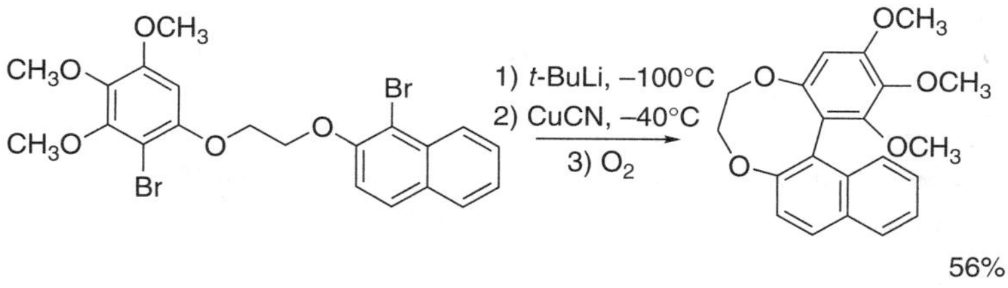

56%   Ref. 97

8.1.2.10. Summary of Synthetic Reactions of Organocopper Reagents and Intermediates. The synthetic procedures involving organocopper reagents and intermediates offer a wide range of carbon-carbon bond-forming reactions. Coupling of alkyl, alkenyl, and aryl groups and the various mixed combinations can be achieved. The coupling of allylic reagents encompasses acetates, sulfonates, and phosphates, as well as halides. These reactions often occur with allylic transposition. Both direct and vinylogous

94. T. Cohen and I. Cristea, *J. Am. Chem. Soc.*, **98**, 748 (1976).

95. F. E. Ziegler, I. Chliwner, K. W. Fowler, S. J. Kanfer, S. J. Kuo, and N. D. Sinha, *J. Am. Chem. Soc.*, **102**, 790 (1980).

96. B. H. Lipshutz, K. Siegmann, and E. Garcia, *Tetrahedron*, **48**, 2579 (1992); B. H. Lipshutz, K. Siegmann, E. Garcia, and F. Kayser, *J. Am. Chem. Soc.*, **115**, 9276 (1993).

97. B. H. Lipshutz, F. Kayser, and N. Maullin, *Tetrahedron Lett.*, **35**, 815 (1994).

epoxide ring-opening reactions are available for the synthesis of alcohols. The reactants for conjugate addition include α, β-unsaturated ketones, esters, amides, and nitriles, and these reactions can be combined with tandem alkylation. These synthetic transformations are summarized below.

8.2. Reactions Involving Organopalladium Intermediates

Organopalladium intermediates are very important in synthetic organic chemistry. Usually, organic reactions involving palladium do not involve the preparation of stoichiometric organopalladium reagents. Rather, organopalladium species are generated in situ during the course of the reaction. In the most useful processes only a *catalytic amount* of palladium is used. The overall reaction mechanisms typically involve several steps in which organopalladium species are formed, react with other reagents, give product, and are regenerated in a catalytically active form. Catalytic processes have both economic and environmental advantages. Since, in principle, the catalyst is not consumed, it can be used to make product without generating by-products. Some processes use *solid phase catalysts*, which further improve the economic and environmental advantages of catalyst recovery. Reactions that involve chiral catalysts can generate enantiomerically enriched or pure materials from achiral starting materials. In this section we focus on carbon-carbon bond formation, but in Chapter 11 we will see that palladium can also catalyze aromatic substitution reactions.

Several types of organopalladium intermediates are of primary importance in the reactions that have found synthetic applications. Alkenes react with Pd(II) to give π complexes that are subject to *nucleophilic attack*. These reactions are closely related to the solvomercuration reactions discussed in Section 4.1.3. The products that are formed from the resulting intermediates depend upon specific reaction conditions. The palladium can be replaced by hydrogen under reductive conditions (path a). In the absence of a reducing agent, elimination of Pd(0) and a proton occurs, leading to net substitution of a vinyl hydrogen by the nucleophile (path b). We return to specific examples of these reactions shortly.

$$RCH{=}CH_2 + Pd(II) \;\rightleftharpoons\; RCH{\overset{Pd^{II}}{\underset{\shortmid}{=}}}CH_2$$

$$Nu + RCH{\overset{Pd^{II}}{\underset{\shortmid}{=}}}CH_2 \;\longrightarrow\; Nu{-}\underset{\underset{R}{\shortmid}}{C}HCH_2Pd^{II}$$

$$Nu{-}\underset{\underset{R}{\shortmid}}{C}HCH_2Pd^{II} \;\begin{array}{l} \xrightarrow{[H]} \; Nu{-}\underset{\underset{R}{\shortmid}}{C}HCH_3 \quad \text{(path a)} \\[2em] \xrightarrow[-H^+]{-Pd(0)} \; Nu{-}\underset{\underset{R}{\shortmid}}{C}{=}CH_2 \quad \text{(path b)} \end{array}$$

A second major group of organopalladium intermediates are π-allyl complexes, which can be obtained from Pd(II) salts, allylic acetates, and other compounds having potential leaving groups in an allylic position.[98] The same type of π-allyl complex can be prepared directly from alkenes by reaction with $PdCl_2$ or $Pd(O_2CCF_3)_2$.[99] The reaction with alkenes occurs by electrophilic attack on the π electrons followed by loss of a proton. The proton loss probably proceeds via an unstable species in which the hydrogen is bound to palladium.[100]

These π-allyl complexes are moderately *electrophilic*[101] in character and react with a variety of nucleophiles, usually at the less-substituted allylic terminus. After nucleophilic addition occurs, the resulting organopalladium intermediate usually breaks down by elimination of Pd(0) and H^+. The overall transformation is an allylic substitution.

Another general process involves the reaction of Pd(0) species with halides or sulfonates by *oxidative addition*, generating reactive intermediates having the organic group attached to Pd(II) by a σ bond. The oxidative addition reaction is very useful for aryl and alkenyl halides, but the products from saturated alkyl halides often decompose by β-elimination. The σ-bonded species formed by oxidative addition can react with alkenes and other unsaturated compounds to form new carbon-carbon bonds. The

98. R. Huttel, *Synthesis*, 225 (1970); B. M. Trost, *Tetrahedron*, **33**, 2615 (1977).
99. B. M. Trost and P. J. Metzner, *J. Am. Chem. Soc.*, **102**, 3572 (1980); B. M. Trost, P. E. Strege, L. Weber, T. J. Fullerton, and T. J. Dietsche, *J. Am. Chem. Soc.*, **100**, 3407 (1978).
100. D. R. Chrisope, P. Beak, and W. H. Saunders, Jr., *J. Am. Chem. Soc.*, **110**, 230 (1988).
101. O. Kuhn and H. Mayr, *Angew. Chem. Int. Ed. Engl.*, **38**, 343 (1998).

σ-bound species also react with a variety of organometallic reagents to give coupling products.

$$CH_2\!=\!CHR$$
$$RCH\!=\!CH_2 \quad Ar\!-\!Pd^{II}\!-\!X \longrightarrow RCH\!=\!CHAr$$
$$Ar\!-\!X + Pd^0 \longrightarrow Ar\!-\!Pd^{II}\!-\!X$$
$$R'\!-\!M \quad R'$$
$$Ar\!-\!Pd^{II}\!-\!X \longrightarrow Ar\!-\!R'$$

These are called *cross-coupling reactions* and usually involve three basic steps: oxidative addition, transmetallation, and reductive elimination. In the transmetallation step an organic group is transferred from the organometallic reagent to palladium.

oxidative addition $\qquad R\!-\!X + Pd^0 \longrightarrow R\!-\!Pd^{II}\!-\!X$

transmetallation $\qquad R\!-\!Pd^{II}\!-\!X + R'\!-\!M \longrightarrow R\!-\!Pd^{II}\!-\!R' + M\!-\!X$

reductive elimination $\qquad R\!-\!Pd^{II}\!-\!R' \longrightarrow R\!-\!R' + Pd^0$

The organometallic reagents that give such reactions include organomagnesium, organolithium, and organozinc compounds, stannanes, and even organoboron compounds. The reactions are very general for sp^2-sp^2 and sp^2-sp coupling and in some systems can also be applied to sp^2-sp^3 coupling. Most of these procedures involve phosphine or related ligands.

$$R\!-\!M + R'\!-\!X \xrightarrow{\ Pd(L)_y\ } R\!-\!R'$$

$$M = Li,\ MgX,\ ZnX,\ SnR_3,\ BR_2$$

Organopalladium intermediates are also involved in the synthesis of ketones and other carbonyl compounds. These reactions involve acylpalladium intermediates, which can be made from acyl halides or by reaction of an organopalladium species with carbon monoxide. A second organic group, usually arising from an organometallic reagent, can then form a ketone. Alternatively, the acylpalladium intermediate may react with nucleophilic solvents such as alcohols to form esters.

$$R\!-\!\overset{O}{\overset{\|}{C}}\!-\!X \xrightarrow{Pd^0} R\!-\!\overset{O}{\overset{\|}{C}}\!-\!Pd^{II} \xleftarrow{\ C\!\equiv\!O\ }{R\!-\!Pd^{II}}$$

$$R\!-\!\overset{O}{\overset{\|}{C}}\!-\!R' \xleftarrow{} R\!-\!\overset{O}{\overset{\|}{C}}\!-\!\overset{R'}{Pd^{II}} \xleftarrow{R'\!-\!M} \qquad \xrightarrow{R'OH} R\!-\!\overset{O}{\overset{\|}{C}}\!-\!OR'$$

In considering the mechanisms involved in organopalladium chemistry, several general points should be kept in mind. Frequently, reactions involving organopalladium intermediates are done in the presence of phosphine ligands, which play a key role by influencing the reactivity at palladium. Another general point concerns the relative weakness of the C–Pd bond and, especially, the instability of alkylpalladium species in which there is a β-hydrogen.

$$R-\underset{\underset{H}{|}}{\overset{\overset{H}{\curvearrowright}}{C}}-CH_2\overset{\curvearrowleft}{-}\underset{|}{\overset{|}{Pd^{II}}}- \longrightarrow H^+ + RCH{=}CH_2 + Pd^0$$

The final stage in many palladium-mediated reactions is the elimination of Pd(0) and H$^+$ to generate a carbon-carbon double bond. This tendency toward elimination distinguishes organopalladium species from most of the organometallic species we have discussed up to this point. Finally, organopalladium(II) species with two organic substituents show the same tendency to react with recombination of the organic groups by reductive elimination that is exhibited by copper(III) intermediates. This reductive elimination generates the new carbon-carbon bond.

$$R-\underset{|}{\overset{\overset{R'}{|}}{Pd^{II}}}- \longrightarrow R-R' + Pd^0$$

8.2.1. Palladium-Catalyzed Nucleophilic Addition and Substitution

8.2.1.1. The Wacker Reaction and Related Oxidations. An important industrial process based on Pd-alkene complexes is the *Wacker reaction*, a catalytic method for conversion of ethene to acetaldehyde. The first step is addition of water to the Pd(II)-activated alkene. The addition intermediate undergoes the characteristic elimination of Pd(0) and H$^+$ to generate the enol of acetaldehyde.

$$CH_2{=}CH_2 + Pd(II) \longrightarrow CH_2\overset{\underset{|}{Pd^{II}}}{\underset{\dot=}{}}CH_2 \xrightarrow{H_2O} HO{-}CH_2CH_2\overset{|}{\underset{|}{-}}Pd^{II} \longrightarrow \underset{H}{\overset{HO}{\diagdown}}C{=}CH_2 + Pd^0 + H^+$$

$$\underset{H}{\overset{HO}{\diagdown}}C{=}CH_2 \longrightarrow CH_3CH{=}O$$

The reaction is run with only a catalytic amount of Pd. The co-reagents CuCl$_2$ and O$_2$ serve to reoxidize the Pd(0) to Pd(II). The net reaction consumes only alkene and oxygen.

$$
\begin{array}{c}
2H^+ + \tfrac{1}{2}O_2 \quad\quad 2\,Cu^I \quad\quad Pd^{II} \quad CH_2{=}CH_2 \\
\\
2\,{}^-OH \quad\quad 2\,Cu^{II} \quad\quad\quad\quad Pd^{II} \\
\quad\quad\quad\quad Pd^0 \quad\quad CH_2\dot{=}CH_2 \\
\\
H^+ + HOCH{=}CH_2 \quad\quad HOCH_2CH_2Pd^{II} \quad H^+ \quad H_2O
\end{array}
$$

The relative reactivity profile of the simple alkenes toward Wacker oxidation is quite shallow and in the order ethene > propene > 1-butene > *E*-2-butene > *Z*-2-butene.[102] This order indicates that steric factors outweigh electronic effects and is consistent with substantial nucleophilic character in the rate-determining step. (Compare with oxymercuration; see Part A, Section 5.8.) The addition step is believed to occur by an internal ligand transfer through a four-center mechanism, leading to *syn* addition.

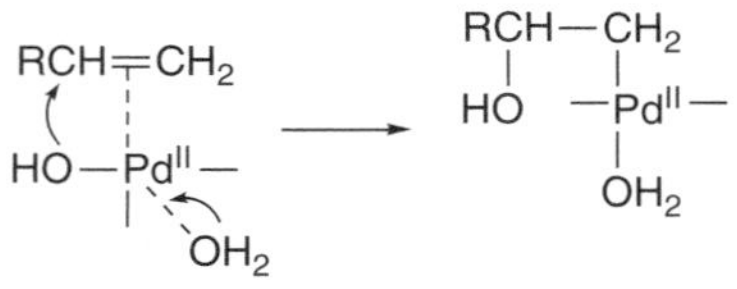

The stereochemistry, however, is sensitive to the concentration of chloride ion, shifting to *anti* when chloride is present.[103]

The Wacker reaction can also be applied to laboratory-scale syntheses.[104] When the Wacker conditions are applied to terminal alkenes, methyl ketones are formed.[105]

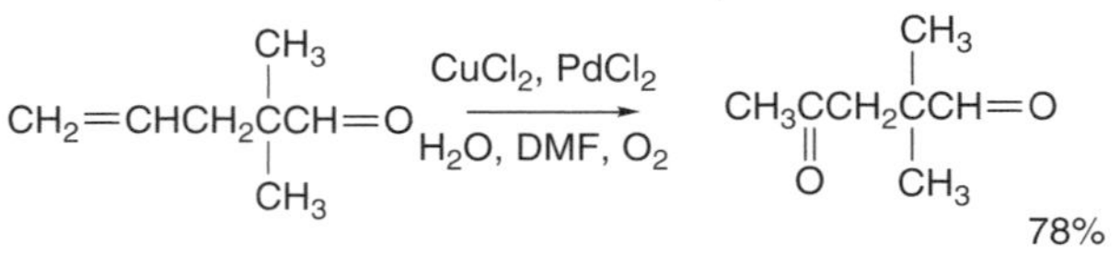

This regiochemistry is consistent with the electrophilic character of Pd(II) in the addition step. Solvent and catalyst composition can affect the regiochemistry of the Wacker reaction. Use of *t*-butanol as the solvent was found to increase the amount of aldehyde formed from terminal alkenes, and is attributed to the greater steric requirement of *t*-butanol. Hydrolysis of the enol ether then leads to the aldehyde.

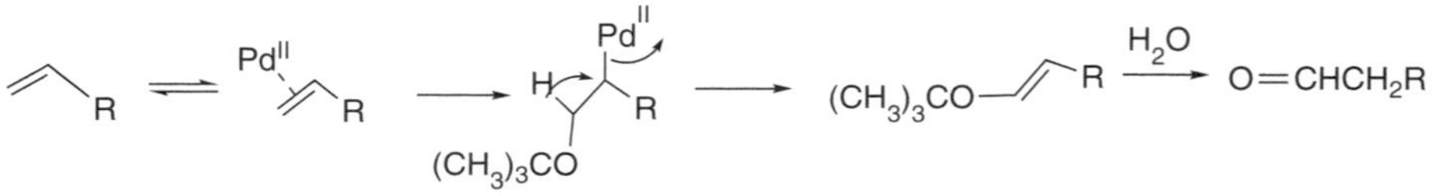

These conditions are particularly effective for allyl acetate.[106]

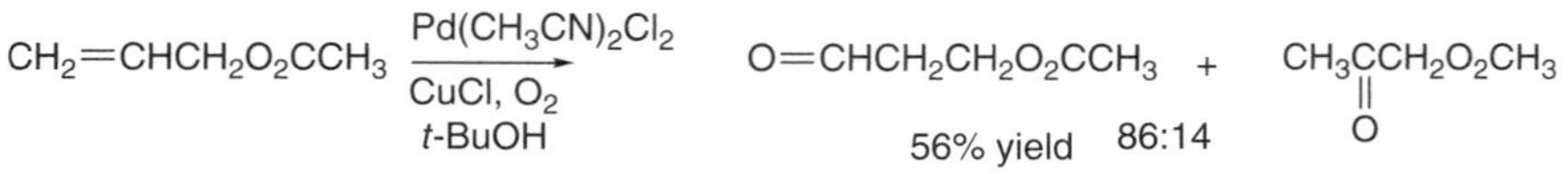

102. K. Zaw and P. M. Henry, *J. Org. Chem.*, **55**, 1842 (1990); A. Lambert, E. G. Derouane, and I. V. Kozhevnikov, *J. Catal.*, **211**, 445 (2002).
103. O. Hamed, P. M. Henry, and C. Thompson, *J. Org. Chem.*, **64**, 7745 (1999).
104. J. M. Takacs and X.-T. Jiang, *Current Org. Chem.*, **7**, 369 (2003).
105. (a) J. Tsuji, I. Shimizu, and K. Yamamoto, *Tetrahedron Lett.*, 2975 (1976); J. Tsuji, H. Nagashima, and H. Nemoto, *Org. Synth.*, **62**, 9 (1984); (c) D. Pauley, F. Anderson, and T. Hudlicky, *Org. Synth.*, **67**, 121 (1988); (d) K. Januszkiewicz and H. Alper, *Tetrahedron Lett.*, **25**. 5159 (1983); (e) K. Januszkiewicz and D. J. H. Smith, *Tetrahedron Lett.*, **26**, 2263 (1985).
106. B. L. Feringa, *J. Chem. Soc., Chem. Commun.*, 909 (1986); T. T. Wenzel, *J. Chem. Soc., Chem. Commun.*, 862 (1993).

Both the regiochemistry and stereochemistry of Wacker oxidation can be influenced by substituents that engage in chelation with Pd. Whereas a single γ-alkoxy function leads to a mixture of aldehyde and ketone, more highly oxygenated systems such as the acetonide or carbonate of the diol **1** lead to dominant aldehyde formation.[107] The diol itself gives only ketone, which perhaps indicates that steric factors are also important.

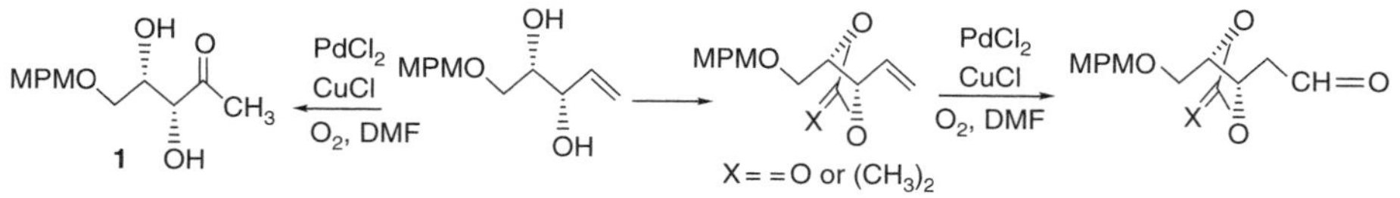

X = =O or (CH$_3$)$_2$

The two reactions shown below are examples of the use of the Wacker reaction in multistep synthesis. In the first case, selectivity is achieved between two terminal alkene units on the basis of a difference in steric accessibility. Both reactions use a reduced amount of Cu(I) salt. In the second reaction this helps to minimize hydrolysis of the acid-sensitive dioxane ring.

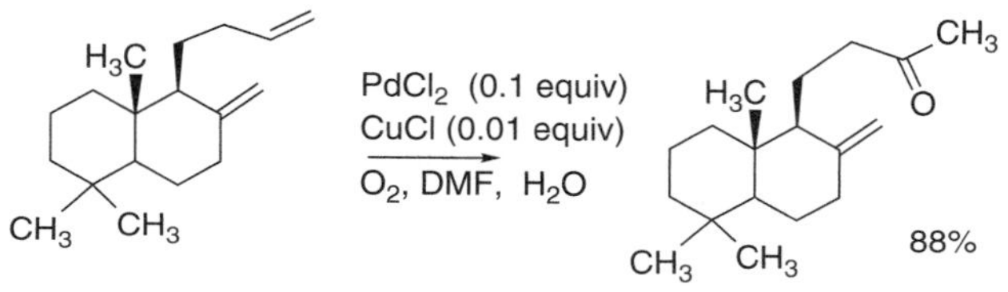

Ref. 108

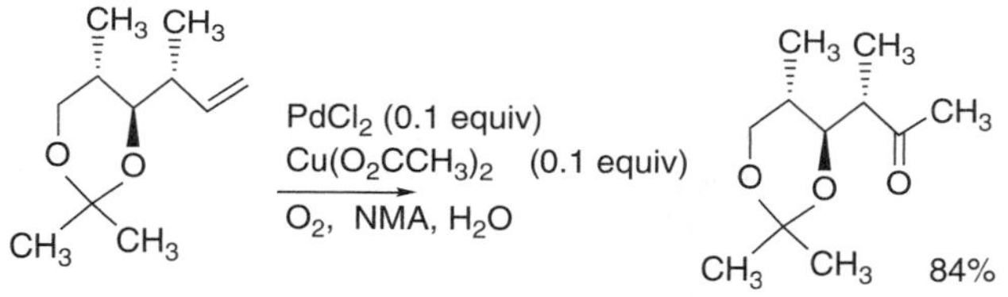

Ref. 109

Palladium(II) like Hg(II) can induce intramolecular nucleophilic addition, but this is followed by elimination of Pd(0) and H$^+$. For example, γ, δ- and δ, ε-unsaturated carboxylic acids can be cyclized to unsaturated lactones by Pd(OAc)$_2$ in DMSO in the presence of O$_2$. Although Cu(OAc)$_2$ can be included as a catalyst for reoxidation of the Pd(0), it is not necessary.[110]

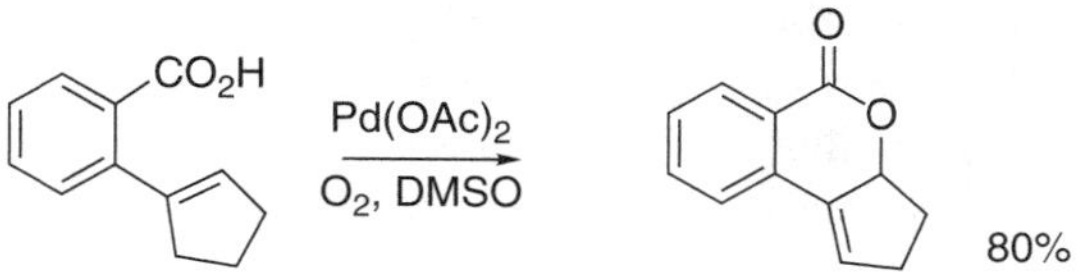

Similarly, phenols with unsaturated side chains can form five- and six-membered rings. In these systems the quaternary carbon imposes the β, γ-elimination. As with the above

107. S.-K. Kang, K.-Y. Jung, J.-U. Chung, E.-Y. Namkoong, and T.-H. Kim, *J. Org. Chem.*, **60**, 4678 (1995).
108. H. Toshima, H. Oikawa, T. Toyomasu, and T. Sassa, *Tetrahedron*, **56**, 8443 (2000).
109. A. B. Smith, III, Y. S. Cho, and G. K. Friestad, *Tetrahedron Lett.*, **39**, 8765 (1998).
110. R. C. Larock and T. R. Hightower, *J. Org. Chem.*, **58**, 5298 (1993).

cyclization, a copper co-oxidant is not needed. The pyridine is evidently involved in accelerating the oxidation of Pd(0) by O_2.[111]

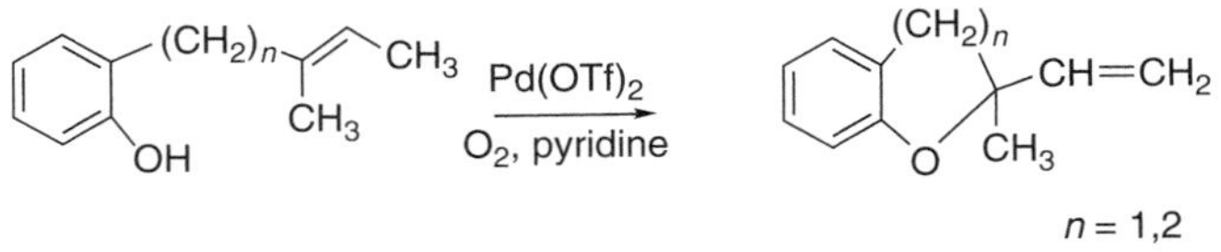

$n = 1,2$

Cyclizations of this type can be carried out with high enantioselectivity using a chiral *bis*-oxazoline catalyst.

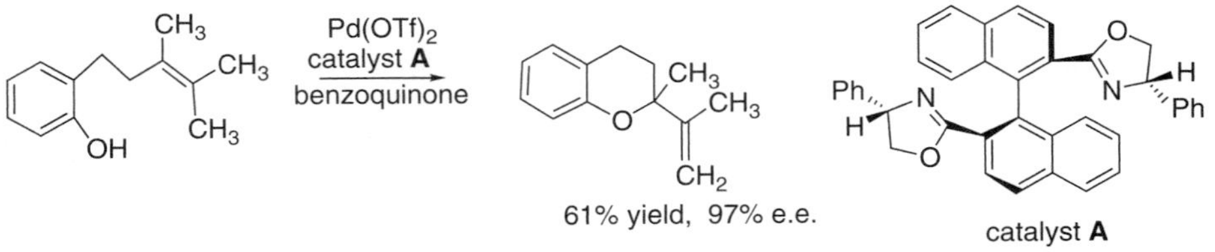

61% yield, 97% e.e.

catalyst **A**

Ref. 112

A deuterium-labeling study of a reaction of this type demonstrated *syn* stereoselectivity in both the oxypalladation and β-elimination, which indicates that the cyclization occurs by internal migration, rather than by an *anti* nucleophilic capture.[113] This particular system also gives products from double-bond migration that occurs by reversible Pd(II)–D addition-elimination.

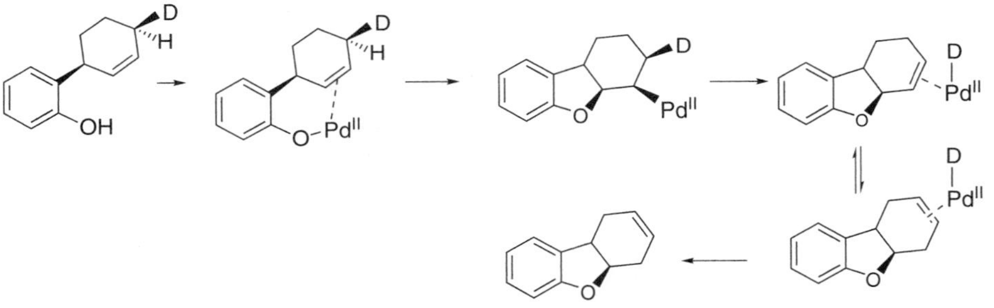

8.2.1.2. Nucleophilic Substitution of π-Allyl Palladium Complexes. π-Allyl palladium species are subject to a number of useful reactions that result in allylation of nucleophiles.[114] The reaction can be applied to carbon-carbon bond formation using relatively stable carbanions, such as those derived from malonate esters and β-sulfonyl esters.[115] The π-allyl complexes are usually generated in situ by reaction of an allylic acetate with a catalytic amount of *tetrakis*-(triphenylphosphine)palladium

111. R. M. Trend, Y. K. Ramtohul, E. M. Ferreira, and B. M. Stoltz, *Angew. Chem. Int. Ed. Engl.*, **42**, 2892 (2003).

112. Y. Uozumi, K. Kato, and T. Hayashi, *J. Am. Chem. Soc.*, **119**, 5063 (1997).

113. T. Hayashi, K. Yamasaki, M. Mimura, and Y. Uozumi, *J. Am. Chem. Soc.*, **126**, 3036 (2004).

114. G. Consiglio and R. M. Waymouth, *Chem. Rev.*, **89**, 257 (1989).

115. B. M. Trost, W. P. Conway, P. E. Strege, and T. J. Dietsche, *J. Am. Chem. Soc.*, **96**, 7165 (1974); B. M. Trost, L. Weber, P. E. Strege, T. J. Fullerton, and T. J. Dietsche, *J. Am. Chem. Soc.*, **100**, 3416 (1978); B. M. Trost, *Acc. Chem. Res.*, **13**, 385 (1980).

or a chelated diphosphine complex.[116] The reactive Pd(0) species is regenerated in an elimination step.

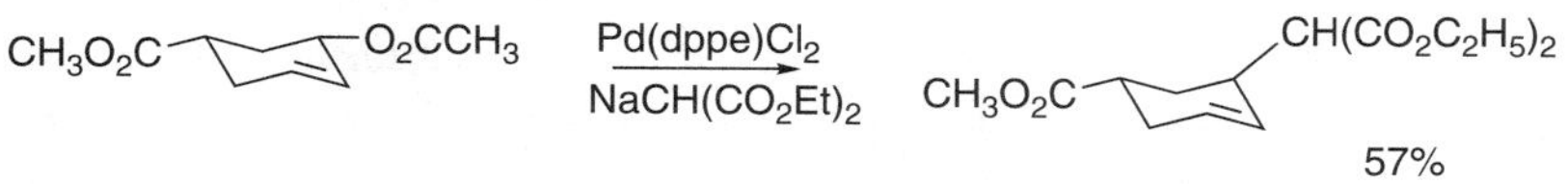

Ref. 117

For unsymmetrical allylic systems both the regiochemistry and stereochemistry of the substitution are critical issues. The palladium normally bonds *anti* to the acetate leaving group. The same products are obtained from 2-acetoxy-4-phenyl-3-butene and 1-acetoxy-1-phenyl-2-butene, indicating a common intermediate. The same product mixture is also obtained from the *Z*-reactants, indicating rapid *E,Z*-equilibration in the allylpalladium intermediate.[118]

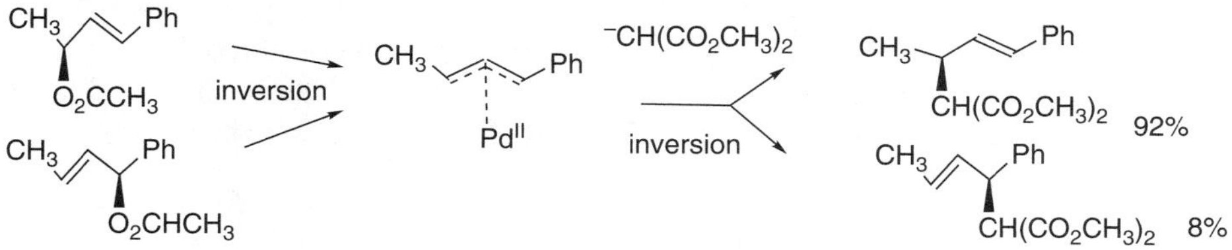

In the presence of chiral phosphine ligands, there is also rapid epimerization to the most stable diastereomeric π-allyl complex. The stereoselectivity arises in the reaction with the nucleophile.[119]

Mechanistically, the nucleophilic addition can occur either by internal ligand transfer or by external attack. Generally, softer more stable nucleophiles (e.g., malonate enolates) are believed to react by the external mechanism and give *anti* addition, whereas harder nucleophiles (e.g., hydroxide) are delivered by internal ligand transfer with *syn* stereochemistry.[120]

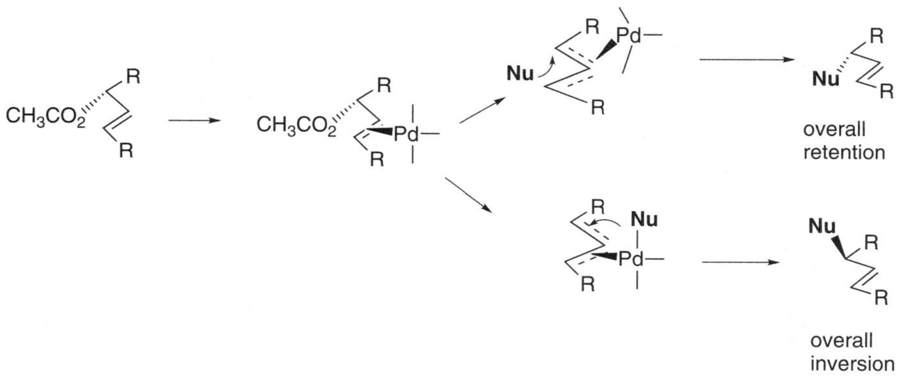

Both the regiochemistry and stereochemistry are influenced by reaction conditions. A striking example is a complete switch to 3-alkylation of dimethyl malonate

116. B. M. Trost and T. R. Verhoeven, *J. Am. Chem. Soc.*, **102**, 4730 (1980).

117. B. M. Trost and P. E. Strege, *J. Am. Chem. Soc.*, **99**, 1649 (1977).

118. T. Hayashi, A. Yamamoto, and T. Hagihara, *J. Org. Chem.*, **51**, 723 (1986).

119. P. B. Mackenzie, J. Whelan, and B. Bosnich, *J. Am. Chem. Soc.*, **107**, 2046 (1985).

120. A. Heumann and M. Reglier, *Tetrahedron*, **51**, 975 (1995).

anion by 1-phenylprop-2-enyl acetate in the presence of iodide ion. In the absence of iodide, using 2 mol % catalyst, the ratio of **2** to **3** is about 4:1. When 2 mol % iodide is added, only **2** is formed. This change is attributed to the involvement of a catalytic species in which I^- is present as a Pd ligand. The effect is diminished when a chelating diphosphine ligand is used, presumably because addition of I^- to the Pd ligand sphere is prevented by the chelate.

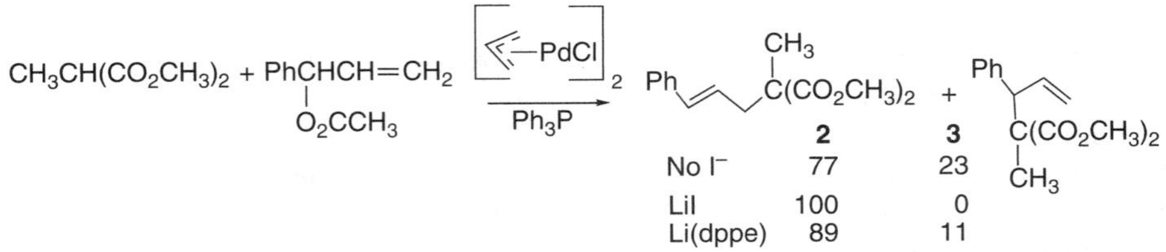

CH3CH(CO2CH3)2 + PhCHCH=CH2
O2CCH3
PdCl
2
Ph3P
Ph
CH3
C(CO2CH3)2
+
Ph
C(CO2CH3)2
CH3
2
3
No I−
77
23
LiI
100
0
Li(dppe)
89
11

Ref. 121

The allylation reaction has also been used to form rings. β-Sulfonyl esters have proven particularly useful in this application for formation of both medium and large rings.[122] In some cases medium-sized rings are formed in preference to six- and seven-membered rings.[123]

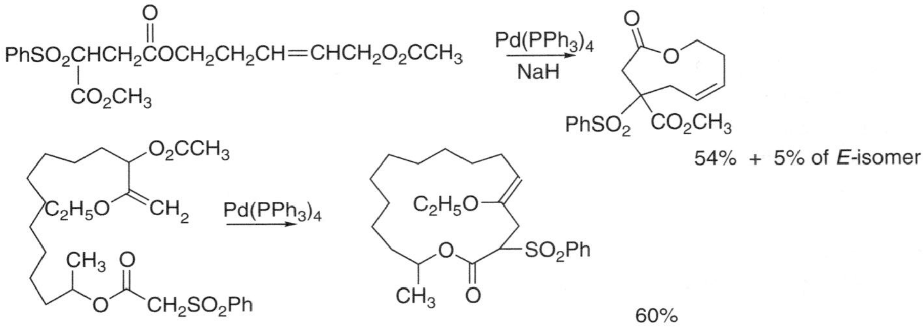

PhSO2CHCH2COCH2CH2CH=CHCH2O2CCH3
CO2CH3
Pd(PPh3)4
NaH
PhSO2
CO2CH3
54% + 5% of E-isomer
O2CCH3
C2H5O
CH2
CH3
CH2SO2Ph
Pd(PPh3)4
C2H5O
SO2Ph
CH3
60%

The sulfonyl substituent can be removed by reduction after the ring closure (see Section 5.6.2). Other appropriate reactants are α-phenylthio nitriles, which can be hydrolyzed to lactones.[124]

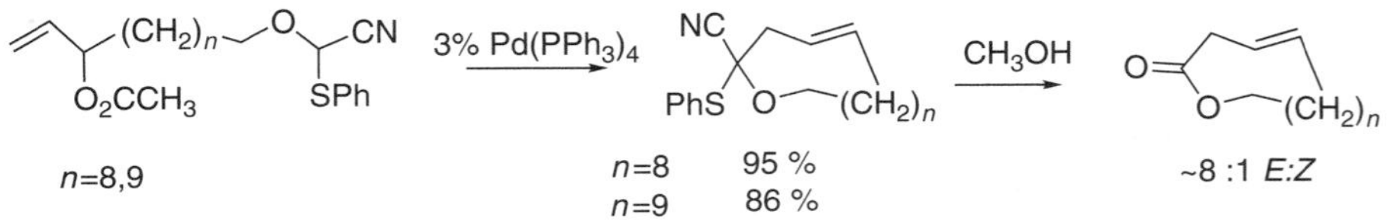

(CH2)n
CN
O2CCH3
SPh
n=8,9
3% Pd(PPh3)4
NC
PhS
(CH2)n
n=8 95 %
n=9 86 %
CH3OH
(CH2)n
~8 :1 E:Z

Allylation reactions can be made highly enantioselective by the use of various chiral phosphine ligands.[125] Examples are included in Scheme 8.8.

121. M. Kawatsura, Y. Uozumi, and T. Hayashi, *J. Chem. Soc., Chem. Commun.*, 217 (1998).
122. B. M. Trost, *Angew. Chem. Int. Ed. Engl.*, **28**, 1173 (1989).
123. B. M. Trost and T. R. Verhoeven, *J. Am. Chem. Soc.*, **102**, 4743 (1980); B. M. Trost and S. J. Brickner, *J. Am. Chem. Soc.*, **105**, 568 (1983); B. M. Trost, B. A. Vos, C. M. Brzezowski, and D. P. Martina, *Tetrahedron Lett.*, **33**, 717 (1992).
124. B. M. Trost and J. R. Granja, *J. Am. Chem. Soc.*, **113**, 1044 (1991).
125. S. J. Sesay and J. M. J. Williams, in *Advances in Asymmetric Synthesis*, Vol. 3, A. Hassner, ed., JAI Press, Stamford, CT, 1998, pp. 235–271; G. Helmchen, *J. Organomet. Chem.*, **576**, 203 (1999).

Scheme 8.8. Enantioselective Allylation of Diethyl Malonate

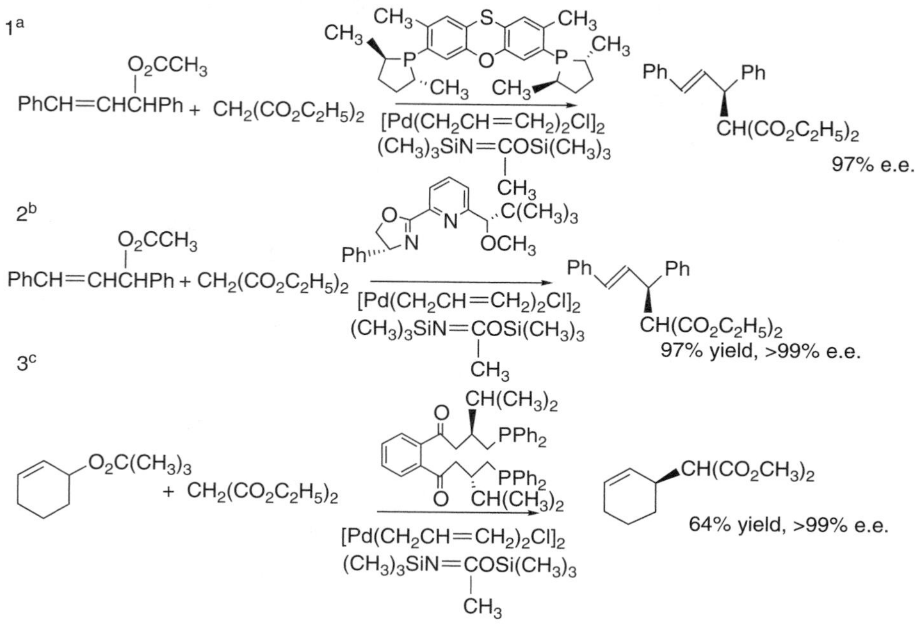

1[a]

$$PhCH=CHCHPh + CH_2(CO_2C_2H_5)_2$$

with O$_2$CCH$_3$ group

97% e.e.

2[b]

$$PhCH=CHCHPh + CH_2(CO_2C_2H_5)_2$$

with O$_2$CCH$_3$ group

$[Pd(CH_2CH=CH_2)_2Cl]_2$
$(CH_3)_3SiN=COSi(CH_3)_3$

97% yield, >99% e.e.

3[c]

with O$_2$C(CH$_3$)$_3$ + CH$_2$(CO$_2$C$_2$H$_5$)$_2$

$[Pd(CH_2CH=CH_2)_2Cl]_2$
$(CH_3)_3SiN=COSi(CH_3)_3$

64% yield, >99% e.e.

a. P. Dierks, S. Ramdeehul, L. Barley, A. DeCian, J. Fischer, P. C. J. Kramer, P. W. N. M. van Leeuwen, and J. A. Osborne, *Angew. Chem. Int. Ed. Engl.*, **37**, 3116 (1998).
b. K. Nordstrom, E. Macedo, and C. Moberg, *J. Org. Chem.*, **62**, 1604 (1997); U. Bremberg, F. Rahm, and C. Moberg, *Tetrahedron: Asymmetry*, **9**, 3437 (1998).
c. A. Saitoh, M. Misawa, and T. Morimoto, *Tetrahedron: Asymmetry*, **10**, 1025 (1999).

8.2.2. The Heck Reaction

Another important type of reactivity of palladium, namely oxidative addition to Pd(0), is the foundation for several methods of forming carbon-carbon bonds. Aryl[126] and alkenyl[127] halides react with alkenes in the presence of catalytic amounts of palladium to give net substitution of the halide by the alkenyl group. The reaction, known as the *Heck reaction*,[128] is quite general and has been observed for simple alkenes, aryl-substituted alkenes, and substituted alkenes such as acrylate esters, vinyl ethers, and *N*-vinylamides.[129]

126. H. A. Dieck and R. F. Heck, *J. Am. Chem. Soc.*, **96**, 1133 (1974); R. F. Heck, *Acc. Chem. Res.*, **12**, 146 (1979); R. F. Heck, *Org. React.*, **27**, 345 (1982).
127. B. A. Patel and R. F. Heck, *J. Org. Chem.*, **43**, 3898 (1978); B. A. Patel, J. I. Kim, D. D. Bender, L. C. Kao, and R. F. Heck, *J. Org. Chem.*, **46**, 1061 (1981); J. I. Kim, B. A. Patel, and R. F. Heck, *J. Org. Chem.*, **46**, 1067 (1981).
128. I. P. Beletskaya and A. V. Cheprakov, *Chem. Rev.*, **100**, 3009 (2000); B. C. G. Soderberg, *Coord. Chem. Rev.*, **224**, 171 (2002); G. T. Crisp, *Chem. Soc. Rev.*, **27**, 427 (1998).
129. C. B. Ziegler, Jr., and R. F. Heck, *J. Org. Chem.*, **43**, 2941 (1978); W. C. Frank, Y. C. Kim, and R. F. Heck, *J. Org. Chem.*, **43**, 2947 (1978); C. B. Ziegler, Jr., and R. F. Heck, *J. Org. Chem.*, **43**, 2949 (1978); H. A. Dieck and R. F. Heck, *J. Am. Chem. Soc.*, **96**, 1133 (1974); C. A. Busacca, R. E. Johnson, and J. Swestock, *J. Org. Chem.*, **58**, 3299 (1993).

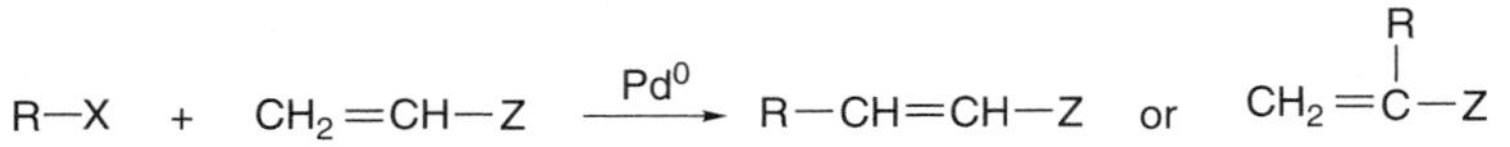
R—X
+
CH2=CH—Z
Pd0
R—CH=CH—Z
or
R
CH2=C—Z

R = alkenyl, aryl
X = halide, sulfonate

Many procedures use $Pd(OAc)_2$ or other Pd(II) salts as catalysts with the catalytically active Pd(0) being generated in situ. The reactions are usually carried out in the presence of a phosphine ligand, with *tris-o*-tolylphosphine being preferred in many cases. *Tris*-(2-furyl)phosphine (tfp) is also used frequently. Several chelating diphosphines, shown below with their common abbreviations, are also effective. Phosphites are also good ligands.[130]

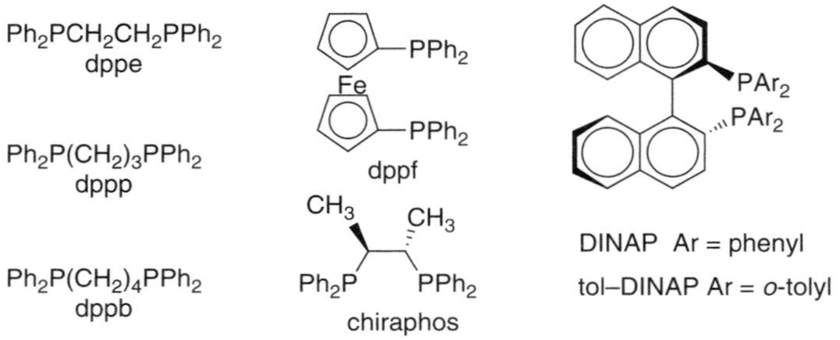
Ph2PCH2CH2PPh2
dppe
Ph2P(CH2)3PPh2
dppp
Ph2P(CH2)4PPh2
dppb
PPh2
Fe
PPh2
dppf
CH3
CH3
Ph2P
PPh2
chiraphos
PAr2
PAr2
DINAP Ar = phenyl
tol–DINAP Ar = o-tolyl

The reaction is initiated by oxidative addition of the halide or sulfonate to a Pd(0) species generated in situ from the Pd(II) catalyst. The arylpalladium(II) intermediate then forms a π complex with the alkene, which rearranges to a σ complex by carbon-carbon bond formation. The σ complex decomposes by β-elimination with regeneration of Pd(0). Both of these reactions occur with *syn* stereoselectivity. The Heck reaction often uses of $Pd(OAc)_2$ as the palladium source along with a triarylphosphine and a tertiary amine. Under these conditions it has been proposed that the initiation of the reaction involves formation of an anionic complex $[Pd(L)_2OAc]^-$.[131] This is a 16-electron species and is considered to be the active form of Pd for the oxidative addition. The base is crucial in maintaining the equilibrium in favor of the active anionic form after the reductive elimination. This is called the *anionic mechanism*. Note that the phosphine ligand is also the reducing agent for formation of the active Pd(0) species.

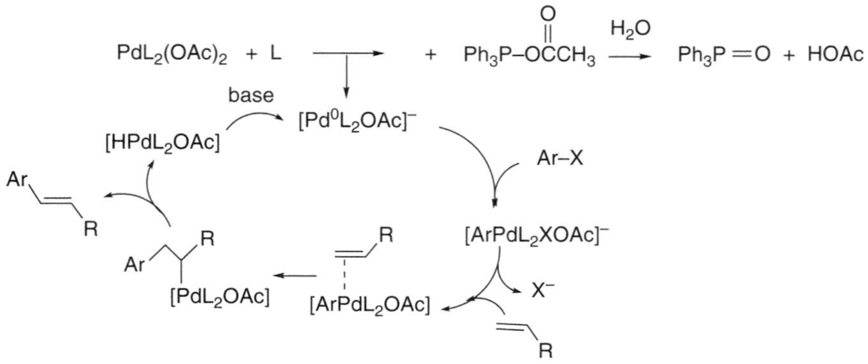
PdL2(OAc)2 + L
+ Ph3P–OCCH3
O
H2O
Ph3P=O + HOAc
base
[Pd0L2OAc]–
[HPdL2OAc]
Ar–X
Ar
R
R
Ar
[PdL2OAc]
R
[ArPdL2OAc]
[ArPdL2XOAc]–
X–
R

Anionic mechanism for Heck reaction

[130] M. Beller and A. Zapt, *Synlett*, 792 (1998).
[131] A. Amatore and A. Jutand, *Acc. Chem. Res.*, **33**, 314 (2000).

Several different Pd(0) species can be involved in both the oxidative addition and π-coordination steps, depending on the anions and ligands present. Because of the equilibria involving dissociation of phosphine ligands and anions, there is dependence on their identity and concentration.[132] High halide concentration promotes formation of the anionic species $[PdL_2X]^-$ by addition of a halide ligand. Use of trifluoromethanesulfonate anions promotes dissociation of the anion from the Pd(II) adduct and accelerates complexation with electron-rich alkenes. The presence of metal ions that bind the halide, e.g., Ag^+, also promotes dissociation. Reactions that proceed through a dissociated species are called *cationic* and are expected to have a more electrophilic interaction with the alkene. A base is included to neutralize the proton released in the β-elimination step. The catalytic cycle under these conditions is shown below.

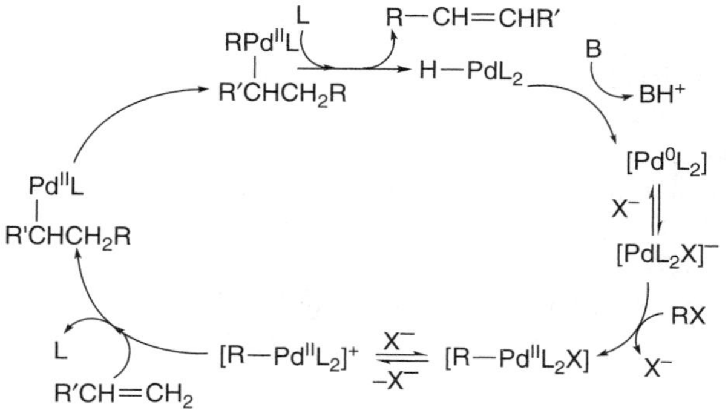

It appears that a modified mechanism operates when *tris-(o-tolyl)phosphine* is used as the ligand,[133] and this phosphine has been found to form a *palladacycle*. Much more stable than noncyclic Pd(0) complexes, this compound is also more reactive toward oxidative addition. As with the other mechanisms, various halide adducts or halide-bridged compounds may enter into the overall mechanism.

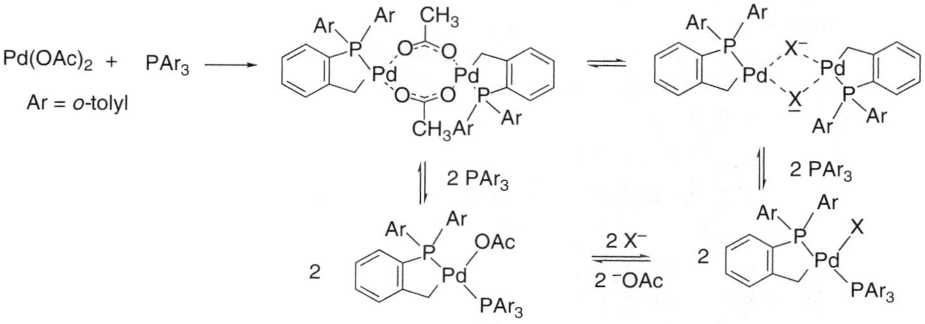

132. W. Cabri, I. Candiani, S. De Bernardinis, F. Francalanci, S. Penco, and R. Santi, *J. Org. Chem.*, **56**, 5796 (1991); F. Ozawa, A. Kubo, and T. Hayashi, *J. Am. Chem. Soc.*, **113**, 1417 (1991).

133. W. A. Hermann, C. Brossmer, K. Ofele, C.-P. Reisinger, T. Priermeier, M. Beller, and H. Fischer, *Angew. Chem. Int. Ed. Engl.*, **34**, 1844 (1995).

Several modified reaction conditions have been developed. One involves addition of silver salts, which activate the halide toward displacement.[134] Use of sodium bicarbonate or sodium carbonate in the presence of a phase transfer catalyst permits especially mild conditions to be used for many systems.[135] Tetraalkylammonium salts also often accelerate reaction.[136] Solid phase catalysts in which the palladium is complexed by polymer-bound phosphine groups have also been developed.[137]

Aryl chlorides are not very reactive under normal Heck reaction conditions but reaction can be achieved by inclusion of tetraphenylphosphonium salts with Pd(OAc)$_2$ or PdCl$_2$ as the catalysts.[138]

Pretreatment with nickel bromide causes normally unreactive aryl chlorides to undergo Pd-catalyzed substitution,[139] and aryl and vinyl triflates have been found to be excellent substrates for Pd-catalyzed alkenylations.[140]

Heck reactions can be carried out in the absence of phosphine ligands.[141] These conditions usually involve Pd(OAc)$_2$ as a catalyst, along with a base and a phase transfer salt such as tetra-*n*-butylammonium bromide. These conditions were originally applied to stereospecific coupling of vinyl iodides with ethyl acrylate and methyl vinyl ketone.

Several optimization studies have been carried out under these phosphine-free conditions. The reaction of bromobenzene and styrene was studied using Pd(OAc)$_2$ as the catalyst, and potassium phosphate and *N,N*-dimethylacetamide (DMA) were found to be the best base and solvent. Under these conditions, the Pd content can be reduced to as low as 0.025 mol %.[142] The reaction of substituted bromobenzenes with methyl α-acetamidoacrylate has also been studied carefully, since the products are potential precursors of modified amino acids. Good results were obtained using either *N, N*-diisopropylethylamine or NaOAc as the base.

134. M. M. Abelman, T. Oh, and L. E. Overman, *J. Org. Chem.*, **52**, 4130 (1987); M. M. Abelman and L. E. Overman, *J. Am. Chem. Soc.*, **110**, 2328 (1988).

135. T. Jeffery, *J. Chem. Soc., Chem. Commun.*, 1287 (1984); T. Jeffery, *Tetrahedron Lett.*, **26**, 2667 (1985); T. Jeffery, *Synthesis*, 70 (1987); R. C. Larock and S. Babu, *Tetrahedron Lett.*, **28**, 5291 (1987).

136. A. de Meijere and F. E. Meyer, *Angew. Chem. Int. Ed. Engl.*, **33**, 2379 (1994); R. Grigg, *J. Heterocycl. Chem.*, **31**, 631 (1994); T. Jeffery, *Tetrahedron*, **52**, 10113 (1996).

137. C.-M. Andersson, K. Karabelas, A. Hallberg, and C. Andersson, *J. Org. Chem.*, **50**, 3891 (1985).

138. M. T. Reetz, G. Lehmer, and R. Schwickard, *Angew. Chem. Int. Ed.*, **37**, 481 (1998).

139. J. J. Bozell and C. E. Vogt, *J. Am. Chem. Soc.*, **110**, 2655 (1988).

140. A. M. Echavarren and J. K. Stille, *J. Am. Chem. Soc.*, **109**, 5478 (1987); K. Karabelas and A. Hallberg, *J. Org. Chem.*, **53**, 4909 (1988).

141. T. Jeffery, *Tetrahedron Lett.*, **26**, 2667 (1985); T. Jeffery, *Synthesis*, 70 (1980).

142. Q. Yao, E. P. Kinney, and Z. Yang, *J. Org. Chem.*, **68**, 7528 (2003).

Ref. 143

Low Pd concentrations are beneficial in preventing precipitation of inactive Pd metal.[144] Small Pd clusters can be observed in phosphine-free systems,[145] and these particles may serve as catalysts or, alternatively, as reservoirs of Pd for formation of soluble reactive species.

The regiochemistry of the Heck reaction is determined by the competitive removal of the β-proton in the elimination step. Mixtures are usually obtained if more than one type of β-hydrogen is present. Often there is also double-bond migration that occurs by reversible Pd-H elimination-addition sequences. For example, the reaction of cyclopentene with bromobenzene leads to all three possible double-bond isomers.[146]

Substituents with stronger electronic effects can influence the competition between α- and β-arylation. Alkenes having EWG substituents normally result in β-arylation. However, alkenes with donor substituents give a mixture of α- and β-regioisomers. The regiochemistry can be controlled to some extent by specific reaction conditions. Bidentate phosphines such as dppp and dppf promote α-arylation of alkenes with donor substituents such as alkoxy, acetoxy, and amido. These reactions are believed to occur through the more electrophilic form of Pd(II) generated by dissociation of the triflate anion (cationic mechanism).[147] Electronic factors favor migration of the aryl group to the α-carbon. The combination of the bidentate ligand and triflate leaving group increases the importance of electronic effects on the regiochemistry.

Substituents without strong donor or acceptor character (e.g., phenyl, succinimido) give mixtures. The reason for the increased electronic sensitivity is thought to be the

---

[143.] C. E. Williams, J. M. C. A. Mulders, J. G. de Vries, and A. H. M. de Vries, *J. Organomet. Chem.*, **687**, 494 (2003).

[144.] A. H. M. de Vries, J. M. C. A. Mulders, J. H. M. Mommers, H. J. W. Hendrickx, and J. G. de Vries, *Org. Lett.*, **5**, 3285 (2003).

[145.] M. T. Reetz and E. Westermann, *Angew. Chem. Int. Ed. Engl.*, **39**, 165 (2000).

[146.] C. G. Hartung, K. Kohler, and M. Beller, *Org. Lett.*, **1**, 709 (1999).

[147.] W. Cabri, I. Cardiani, A. Bedeschi, and R. Santi, *J. Org. Chem.*, **55**, 3654 (1990); W. Cabri, I. Candiani, A. Bedeschi, and R. Santi, *J. Org. Chem.*, **57**, 3558 (1992). W. Cabri, I. Candiani, A. Bedeschi, S. Penco, and R. Santi, *J. Org. Chem.*, **57**, 1481 (1992).

involvement of a cationic, as opposed to a neutral, complex. The triflate anion is more likely to be dissociated than a halide.

Allylic silanes show a pronounced tendency to react at the α-carbon in the presence of bidentate ligands.[148] This regiochemistry is attributed to the preferential stabilization of cationic character by the silyl substituent. The bidentate ligands enhance the electrophilic character of the TS, and the cation stabilization of the silyl group becomes the controlling factor.

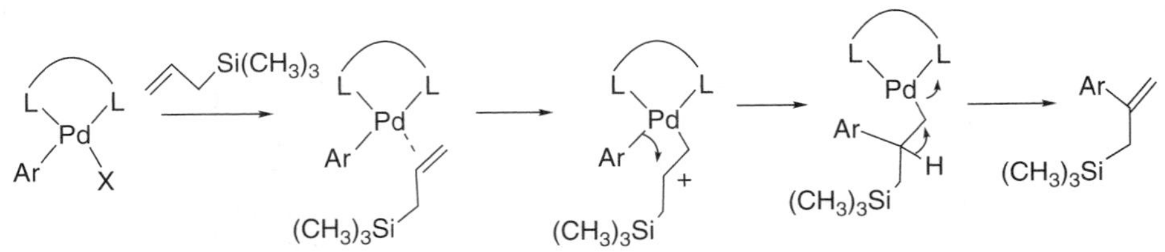

There have been several computational studies of electronic effects on the regioselectivity of the Heck reaction. Vinyl migration was studied for $X = CH_3$ CN, and OCH_3 using PH_3 as the ligand model.[149] Differences were calculated for the best α- and β-migration TS for each substituent. The differences were as follows: CH_3: α-migration favored by 0.1 kcal/mol; CN: β-migration favored by 4 kcal/mol; OCH_3: α-migration favored by 2 kcal/mol.

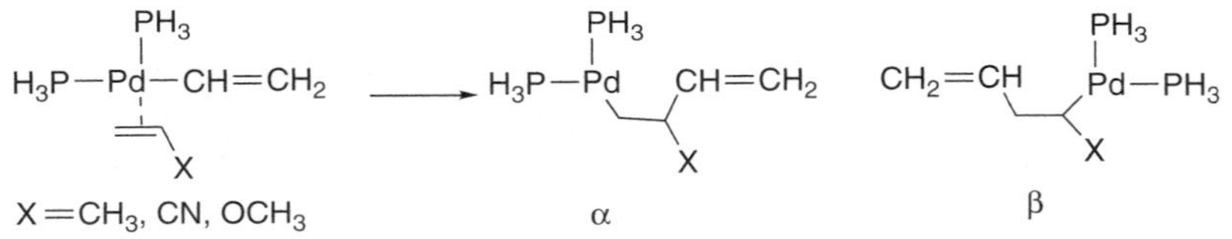

Examination of the HOMO and LUMO orbitals in these TSs indicates that the electronic effect operates mainly through the LUMO. The EWG cyano tends to localize the LUMO on the β-carbon, whereas ERG substituents have the opposite effect. Similar trends were found for Pd coordinated by diimine ligands.[150] These results indicate that the Markovnikov rule applies with the more electrophilic Pd complexes. When steric effects become dominant, the Pd adds to the less hindered position.

The Heck reaction has been applied to synthesis of intermediates and in multistage syntheses. Some examples are given in Scheme 8.9. Entries 1 and 2 illustrate both the β-regioselectivity and selectivity for aryl iodides over bromides. Entries 3 and 4 show conditions that proved favorable for cyclohexene. These examples also indicate preferential *syn* Pd-H elimination, since this accounts for formation of the 3-substituted cyclohexene as the major product.

148. K. Olofsson, M. Larhed, and A. Hallberg, *J. Org. Chem.*, **63**, 5076 (1998).
149. R. J. Deeth, A. Smith, and J. M. Brown, *J. Am. Chem. Soc.*, **126**, 7144 (2004).
150. H. V. Schenck, B. Akermark, and M. Svensson, *J. Am. Chem. Soc.*, **125**, 3503 (2003).

Scheme 8.9. Palladium-Catalyzed Alkenylation of Aryl and Alkenyl Systems

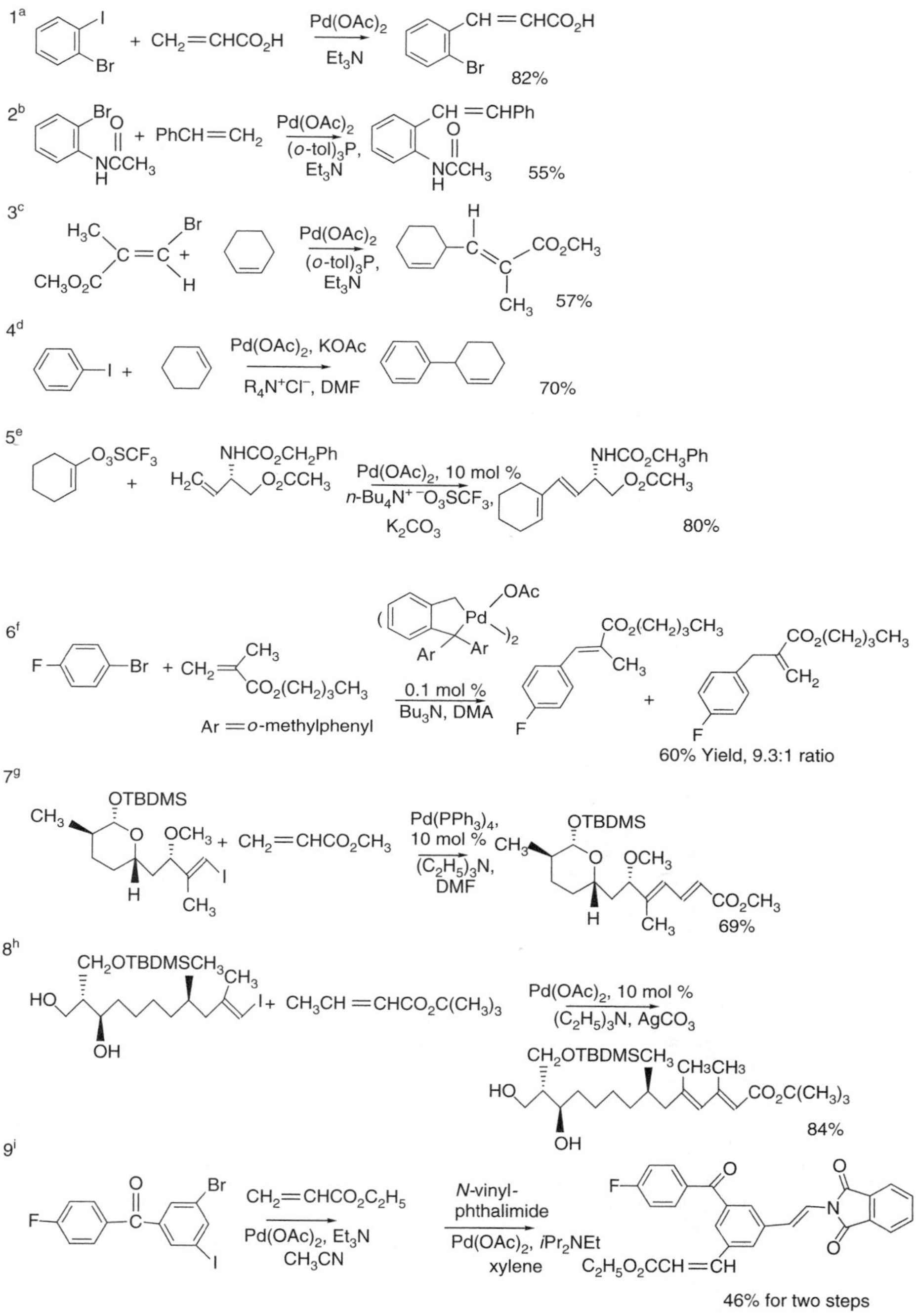

1[a]

2[b]

3[c]

4[d]

5[e]

6[f]

7[g]

8[h]

9[i]

(Continued)

Scheme 8.9. (*Continued*)

Intramolecular reactions

10^j — Pd(OAc)$_2$, PPh$_3$, Et$_3$N — 85%

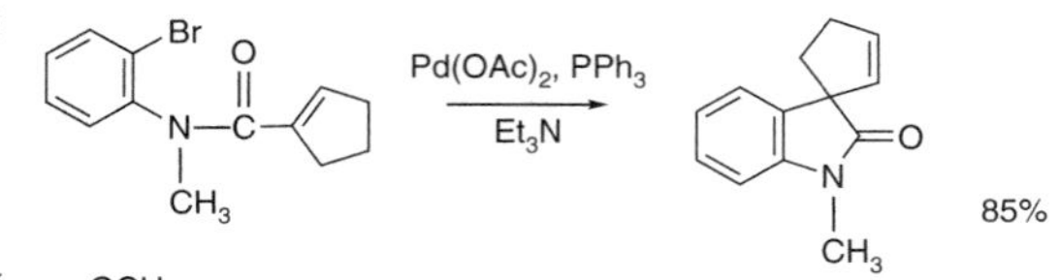

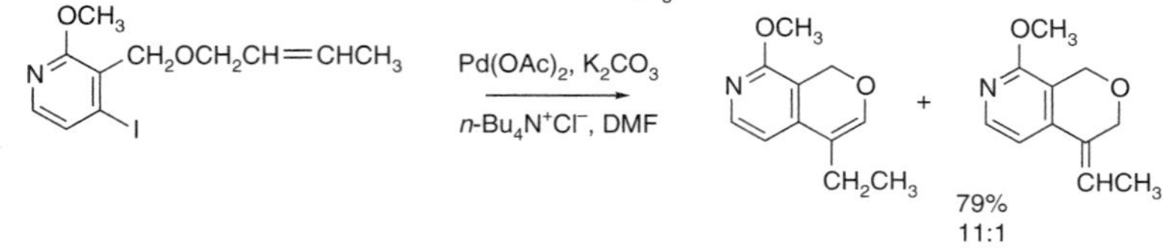

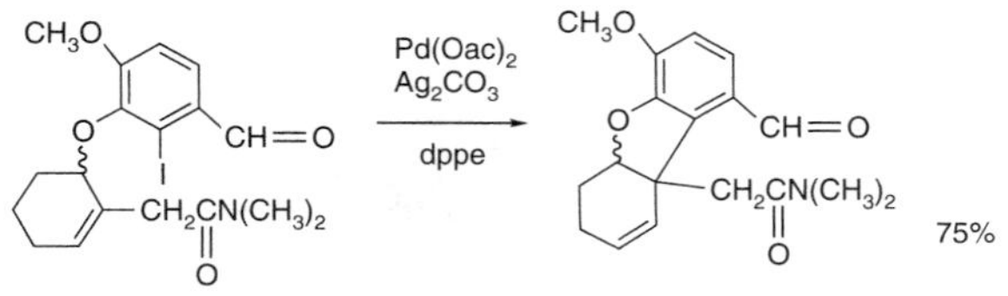

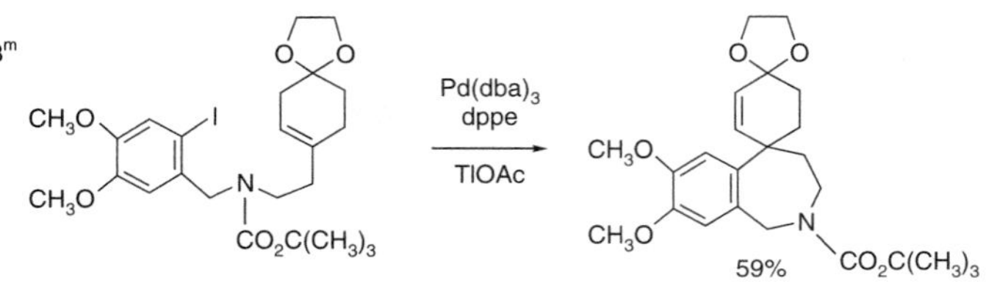

11^k — Pd(OAc)$_2$, K$_2$CO$_3$, n-Bu$_4$N$^+$Cl$^-$, DMF — 79%, 11:1

12^l — Pd(Oac)$_2$, Ag$_2$CO$_3$, dppe — 75%

13^m — Pd(dba)$_3$, dppe, TlOAc — 59%

14^n — 5 mol % Pd$_2$(dba)$_3$, 20 mol % (o-tol)$_3$P, Et$_3$N — 83%

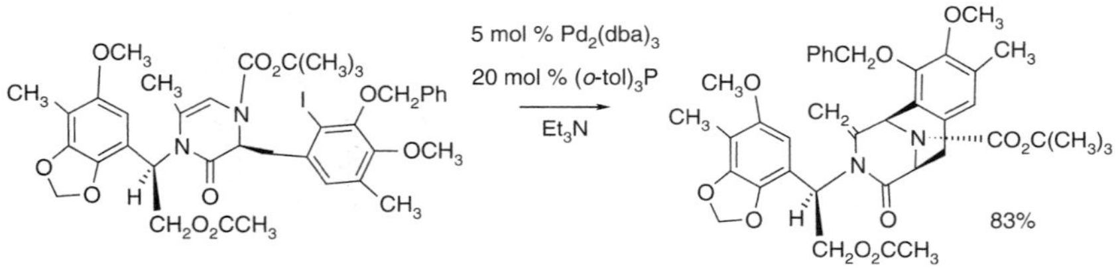

a. J. E. Plevyak, J. E. Dickerson, and R. F. Heck, *J. Org. Chem.*, **44**, 4078 (1979).
b. P. de Mayo, L. K. Sydnes, and G. Wenska, *J. Org. Chem.*, **45**, 1549 (1980).
c. J.-I. Kim, B. A. Patel, and R. F. Heck, *J. Org. Chem.*, **46**, 1067 (1981).
d. R. C. Larock and B. E. Baker, *Tetrahedron Lett.*, **29**, 905 (1988).
e. G. T. Crisp and M. G. Gebauer, *Tetrahedron*, **52**, 12465 (1996).
f. M. Beller and T. H. Riermeier, *Tetrahedron Lett.*, **37**, 6535 (1996).
g. L. Harris, K. Jarowicki, P. Kocienski, and R. Bell, *Synlett*, 903 (1996).
h. P. M. Wovkulich, K. Shankaran, J. Kiegel, and M. R. Uskokovic, *J. Org. Chem.*, **58**, 832 (1993); T. Jeffery and J.-C. Galland, *Tetrahedron Lett.*, **35**, 4103 (1994).
i. D. C. Waite and C. P. Mason, *Org. Proc. Res. Devel.*, **2**, 116 (1998).
j. M. M. Abelman, T. Oh, and L. E. Overman, *J. Org. Chem.*, **59**, 4130 (1987).
k. F. G. Fang, S. Xie, and M. W. Lowery, *J. Org. Chem.*, **59**, 6142 (1994).
l. P. J. Parsons, M. D. Charles, D. M. Harvey, L. R. Sumoreeah, A. Skell, G. Spoors, A. L. Gill, and S. Smith, *Tetrahedron Lett.*, **42**, 2209 (2001).
m. C. Bru, C. Thal, and C. Guillou, *Org. Lett.*, **5**, 1845 (2003).
n. A. Endo, A. Yanagisawa, M. Abe, S. Tohma, T. Kan, and T. Fukuyama, *J. Am. Chem. Soc.*, **124**, 6552 (2002).

syn-arylpalladation   *syn*-β–elimination

Entry 5 illustrates use of a vinyl triflate under the "phosphine-free" conditions. Entry 6 achieved exceptionally high catalyst efficiency by using a palladacycle-type catalyst. Entries 7 and 8 show the introduction of acrylate ester groups using functionalized alkenyl iodides. Entry 9 demonstrates two successive Heck reactions employed in a large-scale synthesis of a potential thromboxane receptor antagonist. These reactions were carried out in the absence of any phosphine ligand. The greater reactivity of the iodide over the bromide permits the sequential introduction of the two substituents.

There are numerous examples of intramolecular Heck reactions,[151] such as in Entries 10 to 14. Entry 11 is part of a synthesis of the antitumor agent camptothecin. The Heck reaction gives an 11:1 endocyclic-exocyclic mixture. Entries 12–14 are also steps in syntheses of biologically active substances. Entry 12 is part of a synthesis of maritidine, an alkaloid with cytotoxic properties; the reaction in Entry 13 is on a route to galanthamine, a potential candidate for treatment of Alzheimer's disease; and Entry 14 is a key step in the synthesis of a potent antitumor agent isolated from a marine organism.

8.2.3. Palladium-Catalyzed Cross Coupling

Palladium can catalyze carbon-carbon bond formation between aryl or vinyl halides and sulfonates and a wide range of organometallic reagents *in cross-coupling reactions*.[152] The organometallic reagents used include organolithium, organomagnesium, and organozinc reagents, as well as cuprates, stannanes, and organoboron compounds. The reaction is quite general for formation of sp^2-sp^2 and sp^2-sp bonds in biaryls, dienes, polyenes, and enynes. There are also some reactions that can couple alkyl organometallic reagents, but these are less general because of the tendency of alkylpalladium intermediates to decompose by β-elimination. Arylation of enolates also can be effected by palladium catalysts.

The basic steps in the cross-coupling reaction include oxidative addition of the aryl or vinyl halide (or sulfonate) to Pd(0), followed by transfer of an organic group from the organometallic to the resulting Pd(II) intermediate (*transmetallation*). The disubstituted Pd(II) intermediate then undergoes reductive elimination, which gives the product by carbon bond formation and regenerates the catalytically active Pd(0) oxidation level.

oxidative addition $R-X + Pd^0 \longrightarrow R-Pd^{II}-X$

transmetallation $R-Pd^{II}-X + R'-M \longrightarrow R-Pd^{II}-R' + M-X$

reductive elimination $R-Pd^{II}-R' \longrightarrow R-R' + Pd^0$

[151.] J. Link, *Org. React.*, **60**, 157 (2002).
[152.] F. Diederich and P. J. Stang, *Metal-Catalyzed Cross-Coupling Reactions*, Wiley-VCH, New York, 1998; S. P. Stanforth, *Tetrahedron*, **54**, 263 (1998).

Ligands and anions play a crucial role in determining the rates and equilibria of the various steps by controlling the detailed coordination environment at palladium.[153] In the next section we discuss coupling reactions involving organolithium, organo-magnesium, organozinc, and organocopper reagents. We then proceed to arylation of enolates mediated by palladium catalysts. Subsequent sections consider cross coupling with stannanes (*Stille reaction*) and boron compounds (*Suzuki reaction*).

8.2.3.1. Coupling with Organometallic Reagents. Tetrakis-(triphenylphosphine) palladium catalyzes coupling of alkenyl halides with Grignard reagents and organo-lithium reagents. The reactions proceed with retention of configuration at the double bond.

Ref. 154

Ref. 155

Organozinc compounds are also useful in palladium-catalyzed coupling with aryl and alkenyl halides. Procedures for arylzinc,[156] alkenylzinc,[157] and alkylzinc[158] reagents have been developed. The ferrocenyldiphosphine dppf has been found to be an especially good Pd ligand for these reactions.[159]

Ref. 156

Ref. 158

153. P. J. Stang, M. H. Kowalski, M. D. Schiavelli, and D. Longford, *J. Am. Chem. Soc.*, **111**, 3347 (1989); P. J. Stang and M. H. Kowalski, *J. Am. Chem. Soc.*, **111**, 3356 (1989); M. Portnoy and D. Milstein, *Organometallics*, **12**, 1665 (1993).

154. M. P. Dang and G. Linstrumelle, *Tetrahedron Lett.*, 191 (1978).

155. M. Yamamura, I. Moritani, and S. Murahashi, *J. Organometal. Chem.*, **91**, C39 (1975).

156. E. Negishi, A. O. King, and N. Okukado, *J. Org. Chem.*, **42**, 1821 (1977); E. Negishi, T. Takahashi, and A. O. King, *Org. Synth.*, **66**, 67 (1987).

157. U. H. Lauk, P. Skrabal, and H. Zollinger, *Helv. Chim. Acta*, **68**, 1406 (1985); E. Negishi, T. Takahashi, S. Baba, D. E. Van Horn, and N. Okukado, *J. Am. Chem. Soc.*, **109**, 2393 (1987); J.-M. Duffault, J. Einhorn, and A. Alexakis, *Tetrahedron Lett.*, **32**, 3701 (1991).

158. E. Negishi, L. F. Valente, and M. Kobayashi, *J. Am. Chem. Soc.*, **102**, 3298 (1980).

159. T. Hayashi, M. Konishi, Y. Kobori, M. Kumada, T. Higuchi, and K. Hirotsu, *J. Am. Chem. Soc.*, **106**, 158 (1984).

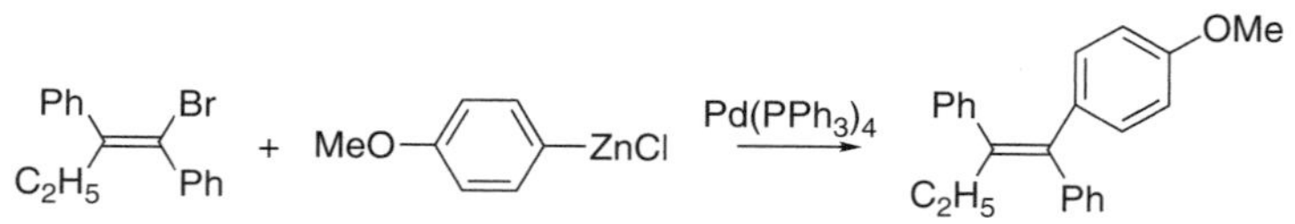

Ref. 160

Scheme 8.10 shows some representative coupling reactions with organomagnesium and organozinc reagents. Entry 1 shows a biaryl coupling accomplished using an arylzinc reagent. Entry 2 involves the use of a chelating ligand with an aryl triflate. The *bis*-phosphines dppe, dppp, and dppb were also effective for this coupling. Entry 3 is an example of use of a vinyl triflate. Entries 4 and 5 illustrate the use of perfluorobutanesulfonate (nonaflate) as an alternative leaving group to triflate. The organozinc

Scheme 8.10. Palladium-Catalyzed Cross Coupling of Organometallic Reagents with Halides and Sulfonates

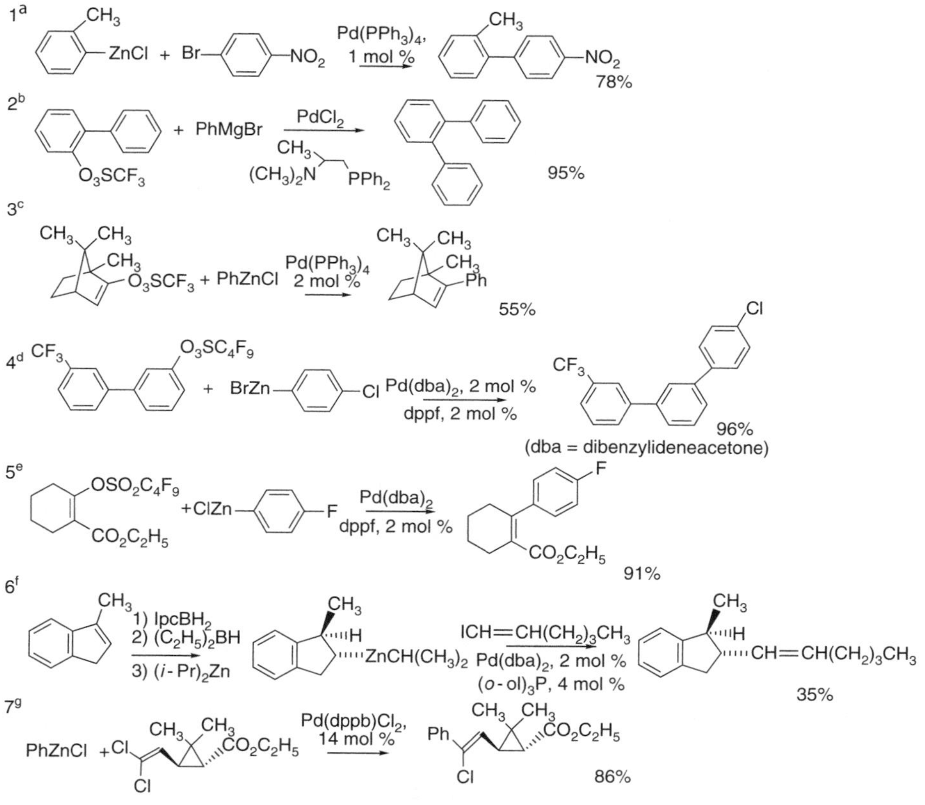

1[a] The reaction of an o-tolylZnCl with Br-C_6H_4-NO_2 using $Pd(PPh_3)_4$, 1 mol %, to give the biaryl coupling product in 78%.

2[b] A 2-aryl aryl triflate (O_3SCF_3) with PhMgBr, $PdCl_2$ and the ligand $(CH_3)_2N$-$CH(CH_3)CH_2$-PPh_2, giving the terphenyl in 95%.

3[c] A camphene-derived vinyl triflate (O_3SCF_3) + PhZnCl, $Pd(PPh_3)_4$ 2 mol %, giving the phenyl-substituted product in 55%.

4[d] A CF_3-biaryl nonaflate ($O_3SC_4F_9$) + BrZn-C_6H_4-Cl, $Pd(dba)_2$, 2 mol %, dppf, 2 mol %, giving the terphenyl in 96%. (dba = dibenzylideneacetone)

5[e] A cyclohexene nonaflate ($OSO_2C_4F_9$) bearing $CO_2C_2H_5$ + ClZn-C_6H_4-F, $Pd(dba)_2$, dppf, 2 mol %, giving the aryl-substituted product in 91%.

6[f] An indene: 1) $IpcBH_2$; 2) $(C_2H_5)_2BH$; 3) $(i\text{-}Pr)_2Zn$ giving $ZnCH(CH_3)_2$; then $ICH=CH(CH_2)_3CH_3$, $Pd(dba)_2$, 2 mol %, $(o\text{-}ol)_3P$, 4 mol %, giving $CH=CH(CH_2)_3CH_3$ product in 35%.

7[g] PhZnCl + a dichlorovinyl cyclopropane ester ($CO_2C_2H_5$), $Pd(dppb)Cl_2$, 14 mol %, giving the Ph-vinyl cyclopropane product in 86%.

a. E. Negishi, T. Takahashi, and A. O. King, *Org. Synth.*, **VIII**, 430 (1993).
b. T. Kamikawa and T. Hayashi, *Synlett*, 163 (1997).
c. G. Stork and R. C. A. Issacs, *J. Am. Chem. Soc.*, **112**, 7399 (1990).
d. M. Rottlander and P. Knochel, *J. Org. Chem.*, **63**, 203 (1998).
e. F. Bellina, D. Ciucci, R. Rossi, and P. Vergamini, *Tetrahedron*, **55**, 2103 (1999).
f. A. Boudier and P. Knochel, *Tetrahedron Lett.*, **40**, 687 (1999).
g. A. Minato, *J. Org. Chem.*, **56**, 4052 (1991).

160. R. B. Miller and M. I. Al-Hassan, *J. Org. Chem.*, **50**, 2121 (1985).

reagent in Entry 4 was prepared by the hydroboration route (see Section 7.3.1.1). The reaction in Entry 7 was used to prepare analogs of the pyrethrin insecticides. There was a substantial difference in the reactivity of the two chlorides, permitting the stereoselective synthesis.

There are a number of procedures for coupling of terminal alkynes with halides and sulfonates, a reaction that is known as the *Sonogashira reaction*.[161] A combination of $Pd(PPh_3)_4$ and $Cu(I)$ effects coupling of terminal alkynes with vinyl or aryl halides.[162] The reaction can be carried out directly with the alkyne, using amines for deprotonation. The alkyne is presumably converted to the copper acetylide, and the halide reacts with Pd(0) by oxidative addition. Transfer of the acetylide group to Pd results in reductive elimination and formation of the observed product.

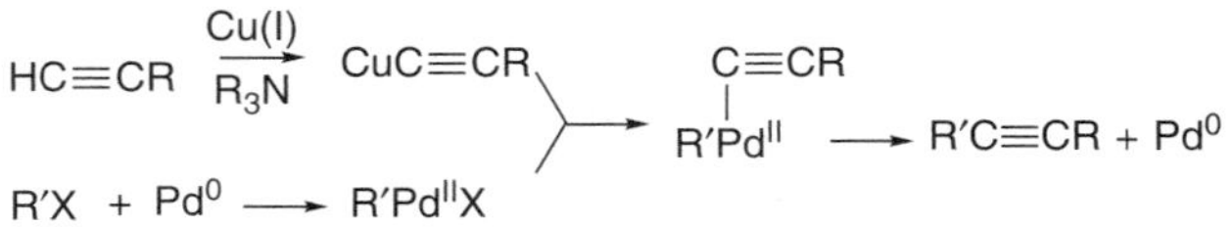

$$HC\equiv CR \xrightarrow[R_3N]{Cu(I)} CuC\equiv CR$$

$$R'X + Pd^0 \longrightarrow R'Pd^{II}X$$

$$\begin{array}{c} C\equiv CR \\ | \\ R'Pd^{II} \end{array} \longrightarrow R'C\equiv CR + Pd^0$$

The original conditions used amines as solvents or cosolvents. Several other bases can replace the amine. Tetrabutylammonium hydroxide or fluoride can be used in THF (see Entry 1 in Scheme 8.11).[163] Tetrabutylammonium acetate is also effective with aryl iodides and EWG-substituted aryl bromides (Entry 2).[164] Use of alkenyl halides in this reaction has proven to be an effective method for the synthesis of enynes[165] (see also Entries 5 and 6 in Scheme 8.11).

$$CH_3(CH_2)_4CH=CHI + HC\equiv C(CH_2)_2OH \xrightarrow[\text{pyrrolidine}]{\substack{Pd(PPh_3)_4, 5 \text{ mol }\% \\ CuI, 10 \text{ mol }\%}} CH_3(CH_2)_4CH=CHC\equiv C(CH_2)_2OH$$

90%

Ref. 166

Several hindered phosphine ligands give enhanced reactivity. Aryl iodides can be coupled at low temperature using $Pd_2(dba)_3$ and *tris*-(mesityl)phosphine.

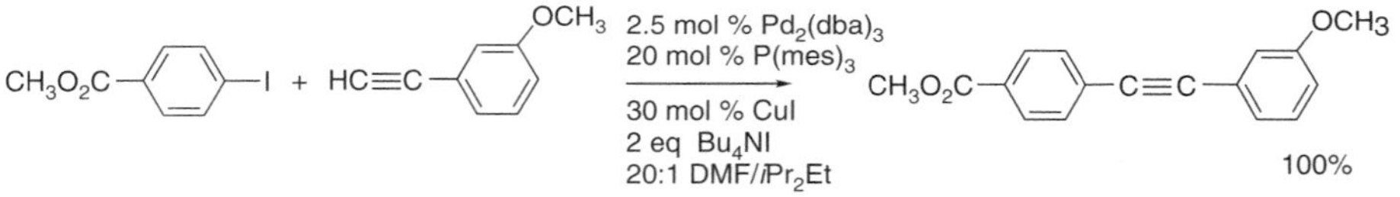

100%

Ref. 167

$Pd_2(dba)_3$ with *tris-t*-butylphosphine is an effective catalyst and functions in the absence of copper.[168]

161. R. R. Tykwinski, *Angew. Chem. Int. Ed. Engl.*, **42**, 1566 (2003).
162. K. Sonogashira, Y. Tohda, and N. Hagihara, *Tetrahedron Lett.*, 4467 (1975).
163. A. Mori, T. Shimada, T. Kondo, and A. Sekiguchi, *Synlett*, 649 (2001).
164. S. Urgaonkar and J. G. Verkade, *J. Org. Chem.*, **69**, 5752 (2004).
165. V. Ratovelomana and G. Linstrumelle, *Synth. Commun.*, **11**, 917 (1981); L. Crombie and M. A. Horsham, *Tetrahedron Lett.*, **28**, 4879 (1987); G. Just and B. O'Connor, *Tetrahedron Lett.*, **29**, 753 (1988); D. Guillerm and G. Linstrumelle, *Tetrahedron Lett.*, **27**, 5857 (1986).
166. M. Alami, F. Ferri, and G. Linstrumelle, *Tetrahedron Lett.*, **34**, 6403 (1993).
167. K. Nakamura, H. Okubo, and M. Yamaguchi, *Synlett*, 549 (1999).
168. V. P. W. Bohm and W. A. Herrmann, *Eur. J. Org. Chem.*, 3679 (2000).

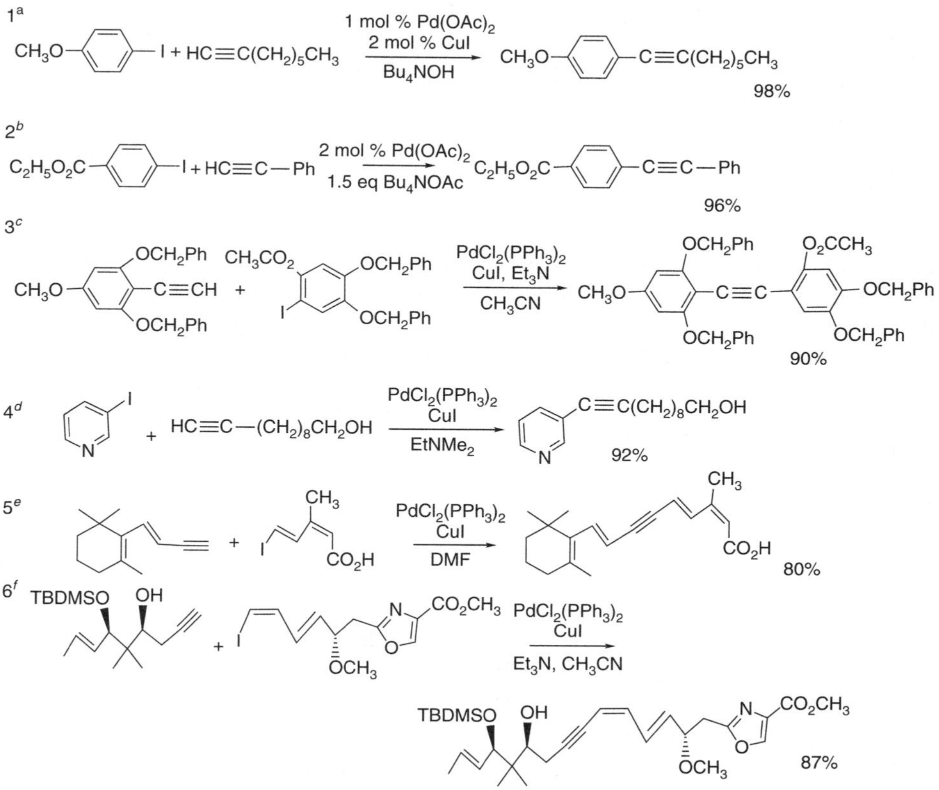

a. S. Urgaonkar and J. G. Verkade, *J. Org. Chem.*, **69**, 5752 (2004).
b. A. Mori, T. Shimada, T. Kondo, and A. Sekiguchi, *Synlett*, 649 (2001)
c. C. C. Li, Z. X. Xie, Y. D. Zhang, J. H. Chen, and Z. Yang, *Org. Lett.*, **5**, 3919 (2003).
d. J. Krauss and F. Bracher, *Arch. Pharm.*, **337**, 371 (2004).
e. M. Abarbri, J. Thibonnet, J.-L. Parrain, and A. Duchene, *Tetrahedron Lett.*, **43**, 4703 (2002).
f. M. C. Hillier, A. T. Price, and A. I. Meyers, *J. Org. Chem.*, **66**, 6037 (2001).

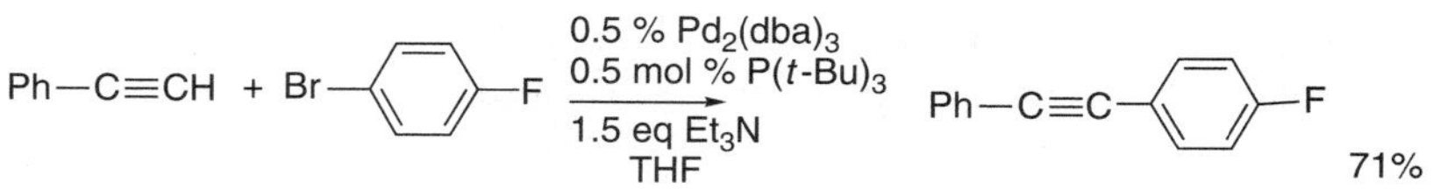

Various aminophosphines have also been found to catalyze coupling in the absence of copper.

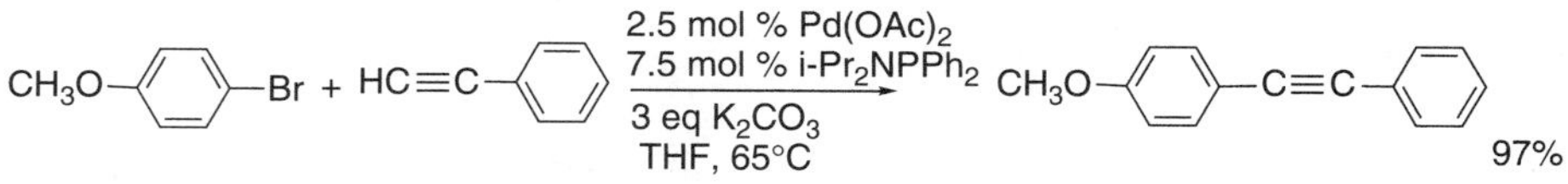

Ref. 169

169. J. Cheng, Y. Sun, F. Wang, M. Guo, J.-H. Xu, Y. Pan, and Z. Zhang, *J. Org. Chem.*, **69**, 5428 (2004).

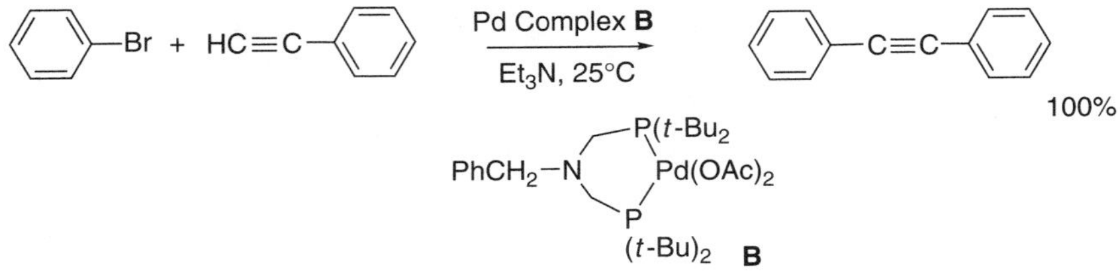

Ref. 170

8.2.3.2. Palladium-Catalyzed Arylation of Enolates. Very substantial progress has been made in the use of Pd-catalyzed cross coupling for arylation of enolates and enolate equivalents. This reaction provides an important method for arylation of enolates, which is normally a difficult transformation to accomplish.[171] A number of phosphine ligands have been found to promote these reactions. Bulky trialkyl phosphines such as *tris*-(*t*-butyl)phosphine with a catalytic amount of $Pd(OAc)_2$ results in phenylation of the enolates of aromatic ketones and diethyl malonate.[172]

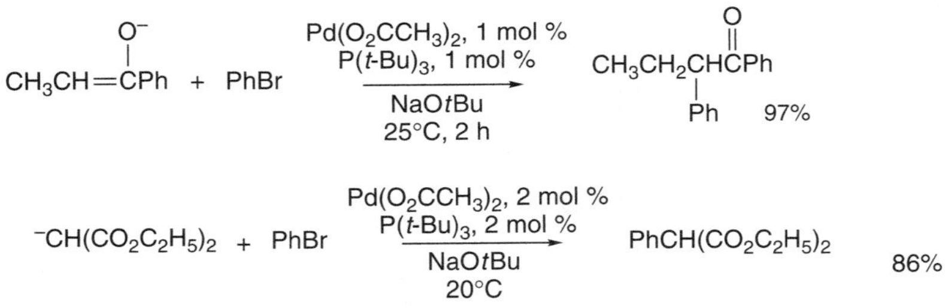

Phenylation has also been achieved with the diphosphine ligands BINAP and tol-BINAP.

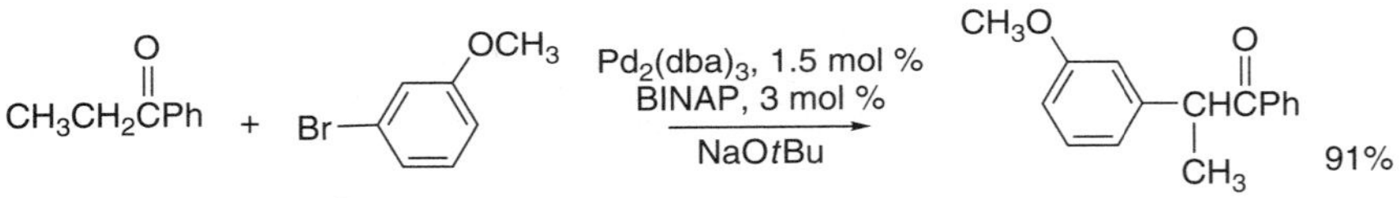

Ref. 173

Several biphenylphosphines with 2'-amino substituents are also effective in arylation of ester enolates.[174] Among the esters that were successfully arylated were *t*-butyl acetate, *t*-butyl propanoate, and ethyl phenylacetate. The ester enolates were generated with LiHMDS.

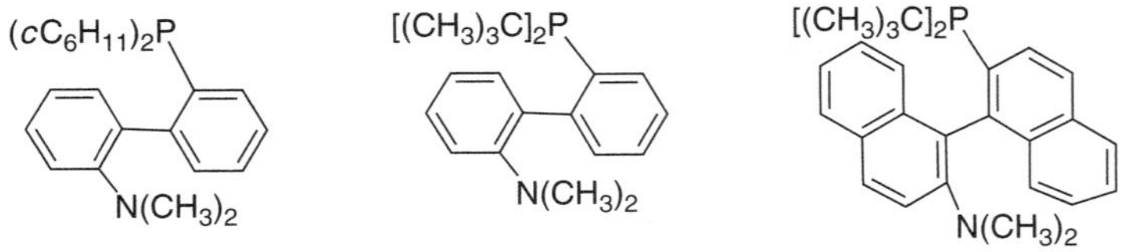

170. D. Mery, K. Heuze, and D. Astruc, *Chem. Commun.*, 1934 (2003).
171. D. A. Culkin and J. F. Hartwig, *Acc. Chem. Res.*, **36**, 234 (2003).
172. M. Kawatsura and J. E. Hartwig, *J. Am. Chem. Soc.*, **121**, 1473 (1999).
173. M. Palucki and S. L. Buchwald, *J. Am. Chem. Soc.*, **119**, 11108 (1997).
174. W. A. Moradi and S. L. Buchwald, *J. Am. Chem. Soc.*, **123**, 7996 (2001).

Carbenoid imidazolidene ligands such as **C** can also be used in conjunction with Pd(dba)$_2$, and this method has been applied to α-arylpropanoic acids (NSAIDS) such as naproxen.[175]

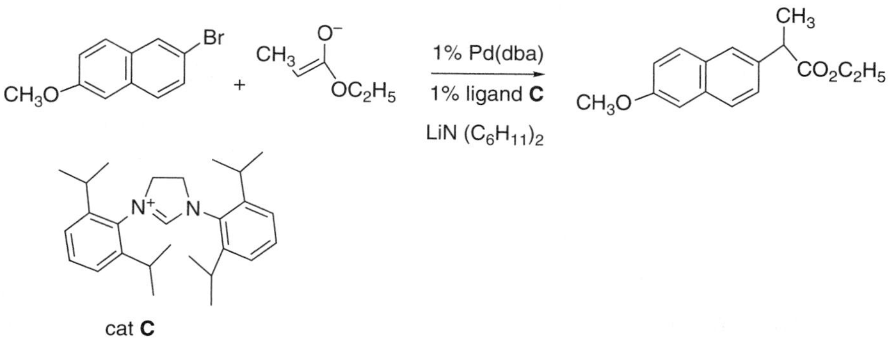

cat **C**

Highly arylated ketones have been prepared successfully. For example, arylation of the enolate of the deoxybenzoin **4** gives 1,1,2-triarylethanones that are related to substances such as tamoxifen.[176]

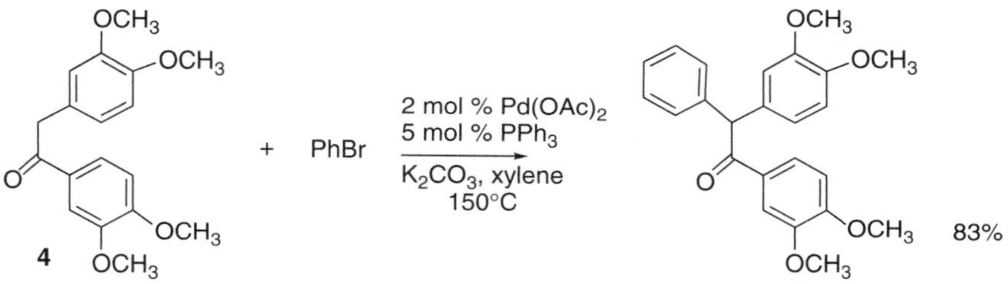

Similar reactions have been carried out using polymer-supported catalysts.[177]

Arylations have also been extended to zinc enolates of esters (Reformatsky reagents).[178]

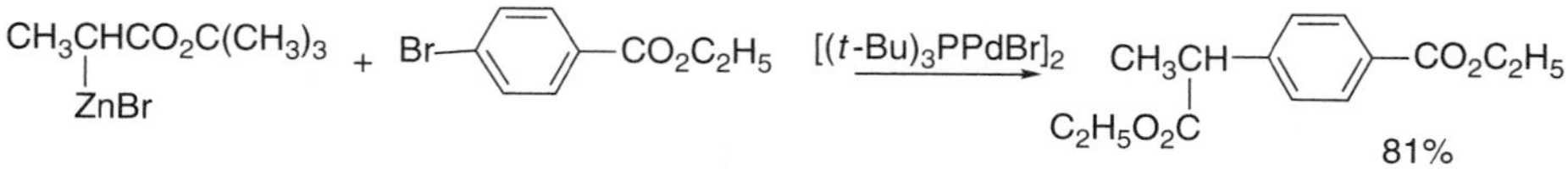

These conditions can also be applied to enolates prepared from α-halo amides.

175. M. Jorgensen, S. Lee, X. Liu, J. P. Wolkowski, and J. F. Hartwig, *J. Am. Chem. Soc.*, **124**, 12557 (2002).
176. F. Churruca, R. SanMartin, I. Tellitu, and E. Dominguez, *Org. Lett.*, **4**, 1591 (2002).
177. F. Churruca, R. SanMartin, M. Carrill, I. Tellitu, and E. Dominguez, *Tetrahedron*, **60**, 2393 (2004).
178. T. Hama, X. Liu, D. A. Culkin, and J. F. Hartwig, *J. Am. Chem. Soc.*, **125**, 11176 (2003).

Enolate arylation has also been extended to aryl tosylates. The preferred catalyst includes a very bulky biphenyl phosphine **D**.[179]

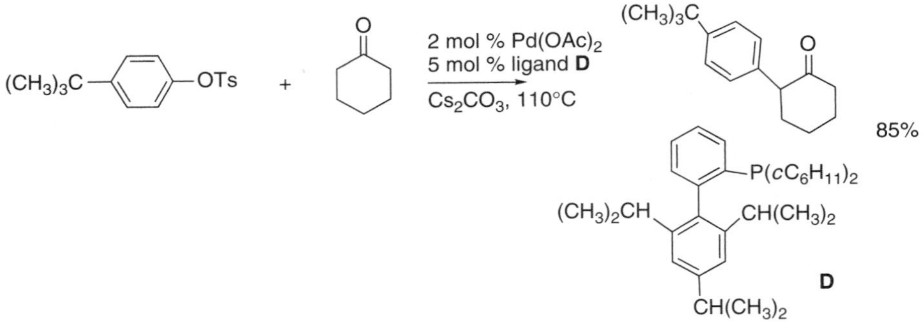

85%

Conditions for arylation of enolate equivalents have also been developed. In the presence of ZnF_2, silyl enol ethers, silyl ketene acetals, and similar compounds react. For example, the TMS derivatives of *N*-acyl oxazolidinones can be arylated.

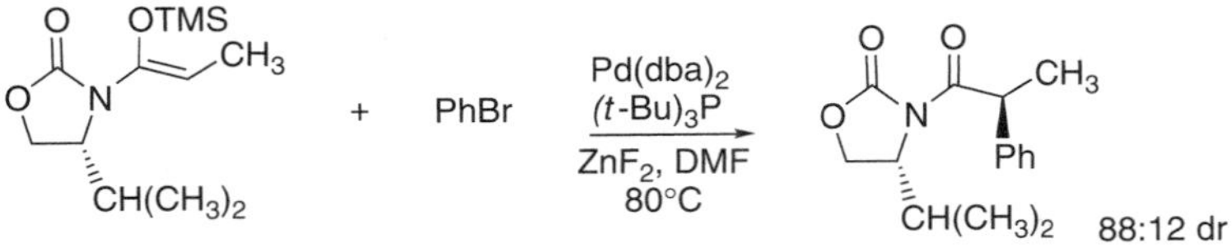

88:12 dr

Arylacetate esters have been generated by coupling aryl bromides with stannyl enolates generated from silyl ketene acetals.

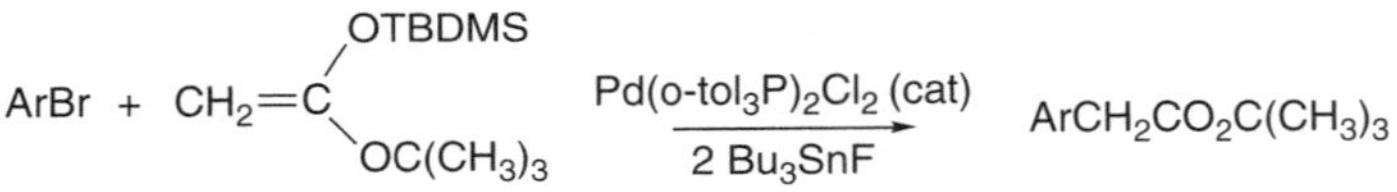

Ref. 180

Intramolecular arylations are possible and several studies have examined the synthesis of biologically active compounds such as oxindoles.[181] For example, a synthesis of physovenine has been reported using this methodology.

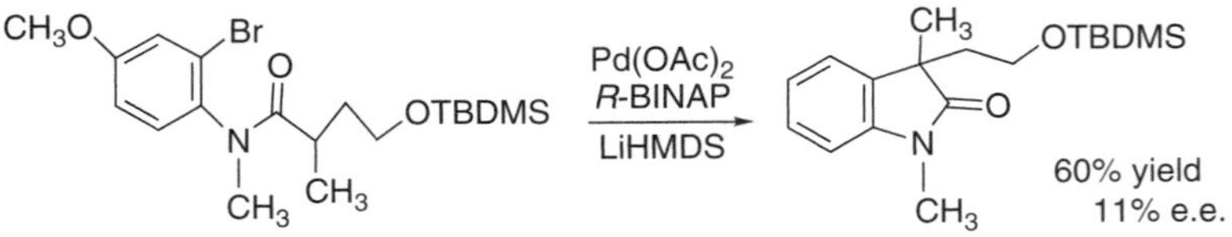

60% yield
11% e.e.

Ref. 182

[179] H. N. Nguyen, X. Huang, and S. L. Buchwald, *J. Am. Chem. Soc.*, **125**, 11818 (2003).
[180] F. Agnelli and G. A. Sulikowski, *Tetrahedron Lett.*, **39**, 8807 (1998).
[181] S. Lee and J. F. Hartwig, *J. Org. Chem.*, **66**, 3402 (2001).
[182] T. Y. Zhang and H. Zhong, *Tetrahedron Lett.*, **43**, 1363 (2002).

8.2.3.3. Coupling with Stannanes. Another important group of cross-coupling reactions, known as *Stille reactions*, uses aryl and alkenyl stannanes as the organometallic component.[183] The reactions are carried out with Pd(0) catalysts in the presence of phosphine ligands and have proven to be very general with respect to the halides that can be used. Benzylic, aryl, alkenyl, and allylic halides can all be utilized,[184] and the groups that can be transferred from tin include alkyl, alkenyl, aryl, and alkynyl. The approximate order of the effectiveness of transfer of groups from tin is alkynyl > alkenyl > aryl > methyl > alkyl, so unsaturated groups are normally transferred selectively.[185] Subsequent studies have found better ligands, including *tris*-(2-furyl)phosphine[186] and triphenylarsine.[187] Aryl-aryl coupling rates are increased by the presence of a Cu(I) cocatalyst,[188] which has led to a simplified protocol in which Pd-C catalyst, along with CuI and Ph$_3$As, gives excellent yields of biaryls.

Ref. 189

The general catalytic cycle of the Stille reaction involves oxidative addition, transmetallation, and reductive elimination.

The role of the ligands is both to stabilize the Pd(0) state and to "tune" the reactivity of the palladium. The outline mechanism above does not specify many detailed aspects of the reaction that are important to understanding the effect of ligands, added salts, and solvents. Moreover, it does not address the stereochemistry, either in terms of the Pd center (tetracoordinate? pentacoordinate?, *cis*?, *trans*?) or of the reacting carbon groups (inversion?, retention?). Some of these issues are addressed by a more detailed mechanism.[190]

[183.] J. K. Stille, *Angew. Chem. Int. Ed. Engl.*, **25**, 508 (1986); T. N. Mitchell, *Synthesis*, 803 (1992); V. Farina, V. Krishnamurthy, and W. J. Scott, *Org. React.*, **50**, 1 (1998).
[184.] F. K. Sheffy, J. P. Godschalx, and J. K. Stille, *J. Am. Chem. Soc.*, **106**, 4833 (1984); I. P. Beltskaya, *J. Organomet. Chem.*, **250**, 551 (1983); J. K. Stille and B. L. Groth, *J. Am. Chem. Soc.*, **109**, 813 (1987).
[185.] J. W. Labadie and J. K. Stille, *J. Am. Chem. Soc.*, **105**, 6129 (1983).
[186.] V. Farina and B. Krishnan, *J. Am. Chem. Soc.*, **113**, 9585 (1991).
[187.] V. Farina, B. Krishnan, D. R. Marshall, and G. P. Roth, *J. Org. Chem.*, **58**, 5434 (1993).
[188.] V. Farina, S. Kapadia, B. Krishnan, C. Wang, and L. S. Liebskind, *J. Org. Chem.*, **59**, 5905 (1994).
[189.] G. P. Roth, V. Farina, L. S. Liebeskind, and E. Pena-Cabrera, *Tetrahedron Lett.*, **36**, 2191 (1995).
[190.] P. Espinet and A. Echavarren, *Angew. Chem. Int. Ed. Engl.*, **43**, 4704 (2004).

The oxidative addition is considered to give a *cis* Pd complex that can rearrange to the more stable *trans* complex. The mechanism also takes account of the possibility of exchange of the ligands by solvent (or anions that may be present). This mechanism suggests that the transmetallation can occur either with retention (TS-A) or inversion (TS-B), which is consistent with experimental observations of both outcomes. The reductive elimination is believed to occur from a *cis* complex, and the ligands can play a role in promoting this configuration. The ligands can also affect the rate and position of the off-on equilibria. Thus there are several factors that affect the detailed kinetics of the reaction and these can be manipulated in optimization of the reaction conditions. Especially when triflates are used as the electrophilic reactant, added LiCl can have a beneficial effect. The chloride is believed to facilitate the oxidative addition step by reversible formation of an anionic complex that is more nucleophilic than the neutral species. (Compare with the anionic mechanisms for the Heck reaction on p. 716.)[191] The harder triflate does not have this effect. Acetate ions can also accelerate the reaction.[192] Copper salts are believed to shift the extent of ligation at the palladium by competing for the phosphine ligand.[193] The kinetics of Stille reactions catalyzed by triphenylarsine have been studied in some detail.[194] In this system, displacement of an arsine ligand by solvent DMF precedes the transmetallation step.

Various phosphine ligands have been employed. *Tris*-(*t*-butyl)phosphine is an excellent ligand and is applicable to both vinyl and arylstannanes, including sterically hindered ones. Aryl chlorides are reactive under these conditions.[195]

191. C. Amatore, A. Jutand, and A. Suarez, *J. Am. Chem. Soc.*, **115**, 9531 (1993); C. Amatore and A. Jutand, *Acc. Chem. Res.*, **33**, 314 (2000).
192. C. Amatore, E. Carre, A. Jutland, M. M'Barki, and G. Meyer, *Organometallics*, **14**, 5605 (1995).
193. A. L. Casado and P. Espinet, *Organometallics*, **22**, 1305 (2003).
194. C. Amatore, A. A. Bahsoun, A. Jutand, G. Meyer, N. A. Ndedi, and L. Ricard, *J. Am. Chem. Soc.*, **125**, 4212 (2003).
195. A. F. Littke, L. Schwarz, and G. C. Fu, *J. Am. Chem. Soc.*, **124**, 6343 (2002).

The Stille reaction can be used with alkenyl stannanes, alkenyl halides, and triflates,[196] and the reactions occur with retention of configuration at both the halide and stannane. These methods are applicable to stereospecific syntheses of materials such as the retinoids.[197]

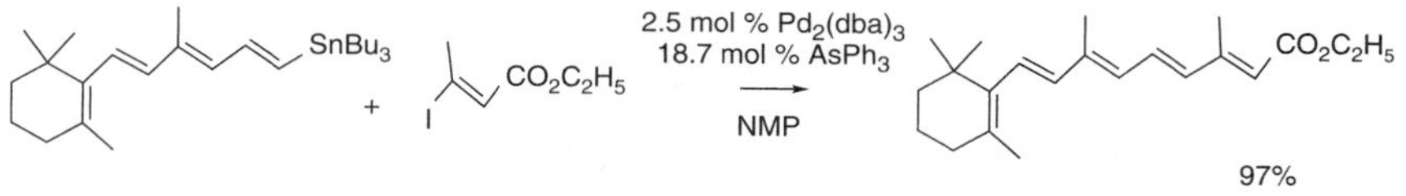

97%

Carotene has been synthesized from a symmetrical 1,10-*bis*-(tri-*n*-butyl stannyl) decapentaene.[198]

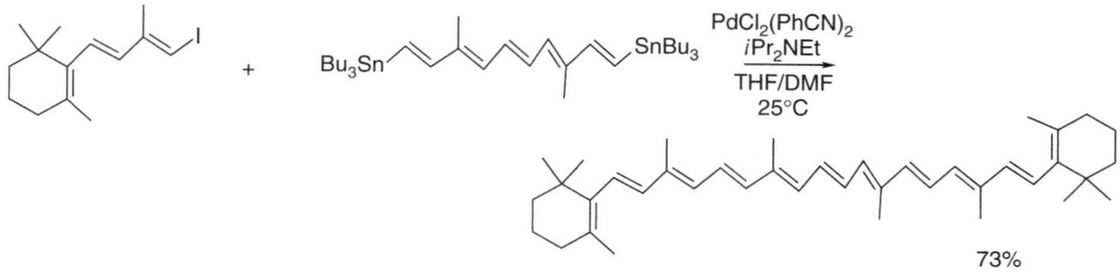

73%

The versatility of Pd-catalyzed coupling of stannanes has been extended by the demonstration that alkenyl triflates are also reactive.[199]

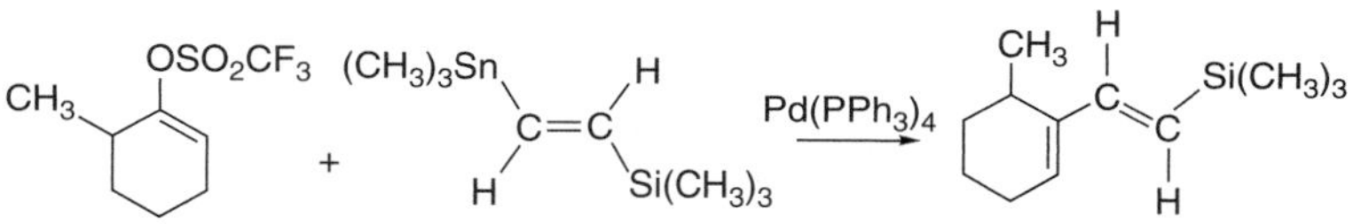

The alkenyl triflates can be prepared from ketones,[200] and methods are available for regioselective preparation of alkenyl triflates from unsymmetrical ketones.[201]

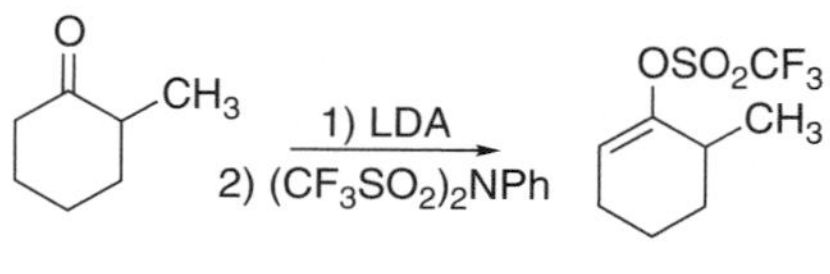

The coupling reaction can tolerate a number of functional groups, as illustrated by a step in the synthesis of the antibiotic nisamycin.

196. W. J. Scott and J. K. Stille, *J. Am. Chem. Soc.*, **108**, 3033 (1986).
197. B. Dominguez, B. Iglesias, and A. R. de Lera, *Tetrahedron*, **55**, 15071 (1999).
198. B. Vaz, R. Alvarez, and A. R. de Lera, *J. Org. Chem.*, **67**, 5040 (2002).
199. W. J. Scott, G. T. Crisp, and J. K. Stille, *J. Am. Chem. Soc.*, **106**, 4630 (1984); W. J. Scott and J. E. McMurry, *Acc. Chem. Res.*, **21**, 47 (1988).
200. P. J. Stang, M. Hanack, and L. R. Subramanian, *Synthesis*, 85 (1982).
201. J. E. McMurry and W. J. Scott, *Tetrahedron Lett.*, **24**, 979 (1983).

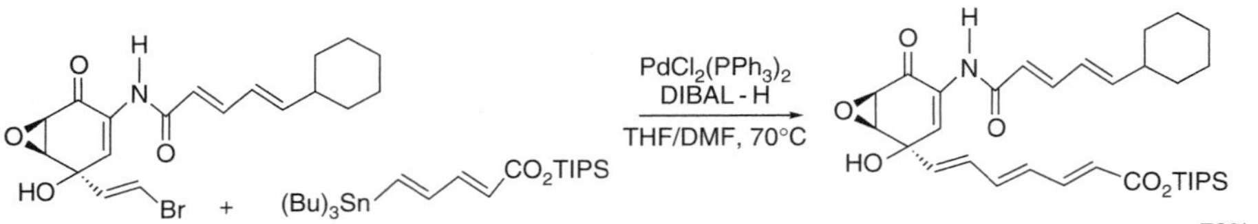

70%

Ref. 202

The Stille coupling reaction is very versatile with respect to the functionality that can be carried in both the halide and the tin reagent. Groups such as ester, nitrile, nitro, cyano, and formyl can be present, which permits applications involving "masked functionality." For example, when the coupling reaction is applied to 1-alkoxy-2-butenylstannanes, the double-bond shift leads to a vinyl ether that can be hydrolyzed to an aldehyde.

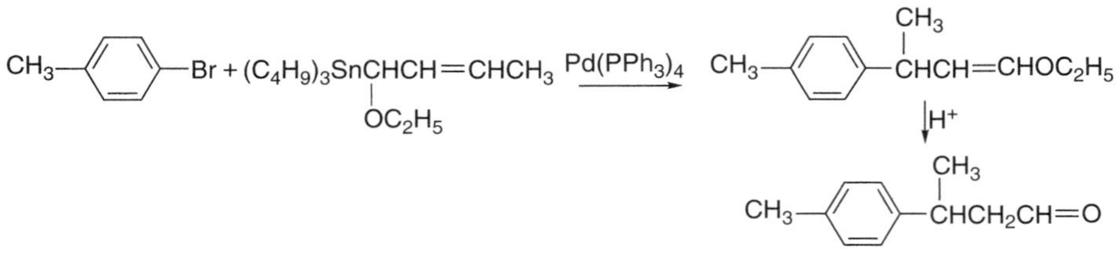

Ref. 203

Alkenylstannanes react with 1,1-dibromoalkenes to give enynes.[204] These reactions are thought to involve elimination of the elements of HBr prior to reductive elimination.

$$PhCH{=}CBr_2 \quad + \quad CH_2{=}CHSnBu_3 \xrightarrow[i\text{-}Pr_2NEt]{\text{2.5\% }Pd_2(dba)_3,\ (MeOPh)_3P} PhC{\equiv}CCH{=}CH_2 \quad 89\%$$

This reaction has been used in the synthesis of a portion of callipeltoside, a substance with anticancer activity.

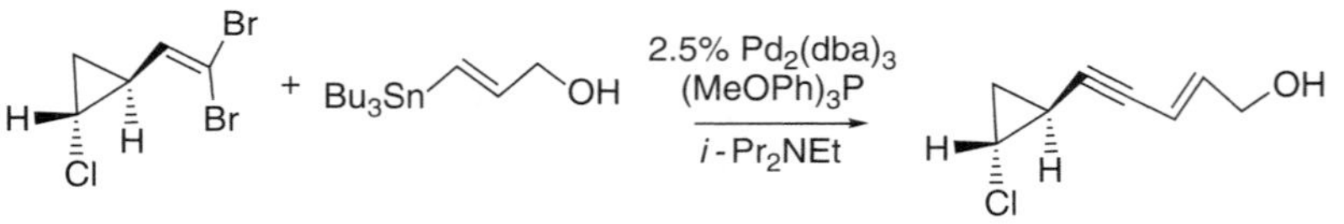

Ref. 205

The most problematic cases for the Stille reaction involve coupling saturated systems. The tendency for β-elimination of alkylpalladium compounds requires special conditions. *Bis*-(dialkylamino)cyclohexylphosphines have shown considerable success

202. P. Wipf and P. D. G. Coish, *J. Org. Chem.*, **64**, 5053 (1999).
203. A. Duchene and J.-P. Quintard, *Synth. Commun.*, **15**, 873 (1987).
204. W. Shen and L. Wang, *J. Org. Chem.*, **64**, 8873 (1999).
205. H. F. Olivo, F. Velazquez, and H. C. Trevisan, *Org. Lett.*, **2**, 4055 (2000).

in promoting coupling of saturated primary bromides and iodides with alkenyl and aryl stannanes.[206]

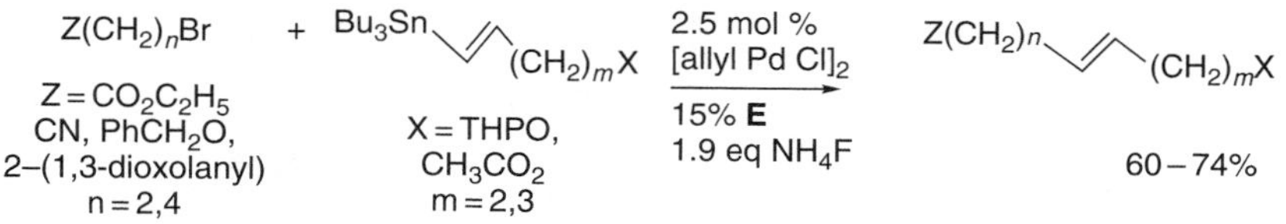

The Stille reaction has been successfully applied to a number of macrocyclic ring closures.[207] In a synthesis of amphidinolide A, the two major fragments were coupled via a selective Stille reaction, presumably governed by steric factors. After deprotection the ring was closed by coupling the second vinyl stannane group with an allylic acetate.[208]

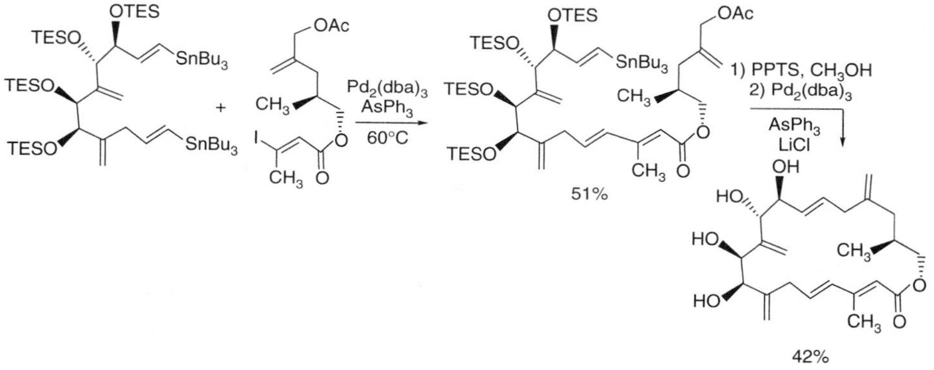

A similar cross-coupling reaction was used for macrocylization in the synthesis of rhizoxin A.[209]

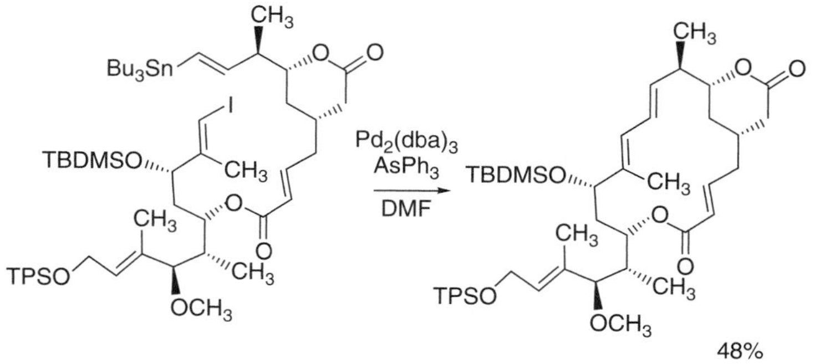

206. H. Tang, K. Menzel, and G. C. Fu, *Angew. Chem. Int. Ed. Engl.*, **42**, 5079 (2003).
207. M. A. J. Duncton and G. Pattenden, *J. Chem. Soc., Perkin Trans. 1*, 1235 (1999).
208. H. W. Lam and G. Pattenden, *Angew. Chem. Int. Ed. Engl.*, **41**, 508 (2002).
209. I. S. Mitchell, G. Pattenden, and J. P. Stonehouse, *Tetrahedron Lett.*, **43**, 493 (2002).

A striking example of a macrocyclic closure is found in the double "stitching" done in the final step of the synthesis of the immunosuppressant rapamycin. *Bis*-1,2-(tri-*n*-butylstannyl)ethene reacted with the diiodide to close a 31-membered ring in 28% yield at 70% conversion. The intermediate iodostannane (from a single coupling) was also isolated in about 30% yield and could be cyclized in a second step.[210]

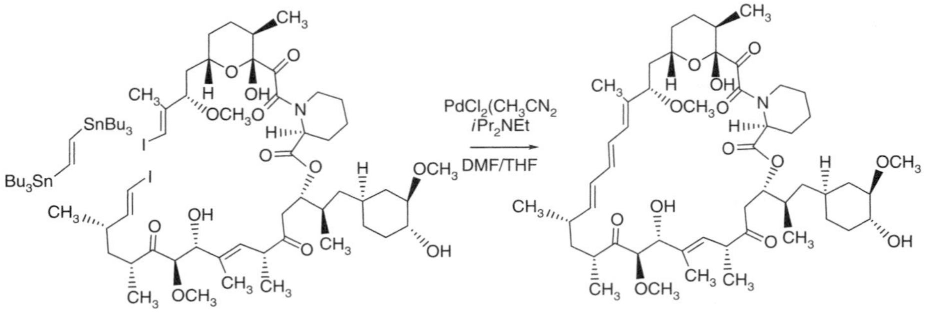

Some other examples of Pd-catalyzed coupling of organostannanes with halides and triflates are given in Scheme 8.12. Entries 1 and 2 are early examples that show that the reaction can be done with either ERG or EWG substituents on the aromatic ring. Entry 3 is an example of the use of an aryl triflate. Entry 5 was developed in the exploration of the synthetic potential of cyclobutendiones. Entries 6 to 11 are various alkenyl-alkenyl and alkenyl-aryl couplings using iodides and triflates. Entries 12 to 14 involve heterocyclic structures in the synthesis of several antibiotics. Entry 15 involves coupling of a protected glycoside with a vinyl triflate and an α-oxystannane. Entry 16 involves an alkynylstannane and generates a deca-1,6-diyne ring. Entries 17 and 18 show the use of allylic and benzylic bromides.

Procedures for the synthesis of ketones based on coupling of organostannanes with acyl halides have also been developed.[211] The catalytic cycle is similar to that involved in coupling with aryl halides. The scope of compounds to which the reaction is applicable includes tetra-*n*-butylstannane. This example indicates that the reductive elimination step competes successfully with β-elimination.

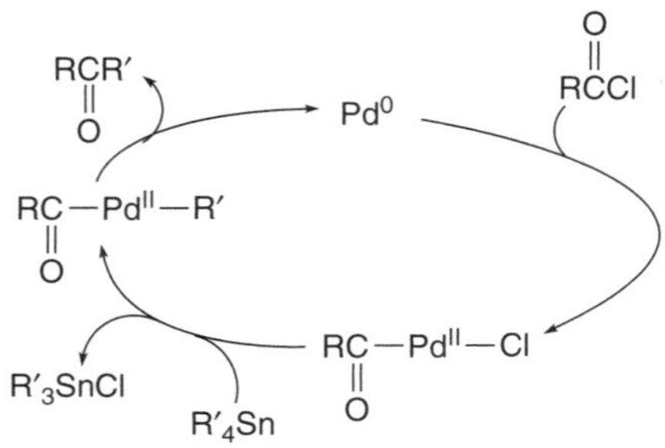

Scheme 8.13 gives some examples of these reactions.

210. K. C. Nicolaou, T. K. Chakraborty, A. D. Piscopio, N. Minowa, and P. Bertinato, *J. Am. Chem. Soc.*, **115**, 4419 (1993).
211. D. Milstein and J. K. Stille, *J. Org. Chem.*, **44**, 1613 (1979); J. W. Labadie and J. K. Stille, *J. Am. Chem. Soc.*, **105**, 6129 (1983).

Scheme 8.12. Palladium-Catalyzed Coupling of Stannanes with Halides and Sulfonates

A. Aryl halides

1[a]

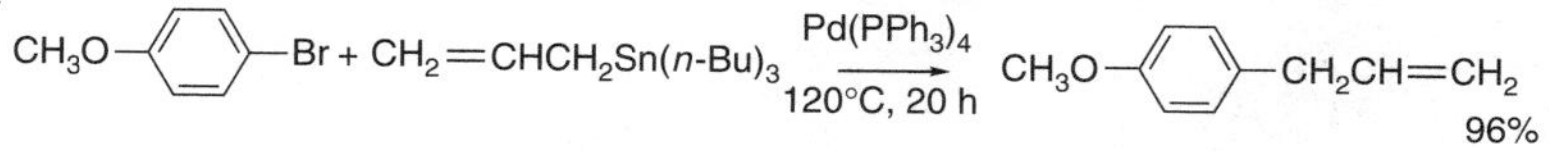

$$CH_3O\!-\!\!\langle\ \rangle\!\!-\!Br + CH_2\!\!=\!\!CHCH_2Sn(n\text{-}Bu)_3 \xrightarrow[120°C, 20\ h]{Pd(PPh_3)_4} CH_3O\!-\!\!\langle\ \rangle\!\!-\!CH_2CH\!\!=\!\!CH_2 \quad 96\%$$

2[b]

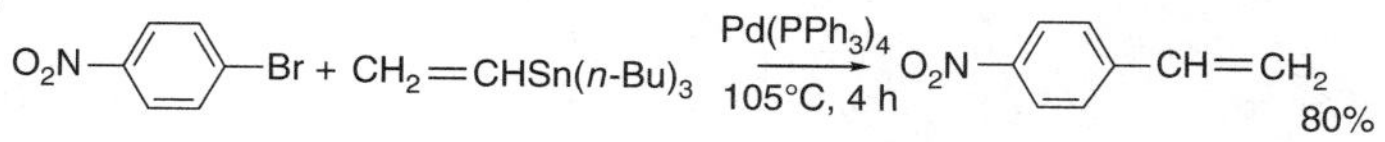

$$O_2N\!-\!\!\langle\ \rangle\!\!-\!Br + CH_2\!\!=\!\!CHSn(n\text{-}Bu)_3 \xrightarrow[105°C, 4\ h]{Pd(PPh_3)_4} O_2N\!-\!\!\langle\ \rangle\!\!-\!CH\!\!=\!\!CH_2 \quad 80\%$$

3[c]

$$CH_3O\!-\!\!\langle\ \rangle\!\!-\!SnBu_3 + CF_3SO_2O\!-\!\!\langle\ \rangle\!\!-\!NO_2 \xrightarrow[LiCl_2, DMF]{PdCl_2(PPh_3)_2} CH_3O\!-\!\!\langle\ \rangle\!\!-\!\!\langle\ \rangle\!\!-\!OCH_3 \quad 48\%$$

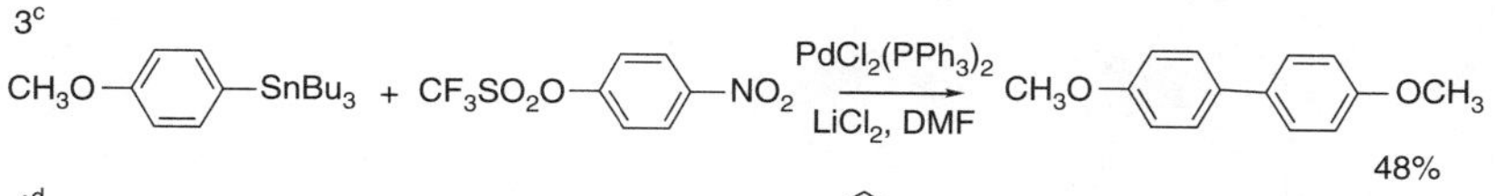

4[d]

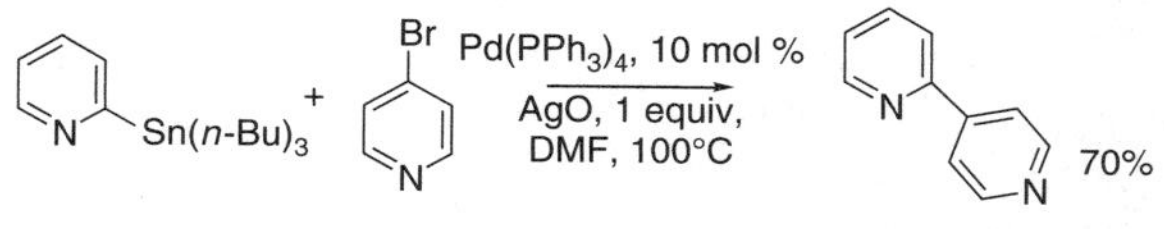

5[e]

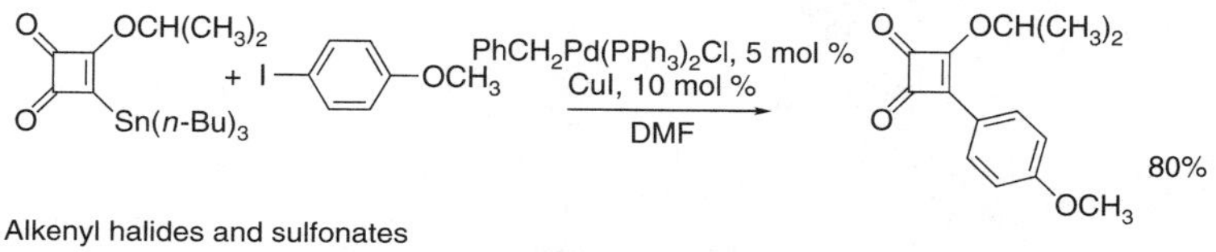

B. Alkenyl halides and sulfonates

6[f]

$$PhCH\!\!=\!\!CHI + CH_2\!\!=\!\!CHSn(n\text{-}Bu)_3 \xrightarrow[25°C, 0.1\ h]{PdCl_2(CH_3CN)_2} PhCH\!\!=\!\!CHCH\!\!=\!\!CH_2 \quad 85\%$$

7[g]

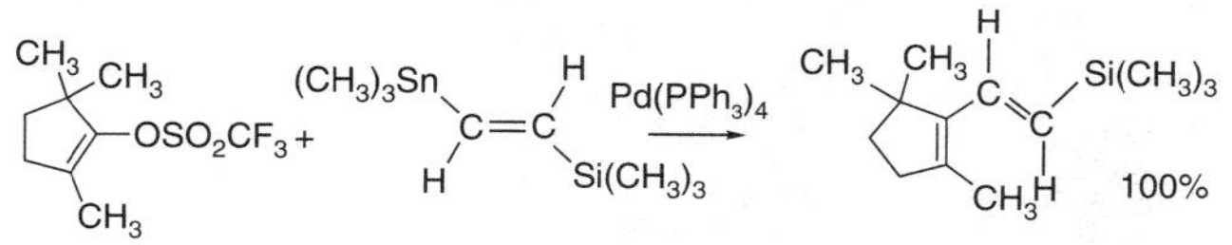

8[h]

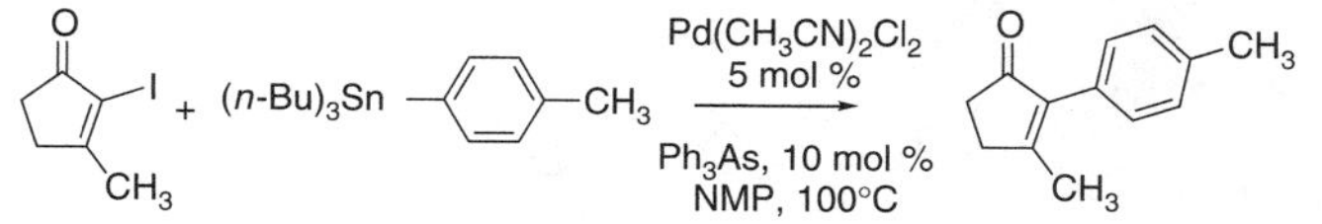

9[i]

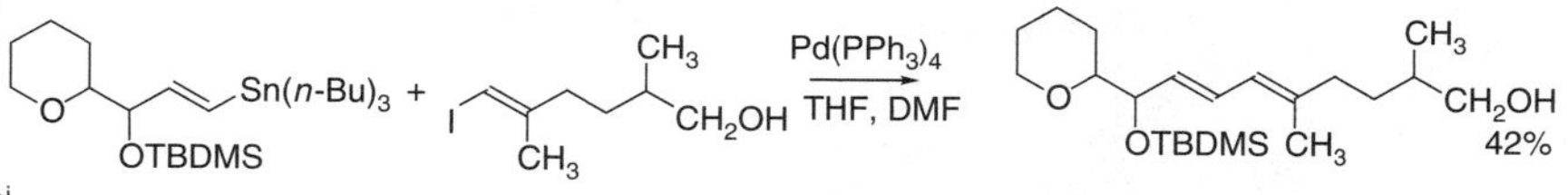

10[j]

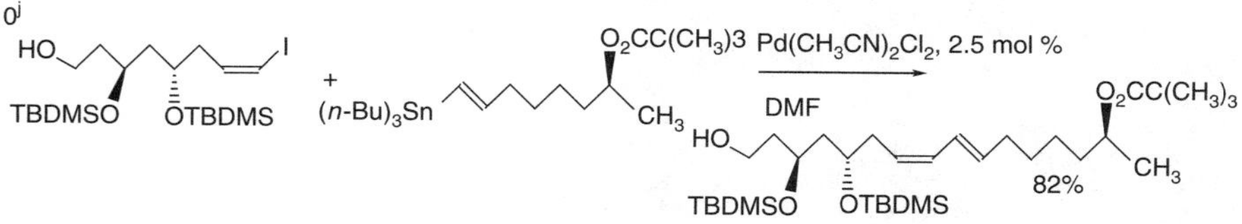

11[k]

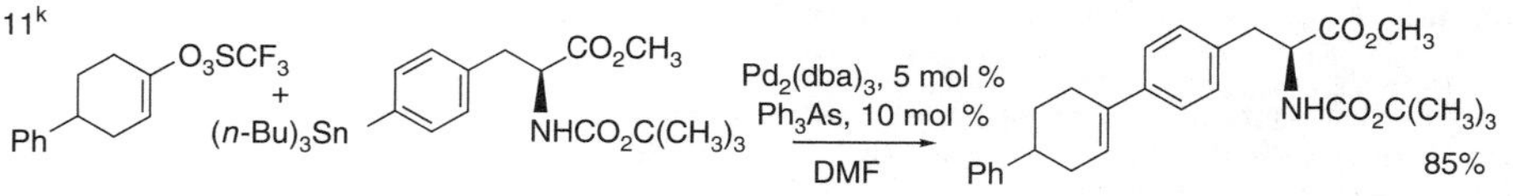

(Continued)

Scheme 8.12. (*Continued*)

12[l]

13[m]

14[n]

15[o]

16[p]

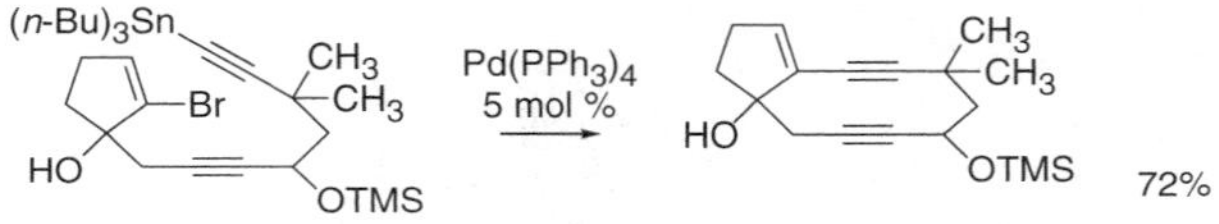

C. Allylic and benzylic halides

17[q]

18[r]

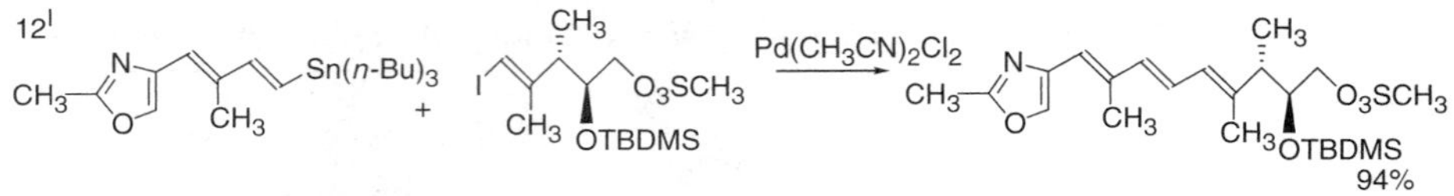

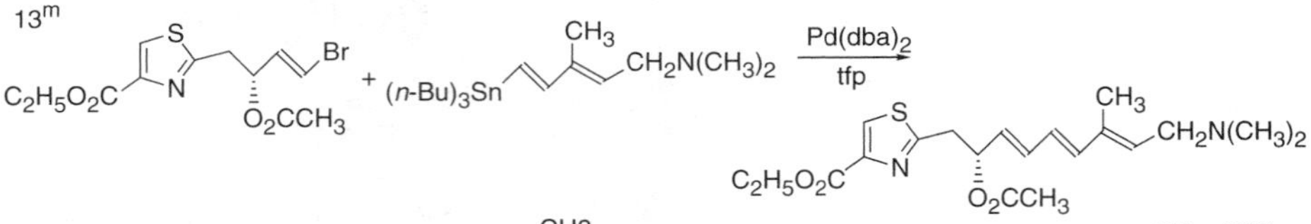

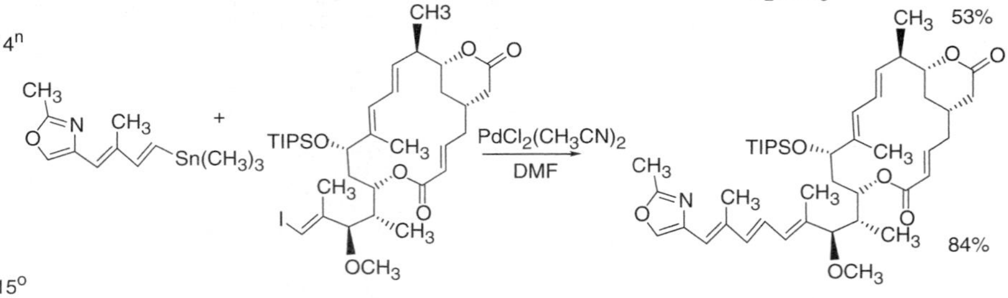

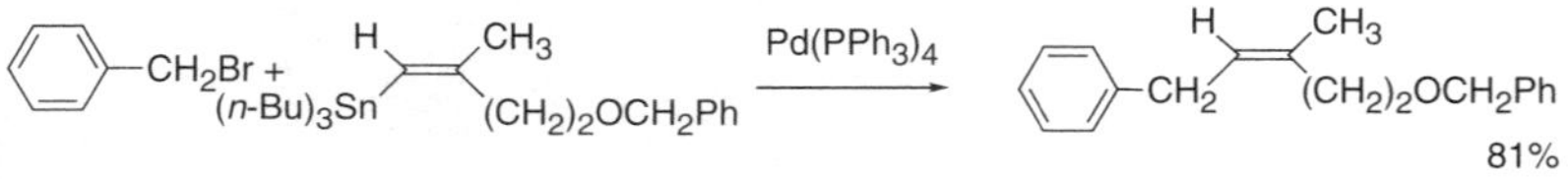

a. M. Kosugi, K. Sasazawa, Y. Shimizu, and T. Migata, *Chem. Lett.*, 301 (1977).
b. D. R. McKean, G. Parrinello, A. F. Renaldo, and J. K. Stille, *J. Org. Chem.*, **52**, 422 (1987).
c. J. K. Stille, A. M. Echavarren, R. M. Williams, and J. A. Hendrix, *Org. Synth.*, **IX**, 553 (1998).
d. J. Malm, P. Bjork, S. Gronowitz, and A.-B. Hornfeldt, *Tetrahedron Lett.*, **33**, 2199 (1992).
e. L. S. Liebeskind and R. W. Fengl, *J. Org. Chem.*, **55**, 5359 (1990).
f. J. K. Stille and B. L. Groh, *J. Am. Chem. Soc.*, **109**, 813 (1987).
g. W. J. Scott, G. T. Crisp, and J. K. Stille, *J. Am. Chem. Soc.*, **106**, 4630 (1984).
h. C. R. Johnson, J. P. Adams, M. P. Braun, and C. B. W. Senanayake, *Tetrahedron Lett.*, **33**, 919 (1992).
i. E. Claus and M. Kalesse, *Tetrahedron Lett.*, **40**, 4157 (1999).
j. A. B. Smith, III, and G. R. Ott, *J. Am. Chem. Soc.*, **120**, 3935 (1998).
k. E. Morera and G. Ortar, *Synlett*, 1403 (1997).
l. J. D. White, M. A. Holoboski, and N. J. Green, *Tetrahedron Lett.*, **38**, 7333 (1997).
m. D. Romo, R. M. Rzasa, H. E. Shea, K. Park, J. M. Langenhan, L. Sun, A. Akhiezer, and J. O. Liu, *J. Am. Chem. Soc.*, **120**, 12237 (1998).

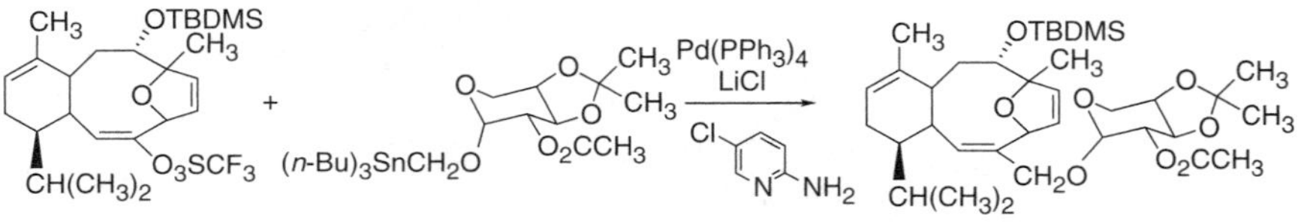

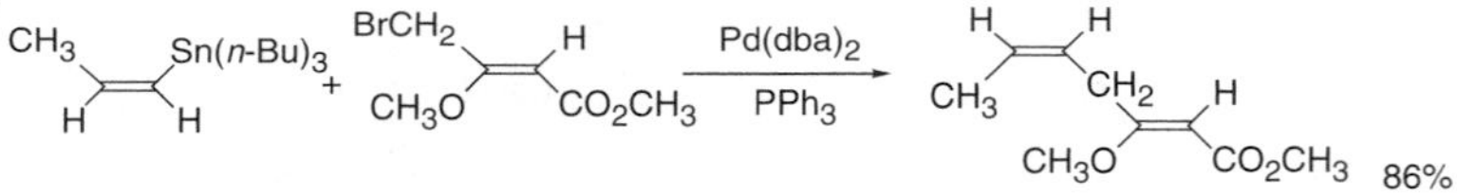

(*Continued*)

Scheme 8.12. (*Continued*)

739

SECTION 8.2

*Reactions Involving
Organopalladium
Intermediates*

n. J. D. White, P. R. Blakemore, N. J. Green, E. B. Hauser, M. A. Holoboski, L. E. Keown, C. S. N. Kolz, and B. W. Phillips, *J. Org. Chem.*, **67**, 7750 (2002).
o. X.-T. Chen, B. Zhou, S. K. Bhattacharya, C. E. Gutteridge, T. R. R. Pettus, and S. Danishefsky, *Angew. Chem. Int. Ed. Engl.*, **37**, 789 (1999).
p. M. Hirama, K. Fujiwara, K. Shigematu, and Y. Fukazawa, *J. Am. Chem. Soc.*, **111**, 4120 (1989).
q. F. K. Sheffy, J. P. Godschalx, and J. K. Stille, *J. Am. Chem. Soc.*, **106**, 4833 (1984).
r. J. Hibino, S. Matsubara, Y. Morizawa, K. Oshima, and H. Nozaki, *Tetrahedron Lett.*, **25**, 2151 (1984).

8.2.3.4. Coupling with Organoboron Reagents.

The *Suzuki reaction* is a palladium-catalyzed cross-coupling reaction in which the organometallic component is a boron compound.[212] The organoboron compounds that undergo coupling include boronic acids,[213] boronate esters,[214] and boranes.[215] The overall mechanism is closely related to that of the other cross-coupling methods. The aryl halide or triflate reacts with the Pd(0) catalyst by oxidative addition. The organoboron compound serves as the source of the

Scheme 8.13. Synthesis of Ketones from Acyl Chlorides and Stannanes

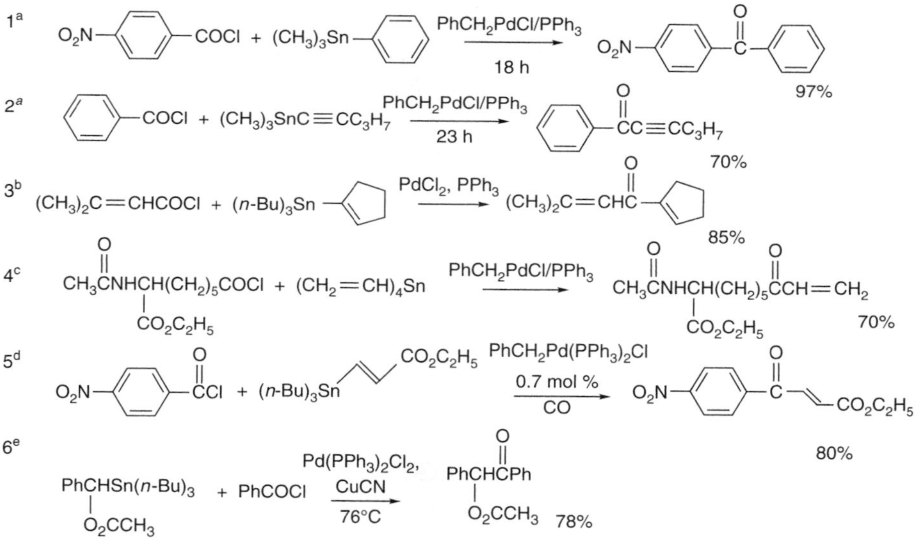

a. J. W. Labadie, D. Tueting, and J. K. Stille, *J. Org. Chem.*, **48**, 4634 (1983).
b. W. F. Goure, M. E. Wright, P. D. Davis, S. S. Labadie, and J. K. Stille, *J. Am. Chem. Soc.*, **106**, 6417 (1984).
c. D. H. Rich, J. Singh, and J. H. Gardner, *J. Org. Chem.*, **48**, 432 (1983).
d. A. F. Renaldo, J. W. Labadie, and J. K. Stille, *Org. Synth.*, **67**, 86 (1988).
e. J. Ye, R. K. Bhatt, and J. R. Falck, *J. Am. Chem. Soc.*, **116**, 1 (1994).

212. N. Miyaura, T. Yanagi, and A. Suzuki, *Synth. Commun.*, **11**, 513 (1981); A. Miyaura and A. Suzuki, *Chem. Rev.*, **95**, 2457 (1995); A. Suzuki, *J. Organomet. Chem.*, **576**, 147 (1999).
213. W. R. Roush, K. J. Moriarty, and B. B. Brown, *Tetrahedron Lett.*, **31**, 6509 (1990); W. R. Roush, J. S. Warmus, and A. B. Works, *Tetrahedron Lett.*, **34**, 4427 (1993); A. R. de Lera, A. Torrado, B. Iglesias, and S. Lopez, *Tetrahedron Lett.*, **33**, 6205 (1992).
214. T. Oh-e, N. Miyaura, and A. Suzuki, *Synlett*, 221 (1990); J. Fu, B. Zhao, M. J. Sharp, and V. Sniekus, *J. Org. Chem.*, **56**, 1683 (1991).
215. T. Oh-e, N. Miyauara, and A. Suzuki, *J. Org. Chem.*, **58**, 2201 (1993); Y. Kobayashi, T. Shimazaki, H. Taguchi, and F. Sato, *J. Org. Chem.*, **55**, 5324 (1990).

second organic group by transmetallation, and the disubstituted Pd(II) intermediate then undergoes reductive elimination. It appears that either the oxidative addition or the transmetallation can be rate determining, depending on reaction conditions.[216] With boronic acids as reactants, base catalysis is normally required and is believed to involve the formation of the more reactive boronate anion in the transmetallation step.[217]

$$ArX + Pd^0 \longrightarrow Ar-Pd^{II}-X$$

$$Ar'B(OH)_2 + {}^-OH \rightleftharpoons [Ar'B(OH)_3]^-$$

$$[Ar'B(OH)_3]^- + Ar-Pd^{II}-X \longrightarrow Ar-Pd^{II}-Ar' + B(OH)_3 + X^-$$

$$Ar-Pd^{II}-Ar' \longrightarrow Ar-Ar' + Pd^0$$

In some synthetic applications, specific bases such as Cs_2CO_3[218] or TlOH[219] have been found preferable to NaOH. Cesium fluoride can play a similar function by forming fluoroborate anions.[220] In addition to aryl halides and triflates, aryldiazonium ions can be the source of the electrophilic component in coupling with arylboronic acids.[221] Conditions for effecting Suzuki coupling in the absence of phosphine ligands have been developed.[222] One of the potential advantages of the Suzuki reaction, especially when boronic acids are used, is that the boric acid is a more innocuous by-product than the tin-derived by-products generated in Stille-type couplings.

Alkenylboronic acids, alkenyl boronate esters, and alkenylboranes can be coupled with alkenyl halides by palladium catalysts to give dienes.[223]

X = OH, OR, R
Y = Br. I

These reactions proceed with retention of double-bond configuration in both the boron derivative and the alkenyl halide. The oxidative addition by the alkenyl halide, transfer

216. G. B. Smith, G. C. Dezeny, D. L. Hughes, A. D. King, and T. R. Verhoeven, *J. Org. Chem.*, **59**, 8151 (1994).

217. K. Matos and J. B. Soderquist, *J. Org. Chem.*, **63**, 461 (1998).

218. A. F. Littke and G. C. Fu, *Angew. Chem. Int. Ed. Engl.*, **37**, 3387 (1998).

219. J. Uenishi, J.-M. Beau, R. W. Armstrong, and Y. Kishi, *J. Am. Chem. Soc.*, **109**, 4756 (1987); J. C. Anderson, H. Namli, and C. A. Roberts, *Tetrahedron*, **53**, 15123 (1997).

220. S. W. Wright, D. L. Hageman, and L. D. McClure, *J. Org. Chem.*, **59**, 6095 (1994).

221. S. Darses, T. Jeffery, J.-P. Genet, J.-L. Brayer, and J.-P. Demoute, *Tetrahedron Lett.*, **37**, 3857 (1996); S. Darses, T. Jeffery, J.-L. Brayer, J.-P. Demoute, and J.-P. Genet, *Bull. Soc. Chim. Fr.*, **133**, 1095 (1996); S. Sengupta and S. Bhattacharyya, *J. Org. Chem.*, **62**, 3405 (1997).

222. T. L. Wallow and B. M. Novak, *J. Org. Chem.*, **59**, 5034 (1994); D. Badone, M. B. R. Cardamone, A. Ielmini, and U. Guzzi, *J. Org. Chem.*, **62**, 7170 (1997).

223. (a) N. Miyaura, K. Yamada, H. Suginome, and A. Suzuki, *J. Am. Chem. Soc.*, **107**, 972 (1985); (b) N. Miyaura, M. Satoh, and A. Suzuki, *Tetrahedron Lett.*, **27**, 3745 (1986); (c) F. Bjorkling, T. Norin, C. R. Unelius, and R. B. Miller, *J. Org. Chem.*, **52**, 292 (1987).

of an alkenyl group from boron to palladium, and reductive elimination all occur with retention of configuration.

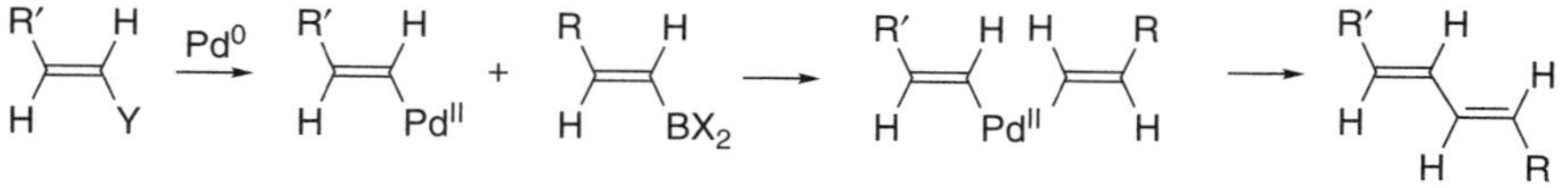

Both alkenyl disiamylboranes and *B*-alkenylcatecholboranes also couple stereo-specifically with alkenyl bromides.[224]

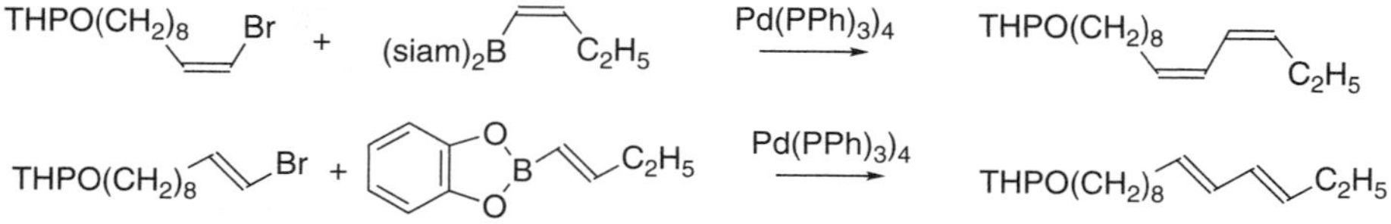

Boronate esters have been used for the preparation of polyunsaturated systems such as retinoic acid esters.

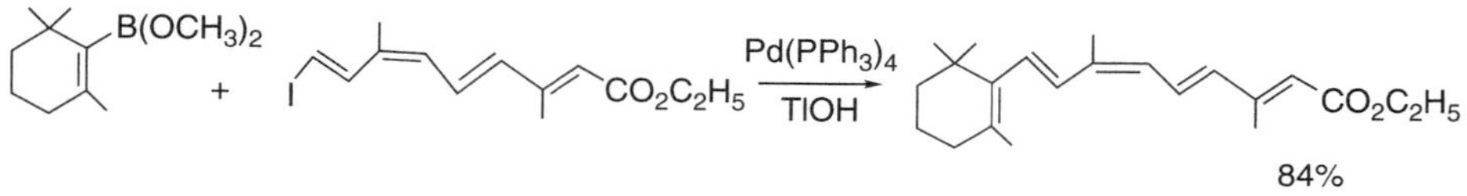

84%

Ref. 225

Intramolecular Suzuki reactions have been done by hydroboration followed by coupling.

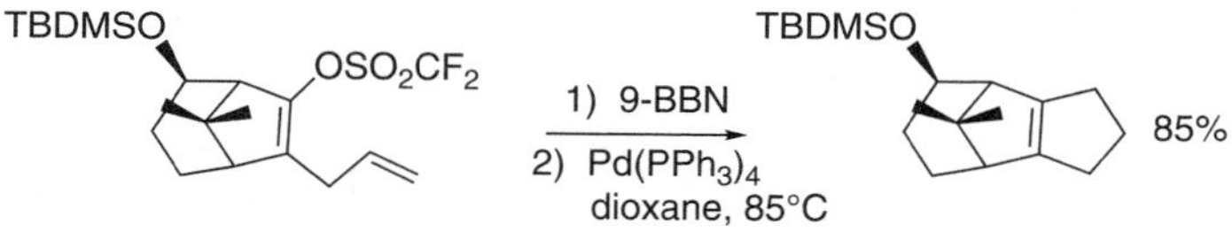

85%

Ref. 226

Triflates prepared from *N*-alkoxycarbonyllactams can be coupled with aryl and alkenylboronic acids.[227]

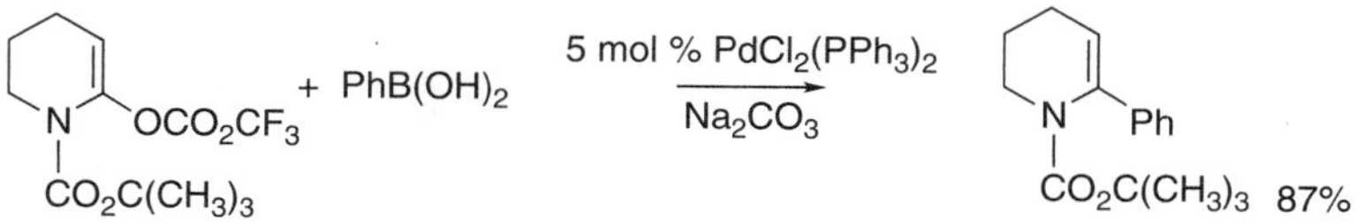

87%

224. (a) N. Miyaura, K. Yamada, H. Suginome, and A. Suzuki, *J. Am. Chem. Soc.*, **107**, 972 (1985); (b) N. Miyaura, T. Ishiyama, M. Ishikawa, and A. Suzuki, *Tetrahedron Lett.*, **27**, 6369 (1986); (c) N. Miyaura, M. Satoh, and A. Suzuki, *Tetrahedron Lett.*, **27**, 3745 (1986); (d) Y. Satoh, H. Serizawa, N. Miyaura, S. Hara, and A. Suzuki, *Tetrahedron Lett.*, **29**, 1811 (1988).
225. Y. Pazos, B. Iglesias, and A. R. de Lera, *J. Org. Chem.*, **66**, 8483 (2001).
226. K. Shimada, M. Nakamura, T. Suzuka, J. Matsui, R. Tatsumi, K. Tsutsumi, T. Morimoto, H. Kurosawa, and K. Kakiuchi, *Tetrahedron Lett.*, **44**, 1401 (2003).
227. E. G. Occhiato, A. Trabocchi, and A. Guarna, *J. Org. Chem.*, **66**, 2459 (2001).

Alkyl substituents on boron in 9-BBN derivatives can be coupled with either vinyl or aryl halides through Pd catalysts.[224b] This is an especially interesting reaction because of its ability to effect coupling of saturated alkyl groups. Palladium-catalyzed couplings of alkyl groups by most other methods often fail because of the tendency for β-elimination

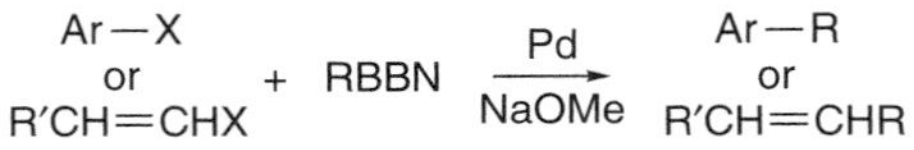

Ar—X
or
R′CH=CHX
+ RBBN
Pd
NaOMe
Ar—R
or
R′CH=CHR

One catalyst that has been found amenable to alkyl systems is $CH_3P(t\text{-}Bu)_2$ or the corresponding phosphonium salt.[228] A range of substituted alkyl bromides were coupled with arylboronic acids.

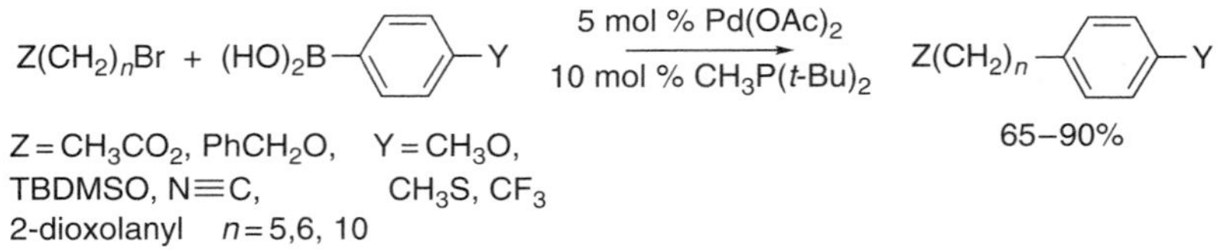

Z(CH2)nBr + (HO)2B
Y
5 mol % Pd(OAc)2
10 mol % CH3P(t-Bu)2
Z(CH2)n
Y
65–90%
Z = CH3CO2, PhCH2O, TBDMSO, N≡C, 2-dioxolanyl
Y = CH3O, CH3S, CF3
n = 5,6, 10

Suzuki couplings have been used in the synthesis of complex molecules. For example, coupling of two large fragments of the epothilone A structure was accomplished in this way.[229]

S
CH3
CH3
O2CCH3
I
+
PhCH2O
OTHP
9BBN
OTBDMS
CH3 CH3
Pd(dppf)Cl2
Cs2CO3
DMF, H2O, 25°C
60%

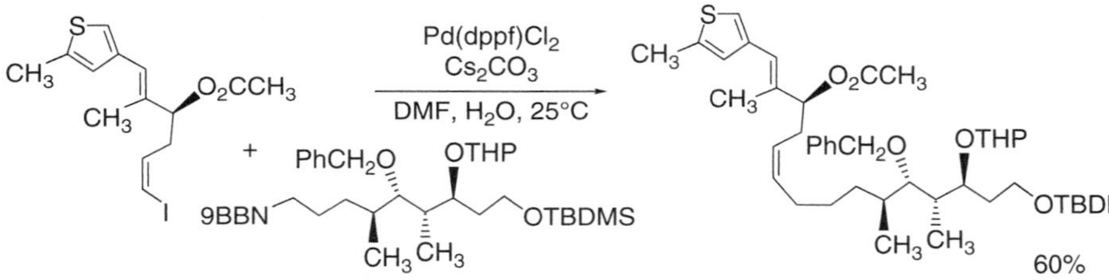

A portion of the side chain of calyculin was prepared by a tandem reaction sequence that combined an alkenylzinc reagent with 2-bromoethenylboronate, followed by Suzuki coupling with a vinyl iodide in the same pot.[230]

H2C
ZnCl
OC2H5
+ Br
B(OiPr)2
Pd(PPh)3
H2C
B(OiPr)2
OC2H5
I
P(OC2H5)2
CH3
O
Ag2O, H2O
64%

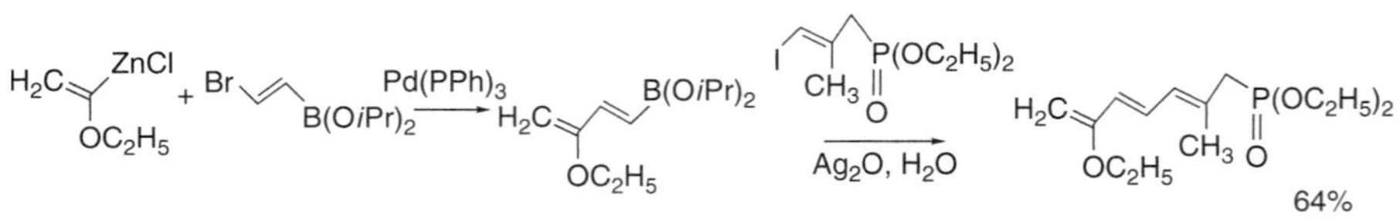

There are also several examples of the use of Suzuki reactions in scale-up synthesis of drug candidates. In the synthesis of CI-1034, an endothelin antagonist, a triflate,

228. J. H. Kirchoff, M. R. Netherton, I. D. Hills, and G. C. Fu, *J. Am. Chem. Soc.*, **124**, 13662 (2002).
229. B. Zhu and J. S. Panek, *Org. Lett.*, **2**, 2575 (2000); see also A. Balog, D. Meng, T. Kamenecka, P. Bertinato, D.-S. Su, E. J. Sorensen, and S. J. Danishefsky, *Angew. Chem. Int. Ed. Engl.*, **35**, 2801 (1996).
230. A. B. Smith, III, G. K. Friestad, J. Barbosa, E. Bertounesque, J. J.-W. Duan, K. G. Hull, M. Iwashima, Y. Qui, P. G. Spoors, and B. A. Salvatore, *J. Am. Chem. Soc.*, **121**, 10478 (1999).

and boronic acid were coupled in 95% yield on an 80-kg scale.[231] For reasons of cost, a replacement was sought for the triflate group, and the most promising was the 4-fluorobenzenesulfonate.

R = CF$_3$, 4-Fluorophenyl

A coupling of a 3-pyridylborane was used in the synthesis of a potential CNS agent.[232] The product (278 kg) was isolated in 92.5% yield as the methanesulfonate salt.

Scheme 8.14 gives some examples of cross coupling using organoboron reagents. Entries 1 to 3 illustrate biaryl coupling. The conditions in Entry 1 are appropriate for the relatively unreactive chlorides. The conditions in Entries 2 and 3 involve no phosphine ligands. The reactions in Entries 4 and 5 illustrate the use of diazonium ions as reactants. Entry 6 illustrates the use of highly substituted reactants. Entry 7 involves use of a cyclic boronate ester. Entries 8 and 9 pertain to heteroaromatic rings. Entry 10 shows the use of a solid-supported reactant. Part B of Scheme 8.14 illustrates several couplings of alkenylboron reagents including catecholboranes (Entries 11 and 12), boronate esters (Entry 13), and boronic acids (Entries 14 and 15). The latter reaction was applied to the synthesis of a retinoate ester. Entry 16 employs a lactone-derived triflate. Entries 17 to 20 are examples of the use of Suzuki couplings in multistage synthesis. Entries 21 to 24 illustrate the applicability of the reaction to alkylboranes. Entry 25 applies phosphine-free conditions to an allylic bromide.

Ketones can also be prepared by palladium-catalyzed reactions of boranes or boronic acids with acyl chlorides. Both saturated and aromatic acyl chlorides react with trialkylboranes in the presence of Pd(PPh$_3$)$_4$.[233]

231. T. E. Jacks, D. T. Belmont, C. A. Briggs, N. M. Horne, G. D. Kanter, G. L. Karrick, J. J. Krikke, R. J. McCabe, J. G. Mustakis, T. N. Nanniga, G. S. Risedorph, R. E. Seamans, R. Skeean, D. D. Winkle, and T. M. Zennie, *Org. Proc. Res. Dev.*, **8**, 201 (2004).

232. M. F. Lipton, M. A. Mauragis, M. T. Maloney, M. F. Veley, D. W. Vander Bor, J. J. Newby, R. B. Appell, and E. D. Daugs, *Org. Proc. Res. Dev.*, **7**, 385 (2003).

233. G. W. Kabalka, R. R. Malladi, D. Tejedor, and S. Kelly, *Tetrahedron Lett.*, **41**, 999 (2000).

Scheme 8.14. Palladium-Catalyzed Cross Coupling of Organoboron Reagents

A. Biaryl formation

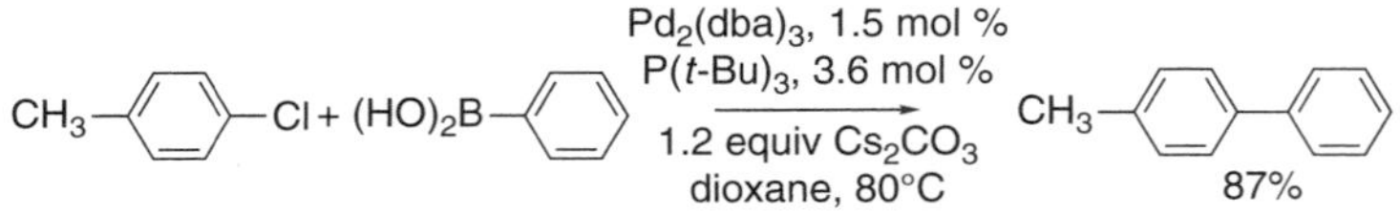

1.

$$CH_3-\langle\rangle-Cl + (HO)_2B-\langle\rangle \xrightarrow[\substack{1.2 \text{ equiv } Cs_2CO_3 \\ \text{dioxane, 80°C}}]{\substack{Pd_2(dba)_3, 1.5 \text{ mol }\% \\ P(t\text{-}Bu)_3, 3.6 \text{ mol }\%}} CH_3-\langle\rangle-\langle\rangle \quad 87\%$$

2[b]

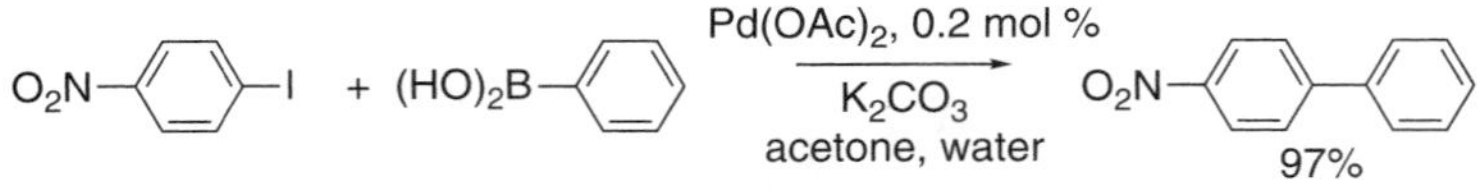

$$O_2N-\langle\rangle-I + (HO)_2B-\langle\rangle \xrightarrow[\substack{K_2CO_3 \\ \text{acetone, water}}]{Pd(OAc)_2, 0.2 \text{ mol }\%} O_2N-\langle\rangle-\langle\rangle \quad 97\%$$

3[c]

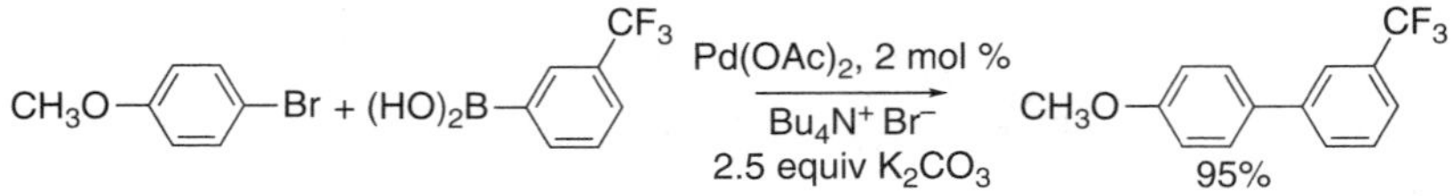

$$CH_3O-\langle\rangle-Br + (HO)_2B-\langle\rangle^{CF_3} \xrightarrow[\substack{Bu_4N^+Br^- \\ 2.5 \text{ equiv } K_2CO_3}]{Pd(OAc)_2, 2 \text{ mol }\%} CH_3O-\langle\rangle-\langle\rangle^{CF_3} \quad 95\%$$

4[d]

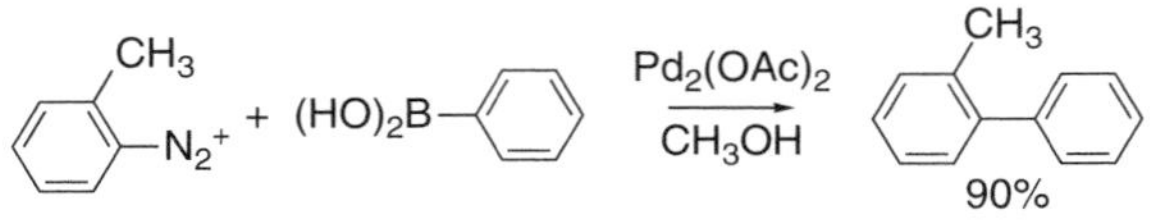

$$\langle\rangle\substack{CH_3 \\ -N_2^+} + (HO)_2B-\langle\rangle \xrightarrow[CH_3OH]{Pd_2(OAc)_2} \langle\rangle\substack{CH_3 \\ -\langle\rangle} \quad 90\%$$

5[e]

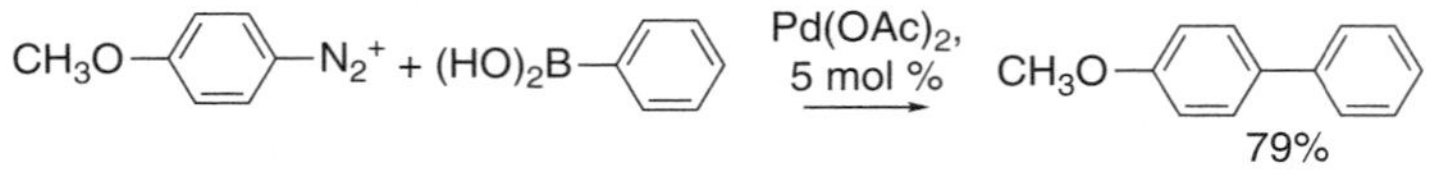

$$CH_3O-\langle\rangle-N_2^+ + (HO)_2B-\langle\rangle \xrightarrow[5 \text{ mol }\%]{Pd(OAc)_2,} CH_3O-\langle\rangle-\langle\rangle \quad 79\%$$

6[f]

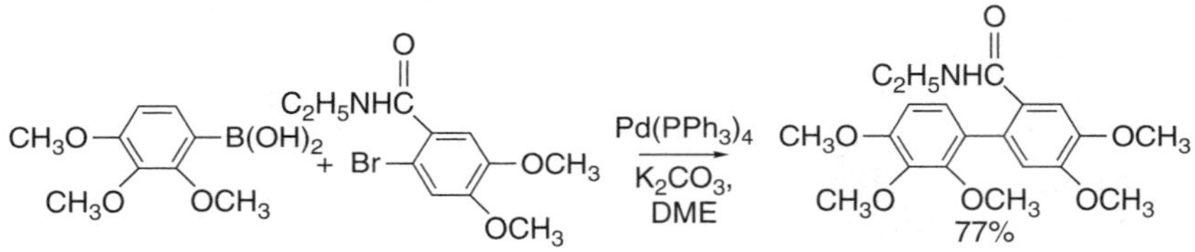

$$77\%$$

7[b]

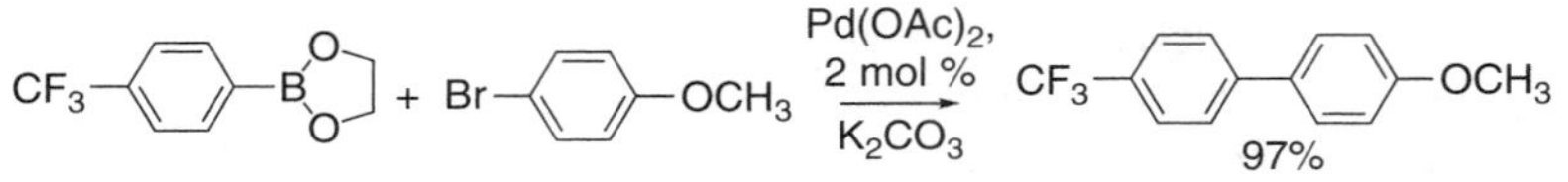

$$CF_3-\langle\rangle-B\langle\substack{O \\ O} + Br-\langle\rangle-OCH_3 \xrightarrow[K_2CO_3]{\substack{Pd(OAc)_2, \\ 2 \text{ mol }\%}} CF_3-\langle\rangle-\langle\rangle-OCH_3 \quad 97\%$$

8[g]

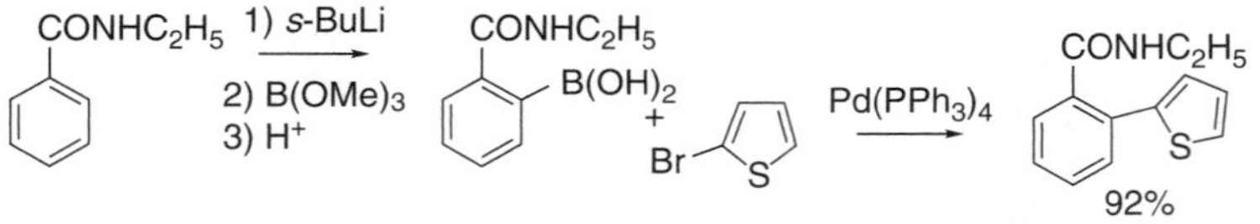

$$92\%$$

9[h]

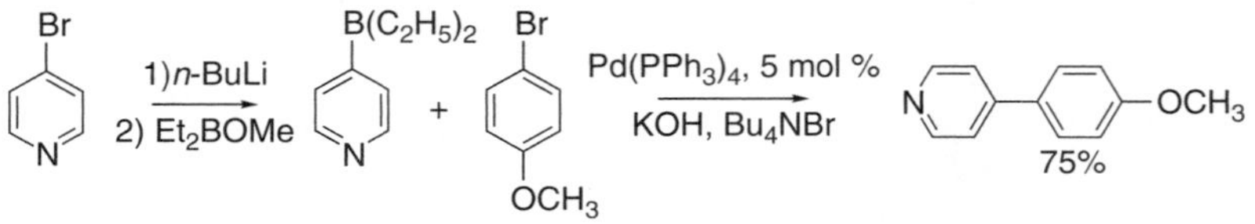

$$75\%$$

10[i]

$$polystyrene-O_2C-\langle\rangle-I + (HO)_2B-\langle\rangle_S \xrightarrow[\substack{1) Pd_2(dba)_3, \\ K_2CO_3 \\ 2) TFA/CH_2Cl_2}]{} HO_2C-\langle\rangle-\langle\rangle_S \quad 91\%$$

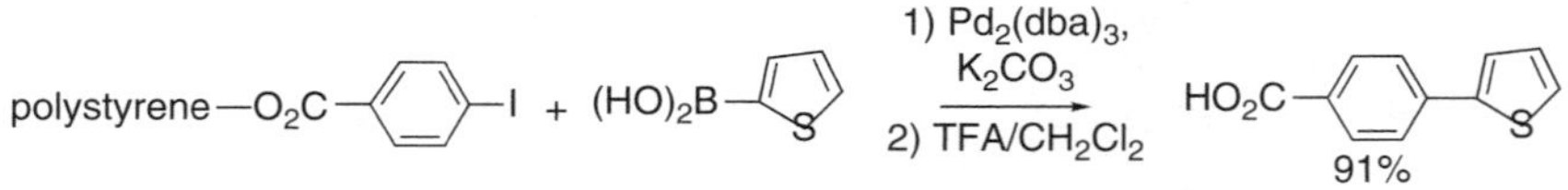

(Continued)

Scheme 8.14. (*Continued*)

745

SECTION 8.2

*Reactions Involving
Organopalladium
Intermediates*

B. Alkenylboranes and alkylboronic acids

11[j]

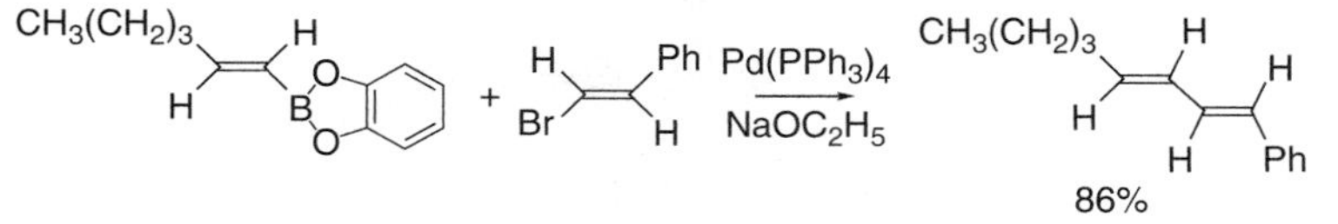

86%

12[k]

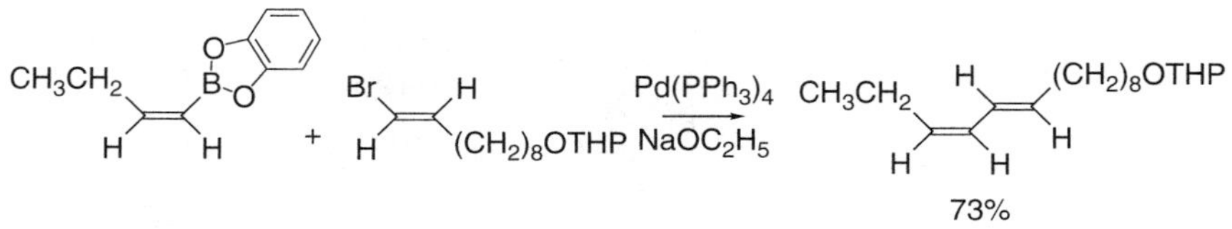

73%

13[l]

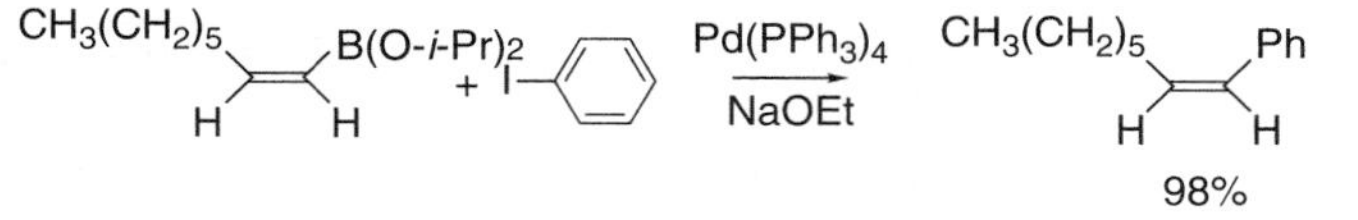

98%

14[m]

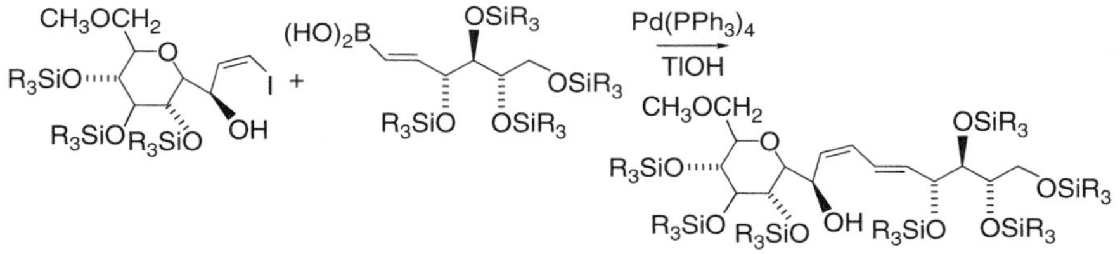

15[n]

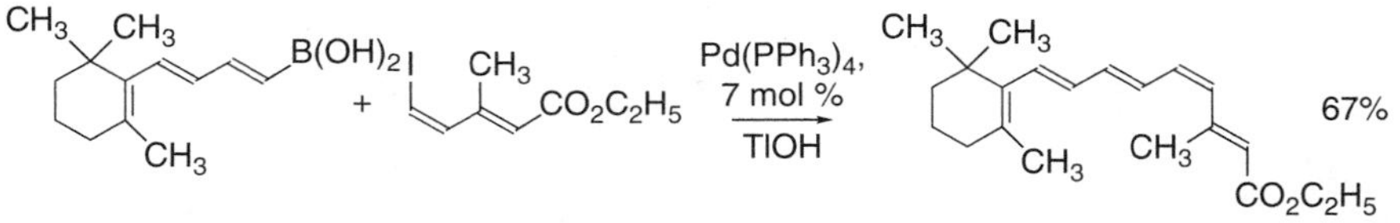

67%

16[o]

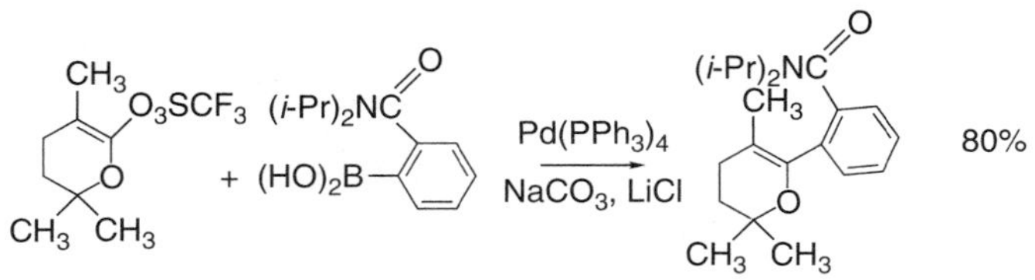

80%

17[p]

62%

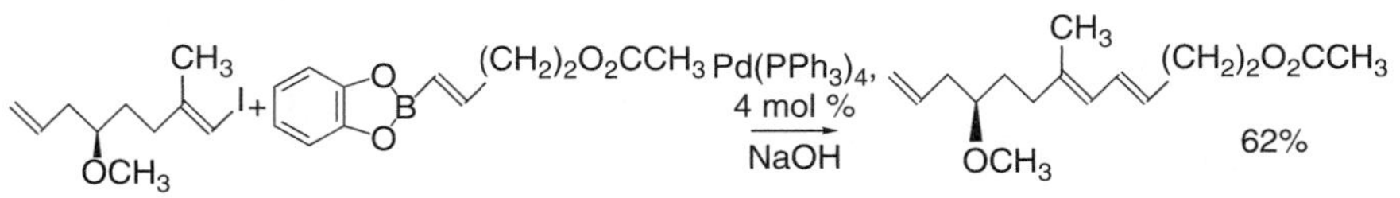

(*Continued*)

Scheme 8.14. (*Continued*)

18[q]

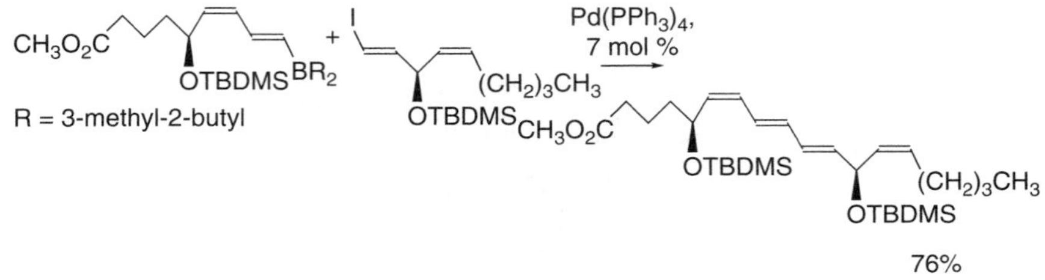

R = 3-methyl-2-butyl

76%

19[r]

72%

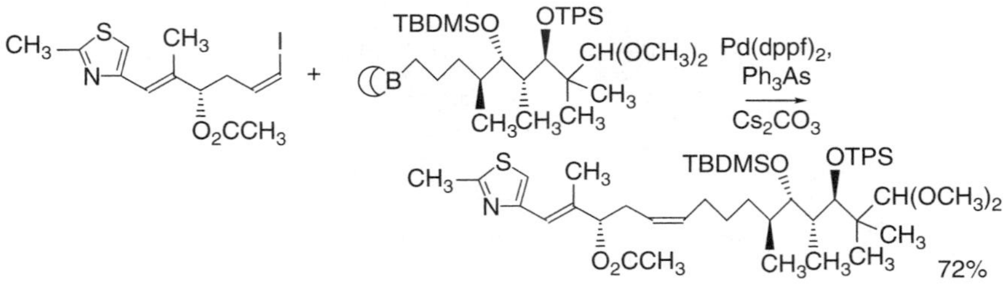

20[s]

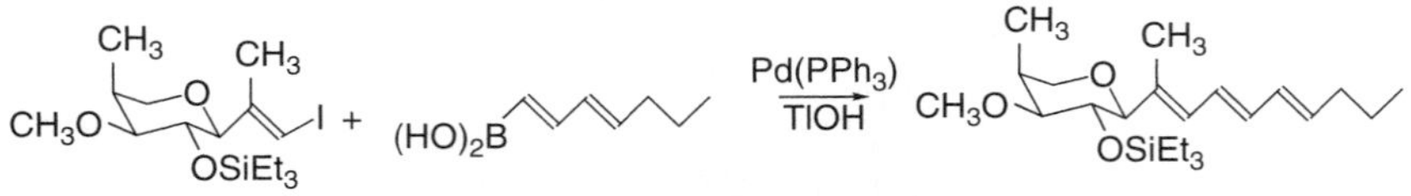

C. Alkyl–aryl coupling

21[t]

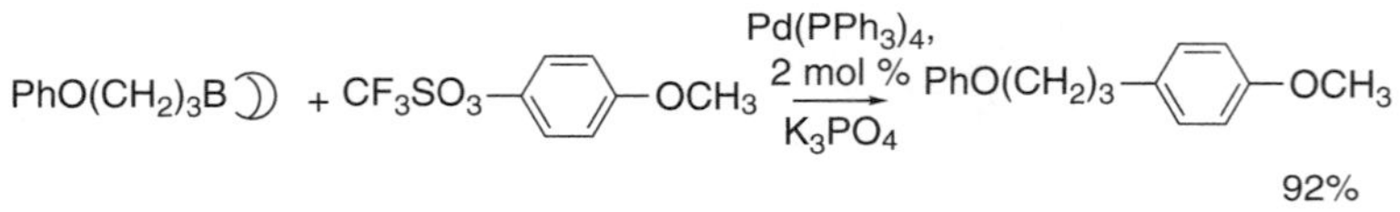

92%

22[u]

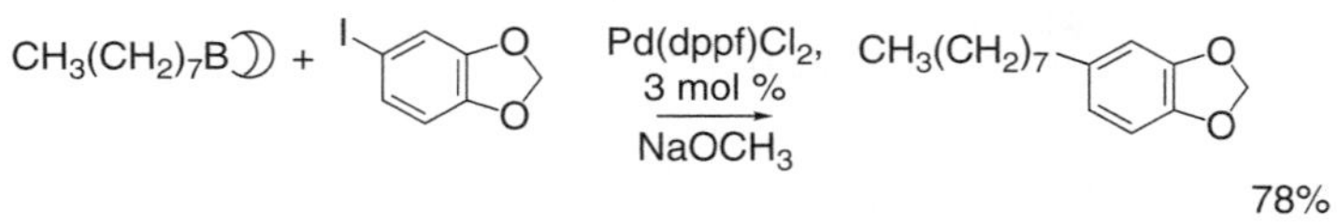

78%

23[v]

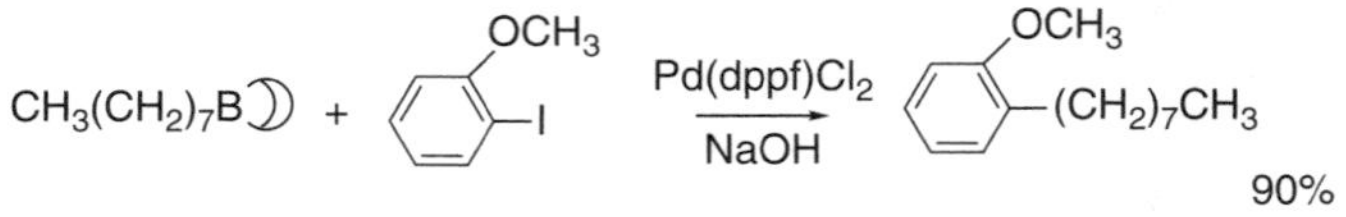

90%

24[w]

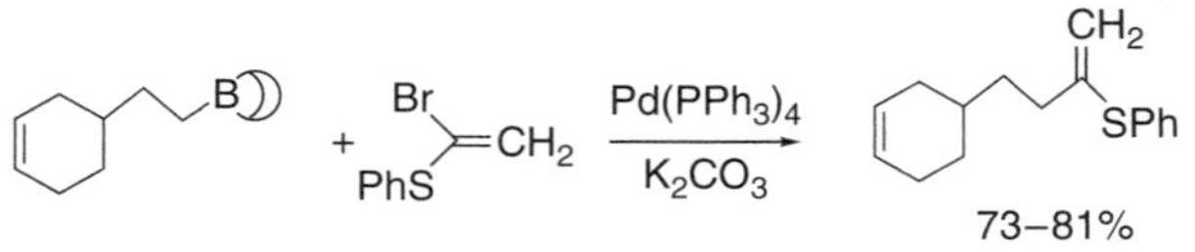

73–81%

25[x]

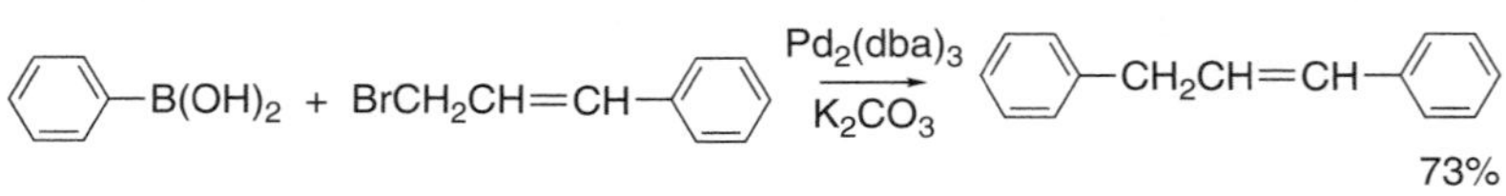

73%

Scheme 8.14. (*Continued*)

747

a. F. Little and G. C. Fu, *Angew. Chem. Int. Ed. Engl.*, **37**, 3387 (1998).
b. T. L. Wallow and B. M. Novak, *J. Org. Chem.*, **59**, 5034 (1994).
c. D. Badone, M. Baroni, R. Cardamone, A Ielmini, and U. Guzzi, *J. Org. Chem.*, **62**, 7170 (1997).
d. S. Darses, T. Jeffery, J.-L. Brayer, J.-P. Demoute, and J.-P. Genet, *Bull. Soc. Chim. Fr.* **133**, 1095 (1996); S. Sengupta and S. Bhattacharyya, *J. Org. Chem.*, **62**, 3405 (1997).
e. S. Darses, T. Jeffery, J.-P. Genet, J.-L. Brayer, and J.-P. Demoute, *Tetrahedron Lett.*, **37**, 3857 (1996).
f. B. I. Alo, A. Kandil, P. A. Patil, M. J. Sharp, M. A. Siddiqui, and V. Snieckus, *J. Org. Chem.*, **56**, 3763 (1991).
g. J. Sharp and V. Snieckus, *Tetrahedron Lett.*, **26**, 5997 (1985).
h. M. Ishikura, T. Ohta, and M. Terashima, *Chem. Pharm. Bull.*, **33**, 4755 (1985).
i. J. W. Guiles, S. G. Johnson, and W. V. Murray, *J. Org. Chem.*, **61**, 5169 (1996).
j. N. Miyaura, K. Yamada, H. Suginome, and A. Suzuki, *J. Am. Chem. Soc.*, **107**, 972 (1985).
k. F. Bjorkling, T. Norin, C. R. Unelius, and R. B. Miller, *J. Org. Chem.*, **52**, 292 (1987).
l. N. Miyaura, M. Satoh, and A. Suzuki, *Tetrahedron Lett.*, **27**, 3745 (1986).
m. J. Uenishi, J.-M. Beau, R. W. Armstrong, and Y. Kishi, *J. Am. Chem. Soc.*, **109**, 4756 (1987).
n. A. R. de Lera, A. Torrado, B. Iglesias, and S. Lopez, *Tetrahedron Lett.*, **33**, 6205 (1992).
o. M. A. F. Brandao, A. B. de Oliveira, and V. Snieckus, *Tetrahedron Lett.*, **34**, 2437 (1993).
p. J. D. White, T. S. Kim, and M. Nambu, *J. Am. Chem. Soc.*, **119**, 103 (1997).
q. Y. Kobayashi, T. Shimazaki, H. Taguchi, and F. Sato, *J. Org. Chem.*, **55**, 5324 (1990).
r. D. Meng, P. Bertinato, A. Balog, D.-S. Su, T. Kamenecka, E. J. Sorensen, and S. J. Danishefsky, *J. Am. Chem. Soc.*, **119**, 10073 (1997).
s. A. G. M. Barrett, A. J. Bennett, S. Menzer, M. L. Smith, A. J. P. White, and D. J. Williams, *J. Org. Chem.*, **64**, 162 (1999).
t. T. Oh-e, N. Miyaura, and A. Suzuki, *J. Org. Chem.*, **58**, 2201 (1993).
u. N. Miyaura, T. Ishiyama, H. Sasaki, M. Ishikawa, M. Satoh, and A. Suzuki, *J. Am. Chem. Soc.*, **111**, 314 (1989).
v. N. Miyaura, T. Ishiyama, M. Ishikawa, and A. Suzuki, *Tetrahedron Lett.*, **27**, 6369 (1986).
w. T. Ishiyama, N. Miyaura, and A. Suzuki, *Org. Synth.*, **71**, 89 (1993).
x. M. Moreno-Manas, F. Pajuelo, and R. Pleixarts, *J. Org. Chem.*, **60**, 2396 (1995).

Aromatic acyl chlorides also react with arylboronic acids to give ketones.[234]

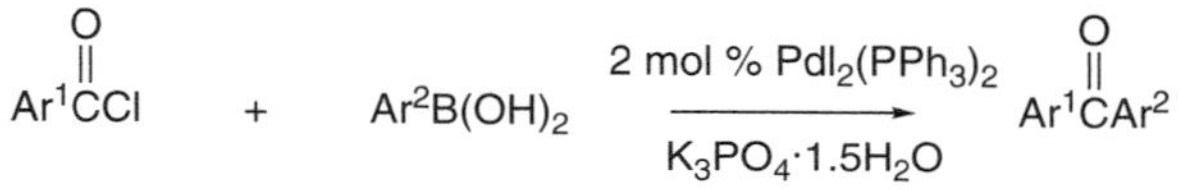

$$Ar^1CCl \quad + \quad Ar^2B(OH)_2 \quad \xrightarrow[\text{K}_3\text{PO}_4\cdot1.5\text{H}_2\text{O}]{\text{2 mol \% PdI}_2(\text{PPh}_3)_2} \quad Ar^1CAr^2$$

α, β-Unsaturated acyl chlorides can also be converted to ketones by reaction with arylboronic acids.[235]

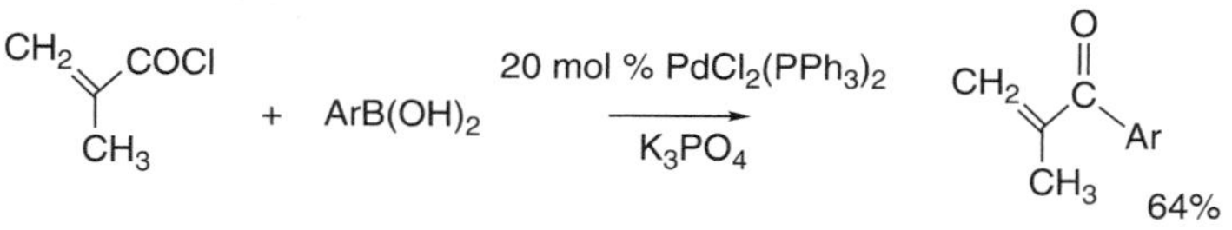

$$\underset{CH_3}{\overset{CH_2}{\diagdown}}COCl \quad + \quad ArB(OH)_2 \quad \xrightarrow[\text{K}_3\text{PO}_4]{\text{20 mol \% PdCl}_2(\text{PPh}_3)_2} \quad \underset{CH_3}{\overset{CH_2}{\diagdown}}\overset{O}{\underset{}{C}}Ar \quad 64\%$$

Ketones can also be prepared directly from carboxylic acids by activation as mixed anhydrides by dimethyl dicarbonate.[236] These conditions were used successfully with alkanoic and alkanedioic acids, as well as aromatic acids.

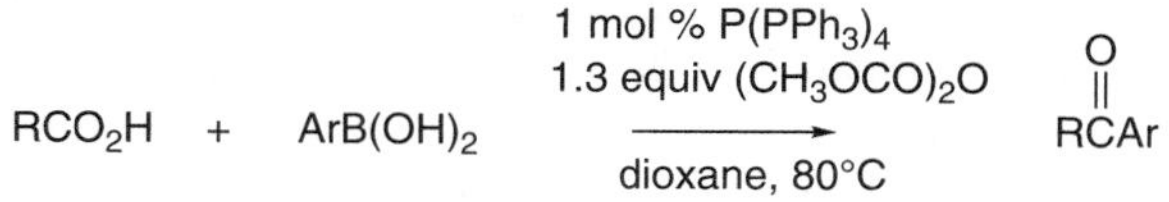

$$RCO_2H \quad + \quad ArB(OH)_2 \quad \xrightarrow[\text{dioxane, 80°C}]{\substack{\text{1 mol \% P(PPh}_3)_4 \\ \text{1.3 equiv (CH}_3\text{OCO)}_2\text{O}}} \quad \overset{O}{\underset{}{R}}Ar$$

In all these reactions, the acylating reagent reacts with the active Pd(0) catalyst to give an acyl Pd(II) intermediate. Transmetallation by the organoboron derivative and reductive elimination generate the ketone.

234. Y. Urawa and K. Ogura, *Tetrahedron Lett.*, **44**, 271 (2003).
235. Y. Urawa, K. Nishiura, S. Souda, and K. Ogura, *Synthesis*, 2882 (2003).
236. R. Kakino, H. Narahashi, I. Shimizu, and A. Yamamoto, *Bull. Chem. Soc. Jpn.*, **75**, 1333 (2002).

Ketones can also be prepared from 4-methylphenylthiol esters. These reactions require a stoichiometric amount of a Cu(I) salt and the thiophene-2-carboxylate was used.[237]

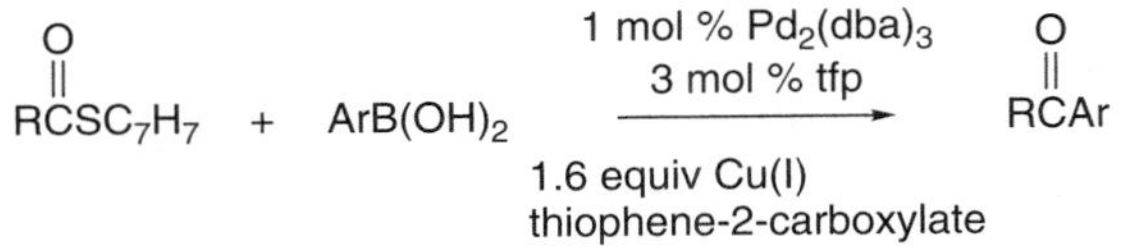

The copper salt is believed to function by promoting the transmetallation stage.

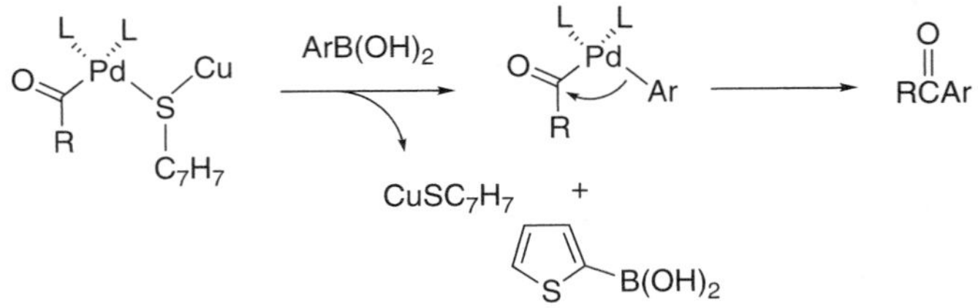

These reaction conditions were applicable to the thiol esters of alkanoic, heteroaromatic, and halogenated acetic acids.

8.2.4. Carbonylation Reactions

Carbonylation reactions involve coordination of carbon monoxide to palladium and a transfer of an organic group from palladium to the coordinated carbon monoxide.

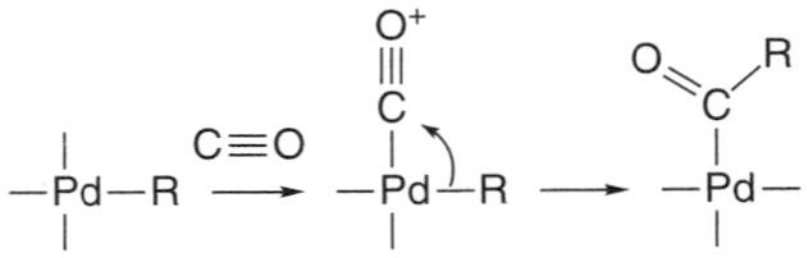

Carbonylation reactions have been observed using both Pd(II)-alkene complexes and σ-bonded Pd(II) species formed by oxidative addition. Under reductive conditions, the double bond can be *hydrocarbonylated*, resulting in the formation of a carboxylic acid or ester.[238] In nucleophilic solvents, the intermediate formed by *solvopalladation* is intercepted by carbonylation and addition of nucleophilic solvent. In both types of reactions, regioisomeric products are possible.

[237.] L. S. Liebeskind and J. Srogl, *J. Am. Chem. Soc.*, **122**, 11260 (2000).

[238.] B. El Ali and H. Alper, in *Handbook of Organopalladium Chemistry for Organic Synthesis*, Vol. 2, E. Negishi and A. de Meijere, eds., Wiley-Interscience, New York, 2000, pp. 2333–2349.

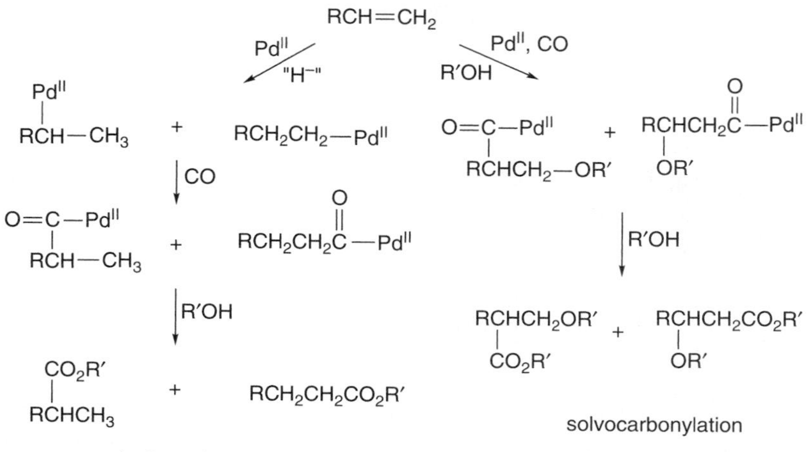

$$RCH=CH_2$$

$$
\begin{array}{ccc}
Pd^{II} & & Pd^{II}, CO \\
RCH-CH_3 & + & RCH_2CH_2-Pd^{II} \\
\end{array}
$$

$$O=C-Pd^{II} \quad + \quad RCHCH_2\overset{O}{\overset{\|}{C}}-Pd^{II}$$
$$RCHCH_2-OR' \qquad OR'$$

$$\downarrow CO$$

$$O=C-Pd^{II} \quad + \quad RCH_2CH_2\overset{O}{\overset{\|}{C}}-Pd^{II}$$
$$RCH-CH_3$$

$$\downarrow R'OH$$

$$\downarrow R'OH$$

$$
\begin{array}{ccc}
CO_2R' & & \\
RCHCH_3 & + & RCH_2CH_2CO_2R' \\
\end{array}
$$

$$RCHCH_2OR' \quad + \quad RCHCH_2CO_2R'$$
$$CO_2R' \qquad\qquad OR'$$

hydrocarbonylation

solvocarbonylation

8.2.4.1. Hydrocarbonylation.

The hydrocarbonylation reaction can be applied to the synthesis of α-arylpropanoic acids of the NSAIDS type.[239] For this synthesis to be effective, selective carbonylation of the more-substituted sp^2 carbon is required. Although many carbonylation conditions are unselective, $PdCl_2(PPh_3)_2$ with *p*-toluenesulfonic acid and LiCl achieves excellent selectivity, which is thought to involve the formation of a benzylic chloride intermediate.

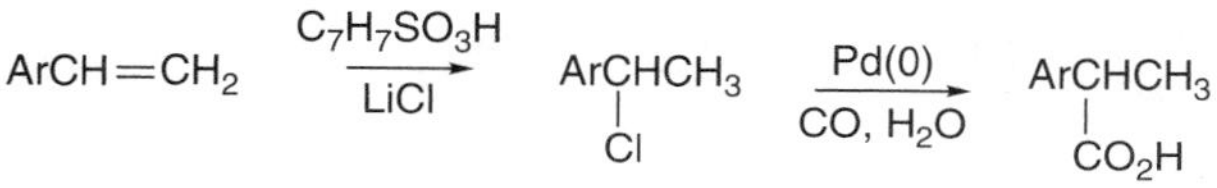

$$ArCH=CH_2 \xrightarrow[\text{LiCl}]{\text{C}_7\text{H}_7\text{SO}_3\text{H}} \underset{Cl}{ArCHCH_3} \xrightarrow[\text{CO, H}_2\text{O}]{Pd(0)} \underset{CO_2H}{ArCHCH_3}$$

Naproxen can be synthesized in 89% yield with 97.5% regioselectivity under these conditions.

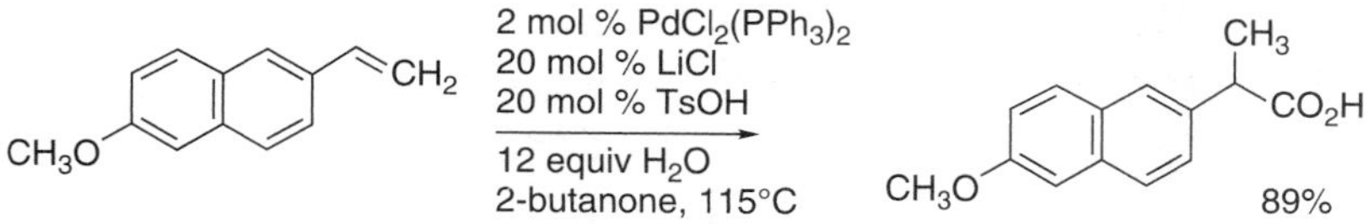

This reaction has been done with good enantioselectivity using 1, 1'-binaphthyl-2, 2'-diyl hydrogen phosphate (BNPPA) as a chiral ligand.[240]

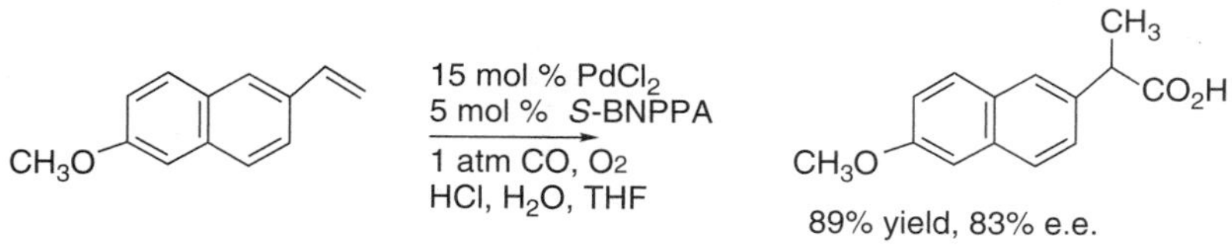

When conducting hydrocarbonylations with dienes, it was found that a mixture of nonchelating and bidentate phosphine ligands was beneficial.[241]

239. A Seayad, S. Jayasree, and R. V. Chaudhari, *Org. Lett.*, **1**, 459 (1999).
240. H. Alper and N. Hamel, *J. Am. Chem. Soc.*, **112**, 2803 (1990).
241. G. Vasapollo, A. Somasunderam, B. El Ali, and H. Alper, *Tetrahedron Lett.*, **35**, 6203 (1994).

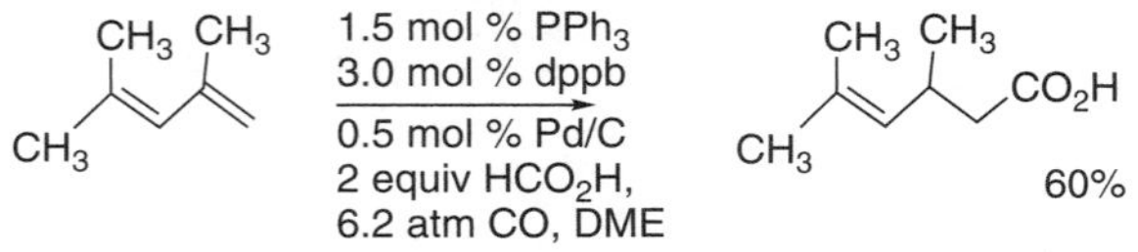

In some cases double-bond migration was noted, as for isoprene.

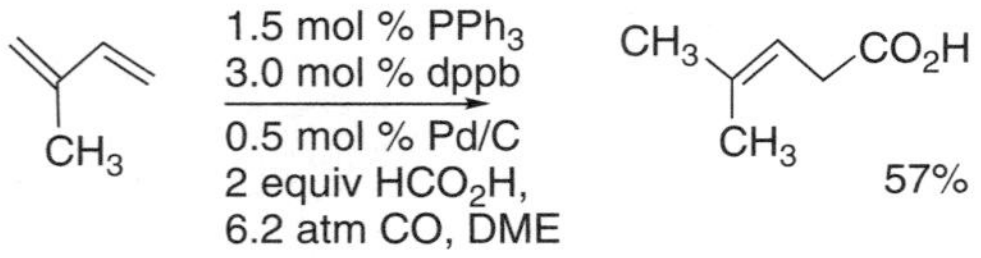

Esters can be formed when the hydrocarbonylation reaction is carried out in an alcohol.[242] Although hydrocarbonylation is the basis for conversion of alkenes to carboxylic acids on an industrial scale, it has seen only limited application in laboratory synthesis.

Olefin hydrocarbonylation can be used in conjunction with oxidative addition to prepare indanones and cyclopentenones, but the reaction is limited to terminal alkenes.[243]

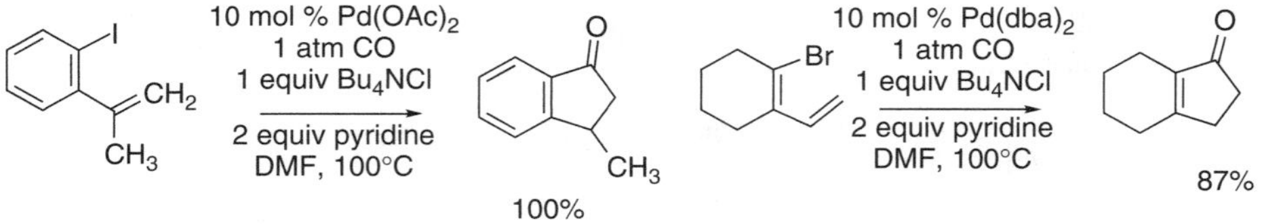

8.2.4.2. Solvocarbonylation. In solvocarbonylation, a substituent is introduced by a nucleophilic addition to a π complex of the alkene. The acylpalladium intermediate is then captured by a nucleophilic solvent such as an alcohol. A catalytic process that involves Cu(II) reoxidizes Pd(0) to the Pd(II) state.[244]

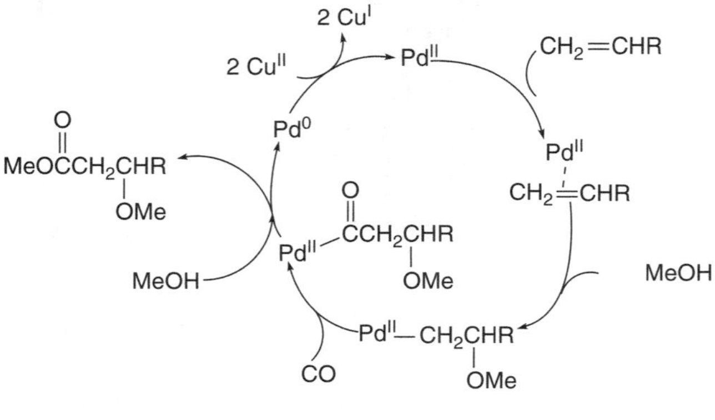

242. S. Oi, M. Nomura, T. Aiko, and Y. Inoue, *J. Mol. Catal. A.*, **115**, 289 (1997).
243. S. V. Gagnier and R. C. Larock, *J. Am. Chem. Soc.*, **125**, 4804 (2003).
244. D. E. James and J. K. Stille, *J. Am. Chem. Soc.*, **98**, 1810 (1976).

This reaction has been shown to proceed with overall *anti* addition in the case of *E*- and *Z*-butene.[245]

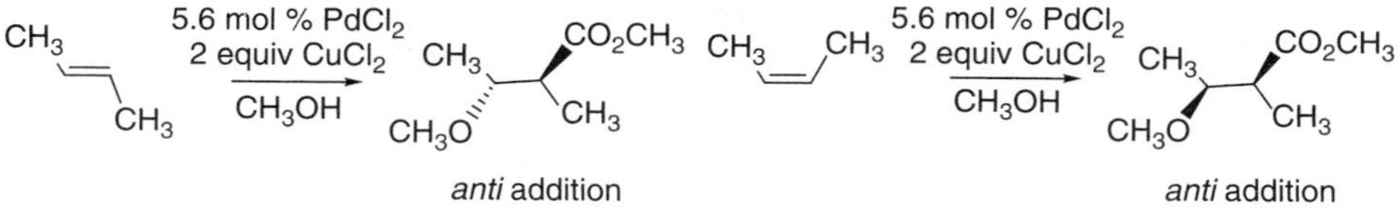

anti addition   *anti* addition

Organopalladium(II) intermediates generated from halides or triflates by oxidative addition react with carbon monoxide in the presence of alcohols to give carboxylic acids[246] or esters.[247]

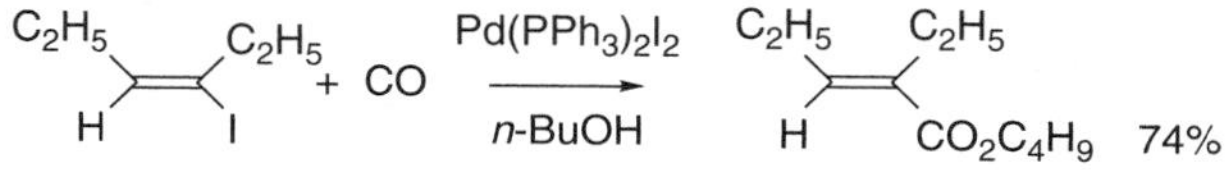

The carbonyl insertion step takes place by migration of the organic group from the metal to the coordinated carbon monoxide, generating an acylpalladium species. This intermediate can react with nucleophilic solvent, releasing catalytically active Pd(0).

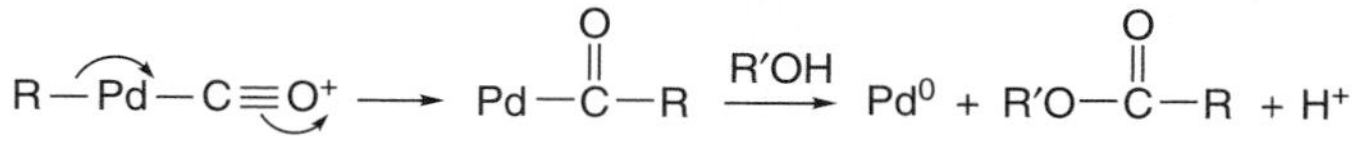

The detailed mechanisms of such reactions have been shown to involve addition and elimination of phosphine ligands. The efficiency of individual reactions can often be improved by careful choice of added ligands.

Allylic acetates and phosphates can be readily carbonylated.[248] Carbonylation usually occurs at the less-substituted end of the allylic system and with inversion of configuration in cyclic systems.

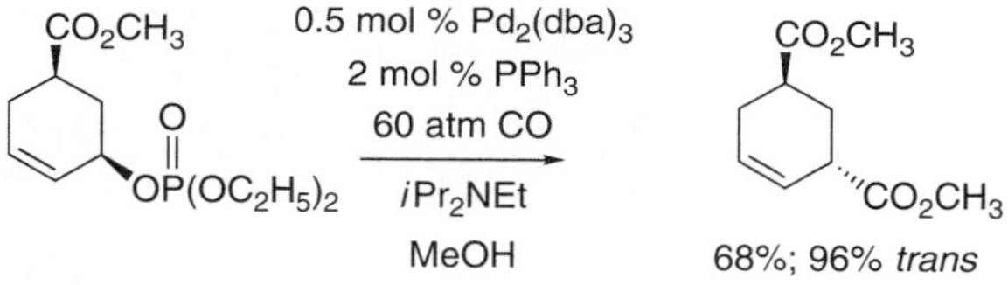

The reactions are accelerated by bromide salts, which are thought to exchange for acetate in the π-allylic complex. The reactions of acyclic compounds occur with minimal $E{:}Z$ isomerization. This result implies that the π-allyl intermediate is captured by carbonylation faster than $E{:}Z$ isomerization occurs.

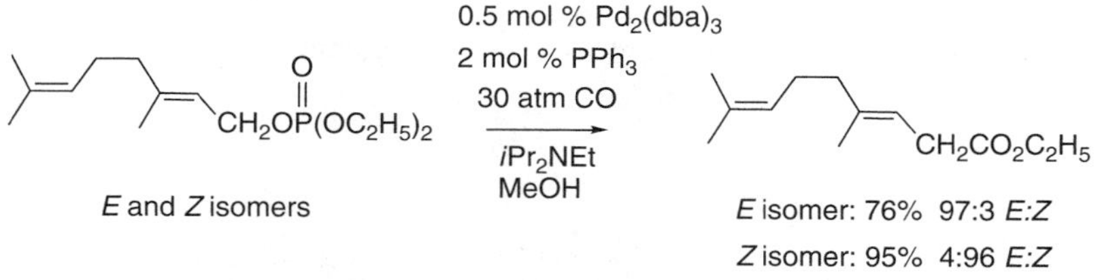

E isomer: 76% 97:3 *E:Z*
Z isomer: 95% 4:96 *E:Z*

245. D. E. James, L. F. Hines, and J. K. Stille, *J. Am. Chem. Soc.*, **98**, 1806 (1976).
246. S. Cacchi and A. Lupi, *Tetrahedron Lett.*, **37**, 3939 (1992).
247. A. Schoenberg, I. Bartoletti, and R. F. Heck, *J. Org. Chem.*, **39**, 3318 (1974); S. Cacchi, E. Morena, and G. Ortar, *Tetrahedron Lett.*, **26**, 1109 (1985).
248. S. Murahashi, Y. Imada, Y. Taniguchi, and S. Higashiura, *J. Org. Chem.*, **58**, 1538 (1993).

Coupling of organostannanes with halides in a carbon monoxide atmosphere leads to ketones by incorporation of a carbonylation step.[249] The catalytic cycle is similar to that involved in the coupling of alkyl or aryl halides. These reactions involve a migration of one of the organic substituents to the carbonyl carbon, followed by reductive elimination.

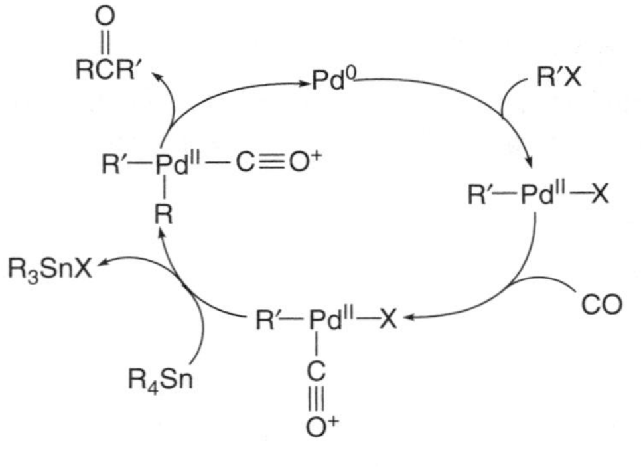

This method can also be applied to alkenyl triflates.

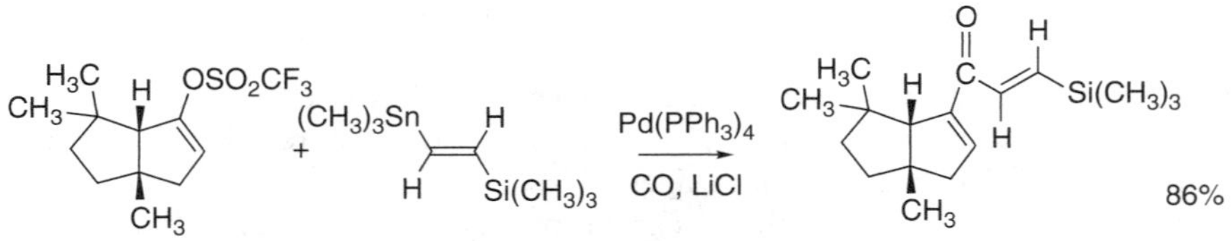

Ref. 250

Carbonylation reactions can be carried out with a boronic acid as the nucleophilic component.[251]

Application of the carbonylation reaction to halides with appropriately placed hydroxy groups leads to lactone formation. In this case the acylpalladium intermediate is trapped intramolecularly.

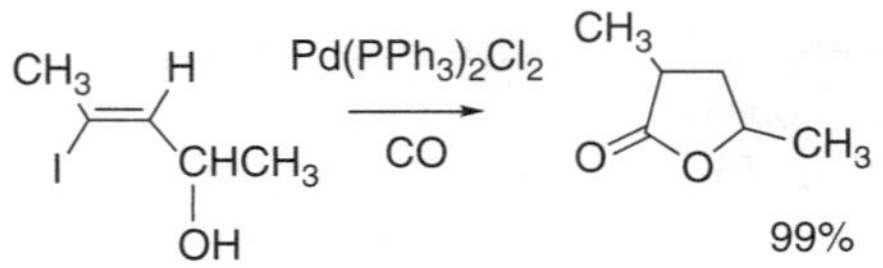

99%

Ref. 252

Carbonylation can also be carried out as a tandem reaction in intramolecular Heck reactions.

249. M. Tanaka, *Tetrahedron Lett.*, 2601 (1979); D. Milstein and J. K. Stille, *J. Org. Chem.*, **44**, 1613 (1979); J. W. Labadie and J. K. Stille, *J. Am. Chem. Soc.*, **105**, 6129 (1983); A. M. Echavarren and J. K. Stille, *J. Am. Chem. Soc.*, **110**, 1557 (1988).

250. G. T. Crisp, W. J. Scott, and J. K. Stille, *J. Am. Chem. Soc.*, **106**, 7500 (1984).

251. T. Ishiyama, H. Kizaki, N. Miyaura, and A. Suzuki, *Tetrahedron Lett.*, **34**, 7595 (1993); T. Ishiyama, H. Kizaki, T. Hayashi, A. Suzuki, and N. Miyaura, *J. Org. Chem.*, **63**, 4726 (1998).

252. A. Cowell and J. K. Stille, *J. Am. Chem. Soc.*, **102**, 4193 (1980).

Scheme 8.15. Synthesis of Ketones, Esters, Carboxylic Acids, and Amides by Palladium-Catalyzed Carbonylation and Acylation

A. Ketones by carbonylation

1[a]

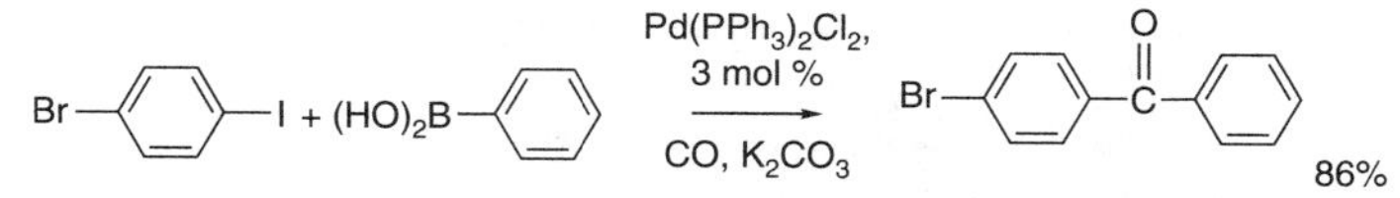

86%

2[b]

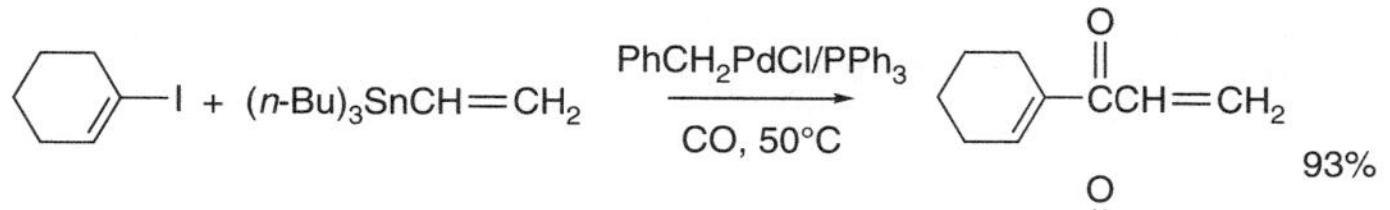

93%

3[c]

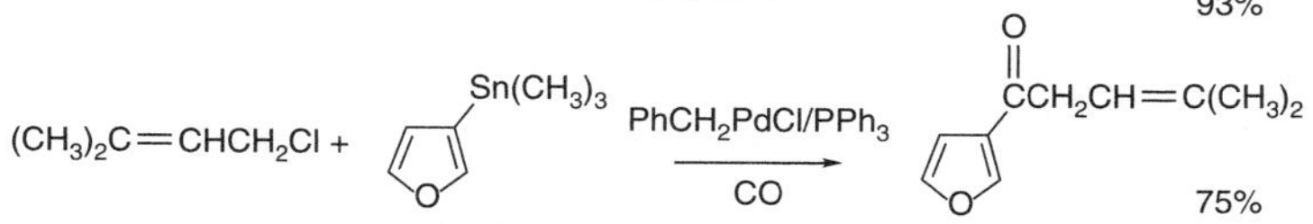

75%

4[d]

85%

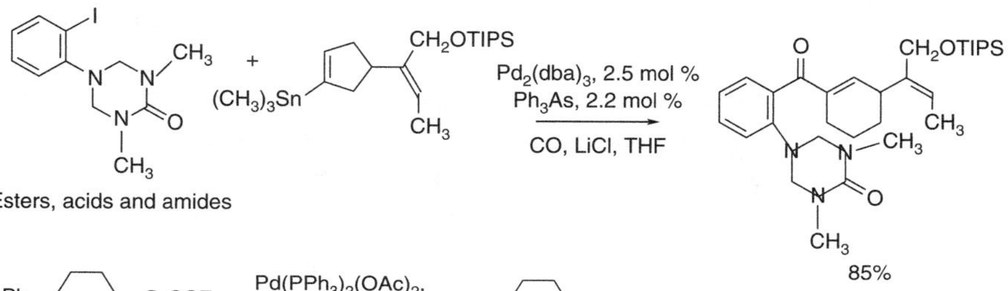

B. Esters, acids and amides

5[e]

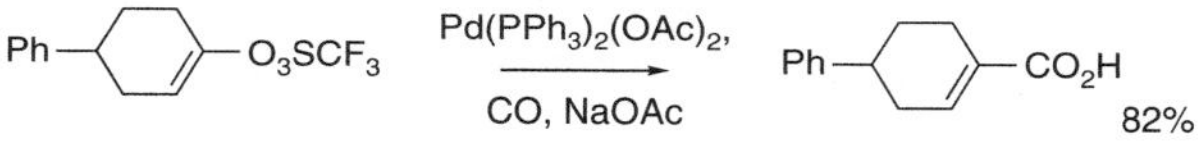

82%

6[f]

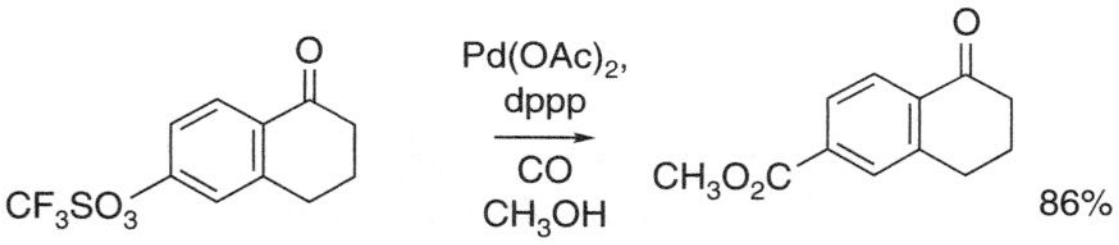

86%

7[g]

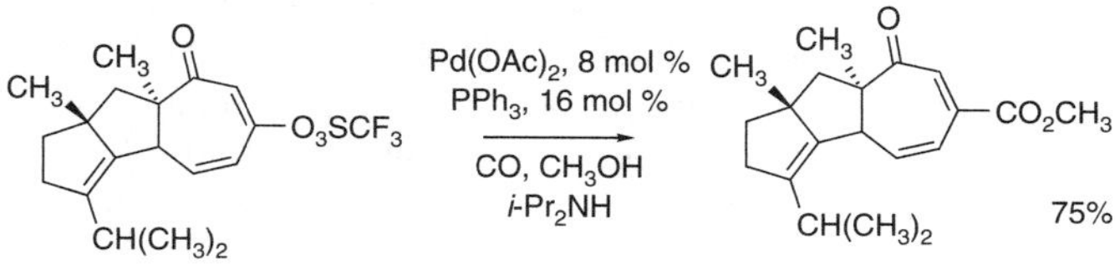

75%

8[h]

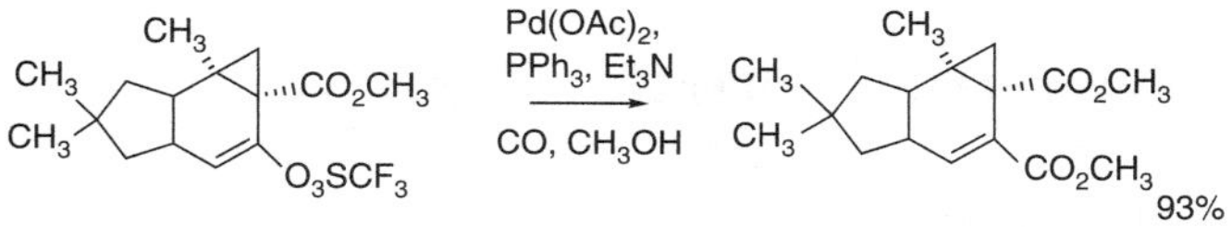

93%

9[i]

83%

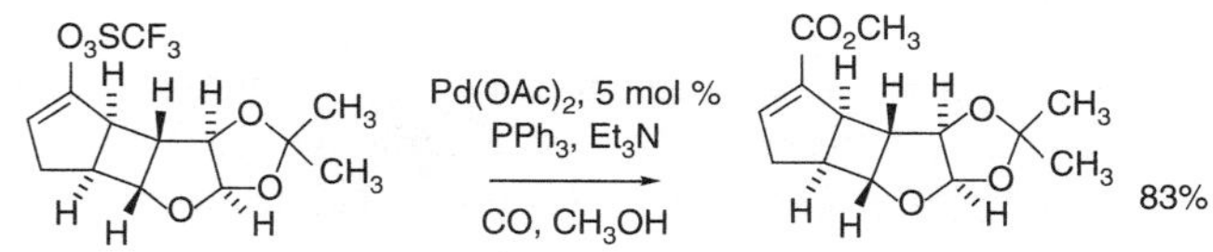

10[j]

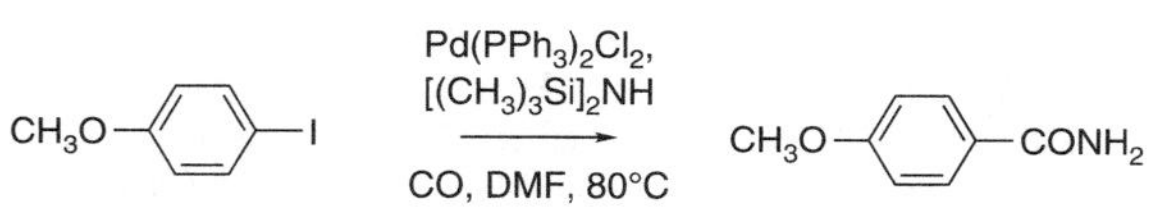

(*Continued*)

a. T. Ishiyama, H. Kizaki, T. Hayashi, A. Suzuki, and N. Miyaura, *J. Org. Chem.*, **63**, 4726 (1998).
b. W. F. Goure, M. E. Wright, P. D. Davis, S. S. Labadie, and J. K. Stille, *J. Am. Chem. Soc.*, **106**, 6417 (1984).
c. F. K. Sheffy, J. P. Godschalx, and J. K. Stille, *J. Am. Chem. Soc.*, **106**, 4833 (1984).
d. S. R. Angle, J. M. Fervig, S. D. Knight, R. W. Marquis, Jr., and L. E. Overman, *J. Am. Chem. Soc.*, **115**, 3966 (1993).
e. S. Cacchi and A. Lupi, *Tetrahedron Lett.*, **33**, 3939 (1992).
f. U. Gerlach and T. Wollmann, *Tetrahedron Lett.*, **33**, 5499 (1992).
g. B. B. Snider, N. H. Vo, and S. V. O'Neill, *J. Org. Chem.*, **63**, 4732 (1998).
h. S. K. Thompson and C. H. Heathcock, *J. Org. Chem.*, **55**, 3004 (1990).
i. A. B. Smith III, G. A. Sulikowski, M. M. Sulikowski, and K. Fujimoto, *J. Am. Chem. Soc.*, **114**, 2567 (1992).
j. E. Morea and G. Ortar, *Tetrahedron Lett.*, **39**, 2835 (1998).

Ref. 253

It can also be done by in situ generation of other types of electrophiles. For example, good yields of *N*-acyl α-amino acids are formed in a process in which an amide and aldehyde combine to generate a carbinolamide and, presumably, an acyliminium ion. The organopalladium intermediate is then carbonylated prior to reaction with water.[254]

Scheme 8.15 gives some examples of carbonylations and acylations involving stannane reagents. Entry 1 illustrates synthesis of diaryl ketones from aryl halides and arylboronic acids. Entries 2 and 3 use stannanes as the nucleophilic reactant. Entry 4 was carried out as part of the synthesis of the *Strychnos* alkaloid akuammicine. The triazinone ring serves to protect the aromatic amino group. Entries 5 and 6 introduce carboxy groups using vinyl and aryl triflates, respectively. Entries 8 and 9 are similar reactions carried out during the course of multistage syntheses. Entry 10 illustrates direct formation of an amide by carbonylation.

8.3. Reactions Involving Other Transition Metals

8.3.1. Organonickel Compounds

The early synthetic processes using organonickel compounds involved the coupling of allylic halides, which react with nickel carbonyl, $Ni(CO)_4$, to give π-allyl complexes. These complexes react with a variety of halides to give coupling products.[255]

253. R. Grigg, P. Kennewall, and A. J. Teasdale, *Tetrahedron Lett.*, **33**, 7789 (1992).
254. M. Beller, M. Eckert, F. M. Vollmuller, S. Bogdanovic, and H. Geissler, *Angew. Chem. Int. Ed. Engl.*, **36**, 1494 (1997); M. Beller, W. A. Maradi, M. Eckert, and H. Neumann, *Tetrahedron Lett.*, **40**, 4523 (1999).
255. M. F. Semmelhack, *Org. React.*, **19**, 115 (1972).

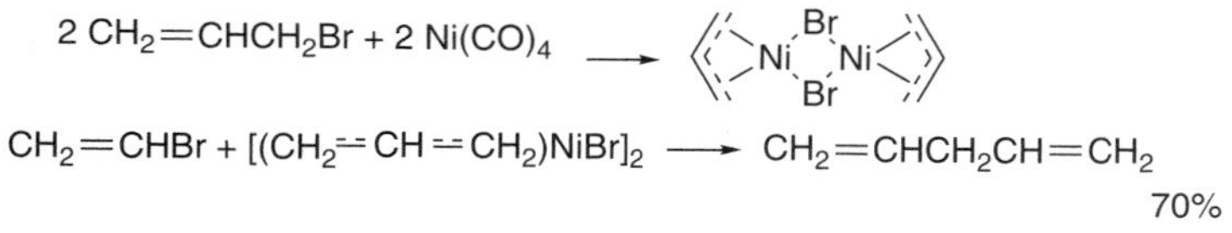

$$2\ CH_2{=}CHCH_2Br + 2\ Ni(CO)_4 \longrightarrow$$

$$CH_2{=}CHBr + [(CH_2{=}CH{=}CH_2)NiBr]_2 \longrightarrow CH_2{=}CHCH_2CH{=}CH_2$$

70% Ref. 256

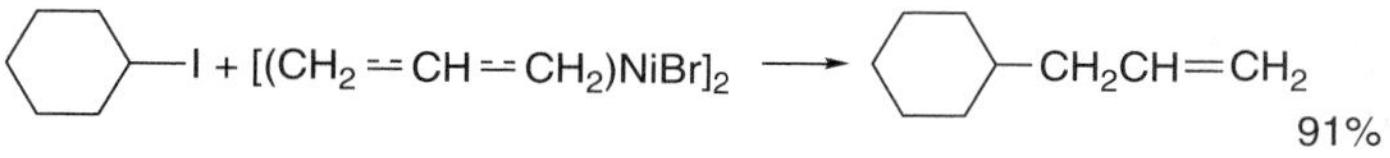

91%

Nickel carbonyl effects coupling of allylic halides when the reaction is carried out in very polar solvents such as DMF or DMSO. This coupling reaction has been used intramolecularly to bring about cyclization of *bis*-allylic halides and was found useful in the preparation of large rings.

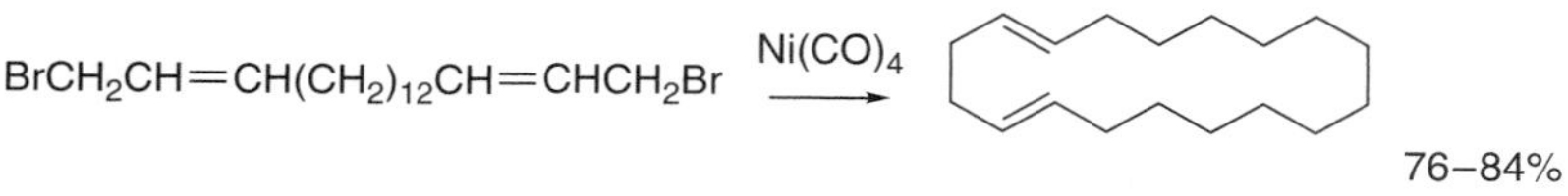

$$BrCH_2CH{=}CH(CH_2)_{12}CH{=}CHCH_2Br \xrightarrow{Ni(CO)_4}$$

76–84%

Ref. 257

$$BrCH_2CH{=}CHCH_2CH_2C \atop BrCH_2CH{=}CHCH_2CH_2CH_2 \xrightarrow{Ni(CO)_4}$$

70–75%

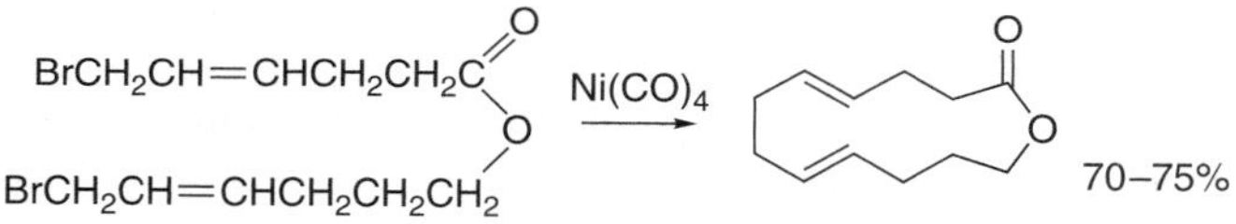

Ref. 258

Nickel carbonyl is an extremely toxic substance, but a number of other nickel reagents with generally similar reactivity can be used in its place. The Ni(0) complex of 1,5-cyclooctadiene, Ni(COD)$_2$, can effect coupling of allylic, alkenyl, and aryl halides.

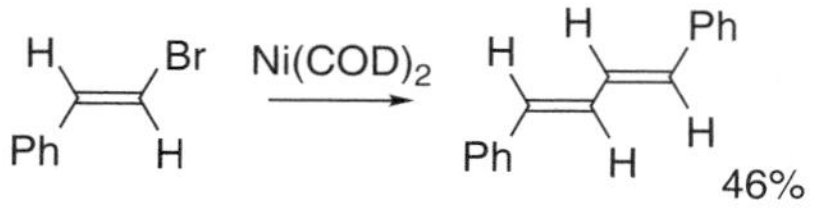

$$\xrightarrow{Ni(COD)_2}$$

46%

Ref. 259

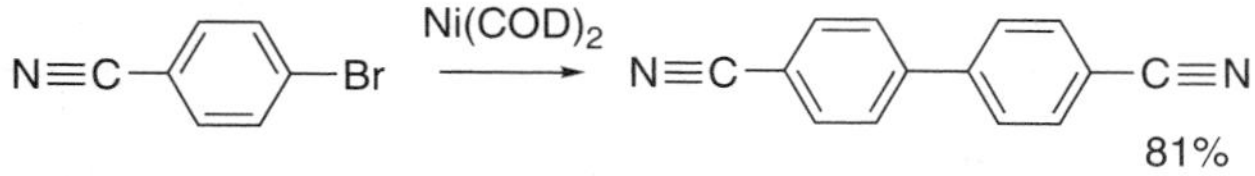

$$N{\equiv}C{-}\!\!\!\bigcirc\!\!\!{-}Br \xrightarrow{Ni(COD)_2} N{\equiv}C{-}\!\!\!\bigcirc\!\!\!{-}\!\!\!\bigcirc\!\!\!{-}C{\equiv}N$$

81% Ref. 260

Tetrakis-(triphenylphosphine)nickel(0) is an effective reagent for coupling aryl halides,[261] and medium rings can be formed in intramolecular reactions.

256. E. J. Corey and M. F. Semmelhack, *J. Am. Chem. Soc.*, **89**, 2755 (1967).
257. E. J. Corey and E. K. W. Wat, *J. Am. Chem. Soc.*, **89**, 2757 (1967).
258. E. J. Corey and H. A. Kirst, *J. Am. Chem. Soc.*, **94**, 667 (1972).
259. M. F. Semmelhack, P. M. Helquist, and J. D. Gorzynski, *J. Am. Chem. Soc.*, **94**, 9234 (1972).
260. M. F. Semmelhack, P. M. Helquist, and L. D. Jones, *J. Am. Chem. Soc.*, **93**, 5908 (1971).
261. A. S. Kende, L. S. Liebeskind, and D. M. Braitsch, *Tetrahedron Lett.*, 3375 (1975).

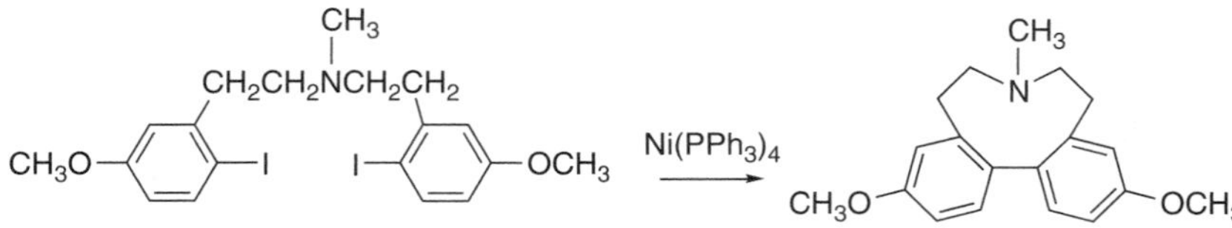

Ref. 262

The homocoupling of aryl halides and triflates can be made catalytic in nickel by using zinc as a reductant for in situ regeneration of the active Ni(0) species.

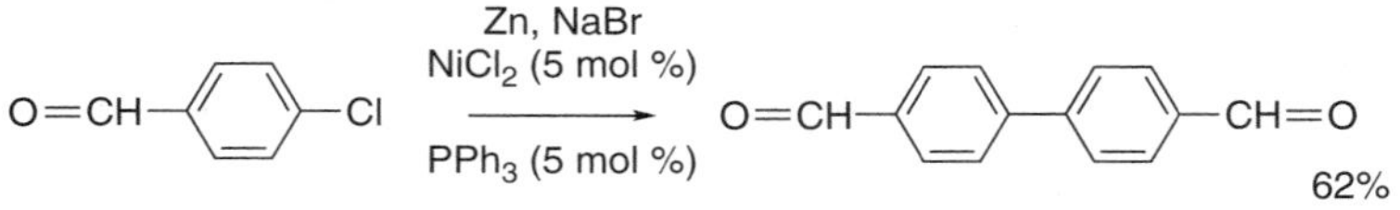

62%

Ref. 263

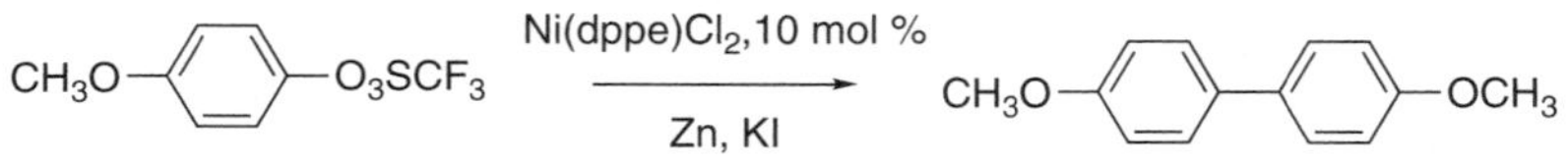

Ref. 265

Mechanistic study of the aryl couplings has revealed the importance of the changes in redox state that are involved in the reaction.[265] Ni(I), Ni(II), and Ni(III) states are believed to be involved. Changes in the degree of coordination by phosphine ligands are also thought to be involved, but these have been omitted in the mechanism shown here. The detailed kinetics of the reaction are inconsistent with a mechanism involving only formation and decomposition of a biarylnickel(II) intermediate. The key aspects of the mechanism are: (1) the oxidative addition involving a Ni(I) species, and (2) the reductive elimination that occurs via a diaryl Ni(III) intermediate and regenerates Ni(I).

initiation by
electron transfer $ArNi(II)X + ArX \longrightarrow ArNi(III)X^+ + Ar\cdot + X^-$

propagation
$$ArNi(III)X^+ + ArNi(II)X \longrightarrow Ar_2Ni(III)X + Ni(II)^+ + X^-$$
$$Ar_2Ni(III)X \longrightarrow Ar\text{-}Ar + Ni(I)X$$
$$Ni(I)X + ArX \longrightarrow ArNi(III)X^+ + X^-$$

Nickel(II) salts are able to catalyze the coupling of Grignard reagents with alkenyl and aryl halides. A soluble *bis*-phosphine complex, $Ni(dppe)_2Cl_2$, is a particularly effective catalyst.[266] The main distinction between this reaction and Pd-catalyzed cross

262. S. Brandt, A. Marfat, and P. Helquist, *Tetrahedron Lett.*, 2193 (1979).

263. M. Zembayashi, K. Tamao, J. Yoshida, and M. Kumada, *Tetrahedron Lett.*, 4089 (1977); I. Colon and D. R. Kelly, *J. Org. Chem.*, **51**, 2627 (1986).

265. A. Jutand and A. Mosleh, *J. Org. Chem.*, **62**, 261 (1997).

265. T. T. Tsou and J. K. Kochi, *J. Am. Chem. Soc.*, **101**, 7547 (1979); L. S. Hegedus and D. H. P. Thompson, *J. Am. Chem. Soc.*, **107**, 5663 (1985); C. Amatore and A. Jutand, *Organometallics*, **7**, 2203 (1988).

266. K. Tamao, K. Sumitani, and M. Kumada, *J. Am. Chem. Soc.*, **94**, 4374 (1972).

coupling is that the nickel reaction can be more readily extended to saturated alkyl groups because of a reduced tendency toward β-elimination.

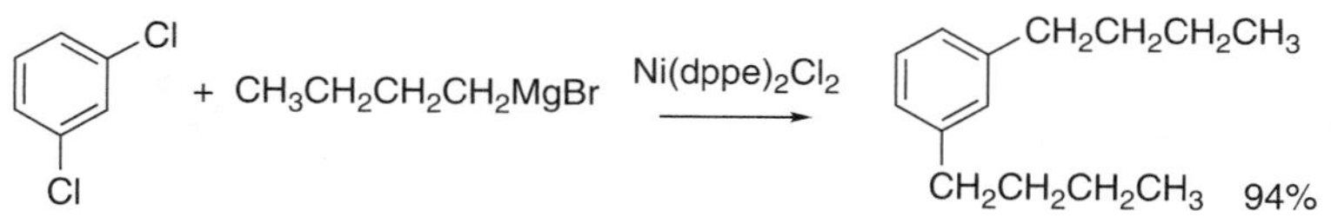

The reaction has been applied to the synthesis of cyclophane-type structures by use of dihaloarenes and Grignard reagents from α, ω-dihalides.

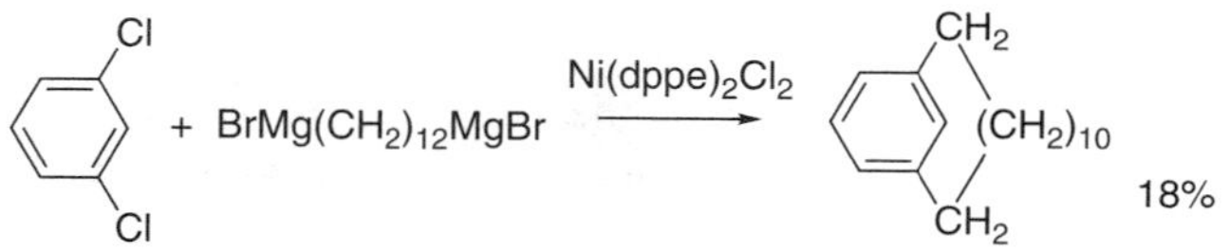

Ref. 267

Recent discoveries have expanded the utility of nickel-catalyzed coupling reactions. Inclusion of butadiene greatly improves the efficiency of the reactions.[268]

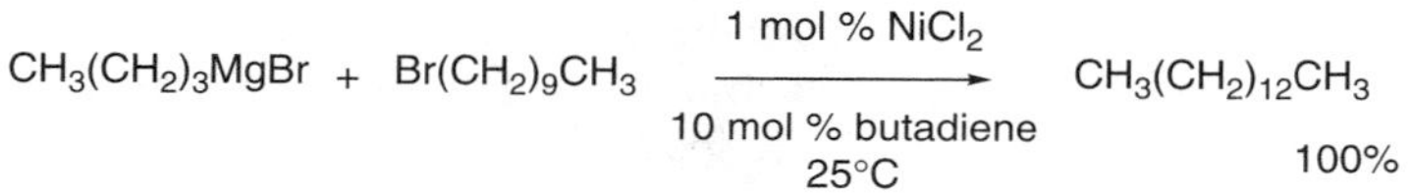

These reaction conditions are applicable to primary chlorides, bromides, and tosylates. The active catalytic species appears to be a *bis-π-allyl* complex formed by dimerization of butadiene.

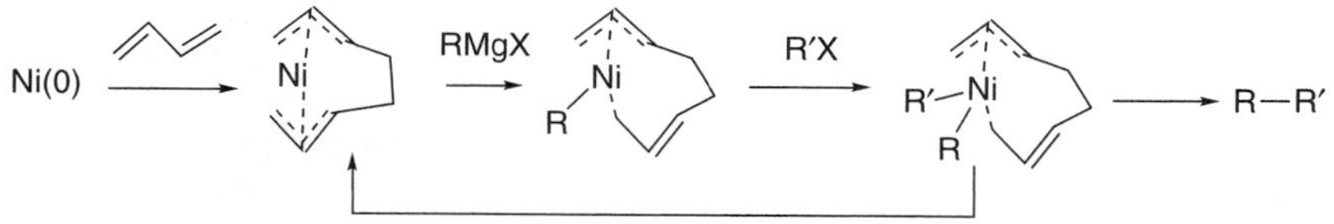

A preparation of Ni(II) on charcoal can also be used as the catalyst. It serves as a reservoir of active Ni(0) formed by reduction by the Grignard reagent.[269]

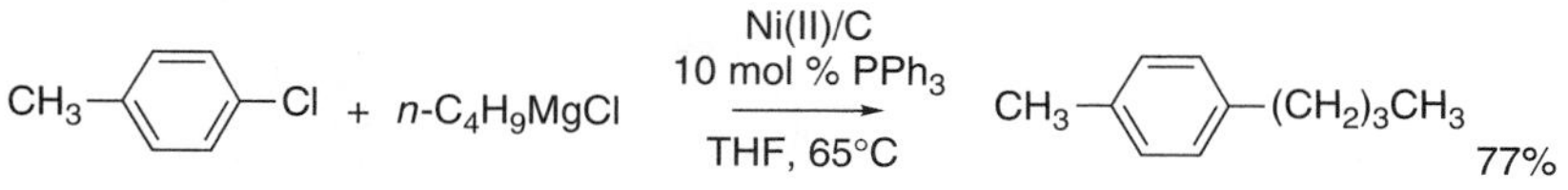

Aryl carbamates are also reactive toward nickel-catalyzed coupling.[270] Since the carbamates can be readily prepared from phenols, they are convenient starting materials.

267. K. Tamno, S. Kodama, T. Nakatsuka, Y. Kiso, and A. Kumada, *J. Am. Chem. Soc.*, **97**, 4405 (1975).
268. J. Terao, H. Watanabe, A. Ikumi, H. Kuniyasu, and N. Kambe, *J. Am. Chem. Soc.*, **124**, 4222 (2002).
269. S. Tasler and R. H. Lipshutz, *J. Org. Chem.*, **68**, 1190 (2003).
270. S. Sengupta, M. Leite, D. S. Raslan, C. Quesnelle, and V. Snieckus, *J. Org. Chem.*, **57**, 4066 (1992).

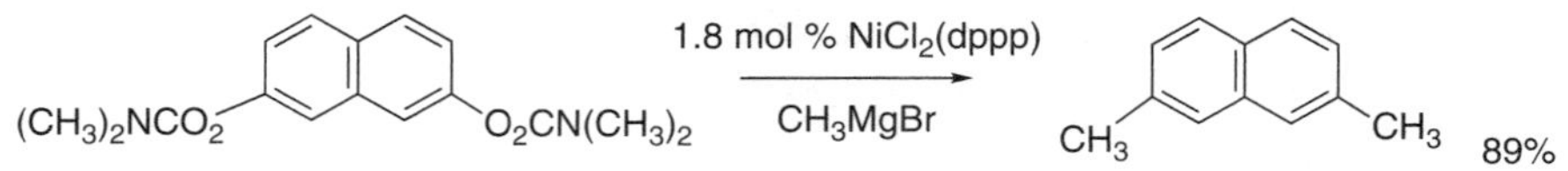

Ref. 271

Vinyl carbamates are also reactive.

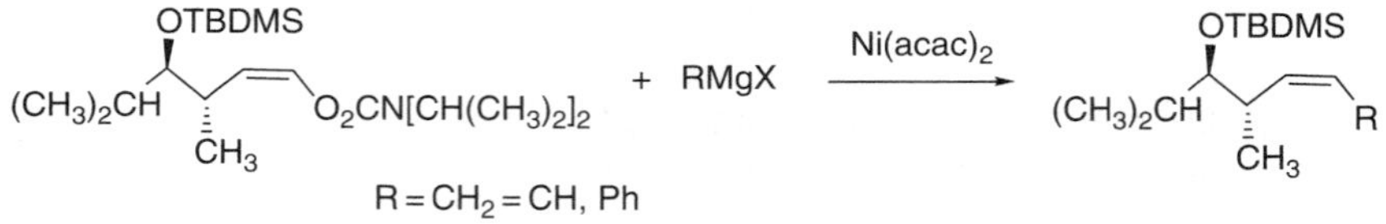

Ref. 272

Similarly, nickel catalysis permits the extension of cross coupling to vinyl phosphates, which are in some cases more readily obtained and handled than vinyl triflates.[273]

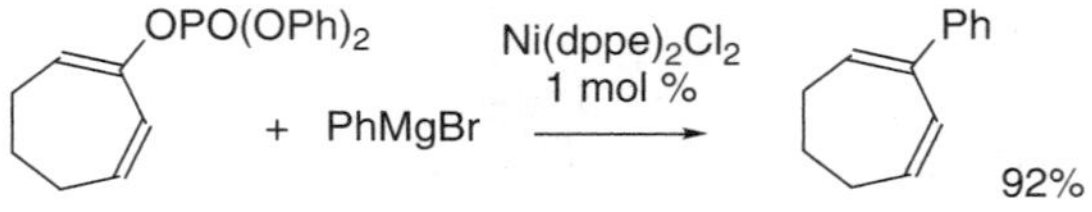

Nickel acetylacetonate, $Ni(acac)_2$, in the presence of a styrene derivative promotes coupling of primary alkyl iodides with organozinc reagents. The added styrene serves to stabilize the active catalytic species, and of the derivatives examined, m-trifluoromethylstyrene was the best.[274]

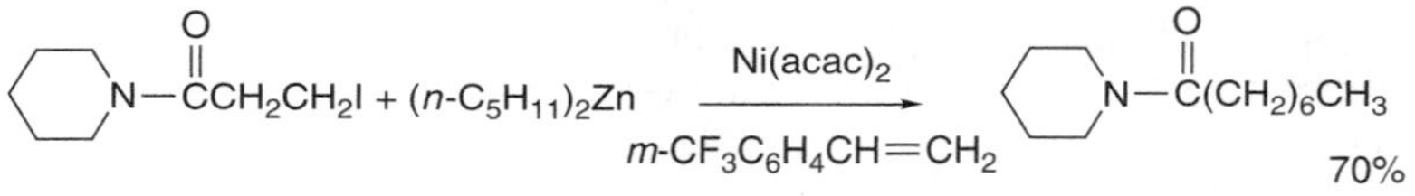

This method can extend Ni-catalyzed cross coupling to functionalized organometallic reagents.

Nickel can also be used in place of Pd in Suzuki-type couplings of boronic acids. The main advantage of nickel in this application is that it reacts more readily with aryl chlorides[275] and methanesulfonates[276] than do the Pd systems. These reactants may be more economical than iodides or triflates in large-scale syntheses.

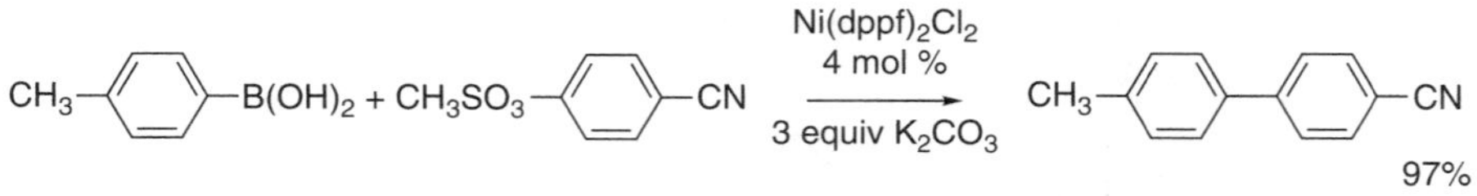

271. C. Dallaire, I. Kolber, and M. Gringas, Org. Synth., 78, 42 (2002).

272. F.-H. Poree, A. Clavel, J.-F. Betzer, A. Pancrazi, and J. Ardisson, Chem. Eur. J., 7553 (2003).

273. A. Sofia, E. Karlstom, K. Itami, and J.-E. Backvall, J. Org. Chem., 64, 1745 (1999); Y. Nan and Z. Yang, Tetrahedron Lett., 40, 3321 (1999).

274. R. Giovannini, T. Studemann, G. Dussin, and P. Knochel, Angew. Chem. Int. Ed. Engl., 37, 2387 (1998); R. Giovannini, T. Studemann, A. Devasagayaraj, G. Dussin, and P. Knochel, J. Org. Chem., 64, 3544 (1999).

275. S. Saito, M. Sakai, and N. Miyaura, Tetrahedron Lett., 37, 2993 (1996); S. Sato, S. Oh-tani, and N. Miyaura, J. Org. Chem., 62, 8024 (1997).

276. V. Percec, J.-Y. Bae, and D. H. Hill, J. Org. Chem., 60, 1060 (1995); M. Ueda, A. Saitoh, S. Oh-tani, and N. Miyaura, Tetrahedron, 54, 13079 (1998).

$NiCl_2[P(c\text{-}C_6H_{11})_3]_2$ is an effective catalyst for coupling aryl tosylates with arylboronic acids.[277]

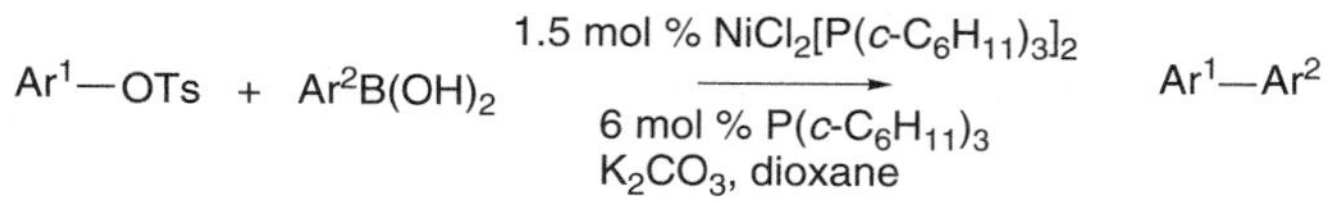

Nickel catalysis has been used in a sequential synthesis of terphenyls, starting with 2-, 3-, or 4-bromophenyl neopentanesulfonates. Conventional Pd-catalyzed Suzuki conditions were used for the first step involving coupling of the bromide and then nickel catalysis was utilized for coupling the sulfonate.

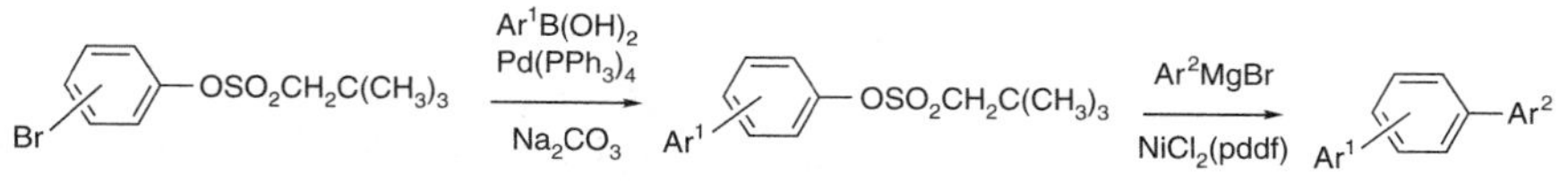

Ref. 278

These coupling reactions can also be done with boronate esters activated by conversion to "ate" reagents by reaction with alkyllithium compounds.[279] For example, analogs of leukotrienes have been synthesized in this way.

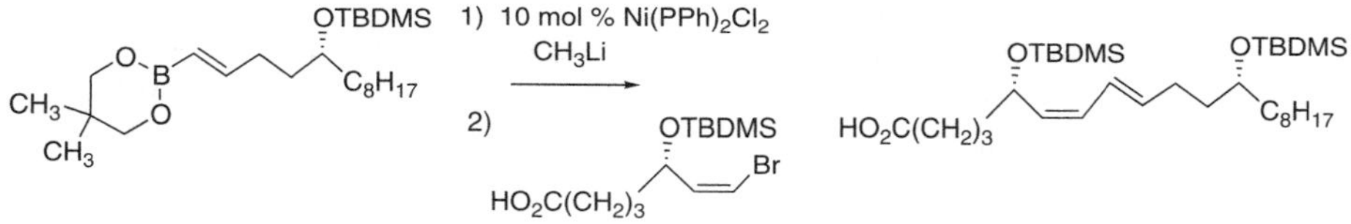

Ref. 280

8.3.2. Reactions Involving Rhodium and Cobalt

Rhodium and cobalt participate in several reactions that are of value in organic syntheses. Rhodium and cobalt are active catalysts for the reaction of alkenes with hydrogen and carbon monoxide to give aldehydes, known as *hydroformylation*.[281]

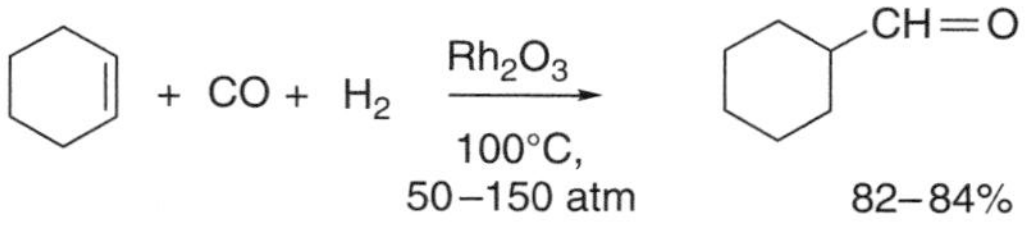

Ref. 282

277. D. Zim, V. R. Lando, J. Dupont, and A. L. Monteiro, *Org. Lett.*, **3**, 3049 (2001).
278. C.-H. Cho, I.-S. Kim, and K. Park, *Tetrahedron*, **60**, 4589 (2004).
279. Y. Kobayashi, Y. Nakayama, and R. Mizojiri, *Tetrahedron*, **54**, 1053 (1998).
280. Y. Nakayama, G. B. Kumar, and Y. Kobayashi, *J. Org. Chem.*, **65**, 707 (2000).
281. R. L. Pruett, *Adv. Organometal. Chem.*, **17**, 1 (1979); H. Siegel and W. Himmele, *Angew. Chem. Int. Ed. Engl.*, **19**, 178 (1980); J. Falbe, *New Syntheses with Carbon Monoxide*, Springer Verlag, Berlin, 1980.
282. P. Pino and C. Botteghi, *Org. Synth.*, **57**, 11 (1977).

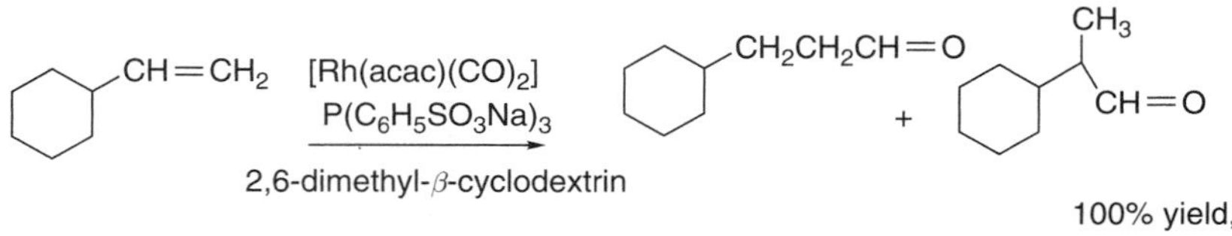

100% yield, 3.2:1 ratio

Ref. 283

The key steps in the reaction are addition of hydridorhodium to the double bond of the alkene and migration of the alkyl group to the complexed carbon monoxide. Hydrogenolysis then leads to the aldehyde.

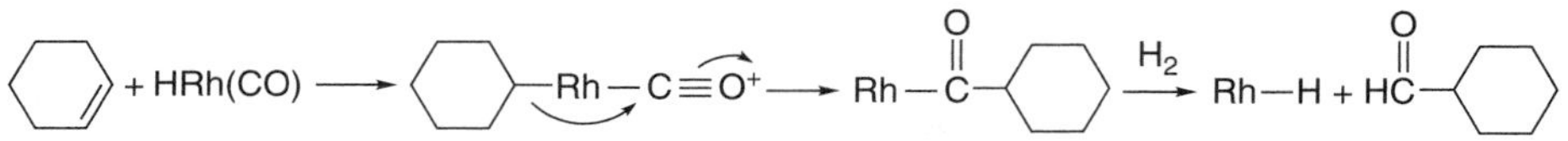

Carbonylation can also be carried out under conditions in which the acylrhodium intermediate is trapped by internal nucleophiles.

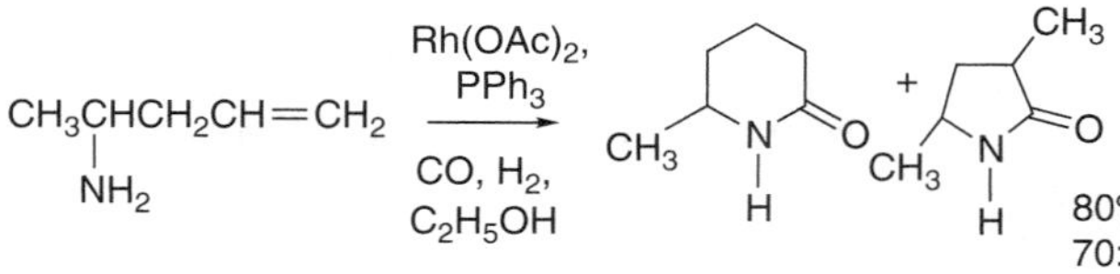

80% yield, 70:30 ratio

Ref. 284

The steps in the hydroformylation reaction are closely related to those that occur in the *Fischer-Tropsch process*, which is the reductive conversion of carbon monoxide to alkanes and occurs by a repetitive series of carbonylation, migration, and reduction steps that can build up a hydrocarbon chain.

$$M + CO \longrightarrow M{-}CO \xrightarrow{+H_2} M{-}CH_3 \xrightarrow{+CO} OC{-}M{-}CH_3$$

$$OC{-}M{-}CH_3 \longrightarrow M{-}\overset{\displaystyle O}{\overset{\|}{C}}{-}CH_3 \xrightarrow{+H_2} M{-}CH_2CH_3$$

$$M{-}CH_2CH_3 \xrightarrow{CO} M{-}\overset{\displaystyle O}{\overset{\|}{C}}{-}CH_2CH_3 \xrightarrow{+H_2} M{-}CH_2CH_2CH_3 \text{ etc.}$$

The Fischer-Tropsch process is of considerable economic interest because it is the basis of conversion of carbon monoxide to synthetic hydrocarbon fuels, and extensive work has been done on optimization of catalyst systems.

The carbonylation step that is involved in both hydroformylation and the Fischer-Tropsch reaction can be reversible. Under appropriate conditions, rhodium catalyst can be used for the decarbonylation of aldehydes[285] and acyl chlorides.[286]

$$R\overset{\displaystyle O}{\overset{\|}{C}}H + Rh(PPh_3)_3Cl \longrightarrow RH \qquad R\overset{\displaystyle O}{\overset{\|}{C}}Cl + Rh(PPh_3)_3Cl \longrightarrow RCl$$

283. E. Monflier, S. Tilloy, G. Fremy, Y. Castanet, and A. Mortreux, *Tetrahedron Lett.*, **36**, 9481 (1995).
284. D. Anastasiou and W. R. Jackson, *Tetrahedron Lett.*, **31**, 4795 (1990).
285. J. A. Kampmeier, S. H. Harris, and D. K. Wedgaertner, *J. Org. Chem.*, **45**, 315 (1980); J. M. O'Connor and J. Ma, *J. Org. Chem.*, **57**, 5074 (1992).
286. J. K. Stille and M. T. Regan, *J. Am. Chem. Soc.*, **96**, 1508 (1974); J. K. Stille and R. W. Fries, *J. Am. Chem. Soc.*, **96**, 1514 (1974).

An acylrhodium intermediate is involved in both cases. The elimination of the hydrocarbon or halide occurs by reductive elimination.[287]

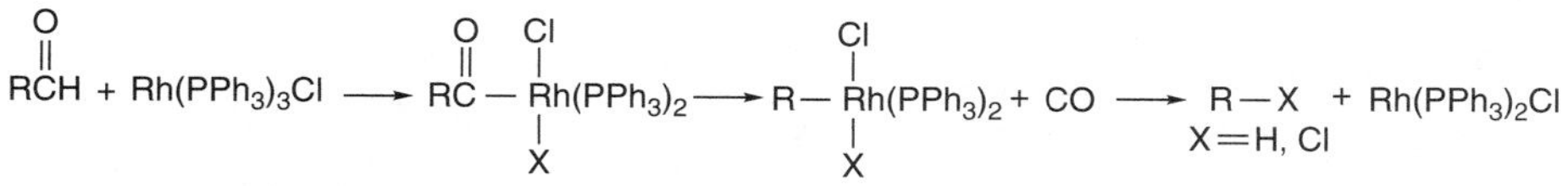

Although the very early studies of transition metal–catalyzed coupling of organometallic reagents included cobalt salts, the use of cobalt for synthetic purposes is quite limited. Vinyl bromide and iodides couple with Grignard reagents in good yield, but a good donor ligand such as NMP or DMPU is required as a cocatalyst.

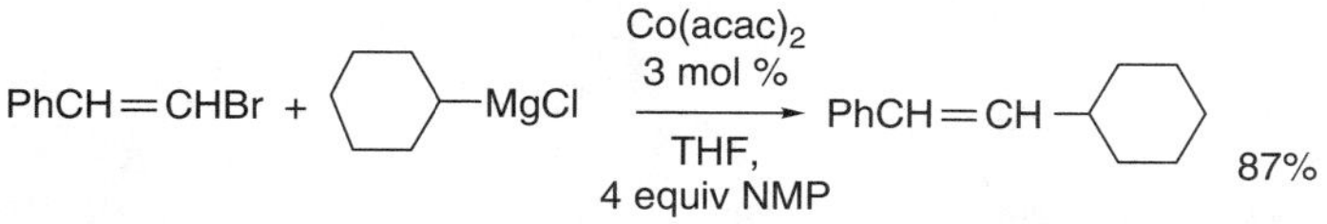

Ref. 288

$Co(acac)_2$ also catalyzes cross coupling of organozinc reagents under these conditions.[289]

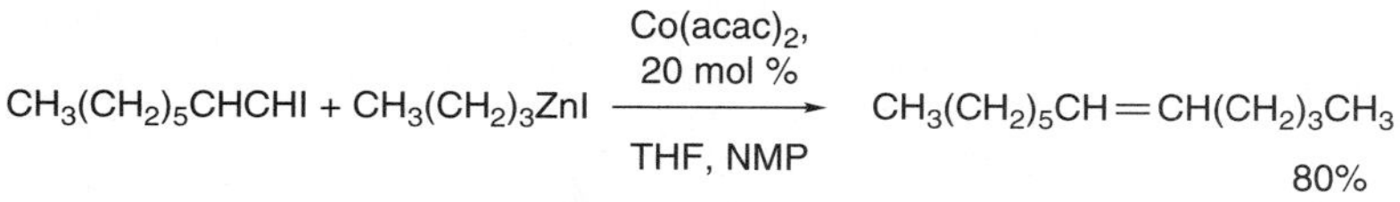

8.4. The Olefin Metathesis Reaction

Several transition metal complexes can catalyze the exchange of partners of two double bonds. Known as the *olefin metathesis reaction*, this process can be used to close or open rings, as well to interchange double-bond components.

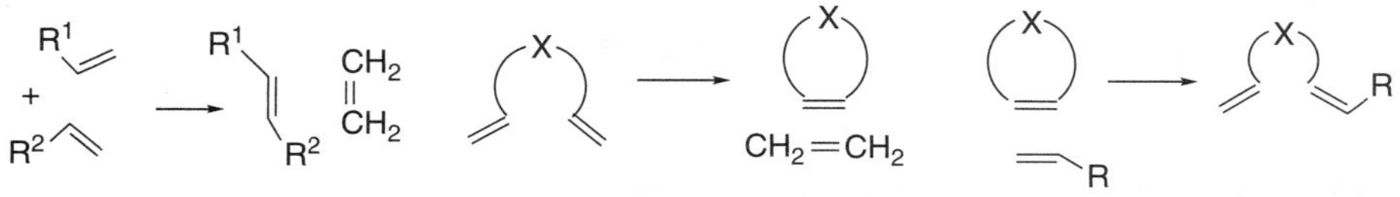

Intermolecular metathesis   Ring-closing metathesis   Ring-opening metathesis

[287.] J. E. Baldwin, T. C. Barden, R. L. Pugh, and W. C. Widdison, *J. Org. Chem.*, **52**, 3303 (1987).
[288.] G. Cahiez and H. Avedissian, *Tetrahedron Lett.*, **39**, 6159 (1998).
[289.] H. Avedissian, L. Berillon, G. Cahiez, and P. Knochel, *Tetrahedron Lett.*, **39**, 6163 (1998).

The catalysts are metal-carbene complexes that react with the alkene to form a metal-locyclobutane intermediate.[290] If the metallocyclobutane breaks down in the alternative path from its formation, an exchange of the double-bond components occurs.

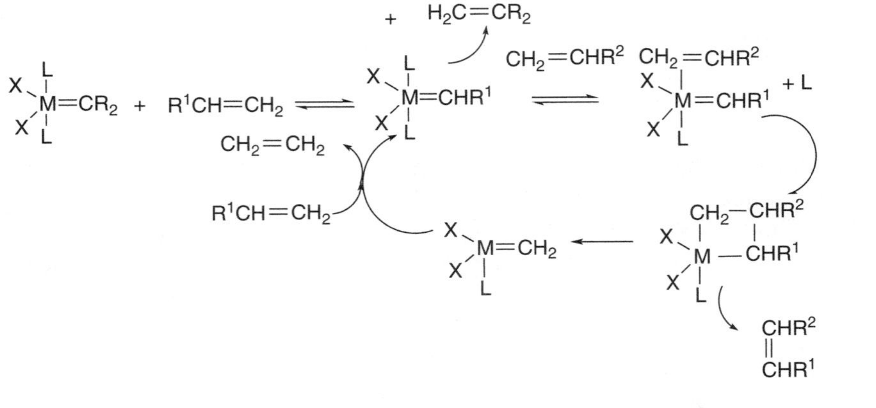

The most commonly used catalyst is the benzylidene complex of $RuCl_2[P(c-C_6H_{11})_3]_2$, **F**, which is called the *Grubbs catalyst*, but several other catalysts are also reactive. Catalyst **H**, which is known as the *second-generation Grubbs catalyst*, is used extensively.

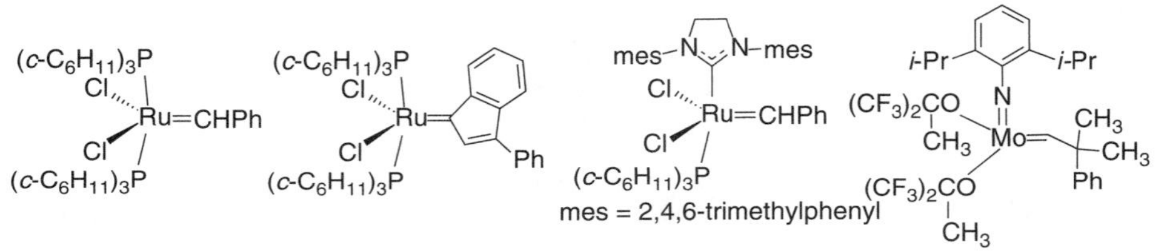

mes = 2,4,6-trimethylphenyl

F[291]   **G**[292]   **H**[293]   **I**[294]

In laboratory synthesis, these catalysts have been utilized primarily to form both common and large rings by coupling two terminal alkenes.[295] For example, catalyst **H** has been used to synthesize the highly oxygenated cyclohexenes known as conduritols.

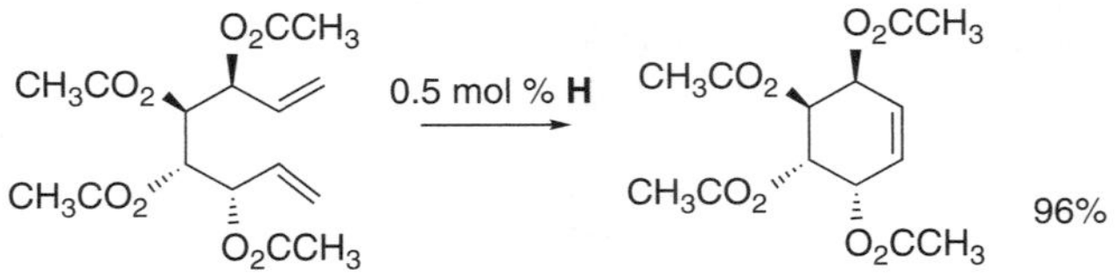

0.5 mol % **H**

96%

290. J.-L. Herisson and Y. Chauvin, *Makromol. Chem.*, **141**, 161 (1971).

291. P. Schwab, R. H. Grubbs, and J. W. Ziller, *J. Am. Chem. Soc.*, **118**, 100 (1996).

292. A. Furstner, M. Liebl, A. F. Hill, and J. D. E. T. Winton-Ely, *Chem. Commun.*, 601 (1999); A. Furstner, O. Guth, A. Duffels, G. Seidel, M. Liebl, B. Gabor, and R. Mynott, *Chem. Eur. J.*, **7**, 4811 (2001).

293. M. Scholl, T. M. Trnka, J. P. Morgan, and R. H. Grubbs, *Tetrahedron Lett.*, **40**, 2247 (1999); J. A. Love, M. S. Sanford, M. W. Day, and R. H. Grubbs, *J. Am. Chem. Soc.*, **125**, 10103 (2003).

294. R. R. Schrock, J. S. Murdzek, G. C. Bazan, J. Robbins, M. Di Mare, and M. O'Regan, *J. Am. Chem. Soc.*, **112**, 3875 (1999).

295. D. L. Wright, *Curr. Org. Chem.*, **3**, 211 (1999); A. Deiters and S. F. Martin, *Chem. Rev.*, **104**, 2199 (2004).

Various heterocyclic rings can be closed, as in the formation of an α, β-unsaturated lactone ring in the synthesis of peloruside A (see also Entries 3 and 5 of Scheme 8.16).

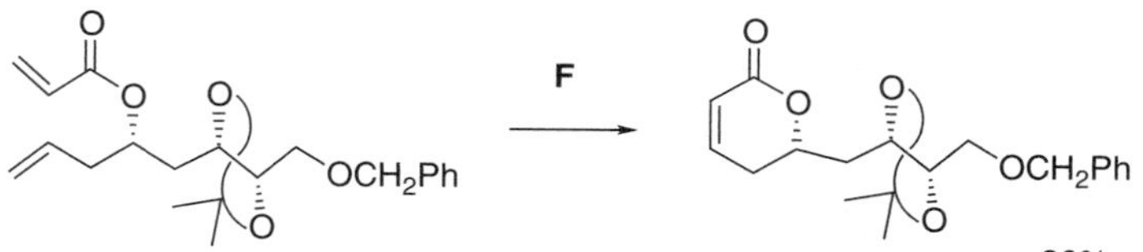

90%　　Ref. 296

Some of the most impressive successes have come in the synthesis of large rings. Several research groups employed the ring-closing metathesis reaction in the synthesis of epothilone and analogs (see Entry 8 of Scheme 8.14).[297] A large ring incorporating a tetrasaccharide unit was synthesized in essentially quantitative yield using either catalyst **F** or **G**. The newly formed double bond is 9:1 *E:Z*.

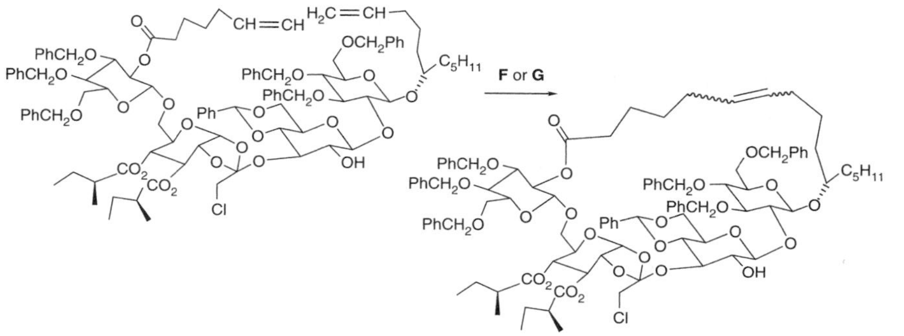

Ref. 298

Olefin metathesis can also be used in intermolecular reactions.[299] For example, a variety of functionally substituted side chains were introduced by exchange with the terminal double bond in **5**.[300] These reactions gave *E:Z* mixtures.

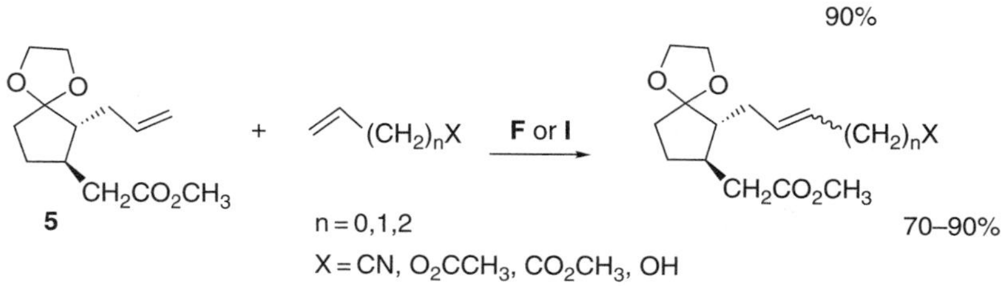

The effectiveness of these intermolecular reactions depends on the relative reactivity of the two components, since self-metathesis leading to dimeric products will occur if one compound is more reactive than the other.

[296.] A. K. Ghosh and J.-H. Kim, *Tetrahedron Lett.*, **44**, 3967 (2003).
[297.] K. C. Nicolaou, H. Vallberg, N. P. King, F. Roschangar, Y. He, D. Vourloumis, and C. G. Nicoloau, *Chem. Eur. J.*, **3**, 1957 (1997); D. Meng, P. Bertinato, A. Balog, D.-S. Su, T. Kamenecka, E. J. Sorensen, and S. J. Danishefsky, *J. Am. Chem. Soc.*, **119**, 10073 (1997); K. Biswas, H. Lin, J. T. Nijardarson, M. D. Chappell, T.-C. Chou, Y. Guan, W. P. Tong, L. He, S. B. Horwitz, and S. J. Danishefsky, *J. Am. Chem. Soc.*, **124**, 9825 (2002).
[298.] A. Furstner, F. Jeanjean, P. Razon, C. Wirtz, and P. Mynott, *Chem. Eur. J.*, 320 (2003).
[299.] S. J. Connon and S. Blechert, *Angew. Chem. Int. Ed. Engl.*, **42**, 1900 (2003).
[300.] O. Brummer, A. Ruckert, and S. Blechert, *Chem. Eur. J.*, **3**, 441 (1997).

Triple bonds can also participate in the metathesis reaction. Intramolecular reactions give vinylcycloalkenes, whereas intermolecular reactions provide conjugated dienes.[301] The mechanism is similar to that for α, ω-diene metathesis, but in contrast to diene cyclization, no carbon atoms are lost.[302]

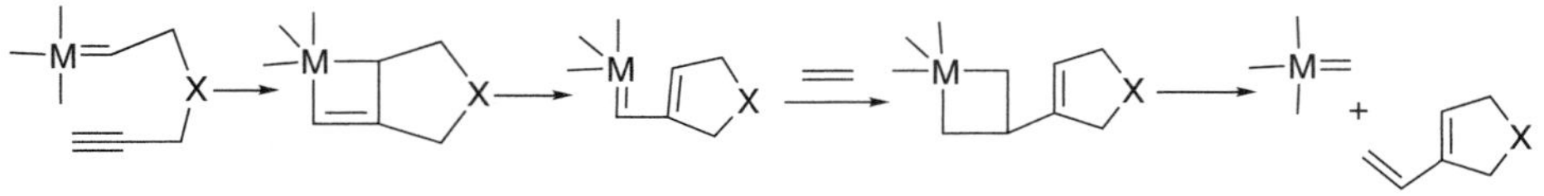

Intramolecular alkene-alkyne metathesis

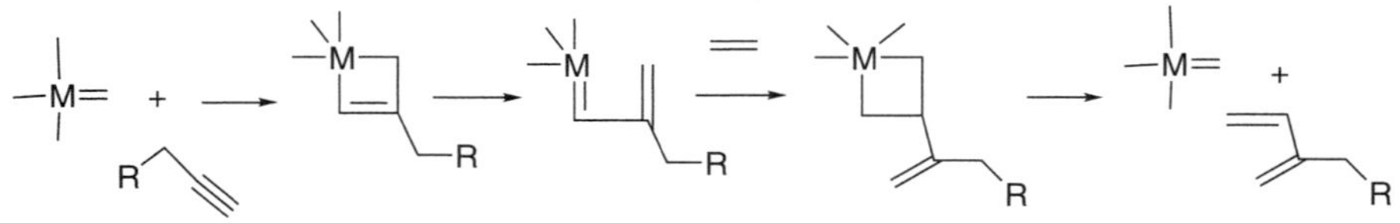

Intermolecular alkene-alkyne metathesis

The reaction has been applied in several synthetic contexts. The intermolecular reaction has been used to construct the conjugated diene side chain of mycothiazole, an antibiotic isolated from a sponge.

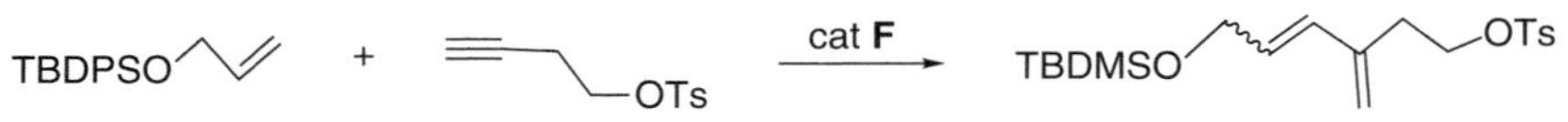

1.1:1 *E:Z* mixture

Ref. 303

The intermolecular version has been used in alkaloid synthesis.

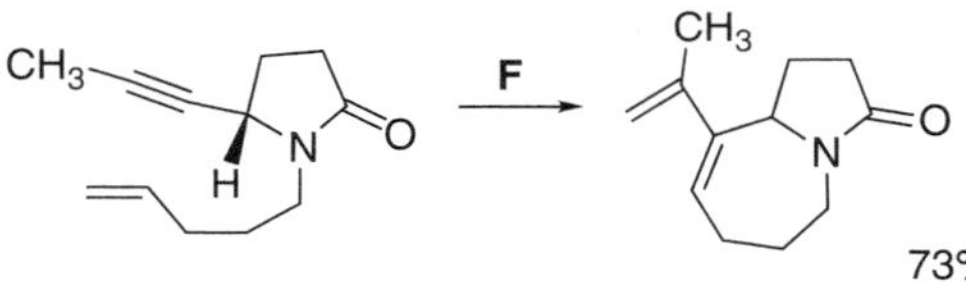

73%   Ref. 304

When the intramolecular version is applied to silyloxyalkynes, the ultimate products are acetyl cycloalkenes.[305]

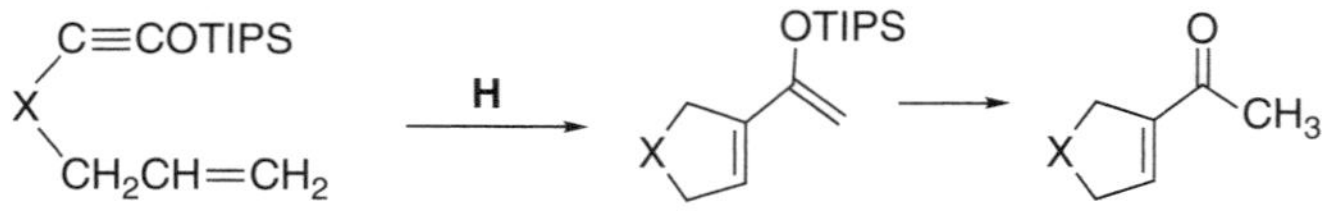

$X = (CH_2)_n, (CH_2)_nO, (CH_2)_nNCO_2CH_3$
or fused ring

This reaction was used to prepare an intermediate suitable for synthesis of the sesquiterpenes α- and β-eremophilane and related structures.[306]

301. S. T. Diver and A. J. Giessert, *Synthesis*, 466 (2004).
302. R. Stragies, M. Schuster, and S. Blechert, *Angew. Chem. Int. Ed. Engl.*, **36**, 2518 (1997).
303. S. Rodriguez-Conesa, P. Candal, C. Jimenez, and J. Rodriguez, *Tetrahedron Lett.*, **42**, 6699 (2001).
304. A. Kinoshita and M. Mori, *J. Org. Chem.*, **61**, 8356 (1996).
305. M. P. Schramm, D. S. Reddy, and S. A. Kozmin, *Angew. Chem. Int. Ed. Engl.*, **40**, 4274 (2001).
306. D. S. Reddy and S. A. Kozmin, *J. Org. Chem.*, **69**, 4860 (2004).

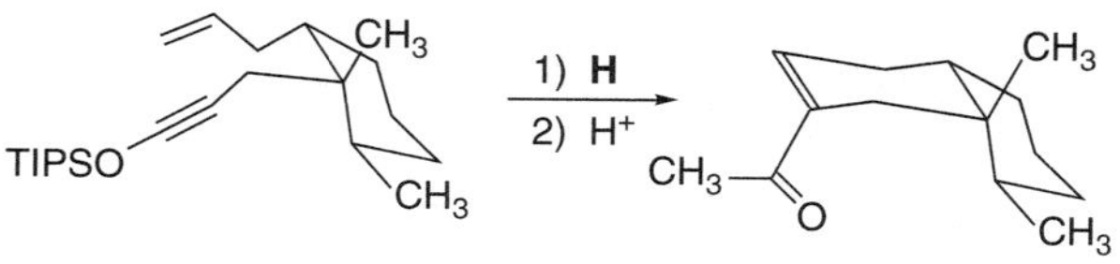

Diynes can be employed in intramolecular ring-closing metathesis. Several catalysts involving Mo and W have been investigated. These cyclizations can be combined with semihydrogenation to give macrocycles with *Z*-double bonds.

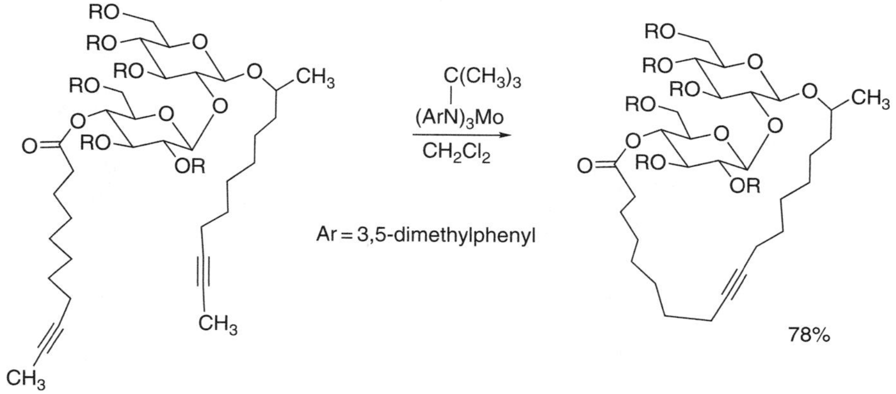

Ar = 3,5-dimethylphenyl

78%

Ref. 307

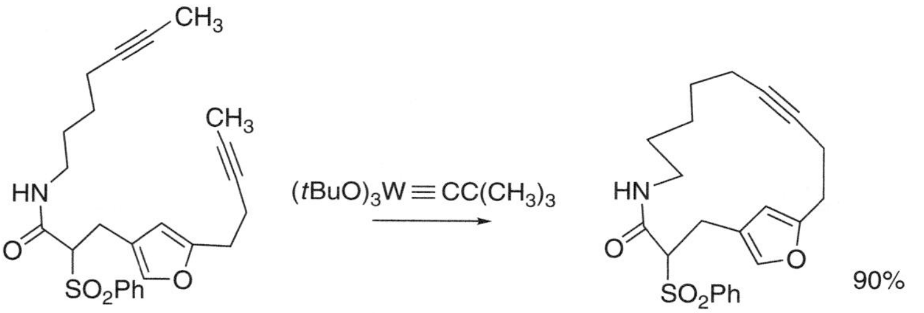

90%   Ref. 308

Scheme 8.16 gives some examples of the synthetic application of the olefin metathesis reaction. Entry 1 is the synthesis of a structure related to a flour beetle aggregation pheromone. Entry 2 was used in the synthesis of a component of sandalwood oil. These two examples illustrate use of the ring-closing metathesis in the synthesis of common rings. Entry 3 forms an α,β-unsaturated lactone and was used in the synthesis of fostriecin, which has anticancer activity. Entry 4 forms a cyclohexenone. Generally, alkenes with EWG substituents have somewhat reduced reactivity and in this case a mild Lewis acid cocatalyst was required. Entry 5 illustrates the synthesis of a medium-sized ring. In this case, catalyst **G** showed a preference for the *E*-double bond but a catalyst similar to **H** formed the *Z*-isomer. This difference was attributed to more rapid reversibility and thermodynamic control in the latter case. Entry 6 also shows the formation of a medium-size ring. Entries 7 and 8 illustrate the application of the ring-closing metathesis to large rings, with Entry 8 being an example of the synthesis of epothilone by this method.

[307.] A. Furstner, O. Guth, A. Rumbo, and S. Seidel, *J. Am. Chem. Soc.*, **121**, 11108 (1999).
[308.] A. Furstner, K. Radkowski, J. Grabowski, C. Wirtz, and R. Mynott, *J. Org. Chem.*, **65**, 8758 (2000).

Scheme 8.16. Examples of the Ring-Closing Olefin Metathesis Reaction

1[a]

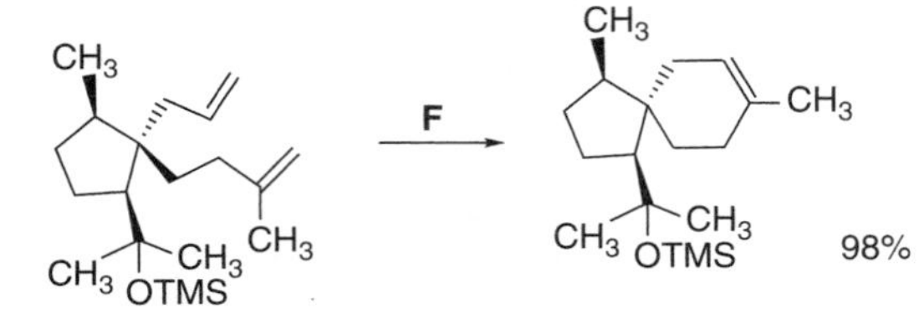

2[b]

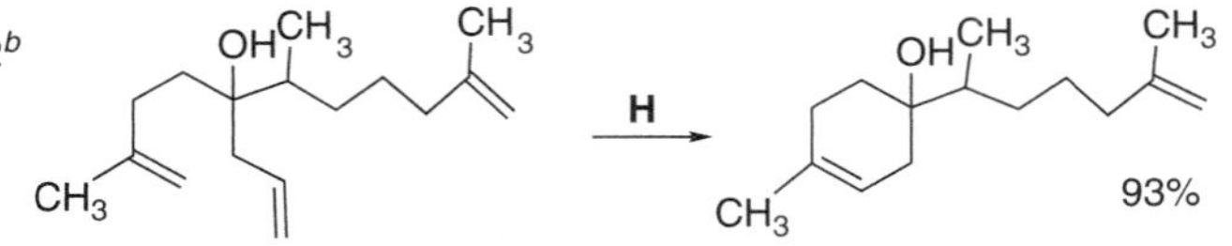

3[c]

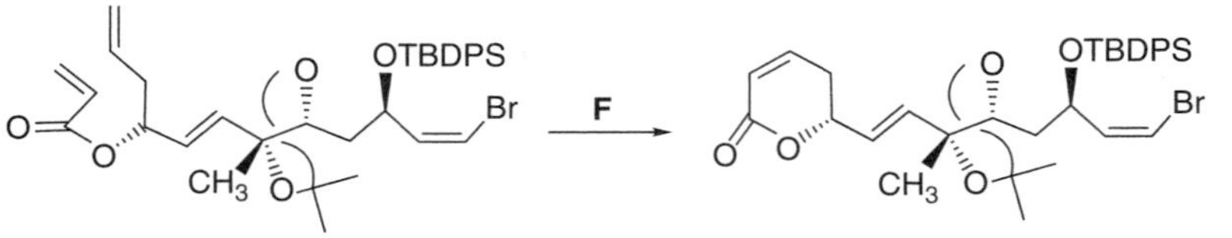

4[d]

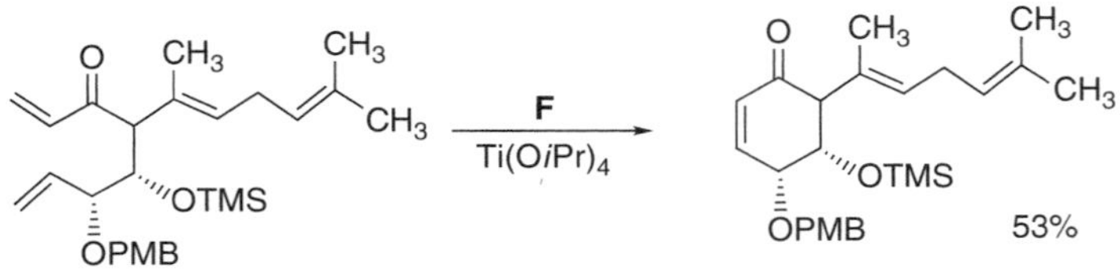

5[e]

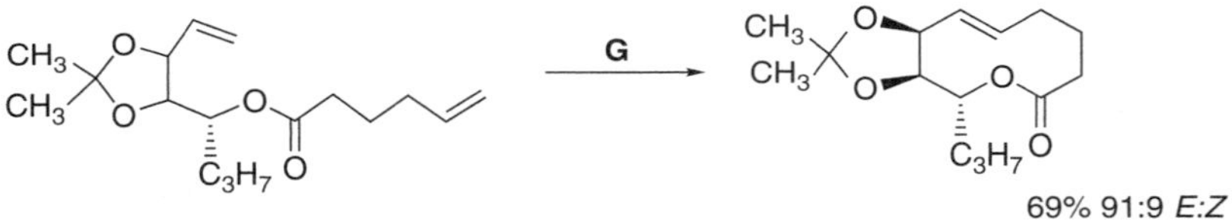

6[f]

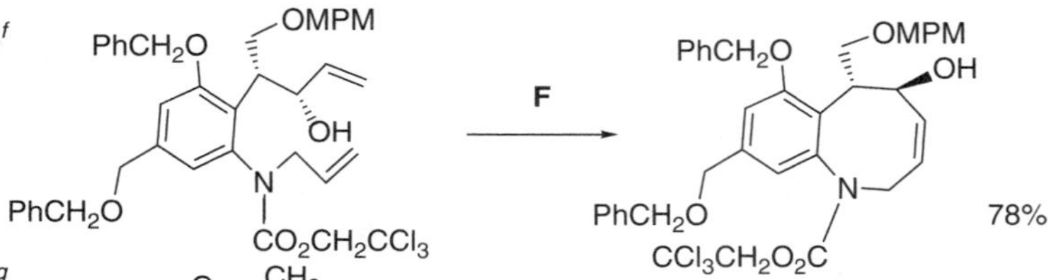

7[g]

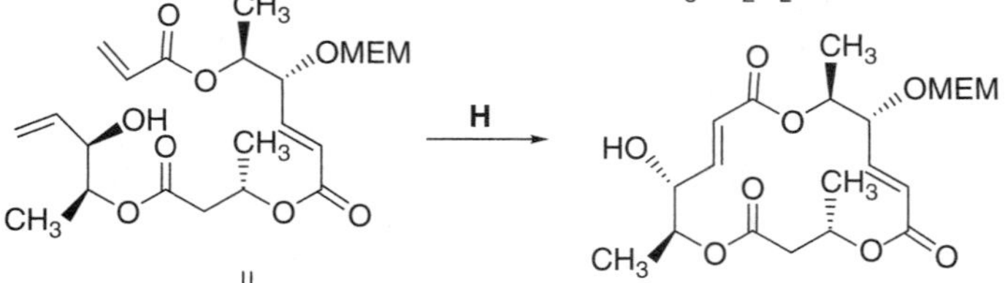

8[h]

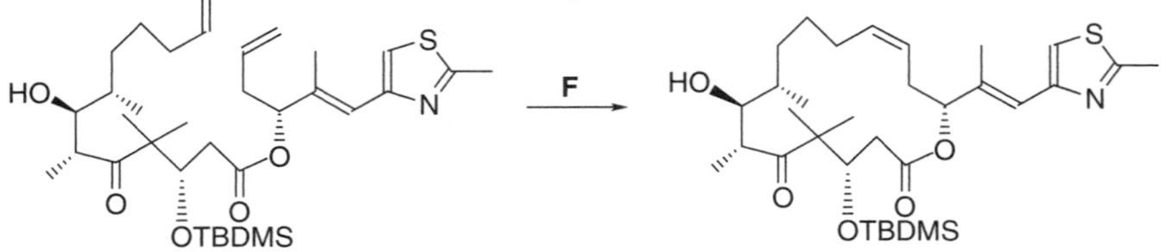

(*Continued*)

a. S. Kurosawa, M. Bando, and K. Mori, *Eur. J. Org. Chem.*, 4395 (2001).
b. J. M. Mörgenthaler and D. Spitzner, *Tetrahedron Lett.*, **45**, 1171 (2004).
c. Y. K. Reddy and J. R. Falck, *Org. Lett.*, **4**, 969 (2002).
d. J.-G. Boiteau, P. Van de Weghe, and J. Eustache, *Org. Lett.*, **3**, 2737 (2001).
e. A. Furstner, K. Radkowski, C. Wirtz, R. Goddard, C. W. Lehmann, and R. Mynott, *J. Am. Chem. Soc.*, **124**, 7061 (2002).
f. I. M. Fellows, D. E. Kaelin, Jr., and S. F. Martin, *J. Am. Chem. Soc.*, **122**, 10781 (2000).
g. Y. Matsuya, T. Kawaguchi, and H. Nemoto, *Org. Lett.*, **5**, 2939 (2003).
h. Z. Yang, Y. He, D. Vourloumis, H. Vallberg, and K. C. Nicolaou, *Angew. Chem. Int. Ed. Engl.*, **36**, 166 (1997).

8.5. Organometallic Compounds with π-Bonding

The organometallic reactions discussed in the previous sections in most cases involved intermediates carbon-metal with σ bonds, although examples of π bonding with alkenes and allyl groups were also encountered. The reactions emphasized in this section involve compounds in which organic groups are bound to the metal through delocalized π systems. Among the classes of organic compounds that can serve as π ligands are alkenes, allyl groups, dienes, the cyclopentadienide anion, and aromatic compounds. There are many such compounds, and we illustrate only a few examples. The bonding of polyenes in π complexes is the result of two major contributions. The filled π orbital acts as an electron donor to empty d orbitals of the metal ion. There is also a contribution to bonding, called "back bonding," from a filled metal orbital interacting with ligand π^* orbitals. These two types of bonding are illustrated in Figure 8.6. These same general bonding concepts apply to all the other π organometallics. The details of structure and reactivity of the individual compound depend on such factors as: (a) the number of electrons that can be accommodated by the metal; (b) the oxidation level of the metal; and (c) the electronic character of other ligands on the metal.

Alkene-metal complexes are usually prepared by a process by which some other ligand is dissociated from the metal. Both thermal and photochemical reactions are used.

$$(C_6H_5CN)_2PdCl_2 + 2\ RCH{=}CH_2 \longrightarrow$$

Ref. 309

Fig. 8.6. Representation of π bonding in a alkene-metal cation complex.

309. M. S. Kharasch, R. C. Seyler, and F. R. Mayo, *J. Am. Chem. Soc.*, **60**, 882 (1938).

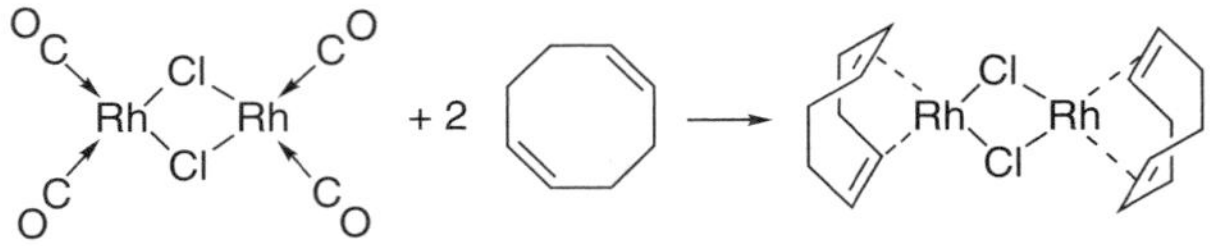

Ref. 310

π-Allyl complexes of palladium were described in Section 8.2.1. Similar π-allyl complexes of nickel can be prepared either by oxidative addition on Ni(0) or by transmetallation of a Ni(II) salt. Some reactions of these allyl nickel species are discussed in Section 8.3.1.

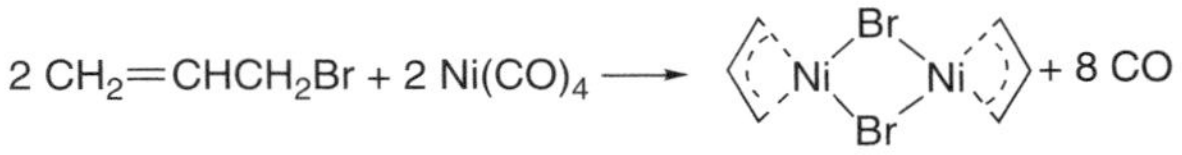

Ref. 311

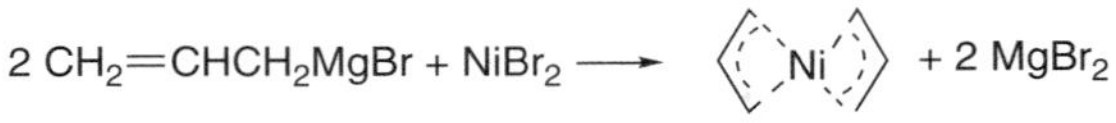

Ref. 312

Organic ligands having a cyclic array of four carbon atoms have been of particular interest in connection with the chemistry of cyclobutadiene. Organometallic compounds containing cyclobutadiene as a ligand were first prepared in 1965.[313] The carbocyclic ring in the cyclobutadiene–iron tricarbonyl complex reacts as an aromatic ring and can undergo electrophilic substitutions.[314] Subsequent studies showed that oxidative decomposition of the complex can liberate cyclobutadiene, which is trapped by appropriate reactants.[315] Some examples of these reactions are given in Scheme 8.17.

One of the most familiar of the π-organometallic compounds is ferrocene, a neutral compound that is readily prepared from cyclopentadienide anion and iron(II).[316]

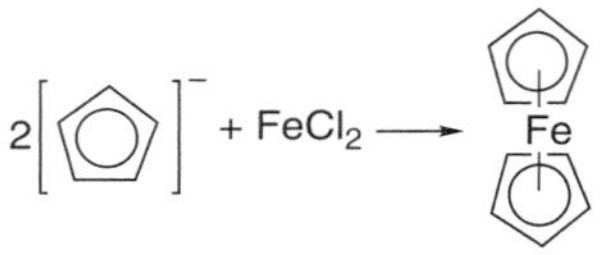

Numerous chemical reactions have been carried out on ferrocene and its derivatives.[317] The molecule behaves as an electron-rich aromatic system, and electrophilic substitution reactions occur readily. Reagents that are relatively strong oxidizing agents, such as the halogens, effect oxidation at iron and destroy the compound.

310. J. Chatt and L. M. Venanzi, *J. Chem. Soc.*, 4735 (1957).
311. E. J. Corey and M. F. Semmelhack, *J. Am. Chem. Soc.*, **89**, 2755 (1967).
312. D. Walter and G. Wilke, *Angew. Chem. Int. Ed. Engl.*, **5**, 151 (1966).
313. G. F. Emerson, L. Watts, and R. Pettit, *J. Am. Chem. Soc.*, **87**, 131 (1965); R. Pettit and J. Henery, *Org. Synth.*, **50**, 21 (1970).
314. J. D. Fitzpatrick, L. Watts, G. F. Emerson, and R. Pettit, *J. Am. Chem. Soc.*, **87**, 3254 (1965).
315. R. H. Grubbs and R. A. Grey, *J. Am. Chem. Soc.*, **95**, 5765 (1973).
316. G. Wilkinson, *Org. Synth.*, **IV**, 473, 476 (1963).
317. A. Federman Neto, A. C. Pelegrino, and V. A. Darin, *Trends in Organometallic Chem.*, **4**, 147 (2002).

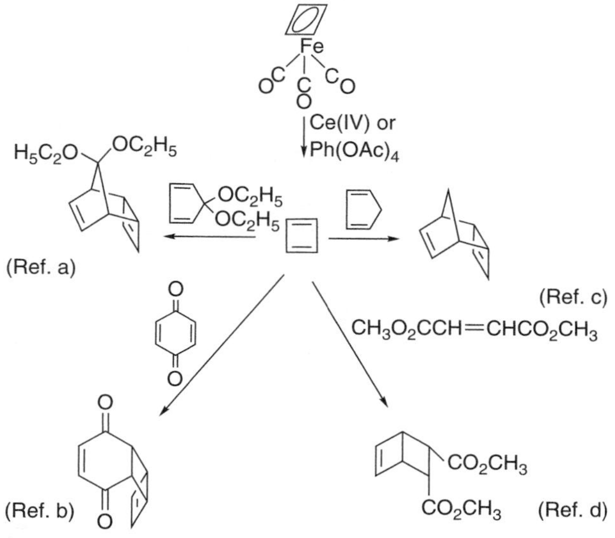

a. J. C. Barborak and R. Pettit, *J. Am. Chem. Soc.*, **89**, 3080 (1967).
b. J. C. Barborak, L. Watts, and R. Pettit, *J. Am. Chem. Soc.*, **88**, 1328 (1966).
c. L. Watts, J. D. Fitzpatrick, and R. Pettit, *J. Am. Chem. Soc.*, **88**, 623 (1966).
d. P. Reeves, J. Henery, and R. Pettit, *J. Am. Chem. Soc.*, **91**, 3889 (1969).

Many other π-organometallic compounds have been prepared. In the most stable of these, the total number of electrons contributed by the ligands (e.g., four for allyl anions and six for cyclopentadiene anion) plus the valence electrons on the metal atom or ion is usually 18, to satisfy the *effective atomic number rule*.[318]

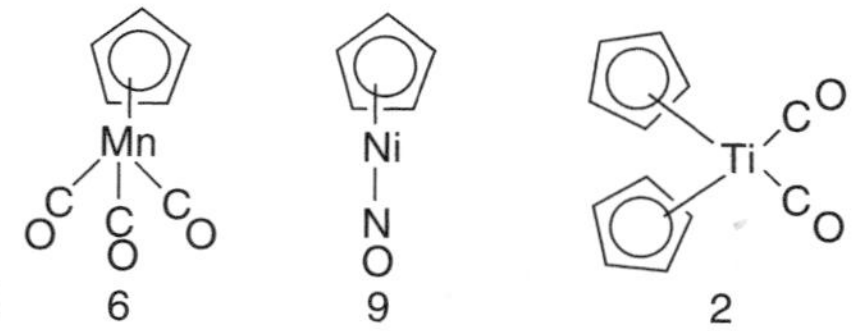

	Mn	Ni	Ti
Metal	6	9	2
Ligands	12	9	16
Total	18	18	18

One of the most useful types of π complexes of aromatic compounds from the synthetic point of view are chromium tricarbonyl complexes obtained by heating benzene or other aromatics with $Cr(CO)_6$.

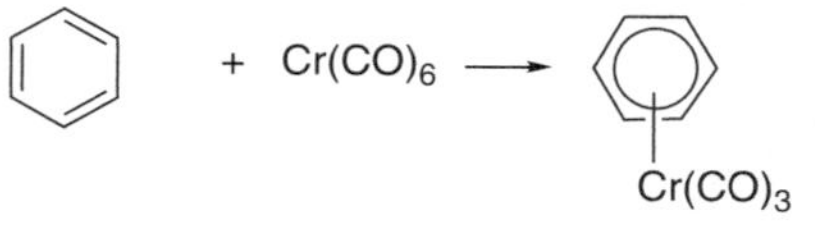

Ref. 319

[318.] M. Tsutsui, M. N. Levy, A. Nakamura, M. Ichikawa, and K. Mori, *Introduction to Metal π-Complex Chemistry*, Plenum Press, New York, 1970, pp. 44–45; J. P. Collman, L. S. Hegedus, J. R. Norton, and R. G. Finke, *Principles and Applications of Organotransition Metal Chemistry*, University Science Books, Mill Valley, CA, 1987, pp. 166–173.

[319.] W. Strohmeier, *Chem. Ber.*, **94**, 2490 (1961).

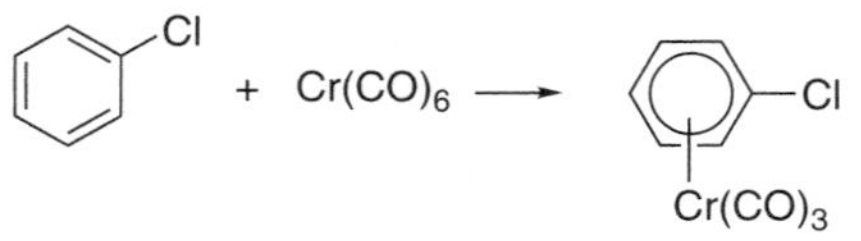

Ref. 320

The $Cr(CO)_3$ unit in these compounds is strongly electron withdrawing and activates the ring to nucleophilic attack. Reactions with certain carbanions results in arylation.[321]

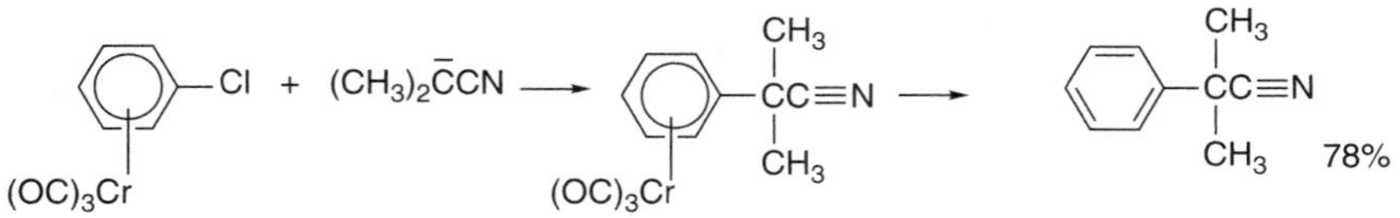

In compounds in which the aromatic ring does not have a leaving group, addition occurs. The intermediate can by oxidized by I_2.

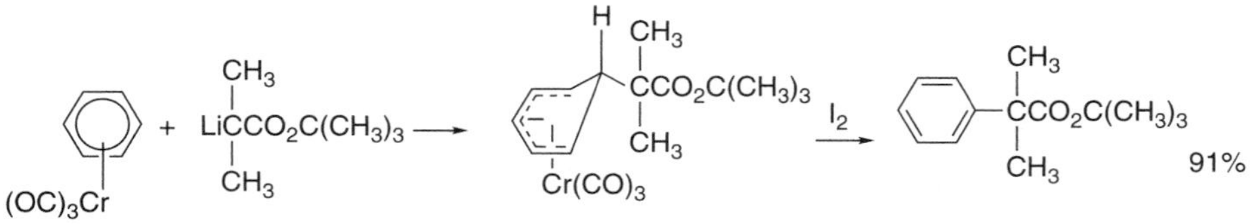

Ref. 322

Existing substituent groups such as CH_3, OCH_3, and $^+N(CH_3)_3$ exert a directive effect, often resulting in a major amount of the *meta* substitution product.[323] The intermediate adducts can be converted to cyclohexadiene derivatives if the adduct is protonolyzed.[324]

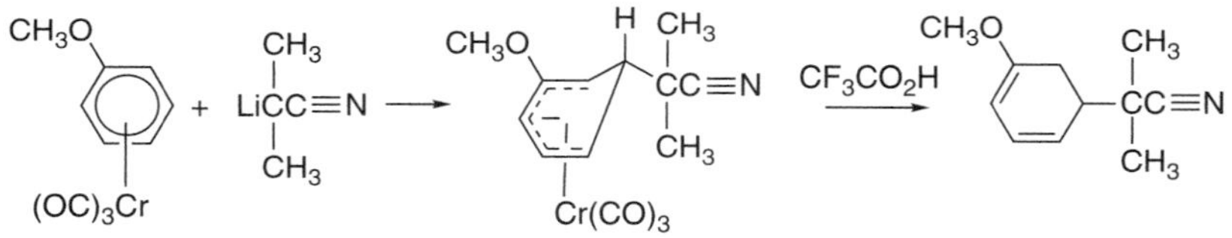

Not all carbon nucleophiles will add to arene chromium tricarbonyl complexes. For example, alkyllithium reagents and simple ketone enolates do not give adducts.[325]

Organometallic chemistry is a very large and active field of research and new compounds, reactions, and useful catalysts are being discovered at a rapid rate. These developments have had a major impact on organic synthesis and future developments can be expected.

[320.] J. F. Bunnett and H. Hermann, *J. Org. Chem.*, **36**, 4081 (1971).

[321.] M. F. Semmelhack and H. T. Hall, *J. Am. Chem. Soc.*, **96**, 7091 (1974).

[322.] M. F. Semmelhack, H. T. Hall, M. Yoshifuji, and G. Clark, *J. Am. Chem. Soc.*, **97**, 1247 (1975); M. F. Semmelhack, H. T. Hall, Jr., R. Farina, M. Yoshifuji, G. Clark, T. Bargar, K. Hirotsu, and J. Clardy, *J. Am. Chem. Soc.*, **101**, 3535 (1979).

[323.] M. F. Semmelhack, G. R. Clark, R. Farina, and M. Saeman, *J. Am. Chem. Soc.*, **101**, 217 (1979).

[324.] M. F. Semmelhack, J. J. Harrison, and Y. Thebtaranonth, *J. Org. Chem.*, **44**, 3275 (1979).

[325.] R. J. Card and W. S. Trahanovsky, *J. Org. Chem.*, **45**, 2555, 2560 (1980).

General References

J. P. Collman, L. S. Hegedus, J. R. Norton, and R. G. Finke, *Principles and Applications of Organotransition Metal Chemistry*, University Science Books, Mill Valley, CA, 1987.

H. M. Colquhoun, J. Holton, D. J. Thomson, and M. V. Twigg, *New Pathways for Organic Synthesis*, Plenum Press, New York, 1984.

R. M. Crabtree *The Organometallic Chemistry of the Transition Metals*, Wiley-Interscience, New York, 2005.

S. G. Davies, *Organo-Transition Metal Chemistry: Applications in Organic Synthesis*, Pergamon Press, Oxford, 1982.

F. Diederich and P. J. Stang, *Metal-Catalyzed Cross-Coupling Reactions*, Wiley-VCH, New York, 1998.

J. K. Kochi, *Organometallic Mechanisms and Catalysis*, Academic Press, New York, 1979.

E. Negishi, *Organometallics in Organic Synthesis*, Wiley, New York, 1980.

M. Schlosser, ed., *Organometallics in Synthesis: A Manual*, Wiley, Chichester, 1994.

Organocopper Reactions

G. Posner, *An Introduction to Synthesis Using Organocopper Reagents*, Wiley-Interscience, New York, 1975.

R. J. K. Taylor, ed., *Organocopper Reagents*, Oxford University Press, Chichester, 1994.

Organopalladium Reactions

R. F. Heck, *Palladium Reagents in Organic Synthesis*, Academic Press, Orlando, FL, 1985.

R. F. Heck, *Org. React.*, **27**, 345 (1982).

E. Negishi and A. de Mejeire, eds., *Handbook of Organopalladium Chemistry for Organic Synthesis*, Vol. 1 and 2, Wiley-Interscience, New York, 2002.

J. Tsuji, *Palladium Reagents and Catalysts: Innovations in Organic Synthesis*, Wiley, New York, 1996.

Problems

(References for these problems will be found on page 1284.)

8.1. Predict the product of the following reactions. Be sure to specify all elements of regiochemistry and stereochemistry.

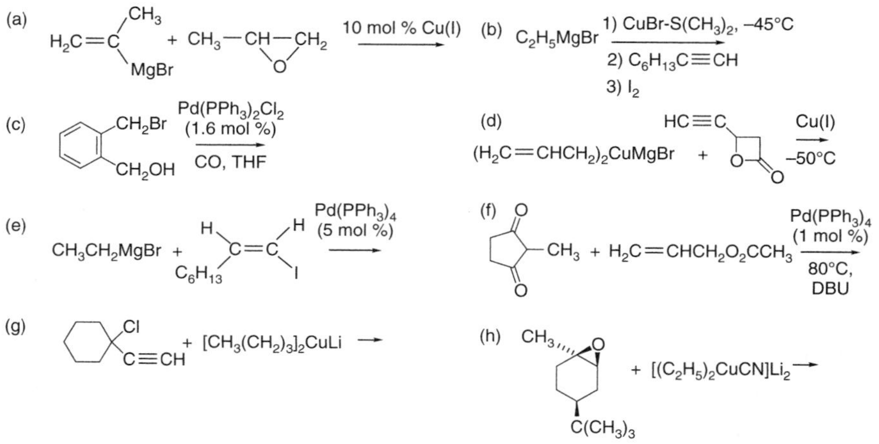

(i)

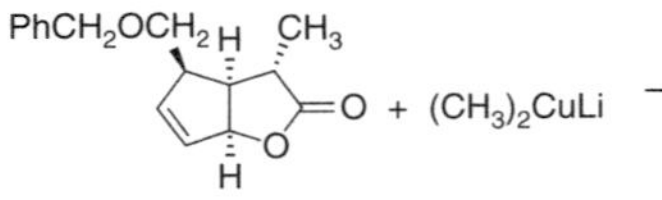

(j)

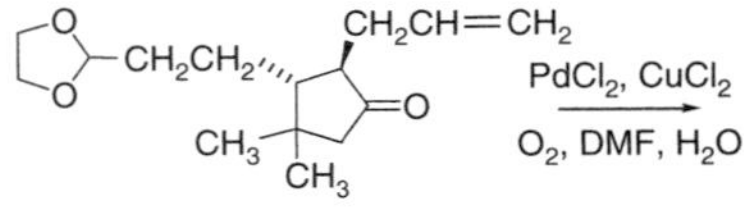

(k)

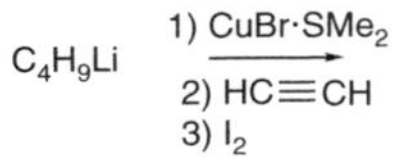

C_4H_9Li 1) CuBr·SMe$_2$
             2) HC≡CH
             3) I$_2$

(l)

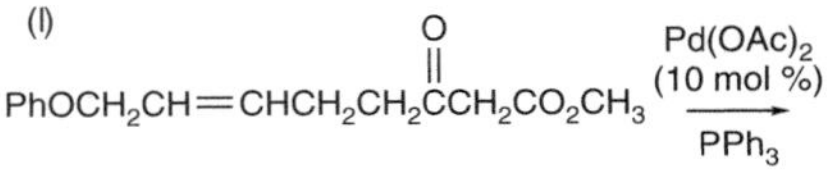

$PhOCH_2CH=CHCH_2CH_2CCH_2CO_2CH_3$ $\xrightarrow[PPh_3]{Pd(OAc)_2 \ (10 \ mol \ \%)}$

(m)

+ CH$_2$=CHMgBr $\xrightarrow[THF]{Ni(dmpe)Cl_2 \ (1 \ mol \ \%)}$

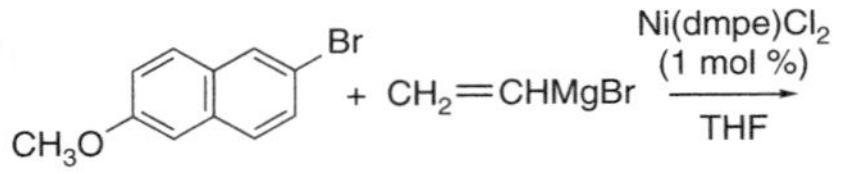

dmpe = 1,2-bis(dimethylphosphino)ethane

(n)

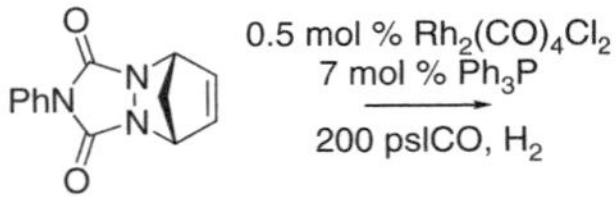

$\xrightarrow[200 \ psiCO, \ H_2]{0.5 \ mol \ \% \ Rh_2(CO)_4Cl_2 \ \ 7 \ mol \ \% \ Ph_3P}$

(o)

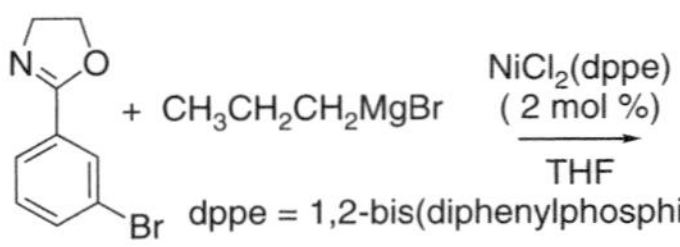

+ CH$_3$CH$_2$CH$_2$MgBr $\xrightarrow[THF]{NiCl_2(dppe) \ (2 \ mol \ \%)}$ dppe = 1,2-bis(diphenylphosphino)ethane

(p)

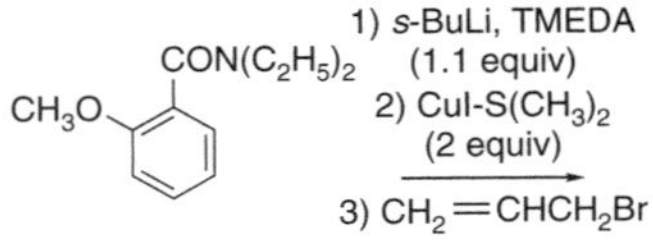

1) s-BuLi, TMEDA
   (1.1 equiv)
2) CuI-S(CH$_3$)$_2$
   (2 equiv)
3) CH$_2$=CHCH$_2$Br

(q)

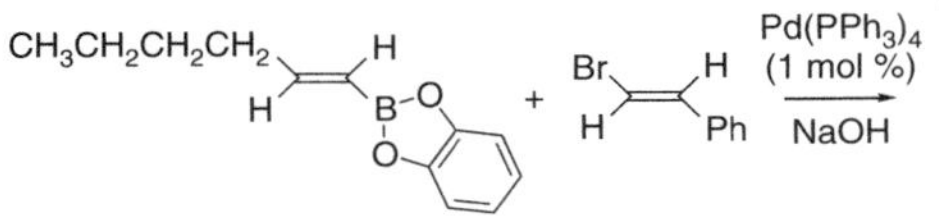

+

$\xrightarrow[NaOH]{Pd(PPh_3)_4 \ (1 \ mol \ \%)}$

(r)

+

$\xrightarrow[Pd(dba)_2]{CO, \ 55 \ psi}$

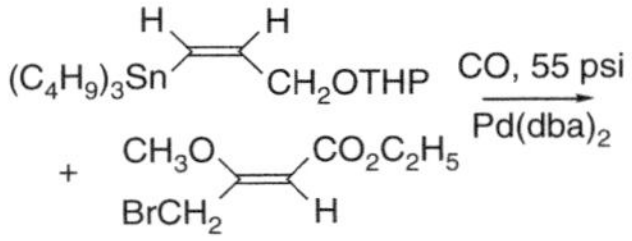

dba = dibenzylideneacetonate

8.2. Give the products expected from each of the following reactions involving mixed cuprate reagents.

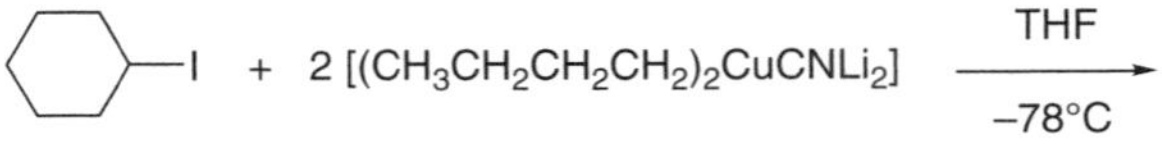

(a)

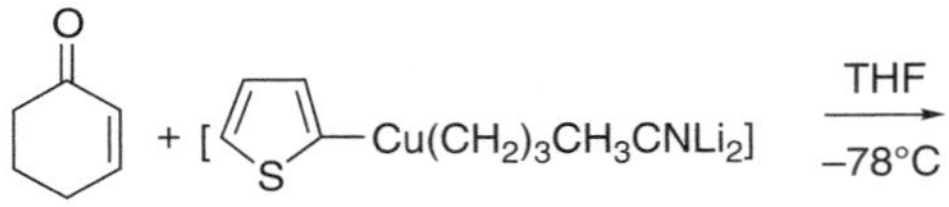

+ [—Cu(CH$_2$)$_3$CH$_3$CNLi$_2$] $\xrightarrow[-78°C]{THF}$

(b)

—I   +   2 [(CH$_3$CH$_2$CH$_2$CH$_2$)$_2$CuCNLi$_2$] $\xrightarrow[-78°C]{THF}$

(c)

+ [(CH$_3$)$_3$CCuCN]Li $\longrightarrow$

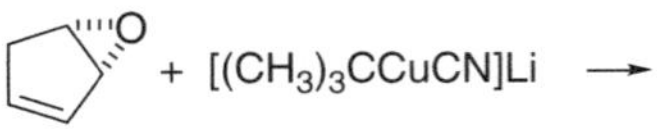

(d)

+ [(CH$_3$)$_3$CCuCH$_2$SCH$_3$]Li $\longrightarrow$

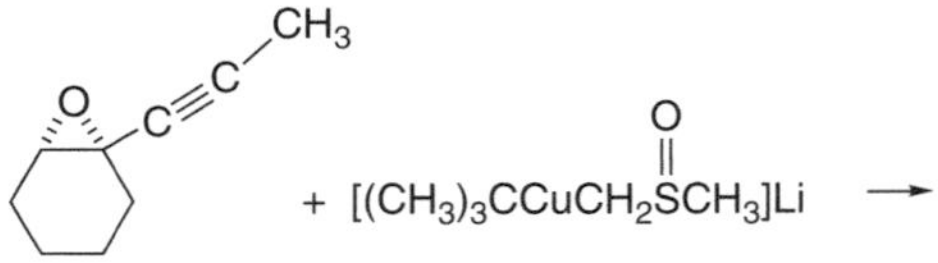

8.3. Write a mechanism for each of the following reactions that accounts for the observed product and is in accord with other information that is available concerning the reaction.

(a)

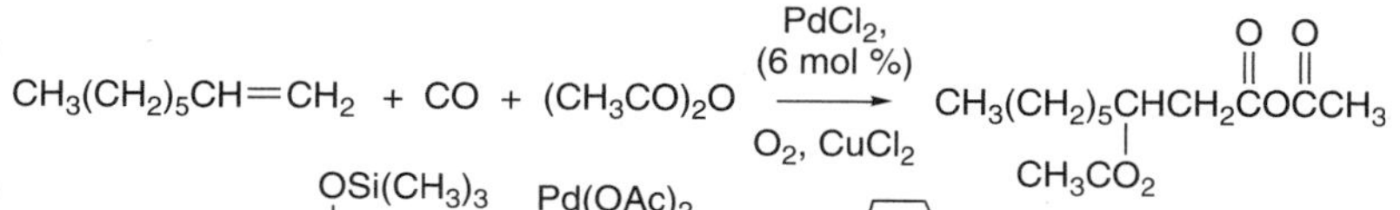

$$CH_3(CH_2)_5CH=CH_2 + CO + (CH_3CO)_2O \xrightarrow[O_2,\ CuCl_2]{\begin{array}{c}PdCl_2,\\(6\ mol\ \%)\end{array}} CH_3(CH_2)_5\underset{CH_3CO_2}{\overset{}{CH}}CH_2\overset{O}{\overset{\|}{C}}\overset{O}{\overset{\|}{C}}CH_3$$

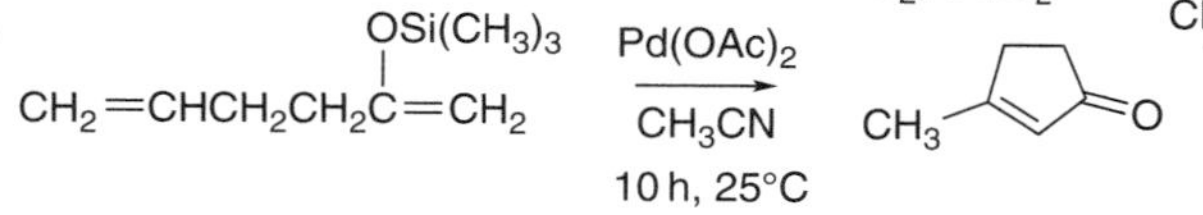

(b)

$$CH_2=CHCH_2CH_2\underset{}{\overset{OSi(CH_3)_3}{C}}=CH_2 \xrightarrow[\begin{array}{c}CH_3CN\\10\ h,\ 25°C\end{array}]{Pd(OAc)_2}$$

(c)

$$PhCH_2O-\underset{\underset{OH}{}}{\overset{CH_3\ CH_3}{}} \xrightarrow[\begin{array}{c}Rh_2(OAc)_2,\\PPh_3,\ 100°C\end{array}]{CO,\ H_2} PhCH_2O-$$

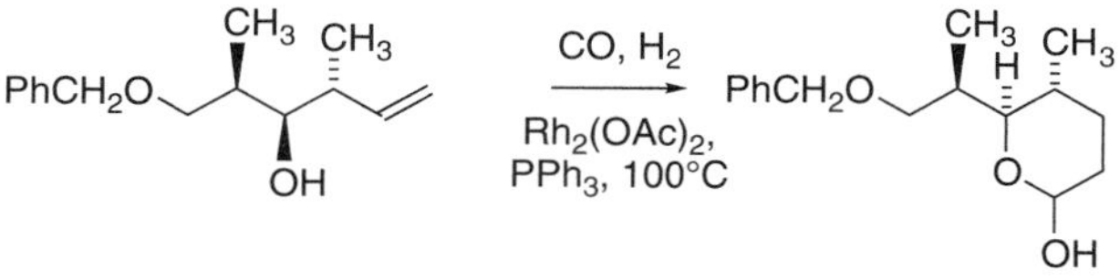

(d)

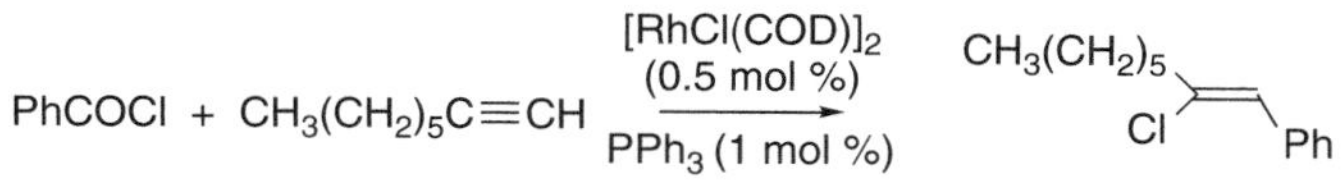

$$PhCOCl + CH_3(CH_2)_5C\equiv CH \xrightarrow[\begin{array}{c}PPh_3\ (1\ mol\ \%)\end{array}]{\begin{array}{c}[RhCl(COD)]_2\\(0.5\ mol\ \%)\end{array}} \underset{Cl}{\overset{CH_3(CH_2)_5}{}}=\underset{Ph}{}$$

(e)

$$PhCO_2CH_2C\equiv CH + \bigcirc \xrightarrow[(see\ 762)]{cat\ G} PhCO_2CH_2$$

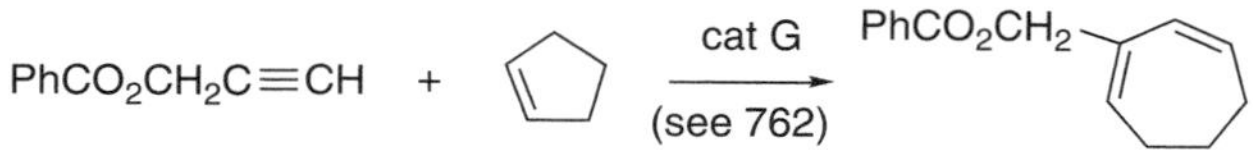

8.4. Indicate appropriate conditions and reagents for effecting the following transformations. Identify necessary co-reactants, reagents, and catalysts. One-pot processes are possible in all cases.

(a)

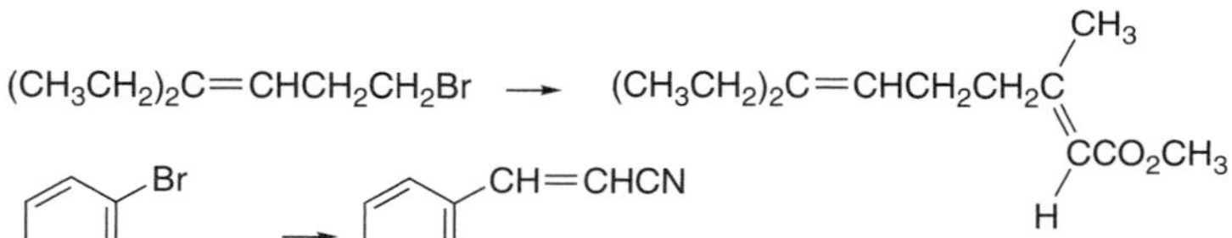

$$(CH_3CH_2)_2C=CHCH_2CH_2Br \longrightarrow (CH_3CH_2)_2C=CHCH_2CH_2\overset{CH_3}{\underset{H}{\overset{}{C}}}CCO_2CH_3$$

(b)

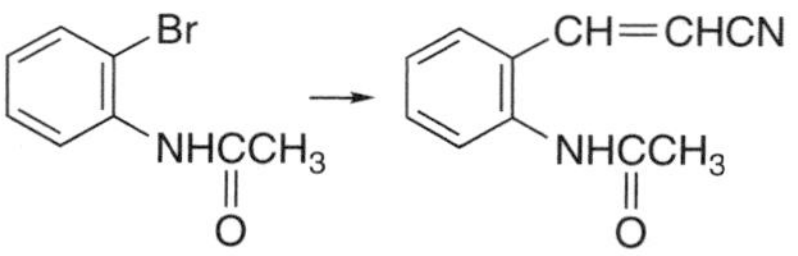

(c)

$$CH_3(CH_2)_3Br + CH_3O_2CC\equiv CCO_2CH_3 \longrightarrow \underset{CH_3(CH_2)_3}{\overset{CH_3O_2C}{}}=\underset{H}{\overset{CO_2CH_3}{}}$$

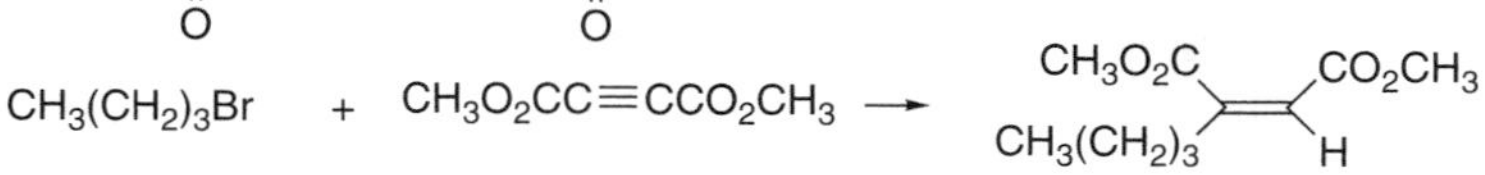

(d)

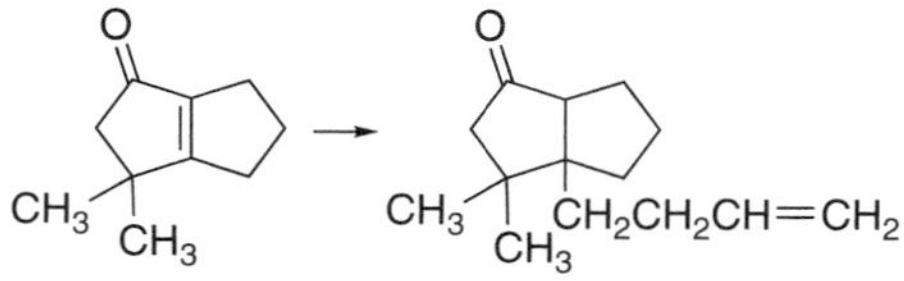

(e)

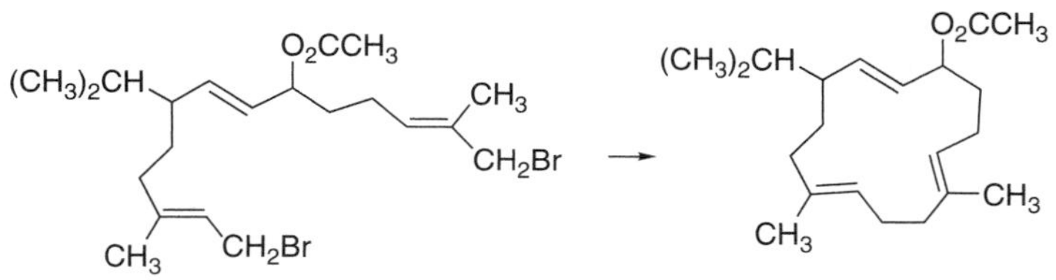

(f)

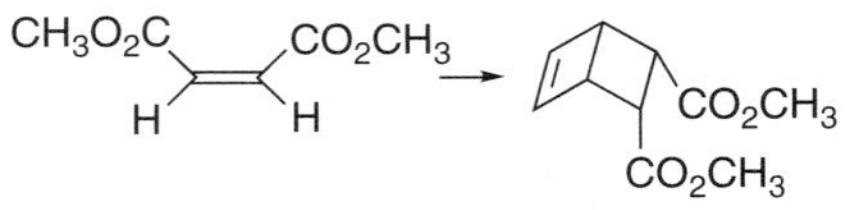

(g) $(CH_3)_2C=CCH_3 \longrightarrow (CH_3)_2C=CCH=CHCO_2H$

with Br below the first structure and CH$_3$ above the product

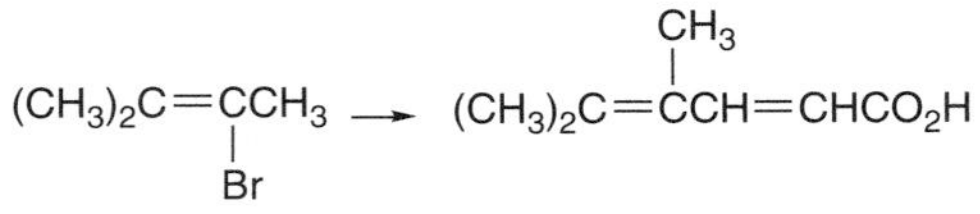

(h)

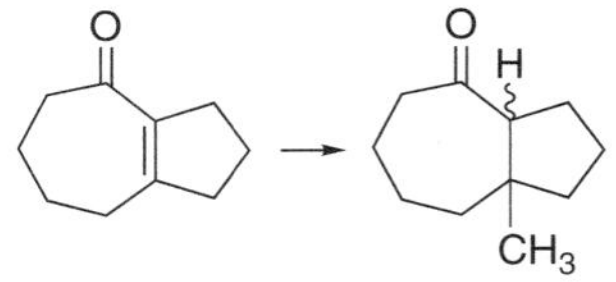

(i)

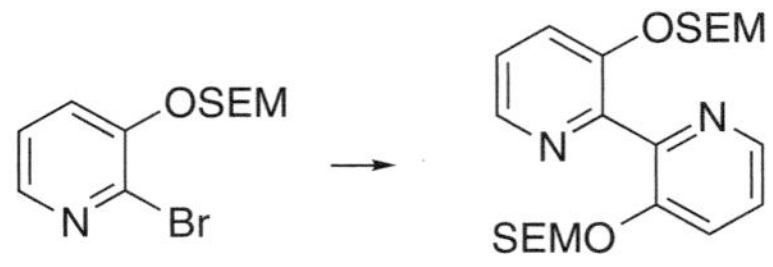

(j)

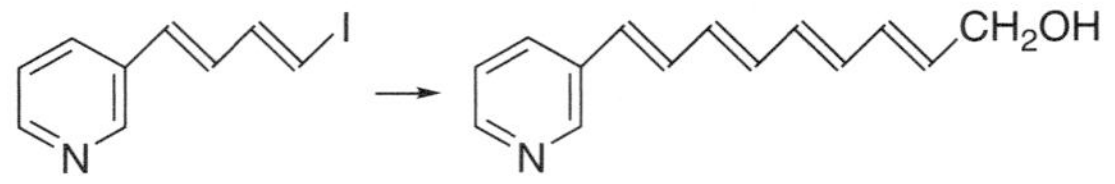

(k)

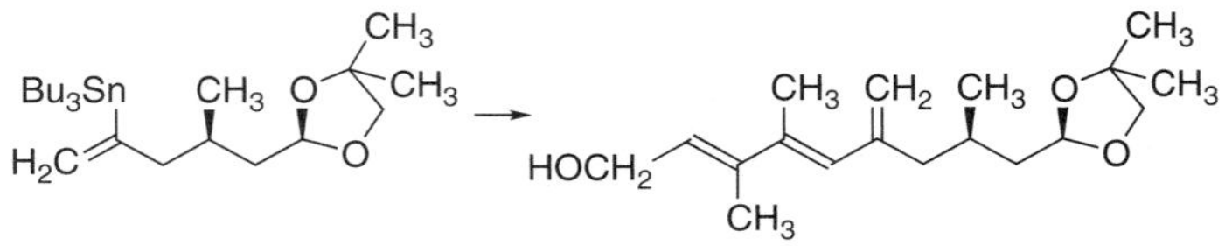

(l)

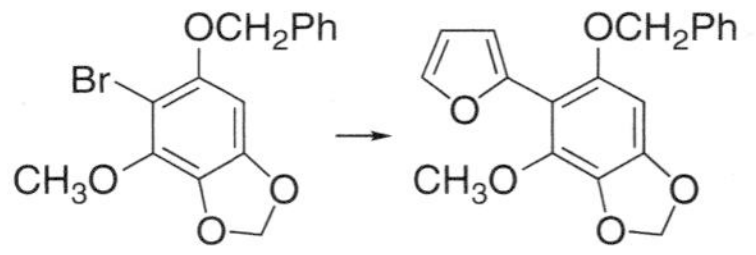

(m)

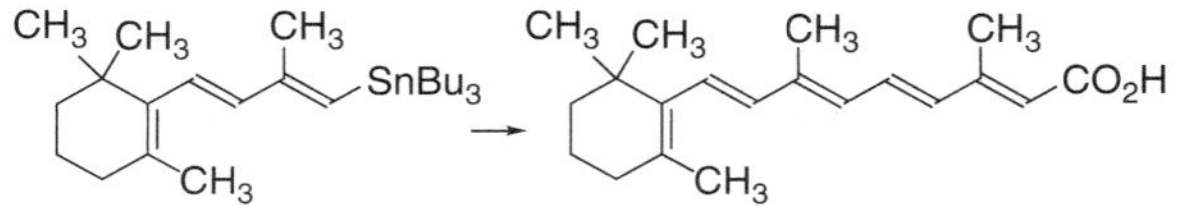

(n) $PhCH=CHCH_2O_2CH_3 \longrightarrow PhCH=CHCH_2CHCCH_3$

with O above CCH$_3$ and CO$_2$C$_2$H$_5$ below

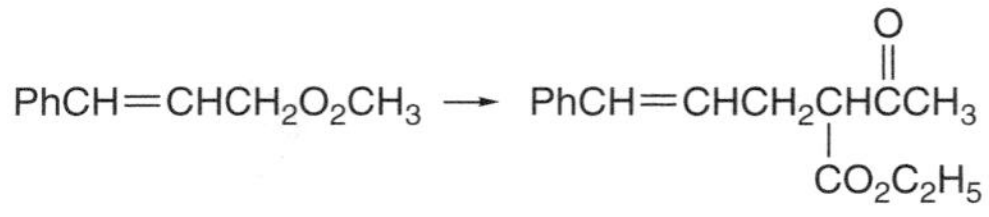

(o) $CH_3(CH_2)_3C\equiv CH \longrightarrow$

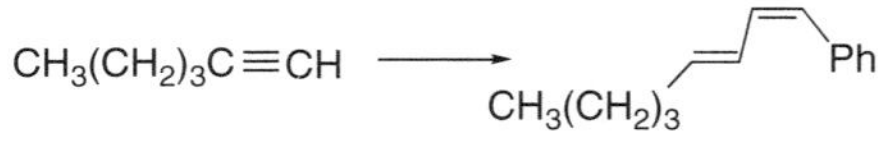

(p)

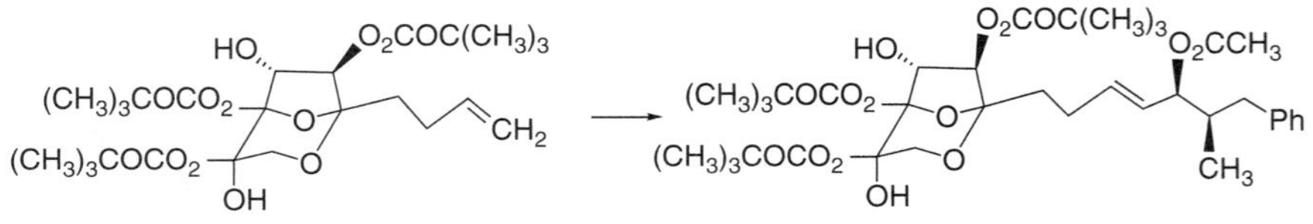

8.5. Vinyltriphenylphosphonium ion has been found to react with cuprate reagents by nucleophilic addition, generating an ylide that can react with aldehydes to give alkenes. In another version of the reaction, an intermediate formed by the reaction of the cuprate with acetylene adds to vinyltriphenylphosphonium ion to generate an ylide intermediate. Show how these reactions can be used to prepare the following products from the specified starting materials.

(a)

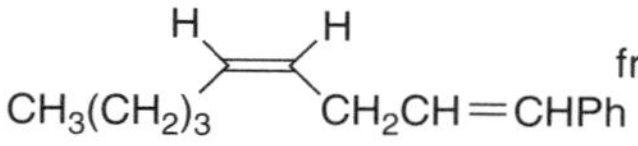

$$H_2C=CH-CH_2CH=CHPh$$

CH₃(CH₂)₃ CH₂CH=CHPh

from

CH₃(CH₂)₃ I

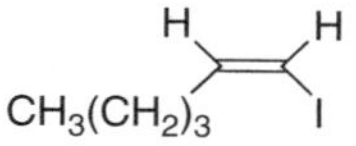

(b)

PhCH₂CH=CH(CH₂)₃CH₃   from   —I

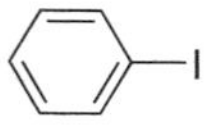

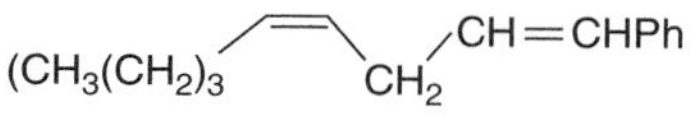

(c)

(CH₃(CH₂)₃   CH=CHPh   from   CH₃(CH₂)₃Br
          CH₂

8.6. It has been observed that the reaction of $[(C_2H_5)_2Cu]Li$ or $[(C_2H_5)_2CuCNLi_2]$ with 2-iodooctane proceeds with racemization in both cases. On the other hand, the corresponding bromide reacts with nearly complete inversion of configuration with both reagents. When 6-halo-2-heptenes are used in similar reactions with $[(CH_3)_2Cu]Li$, the iodide gives a cyclic product 1-ethyl-2-methylcyclopentane, whereas the bromide gives mainly 6-methyl-1-heptene. Propose a mechanism that accounts for the different behavior of the iodides as compared to the bromides.

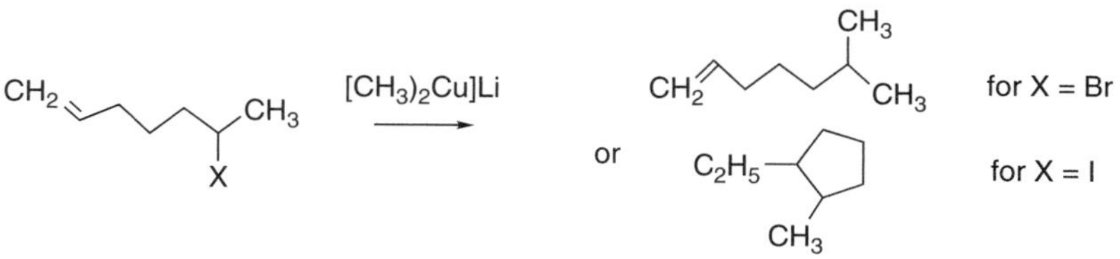

8.7. Short synthetic sequences involving no more than three steps can be used to prepare the compound shown on the left from the potential starting materials on the right. Suggest an appropriate series of reactions involving one or more organometallic reagent for each transformation.

(a)

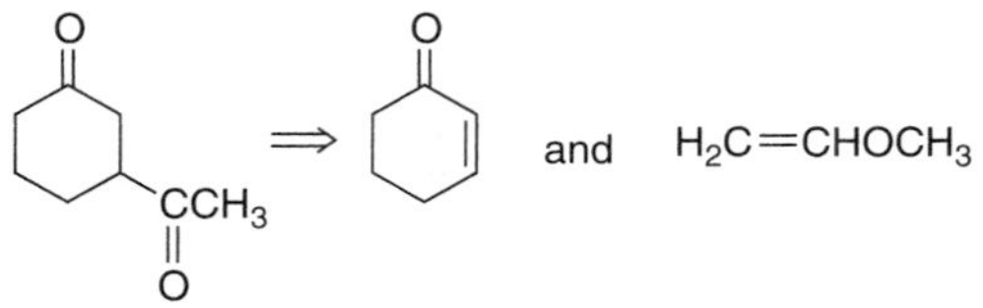

and H₂C=CHOCH₃

(b)

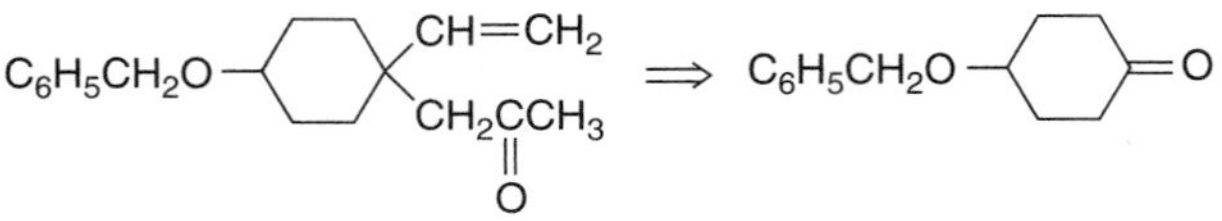

(c)

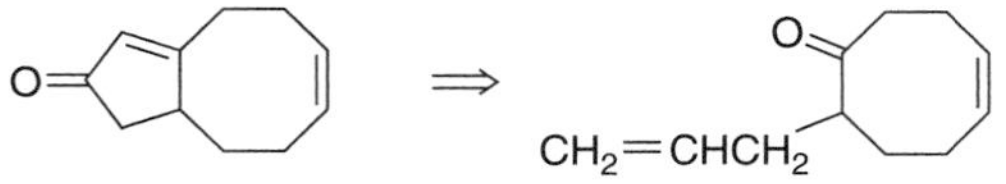

(d)

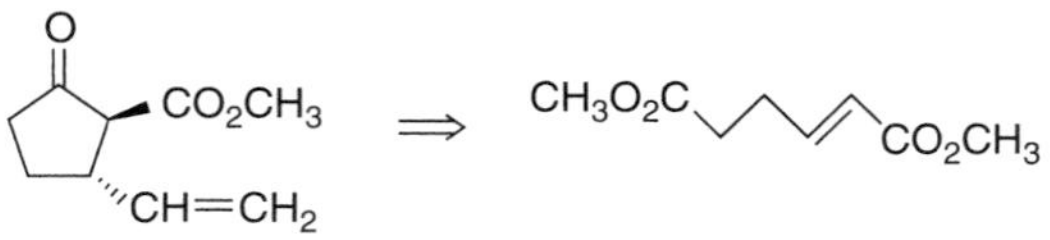

(e)

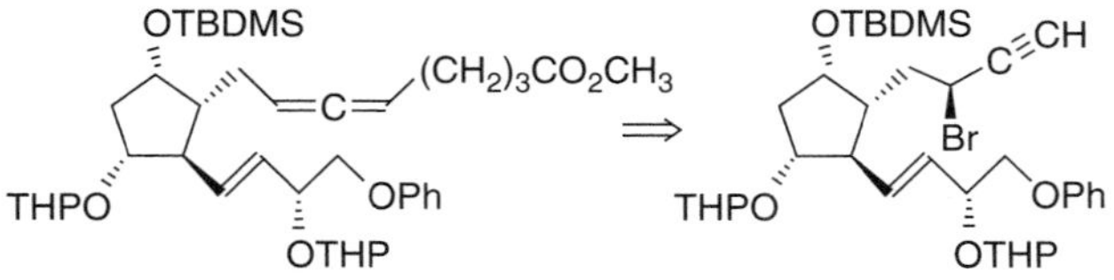

(f)

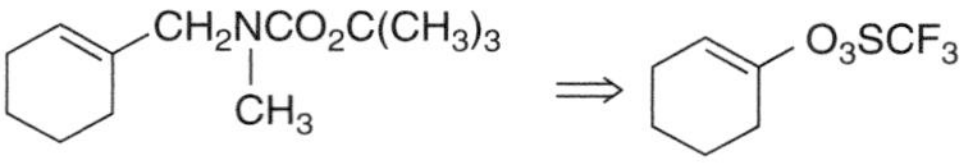

(g)

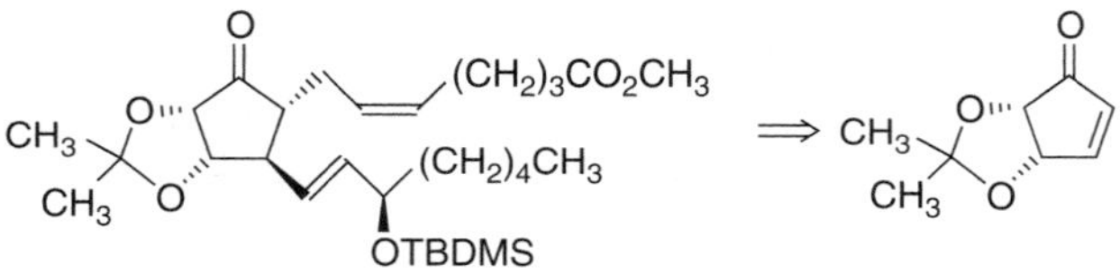

(h)

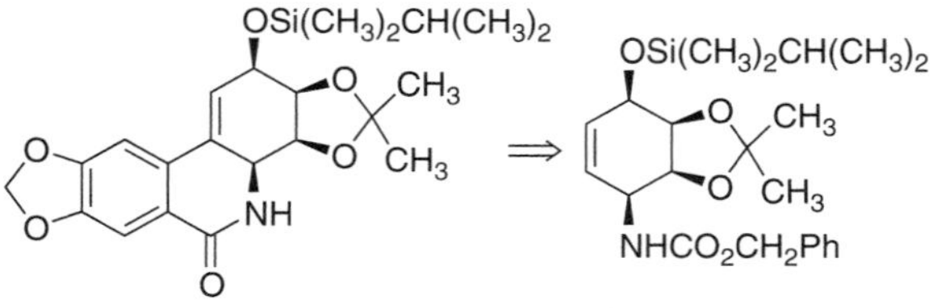

8.8. The conversions shown below can be carried out in multistep, but one-pot, reactions in which none of the intermediates needs to be isolated. Show how you would perform the transformations by suggesting a sequence of reagents and the approximate reaction conditions.

(a)

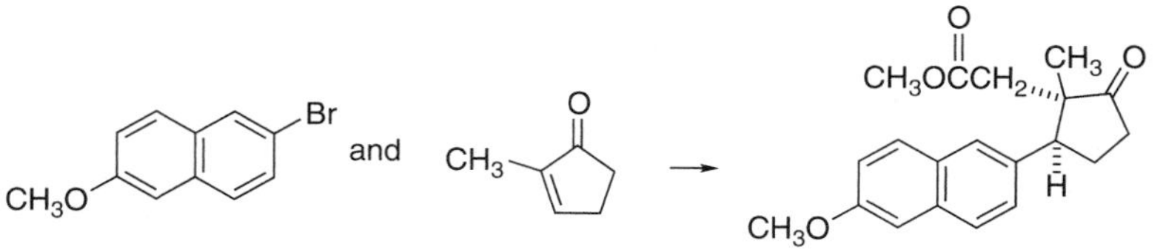

(b)

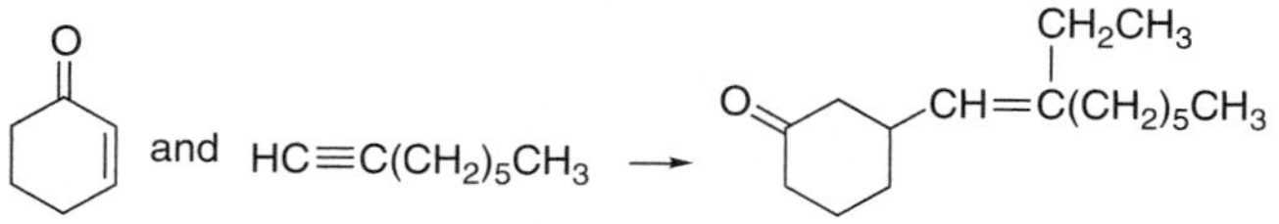

(c)

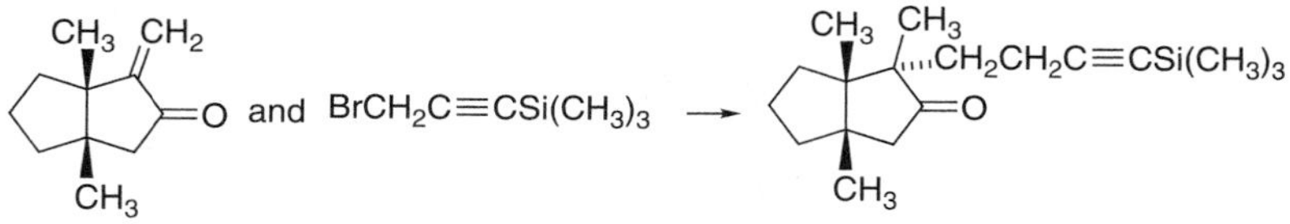

(d)

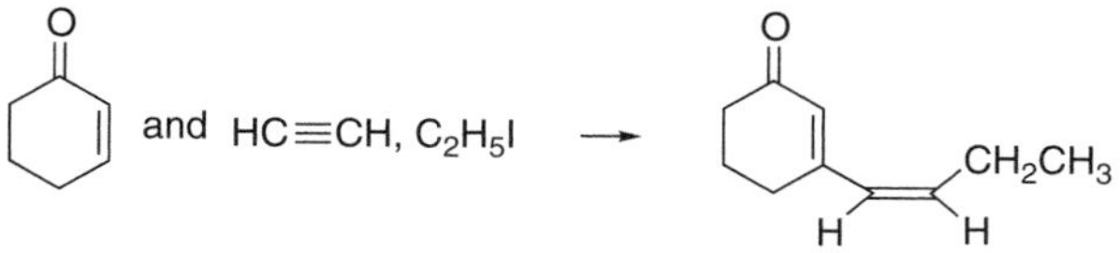

(e)

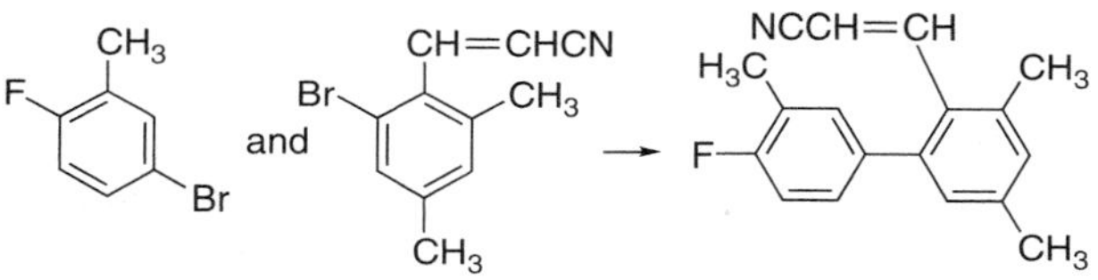

8.9. A number of syntheses of medium- and large-ring compounds that involve transition metal reagents or catalysts have been described. Suggest an organometallic reagent or catalyst that could bring about each of the following transformations.

(a)

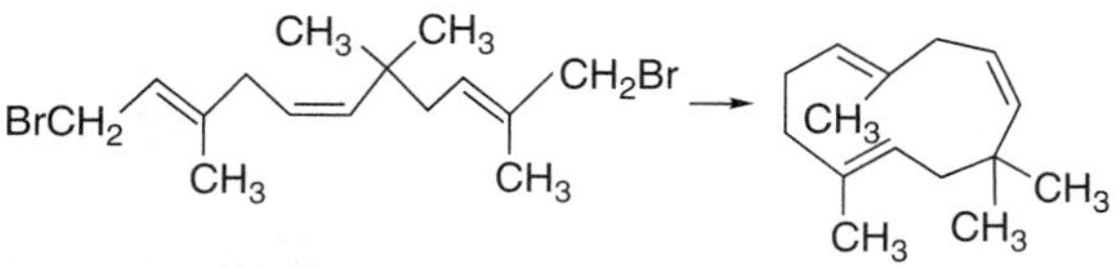

(b)

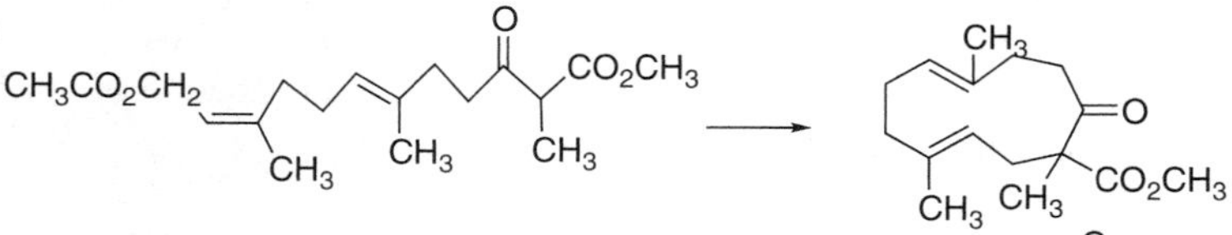

(c)

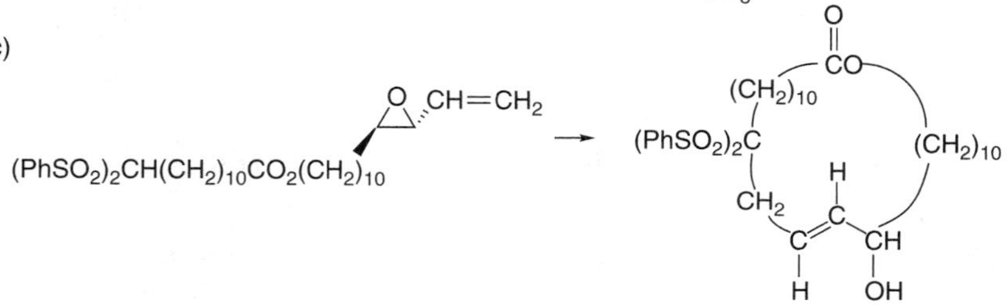

(d)

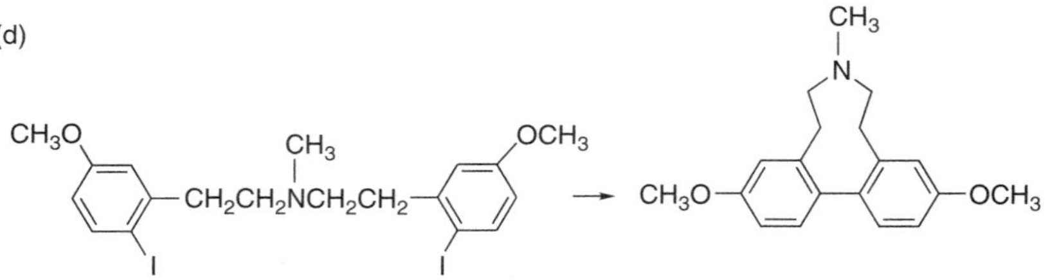

(e)

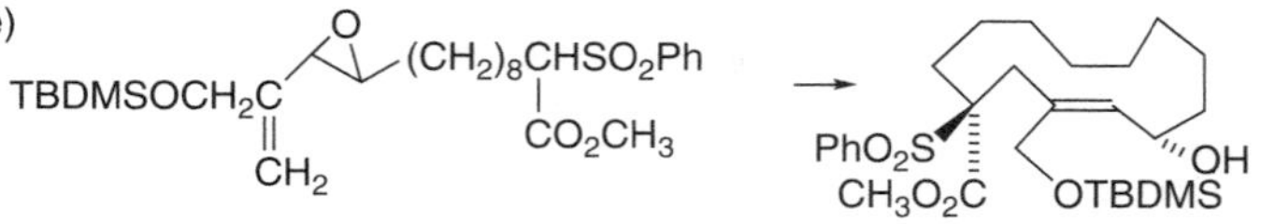

(f)

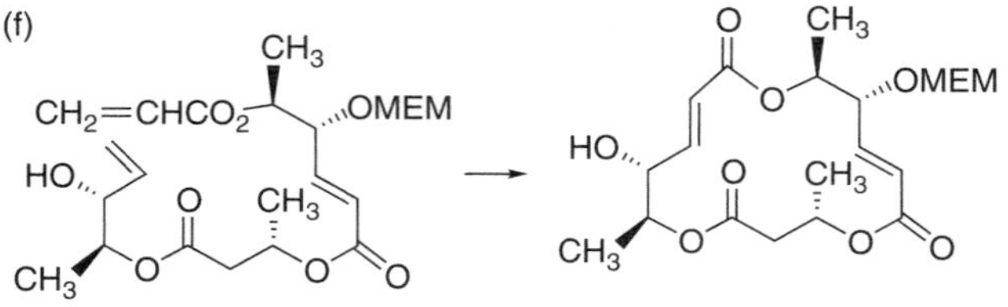

8.10. The cyclobutadiene complex **10-A** can be prepared in enantiomerically pure form. When the complex is decomposed by an oxidizing reagent in the presence of a potential trapping agent, the products are racemic. When the reaction is carried out only to partial completion, the unreacted complex remains enantiomerically pure. Discuss the relevance of these results to the following question: "In oxidative decomposition of cyclobutadiene–iron tricarbonyl complexes, is the cyclobutadiene released from the complex before or after it has reacted with the trapping reagent?"

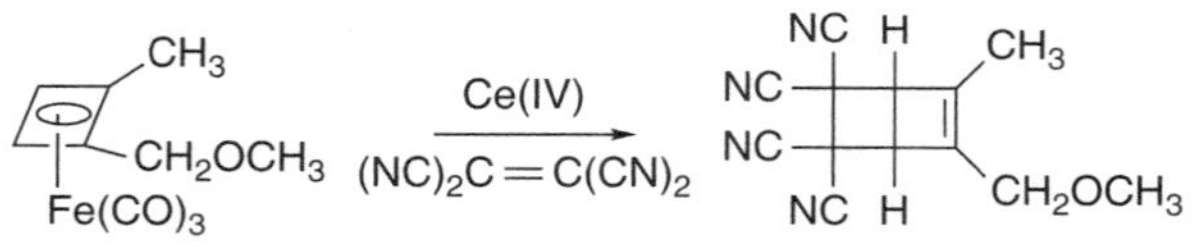

10 – A

8.11. When the isomeric allylic acetates **11-A** and **11-B** react with dialkylcuprates, they give very similar product mixtures that contain mainly **11-C** with a small amount of **11-D**. Discuss the mechanistic implications of the formation of essentially the same product mixture from both reactants.

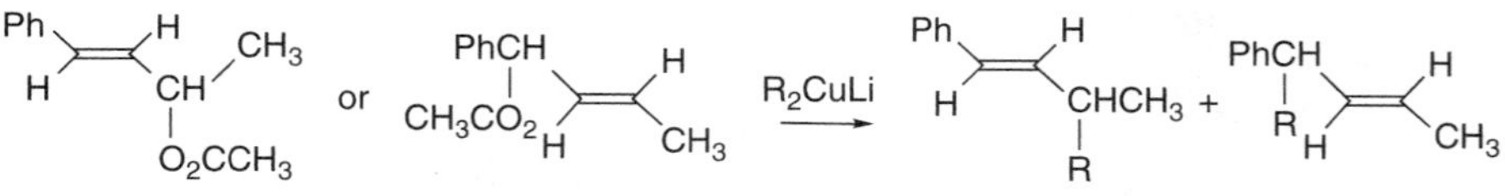

11-A   **11-B**   **11-C**   **11-D**

8.12. The compound shown below is a constituent of the pheromone of the codling moth. It has been synthesized using *n*-propyl bromide, propyne, 1-pentyne, ethylene oxide, and CO_2 as the source of the carbon atoms. Devise a route for such a synthesis. Hint: Extensive use of organocopper reagents is the basis for the synthesis.

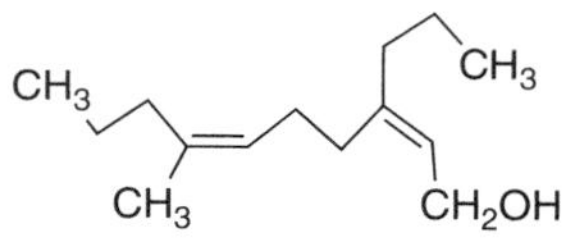

8.13. (S)-3-Hydroxy-2-methylpropanoic acid, **13-A**, can be obtained in enantiomerically pure form from isobutyric acid by a microbiological oxidation. The aldehyde **13-B** is available from a natural product, pulegone, also in enantiomerically pure form. Devise a synthesis of enantiomerically pure **13-C**, a compound of interest as a starting material for the synthesis of α-tocopherol (vitamin E).

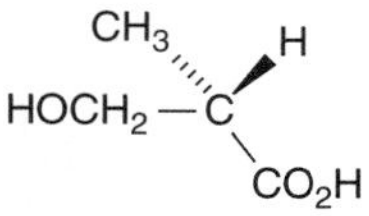

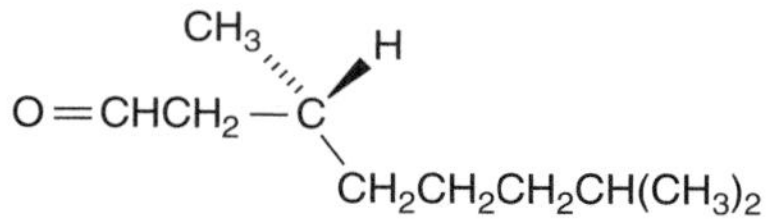

8.14. Each of the following conjugate additions can be carried out in good yield under optimized conditions. Consider the special factors in each case and suggest a reagent and reaction conditions that would be expected to give good yields.

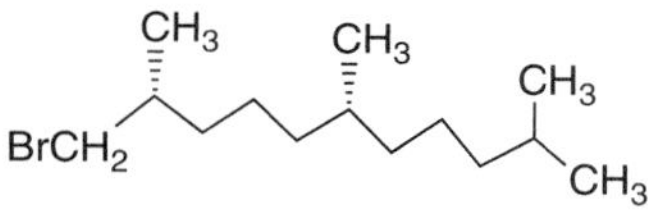

(a)

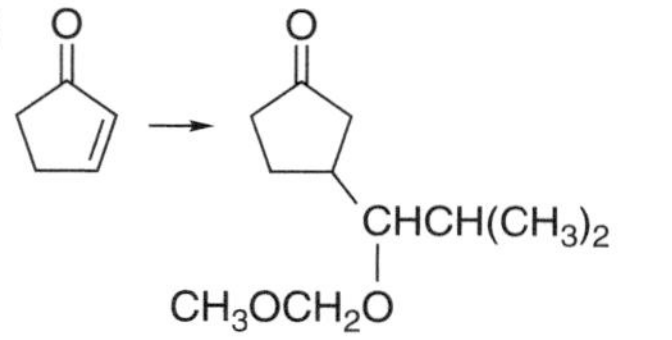

(b)

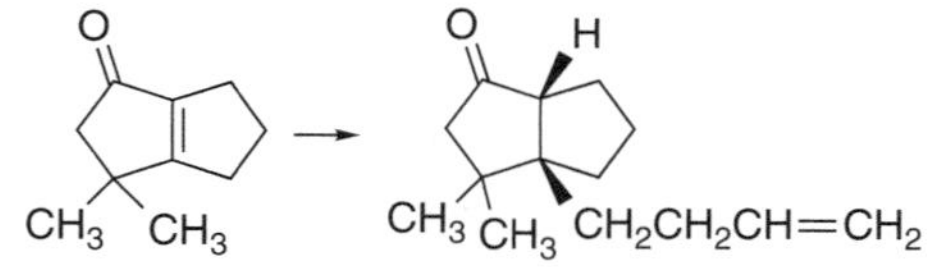

(c) $(CH_3)_2C=CHCO_2C_2H_5 \longrightarrow CH_3(CH_2)_3C(CH_3)_2CH_2CO_2C_2H_5$

(d)

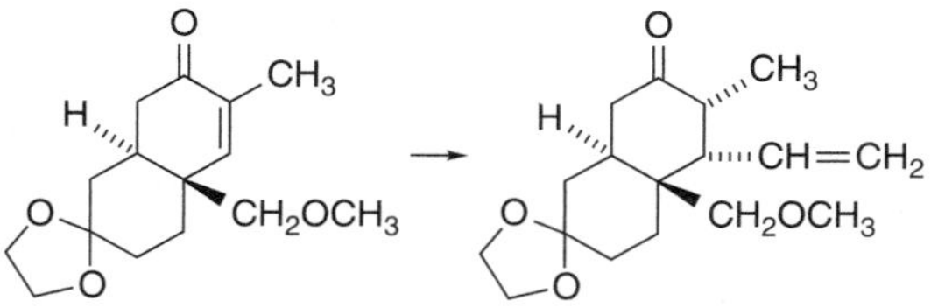

8.15. Each of the following synthetic transformations can be accomplished by use of organometallic reagents and/or catalysts. Indicate a sequence of reactions that will permit each of the syntheses to be completed.

(a)

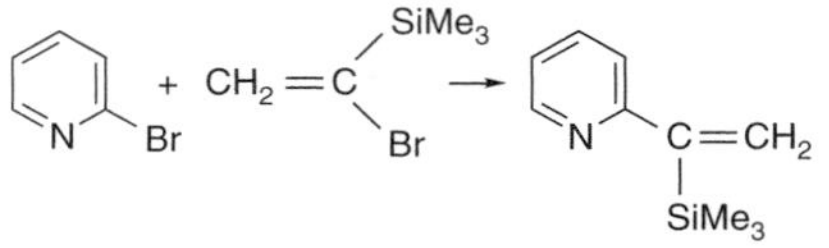

(b)

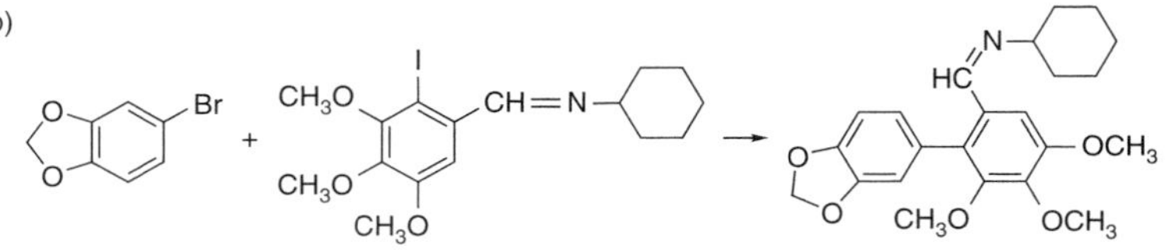

(c)

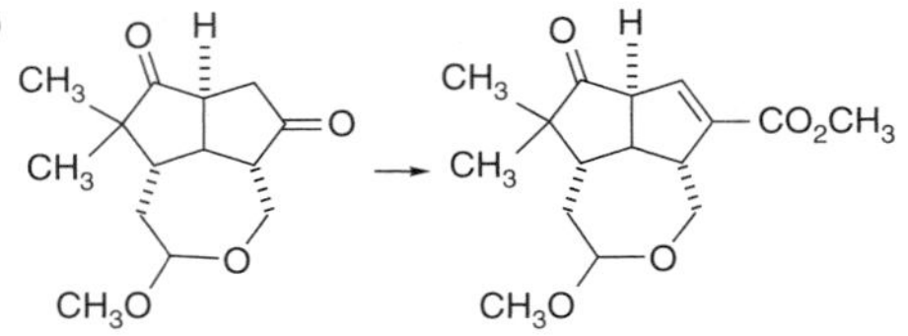

(d)

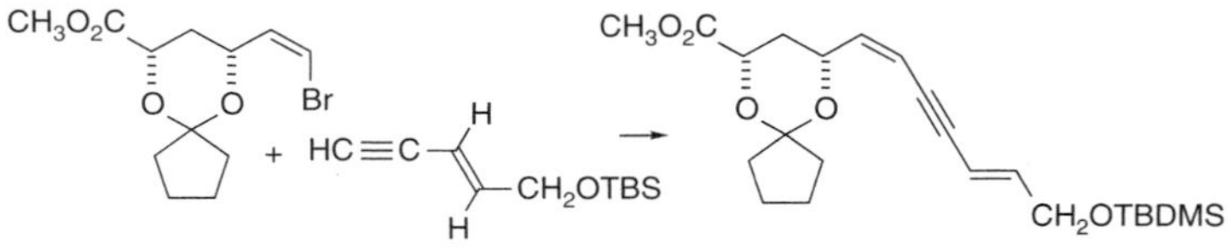

(e)

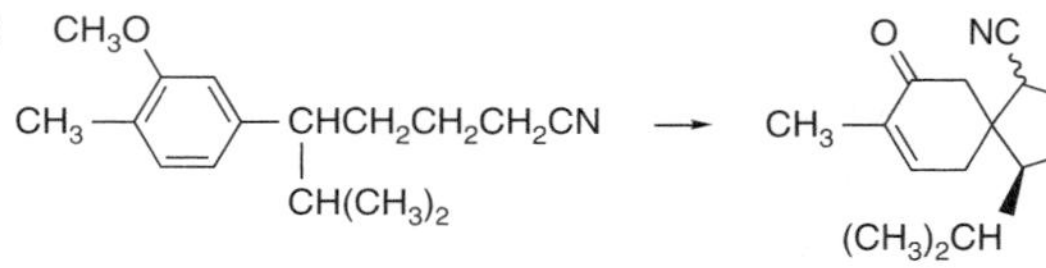

(f)

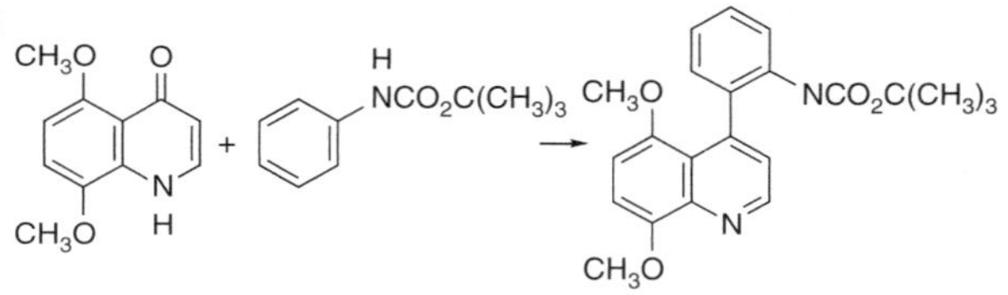

(g)

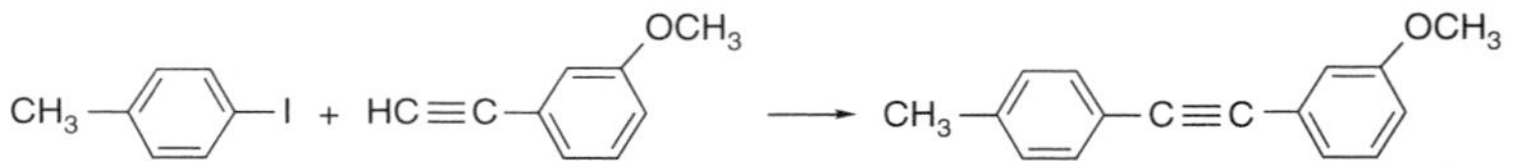

(h)

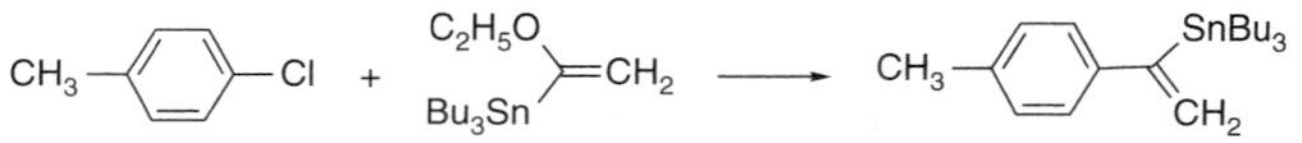

(i)

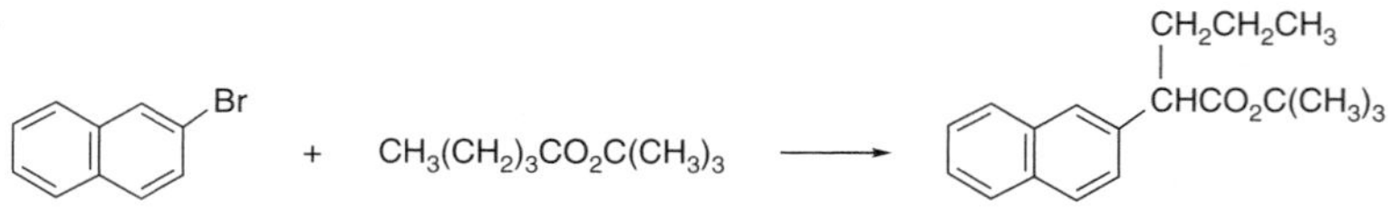

(j)

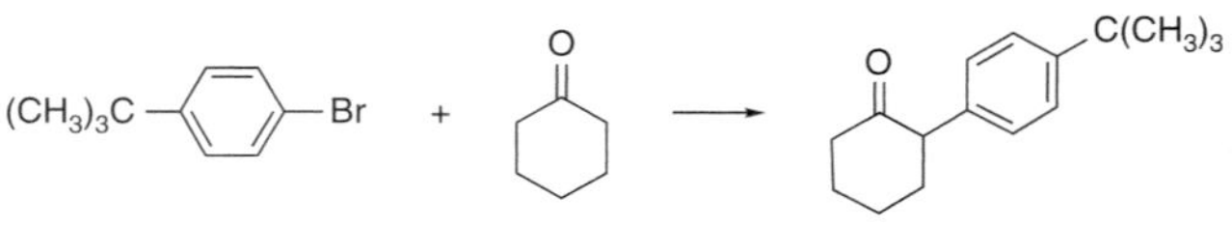

(k)

(l)

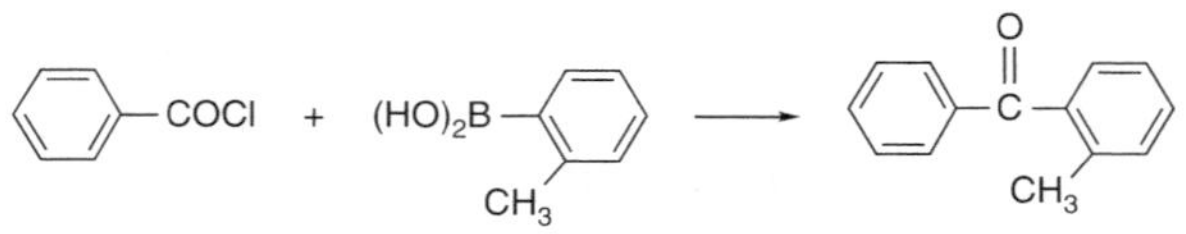

8.16. Each of the following reactions can be accomplished with a palladium reagent or catalyst. Write a detailed mechanism for each reaction. The number of equivalents of each reagent is given in parentheses. Specify the oxidation state of Pd in the intermediates. Be sure your mechanism accounts for the regeneration of catalytically active species in those reactions that are catalytic in palladium.

(a)

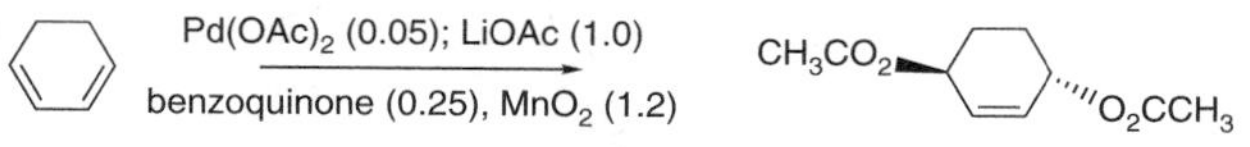

(b)

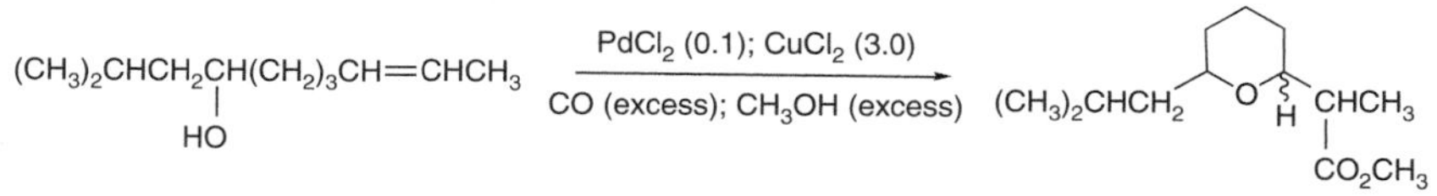

(c)

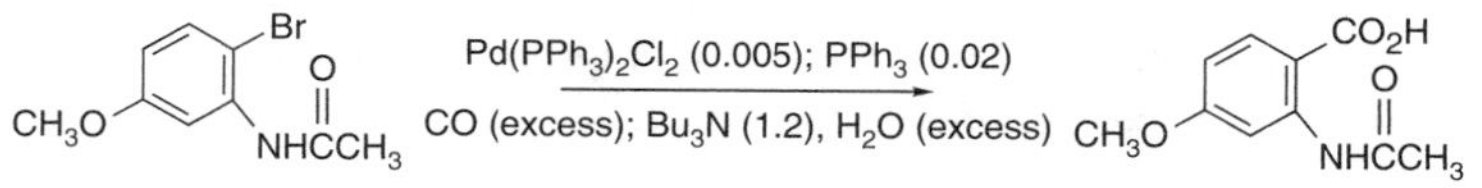

(d)

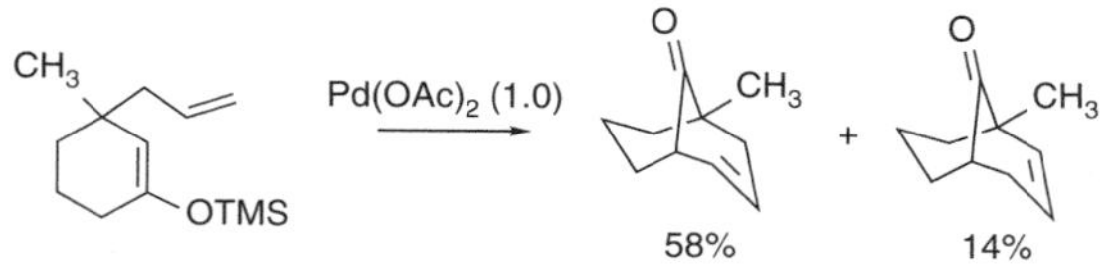

58%   14%

8.17. The reaction of lithium dimethylcuprate with **17-A** shows considerable 1,4-diastereoselectivity. Offer an explanation, including a transition structure.

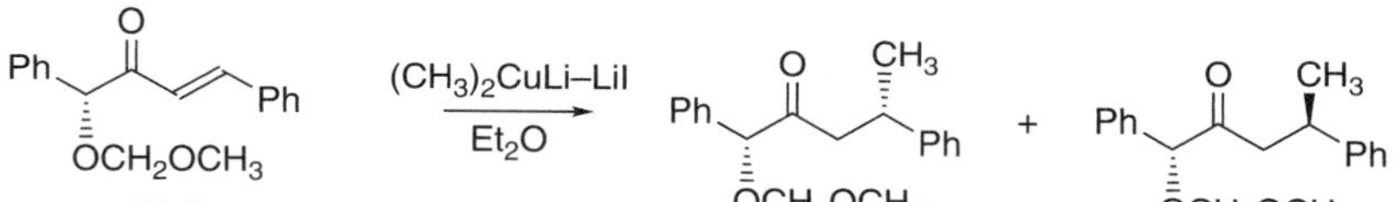

13:1 ratio

8.18. The following transformations have been carried out to yield a specific enantiomer using organometallic reagents. Devise a strategy by which organometallic reagents or catalysts can be used to prepare the desired compound from the specified starting material.

(a)

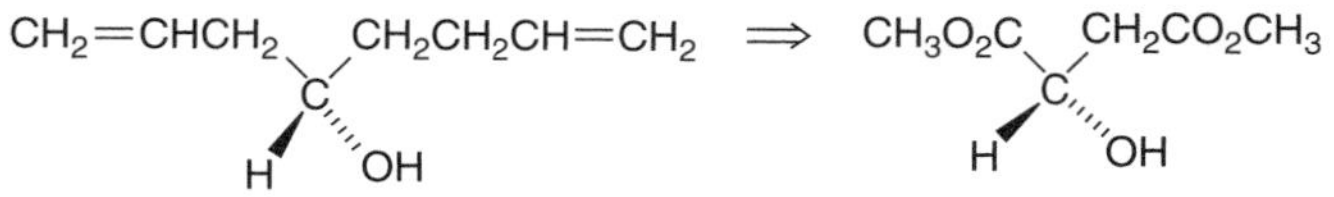

(b)

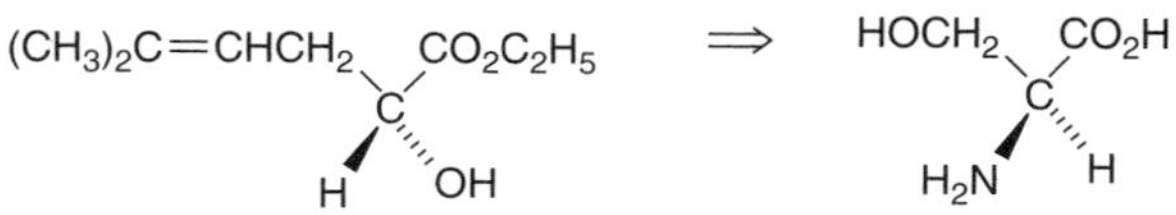

(c)

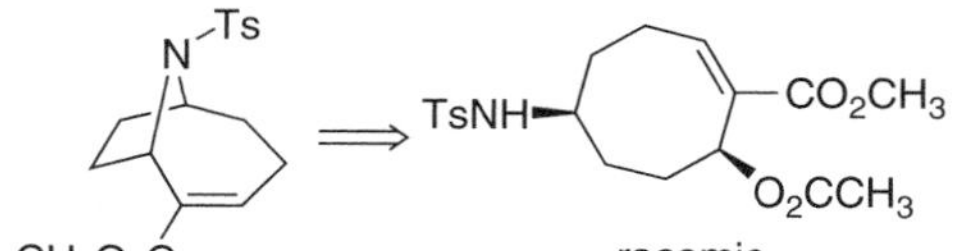

racemic

8.19. Under the conditions of the Wacker oxidation, 4-trimethylsilyl-3-alkyn-1-ols give γ-lactones. Similarly, *N*-carbamoyl or *N*-acetyl 4-trimethylsilyl-3-alkynamines cyclize to γ-lactams. Formulate a mechanism for these reactions. (Hint: In D_2O, the reaction gives 3,3-dideuterated products.)

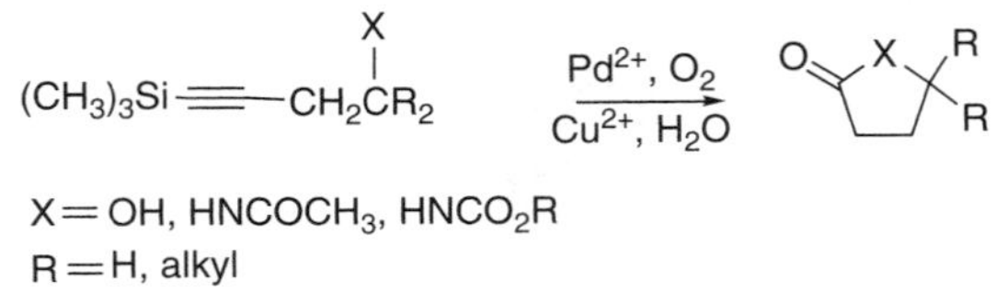

$$(CH_3)_3Si \equiv\!\!\!=\!\!\!-CH_2\overset{\overset{\textstyle X}{\textstyle |}}{C}R_2 \quad \xrightarrow[Cu^{2+},\ H_2O]{Pd^{2+},\ O_2} \quad$$

X = OH, HNCOCH₃, HNCO₂R
R = H, alkyl

8.20. The tricyclic compound **20-C**, a potential intermediate for alkaloid synthesis, has been prepared by an intramolecular Diels-Alder reaction of the ketone obtained by deprotection and oxidation of **20-B**. Compound **20-B** was prepared from **20-A** using alkyne-ethene metathesis chemistry. Show the mechanistic steps involved in conversion of **20-A** to **20-B**.

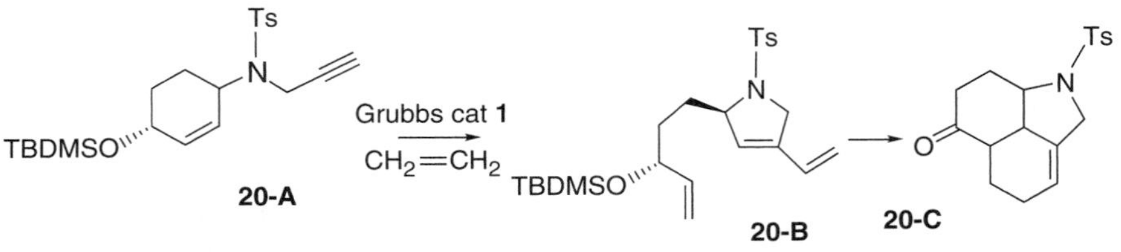

Grubbs cat **1**
CH₂=CH₂

20-A   **20-B**   **20-C**

9

Carbon-Carbon Bond-Forming Reactions of Compounds of Boron, Silicon, and Tin

Introduction

In this chapter we discuss the use of boron, silicon, and tin compounds to form carbon-carbon bonds. These elements are at the metal-nonmetal boundary, with boron being the most and tin the least electronegative of the three. The neutral alkyl derivatives of boron have the formula R_3B, whereas silicon and tin are tetravalent compounds, R_4Si and R_4Sn. These compounds are relatively volatile nonpolar substances that exist as discrete molecules and in which the carbon-metal bonds are largely covalent. By virtue of the electron deficiency at boron, the boranes are Lewis acids. Silanes do not have strong Lewis acid character but can form pentavalent adducts with hard bases such as alkoxides and especially fluoride. Silanes with halogen or sulfonate substituents are electrophilic and readily undergo nucleophilic displacement. Stannanes have the potential to act as Lewis acids when substituted by electronegative groups such as halogens. Either displacement of a halide or expansion to pentacoordinate or hexacoordinate structures is possible.

In contrast to the transition metals, where there is often a change in oxidation level at the metal during the reaction, there is usually no change in oxidation level for boron, silicon, and tin compounds. The synthetically important reactions of these three groups of compounds involve transfer of a carbon substituent with one (radical equivalent) or two (carbanion equivalent) electrons to a reactive carbon center. Here we focus on the nonradical reactions and deal with radical reactions in Chapter 10. We have already introduced one important aspect of boron and tin chemistry in the transmetallation reactions involved in Pd-catalyzed cross-coupling reactions, discussed

783

784

CHAPTER 9

*Carbon-Carbon
Bond-Forming Reactions
of Compounds of Boron,
Silicon, and Tin*

in Section 8.2.3. This chapter emphasizes the use of boranes, silanes, and stannanes as sources of nucleophilic carbon groups toward a variety of electrophiles, especially carbonyl compounds.

Allylic derivatives are particularly important in the case of boranes, silanes, and stannanes. Allylic boranes effect nucleophilic addition to carbonyl groups via a cyclic TS that involves the Lewis acid character of the borane. 1,3-Allylic transposition occurs through the cyclic TS.

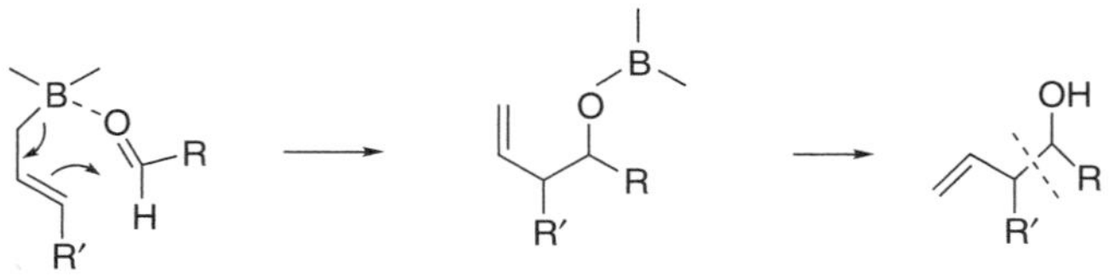

Allylic silanes and stannanes react with various electrophiles with demetallation. These reactions can occur via several related mechanisms. Both types of reagents can deliver allylic groups to electrophilic centers such as carbonyl and iminium.

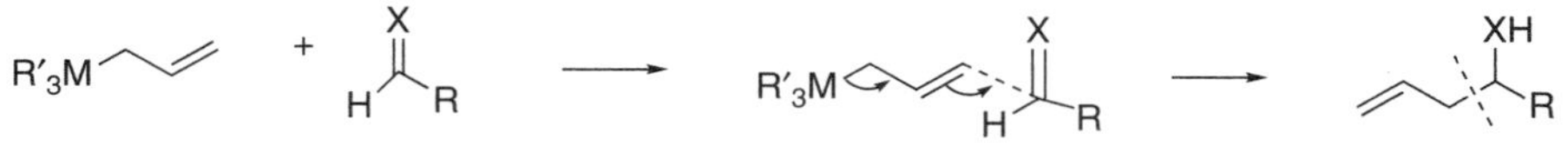

M = Si, Sn   X = O, NY

Alkenyl silanes and stannanes have the potential for nucleophilic delivery of vinyl groups to a variety of electrophiles. Demetallation also occurs in these reactions, so the net effect is substitution for the silyl or the stannyl group.

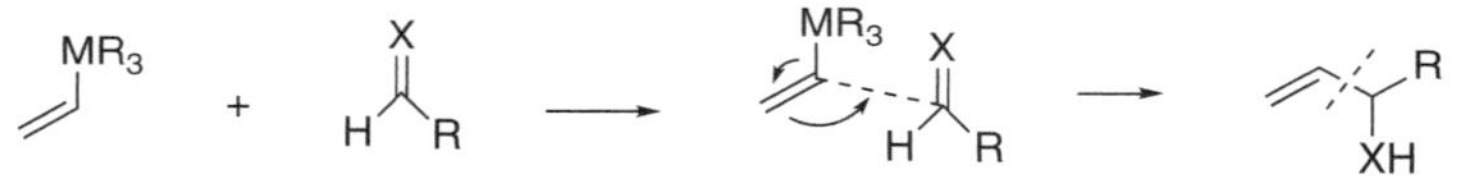

M = Si, Sn   X = O, NY

9.1. Organoboron Compounds

9.1.1. Synthesis of Organoboranes

The most widely used route to organoboranes is hydroboration, introduced in Section 4.5.1, which provides access to both alkyl- and alkenylboranes. Aryl-, methyl-, allylic, and benzylboranes cannot be prepared by hydroboration, and the most general route to these organoboranes is by reaction of an organometallic compound with a halo- or alkoxyboron derivative.[1]

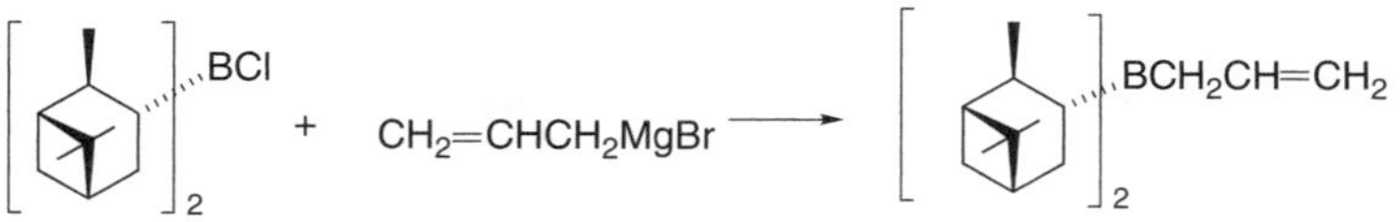

[1] H. C. Brown and P. K. Jadhar, *J. Am. Chem. Soc.*, **105**, 2092 (1983).

Alkyl, aryl, and allyl derivatives of boron can be prepared directly from the corresponding halides, BF_3, and magnesium metal. This process presumably involves in situ generation of a Grignard reagent, which then displaces fluoride from boron.[2]

$$3\ R-X\ +\ BF_3\ +\ 3\ Mg\ \longrightarrow\ R_3B\ +\ 3\ MgXF$$

Alkoxy groups can be displaced from boron by alkyl- or aryllithium reagents. The reaction of diisopropoxy boranes with an organolithium reagent, for example, provides good yields of unsymmetrically disubstituted isopropoxyboranes.[3]

$$RB(Oi\text{-}Pr)_2 + R'Li \longrightarrow \underset{R'}{\overset{R}{\diagup}} B-O-i\text{-}Pr$$

Organoboranes can also be made using organocopper reagents. One route to methyl and aryl derivatives is by reaction of a dialkylborane, such as 9-BBN, with a cuprate reagent.[4]

$$BH + R_2CuLi \longrightarrow B-R + [RCuH]^-Li^+$$

These reactions occur by oxidative addition at copper, followed by decomposition of the Cu(III) intermediate.

$$R'_2B-H + {}^-[Cu^IR_2] \longrightarrow R'-\overset{\overset{\displaystyle H}{|}}{\underset{\underset{\displaystyle R'}{|}}{B}}\overset{\diagup R}{\underset{\diagdown R}{\cdots Cu}} \longrightarrow \left[R'-\overset{\overset{\displaystyle R'\ H}{|\ \ |}}{\underset{\underset{\displaystyle R}{|}}{B}}=Cu^{III}R \right]^- \longrightarrow \underset{R'}{\overset{R'}{\diagup}}B-R\ +\ [RCu^IH]^-$$

Two successive reactions with different organocuprates can convert thexylborane to an unsymmetrical trialkylborane.[5]

$$\overset{|}{\underset{|}{\diagup}}-BH_2 \xrightarrow{R_2^1CuLi} \xrightarrow{R_2^2CuLi} \overset{|}{\underset{|}{\diagup}}-B\overset{\diagup R^1}{\underset{\diagdown R^2}{}}$$

In addition to trialkylboranes, various alkoxyboron compounds have prominent roles in synthesis. Some of these, such as catecholboranes (see. p. 340) can be made by hydroboration. Others are made by organometallic or related substitution reactions. Alkoxyboron compounds are usually named as esters. Compounds with one alkoxy group are esters of borinic acids and are called *borinates*. Compounds with two alkoxy groups are called *boronates*. Trialkoxyboron compounds are *borates*.

R_2BOH	R_2BOR'	$RB(OH)_2$	$RB(OR')_2$	$B(OH)_3$	$B(OR')_3$
borinic acid	borinate	boronic acid	boronate	boric acid	borate

[2] H. C. Brown and U. S. Racherla, *J. Org. Chem.*, **51**, 427 (1986).
[3] H. C. Brown, T. E. Cole, and M. Srebnik, *Organometallics*, **4**, 1788 (1985).
[4] C. G. Whiteley and I. Zwane, *J. Org. Chem.*, **50**, 1969 (1985).
[5] C. G. Whiteley, *Tetrahedron Lett.*, **25**, 5563 (1984).

The cyclic five- and six-membered boronate esters are used frequently. Their systematic names are 1,3,2-dioxaborolane and 1,3,2-dioxaborinanes, respectively.

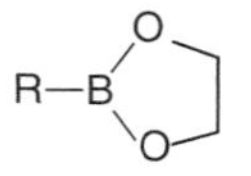

1,3,2-dioxaborolane 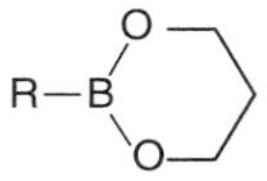   1,3,2-dioxaborinane

9.1.2. Carbonylation and Other One-Carbon Homologation Reactions

The reactions of organoboranes that we discussed in Chapter 4 are valuable methods for introducing functional groups such as hydroxy, amino, and halogen into alkenes. In this section we consider carbon-carbon bond-forming reactions of boron compounds.[6] Trivalent organoboranes are not very nucleophilic but they are moderately reactive Lewis acids. Most reactions in which carbon-carbon bonds are formed involve a tetracoordinate intermediate that has a negative charge on boron. Adduct formation weakens the boron-carbon bonds and permits a transfer of a carbon substituent with its electrons. The general mechanistic pattern is shown below.

The electrophilic center is sometimes generated from the Lewis base by formation of the adduct, and the reaction proceeds by migration of a boron substituent.

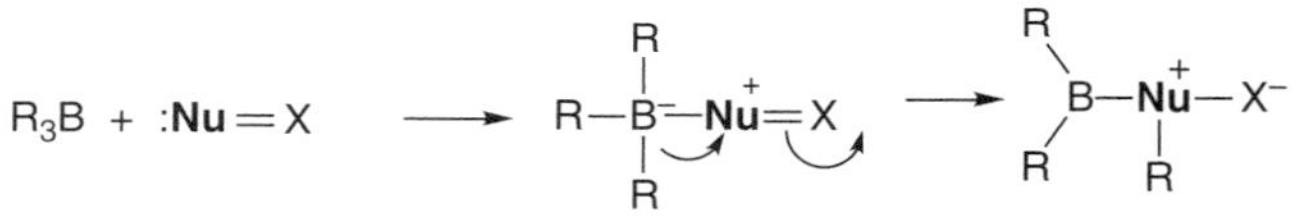

A significant group of reactions of this type involves the reactions of organoboranes with carbon monoxide, which forms Lewis acid-base complexes with the organoboranes. In these adducts the boron bears a formal negative charge and carbon is electrophilic because the triply bound oxygen bears a formal positive charge. The adducts undergo boron to carbon migration of the alkyl groups. The reaction can be controlled so that it results in the migration of one, two, or all three of the boron substituents.[7] If the organoborane is heated with carbon monoxide to 100°–125 °C, all of the groups migrate and a tertiary alcohol is obtained after workup by oxidation. The presence of water causes the reaction to cease after migration of two groups from boron to carbon. Oxidation of the reaction mixture at this stage gives a ketone.[8] Primary alcohols are obtained when the carbonylation is carried out in the presence of

---

[6.] For a review of this topic, see E. Negishi and M. Idacavage, *Org. React.*, **33**, 1 (1985).

[7.] H. C. Brown and M. W. Rathke, *J. Am. Chem. Soc.*, **89**, 2737 (1967).

[8.] H. C. Brown and M. W. Rathke, *J. Am. Chem. Soc.*, **89**, 2738 (1967).

sodium borohydride or lithium borohydride.[9] The product of the first migration step is reduced and subsequent hydrolysis gives a primary alcohol.

$$R_3B^=C\equiv O^+$$

NaBH$_4$ 100–125°C

100°C | H$_2$O

   OH
   |
R$_2$B—CHR

H$_2$O, | OH

RCH$_2$OH

   OH
   |
RB—CR$_2$
|
HO

| H$_2$O$_2$

  O
  ||
 RCR

"O=B—CR$_3$"

| H$_2$O$_2$, OH

R$_3$COH

In this synthesis of primary alcohols, only one of the three groups in the organoborane is converted to product. This disadvantage can be overcome by using a dialkylborane, particularly 9-BBN, in the initial hydroboration. (See p. 338 to review the abbreviations of some of the common boranes.) After carbonylation and B → C migration, the reaction mixture can be processed to give an aldehyde, an alcohol, or the homologated 9-alkyl-BBN.[10] The utility of 9-BBN in these procedures is the result of the minimal tendency of the bicyclic ring to undergo migration.

$^-$OH → HOCH$_2$CH$_2$CH$_2$R

B—CH$_2$CH$_2$R   RCH=CH$_2$ B—CH$_2$CH$_2$R   KBH(O-*i*-Pr)$_3$ / CO   B—CHCH$_2$CH$_2$R

LiAlH / −20°C → B—CH$_2$CH$_2$CH$_2$R

H$_2$O$_2$ → O=CHCH$_2$CH$_2$R

Several alternative procedures have been developed in which other reagents replace carbon monoxide as the migration terminus.[11] The most generally applicable of these methods involves the use of cyanide ion and trifluoroacetic anhydride (TFAA). In this reaction the borane initially forms an adduct with cyanide ion. The migration is induced by N-acylation of the cyano group by TFAA. Oxidation and hydrolysis then give a ketone.

$$R_3B + {}^-CN \longrightarrow R_3B^- —C\equiv N \xrightarrow{(CF_3CO)_2O} R_3B^- —C\equiv N^+ —\overset{\overset{O}{||}}{C}CF_3$$

$$R_3B^- —C\equiv N^+ —\overset{\overset{O}{||}}{C}CF_3 \longrightarrow R_2B—\underset{R}{\overset{\overset{O}{||}}{C}}=N—\overset{\overset{O}{||}}{C}CF_3 \longrightarrow R—B\underset{R\quad R}{\overset{O—CCF_3}{\diagdown \ \diagup N}}C \xrightarrow{H_2O_2} \overset{\overset{O}{||}}{R}CR$$

9. M. W. Rathke and H. C. Brown, *J. Am. Chem. Soc.*, **89**, 2740 (1967).
10. H. C. Brown, E. F. Knights, and R. A. Coleman, *J. Am. Chem. Soc.*, **91**, 2144 (1969); H. C. Brown, T. M. Ford, and J. L. Hubbard, *J. Org. Chem.*, **45**, 4067 (1980).
11. H. C. Brown and S. M. Singh, *Organometallics*, **5**, 998 (1986).

788

CHAPTER 9

*Carbon-Carbon
Bond-Forming Reactions
of Compounds of Boron,
Silicon, and Tin*

Another useful reagent for introduction of the carbonyl carbon is dichloromethyl methyl ether. In the presence of a hindered alkoxide base, it is deprotonated and acts as a nucleophile toward boron. Rearrangement then ensues with migration of two boron substituents. Oxidation gives a ketone.

$$R_3B + {}^-:CCl_2OCH_3 \longrightarrow R_3B^- {-} CCl_2OCH_3$$

$$R_3B^- {-} CCl_2OCH_3 \longrightarrow \underset{\underset{R}{|}}{R_2}B^- {-} \overset{\overset{Cl}{|}}{C}ClOCH_3 \longrightarrow RB^- {-} \underset{\underset{R}{|}}{C}OCH_3$$

$$R {-} \overset{\overset{Cl}{|}}{\underset{\underset{Cl}{|}}{B^-}} {-} \overset{\overset{R}{|}}{\underset{\underset{R}{|}}{C}}OCH_3 \xrightarrow{H_2O_2} R_2C{=}O$$

Unsymmetrical ketones can be made by using either thexylborane or thexylchloroborane.[12] Thexylborane works well when one of the desired carbonyl substituents is derived from a moderately hindered alkene. Under these circumstances, a clean monoalkylation of thexylborane can be accomplished, which is then followed by reaction with a second alkene and carbonylation.

$$(CH_3)_2CHC\underset{\underset{CH_3}{|}}{\overset{\overset{CH_3}{|}}{}}{-}BH_2 \xrightarrow[\text{RCH=CHR} \ \ \text{R'CH=CH}_2]{} \xrightarrow[\substack{2) \ H_2O, \ 100°C \\ 3) \ H_2O_2}]{1) \ CO} RCH_2\underset{\underset{R}{|}}{C}H\overset{\overset{O}{\|}}{C}CH_2CH_2R'$$

Thexylchloroborane can be alkylated and then converted to a dialkylborane by a reducing agent such as $KBH[OCH(CH_3)_2]_3$, an approach that is preferred for terminal alkenes.

$$(CH_3)_2CHC\underset{\underset{CH_3}{|}}{\overset{\overset{CH_3}{|}}{}}{-}BHCl \xrightarrow[\substack{2) \ KBH[OCH(CH_3)_2]_3}]{1) \ CH_3CH_2CH=CH_2} (CH_3)_2CHC\underset{\underset{CH_3}{|}}{\overset{\overset{CH_3}{|}}{}}{-}BH{-}CH_2CH_2CH_2CH_3$$

The success of both of these methods depends upon the thexyl group being noncompetitive with the other groups in the migration steps.

The formation of unsymmetrical ketones can also be done starting with $IpcBCl_2$. Sequential reduction and hydroboration are carried out with two different alkenes. The first reduction can be done with $(CH_3)_3SiH$, but the second stage requires $LiAlH_4$.

12. H. C. Brown and E. Negishi, *J. Am. Chem. Soc.*, **89**, 5285 (1967); S. U. Kulkarni, H. D. Lee, and H. C. Brown, *J. Org. Chem.*, **45**, 4542 (1980).

In this procedure, dichloromethyl methyl ether is used as the source of the carbonyl carbon.[13]

Scheme 9.1 shows several examples of one-carbon homologations involving boron to carbon migration. Entry 1 illustrates the synthesis of a symmetrical tertiary alcohol. Entry 2 involves interception of the intermediate after the first migration by reduction. Acid then induces a second migration. This sequence affords secondary alcohols.

Entries 3 to 5 show the use of alternative sources of the one carbon unit. In Entry 3, a tertiary alcohol is formed with one of the alkyl groups being derived from the dithioacetal reagent. Related procedures have been developed for ketones and tertiary alcohols using 2-lithio-2-alkyl-1,3-benzodithiole as the source of the linking carbon.[14] Problem 9.3 deals with the mechanisms of these reactions.

Section B of the Scheme 9.1 shows several procedures for the synthesis of ketones. Entry 6 is the synthesis of a symmetrical ketone by carbonylation. Entry 7 illustrates the synthesis of an unsymmetrical ketone by the thexylborane method and also demonstrates the use of a functionalized olefin. Entries 8 to 10 illustrate synthesis of ketones by the cyanide-TFAA method. Entry 11 shows the synthesis of a bicyclic ketone involving intramolecular hydroboration of 1,5-cyclooctadiene. Entry 12 is another ring closure, generating a potential steroid precursor.

Section C illustrates the synthesis of aldehydes by boron homologation. Entry 13 is an example of synthesis of an aldehyde from an alkene using 9-BBN for hydroboration. Entry 14 illustrates an efficient process for one-carbon homologation to aldehydes that is based on cyclic boronate esters. These can be prepared by hydroboration of an alkene with dibromoborane, followed by conversion of the dibromoborane to the cyclic boronate. The homologation step is carried out by addition of methoxy(phenylthio)methyllithium to the boronate. The migration step is induced by mercuric ion. Use of chiral boranes and boronates leads to products containing groups of retained configuration.[15]

13. H. C. Brown, S. V. Kulkarni, U. S. Racherla, and U. P. Dhokte, *J. Org. Chem.*, **63**, 7030 (1998).
14. S. Ncube, A. Pelter, and K. Smith, *Tetrahedron Lett.*, 1893, 1895 (1979).
15. M. V. Rangaishenvi, B. Singaram, and H. C. Brown, *J. Org. Chem.*, **56**, 3286 (1991).

Scheme 9.1. Homologation and Coupling of Organoboranes by Carbon Monoxide and Other One-Carbon Donors

A. Formation of alcohols

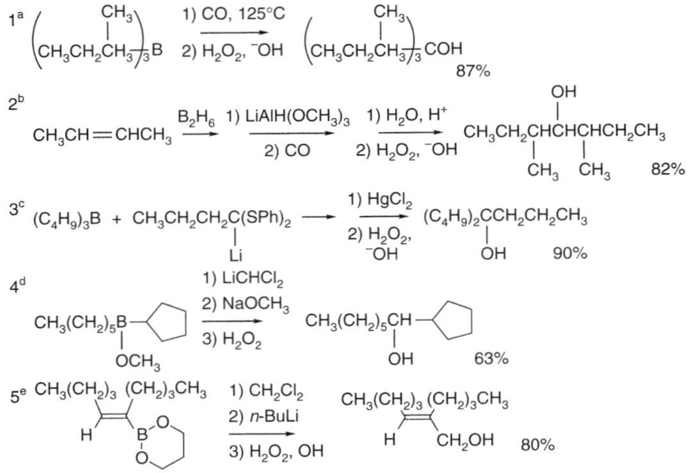

B. Formation of ketones

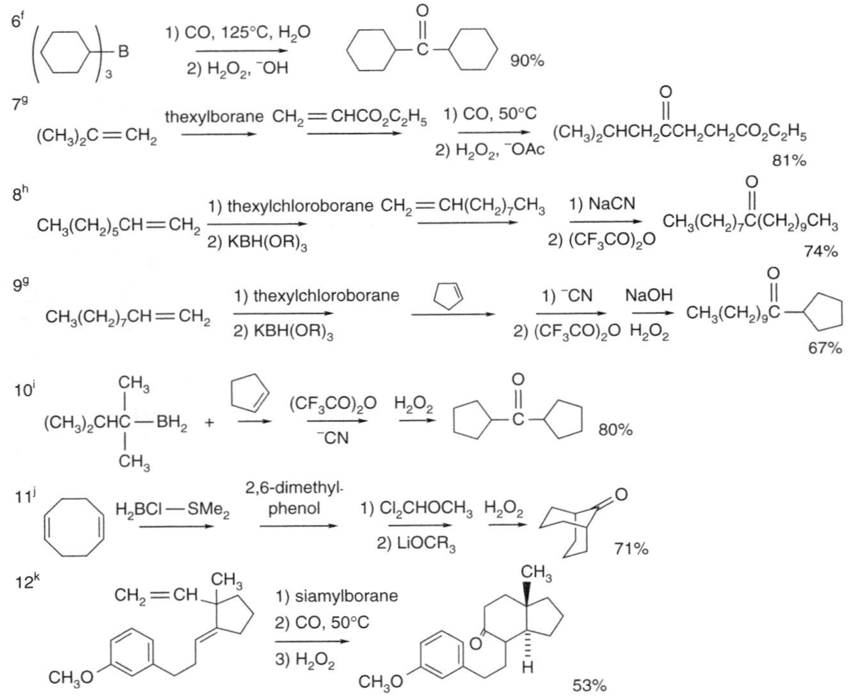

(Continued)

Scheme 9.1. (*Continued*) 791

C. Formation of aldehydes

13[l]

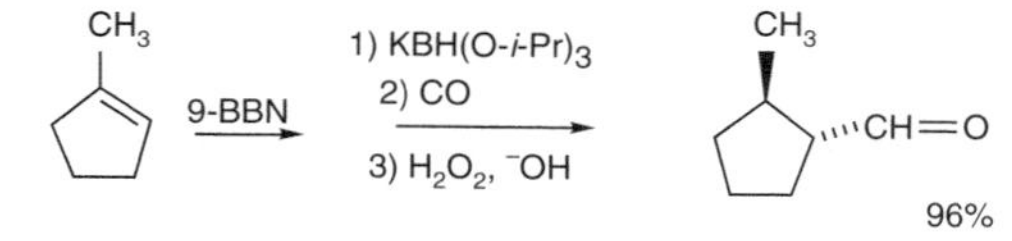

1) KBH(O-*i*-Pr)$_3$
2) CO
3) H$_2$O$_2$, $^-$OH

96%

14[m]

1) LiCHOCH$_3$
   |
   SPh
2) HgCl$_2$
3) H$_2$O$_2$, pH 8

64%

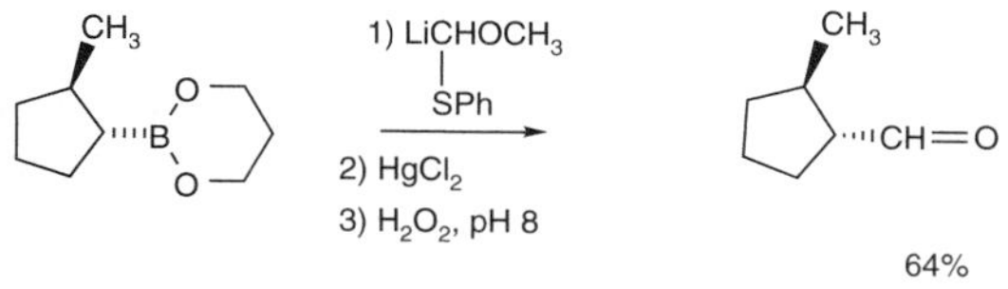

a. H. C. Brown and M. W. Rathke, *J. Am. Chem. Soc.*, **89**, 2737 (1967).
b. J. L. Hubbard and H. C. Brown, *Synthesis*, 676 (1978).
c. R. J. Hughes, S. Ncube, A. Pelter, K. Smith, E. Negishi, and T. Yoshida, *J. Chem. Soc., Perkin Trans. 1*, 1172 (1977); S. Ncube, A. Pelter, and K. Smith, *Tetrahedron Lett.*, 1893, 1895 (1979).
d. H. C. Brown, T. Imai, P. T. Perumal, and B. Singaram, *J. Org. Chem.*, **50**, 4032 (1985).
e. H. C. Brown, A. S. Phadke, and N. G. Bhat, *Tetrahedron Lett.*, **34**, 7845 (1993).
f. H. C. Brown and M. W. Rathke, *J. Am. Chem. Soc.*, **89**, 2738 (1967).
g. H. C. Brown and E. Negishi, *J. Am. Chem. Soc.*, **89**, 5285 (1967).
h. S. U. Kulkarni, H. D. Lee, and H. C. Brown, *J. Org. Chem.*, **45**, 4542 (1980).
i. A. Pelter, K. Smith, M. G. Hutchings, and K. Rowe, *J. Chem. Soc., Perkin Trans. 1*, 129 (1975).
j. H. C. Brown and S. U. Kulkarni, *J. Org. Chem.*, **44**, 2422 (1979).
k. T. A. Bryson and W. E. Pye, *J. Org. Chem.*, **42**, 3214 (1977).
l. H. C. Brown, J. L. Hubbard, and K. Smith, *Synthesis*, 701 (1979).
m. H. C. Brown and T. Imai, *J. Am. Chem. Soc.*, **105**, 6285 (1983).

As can be judged from the preceding discussion, organoboranes are versatile intermediates for formation of carbon-carbon bonds. An important aspect of all of these synthetic procedures involving boron to carbon migration is that they occur with *retention of the configuration of the migrating group.* Since effective procedures for enantioselective hydroboration have been developed (see Section 4.5.3), these reactions offer the opportunity for enantioselective synthesis. A sequence for enantioselective formation of ketones starts with hydroboration by mono(isopinocampheyl)borane, (IpcBH$_2$), which can be obtained in high enantiomeric purity.[16] The hydroboration of a prochiral alkene establishes a new stereocenter. A third alkyl group can be introduced by a second hydroboration step.

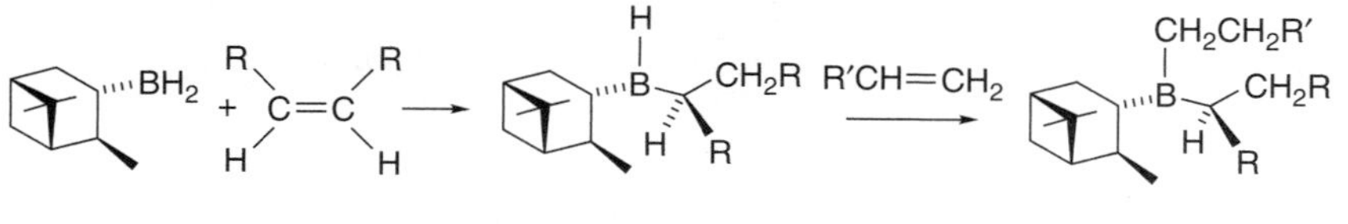

The trialkylborane can be transformed to a dialkyl(ethoxy)borane by heating with acetaldehyde, which releases the original chiral α-pinene. Finally application of one of the carbonylation procedures outlined in Scheme 9.1 gives a chiral ketone.[17] The enantiomeric excess observed for ketones prepared in this way ranges from 60–90%.

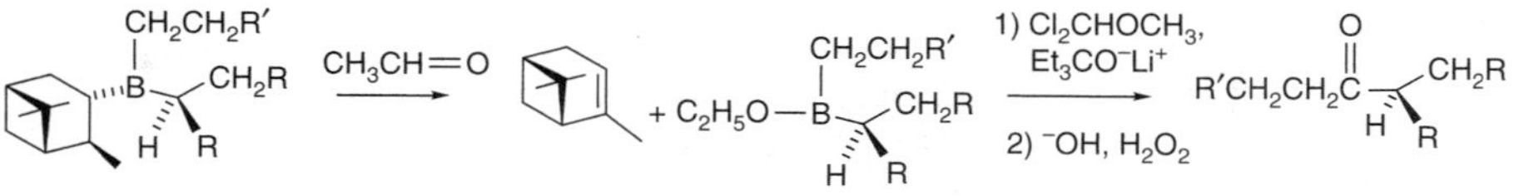

[16.] H. C. Brown, P. K. Jadhav, and A. K. Mandal, *J. Org. Chem.*, **47**, 5074 (1982).
[17.] H. C. Brown, R. K. Jadhav, and M. C. Desai, *Tetrahedron*, **40**, 1325 (1984).

792

CHAPTER 9

*Carbon-Carbon
Bond-Forming Reactions
of Compounds of Boron,
Silicon, and Tin*

Higher enantiomeric purity can be obtained by a modified procedure in which the monoalkylborane intermediate is prepared by reduction of a cyclic boronate.[18]

Subsequent steps involve introduction of a thexyl group and then the second ketone substituent. Finally, the ketone is formed by the cyanide-TFAA method.

By starting with enantiomerically enriched IpcBHCl, it is possible to construct chiral cyclic ketones. For example, stepwise hydroboration of 1-allylcyclohexene and ring construction provides *trans*-1-decalone in greater than 99% e.e.[19]

9.1.3. Homologation via α-Haloenolates

Organoboranes can also be used to construct carbon-carbon bonds by several other types of reactions that involve migration of a boron substituent to carbon. One such reaction involves α-halo carbonyl compounds.[20] For example, ethyl bromoacetate reacts with trialkylboranes in the presence of base to give alkylated acetic acid derivatives in excellent yield. The reaction is most efficiently carried out with a 9-BBN derivative. These reactions can also be effected with β-alkenyl derivatives of 9-BBN to give β,γ-unsaturated esters.[21]

18. H. C. Brown, R. K. Bakshi, and B. Singaram, *J. Am. Chem. Soc.*, **110**, 1529 (1988); H. C. Brown, M. Srebnik, R. K. Bakshi, and T. E. Cole, *J. Am. Chem. Soc.*, **109**, 5420 (1987).
19. H. C. Brown, V. K. Mahindroo, and U. P. Dhokte, *J. Org. Chem.*, **61**, 1906 (1996); U. P. Dhokte, P. M. Pathare, V. K. Mahindroo, and H. C. Brown, *J. Org. Chem.*, **63**, 8276 (1998).
20. H. C. Brown, M. M. Rogic, M. W. Rathke, and G. W. Kabalka, *J. Am. Chem. Soc.*, **90**, 818 (1968); H. C. Brown and M. M. Rogic, *J. Am. Chem. Soc.*, **91**, 2146 (1969).
21. H. C. Brown, N. G. Bhat, and J. B. Cambell, Jr., *J. Org. Chem.*, **51**, 3398 (1986).

The reactions can be made enantioselective by using enantiomerically pure IpcBH$_2$ for hydroboration of alkenes and then transforming the products to enantiomerically pure derivatives of 9-BBN by reaction with 1,5-cyclooctadiene.[22]

The mechanism of these alkylations involves a tetracoordinate boron intermediate formed by addition of the enolate of the α-bromo ester to the organoborane. The migration then occurs with displacement of bromide ion. In agreement with this mechanism, retention of configuration of the migrating group is observed.[23]

α-Halo ketones and α-halo nitriles undergo similar reactions.[24]

A closely related reaction employs α-diazo esters or α-diazo ketones.[25] With these compounds, molecular nitrogen acts as the leaving group in the migration step. The best results are achieved using dialkylchloroboranes or monoalkyldichloroboranes.

$$RBCl_2 + N_2CHCO_2CH_3 \longrightarrow RCH_2CO_2CH_3$$

A number of these alkylation reactions are illustrated in Scheme 9.2. Entries 1 and 2 are typical examples of α-halo ester reactions. Entry 3 is a modification in which the highly hindered base potassium 2,6-di-*t*-butylphenoxide is used. Similar reaction conditions can be used with α-halo ketones (Entries 4 and 5) and nitriles (Entry 6). Entries 7 to 9 illustrate the use of diazo esters and diazo ketones. Entry 10 shows an application of the reaction to the synthesis of an amide.

9.1.4. Stereoselective Alkene Synthesis

Several methods for stereoselective alkene synthesis are based on boron intermediates. One approach involves alkenylboranes, which can be prepared from terminal alkynes. Procedures have been developed for the synthesis of both *Z*- and *E*-alkenes.

22. H. C. Brown, N. N. Joshi, C. Pyun, and B. Singaram, *J. Am. Chem. Soc.*, **111**, 1754 (1989).
23. H. C. Brown, M. M. Rogic, M. W. Rathke, and G. W. Kabalka, *J. Am. Chem. Soc.*, **91**, 2151 (1969).
24. H. C. Brown, M. M. Rogic, H. Nambu, and M. W. Rathke, *J. Am. Chem. Soc.*, **91**, 2147 (1969); H. C. Brown, H. Nambu, and M. M. Rogic, *J. Am. Chem. Soc.*, **91**, 6853, 6855 (1969).
25. H. C. Brown, M. M. Midland, and A. B. Levy, *J. Am. Chem. Soc.*, **94**, 3662 (1972); J. Hooz, J. N. Bridson, J. G. Calzada, H. C. Brown, M. M. Midland, and A. B. Levy, *J. Org. Chem.*, **38**, 2574 (1973).

794

CHAPTER 9

*Carbon-Carbon
Bond-Forming Reactions
of Compounds of Boron,
Silicon, and Tin*

Scheme 9.2. Homologation of Boranes by α-Halocarbonyl and Related Compounds

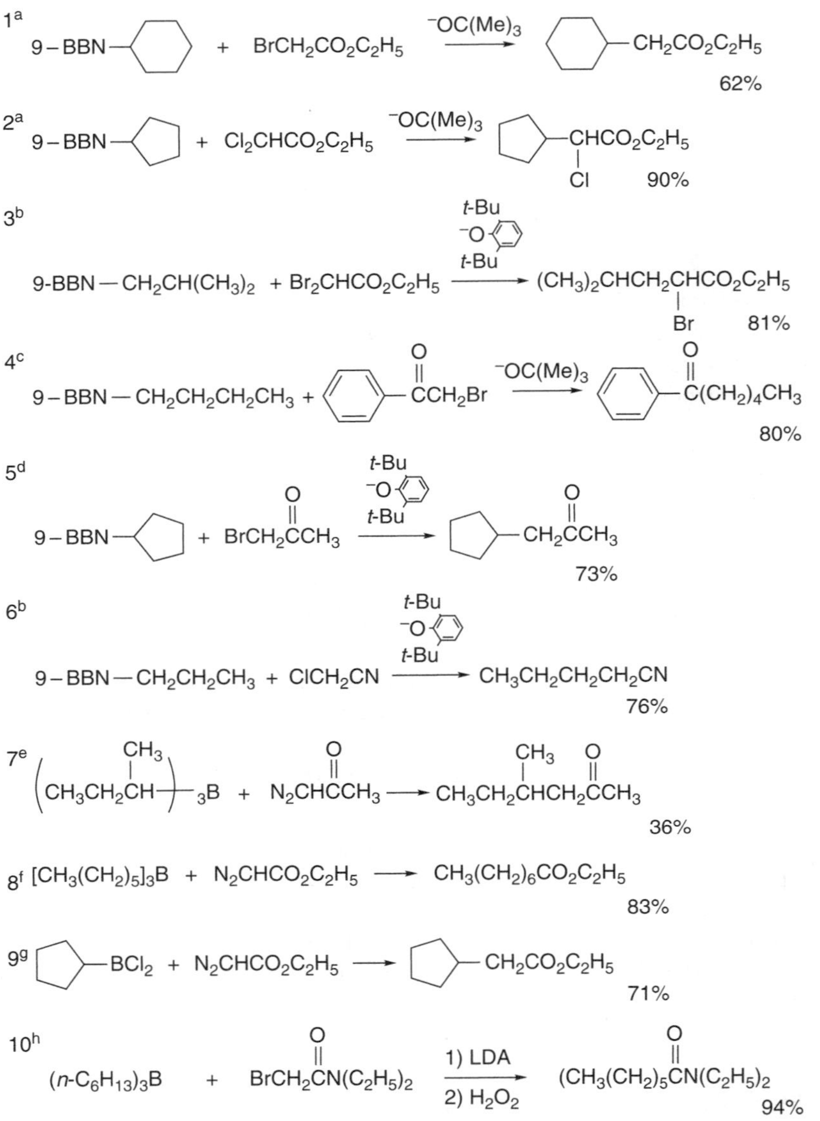

1[a]

9–BBN—⬡ + $BrCH_2CO_2C_2H_5$ $\xrightarrow{^-OC(Me)_3}$ ⬡—$CH_2CO_2C_2H_5$

62%

2[a]

9–BBN—⬠ + $Cl_2CHCO_2C_2H_5$ $\xrightarrow{^-OC(Me)_3}$ ⬠—CHCO$_2$C$_2$H$_5$ | Cl

90%

3[b]

9-BBN—$CH_2CH(CH_3)_2$ + $Br_2CHCO_2C_2H_5$ $\xrightarrow[\text{\emph{t}-Bu}]{^-O\langle\rangle\text{\emph{t}-Bu}}$ $(CH_3)_2CHCH_2CHCO_2C_2H_5$ | Br

81%

4[c]

9–BBN—$CH_2CH_2CH_2CH_3$ + Ph—$\overset{O}{\overset{\|}{C}}CH_2Br$ $\xrightarrow{^-OC(Me)_3}$ Ph—$\overset{O}{\overset{\|}{C}}(CH_2)_4CH_3$

80%

5[d]

9–BBN—⬠ + $BrCH_2\overset{O}{\overset{\|}{C}}CH_3$ $\xrightarrow[\text{\emph{t}-Bu}]{^-O\langle\rangle\text{\emph{t}-Bu}}$ ⬠—$CH_2\overset{O}{\overset{\|}{C}}CH_3$

73%

6[b]

9–BBN—$CH_2CH_2CH_3$ + $ClCH_2CN$ $\xrightarrow[\text{\emph{t}-Bu}]{^-O\langle\rangle\text{\emph{t}-Bu}}$ $CH_3CH_2CH_2CH_2CN$

76%

7[e]

$\left(CH_3CH_2\underset{CH_3}{\overset{CH_3}{CH}}\right)_3B$ + $N_2CH\overset{O}{\overset{\|}{C}}CH_3$ $\longrightarrow$ $CH_3CH_2\underset{CH_3}{\overset{CH_3}{CH}}CH_2\overset{O}{\overset{\|}{C}}CH_3$

36%

8[f] $[CH_3(CH_2)_5]_3B$ + $N_2CHCO_2C_2H_5$ $\longrightarrow$ $CH_3(CH_2)_6CO_2C_2H_5$

83%

9[g] ⬠—BCl_2 + $N_2CHCO_2C_2H_5$ $\longrightarrow$ ⬠—$CH_2CO_2C_2H_5$

71%

10[h]

$(n\text{-}C_6H_{13})_3B$ + $BrCH_2\overset{O}{\overset{\|}{C}}N(C_2H_5)_2$ $\xrightarrow[\text{2) }H_2O_2]{\text{1) LDA}}$ $(CH_3(CH_2)_5\overset{O}{\overset{\|}{C}}N(C_2H_5)_2$

94%

a. H. C. Brown and M. M. Rogic, *J. Am. Chem. Soc.*, **91**, 2146 (1969).
b. H. C. Brown, H. Nambu, and M. M. Rogic, *J. Am. Chem. Soc.*, **91**, 6855 (1969).
c. H. C. Brown, M. M. Rogic, H. Nambu, and M. W. Rathke, *J. Am. Chem. Soc.*, **91**, 2147 (1969).
d. H. C. Brown, H. Nambu, and M. M. Rogic, *J. Am. Chem. Soc.*, **91**, 6853 (1969).
e. J. Hooz and S. Linke, *J. Am. Chem. Soc.*, **90**, 5936 (1968).
f. J. Hooz and S. Linke, *J. Am. Chem. Soc.*, **90**, 6891 (1968).
g. J. Hooz, J. N. Bridson, J. G. Caldaza, H. C. Brown, M. M. Midland, and A. B. Levy, *J. Org. Chem.*, **38**, 2574 (1973).
h. N.-S. Li, M.-Z. Deng, and Y.-Z. Huang, *J. Org. Chem.*, **58**, 6118 (1993).

Treatment of alkenyldialkylboranes with iodine results in the formation of the Z-alkene with migration of one boron substituent.[26]

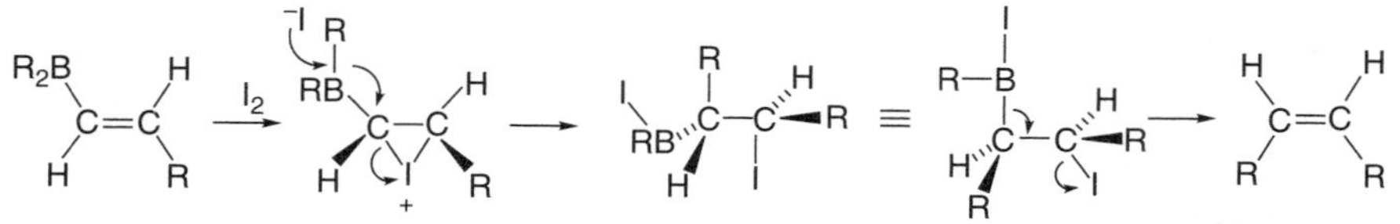

Similarly, alkenyllithium reagents add to dimethyl boronate to give adducts that decompose to Z-alkenes on treatment with iodine.[27]

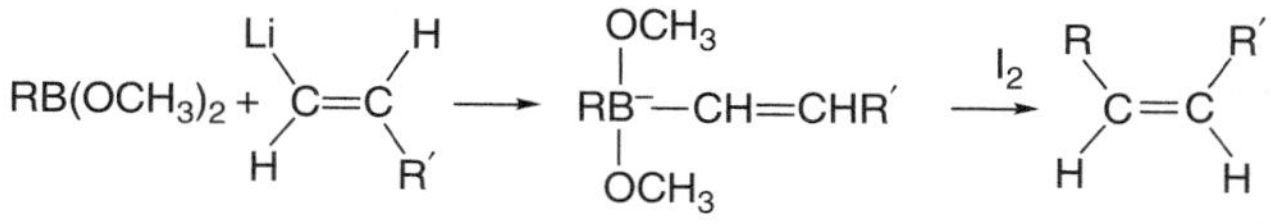

The synthesis of Z-alkenes can also be carried out starting with an alkylbromoborane, in which case migration presumably follows replacement of the bromide by methoxide.[28]

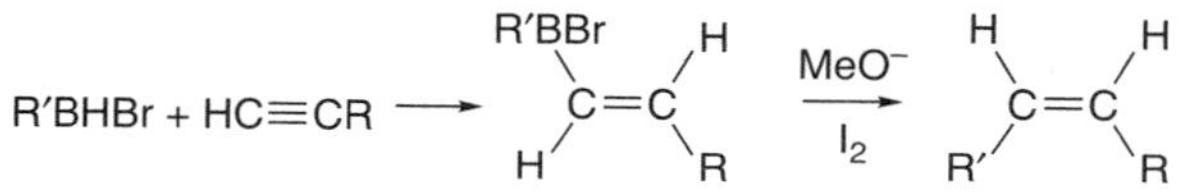

The stereoselectivity of these reactions arises from a base-induced *anti* elimination after the migration. The elimination is induced by addition of methoxide to the boron, generating an anionic center.

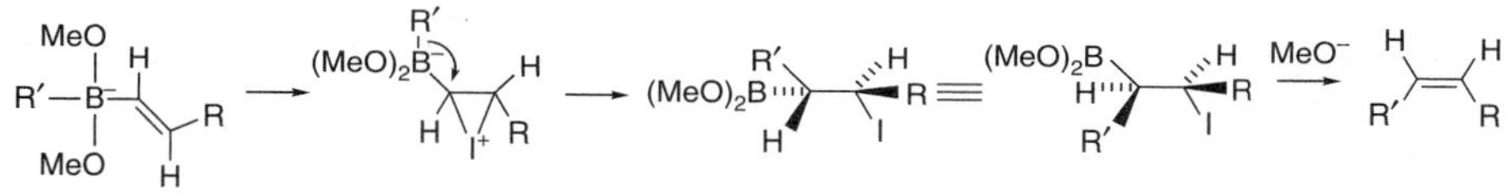

E-Alkenes can be prepared by several related reactions.[29] Hydroboration of a bromoalkyne generates an α-bromoalkenylborane. On treatment with methoxide ion these intermediates undergo B → C migration to give an alkyl alkenylborinate. Protonolysis generates an *E*-alkene.

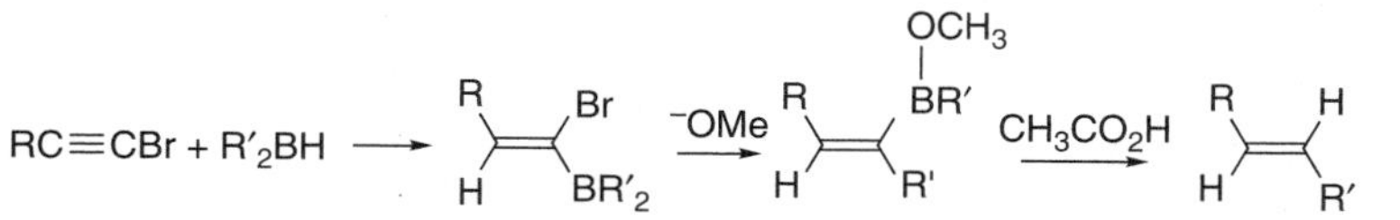

[26.] G. Zweifel, H. Arzoumanian, and C. C. Whitney, *J. Am. Chem. Soc.*, **89**, 3652 (1967); G. Zweifel, R. P. Fisher, J. T. Snow, and C. C. Whitney, *J. Am. Chem. Soc.*, **93**, 6309 (1971).

[27.] D. A. Evans, T. C. Crawford, R. C. Thomas, and J. A. Walker, *J. Org. Chem.*, **41**, 3947 (1976).

[28.] H. C. Brown, D. Basavaiah, S. U. Kulkarni, N. G. Bhat, and J. V. N. Vara Prasad, *J. Org. Chem.*, **53**, 239 (1988).

[29.] H. C. Brown, D. Basavaiah, S. U. Kulkarni, H. P. Lee, E. Negishi, and J.-J. Katz, *J. Org. Chem.*, **51**, 5270 (1986).

796

CHAPTER 9

*Carbon-Carbon
Bond-Forming Reactions
of Compounds of Boron,
Silicon, and Tin*

The dialkylboranes can be prepared from thexylchloroborane. The thexyl group does not normally migrate.

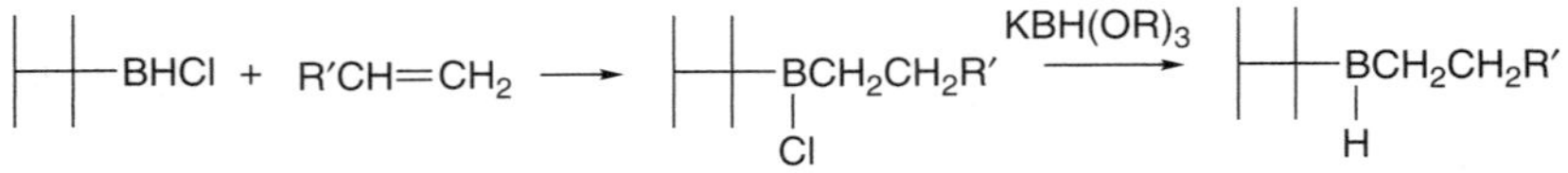

A similar strategy involves initial hydroboration by $BrBH_2$.[30]

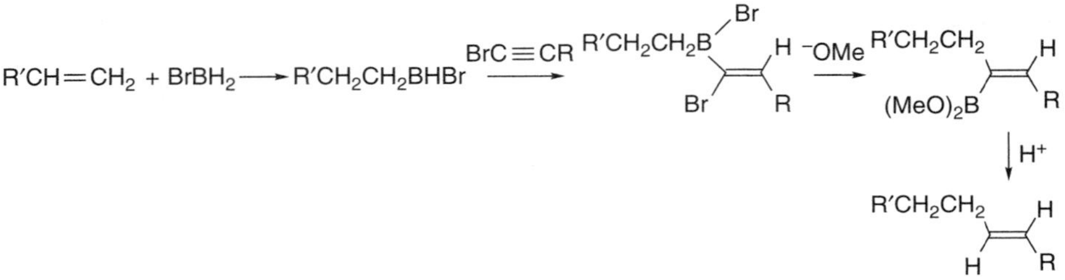

Stereoselective syntheses of trisubstituted alkenes are based on *E*- and *Z*-alkenyldioxaborinanes. Reaction with an alkyllithium reagent forms an "ate" adduct that rearranges on treatment with iodine in methanol.[31]

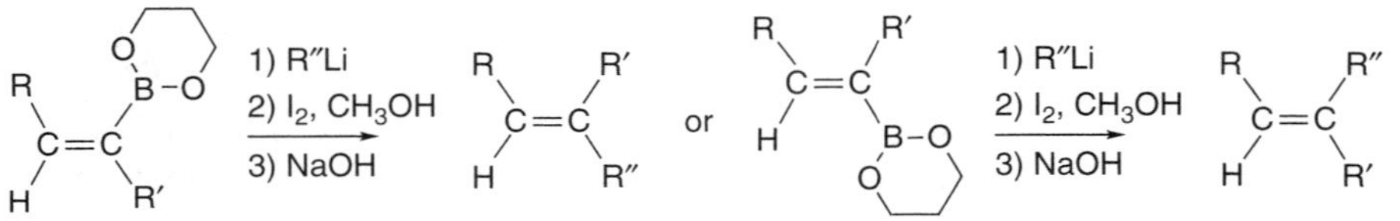

Both alkynes and alkenes can be obtained from adducts of terminal alkynes and boranes. Reaction with iodine induces migration and results in the formation of the alkylated alkyne.[32]

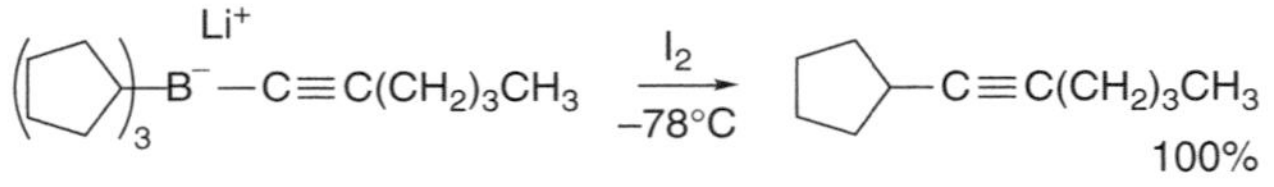

The mechanism involves electrophilic attack by iodine at the triple bond, which induces migration of an alkyl group from boron. This is followed by elimination of dialkyliodoboron.

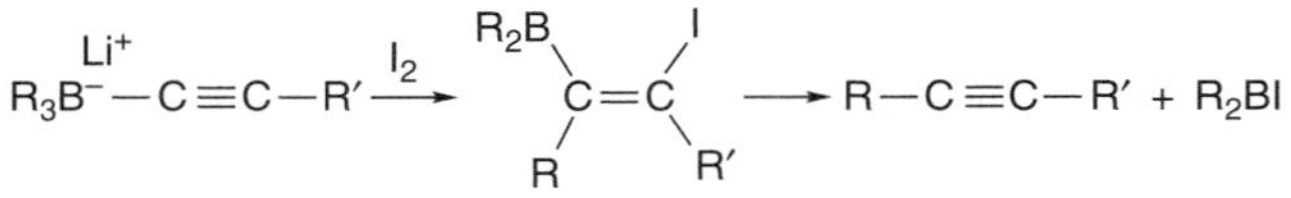

30. H. C. Brown, T. Imai, and N. G. Bhat, *J. Org. Chem.*, **51**, 5277 (1986); H. C. Brown, D. Basavaiah, and S. U. Kulkarni, *J. Org. Chem.*, **47**, 3808 (1982).

31. H. C. Brown and N. G. Bhat, *J. Org. Chem.*, **53**, 6009 (1988).

32. A. Suzuki, N. Miyaura, S. Abiko, M. Itoh, H. C. Brown, J. A. Sinclair, and M. M. Midland, *J. Am. Chem. Soc.*, **95**, 3080 (1973); A. Suzuki, N. Miyaura, S. Abiko, M. Itoh, M. M. Midland, J. A. Sinclair, and H. C. Brown, *J. Org. Chem.*, **51**, 4507 (1986).

If the alkyne is hydroborated and then protonolyzed a *Z*-alkene is formed. This method was used to prepare an insect pheromone containing a *Z*-double bond.

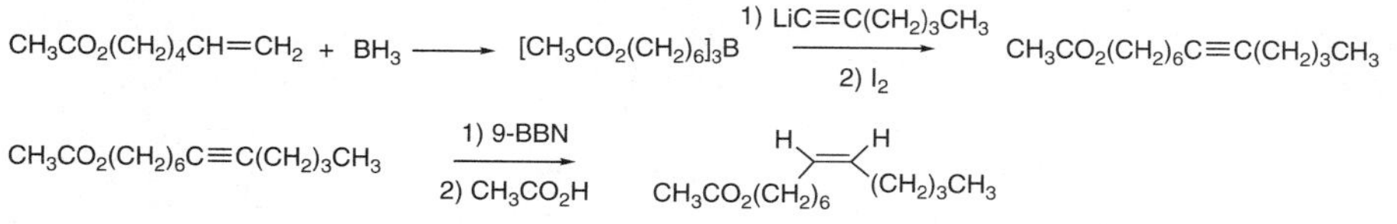

Ref. 33

The B → C migration can also be induced by other types of electrophiles. Trimethylsilyl chloride or trimethylsilyl triflate induces a stereospecific migration to form β-trimethylsilyl alkenylboranes having *cis* silicon and boron substituents.[34] It has been suggested that this stereospecificity arises from a silicon-bridged intermediate.

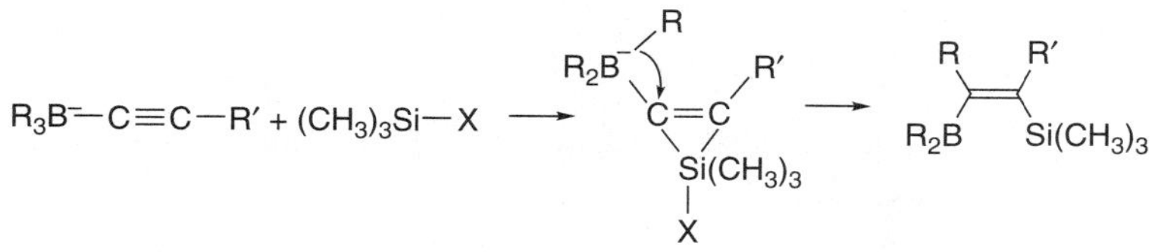

Tributyltin chloride also induces migration and gives the product in which the C–Sn bond is *cis* to the C–B bond. Protonolysis of both the C–Sn and C–B bonds by acetic acid gives the corresponding *Z*-alkene.[35]

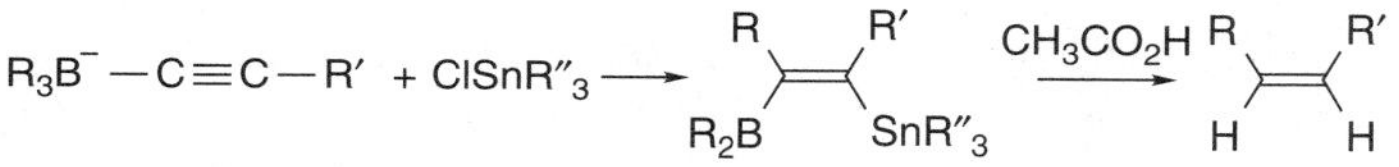

9.1.5. Nucleophilic Addition of Allylic Groups from Boron Compounds

Allylic boranes such as 9-allyl-9-BBN react with aldehydes and ketones to give allylic carbinols. The reaction begins by Lewis acid-base coordination at the carbonyl oxygen, which both increases the electrophilicity of the carbonyl group and weakens the C–B bond to the allyl group. The dipolar adduct then reacts through a cyclic TS. Bond formation takes place at the γ-carbon of the allyl group and the double bond shifts.[36] After the reaction is complete, the carbinol product is liberated from the borinate ester by displacement with ethanolamine. Yields for a series of aldehydes and ketones were usually above 90% for 9-allyl-9-BBN.

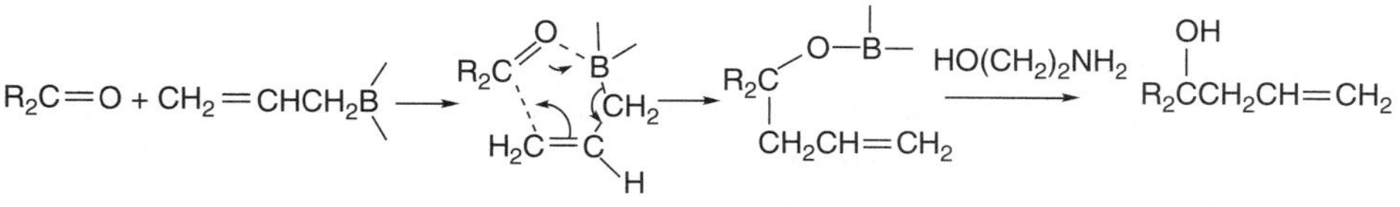

33. H. C. Brown and K. K. Wang, *J. Org. Chem.*, **51**, 4514 (1986).
34. P. Binger and R. Koester, *Synthesis*, 309 (1973); E. J. Corey and W. L. Seibel, *Tetrahedron Lett.*, **27**, 905 (1986).
35. K. K. Wang and K.-H. Chu, *J. Org. Chem.*, **49**, 5175 (1984).
36. G. W. Kramer and H. C. Brown, *J. Org. Chem.*, **42**, 2292 (1977).

798

CHAPTER 9

*Carbon-Carbon
Bond-Forming Reactions
of Compounds of Boron,
Silicon, and Tin*

The cyclic mechanism predicts that the addition reaction will be stereospecific with respect to the geometry of the double bond in the allylic group, and this has been demonstrated to be the case. The *E*- and *Z*-2-butenyl cyclic boronates **1** and **2** were synthesized and allowed to react with aldehydes. The *E*-boronate gave the carbinol with *anti* stereochemistry, whereas the *Z*-boronate resulted in the *syn* product.[37]

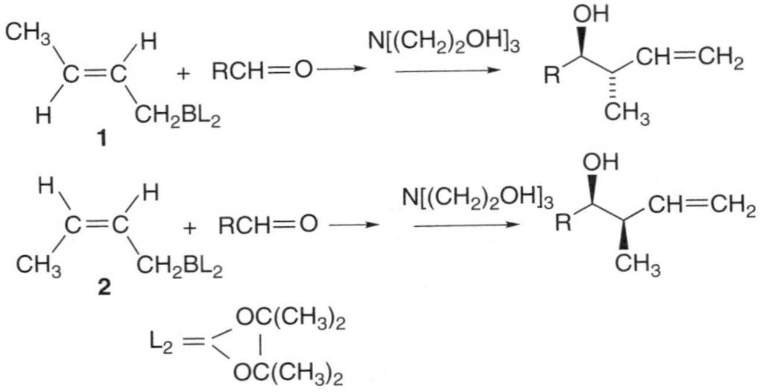

This stereochemistry is that predicted by a cyclic TS in which the aldehyde substituent occupies an equatorial position.

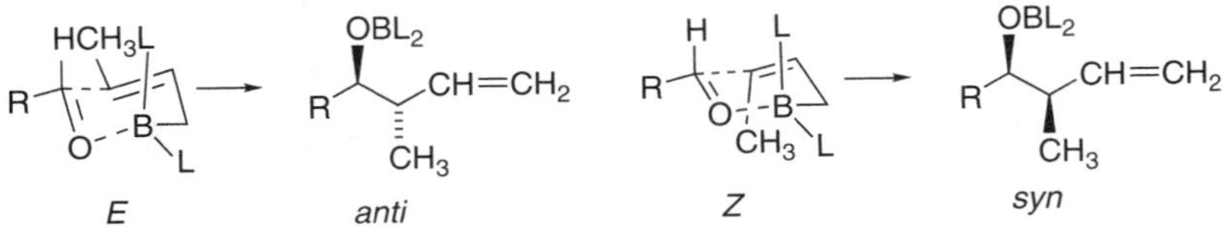

The diastereoselectivity observed in simple systems led to investigation of enantiomerically pure aldehydes. It was found that the *E*- and *Z*-2-butenylboronates both exhibit high *syn-anti* diastereoselectivity with chiral α-substituted aldehydes. However, only the *Z*-isomer also exhibited high selectivity toward the diastereotopic faces of the aldehyde.[38]

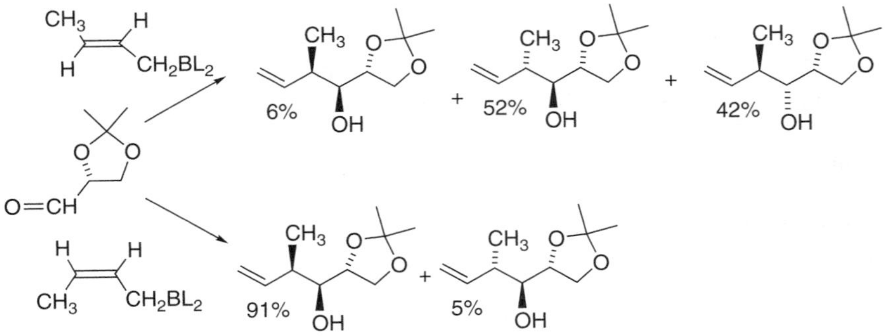

The allylation reaction has been extended to enantiomerically pure allylic boranes and borinates. For example, the 3-methyl-2-butenyl derivative of $(Ipc)_2BH$ reacts with aldehydes to give carbinols of greater than 90% e.e. in most cases.[39]

37. R. W. Hoffmann and H.-J. Zeiss, *J. Org. Chem.*, **46**, 1309 (1981); K. Fujita and M. Schlosser, *Helv. Chim. Acta*, **65**, 1258 (1982).

38. W. R. Roush, M. A. Adam, A. E. Walts, and D. J. Harris, *J. Am. Chem. Soc.*, **108**, 3422 (1986).

39. H. C. Brown and P. K. Jadhav, *Tetrahedron Lett.*, **25**, 1215 (1984); H. C. Brown, P. K. Jadhav, and K. S. Bhat, *J. Am. Chem. Soc.*, **110**, 1535 (1988).

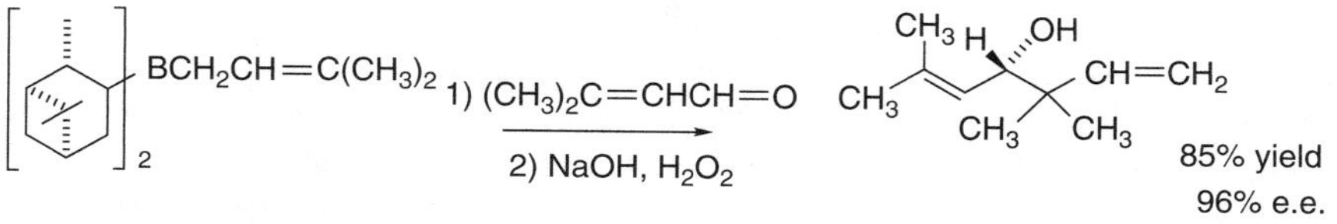

$$\left[\begin{array}{c} \\ \end{array}BCH_2CH=C(CH_3)_2\right]_2 \quad \xrightarrow[\text{2) NaOH, H}_2O_2]{\text{1) (CH}_3)_2C=CHCH=O}$$

85% yield
96% e.e.

β-Allyl-*bis*-(isopinocampheyl)borane exhibits high stereoselectivity in reactions with chiral α-substituted aldehydes.[40] The stereoselectivity is *reagent controlled*, in that there is no change in stereoselectivity between the two enantiomeric boranes in reaction with a chiral aldehyde. Rather, the configuration of the product is determined by the borane. Both enantiomers of $(Ipc)_2BH$ are available, so either enantiomer can be prepared from a given aldehyde.

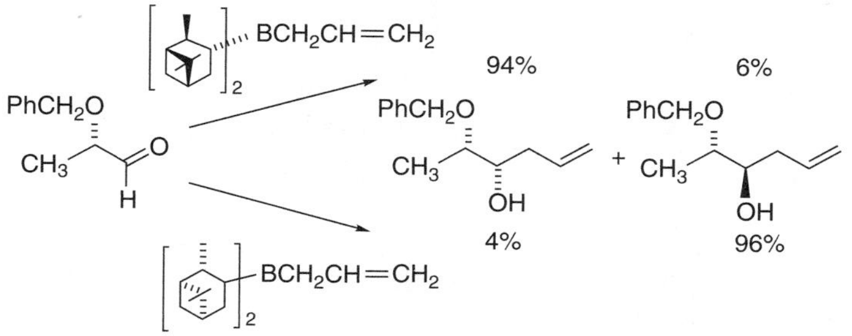

94%   6%

+

4%   96%

It has been found that conditions in which purified allylic boranes are used give even higher enantioselectivity and faster reactions than the reagents prepared and used in situ. The boranes are prepared from Grignard reagents and evidently the residual Mg^{2+} salts inhibit the addition reaction. Magnesium-free borane solutions can be obtained by precipitation and extracting the borane into pentane. These purified reagents react essentially instantaneously with typical aldehydes at $-100\,°C$.[41]

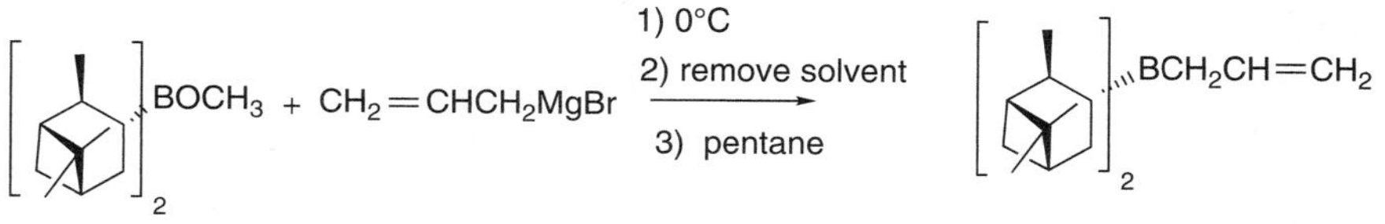

$$\left[\begin{array}{c} \\ \end{array}BOCH_3\right]_2 + CH_2=CHCH_2MgBr \quad \xrightarrow[\text{3) pentane}]{\begin{array}{c}\text{1) 0°C}\\\text{2) remove solvent}\end{array}} \quad \left[\begin{array}{c} \\ \end{array}BCH_2CH=CH_2\right]_2$$

Another extensively developed group of allylic boron reagents for enantioselective synthesis is derived from tartrates.[42]

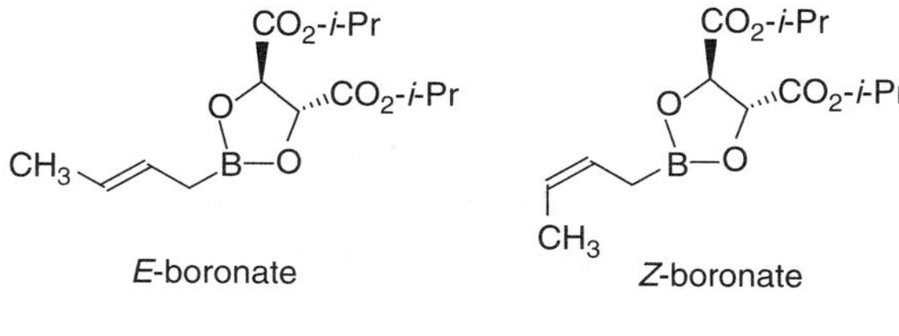

E-boronate   *Z*-boronate

40. H. C. Brown, K. S. Bhat, and R. S. Randad, *J. Org. Chem.*, **52**, 319 (1987); H. C. Brown, K. S. Bhat, and R. S. Randad, *J. Org. Chem.*, **54**, 1570 (1989).
41. U. S. Racherla and H. C. Brown, *J. Org. Chem.*, **56**, 401 (1991).
42. W. R. Roush, K. Ando, D. B. Powers, R. L. Halterman, and A. Palkowitz, *Tetrahedron Lett.*, **29**, 5579 (1988); W. R. Roush, L. Banfi, J. C. Park, and L. K. Hong, *Tetrahedron Lett.*, **30**, 6457 (1989).

800

CHAPTER 9

Carbon-Carbon
Bond-Forming Reactions
of Compounds of Boron,
Silicon, and Tin

With unhindered aldehydes such as cyclohexanecarboxaldehyde, the diastereoselectivity is higher than 95%, with the *E*-boronate giving the *anti* adduct and the *Z*-boronate giving the *syn* adduct. Enantioselectivity is about 90% for the *E*-boronate and 80% for the *Z*-boronate. With more hindered aldehydes, such as pivaldehyde, the diastereoselectivity is maintained but the enantioselectivity drops somewhat. These reagents also give excellent double stereodifferentiation when used with chiral aldehydes. For example, the aldehydes **3** and **4** give at least 90% enantioselection with both the *E*- and *Z*-boronates.[43]

These reagents exhibit reagent control of stereoselectivity and have proven to be very useful in stereoselective synthesis of polyketide natural products, which frequently contain arrays of alternating methyl and oxygen substituents.[44]

The enantioselectivity is consistent with cyclic TSs. The key element determining the orientation of the aldehyde within the TS is the interaction of the aldehyde group with the tartrate ligand.

The preferred orientation results from the greater repulsive interaction between the carbonyl groups of the aldehyde and ester in the disfavored orientation.[45] There is also an attractive electrostatic interaction between the ester carbonyl and the aldehyde

43. W. R. Roush, A. D. Palkowitz, and M. A. J. Palmer, *J. Org. Chem.*, **52**, 316 (1987); W. R. Roush, K. Ando, D. B. Powers, A. D. Palkowitz, and R. L. Halterman, *J. Am. Chem. Soc.*, **112**, 6339 (1990); W. R. Roush, A. D. Palkowitz, and K. Ando, *J. Am. Chem. Soc.*, **112**, 6348 (1990).
44. W. R. Roush and A. D. Palkowitz, *J. Am. Chem. Soc.*, **109**, 953 (1987).
45. W. R. Roush, A. E. Walts, and L. K. Hoong, *J. Am. Chem. Soc.*, **107**, 8186 (1985); W. R. Roush, L. K. Hoong, M. A. J. Palmer, and J. C. Park, *J. Org. Chem.*, **55**, 4109 (1990).

carbon.[46] This orientation and the E- or Z-configuration of the allylic group as part of a chair TS determine the stereochemistry of the product.

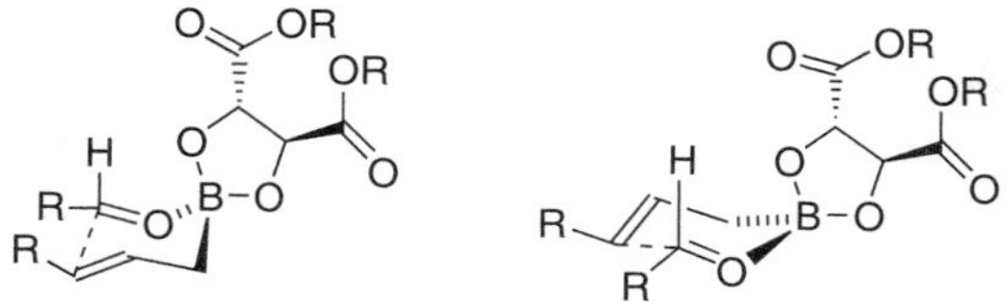

favored   disfavored

Detailed studies have been carried out on the stereoselectivity of α- and β-substituted aldehydes toward the tartrate boronates.[47] α-Benzyloxy and β-benzyloxy-α-methylpropionaldehyde gave approximately 4:1 diastereoselectivity with both the *R, R*- and *S, S*- enantiomers. The stereoselectivity is reagent (tartrate) controlled. The acetonide of glyceraldehydes showed higher stereoselectivity.

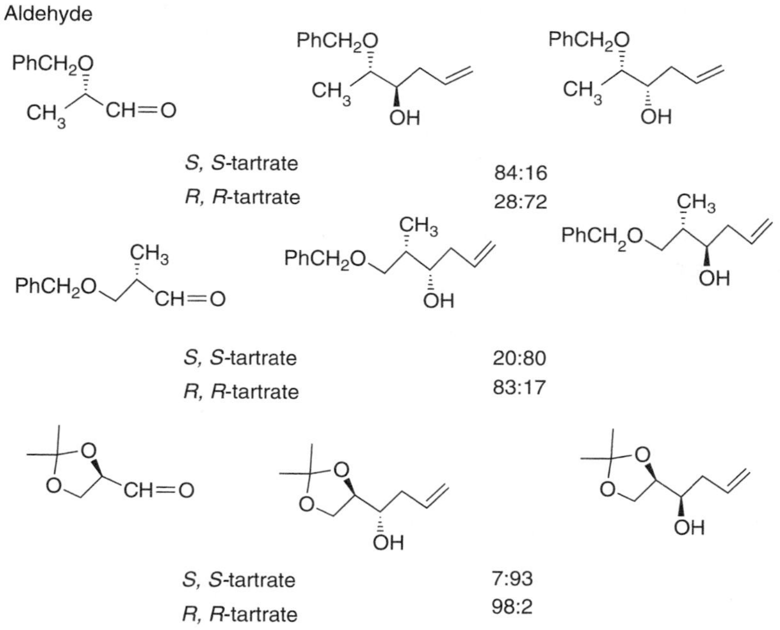

Aldehyde

S, S-tartrate	84:16	
R, R-tartrate	28:72	

S, S-tartrate	20:80	
R, R-tartrate	83:17	

S, S-tartrate	7:93	
R, R-tartrate	98:2	

The tartrate-based allylboration reaction has been studied computationally using B3LYP/6-31G* calculations.[46] The ester groups were modeled by formyl. It was concluded that the major factor in determining enantioselectivity is a favorable electrostatic interaction between a formyl oxygen lone pair and the positively polarized carbon of the reacting aldehyde. This gives rise to a calculated energy difference of 1.6 kcal/mol between the best *si* and the best *re* TS (see Figure 9.1). In the preferred conformation of the TS, the formyl carbonyl is nearly in the plane of the dioxaborolane ring. This orientation has been calculated to be optimal for α-oxy esters[48] and is observed in the crystal structure of the tartrate ligands.[49]

[46] B. W. Gung, X. Xue, and W. R. Roush, *J. Am. Chem. Soc.*, **124**, 10692 (2002).
[47] W. R. Roush, L. K. Hoong, M. A. J. Palmer, J. A. Straub, and A. D. Palkowitz, *J. Org. Chem.*, **55**, 4117 (1990).
[48] K. B. Wiberg and K. E. Laiding, *J. Am. Chem. Soc.*, **109**, 5935 (1987).
[49] W. R. Roush, A. M. Ratz, and J. A. Jablonowski, *J. Org. Chem.*, **57**, 2047 (1992).

802

CHAPTER 9

*Carbon-Carbon
Bond-Forming Reactions
of Compounds of Boron,
Silicon, and Tin*

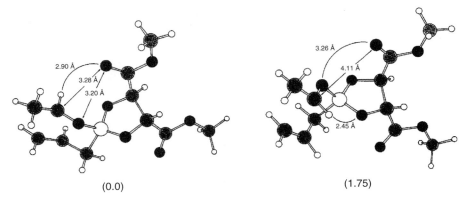

Fig. 9.1. Most favorable *si* and *re* transition structures for allylboration of acetaldehyde. The *si* TS is favored by 1.75 kcal/mol, which is attributed to an electrostatic attraction between a formyl carbonyl oxygen lone pair and the acetaldehyde carbonyl carbon. In the *re* TS, there is a repulsive interaction between lone pairs on the formyl and acetaldehyde carbonyl oxygens. Reproduced from *J. Am. Chem. Soc.*, **124**, 10692 (2002), by permission of the American Chemical Society.

Visual models, additional information and exercises on Allylboration can be found in the Digital Resource available at: Springer.com/carey-sundberg.

Another computational study examined the effect that the boron ligands might have on the reactivity of allyl derivatives.[50] The order found is shown below and is related to the level of the boron LUMO. The dominant factor seems to be the π-donor capacity of the ligands. The calculated order is consistent with experimental data.[51]

Recently the scope of the allylboration has been expanded by the discovery that it is catalyzed by certain Lewis acids, especially Sc(OTf)$_3$.[52] The catalyzed reaction exhibits the same high diastereoselectivity as the uncatalyzed reaction, which indicates that it proceeds through a cyclic TS.

Ref. 52b

50. K. Omoto and H. Fujimoto, *J. Org. Chem.*, **63**, 8331 (1998).
51. H. C. Brown, U. S. Racherla, and P. J. Pellechia, *J. Org. Chem.*, **55**, 1868 (1990).
52. (a) J. W. J. Kennedy and D. G. Hall, *J. Am. Chem. Soc.*, **124**, 11586 (2002); (b) T. Ishiyama, T.-A. Ahiko, and N. Miyaura, *J. Am. Chem. Soc.*, **124**, 12414 (2002).

The catalysis has made reactions of certain functionalized boronates possible. For example, a carbocupration and alkylation allowed the synthesis of boronate **5**. Reaction with aldehydes gave α-methylene lactones with high stereoselectivity.[53]

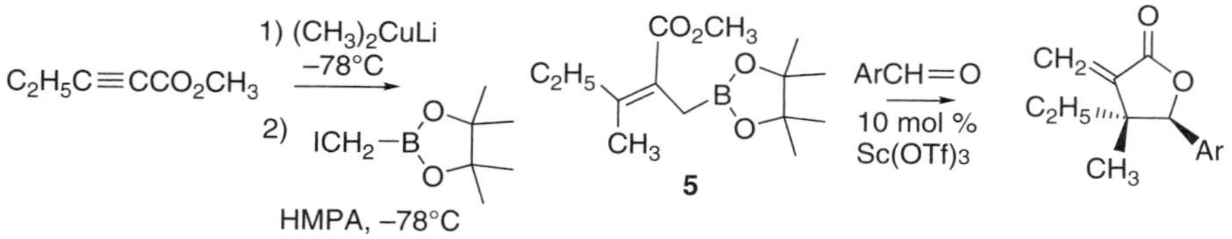

The catalysis has been extended for use with chiral boronates and those from the phenyl-substituted bornane diol derivatives **A** and **B**[54] have been found to be particularly effective.[55]

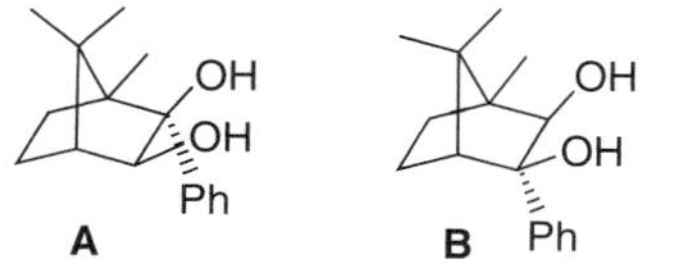

These reagents have been utilized for allyl-, 2-methylallyl-, and E- and Z-2-butenyl derivatives. Enantioselectivity of 90–95% is achieved with alkyl- and aryl-, as well as α- and β-siloxy aldehydes.

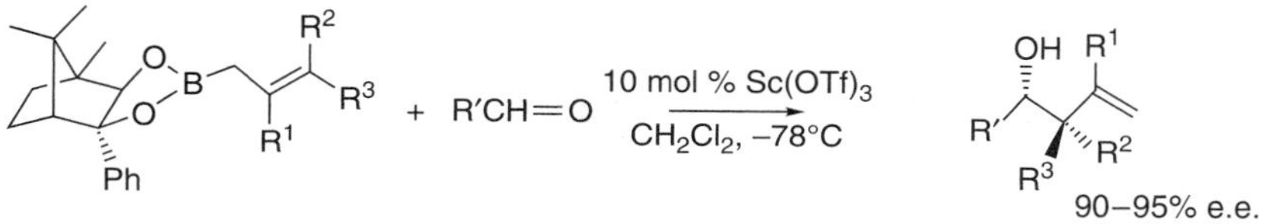

This method has been applied to the synthesis of (S)-2-methyl-4-octanol, an aggregation pheromone of *Metamasius hemipterus*.[56]

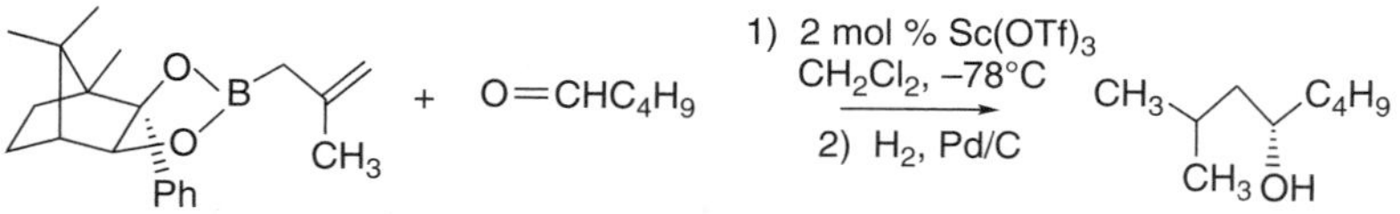

Mechanistic studies have suggested that the TS involves bonding of Sc^{3+} to one of the boronate oxygens,[57] which is consistent with the observation that the catalysts do not have much effect on the rate of allylic boranes. The phenyl substituent on the

[53] J. W. J. Kennedy and D. G. Hall, *J. Org. Chem.*, **69**, 4412 (2004).
[54] T. Herold, U. Schrott, and R. W. Hoffmann, *Chem. Ber.*, **114**, 359 (1981).
[55] H. Lachance, X. Lu, M. Gravel, and D. G. Hall, *J. Am. Chem. Soc.*, **125**, 10160 (2003).
[56] M. Gravel, H. Lachance, X. Lu, and D. G. Hall, *Synthesis*, 1290 (2004).
[57] V. Rauniyar and D. G. Hall, *J. Am. Chem. Soc.*, **126**, 4518 (2004).

804

CHAPTER 9

*Carbon-Carbon
Bond-Forming Reactions
of Compounds of Boron,
Silicon, and Tin*

boronate is thought to assist in the aldehyde binding through a π-π^* interaction with the aromatic ring.

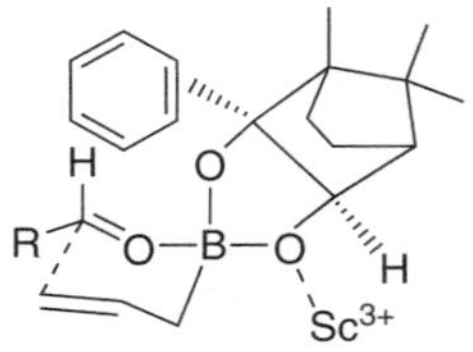

Various functionalized allylic boronates have been prepared.[58] Z-3-Methoxy derivatives can be prepared by lithiation of allyl methyl ether and substitution.[59]

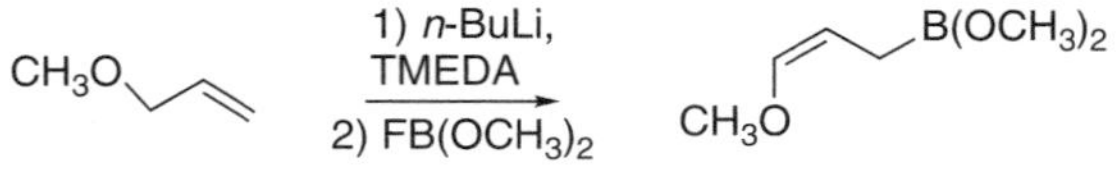

They react with aldehydes to give α-methoxy alcohols.

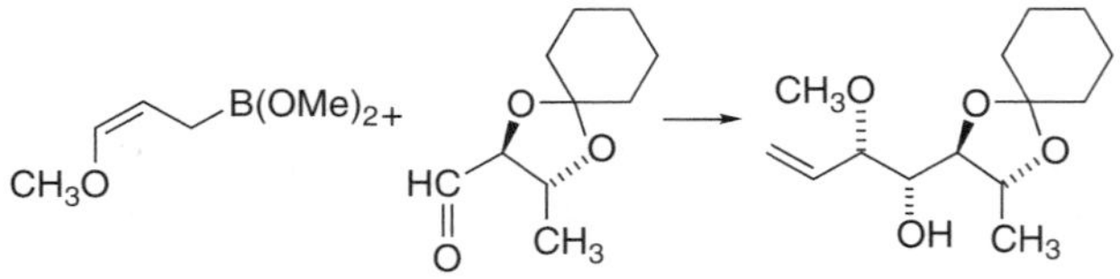

Ref. 60

Oxygenated allylic derivatives of $(Ipc)_2BH$ also show excellent diastereoselectivity.

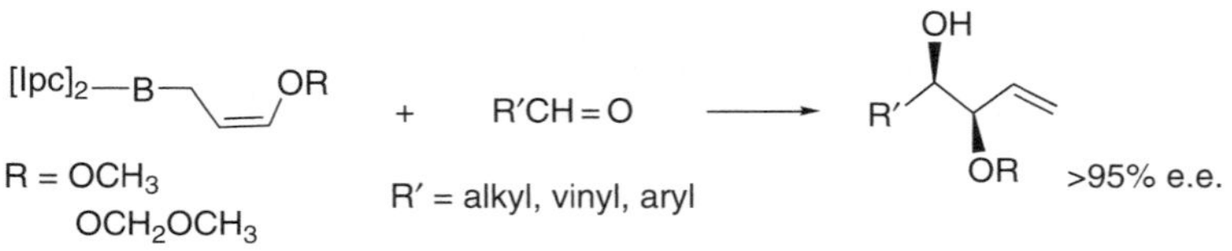

1-Methoxy-2-butenyl pinacol boronates show good stereoselectivity toward achiral aldehydes.[61]

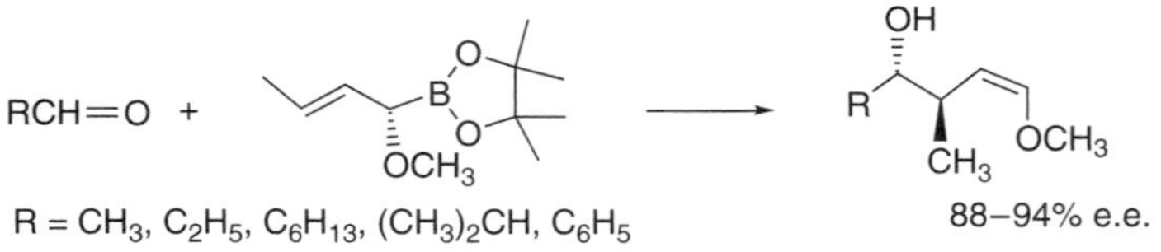

These reagents were also examined with chiral α-substituted aldehydes. The allylboration reagent dominates the enantioselectivity in both matched and mismatched pairs.

58. P. G. M. Wuts, P. A. Thompson, and G. R. Callen, *J. Org. Chem.*, **48**, 5398 (1983); E. Moret and M. Schlosser, *Tetrahedron Lett.*, **25**, 4491 (1984).
59. P. G. M. Wuts and S. S. Bigelow, *J. Org. Chem.*, **47**, 2498 (1982); K. Fujita and M. Schlosser, *Helv. Chim. Acta*, **65**, 1258 (1982).
60. W. R. Roush, M. R. Michaelides, D. F. Tai, and W. K. M. Chong, *J. Am. Chem. Soc.*, **109**, 7575 (1987).
61. R. W. Hoffmann and S. Dresely, *Chem. Ber.*, **122**, 903 (1989).

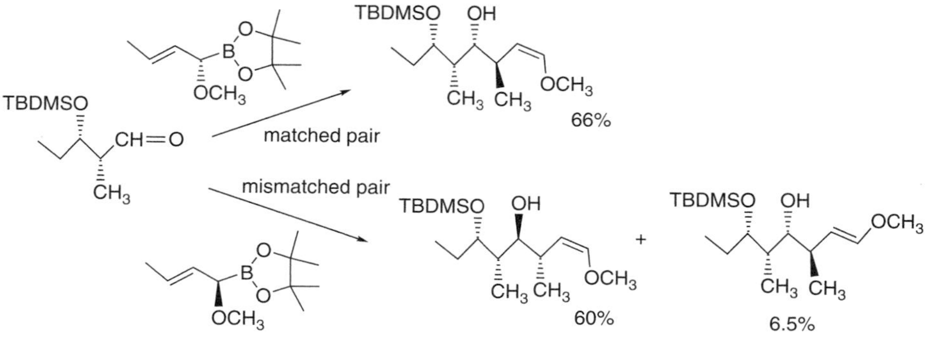

Chloro-substituted [Ipc]$_2$BH derivatives have proven useful for enantioselective synthesis of vinyl epoxides.[62]

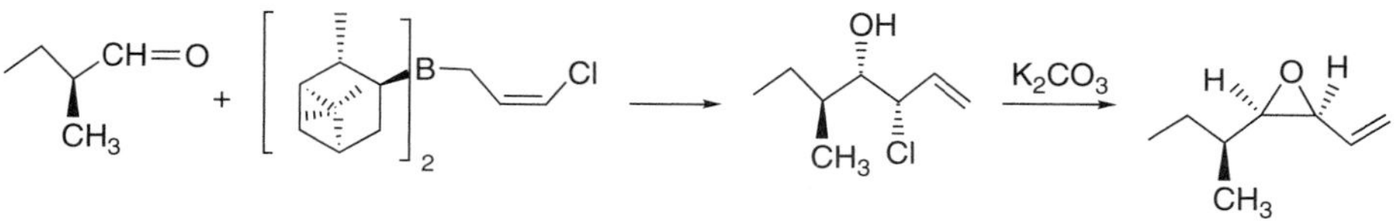

Allyl tetrafluoroborates are also useful allylboration reagents. They can be made from allylic boronic acids and are stable solids.[63] The reaction with aldehydes is mediated by BF$_3$, which is believed to provide the difluoroborane by removing a fluoride. The addition reactions occur with high stereoselectivity, indicating a cyclic TS.

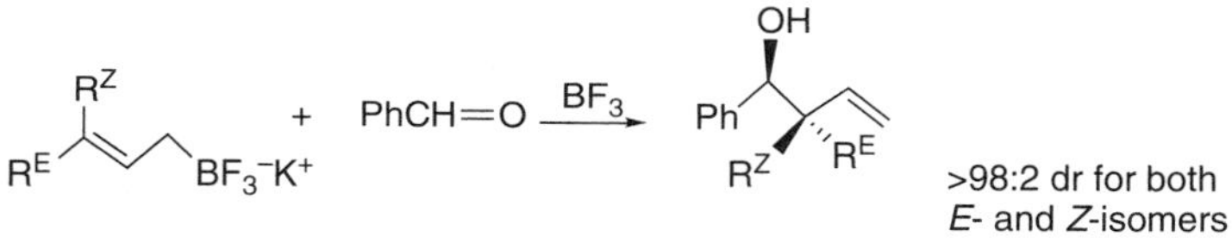

β-Alkynyl derivatives of 9-BBN act as mild sources of nucleophilic acetylenic groups. Reaction occurs with both aldehydes and ketones, but the rate is at least 100 times faster for aldehydes.[64]

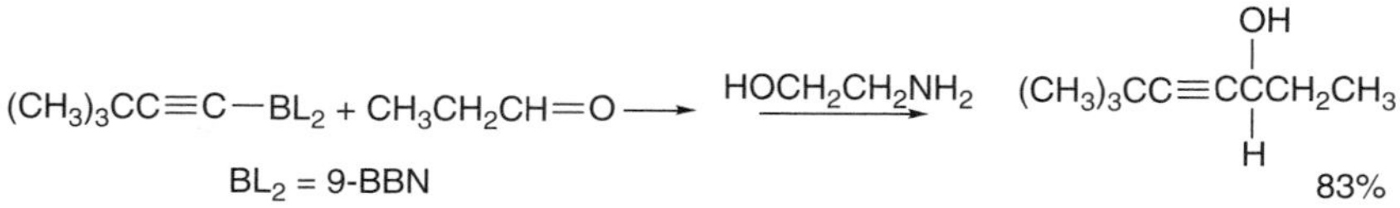

The facility with which the transfer of acetylenic groups occurs is associated with the relative stability of the *sp*-hybridized carbon. This reaction is an alternative to the more common addition of magnesium or lithium salts of acetylides to aldehydes.

Scheme 9.3 illustrates some examples of syntheses of allylic carbinols via allylic boranes and boronate esters. Entries 1 and 2 are among the early examples that

62. S. Hu, S. Jayaraman, and A. C. Oehschlager, *J. Org. Chem.*, **63**, 8843 (1998).
63. R. A. Batey, A. N. Thandani, D. V. Smil, and A. J. Lough, *Synthesis*, 990 (2000).
64. H. C. Brown, G. A. Molander, S. M. Singh, and U. S. Racherla, *J. Org. Chem.*, **50**, 1577 (1985).

806

CHAPTER 9

*Carbon-Carbon
Bond-Forming Reactions
of Compounds of Boron,
Silicon, and Tin*

Scheme 9.3. Addition Reactions of Allylic Boranes and Carbonyl Compounds

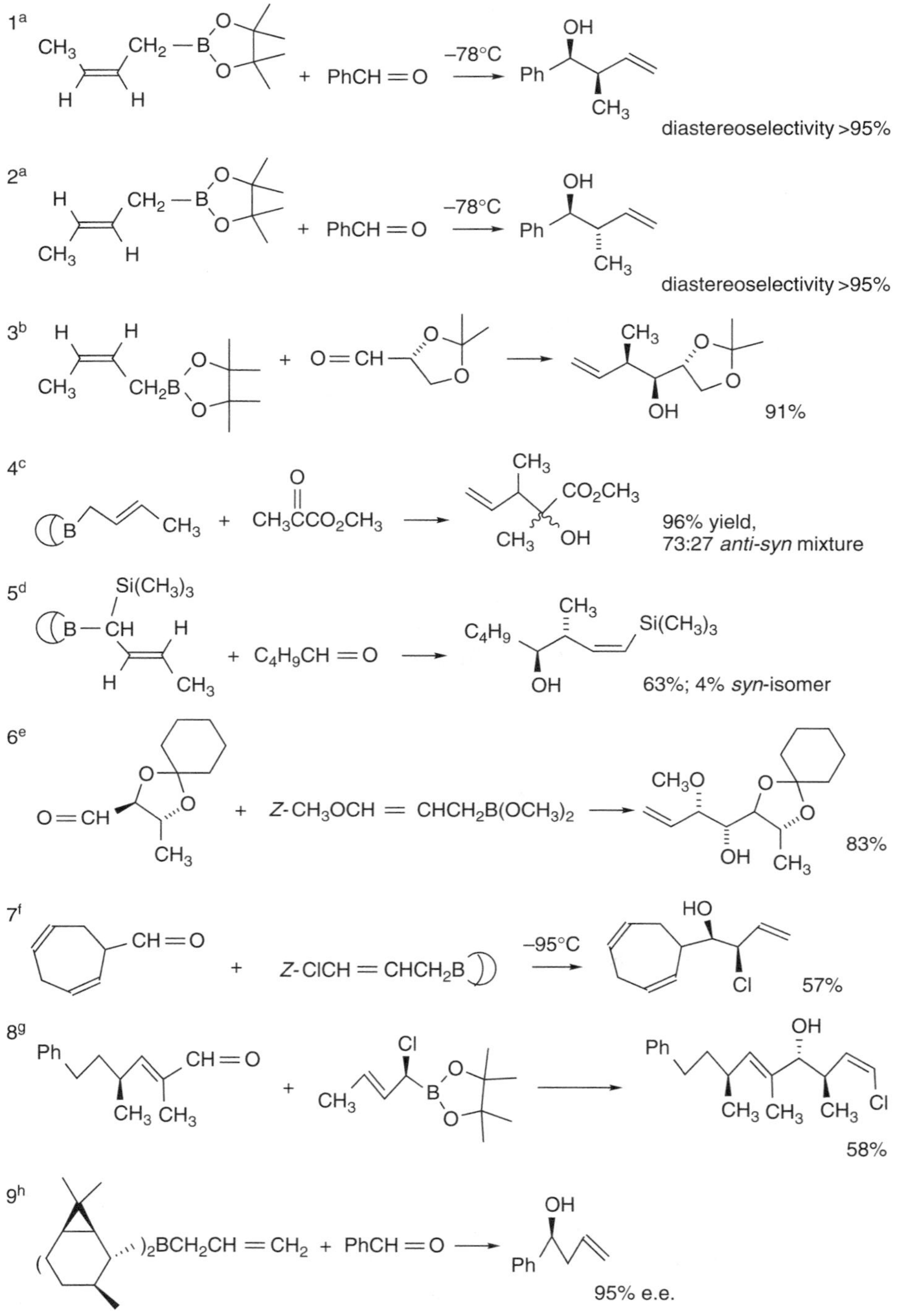

1[a]

diastereoselectivity >95%

2[a]

diastereoselectivity >95%

3[b]

91%

4[c]

96% yield,
73:27 *anti-syn* mixture

5[d]

63%; 4% *syn*-isomer

6[e]

83%

7[f]

57%

8[g]

58%

9[h]

95% e.e.

(Continued)

Scheme 9.3. (*Continued*)

807

SECTION 9.1

*Organoboron
Compounds*

10[i]

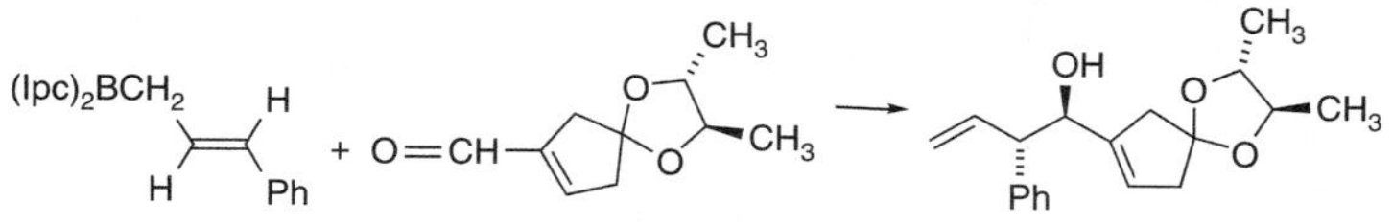

96:4 diastereoselectivity,
89% e.e.

11[j]

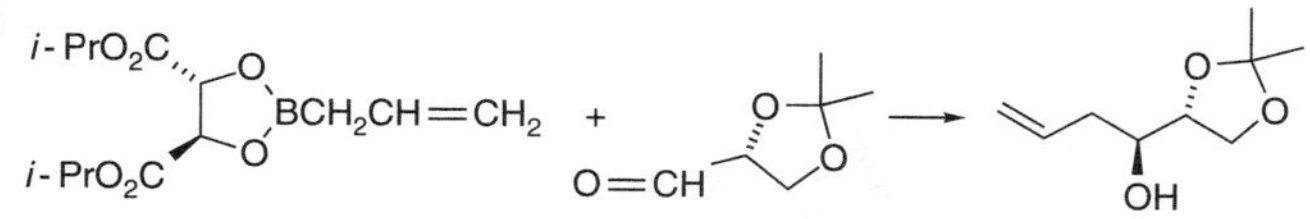

91% yield
96:4 diastereoselectivity

12[k]

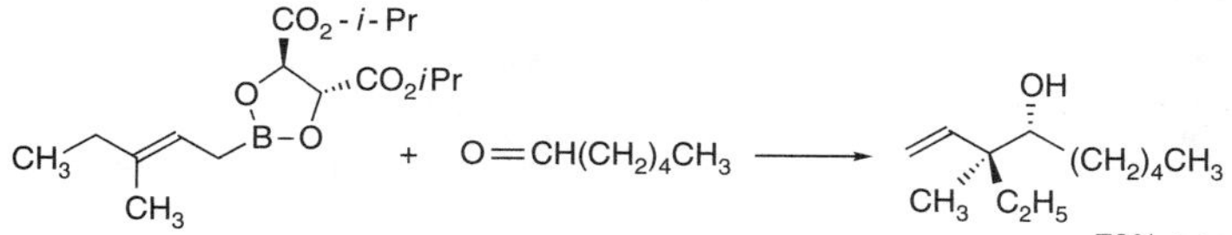

73% e.e.

13[l]

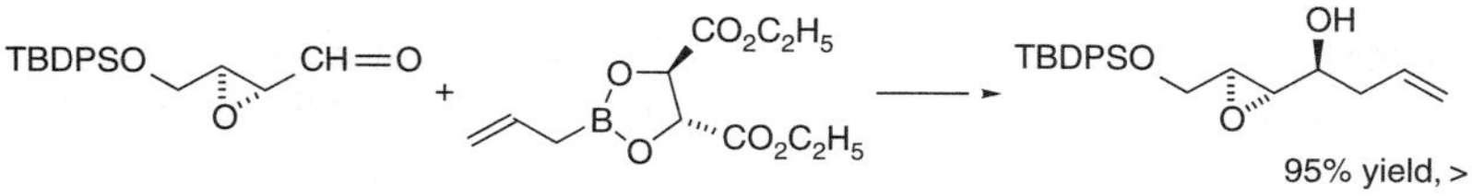

95% yield, > 96% e.e.

14[m]

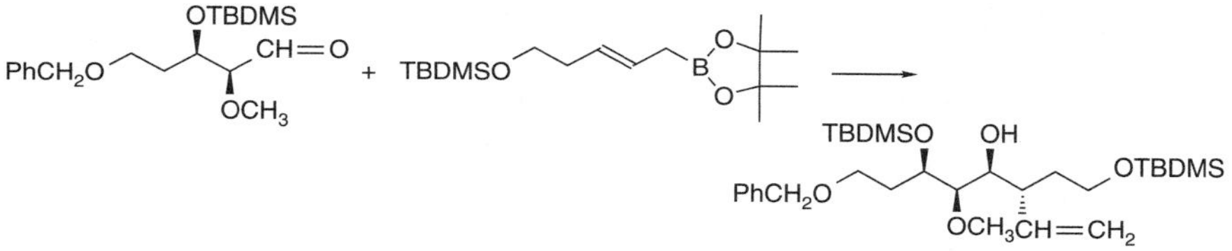

85% total yield,
major isomer of mixture

15[n]

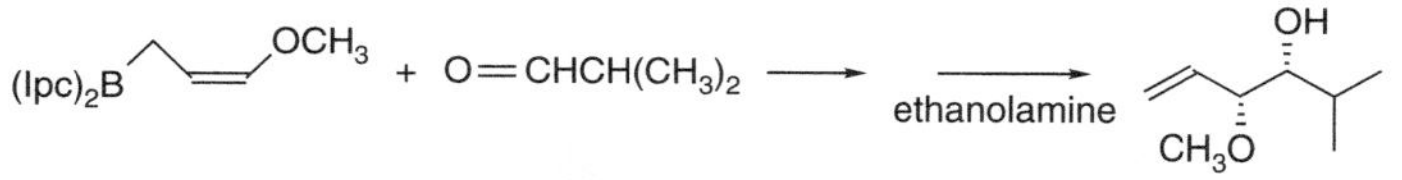

57% yield,
100% *anti*, 88% e.e.

16[o]

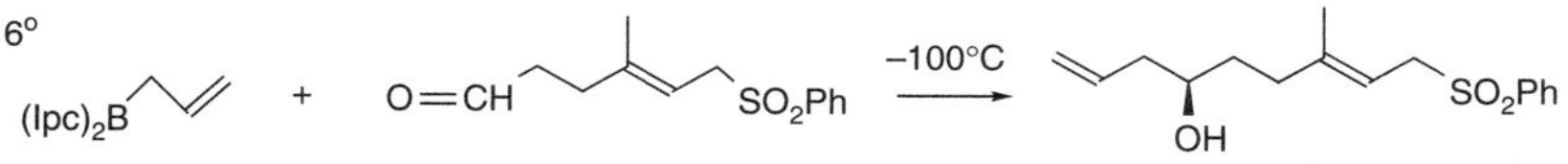

62% yield, 86% e.e.

17[p]

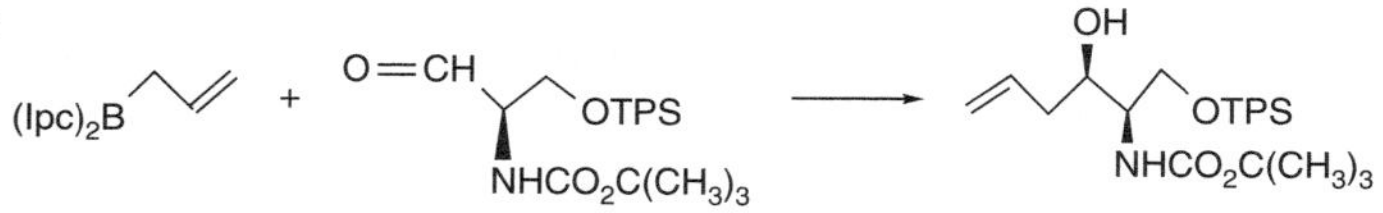

18[q]

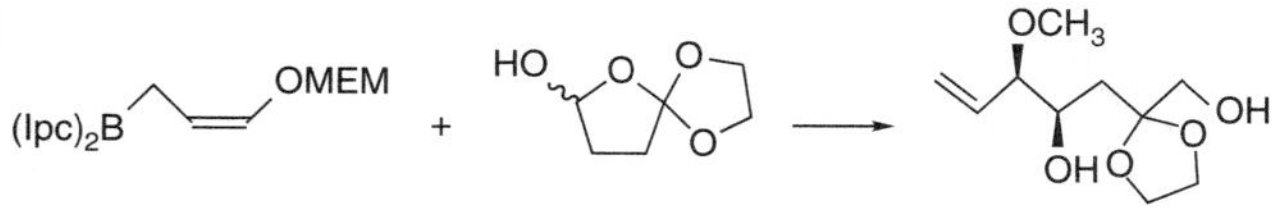

87% >98% e.e.

19[r]

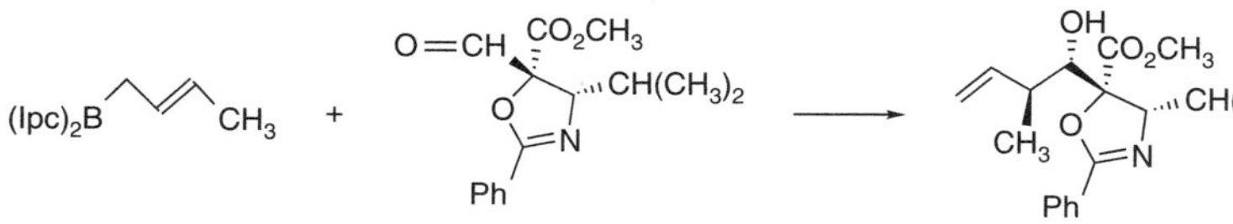

70%

(*Continued*)

a. R. W. Hoffmann and H.-J. Zeiss, *J. Org. Chem.*, **46**, 1309 (1981).
b. W. R. Roush and A. E. Walts, *Tetrahedron Lett.*, **26**, 3427 (1985); W. R. Roush, M. A. Adam, and D. J. Harris, *J. Org. Chem.*, **50**, 2000 (1985).
c. Y. Yamamoto, K. Maruyama, T. Komatsu, and W. Ito, *J. Org. Chem.*, **51**, 886 (1986).
d. Y. Yamamoto, H. Yatagai, and K. Maruyama, *J. Am. Chem. Soc.*, **103**, 3229 (1981).
e. W. R. Roush, M. R. Michaelides, D. F. Tai, and W. K. M. Chong, *J. Am. Chem. Soc.*, **109**, 7575 (1987).
f. C. Hertweck and W. Boland, *Tetrahedron Lett.*, **53**, 14651 (1997).
g. H. C. Brown, R. S. Randad, K. S. Bhat, M. Zaidlewicz, and U. S. Racherla, *J. Am. Chem. Soc.*, **112**, 2389 (1990).
h. R. W. Hoffmann, E. Haeberlin, and T. Rolide, *Synthesis*, 207 (2002).
i. L. K. Truesdale, D. Swanson, and R. C. Sun, *Tetrahedron Lett.*, **26**, 5009 (1985).
j. W. R. Roush, A. E. Walts, and L. K. Hoong, *J. Am. Chem. Soc.*, **107**, 8186 (1985).
k. Y. Yamamoto, S. Hara, and A. Suzuki, *Synlett*, 883 (1996).
l. W. R. Roush, J. A. Straub, and M. S. Van Nieuwenhze, *J. Org. Chem.*, **56**, 1636 (1985).
m. P. G. M. Wuts and S. S. Bigelow, *J. Org. Chem.*, **53**, 5023 (1988).
n. H. C. Brown, P. K. Jadhav, and K. S. Bhat, *J. Am. Chem. Soc.*, **110**, 1535 (1988).
o. M. Z. Hoemann, K. A. Agrios, and J. Aube, *Tetrahedron*, **53**, 11087 (1997).
p. K. C. Nicolaou, M. E. Bunnage, and K. Koide, *Chem. Eur. J.*, **1**, 454 (1995).
q. A. L. Smith, E. N. Pitsinos, C.-K. Hwang, Y. Mizuno, H. Saimoto, G. R. Scarlato, T. Suzuki, and K. C. Nicolaou, *J. Am. Chem. Soc.*, **115**, 7612 (1993).
r. T. Sunazuka, T. Nagamitsu, K. Matsuzaki, H. Tanaka, S. Omura, and A. B. Smith, III, *J. Am. Chem. Soc.*, **115**, 5302 (1993).

demonstrate the high diastereoselectivity of the allylboration reaction. Entry 3 examines the facial selectivity of glyceraldehyde acetonide toward the achiral reagents derived from butenyl pinacol borane. It was found that the reaction with the *Z*-2-butenyl derivative is highly enantioselective, the *E*-isomer was much less so. It was suggested that steric interaction of the *E*-methyl group with the dioxolane in the expected TS ring led to involvement of a second transition structure.

strongly favored
for *Z*-boronate

two competing transition
structures for *E*-boronate

Entry 4 shows the reaction of 9-(*E*-2-butenyl)-9BBN with methyl pyruvate. This reaction is not very stereoselective, which is presumably due to a modest preference for the orientation of the methyl and methoxycarbonyl groups in the TS. Only use of an extremely sterically demanding pyruvic ester achieved high diastereoselectivity.

R	product ratio	
CH₃	73	27
Ph	80	20
2,6-diMePh	75	25
2,4,6-tri-*t*-BuPh	100	0

Entry 5 is an example of use of an α-trimethylsilylallyl group to prepare a vinylsilane. The stereochemistry is consistent with a cyclic TS having the trimethylsilyl substituent in a quasi-axial position to avoid interaction with the bridgehead hydrogen of the bicyclic ring.

Entries 6 and 7 involve functionalized allyl groups, with a Z-γ-methoxy group in Entry 6 and a Z-γ-chloro group in Entry 7. Both give *syn* products; in the case of Entry 7 the chlorohydrin was cyclized to the *cis* epoxide, which is a pheromone (lamoxirene) of a species of algae. Entry 8 is another example of the use of a chloro-substituted allylic borane. Entry 9 involves one of the alternatives to (Ipc)$_2$BH for enantioselective allylation. In Entry 10, both the aldehyde and allyl group contain chiral centers, but the borane is presumably the controlling factor in the stereoselectivity. Entries 11 to 13 demonstrate several enantioselective reactions using the tartrate-derived chiral auxiliaries. Entry 14 is an example of *reactant-controlled* stereochemistry involving the achiral β-allyl pinacol borane. This reaction proceeded with low stereochemical control to give four isomers in a ratio of 18:3.4:1.4:1. Entry 15 shows high diastereoselectivity and enantioselectivity in a reaction with a Z-γ-methoxyallyl-(Ipc)$_2$-borane. Entries 16 to 19 are examples of the use of allylboration in multistage syntheses. Entry 16 involves magnesium-free conditions (see p. 799). Entry 17 was used to construct balanol, a PKC inhibitor, and demonstrates *reagent control* of stereochemistry by allyl-B(Ipc)$_2$ without interference from the protected α-amino and β-hydroxy substituents. Entries 18 and 19 also involve functionalized aldehydes.

9.2. Organosilicon Compounds

9.2.1. Synthesis of Organosilanes

Silicon is similar in electronegativity to carbon. The carbon-silicon bond is quite strong (~75 kcal) and trialkylsilyl groups are stable to many of the reaction conditions that are used in organic synthesis. Much of the repertoire of synthetic organic chemistry can be used for elaboration of organosilanes.[65] For example, the Grignard reagent derived from chloromethyltrimethylsilane is a source of nucleophilic $CH_2Si(CH_3)_3$ units. Two of the most general means of synthesis of organosilanes are nucleophilic displacement of halogen from a halosilane by an organometallic reagent and addition of silanes at multiple bonds (*hydrosilation*). Organomagnesium and organolithium compounds react with trimethylsilyl chloride to give the corresponding tetrasubstituted silanes.

$$CH_2{=}CHMgBr \quad + \quad (CH_3)_3SiCl \quad \longrightarrow \quad CH_2{=}CHSi(CH_3)_3 \qquad \text{Ref. 66}$$

65. L. Birkofer and O. Stuhl, in *The Chemistry of Organic Silicon Compounds*, S. Patai and Z. Rappoport, eds., Wiley-Interscience, 1989, New York, Chap. 10.
66. R. K. Boeckman, Jr., D. M. Blum, B. Ganem, and N. Halvey, *Org. Synth.*, **58**, 152 (1978).

$$CH_2=\overset{|}{\underset{Li}{C}}OC_2H_5 + (CH_3)_3SiCl \longrightarrow CH_2=\overset{|}{\underset{Si(CH_3)_3}{C}}OC_2H_5$$

Ref. 67

Metallation of alkenes with *n*-BuLi-KOC(CH$_3$)$_3$ provides a route that is stereoselective for Z-allylic silanes.[68] (See p. 632 for discussion of this metallation method.)

$$RCH_2CH=CH_2 \xrightarrow[KOC(CH_3)_3]{n\text{-BuLi}} \quad \xrightarrow{(CH_3)_3SiCl}$$

R = alkyl

50 – 75% yield
Z:E = 95 – 98:5 – 2

These conditions are also applicable to functionalized systems that are compatible with metallation by this "superbase."[69]

38%

Silicon substituents can be introduced into alkenes and alkynes by hydrosilation.[70] This reaction, in contrast to hydroboration, does not occur spontaneously, but it can be carried out in the presence of catalysts such as H$_2$PtCl$_6$, hexachloroplatinic acid. Other catalysts are also available.[71] Halosilanes are more reactive than trialkylsilanes.[72]

Alkenylsilanes can be made by Lewis acid–catalyzed hydrosilation of alkynes. Both AlCl$_3$ and C$_2$H$_5$AlCl$_2$ are effective catalysts.[73] The reaction proceeds by net *anti* addition, giving the Z-alkenylsilane. The reaction is regioselective for silylation of the terminal carbon.

$$PhCH_2C\equiv CH + (C_2H_5)_3SiH \xrightarrow{AlCl_3}$$

67. R. F. Cunico and C.-P. Kuan, *J. Org. Chem.*, **50**, 5410 (1985).
68. O. Desponds, L. Franzini, and M. Schlosser, *Synthesis*, 150 (1997).
69. E. Moret, L. Franzini, and M. Schlosser, *Chem. Ber.*, **130**, 335 (1997).
70. J. L. Speier, *Adv. Organomet. Chem.*, **17**, 407 (1979); E. Lukenvics, *Russ. Chem. Rev.* (Engl. Transl.), **46**, 264 (1977); N. D. Smith, J. Mancuso, and M. Lautens, *Chem. Rev.*, **100**, 3257 (2000); M. Brunner, *Angew. Chem. Int. Ed. Engl.*, **43**, 2749 (2004); B. M. Trost and Z. T. Ball, *Synthesis*, 853 (2005).
71. A. Onopchenko and E. T. Sabourin, *J. Org. Chem.*, **52**, 4118 (1987). H. M. Dickens, R. N. Hazeldine, A. P. Mather, and R. V. Parish, *J. Organomet. Chem.*, **161**, 9 (1978); A. J. Cornish and M. F. Lappert, *J. Organomet. Chem.*, **271**, 153 (1984).
72. T. G. Selin and R. West, *J. Am. Chem. Soc.*, **84**, 1863 (1962).
73. N. Asao, T. Sudo, and Y. Yamamoto, *J. Org. Chem.*, **61**, 7654 (1996); T. Sudo, N. Asoa, V. Gevorgyan, and Y. Yamamoto, *J. Org. Chem.*, **64**, 2494 (1999).

These conditions can also be applied to internal alkynes and show a regiochemical preference for silylation β to aryl substituents.

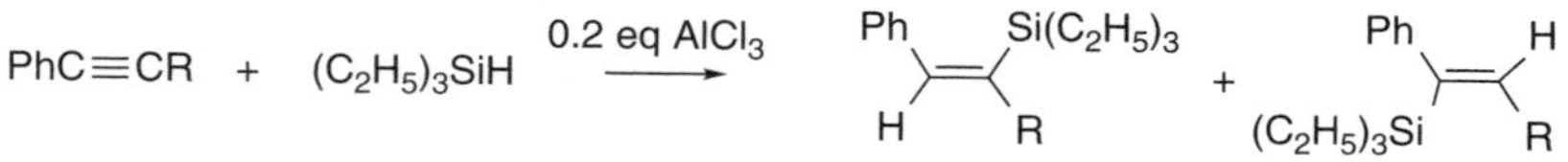

PhC≡CR + (C2H5)3SiH
0.2 eq AlCl3
Ph
Si(C2H5)3
H
R
+
Ph
H
(C2H5)3Si
R

R		
CH_3	76%	10%
C_2H_5	54%	26%

The reaction is formulated as an electrophilic attack by the aluminum halide, followed by hydride abstraction and transmetallation. A vinyl cation intermediate can account for both the regiochemistry and the stereochemistry.

H
R
Ph
SiEt3
—AlCl
Ph—C≡C—R
Ph—C=C
Al
Cl
Et3Si+
Et3SiH

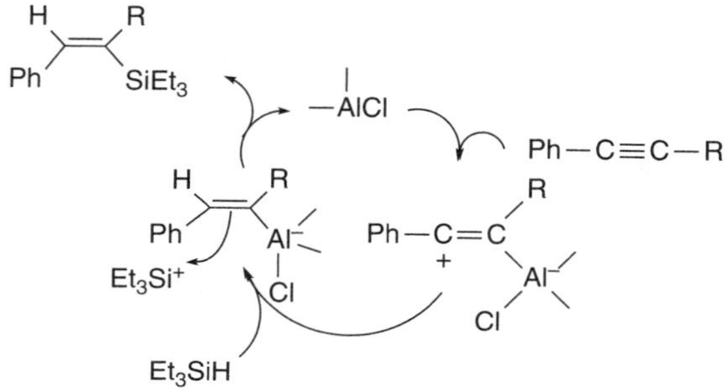

A variety of transition metal complexes catalyze hydrosilylation of alkynes. Catalysis of hydrosilylation by rhodium gives *E*-alkenylsilanes from 1-alkynes.[74]

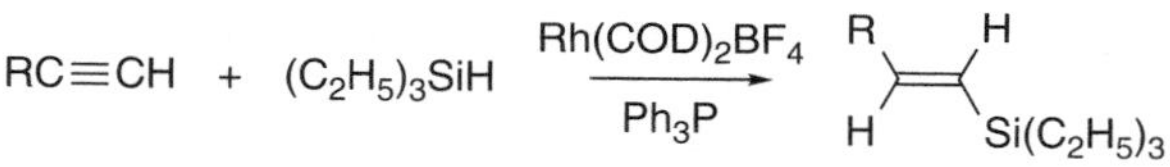

RC≡CH + (C2H5)3SiH
Rh(COD)2BF4
Ph3P
R
H
H
Si(C2H5)3

Ref. 75

$CpRu(CH_3CN)_3PF_6$ catalyzes hydrosilylation of both terminal and internal alkynes. With this catalyst, addition exhibits the opposite regiochemistry.

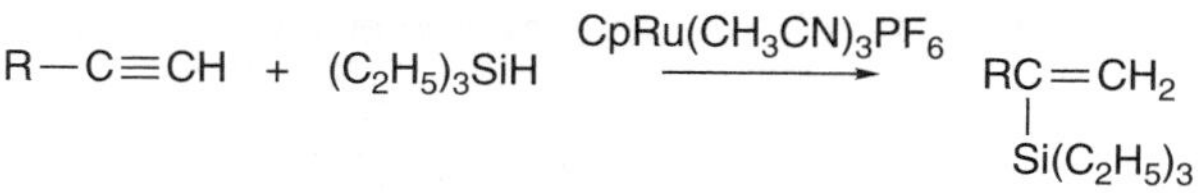

R—C≡CH + (C2H5)3SiH
CpRu(CH3CN)3PF6
RC=CH2
Si(C2H5)3

With internal alkynes, the stereochemistry of addition is *anti*.

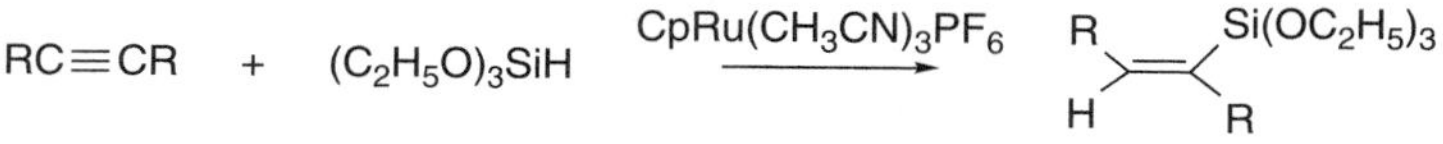

RC≡CR + (C2H5O)3SiH
CpRu(CH3CN)3PF6
R
Si(OC2H5)3
H
R

74. R. Takeuchi, S. Nitta, and D. Watanabe, *J. Org. Chem.*, **60**, 3045 (1995).
75. B. M. Trost and Z. T. Ball, *J. Am. Chem. Soc.*, **123**, 12726 (2001).

812

CHAPTER 9

*Carbon-Carbon
Bond-Forming Reactions
of Compounds of Boron,
Silicon, and Tin*

This method has been used to prepare alkenyl benzyldimethylsilanes.[76] These derivatives are amenable to synthetic transformation involving F^--mediated debenzylation.

$$CH_3O_2C(CH_2)_8C{\equiv}CH \;+\; PhCH_2SiH(CH_3)_2 \;\longrightarrow\; \underset{\underset{PhCH_2(CH_3)_2Si}{}}{\overset{\overset{CH_3O_2C(CH_2)_8}{}}{}}C{=}CH_2$$

Other ruthenium-based catalysts are also active. Ruthenium dichloride–cymene complex is stereoselective for formation of the Z-vinyl silanes from terminal alkynes.

$$RC{\equiv}CH \;+\; Ph_3SiH \;\xrightarrow[\text{(5 mol \%)}]{RuCl_2(cymene)_2}\; \underset{\underset{H}{}\ \underset{H}{}}{\overset{\overset{R}{}\ \overset{SiPh_3}{}}{}} \quad {>}95\%\ Z$$

R = alkyl, aryl, alkoxyalkyl, acyloxyalkyl

Palladium-phosphine catalysts have also been used in the addition of triphenylsilane.[77] In this case, the E-silane is formed.

$$RC{\equiv}CH \;+\; Ph_3SiH \;\xrightarrow[\text{0.5 mol \%}]{Pd_2(dba)_3}\; \underset{\underset{H}{}\ \underset{SiPh_3}{}}{\overset{\overset{R}{}\ \overset{H}{}}{}}$$

High stereoselectivity was noted with Wilkinson's catalyst in the reaction of arylalkynes with diethoxymethylsilane. Interestingly, the stereoselectivity was dependent on the order of mixing of the reagents and the catalyst. When the alkyne was added to a mixture of catalyst and silane, the Z-isomer was formed. Reversing the order and adding the silane to an alkyne-catalyst mixture led to formation of the E-product.[78]

$$ArC{\equiv}CH \;+\; CH_3SiH(OC_2H_5)_2 \;\xrightarrow[\text{5 mol \% NaI}]{\substack{RhCl(PPh_3)_3\\(0.1\ mol\ \%)}}\; \underset{\underset{Ar}{}\ \underset{H}{}}{\overset{\overset{H}{}\ \overset{Si(OC_2H_5)_2CH_3}{}}{}}$$

Tandem *syn* addition of alkyl and trimethylsilyl groups can be accomplished with dialkylzinc and trimethylsilyl iodide in the presence of a Pd(0) catalyst.[79]

$$RC{\equiv}CH \;+\; R'_2Zn \;+\; (CH_3)_3SiI \;\xrightarrow{Pd(PPh_3)_4}\; \underset{\underset{R'}{}\ \underset{Si(CH_3)_3}{}}{\overset{\overset{R}{}\ \overset{H}{}}{}}C{=}C$$

76. B. M. Trost, M. R. Machacek, and Z. T. Ball, *Org. Lett.*, **5**, 1895 (2003).
77. D. Motoda, H. Shinokubo, and K. Oshima, *Synlett*, 1529 (2002).
78. A. Mori, E. Takahisa, H. Kajiro, K. Hirabayashi, Y. Nishihara, and T. Hiyama, *Chem. Lett.*, 443 (1998).
79. N. Chatani, N. Amishiro, T. Morii, T. Yamashita, and S. Murai, *J. Org. Chem.*, **60**, 1834 (1995).

A possible mechanism involves formation of a Pd(II) intermediate that can undergo cross coupling with the zinc reagent.

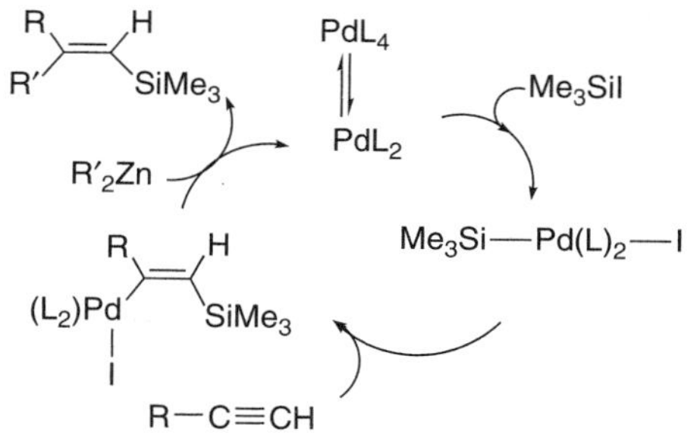

Several variations of the Peterson reaction have been developed for synthesis of alkenylsilanes.[80] E-β-Arylvinylsilanes can be obtained by dehydration of β-silyloxy alkoxides formed by addition of lithiomethyl trimethylsilane to aromatic aldehydes. Specific Lewis acids have been found to be advantageous for the elimination step.[81]

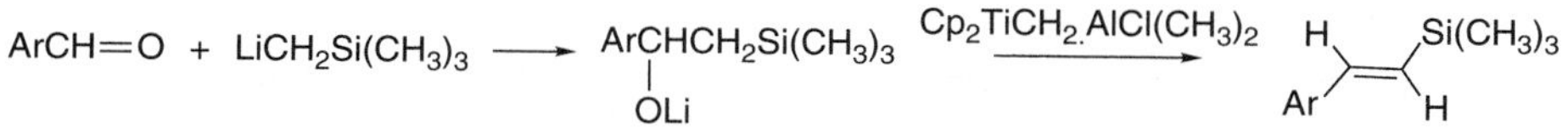

Alkenylsilanes can be prepared from aldehydes and ketones using lithio(chloromethyl)trimethylsilane. The adducts are subjected to a reductive elimination by lithium naphthalenide. This procedure is stereoselective for the E-isomer with both alkyl and aryl aldehydes.[82]

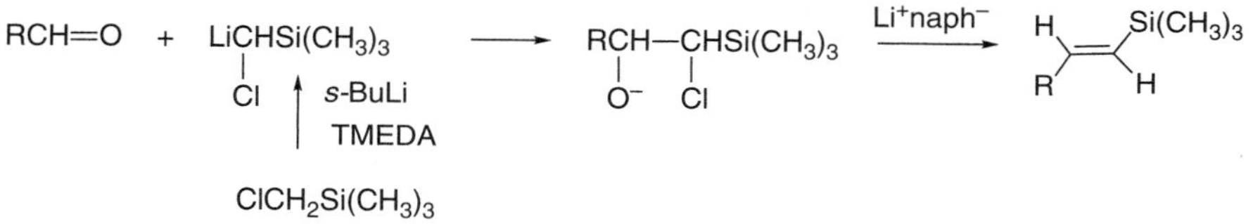

The adducts can be directed toward Z-alkenylsilanes by acetylation and reductive elimination using SmI_2.[83]

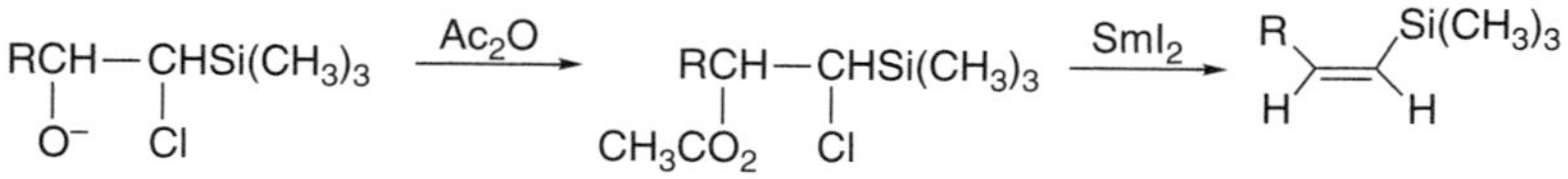

The stereoselectivity in this case is attributed to elimination through a cyclic TS, but is considerably reduced with aryl aldehydes.

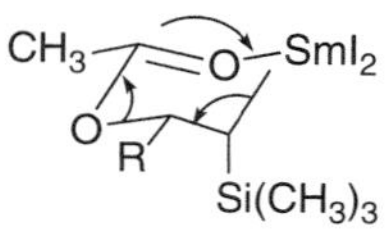

80. C. Trindle, J.-T. Hwang, and F. A. Carey, *J. Org. Chem.*, **38**, 2664 (1973); P. F. Hudrlik, E. L. Agwaramgbo, and A. M. Hudrlik, *J. Org. Chem.*, **54**, 5613 (1989).
81. M. L. Kwan, C. W. Yeung, K. L. Breno, and K. M. Doxsee, *Tetrahedron Lett.*, **42**, 1411 (2001).
82. J. Barluenga, J. L. Fernandez-Simon, J. M. Concellon, and M. Yus, *Synthesis*, 234 (1988).
83. J. M. Concellon, P. L. Bernad, and E. Bardales, *Org. Lett.*, **3**, 937 (2001).

814

CHAPTER 9

*Carbon-Carbon
Bond-Forming Reactions
of Compounds of Boron,
Silicon, and Tin*

Specialized silyl substituents have been developed. High yields of
E-alkenylsilanes were obtained using *bis*-(dimethyl-2-pyridyl)silylmethyllithium.[84]
The stereoselectivity is attributed to a cyclic TS for the addition step.

If necessary for further applications, the 2-pyridyl group can be exchanged by alkyl in
a two-step sequence that takes advantage of the enhanced leaving-group ability of the
2-pyridyl group.

9.2.2. General Features of Carbon-Carbon Bond-Forming Reactions of Organosilicon Compounds

Alkylsilanes are not very nucleophilic because there are no high-energy electrons
in the sp^3-sp^3 carbon-silicon bond. Most of the valuable synthetic procedures based
on organosilanes involve either alkenyl or allylic silicon substituents. The dominant
reactivity pattern involves attack by an electrophilic carbon intermediate at the double
bond that is followed by desilylation. Attack on alkenylsilanes takes place at the
α-carbon and results in overall replacement of the silicon substituent by the
electrophile. Attack on allylic groups is at the γ-carbon and results in loss of the silicon
substituent and an allylic shift of the double bond.

The crucial influence on the reactivity pattern in both cases is the *very high stabilization that silicon provides for carbocationic character at the β-carbon atom.* This
stabilization is attributed primarily to hyperconjugation with the C–Si bond (see Part A,
Section 3.4.1).[85]

84. K. Itami, T. Nokami, and J. Yoshida, *Org. Lett.*, **2**, 1299 (2000).

85. S. G. Wierschke, J. Chandrasekhar, and W. L. Jorgensen, *J. Am. Chem. Soc.*, **107**, 1496 (1985);
J. B. Lambert, G. Wang, R. B. Finzel, and D. H. Teramura, *J. Am. Chem. Soc.*, **109**, 7838 (1987).

Most reactions of alkenyl and allylic silanes require strong carbon electrophiles and Lewis acid catalysts are often involved. The most useful electrophiles from a synthetic standpoint are carbonyl compounds, iminium ions, and electrophilic alkenes.

There are also some reactions of allylic silanes that proceed through anionic silicate species. These reactions usually involve activation by fluoride and result in transfer of an allylic anion.

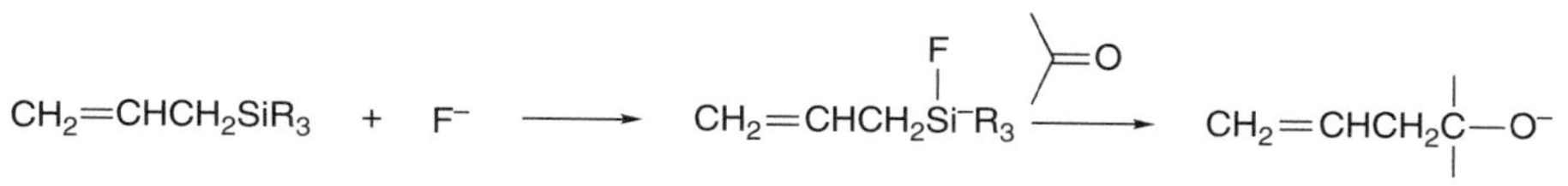

$$CH_2=CHCH_2SiR_3 \ + \ F^- \ \longrightarrow \ CH_2=CHCH_2\overset{F}{\underset{|}{Si^-}}R_3 \ \overset{\overset{\displaystyle >=O}{}}{\longrightarrow} \ CH_2=CHCH_2\overset{|}{\underset{|}{C}}-O^-$$

Trichloro- and trifluorosilanes introduce another dimension into the reactivity of allylic silanes. The silicon in these compounds is electrophilic and can expand to pentacoordinate and hexacoordinate structures. These reactions can occur through a cyclic or chelated TS.

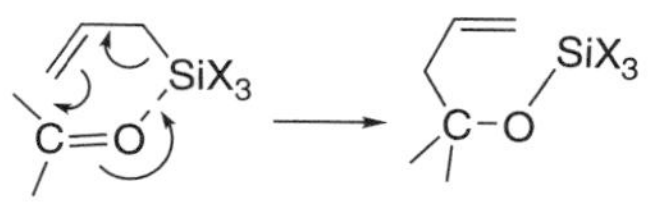

9.2.3. Addition Reactions with Aldehydes and Ketones

A variety of electrophilic catalysts promote the addition of allylic silanes to carbonyl compounds.[86] The original catalysts included typical Lewis acids such as $TiCl_4$ or BF_3.[87] This reaction is often referred to as the *Sakurai reaction*.

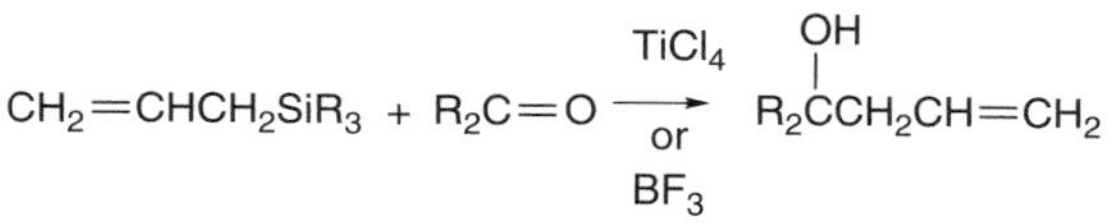

$$CH_2=CHCH_2SiR_3 \ + \ R_2C=O \ \underset{\underset{BF_3}{or}}{\overset{TiCl_4}{\longrightarrow}} \ R_2\overset{OH}{\underset{|}{C}}CH_2CH=CH_2$$

These reactions involve activation of the carbonyl group by the Lewis acid. A nucleophile, either a ligand from the Lewis acid or the solvent, assists in the desilylation step.

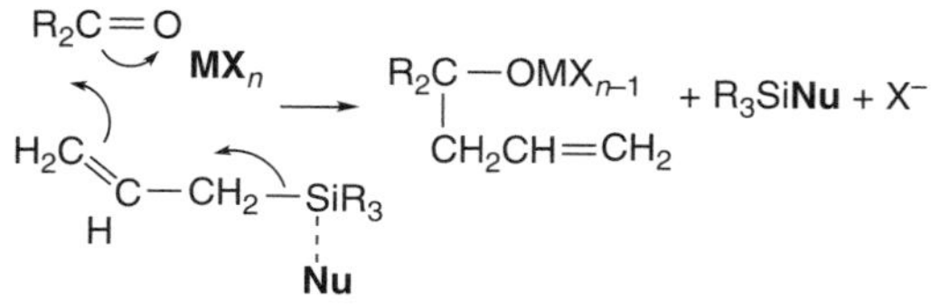

Various other Lewis acids have been explored as catalysts, and the combination $InCl_3$-$(CH_3)_3SiCl$ has been found to be effective.[88] The catalysis requires both components and is attributed to assistance from O-silylation of the carbonyl compound.

[86.] A. Hosomi, *Acc. Chem. Res.*, **21**, 200 (1988); I. Fleming, J. Dunoques, and R. Smithers, *Org. React.*, **37**, 57 (1989).

[87.] A. Hosomi and H. Sakurai, *Tetrahedron Lett.*, 1295 (1976).

[88.] Y. Onishi, T. Ito, M. Yasuda, and A. Baba, *Eur. J. Org. Chem.*, 1578 (2002); Y. Onishi, T. Ito, M. Yasuda, and A. Baba, *Tetrahedron*, **58**, 8227 (2002).

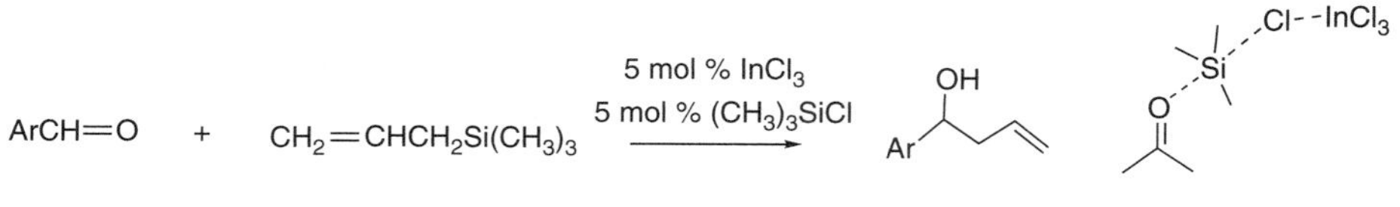

ArCH=O
+
$CH_2=CHCH_2Si(CH_3)_3$
5 mol % $InCl_3$
5 mol % $(CH_3)_3SiCl$
OH
Ar
Cl- -$InCl_3$
Si
O

Lanthanide salts, such as $Sc(O_3SCF_3)_3$, are also effective catalysts.[89]

Silylating reagents such as TMSI and TMS triflate have only a modest catalytic effect, but the still more powerful silylating reagent $(CH_3)_3SiB(O_3SCF_3)_4$ does induce addition to aldehydes.[90]

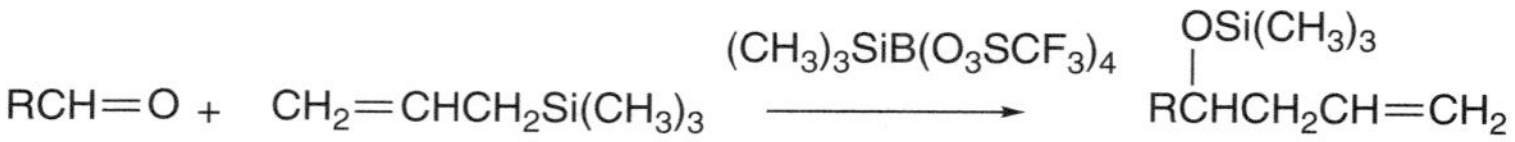

RCH=O +
$CH_2=CHCH_2Si(CH_3)_3$
$(CH_3)_3SiB(O_3SCF_3)_4$
$OSi(CH_3)_3$
$RCHCH_2CH=CH_2$

In another procedure, $(CH_3)_3SiN(O_3SCF_3)$ is generated in situ from triflimide.[91]

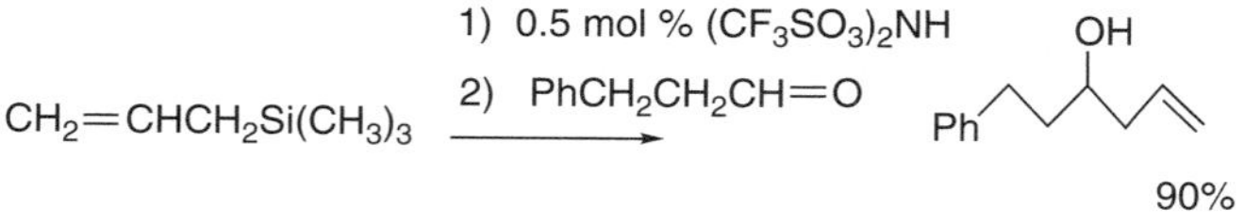

$CH_2=CHCH_2Si(CH_3)_3$
1) 0.5 mol % $(CF_3SO_3)_2NH$
2) $PhCH_2CH_2CH=O$
OH
Ph
90%

These reagents initiate a catalytic cycle that regenerates the active silyation species.[92] (See p. 83 for a similar cycle in the Mukaiyama reaction.)

RCH=O
$Si(CH_3)_3$
$RCH=O^+$
$Si(CH_3)_3$
$(CH_3)_3SiX$
$OSi(CH_3)_3$
R
+
$Si(CH_3)_3$ X^-
$OSi(CH_3)_3$
R

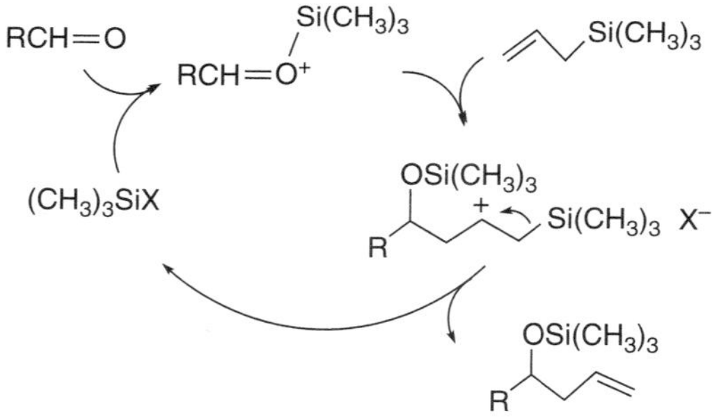

Although the allylation reaction is formally analogous to the addition of allylic boranes to carbonyl derivatives, it does not normally occur through a cyclic TS. This is because, in contrast to the boranes, the silicon in allylic silanes has little Lewis acid character and does not coordinate at the carbonyl oxygen. The stereochemistry of addition of allylic silanes to carbonyl compounds is consistent with an acyclic TS. The *E*-stereoisomer of 2-butenyl(trimethyl)silane gives nearly exclusively the product in

89. V. K. Aggarwal and G. P. Vennall, *Tetrahedron Lett.*, **37**, 3745 (1996).
90. A. P. Davis and M. Jaspars, *Angew. Chem. Int. Ed. Engl.*, **31**, 470 (1992).
91. K. Ishihara, Y. Hiraiwa, and H. Yamamoto, *Synlett*, 1851 (2001).
92. T. K. Hollis and B. Bosnich, *J. Am. Chem. Soc.*, **117**, 4570 (1995).

which the newly formed hydroxyl group is *syn* to the methyl substituent; the *Z*-isomer is also modestly selective for the *syn* isomer.[93]

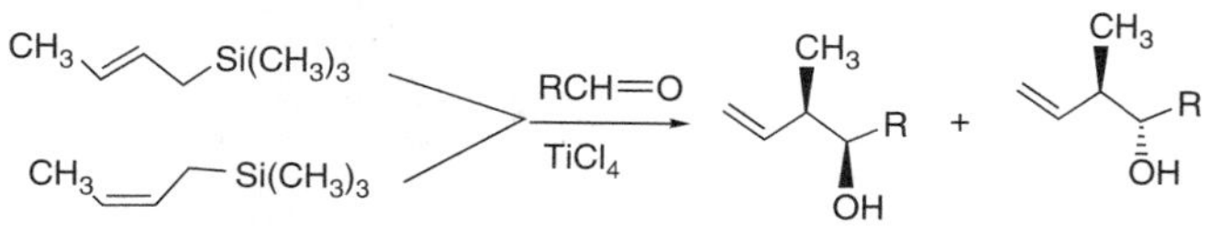

R	E-silane syn:anti	Z-silane syn:anti
Et	>95:5	65:35
i-Pr	>97:3	64:36
t-Bu	>99:1	69:31

Both *anti*-synclinal and *anti*-periplanar TSs are considered to be feasible. These differ in the relative orientation of the C=C and C=O bonds. The *anti*-synclinal arrangement is usually preferred.[94]

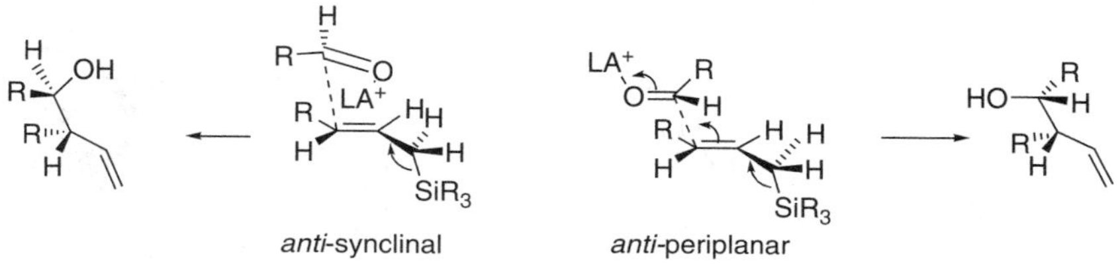

anti-synclinal   anti-periplanar

The addition reaction of allylsilane to acetaldehyde with BF_3 as the Lewis acid has been modeled computationally.[95] The lowest-energy TSs found, which are shown in Figure 9.2, were of the synclinal type, with dihedral angles near 60°. Although the structures are acyclic, there is an apparent electrostatic attraction between the fluorine and the silicon that imparts some cyclic character to the TS. Both *anti* and *syn* structures were of comparable energy for the model. However, steric effects that arise by replacement of hydrogen on silicon with methyl are likely to favor the *anti* TS.

When chiral aldehydes such as **6** are used, there is a modest degree of diastereoselectivity in the direction predicted by an open Felkin TS.[96]

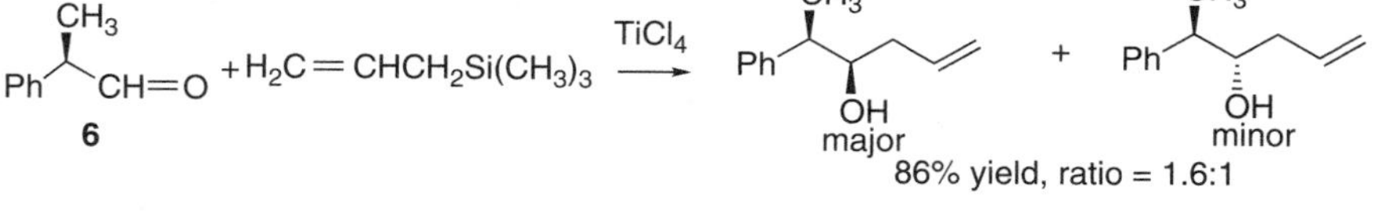

86% yield, ratio = 1.6:1

93. T. Hayashi, K. Kabeta, I. Hamachi, and K. Kumada, *Tetrahedron Lett.*, **24**, 2865 (1983).
94. S. E. Denmark and N. G. Almstead, *J. Org. Chem.*, **59**, 5130 (1994).
95. A. Bottoni, A. L. Costa, D. Di Tommaso, I. Rossi, and E. Tagliavini, *J. Am. Chem. Soc.*, **119**, 12131 (1997).
96. M. Nakada, Y. Urano, S. Kobayashi, and M. Ohno, *J. Am. Chem. Soc.*, **110**, 4826 (1988).

818

CHAPTER 9

*Carbon-Carbon
Bond-Forming Reactions
of Compounds of Boron,
Silicon, and Tin*

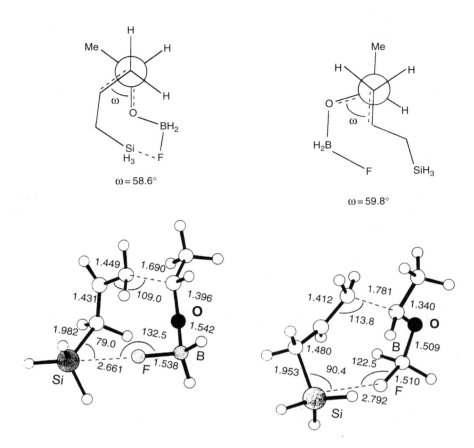

Fig. 9.2. Most favorable transition structures for reaction of allylsilane with acetaldehyde-fluoroborane: (left) *anti* synclinal; (right) *syn* synclinal. Reproduced from *J. Am. Chem. Soc.*, **119**, 12131 (1997), by permission of the American Chemical Society.

Aldehydes with α- or β-benzyloxy substituents react with allyltrimethylsilane in the presence of SnCl$_4$ to give high yields of product resulting from chelation control.[97]

97. C. H. Heathcock, S. Kiyooka, and T. Blumenkopf, *J. Org. Chem.*, **49**, 4214 (1984).

The stereochemistry is consistent with approach of the silane *anti* to the methyl substituent.

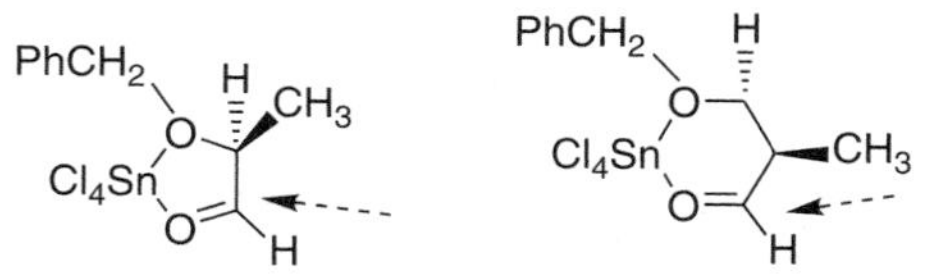

In contrast, BF_3 showed very low stereoselectivity, consistent with its inability to form a chelate.

Intramolecular reactions can also occur between carbonyl groups and allylic silanes. These reactions frequently show good stereoselectivity. For example, **7** cyclizes primarily to **8** with 4% of **9** as a by-product. The two other possible stereoisomers are not observed.[98] The stereoselectivity is attributed to a preference for TS **7A** over TS **7B**. These are both synclinal structures but differ stereoelectronically. In **7A**, the electron flow is approximately *anti* parallel, whereas in **7B** it is skewed. It was suggested that this difference may be the origin of the stereoselectivity.

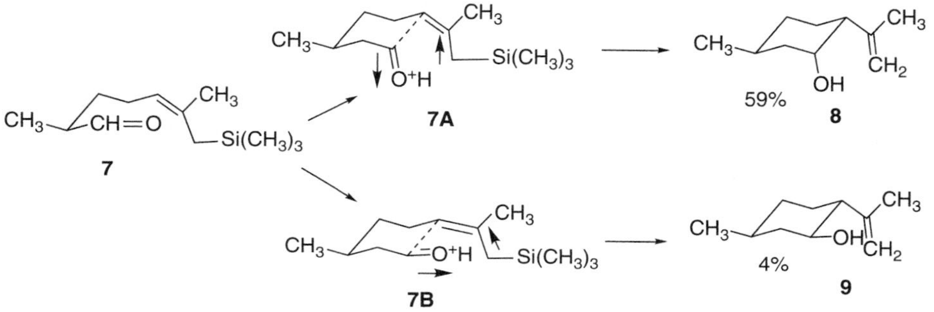

The differential in chelation capacity between BF_3 and $SnCl_4$ was used to control the stereochemistry of the cyclization of the vinyl silane **10**.[99] With BF_3, the reaction proceeds through a nonchelated TS and the stereochemistry at the new bond is *trans*. With $SnCl_4$, a chelated TS leads to the *cis* diastereomer.

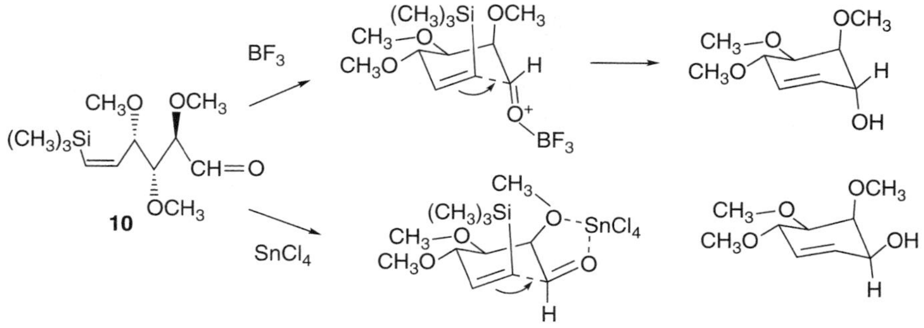

Both ketals[100] and enol ethers[101] can be used as electrophiles in place of aldehydes with appropriate catalysts. Trimethylsilyl iodide can be used in catalytic quantities

[98.] M. Schlosser, L. Franzini, C. Bauer, and F. Leroux, *Chem. Eur. J.*, **7**, 1909 (2001).
[99.] M. C. McIntosh and S. M. Weinreb, *J. Org. Chem.*, **56**, 5010 (1991).
[100.] T. K. Hollis, N. P. Robinson, J. Whelan, and B. Bosnich, *Tetrahedron Lett.*, **34**, 4309 (1993).
[101.] T. Yokozawa, K. Furuhashi, and H. Natsume, *Tetrahedron Lett.*, **36**, 5243 (1995).

820

CHAPTER 9

*Carbon-Carbon
Bond-Forming Reactions
of Compounds of Boron,
Silicon, and Tin*

because it is regenerated by recombination of iodide ion with silicon in the desilylation step.[102]

$$R_2C(OCH_3)_2 \quad + \quad \xrightarrow{TMS-I} \quad R_2C\overset{+}{=}OCH_3 \quad \xrightarrow{CH_2=CH-CH_2-Si(CH_3)_3 \ I^-} \quad R_2CCH_2CH=CH_2 \atop OCH_3$$

This type of reaction has been used for the extension of the carbon chain of protected carbohydrate acetals.[103]

Reaction of allylic silanes with enantiomerically pure 1,3-dioxanes has been found to proceed with moderate enantioselectivity.[104] The homoallylic alcohol can be liberated by oxidation followed by base-catalyzed β-elimination. The alcohols obtained in this way are formed in $70\pm5\%$ e.e.

The enantioselectivity is dependent on several reaction variables, including the Lewis acid and the solvent. The observed stereoselectivity appears to reflect differences in the precise structure of the electrophilic species generated. Mild Lewis acids tend to react with inversion of configuration at the reaction site, whereas very strong Lewis acids cause loss of enantioselectivity. The strength of the Lewis acid, together with related effects of solvent and other experimental variables, determines the nature of the electrophile. With mild Lewis acids, a tight ion pair favors inversion, whereas stronger Lewis acids cause complete dissociation to an acyclic species. These two species represent extremes of behavior and intermediate levels of enantioselectivity are also observed.[105]

102. H. Sakurai, K. Sasaki, and A. Hosomi, *Tetrahedron Lett.*, **22**, 745 (1981).
103. A. P. Kozikowski, K. L. Sorgi, B. C. Wang, and Z. Xu, *Tetrahedron Lett.*, **24**, 1563 (1983).
104. P. A. Bartlett, W. S. Johnson, and J. D. Elliott, *J. Am. Chem. Soc.*, **105**, 2088 (1983).
105. S. E. Denmark and N. G. Almstead, *J. Am. Chem. Soc.*, **113**, 8089 (1991).

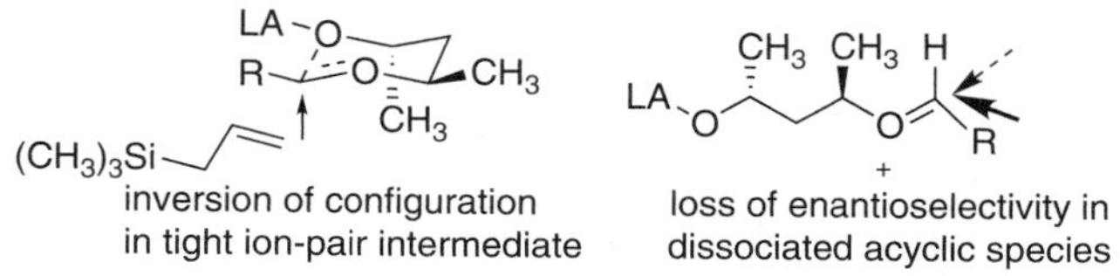

inversion of configuration
in tight ion-pair intermediate

loss of enantioselectivity in
dissociated acyclic species

Although most studies of alkenyl and allylic silanes have been done with trialkylsilyl analogs, the reactivity of the system can be adjusted by varying the silicon substituents. Allylic trichlorosilanes react with aldehydes in DMF to give homoallylic alcohols.[106] The reactions are highly stereoselective with respect to the silane geometry and give the product expected for a cyclic TS. The reaction is thought to proceed through a hexacoordinate silicon intermediate.

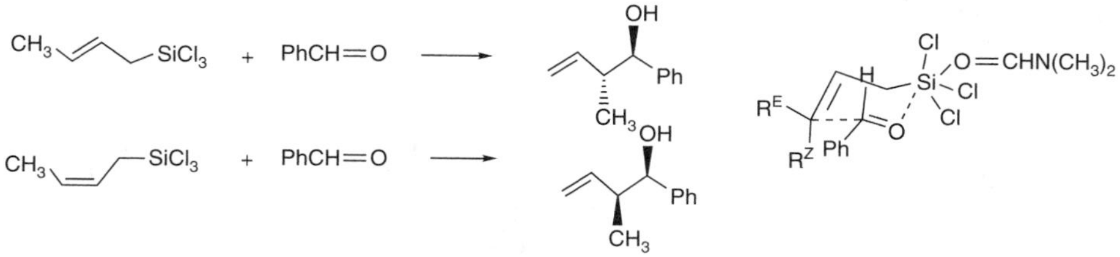

Allylic trichlorosilanes have shown promise in the development of methods for enantioselective reactions by use of chiral phosphoramides such as **C**.

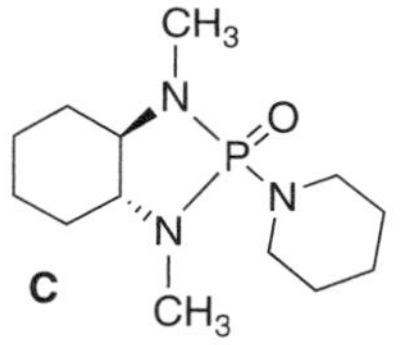

Mechanistic studies suggested that two phosphoramide molecules were involved.[107] This led to the development of linked phosphoramides such as **D**.[108]

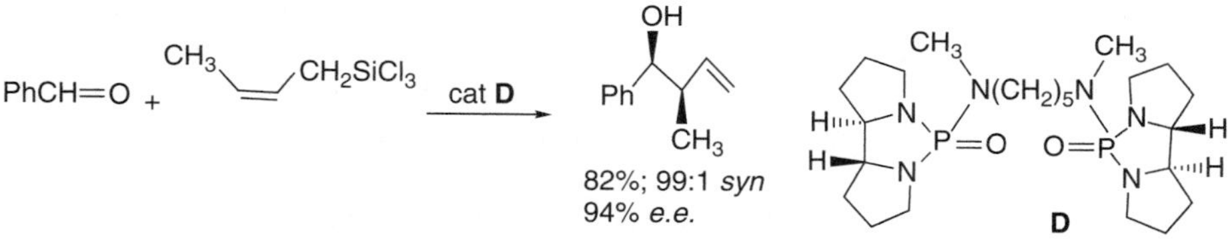

82%; 99:1 *syn*
94% e.e.

The axially chiral 2, 2′-bipyridine **E** is also an effective enantioselective catalyst for addition of allyltrichlorosilane to aldehydes.[109]

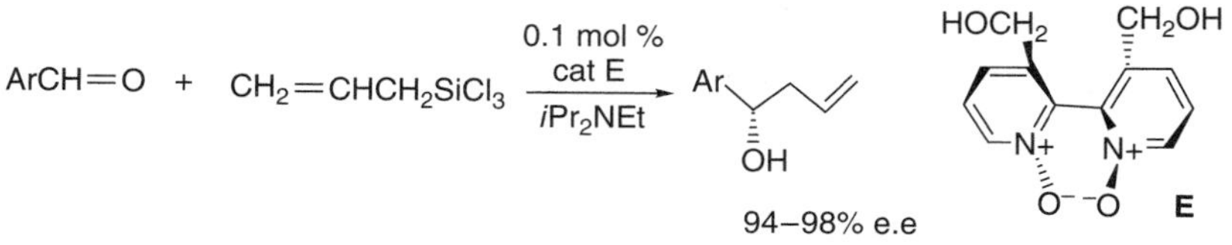

94–98% e.e

106. S. Kobayashi and K. Nishio, *J. Org. Chem.*, **59**, 6620 (1994).
107. S. E. Denmark and J. Fu, *J. Am. Chem. Soc.*, **123**, 9488 (2001).
108. S. E. Denmark and J. Fu, *J. Am. Chem. Soc.*, **125**, 2208 (2003).
109. T. Shimada, A. Kina, S. Ikeda, and T. Hayashi, *Org. Lett.*, **4**, 2799 (2002).

822

CHAPTER 9

*Carbon-Carbon
Bond-Forming Reactions
of Compounds of Boron,
Silicon, and Tin*

The use of trifluorosilanes permits reactions through hexacoordinate silicon, which presents an opportunity for chelation control. For example, α-hydroxy ketones give *syn* diols.[110]

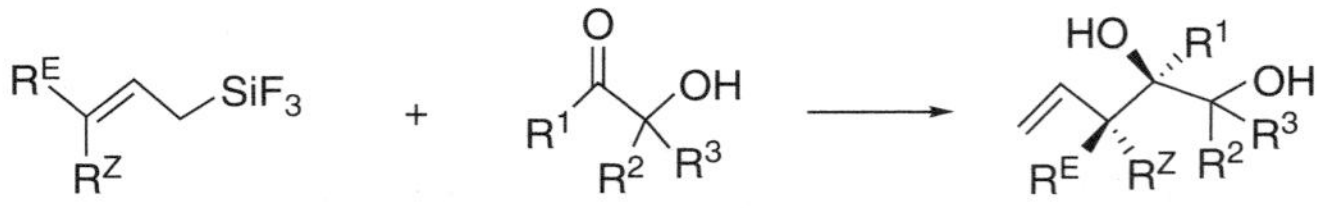

Advantage of this chelation has been taken in the construction of compounds with several contiguous chiral centers. *Z*-2-Butenyl trifluorosilanes give *syn*-1,3-diols on reaction with *anti*-β-hydroxy-α-methyl aldehydes.[111] The stereoselectivity is consistent with a chelated bicyclic TS.

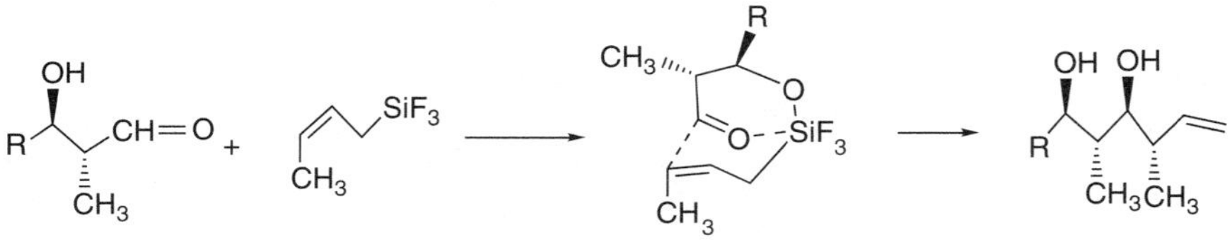

This methodology was applied to construct the all *anti* stereochemistry for a segment of the antibiotic zincophorin.

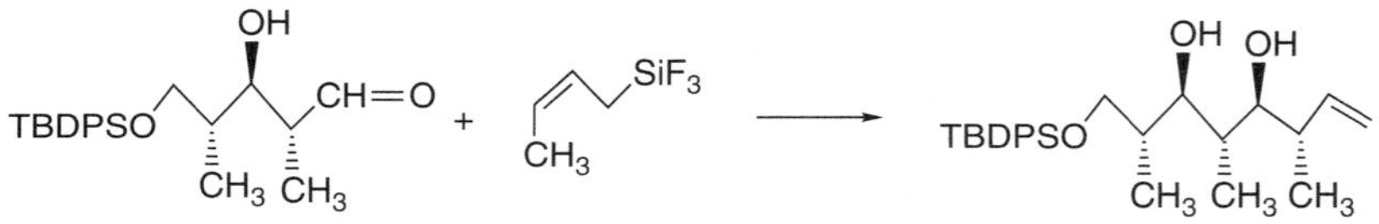

The corresponding *syn*-β-hydroxy-α-methyl aldehydes do not react through a chelated TS,[112] which appears to be due to steric factors that raise the bicyclic TS by several kcal relative to the *anti* isomers. The monocyclic six-membered TS does not incorporate these factors and the *syn* isomer reacts through a monocyclic TS. Figure 9.3 depicts the competing TSs and their relative energies as determined by MNDO calculations.

The electrophilicity of silicon is enhanced in five-membered ring structures. Chloro dioxasilolanes, oxazasilolidines, and diazasilolidines react with aldehydes in the absence of an external Lewis acid catalyst.[113]

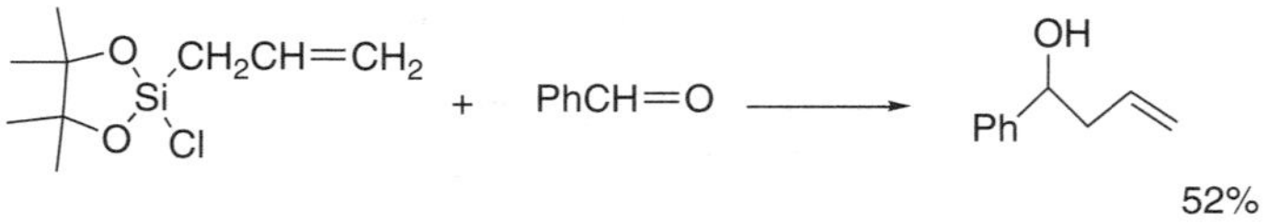

52%

110. K. Sato, M. Kira, and H. Sakurai, *J. Am. Chem. Soc.*, **111**, 6429 (1989).
111. S. R. Chemler and W. R. Roush, *J. Org. Chem.*, **63**, 3800 (1998).
112. S. R. Chemler and W. R. Roush, *J. Org. Chem.*, **68**, 1319 (2003).
113. J. W. A. Kinnaird, P. Y. Ng, K. Kubota, X. Wang, and J. L. Leighton, *J. Am. Chem. Soc.*, **124**, 7920 (2002).

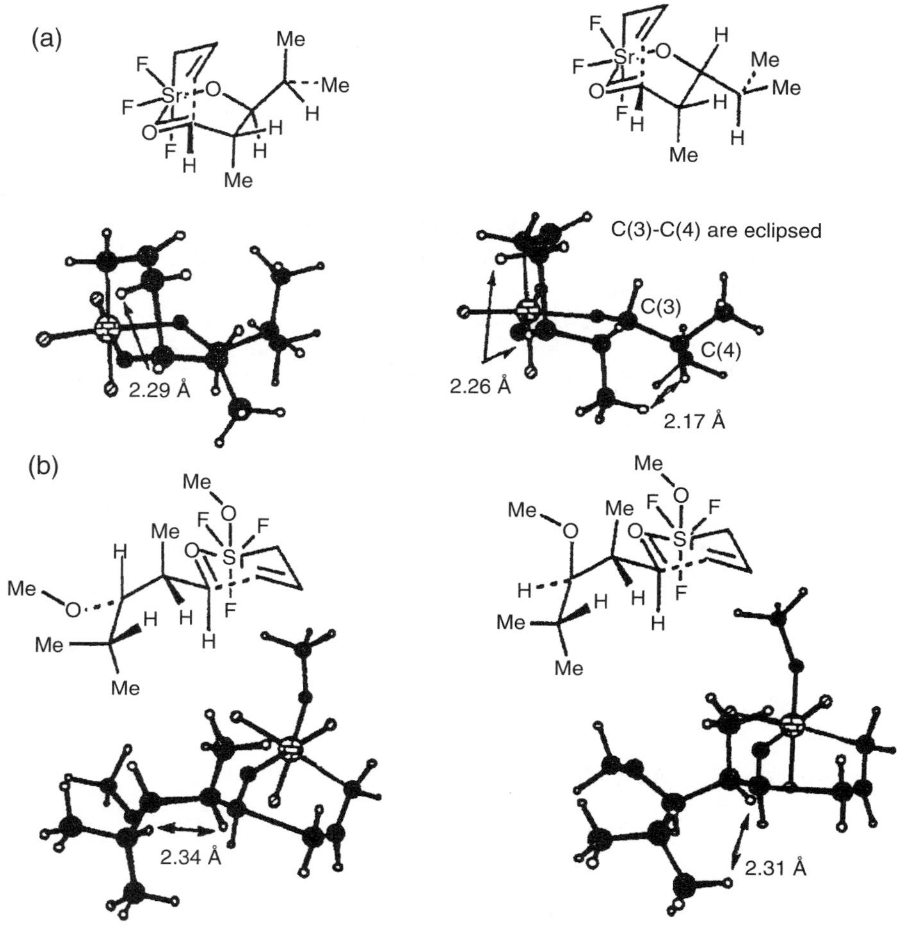

Fig. 9.3. Comparison of chelated bicyclic and nonchelated monocyclic transition structures for addition of allyl trifluorosilane to *syn-* and *anti-*3-methoxy-2,4-dimethylpentanal based on MNDO computations: (a) chelated bicyclic transition structures differ by 6 kcal/mol owing to nonbonded interactions in the *syn* case; (b) nonchelated monocyclic transition structures are of comparable energy for both isomers. Reproduced from *J. Org. Chem.*, **68**, 1319 (2003), by permission of the American Chemical Society.

The oxazasilolidine derived from pseudoephedrine incorporates chirality around the silicon and leads to enantioselective addition.

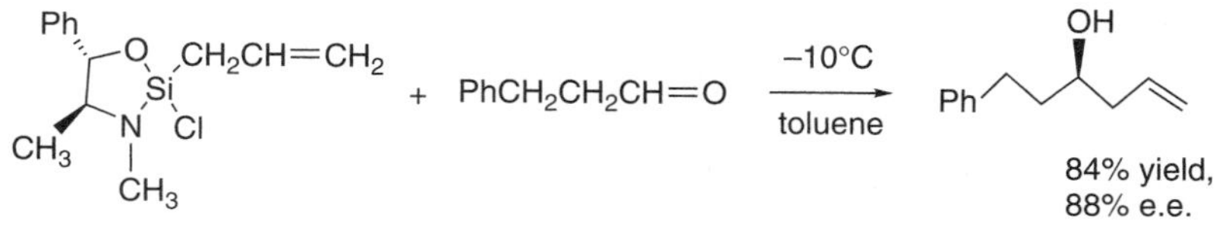

While trifluoro and other halosilanes function by increased *electrophilicity* at silicon, *nucleophilic* reactivity of allylic silanes can be enhanced by formation of anionic adducts (silicates). Reaction of allylic silanes with aldehydes and ketones can

824

CHAPTER 9

*Carbon-Carbon
Bond-Forming Reactions
of Compounds of Boron,
Silicon, and Tin*

be induced by fluoride ion. Fluoride adds at silicon to form a hypervalent anion having enhanced nucleophilicity.[114]

$$CH_2{=}CHCH_2SiR_3 \ + \ F^- \ \longrightarrow \ CH_2{=}CHCH_2{-}\overset{|}{\underset{/\,\backslash}{Si^-}}{-}F$$

The THF-soluble salt tetrabutylammonium fluoride (TBF) is a common source of fluoride. An alternative reagent is tetrabutylammonium triphenyldifluorosilicate (TBAF).[115] Unsymmetrical allylic anions generated in this way react with ketones at their less-substituted terminus.

$$CH_2{=}CHCH_2Si(CH_3)_3 \ + \ F^- \ \longrightarrow \ CH_2{=}CHCH_2\overset{\overset{F}{|}}{Si}(CH_3)_3 \ \xrightarrow{RCH=O} \ RCH\overset{\overset{OH}{|}}{CH_2}CH{=}CH_2$$

$$(CH_3)_2C{=}CHCH_2Si(CH_3) \ + Ph_2C{=}O \ \xrightarrow{TBAF} \ \xrightarrow{H_2O} \ (CH_3)_2C{=}CHCH_2\underset{\underset{OH}{|}}{C}Ph_2 \quad 87\%$$

An allylic silane of this type serves as a reagent for the introduction of isoprenoid structures.[116]

$$(CH_3)_2C{=}CHCH{=}O \ + \ (CH_3)_3SiCH_2\underset{\underset{CH_2}{\|}}{C}CH{=}CH_2 \ \xrightarrow{R_4N^+F^-} \ (CH_3)_2C{=}CHCHCH_2\underset{\underset{CH_2}{\|}}{C}CH{=}CH_2$$
$$\underset{OH \quad\quad CH_2}{} \quad 70\%$$

Fluoride-induced desilylation has also been used to effect ring closures.[117]

Allylic trimethoxysilanes are activated by a catalytic combination of CuCl and TBAF.[118] The mechanism of this reaction is not entirely clear, but it seems to involve fluoride activation of the silane. These reactions are stereoconvergent for the isomeric 2-butenyl silanes, indicating that reaction occurs through an acyclic TS.

2.6:1 *syn:anti*

114. A. Hosomi, A. Shirahata, and H. Sakurai, *Tetrahedron Lett.*, 3043 (1978); G. G. Furin, O. A. Vyazankina, B. A. Gostevsky, and N. S. Vyazankin, *Tetrahedron*, **44**, 2675 (1988).

115. A. S. Pilcher and P. De Shong, *J. Org. Chem.*, **61**, 6901 (1996).

116. A. Hosomi, Y. Araki, and H. Sakurai, *J. Org. Chem.*, **48**, 3122 (1983).

117. B. M. Trost and J. E. Vincent, *J. Am. Chem. Soc.*, **102**, 5680 (1980); B. M. Trost and D. P. Curran, *J. Am. Chem. Soc.*, **103**, 7380 (1981).

118. S. Yamasaki, K. Fujii, R. Wada, M. Kanai, and M. Shibasaki, *J. Am. Chem. Soc.*, **124**, 6536 (2002).

p-Tol-BINAP-AgF effects enantioselective additions with trimethoxysilanes.[119] These reactions give *anti* products, regardless of the configuration of the allylic silane.

$$ArCH=O \ + \ CH_3 \diagdown\!\!\!\diagup Si(OCH_3)_3 \xrightarrow[\text{10 mol \% AgF}]{\text{6 mol \% } p\text{-tol-BINAP}} \underset{CH_3}{\overset{OH}{Ar\diagdown\!\!\!\diagup\!\!\!\diagdown}}$$

96% e.e.

The combination BINAP-Ag$_2$O-KF with 18-crown-6 also leads to high enantioselectivity.[120]

9.2.4. Reactions with Iminium Ions

Iminium ions are reactive electrophiles toward both alkenyl and allylic silanes. Useful techniques for closing nitrogen-containing rings are based on in situ generation of iminium ions from amines and formaldehyde.[121]

$$\underset{H}{\overset{CH_2}{PhCH_2NCH_2CH_2CCH_2Si(CH_3)_3}} \xrightarrow[H_2C=O]{CF_3CO_2H} \underset{CH_2 \quad CH_2}{PhCH_2\overset{+}{N}CH_2CH_2CCH_2Si(CH_3)_3} \longrightarrow PhCH_2N\!\!\!\bigcirc\!\!\!=CH_2$$

73%

When primary amines are employed, the initially formed 3-butenylamine undergoes a further reaction forming 4-piperidinols.[122]

$$PhCH_2\overset{+}{N}H_3 + CH_2=CHCH_2Si(CH_3)_3 + CH_2=O \longrightarrow PhCH_2N\!\!\!\bigcirc\!\!\!-OH$$

Reactions of this type can also be observed with 4-(trimethylsilyl)-3-alkenylamines.[123]

$$\underset{H}{\overset{}{R^1NCH_2CH_2CH=CR^3}} \ \underset{Si(CH_3)_3}{} \xrightarrow[H^+]{CH_2=O} \text{(ring product with } R_3 \text{, N-}R^1\text{)}$$

Mechanistic investigation in this case has shown that there is an equilibrium between an alkenyl silane and an allylic silane by a rapid 3,3-sigmatropic process. The cyclization occurs through the more reactive allylic silane.

119. A. Yanagisawa, H. Kageyama, Y. Nakatsuka, K. Asakawa, Y. Matsumoto, and H. Yamamoto, *Angew. Chem. Int. Ed. Engl.*, **38**, 3701 (1999).
120. M. Wadamoto, N. Ozasa, A. Yanagisawa, and H. Yamamoto, *J. Org. Chem.*, **68**, 5593 (2003).
121. P. A. Grieco and W. F. Fobare, *Tetrahedron Lett.*, **27**, 5067 (1986).
122. S. D. Larsen, P. A. Grieco, and W. F. Fobare, *J. Am. Chem. Soc.*, **108**, 3512 (1986).
123. C. Flann, T. C. Malone, and L. E. Overman, *J. Am. Chem. Soc.*, **109**, 6097 (1987).

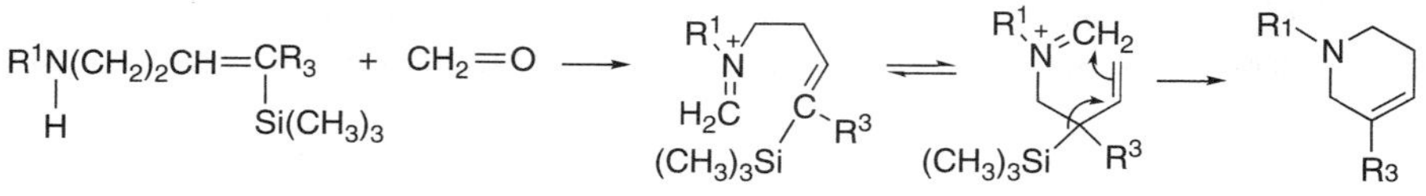

N-Acyliminium ions, which are even more reactive toward allylic and alkenylsilanes, are usually obtained from imides by partial reduction (see Section 2.2.2). The partially reduced *N*-acylcarbinolamines can then generate acyliminium ions. Such reactions have been employed in intramolecular situations with both allylic and vinyl silanes.

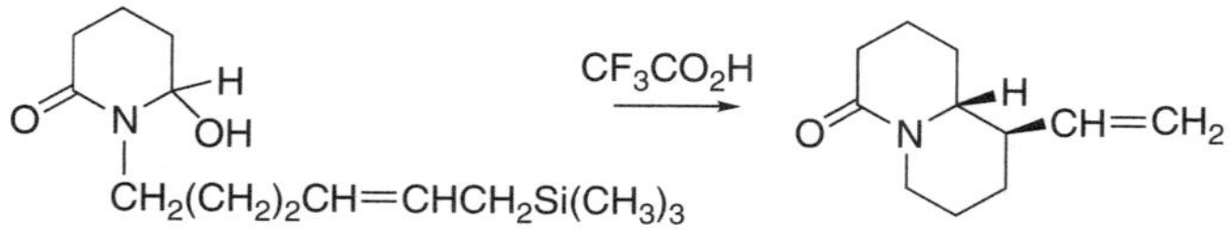

Ref. 124

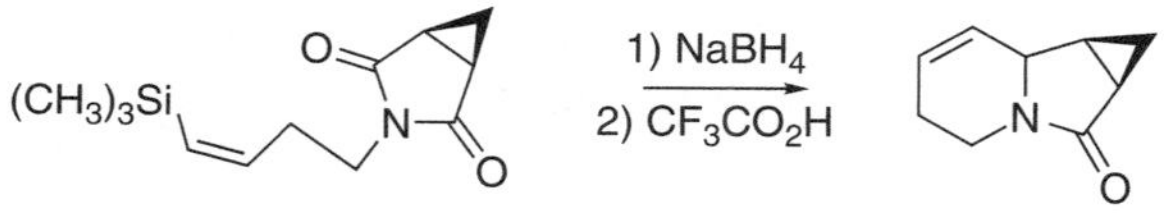

Ref. 125

9.2.5. Acylation Reactions

Reaction of alkenyl silanes with acid chlorides is catalyzed by aluminum chloride or stannic chloride.[126]

$$\mathrm{RCH{=}CHSi(CH_3)_3 \;+\; RCOCl \;\xrightarrow[\text{or } SnCl_4]{AlCl_3}\; RCH{=}CH\overset{O}{\overset{\|}{C}}R}$$

Titanium tetrachloride induces reaction with dichloromethyl methyl ether to give α, β-unsaturated aldehydes.[127]

$$\mathrm{RCH{=}CHSi(CH_3)_3 \;+\; Cl_2CHOCH_3 \;\xrightarrow{TiCl_4}\; RCH{=}CHCH{=}O}$$

Similar conditions are used to effect reactions of allylsilanes with acyl halides, resulting in β, γ-unsaturated ketones.[128]

$$\mathrm{Ph\overset{O}{\overset{\|}{C}}Cl \;+\; CH_2{=}CHCH_2Si(CH_3)_3 \;\xrightarrow{AlCl_3}\; Ph\overset{O}{\overset{\|}{C}}CH_2CH{=}CH_2}$$

124. H. Hiemstra, M. H. A. M. Sno, R. J. Vijn, and W. N. Speckamp, *J. Org. Chem.*, **50**, 4014 (1985).
125. G. Kim, M. Y. Chu-Moyer, S. J. Danishefsky, and G. K. Schulte, *J. Am. Chem. Soc.*, **115**, 30 (1993).
126. I. Fleming and A. Pearce, *J. Chem. Soc., Chem. Commun.*, 633 (1975); W. E. Fristad, D. S. Dime, T. R. Bailey, and L. A. Paquette, *Tetrahedron Lett.*, 1999 (1979).
127. K. Yamamoto, O. Nunokawa, and J. Tsuji, *Synthesis*, 721 (1977).
128. J.-P. Pillot, G. Deleris, J. Dunogues, and R. Calas, *J. Org. Chem.*, **44**, 3397 (1979); R. Calas, J. Dunogues, J.-P. Pillot, C. Biran, F. Pisciotti, and B. Arreguy, *J. Organomet. Chem.*, **85**, 149 (1975).

Indium tribromide also gives good yields, with minor isomerization to the α, β-isomers.[129]

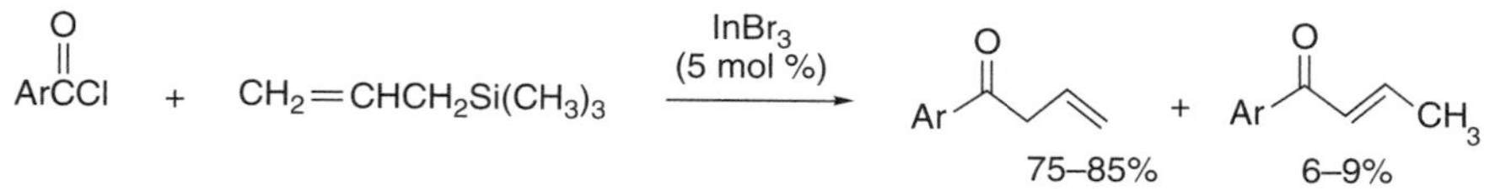

These reactions probably involve acylium ions as the electrophiles.

Scheme 9.4 shows some representative reactions of allylic and alkenyl silanes. Entry 1 involves 3-trimethylsilylcyclopentene, which can be made by hydrosilylation of cyclopentadiene by chlorodimethylsilane, followed by reaction with methylmagnesium bromide.

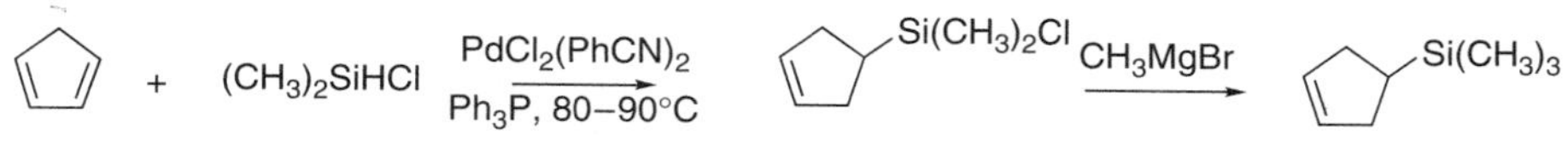

Entry 2 was reported as part of a study of the stereochemistry of addition of allyltrimethylsilane to protected carbohydrates. Use of BF_3 as the Lewis acid, as shown, gave the product from an open TS, whereas $TiCl_4$ led to the formation of the alternate stereoisomer through chelation control. Similar results were reported for a protected galactose.

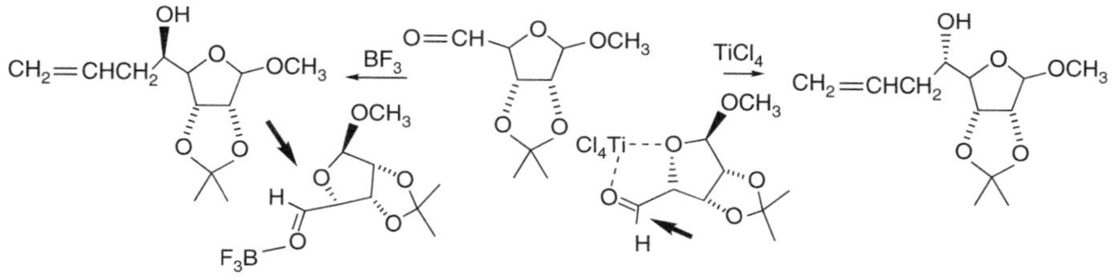

In Entry 3, BF_3-mediated addition exhibits a preference for the Felkin stereochemistry.

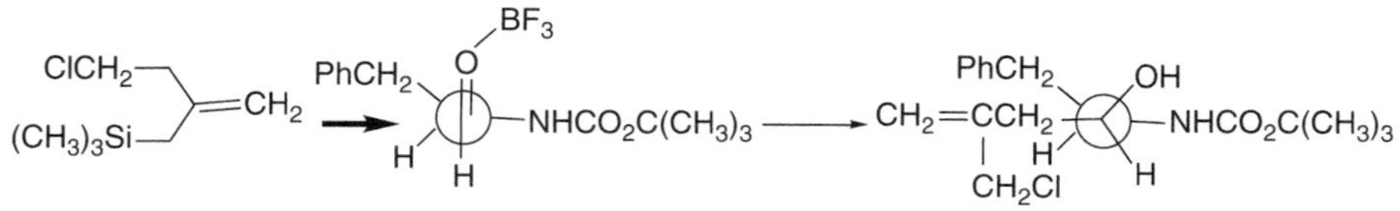

Entries 4 and 5 are examples of use of the Sakurai reaction to couple major fragments in multistage synthesis. In Entry 4 an unusual catalyst, a chiral acyloxyboronate (see p. 126) was used to effect an enantioselective coupling. (See p. 847 for another application of this catalyst.) Entry 5 was used in the construction of amphidinolide P, a compound with anticancer activity.

Entries 6 to 8 demonstrate addition of allyl trimethylsilane to protected carbohydrate acetals. This reaction can be a valuable method for incorporating the chirality of carbohydrates into longer carbon chains. In cases involving cyclic acetals, reactions occur through oxonium ions and the stereochemistry is governed by steric and stereoelectronic effects of the ring. Note that Entry 8 involves the use of trimethylsilyl

[129.] J. S. Yadav, B. V. S. Reddy, M. S. Reddy, and G. Parimala, *Synthesis*, 2390 (2003).

828

CHAPTER 9

*Carbon-Carbon
Bond-Forming Reactions
of Compounds of Boron,
Silicon, and Tin*

Scheme 9.4. Reactions of Alkenyl and Allylic Silanes with Aldehydes, Ketones, Acetals, Iminium Ions, and Acyl Halides

A. Reactions with carbonyl compounds

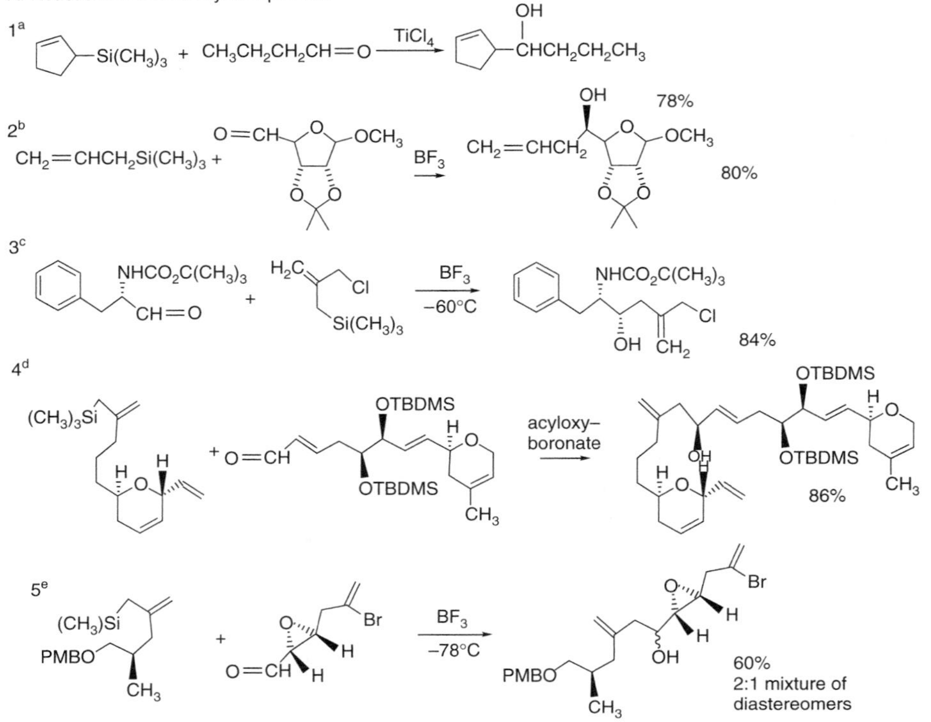

B. Reactions with acetals and related compounds

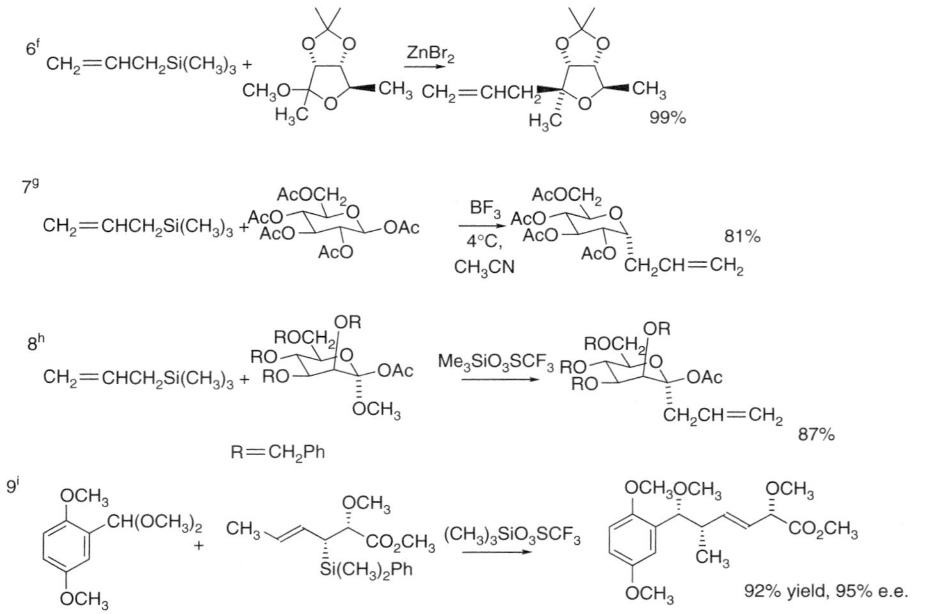

Scheme 9.4. *(Continued)*

829

SECTION 9.2

*Organosilicon
Compounds*

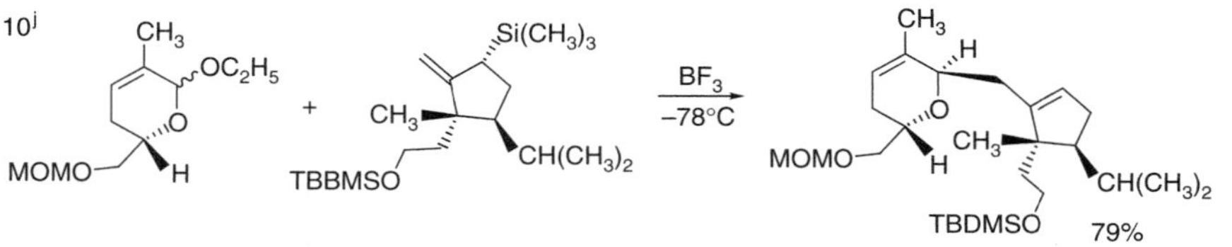

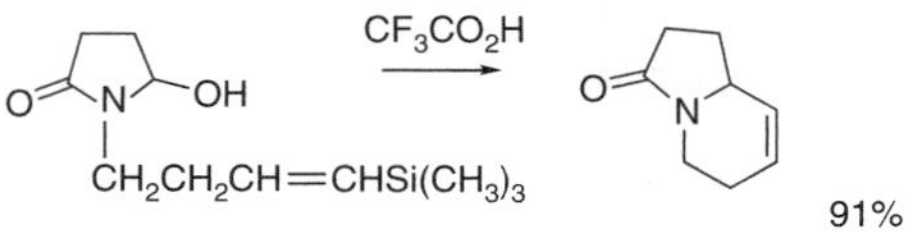

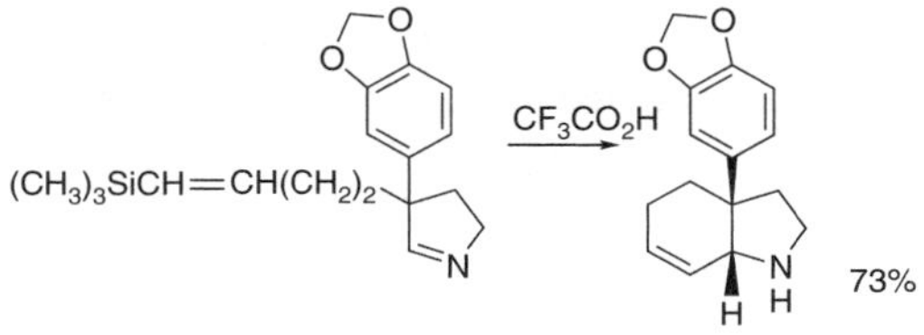

C. Reactions with Iminium ions

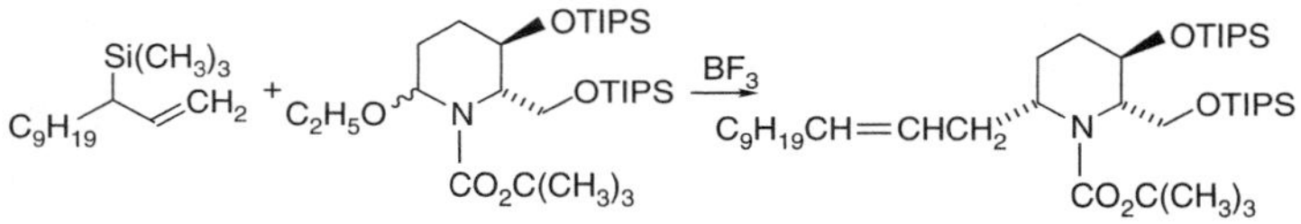

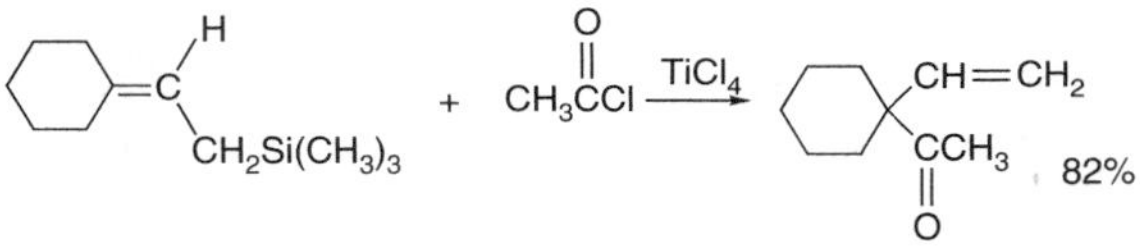

D. Acylation reactions

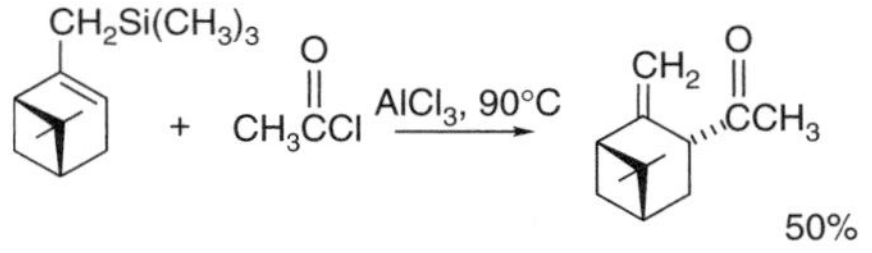

a. I. Ojima, J. Kumagai, and Y. Miyazawa, *Tetrahedron Lett.*, 1385 (1977).
b. S. Danishefsky and M. De Ninno, *Tetrahedron Lett.*, **26**, 823 (1985).
c. F. D'Aniello and M. Taddei, *J. Org. Chem.*, **57**, 5247 (1992).
d. P. A. Wender, S. G. Hegde, R. D. Hubbard, and L. Zhang, *J. Am. Chem. Soc.*, **124**, 4956 (2002).
e. D. R. Williams, B. J. Myers, and L. Mi, *Org. Lett.*, **2**, 945 (2000).
f. H. Suh and C. S. Wilcox, *J. Am. Chem. Soc.*, **110**, 470 (1988).
g. A. Giannis and K. Sanshoff, *Tetrahedron Lett.*, **26**, 1479 (1985).
h. A. Hosomi, Y. Sakata, and H. Sakurai, *Tetrahedron Lett.*, **25**, 2383 (1984).
i. J. S. Panek and M. Yang, *J. Am. Chem. Soc.*, **113**, 6594 (1991).
j. D. R. Williams and R. W. Heidebrecht, Jr., *J. Am. Chem. Soc.*, **125**, 1843 (2003).
k. C. Flann, T. C. Malone, and L. E. Overman, *J. Am. Chem. Soc.*, **109**, 6097 (1987).
l. L. E. Overman and R. M. Burk, *Tetrahedron Lett.*, **25**, 5739 (1984).
m. I. Ojima and E. S. Vidal, *J. Org. Chem.*, **63**, 7999 (1998).
n. I. Fleming and I. Paterson, *Synthesis*, 446 (1979).
o. J. P. Pillot, G. Deleris, J. Dunogues, and R. Calas, *J. Org. Chem.*, **44**, 3397 (1979).

830

CHAPTER 9

*Carbon-Carbon
Bond-Forming Reactions
of Compounds of Boron,
Silicon, and Tin*

triflate as the catalyst. Entry 9 is a case of substrate control of enantioselectivity. Both high diastereoselectivity and enantioselectivity at the new chiral center were observed. The reaction is believed to proceed through an *O*-methyloxonium and to involve an open TS. Entry 10 involves generation of a cyclic oxonium ion. The observed stereochemistry is consistent with a synclinal orientation in the TS.

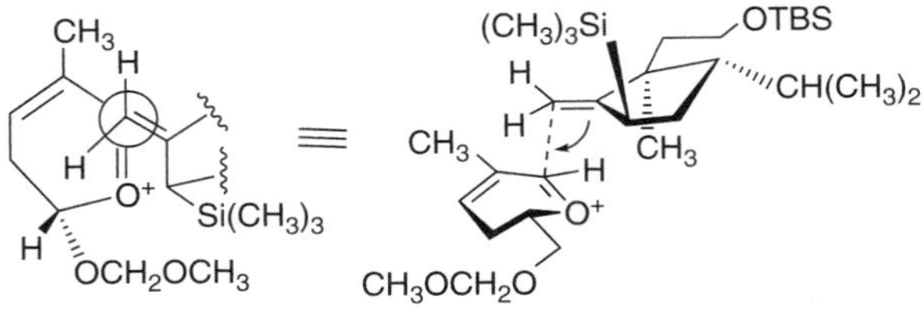

Entries 11 to 13 are examples of iminium ion and acyliminium ion reactions. Note that in Entries 11 and 12, vinyl, rather than allylic, silane moieties are involved. Entries 14 and 15 illustrate the synthesis of β, γ-unsaturated ketones by acylation of allylic silanes.

9.2.6. Conjugate Addition Reactions

Allylic silanes act as nucleophilic species toward α, β-unsaturated ketones in the presence of Lewis acids such as $TiCl_4$.[130]

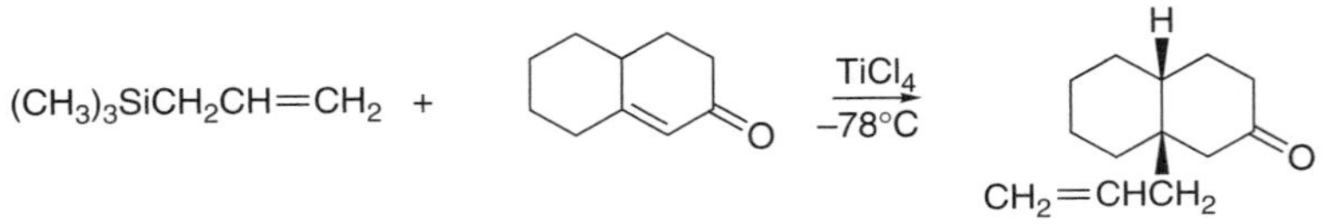

The stereochemistry of this reaction in cyclic systems is in accord with expectations for stereoelectronic control. The allylic group approaches from a trajectory that is appropriate for interaction with the LUMO of the conjugated system.[131]

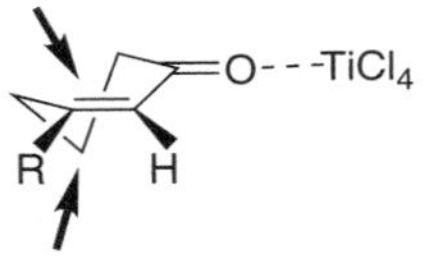

The stereoselectivity then depends on the conformation of the enone and the location of substituents that establish a steric bias for one of the two potential directions of approach. In the ketone **11**, the preferred approach is from the β-face, since this permits maintaining a chair conformation as the reaction proceeds.[132]

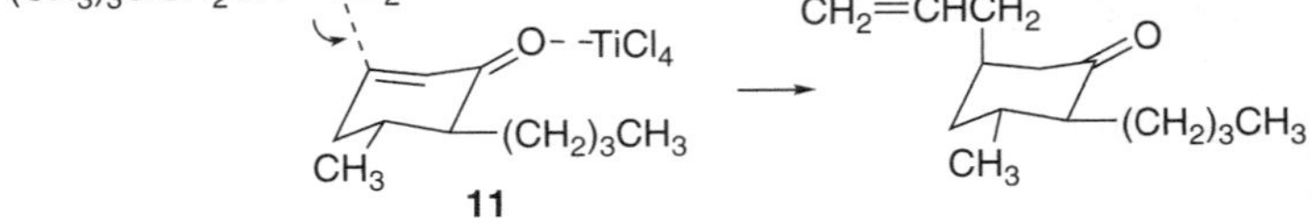

130. A. Hosomi and H. Sakurai, *J. Am. Chem. Soc.*, **99**, 1673 (1977).
131. T. A. Blumenkopf and C. H. Heathcock, *J. Am. Chem. Soc.*, **105**, 2354 (1983).
132. W. R. Roush and A. E. Walts, *J. Am. Chem. Soc.*, **106**, 721 (1984).

Conjugate addition to acyclic enones is subject to chelation control when $TiCl_4$ is used as the Lewis acid. Thus, whereas the *E*-enone **12** gives *syn* product **13** via an acyclic TS, the *Z*-isomer **14** reacts through a chelated TS to give **15**.[133]

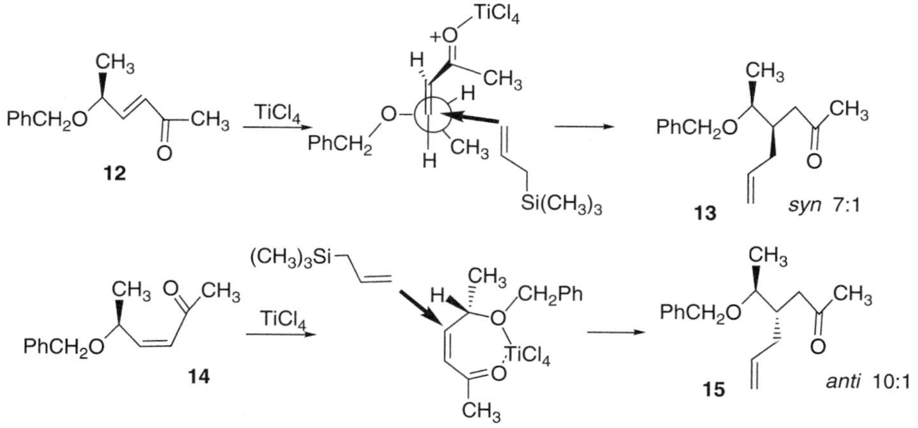

Conjugate additions of allylic silanes to enones are also catalyzed by $InCl_3$-TMSCl.[134]

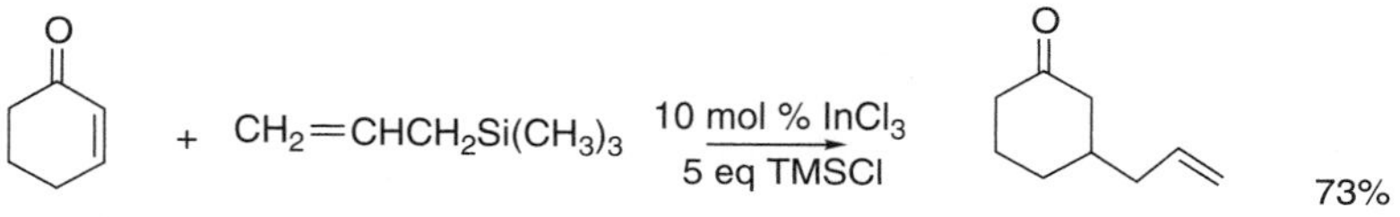

The reaction can also be carried out using indium metal. Under these conditions $InCl_3$ is presumably generated in situ.[135]

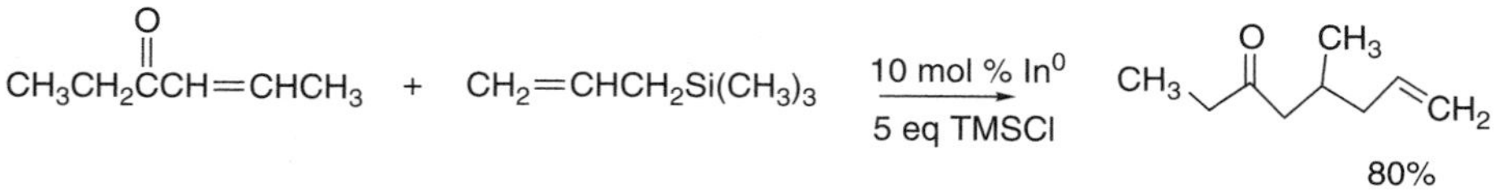

Conjugate addition can also be carried out by fluoride-mediated disilylation. A variety of α,β-unsaturated esters and amides have been found to undergo this reaction.[136]

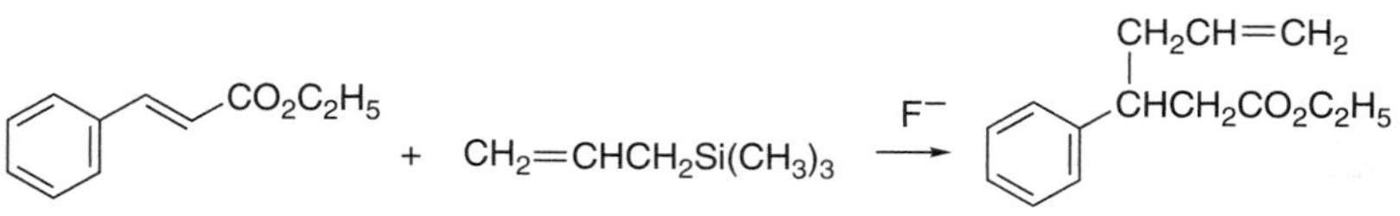

133. C. H. Heathcock, S. Kiyooka, and T. A. Blujenkopf, *J. Org. Chem.*, **49**, 4214 (1984).
134. P. H. Lee, K. Lee, S.-Y. Sung, and S. Chang, *J. Org. Chem.*, **66**, 8646 (2001); Y. Onishi, T. Ito, M. Yasuda, and A. Baba, *Eur. J. Org. Chem.*, 1578 (2002).
135. P. H. Lee, D. Seomoon, S. Kim, K. Nagaiah, S. V. Damle, and K. Lee, *Synthesis*, 2189 (2003).
136. G. Majetich, A. Casares, D. Chapman, and M. Behnke, *J. Org. Chem.*, **51**, 1745 (1986).

Scheme 9.5. Conjugate Addition of Allylic Silanes to α, β-Unsaturated Enones

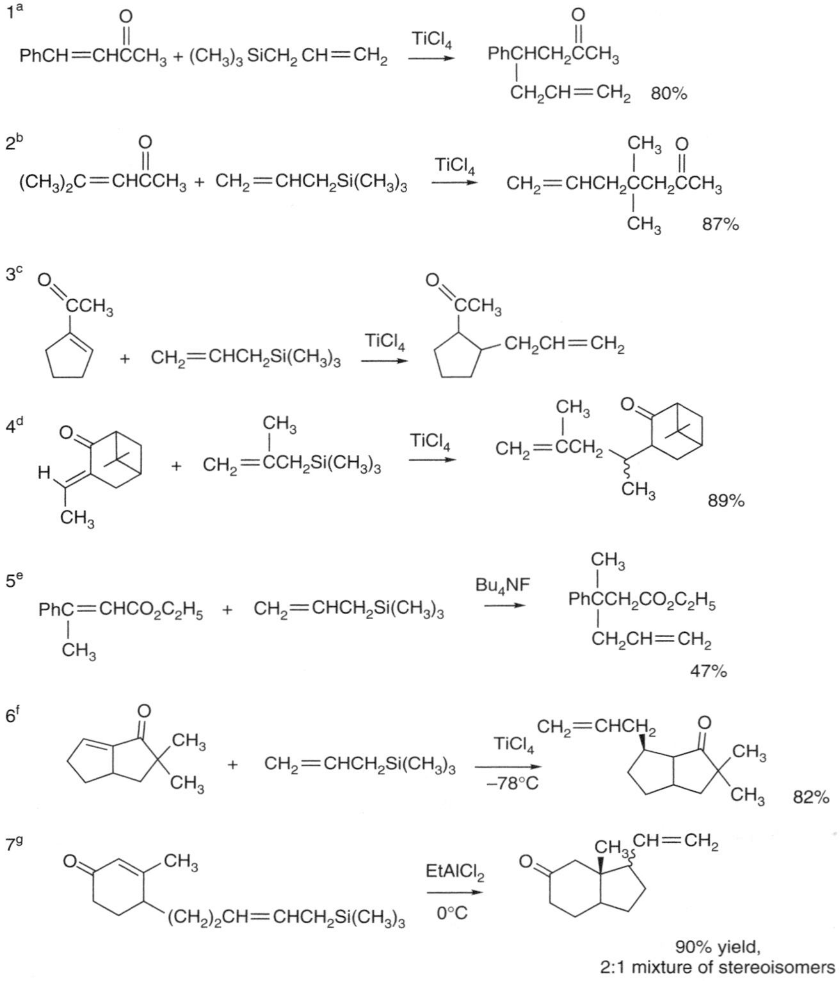

1[a]

PhCH=CHCCH$_3$ + (CH$_3$)$_3$SiCH$_2$CH=CH$_2$ $\xrightarrow{\text{TiCl}_4}$ PhCHCH$_2$CCH$_3$ 80%
CH$_2$CH=CH$_2$

2[b]

(CH$_3$)$_2$C=CHCCH$_3$ + CH$_2$=CHCH$_2$Si(CH$_3$)$_3$ $\xrightarrow{\text{TiCl}_4}$ CH$_2$=CHCH$_2$CCH$_2$CCH$_3$ 87%

3[c]

+ CH$_2$=CHCH$_2$Si(CH$_3$)$_3$ $\xrightarrow{\text{TiCl}_4}$

4[d]

+ CH$_2$=CCH$_2$Si(CH$_3$)$_3$ $\xrightarrow{\text{TiCl}_4}$ 89%

5[e]

PhC=CHCO$_2$C$_2$H$_5$ + CH$_2$=CHCH$_2$Si(CH$_3$)$_3$ $\xrightarrow{\text{Bu}_4\text{NF}}$ PhCCH$_2$CO$_2$C$_2$H$_5$ 47%
CH$_3$

6[f]

+ CH$_2$=CHCH$_2$Si(CH$_3$)$_3$ $\xrightarrow[-78°\text{C}]{\text{TiCl}_4}$ 82%

7[g]

(CH$_2$)$_2$CH=CHCH$_2$Si(CH$_3$)$_3$ $\xrightarrow[0°\text{C}]{\text{EtAlCl}_2}$ 90% yield, 2:1 mixture of stereoisomers

a. H. Sakurai, A. Hosmoni, and J. Hayashi, *Org. Synth.*, **62**, 86 (1984).
b. D. H. Hua, *J. Am. Chem. Soc.*, **108**, 3835 (1986).
c. H. O. House, P. C. Gaa, and D. Van Derveer, *J. Org. Chem.*, **48**, 1661 (1983).
d. T. Yanami, M. Miyashita, and A. Yoshikoshi, *J. Org. Chem.*, **45**, 607 (1980).
e. G. Majetich, A Casares, D. Chapman, and M. Behnke, *J. Org. Chem.*, **51**, 1745 (1986).
f. C. E. Davis, B. C. Duffy, and R. M. Coates, *Org. Lett.*, **2**, 2717 (2000).
g. D. Schinzer, S. Solyom, and M. Becker, *Tetrahedron Lett.*, **26**, 1831 (1985).

With unsaturated aldehydes, 1,2-addition occurs and with ketones both the 1,2- and 1,4-products are formed.

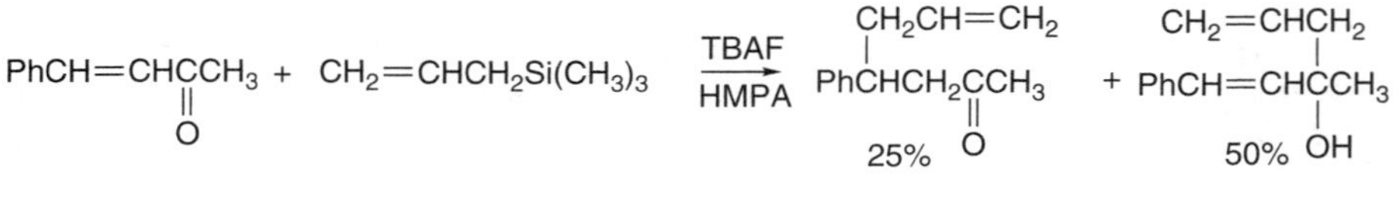

PhCH=CHCCH$_3$ + CH$_2$=CHCH$_2$Si(CH$_3$)$_3$ $\xrightarrow[\text{HMPA}]{\text{TBAF}}$ PhCHCH$_2$CCH$_3$ + PhCH=CHCCH$_3$
25% 50%

Some examples of conjugate addition reactions of allylic silanes are given in Scheme 9.5. Entries 1 to 3 illustrate the synthesis of several β-allyl ketones. Note that Entry 2 involves the creation of a quaternary carbon. Entry 4 was used in the synthesis of a terpenoid ketone, (+)-nootkatone. Entry 5 illustrates fluoride-mediated addition using tetrabutylammonium fluoride. These conditions were found to be especially effective for unsaturated esters. In Entry 6, the addition is from the convex face of the ring system. Entry 7 illustrates a ring closure by intramolecular conjugate addition.

9.3. Organotin Compounds

9.3.1. Synthesis of Organostannanes

The readily available organotin compounds include tin hydrides (stannanes) and the corresponding chlorides, with the tri-*n*-butyl compounds being the most common. Trialkylstannanes can be added to carbon-carbon double and triple bonds. The reaction is usually carried out by a radical chain process,[137] and the addition is facilitated by the presence of radical-stabilizing substituents.

$$(C_2H_5)_3SnH \ + \ CH_2{=}CHCN \ \xrightarrow{\text{AIBN}} \ (C_2H_5)_3SnCH_2CH_2CN$$

Ref. 138

$$(C_4H_9)_3SnH \quad CH_2{=}C{\overset{\displaystyle CO_2CH_3}{\underset{\displaystyle Ph}{}}} \ \longrightarrow \ (C_4H_9)_3SnCH_2\underset{\displaystyle Ph}{CH}CO_2CH_3$$

Ref. 139

With terminal alkynes, the stannyl group is added at the unsubstituted carbon and the Z-stereoisomer is initially formed but is readily isomerized to the *E*-isomer.[140]

$$HC{\equiv}CCH_2OTHP \ \xrightarrow[\text{AIBN}]{(C_4H_9)SnH} \ \underset{(C_4H_9)_3Sn}{\overset{H}{\diagup}}{=}\underset{CH_2OTHP}{\overset{H}{\diagdown}} \ \longrightarrow \ \underset{H}{\overset{(C_4H_9)_3Sn}{\diagup}}{=}\underset{CH_2OTHP}{\overset{H}{\diagdown}}$$

The reaction with internal acetylenes leads to a mixture of both regioisomers and stereoisomers.[141]

Lewis acid–catalyzed hydrostannylation has been observed using $ZrCl_4$. With terminal alkynes the Z-alkenylstannane is formed.[142] These reactions are probably similar in mechanism to Lewis acid–catalyzed additions of silanes (see p. 811).

$$RC{\equiv}CH \ + \ (n{-}C_4H_9)_3SnH \ \xrightarrow{ZrCl_4} \ \underset{H \quad H}{\overset{R \qquad Sn(n{-}C_4H_9)_3}{\diagup\diagdown}}$$

137. H. G. Kuivila, *Adv. Organomet. Chem.*, **1**, 47 (1964).
138. A. J. Leusinsk and J. G. Noltes, *Tetrahedron Lett.*, 335 (1966).
139. I. Fleming and C. J. Urch, *Tetrahedron Lett.*, **24**, 4591 (1983).
140. E. J. Corey and R. H. Wollenberg, *J. Org. Chem.*, **40**, 2265 (1975).
141. H. E. Ensley, R. R. Buescher, and K. Lee, *J. Org. Chem.*, **47**, 404 (1982).
142. N. Asao, J.-X. Liu, T. Sudoh, and Y. Yamamoto, *J. Org. Chem.*, **61**, 4568 (1996).

834

CHAPTER 9

*Carbon-Carbon
Bond-Forming Reactions
of Compounds of Boron,
Silicon, and Tin*

Palladium-catalyzed procedures have also been developed for addition of stannanes to alkynes,[143] and these reactions usually occur by *syn* addition.

$$PhC{\equiv}CCH_3 \ + \ (C_4H_9)SnH \ \xrightarrow{PdCl_2\text{-}PPh_3} \ \underset{Ph \quad\ H}{\overset{(C_4H_9)_3Sn \quad CH_3}{\diagdown=\diagup}}$$

Hydrostannylation of terminal alkynes can also be achieved by reaction with stannyl-cyanocuprates.

$$HOCH_2CH_2C{\equiv}CH \ + \ \underset{\underset{CH_3}{|}}{(CH_3)_3SnCu(CN)Li_2} \ \longrightarrow \ \underset{H \qquad Sn(CH_3)_3}{\overset{HOCH_2CH_2 \quad H}{\diagdown=\diagup}}$$

Ref. 144

$$(C_2H_5O)_2CHC{\equiv}CH \ + \ \underset{\underset{C_4H_9}{|}}{(n\text{-}C_4H_9)_3SnCu(CN)Li_2} \ \longrightarrow \ \underset{H \qquad Sn(n\text{-}C_4H_9)_3}{\overset{(C_2H_5O)_2CH \quad H}{\diagdown=\diagup}}$$

Ref. 145

These reactions proceed via a *syn* addition followed by protonolysis.

$$RC{\equiv}CH \ + \ \underset{\underset{R''}{|}}{R'_3SnCu(CN)Li_2} \ \longrightarrow \ \underset{\underset{''R}{\overset{|}{-Cu-}}\quad SnR'_3}{\overset{R \qquad H}{\diagdown=\diagup}} \ \xrightarrow{SOH} \ \underset{H \qquad SnR'_3}{\overset{R \qquad H}{\diagdown=\diagup}}$$

Allylic stannanes can be prepared from allylic halides and sulfonates by displacement with or $LiSnMe_3$ or $LiSnBu_3$.[146] They can also be prepared by Pd-catalyzed substitution of allylic acetates and phosphates using $(C_2H_5)_2AlSn(n\text{-}C_4H_9)_3$.[147]

Another major route for synthesis of stannanes is reaction of an organometallic reagent with a trisubstituted halostannane, which is the normal route for the preparation of aryl stannanes.

$$CH_3O{-}\!\!\left\langle\!\!\bigcirc\!\!\right\rangle\!\!{-}MgBr \ + \ BrSn(CH_3)_3 \ \longrightarrow \ CH_3O{-}\!\!\left\langle\!\!\bigcirc\!\!\right\rangle\!\!{-}Sn(CH_3)_3$$

Ref. 148

[143]. H. X. Zhang, F. Guibe, and G. Balavoine, *Tetrahedron Lett.*, **29**, 619 (1988); M. Benechie, T. Skrydstrup, and F. Khuong-Huu, *Tetrahedron Lett.*, **32**, 7535 (1991); N. D. Smith, J. Mancuso, and M. Lautens, *Chem. Rev.*, **100**, 3257 (2000).

[144]. I. Beaudet, J.-L. Parrain, and J.-P. Quintard, *Tetrahedron Lett.*, **32**, 6333 (1991).

[145]. A. C. Oehlschlager, M. W. Hutzinger, R. Aksela, S. Sharma, and S. M. Singh, *Tetrahedron Lett.*, **31**, 165 (1990).

[146]. E. Winter and R. Bruckner, *Synlett*, 1049 (1994); G. Naruta and K. Maruyama, *Chem. Lett.*, 881 (1979); G. E. Keck and S. D. Tonnies, *Tetrahedron Lett.*, **34**, 4607 (1993); S. Weigand and R. Bruckner, *Synthesis*, 475 (1996).

[147]. B. M. Trost and J. W. Herndon, *J. Am. Chem. Soc.*, **106**, 6835 (1984); S. Matsubara, K. Wakamatsu, J. Morizawa, N. Tsuboniwa, K. Oshima, and H. Nozaki, *Bull. Chem. Soc. Jpn.*, **58**, 1196 (1985).

[148]. C. Eaborn, A. R. Thompson, and D. R. M. Walton, *J. Chem. Soc. C*, 1364 (1967); C. Eaborn, H. L. Hornfeld, and D. R. M. Walton, *J. Chem. Soc. B*, 1036 (1967).

$$\text{(H, Li on Ph, OCH}_3\text{ alkene)} \quad + \quad (CH_3)_3SnCl \quad \longrightarrow \quad \text{(H, Sn(CH}_3)_3 \text{ on Ph, OCH}_3\text{ alkene)}$$

Ref. 149

There are several procedures for synthesis of terminal alkenyl stannanes that involve addition to aldehydes. A well-established three-step sequence culminates in a radical addition to a terminal alkyne.[150]

$$RCH{=}O + CBr_4 \xrightarrow[\text{Zn}]{P(Ph)_3} RCH{=}CBr_2 \xrightarrow[\text{2) H}_2O]{\text{1) } n\text{-BuLi}} RC{\equiv}CH \xrightarrow[\text{AIBN}]{(n\text{-}C_4H_9)_3SnH} RCH{=}CHSn(n\text{-}C_4H_9)_3$$

Another sequence involves a dibromomethyl(trialkyl)stannane as the starting material. On reaction with $CrCl_2$, addition to the aldehyde is followed by reductive elimination.[151]

$$RCH{=}O + R'_3SnCHBr_2 \xrightarrow[\text{LiI}]{CrCl_2} RCH{=}CHSnR'_3$$

Deprotonated trialkylstannanes are potent nucleophiles. Addition to carbonyl groups or iminium intermediates provides routes to α-alkoxy- and α-amino-alkylstannanes.

$$RCH{=}O + (C_4H_9)_3SnLi \longrightarrow \overset{O^-}{\underset{}{RCHSn(C_4H_9)_3}} \xrightarrow{R'X} \overset{OR'}{\underset{}{RCHSn(C_4H_9)_3}}$$

Ref. 152

$$R_2NCH_2SPh + (C_4H_9)_3SnLi \longrightarrow R_2NCH_2Sn(C_4H_9)_3$$

Ref. 153

α-Silyoxystannanes can be prepared directly from aldehydes and tri-*n*-butyl (trimethylsilyl)stannane.[154]

$$RCH{=}O + (n\text{-}C_4H_9)_3SnSi(CH_3)_3 \xrightarrow{R'_4N^+CN^-} \overset{OSi(CH_3)_3}{\underset{}{RCHSn(n\text{-}C_4H_9)_3}}$$

Addition of tri-*n*-butylstannyllithium to aldehydes followed by iodination and dehydrohalogenation gives primarily *E*-alkenylstannanes.[155]

$$RCH_2CH{=}O + (n\text{-}C_4H_9)_3SnLi \xrightarrow{Ph_3P\text{-}I_2} \overset{I}{\underset{}{RCH_2CHSn(n\text{-}C_4H_9)_3}} \xrightarrow{DBU} \text{(R, H on alkene; H, Sn(n-C}_4H_9)_3\text{)}$$

149. J. A. Soderquist and G. J.-H. Hsu, *Organometallics*, **1**, 830 (1982).
150. E. J. Corey and P. L. Fuchs, *Tetrahedron Lett.*, 3769 (1972).
151. M. D. Cliff and S. G. Payne, *Tetrahedron Lett.*, **36**, 763 (1995); D. M. Hodgson, *Tetrahedron Lett.*, **33**, 5603 (1992); D. M. Hodgson, L. T. Boulton, and G. N. Maw, *Tetrahedron Lett.*, **35**, 2231 (1994).
152. W. C. Still, *J. Am. Chem. Soc.*, **100**, 1481 (1978).
153. D. J. Peterson, *J. Am. Chem. Soc.*, **93**, 4027 (1971).
154. R. M. Bhatt, J. Ye, and J. R. Falck, *Tetrahedron Lett.*, **35**, 4081 (1994).
155. J. M. Chong and S. B. Park, *J. Org. Chem.*, **58**, 523 (1993).

9.3.2. Carbon-Carbon Bond-Forming Reactions

As with the silanes, the most useful synthetic procedures involve electrophilic attack on alkenyl and allylic stannanes. The stannanes are considerably more reactive than the corresponding silanes because there is more anionic character on carbon in the C–Sn bond and it is a weaker bond.[156] The most useful reactions in terms of syntheses involve the Lewis acid–catalyzed addition of allylic stannanes to aldehydes.[157] The reaction occurs with allylic transposition.

$$RCH=O \quad + \quad R^3CH=CHCHSnBu_3 \quad \longrightarrow \quad \underset{HO}{RCHCHCH}=CHR^1$$

There are also useful synthetic procedures in which organotin compounds act as carbanion donors in transition metal–catalyzed reactions, as discussed in Section 8.2.3.3. Organotin compounds are also very important in free radical reactions, as is discussed in Chapter 10.

9.3.2.1. Reactions of Allylic Trialkylstannnanes. Allylic organotin compounds are not sufficiently reactive to add directly to aldehydes or ketones, although reactions with aldehydes do occur with heating.

$$Cl-C_6H_4-CH=O \quad + \quad CH_2=CHCH_2Sn(C_2H_5)_3 \quad \xrightarrow[4\,h]{100°C} \quad Cl-C_6H_4-CHCH_2CH=CH_2$$

90%

Ref. 158

Use of Lewis acid catalysts allows allylic stannanes to react under mild conditions. As is the case with allylic silanes, a double-bond transposition occurs in conjunction with destannylation.[159]

$$PhCH=O \quad + \quad \underset{H \quad H}{\overset{CH_3 \quad CH_2Sn(C_4H_9)_3}{\diagup}} \quad \xrightarrow{BF_3} \quad \underset{OH}{PhCHCHCH}=CH_2$$

92%

The stereoselectivity of addition to aldehydes has been of considerable interest.[160] With benzaldehyde the addition of 2-butenylstannanes catalyzed by BF_3 gives the *syn* isomer, irrespective of the stereochemistry of the butenyl group.[161]

[156.] J. Burfeindt, M. Patz, M. Mueller, and H. Mayr, *J. Am. Chem. Soc.*, **120**, 3629 (1998).

[157.] B. W. Gung, *Org. React.*, **64**, 1 (2004).

[158.] K. König and W. P. Neumann, *Tetrahedron Lett.*, 495 (1967).

[159.] H. Yatagai, Y. Yamamoto, and K. Maruyama, *J. Am. Chem. Soc.*, **102**, 4548 (1980); Y. Yamamoto, H. Yatagai, Y. Naruta, and K. Maruyama, *J. Am. Chem. Soc.*, **102**, 7107 (1989).

[160.] Y. Yamomoto, *Acc. Chem. Res.*, **20**, 243 (1987); Y. Yamoto and N. Asao, *Chem. Rev.*, **93**, 2207 (1993).

[161.] (a) Y. Yamamoto, H. Yatagai, H. Ishihara, and K. Maruyama, *Tetrahedron*, **40**, 2239 (1984); (b) G. E. Keck, K. A. Savin, E. N. K. Cressman, and D. E. Abbott, *J. Org. Chem.*, **59**, 7889 (1994).

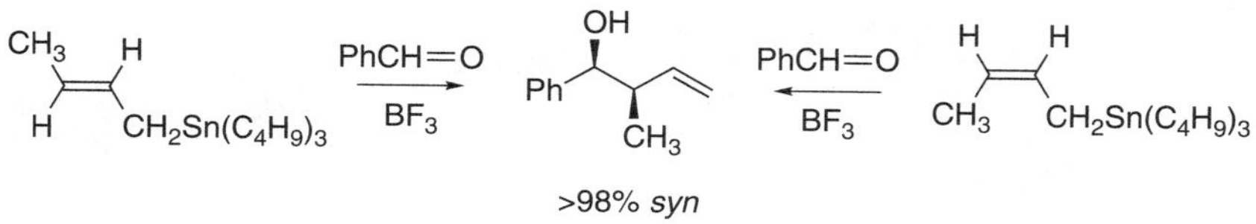

Synclinal and antiperiplanar conformations of the TS are possible. The two TSs are believed to be close in energy and either may be involved in individual systems. An electronic π interaction between the stannane HOMO and the carbonyl LUMO, as well as polar effects appear to favor the synclinal TS and can overcome the unfavorable steric effects.[161b, 162] Generally the synclinal TS seems to be preferred for intramolecular reactions. The steric effects that favor the antiperiplanar TS are not present in intramolecular reactions, since the aldehyde and the stannane substituents are then part of the intramolecular linkage.

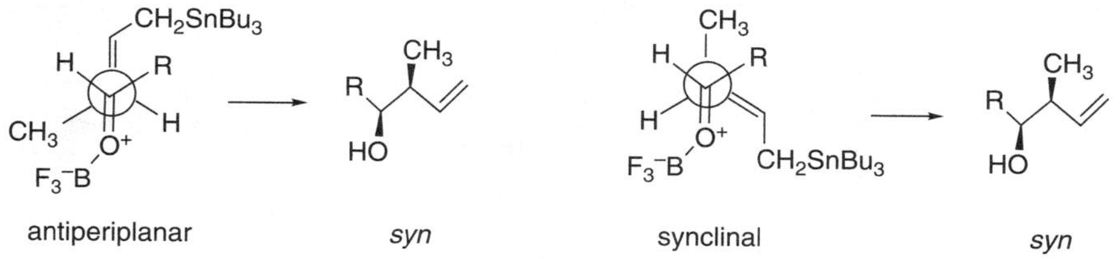

antiperiplanar   *syn*   synclinal   *syn*

With chiral aldehydes, reagent approach is generally consistent with a Felkin model.[163] This preference can be reinforced or opposed by the effect of other stereocenters. For example, the addition of allyl stannane to 1,4-dimethyl-3-(4-methoxybenzyloxy)pentanal is strongly in accord with the Felkin model for the *anti* stereoisomer but is anti-Felkin for the *syn* isomer.

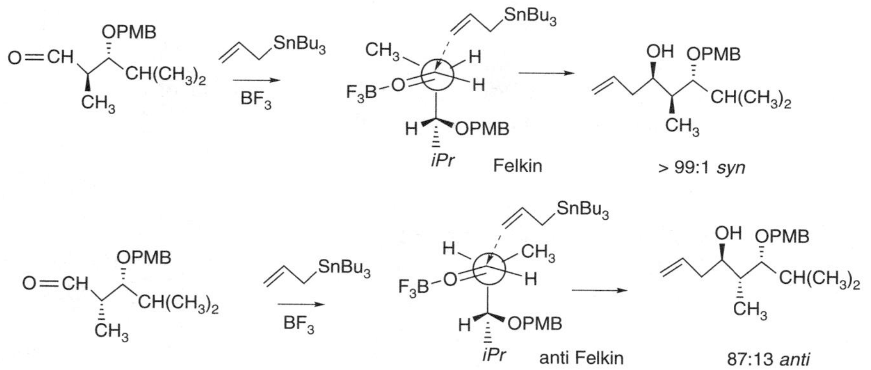

When an aldehyde subject to chelation control is used, the *syn* stereoisomer dominates, with $MgBr_2$ as the Lewis acid.[164]

162. S. E. Denmark, E. J. Weber, T. Wilson, and T. M. Willson, *Tetrahedron*, **45**, 1053 (1989).
163. D. A. Evans, M. J. Dart, J. L. Duffy, M. G. Yang, and A. B. Livingston, *J. Am. Chem. Soc.*, **117**, 6619 (1995); D. A. Evans, M. J. Dart, J. L. Duffy, and M. G. Yang, *J. Am. Chem. Soc.*, **118**, 4322 (1996).
164. G. E. Keck and E. P. Boden, *Tetrahedron Lett.*, **25**, 265 (1984); G. E. Keck, D. E. Abbott, and M. R. Wiley, *Tetrahedron Lett.*, **28**, 139 (1987).

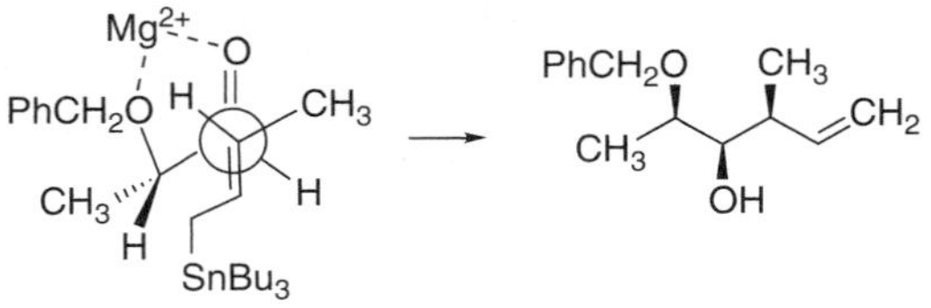

838

CHAPTER 9

*Carbon-Carbon
Bond-Forming Reactions
of Compounds of Boron,
Silicon, and Tin*

The introduction of a β-methyl group shifts the stereoselectivity to *anti*, indicating a preference for TS **E**. There is some dependence on the Lewis acid. For example, the reaction below gives a high ratio of chelation control with $MgBr_2$ and $SnCl_4$, but not with $TiCl_4$.[165]

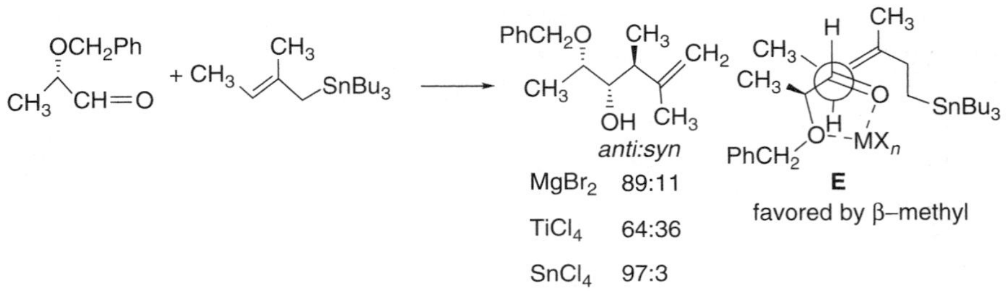

β-Oxy substituents can also lead to chelation control. Excellent stereoselectivity is observed using $SnCl_4$ at low temperature.[166]

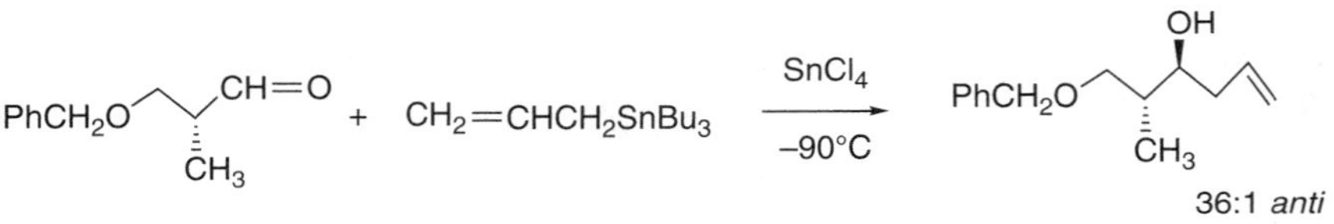

The chelation control approach has been used during the synthesis of the C(13)–C(19) fragment of a marine natural product called calculin A-D.[167]

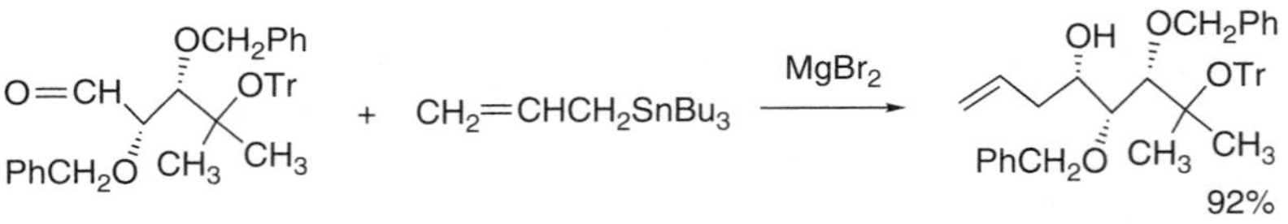

9.3.2.2. Reactions of Allylic Halostannanes. Various allyl halostannanes can transfer allyl groups to carbonyl compounds. In this case the reagent acts both as a Lewis acid and as the source of the nucleophilic allyl group. Reactions involving halostannanes are believed to proceed through cyclic TSs.

165. K. Mikami, K. Kawamoto, T.-P. Loh, and T. Nakai, *J. Chem. Soc., Chem. Commun.*, 1161 (1990).
166. G. E. Keck and D. E. Abbott, *Tetrahedron Lett.*, **25**, 1883 (1984); R. J. Linderman, K. P. Cusack, and M. R. Jaber, *Tetrahedron Lett.*, **37**, 6649 (1996).
167. O. Hara, Y. Hamada, and T. Shiori, *Synlett*, 283 285 (1991).

$$\text{PhCH}_2\text{CH}_2\overset{\overset{\displaystyle O}{\|}}{\text{C}}\text{CH}_3 + (\text{CH}_2{=}\text{CHCH}_2)_2\text{SnBr}_2 \longrightarrow \text{PhCH}_2\text{CH}_2\overset{\overset{\displaystyle OH}{|}}{\underset{\underset{\displaystyle CH_3}{|}}{\text{C}}}\text{CH}_2\text{CH}{=}\text{CH}_2$$

$$\text{RCH}{=}\text{O} + \text{CH}_2{=}\text{CHCH}_2\text{Sn}(n\text{-C}_4\text{H}_9)_3 \xrightarrow[\substack{\text{RCOCl or} \\ (\text{CH}_3)_3\text{SiCl}}]{(n\text{-C}_4\text{H}_9)_2\text{SnCl}_2} \overset{\overset{\displaystyle OX}{|}}{\text{R}}\text{CHCH}_2\text{CH}{=}\text{CH}_2$$

$$X = \text{RCO or } (\text{CH}_3)_3\text{Si}$$

Ref. 168

The halostannanes can also be generated in situ by reactions of allylic halides with tin metal or stannous halides.

$$\text{PhCH}{=}\text{O} + \text{CH}_2{=}\text{CHCH}_2\text{I} \xrightarrow{\text{Sn}} \xrightarrow{\text{H}_2\text{O}} \overset{\overset{\displaystyle OH}{|}}{\text{Ph}}\text{CHCH}_2\text{CH}{=}\text{CH}_2$$

Ref. 169

$$\text{PhCH}{=}\text{CHCH}{=}\text{O} + \text{CH}_2{=}\text{CHCH}_2\text{I} \xrightarrow{\text{SnF}_2} \text{PhCH}{=}\text{CH}\overset{\overset{\displaystyle OH}{|}}{\text{C}}\text{HCH}_2\text{CH}{=}\text{CH}_2$$

Ref. 169

The allylation reaction can be adapted to the synthesis of terminal dienes by using 1-bromo-3-iodopropene and stannous chloride. The elimination step is a reductive elimination of the type discussed in Section 5.8. Excess stannous chloride acts as the reducing agent.

$$\text{PhCH}{=}\text{O} + \text{ICH}_2\text{CH}{=}\text{CHBr} \xrightarrow{\text{SnCl}_2} \text{Ph}\overset{\overset{\displaystyle Br}{|}}{\underset{\underset{\displaystyle OH}{|}}{\text{C}}}\text{H}\text{CHCH}{=}\text{CH}_2 \longrightarrow \text{PhCH}{=}\text{CHCH}{=}\text{CH}_2$$

Ref. 170

Allylic Sn(II) species are believed to be involved in reactions of allylic trialkyl stannanes in the presence of SnCl$_2$. These reactions are particularly effective in acetonitrile, which appears to promote the exchange reaction. Ketones as well as aldehydes are reactive under these conditions.[171]

168. T. Mukaiyama and T. Harada, *Chem. Lett.*, 1527 (1981).

169. T. Mukaiyama, T. Harada, and S. Shoda, *Chem. Lett.*, 1507 (1980).

170. J. Auge, *Tetrahedron Lett.*, **26**, 753 (1985).

171. (a) M. Yasuda, Y. Sugawa, A. Yamamoto, I. Shibata, and A. Baba, *Tetrahedron Lett.*, **37**, 5951 (1996); (b) M. Yasuda, K. Hirata, M. Nishino, A. Yamamoto, and A. Baba, *J. Am. Chem. Soc.*, **124**, 13442 (2002).

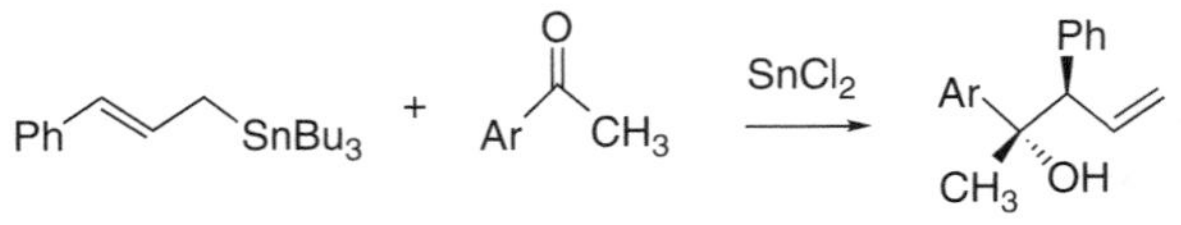

840

CHAPTER 9

*Carbon-Carbon
Bond-Forming Reactions
of Compounds of Boron,
Silicon, and Tin*

The *anti* stereochemistry is consistent with a cyclic TS, but the reaction is stereoconvergent for the *E*- and *Z*-2-butenylstannanes, indicating that isomerization must occur at the transmetallation stage. The adducts are equilibrated at 82°C and under these conditions the *anti* product is isolated on workup.

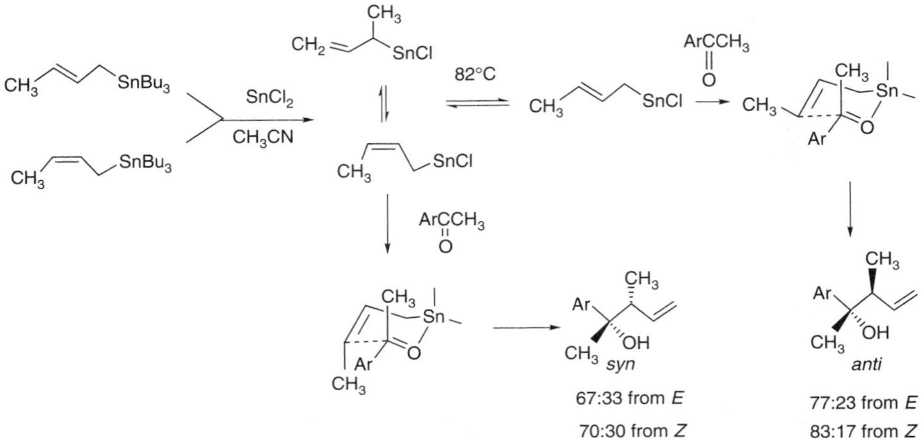

Cyclic allylstannanes give *syn* products with high selectivity.

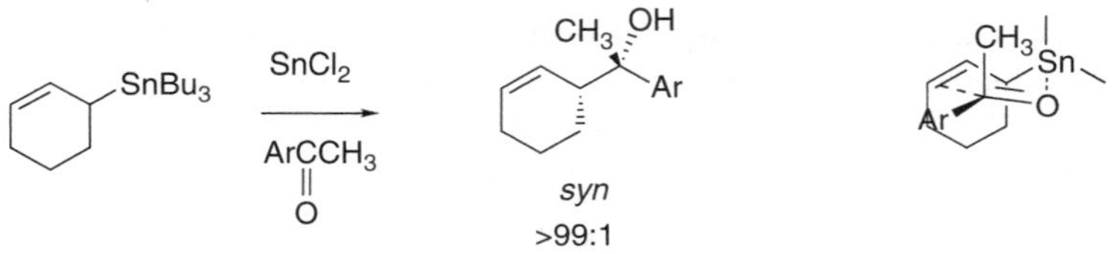

The reaction with α-hydroxy and α-methoxy ketones under these conditions are chelation controlled.

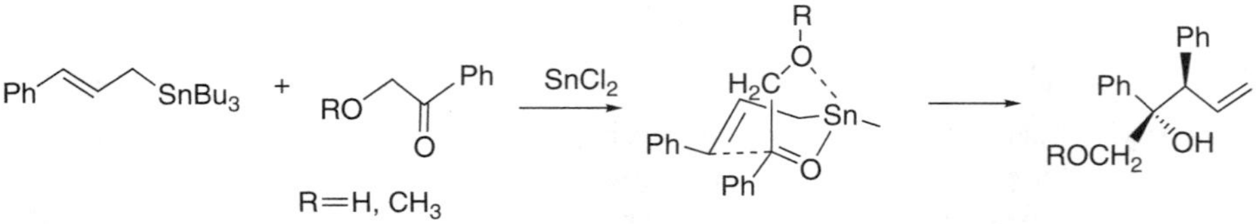

Use of di-(*n*-butyl)stannyl dichloride along with an acyl or silyl halide leads to addition of allylstannanes to the aldehydes.[172a, 172] Reaction is also promoted by butylstannyl trichloride.[173] Both SnCl$_4$ and SnCl$_2$ also catalyze this kind of addition.

172. J. K. Whitesell and R. Apodaca, *Tetrahedron Lett.*, **37**, 3955 (1996).
173. H. Miyake and K. Yamamura, *Chem. Lett.*, 1369 (1992); H. Miyake and K. Yamamura, *Chem. Lett.*, 1473 (1993).

Reactions of tetraallylstannanes with aldehydes catalyzed by $SnCl_4$ also appear to involve a halostannane intermediate. It can be demonstrated by NMR that there is a rapid redistribution of the allyl group.[174] Reactions with these halostannanes are believed to proceed through a cyclic TS.

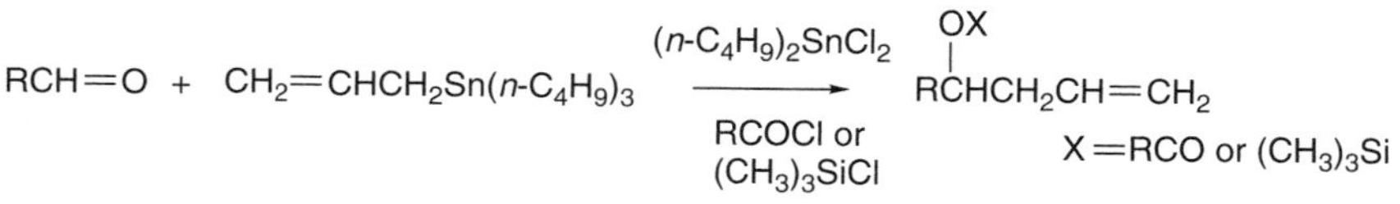

9.3.2.3. Reactions Involving Transmetallation. With certain Lewis acids, the reaction may involve a prior transmetallation. This introduces several additional factors into the analysis of the stereoselectivity, as the stereochemistry of the transmetallation has to be considered. Reactions involving halo titanium and halo tin intermediates formed by transmetallation can proceed through a cyclic TS. When $TiCl_4$ is used as the catalyst, the stereoselectivity depends on the order of addition of the reagents. When *E*-2-butenylstannane is added to a $TiCl_4$-aldehyde mixture, *syn* stereoselectivity is observed. When the aldehyde is added to a premixed solution of the 2-butenylstannane and $TiCl_4$, the *anti* isomer predominates.[175]

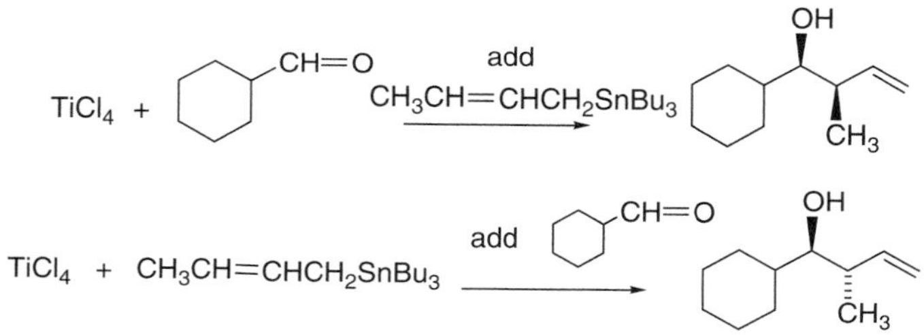

The formation of the *anti* stereoisomer is attributed to involvement of a butenyltitanium intermediate formed by rapid exchange with the butenylstannane. This intermediate then reacts through a cyclic TS.

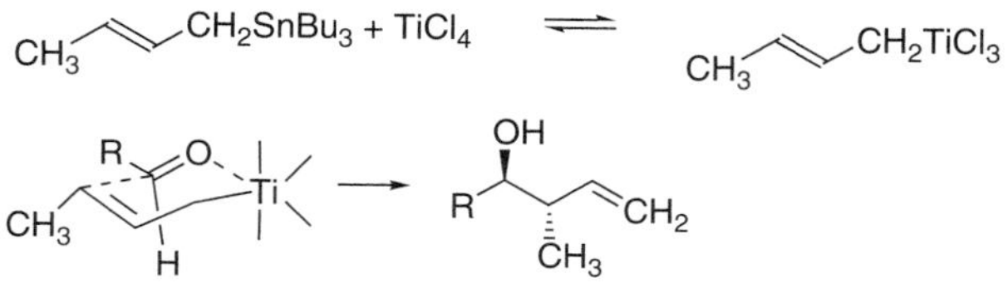

Indium chloride in polar solvents such as acetone or acetonitrile leads to good diastereoselectivity with cyclohexanecarboxaldehyde and other representative aldehydes.[176]

[174.] S. E. Denmark, T. Wilson, and T. M. Willson, *J. Am. Chem. Soc.*, **110**, 984 (1988); G. E. Keck, M. B. Andrus, and S. Castellino, *J. Am. Chem. Soc.*, **111**, 8136 (1989).
[175.] G. E. Keck, D. E. Abbott, E. P. Boden, and E. J. Enholm, *Tetrahedron Lett.*, **25**, 3927 (1984).
[176.] J. A. Marshall and K. W. Hinkle, *J. Org. Chem.*, **60**, 1920 (1995).

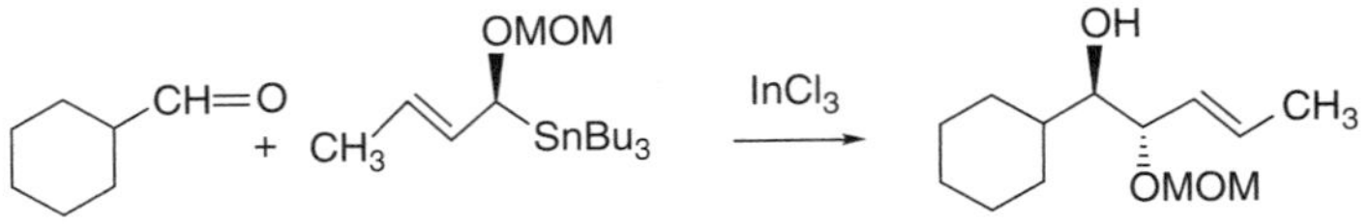

842

CHAPTER 9

Carbon-Carbon
Bond-Forming Reactions
of Compounds of Boron,
Silicon, and Tin

These reactions are believed to proceed via transmetallation. Configurational inversion occurs at both the transmetallation and addition steps, leading to overall retention of the allylic stereochemistry.

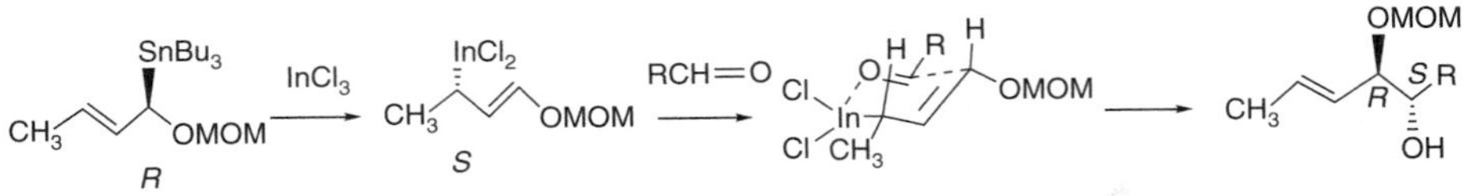

These reagents are useful in enantioselective synthesis and are discussed further in the following section.

9.3.2.4. γ-Oxygen-Substituted Stannanes. Oxygenated allylic stannanes have been synthesized and used advantageously in several types of syntheses. Both α- and γ-alkoxy and silyloxy stannane can be prepared by several complementary methods.[177] *E*-γ-Alkoxy and silyloxy allylic stannanes react with aldehydes to give primarily *syn* adducts.[178]

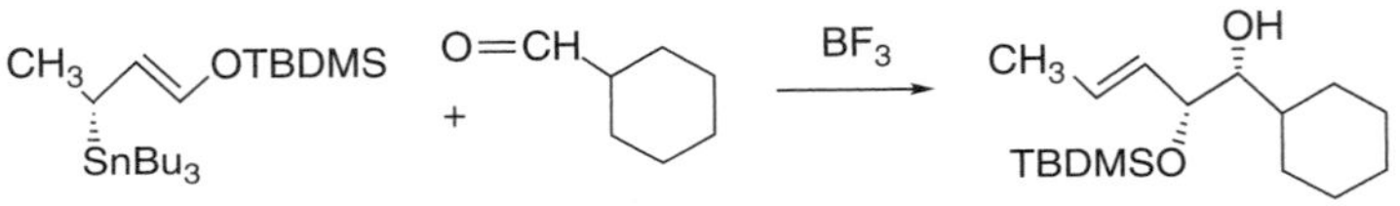

79% yield, 98:2 *syn*

Allylic silanes with γ-alkoxy substituents also give a preference for the *syn* stereo-chemistry.[179]

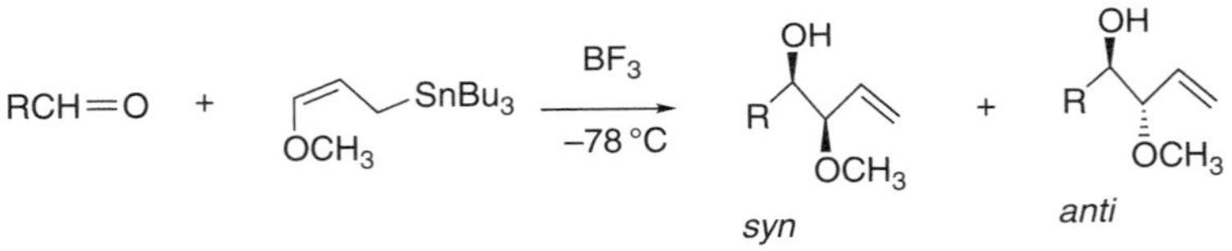

R	syn:anti
Ph	10:1
i-Pr	25:1
c-C$_6$H$_{11}$	5:1

Improved stereoselectivity is observed with methoxymethoxy (MOM) and TBDMSO substituents.[180]

177. J. A. Marshall, *Chem. Rev.*, **96**, 31 (1996).
178. J. A. Marshall, J. A. Jablonowski, and L. M. Elliott, *J. Org. Chem.*, **60**, 2662 (1995).
179. M. Koreeda and Y. Tanaka, *Tetrahedron Lett.*, **28**, 143 (1987).
180. J. A. Marshall and J. A. Welmaker, *J. Org. Chem.*, **57**, 7158 (1992).

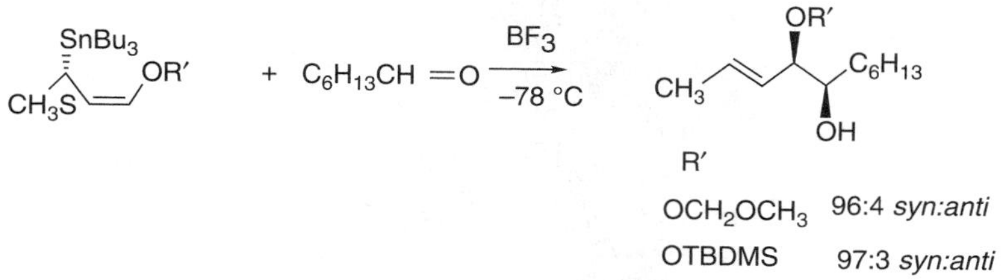

SnBu₃ ... OR'

CH₃S ... + C₆H₁₃CH=O →(BF₃, −78 °C)→ CH₃ ... OR' ... C₆H₁₃ ... OH

R'

OCH₂OCH₃   96:4 *syn:anti*

OTBDMS   97:3 *syn:anti*

Use of oxygenated stannanes with α-substituted aldehydes leads to matched and mismatched combinations.[181] For example, with the γ-MOM derivative and α-benzyloxypropanal, the matched pair gives a single stereoisomer of the major product, whereas the mismatched pair gives a 67:33 *syn:anti* mixture. The configuration at the alkoxy-substituted center is completely controlled by the chirality of the stannane.

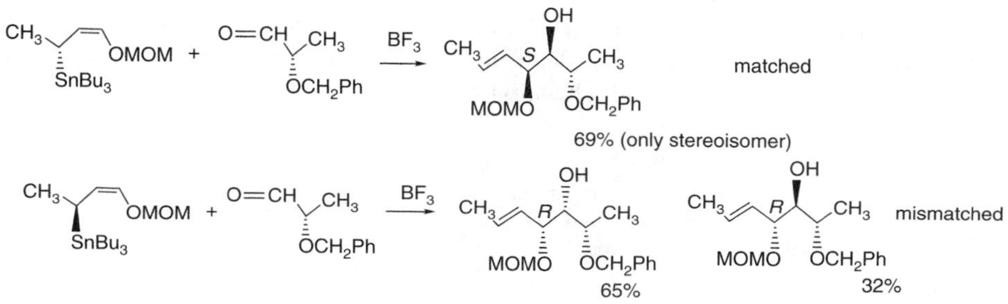

CH₃ ... OMOM ... SnBu₃ + O=CH ... CH₃ ... OCH₂Ph →(BF₃)→ CH₃ ... S ... OH ... CH₃ ... MOMO ... OCH₂Ph   matched

69% (only stereoisomer)

CH₃ ... OMOM ... SnBu₃ + O=CH ... CH₃ ... OCH₂Ph →(BF₃)→ CH₃ ... R ... OH ... CH₃ ... MOMO ... OCH₂Ph   65% ... CH₃ ... R ... OH ... CH₃ ... MOMO ... OCH₂Ph   mismatched   32%

Use of MgBr₂, which results in chelation control, reverses the matched and mismatched combinations.

CH₃ ... OMOM ... SnBu₃ + O=CH ... CH₃ ... OCH₂Ph →(MgBr₂)→ CH₃ ... S ... OH ... CH₃ ... MOMO ... OCH₂Ph 56% + CH₃ ... R ... OH ... CH₃ ... MOMO ... OCH₂Ph   mismatched   18%

CH₃ ... OMOM ... SnBu₃ + O=CH ... CH₃ ... OCH₂Ph →(MgBr₂)→ CH₃ ... R ... OH ... CH₃ ... MOMO ... OCH₂Ph   matched

83% (only stereoisomer)

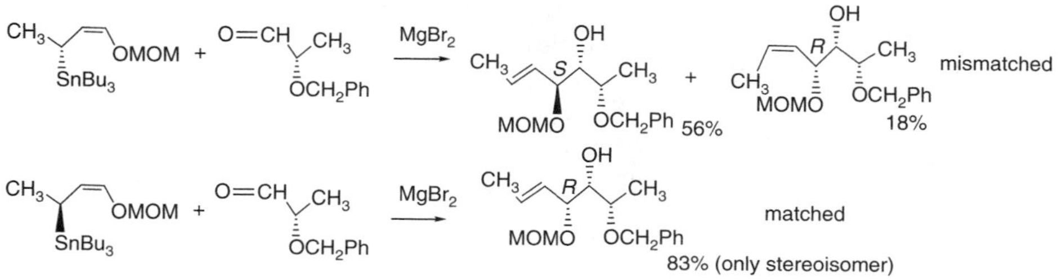

9.3.2.5. Enantioselective Addition Reactions of Allylic Stannanes. There have been several studies of the enantiomers of α-oxygenated alkenyl stannanes. The chirality of the α-carbon exerts powerful control on enantioselectivity with the preference for the stannyl group to be *anti* to the forming bond. This is presumably related to the stereoelectronic effect that facilitates the transfer of electron density from the tin to the forming double bond.[182]

[181.] J. A. Marshall, J. A. Jablonowski, and G. P. Luke, *J. Org. Chem.*, **59**, 7825 (1994).
[182.] J. A. Marshall and W. Y. Gung, *Tetrahedron*, **45**, 1043 (1989).

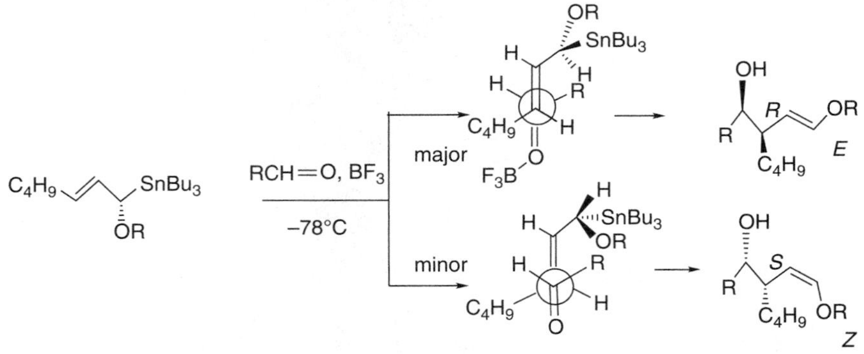

anti relationship between stannyl substituent
and developing bond exerts control on double
bond configuration

Allylic stannanes with γ-oxygen substituents have been used to build up polyoxy-genated carbon chains. For example, **16** reacts with the stannane **17** to give a high preference for the stereoisomer in which the two oxygen substituents are *anti*. This stereoselectivity is consistent with chelation control.[183]

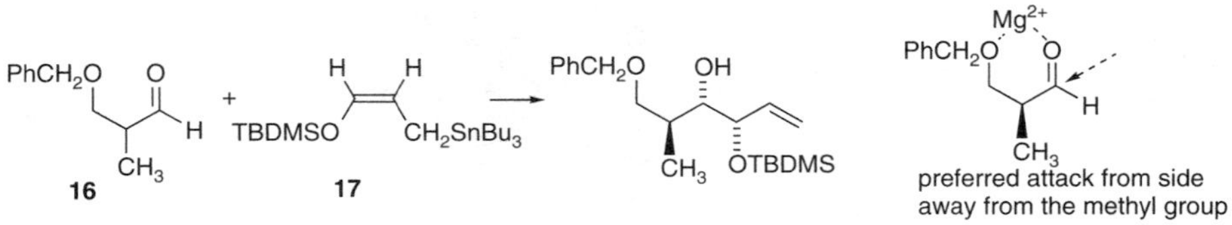

preferred attack from side
away from the methyl group

The substrate-controlled addition of **18** to **19** proceeded with good enantioselectivity and was used to prepare the epoxide (+)-disparlure, a gypsy moth pheromone.[184]

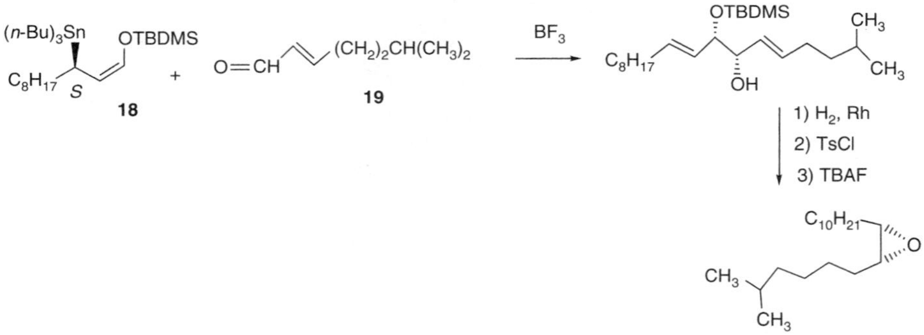

Reagent-controlled stereoselectivity can provide stereochemical relationships over several centers when a combination of acyclic and chelation control and cyclic TS resulting from transmetallation is utilized. In reactions mediated by BF_3 or $MgBr_2$ the new centers are *syn*. Indium reagents can be used to create an *anti* relationship between two new chiral centers. The indium reagents are formed by transmetallation and react

183. G. E. Keck, K. A. Savin, E. N. K. Cressman, and D. E. Abbott, *J. Org. Chem.*, **59**, 7889 (1994).
184. J. A. Marshall, J. A. Jablonowski, and H. Jiang, *J. Org. Chem.*, **64**, 2152 (1999).

through cyclic TSs leading to *anti* stereochemistry at the new bond. The complementary relationship has been used to construct all eight possible hexose configurations.[185]

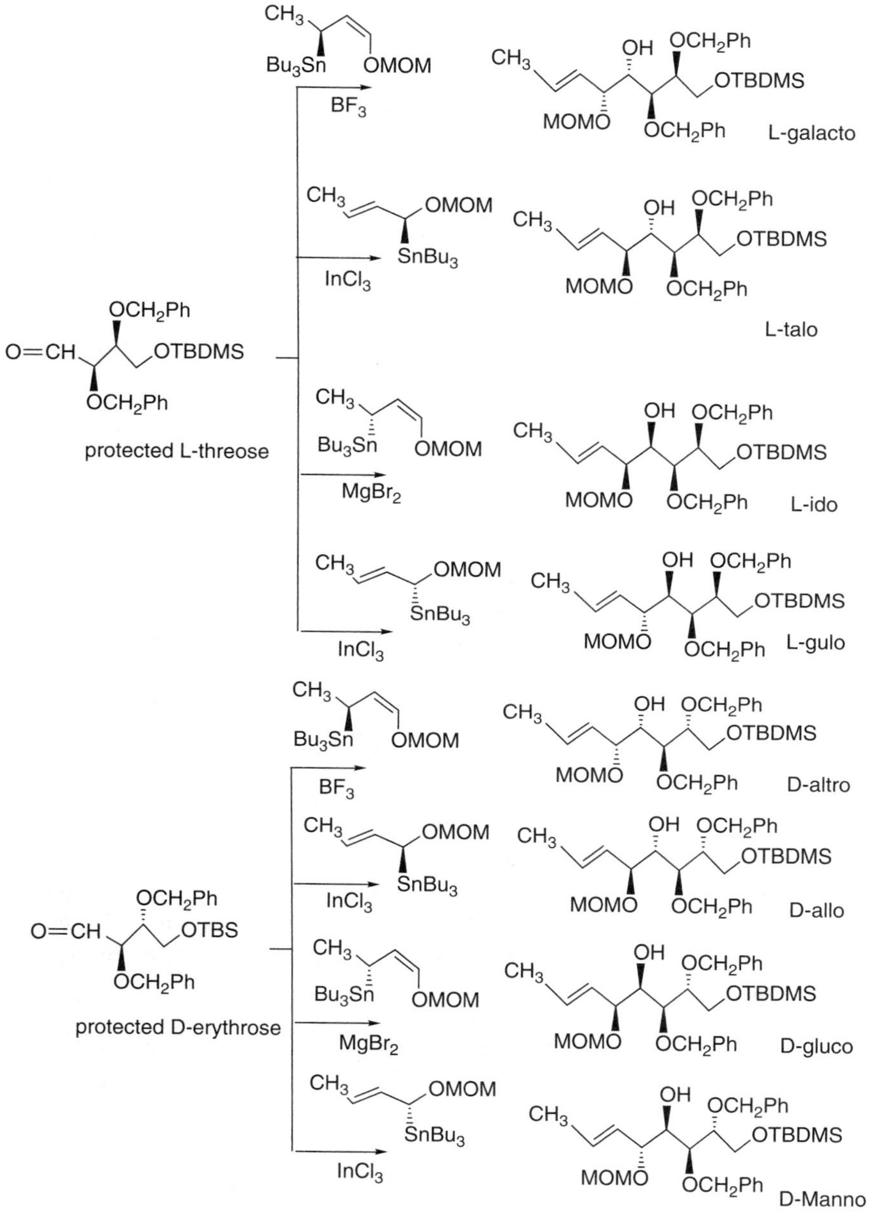

More remote oxygen substituents can also influence stereochemistry. 4-Benzyloxy-2-pentenyl tri-*n*-butylstannane exhibits excellent enantioselectivity in reactions with aldehydes.[186] This reaction is believed to involve chelation of the

185. J. A. Marshall and K. W. Hinkle, *J. Org. Chem.*, **61**, 105 (1996).

186. E. J. Thomas, *J. Chem. Soc. Chem. Commun.*, 411 (1997); A. H. McNeill and E. J. Thomas, *Synthesis*, 322 (1998).

846

CHAPTER 9

*Carbon-Carbon
Bond-Forming Reactions
of Compounds of Boron,
Silicon, and Tin*

benzyloxy group in both the transmetallation and addition steps. The transmetallation is thought to involve coordination with $SnCl_4$ through the benzyloxy group that is maintained in the addition step.

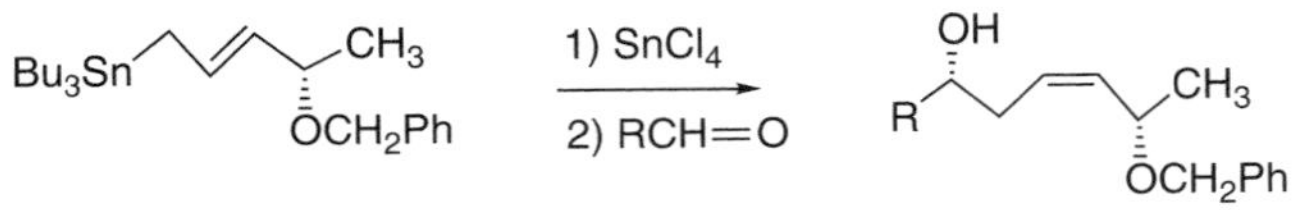

>90% 1,5-*syn*

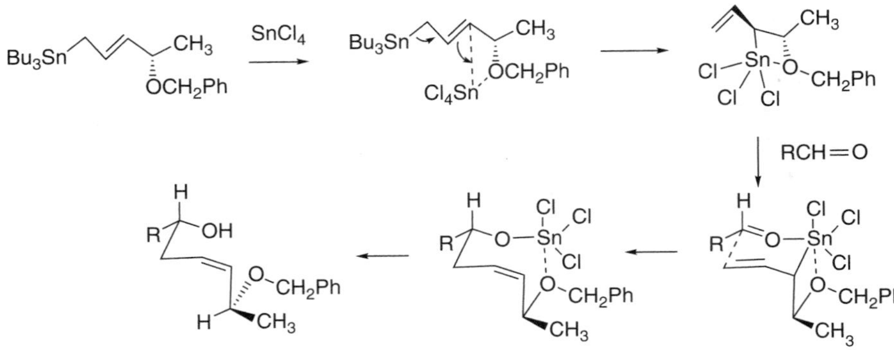

Allylstannane additions to aldehydes can be made enantioselective by use of chiral catalysts. A catalyst prepared from the chiral binaphthols *R*- or *S*-BINOL and $Ti(O\text{-}i\text{-}Pr)_4$ achieves 85–95% enantioselectivity.[187]

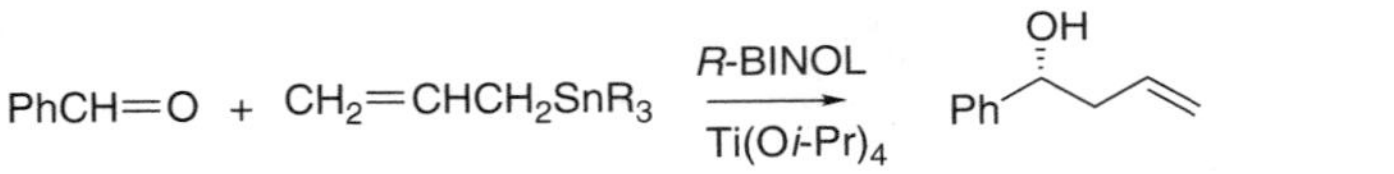

$$PhCH{=}O \ + \ CH_2{=}CHCH_2SnR_3 \xrightarrow[\;Ti(Oi\text{-}Pr)_4\;]{R\text{-BINOL}}$$

87–96% e.e.

BINAP-AgF gives good enantioselectivity, especially for the major *anti* product in the addition of 2-butenylstannanes to benzaldehyde.[188] This system appears to be stereoconvergent, suggesting that isomerization of the 2-butenyl system occurs, perhaps by transmetallation.

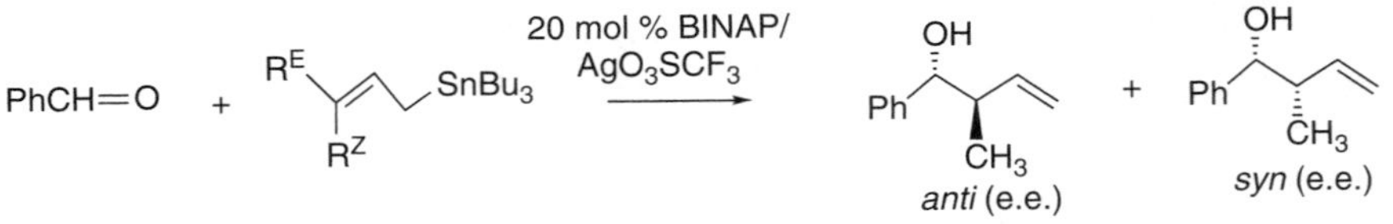

	anti (e.e.)	*syn* (e.e.)
E	85 (94)	15 (64)
Z	85 (91)	15 (50)

187. G. E. Keck, K. H. Tarbet, and L. S. Geraci, *J. Am. Chem. Soc.*, **115**, 8467 (1993); A. L. Costa, M. G. Piazza, E. Tagliavini, C. Trombini, and A. Umani-Ronchi, *J. Am. Chem. Soc.*, **115**, 7001 (1993); G. E. Keck and L. S. Geraci, *Tetrhahedron Lett.*, **34**, 7827 (1993); G. E. Keck, D. Krishnamurthy, and M. C. Grier, *J. Org. Chem.*, **58**, 6543 (1993).

188. A. Yanagisawa, H. Nakashima, Y. Nakatsuka, A. Ishiba, and H. Yamamoto, *Bull. Chem. Soc. Jpn.*, **74**, 1129 (2001).

The coupling of the achiral stannane **20** and aldehyde **21** was achieved with fair to good enantioselectivity and fair yield using chiral catalysts. Ti-BINOL gave 52% e.e. and 31% yield, whereas an acyloxyborane catalyst (see p. 127) gave 90% e.e. and 24% yield.[189]

cat **E**   1 equiv Ti(O*i*Pr)$_4$; 1 equiv BINOL   cat **F**   1 equiv 2,6-diMeOPhCO$_2$

1 equiv BH$_3$-THF; 2 equiv (CF$_3$SO$_2$)$_2$O

Lewis acid–mediated ionization of acetals also generates electrophilic carbon intermediates that react readily with allylic stannanes.[190] Dithioacetals can be activated by the sulfonium salt $[(CH_3)_2SSCH_3]^+BF_4^-$.[191]

$PhCH_2CH_2CH(OCH_3)_2 + CH_2=CHCH_2Sn(CH_3)_3 \xrightarrow{(R_2AlO)_2SO_2} PhCH_2CH_2CHCH_2CH=CH_2$ (OCH$_3$)

$PhCH_2CH(OCH_3)_2 + Sn(CH_2CH=CH_2)_4 \xrightarrow[CH_3OH]{\substack{CF_3CO_2H \\ \text{silica gel}}} PhCH_2CHCH_2CH=CH_2$ (OCH$_3$)

$CH_3(CH_2)_4C(SCH_3)_2 + CH_2=CHCH_2Sn(C_4H_9)_3 \xrightarrow{[(CH_3)_2SSCH_3]^+BF_4^-} CH_3(CH_2)_4CCH_2CH=CH_2$ (CH$_3$, SCH$_3$)

Scheme 9.6 gives some other examples of Lewis acid–catalyzed reactions of allylic stannanes with carbonyl compounds. Entry 1 demonstrates the *syn* stereoselectivity observed with *E*-allylic systems. Entries 2 and 3 illustrate the use of mono- and dihalostannanes in reactions with acetone. Entry 4 involves addition to acrolein, using Bu$_2$SnCl$_2$ as the catalyst. This reaction was run at room temperature for 24 h and gave exclusively the *Z*-configuration of the new double bond. It seems likely that this is the result of thermodynamic control. Entry 5 involves an α-ethoxyallylstannane and shows *syn* stereoselectivity. Entry 6 involving an α-benzyloxy aldehyde occurred with high chelation control. The addition in Entry 7 involves in situ generation of an allylic stannane and favored the *anti* stereoisomer by about 4:1. Entry 8 was used to establish relative stereochemistry in a short synthesis of racemic Prelog-Djerassi lactone. Although the methoxycarbonyl group is a potential chelating ligand, the use of

189. J. A. Marshall and J. Liao, *J. Org. Chem.*, **63**, 5962 (1998).
190. A. Hosomi, H. Iguchi, M. Endo, and H. Sakurai, *Chem. Lett.*, 977 (1979).
191. B. M. Trost and T. Sato, *J. Am. Chem. Soc.*, **107**, 719 (1985).

Scheme 9.6. Reactions of Allylic Stannanes with Carbonyl Compounds

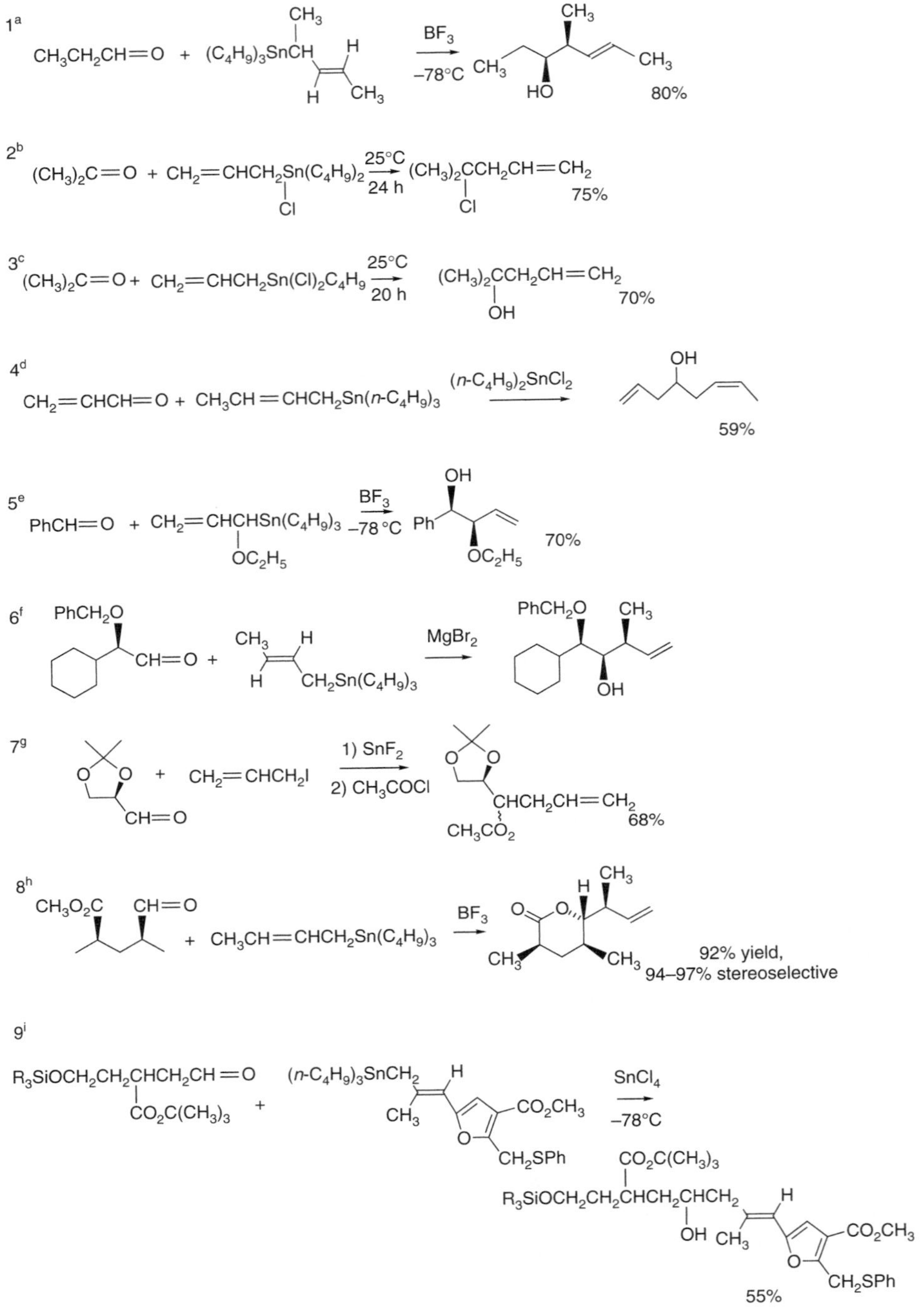

(Continued)

Scheme 9.6. (*Continued*)

849

SECTION 9.3

Organotin Compounds

10^j

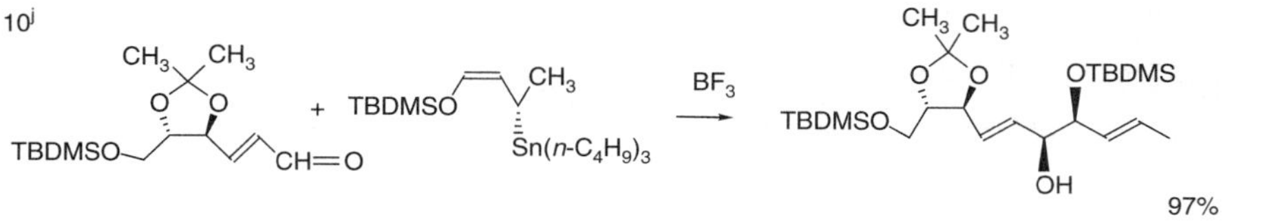

97%

11^k

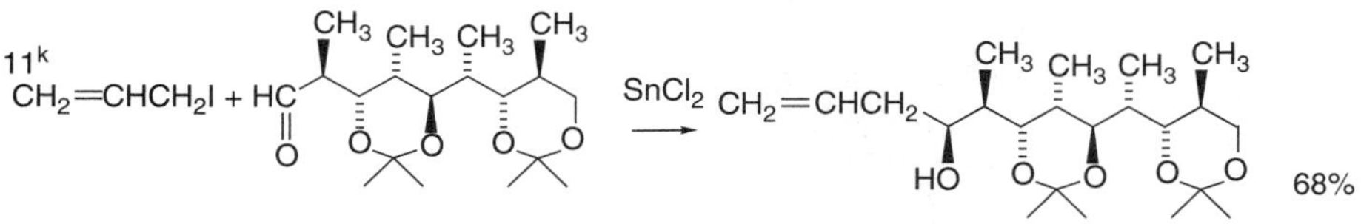

68%

12^l

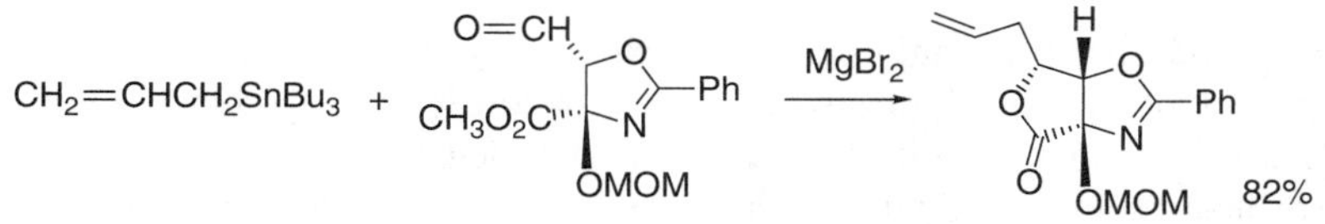

82%

13^m

78%

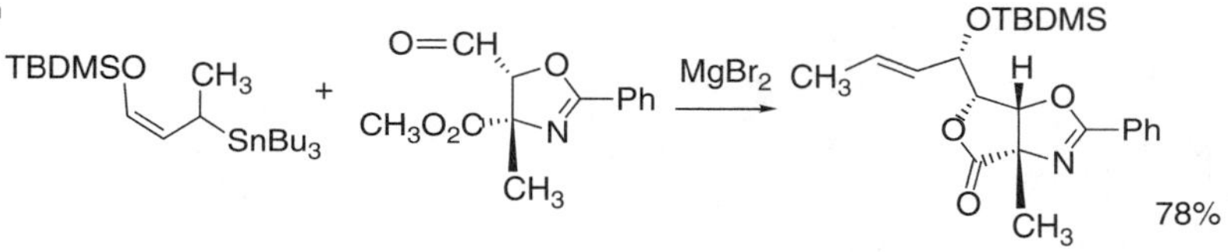

a. M. Koreeda and Y. Tanaka, *Chem. Lett.*, 1297 (1982).
b. V. Peruzzo and G. Tagliavini, *J. Organomet. Chem.*, **162**, 37 (1978).
c. A. Gambaro, V. Peruzzo, G. Plazzogna, and G. Tagliavini, *J. Organomet. Chem.*, **197**, 45 (1980).
d. L. A. Paquette and G. D. Maynard, *J. Am. Chem. Soc.*, **114**, 5018 (1992).
e. D.-P. Quintard, B. Elissondo, and M. Pereyre, *J. Org. Chem.*, **48**, 1559 (1983).
f. G. E. Keck and E. P. Boden, *Tetrahedron Lett.*, **25**, 1879 (1984).
g. T. Harada and T. Mukaiyama, *Chem. Lett.*, 1109 (1981).
h. K. Maruyama, Y. Ishiara, and Y. Yamamoto, *Tetrahedron Lett.*, **22**, 4235 (1981).
i. L. A. Paquette and P. C. Astles, *J. Org. Chem.*, **58**, 165 (1993).
j. J. A. Marshall, S. Beaudoin, and K. Lewinski, *J. Org. Chem.* **58**, 5876 (1993).
k. H. Nagaoka, and Y. Kishi, *Tetrahedron*, **37**, 3873 (1981).
l. K.-Y. Lee, C.-Y. Oh, Y.-H. Kim, J. E. Joo, and W.-H. Ham, *Tetrahedron Lett.*, **43**, 9361 (2002).
m. K.-Y. Lee, C.-Y. Oh, and W.-H. Ham, *Org. Lett.*, **4**, 4403 (2002).

BF_3 should involve an open TS. The observed stereochemistry is *syn* but the approach is anti-Felkin.

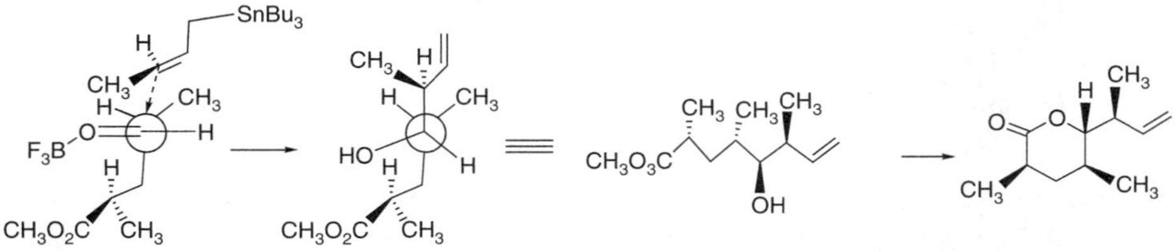

Entry 9 was used in the synthesis of a furanocembranolide. This reaction presumably proceeds through a trichlorostannane intermediate and involves allylic

850

CHAPTER 9

*Carbon-Carbon
Bond-Forming Reactions
of Compounds of Boron,
Silicon, and Tin*

shift at both the transmetallation and addition steps, resulting in restoration of the original allylic structure.

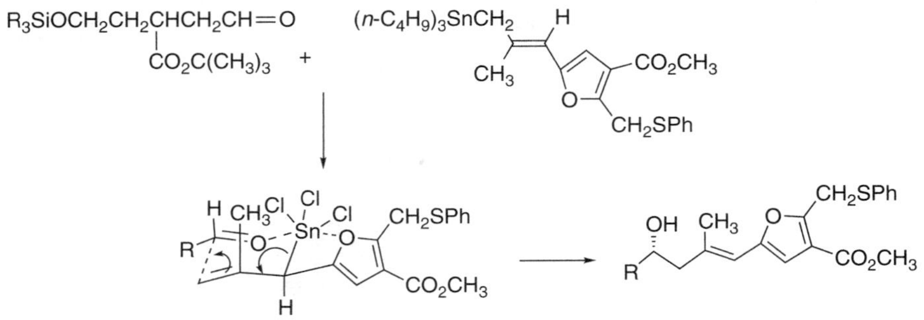

Entry 10 was used in conjunction with dihydroxylation in the enantiospecific synthesis of polyols. Entry 11 illustrates the use of $SnCl_2$ with a protected polypropionate. Entries 12 and 13 result in the formation of lactones, after $MgBr_2$-catalyzed additions to heterocyclic aldehyde having ester substituents. The stereochemistry of both of these reactions is consistent with approach to a chelate involving the aldehyde oxygen and oxazoline oxygen.

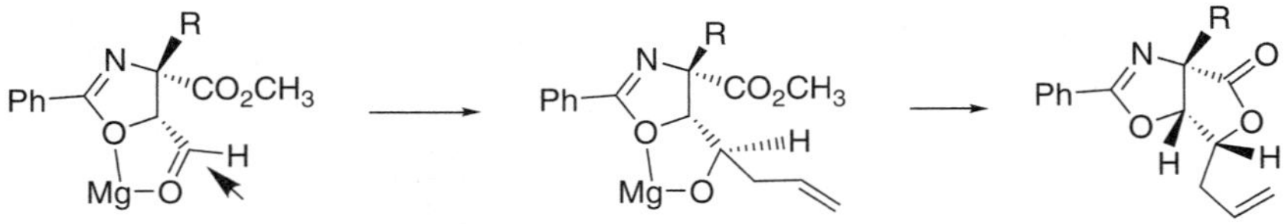

9.3.2.6. Allenyl Stannanes. Allenyl stannanes are a useful variation of the allylic stannanes.[192] They can be made in enantiomerically pure form by S_N2' displacements on propargyl tosylates.[193]

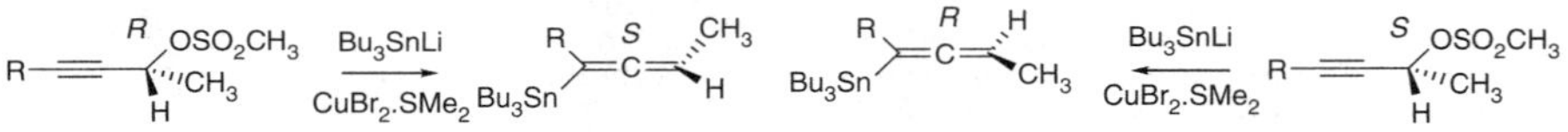

The allenic stannanes react with aldehydes under the influence of Lewis acids such as BF_3 and $MgBr_2$. Unbranched aldehydes are not very stereoselective, but branched aldehydes show a strong preference for the *syn* adduct.

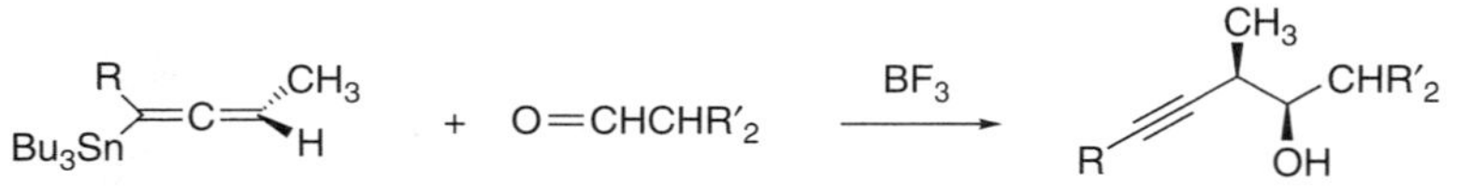

With α-benzyloxypropanal, using $MgBr_2$ as the Lewis acid, chelation control is observed. The stereospecificity is determined by an *anti* orientation of the C–Sn bond

192. J. A. Marshall, *Chem. Rev.*, **96**, 31 (1996).
193. J. A. Marshall and X. Wang, *J. Org. Chem.*, **56**, 3211 (1991).

and the forming C–C bond. As a result, the (S) reactant gives a *syn* adduct, whereas the (R) reactant gives the *anti* isomer.

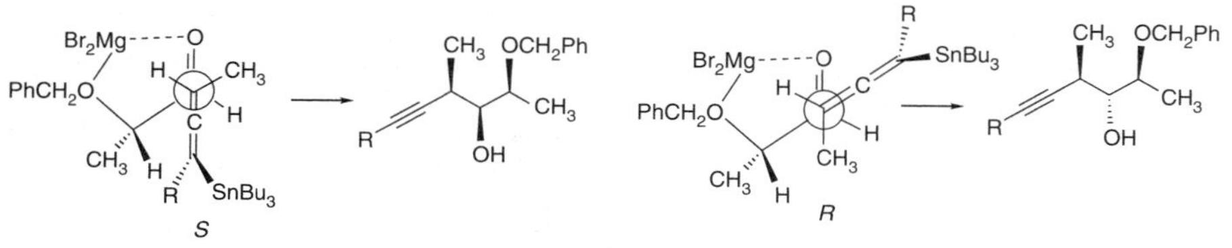

The allenic stannanes can be transmetallated by treatment with $SnCl_4$, a reaction that results in the formation of the a propargyl stannane. If the transmetallation reaction is allowed to equilibrate at $0\,°C$, an allenic structure is formed. These reagents add stereospecifically to the aldehyde through cyclic TSs.[194]

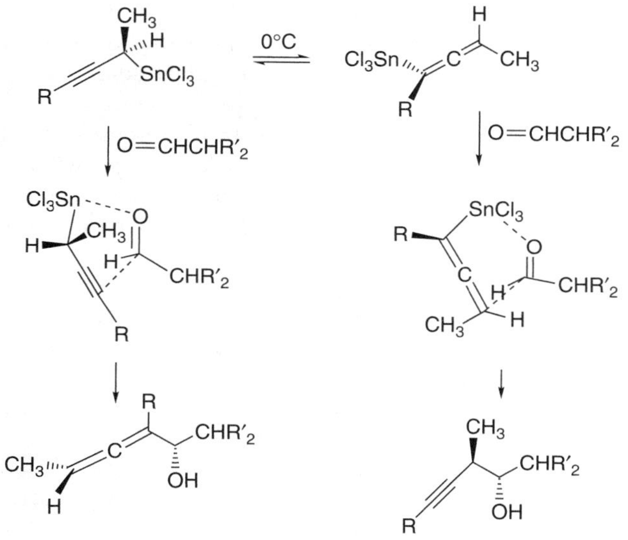

The combination of reagents and methods can provide for stereochemical control of addition to α-substituted aldehydes.[195] An application of the methodology can be found in the synthesis of (+)-discodermolide that was carried out by J. A. Marshall and co-workers and is described in Scheme 13.69.

9.4. Summary of Stereoselectivity Patterns

In this chapter, we have seen a number of instances of stereoselectivity. Although they are affected by specific substitution patterns, every case can be recognized as conforming to one of several general patterns.

1. Reactions proceeding through a monocyclic TS with substrate control: These reactions exhibit predictable stereoselectivity determined by the monocyclic

[194] J. A. Marshall and J. Perkins, *J. Org. Chem.*, **60**, 3509 (1995).
[195] J. A. Marshall, J. F. Perkins, and M. A. Wolf, *J. Org. Chem.*, **60**, 5556 (1995).

Scheme 9.7. Summary of Stereoselectivity of Allylic Reagents in Carbonyl Addition Reactions

Monocyclic TS	Open TS	Chelation TS	Stereoconvergent
Allylboration with β-allylic boranes and boronates	Lewis acid–catalyzed addition of allylic silanes	Lewis acid–catalyzed addition of allylic silanes and stannanes α- and β-oxy aldehydes	$SnCl_2$- mediated addition of allylic to stannanes aryl methyl ketones
Addition of allylic trihalo stannanes to aldehydes	Lewis Acid-catalyzed addition of allylic stannanes		

TS, which is usually based on the chair (Zimmerman-Traxler) model. This pattern is particularly prevalent for the allylic borane reagents, where the Lewis acidity of boron promotes a tight cyclic TS, but at the same time limits the possibility of additional chelation. The dominant factors in these cases are the *E*- or *Z*-configuration of the allylic reagent and the conformational preferences of the reacting aldehyde (e.g., a Felkin-type preference.)

2. Reactions proceeding through open TS: In this group, exemplified by BF_3-catalyzed additions of allylic silanes and stannanes, the degree of stereochemical control is variable and often moderate. The stereoselectivity depends on steric factors in the open TS and can differ significantly for the *E*- and *Z*-isomers of the allylic reactant.

3. Reactions through chelated TS: Reactions of α- or β-oxy-substituted aldehydes often show chelation-controlled stereoselectivity with Lewis acids that can accommodate five or six ligands. Chelation with substituents in the allylic reactant can also occur. The overall stereoselectivity depends on steric and stereoelectronic effects in the chelated TS.

4. Stereoconvergence owing to reactant or product equilibration: We also saw several cases where the product composition was the same for stereoisomeric reactants, e.g., for *E*- and *Z*-allylic reactants. This can occur if there is an intermediate step in the mechanism that permits *E*- and *Z*-equilibration or if the final stereoisomeric product can attain equilibrium.

Scheme 9.7 gives examples of each of these types of stereoselectivities. The analysis of any particular system involves determination of the nature of the reactant, e.g., has transmetallation occurred, the coordination capacity of the Lewis acid, and the specific steric and stereoelectronic features of the two reactants.

General References

Organoborane Compounds

H. C. Brown, *Organic Synthesis via Boranes*, Wiley, New York, 1975.
A. Pelter, K. Smith, and H. C. Brown, *Borane Reagents*, Academic Press, New York, 1988.
A. Pelter, in *Rearrangements in Ground and Excited States*, Vol. 2, P. de Mayo, ed., Academic Press, New York, 1980, Chap. 8.
B. M. Trost, ed., *Stereodirected Synthesis with Organoboranes*, Springer, Berlin, 1995.

E. W. Colvin, *Silicon Reagents in Organic Synthesis*, Academic Press, London, 1988.
I. Fleming, J. Dunogves, and R. Smithers, *Org. React.*, **37**, 57 (1989).
W. Weber, *Silicon Reagents for Organic Synthesis*, Springer, Berlin, 1983.

Organotin Compounds

A. G. Davies, *Organotin Chemistry*, VCH, Weinheim, 1997.
S. Patai, ed., *The Chemistry of Organic Germanium, Tin and Lead Compounds*, Wiley-Interscience, New York, 1995.
M. Pereyre, J.-P. Quintard, and A. Rahm, *Tin in Organic Synthesis*, Butterworths, London, 1983.

Problems

(References for these problems will be found on page 1286.)

9.1. Give the expected product(s) for the following reactions:

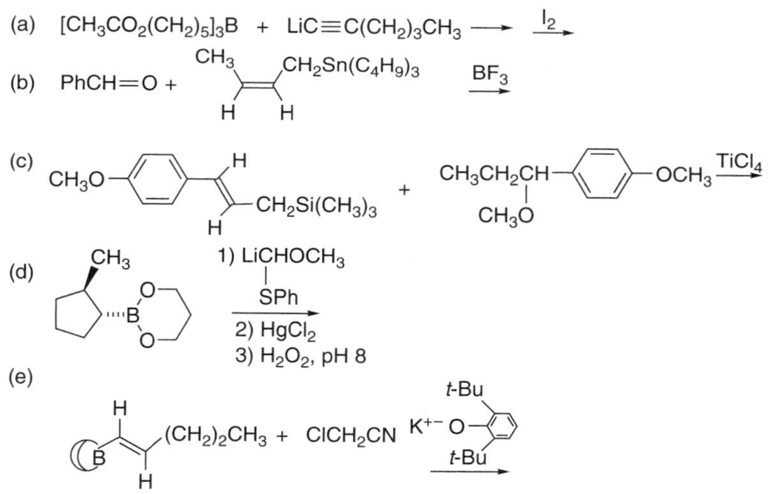

(a) $[CH_3CO_2(CH_2)_5]_3B$ + $LiC \equiv C(CH_2)_3CH_3 \longrightarrow \xrightarrow{I_2}$

(b) $PhCH = O$ +

(c)

(d)

(e)

9.2. Starting with an alkene $RCH = CH_2$, indicate how an organoborane intermediate could be used for each of the following synthetic transformations:

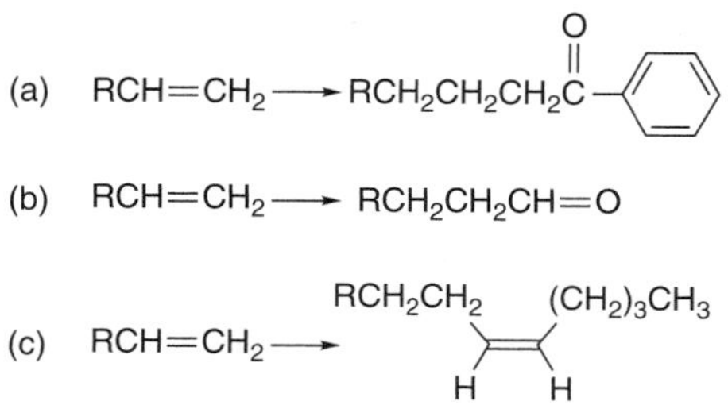

(a) $RCH = CH_2 \longrightarrow RCH_2CH_2CH_2C(=O)$—Ph

(b) $RCH = CH_2 \longrightarrow RCH_2CH_2CH = O$

(c) $RCH = CH_2 \longrightarrow$

(d) $RCH{=}CH_2 \longrightarrow RCH_2CH_2\overset{\overset{\displaystyle O}{\|}}{C}CH_2CH_2R$

(e) $RCH{=}CH_2 \longrightarrow RCH_2CH_2CH_2CO_2C_2H_5$

9.3. Scheme 9.1 describes reactions with several lithiated compounds, including dichloromethane, dichloromethyl methyl ether, phenylthiomethyl methyl ether, and phenylthioacetals. Compare the structure of these reagents and the final products for these reactions. Develop a mechanistic outline that encompasses these reactions. Discuss the features that these reagents have in common with one another and with carbon monoxide.

9.4. Each of the following transformations was performed advantageously with a thexylborane derivative. Give appropriate reactants, reagents, and reaction conditions for effecting the following syntheses in a one-pot" process.

(a) 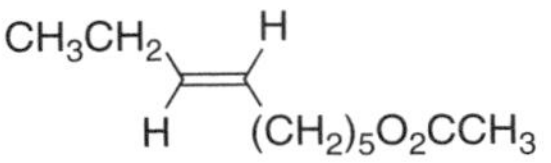   from $IC{\equiv}CCH_2CH_3$ and $CH_2{=}CH(CH_2)_3O_2CCH_3$

(b) $CH_3(CH_2)_3C{\equiv}C(CH_2)_7CH_3$ from $CH_2{=}CH(CH_2)_5CH_3$ and $HC{\equiv}C(CH_2)_3CH_3$

(c) 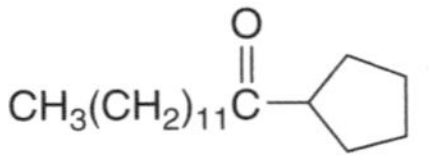   from $CH_2{=}CH(CH_2)_9CH_3$ and

(d) 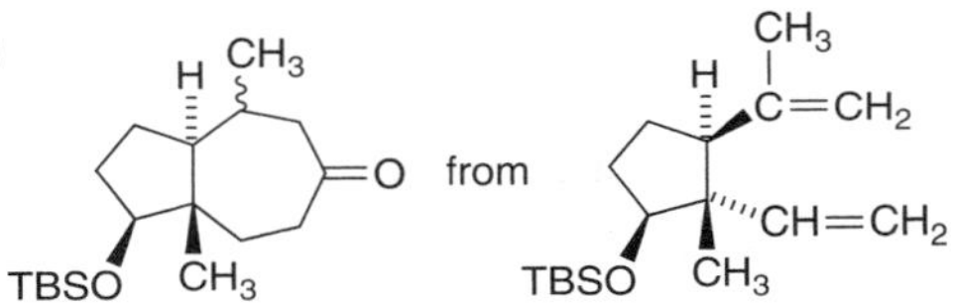   from

9.5. Provide mechanisms for the formation of the new carbon-carbon bonds in each of the following reactions:

(a)

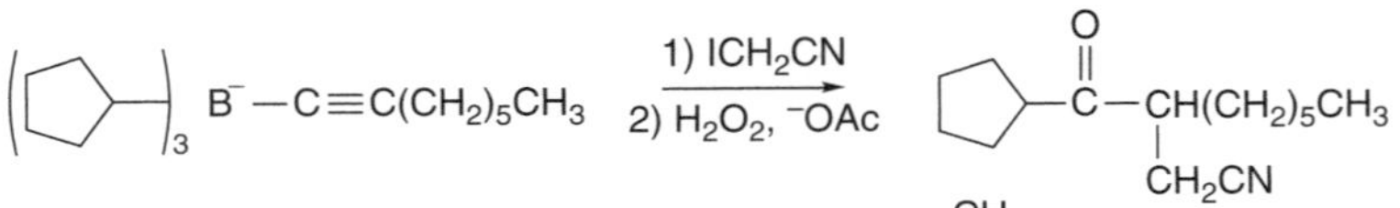

(b)

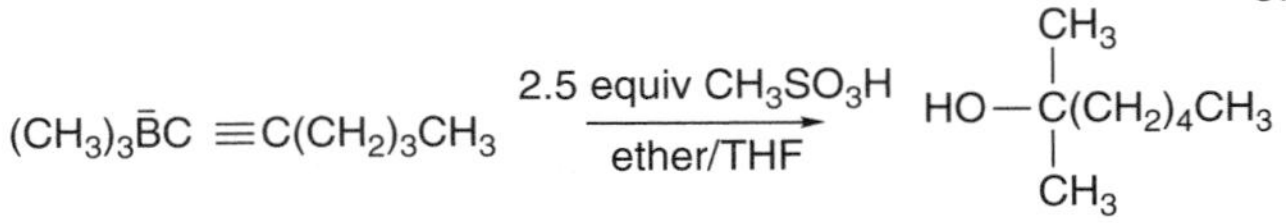

(c)

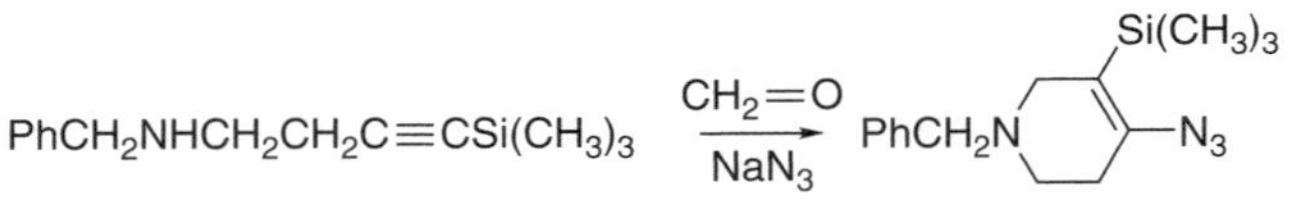

(d)

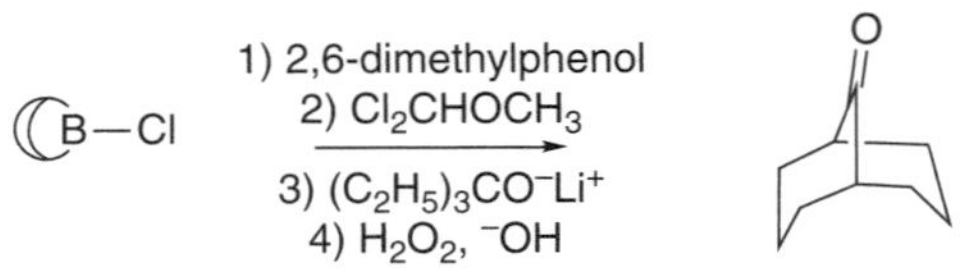

9.6. Offer a detailed mechanistic explanation for the following observations.

a. When the *E*- and *Z*-isomers of 2-butenyl-1,3,2-dioxaborolane **6-A** react with aldehyde **6-B**, the *Z*-isomer gives *syn* product **6-C** with greater than 90% stereoselectivity. The *E*-isomer, however, gives a nearly 1:1 mixture of two *anti* products **6-D** and **6-E**.

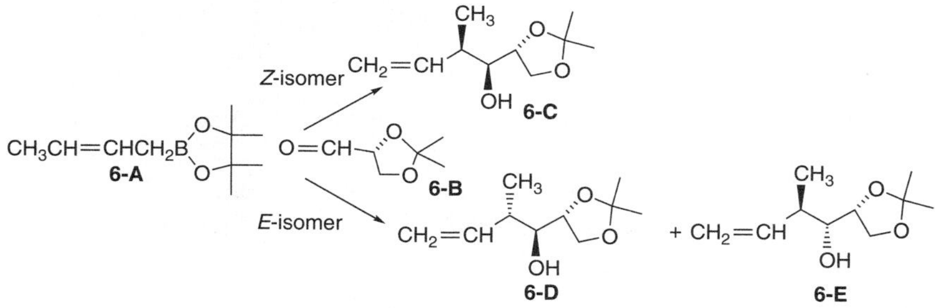

b. The reaction of several $\Delta^{2,3}$-pyranyl acetates with allyl trimethylsilane under the influence of Lewis acids gives 2-allyl-$\Delta^{3,4}$-pyrans. The stereochemistry depends on whether the *E*- or *Z*-allylsilane is used. There is a preference for *anti* stereochemistry at the new bond with the *E*-silane but *syn* stereochemistry with the *Z*-silane. The preference for the *syn* stereochemistry is increased by use of a more bulky silyl substituent. Analyze the competing transition structures for the *E*- and *Z*-silanes and suggest an explanation for the observed stereoselectivity.

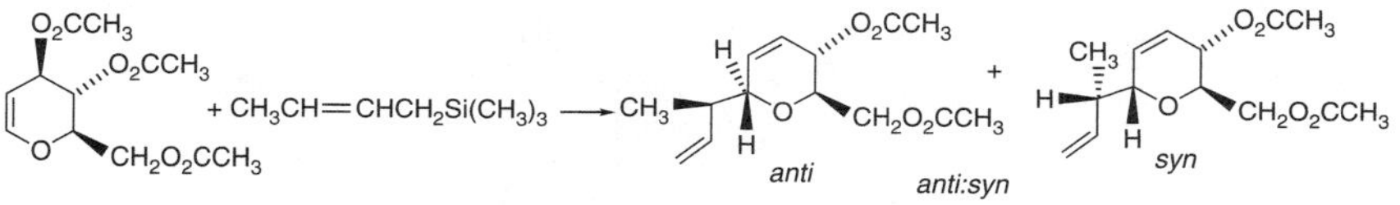

		anti:syn
E	SiMe$_3$	3:1
Z	SiMe$_3$	1:3
Z	SiMe$_2$Ph	1:3.2
Z	SiMePh$_2$	1.4.5
Z	Si(*t*Bu)Ph$_2$	1:7

c. In the reaction of 2-pentenyl tri-*n*-butylstannanes with benzaldehyde and BF$_3$, the diastereoselectivity is dependent on the identity of the 3-substituent group. Offer an explanation in terms of possible transition structures.

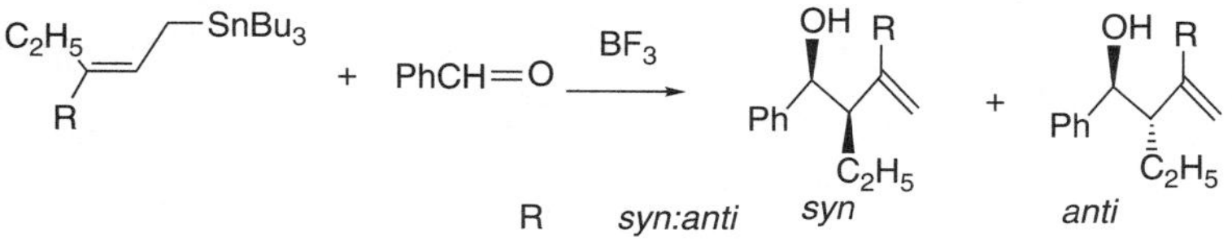

R	syn:anti
H	78:22
CH$_3$	91:9
i-Pr	84:16
t-Bu	13:87

856

CHAPTER 9

*Carbon-Carbon
Bond-Forming Reactions
of Compounds of Boron,
Silicon, and Tin*

d. It is observed that the stereoselectivity of cyclizative condensation of aminoalkyl silane **6-F** depends on the steric bulk of the amino substituent. Offer an explanation for this observation in terms of the transition structure for the addition reaction.

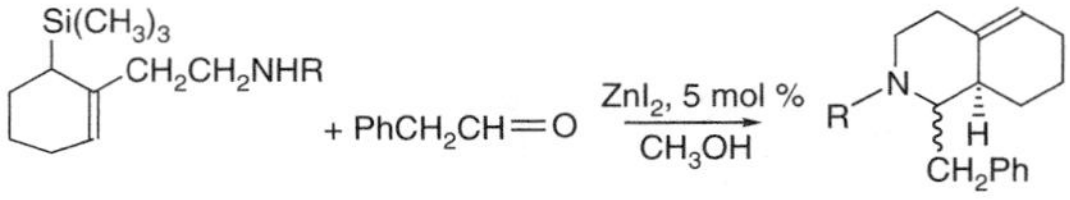

6-F

R	Yield (%)	*trans:cis* ratio
CH₃ᵃ	68	20:80
PhCH₂	88	58:42
Ph₂CH	73	>99:1
Dibenzocycloheptyl	67	>99:1

ᵃ Ph(CH₃)₂Si instead of (CH₃)₃Si.

9.7. A number of procedures for stereoselective synthesis of alkenes involving alkenylboranes have been developed. For each of the reactions given below, show the structure of the intermediates and outline the mechanism in sufficient detail to account for the observed stereoselectivity.

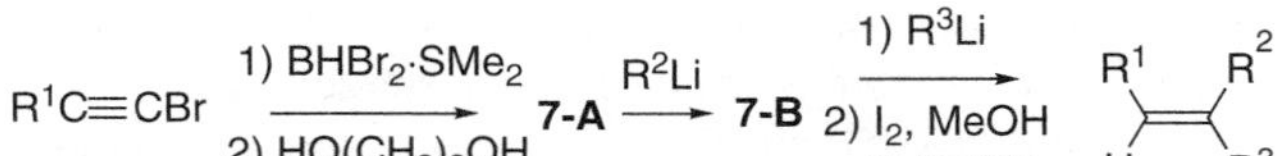

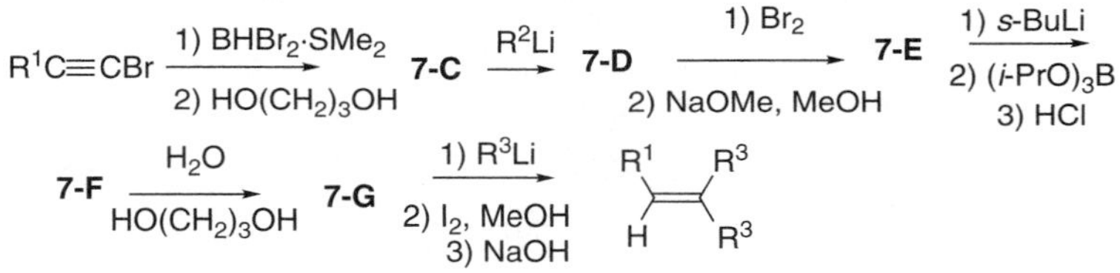

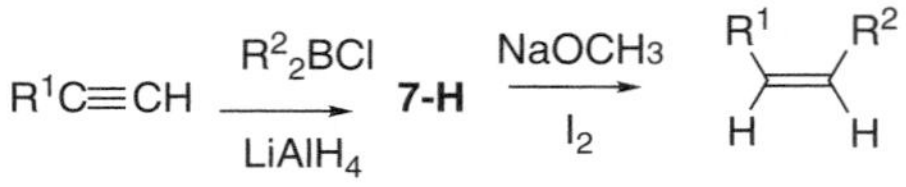

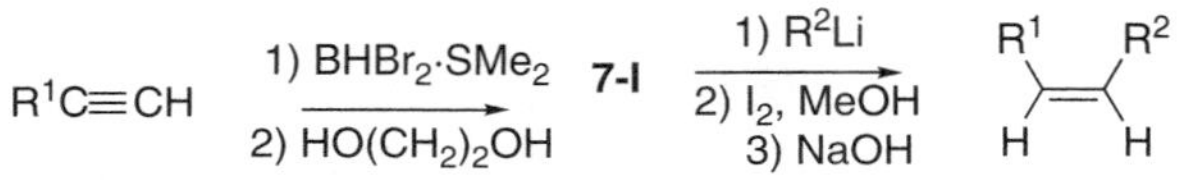

9.8. Suggest reagents and reaction conditions that would be effective for the following cyclization reactions:

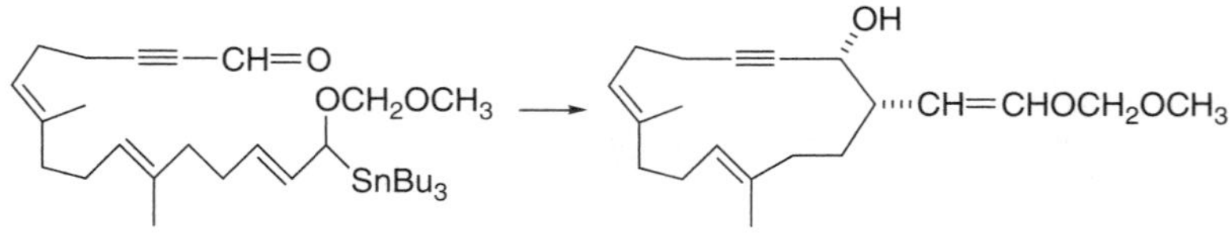

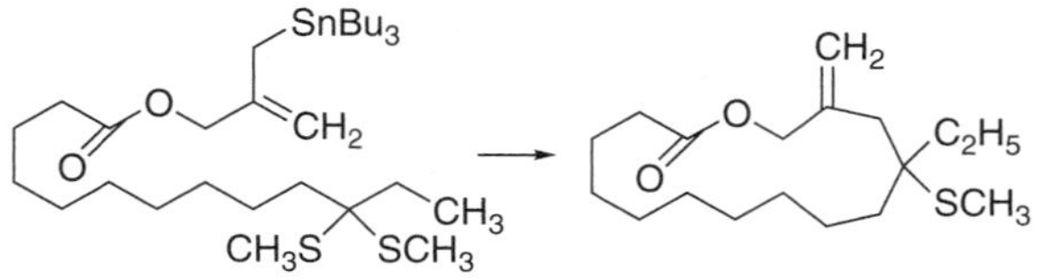

9.9. Show how the following silanes and stannanes can be synthesized from the suggested starting material.

(a)

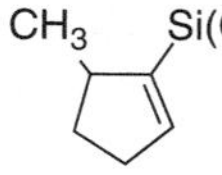

from

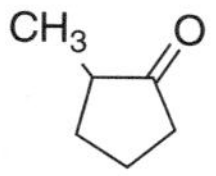

(b)

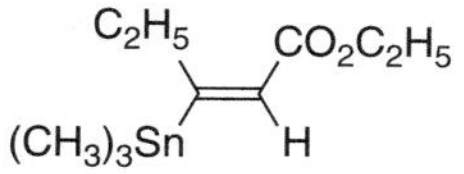

from $CH_3CH_2C\equiv CCO_2C_2H_5$

(c) $Bu_3SnCH=CHSnBu_3$ from $HC\equiv CH$, Bu_3SnCl, and Bu_3SnH

(d)

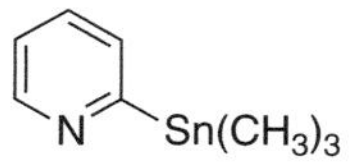

from

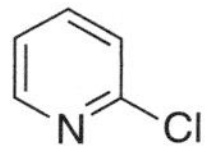

(e) $RCCH_2Si(CH_3)_3$ from $RCOCl$ or RCO_2R'
     $\parallel$
     CH_2

(f) $CH_2Si(CH_3)_3$
      $\mid$
     $CH_2=CCH=CH_2$ from $CH_2=CCH=CH_2$
                                    $\mid$
                                    Cl

9.10. Each of the unsaturated cyclic amines shown below has been synthesized by reaction of an amino-substituted allylic silane under iminium ion cyclization conditions ($CH_2=O$, TFA). By retrosynthetic analysis, identify the appropriate precursor for each cyclization. Suggest a method of synthesis of each of the required amines.

(a)

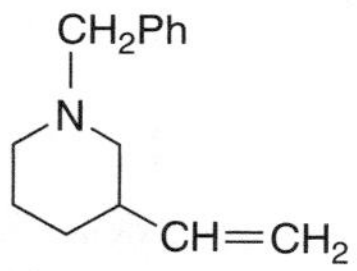

(b)

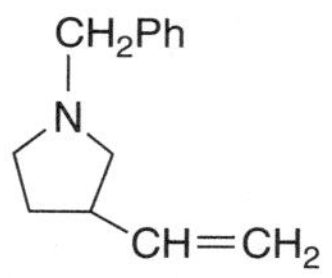

(c)

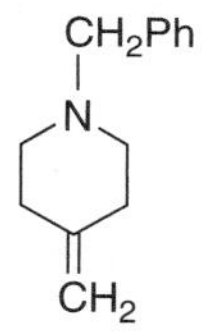

9.11. Both E- and Z-isomers of the terpene γ-bisabolene have been isolated from natural sources. The synthesis of these compounds can be achieved by stereoselective alkene syntheses using borane intermediates. An outline of each synthesis is given below. Indicate the reaction conditions that would permit the stereoselective synthesis of each isomer.

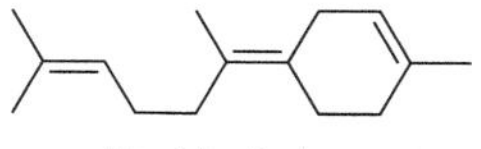

E-γ-bisabolene

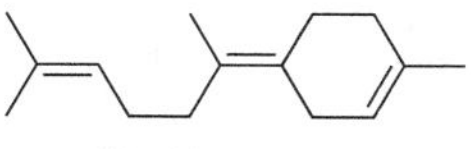

Z-γ-bisabolene

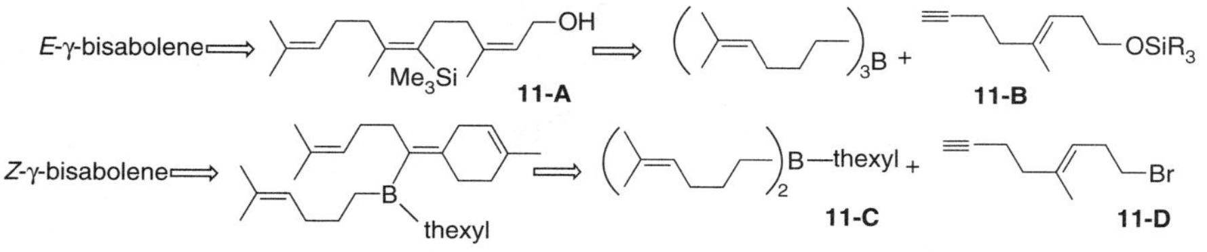

858

CHAPTER 9

*Carbon-Carbon
Bond-Forming Reactions
of Compounds of Boron,
Silicon, and Tin*

9.12. By retrosynthetic analysis, devise a sequence of reactions that would provide the desired compound from the indicated starting materials.

(a)

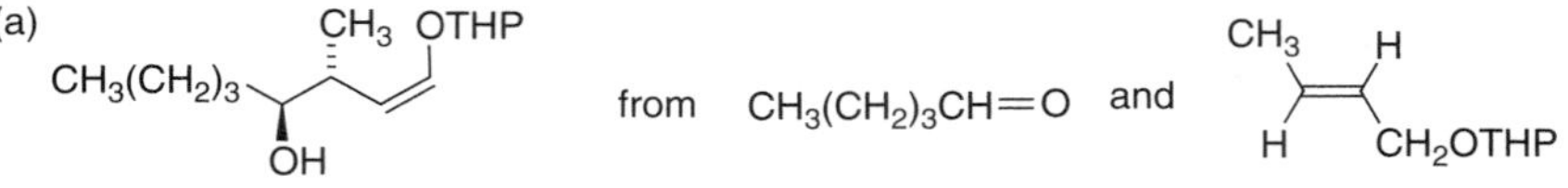

from $CH_3(CH_2)_3CH=O$ and

(b)

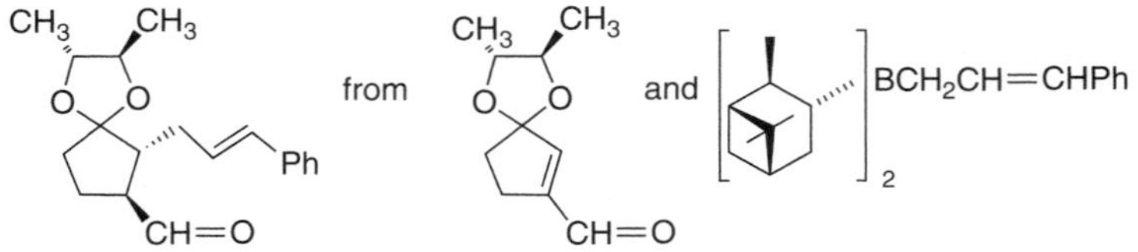

from

and

(c)

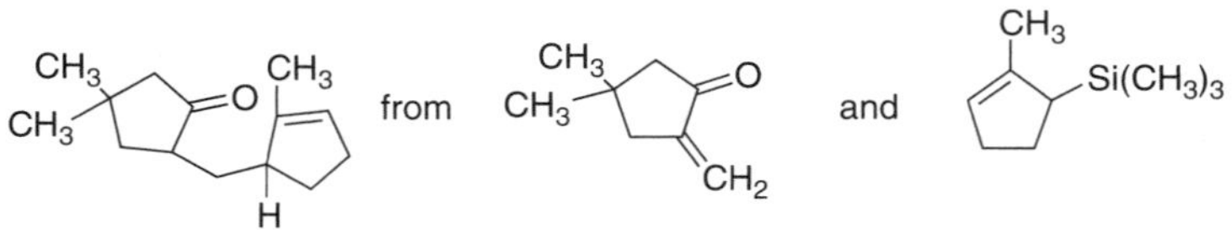

from

and

(d)

from

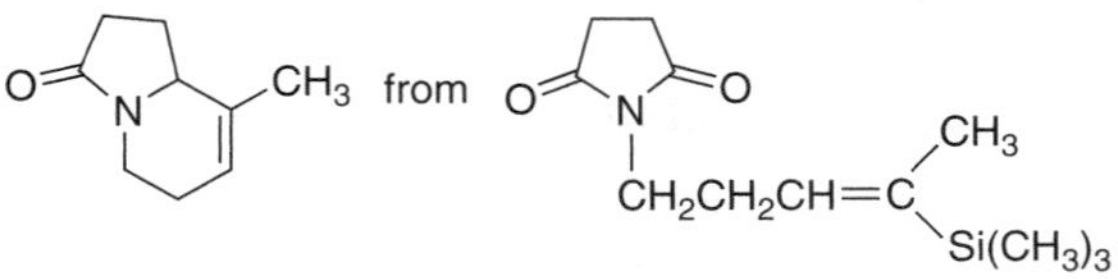

9.13. Show how the following compounds could be prepared in high enantiomeric purity using enantiopure boranes as reactants.

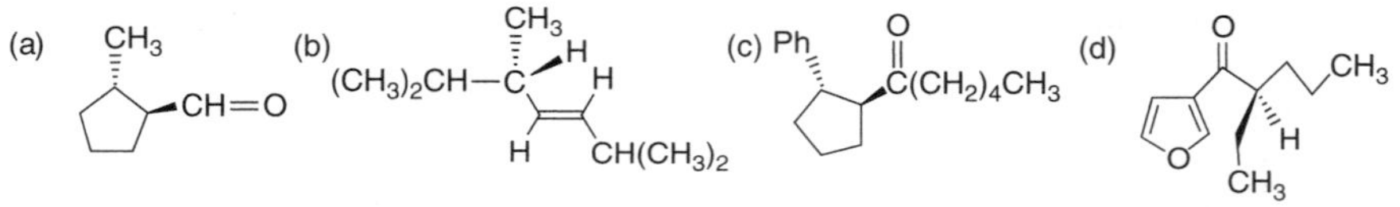

9.14. Show how organoborane intermediates can be used to synthesize the gypsy moth pheromone $E, Z\text{-}CH_3CO_2(CH_2)_4CH=CH(CH_2)_2CH=CH(CH_2)_3CH_3$ from hept-6-ynyl acetate, allyl bromide, and 1-hexyne.

9.15. Predict the major stereoisomer that will be formed in the following reactions. Show the transition structure that is the basis for your response.

(a)

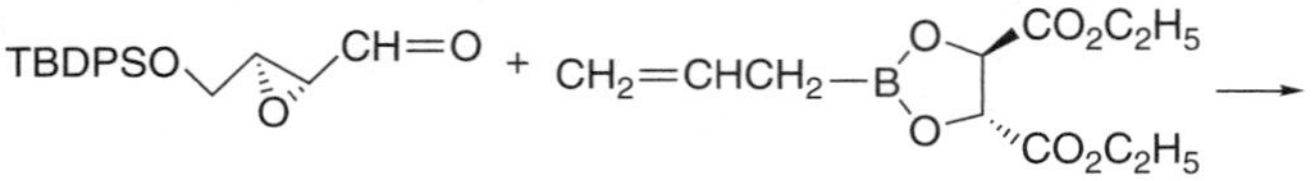

(b) $CH_2=CHCH=O + CH_3CH=CHCH_2Sn(n\text{-}C_4H_9)_3 \xrightarrow[\text{25°C, 24 h}]{(n\text{-}C_4H_9)_2SnCl_2}$

(3:1 *E:Z*-mixture)

(c)

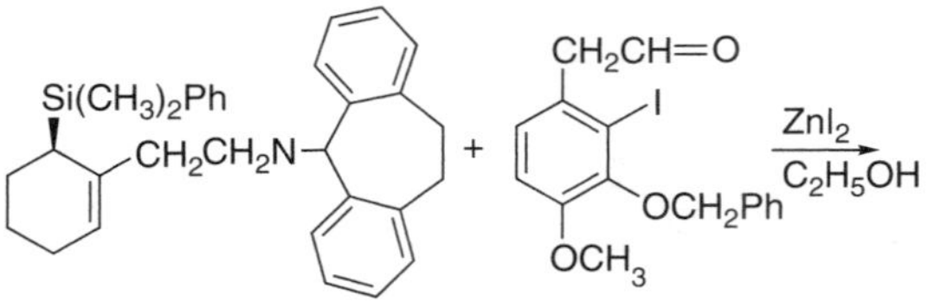

(d)

(e)

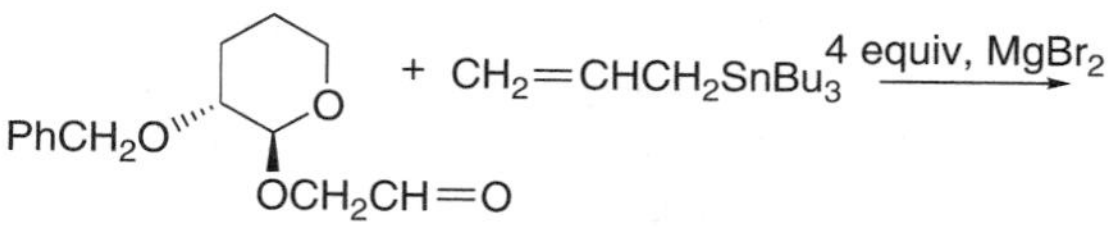

(f)

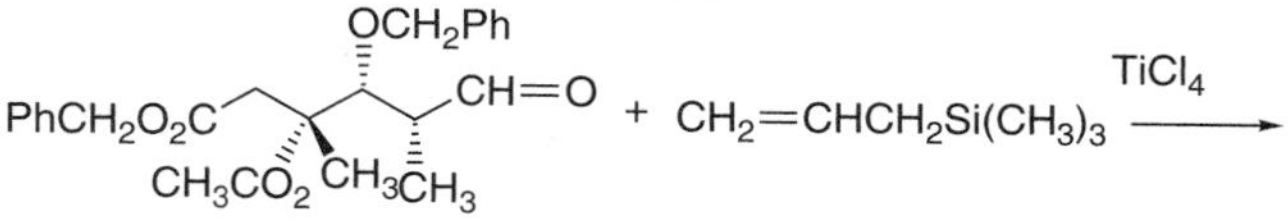

9.16. The stereoselectivity of the β-carboethoxyallylic boronate derived from the *endo*-phenyl auxiliary **A** (p. 803) toward *R*- and *S*-glyceraldehyde acetonide has been investigated. One enantiomer gives the *anti* product in 98:2 ratio, whereas the other favors the *syn* product by a 65:35 ratio. Based on the proposed transition structure for this boronate, determine which combination leads to the higher stereoselectivity and which to the lower. Propose the favored transition structure in each case.

9.17. The *R*- and *S*-enantiomers of *Z*-3-methoxymethyl-1-methylpropenylstannane have been allowed to react with the protected erythrose- and threose-derived aldehydes **17-A** and **17-B**. The products are shown below. Indicate the preferred transition structure for each combination.

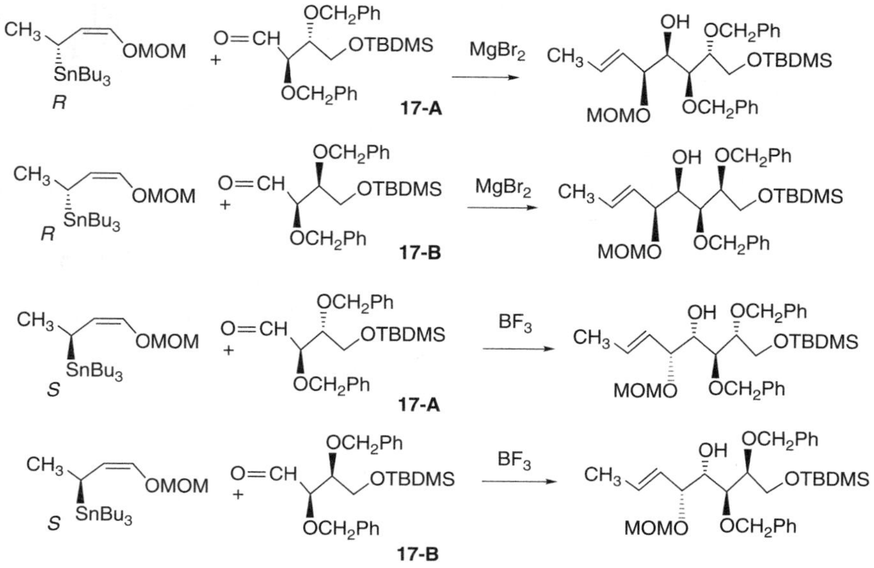

860

CHAPTER 9

*Carbon-Carbon
Bond-Forming Reactions
of Compounds of Boron,
Silicon, and Tin*

9.18. In the original report of the reaction in Entry 8 of Scheme 9.6, it was found that use of three equivalents of BF_3 led to loss of stereoselectivity, but not yield.

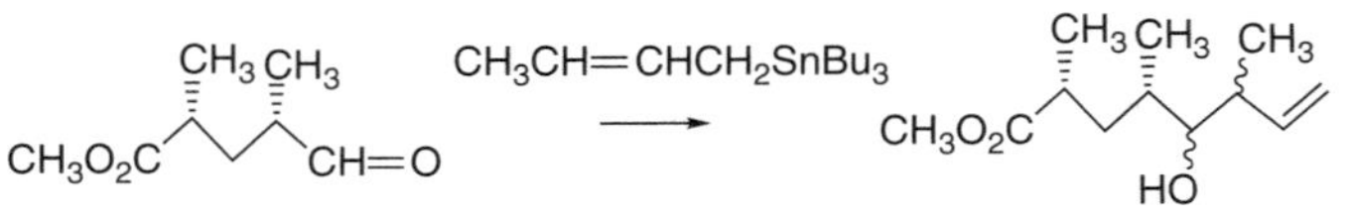

Product Composition

Equiv BF_3	Total Yield	*anti* anti-Felkin	*anti* Felkin	*syn* Felkin	*syn* anti-Felkin
1	92	94–97	3–4	1	1
2	90	83–91	5–9	1–3	2–5
3	90	41	10	17	32

These results were attributed to a preference for an eight-membered chelated transition structure that was lost in the presence of excess BF_3 because of coordination of a second BF_3 at the ester group. What objections would you raise to this explanation? What alternative would you propose?

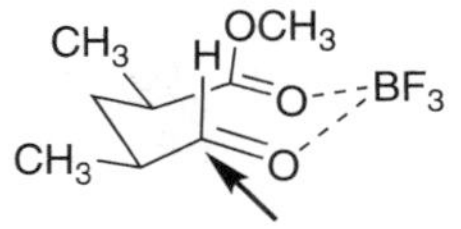

9.19. The aldehyde **19-A** shows differential stereoselectivity toward the enantiomeric stannanes (*S*)-**19-B** and (*R*)-**19-B**. The former aldehyde gives a single product in high yield, whereas the latter gives a somewhat lower yield and a mixture of two stereoisomers under the same conditions and is a mixture of two stereoisomers. Propose TSs to account for each product and indicate the reasons for the enhanced stereoselectivity of (*S*)-**19-B**.

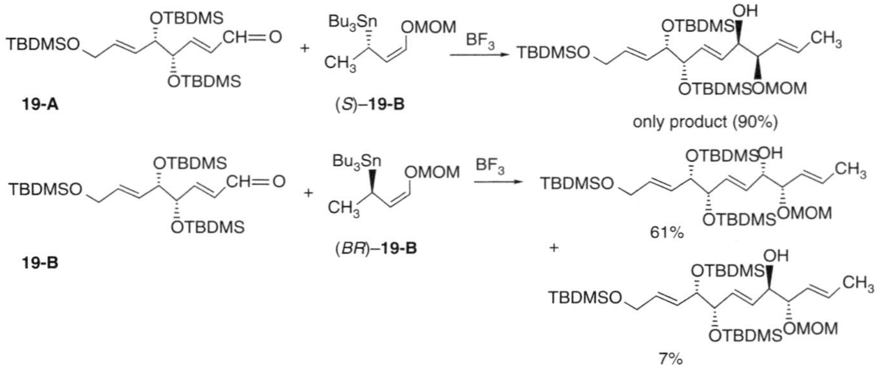

Reactions Involving Carbocations, Carbenes, and Radicals as Reactive Intermediates

Introduction

Trivalent carbocations, carbanions, and radicals are the most fundamental classes of reactive intermediates. The basic aspects of the structural and reactivity features of these intermediates were introduced in Chapter 3 of Part A. Discussion of carbanion intermediates in synthesis began in Chapter 1 of the present volume and continued through several further chapters. The focus in this chapter is on *electron-deficient reactive intermediates*, including carbocations, carbenes, and carbon-centered radicals. Both carbocations and carbenes have a carbon atom with *six valence electrons* and are therefore *electron-deficient* and *electrophilic* in character, and they have the potential for skeletal rearrangements. We also discuss the use of carbon radicals to form carbon-carbon bonds. Radicals react through homolytic bond-breaking and bond-forming reactions involving intermediates with *seven valence electrons*.

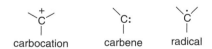

A common feature of these intermediates is that they are of high energy, compared to structures with completely filled valence shells. Their lifetimes are usually very short. Bond formation involving carbocations, carbenes, and radicals often occurs with low activation energies. This is particularly true for addition reactions with alkenes and other systems having π bonds. These reactions replace a π bond with a σ bond and are usually exothermic.

$$\overset{+}{C} + \overset{}{C}=\overset{}{C} \longrightarrow -\overset{|}{\underset{|}{C}}-\overset{|}{\underset{|}{C}}-\overset{+}{C} \quad \text{or} \quad \overset{\cdot}{C} + \overset{}{C}=\overset{}{C} \longrightarrow -\overset{|}{\underset{|}{C}}-\overset{|}{\underset{|}{C}}-\overset{\cdot}{C}$$

Owing to the low barriers to bond formation, *reactant conformation* often plays a decisive role in the outcome of these reactions. Carbocations, carbene, and radicals frequently undergo very efficient *intramolecular reactions* that depend on the proximity of the reaction centers. Conversely, because of the short lifetimes of the intermediates, reactions through unfavorable conformations are unusual. Mechanistic analyses and synthetic designs that involve carbocations, carbenes, and radicals must pay particularly close attention to conformational factors.

10.1. Reactions and Rearrangement Involving Carbocation Intermediates

In this section, the emphasis is on carbocation reactions that modify the carbon skeleton, including carbon-carbon bond formation, rearrangements, and fragmentation reactions. The fundamental structural and reactivity characteristics of carbocations toward nucleophilic substitution were explored in Chapter 4 of Part A.

10.1.1. Carbon-Carbon Bond Formation Involving Carbocations

10.1.1.1. Intermolecular Alkylation by Carbocations. The formation of carbon-carbon bonds by electrophilic attack on the π system is a very important reaction in aromatic chemistry, with both Friedel-Crafts alkylation and acylation following this pattern. These reactions are discussed in Chapter 11. There also are useful reactions in which carbon-carbon bond formation results from electrophilic attack by a carbocation on an alkene. The reaction of a carbocation with an alkene to form a new carbon-carbon bond is both kinetically accessible and thermodynamically favorable.

$$\overset{+}{C} + \overset{}{C}=\overset{}{C} \longrightarrow -\overset{|}{\underset{|}{C}}-\overset{|}{\underset{|}{C}}-\overset{+}{C}$$

There are, however, serious problems that must be overcome in the application of this reaction to synthesis. The product is a new carbocation that can react further. Repetitive addition to alkene molecules leads to polymerization. Indeed, this is the mechanism of acid-catalyzed polymerization of alkenes. There is also the possibility of rearrangement. A key requirement for adapting the reaction of carbocations with alkenes to the synthesis of small molecules is control of the reactivity of the newly formed carbocation intermediate. Synthetically useful carbocation-alkene reactions require a suitable termination step. We have already encountered one successful strategy in the reaction of alkenyl and allylic silanes and stannanes with electrophilic carbon (see Chapter 9). In those reactions, the silyl or stannyl substituent is eliminated and a stable alkene is formed. The increased reactivity of the silyl- and stannyl-substituted alkenes is also favorable to the synthetic utility of carbocation-alkene reactions because the reactants are more nucleophilic than the product alkenes.

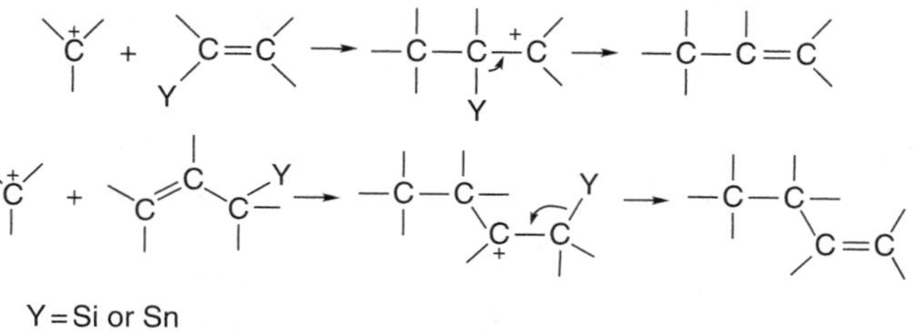

863

SECTION 10.1

Reactions and
Rearrangement
Involving Carbocation
Intermediates

Y = Si or Sn

Silyl enol ethers and silyl ketene acetals also offer both enhanced reactivity and a favorable termination step. Electrophilic attack is followed by desilylation to give an α-substituted carbonyl compound. The carbocations can be generated from tertiary chlorides and a Lewis acid, such as $TiCl_4$. This reaction provides a method for introducing tertiary alkyl groups α to a carbonyl, a transformation that cannot be achieved by base-catalyzed alkylation because of the strong tendency for tertiary halides to undergo elimination.

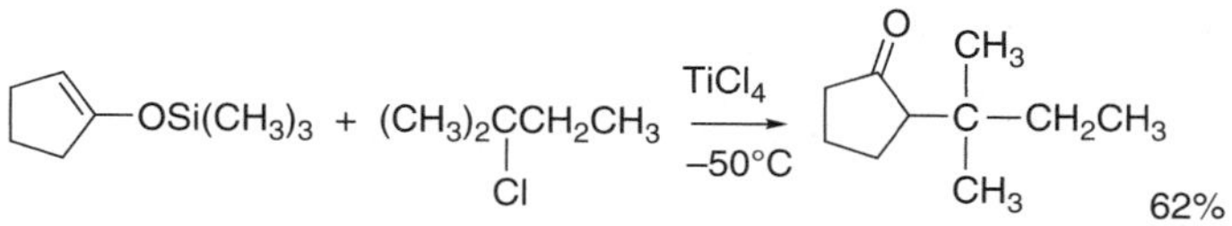

Ref. 1

Secondary benzylic bromides, allylic bromides, and α-chloro ethers can undergo analogous reactions using $ZnBr_2$ as the catalyst.[2] Primary iodides react with silyl ketene acetals in the presence of AgO_2CCF_3.[3]

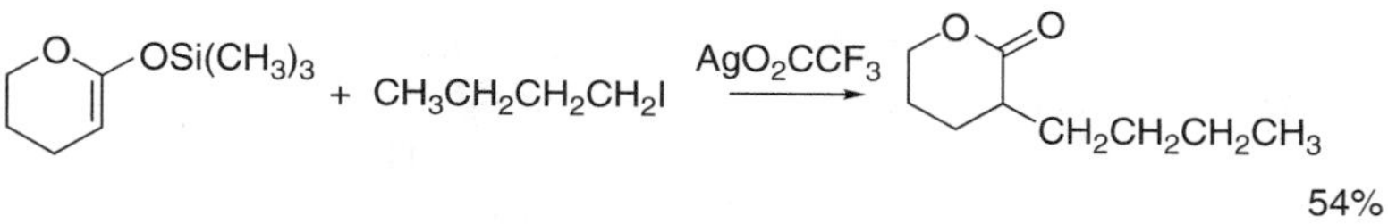

Alkylations via an allylic cation have been observed using $LiClO_4$ to promote ionization.[4]

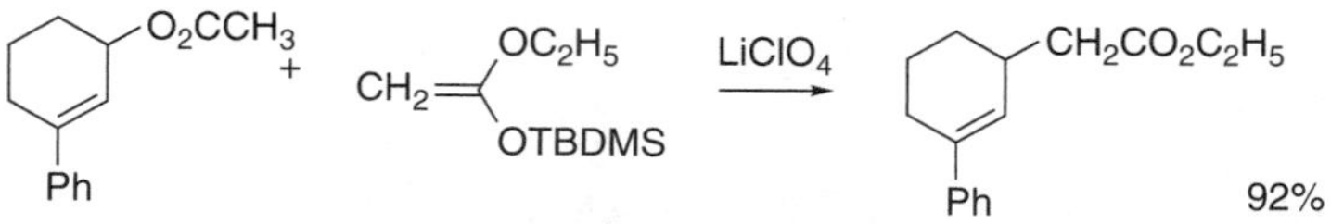

These reactions provide examples of intermolecular carbocation alkylations. Despite the feasibility of this type of reaction, the requirements for good yields are stringent and the number of its synthetic applications is limited.

1. M. T. Reetz, I. Chatziiosifidis, U. Loewe, and W. F. Maier, *Tetrahedron Lett.*, 1427 (1979); M. T. Reetz, I. Chatziiosifidis, F. Huebner, and H. Heimbach, *Org. Synth.*, **62**, 95 (1984).
2. I. Paterson, *Tetrahedron Lett.*, 1519 (1979).
3. C. W. Jefford, A. W. Sledeski, P. Lelandais, and J. Boukouvalas, *Tetrahedron Lett.*, **33**, 1855 (1992).
4. W. H. Pearson and J. M. Schkeryantz, *J. Org. Chem.*, **57**, 2986 (1992).

864

CHAPTER 10

*Reactions Involving
Carbocations, Carbenes,
and Radicals as Reactive
Intermediates*

10.1.1.2. Polyene Cyclization. Perhaps the most synthetically useful of the carbocation alkylation reactions is the cyclization of polyenes having two or more double bonds positioned in such a way that successive bond-forming steps can occur. This process, called *polyene cyclization*, has proven to be an effective way of making polycyclic compounds containing six-membered and, in some cases, five-membered rings. The reaction proceeds through an electrophilic attack and requires that the double bonds that participate in the cyclization be properly positioned. For example, compound **1** is converted quantitatively to **2** on treatment with formic acid. The reaction is initiated by protonation and ionization of the allylic alcohol and is terminated by nucleophilic capture of the cyclized secondary carbocation.

Ref. 5

More extended polyenes can cyclize to tricyclic systems.

(Product is a mixture of four diene isomers indicated by dotted lines)   Ref. 6

These cyclizations are usually highly stereoselective, with the stereochemical outcome being determined by the reactant conformation.[7] The stereochemistry of the products in the decalin system can be predicted by assuming that cyclization occurs through conformations that resemble chair cyclohexane rings. The stereochemistry at ring junctures is that resulting from *anti* attack at the participating double bonds.

trans   *cis*

   To be of maximum synthetic value, the generation of the cationic site that initiates cyclization must involve mild reaction conditions. Formic acid and stannic chloride are effective reagents for cyclization of polyunsaturated allylic alcohols. Acetals generate oxonium ions in acidic solution and can also be used to initiate the cyclization of polyenes.[8]

5. W. S. Johnson, P. J. Neustaedter, and K. K. Schmiegel, *J. Am. Chem. Soc.*, **87**, 5148 (1965).
6. W. J. Johnson, N. P. Jensen, J. Hooz, and E. J. Leopold, *J. Am. Chem. Soc.*, **90**, 5872 (1968).
7. W. S. Johnson, *Acc. Chem. Res.*, **1**, 1 (1968); P. A. Bartlett, in *Asymmetric Synthesis*, Vol. 3, J. D. Morrison, ed., Academic Press, New York, 1984, Chap. 5.
8. A van der Gen, K. Wiedhaup, J. J. Swoboda, H. C. Dunathan, and W. S. Johnson, *J. Am. Chem. Soc.*, **95**, 2656 (1973).

865

SECTION 10.1

*Reactions and
Rearrangement
Involving Carbocation
Intermediates*

(Dotted lines indicate mixture
of unsaturated products)

Another significant method for generating the electrophilic site is acid-catalyzed epoxide ring opening.[9] Lewis acids such as BF_3, $SnCl_4$, CH_3AlCl_2, or $TiCl_3(O\text{-}i\text{-Pr})$ can be used,[10] as illustrated by Entries 4 to 7 in Scheme 10.1.

Mercuric ion is capable of inducing cyclization of polyenes.

Ref. 11

The particular example shown also has a special mechanism for stabilization of the cyclized carbocation. The adjacent acetoxy group is captured to form a stabilized dioxanylium cation. After reductive demercuration (see Section 4.1.3) and hydrolysis, a diol is isolated.

As the intermediate formed in a polyene cyclization is a carbocation, the isolated product is often found to be a mixture of closely related compounds resulting from competing modes of reaction. The products result from capture of the carbocation by solvent or other nucleophile or by deprotonation to form an alkene. Polyene cyclizations can be carried out on reactants that have structural features that facilitate transformation of the carbocation to a stable product. Allylic silanes, for example, are stabilized by desilylation.[12]

The incorporation of silyl substituents not only provides for specific reaction products but can also improve the effectiveness of polyene cyclization. For example, although cyclization of **2a** gave a mixture containing at least 17 products, the allylic silane **2b** gave a 79% yield of a 1:1 mixture of stereoisomers.[13] This is presumably due to the enhanced reactivity and selectivity of the allylic silane.

9. E. E. van Tamelen and R. G. Nadeau, *J. Am. Chem. Soc.*, **89**, 176 (1967).
10. E. J. Corey and M. Sodeoka, *Tetrahedron Lett.*, **33**, 7005 (1991); P. V. Fish, A. R. Sudhakar, and W. S. Johnson, *Tetrahedron Lett.*, **34**, 7849 (1993).
11. M. Nishizawa, H. Takenaka, and Y. Hayashi, *J. Org. Chem.*, **51**, 806 (1986); E. J. Corey, J. G. Reid, A. G. Myers, and R. W. Hahl, *J. Am. Chem. Soc.*, **109**, 918 (1987).
12. W. S. Johnson, Y.-Q. Chen, and M. S. Kellogg, *J. Am. Chem. Soc.*, **105**, 6653 (1983).
13. P. V. Fish, *Tetrahedron Lett.*, **35**, 7181 (1994).

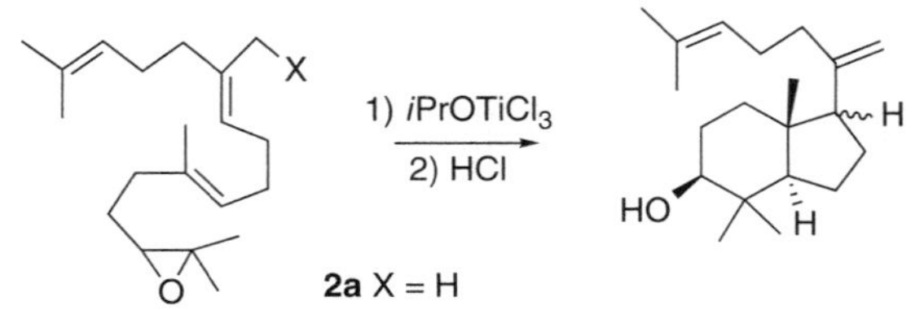

866

CHAPTER 10

*Reactions Involving
Carbocations, Carbenes,
and Radicals as Reactive
Intermediates*

2a X = H

2b X = Si(CH₃)₃

The efficiency of cyclization can also be affected by stereoelectronic factors. For example, there is a significant difference in the efficiency of the cyclization of the *Z*- and *E*-isomers of **3**. Only the *Z*-isomer presents an optimal alignment for electronic stabilization.[14] These effects of the terminating substituent point to considerable concerted character for the cyclizations.

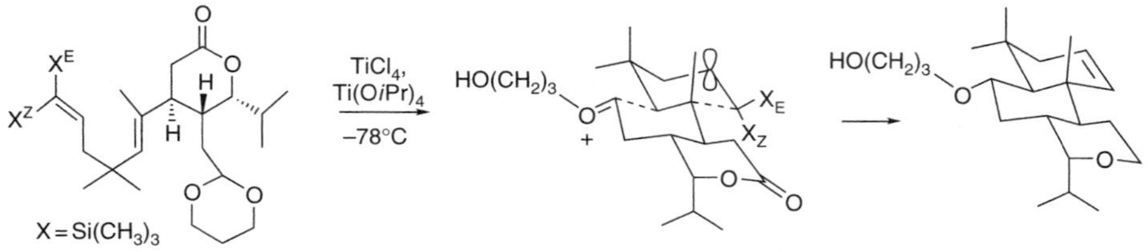

X = Si(CH₃)₃

3

30–40% for *E*-isomer

85–90% for *Z*-isomer

When a cyclization sequence is terminated by an alkyne, vinyl cations are formed. Capture of water leads to formation of a ketone.[15]

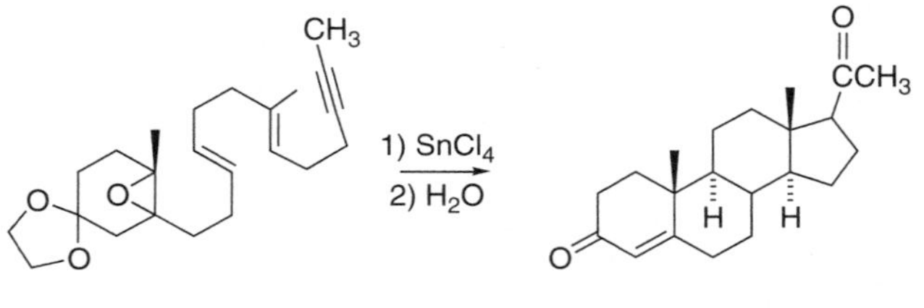

Use of chiral acetal groups can result in enantioselective cyclization.[16]

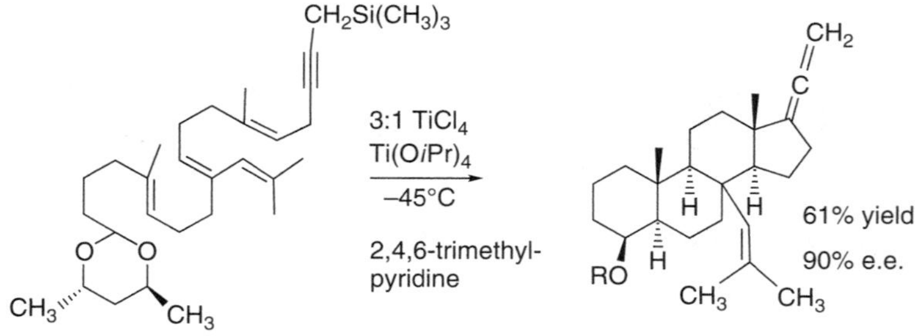

61% yield

90% e.e.

14. S. D. Burke, M. E. Kort, S. M. S. Strickland, H. M. Organ, and L. A. Silks, III, *Tetrahedron Lett.*, **35**, 1503 (1994).
15. E. E. van Tamelen and J. R. Hwu, *J. Am. Chem. Soc.*, **105**, 2490 (1983).
16. D. Guay, W. S. Johnson, and U. Schubert, *J. Org. Chem.*, **54**, 4731 (1989).

Polyene cyclizations are of substantial value in the synthesis of polycyclic terpene natural products. These syntheses resemble the processes by which the polycyclic compounds are assembled in nature. The most dramatic example of biosynthesis of a polycyclic skeleton from a polyene intermediate is the conversion of squalene oxide to the steroid lanosterol. In the biological reaction, an enzyme not only to induces the cationic cyclization but also holds the substrate in a conformation corresponding to stereochemistry of the polycyclic product.[17] In this case, the cyclization is terminated by a series of rearrangements.

Scheme 10.1 gives some representative examples of laboratory syntheses involving polyene cyclization. The cyclization in Entry 1 is done in anhydrous formic acid and involves the formation of a symmetric tertiary allylic carbocation. The cyclization forms a six-membered ring by attack at the terminal carbon of the vinyl group. The bicyclic cation is captured as the formate ester. Entry 2 also involves initiation by a symmetric allylic cation. In this case, the triene unit cyclizes to a tricyclic ring system. Entry 3 results in the formation of the steroidal skeleton with termination by capture of the alkynyl group and formation of a ketone. The cyclization in Entry 4 is initiated by epoxide opening.

Entries 5 and 6 also involve epoxide ring opening. In Entry 5 the cyclization is terminated by electrophilic substitution on the highly reactive furan ring. In Entry 6 a silyl enol ether terminates the cyclization sequence, leading to the formation of a ketone. Entry 7 incorporates two special features. The terminal propargylic silane generates an allene. The fluoro substituent was found to promote the formation of the six-membered D ring by directing the regiochemistry of formation of the C(8)−C(14) bond. After the cyclization, the five-membered A ring was expanded to a six-membered ring by oxidative cleavage and aldol condensation. The final product of this synthesis was β-amyrin. Entry 8 also led to the formation of β-amyrin and was done using the enantiomerically pure epoxide.

β-Amyrin

[17.] D. Cane, *Chem. Rev.*, **90**, 1089 (1990); I. Abe, M. Rohmer, and G. D. Prestwich, *Chem. Rev.*, **93**, 2189 (1993); K. U. Wendt and G. E. Schulz, *Structure*, **6**, 127 (1998).

Scheme 10.1. Polyene Cyclizations

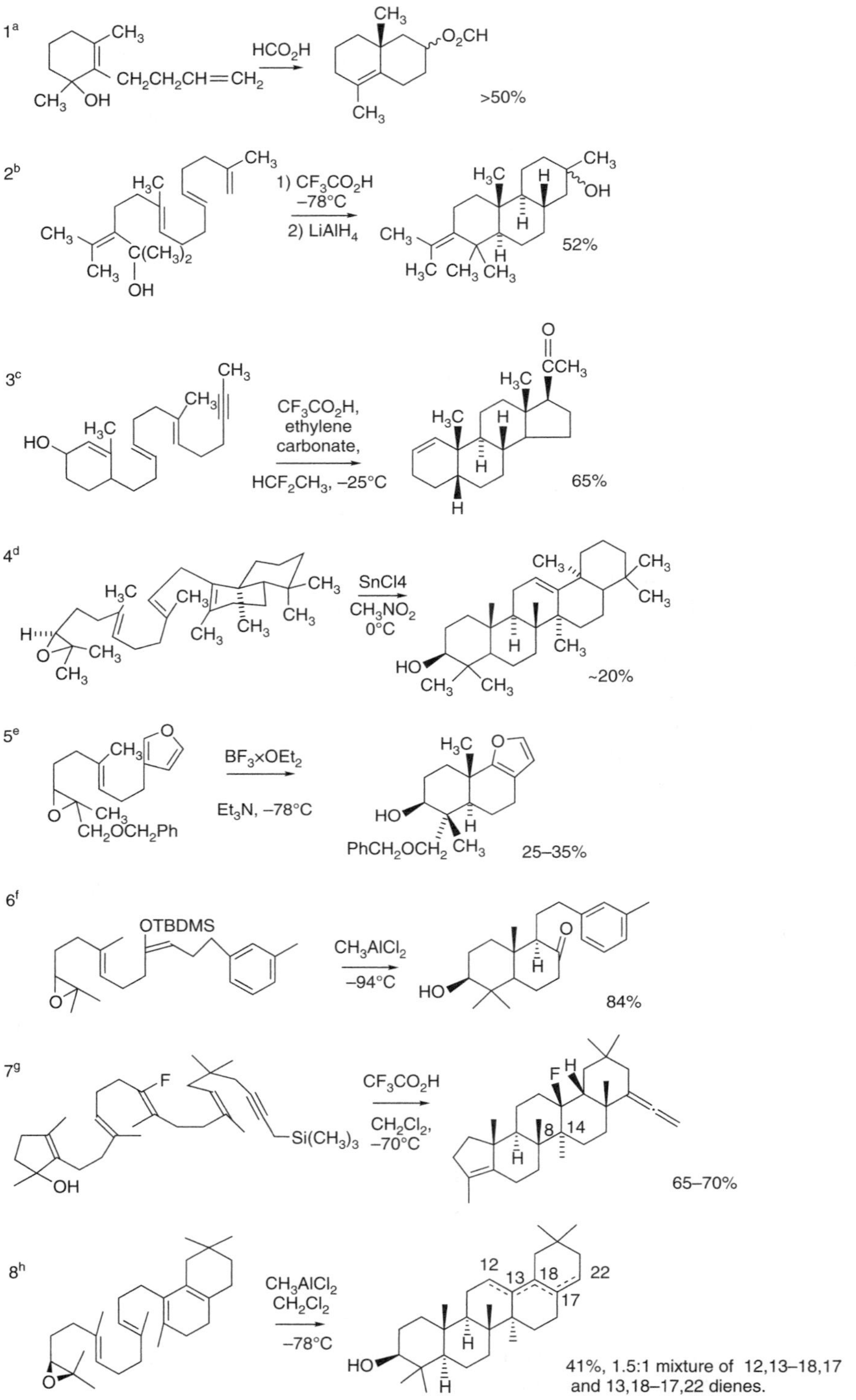

(*Continued*)

Scheme 10.1. (*Continued*)

869

SECTION 10.1

*Reactions and
Rearrangement
Involving Carbocation
Intermediates*

a. J. A. Marshall, N. Cohen, and A. R. Hochstetler, *J. Am. Chem. Soc.*, **88**, 3408 (1966).
b. W. S. Johnson and T. K. Schaaf, *J. Chem. Soc., Chem. Commun.*, 611 (1969).
c. B. E. McCarry, R. L. Markezich, and W. S. Johnson, *J. Am. Chem. Soc.*, **95**, 4416 (1973).
d. E. E. van Tamelen, R. A. Holton, R. E. Hopla, and W. E. Konz, *J. Am. Chem. Soc.*, **94**, 8228 (1972).
e. S. P. Tanis, Y.-H. Chuang, and D. B. Head, *J. Org. Chem.*, **53**, 4929 (1988).
f. E. J. Corey, G. Luo, and L. S. Lin, *Angew. Chem. Int. Ed. Engl.*, **37**, 1126 (1998).
g. W. S. Johnson, M. S. Plummer, S. P. Reddy, and W. R. Bartlett, *J. Am. Chem. Soc.*, **115**, 515 (1993).
h. E. J. Corey and J. Lee, *J. Am. Chem. Soc.*, **115**, 8873 (1993).

10.1.1.3. Ene and Carbonyl-Ene Reactions. Certain double bonds undergo electrophilic addition reactions with alkenes in which an allylic hydrogen is transferred to the reactant. This process is called the *ene reaction* and the electrophile is known as an *enophile*.[18] When a carbonyl group serves as the enophile, the reaction is called a *carbonyl-ene reaction* and leads to β,γ-unsaturated alcohols. The reaction is also called the *Prins reaction*.

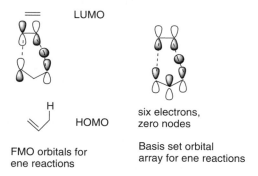

A variety of double bonds give reactions corresponding to the pattern of the ene reaction. Those that have been studied from a mechanistic and synthetic perspective include alkenes, aldehydes and ketones, imines and iminium ions, triazoline-2,5-diones, nitroso compounds, and singlet oxygen, $^1O{=}O$. After a mechanistic overview of the reaction, we concentrate on the carbon-carbon bond-forming reactions. The important and well-studied reaction with $^1O{=}O$ is discussed in Section 12.3.2.

The concerted mechanism shown above is allowed by the Woodward-Hoffmann rules. The TS involves the π electrons of the alkene and enophile and the σ electrons of the allylic C—H bond. The reaction is classified as a $[\pi 2 + \pi 2 + \sigma 2]$ and either an FMO or basis set orbital array indicates an allowed concerted process.

LUMO

HOMO

FMO orbitals for
ene reactions

six electrons,
zero nodes

Basis set orbital
array for ene reactions

Because the enophiles are normally the electrophilic reagent, their reactivity increases with addition of EWG substituents. Ene reactions between unsubstituted alkenes have high-energy barriers, but compounds such as acrylate or propynoate esters

---

[18.] For reviews of the ene reaction, see H. M. R. Hoffmann, *Angew. Chem. Int. Ed. Engl.*, **8**, 556 (1969); W. Oppolzer, *Pure Appl. Chem.*, **53**, 1181 (1981); K. Mikami and M. Shimizu, *Chem. Rev.*, **92**, 1020 (1992).

or, especially, maleic anhydride are more reactive. Similarly, for carbonyl compounds, glyoxylate, oxomalonate, and dioxosuccinate esters are among the typical reactants under thermal conditions.

glyoxylate
ester

oxomalonate
ester

dioxosuccinate
ester

Mechanistic studies have been designed to determine if the concerted cyclic TS provides a good representation of the reaction. A systematic study of all the *E*- and *Z*-decene isomers with maleic anhydride showed that the stereochemistry of the reaction could be accounted for by a concerted cyclic mechanism.[19] The reaction is only moderately sensitive to electronic effects or solvent polarity. The ρ value for reaction of diethyl oxomalonate with a series of 1-arylcyclopentenes is -1.2, which would indicate that there is little charge development in the TS.[20] The reaction shows a primary kinetic isotope effect indicative of C$-$H bond breaking in the rate-determining step.[21] There is good agreement between measured isotope effects and those calculated on the basis of TS structure.[22] These observations are consistent with a concerted process.

The carbonyl-ene reaction is strongly catalyzed by Lewis acids,[23] such as BF_3, $SnCl_4$, and $(CH_3)_2AlCl$.[24,25] Coordination of a Lewis acid at the carbonyl group increases its electrophilicity and allows reaction to occur at or below room temperature. The reaction becomes much more polar under Lewis acid catalysis and is more sensitive to solvent polarity[26] and substituent effects. For example, the ρ for 1-arylcyclopentenes with diethyl oxomalonate goes from -1.2 for the thermal reaction to -3.9 for a $SnCl_4$-catalyzed reaction. Mechanistic analysis of Lewis acid–catalyzed reactions indicates they are electrophilic substitution processes. At one mechanistic extreme, this might be a concerted reaction. At the other extreme, the reaction could involve formation of a carbocation. In synthetic practice, the reaction is often carried out using Lewis acid catalysts and probably is a stepwise process.

concerted carbonyl–ene reaction

stepwise mechanism

[19] S. H. Nahm and H. N. Cheng, *J. Org. Chem.*, **57** 5093 (1996).

[20] H. Kwart and M. Brechbiel, *J. Org. Chem.*, **47**, 3353 (1982).

[21] F. R. Benn and J. Dwyer, *J. Chem. Soc., Perkin Trans. 2*, 533 (1977); O. Achmatowicz and J. Szymoniak, *J. Org. Chem.*, **45**, 4774 (1980); H. Kwart and M. Brechbiel, *J. Org. Chem.*, **47**, 3353 (1982).

[22] D. A. Singleton and C. Hang, *Tetrahedron Lett.*, **40**, 8939 (1999).

[23] B. B. Snider, *Acc. Chem. Res.*, **13**, 426 (1980).

[24] K. Mikami and M. Shimizu, *Chem. Rev.*, **92**, 1020 (1992).

[25] M. F. Salomon, S. N. Pardo, and R. G. Salomon, *J. Org. Chem.*, **49**, 2446 (1984); *J. Am. Chem. Soc.*, **106**, 3797 (1984).

[26] P. Laszlo and M. Teston-Henry, *J. Phys. Org. Chem.*, **4** 605 (1991).

The experimental isotope effects have been measured for the reaction of 2-methylbutene with formaldehyde with diethylaluminum chloride as the catalyst,[27] and are consistent with a stepwise mechanism or a concerted mechanism with a large degree of bond formation at the TS. B3LYP/6-31G* computations using H^+ as the Lewis acid favored a stepwise mechanism.

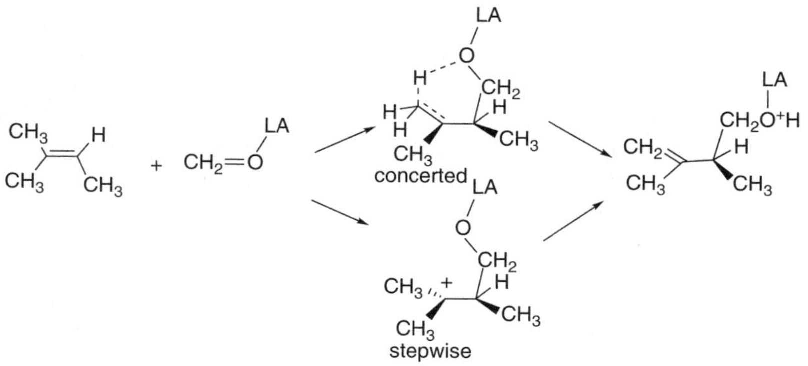

The best carbonyl components for these reactions are highly electrophilic compounds such as glyoxylate, pyruvate, and oxomalonate esters, as well as chlorinated and fluorinated aldehydes. Most synthetic applications of the carbonyl-ene reaction utilize Lewis acids. Although such reactions may be stepwise in character, the stereo-chemical outcome is often consistent with a cyclic TS. It was found, for example, that steric effects of trimethylsilyl groups provide a strong stereochemical influence.[28]

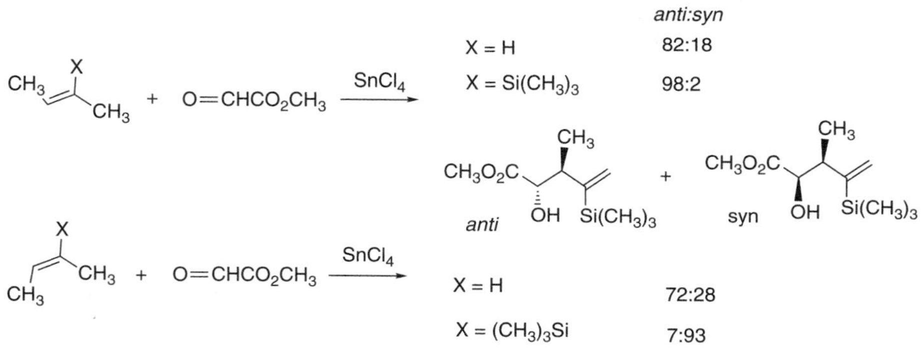

These results are consistent with two competing TSs differing in the facial orientation of the glyoxylate ester group. When X=H, the interaction with the ester group is small and the R^Z-ester interaction controls the stereochemistry. When the silyl group is present, there is a strong preference for TS **A**, which avoids interaction of the silyl group with the ester substituents.

27. D. A. Singleton and C. Hang, *J. Org. Chem.*, **65**, 895 (2000).
28. K. Mikami, T. P. Loh, and T. Nakai, *J. Am. Chem. Soc.*, **112**, 6737 (1990).

872

CHAPTER 10

*Reactions Involving
Carbocations, Carbenes,
and Radicals as Reactive
Intermediates*

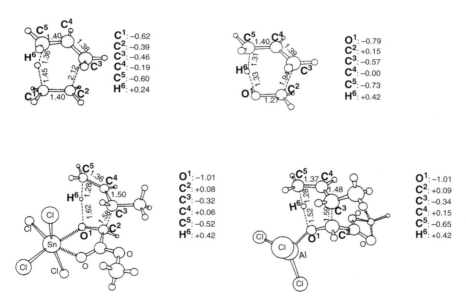

The mechanisms of simple ene reactions, such as those involving propene with ethene and formaldehyde, have been explored computationally. Concerted mechanisms and E_a values in general agreement with experiment are found using B3LYP/6-31G*,[29] MP2/6-31G*,[30] and MP4/6-31G*[31] computations. Yamanaka and Mikami used HF/6-31G* computations to compare the TS for ene reactions of propene with ethene and formaldehyde, and also for SnCl$_4$- and AlCl$_3$-catalyzed reactions with methyl glyoxylate.[32] The TS geometries and NPA charges are given in Figure 10.1. The ethene and formaldehyde TSs are rather similar, with the transferring hydrogen being positive in character, more so with formaldehyde than ethene. The catalyzed reactions are much more asynchronous, with C—C bond formation quite advanced. The two catalyzed reaction TSs correlate nicely with the observed stereoselectivity of the reaction. The stereochemistry of the 2-butene-methyl glyoxylate reaction shows a strong dependence on the Lewis acid that is used. The SnCl$_4$-catalyzed reaction gives the *anti* product via an *exo* TS, whereas AlCl$_3$ gives the *syn* product via an *endo* TS. The glyoxylate is chelated with SnCl$_4$, but not with AlCl$_3$, which leads to a difference in the orientation

Fig. 10.1. Minimum-energy transition structures for ene reactions: (a) propene and ethene; (b) propene and formaldehyde; (c) butene and methyl glyoxylate–SnCl$_4$; (d) butene and methyl glyoxylate–AlCl$_3$. Reproduced from *Helv. Chim. Acta*, **85**, 4264 (2002), by permission of Wiley-VCH.

29. Q. Deng, B. E. Thomas, IV, K. N. Houk, and P. Dowd, *J. Am. Chem. Soc.*, **119**, 6902 (1997).

30. J. Pranata, *Int. J. Quantum Chem.*, **62**, 509 (1997).

31. S. M. Bachrach and S. Jiang, *J. Org. Chem.*, **62**, 8319 (1997).

32. M. Yamanaka and K. Mikami, *Helv. Chim. Acta*, **85**, 4264 (2002).

of the unshared electrons on the ester oxygen. The *exo* TS is believed to be favored by an electrostatic interaction between the oxygen and C(4).

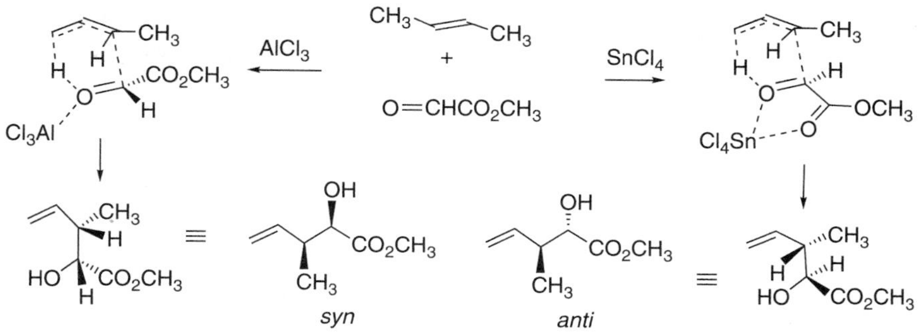

Despite the cyclic character of these TSs, both the bond distances and charge distribution are characteristic of a high degree of charge separation, with the butenyl fragment assuming the character of an allylic carbocation.

Visual models, additional information and exercises on the Carbonyl-Ene Reaction can be found in the Digital Resource available at: Springer.com/carey-sundberg.

Examples of catalyst control of stereoselectivity have been encountered in the course of the use of the ene reaction to elaborate a side chain on the steroid nucleus. The steroid **4** gave stereoisomeric products, depending on the catalysts and specific aldehyde that were used.[33] This is attributed to the presence of a chelated structure in the case of the SnCl$_4$ catalyst.

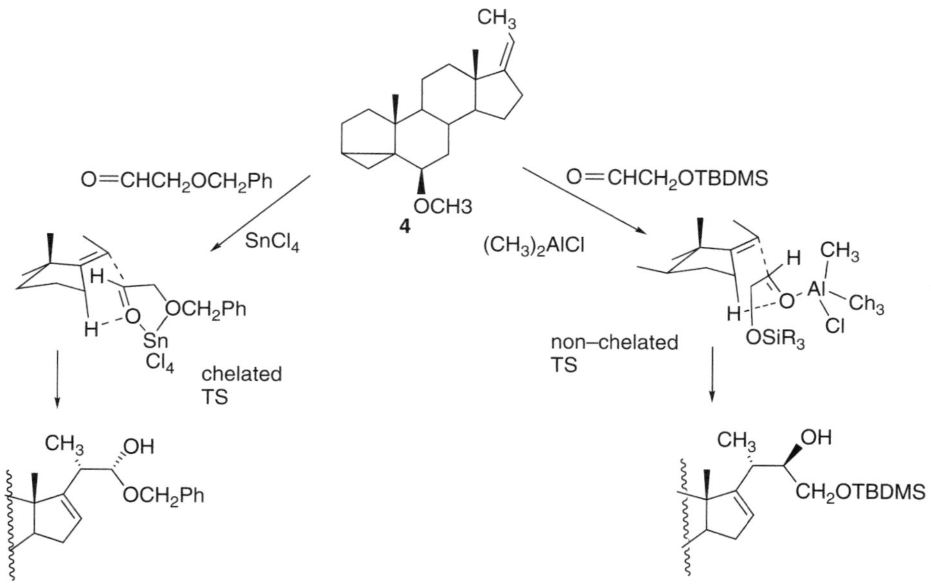

[33.] K. Mikami, H. Kishino, and T.-P. Loh, *J. Chem. Soc., Chem. Commun.*, 495 (1994).

The stereoselectivity of the $(CH_3)_2AlCl$-catalyzed reaction has also been found to be sensitive to the steric bulk of the aldehyde.[34]

The use of Lewis acid catalysts greatly expands the synthetic utility of the carbonyl-ene reaction. Aromatic aldehydes and acrolein undergo the ene reaction with activated alkenes such as enol ethers in the presence of $Yb(fod)_3$.[35] $Sc(O_3SCF_3)_3$ has also been used to catalyze carbonyl-ene reactions.[36]

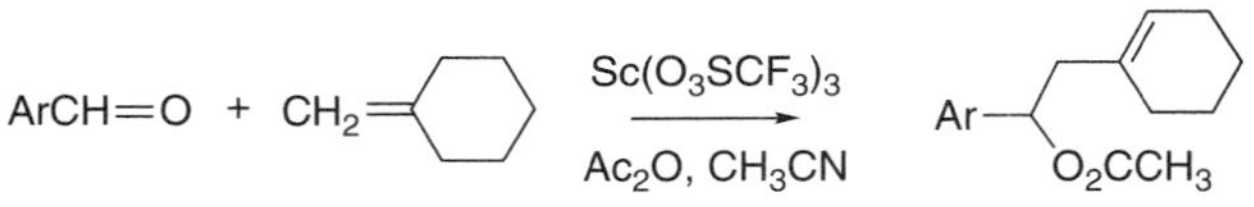

Among the more effective conditions for reaction of formaldehyde with α-methylstyrenes is BF_3 in combination with 4A molecular sieves.[37]

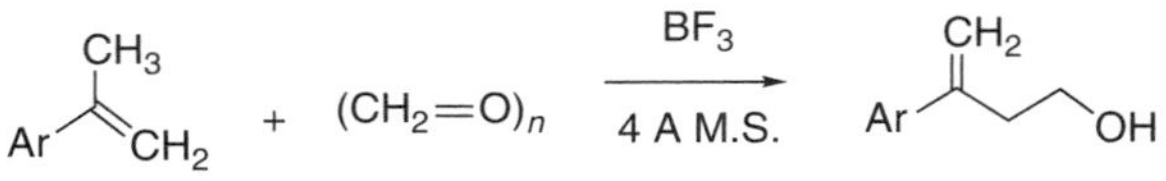

The function of the molecular sieves in this case is believed to be as a base that sequesters the protons, which otherwise would promote a variety of side reactions. With chiral catalysts, the carbonyl ene reaction becomes enantioselective. Among the successful catalysts are diisopropoxyTi(IV)BINOL and copper-BOX complexes.

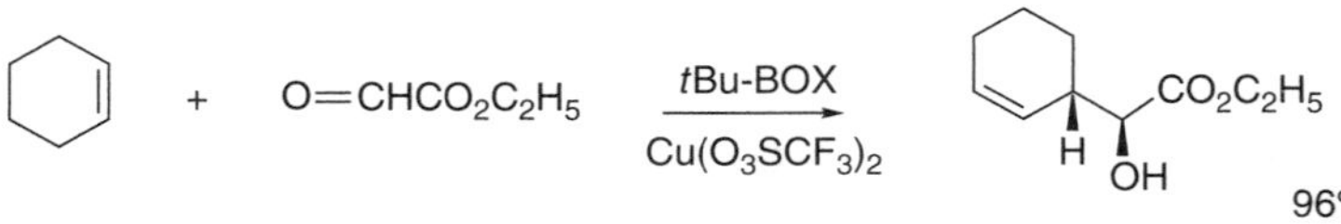

96% e.e.

Ref. 38

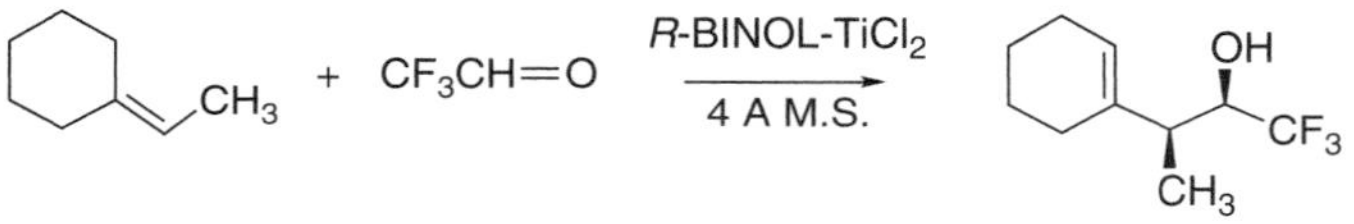

94% yield, 98% syn, 96% e.e.

Ref. 39

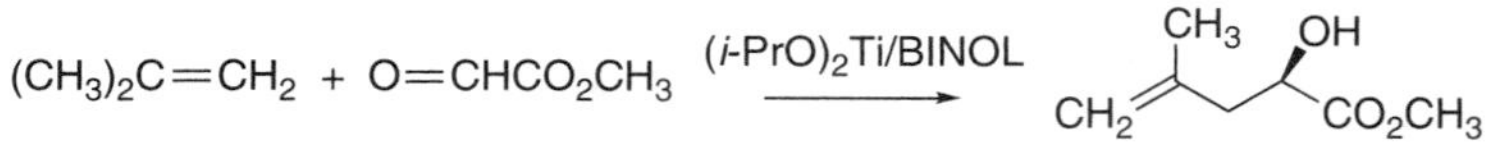

72% yield, 95% e.e.

Ref. 40

34. T. A. Houston, Y. Tanaka, and M. Koreeda, *J. Org. Chem.*, **58**, 4287 (1993).
35. M. A. Ciufolini, M. V. Deaton, S. R. Zhu, and M. Y. Chen, *Tetrahedron*, **53**, 16299 (1997); M. A. Ciufolini and S. Zhu, *J. Org. Chem.*, **63**, 1668 (1998).
36. V. K. Aggarawal, G. P. Vennall, P. N. Davey, and C. Newman, *Tetrahedron Lett.*, **39**, 1997 (1998).
37. T. Okachi, K. Fujimoto, and M. Onaka, *Org. Lett.*, **4**, 1667 (2002).
38. D. A. Evans, C. S. Burgey, N. A. Paras, T. Vojkovsky, and S. W. Tregay, *J. Am. Chem. Soc.*, **120**, 5824 (1998).
39. K. Mikami, T. Yajima, T. Takasaki, S. Matsukawa, M. Terada, T. Uchimaru, and M. Maruta, *Tetrahedron*, **52**, 85 (1996).
40. K. Mikami, M. Terada, and T. Nakai, *J. Am. Chem. Soc.*, **112**, 3949 (1990).

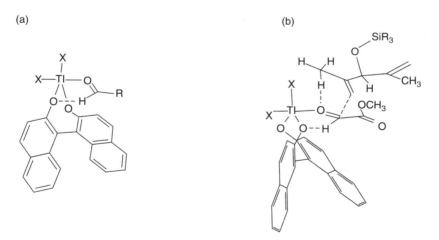

95% yield, 96% e.e.

Ref. 41

The enantioselectivity of the BINOL-Ti(IV)-catalyzed reactions can be interpreted in terms of several fundamental structural principles.[42] The aldehyde is coordinated to Ti through an apical position and there is also a O—HC=O hydrogen bond involving the formyl group. The most sterically favored approach of the alkene toward the complexed aldehyde then leads to the observed product. Figure 10.2 shows a representation of the complexed aldehyde and the TS structure for the reaction.

Most carbonyl-ene reactions used in synthesis are intramolecular and can be carried out under either thermal or catalyzed conditions,[43] but generally Lewis acids are used. Stannic chloride catalyzes cyclization of the unsaturated aldehyde **5**.

Ref. 44

(a)

(b)

Fig. 10.2. Structures of complexed aldehyde reagent (a) and transition structure (b) for enantio-selective catalysis of the carbonyl-ene reaction by BINOL-Ti(IV). Reproduced from *Tetrahedron Lett.*, **38**, 6513 (1997), by permission of Elsevier.

41. D. A. Evans, S. W. Tregay, C. S. Burgey, N. A. Paras, and T. Vojkovsky, *J. Am. Chem. Soc.*, **122**, 7936 (2000).
42. E. J. Corey, D. L. Barnes-Seeman, T. W. Lee and S. N. Goodman, *Tetrahedron Lett.*, **38**, 6513 (1997).
43. W. Oppolzer and V. Snieckus, *Angew. Chem. Int. Ed. Engl.*, **17**, 476 (1978).
44. L. A. Paquette and Y.-K. Han, *J. Am. Chem. Soc.*, **103**, 1835 (1981).

876

CHAPTER 10

Reactions Involving
Carbocations, Carbenes,
and Radicals as Reactive
Intermediates

The cyclization of the α-ketoester **6** can be effected by $Mg(ClO_4)_2$, $Yb(OTf)_3$, $Cu(OTf)_2$, or $Sc(OTf)_3$.[45] The reaction exhibits a 20:1 preference for formation of the *trans*-2-(1-methylpropenyl) isomer. The reaction can be conducted with greater than 90% e.e. using $Cu(OTf)_2$ or $Sc(OTf)_3$ with the *t*-Bu-BOX ligand.

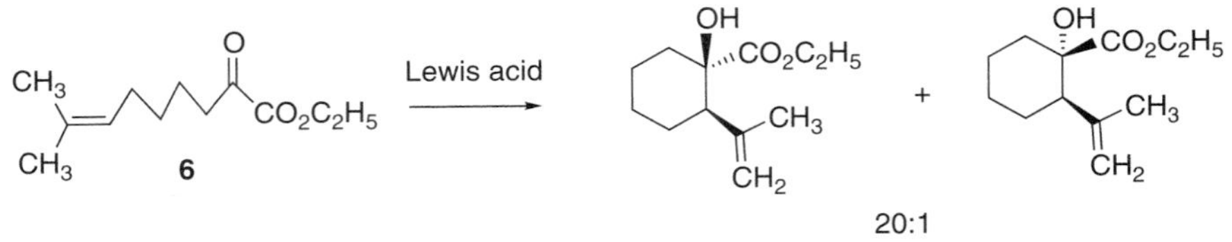

20:1

As an example of a thermal reaction, **7** cyclizes at 180°C. The reaction is stereoselective and the two stereoisomers can be formed from competing cyclic TSs.[46]

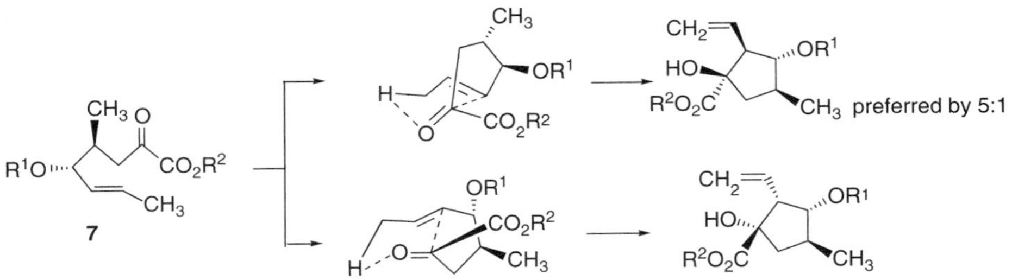

Carbonyl-ene reactions can be carried out in combination with other kinds of reactions. Mixed acetate acetals of γ,δ-enols, which can be prepared from the corresponding acetate esters, undergo cyclization with nucleophilic capture. When $SnBr_4$ is used for cyclization, the 4-substituent is bromine, whereas BF_3 in acetic acid gives acetates.[47]

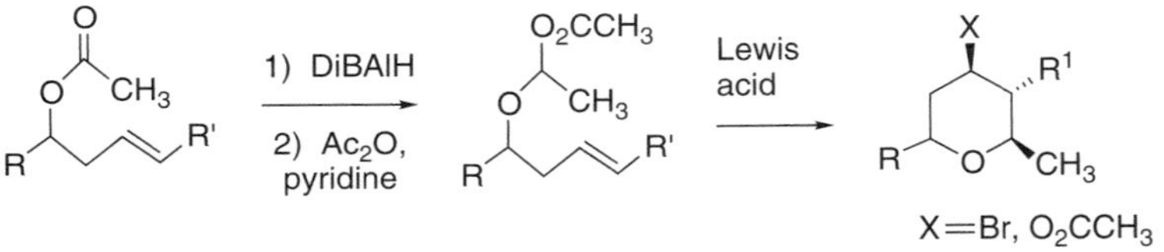

$X = Br, O_2CCH_3$

The reaction stereochemistry is consistent with a cyclic TS.

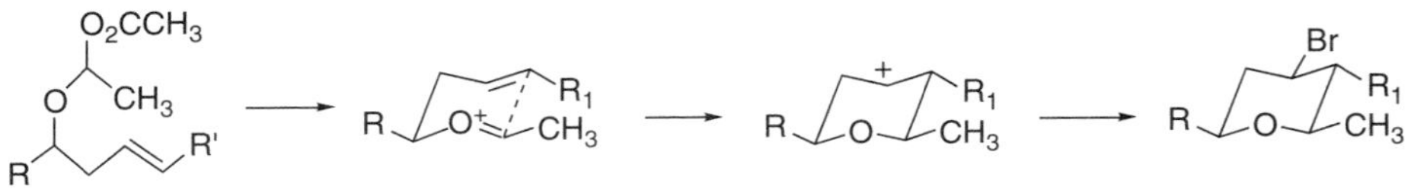

A tandem combination initiated by a Mukaiyama reaction generates an oxonium ion that cyclizes to give a tetrahydropyran rings.[48]

45. D. Yang, M. Yang, and N.-Y. Zhu, *Org. Lett.*, **5**, 3749 (2003).
46. H. Helmboldt, J. Rehbein, and M. Hiersemann, *Tetrahedron Lett.*, **45**, 289 (2004).
47. J. J. Jaber, K. Mitsui, and S. D. Rychnovsky, *J. Org. Chem.*, **66**, 4679 (2001).
48. B. Patterson and S. D. Rychnovsky, *Synlett*, 543 (2004).

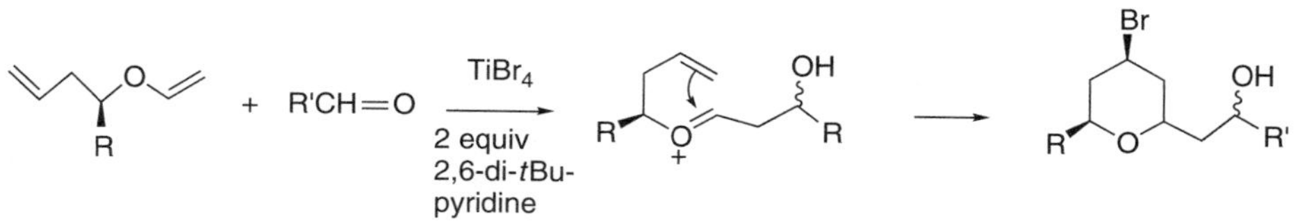

877

SECTION 10.1

*Reactions and
Rearrangement
Involving Carbocation
Intermediates*

This reaction has been used in coupling two fragments in a synthesis of leucascandrolide, a cytotoxic substance isolated from a sponge.[49]

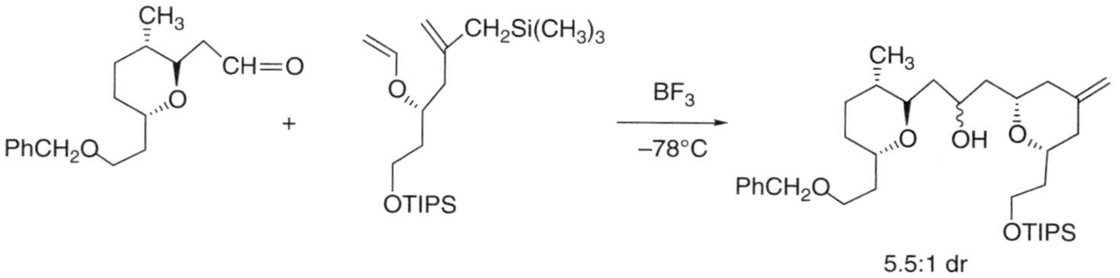

5.5:1 dr

A tandem Sakurai-carbonyl-ene sequence was used to create a tricyclic skeleton in the synthesis of a steroidal structure.[50]

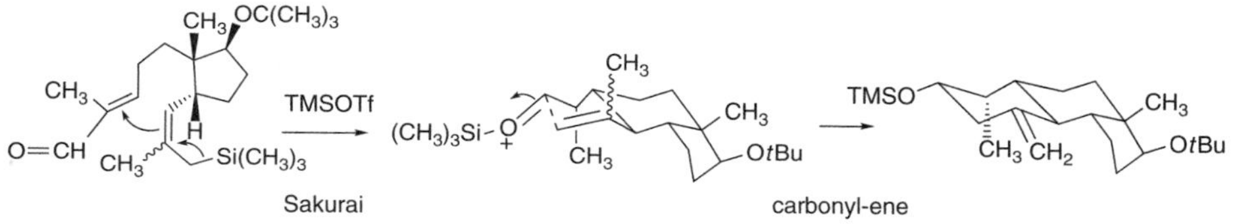

Sakurai   carbonyl-ene

Section 10.1.2.2 describes another tandem reaction sequence involving a carbonyl-ene reaction.

Scheme 10.2 gives some examples of ene and carbonyl-ene reactions. Entries 1 and 2 are thermal ene reactions. Entries 3 to 7 are intermolecular ene and carbonyl-ene reactions involving Lewis acid catalysts. Entry 3 is interesting in that it exhibits a significant preference for the terminal double bond. Entry 4 demonstrates the reactivity of methyl propynoate as an enophile. Nonterminal alkenes tend to give cyclobutenes with this reagent combination. The reaction in Entry 5 uses an acetal as the reactant, with an oxonium ion being the electrophilic intermediate.

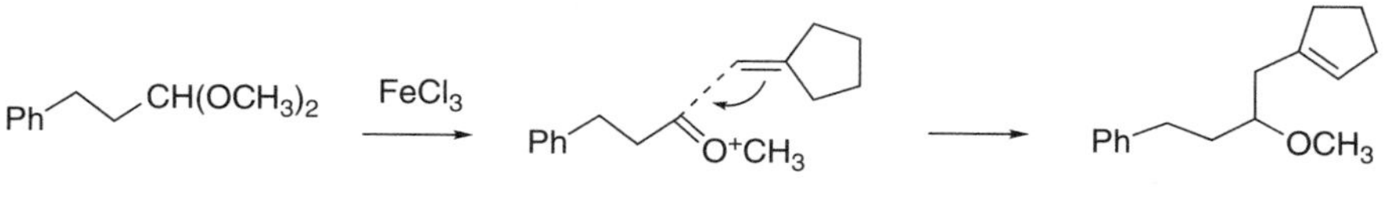

Entry 6 uses diisopropoxytitanium with racemic BINOL as the catalyst. Entry 7 shows the use of $(CH_3)_2AlCl$ with a highly substituted aromatic aldehyde. The product

[49.] D. J. Kopecky and S. D. Rychnovsky, *J. Am. Chem. Soc.*, **123**, 8420 (2001).
[50.] L. F. Tietze and M. Rischer, *Angew. Chem. Int. Ed. Engl.*, **31**, 1221 (1992).

Scheme 10.2. Ene and Carbonyl-Ene Reactions

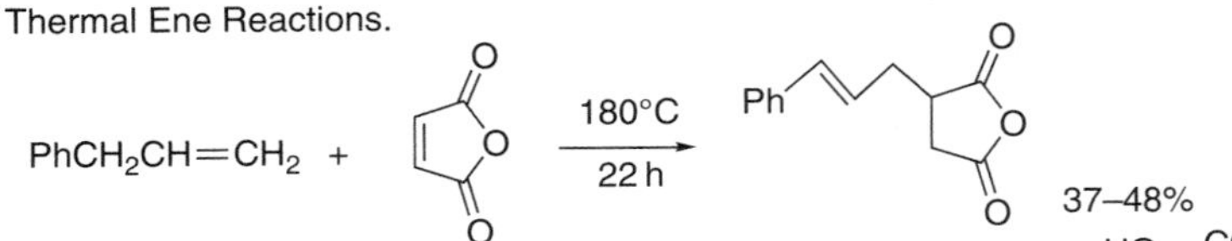

A. Thermal Ene Reactions.

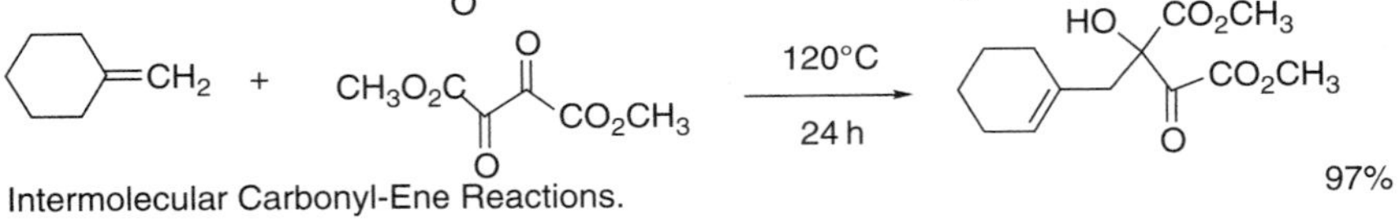

B. Intermolecular Carbonyl-Ene Reactions.

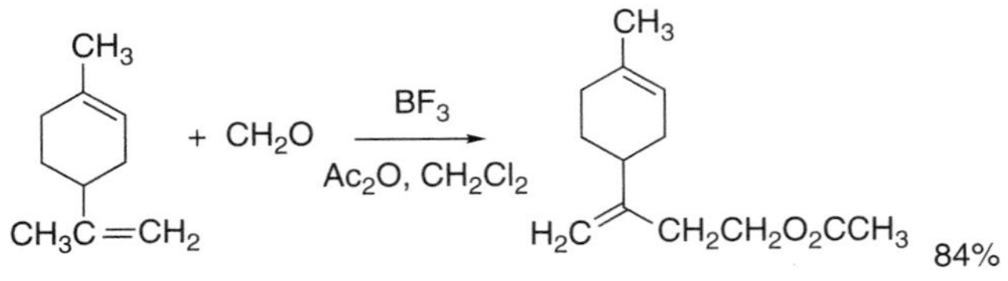

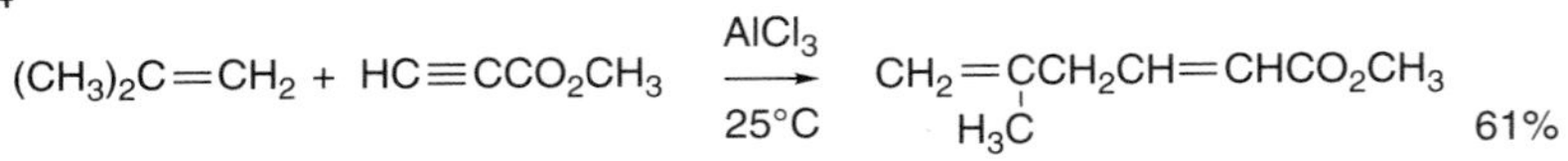

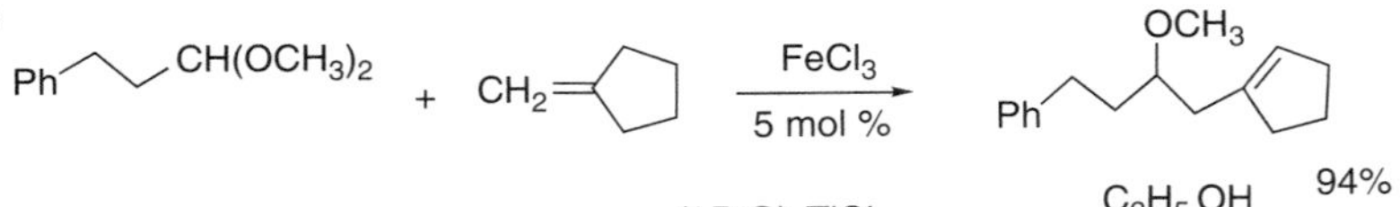

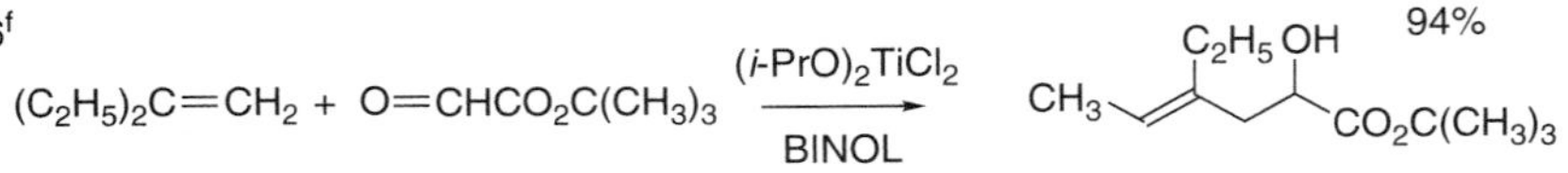

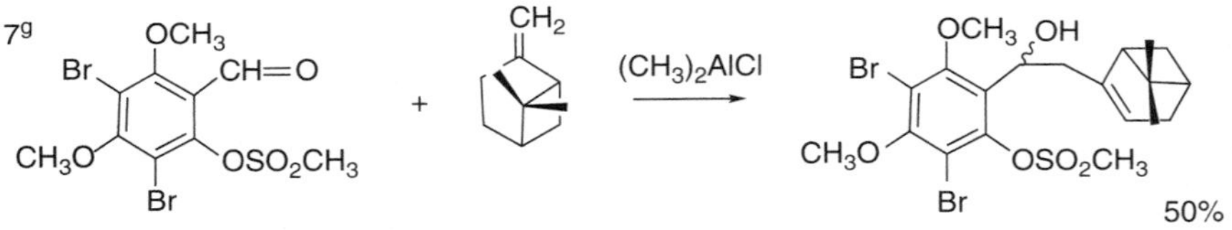

C. Intramolecular Ene Reactions.

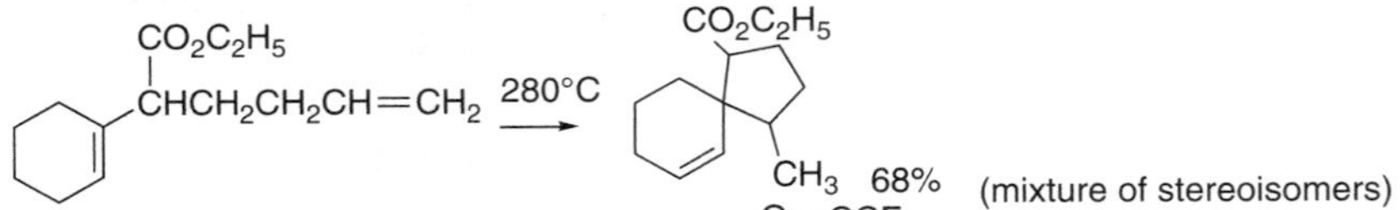

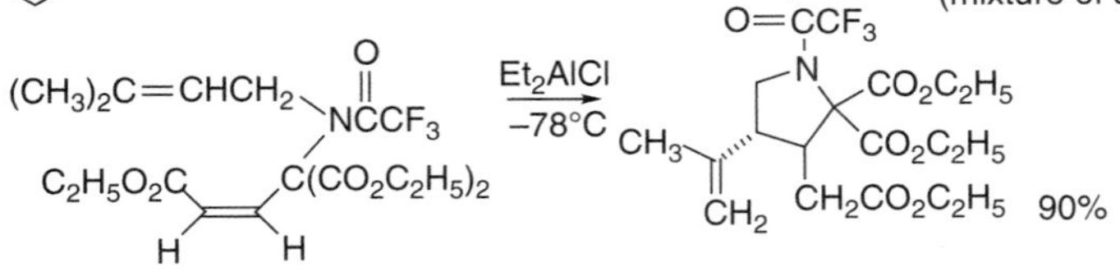

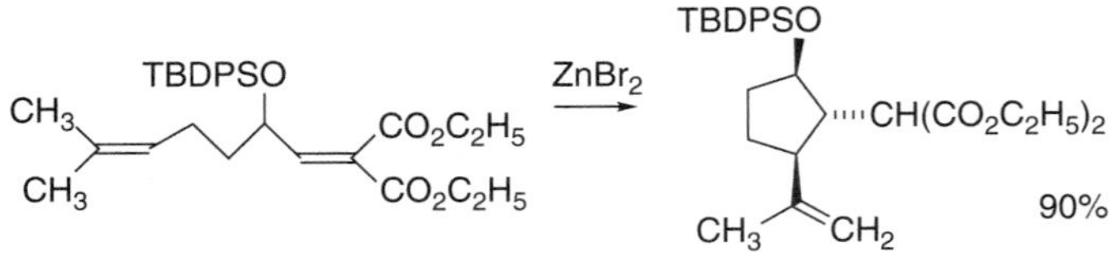

(Continued)

Scheme 10.2. (*Continued*)

879

SECTION 10.1

*Reactions and
Rearrangement
Involving Carbocation
Intermediates*

11[k]

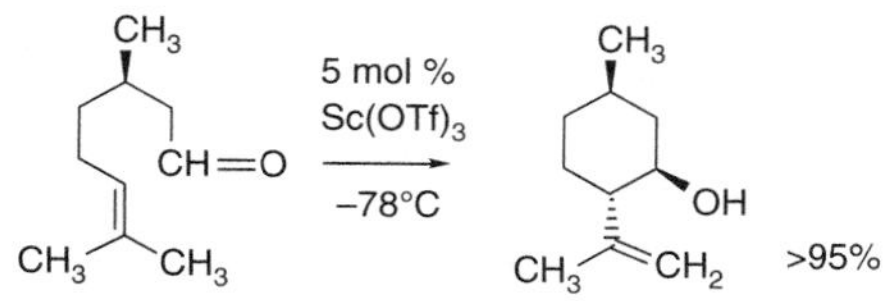

5 mol %
Sc(OTf)$_3$

−78°C

>95%

12[l]

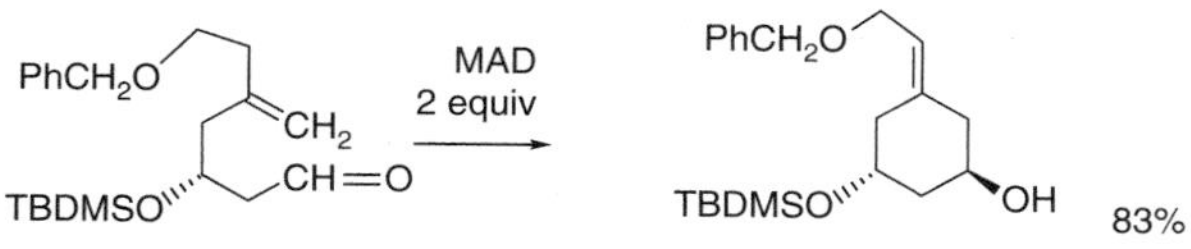

MAD
2 equiv

83%

MAD = methyl-*bis*-(2,6-di-*t*-butylphenoxy)aluminum

13[m]

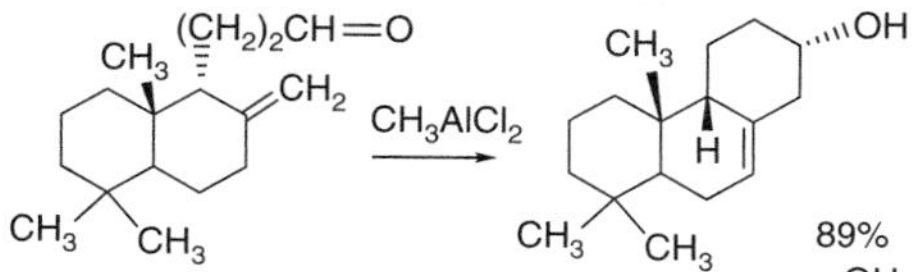

CH$_3$AlCl$_2$

89%

14[n]

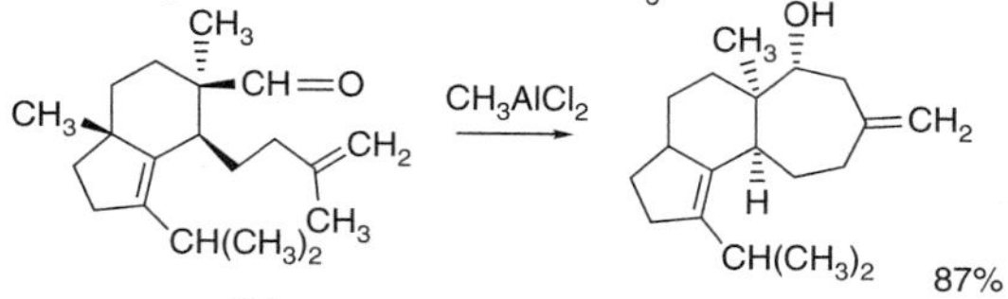

CH$_3$AlCl$_2$

87%

15[o]

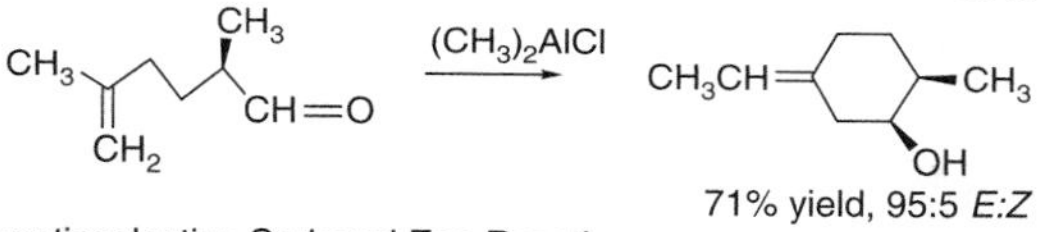

(CH$_3$)$_2$AlCl

71% yield, 95:5 *E:Z*

D. Enantioselective Carbonyl Ene Reactions.

16[p]

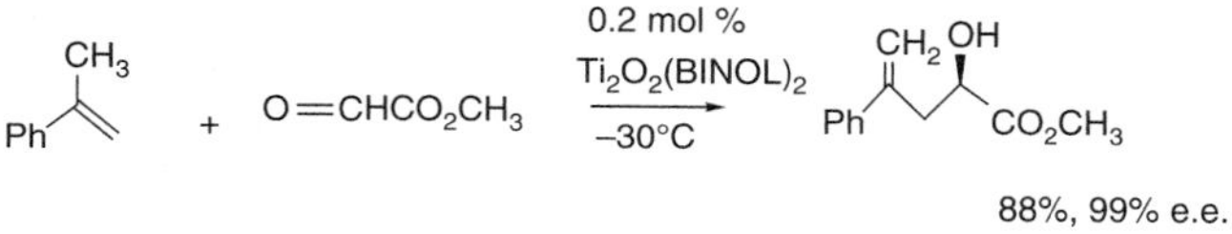

0.2 mol %
Ti$_2$O$_2$(BINOL)$_2$

−30°C

88%, 99% e.e.

17[q]

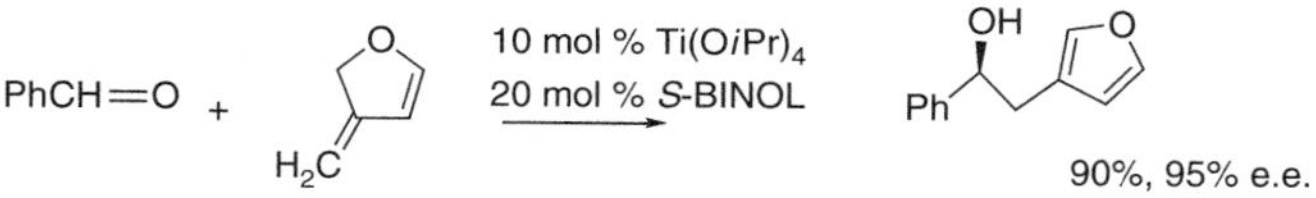

PhCH=O +

10 mol % Ti(O*i*Pr)$_4$
20 mol % *S*-BINOL

90%, 95% e.e.

18[r]

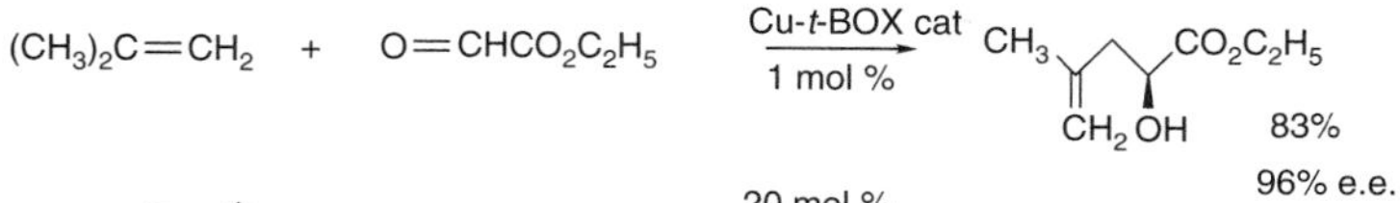

(CH$_3$)$_2$C=CH$_2$ + O=CHCO$_2$C$_2$H$_5$

Cu-*t*-BOX cat

1 mol %

83%

96% e.e.

19[s]

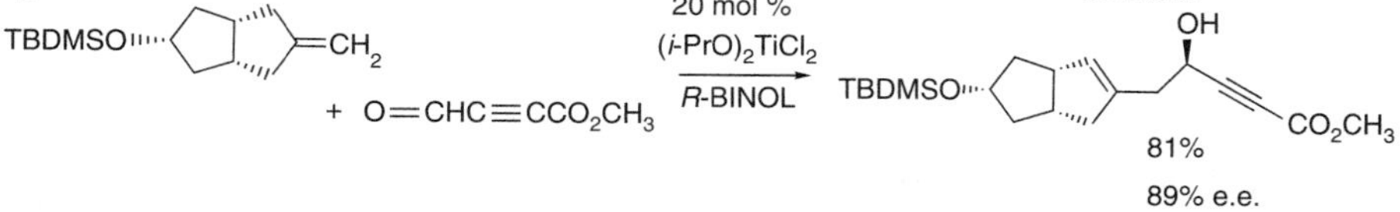

+ O=CHC≡CCO$_2$CH$_3$

20 mol %
(*i*-PrO)$_2$TiCl$_2$

R-BINOL

81%

89% e.e.

(*Continued*)

880

CHAPTER 10

*Reactions Involving
Carbocations, Carbenes,
and Radicals as Reactive
Intermediates*

Scheme 10.2. (*Continued*)

a. C. S. Rondestvedt, Jr., *Org. Synth.*, **IV**, 766 (1963).
b. P. Beak, Z. Song, and J. E. Resek, *J. Org. Chem.*, **57**, 944 (1992).
c. A. T. Blomquist and R. J. Himics, *J. Org. Chem.*, **33**, 1156 (1968).
d. B. B. Snider, D. J. Rodini, R. S. E. Conn, and S. Sealfon, *J. Am. Chem. Soc.*, **101**, 5283 (1979).
e. A. Ladepeche, E. Tam, J.-E. Arcel, and L. Ghosez, *Synthesis*, 1375 (2004).
f. M. A. Brimble and M. K. Edmonds, *Synth. Commun.*, **26**, 243 (1996).
g. M. Majewski and G. W. Bantle, *Synth. Commun.*, **20**, 2549 (1990); M. Majewski, N. M. Irvine, and G. W. Bantle, *J. Org. Chem.*, **59**, 6697 (1994).
h. W. Oppolzer, K. K. Mahalanabis, and K. Battig, *Helv. Chim. Acta*, **60**, 2388 (1977).
i. W. Oppolzer and C. Robbiani, *Helv. Chim. Acta*, **63**, 2010 (1980).
j. T. K. Sarkar, B. K. Ghorai, S. K. Nandy, B. Mukherjee, and A. Banerji, *J. Org. Chem.*, **62**, 6006 (1997).
k. V. K. Aggarwal, G. P Vennall, P. N. Davey, and C. Newman, *Tetrahedron Lett.*, **39**, 1997 (1998).
l. L. F. Courtney, M. Lange, M. R. Uskokovics, and P. M. Wovkulich, *Tetrahedron Lett.*, **39**, 3363 (1998).
m. J.-M. Weibel and D. Heissler, *Synlett*, 391 (1993).
n. B. B. Snider, N. H. Vo, and S. V. O'Neill, *J. Org. Chem.*, **63**, 4732 (1998).
o. J. A. Marshall and M. W. Andersen, *J. Org. Chem.*, **57**, 5851 (1992).
p. M. Terada and K. Mikami, *J. Chem. Soc., Chem. Commun.*, 833 (1994).
q. W. H. Miles, E. J. Fialcowitz, and E. S. Halstead, *Tetrahedron*, **57**, 9925 (2001).
r. D. A. Evans, S. W. Tregay, C. S. Burgey, N. A. Paras, and T. Vojkovsky, *J. Am. Chem. Soc.*, **122**, 7936 (2000).
s. K. Mikami, A. Yoshida, and Y. Matsumoto, *Tetrahedron Lett.*, **37**, 8515 (1996).

was used in syntheses of derivatives of robustadial, which are natural products from *Eucalyptus* that have antimalarial activity.

Entries 8 to 15 are examples of intramolecular reactions. Entry 8 involves two unactivated double bonds and was carried out at a temperature of 280°C. The product was a mixture of epimers at the ester site but the methyl group and cyclohexenyl double bond are *cis*, which indicates that the reaction occurred entirely through an *endo* TS.

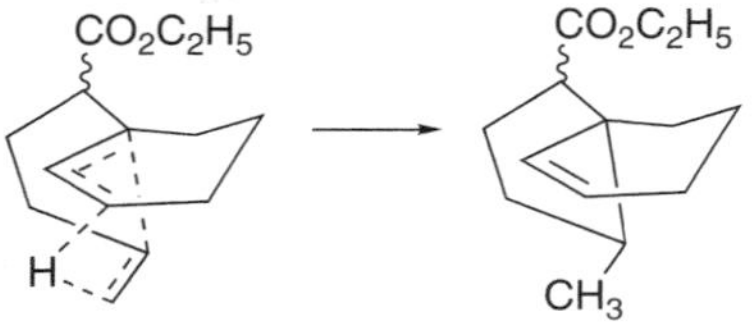

The reaction in Entry 9 was completely stereospecific. The corresponding *E*-isomer gave mainly the *cis* isomer. These results are consistent with a cyclic TS for the hydrogen transfer.

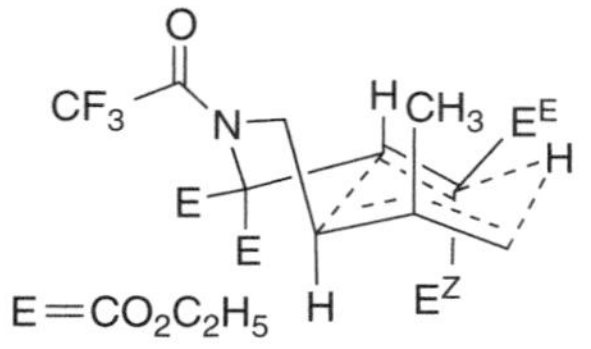

The stereoselectivity of the reaction in Entry 10 is also consistent with a TS in which the hydrogen is transferred through a chairlike TS.

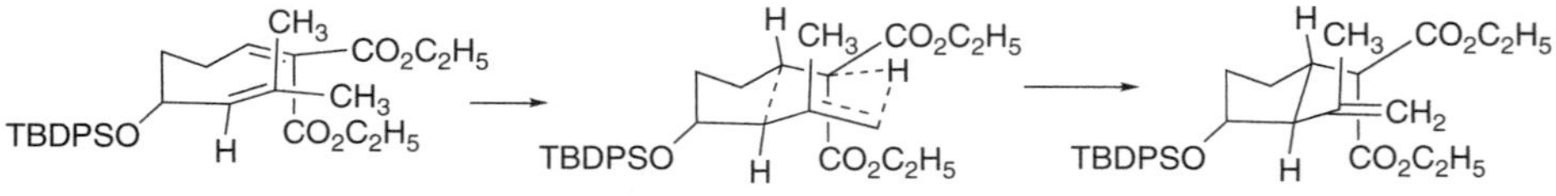

Entry 11 illustrates the facility of a $Sc(OTf)_3$-mediated reaction. The catalyst in Entry 12 is a hindered *bis*-phenoxyaluminum compound. The proton removal in Entry 12 is highly stereoselective, giving rise to a single exocyclic double-bond isomer. This stereochemistry is consistent with a TS that incorporates the six-membered hydrogen transfer TS into a bicyclic framework.

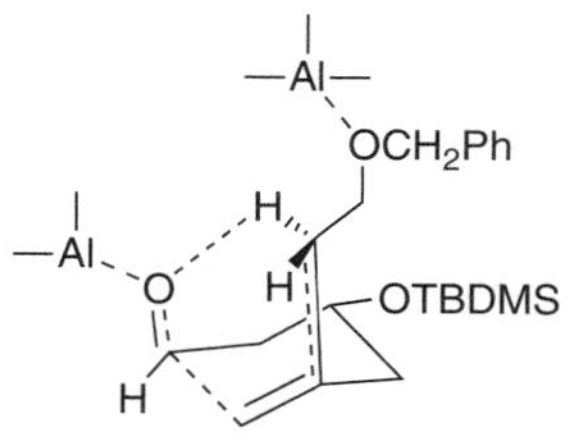

Entries 13 to 15 are examples of high-yield cyclizations of aldehydes effected by CH_3AlCl_2.

Section D of Scheme 10.2 shows some enantioselective reactions. Entry 16 illustrates the enantioselective reaction of methyl glyoxylate with a simple alkene. The catalyst is a dioxido-bridged dimer of Ti(BINOL) prepared azeotropically from BINOL and $TiCl_2(O\text{-}i\text{-Pr})_2$. Entry 17 also uses a Ti(BINOL) catalyst. The methylenedihydrofuran substrate is highly reactive owing to the donor effect of the vinyl ether and the stabilization provided by formation of the aromatic furan ring. Entry 18 shows the use of a Cu-BOX catalysts to achieve a highly enantioselective reaction between isobutene and ethyl glyoxylate. The reaction in Entry 19 was done with a $(i\text{-PrO})_2TiCl_2\text{-}(R)$-BINOL and the product had an e.e. of 89%.

10.1.1.4. Reactions with Acylium Ions. Alkenes react with acyl halides or acid anhydrides in the presence of a Lewis acid catalyst to give β,γ-unsaturated ketones. The reactions generally work better with cyclic than acyclic alkenes.

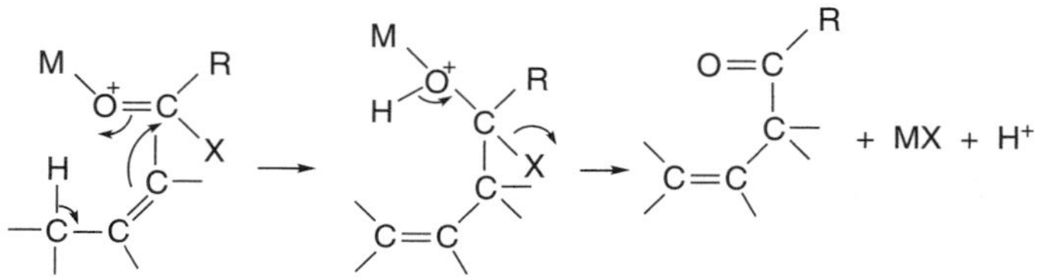

It has been suggested that the kinetic preference for formation of β,γ-unsaturated ketones results from an intramolecular deprotonation, as shown in the mechanism above.[51] The carbonyl-ene and alkene acylation reactions have several similarities. Both reactions occur most effectively in intramolecular circumstances and provide a useful method for ring closure. Although both reactions appear to occur through highly polarized TSs, there is a strong tendency toward specificity in the proton abstraction step. This specificity and other similarities in the reaction are consistent with a cyclic formulation of the mechanism.

A variety of reaction conditions have been examined for acylation of alkenes by acyl chlorides. With the use of Lewis acid catalysts, reaction typically occurs

[51] P. Beak and K. R. Berger, *J. Am. Chem. Soc.*, **102**, 3848 (1980).

882

CHAPTER 10

*Reactions Involving
Carbocations, Carbenes,
and Radicals as Reactive
Intermediates*

to give both β,γ-enones and β-haloketones.[52] One of the more effective catalysts is ethylaluminum dichloride.[53]

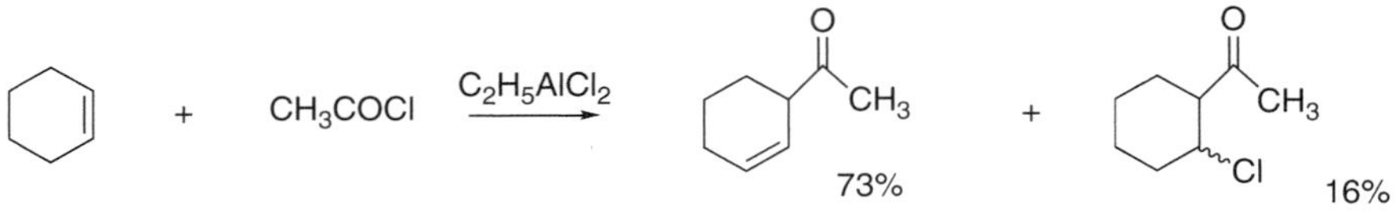

73%   16%

Zinc chloride also gives good results, especially with cyclic alkenes.[51]

A similar reaction occurs between alkenes and acylium ions, as in the reaction between 2-methylpropene, and the acetylium ion leads regiospecifically to β,γ-enones.[54] A concerted mechanism has been suggested to account for this regiochemical preference.

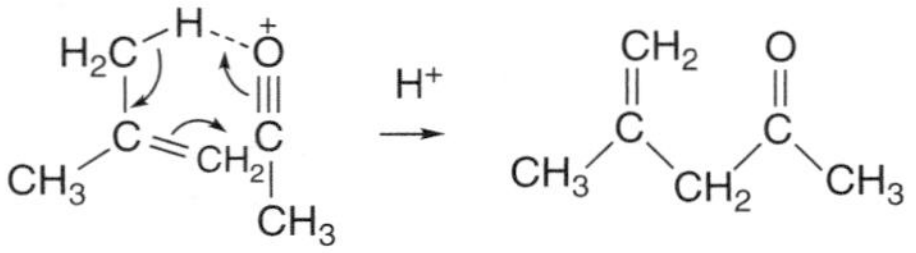

Highly reactive mixed anhydrides can also promote acylation. Phenylacetic acid reacts with alkenes to give 2-tetralones in TFAA-H_3PO_4.[55] This reaction involves an intramolecular Friedel-Crafts alkylation subsequent to the acylation.

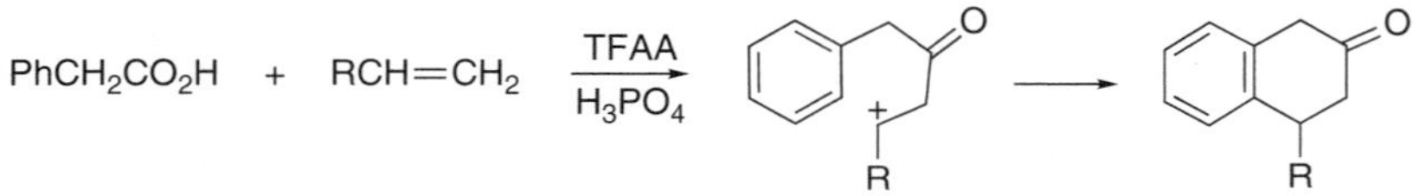

The acylation reaction has been most synthetically useful in intramolecular reactions. The following examples are illustrative.

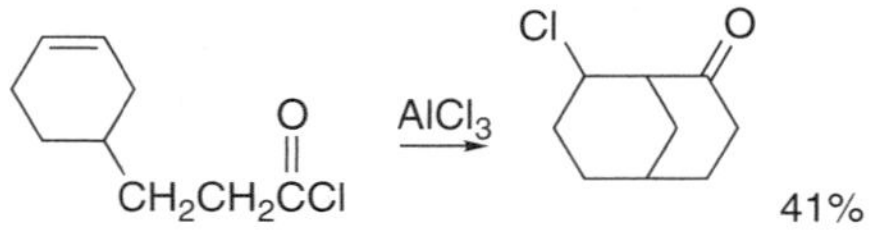

41%   Ref. 56

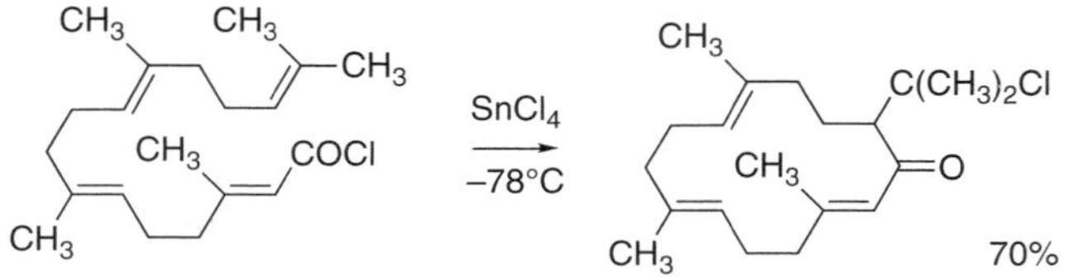

70%   Ref. 57

52. See, e.g., T. S. Cantrell, J. M. Harless, and B. L. Strasser, *J. Org. Chem.*, **36**, 1191 (1971); L. Rand and R. J. Dolinski, *J. Org. Chem.*, **31**, 3063 (1966).
53. B. B. Snider and A. C. Jackson, *J. Org. Chem.*, **47**, 5393 (1982).
54. H. M. R. Hoffmann and T. Tsushima, *J. Am. Chem. Soc.*, **99**, 6008 (1977).
55. A. D. Gray and T. P. Smyth, *J. Org. Chem.*, **66**, 7113 (2001).
56. E. N. Marvell, R. S. Knutson, T. McEwen, D. Sturmer, W. Federici, and K. Salisbury, *J. Org. Chem.*, **35**, 391 (1970).
57. T. Kato, M. Suzuki, T. Kobayashi, and B. P. Moore, *J. Org. Chem.*, **45**, 1126 (1980).

Several successful cyclizations of quite complex structures were achieved using polyphosphoric acid trimethylsilyl ester, a viscous material that contains reactive anhydrides of phosphoric acid.[58] Presumably the reactive acylating agent is a mixed phosphoric anhydride of the carboxylic acid.

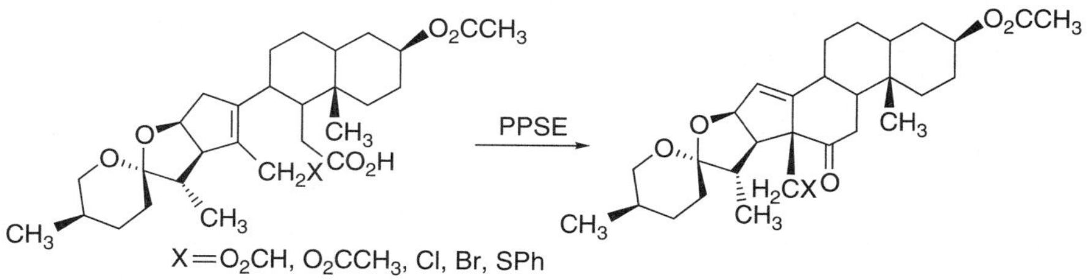

X=O₂CH, O₂CCH₃, Cl, Br, SPh

Ref. 59

10.1.2. Rearrangement of Carbocations

Carbocations, as we learned in Chapter 4 of Part A, can readily rearrange to more stable isomers. To be useful in synthesis, such reactions must be controlled and predictable. This goal can be achieved on the basis of substituent effects and stereoelectronic factors. Among the most important rearrangements in synthesis are those directed by oxygen substituents, which can provide predictable outcomes on the basis of electronic and stereoelectronic factors.

10.1.2.1. Pinacol Rearrangement. Carbocations can be stabilized by the migration of hydrogen, alkyl, alkenyl, or aryl groups, and, occasionally, even functional groups can migrate. A mechanistic discussion of these reactions is given in Section 4.4.4 of Part A. Reactions involving carbocation rearrangements can be complicated by the existence of competing rearrangement pathways. Rearrangements can be highly selective and, therefore, reliable synthetic reactions when the structural situation is such as to strongly favor a particular reaction path. One example is the reaction of carbocations having a hydroxy group on an adjacent carbon, which leads to the formation of a carbonyl group.

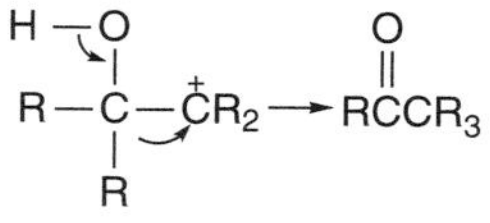

A reaction that follows this pattern is the acid-catalyzed conversion of diols to ketones, which is known as the *pinacol rearrangement*.[60] The classic example of this reaction is the conversion of 2,3-dimethylbutane-2,3-diol(pinacol) to methyl *t*-butyl ketone (pinacolone).[61]

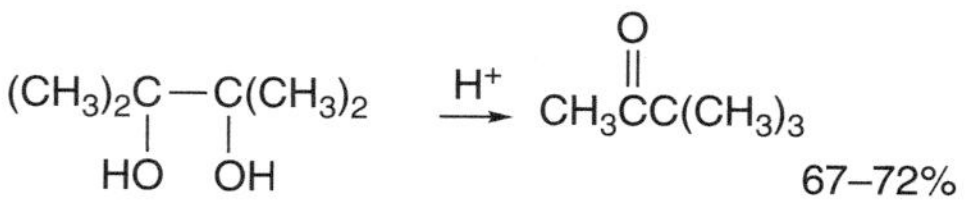

67–72%

58. K. Yamamoto and H. Watanabe, *Chem. Lett.*, 1225 (1982).
59. W. Li and P. L. Fuchs, *Org. Lett.*, **5**, 4061 (2003).
60. C. J. Collins, *Q. Rev.*, **14**, 357 (1960).
61. G. A. Hill and E. W. Flosdorf, *Org. Synth.*, **I**, 451 (1932).

884

CHAPTER 10

*Reactions Involving
Carbocations, Carbenes,
and Radicals as Reactive
Intermediates*

The acid-catalyzed mechanism involves carbocation formation and substituent migration assisted by the hydroxy group.

Under acidic conditions, the more easily ionized $C-O$ bond generates the carbocation, and migration of one of the groups from the adjacent carbon ensues. Both stereochemistry and "migratory aptitude" are factors in determining the extent of migration of the different groups. The issue of the electronic component in migratory aptitude has been examined by calculating (MP2/6-31G*) the relative energy for several common groups in a prototypical TS for migration. The order is vinyl > cyclopropyl > alkynyl > methyl ~ hydrogen.[62] The tendency for migration of alkenyl groups is further enhanced by ERG substituents and selective migration of trimethylsilyl-substituted groups has been exploited in pinacol rearrangements.[63] In the example shown, the triethylsilane serves to reduce the intermediate silyloxonium ion and generate a primary alcohol.

Another method for achieving selective pinacol rearrangement involves synthesis of a glycol monosulfonate ester. These compounds rearrange under the influence of base.

Rearrangements of monosulfonates permit greater control over the course of the rearrangement because ionization occurs only at the sulfonylated alcohol. These reactions have been of value in the synthesis of ring systems, especially terpenes, as illustrated by Entries 3 and 4 in Scheme 10.3.

In cyclic systems that enforce structural rigidity or conformational bias, the course of the rearrangement is controlled by stereoelectronic factors. The carbon substituent that is *anti* to the leaving group is the one that undergoes migration. In cyclic systems such as **8**, for example, selective migration of the ring fusion bond occurs because

[62] K. Nakamura and Y. Osamura, *J. Am. Chem. Soc.*, **115**, 9112 (1993).

[63] K. Suzuki, T. Ohkuma, and G. Tsuchihashi, *Tetrahedron Lett.*, **26**, 861 (1985); K. Suzuki, M. Shimazaki, and G. Tsuchihashi, *Tetrahedron Lett.*, **27**, 6233 (1986); M. Shimazaki, M. Morimoto, and K. Suzuki, *Tetrahedron Lett.*, **31**, 3335 (1990).

of this stereoelectronic effect. In both cyclic and acyclic systems, the rearrangement takes place with *retention of configuration* at the migrating group.

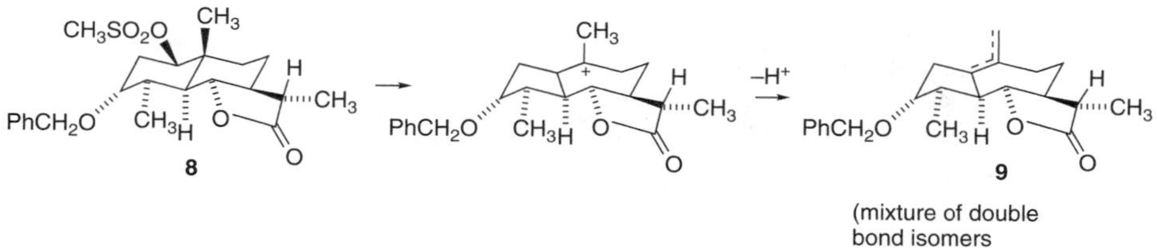

8

9

(mixture of double bond isomers)

Ref. 64

Similarly, **10** gives **11** by antiperiplanar migration.

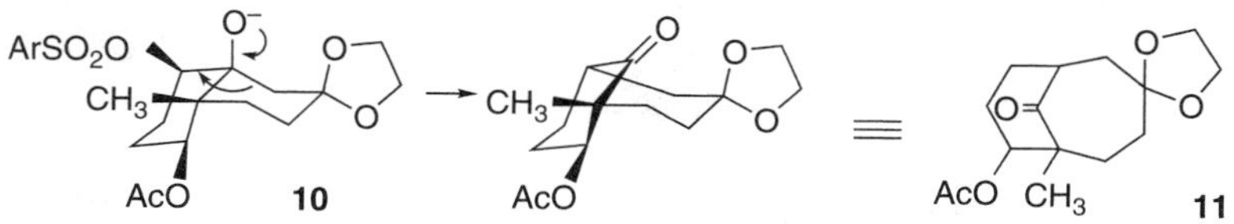

10   **11**

Ref. 65

Rearrangement of diol monosulfonates can also be done using Lewis acids. These conditions lead to *inversion of configuration* at the migration terminus, as would be implied by a concerted mechanism.[66]

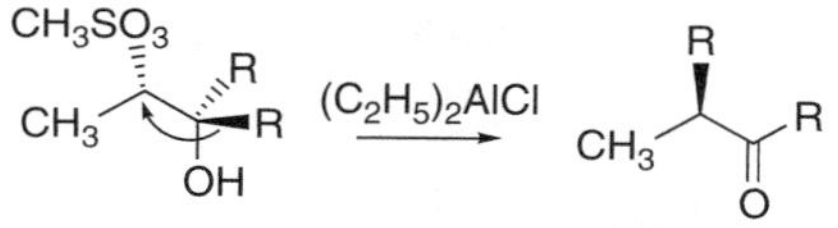

Triethylaluminum is also effective in catalyzing rearrangement of monosulfonate with high stereospecificity. The reactions are believed to proceed through a cyclic TS.[67]

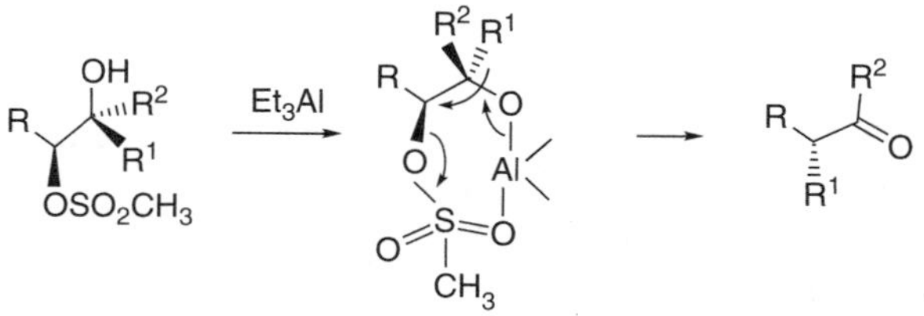

The reactants can be prepared by chelation-controlled addition of organometallic reagents to α-(1-ethoxyethoxy)methyl ketones. Selective sulfonylation occurs at the

64. M. Ando, A. Akahane, H. Yamaoka, and K. Takase, *J. Org. Chem.*, **47**, 3909 (1982).
65. C. H. Heathcock, E. G. Del Mar, and S. L. Graham, *J. Am. Chem. Soc.*, **104**, 1907 (1982).
66. G. Tsuchihashi, K. Tomooka, and K. Suzuki, *Tetrahedron Lett.*, **25**, 4253 (1984).
67. K. Suzuki, E. Katayama, and G. Tsuchihashi, *Tetrahedron Lett.*, **24**, 4997 (1983); K. Suzuki, E. Katayama, and G. Tsuchihashi, *Tetrahedron Lett.*, **25**, 1817 (1984); T. Shinohara and K. Suzuki, *Synthesis*, 141 (2003).

886

CHAPTER 10

*Reactions Involving
Carbocations, Carbenes,
and Radicals as Reactive
Intermediates*

less hindered secondary hydroxy group. The rearranged ketones were obtained in greater than 99% e.e.

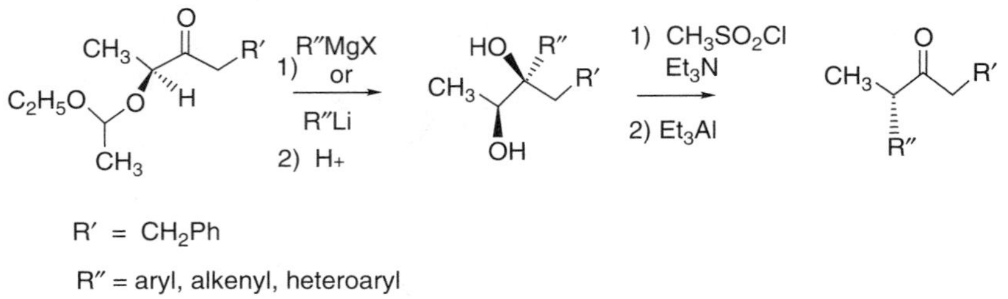

R′ = CH₂Ph

R″ = aryl, alkenyl, heteroaryl

A related method was applied in the course of synthesis of a precursor of a macrolide antibiotic, protomycinolide IV. The migrating group was an α-trimethylsilylalkenyl group.[68] In this procedure, the DiBAlH first reduces the ketone and then, after rearrangement, reduces the aldehyde to a primary alcohol.

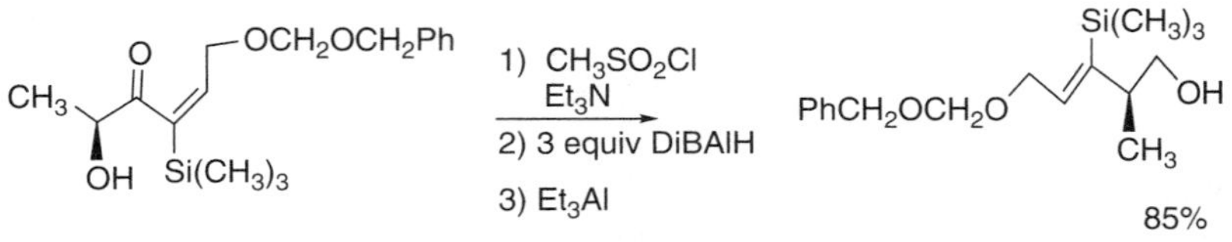

Stereospecfic ring expansion can be done by taking advantage of the hydroxy-directed epoxidation and SnCl₄-mediated rearrangement of 1-hydroxycycloalkyl epoxides.[69]

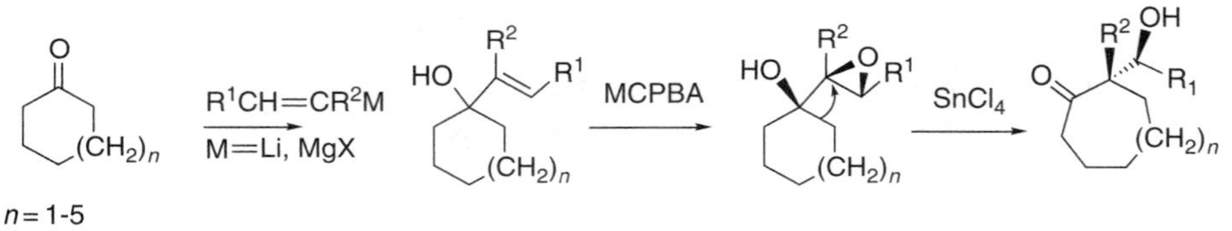

The overall transformation of this sequence corresponds to the aldol addition of an aldehyde with a cyclic ketone. The actual aldol addition frequently proceeds with low stereocontrol, so this sequence constitutes a method for stereoselective synthesis of the aldol adducts. The reaction has been done with several Lewis acids, including SnCl₄, BF₃, and Ti(O-*i*-Pr)₃Cl.

10.1.2.2. Pinacol Rearrangement in Tandem with the Carbonyl-Ene Reaction.
Overman and co-workers have developed protocols in which pinacol rearrangement

68. K. Suzuki, K. Tomooka, E. Katayama, T. Matsumoto, and G. Tsuchihashi, *J. Am. Chem. Soc.*, **108**, 5221 (1986).
69. S. W. Baldwin, P. Chen, N. Nikolic, and D. C. Weinseimer, *Org. Lett.*, **2**, 1193 (2000); C. M. Marson, A. Khan, R. A. Porter, and A. J. A. Cobb, *Tetrahedron Lett.*, **43**, 6637 (2002).

occurs in tandem with a carbonyl-ene reaction and results in both a ring closure and ring expansion.[70]

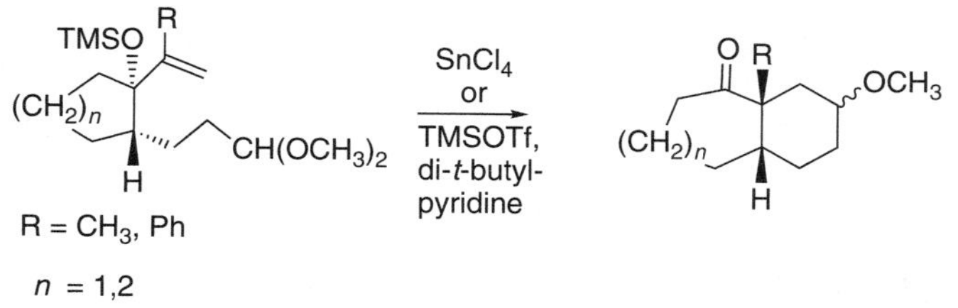

R = CH$_3$, Ph

n = 1,2

These reactions appear to proceed through the sequence **C → D → E** . When the seven-membered analog (*n* = 3) reacts, two products are formed. The more flexible seven-membered ring accommodates the competing sequence. **F → G → H.**

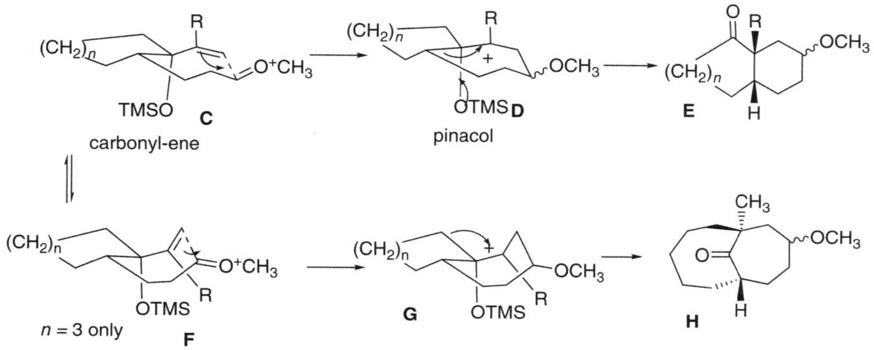

The carbonyl-ene–pinacol sequence has also been observed in reactions leading to the formation of tetrahydrofurans.[71]

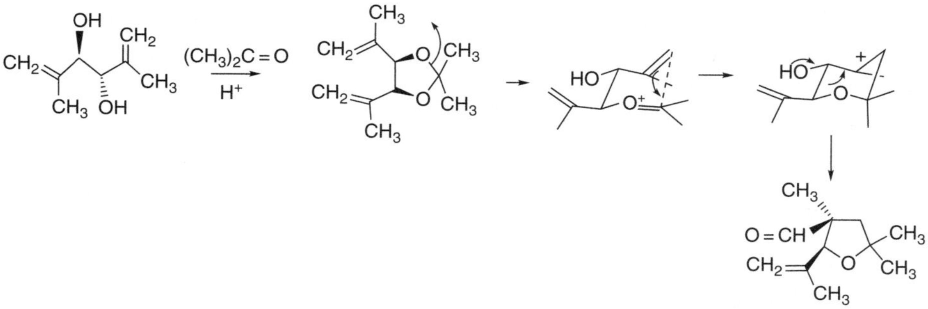

The reaction has been developed for the synthesis of both oxygen heterocycles and carbocyclic compounds.[72]

70. S. Ando, K. P. Minor, and L. E. Overman, *J. Org. Chem.*, **62**, 6379 (1997).

71. P. Martinet and G. Moussel, *Bull. Soc. Chim. Fr.*, 4093 (1971); C. M. Gasparski, P. M. Herrinton, L. E. Overman, and J. P. Wolfe, *Tetrahedron Lett.*, **41**, 9431 (2000).

72. L. E. Overman, *Acc. Chem. Res.*, **25**, 352 (1992); L. E. Overman and L. D. Pennington, *J. Org. Chem.*, **68**, 7143 (2003).

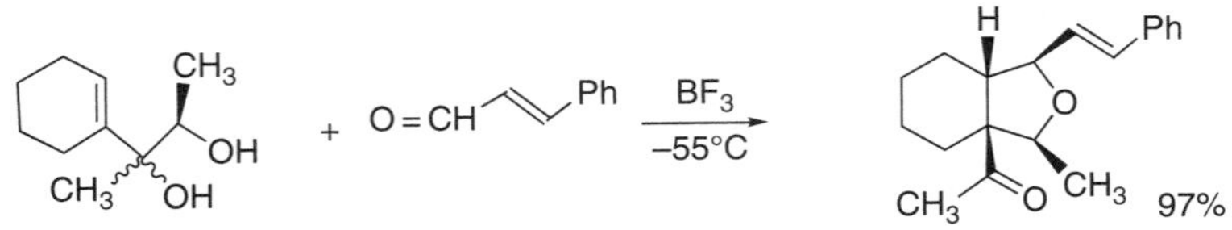

888

CHAPTER 10

*Reactions Involving
Carbocations, Carbenes,
and Radicals as Reactive
Intermediates*

Ref. 73

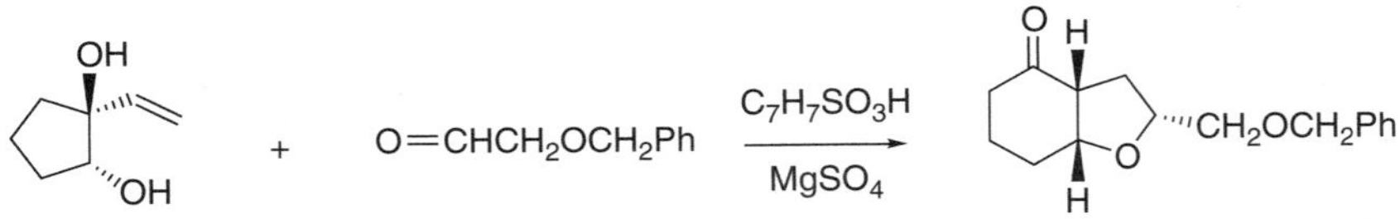

Ref. 74

These reactions can also be adapted to carbocyclic ring formation and expansion.

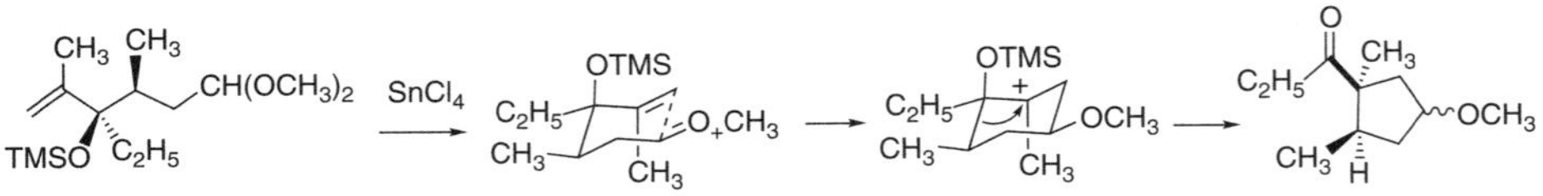

Ref. 75

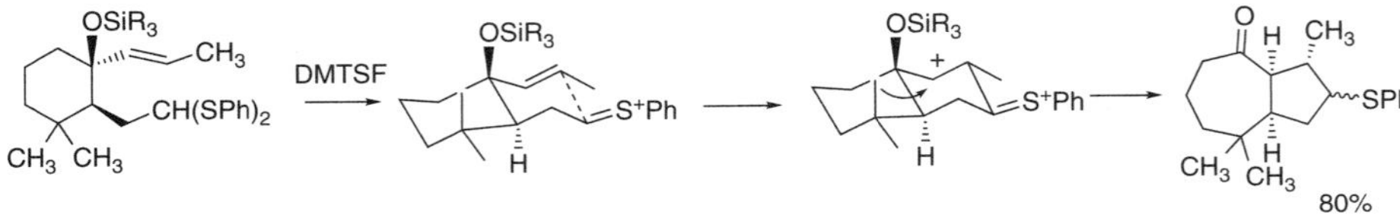

Ref. 76

Scheme 10.3 gives some examples of pinacol and related rearrangements. Entry 1 is a rearrangement done under strongly acidic conditions. The selectivity leading to ring expansion results from the preferential ionization of the diphenylcarbinol group. Entry 2, a preparation of 2-indanone, involves selective ionization at the benzylic alcohol, followed by a hydride shift.

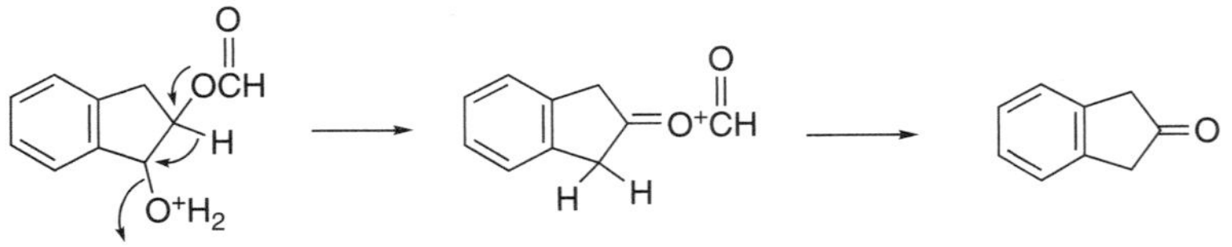

Entries 3 and 4 are examples of stereospecific *anti* migrations governed by the stereochemistry of the sulfonate leaving group. These transformations are parts of synthetic schemes that use available terpene starting materials for synthesis of more complex natural products. The ring expansion in Entry 5 was used to form an eight-membered ring found in certain diterpenes. This highly efficient and selective rearrangement

73. D. W. C. MacMillan, L. E. Overman, and L. D. Pennington, *J. Am. Chem. Soc.*, **123**, 9033 (2001).

74. M. J. Brown, T. Harrison, P. M. Herrinton, M. H. Hopkins, K. D. Hutchinson, P. Mishra, and L. E. Overman, *J. Am. Chem. Soc.*, **113**, 5365 (1991).

75. T. C. Gahman and L. E. Overman, *Tetrahedron*, **58**, 6473 (2002).

76. A. D. Lebsack, L. E. Overman, and R. J. Valentekovich, *J. Am. Chem. Soc.*, **123**, 4851 (2001).

889

SECTION 10.1

*Reactions and
Rearrangement
Involving Carbocation
Intermediates*

Scheme 10.3. Rearrangements Promoted by Adjacent Heteroatoms

A. Pinacol-type rearrangements

1[a]

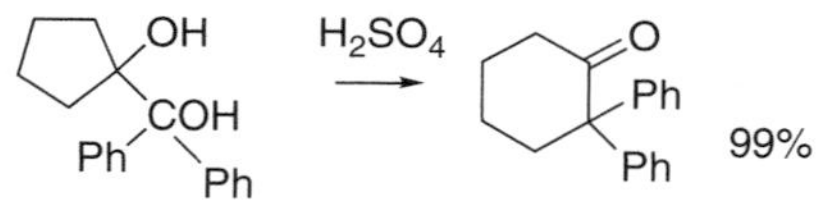

99%

2[b]

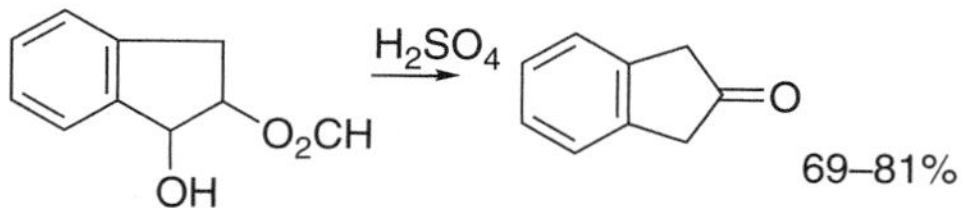

69–81%

3[c]

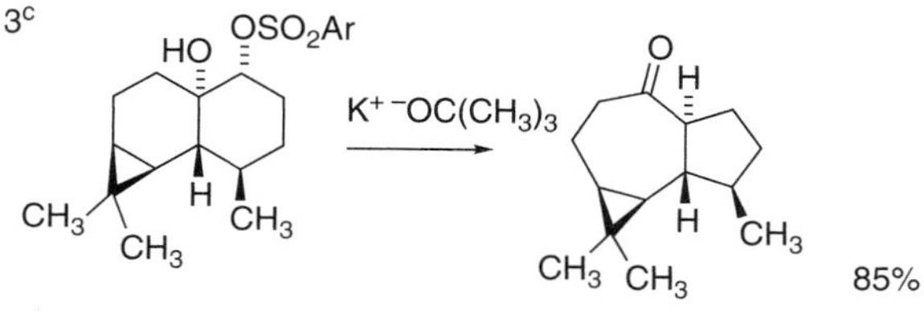

85%

4[d]

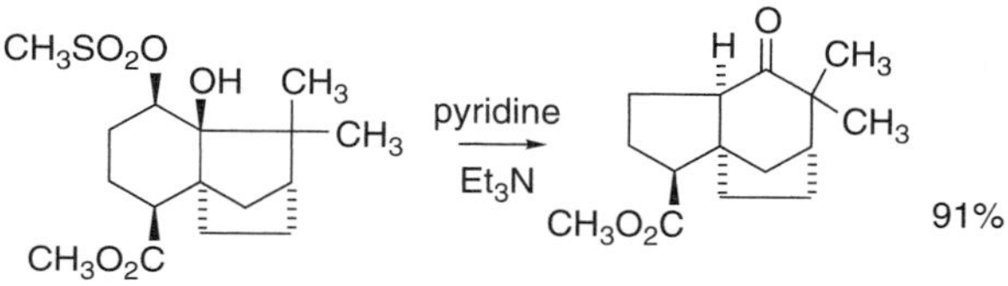

91%

5[e]

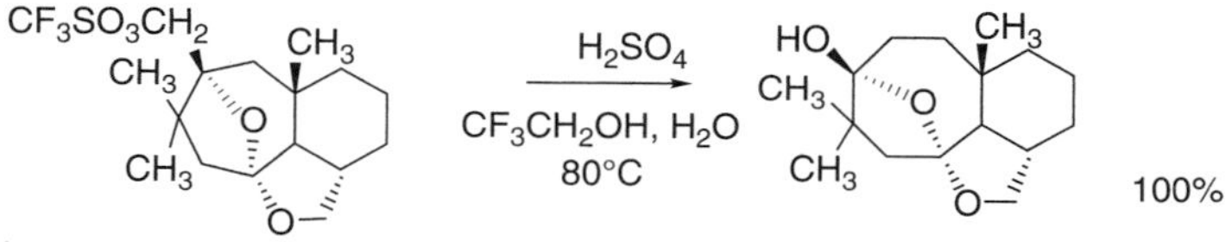

100%

6[f]

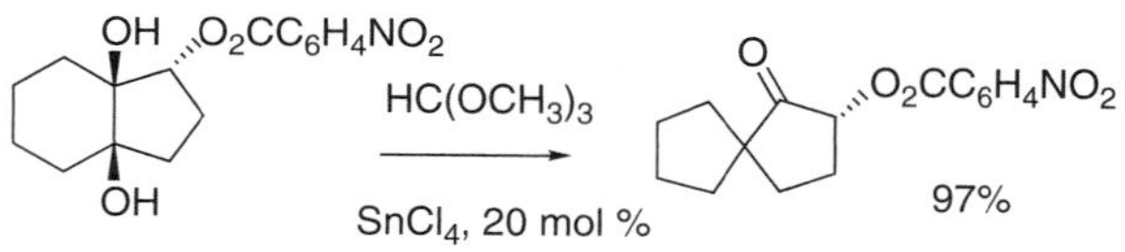

97%

7[g]

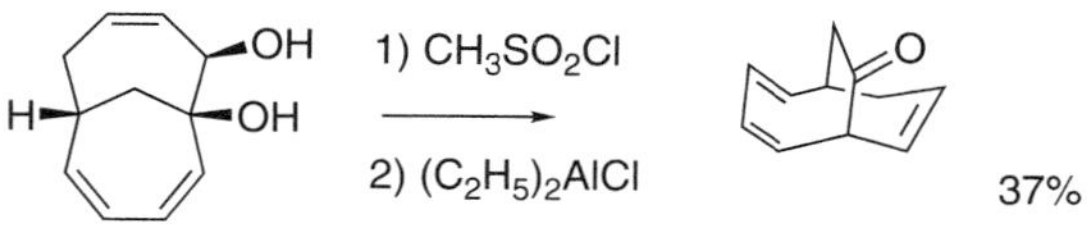

37%

8[h]

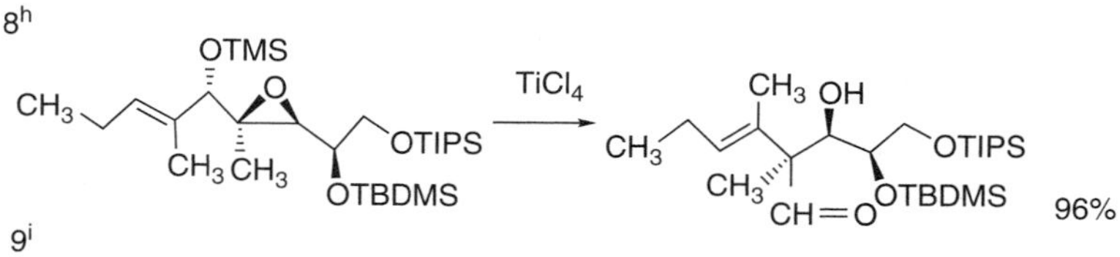

96%

9[i]

63%

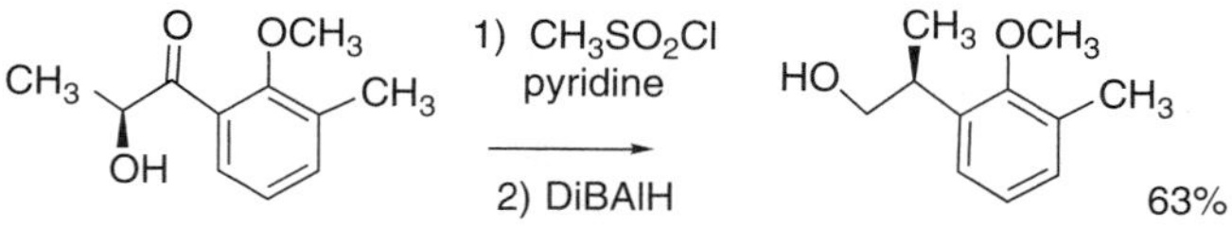

(*Continued*)

Scheme 10.3. (*Continued*)

a. H. E. Zaugg, M. Freifelder, and B. W. Horrom, *J. Org. Chem.*, **15**, 1191 (1950).
b. J. E. Horan and R. W. Schliessler, *Org. Synth.*, **41**, 53 (1961).
c. G. Buchi, W. Hofheinz, and J. V. Paukstelis, *J. Am. Chem. Soc.*, **91**, 6473 (1969).
d. D. F. MacSweeney and R. Ramage, *Tetrahedron*, **27**, 1481 (1971).
e. P. Magnus, C. Diorazio, T. J. Donohoe, M. Giles, P. Pye, J. Tarrant, and S. Thom, *Tetrahedron*, **52**, 14147 (1996).
f. Y. Kita, Y. Yoshida, S. Mihara, D.-F. Fang, K. Higuchi, A. Furukawa, and H. Fujioka, *Tetrahedron Lett.*, **38**, 8315 (1997).
g. J. H. Rigby and K. R. Fales, *Tetrahedron Lett.*, **39**, 1525 (1998).
h. K. D. Eom, J. V. Raman, H. Kim, and J. K. Cha, *J. Am. Chem. Soc.*, **125**, 5415 (2003).
i. H. Arimoto, K. Nishimura, M. Kuramoto, and D. Uemura, *Tetrahedron Lett.*, **39**, 9513 (1998).

presumably proceeds with participation of the adjacent oxygen, which accounts for the specific migration of bond *a* over bond *b*.

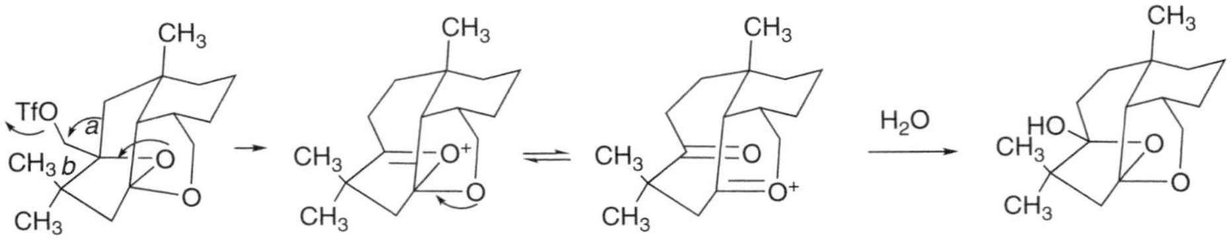

Entry 6 illustrates a significant regioselectivity in that two tertiary alcohol groups are present in the reactant. This reaction is thought to involve a cyclic orthoester. The preferred rupture of the C—O bond distal to the *p*-nitrobenzoyloxy group is likely due to the dipolar effect of the C—O bond on ionization. No migration of the oxy-substituted ring is observed, indicating that the *p*-nitrobenzoyloxy group minimizes any potential electron donation by the oxygen.

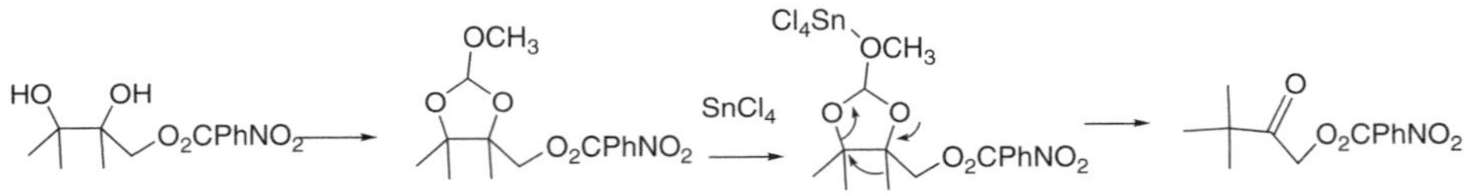

Entry 7 involves formation and ionization of a secondary allylic sulfonate and migration of a dienyl group.

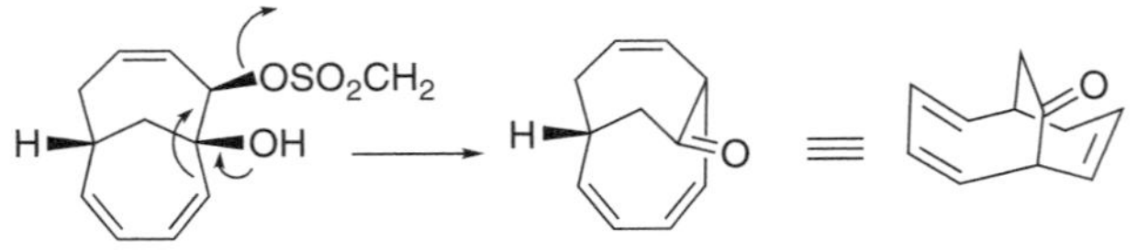

Entry 8 involves a migration initiated by epoxide ring opening. This reaction involves migration of a vinyl substituent. Entry 9 is a stereospecific migration of the aryl group. The DiBAlH both promotes the rearrangement and reduces the product aldehyde.

10.1.2.3. Rearrangements Involving Diazonium Ions. Aminomethyl carbinols yield ketones when treated with nitrous acid. The reaction proceeds by formation and rearrangement of diazonium ions. The diazotization reaction generates the same type of β-hydroxycarbocation that is involved in the pinacol rearrangement.

891

SECTION 10.1

*Reactions and
Rearrangement
Involving Carbocation
Intermediates*

This reaction has been used to form ring-expanded cyclic ketones, a procedure known as the *Tiffeneau-Demjanov reaction.*[77]

Ref. 78

The reaction of ketones with diazomethane sometimes leads to a ring-expanded ketone in synthetically useful yields.[79] The reaction occurs by addition of the diazomethane, followed by elimination of nitrogen and migration.

The rearrangement proceeds via essentially the same intermediate that is involved in the Tiffeneau-Demjanov reaction. Since the product is also a ketone, subsequent addition of diazomethane can lead to higher homologs. The best yields are obtained when the starting ketone is substantially more reactive than the product. For this reason, strained ketones work especially well. Higher diazoalkanes can also be used in place of diazomethane. The reaction is found to be accelerated by alcoholic solvents. This effect probably involves the hydroxy group being hydrogen bonded to the carbonyl oxygen and serving as a proton donor in the addition step.[80]

Trimethylaluminum also promotes ring expansion by diazoalkanes.[81]

Ketones react with esters of diazoacetic acid in the presence of Lewis acids such as BF_3 and $SbCl_5$.[82]

77. P. A. S. Smith and D. R. Baer, *Org. React.*, **11**, 157 (1960).

78. F. F. Blicke, J. Azuara, N. J. Dorrenbos, and E. B. Hotelling, *J. Am. Chem. Soc.*, **75**, 5418 (1953).

79. C. D. Gutsche, *Org. React.*, **8**, 364 (1954).

80. J. N. Bradley, G. W. Cowell, and A. Ledwith, *J. Chem. Soc.*, 4334 (1964).

81. K. Maruoka, A. B. Concepcion, and H. Yamamoto, *J. Org. Chem.*, **59**, 4725 (1994).

82. H. J. Liu and T. Ogino, *Tetrahedron Lett.*, 4937 (1973); W. T. Tai and E. W. Warnhoff, *Can. J. Chem.*, **42**, 1333 (1964); W. L. Mock and M. E. Hartman, *J. Org. Chem.*, **42**, 459 (1977); V. Dave and E. W. Warnhoff, *J. Org. Chem.*, **48**, 2590 (1983).

892

CHAPTER 10

*Reactions Involving
Carbocations, Carbenes,
and Radicals as Reactive
Intermediates*

These reactions involve addition of the diazo ester to an adduct of the carbonyl compound and the Lewis acid. Elimination of nitrogen then triggers migration. Triethyloxonium tetrafluoroborate also effects ring expansion of cyclic ketones by ethyl diazoacetate.[83]

Scheme 10.4 gives some examples of synthetic applications of rearrangements of diazonium ions. The diazotization rearrangement in Entry 1 was used to assemble the four contiguous stereogenic centers of the oxygenated cyclopentane ring found in prostaglandins. The synthesis started with *cis,cis*-1,3,5-cyclohexanetriol. Entry 2 uses trimethylsilyl cyanide addition, followed by LiAlH$_4$ reduction to generate the amino alcohol. The minor product in this reaction is formed by competing migration of the bridgehead carbon. The reaction was part of a synthesis of the terpene cedrene. Entry 3 is an example of the use of diazomethane to effect ring expansion of a strained ketone. The reaction was carried out by generating the diazomethane in situ. Entry 4 is an example of BF$_3$-mediated addition and rearrangement using ethyl diazoacetate. In Entry 5, the diazo group was generated in situ, and the intramolecular addition-rearrangement occurs at 25°C and under alkaline conditions. In this case there is little selectivity between the two competing migration possibilities.

10.1.3. Related Rearrangements

The subjects of this section are two reactions that do not actually involve carbocation intermediates. They do, however, result in carbon to carbon rearrangements that are structurally similar to the pinacol rearrangement. In both reactions cyclic intermediates are formed, at least under some circumstances. In the *Favorskii rearrangement*, an α-halo ketone rearranges to a carboxylic acid or ester. In the *Ramberg-Backlund reaction*, an α-halo sulfone gives an alkene.

10.1.3.1. The Favorskii Rearrangement. When treated with base, α-halo ketones undergo a skeletal change that is similar to the pinacol rearrangement. The most commonly used bases are alkoxide ions, which lead to esters as the reaction products. This reaction is known as the *Favorskii rearrangement*.[84]

83. L. J. MacPherson, E. K. Bayburt, M. P. Capparelli, R. S. Bohacek, F. H. Clarke, R. D. Ghai, Y. Sakane, C. J. Berry, J. V. Peppard, and A. J. Trapani, *J. Med. Chem.*, **36**, 3821 (1993).
84. A. S. Kende, *Org. React.*, **11**, 261 (1960); A. A. Akhrem, T. K. Ustynyuk, and Y. A. Titov, *Russ. Chem. Rev.* (English Transl.), **39**, 732 (1970).

Scheme 10.4. Rearrangement Involving Diazonium Ions

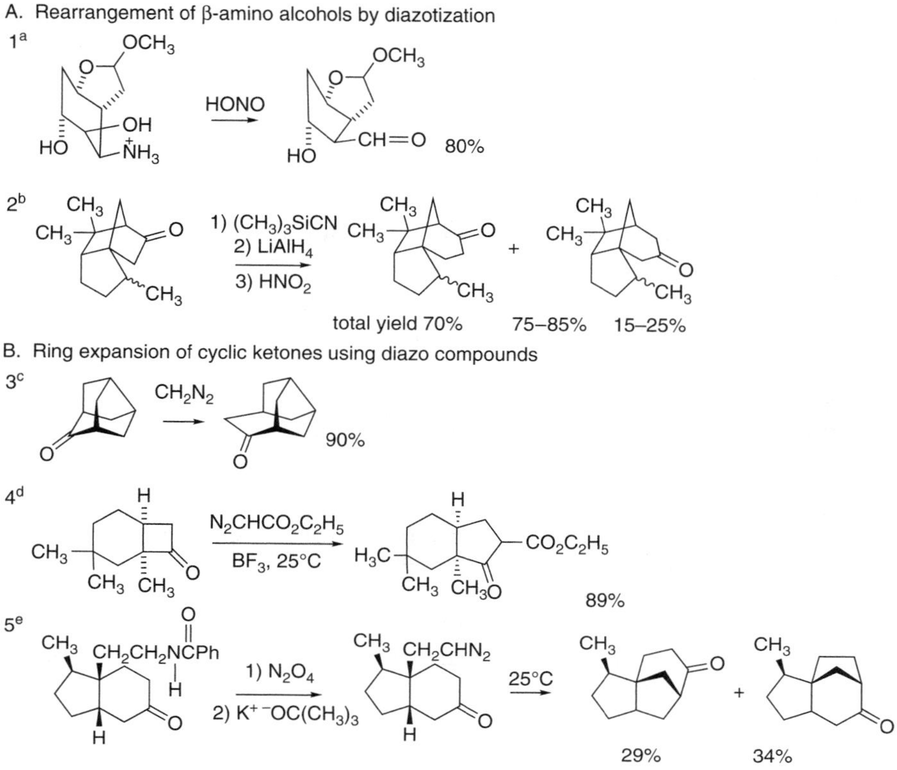

893

SECTION 10.1

*Reactions and
Rearrangement
Involving Carbocation
Intermediates*

A. Rearrangement of β-amino alcohols by diazotization

1[a]

HONO

80%

2[b]

1) $(CH_3)_3SiCN$
2) $LiAlH_4$
3) HNO_2

+

total yield 70%   75–85%   15–25%

B. Ring expansion of cyclic ketones using diazo compounds

3[c]

CH_2N_2

90%

4[d]

$N_2CHCO_2C_2H_5$
BF_3, 25°C

$CO_2C_2H_5$

89%

5[e]

1) N_2O_4
2) $K^+ {}^-OC(CH_3)_3$

25°C

+

29%   34%

a. R. B. Woodward, J. Gosteli, I. Ernest, R. J. Friary, G. Nestler, H. Raman, R. Sitrin, C. Suter, and J. K. Whitesell, *J. Am. Chem. Soc.*, **95**, 6853 (1973).
b. E. G. Breitholle and A. G. Fallis, *J. Org. Chem.*, **43**, 1964 (1978).
c. Z. Majerski, S.Djigas, and V. Vinkovic, *J. Org. Chem.*, **44**, 4064 (1979).
d. H. J. Liu and T. Ogina, *Tetrahedron Lett.*, 4937 (1973).
e. P. R. Vettel and R. M. Coates, *J. Org. Chem.*, **45**, 5430 (1980).

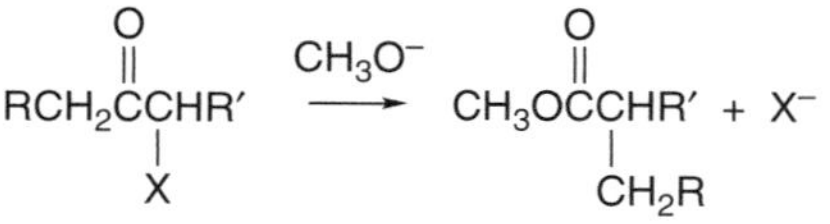

$$RCH_2\overset{O}{\overset{\|}{C}}CHR' \quad \xrightarrow{CH_3O^-} \quad CH_3O\overset{O}{\overset{\|}{C}}CHR' + X^-$$
$$\underset{X}{\mid} \qquad\qquad\qquad \underset{CH_2R}{\mid}$$

If the ketone is cyclic, a ring contraction occurs.

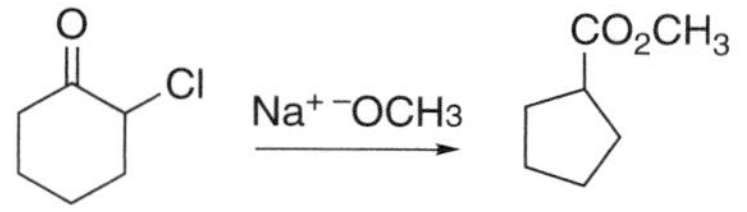

$Na^+ {}^-OCH3$

CO_2CH_3

Ref. 85

There is evidence that the rearrangement involves cyclopropanones or their open 1,3-dipolar equivalents as reaction intermediates.[86]

85. D. W. Goheen and W. R. Vaughan, *Org. Synth.*, **IV**, 594 (1963).
86. F. G. Bordwell, T. G. Scamehorn, and W. R. Springer, *J. Am. Chem. Soc.*, **91**, 2087 (1969); F. G. Bordwell and J. G. Strong, *J. Org. Chem.*, **38**, 579 (1973).

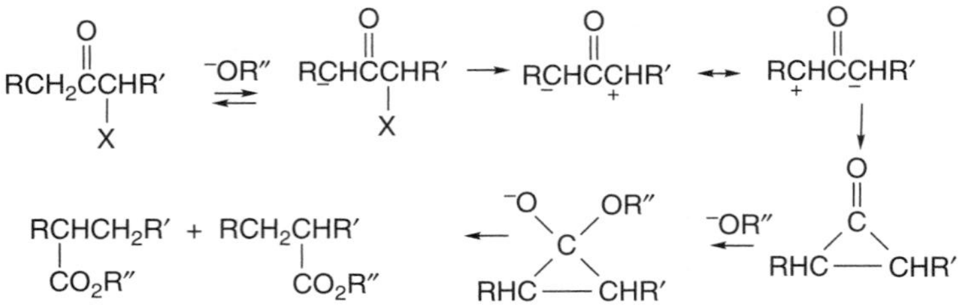

894

CHAPTER 10

*Reactions Involving
Carbocations, Carbenes,
and Radicals as Reactive
Intermediates*

There is also a mechanism that can operate in the absence of an acidic α-hydrogen. This process, called the *semibenzilic rearrangement*, is closely related to the pinacol rearrangement. A tetrahedral intermediate is formed by nucleophilic addition to the carbonyl group and the halide serves as the leaving group.

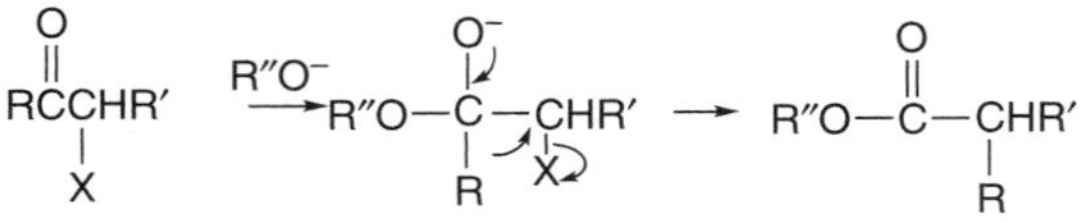

The net structural change is the same for both mechanisms. The energy requirements of the cyclopropanone and semibenzilic mechanism may be fairly closely balanced.[87] Cases of operation of the semibenzilic mechanism have been reported even for compounds having a hydrogen available for enolization.[88] Among the evidence that the cyclopropanone mechanism operates is the demonstration that a symmetrical intermediate is involved. The isomeric chloro ketones **12** and **13**, for example, lead to the same ester.

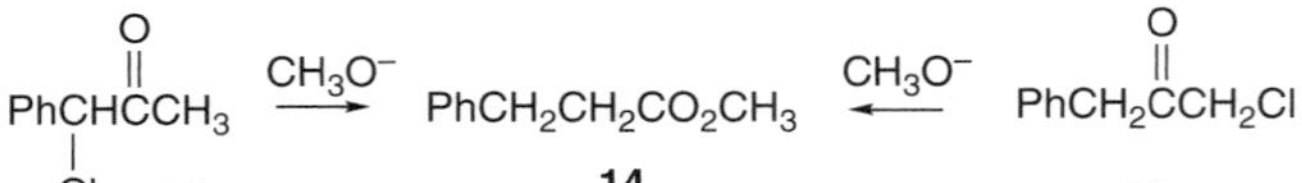

Ref. 37

The occurrence of a symmetrical intermediate has also been demonstrated by ^{14}C labeling in the case of α-chlorocyclohexanone.[89]

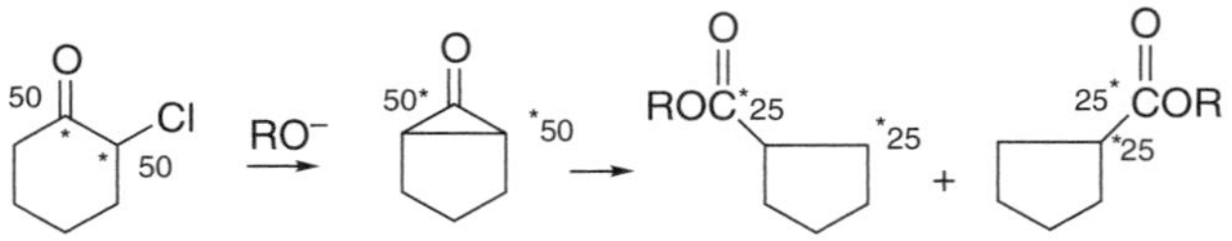

* = ^{14}C label

Numbers refer to percentage of label at each carbon.

When the two carbonyl substituents are identical, either the cyclopropanone or the dipolar equivalent is symmetric. As the α- and α′-carbons are electronically similar (identical in symmetrical cases) in these intermediates, the structure of the ester product

87. V. Moliner, R. Castillo, V. S. Safont, M. Oliva, S. Bohn, I. Tunon, and J. Andres, *J. Am. Chem. Soc.*, **119**, 1941 (1997).
88. E. W. Warnhoff, C. M. Wong, and W. T. Tai, *J. Am. Chem. Soc.*, **90**, 514 (1968).
89. R. B. Loftfield, *J. Am. Chem. Soc.*, **73**, 4707 (1951).

895

SECTION 10.1

*Reactions and
Rearrangement
Involving Carbocation
Intermediates*

cannot be predicted directly from the structure of the reacting haloketone. Instead, the identity of the product is governed by the direction of ring opening of the cyclopropanone intermediate. The dominant mode of ring opening is expected to be the one that forms the more stable of the two possible ester enolates. For this reason, a phenyl substituent favors breaking the bond to the substituted carbon, but an alkyl group directs the cleavage to the less-substituted carbon.[90] That both **12** and **13** above give the same ester, **14**, is illustrative of the directing effect that the phenyl group has on the ring-opening step.

Scheme 10.5 gives some examples of Favorskii rearrangements. Entries 1 and 2 are examples of classical reaction conditions, the latter involving a ring contraction. Entry 3 is an interesting ring contraction-elimination. The reaction was shown to be highly stereospecific, with the *cis*-dibromide giving exclusively the *E*-double bond, whereas the *trans*-dibromide gave mainly the *Z*-double bond. Entry 4 is a ring contraction leading to the formation of an interesting strained-cage hydrocarbon skeleton. Entry 5 is a step in the synthesis of the natural analgesic epibatidine.

10.1.3.2. The Ramberg-Backlund Reaction. α-Halosulfones undergo a related rearrangement known as the *Ramberg-Backlund reaction*.[91] The carbanion formed by deprotonation gives an unstable thiirane dioxide that decomposes with elimination of sulfur dioxide. This elimination step is considered to be a concerted cycloelimination.

The overall transformation is the conversion of the carbon-sulfur bonds to a carbon-carbon double bond. The original procedure involved halogenation of a sulfide, followed by oxidation to the sulfone. Recently, the preferred method has reversed the order of the steps. After the oxidation, which is normally done with a peroxy acid, halogenation is done under basic conditions by use CBr_2F_2 or related polyhalomethanes for the halogen transfer step.[92] This method was used, for example, to synthesize 1,8-diphenyl-1,3,5,7-octatetraene.

90. C. Rappe, L. Knutsson, N. J. Turro, and R. B. Gagosian, *J. Am. Chem. Soc.*, **92**, 2032 (1970).
91. L. A. Paquette, *Acc. Chem. Res.*, **1**, 209 (1968); L. A. Paquette, in *Mechanism of Molecular Migrations*, Vol. 1, B. S. Thyagarajan, ed., Wiley-Interscience, New York, 1968, Chap. 3; L. A. Paquette, *Org. React.*, **25**, 1 (1977); R. J. K. Taylor, *J. Chem. Soc.,Chem. Commun.*, 217 (1999); R. J. K. Taylor and G. Casy, *Org. React.*, **62**, 357 (2003).
92. T.-L. Chan, S. Fong, Y. Li, T.-O. Mau, and C.-D. Poon, *J. Chem. Soc., Chem. Commun.*, 1771 (1994); X.-P. Cao, *Tetrahedron*, **58**, 1301 (2002).

Scheme 10.5. Base-Mediated Rearrangements of α-Haloketones

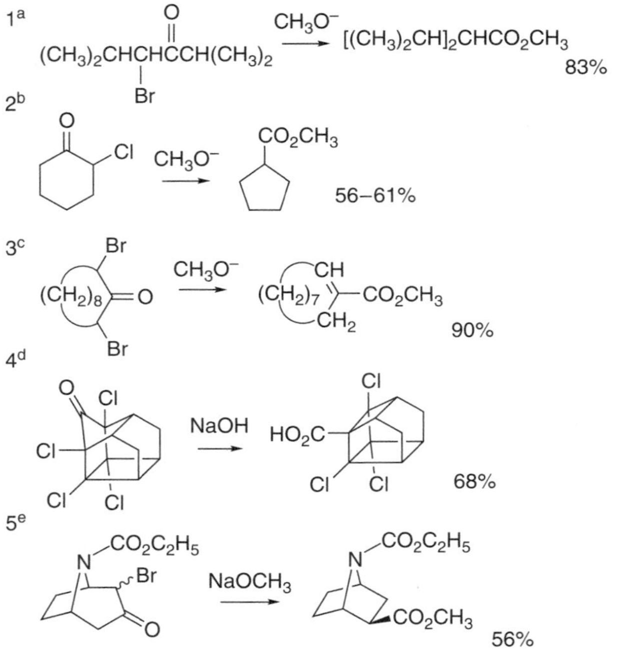

1[a]

$(CH_3)_2CHCHCCH(CH_3)_2$ (with O above, Br below) $\xrightarrow{CH_3O^-}$ $[(CH_3)_2CH]_2CHCO_2CH_3$

83%

2[b]

56–61%

3[c]

90%

4[d]

$\xrightarrow{NaOH}$

68%

5[e]

$\xrightarrow{NaOCH_3}$

56%

a. S. Sarel and M. S. Newman, *J. Am. Chem. Soc.*, **78**, 5416 (1956).
b. D. W. Goheen and W. R. Vaughan, *Org. Synth.*, **IV**, 594 (1963).
c. E. W. Garbisch, Jr., and J. Wohllebe, *J. Org. Chem.*, **33**, 2157 (1968).
d. R. J. Stedman, L. S. Miller, L. D. Davis, and J. R. E. Hoover, *J. Org. Chem.*, **35**, 4169 (1970).
e. D. Bai, R. Xu, G. Chu, and X. Zhu, *J. Org. Chem.*, **61**, 4600 (1996).

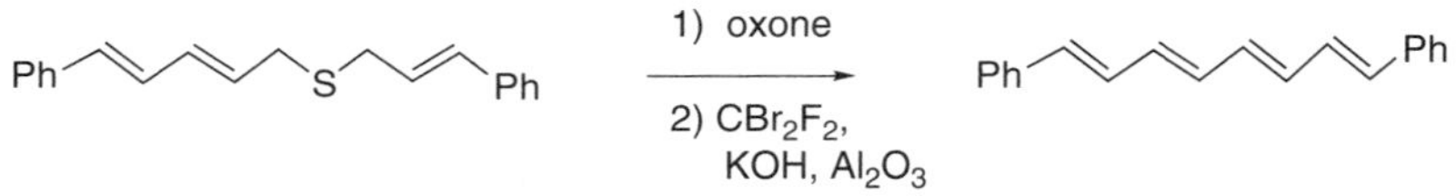

The Ramberg-Backlund reaction has found several applications. Owing to the concerted nature of the elimination, it can applied to both small and large rings containing a double bond.

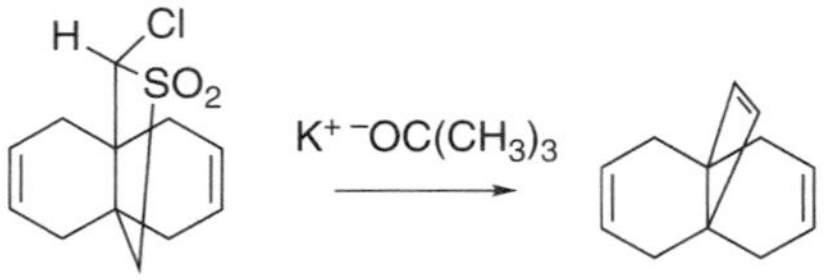

Ref. 93

93. L. A. Paquette, J. C. Philips, and R. E. Wingard, Jr., *J. Am. Chem. Soc.*, **93**, 4516 (1971).

897

SECTION 10.1

*Reactions and
Rearrangement
Involving Carbocation
Intermediates*

Ref. 94

A recently developed application of the Ramberg-Backlund reaction is the synthesis of C-glycosides. The required thioethers can be prepared easily by exchange with a thiol. The application of the Ramberg-Backlund conditions then leads to an exocyclic vinyl ether that can be reduced to the C-nucleoside.[95] Entries 3 and 4 in Scheme 10.6 are examples. The vinyl ether group can also be transformed in other ways. In the synthesis of partial structures of the antibiotic altromycin, the vinyl ether product was subjected to diastereoselective hydroboration.

Scheme 10.6 gives some examples of the Ramberg-Backlund reaction. Entry 1 was used to prepare analogs of the antimalarial compound artemisinin for biological evaluation. The reaction in Entry 2 was used to install the side chain in a synthesis of the chrysomycin type of antibiotic. Entries 3 and 4 are examples of formation of C-glycosides.

10.1.4. Fragmentation Reactions

The classification *fragmentation* applies to reactions in which a carbon-carbon bond is broken. One structural feature that permits fragmentation to occur readily is the presence of a carbon that can accommodate carbocationic character β to a developing electron deficiency. This type of reaction, known as the *Grob fragmentation*, occurs particularly readily when the γ-atom is a heteroatom, such as nitrogen or oxygen, that has an unshared electron pair that can stabilize the new cationic center.[96]

The fragmentation can be concerted or stepwise. The concerted mechanism is restricted to molecular geometry that is appropriate for continuous overlap of the participating

94. I. MaGee and E. J. Beck, *Can. J. Chem.*, **78**, 1060 (2000).
95. F. K. Griffin, D. E. Paterson, P. V. Murphy, and R. J. K. Taylor, *Eur. J. Org. Chem.*, 1305 (2002).
96. C. A. Grob, *Angew. Chem. Int. Ed. Engl.*, **8**, 535 (1969).

Scheme 10.6. Ramberg-Backlund Reaction

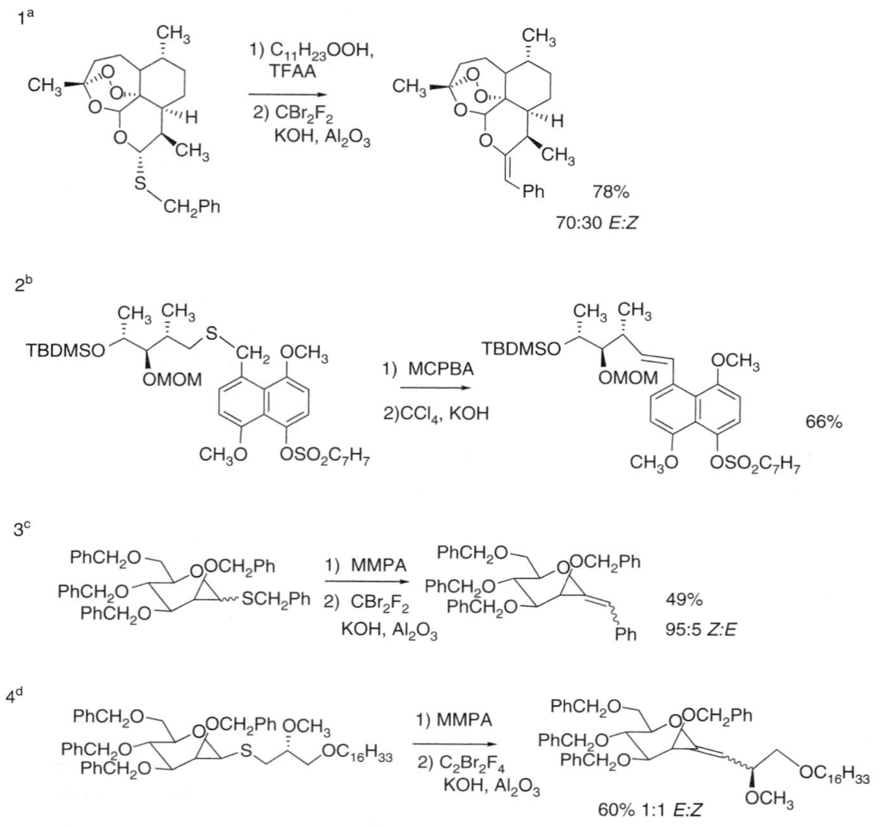

a. S. Oh, I. H. Jeong, W.-S. Shin, and S. Lee, *Biorg. Med. Chem. Lett.*, **14**, 3683 (2004).
b. D. J. Hart, G. H. Merriman, and D. G. J. Young, *Tetrahedron*, **52**, 14437 (1996).
c. P. S. Belica and R. W. Franck, *Tetrahedron Lett.*, **39**, 8225 (1998).
d. G. Yang, R. W. Franck, H. S. Byun, R. Bittman, P. Samadder, and G. Arthur, *Org. Lett.*, **1**, 2149 (1999).

orbitals. An example is the solvolysis of 4-chloropiperidine, which is faster than the solvolysis of chlorocyclohexane and occurs by fragmentation of the C(2)−C(3) bond.[97]

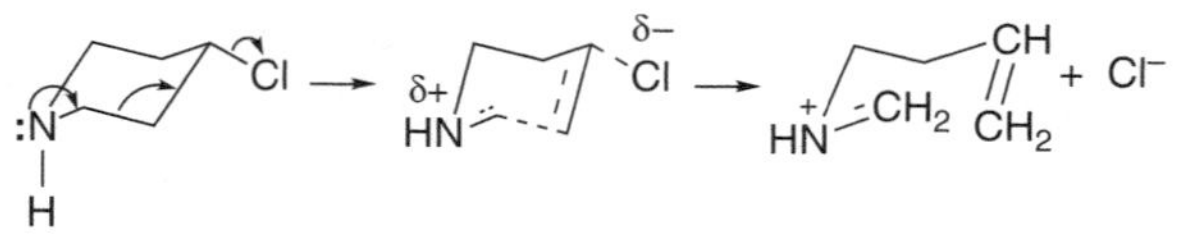

1,3-Diols or β-hydroxy ethers are particularly useful substrates for fragmentation. If the diol or hydroxy ether is converted to a monotosylate, the remaining oxy group can promote fragmentation.

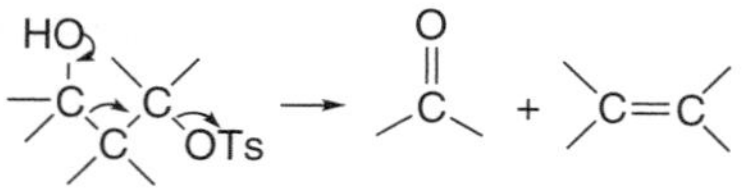

97. R. D'Arcy, C. A. Grob, T. Kaffenberger, and V. Krasnobajew, *Helv. Chim. Acta*, **49**, 185 (1966).

This reaction can be used in synthesis of medium-sized rings by cleavage of specific bonds. An example of this reaction pattern can be seen in a fragmentation used to construct the ring structure found in the taxane group of diterpenes.

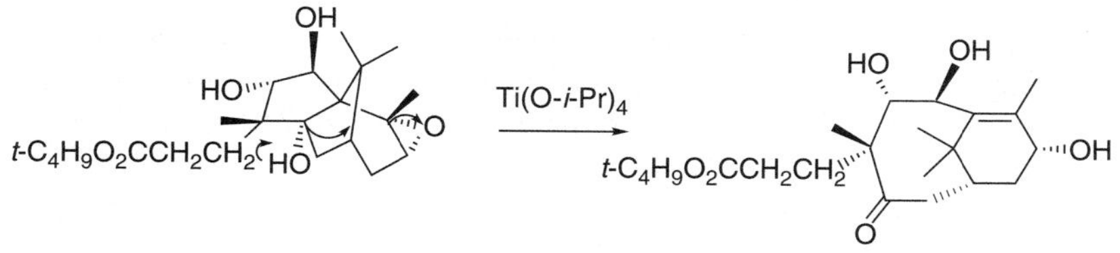

Ref. 98

Similarly, a carbonyl group at the fifth carbon from a leaving group, reacting as the enolate, promotes fragmentation with formation of an enone.[99] This is a *vinylogous* analog of the Grob fragmentation.

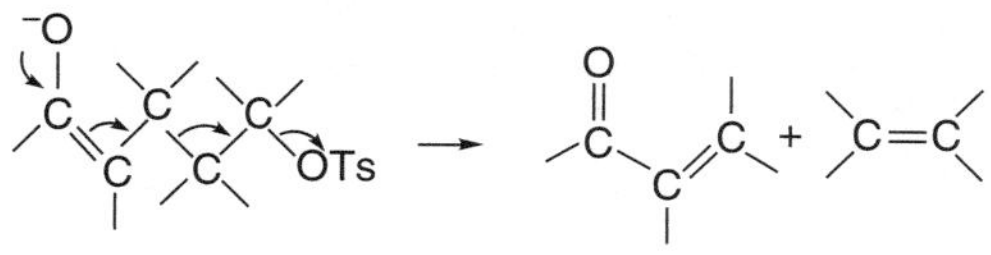

β-Hydroxyketones are also subject to fragmentation. Lewis acids promote fragmentation of mixed aldol products derived from aromatic aldehydes.[100]

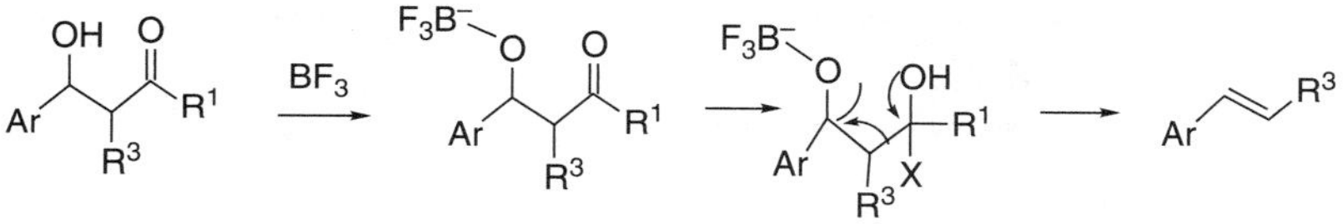

The same fragmentation is effected by Yb(OTf)$_3$ on heating with the aldol adduct in the absence of solvent.[101]

Organoboranes undergo fragmentation if a good leaving group is present on the δ-carbon.[102] The reactive intermediate is the tetrahedral borate formed by addition of hydroxide ion at boron.

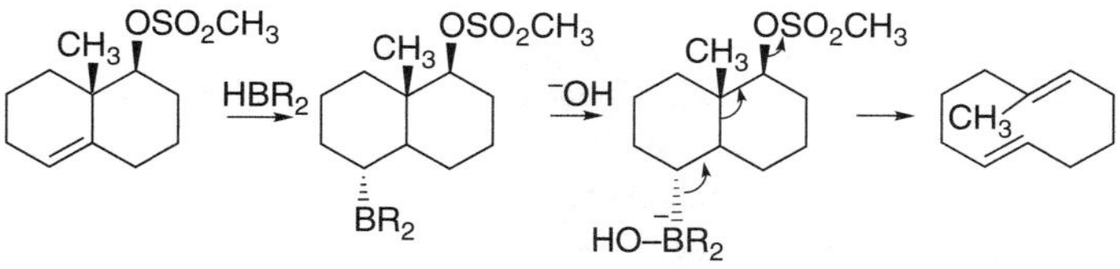

Ref. 103

[98] R. A. Holton, R. R. Juo, H. B. Kim, A. Q. Williams, S. Harusawa, P. E. Lowenthal, and S. Yogai, *J. Am. Chem. Soc.*, **110**, 6558 (1988).

[99] J. M. Brown, T. M. Cresp, and L. N. Mander, *J. Org. Chem.*, **42**, 3984 (1977); D. A. Clark and P. L. Fuchs, *J. Am. Chem. Soc.*, **101**, 3567 (1979).

[100] G. W. Kabalka, N.-S. Li, D. Tejedor, R. R. Malladi, and S. Trotman, *J. Org. Chem.*, **64**, 3157 (1999).

[101] M. Curini, F. Epifano, F. Maltese, and M. C. Marcotullio, *Chem. Eur. J.*, 1631 (2003).

[102] J. A. Marshall, *Synthesis*, 229 (1971); J. A. Marshall and G. L. Bundy, *J. Chem. Soc., Chem. Commun.*, 854 (1967); P. S. Wharton, C. E. Sundin, D. W. Johnson, and H. C. Kluender, *J. Org. Chem.*, **37**, 34 (1972).

[103] J. A. Marshall and G. L. Bundy, *J. Am. Chem. Soc.*, **88**, 4291 (1966).

900

CHAPTER 10

*Reactions Involving
Carbocations, Carbenes,
and Radicals as Reactive
Intermediates*

The usual synthetic objective of a fragmentation reaction is the construction of a medium-sized ring from a fused ring system. As the fragmentation reactions are usually concerted stereoselective processes, the stereochemistry is predictable. In 3-hydroxy tosylates, the fragmentation is most favorable for a geometry in which the carbon-carbon bond being broken is in an *anti*-periplanar relationship to the leaving group.[104] Other stereochemical relationships in the molecule are retained during the concerted fragmentation. In the case below, for example, the newly formed double bond has the *E*-configuration.

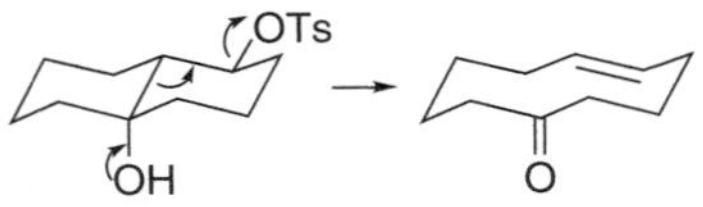

Fragmentation reactions can also be used to establish stereochemistry of acyclic systems based on stereochemical relationships built into cyclic reactants. In both the examples shown below, the aldehyde group generated by fragmentation was reduced in situ.

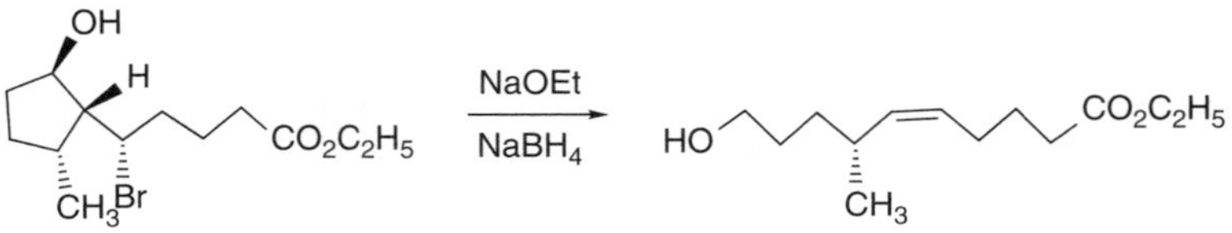

Ref. 105

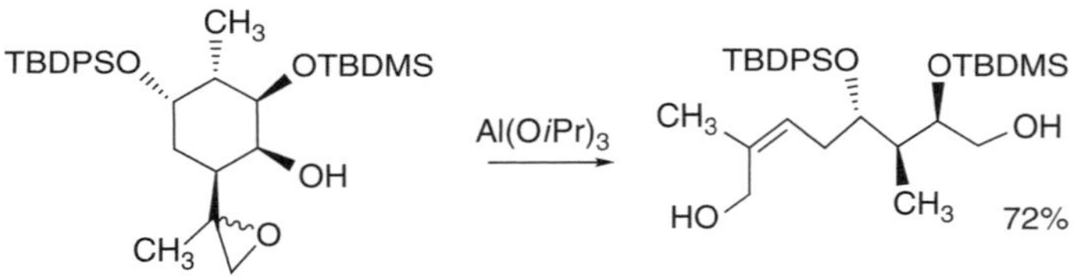

Ref. 106

Scheme 10.7 provides some additional examples of fragmentation reactions that have been employed in a synthetic context. Entry 1 was used in the late stages of the synthesis of (±)-hinesol, an example of a terpene possessing a spiro[4,5]decane skeleton. The fragmentation provides the spiro ring system with a vinyl side chain. Entry 2 illustrates the formation of a medium ring by fragmentation of a bicyclic system. In this case LiAlH$_4$ serves as a base and also reduces the carbonyl group in the product, but closely related reactions were carried out with the more usual alkoxide bases. The reaction in Entry 3 was developed during exploration of the

104. P. S. Wharton and G. A. Hiegel, *J. Org. Chem.*, **30**, 3254 (1965); C. H. Heathcock and R. A. Badger, *J. Org. Chem.*, **37**, 234 (1972).
105. Y. M. A. W. Lamers, G. Rusu, J. B. P. A. Wijnberg, and A. de Groot, *Tetrahedron*, **59**, 9361 (2003).
106. X. Z. Zhao, Y. Q. Tu, L. Peng, X. Q. Li, and Y. X. Jia, *Tetrahedron Lett.*, **45**, 3213 (2004).

901

SECTION 10.1

*Reactions and
Rearrangement
Involving Carbocation
Intermediates*

Scheme 10.7. Synthetic Applications of Fragmentation Reactions

A. Heteroatom-promoted fragmentation

1[a]

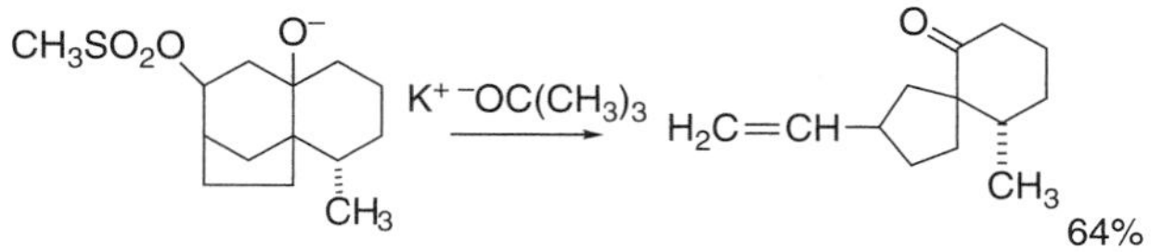

64%

2[b]

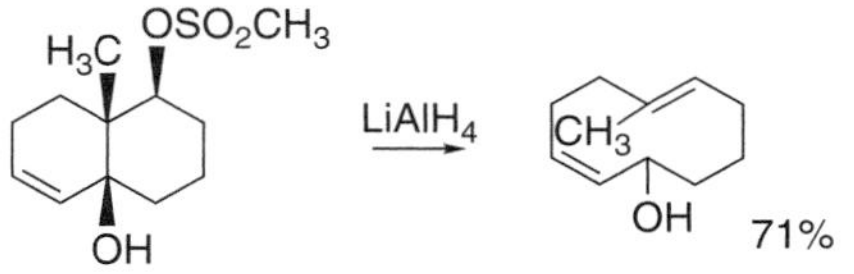

71%

3[c]

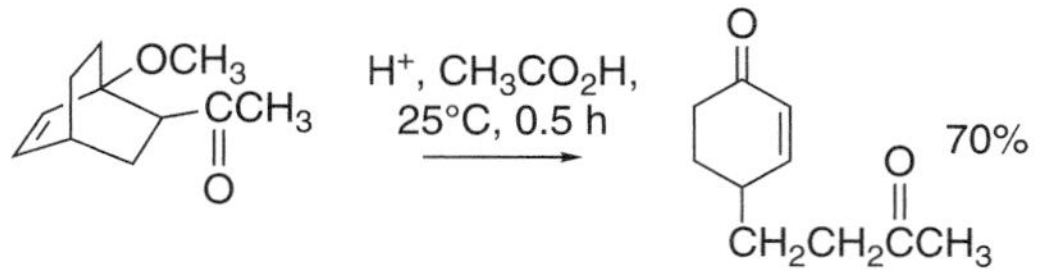

70%

4[d]

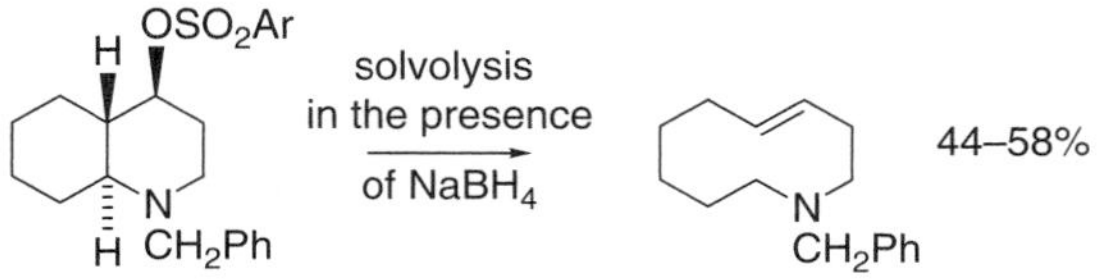

44–58%

5[e]

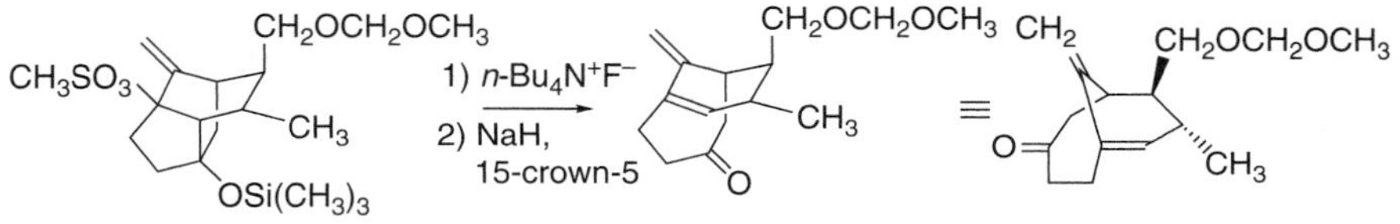

6[f]

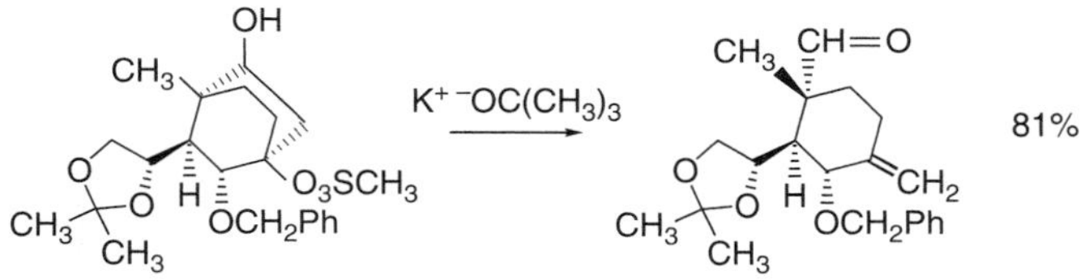

81%

7[g]

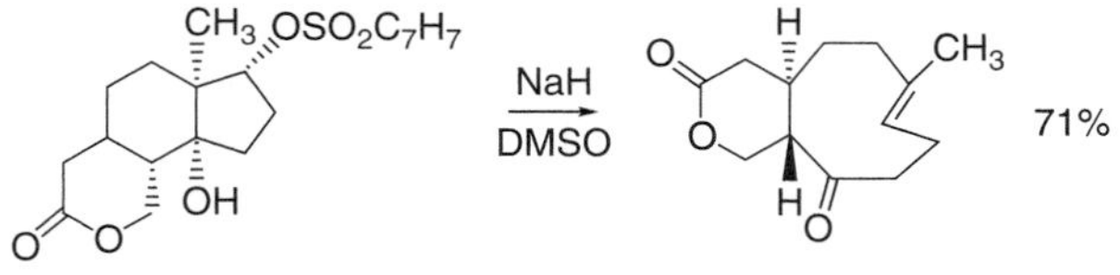

71%

8[h]

98%

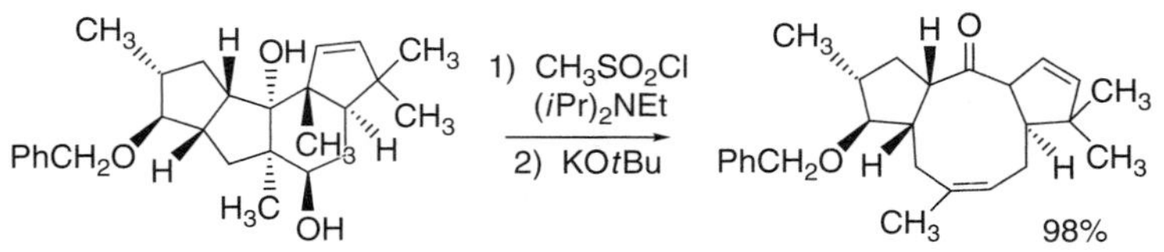

(*Continued*)

B. Boronate fragmentation

9[i]

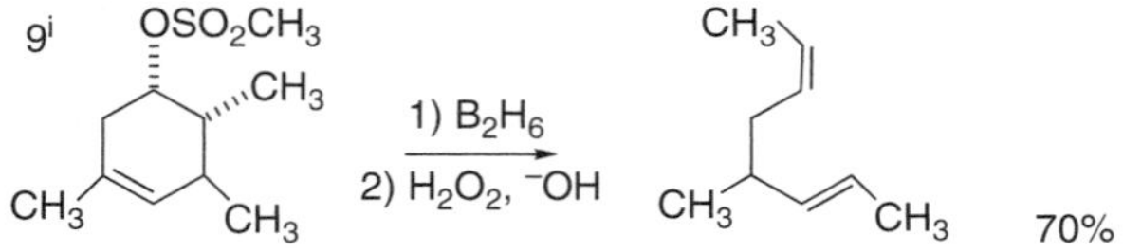

1) B_2H_6
2) H_2O_2, ^-OH

70%

C. δ-Tosyloxy fragmentation

10[j]

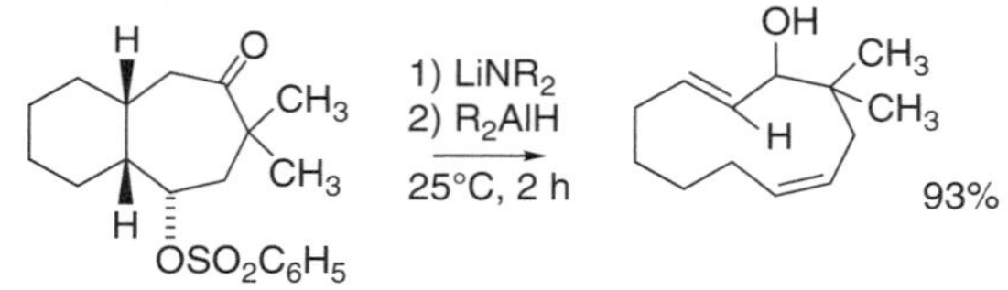

1) $LiNR_2$
2) R_2AlH
25°C, 2 h

93%

a. J. A. Marshall and S. F. Brady, *J. Org. Chem.*, **35**, 4068 (1970).
b. J. A. Marshall, W. F. Huffman, and J. A. Ruth, *J. Am. Chem. Soc.*, **94**, 4691 (1972).
c. A. J. Birch and J. S. Hill, *J. Chem. Soc., C*, 419 (1966).
d. J. A. Marshall and J. H. Babler, *J. Org. Chem.*, **34**, 4186 (1969).
e. T. Yoshimitsu, M Yanagiya, and H. Nagoka, *Tetrahedron Lett..*, **40**, 5215 (1999).
f. Y. Hirai, T. Suga, and H. Nagaoka, *Tetrahedron Lett.*, **38**, 4997 (1997).
g. D. Rennenberg, H. Pfander, and C. J. Leumann, *J. Org. Chem.*, **65**, 9069 (2000).
h. L. A. Paquette, J. Yang, and Y. O. Long, *J. Am. Chem. Soc.*, **124**, 6542 (2002).
i. J. A. Marshall and J. H. Babler, *Tetrahedron Lett.*, 3861 (1970).
j. D. A. Clark and P. L. Fuchs, *J. Am. Chem. Soc.*, **101**, 3567 (1979).

chemistry of the reactant, which is readily available by a Diels-Alder reaction of 1-methoxycyclohexadiene. This acid-catalyzed fragmentation is induced by protonation of the acetyl group.

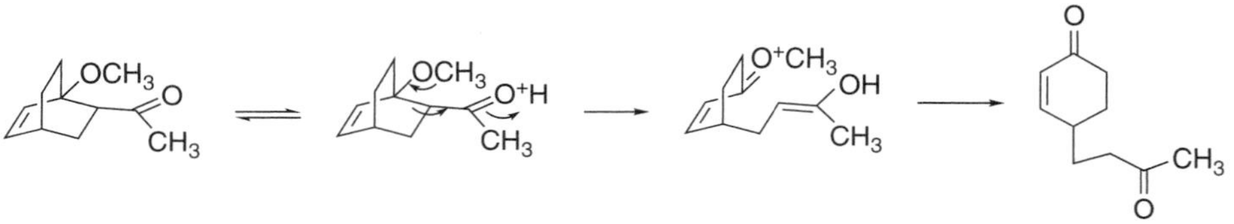

Entry 4 involves nitrogen participation and formation of an iminium ion that is reduced by $NaBH_4$. The reaction in Entry 5 creates an 11-methylenebicyclo[4.3.1]undecen-3-one structure found in a biologically active natural product. Note that this fragmentation creates a bridgehead double bond. Entry 6 involves construction of a portion of the taxol structure. The reaction in Entry 7 is stereospecific, leading to the *E*-double bond.

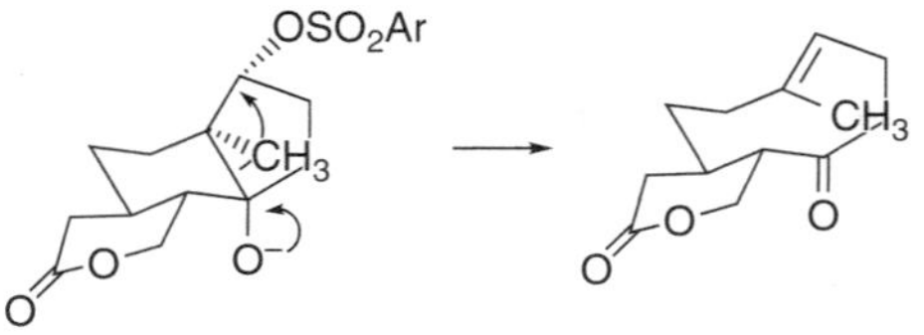

Entry 8 was used to create the central nine-membered ring system found in the diterpene jatrophatrione. Entry 9 is an example of a boronate fragmentation (see p. 899). Entry 10 illustrates enolate fragmentation. The reaction presumably proceeds through an extended conformation that aligns the enolate and sulfonate leaving group advantageously and results in an *E*-double bond.

10.2. Reactions Involving Carbenes and Related Intermediates

Carbenes can be included with carbanions, carbocations, and carbon-centered radicals as being among the fundamental intermediates in the reactions of carbon compounds. Carbenes are neutral divalent derivatives of carbon. As would be expected from their electron-deficient nature, most carbenes are highly reactive. Depending upon the mode of generation, a carbene can be formed in either the singlet or the triplet state, no matter which is lower in energy. The two electronic configurations have different geometry and reactivity. A conceptual picture of the bonding in the singlet assumes sp^2 hybridization at carbon, with the two unshared electrons in an sp^2 orbital. The p orbital is unoccupied. The R−C−R angle would be expected to be contracted slightly from 120° because of the electronic repulsions between the unshared electron pair and the electrons in the two bonding σ orbitals. The bonds in a triplet carbene are considered to be formed from sp orbitals with the unpaired electrons being in two equivalent p orbitals. This bonding arrangement corresponds to a linear structure.

Both theoretical and experimental studies have provided more detailed information about carbene structure. Molecular orbital calculations lead to the prediction of H−C−H angles for methylene of roughly 135° for the triplet and about 105° for the singlet. The triplet is calculated to be about 8 kcal/mol lower in energy than the singlet.[107] Experimental determinations of the geometry of CH_2 accord with the theoretical results. The H−C−H angle of the triplet state, as determined from the ESR spectrum is 125°–140°. The H−C−H angle of the singlet state is found to be 102° by electronic spectroscopy. The available evidence is consistent with the triplet being the ground state species.

Substituents perturb the relative energies of the singlet and triplet states. In general, alkyl groups resemble hydrogen as a substituent and dialkylcarbenes are ground state

107. J. F. Harrison, *Acc. Chem. Res.*, **7**, 378 (1974); P. Saxe, H. F. Shaefer, and N. C. Hardy, *J. Phys. Chem.*, **85**, 745 (1981); C. C. Hayden, M. Newmark, K. Shobatake, R. K. Sparks, and Y. T. Lee, *J. Chem. Phys.*, **76**, 3607 (1982); R. K. Lengel and R. N. Zare, *J. Am. Chem. Soc.*, **100**, 739 (1978); C. W. Bauschlicher, Jr., and I. Shavitt, *J. Am. Chem. Soc.*, **100**, 739 (1978); A. R. W. M. Kellar, P. R. Bunker, T. J. Sears, K. M. Evenson, R. Saykally, and S. R. Langhoff, *J. Chem. Phys.*, **79**, 5251 (1983).

904

CHAPTER 10

*Reactions Involving
Carbocations, Carbenes,
and Radicals as Reactive
Intermediates*

triplets. Substituents that act as electron-pair donors stabilize the singlet state more than the triplet state by delocalization of an electron pair into the empty *p* orbital.[108]

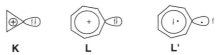

$$X = F, Cl, OR, NR_2$$

The presence of more complex substituent groups complicates the description of carbene structure. Furthermore, since carbenes are high-energy species, structural entities that would be unrealistic for more stable species must be considered. As an example, one set of MO calculations[109] arrives at structure **I** as a better description of carbomethoxycarbene than the conventional structure **J**.

π-Delocalization involving divalent carbon in conjugated cyclic systems has been studied in the interesting species cyclopropenylidene (**K**)[110] and cycloheptatrienylidene (**L**).[111] In these molecules the empty *p* orbital on the carbene carbon can be part of the aromatic π system and be delocalized over the entire ring. Currently available data indicate that the ground state structures for both **K** and **L** are singlets, but for **L**, the most advanced theoretical calculations indicate that the most stable singlet structure has an electronic configuration in which one of the nonbonded electrons is in the π orbital.[112]

There are a number of ways of generating carbenes that will be discussed shortly. In some cases, the reactions involve complexes or precursors of carbenes rather than the carbene per se. For example, carbenes can be generated by α-elimination reactions. Under some circumstances the question arises as to whether the carbene has a finite lifetime, and in some cases a completely free carbene structure is never attained.

108. N. C. Baird and K. F. Taylor, *J. Am. Chem. Soc.*, **100**, 1333 (1978); J. F. Harrison, R. C. Liedtke, and J. F. Liebman, *J. Am. Chem. Soc.*, **101**, 7162 (1979); P. H. Mueller, N. G. Rondan, K. N. Houk, J. F. Harrison, D. Hooper, B. H. Willen, and J. F. Liebman, *J. Am. Chem. Soc.*, **103**, 5049 (1981).
109. R. Noyori and M. Yamanaka, *Tetrahedron Lett.*, 2851 (1980).
110. H. P. Reisenauer, G. Maier, A. Reimann, and R. W. Hoffmann, *Angew. Chem. Int. Ed. Engl.*, **23**, 641 (1984); T. J. Lee, A. Bunge, and H. F. Schaefer, III, *J. Am. Chem. Soc.*, **107**, 137 (1985); J. M. Bofill, J. Farras, S. Olivella, A. Sole, and J. Vilarrasa, *J. Am. Chem. Soc.*, **110**, 1694 (1988).
111. R. J. McMahon and O. L. Chapman, *J. Am. Chem. Soc.*, **108**, 1713 (1986); M. Kusaz, H. Luerssen, and C. Wentrup, *Angew. Chem. Int. Ed. Engl.*, **25**, 480 (1986); C. L. Janssen and H. F. Schaefer, III, *J. Am. Chem. Soc.*, **109**, 5030 (1987); M. W. Wong and C. Wentrup, *J. Org. Chem.*, **61**, 7022 (1996).
112. S. Matzinger, T. Bally, E. V. Patterson, and R. J. McMahon, *J. Am. Chem. Soc.*, **118**, 1535 (1996); P. R. Schreiner, W. L. Karney, P. v. R. Schleyer, W. T. Borden, T. P. Hamilton, and H. F. Schaefer, III, *J. Org. Chem.*, **61**, 7030 (1996).

When a reaction appears to involve a species that reacts as expected for a carbene but must still be at least partially bound to other atoms, the term *carbenoid* is used. Some carbenelike processes involve transition metal ions. In many of these reactions, the divalent carbene is bound to the metal. Some compounds of this type are stable, whereas others exist only as transient intermediates. In most cases, the reaction involves the metal-bound carbene, rather than a free carbene.

$$M=C\diagup_{\diagdown}$$

metal-bound carbene

The stability and reactivity of metallocarbenes depends on the degree of back donation from the metal to the carbene. If this is small, the metallocarbenes are highly reactive and electrophilic in character. If back bonding is substantial, the carbon will be less electrophilic, and the reactions are more likely to involve the metal.

$$\overset{..}{M}-C\diagup_{\diagdown} + \qquad\qquad M=C\diagup_{\diagdown}$$

dominant electron distribution in electrophilic metallo-carbenes

dominant electron distribution in metallo-carbenes with strong back-bonding

Carbenes and carbenoids can add to double bonds to form cyclopropanes or insert into C—H bonds.

$$R_3C-\overset{\overset{\displaystyle X}{|}}{\underset{\underset{\displaystyle H}{|}}{C}}-Y \xleftarrow{R_3C-H} \overset{\overset{\displaystyle X}{|}}{:C}-Y \xrightarrow{R_2C=CR_2} \overset{X\quad Y}{\underset{\underset{R\quad R}{R\diagup\quad\diagdown R}}{\triangle}}$$

insertion   addition

These reactions have *very low activation energies* when the intermediate is a "free" carbene. Intermolecular insertion reactions are inherently nonselective. The course of intramolecular reactions is frequently controlled by the proximity of the reacting groups.[113] Carbene intermediates can also be involved in rearrangement reactions. In the sections that follow we also consider a number of rearrangement reactions that probably do not involve carbene intermediates, but lead to transformations that correspond to those of carbenes.

10.2.1. Reactivity of Carbenes

From the point of view of both synthetic and mechanistic interest, much attention has been focused on the addition reaction between carbenes and alkenes to give cyclopropanes. Characterization of the reactivity of substituted carbenes in addition reactions has emphasized stereochemistry and selectivity. The reactivities of singlet and triplet states are expected to be different. The triplet state is a diradical, and would be expected to exhibit a selectivity similar to free radicals and other species with unpaired electrons. The singlet state, with its unfilled *p* orbital, should be electrophilic and exhibit reactivity patterns similar to other electrophiles. Moreover, a triplet addition

[113.] S. D. Burke and P. A. Grieco, *Org. React.*, **26**, 361 (1979).

process must go through a 1,3-diradical intermediate that has two unpaired electrons of the same spin. In contrast, a singlet carbene can go to a cyclopropane in a single concerted step.[114] As a result, it was predicted that additions of singlet carbenes would be stereospecific, whereas those of triplet carbenes would not be.[115] This expectation has been confirmed and the stereoselectivity of addition reactions with alkenes is used as a test for the involvement of the singlet versus triplet carbene in specific reactions.[116]

Transition structure for
concerted singlet carbene
addition

Diradical intermediate in
triplet carbene addition

The radical versus electrophilic character of triplet and singlet carbenes also shows up in relative reactivity patterns given in Table 10.1. The relative reactivity of singlet dibromocarbene toward alkenes is more similar to electrophiles (bromination, epoxidation) than to radicals ($\cdot CCl_3$).

Carbene reactivity is strongly affected by substituents.[117] Various singlet carbenes have been characterized as nucleophilic, ambiphilic, and electrophilic as shown in Table 10.2 This classification is based on relative reactivity toward a series of both nucleophilic alkenes, such as tetramethylethylene, and electrophilic ones, such as acrylonitrile. The principal structural feature that determines the reactivity of the carbene is the ability of the substituent to act as an electron donor. For example, dimethoxycarbene is devoid of electrophilicity toward alkenes because of electron donation by the methoxy groups.[118]

Table 10.1. Relative Rates of Addition to Alkenes[a]

Alkene	$\cdot CCl_3$	$:CBr_2$	Br_2	Epoxidation
2-Methylpropene	1.0	1.0	1.0	1.0
Styrene	>19	0.4	0.6	0.1
2-Methyl-2-butene	0.17	3.2	1.9	13.5

a. P. S. Skell and A. Y. Garner, *J. Am. Chem. Soc.*, **78**, 5430 (1956).

114. A. E. Keating, S. R. Merrigan, D. A. Singleton, and K. N. Houk, *J. Am. Chem. Soc.*, **121**, 3933 (1999).

115. P. S. Skell and A. Y. Garner, *J. Am. Chem. Soc.*, **78**, 5430 (1956).

116. R. C. Woodworth and P. S. Skell, *J. Am. Chem. Soc.*, **81**, 3383 (1959); P. S. Skell, *Tetrahedron*, **41**, 1427 (1985).

117. A comprehensive review of this topic is given by R. A. Moss, in *Carbenes*, M. Jones, Jr., and R. A. Moss, eds., John Wiley & Sons, New York, 1973, pp. 153–304; R. A. Moss, *Acc. Chem. Res.*, **22**, 15 (1989). More recent work is reviewed in the series *Reactive Intermediates*, R. A. Moss, M. S. Platz, and M. Jones, Jr., eds., Wiley, New York, 2004, Chap. 7.

118. D. M. Lemal, E. P. Gosselink, and S. D. McGregor, *J. Am. Chem. Soc.*, **88**, 582 (1966).

Table 10.2. Classification of Carbenes on the Basis of Reactivity toward Alkenes[a]

Nucleophilic	Ambiphilic	Electrophilic
$(CH_3O)_2C$	CH_3OCCl	Cl_2C
$CH_3OCN(CH_3)_2$	CH_3OCF	$PhCCl$
		CH_3CCl
		$BrCCO_2C_2H_5$

a. R. A. Moss and R. C. Munjai, *Tetrahedron Lett.*, 4721 (1979); R. A. Moss, *Acc. Chem. Res.*, **13**, 58 (1980); R. A. Moss, *Acc. Chem. Res.*, **22**, 15 (1989).

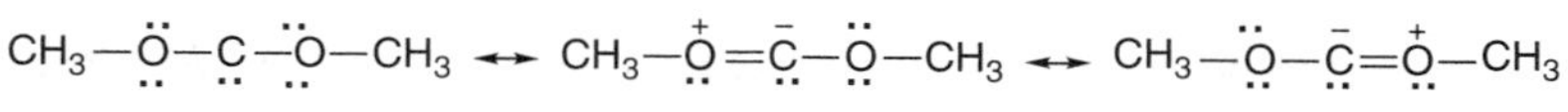

$$CH_3-\overset{..}{\underset{..}{O}}-C-\overset{..}{\underset{..}{O}}-CH_3 \leftrightarrow CH_3-\overset{+}{\underset{..}{O}}=\overset{..}{\underset{-}{C}}-\overset{..}{\underset{..}{O}}-CH_3 \leftrightarrow CH_3-\overset{..}{\underset{..}{O}}-\overset{-}{C}=\overset{+}{\underset{..}{O}}-CH_3$$

Absolute rates have been measured for some carbene reactions. The rate of addition of phenylchlorocarbene shows a small dependence on alkene substituents, but as expected for a very reactive species, the range of reactivity is quite narrow.[119] The rates are comparable to moderately fast bimolecular addition reactions of radicals (see Part A, Table 11.3).

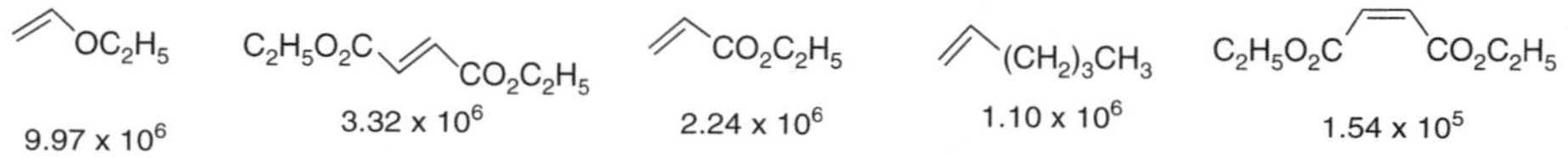

9.97 x 10^6   3.32 x 10^6   2.24 x 10^6   1.10 x 10^6   1.54 x 10^5

Absolute rate of addition of phenylchlorocarbene, k $M^{-1}s^{-1}$

The rates of phenylchlorocarbene have also been compared with the fluoro and bromo analogs.[120] The data show slightly decreased rates in the order Br > Cl > F. The alkene reactivity difference is consistent with an electrophilic attack. These reactions have low activation barriers and the reactivity differences are dominated by entropy effects.

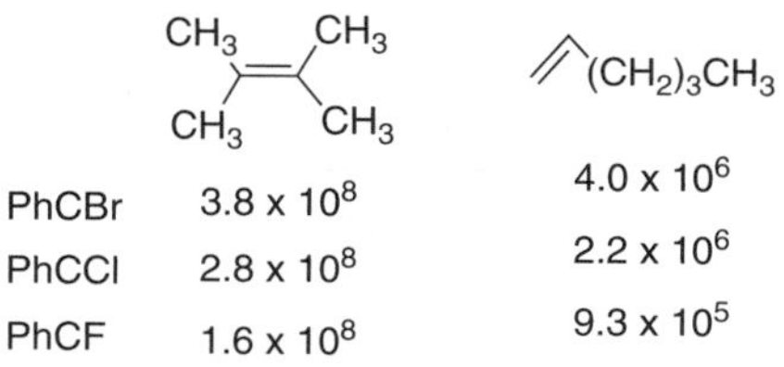

	$\underset{CH_3}{\overset{CH_3}{>}}=\underset{CH_3}{\overset{CH_3}{<}}$	$(CH_2)_3CH_3$
PhCBr	3.8 x 10^8	4.0 x 10^6
PhCCl	2.8 x 10^8	2.2 x 10^6
PhCF	1.6 x 10^8	9.3 x 10^5

Absolute rate of addition, k $M^{-1}s^{-1}$

[119]. N. Soundararajan, M. S. Platz, J. E. Jackson, M. P. Doyle, S.-M. Oon, M. J. H. Liu, and S. M. Anand, *J. Am. Chem. Soc.*, **110**, 7143 (1988).

[120]. R. A. Moss, W. Lawrynowicz, N. J. Turro, I. R. Gould, and Y. Cha, *J. Am. Chem. Soc.*, **108**, 7028 (1986).

908

CHAPTER 10

*Reactions Involving
Carbocations, Carbenes,
and Radicals as Reactive
Intermediates*

There is a small dependence on the rate of solvent insertion reactions for saturated hydrocarbons.[121] Benzene is much less reactive.

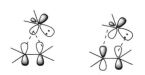

$7.1 \times 10^4 s^{-1}$	$1.9 \times 10^4 s^{-1}$	$2.8 \times 10^4 s^{-1}$

Absolute rate for solvent insertion by 4-methylphenylchlorocarbene

An HSAB analysis of singlet carbene reactivity based on B3LYP/6-31G* computations has calculated the extent of charge transfer for substituted alkenes,[122] and the results are summarized in Figure 10.3 The trends are as anticipated for changes in structure of both the carbene and alkene. The charge transfer interactions are consistent with HOMO-LUMO interactions between the carbene and alkene. Similarly, a correlation was found for the global electrophilicity parameter, ω, and the ΔN_{max} parameters (see Topic 1.5, Part A for definition of these DFT-based parameters).[123]

HOMO-LUMO interactions
in carbene alkene addition

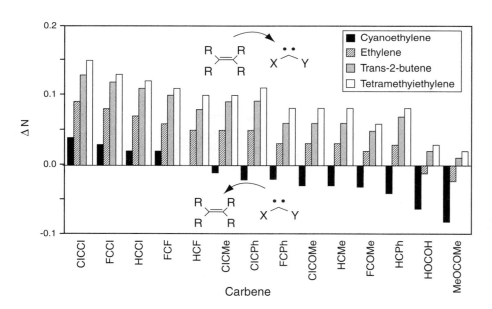

Fig. 10.3. Net charge transfer (ΔN) calculated for substituted carbenes with several alkenes. Reproduced from *J. Org. Chem.*, **64**, 7061 (1999), by permission of the American Chemical Society.

[121.] R. Bonneau and M. T. H. Liu, *J. Photochem. Photobiol. A*, **68**, 97 (1992).
[122.] F. Mendez and M. A. Garcia-Garibay, *J. Org. Chem.*, **64**, 7061 (1999).
[123.] P. Perez, *J. Phys. Chem. A*, **107**, 522 (2003).

10.2.2. Generation of Carbenes

There are several ways of generating carbene intermediates. Some of the most general routes are summarized in Scheme 10.8 and are discussed in the succeeding paragraphs.

10.2.2.1. Carbenes from Diazo Compounds. Decomposition of diazo compounds to form carbenes is a quite general reaction that is applicable to diazomethane and other diazoalkanes, diazoalkenes, and diazo compounds with aryl and acyl substituents. The main restrictions on this method are the limitations on synthesis and limited stability of the diazo compounds. The smaller diazoalkanes are toxic and potentially explosive, and they are usually prepared immediately before use. The most general synthetic routes involve base-catalyzed decomposition of *N*-nitroso derivatives of amides, ureas, or sulfonamides, as illustrated by several reactions used for the preparation of diazomethane.

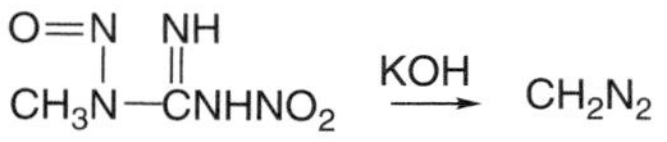

$$\underset{\underset{\displaystyle CH_3N-CNHNO_2}{|\quad\;\;||}}{O=N\;\;NH}\quad\xrightarrow{KOH}\quad CH_2N_2$$

Ref. 124

Scheme 10.8. General Methods for Generation of Carbenes

Precursor	Conditions	Products	Ref.
Diazoalkanes $R_2C=N^+=N^-$	Photolysis, thermolysis or metal catalysis	$R_2C: + N_2$	a
Salts of sulfonylhydrazones $[R_2C=NNSO_2Ar]^-$	Photolysis or thermolysis; diazoalkanes are intermediates	$R_2C: + N_2 + ArSO_2^-$	b
Diazirines 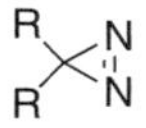 $\underset{R}{\overset{R}{>}}\!\!\times\!\!\underset{N}{\overset{N}{\vert\vert}}$	Photolysis	$R_2C: + N_2$	c
Alkyl halides R_2CH-X	Strong base, including metalation	$R_2C: + X^- + B-H$	d
α-Haloalkylmercury compounds $\underset{\underset{\displaystyle X}{\vert}}{R_2CHgZ}$	Thermolysis	$R_2C: + HgXZ$	e

a. W. J. Baron, M. R. DeCamp, M. E. Hendrick, M. Jones, Jr., R. H. Levin, and M. B. Sohn, in *Carbenes*, M. Jones, Jr., and R. A. Moss, eds. John Wiley & Sons, New York, 1973, pp. 1–151.
b. W. B. Bamford and T. S. Stevens, *J. Chem. Soc.*, 4735 (1952).
c. H. M. Frey, *Adv. Photochem.*, **4**, 225 (1966); R. A. G. Smith and J. R. Knowles, *J. Chem. Soc., Perkin Trans. 2*, 686 (1975); T. C. Celius and J. P. Toscano, *CRC Handbook of Organo Photochemistry and Photobiology*, 2nd Edition, 2004, pp 92/1–92/10.
d. W. Kirmse, *Carbene Chemistry*, Academic Press, New York, 1971, pp. 96–109, 129–149.
e. D. Seyferth, *Acc. Chem. Res.* **5**, 65 (1972).

124. M. Neeman and W. S. Johnson, *Org. Synth.*, **V**, 245 (1973).

910

CHAPTER 10

*Reactions Involving
Carbocations, Carbenes,
and Radicals as Reactive
Intermediates*

$$\underset{\overset{\displaystyle O=N}{|}\ \overset{\displaystyle O}{||}}{CH_3N-CNH_2} \xrightarrow[RO^-]{HO^-} CH_2N_2 \qquad \text{Ref. 125}$$

$$\underset{CH_3N-C}{\overset{O=N}{|}\ \overset{O}{||}}\!\!-\!\!\!\bigcirc\!\!\!-\!\!\underset{}{\overset{O}{||}\ \overset{N=O}{|}}{C-NCH_3} \xrightarrow{NaOH} CH_2N_2 \qquad \text{Ref. 126}$$

$$\underset{\overset{\displaystyle O=N}{|}}{CH_3N-SO_2Ph} \xrightarrow{KOH} CH_2N_2 \qquad \text{Ref. 127}$$

The details of the base-catalyzed decompositions vary somewhat but the mechanisms involve two essential steps.[128] The initial reactants undergo a base-catalyzed addition-elimination to form an alkyl diazoate. This is followed by a deprotonation of the α-carbon and elimination of hydroxide.

$$RCH_2N\overset{N=O}{\underset{Z}{\overset{|}{\diagdown}X}} \xrightarrow{^-OH} RCH_2N\overset{N=O}{\underset{OH}{\overset{|}{-}}C\diagdown X^-}{\underset{HO^-}{\diagup Z}} \longrightarrow \underset{\overset{|}{H}}{RCH-N=N-O-H} \xrightarrow{-H_2O} RCH=\overset{+}{N}=\ddot{N}^-$$

Diazo compounds can also be obtained by oxidation of the corresponding hydrazone,[129] the route that is most common when one of the substituents is an aromatic ring.

$$Ph_2C=NNH_2 \xrightarrow{HgO} Ph_2C=\overset{+}{N}=\ddot{N}^- \qquad \text{Ref. 130}$$

The higher diazoalkanes can be made by $Pb(O_2CCH_3)_4$ oxidation of hydrazones.[129]

α-Diazoketones are especially useful in synthesis.[131] There are several methods of preparation. Reaction of diazomethane with an acyl chloride results in formation of a diazomethyl ketone.

$$\underset{RCCl}{\overset{O}{||}} + H_2C=\overset{+}{N}=N^- \longrightarrow \underset{RCCH}{\overset{O}{||}}=\overset{+}{N}=N^-$$

The HCl generated in this reaction destroys one equivalent of diazomethane, but this can be avoided by including a base, such as triethylamine, to neutralize the acid.[132]

125. F. Arndt, *Org. Synth.*, **II**, 165 (1943).
126. T. J. de Boer and H. J. Backer, *Org. Synth.*, **IV**, 250 (1963).
127. J. A. Moore and D. E. Reed, *Org. Synth.*, **V**, 351 (1973).
128. W. M. Jones, D. L. Muck, and T. K. Tandy, Jr., *J. Am. Chem. Soc.*, **88**, 3798 (1966); R. A. Moss, *J. Org. Chem.*, **31**, 1082 (1966); D. E. Applequist and D. E. McGreer, *J. Am. Chem. Soc.*, **82**, 1965 (1960); S. M. Hecht and J. W. Kozarich, *J. Org. Chem.*, **38**, 1821 (1973); E. H. White, J. T. DePinto, A. J. Polito, I. Bauer, and D. F. Roswell, *J. Am. Chem. Soc.*, **110**, 3708 (1988).
129. T. L. Holton and H. Shechter, *J. Org. Chem.*, **60**, 4725 (1995).
130. L. I. Smith and K. L. Howard, *Org. Synth.*, **III**, 351 (1955).
131. T. Ye and M. A. McKervey, *Chem. Rev.*, **94**, 1091 (1994).
132. M. S. Newman and P. Beall, III, *J. Am. Chem. Soc.*, **71**, 1506 (1949); M. Berebom and W. S. Fones, *J. Am. Chem. Soc.*, **71**, 1629 (1949); L. T. Scott and M. A. Minton, *J. Org. Chem.*, **42**, 3757 (1977).

Cyclic α-diazoketones, which are not available from acyl chlorides, can be prepared by reaction of an enolate equivalent with a sulfonyl azide, in a reaction known as *diazo transfer*.[133] Various arenesulfonyl azides[134] and methanesulfonyl azide[135] are used most frequently. Because of the potential explosion hazard of sulfonyl azides, safety is a factor in choosing the reagent. 4-Dodecylbenzenesulfonyl azide has been recommended on the basis of relative thermal stability.[136] This reagent has been used in an *Organic Synthesis* preparation of 1-diazo-4-phenylbut-3-en-2-one.[137]

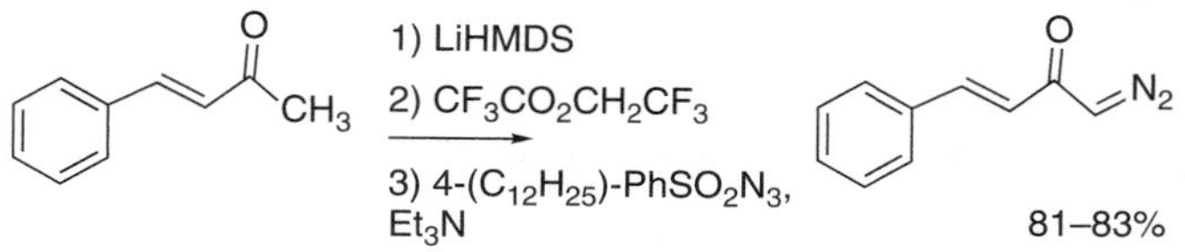

A polymer bound arenesulfonyl azide can be prepared from polystyrene.[138]

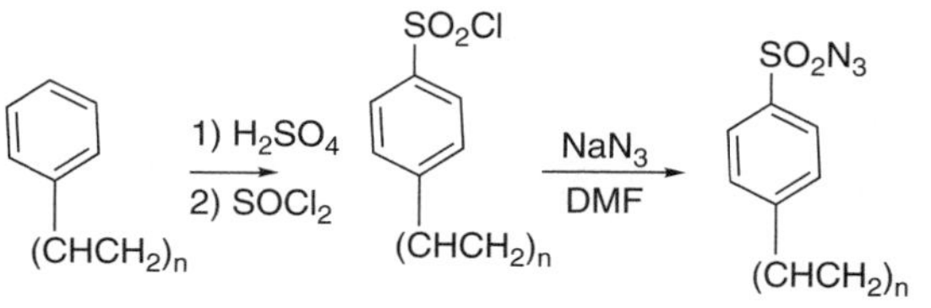

This reagent effects diazo transfer in good yield.

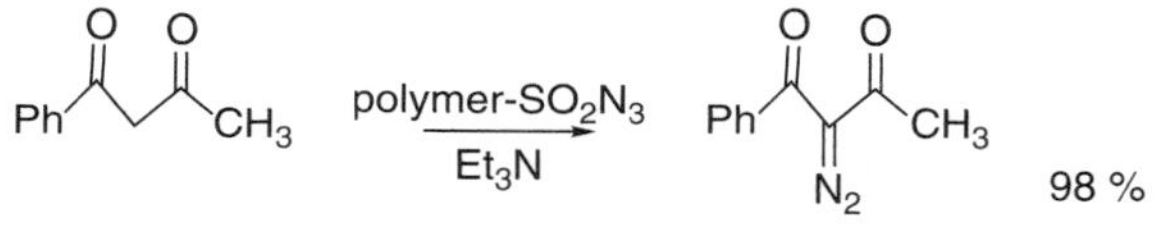

Several types of compounds can act as the carbon nucleophile in diazo transfer, including the oxymethylene[139] or dialkylaminomethylene[140] derivatives of the ketone. These activating substituents are lost during these reactions.

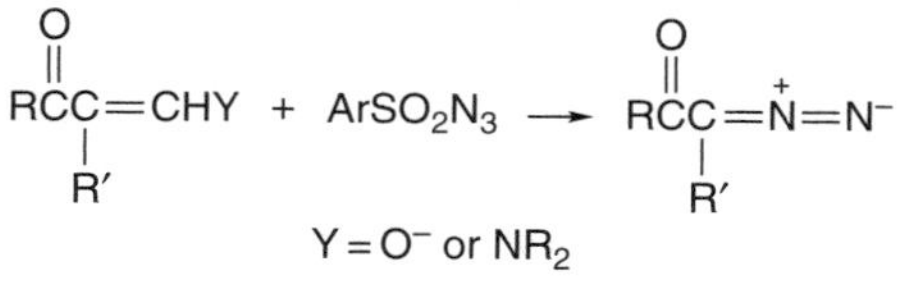

133. F. W. Bollinger and L. D. Tuma, *Synlett*, 407 (1996).

134. J. B. Hendrickson and W. A. Wolf, *J. Org. Chem.*, **33**, 3610 (1968); J. S. Baum, D. A. Shook, H. M. L. Davies, and H. D. Smith, *Synth. Commun.*, **17**, 1709 (1987); L. Lombardo and L. N. Mander, *Synthesis*, 368 (1980).

135. D. F. Taber, R. E. Ruckle, and M. J. Hennessy, *J. Org. Chem.*, **57**, 4077 (1986); R. L. Danheiser, D. S. Casebier, and F. Firooznia, *J. Org. Chem.*, **60**, 8341 (1995).

136. L. D. Tuma, *Thermochimica Acta*, **243**, 161 (1994).

137. R. L. Danheiser, R. F. Miller, and R. G. Brisbois, *Org. Synth.*, **73**, 134 (1996).

138. G. M. Green, N. P. Peet, and W. A. Metz, *J. Org. Chem.*, **66**, 2509 (2001).

139. M. Regitz and G. Heck, *Chem. Ber.*, **97**, 1482 (1964); M. Regitz, *Angew. Chem. Int. Ed. Engl.*, **6**, 733 (1967).

140. M. Rosenberger, P. Yates, J. B. Hendrickson, and W. Wolf, *Tetrahedron Lett.*, 2285 (1964); K. B. Wiberg, B. L. Furtek, and L. K. Olli, *J. Am. Chem. Soc.*, **101**, 7675 (1979).

α-Trifluoroacetyl derivatives of ketones are also useful substrates for diazo transfer reactions.[141] They are made by enolate acylation using 2,2,2-trifluoroethyl trifluoroacetate. The trifluoroacetyl group is cleaved during diazo transfer.

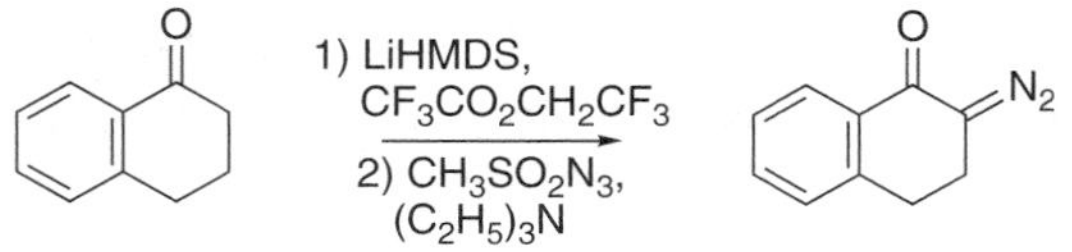

Benzoyl groups are also selectively cleaved during diazo transfer. This method has been used to prepare diazo ketones and diazo esters.[142]

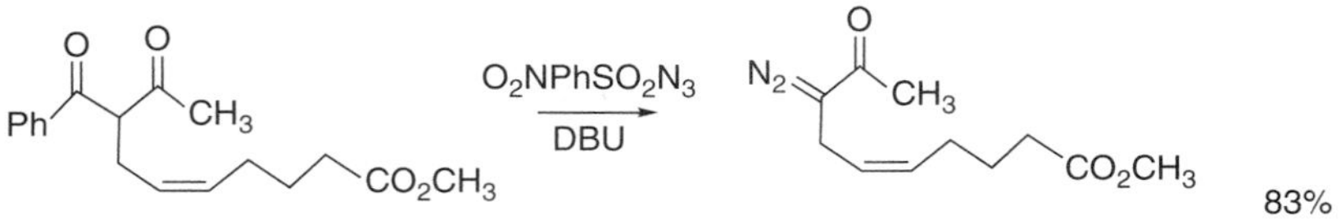

83%

α-Diazo ketones can also be made by first converting the ketone to an α-oximino derivative by nitrosation and then oxidizing the oximino ketone with chloramine.[143]

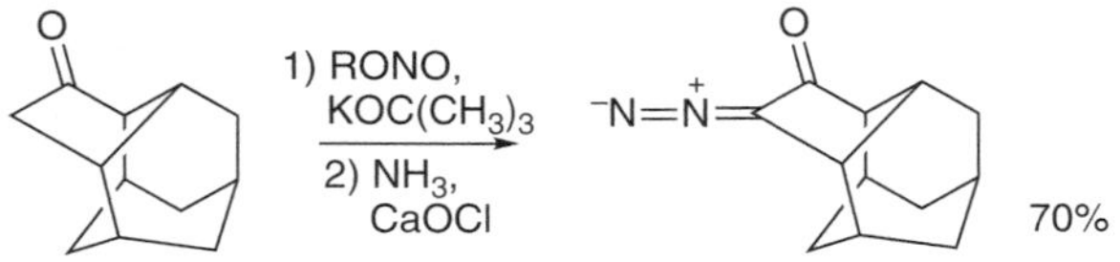

70%

Ref. 144

α-Diazo esters can be prepared by esterification of alcohols with the tosylhydrazone of glyoxyloyl chloride, followed by reaction with triethylamine.[145]

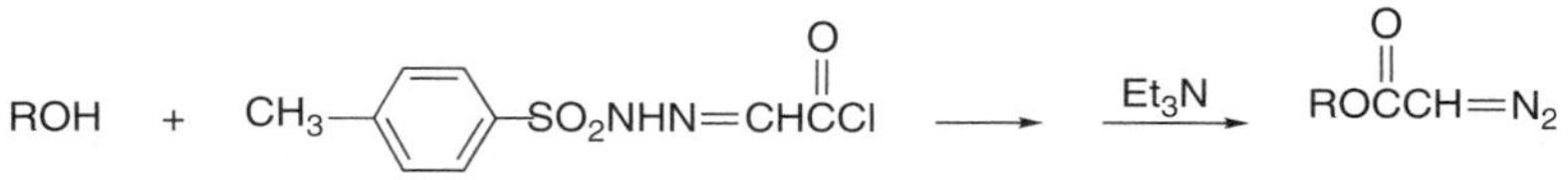

The driving force for decomposition of diazo compounds to carbenes is the formation of the very stable nitrogen molecule. Activation energies for decomposition of diazoalkanes in the gas phase are about 30 kcal/mol. The requisite energy can also be supplied by photochemical excitation. It is often possible to control the photochemical process to give predominantly singlet or triplet carbene. Direct photolysis leads to the singlet intermediate when the dissociation of the excited diazoalkene is faster than intersystem crossing to the triplet state. The triplet carbene is the principal intermediate in photosensitized decomposition of diazoalkanes. (See Part A, Chapter 12 to review photosensitization.)

Reaction of diazo compounds with a variety of transition metal compounds leads to evolution of nitrogen and formation of products of the same general type as those formed by thermal and photochemical decomposition of diazoalkanes. These transition

141. R. L. Danheiser, R. F. Miller, R. G. Brisbois, and S. Z. Park, *J. Org. Chem.*, **55**, 1959 (1990).
142. D. F. Taber, D. M. Gleave, R. J. Herr, K. Moody, and M. J. Hennessy, *J. Org. Chem.*, **60**, 1093 (1995).
143. T. N. Wheeler and J. Meinwald, *Org. Synth.*, **52**, 53 (1972).
144. T. Sasaki, S. Eguchi, and Y. Hirako, *J. Org. Chem.*, **42**, 2981 (1977).
145. E. J. Corey and A. G. Myers, *Tetrahedron Lett.*, **25**, 3559 (1984).

metal–catalyzed reactions involve carbenoid intermediates in which the carbene is bound to the metal.[146] The metals that have been used most frequently in synthetic reactions are copper and rhodium, and these reactions are discussed in Section 10.2.3.2

10.2.2.2. Carbenes from Sulfonylhydrazones. The second method listed in Scheme 10.8, thermal or photochemical decomposition of salts of arenesulfonylhydrazones, is actually a variation of the diazoalkane method, since diazo compounds are intermediates. The conditions of the decomposition are usually such that the diazo compound reacts immediately on formation.[147] The nature of the solvent plays an important role in the outcome of sulfonylhydrazone decompositions. In protic solvents, the diazoalkane can be diverted to a carbocation by protonation.[148] Aprotic solvents favor decomposition via the carbene pathway.

$$\underset{\substack{\| \\ RCR}}{\overset{O}{}} + NH_2NHSO_2Ar \longrightarrow R_2C{=}NNSO_2Ar \overset{base}{\longrightarrow} R_2C{=}N{-}\bar{N}SO_2Ar$$

$$R_2C{=}N{-}\bar{N}SO_2Ar \longrightarrow R_2\overset{+}{C}{=}\overset{}{N}{=}N^- \overset{h\nu}{\underset{or\ \Delta}{\longrightarrow}} R_2C:$$

$$R_2\overset{+}{C}{=}N{=}N^- \overset{SOH}{\longrightarrow} R_2\overset{H}{\underset{}{C}}{-}\overset{+}{N}{\equiv}N \longrightarrow R_2\overset{+}{C}H + N_2$$

10.2.2.3. Carbenes from Diazirines. The diazirine precursors of carbenes (Scheme 10.8, Entry 3) are cyclic isomers of diazo compounds. The strain of the small ring and the potential for formation of nitrogen make them highly reactive toward loss of nitrogen on photoexcitation. Diazirines have been used mainly in mechanistic investigations of carbenes. They are, in general, somewhat less easily available than diazo compounds or arenesulfonylhydrazones. However, there are several useful synthetic routes.[149]

Ref. 150

Ref. 151

146. W. R. Moser, *J. Am. Chem. Soc.*, **91**, 1135, 1141 (1969); M. P. Doyle, *Chem. Rev.*, **86**, 919 (1986); M. Brookhart, and H. B. Studabaker *Chem. Rev.*, **87**, 411 (1987).

147. G. M. Kaufman, J. A. Smith, G. G. Vander Stouw, and H. Shechter, *J. Am. Chem. Soc.*, **87**, 935 (1965).

148. J. H. Bayless, L. Friedman, F. B. Cook, and H. Shechter, *J. Am. Chem. Soc.*, **90**, 531 (1968).

149. For reviews of synthesis of diazirines, see E. Schmitz, *Dreiringe mit Zwei Heteroatomen*, Springer Verlag, Berlin, 1967, pp. 114–121; E. Schmitz, *Adv. Heterocycl. Chem.*, **24**, 63 (1979); H. W. Heine, in *Chem. Heterocycl. Compounds*, Vol. 42, Part 2, A. Hassner, ed., Wiley-Interscience, New York, 1983, pp. 547–628.

150. G. Kurz, J. Lehmann, and R. Thieme, *Carbohydrate Res.*, **136**, 125 (1983).

151. D. F. Johnson and R. K. Brown, *Photochem. Photobiol.*, **43**, 601 (1986).

914

CHAPTER 10

*Reactions Involving
Carbocations, Carbenes,
and Radicals as Reactive
Intermediates*

10.2.2.4. Carbenes from Halides by α-Elimination. The α-elimination of hydrogen halide induced by strong base (Scheme 10.8, Entry 4) is restricted to reactants that do not have β-hydrogens, because dehydrohalogenation by β-elimination dominates when it can occur. The classic example of this method of carbene generation is the generation of dichlorocarbene by base-catalyzed decomposition of chloroform.[152]

$$HCCl_3 \ + \ {}^-OR \ \rightleftharpoons \ {}^-:CCl_3 \ \longrightarrow \ :CCl_2 \ + \ Cl^-$$

Both phase transfer and crown ether catalysis have been used to promote α-elimination reactions of chloroform and other haloalkanes.[153] The carbene can be trapped by alkenes to form dichlorocyclopropanes.

$$Ph_2C{=}CH_2 \ + \ CHCl_3 \ \xrightarrow[\text{50\% NaOH}]{PhCH_2\overset{+}{N}(C_2H_5)_3} \ \underset{Ph}{\overset{Ph}{\diagdown}}{\triangle}\overset{Cl}{\underset{}{\diagup}}Cl$$

Ref. 154

Dichlorocarbene can also be generated by sonication of a solution of chloroform with powdered KOH.[155]

α-Elimination also occurs in the reaction of dichloromethane and benzyl chlorides with alkyllithium reagents. The carbanion stabilization provided by the chloro and phenyl groups makes the lithiation feasible.

$$H_2CCl_2 \ + \ RLi \ \longrightarrow \ RH \ + \ LiCHCl_2 \ \longrightarrow \ :CHCl \ + \ LiCl$$

Ref. 156

$$ArCH_2X \ + \ RLi \ \longrightarrow \ RH \ + \ Ar\overset{\overset{\displaystyle Li}{|}}{C}HX \ \longrightarrow \ Ar\overset{..}{C}H \ + \ LiX$$

Ref. 157

The reactive intermediates under some conditions may be the carbenoid α-haloalkyllithium compounds or carbene-lithium halide complexes.[158] In the case of the trichloromethyllithium to dichlorocarbene conversion, the equilibrium lies heavily to the side of trichloromethyllithium at $-100°C$.[159] The addition reaction with alkenes seems to involve dichlorocarbene, however, since the pattern of reactivity toward different alkenes is identical to that observed for the free carbene in the gas phase.[160]

152. J. Hine, *J. Am. Chem. Soc.*, **72**, 2438 (1950); J. Hine and A. M. Dowell, Jr., *J. Am. Chem. Soc.*, **76**, 2688 (1954).
153. W. P. Weber and G. W. Gokel, *Phase Transfer Catalysis in Organic Synthesis*, Springer Verlag, New York, 1977, Chaps. 2–4.
154. E. V. Dehmlow and J. Schoenefeld, *Liebigs Ann. Chem.*, **744**, 42 (1971).
155. S. L. Regen and A. Singh, *J. Org. Chem.*, **47**, 1587 (1982).
156. G. Köbrich, H. Trapp, K. Flory, and W. Drischel, *Chem. Ber.*, **99**, 689 (1966); G. Kobrich and H. R. Merkle, *Chem. Ber.*, **99**, 1782 (1966).
157. G. L. Closs and L. E. Closs, *J. Am. Chem. Soc.*, **82**, 5723 (1960).
158. G. Kobrich, *Angew. Chem. Int. Ed. Engl.*, **6**, 41 (1967).
159. W. T. Miller, Jr., and D. M. Whalen, *J. Am. Chem. Soc.*, **86**, 2089 (1964); D. F. Hoeg, D. I. Lusk, and A. L. Crumbliss, *J. Am. Chem. Soc.*, **87**, 4147 (1965).
160. P. S. Skell and M. S. Cholod, *J. Am. Chem. Soc.*, **91**, 6035, 7131 (1969); P. S. Skell and M. S. Cholod, *J. Am. Chem. Soc.*, **92**, 3522 (1970).

A method that provides an alternative route to dichlorocarbene is the decarboxylation of trichloroacetic acid.[161] The decarboxylation generates the trichloromethyl anion, which decomposes to the carbene. Treatment of alkyl trichloroacetates with an alkoxide also generates dichlorocarbene.

$$^-O-\overset{\overset{\displaystyle O}{\|}}{C}-CCl_3 \xrightarrow{-CO_2} \quad :\bar{C}Cl_3 \quad \longleftarrow \quad Cl_3\overset{\overset{\displaystyle O}{\|}}{C}-COR \quad \longleftarrow \quad Cl_3C\overset{\overset{\displaystyle O}{\|}}{C}OR$$

$$\downarrow \qquad\qquad\qquad\qquad\qquad |\qquad\qquad\qquad |$$
$$:CCl_2 + Cl^- \qquad\qquad\qquad OR' \qquad\qquad\qquad \bar{O}R'$$

The applicability of these methods is restricted to polyhalogenated compounds, since the inductive effect of the halogen atoms is necessary for facilitating formation of the carbanion.

Hindered lithium dialkylamides can generate aryl-substituted carbenes from benzyl halides.[162] Reaction of α,α-dichlorotoluene or α,α-dibromotoluene with potassium *t*-butoxide in the presence of 18-crown-6 generates the corresponding α-halophenylcarbene.[163] The relative reactivity data for carbenes generated under these latter conditions suggest that they are "free." The potassium cation would be expected to be strongly solvated by the crown ether and it is evidently not involved in the carbene-generating step.

10.2.2.5. Carbenes from Organomercury Compounds.

The α-elimination mechanism is also the basis for the use of organomercury compounds for carbene generation (Scheme 10.8 , Entry 5). The carbon-mercury bond is much more covalent than the C—Li bond, however, so the mercury reagents are generally stable at room temperature and can be isolated. They decompose to the carbene on heating.[164] Addition reactions occur in the presence of alkenes. The decomposition rate is not greatly influenced by the alkene. This observation implies that the rate-determining step is generation of the carbene from the organomercury precursor.[165]

$$PhHg-\overset{\overset{\displaystyle Cl}{|}}{\underset{\underset{\displaystyle Cl}{|}}{C}}-Br \longrightarrow :CCl_2 + PhHgBr$$

A variety of organomercury compounds that can serve as precursors of substituted carbenes have been synthesized. For example, carbenes with carbomethoxy or trifluoromethyl substituents can be generated in this way.[166]

$$PhHg-\overset{\overset{\displaystyle Cl}{|}}{\underset{\underset{\displaystyle CCF_3}{|}}{C}}-Br \longrightarrow Cl\ddot{C}CF_3$$

$$PhHgCCl_2CO_2CH_3 \longrightarrow Cl\ddot{C}CO_2CH_3$$

161. W. E. Parham and E. E. Schweizer, *Org. React.*, **13**, 55 (1963).
162. R. A. Olofson and C. M. Dougherty, *J. Am. Chem. Soc.*, **95**, 581 (1973).
163. R. A. Moss and F. G. Pilkiewicz, *J. Am. Chem. Soc.*, **96**, 5632 (1974).
164. D. Seyferth, J. M. Burlitch, R. J. Minasz, J. Y.-P. Mui, H. D. Simmons, Jr., A. J. H. Treiber, and S. R. Dowd, *J. Am. Chem. Soc.*, **87**, 4259 (1965).
165. D. Seyferth, J. Y.-P. Mui, and J. M. Burlitch, *J. Am. Chem. Soc.*, **89**, 4953 (1967).
166. D. Seyferth, D. C. Mueller, and R. L. Lambert, Jr., *J. Am. Chem. Soc.*, **91**, 1562 (1969).

916

CHAPTER 10

*Reactions Involving
Carbocations, Carbenes,
and Radicals as Reactive
Intermediates*

The addition reaction of alkenes and phenylmercuric bromide typically occurs at about 80°C. Phenylmercuric iodides are somewhat more reactive and may be advantageous in reactions with relatively unstable alkenes.[167]

10.2.3. Addition Reactions

Addition reactions with alkenes to form cyclopropanes are the most studied reactions of carbenes, both from the point of view of understanding mechanisms and for synthetic applications. A concerted mechanism is possible for singlet carbenes. As a result, the stereochemistry present in the alkene is retained in the cyclopropane. With triplet carbenes, an intermediate 1,3-diradical is involved. Closure to cyclopropane requires spin inversion. The rate of spin inversion is slow relative to rotation about single bonds, so mixtures of the two possible stereoisomers are obtained from either alkene stereoisomer.

Reactions involving free carbenes are very exothermic since two new σ bonds are formed and only the alkene π bond is broken. The reactions are very fast and, in fact, theoretical treatment of the addition of singlet methylene to ethylene suggests that there is no activation barrier.[168] The addition of carbenes to alkenes is an important method for synthesis of many types of cyclopropanes and several of the methods for carbene generation listed in Scheme 10.8 have been adapted for use in synthesis. Scheme 10.9, at the end of this section, gives a number of specific examples.

10.2.3.1. Cyclopropanation with Halomethylzinc Reagents. A very effective means for conversion of alkenes to cyclopropanes by transfer of a CH_2 unit involves reaction with methylene iodide and a zinc-copper couple, referred to as the *Simmons-Smith reagent*.[169] The reactive species is iodomethylzinc iodide.[170] The transfer of methylene occurs stereospecifically. Free :CH_2 is not an intermediate. Entries 1 to 3 in Scheme 10.9 are typical examples.

[167] D. Seyferth and C. K. Haas, *J. Org. Chem.*, **40**, 1620 (1975).
[168] B. Zurawski and W. Kutzelnigg, *J. Am. Chem. Soc.*, **100**, 2654 (1978).
[169] H. E. Simmons and R. D. Smith, *J. Am. Chem. Soc.*, **80**, 5323 (1958); H. E. Simmons and R. D. Smith, **81**, 4256 (1959); H. E. Simmons, T. L. Cairns, S. A. Vladuchick, and C. M. Hoiness, *Org. React.*, **20**, 1 (1973); W. B. Motherwell and C. J. Nutley, *Contemporary Org. Synth.*, **1**, 219 (1994); A. B. Charette and A. Beauchemin, *Org. React.*, **58**, 1 (2001).
[170] A. B. Charette and J.-F. Marcoux, *J. Am. Chem. Soc.*, **118**, 4539 (1996).

A modified version of the Simmons-Smith reaction uses dibromomethane and in situ generation of the Cu-Zn couple.[171] Sonication is used in this procedure to promote reaction at the metal surface.

Ref. 172

Another useful reagent combination involves diethylzinc and diiodomethane or chloroiodomethane.

Ref. 173

Several modifications of the Simmons-Smith procedure have been developed in which an electrophile or Lewis acid is included. Inclusion of acetyl chloride accelerates the reaction and permits the use of dibromomethane.[174] Titanium tetrachloride has similar effects in the reactions of unfunctionalized alkenes.[175] Reactivity can be enhanced by inclusion of a small amount of trimethylsilyl chloride.[176] The Simmons-Smith reaction has also been found to be sensitive to the purity of the zinc used. Electrolytically prepared zinc is much more reactive than zinc prepared by metallurgic smelting, and this has been traced to small amounts of lead in the latter material.

The nature of reagents prepared under different conditions has been explored both structurally and spectroscopically.[177] $C_2H_5ZnCH_2I$, $Zn(CH_2I)_2$, and ICH_2ZnI are all active methylene transfer reagents.

$$(C_2H_5)_2Zn + CH_2I_2 \longrightarrow C_2H_5ZnCH_2I + C_2H_5I$$

$$C_2H_5ZnI + CH_2I_2 \longrightarrow ICH_2ZnI + C_2H_5I$$

A crystal structure has been obtained for $Zn(CH_2I)_2$ complexed with *exo,exo*-2,3-dimethoxybornane and is shown in Figure 10.4.

Computational studies were done on several $ClZnCH_2Cl$ models, and the results are summarized in Figure 10.5.[178] A minimal TS consisting of $ClZnCH_2Cl$ and ethene shows charge transfer mainly to the departing Cl; that is, the ethene displaces chloride in the zinc coordination sphere. The model can be elaborated by inclusion of $ZnCl_2$,

171. E. C. Friedrich, J. M. Demek, and R. Y. Pong, *J. Org. Chem.*, **50**, 4640 (1985).
172. S. Sawada and Y. Inouye, *Bull. Chem. Soc. Jpn.*, **42**, 2669 (1969); N. Kawabata, T. Nakagawa, T. Nakao, and S. Yamashita, *J. Org. Chem.*, **42**, 3031 (1977); J. Furukawa, N. Kawabata, and J. Nishimura, *Tetrahedron*, **24**, 53 (1968).
173. J. Furukawa, N. Kawabata, and J. Nishimura, *Tetrahedron*, **24**, 53 (1968); S. Miyano and H. Hashimoto, *Bull. Chem. Soc. Jpn.*, **46**, 892 (1973); S. E. Denmark and J. P. Edwards, *J. Org. Chem.*, **56**, 6974 (1991).
174. E. C. Friedrich and E. J. Lewis, *J. Org. Chem.*, **55**, 2491 (1990).
175. E. C. Friedrich, S. E. Lunetta, and E. J. Lewis, *J. Org. Chem.*, **54**, 2388 (1989).
176. K. Takai, T. Kakiuchi, and K. Utimoto, *J. Org. Chem.*, **59**, 2671 (1994).
177. S. E. Denmark, J. P. Edwards, and S. R. Wilson, *J. Am. Chem. Soc.*, **114**, 2592 (1992); A. B. Charette and J.-F. Marcoux, *J. Am. Chem. Soc.*, **118**, 4539 (1996).
178. M. Nakamura, A. Hirai, and E. Nakamura, *J. Am. Chem. Soc.*, **125**, 2341 (2003).

918

CHAPTER 10

*Reactions Involving
Carbocations, Carbenes,
and Radicals as Reactive
Intermediates*

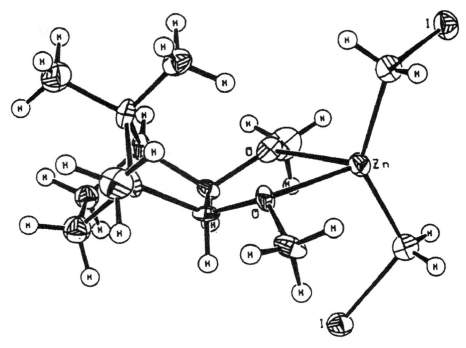

Fig. 10.4. Crystal structure of one molecule of $Zn(CH_2I)_2$ complexed with *exo,exo*-2,3-dimethoxybornane. Reproduced from *J. Am. Chem. Soc.*, **114**, 2592 (1992), by permission of the American Chemical Society.

which is present under most experimental conditions and can have an accelerating effect. Models were also calculated for the directing and activating effect of allylic hydroxy groups. Definitive results were not obtained for this case, but an aggregated structure with the oxygen coordinated to zinc is plausible.

Other reagents have been developed in which one of the zinc ligands is an oxy anion. Compounds with trifluoroacetate anions are prepared by protonolysis of C_2H_5 or CH_2I groups on zinc.[179]

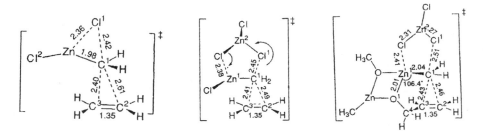

Fig. 10.5. Transition structures for CH_2 transfer from $ClCH_2ZnCl_2$ and $ClZnCH_2Cl-ZnCl_2$ to ethene and to coordinated allyl alcohol. Reproduced from *J. Am. Chem. Soc.*, **125**, 2341 (2003), by permission of the American Chemical Society.

[179.] J. C. Lorenz, J. Long, Z. Yang, S. Xue, Y. Xie, and Y. Shi, *J. Org. Chem.*, **69**, 327 (2004).

$$(C_2H_5)_2Zn + CF_3CO_2H \longrightarrow CF_3CO_2ZnC_2H_5 \xrightarrow{CH_2I_2} CF_3CO_2ZnCH_2I$$

Iodomethylzinc phenoxides can be prepared in a similar fashion. The best phenols are the 2,4,6-trihalophenols and the readily available 2,4,6-trichlorophenol was examined most thoroughly.[180]

$$(C_2H_5)_2Zn + ArOH \longrightarrow ArOZnC_2H_5 \xrightarrow{CH_2I_2} ArOZnCH_2I$$

This reagent can achieve better than 90% yields for a variety of unactivated alkenes.

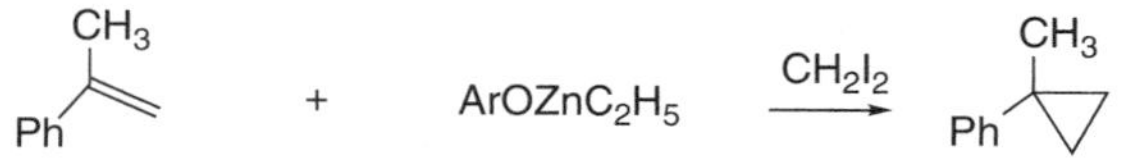

The reactivity of the oxy anions is in the order $CF_3CO_2^- > ArO^- >> RO^-$.

In molecules containing hydroxy groups, the CH_2 unit is selectively introduced on the side of the double bond *syn* to the hydroxy group in the Simmons-Smith reaction and related cyclopropanations. This indicates that the reagent is complexed to the hydroxy group and that the complexation facilitates the addition. Entries 3 and 4 in Scheme 10.9 illustrate the stereodirective effect of the hydroxy group. It is evidently the Lewis base character of the group that is important, in contrast to the hydrogen bonding that is involved in epoxidation. The lithium salts of allylic alcohols are also strongly activated, even more so than the alcohols. This reactivity has been used to advantage in the preparation of relatively unstable products.[181]

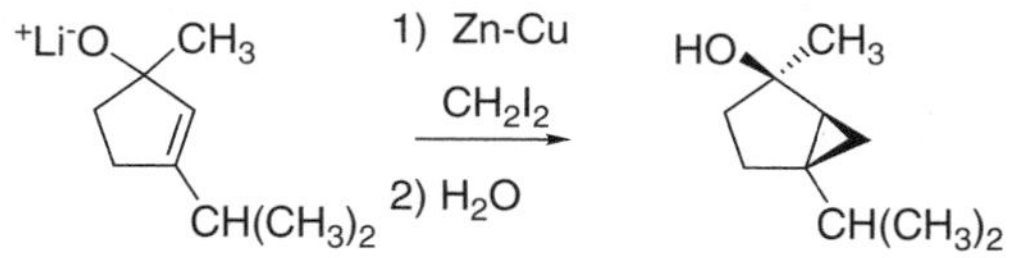

While amino groups alone are not effective directing groups, both ephedrine and pseudoephedrine derivatives give high diastereoselectivity. This is evidently due to chelation by the hydroxy group, as both auxiliaries give the same facial selectivity despite differing in configuration at the nitrogen position.[182]

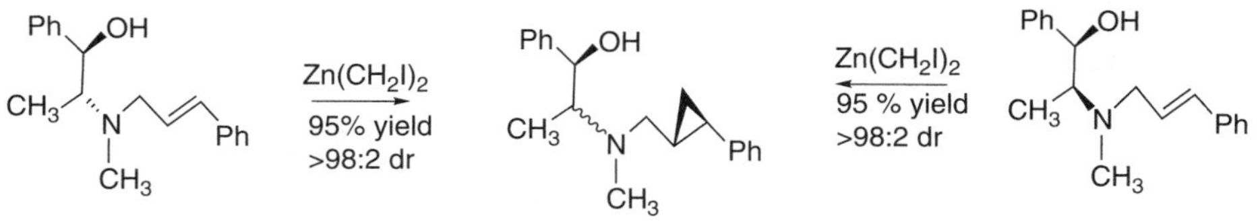

Dioxolanyl oxygens are also effective directing groups.[183]

180. A. B. Charette, S. Francouer, J. Martel, and N. Wilb, *Angew. Chem. Int. Ed. Engl.*, **39**, 4539 (2000).
181. D. Chang, T. Kreethadumrongdat, and T. Cohen, *Org. Lett.*, **3**, 2121 (2001).
182. V. K. Aggarwal, G. Y. Fang, and G. Meek, *Org. Lett.*, **5**, 4417 (2003).
183. A. G. M. Barrett, K. Kasdorf, and D. J. Williams, *J. Chem. Soc., Chem. Commun.*, 1781 (1994).

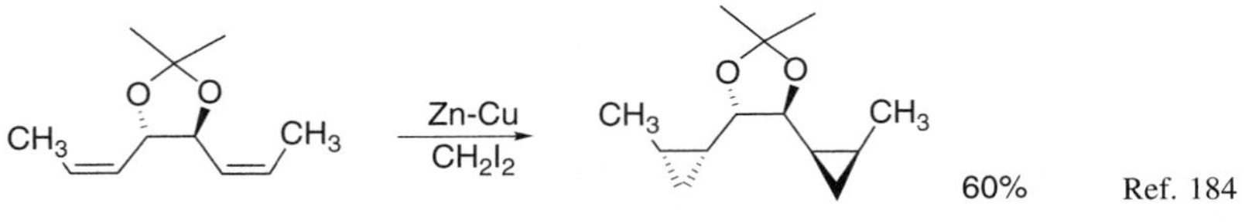

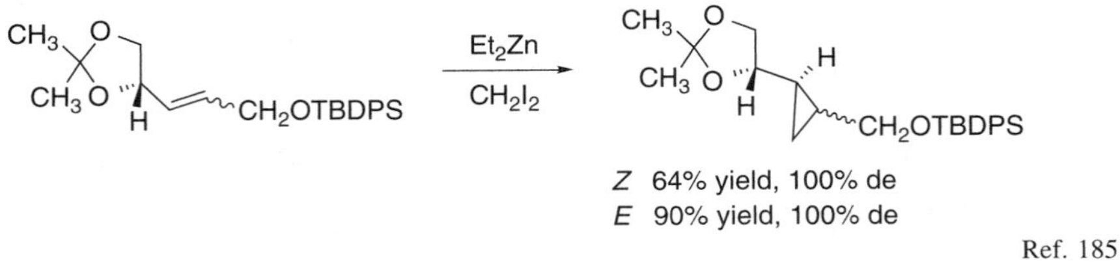

920

CHAPTER 10

*Reactions Involving
Carbocations, Carbenes,
and Radicals as Reactive
Intermediates*

60%   Ref. 184

Z   64% yield, 100% de
E   90% yield, 100% de

Ref. 185

The stereoselectivity is accounted for by a TS in which the allylic oxygen is coordinated to the zinc.

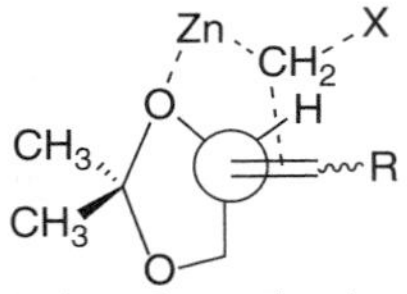

preferred conformation for directing
effect of dioxolanyl substituents

The directive effect of allylic hydroxy groups can be used in conjunction with chiral catalysts to achieve enantioselective cyclopropanation. The chiral ligand used is a boronate ester derived from the N,N,N',N'-tetramethyl amide of tartaric acid.[186] Similar results are obtained using the potassium alkoxide, again indicating the Lewis base character of the directive effect.

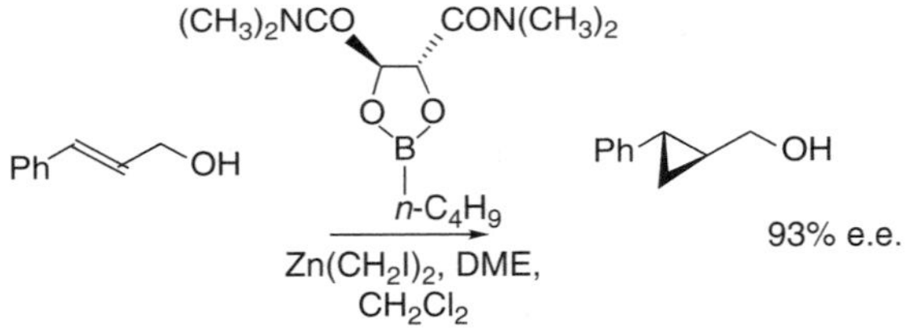

93% e.e.

These conditions were used to make natural products containing several successive cyclopropane rings.[187]

U-106305

184. T. Onoda, R. Shirai, Y. Koiso, and S. Iwasaki, *Tetrahedron Lett.*, **37**, 4397 (1996).
185. T. Morikawa, H. Sasaki, R. Hanai, A. Shibuya, and T. Taguchi, *J. Org. Chem.*, **59**, 97 (1994).
186. A. B. Charette and H. Juteau, *J. Am. Chem. Soc.*, **116**, 2651 (1994); A. B. Charette, S. Prescott, and C. Brochu, *J. Org. Chem.*, **60**, 1081 (1995).
187. A. B. Charette and H. Lebel, *J. Am. Chem. Soc.*, **118**, 10327 (1996).

The starting material was *trans*-cyclopropane-1,2-dimethanol. The contiguous cyclopropane units were added by two iterative sequences of oxidation–Wadsworth-Emmons–reduction–cyclopropanation.

10.2.3.2. Metal-Catalyzed Cyclopropanation. Carbene addition reactions can be catalyzed by several transition metal complexes. Most of the synthetic work has been done using copper or rhodium complexes and we focus on these. The copper-catalyzed decomposition of diazo compounds is a useful reaction for formation of substituted cyclopropanes.[188] The reaction has been carried out with several copper salts,[189] and both Cu(I) and Cu(II) triflate are useful.[190] Several Cu(II)salen complexes, such as the *N-t*-butyl derivative, which is called Cu(TBS)$_2$, have become popular catalysts.[191]

Ref. 192

An NMR and structural study characterized the intermediates generated from diimine catalysts on reaction with diazodiphenylmethane.[193] The dominant species in solution is dinuclear, but a monomeric metallocarbene species can be detected.

The monomeric species can be isolated as a solid in the case of the *N,N'*-dimesityl derivative. The crystal structures of both dimeric and monomeric structures are shown in Figure 10.6.

[188.] W. Kirmse, *Angew. Chem. Int. Ed. Engl.*, **42**, 1088 (2003).

[189.] W. von E. Doering and W. R. Roth, *Tetrahedron*, **19**, 715 (1963); J. P. Chesick, *J. Am. Chem. Soc.*, **84**, 3250 (1962); H. Nozaki, H. Takaya, S. Moriuti, and R. Noyori, *Tetrahedron*, **24**, 3655 (1968); R. G. Salomon and J. K. Kochi, *J. Am. Chem. Soc.*, **95**, 3300 (1973); M. E. Alonso, P. Jano, and M. I. Hernandez, *J. Org. Chem.*, **45**, 5299 (1980); T. Hudlicky, F. J. Koszyk, T. M. Kutchan, and J. P. Sheth, *J. Org. Chem.*, **45**, 5020 (1980); M. P. Doyle and M. L. Truell, *J. Org. Chem.*, **49**, 1196 (1984); E. Y. Chen, *J. Org. Chem.*, **49**, 3245 (1984).

[190.] R. T. Lewis and W. B. Motherwell, *Tetrahedron Lett.*, **29**, 5033 (1988).

[191.] E. J. Corey and A. G. Myers, *Tetrahedron Lett.*, **25**, 3559 (1984); J. D. Winkler and E. Gretler, *Tetrahedron Lett.*, **32**, 5733 (1991).

[192.] S. F. Martin, R. E. Austin, and C. J. Oalmann, *Tetrahedron Lett.*, **31**, 4731 (1990).

[193.] X. Dai and T. H. Warren, *J. Am. Chem. Soc.*, **126**, 10085 (2004).

922

CHAPTER 10

*Reactions Involving
Carbocations, Carbenes,
and Radicals as Reactive
Intermediates*

(a)   (b)

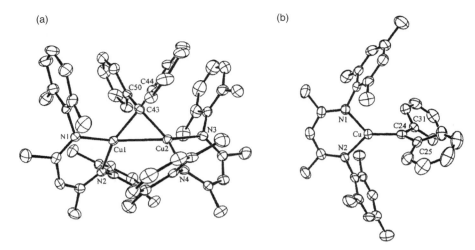

Fig. 10.6. Dimeric (Ar = 2,6-dimethylphenyl) (a) and monomeric (Ar = 2,4,6-trimethylphenyl) (b) copper complexes with diphenylcarbene. Reproduced from *J. Am. Chem. Soc.*, **126**, 10085 (2004), by permission of the American Chemical Society.

There has also been computational investigation of copper-catalyzed carbenoid addition reactions, as shown in Figure 10.7.[194] These computational studies agree with experimental investigations in identifying nitrogen extrusion as the rate-determining step. The addition step is a direct carbene transfer, as opposed to involving a metallo-cyclobutane intermediate.

Various other transition metal complexes are also useful, including rhodium,[195] palladium,[196] and molybdenum[197] compounds. The catalytic cycle can generally be represented as shown below.[198]

$$L_nM \quad\quad R_2CN_2$$
$$L_nM=CR_2 \quad N_2$$

194. J. M. Fraile, J. I. Garcia, V. Martinez-Merino, J. A. Mayoral, and L. Salvatella, *J. Am. Chem. Soc.*, **123**, 7616 (2001); T. Rasmussen, J. F. Jensen, N. Ostergaard, D. Tanner, T. Ziegler, and P.-O. Norrby, *Chem. Eur. J.*, **8**, 177 (2002).

195. S. Bien and Y. Segal, *J. Org. Chem.*, **42**, 1685 (1977); A. J. Anciaux, A. J. Hubert, A. F. Noels, N. Petiniot, and P. Teyssie, *J. Org. Chem.*, **45**, 695 (1980); M. P. Doyle, W. H. Tamblyn, and V. Baghari, *J. Org. Chem.*, **46**, 5094 (1981); D. F. Taber and R. E. Ruckle, Jr., *J. Am. Chem. Soc.*, **108**, 7686 (1986).

196. R. Paulissen, A. J. Hubert, and P. Teyssie, *Tetrahedron Lett.*, 1465 (1972); U. Mende, B. Raduchel, W. Skuballa, and H. Vorbruggen, *Tetrahedron Lett.*, 629 (1975); M. Suda, *Synthesis*, 714 (1981); M. P. Doyle, L. C. Wang, and K.-L. Loh, *Tetrahedron Lett.*, **25**, 4087 (1984); L. Strekowski, M. Visnick, and M. A. Battiste, *J. Org. Chem.*, **51**, 4836 (1986).

197. M. P. Doyle and J. G. Davidson, *J. Org. Chem.*, **45**, 1538 (1980); M. P. Doyle, R. L. Dorow, W. E. Buhro, J. H. Tamblyn, and M. L. Trudell, *Organometallics*, **3**, 44 (1984).

198. M. P. Doyle, *Chem. Rev.*, **86**, 919 (1986).

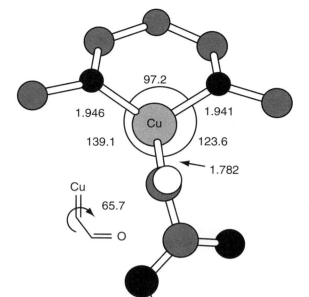

Fig. 10.7. Computational (B3LYP/6-31G(d)) minimum-energy structure of carbomethoxycarbene derivative of copper N,N'-dimethylpropane-1,3-diimine. Reproduced from *J. Am. Chem. Soc.*, **123**, 7616 (2001), by permission of the American Chemical Society.

The metal-carbene complexes are electrophilic in character. They can, in fact, be represented as metal-stabilized carbocations.

$$\ddot{M} + R_2\bar{C}-\overset{+}{N}\equiv N \xrightarrow{-N_2} \ddot{M}-\overset{R}{\underset{R}{C^+}} \longleftrightarrow M=\overset{R}{\underset{R}{C}}$$

In most transition metal–catalyzed reactions, one of the carbene substituents is a carbonyl group, which further enhances the electrophilicity of the intermediate. There are two general mechanisms that can be considered for cyclopropane formation. One involves formation of a four-membered ring intermediate that incorporates the metal. The alternative represents an electrophilic attack giving a polar species that undergoes 1,3-bond formation.

Since the additions are normally stereospecific with respect to the alkene, if an open-chain intermediate is involved it must collapse to product more rapidly than single-bond rotations that would destroy the stereoselectivity.

In recent years, much attention has been focused on rhodium-mediated carbenoid reactions. One goal has been to understand how the rhodium ligands control reactivity and selectivity, especially in cases in which both addition and insertion reactions are possible. These catalysts contain Rh−Rh bonds but function by mechanisms similar to other transition metal catalysts.

The original catalyst was $Rh_2(O_2CCH_3)_4$, but other carboxylates such as nonafluorobutanoate and amide anions, such as those from acetamide and caprolactam, also have good catalytic activity.[199]

rhodium carboxylates
R = CH_3, (CF_2)_3CF_3

rhodium acetamidate
Rh_2(acam)_4

Rh_2(caprolactamate)_4
(two ligands not shown)

The ligands adjust the electrophilicity of the catalyst with the nonafluorobutanoate being more electrophilic and the amido ligands less electrophilic than the acetate. These catalysts show differing reactivity. For example, $Rh_2(O_2C_4F_9)_4$ was found to favor aromatic substitution over cyclopropanation, whereas $Rh_2(caprolactamate)_4$ was selective for cyclopropanation.[200] In competition between tertiary alkyl insertion versus cyclopropanation, the order in favor of cyclopropanation is also $Rh_2(caprolactamate)_4 > Rh_2(O_2CCH_3)_4 > Rh_2(O_2CC_4F_9)_4$. These predictable selectivity patterns have made the rhodium catalysts useful in a number of synthetic applications.[201] For example, $Rh_2(O_2C_4F_9)_4$ gave exclusively insertion, whereas $Rh_2(caprolactamate)_4$ gave exclusively cyclopropanation. $Rh_2(O_2CCH_3)_4$ gave a mixture of the two products.[202]

199. M. P. Doyle, V. Bagheri, T. J. Wandless, N. K. Harn, D. B. Brinker, C. T. Eagle, and K.-L. Loh, *J. Am. Chem. Soc.*, **112**, 1906 (1990).

200. A. Padwa, D. J. Austin, A. T. Price, M. A. Semones, M. P. Doyle, M. N. Protopova, W. R. Winchester, and A. Tran, *J. Am. Chem. Soc.*, **115**, 8669 (1993).

201. M. P. Doyle and D. Forbes, *Chem. Rev.*, **98**, 911 (1998); C. A. Merlic and A. L. Zechman, *Synthesis*, 1137 (2003).

202. A. Padwa, D. J. Austin, S. F. Hornbuckle, and M. A. Semones, *J. Am. Chem. Soc.*, **114**, 1874 (1992).

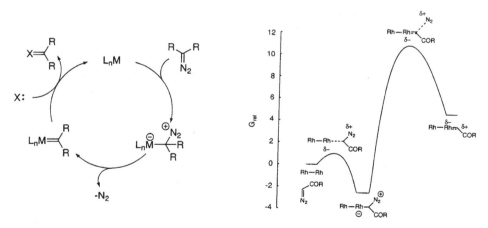

	Rh$_2$(O$_2$CC$_4$F$_9$)$_4$	100%	0%
	Rh$_2$(O$_2$CCH$_3$)$_4$	56%	44%
	Rh$_2$(caprolactamate)$_4$	0%	100%

    Mechanistic and computational studies have elucidated some of the key details of the reactions. A kinetic study of Rh$_2$[O$_2$CC(CH$_3$)$_3$]$_4$ involving several different reaction types established that the rate-determining step in the rhodium-catalyzed reactions is loss of nitrogen.[203] The basic mechanism and reaction energy profile are given in Figure 10.8. In addition, certain reactants and solvents were shown to have an inhibitory effect by competing with the diazo compound for coordination at the rhodium center. For example, anisole has such an effect.

    Another study combined measurement of kinetic isotope effects with computational modeling of the TS.[204] The computed energy profile suggests that there is no barrier for the reaction of styrene with the carbene complex from methyl diazoacetate. In contrast, a barrier of about 12 kcal/mol is found for methyl 2-diazobut-3-enoate. This is consistent with experimental work showing that alkenyl and aryl-substituted diazo esters have greater selectivity. Figure 10.9 shows the computed TS for the reaction of the phenyl-substituted ester with styrene. The addition is highly asynchronous and has an early TS. The kinetic isotope effects calculated for this model are in excellent agreement with the experimental values.

    This study also gives a good account of the stereoselectivity of the 2-diazobut-3-enoate addition reaction with styrene. There is a preference for the ester group

Fig. 10.8. Basic catalytic cycle and energy profile for rhodium-catalyzed carbenoid reactions. Reproduced from *J. Am. Chem. Soc.*, **124**, 1014 (2002), by permission of the American Chemical Society.

[203.] M. C. Pirrung, H. Liu, and A. T. Morehead, Jr., *J. Am. Chem. Soc.*, **124**, 1014 (2002).
[204.] D. T. Nowlan, III, T. M. Gregg, H. M. L. Davies, and D. A. Singleton, *J. Am. Chem. Soc.*, **125**, 15902 (2003).

926

CHAPTER 10

*Reactions Involving
Carbocations, Carbenes,
and Radicals as Reactive
Intermediates*

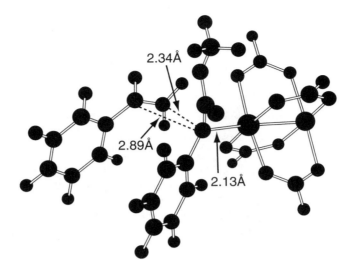

Fig. 10.9. Computed transition structure for addition of methyl phenyl-diazoacetate to styrene from B3LYP/6-31G*/LANL2DZ computations. Reproduced from *J. Am. Chem. Soc.*, **125**, 15902 (2003), by permission of the American Chemical Society.

to be *trans* to the phenyl group. The calculated difference between the two TSs is 1.7 kcal/mol. The main difference is the closer approach of the phenyl group to the ester oxygen in the disfavored TS. Steric interactions with the ester group also explain why *trans*-disubstituted alkenes are unreactive with this catalyst, whereas *cis*-alkenes are reactive (see Figure 10.10). We will see shortly that the same TS feature can account for the enantioselectivity of chiral rhodium catalysts.

As would be expected for a highly electrophilic species, rhodium-catalyzed carbenoid additions are accelerated by aryl substituents, as well as by other cation-stabilizing groups on the alkene reactant.[205] When applied to 1,1-diarylethenes, ERG substituents favor the position *trans* to the ester group.[206] This can be understood in terms of maximizing the interaction between this ring and the reacting double bond.

[205.] H. M. L. Davies and S. A. Panaro, *Tetrahedron*, **56**, 4871 (2000).
[206.] H. M. L. Davies, T. Nagashima, and J. L. Klino, III, *Org. Lett.*, **2**, 823 (2000).

(a)

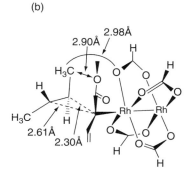

$E_{rel} = -12.1$ $E_{rel} = -10.4$

(b)

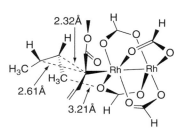

Fig. 10.10. Steric interactions in rhodium-catalyzed addition of methyl 2-diazobut-3-enoate to styrene (a) and *cis* and *trans* butene (b). Reproduced from *J. Am. Chem. Soc.*, **125**, 15902 (2003), by permission of the American Chemical Society.

10.2.3.3. Other Cyclopropanation Methods. Haloalkylmercury compounds are also useful in synthesis. The addition reactions are usually carried out by heating the organomercury compound with the alkene. Two typical examples are given in Section C of Scheme 10.9.

The addition of dichlorocarbene, generated from chloroform, to alkenes gives dichlorocyclopropanes. The procedures based on lithiated halogen compounds have been less generally used in synthesis. Section D of Scheme 10.9 gives a few examples of addition reactions of carbenes generated by α-elimination.

10.2.3.4. Examples of Cyclopropanations. Scheme 10.9 illustrates some of these cyclopropanation methods. Section A pertains to the Simmons-Smith type of cyclopropanation. Entry 1 is an example using readily available sources of the of cyclopropanation reagent. Only a modest excess of the reagents was needed, and good yields were obtained from several unfunctionalized cycloalkenes under these conditions. Entry 2 is a case of an allylic alcohol and illustrates the hydroxy-directing effect. Entries 3 to 6 are also examples of the directive effect of hydroxy groups in ring systems. Entry 4 was done using the diethylzinc-diiodomethane conditions. The vinyl ether group is expected to be quite reactive because of the electrophilic character of the methylene transfer reaction. Entry 5 illustrates the application of the hydroxy-directing

Scheme 10.9. Cyclopropane Formation by Carbenoid Addition

A. Cyclopropanes by methylene transfer

1[a]

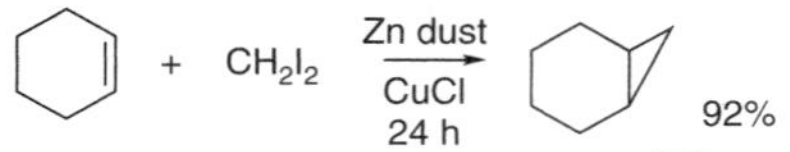

92%

2[b]

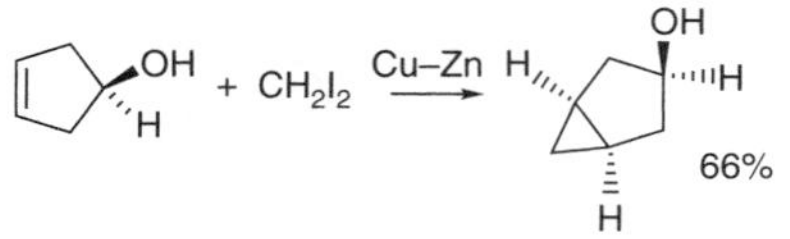

66%

3[c]

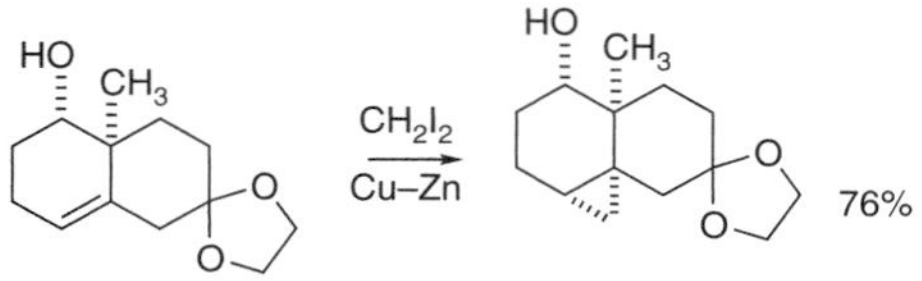

76%

4[d]

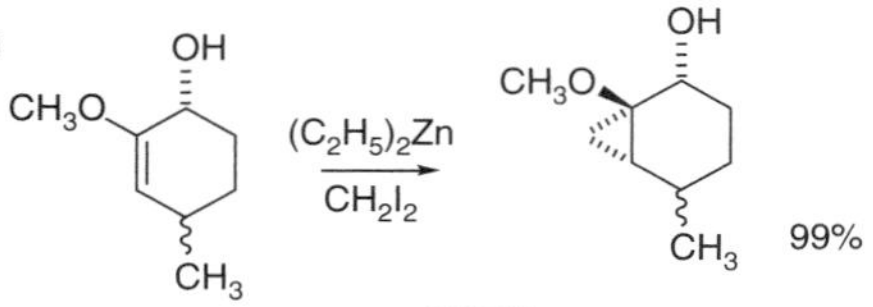

99%

5[e]

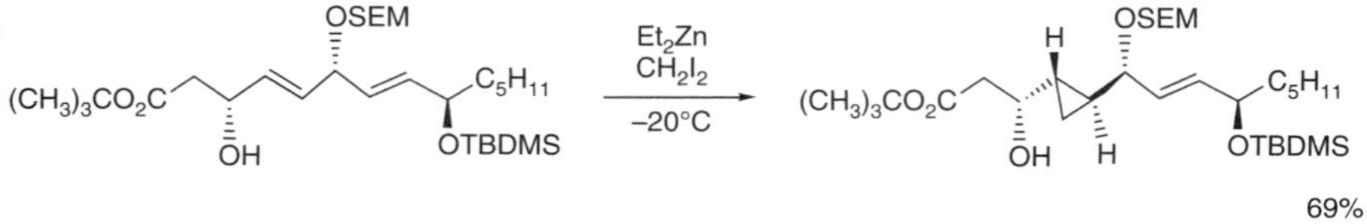

69%

6[f]

73%

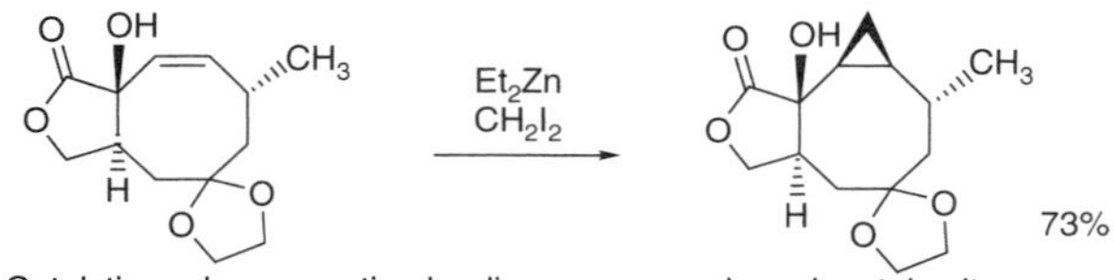

B. Catalytic cyclopropanation by diazo compounds and metal salts

7[g]

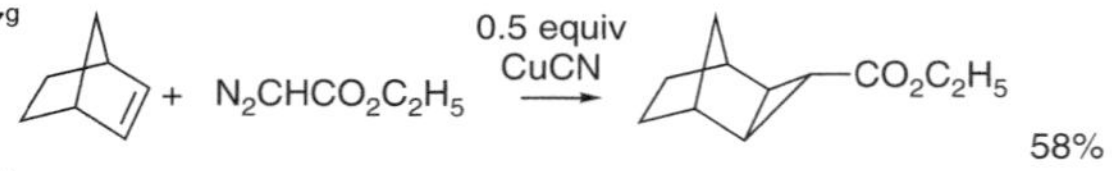

58%

8[h]

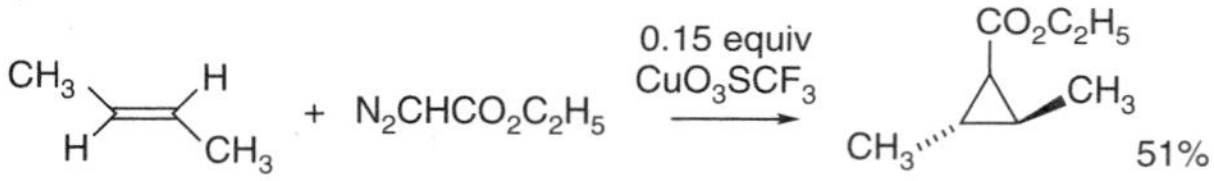

51%

9[i]

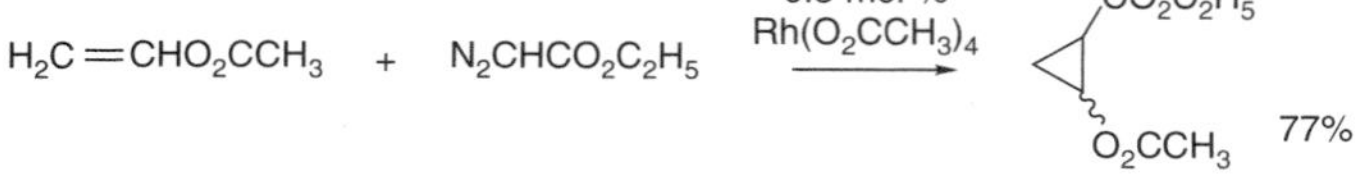

77%

10[j]

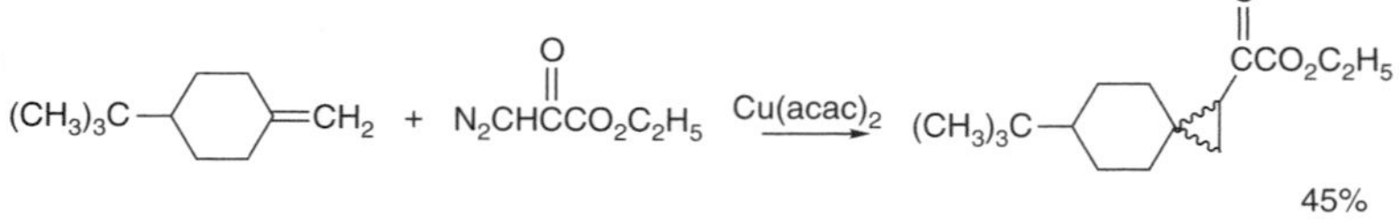

45%

11[k]

96%

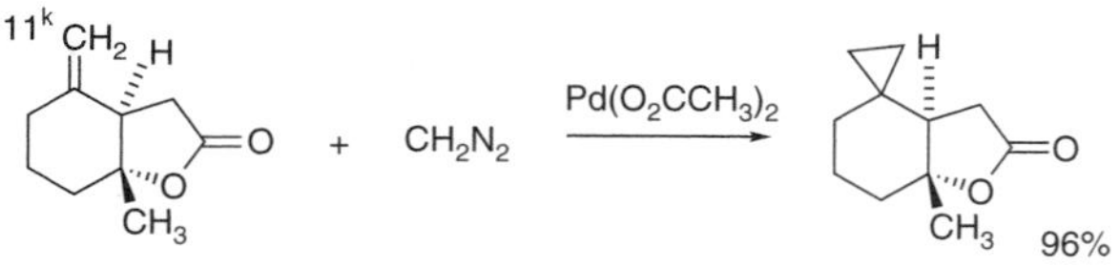

(*Continued*)

C. Cyclopropane formation using haloalkylmercurials

12[l]

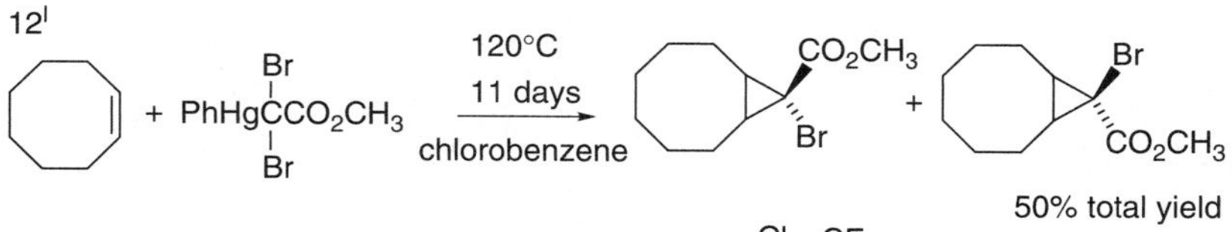

50% total yield

13[m]

58%

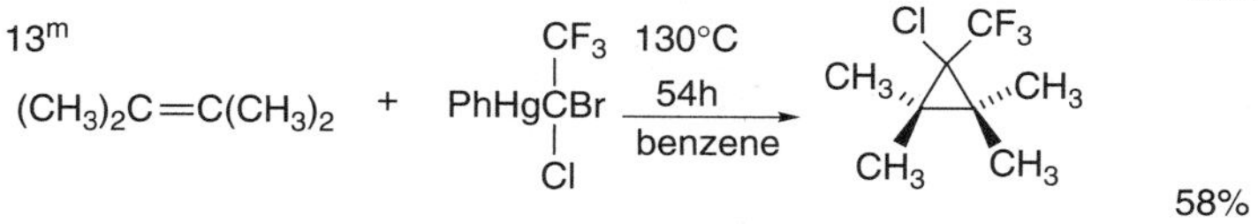

D. Reactions of carbenes generated by α-elimination

14[n]

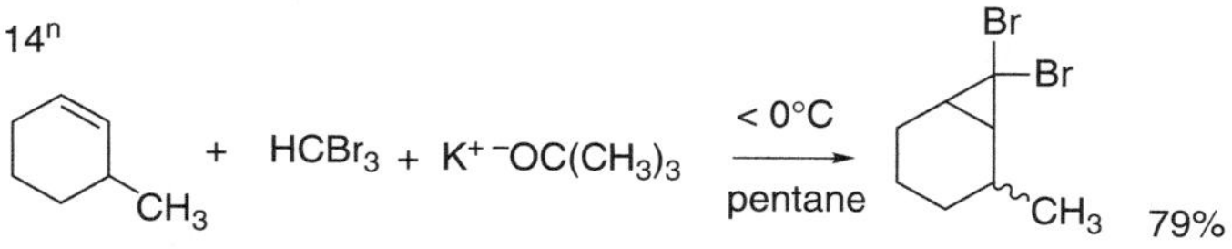

79%

15[o]

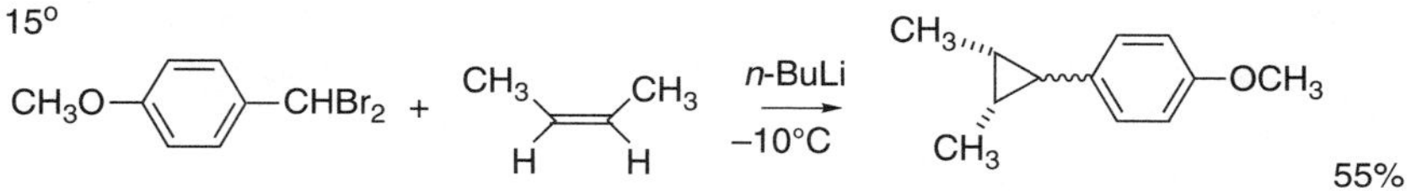

55%

16[p]

55%

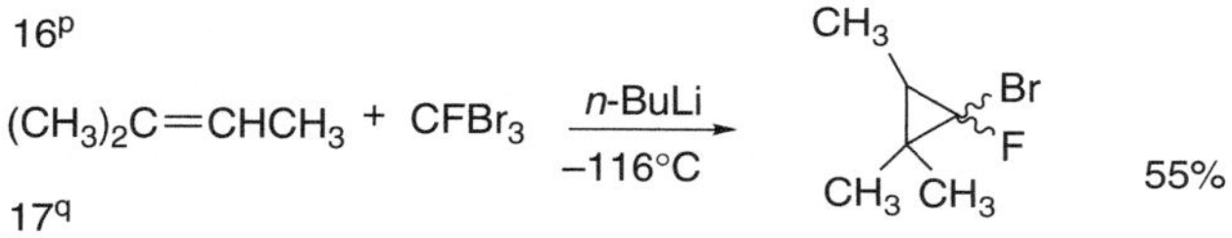

17[q]

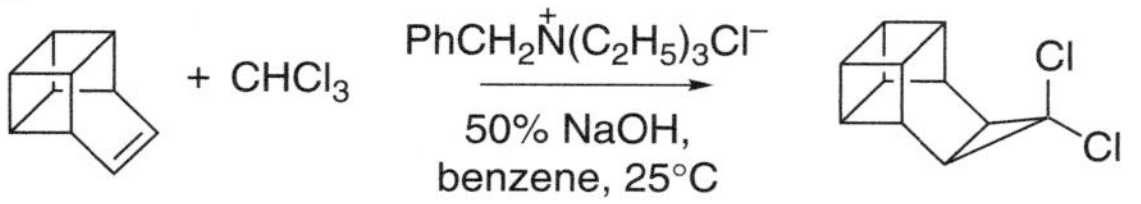

E. Intramolecular cyclopropanation reactions

18[r]

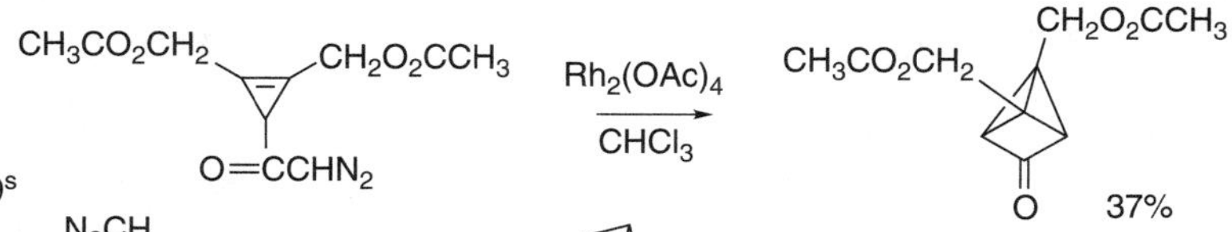

37%

19[s]

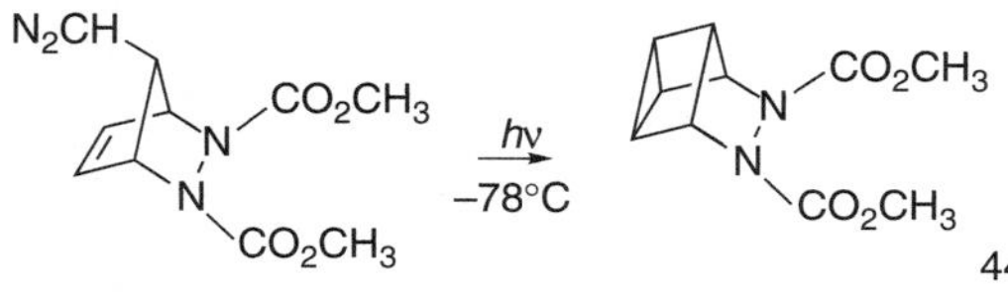

44%

20[t]

87% 6.7:1 dr

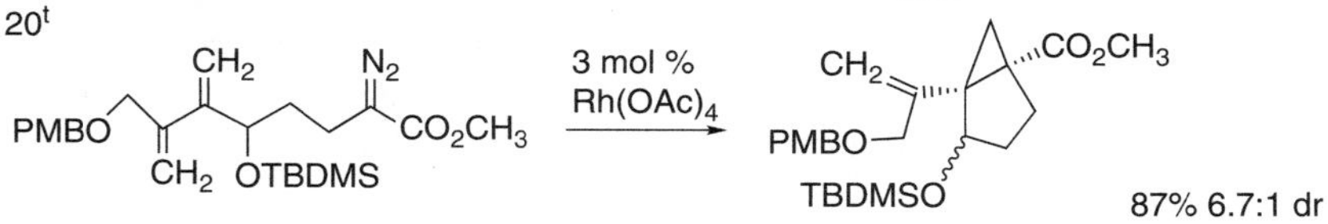

Scheme 10.9. (*Continued*)

a. R. J. Rawson and I. T. Harrison, *J. Org. Chem.*, **35**, 2057 (1970).
b. S. Winstein and J. Sonnenberg, *J. Am. Chem. Soc.*, **83**, 3235 (1961).
c. P. A. Grieco, T. Oguir, C.-L. J. Wang, and E. Williams, *J. Org. Chem.*, **42**, 4113 (1977).
d. R. C. Gadwood, R. M. Lett, and J. E. Wissinger, *J. Am. Chem. Soc.*, **108**, 6343 (1986).
e. Y. Baba, G. Saha, S. Nakao, C. Iwata, T. Tanaka, T. Ibuka, H. Ohishi, and Y. Takemoto, *J. Org. Chem.*, **66**, 81 (2001).
f. L. A. Paquette, J. Ezquerra, and W. He, *J. Org. Chem.*, **60**, 1435 (1995).
g. R. R. Sauers and P. E. Sonnett, *Tetrahedron*, **20**, 1029 (1964).
h. R. G. Salomon and J. K. Kochi, *J. Am. Chem. Soc.*, **95**, 3300 (1973).
i. A. J. Anciaux, A. J. Hubert, A. F. Noels, N. Petiniot, and P. Teyssie, *J. Org. Chem.*, **45**, 695 (1980).
j. M. E. Alonso, P. Jano, and M. I. Hernandez, *J. Org. Chem.*, **45**, 5299 (1980).
k. L. Stekowski, M. Visnick, and M. A. Battiste, *J. Org. Chem.*, **51**, 4836 (1986).
l. D. Seyferth, D. C. Mueller, and R. L. Lambert, Jr., *J. Am. Chem. Soc.*, **91**, 1562 (1969).
m. D. Seyferth and D. C. Mueller, *J. Am. Chem. Soc.*, **93**, 3714 (1971).
n. L. A. Paquette, S. E. Wilson, R. P. Henzel, and G. R. Allen, Jr., *J. Am. Chem. Soc.*, **94**, 7761 (1972).
o. G. L. Closs and R. A. Moss, *J. Am. Chem. Soc.*, **86**, 4042 (1964).
p. D. J. Burton and J. L. Hahnfeld, *J. Org. Chem.*, **42**, 828 (1977).
q. T. T. Sasaki, K. Kanematsu, and N. Okamura, *J. Org. Chem.*, **40**, 3322 (1975).
r. P. Dowd, P. Garner, R. Schappert, H. Irngartiner, and A. Goldman, *J. Org. Chem.*, **47**, 4240 (1982).
s. B. M. Trost, R. M. Cory, P. H. Scudder, and H. B. Neubold, *J. Am. Chem. Soc.*, **95**, 7813 (1973).
t. K. C. Nicolaou, M. H. D. Postema, N. D. Miller, and G. Yang, *Angew. Chem. Int. Ed. Engl.*, **36**, 2821 (1997).

effect in an acyclic system. Not only is the hydroxy group stereodirective, but it also provides selectivity with respect to the two double bonds. The reaction in Entry 6 was carried out in the course of synthesis of crenulide derivatives, which are obtained from seaweed.

Section B gives some examples of metal-catalyzed cyclopropanations. In Entries 7 and 8, Cu(I) salts are used as catalysts for intermolecular cyclopropanation by ethyl diazoacetate. The *exo* approach to norbornene is anticipated on steric grounds. In both cases, the Cu(I) salts were used at a rather high ratio to the reactants. Entry 9 illustrates use of $Rh_2(O_2CCH_3)_4$ as the catalyst at a much lower ratio. Entry 10 involves ethyl diazopyruvate, with copper acetylacetonate as the catalyst. The stereoselectivity of this reaction was not determined. Entry 11 shows that $Pd(O_2CCH_3)$ is also an active catalyst for cyclopropanation by diazomethane.

Section C shows cases involving organomercury reagents, which are useful for introducing functionalized cyclopropane rings when the necessary reagents can be obtained as mercury compounds. The very vigorous conditions needed for these reactions indicate the relatively low reactivity of the organomercury compounds toward α-elimination.

Section D illustrates formation of carbenes from halides by α-elimination. The carbene precursors are formed either by deprotonation (Entries 14 and 17) or halogen-metal exchange (Entries 15 and 16). The carbene additions can take place at low temperature. Entry 17 is an example of generation of dichlorocarbene from chloroform under phase transfer conditions.

Intramolecular carbene addition reactions have a special importance in the synthesis of strained-ring compounds. Because of the high reactivity of carbene or carbenoid species, the formation of highly strained bonds is possible. The strategy for synthesis is to construct a potential carbene precursor, such as a diazo compound or di- or trihalo compound that can undergo intramolecular addition to give the desired structure. Section E of Scheme 10.9 gives some representative examples. Entries 18 and 19 are cases of formation of strained compounds. The reaction in Entry 20 shows a preference between the two double bonds, based on proximity, and establishes a ring system that subsequently undergoes a divinylcyclopropane rearrangement to generate a nine-membered ring.

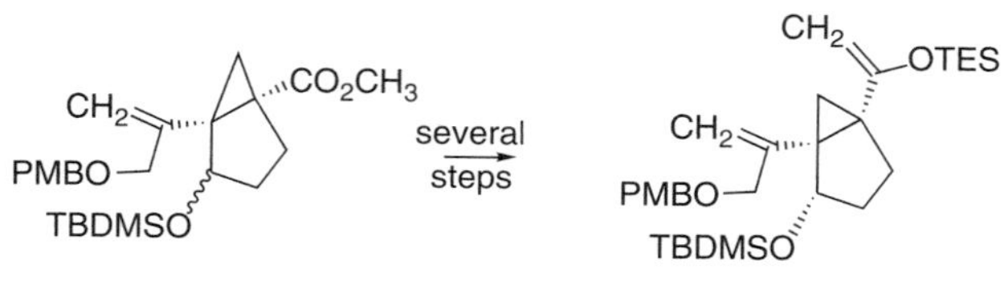

10.2.3.5. Enantioselective Cyclopropanation. Enantioselective versions of both copper and rhodium cyclopropanation catalysts are available. The copper-imine class of catalysts is enantioselective when chiral imines are used. Some of the chiral ligands that have been utilized in conjunction with copper salts are shown in Scheme 10.10.

Several chiral ligands have been developed for use with the rhodium catalysts, among them are pyrrolidinones and imidazolidinones.[207] For example, the lactamate of pyroglutamic acid gives enantioselective cyclopropanation reactions.

Scheme 10.10. Chiral Copper Catalysts Used in Enantioselective Cyclopropanation

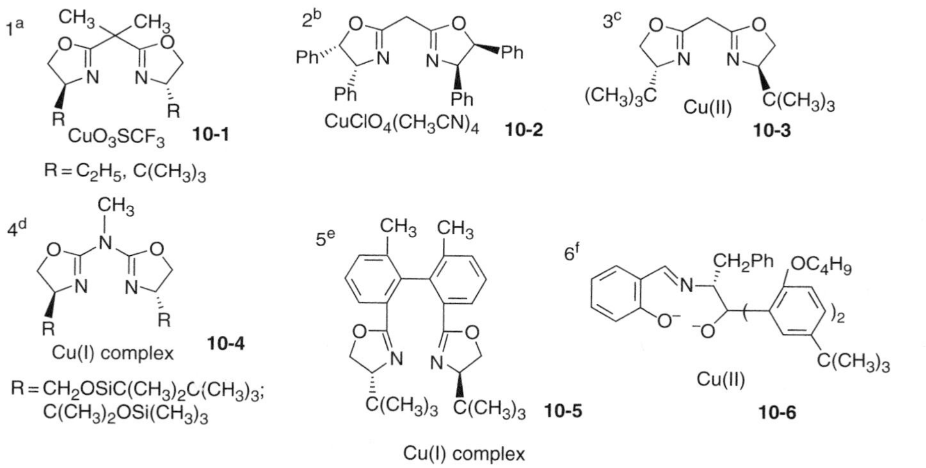

a. D. A. Evans, K. A. Woerpel, M. M. Hinman, and M. M. Faul, *J. Am. Chem. Soc.*, **113**, 726 (1991); D. A. Evans, K. A. Woerpel, and M. I. Scott, *Angew. Chem. Int. Ed. Engl.*, **31**, 430 (1992).
b. R. E. Lowenthal and S. Masamune, *Tetrahedron Lett.*, **32**, 7373 (1991).
c. R. E. Lowenthal, A. Abiko, and S. Masamune, *Tetrahedron Lett.*, **31**, 6005 (1990).
d. A. Pfaltz, *Acc. Chem. Res.*, **26**, 339 (1993).
e. T. G. Gant, M. C. Noe, and E. J. Corey, *Tetrahedron Lett.*, **36**, 8745 (1995).
f. T. Aratani, Y. Yoneyoshi, and T. Nagase, *Tetrahedron Lett.*, **23**, 685 (1982).

207. M. P. Doyle, R. E. Austin, A. S. Bailey, M. P. Dwyer, A. B. Dyatkin, A. V. Kalinin, M. M.-Y. Kwan, S. Liras, C. J. Oalmann, R. J. Pieters, M. N. Protopopova, C. E. Raab, G. H. P. Roos, Q. L. Zhou, and S. F. Martin, *J. Am. Chem. Soc.*, **117**, 5763 (1995); M. P. Doyle, A. B. Dyatkin, M. N. Protopopova, C. I. Yang, G. S. Miertschin, W. R. Winchester, S. H. Simonsen, V. Lynch, and R. Ghosh, *Rec. Trav. Chim. Pays-Bas*, **114**, 163 (1995); M. P. Doyle, *Pure Appl. Chem.*, **70**, 1123 (1998); M. P. Doyle and M. N. Protopopova, *Tetrahedron*, **54**, 7919 (1998); M. P. Doyle and D. C. Forbes, *Chem. Rev.*, **98**, 911 (1998).

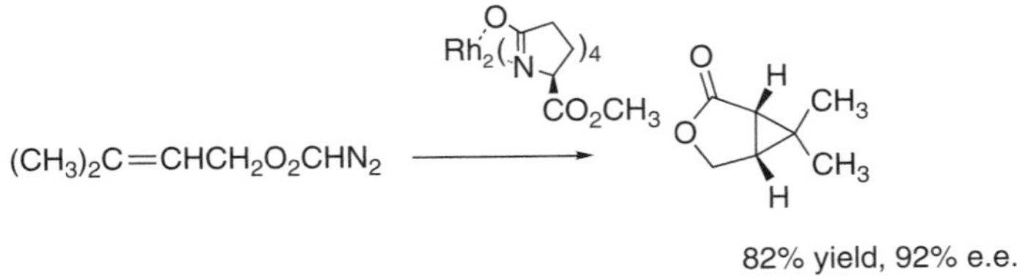

932

CHAPTER 10

*Reactions Involving
Carbocations, Carbenes,
and Radicals as Reactive
Intermediates*

$(CH_3)_2C=CHCH_2O_2CHN_2$

82% yield, 92% e.e.

The 1-acetyl and 1-benzoyl derivatives of 4-carbomethoxyimidazolinone are also effective catalysts. Another group of catalysts is made up of *N*-arenesulfonylprolinates. The structures and abbreviations are given in Scheme 10.11. The PY series of catalysts is derived from pyroglutamic acid, whereas the IM and OX designations apply to imidazolines and oxazolines, respectively. The designations ME and NE refer to methyl and neopentyl *esters*, and MA and PA indicate *amide*s of acetic acid and phenylacetic acid, respectively. Only two of the four ligands that are present are shown.

A comparison of several of the PY and IM types of catalysts in intramolecular reactions of allylic diazoacetates led to a consistent model for the enantioselectivity. The highest e.e. values are observed for *cis*-substituted allylic esters. Both R^t and R^i are directed toward the catalyst and introduce steric interactions that detract from enantioselectivity.[208]

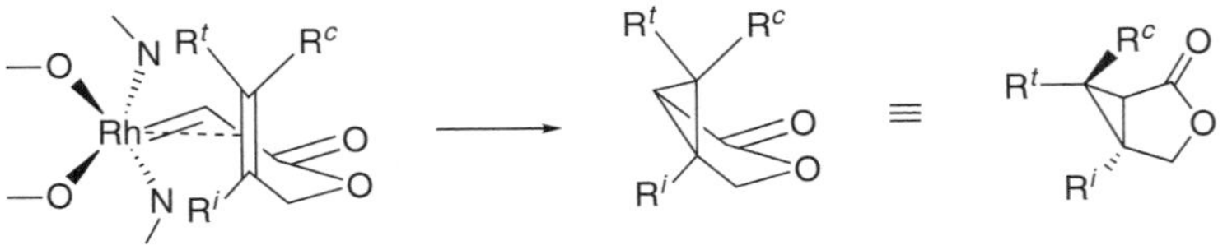

The 1-arenesulfonylprolinate catalysts have been studied computationally.[209] A computed TS and conceptual model that is consistent with experimentally observed enantioselectivity is shown in Figure 10.11. The arenesulfonyl groups block one of the directions of approach to the carbene catalyst and also orient the alkene substituent away from the metal center.

Several of the copper and rhodium catalysts were compared in an intramolecular cyclopropanation.[210] For the reaction leading to formation of a 10-membered ring, shown below, the copper catalysts gave higher enantioselectivity, but there were many subtleties, depending on ring size and other structural features in related systems.

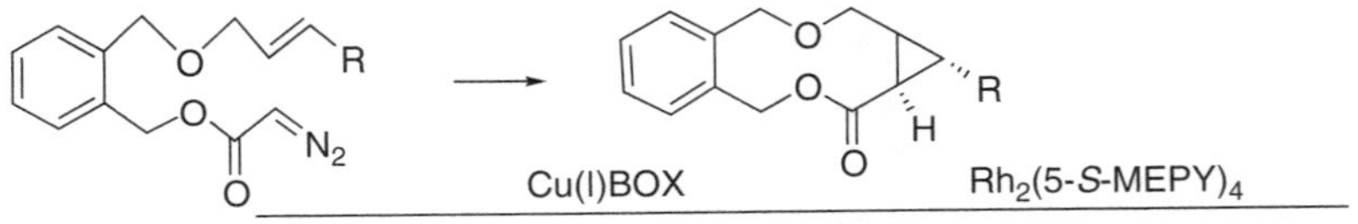

	Cu(I)BOX	Rh$_2$(5-*S*-MEPY)$_4$
$R = CH_3$	82% yield, 90% e.e.	81% yield, 45% e.e.
$R = CH(CH_3)_2$	93% yield, 84% e.e.	80% yield, 19% e.e.

208. M. P. Doyle, R. E. Austin, A. S. Bailey, M. P. Dwyer, A. B. Dyatkin, A. V. Kalinin, M. M. Y. Kwan, S. Liras, C. J. Oalmann, R. J. Pieters, M. N. Protopopova, C. E. Raab, G. H. P. Roos, Q.-L. Zhou, and S. F. Martin, *J. Am. Chem. Soc.*, **117**, 5763 (1995).
209. D. T. Nowlan, III, T. M. Gregg, H. M. L. Davies, and D. A. Singleton, *J. Am. Chem. Soc.*, **125**, 15902 (2003).
210. M. P. Doyle, W. Hu, B. Chapman, A. B. Marnett, C. S. Peterson, J. P. Vitale, and S. A. Stanley, *J. Am. Chem. Soc.*, **122**, 5718 (2000).

Scheme 10.11. Chiral Dirhodium Catalysts

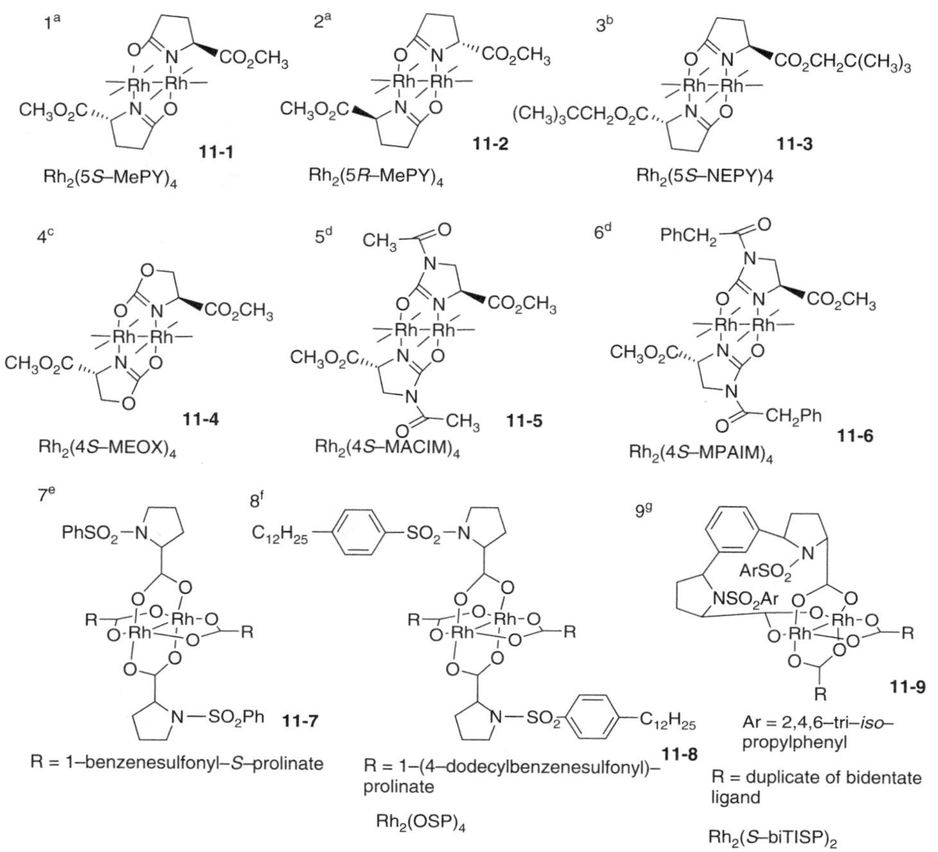

1[a]

11-1

$Rh_2(5S–MePY)_4$

2[a]

11-2

$Rh_2(5R–MePY)_4$

3[b]

11-3

$Rh_2(5S–NEPY)4$

4[c]

11-4

$Rh_2(4S–MEOX)_4$

5[d]

11-5

$Rh_2(4S–MACIM)_4$

6[d]

11-6

$Rh_2(4S–MPAIM)_4$

7[e]

11-7

R = 1–benzenesulfonyl–*S*–prolinate

8[f]

11-8

R = 1–(4–dodecylbenzenesulfonyl)–prolinate

$Rh_2(OSP)_4$

9[g]

11-9

Ar = 2,4,6–tri–*iso*–propylphenyl

R = duplicate of bidentate ligand

$Rh_2(S–biTISP)_2$

a. M. P. Doyle, R. J. Pieters, S. F. Martin, R. E. Austin, P.. J. Oalmann, and P. Mueller, *J. Am. Chem. Soc.*, **113**, 1423 (1991); M. P. Doyle, W. R. Winchester, J. A. A. Hoorn, V. Lynch, S. H. Simonsen, and R. Ghosh, *J. Am. Chem. Soc.*, **115**, 9968 (1993).
b. M. P. Doyle, A. van Oeveren, L. J. Westrum, M. N. Protopopova, and W. T. Clayton, Jr., *J. Am. Chem. Soc.*, **113**, 8982 (1991).
c. M. P. Doyle, A. B. Dyatkin, M. N. Protopopova, C. I. Yang, C. S. Miertschin, W. R. Winchester, S. H. Simonsen, V. Lynch, and R. Ghosh, *Recl. Trav. Chim. Pays-Bas*, **114**, 163 (1995).
d. M. P. Doyle, A. B. Dyatkin, G. H. P. Roos, F. Canas, D. A. Pierson, A. van Basten, P. Mueller, and P. Polleux, *J. Am. Chem. Soc.*, **116**, 4507 (1994).
e. M. A. McKervey and T. Ye, *J. Chem. Soc., Chem. Commun.*, 823 (1992).
f. H. M. L. Davies and D. K. Hutcheson, *Tetrahedron Lett.*, **34**, 7243 (1993); H. M. L. Davies, P. R. Bruzinski, D. H. Lake, N. Kong, and M. J. Fall, *J. Am. Chem. Soc.*, **118**, 6897 (1996).
g. H. M. L. Davies and S. A. Panaro, *Tetrahedron Lett.*, **40**, 5287 (1999).

Scheme 10.12 gives some examples of enantioselective cyclopropanations. Entry 1 uses the *bis-t*-butyloxazoline (BOX) catalyst. The catalytic cyclopropanation in Entry 2 achieves both stereo- and enantioselectivity. The electronic effect of the catalysts (see p. 926) directs the alkoxy-substituted ring *trans* to the ester substituent (87:13 ratio), and very high enantioselectivity was observed. Entry 3 also used the *t*-butyl-BOX catalyst. The product was used in an enantioselective synthesis of the alkaloid quebrachamine. Entry 4 is an example of enantioselective methylene transfer using the tartrate-derived dioxaborolane catalyst (see p. 920). Entry 5 used the $Rh_2[5(S)-MePY]_4$

934

CHAPTER 10

*Reactions Involving
Carbocations, Carbenes,
and Radicals as Reactive
Intermediates*

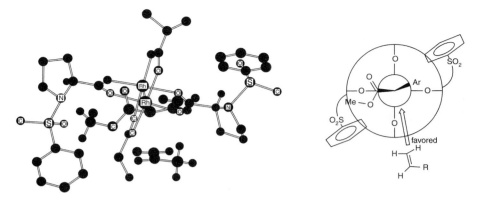

Fig. 10.11. General schematic model for favored approach of alkenes to 1-arenesulfonylprolinate catalysts (right); and B3LYP/6-31G*/LANL2DZ computational model of preferred approach of propene to 1-carbomethoxyprop-2-enylidene complex with Rh_2(1-benzenesulfonylprolinate)$_2$(isobutyrate)$_2$ (left). Reproduced from *J. Am. Chem. Soc.*, **125**, 15902 (2003), by permission of the American Chemical Society.

catalyst. Entry 6 is an intramolecular cyclopropanation done using a *bis*-(oxazolinyl) biphenyl catalyst (see Scheme 10.10, Entry 5).

10.2.4. Insertion Reactions

Insertion reactions are processes in which a reactive intermediate, in this case a carbene, interposes itself into an existing bond. In terms of synthesis, this usually involves C−H bonds. Many singlet carbenes are sufficiently reactive that insertion can occur as a one-step process.

$$CH_3-CH_2-CH_3 \ + \ :CH_2 \ \longrightarrow \ CH_3-\underset{\underset{CH_3}{|}}{CH}-CH_3$$

The same products can be formed by a two-step hydrogen abstraction and recombination involving a triplet carbene.

$$CH_3-CH_2-CH_3 \ + \ \overset{\cdot}{C}H_2 \ \longrightarrow \ CH_3-\underset{\underset{\overset{\cdot}{C}H_3}{|}}{\overset{\cdot}{C}}-CH_3 \ \longrightarrow \ CH_3-\underset{\underset{CH_3}{|}}{CH}-CH_3$$

It is sometimes difficult to distinguish clearly between these mechanisms, but determination of reaction stereochemistry provides one approach. The true one-step insertion must occur with complete *retention of configuration*. The results for the two-step process will depend on the rate of recombination in competition with stereorandomization of the radical pair intermediate.

Owing to the high reactivity of the intermediates involved, intermolecular carbene insertion reactions are not very selective. The distribution of products from the photolysis of diazomethane in heptane, for example, is almost exactly that expected on a statistical basis.[211]

[211.] D. B. Richardson, M. C. Simmons, and I. Dvoretzky, *J. Am. Chem. Soc.*, **83**, 1934 (1961).

Scheme 10.12. Enantioselective Cyclopropanation

1[a]

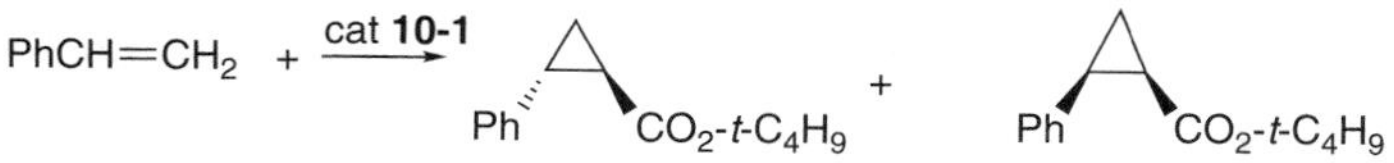

catalyst **10-1**, R = *t*-Bu, Scheme 10.10

61% yield, 96% e.e.   14% yield, 93% e.e.

2[b]

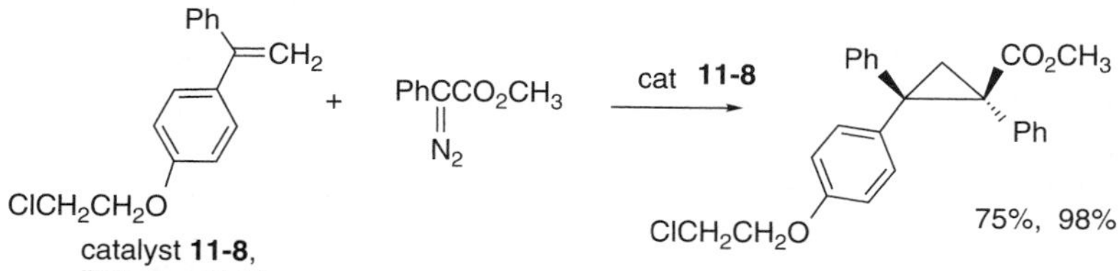

catalyst **11-8**, Scheme 10.11

75%, 98% e.e.

3[c]

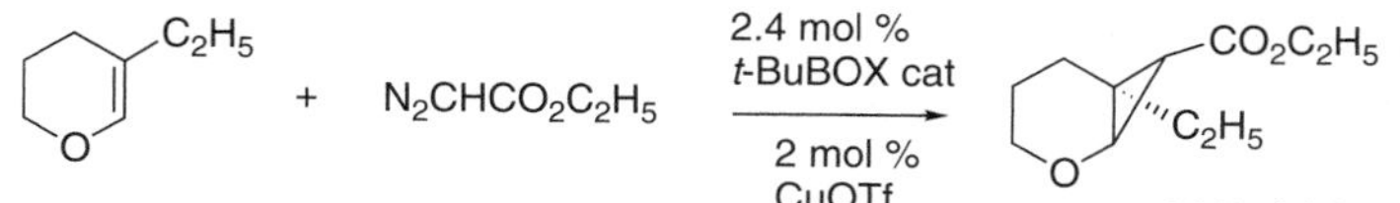

catalyst **10-1**, R = *t*-Bu Scheme 10,10

52% yield, >95% e.e.

4[d]

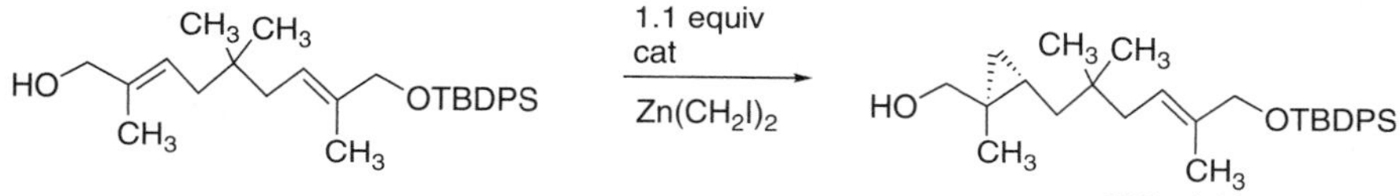

catalyst is *B*-butyl 1,3,2-dioxaborolane 4,5-*bis*-(*N,N*-dimethylcarboxamide).

>95% yield
86% e.e.

5[e]

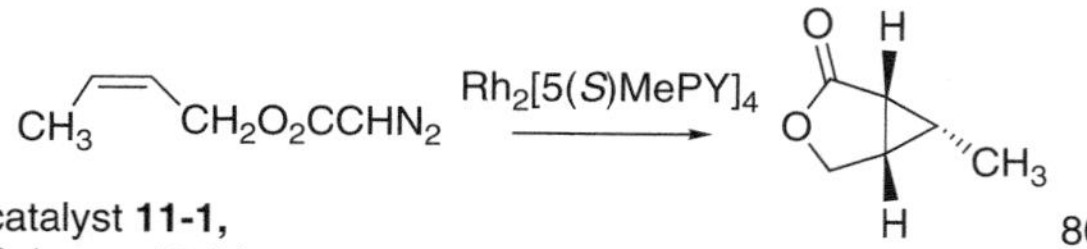

catalyst **11-1**, Scheme 10.11

80% yield, 92% e.e.

6[f]

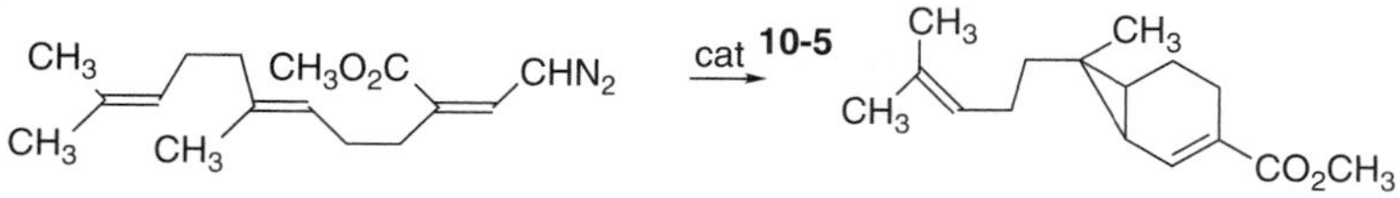

catalyst **10-5**, Scheme 10.10

77% yield, 90% e.e.

a. D. A. Evans, K. A. Woerpel, M. M. Hinman, and M. M. Faul, *J. Am. Chem. Soc.*, **113**, 726 (1991).
b. H. M. L. Davies, T. Nagashima, and J. L. Klino, III, *Org. Lett.*, **2**, 823 (2000).
c. O. Temme, S.-A. Taj, and P. G. Andersson, *J. Org. Chem.*, **63**, 6007 (1998).
d. A. B. Charette and H. Juteau, *Tetrahedron*, **53**, 16277 (1997).
e. S. M. Berberich, R. J. Cherney, J. Colucci, C. Courillon, L. S. Geraci, T. A. Kirkland, M. A. Marx, M. Schneider, and S. F. Martin, *Tetrahedron*, **59**, 6819 (2003).
f. T. G. Grant, M. C. Noe, and E. J. Corey, *Tetrahedron Lett.*, **36**, 8745 (1995).

$$CH_3CH_2CH_2CH_2CH_2CH_2CH_3 \xrightarrow[h\nu]{CH_2N_2} CH_3(CH_2)_6CH_3 + CH_3\underset{\underset{CH_3}{|}}{C}H(CH_2)_4CH_3$$

38%   25%

$$+ \quad CH_3CH_2\underset{\underset{CH_3}{|}}{C}H(CH_2)_3CH_3 + (CH_3CH_2CH_2)_2CHCH_3$$

24%   13%

There is some increase in selectivity with functionally substituted carbenes, but it is still not high enough to prevent formation of mixtures. Phenylchlorocarbene gives a relative reactivity ratio of 2.1:1:0.09 in insertion reactions with *i*-propylbenzene, ethylbenzene, and toluene.[212] For cycloalkanes, tertiary positions are about 15 times more reactive than secondary positions toward phenylchlorocarbene.[213] Carbethoxycarbene inserts at tertiary C−H bonds about three times as fast as at primary C−H bonds in simple alkanes.[214] Owing to low selectivity, intermolecular insertion reactions are seldom useful in syntheses. Intramolecular insertion reactions are of considerably more value. Intramolecular insertion reactions usually occur at the C−H bond that is closest to the carbene and good yields can frequently be achieved. Intramolecular insertion reactions can provide routes to highly strained structures that would be difficult to obtain in other ways.

Rhodium carboxylates have been found to be effective catalysts for intramolecular C−H insertion reactions of α-diazo ketones and esters.[215] In flexible systems, five-membered rings are formed in preference to six-membered ones. Insertion into methine hydrogen is preferred to a methylene hydrogen. Intramolecular insertion can be competitive with intramolecular addition. Product ratios can to some extent be controlled by the specific rhodium catalyst that is used.[216] In the example shown, insertion is the exclusive reaction with $Rh_2(O_2CC_4F_9)_4$, whereas only addition occurs with $Rh_2(caprolactamate)_4$, which indicates that the more electrophilic carbenoids favor insertion.

$Rh_2(X^-)_4$	Yield (%)	Ratio
$Rh_2(O_2CCH_3)_4$	99	67:33
$Rh_2(O_2CC_4F_9)_4$	95	0:100
$Rh_2(caprolactamate)_4$	72	100:0

The insertion reaction can be used to form lactones from α-diazo-β-keto esters.

212. M. P. Doyle, J. Taunton, S.-M. Oon, M. T. H. Liu, N. Soundararajan, M. S. Platz, and J. E. Jackson, *Tetrahedron Lett.*, **29**, 5863 (1988).

213. R. M. Moss and S. Yan, *Tetrahedron Lett.*, **39**, 9381 (1998).

214. W. von E. Doering and L. H. Knox, *J. Am. Chem. Soc.*, **83**, 1989 (1961).

215. D. F. Taber and E. H. Petty, *J. Org. Chem.*, **47**, 4808 (1982); D. F. Taber and R. E. Ruckle, Jr., *J. Am. Chem. Soc.*, **108**, 7686 (1986).

216. (a) M. P. Doyle, L. J. Westrum, W. N. E. Wolthuis, M. M. See, W. P. Boone, V. Bagheri, and M. M. Pearson, *J. Am. Chem. Soc.*, **115**, 958 (1993); (b) A. Padwa and D. J. Austin, *Angew. Chem. Int. Ed. Engl.*, **33**, 1797 (1994).

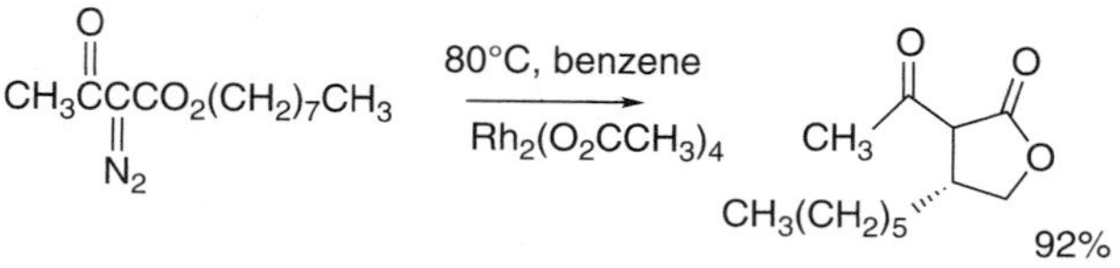

When the reactant provides more than one kind of hydrogen for insertion, the catalyst can influence selectivity. For example, $Rh_2(acam)_4$ gives exclusively insertion at a tertiary position, whereas $Rh_2(O_2CC_4F_9)_4$ leads to nearly a statistical mixture.[217a] The attenuated reactivity of the amidate catalyst enhances selectivity.

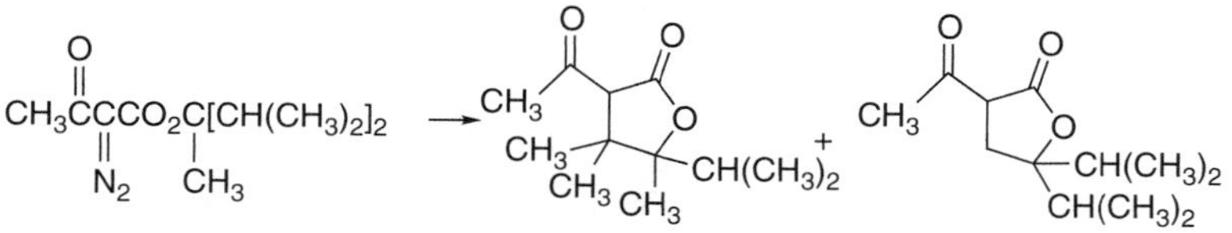

$Rh_2(X^-)_4$	Ratio
$Rh_2(O_2CCH_3)_4$	90:10
$Rh_2(O_2CC_4F_9)_4$	39:61
$Rh_2(NHCOCH_3)_4$	>99:1

Stereoselectivity is also influenced by the catalysts. For example, **16** can lead to either *cis* or *trans* products. Although $Rh_2(O_2CCH_3)_4$ is unselective, the $Rh_2(MACIM)_4$ catalyst **11-5** (Scheme 10.11) is selective for the *cis* isomer and also gives excellent enantioselectivity in the major product.[217]

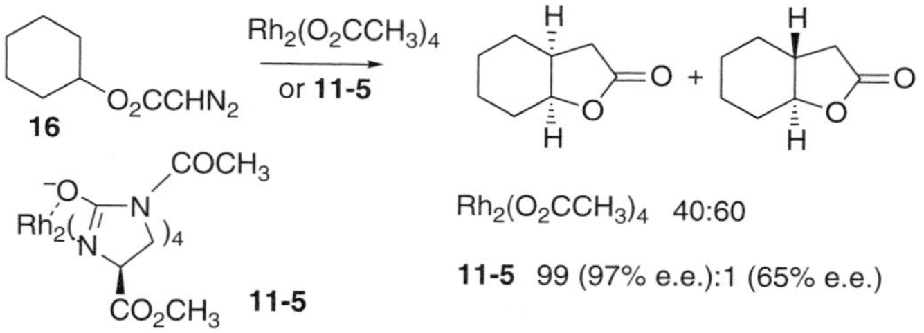

$Rh_2(O_2CCH_3)_4$ 40:60

11-5 99 (97% e.e.):1 (65% e.e.)

Certain sterically hindered rhodium catalysts also lead to improved selectivity. For example, rhodium triphenylacetate improves the selectivity for **17** over **18** from 5:1 to 99:1.[218]

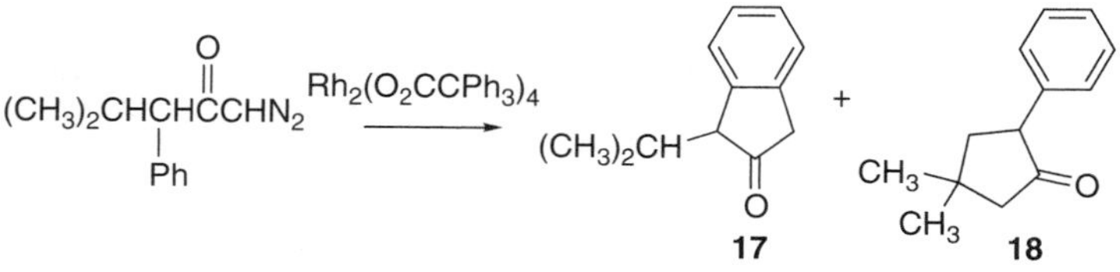

217. M. P. Doyle, A. B. Dyatkin, G. H. P. Roos, F. Canas, D. A. Pierson, and A. van Basten, *J. Am. Chem. Soc.*, **116**, 4507 (1994).
218. S. Hashimoto, N. Watanabe, and S. Ikegami, *J. Chem. Soc., Chem. Commun.*, 1508 (1992); S. Hashimoto, N. Watanabe, and S. Ikegami, *Tetrahedron Lett.*, **33**, 2709 (1992).

Intramolecular insertion reactions show a strong preference for formation of five-membered rings.[219] This was seen in a series of α-diazomethyl ketones of increasing chain length. With only one exception, all of the products were five-membered lactones.[220] In the case of $n = 3$, the cyclization occurs in the side chain, again forming a five-membered ring.

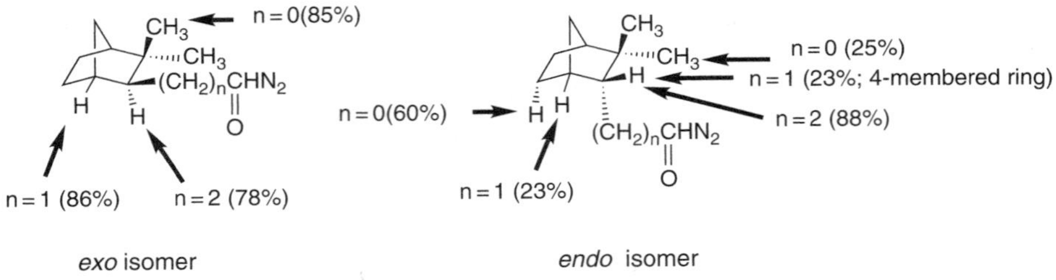

Scheme 10.13 gives some additional examples of intramolecular insertion reactions. Entries 1 and 2 were done under the high-temperature conditions of the Bamford-Stevens reaction (see p. 913). Entries 3 to 5 are metal-catalyzed intramolecular reactions in which 5-membered rings are formed. Entries 6 and 7 result in generation of strained rings by insertion into proximate C—H bonds. The insertion in Entry 6 via a diazirine was done in better yield (92%) by thermolysis (200°C) of the corresponding tosylhydrazone salt. Entry 8 is a case of enantioselective insertion, using one of the *N*-acyl methoxycarbonylimidazolonato rhodium catalysts.

10.2.5. Generation and Reactions of Ylides by Carbenoid Decomposition

Compounds in which a carbonyl or other nucleophilic functional group is close to a carbenoid carbon can react to give ylide intermediate.[221] One example is the formation of carbonyl ylides that go on to react by 1,3-dipolar addition. Both intramolecular and intermolecular cycloadditions have been observed.

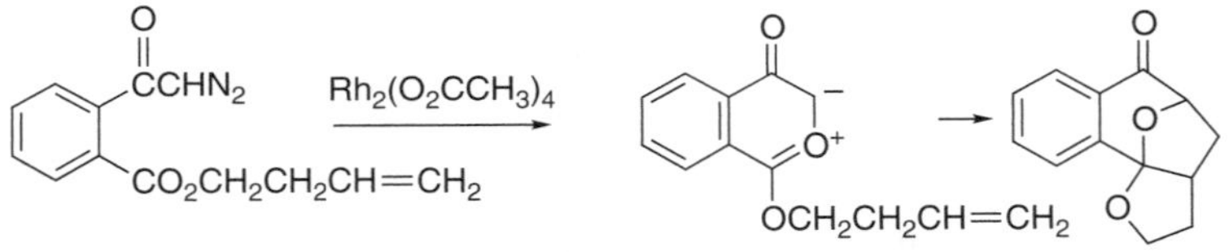

Ref. 222

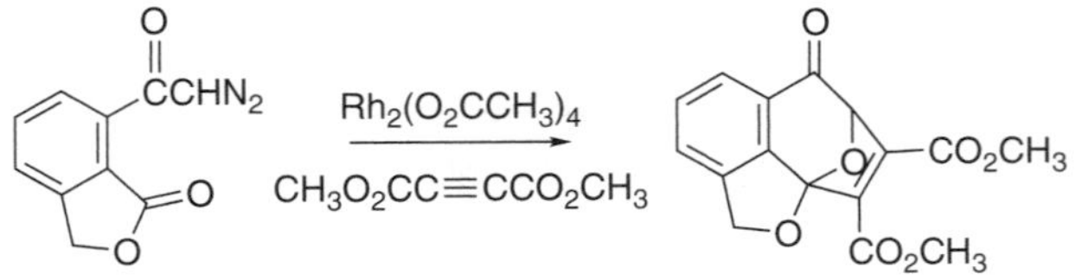

Ref. 221

219. D. F. Taber and R. E. Ruckle, Jr., *J. Am. Chem. Soc.*, **108**, 7686 (1986).
220. H. R. Sonawane, N. S. Bellur, J. R. Ahuja, and D. G. Kulkarni, *J. Org. Chem.*, **56**, 1434 (1991).
221. A. Padwa and S. F. Hornbuckle, *Chem. Rev.*, **91**, 263 (1991).
222. A. Padwa, S. P. Carter, H. Nimmesgern, and P. D. Stull, *J. Am. Chem. Soc.*, **110**, 2894 (1988).

Scheme 10.13. Intramolecular Carbene-Insertion Reactions

1[a]

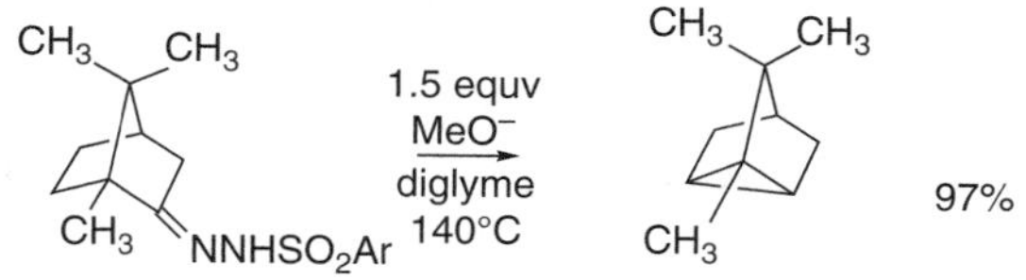

1.5 equv
MeO⁻
diglyme
140°C

97%

2[b]

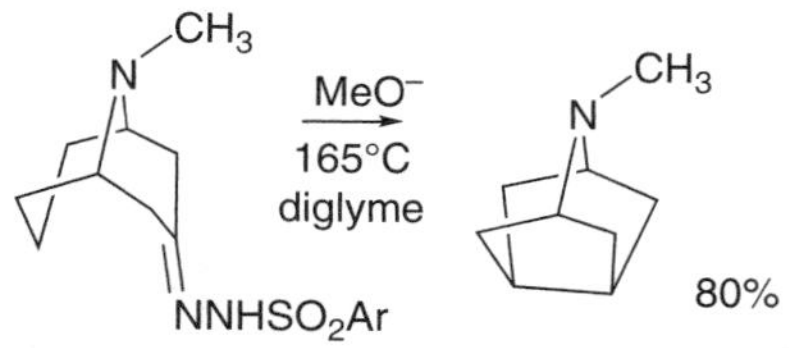

MeO⁻
165°C
diglyme

80%

3[c]

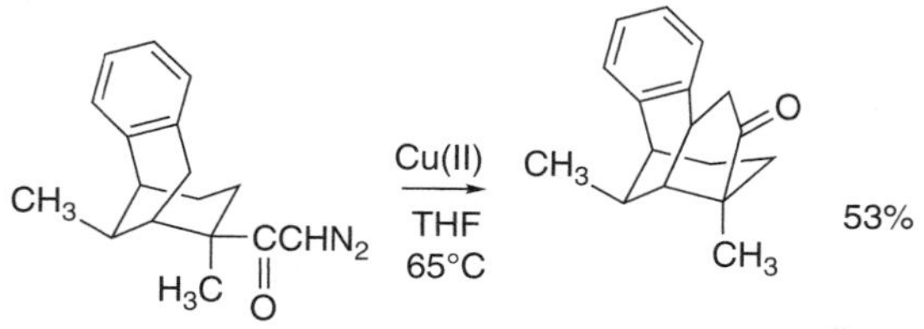

Cu(II)
THF
65°C

53%

4[d]

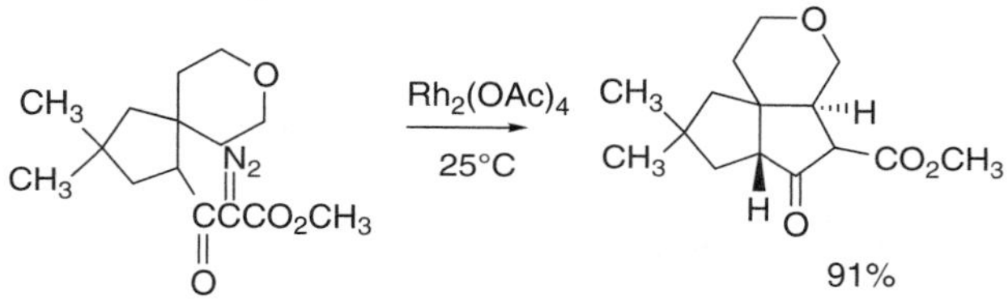

Rh₂(OAc)₄
25°C

91%

5[e]

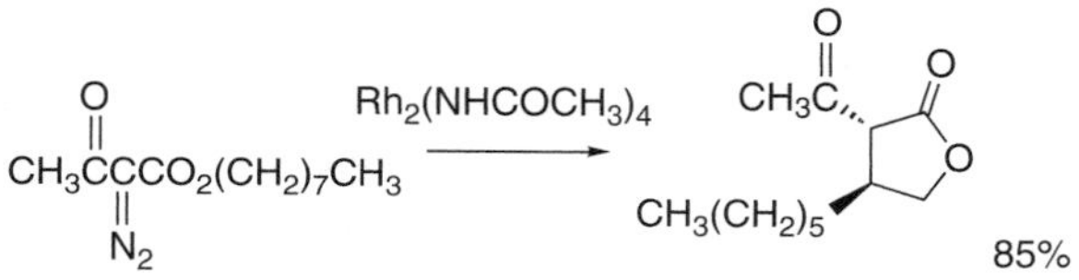

$CH_3CCCO_2(CH_2)_7CH_3$

Rh₂(NHCOCH₃)₄

85%

6[f]

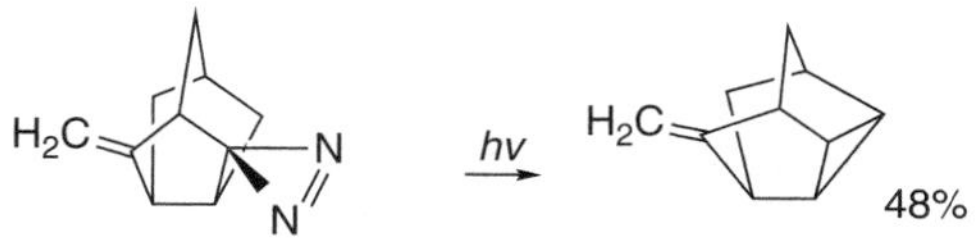

hv

48%

7[g]

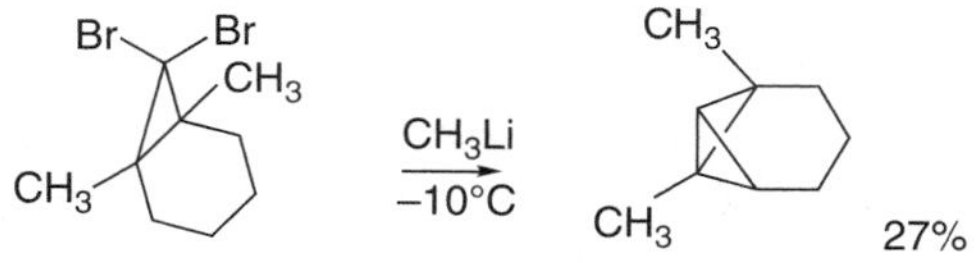

CH₃Li
−10°C

27%

8[h]

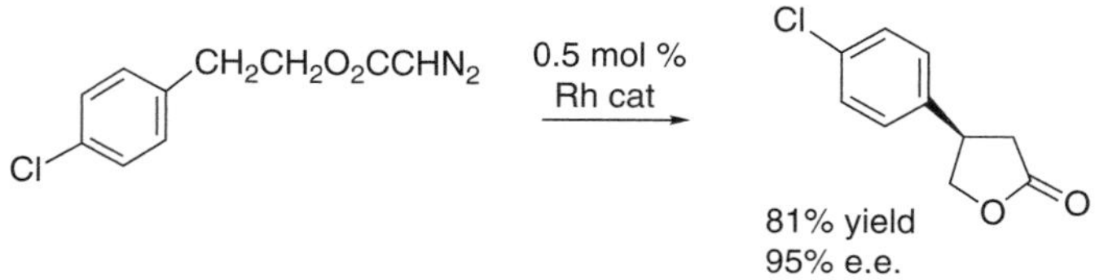

CH₂CH₂O₂CCHN₂

0.5 mol %
Rh cat

81% yield
95% e.e.

catalyst is *tetrakis*-[*N*-phenylpropanoyl-4-methoxycarbonylimidazolonato] dirhodium

(Continued)

Scheme 10.13. (*Continued*)

a. R. H. Shapiro, J. H. Duncan, and J. C. Clopton, *J. Am. Chem. Soc.*, **89**, 1442 (1967).
b. T. Sasaki, S. Eguchi, and T. Kiriyama, *J. Am. Chem. Soc.*, **91**, 212 (1969).
c. U. R. Ghatak and S. Chakrabarty, *J. Am. Chem. Soc.*, **94**, 4756 (1972).
d. D. F. Taber and J. L. Schuchardt, *J. Am. Chem. Soc.*, **107**, 5289 (1985).
e. M. P. Doyle, V. Bagheri, M. M. Pearson, and J. D. Edwards, *Tetrahedron Lett.*, **30**, 7001 (1989).
f. Z. Majerski, Z. Hamersak, and R. Sarac-Arneri, *J. Org. Chem.*, **53**, 5053 (1988).
g. L. A. Paquette, S. E. Williams, R. P. Henzel, and G. R. Allen, Jr., *J. Am. Chem. Soc.*, **94**, 7761 (1972).
h. M. P. Doyle and W. Hu, *Chirality*, **14**, 169 (2002).

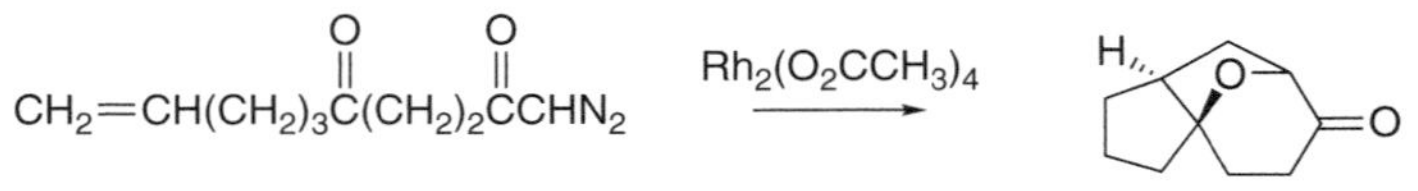

$$CH_2=CH(CH_2)_3\overset{O}{\underset{||}{C}}(CH_2)_2\overset{O}{\underset{||}{C}}CHN_2 \xrightarrow{Rh_2(O_2CCH_3)_4}$$

Ref. 223

Allylic ethers and acetals can react with carbenoid reagents to generate oxonium ylides that undergo [2,3]-sigmatropic shifts.[224]

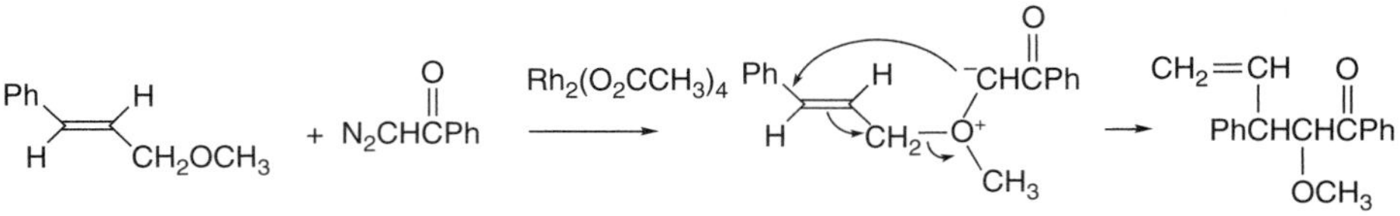

Ref. 224

10.2.6. Rearrangement Reactions

The most common rearrangement reaction of alkyl carbenes is the shift of hydrogen, generating an alkene. This mode of stabilization predominates to the exclusion of most intermolecular reactions of aliphatic carbenes and often competes with intramolecular insertion reactions. For example, the carbene generated by decomposition of the tosylhydrazone of 2-methylcyclohexanone gives mainly 1- and 3-methylcyclohexene rather than the intramolecular insertion product.

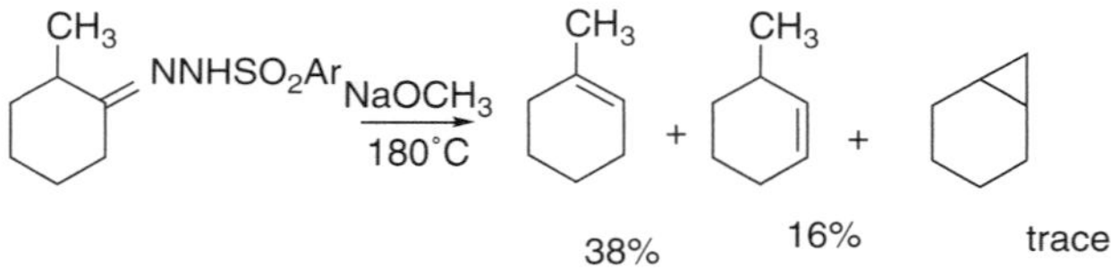

$$\xrightarrow[180°C]{NaOCH_3}$$

38%   16%   trace   Ref. 225

Carbenes can also be stabilized by migration of alkyl or aryl groups. 2-Methyl-2-phenyl-1-diazopropane provides a case in which products of both phenyl and methyl migration, as well as intramolecular insertion, are observed.

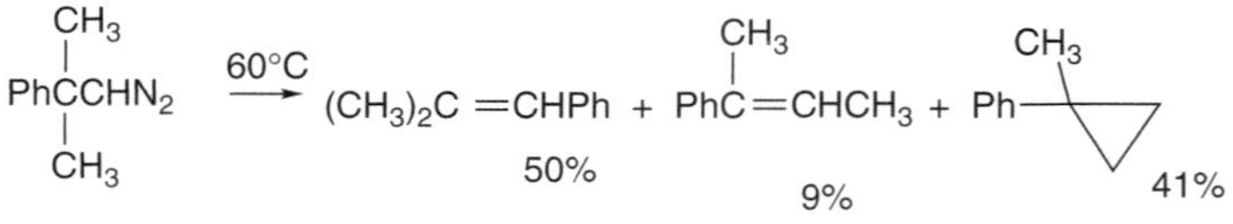

$$\underset{\underset{CH_3}{|}}{\overset{\overset{CH_3}{|}}{PhCCHN_2}} \xrightarrow{60°C} (CH_3)_2C=CHPh + PhC=CHCH_3 + Ph—\triangleleft$$

50%   9%   41%   Ref. 226

223. A. Padwa, S. F. Hornbuckle, G. E. Fryxell, and P. D. Stull, *J. Org. Chem.*, **54**, 819 (1989).
224. M. P. Doyle, V. Bagheri, and N. K. Harn, *Tetrahedron Lett.*, **29**, 5119 (1988).
225. J. W. Wilt and W. J. Wagner, *J. Org. Chem.*, **29**, 2788 (1964).
226. H. Philip and J. Keating, *Tetrahedron Lett.*, 523 (1961).

Bicyclo[3.2.2]non-1-ene, a strained bridgehead alkene, is generated by rearrangement when bicyclo[2.2.2]octyldiazomethane is photolyzed.[227]

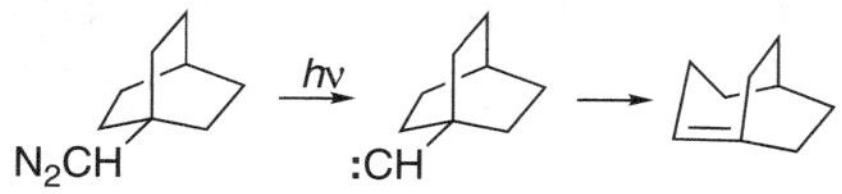

Carbene centers adjacent to double bonds (vinyl carbenes) usually cyclize to cyclopropenes.[228]

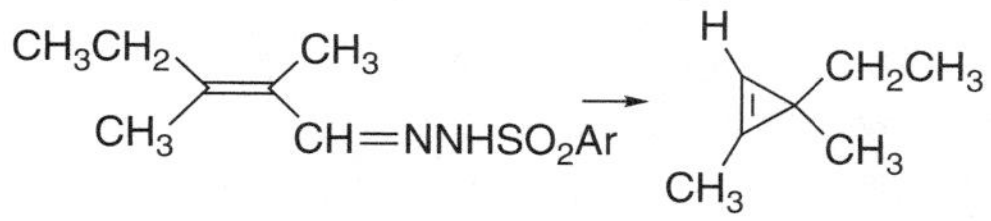

Ref. 229

Cyclopropylidenes undergo ring opening to give allenes. Reactions that would be expected to generate a cyclopropylidene therefore lead to allene, often in preparatively useful yields.

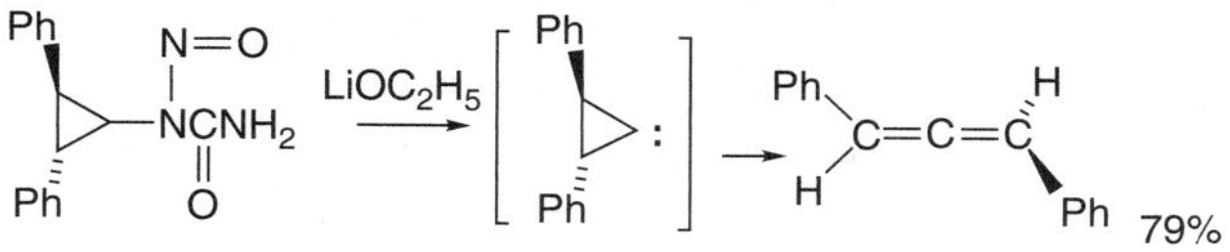

Ref. 230

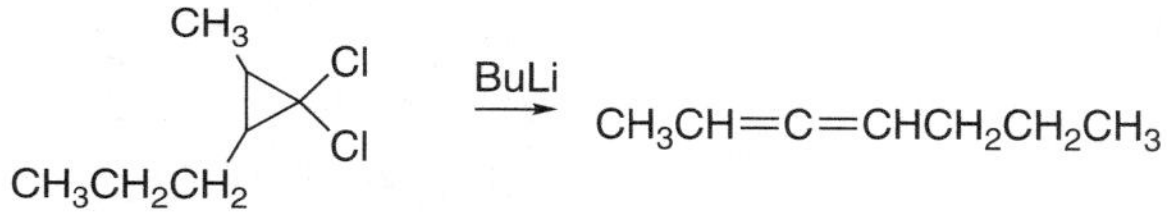

Ref. 231

10.2.7. Related Reactions

There are several reactions that are conceptually related to carbene reactions but do not involve carbene, or even carbenoid, intermediates. Usually, these are reactions in which the generation of a carbene is circumvented by a concerted rearrangement process. Important examples of this type are the thermal and photochemical reactions of α-diazo ketones. When α-diazo ketones are decomposed thermally or photochemically, they usually rearrange to ketenes, in a reaction known as the *Wolff rearrangement*.[232]

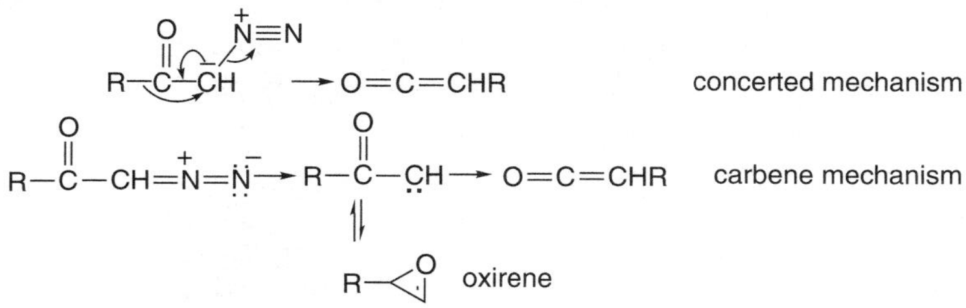

concerted mechanism

carbene mechanism

oxirene

227. M. S. Gudipati, J. G. Radziszewski, P. Kaszynski, and J. Michl, *J. Org. Chem.*, **58**, 3668 (1993).
228. G. L. Closs, L. E. Closs, and W. A. Böll, *J. Am. Chem. Soc.*, **85**, 3796 (1963).
229. E. J. York, W. Dittmar, J. R. Stevenson, and R. G. Bergman, *J. Am. Chem. Soc.*, **95**, 5680 (1973).
230. W. M. Jones, J. W. Wilson, Jr., and F. B. Tutwiler, *J. Am. Chem. Soc.*, **85**, 3309 (1963).
231. W. R. Moore and H. R. Ward, *J. Org. Chem.*, **25**, 2073 (1960).
232. W. Kirmse, *Eur. J. Org. Chem.*, 2193 (2002); T. Ye and M. A. McKervey, *Chem. Rev.*, **94**, 1091 (1994).

If this reaction proceeds in a concerted fashion, a carbene intermediate is avoided. Mechanistic studies have been aimed at determining whether migration is concerted with the loss of nitrogen. The conclusion that has emerged is that a carbene is generated in photochemical reactions but that the reaction can be concerted under thermal conditions.

A related issue is whether the carbene, when it is involved, is in equilibrium with a ring-closed isomer, an oxirene.[233] This aspect of the reaction has been probed using isotopic labeling. If a symmetrical oxirene is formed, the label should be distributed to both the carbonyl and α-carbon. A concerted reaction or a carbene intermediate that did not equilibrate with the oxirene should have label only in the carbonyl carbon. The extent to which the oxirene is formed depends on the structure of the diazo compound. For diazoacetaldehyde, photolysis leads to only 8% migration of label, which would correspond to formation of 16% of the product through the oxirene.[234]

8%   92%
distribution of label

The diphenyl analog shows about 20–30% rearrangement.[235] α-Diazocyclohexanone gives no evidence of an oxirene intermediate, since all the label remains at the carbonyl carbon.[236]

100%

The reactivity of diazo carbonyl compounds appears to be related to the conformational equilibria between *s-cis* and *s-trans* conformations. A concerted rearrangement is favored by the *s-cis* conformation.[237] The *t*-butyl compound **19**, which exists in the *s-trans* conformation, gives very little di-*t*-butylketene on photolysis.[238] A similarly

[233.] M. Torres, E. M. Lown, H. E. Gunning, and O. P. Strausz, *Pure Appl. Chem.*, **52**, 1623 (1980); E. G. Lewars, *Chem. Rev.*, **83**, 519 (1983); M. A. Blaustein and J. A. Berson, *Tetrahedron Lett.*, **22**, 1081 (1981); A. P. Scott, R. H. Nobes, H. F. Schaeffer, III, and L. Radom, *J. Am. Chem. Soc.*, **116**, 10159 (1994).

[234.] K.-P. Zeller, *Tetrahedron Lett.*, 707 (1977); see also Y. Chiang, A. J. Kresge, and V. V. Popik, *J. Chem. Soc., Perkin Trans. 2*, 1107 (1999).

[235.] K.-P. Zeller, H. Meier, H. Kolshorn, and E. Mueller, *Chem. Ber.*, **105**, 1875 (1972).

[236.] U. Timm, K.-P. Zeller, and H. Meier, *Tetrahedron*, **33**, 453 (1977).

[237.] F. Kaplan and G. K. Meloy, *J. Am. Chem. Soc.*, **88**, 950 (1966).

[238.] M. S. Newman and A. Arkell, *J. Org. Chem.*, **24**, 385 (1959).

substituted cyclic diazoketone **20**, which is in the *s-cis* conformation, gives a high yield of the ring-contracted ketene.[239]

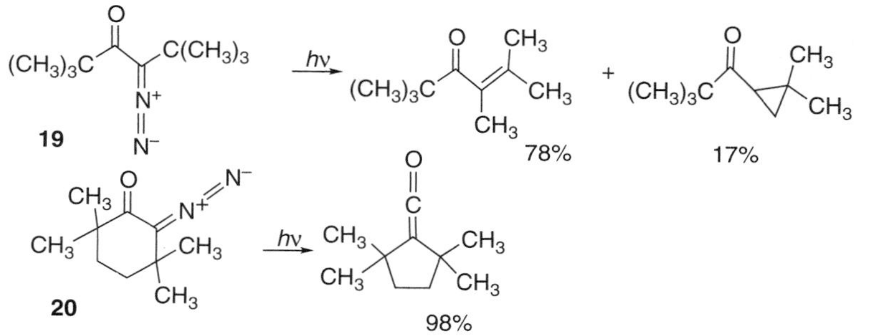

In a flash photolysis study of a series of diazo carbonyl compounds, a correlation was found between the amount of carbene that could be trapped by pyridine and the amount of *s-trans* ketone.[240]

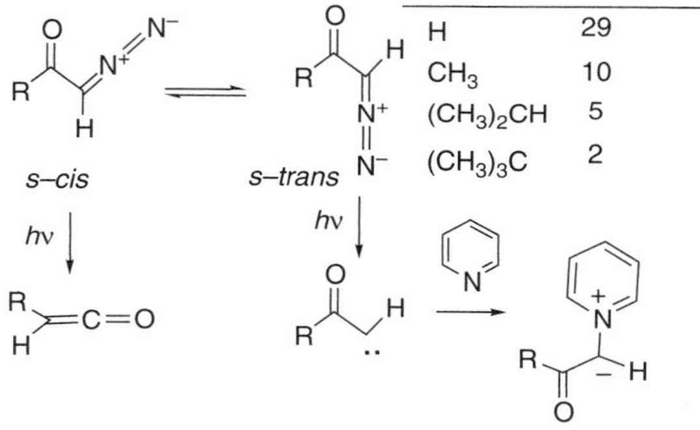

R	% s–trans	% trapped as ylide
H	29	42
CH₃	10	15
(CH₃)₂CH	5	13
(CH₃)₃C	2	9

Flash photolysis of benzoyl and naphthoyl diazomethane, which should exist in the *s-cis* conformation, led to ketene intermediates within the duration of the pulse (~ 20 ns).[241]

The main synthetic application of the Wolff rearrangement is for the one-carbon homologation of carboxylic acids.[242] In this procedure, a diazomethyl ketone is synthesized from an acyl chloride. The rearrangement is then carried out in a nucleophilic solvent that traps the ketene to form a carboxylic acid (in water) or an ester (in alcohols). Silver oxide is often used as a catalyst, since it seems to promote the rearrangement over carbene formation.[243]

The photolysis of cyclic α-diazoketones results in ring contraction to a ketene, which can be isolated as the corresponding ester.

239. F. Kaplan and M. L. Mitchell, *Tetrahedron Lett.*, 759 (1979).
240. J. P. Toscano and M. S. Platz, *J. Am. Chem. Soc.*, **117**, 4712 (1995).
241. Y. Chiang, A. J. Kresge, and V. V. Popik, *J. Am. Chem. Soc.*, **121**, 5930 (1999).
242. W. E. Bachmann and W. S. Stuve, *Org. React.*, **1**, 38 (1942); L. L. Rodina and I. K. Korobitsyna, *Russ. Chem. Rev.* (English Transl.), **36**, 260 (1967); W. Ando, in *Chemistry of Diazonium and Diazo Groups*, S. Patai, ed., John Wiley, New York (1978), pp. 458–475; H. Meier and K.-P. Zeller, *Angew. Chem. Int. Ed. Engl.*, **14**, 32 (1975).
243. T. Hudlicky and J. P. Sheth, *Tetrahedron Lett.*, 2667 (1979).

944

CHAPTER 10

*Reactions Involving
Carbocations, Carbenes,
and Radicals as Reactive
Intermediates*

42%   Ref. 244

60%   Ref. 245

Scheme 10.14 gives some other examples of Wolff rearrangement reactions. Entries 1 and 2 are reactions carried out under the classical silver ion catalysis conditions. Entry 3 is an example of a thermolysis. Entries 4 to 7 are ring contractions done under photolytic conditions. Entry 8, done using a silver catalyst, was a step in the synthesis of macbecin, an antitumor antibiotic. Entry 9, a step in the synthesis of a drug candidate, illustrates direct formation of an amide by trapping the ketene intermediate with an amine.

10.2.8. Nitrenes and Related Intermediates

The nitrogen analogs of carbenes are called nitrenes. As with carbenes, both singlet and triplet electronic states are possible.

singlet   triplet
nitrene   nitrene

The triplet state is usually the ground state for non-conjugated structures, but either species can be involved in reactions. The most common method for generating nitrene intermediates, analogous to formation of carbenes from diazo compounds, is by thermolysis or photolysis of azides.[246]

$$R\overset{..}{-}\overset{-}{N}\overset{+}{-}N\equiv N \xrightarrow[\text{or } h\nu]{\Delta} R\overset{..}{-}\overset{..}{N} + N_2$$

The types of azides that have been used for generation of nitrenes include alkyl,[247] aryl,[248] acyl,[249] and sulfonyl[250] derivatives.

244. K. B. Wiberg, L. K. Olli, N. Golembeski, and R. D. Adams, *J. Am. Chem. Soc.*, **102**, 7467 (1980).

245. K. B. Wiberg, B. L. Furtek, and L. K. Olli, *J. Am. Chem. Soc.*, **101**, 7675 (1979).

246. E. F. V. Scriven, ed., *Azides and Nitrenes: Reactivity and Utility*, Academic Press, Orlando, FL, 1984.

247. F. D. Lewis and W. H. Saunders, Jr., in *Nitrenes*, W. Lwowski, ed., Interscience, New York, 1970, pp. 47–98; E. P. Kyba, in *Azides and Nitrenes*, E. F. V. Scriven, ed., Academic Press, Orlando, FL, 1984, pp. 2–34.

248. P. A. Smith, in *Nitrenes*, W. Lwowski, ed., Interscience, New York, 1970, pp. 99–162; P. A. S. Smith, in *Azides and Nitrenes*, E. F. V. Scriven, ed., Academic Press, Orlando, FL, 1984, pp. 95–204.

249. W. Lwowski, in *Nitrenes*, W. Lwowski, ed., Interscience, New York, 1970, pp. 185–224; W. Lwowski, in *Azides and Nitrenes*, E. F. V. Scriven, ed., Academic Press, Orlando, FL, 1984, pp. 205–246.

250. D. S. Breslow, in *Nitrenes*, W. Lwowski, ed., Interscience, New York, 1970, pp. 245–303; R. A. Abramovitch and R. G. Sutherland, *Fortshr. Chem. Forsch.*, **16**, 1 (1970).

Scheme 10.14. Wolff Rearrangements of α-Diazoketones

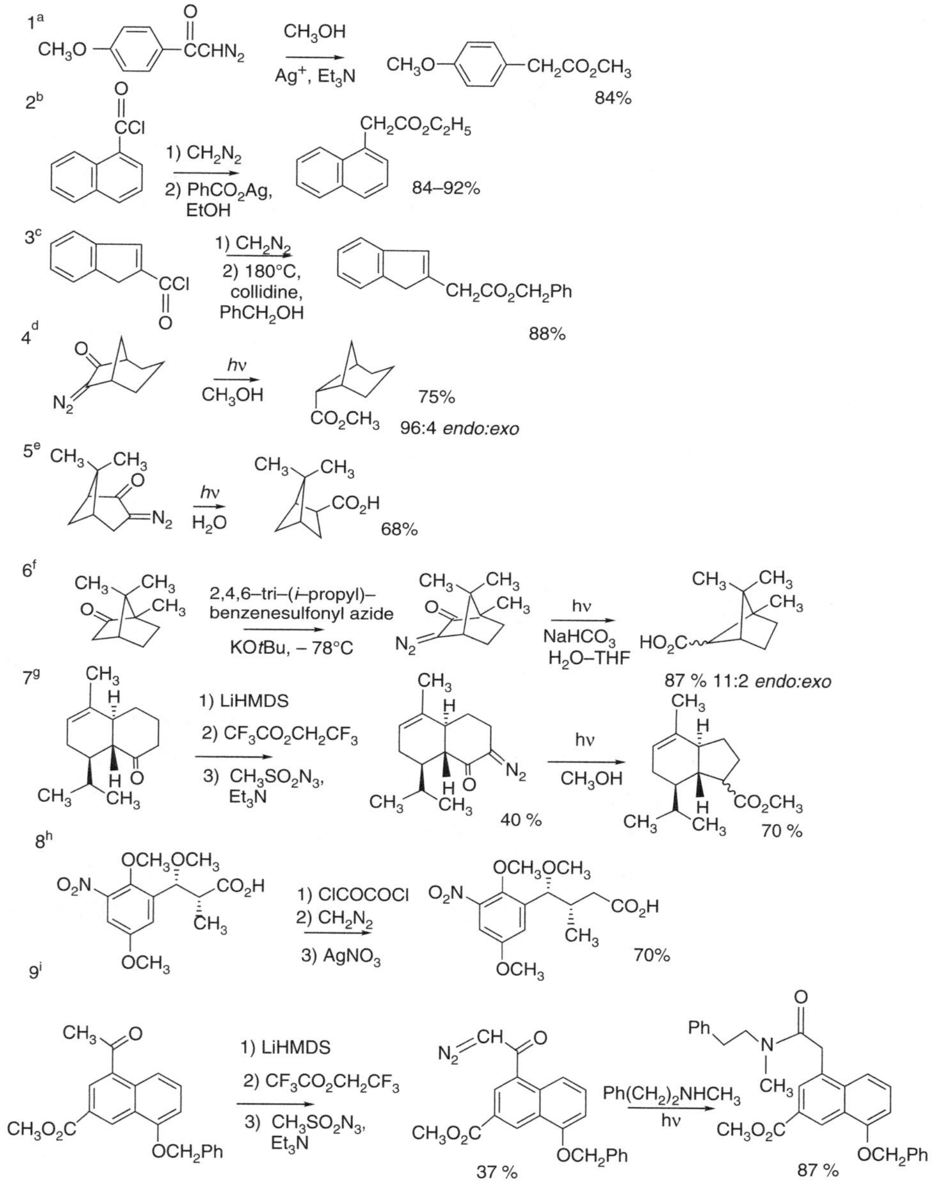

a. M. S. Newman and P. F. Beal, III, *J. Am. Chem. Soc.*, **72**, 5163 (1950).
b. V. Lee and M. S. Newman, *Org. Synth.*, **50**, 77 (1970).
c. E. D. Bergmann and E. Hoffmann, *J. Org. Chem.*, **26**, 3555 (1961).
d. K. B. Wiberg and B. A. Hess, Jr., *J. Org. Chem.*, **31**, 2250 (1966).
e. J. Meinwald and P. G. Gassman, *J. Am. Chem. Soc.*, **82**, 2857 (1960).
f. T. Uyehara, N. Takehara, M. Ueno, and T. Sato, *Bull. Chem. Soc. Jpn.*, **68**, 2687 (1995).
g. D. F. Taber, S. Kong, and S. C. Malcolm, *J. Org. Chem.*, **63**, 7953 (1998).
h. D. A. Evans, S. J. Miller, M. D. Ennis, and P. L. Ornstein, *J. Org. Chem.*, **57**, 1067 (1992); D. A. Evans, S. J. Miller, and M. D. Ennis, *J. Org. Chem.*, **58**, 471 (1993).
i. I. Pendrak and P. A. Chambers, *J. Org. Chem.*, **60**, 3249 (1995).

946

CHAPTER 10

*Reactions Involving
Carbocations, Carbenes,
and Radicals as Reactive
Intermediates*

The characteristic reaction of an alkyl nitrene is migration of one of the substituents to nitrogen, giving an imine.

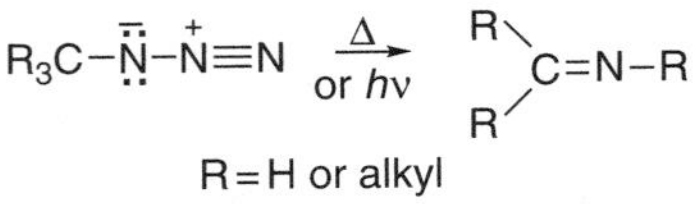

$$R = H \text{ or alkyl}$$

Intramolecular insertion and addition reactions are very rare for alkyl nitrenes. In fact, it is not clear that the nitrenes are formed as discrete species. The migration may be concerted with elimination, as is often the case in the Wolff rearrangement.[251]

Aryl nitrenes also generally rearrange rather than undergo addition or insertion reactions.[252]

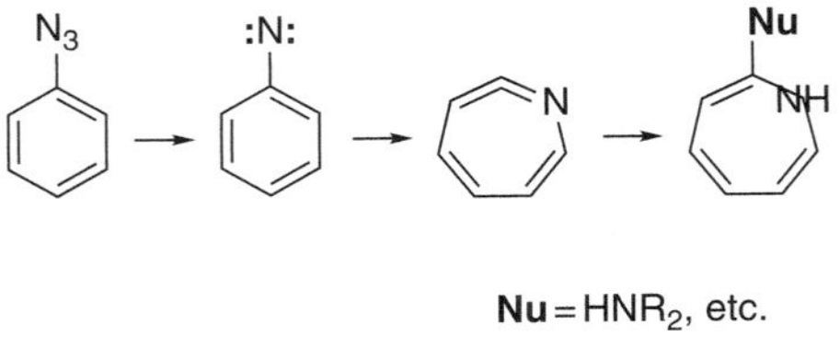

$$Nu = HNR_2, \text{ etc.}$$

A few intramolecular insertion reactions, especially in aromatic systems, go in good yield.[253]

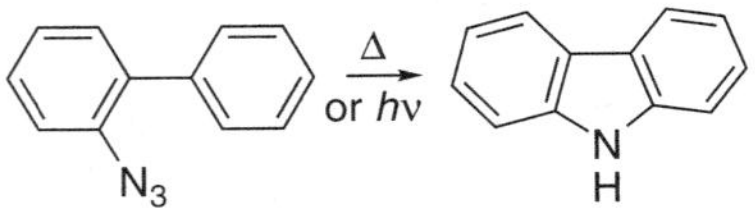

The nitrenes that most consistently give addition and insertion reactions are carboalkoxynitrenes generated from alkyl azidoformates.

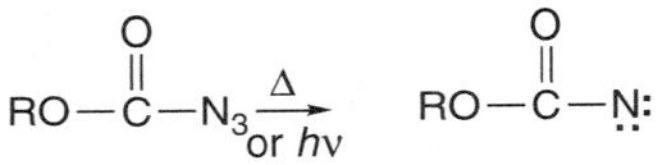

These intermediates undergo addition reactions with alkenes and aromatic compounds and insertion reactions with saturated hydrocarbons.[254]

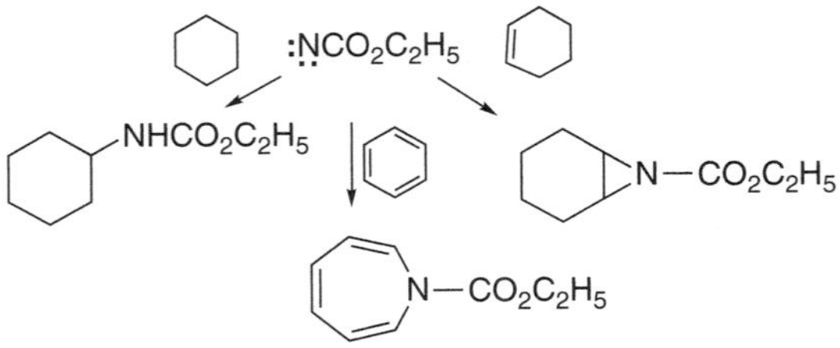

251. R. M. Moriarty and R. C. Reardon, *Tetrahedron*, **26**, 1379 (1970); R. A. Abramovitch and E. P. Kyba, *J. Am. Chem. Soc.*, **93**, 1537 (1971); R. M. Moriarty and P. Serridge, *J. Am. Chem. Soc.*, **93**, 1534 (1971).

252. O. L. Chapman and J.-P. LeRoux, *J. Am. Chem. Soc.*, **100**, 282 (1978); O. L. Chapman, R. S. Sheridan, and J.-P. LeRoux, *Rec. Trav. Chim. Pays-Bas*, **98**, 334 (1979); R. J. Sundberg, S. R. Suter, and M. Brenner, *J. Am. Chem. Soc.*, **94**, 573 (1972).

253. P. A. S. Smith and B. B. Brown, *J. Am. Chem. Soc.*, **73**, 2435, 2438 (1951); J. S. Swenton, T. J. Ikeler, and B. H. Williams, *J. Am. Chem. Soc.*, **92**, 3103 (1970).

254. W. Lwowski, *Angew. Chem. Int. Ed. Engl.*, **6**, 897 (1967).

Carboalkoxynitrenes are somewhat more selective than the corresponding carbenes, showing selectivities of roughly 1:10:40 for the primary, secondary, and tertiary positions in 2-methylbutane in insertion reactions.

Sulfonylnitrenes are formed by thermal decomposition of sulfonyl azides. Insertion reactions occur with saturated hydrocarbons.[255] With aromatic compounds the main products are formally insertion products, but they are believed to be formed through addition intermediates.

Ref. 256

Aziridination of alkenes can be carried out using *N*-(*p*-toluenesulfonylimino) phenyliodinane and copper triflate or other copper salts.[257] These reactions are mechanistically analogous to metal-catalyzed cyclopropanation. Rhodium acetate also acts as a catalyst.[258] Other arenesulfonyliminoiodinanes can be used,[259] as can chloroamine T[260] and bromoamine T.[261] The range of substituted alkenes that react includes acrylate esters.[262]

10.2.9. Rearrangements to Electron-Deficient Nitrogen

In contrast to the rather limited synthetic utility of nitrenes, there is an important group of reactions in which migration occurs to electron-deficient nitrogen. One of the most useful of these reactions is the *Curtius rearrangement*,[263] which has the same relationship to acyl nitrene intermediates that the Wolff rearrangment has to acyl carbenes. This reaction is usually considered to be a concerted process in which migration accompanies loss of nitrogen.[264] The temperature required for reaction is in the vicinity of 100°C. The initial product is an isocyanate that can be isolated or trapped by a nucleophilic solvent. The migrating group retains its stereochemical configuration.

255. D. S. Breslow, M. F. Sloan, N. R. Newburg, and W. B. Renfrow, *J. Am. Chem. Soc.*, **91**, 2273 (1969).
257. R. A. Abramovitch, G. N. Knaus, and V. Uma, *J. Org. Chem.*, **39**, 1101 (1974).
257. D. A. Evans, M. M. Faulk, and M. T. Bilodeau, *J. Am. Chem. Soc.*, **116**, 2742 (1994).
258. P. Mueller, C. Baud, and Y. Jacquier, *Tetrahedron*, **52**, 1543 (1996).
259. M. J. Sodergren, D. A. Alonso, and P. G. Andersson, *Tetrahedron: Asymmetry*, **8**, 3563 (1991); M. J. Sodergren, D. A. Alonso, A. V. Bedekar, and P. G. Andersson, *Tetrahedron Lett.*, **38**, 6897 (1997).
260. D. P. Albone, P. S. Aujla, P. C. Taylor, S. Challenger, and A. M. Derrick, *J. Org. Chem.*, **63**, 9569 (1998).
261. R. Vyas, B. M. Chandra, and A. V. Bedekar, *Tetrahedron Lett.*, **39**, 4715 (1998).
262. P. Dauban and R. H. Dodd, *Tetrahedron Lett.*, **39**, 5739 (1998).
263. P. A. S. Smith, *Org. React.*, **3**, 337 (1946).
264. S. Linke, G. T. Tisue, and W. Lwowski, *J. Am. Chem. Soc.*, **89**, 6308 (1967).

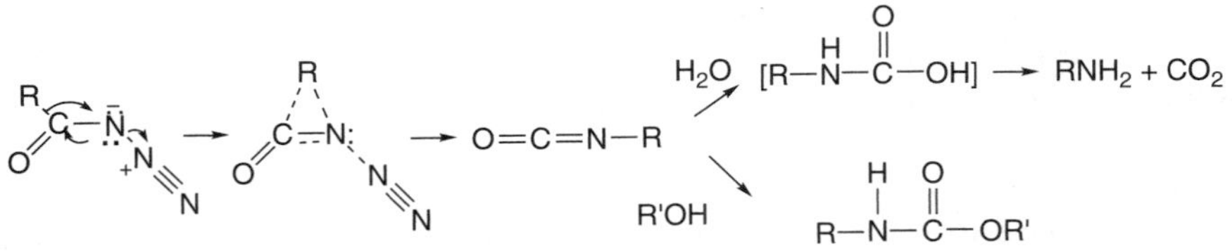

948

CHAPTER 10

*Reactions Involving
Carbocations, Carbenes,
and Radicals as Reactive
Intermediates*

The acyl azide intermediates are prepared either by reaction of sodium azide with a reactive acylating agent or by diazotization of an acyl hydrazide. An especially convenient version of the former process is treatment of the carboxylic acid with ethyl chloroformate to form a mixed anhydride, which then reacts with azide ion.[265]

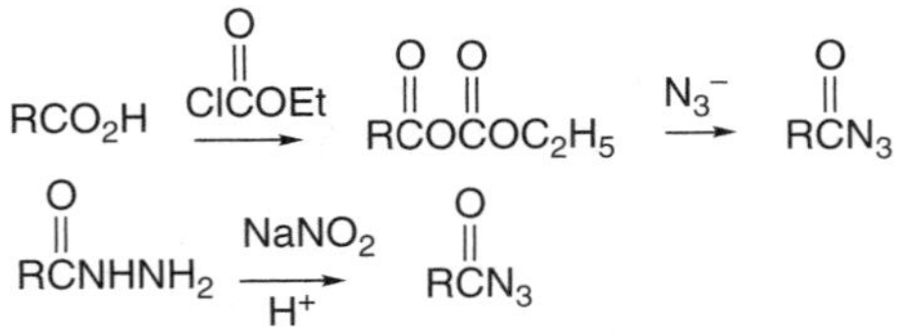

The transformation can also be carried out on the acid using diphenylphosphoryl azide (DPPA).[266]

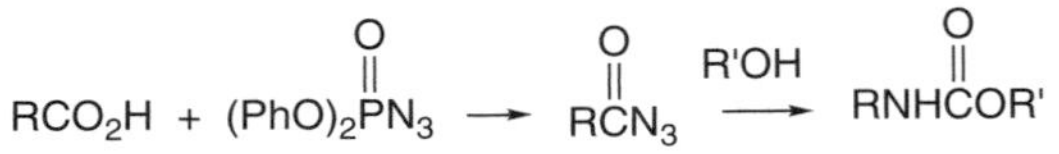

This version of the Curtius rearrangement has been applied to the synthesis of amino acid analogs and structures containing amino acids. Several *cis*-2-aminocyclopropane carboxylate esters were prepared by selective hydrolysis of cyclopropane-1,2-dicarboxylates, followed by reaction with DPPA.[267]

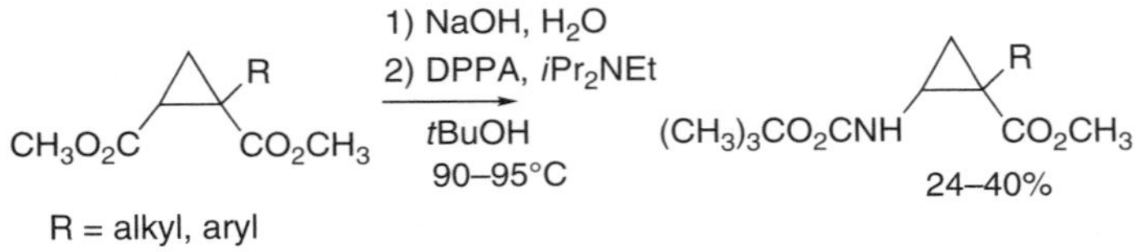

The Curtius reaction has occasionally been used in formation of medium[268] and large[269] rings, usually in modest yield.

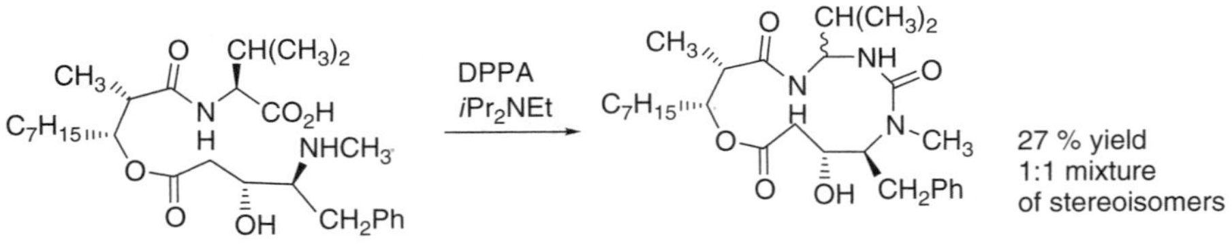

265. J. Weinstock, *J. Org. Chem.*, **26**, 3511 (1961).
266. D. Kim and S. M. Weinreb, *J. Org. Chem.*, **43**, 125 (1978).
267. S. Mangelinckx and N. De Kimpe, *Tetrahedron Lett.*, **44**, 1771 (2003).
268. C. Hermann, G. C. G. Pais, A Geyer, S. M. Kuhnert, and M. E. Maier, *Tetrahedron*, **56**, 8461 (2000).
269. Y. Hamada, M. Shibata, and T. Shioiri, *Tetrahedron Lett.*, **26**, 5155, 5159 (1985).

Another reaction that can be used for conversion of carboxylic acids to the corresponding amines with loss of carbon dioxide is the *Hofmann rearrangement*. The classic reagent is hypobromite ion, which reacts to form an *N*-bromoamide intermediate. Like the Curtius reaction, this rearrangement is believed to be a concerted process and proceeds through an isocyanate intermediate.

$$RCNH_2 + {}^-OBr \longrightarrow RCNHBr + {}^-OH \rightleftharpoons RCNBr + H_2O$$

$$R-C-N-Br \longrightarrow O=C=N-R + Br^- \xrightarrow{H_2O} H_2NR + CO_2$$

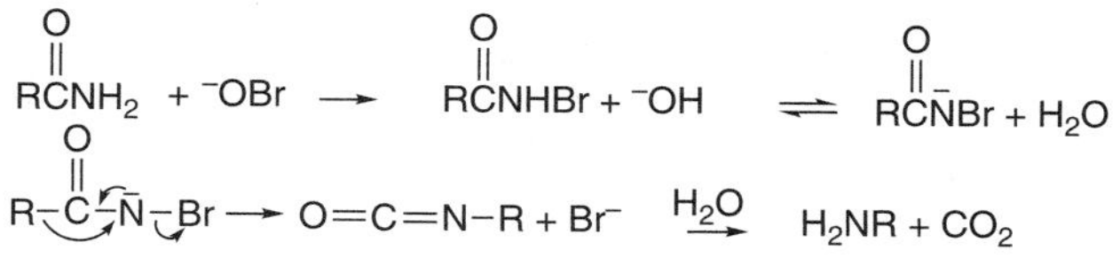

The reaction is useful in the conversion of aromatic carboxylic acids to aromatic amines.

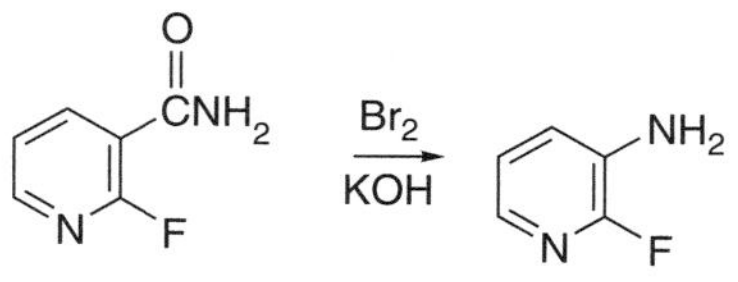

Ref. 270

Use of *N*-bromosuccinimide in the presence of sodium methoxide or DBU in methanol traps the isocyanate intermediate as a carbamate.[271]

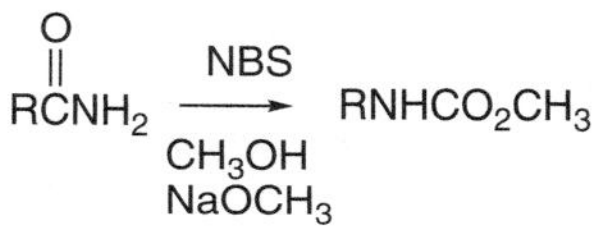

$$RCNH_2 \xrightarrow[\substack{CH_3OH \\ NaOCH_3}]{NBS} RNHCO_2CH_3$$

Direct oxidation of amides can also lead to Hofmann-type rearrangement with formation of amines or carbamates. One reagent that is used is $Pb(O_2CCH_3)_4$.

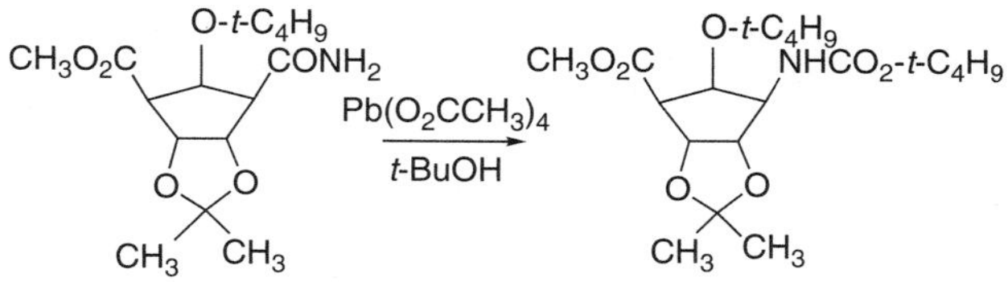

Ref. 272

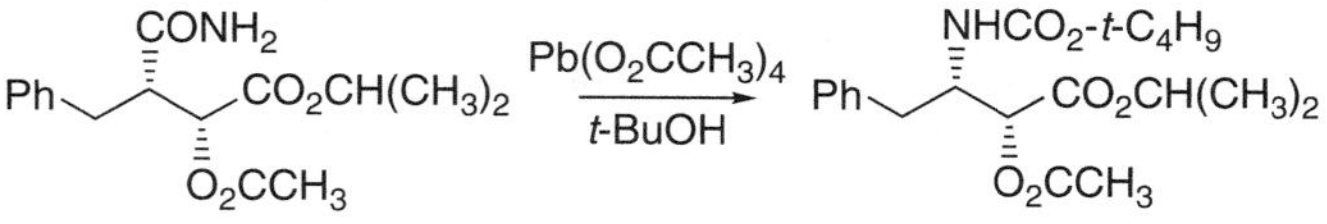

Ref. 273

270. G. C. Finger, L. D. Starr, A. Roe, and W. J. Link, *J. Org. Chem.*, **27**, 3965 (1962).
271. X. Huang and J. W. Keillor, *Tetrahedron Lett.*, **38**, 313 (1997); X. Huang, M. Said, and J. W. Keillor, *J. Org. Chem.*, **62**, 7495 (1997); J. W. Keillor and X. Huang, *Org. Synth.*, **78**, 234 (2002).
272. A. Ben Cheikh, L. E. Craine, S. G. Recher, and J. Zemlicka, *J. Org. Chem.*, **53**, 929 (1988).
273. R. W. Dugger, J. L. Ralbovsky, D. Bryant, J. Commander, S. S. Massett, N. A. Sage, and J. R. Selvidio, *Tetrahedron Lett.*, **33**, 6763 (1992).

950

CHAPTER 10

*Reactions Involving
Carbocations, Carbenes,
and Radicals as Reactive
Intermediates*

Phenyliodonium diacetate,[274,275] and phenyliodonium *bis*-trifluoroacetate,[276] are also useful oxidants for converting amides to carbamates.

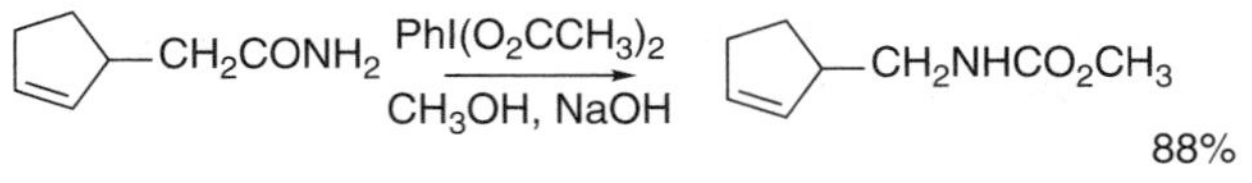

Among the recent applications of the Hofmann reaction has been the preparation of relatively unstable geminal diamides and carbinolamides. For example, 1,1-diacetamidocyclohexane can be prepared in this way.[277]

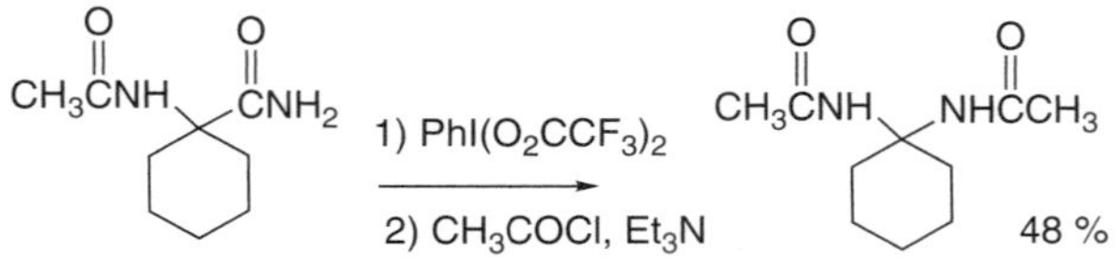

Carboxylic acids and esters can also be converted to amines with loss of the carbonyl group by reaction with hydrazoic acid, HN_3, which is known as the *Schmidt reaction*.[278] The mechanism is related to that of the Curtius reaction. An azido intermediate is generated by addition of hydrazoic acid to the carbonyl group. The migrating group retains its stereochemical configuration.

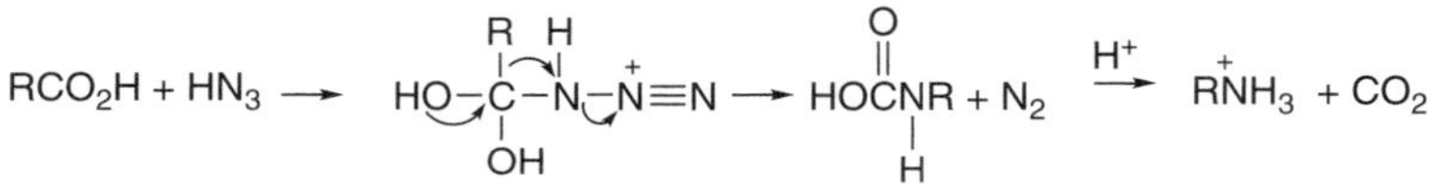

Reaction with hydrazoic acid converts ketones to amides.

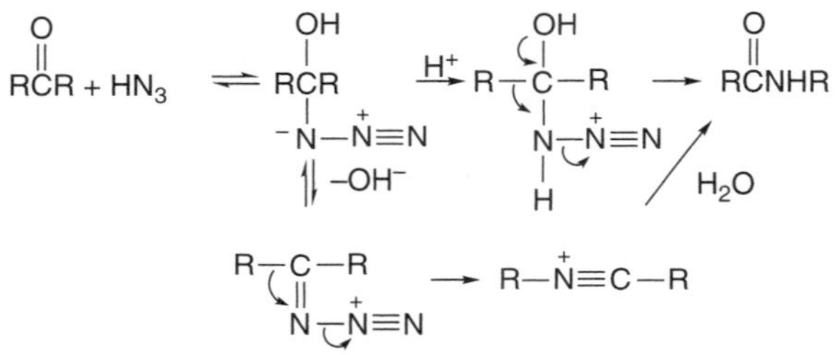

Unsymmetrical ketones can give mixtures of products because it is possible for either group to migrate.

274. R. M. Moriarty, C. J. Chany, II, R. K. Vaid, O. Prakash, and S. M. Tuladar, *J. Org. Chem.*, **58**, 2478 (1993).
275. L.-H. Zhang, G. S. Kaufman, J. A. Pesti, and J. Yin, *J. Org. Chem.*, **62**, 6918 (1997).
276. G. M. Loudon, A. S. Radhakrishna, M. R. Almond, J. K. Blodgett, and R. H. Boutin, *J. Org. Chem.*, **49**, 4272 (1984).
277. M. C. Davis, D. Stasko, and R. D. Chapman, *Synth. Commun.*, **33**, 2677 (2003).
278. H. Wolff, *Org. React.*, **3**, 307 (1946); P. A. S. Smith, in *Molecular Rearrangements*, P. de Mayo ed., Vol. 1, Interscience, New York, 1963, pp. 507–522.

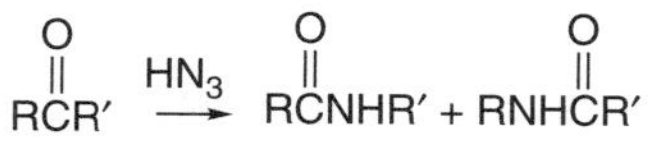

$$RCR' \xrightarrow{HN_3} RCNHR' + RNHCR'$$

Both inter- and intramolecular variants of the Schmidt reaction in which an alkyl azide effects overall insertion have been observed.

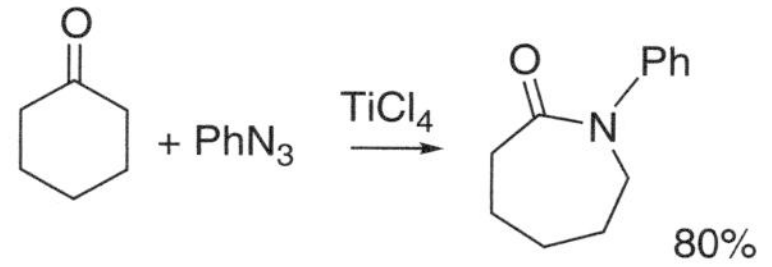

80%   Ref. 279

91%   Ref. 280

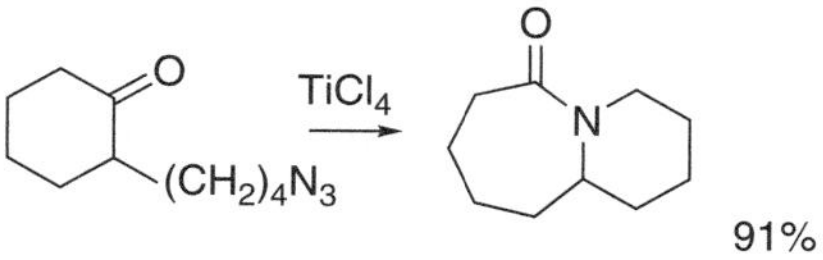

These reactions are especially favorable for β- and γ-hydroxy azides, where reaction can proceed through a hemiketal intermediate.

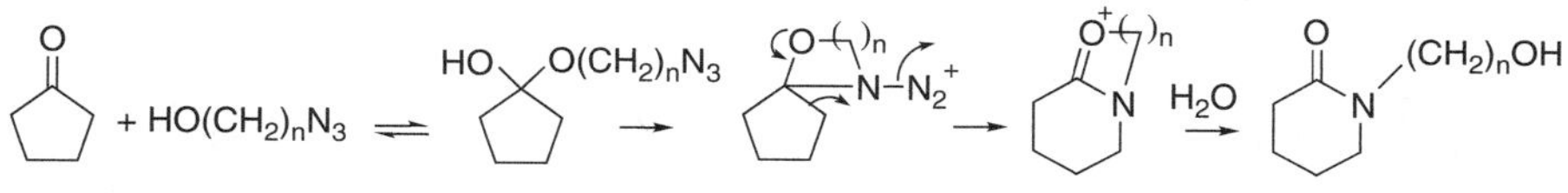

Ref. 281

Another important reaction involving migration to electron-deficient nitrogen is the *Beckmann rearrangement*, in which oximes are converted to amides.[282]

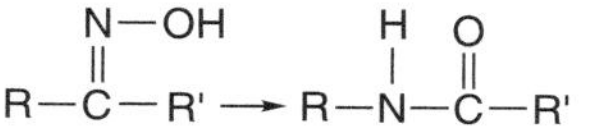

$$\underset{R-C-R'}{\overset{N-OH}{\|}} \rightarrow \underset{R-N-C-R'}{\overset{H \quad O}{\underset{|}{\|}}}$$

A variety of protic acids, Lewis acids, acid anhydrides, or acyl and sulfonyl halides can cause the reaction to occur. The mechanism involves conversion of the oxime hydroxy group to a leaving group. Ionization and migration then occur as a concerted process, with the group that is *anti* to the oxime leaving group migrating. The migration results in formation of a nitrilium ion, which captures a nucleophile. Eventually hydrolysis leads to the amide.

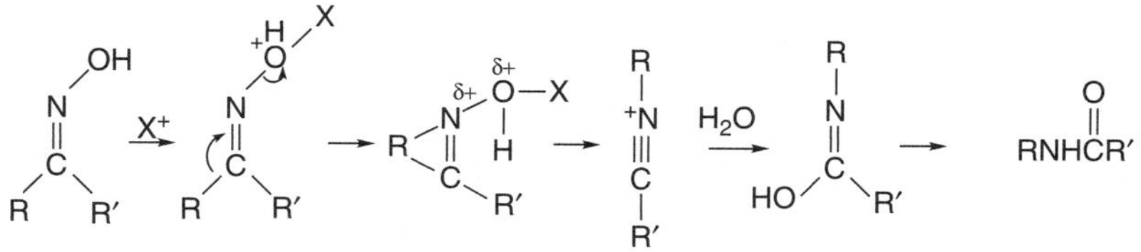

279. J. Aube and G. L. Milligan, *J. Org. Chem.*, **57**, 1635 (1992).
280. J. Aube and G. L. Milligan, *J. Am. Chem. Soc.*, **113**, 8965 (1991).
281. V. Gracias, K. E. Frank, G. L. Milligan, and J. Aube, *Tetrahedron*, **53**, 16241 (1997).
282. L. G. Donaruma and W. Z. Heldt, *Org. React.*, **11**, 1 (1960); P. A. S. Smith, *Open Chain Nitrogen Compounds*, Vol. II, W. A. Benjamin, New York, 1966, pp. 47–54; P. A. S. Smith, in *Molecular Rearrangements*, Vol. 1, P. de Mayo, ed., Interscience, New York, 1973, pp. 483–507; G. R. Krow, *Tetrahedron*, **37**, 1283 (1981); R. E. Gawley, *Org. React.*, **35**, 1 (1988).

952

CHAPTER 10

*Reactions Involving
Carbocations, Carbenes,
and Radicals as Reactive
Intermediates*

The migrating group retains its configuration. Some reaction conditions can lead to *syn-anti* isomerization at a rate exceeding rearrangement, and when this occurs, a mixture of products is formed. The reagents that have been found least likely to cause competing isomerization are phosphorus pentachloride and *p*-toluenesulfonyl chloride.[283]

A fragmentation reaction occurs if one of the oxime substituents can give rise to a relatively stable carbocation. Fragmentation is very likely to occur if a nitrogen, oxygen, or sulfur atom is present α to the oximino group.

$$X-C-\underset{\underset{R}{|}}{C}=N-OY \longrightarrow X^+=C + RC\equiv N + {}^-OY$$

Fragmentation can also occur when the α-carbon can support cationic character.

Ref. 284

Section D of Scheme 10.15 provides some examples of the Beckmann rearrangement.

Section A of Scheme 10.15 contains a number of examples of Curtius rearrangements. Entry 1 is an example carried out in a nonnucleophilic solvent, permitting isolation of the isocyanate. Entries 2 and 3 involve isolation of the amine after hydrolysis of the isocyanate. In Entry 2, the dihydrazide intermediate is isolated as a solid and diazotized in aqueous solution, from which the amine is isolated as the dihydrochloride. Entry 3 is an example of the mixed anhydride procedure (see p. 948). The first stage of the reaction is carried out in acetone and the thermolysis of the acyl azide is done in refluxing toluene. The crude isocyanate is then hydrolyzed in acidic water. Entry 4 is a reaction that demonstrates the retention of configuration during rearrangement.

Entries 5 to 8 are synthetic applications in more complex molecules. Entries 5 and 6 illustrate the diphenylphosphoroyl azide method. Entry 7 was used in the late stages of the synthesis of an antitumor macrolide, zampanolide, to introduce the amino group. The ultimate target molecule in Entry 8 is himandrine, one of several polycyclic alkaloids isolated from an ancient plant species.

himandrine

283. R. F. Brown, N. M. van Gulick, and G. H. Schmid, *J. Am. Chem. Soc.*, **77**, 1094 (1955); J. C. Craig and A. R. Naik, *J. Am. Chem. Soc.*, **84**, 3410 (1962).
284. R. T. Conley and R. J. Lange, *J. Org. Chem.*, **28**, 210 (1963).

Scheme 10.15. Rearrangement to Electron-Deficient Nitrogen

A. Curtius Rearrangements

1^a

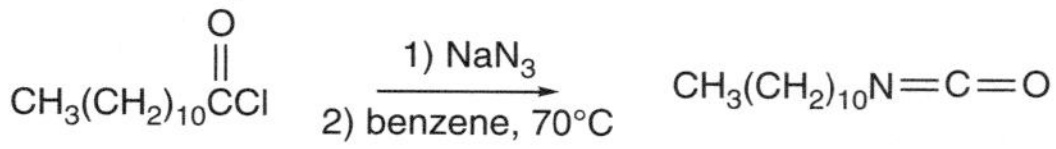

$$CH_3(CH_2)_{10}CCl \xrightarrow[\text{2) benzene, 70°C}]{\text{1) NaN}_3} CH_3(CH_2)_{10}N=C=O$$

2^b

$$C_2H_5O_2C(CH_2)_4CO_2C_2H_5 \xrightarrow[\substack{\text{3) }\Delta \\ \text{4) H}^+, \text{H}_2\text{O}}]{\substack{\text{1) N}_2\text{H}_4 \\ \text{2) HNO}_2}} Cl^- \overset{+}{H_3N}(CH_2)_4\overset{+}{NH_3} Cl^-$$

3^c

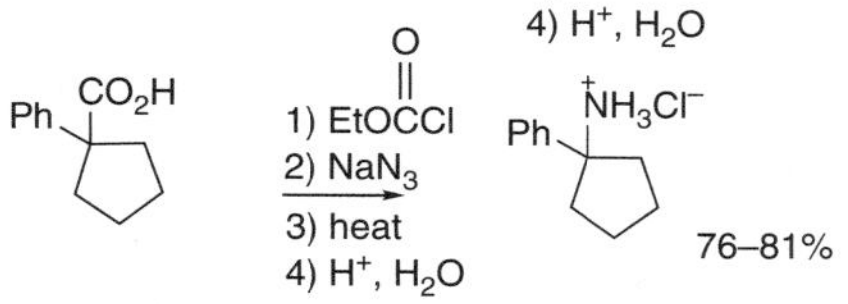

1) EtOCCl
2) NaN$_3$
3) heat
4) H$^+$, H$_2$O
76–81%

4^d

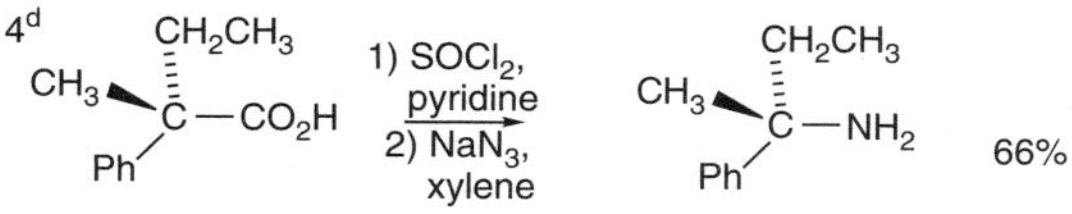

1) SOCl$_2$, pyridine
2) NaN$_3$, xylene
66%

5^e

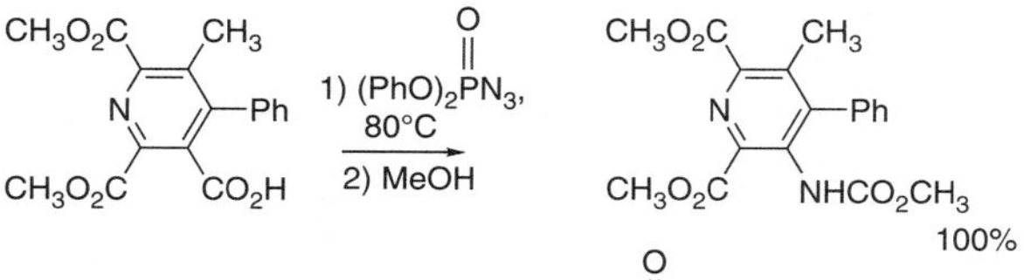

1) (PhO)$_2$PN$_3$, 80°C
2) MeOH
100%

6^f

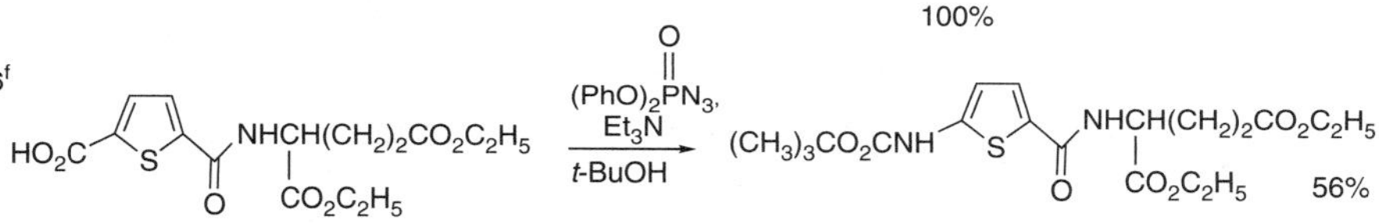

(PhO)$_2$PN$_3$, Et$_3$N, t-BuOH
56%

7^g

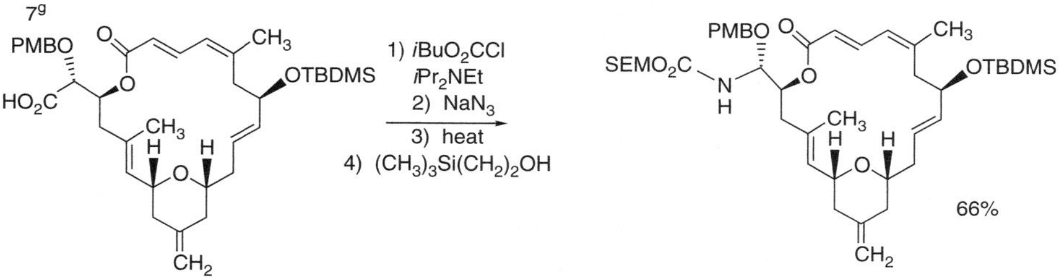

1) iBuO$_2$CCl, iPr$_2$NEt
2) NaN$_3$
3) heat
4) (CH$_3$)$_3$Si(CH$_2$)$_2$OH
66%

8^h

1) (COCl)$_2$, pyridine
2) NaN$_3$, H$_2$O
3) heat
4) CH$_3$OH, NaOCH$_3$
77%

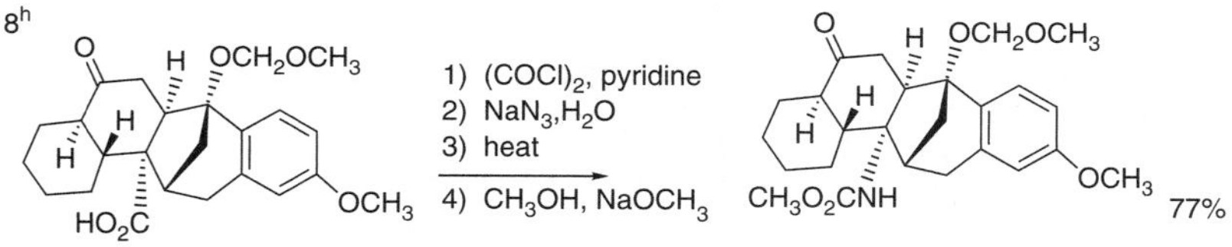

B. Hofmann Rearrangements.

9^i

$$N{\equiv}C(CH_2)_4CONH_2 \xrightarrow[\text{NaOCH}_3]{\text{Br}_2} N{\equiv}C(CH_2)_4NHCO_2CH_3 \quad 94\%$$

10^j

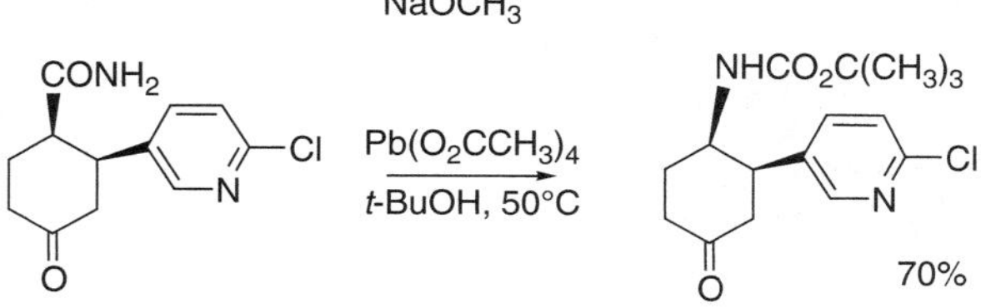

Pb(O$_2$CCH$_3$)$_4$, t-BuOH, 50°C
70%

(Continued)

Scheme 10.15. (*Continued*)

11[k]

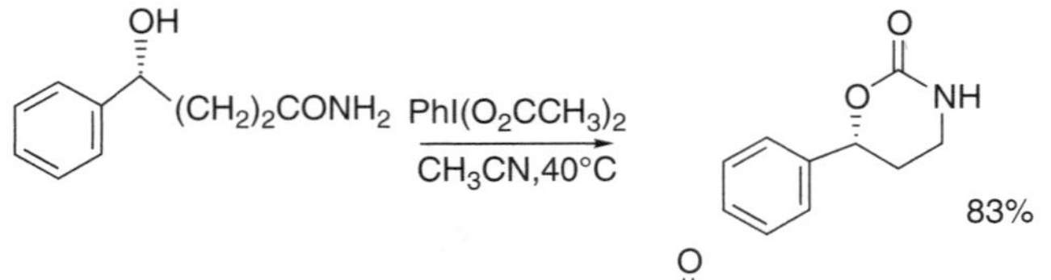

83%

12[l]

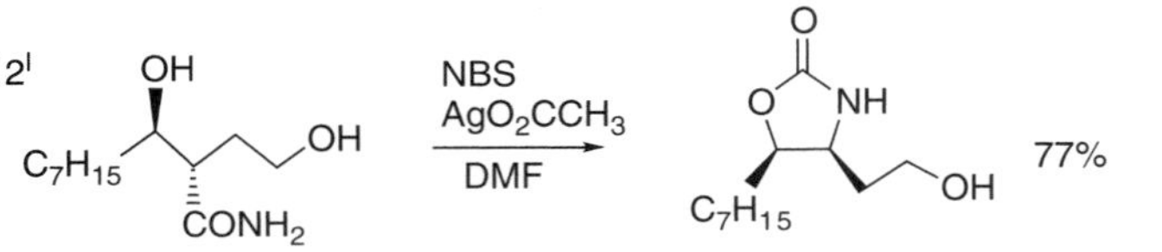

77%

13[m]

50%

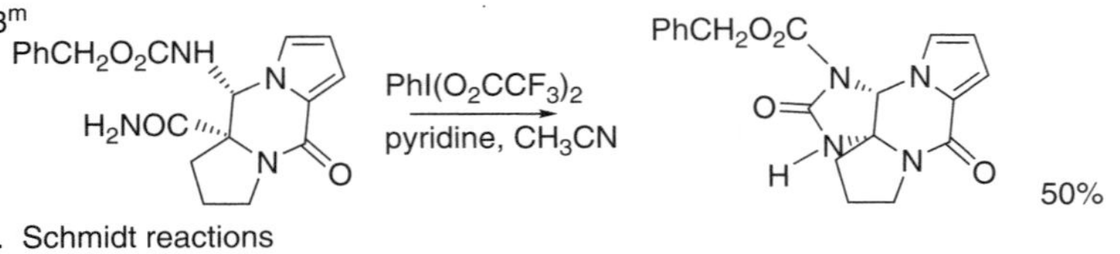

C. Schmidt reactions

14[n]

$$PhCH_2CO_2H \xrightarrow[\text{acid}]{\substack{NaN_3 \\ \text{polyphosphoric}}} PhCH_2NH_2 \quad 67\%$$

15[o]

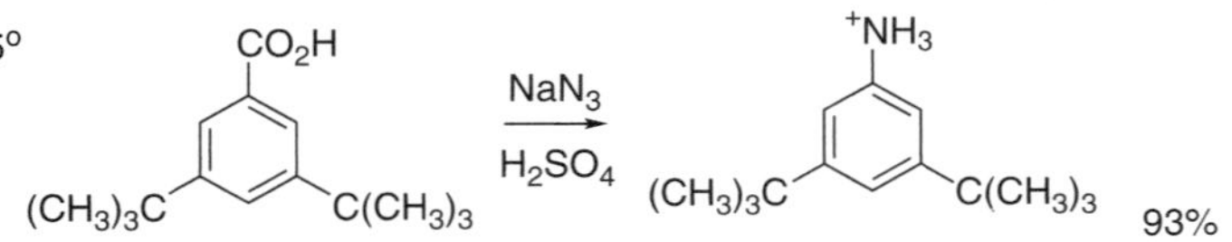

93%

16[p]

59%

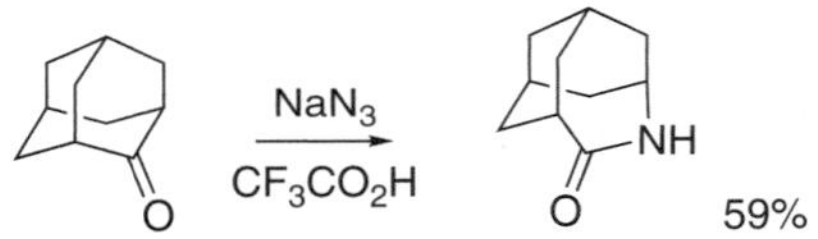

D. Beckmann Rearrangements

17[q]

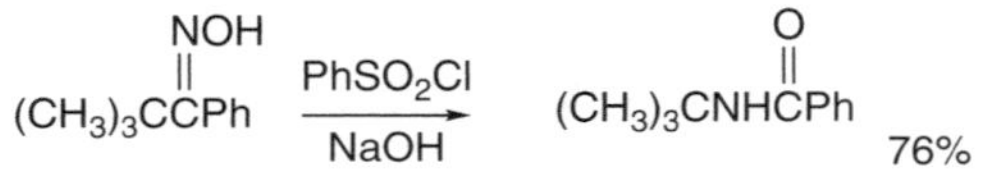

76%

18[r]

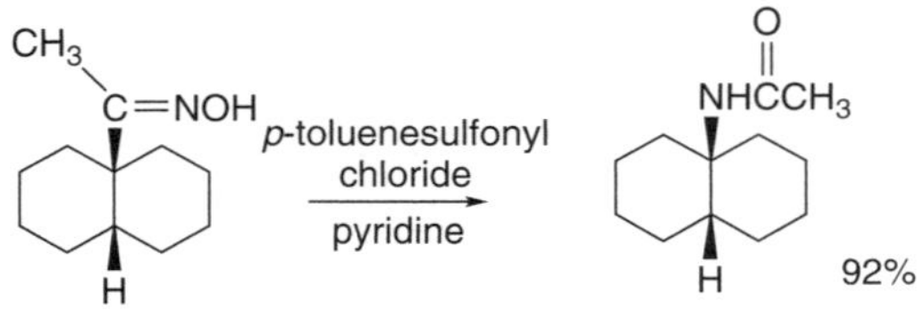

92%

19[s]

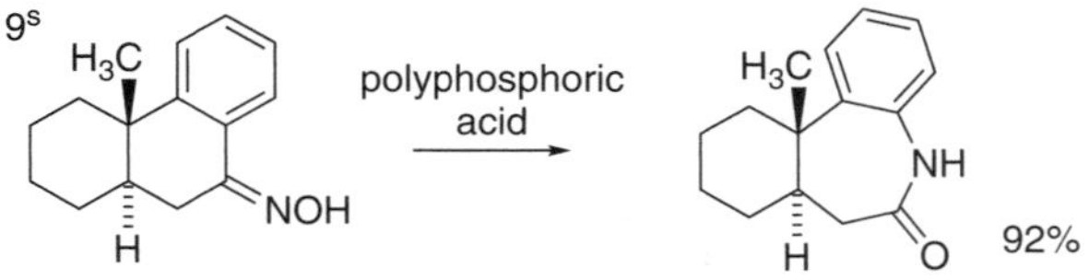

92%

20[t]

91%

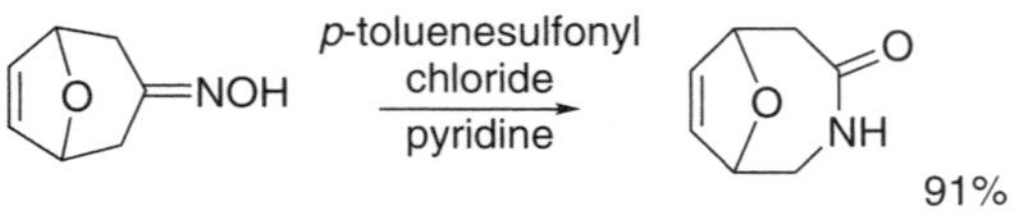

(*continued*)

a. C. F. H. Allen and A. Bell, *Org. Synth.*, **III**, 846 (1955).
b. P. A. S. Smith, *Org. Synth.*, **IV**, 819 (1963).
c. C. Kaiser and J. Weinstock, *Org. Synth.*, **51**, 48 (1971).
d. D. J. Cram and J. S. Bradshaw, *J. Am. Chem. Soc.*, **85**, 1108 (1963).
e. D. Kim and S. M. Weinreb, *J. Org. Chem.*, **43**, 125 (1975).
f. S. L. Cao, R. Wan, and Y.-P. Feng, *Synth. Commun.* **33**, 3519 (2003).
g. A. B. Smith, III, I. G. Safonov, and R. M. Corbett, *J. Am. Chem. Soc.*, **124**, 11102 (2002).
h. P. D. O'Connor, L. N. Mander, and M. M. W. McLachlan, *Org. Lett.*, **6**, 703 (2004).
i. R. Shapiro, R. DiCosimo, S. M. Hennessey, B. Stieglitz, O. Campopiano, and G. C. Chiang, *Org. Process Res. Dev.*, **5**, 593 (2001).
j. D. A. Evans, K. A. Scheidt, and C. W. Downey, *Org. Lett*, **3**, 3009 (2001).
k. J. W. Hilborn, Z.-H. Lu, A. R. Jurgens, Q. K. Fang, P. Byers, S. A. Wald, and C. H. Senanayake, *Tetrahedron Lett.*, **42**, 8919 (2001).
l. T. Hakogi, Y. Monden, M. Taichi, S. Iwama, S. Fujii, K. Ikeda, and S. Katsumura, *J. Org. Chem.*, **67**, 4839 (2002).
m.. K. G. Poullennec and D. Romo, *J. Am. Chem. Soc.*, **125**, 6344 (2003).
n. R. M. Palmere and R. T. Conley, *J. Org. Chem.*, **35**, 2703 (1970).
o. J. W. Elder and R. P. Mariella, *Can. J. Chem.*, **41**, 1653 (1963).
p. T. Sasaki, S. Eguchi, and T. Toru, *J. Org. Chem.*, **35**, 4109 (1970).
q. R. F. Brown, N. M. van Gulick, and G. H. Schmid, *J. Am. Chem. Soc.*, **77**, 1094 (1955).
r. R. K. Hill and O. T. Chortyk, *J. Am. Chem. Soc.*, **84**, 1064 (1962).
s. R. A. Barnes and M. T. Beachem, *J. Am. Chem. Soc.*, **77**, 5388 (1955).
t. S. R. Wilson, R. A. Sawicki, and J. C. Huffman, *J. Org. Chem.*, **46**, 3887 (1981).

Section B shows some Hofmann rearrangements. Entry 9, using basic conditions with bromine, provided an inexpensive route to an intermediate for a commercial synthesis of an herbicide. Entry 10, which uses the Pb(OAc)$_4$ conditions (see p. 949), was utilized in an enantiospecific synthesis of the naturally occurring analagesic (–)-epibatidine. Entry 11 uses phenyliodonium diacetate as the reagent. The product is the result of cyclization of the intermediate isocyanate and was used in an enantioselective synthesis of the antianxiety drug (*R*)-fluoxetine.

(*R*)-Fluoxetine

Entries 12 and 13 also involve cyclization of the isocyanate intermediates.

Section C of Scheme 10.15 shows some Schmidt reactions. Entry 14 is a procedure using polyphosphoric acid, whereas Entry 15 was done in H$_2$SO$_4$. Entry 16 is a case of conversion of a cyclic ketone, adamantanone, to the corresponding lactam.

Section D shows some representative Beckmann rearrangements. Entry 17 shows a selective migration of a *t*-butyl group and illustrates the use of oxime sulfonates to control regioselectivity. The opposite regioisomer, resulting from migration of the phenyl group, was observed using HCl in acetic acid. Entry 18 illustrates another aspect of the stereochemistry of the Beckmann rearrangement. As shown, use of the benzenesulfonate led to retention of the *cis* ring juncture. When the reaction was done in H$_2$SO$_4$ or polyphosphoric acid, the *trans* isomer was formed, presumably as the result of fragmentation to a tertiary carbocation.

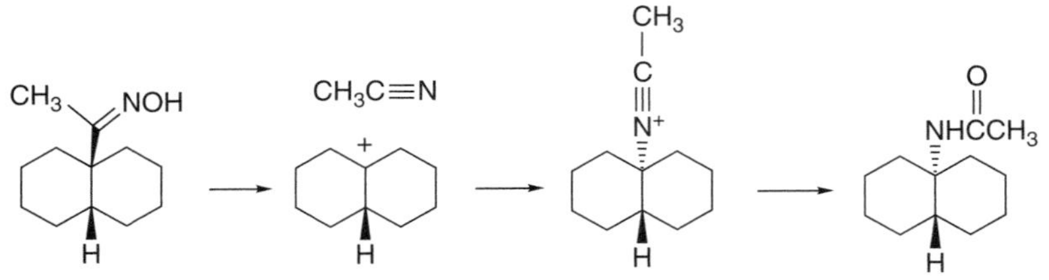

Entries 19 and 20 are examples of lactam formation by ring expansion of cyclic oximes.

10.3. Reactions Involving Free Radical Intermediates

The fundamental mechanisms of free radical reactions were considered in Chapter 11 of Part A. Several mechanistic issues are crucial in development of free radical reactions for synthetic applications.[285] Free radical reactions are usually chain processes, and the lifetimes of the intermediate radicals are very short. To meet the synthetic requirements of high selectivity and efficiency, all steps in a desired sequence must be fast in comparison with competing reactions. Owing to the requirement that all the steps be fast, only steps that are exothermic or very slightly endothermic can participate in chain processes. Comparison between addition of a radical to a carbon-carbon double bond and addition to a carbonyl group can illustrate this point.

$$\Delta H = (C-C) - (C^\pi - C^\pi)$$
$$= -81 - (-64) = -17$$

$$\Delta H = (C-C) - (C^\pi - O^\pi)$$
$$= -81 - (-94) = +13$$

This comparison suggests that of these two similar reactions, only alkene additions are likely to be a part of an efficient radical chain sequence. Radical additions to carbon-carbon double bonds can be further enhanced by radical stabilizing groups. Addition to a carbonyl group, in contrast, is endothermic. In fact, the reverse fragmentation reaction is commonly observed (see Section 10.3.6) A comparison can also be made between abstraction of hydrogen from carbon as opposed to oxygen.

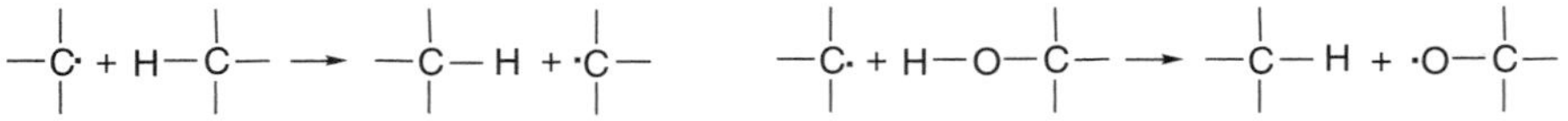

$$\Delta H = 0$$

$$\Delta H = (C-H) - (O-H) = -98 - (-109) = +11$$

The reaction endothermicity establishes a *minimum* for the activation energy; whereas abstraction of a hydrogen atom from carbon is a feasible step in a chain process, abstraction of a hydrogen atom from a hydroxy group is unlikely. Homolytic cleavage of an O—H bond is likely only if the resulting oxygen radical is stabilized, such as in phenoxy radicals formed from phenols.

285. C. Walling, *Tetrahedron*, **41**, 3887 (1985).

There is a good deal of information available about the absolute rates of free radical reactions. A selection from these data is given in Table 11.3 of Part A. If the steps in a projected reaction sequence correspond to reactions for which absolute rates are known, this information can allow evaluation of the kinetic feasibility of the reaction sequence.

10.3.1. Sources of Radical Intermediates

There is a discussion of some of the sources of radicals for mechanistic studies in Section 11.1.4 of Part A. Some of the reactions discussed there, particularly the use of azo compounds and peroxides as initiators, are also important in synthetic chemistry. One of the most useful sources of free radicals in preparative chemistry is the reaction of halides with stannyl radicals. Stannanes undergo hydrogen abstraction reactions and the stannyl radical can then abstract halogen from the alkyl group. For example, net addition of an alkyl group to a reactive double bond can follow halogen abstraction by a stannyl radical.

$$
\begin{aligned}
\text{initiation} \quad & \text{In·} + R'_3\text{SnH} \longrightarrow R'_3\text{Sn·} + \text{In—H} \\
\text{propagation} \quad & R—X + R'_3\text{Sn·} \longrightarrow R· + R'_3\text{Sn—X} \\
& R· + X{=}Y \longrightarrow R—X—Y· \\
& R—X—Y· + R'_3\text{Sn—H} \longrightarrow R—X—Y—H + R'_3\text{Sn·}
\end{aligned}
$$

This generalized reaction sequence consumes the halide, the stannane, and the reactant $X{=}Y$, and effects addition to the organic radical and a hydrogen atom to the $X{=}Y$ bond. The order of reactivity of organic halides toward stannyl radicals is iodides > bromides > chlorides.

Esters of N-hydroxypyridine-2-thione are another versatile source of radicals,[286] where the radical is formed by decarboxylation of an adduct formed by attack at sulfur by the chain-carrying radical.[287] The generalized chain sequence is as follows.

$$R· + X—Y \longrightarrow R—Y + X·$$

When $X—Y$ is $R_3\text{Sn—H}$ the net reaction is decarboxylation and reduction of the original acyloxy group. Halogen atom donors can also participate in such reactions.

[286.] D. Crich, *Aldrichimica Acta*, **20**, 35 (1987); D. H. R. Barton, *Aldrichimica Acta*, **23**, 3 (1990).

[287.] D. H. R. Barton, D. Crich, and W. B. Motherwell, *Tetrahedron*, **41**, 3901 (1985); D. H. R. Barton, D. Crich, and G. Kretzschmar, *J. Chem. Soc., Perkin Trans. 1*, 39 (1986); D. H. R. Barton, D. Bridson, I. Fernandez-Picot, and S. Z. Zard, *Tetrahedron*, **43**, 2733 (1987).

958

CHAPTER 10

*Reactions Involving
Carbocations, Carbenes,
and Radicals as Reactive
Intermediates*

When X—Y is Cl_3C—Cl, the final product is a chloride.[288] Use of Cl_3C—Br gives the corresponding bromide.[289]

The precise reaction conditions for optimal yields depend upon the specific reagents and both thermal[290] and photochemical[291] conditions have been developed. Phenyl thionocarbonates are easily prepared and are useful in radical generating reactions.[292] A variety of other thiono esters, including xanthates and imidazolyl thionocarbonates also can be used.[293]

Selenyl groups can be abstracted by stannyl radicals from alkyl and acyl selenides to generate the corresponding radicals.[294] Among the types of compounds that react by selenyl transfer are α-selenylphosphonates[295] and α-selenylcyanides.[296] The radicals generated can undergo addition and/or cyclization. The chain reaction is propagated by abstraction of hydrogen from the stannane.

Trialkylboranes, especially triethylborane, are used in conjunction with O_2 to generate radicals.[297] The alkyl radicals are generated by breakdown of a borane-oxygen adduct. An advantage this method has over many other radical initiation systems is that it proceeds at low temperature, e.g., $-78°C$.

$$R_3B + O_2 \longrightarrow \cdot O-OBR_2 + R\cdot$$
$$R\cdot + O_2 \longrightarrow RO_2\cdot$$
$$RO_2\cdot + R_3B \longrightarrow RO_2BR_2 + R\cdot$$

288. D. H. R. Barton, D. Crich, and W. B. Motherwell, *Tetrahedron Lett.*, **24**, 4979 (1983).

289. D. H. R. Barton, R. Lacher, and S. Z. Zard, *Tetrahedron Lett.*, **26**, 5939 (1983).

290. D. H. R. Barton, J. L. Jaszberenyi, and D. Tang, *Tetrahedron Lett.*, **54**, 3381 (1993).

291. J. Bouivin, E. Crepon, and S. Z. Zard, *Tetrahedron Lett.*, **32**, 199 (1991).

292. M. J. Robins, J. S. Wilson, and F. Hansske, *J. Am. Chem. Soc.*, **105**, 4059 (1983).

293. D. H. R. Barton and S. W. McCombie, *J. Chem. Soc., Perkin Trans. 1*, 1574 (1975).

294. J. Pfenninger, C. Heuberger, and W. Graf, *Helv. Chim. Acta*, **63**, 2328 (1980); D. L. Boger and R. J. Mathvink, *J. Org. Chem.*, **53**, 3377 (1988); D. L. Boger and R. J. Mathvink, *J. Org. Chem.*, **57**, 1429 (1992).

295. P. Balczewski, W. M. Pietrzykowski, and M. Mikolajczyk, *Tetrahedron*, **51**, 7727 (1995).

296. D. L. J. Clive, T. L. B. Boivin, and A. G. Angoh, *J. Org. Chem.*, **52**, 4943 (1987).

297. C. Ollivier and P. Renaud, *Chem. Rev.*, **101**, 3415 (2001).

The radicals generated in this way can initiate a variety of chain processes. Alkyl radicals can be generated from alkyl iodides.[298] For example, addition of alkyl radicals to alkynes can be accomplished under these conditions.

Ref. 299

These reactions result in *iodine atom transfer* and introduce a potential functional group into the product. The trialkylborane method of radical generation can also be used in conjunction with either tri-*n*-butyl stannane or *tris*-(trimethylsilyl)silane, in which case the product is formed by hydrogen atom transfer.

The reductive decomposition of alkylmercury compounds is also a useful source of radicals.[300] The organomercury compounds are available by oxymercuration (see Section 4.1.3) or from organometallic compounds as a result of metal-metal exchange (see Section 7.3.3).

$$RCH=CH_2 + HgX_2 \xrightarrow{SH} RCHCH_2HgX$$
$$\underset{S}{|} \quad (SH = solvent)$$

$$RLi + HgX_2 \longrightarrow RHgX + LiX$$

Alkylmercury reagents can also be prepared from alkyl boranes.

$$R_3B + 3\,Hg(OAc)_2 \longrightarrow 3\,RHgOAc$$

Ref. 301

The mercuric hydride formed by reduction undergoes chain decomposition to generate alkyl radicals.

reduction $\quad RHgX + \tfrac{1}{4}\,NaBH_4 \longrightarrow RHgH + \tfrac{1}{4}\,NaBX_4$

initiation $\quad RHgH \longrightarrow R^{\cdot} + HgH$

propagation $\quad R\cdot + RHgH \longrightarrow R{-}H + RHg$

$\quad RHg \longrightarrow R^{\cdot} + Hg^0$

overall reaction $\quad RHgX + \tfrac{1}{4}\,NaBH_4 \longrightarrow R{-}H + \tfrac{1}{4}\,NaBX_4$

10.3.2. Addition Reactions of Radicals with Substituted Alkenes

The most general method for formation of new carbon-carbon bonds via radical intermediates involves addition of the radical to an alkene. The reaction generates a new radical that can propagate a chain sequence. The preferred alkenes for trapping alkyl

[298] H. C. Brown and M. M. Midland, *Angew. Chem. Int. Ed. Engl.*, **11**, 692 (1972); K. Nozaki, K. Oshima, and K. Utimoto, *Tetrahedron Lett.*, **29**, 1041 (1988).

[299] Y. Ichinose, S. Matsunaga, K. Fugami, K. Oshima, and K. Utimoto, *Tetrahedron Lett.*, **30**, 3155 (1989).

[300] G. A. Russell, *Acc. Chem. Res.*, **22**, 1 (1989).

[301] R. C. Larock and H. C. Brown, *J. Am. Chem. Soc.*, **92**, 2467 (1976).

960

CHAPTER 10

*Reactions Involving
Carbocations, Carbenes,
and Radicals as Reactive
Intermediates*

radicals are ethene derivatives with electron-attracting groups, such as cyano, ester, or other carbonyl substituents.[302] There are three factors that make such compounds particularly useful: (1) alkyl radicals are relatively *nucleophilic* and react at enhanced rates with alkenes having EWG substituents; (2) alkenes with such substituents exhibit a good degree of regioselectivity, resulting from a combination of steric and radical-stabilizing effects of the substituent; (3) the EWG substituent makes the adduct radical *more electrophilic* and increases the rate of the subsequent hydrogen abstraction step. The "nucleophilic" versus "electrophilic" character of radicals can be understood in terms of the FMO description of substituent effects on radicals. The three most important cases are outlined in Figure 10.12. An ERG in the radical raises the energy of the SOMO, which increases the stabilizing interaction with the LUMO of alkenes having EWG substituents. In the opposite combination, an EWG substituent on the radicals lowers the SOMO and the strongest interaction is with the alkene HOMO. This interaction is stabilizing because of lowering of the alkene HOMO.

Radicals for addition reactions can be generated by halogen atom abstraction by stannyl radicals. The chain mechanism for alkylation of alkyl halides by reaction with a substituted alkene is outlined below. There are three reactions in the propagation cycle of this chain mechanism: addition, hydrogen atom abstraction, and halogen atom transfer.

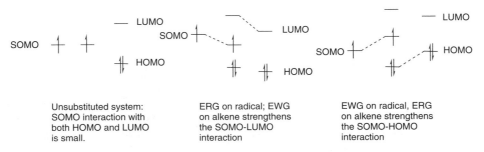

The rates of each of these steps must exceed competing chain termination reactions in order for good yields to be obtained. The most important competitions are between: (a) the addition step k_1 and reaction of the intermediate R· with Bu$_3$SnH, and (b) between the H abstraction step k_2 and addition to another molecule of the alkene. If

| SOMO interaction with both HOMO and LUMO is small. | ERG on radical; EWG on alkene strengthens the SOMO-LUMO interaction | EWG on radical, ERG on alkene strengthens the SOMO-HOMO interaction |

Unsubstituted system: SOMO interaction with both HOMO and LUMO is small.

ERG on radical; EWG on alkene strengthens the SOMO-LUMO interaction

EWG on radical, ERG on alkene strengthens the SOMO-HOMO interaction

Fig. 10.12. Frontier orbital interpretation of radical substituent effects.

[302] B. Giese, *Angew. Chem. Int. Ed. Engl.*, **22**, 753 (1983); B. Giese, *Angew. Chem. Int. Ed. Engl.*, **24**, 553 (1985).

the addition step k_1 is not fast enough, the radical R· will abstract H from the stannane and the overall reaction will simply be dehalogenation. If step k_2 is not fast relative to a successive addition step, formation of oligomers containing several alkene units will occur. For good yields R· must be more reactive to the substituted alkene than is $RCH_2C·HZ$ and $RCH_2C·HZ$ must be more reactive toward Bu_3SnH than is R·. These requirements are met when Z is an electron-attracting group. Yields are also improved if the concentration of Bu_3SnH is kept low to minimize the reductive dehalogenation, which can be done by adding the stannane slowly as the reaction proceeds. Another method is to use only a small amount of the trialkyltin hydride along with a reducing agent, such as $NaBH_4$ or $NaBH_3CN$, that can regenerate the reactive stannane.[303] Radicals formed by fragmentation of thionocarbonates and related thiono esters can also be trapped by reactive alkenes. The mechanism of radical generation from thiono esters was discussed in connection with the Barton deoxygenation method in Section 5.5.

Although most radical reactions involving chain propagation by hydrogen atom transfer can be done using trialkylstannanes, several silanes have been investigated as alternatives.[304] *Tris*-(trimethylsilyl)silane reacts with alkyl radicals at about one-tenth the rate of tri-*n*-butylstannane. The *tris*-(trimethylsilyl)silyl radical is reactive toward iodides, sulfides, selenides, and thiono esters, permitting chain transfer. Thus it is possible to substitute *tris*-(trimethylsilyl)silane for tri-*n*-butylstannane in reactions such as dehalogenations, radical additions, and cyclizations. A virtue of the silane donors is that they avoid the tin-containing by-products of stannane reactions that can cause purification problems.

$$CH_3(CH_2)_{15}I \xrightarrow[\text{AIBN}]{[(CH_3)_3Si]_3SiH} CH_3(CH_2)_{14}CH_3$$

Ref. 305

Ref. 306

Ref. 306

Alkyl radicals generated by reduction of organomercury compounds can also add to alkenes having EWG groups. Radicals are generated by reduction of the organomercurial by $NaBH_4$ or a similar reductant. These techniques have been

303. B. Giese, J. A. Gonzalez-Gomez, and T. Witzel, *Angew. Chem. Int. Ed. Engl.*, **23**, 69 (1984).
304. C. Chatgilialoglu, *Acc. Chem. Res.*, **25**, 188 (1991).
305. C. Chatgilialoglu, A. Guerrini, and G. Sesoni, *Synlett*, 219 (1990).
306. B. Giese, B. Kopping, and C. Chatgilialoglu, *Tetrahedron Lett.*, **30**, 681 (1989).

962

CHAPTER 10

Reactions Involving
Carbocations, Carbenes,
and Radicals as Reactive
Intermediates

applied to β-hydroxy-,[307] β-alkoxy-,[308] and β-amido-[309] alkylmercury derivatives. α-Acetoxyalkylmercury compounds can be prepared from hydrazones by mercuric oxide and mercuric acetate.

Ref. 310

Several other examples of addition reactions involving organomercury compounds are given in Section B of Scheme 10.16 at the end of this section.

There are also reactions in which electrophilic radicals react with relatively nucleophilic alkenes. These reactions are exemplified by a group of procedures in which a radical intermediate is formed by oxidation of readily enolizable compounds. This reaction was initially developed for β-ketoacids,[311] and the method has been extended to β-diketones, malonic acids, and cyanoacetic acid.[312] The radicals formed by the addition step are rapidly oxidized to cations, which give rise to the final product by intramolecular capture of a carboxylate group.

Phenacyl radicals can be generated from the corresponding xanthates and add in good yield to various substituted propenes. The products of the reaction can then be cyclized to tetralones using an equivalent of a peroxide.[313]

307. A. P. Kozikowski, T. R. Nieduzak, and J. Scripko, *Organometallics*, **1**, 675 (1982).
308. B. Giese and K. Heuck, *Chem. Ber.*, **112**, 3759 (1979); B. Giese and U. Luening, *Synthesis*, 735 (1982).
309. A. P. Kozikowski and J. Scripko, *Tetrahedron Lett.*, **24**, 2051 (1983).
310. B. Giese and U. Erfort, *Chem. Ber.*, **116**, 1240 (1983).
311. E. Heiba and R. M. Dessau, *J. Org. Chem.*, **39**, 3456 (1974).
312. E. J. Corey and M. C. Kang, *J. Am. Chem. Soc.*, **106**, 5384 (1984); E. J. Corey and A. W. Gross, *Tetrahedron Lett.*, **26**, 4291 (1985); W. E. Fristad and S. S. Hershberger, *J. Org. Chem.*, **50**, 1026 (1985).
313. A. Liard, B. Quiclet-Sire, R. N. Saicic, and S. Z. Zard, *Tetrahedron Lett.*, **38**, 1759 (1997).

This methodology has been applied to carbohydrate derivatives and provides a route to certain C-aryl glycosides.

Ar = 4-chlorophenyl
4-fluorophenyl

Ref. 314

Scheme 10.16 gives some examples of radical addition reactions. Entry 1 is a typical alkylation reaction using Bu_3SnH as the chain carrier and hydrogen atom donor. The reaction was done at 100°C in toluene by slow (syringe pump) addition of one equivalent of Bu_3SnH. Five equivalents of methyl acrylate was used. Entry 2 utilized *in situ* generation of Bu_3SnH. This carbohydrate-derived bromide could not be added successfully to acrylonitrile or methyl acrylate under standard conditions. A tenfold excess of phenyl vinyl sulfone was used. In Entry 3, a carbohydrate-derived acrylate is the reactant. The stannane was added by syringe pump and a 20-fold excess of the iodoacetamide was used. In Entry 4, the unprotected carbohydrate hydroxy group was converted to a xanthate ester and then added to acrylonitrile. The stereoselectivity is determined by conformational factors that establish a preference for the direction of reagent approach. Radicals with a large bias can give highly stereoselective reactions.

Entry 5 is an example of the use of *tris*-(trimethylsilyl)silane as the chain carrier. Entries 6 to 11 show additions of radicals from organomercury reagents to substituted alkenes. In general, the stereochemistry of these reactions is determined by reactant conformation and steric approach control. In Entry 9, for example, addition is from the *exo* face of the norbornyl ring. Entry 12 is an example of addition of an acyl radical from a selenide. These reactions are subject to competition from decarbonylation, but the relatively slow decarbonylation of aroyl radicals (see Part A, Table 11.3) favors addition in this case.

Allylic stannanes are an important class of compounds that undergo substitution reactions with alkyl radicals. The chain is propagated by elimination of the trialkyl-stannyl radical.[315] The radical source must have some functional group that can be abstracted by trialkylstannyl radicals. In addition to halides, both thiono esters[316] and selenides[317] are reactive.

314. A. Cordero-Vargus, B. Quiclet-Sire, and S. Z. Zard, *Tetrahedron Lett.*, **45**, 7335 (2004).
315. G. E. Keck and J. B. Yates, *J. Am. Chem. Soc.*, **104**, 5829 (1982).
316. G. E. Keck, D. F. Kachensky, and E. J. Enholm, *J. Org. Chem.*, **49**, 1462 (1984).
317. R. R. Webb and S. Danishefsky, *Tetrahedron Lett.*, **24**, 1357 (1983); T. Toru, T. Okumura, and Y. Ueno, *J. Org. Chem.*, **55**, 1277 (1990).

Scheme 10.16. Addition of Alkyl Radicals to Alkenes

A. With radical generation using trisubstituted stannanes

1[a] Bu$_3$SnH, AIBN

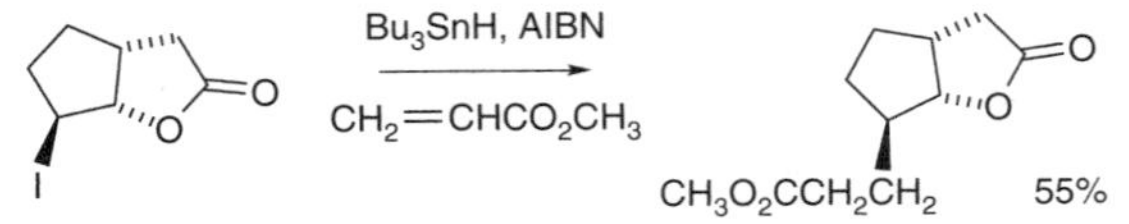

CH$_2$=CHCO$_2$CH$_3$

CH$_3$O$_2$CCH$_2$CH$_2$ 55%

2[b] Bu$_3$SnCl, NaBH$_3$CN

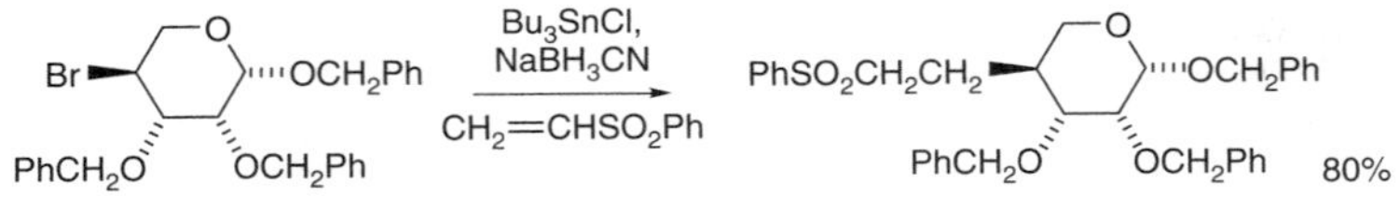

CH$_2$=CHSO$_2$Ph

PhSO$_2$CH$_2$CH$_2$ 80%

3[c] CH$_3$O$_2$CCH=CH

Bu$_3$SnH, hv

ICH$_2$CONH$_2$

H$_2$NOCCH$_2$ 64%

4[d] 1) CS$_2$, NaH
2) CH$_3$I
3) Bu$_3$SnH, CH$_2$=CHCN

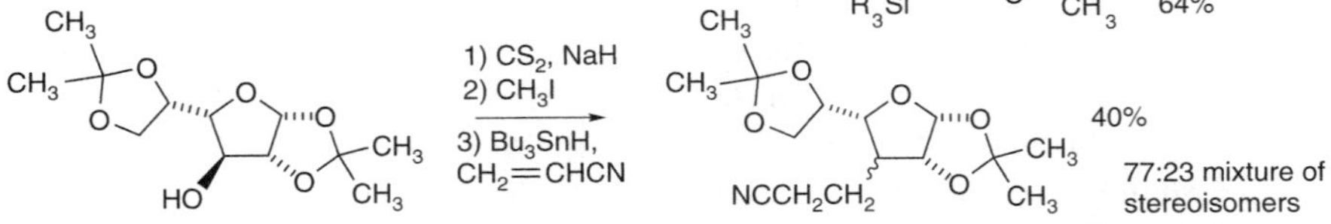

40%

77:23 mixture of stereoisomers

5[e] + CH$_2$=CHCCH$_3$

[(CH$_3$)$_3$Si]$_3$SiH

AIBN

72 % yield;
82:18 β:α

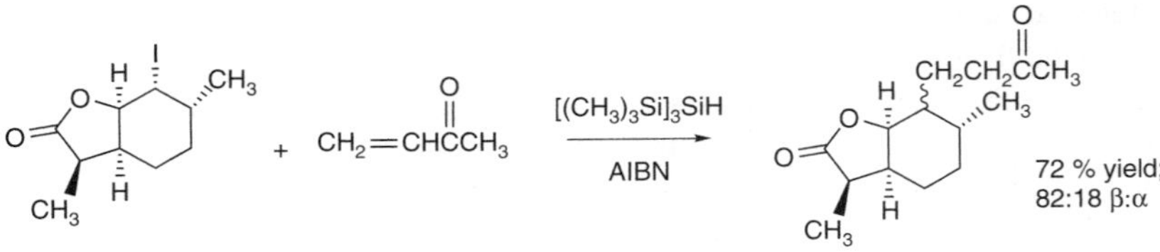

B. Using other methods of radical generation

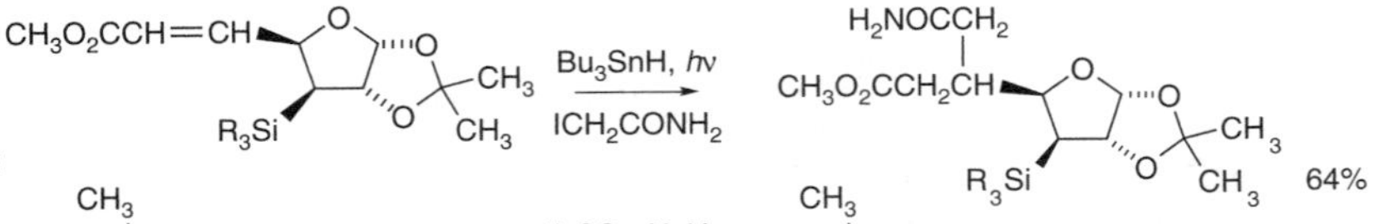

6[f] NaBH(OCH$_3$)$_3$

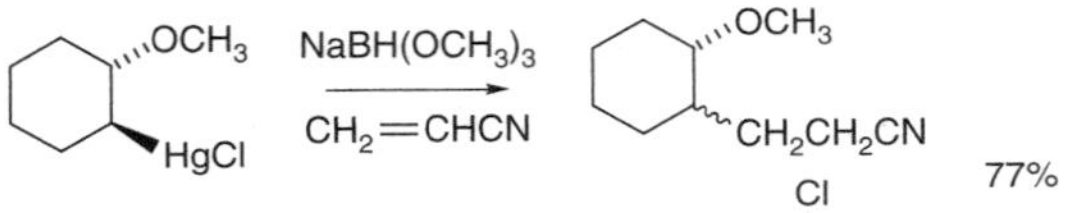

CH$_2$=CHCN

77%

7[g] NaBH$_4$

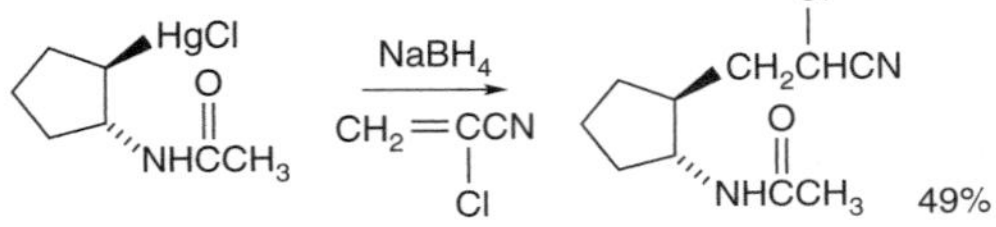

CH$_2$=CCN, Cl

49%

8[h] CH$_3$(CH$_2$)$_8$CCH$_2$HgBr

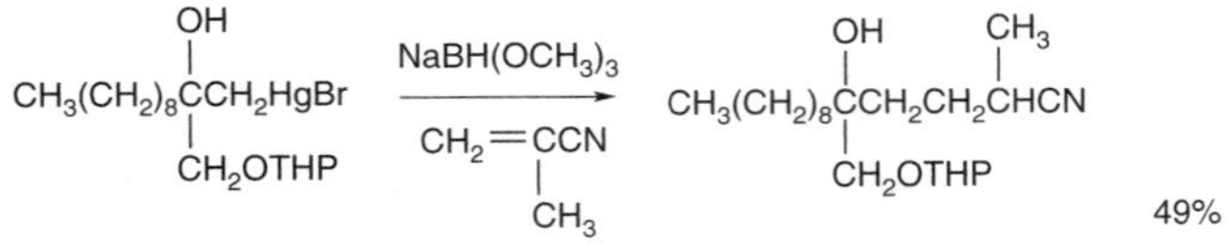

NaBH(OCH$_3$)$_3$

CH$_2$=CCN, CH$_3$

CH$_3$(CH$_2$)$_8$CCH$_2$CH$_2$CHCN   49%

9[g] HgOAc

NaBH$_4$

CH$_2$=CHCO$_2$CH$_3$

CH$_2$CH$_2$CO$_2$CH$_3$ 75%

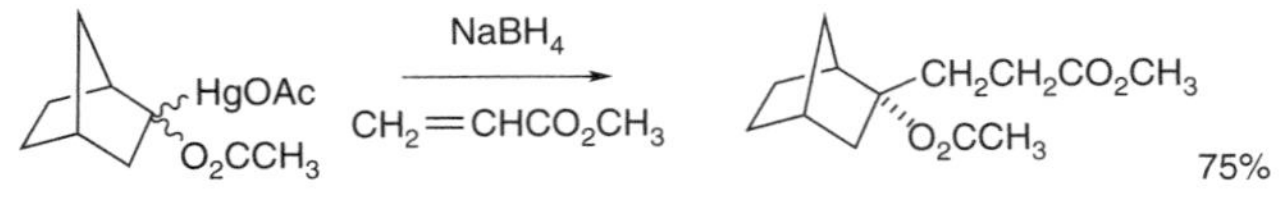

10[i] BrHg

NaBH(OCH$_3$)$_3$

CH$_2$=CCN, CH$_3$

49%

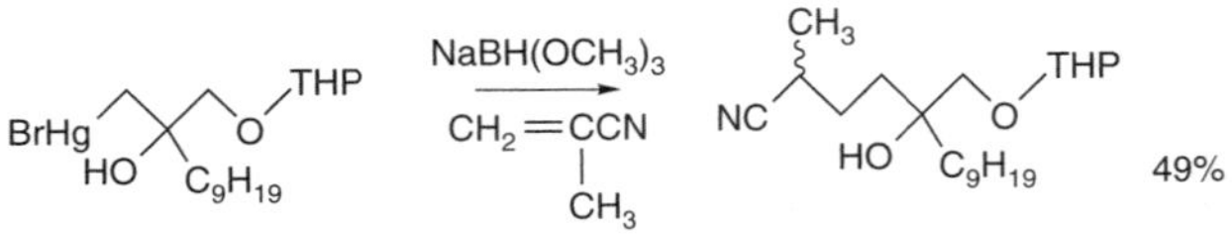

(Continued)

11[j]

12[k]

a. S. D. Burke, W. B. Fobare, and D. M. Arminsteadt, *J. Org. Chem.*, **47**, 3348 (1982).
b. M. V. Rao and M. Nagarajan, *J. Org. Chem.*, **53**, 1432 (1988).
c. G. Sacripante, C. Tan, and G. Just, *Tetrahedron Lett.*, **26**, 5643 (1985).
d. B. Giese, J. A. Gonzalez-Gomez, and T. Witzel, *Angew. Chem. Int. Ed. Engl.*, **23**, 69 (1984).
e. J. S. Yadav, R. S. Babu, and G. Sabitha, *Tetrahedron Lett.*, **44**, 387 (2003).
f. B. Giese and K. Heuck, *Chem. Ber.*, **112**, 3759 (1979).
g. R. Henning and H. Urbach, *Tetrahedron Lett.*, **24**, 5343 (1983).
h. A. P. Kozikowski, T. R. Nieduzak, and J. Scripko, *Organometallics*, **1**, 675 (1982).
i. B. Giese and U. Erfort, *Chem. Ber.*, **116**, 1240 (1983).
j. S. Danishefsky, E. Taniyama, and R. P. Webb, II, *Tetrahedron Lett.*, **24**, 11 (1983).
k. D. L. Boger and R. J. Mathvink, *J. Org. Chem.*, **57**, 1429 (1992).

Allyl *tris*-(trimethylsilyl)silane can react similarly.[318]

Allylation reactions can be initiated by triethylboron. This procedure has been found to give improved stereoselectivity in acyclic allylations.[319]

89 %

22:1 *erythro*

Scheme 10.17 illustrates allylation by reaction of radical intermediates with allyl stannanes. The first entry uses a carbohydrate-derived xanthate as the radical source. The addition in this case is highly stereoselective because the shape of the bicyclic ring system provides a steric bias. In Entry 2, a primary phenylthiocarbonate ester is used as the radical source. In Entry 3, the allyl group is introduced at a rather congested carbon. The reaction is completely stereoselective, presumably because of steric features of the tricyclic system. In Entry 4, a primary selenide serves as the radical source. Entry 5 involves a tandem alkylation-allylation with triethylboron generating the ethyl radical that initiates the reaction. This reaction was done in the presence of a Lewis acid, but lanthanide salts also give good results.

318. C. Chatgilialoglu, C. Ferreri, M. Ballestri, and D. P. Curran, *Tetrahedron Lett.*, **37**, 6387 (1996).
319. Y. Guindon, J. F. Lavallee, L. Boisvert, C. Chabot, D. Delorme, C. Yoakim, D. Hall, R. Lemieux, and B. Simoneau, *Tetrahedron Lett.*, **32**, 27 (1991).

Scheme 10.17. Allylation of Radical Centers

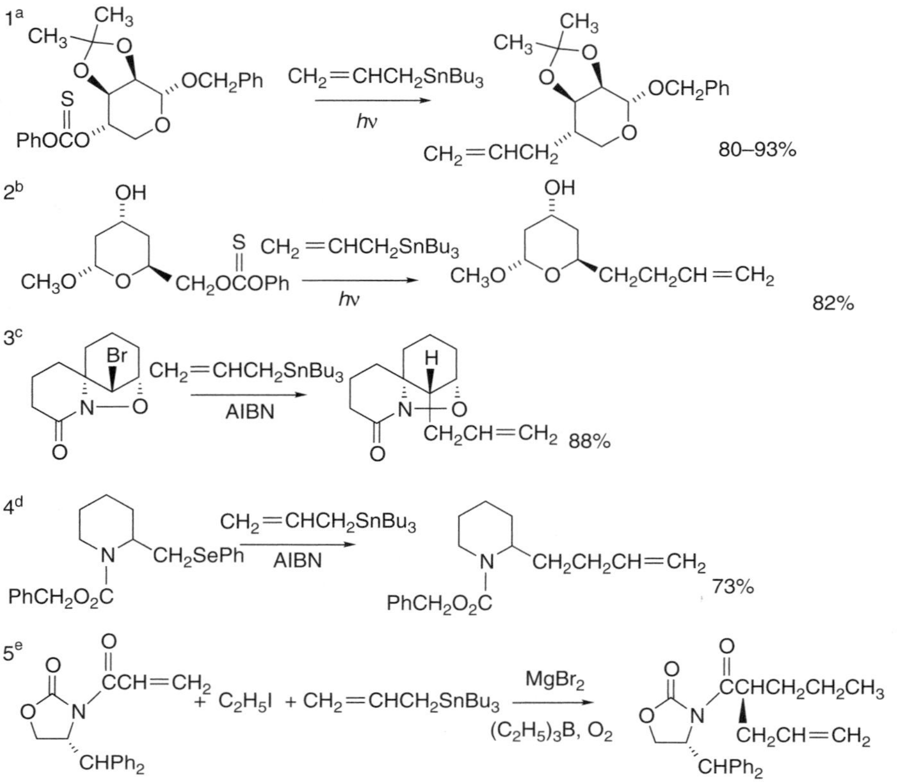

a. G. E. Keck, D. F. Kachensky, and E. J. Enholm, *J. Org. Chem.*, **50**, 4317 (1985).
b. G. E. Keck and D. F. Kachensky, *J. Org. Chem.*, **51**, 2487 (1986).
c. G. E. Keck and J. B. Yates, *J. Org. Chem.*, **47**, 3590 (1982).
d. R. R. Webb, II, and S. Danishefsky, *Tetrahedron Lett.*, **24**, 1357 (1983).
e. M. P. Sibi and J. Ji, *J. Org. Chem.*, **61**, 6090 (1996).

These reactions exhibit excellent diastereoselectivity derived from the chiral oxazolidinone auxiliary. The Lewis acid forms a chelate with the oxazoline and presumably also serves to enhance reactivity. In addition to ethyl, other primary, secondary, and tertiary alkyl radicals, as well as acetyl and benzoyl radicals were used successfully in analogous reactions.

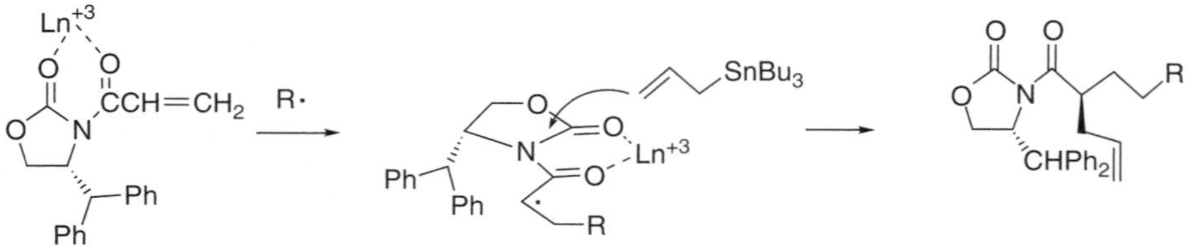

10.3.3. Cyclization of Free Radical Intermediates

Cyclization of radical intermediates is an important method for ring synthesis.[320] The key step involves addition of a radical center to an unsaturated functional group. Many of these reactions involve halides as the source of the radical intermediate. The radicals are normally generated by halogen atom abstraction using a trialkylstannane as the reagent and AIBN as the initiator. The cyclization step must be fast relative to hydrogen abstraction from the stannane. The chain is propagated when the cyclized radical abstracts hydrogen from the stannane.

initiation $\qquad$ In· + Bu$_3$Sn—H ⟶ In—H + Bu$_3$Sn·

propagation $\;$ Bu$_3$Sn· + X—CH$_2$ CH=CH$_2$ ⟶ Bu$_3$Sn—X + ·CH$_2$ $\;$ CH=CH$_2$ ⟶ CH$_2$–CH–CH$_2$·

$\qquad$ CH$_2$–CH–CH$_2$· + Bu$_3$Sn—H ⟶ CH$_2$–CH–CH$_3$ + Bu$_3$Sn·

From a synthetic point of view, the regioselectivity and stereoselectivity of the cyclization are of paramount importance. As discussed in Section 11.2.3.3 of Part A, the order of preference for cyclization of alkyl radicals is 5-*exo* > 6-*endo*; 6-*exo* > 7-*endo*; 8-*endo* > 7-*exo* because of stereoelectronic preferences. For relatively rigid cyclic structures, proximity and alignment factors determined by the specific geometry of the ring system are of major importance. Theoretical analysis of radical addition indicates that the major interaction of the attacking radical is with the alkene LUMO.[321] The preferred direction of attack is not perpendicular to the π system, but rather at an angle of about 110°.

Figure 10.13 shows the preferred geometries and calculated energy differences based on MM2 modeling.

Another major influence on the direction of cyclization is the presence of substituents. Attack at a less hindered position is favored by both steric effects and the stabilizing effect that most substituents have on a radical center. These have been examined by DFT (UB3LYP/6-31+G**) calculations, and the results for 5-hexenyl radicals are shown in Figure 10.14. For the unsubstituted system, the 5-*exo* chair TS is favored over the 6-*endo* chair by 2.7 kcal/mol. A 5-methyl substituent disfavors the 5-*exo* relative to the 6-*endo* mode by 0.7 kcal/mol, whereas a 6-methyl substituent increases the preference for the 5-*exo* TS to 3.3 kcal/mol.[322]

[320] D. P. Curran, *Synthesis*, 417 (1988); *Synthesis*, 489 (1988); C. P. Jasperse, D. P. Curran, and T. L. Fervig, *Chem. Rev.*, **91**, 1237 (1991); K. C. Majumdar, P. K. Basu, and P. P. Mukhopadhyay, *Tetrahedron*, **60**, 6239 (2004).

[321] A. L. J. Beckwith and C. H. Schiesser, *Tetrahedron*, **41**, 3925 (1985); D. C. Spellmeyer and K. N. Houk, *J. Org. Chem.*, **52**, 959 (1987).

[322] A. G. Leach, R. Wang, G. E. Wohlhieter, S. I. Khan, M. E. Jung, and K. N. Houk, *J. Am. Chem. Soc.*, **125**, 4271 (2003).

968

CHAPTER 10

*Reactions Involving
Carbocations, Carbenes,
and Radicals as Reactive
Intermediates*

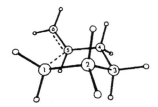

5-hexenyl, <u>exo</u>   5-hexenyl, <u>endo</u>

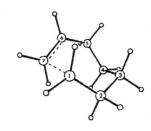

6-heptenyl, <u>exo</u>   6-heptenyl, <u>endo</u>

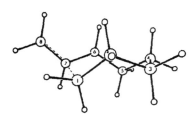
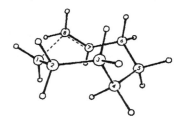

7-octenyl, <u>exo</u>   7-octenyl, <u>endo</u>

Calculated Energy (kcal/mol) of Transition Structures

Ring closure	exo	endo
5/6	7.5	10.3
6/7	9.1	10.8
7/8	15.0	13.0

Fig. 10.13. MM2 models of *exo* and *endo* cyclization transition structures for 5-hexenyl, 6-heptenyl, and 7-octenyl radicals. Reproduced from *Tetrahedron*, **41**, 3925 (1985), by permission of Elsevier.

Radical cyclization reactions have been extensively applied in synthesis. Among the first systems to be studied were unsaturated mixed acetals of bromoacetaldehyde.[323]

[323.] G. Stork, R. Mook, Jr., S. A. Biller, and S. D. Rychnovsky, *J. Am. Chem. Soc.*, **105**, 3741 (1983).

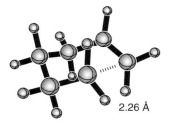

6-endo chair   6-endo boat

2.26 Å   2.24 Å

5-exo chair   5-exo boat

2.19 Å   2.19 Å

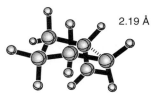

	6-endo pathway		5-exo pathway	
reactant	chair TS	boat TS	chair TS	boat TS
R^1=H, R^2=H, R^3=H,	9.1	11.6	6.4	8.1
R^1=Me, R^2=H, R^3=H,	9.6	12.2	7.0	8.7
R^1=H, R^2=Me, R^3=H,	8.4	10.7	9.1	10.3
R^1=H, R^2=H, R^3=Me,	9.8	12.5	6.5	8.1

Fig. 10.14. Relative energies of 5-*exo* and 6-*endo* transition structures. Insert shows the effect of methyl substituents. Reproduced from *J. Am. Chem. Soc.*, **125**, 4271 (2003), by permission of the American Chemical Society.

This reaction has subsequently been used in a number of other cases.[324] The five-membered rings are usually fused in a *cis* manner, minimizing strain. When cyclization is followed by hydrogen abstraction, the hydrogen atom is normally delivered from the less hindered side of the molecule. The following example illustrates these generalizations. The initial tetrahydrofuran ring closure gives the *cis*-fused ring. The subsequent hydrogen abstraction is from the less hindered axial direction.[325]

[324] X. J. Salom-Roig, F. Denes, and P. Renaud, *Synthesis*, 1903 (2004).
[325] M. J. Begley, H. Bhandal, J. H. Hutchinson, and G. Pattenden, *Tetrahedron Lett.*, **28**, 1317 (1987).

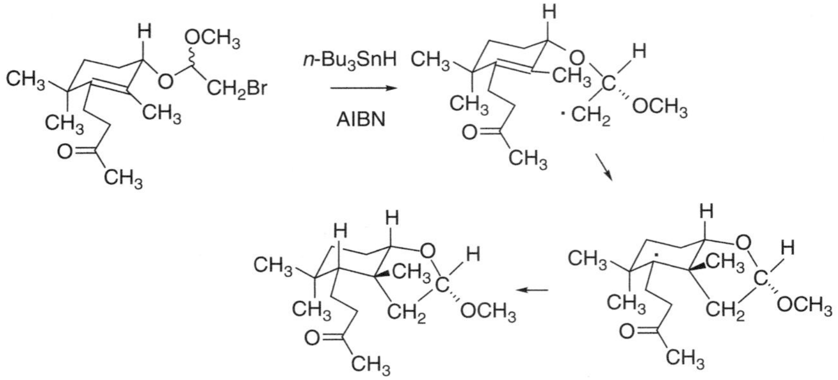

970

CHAPTER 10

*Reactions Involving
Carbocations, Carbenes,
and Radicals as Reactive
Intermediates*

Reaction conditions have been developed in which the cyclized radical can react in some manner other than hydrogen atom abstraction. One such reaction is an iodine atom transfer. The cyclization of 2-iodo-2-methyl-6-heptyne is a structurally simple example.

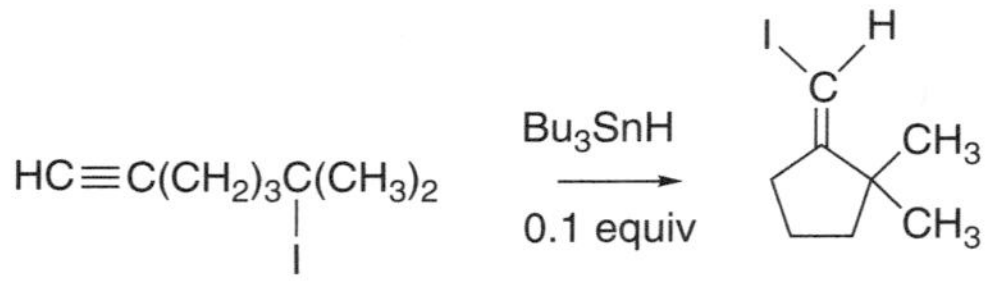

Ref. 326

In this reaction, the trialkylstannane serves to initiate the chain sequence but it is present in low concentration to minimize the rate of hydrogen atom abstraction from the stannane. Under these conditions, the chain is propagated by iodine atom abstraction.

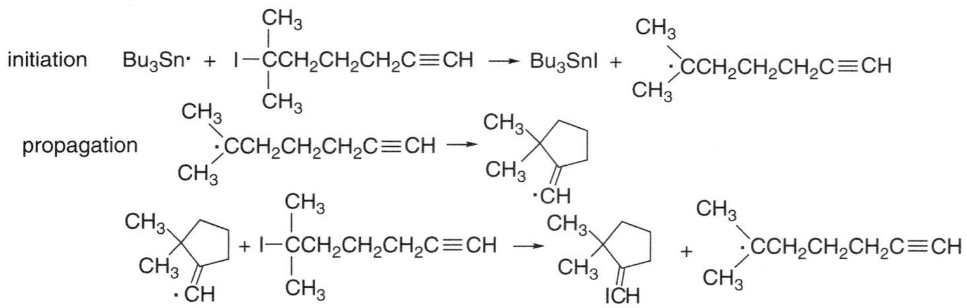

The fact that the cyclization is directed toward an acetylenic group and leads to formation of an alkenyl radical is significant. Formation of a saturated iodide could lead to a more complex product mixture because the cyclized product could undergo iodine atom transfer and proceed to add to a second unsaturated center. Vinyl iodides are much less reactive and the reaction product is unreactive. Owing to the potential

326. D. P. Curran, M.-H. Chen, and D. Kim, *J. Am. Chem. Soc.*, **108**, 2489 (1986); D. P. Curran, M.-H. Chen, and D. Kim, *J. Am. Chem. Soc.*, **111**, 6265 (1989).

for competition from reduction by the stannane, other reaction conditions have been developed to promote cyclization. Hexabutylditin can be used.[327]

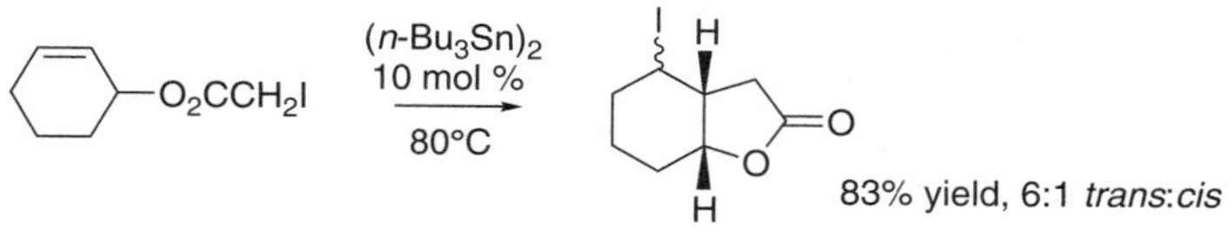

83% yield, 6:1 *trans:cis*

Alkenyl radicals generated by addition of trialkylstannyl radicals to terminal alkynes can undergo cyclization with a nearby double bond.

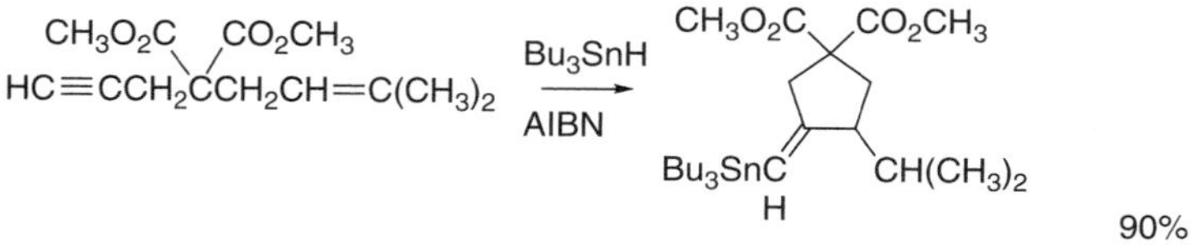

90%

Ref. 328

The addition of a vinyl radical to a double bond is usually favorable thermodynamically because a more stable alkyl radical is formed. The vinyl radical can be generated by dehalogenation of vinyl bromides or iodides. An early study provided examples of both five-and six-membered rings being formed.[329] The six-membered ring is favored when a branching substituent is introduced.

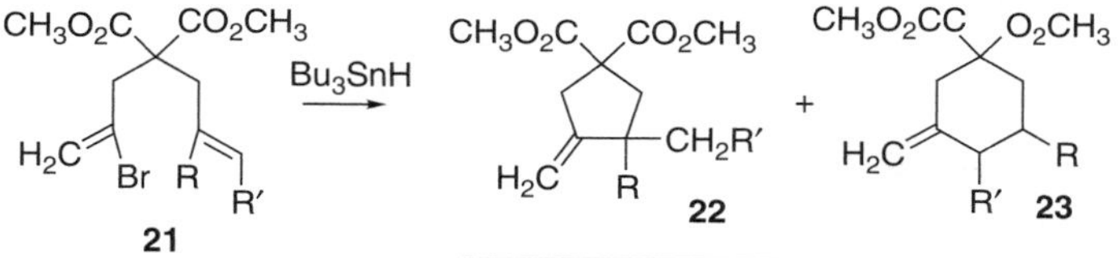

R	R'	Product ratio **22:23**
H	H	3:1
CH3	H	**23** exclusively
H	CH3	2:1

An alternative system for initiating radical cyclization uses triethylborane and oxygen. Under these conditions, *tris*-(trimethylsilyl)silane is an effective hydrogen donor.[330]

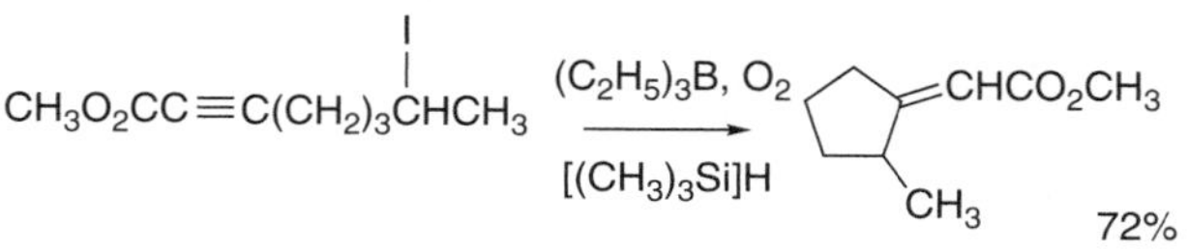

72%

327. D. P. Curran and J. Tamine, *J. Org. Chem.*, **56**, 2746 (1991).
328. G. Stork and R. Mook, Jr., *J. Am. Chem. Soc.*, **109**, 2829 (1987).
329. G. Stork and N. H. Baine, *J. Am. Chem. Soc.*, **104**, 2321 (1982).
330. (a) T. B. Lowinger and L. Weiler, *J. Org. Chem.*, **57**, 6099 (1992); (b) P. A. Evans and J. D. Roseman, *J. Org. Chem.*, **61**, 2252 (1996).

972

CHAPTER 10

Reactions Involving
Carbocations, Carbenes,
and Radicals as Reactive
Intermediates

These cyclizations can also be carried out without a hydrogen donor, in which case the chain is propagated by iodine atom transfer.[331] If necessary, ethyl iodide can be added to facilitate iodine atom transfer.

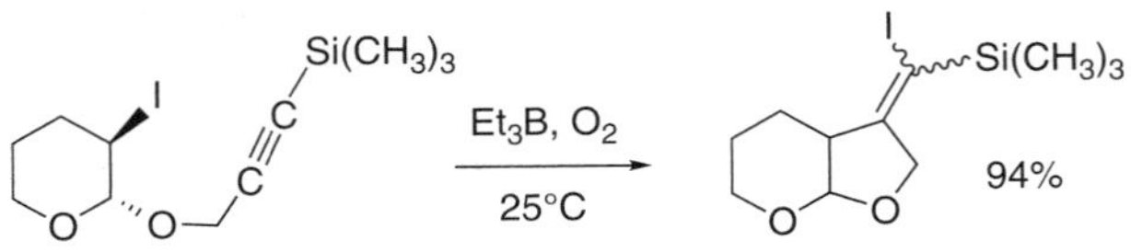

Ref. 332

Intramolecular additions have also been accomplished using xanthate and thiono-carbonates.

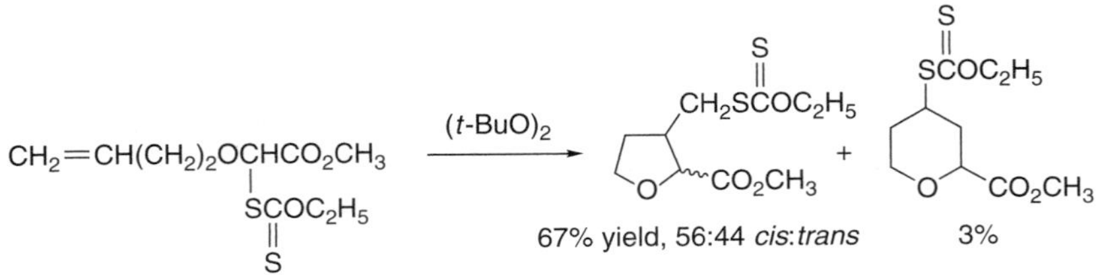

Ref. 333

When a hydrogen donor is present, the product results from reduction.

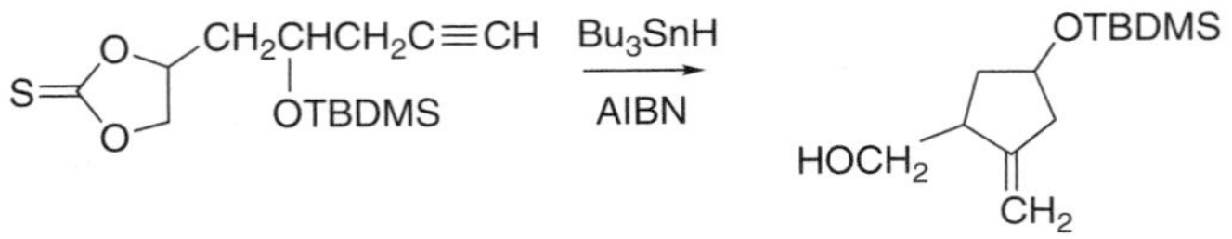

Ref. 334

Cyclization of both alkyl and acyl radicals generated by selenide abstraction have also been observed.

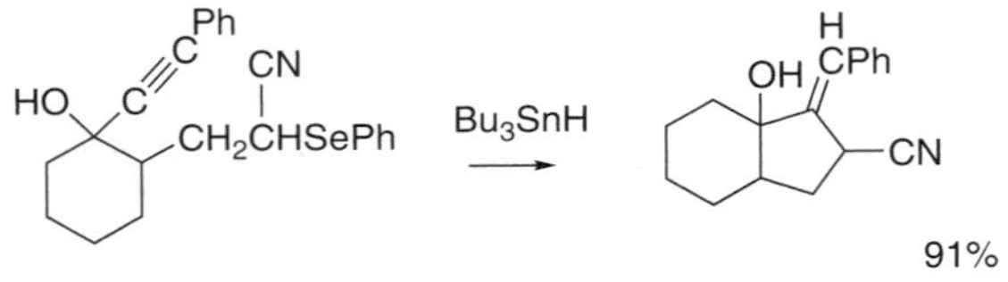

Ref. 335

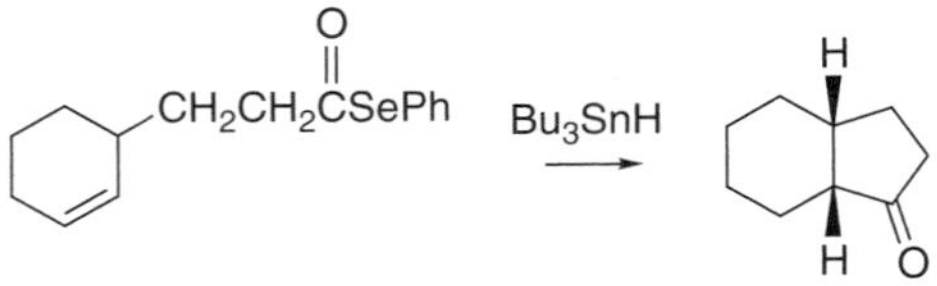

Ref. 336

331. T. J. Woltering and H. M. R. Hoffman, *Tetrahedron*, **51**, 7389 (1995).
332. Y. Ichinose, S. J. Matsunaga, K. Fugami, K. Oshima, and K. Utimoto, *Tetrahedron Lett.*, **30**, 3155 (1989).
333. J. H. Udding, J. P. M. Giesselink, H. Hiemstra, and W. N. Speckamp, *J. Org. Chem.*, **59**, 6671 (1994).
334. F. E. Ziegler and C. A. Metcalf, III, *Tetrahedron Lett.*, **33**, 3117 (1992).
335. D. L. J. Clive, T. L. B. Boivin, and A. G. Angoh, *J. Org. Chem.*, **52**, 4943 (1987).
336. D. L. Boger and R. J. Mathvink, *J. Org. Chem.*, **53**, 3377 (1988).

Triethylborane can also be used for radical initiation and the low temperature can lead to improved yields and stereoselectivity.

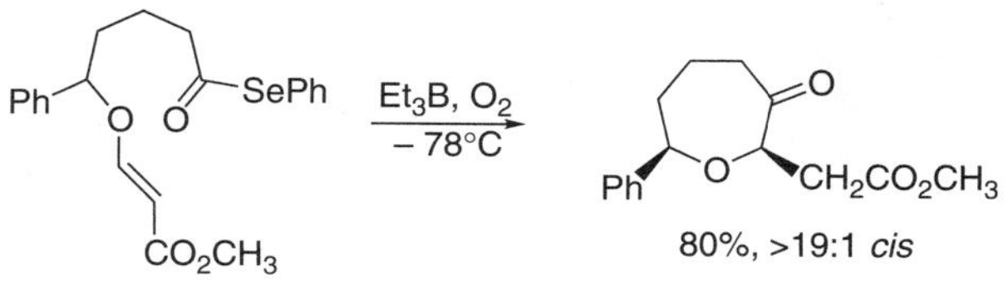

80%, >19:1 *cis*

Ref. 330b

10.3.4. Additions to C=N Double Bonds

Several functional groups containing carbon-nitrogen double bonds can participate in radical cyclizations. Among these are oxime ethers, imines, and hydrazones.[337] Hydrazones and oximes are somewhat more reactive than imines, evidently because the adjacent substituents can stabilize the radical center at nitrogen.[338] Cyclization at these functional groups leads to amino- substituted products.

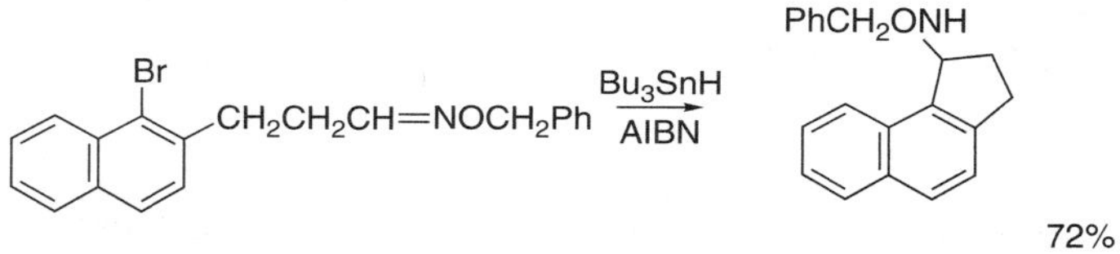

72%

Ref. 339

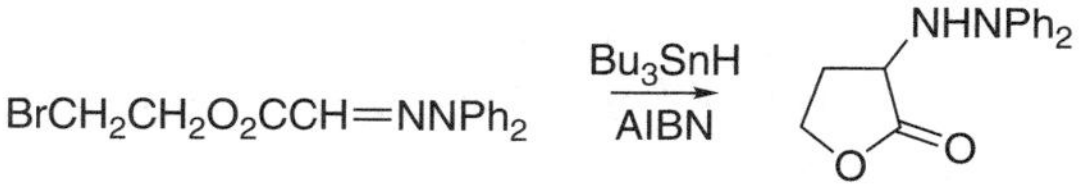

Ref. 340

Ph   70%

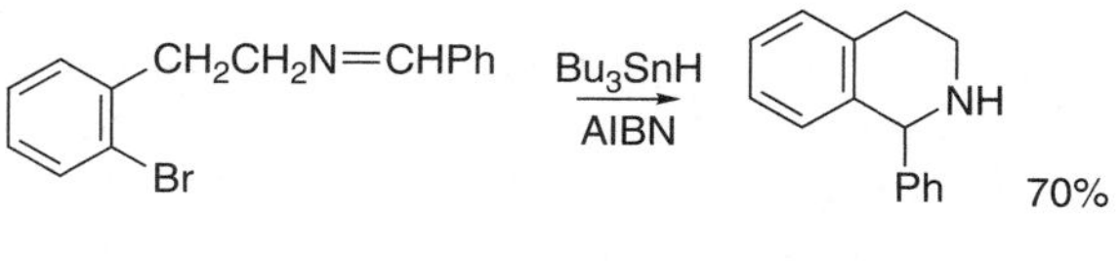

Ref. 341

A radical cyclization of this type was used to synthesize the 4-amino-5-hydroxyhexahydroazepine group found in the PKC inhibitor balanol. The cyclization involves an α-stannyloxy radical formed by addition of the stannyl radical to the aldehyde oxygen.

337. G. K. Friestad, *Tetrahedron*, **57**, 5461 (2001).
338. A. G. Fallis and I. M. Brinza, *Tetrahedron*, **53**, 17543 (1997).
339. J. W. Grissom, D. Klingberg, S. Meyenburg, and B. L. Stallman, *J. Org. Chem.*, **59**, 7876 (1994).
340. D. L. J. Clive and J. Zhang, *J. Chem. Soc.,Chem. Commun.*, 549 (1997).
341. M. J. Tomaszewski and J. Warkentin, *Tetrahedron Lett.*, **33**, 2123 (1992).

974

CHAPTER 10

*Reactions Involving
Carbocations, Carbenes,
and Radicals as Reactive
Intermediates*

$$O=CH(CH_2)_3NCH_2CH=NOCH_2Ph \xrightarrow{Bu_3SnH}$$

with $CO_2\text{-}t\text{-}C_4H_9$ on the nitrogen

Product: seven-membered ring with OH, NHOCH$_2$Ph substituents, N-CO$_2$-t-C$_4$H$_9$

50% 1:2.6 *cis.trans*

Ref. 342

The reactivity of oxime ethers as radical acceptors is enhanced by Lewis acids, BF_3 being the most effective.[343]

$$C_2H_5CH=NOCH_2Ph \xrightarrow[\substack{2)\ Et_3B \\ 3)\ BF_3}]{1)\ RI,\ Bu_3SnH} C_2H_5\underset{R}{CHNHOCH_2Ph}$$

R = alkyl

Addition to oxime ethers of glyoxylic acid generates *N*-benzyloxyamino acids. These reactions have been done in both organic solvents[344] and aqueous mixtures.[345] The reactions can be done with or without Bu_3SnH as a chain carrier.

$$HO_2CCH=NOCH_2Ph \ +\ RI \xrightarrow[\substack{2)\ Et_3B}]{1)\ RI,\ (Bu_3SnH)} HO_2C\underset{R}{CHNHOCH_2Ph}$$

Scheme 10.18 gives some additional examples of cyclization reactions involving radical intermediates. Section A pertains to reactions of alkyl halides. Entry 1 is an early example of the application of a radical cyclization and was used in the synthesis of the terpenes sativene and copacamphene. Entry 2 is an example of the use of the β-bromo-α-ethoxyethyl group in radical cyclization. Ring strain effects dictate the formation of the *cis*-fused five-membered ring, and the stereochemistry of the decalin ring junction is then controlled by the shape of the tricyclic radical intermediate, resulting in good stereochemical control. Entry 3 involves addition of an alkenyl radical. Entry 4 involves generation of a vinyl radical that undergoes stereoequilibration faster than cyclization. The 6-*endo* mode of cylization is favored by both steric and radical stabilization effects. Entry 5 is an 5-*exo* cyclization. Several similar reactions showed a preference of about 8:1 for generation of the *anti* stereochemical relationship at the two new stereocenters. Another noteworthy feature of this reaction is the successful reaction between a relatively electrophilic radical and the acrylate moiety. Entry 6 has several interesting aspects. The reaction proceeds by iodine atom transfer and the cyclization mode is 9-*endo*. The initiation is by triethylborane and the reaction gives much higher yields in water than in benzene. The efficiency of the cyclization and the solvent sensitivity are probably related to reactant conformation. Entry 7 is another iodine atom transfer cyclization initiated by triethylboron. Entry 8 involves 5-*exo* addition to a alkynylsilane.

342. H. Miyabe, M. Torieda, K. Inoue, K. Tajiri, T. Kiguchi, and T. Naito, *J. Org. Chem.*, **63**, 4397 (1998).

343. H. Miyabe, M. Ueda, and T. Naito, *Synlett*, 1140 (2004).

344. H. Miyabe, M. Ueda, N. Yoshioka, and T. Naito, *Synlett*, 465 (1999); H. Miyabe, M. Ueda, N. Yoshioka, K. Yamakawa, and T. Naito, *Tetrahedron*, **56**, 2413 (2000).

345. H. Miyabe, M. Ueda, and T. Naito, *J. Org. Chem.*, **65**, 5043 (2000).

A. Cyclizations of halides terminated by hydrogen atom abstraction or halogen atom transfer

1[a]

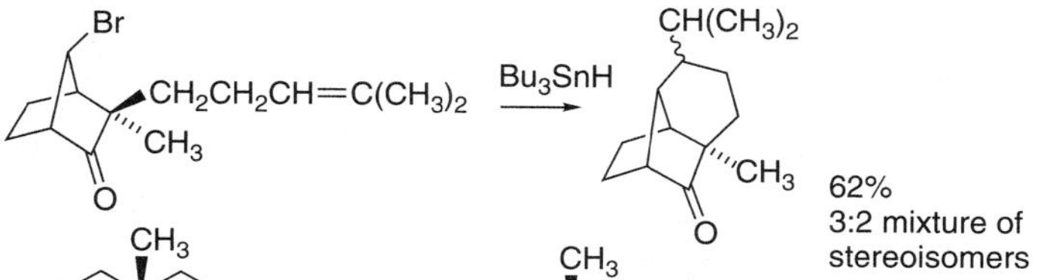

62%
3:2 mixture of
stereoisomers

2[b]

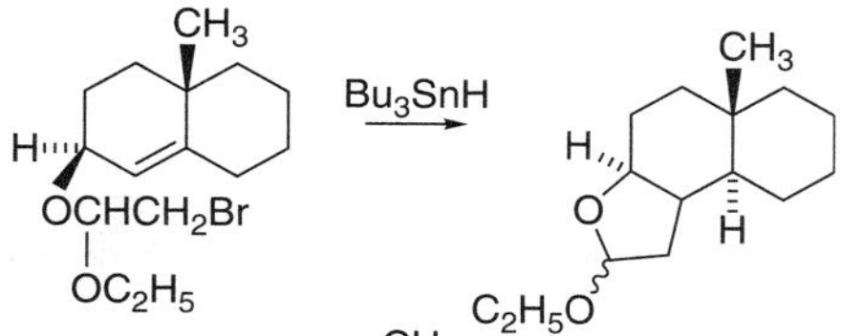

3[c]

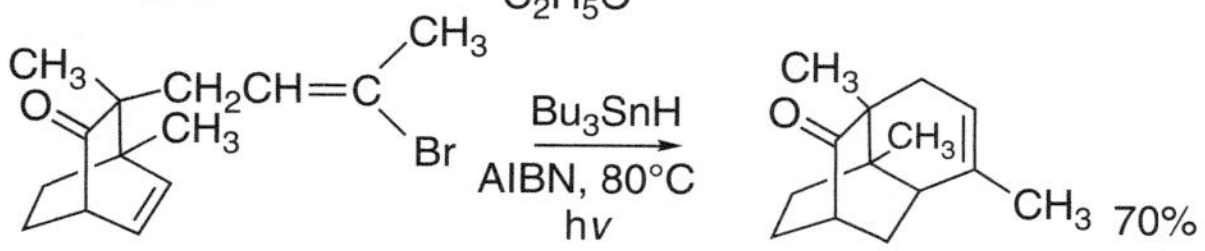

70%

4[d]

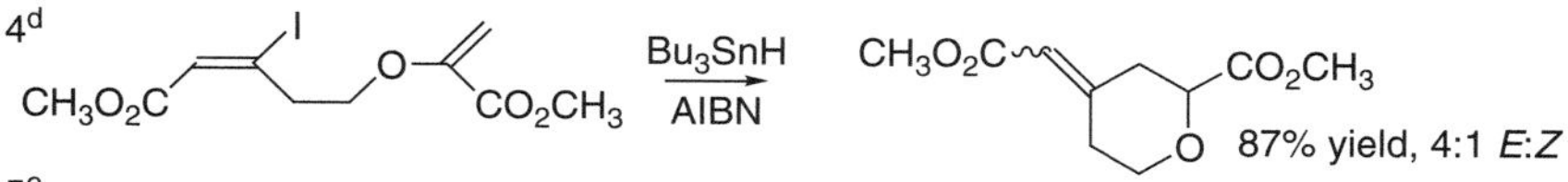

87% yield, 4:1 *E:Z*

5[e]

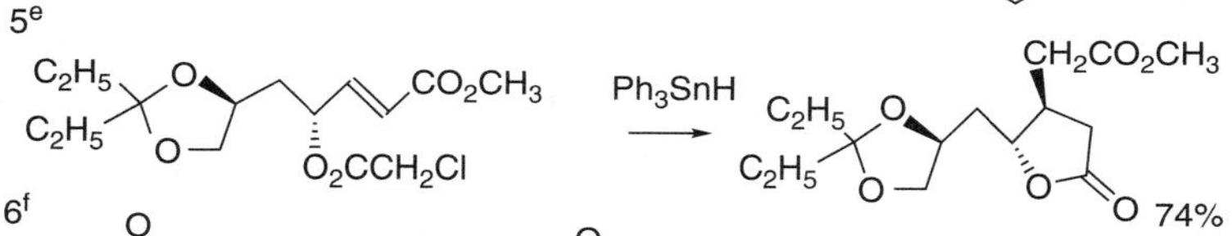

74%

6[f]

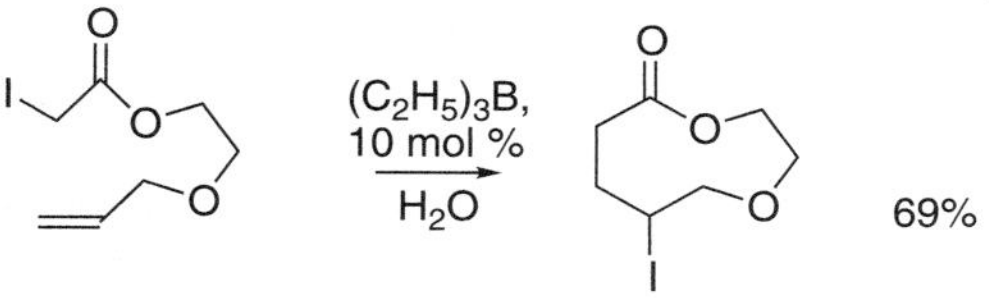

69%

7[g]

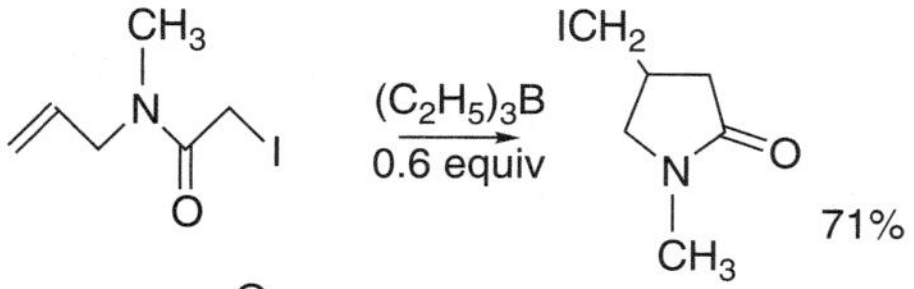

71%

8[h]

69% yield, 1:9 *E:Z*

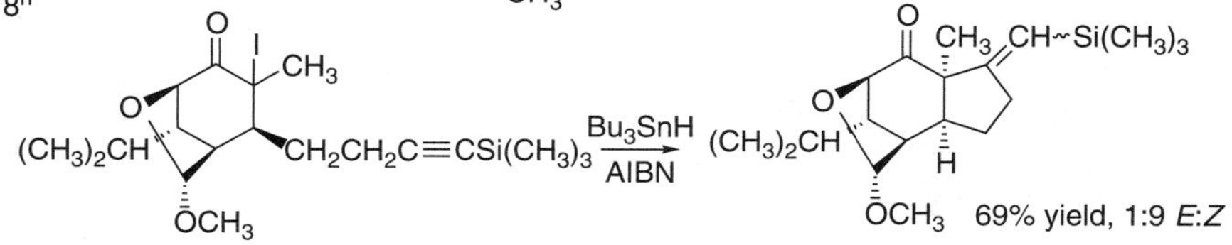

B. Cyclization of thiono esters, sulfides, and selenides

9[i]

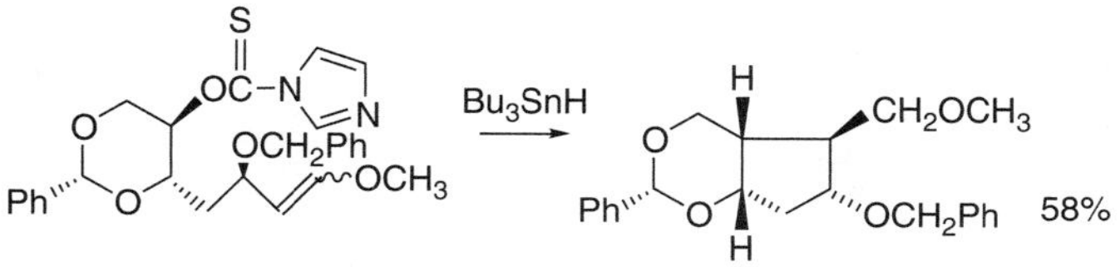

58%

(*Continued*)

976

CHAPTER 10

*Reactions Involving
Carbocations, Carbenes,
and Radicals as Reactive
Intermediates*

Scheme 10.18. (*Continued*)

10[j]

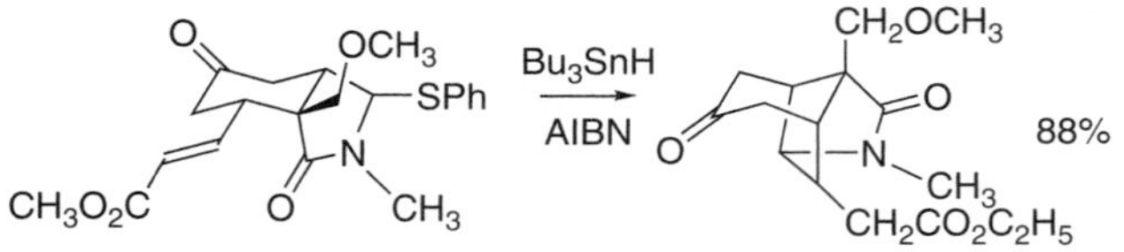

Bu₃SnH / AIBN → 88%

11[k]

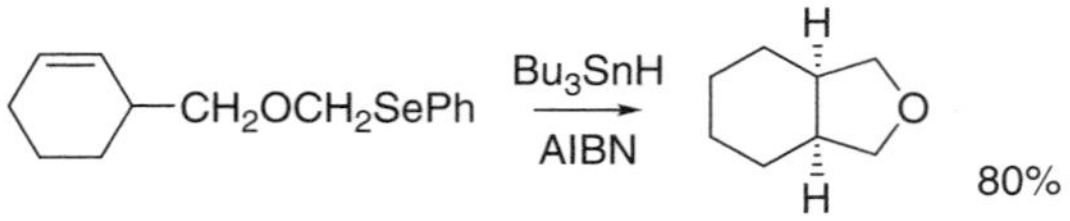

Bu₃SnH / AIBN → 80%

12[l]

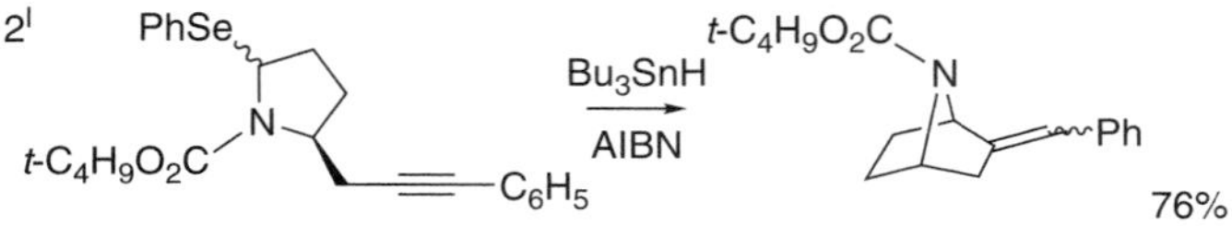

Bu₃SnH / AIBN → 76%

13[m]

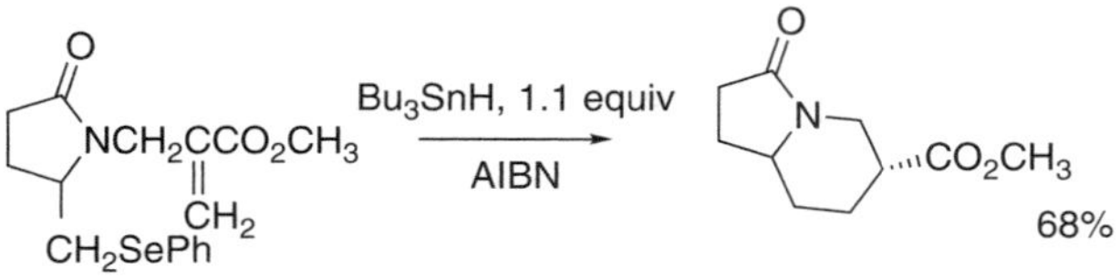

Bu₃SnH, 1.1 equiv / AIBN → 68%

14[n]

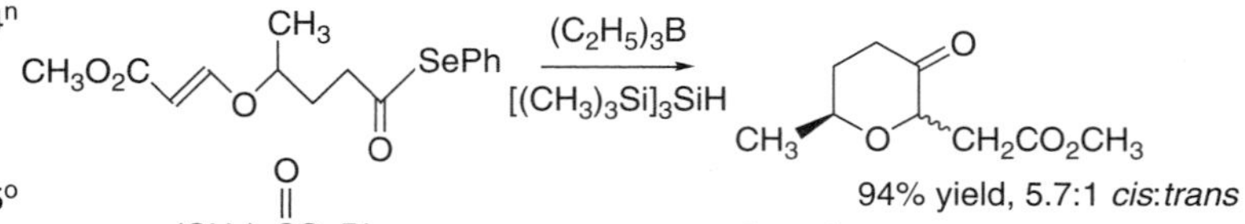

(C₂H₅)₃B / [(CH₃)₃Si]₃SiH → 94% yield, 5.7:1 *cis:trans*

15[o]

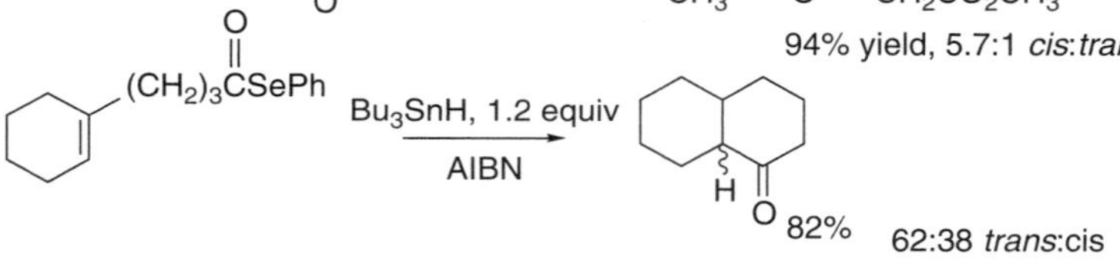

Bu₃SnH, 1.2 equiv / AIBN → 82%   62:38 *trans*:cis

16[p]

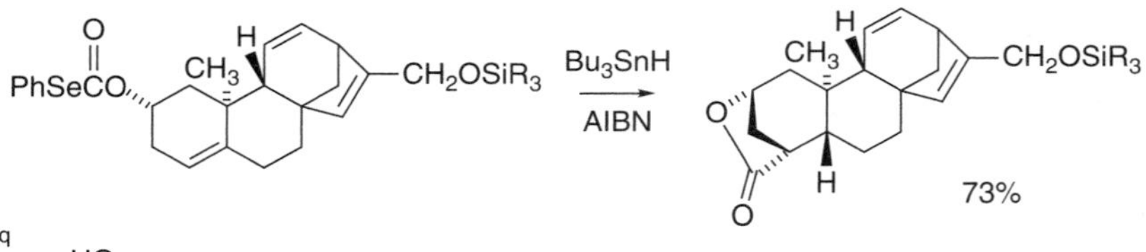

Bu₃SnH / AIBN → 73%

17[q]

Ph₃SnH, 15 equiv / AIBN

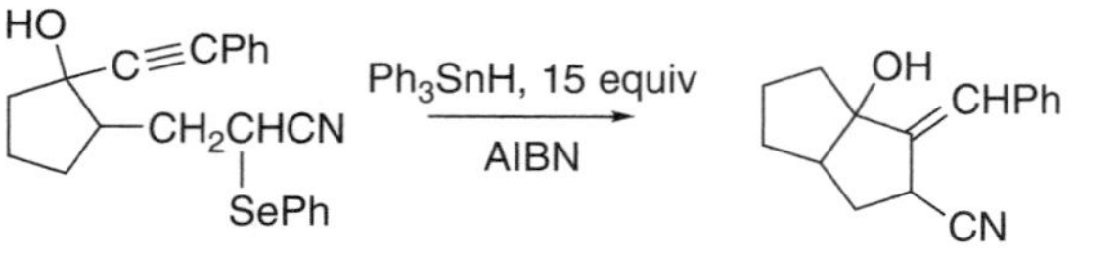

C. Oxidative cyclization with Mn(O₂CCH₃)₃

18[r]

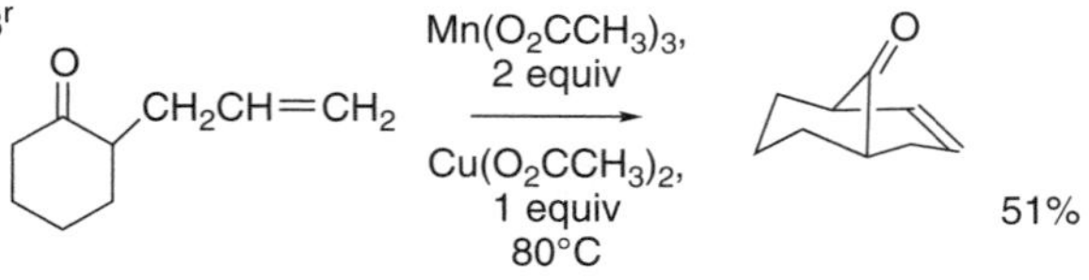

Mn(O₂CCH₃)₃, 2 equiv / Cu(O₂CCH₃)₂, 1 equiv / 80°C → 51%

(*continued*)

19[s]

Mn(O$_2$CCH$_3$)$_3$,
15 equiv

EtOH/HOAc
90°C

58% yield, 1:1.4 *E:Z*

D. Additions to C=N bonds

20[t]

Ph$_3$SnH

(C$_2$H$_5$)$_3$B

91% 1:14 *E:Z*

21[u]

Ph$_3$SnH

AIBN

79% 1:1.1 *trans:cis*

22[v]

O=CH(CH$_2$)$_2$NCH$_2$CH=NOCH$_3$
|
CO$_2$CH$_2$Ph

Bu$_3$SnH,
2 equiv

62% yield, 1:1.3 *cis:trans*

23[w]

Ph$_3$SnH

AIBN

70%

a. P. Bakuzis, O. O. S. Campos, and M. L. F. Bakuzis, *J. Org. Chem.*, **41**, 3261 (1976).
b. G. Stork and M. Kahn, *J. Am. Chem. Soc.*, **107**, 500 (1985).
c. G. Stork and N. H. Baine, *Tetrahedron Lett.*, **26**, 5927 (1985).
d. R. J. Maguire, S. P. Munt, and E. J. Thomas, *J. Chem. Soc., Perkin Trans. 1*, 2853 (1998).
e. S. Hanessian, R. DiFabio, J.-F. Marcoux, and M. Prud'homme, *J. Org. Chem.*, **55**, 3436 (1990).
f. H. Yorimitsu, T. Nakamura, H. Shinokubo, and K. Oshima, *J. Org. Chem.*, **63**, 8604 (1998).
g. M. Ikeda, H. Teranishi, K. Nozaki, and H. Ishibashi, *J. Chem. Soc., Perkin Trans. 1*, 1691 (1998).
h. C.-K. Sha, R.-T. Chiu, C.-F. Yang, N.-T. Yao, W.-H. Tseng, F.-L. Liao, and S.-L. Wang, *J. Am. Chem. Soc.*, **119**, 4130 (1997).
i. T. V. RajanBabu, *J. Org. Chem.*, **53**, 4522 (1988).
j. J.-K. Choi, D.-C. Ha, D. J. Hart, C.-S. Lee, S. Ramesh, and S. Wu, *J. Org. Chem.*, **54**, 279 (1989).
k. V. H. Rawal, S. P. Singh, C. Dufour, and C. Michoud, *J. Org. Chem.*, **56**, 5245 (1991).
l. D. L. J. Clive and V. S. C. Yeh, *Tetrahedron Lett.*, **39**, 4789 (1998).
m. S. Knapp and F. S. Gibson, *J. Org. Chem.*, **57**, 4802 (1992).
n. P. A. Evans and J. D. Roseman, *J. Org. Chem.*, **61**, 2252 (1996).
o. D. L. Boger and R. J. Mathvink, *J. Org. Chem.*, **57**, 1429 (1992).
p. A. K. Singh, R. K. Bakshi, and E. J. Corey, *J. Am. Chem. Soc.*, **109**, 6187 (1987).
q. D. L. J. Clive, T. L. B. Boivin, and A. G. Angoh, *J. Org. Chem.*, **52**, 4943 (1987).
r. B. McC. Cole, L. Han, and B. B. Snider, *J. Org. Chem.*, **61**, 7832 (1996).
s. S. V. O'Neill, C. A. Quickley, and B. B. Snider, *J. Org. Chem.* **62**, 1970 (1997).
t. J. Marco-Contelles, C. Destabel, P. Gallego, J. L. Chiara, and M. Bernabe, *J. Org. Chem.*, **61**, 1354 (1996).
u. J. Zhang and D. L. J. Clive, *J. Org. Chem.*, **64**, 1754 (1999).
v. T. Naito, K. Nakagawa, T. Nakamura, A. Kasei, I. Ninomiya, and T. Kiguchi, *J. Org. Chem.*, **64**, 2003 (1999).
w. G. E. Keck, S. F. McHardy, and J. A. Murry, *J. Org. Chem.*, **64**, 4465 (1999)

978

CHAPTER 10

*Reactions Involving
Carbocations, Carbenes,
and Radicals as Reactive
Intermediates*

Section B of Scheme 10.18 shows examples of the use of sulfides, thiono esters, and selenides as radical sources. The imidazolyl thionocarbamate group used in Entry 9 is one of the thioester groups developed as a source of radicals. In this particular reaction, the phenylthionocarbonate group is even more effective. The ring closure generates an *anti* relationship between the benzyloxy and methoxymethyl substituents. This stereochemistry is consistent with a boatlike TS that may be preferred in order to maintain the preferred conformation of the dioxane ring while avoiding allylic strain in the side chain.

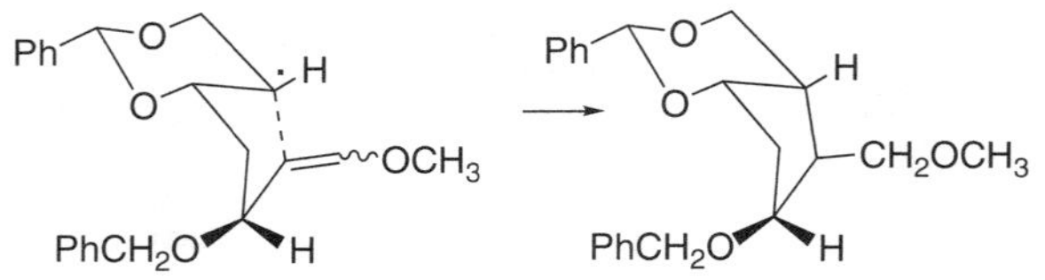

Entry 10 shows the occurrence of 5-*exo* cyclization. The radical in this case is generated from an amino sulfide. This reaction requires a specific, somewhat disfavored conformation of the reactant in order for cyclization to occur. When the unsubstituted vinyl substituent was used, no cyclization occurred. However, increasing the reactivity of the double bond by adding the ester substituent led to successful cyclization.

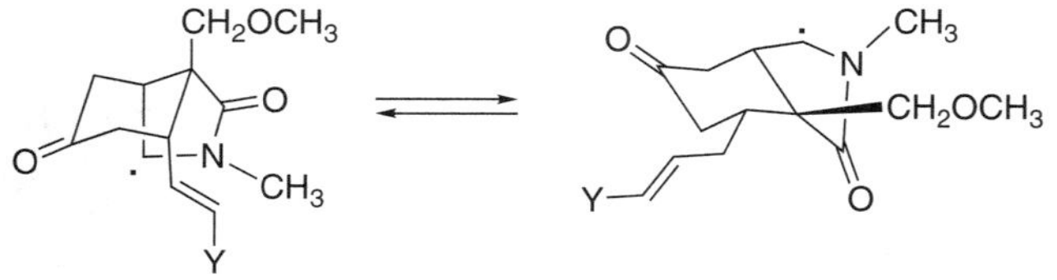

Entry 11 involves generation and cyclization of an alkoxymethyl radical from a selenide. The cyclization mode is the anticipated 5-*exo* with a *cis* ring juncture. This is a case in which the electronic characteristics of the radical are not particularly favorable (ERG oxygen in the radical), but cyclization nevertheless proceeds readily. The reaction in Entry 12 was used to prepare a precursor of epibatidine. Entry 13 shows a 6-*endo* cyclization that is favored by steric factors. The 6-*endo* cyclization is also favored with a tetrahydropyranyloxy substituent in place of the ester, indicating that the electronic effect is not important. Entries 14 to 16 involve acyl radicals generated from selenides. The preferred 6-*endo* cyclization in Entry 15 is thought to be due to the preference for the less-substituted end of the double bond. Entry 17 is an example of a 5-*exo-dig* cyclization.

Entries 18 to 19 pertain to cyclizations of electrophilic radicals generated by oxidations. Entry 18 is the prototype for cyclization of a number of more highly substituted systems. The reaction outcome is consistent with oxidation of the less-substituted enolic position followed by a 6-*endo* cyclization. The cyclized radical is then oxidized and deprotonated. In Entry 19, the vinyl radical formed by cyclization is reduced by hydrogen abstraction from the solvent ethanol.

Entries 20 to 23 involve additions to C=N double bonds in oxime ethers and hydrazones. These reactions result in installation of a nitrogen substituent on the newly formed rings. Entry 20 involves the addition of the triphenylstannyl radical to the terminal alkyne followed by cyclization of the resulting vinyl radical. The product can be proto-destannylated in good yield. The ring closure generates an *anti* relationship for the amino substituent, which is consistent with the TS shown below.

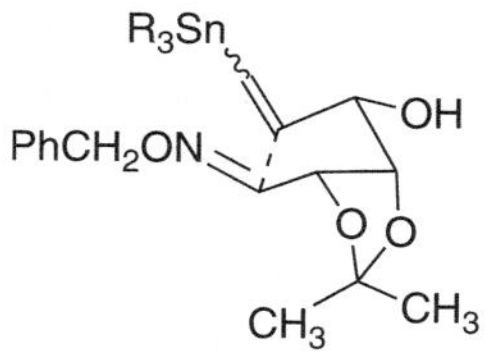

Entry 21 involves addition to a glyoxylic hydrazone and the *cis* ring junction is dictated by strain effects. The primary phenylselenyl group is reductively removed under the reaction conditions. Entry 22 involves generation of a stannyloxy radical by addition of the stannyl radical at the carbonyl *oxygen*. Cyclization then ensues, with the *cis-trans* ratio being determined by the conformation of the cyclization TS.

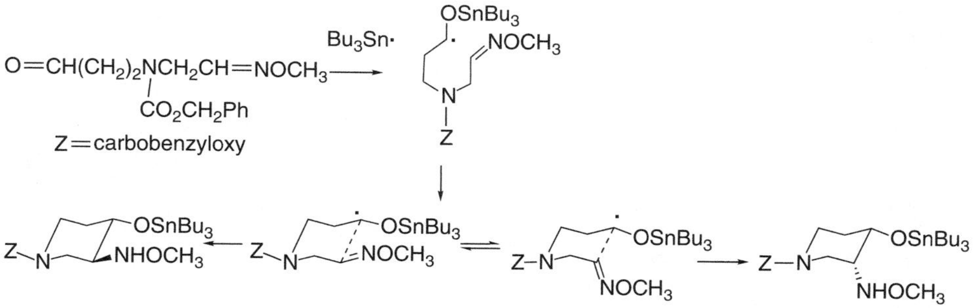

Entry 23 was part of a synthesis of the pancratistatin structure. The lactone ring was used to control the stereochemistry at the cyclization center. Noncyclic analogs gave a mixture of stereoisomers at this center. In this reaction, triphenylstannane gave much better yields than tri-*n*-butylstannane.

10.3.5. Tandem Radical Cyclizations and Alkylations

The synthetic scope of radical cyclizations can be further extended by tandem trapping by an electrophilic alkene.

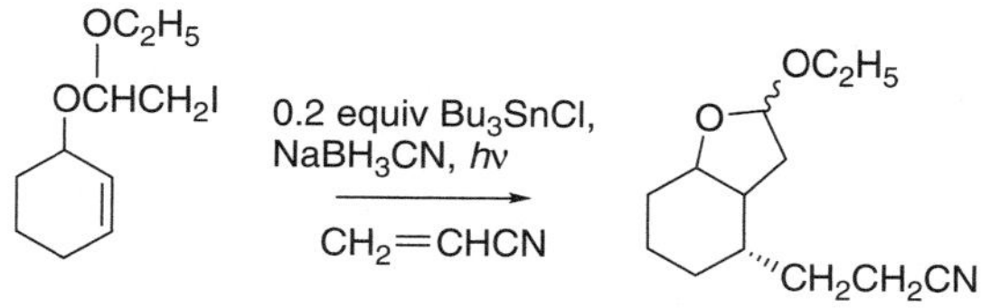

Ref.346

Alkenyl radicals generated by intramolecular addition to a triple bond can add to a nearby double bond, resulting in a tandem cyclization process.

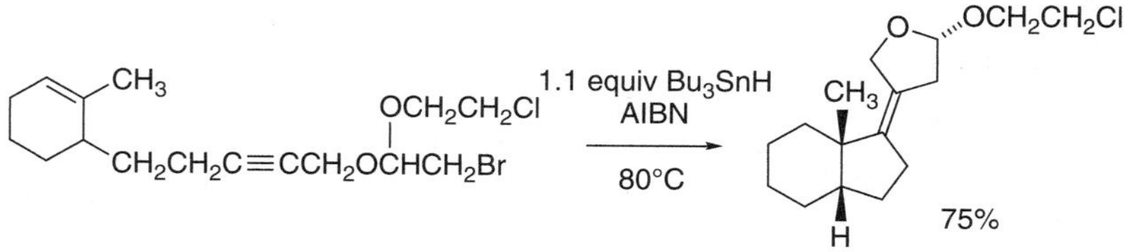

Ref. 347

346. G. Stork and P. M. Sher, *J. Am. Chem. Soc.*, **108**, 303 (1986).
347. G. Stork and R. Mook, Jr., *J. Am. Chem. Soc.*, **105**, 3720 (1983).

980

CHAPTER 10

*Reactions Involving
Carbocations, Carbenes,
and Radicals as Reactive
Intermediates*

As with carbocation-initiated polyene cyclizations, radical cyclizations can proceed through several successive steps if the steric and electronic properties of the reactant provide potential reaction sites. Cyclization may be followed by a second intramolecular step or by an intermolecular addition or alkylation. Intermediate radicals can be constructed so that hydrogen atom transfer can occur as part of the overall process. For example, 2-bromohexenes having radical stabilizing substituents at C(6) can undergo cyclization after a hydrogen atom transfer step.[348]

E = CO_2CH_3; X,Y = TBDMSO, H; Ph, H; CO_2CH_3, H; CO_2CH_3, CO_2CH_3; 2-dioxolanyl

The success of such reactions depends on the intramolecular hydrogen transfer being faster than hydrogen atom abstraction from the stannane reagent. In the example shown, hydrogen transfer is favored by the thermodynamic driving force of radical stabilization, by the intramolecular nature of the hydrogen transfer, and by the steric effects of the central quaternary carbon. This substitution pattern often favors intramolecular reactions as a result of conformational effects.

This type of cyclization has also been carried out using thiophenol to generate the reactive radicals. Good yields were obtained for both EWG and ERG substituents.[349]

X	Y	% yield
Ph	H	85
CN	H	70
$CO_2C_2H_5$	H	57
TBDMSO	H	89
—O$(CH_2)_2$O—		90
CH_3	CH_3	83

Scheme 10.19 gives some other examples of tandem radical reactions. Entry 1 was used to construct the disubstituted cyclopentane system found in the prostaglandins. The first 5-*exo* cyclization to generate the tetrahydrofuran ring is followed by intermolecular trapping of the radical by the α-(trimethylsilyl)enone. In Entry 2, a primary radical was generated and adds to the cyclopentene, generating a tertiary radical that adds to the terminal alkyne. Both ring junctions are *cis*. In Entry 3, a reactive radical is generated from the xanthate groups, and it adds to the styrene double bond faster than

[348.] D. P. Curran, D. Kim, H. T. Liu, and W. Shen, *J. Am. Chem. Soc.*, **110**, 5900 (1988).
[349.] F. Beaufils, F. Denes, and P. Renaud, *Org. Lett.*, **6**, 2563 (2004).

Scheme 10.19. Radical Cyclizations with Tandem Alkylation

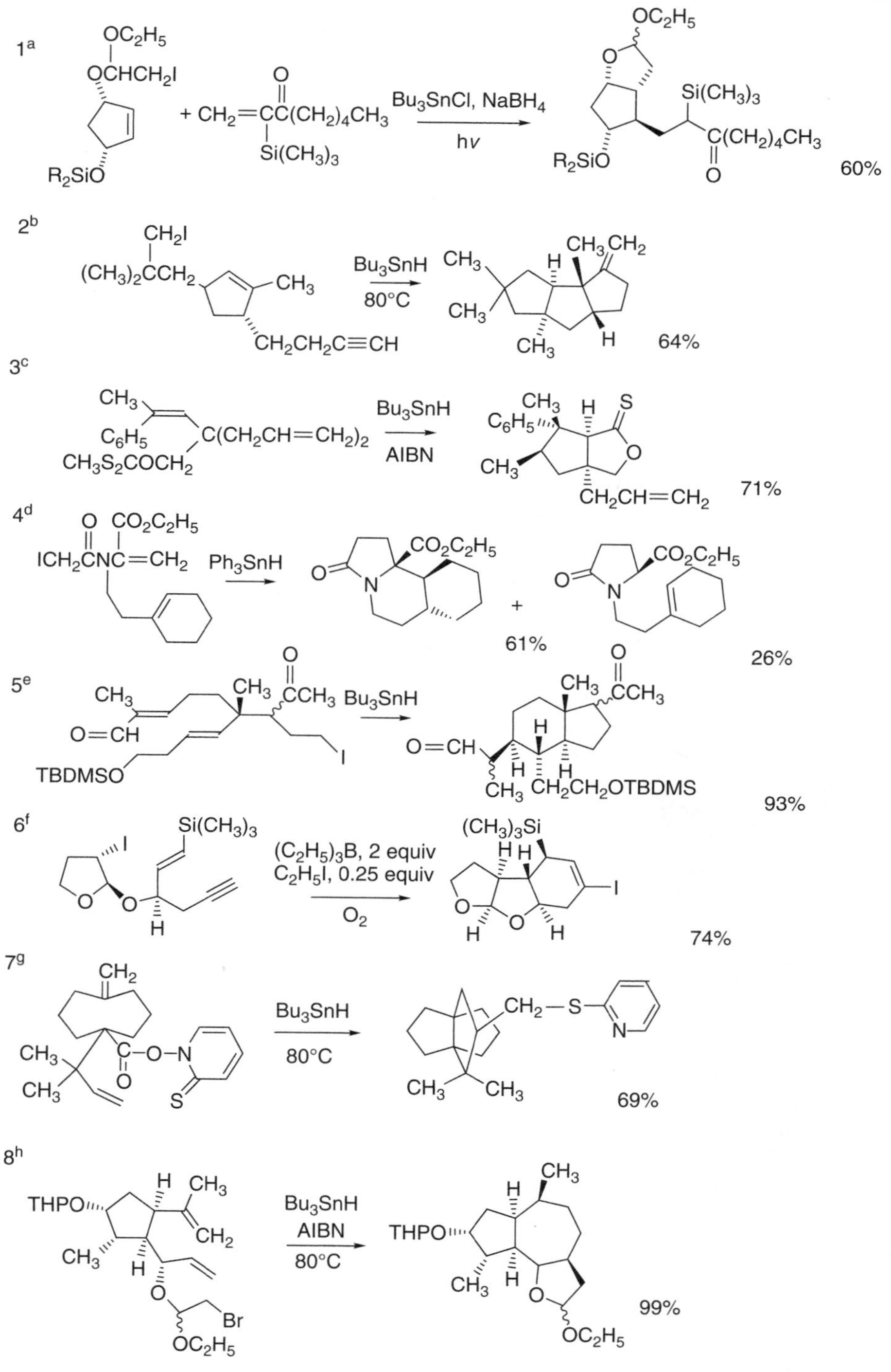

1[a]

60%

2[b]

64%

3[c]

71%

4[d]

61%   26%

5[e]

93%

6[f]

74%

7[g]

69%

8[h]

99%

(Continued)

9[i]

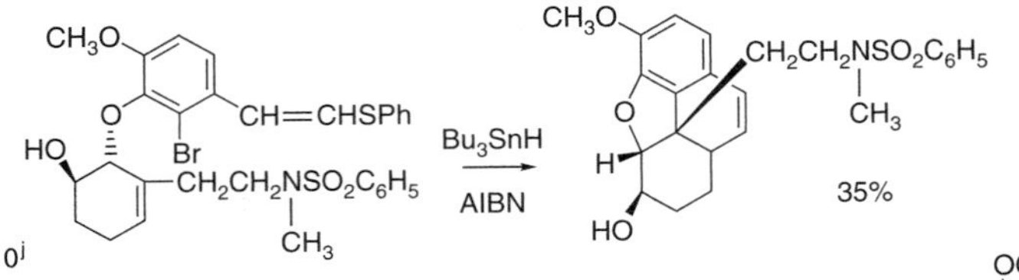

Bu₃SnH / AIBN → 35%

10[j]

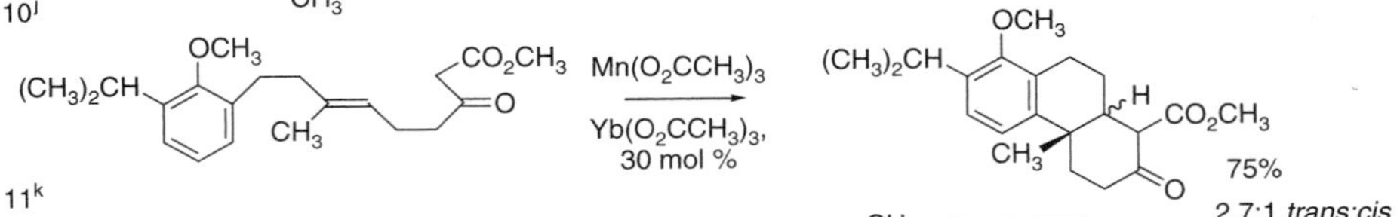

Mn(O₂CCH₃)₃ / Yb(O₂CCH₃)₃, 30 mol % → 75% 2.7:1 *trans:cis*

11[k]

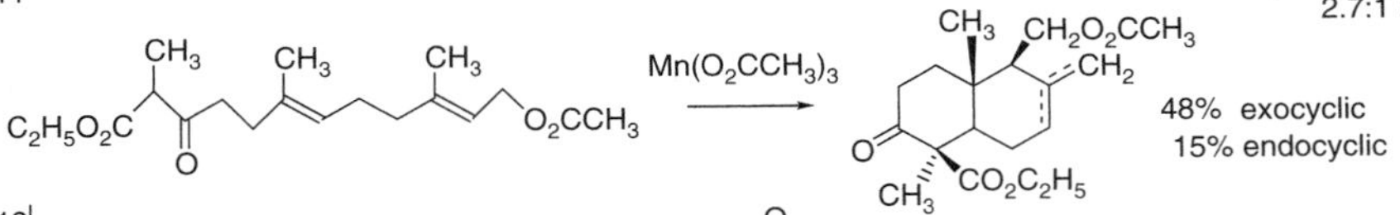

Mn(O₂CCH₃)₃ → 48% exocyclic 15% endocyclic

12[l]

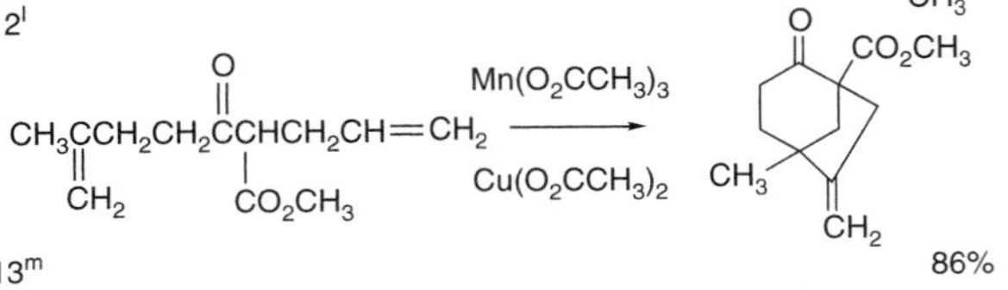

Mn(O₂CCH₃)₃ / Cu(O₂CCH₃)₂ → 86%

13[m]

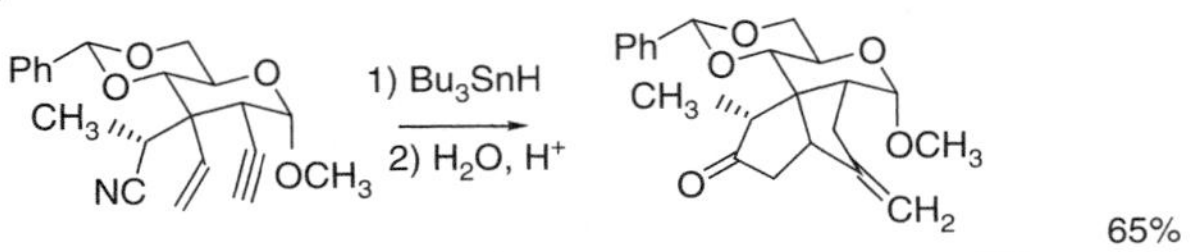

1) Bu₃SnH 2) H₂O, H⁺ → 65%

14[n]

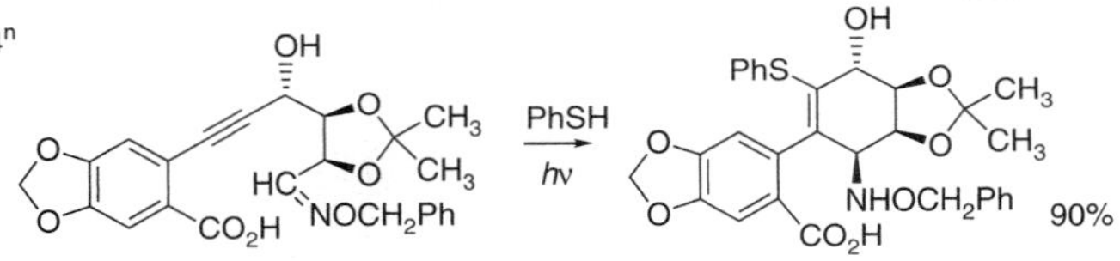

PhSH / hv → 90%

15[o]

Et₃B, O₂ / −78°C → 46% 37:1 *cis*

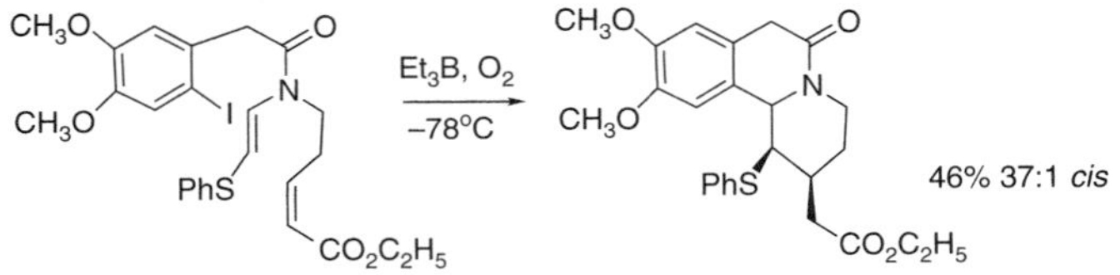

a. G. Stork, P. M. Sher, and H.-L. Chen, *J. Am. Chem. Soc.*, **108**, 6384 (1986).
b. D. P. Curran and D. W. Rakiewicz, *Tetrahedron*, **41**, 3943 (1985).
c. S. Isawa, M. Yamamoto, S. Kohmoto, and K. Yamada, *J. Org. Chem.*, **56**, 2849 (1991).
d. S. R. Baker, A. F. Parsons, J.-F. Pons, and M. Wilson, *Tetrahedron Lett.*, **39**, 7197 (1998); S. R. Baker, K. I. Burton, A. F. Parsons, J.-F. Pons, and M. Wilson, *J. Chem. Soc., Perkin Trans. 1*, 427 (1999).
e. T. Takahshi, S. Tomida, Y. Sakamoto, and H. Yamada, *J. Org. Chem.*, **62**, 1912 (1997).
f. M. Breithor, U. Herden, and H. M. R. Hoffmann, *Tetrahedron*, **53**, 8401 (1997).
g. D. P. Curran and W. Shen, *Tetrahedron*, **49**, 755 (1993).
h. E. Lee, J. W. Lim, C. H. Yoon, Y.-S. Sung, Y. K. Kim, M. Yun, and S. Kim, *J. Am. Chem. Soc.*, **119**, 8391 (1995)
i. K. A. Parker and D. Fokas, *J. Am. Chem. Soc.*, **114**, 9688 (1992).
j. D. Yang, X.-Y. Ye, S. Gu, and M. Xu, *J. Am. Chem. Soc.*, **121**, 5579 (1999).
k. M. A. Dombroski, S. A. Kates, and B. B. Snider, *J. Am. Chem. Soc.*, **112**, 2759 (1990).
l. B. B. Snider, R. Mohan, and S. A. Kates, *Tetrahedron Lett.*, **28**, 841 (1987).
m. H. Pak, I. I. Canalda, and B. Fraser-Reid, *J. Org. Chem.*, **55**, 3009 (1990).
n. G. E. Keck, T. T. Wager, and J. F. D. Rodriquez, *J. Am. Chem. Soc.*, **121**, 5176 (1999).
o. H. Ishibashi, M. Inomata, M Ohba, and M. Ikeda, *Tetrahedron Lett.*, **40**, 1149 (1999).

it fragments. The benzylic radical that is generated by cyclization adds to one of the allyl groups. The chain is then propagated by hydrogen abstraction from the stannane.

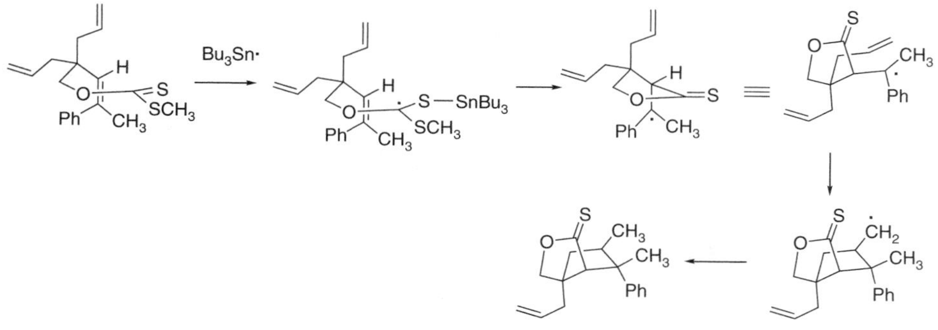

In Entry 4, the initial cyclization is evidently a 5-*endo* process, which in this case is strongly favored by the substitution pattern (capto-dative substituents; see Part A, Section 11.1.6). Most of the cyclized radical then undergoes addition to the cyclohexene ring, generating the major product. In this step, the 6-*endo* process is favored both thermodynamically (5,6- versus 5,5-ring fusion) and by the less-substituted nature of the double bond in this mode. Entry 5 illustrates creation of a CD fragment of the steroid ring system, with side chains in place to create the B ring. The stereochemistry at the ring junction and substitution sites was highly selective. Entry 6 involves a 5-*exo* cyclization followed by a 6-*endo-dig* cyclization. It was found that the selectivity of the tandem sequence was improved by the trimethylsilyl substituent. Entry 7 was used in the synthesis of the carbon skeleton of the terpene modhephene. The sequence consists of two 5-*exo* cyclizations, the first of which is transannular. In Entry 8, the first step is a 5-*exo* cyclization of a bromoacetaldehyde acetal. This is followed by a 7-*endo* cyclization that is favored by the steric and substituent effects of the isopropenyl group. The hydrogen abstraction at the terminal tertiary radical site is highly stereoselective because of ring geometry.

In Entry 9, the initial reaction involves 5-*exo* addition of the aryl radical to the more-substituted end of the cyclohexene double bond, followed by a 6-*endo* addition to the phenylthiovinyl group. The reaction is completed by elimination of the phenylthio radical. The product is an intermediate in the synthesis of morphine.

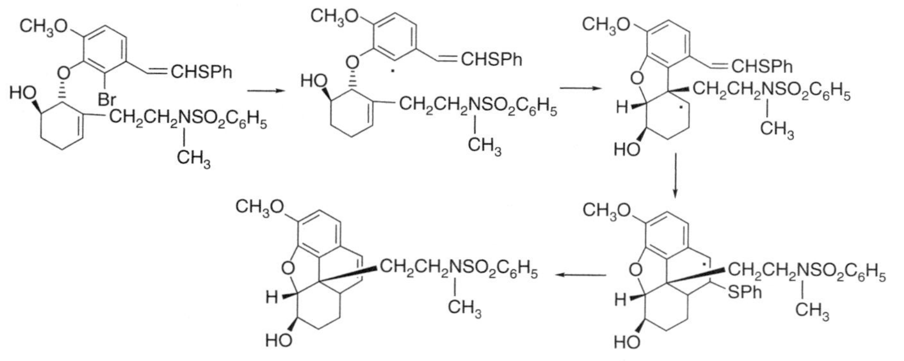

Entries 10 to 12 are examples of oxidative generation of radicals, followed by tandem cyclization. The reaction in Entry 10 includes a lanthanide catalyst. Entry 11

984

CHAPTER 10

*Reactions Involving
Carbocations, Carbenes,
and Radicals as Reactive
Intermediates*

results in the formation of the *trans* decalin product. The by-products of this reaction suggest that the first cyclization is a radical reaction but that oxidation to the tertiary carbocation occurs prior to the second cyclization. Entry 12 involves a tandem process in which the intermediate radical is captured by the second double bond. The presence of Cu(II) results in oxidation of the cyclized radical to an alkene.

Entries 13 to 15 involve adding to carbon-nitrogen multiple bonds. The reaction in Entry 13 is initiated by addition of the stannyl radical to the terminal alkyne. Cyclization generates a primary radical that adds to the cyano group. Cyano groups are not particularly good radical traps, but in this case the group is in close proximity to the radical center. The imine formed by the addition is hydrolyzed and the vinylstannane undergoes proto-destannylation on exposure to silica. In Entry 14, a vinyl radical is generated by thiyl radical addition, followed by cyclization with the oximino ether. Entry 15 involves generation of an aryl radical using the triethylborane system. The low temperature available under these conditions results in much higher stereoselectivity at the acetate side chain than the reaction initiated by a stannyl radical.

10.3.6. Fragmentation and Rearrangement Reactions

Fragmentation is the reverse of radical addition. Fragmentation of radicals is often observed to be fast when the overall transformation is exothermic.

The fragmentation of alkoxyl radicals is especially favorable because the formation of a carbonyl bond makes such reactions exothermic. Rearrangements of radicals frequently occur by a series of addition-fragmentation steps. The following two reactions involve radical rearrangements that proceed through addition-elimination sequences.

Ref. 350

350. P. Dowd and S.-C. Choi, *J. Am. Chem. Soc.*, **109**, 6548 (1987).

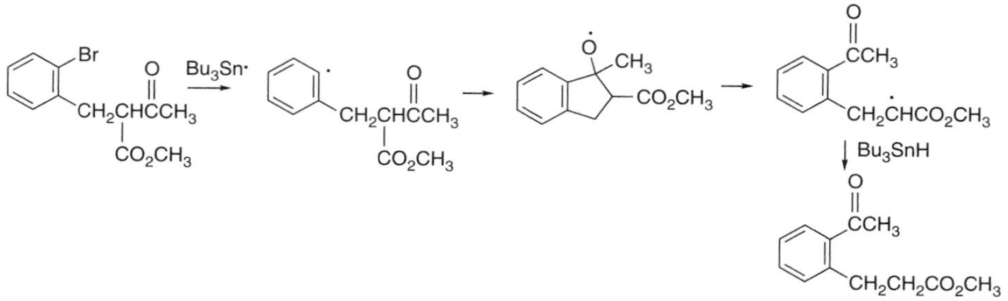

Ref. 351a

Both of these transformations feature addition of a carbon-centered radical to a carbonyl group, followed by fragmentation to a more stable radical. The rearranged radical then abstracts hydrogen from the co-reactant n-Bu$_3$SnH. The addition step must be fast relative to hydrogen abstraction because if this is not the case, simple reductive dehalogenation will occur. The fragmentation step is usually irreversible for two reasons: (1) the reverse addition is endothermic; (2) the product radical is substituted by the electron-withdrawing alkoxycarbonyl group and is unreactive to addition to carbonyl bonds.

The two reactions above are examples of a more general reactivity pattern.[351]

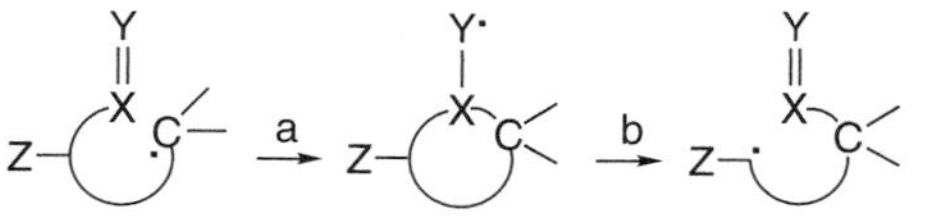

The unsaturated group X=Y that is formally "transferred" by the rearrangement process can be C=C, C=O, C=N, or any other group that fulfills the following general criteria: (1) the addition step **a** must be fast relative to other potentially competing reactions; and (2) the group Z must stabilize the product radical so that the overall process is energetically favorable. A direct comparison of the ease with which unsaturated groups migrate by cyclization-fragmentation has been made for the case of 1,2-migration.

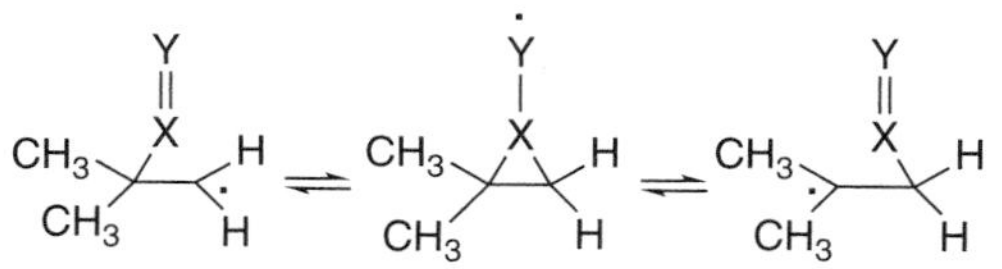

In this system, the overall driving force is the conversion of a primary radical to a tertiary one ($\Delta H \sim -5$ kcal) and the activation barrier incorporates strain associated with formation of the three-membered ring. Rates and activation energies for several migrating groups were determined.[352] A noteworthy feature is the low reactivity of

351. (a) A. L. J. Beckwith, D. M. O'Shea, and S. W. Westwood, *J. Am. Chem. Soc.*, **110** 2565 (1988); (b) R. Tsang, J. K. Pickson, Jr., H. Pak, R. Walton, and B. Fraser-Reid, *J. Am. Chem. Soc.*, **109**, 3484 (1987).
352. D. A. Lindsay, J. Lusztyk, and K. U. Ingold, *J. Am. Chem. Soc.*, **106**, 7087 (1984).

986

CHAPTER 10

*Reactions Involving
Carbocations, Carbenes,
and Radicals as Reactive
Intermediates*

alkyne and cyano groups, which is due to the additional strain introduced in the three-membered ring by the sp^2 carbon. Aryl groups are also relatively unreactive because of the loss of aromaticity in the cyclic intermediate.

		X=Y			
	$HC=CH_2$	$(CH_3)_3C$ $C=O$	(phenyl)	$-C\equiv CC(CH_3)_3$	$C\equiv N$
$kr\,(s^{-1})$	10^7	1.7×10^5	7.6×10^3	93	0.9
E_a (kcal/mol)	5.7	7.8	11.8	12.8	16.4

Among the most useful radical fragmentation reactions from a synthetic point of view are decarboxylations and fragmentations of alkoxyl radicals. The use of N-hydroxy-2-thiopyridine esters for decarboxylation is quite general. Several procedures and reagents are available for preparation of the esters,[353] and the reaction conditions are compatible with many functional groups.[354] t-Butyl mercaptan and thiophenol can serve as hydrogen atom donors.

Ref. 355

Esters of N-hydroxyphthalimide can also be used for decarboxylation. Photolysis in the presence of an electron donor and a hydrogen atom donor leads to decarboxylation. Carboxyl radicals are formed by one-electron reduction of the phthalimide ring.

Ref. 356

Fragmentation of cyclopropylcarbinyl radicals has been incorporated into several synthetic schemes.[357] For example, 2-dienyl-1,1-(dimethoxycarbonyl)-cyclopropanes undergo ring expansion to cyclopentenes.

353. F. J. Sardina, M. H. Howard, M. Morningstar, and H. Rapoport, *J. Org. Chem.*, **55**, 5025 (1990); D. Bai, R. Xu, G. Chu, and X. Zhu, *J. Org. Chem.*, **61**, 4600 (1996).
354. D. H. R. Barton, D. Crich, and W. B. M. Motherwell, *Tetrahedron*, **41**, 3901 (1985).
355. M. Bruncko, D. Crich, and R. Samy, *J. Org. Chem.*, **59**, 5543 (1994).
356. K. Okada, K. Okamoto, and M. Oda, *J. Am. Chem. Soc.*, **110**, 8736 (1988).
357. P. Dowd and W. Zhang, *Chem. Rev.*, **93**, 2091 (1993).

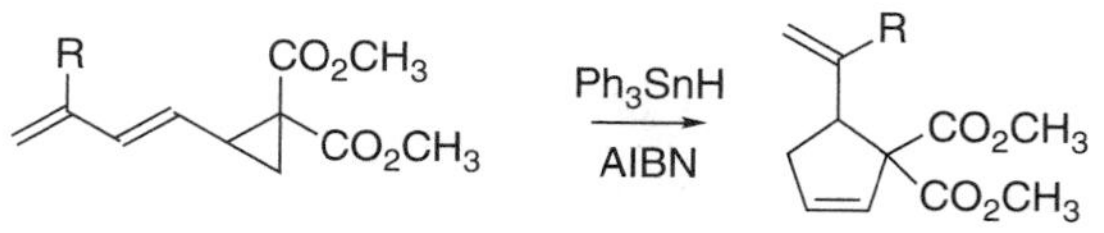

Ref. 358

These reactions presumably involve terminal addition of the chain-carrying radical, followed by fragmentation and recyclization.

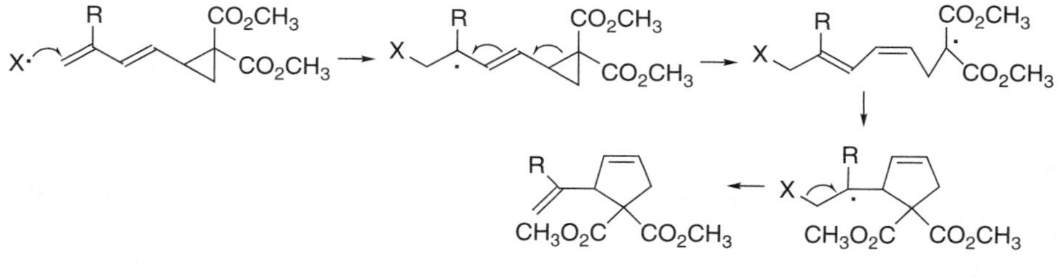

Other intramolecular cyclizations can follow generation and fragmentation of cyclopropylcarbinyl radicals. In the example below, the fragmented radical adds to the alkyne.

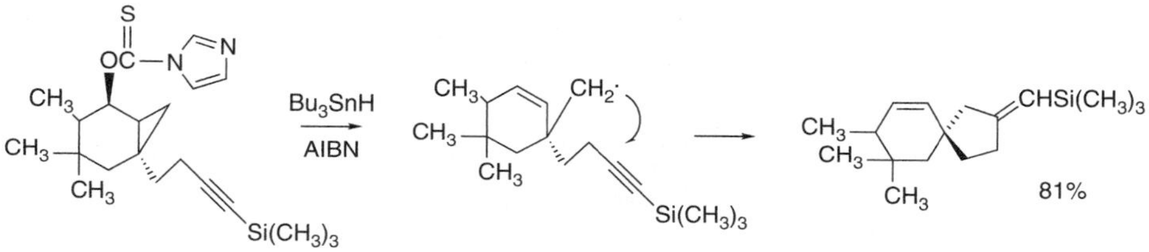

Ref. 359

Cyclic α-halomethyl or α-phenylselenenylmethyl β-ketoesters undergo one-carbon ring expansion via transient cyclopropylalkoxy radicals.[360]

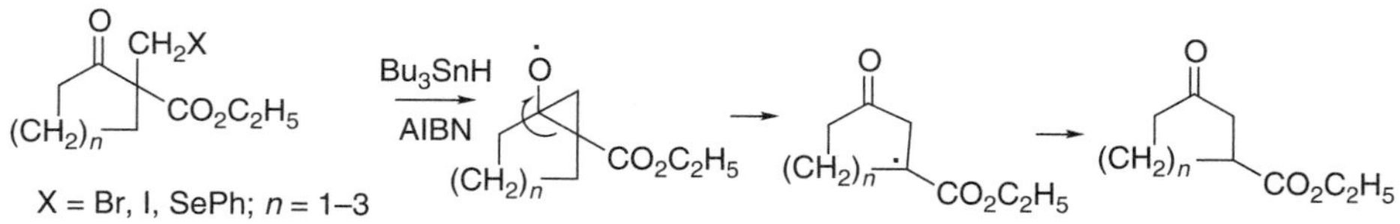

X = Br, I, SePh; n = 1–3

Comparable cyclization-fragmentation sequences have been developed for acyclic and heterocyclic systems.

358. K. Miura, K. Fagami, K. Oshima, and K. Utimoto, *Tetrahedron Lett.*, **29**, 1543 (1988).
359. R. A. Batey, J. D. Harling, and W. R. Motherwell, *Tetrahedron*, **46**, 8031 (1992).
360. P. Dowd and S.-C. Choi, *Tetrahedron*, **45**, 77 (1989); A. L. J. Beckwith, D. M. O'Shea, and S. W. Westwood, *J. Am. Chem. Soc.*, **110**, 2565 (1988), P. Dowd and S.-C. Choi, *Tetrahedron*, **48**, 4773 (1992).

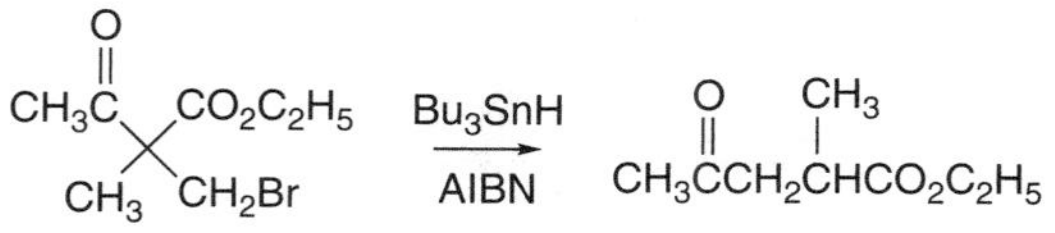

64% Ref. 361

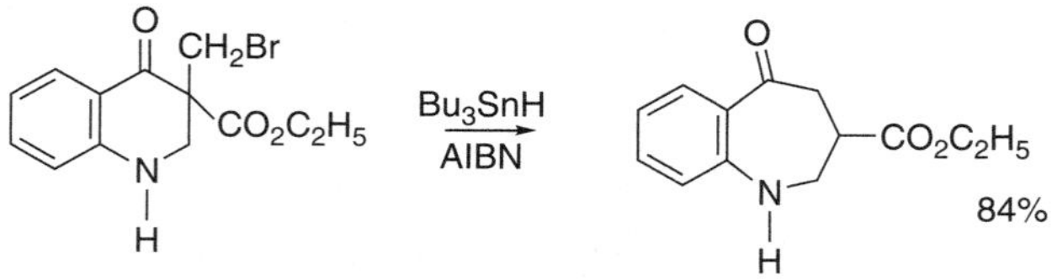

84% Ref. 362

Similar reactions can be conducted using *tris*-(trimethylsilyl)silane as the hydrogen atom donor.[363]

Fragmentation of alkoxy radicals finds use in construction of medium-size rings.[364] One useful reagent combination is phenyliodonium diacetate and iodine.[365] The radical formed by fragmentation is normally oxidized to the corresponding carbocation and trapped by iodide or another nucleophile.

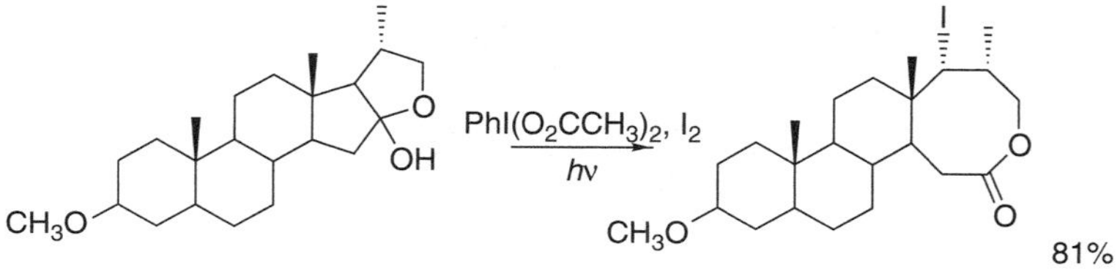

81%

This reagent also can cleave the C(1)−C(2) bond in furanose carbohydrates.

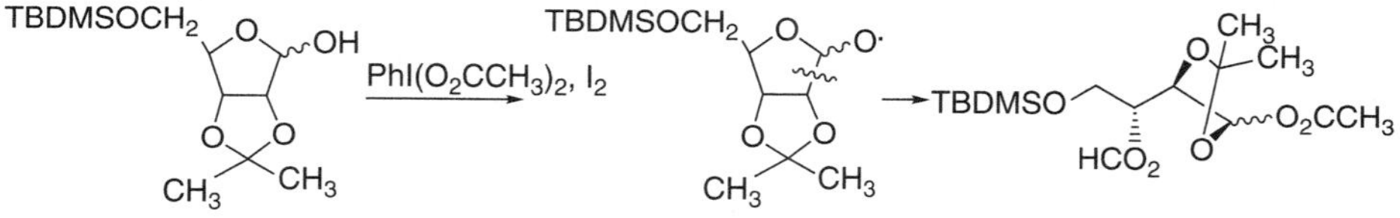

Ref. 366

When the 5-hydroxy group is unprotected, it can capture the fragmented intermediate.[367]

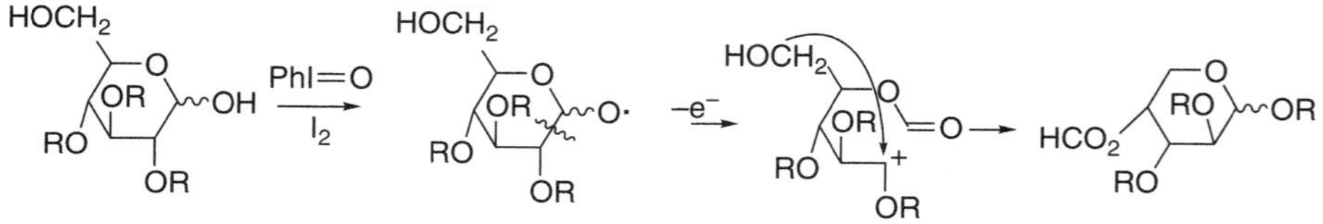

361. P. Dowd and S.-C. Choi, *Tetrahedron*, **45**, 77 (1989).

362. Z. B. Zheng and P. Dowd, *Tetrahedron Lett.*, **34**, 7709 (1993); P. Dowd and S.-C. Choi, *Tetrahedron*, **47**, 4847 (1991).

363. M. Sugi and H. Togo, *Tetrahedron*, **58**, 3171 (2002).

364. L. Yet, *Tetrahedron*, **55**, 9349 (1999).

365. R. Freire, J. J. Marrero, M. S. Rodriquez, and E. Suarez, *Tetrahedron Lett.*, **27**, 383 (1986); M. T. Arencibia, R. Freire, A. Perales, M. S. Rodriguez, and E. Suarez, *J. Chem. Soc., Perkin Trans. 1*, 3349 (1991).

366. P. de Armas, C. G. Francisco, and E. Suarez, *Angew. Chem. Intl. Ed. Engl.*, **31**, 772 (1992).

367. P. de Armas, C. G. Francisco, and E. Suarez, *J. Am. Chem. Soc.*, **115**, 8865 (1993).

Bicyclic lactols afford monocyclic iodolactones.

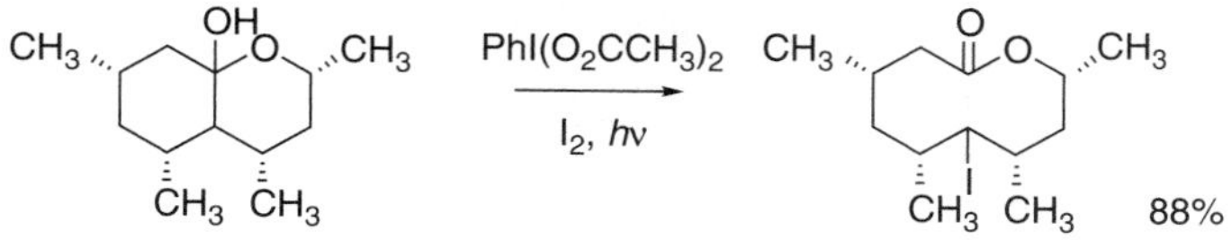

Ref. 368

Similarly, bicyclic hemiacetals fragment to medium-size lactones.

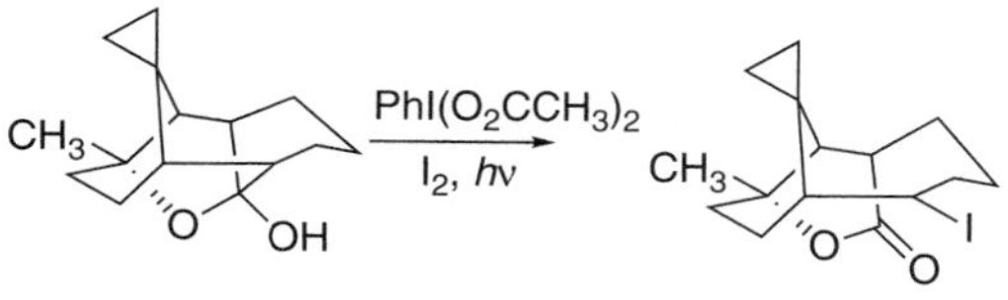

Ref. 369

These reactions are believed to proceed through hypoiodite intermediates.

Alkoxy radical fragmentation is also involved in ring expansion of 3- and 4-haloalkyl cyclohexanones. The radical formed by halogen atom abstraction adds to the carbonyl group, after which fragmentation to the carboethoxy-stabilized radical occurs.[370]

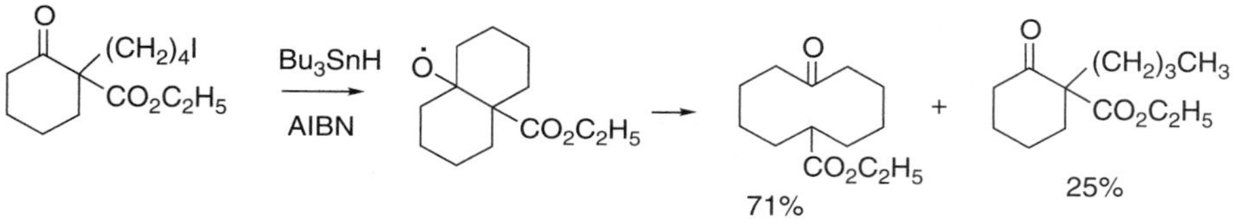

The by-product results from competing reduction of the radical by hydrogen atom abstraction.

10.3.7. Intramolecular Functionalization by Radical Reactions

In this section we focus on intramolecular functionalization. Such reactions normally achieve selectivity on the basis of proximity of the reacting centers. In acyclic molecules, intramolecular functionalization normally involves hydrogen atom abstraction via a six-membered cyclic TS. The net result is introduction of functionality at the δ-atom in relation to the radical site.

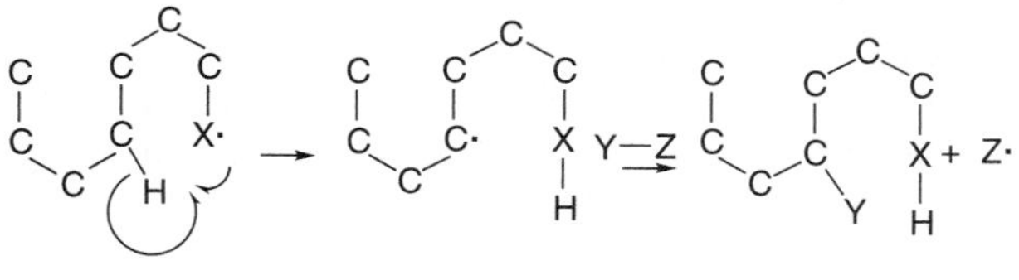

[368.] M. Kaino, Y. Naruse, K. Ishihara, and H. Yamamoto, *J. Org. Chem.*, **55**, 5814 (1990).
[369.] J. Lee, J. Oh, S. Jin, J.-R. Choi, J. L. Atwood, and J. K. Cha, *J. Org. Chem.*, **59**, 6955 (1994).
[370.] P. Dowd and S.-C. Choi, *Tetrahedron*, **45**, 77 (1989); P. Dowd and S.-C. Choi, *J. Am. Chem. Soc.*, **109**, 6548 (1987).

One example of this type of reaction is the photolytically initiated decomposition of N-chloroamines in acidic solution, which is known as the *Hofmann-Loeffler-Freytag reaction*.[371] The initial products are δ-chloroamines, but these are usually converted to pyrrolidines by intramolecular nucleophilic substitution.

initiation

$$RCH_2CH_2CH_2CH_2\overset{+}{N}HCH_3 \xrightarrow{h\nu} RCH_2CH_2CH_2CH_2\overset{+}{N}HCH_3 + Cl\cdot$$
$$\underset{|}{Cl} \qquad\qquad\qquad \underset{\cdot}{}$$

propagation

$$RCH_2CH_2CH_2CH_2\overset{+}{N}HCH_3 \longrightarrow R\underset{\cdot}{C}HCH_2CH_2CH_2\overset{+}{N}H_2CH_3$$
$$\underset{|}{Cl}$$

$$R\underset{\cdot}{C}HCH_2CH_2CH_2\overset{+}{N}H_2CH_3 + RCH_2CH_2CH_2CH_2\overset{+}{N}HCH_3 \longrightarrow$$
$$\underset{|}{Cl}$$

$$RCHCH_2CH_2CH_2\overset{+}{N}H_2CH_3 + RCH_2CH_2CH_2CH_2\overset{+}{N}HCH_3$$
$$\underset{|}{Cl}$$

base-catalyzed cyclization

$$RCHCH_2CH_2CH_2\overset{+}{N}H_2CH_3 \xrightarrow{NaOH} R-\underset{\underset{CH_3}{|}}{\boxed{N}}$$
$$\underset{|}{Cl}$$

A closely related procedure results in formation of γ-lactones. Amides are converted to N-iodoamides by reaction with iodine and t-butyl hypochlorite. Photolysis of the N-iodoamides gives lactones via iminolactone intermediates.[372]

$$RCH_2(CH_2)_2\overset{O}{\overset{||}{C}}NHI \xrightarrow{h\nu} RCH(CH_2)_2\overset{O}{\overset{||}{C}}NH_2 \longrightarrow$$
$$\underset{|}{I}$$

Steps similar to the Hofmann-Loeffler reaction are also involved in cyclization of N-alkylmethanesulfonamides by oxidation with $Na_2S_2O_4$ in the presence of cupric ion.[373]

$$RCH_2(CH_2)_3NHSO_2CH_3 \xrightarrow[-H^+]{-e^-} RCH_2(CH_2)_3\overset{\cdot}{N}SO_2CH_3 \longrightarrow R\overset{\cdot}{C}H(CH_2)_3\underset{\underset{H}{|}}{N}SO_2CH_3$$

$$R\overset{\cdot}{C}H(CH_2)_3NHSO_2CH_3 \xrightarrow{Cu^{2+}} R\overset{+}{C}H(CH_2)_3NHSO_2CH_3 \longrightarrow R-\underset{\underset{SO_2CH_3}{|}}{\boxed{N}}$$

There are also useful intramolecular functionalization methods that involve hydrogen atom abstraction by oxygen radicals. The conditions that were originally developed involved thermal or photochemical dissociation of alkoxy derivative of Pb(IV) generated by exchange with $Pb(OAc)_4$.[374] These decompose, giving alkoxy

[371.] M. E. Wolff, *Chem. Rev.*, **63**, 55 (1963).
[372.] D. H. R. Barton, A. L. J. Beckwith, and A. Goosen, *J. Chem. Soc.*, 181 (1965).
[373.] G. I. Nikishin, E. I. Troyansky, and M. Lazareva, *Tetrahedron Lett.*, **26**, 1877 (1985).
[374.] K. Heusler, *Tetrahedron Lett.*, 3975 (1964).

radicals with reduction to Pb(III). The subsequent oxidation of the radical to a carbocation is effected by Pb(IV) or Pb(III).

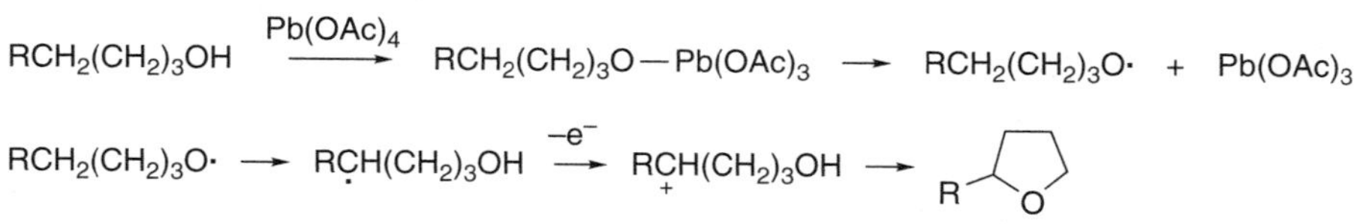

$$RCH_2(CH_2)_3OH \xrightarrow{Pb(OAc)_4} RCH_2(CH_2)_3O-Pb(OAc)_3 \longrightarrow RCH_2(CH_2)_3O\cdot + Pb(OAc)_3$$

$$RCH_2(CH_2)_3O\cdot \longrightarrow R\dot{C}H(CH_2)_3OH \xrightarrow{-e^-} R\overset{+}{C}H(CH_2)_3OH \longrightarrow$$

Current procedures include iodine and are believed to involve a hypoiodite intermediate.[375]

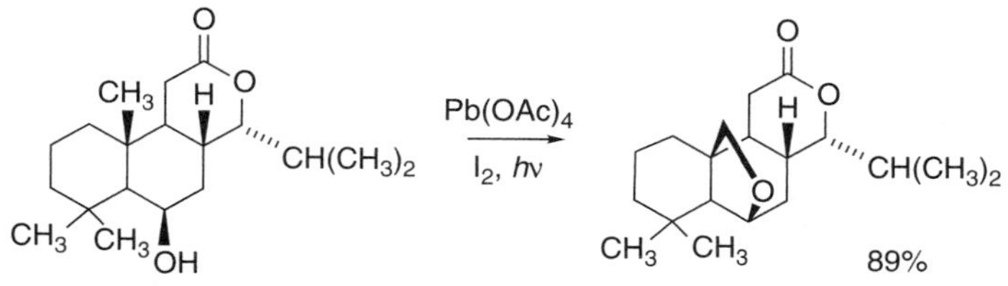

Ref. 376

The reactions can also be effected by phenyliodonium diacetate.[377] A mechanistic prototype can be found in the conversion of pentanol to 2-methyltetrahydrofuran. The secondary radical is most likely captured by iodine or oxidized to the carbocation prior to cyclization.[378]

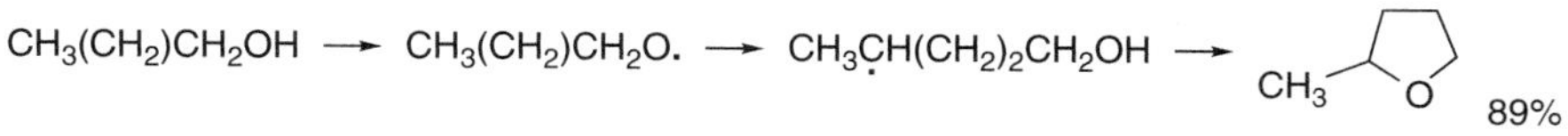

$$CH_3(CH_2)CH_2OH \longrightarrow CH_3(CH_2)CH_2O\cdot \longrightarrow CH_3\dot{C}H(CH_2)_2CH_2OH \longrightarrow$$

89%

Alkoxy radicals are also the active hydrogen-abstracting species in a procedure that involves photolysis of nitrite esters. This reaction was originally developed as a method for functionalization of methyl groups in steroids. [379]

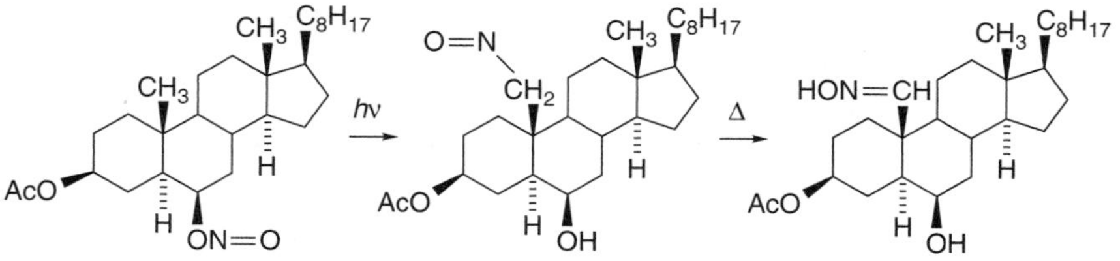

It has found other synthetic applications.

375. K. Heusler, P. Wieland, and C. Meystre, *Org. Synth.*, **V**, 692 (1973); K. Heusler and J. Kalvoda, *Angew. Chem. Int. Ed. Engl.*, **3**, 525 (1964).
376. S. D. Burke, L. A. Silks, III, and S. M. S. Strickland, *Tetrahedron Lett.*, **29**, 2761 (1988).
377. J. I. Concepcion, C. G. Francisco, R. Hernandez, J. A. Salazar, and E. Suarez, *Tetrahedron Lett.*, **25**, 1953 (1984).
378. J. L. Courtneidge, J. Lusztyk, and D. Page, *Tetrahedron Lett.*, **35**, 1003 (1994).
379. D. H. R. Barton, J. M. Beaton, L. E. Geller, and M. M. Pechet, *J. Am. Chem. Soc.*, **83**, 4076 (1961).

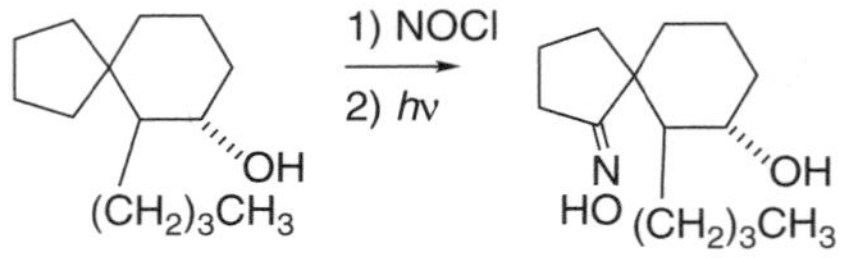

Ref. 380

These reactions depend on the proximity of the alkoxy radical to a particular hydrogen for selectivity.

Problems

(*References for these problems will be found on page 1287.*)

10.1. Indicate the major product to be expected in the following reactions:

(a)

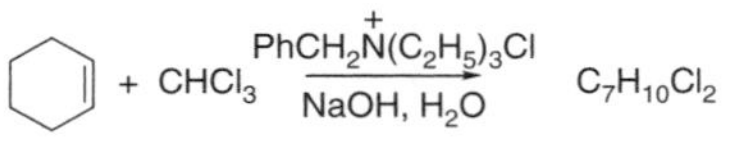

(b)

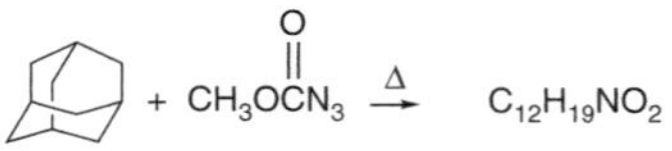

(c)

(d)

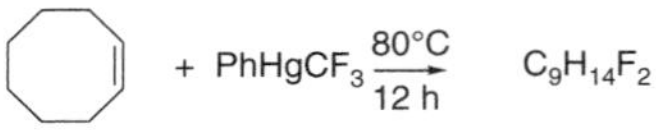

(e)

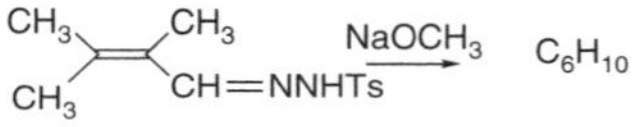

(f)

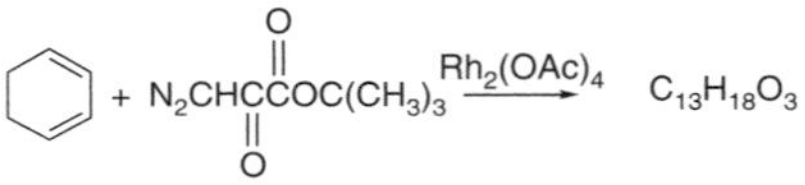

(g)

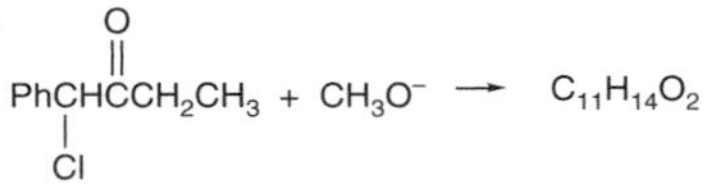

(h)

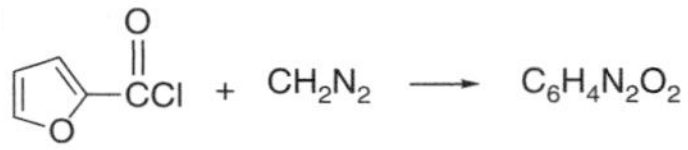

(i)

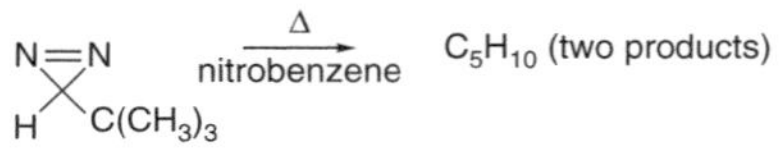

(j)

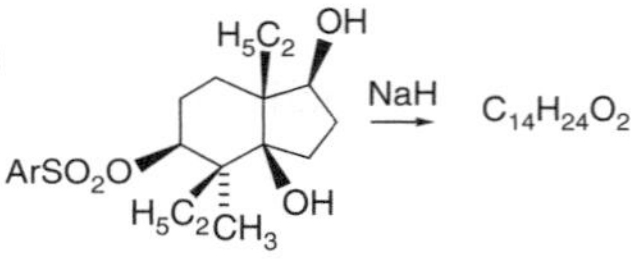

(k)

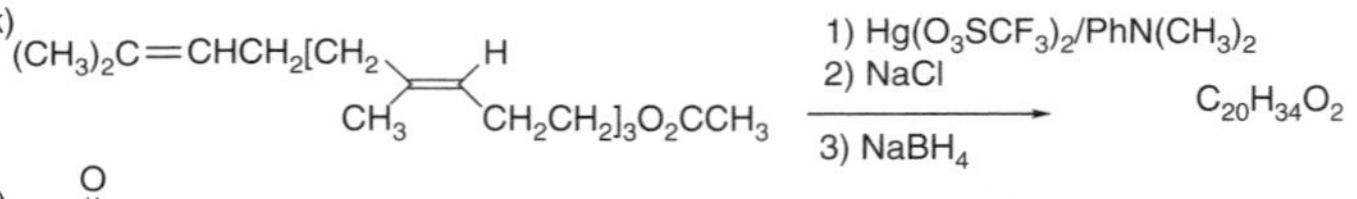

(l)

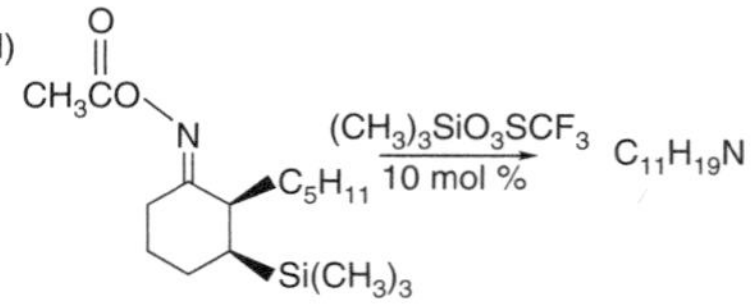

(m)

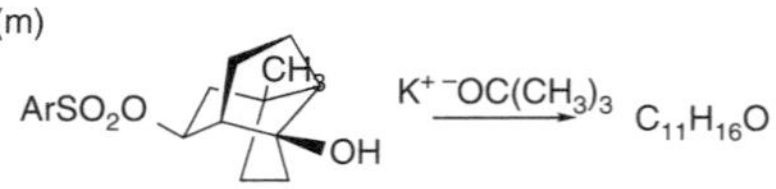

(n)

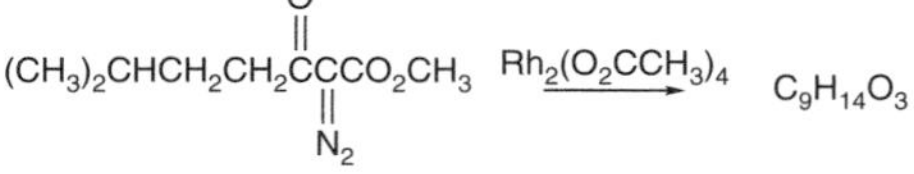

(o)

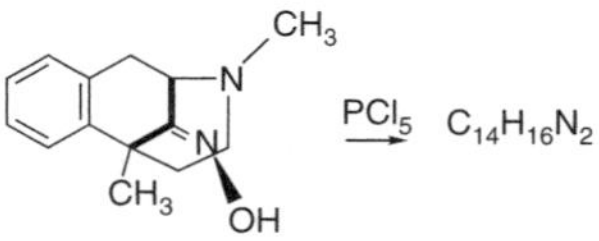

380. E. J. Corey, J. F. Arnett, and G. N. Widiger, *J. Am. Chem. Soc.*, **97**, 430 (1975).

(p) 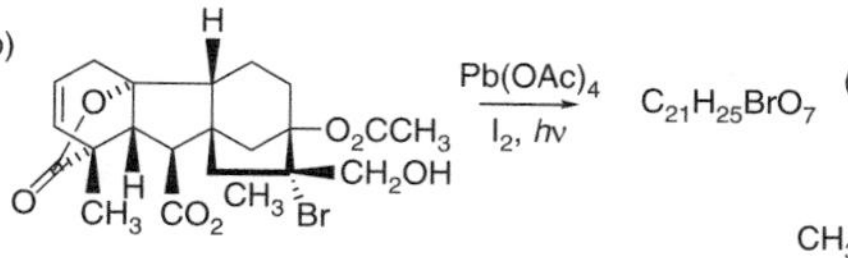 $\xrightarrow[I_2,\ h\nu]{Pb(OAc)_4}$ $C_{21}H_{25}BrO_7$

(q) 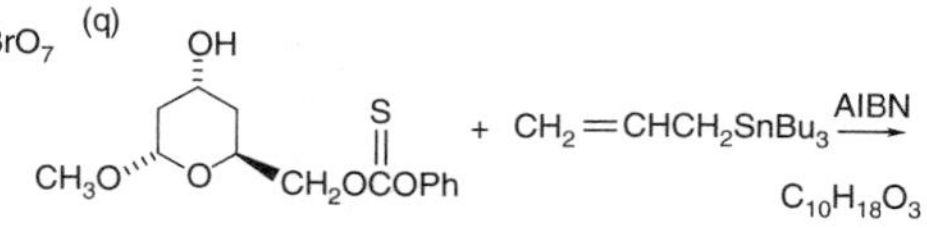 $+\ CH_2{=}CHCH_2SnBu_3 \xrightarrow{AIBN}$ $C_{10}H_{18}O_3$

(r)

$$CH_2{=}CHCH_2CH_3\ +\ C_2H_5O_2CN{=}NCO_2C_2H_5 \longrightarrow C_{10}H_{18}N_2O_4$$

(s) $(CH_3)_2C{=}CHCH_3\ +\ HC{\equiv}CCO_2CH_3 \xrightarrow{AlCl_3} C_9H_{14}O_2$

(t) 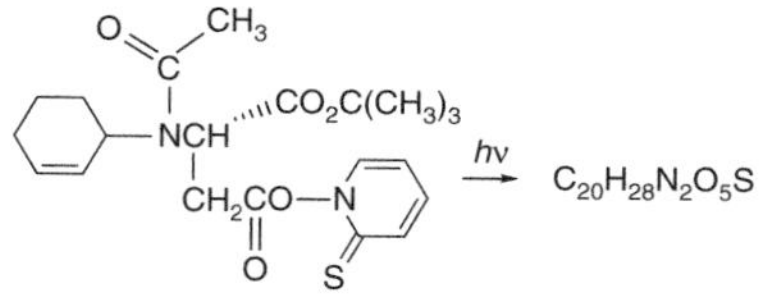 $\xrightarrow{h\nu}$ $C_{20}H_{28}N_2O_5S$

(u) 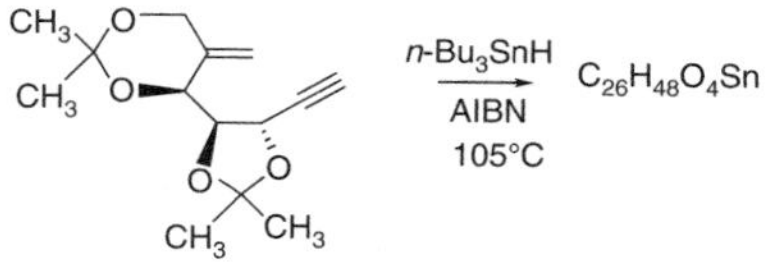 $\xrightarrow[\substack{AIBN \\ 105°C}]{n\text{-}Bu_3SnH}$ $C_{26}H_{48}O_4Sn$

(v) 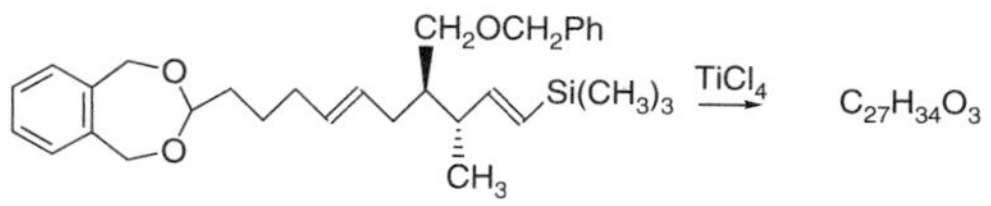 $\xrightarrow{TiCl_4}$ $C_{27}H_{34}O_3$

(w) 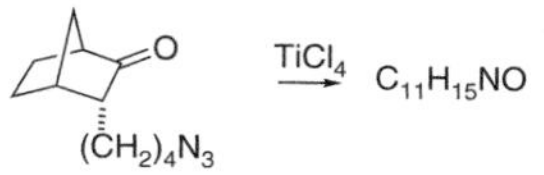 $\xrightarrow{TiCl_4}$ $C_{11}H_{15}NO$

(x) 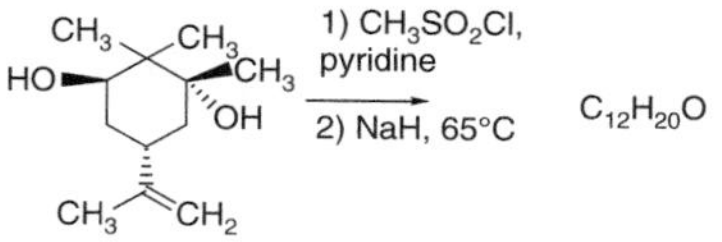 $\xrightarrow[\substack{2)\ NaH,\ 65°C}]{1)\ CH_3SO_2Cl,\ pyridine}$ $C_{12}H_{20}O$

10.2. Indicate appropriate reagents and reaction conditions or a short reaction sequence that could be expected to effect the following transformations:

(a)

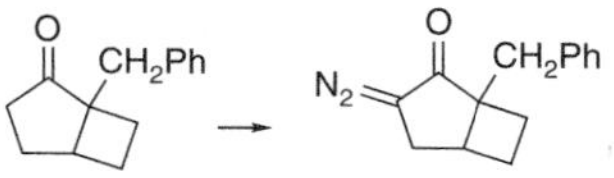

(b)

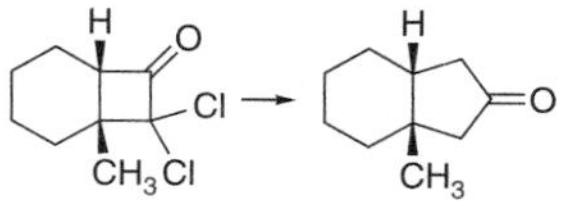

(c)

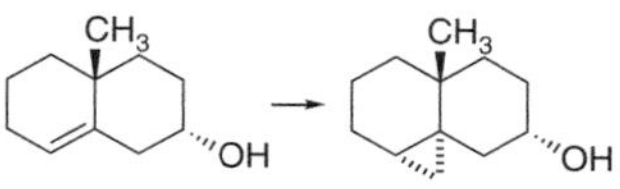

(d)

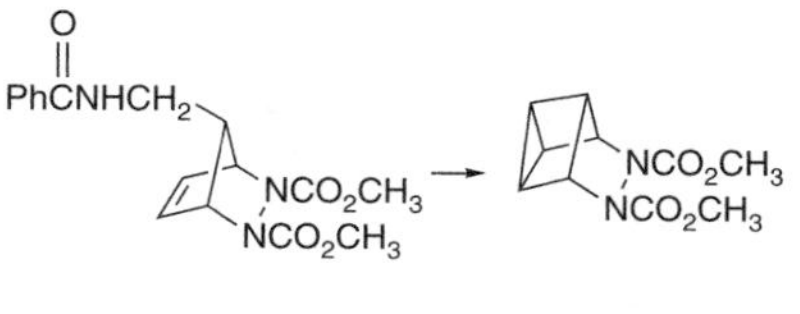

(e)

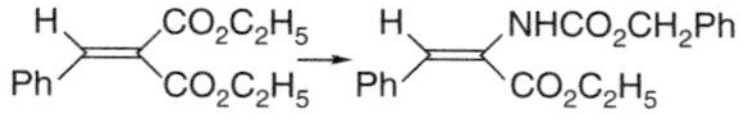

(f)

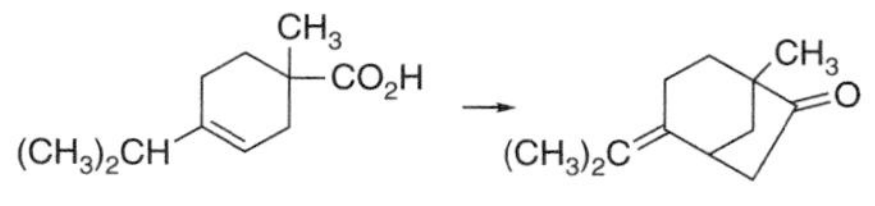

(g)

$$CH_2{=}CHCH{=}CHCO_2H \longrightarrow CH_2{=}CHCH{=}CHNHCO_2CH_2Ph$$

(h)

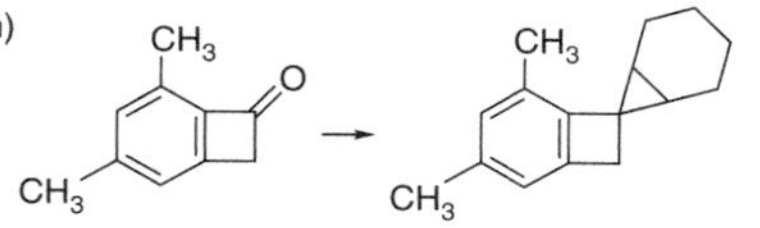

(i)

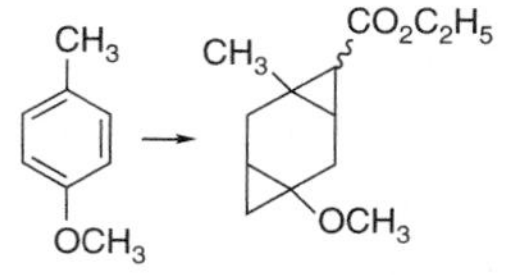

994

CHAPTER 10

*Reactions Involving
Carbocations, Carbenes,
and Radicals as Reactive
Intermediates*

(j)

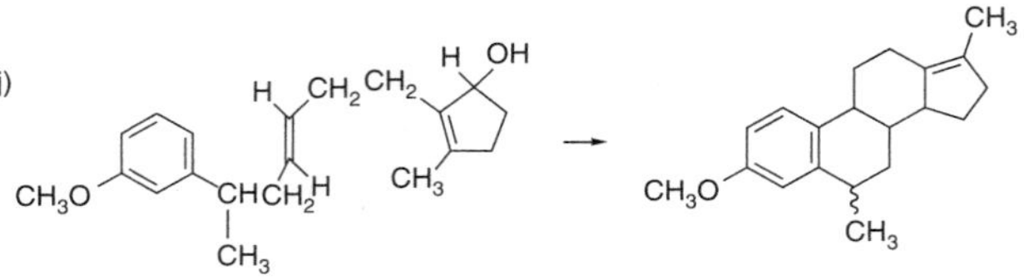

(k)

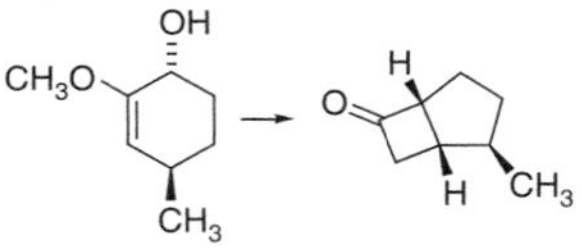

(l)

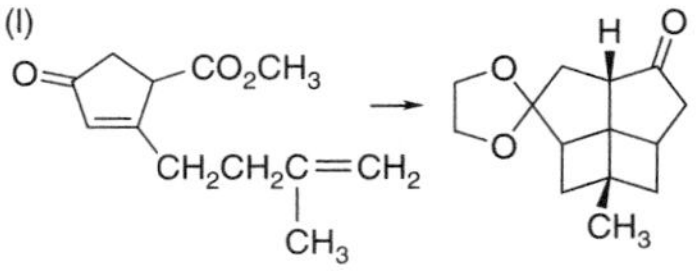

(m)

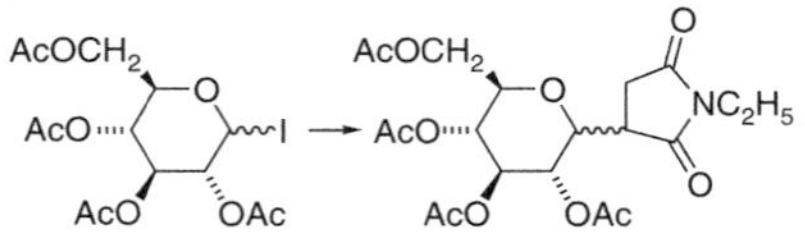

(n)

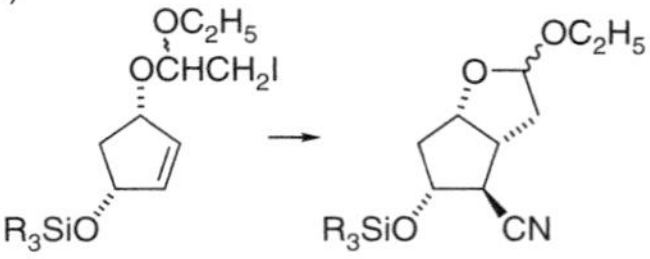

(o)

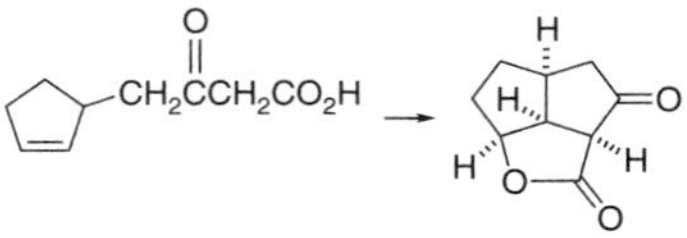

(p)

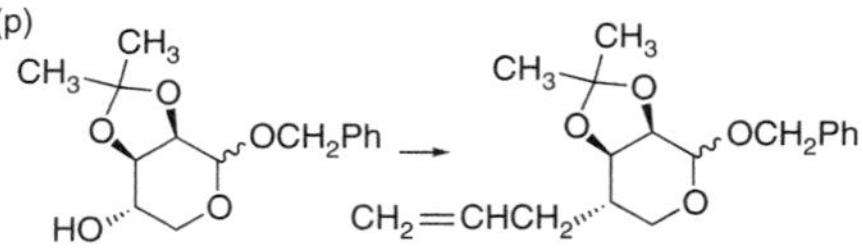

(q)

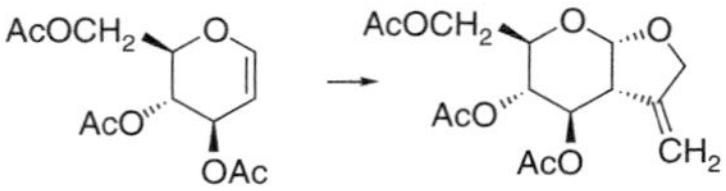

(r)

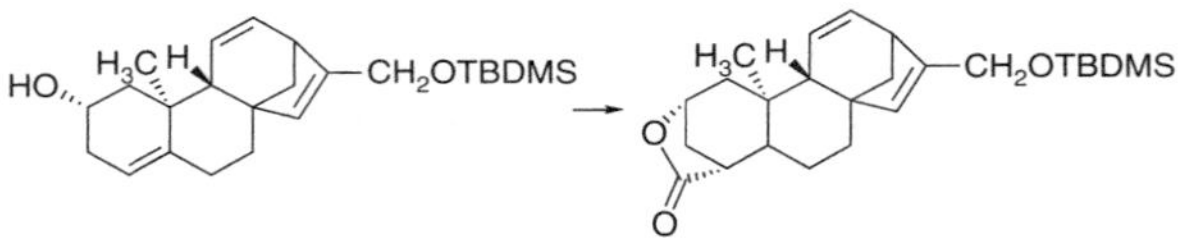

(s)

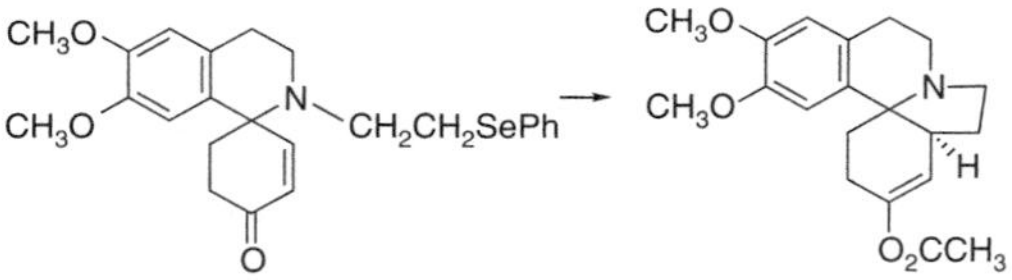

(t)

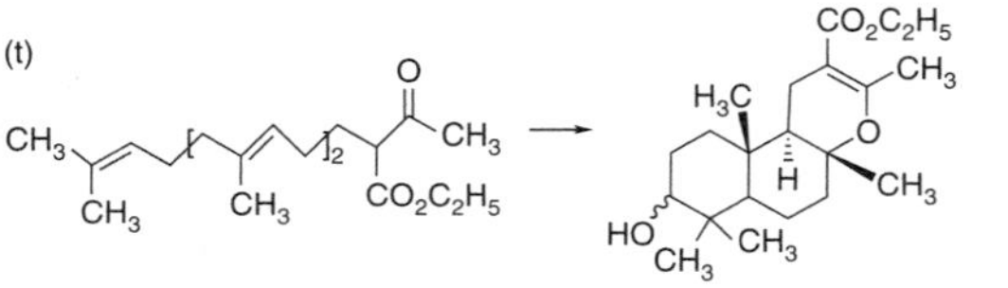

10.3. Each of the following carbenes has been predicted to have a singlet ground state, either as the result of qualitative structural considerations or theoretical calculations. Indicate what structural features might stabilize the singlet state in each case.

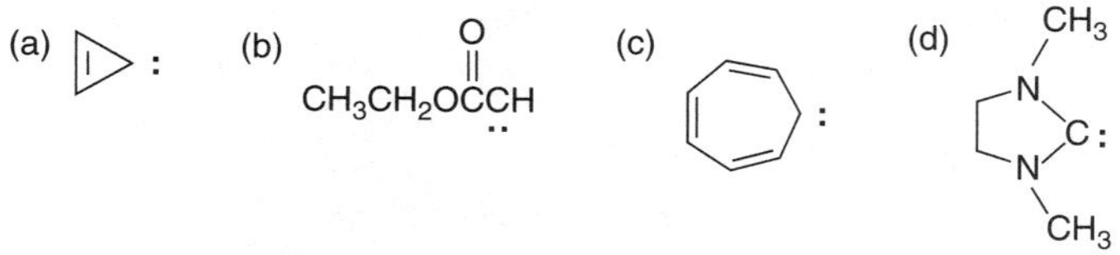

(a) ▷ :

(b)
$$CH_3CH_2O\overset{\overset{\displaystyle O}{\|}}{C}CH\ddot{\,}$$

(c) ⬡ :

(d)
$$\begin{array}{c} CH_3 \\ | \\ N \\ \diagdown \\ C: \\ \diagup \\ N \\ | \\ CH_3 \end{array}$$

10.4. The hydroxy group in *E*-cycloocten-3-ol determines the stereochemistry of the reaction with the Simmons-Smith reagent. By examining a model, predict the stereochemistry of the product.

10.5. Discuss the significance of the relationship between reactant stereochemistry and product composition exhibited in the reactions shown below.

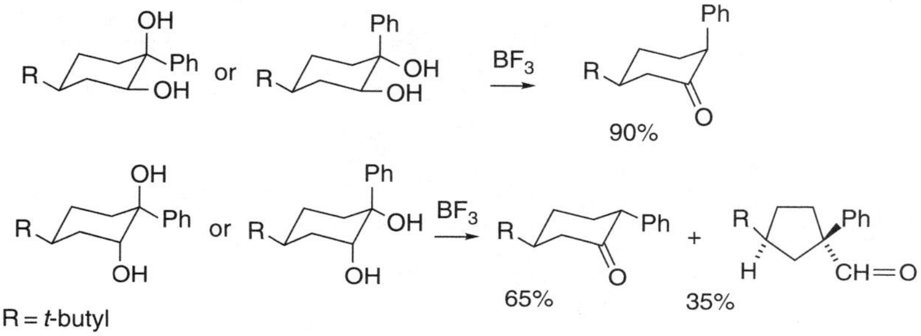

R = *t*-butyl

10.6. Suggest a mechanistic rationalization for the following reactions. Point out the structural features that contribute to the unusual or abnormal course of the reaction. What product would have been expected if the reaction followed a "normal" course.

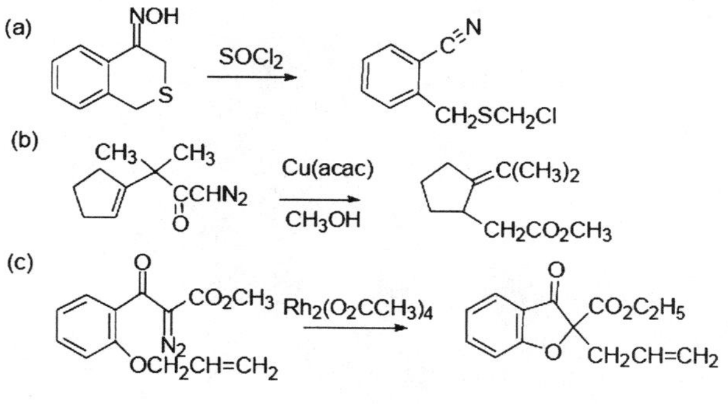

.

10.7. It has been found that the bromo ketones **10-7a-c** can rearrange by either the cyclopropanone or the semibenzilic mechanism, depending on the size of the ring and the reaction conditions. Suggest two experiments that would permit you to distinguish between the two mechanisms under a given set of circumstances.

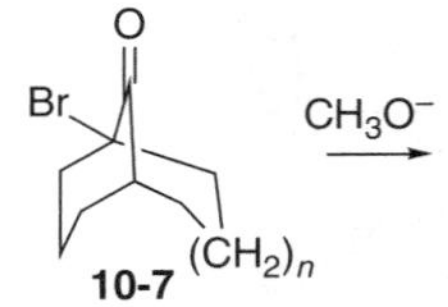

996

CHAPTER 10

Reactions Involving
Carbocations, Carbenes,
and Radicals as Reactive
Intermediates

10-7 $(CH_2)_n$

a–c; n = 1–3

10.8. Predict the major product of the following reactions:

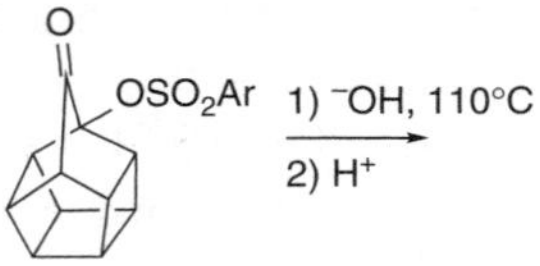

(a)

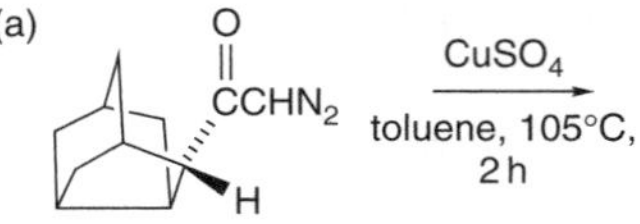

$\xrightarrow[\substack{\text{toluene, 105°C,} \\ 2\,h}]{\text{CuSO}_4}$

(b)

$\xrightarrow[\text{2) H}^+]{\text{1) }^-\text{OH, 110°C}}$

(c)

$\xrightarrow{t\text{-AmO}^-}$

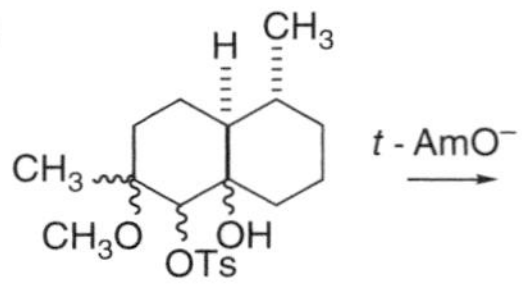

(d)

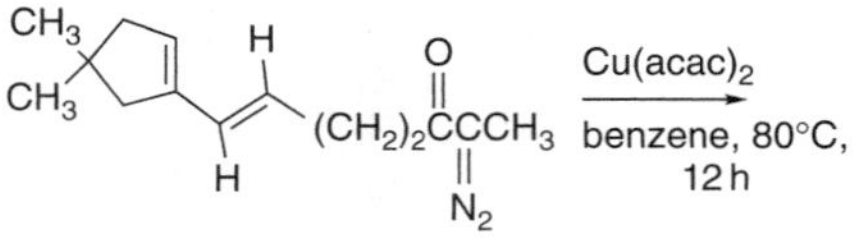

$\xrightarrow[\substack{\text{benzene, 80°C,} \\ 12\,h}]{\text{Cu(acac)}_2}$

(e)

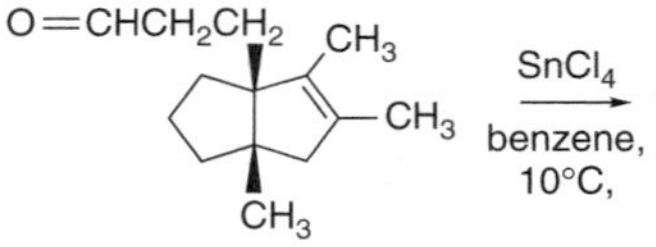

$\xrightarrow[\substack{\text{benzene,} \\ 10°C,}]{\text{SnCl}_4}$

(f)

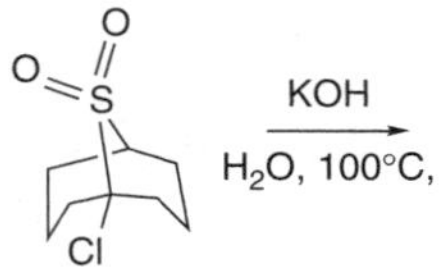

$\xrightarrow[\text{H}_2\text{O, 100°C,}]{\text{KOH}}$

(g)

1) CH₃SO₂Cl,
   (C₂H₅)₃N

2) K⁺⁻O-*t*-C₄H₉

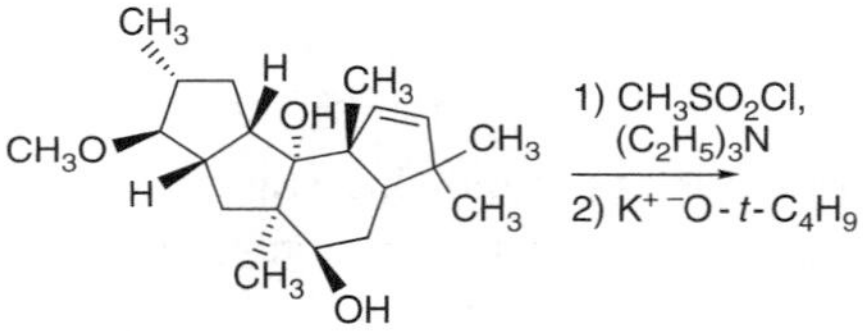

10.9. Short reaction series can effect formation of the desired material on the left from the starting material on the right. Devise an appropriate reaction sequence.

(a)

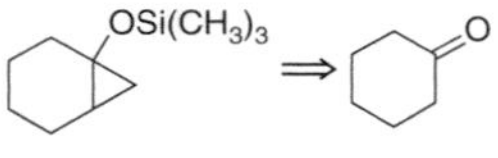

(b)

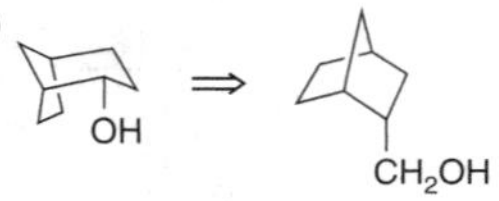

(c)

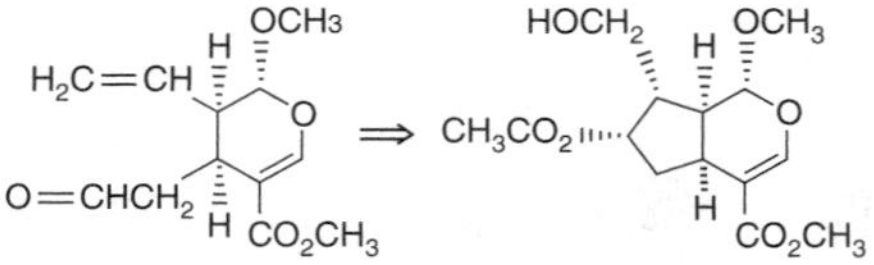

(d)

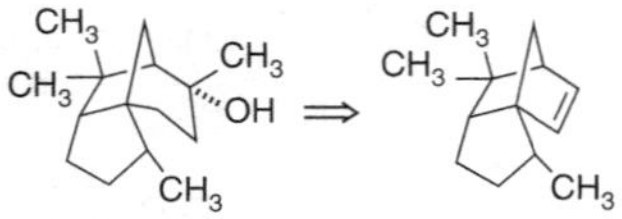

(e)

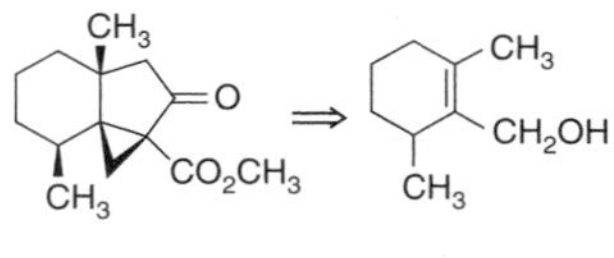

(f)

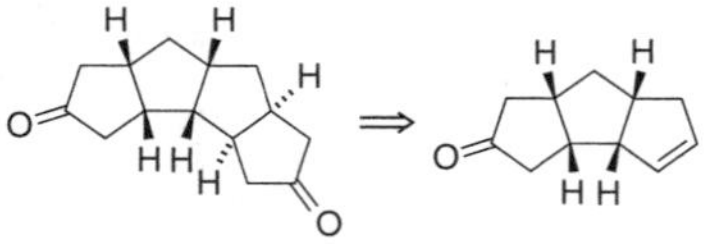

(g)

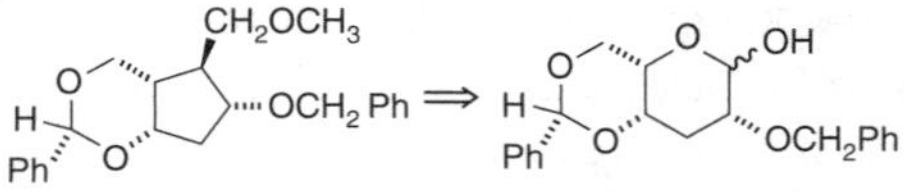

(h)

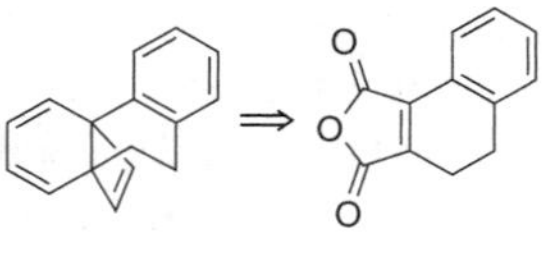

(i)

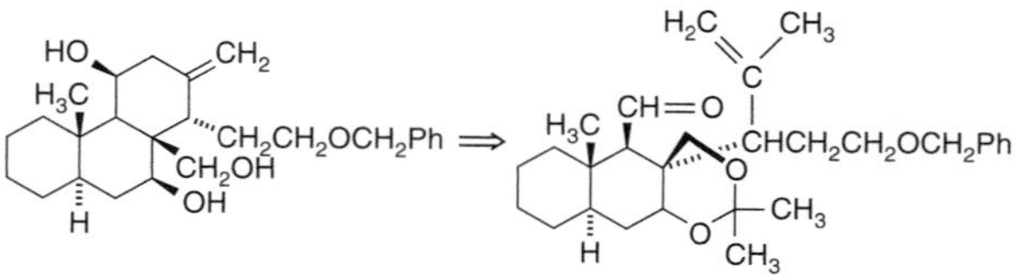

(j)

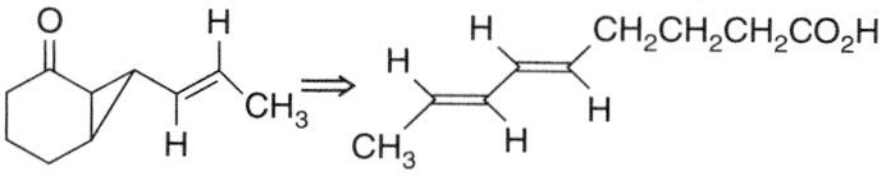

(k) 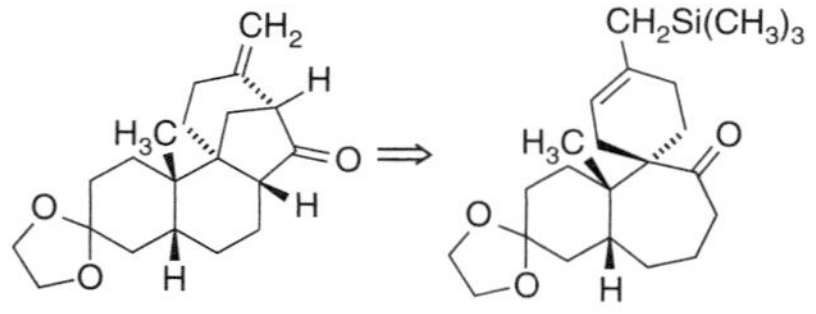   (l)

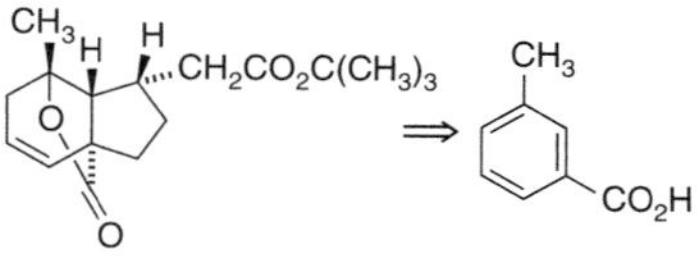

(m)

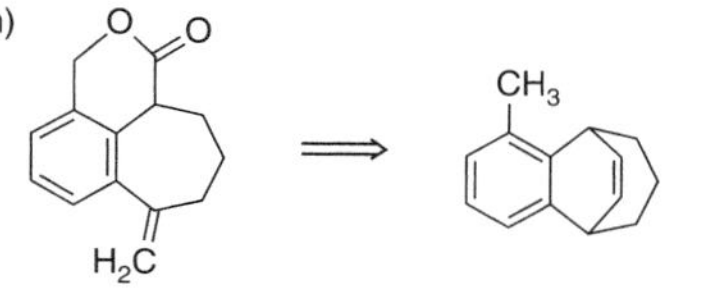

10.10. Formulate mechanisms for the following reactions:

(a)

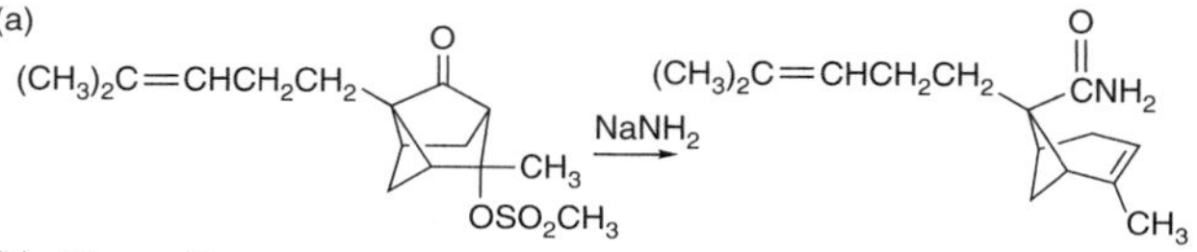

$(CH_3)_2C=CHCH_2CH_2$ — $\xrightarrow{\text{NaNH}_2}$ — $(CH_3)_2C=CHCH_2CH_2$ — $\overset{O}{\underset{}{C}}NH_2$

(b)

$\xrightarrow[\text{2) H}^+]{\text{1) KOH}}$ CO_2H

(c)

$\xrightarrow[\text{2) H}^+]{\text{1) KOH}}$ $H_2C=CHCH_2CCO_2H$, CH_2

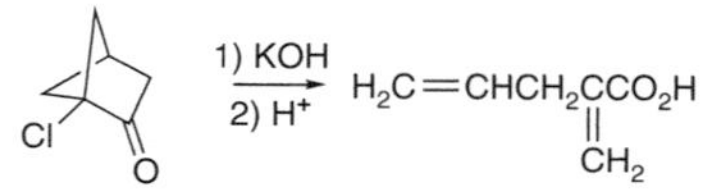

(d)

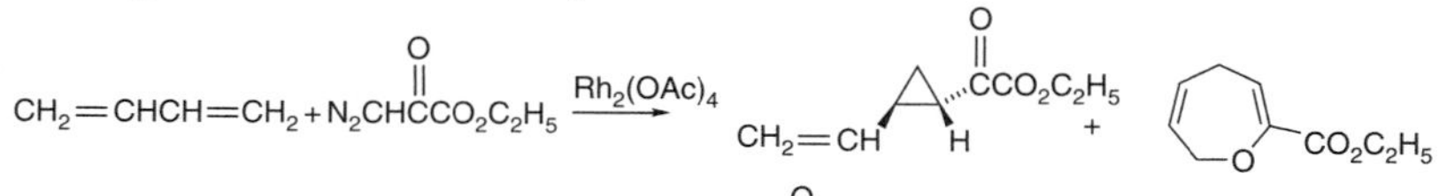

$CH_2=CHCH=CH_2 + N_2CHCCO_2C_2H_5 \xrightarrow{\text{Rh}_2(\text{OAc})_4} CH_2=CH$ ... $CO_2C_2H_5$ + $CO_2C_2H_5$

(e)

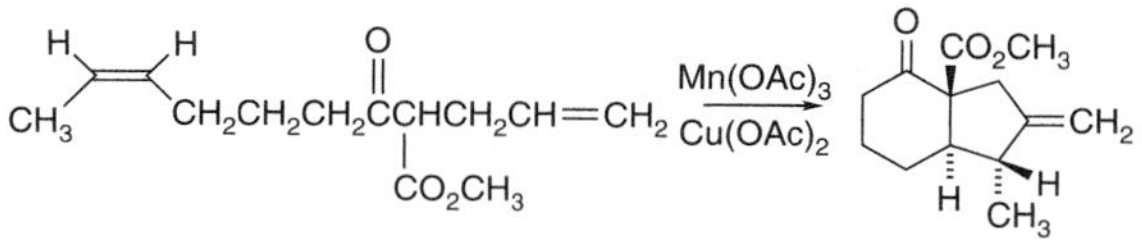

$\xrightarrow[\text{Cu(OAc)}_2]{\text{Mn(OAc)}_3}$ CO_2CH_3

(f)

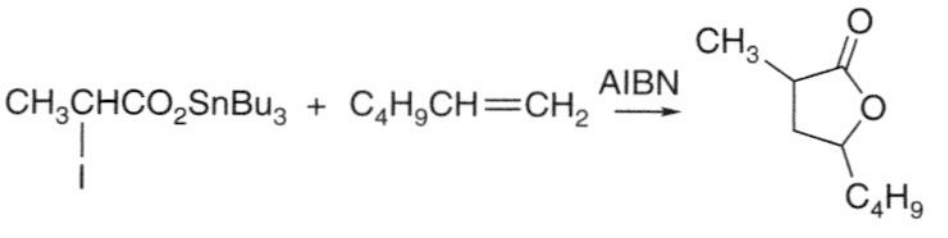

$CH_3CHCO_2SnBu_3 + C_4H_9CH=CH_2 \xrightarrow{\text{AIBN}}$ C_4H_9

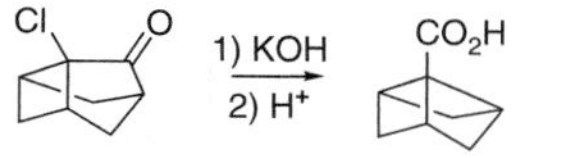

(g)

$\xrightarrow[\text{AIBN}]{\text{Bu}_3\text{SnH}}$

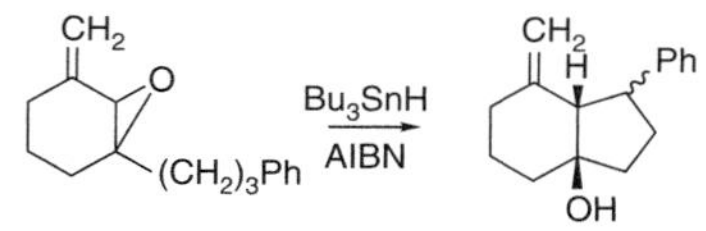

998

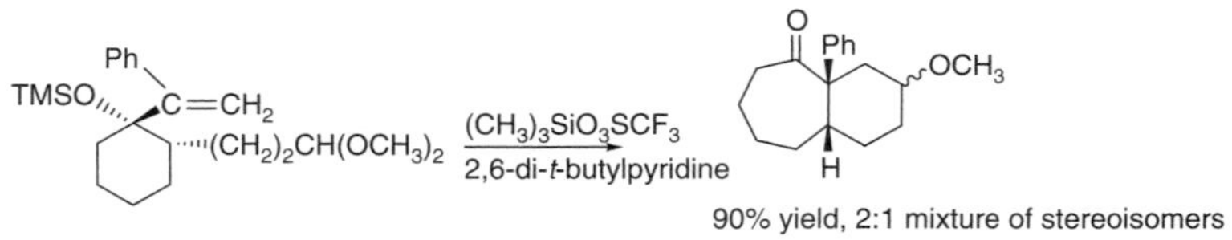

CHAPTER 10

Reactions Involving
Carbocations, Carbenes,
and Radicals as Reactive
Intermediates

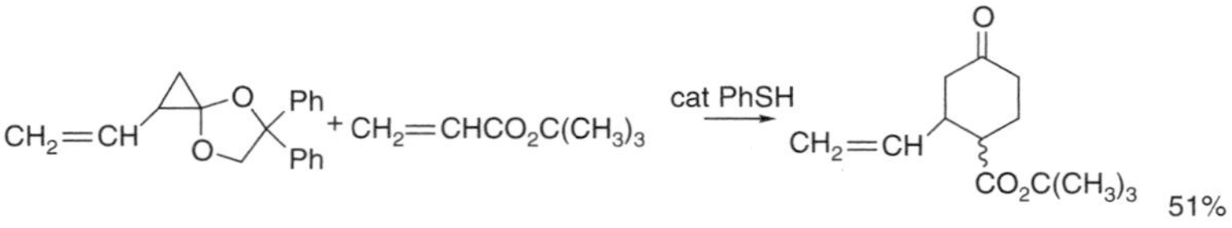

(h)

$(CH_3)_3SiO_3SCF_3$
2,6-di-*t*-butylpyridine

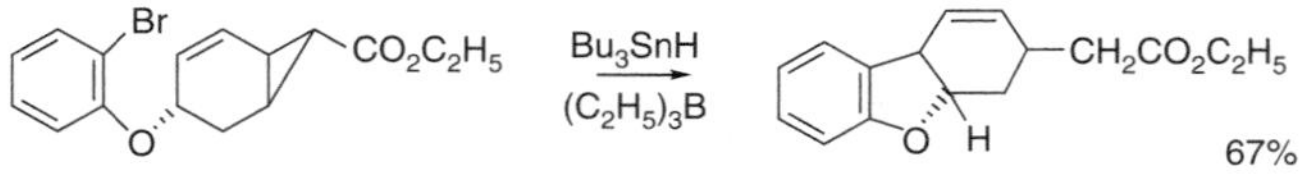

90% yield, 2:1 mixture of stereoisomers

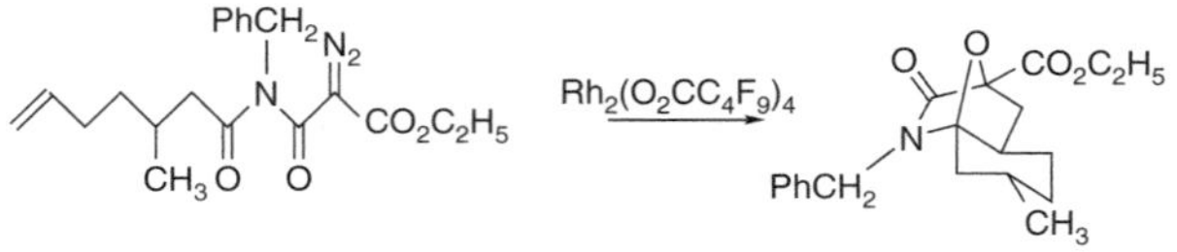

(i)

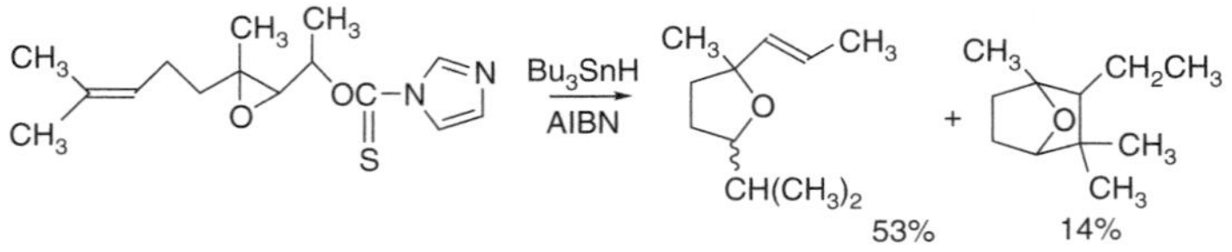

Bu_3SnH
AIBN

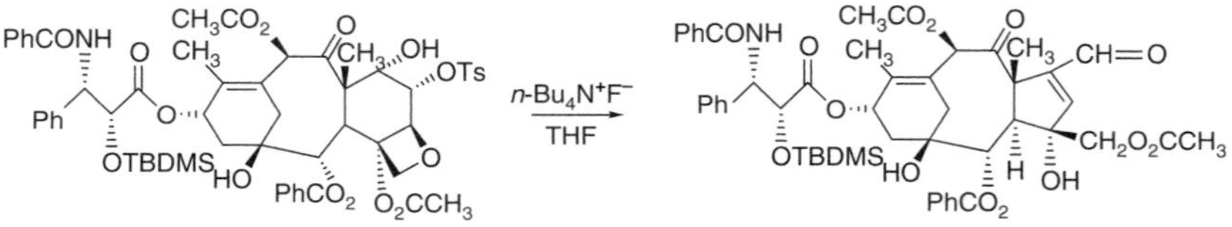

53%   14%

(j)

cat PhSH

51%

(k)

Bu_3SnH
$(C_2H_5)_3B$

67%

(l)

$Rh_2(O_2CC_4F_9)_4$

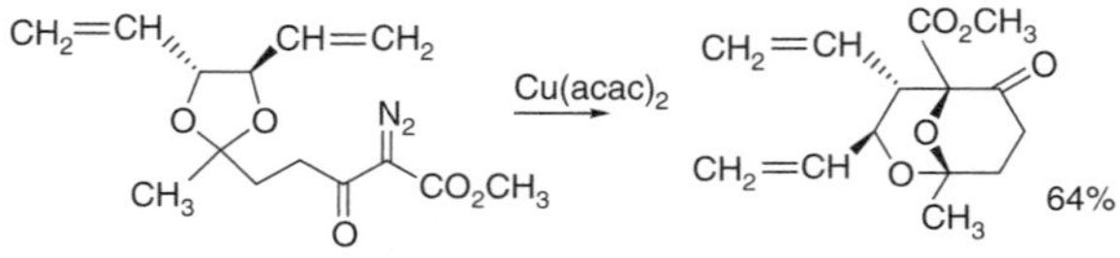

(m)

n-$Bu_4N^+F^-$
THF

(n)

$Cu(acac)_2$

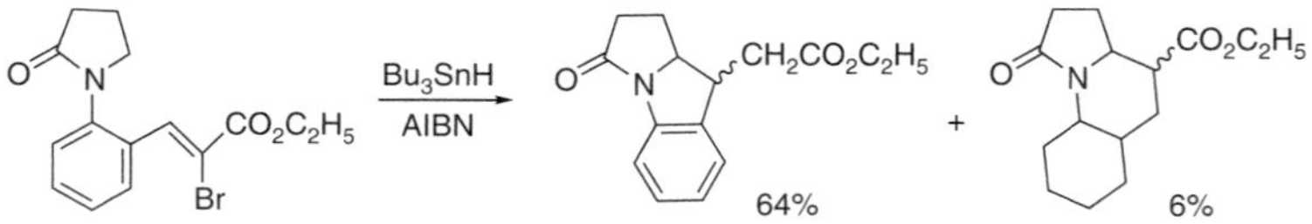

64%

(o)

Bu_3SnH
AIBN

64%   6%

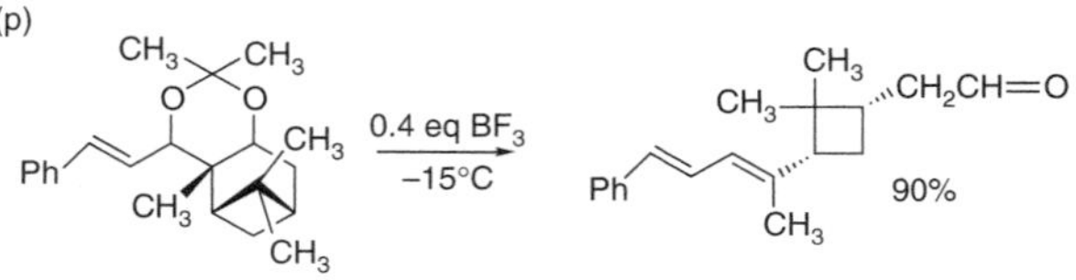

(p)

0.4 eq BF_3
−15°C

90%

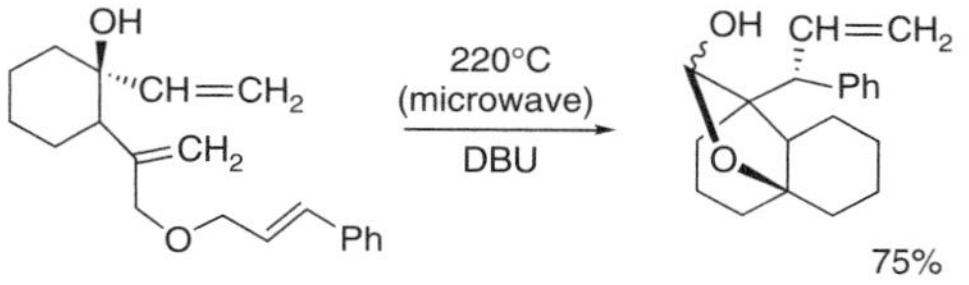

10.11. A sequence of reactions for conversion of acyclic and cyclic ketones into α,β-unsaturated ketones with insertion of a =CHCH$_3$ unit has been developed. The method uses 1-lithio-1,1-dichloroethane as a key carbenoid reagent. The overall sequence involves three steps, one of them before and one after the carbenoid reaction. By analysis of the bonding changes and application of your knowledge of carbene reactions, devise a reaction sequence that would accomplish the transformation.

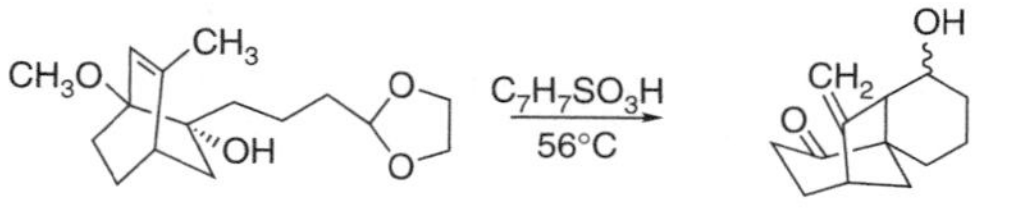

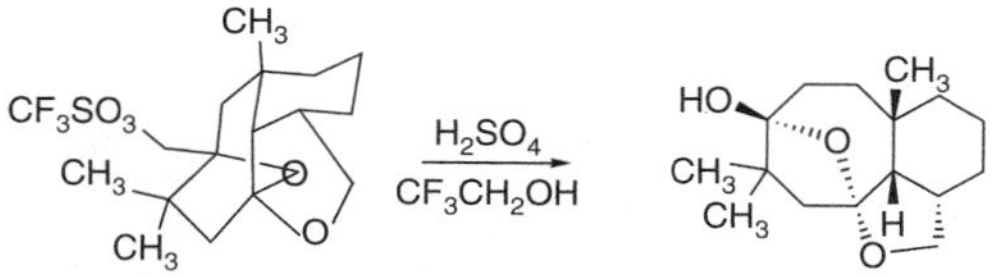

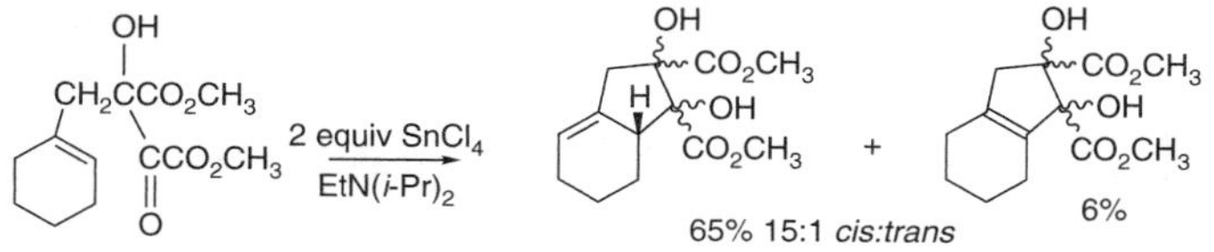

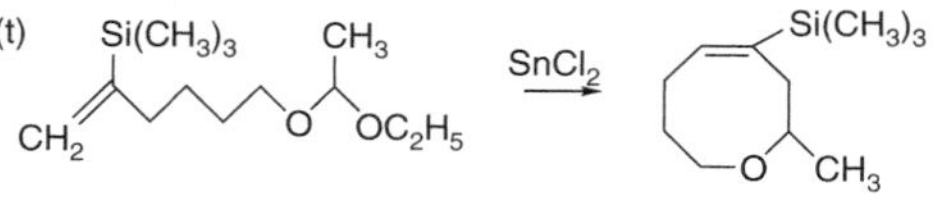

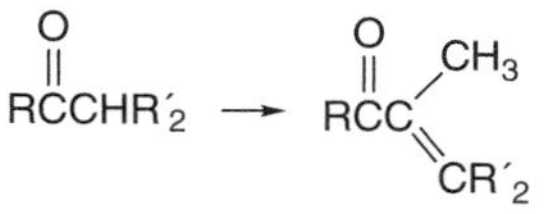

10.12. The synthesis of globulol from the octalin derivative shown proceeds in four stages. These include, not necessarily in sequence, addition of a carbene, a fragmentation reaction, and acid-catalyzed cyclization of a cyclodeca-2,7-dienol. The final step of the synthesis converts a dibromocyclopropane to the dimethyl-cyclopropane structure using dimethylcuprate. Using retrosynthetic analysis, devise an appropriate sequence of reactions and suggest reagents for each step.

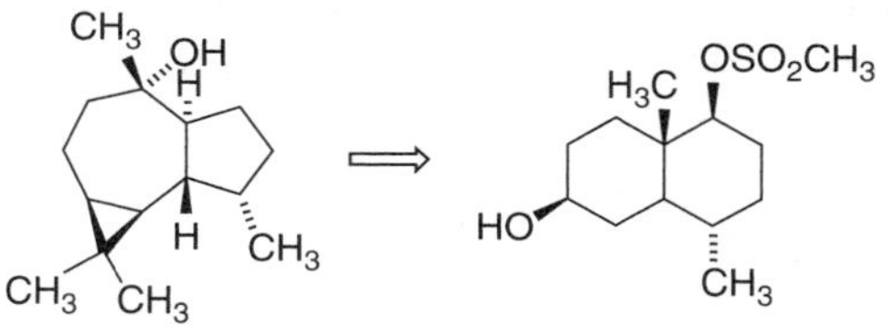

globulol

1000

CHAPTER 10

*Reactions Involving
Carbocations, Carbenes,
and Radicals as Reactive
Intermediates*

10.13. Both the *E*- and *Z*-isomers of vinylsilane **13-A** have been subjected to polyene cyclization using $TiCl_4$-$Ti(O\text{-}i\text{-}Pr)_4$. Although the *Z*-isomer gives an 85–90% yield, the *E*-isomer affords only a 30–40% yield. Offer an explanation.

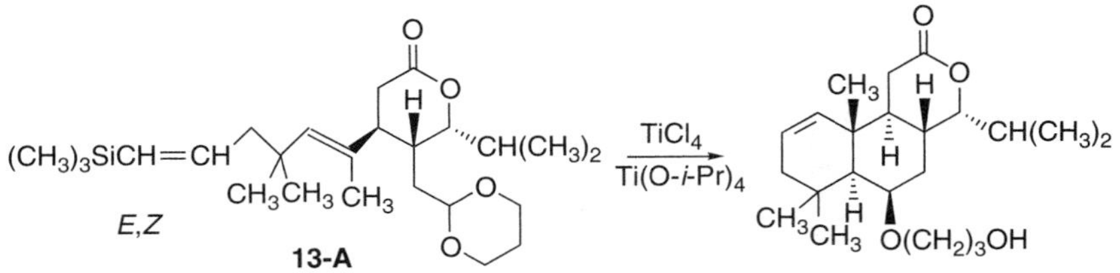

10.14. Each of the three decahydroquinoline sulfonates shown below gives a different product composition on solvolysis. One gives 9-methylamino-*E*-non-5-enal, one gives 9-methylamino-*Z*-non-5-enal, and one gives a mixture of the two quinoline derivatives **14-D** and **14-E**. Deduce which compound gives rise to which product. Explain your reasoning.

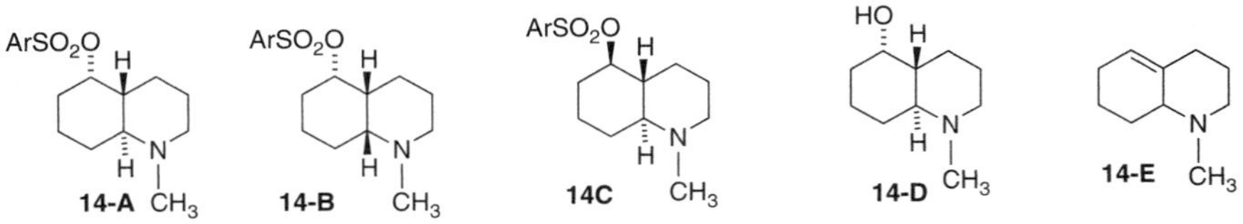

10.15. Normally, the dominant reaction between acyl diazo compounds and simple α,β-unsaturated carbonyl compounds is a cycloaddition.

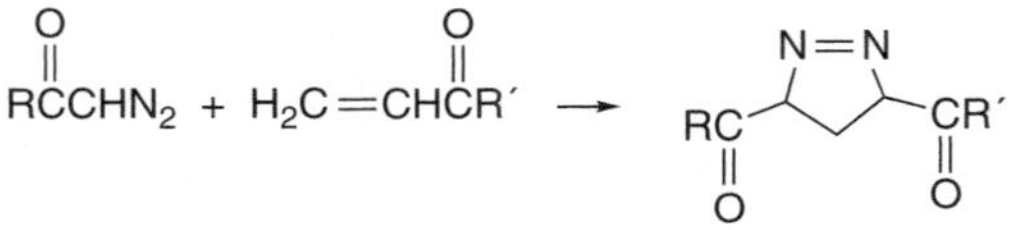

If, however, the reaction is run in the presence of a Lewis acid, particularly SbF_5, the reaction takes a different course, giving a diacyl cyclopropane.

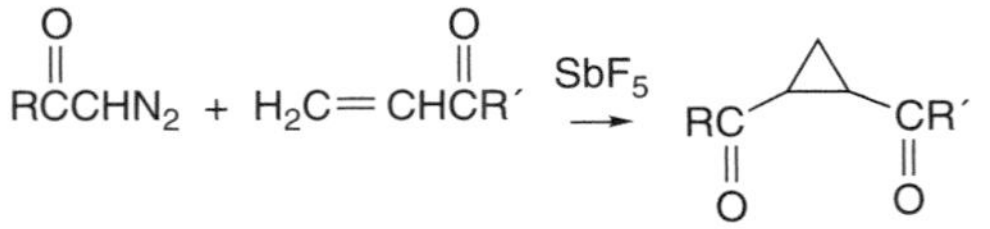

Formulate a mechanism to account for the altered course of the reaction in the presence of SbF_5.

10.16. Compound **16-A** on reaction with Bu_3SnH in the presence of AIBN gives **16-B** rather than **16-C**. How is **16-B** formed? Why is **16-C** not formed? What relationship do these results have to the rate data given on p. 986?

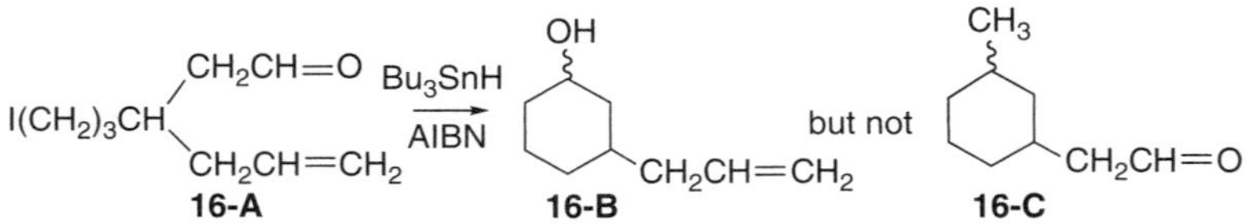

10.17. The following molecules have been synthesized by radical cyclization and tandem radical cyclizations. Identify the bond or bonds that could be formed by radical cyclizations and suggest an appropriate reactant and reaction conditions that would lead to the specified products.

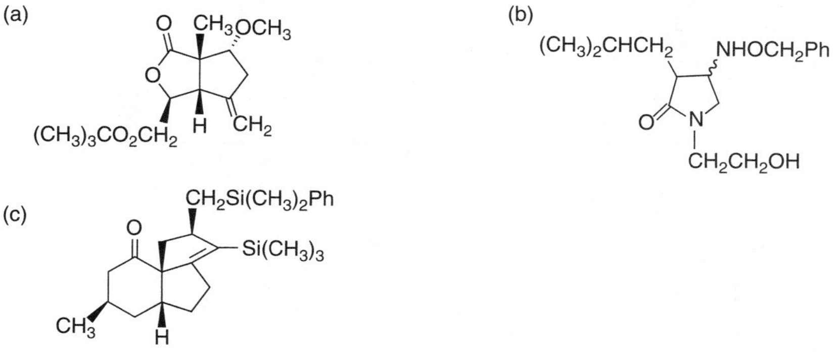

(a)

(b)

(c)

10.18. Attempted deoxygenation of several β-aryl thiono carbonates gave the unexpected product shown. In contrast, the corresponding α-isomers gave the desired deoxygenation product. Account for the formation of the observed products, and indicate why these products are not formed from the α-stereoisomers.

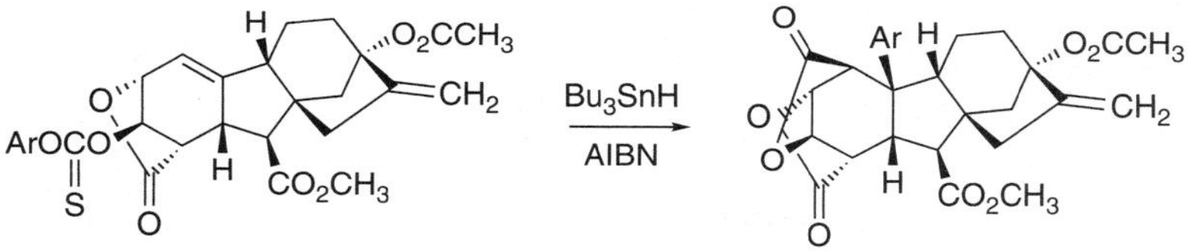

10.19. *cis*-Chrysanthemic acid has been synthesized through three intermediates using the reaction conditions shown. Assign structures to the intermediates and indicate the nature of each of the reactions.

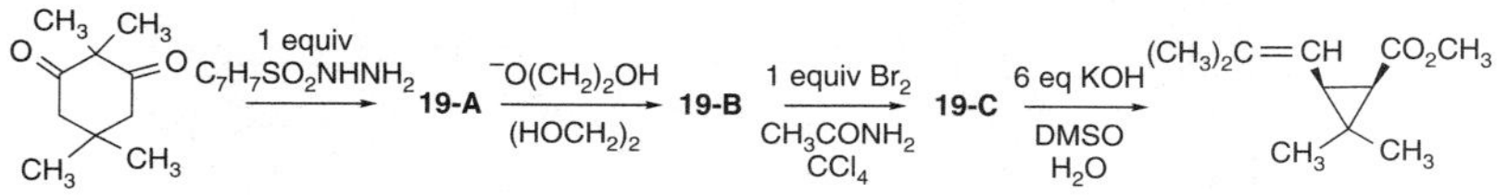

10.20. The photolysis of alkoxy chlorodiazirines generates carbenes. The reaction has been examined in pentane and CH_2Cl_2 with increasing amounts of methanol. Three products, the bridgehead chloride, bridgehead ether, and bridgehead alcohol are formed. The former two products arise from fragmentation of the carbene. The last results from trapping of the carbene prior to fragmentation.

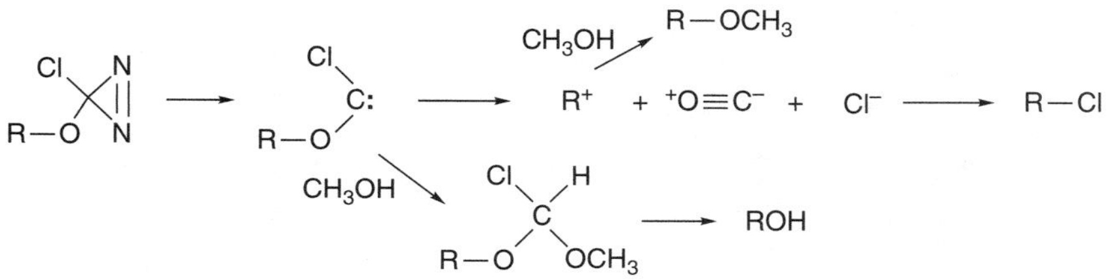

1002

CHAPTER 10

Reactions Involving
Carbocations, Carbenes,
and Radicals as Reactive
Intermediates

The activation energies for the fragmentation of the carbene in CH_2Cl_2 were calculated by the B3LYP/6-31G* method to be 14.6, 2.2, and −0.95 for the bicyclo[2.2.1]heptyl, bicyclo[2.2.2]octyl, and adamantyl systems, respectively. Are the product trends consistent with these computational results, which presumably reflect the relative stability of the carbocation formed by the fragmentation?

[MeOH]	pentane			CH_2Cl_2		
	R−Cl	R−OCH$_3$	R−OH	R−Cl	R−OCH$_3$	R−OH
R = bicyclo[2.2.1]heptyl						
0				100		
0.25	15	3	82	64	trace	35
0.50	23	trace	77	59	1	40
1.00	45	trace	55	57	2	41
R = bicyclo[2.2.2]octyl						
0				100		
0.25	38	19	43	60	8	32
0.50	34	19	47	52	13	35
1.00	40	20	40	45	21	34
R = adamantyl						
0				100		
0.25	81	trace	19	93	trace	7
0.50	83	trace	17	91	trace	9
1.00	83	trace	17	79	10	11

10.21 a. The oxidation of norbornadiene by *t*-butyl perbenzoate and Cu(I) leads to 7-*t*-butoxynorbornadiene. Similarly, oxidation with dibenzoyl peroxide and CuBr leads to 7-benzoyloxynorbornadiene. In both reactions, when a 2-deuterated sample of norbornadiene is used, the deuterium is found distributed among all positions in the product in approximately equal amounts. Provide a mechanism that can account for this result.

b. A very direct synthesis of certain lactones involves heating an alkene with a carboxylic acid and the Mn(III) salt of the acid. Suggest a mechanism by which this reaction might occur.

$$CH_3(CH_2)_5CH{=}CH_2 \quad + \quad Mn(O_2CCH_3)_3 \quad \xrightarrow{CH_3CO_2H} \quad$$

Aromatic Substitution Reactions

Introduction

This chapter is concerned with reactions that introduce or replace substituent groups on aromatic rings. The synthetic methods for aromatic substitution were among the first to be developed. The basic mechanistic concepts for electrophilic aromatic substitution and some of the fundamental reactions are discussed in Chapter 9 of Part A. These reactions provide methods for introduction of nitro groups, the halogens, sulfonic acids, and alkyl and acyl groups. The regioselectivity of these reactions depends upon the nature of the existing substituent and can be *ortho*, *meta*, or *para* selective.

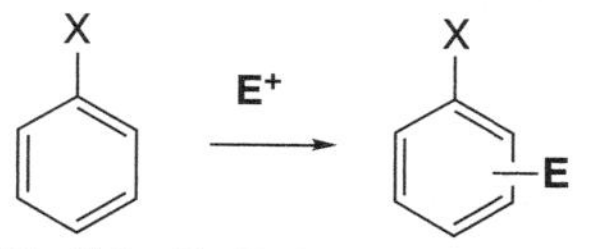

$E = NO_2$, F, Cl, Br. I, SO_3H, SO_2Cl, R, $RC=O$

A second group of aromatic substitution reactions involves aryl diazonium ions. As for electrophilic aromatic substitution, many of the reactions of aromatic diazonium ions date to the nineteenth century. There have continued to be methodological developments for substitution reactions of diazonium intermediates. These reactions provide routes to aryl halides, cyanides, and azides, phenols, and in some cases to alkenyl derivatives.

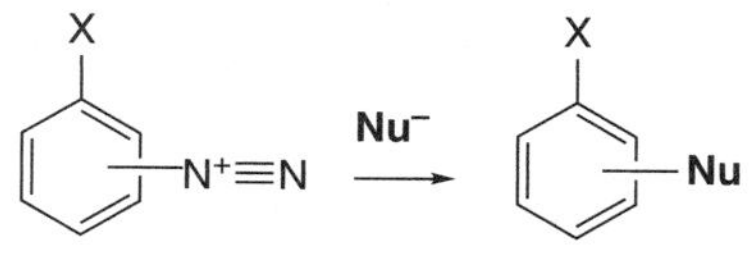

$Nu = $ F, Cl, Br, I, CN, N_3, OH, $CH=CHR$

Direct nucleophilic displacement of halide and sulfonate groups from aromatic rings is difficult, although the reaction can be useful in specific cases. These reactions can occur by either addition-elimination (Section 11.2.2) or elimination-addition (Section 11.2.3). Recently, there has been rapid development of metal ion catalysis, and old methods involving copper salts have been greatly improved. Palladium catalysts for nucleophilic substitutions have been developed and have led to better procedures. These reactions are discussed in Section 11.3.

$$Z = I, Br, Cl, O_3SAr \qquad Nu = CN, R_2N, RO$$

Several radical reaction have some synthetic application, including radical substitution (Section 11.4.1) and the $S_{RN}1$ reaction (Section 11.4.2).

11.1. Electrophilic Aromatic Substitution

The basic mechanistic concepts and typical electrophilic aromatic substitution reactions are discussed in Sections 9.1 and 9.4 of Part A. In the present section, we expand on that material, with particular emphasis on synthetic methodology.

11.1.1. Nitration

Nitration is the most important method for introduction of nitrogen functionality on aromatic rings. Nitro compounds can be reduced easily to the corresponding amino derivatives, which can provide access to diazonium ions. There are several reagent systems that are useful for nitration. A major factor in the choice of reagent is the reactivity of the ring to be nitrated. Nitration is a very general reaction and satisfactory conditions can normally be developed for both activated and deactivated aromatic compounds. Since each successive nitro group reduces the reactivity of the ring, it is easy to control conditions to obtain a mononitration product. If polynitration is desired, more vigorous conditions are used.

Concentrated nitric acid can effect nitration but it is not as reactive as a mixture of nitric acid with sulfuric acid. The active nitrating species in both media is the nitronium ion, NO_2^+, which is formed by protonation and dissociation of nitric acid. The concentration of NO_2^+ is higher in the more strongly acidic sulfuric acid than in nitric acid.

$$HNO_3 + 2\,H^+ \rightleftharpoons H_3O^+ + NO_2^+$$

Nitration can also be carried out in organic solvents, with acetic acid and nitromethane being common examples. In these solvents the formation of the NO_2^+ is often the rate-controlling step.[1]

[1] E. D. Hughes, C. K. Ingold, and R. I. Reed, *J. Chem. Soc.*, 2400 (1950); J. G. Hoggett, R. B. Moodie, and K. Schofield, *J. Chem. Soc. B*, 1 (1969); K. Schofield, *Aromatic Nitration*, Cambridge University Press, Cambridge, 1980, Chap. 2.

$$2\ HNO_3 \rightleftharpoons H_2NO_3^+ + NO_3^-$$

$$H_2NO_3^+ \xrightarrow{\text{slow}} NO_2^+ + H_2O$$

$$ArH + NO_2^+ \xrightarrow{\text{fast}} ArNO_2 + H^+$$

Another useful medium for nitration is a solution prepared by dissolving nitric acid in acetic anhydride, which generates acetyl nitrate. This reagent tends to give high *ortho:para* ratios for some nitrations.[2]

$$HNO_3 + (CH_3CO)_2O \rightleftharpoons CH_3\overset{\overset{\displaystyle O}{\|}}{C}ONO_2 + CH_3CO_2H$$

A convenient procedure involves reaction of the aromatic in chloroform or dichloromethane with a nitrate salt and trifluoroacetic anhydride.[3] Presumably trifluoroacetyl nitrate is generated under these conditions.

$$NO_3^- + (CF_3CO)_2O \longrightarrow CF_3\overset{\overset{\displaystyle O}{\|}}{C}ONO_2 + CF_3CO_2^-$$

Acetic anhydride and trifluoroacetic anhydride have both been used in conjunction with nitric acid and zeolite β. This system gives excellent *para* selectivity in many cases.[4] The improved selectivity is thought to occur as a result of nitration within the zeolite pores, which may restrict access to the *ortho* position; see, e.g., Entry 7 in Scheme 11.1.

Nitration can be catalyzed by lanthanide salts. For example, the nitration of benzene, toluene, and naphthalene by aqueous nitric acid proceeds in good yield in the presence of $Yb(O_3SCF_3)_3$.[5] The catalysis presumably results from an oxyphilic interaction of nitrate ion with the cation, which generates or transfers the NO_2^+ ion.[6] This catalytic procedure uses a stoichiometric amount of nitric acid and avoids the excess strong acidity associated with conventional nitration conditions.

$$Ln^{3+} \cdots O-N\overset{\overset{\displaystyle O^-}{\diagup}}{\underset{\displaystyle O}{\diagdown}} \longrightarrow [O=N^+=O]$$

A variety of aromatic compounds can be nitrated using $Sc(O_3SCF_3)_3$, with $LiNO_3$ or $Al(NO_3)_3$ and acetic anhydride (see Scheme 11.1, Entry 9).[7]

Salts containing the nitronium ion can be prepared and are reactive nitrating agents. The tetrafluoroborate salt has been used most frequently,[8] but the

2. A. K. Sparks, *J. Org. Chem.*, **31**, 2299 (1966).
3. J. V. Crivello, *J. Org. Chem.*, **46**, 3056 (1981).
4. K. Smith, T. Gibbins, R. W. Millar, and R. P. Claridge, *J. Chem. Soc., Perkin Trans. 1*, 2753 (2000); K. Smith, A. Musson, and G. A. DeBoos, *J. Org. Chem.*, **63**, 8448 (1998).
5. F. J. Walker, A. G. M. Barrett, D. C. Braddock, and D. Ramprasad, *J. Chem. Soc., Chem. Commun.*, 613 (1997).
6. F. J. Walker, A. G. M. Barrett, D. C. Braddock, R. M. McKinnell, and D. Ramprasad, *J. Chem. Soc., Perkin Trans. 1*, 867 (1999).
7. A. Kawada, S. Takeda, K. Yamashita, H. Abe, and T. Harayama, *Chem. Pharm. Bull.*, **50**, 1060 (2002).
8. S. J. Kuhn and G. A. Olah, *J. Am. Chem. Soc.*, **83**, 4564 (1961); G. A. Olah and S. J. Kuhn, *J. Am. Chem. Soc.*, **84**, 3684 (1962); G. A. Olah, S. C. Narang, J. A. Olah, and K. Lammertsma, *Proc. Natl. Acad. Sci., USA*, **79**, 4487 (1982); C. L. Dwyer and C. W. Holzapfel, *Tetrahedron*, **54**, 7843 (1998).

trifluoromethansulfonate can also be prepared readily.[9] Nitrogen heterocycles such as pyridine and quinoline form *N*-nitro salts on reaction with NO_2BF_4.[10] These *N*-nitro heterocycles in turn can act as nitrating reagents, in a reaction called *transfer nitration* (see Scheme 11.1, Entry 10).

Another nitration procedure uses ozone and nitrogen dioxide.[11] With aromatic hydrocarbons and activated derivatives, this nitration is believed to involve the radical cation of the aromatic reactant.

$$NO_2 + O_3 \longrightarrow NO_3 + O_2$$
$$ArH + NO_3 \longrightarrow [ArH]^{\cdot+} + NO_3^-$$
$$[ArH]^{\cdot+} + NO_2 \longrightarrow [Ar\underset{NO_2}{\overset{H}{<}}]^+ \longrightarrow ArNO_2 + H^+$$

Compounds such as phenylacetate esters and phenylethyl ethers, which have oxygen substituents that can serve as directing groups, show high *ortho:para* ratios under these conditions.[12] These reactions are believed to involve coordination of the NO_2^+ at the substituent oxygen, followed by intramolecular transfer.

$$o:m:p = 81:2:17$$

Scheme 11.1 gives some examples of nitration reactions. Entries 1 to 3 are cases involving mixed nitric and sulfuric acids. Entry 2 illustrates the *meta*-directing effect of the protonated amino substituent. Entry 3 is an example of dinitration. Entry 4 involves an activated ring, and nitric acid suffices for nitration. At first glance, the position of substitution might seem surprising, but it may be that the direct resonance interaction of the 4-methoxy group with the formyl group attenuates its donor effect, leading to dominance of the 3-methoxy group.

Entry 5 is an example of nitration in acetic anhydride. An interesting aspect of this reaction is its high selectivity for the *ortho* position. Entry 6 is an example of the use of trifluoroacetic anhydride. Entry 7 illustrates the use of a zeolite catalyst with improved *para* selectivity. With mixed sulfuric and nitric acids, this reaction gives a 1.8:1 *para:ortho* ratio. Entry 8 involves nitration using a lanthanide catalyst, whereas Entry 9 illustrates catalysis by $Sc(O_3SCF_3)_3$. Entry 10 shows nitration done directly with $NO_2^+BF_4^-$, and Entry 11 is also a transfer nitration. Entry 12 is an example of the use of the NO_2-O_3 nitration method.

9. C. L. Coon, W. G. Blucher, and M. E. Hill, *J. Org. Chem.*, **38**, 4243 (1973).
10. G. A. Olah, S. C. Narang, J. A. Olah, R. L. Pearson, and C. A. Cupas, *J. Am. Chem. Soc.*, **102**, 3507 (1980).
11. H. Suzuki and T. Mori, *J. Chem. Soc., Perkin Trans. 2*, 677 (1996); N. Noryama, T. Mori, and H. Suzuki, *Russ. J. Org. Chem.*, **34**, 1521 (1998).
12. H. Suzuki, T. Takeuchi, and T. Mori, *J. Org. Chem.*, **61**, 5944 (1996).

Scheme 11.1. Aromatic Nitration

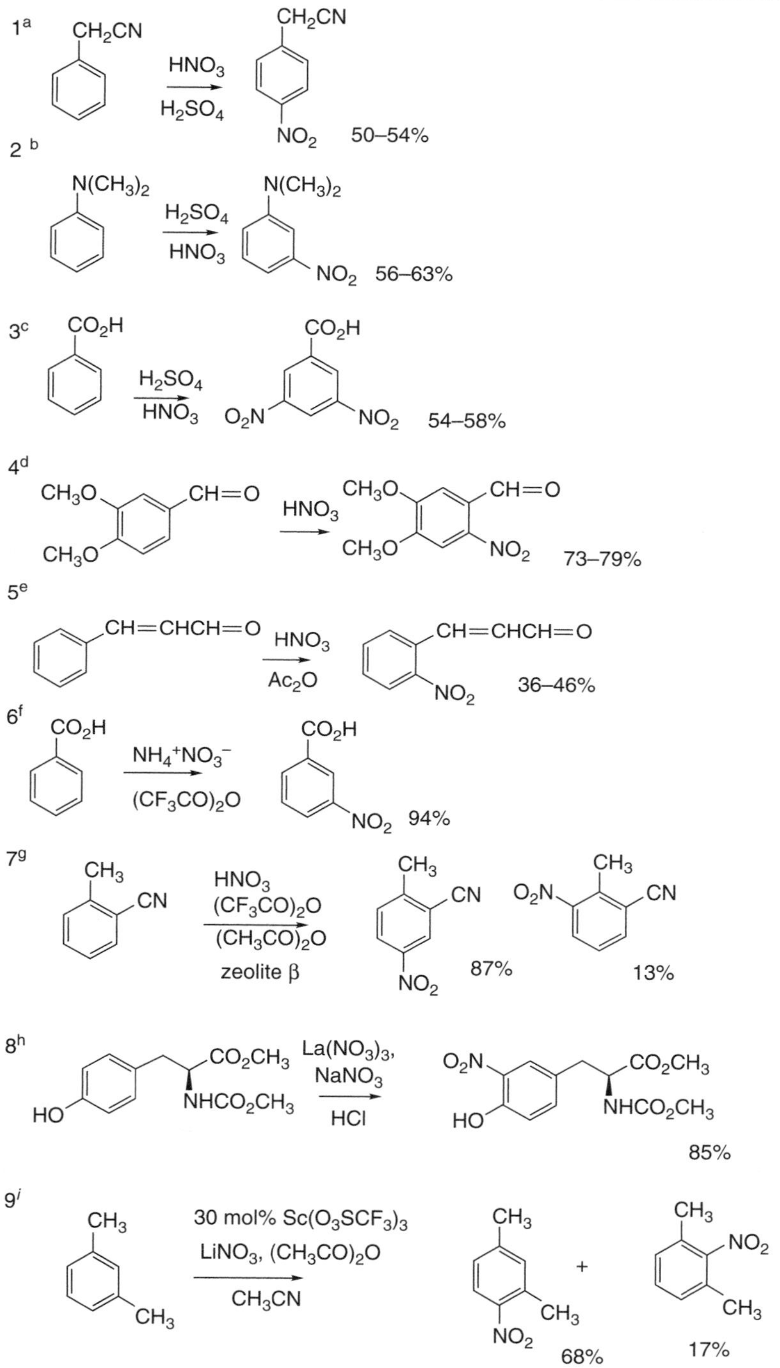

(Continued)

Scheme 11.1. (*Continued*)

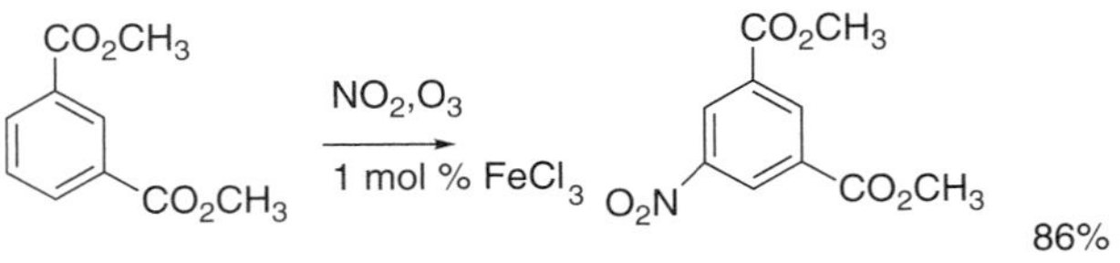

a. G. R. Robertson, *Org. Synth.*, **I**, 389 (1932).
b. H. M. Fitch, *Org. Synth.*, **III**, 658 (1955).
c. R. Q. Brewster, B. Williams, and R. Phillips, *Org. Synth.*, **III**, 337 (1955).
d. C. A. Fetscher, *Org. Synth.*, **IV**, 735 (1963).
e. R. E. Buckles and M. P. Bellis, *Org. Synth.*, **IV**, 722 (1963).
f. J. V. Crievello, *J. Org. Chem.*, **46**, 3056 (1981).
g. K. Smith, T. Gibbins, R. W. Millar, and R. P. Claridge, *J. Chem. Soc., Perkin Trans. 1*, 2753 (2000).
h. D. Ma and W. Tang, *Tetrahedron Lett.*, **39**, 7369 (1998).
i. A. Kawada, S. Takeda, K. Yamashita, H. Abe, and T. Harayama, *Chem. Pharm. Bull.*, **50**, 1060 (2002).
j. C. L. Dwyer and C. W. Holzpael, *Tetrahedron*, **54**, 7843 (1998).
k. C. A. Cupas and R. L. Pearson, *J. Am. Chem. Soc.*, **90**, 4742 (1968).
l. M. Nose, H. Suzuki, and H. Suzuki, *J. Org. Chem.*, **66**, 4356 (2001).

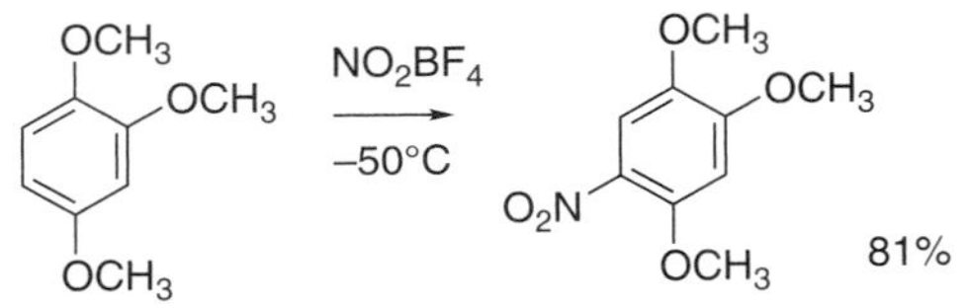

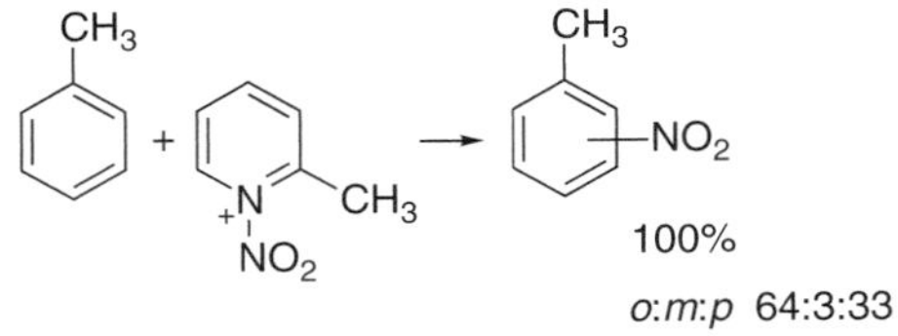

11.1.2. Halogenation

The introduction of the halogens onto aromatic rings by electrophilic substitution is an important synthetic procedure. Chlorine and bromine are reactive toward aromatic hydrocarbons, but Lewis acid catalysts are normally needed to achieve desirable rates. Elemental fluorine reacts very exothermically and careful control of conditions is required. Molecular iodine can effect substitution only on very reactive aromatics, but a number of more reactive iodination reagents have been developed.

Rate studies show that chlorination is subject to acid catalysis, although the kinetics are frequently complex.[13] The proton is believed to assist Cl–Cl bond breaking in a reactant-Cl_2 complex. Chlorination is much more rapid in polar than in nonpolar solvents.[14] Bromination exhibits similar mechanistic features.

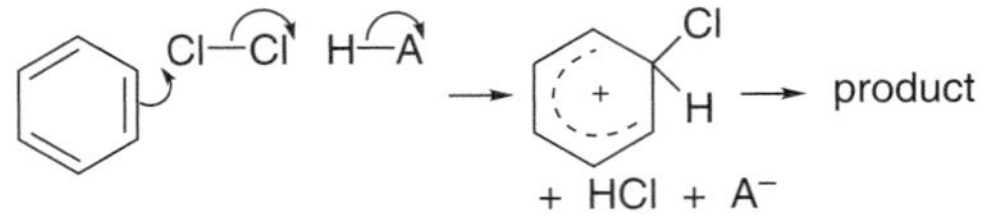

13. L. M. Stock and F. W. Baker, *J. Am. Chem. Soc.*, **84**, 1661 (1962); L. J. Andrews and R. M. Keefer, *J. Am. Chem. Soc.*, **81**, 1063 (1959); R. M. Keefer and L. J. Andrews, *J. Am. Chem. Soc.*, **82**, 4547 (1960); L. J. Andrews and R. M. Keefer, *J. Am. Chem. Soc.*, **79**, 5169 (1957).
14. L. M. Stock and A. Himoe, *J. Am. Chem. Soc.*, **83**, 4605 (1961).

For preparative reactions, Lewis acid catalysts are used. Zinc chloride or ferric chloride can be used in chlorination, and metallic iron, which generates ferric bromide, is often used in bromination. The Lewis acid facilitates cleavage of the halogen-halogen bond.

N-Bromosuccinimide (NBS) and N-chlorosuccinimide (NCS) are alternative halogenating agents. Activated aromatics, such as 1,2,4-trimethoxybenzene, are brominated by NBS at room temperature.[15] Both NCS and NBS can halogenate moderately active aromatics in nonpolar solvents by using HCl[16] or HClO$_4$[17] as a catalyst. Many other "positive halogen" compounds can act as halogenating agents. (See Table 4.2 for examples of such reagents.)

A wide variety of aromatic compounds can be brominated. Highly reactive ones, such as anilines and phenols, may undergo bromination at all activated positions. More selective reagents such as pyridinium bromide perbromide or tetraalkylammonium tribromides can be used in such cases.[18] Moderately reactive compounds such as anilides, haloaromatics, and hydrocarbons can be readily brominated and the usual directing effects control the regiochemistry. Use of Lewis acid catalysts permits bromination of rings with deactivating substituents, such as nitro and cyano.

Halogenations are strongly catalyzed by mercuric acetate or trifluoroacetate. These conditions generate acyl hypohalites, which are the active halogenating agents. The trifluoroacetyl hypohalites are very reactive reagents. Even nitrobenzene, for example, is readily brominated by trifluoroacetyl hypobromite.[19]

$$Hg(O_2CR)_2 + X_2 \; \rightleftharpoons \; HgX(O_2CR) + RCO_2X$$

A solution of bromine in CCl$_4$ containing sulfuric acid and mercuric oxide is also a reactive brominating agent.[20]

Fluorination can be carried out using fluorine diluted with an inert gas. However, great care is necessary to avoid uncontrolled reaction.[21] Several other reagents have been devised that are capable of aromatic fluorination.[22] Acetyl hypofluorite can be prepared in situ from fluorine and sodium acetate.[23] This reagent effects fluorination

15. M. C. Carreno, J. L. Garcia Ruano, G. Sanz, M. A. Toledo, and A. Urbano, *J. Org. Chem.*, **60**, 5328 (1995).

16. B. Andersh, D. L. Murphy, and R. J. Olson, *Synth. Commun.*, **30**, 2091 (2000).

17. Y. Goldberg and H. Alper, *J. Org. Chem.*, **58**, 3072 (1993).

18. W. P. Reeves and R. M. King, II, *Synth. Commun.*, **23**, 855 (1993); J. Berthelot, C. Guette, P. L. Desbene, and J. J. Basselier, *Can. J. Chem.*, **67**, 2061 (1989); S. Kajgaeshi, T. Kakinami, T. Inoue, M. Kondo, H. Nakamura, M. Fujikawa, and T. Okamoto, *Bull. Chem. Soc. Jpn.*, **61**, 597 (1988); S. Kajigaeshi, T. Kakinami, T. Yamasaki, S. Fujisaki, M. Fujikawa, and T. Okamoto, *Bull. Chem. Soc. Jpn.*, **61**, 2681 (1988); S. Gervat, E. Leonel, J.-Y. Barraud, and V. Ratovelomanana, *Tetrahedron Lett.*, **34**, 2115 (1993). M. K. Chaudahuri, A. J. Khan, B. K. Patel, D. Dey, W. Kharmawophlang, T. R. Lakshimprabha, and G. C. Mandal, *Tetrahedron Lett.*, **39**, 8163 (1998).

19. J. R. Barnett, L. J. Andrews, and R. M. Keefer, *J. Am. Chem. Soc.*, **94**, 6129 (1972).

20. S. A. Khan, M. A. Munawar, and M. Siddiq, *J. Org. Chem.*, **53**, 1799 (1988).

21. F. Cacace, P. Giacomello, and A. P. Wolf, *J. Am. Chem. Soc.*, **102**, 3511 (1980).

22. S. T. Purrington, B. S. Kagan, and T. B. Patrick, *Chem. Rev.*, **86**, 997 (1986).

23. O. Lerman, Y. Tor, and S. Rozen, *J. Org. Chem.*, **46**, 4629 (1981); O. Lerman, Y. Tor, D. Hebel, and S. Rozen, *J. Org. Chem.*, **49**, 806 (1984); G. W. M. Visser, C. N. M. Bakker, B. W. v. Halteren, J. D. M. Herscheid, G. A. Brinkman, and A. Hoekstra, *J. Org. Chem.*, **51**, 1886 (1986).

of activated aromatics. Although this procedure does not avoid the special precautions necessary for manipulation of elemental fluorine, it does provide a system with much greater selectivity. Acetyl hypofluorite shows a strong preference for *o*-fluorination of alkoxy and acetamido-substituted rings. *N*-Fluoro-*bis*-(trifluoromethansulfonyl)amine (*N*-fluorotriflimide) displays similar reactivity and can fluorinate benzene and activated aromatics.[24]

Several *N*-fluoro derivatives of 1,4-diazabicyclo[2.2.2]octane are useful for aromatic fluorination.[25]

Iodinations can be carried out by mixtures of iodine and various oxidants such as periodic acid,[26] I_2O_5,[27] NO_2,[28] and $Ce(NH_3)_2(NO_3)_6$.[29] A mixture of cuprous iodide and a cupric salt can also effect iodination.[30]

Iodination of moderately reactive aromatics can be effected by mixtures of iodine and silver or mercuric salts.[31] Hypoiodites are presumably the active iodinating species. *Bis*-(pyridine)iodonium salts can iodinate benzene and activated derivatives in the presence of strong acids such as HBF_4 or CF_3SO_3H.[32]

Scheme 11.2 shows some representative halogenation reactions. Entries 1 and 2 involve Lewis acid–catalyzed chlorination. Entry 3 is an acid-catalyzed chlorination using NCS as the reagent. Entry 4 shows a high-yield chlorination of acetanilide by *t*-butyl hypochlorite. This seems to be an especially facile reaction, since anisole is not chlorinated under these conditions, and may involve the *N*-chloroamide as an intermediate. Entry 5 describes a large-scale chlorination done with NCS. The product was used for the synthesis of sulamserod, a drug candidate.

24. S. Singh, D. D. DesMarteau, S. S. Zuberi, M. Whitz, and H.-N. Huang, *J. Am. Chem. Soc.*, **109**, 7194 (1987).

25. T. Shamma, H. Buchholz, G. K. S. Prakash, and G. A. Olahn, *Israel J. Chem.*, **39**, 207 (1999); A. J. Poss and G. A. Shia, *Tetrahedron Lett.*, **40**, 2673 (1999); T. Umemoto and M. Nagayoshi, *Bull. Chem. Soc. Jpn.*, **69**, 2287 (1996).

26. H. Suzuki, *Org. Synth.*, **VI**, 700, (1988).

27. L. C. Brazdil and C. J. Cutler, *J. Org. Chem.*, **61**, 9621 (1996).

28. Y. Noda and M. Kashima, *Tetrahedron Lett.*, **38**, 6225 (1997).

29. T. Sugiyama, *Bull. Chem. Soc. Jpn.*, **54**, 2847 (1981).

30. W. C. Baird, Jr., and J. H. Surridge, *J. Org. Chem.*, **35**, 3436 (1970).

31. Y. Kobayashi, I. Kumadaki, and T. Yoshida, *J. Chem. Res.* (Synopses), 215 (1977); R. N. Hazeldine and A. G. Sharpe, *J. Chem. Soc.*, 993 (1952); W. Minnis, *Org. Synth.*, **II**, 357 (1943); D. E. Janssen and C. V. Wilson, *Org. Synth.*, **IV**, 547 (1963); N.-W. Sy and B. A. Lodge, *Tetrahedron Lett.*, **30**, 3769 (1989).

32. J. Barluenga, J. M. Gonzalez, M. A. Garcia-Martin, P. J. Campos, and G. Asensio, *J. Org. Chem.*, **58**, 2058 (1993).

Scheme 11.2. Aromatic Halogenation

A. Chlorination

1[a]

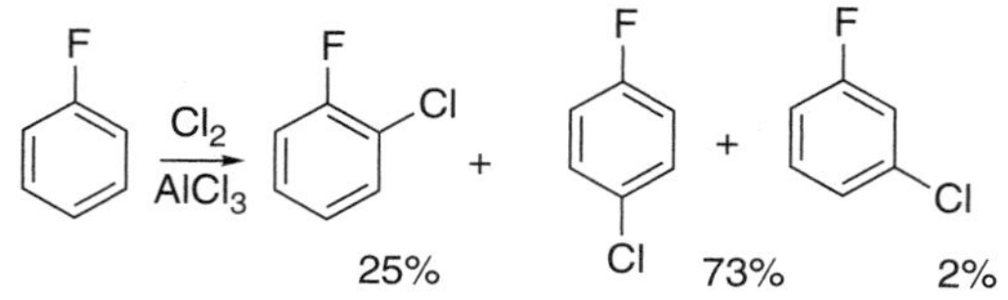

25%   73%   2%

2[b]

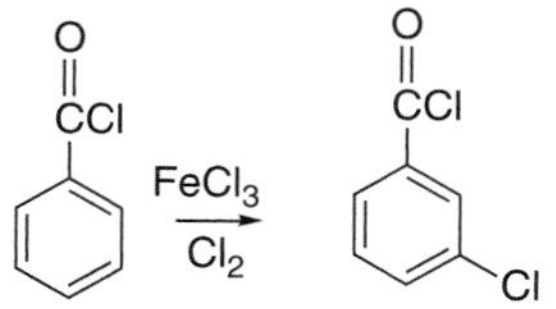

3[c]

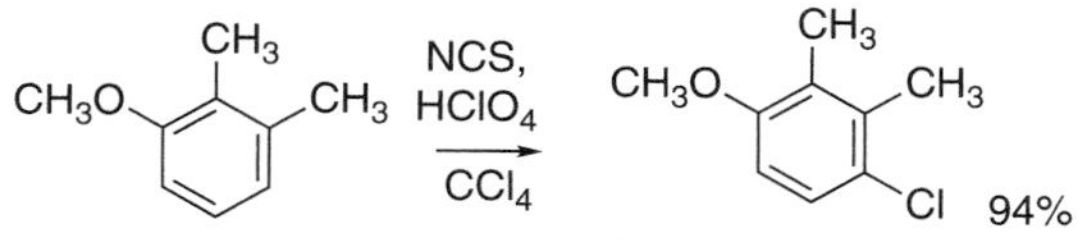

94%

4[d]

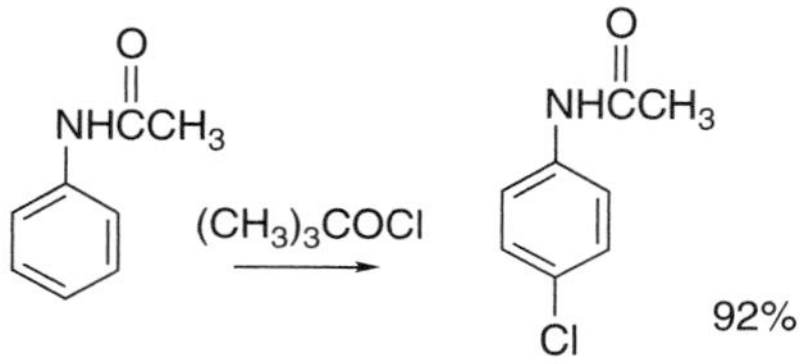

92%

5[e]

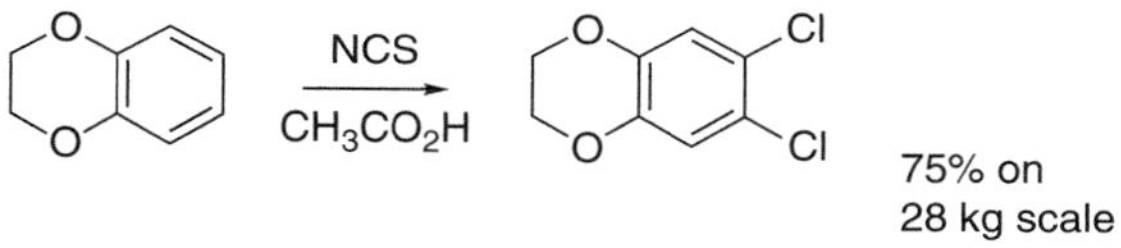

75% on
28 kg scale

B. Bromination

6[f]

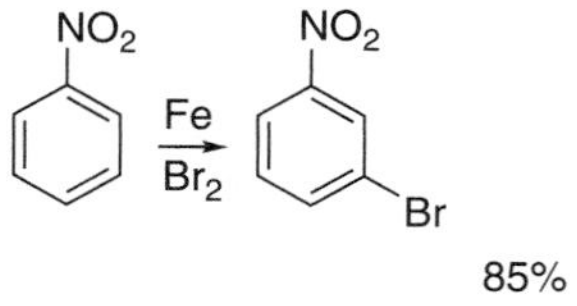

85%

7[g]

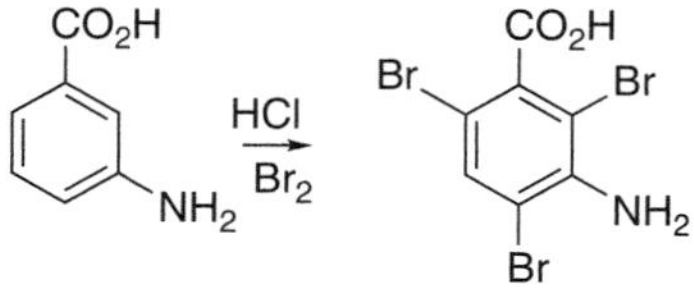

8[h]

98%

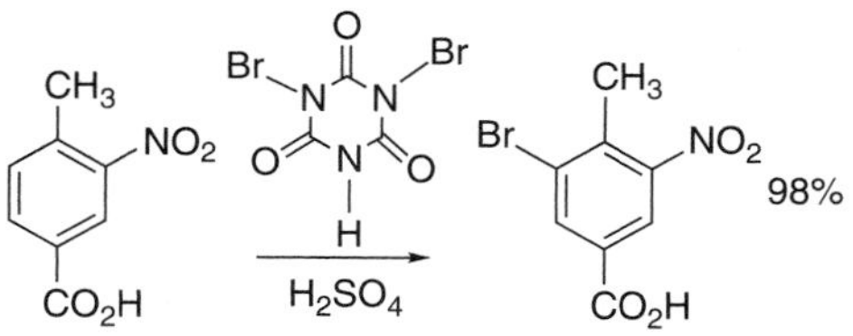

(Continued)

Scheme 11.2. (*Continued*)

9[i]

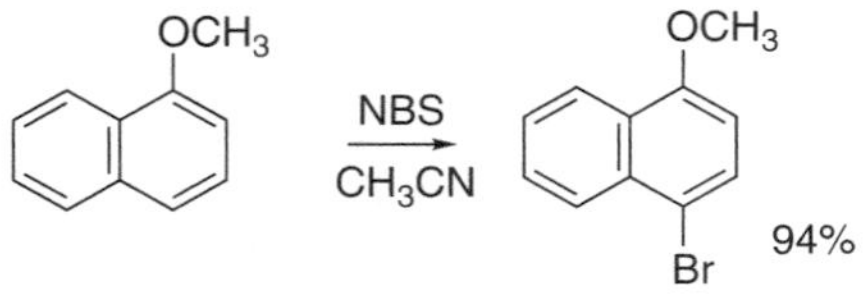

$$\xrightarrow[\text{CH}_3\text{CN}]{\text{NBS}}$$

94%

10[j]

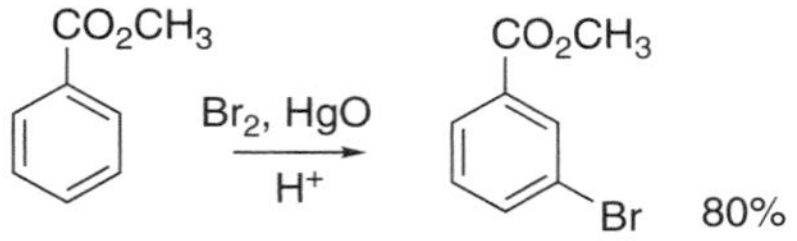

$$\xrightarrow[\text{H}^+]{\text{Br}_2, \text{HgO}}$$

80%

11[k]

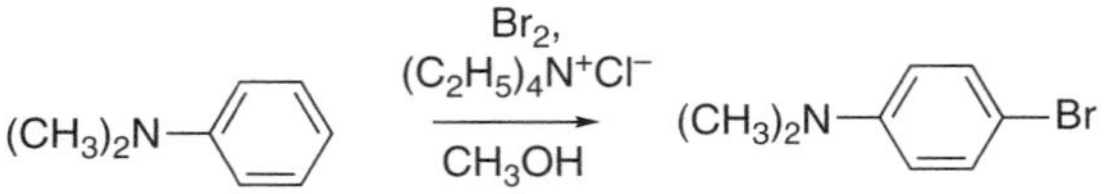

$$(\text{CH}_3)_2\text{N}\text{—}\text{C}_6\text{H}_5 \xrightarrow[\text{CH}_3\text{OH}]{\begin{array}{c}\text{Br}_2,\\(\text{C}_2\text{H}_5)_4\text{N}^+\text{Cl}^-\end{array}} (\text{CH}_3)_2\text{N}\text{—}\text{C}_6\text{H}_4\text{—Br}$$

C. Iodination

12[l]

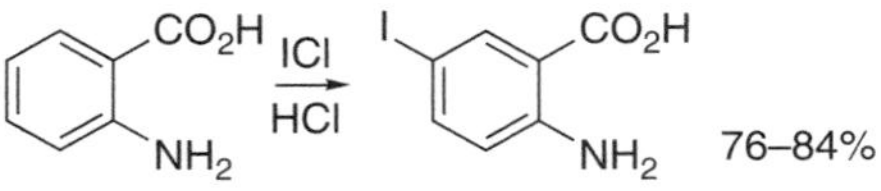

$$\xrightarrow[\text{HCl}]{\text{ICl}}$$

76–84%

13[m]

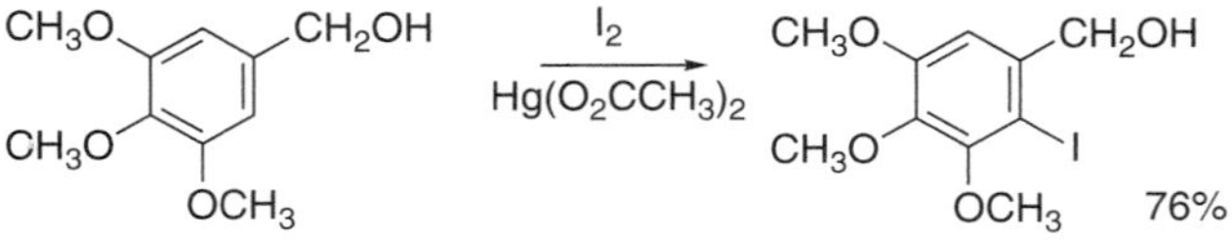

$$\xrightarrow[\text{Hg}(\text{O}_2\text{CCH}_3)_2]{\text{I}_2}$$

76%

14[n]

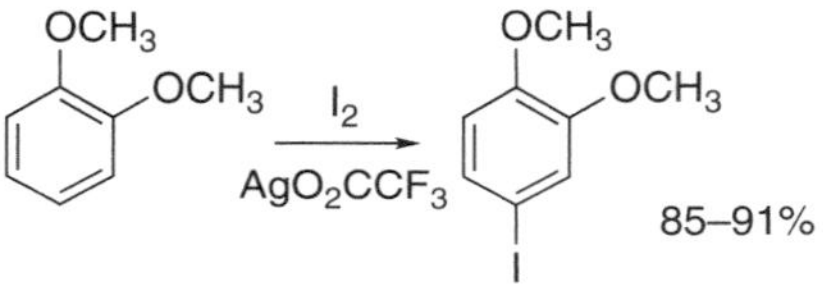

$$\xrightarrow[\text{AgO}_2\text{CCF}_3]{\text{I}_2}$$

85–91%

15[o]

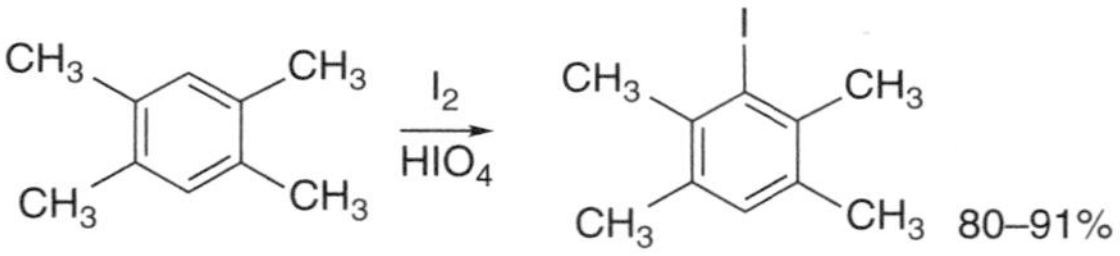

$$\xrightarrow[\text{HIO}_4]{\text{I}_2}$$

80–91%

16[p]

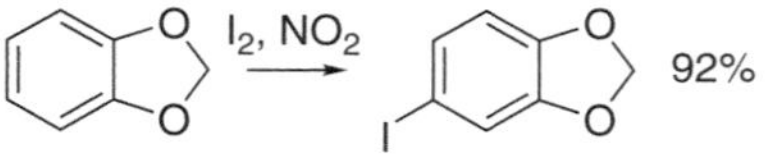

$$\xrightarrow{\text{I}_2, \text{NO}_2}$$

92%

17[q]

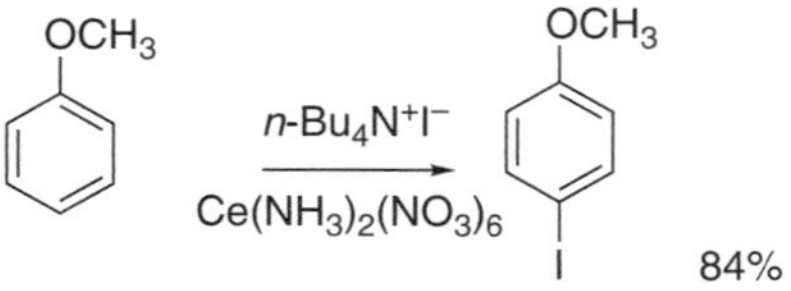

$$\xrightarrow[\text{Ce}(\text{NH}_3)_2(\text{NO}_3)_6]{n\text{-Bu}_4\text{N}^+\text{I}^-}$$

84%

18[r]

$$\text{Br}\text{—}\text{C}_6\text{H}_5 \xrightarrow[\text{CF}_3\text{SO}_3\text{H}]{\text{I}^+(\text{pyridine})_2} \text{Br}\text{—}\text{C}_6\text{H}_4\text{—I}$$

90%

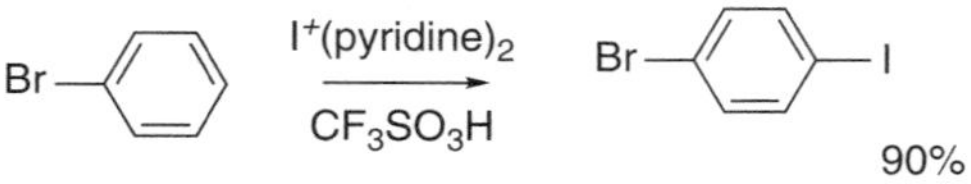

(*Continued*)

D. Fluorination

19[s]

60%

20[t]

21%   24%   2.4%

a. G. A. Olah, S. J. Kuhn, and B. A. Hardi, *J. Am. Chem. Soc.*, **86**, 1055 (1964).
b. E. Hope and G. F. Riley, *J. Chem. Soc.*, **121**, 2510 (1922).
c. V. Goldberg and H. Alper, *J. Org. Chem.*, **58**, 3072 (1993).
d. I. Lengyel, V. Cesare, and R. Stephani, *Synth. Commun.*, **28**, 1891 (1998).
e. B. A. Kowalczyk, J. Robinson, III, and J. O. Gardner, *Org. Proc. Res. Dev.*, **5**, 116 (2001).
f. J. R. Johnson and C. G. Gauerke, *Org. Synth.*, **I**, 123 (1941).
g. M. M. Robison and B. L. Robison, *Org. Synth.*, **IV**, 947 (1963).
h. A. R. Leed, S. D. Boettger, and B. Ganem, *J. Org. Chem.*, **45**, 1098 (1980).
i. M. C. Carreno, J. L. Garcia Russo, G. Sanz, M. A. Toledo, and A Urbano, *J. Org. Chem.*, **60**, 5328 (1995).
j. S. A. Khan, M. A. Munawar, and M. Siddiq, *J. Org. Chem.*, **53**, 1799 (1988).
k. S. Gervat, E. Leonel, J.-Y. Barraud, and V. Ratovelomanana, *Tetrahedron Lett.*, **34**, 2115 (1993).
l. V. H. Wallingford and P. A. Krueger, *Org. Synth.*, **II**, 349 (1943).
m. F. E. Ziegler and J. A. Schwartz, *J. Org. Chem.*, **43**, 985 (1978).
n. D. E. Janssen and C. V. Wilson, *Org. Synth.*, **IV**, 547 (1963).
o. H. Suzuki, *Org. Synth.*, **51**, 94 (1971).
p. Y. Noda and M. Kashima, *Tetrahedron Lett.*, **38**, 6225 (1997).
q. T. Sugiyama, *Bull. Chem. Soc. Jpn.*, **54**, 2847 (1981).
r. J. Barluenga, J. M. Gonzalez, M. A. Garcia-Martin, P. J. Campos, and G. Asensio, *J. Org. Chem.*, **58**, 2058 (1993).
s. M. A. Tius, J. K. Kawakami, W. A. G. Hill, and A. Makriyannis, *J. Chem. Soc., Chem. Commun.*, 2085 (1996).
t. T. Umemoto and M. Nagayoshi, *Bull. Chem. Soc. Jpn.*, **69**, 2287 (1996).

Entry 6 is a case of *meta* bromination of a deactivated aromatic. Entry 7 is a case in which all activated positions are brominated. It is interesting that the reaction occurs in acidic solution. It may be that each successive bromine addition accelerates the reaction by decreasing the basicity of the aniline and increasing the amount that is present in the neutral form. Entry 8 employs dibromoisocyanuric acid in concentrated H_2SO_4 as a brominating reagent. These conditions have been found useful for unreactive aromatics. Entry 9 is an example of bromination using NBS. Entry 10 uses bromine and mercuric oxide under conditions that were found effective for deactivated aromatics. Entry 11 describes conditions that are applicable for bromination of anilines. It is suggested that the reaction may involve formation of methyl hypobromite as the active bromination reagent.

Entries 12 to 18 show iodinations under various conditions. The reaction in Entry 12, using iodine monochloride, is done in concentrated HCl, but presumably occurs through the neutral form of the reactant ($pK_1 = 2.17$). Entries 13 and 14 involve reactions activated by mercuric and silver salts, respectively, and probably involve the hypoiodites as the active reagents. Entry 15 uses iodine and periodic acid, a reagent combination that was found effective for moderately activated aromatics. The I_2-NO_2 combination illustrated in Entry 16 is also applicable to activated species. Entry 17 illustrates an oxidative procedure that can be used with moderately activated aromatics such as the methyl and methoxy derivatives of benzene. The *bis*-pyridine-iodonium reagent shown in Entry 18 was used with two equivalents of a strong acid, either HBF_4 or CF_3SO_3H, in dichloromethane. These conditions were applicable even to deactivated aromatics, such as methyl benzoate and nitrobenzene.

Entries 19 and 20 are fluorinations. In Entry 19, the fluorination is on an activated ring in the antinausea drug nabilone. Entry 20 illustrates the use N,N'-difluoro-1,4-diazabicyclo[2.2.2]octane ditriflate.

11.1.3. Friedel-Crafts Alkylation

Friedel-Crafts alkylation reactions are an important method for introducing carbon substituents on aromatic rings. The reactive electrophiles can be either discrete carbocations or polarized complexes that contain a reactive leaving group. Various combinations of reagents can be used to generate alkylating species. Alkylations usually involve alkyl halides and Lewis acids or reactions of alcohols or alkenes with strong acids.

$$R-X + AlCl_3 \rightleftharpoons R-\overset{+}{X}-\bar{A}lCl_3 \rightleftharpoons R^+ + X\bar{A}lCl_3$$

$$R-OH + H^+ \rightleftharpoons R-\overset{+}{\underset{H}{O}}H \rightleftharpoons R^+ + H_2O$$

$$RCH=CH_2 + H^+ \rightleftharpoons R\overset{+}{C}HCH_3$$

Owing to the involvement of carbocations, Friedel-Crafts alkylations can be accompanied by rearrangement of the alkylating group. For example, isopropyl groups are often introduced when *n*-propyl reactants are used.[33]

Similarly, under a variety of reaction conditions, alkylation of benzene with either 2-chloro or 3-chloropentane gives rise to a mixture of both 2-pentyl- and 3-pentylbenzene.[34]

Rearrangement can also occur after the initial alkylation. The reaction of 2-chloro-2-methylbutane with benzene is an example of this behavior.[35] With relatively mild Friedel-Crafts catalysts such as BF_3 or $FeCl_3$, the main product is **1**. With $AlCl_3$, equilibration of **1** and **2** occurs and the equilibrium favors **2**. The rearrangement is the result of product equilibration via reversibly formed carbocations.

33. S. H. Sharman, *J. Am. Chem. Soc.*, **84**, 2945 (1962).

34. R. M. Roberts, S. E. McGuire, and J. R. Baker, *J. Org. Chem.*, **41**, 659 (1976).

35. A. A. Khalaf and R. M. Roberts, *J. Org. Chem.*, **35**, 3717 (1970); R. M. Roberts and S. E. McGuire, *J. Org. Chem.*, **35**, 102 (1970).

Alkyl groups can also migrate from one position to another on the ring.[36] Such migrations are also thermodynamically controlled and proceed in the direction of minimizing steric interactions between substituents.

The relative reactivity of Friedel-Crafts catalysts has not been described in a quantitative way, but comparative studies using a series of benzyl halides has resulted in the qualitative groupings shown in Table 11.1. Proper choice of catalyst can minimize subsequent product equilibrations.

The Friedel-Crafts alkylation reaction does not proceed successfully with aromatic reactants having EWG substituents. Another limitation is that each alkyl group that is introduced *increases the reactivity of the ring toward further substitution*, so polyalkylation can be a problem. Polyalkylation can be minimized by using the aromatic reactant in excess.

Apart from the alkyl halide–Lewis acid combination, two other sources of carbocations are often used in Friedel-Crafts reactions. Alcohols can serve as carbocation precursors in strong acids such as sulfuric or phosphoric acid. Alkylation can also be effected by alcohols in combination with BF_3 or $AlCl_3$.[37] Alkenes can serve as alkylating agents when a protic acid, especially H_2SO_4, H_3PO_4, and HF, or a Lewis acid, such as BF_3 and $AlCl_3$, is used as a catalyst.[38]

Stabilized carbocations can be generated from allylic and benzylic alcohols by reaction with $Sc(O_3SCF_3)_3$ and results in formation of alkylation products from benzene and activated derivatives.[39]

Table 11.1. Relative Activity of Friedel-Crafts Catalysts[a]

Very active	Moderately active	Mild
$AlCl_3$, $AlBr_3$, $GaCl_3$, $GaCl_2$, SbF_5, $MoCl_5$,	$InCl_3$, $InBr_3$, $SbCl_4$, $FeCl_3$, $AlCl_3-CH_3NO_2$, $SbF_5-CH_3NO_2$	BCl_3, $SnCl_4$, $TiCl_4$, $TiBr_4$, $FeCl_2$

a. G. A. Olah, S. Kobayashi, and M. Tashiro, *J. Am. Chem. Soc.*, **94**, 7448 (1972).

36. R. M. Roberts and D. Shiengthong, *J. Am. Chem. Soc.*, **86**, 2851 (1964).
37. A. Schriesheim, in *Friedel-Crafts and Related Reactions*, Vol. II, G. Olah, ed., Interscience, New York, 1964, Chap. XVIII.
38. S. H. Patinkin and B. S. Friedman, in *Friedel-Crafts and Related Reactions*, Vol. II, G. Olah, ed., Interscience, New York, 1964, Chap. XIV.
39. T. Tsuchimoto, K. Tobita, T. Hiyama, and S. Fukuzawa, *Synlett*, 557 (1996); T. Tsuchimoto, K. Tobita, T. Hiyama, and S. Fukuzawa, *J. Org. Chem.*, **62**, 6997 (1997).

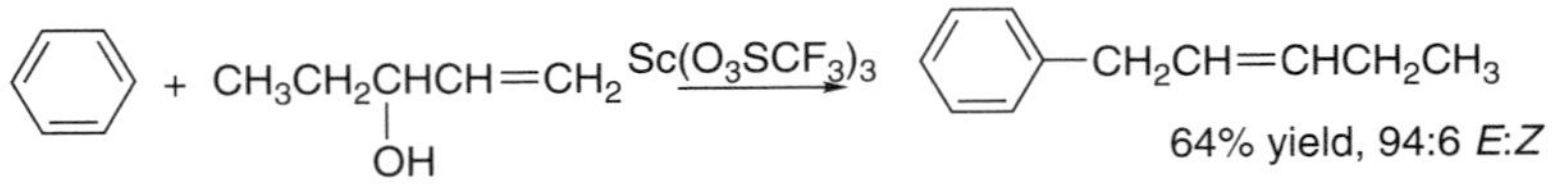

64% yield, 94:6 *E:Z*

This kind of reaction has been used to synthesize α-tocopherol, in a reaction that involves alkylation, followed by cyclization involving the phenyl hydroxy group.

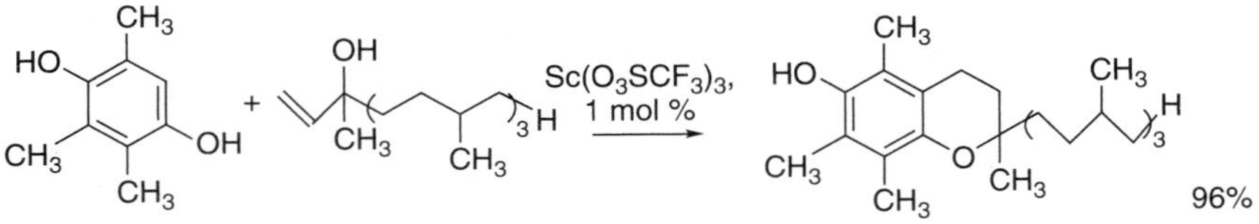

96%

Ref. 40

Methanesulfonate esters of secondary alcohols also give Friedel-Crafts products in the presence of $Sc(O_3SCF_3)_3$[41] or $Cu(O_3SCF_3)_2$.[42]

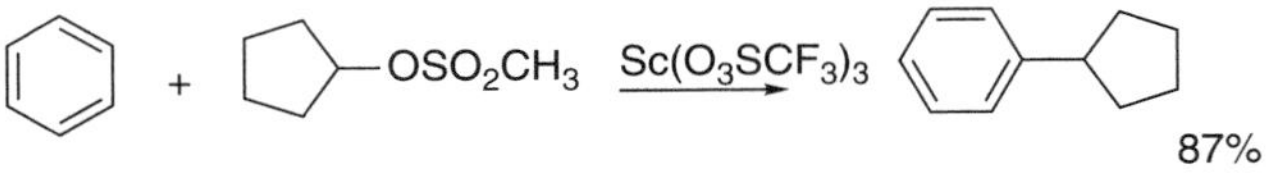

87%

Friedel-Crafts alkylation can occur intramolecularly to form a fused ring. Intramolecular Friedel-Crafts reactions provide an important method for constructing polycyclic hydrocarbon frameworks. It is somewhat easier to form six-membered than five-membered rings in such reactions. Thus, whereas 4-phenyl-1-butanol gives a 50% yield of a cyclized product in phosphoric acid, 3-phenyl-1-propanol is mainly dehydrated to alkenes.[43]

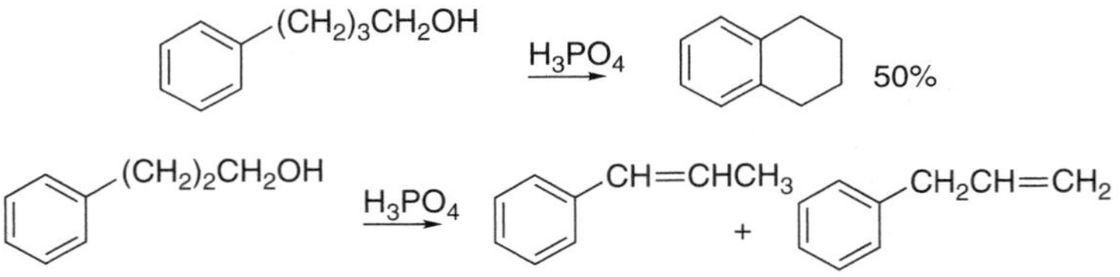

If a potential carbocation intermediate can undergo a hydride or alkyl shift, this shift occurs in preference to closure of the five-membered ring.

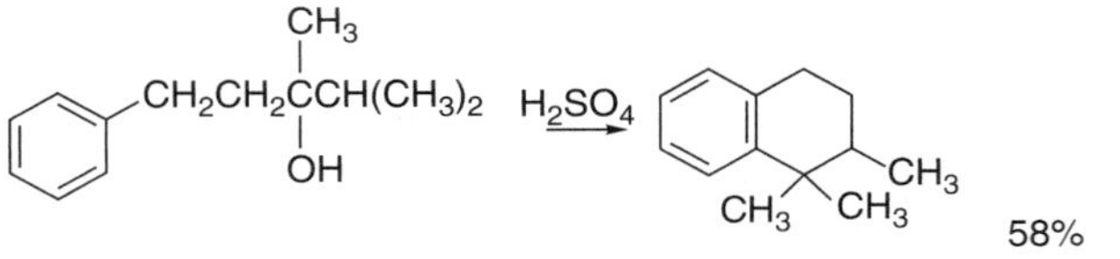

58%   Ref. 44

40. M. Matsui, N. Karibe, K. Hayashi, and H. Yamamoto, *Bull. Chem. Soc. Jpn.*, **68**, 3569 (1995).

41. H. Kotsuki, T. Ohishi, and M. Inoue, *Synlett*, 255 (1998); H. Kotsuki, T. Ohishi, M. Inoue, and T. Kojima, *Synthesis*, 603 (1999).

42. R. P. Singh, R. M. Kamble, K. L. Chandra, P. Saravaran, and V. K. Singh, *Tetrahedron*, **57**, 241 (2001).

43. A. A. Khalaf and R. M. Roberts, *J. Org. Chem.*, **34**, 3571 (1969).

44. A. A. Khalaf and R. M. Roberts, *J. Org. Chem.*, **37**, 4227 (1972).

These results reflect a rather general tendency for 6 > 5, 7 in ring closure by intramolecular Friedel-Crafts reactions.[44,45] The difficulty in forming five-membered rings may derive from steric and electronic factors. Some strain must develop because of the sp^2 carbons included in the ring. Perhaps more important is the need for approach perpendicular to the ring. With three of the five carbons coplanar, it is difficult to align the empty p orbital of the carbocation with the π system.

Scheme 11.3 gives some examples of both inter- and intramolecular Friedel-Crafts alkylations. Entry 1 is carried out using $AlCl_3$ in an excess of refluxing benzene. Entry 2 was also done using benzene as the solvent, but this reaction is done at 0°C. A tertiary carbocation is generated by protonation of the double bond. Entry 3 involves alkylation by both bromo substituents in the reactant. The reaction is carried out in excess benzene, using $AlBr_3$. Entry 4 demonstrates the ability of a typical aromatic sulfonic acid to generate a reactive carbocation by alkene protonation. The reaction was carried out in excess toluene at 105°C. Note the relatively weak position selectivity (see also Part A, Section 9.4.4). Secondary alkyl tosylates are also sources of reactive carbocations under these conditions.

Entries 5 to 7 show intramolecular reactions. Entry 5 is an example of formation of a polycyclic ring system. The product is a 3:1 mixture of β:α methyl isomers at the new ring junction, and reflects a preference for TS **A** over TS **B**.

Entry 6 involves formation of a stabilized benzylic carbocation and results in a very efficient closure of a six-membered ring. Entry 7 involves an activated ring. The reaction was done using enantiomerically pure alcohol, but, as expected for a carbocation intermediate, the product was nearly racemic (6% e.e.). This cyclization was done enantiospecifically by first forming the $Cr(CO)_3$ complex (see Section 8.5).

11.1.4. Friedel-Crafts Acylation

Friedel-Crafts acylation generally involves reaction of an acyl halide and Lewis acid such as $AlCl_3$, SbF_5, or BF_3. Bismuth(III) triflate is also a very active acylation catalyst.[46] Acid anhydrides can also be used in some cases. For example, a combination

45. R. J. Sundberg and J. P. Laurino, *J. Org. Chem.*, **49**, 249 (1984); S. R. Angle and M. S. Louie, *J. Org. Chem.*, **56**, 2853 (1991).
46. C. Le Roux and J. Dubac, *Synlett*, 181 (2002); J. R. Desmurs, M. Labrouillere, C. Le Roux, H. Gaspard, A. Laporterie, and J. Dubac, *Tetrahedron Lett.*, **38**, 8871 (1997); S. Repichet, C. LeRoux, J. Dubac, and J.-R. Desmurs, *Eur. J. Org. Chem.*, 2743 (1998).

Scheme 11.3. Friedel-Crafts Alkylation Reactions

A. Intermolecular reactions

1[a]

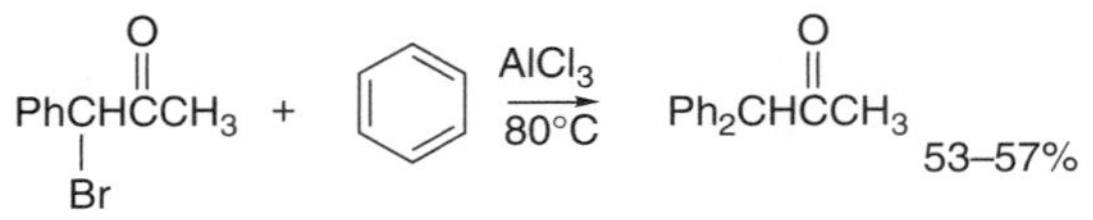

53–57%

2[b]

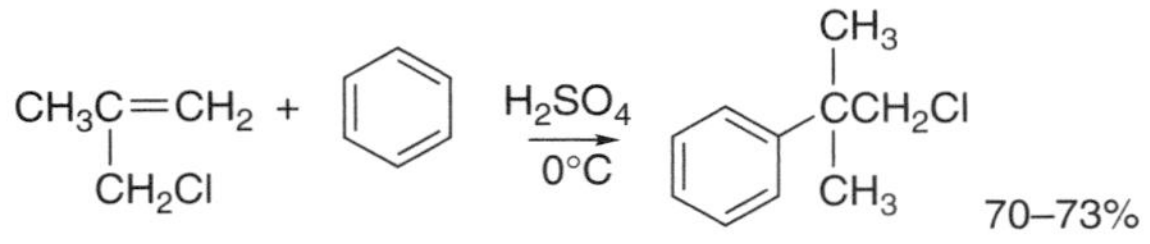

70–73%

3[c]

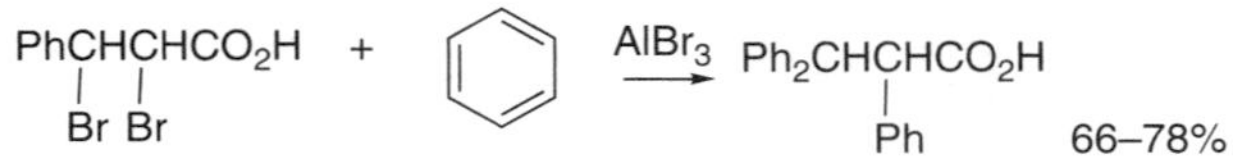

66–78%

4[d]

98% yield, *o:m:p* = 29:18:53

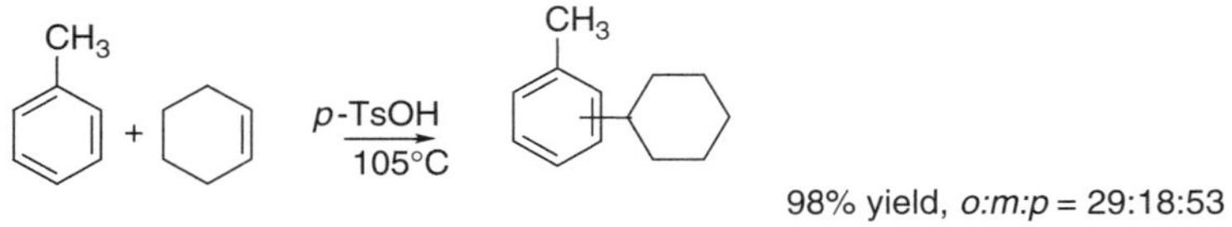

B. Intramolecular Friedel–Crafts cyclizations

5[e]

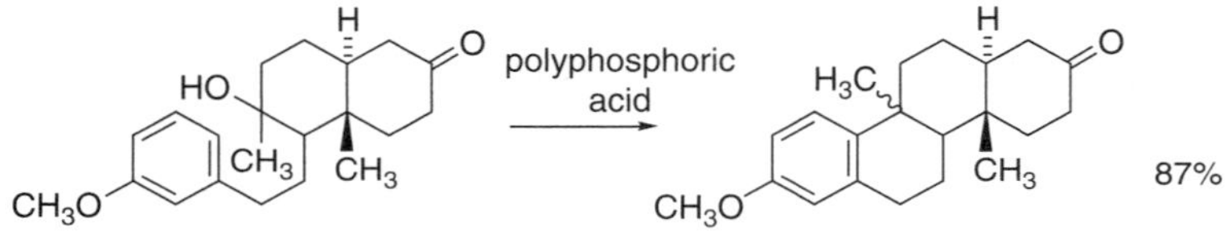

87%

6[f]

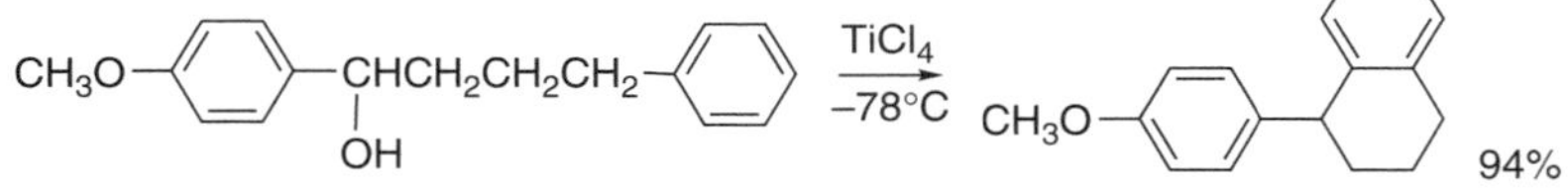

94%

7[g]

85%

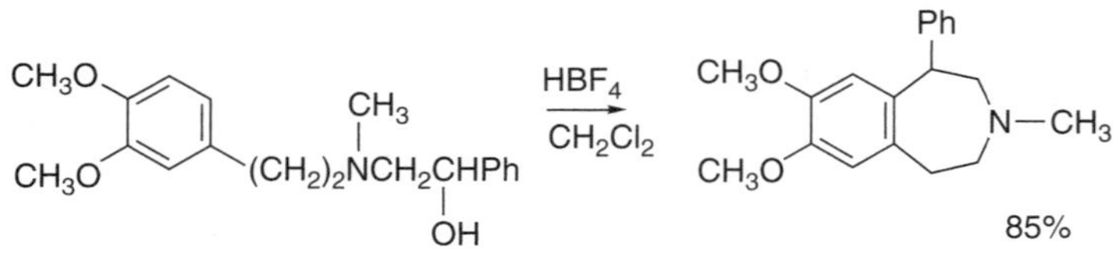

a. E. M. Shultz and S. Mickey, *Org. Synth.*, **III**, 343 (1955).
b. W. T. Smith, Jr., and J. T. Sellas, *Org. Synth.*, **IV**, 702 (1963).
c. C. P. Krimmol, L. E. Thielen, E. A. Brown, and W. J. Heidtke, *Org. Synth.*, **IV**, 960 (1963).
d. M. P. D. Mahindaratne and K. Wimalasena, *J. Org. Chem.*, **63**, 2858 (1998).
e. R. E. Ireland, S. W. Baldwin, and S. C. Welch, *J. Am. Chem. Soc.*, **94**, 2056 (1972).
f. S. R. Angle and M. S. Louie, *J. Org. Chem.*, **56**, 2853 (1991).
g. S. J. Coote, S. G. Davies, D. Middlemiss, and A. Naylor, *Tetrahedron Lett.*, **30**, 3581 (1989).

of hafnium(IV) triflate and $LiClO_4$ in nitromethane catalyzes acylation of moderately reactive aromatics by acetic anhydride.

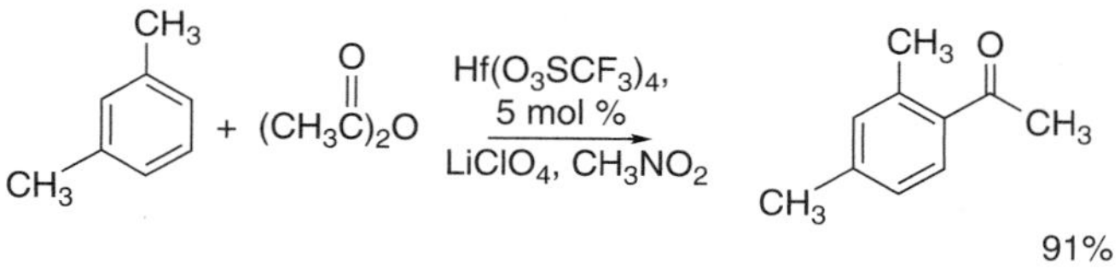

91%

Ref. 47

Mixed anhydrides with trifluoroacetic acid are particularly reactive acylating agents.[48] For example, Entry 5 in Scheme 11.4 shows the use of a mixed anhydride in the course of synthesis of the anticancer agent tamoxifen.

As in the alkylation reaction, the reactive intermediate in Friedel-Crafts acylation can be a dissociated acylium ion or a complex of the acyl chloride and Lewis acid.[49] Recent mechanistic studies have indicated that with benzene and slightly deactivated derivatives, it is the *protonated acylium ion* that is the kinetically dominant electrophile.[50]

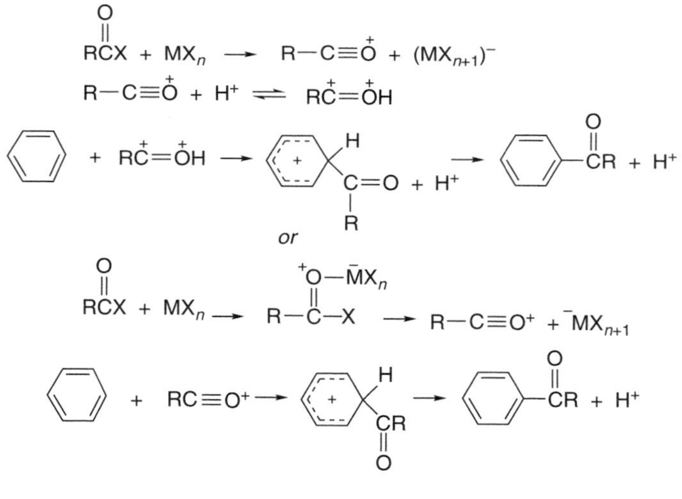

Regioselectivity in Friedel-Crafts acylations can be quite sensitive to the reaction solvent and other procedural variables.[51] In general, *para* attack predominates for

[47] I. Hachiya, M. Moriwaki, and S. Kobayashi, *Tetrahedron Lett.*, **36**, 409 (1995); A. Kawada, S. Mitamura, and S. Kobabyashi, *J. Chem. Soc., Chem. Commun.*, 183 (1996); I. Hachiya, M. Moriwaki, and S. Kobayashi, *Bull. Chem. Soc. Jpn.*, **68**, 2053 (1995).

[48] E. J. Bourne, M. Stacey, J. C. Tatlow, and J. M. Teddar, *J. Chem. Soc.*, 719 (1951); C. Galli, *Synthesis*, 303 (1979); B. C. Ranu, K. Ghosh, and U. Jana, *J. Org. Chem.*, **61**, 9546 (1996).

[49] F. R. Jensen and G. Goldman, in *Friedel-Crafts and Related Reactions*, Vol. III, G. Olah, ed., Interscience, New York, 1964, Chap. XXXVI.

[50] Y. Sato, M. Yato, T. Ohwada, S. Saito, and K. Shudo, *J. Am. Chem. Soc.*, **117**, 3037 (1995).

[51] For example, see L. Friedman and R. J. Honour, *J. Am. Chem. Soc.*, **91**, 6344 (1969).

alkylbenzenes.[52] The percentage of *ortho* attack increases with the electrophilicity of the acylium ion and as much as 50% *ortho* product is observed with the formylium and 2,4-dinitrobenzoylium ions.[53] Rearrangement of the acyl group is not a problem in Friedel-Craft acylation. Neither is polyacylation, because the first acyl group serves to deactivate the ring to further attack. For these reasons, it is often preferable to introduce primary alkyl groups by a sequence of acylation followed by reduction of the acyl group (see Section 5.7.1).

Intramolecular acylations are very common, and the normal conditions involving an acyl halide and Lewis acid can be utilized. One useful alternative is to dissolve the carboxylic acid in polyphosphoric acid (PPA) and heat to effect cyclization. This procedure probably involves formation of a mixed phosphoric-carboxylic anhydride.[54]

Cyclizations can also be carried out with an esterified oligomer of phosphoric acid called "polyphosphate ester," which is chloroform soluble.[55] Another reagent of this type is trimethylsilyl polyphosphate (Scheme 11.4, Entry 13).[56] Neat methanesulfonic acid is also an effective reagent for intramolecular Friedel-Crafts acylation (Scheme 11.4, Entry 14).[57]

A classical procedure for fusing a six-membered ring to an aromatic ring uses succinic anhydride or a derivative. An intermolecular acylation is followed by reduction and an intramolecular acylation. The reduction step is necessary to provide a more reactive ring for the second acylation.

Ref. 58

Scheme 11.4 shows some other representative Friedel-Crafts acylation reactions. Entries 1 and 2 show typical Friedel-Crafts acylation reactions using $AlCl_3$. Entries 3 and 4 are similar, but include some functionality in the acylating reagents. Entry 5 involves formation of a mixed trifluoroacetic anhydride, followed by acylation in 85% H_3PO_4. The reaction was conducted on a kilogram scale and provides a starting material for the synthesis of tamoxifen. Entry 6 illustrates the use of bismuth triflate as

52. H. C. Brown, G. Marino, and L. M. Stock, *J. Am. Chem. Soc.*, **81**, 3310 (1959); H. C. Brown and G. Marino, *J. Am. Chem. Soc.*, **81**, 5611 (1959); G. A. Olah, M. E. Moffatt, S. J. Kuhn, and B. A. Hardie, *J. Am. Chem. Soc.*, **86**, 2198 (1964).
53. G. A. Olah and S. Kobayashi, *J. Am. Chem. Soc.*, **93**, 6964 (1971).
54. W. E. Bachmann and W. J. Horton, *J. Am. Chem. Soc.*, **69**, 58 (1947).
55. Y. Kanaoka, O. Yonemitsu, K. Tanizawa, and Y. Ban, *Chem. Pharm. Bull.*, **12**, 773 (1964); T. Kametani, S. Takano, S. Hibino, and T. Terui, *J. Heterocycl. Chem.*, **6**, 49 (1969).
56. E. M. Berman and H. D. H. Showalter, *J. Org. Chem.*, **54**, 5642 (1989).
57. V. Premasagar, V. A. Palaniswamy, and E. J. Eisenbraun, *J. Org. Chem.*, **46**, 2974 (1981).
58. E. J. Eisenbraun, C. W. Hinman, J. M. Springer, J. W. Burnham, T. S. Chou, P. W. Flanagan, and M. C. Hamming, *J. Org. Chem.*, **36**, 2480 (1971).

Scheme 11.4. Friedel-Crafts Acylation Reactions

A. Intermolecular reactions

1[a]

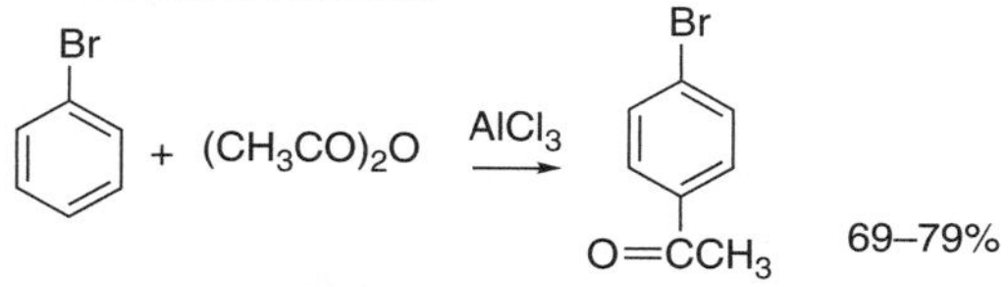

69–79%

2[b]

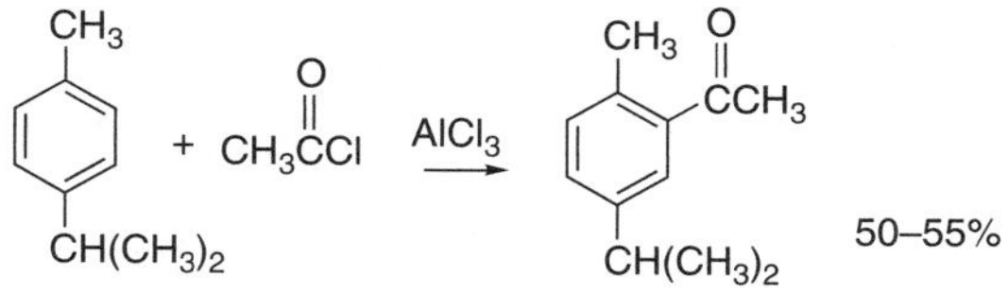

50–55%

3[c]

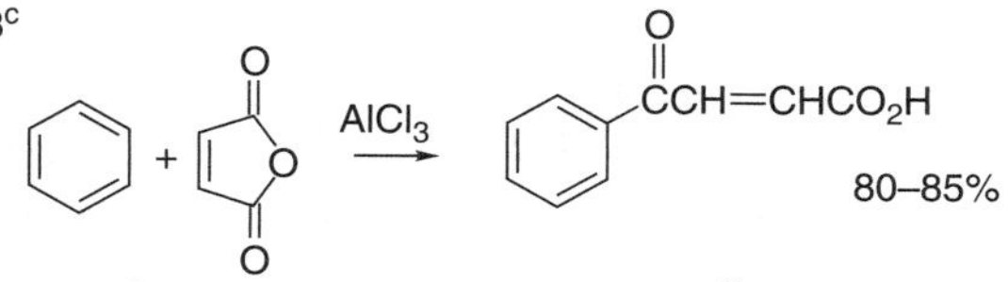

80–85%

4[d]

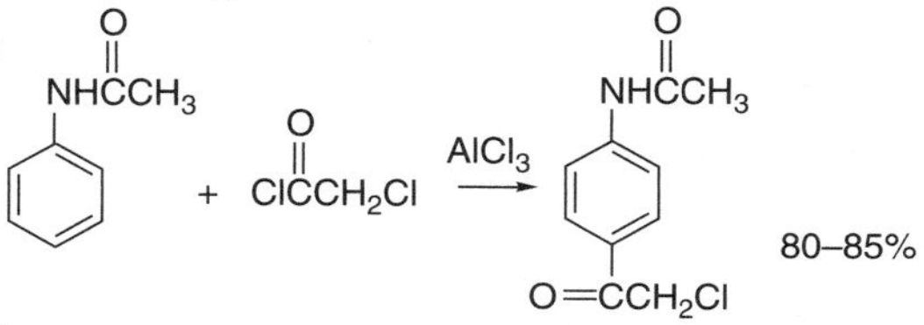

80–85%

5[e]

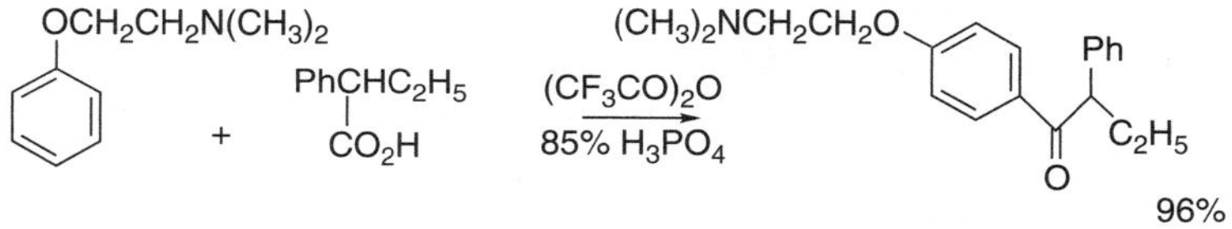

96%

6[f]

86%

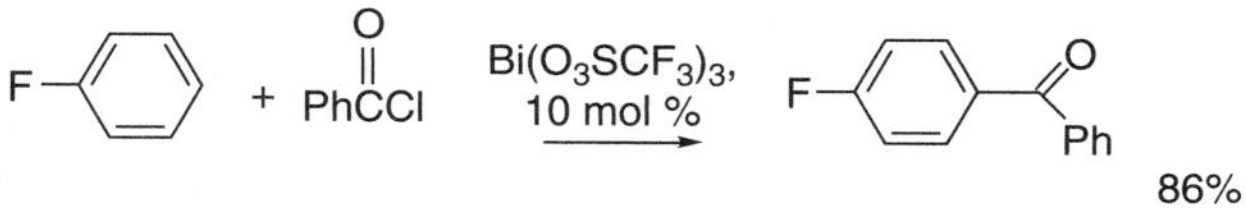

B. Intramolecular friedel–crafts acylations

7[g]

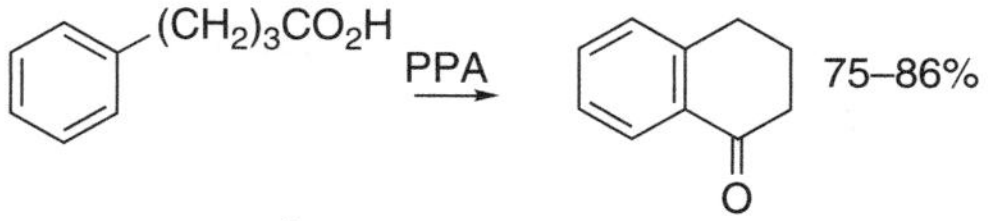

75–86%

8[h]

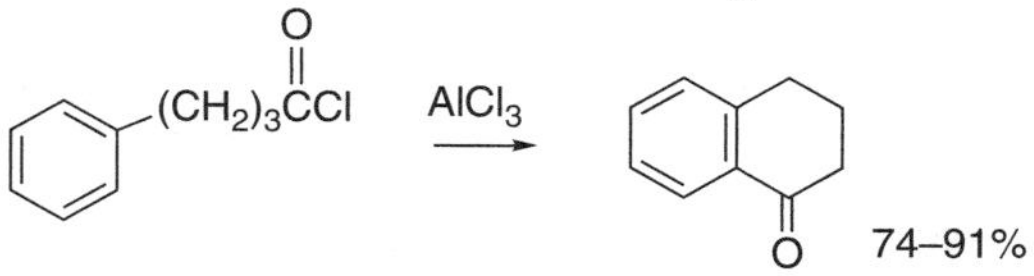

74–91%

9[i]

91–96%

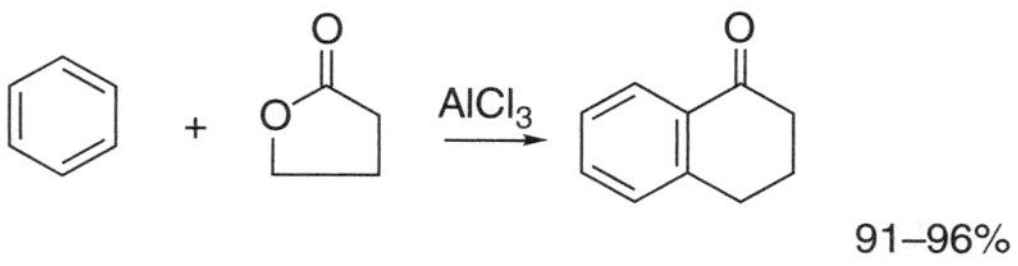

(Continued)

Scheme 11.4. (*Continued*)

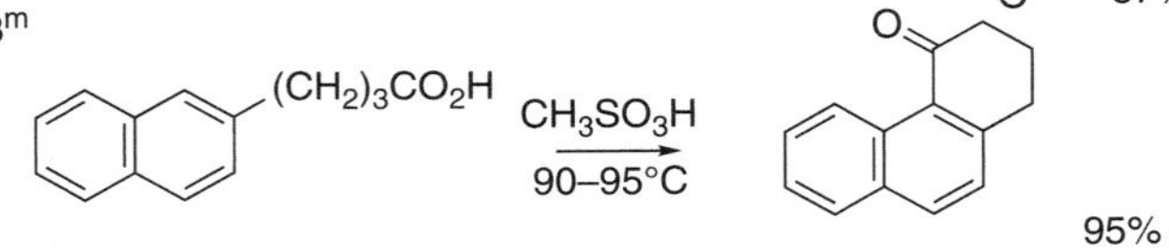

a. R. Adams and C. R. Noller, *Org. Synth.*, **I**, 109 (1941).
b. C. F. H. Allen, *Org. Synth.*, **II**, 3 (1943).
c. O. Grummitt, E. I. Becker, and C. Miesse, *Org. Synth.*, **III**, 109 (1955).
d. J. L. Leiserson and A. Weissberger, *Org. Synth.*, **III**, 183 (1955).
e. T. P. Smythe and B. W. Corby, *Org. Process Res. Dev.*, **1**, 264 (1997).
f. J. R. Desmurs, M. Labrouillere, C. Le Roux, H. Gaspard, A. Laporterie, and J. Dubac, *Tetrahedron Lett.*, **38**, 8871 (1997).
g. L. Arsnijevic, V. Arsenijevic, A. Horeua, and J. Jaques, *Org. Synth.*, **53**, 5 (1973).
h. E. L. Martin and L. F. Fieser, *Org. Synth.*, **II**, 569 (1943).
i. C. E. Olson and A. F. Bader, *Org. Synth.*, **IV**, 898 (1963).
j. M. B. Floyd and G. R. Allen, Jr., *J. Org. Chem.*, **35**, 2647 (1970).
k. M. C. Venuti, *J. Org. Chem.*, **46**, 3124 (1981).
l. G. Esteban, M. A. Lopez-Sanchez, E. Martinez, and J. Plumet, *Tetrahedron*, **54**, 197 (1998).
m. V. Premasagar, V. A. Palaniswamy, and E. J. Eisenbraun, *J. Org. Chem.*, **46**, 2974 (1981).

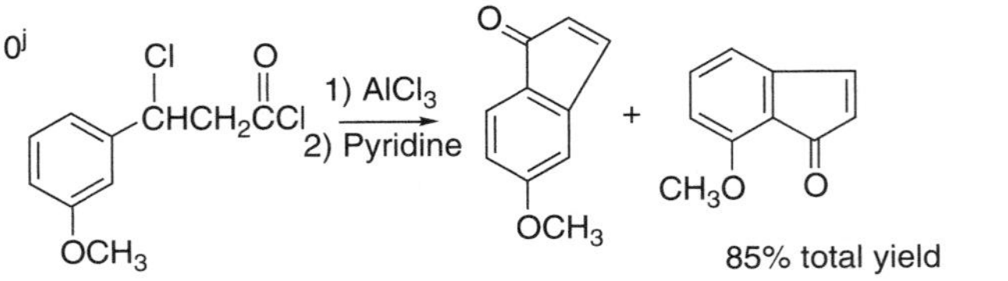

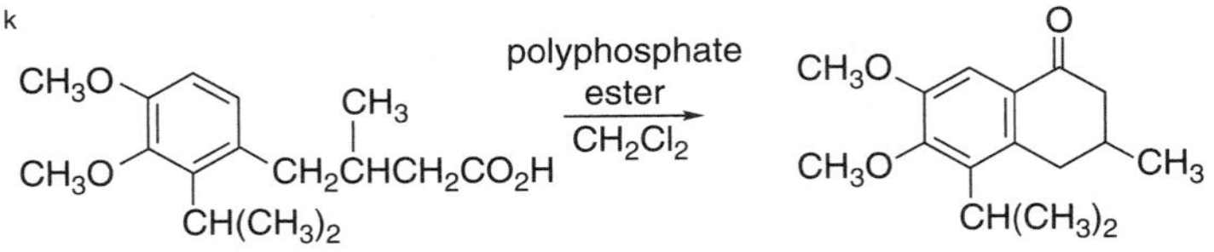

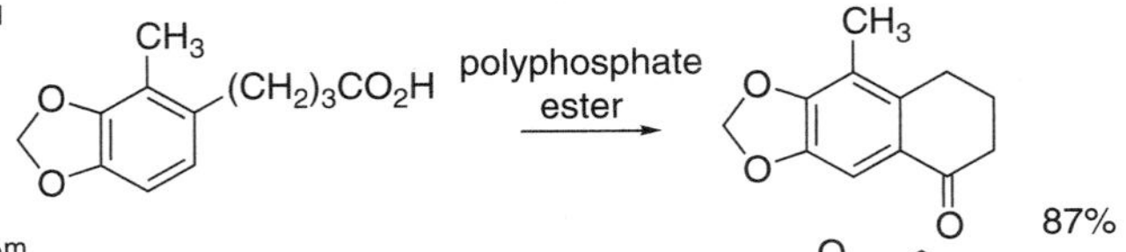

a Lewis acid. Entries 7 and 8 exemplify typical conditions for intramolecular Friedel-Crafts reactions. In Entry 9, both alkylation and acylation occur, presumably in that order.

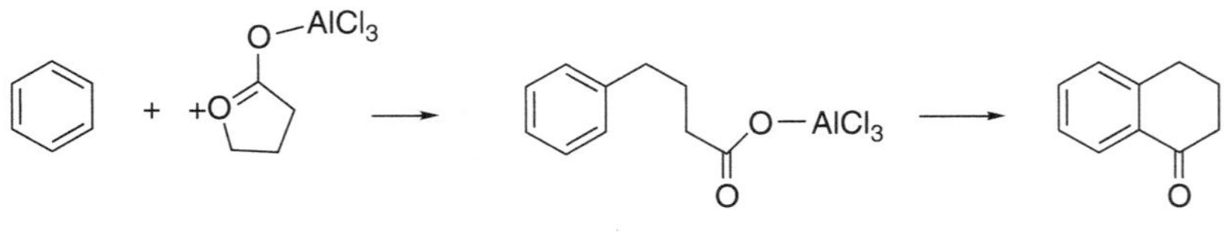

In Entry 10, intramolecular acylation is followed by dehydrohalogenation. Entries 11 and 12 illustrate the use of polyphosphate ester. The cyclization in Entry 13 is done in neat methanesulfonic acid.

A special case of aromatic acylation is the *Fries rearrangement*, which is the conversion of an ester of a phenol to an *o*-acyl phenol by a Lewis acid.

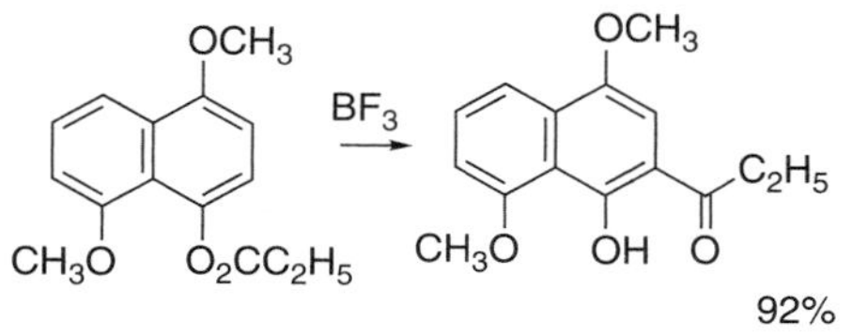

Ref. 59

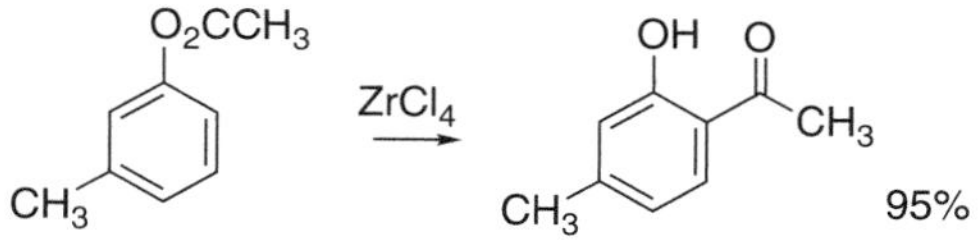

Ref. 60

Lanthanide triflates are also good catalysts for Fries rearrangements.[61]

11.1.5. Related Alkylation and Acylation Reactions

There are a number of variations of the Friedel-Crafts reactions that are useful in synthesis. The introduction of chloromethyl substituents is brought about by reaction with formaldehyde in concentrated hydrochloric acid and halide salts, especially zinc chloride.[62] The reaction proceeds with benzene and activated derivatives. The reactive electrophile is probably the chloromethylium ion.

$$CH_2=O + HCl + H^+ \rightleftharpoons H_2\overset{+}{O}CH_2Cl \longrightarrow CH_2=Cl^+$$

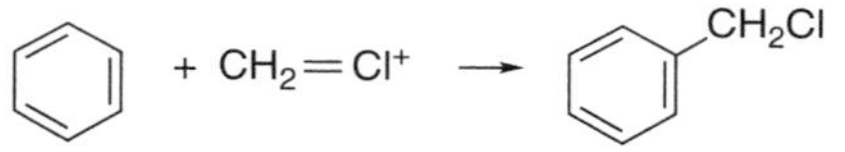

Chloromethylation can also be carried out using various chloromethyl ethers and $SnCl_4$.[63]

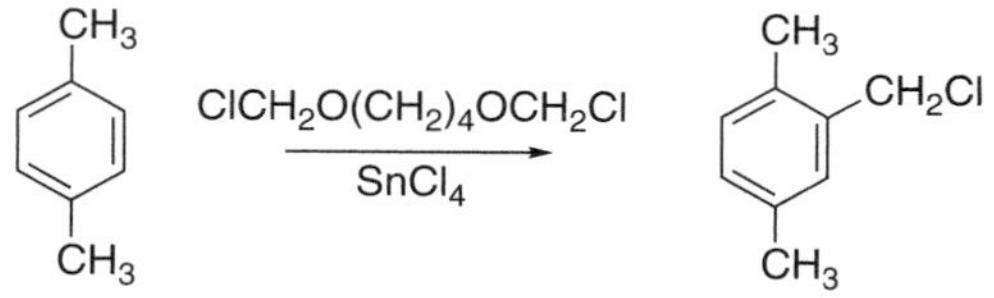

Carbon monoxide, hydrogen cyanide, and nitriles also react with aromatic compounds in the presence of strong acids or Friedel-Crafts catalysts to introduce formyl or acyl substituents. The active electrophiles are believed to be *dications* resulting from diprotonation of CO, HCN, or the nitrile.[64] The general outlines of the mechanisms of these reactions are given below.

[59.] Y. Naruta, Y. Nishgaichi, and K. Maruyama, *J. Org. Chem.*, **53**, 1192 (1988).

[60.] D. C. Harrowven and R. F. Dainty, *Tetrahedron Lett.*, **37**, 7659 (1996).

[61.] S. Kobayahis, M. Moriwaki, and J. Hachiya, *Bull. Chem. Soc. Jpn.*, **70**, 267 (1997).

[62.] R. C. Fuson and C. H. McKeever, *Org. React.*, **1**, 63 (1942); G. A. Olah and S. H. Yu, *J. Am. Chem. Soc.*, **97**, 2293 (1975).

[63.] G. A. Olah, D. A. Beal, and J. A. Olah, *J. Org. Chem.*, **41**, 1627 (1976); G. A. Olah, D. A. Bell, S. H. Yu, and J. A. Olah, *Synthesis*, 560 (1974).

[64.] M. Yato, T. Ohwada, and K. Shudo, *J. Am. Chem. Soc.*, **113**, 691 (1991); Y. Sato, M. Yato, T. Ohwada, S. Saito, and K. Shudo, *J. Am. Chem. Soc.*, **117**, 3037 (1995).

a. Formylation with carbon monoxide:

$$\bar{C}\equiv\overset{+}{O} + H^+ \rightleftharpoons H-C\equiv\overset{+}{O} \overset{H^+}{\rightleftharpoons} H-\overset{+}{C}=\overset{+}{O}-H$$

$$ArH + H\overset{+}{C}=\overset{\cdot\cdot}{O}-H \longrightarrow ArCH=O + 2H^+$$

b. Formylation with hydrogen cyanide:

$$H-C\equiv N + H^+ \rightleftharpoons H-C\equiv\overset{+}{N}-H \overset{H^+}{\rightleftharpoons} H\overset{+}{C}=\overset{+}{N}H_2$$

$$ArH + H\overset{+}{C}=\overset{+}{N}H_2 \longrightarrow ArCH=\overset{+}{N}H_2 \overset{H_2O}{\longrightarrow} ArCH=O$$

c. Acylation with nitriles:

$$R-C\equiv N + H^+ \rightleftharpoons R-C\equiv\overset{+}{N}-H \overset{H^+}{\rightleftharpoons} R\overset{+}{C}=\overset{+}{N}H_2$$

$$ArH + R\overset{+}{C}=\overset{+}{N}H_2 \longrightarrow \underset{Ar}{R-C}=\overset{+}{N}H_2 \overset{H_2O}{\longrightarrow} \overset{O}{\underset{\|}{ArCR}}$$

Many specific examples of these reactions can be found in reviews in the *Organic Reactions* series.[65] Dichloromethyl ethers are also precursors of the formyl group via alkylation catalyzed by $SnCl_4$ or $TiCl_4$.[66] The dichloromethyl group is hydrolyzed to a formyl group.

$$Ar-H \overset{Cl_2CHOR}{\underset{SnCl_4}{\longrightarrow}} ArCHCl_2 \overset{H_2O}{\longrightarrow} ArCH=O$$

Another useful method for introducing formyl and acyl groups is the *Vilsmeier-Haack reaction*.[67] *N,N*-dialkylamides react with phosphorus oxychloride or oxalyl chloride[68] to give a chloroiminium ion, which is the reactive electrophile.

$$\overset{O}{\underset{\|}{RCN(CH_3)_2}} + POCl_3 \longrightarrow \overset{Cl}{\underset{|}{RC}}=\overset{+}{N}(CH_3)_2$$

This species acts as an electrophile in the absence of any added Lewis acid, but only rings with ERG substituents are reactive.

Scheme 11.5 gives some examples of these acylation reactions. Entry 1 is an example of a chloromethylation reaction. Entry 2 is a formylation using carbon monoxide. Entry 3 is an example of formylation via *bis*-chloromethyl ether. A cautionary note on this procedure is the potent carcinogenicity of this reagent. Entries 4 and 5 are examples of formylation and acetylation, using HCN and acetonitrile, respectively. Entries 6 to 8 are examples of Vilsmeier-Haack reactions, all of which are conducted on strongly activated aromatics.

[65] N. N. Crounse, *Org. React.*, **5**, 290 (1949); W. E. Truce, *Org. React.*, **9**, 37 (1957); P. E. Spoerri and A. S. DuBois, *Org. React.*, **5**, 387 (1949); see also G. A. Olah, L. Ohannesian, and M. Arvanaghi, *Chem. Rev.*, **87**, 671 (1987).

[66] P. E. Sonnet, *J. Med. Chem.*, **15**, 97 (1972); C. H. Hassall and B. A. Morgan, *J. Chem. Soc., Perkin Trans. 1*, 2853 (1973); R. Halterman and S.-T. Jan, *J. Org. Chem.*, **56**, 5253 (1991).

[67] G. Martin and M. Martin, *Bull. Soc. Chim. Fr.*, 1637 (1963); S. Seshadri, *J. Sci. Ind. Res.*, **32**, 128 (1973); C. Just, in *Iminium Salts in Organic Chemistry*, H. Bohme and H. G. Viehe, eds., Vol. 9 in *Advances in Organic Chemistry: Methods and Results*, Wiley-Interscience, 1976, pp. 225–342.

[68] J. N. Frekos, G. W. Morrow, and J. S. Swenton, *J. Org. Chem.*, **50**, 805 (1985).

Scheme 11.5. Other Electrophilic Aromatic Substitutions Related to Friedel-Crafts Reactions

A. Chloromethylation

1[a]

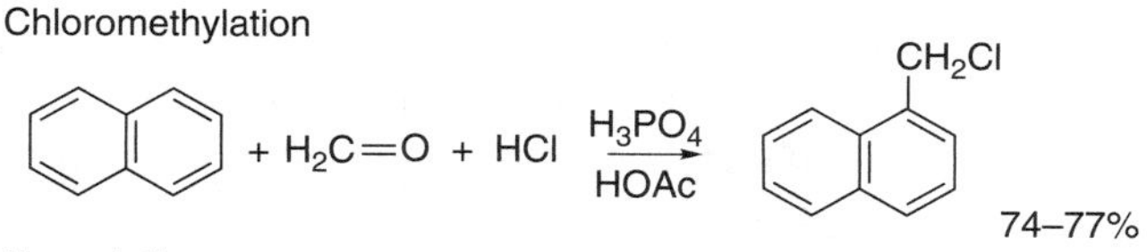

74–77%

B. Formylation

2[b]

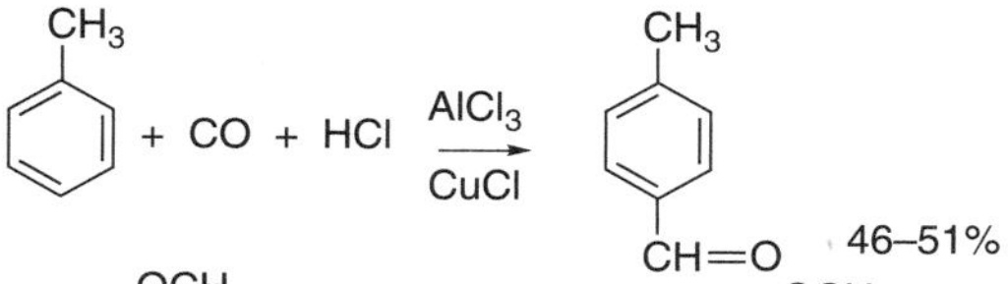

46–51%

3[c]

99%

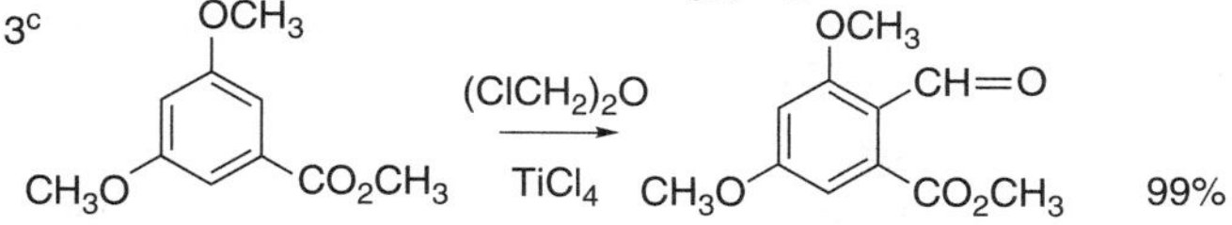

C. Acylation with cyanide and nitriles

4[d]

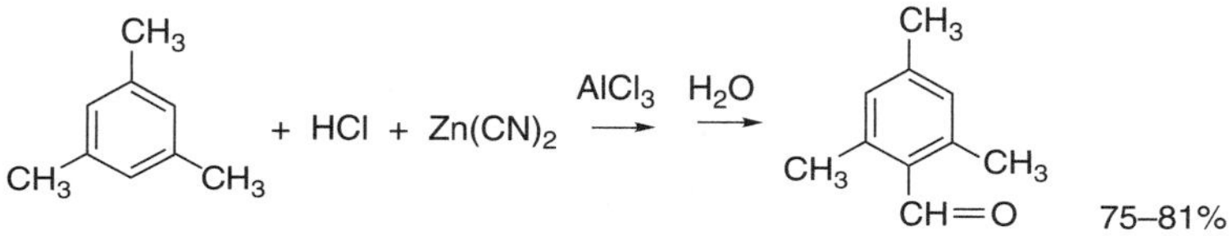

75–81%

5[e]

74–87%

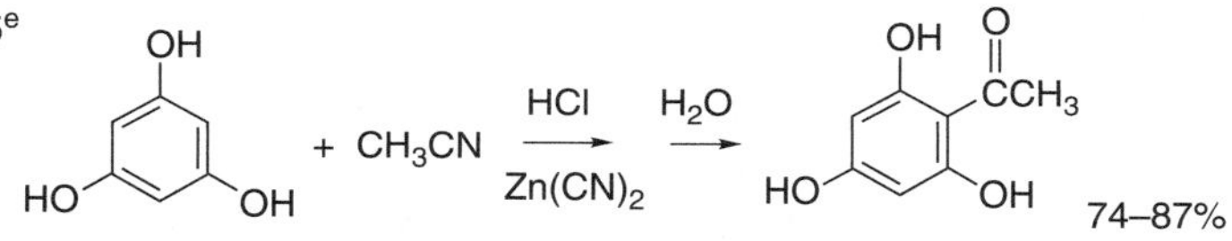

D. Vilsmeier–Haack acylation

6[f]

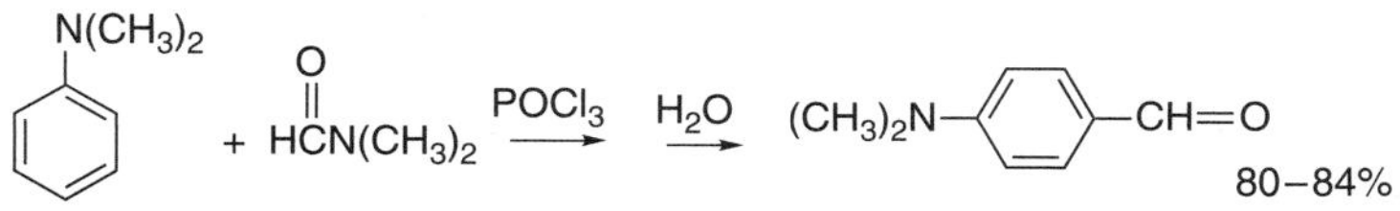

80–84%

7[g]

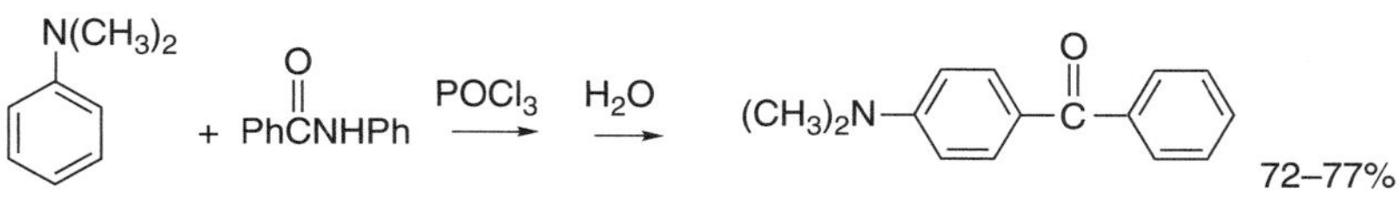

72–77%

8[h]

74–84%

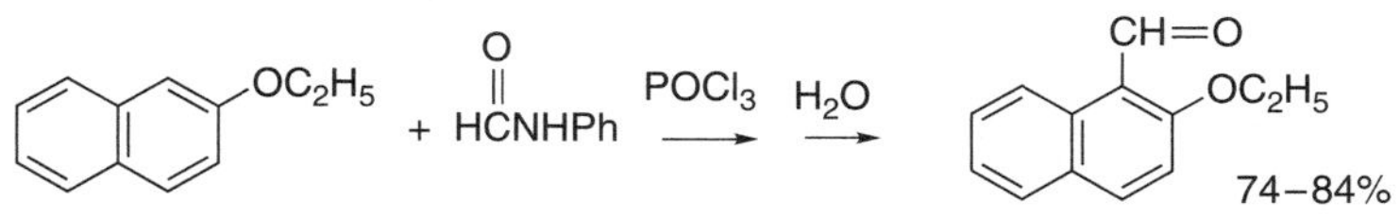

a. C. Grummitt and A. Buck, *Org. Synth.*, **III**, 195 (1955).
b. G. H. Coleman and D. Craig, *Org. Synth.*, **II**, 583 (1943).
c. C. H. Hassall and B. A. Morgan, *J. Chem. Soc., Perkin Trans. 1*, 2853 (1973).
d. R. C. Fuson, E. C. Horning, S. P. Rowland, and M. L. Ward, *Org. Synth.*, **III**, 549 (1955).
e. K. C. Gulati, S. R. Seth, and K. Venksataraman, *Org. Synth.*, **II**, 522 (1943).
f. E. Campaigne and W. L. Archer, *Org. Synth.*, **IV**, 331 (1963).
g. C. D. Hurd and C. N. Webb, *Org. Synth.*, **I**, 217 (1941).
h. J. H. Wood and R. W. Bost, *Org. Synth.*, **III**, 98 (1955).

11.1.6. Electrophilic Metallation

Aromatic compounds react with mercuric salts to give arylmercury compounds.[69] Mercuric acetate or mercuric trifluoroacetate are the usual reagents.[70] The reaction shows substituent effects that are characteristic of electrophilic aromatic substitution.[71] Mercuration is one of the few electrophilic aromatic substitutions in which proton loss from the σ complex is rate determining. Mercuration of benzene shows an isotope effect $k_H/k_D = 6$,[72] which indicates that the σ complex must be formed reversibly.

The synthetic utility of the mercuration reaction derives from subsequent transformations of the arylmercury compounds. As indicated in Section 7.3.3, these compounds are only weakly nucleophilic, but the carbon-mercury bond is reactive to various electrophiles. They are particularly useful for synthesis of nitroso compounds. The nitroso group can be introduced by reaction with nitrosyl chloride[73] or nitrosonium tetrafluoroborate[74] as the electrophile. Arylmercury compounds are also useful in certain palladium-catalyzed reactions, as discussed in Section 8.2.

Thallium(III), particularly as the trifluoroacetate salt, is also a reactive electrophilic metallating species, and a variety of synthetic schemes based on arylthallium intermediates have been devised.[75] Arylthallium compounds are converted to chlorides or bromides by reaction with the appropriate cupric halide.[76] Reaction with potassium iodide gives aryl iodides.[77] Fluorides are prepared by successive treatment with potassium fluoride and boron trifluoride.[78] Procedures for converting arylthallium compounds to nitriles and phenols have also been described.[79]

The thallium intermediates can be useful in directing substitution to specific positions when the site of thallation can be controlled in an advantageous way. The two principal means of control are chelation and the ability to effect thermal equilibration of arylthallium intermediates. Oxygen-containing groups normally direct thallation to the *ortho* position by a chelation effect. The thermodynamically favored position is

69. W. Kitching, *Organomet. Chem. Rev.*, **3**, 35 (1968).

70. A. J. Kresge, M. Dubeck, and H. C. Brown, *J. Org. Chem.*, **32**, 745 (1967); H. C. Brown and R. A. Wirkkala, *J. Am. Chem. Soc.*, **88**, 1447, 1453, 1456 (1966).

71. H. C. Brown and C. W. McGary, Jr., *J. Am. Chem. Soc.*, **77**, 2300, 2310 (1955); A. J. Kresge and H. C. Brown, *J. Org. Chem.*, **32**, 756 (1967); G. A. Olah, I. Hashimoto, and H. C. Lin, *Proc. Natl. Acad. Sci., USA*, **74**, 4121 (1977).

72. C. Perrin and F. H. Westheimer, *J. Am. Chem. Soc.*, **85**, 2773 (1963); A. J. Kresge and J. F. Brennan, *J. Org. Chem.*, **32**, 752 (1967); C. W. Fung, M. Khorramdel-Vahad, R. J. Ranson, and R. M. G. Roberts, *J. Chem. Soc., Perkin Trans. 2*, 267 (1980).

73. L. I. Smith and F. L. Taylor, *J. Am. Chem. Soc.*, **57**, 2460 (1935); S. Terabe, S. Kuruma, and R. Konaka, *J. Chem. Soc., Perkin Trans. 2*, 1252 (1973).

74. L. M. Stock and T. L. Wright, *J. Org. Chem.*, **44**, 3467 (1979).

75. E. C. Taylor and A. McKillop, *Acc. Chem. Res.*, **3**, 338 (1970).

76. S. Uemura, Y. Ikeda, and K. Ichikawa, *Tetrahedron*, **28**, 5499 (1972).

77. A. McKillop, J. D. Hunt, M. J. Zelesko, J. S. Fowler, E. C. Taylor, G. McGillivray, and F. Kienzle, *J. Am. Chem. Soc.*, **93**, 4841 (1971); M. L. dos Santos, G. C. de Magalhaes, and R. Braz Filhe, *J. Organomet. Chem.*, **526**, 15 (1996).

78. E. C. Taylor, E. C. Bigham, and D. K. Johnson, *J. Org. Chem.*, **42**, 362 (1977).

79. S. Uemura, Y. Ikeda, and K. Ichikawa, *Tetrahedron*, **28**, 3025 (1972); E. C. Taylor, H. W. Altland, R. H. Danforth, G. McGillivray, and A. McKillop, *J. Am. Chem. Soc.*, **92**, 3520 (1970).

normally the *meta* position, and heating the thallium derivatives of alkylbenzenes gives a predominance of the *meta* isomer.[80] Both mercury and thallium compounds are very toxic, so special care is needed in their manipulation.

11.2. Nucleophilic Aromatic Substitution

Synthetically important substitutions of aromatic compounds can also be done by nucleophilic reagents. There are several general mechanism for substitution by nucleophiles. Unlike nucleophilic substitution at saturated carbon, aromatic nucleophilic substitution does not occur by a single-step mechanism. The broad mechanistic classes that can be recognized include addition-elimination, elimination-addition, and metal-catalyzed processes. (See Section 9.5 of Part A to review these mechanisms.) We first discuss diazonium ions, which can react by several mechanisms. Depending on the substitution pattern, aryl halides can react by either addition-elimination or elimination-addition. Aryl halides and sulfonates also react with nucleophiles by metal-catalyzed mechanisms and these are discussed in Section 11.3.

11.2.1. Aryl Diazonium Ions as Synthetic Intermediates

The first widely used intermediates for nucleophilic aromatic substitution were the aryl diazonium salts. Aryl diazonium ions are usually prepared by reaction of an aniline with nitrous acid, which is generated in situ from a nitrite salt.[81] Unlike aliphatic diazonium ions, which decompose very rapidly to molecular nitrogen and a carbocation (see Part A, Section 4.1.5), aryl diazonium ions are stable enough to exist in solution at room temperature and below. They can also be isolated as salts with nonnucleophilic anions, such as tetrafluoroborate or trifluoroacetate.[82] Salts prepared with *o*-benzenedisulfonimidate also appear to have potential for synthetic application.[83]

benzenedisulfonimidate anion

The steps in forming a diazonium ion are addition of the nitrosonium ion, ^+NO, to the amino group, followed by elimination of water.

$$ArNH_2 + HONO \xrightarrow{H^+} ArN(H)-N{=}O + H_2O$$

$$ArN(H)-N{=}O \longrightarrow ArN{=}N-OH \xrightarrow{H^+} Ar\overset{+}{N}{\equiv}N + H_2O$$

---

[80] A. McKillop, J. D. Hunt, M. J. Zelesko, J. S. Fowler, E. C. Taylor, G. McGillivray, and F. Kienzle, *J. Am. Chem. Soc.*, **93**, 4841 (1971); M. L. dos Santos, G. C. de Mangalhaes, and R. Braz Filho, *J. Organomet. Chem.*, **526**, 15 (1996).

[81] H. Zollinger, *Azo and Diazo Chemistry*, Interscience, New York, 1961; S. Patai, ed., *The Chemistry of Diazonium and Diazo Groups*, Wiley, New York, 1978, Chaps. 8, 11, and 14; H. Saunders and R. L. M. Allen, *Aromatic Diazo Compounds*, 3rd Edition, Edward Arnold, London, 1985.

[82] C. Colas and M. Goeldner, *Eur. J. Org. Chem.*, 1357 (1999).

[83] M. Barbero, M. Crisma, I. Degani, R. Fochi, and P. Perracino, *Synthesis*, 1171 (1998); M. Babero, I. Degani, S. Dughera, and R. Fochi, *J. Org. Chem.*, **64**, 3448 (1999).

In alkaline solution, diazonium ions are converted to diazoate anions, which are in equilibrium with diazo oxides.[84]

$$ArN^+\!\!\equiv\!\!N \ + \ 2\ ^-OH \longrightarrow ArN\!\!=\!\!N\!\!-\!\!O^- + H_2O$$
$$\text{diazoate anion}$$

$$ArN\!\!=\!\!N\!\!-\!\!O^- + \ ArN^+\!\!\equiv\!\!N \ \rightleftharpoons \ ArN\!\!=\!\!N\!\!-\!\!O\!\!-\!\!N\!\!=\!\!NAr$$
$$\text{diazo oxide}$$

In addition to the aqueous method for diazotization, diazonium ions can be generated in organic solvents by reaction with alkyl nitrites.

$$RO\!\!-\!\!N\!\!=\!\!O \ + \ ArNH_2 \longrightarrow \overset{\overset{\displaystyle H}{\displaystyle |}}{ArN}\!\!-\!\!N\!\!=\!\!O \ + \ ROH$$

$$\overset{\overset{\displaystyle H}{\displaystyle |}}{ArN}\!\!-\!\!N\!\!=\!\!O \ \rightleftharpoons \ ArN\!\!=\!\!N\!\!-\!\!OH \ \xrightarrow{H^+} \ ArN^+\!\!\equiv\!\!N \ + \ H_2O$$

Diazonium ions form stable adducts with certain nucleophiles such as secondary amines and sulfide anions.[85] These compounds can be used as precursors of diazonium ion intermediates.

$$ArN^+\!\!\equiv\!\!N \ + \ HNR_2 \longrightarrow ArN\!\!=\!\!NNR_2$$

$$ArN^+\!\!\equiv\!\!N \ + \ ^-SR \longrightarrow ArN\!\!=\!\!NSR$$

The wide utility of aryl diazonium ions as synthetic intermediates results from the excellence of N_2 as a leaving group. There are several general mechanisms by which substitution can occur. One involves unimolecular thermal decomposition of the diazonium ion, followed by capture of the resulting aryl cation by a nucleophile. The phenyl cation is very unstable (see Part A, Section 3.4.1.1) and therefore highly unselective.[86] Either the solvent or an anion can act as the nucleophile.

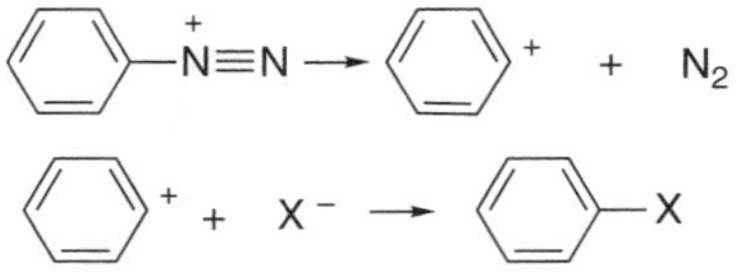

Another general mechanism for substitution is adduct formation followed by collapse of the adduct with loss of nitrogen.

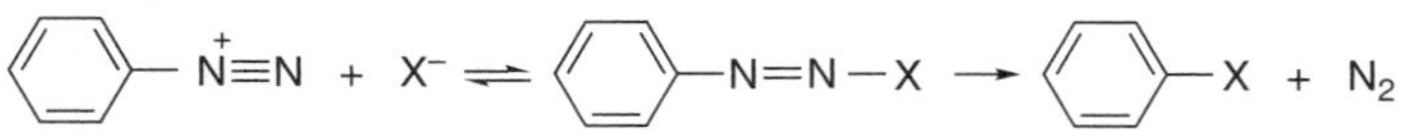

84. E. S. Lewis and M. P. Hanson, *J. Am. Chem. Soc.*, **89**, 6268 (1967).
85. M. L. Gross, D. H. Blank, and W. M. Welch, *J. Org. Chem.*, **58**, 2104 (1993); S. A. Haroutounian, J. P. DiZio, and J. A. Katzenellenbogen, *J. Org. Chem.*, **56**, 4993 (1991).
86. C. G. Swain, J. E. Sheats, and K. G. Harbison, *J. Am. Chem. Soc.*, **97**, 783 (1975).

A third mechanism involves redox processes,[87] and is particularly likely to operate in reactions in which copper salts are used as catalysts.[88]

$$ArN^+ \equiv N + [Cu(I)X_2]^- \longrightarrow Ar-Cu(III)X_2 + N_2$$

$$Ar-Cu(III)X_2 \longrightarrow ArX + Cu(I)X$$

Examples of the three mechanistic types are, respectively: (a) hydrolysis of diazonium salts to phenols[89]; (b) reaction with azide ion to form aryl azides[90]; and (c) reaction with cuprous halides to form aryl chlorides or bromides.[91] In the paragraphs that follow, these and other synthetically useful reactions of diazonium intermediates are considered. The reactions are organized on the basis of the group that is introduced, rather than on the mechanism involved. It will be seen that the reactions that are discussed fall into one of the three general mechanistic types.

11.2.1.1. Reductive Dediazonization. Replacement of a nitro or amino group by hydrogen is sometimes required as a sequel to a synthetic operation in which the substituent was used to control the position selectivity of a prior transformation. The best reagents for reductive dediazonation are hypophosphorous acid, H_3PO_2,[92] and $NaBH_4$.[93] The reduction by H_3PO_2 is substantially improved by catalysis with cuprous oxide.[94] The reduction by H_3PO_2 proceeds by one-electron reduction followed by loss of nitrogen and formation of the phenyl radical.[95] The hypophosphorous acid then serves as a hydrogen atom donor.

$$\begin{aligned} \text{initiation} \quad & ArN^+ \equiv N + e^- \longrightarrow Ar\cdot + N_2 \\ \text{propagation} \quad & Ar\cdot + H_3PO_2 \longrightarrow Ar-H + [H_2PO_2\cdot] \\ & ArN^+ \equiv N + [H_2PO_2\cdot] \longrightarrow Ar\cdot + N_2 + [H_2PO_2^+] \\ & [H_2PO_2^+] + H_2O \longrightarrow H_3PO_3 + H^+ \end{aligned}$$

An alternative method for reductive dediazonation involves in situ diazotization by an alkyl nitrite in dimethylformamide.[96] This reduction is a chain reaction with the solvent acting as the hydrogen atom donor.

87. C. Galli, *Chem. Rev.*, **88**, 765 (1988).
88. T. Cohen, R. J. Lewarchik, and J. Z. Tarino, *J. Am. Chem. Soc.*, **97**, 783 (1975).
89. E. S. Lewis, L. D. Hartung, and B. M. McKay, *J. Am. Chem. Soc.*, **91**, 419 (1969).
90. C. D. Ritchie and D. J. Wright, *J. Am. Chem. Soc.*, **93**, 2429 (1971); C. D. Ritchie and P. O. I. Virtanen, *J. Am. Chem. Soc.*, **94**, 4966 (1972).
91. J. K. Kochi, *J. Am. Chem. Soc.*, **79**, 2942 (1957); S. C. Dickerman, K. Weiss, and A. K. Ingberman, *J. Am. Chem. Soc.*, **80**, 1904 (1958).
92. N. Kornblum, *Org. React.*, **2**, 262 (1944).
93. J. B. Hendrickson, *J. Am. Chem. Soc.*, **83**, 1251 (1961).
94. S. Korzeniowski, L. Blum, and G. W. Gokel, *J. Org. Chem.*, **42**, 1469 (1977).
95. N. Kornblum, G. D. Cooper, and J. E. Taylor, *J. Am. Chem. Soc.*, **72**, 3013 (1950).
96. M. P. Doyle, J. F. Dellaria, Jr., B. Siegfried, and S. W. Bishop, *J. Org. Chem.*, **42**, 3494 (1977); J. H. Markgraf, R. Chang, J. R. Cort, J. L. Durant, Jr., M. Finkelstein, A. W. Gross, M. H. Lavyne, W. M. Moore, R. C. Peterson, and S. D. Ross, *Tetrahedron*, **53**, 10009 (1997).

initiation $\quad ArN^+\equiv N + e^- \longrightarrow Ar\cdot + N_2$

propagation $\quad Ar\cdot + HCN(CH_3)_2 \longrightarrow Ar-H + \cdot CN(CH_3)_2$
$$\overset{\|}{O} \qquad\qquad\qquad \overset{\|}{O}$$

$$ArN^+\equiv N + \cdot CN(CH_3)_2 \longrightarrow Ar\cdot + N_2 + C\equiv O + CH_3N^+H=CH_2$$
$$\overset{\|}{O}$$

This reaction can be catalyzed by $FeSO_4$.[97]

11.2.1.2. Phenols from Diazonium Ion Intermediates. Aryl diazonium ions can be converted to phenols by heating in water. Under these conditions, there is probably formation of a phenyl cation.

$$ArN^+\equiv N \longrightarrow Ar^+ + N_2 \xrightarrow{H_2O} ArOH + H^+$$

By-products from capture of nucleophilic anions may be observed.[53] Phenols can be formed under milder conditions by an alternative redox mechanism.[98] The reaction is initiated by cuprous oxide, which effects reduction and decomposition to an aryl radical, and is run in the presence of Cu(II) salts. The radical is captured by Cu(II) and converted to the phenol by reductive elimination. This procedure is very rapid and gives good yields of phenols over a range of structural types.

$$ArN^+\equiv N + Cu(I) \longrightarrow Ar\cdot + N_2 + Cu(II)$$

$$Ar\cdot + Cu(II) \longrightarrow [Ar-Cu^{III}]^{2+} \xrightarrow{H_2O} ArOH + Cu(I) + H^+$$

11.2.1.3. Aryl Halides from Diazonium Ion Intermediates. Replacement of diazonium groups by halides is a valuable alternative to direct halogenation for the preparation of aryl halides. Aryl bromides and chlorides are usually prepared by a reaction using the appropriate Cu(I) salt, which is known as the *Sandmeyer reaction.* Under the classic conditions, the diazonium salt is added to a hot acidic solution of the cuprous halide.[99] The Sandmeyer reaction occurs by an oxidative addition reaction of the diazonium ion with Cu(I) and halide transfer from a Cu(III) intermediate.

$$ArN^+\equiv N + [Cu^I X_2]^- \longrightarrow Ar-Cu^{III} X_2 + N_2$$

$$Ar-Cu^{III}X_2 \longrightarrow ArX + Cu^I X$$

Good yields of chlorides have also been obtained for reaction of isolated diazonium tetrafluoroborates with $FeCl_2$-$FeCl_3$ mixtures.[100] It is also possible to convert anilines to aryl halides by generating the diazonium ion in situ. Reaction of anilines with alkyl nitrites and Cu(II) halides in acetonitrile gives good yields of aryl chlorides and bromides.[101]

97. F. W. Wassmundt and W. F. Kiesman, *J. Org. Chem.*, **60**, 1713 (1995).
98. T. Cohen, A. G. Dietz, Jr., and J. R. Miser, *J. Org. Chem.*, **42**, 2053 (1977).
99. W. A. Cowdrey and D. S. Davies, *Q. Rev. Chem. Soc.*, **6**, 358 (1952); H. H. Hodgson, *Chem. Rev.*, **40**, 251 (1947).
100. K. Daasbjerg and H. Lund, *Acta Chem. Scand.*, **46**, 157 (1992).
101. M. P. Doyle, B. Sigfried, and J. F. Dellaria, Jr., *J. Org. Chem.*, **42**, 2426 (1977).

Diazonium salts can also be converted to halides by processes involving aryl free radicals. In basic solutions, aryl diazonium ions are converted to radicals via the diazo oxide.[102]

$$2\ Ar\overset{+}{N}{\equiv}N\ +\ 2\ ^-OH\ \longrightarrow\ ArN{=}N{-}O{-}N{=}NAr\ +\ H_2O$$

$$ArN{=}N{-}O{-}N{=}NAr\ \longrightarrow\ ArN{=}N{-}O{\cdot}\ +\ Ar{\cdot}\ +\ N_2$$

The reaction can be carried out efficiently using aryl diazonium tetrafluoroborates with crown ethers, polyethers, or phase transfer catalysts.[103] In solvents that can act as halogen atom donors, the radicals react to give aryl halides. Bromotrichloromethane gives aryl bromides, whereas methyl iodide and diiodomethane give iodides.[104] The diazonium ions can also be generated by in situ methods. Under these conditions bromoform and bromotrichloromethane have been used as bromine donors and carbon tetrachloride is the best chlorine donor.[105] This method was used successfully for a challenging chlorodeamination in the vancomycin system (Entry 6, Scheme 11.6).

Fluorine substituents can also be introduced via diazonium ions. One procedure is to isolate aryl diazonium tetrafluoroborates. These decompose thermally to give aryl fluorides.[106] Called the *Schiemann reaction*, it probably involves formation of an aryl cation that abstracts fluoride ion from the tetrafluoroborate anion.[107]

$$Ar\overset{+}{N}{\equiv}N\ +\ BF_4^-\ \longrightarrow\ ArF\ +\ N_2\ +\ BF_3$$

Hexafluorophosphate salts behave similarly.[108] The diazonium tetrafluoroborates can be prepared either by precipitation from an aqueous solution by fluoroboric acid[109] or by anhydrous diazotization in ether, THF, or acetonitrile using *t*-butyl nitrite and boron trifluoride.[110] Somewhat milder reaction conditions can be achieved by reaction of aryl diazo sulfide adducts with pyridine-HF in the presence of AgF or AgNO$_3$.

39%

Ref. 111

Aryl diazonium ions are converted to iodides in high yield by reaction with iodide salts. This reaction is initiated by reduction of the diazonium ion by iodide. The aryl radical then abstracts iodine from either I$_2$ or I$_3^-$. A chain mechanism then proceeds

102. C. Rüchardt and B. Freudenberg, *Tetrahedron Lett.*, 3623 (1964); C. Rüchardt and E. Merz, *Tetrahedron Lett.*, 2431 (1964).
103. S. H. Korzeniowski and G. W. Gokel, *Tetrahedron Lett.*, 1637 (1977).
104. S. H. Korzeniowski and G. W. Gokel, *Tetrahedron Lett.*, 3519 (1977); R. A. Bartsch and I. W. Wang, *Tetrahedron Lett.*, 2503 (1979); W. C. Smith and O. C. Ho, *J. Org. Chem.*, **55**, 2543 (1990).
105. J. I. G. Cadogan, D. A. Roy, and D. M. Smith, *J. Chem. Soc. C*, 1249 (1966).
106. A. Roe, *Org. React.*, **5**, 193 (1949).
107. C. G. Swain and R. J. Rogers, *J. Am. Chem. Soc.*, **97**, 799 (1975).
108. M. S. Newman and R. H. B. Galt, *J. Org. Chem.*, **25**, 214 (1960).
109. E. B. Starkey, *Org. Synth.*, **II**, 225 (1943); G. Schiemann and W. Winkelmuller, *Org. Synth.*, **II**, 299 (1943).
110. M. P. Doyle and W. J. Bryker, *J. Org. Chem.*, **44**, 1572 (1979).
111. S. A. Haroutounian, J. P. DiZio, and J. A. Katzenellenbogen, *J. Org. Chem.*, **56**, 4993 (1991).

and consumes I^- and ArN_2^+.[112] Evidence for the involvement of radicals includes the isolation of cyclized products from *o*-allyl derivatives.

$$Ar\overset{+}{N}\equiv N + I^- \longrightarrow Ar\cdot + N_2 + I\cdot \qquad 2I \longrightarrow I_2$$

$$Ar\cdot + I_3^- \longrightarrow ArI + I_2^{-\cdot}$$

$$Ar\overset{+}{N}\equiv N + I_2^{-\cdot} \longrightarrow Ar\cdot + N_2 + I_2 \qquad I_2 + I^- \longrightarrow I_3^-$$

11.2.1.4. Introduction of Other Nucleophiles Using Diazonium Ion Intermediates. Cyano and azido groups are also readily introduced via diazonium intermediates. The former involves a copper-catalyzed reaction analogous to the Sandmeyer reaction. Reaction of diazonium salts with azide ion gives adducts that smoothly decompose to nitrogen and the aryl azide.[56]

$$Ar\overset{+}{N}\equiv N + {}^-N\overset{+}{=}N=N^- \longrightarrow ArN=N-N\overset{+}{=}N=N^- \longrightarrow ArN\overset{+}{=}N=N^- + N_2$$

Aryl thiolates react with aryl diazonium ions to give diaryl sulfides. This reaction is believed to be a radical chain process, similar to the mechanism for reaction of diazonium ions with iodide ion.[113]

initiation $Ar\overset{+}{N}\equiv N + PhS^- \longrightarrow ArN=NSPh$

$ArN=NSPh \longrightarrow Ar\cdot + N_2 + PhS\cdot$

propagation $Ar\cdot + PhS^- \longrightarrow Ar\overset{-\cdot}{S}Ph$

$Ar\overset{-\cdot}{S}Ph + Ar\overset{+}{N}\equiv N \longrightarrow ArSPh + Ar\cdot + N_2$

Scheme 11.6 gives some examples of the various substitution reactions of aryl diazonium ions. Entries 1 to 6 are examples of reductive dediazonization. Entry 1 is an older procedure that uses hydrogen abstraction from ethanol for reduction. Entry 2 involves reduction by hypophosphorous acid. Entry 3 illustrates use of copper catalysis in conjunction with hypophosphorous acid. Entries 4 and 5 are DMF-mediated reductions, with ferrous catalysis in the latter case. Entry 6 involves reduction by $NaBH_4$.

Entries 7 and 8 illustrate conversion of diazonium salts to phenols. Entries 9 and 10 use the traditional conditions for the Sandmeyer reaction. Entry 11 is a Sandmeyer reaction under in situ diazotization conditions, whereas Entry 12 involves halogen atom transfer from solvent. Entry 13 is an example of formation of an aryl iodide. Entries 14 and 15 are Schiemann reactions. The reaction in Entry 16 was used to introduce a chlorine substituent on vancomycin. Of several procedures investigated, the $CuCl-CuCl_2$ catalysis of chlorine atom transfer form CCl_4 proved to be the best. The diazonium salt was isolated as the tetrafluoroborate after in situ diazotization. Entries 17 and 18 show procedures for introducing cyano and azido groups, respectively.

112. P. R. Singh and R. Kumar, *Aust. J. Chem.*, **25**, 2133 (1972); A. Abeywickrema and A. L. J. Beckwith, *J. Org. Chem.*, **52**, 2568 (1987).

113. A. N. Abeywickrema and A. L. J. Beckwith, *J. Am. Chem. Soc.*, **108**, 8227 (1986).

Scheme 11.6. Aromatic Substitution via Diazonium Ions

A. Replacement by hydrogen

1[a]

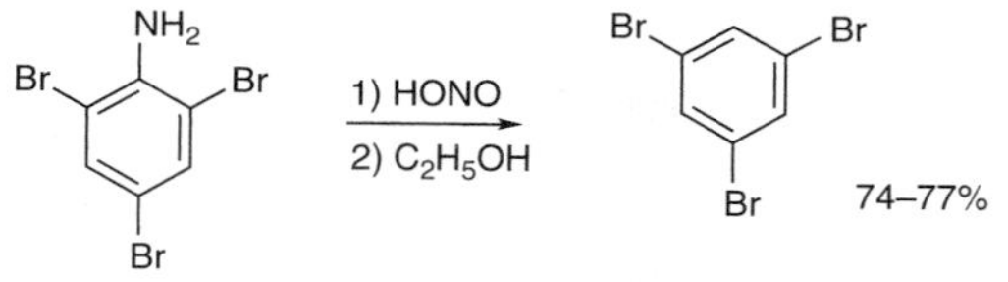

1) HONO
2) C₂H₅OH

74–77%

2[b]

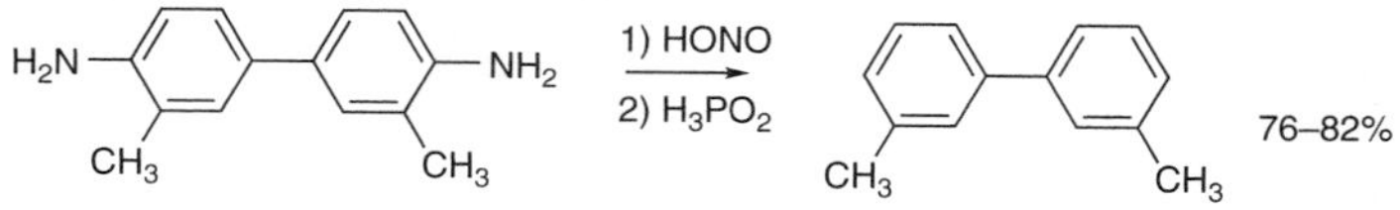

1) HONO
2) H₃PO₂

76–82%

3[c]

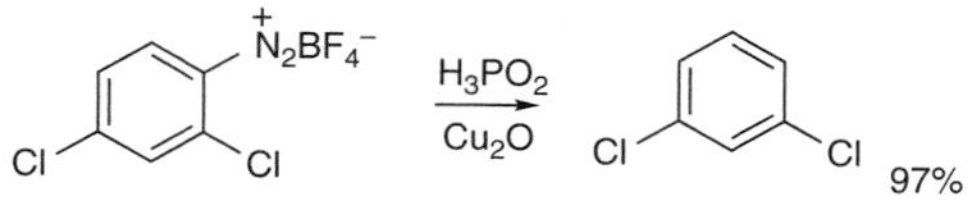

H₃PO₂
Cu₂O

97%

4[d]

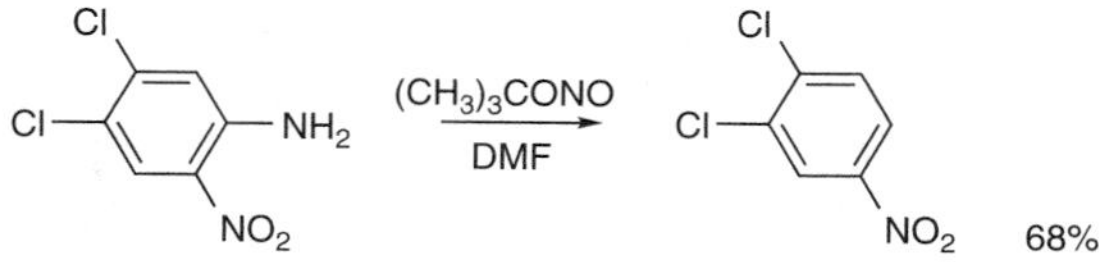

(CH₃)₃CONO
DMF

68%

5[e]

1) NaNO₂, CH₃CO₂H
2) FeSO₄, DMF

76%

6[f]

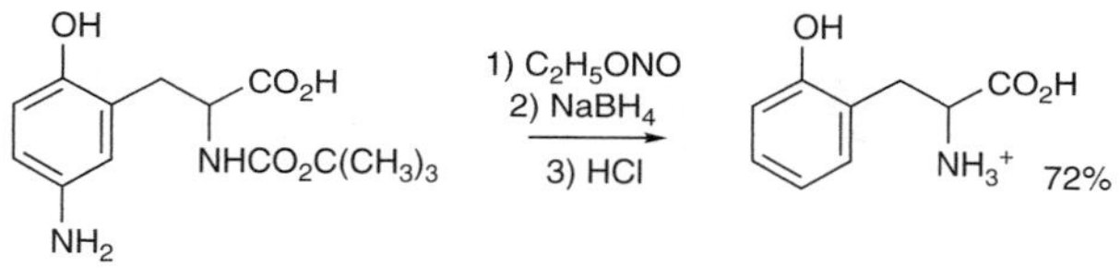

1) C₂H₅ONO
2) NaBH₄
3) HCl

72%

B. Replacement by hydroxy

7[g]

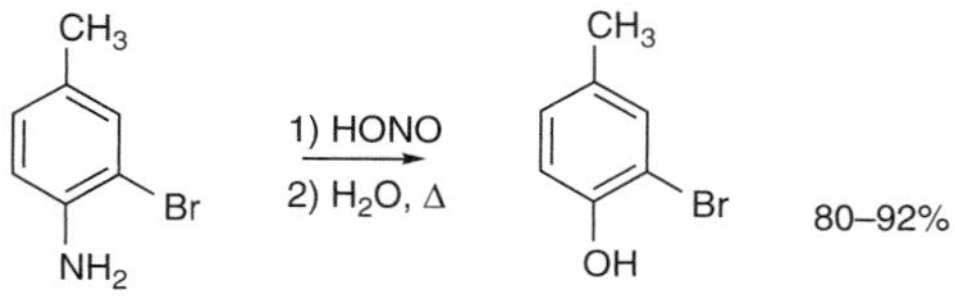

1) HONO
2) H₂O, Δ

80–92%

8[h]

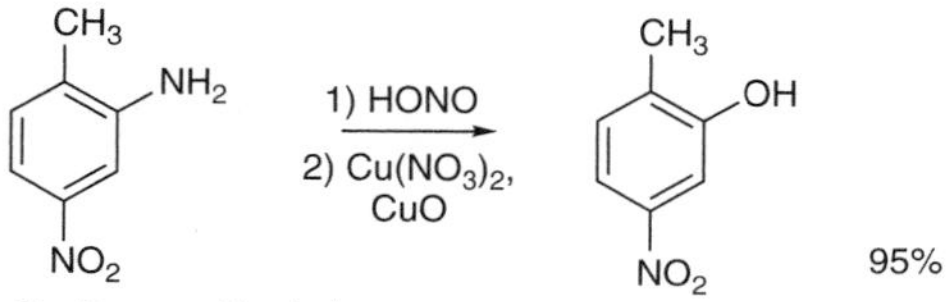

1) HONO
2) Cu(NO₃)₂, CuO

95%

C. Replacement by halogen

9[i]

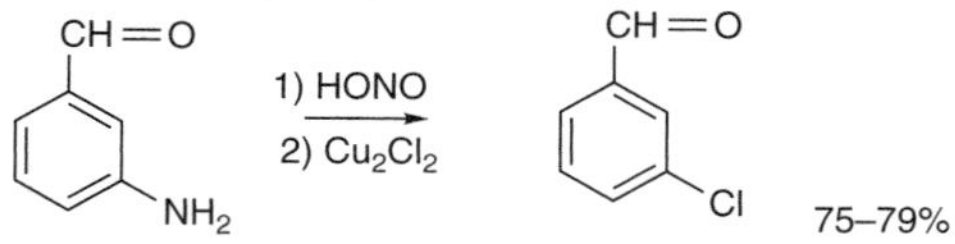

1) HONO
2) Cu₂Cl₂

75–79%

10[j]

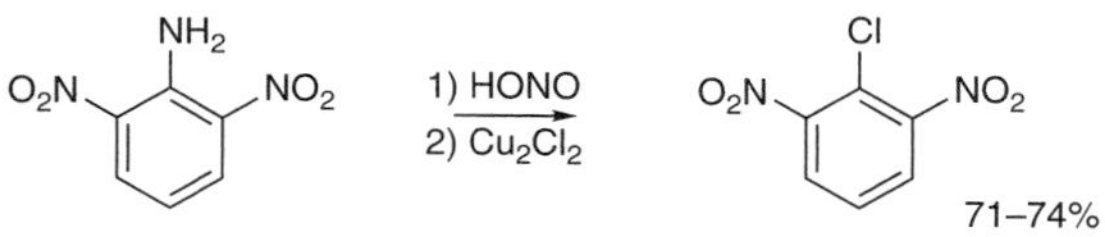

1) HONO
2) Cu₂Cl₂

71–74%

(*Continued*)

Scheme 11.6. (*Continued*)

11[k]

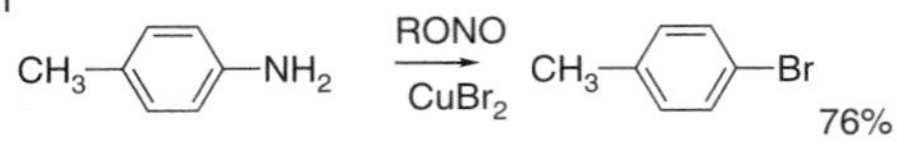

CH_3—〈〉—NH_2 $\xrightarrow[\text{CuBr}_2]{\text{RONO}}$ CH_3—〈〉—Br 76%

12[l]

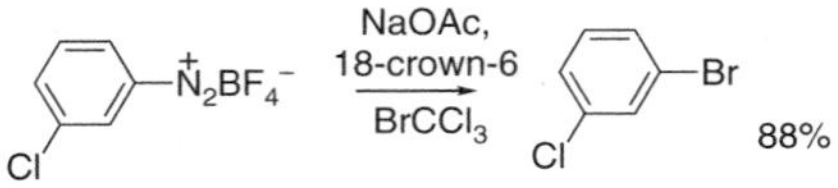

$\xrightarrow[\text{BrCCl}_3]{\text{NaOAc, 18-crown-6}}$ 88%

13[m]

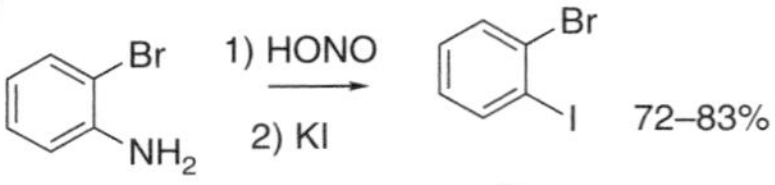

1) HONO 2) KI 72–83%

14[n]

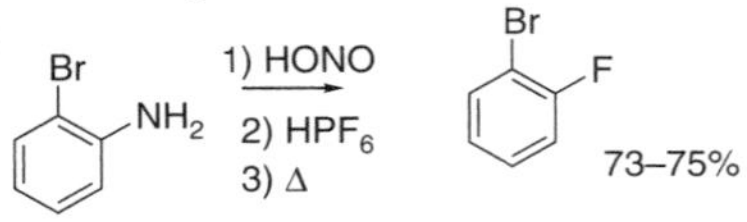

1) HONO 2) HPF_6 3) Δ 73–75%

15[o]

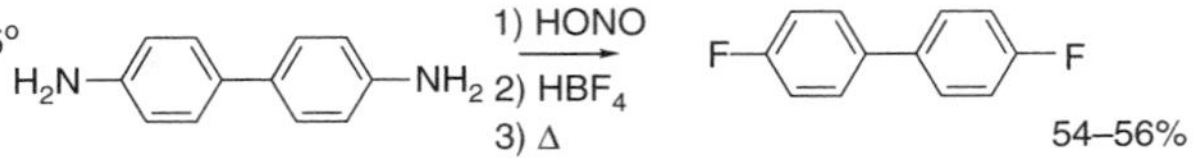

H_2N—〈〉—〈〉—NH_2 1) HONO 2) HBF_4 3) Δ F—〈〉—〈〉—F 54–56%

16[p]

1) $SnCl_2$, DMF
2) $t\text{-}C_4H_9ONO$, BF_3
3) CuCl, $CuCl_2$, CCl_4

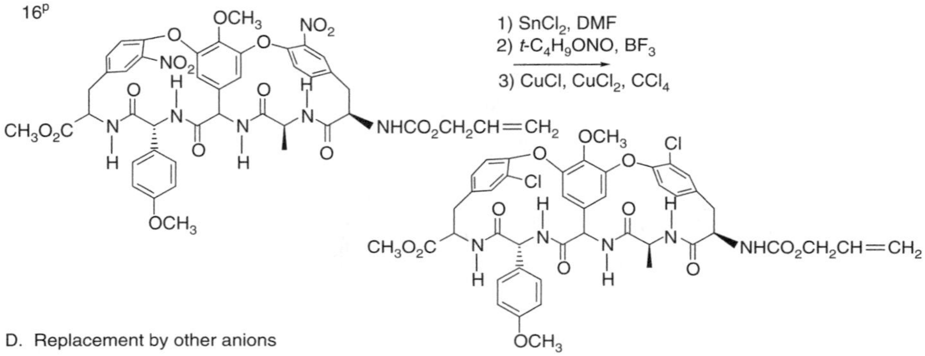

D. Replacement by other anions

17[q]

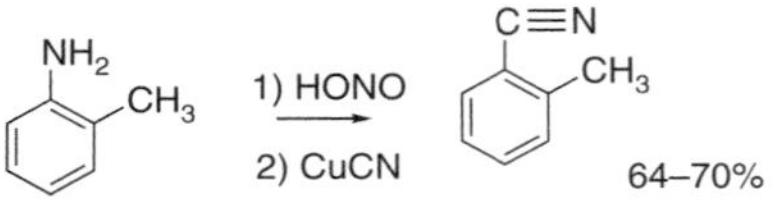

1) HONO 2) CuCN 64–70%

18[r]

1) HONO 2) NaN_3 88%

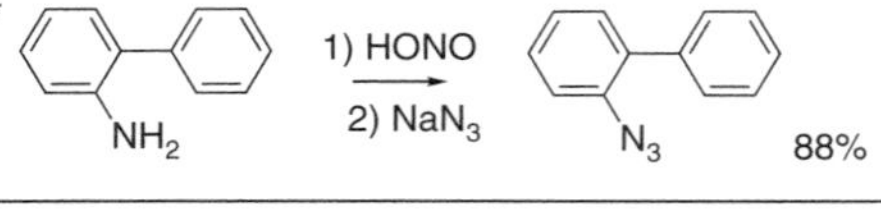

a. G. H. Coleman and W. F. Talbot, *Org. Synth.*, **II**, 592 (1943).
b. N. Kornblum, *Org. Synth.*, **III**, 295 (1955).
c. S. H. Korzeniowski, L. Blum, and G. W. Gokel, *J. Org. Chem.*, **42**, 1469 (1977).
d. M. P. Doyle, J. F. Dellaria, Jr., B. Siegfried, and S. W. Bishop, *J. Org. Chem.*, **42**, 3494 (1977).
e. F. W. Wassmundt and W. F. Kiesman, *J. Org. Chem.*, **60**, 1713 (1995).
f. C. Dugave, *J. Org. Chem.*, **60**, 601 (1995).
g. H. E. Ungnade and E. F. Orwoll, *Org. Synth.*, **III**, 130 (1943).
h. T. Cohen, A. G. Dietz, Jr., and J. R. Miser, *J. Org. Chem.*, **42**, 2053 (1977).
i. J. S. Buck and W. S. Ide, *Org. Synth.*, **II**, 130 (1943).
j. F. D. Gunstone and S. H. Tucker, *Org. Synth.*, **1V**, 160 (1963).
k. M. P. Doyle, B. Siegfried, and J. F. Dellaria, Jr., *J. Org. Chem.*, **42**, 2426 (1977).
l. S. H. Korzeniowsky and G. W. Gokel, *Tetrahedron Lett.*, 3519 (1977).
m. H. Heaney and I. T. Millar, *Org. Synth.*, **40**, 105 (1960).
n. K. G. Rutherford and W. Redmond, *Org. Synth.*, **43**, 12 (1963).
o. G. Schiemann and W. Winkelmuller, *Org. Synth.*, **II**, 188 (1943).
p. C. Vergne, M. Bois-Choussy, and J. Zhu, *Synlett*, 1159 (1998).
q. H. T. Clarke and R. R. Read, *Org. Synth.*, **I**, 514 (1941).
r. P. A. S. Smith and B. B. Brown, *J. Am. Chem. Soc.*, **73**, 2438 (1951).

11.2.1.5. Meerwein Arylation Reactions. Aryl diazonium ions can also be used to form certain types of carbon-carbon bonds. The copper-catalyzed reaction of diazonium ions with conjugated alkenes results in arylation of the alkene, known as the *Meerwein arylation reaction.*[114] The reaction sequence is initiated by reduction of the diazonium ion by Cu(I). The aryl radical adds to the alkene to give a new β-aryl radical. The final step is a ligand transfer that takes place in the copper coordination sphere. An alternative course is oxidation-deprotonation, which gives a styrene derivative.

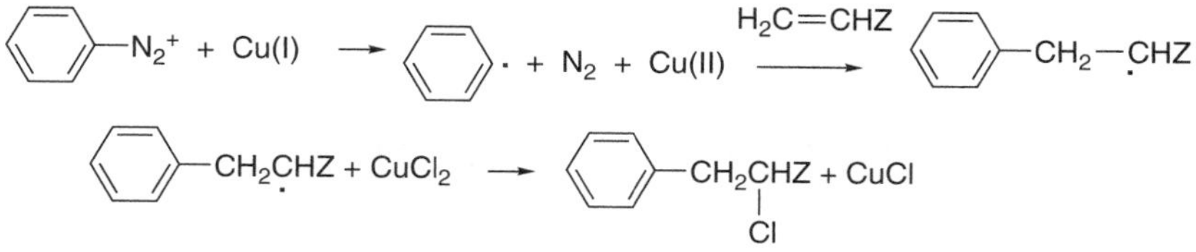

$$\text{Ph-}N_2^+ + \text{Cu(I)} \longrightarrow \text{Ph} \cdot + N_2 + \text{Cu(II)} \xrightarrow{H_2C=CHZ} \text{Ph-}CH_2\text{-}\overset{\cdot}{C}HZ$$

$$\text{Ph-}CH_2\overset{\cdot}{C}HZ + CuCl_2 \longrightarrow \text{Ph-}CH_2\underset{\underset{Cl}{|}}{C}HZ + CuCl$$

The reaction gives better yield with dienes, styrenes, or alkenes substituted with EWGs than with simple alkenes. These groups increase the rate of capture of the aryl radical. The standard conditions for the Meerwein arylation employ aqueous solutions of diazonium ions. Conditions for in situ diazotization by *t*-butyl nitrite in the presence of $CuCl_2$ and acrylonitrile or styrene are also effective.[115]

Reduction of aryl diazonium ions by Ti(III) in the presence of α,β-unsaturated ketones and aldehydes leads to β-arylation and formation of the saturated ketone or aldehyde. The early steps in this reaction parallel the copper-catalyzed reaction. However, rather than being oxidized, the radical formed by the addition step is reduced by Ti(III).[116]

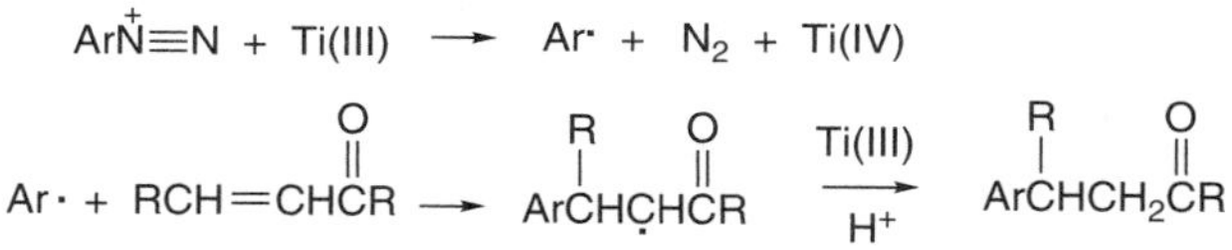

$$\text{Ar}\overset{+}{N}\equiv N + \text{Ti(III)} \longrightarrow \text{Ar} \cdot + N_2 + \text{Ti(IV)}$$

$$\text{Ar} \cdot + \text{RCH}=\text{CHCR} \longrightarrow \text{ArCHCHCR} \xrightarrow[H^+]{\text{Ti(III)}} \text{ArCHCH}_2\text{CR}$$

Scheme 11.7 illustrates some arylation of alkenes by diazonium ions. Entries 1 to 4 are typical conditions. Entry 5 illustrates generation of the diazonium ion under in situ conditions. Entry 6 is an example of the reductive conditions using Ti(III).

11.2.2. Substitution by the Addition-Elimination Mechanism

The addition of a nucleophile to an aromatic ring, followed by elimination of a substituent, results in nucleophilic substitution. The major energetic requirement for this mechanism is formation of the addition intermediate. The addition step is greatly facilitated by strongly electron-attracting substituents, and nitroaromatics are the best reactants for nucleophilic aromatic substitution. Other EWGs such as cyano, acetyl, and trifluoromethyl also enhance reactivity.

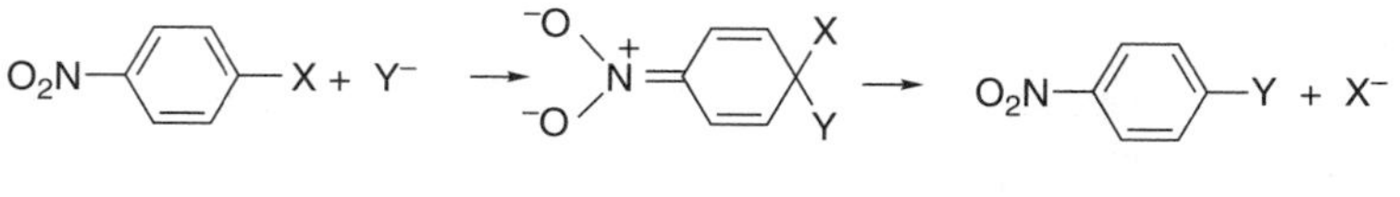

$$O_2N\text{-}C_6H_4\text{-}X + Y^- \longrightarrow \left[\text{intermediate} \right] \longrightarrow O_2N\text{-}C_6H_4\text{-}Y + X^-$$

114. C. S. Rondestvedt, Jr., *Org. React.*, **11**, 189 (1960); C. S. Rondestvedt, *Org. React.*, **24**, 225 (1976); A. V. Dombrovskii, *Russ. Chem. Rev.* (Engl. Transl.), **53**, 943 (1984).
115. M. P. Doyle, B. Siegfried, R. C. Elliot, and J. F. Dellaria, Jr., *J. Org. Chem.*, **42**, 2431 (1977).
116. A. Citterio and E. Vismara, *Synthesis*, 191 (1980); A. Citterio, A. Cominelli, and F. Bonavoglia, *Synthesis*, 308 (1986).

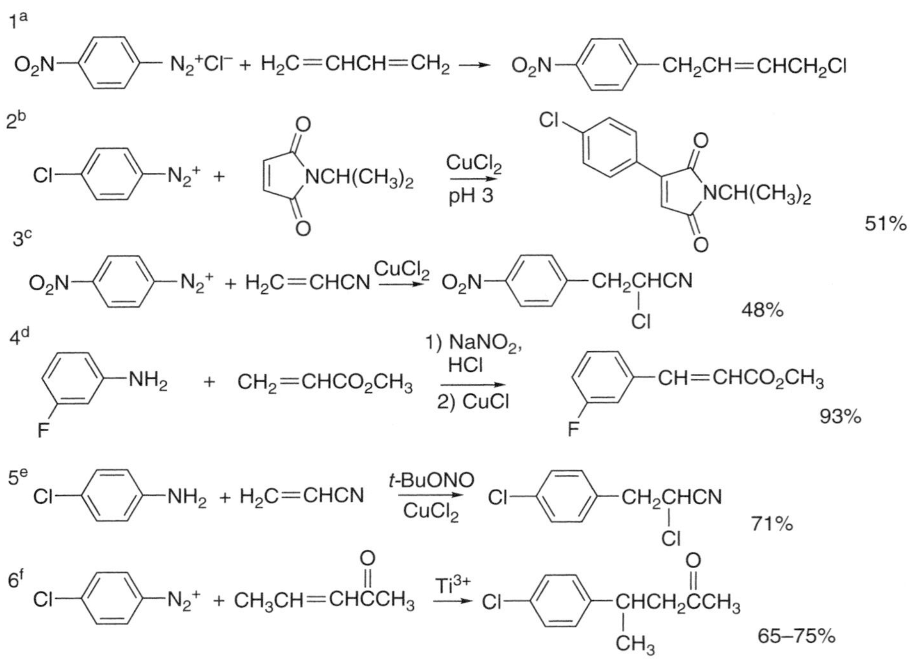

a. G. A. Ropp and E. C. Coyner, *Org. Synth.*, **IV**, 727 (1963).
b. C. S. Rondestvedt, Jr., and O. Vogel, *J. Am. Chem. Soc.*, **77**, 2313 (1955).
c. C. F. Koelsch, *J. Am. Chem. Soc.*, **65**, 57 (1943).
d. G. Theodoridis and P. Malamus, *J. Heterocycl. Chem.*, **28**, 849 (1991).
e. M. P. Doyle, B. Siegfried, R. C. Elliott, and J. F. Dellaria, Jr., *J. Org. Chem.*, **42**, 2431 (1977).
f. A. Citterio and E. Vismara, *Synthesis*, 191 (1980); A. Citterio, *Org. Synth.*, **62**, 67 (1984).

Nucleophilic substitution occurs when there is a potential leaving group present at the carbon at which addition occurs. Although halides are the most common leaving groups, alkoxy, cyano, nitro, and sulfonyl groups can also be displaced. The leaving group ability does not necessarily parallel that found for nucleophilic substitution at saturated carbon. As a particularly striking example, fluoride is often a better leaving group than the other halogens in nucleophilic aromatic substitution. The relative reactivity of the *p*-halonitrobenzenes toward sodium methoxide at 50°C is F(312) >> Cl(1) > Br (0.74) > I (0.36).[117] A principal reason for the order I > Br > Cl > F in S_N2 reactions is the carbon-halogen bond strength, which increases from I to F. The carbon-halogen bond strength is not so important a factor in nucleophilic aromatic substitution because bond breaking is not ordinarily part of the rate-determining step. Furthermore, the highly electronegative fluorine favors the addition step more than the other halogens.

The addition-elimination mechanism has been used primarily for arylation of oxygen and nitrogen nucleophiles. There are not many successful examples of arylation of carbanions by this mechanism. A major limitation is the fact that aromatic nitro

[117] G. P. Briner, J. Mille, M. Liveris, and P. G. Lutz, *J. Chem. Soc.*, 1265 (1954).

compounds often react with carbanions by electron transfer processes.[118] However, substitution by carbanions can be carried out under the conditions of the $S_{RN}1$ reaction (see Section 11.4).

The pyridine family of heteroaromatic nitrogen compounds is reactive toward nucleophilic substitution at the C(2) and C(4) positions. The nitrogen atom serves to activate the ring toward nucleophilic attack by stabilizing the addition intermediate. This kind of substitution reaction is especially important in the chemistry of pyrimidines.

Ref. 119

Ref. 120

A variation of the aromatic nucleophilic substitution process in which the leaving group is part of the entering nucleophile has been developed and is known as *vicarious nucleophilic aromatic substitution*. These reactions require a strong EWG substituent such as a nitro group but require no halide or other leaving group. The reactions proceed through addition intermediates.[121]

The combinations $Z = CN$, RSO_2, CO_2R, and SR and $X = F$, Cl, Br, I, ArO, ArS, and $(CH_3)_2NCS_2$ are among those that have been demonstrated.[122]

Scheme 11.8 gives some examples of addition-elimination reactions. Entries 1 and 2 illustrate typical *o*- and *p*-nitrophenylations of amines. Note the rather vigorous conditions that are required. Entry 3 shows a rather unusual case in which an acetyl group is the activating substituent. Good yields were obtained for a number of amines in polar aprotic solvents. The corresponding chloro and bromo derivative were much less reactive. Entry 4 represents a case of a very electrophilic aromatic ring, but

[118]. R. D. Guthrie, in *Comprehensive Carbanion Chemistry*, Part A, E. Buncel and T. Durst, eds., Elsevier, Amsterdam, 1980, Chap. 5.

[119]. J. A. Montgomery and K. Hewson, *J. Med. Chem.*, **9**, 354 (1966).

[120]. D. J. Brown, B. T. England, and J. M. Lyall, *J. Chem. Soc. C*, 226 (1966).

[121]. M. Makosza, T. Lemek, A. Kwast, and F. Terrier, *J. Org. Chem.*, **67**, 394 (2002); M. Makosza and A. Kwast, *J. Phys. Org. Chem.*, **11**, 341 (1998).

[122]. M. Makosza and J. Winiarski, *J. Org. Chem.*, **45**, 1534 (1980); M. Makosza, J. Golinski, and J. Baran, *J. Org. Chem.*, **49**, 1488 (1984); M. Makosza and J. Winiarski, *J. Org. Chem.*, **49**, 1494 (1984); M. Makosza and J. Winiarski, *J. Org. Chem.*, **49**, 5272 (1984); M. Makosza and J. Winiarski, *Acc. Chem. Res.*, **20**, 282 (1987); M. Makosza and K. Wojciechowski, *Liebigs Ann. Chem./Recueil*, 1805 (1997).

Scheme 11.8. Nucleophilic Aromatic Substitution

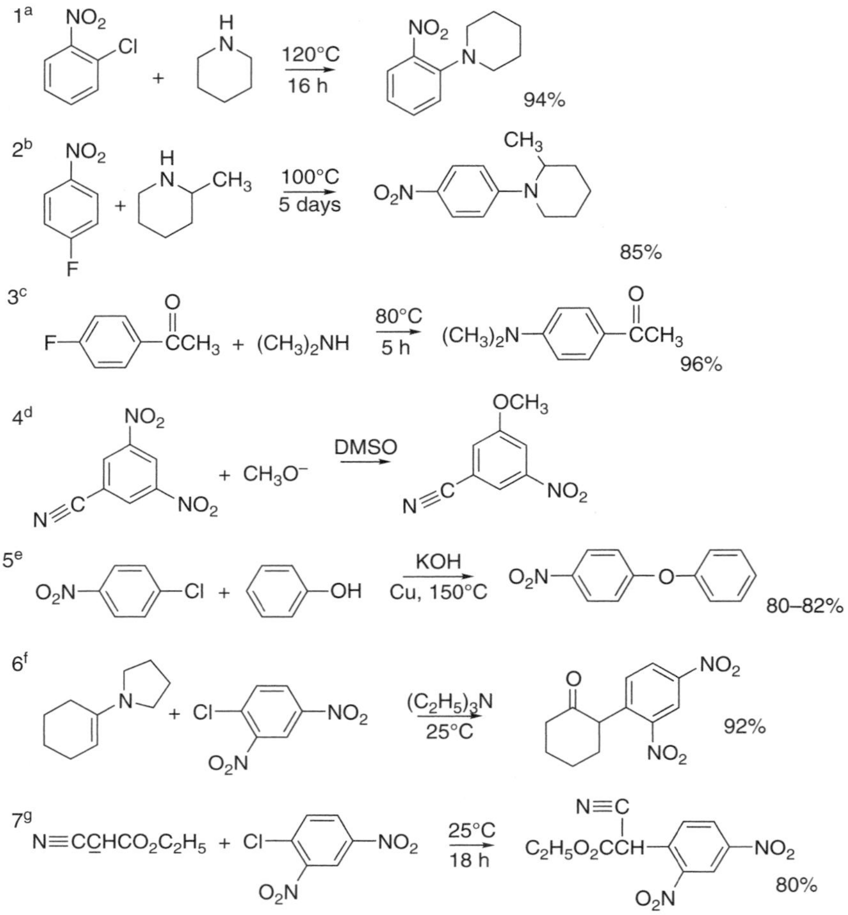

a. S. D. Ross and M. Finkelstein, *J. Am. Chem. Soc.*, **85**, 2603 (1963).
b. F. Pietra and F. Del Cima, *J. Org. Chem.*, **33**, 1411 (1968).
c. H. Bader, A. R. Hansen, and F. J. McCarty, *J. Org. Chem.*, **31**, 2319 (1966).
d. E. J. Fendler, J. H. Fendler, N. I. Arthur, and C. E. Griffin, *J. Org. Chem.*, **37**, 812 (1972).
e. R. O. Brewster and T. Groening, *Org. Synth.*, **II**, 445 (1943).
f. M. E. Kuehne, *J. Am. Chem. Soc.*, **84**, 837 (1962).
g. H. R. Snyder, E. P. Merica, C. G. Force, and E. G. White, *J. Am. Chem. Soc.*, **80**, 4622 (1958).

the favored addition intermediate does not have a potential leaving group. Reaction evidently occurs through a minor adduct.

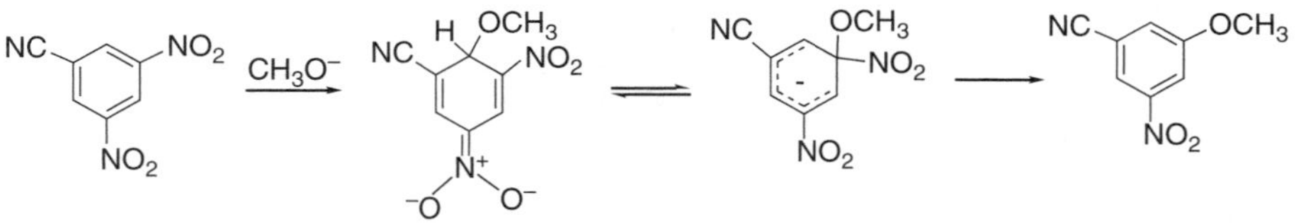

Entry 5 involves metallic copper as a catalyst and is probably a metal-catalyzed reaction (see Section 11.3). The reaction is carried out with excess phenol without solvent. Entries 6 and 7 are cases of C-arylation, both using 2,4-dinitrochlorobenzene.

11.2.3. Substitution by the Elimination-Addition Mechanism

The elimination-addition mechanism involves a highly unstable intermediate called *dehydrobenzene or benzyne*.[123] (See Section 10.6 of Part A for a discussion of the structure of benzyne.)

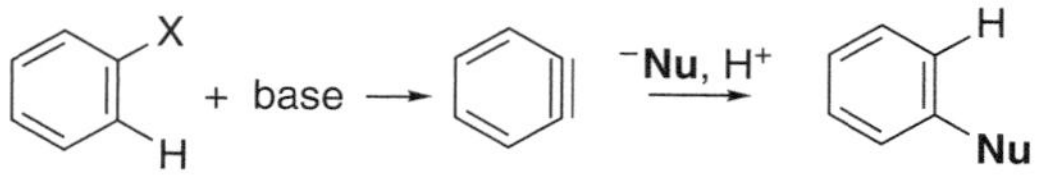

A unique feature of this mechanism is that the entering nucleophile does not necessarily become bound to the carbon to which the leaving group was attached.

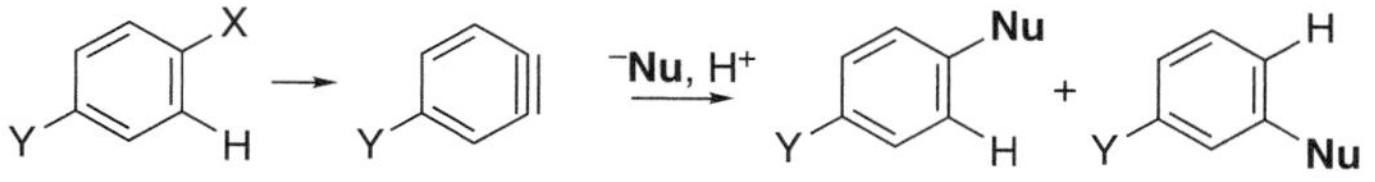

The elimination-addition mechanism is facilitated by electronic effects that favor removal of a hydrogen from the ring as a proton. Relative reactivity also depends on the halide. The order Br > I > Cl >> F has been established in the reaction of aryl halides with KNH_2 in liquid ammonia[124] and has been interpreted as representing a balance of two effects. The polar order favoring proton removal would be F > Cl > Br > I, but this is largely overwhelmed by the ease of bond breaking, which is I > Br > Cl > F. With organolithium reagents in ether solvents, the order of reactivity is F > Cl > Br > I, which indicates that the acidity of the ring hydrogen is the dominant factor governing reactivity.[125]

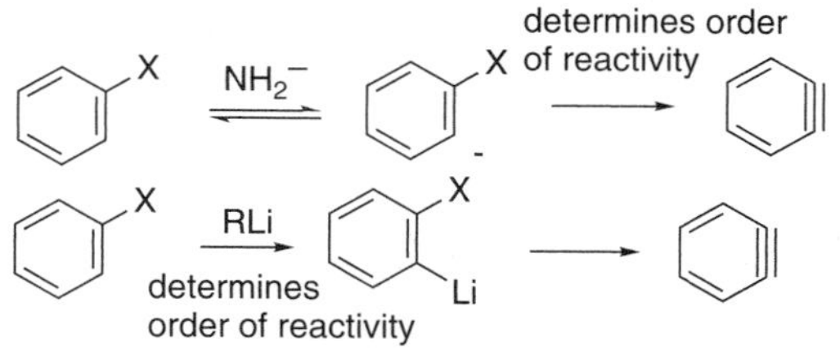

Benzyne can also be generated from *o*-dihaloaromatics. Reaction with lithium amalgam or magnesium results in the formation of transient organometallic compounds that decompose with elimination of lithium halide. *o*-Fluorobromobenzene is the usual starting material in this procedure.[126]

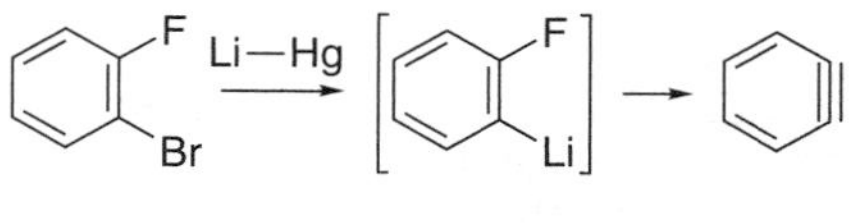

123. R. W. Hoffmann, *Dehydrobenzene and Cycloalkynes*, Academic Press, New York, 1967.
124. F. W. Bergstrom, R. E. Wright, C. Chandler, and W. A. Gilkey, *J. Org. Chem.*, **1**, 170 (1936).
125. R. Huisgen and J. Sauer, *Angew. Chem.*, **72**, 91 (1960).
126. G. Wittig and L. Pohmer, *Chem. Ber.*, **89**, 1334 (1956); G. Wittig, *Org. Synth.*, **IV**, 964 (1963).

There are several methods for generation of benzyne in addition to base-catalyzed elimination of hydrogen halide from a halobenzene and some of these are more generally applicable for preparative work. Probably the most useful method is diazotization of *o*-aminobenzoic acids.[127] Loss of nitrogen and carbon dioxide follows diazotization and generates benzyne. This method permits generation of benzyne in the presence of a number of molecules with which it can react.

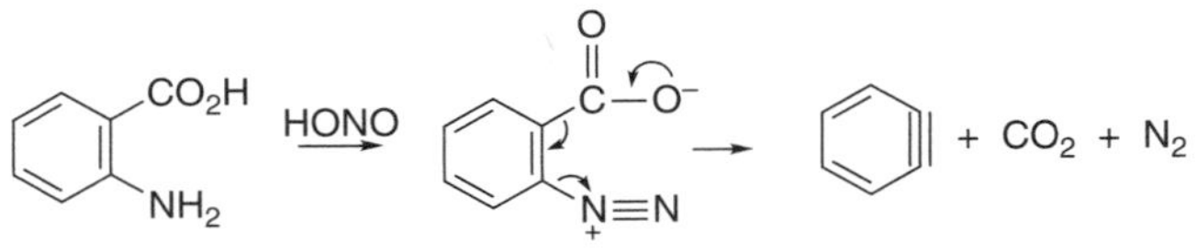

Oxidation of 1-aminobenzotriazole also serves as a source of benzyne under mild conditions. An oxidized intermediate decomposes with loss of two molecules of nitrogen.[128]

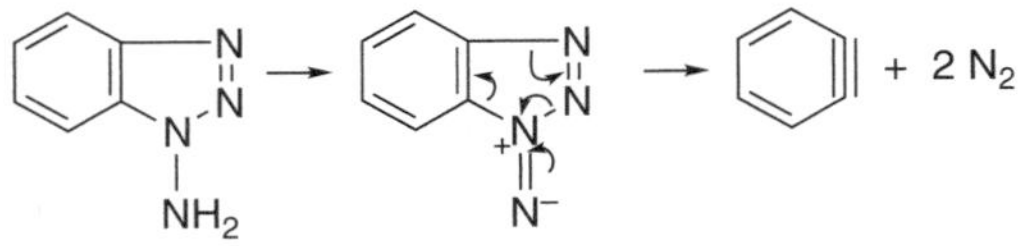

Another heterocyclic molecule that can serve as a benzyne precursor is benzothiadiazole-1,1-dioxide, which decomposes with elimination of nitrogen and sulfur dioxide.[129]

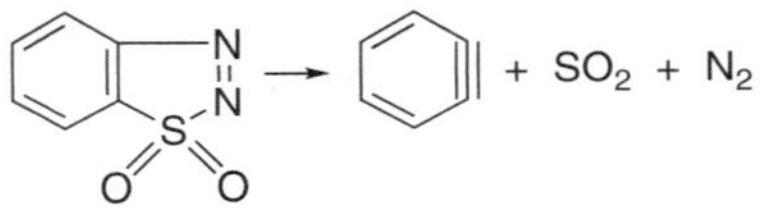

Addition of nucleophiles such as ammonia or alcohols, or their conjugate bases, to benzynes takes place very rapidly. The addition is believed to involve capture of the nucleophile by benzyne, followed by protonation to give the substitution product.[130] Electronegative groups tend to favor addition of the nucleophile at the more distant end of the "triple bond," since this permits stabilization of the developing negative charge. Selectivity is usually not high, however, and formation of both possible products from monosubstituted benzynes is common.[131]

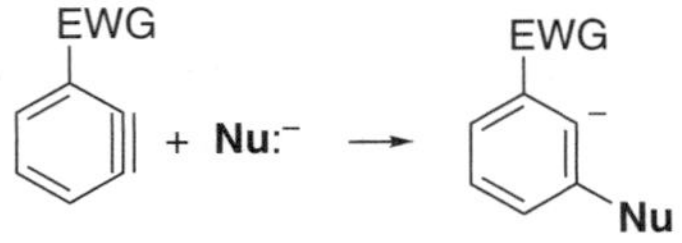

127. M. Stiles, R. G. Miller, and U. Burckhardt, *J. Am. Chem. Soc.*, **85**, 1792 (1963); L. Friedman and F. M. Longullo, *J. Org. Chem.*, **34**, 3089 (1969).

128. C. D. Campbell and C. W. Rees, *J. Chem. Soc. C*, 742, 752 (1969); S. E. Whitney and B. Rickborn, *J. Org. Chem.*, **53**, 5595 (1988); H. Hart and D. Ok, *J. Org. Chem.*, **52**, 3835 (1987).

129. G. Wittig and R. W. Hoffmann, *Org. Synth.*, **47**, 4 (1967); G. Wittig and R. W. Hoffmann, *Chem. Ber.*, **95**, 2718, 2729 (1962).

130. J. F. Bunnett, D. A. R. Happer, M. Patsch, C. Pyun, and H. Takayama, *J. Am. Chem. Soc.*, **88**, 5250 (1966); J. F. Bunnett and J. K. Kim, *J. Am. Chem. Soc.*, **95**, 2254 (1973).

131. E. R. Biehl, E. Nieh, and K. C. Hsu, *J. Org. Chem.*, **34**, 3595 (1969).

When benzyne is generated in the absence of another reactive molecule it dimerizes to biphenylene.[132] In the presence of dienes, benzyne is a very reactive dienophile and [4+2] cycloaddition products are formed. The adducts with furans can be converted to polycyclic aromatic compounds by elimination of water. Similarly, cyclopentadienones can give a new aromatic ring by loss of carbon monoxide. Pyrones give adducts that can aromatize by loss of CO_2, as illustrated by Entry 7 in Scheme 11.9.

Ref. 133

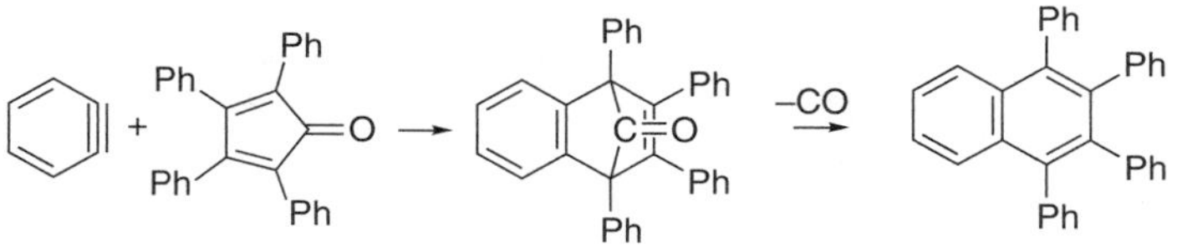

Ref. 134

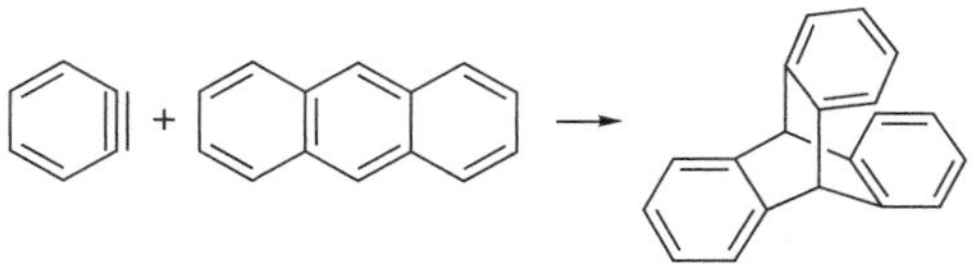

Ref. 135

Benzyne gives both [2+2] cycloaddition and ene reaction products with simple alkenes.[136]

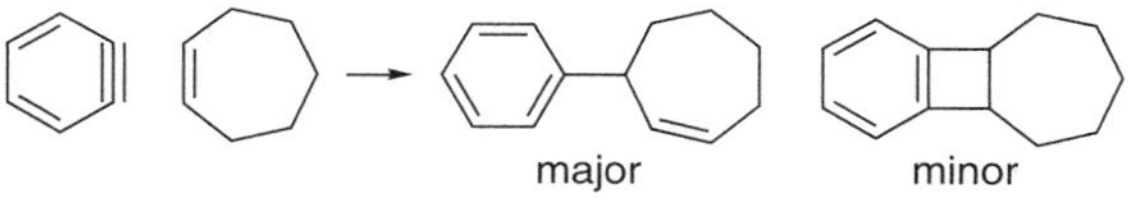

Scheme 11.9 illustrates some of the types of compounds that can be prepared via benzyne intermediates. Entry 1 is an example of the generation of benzyne in a strongly basic DMSO solution. Entry 2 is a Diels-Alder reaction involving in situ generation of benzyne. The adduct was used to synthesize several polycyclic strained-ring systems having fused benzene rings. Entry 3 illustrates the formation of benzyne from *o*-bromofluorobenzene by reaction with magnesium. The benzyne undergoes a Diels-Alder reaction with anthracene. Entry 4 also uses this method of benzyne generation and results in a [2+2] cycloaddition with an enamine. Entry 5 is photolytic generation of benzyne employing phthaloyl peroxide. This method seems to have been used only rarely. Entry 6 shows a case of intramolecular trapping of benzyne by a nitrile-stabilized carbanion. Entry 7 is a Diels-Alder reaction with a pyrone, in which the adduct undergoes decarboxylation under the reaction conditions.

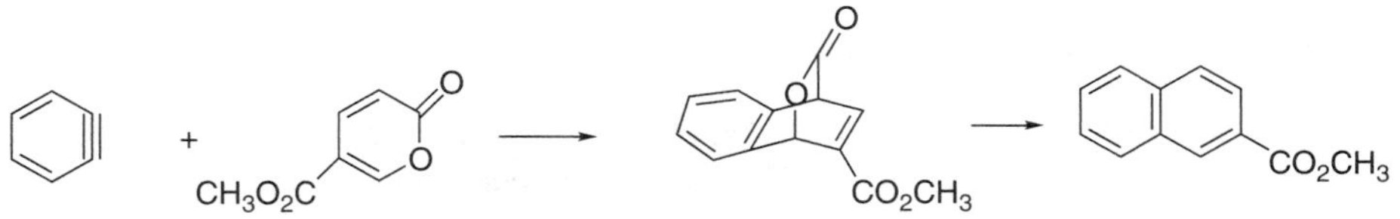

132. F. M. Logullo, A. H. Seitz, and L. Friedman, *Org. Synth.*, **V**, 54 (1973).
133. G. Wittig and L. Pohmer, *Angew. Chem.*, **67**, 348 (1955).
134. L. F. Fieser and M. J. Haddadin, *Org. Synth.*, **V**, 1037 (1973).
135. L. Friedman and F. M. Logullo, *J. Org. Chem.*, **34**, 3089 (1969).
136. P. Crews and J. Beard, *J. Org. Chem.*, **38**, 522 (1973).

Scheme 11.9. Syntheses via Benzyne Intermediates

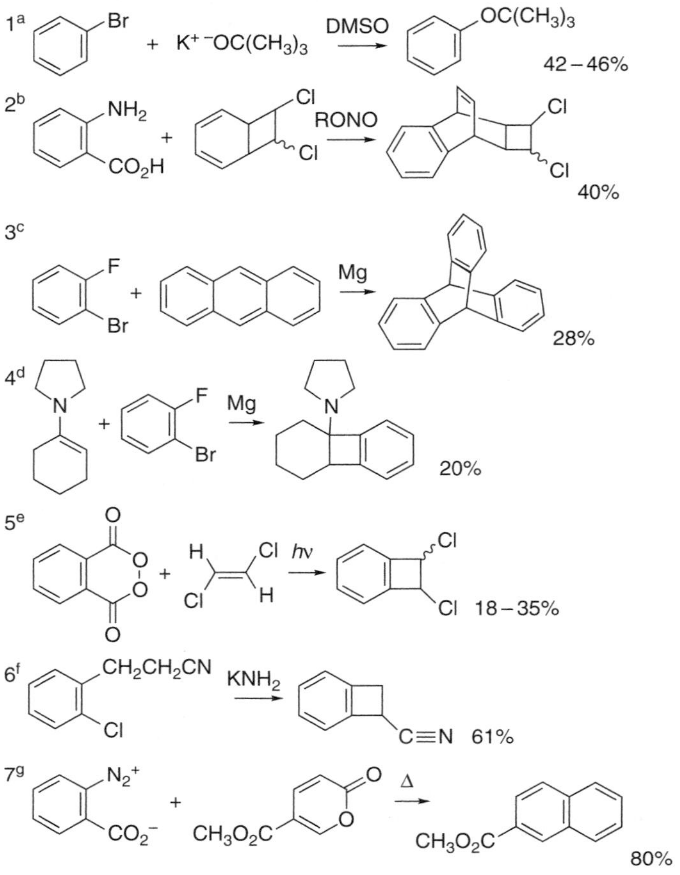

a. M. R. V. Sahyun and D. J. Cram, *Org. Synth.*, **45**, 89 (1965).
b. L. A. Paquette, M. J. Kukla, and J. C. Stowell, *J. Am. Chem. Soc.*, **94**, 4920 (1972).
c. G. Wittig, *Org. Synth.*, **IV**, 964 (1963).
d. M. E. Kuehne, *J. Am. Chem. Soc.*, **84**, 837 (1962).
e. M. Jones, Jr., and M. R. DeCamp, *J. Org. Chem.*, **36**, 1536 (1971).
f. J. F. Bunnett and J. A. Skorcz, *J. Org. Chem.*, **27**, 3836 (1962).
g. S. Escudero, D. Perez, E. Guitan, and L. Castedo, *Tetrahedron Lett.*, **38**, 5375 (1997).

11.3. Transition Metal–Catalyzed Aromatic Substitution Reactions

11.3.1. Copper-Catalyzed Reactions

As noted in Section 11.2.2, nucleophilic substitution of aromatic halides lacking activating substituents is generally difficult. It has been known for a long time that the nucleophilic substitution of aromatic halides can be catalyzed by the presence of copper metal or copper salts.[137] Synthetic procedures based on this observation are used to prepare aryl nitriles by reaction of aryl bromides with Cu(I)CN. The reactions are usually carried out at elevated temperature in DMF or a similar solvent.

[137.] J. Lindley, *Tetrahedron*, **40**, 1433 (1984).

1043

SECTION 11.3

*Transition
Metal–Catalyzed
Aromatic Substitution
Reactions*

Ref. 138

Ref. 139

A general mechanistic description of the copper-promoted nucleophilic substitution involves an oxidative addition of the aryl halide to Cu(I) followed by collapse of the arylcopper intermediate with a ligand transfer (reductive elimination).[140]

$$Ar-X + Cu(I)Z \longrightarrow Ar-\underset{\underset{X}{|}}{Cu(III)}-Z \longrightarrow Ar-Z + CuX$$

X = halide
Z = nucleophile

Several other kinds of nucleophiles can be arylated by copper-catalyzed substitution. Among the reactive nucleophiles are carboxylate ions,[141] alkoxide ions,[142] amines,[143] phthalimide anions,[144] thiolate anions,[145] and acetylides.[146] In some of these reactions there is competitive reduction of the aryl halide to the dehalogenated arene, which is attributed to protonolysis of the arylcopper intermediate. Most of these reactions are carried out at high temperature under heterogeneous conditions using copper powder or copper bronze as the catalyst. The general mechanism suggests that these catalysts act as sources of Cu(I) ions. Homogeneous reactions can be carried out using soluble Cu(I) salts, particularly Cu(I)O$_3$SCF$_3$.[147] These reactions occur under milder conditions than those using other sources of copper. The range and effectiveness of coupling aryl halides and phenolates to give diaryl ethers is improved by use of with CsCO$_3$.[148] Reaction occurs in refluxing toluene.

Some reactions of this type are accelerated further by use of naphthoic acid as an additive. This effect is believed to result from formation of a mixed anionic cuprate

138. L. Friedman and H. Shechter, *J. Org. Chem.*, **26**, 2522 (1961).
139. M. S. Newman and H. Bode, *J. Org. Chem.*, **26**, 2525 (1961).
140. T. Cohen, J. Wood, and A. G. Dietz, *Tetrahedron Lett.*, 3555 (1974).
141. T. Cohen and A. H. Lewin, *J. Am. Chem. Soc.*, **88**, 4521 (1966).
142. R. G. R. Bacon and S. C. Rennison, *J. Chem. Soc. C*, 312 (1969).
143. A. J. Paine, *J. Am. Chem. Soc.*, **109**, 1496 (1987).
144. R. G. R. Bacon and A. Karim, *J. Chem. Soc., Perkin Trans. 1*, 272 (1973).
145. H. Suzuki, H. Abe, and A. Osuka, *Chem. Lett.*, 1303 (1980); R. G. R. Bacon and H. A. O. Hill, *J. Chem. Soc.*, 1108 (1964).
146. C. E. Castro, R. Havlin, V. K. Honwad, A. Malte, and S. Moje, *J. Am. Chem. Soc.*, **91**, 6464 (1969).
147. T. Cohen and J. G. Tirpak, *Tetrahedron Lett.*, 143 (1975).
148. J. F. Marcoux, S. Doye, and S. L. Buchwald, *J. Am. Chem. Soc.*, **119**, 10539 (1997).

having naphthoate as one of the ligands. The Cs^+ salts are beneficial in maximizing the solubility of the phenolate and naphthoates.

It has been found that a number of bidentate ligands greatly expand the scope of copper catalysis. Copper(I) iodide used in conjunction with a chelating diamine is a good catalyst for amidation of aryl bromides. Of several diamines that were examined, *trans-N,N'*-dimethylcyclohexane-1,2-diamine was among the best. These conditions are applicable to aryl bromides and iodides with either ERG or EWG substituents, as well as to relatively hindered halides. The nucleophiles that are reactive under these conditions include acyclic and cyclic amides.[149]

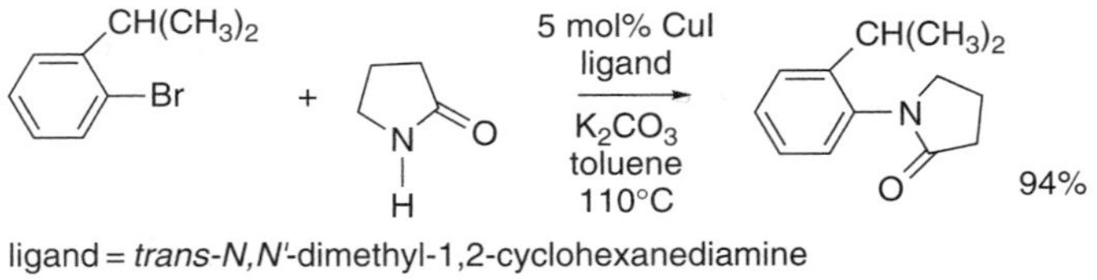

ligand = *trans-N,N'*-dimethyl-1,2-cyclohexanediamine

This catalytic system also promotes exchange of iodide for bromide on aromatic rings.[150] The reaction is an equilibrium process that is driven forward by the low solubility of NaBr in the solvent, dioxane.

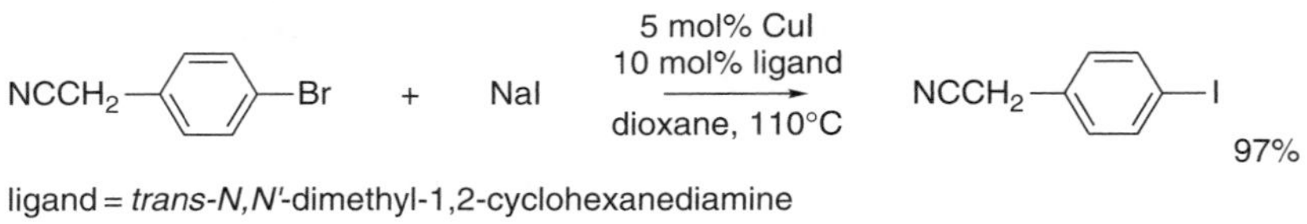

ligand = *trans-N,N'*-dimethyl-1,2-cyclohexanediamine

The *N,N*-diethylamide of salicylic acid is a useful ligand in conjunction with CuI and permits amination of aryl bromides by primary alkylamines.[151]

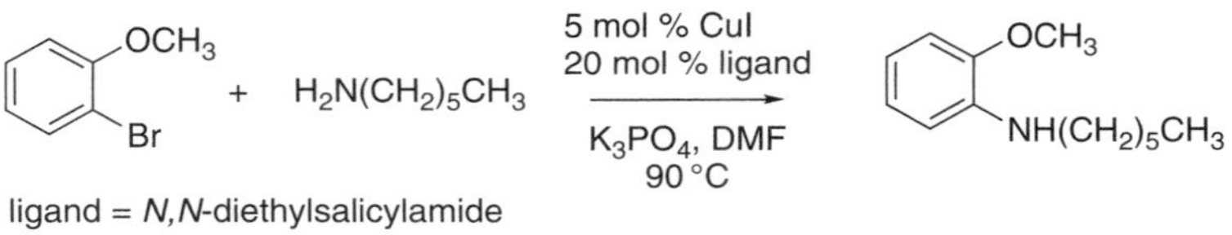

ligand = *N,N*-diethylsalicylamide

Copper(I) iodide with 1,10-phenanthroline catalyzes substitution of aryl iodides by alcohols. The reaction can be done either in excess alcohol or in toluene.[152]

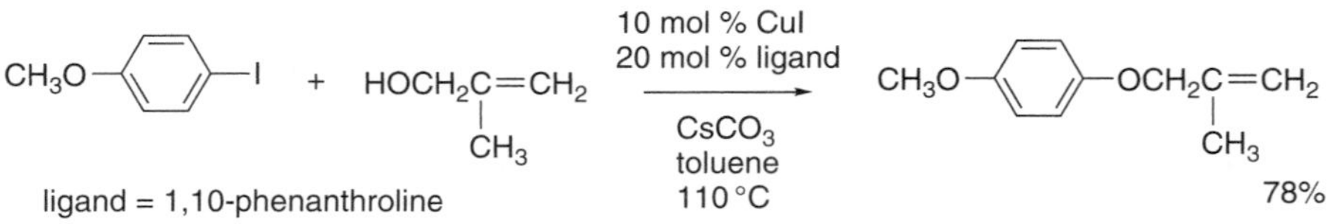

ligand = 1,10-phenanthroline

These copper-catalyzed reactions are generally applicable to aryl halides with either EWG or ERG substituents. The order of reactivity is I > Br> Cl > OSO_2R, which is consistent with an oxidative addition mechanism.

149. A. Klapars, X. Huang, and S. L. Buchwald, *J. Am. Chem. Soc.*, **124**, 7421 (2002).

150. A. Klapars and S. L. Buchwald, *J. Am. Chem. Soc.*, **124**, 14844 (2002).

151. F. Y. Kwong and S. L. Buchwald, *Org. Lett.*, **5**, 793 (2003).

152. M. Wolter, G. Nordmann, G. E. Job, and S. L. Buchwald, *Org. Lett.*, **4**, 973 (2002).

One aspect of the copper catalytic system that has received attention is the identity of the active catalytic species. In the case of displacement of aryl bromides by methoxide ion in the presence of CuBr, it has been suggested that the active species is $Cu(I)(OCH_3)_2$, an anionic cuprate.[153]

$$CuBr + 2\,NaOCH_3 \rightleftharpoons [Cu^I(OCH_3)_2]^-$$

$$[Cu^I(OCH_3)_2]^- + ArBr \xrightarrow[\text{addition}]{\text{oxidative}} [Ar-\underset{\underset{Br}{|}}{Cu^{III}}(OCH_3)_2]^- \xrightarrow[\text{elimination}]{\text{reductive}} ArOCH_3 + [Cu^IBr(OCH_3)]^-$$

11.3.2. Palladium-Catalyzed Reactions

In Section 8.2.3.2, we discussed arylation of enolates and enolate equivalents using palladium catalysts. Related palladium-phosphine combinations are very effective catalysts for aromatic nucleophilic substitution reactions. For example, conversion of aryl iodides to nitriles can be done under mild conditions with $Pd(PPh_3)_4$ as a catalyst.

$$CH_3O-\!\!\left\langle\;\right\rangle\!\!-I \xrightarrow[\substack{(CH_3)_3SiCN \\ 80\,°C}]{\substack{Pd(PPh_3)_4, \\ (C_2H_5)_3N}} CH_3O-\!\!\left\langle\;\right\rangle\!\!-CN$$

89%

Ref. 154

A great deal of effort has been devoted to finding efficient catalysts for substitution by oxygen and nitrogen nucleophiles.[155] These studies have led to optimization of the catalysis with ligands such as triarylphosphines,[156] *bis*-phosphines such as BINAP,[157] dppf,[158] and phosphines with additional chelating substituents.[159] Among the most effective catalysts are highly hindered trialkyl phosphines such as tri-*t*-butyl and tricyclohexylphosphine.[160] A series of 2-biphenylphosphines **3–6** has also been found to have excellent activity.[161]

153. H. L. Aalten, C. van Koten, D. M. Grove, T. Kuilman, O. G. Piekstra, L. A. Hulshof, and R. A. Sheldon, *Tetrahedron*, **45**, 5565 (1989).
154. N. Chatani and T. Hanafusa, *J. Org. Chem.*, **51**, 4714 (1986).
155. S. L. Buchwald, A. S. Guram, and R. A. Rennels, *Angew. Chem. Intl. Ed. Engl.*, **34**, 1348 (1995); J. F. Hartwig, *Synlett*, 329 (1997); J. F. Hartwig, *Angew. Chem. Intl. Ed. Engl.*, **37**, 2047 (1998); J. P. Wolfe, S. Wagaw, J. F. Marcoux, and S. L. Buchwald, *Acc. Chem. Res.*, **31**, 805 (1998); J. F. Hartwig, *Acc. Chem. Res.*, **31**, 852 (1998); B. H. Yang and S. L. Buchwald, *J. Organomet. Chem.*, **576**, 125 (1999).
156. J. P. Wolfe and S. L. Buchwald, *J. Org. Chem.*, **61**, 1133 (1996); J. Louie and J. F. Hartwig, *Tetrahedron Lett.*, **36**, 3609 (1995).
157. J. P. Wolfe, S. Wagaw, and S. L. Buchwald, *J. Am. Chem. Soc.*, **118**, 7215 (1996).
158. M. S. Driver and J. F. Hartwig, *J. Am. Chem. Soc.*, **118**, 7217 (1996).
159. D. W. Old, J. P. Wolfe, and S. L. Buchwald, *J. Am. Chem. Soc.*, **120**, 9722 (1998); B. C. Hamann and J. F. Hartwig, *J. Am. Chem. Soc.*, **120**, 7369 (1998); S. Vyskocil, M. Smrcina, and P. Kocovsky, *Tetrahedron Lett.*, **39**, 9289 (1998).
160. M. Nishiyama, T. Yamamoto, and Y. Koie, *Tetrahedron Lett.*, **39**, 617 (1998); N. P. Reddy and M. Tanaka, *Tetrahedron Lett.*, **38**, 4807 (1997).
161. M. C. Harris, X. Huang, and S. L. Buchwald, *Org. Lett.*, **4**, 2885 (2002); D. W. Old, J. P. Wolfe, and S. L. Buchwald, *J. Am. Chem. Soc.*, **120**, 9722 (1998); H. Tomori, J. M. Fox, and S. L. Buchwald, *J. Org. Chem.*, **65**, 5334 (2000).

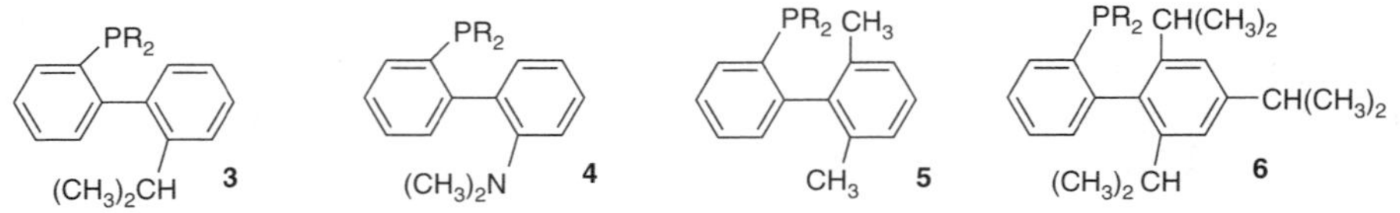

$R = t\text{-Bu}, c\text{-C}_6H_{11}$

A stable palladacycle **7** derived from biphenyl is also an active catalyst.[162]

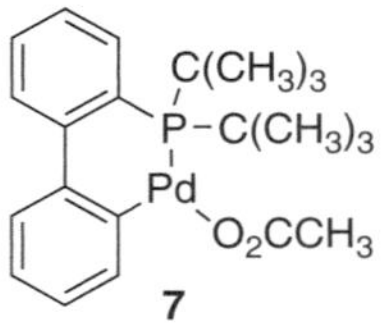

In addition to bromides and iodides, the reaction has been successfully extended to chlorides,[163] triflates,[164] and nonafluorobutanesulfonates (nonaflates).[165] These reaction conditions permit substitution in both electron-poor and electron-rich aryl systems by a variety of nitrogen nucleophiles, including alkyl or aryl amines and heterocycles. These reactions proceed via a catalytic cycle involving Pd(0) and Pd(II) intermediates.

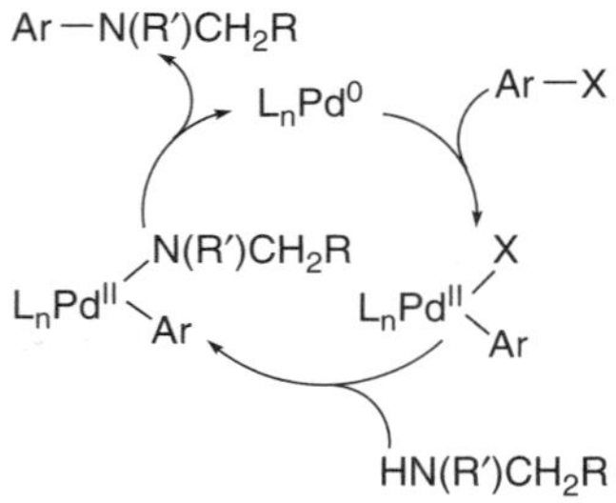

Some of the details of the mechanism may differ for various catalytic systems. There have been kinetic studies on two of the amination systems discussed here. The results of a study of the kinetics of amination of bromobenzene using Pd$_2$(dba)$_3$, BINAP, and sodium *t*-amyloxide in toluene were consistent with the oxidative addition occurring *after* addition of the amine at Pd. The reductive elimination is associated with *deprotonation of the aminated palladium complex.*[166]

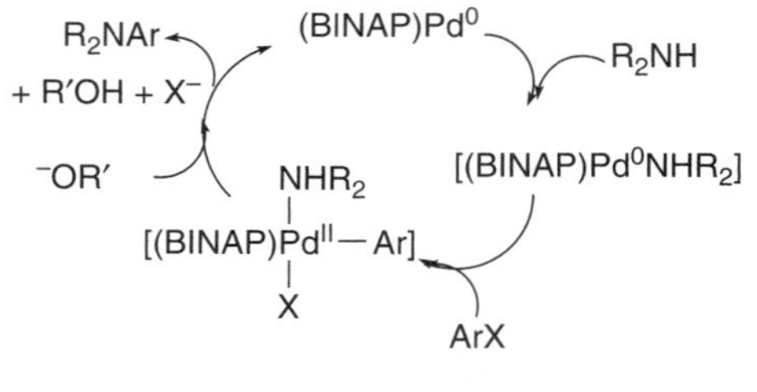

162. D. Zim and S. L. Buchwald, *Org. Lett.*, **5**, 2413 (2003).

163. X. Bei, A. S. Guram, H. W. Turner, and W. H. Weinberg, *Tetrahedron Lett.*, **40**, 1237 (1999).

164. J. P. Wolfe and S. L. Buchwald, *J. Org. Chem.*, **62**, 1264 (1997); J. Louie, M. S. Driver, B. C. Hamann, and J. T. Hartwig, *J. Org. Chem.*, 62, 1268 (1997).

165. K. W. Anderson, M. Mendez-Perez, J. Priego, and S. L. Buchwald, *J. Org. Chem.*, **68**, 9563 (2003).

166. U. K. Singh, E. R. Strieter, D. G. Blackmond, and S. L. Buchwald, *J. Am. Chem. Soc.*, **124**, 14104 (2002).

1047

SECTION 11.3

*Transition
Metal–Catalyzed
Aromatic Substitution
Reactions*

A study of the reaction of chlorobenzene with *N*-methylaniline in the presence of Pd[P(*t*-Bu)$_3$]$_2$ and several different bases indicated that two mechanisms may occur concurrently, with their relative importance depending on the base, as indicated in the catalytic cycle below. The cycle on the right depicts oxidative addition followed by ligation by the deprotonated amine. The cycle on the left suggests that oxidative addition occurs on an anionic adduct of the catalyst and the base, followed by exchange with the amine ligand.[167]

A comparison of several of the biphenylphosphine ligands has provided some insight into the mechanism of catalyst activation.[168] The results of this study suggest that dissociation of the diphosphino to a monophosphino complex is an essential step in catalyst activation, which would explain why some of the most hindered phosphines are among the best catalyst ligands. This study also indicated that deprotonation of the amine ligand is an essential step. Finally, in catalyst systems that are based on Pd(II) salts, there must be a mechanism for reduction to the active Pd(0) species. In the case of amines, this may occur by reduction by the amine ligand.

Steps in Catalyst Activation

The various palladium species can be subject to decomposition and deposition of palladium metal, which generally leads to catalyst inactivation. Apart from their effect on the catalyst activity, the ligands and bases also affect catalyst longevity.

Most of the synthetic applications to date have been based on empirical screening and comparison of ligand systems for effectiveness. A number of useful procedures have been developed. Aryl chlorides are generally less reactive than iodides and

[167.] L. M. Alcazar-Roman and J. F. Hartwig, *J. Am. Chem. Soc.*, **123**, 12905 (2001).

[168.] E. R. Strieter, D. G. Blackmond, and S. L. Buchwald, *J. Am. Chem. Soc.*, **125**, 13978 (2003).

bromides. The palladacycle **7** (see p. 1046), was used successfully in the amination of aryl chlorides.[169]

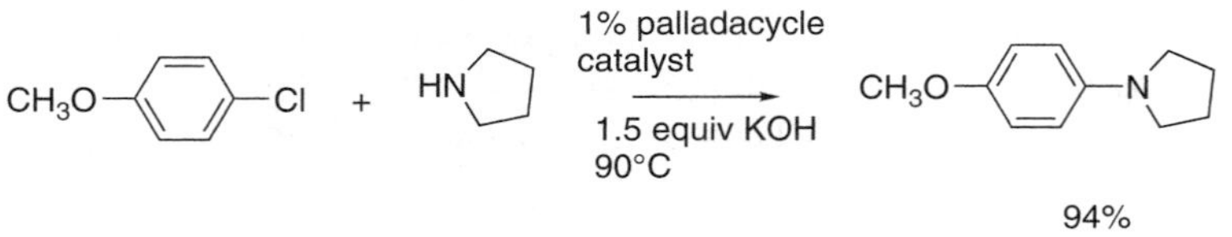

94%

Palladium-catalyzed substitution can also be applied to nonbasic nitrogen heterocycles, such as indoles, in the absence of strong bases.

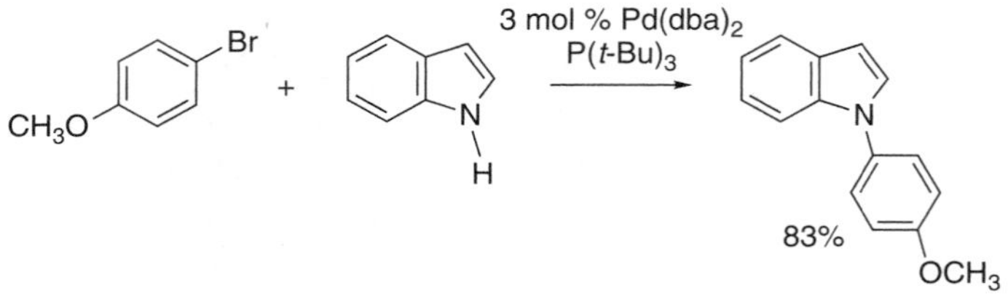

83%

Ref. 170

Except for the perfluoro cases, aryl sulfonates are generally less reactive than the halides. However certain catalyst systems can achieve reactions with benzenesulfonates and tosylates. The hindered biphenyphosphines are the most effective ligands.

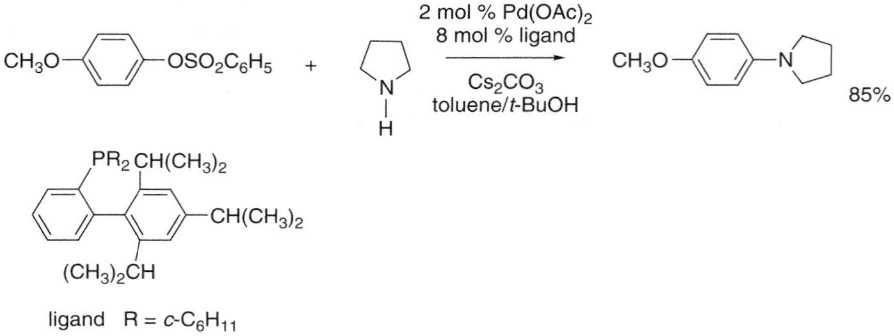

85%

ligand   R = c-C$_6$H$_{11}$

Ref. 171

These conditions were also successfully applied to arylation of amides and carbamates.

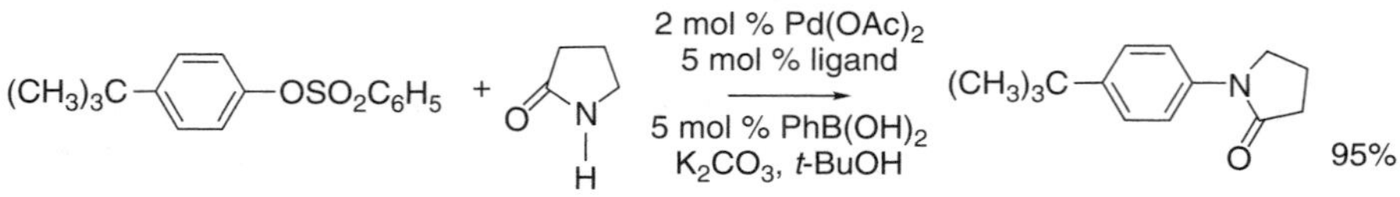

95%

169. D. Zim and S. L. Buchwald, *Org. Lett.*, **5**, 2413 (2003).
170. J. F. Hartwig, M. Kawatsura, S. I. Hauck, K. H. Shaughnessy, and L. M. Alcazar-Roman, *J. Org. Chem.*, **64**, 5575 (1999).
171. X. Huang, K. W. Anderson, D. Zim, L. Jiang, A. Klapars, and S. L. Buchwald, *J. Am. Chem. Soc.*, **125**, 6653 (2003).

1049

SECTION 11.3

*Transition
Metal–Catalyzed
Aromatic Substitution
Reactions*

Amination of tosylates has been achieved using a hindered ferrocenyldiphosphine ligand.[172]

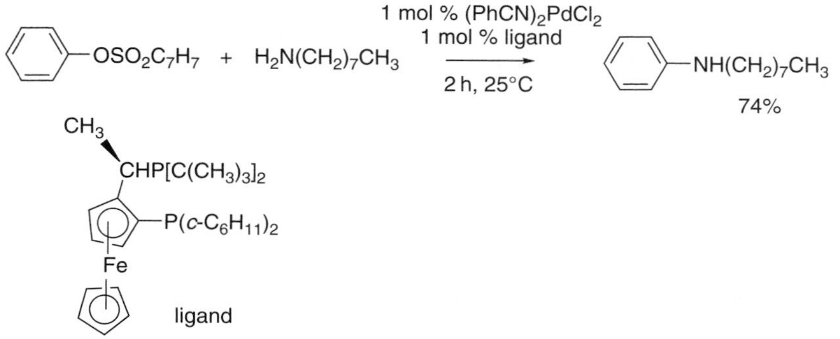

Similar reactions have been used for substitution by alkoxide and phenoxide nucleophiles. Hindered binaphthyl ligands have proven useful in substitutions by alcohols.[173]

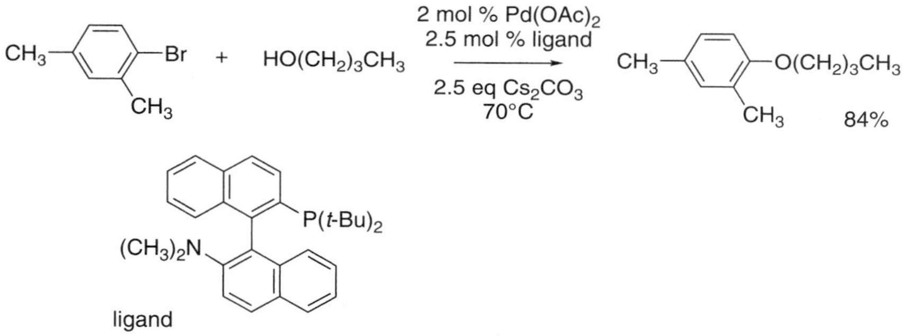

Palladium acetate in conjunction with a diphosphine ligand, xantphos, is active for arylation of amides, ureas, oxazolidinones and sulfonamides.[174]

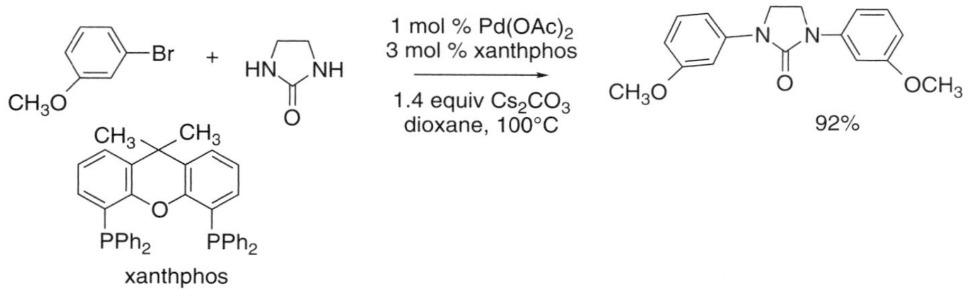

172. A. H. Roy and J. F. Hartwig, *J. Am. Chem. Soc.*, **125**, 8704 (2003).
173. K. E. Torraca, X. Huang, C. A. Parrish, and S. L. Buchwald, *J. Am. Chem. Soc.*, **123**, 10770 (2001).
174. J. Yin and S. L. Buchwald, *J. Am. Chem. Soc.*, **124**, 6043 (2002).

Scheme 11.10. Copper- and Palladium-Catalyzed Aromatic Substitution

A. Copper-catalyzed substitution

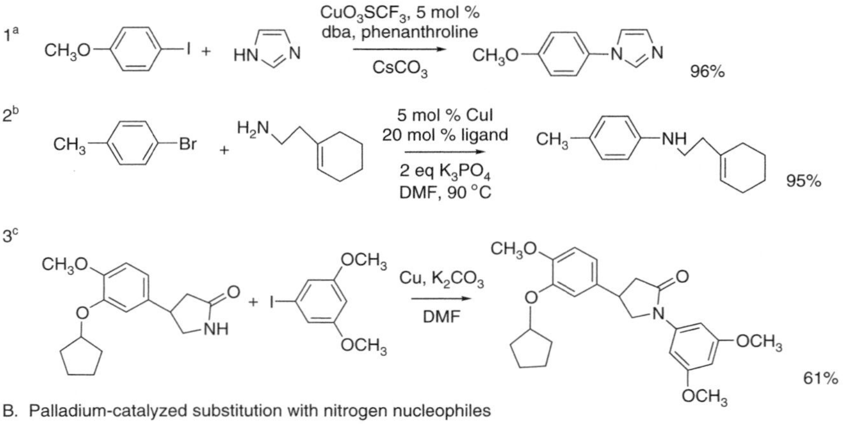

B. Palladium-catalyzed substitution with nitrogen nucleophiles

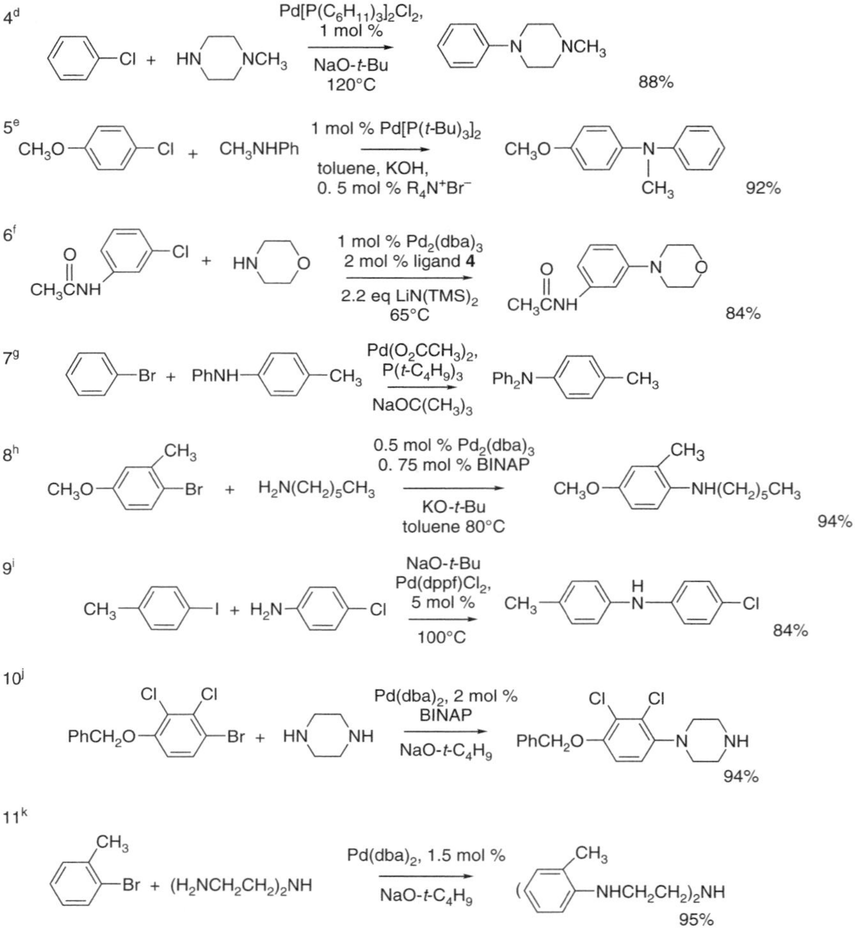

(Continued)

Scheme 11.10. (Continued)

1051

SECTION 11.3

*Transition
Metal–Catalyzed
Aromatic Substitution
Reactions*

12[l]

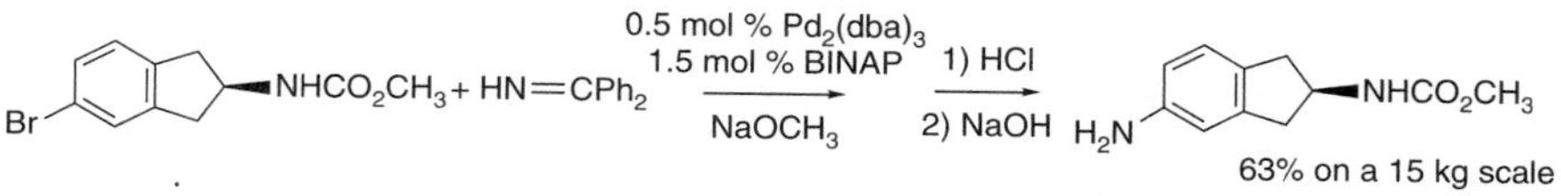

$$\text{Br}\text{—indanyl—}NHCO_2CH_3 + HN{=}CPh_2 \xrightarrow[\text{NaOCH}_3]{\substack{0.5 \text{ mol } \% \text{ Pd}_2(\text{dba})_3 \\ 1.5 \text{ mol } \% \text{ BINAP}}} \xrightarrow{\substack{1) \text{ HCl} \\ 2) \text{ NaOH}}} H_2N\text{—indanyl—}NHCO_2CH_3$$

63% on a 15 kg scale

13[m]

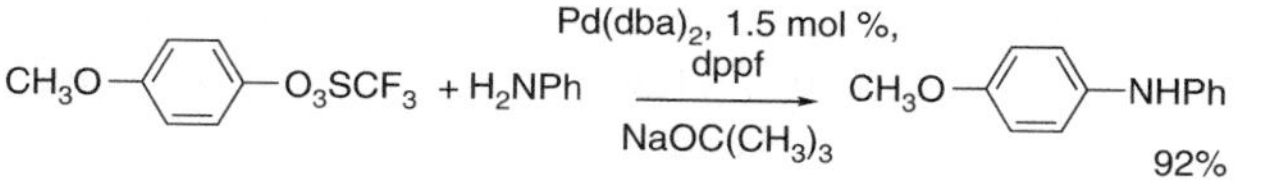

$$CH_3O{-}\langle\rangle{-}O_3SCF_3 + H_2NPh \xrightarrow[\text{NaOC(CH}_3)_3]{\substack{\text{Pd(dba)}_2, 1.5 \text{ mol } \%, \\ \text{dppf}}} CH_3O{-}\langle\rangle{-}NHPh$$

92%

14[n]

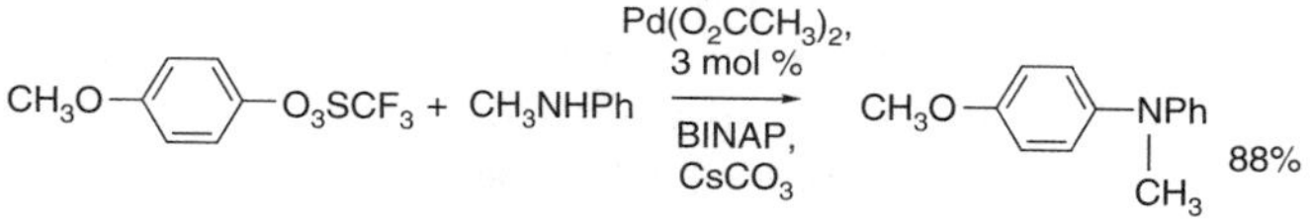

$$CH_3O{-}\langle\rangle{-}O_3SCF_3 + CH_3NHPh \xrightarrow[\substack{\text{BINAP,} \\ \text{CsCO}_3}]{\substack{\text{Pd(O}_2\text{CCH}_3)_2, \\ 3 \text{ mol } \%}} CH_3O{-}\langle\rangle{-}N(CH_3)Ph$$

88%

15[o]

$$\text{(piperidinone)} + \text{BrPh} \xrightarrow[\text{NaO-}t\text{-C}_4\text{H}_9]{\substack{\text{Pd(O}_2\text{CCH}_3)_2, 5 \text{ mol } \%, \\ \text{dppf}}} \text{(N-Ph piperidinone)}$$

95%

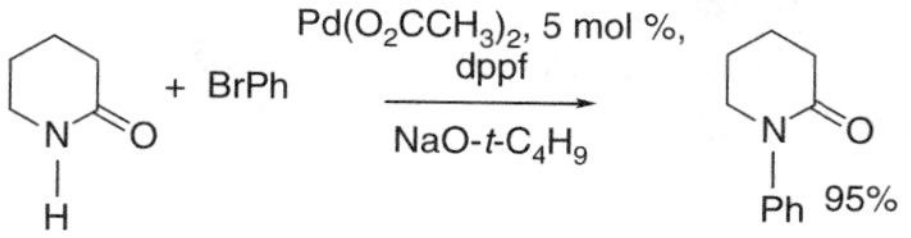

C. Palladium-catalyzed reactions with oxygen nucleophiles.

16[p]

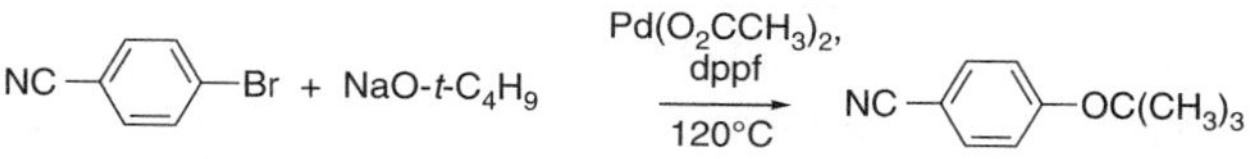

$$NC{-}\langle\rangle{-}Br + NaO\text{-}t\text{-}C_4H_9 \xrightarrow[120°C]{\substack{\text{Pd(O}_2\text{CCH}_3)_2, \\ \text{dppf}}} NC{-}\langle\rangle{-}OC(CH_3)_3$$

17[q]

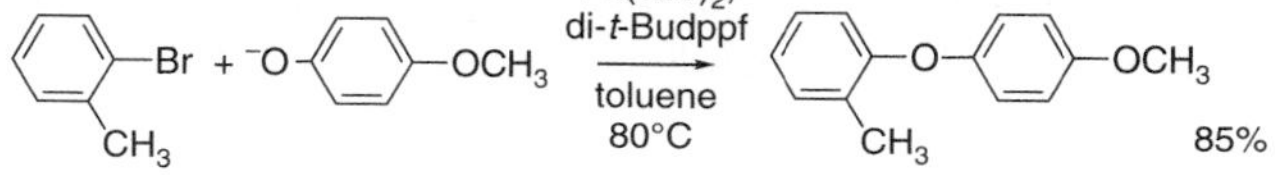

$$\text{(o-tolyl)}\text{—Br} + {}^-O{-}\langle\rangle{-}OCH_3 \xrightarrow[\substack{\text{toluene} \\ 80°C}]{\substack{\text{Pd(dba)}_2, \\ \text{di-}t\text{-Budppf}}} \text{(o-tolyl)}{-}O{-}\langle\rangle{-}OCH_3$$

85%

18[r]

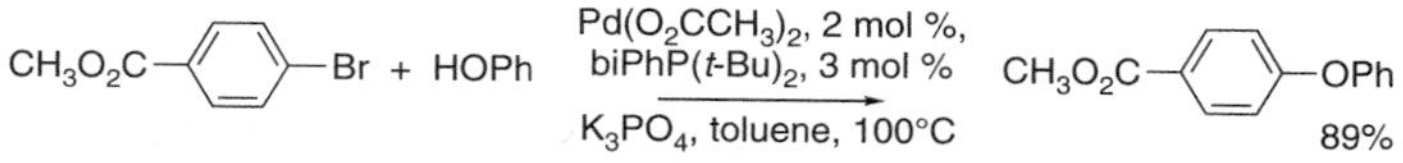

$$CH_3O_2C{-}\langle\rangle{-}Br + HOPh \xrightarrow[\text{K}_3\text{PO}_4, \text{ toluene, } 100°C]{\substack{\text{Pd(O}_2\text{CCH}_3)_2, 2 \text{ mol } \%, \\ \text{biPhP(}t\text{-Bu)}_2, 3 \text{ mol } \%}} CH_3O_2C{-}\langle\rangle{-}OPh$$

89%

19[s]

$$CH_3(CH_2)_3{-}\langle\rangle{-}Cl + NaOC(CH_3)_3 \xrightarrow[\text{toluene, } 100°C]{\substack{2.5 \text{ mol } \% \text{ Pd(OAc)}_2 \\ 3 \text{ mol } \% \text{ MebiPhP(}t\text{-Bu)}_2}} CH_3(CH_2)_3{-}\langle\rangle{-}OC(CH_3)_3$$

92%

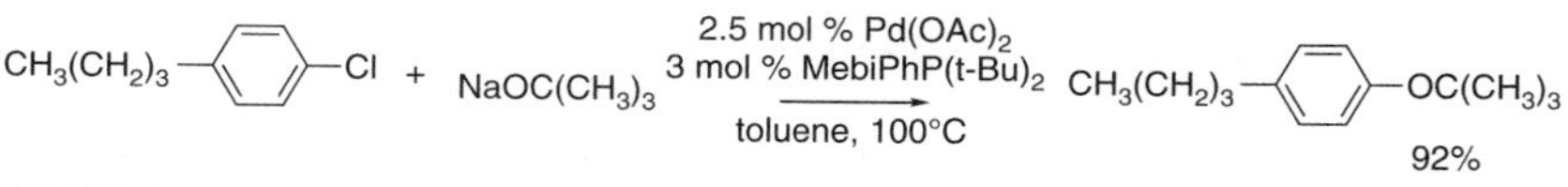

a. A. Kiyomori, J.-F. Marcoux, and S. L. Buchwald, *Tetrahedron Lett.*, **40**, 2657 (1999).
b. F. Y. Kwong and S. L. Buchwald, *Org. Lett.*, **5**, 793 (2003).
c. E. Aebischer, E. Bacher, F. W. J. Demnitz, T. H. Keller, M. Kurzmeyer, M. L. Ortiz, E. Pombo-Villar, and H.-P. Weber, *Hetereocycles*, **48**, 2225 (1998).
d. N. P. Reddy and M. Tanaka, *Tetrahedron Lett.*, **38**, 4807 (1997).
e. R. Kuwano, M. Utsunomiya, and J. F. Hartwig, *J. Org. Chem.*, **67**, 6479 (2002).
f. M. C. Harris, X. Huang, and S. L. Buchwald, *Org. Lett.*, **4**, 2885 (2002).
g. T. Yamamoto, M. Nishiyama, and Y. Koie, *Tetrahedron Lett.*, **39**, 2367 (1998).
h. K. E. Torraca, X. Huang, C. A. Parrish, and S. L. Buchwald, *J. Am. Chem. Soc.*, **123**, 10770 (2001).
i. M. S. Driver and J. F. Hartwig, *J. Am. Chem. Soc.*, **118**, 7217 (1996).
j. S. Morita, K. Kitano, J. Matsubara, T. Ohtani, Y. Kawano, K. Otsubo, and M. Uchida, *Tetrahedron*, **54**, 4811 (1998).
k. Y. Hong, C. H. Senanayake, T. Xiang, C. P. Vandenbossche, G. J. Tanoury, R. P. Bakale, and S. A. Wald, *Tetrahedron Lett.*, **39**, 3121 (1998).
l. M. Prashad, B. Hu, D. Har, O. Repic, T. J. Blacklock, and M. Avemoglu, *Adv. Synth. Catal.*, **343**, 461 (2001).
m. J. Louie, M. S. Driver, B. C. Hamann, and J. F. Hartwig, *J. Org. Chem.*, **62**, 1268 (1997).
n. J. Ahman and S. L. Buchwald, *Tetrahedron Lett.*, **38**, 6363 (1997).
o. W. C. Shakespeare, *Tetrahedron Lett.*, **40**, 2035 (1999).
p. G. Mann and J. F. Hartwig, *J. Org. Chem.*, **62**, 5413 (1997).
q. G. Mann, C. Incarvito, A. L. Rheingold, and J. F. Hartwig, *J. Am. Chem. Soc.*, **121**, 3224 (1999).
r. A. Aranyos, D. W. Old, A. Kiyomori, J. P. Wolfe, J. P. Sadighi, and S. L. Buchwald, *J. Am. Chem. Soc.*, **121**, 4369 (1999).
s. C. A. Parrish and S. L. Buchwald, *J. Org. Chem.*, **66**, 2498 (2001).

Some other examples of metal-catalyzed substitutions are given in Scheme 11.10. Entries 1 to 3 are copper-catalyzed reactions. Entry 1 is an example of arylation of imidazole. Both dibenzylideneacetone and 1,10-phenanthroline were included as ligands and Cs_2CO_3 was used as the base. Entry 2 is an example of amination by a primary amine. The ligand used in this case was *N,N*-diethylsalicylamide. These conditions proved effective for a variety of primary amines and aryl bromides with both ERG and EWG substituents. Entry 3 is an example of more classical conditions. The target structure is a phosphodiesterase inhibitor of a type used in treatment of asthma. Copper powder was used as the catalyst.

The remainder of the entries in Scheme 11.10 depict palladium-catalyzed reactions. Entries 4 to 6 are examples of aminations of aryl chlorides. In Entry 4, a Pd(II) salt with a hindered phosphine ligand was used as the catalyst. Entry 5 uses the Pd(0)-tri-(*t*-butyl)phosphine complex as the catalyst in conjunction with a phase transfer salt. The reaction was done in a water-toluene mixture and these conditions were applicable to chlorides with both ERG and EWG substituents. Entry 6 used the biphenyl ligand **4** (see p. 1046). LiHMDS was a particularly good base in this case. Entries 7 to 11 use bromides (or iodides) as reactants and *t*-alkoxides as bases. In cases where the catalyst source is a Pd(II) salt, catalyst activation by reduction is necessary. Entry 12 is a large-scale amination carried out using the imine of benzophenone as the nucleophile, with subsequent hydrolysis to provide the amine. Entries 13 and 14 use aryl triflates as reactants. Again, the palladium sources must be reduced as part of catalyst activation. Entry 15 is an example of arylation of an amide. The conditions are similar to those for amination, and subsequent studies have shown that many other nonbasic nitrogen compounds can be arylated (e.g. see p. 1049). Entries 16 to 19 involve alkoxide and phenoxide nucleophiles. The best ligands for these reactions seem to be highly hindered phosphines.

11.4. Aromatic Substitution Reactions Involving Radical Intermediates

11.4.1. Aromatic Radical Substitution

Aromatic rings are moderately reactive toward addition of free radicals (see Part A, Section 12.2) and certain synthetically useful substitution reactions involve free radical substitution. One example is the synthesis of biaryls.[175]

There are some inherent limits to the usefulness of such reactions. Radical substitutions are only moderately sensitive to substituent directing effects, so substituted reactants usually give a mixture of products. This means that the practical utility is limited to symmetrical reactants, such as benzene, where the position of attack

---

[175.] W. E. Bachmann and R. A. Hoffman, *Org. React.*, **2**, 224 (1944); D. H. Hey, *Adv. Free Radical Chem.*, **2**, 47 (1966).

is immaterial. The best sources of aryl radicals are aryl diazonium ions and *N*-nitrosoacetanilides. In the presence of base, diazonium ions form diazooxides, which decompose to aryl radicals.[176]

$$ArN \equiv N + 2\ {}^-OH \longrightarrow ArN = N - O - N = NAr + H_2O$$
$$ArN = N - O - N = NAr \longrightarrow Ar\cdot\ + N_2 +\ \cdot O - N = NAr$$

In the classical procedure, base is added to a two-phase mixture of the aqueous diazonium salt and an excess of the aromatic that is to be substituted. Improved yields can be obtained by using polyethers or phase transfer catalysts with solid aryl diazonium tetrafluoroborate salts in an excess of the aromatic reactant.[177] Another source of aryl radicals is *N*-nitrosoacetanilides, which rearrange to diazonium acetates and give rise to aryl radicals via diazo oxides.[178]

$$\underset{\underset{O}{\overset{\displaystyle N=O}{\underset{\|}{ArNCCH_3}}}}{} \longrightarrow \underset{\overset{\|}{O}}{ArN=N-OCCH_3}$$

$$2\ \underset{\overset{\|}{O}}{ArN=N-OCCH_3} \longrightarrow ArN=N-O-N=NAr + (CH_3CO)_2O$$

A procedure for arylation involving in situ diazotization has also been developed.[179]

Scheme 11.11 gives some representative preparative reactions based on these methods. Entry 1 is an example of the classical procedure. Entry 2 uses crown-ether catalysis. These reactions were conducted in the aromatic reactant as the solvent. In the study cited for Entry 2, it was found that substituted aromatic reactants such as toluene, anisole, and benzonitrile tended to give more *ortho* substitution product than expected on a statistical basis.[180] The nature of this directive effect does not seem to have been studied extensively. Entries 3 and 4 involve in situ decomposition of *N*-nitrosoamides. Entry 5 is a case of in situ nitrosation.

11.4.2. Substitution by the $S_{RN}1$ Mechanism

The mechanistic aspects of the $S_{RN}1$ reaction were discussed in Section 11.6 of Part A. The distinctive feature of the $S_{RN}1$ mechanism is an electron transfer between the nucleophile and the aryl halide.[181] The overall reaction is normally a chain process.

[176] C. Rüchardt and B. Freudenberg, *Tetrahedron Lett.*, 3623 (1964); C. Rüchardt and E. Merz, *Tetrahedron Lett.*, 2431 (1964); C. Galli, *Chem. Rev.*, **88**, 765 (1988).

[177] J. R. Beadle, S. H. Korzeniowski, D. E. Rosenberg, G. J. Garcia-Slanga, and G. W. Gokel, *J. Org. Chem.*, **49**, 1594 (1984).

[178] J. I. G. Cadogan, *Acc. Chem. Res.*, **4**, 186 (1971); *Adv. Free Radical Chem.*, **6**, 185 (1980).

[179] J. I. G. Cadogan, *J. Chem. Soc.*, 4257 (1962).

[180] See also T. Inukai, K. Kobayashi, and O. Shinmura, *Bull. Chem. Soc. Jpn.*, **35**, 1576 (1962).

[181] J. F. Bunnett, *Acc. Chem. Res.*, **11**, 413 (1978); R. A. Rossi and R. H. de Rossi, *Aromatic Substitution by the $S_{RN}1$ Mechanism*, ACS Monograph Series, No. 178, American Chemical Society, Washington, DC, 1983.

Scheme 11.11. Synthesis of Biaryls by Radical Substitution

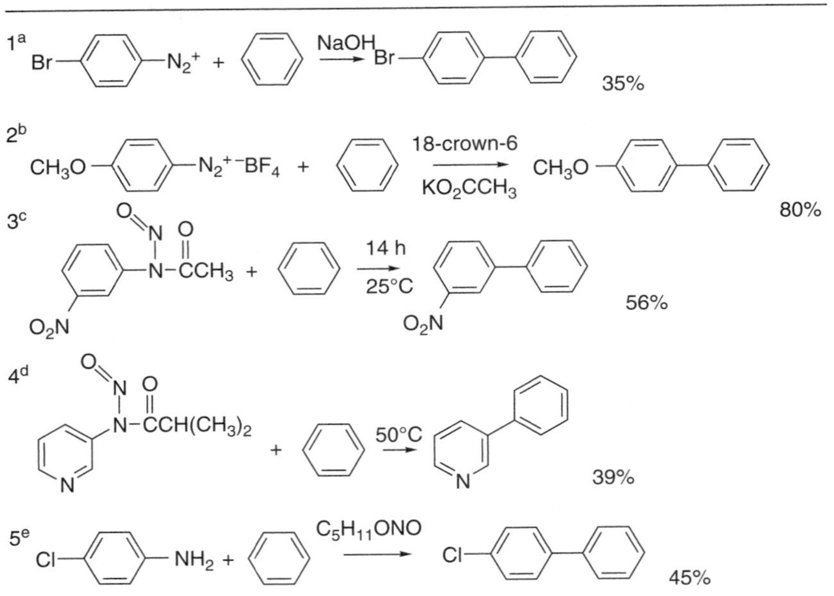

1[a]

$$Br-\phi-N_2^+ + \phi \xrightarrow{NaOH} Br-\phi-\phi \quad 35\%$$

2[b]

$$CH_3O-\phi-N_2^+ \cdot BF_4 + \phi \xrightarrow[KO_2CCH_3]{18\text{-crown-}6} CH_3O-\phi-\phi \quad 80\%$$

3[c]

$$O_2N-\phi-N(N=O)-CCH_3 + \phi \xrightarrow[25°C]{14\text{ h}} O_2N-\phi-\phi \quad 56\%$$

4[d]

$$\text{pyridyl}-N(N=O)-CCH(CH_3)_2 + \phi \xrightarrow{50°C} \text{pyridyl}-\phi \quad 39\%$$

5[e]

$$Cl-\phi-NH_2 + \phi \xrightarrow{C_5H_{11}ONO} Cl-\phi-\phi \quad 45\%$$

a. M. Gomberg and W. E. Bachman, *Org. Synth.*, **I**, 113 (1941).
b. S. H. Korzeniowski, L. Blum, and G. W. Gokel, *Tetrahedron Lett.*, 1871 (1977); J. R. Beadle, S. H. Korzeniowski, D. E. Rosenberg, B. J. Garcia-Slanga, and G. W. Gokel, *J. Org. Chem.*, **49**, 1594 (1984).
c. W. E. Bachmann and R. A. Hoffman, *Org. React.*, **2**, 249 (1944).
d. H. Rapoport, M. Lick, and G. J. Kelly, *J. Am. Chem. Soc.*, **74**, 6293 (1952).
e. J. I. G. Cadogan, *J. Chem. Soc.*., 4257 (1962).

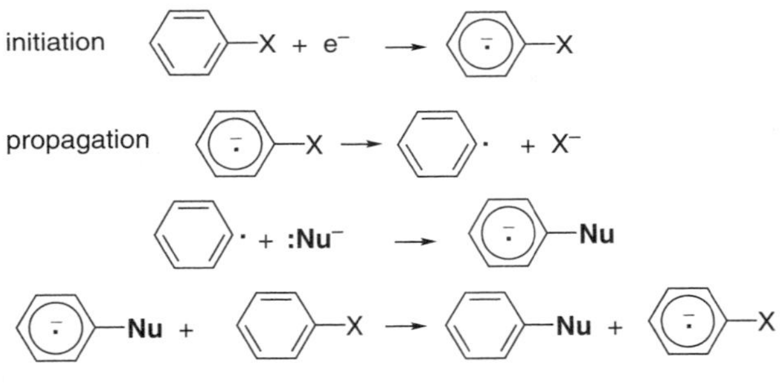

initiation

propagation

A potential advantage of the $S_{RN}1$ mechanism is that it is not particularly sensitive to the nature of other aromatic ring substituents, although EWG substituents favor the nucleophilic addition step. For example, chloropyridines and chloroquinolines are excellent reactants.[182] A variety of nucleophiles undergo the reaction, although not always in high yield. The nucleophiles that have been found to participate in

---

182. J. V. Hay, T. Hudlicky, and J. F. Wolfe, *J. Am. Chem. Soc.*, **97**, 374 (1975); J. V. Hay and J. F. Wolfe, *J. Am. Chem. Soc.*, **97**, 3702 (1975); A. P. Komin and J. F. Wolfe, *J. Org. Chem.*, **42**, 2481 (1977); R. Beugelmans, M. Bois-Choussy, and B. Boudet, *Tetrahedron*, **24**, 4153 (1983).

$S_{RN}1$ substitution include ketone enolates,[183] ester enolates,[184] amide enolates,[185] 2,4-pentanedione dianion,[186] pentadienyl and indenyl carbanions,[187] phenolates,[188] diethyl phosphite anion,[189] phosphides,[190] and thiolates.[191] The reactions are frequently initiated by light, which promotes the initiating electron transfer. As for other radical chain processes, the reaction is sensitive to substances that can intercept the propagation intermediates.

Scheme 11.12 provides some examples of the preparative use of the $S_{RN}1$ reaction. Entries 1 and 2 involve arylations of ketone enolates, whereas Entry 3 involves a dianion. Entry 4 is an example of a convenient preparation of arylphosphonates. Entry 5 is an example of application of the $S_{RN}1$ reaction to a chloropyridine.

Scheme 11.12. Aromatic Substitution by the $S_{RN}1$ Mechanism

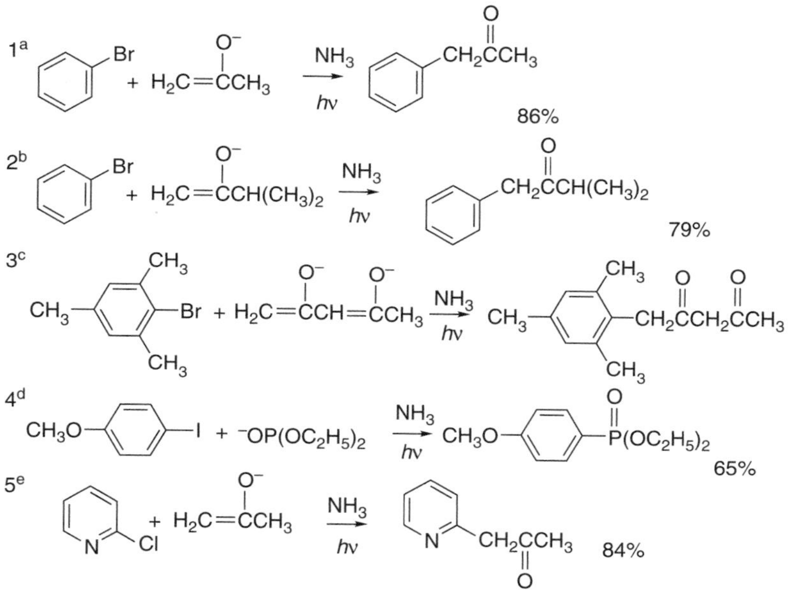

a. R. A. Rossi and J. F. Bunnett, *J. Org. Chem.*, **38**, 1407 (1973).
b. M. F. Semmelhack and T. Bargar, *J. Am. Chem. Soc.*, **102**, 7765 (1980).
c. J. F. Bunnett and J. E. Sundberg, *J. Org. Chem.*, **41**, 1702 (1976).
d. J. F. Bunnett and X. Creary, *J. Org. Chem.*, **39**, 3612 (1974).
e. A. P. Komin and J. F. Wolfe, *J. Org. Chem.*, **42**, 2481 (1977).

[183.] M. F. Semmelhack and T. Bargar, *J. Am. Chem. Soc.*, **102**, 7765 (1980).
[184.] J.-W. Wong, K. J. Natalie, Jr., G. C. Nwokogu, J. S. Pisipati, S. Jyothi, P. T. Flaherty, T. D. Greenwood, and J. F. Wolfe, *J. Org. Chem.*, **62**, 6152 (1997).
[185.] R. A. Rossi and R. A. Alonso, *J. Org. Chem.*, **45**, 1239 (1980).
[186.] J. F. Bunnett and J. E. Sundberg, *J. Org. Chem.*, **41**, 1702 (1976).
[187.] R. A. Rossi and J. F. Bunnett, *J. Org. Chem.*, **38**, 3020 (1973).
[188.] A. B. Pierini, M. T. Baumgartner, and R. A. Rossi, *Tetrahedron Lett.*, **29**, 3429 (1988).
[189.] J. F. Bunnett and X. Creary, *J. Org. Chem.*, **39**, 3612 (1974); A. Boumekouez, E. About-Jaudet, N. Collignon, and P. Savignac, *J. Organomet. Chem.*, **440**, 297 (1992).
[190.] E. Austin, R. A. Alonso, and R. A. Rosi, *J. Org. Chem.*, **56**, 4486 (1991).
[191.] J. F. Bunnett and X. Creary, *J. Org. Chem.*, **39**, 3173, 3611 (1974); J. F. Bunnett and X. Creary, *J. Org. Chem.*, **40**, 3740 (1975).

Problems

(*References for these problems will be found on page 1289.*)

11.1. Give reagents and reaction conditions that would accomplish each of the following transformations. Multistep schemes are not necessary. Be sure to choose conditions that would lead to the desired isomer as the major product.

(a)

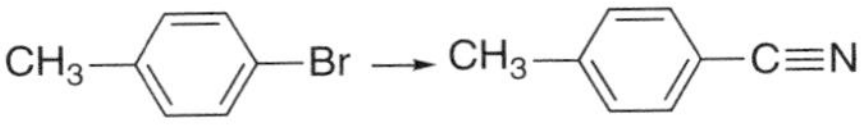

(b)

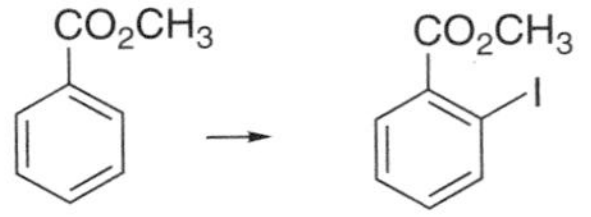

(c)

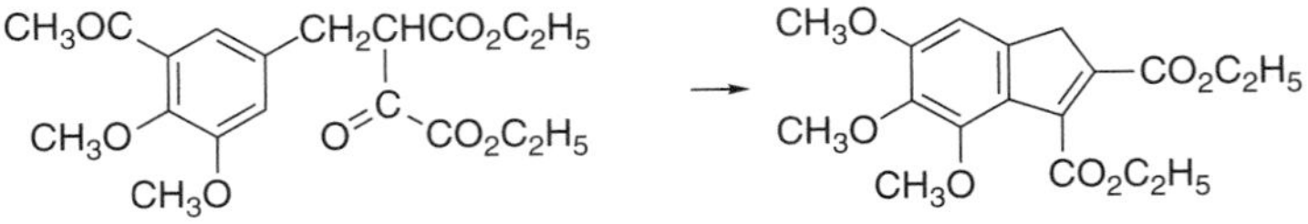

(d)

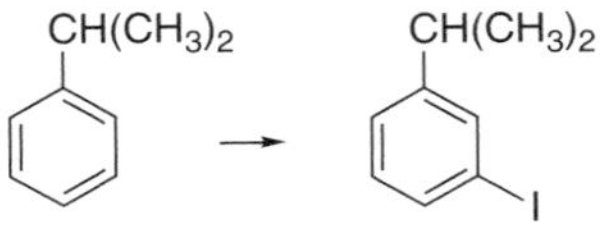

(e)

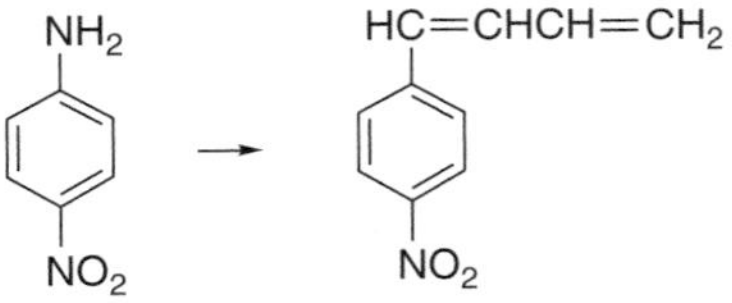

(f)

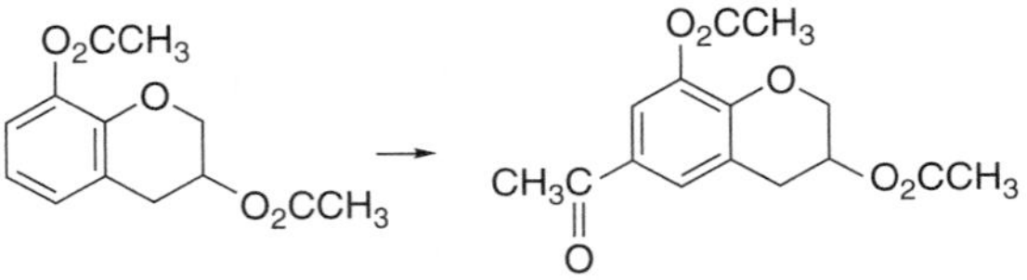

(g)

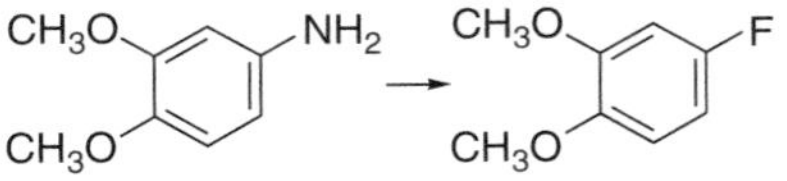

(h)

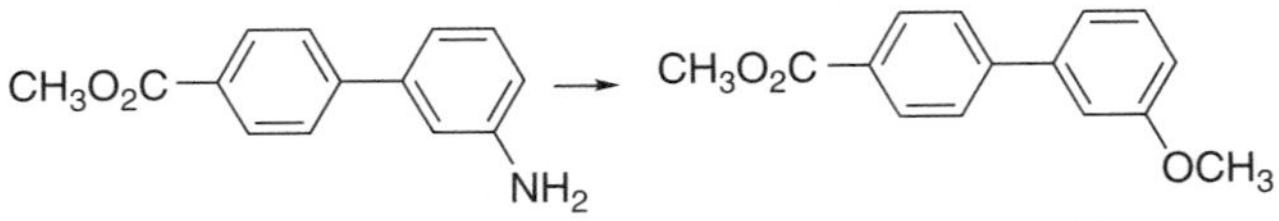

(i)

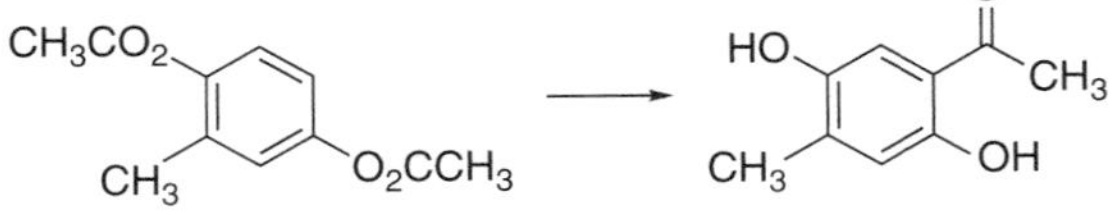

11.2. Suggest a short series of reactions that would be expected to transform the material on the right into the desired product shown on the left.

(a)

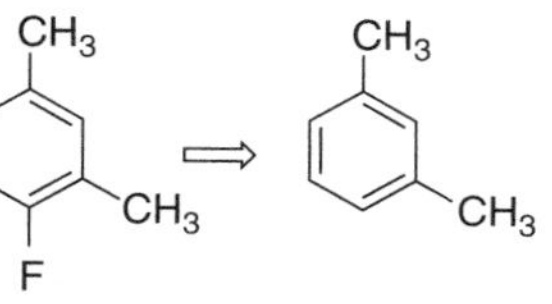

(b)

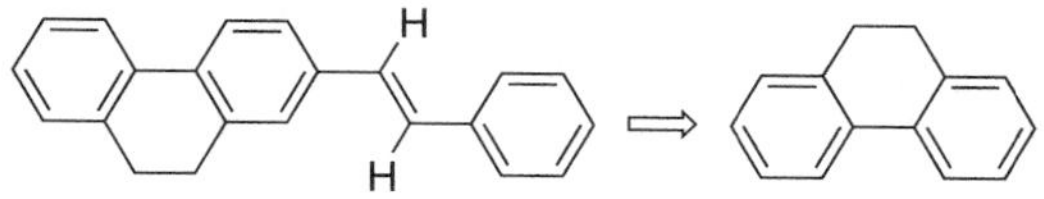

(c)

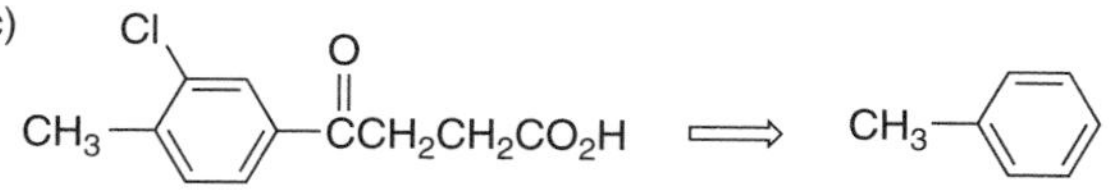

(d)

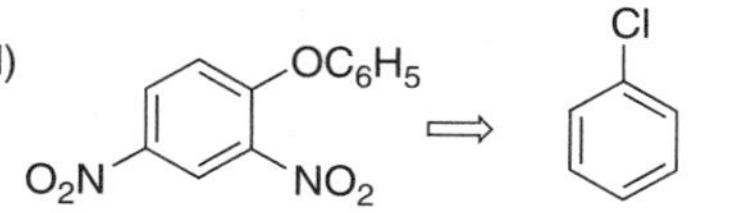

(e)

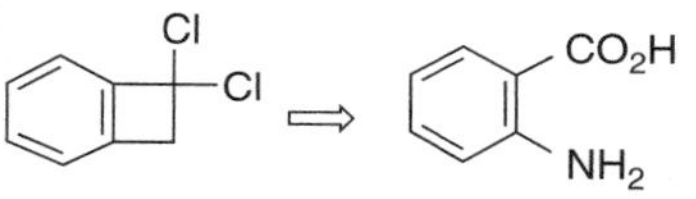

11.3. Write mechanisms that would account for the following reactions:

(a)

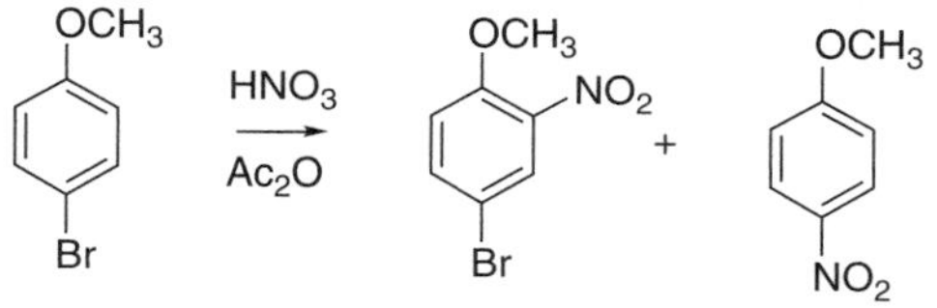

(b)

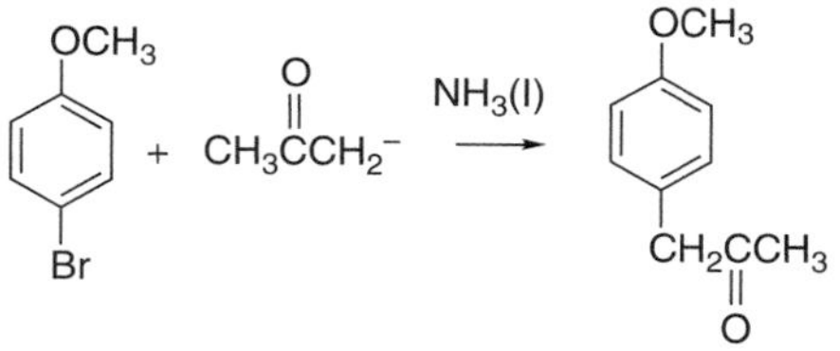

(c)

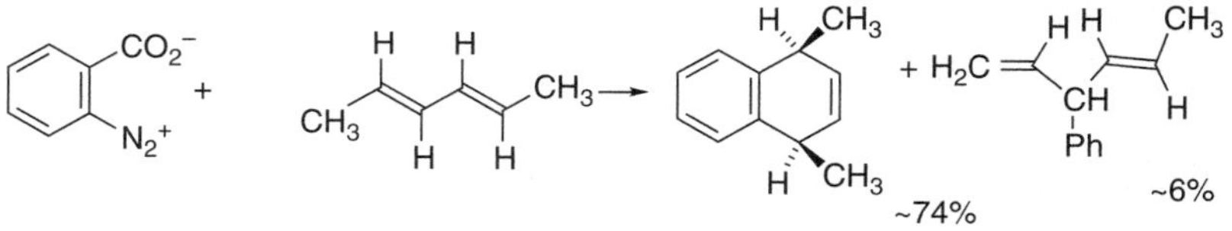

(d)

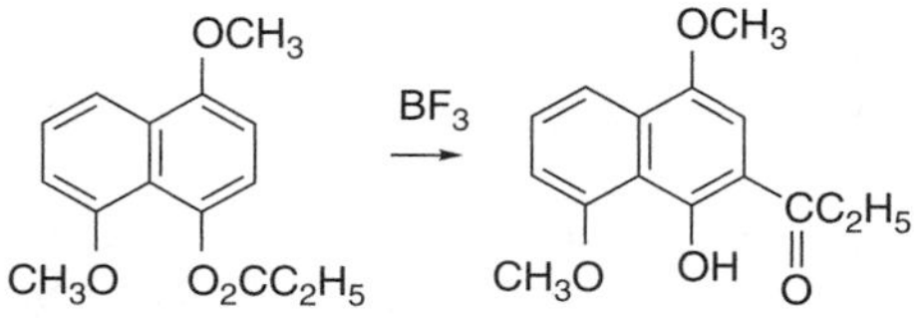

(e)

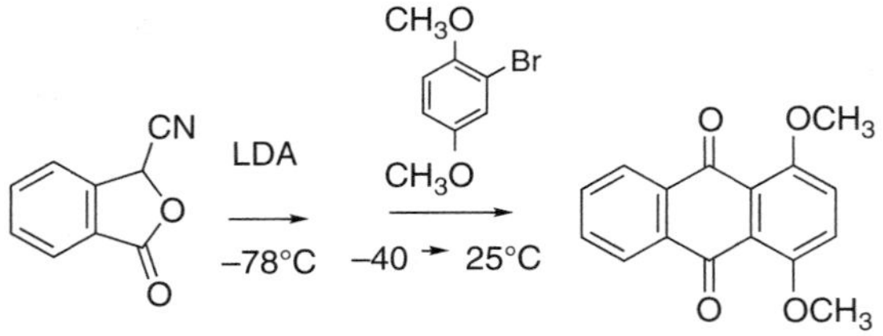

(f)

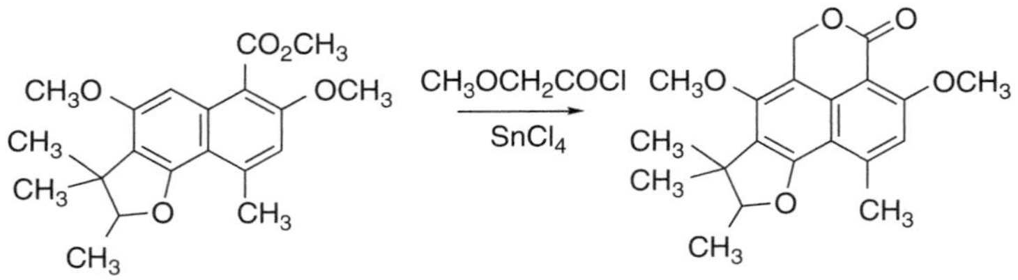

11.4. Predict the product(s) of the following reactions. If more than one product is expected, indicate which will be major and which will be minor.

(a)

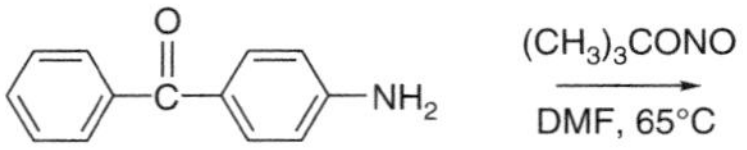

(b)

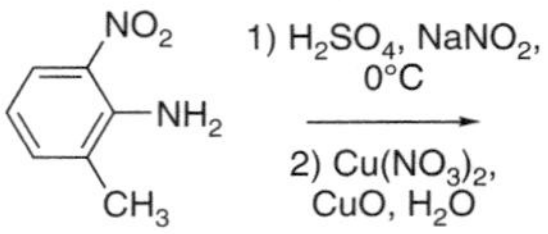

(c)

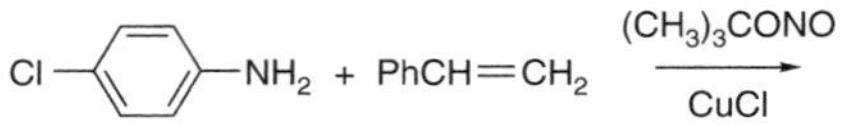

(d)

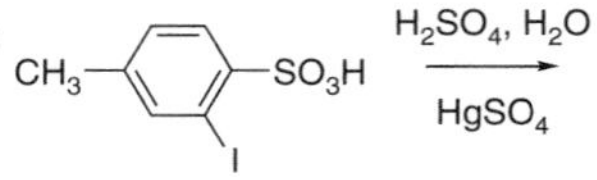

(e)

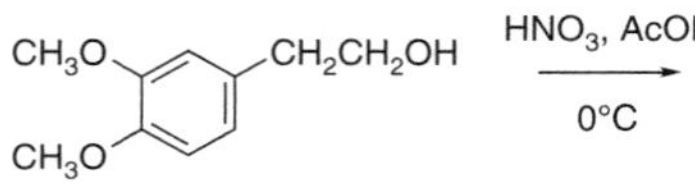

(f)

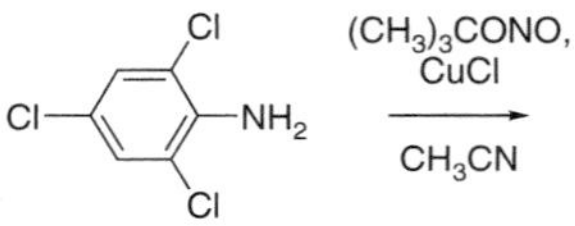

(g)

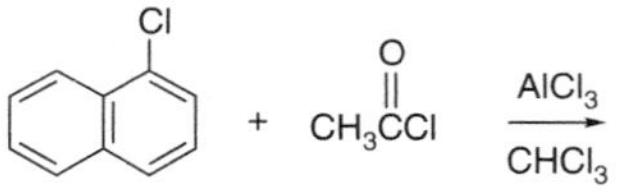

(h)

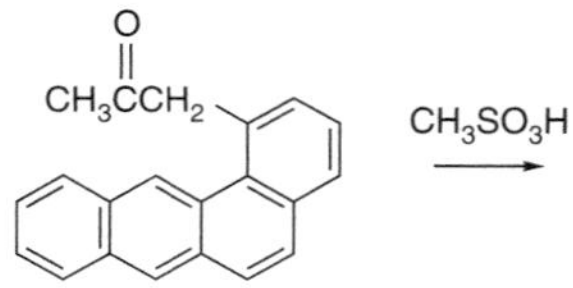

(i)

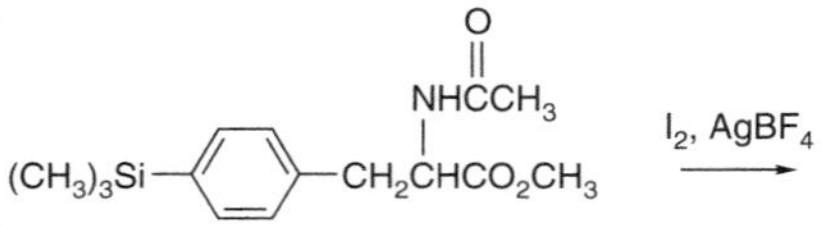

(j)

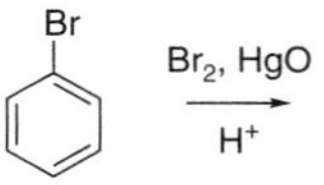

(k)

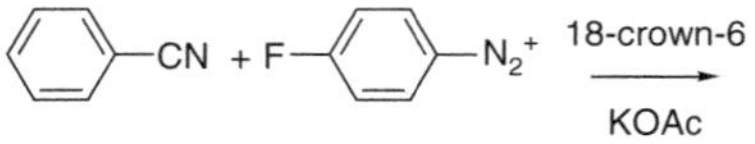

11.5. Suggest efficient syntheses of *o*-, *m*-, and *p*-fluoropropiophenone from benzene and other necessary reagents.

11.6. Treatment of compound **6-1** in dibromethane with one equivalent of aluminum bromide yields **6-2** as the only product in 78% yield. When three equivalents of aluminum bromide are used, compounds **6-3** and **6-4** are obtained in a combined yield of 97%. Suggest an explanation for these observations.

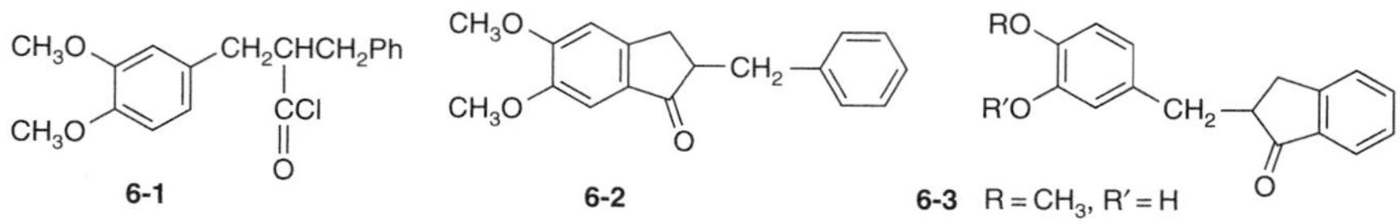

6-1

6-2

6-3 R = CH$_3$, R′ = H

6-4 R = H, R′ = CH$_3$

11.7. Some data on the alkylation of naphthalene by 2-bromopropane using AlCl$_3$ under different conditions are given below. What factors are responsible for the differing product ratios for the two solvents, and why does the product ratio change with time?

	α:β Product ratio	
	Solvent	
Time (min)	CS$_2$	CH$_3$NO$_2$
5	4:96	83:17
15	2.5:97.5	74:26
45	2:98	70:30

11.8. Addition of a solution of bromine and potassium bromide to a solution of the carboxylate salt **8-1** results in the precipitation of a neutral compound having the formula C$_{11}$H$_{13}$BrO$_3$. Spectroscopic data show that the compound is nonaromatic. Suggest a structure and discuss the mechanistic significance of its formation.

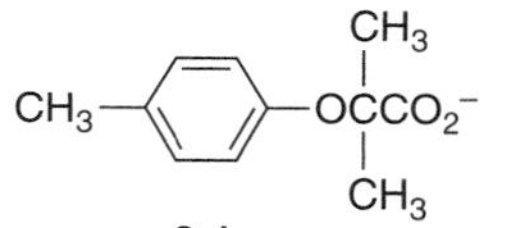

8-1

11.9. Benzaldehyde, benzyl methyl ether, benzoic acid, methyl benzoate, and phenylacetic acid all undergo thallation initially in the *ortho* position. Explain this observation.

11.10. Reaction of 3,5,5-trimethyl-2-cyclohexenone with three equivalents of NaNH$_2$ in THF generates the corresponding enolate. When bromobenzene is added and the solution stirred for 4 h, the product **10-1** is isolated in 30% yield. Formulate a mechanism for this transformation.

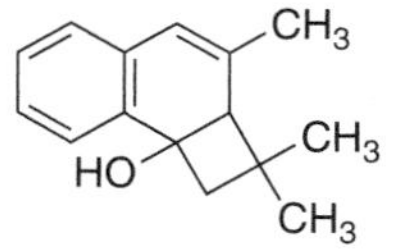

10-1

11.11. When phenylacetonitrile is converted to its anion in the presence of excess LDA and then allowed to react with 2-bromo-4-methyl-1-methoxybenzene, the product contains both a benzyl and cyano substituent. Propose a mechanism for this reaction.

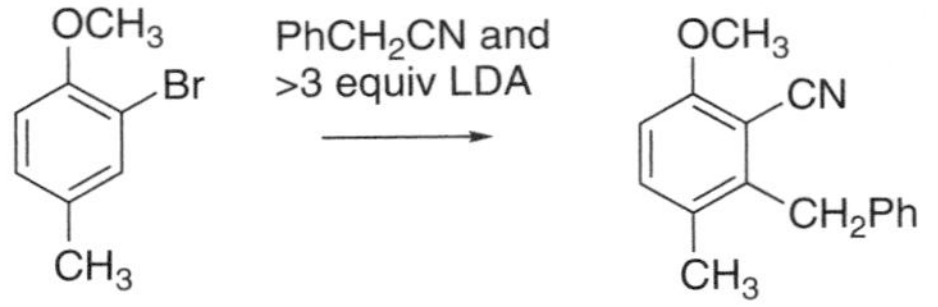

11.12. Suggest a reaction sequence that would permit synthesis of the following aromatic compounds from the starting material indicated on the right.

(a)

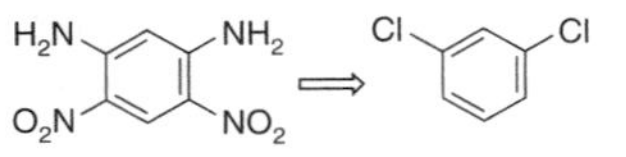

(b)

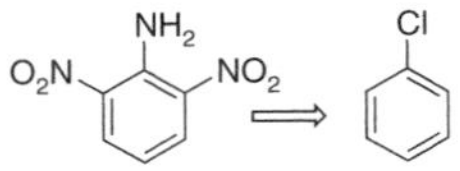

(c)

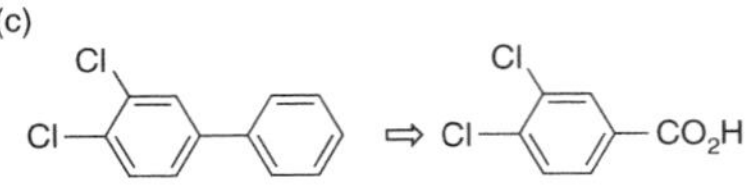

(d)

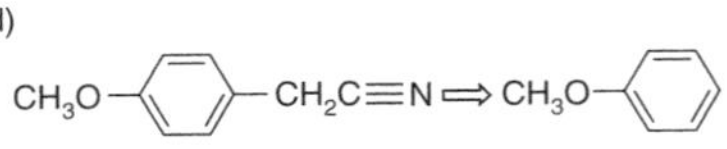

(e)

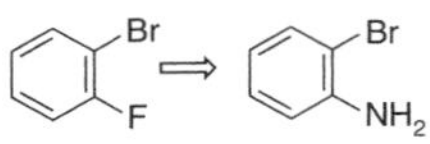

(f)

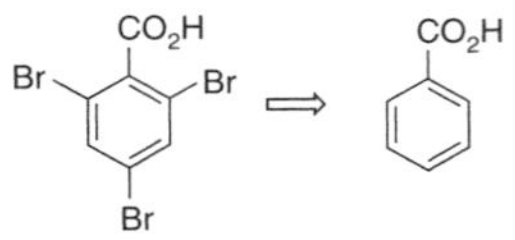

(g)

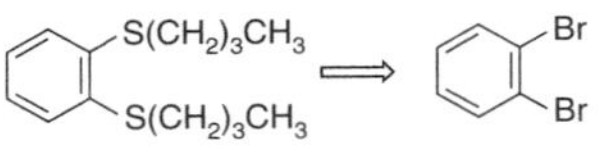

(h)

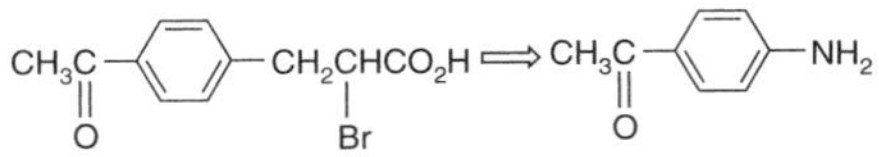

11.13. Aromatic substitution reactions are key steps in the multistep synthetic sequences that effect the following transformations. Suggest a sequence of reactions that could effect the desire syntheses.

(a)

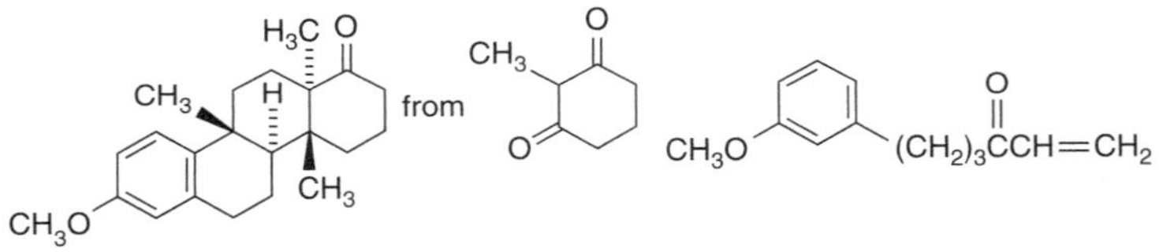

(b)

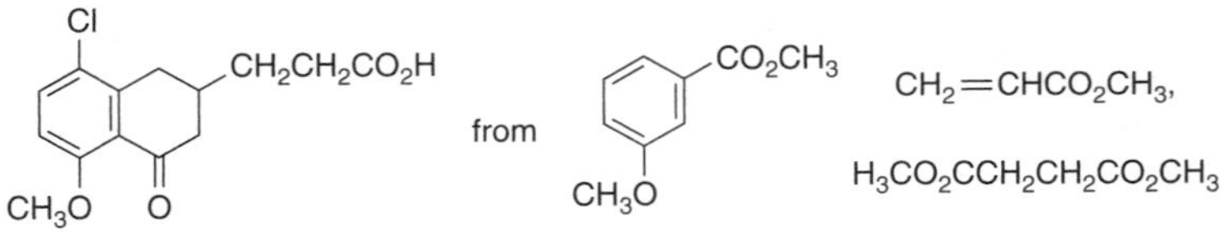

(c)

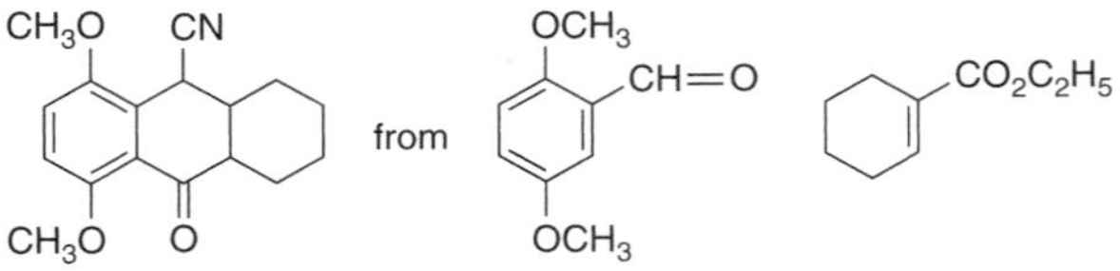

(d)

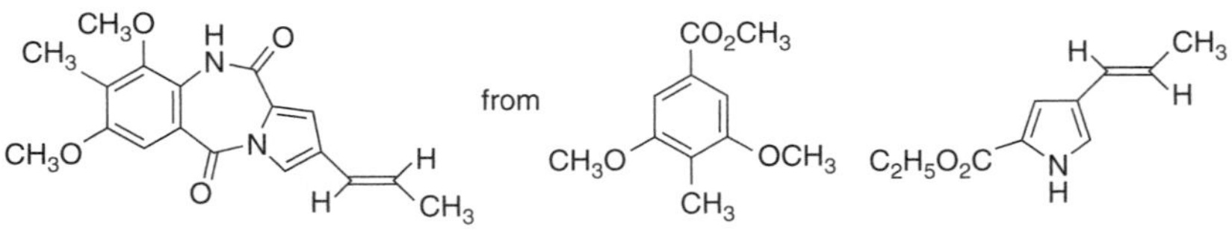

(e)

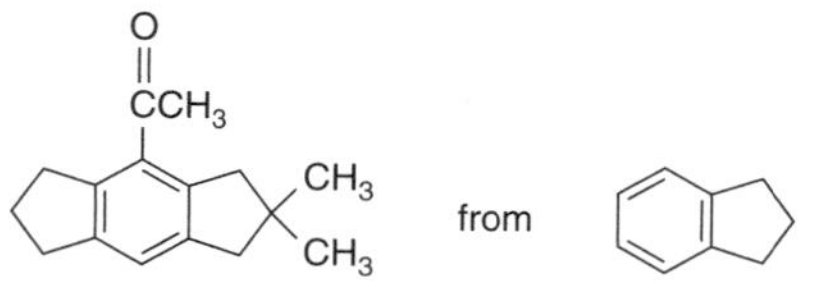

(f)

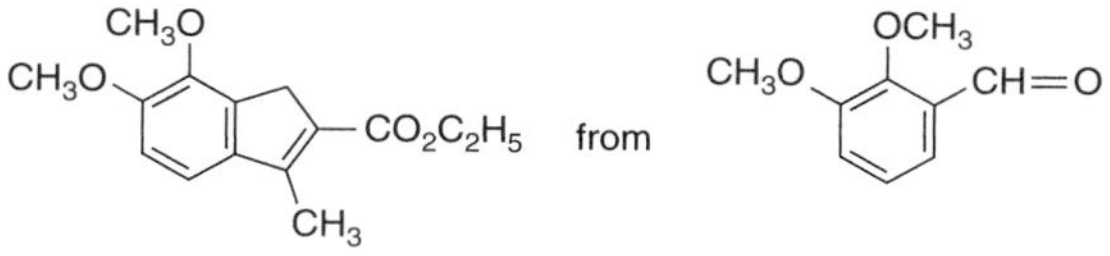

11.14. The following intermediates in the synthesis of naturally occurring materials have been synthesized by reactions based on a benzyne intermediate. The benzyne precursor is shown. By retrosynthetic analysis identify an appropriate co-reactant that would form the desired compound.

a)

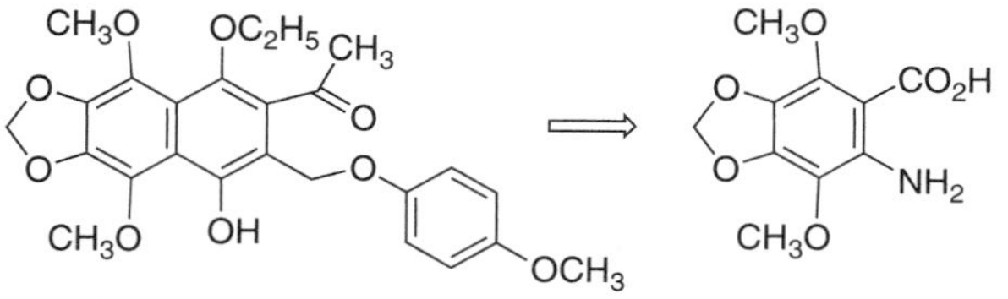

b)

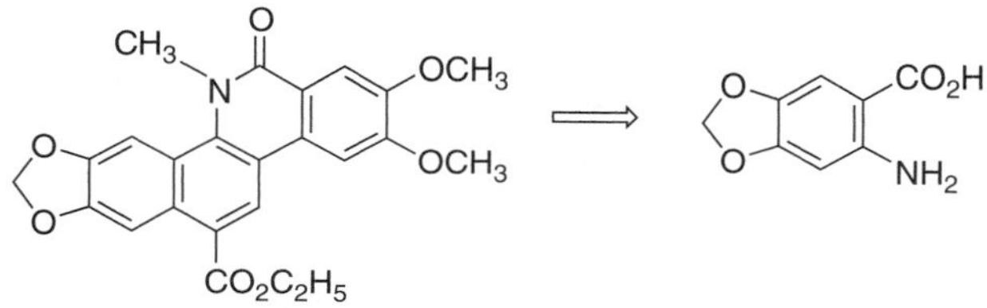

11.15. Aryltrimethylsilanes has been found to be a useful complement to direct thallation in the preparation of arylthallium(III) intermediates. The thallium(III) replaces the silyl substituent and the scope of the reaction is expanded to include some EWGs, such as trifluoromethyl. How does the silyl group function in these systems?

11.16. The Pschorr reaction is a method of synthesis of phenanthrenes from diazotized Z-2-aminostilbenes. A traditional procedure involves heating with a copper catalyst. Improved yields are often observed, however, if the diazonium ion is treated with iodide ion. Suggest a mechanism for the iodide-catalyzed reaction.

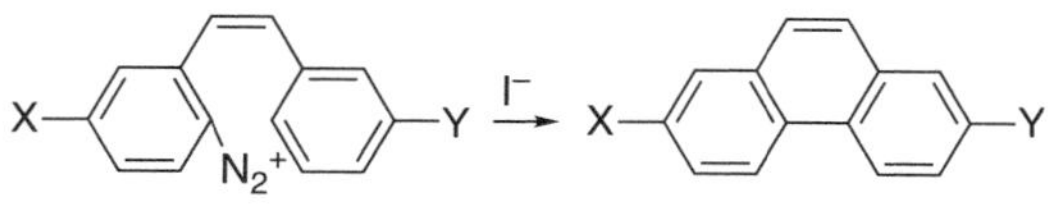

11.17. When compound **17-1** is dissolved in FSO_3H at $-78°C$, NMR spectroscopy shows that a carbocation is formed. If the solution is then allowed to warm to $-10°C$, a different ion forms. The first ion gives compound **17-2** when quenched with base, whereas the second ion gives **17-3**. What are the structures of the two carbocations, and why do they give different products on quenching?

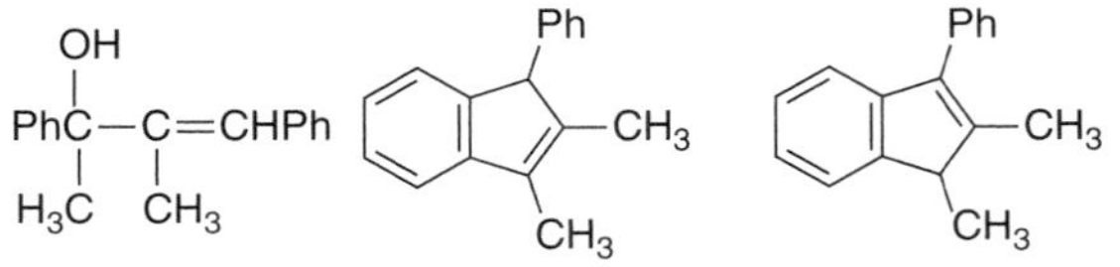

17-1	**17-2**	**17-3**

11.18. Various phenols can be selectively hydroxymethylated at the *ortho* position by heating with paraformaldehyde and phenylboronic acid. An intermediate **18-1** having the formula $C_{14}H_{13}O_2B$ for the case shown can be isolated prior to the oxidation. Suggest a structure for the intermediate and comment on its role in the reaction.

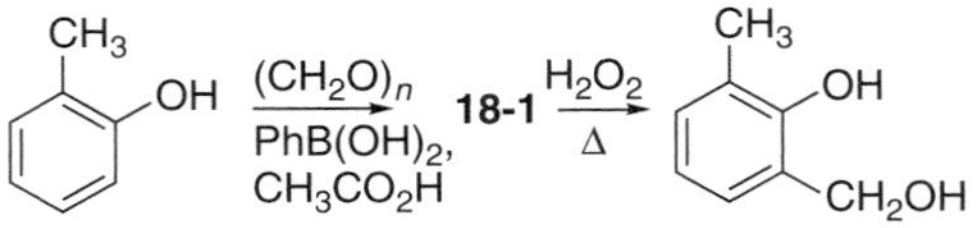

11.19. The electrophilic cyclization of **19-1** and **19-2** gives two isomers, but with the unsubstituted reactant **19-3**, only a single stereoisomer is formed. Explain the origin of the isomers and the absence of isomer formation in the case of **19-3**.

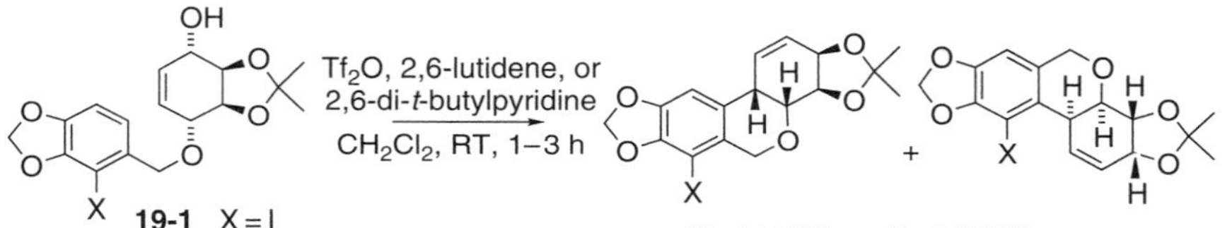

19-1 X = I
19-2 X = Br
19-3 X = H

X = I (42%)   X = I (21%)
X = Br (50%)   X = Br (16%)
X = H (73%)

11.20. Entry 5 in Scheme 11.4 is a step in the synthesis of the anticancer drug tamoxifen. Explain why the 2-phenylbutanoyl group is introduced in preference to a trifluoroacetyl group.

Oxidations

Introduction

This chapter is concerned with reactions that transform a functional group to a more highly oxidized derivative by removal of hydrogen and/or addition of oxygen. There are a great many oxidation methods, and we have chosen the reactions for discussion on the basis of their utility in synthesis. As the reactions are considered, it will become evident that the material in this chapter spans a broader range of mechanisms than most of the previous chapters. Owing to this range, the chapter is organized according to the functional group transformation that is accomplished. This organization facilitates comparison of the methods available for effecting a given synthetic transformation. The major sections consider the following reactions: (1) oxidation of alcohols; (2) addition of oxygen at double bonds; (3) allylic oxidation; (4) oxidative cleavage of double bonds; (5) oxidative cleavage of other functional groups; (6) oxidations of aldehydes and ketones; and (7) oxidation at unfunctionalized positions. The oxidants are grouped into three classes: transition metal derivatives; oxygen, ozone, and peroxides; and other reagents.

12.1. Oxidation of Alcohols to Aldehydes, Ketones, or Carboxylic Acids

12.1.1. Transition Metal Oxidants

The most widely employed transition metal oxidants for alcohols are based on Cr(VI). The specific reagents are generally prepared from chromic trioxide, CrO_3, or a dichromate salt, $[Cr_2O_7]^{2-}$. The form of Cr(VI) in aqueous solution depends upon concentration and pH; the pK_1 and pK_2 of H_2CrO_4 are 0.74 and 6.49, respectively. In dilute solution, the monomeric acid chromate ion $[HCrO_3]^-$ is the main species present; as concentration increases, the dichromate ion dominates.

$$2 \ HO-\overset{\overset{O}{\|}}{\underset{\underset{O}{\|}}{Cr}}-O^- \ \rightleftharpoons \ ^-O-\overset{\overset{O}{\|}}{\underset{\underset{O}{\|}}{Cr}}-O-\overset{\overset{O}{\|}}{\underset{\underset{O}{\|}}{Cr}}-O^- \ + \ H_2O$$

In acetic acid, Cr(VI) is present as mixed anhydrides of acetic acid and chromic acid.[1]

$$2 \ CH_3CO_2H \ + \ CrO_3 \ \rightleftharpoons \ CH_3CO_2\overset{\overset{O}{\|}}{\underset{\underset{O}{\|}}{Cr}}-OH \ \rightleftharpoons \ CH_3CO_2\overset{\overset{O}{\|}}{\underset{\underset{O}{\|}}{Cr}}O_2CCH_3 \ + \ H_2O$$

In pyridine, an adduct involving Cr–N bonding is formed.

$$\text{(pyridine)} N \ + \ CrO_3 \ \longrightarrow \ \text{(pyridine)} \overset{+}{N}-\overset{\overset{O}{\|}}{\underset{\underset{O}{\|}}{Cr}}-O^-$$

The oxidation state of Cr in each of these species is (VI) and they are all powerful oxidants. The precise reactivity depends on the solvent and the chromium ligands, so substantial selectivity can be achieved by the choice of the particular reagent and conditions.

The general mechanism of alcohol oxidation involves coordination of the alcohol at chromium and a rate-determining deprotonation.

$$R_2CHOH + HO-\overset{\overset{O}{\|}}{\underset{\underset{O}{\|}}{Cr^{(VI)}}}O^- + H^+ \longrightarrow R_2CHO-\overset{\overset{O}{\|}}{\underset{\underset{O}{\|}}{Cr^{(VI)}}}OH + H_2O$$

$$R_2\overset{\displaystyle }{\underset{\underset{H}{|}}{C}}{\longrightarrow}O\overset{}{\overset{\overset{O}{\|}}{\underset{\underset{O}{\|}}{Cr^{(VI)}}}}OH \longrightarrow R_2C=O + O=\overset{}{\underset{\underset{O^-}{|}}{Cr^{(IV)}}}OH + H^+$$

An important piece of evidence for this mechanism is the fact that a primary isotope effect is observed when the α-hydrogen is replaced by deuterium.[2] The Cr(IV) that is produced in the initial step is not stable and is capable of a further oxidation. It is believed that Cr(IV) is reduced to Cr(II), which is then oxidized by Cr(VI) generating Cr(V). This mechanism accounts for the overall stoichiometry of the reaction.[3]

$$
\begin{array}{rcl}
R_2CHOH \ + \ Cr(VI) & \longrightarrow & R_2C{=}O \ + \ Cr(IV) \ + \ 2H^+ \\
R_2CHOH \ + \ Cr(IV) & \longrightarrow & R_2C{=}O \ + \ Cr(II) \ + \ 2H^+ \\
Cr(II) \ + \ Cr(VI) & \longrightarrow & Cr(III) \ + \ Cr(V) \\
R_2CHOH \ + \ Cr(V) & \longrightarrow & R_2C{=}O \ + \ Cr(III) \ + \ 2H^+ \\
\hline
3 \ R_2CHOH \ + \ 2 \ Cr(VI) & \longrightarrow & 3 \ R_2C{=}O \ + \ 2 \ Cr(III) \ + \ 6 \ H^+
\end{array}
$$

[1]. K. B. Wiberg, *Oxidation in Organic Chemistry*, Part A, Academic Press, New York, 1965, pp. 69–72.
[2]. F. H. Westheimer and N. Nicolaides, *J. Am. Chem. Soc.*, **71**, 25 (1949).
[3]. S. L. Scott, A. Bakac, and J. H. Esperson, *J. Am. Chem. Soc.*, **114**, 4205 (1992); J. F. Perez-Benito and C. Arias, *Can. J. Chem.*, **71**, 649 (1993).

Various experimental conditions have been used for oxidations of alcohols by Cr(VI) on a laboratory scale, and several examples are shown in Scheme 12.1. Entry 1 is an example of oxidation of a primary alcohol to an aldehyde. The propanal is distilled from the reaction mixture as oxidation proceeds, which minimizes overoxidation. For secondary alcohols, oxidation can be done by addition of an acidic aqueous solution containing chromic acid (known as *Jones' reagent*) to an acetone solution of the alcohol. Oxidation normally occurs rapidly, and overoxidation is minimal. In acetone solution, the reduced chromium salts precipitate and the reaction solution can be decanted. Entries 2 to 4 in Scheme 12.1 are examples of this method.

The chromium trioxide-pyridine complex is useful in situations when other functional groups might be susceptible to oxidation or the molecule is sensitive to acid.[4] A procedure for utilizing the CrO_3-pyridine complex, which was developed by Collins,[5] has been widely adopted. The CrO_3-pyridine complex is isolated and dissolved in dichloromethane. With an excess of the reagent, oxidation of simple alcohols is complete in a few minutes, giving the aldehyde or ketone in good yield. A procedure that avoids isolation of the complex can further simplify the experimental operations.[6] Chromium trioxide is added to pyridine in dichloromethane. Subsequent addition of the alcohol to this solution results in oxidation in high yield. Other modifications for use of the CrO_3-pyridine complex have been developed.[7] Entries 5 to 9 in Scheme 12.1 demonstrate the excellent results that have been reported using the CrO_3-pyridine complex in dichloromethane. Entries 5 and 6 involve conversion of primary alcohols to aldehydes, Entry 7 describes preparation of the reagent in situ, and Entry 8 is an example of application of these conditions to a primary alcohol. The conditions described in Entry 9 were developed to optimize the oxidation of sensitive carbohydrates. It was found that inclusion of 4A molecular sieves and a small amount of acetic acid accelerated the reaction.

Another very useful Cr(VI) reagent is pyridinium chlorochromate (PCC), which is prepared by dissolving CrO_3 in hydrochloric acid and adding pyridine to obtain a solid reagent having the composition $CrO_3Cl \cdot pyrH$.[8] This reagent can be used in amounts close to the stoichiometric ratio. Entries 10 and 11 are examples of the use of PCC. Reaction of pyridine with CrO_3 in a small amount of water gives pyridinium dichromate (PDC), which is also a useful oxidant.[9] As a solution in DMF or a suspension in dichloromethane, this reagent oxidizes secondary alcohols to ketones. Allylic primary alcohols give the corresponding aldehydes. Depending upon the conditions, saturated primary alcohols give either an aldehyde or the corresponding carboxylic acid.

$$CH_3(CH_2)_8CH_2OH \xrightarrow[\text{DMF, 25°C}]{\text{PDC}} CH_3(CH_2)_8CH{=}O$$
$$98\%$$

4. G. I. Poos, G. E. Arth, R. E. Beyler, and L. H. Sarett, *J. Am. Chem. Soc.*, **75**, 422 (1953); W. S. Johnson, W. A. Vredenburgh, and J. E. Pike, *J. Am. Chem. Soc.*, **82**, 3409 (1960); W. S. Allen, S. Bernstein, and R. Little, *J. Am. Chem. Soc.*, **76**, 6116 (1954).

5. J. C. Collins, W. W. Hess, and F. J. Frank, *Tetrahedron Lett.*, 3363 (1968).

6. R. Ratcliffe and R. Rodehorst, *J. Org. Chem.*, **35**, 4000 (1970).

7. J. Herscovici, M.-J. Egron, and K. Antonakis, *J. Chem. Soc., Perkin Trans. 1*, 1967 (1982); E. J. Corey and G. Schmidt, *Tetrahedron Lett.*, 399 (1979); S. Czernecki, C. Georgoulis, C. L. Stevens, and K. Vijayakumaran, *Tetrahedron Lett.*, **26**, 1699 (1985).

8. E. J. Corey and J. W. Suggs, *Tetrahedron Lett.*, 2647 (1975); G. Piancatelli, A. Scettri, and M. D'Auria, *Synthesis*, 245 (1982).

9. E. J. Corey and G. Schmidt, *Tetrahedron Lett.*, 399 (1979).

Scheme 12.1. Oxidation with Chromium(VI) Reagents

A. Chromic acid solutions

1[a] $CH_3CH_2CH_2OH$ $\xrightarrow[H_2O]{H_2CrO_4}$ $CH_3CH_2CH{=}O$

45–49%

2[b]

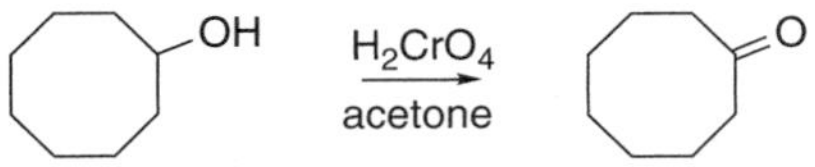

92–96%

3[c]

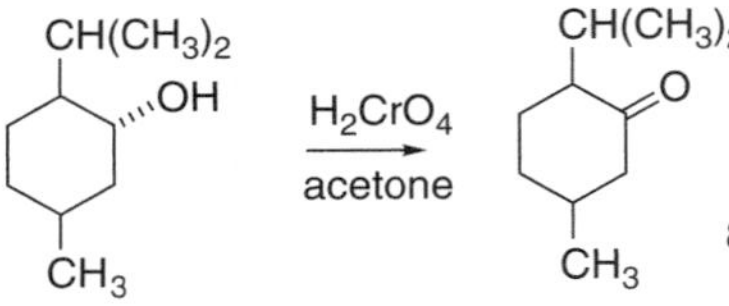

84%

4[d]

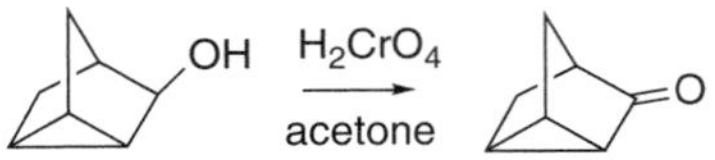

79–88%

B. Chromium trioxide–pyridine

5[e] $CH_3(CH_2)_5CH_2OH$ $\xrightarrow[CH_2Cl_2]{CrO_3-pyridine}$ $CH_3(CH_2)_5CH{=}O$

70–84%

6[f]

$CH_3CH_2CH(CH_2)_4CH_2OH$ $\xrightarrow[CH_2Cl_2]{CrO_3-pyridine}$ $CH_3CH_2CH(CH_2)_4CH{=}O$

69%

7[g]

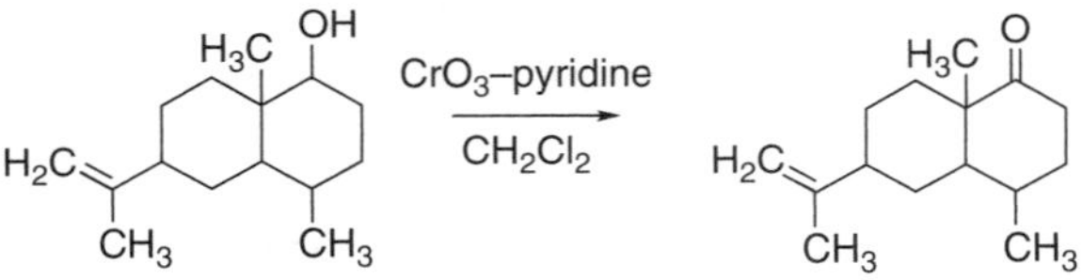

95%

8[h]

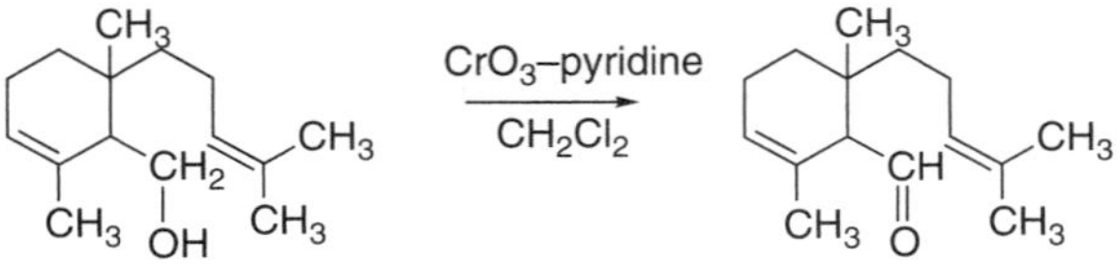

9[i]

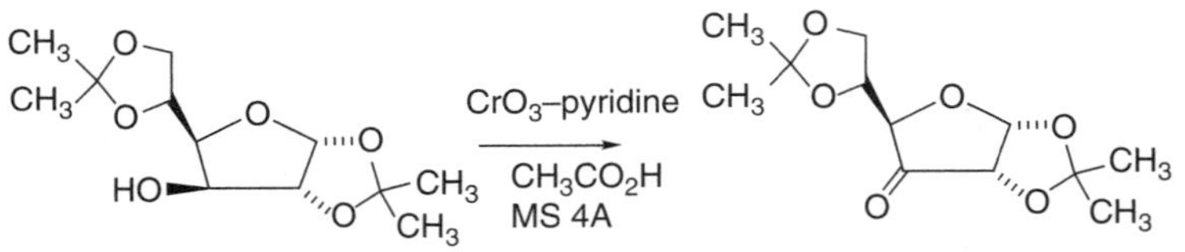

96%

(Continued)

C. Pyridinium chlorochromate

10^j (CH$_3$)$_2$C=CHCH$_2$CH$_2$CHCH$_2$CH$_2$OH $\xrightarrow{\text{PCC}}$ (CH$_3$)$_2$C=CHCH$_2$CH$_2$CHCH$_2$CH=O

with CH$_3$ substituent on both sides   82%

11^k HOCH$_2$CH$_2$CCH$_2$CH=CHCO$_2$CH$_3$ $\xrightarrow{\text{PCC}}$ O=CHCH$_2$CCH$_2$CH=CHCO$_2$CH$_3$

with CH$_3$ above and below on both sides   83%

a. C. D. Hurd and R. N. Meinert, *Org. Synth.*, **II**, 541 (1943).
b. E. J. Eisenbraun, *Org. Synth.*, **V**, 310 (1973).
c. H. C. Brown, C. P. Garg, and K.-T. Liu, *J. Org. Chem.*, **36**, 387 (1971).
d. J. Meinwald, J. Crandall, and W. E. Hymans, *Org. Synth.*, **V**, 866 (1973).
e. J. C. Collins and W. W. Hess, *Org. Synth.*, **52**, 5 (1972).
f. J. I. DeGraw and J. O. Rodin, *J. Org. Chem.*, **36**, 2902 (1971).
g. R. Ratcliffe and R. Rodehorst, *J. Org. Chem.*, **35**, 4000 (1970).
h. M. A. Schwartz, J. D. Crowell, and J. H. Musser, *J. Am. Chem. Soc.*, **94**, 4361 (1972).
i. C. Czernecki, C. Gerogoulis, C. L. Stevens, and K. Vijayakumaran, *Tetrahedron Lett.*, **26**, 1699 (1985).
j. E. J. Corey and J. W. Suggs, *Tetrahedron Lett.*, 2647 (1975).
k. R. D. Little and G. W. Muller, *J. Am. Chem. Soc.*, **103**, 2744 (1981).

Although Cr(VI) oxidants are very versatile and efficient, they have one drawback, which becomes especially serious in larger-scale work: the toxicity and environmental hazards associated with chromium compounds. The reagents are used in stoichiometric or excess amount and the Cr(III) by-products must be disposed of safely.

Potassium permanganate, KMnO$_4$, is another powerful transition metal oxidant, but it has found relatively little application in the oxidation of alcohols to ketones and aldehydes. The reagent is less selective than Cr(VI), and overoxidation is a problem. On the other hand, manganese(IV) dioxide is quite useful.[10] This reagent, which is selective for allylic and benzylic alcohols, is prepared by reaction of Mn(II)SO$_4$ with KMnO$_4$ and sodium hydroxide. The precise reactivity of MnO$_2$ depends on its mode of preparation and the extent of drying.[11]

Scheme 12.2 shows various types of alcohols that are most susceptible to MnO$_2$ oxidation. Entries 1 and 2 illustrate the application of MnO$_2$ to simple benzylic and allylic alcohols. In Entry 2, the MnO$_2$ was activated by azeotropic drying. Entry 3 demonstrates the application of the reagent to cyclopropylcarbinols. Entry 4 is an application to an acyloin. Entry 5 involves oxidation of a sensitive conjugated system.

A reagent system that is selective for allylic, benzylic, and cyclopropyl alcohols uses iodosobenzene in conjunction with a Cr(III)(salen) complex.[12]

15 mol % Cr^{III}salen
1.5 equiv PhI=O
30 mol % 4-phenyl-pyridine-*N*-oxide

10. D. G. Lee, in *Oxidation*, Vol. 1, R. L. Augustine, ed., Marcel Dekker, New York, 1969, pp. 66–70; A. J. Fatiadi, *Synthesis*, 65 (1976); A. J. Fatiadi, *Synthesis*, 133 (1976).

11. J. Attenburrow, A. F. B. Cameron, J. H. Chapman, R. M. Evans, A. B. A. Jansen, and T. Walker, *J. Chem. Soc.*, 1094 (1952); I. M. Goldman, *J. Org. Chem.*, **34**, 1979 (1969).

12. W. Adam, F. G. Gelacha, C. R. Saha-Moeller, and V. R. Stegmann, *J. Org. Chem.*, **65**, 1915 (2000); see also S. S. Kim and D. W. Kim, *Synlett*, 1391 (2003).

Scheme 12.2. Oxidation of Alcohols with Manganese Dioxide

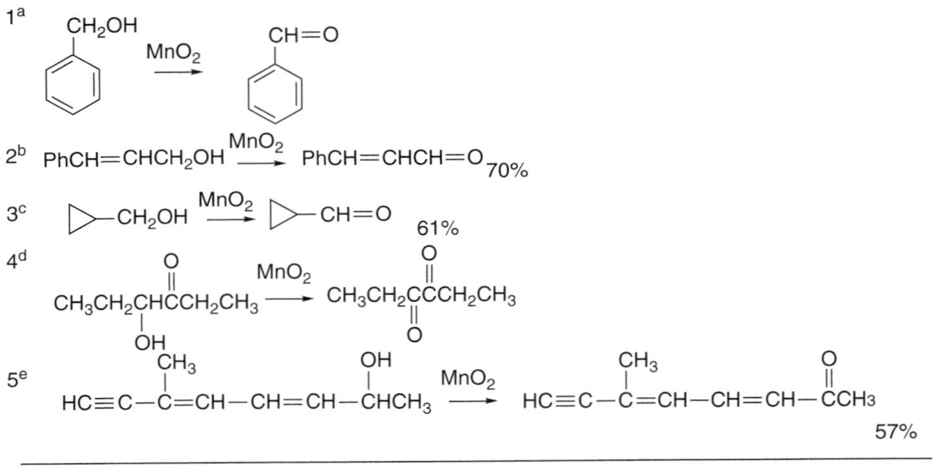
1a
CH_2OH
MnO_2
CH=O
2b PhCH=CHCH$_2$OH
MnO_2
PhCH=CHCH=O
70%
3c
CH_2OH
MnO_2
CH=O
61%
4d
$CH_3CH_2CHCCH_2CH_3$
OH
MnO_2
$CH_3CH_2CCCH_2CH_3$
5e
CH_3
OH
HC≡C—C=CH—CH=CH—CHCH$_3$
MnO_2
HC≡C—C=CH—CH=CH—CCH$_3$
57%

a. E. F. Pratt and J. F. Van De Castle, *J. Org. Chem.*, **26**, 2973 (1961).
b. I. M. Goldman, *J. Org. Chem.*, **34**, 1979 (1969).
c. L. Crombie and J. Crossley, *J. Chem. Soc.*, 4983 (1963).
d. E. P. Papadopoulos, A. Jarrar, and C. H. Issidorides, *J. Org. Chem.*, **31**, 615 (1966).
e. J. Attenburrow, A. F. B. Cameron, J. H. Chapman, R. M. Evans, B. A. Hems, A. B. A. Janssen, and T. Walker, *J. Chem. Soc.*, 1094 (1952).

Another recently developed oxidant is CrO_2, a solid known as Magtrieve™ that is prepared commercially (for other purposes), which oxidizes allylic and benzylic alcohols in good yield.[13] It is also reactive toward saturated alcohols. Because the solid remains ferromagnetic, it can be recovered by use of a magnet and can be reactivated by exposure to air at high temperature, making it environmentally benign.

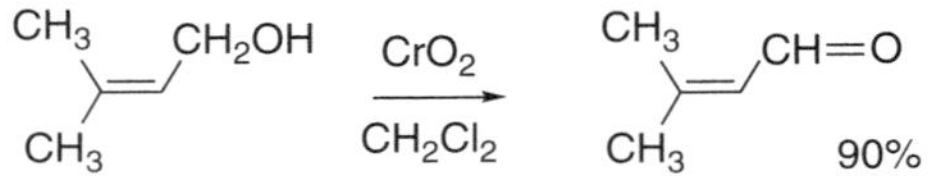
CH_3
CH_2OH
CH_3
CrO_2
CH_2Cl_2
CH_3
CH=O
CH_3
90%

Another possible alternative oxidant that has recently been investigated is an Fe(VI) species, potassium ferrate, K_2FeO_4, supported on montmorillonite clay.[14] This reagent gives clean, high-yielding oxidation of benzylic and allylic alcohols, but saturated alcohols are less reactive.

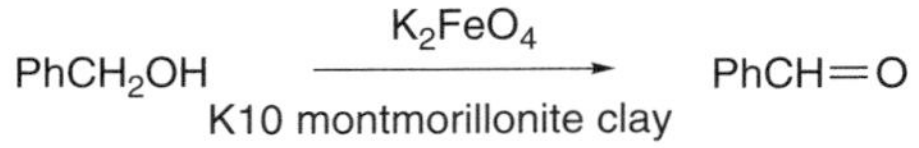
$PhCH_2OH$
K_2FeO_4
K10 montmorillonite clay
PhCH=O

A catalytic system that extends the reactivity of MnO_2 to saturated secondary alcohols has been developed.[15] This system consists of a Ru(II) salt, $RuCl_2(p\text{-cymene})_2$, and 2,6-di-*t*-butylbenzoquinone.

[13] R. A. Lee and D. S. Donald, *Tetrahedron Lett.*, **38**, 3857 (1997).
[14] L. Delaude and P. Laszlo, *J. Org. Chem.*, **61**, 6360 (1996).
[15] U. Karlsson, G. Z. Wang, and J.-E. Backvall, *J. Org. Chem.*, **59**, 1196 (1994).

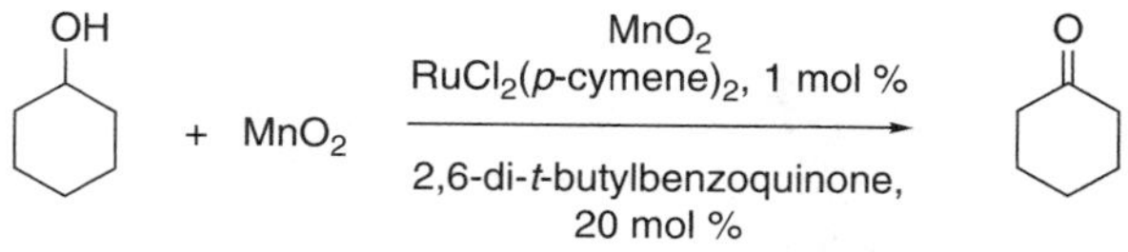

Ruthenium is the active oxidant and benzoquinone functions as an intermediary hydride transfer agent.

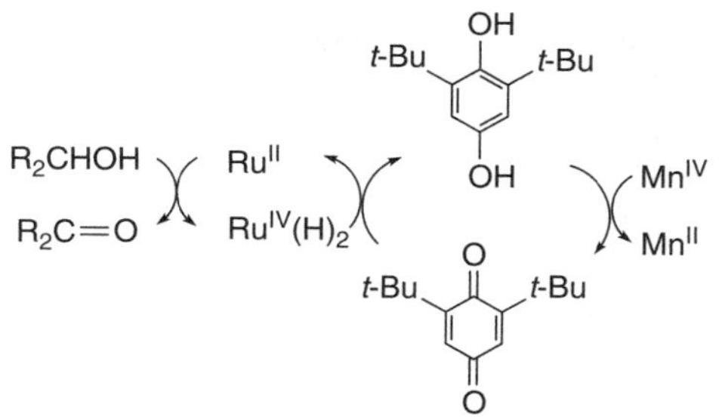

Another reagent that finds application of oxidations of alcohols to ketones is ruthenium tetroxide. The oxidations are typically carried out using a catalytic amount of the ruthenium source, e.g., $RuCl_3$, with $NaIO_4$ or $NaOCl$ as the stoichiometric oxidant.[16] Acetonitrile is a favorable solvent because of its ability to stabilize the ruthenium species that are present.[17] For example, the oxidation of **1** to **2** was successfully achieved with this reagent after a number of other methods failed.

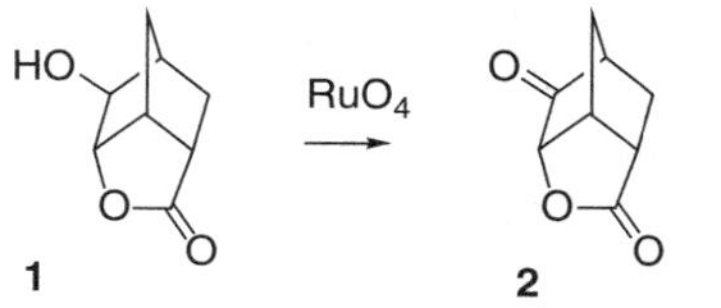

Ref. 18

Ruthenium tetroxide is a potent oxidant, however, and it readily attacks carbon-carbon double bonds.[19] Primary alcohols are oxidized to carboxylic acids, methyl ethers give methyl esters, and benzyl ethers are oxidized to benzoate esters.

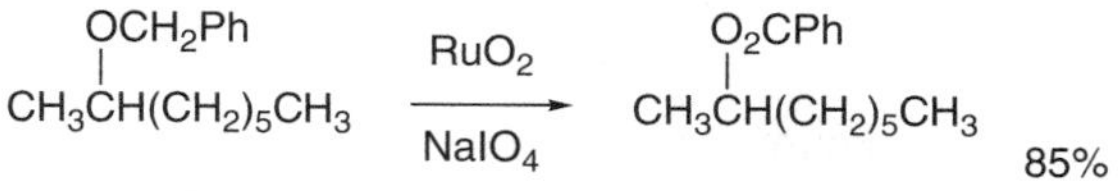

85%   Ref. 20

This reagent has been used in multistep syntheses to convert a tetrahydrofuran ring into a γ-lactone.

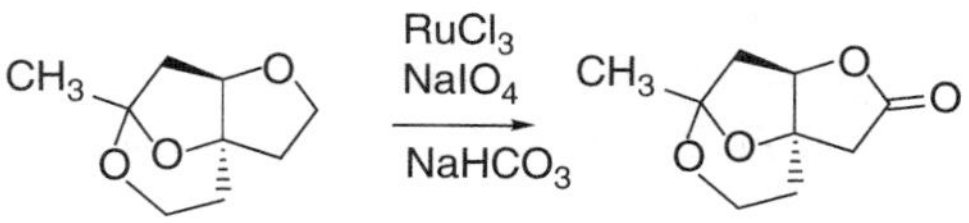

Ref. 21

16. P. E. Morris, Jr., and D. E. Kiely, *J. Org. Chem.*, **52**, 1149 (1987).
17. P. H. J. Carlsen, T. Katsuki, V. S. Martin, and K. B. Sharpless, *J. Org. Chem.*, **46**, 3936 (1981).
18. R. M. Moriarty, H. Gopal, and T. Adams, *Tetrahedron Lett.*, 4003 (1970).
19. J. L. Courtney and K. F. Swansborough, *Rev. Pure Appl. Chem.*, **22**, 47 (1972); D. G. Lee and M. van den Engh, in *Oxidation*, Part B, W. S. Trahanovsky, ed., Academic Press, New York, 1973, Chap. IV.
20. P. F. Schuda, M. B. Cichowitz, and M. P. Heinmann, *Tetrahedron Lett.*, **24**, 3829 (1983).
21. J.-S. Han and T. L. Lowary, *J. Org. Chem.*, **68**, 4116 (2003).

12.1.2. Other Oxidants

12.1.2.1. Oxidations Based on Dimethyl Sulfoxide. A very useful group of procedures for oxidation of alcohols to ketones employs dimethyl sulfoxide (DMSO) and any one of several electrophilic reagents, such as dicyclohexylcarbodiimide (DCCI), acetic anhydride, trifluoroacetic anhydride (TFAA), oxalyl chloride, or sulfur trioxide.[22] The original procedure involved DMSO and DCCI.[23] The mechanism of the oxidation involves formation of intermediate **A** by nucleophilic attack by DMSO on the carbodiimide, followed by reaction of the intermediate with the alcohol.[24] A proton transfer leads to an alkoxysulfonium ylide that is converted to product by an intramolecular proton transfer and elimination.

The activation of DMSO toward the addition step can be accomplished by other electrophiles. All of these reagents are believed to form a sulfoxonium species by electrophilic attack at the sulfoxide oxygen. The addition of the alcohol and the departure of the sulfoxide oxygen as part of a leaving group generates an intermediate comparable to **C** in the carbodiimide mechanism.

Preparatively useful procedures based on acetic anhydride,[25] trifluoroacetic anhydride,[26] and oxalyl chloride[27] have been developed. The last method, known as the *Swern oxidation*, is currently the most popular.

Scheme 12.3 gives some representative examples of these methods. Entry 1 is an example of the original procedure using DCCI. Entries 2 and 3 use SO_3 and $(CH_3CO)_2O$, respectively, as the electrophilic reagents. Entry 3 is noteworthy in successfully oxidizing an alcohol without effecting the sensitive indole ring. Entry 4 is

22. A. J. Mancuso and D. Swern, *Synthesis*, 165 (1981); T. T. Tidwell, *Synthesis*, 857 (1990).
23. K. E. Pfitzner and J. G. Moffatt, *J. Am. Chem. Soc.*, **87**, 5661, 5670 (1965).
24. J. G. Moffatt, *J. Org. Chem.*, **36**, 1909 (1971).
25. J. D. Albright and L. Goldman, *J. Am. Chem. Soc.*, **89**, 2416 (1967).
26. J. Yoshimura, K. Sato, and H. Hashimoto, *Chem. Lett.*, 1327 (1977); K. Omura, A. K. Sharma, and D. Swern, *J. Org. Chem.*, **41**, 957 (1976); S. L. Huang, K. Omura, and D. Swern, *J. Org. Chem.*, **41**, 3329 (1976).
27. A. J. Mancuso, S.-L. Huang, and D. Swern, *J. Org. Chem.*, **43**, 2480 (1978).

Scheme 12.3. Oxidation of Alcohols Using Dimethyl Sulfoxide

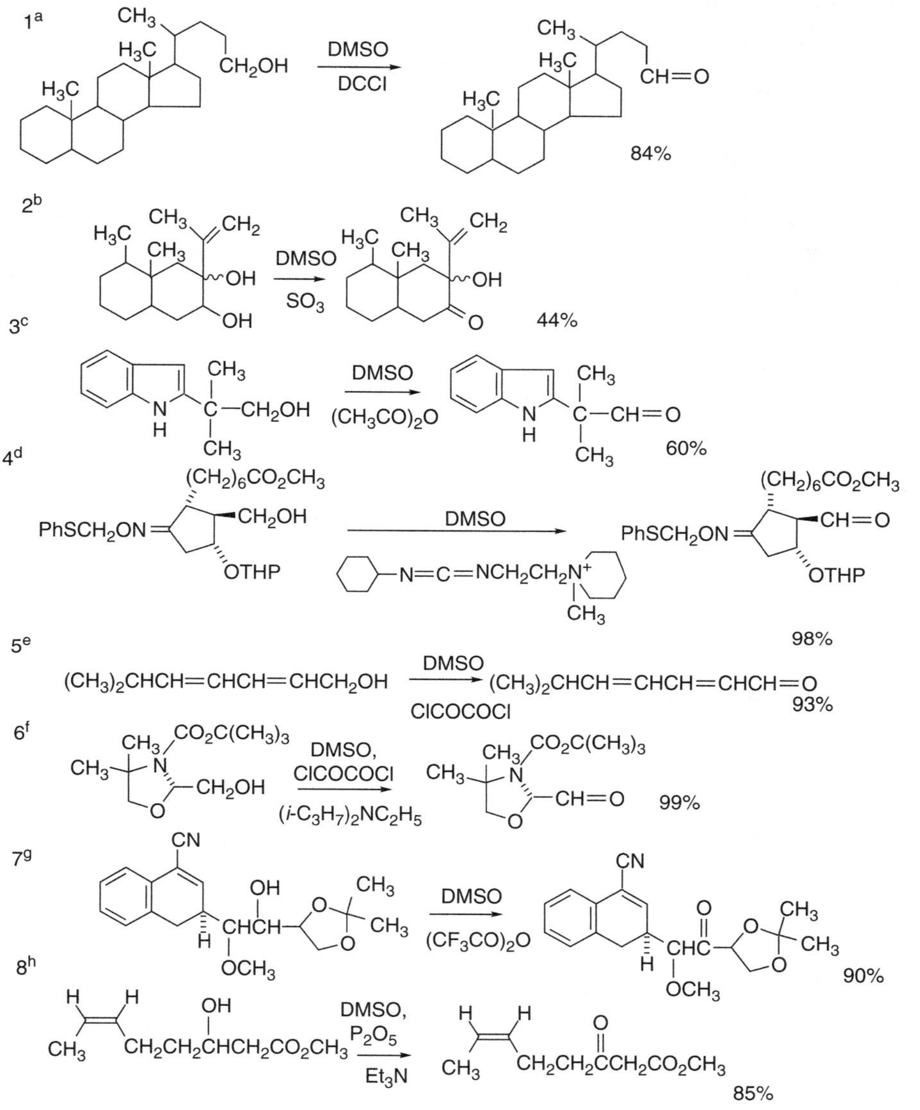

1[a]

DMSO / DCCI → 84%

2[b]

DMSO / SO₃ → 44%

3[c]

DMSO / (CH₃CO)₂O → 60%

4[d]

DMSO → 98%

5[e]

$(CH_3)_2CHCH=CHCH=CHCH_2OH$ — DMSO / ClCOCOCl → $(CH_3)_2CHCH=CHCH=CHCH=O$ — 93%

6[f]

DMSO, ClCOCOCl / $(i\text{-}C_3H_7)_2NC_2H_5$ → 99%

7[g]

DMSO / $(CF_3CO)_2O$ → 90%

8[h]

DMSO, P_2O_5 / Et_3N → 85%

a. J. G. Moffat, *Org. Synth.*, **47**, 25 (1967).
b. J. A. Marshall and G. M. Cohen, *J. Org. Chem.*, **36**, 877 (1971).
c. E. Houghton and J. E. Saxton, *J. Chem. Soc. C*, 595 (1969).
d. N. Finch, L. D. Veccia, J. J. Fitt, R. Stephani, and I. Vlatta, *J. Org. Chem.*, **38**, 4412 (1973).
e. W. R. Roush, *J. Am. Chem. Soc.*, **102**, 1390 (1980).
f. A. Dondoni and D. Perrone, *Synthesis*, 527 (1997).
g. R. W. Franck and T. V. John, *J. Org. Chem.*, **45**, 1170 (1987).
h. D. F. Taber, J. C. Amedio, Jr., and K.-Y. Jung, *J. Org. Chem.*, **52**, 5621 (1987).

an example of the use of a water-soluble carbodiimide as the activating reagent. The modified carbodiimide facilitates product purification by providing for easy removal of the urea by-product. Entries 5 and 6 are examples of the Swern procedure. Entry 7 uses TFAA as the electrophile. Entry 8, which uses the inexpensive reagent P_2O_5 as the electrophile, was conducted on a 60-g scale.

12.1.2.2. Oxidation by the Dess-Martin Reagent. Another reagent that has become important for laboratory synthesis is known as the *Dess-Martin reagent*,[28] which is a hypervalent iodine(V) compound.[29] The reagent is used in inert solvents such as chloroform or acetonitrile and gives rapid oxidation of primary and secondary alcohols. The by-product, *o*-iodosobenzoic acid, can be extracted with base and recycled.

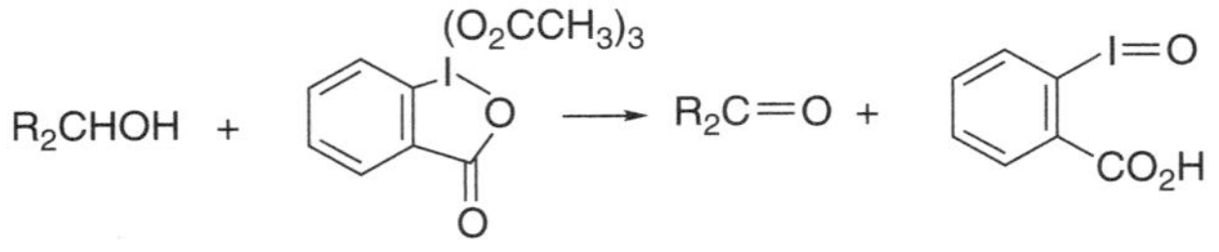

Scheme 12.4. Oxidation by the Dess-Martin Reagent

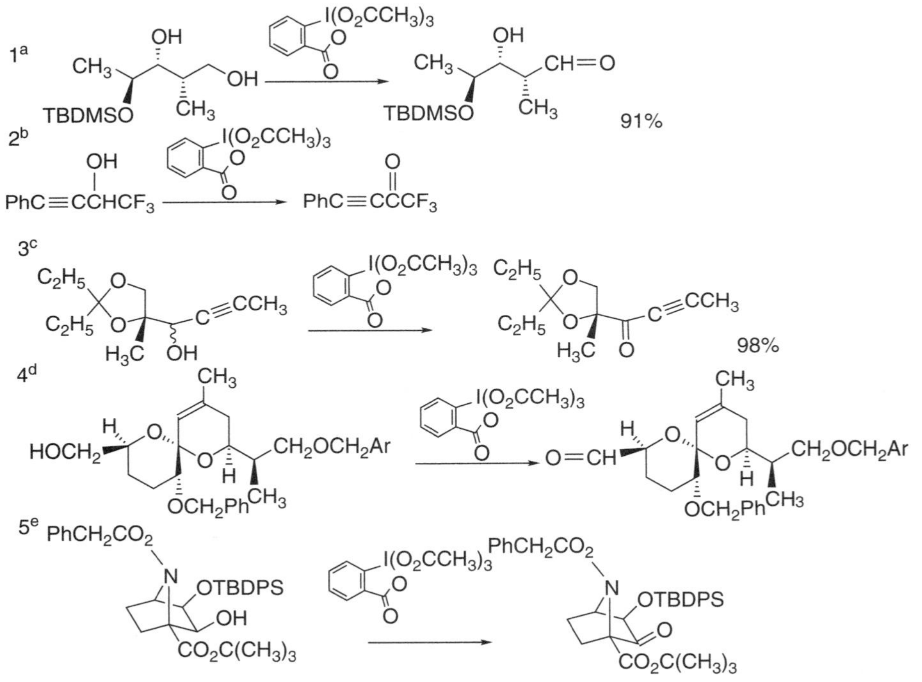

a. P. R. Blakemore, P. J. Kocienski, A. Morley, and K. Muir, *J. Chem. Soc., Perkin Trans. 1*, 955 (1999).
b. R. J. Linderman and D. M. Graves, *Tetrahedron Lett.*, **28**, 4259 (1987).
c. S. D. Burke, J. Hong, J. R. Lennox, and A. P. Mongin, *J. Org. Chem.*, **63**, 6952 (1998).
d. S. F. Sabes, R. A. Urbanek, and C. J. Forsyth, *J. Am. Chem. Soc.*, **120**, 2534 (1998).
e. B. P. Hart and H. Rapoport, *J. Org. Chem.*, **64**, 2050 (1999).

28. D. B. Dess and J. C. Martin, *J. Org. Chem.*, **48**, 4155 (1983); R. E. Ireland and L. Liu, *J. Org. Chem.*, **58**, 2899 (1993); S. D. Meyer and S. L. Schreiber, *J. Org. Chem.*, **59**, 7549 (1994).
29. T. Wirth and U. H. Hirt, *Synthesis*, 471 (1999).

The mechanism of the Dess-Martin oxidation involves exchange of the alcohol for acetate, followed by proton removal.[30]

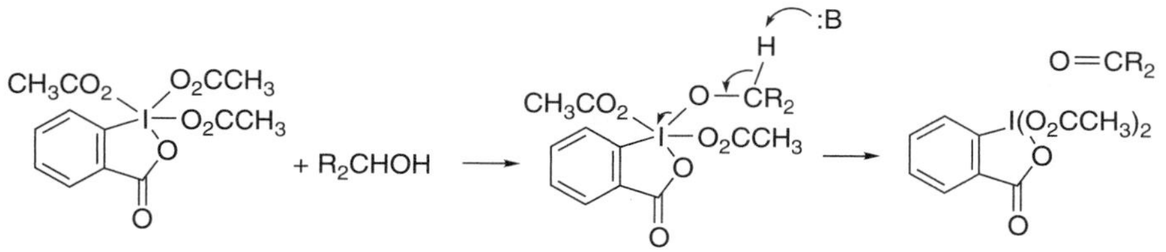

Scheme 12.4 shows several examples of the use of the Dess-Martin reagent.

Scheme 12.5. Oxidations Using TEMPO

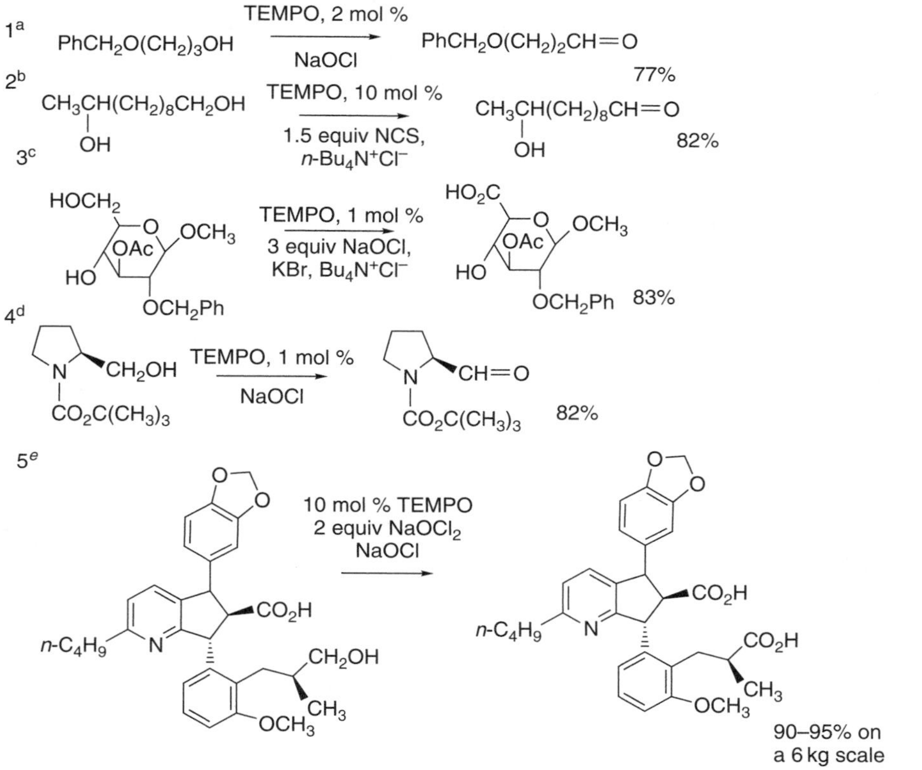

a. B. G. Szczepankiewicz and C. H. Heathcock, *Tetrahedron*, **53**, 8853 (1997).
b. J. Einhorn, C. Einhorn, F. Ratajczak, and J.-L. Pierre, *J. Org. Chem.*, **61**, 7452 (1996).
c. N. J. Davis and S. L. Flitsch, *Tetrahderon Lett.*, **34**, 1181 (1993).
d. M. R. Leanna, T. J. Sowin, and H. E. Morton, *Tetrahedron Lett.*, **33**, 5029 (1992).
e. Z. J. Song, M. Zhao, R. Desmond, P. Devine, D. M. Tscaen, R. Tillyer, L. Frey, R. Heid, F. Xu, B. Foster, J. Li, R. Reamer, R. Volante, E. J. Grabowski, U. H. Dolling, P. J. Reider, S. Okada, Y. Kato, and E. Mano, *J. Org. Chem.*, **64**, 9658 (1999).

[30.] S. De Munari, M. Frigerio, and M. Santagostino, *J. Org. Chem.*, **61**, 9272 (1996).

12.1.2.3. Oxidations Using Oxoammonium Ions. Another oxidation procedure uses an oxoammonium ion, usually derived from the stable nitroxide tetramethylpiperidine nitroxide, TEMPO, as the active reagent.[31] It is regenerated in a catalytic cycle using hypochlorite ion[32] or NCS[33] as the stoichiometric oxidant. These reactions involve an intermediate adduct of the alcohol and the oxoammonium ion.

One feature of this oxidation system is that it can selectively oxidize primary alcohols in preference to secondary alcohols, as illustrated by Entry 2 in Scheme 12.5. The reagent can also be used to oxidize primary alcohols to carboxylic acids by a subsequent oxidation with sodium chlorite.[34] Entry 3 shows the selective oxidation of a primary alcohol in a carbohydrate to a carboxylic acid without affecting the secondary alcohol group. Entry 5 is a large-scale preparation that uses $NaClO_2$ in conjunction with bleach as the stoichiometric oxidant.

12.2. Addition of Oxygen at Carbon-Carbon Double Bonds

12.2.1. Transition Metal Oxidants

12.2.1.1. Dihydroxylation of Alkenes. The higher oxidation states of certain transition metals, particularly the permanganate ion and osmium tetroxide, are effective reagents for addition of two oxygen atoms at a carbon-carbon double bond. Under carefully controlled reaction conditions, potassium permanganate can effect conversion of alkenes to glycols. However, this oxidant is capable of further oxidizing the glycol with cleavage of the carbon-carbon bond. A cyclic manganese ester is an intermediate in these oxidations. Owing to the cyclic nature of this intermediate, the glycols are formed by *syn* addition.

31. N. Merbouh, J. M. Bobbitt, and C. Brueckner, *Org. Prep. Proced. Int.*, **36**, 3 (2004).
32. R. Siedlecka, J. Skarzewski, and J. Mlochowski, *Tetrahedron Lett.*, **31**, 2177 (1990); T. Inokuchi, S. Matsumoto, T. Nishiyama, and S. Torii, *J. Org. Chem.*, **55**, 462 (1990); P. L. Anelli, S. Banfi, F. Montanari, and S. Quici, *J. Org. Chem.*, **54**, 2970 (1989); M. R. Leanna, T. J. Sowin, and H. E. Morton, *Tetrahedron Lett.*, **33**, 5029 (1992).
33. J. Einhorn, C. Einhorn, F. Ratajczak, and J.-L. Pierre, *J. Org. Chem.*, **61**, 7452 (1996).
34. P. M. Wovkulich, K. Shankaran, J. Kiegel, and M. R. Uskokovic, *J. Org. Chem.*, **58**, 832 (1993).

Ketols are also observed as products of permanganate oxidation of alkenes. The ketols are believed to be formed as a result of oxidation of the cyclic intermediate.[35]

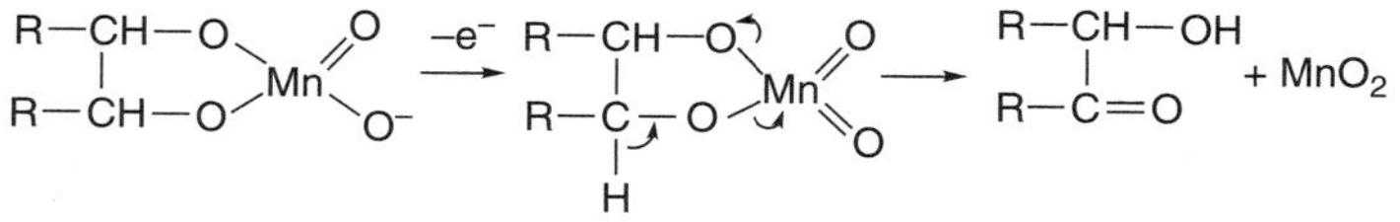

Ruthenium tetroxide can also be used in the oxidation of alkenes. Conditions that are selective for formation of ketols have been developed.[36] Use of 1 mol % of $RuCl_3$ and five equivalents of $KHSO_5$ (Oxone®) in an ethyl acetate-acetonitrile-water mixture gives mainly hydroxymethyl ketones from terminal alkenes.

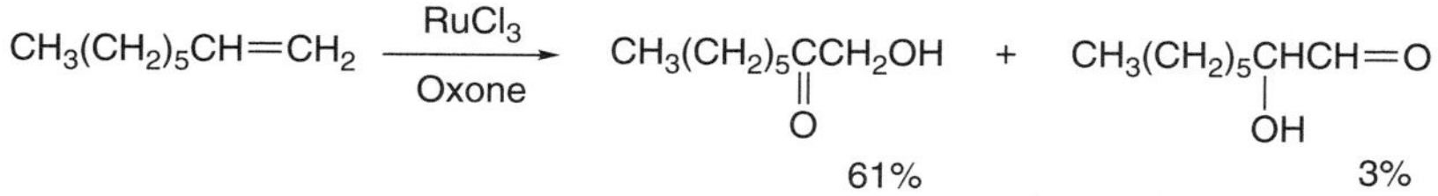

With aryl-substituted alkenes, the aryl ketone is the major product.

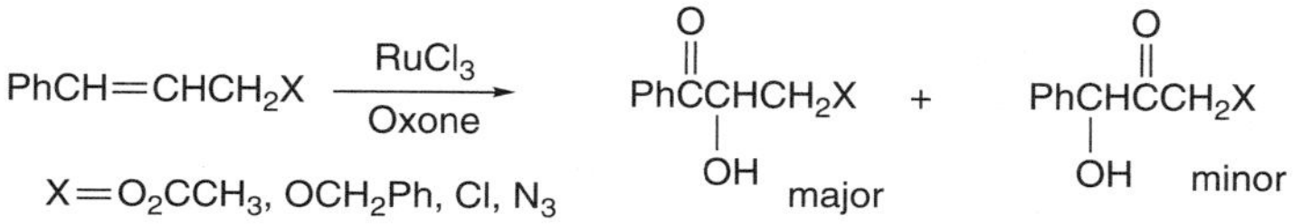

The mechanistic basis of this method depends on the use of excess peroxysulfate so that the major pathway leads to ketol rather than diol.

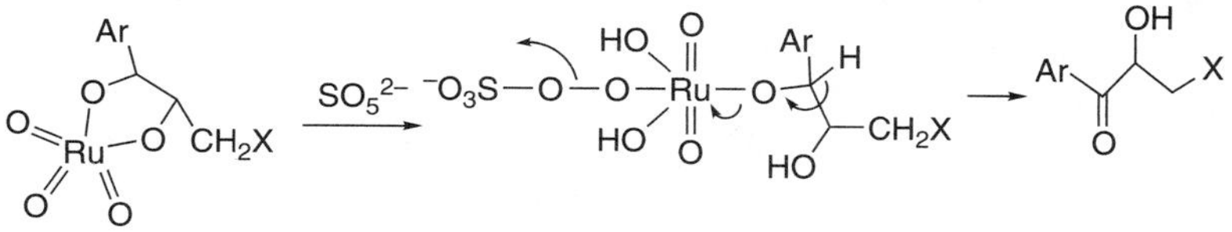

Permanganate ion can be used to oxidize acetylenes to diones.

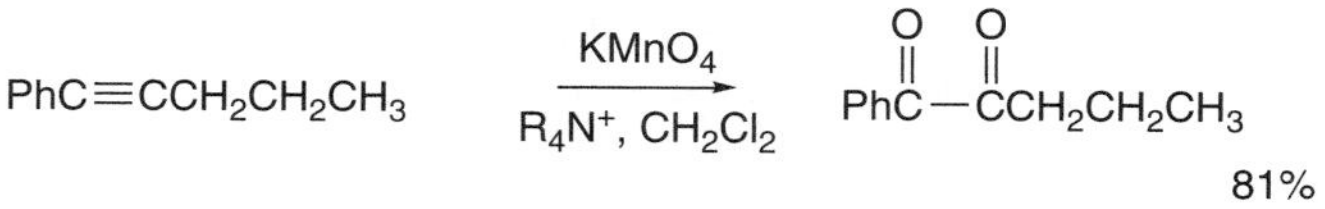

A mixture of $NaIO_4$ and RuO_2 in a heterogeneous solvent system is also effective for this transformation.

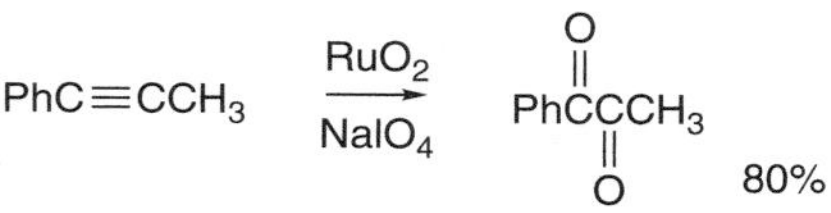

35. S. Wolfe, C. F. Ingold, and R. U. Lemieux, *J. Am. Chem. Soc.*, **103**, 938 (1981); D. G. Lee and T. Chen, *J. Am. Chem. Soc.*, **111**, 7534 (1989).
36. B. Plietker, *J. Org. Chem.*, **69**, 8287 (2004).
37. D. G. Lee and V. S. Chang, *J. Org. Chem.*, **44**, 2726 (1979).
38. R. Zibuck and D. Seebach, *Helv. Chim. Acta*, **71**, 237 (1988).

The most widely used reagent for oxidation of alkenes to glycols is osmium tetroxide. Osmium tetroxide is a highly selective oxidant that gives glycols by a stereospecific *syn* addition.[39] The reaction occurs through a cyclic osmate ester that is formed by a $[3+2]$ cycloaddition.[40]

The reagent is toxic and expensive but these disadvantages are minimized by methods that use only a catalytic amount of osmium tetroxide. A very useful procedure involves an amine oxide such as morpholine-*N*-oxide as the stoichiometric oxidant.[41]

t-Butyl hydroperoxide,[42] barium chlorate,[43] or potassium ferricyanide[44] can also be used as oxidants in catalytic procedures.

Scheme 12.6 provides some examples of oxidations of alkenes to glycols by both permanganate and osmium tetroxide. The oxidation by $KMnO_4$ in Entry 1 is done in cold aqueous solution. The reaction is very sensitive to the temperature control during the reaction. The reaction in Entry 2 was also done by the catalytic OsO_4 method using *N*-methylmorpholine-*N*-oxide in better (80%) yield. Note that the hydroxy groups are introduced from the less hindered face of the double bond. Entries 3 to 5 illustrate several of the catalytic procedures for OsO_4 oxidation. In each case the reaction is a stereospecific *syn* addition. Note also that in Entries 4 and 5 the double bond is conjugated with an EWG substituent, so the range of the reaction includes deactivated alkenes.

Osmium tetroxide oxidations can be highly enantioselective in the presence of chiral ligands. The most highly developed ligands are derived from the cinchona alkaloids dihydroquinine (DHQ) and dihydroquinidine (DHQD).[45] The most effective

39. M. Schroeder, *Chem. Rev.*, **80**, 187 (1980).

40. A. J. DelMonte, J. Haller, K. N. Houk, K. B. Sharpless, D. A. Singleton, T. Strassner, and A. A. Thomas, *J. Am. Chem. Soc.*, **119**, 9907 (1997); U. Pidun, C. Boehme, and G. Frenking, *Angew. Chem. Intl. Ed. Engl.*, **35**, 2817 (1997).

41. V. Van Rheenen, R. C. Kelly, and D. Y. Cha, *Tetrahedron Lett.*, 1973 (1976).

42. K. B. Sharpless and K. Akashi, *J. Am. Chem. Soc.*, **98**, 1986 (1976); K. Akashi, R. E. Palermo, and K. B. Sharpless, *J. Org. Chem.*, **43**, 2063 (1978).

43. L. Plaha, J. Weichert, J. Zvacek, S. Smolik, and B. Kakac, *Collect. Czech. Chem. Commun.*, **25**, 237 (1960); A. S. Kende, T. V. Bentley, R. A. Mader, and D. Ridge, *J. Am. Chem. Soc.*, **96**, 4332 (1974).

44. M. Minato, K. Yamamoto, and J. Tsuji, *J. Org. Chem.*, **55**, 766 (1990); K. B. Sharpless, W. Amberg, Y. L. Bennani, G. A. Crispino, J. Hartung, K.-S. Jeong, H.-L. Kwong, K. Morikawa, Z.-M. Wang, D. Xu, and X.-L. Zhang, *J. Org. Chem.*, **57**, 2768 (1992); J. Eames, H. J. Mitchell, A. Nelson, P. O'Brien, S. Warren, and P. Wyatt, *Tetrahedron Lett.*, **36**, 1719 (1995).

45. H. C. Kolb, M. S. VanNieuwenhze, and K. B. Sharpless, *Chem. Rev.*, **94**, 2483 (1994).

Scheme 12.6. Examples of *syn* Dihydroxylation of Alkenes

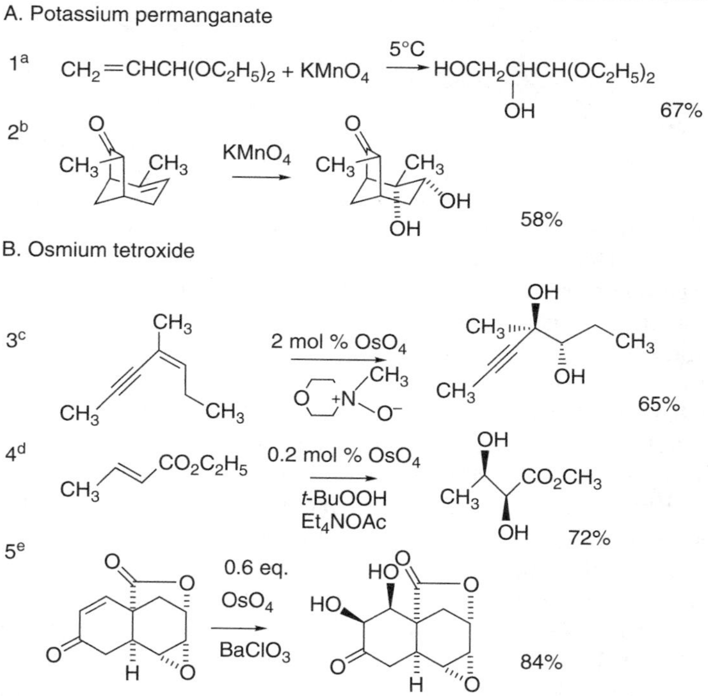

A. Potassium permanganate

B. Osmium tetroxide

a. E. J. Witzeman, W. L. Evans, H. Haas, and E. F. Schroeder, *Org. Synth.*, **II**, 307 (1943).
b. S. D. Larsen and S. A. Monti, *J. Am. Chem. Soc.*, **99**, 8015 (1977).
c. E. J. Corey, P. B. Hopkins, S. Kim, S. Yoo, K. P. Nambiar, and J. R. Falck, *J. Am. Chem. Soc.*, **101**, 7131 (1979).
d. K. Akashi, R. E. Palermo, and K. B. Sharpless, *J. Org. Chem.*, **43**, 2063 (1978).
e. S. Danishefsky, P. F. Schuda, T. Kitahara, and S. J. Etheredge, *J. Am. Chem. Soc.*, **99**, 6066 (1977).

ligands are dimeric derivatives of these alkaloids.[46] These ligands both induce high enantioselectivity and accelerate the reaction.[47] Potassium ferricyanide is usually used as the stoichiometric oxidant. Optimization of the reaction conditions permits rapid and predictable dihydroxylation of many types of alkenes.[48] The premixed catalysts are available commercially and are referred to by the trade name AD-mix™. Several heterocyclic compounds including phthalazine (PHAL), pyrimidine (PYR), pyridazine (PYDZ), and diphenylpyrimidine (DPPYR) have been used as linking groups for the alkaloids.

[46.] (a) G. A. Crispino, K. S. Jeong, H. C. Kolb, Z.-M. Wang, D. Xu, and K. B. Sharpless, *J. Org. Chem.*, **58**, 3785 (1993); (b) G. A. Crispino, A. Makita, Z.-M. Wang, and K. B. Sharpless, *Tetrahedron Lett.*, **35**, 543 (1994); (c) K. B. Sharpless, W. Amberg, Y. L. Bennani, G. A. Crispino, J. Hartung, K.-S. Jeong, H.-L. Kwong, K. Morikawa, Z.-M. Wang, D. Xu, and X.-L. Zhang, *J. Org. Chem.*, **57**, 2768 (1992); (d) W. Amberg, Y. L. Bennani, R. K. Chadha, G. A. Crispino, W. D. Davis, J. Hartung, K. S. Jeong, Y. Ogino, T. Shibata, and K. B. Sharpless, *J. Org. Chem.*, **58**, 844 (1993); (e) H. Becker, S. B. King, M. Taniguchi, K. P. M. Vanhessche, and K. B. Sharpless, *J. Org. Chem.*, **60**, 3940 (1995).
[47.] P. G. Anderson and K. B. Sharpless, *J. Am. Chem. Soc.*, **115**, 7047 (1993).
[48.] T. Gobel and K. B. Sharpless, *Angew. Chem. Int. Ed. Engl.*, **32**, 1329 (1993).

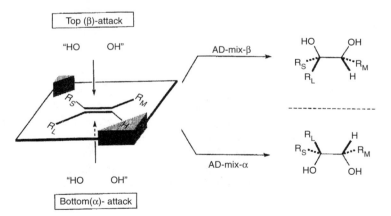

(DHQ)₂-PHAL

(DHQD)₂-DPPYR

Empirical analysis led to the predictive model for enantioselectivity shown in Figure 12.1.[46c, 49] The two alkaloids are of opposite chirality and give enantiomeric products. The commercial reagents are designated AD-mix-α and AD-mix-β. The configuration of the products can be predicted by a model based on the relative size of the substituent groups. *E*-Alkenes give the best fit to the binding pocket and give the highest reactivity and enantioselectivity.

There have been two computational studies of the basis for the catalysis and enantioselectivity. A study of the reaction of styrene with the (DHQD)₂PYDZ ligand was done using a hybrid DFT/MM protocol.[50] Two orientations of the styrene molecule were found that were about 3.0 kcal/mol more favorable than any of the others. These TSs are shown in Figure 12.2. Both these structures predict the observed *R*-configuration for the product. Most of the difference among the various structures is found in the MM terms and they are exothermic, that is, there are *net attractive forces involved in the binding of the reactant*. The second study used stilbene as the reactant and (DHQD)₂PHAL as the catalyst ligand.[51] This study arrives at the TS shown in Figure 12.3. The two phenyl groups of stilbene occupy *both* of the sites found for the two low-energy TSs for styrene.

Fig. 12.1. Predictive model for enantioselective dihydroxylation by dimeric alkaloid catalysts. (DHQD)₂ catalysts give β-approach; (DHQ)₂ catalysts give α-approach. Reproduced from *J. Org. Chem.*, **57**, 2768 (1992), by permission of the American Chemical Society.

49. H. C. Kolb, M. S. VanNieuwenhze, and K. B. Sharpless, *Chem. Rev.*, **94**, 2483 (1994).
50. G. Ujaque, F. Maseras, and A. Lledos, *J. Am. Chem. Soc.*, **121**, 1317 (1999).
51. P.-O. Norrby, T. Rasmussen, J. Haller, T. Strassner, and K. N. Houk, *J. Am. Chem. Soc.*, **121**, 10186 (1999).

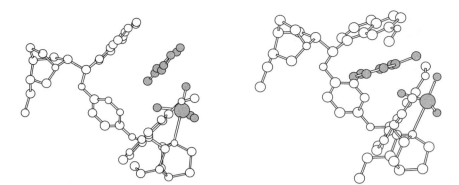

Fig. 12.2. Two lowest-energy transition structures for oxidation of styrene by $(DHQD)_2PYDZ-OsO_4$ catalysts. The structure on the left is about 0.4 kcal more stable than the one on the right. Both structures predict the formation of *R*-styrene oxide. Reproduced from *J. Am. Chem. Soc.*, **121**, 1317 (1999), by permission of the American Chemical Society.

Visual models, additional information and exercises on Dihydroxylation can be found in the Digital Resource available at: Springer.com/carey-sundberg.

Scheme 12.7 gives some examples of enantioselective hydroxylations using these reagents. Entry 1 is an allylic ether with a terminal double bond. *para*-Substituted derivatives also gave high e.e. values, but some *ortho* substituents led to lower e.e. values. Entry 2 is one of several tertiary allylic alcohols that gave excellent results. Entry 3 is a *trans*-substituted alkene with rather large (but unbranched) substituents. The inclusion of methanesulfonamide, as in this example, has been found to be beneficial for di- and trisubstituted alkenes. It functions by speeding the hydrolysis of the osmate ester intermediate. The product in this case goes on to cyclize to the

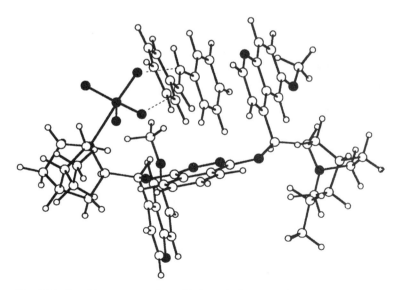

Fig. 12.3. Transition structure for oxidation of stilbene by $(DHQD)_2PHAL-OsO_4$ catalyst. Reproduced from *J. Am. Chem. Soc.*, **121**, 10186 (1999), by permission of the American Chemical Society.

Scheme 12.7. Enantioselective Osmium-Catalyzed Dihydroxylation of Alkenes

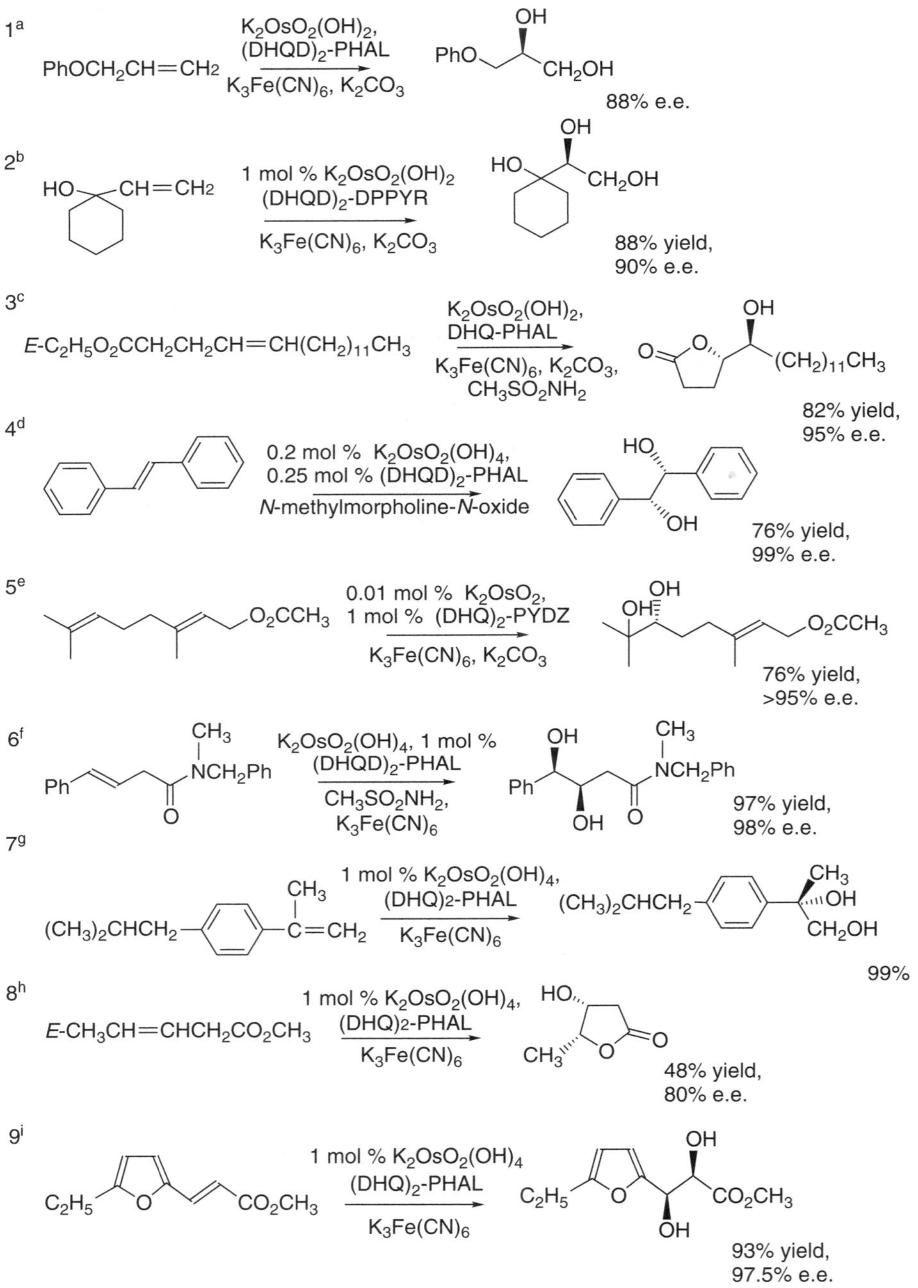

1[a]

PhOCH₂CH=CH₂

$\xrightarrow[\text{K}_3\text{Fe(CN)}_6, \text{K}_2\text{CO}_3]{\substack{\text{K}_2\text{OsO}_2\text{(OH)}_2, \\ \text{(DHQD)}_2\text{-PHAL}}}$

PhO—CH(OH)—CH₂OH

88% e.e.

2[b]

HO—C(cyclohexyl)—CH=CH₂

$\xrightarrow[\text{K}_3\text{Fe(CN)}_6, \text{K}_2\text{CO}_3]{\substack{\text{1 mol \% K}_2\text{OsO}_2\text{(OH)}_2 \\ \text{(DHQD)}_2\text{-DPPYR}}}$

HO—C—CH(OH)—CH₂OH

88% yield,
90% e.e.

3[c]

E-C₂H₅O₂CCH₂CH₂CH=CH(CH₂)₁₁CH₃

$\xrightarrow[\substack{\text{K}_3\text{Fe(CN)}_6, \text{K}_2\text{CO}_3, \\ \text{CH}_3\text{SO}_2\text{NH}_2}]{\substack{\text{K}_2\text{OsO}_2\text{(OH)}_2, \\ \text{DHQ-PHAL}}}$

lactone—CH(OH)—(CH₂)₁₁CH₃

82% yield,
95% e.e.

4[d]

stilbene (PhCH=CHPh)

$\xrightarrow[\text{N-methylmorpholine-N-oxide}]{\substack{\text{0.2 mol \% K}_2\text{OsO}_2\text{(OH)}_4, \\ \text{0.25 mol \% (DHQD)}_2\text{-PHAL}}}$

PhCH(OH)—CH(OH)Ph

76% yield,
99% e.e.

5[e]

(CH₃)₂C=CH...C(CH₃)=CH—CH₂O₂CCH₃

$\xrightarrow[\text{K}_3\text{Fe(CN)}_6, \text{K}_2\text{CO}_3]{\substack{\text{0.01 mol \% K}_2\text{OsO}_2, \\ \text{1 mol \% (DHQ)}_2\text{-PYDZ}}}$

(CH₃)₂C(OH)—CH(OH)...C(CH₃)=CH—CH₂O₂CCH₃

76% yield,
>95% e.e.

6[f]

Ph—CH=CH—CH₂—C(=O)—N(CH₃)CH₂Ph

$\xrightarrow[\substack{\text{CH}_3\text{SO}_2\text{NH}_2, \\ \text{K}_3\text{Fe(CN)}_6}]{\substack{\text{K}_2\text{OsO}_2\text{(OH)}_4, \text{1 mol \%} \\ \text{(DHQD)}_2\text{-PHAL}}}$

Ph—CH(OH)—CH(OH)—CH₂—C(=O)—N(CH₃)CH₂Ph

97% yield,
98% e.e.

7[g]

(CH₃)₂CHCH₂—C₆H₄—C(CH₃)=CH₂

$\xrightarrow[\text{K}_3\text{Fe(CN)}_6]{\substack{\text{1 mol \% K}_2\text{OsO}_2\text{(OH)}_4, \\ \text{(DHQ)}_2\text{-PHAL}}}$

(CH₃)₂CHCH₂—C₆H₄—C(CH₃)(OH)—CH₂OH

99%

8[h]

E-CH₃CH=CHCH₂CO₂CH₃

$\xrightarrow[\text{K}_3\text{Fe(CN)}_6]{\substack{\text{1 mol \% K}_2\text{OsO}_2\text{(OH)}_4, \\ \text{(DHQ)}_2\text{-PHAL}}}$

CH₃—CH(OH) lactone

48% yield,
80% e.e.

9[i]

C₂H₅—furan—CH=CH—CO₂CH₃

$\xrightarrow[\text{K}_3\text{Fe(CN)}_6]{\substack{\text{1 mol \% K}_2\text{OsO}_2\text{(OH)}_4, \\ \text{(DHQ)}_2\text{-PHAL}}}$

C₂H₅—furan—CH(OH)—CH(OH)—CO₂CH₃

93% yield,
97.5% e.e.

a. Z.-M. Wang, X.-L. Zhang, and K. B. Sharpless, *Tetrahedron Lett.*, **34**, 2267 (1993).

b. Z.-M. Wang and K. B. Sharpless, *Tetrahedron Lett.*, **34**, 8225 (1993).

c. Z.-M. Wang, X.-L. Zhang, K. B. Sharpless, S. C. Sinha, A. Sinha-Bagchi, and E. Keinan, *Tetrahedron Lett.*, **33**, 6407 (1992).

d. H. T. Chang and K. B. Sharpless, *J. Org. Chem.*, **61**, 6456 (1996).

e. E. J. Corey, M. C. Noe, and W.-C. Shieh, *Tetrahedron Lett.*, **34**, 5995 (1993).

f. Y. L. Bennani and K. B. Sharpless, *Tetrahedron Lett.*, **34**, 2079 (1993).

g. H. Ishibashi, M. Maeki, J. Yagi, M. Ohba, and T. Kanai, *Tetrahedron*, **55**, 6075 (1999).

h. T. Berkenbusch and R. Bruckner, *Tetrahedron*, **54**, 11461 (1998).

i. T. Taniguchi, M. Takeuchi, and K. Ogasawara, *Tetrahedron: Asymmetry*, **9**, 1451 (1998).

observed lactone. This particular oxidation was also carried out with (DHQD)$_2$-PHAL, which gave the enantiomeric lactone. Entry 4 is an optimized oxidation of stilbene that was done on a 1-kg scale. Entry 5 is the dihydroxylation of geranyl acetate that shows selectivity for the 6,7-double bond. Entry 6 involves an unsaturated amide and required somewhat higher catalyst loading than normal. Entry 7 provided a starting material for the enantioselective synthesis of S-ibuprofen. The reaction in Entry 8 was used to prepare the lactone shown (and its enantiomer) as starting materials for enantioselective synthesis of several natural products. The furan synthesized in Entry 9 was used to prepare a natural material by a route involving eventual oxidation of the furan ring.

Various other chiral diamines have also been explored for use with OsO$_4$, some of which are illustrated in Scheme 12.8. They presumably function by forming hexaco-ordinate chelates with OsO$_4$. The reactant in Entry 3 also raises the issue of diastereo-selectivity with respect to the allylic substituent. Normally, the dihydroxylation is *anti* toward such substituents.[52] There are thus matched and mismatched combinations with the chiral osmium ligand. The R,R-diamine shown gives the matched combination and leads to high diastereoselectivity, as well as high enantioselectivity.

12.2.1.2. Transition Metal–Catalyzed Epoxidation of Alkenes. Other transition metal oxidants can convert alkenes to epoxides. The most useful procedures involve *t*-butyl hydroperoxide as the stoichiometric oxidant in combination with vanadium or

Scheme 12.8. Enantioselective Hydroxylation Using Chiral Diamines

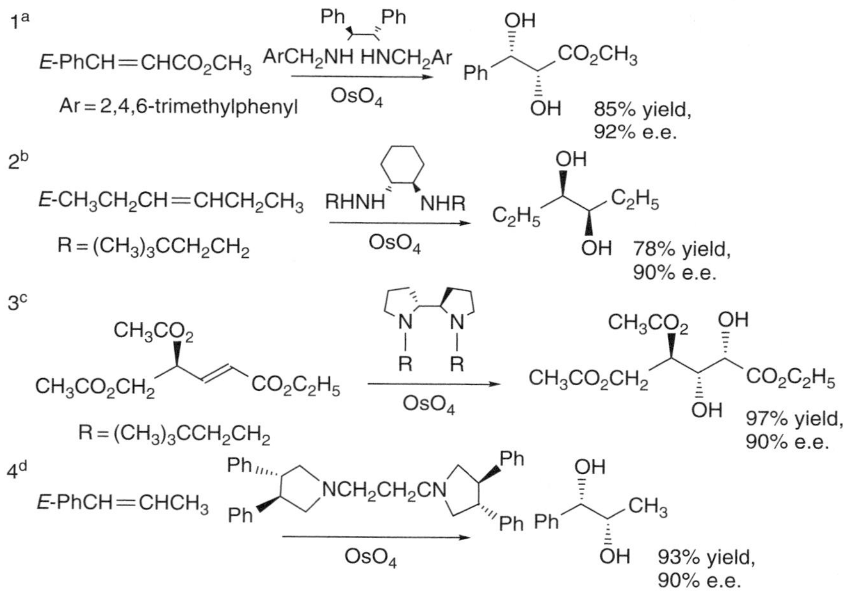

a. E. J. Corey, P. D. Jardine, S. Virgil, P.-W. Yuen, and R. D. Connell, *J. Am. Chem. Soc.*, **111**, 9243 (1989).
b. S. Hannessian, P. Meffre, M. Girard, S. Beaudoin, J.-Y. Sanceau, and Y. Bennani, *J. Org. Chem.*, **58**, 1991 (1993).
c. T. Oishi, K. Iida, and M. Hirama, *Tetrahedron Lett.*, **34**, 3573 (1993).
d. K. Tomioka, M. Nakajima, and K. Koga, *Tetrahedron Lett.*, **31**, 1741 (1990).

[52.] J. K. Cha, W. J. Christ, and Y. Kishi, *Tetrahedron*, **40**, 2247 (1984).

titanium compounds. The most reliable substrates for oxidation are allylic alcohols. The hydroxy group of the alcohol plays both an activating and stereodirecting role in these reactions. *t*-Butyl hydroperoxide and a catalytic amount of VO(acac) convert allylic alcohols to the corresponding epoxides in good yields.[53] The reaction proceeds through a complex in which the allylic alcohol is coordinated to vanadium by the hydroxy group. In cyclic alcohols, this results in epoxidation *cis* to the hydroxy group. In acyclic alcohols the observed stereochemistry is consistent with a TS in which the double bond is oriented at an angle of about 50° to the coordinated hydroxy group. This TS leads to diastereoselective formation of the *syn*-alcohol. This stereoselectivity is observed for both *cis*- and *trans*-disubstituted allylic alcohols.[54]

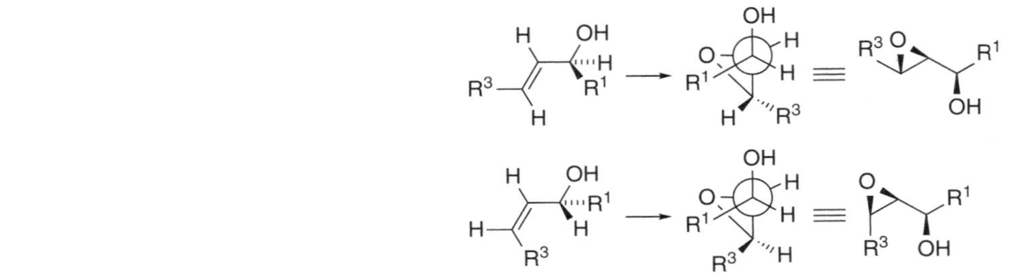

The epoxidation of allylic alcohols can also be effected by *t*-butyl hydroperoxide and titanium tetraisopropoxide. When enantiomerically pure tartrate ligands are included, the reaction is highly enantioselective. This reaction is called the *Sharpless asymmetric epoxidation*.[55] Either the (+) or (−) tartrate ester can be used, so either enantiomer of the desired product can be obtained.

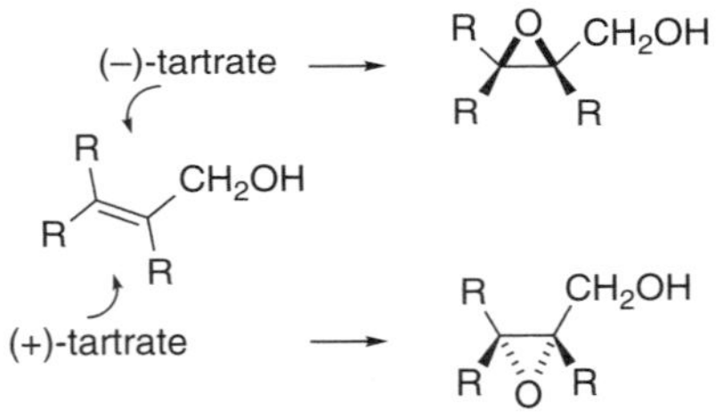

The mechanism by which the enantioselective oxidation occurs is generally similar to that for the vanadium-catalyzed oxidations. The allylic alcohol serves to coordinate the substrate to titanium. The tartrate esters are also coordinated at titanium, creating a chiral environment. The active catalyst is believed to be a dimeric species, and the mechanism involves rapid exchange of the allylic alcohol and *t*-butylhydroperoxide at the titanium ion.

53. K. B. Sharpless and R. C. Michaelson, *J. Am. Chem. Soc.*, **95**, 6136 (1973).
54. E. D. Mihelich, *Tetrahedron Lett.*, 4729 (1979); B. E. Rossiter, T. R. Verhoeven, and K. B. Sharpless, *Tetrahedron Lett.*, 4733 (1979).
55. For reviews, see A. Pfenninger, *Synthesis*, 89 (1986); R. A. Johnson and K. B. Sharpless, in *Catalytic Asymmetric Synthesis*, I. Ojima, ed., VCH Publishers, New York, 1993, pp. 103–158.

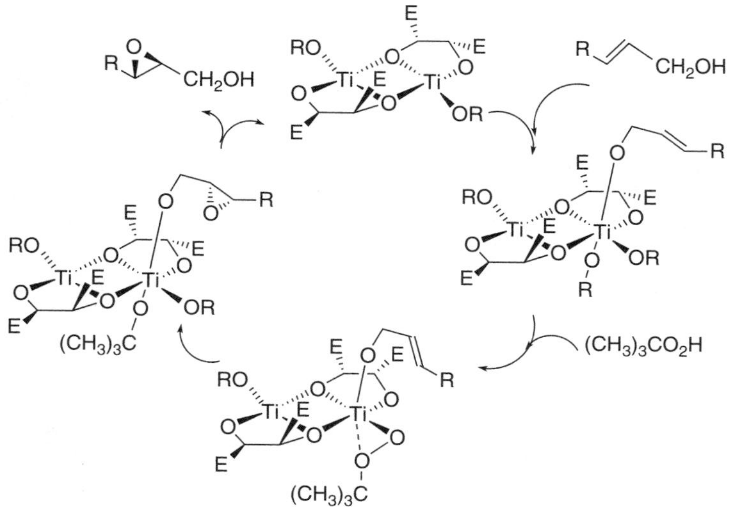

This method has proven to be an extremely useful means of synthesizing enantiomerically enriched compounds. Various improvements in the methods for carrying out the Sharpless oxidation have been developed.[56] The reaction can be done with catalytic amounts of titanium isopropoxide and the tartrate ligand.[57] This procedure uses molecular sieves to sequester water, which has a deleterious effect on both the rate and enantioselectivity of the reaction.

The orientation of the reactants is governed by the chirality of the tartrate ligand. In the TS an oxygen atom from the peroxide is transferred to the double bond. The enantioselectivity is consistent with a TS such as that shown below.[58]

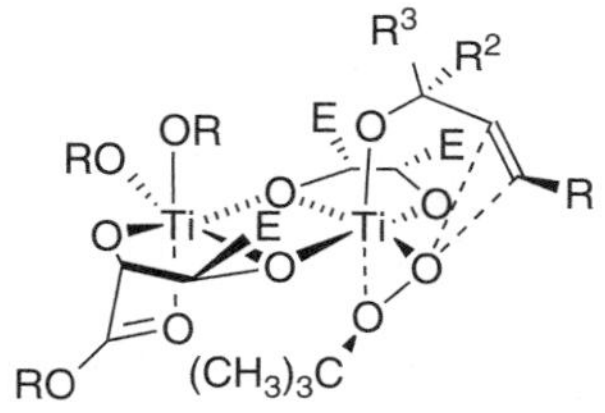

There has been a DFT (BLYP/6-31G*) study of the TS and its relationship to the enantioselectivity of the reaction.[59] The strategy used was to build up the model by successively adding components. First the titanium coordination sphere, including an alkene and peroxide group, was modeled (Figure 12.4a). In Figure 12.4b, the diol

56. J. G. Hill, B. E. Rossiter, and K. B. Sharpless, *J. Org. Chem.*, **48**, 3607 (1983); L. A. Reed, III, S. Masamune, and K. B. Sharpless, *J. Am. Chem. Soc.*, **104**, 6468 (1982).

57. R. M. Hanson and K. B. Sharpless, *J. Org. Chem.*, **51**, 1922 (1986); Y. Gao, R. M. Hanson, J. M. Klunder, S. Y. Ko, H. Masamune, and K. B. Sharpless, *J. Am. Chem. Soc.*, **109**, 5765 (1987).

58. V. S. Martin, S. S. Woodard, T. Katsuki, Y. Yamada, M. Ikeda, and K. B. Sharpless, *J. Am. Chem. Soc.*, **103**, 6237 (1981); K. B. Sharpless, S. S. Woodard, and M. G. Finn, *Pure Appl. Chem.*, **55**, 1823 (1983); M. G. Finn and K. B. Sharpless, in *Asymmetric Synthesis*, Vol. 5, J. D. Morrison, ed., Academic Press, New York, 1985, Chap 8; M. G. Finn and K. B. Sharpless, *J. Am. Chem. Soc.*, **113**, 113 (1991); B. H. McKee, T. H. Kalantar, and K. B. Sharpless, *J. Org. Chem.*, **56**, 6966 (1991); For an alternative description of the origin of enantioselectivity, see E. J. Corey, *J. Org. Chem.*, **55**, 1693 (1990).

59. Y.-D. Wu and D. F. W. Lai, *J. Am. Chem. Soc.*, **117**, 11327 (1995).

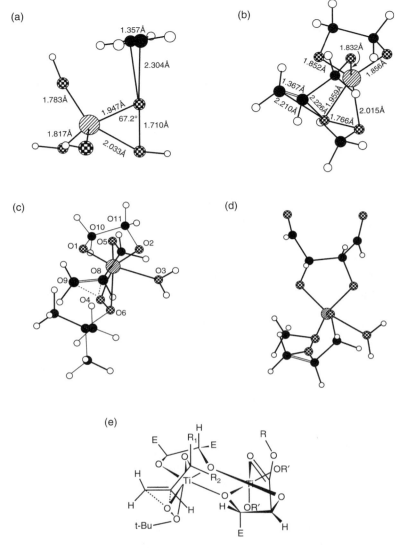

Fig. 12.4. Successive models of the transition state for Sharpless epoxidation. (a) the hexacoordinate Ti core with uncoordinated alkene; (b) Ti with methylhydroperoxide, allyl alcohol, and ethanediol as ligands; (c) monomeric catalytic center incorporating *t*-butylhydroperoxide as oxidant; (d) monomeric catalytic center with formyl groups added; (e) dimeric transition state with chiral tartrate model (E = CH = O). Reproduced from *J. Am. Chem. Soc.*, **117**, 11327 (1995), by permission of the American Chemical Society.

ligand and allylic alcohol were added to the coordination sphere. Then the steric bulk associated with the hydroperoxide was added (Figure 12.4c), and finally the tartrate ligands were added (using formyl groups as surrogates; Figure 12.4d) This led successively to TSs of increasingly detailed structure. The energies were minimized to identify the most stable structure at each step. The key features of the final TS model are the following: (1) The peroxide-titanium interaction has a spiro, rather than planar, arrangement in the TS for oxygen transfer. (2) The orientation of the alkyl group of the peroxide plays a key role in the enantioselectivity, which is consistent with the experimental observation that less bulky hydroperoxides give much lower enantioselectivity. (3) The C–O bond of the allylic alcohol bisects the Ti–O bond formed by the water and peroxy ligands. (4) The tartrate groups at the active catalytic center are in equatorial positions and do not coordinate to titanium. This implies a conformation flip of the diolate ring as part of the activation process, since the ester groups are in axial positions in the dimeric catalyst.

Visual models, additional information and exercises on Sharpless Epoxidation can be found in the Digital Resource available at: Springer.com/carey-sundberg.

Owing to the importance of the allylic hydroxy group in coordinating the reactant to the titanium, the structural relationship between the double bond and the hydroxy group is crucial. Homoallylic alcohols can be oxidized but the degree of enantioselectivity is reduced. Interestingly, the facial selectivity is reversed from that observed with allylic alcohols.[60] Compounds lacking a coordinating hydroxy group are not reactive under the standard reaction conditions.

Substituted allylic alcohols also exhibit diastereoselectivity. A DFT study has examined the influence of alkyl substituents in the allylic alcohol on the stereoselectivity.[61] Alcohols **3a**, **3b**, and **3c** were studied. The catalytic entity was modeled by $Ti(OH)_4$-CH_3OOH. This approach neglects the steric influence of the *t*-butyl and tartrate ester groups and focuses on the structural features of the allylic alcohols, which are placed on the catalytic core in their minimum energy conformation. Figure 12.5 shows these conformations. The TS structural parameters were derived from the Wu-Lai TS model (see Figure 12.4). The relative energies of the TSs leading to the *erythro* and *threo* products for each alcohol were compared (Figure 12.6). A solvent dielectric chosen to simulate CH_2Cl_2 was used. The general conclusion drawn from this study is that the reactant conformation is the critical feature determining the diastereoselectivity of the epoxidation.

In allylic alcohols with $A^{1,3}$ strain, the main product is *syn*. A methyl substituent at R^4 leads to the methyl group being positioned *anti* to the complexed oxidant. If R^4 is hydrogen, a TS with the methyl group in an "inside" position is favored, as shown in Figure 12.6.

The two TSs for **3a** are shown in Figure 12.7. **TS A** also has a more favorable orientation of the spiro ring structure. The ideal angle is 90°, at which point the two rings are perpendicular. This angle is 78.2° in **TS A** and 36.2° in **TS B**. **TS A** has a O(1)−C(2)−C(3)−C(4) angle of 35.6°, **TS B** has a corresponding angle of 96.1°. Based on the reactant conformational profile, this will introduce about 0.7 kcal more

[60.] B. E. Rossiter and K. B. Sharpless, *J. Org. Chem.*, **49**, 3707 (1984).
[61.] M. Cui, W. Adam, J. H. Shen, X. M. Luo, X. J. Tan, K. X. Chen, R. Y. Ji, and H. L. Jiang, *J. Org. Chem.*, **67**, 1427 (2002).

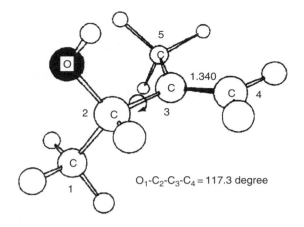

$O_1\text{-}C_2\text{-}C_3\text{-}C_4 = 117.3$ degree

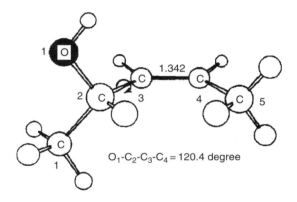

$O_1\text{-}C_2\text{-}C_3\text{-}C_4 = 120.4$ degree

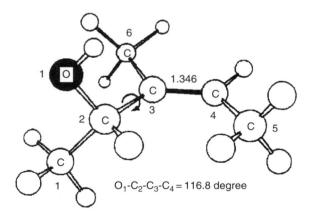

$O_1\text{-}C_2\text{-}C_3\text{-}C_4 = 116.8$ degree

Fig. 12.5. Minimum energy conformations for allylic alcohol. **3a,**
3b, and **3c**. Reproduced from *J. Org. Chem.*, **67**, 1427 (2002), by
permission of the American Chemical Society.

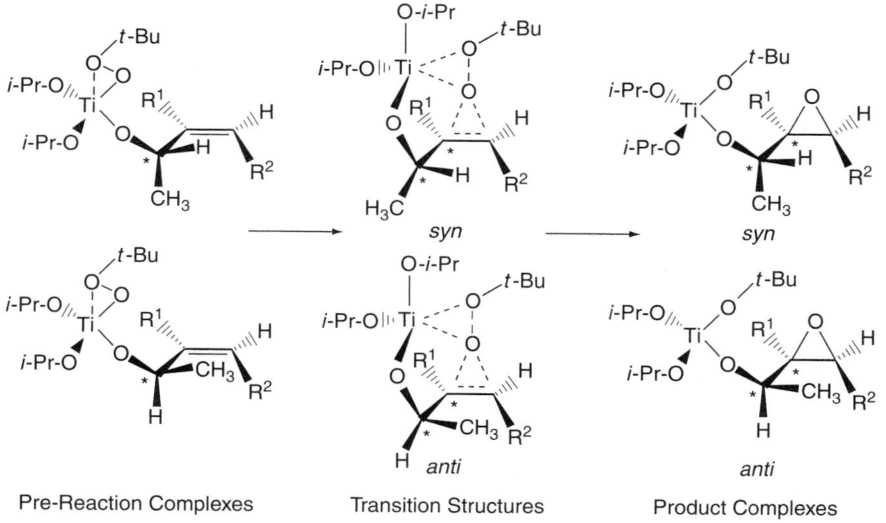

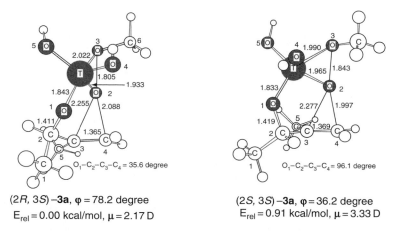

Fig. 12.6. Conformational factors affecting *syn* and *anti* diastereoselectivity in Sharpless epoxidation. If substituent $R^4 > H$, $A^{1,3}$ strain favors the *syn* product. If $R^4 = H$, the preferred transition structure leads to *anti* product. Reproduced from *J. Org. Chem.*, **67**, 1427 (2002), by permission of the American Chemical Society.

strain in **TS B** than in **TS A**. Similar analyses were done on the two TSs for **3b** and **3c**. The TS energies were used to compare computational ΔE_a with experimental diastereoselectivity. Whereas **TS A** is favored for **3a**, **TS B** is favored for **3b** and **3c**, in agreement with the experimental stereoselectivity.

Fig. 12.7. Alternate orientations of 3-methylbut-3-en-1-ol (**3a**) in the transition state for Ti-mediated epoxidation. Angle φ is the inter-ring angle of the spiro rings. Reproduced from *J. Org. Chem.*, **67**, 1427 (2002), by permission of the American Chemical Society.

	R^3	R^4	predicted	observed
3a	CH_3	H	12:88	22:78
3b	H	CH_3	92:8	91:9
3c	CH_3	CH_3	77:23	83:17

Visual models, additional information and exercises on Sharpless Epoxidation can be found in the Digital Resource available at: Springer.com/carey-sundberg.

Scheme 12.9 gives some examples of enantioselective oxidation of allylic alcohols. Entry 1 is a representative procedure, as documented in an *Organic Syntheses* preparation. The reaction in Entry 2 was used to prepare a starting material for synthesis of leukotriene C-1. Entry 3 is an example incorporating the use of molecular sieves. The reaction in Entry 4 was the departure point in a synthesis of part of the polyether antibiotic X-206. Entry 5 is another example of the procedure using molecular sieves. The catalyst loading in this reaction is 5%. The reaction in Entry 6 is diastereoselective for the *anti* isomer. Entry 7 also shows a case of diastereoselectivity, in this instance with respect to the 4-methyl group. Note that both of these reactions involve oxidation of the alkene from the same face, although they differ in configuration at C(4). Thus, the enantioselectivity is under reagent control.

Several catalysts that can effect enantioselective epoxidation of unfunctionalized alkenes have been developed, most notably manganese complexes of diimines derived from salicylaldehyde and chiral diamines (salens).[62]

These catalysts are used in conjunction with a stoichiometric amount of an oxidant and the active oxidant is believed to be an oxo Mn(V) species. The stoichiometric oxidants that have been used include NaOCl,[63] periodate,[64] and amine oxides.[65] Various other

62. W. Zhang, J. L. Loebach, S. R. Wilson, and E. N. Jacobsen, *J. Am. Chem. Soc.*, **112**, 2801 (1990); E. N. Jacobsen, W. Zhang, A. R. Muci, J. R. Ecker, and L. Deng, *J. Am. Chem. Soc.*, **113**, 7063 (1991).
63. W. Zhang and E. N. Jacobsen, *J. Org. Chem.*, **56**, 2296 (1991); B. D. Brandes and E. N. Jacobsen, *J. Org. Chem.*, **59**, 4378 (1994).
64. P. Pietikainen, *Tetrahedron Lett.*, **36**, 319 (1995).
65. M. Palucki, P. J. Pospisil, W. Zhang, and E. N. Jacobsen, *J. Am. Chem. Soc.*, **116**, 9333 (1994).

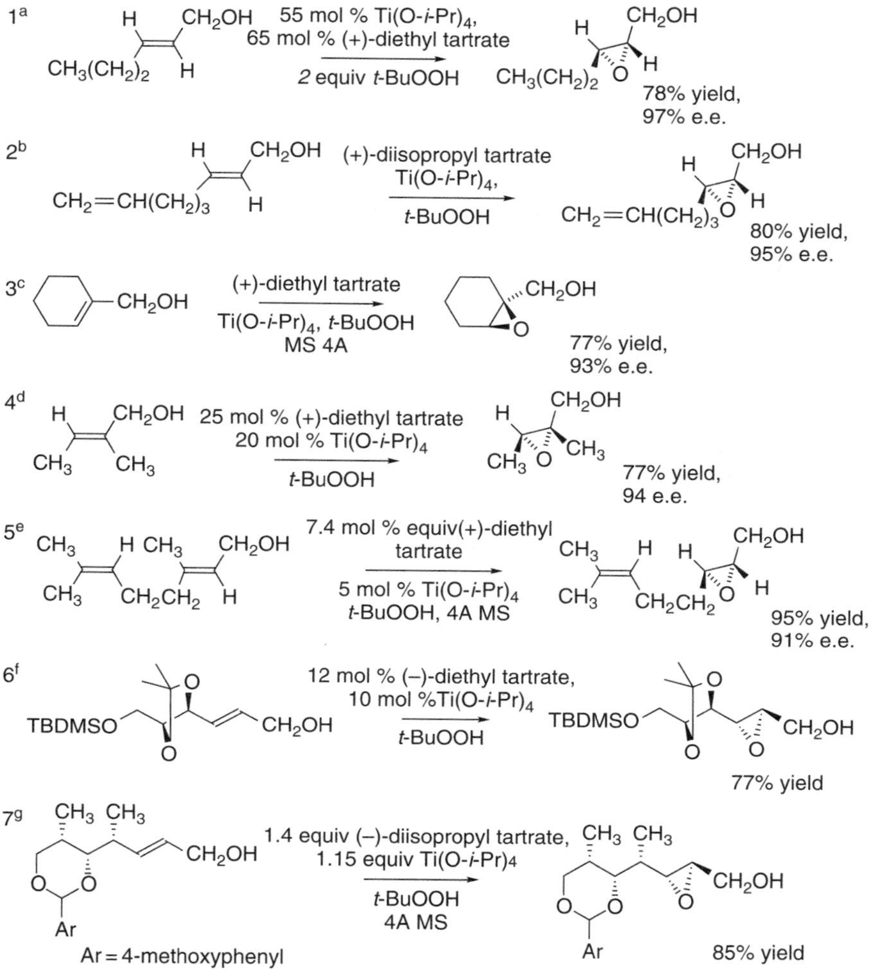

Scheme 12.9. Enantioselective Epoxidation of Allylic Alcohols

[1a] CH₃(CH₂)₂ ... 55 mol % Ti(O-i-Pr)₄, 65 mol % (+)-diethyl tartrate, 2 equiv t-BuOOH → 78% yield, 97% e.e.

[2b] CH₂=CH(CH₂)₃ ... (+)-diisopropyl tartrate Ti(O-i-Pr)₄, t-BuOOH → 80% yield, 95% e.e.

[3c] (+)-diethyl tartrate, Ti(O-i-Pr)₄, t-BuOOH, MS 4A → 77% yield, 93% e.e.

[4d] 25 mol % (+)-diethyl tartrate, 20 mol % Ti(O-i-Pr)₄, t-BuOOH → 77% yield, 94 e.e.

[5e] 7.4 mol % equiv(+)-diethyl tartrate, 5 mol % Ti(O-i-Pr)₄, t-BuOOH, 4A MS → 95% yield, 91% e.e.

[6f] TBDMSO ... 12 mol % (−)-diethyl tartrate, 10 mol % Ti(O-i-Pr)₄, t-BuOOH → 77% yield

[7g] 1.4 equiv (−)-diisopropyl tartrate, 1.15 equiv Ti(O-i-Pr)₄, t-BuOOH, 4A MS → 85% yield. Ar = 4-methoxyphenyl

a. J. G. Hill and K. B. Sharpless, *Org. Synth.*, **63**, 66 (1985).
b. B. E. Rossiter, T. Katsuki, and K. B. Sharpless, *J. Am. Chem. Soc.*, **103**, 464 (1981).
c. Y. Gao, R. M. Hanson. J. M. Klunder, S. Y. Ko, H. Masamune, and K. B. Sharpless, *J. Am. Chem. Soc.*, **109**, 5765 (1987).
d. D. A. Evans, S. L. Bender, and J. Morris, *J. Am. Chem. Soc.*, **110**, 2506 (1988).
e. R. M. Hanson and K. B. Sharpless, *J. Org. Chem.*, **51**, 1922 (1986).
f. A. K. Ghosh and Y. Wang, *J. Org. Chem.*, **64**, 2789 (1999).
g. J. A. Marshall, Z.-H. Lu, and B. A. Johns, *J. Org. Chem.*, **63**, 817 (1998).

chiral salen-type ligands have also been explored.[66] These epoxidations are not always stereospecific with respect to the alkene geometry, which is attributed to an electron transfer mechanism that involves a radical intermediate.

[66.] N. Hosoya, R. Irie, and T. Katsuki, *Synlett*, 261 (1993); S. Chang, R. M. Heid, and E. N. Jacobsen, *Tetrahedron Lett.*, **35**, 669 (1994).

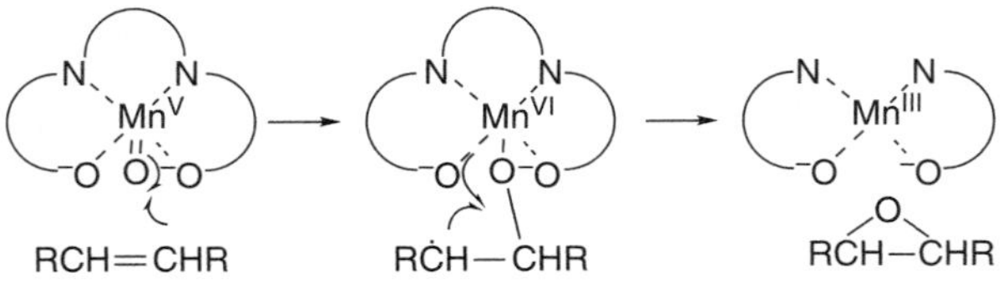

Scheme 12.10 gives some examples of these oxidations. Entry 1 is one of several aryl-conjugated alkenes that were successfully epoxidized. Entry 2 is a reaction that was applied to enantioselective synthesis of the taxol side chain. Entry 3 demonstrates

Scheme 12.10. Enantioselective Epoxidation with Chiral Manganese Catalysts[a]

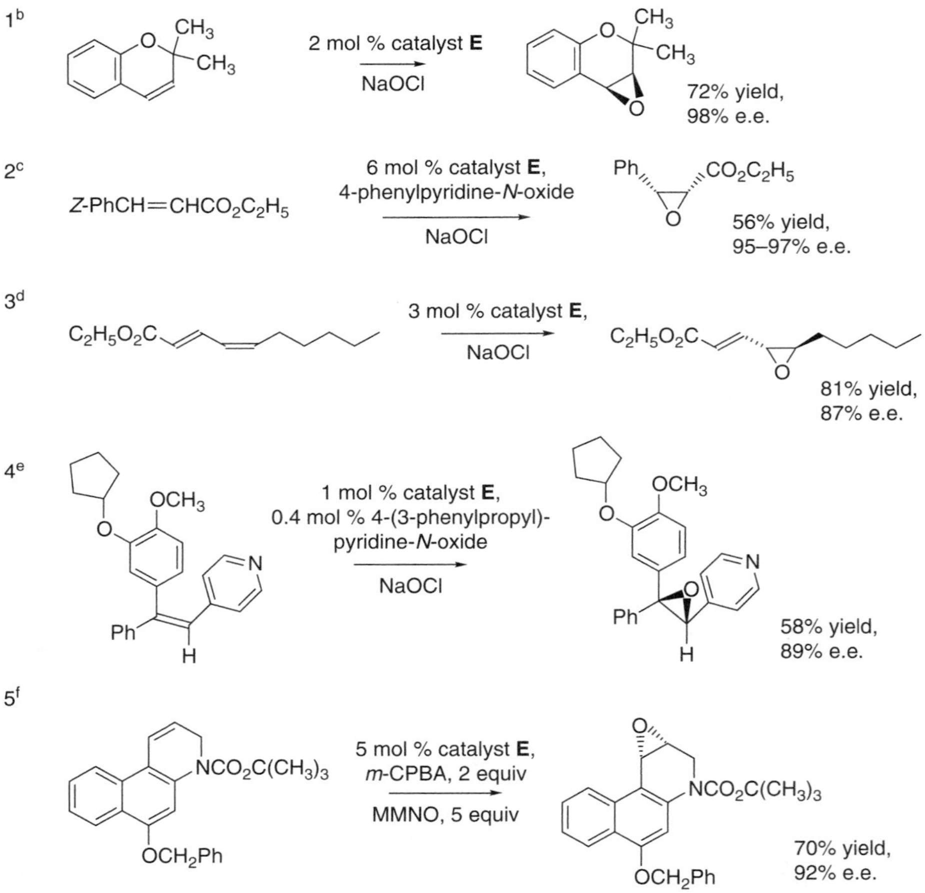

a. The structure of catalyst **E** is shown on p. 1088.
b. E. N. Jacobsen, W. Zhang, A. R. Muci, J. R. Ecker, and L. Deng, *J. Am. Chem. Soc.*, **113**, 7063 (1991).
c. L. Deng and E. N. Jacobsen, *J. Org. Chem.*, **57**, 4320 (1992).
d. S. Chang, N. H. Lee, and E. N. Jacobsen, *J. Org. Chem.*, **58**, 6939 (1993).
e. J. E. Lynch, W.-B. Choi, H. R. O. Churchill, R. P. Volante, R. A. Reamer, and R. G. Ball, *J. Org. Chem.*, **62**, 9223 (1997).
f. D. L. Boger, J. A. McKie, and C. W. Boyce, *Synlett*, 515 (1997).

chemoselectivity for the 4,5-double bond in a dienoate ester. This case also illustrates the occurrence of isomerization during the epoxidation. Entry 4 is a step in the enantioselective synthesis of CDP840, a phosphodiesterase inhibitor. The reaction in Entry 5 provided a starting material for the synthesis of the DNA-alkylating antitumor agent CC-1065.

12.2.2. Epoxides from Alkenes and Peroxidic Reagents

12.2.2.1. Epoxidation by Peroxy Acids and Related Reagents. The most general reagents for conversion of simple alkenes to epoxides are peroxycarboxylic acids.[67] *m*-Chloroperoxybenzoic acid[68] (MCPBA) is a particularly convenient reagent. The magnesium salt of monoperoxyphthalic acid is an alternative.[69] Potassium hydrogen peroxysulfate, which is sold commercially as Oxone®, is a convenient reagent for epoxidations that can be done in aqueous methanol.[70] Peroxyacetic acid, peroxybenzoic acid, and peroxytrifluoroacetic acid have also been used frequently for epoxidation. All of the peroxycarboxylic acids are potentially hazardous materials and require appropriate precautions.

It has been demonstrated that ionic intermediates are not involved in the epoxidation reaction. The reaction rate is not very sensitive to solvent polarity.[71] Stereospecific *syn* addition is consistently observed. The oxidation is therefore believed to be a concerted process. A representation of the transition structure is shown below.

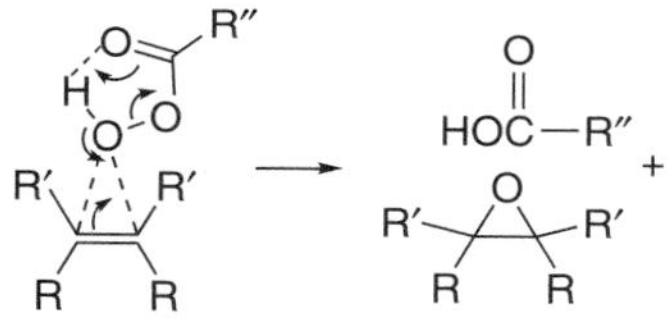

The rate of epoxidation of alkenes is increased by alkyl groups and other ERG substituents and the reactivity of the peroxy acids is increased by EWG substituents.[72] These structure-reactivity relationships demonstrate that the peroxyacid acts as an electrophile in the reaction. Decreased reactivity is exhibited by double bonds that are conjugated with strongly electron-attracting substituents, and more reactive peroxyacids, such as trifluoroperoxyacetic acid, are required for oxidation of such compounds.[73] Electron-poor alkenes can also be epoxidized by alkaline solutions of

[67.] D. Swern, *Organic Peroxides*, Vol. II, Wiley-Interscience, New York, 1971, pp. 355–533; B. Plesnicar, in *Oxidation in Organic Chemistry*, Part C, W. Trahanovsky, ed., Academic Press, New York, 1978, pp. 211–253.

[68.] R. N. McDonald, R. N. Steppel, and J. E. Dorsey, *Org. Synth.*, **50**, 15 (1970).

[69.] P. Brougham, M. S. Cooper, D. A. Cummerson, H. Heaney, and N. Thompson, *Synthesis*, 1015 (1987).

[70.] R. Bloch, J. Abecassis, and D. Hassan, *J. Org. Chem.*, **50**, 1544 (1985).

[71.] N. N. Schwartz and J. N. Blumbergs, *J. Org. Chem.*, **29**, 1976 (1964).

[72.] B. M. Lynch and K. H. Pausacker, *J. Chem. Soc.*, 1525 (1955).

[73.] W. D. Emmons and A. S. Pagano, *J. Am. Chem. Soc.*, **77**, 89 (1955).

hydrogen peroxide or *t*-butyl hydroperoxide. A quite different mechanism, involving conjugate nucleophilic addition, operates in this case.[74]

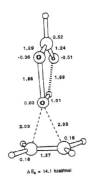

There have been a number of computational studies of the epoxidation reaction. These studies have generally found that the hydrogen-bonded peroxy acid is approximately perpendicular to the axis of the double bond, giving a spiro structure.[75] Figure 12.8 shows TS structures and E_a values based on B3LYP/6-31G* computations. The E_a trend is as expected for an electrophilic process: $OCH_3 < CH_3 \sim CH=CH_2 < H < CN$. Similar trends were found in MP4/6-31G* and QCISD/6-31G* computations.

The stereoselectivity of epoxidation with peroxycarboxylic acids has been well studied. Addition of oxygen occurs preferentially from the less hindered side of the molecule. Norbornene, for example, gives a 96:4 *exo:endo* ratio.[76] In molecules where two potential modes of approach are not very different, a mixture of products is formed.

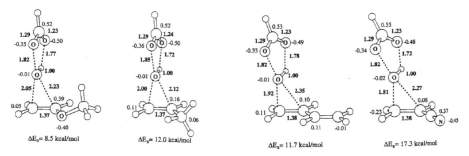

Fig. 12.8. Comparison of epoxidation transition structures and activation energies for ethene and substituted ethenes. Reproduced from *J. Am. Chem. Soc.*, **119**, 10147 (1997), by permission of the American Chemical Society.

74. C. A. Bunton and G. J. Minkoff, *J. Chem. Soc.*, 665 (1949).
75. R. D. Bach, M. N. Glukhovtsev, and C. Gonzalez, *J. Am. Chem. Soc.*, **120**, 9902 (1998); K. N. Houk, J. Liu, N. C. DeMello, and K. R. Condroski, *J. Am. Chem. Soc.*, **119**, 10147 (1997).
76. H. Kwart and T. Takeshita, *J. Org. Chem.*, **28**, 670 (1963).

For example, the unhindered exocyclic double bond in 4-*t*-butylmethylenecyclohexane gives both stereoisomeric products.[77]

Hydroxy groups exert a directive effect on epoxidation and favor approach from the side of the double bond closest to the hydroxy group.[78] Hydrogen bonding between the hydroxy group and the reagent evidently stabilizes the TS.

This is a strong directing effect that can exert stereochemical control even when steric effects are opposed. Entries 4 and 5 in Scheme 12.11 illustrate the hydroxy-directing effect. Other substituents capable of hydrogen bonding, in particular amides, also can exert a *syn*-directing effect.[79]

The hydroxy-directing effect has been studied computationally, as the hydrogen bond can have several possible orientations.[80] Studies on 2-propen-1-ol show the same preference for the spiro TS as for unfunctionalized alkenes. There is a small preference for hydrogen bonding to a peroxy oxygen, as opposed to the carbonyl oxygen. The TSs for conformations of 2-propen-1-ol that are not hydrogen-bonded are 2–3 kcal/mol higher in energy than the best of the hydrogen-bonded structures. For substituted allylic alcohols, $A^{1,2}$ and $A^{1,3}$ strain comes into play. Figure 12.9 shows the structures and relative energies of the four possible TSs for prop-2-en-1-ol. The *syn,exo* structure with hydrogen-bonding to the transferring oxygen is preferred to the *endo* structure, in which the hydrogen-bonding is to the carbonyl oxygen.

Torsional effects are important in cyclic systems. A PM3 study of the high stereoselectivity of compounds **4a-d** found torsional effects to be the major difference between the diastereomeric TSs.[81] The computed TSs for **4a** are shown in Figure 12.10. The structures all show similar stereoselectivity, regardless of the presence and nature of a 3-substituent.

[77.] R. G. Carlson and N. S. Behn, *J. Org. Chem.*, **32**, 1363 (1967).

[78.] H. B. Henbest and R. A. L. Wilson, *J. Chem. Soc.*, 1958 (1957).

[79.] F. Mohamadi and M. M. Spees, *Tetrahedron Lett.*, **30**, 1309 (1989); P. G. M. Wuts, A. R. Ritter, and L. E. Pruitt, *J. Org. Chem.*, **57**, 6696 (1992); A. Jemmalm, W. Bets, K. Luthman, I. Csoregh, and U. Hacksell, *J. Org. Chem.*, **60**, 1026 (1995); P. Kocovsky and I. Stary, *J. Org. Chem.*, **55**, 3236 (1990); A. Armstrong, P. A. Barsanti, P. A. Clarke, and A. Wood, *J. Chem. Soc., Perkin Trans. 1*, 1373 (1996).

[80.] M. Freccero, R. Gandolfi, M. Sarzi-Amade, and A. Rastelli, *J. Org. Chem.*, **64**, 3853 (1999); M. Freccero, R. Gandolfi, M. Sarzi-Amade, and A. Rastelli, *J. Org. Chem.*, **65**, 2030 (2000).

[81.] M. J. Lucero and K. N. Houk, *J. Org. Chem.*, **63**, 6973 (1998).

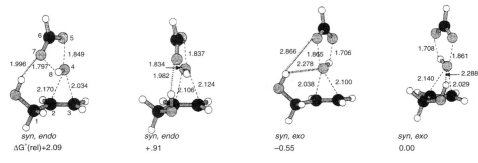

syn, endo   syn, endo   syn, exo   syn, exo
$\Delta G^{\ddagger}$(rel)+2.09   +.91   −0.55   0.00

Fig. 12.9. Structure and relative energies of four modes of hydrogen bonding in transition structures for epoxidation of 2-propen-1-ol by peroxyformic acid. Relative energies are from B3LYP/6-311G*-level computations with a solvation model for CH_2Cl_2, $\epsilon = 8.9$. Reproduced from *J. Org. Chem.*, **64**, 3853 (1999), by permission of the American Chemical Society.

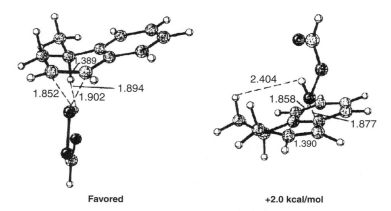

	R	X	anti	syn
4a	CH$_3$	H	85:15	
4b	Ph	H	high	
4c	Ph	CO$_2$CH$_3$	>95:5	
4d	Ph	CH$_2$OH	>95:5	

Even in the absence of a 3-substituent (**4a**, **4b**) and with only a small 4-methyl group (**4a**), the stereoselectivity is high. The preference arises from the staggered relationship between the forming C–O bond and the axial allylic hydrogen.

Favored   **+2.0 kcal/mol**

Fig. 12.10. Comparison of *trans*- and *cis*-oriented transition structures for epoxidation of 1-methyl-1,2-dihydronapththalene. Reproduced from *J. Org. Chem.*, **63**, 6973 (1998), by permission of the American Chemical Society.

A process that is effective for epoxidation and avoids acidic conditions involves reaction of an alkene, a nitrile, and hydrogen peroxide.[82] The nitrile and hydrogen peroxide react, forming a peroxyimidic acid, which epoxidizes the alkene, by a mechanism similar to that for peroxyacids. An important contribution to the reactivity of the peroxyimidic acid comes from the formation of the stable amide carbonyl group.

At least in some cases, the hydroxy-directing effect also operates for this version of the reaction.

Ref. 83

Scheme 12.11 gives some examples of epoxidation using peroxyacids and related reagents. Entry 1 shows standard epoxidation conditions applied to styrene. The reaction in Entry 2 uses typical epoxidation conditions and also illustrates the approach from the less hindered face of the molecule. In Entry 3, the selectivity for the more-substituted double bond was used to achieve regioselectivity. Entries 4 and 5 illustrate stereochemical control by hydroxy participation. The reaction in Entry 6 is an example of diastereoselectivity, most likely due to hydrogen bonding by the amide group. Entries 7 and 8 are cases of application of nucleophilic peroxidation conditions to alkenes conjugated with EWG substituents. In Entry 9, the more reactive trifluoroperoxyacetic acid was used to oxidize a deactivated double bond. Entry 10 is an example of use of the peroxyimidic acid conditions.

There is interest in being able to use H_2O_2 directly as an epoxidizing reagent because it is the ultimate source of most peroxides. The reactivity of H_2O_2 is substantially enhanced in hexafluoro-2-propanol (HFIP) and other polyfluorinated alcohols such as nonafluoro-*t*-butanol.[84] Either 30 or 60% H_2O_2 can oxidize alkenes to epoxides in these solvents. The system shows the normal trend of higher reactivity for more-substituted alkenes. The activation is attributed to polarization of the H_2O_2 by hydrogen bonding with the β-fluoroalcohols. The fluoro substituents also increase the acidity of the hydroxy group.

[82] G. B. Payne, *Tetrahedron*, **18**, 763 (1962); R. D. Bach and J. W. Knight, *Org. Synth.*, **60**, 63 (1981); L. A. Arias, S. Adkins, C. J. Nagel, and R. D. Bach, *J. Org. Chem.*, **48**, 888 (1983).

[83] W. C. Frank, *Tetrahedron: Asymmetry*, **9**, 3745 (1998).

[84] K. Neimann and R. Neumann, *Org. Lett.*, **2**, 2861 (2000).

Scheme 12.11. Synthesis of Epoxides from Alkenes Using Peroxy Acids

A. Oxidation of alkenes with peroxy acids

1[a]

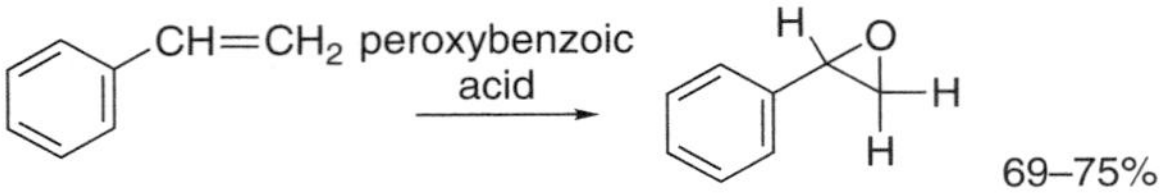

69–75%

2[b]

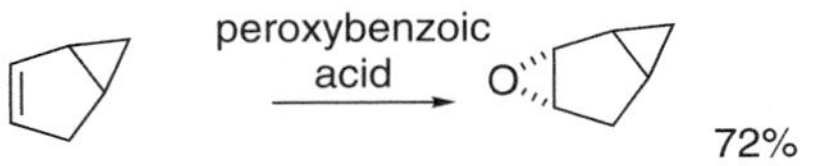

72%

3[c]

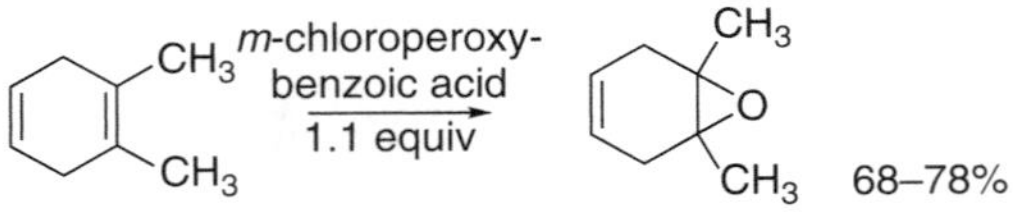

68–78%

4[d]

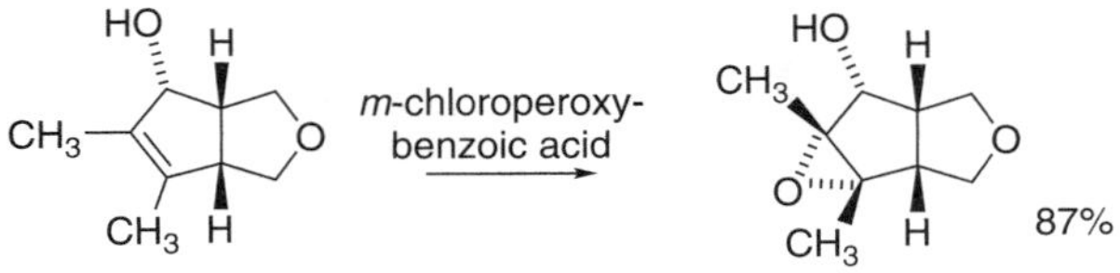

87%

5[e]

78%

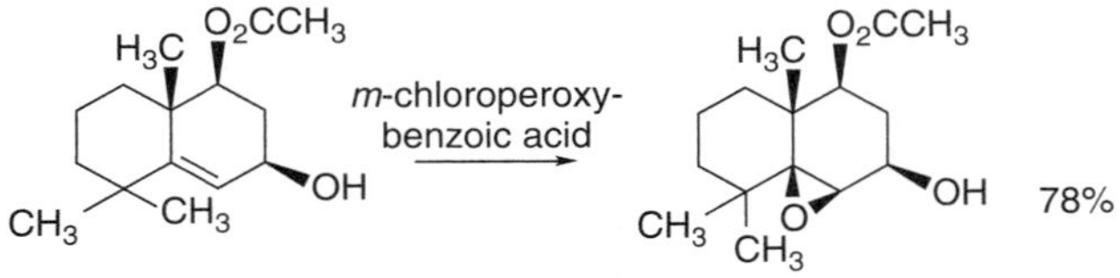

6[f]

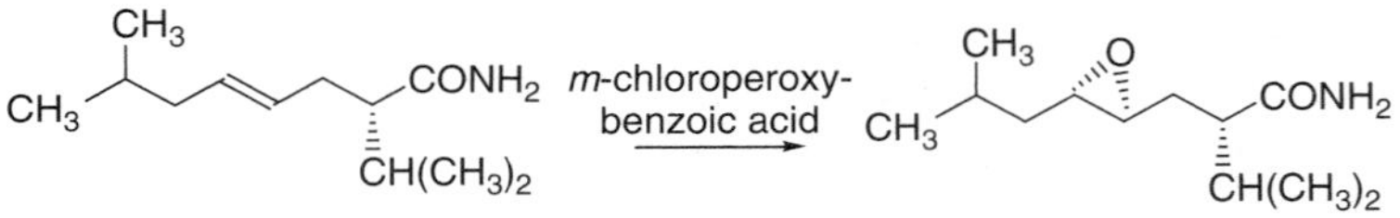

12:1 diastereoselectivity

B. Epoxidation of electrophilic alkenes

7[g]

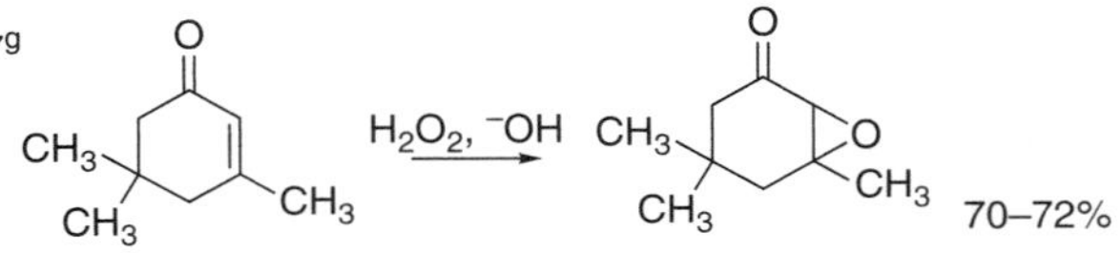

70–72%

8[h]

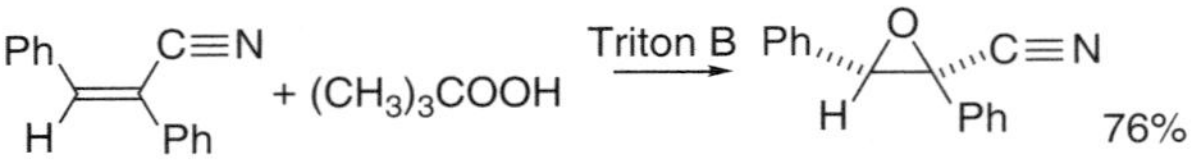

76%

9[i]

$CH_3CH=CHCO_2C_2H_5$ CF_3CO_3H

73%

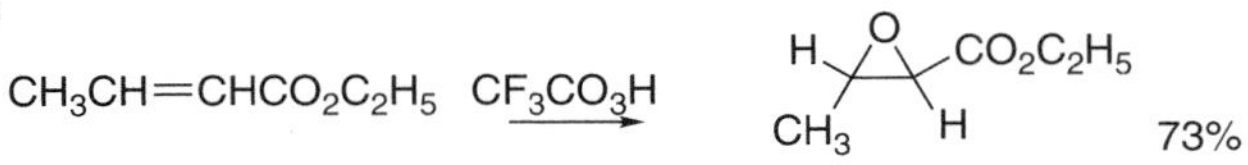

C. Epoxidation with peroxyimidic Acids

10[j]

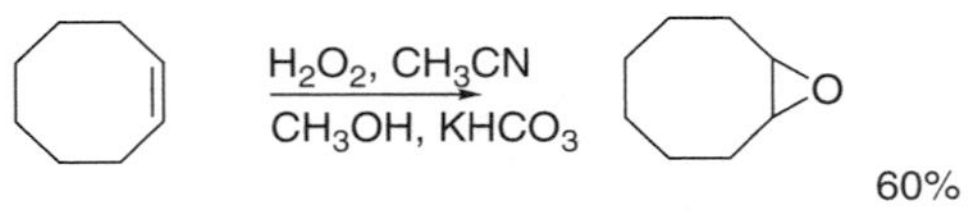

H_2O_2, CH_3CN / CH_3OH, $KHCO_3$

60%

(Continued)

a. H. Hibbert and P. Burt, *Org. Synth.*, **I**, 481 (1932).
b. E. J. Corey and R. L. Dawson, *J. Am. Chem. Soc.*, **85**, 1782 (1963).
c. L. A. Paquette and J. H. Barrett, *Org. Synth.*, **49**, 62 (1969).
d. R. M. Scarborough, Jr., B. H. Toder, and A. B. Smith, III, *J. Am. Chem. Soc.*, **102**, 3904 (1980).
e. M. Miyashita and A. Yoshikoshi, *J. Am. Chem. Soc.*, **96**, 1917 (1974).
f. P. G. M. Wuts, A. R. Ritter, and L. E. Pruitt, *J. Org. Chem.*, **57**, 6696 (1992).
g. R. L. Wasson and H. O. House, *Org. Synth.*, **IV**, 552 (1963).
h. G. B. Payne and P. H. Williams, *J. Org. Chem.*, **26**, 651 (1961).
i. W. D. Emmons and A. S. Pagano, *J. Am. Chem. Soc.*, **77**, 89 (1955).
j. R. D. Bach and J. W. Knight, *Org. Synth.*, **60**, 63 (1981).

A variety of electrophilic reagents have been examined with the objective of activating H_2O_2 to generate a good epoxidizing agent. In principle, any species that can convert one of the hydroxy groups to a good leaving group can generate a reactive epoxidizing reagent.

In practice, promising results have been obtained for several systems. For example, fair to good yields of epoxides are obtained when a two-phase system consisting of alkene and ethyl chloroformate is stirred with a buffered basic solution of hydrogen peroxide. The active oxidant is presumed to be *O*-ethyl peroxycarbonic acid.[85]

Although these reagent combinations are not as generally useful as the peroxycarboxylic acids, they serve to illustrate that epoxidizing activity is not unique to the peroxyacids.

12.2.2.2. Epoxidation by Dioxirane Derivatives.

Another useful epoxidizing agent is dimethyldioxirane (DMDO),[86] which is generated by in situ reaction of acetone and peroxymonosulfate in buffered aqueous solution. Distillation gives about a 0.1 *M* solution of DMDO in acetone.[87]

[85.] R. D. Bach, M. W. Klein, R. A. Ryntz, and J. W. Holubka, *J. Org. Chem.*, **44**, 2569 (1979).
[86.] R. W. Murray, *Chem. Rev.*, **89**, 1187 (1989); W. Adam and L. P. Hadjiarapoglou, *Topics Current Chem.*, **164**, 45 (1993); W. Adam, A. K. Smerz, and C. G. Zhao, *J. Prakt. Chem., Chem. Zeit.*, **339**, 295 (1997).
[87.] R. W. Murray and R. Jeyaraman, *J. Org. Chem.*, **50**, 2847 (1985); W. Adam, J. Bialas, and L. Hadjiara-paglou, *Chem. Ber.*, **124**, 2377 (1991).

Higher concentrations of DMDO can be obtained by extraction of a 1:1 aqueous dilution of the distillate by CH_2Cl_2, $CHCl_3$, or CCl_4.[88] Another method involves in situ generation of DMDO under phase transfer conditions.[89]

$$CH_3CH{=}CH(CH_2)_3OCH_2Ph \xrightarrow[\substack{pH\ 7.8\ buffer,\\ n\text{-}Bu_4N^+\ HSO_4^-,}]{\substack{O\\ \parallel\\ CH_3CCH_3,\\ KOSO_2OOH}} CH_3\overset{O}{\overset{\diagup\backslash}{CH}}{-}CH(CH_2)_3OCH_2Ph$$

The yields and rates of oxidation by DMDO under these in situ conditions depend on pH and other reaction parameters.[90]

Various computational models agree that the reaction occurs by a concerted mechanism.[91] Comparison between epoxidation by peroxy acids and dioxiranes suggests that they have similar transition structures.

Kinetics and isotope effects are consistent with this mechanism.[92] The reagent is electrophilic in character and reaction is facilitated by ERG substituents in the alkene. A B3LYP/6-31G* computation found the transition structures and E_a values shown in Figure 12.11.

Similarly to peroxycarboxylic acids, DMDO is subject to *cis or syn* stereoselectivity by hydroxy and other hydrogen-bonding functional groups.[93] However a study of several substituted cyclohexenes in $CH_3CN - H_2O$ suggested a dominance by steric effects. In particular, the hydroxy groups in cyclohex-2-enol and

88. M. Gilbert. M. Farrert, F. Sanchez-Baeza, and A. Messeguer, *Tetrahedron*, **53**, 8643 (1997).

89. S. E. Denmark, D. C. Forbes, D. S. Hays, J. S. DePue, and R. G. Wilde, *J. Org. Chem.*, **60**, 1391 (1995).

90. M. Frohn, Z.-X. Wang, and Y. Shi, *J. Org. Chem.*, **63**, 6425 (1998); A. O'Connell, T. Smyth, and B. K. Hodnett, *J. Chem. Technol. Biotech.*, **72**, 60 (1998).

91. R. D. Bach, M. N. Glukhovtsev, C. Gonzalez, M. Marquez, C. M. Estevez, A. G. Baboul, and H. Schlegel, *J. Phys. Chem.*, **101**, 6092 (1997); K. N. Houk, J. Liu, N. C. DeMello, and K. R. Condroski, *J. Am. Chem. Soc.*, **119**, 10147 (1997); C. Jenson, J. Liu, K. N. Houk, and W. L. Jorgensen, *J. Am. Chem. Soc.*, **119**, 12982 (1987); R. D. Bach, M. N. Glukhovtsev, and C. Canepa, *J. Am. Chem. Soc.*, **120**, 775 (1998); M. Freccero, R. Gandolfi, M. Sarzi-Amade, and A. Rastelli, *Tetrahedron*, **54**, 6123 (1998); J. Liu, K. N. Houk, A. Dinoi, C. Fusco, and R. Curci, *J. Org. Chem.*, **63**, 8565 (1998); R. D. Bach, O. Dmitrenko, W. Adam, and S. Schambony, *J. Am. Chem. Soc.*, **125**, 924 (2003).

92. W. Adam, R. Paredes, A. K. Smerz, and L. A. Veloza, *Liebigs Ann. Chem.*, 547 (1997); A. L. Baumstark, E. Michalenabaez, A. M. Navarro, and H. D. Banks, *Heterocycl. Commun.*, **3**, 393 (1997); Y. Angelis, X. Zhang, and M. Orfanopoulos, *Tetrahedron Lett.*, **37**, 5991 (1996).

93. R. W. Murray, M. Singh, B. L. Williams, and H. M. Moncrief, *J. Org. Chem.*, **61**, 1830 (1996); G. Asensio, C. Boix-Bernardini, C. Andreu, M. E. Gonzalez-Nunez, R. Mello, J. O. Edwards, and G. B. Carpenter, *J. Org. Chem.*, **64**, 4705 (1999).

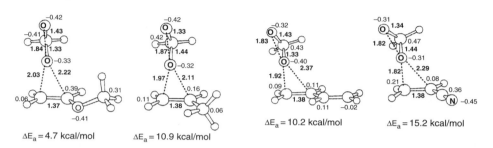

Fig. 12.11. Transition structures and E_a values for epoxidation of ethene and substituted derivatives by dimethyloxirane. Reproduced from *J. Am. Chem. Soc.*, **119**, 10147 (1997), by permission of the American Chemical Society.

3-methylcyclohex-2-enol were not very strongly *syn* directing.[94] The hydroxylic solvent may minimize any directive effect by competing hydrogen bonding.[95]

$$\begin{array}{cccc}
\text{OH} & \text{OH} & \text{OTBDMS} & \text{OTBDMS} \\
1.2{:}1 & 1.4{:}1 & 4.8{:}1 & 13.6{:}1
\end{array}$$

trans:cis ratio for epoxidation by DMDO

Directing effects have also been attributed to more remote substituents, as, e.g., a urea NH.

Ref. 96

94. D. Yang, G.-S. Jiao, Y.-C. Yip, and M.-K. Wong, *J. Org. Chem.*, **64**, 1635 (1999).
95. W. Adam, R. Paredes, A. K. Smerz, and L. A. Veloza, *Eur. J. Org. Chem.*, 349 (1998).
96. W. Adam, K. Peters, E.-M. Peters, and S. B. Schambony, *J. Am. Chem. Soc.*, **123**, 7228 (2001).

Several disubstituted 3,4-dimethylcyclobutenes show *syn* selectivity. The mesylate groups were strongly *syn* directive, with the hydroxy, methoxy, and acetoxy groups being somewhat less so.[97] The same groups were even more strongly *syn* directing with MCPBA. The effects are attributed to an attractive electrostatic interaction of the relatively positive methylene hydrogens and the oxygens of the dioxirane and peroxy acid.

X	DMDO	MCPBA
OH	67:33	82:18
OCH$_3$	62:38	76:24
O$_2$CCH$_3$	68:32	69:31
OSO$_2$CH$_3$	79:21	87:13

For other substituents, both steric and dipolar factors seem to have an influence and several complex reactants have shown good stereoselectivity, although the precise origin of the stereoselectivity is not always evident.[98]

Other ketones besides acetone can be used for in situ generation of dioxiranes by reaction with peroxysulfate or another suitable peroxide. More electrophilic ketones give more reactive dioxiranes. 3-Methyl-3-trifluoromethyldioxirane is a more reactive analog of DMDO.[99] This reagent, which is generated in situ from 1,1,1-trifluoroacetone, can oxidize less reactive compounds such as methyl cinnamate.

97%

Ref. 100

Hexafluoroacetone and hydrogen peroxide in buffered aqueous solution can epoxidize alkenes and allylic alcohols.[101] *N, N*-Dialkylpiperidin-4-one salts are also good catalysts for epoxidation.[102] The polar effect of the quaternary nitrogen enhances the

97. M. Freccero, R. Gandolfi, and M. Sarzi-Amade, *Tetrahedron*, **55**, 11309 (1999).

98. R. C. Cambie, A. C. Grimsdale, P. S. Rutledge, M. F. Walker, and A. D. Woodgate, *Austr. J. Chem.*, **44**, 1553 (1991); P. Boricelli and P. Lupattelli, *J. Org. Chem.*, **59**, 4304 (1994); R. Curci, A. Detomaso, T. Prencipe, and G. B. Carpenter, *J. Am. Chem. Soc.*, **116**, 8112 (1994); T. C. Henninger, M. Sabat, and R. J. Sundberg, *Tetrahedron*, **52**, 14403 (1996).

99. R. Mello, M. Fiorentino, O. Sciacevolli, and R. Curci, *J. Org. Chem.*, **53**, 3890 (1988).

100. D. Yang, M.-K. Wong, and Y.-C. Yie, *J. Org. Chem.*, **60**, 3887 (1995).

101. R. P. Heggs and B. Ganem, *J. Am. Chem. Soc.*, **101**, 2484 (1979); A. J. Biloski, R. P. Hegge, and B. Ganem, *Synthesis*, 810 (1980); W. Adam, H.-G. Degen, and C. R. Saha-Moller, *J. Org. Chem.*, **64**, 1274 (1999).

102. S. E. Denmark, D. C. Forbes, D. S. Hays, J. S. DePue, and R. G. Wilde, *J. Org. Chem.*, **60**, 1391 (1995).

reactivity of the ketone toward nucleophilic addition and also makes the dioxirane intermediate more reactive.

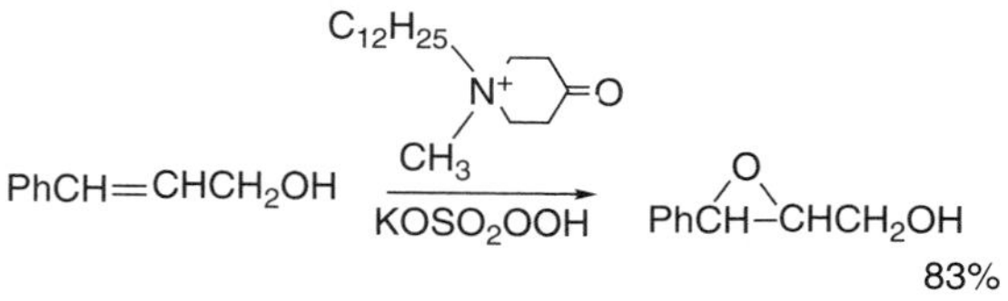
$C_{12}H_{25}$
N+
O
CH_3
$PhCH{=}CHCH_2OH$
$KOSO_2OOH$
O
$PhCH{-}CHCH_2OH$
83%

The cyclic sulfone 4-thiopyrone-*S*, *S*-dioxide also exhibits enhanced reactivity as a result of the effect of the sulfone dipole.[103]

Scheme 12.12 gives some examples of epoxidations involving dioxiranes. Entry 1 indicates the ability of the reagent to expoxidize deactivated double bonds. Entry 2

Scheme 12.12. Epoxidation by Dioxiranes

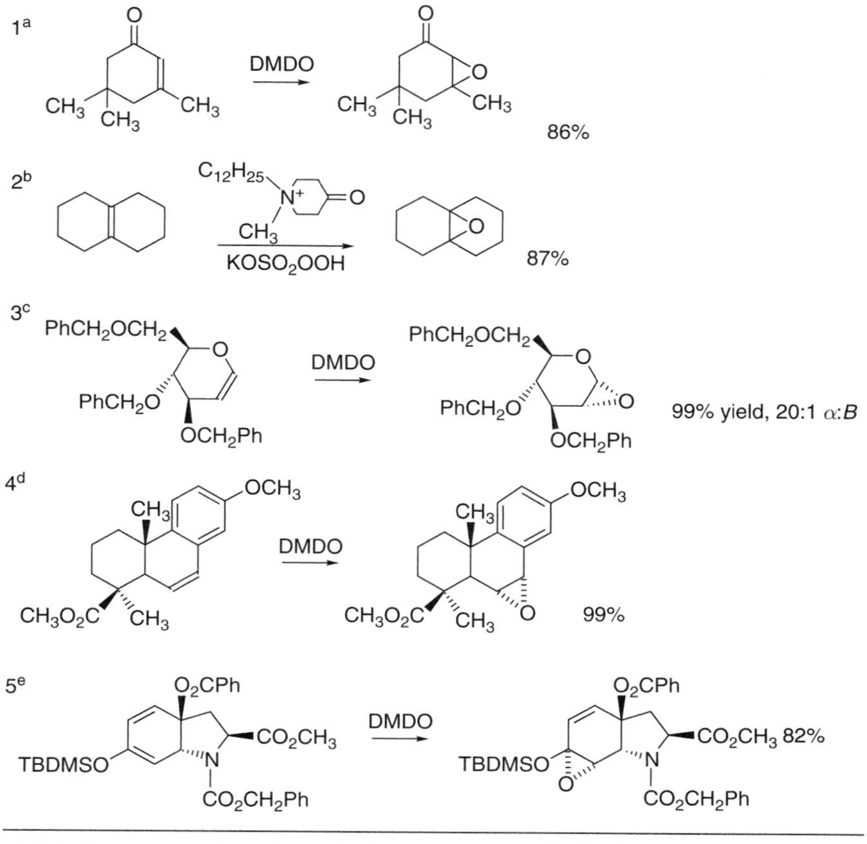
1a
DMDO
CH_3
CH_3
CH_3
86%
2b
$C_{12}H_{25}$
N+
CH_3
$KOSO_2OOH$
87%
3c
$PhCH_2OCH_2$
$PhCH_2O$
OCH_2Ph
DMDO
99% yield, 20:1 α:B
4d
OCH_3
CH_3
CH_3O_2C
DMDO
99%
5e
O_2CPh
CO_2CH_3
TBDMSO
N
CO_2CH_2Ph
DMDO
82%

a. W. Adam, L. Hadjarapaglou, and B. Nestler, *Tetrahedron Lett.*, **31**, 331 (1990).
b. S. E. Denmark, D. C. Forbes, D. S. Hays, J. S. DePue, and R. G. Wilde, *J. Org. Chem.*, **60**, 1391 (1995).
c. R. L. Halcomb and S. J. Danishefsky, *J. Am. Chem. Soc.*, **111**, 6661 (1989).
d. R. C. Cambie, A. C. Grimsdale, P. S. Rutledge, M. F. Walker, and P. D. Woodgate, *Aust. J. Chem.*, **44**, 1553 (1991).
e. T. C. Henninger, M. Sabat, and R. J. Sundberg, *Tetrahedron*, **52**, 14403 (1996).

103. D. Yang, Y.-C. Yip, G.-S. Jiao, and M.-K. Wong, *J. Org. Chem.*, **63**, 8952 (1998).

illustrates the use of a piperidone salt for in situ generation of a dioxirane. The long alkyl chain imparts phase transfer capability to the ketone. The dioxirane is generated in the aqueous phase but can carry out the epoxidation in the organic phase. Entries 3 to 5 are examples of stereoselective epoxidations. In each case, high stereoselectivity is observed in the presence of nearby functional groups. The exact origins of the stereoselectivity are not clear.

A number of chiral ketones have been developed that are capable of enantiose-lective epoxidation via dioxirane intermediates.[104] Scheme 12.13 shows the structures of some chiral ketones that have been used as catalysts for enantioselective epoxidation. The BINAP-derived ketone shown in Entry 1, as well as its halogenated derivatives, have shown good enantioselectivity toward di- and trisubstituted alkenes.

87%, 78% e.e.

Ref. 105

Scheme 12.13. Chiral Ketones Used for Enantioselective Epoxidation

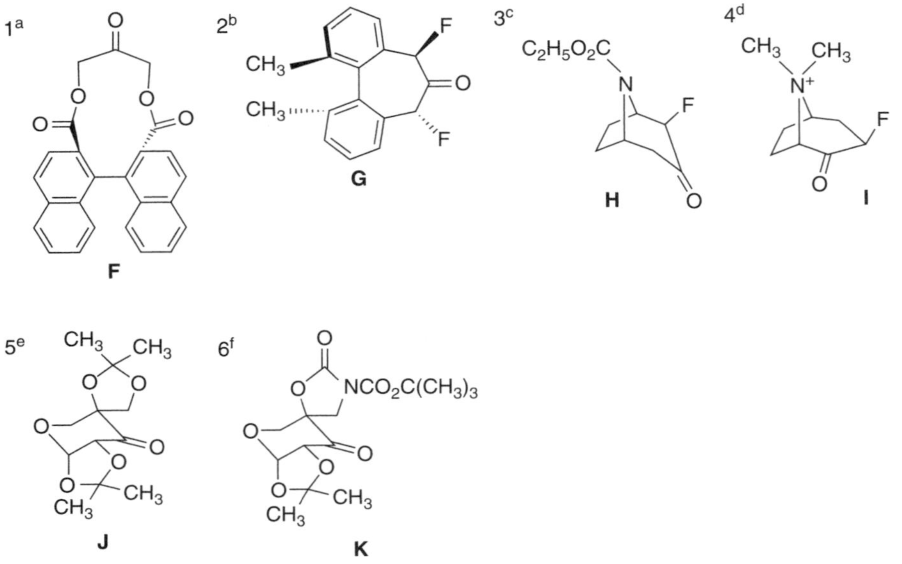

a. D. Yang, M.-K. Wong, Y.-C. Yip, X.-C. Wang, M.-W. Tang, J.-H. Zheng, and K. K. Cheung, *J. Am. Chem. Soc.*, **120**, 5943 (1998).
b. S. E. Denmark and Z. C. Wu, *Synlett*, 847 (1999); M. Frohn and Y. Shi, *Synthesis*, 1979 (2000).
c. A. Armstrong, G. Ahmed, B. Dominguez-Fernandez, B. R. Hayter, and J. S. Wailes, *J. Org. Chem.*, **67**, 8610 (2002).
d. S. E. Denmark and H. Matsuhashi, *J. Org. Chem.*, **67**, 3479 (2002).
e. Z.-X. Wang, Y. Tu, M. Frohn, J.-R. Zhang, and Y. Shi, *J. Am. Chem. Soc.*, **119**, 11224 (1997).
f. H. Tian, X. She, H. Yu, L. Shu, and Y. Shi, *J. Org. Chem.*, **67**, 2435 (2002).

104. D. Yang, *Acc. Chem. Res.*, **37**, 497 (2004); Y. Shi, *Acc. Chem. Res.*, **37**, 488 (2004).
105. T. Furutani, R. Imashiro, M. Hatsuda, and M. Seki, *J. Org. Chem.*, **67**, 4599 (2002).

The use of chiral α-fluoro ketone **G** can lead to enantioselective epoxidation.[106]

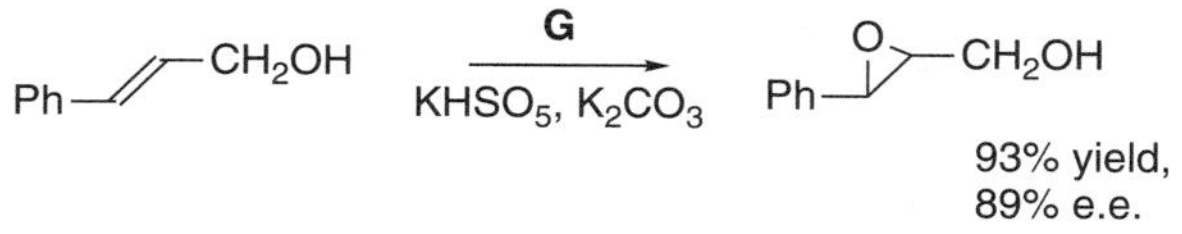

$$Ph\text{—}CH=CH\text{—}CH_2OH \xrightarrow[\text{KHSO}_5,\ K_2CO_3]{\textbf{G}} Ph\text{—}\overset{O}{\triangle}\text{—}CH_2OH$$

93% yield,
89% e.e.

The fluorinated tropones **H** and **I** also show good reactivity and are enantioselective in favorable cases, but show considerable dependence on reactant structure. The carbohydrate structures **J** and **K** also benefit from a polar effect of the adjacent oxygens and give good enantioselectivity with a variety of *trans* di- and trisubstituted alkenes. The oxazolidinone derivative **K** also shows good enantioselectivity toward *cis*-substituted and terminal alkenes. Transition structures **TS J** and **TS K** have been suggested for epoxidation by these ketones. It has been noted that alkenes with conjugated π systems have a preferred orientation toward the oxazolidinone ring.

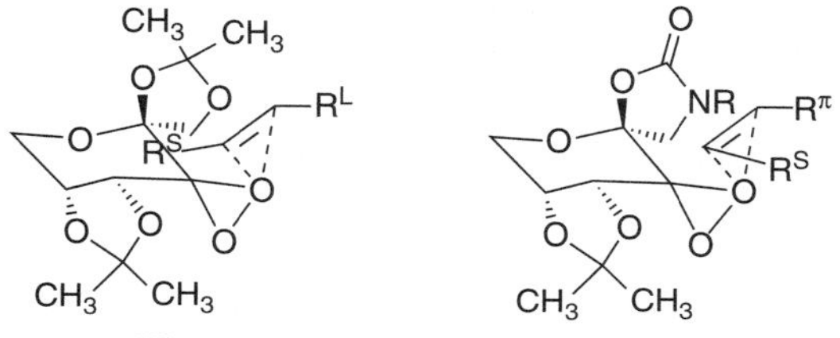

TS–J

TS–K

These ketones can also be used in kinetic resolutions.[107] The carbohydrate-derived ketones have been used in conjunction with acetonitrile and H_2O_2. The reactions are believed to proceed through dioxiranes generated by a catalytic cycle involving a peroxyimidic acid.[108]

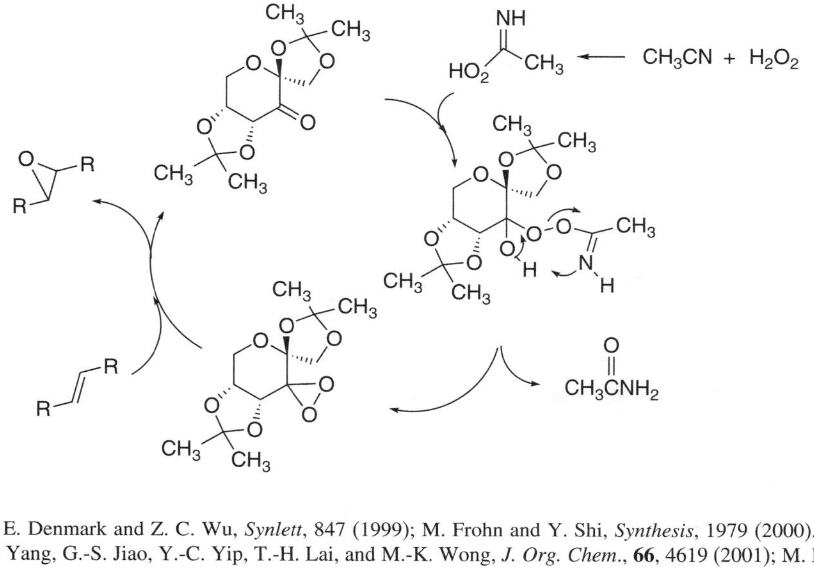

$$CH_3CN + H_2O_2$$

106. S. E. Denmark and Z. C. Wu, *Synlett*, 847 (1999); M. Frohn and Y. Shi, *Synthesis*, 1979 (2000).
107. D. Yang, G.-S. Jiao, Y.-C. Yip, T.-H. Lai, and M.-K. Wong, *J. Org. Chem.*, **66**, 4619 (2001); M. Frohn, X. Zhou, J.-R. Zhang, Y. Tang, and Y. Shi, *J. Am. Chem. Soc.*, **121**, 7718 (1999).
108. L. Shu and Y. Shi, *Tetrahedron*, **57**, 5213 (2001).

12.2.3. Subsequent Transformations of Epoxides

Epoxides are useful synthetic intermediates and the conversion of an alkene to an epoxide is often part of a more extensive molecular transformation.[109] In many instances advantage is taken of the reactivity of the epoxide ring toward nucleophiles to introduce additional functionality. Since epoxide ring opening is usually stereospecific, such reactions can be used to establish stereochemical relationships between adjacent substituents. Such two- or three-step operations can accomplish specific oxidative transformations of an alkene that may not be easily accomplished in a single step. Scheme 12.14 provides a preview of the type of reactivity to be discussed.

12.2.3.1. Nucleophilic and Solvolytic Ring Opening. Epoxidation may be preliminary to solvolytic or nucleophilic ring opening in synthetic sequences. Epoxides can undergo ring opening under either basic or acidic conditions. Base-catalyzed reactions, in which the nucleophile provides the driving force for ring opening, usually involve breaking the epoxide bond at the less-substituted carbon, since this is the position most accessible to nucleophilic attack.[110] These reactions result in an *anti* relationship between the epoxide oxygen and the nucleophile. The situation in acid-catalyzed reactions is more complex. The bonding of a proton to the oxygen weakens the C—O bonds and facilitates rupture by weak nucleophiles. If the C—O bond is largely intact at the TS, the nucleophile becomes attached to the less-substituted position for the same steric reasons that were cited for nucleophilic ring opening. If, on the other hand, C—O rupture is more complete at the TS, the opposite orientation is observed. This change in regiochemistry results from the ability of the more-substituted carbon to better stabilize the developing positive charge.

Scheme 12.14. Synthetic Transformations of Epoxides

A. Epoxidation followed by nucleophilic ring opening

B. Epoxidation followed by reductive ring opening

C. Epoxidation followed by rearrangement to a carbonyl compound

D. Epoxidation followed by ring opening to an allyl alcohol

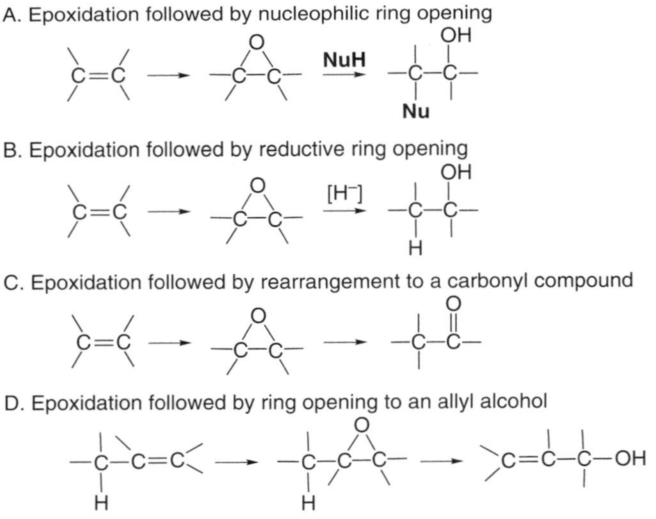

109. J. G. Smith, *Synthesis*, 629 (1984).
110. R. E. Parker and N. S. Isaacs, *Chem. Rev.*, **59**, 737 (1959).

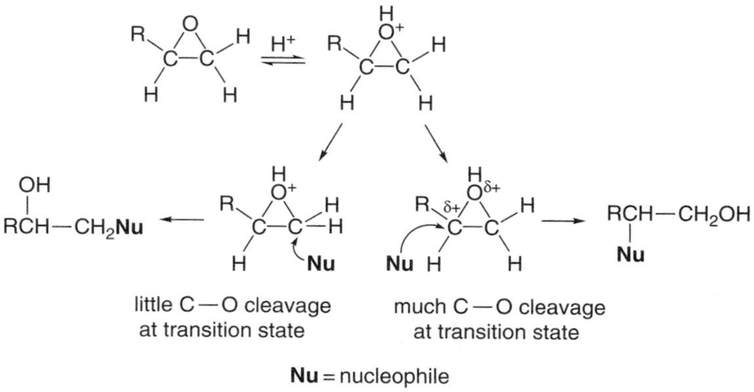

little C—O cleavage
at transition state

much C—O cleavage
at transition state

Nu = nucleophile

When simple aliphatic epoxides such as propylene oxide react with hydrogen halides, the dominant product has the halide at the less-substituted primary carbon.[111]

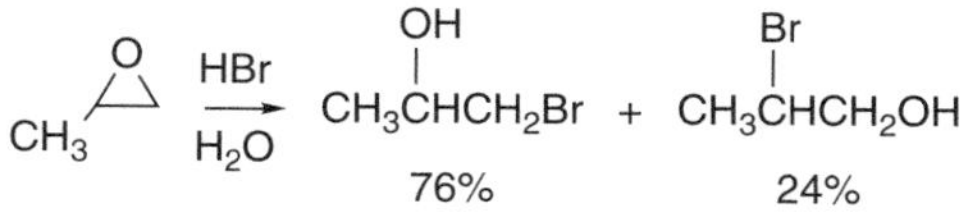

Substituents that further stabilize a carbocation intermediate lead to reversal of the mode of addition.[112] The case of styrene oxide hydrolysis has been carefully examined. Under acidic conditions, the bond breaking is exclusively at the benzylic position. Under basic conditions, ring opening occurs at both epoxide carbons.[113] Styrene also undergoes highly regioselective ring opening in the presence of Lewis acids. For example, methanolysis is catalyzed by $SnCl_4$ and occurs with greater than 95% attack at the benzyl carbon and with high inversion.[114] The stereospecificity indicates a concerted nucleophilic opening of the complexed epoxide.

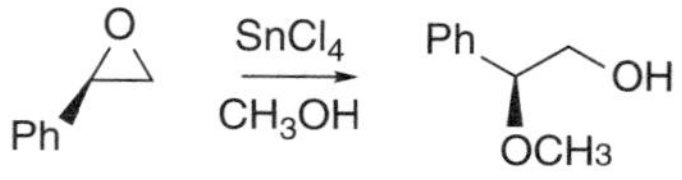

In cyclic systems, ring opening gives the diaxial diol.

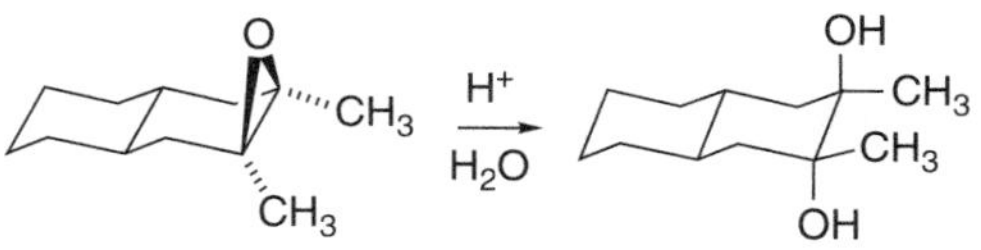

Ref. 115

Under some circumstances, acid-catalyzed ring opening of 2,2-disubstituted epoxides by sulfuric acid in dioxane goes with high *inversion* at the tertiary center.[116]

[111.] C. A. Stewart and C. A. VanderWerf, *J. Am. Chem. Soc.*, **76**, 1259 (1954).

[112.] S. Winstein and L. L. Ingraham, *J. Am. Chem. Soc.*, **74**, 1160 (1952).

[113.] R. Lin and D. L. Whalen, *J. Org. Chem.*, **59**, 1638 (1994); J. J. Blumenstein, V. C. Ukachukwa, R. S. Mohan, and D. Whalen, *J. Org. Chem.*, **59**, 1638 (1994).

[114.] C. Moberg, L. Rakos, and L. Tottie, *Tetrahedron Lett.*, **33**, 2191 (1992).

[115.] B. Rickborn and D. K. Murphy, *J. Org. Chem.*, **34**, 3209 (1969).

[116.] R. V. A. Orru, S. F. Mayer, W. Kroutil, and K. Faber, *Tetrahedron*, **54**, 859 (1998).

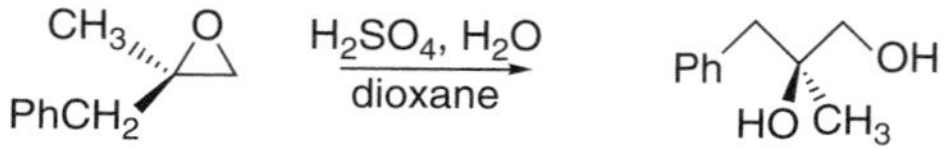

Ref. 117

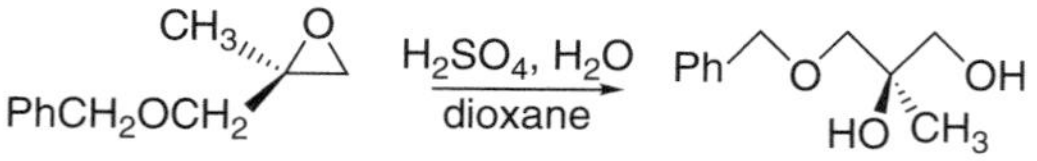

Ref. 118

Under somewhat modified conditions (H_2SO_4 on silica), this reaction has been successfully applied to a complex alkaloid structure.[119]

Recently a number of procedures for epoxide ring opening that feature the oxyphilic Lewis acids, including lanthanides, have been developed. $LiClO_4$, LiO_3SCF_3, $Mg(ClO_4)_2$, $Zn(O_3SCF_3)_2$, and $Yb(O_3SCF_3)_3$ have been shown to catalyze epoxide ring opening.[120] The cations catalyze *anti* addition of amines at the less-substituted carbon, which is consistent with a Lewis acid–assisted nucleophilic ring opening.

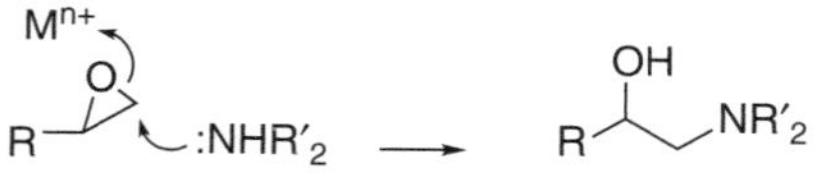

Styrene oxide gives mixtures of C-α and C-β attack, as a result of competition between the activated benzylic site and the primary site.

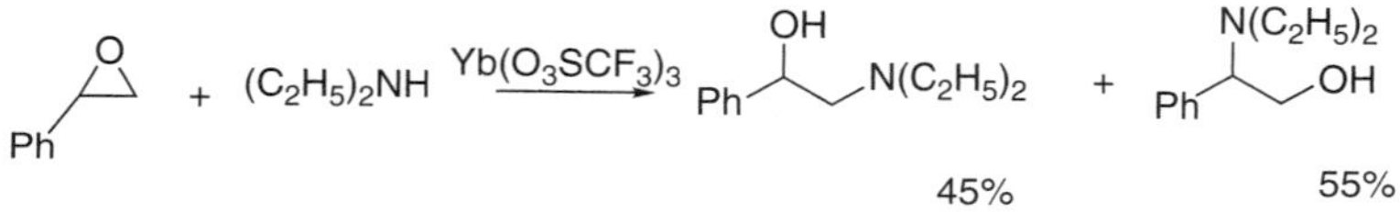

45%   55%

The same salts can be used to catalyze ring opening by other nucleophiles such as azide ion[121] and cyanide ion.[122]

A variety of reaction conditions have been developed for nucleophilic ring opening by cyanide.[123] Heating an epoxide with acetone cyanohydrin (which serves as the cyanide source) and triethylamine leads to ring opening at the less-substituted position.

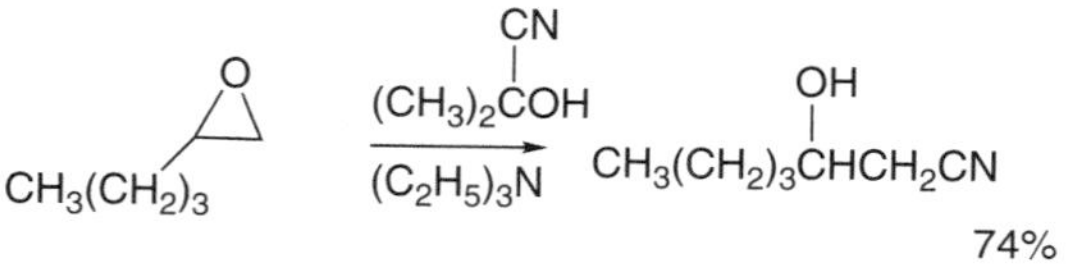

74%   Ref. 124

117. R. V. A. Orru, I. Osprian, W. Kroutil, and K. Faber, *Synthesis*, 1259 (1998).
118. A. Steinreiber, H. Hellstrom, S. F. Mayer, R. V. A. Orru, and K. Faber, *Synlett*, 111 (2001).
119. M. E. Kuehne, Y. Qin, A. E. Huot, and S. L. Bane, *J. Org. Chem.*, **66**, 5317 (2001).
120. M. Chini, P. Crotti, and F. Macchia, *Tetrahedron Lett.*, **31**, 4661 (1990); M. Chini, P. Crotti, L. Favero, F. Macchia, and M. Pineschi, *Tetrahedron Lett.*, **35**, 433 (1994); J. Auge and F. Leroy, *Tetrahedron Lett.*, **37**, 7715 (1996).
121. M. Chini, P. Crotti, and F. Macchia, *Tetrahedron Lett.*, **31**, 5641 (1990); P. Van de Weghe and J. Collin, *Tetrahedron Lett.*, **36**, 1649 (1995).
122. M. Chini, P. Crotti, L. Favera, and F. Macchia, *Tetrahedron Lett.*, **32**, 4775 (1991).
123. R. A. Smiley and C. J. Arnold, *J. Org. Chem.*, **25**, 257 (1960); J. A. Ciaccio, C. Stanescu, and J. Bontemps, *Tetrahedron Lett.*, **33**, 1431 (1992).
124. D. Mitchell and T. M. Koenig, *Tetrahedron Lett.*, **33**, 3281 (1992).

Trimethylsilyl cyanide in conjunction with KCN and a crown ether also results in nucleophilic ring opening.

80%

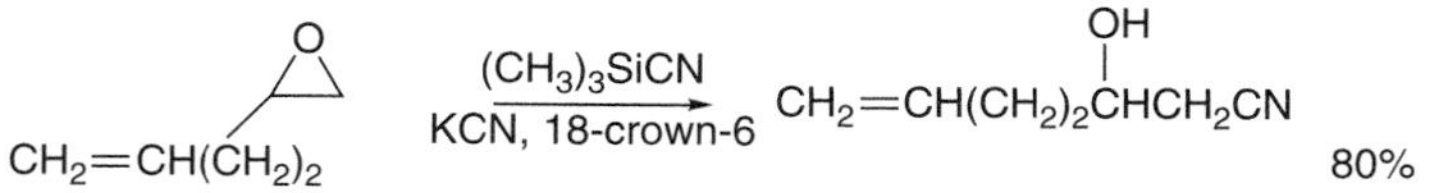

Ref. 125

Diethylaluminum cyanide can also be used for preparation of β-hydroxynitriles.

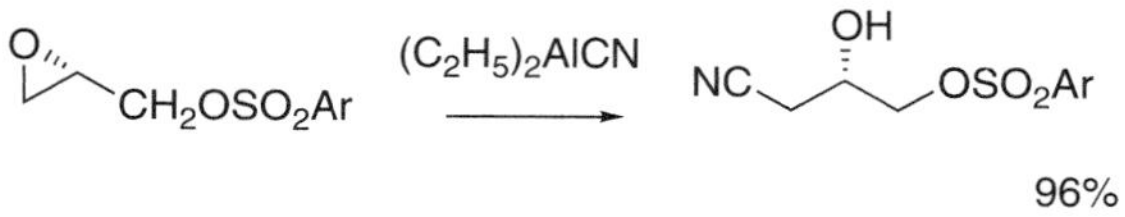

96%

Ref. 126

Similarly, diethylaluminum azide gives β-azido alcohols. The epoxide of 1-methylcyclohexene gives the tertiary azide, indicating that the regiochemistry is controlled by bond cleavage, but with diaxial stereoselectivity.

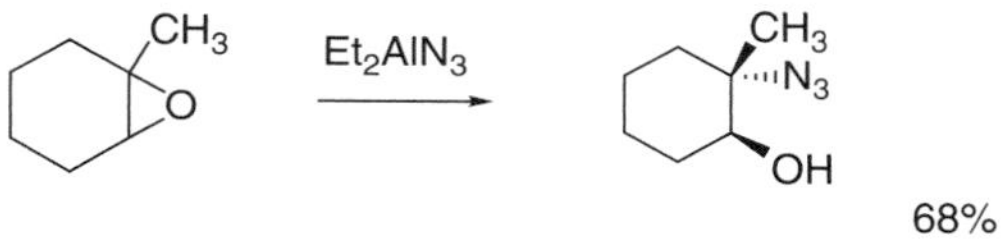

68%

Ref. 127

Epoxides of allylic alcohols exhibit chelation-controlled regioselectivity.[128]

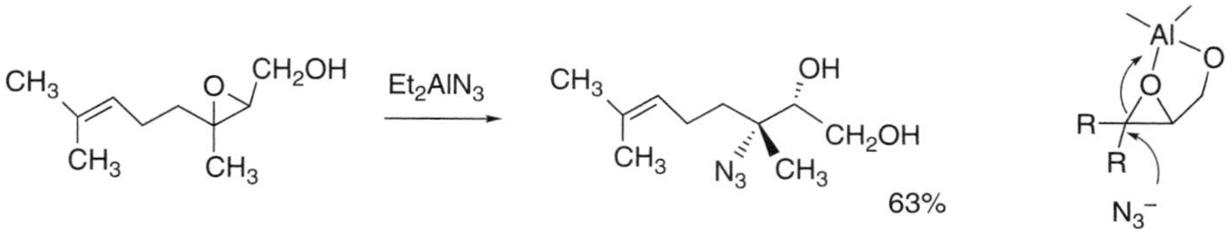

63%

Scheme 12.15 gives some examples of both acid-catalyzed and nucleophilic ring openings of epoxides. Entries 1 and 2 are cases in which epoxidation and solvolysis are carried out without isolation of the epoxide. Both cases also illustrate the preference for *anti* stereochemistry. The regioselectivity in Entry 3 is indicative of dominant bond cleavage in the TS. The reaction in Entry 4 was studied in a number of solvents. The product results from net *syn* addition as a result of phenonium ion participation. The *cis*-epoxide also gives mainly the *syn* product, presumably via isomerization to the

125. M. B. Sassaman, G. K. Surya Prakash, and G. A. Olah, *J. Org. Chem.*, **55**, 2016 (1990).

126. J. M. Klunder, T. Onami, and K. B. Sharpless, *J. Org. Chem.*, **54**, 1295 (1989).

127. H. B. Mereyala and B. Frei, *Helv. Chim. Acta*, **69**, 415 (1986).

128. F. Benedetti, F. Berti, and S. Norbedo, *Tetrahedron Lett.*, **39**, 7971 (1998); C. E. Davis, J. L. Bailey, J. W. Lockner, and R. M. Coates, *J. Org. Chem.*, **68**, 75 (2003).

Scheme 12.15. Nucleophilic and Solvolytic Ring Opening of Epoxides

A. Epoxidation with solvolysis of the intermediate epoxide

1[a] 65–73%

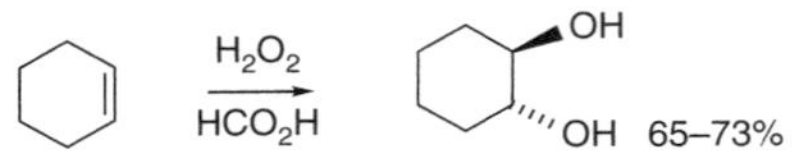

2[b]

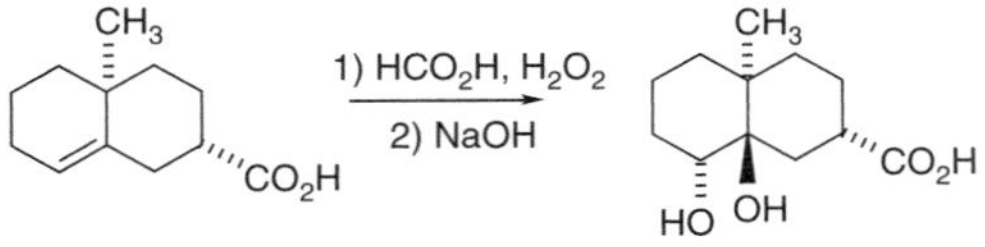

B. Acid-catalyzed solvolytic ring opening

3[c] 76%

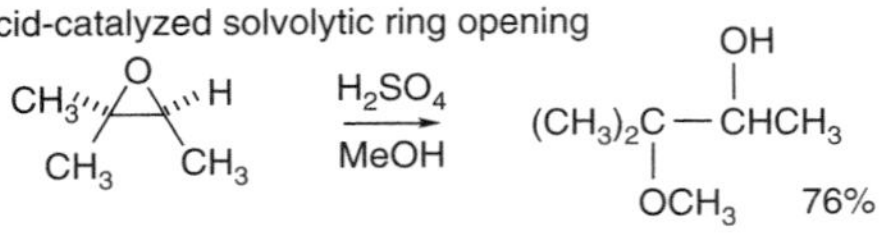

4[d] 93%

5[e] 100%

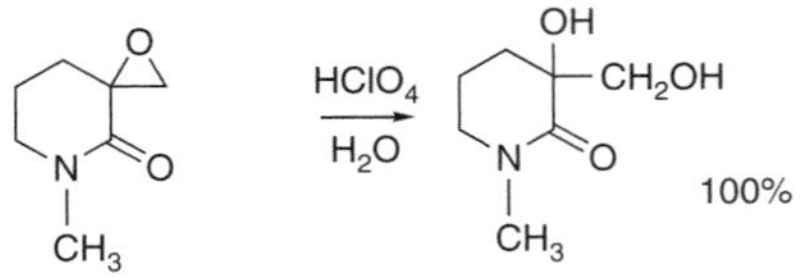

C. Nucleophilic ring-opening reactions

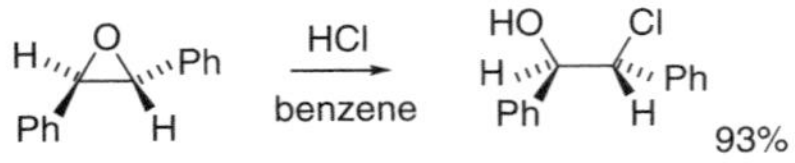

6[c] 53%

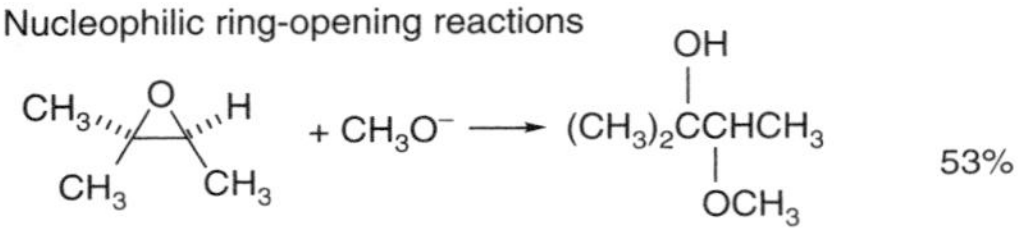

7[f] 100%

8[g] 83%

9[h] 88%

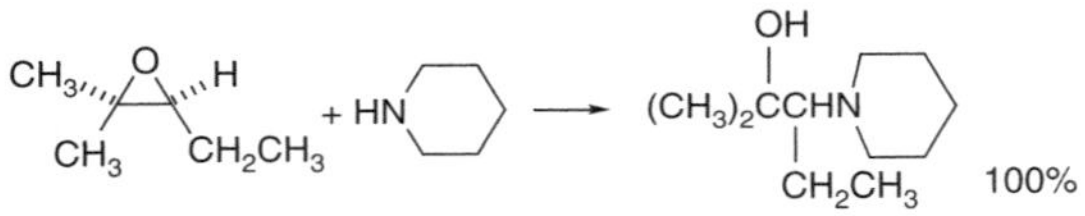

10[i] 63%

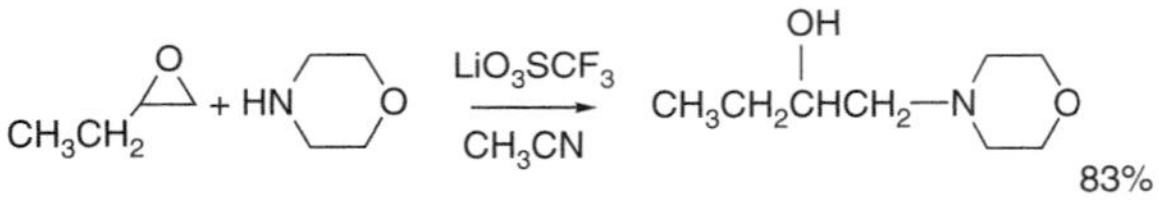

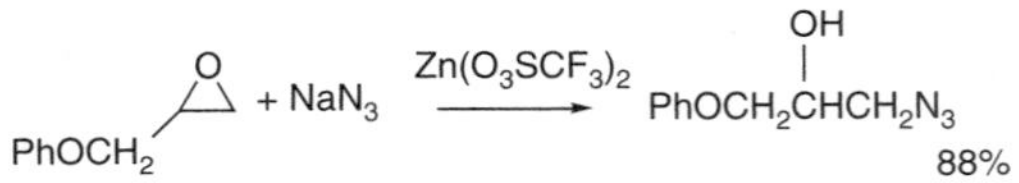

11[j] 84%

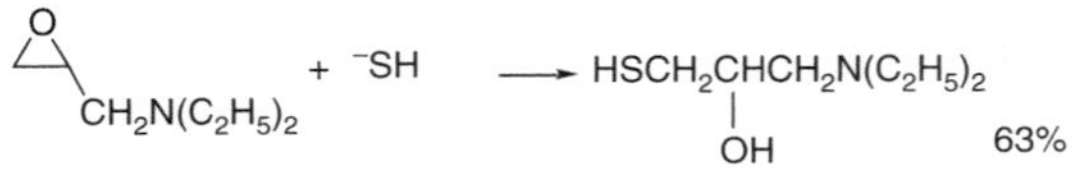

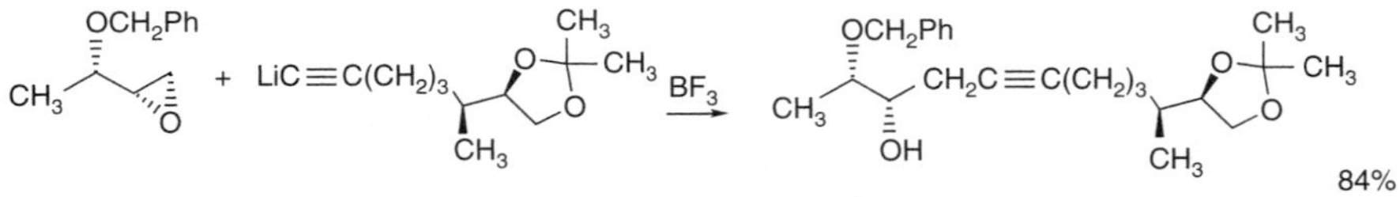

(Continued)

Scheme 12.15. (*Continued*)

1109

SECTION 12.2

*Addition of Oxygen at
Carbon-Carbon Double
Bonds*

a. A. Roebuck and H. Adkins, *Org. Synth.*, **III**, 217 (1955).
b. T. R. Kelly, *J. Org. Chem.*, **37**, 3393 (1972).
c. S. Winstein and L. L. Ingraham, *J. Am. Chem. Soc.*, **74**, 1160 (1952).
d. G. Berti, F. Bottari, P. L. Ferrarini, and B. Macchia, *J. Org. Chem.*, **30**, 4091 (1965).
e. M. L. Rueppel and H. Rapoport, *J. Am. Chem. Soc.*, **94**, 3877 (1972).
f. T. Colclough, J. I. Cunneen, and C. G. Moore, *Tetrahedron*, **15**, 187 (1961).
g. J. Auge and F. Leroy, *Tetrahedron Lett.*, **37**, 7715 (1996).
h. M. Chini, P. Crotti, and F. Macchia, *Tetrahedron Lett.*, **31**, 5641 (1990).
i. D. M. Burness and H. O. Bayer, *J. Org. Chem.*, **28**, 2283 (1963).
j. Z. Liu, C. Yu, R.-F.Wang, and G. Li, *Tetrahedron Lett.*, **39**, 5261 (1998).

more stable *trans* isomer by reversible ring opening and formation of the more stable *trans*-phenonium ion.

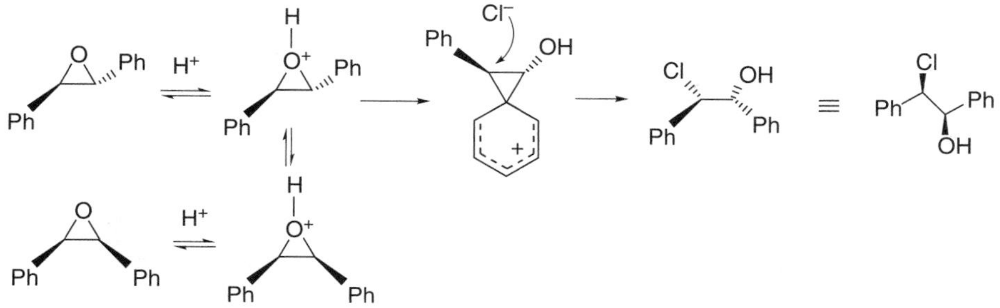

Entry 5 is an example of synthetic application of acid-catalyzed ring opening.

Entries 6 to 11 are examples of nucleophilic ring opening. Each of these entries displays the expected preference for reaction at the less hindered carbon. Entries 8 and 9 involve metal ion catalysis. Entry 11, which involves carbon-carbon bond formation, was part of a synthesis of epothilone A.

12.2.3.2. Reductive Ring Opening. Epoxides can be reduced to saturated alcohols. Lithium aluminum hydride acts as a nucleophilic reducing agent and the hydride is added at the less-substituted carbon atom of the epoxide ring. Substituted cyclohexene oxides prefer diaxial ring opening. A competing process, which accounts for about 10% of the product in the examples shown, involves rearrangement to the cyclohexanone (see below) by hydride shift, followed by reduction.[129]

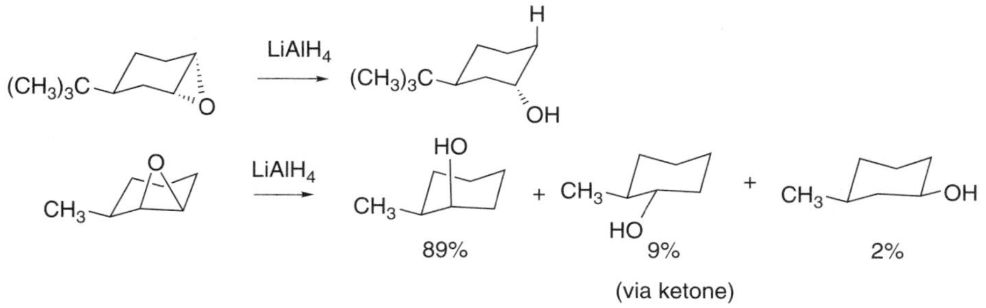

(via ketone)

[129.] B. Rickborn and J. Quartucci, *J. Org. Chem.*, **29**, 3185 (1964); B. Rickborn and W. Z. Lamke, II, *J. Org. Chem.*, **32**, 537 (1967).

The *trans*-3-methyl isomer appears to react through two conformers, with the axial methyl conformer giving *trans*-2-methylcyclohexanol.

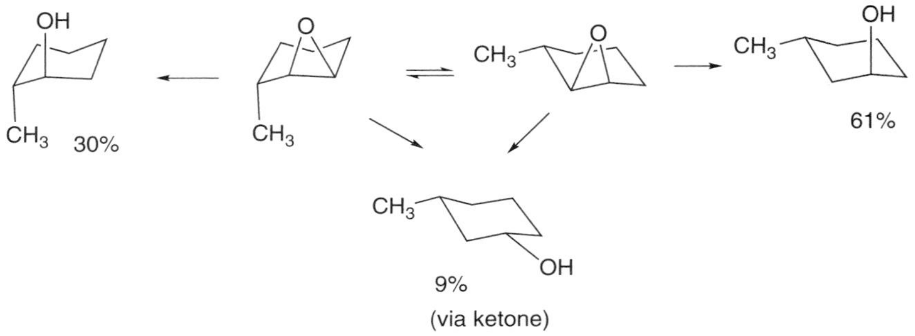

Lithium triethylborohydride is more reactive than LiAlH$_4$ and is superior for epoxides that are resistant to reduction.[130] Reduction by dissolving metals, such as lithium in ethylenediamine,[131] also gives good yields. Di-*i*-butylaluminum hydride also reduces epoxides. 1,2-Epoxyoctane gives 2-octanol in excellent yield, and styrene oxide gives a 1:6 mixture of the secondary and primary alcohols.[132] This relationship indicates that nucleophilic ring opening controls the regiochemistry for 1,2-epoxyoctane but that ring cleavage at the benzylic position is the major factor for styrene oxide.

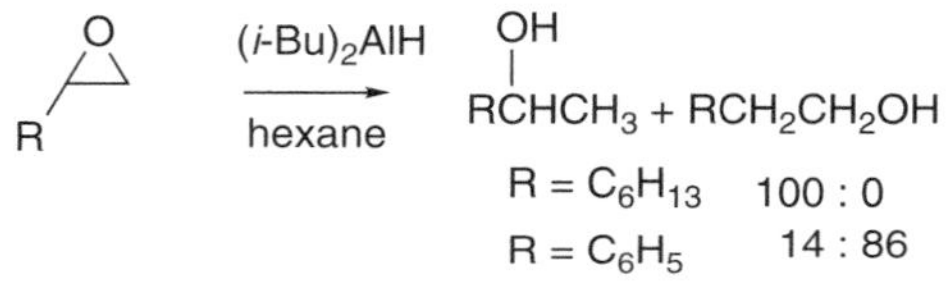

Diborane in THF reduces epoxides, but the yields are low, and other products are formed by pathways that result from the electrophilic nature of diborane.[133] Better yields are obtained when BH$_4^-$ is included in the reaction system, but the electrophilic nature of diborane is still evident because the dominant product results from addition of the hydride at the more-substituted carbon.[134]

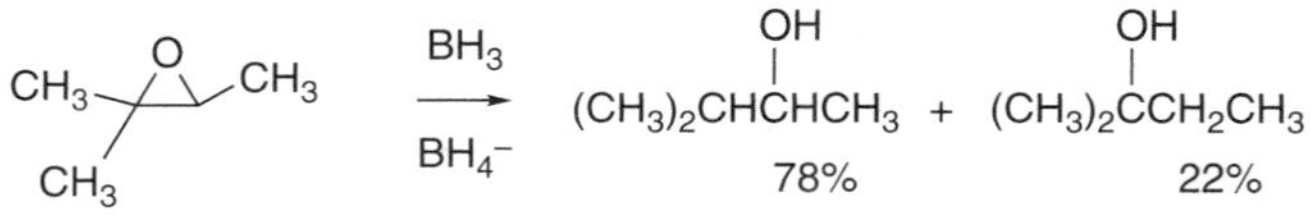

The overall transformation of alkenes to alcohols that is accomplished by epoxidation and reduction corresponds to alkene hydration. Assuming a nucleophilic ring opening by hydride addition at the less-substituted carbon, the reaction corresponds to the Markovnikov orientation. This reaction sequence is therefore an alternative to the hydration methods discussed in Chapter 4 for converting alkenes to alcohols.

130. S. Krishnamurthy, R. M. Schubert, and H. C. Brown, *J. Am. Chem. Soc.*, **95**, 8486 (1973).
131. H. C. Brown, S. Ikegami, and J. H. Kawakami, *J. Org. Chem.*, **35**, 3243 (1970).
132. J. J. Eisch, Z.-R. Liu, and M. Singh, *J. Org. Chem.*, **57**, 1618 (1992).
133. D. J. Pasto, C. C. Cumbo, and J. Hickman, *J. Am. Chem. Soc.*, **88**, 2201 (1966).
134. H. C. Brown and N. M. Yoon, *J. Am. Chem. Soc.*, **90**, 2686 (1968).

12.2.3.3. Rearrangement of Epoxides to Carbonyl Compounds. Epoxides can be isomerized to carbonyl compounds by Lewis acids.[135] This reaction is closely related to the pinacol rearrangement (see p. 883). The epoxide oxygen functions as the leaving group and becomes the oxygen in the new carbonyl group.

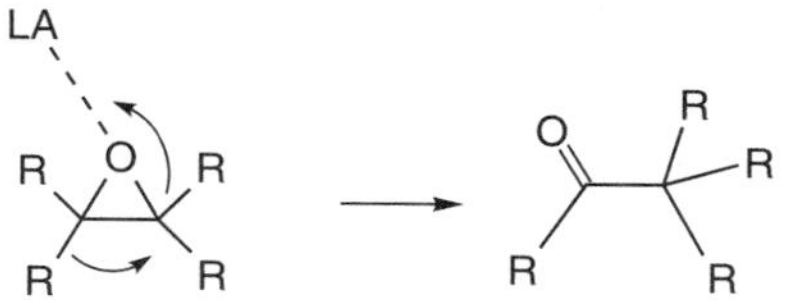

Carbocation intermediates are involved and the structure and stereochemistry of the product are determined by the factors that govern substituent migration in the carbocation. Clean, high-yield reactions can be expected only where structural or conformational factors promote a selective rearrangement. Boron trifluoride is frequently used as the reagent.

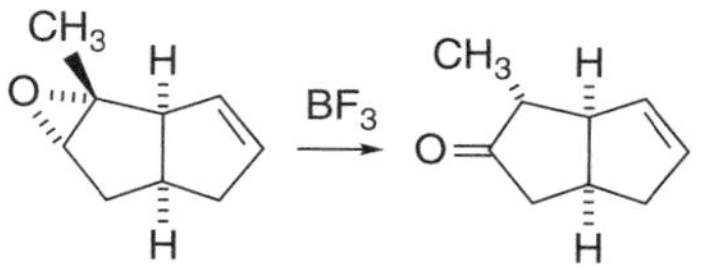

Ref. 136

Catalytic amounts of $Bi(O_3SCF_3)_3$ also promote this rearrangement.[137]

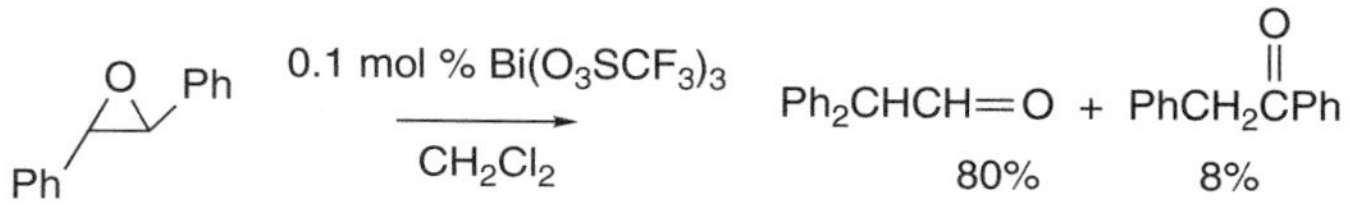

Bulky diaryloxymethylaluminum reagents are also effective for this transformation.

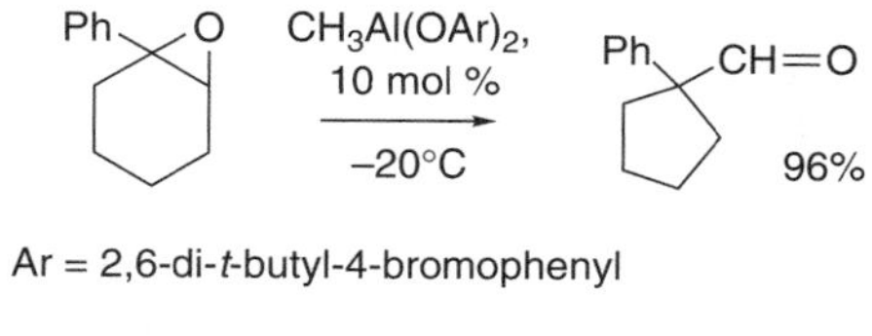

Ar = 2,6-di-*t*-butyl-4-bromophenyl

Ref. 138

This reagent is selective for rearrangement to aldehydes in cases where BF_3, $SnCl_4$, and SbF_5 give mixtures.[139]

135. J. N. Coxon, M. P. Hartshorn, and W. J. Rae, *Tetrahedron*, **26**, 1091 (1970).
136. J. K. Whitesell, R. S. Matthews, M. A. Minton, and A. M. Helbling, *J. Am. Chem. Soc.*, **103**, 3468 (1981).
137. K. A. Bhatia, K. J. Eash, N. M. Leonard, M. C. Oswald, and R. S. Mohan, *Tetrahedron Lett.*, **42**, 8129 (2001).
138. K. Maruoka, S. Nagahara, T. Ooi, and H. Yamamoto, *Tetrahedron Lett.*, **30**, 5607 (1989).
139. K. Maruoka, T. Ooi, and H. Yamamoto, *Tetrahedron*, **48**, 3303 (1992); K. Maruoka, N. Murase, R. Bureau, T. Ooi, and H. Yamamoto, *Tetrahedron*, **50**, 3663 (1994).

	yield	product ratio
$CH_3Al(OAr)_2$	72%	100:0
BF_3	55%	33:67
$SnCl_4$	72%	50:50
SbF_5	79%	15:85

This selectivity is attributed to the steric bulk of the aluminum reagent favoring the migration of the larger alkyl group. The same selectivity pattern is observed with unbranched substituents.

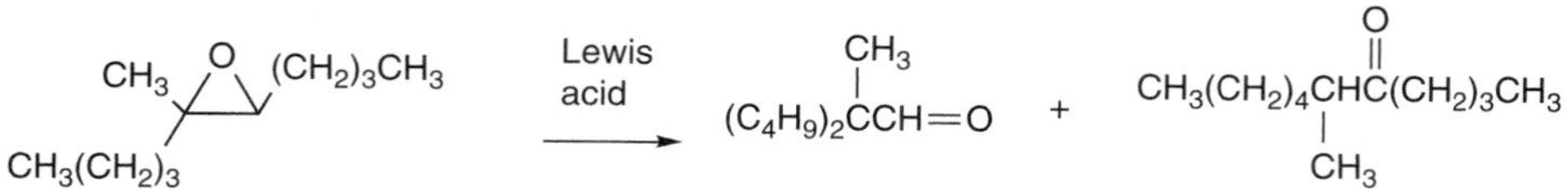

	yield	product ratio
$CH_3Al(OAr)_2$	73%	100:0
BF_3	77%	30:70
SbF_5	86%	82:18

Double bonds having oxygen and halogen substituents are susceptible to epoxidation, and the reactive epoxides that are generated serve as intermediates in some useful synthetic transformations in which the substituent migrates to the other carbon of the original double bond. Vinyl chlorides furnish haloepoxides that can rearrange to α-haloketones.

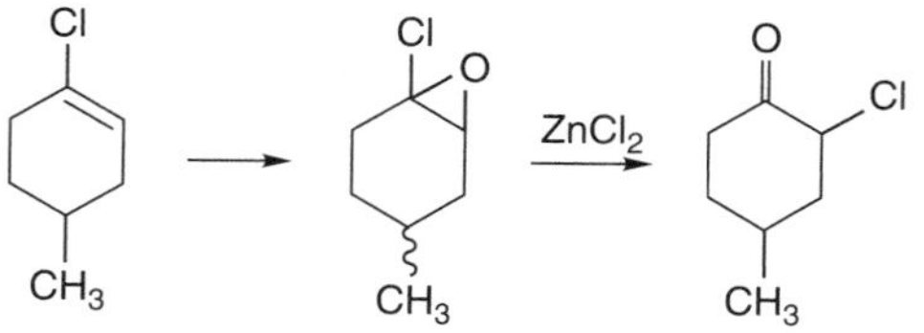

Ref. 140

When this reaction sequence is applied to enol esters or enol ethers, the result is α-oxygenation of the starting carbonyl compound. Enol acetates form epoxides that rearrange to α-acetoxyketones.

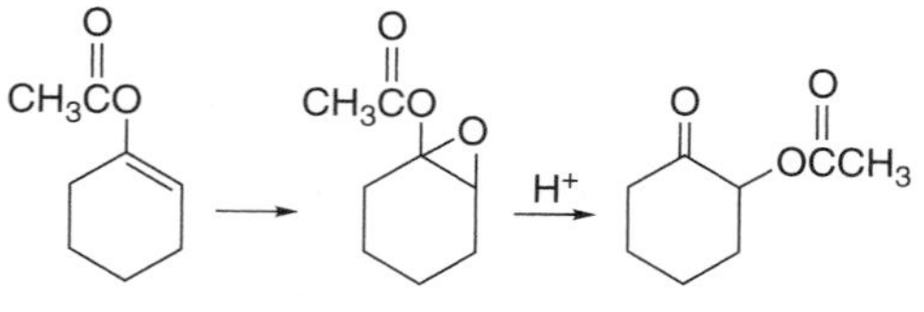

Ref. 141

140. R. N. McDonald and T. E. Tabor, *J. Am. Chem. Soc.*, **89**, 6573 (1967).
141. K. L. Williamson, J. I. Coburn, and M. F. Herr, *J. Org. Chem.*, **32**, 3934 (1967).

The stereochemistry of the reaction depends on the Lewis acid. Protic acids favor retention of configuration, as does TMSOTf. Most metal halides give mixtures of inversion and retention, but $Al(CH_3)_3$ gives dominant inversion.[142] Inversion is suggestive of direct carbonyl group participation.

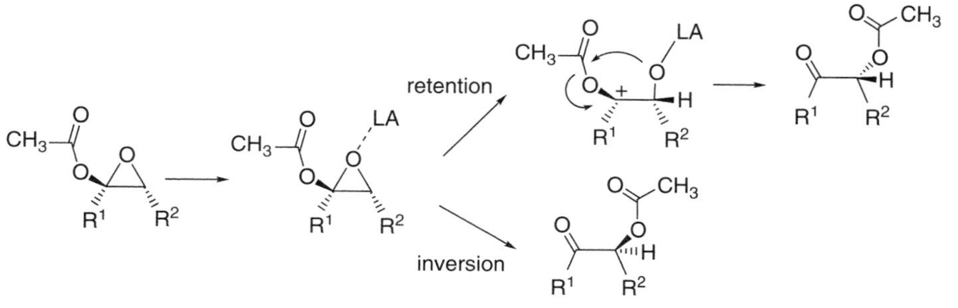

The reaction can also be done thermally. The stereochemistry of the thermal rearrangement of the acetoxy epoxides involves inversion at the carbon to which the acetoxy group migrates,[143] and reaction probably proceeds through a cyclic TS.

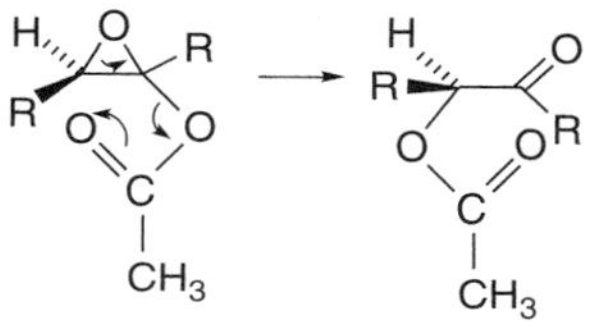

A more synthetically reliable version of this reaction involves epoxidation of silyl enol ethers. Epoxidation of the silyl enol ethers followed by aqueous workup gives α-hydroxyketones and α-hydroxyaldehydes.[144]

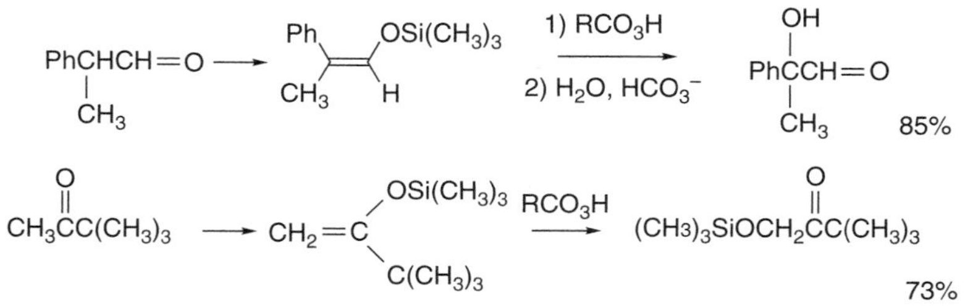

The epoxidation can be done either with peroxy acids or DMDO. In the former case, the rearrangement is catalyzed by the carboxylic acid that is formed, whereas with DMDO, the intermediate epoxides can sometimes be isolated.

142. Y. Zhu, L. Shu., Y. Tu, and Y. Shi, *J. Org. Chem.*, **66**, 1818 (2001).
143. K. L. Williamson and W. S. Johnson, *J. Org. Chem.*, **26**, 4563 (1961).
144. A. Hassner, R. H. Reuss, and H. W. Pinnick, *J. Org. Chem.*, **40**, 3427 (1975).

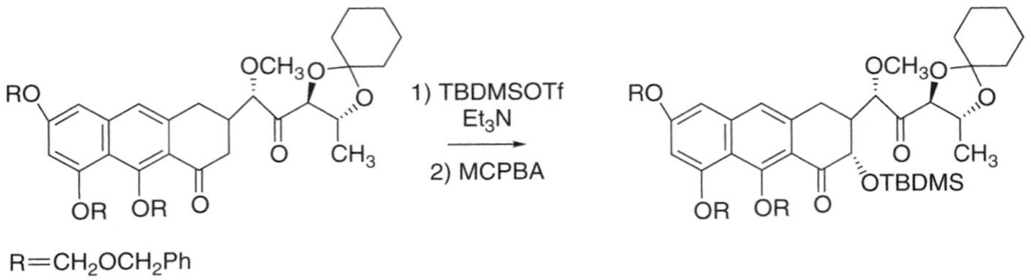

Ref. 145

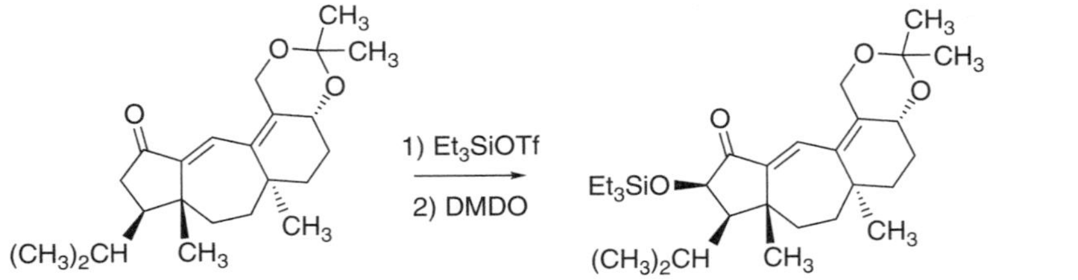

Ref. 146

The oxidation of silyl enol ethers with the osmium tetroxide–amine oxide combination also leads to α-hydroxyketones in generally good yields.[147]

Epoxides derived from vinylsilanes are converted by mildly acidic conditions into ketones or aldehydes.[148]

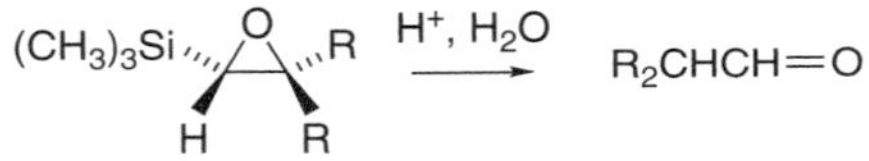

The regioselective ring opening of the silyl epoxides is facilitated by the stabilizing effect that silicon has on a positive charge in the β-position. This facile transformation permits vinylsilanes to serve as the equivalent of carbonyl groups in multistep synthesis.[149]

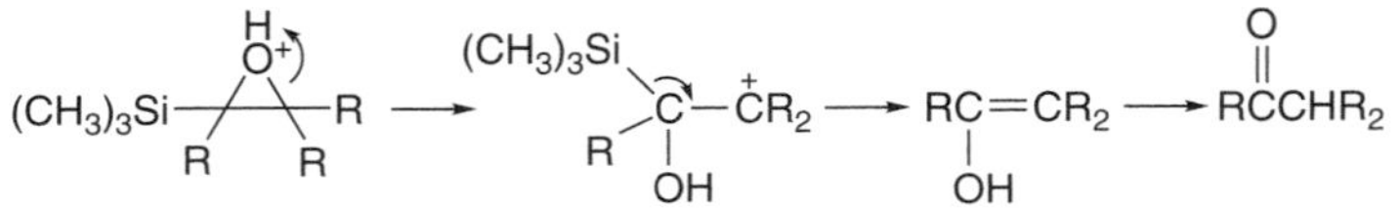

12.2.3.4. Base-Catalyzed Ring Opening of Epoxides. Base-catalyzed ring opening of epoxides provides a route to allylic alcohols.[150]

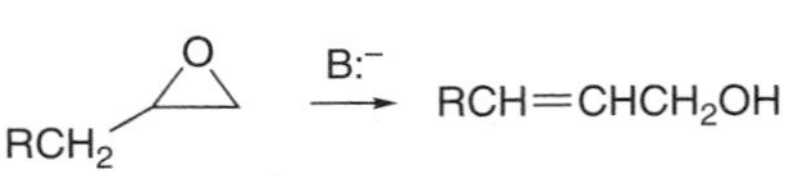

145. W. R. Roush, M. R. Michaelides, D. F. Tai, and W. K. M. Chong, *J. Am. Chem. Soc.*, **109**, 7575 (1987).
146. M. Mandal and S. J. Danishefsky, *Tetrahedron Lett.*, **45**, 3831 (2004).
147. J. P. McCormick, W. Tomasik, and M. W. Johnson, *Tetrahedron Lett.*, 607 (1981).
148. G. Stork and E. Colvin, *J. Am. Chem. Soc.*, **93**, 2080 (1971).
149. G. Stork and M. E. Jung, *J. Am. Chem. Soc.*, **96**, 3682 (1974).
150. J. K. Crandall and M. Apparu, *Org. React.*, **29**, 345 (1983).

Strongly basic reagents, such as the lithium salt of dialkylamines, are required to promote the reaction. The stereochemistry of the ring opening has been investigated by deuterium labeling. A proton *cis* to the epoxide ring is selectively removed.[151]

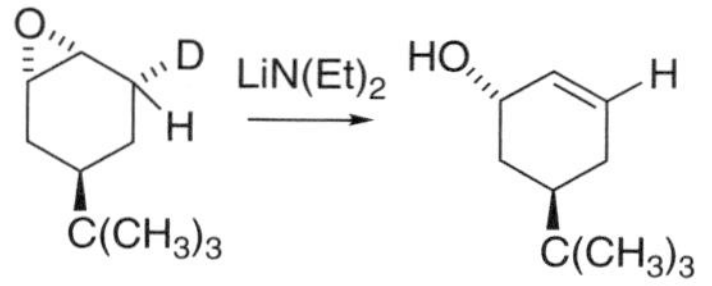

A TS represented by structure **L** accounts for this stereochemistry. Such an arrangement is favored by ion pairing that would bring the amide anion and lithium cation into close proximity. Simultaneous coordination of the lithium ion at the epoxide results in a *syn* elimination.

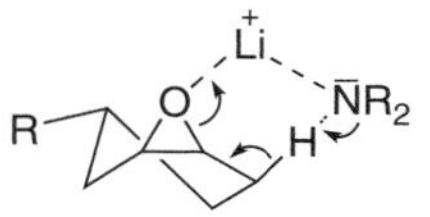

L

Among other reagents that effect epoxide ring opening are diethylaluminum 2,2,6,6-tetramethylpiperidide and magnesium *N*-cyclohexyl-*N*-(*i*-propyl)amide.

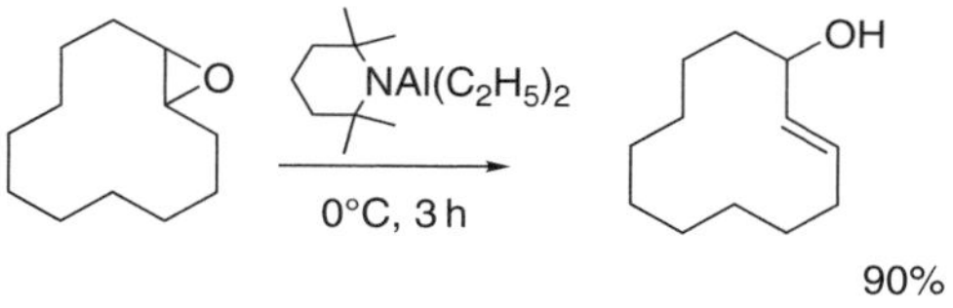

Ref. 152

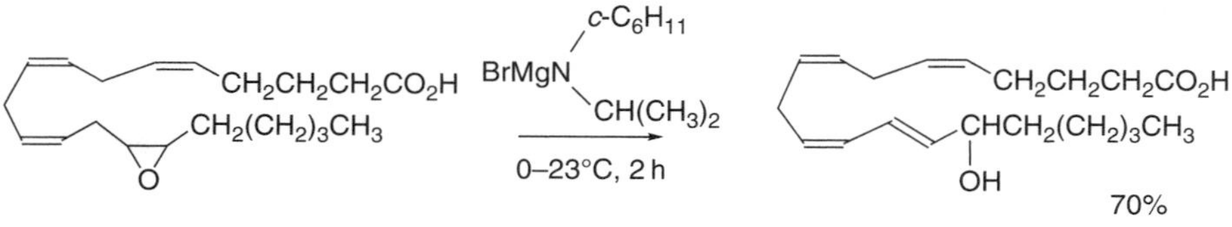

Ref. 153

These reagents are appropriate even for very sensitive molecules. Their efficacy is presumably due to the Lewis acid effect of the aluminum and magnesium ions. The hindered nature of the amide bases also minimizes competition from nucleophilic ring opening.

151. R. P. Thummel and B. Rickborn, *J. Am. Chem. Soc.*, **92**, 2064 (1970).
152. A. Yasuda, S. Tanaka, K. Oshima, H. Yamamoto, and H. Nozaki, *J. Am. Chem. Soc.*, **96**, 6513 (1974).
153. E. J. Corey, A. Marfat, J. R. Falck, and J. O. Albright, *J. Am. Chem. Soc.*, **102**, 1433 (1980).

Epoxides can also be converted to allylic alcohols using electrophilic reagents. The treatment of epoxides with trialkyl silyl iodides and an organic base gives the silyl ether of the corresponding allylic alcohols.[154]

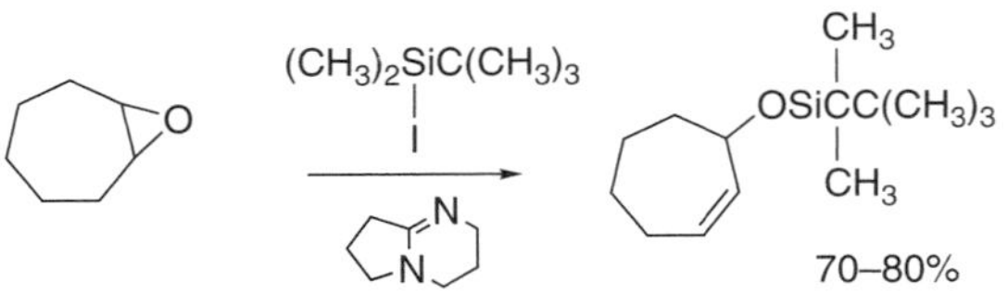

70–80%

Similar ring openings have been achieved using trimethylsilyl triflate and 2,6-di-*t*-butylpyridine.[155]

Each of these procedures for epoxidation and ring opening is the equivalent of an allylic oxidation of a double bond with migration of the double bond.

$$R_2CHCH{=}CHR' \longrightarrow R_2C{=}CH{-}\overset{\overset{\textstyle OH}{|}}{C}HR'$$

In Section 12.3, other means of effecting this transformation are described.

12.3. Allylic Oxidation

12.3.1. Transition Metal Oxidants

Carbon-carbon double bonds, apart from being susceptible to addition of oxygen or cleavage, can also react at allylic positions. Synthetic utility requires that there be good selectivity between the possible reactions. Among the transition metal oxidants, the CrO_3-pyridine reagent in methylene chloride[156] and a related complex in which 3,5-dimethylpyrazole replaces pyridine[157] are the most satisfactory for allylic oxidation.

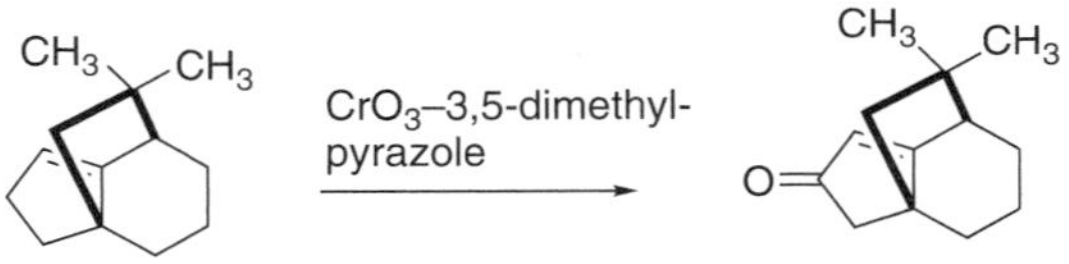

Ref. 158

Several pieces of mechanistic evidence implicate allylic radicals or cations as intermediates in these oxidations. Thus ^{14}C in cyclohexene is distributed in the product cyclohexenone indicating that a symmetrical allylic intermediate is involved at some stage.[159]

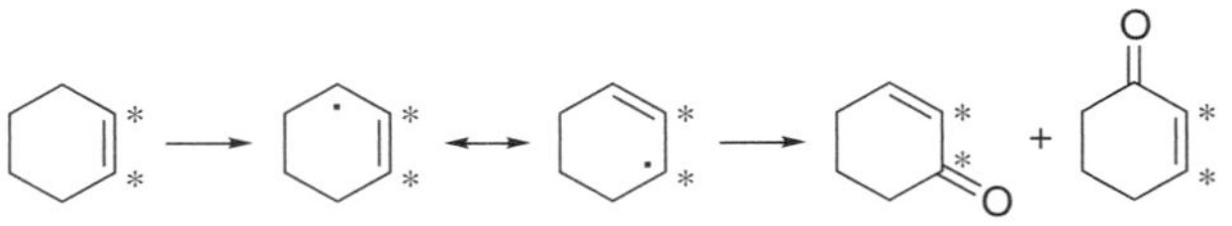

154. M. R. Detty, *J. Org. Chem.*, **45**, 924 (1980); M. R. Detty and M. D. Seiler, *J. Org. Chem.*, **46**, 1283 (1981).
155. S. F. Martin and W. Li, *J. Org. Chem.*, **56**, 642 (1991).
156. W. G. Dauben, M. Lorber, and D. S. Fullerton, *J. Org. Chem.*, **34**, 3587 (1969).
157. W. G. Salmond, M. A. Barta, and J. L. Havens, *J. Org. Chem.*, **43**, 2057 (1978); R. H. Schlessinger, J. L. Wood, A. J. Poos, R. A. Nugent, and W. H. Parson, *J. Org. Chem.*, **48**, 1146 (1983).
158. A. B. Smith, III, and J. P. Konopelski, *J. Org. Chem.*, **49**, 4094 (1984).
159. K. B. Wiberg and S. D. Nielsen, *J. Org. Chem.*, **29**, 3353 (1964).

In many allylic oxidations, the double bond is found in a position indicating that an allylic transposition occurs during the oxidation.

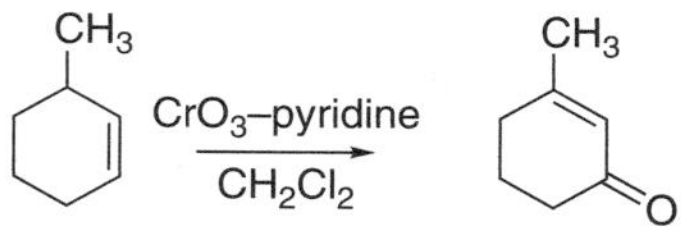

CrO$_3$–pyridine / CH$_2$Cl$_2$ 68%   Ref. 156

Detailed mechanistic understanding of the allylic oxidation has not been developed. One possibility is that an intermediate oxidation state of Cr, specifically Cr(IV), acts as the key reagent by abstracting hydrogen.[160]

Several catalytic systems based on copper can also achieve allylic oxidation. These reactions involve induced decomposition of peroxy esters (see Part A, Section 11.1.4). When chiral copper ligands are used, enantioselectivity can be achieved. Table 12.1 shows some results for the oxidation of cyclohexene under these conditions.

12.3.2. Reaction of Alkenes with Singlet Oxygen

Among the oxidants that add oxygen at carbon-carbon double bonds is singlet oxygen.[161] For most alkenes this reaction proceeds with the removal of an allylic

Table 12.1. Enantioselective Copper-Catalyzed Allylic Oxidation of Cyclohexene

	Catalyst	Yield%	e.e.%
1[a]	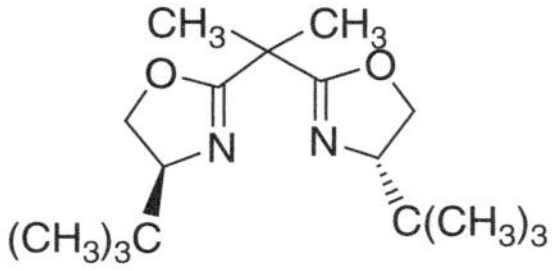	43	80
2[b]	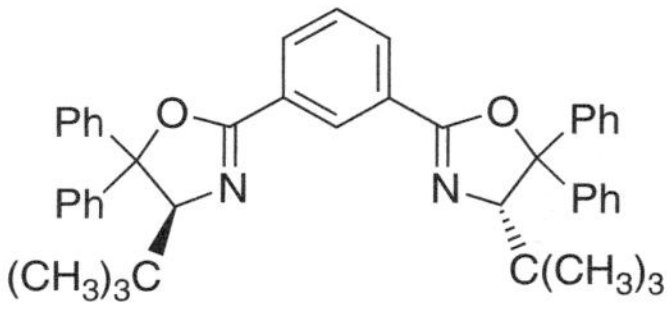	73	75
3[c]	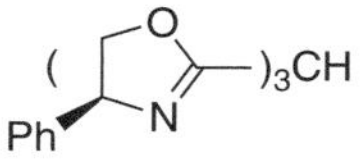	19	42
4[d]	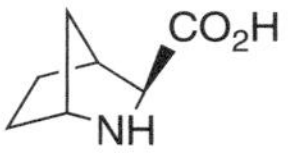	67	50

a. M. B. Andrus and X. Chen, *Tetrahedron*, **53**, 16229 (1997).
b. G. Sekar, A. Datta Gupta, and V. K. Singh, *J. Org. Chem.*, **62**, 2961 (1998).
c. K. Kawasaki and T. Katsuki, *Tetrahedron*, **53**, 6337 (1997).
d. M. J. Sodergren and P. G. Andersson, *Tetrahedron Lett.*, **37**, 7577 (1996).

[160] P. Mueller and J. Rocek, *J. Am. Chem. Soc.*, **96**, 2836 (1974).

[161] H. H. Wasserman and R. W. Murray, eds., *Singlet Oxygen*, Academic Press, New York, 1979; A. A. Frimer, *Chem. Rev.*, **79**, 359 (1979); A. Frimer, ed., *Singlet Oxygen*, CRC Press, Boca Raton, FL, 1985; C. S. Foote and E. L. Clennan, in *Active Oxygen in Chemistry*, C. S. Foote, J. S. Valentine, A. Greenberg, and J. F. Liebman, eds., Blackie Academic & Professional, London, 1995, pp. 105–140; M. Prein and W. Adam, *Angew. Chem. Int. Ed. Engl.*, **35**, 477 (1996); M. Orfanopoulos, *Molec. Supramolec. Photochem.*, **8**, 243 (2001).

hydrogen and shift of the double bond to provide an allylic hydroperoxide as the initial product.

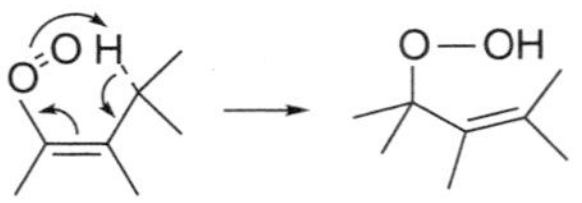

The allylic hydroperoxides generated by singlet oxygen oxidation are normally reduced to the corresponding allylic alcohol. The net synthetic transformation is then formation of an allylic alcohol with transposition of the double bond.

A number of methods of generating singlet oxygen are summarized in Scheme 12.16. Singlet oxygen is usually generated from oxygen by dye-sensitized photoexcitation. Porphyrins are also often used as sensitizers. An alternative chemical means of generating $^1O_2$ involves the reaction of hydrogen peroxide with sodium hypochlorite (Entry 2). The method in Entry 3 involves formation of unstable trioxaphosphetane intermediates from O_3 and phosphine or phosphate esters. The adducts are formed at low temperature (−70 °C) and decomposition with generation of singlet oxygen occurs at about −35 °C. The peroxide intermediate in Entry 4 is formed by photolytic addition of oxygen to diphenylanthracene and reacts at around 80 °C to generate $^1O_2$. The method in Entry 5 involves formation of an unstable precursor of $^1O_2$, a trialkylsilyl hydrotrioxide. The half-life of the adduct is roughly 2.5 min at −60 °C.

$$(C_2H_5)_3SiH + O_3 \longrightarrow (C_2H_5)_3SiOOOH \longrightarrow (C_2H_5)_3SiOH + O{=}O$$

Scheme 12.16. Generation of Singlet Oxygen

1[a]

$$\text{Photosensitizer} + h\nu \longrightarrow {}^1[\text{Photosensitizer}]^*$$

$${}^1[\text{Photosensitizer}]^* \longrightarrow {}^3[\text{Photosensitizer}]^*$$

$${}^3[\text{Photosensitizer}]^* + {}^3O_2 \longrightarrow {}^1O_2 + \text{Photosensitizer}$$

2[b]

$$H_2O_2 + {}^-OCl \longrightarrow {}^1O_2 + H_2O + Cl^-$$

3[c]

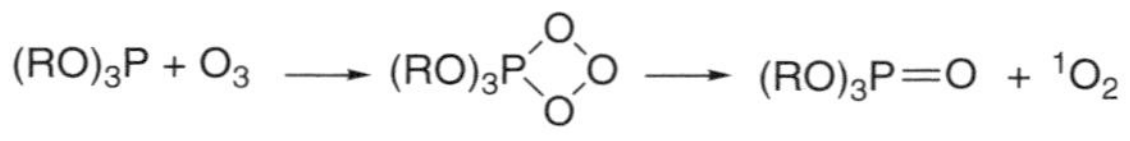

4[d]

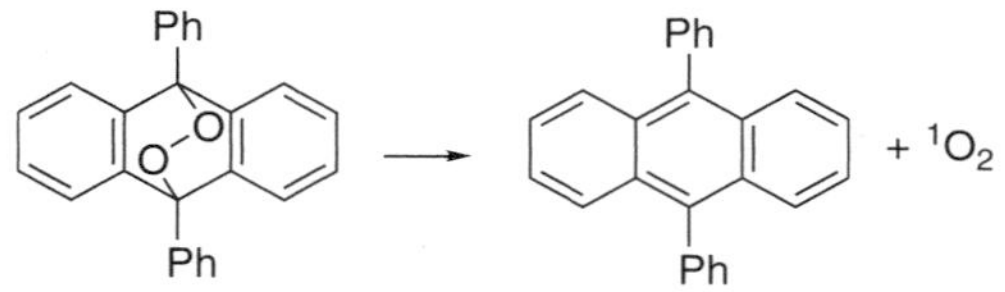

5[e]

$$(C_2H_5)_3SiH + O_3 \longrightarrow (C_2H_5)_3SiOOOH \longrightarrow (C_2H_5)_3SiOH + {}^1O_2$$

a. C. S. Foote and S. Wexler, *J. Am. Chem. Soc.*, **86**, 3880 (1964).
b. C. S. Foote and S. Wexler, *J. Am. Chem. Soc.*, **86**, 3879 (1964).
c. R. W. Murray and M. L. Kaplan, *J. Am. Chem. Soc.*, **90**, 537 (1968).
d. H. H. Wasserman, J. R. Sheffler, and J. L. Cooper, *J. Am. Chem. Soc.*, **94**, 4991 (1972).
e. E. J. Corey, M. M. Mehotra, and A. U. Khan, *J. Am. Chem. Soc.*, **108**, 2472 (1986).

Singlet oxygen decays to the ground state triplet at a rate that is strongly dependent on the solvent.[162] Measured half-lives range from about $700\,\mu s$ in carbon tetrachloride to $2\,\mu s$ in water. The choice of solvent can therefore have a pronounced effect on the efficiency of oxidation; the longer the singlet state lifetime, the more likely it is that reaction with the alkene can occur.

The reactivity order of alkenes is that expected for attack by an electrophilic reagent. Reactivity increases with the number of alkyl substituents.[163] Terminal alkenes are relatively inert. The reaction has a low $\Delta H^\ddagger$ and relative reactivity is dominated by entropic factors.[164] Steric effects govern the direction of approach of the oxygen, so the hydroperoxy group is usually introduced on the less hindered face of the double bond. A key mechanistic issue in singlet oxygen oxidations is whether it is a concerted process or involves an intermediate formulated as a "perepoxide." Most of the available evidence points to the perepoxide mechanism.[165]

concerted mechanism   perepoxide-intermediate mechanism

Many alkenes present several different allylic hydrogens, and in this type of situation it is important to be able to predict the degree of selectivity.[166] A useful generalization is that *there is a preference for removal of a hydrogen from the more congested side of the double bond.*[167]

This "*cis* effect" is ascribed to a more favorable TS when the singlet O_2 can interact with two allylic hydrogens. The stabilizing interaction has been described both in FMO[168] and hydrogen-bonding[169] terminology and can be considered an electrostatic effect. The *cis* effect does not apply to alkene having *t*-butyl substituents.[170] There are

162. P. B. Merkel and D. R. Kearns, *J. Am. Chem. Soc.*, **94**, 1029, 7244 (1972); P. R. Ogilby and C. S. Foote, *J. Am. Chem. Soc.*, **105**, 3423 (1983); J. R. Hurst, J. D. McDonald, and G. B. Schuster, *J. Am. Chem. Soc.*, **104**, 2065 (1982).

163. K. R. Kopecky and H. J. Reich, *Can. J. Chem.*, **43**, 2265 (1965); C. S. Foote and R. W. Denny, *J. Am. Chem. Soc.*, **93**, 5162 (1971); A. Nickon and J. F. Bagli, *J. Am. Chem. Soc.*, **83**, 1498 (1961).

164. J. R. Hurst and G. B. Schuster, *J. Am. Chem. Soc.*, **104**, 6854 (1982).

165. M. Orfanopoulos, I. Smonou, and C. S. Foote, *J. Am. Chem. Soc.*, **112**, 3607 (1990); M. Statakis, M. Orfanopoulos, J. S. Chen, and C. S. Foote, *Tetrahedron Lett.*, **37**, 4105 (1996).

166. M. Stratakis and M. Orfanopoulos, *Tetrahedron*, **56**, 1595 (2000).

167. M. Orfanopoulos, M. B. Grdina, and L. M. Stephenson, *J. Am. Chem. Soc.*, **101**, 275 (1979); K. H. Schulte-Elte, B. L. Muller, and V. Rautenstrauch, *Helv. Chim. Acta*, **61**, 2777 (1978); K. H. Schulte-Elte and V. Rautenstrauch, *J. Am. Chem. Soc.*, **102**, 1738 (1980).

168. L. M. Stephenson, *Tetrahedron Lett.*, 1005 (1980).

169. J. R. Hurst, S. L. Wilson, and G. B. Schuster, *Tetrahedron*, **41**, 2191 (1985).

170. M. Stratakis and M. Orfanopoulos, *Tetrahedron Lett.*, **36**, 4291 (1995).

probably two reasons for this: the *t*-butyl group does not provide any allylic hydrogens and its steric bulk may interfere with approach by $^1O_2$.

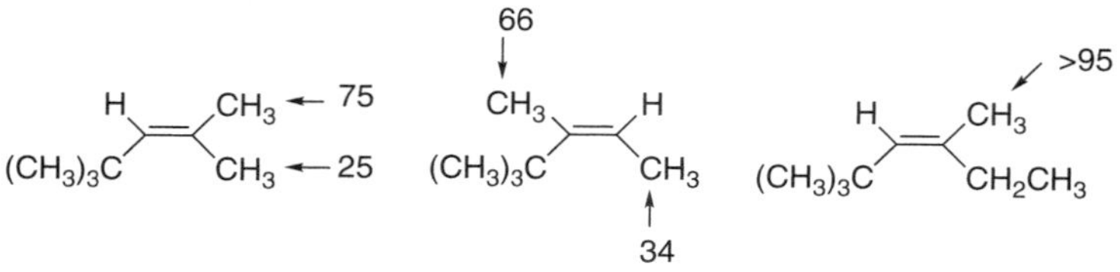

Polar functional groups such as carbonyl, cyano, and sulfoxide, as well as silyl and stannyl groups, exert a strong directing effect, favoring proton removal from the geminal methyl group.[171]

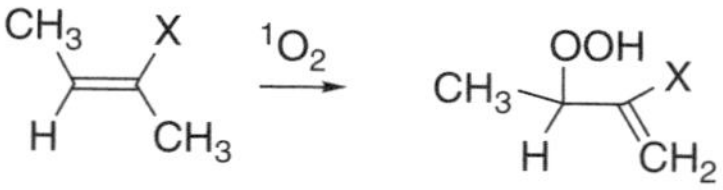

X = CO_2CH_3, CH=O, C≡N,
SOPh, $Si(CH_3)_3$, $Sn(CH_3)_3$

Hydroxy[172] and amino[173] groups favor *syn* stereoselectivity. This is similar to the substituent effects observed for peroxy acids and suggests that the substituents may stabilize the TS by hydrogen bonding.

Recently techniques have been developed for $^1O_2$ oxidations in zeolite cavities.[174] The photosensitizer is absorbed in the zeolite and generation of $^1O_2$ and reaction with the alkene occurs within the cavity. The reactions under these conditions show changes in both regiochemistry[175] and stereoselectivity. The *cis* effect is reduced and there is a preference for hydrogen abstraction from methyl groups.

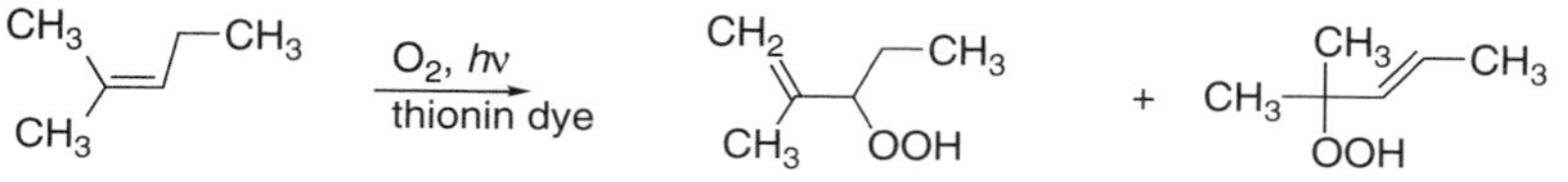

zeolite	100	0
CH_3CN soln	40	60

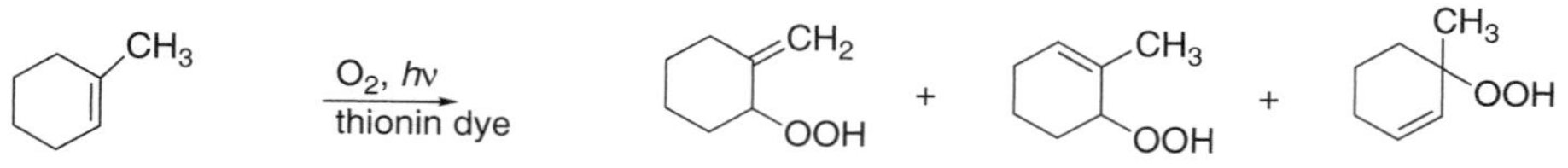

zeolite	88	2	10
CH_3CN soln	40	15	45

171. E. L. Clennan, X. Chen, and J. J. Koola, *J. Am. Chem. Soc.*, **112**, 5193 (1990); M. Orfanopoulos, M. Stratakis, and Y. Elemes, *J. Am. Chem. Soc.*, **112**, 6417 (1990); W. Adam and M. J. Richter, *Tetrahedron Lett.*, **34**, 8423 (1993).

172. W. Adam and B. Nestler, *J. Am. Chem. Soc.*, **114**, 6549 (1992); W. Adam and B. Nestler, *J. Am. Chem. Soc.*, **115**, 5041 (1993); M. Stratakis, M. Orfanopoulos, and C. S. Foote, *Tetrahedron Lett.*, **37**, 7159 (1996).

173. H.-G. Brunker and W. Adam, *J. Am. Chem. Soc.*, **117**, 3976 (1995).

174. X. Li and V. Ramamurthy, *J. Am. Chem. Soc.*, **118**, 10666 (1996).

175. J. Shailaja, J. Sivaguru, R. J. Robbins, V. Ramamurthy, R. B. Sunoj, and J. Chandrasekhar, *Tetrahedron*, **56**, 6927 (2000); E. L. Clennan and J. P. Sram, *Tetrahedron*, **56**, 6945 (2000); M. Stratakis, C. Rabalakos, G. Mpourmpakis, and L. G. Froudakis, *J. Org. Chem.*, **68**, 2839 (2003).

These changes in regio- and stereochemistry are likely due to conformation changes and electrostatic factors within the cavity. The intrazeolite oxidations can be improved by use of fluorocarbon solvents, owing to an enhanced lifetime of $^1O_2$ and to improved occupancy of the cavity by hydrocarbons in this solvent.[176]

The singlet oxidation mechanism has been subject of a comparative study by kinetic isotope effects and computation of the reaction energy surface.[177] The reaction is described as proceeding through the perepoxide structure, but rather than being a distinct intermediate, this structure occurs at a saddle point on the energy surface; that is, there is no barrier to the second stage of the reaction, the hydrogen abstraction. Figure 12.12 is a representation of such a surface and Figure 12.13 shows the computed geometric characteristics for the perepoxides from *Z*-2-butene and 2,3-dimethyl-2-butene. This study also gives a consistent account for the *cis* effect. The perepoxide structure for engagement of the *cis* hydrogens is of lower energy than the corresponding structure involving the *trans* hydrogens. The *cis* transition structure is attained earlier and retains the synchronous character of the TSs from the symmetrical alkenes, as shown in Figure 12.14.

Scheme 12.17 gives some examples of oxidations by singlet oxygen. The reaction in Entry 1 was used to demonstrate that $^1O_2$ can be generated from H_2O_2 and ClO^-. Similarly, the reaction in Entry 2 was used to verify that the phosphite-ozone adducts

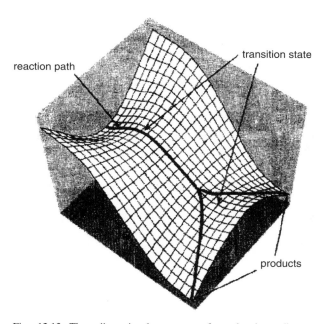

Fig. 12.12. Three-dimensional energy surface showing adjacent transition structures without an intervening intermediate. Reproduced from *J. Am. Chem. Soc.*, **125**, 1319 (2003), by permission of the American Chemical Society.

[176]. A. Pace and E. L. Clennan, *J. Am. Chem. Soc.*, **124**, 11236 (2002).
[177]. D. A. Singleton, C. Hang, M. J. Szymanksi, M. P. Meyer, A. G. Leach, K. T. Kuwata, J. S. Chen, A. Greer, C. S. Foote, and K. N. Houk, *J. Am. Chem. Soc.*, **125**, 1319 (2003).

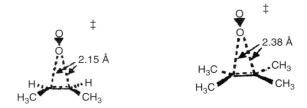

Fig. 12.13. Perepoxide transition structures from *Z*-2-butene and 2,3-dimethyl-2-butene. Reproduced from *J. Am. Chem. Soc.*, **125**, 1319 (2003), by permission of the American Chemical Society.

can serve as a $^1O_2$ source. The reactions in Entries 3 and 4 are representative photo-sensitized procedures with subsequent reduction of the hydroperoxide. Entry 5 used tetra-(perfluorophenyl)phorphyrin as the photosensitizer. This compound, as well as the tetra-(2,6-dichlorophenyl) analog, is reported to have improved stability to degra-dation under the reaction conditions. In this case the intermediate hydroperoxide was dehydrated to an enone using acetic anhydride. This reaction was carried out on a 25-g scale.

Certain compounds react with singlet oxygen in a different manner, giving dioxe-tanes as products.[178]

$$\underset{R}{\overset{R}{>}}{=}\underset{R}{\overset{R}{<}} \ + \ {}^1O_2 \ \longrightarrow \ R\overset{O-O}{\underset{R}{\overline{\quad\quad}}}R$$

This reaction is not usually a major factor with alkenes bearing only alkyl groups, but is important for vinyl ethers and other alkenes with donor substituents. These

Fig. 12.14. Competing *cis* abstraction and *trans* abstraction transition structures for hydroperoxide formation 2-methyl-2-butene. Adapted *J. Am. Chem. Soc.*, **125**, 1319 (2003), by permission of the American Chemical Society.

[178.] W. Fenical, D. R. Kearns, and P. Radlick, *J. Am. Chem. Soc.*, **91**, 3396 (1969); S. Mazur and C. S. Foote, *J. Am. Chem. Soc.*, **92**, 3225 (1970); P. D. Bartlett and A. P. Schaap, *J. Am. Chem. Soc.*, **92**, 3223 (1970).

1[a]

2[b]   −35°C   53%

3[c]   82%

4[d]   63%

5[e]

1) O₂, hν
Ar₄porphyrin
2) Ac₂O, DMAP   70%

Ar = perfluorophenyl

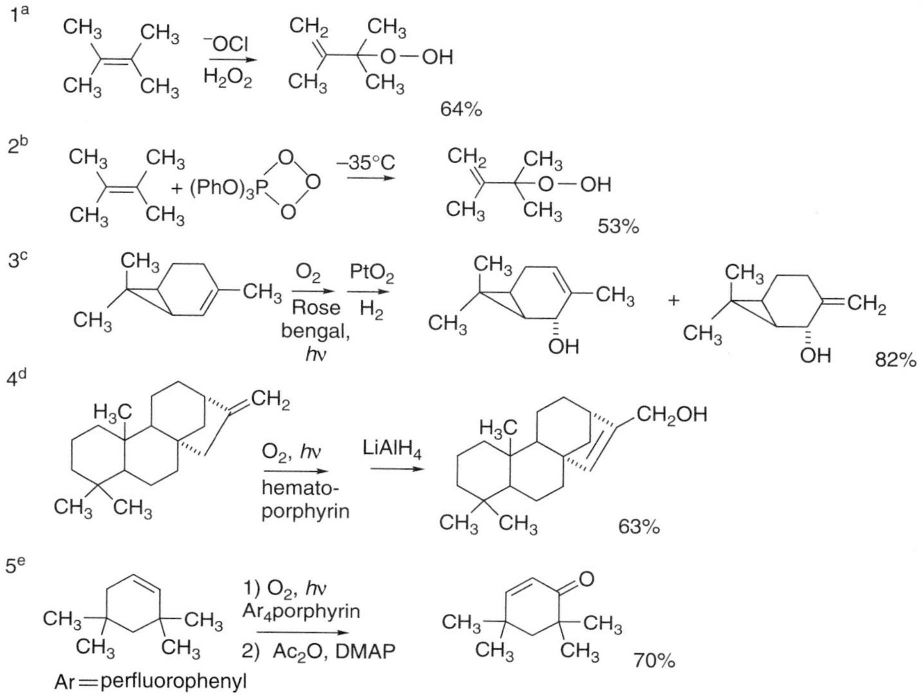

a. C. S. Foote, S. Wexler, W. Ando, and R. Higgins, *J. Am. Chem. Soc.*, **90**, 975 (1968).
b. R. W. Murray and M. L. Kaplan, *J. Am. Chem. Soc.*, **91**, 5358 (1969).
c. K. Gollnick and G. Schade, *Tetrahedron Lett.*, 2335 (1966).
d. R. A. Bell, R. E. Ireland, and L. N. Mander, *J. Org. Chem.*, **31**, 2536 (1966).
e. H. Quast, T. Dietz, and A. Witzel, *Liebigs Ann. Chem.*, 1495 (1995).

reactions are believed to proceed via zwitterionic intermediates that can be diverted by appropriate trapping reagents.[179]

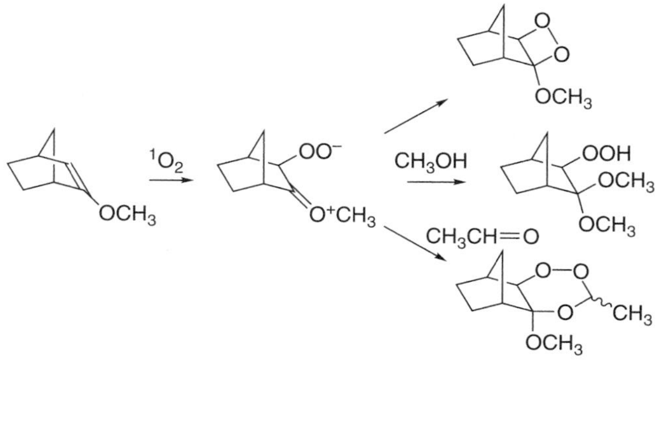

179. C. W. Jefford, S. Kohmoto, J. Boukouvalas, and U. Burger, *J. Am. Chem. Soc.*, **105**, 6498 (1983).

Enaminoketones undergo a clean oxidative cleavage to α-diketones, presumably through a dioxetane intermediate.[180]

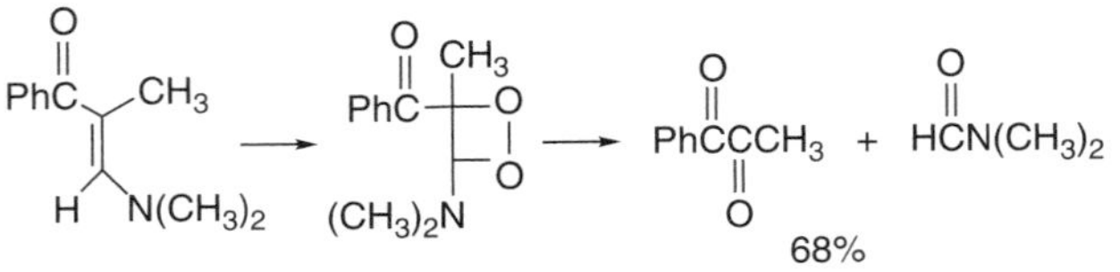

Singlet oxygen undergoes [4 + 2] cycloaddition with dienes.

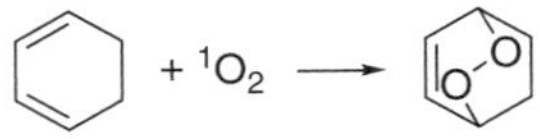

Ref. 181

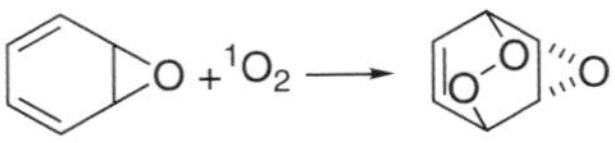

Ref. 182

12.3.3. Other Oxidants

Selenium dioxide is a useful reagent for allylic oxidation of alkenes. The products can include enones, allylic alcohols, or allylic esters, depending on the reaction conditions. The mechanism consists of three essential steps: (a) an electrophilic "ene" reaction with SeO_2, (b) a [2,3]-sigmatropic rearrangement that restores the original location of the double bond, and (c) solvolysis of the resulting selenium ester.[183]

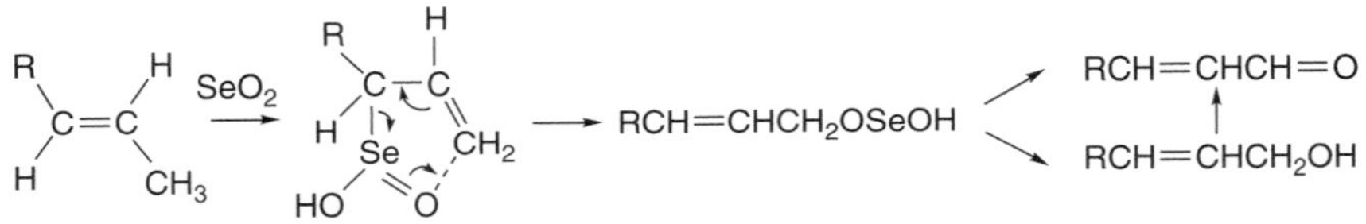

The allylic alcohols that are the initial oxidation products can be further oxidized to carbonyl groups by SeO_2 and the conjugated carbonyl compound is usually isolated. If the alcohol is the desired product, the oxidation can be run in acetic acid, in which case acetate esters are formed.

The mechanism of the reaction has been studied by determining isotope effects for 2-methyl-2-butene and comparing them with predicted values.[184] The isotope effect at the vinyl hydrogen is 0.92 ± 0.01, which is consistent with rehybridization. B3LYP/6-31G* computations located several related TSs with E_a values in the range of 6.0–8.9 kcal/mol. These TSs give calculated isotope effects in good agreement with the experimental values. Although these results are not absolutely definitive, they are consistent with the other evidence for a concerted ene-type mechanism as the first step in SeO_2 oxidation.

180. H. H. Wasserman and J. L. Ives, *J. Am. Chem. Soc.*, **98**, 7868 (1976).
181. C. S. Foote, S. Wexler, W. Ando, and R. Higgins, *J. Am. Chem. Soc.*, **90**, 975 (1968).
182. C. H. Foster and G. A. Berchtold, *J. Am. Chem. Soc.*, **94**, 7939 (1972).
183. K. B. Sharpless and R. F. Lauer, *J. Am. Chem. Soc.*, **94**, 7154 (1972).
184. D. A. Singleton and C. Hang, *J. Org. Chem.*, **65**, 7554 (2000).

Although the traditional conditions for effecting SeO$_2$ oxidations involve use of a stoichiometric or excess amount of SeO$_2$, it is also possible to carry out the reaction with 1.5–2 mol % SeO$_2$, using *t*-butyl hydroperoxide as a stoichiometric oxidant. Under these conditions, the allylic alcohol is the major product and is obtained in good yields, even from alkenes that are poorly reactive under the traditional conditions.[185]

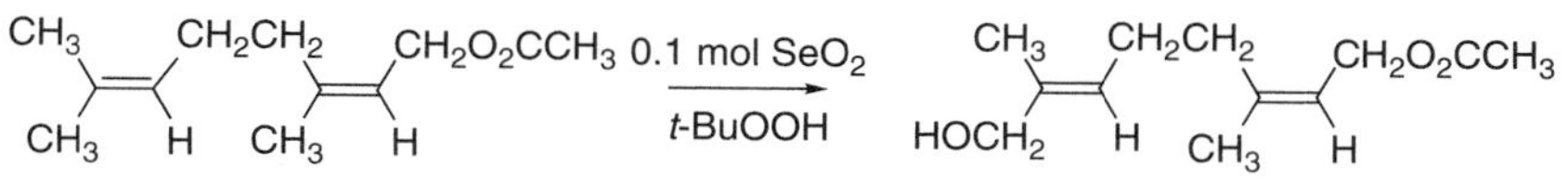

50% + 5% aldehyde

Trisubstituted alkenes are oxidized selectively at the more-substituted end of the carbon-carbon double bond, indicating that the ene reaction step is electrophilic in character.

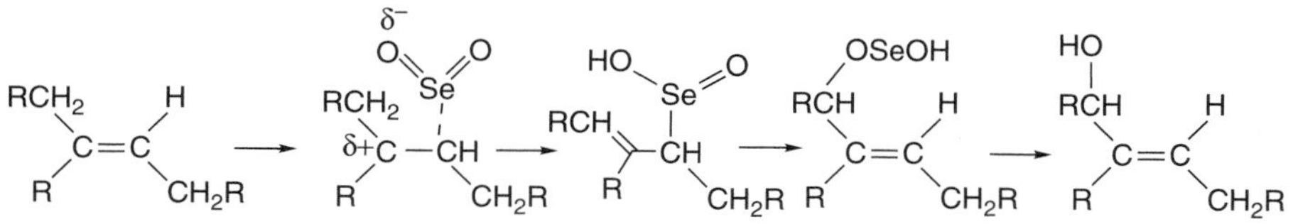

Ref. 186

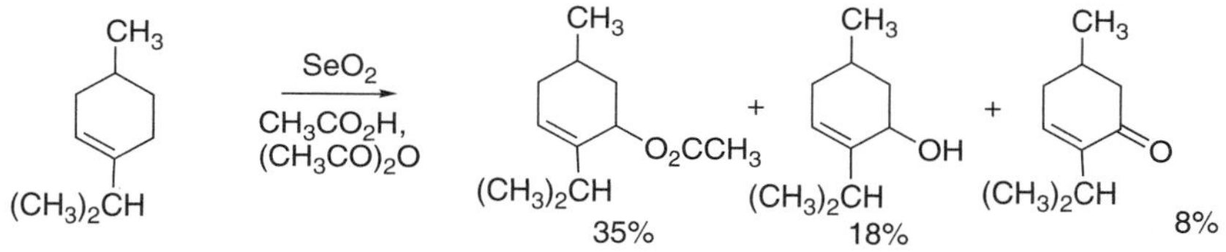

Selenium dioxide reveals a useful stereoselectivity when applied to trisubstituted *gem*-dimethyl alkenes. The products are predominantly the *E*-allylic alcohol or unsaturated aldehyde.[187]

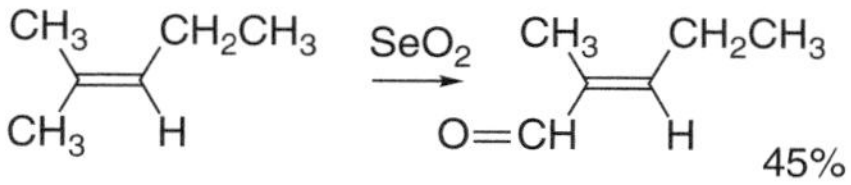

This stereoselectivity can be explained by a five-membered TS for the sigmatropic rearrangement step. The observed *E*-stereochemistry results if the larger alkyl substituent adopts a pseudoequatorial conformation.

185. M. A. Umbreit and K. B. Sharpless, *J. Am. Chem. Soc.*, **99**, 5526 (1977).
186. T. Suga, M. Sugimoto, and T. Matsuura, *Bull. Chem. Soc. Jpn.*, **36**, 1363 (1963).
187. U. T. Bhalerao and H. Rapoport, *J. Am. Chem. Soc.*, **93**, 4835 (1971); G. Buchi and H. Wuest, *Helv. Chim. Acta*, **50**, 2440 (1967).

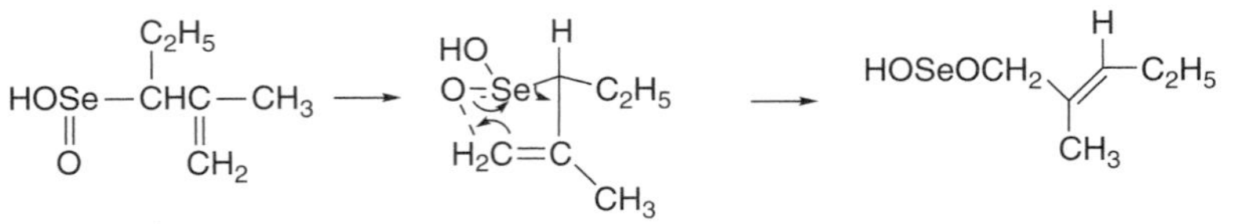

The equivalent to allylic oxidation of alkenes, but with allylic transposition of the carbon-carbon double bond, can be carried out by an indirect oxidative process involving addition of an electrophilic arylselenenyl reagent, followed by oxidative elimination of selenium. In one procedure, addition of an arylselenenyl halide is followed by solvolysis and oxidative elimination.

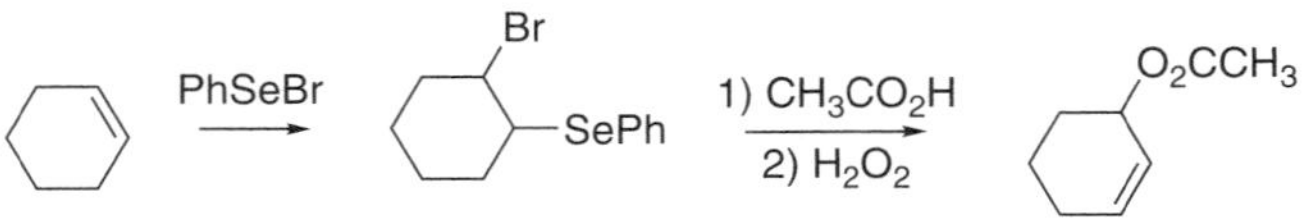

Ref. 188

This reaction depends upon the facile solvolysis of β-haloselenides and the facile oxidative elimination of a selenoxide, which was discussed in Section 6.6.3. An alternative method, which is experimentally simpler, involves reaction of alkenes with a mixture of diphenyl diselenide and phenylseleninic acid.[189] The two selenium reagents generate an electrophilic selenium species, phenylselenenic acid, PhSeOH.

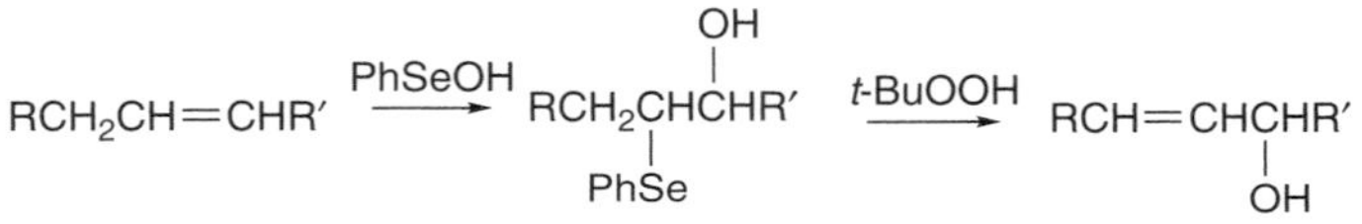

The elimination is promoted by oxidation of the addition product to the selenoxide by *t*-butyl hydroperoxide. The regioselectivity in this reaction is such that the hydroxy group becomes bound at the more-substituted end of the carbon-carbon double bond. The regioselectivity of the addition step follows Markovnikov's rule with PhSe$^+$ acting as the electrophile. The elimination step specifically proceeds away from the oxygen functionality.

12.4. Oxidative Cleavage of Carbon-Carbon Double Bonds

12.4.1. Transition Metal Oxidants

The most selective methods for cleaving organic molecules at carbon-carbon double bonds involve glycols as intermediates. Oxidations of alkenes to glycols was discussed in Section 12.2.1. Cleavage of alkenes can be carried out in one operation under mild conditions by using a solution containing periodate ion and a catalytic

188. K. B. Sharpless and R. F. Lauer, *J. Org. Chem.*, **39**, 429 (1974); D. L. J. Clive, *J. Chem. Soc., Chem. Commun.*, 100 (1974).
189. T. Hori and K. B. Sharpless, *J. Org. Chem.*, **43**, 1689 (1978).

amount of permanganate ion.[190] The permanganate ion effects the hydroxylation and the glycol is then cleaved by reaction with periodate. A cyclic intermediate is believed to be involved in the periodate oxidation. Permanganate is regenerated by the oxidizing action of periodate.

Osmium tetroxide used in combination with sodium periodate can also effect alkene cleavage.[191] Successful oxidative cleavage of double bonds using ruthenium tetroxide and sodium periodate has also been reported.[192] In these procedures the osmium or ruthenium can be used in substoichiometric amounts because the periodate reoxidizes the metal to the tetroxide state. Entries 1 to 4 in Scheme 12.18 are examples of these procedures. Entries 5 and 6 show reactions carried out in the course of multistep syntheses. The reaction in Entry 5 followed a 5-*exo* radical cyclization and served to excise an extraneous carbon. The reaction in Entry 6 followed introduction of the allyl group by enolate alkylation. The aldehyde group in the product was used to introduce an amino group by reductive alkylation (see Section 5.3.1.2).

The strong oxidants Cr(VI) and MnO_4^- can also be used for oxidative cleavage of double bonds, provided there are no other sensitive groups in the molecule. The permanganate oxidation proceeds first to the diols and ketols, as described earlier (see p. 1075), and these are then oxidized to carboxylic acids or ketones. Good yields can be obtained provided care is taken to prevent subsequent oxidative degradation of the products. The oxidation of cyclic alkenes by Cr(VI) reagents can be a useful method for formation of dicarboxylic acids. The initial oxidation step appears to yield an epoxide that undergoes solvolytic ring opening to a glycol or glycol monoester, which is then oxidatively cleaved.[193] Two possible complications that can be encountered are competing allylic attack and skeletal rearrangement. Allylic attack can lead to eventual formation of a dicarboxylic acid that has lost one carbon atom. Pinacol-type rearrangements of the epoxide or glycol intermediates can give rise to rearranged products.

Entries 7 to 9 in Scheme 12.18 are illustrative of these oxidative ring cleavages.

[190] R. U. Lemieux and E. von Rudloff, *Can. J. Chem.*, **33**, 1701, 1710 (1955); E. von Rudloff, *Can. J. Chem.*, **33**, 1714 (1955).

[191] R. Pappo, D. S. Allen, Jr., R. U. Lemieux, and W. S. Johnson, *J. Org. Chem.*, **21**, 478 (1956); H. Vorbrueggen and C. Djerassi, *J. Am. Chem. Soc.*, **84**, 2990 (1962).

[192] W. G. Dauben and L. E. Friedrich, *J. Org. Chem.*, **37**, 241 (1972); B. E. Rossiter, T. Katsuki, and K. B. Sharpless, *J. Am. Chem. Soc.*, **103**, 464 (1981); J. W. Patterson, Jr., and D. V. Krishna Murthy, *J. Org. Chem.*, **48**, 4413 (1983).

[193] J. Rocek and J. C. Drozd, *J. Am. Chem. Soc.*, **92**, 6668 (1970); A. K. Awasthy and J. Rocek, *J. Am. Chem. Soc.*, **91**, 991 (1969).

Scheme 12.18. Oxidative Cleavage of Carbon-Carbon Double Bonds Using Transition Metal Oxidants

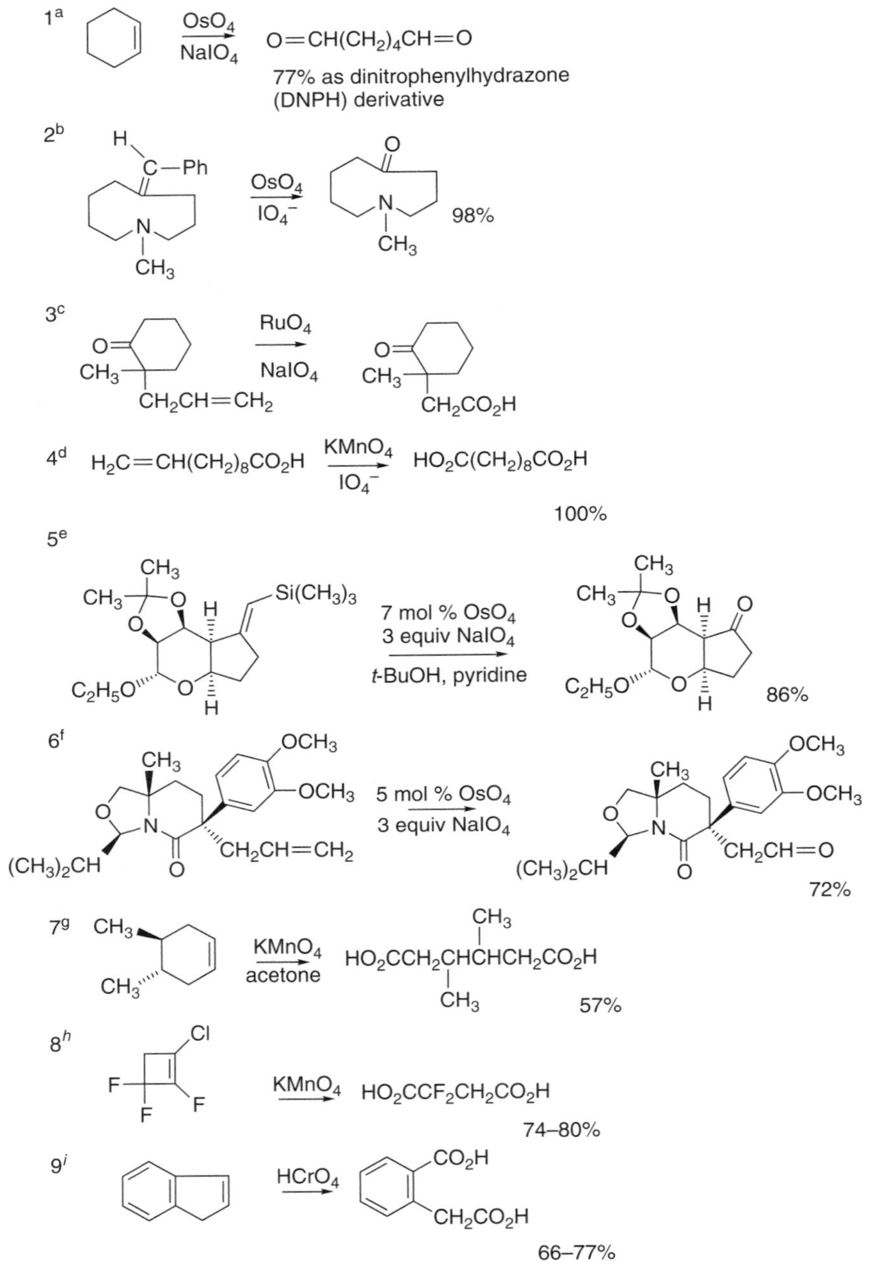

1[a] cyclohexene $\xrightarrow[\text{NaIO}_4]{\text{OsO}_4}$ $O=CH(CH_2)_4CH=O$

77% as dinitrophenylhydrazone (DNPH) derivative

2[b] $\xrightarrow[\text{IO}_4^-]{\text{OsO}_4}$ 98%

3[c] $\xrightarrow[\text{NaIO}_4]{\text{RuO}_4}$

4[d] $H_2C=CH(CH_2)_8CO_2H \xrightarrow[\text{IO}_4^-]{\text{KMnO}_4} HO_2C(CH_2)_8CO_2H$

100%

5[e] $\xrightarrow[\text{$t$-BuOH, pyridine}]{\substack{\text{7 mol \% OsO}_4 \\ \text{3 equiv NaIO}_4}}$ 86%

6[f] $\xrightarrow{\substack{\text{5 mol \% OsO}_4 \\ \text{3 equiv NaIO}_4}}$

72%

7[g] $\xrightarrow[\text{acetone}]{\text{KMnO}_4}$ $HO_2CCH_2CH\overset{\text{CH}_3}{\text{CH}}CHCH_2CO_2H$

57%

8[h] $\xrightarrow{\text{KMnO}_4}$ $HO_2CCF_2CH_2CO_2H$

74–80%

9[i] $\xrightarrow{\text{HCrO}_4}$

66–77%

a. R. U. Lemieux and E. von Rudloff, *Can. J. Chem.*, **33**, 1701 (1955).
b. M. G. Reinecke, L. R. Kray, and R. F. Francis, *J. Org. Chem.*, **37**, 3489 (1972).
c. A. A. Asselin, L. G. Humber, T. A. Dobson, J. Komlossy, and R. R. Martel, *J. Med. Chem.*, **19**, 787 (1976).
d. R. Pappo, D. S. Allen, Jr., R. U. Lemieux, and W. S. Johnson, *J. Org. Chem.*, **21**, 478 (1956).
e. T. Honda, M. Hoshi, K. Kanai, and M. Tsubuki, *J. Chem. Soc., Perkin Trans. 1*, 2091 (1994).
f. A. I. Meyers, R. Hanreich, and K. T. Wanner, *J. Am. Chem. Soc.*, **107**, 7776 (1985).
g. W. C. M. C. Kokke and F. A. Varkvisser, *J. Org. Chem.*, **39**, 1535 (1974).
h. N. S. Raasch and J. E. Castle, *Org. Synth.*, **42**, 44 (1962).
i. O. Grummitt, R. Egan, and A. Buck, *Org. Synth.*, **III**, 449 (1955).

12.4.2. Ozonolysis

The reaction of alkenes with ozone is a general and selective method of cleaving carbon-carbon double bonds.[194] Application of low-temperature spectroscopic techniques has provided information about the rather unstable intermediates in the ozonolysis process. These studies, along with isotopic-labeling results, have provided an understanding of the reaction mechanism.[195] The two key intermediates in ozonolysis are the 1,2,3-trioxolane, or initial ozonide, and the 1,2,4-trioxolane, or ozonide. The first step of the reaction is a 1,3-dipolar cycloaddition to give the 1,2,3-trioxolane. This is followed by a fragmentation and recombination to give the isomeric 1,2,4-trioxolane. Ozone is a very electrophilic 1,3-dipole because of the accumulation of electronegative oxygen atoms in the ozone molecule. The cycloaddition, fragmentation, and recombination are all predicted to be exothermic on the basis of thermochemical considerations.[196]

The products isolated after ozonolysis depend upon the conditions of workup. Simple hydrolysis leads to the carbonyl compounds and hydrogen peroxide, and these can react to give secondary oxidation products. It is usually preferable to include a mild reducing agent that is capable of reducing peroxidic bonds. The current practice is to use dimethyl sulfide, though numerous other reducing agents have been used, including zinc,[197] trivalent phosphorus compounds,[198] and sodium sulfite.[199] If the alcohols resulting from the reduction of the carbonyl cleavage products are desired, the reaction mixture can be reduced with $NaBH_4$.[200] Carboxylic acids are formed in good yields from aldehydes when the ozonolysis reaction mixture is worked up in the presence of excess hydrogen peroxide.[201]

Several procedures that intercept the intermediates have been developed. When ozonolysis is done in alcoholic solvents, the carbonyl oxide fragmentation product can be trapped as an α-hydroperoxy ether.[202] Recombination to the ozonide is then prevented, and the carbonyl compound formed in the fragmentation step can also be

194. P. S. Bailey, *Ozonization in Organic Chemistry*, Vol. 1, Academic Press, New York, 1978.
195. R. P. Lattimer, R. L. Kuckowski, and C. W. Gillies, *J. Am. Chem. Soc.*, **96**, 348 (1974); C. W. Gillies, R. P. Lattimer, and R. L. Kuczkowski, *J. Am. Chem. Soc.*, **96**, 1536 (1974); G. Klopman and C. M. Joiner, *J. Am. Chem. Soc.*, **97**, 5287 (1975); P. S. Bailey and T. M. Ferrell, *J. Am. Chem. Soc.*, **100**, 899 (1978); I. C. Histasune, K. Shinoda, and J. Heicklen, *J. Am. Chem. Soc.*, **101**, 2524 (1979); J.-I. Choe, M. Srinivasan, and R. L. Kuczkowski, *J. Am. Chem. Soc.*, **105**, 4703 (1983). R. L. Kuczkowski, in *1,3-Dipolar Cycloaddition Chemistry*, A. Padwa, ed., Wiley-Interscience, New York, Vol. 2, Chap. 11, 1984; R. L. Kuczkowski, *Chem. Soc. Rev.*, **21**, 79 (1992); C. Geletneky and S. Barger, *Eur. J. Chem.*, 1625 (1998); K. Schank, *Helv. Chim. Acta*, **87**, 2074 (2004).
196. P. S. Nangia and S. W. Benson, *J. Am. Chem. Soc.*, **102**, 3105 (1980).
197. S. M. Church, F. C. Whitmore, and R. V. McGrew, *J. Am. Chem. Soc.*, **56**, 176 (1934).
198. W. S. Knowles and Q. E. Thompson, *J. Org. Chem.*, **25**, 1031 (1960).
199. R. H. Callighan and M. H. Wilt, *J. Org. Chem.*, **26**, 4912 (1961).
200. F. L. Greenwood, *J. Org. Chem.*, **20**, 803 (1955).
201. A. L. Henne and P. Hill, *J. Am. Chem. Soc.*, **65**, 752 (1943).
202. W. P. Keaveney, M. G. Berger, and J. J. Pappas, *J. Org. Chem.*, **32**, 1537 (1967).

Scheme 12.19. Ozonolysis Reactions

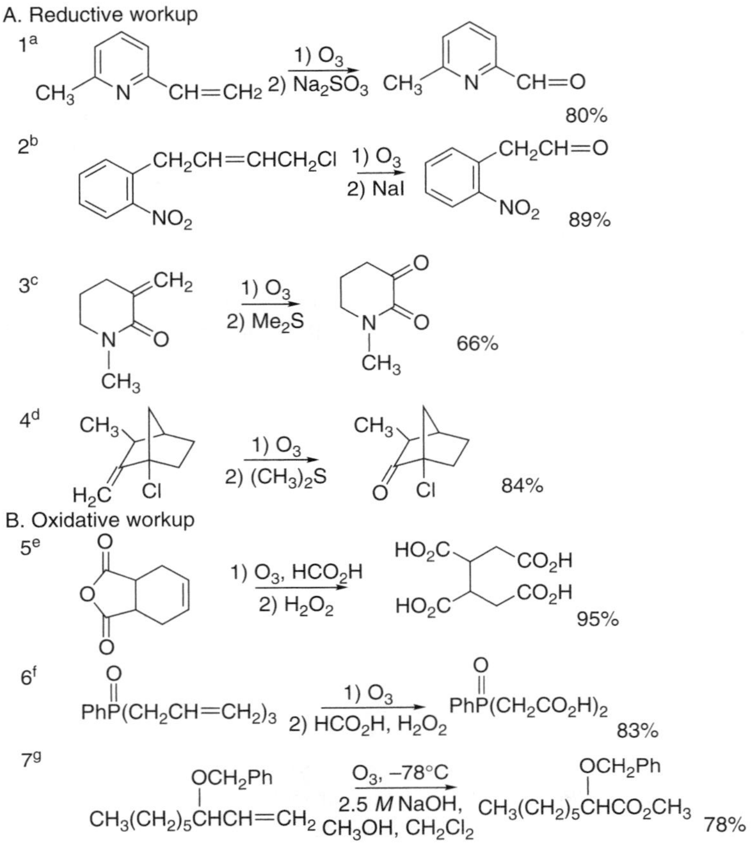

A. Reductive workup

1[a]

80%

2[b]

89%

3[c]

66%

4[d]

84%

B. Oxidative workup

5[e]

95%

6[f]

$$PhP(CH_2CH=CH_2)_3 \xrightarrow[\text{2) } HCO_2H, H_2O_2]{\text{1) } O_3} PhP(CH_2CO_2H)_2$$

83%

7[g]

78%

a. R. H. Callighan and M. H. Wilt, *J. Org. Chem.*, **26**, 4912 (1961).
b. W. E. Noland and J. H. Sellstedt, *J. Org. Chem.*, **31**, 345 (1966).
c. M. L. Rueppel and H. Rapoport, *J. Am. Chem. Soc.*, **94**, 3877 (1972).
d. J. V. Paukstelis and B. W. Macharia, *J. Org. Chem.*, **38**, 646 (1973).
e. J. E. Franz, W. S. Knowles, and C. Ousch, *J. Org. Chem.*, **30**, 4328 (1965).
f. J. L. Eichelberger and J. K. Stille, *J. Org. Chem.*, **36**, 1840 (1971).
g. J. A. Marshall and A. W. Garofalo, *J. Org. Chem.*, **58**, 3675 (1993).

isolated. If the reaction mixture is then treated with dimethyl sulfide, the hydroperoxide is reduced and the second carbonyl compound is also formed in good yield.[203]

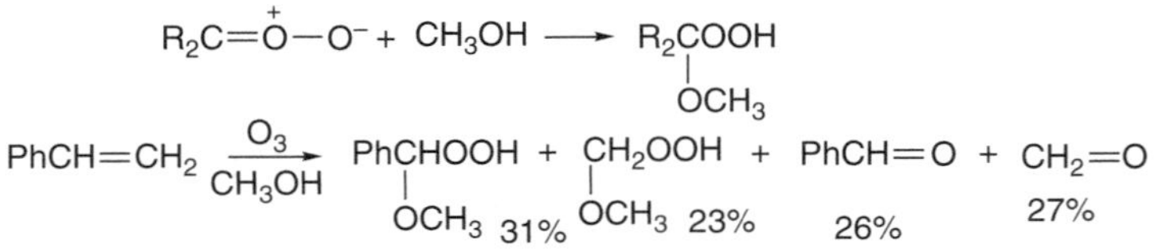

Ozonolysis in the presence of NaOH or NaOCH$_3$ in methanol with CH$_2$Cl$_2$ as a cosolvent leads to formation of esters. This transformation proceeds by trapping both

203. J. J. Pappas, W. P. Keaveney, E. Gancher, and M. Berger, *Tetrahedron Lett.*, 4273 (1966).

the carbonyl oxide and aldehyde products of the fragmentation step.[204] The anionic adducts are then oxidized by O_3.

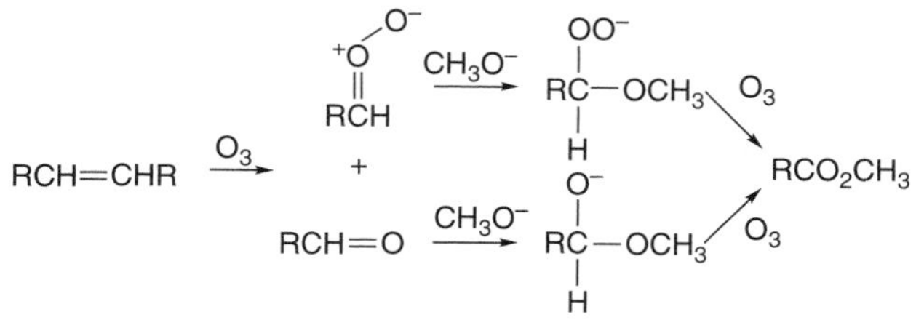

Cyclooctene gives dimethyl octanedioate under these conditions.

Especially reactive carbonyl compounds such as methyl pyruvate can trap the carbonyl oxide component. For example, ozonolysis of cyclooctene in the presence of methyl pyruvate leads to **5**; when treated with triethylamine **5** is converted to **6**, in which the two carbons of the original double bond have been converted to different functionalities.[205]

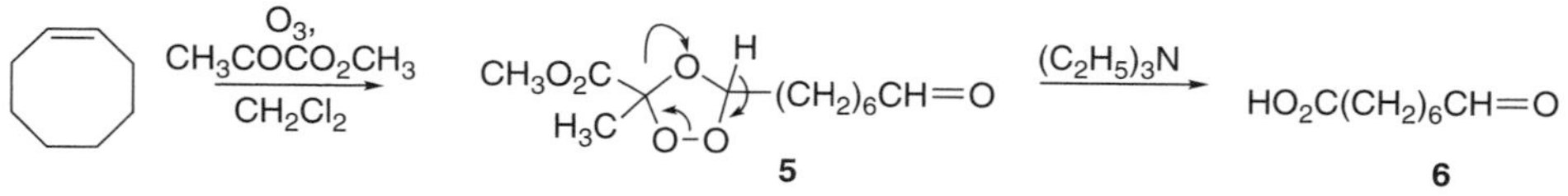

Scheme 12.19 illustrates some cases in which ozonolysis reactions have been used in the course of syntheses. Entries 1 to 4 are examples of use of ozonolysis to introduce carbonyl groups under reductive workup. Entries 5 and 6 involve oxidative workup and give dicarboxylic acid products. The reaction in Entry 7 is an example of direct generation of a methyl ester by methoxide trapping.

12.5. Oxidation of Ketones and Aldehydes

12.5.1. Transition Metal Oxidants

Ketones are oxidatively cleaved by Cr(VI) or Mn(VII) reagents. The reaction is sometimes of utility in the synthesis of difunctional molecules by ring cleavage. The mechanism for both reagents is believed to involve an enol intermediate.[206] A study involving both kinetic data and quantitative product studies has permitted a fairly complete description of the Cr(VI) oxidation of benzyl phenyl ketone.[207] The products include both oxidative-cleavage products and benzil, **7**, which results from oxidation α to the carbonyl. In addition, the dimeric product **8**, which is suggestive of radical intermediates, is formed under some conditions.

[204] J. A. Marshall and A. W. Gordon, *J. Org. Chem.*, **58**, 3675 (1993).

[205] Y.-S. Hon and J.-L. Yan, *Tetrahedron*, **53**, 5217 (1997).

[206] K. B. Wiberg and R. D. Geer, *J. Am. Chem. Soc.*, **87**, 5202 (1965); J. Rocek and A. Riehl, *J. Am. Chem. Soc.*, **89**, 6691 (1967).

[207] K. B. Wiberg, O. Aniline, and A. Gatzke, *J. Org. Chem.*, **37**, 3229 (1972).

$$\text{PhCH}_2\overset{\text{O}}{\underset{\|}{\text{C}}}\text{Ph} \xrightarrow{\text{Cr(VI)}} \text{PhC}\overset{\text{O}}{\underset{\|}{\text{C}}}\text{Ph} + \text{PhCH} + \text{PhCO}_2\text{H} + \text{PhCH}-\text{CHPh}$$

Both the diketone and the cleavage products were shown to arise from an α-hydroxyketone intermediate (benzoin) **9**.

$$\text{PhCH}_2\overset{\text{O}}{\underset{\|}{\text{C}}}\text{Ph} \rightleftharpoons \text{PhCH}=\overset{\text{OH}}{\underset{|}{\text{C}}}\text{Ph} \xrightarrow[\text{H}_2\text{O}]{\text{H}_2\text{CrO}_4} \text{Ph}-\text{CH}=\text{CH}-\text{Ph} \longrightarrow \text{PhCH}-\overset{\text{O}}{\underset{\|}{\text{C}}}\text{Ph} + \text{Cr(IV)} \longrightarrow \text{products}$$

The coupling product is considered to involve a radical intermediate formed by one-electron oxidation, probably effected by Cr(IV). Similarly, the oxidation of cyclohexanone involves 2-hydroxycyclohexanone and 1,2-cyclohexanedione as intermediates.[208]

Owing to the efficient oxidation of alcohols to ketones, alcohols can be used as the starting materials in oxidative cleavages. The conditions required are more vigorous than for the alcohol to ketone transformation (see Section 12.1.1).

Aldehydes can be oxidized to carboxylic acids by both Mn(VII) and Cr(VI). Fairly detailed mechanistic studies have been carried out for Cr(VI). A chromate ester of the aldehyde hydrate is believed to be formed, and this species decomposes in the rate-determining step by a mechanism similar to the one that operates in alcohol oxidations.[209]

$$\text{RCH}=\text{O} + \text{H}_2\text{Cr}^{\text{(VI)}}\text{O}_4 \rightleftharpoons \text{RC}\overset{\text{OH}}{\underset{\text{H}}{|}}-\text{O}-\text{CrO}_3\text{H} \longrightarrow \text{RCO}_2\text{H} + [\text{Cr}^{\text{(IV)}}\text{O}_3\text{H}]^- + \text{H}^+$$

Effective conditions for oxidation of aldehydes to carboxylic acids with $KMnO_4$ involve use of *t*-butanol and an aqueous NaH_2PO_4 buffer as the reaction medium.[210] Buffered sodium chlorite is also a convenient oxidant.[211] Both $KMnO_4$ and $NaClO_2$ can be used in the form of solid-supported materials, using silica and ion exchange resins, respectively,[212] which permits facile workup of the product. Silver oxide is one of the older reagents used for carrying out the aldehyde to carboxylic acid oxidation.

208. J. Rocek and A. Riehl, *J. Org. Chem.*, **32**, 3569 (1967).
209. K. B. Wiberg, *Oxidation in Organic Chemistry*, Part A, Academic Press, New York, 1965, pp. 172–178.
210. A. Abiko, J. C. Roberts, T. Takemasa, and S. Masamune, *Tetrahedron Lett.*, **27**, 4537 (1986).
211. E. Dalcanale and F. Montanari, *J. Org. Chem.*, **51**, 567 (1986); J. P. Bayle, F. Perez, and J. Cortieu, *Bull. Soc. Chim. Fr.*, 565 (1996); E. J. Corey and G. A. Reichard, *Tetrahedron Lett.*, **34**, 6973 (1993); P. M. Wovkulich, K. Shankaran, J. Kiegiel, and M. R. Uskokovic, *J. Org. Chem.*, **58**, 832 (1993); B. R. Babu and K. K. Balasubramaniam, *Org. Prep. Proc. Int.*, **26**, 123 (1994).
212. T. Takemoto, K. Yasuda, and S. V. Ley, *Synlett*, 1555 (2001).

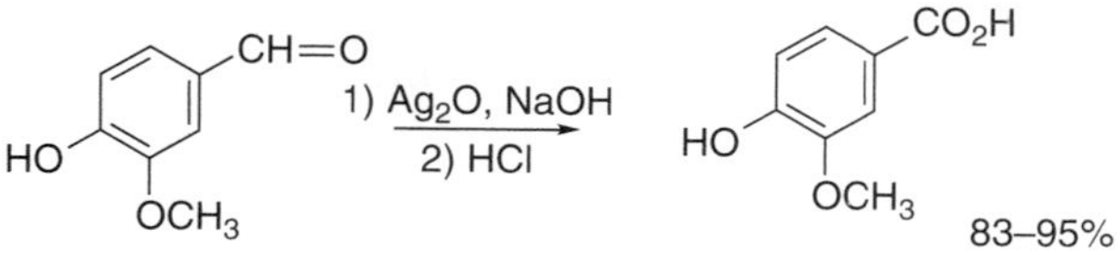

The reaction of aldehydes with MnO_2 in the presence of cyanide ion in an alcoholic solvent is a convenient method of converting aldehydes directly to esters.[214] This reaction involves the cyanohydrin as an intermediate. The initial oxidation product is an acyl cyanide, which is solvolyzed under these reaction conditions.

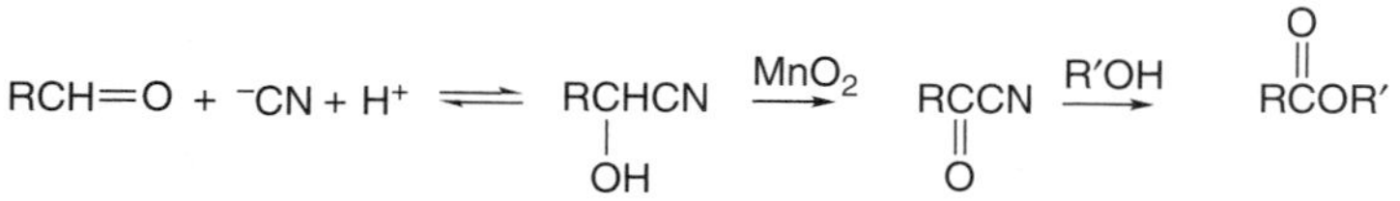

Lead tetraacetate can effect oxidation of carbonyl groups, leading to formation of α-acetoxy ketones,[215] but the yields are seldom high. Boron trifluoride can be used to catalyze these oxidations. It is presumed to function by catalyzing the formation of the enol, which is thought to be the reactive species.[216] With unsymmetrical ketones, products from oxidation at both α-methylene groups are found.[217]

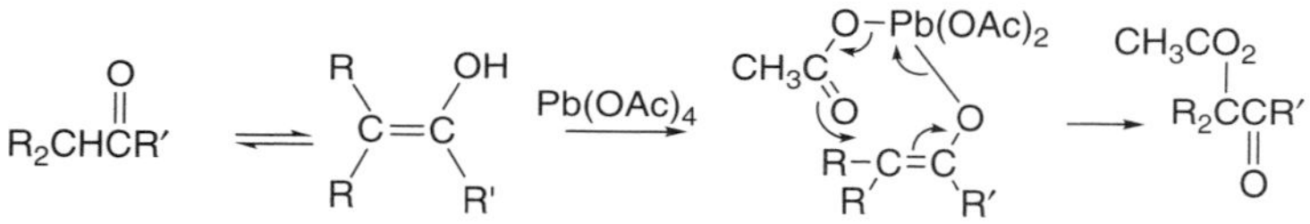

With enol ethers, $Pb(OCCH_3)_4$ gives α-methoxyketones.[218]

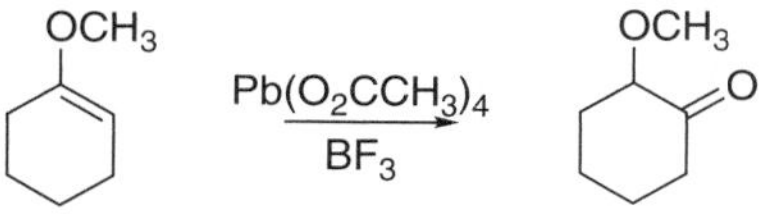

Introduction of oxygen α to a ketone function can also be carried out via the silyl enol ether. Lead tetraacetate gives the α-acetoxy ketone.[219]

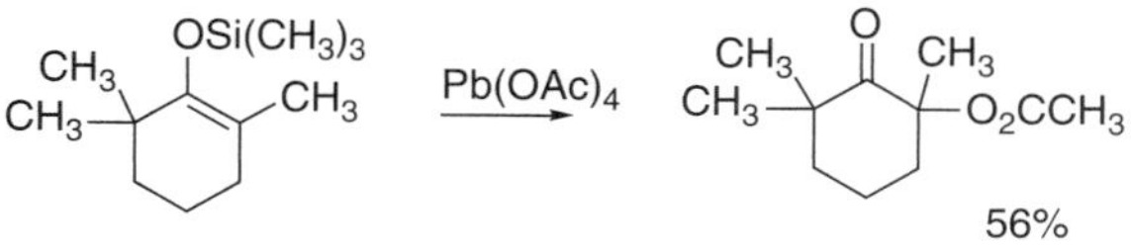

56%

213. I. A. Pearl, *Org. Synth.*, **IV**, 972 (1963).

214. E. J. Corey, N. W. Gilman, and B. E. Ganem, *J. Am. Chem. Soc.*, **90**, 5616 (1968).

215. R. Criegee, in *Oxidation in Organic Chemistry*, Part A, K. B. Wiberg, ed., Academic Press, New York, 1965, pp. 305–312.

216. J. D. Cocker, H. B. Henbest, G. H. Philipps, G. P. Slater, and D. A. Thomas, *J. Chem. Soc.*, 6 (1965).

217. S. Moon and H. Bohm, *J. Org. Chem.*, **37**, 4338 (1972).

218. V. S. Singh, C. Singh, and D. K. Dikshit, *Synth. Commun.*, **28**, 45 (1998).

219. G. M. Rubottom, J. M. Gruber, and K. Kincaid, *Synth. Commun.*, **6**, 59 (1976); G. M. Rubottom and J. M. Gruber, *J. Org. Chem.*, **42**, 1051 (1977); G. M. Rubottom and H. D. Juve, Jr., *J. Org. Chem.*, **48**, 422 (1983).

α-Hydroxyketones can be obtained from silyl enol ethers by oxidation using a catalytic amount of OsO_4 with an amine oxide serving as the stoichiometric oxidant.[220]

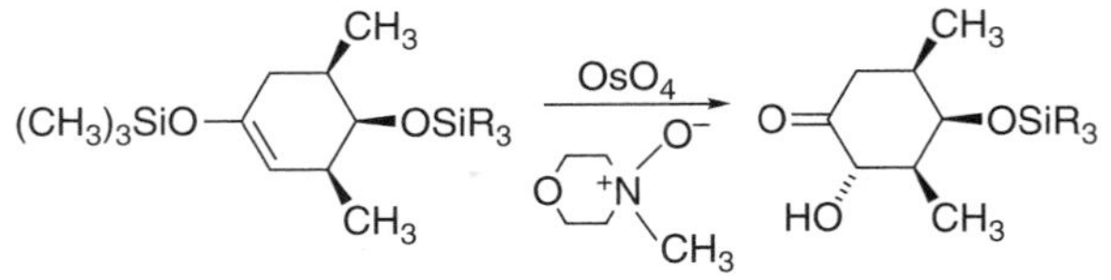

Ref. 221

Other procedures for α-oxidation of ketones are based on prior generation of the enolate. Among the reagents used is a molybdenum compound, MoO_5-pyridine-HMPA, which is prepared by dissolving MoO_3 in hydrogen peroxide, followed by addition of HMPA. This reagent oxidizes the enolates of aldehydes, ketones, esters, and lactones to the corresponding α-hydroxy compound.[222]

Ref. 223

12.5.2. Oxidation of Ketones and Aldehydes by Oxygen and Peroxidic Compounds

12.5.2.1. Baeyer-Villiger Oxidation of Ketones. In the presence of acid catalysts, peroxy compounds are capable of oxidizing ketones by insertion of an oxygen atom into one of the carbon-carbon bonds at the carbonyl group. Known as the *Baeyer-Villiger oxidation*,[224] the mechanism involves a sequence of steps that begins with addition to the carbonyl group, followed by peroxide bond cleavage with migration to oxygen.

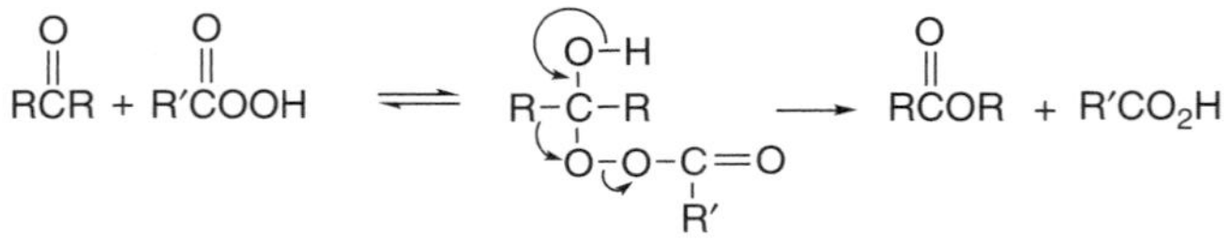

220. J. P. McCormick, W. Tomasik, and M. W. Johnson, *Tetrahedron Lett.*, **22**, 607 (1981).
221. R. K. Boeckman, Jr., J. E. Starrett, Jr., D. G. Nickell, and P.-E. Sun, *J. Am. Chem. Soc.*, **108**, 5549 (1986).
222. E. Vedejs, *J. Am. Chem. Soc.*, **96**, 5945 (1974); E. Vedejs, D. A. Engler, and J. E. Telschow, *J. Org. Chem.*, **43**, 188 (1978); E. Vedejs and S. Larsen, *Org. Synth.*, **64**, 127 (1985).
223. S. P. Tanis and K. Nakanishi, *J. Am. Chem. Soc.*, **101**, 4398 (1979).
224. C. H. Hassall, *Org. React.*, **9**, 73 (1957); G. R. Krow, *Org. React.*, **43**, 252 (1993); M. Renz and B. Beunier, *Eur. J. Org. Chem.*, 737 (1999); G.-J. ten Brink, I. W. C. E. Arends, and R. A. Sheldon, *Chem. Rev.*, **104**, 4105 (2004).

The concerted O—O heterolysis-migration is usually the rate-determining step.[225] The reaction is catalyzed by protic and Lewis acids,[226] including $Sc(O_3SCF_3)_3$[227] and $Bi(O_3SCF_3)_3$.[228]

When the reaction involves an unsymmetrical ketone, the structure of the product depends on which group migrates. A number of studies have been directed at ascertaining the basis of migratory preference in the Baeyer-Villiger oxidation, and a general order of likelihood of migration has been established: *tert*-alkyl, *sec*-alkyl>benzyl, phenyl>*pri*-alkyl>cyclopropyl>methyl.[229] Thus, methyl ketones uniformly give acetate esters resulting from migration of the larger group.[230] A major factor in determining which group migrates is the ability to accommodate partial positive charge. In *para*-substituted phenyl groups, ERG substituents favor migration.[231] Similarly, silyl substituents enhance migratory aptitude of alkyl groups.[232] As is generally true of migration to an electron-deficient center, the configuration of the migrating group is retained in Baeyer-Villiger oxidations.

Steric and conformational factors are also important, especially in cyclic systems.[233] There is a preference for the migration of the group that is antiperiplanar with respect to the peroxide bond. In relatively rigid systems, this effect can outweigh the normal preference for the migration of the more branched group.[234]

This stereoelectronic effect also explains the contrasting regioselectivity of *cis*- and *trans*-2-fluoro-4-*t*-butylcyclohexanone.[235] As a result of a balance between its polar effect and hyperconjugation, the net effect of a fluoro substituent in acyclic systems is small. However, in 2-fluorocyclohexanones an unfavorable dipole-dipole interaction comes into play for the *cis* isomer and preferential migration of the fluoro-substituted carbon is observed.

225. Y. Ogata and Y. Sawaki, *J. Org. Chem.*, **37**, 2953 (1972).

226. G. Stukul, *Angew. Chem. Intl. Ed. Engl.*, **37**, 1199 (1998).

227. H. Kotsuki, K. Arimura, T. Araki, and T. Shinohara, *Synlett*, 462 (1999).

228. M. M. Alam, R. Varala, and S. R. Adapa, *Synth. Commun.*, **33**, 3035 (2003).

229. H. O. House, *Modern Synthetic Reactions*, 2nd Edition, W. A. Benjamin, Menlo Park, CA, 1972, p. 325.

230. P. A. S. Smith, in *Molecular Rearrangements*, P. de Mayo, ed., Interscience, New York, 1963, pp. 457–591.

231. W. E. Doering and L. Speers, *J. Am. Chem. Soc.*, **72**, 5515 (1950).

232. P. F. Hudrlik, A. M. Hudrlik, G. Nagendrappa, T. Yimenu, E. T. Zellers, and E. Chin, *J. Am. Chem. Soc.*, **102**, 6894 (1980).

233. M. F. Hawthorne, W. D. Emmons, and K. S. McCallum, *J. Am. Chem. Soc.*, **80**, 6393 (1958); J. Meinwald and E. Frauenglass, *J. Am. Chem. Soc.*, **82**, 5235 (1960); P. M. Goodman and Y. Kishi, *J. Am. Chem. Soc.*, **120**, 9392 (1998).

234. S. Chandrasekhar and C. D. Roy, *J. Chem. Soc., Perkin Trans. 2*, 2141 (1994).

235. C. M. Crudden, A. C. Chen, and L. A. Calhoun, *Angew. Chem. Int. Ed. Engl.*, **39**, 2852 (2000).

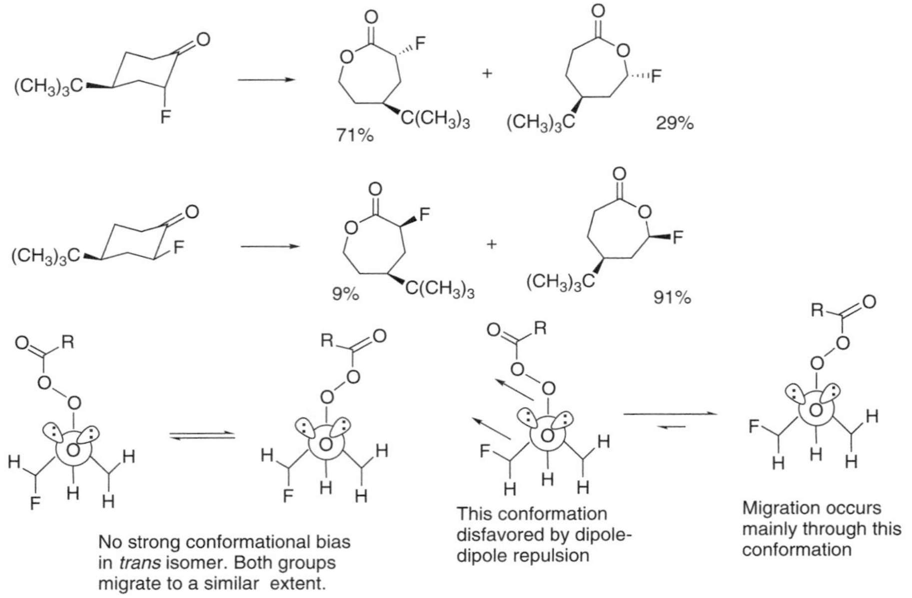
(CH3)3C
F
O
71%
C(CH3)3
+
(CH3)3C
29%
9%
91%
No strong conformational bias in *trans* isomer. Both groups migrate to a similar extent.
This conformation disfavored by dipole-dipole repulsion
Migration occurs mainly through this conformation

In 2-(trifluoromethyl)cyclohexanone, the methylene group migrates in preference to the trifluoromethylmethine group,[236] owing primarily to the EWG effect of the trifluoromethyl group. The computational energy profile, shown in Figure 12.15, indicates that the reaction proceeds through a minor conformation of the adduct in which the trifluoromethyl group is axial. The same regioselectivity is computed for the adduct having the peroxy substituent in an equatorial position, but this adduct is about 1 kcal/mol higher in energy.

The Baeyer-Villiger reaction has found considerable application in the synthesis of prostaglandins. One common pattern involves the use of bicyclo[2.2.1]heptan-2-one derivatives, which are generally obtained by Diels-Alder reactions. For example, compound **10** is known as the *Corey lactone* and has played a prominent role in the synthesis of prostaglandins.[237] This compound was originally prepared by a Baeyer-Villiger oxidation of 7-(methoxymethyl)bicyclo[2.2.1]hept-5-en-2-one.[238]

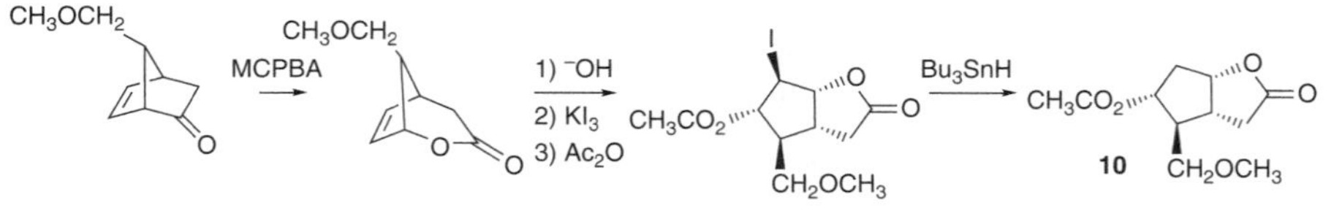
CH3OCH2
MCPBA
1) -OH
2) KI3
3) Ac2O
CH3CO2
CH2OCH3
Bu3SnH
10

[236]. Y. Itoh, M. Yamanaka, and K. Mikami, *Org. Lett.*, **5**, 4803 (2003).

[237]. R. Bansal, G. F. Cooper, and E. J. Corey, *J. Org Chem.*, **56**, 1329 (1991).

[238]. E. J. Corey, N. M. Weinshenker, T. K. Schaaf, and W. Huber, *J. Am. Chem. Soc.*, **91**, 5675 (1969).

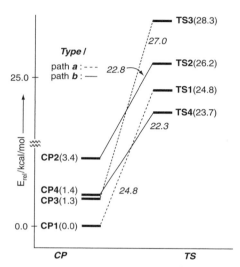

CP1, TS1: $R^1 = CF_3$, $R^2 = H$, *a*
CP2, TS2: $R^1 = CF_3$, $R^2 = H$, *b*
CP3, TS3: $R^1 = H$, $R^2 = CF_3$, *a*
CP4, TS4: $R^1 = H$, $R^2 = CF_3$, *b*

Fig. 12.15. Computational comparison of reactants (adducts) and transition structures for Baeyer-Villiger oxidation of 2-(trifluoromethyl)cyclohexanone by peroxytrifluoroacetic acid. Reproduced from *Org. Lett.*, **5**, 4803 (2003), by permission of the American Chemical Society.

This intermediate has the oxygenation and pattern and *trans*-disubstitution pattern found in the prostaglandins. Several syntheses of similar intermediates have been developed.[239]

In the synthesis of Travoprost, an antiglaucoma agent, a bicyclo[2.2.1]heptan-2-one is converted to a lactone.[240] The commercial process uses peroxyacetic acid as the oxidant and gives a 40% yield. The regioselectivity in this case is only 3:1 but the unwanted isomer can be removed by selective hydrolysis.

239. I. Vesely, V. Kozmik, V. Dedek, J. Palecek, J. Mostecky, and I. Stibor, *Coll. Czech. Chem. Commun.*, **54**, 1683 (1989); J. S. Bindra, A. Grodski, T. K. Schaaf, and E. J. Corey, *J. Am. Chem. Soc.*, **95**, 7522 (1973).
240. L. T. Boulton, D. Brick, M. E. Fox, M. Jackson, I. C. Lennon, R. McCague, N. Parkin, D. Rhodes, and G. Ruecroft, *Org. Proc. Res. Dev.*, **6**, 128 (2002).

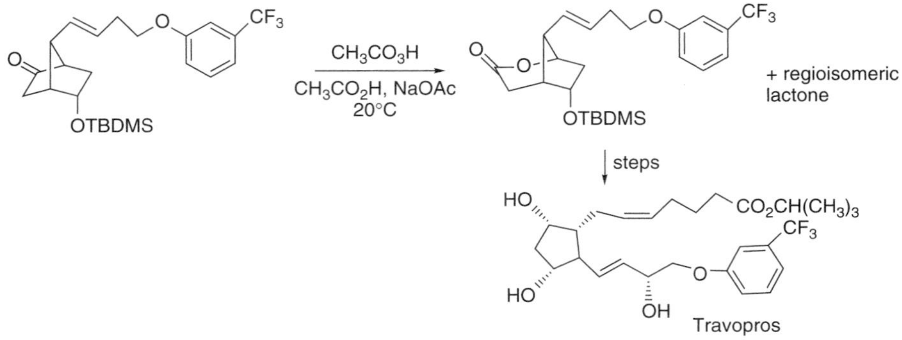

A series of 2-vinyl-3-silyloxybicyclo[3.2.0]heptan-6-ones has also been converted to prostanoid lactones in excellent yield but variable regioselectivity. Some of the best regioselectivity was obtained using H_2O_2 in trifluoroethanol (see p. 1097).[241] The strained cyclobutanone ring and the relatively unreactive terminal vinyl group favor the desired reaction in preference to alkene epoxidation.

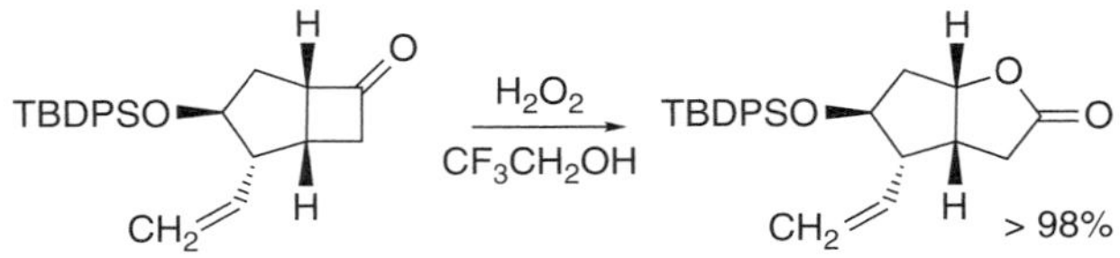

Some typical examples of Baeyer-Villiger oxidations are shown in Scheme 12.20. Entry 1 uses peroxysulfuric acid, the original reagent discovered by Baeyer and Villiger. Entries 2 and 3 generate lactones in good yield from cyclic ketones using peroxyacetic acid. Entry 3 also illustrates the preference for the migration of the more branched group. Entry 4 is a case of formation of an acetate ester from a methyl ketone. Entry 5 illustrates the use of magnesium monoperoxyphthalate and also shows the normal preference for migration of the more branched group. The reaction in Entry 6 exhibits very high regioselectivity. Although this example is consistent with the generalization that the more branched group will migrate, there may be other factors associated with ring geometry that lead to the complete regioselectivity. Entries 7 and 8 use peroxytrifluoroacetic acid and again illustrate the conversion of methyl ketones to acetate esters.

12.5.2.2. Oxidation of Enolates and Enolate Equivalents.   Although ketones are essentially inert to molecular oxygen, enolate anions are susceptible to oxidation. The combination of oxygen and a strong base has found some utility in the introduction of an oxygen function at carbanionic sites.[242] Hydroperoxides are the initial products of such oxidations, but when DMSO or some other substance capable of reducing the hydroperoxide is present, the corresponding alcohol is isolated. A procedure that has met with

[241.] D. Depre, L.-Y. Chen, and L. Ghosez, *Tetrahedron*, **59**, 6797 (2003).
[242.] J. N. Gardner, T. L. Popper, F. E. Carlon, O. Gnoj, and H. L. Herzog, *J. Org. Chem.*, **33**, 3695 (1968).

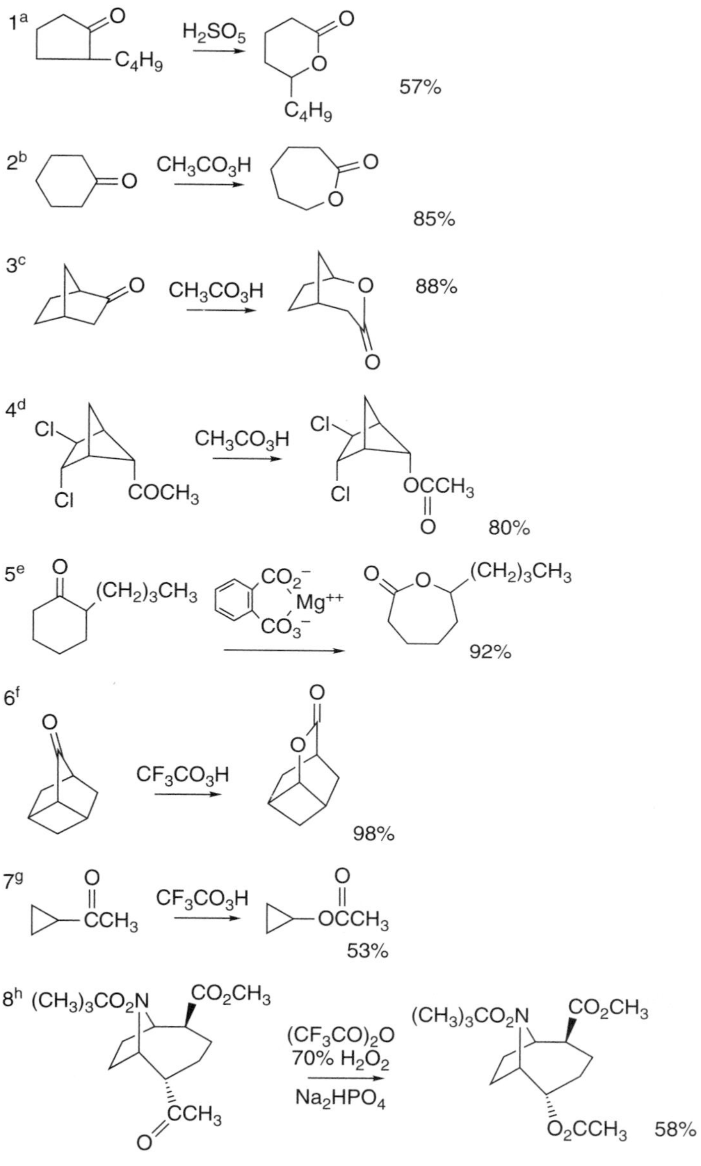

Scheme 12.20. Baeyer-Villiger Oxidation

1[a]
57%

2[b]
85%

3[c]
88%

4[d]
80%

5[e]
92%

6[f]
98%

7[g]
53%

8[h]
58%

a. T. H. Parliament, M. W. Parliament, and J. S. Fagerson, *Chem. Ind.*, 1845 (1966).
b. P. S. Strarcher and B. Phillips, *J. Am. Chem. Soc.*, **80**, 4079 (1958).
c. J. Meinwald and E. Frauenglass, *J. Am. Chem. Soc.*, **82**, 5235 (1960).
d. K. B. Wiberg and R. W. Ubersax, *J. Org. Chem.*, **37**, 3827 (1972).
e. M. Hirano, S. Yakabe, A. Satoh, J. H. Clark, and T. Morimoto, *Synth. Commun.*, **26**, 4591 (1996); T. Mino, S. Masuda, M. Nishio, and M. Yamashita, *J. Org. Chem.*, **62**, 2633 (1997).
f. S. A. Monti and S.-S. Yuan, *J. Org. Chem.*, **36**, 3350 (1971).
g. W. D. Emmons and G. B. Lucas, *J. Am. Chem. Soc.*, **77**, 2287 (1955).
h. F. J. Sardina, M. H. Howard, M. Morningstar, and H. Rapoport, *J. Org. Chem.*, **55**, 5025 (1990).

considerable success involves oxidation in the presence of a trialkyl phosphite.[243] The intermediate hydroperoxide is efficiently reduced by the phosphite ester.

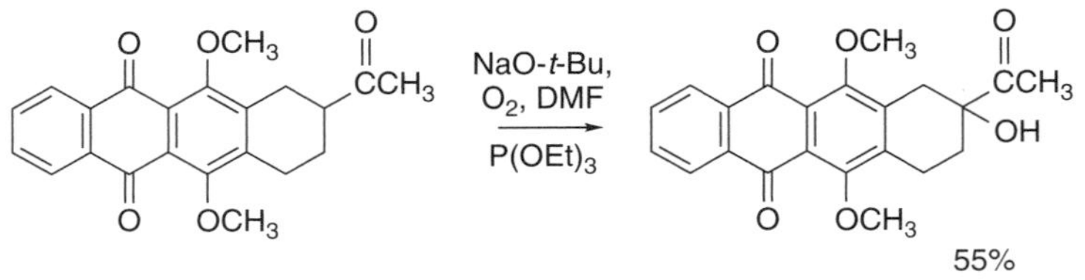

55%   Ref. 144

This oxidative process has been successful with ketones,[244] esters,[245] and lactones.[246] Hydrogen peroxide can also be used as the oxidant, in which case the alcohol is formed directly.[247] The mechanisms for the oxidation of enolates by oxygen is a radical chain autoxidation in which the propagation step involves electron transfer from the carbanion to a hydroperoxy radical.[248]

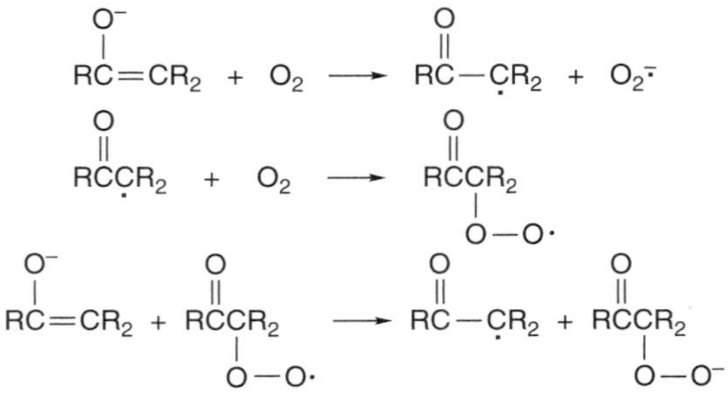

Arguments for a nonchain reaction between the enolate and oxygen to give the hydroperoxide anion directly have been advanced as well.[249]

The silyl enol ethers of ketones are also oxidized to α-hydroxy ketones by *m*-chloroperoxybenzoic acid. If the reaction workup includes acylation, α-acyloxy ketones are obtained.[250] These reactions proceed by initial epoxidation of the silyl enol ether, which then undergoes ring opening. Subsequent transfer of either the *O*-acyl or *O*-TMS substituent occurs, depending on the reaction conditions.

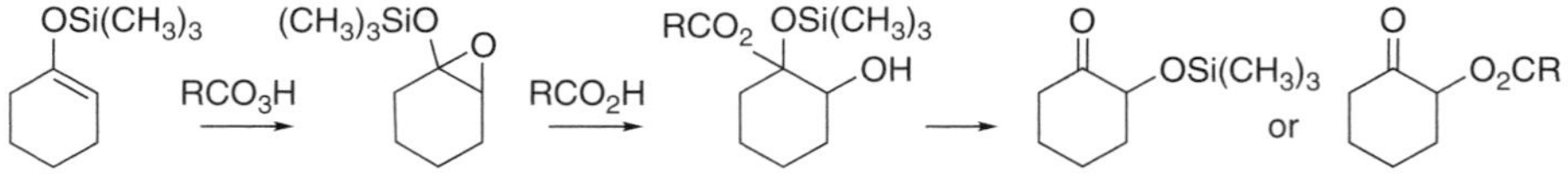

243. J. N. Gardner, F. E. Carlon, and O. Gnoj, *J. Org. Chem.*, **33**, 3294 (1968).

244. F. A. J. Kerdesky, R. J. Ardecky, M. V. Lashmikanthan, and M. P. Cava, *J. Am. Chem. Soc.*, **103**, 1992 (1981).

245. E. J. Corey and H. E. Ensley, *J. Am. Chem. Soc.*, **97**, 6908 (1975).

246. J. J. Plattner, R. D. Gless, and H. Rapoport, *J. Am. Chem. Soc.*, **94**, 8613 (1972); R. Volkmann, S. Danishefsky, J. Eggler, and D. M. Solomon, *J. Am. Chem. Soc.*, **93**, 5576 (1971).

247. G. Buchi, K. E. Matsumoto, and H. Nishimura, *J. Am. Chem. Soc.*, **93**, 3299 (1971).

248. G. A. Russell and A. G. Bemix, *J. Am. Chem. Soc.*, **88**, 5491 (1966).

249. H. R. Gersmann and A. F. Bickel, *J. Chem. Soc. B*, 2230 (1971).

250. G. M. Rubottom, J. M. Gruber, R. K. Boeckman, Jr., M. Ramaiah, and J. B. Medwick, *Tetrahedron Lett.*, 4603 (1978); G. M. Rubottom and J. M. Gruber, *J. Org. Chem.*, **43**, 1599 (1978); G. M. Rubottom, M. A. Vazquez, and D. R. Pelegrina, *Tetrahedron Lett.*, 4319 (1974).

N-Sulfonyloxaziridines are useful reagents for oxidation of enolates to α-hydroxyketones.[251] The best results are frequently achieved by using KHMDS to form the enolate. The hydroxylation occurs preferentially from the less hindered enolate face.

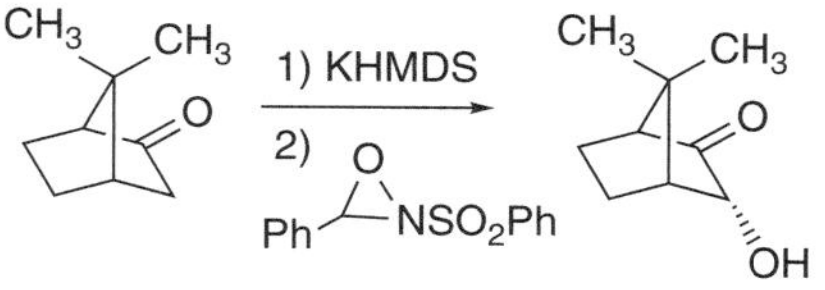

The mechanism of oxygen transfer is believed to involve nucleophilic opening of the oxaziridine, followed by collapse of the resulting *N*-sulfonylcarbinolamine.[252]

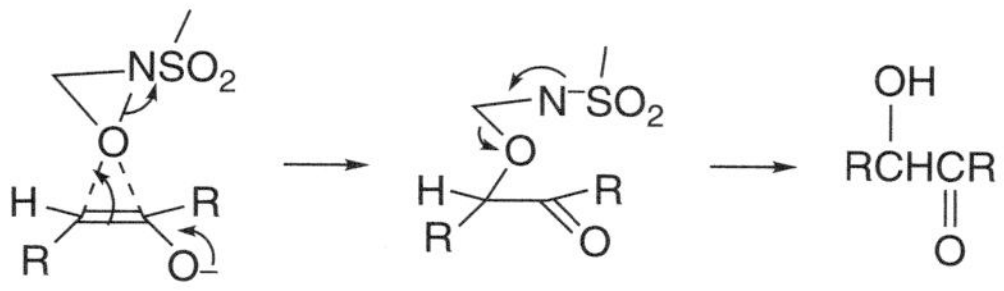

These reagents exhibit good stereoselectivity toward chiral reactants, such as acyloxazolidinones.[253] Chiral oxaziridine reagents have been developed that can achieve enantioselective oxidation of enolates to α-hydroxyketones.[254]

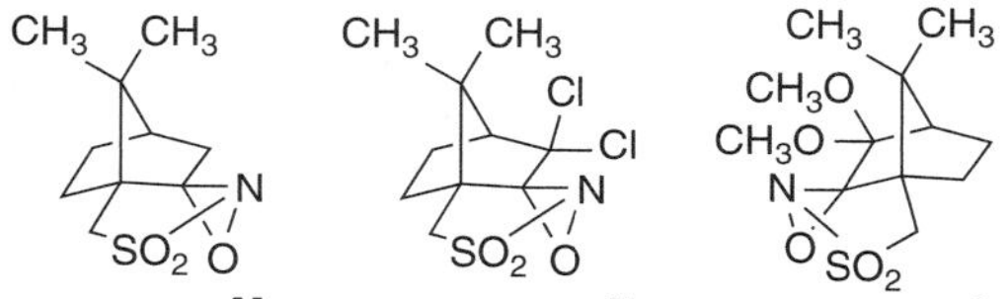

Scheme 12.21 gives some examples of enolate oxidation using *N*-sulfonyloxaziridines. Entries 1 to 3 are examples of enantioselective oxidations using chiral oxaziridines with racemic reactants. In Entry 4, the stereoselectivity is presumably controlled by the reactant shape. The analog with all *cis* stereochemistry at the cyclobutane ring also gave oxidation from the less hindered face of the molecule. Entry 5 is an example of diastereoselective oxidation. The observed *syn* selectivity is consistent with reactant conformation being the controlling factor in reagent approach.

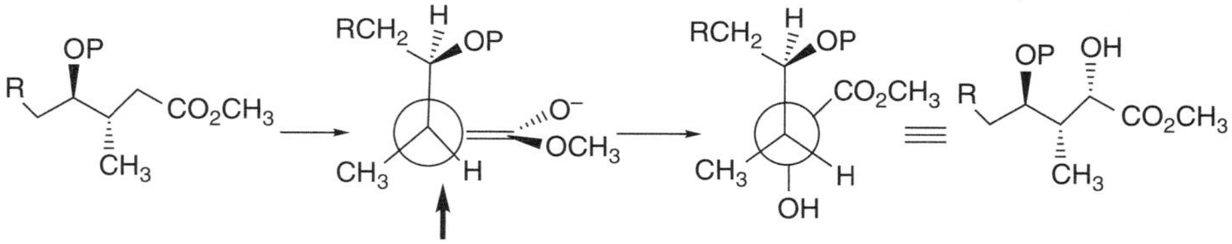

251. F. A. Davis, L. C. Vishwakarma, J. M. Billmers, and J. Finn, *J. Org. Chem.*, **49**, 3241 (1984); L. C. Vishwakarma, O. D. Stringer, and F. A. Davis, *Org. Synth.*, **66**, 203 (1988).
252. F. A. Davis, A. C. Sheppard, B.-C. Chen, and M. S. Haque, *J. Am. Chem. Soc.*, **112**, 6679 (1990).
253. D. A. Evans, M. M. Morrissey, and R. L. Dorow, *J. Am. Chem. Soc.*, **107**, 4346 (1985).
254. F. A. Davis and B.-C. Chen, *Chem. Rev.*, **92**, 919 (1992).

Scheme 12.21. Oxidation of Enolates by Oxaziridines

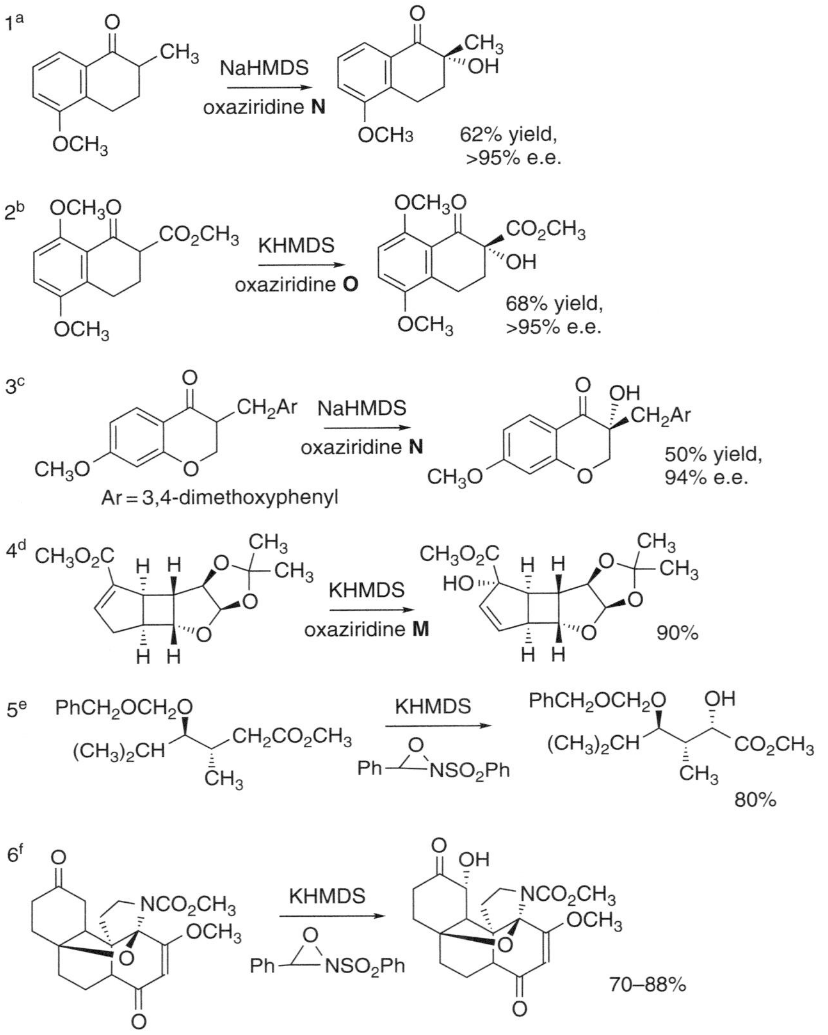

1[a]

NaHMDS / oxaziridine **N**

62% yield, >95% e.e.

2[b]

KHMDS / oxaziridine **O**

68% yield, >95% e.e.

3[c]

NaHMDS / oxaziridine **N**

Ar = 3,4-dimethoxyphenyl

50% yield, 94% e.e.

4[d]

KHMDS / oxaziridine **M**

90%

5[e]

KHMDS

80%

6[f]

KHMDS

70–88%

a. F. A. Davis and M. C. Weismiller, *J. Org. Chem.*, **55**, 3715 (1990).

b. F. A. Davis, A. Kumar, and B.-C. Chen, *Tetrahedron Lett.*, **32**, 867 (1991).

c. F. A. Davis and B.-C. Chen, *J. Org. Chem.*, **58**, 1751 (1993).

d. A. B. Smith, III, G. A. Sulikowski, M. M. Sulikowsii, and K. Fujimoto, *J. Am. Chem. Soc.*, **114**, 2567 (1992).

e. S. Hanessian, Y. Gai, and W. Wang, *Tetrahedron Lett.*, **37**, 7473 (1996).

f. M. A. Tius and M. A. Kerr, *J. Am. Chem. Soc.*, **114**, 5959 (1992).

Both the regio- and stereochemistry of Entry 6 are of interest. The regioselectivity is imposed by the rigid ring geometry, which favors enolization at the observed position. Inspection of a molecular model also shows that α-face of the enolate is more accessible.

12.5.3. Oxidation with Other Reagents

Selenium dioxide can be used to oxidize ketones and aldehydes to α-dicarbonyl compounds. The reaction often gives high yields of products when there is a single type of CH_2 group adjacent to the carbonyl group. In unsymmetrical ketones, oxidation usually occurs at the CH_2 that is most readily enolized.[255]

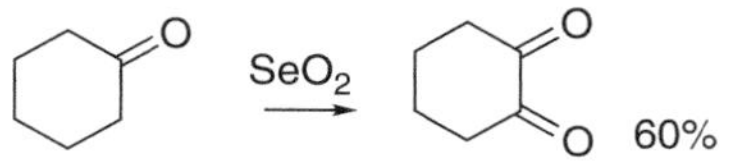

Ref. 256

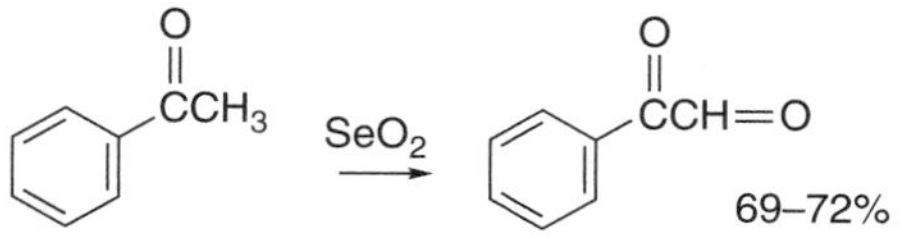

Ref. 257

The oxidation is regarded as taking place by an electrophilic attack of selenium dioxide (or selenous acid, H_2SeO_3, the hydrate) on the enol of the ketone or aldehyde. This is followed by hydrolytic elimination of the selenium.[258]

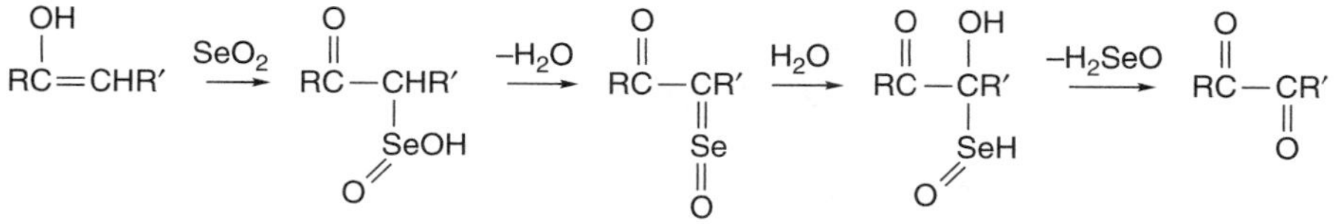

Methyl ketones are degraded to the next lower carboxylic acid by reaction with hypochlorite or hypobromite ions. The initial step in these reactions involves base-catalyzed halogenation. The α-haloketones are more reactive than their precursors, and rapid halogenation to the trihalo compound results. Trihalomethyl ketones are susceptible to alkaline cleavage because of the inductive stabilization provided by the halogen atoms.

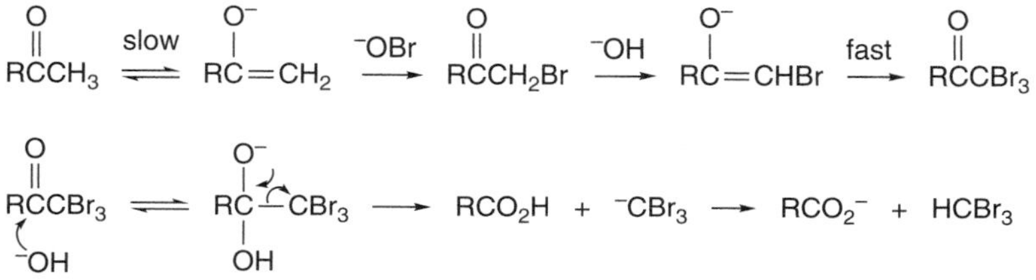

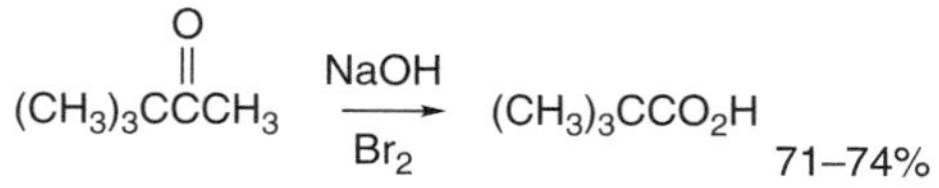

Ref. 259

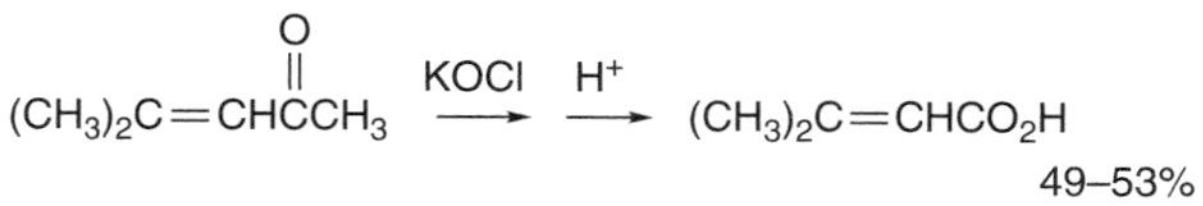

Ref. 260

255. E. N. Trachtenberg, in *Oxidation*, Vol. 1, R. L. Augustine, ed., Marcel Dekker, New York, 1969, Chap. 3.
256. C. C. Hach, C. V. Banks, and H. Diehl, *Org. Synth.*, **IV**, 229 (1963).
257. H. A. Riley and A. R. Gray, *Org. Synth.*, **II**, 509 (1943).
258. K. B. Sharpless and K. M. Gordon, *J. Am. Chem. Soc.*, **98**, 300 (1976).
259. L. T. Sandborn and E. W. Bousquet, *Org. Synth.*, **1**, 512 (1932).
260. L. I. Smith, W. W. Prichard, and L. J. Spillane, *Org. Synth.*, **III**, 302 (1955).

12.6. Selective Oxidative Cleavages at Functional Groups

12.6.1. Cleavage of Glycols

As discussed in connection with cleavage of double bonds by permanganate–periodate or osmium tetroxide–periodate (see p. 1127), the glycol unit is susceptible to mild oxidative cleavage. The most commonly used reagent for this oxidative cleavage is the periodate ion.[261] The fragmentation is believed to occur via a cyclic adduct of the glycol and the oxidant.

Structural features that retard formation of the cyclic intermediate decrease the reaction rate. For example, *cis*-1,2-dihydroxycyclohexane is substantially more reactive than the *trans* isomer.[262] Glycols in which the geometry of the molecule precludes the possibility of a cyclic intermediate are essentially inert to periodate.

Certain other combinations of adjacent functional groups are also cleaved by periodate. Diketones are cleaved to carboxylic acids, and it is proposed that a reactive cyclic intermediate is formed by nucleophilic attack on the diketone.[263]

α-Hydroxy ketones and α-amino alcohols are also subject to oxidative cleavage, presumably by a similar mechanism.

Lead tetraacetate is an alternative reagent to periodate for glycol cleavage. It is particularly useful for glycols that have low solubility in the aqueous media used for periodate reactions. A cyclic intermediate is suggested by the same kind of stereochemistry-reactivity relationship discussed for periodate.[264] Unlike periodate, however, glycols that cannot form cyclic intermediates are eventually oxidized. For example, *trans*-9,10-dihydroxydecalin is oxidized, but the rate is 100 times less than for the *cis* isomer.[265] Thus, whereas a cyclic mechanism appears to provide the lowest-energy pathway for this oxidative cleavage, it is not the only possible mechanism. Both

[261]. C. A. Bunton, in *Oxidation in Organic Chemistry*, Part A, K. B. Wiberg, ed., Academic Press, New York, 1965, pp. 367–388; A. S. Perlin, in *Oxidation*, Vol. 1, R. L. Augustine, ed., Marcel Dekker, New York, 1969, pp. 189–204.

[262]. C. C. Price and M. Knell, *J. Am. Chem. Soc.*, **64**, 552 (1942).

[263]. C. A. Bunton and V. J. Shiner, *J. Chem. Soc.*, 1593 (1960).

[264]. C. A. Bunton, in *Oxidation in Organic Chemistry*, K. Wiberg, ed., Academic Press, New York, 1965, pp. 398–405; W. S. Trahanovsky, J. R. Gilmore, and P. C. Heaton, *J. Org. Chem.*, **38**, 760 (1973).

[265]. R. Criegee, E. Hoeger, G. Huber, P. Kruck, F. Marktscheffel, and H. Schellenberger, *Liebigs Ann. Chem.*, **599**, 81 (1956).

the periodate cleavage and lead tetraacetate oxidation can be applied synthetically to the generation of medium-sized rings when the glycol is at the junction of two rings.

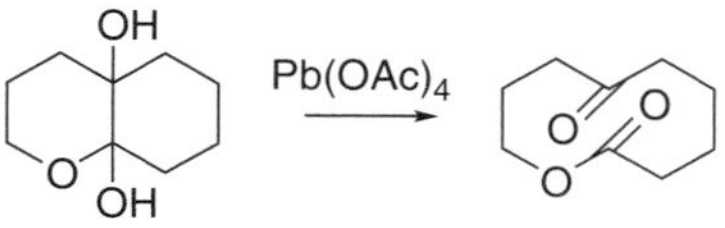

Ref. 266

12.6.2. Oxidative Decarboxylation

Carboxylic acids are oxidized by lead tetraacetate. Decarboxylation occurs and the product may be an alkene, alkane or acetate ester, or under modified conditions a halide. A free radical mechanism operates and the product composition depends on the fate of the radical intermediate.[267] The reaction is catalyzed by cupric salts, which function by oxidizing the intermediate radical to a carbocation (Step 3b in the mechanism). Cu(II) is more reactive than $Pb(OAc)_4$ in this step.

$$Pb(OAc)_4 + RCO_2H \rightleftharpoons RCO_2Pb(OAc)_3 + CH_3CO_2H \quad (1)$$

$$RCO_2Pb(OAc)_3 \longrightarrow R\cdot + CO_2 + Pb(OAc)_3 \quad (2)$$

$$R\cdot + Pb(OAc)_4 \longrightarrow R^+ + Pb(OAc)_3 + CH_3CO_2^- \quad (3a)$$

and

$$R\cdot + Pb(OAc)_3 \xrightarrow{Cu(II)} R^+ + Cu(I) \quad (3b)$$

Alkanes are formed when the radical intermediate abstracts hydrogen from solvent faster than it is oxidized to the carbocation. This reductive step is promoted by good hydrogen donor solvents. It is also more prevalent for primary alkyl radicals because of the higher activation energy associated with formation of primary carbocations. The most favorable conditions for alkane formation involve photochemical decomposition of the carboxylic acid in chloroform, which is a relatively good hydrogen donor.

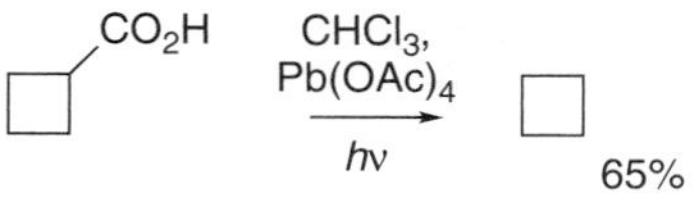

65%

Ref. 268

Normally, the dominant products are the alkene and acetate ester, which arise from the carbocation intermediate by, respectively, elimination of a proton and capture of an acetate ion.[269]

266. T. Wakamatsu, K. Akasaka, and Y. Ban, *Tetrahedron Lett.*, 2751, 2755 (1977).
267. R. A. Sheldon and J. K. Kochi, *Org. React.*, **19**, 279 (1972).
268. J. K. Kochi and J. D. Bacha, *J. Org. Chem.*, **33**, 2746 (1968).
269. J. D. Bacha and J. K. Kochi, *Tetrahedron*, **24**, 2215 (1968).

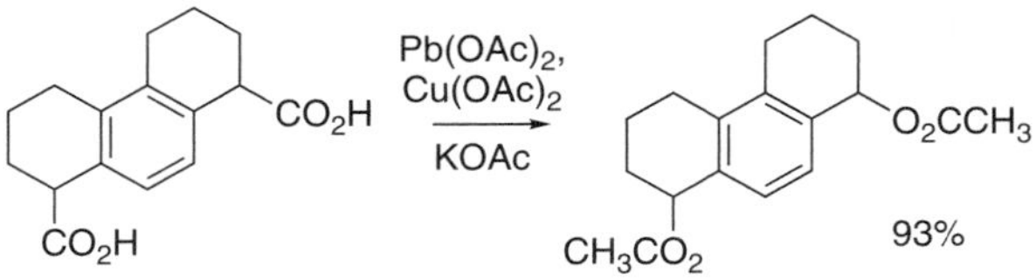

Ref. 270

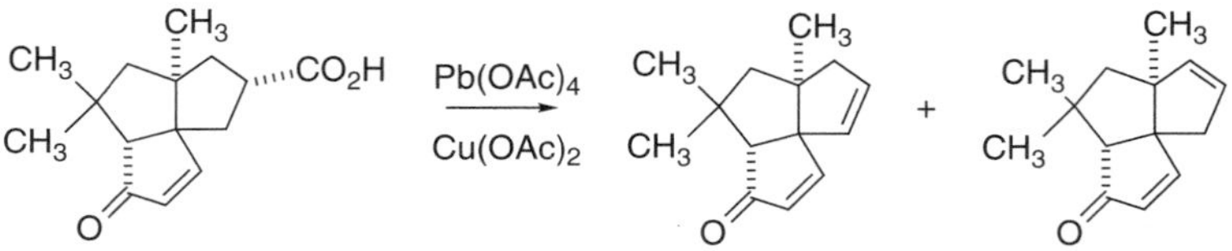

Ref. 271

In the presence of lithium chloride, the product is the corresponding chloride.[272]

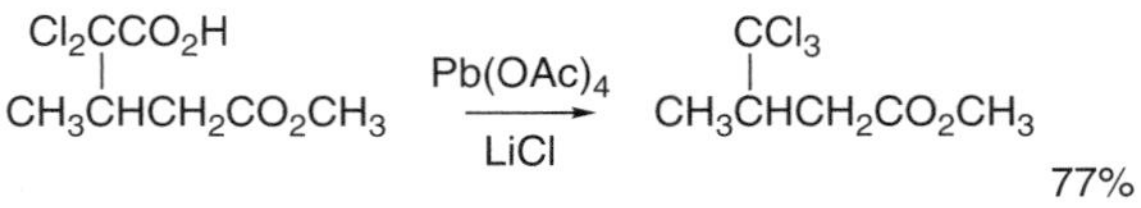

77%   Ref. 273

5-Arylpentanoic acids give tetrahydronaphthalenes, a reaction that is consistent with a radical cyclization.

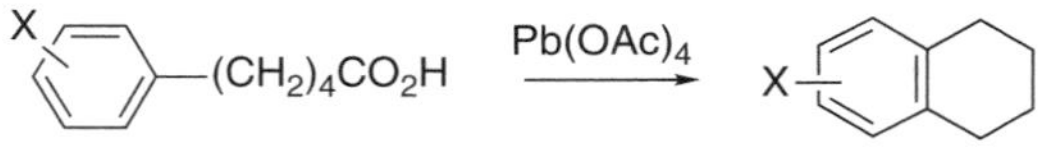

Ref. 274

On the other hand, γ, δ-unsaturated acids give lactones that involve cyclization without decarboxylation.

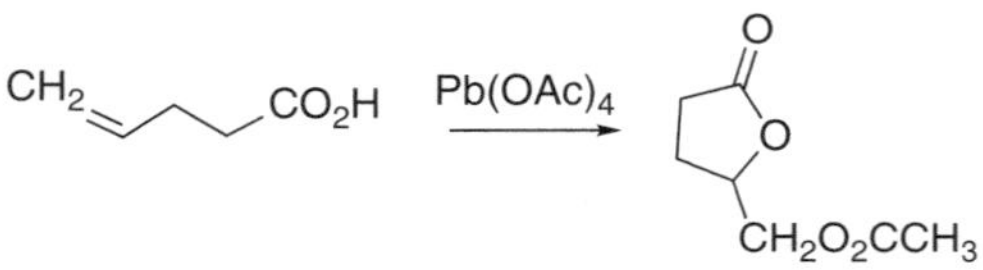

Ref. 275

These products can be formed by a ligand transfer from an intermediate in which the double bond is associated with the Pb.

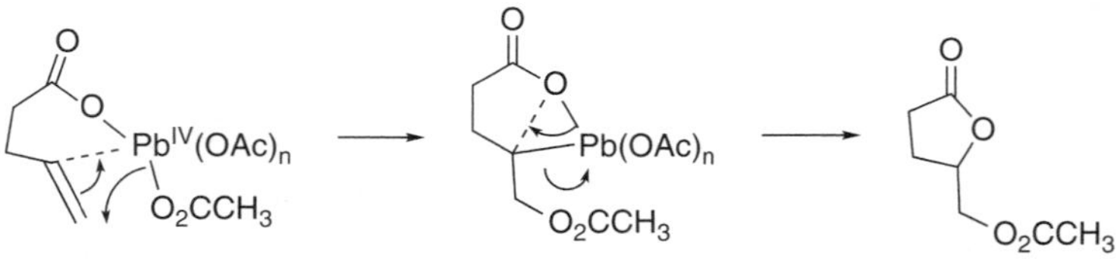

[270.] P. Caluwe and T. Pepper, *J. Org. Chem.*, **53**, 1786 (1988).

[271.] D. D. Sternbach, J. W. Hughes, D. E. Bardi, and B. A. Banks, *J. Am. Chem. Soc.*, **107**, 2149 (1985).

[272.] J. K. Kochi, *J. Org. Chem.*, **30**, 3265 (1965).

[273.] S. E. de Laszlo and P. G. Williard, *J. Am. Chem. Soc.*, **107**, 199 (1985).

[274.] D. I. Davies and C. Waring, *J. Chem. Soc. C*, 1865 (1968).

[275.] M. G. Moloney, E. Nettleton, and K. Smithies, *Tetrahedron Lett.*, **43**, 907 (2002).

A related method for conversion of carboxylic acids to bromides with decarboxylation is the *Hunsdiecker reaction*.[276] The usual method for carrying out this transformation involves heating the carboxylic acid with mercuric oxide and bromine.

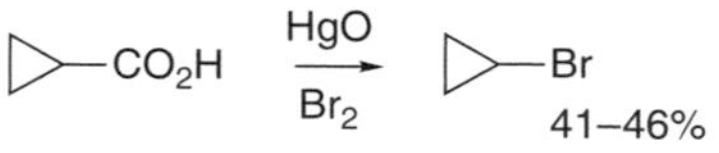

41–46%

Ref. 277

The overall transformation can also be accomplished by reaction of thallium(I) carboxylate with bromine.[278] Phenyliodonium diacetate and bromine also lead to brominative decarboxylation.[279]

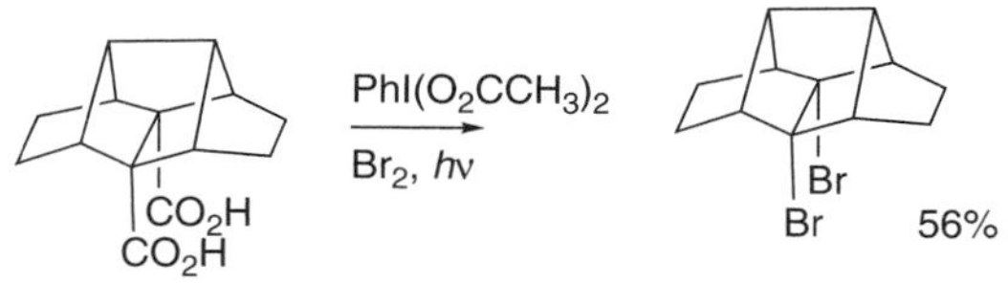

1,2-Dicarboxylic acids undergo *bis*-decarboxylation on reaction with lead tetraacetate to give alkenes. This reaction has been of occasional use for the synthesis of strained alkenes.

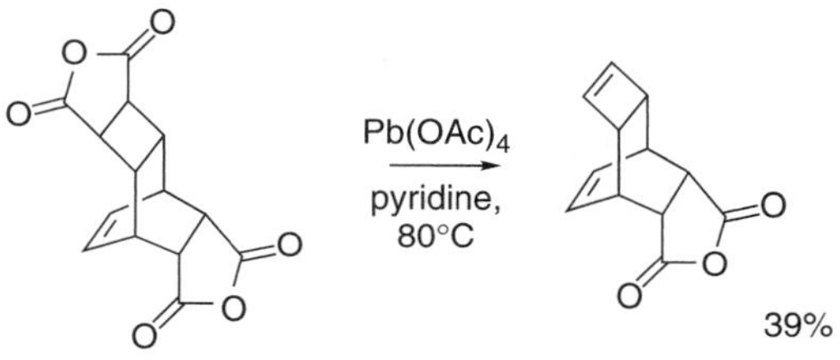

39%

Ref. 280

The reaction can occur by a concerted fragmentation process initiated by a two-electron oxidation.

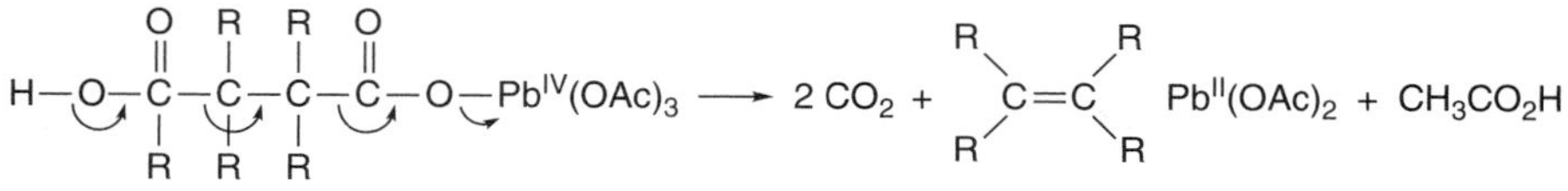

A concerted mechanism is also possible for α-hydroxycarboxylic acids, and these compounds readily undergo oxidative decarboxylation to ketones.[281]

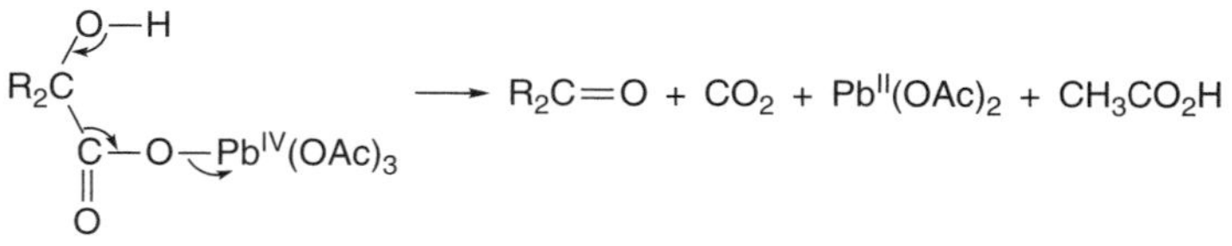

$$R_2C=O + CO_2 + Pb^{II}(OAc)_2 + CH_3CO_2H$$

276. C. V. Wilson, *Org. React.*, **9**, 332 (1957); R. A. Sheldon and J. Kochi, *Org. React.*, **19**, 326 (1972).
277. J. S. Meek and D. T. Osuga, *Org. Synth.*, **V**, 126 (1973).
278. A. McKillop, D. Bromley, and E. C. Taylor, *J. Org. Chem.*, **34**, 1172 (1969).
279. P. Camps, A. E. Lukach, X. Pujol, and S. Vazquez, *Tetrahedron*, **56**, 2703 (2000).
280. E. Grovenstein, Jr., D. V. Rao, and J. W. Taylor, *J. Am. Chem. Soc.*, **83**, 1705 (1961).
281. R. Criegee and E. Büchner, *Chem. Ber.*, **73**, 563 (1940).

γ-Ketocarboxylic acids are oxidatively decarboxylated to enones.[282] This reaction is presumed to proceed through the usual oxidative decarboxylation, with the carbocation intermediate being efficiently deprotonated because of the developing conjugation.

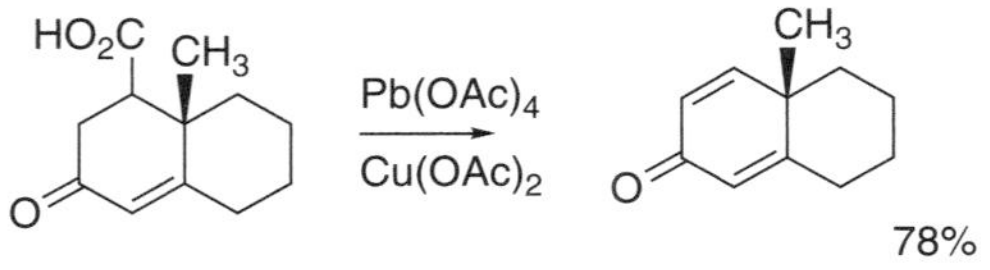

78%   Ref. 119

Oxidation of β-silyl and β-stannyl acids leads to loss of the substituent and alkene formation.[283]

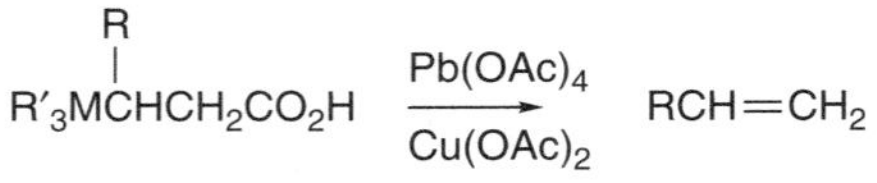

12.7. Oxidations at Unfunctionalized Carbon

Attempts to achieve selective oxidations of hydrocarbons or other compounds when the desired site of attack is remote from an activating functional group are faced with several difficulties. With powerful transition-metal oxidants, the initial oxidation products are almost always more susceptible to oxidation than the starting material. When a hydrocarbon is oxidized, it is likely to be oxidized to a carboxylic acid, with chain cleavage by successive oxidation of alcohol and carbonyl intermediates. There are a few circumstances under which oxidations of hydrocarbons can be synthetically useful processes. One group involves catalytic industrial processes. Much effort has been expended on the development of selective catalytic oxidation processes and several have economic importance. We focus on several reactions that are used on a laboratory scale.

The most general hydrocarbon oxidation is the oxidation of side chains on aromatic rings. Two factors contribute to making this a high-yield procedure, despite the use of strong oxidants. First, the benzylic position is susceptible to hydrogen abstraction by the oxidants.[284] Second, the aromatic ring is resistant to attack by Mn(VII) and Cr(VI) reagents that oxidize the side chain.

Scheme 12.22 provides some examples of the oxidation of aromatic alkyl substituents to carboxylic acid groups. Entries 1 to 3 are typical oxidations of aromatic methyl groups to carboxylic acids. Entries 4 and 5 bring the carbon adjacent to the aromatic ring to the carbonyl oxidation level.

Selective oxidations are possible for certain bicyclic hydrocarbons.[285] Here, the bridgehead position is the preferred site of initial attack because of the order of reactivity of C—H bonds, which is 3° > 2° > 1°. The tertiary alcohols that are the initial oxidation products are not easily further oxidized. The geometry of the bicyclic rings (*Bredt's rule*) prevents both dehydration of the tertiary bridgehead alcohols and further oxidation to ketones. Therefore, oxidation that begins at a bridgehead position

282. J. E. McMurry and L. C. Blaszczak, *J. Org. Chem.*, **39**, 2217 (1974).
283. H. Nishiyama, H. Matsumoto, H. Arai, H. Sakaguchi, and K. Itoh, *Tetrahedron Lett.*, **27**, 1599 (1986).
284. K. A. Gardner, L. L. Kuehnert, and J. M. Mayer, *Inorg. Chem.*, **36**, 2069 (1997).
285. R. C. Bingham and P. v. R. Schleyer, *J. Org. Chem.*, **36**, 1198 (1971).

Scheme 12.22. Side Chain Oxidation of Aromatic Compounds

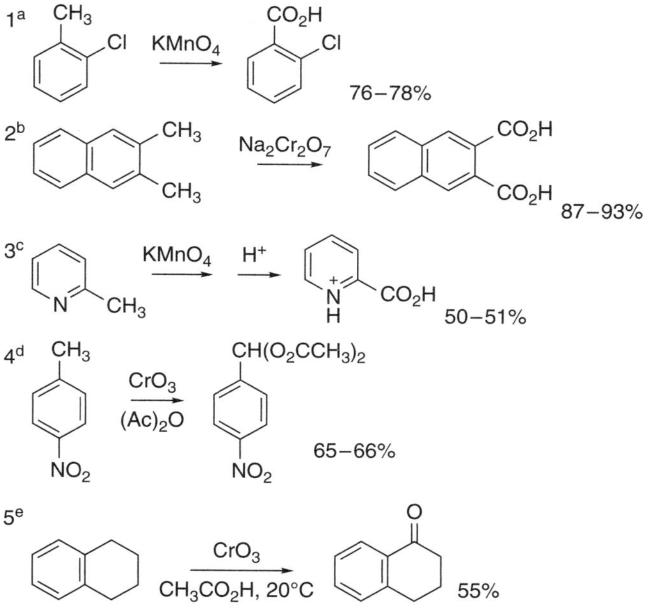

a. H. T. Clarke and E. R. Taylor, *Org. Synth.*, **II**, 135 (1943).
b. L. Friedman, *Org. Synth.*, **43**, 80 (1963); L. Friedman, D. L. Fishel, and H. Shechter, *J. Org. Chem.*, **30**, 1453 (1965).
c. A. W. Singer and S. M. McElvain, *Org. Synth.*, **III**, 740 (1955).
d. T. Nishimura, *Org. Synth.*, **IV**, 713 (1963).
e. J. W. Burnham, W. P. Duncan, E. J. Eisenbraun, G. W. Keen, and M. C. Hamming, *J. Org. Chem.*, **39**, 1416 (1974).

stops at the alcohol stage. Chromic acid oxidation has been the most useful reagent for functionalizing unstrained bicyclic hydrocarbons. The reaction fails for strained bicyclic compounds such as norbornane because the reactivity of the bridgehead position is lowered by the unfavorable energy of radical or carbocation intermediates.

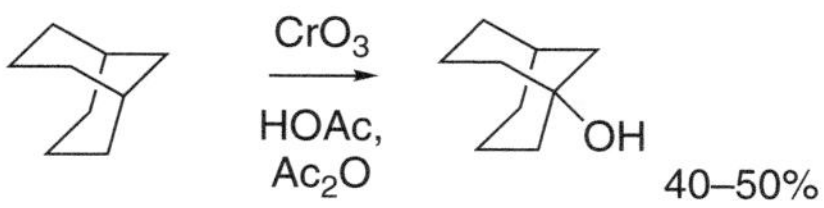

Other successful selective oxidations of hydrocarbons by Cr(VI) have been reported—for example, the oxidation of *cis*-decalin to the corresponding alcohol—but careful attention to reaction conditions is required.

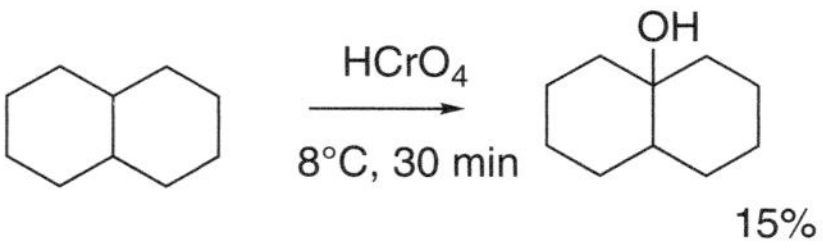

Ref. 286

286. K. B. Wiberg and G. Foster, *J. Am. Chem. Soc.*, **83**, 423 (1961).

Interesting hydrocarbon oxidations have been observed using Fe(II) catalysts with oxygen or hydrogen peroxide as the oxidant. These catalytic systems have become known as "Gif chemistry" after the location of their discovery in France.[287] An improved system involving Fe(III), picolinic acid, and H_2O_2 has been developed. The reactive species generated in these systems is believed to be at the Fe(V)=O oxidation level.[288] The key step is hydrogen abstraction from the hydrocarbon by this Fe(V)=O intermediate.

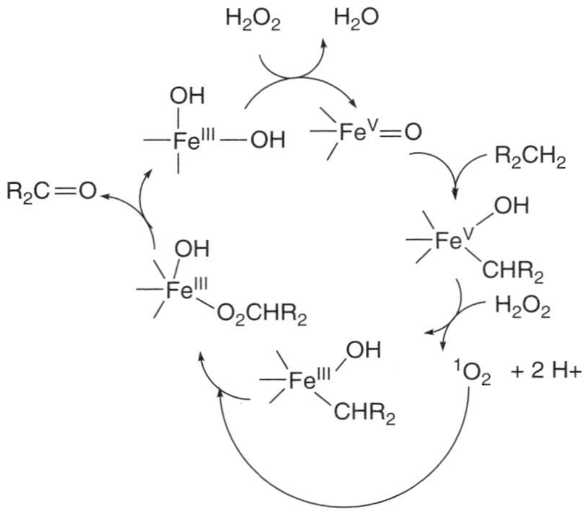

Oxidation of *trans*-decalin leads to a mixture of 1- and 2-*trans*-decalone.[289]

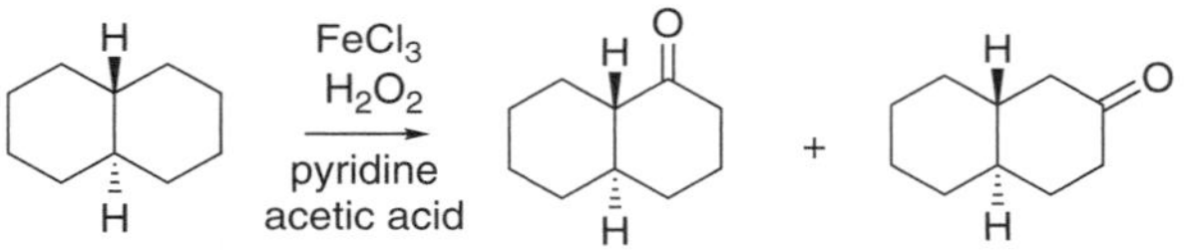

The initial intermediates containing C−Fe bonds can be diverted by reagents such as $CBrCl_3$ or CO, among others.[290]

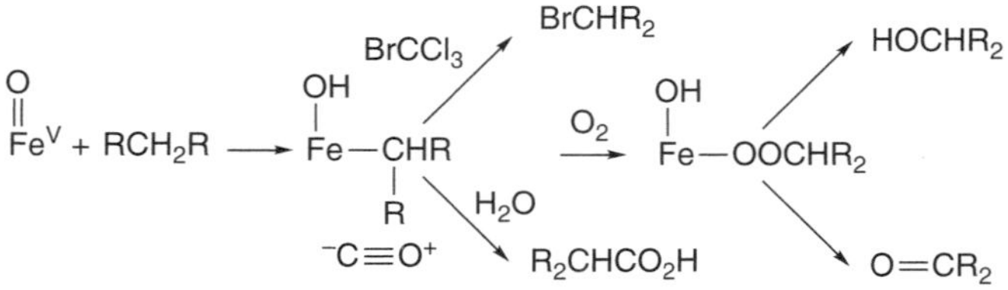

287. D. H. R. Barton and D. Doller, *Acc. Chem. Res.*, **25**, 504 (1992); D. H. R. Barton, *Chem. Soc. Rev.*, **25**, 237 (1996); D. H. R. Barton, *Tetrahedron*, **54**, 5805 (1998).

288. D. H. R. Barton, S. D. Beviere, W. Chavasiri, E. Csuhai, D. Doller, and W. G. Liu, *J. Am. Chem. Soc.*, **114**, 2147 (1992).

289. U. Schuchardt, M. J. D. M. Jannini, D. T. Richens, M. C. Guerreiro, and E. V. Spinace, *Tetrahedron*, **57**, 2685 (2001).

290. D. H. R. Barton, E. Csuhai, and D. Doller, *Tetrahedron Lett.*, **33**, 3413 (1992); D. H. R. Barton, E. Csuhai, and D. Doller, *Tetrahedron Lett.*, **33**, 4389 (1992).

Problems

(References for these problems will be found on page 1290.)

12.1. Indicate an appropriate oxidant for carrying out the following transformations.

(a)

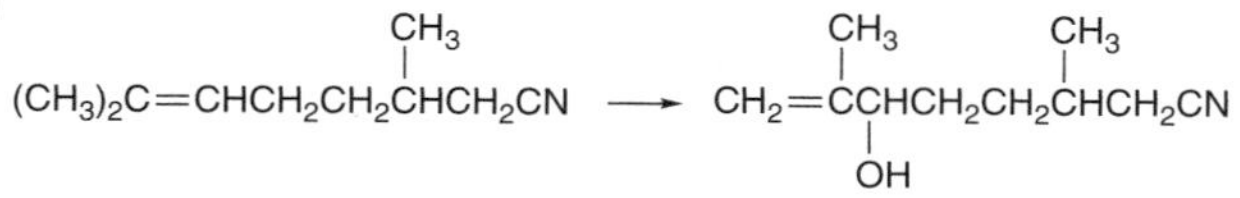

$$(CH_3)_2C = CHCH_2CH_2\overset{\overset{\displaystyle CH_3}{|}}{C}HCH_2CN \longrightarrow CH_2 = \overset{\overset{\displaystyle CH_3}{|}}{C}CHCH_2CH_2\overset{\overset{\displaystyle CH_3}{|}}{C}HCH_2CN$$

with OH

(b)

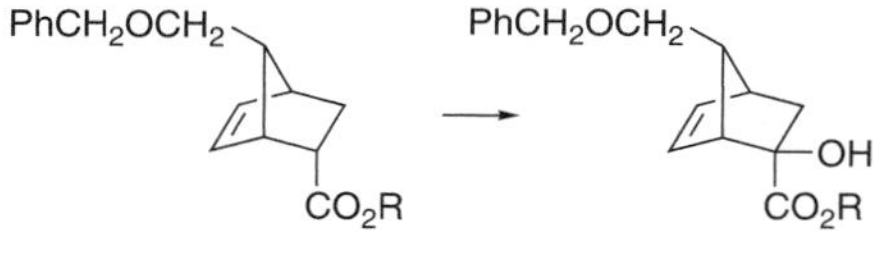

(c)

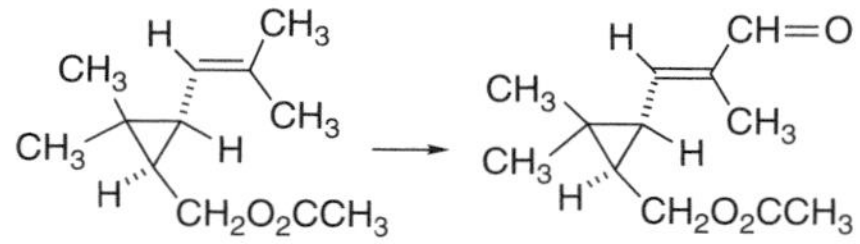

(d)

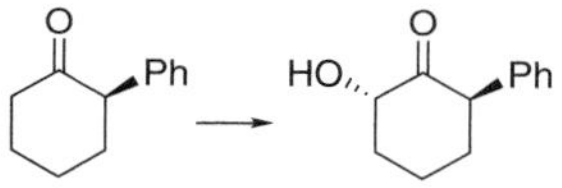

(e)

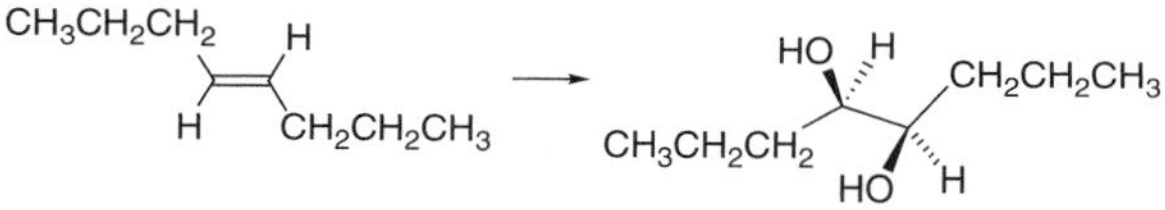

(racemic)

(f)

$$\underset{\overset{\|}{O}}{Ph\overset{}{C}CH_2CH_3} \longrightarrow \underset{\overset{\|}{O}}{Ph\overset{}{C}\overset{}{C}HCH_3}$$

with OH

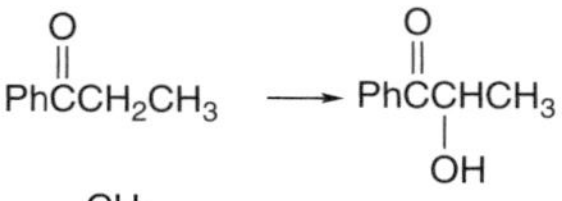

(g)

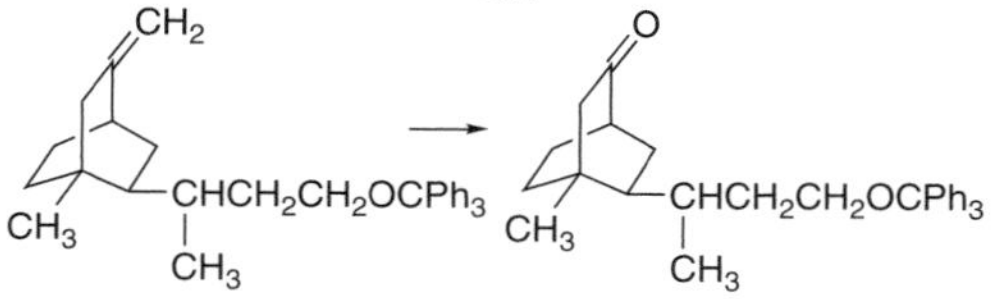

(h)

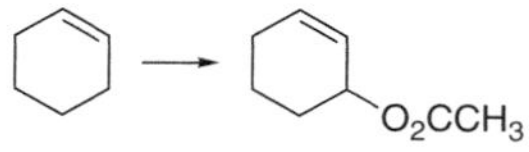

(i)

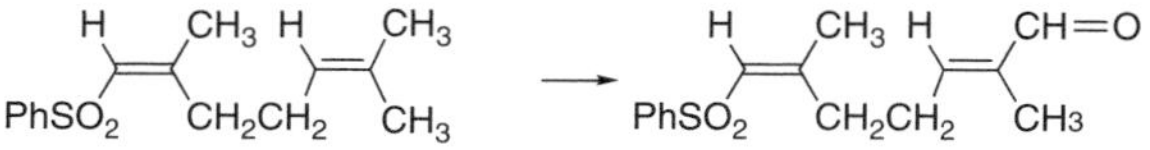

(j)

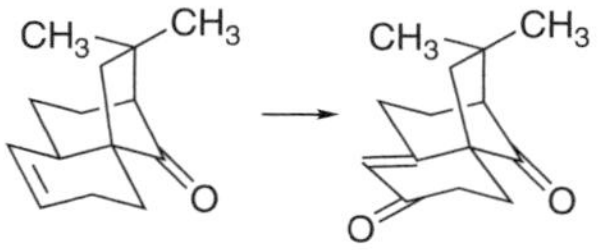

(k)

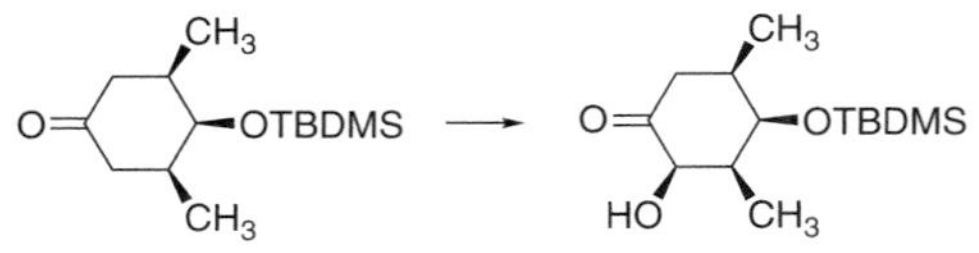

(l)

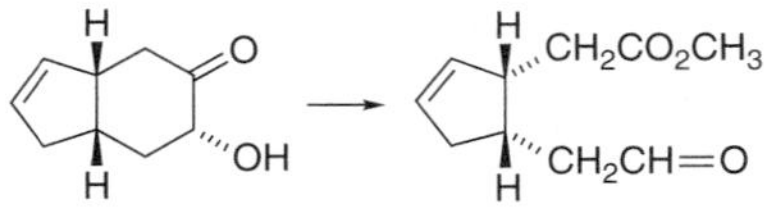

(m)

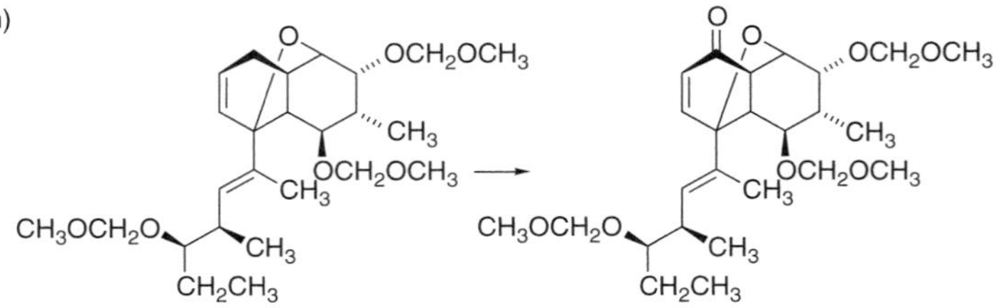

(n)

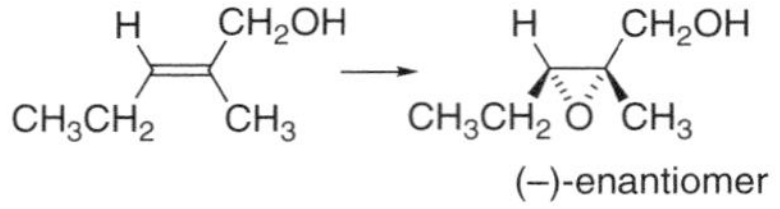

(−)-enantiomer

(o)

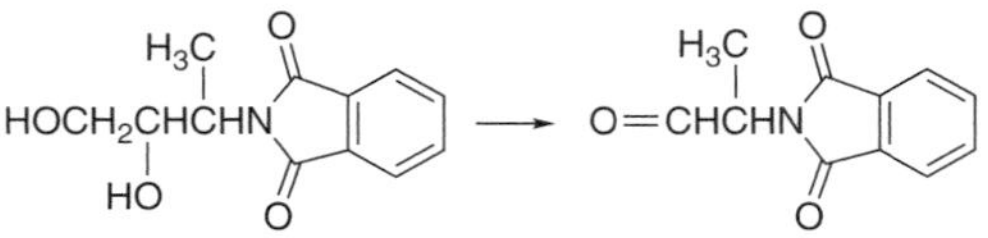

(p)

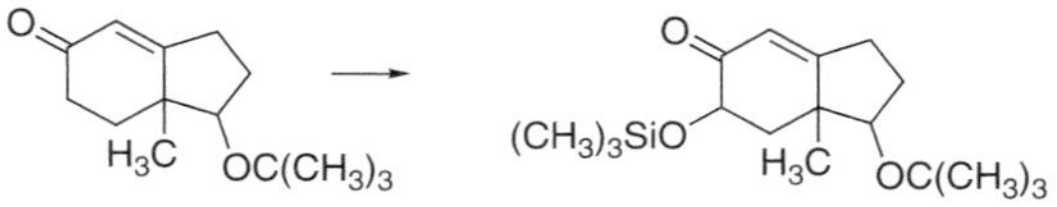

(q)

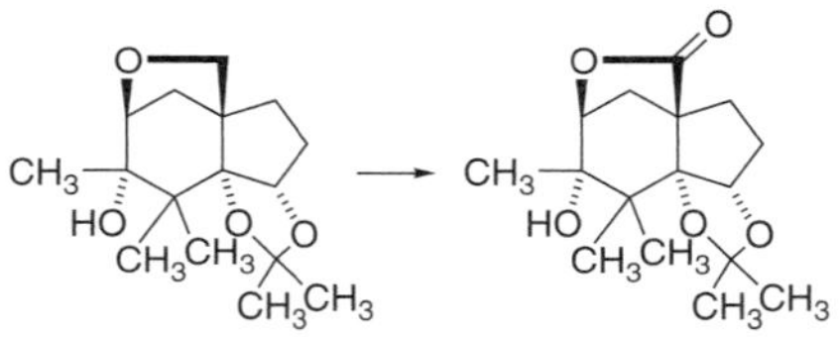

12.2. Predict the products of the following reactions. Be careful to consider all stereochemical aspects.

(a)

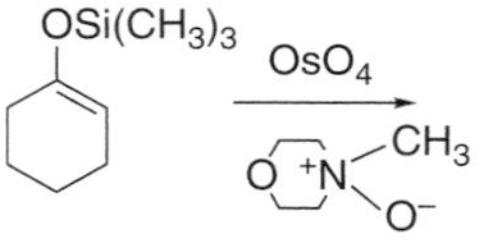

OsO$_4$

(b)

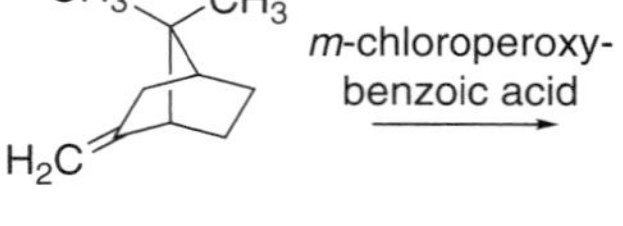

m-chloroperoxy-benzoic acid

(c)

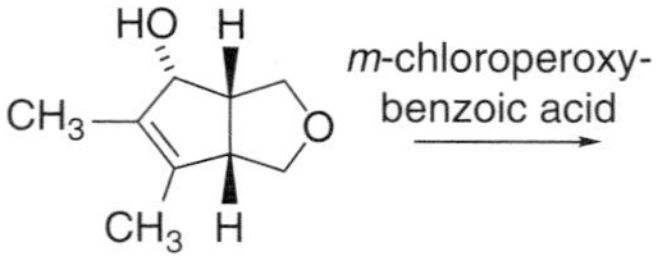

m-chloroperoxy-benzoic acid

(d)

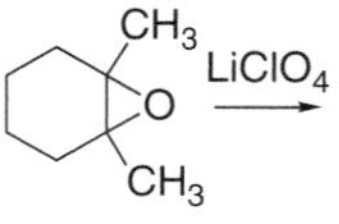

LiClO$_4$

(e)

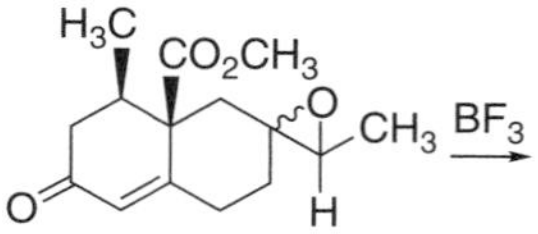

(f)

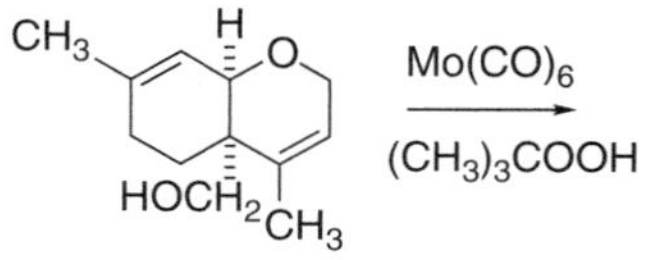

(g)

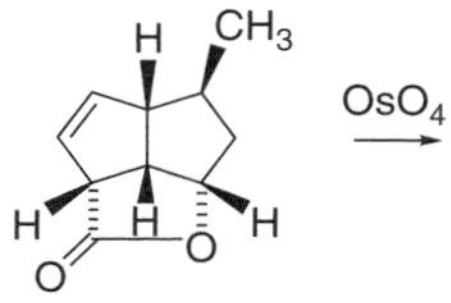

(h)

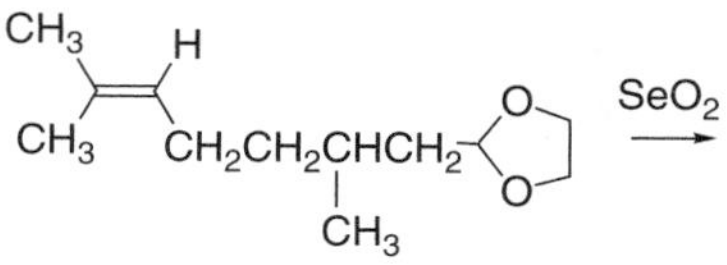

(i)

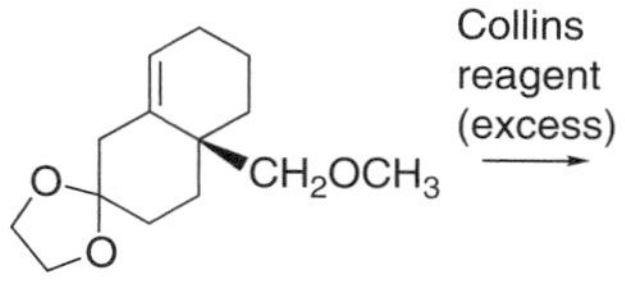

(j)

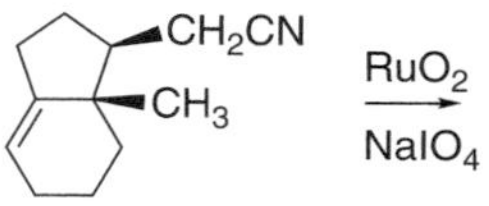

(k)

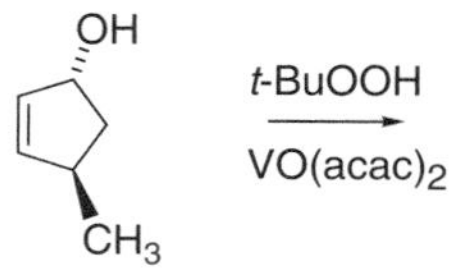

(l)

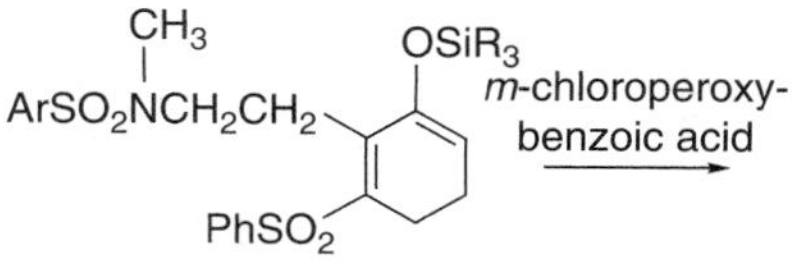

12.3. In chromic acid oxidation of stereoisomeric cyclohexanols, it is usually found that axial hydroxy groups react more rapidly than equatorial groups. For example, *trans*-4-*t*-butylcyclohexanol is less reactive (by a factor of 3.2) than the *cis* isomer. An even larger difference is noted with *cis*- and *trans*-3,3,5-trimethylcyclohexanol. The axial hydroxy in the *trans* isomer is 35 times more reactive than then equatorial hydroxy in the *cis* isomer, even though it is in a more hindered environment. A general relationship is found for pairs of epimeric cyclohexanols in that the ratio of the rates of the isomers is approximately equal to the equilibrium constant for equilibration of the isomers: $k_{ax}/k_{eq} \sim K_{ax/eq}$. Are these data compatible with the mechanism given on p. 1064? What additional details do these data provide about the reaction mechanism? Explain.

12.4. Predict the products from opening of the two stereoisomeric epoxides derived from limonene shown below by reaction with (a) acetic acid and (b) dimethylamine.

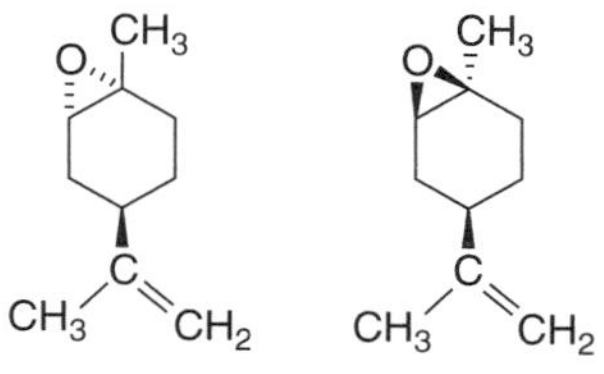

12.5. The direct oxidative conversion of primary halides and sulfonates to aldehydes can be carried out by reaction with DMSO under alkaline conditions. Formulate a mechanism for this reaction.

$$RCH_2X \quad + \quad (CH_3)_2S=O \quad \xrightarrow{\text{base}} \quad RCH=O$$

X = halide or sulfonate

12.6. The following questions pertain to the details of mechanism of ozonolysis under modified conditions.

a. A method for synthesis of ozonides that involves no ozone has been reported. It consists of photosensitized oxidation of diazo compounds in the presence of an aldehyde. Suggest a mechanism for this reaction.

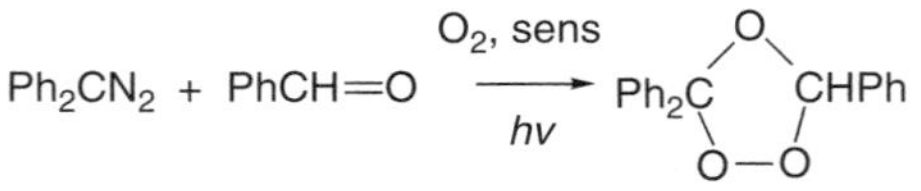

$$Ph_2CN_2 + PhCH=O \quad \xrightarrow[hv]{O_2, \text{ sens}} \quad Ph_2C \underset{O-O}{\overset{O}{\diagdown}} CHPh$$

b. Overoxidation of carbonyl products during ozonolysis can be prevented by addition of tetracyanoethylene to the reaction mixture. The stoichiometry of the reaction is then:

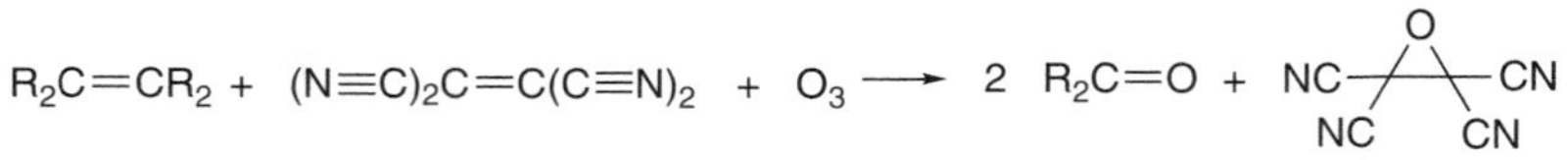

$$R_2C=CR_2 + (N\equiv C)_2C=C(C\equiv N)_2 + O_3 \longrightarrow 2 \ R_2C=O + \ NC\overset{O}{\underset{NC}{\diagup}}\overset{}{\underset{CN}{\diagdown}}CN$$

Propose a mechanism that would account for the effect of tetracyanoethylene. Does your mechanism suggest that tetracyanoethylene would be a particularly effective alkene for this purpose? Explain.

c. It has been found that when unsymmetrical alkenes are ozonized in methanol, there is often a large preference for one cleavage mode of the initial ozonide over the other. For example:

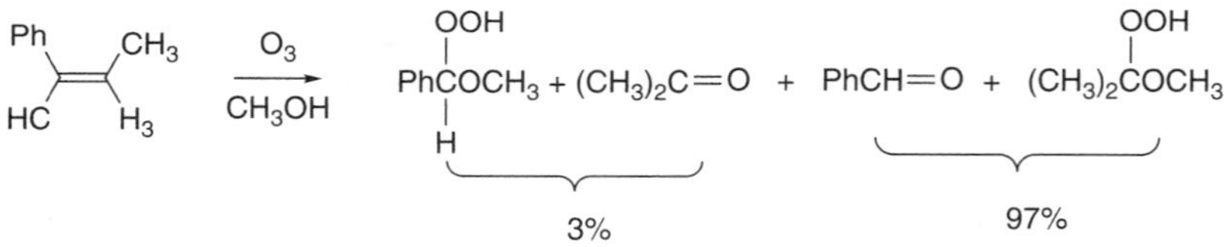

$$\underset{HC}{\overset{Ph}{\diagup}}C=C\underset{H_3}{\overset{CH_3}{\diagup}} \quad \xrightarrow[CH_3OH]{O_3} \quad \underset{H}{\overset{OOH}{\underset{|}{PhCOCH_3}}} + (CH_3)_2C=O \quad + \quad PhCH=O \quad + \quad \overset{OOH}{\underset{|}{(CH_3)_2COCH_3}}$$

$$\underbrace{\hspace{4cm}}_{3\%} \qquad\qquad \underbrace{\hspace{4cm}}_{97\%}$$

Account for this selectivity.

12.7. Suggest mechanisms by which the "abnormal" oxidations shown below could occur.

(a)

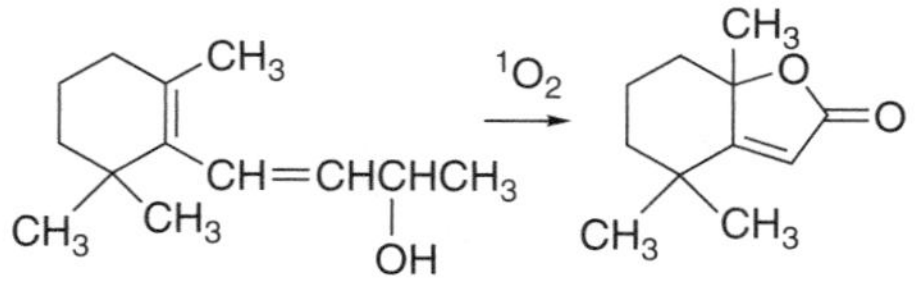

(b)

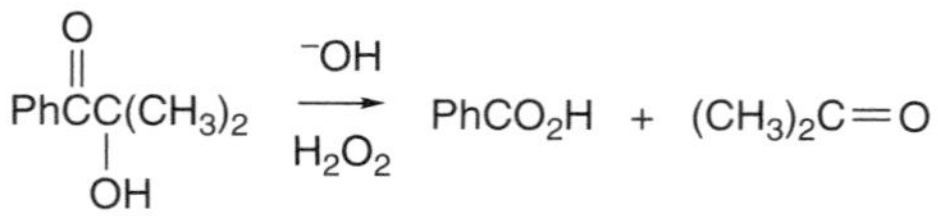

$$\underset{OH}{\overset{O}{\underset{|}{PhCC(CH_3)_2}}} \quad \xrightarrow[H_2O_2]{^-OH} \quad PhCO_2H \quad + \quad (CH_3)_2C=O$$

(c)

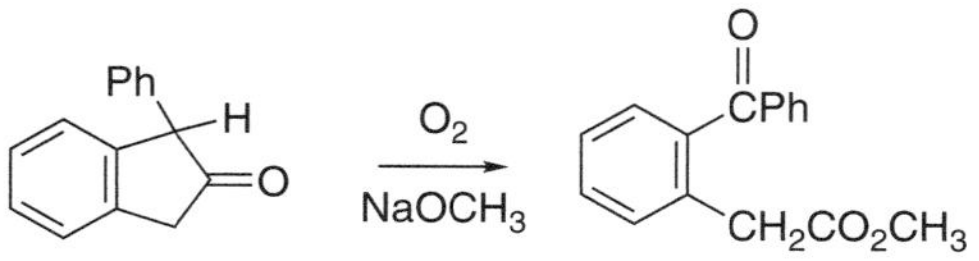

(d)

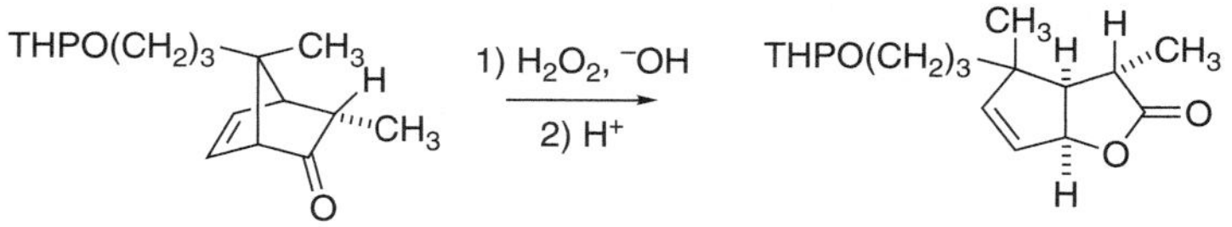

(e)

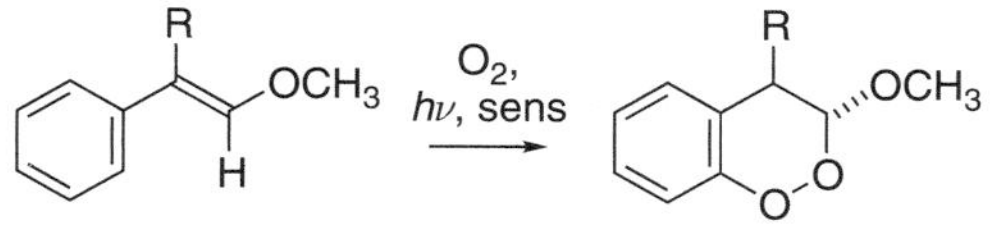

(f)

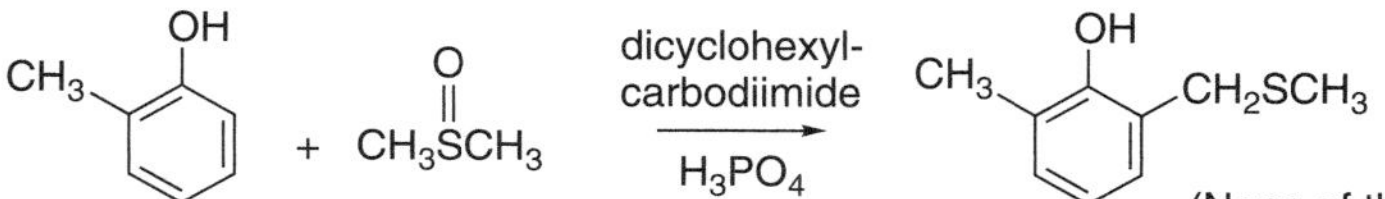

(None of the *para* isomer is formed.)

(g)

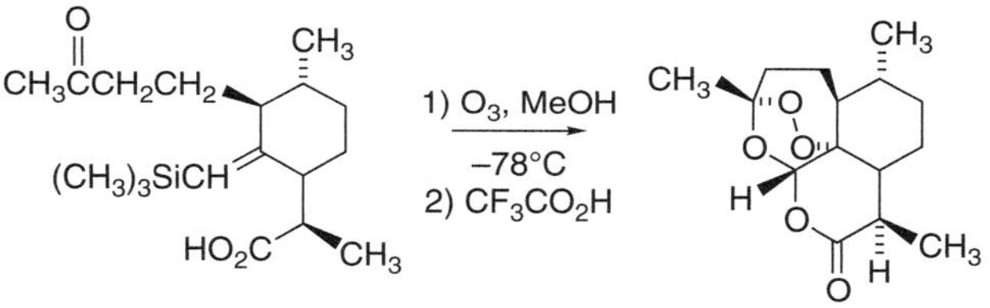

(h)

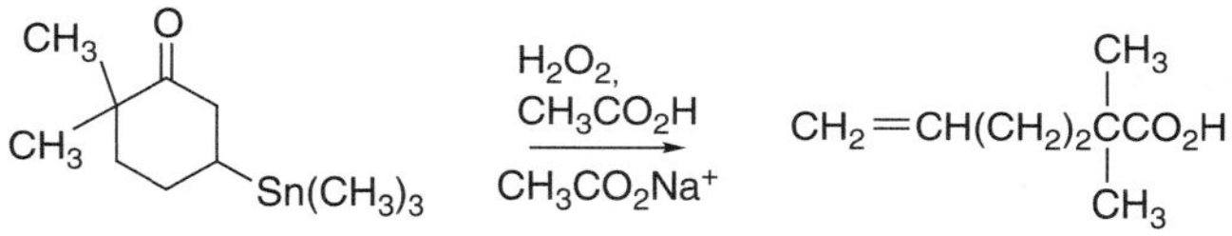

12.8. Indicate one or more satisfactory oxidants for effecting the following transformations. Each molecule poses issues of selectivity or the need to preserve a sensitive functional group. Select oxidants that can avoid the installation of protecting groups. In most cases, a one-pot reaction is possible, and in no case is a sequence of more than three steps required. Explain the reason for your choice of reagent(s).

(a)

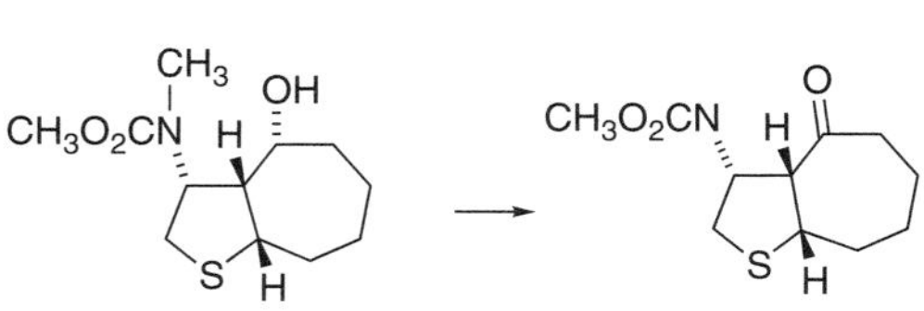

(b)

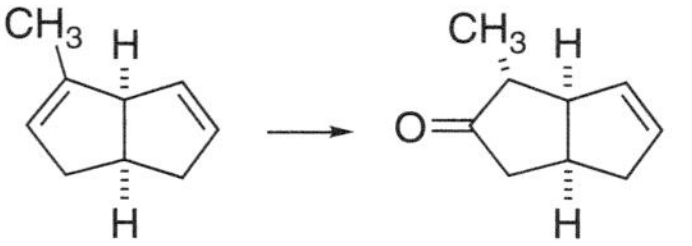

(c)

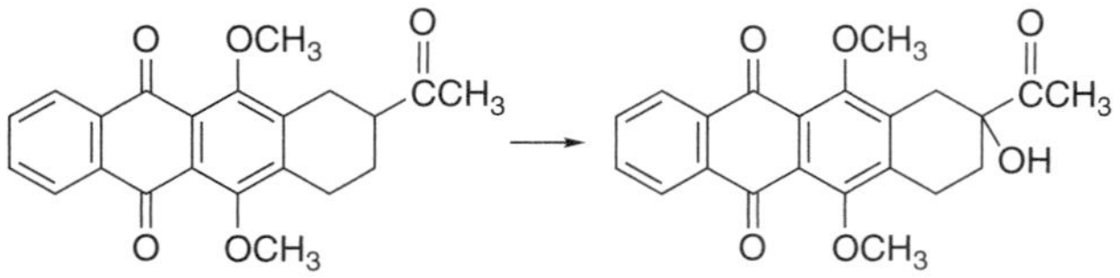

(d)

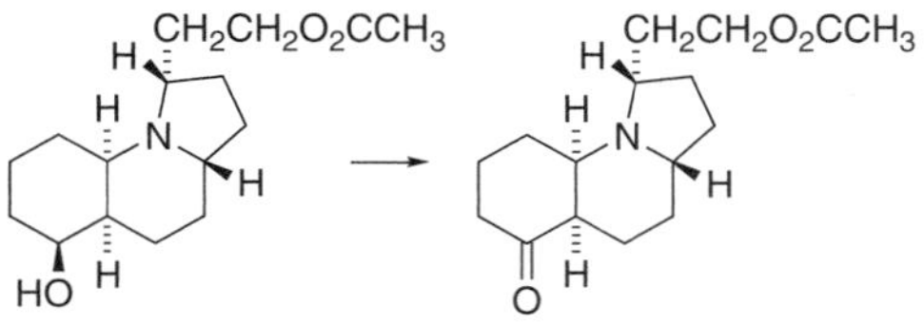

(e)

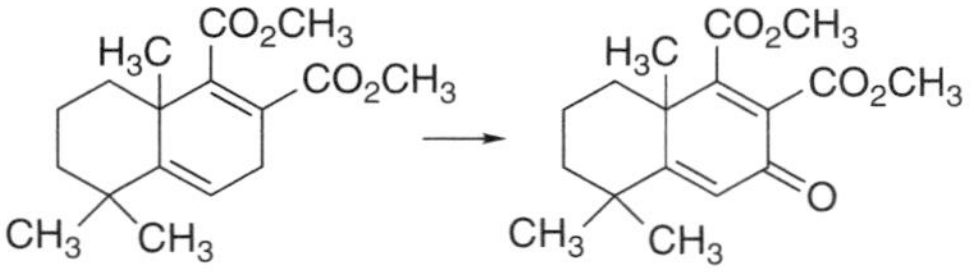

(f)

(g)

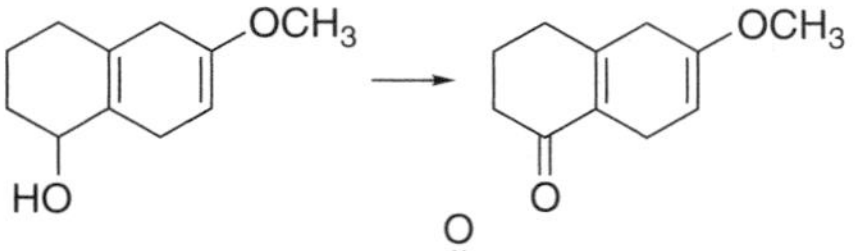

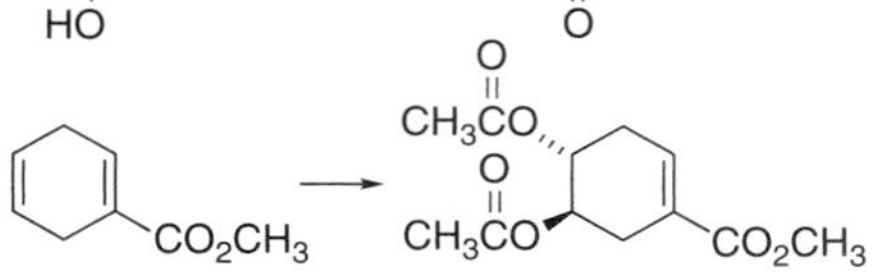

(h)

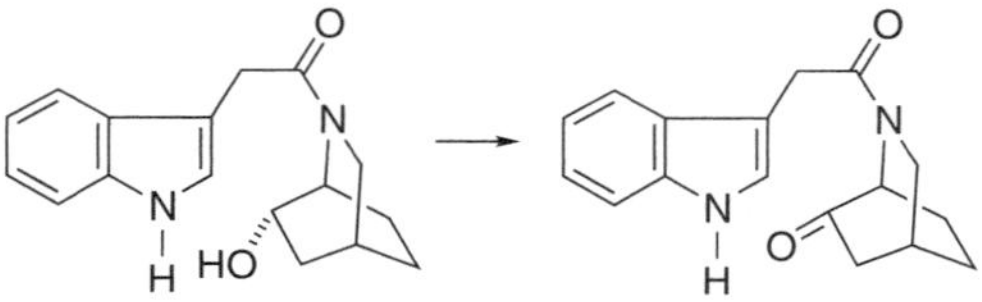

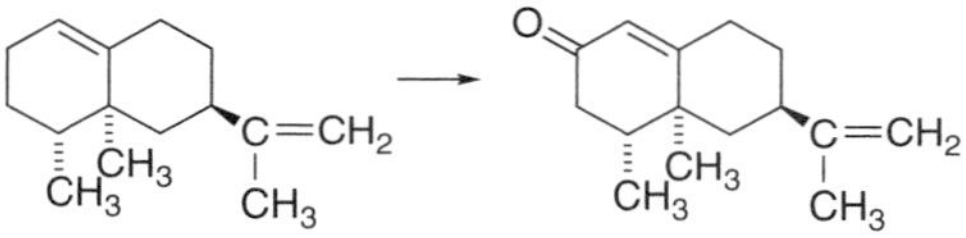

(i)

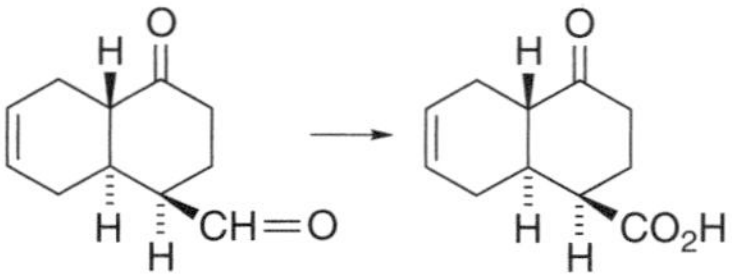

(j)

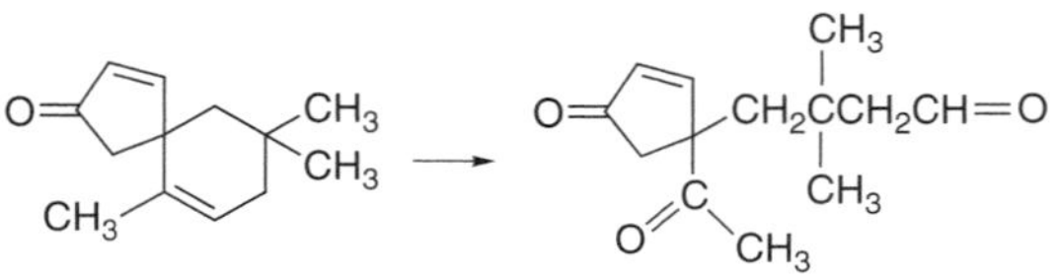

(k)

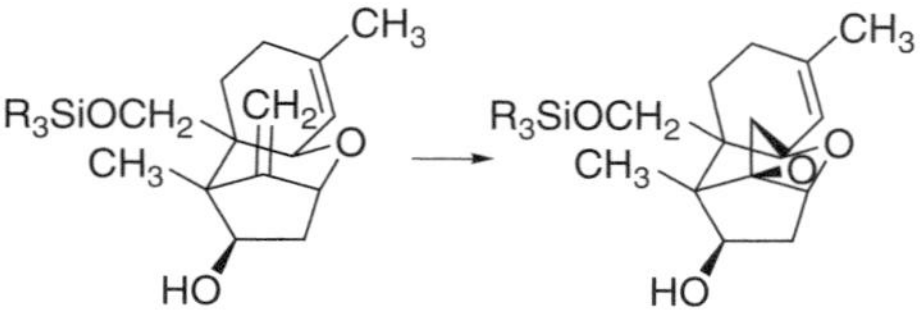

(l)

12.9. A method for oxidative cleavage of cyclic ketones involves a four-stage process. First, the ketone is converted to an α-phenylthio derivative (see Section 4.3.2). The ketone is then converted to an alcohol, either by reduction with NaBH$_4$ or by addition of an organolithium reagent. The alcohol is then treated with Pb(OAc)$_4$ to give an oxidation product in which the hydroxy group has been acetylated and an additional oxygen added to the β-thioalcohol. Aqueous hydrolysis of this intermediate in the presence of Hg^{2+} gives a dicarbonyl compound. Formulate likely structures for the products of each step in this sequence.

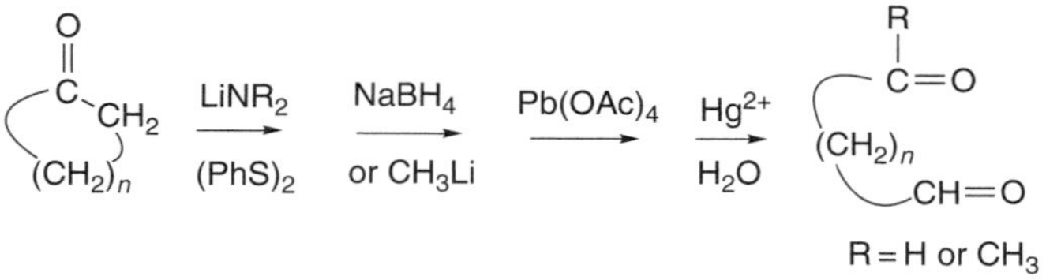

12.10. The transformations shown below have been carried out using reaction sequences involving several oxidation steps. Devise a series of steps that could accomplish these transformations and suggest reagents that would be suitable for each step. Some sequences may also require nonoxidative steps, such as introduction or removal of protecting groups.

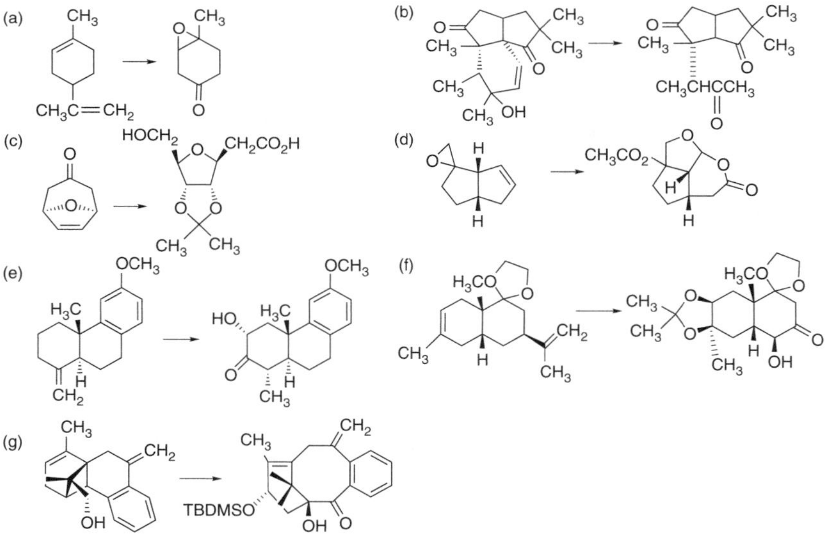

12.11. Provide mechanistic interpretations of the following reactions.

a. Account for the products formed under the following conditions. In particular, why does the inclusion of cupric acetate change the course of the reaction?

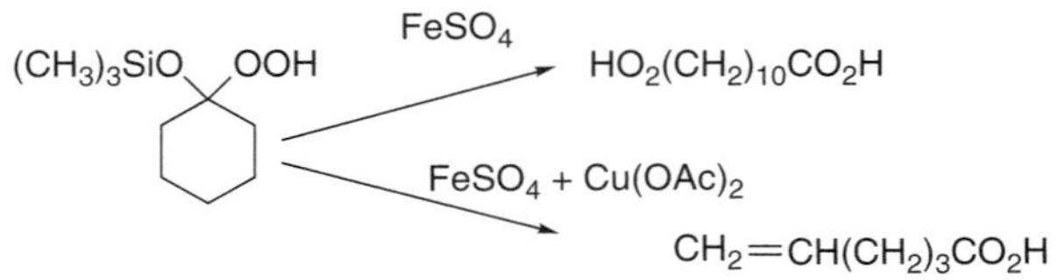

$$(CH_3)_3SiO \quad OOH$$

$$\xrightarrow{FeSO_4} HO_2(CH_2)_{10}CO_2H$$

$$\xrightarrow{FeSO_4 + Cu(OAc)_2} CH_2=CH(CH_2)_3CO_2H$$

b. Account for this oxidative decyanation.

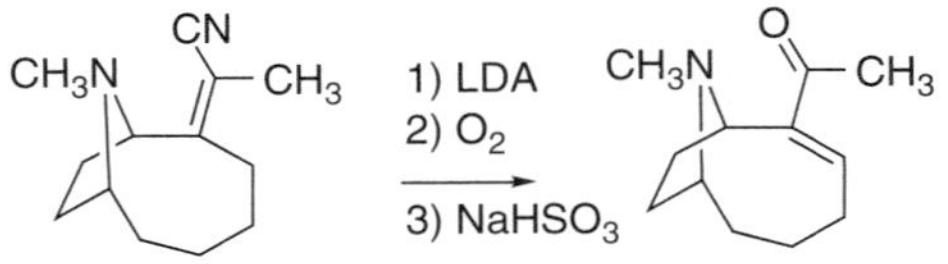

$$\xrightarrow[\text{3) NaHSO}_3]{\text{1) LDA} \quad \text{2) O}_2}$$

c. It is found that the oxidative decarboxylation of β-silyl and β-stannyl carboxylic acids is substantially accelerated by cupric acetate.

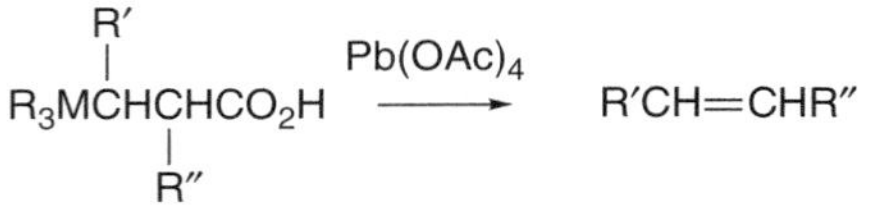

$$R_3MCHCHCO_2H \xrightarrow{Pb(OAc)_4} R'CH=CHR''$$

M = Si, Sn

12.12. Use retrosynthetic analysis to devise a sequence of reactions that could accomplish the formation of the structure on the left from the potential precursor on the right.

(a)

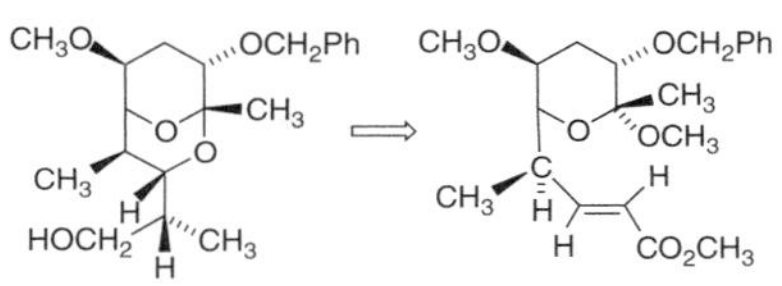

(b)

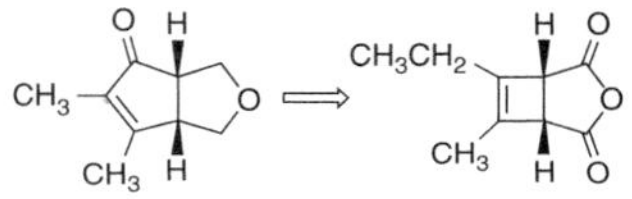

(c)

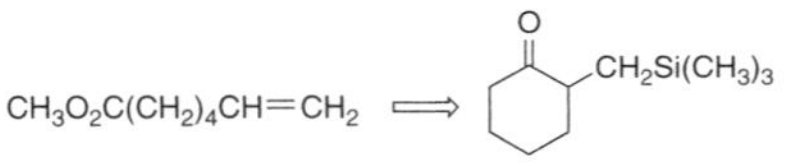

(d)

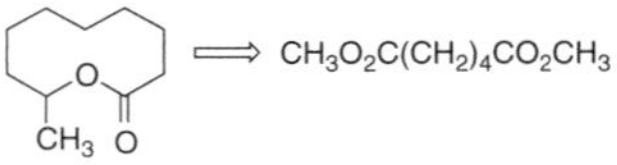

(e)

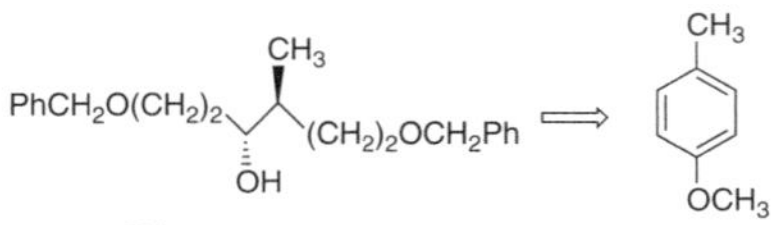

(f)

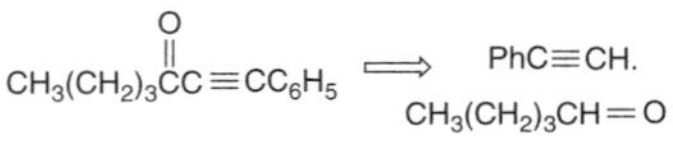

(g)

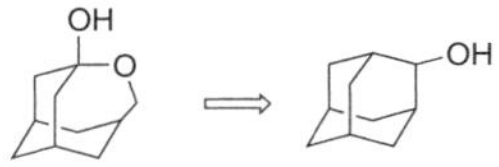

(h)

(i)

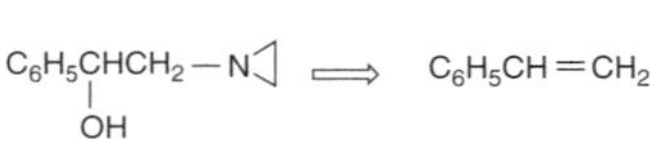

(j)

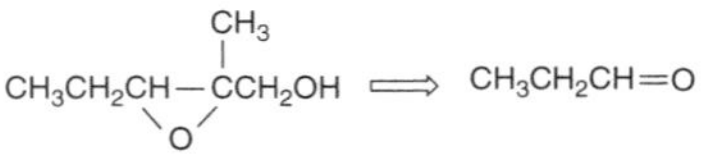

(k)

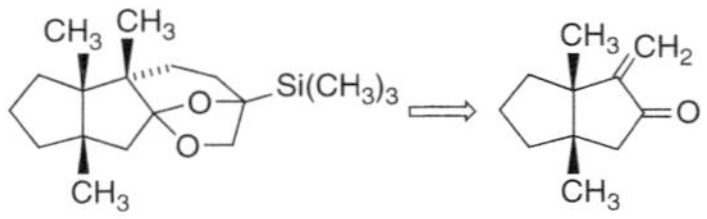

(l)

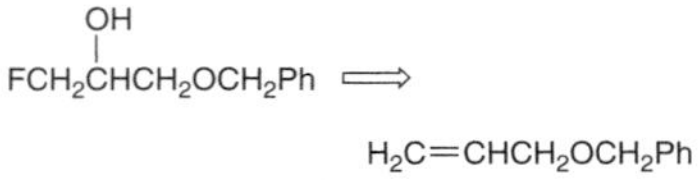

(m) CH₃O₂CC=CHOCH₃ ⟹ CH₃O₂CC=CHOCH₃

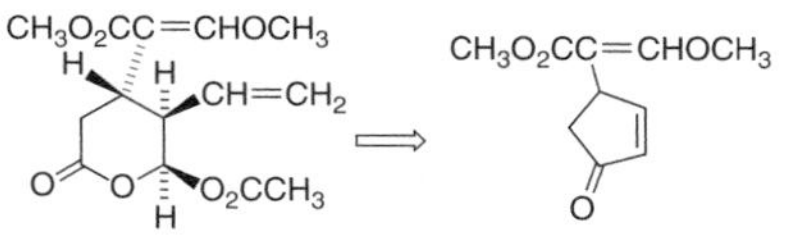

(n) ⟹

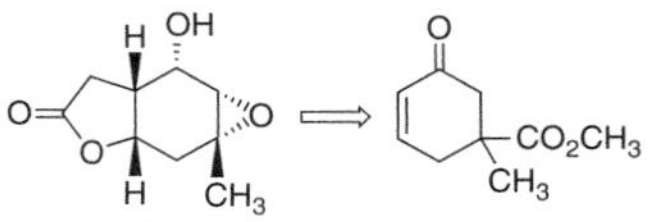

(o) ⟹

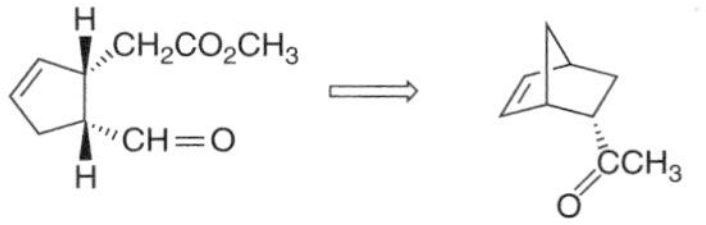

(p) ⟹

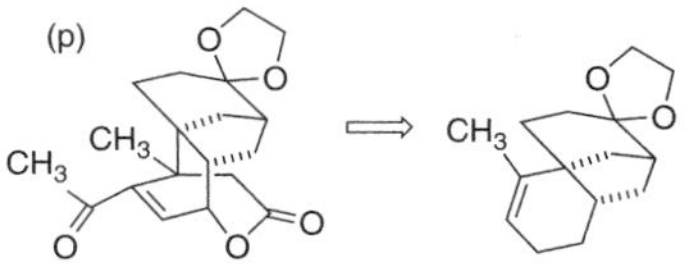

(q) ⟹

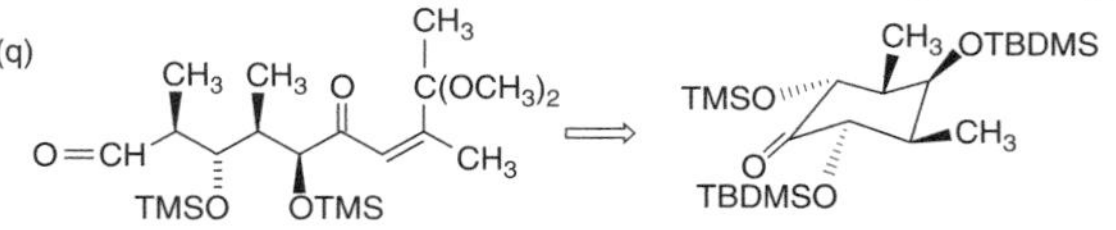

12.13. Tomoxetine and fluoxetine are antidepressants. Both enantiomers of each compound can be prepared enantiospecifically starting from cinnamyl alcohol. Give a reaction sequence that will accomplish this objective.

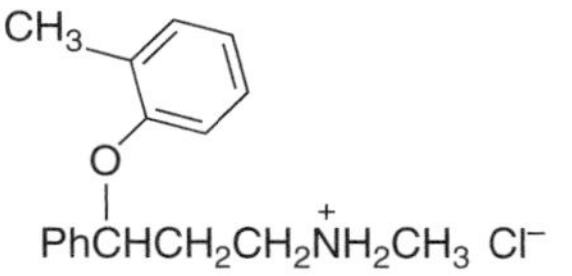

tomoxetine

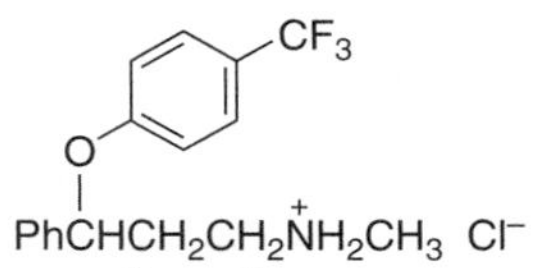

fluoxetine

12.14. The irradiation of **14-A** in the presence of rose bengal and oxygen in methanol gives **14-B** as the only observable product (72% yield). When the irradiation is carried out in acetaldehyde as solvent, the yield of **14-B** is reduced to 54% and two additional products, **14-C** (19%) and **14-D** (17%), are formed. Account for the formation of each product.

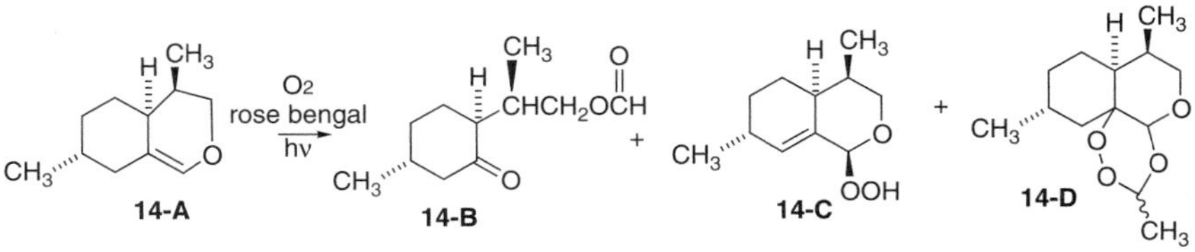

O₂
rose bengal
hv

14-A   **14-B**   **14-C**   **14-D**

12.15. Analyze the following data on the product ratios obtained in the epoxidation of 3-substituted cyclohexenes by dimethyldioxirane. What are the principal factors that determine the stereoselectivity?

Substituent	*trans:cis*[a]
OH	66:34[b]
OH	15:85
OCH₃	85:15
O₂CCH₃	62:38
CO₂CH₃	68:32
CO₂H	84:16

(*Continued*)

Substituent	*trans:cis*[a]
NHCOPh	3:97
Cl	90:10[c]
CF$_3$	90:10[c]
CH$_3$	47:53
(CH$_3$)$_2$CHCH$_2$	54:46
(CH$_3$)$_3$C	95:5[c]
Ph	85:15

a. Solvent is 9:1 CCl$_4$-acetone except as noted otherwise.
b. Solvent is 9:1 methanol-acetone.
c. Solvent is acetone.

12.16. Offer a mechanistic explanation for the following observations.

a. A change from ether as solvent to pentane with 12-crown-4 reverses the stereoselectivity of LiAlH$_4$ reduction of *cis*-3-benzyloxycyclohexene oxide, but not the *trans* isomer.

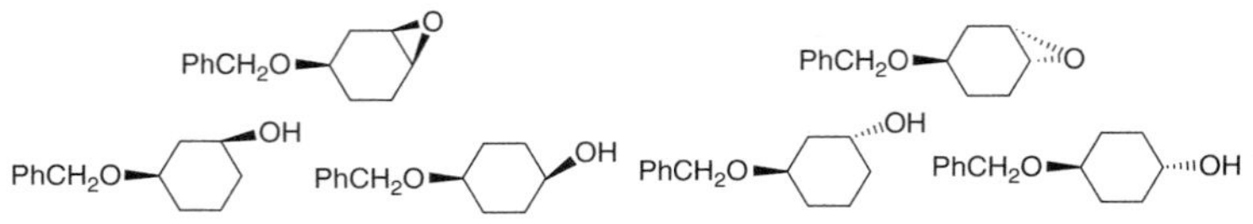

LiAlH$_4$-ether	2	98	97	3
LiAlH$_4$-pentane 12-crown-4	82	18	97	3

b. In the presence of a strong protic acid or a Lewis acid, acetophenones and propiophenones rearrange to arylalkanoic acid on reaction with Pb(OAc)$_4$.

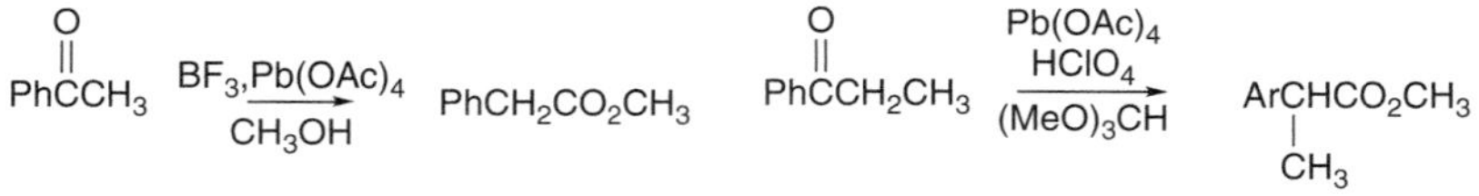

c. Acylation leads to reaction of the hydroperoxides **16α** and **16β**, but the products are different. In **16β**, the vinyl substituent migrates giving ring expansion, whereas with **16α** an enone is formed.

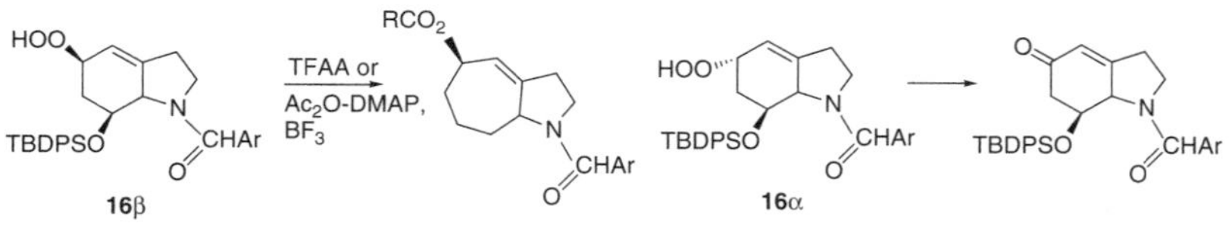

12.17. Various terpene-derived materials are important in the formulation of fragrances and flavors. One example is the tricyclic furan shown below, which is commercially used under the trademark Ambrox.® The synthetic sequences below have been developed to prepare related structures. Suggest reagents for each step in these sequences.

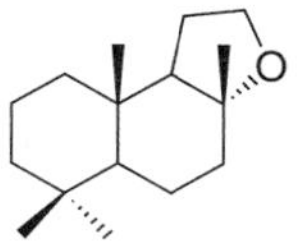

Ambrox^R

a. This sequence was developed to avoid the use of transition metal reagents and minimize by-products.

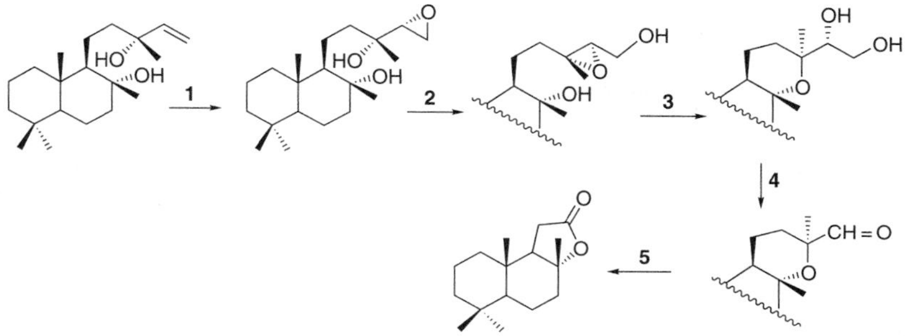

b. The following sequence led to the 6-α–hydroxy derivative.

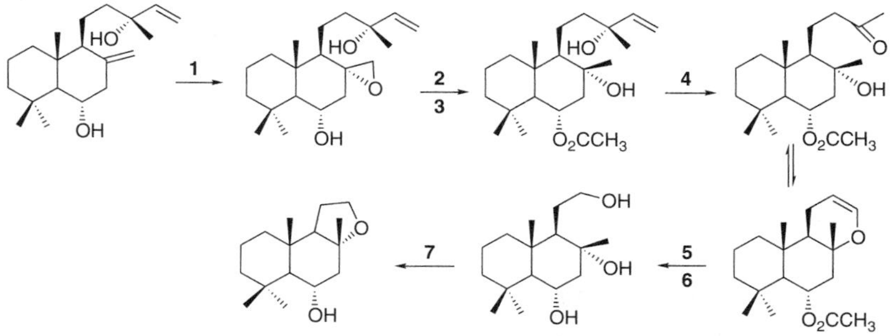

12.18. The closely related enones **18-A** and **18-B** give different products when treated with Pb(OAc)$_4$ in CH$_3$CN. Formulate mechanisms to account for both products and identify the factor(s) that lead to the divergent structures.

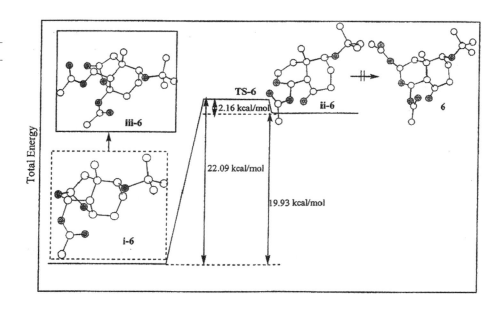

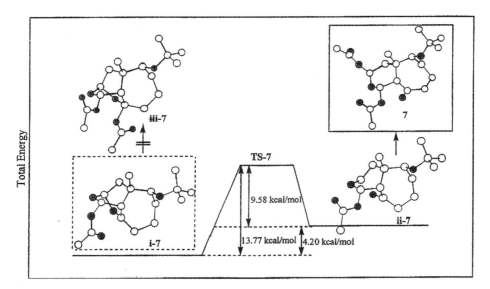

Fig. 12.P18. Comparison of the computed energy profiles for **18-A** and **18-B**. Reproduced from *J. Org. Chem.*, **67**, 2447 (2002), by permission of the American Chemical Society.

12.19. Predict the structure and stereochemistry of the Lewis acid–catalyzed rearrangement of the following epoxides.

13

Multistep Syntheses

Introduction

The reactions discussed in the preceding chapters provide tools for synthesizing new and complex molecules, but a strategy for using these reactions is essential for successful multistep syntheses. The sequence of individual reactions must be planned so that the reactions are mutually compatible with the final synthetic goal. Certain functional groups can interfere with prospective reactions and such problems must be avoided either by a modification of the sequence or by temporarily masking (protecting) the interfering group. *Protective groups* are used to temporarily modify functionality, which is then restored when the protecting group is removed. Another approach is to use a *synthetic equivalent group* in which a particular functionality is introduced as an alternative structure that can subsequently be converted to the desired group.

Protective groups and synthetic equivalent groups are tactical tools of multistep syntheses. They are the means, along with the individual synthetic methods, to reach the goal of a completed synthesis, and these tactical steps must be incorporated into an overall synthetic plan. A synthetic plan is normally created on the basis of a *retrosynthetic analysis*, which involves identification of the particular bonds that can be formed to obtain the desired molecule. Depending on the complexity of the synthetic target, the retrosynthetic analysis may be obvious or intricate. A synthetic plan identifies potential starting materials and reactions that can lead to the desired molecule, and most such plans involve a combination of *linear sequences* and *convergent steps*. Linear sequences construct the target molecule step-by-step by incremental additions and functional group transformations. Convergent steps bring together larger segments of the molecule that have been created by linear sequences. As the overall synthetic yield is the multiplication product of the yield of each of the individual steps in the synthesis, incorporation of a convergent step improves overall yield by reducing the length of the linear sequences. After discussing some general aspects of synthetic analysis and planning, we summarize several syntheses that illustrate application of multistep synthetic methods to representative molecules. In the final sections of the chapter, we consider solid phase synthesis and its application to polypeptide, poly-nucleotide, and combinatorial syntheses.

1163

13.1. Synthetic Analysis and Planning

13.1.1. Retrosynthetic Analysis

The tools available to the synthetic chemist consist of an extensive catalog of reactions and the associated information on such issues as stereoselectivity and mutual reactivity. This knowledge permits a judgment on the applicability of a particular reaction in a synthetic sequence. Broad mechanistic insight is also crucial to synthetic analysis. The relative position of functional groups in a potential reactant may lead to specific interactions or reactions. The ability to recognize such complications enables appropriate adjustments to the synthetic plan. Mechanistic concepts can guide optimization of reaction conditions. They are as well the basis for developing new reactions that may be necessary in a particular situation.

The planning of a synthesis involves a critical comparative evaluation of alternative reaction sequences that could reasonably be expected to lead to the desired structure from appropriate starting materials. In general, the complexity of a synthetic plan increases with the size of the molecule and with increasing numbers of functional groups and stereogenic centers. The goal of synthetic analysis is to recognize possible pathways to the target compound and to develop a suitable sequence of synthetic steps. In general, a large number of syntheses of any given compound are possible. The objective of synthetic analysis and planning is to develop a reaction sequence that will complete the desired synthesis efficiently within the constraints that apply.

The restrictions that apply depend on the purposes for which the synthesis is being done. A synthesis of a material to be prepared in substantial quantity may impose a limitation on the cost of the starting materials. Syntheses for commercial production must meet such criteria as economic feasibility, acceptability of by-products, and safety. Syntheses of structures having several stereogenic centers must deal with the problem of stereoselectivity. If an enantiomerically pure material is to be synthesized, the means of controlling absolute configuration must be considered. The development of a satisfactory plan is the chemist's intellectual challenge and it puts a premium on creativity and ingenuity. There is no single correct solution. Although there is no established routine by which a synthetic plan can be formulated, general principles that can guide synthetic analysis and planning have been described.[1]

The initial step in creating a synthetic plan involves a *retrosynthetic analysis*. The structure of the molecule is dissected step by step along reasonable pathways to successively simpler compounds until molecules that are acceptable as starting materials are identified. Several factors enter into this process, and all are closely interrelated. The recognition of *bond disconnections* allows the molecule to be broken down into *key intermediates*. Such disconnections must be made in such a way that it is feasible to form the bonds by some synthetic process. The relative placement of potential functionality strongly influences which bond disconnections are preferred. To emphasize that these disconnections must correspond to transformations that can be conducted in the synthetic sense, they are sometimes called *antisynthetic transforms*, i.e., the reverse of synthetic steps. An open arrow symbol, $\Rightarrow$, is used to indicate an antisynthetic transform.

Retrosynthetic analysis can identify component segments of a target molecule that can serve as key intermediates, and the subunits that are assembled to construct

---

[1] E. J. Corey and X.-M. Cheng, *The Logic of Chemical Synthesis*, Wiley, New York, 1989.

them are sometimes called *synthons*. Synthons must not only correspond structurally to the desired subunit, but they must also have appropriate reactivity to allow bond formation with adjacent subunits. For example, in the case of aldol reactions, one reagent must serve as the electrophile and the other as the nucleophile. In a ring construction by a Diels-Alder reaction, the diene and dienophile must have compatible reactivity. Similarly, in bond constructions done using organometallic intermediates, the synthons must possess appropriate mutual reactivity.

The overall synthetic plan consists of a sequence of reactions designed to construct the total molecular framework from the key intermediates. The plan should take into account the advantages of a *convergent synthesis*. The purpose of making a synthesis more convergent is to shorten its overall length. In general, it is desirable to construct the molecule from a few key segments that can be combined late in the synthesis rather than build the molecule step-by-step from a single starting material. The overall yield is the multiplication product of the yields for all the individual steps. Overall yields decrease with the increasing number of steps to which the original starting material is subjected.[2] One of the characteristics of a multistep sequence is the *longest linear sequence*, which is the maximum number of steps from an original starting material to the final product. For example, in the case below, a single convergency that reduces the longest linear sequence from six to three improves the overall yield from 53 to 73% if the yield was 90% in each transformation.

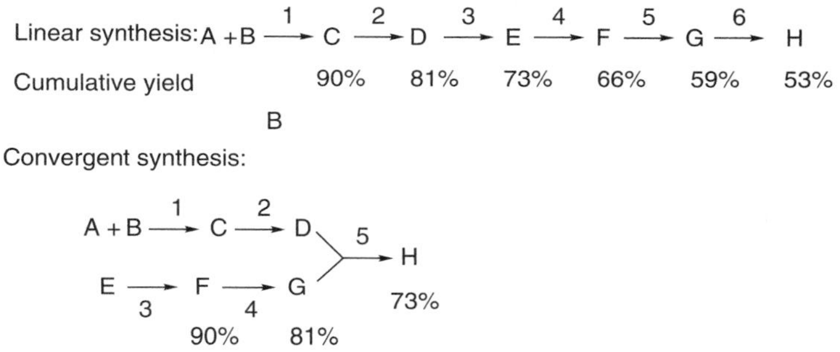

Splitting a 15-step synthesis into three branches of four steps each will improve the yield from 8 to 48% if each step occurs in 90% yield.

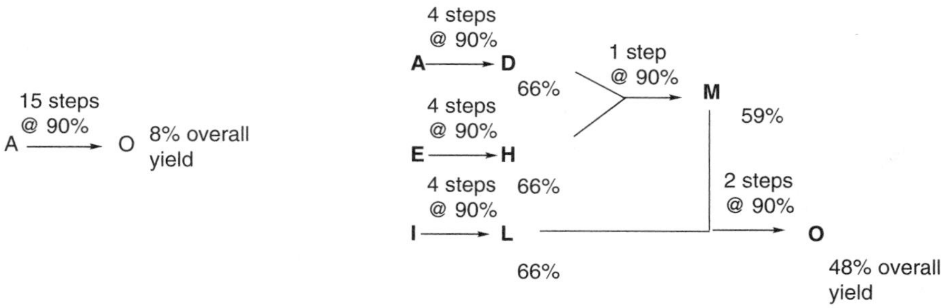

After a plan for assembly of the key intermediates into the molecular framework has been developed, the details of incorporation and transformation of functional

[2.] A formal analysis of the concept of convergency has been presented by J. B. Hendrickson, *J. Am. Chem. Soc.*, **99**, 5439 (1977).

groups are considered. It is frequently necessary to interconvert functional groups, which may be to done to develop a particular kind of reactivity at a center or to avoid interference with a reaction step. Protective groups and synthetic equivalent groups are important for planning of functional group transformations. Owing to the large number of procedures for interconverting the common functional groups, achieving the final array of functionality is often less difficult than establishing the overall molecular skeleton and stereochemistry.

The synthetic plan must also provide for control of stereochemistry. In the case of cyclic compounds, advantage often can be taken of the facial preferences of the rings and the stereoselectivity of reagents to establish the stereochemistry of substituents. For example, the *syn*-directive effect of hydroxy groups in epoxidation (see p. 1093) or the strong preference for *anti* addition in iodolactonization (see p. 311) can be used to determine the configuration of new stereogenic centers. Similarly, the cyclic TS of sigmatropic rearrangements often allows predictable stereoselectivity. Chiral auxiliaries and catalysts provide means of establishing configuration in enantioselective syntheses. A plan for a stereo- or enantioselective synthesis must include the basis for controlling the configuration at each stereocenter.

The care with which a synthesis is analyzed and planned will have a great impact on the likelihood of its success. The investment of material and effort that is made when the synthesis is begun may be lost if the plan is faulty. Even with the best of planning, however, unexpected problems are often encountered. This circumstance again tests the ingenuity of the chemist to devise a modified plan that can overcome the unanticipated obstacle.

13.1.2. Synthetic Equivalent Groups

Retrosynthetic analysis may identify a need to use *synthetic equivalent groups*. These groups are synthons that correspond structurally to a subunit of the target structure, but in which the reactivity of the functionality is masked or modified. As an example, suppose the transformation shown below was to be accomplished.

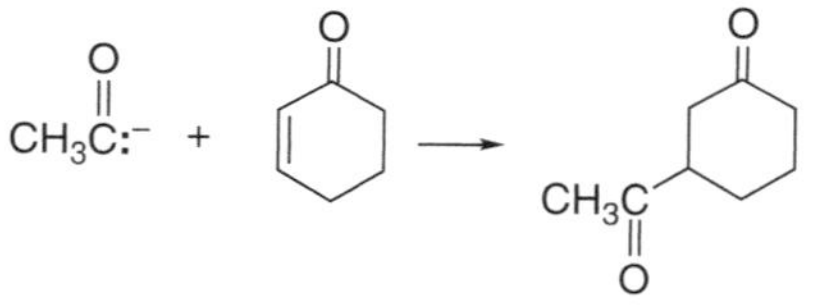

The electrophilic α,β-unsaturated ketone is reactive toward nucleophiles, but the nucleophile that is required, an acyl anion, is not normally an accessible entity. There are several potential reagents that could introduce the desired acyl anion in a masked form. The masked functionality used in place of an inaccessible species is called a synthetically equivalent group. Often the concept of "umpolung" is involved in devising synthetic equivalent groups. The term *umpolung* refers to the formal reversal of the normal polarity of a functional group.[3] Acyl groups are normally *electrophilic*, but a synthetic operation may require the transfer of an acyl group as a *nucleophile*. The *acyl anion* is an umpolung equivalent of the electrophilic acylium cation.

[3.] For a general discussion and many examples of the use of the umpolung concept, see D. Seebach, *Angew. Chem. Int. Ed. Engl.*, **18**, 239 (1979).

Owing to the great importance of carbonyl groups in synthesis, a substantial effort has been devoted to developing nucleophilic equivalents for introduction of acyl groups.[4] One successful method involves a three-step sequence in which an aldehyde is converted to an O-protected cyanohydrin. The α-alkoxynitrile is then deprotonated, generating a nucleophilic carbanion **A**.[5] After carbon-carbon bond formation, the carbonyl group can be regenerated by hydrolysis of the cyanohydrin. This sequence has been used to solve the problem of introducing an acetyl group at the β-position of cyclohexenone.[6]

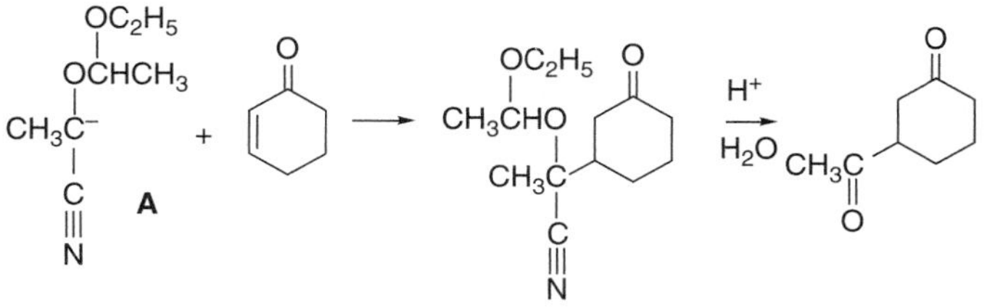

Ref. 5

α-Lithiovinyl ethers and the corresponding cuprates are other examples of acyl anion equivalents.

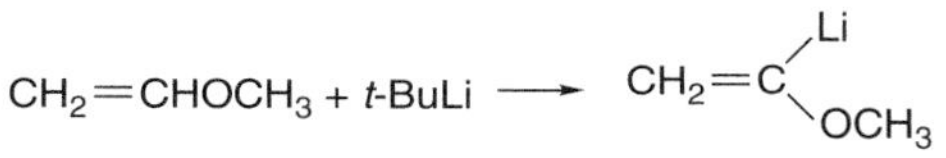

Ref. 7

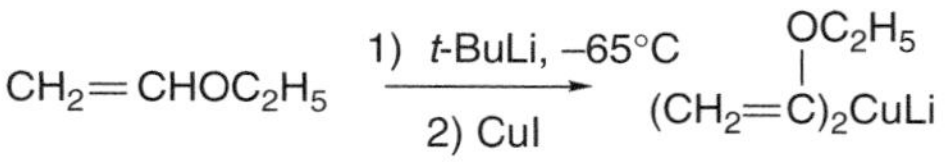

Ref. 8

These reagents are capable of adding the α-alkoxyvinyl group to electrophilic centers. Subsequent hydrolysis can generate the carbonyl group and complete the desired transformation.

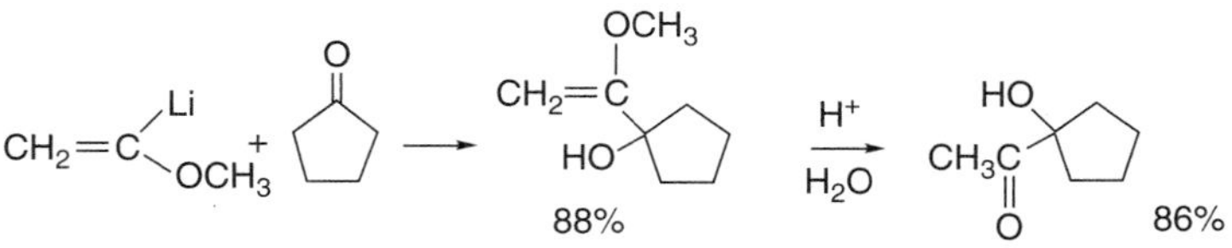

Ref. 7

4. For a review of acyl anion synthons, see T. A. Hase and J. K. Koskimies, *Aldrichica Acta*, **15**, 35 (1982).
5. G. Stork and L. Maldonado, *J. Am. Chem. Soc.*, **93**, 5286 (1971); *J. Am. Chem. Soc.*, **96**, 5272 (1974).
6. For further discussion of synthetic applications of the carbanions of O-protected cyanohydrins, see J. D. Albright, *Tetrahedron*, **39**, 3207 (1983).
7. J. E. Baldwin, G. A. Hoefle, and O. W. Lever, Jr., *J. Am. Chem. Soc.*, **96**, 7125 (1974).
8. R. K. Boeckman, Jr., and K. J. Bruza, *J. Org. Chem.*, **44**, 4781 (1979).

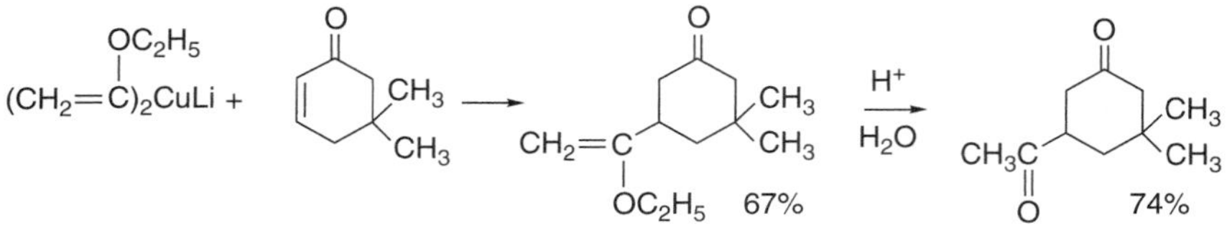

Ref. 8

Lithiation of vinyl thioethers[9] and vinyl carbamates[10] also provides acyl anion equivalents.

Sulfur compounds are useful as nucleophilic acyl equivalents. The most common reagents of this type are 1,3-dithianes, which on lithiation provide a nucleophilic acyl equivalent. In dithianes an umpolung is achieved on the basis of the carbanion-stabilizing ability of the sulfur substituents. The lithio derivative is a reactive nucleophile toward alkyl halides and carbonyl compounds.[11]

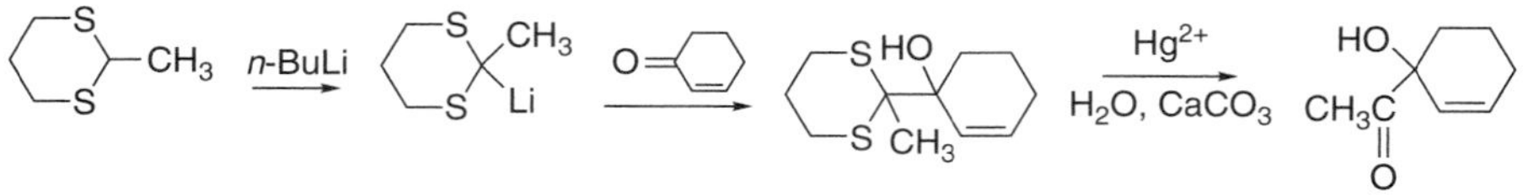

1,3-Dithianes have found considerable application in multistep syntheses.[12] Scheme 13.1 summarizes some examples of synthetic sequences that employ acyl anion equivalents.

Scheme 13.1. Synthetic Sequences Using Acyl Anion Equivalents

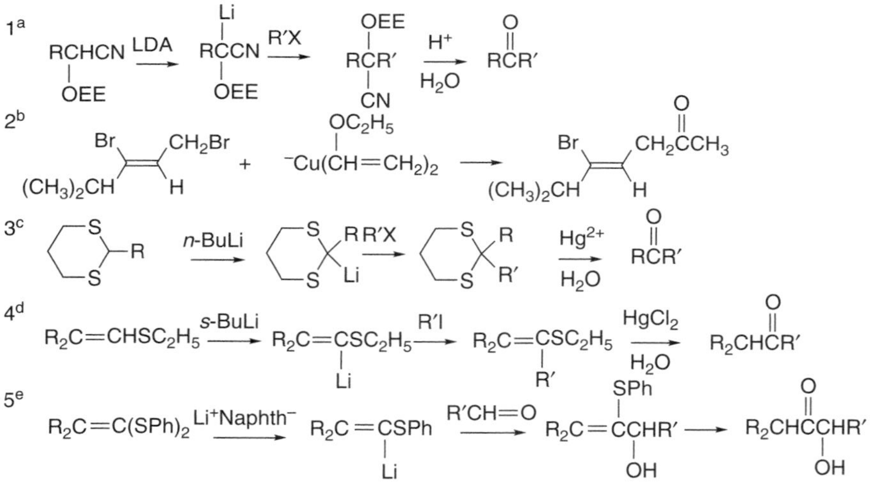

a. G. Stork and L. Maldonado, *J. Am. Chem. Soc.*, **93**, 5236 (1971).
b. P. Canonne, R. Boulanger, and P. Angers, *Tetrahedron Lett.*, **32**, 5861 (1991).
c. D. Seebach and E. J. Corey, *J. Org. Chem.*, **40**, 231 (1975).
d. K. Oshima, K. Shimoji, H. Takahashi, H. Yamamoto, and H. Nozaki, *J. Am. Chem. Soc.*, **95**, 2694 (1973).
e. T. Cohen and R. B. Weisenfeld, *J. Org. Chem.*, **44**, 3601 (1979).

9. K. Oshima, K. Shimoji, H. Takahashi, and H. Nozaki, *J. Am. Chem. Soc.*, **95**, 2694 (1973).
10. S. Sengupta and V. Sniekus, *J. Org. Chem.*, **55**, 5680 (1990).
11. D. Seebach and E. J. Corey, *J. Org. Chem.*, **40**, 231 (1975); B. H. Lipshutz and E. Garcia, *Tetrahedron Lett.*, **31**, 7261 (1990).
12. M. Yus, C. Najera, and F. Foubelo, *Tetrahedron*, **59**, 6147 (2003); A. B. Smith, III, and C. M. Adams, *Acc. Chem. Res.*, **37**, 365 (2004).

Another synthetic equivalent that has been extensively developed corresponds to the propanal "homoenolate," $^-CH_2CH_2CH = O$.[13] This structure is the umpolung equivalent of an important electrophilic reagent, the α,β-unsaturated aldehyde acrolein. Scheme 13.2 illustrates some of the propanal homoenolate equivalents that have been developed. In general, the reagents used for these transformations are reactive toward electrophiles such as alkyl halides and carbonyl compounds. Several general points can be made about the reagents in Scheme 13.2. First, it should be noted that they all deliver the aldehyde functionality in a masked form, such as an acetal or enol ether. The aldehyde is liberated in a final step from the protected precursor. Several of the reagents involve delocalized allylic anions, which gives rise to the possibility of electrophilic attack at either the α- or γ-position of the allylic group. In most cases, the γ-attack that is necessary for the anion to function as a propanal homoenolate is dominant. In Entry 1, the 2-methoxycyclopropyllithium is used to form a cyclopropyl carbinol. The methoxy group serves both to promote fragmentation of the cyclopropyl ring and to establish the aldehyde oxidation level. In Entry 2, the lithiation product of allyl methyl ether serves as a nucleophile and the aldehyde group is liberated by hydrolysis. Entry 3 is similar, but uses a trimethylsilyl ether. In Entry 4, allylic lithiation of an *N*-allylamine provides a nucleophile and can subsequently be hydrolyzed to the aldehyde.

In Entry 5, the carbanion-stabilizing ability of the sulfonyl group enables lithiation and is then reductively removed after alkylation. The reagent in Entry 6 is prepared by dilithiation of allyl hydrosulfide using *n*-butyllithium. After nucleophilic addition and S-alkylation, a masked aldehyde is present in the form of a vinyl thioether. Entry 7 uses the epoxidation of a vinyl silane to form a γ-hydroxy aldehyde masked as a cyclic acetal. Entries 8 and 9 use nucleophilic cuprate reagents to introduce alkyl groups containing aldehydes masked as acetals.

The concept of developing reagents that are the synthetic equivalent of inaccessible species can be taken another step by considering dipolar species. For example, structures **B** and **C** incorporate both electrophilic and nucleophilic centers. Such reagents might be incorporated into ring-forming schemes, since they have the ability, at least formally, of undergoing cycloaddition reactions.

$$C_2H_5OC\overset{-}{C}HCH_2\overset{+}{C}H_2$$
$$\underset{O}{\|}$$
B

$$\overset{O}{\underset{\|}{}}$$
$$^-CCH_2CH_2{}^+$$
C

Among the real chemical species that have been developed along these lines are the cyclopropyl phosphonium ions **1** and **2**.

$$Ph_3P^+ \diagdown \diagup CO_2C_2H_5$$
1

$$Ph_3P^+ \diagdown \diagup SPh$$
2

[13] For reviews of homoenolate anions, see J. C. Stowell, *Chem. Rev.*, **84**, 409 (1984); N. H. Werstiuk, *Tetrahedron*, **39**, 205 (1983).

Scheme 13.2. Synthetic Sequences Using Homoenolate Synthetic Equivalents

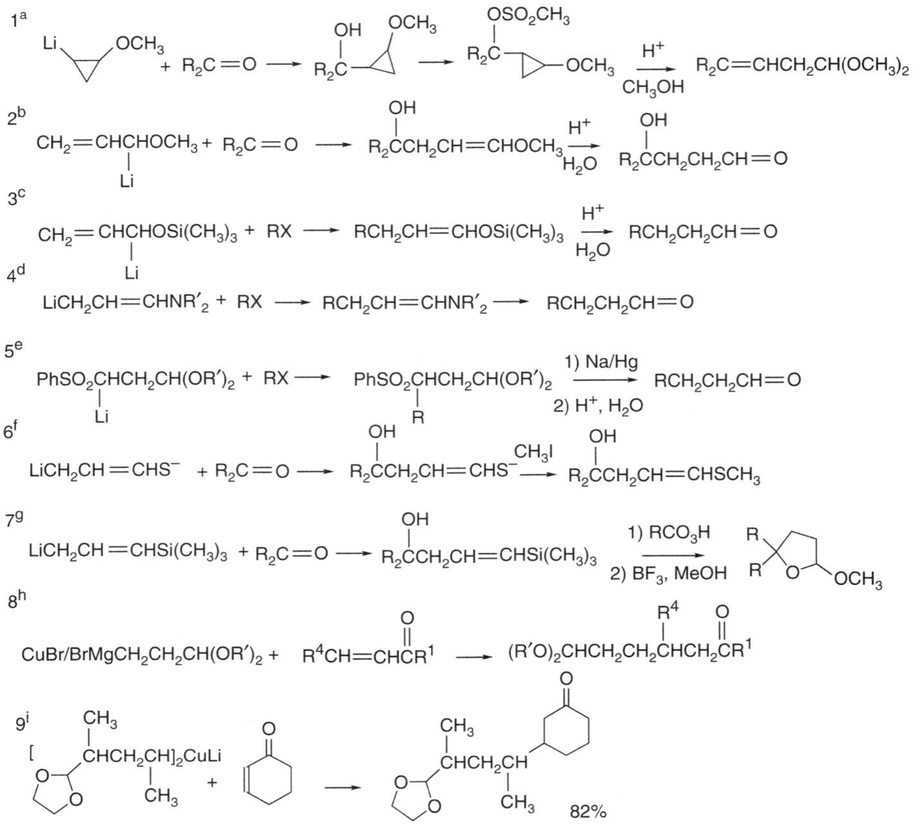

a. E. J. Corey and P. Ulrich, *Tetrahedron Lett.*, 3685 (1975).
b. D. A. Evans, G. C. Andrews, and B. Buckwalter, *J. Am. Chem. Soc.*, **96**, 5560 (1974).
c. W. C. Still and T. L. Macdonald, *J. Am. Chem. Soc.*, **96**, 5561 (1974).
d. H. Ahlbrecht and J. Eichler, *Synthesis*, 672 (1974); S. F. Martin and M. T. DuPriest, *Tetrahedron Lett.*, 3925 (1977); H. Ahlbrecht G. Bonnet, D. Enders, and G. Zimmerman, *Tetrahedron Lett.*, **21**, 3175 (1980). e. M. Julia and B. Badet, *Bull. Soc. Chim. Fr.*, 1363 (1975); K. Kondo and D. Tunemoto, *Tetrahedron Lett.*, 1007 (1975).
f. K.-H. Geiss, B. Seuring, R. Pieter, and D. Seebach, *Angew. Chem. Int. Ed. Engl.*, **13**, 479 (1974); K.-H. Geiss, D. Seebach, and B. Seuring, *Chem. Ber.*, **110**, 1833 (1977).
g. E. Ehlinger and P. Magnus, *J. Am. Chem. Soc.*, **102**, 5004 (1990).
h. A. Marfat and P. Helquist, *Tetrahedron Lett.*, 4217 (1978); A. Leone-Bay and L. A. Paquette, *J. Org. Chem.*, **47**, 4172 (1982).
i. J. P. Cherkaukas and T. Cohen, *J. Org. Chem.*, **57**, 6 (1992).

The phosphonium salt **1** reacts with β-ketoesters and β-ketoaldehydes to give excellent yields of cyclopentenecarboxylate esters.

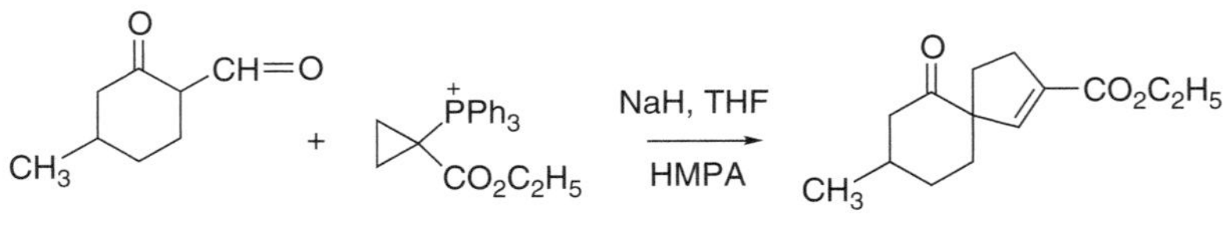

Ref. 14

14. W. G. Dauben and D. J. Hart, *J. Am. Chem. Soc.*, **99**, 7307 (1977).

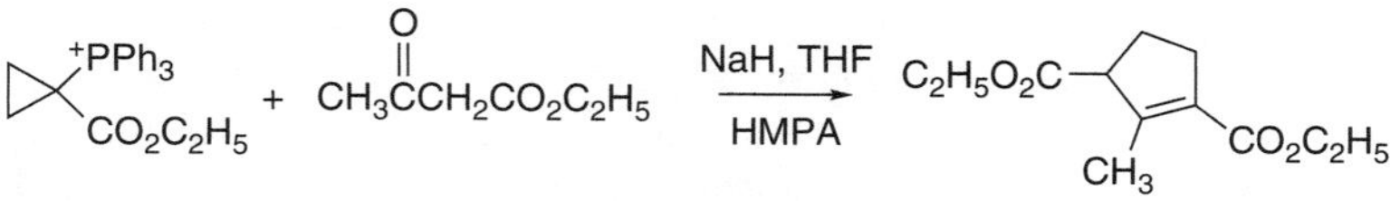

Ref. 15

Several steps are involved in these reactions. First, the enolate of the β-ketoester opens the cyclopropane ring. The polarity of this process corresponds to that in the formal synthon **B** because the cyclopropyl carbons are electrophilic. The product of the ring-opening step is a stabilized Wittig ylide, which can react with the ketone carbonyl to form the carbocyclic ring.

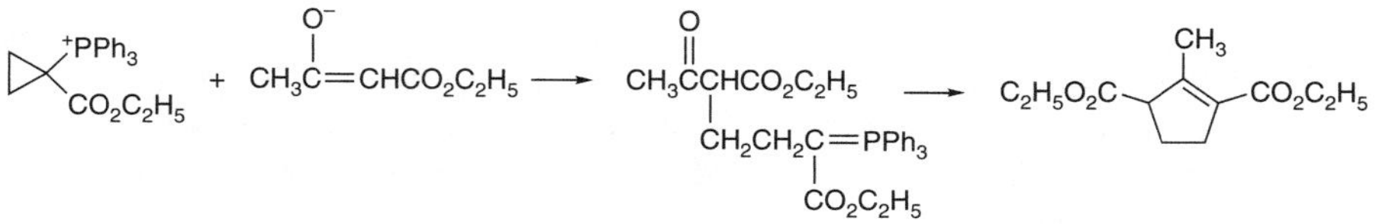

The phosphonium ion **2** reacts similarly with enolates to give vinyl sulfides. The vinyl sulfide group can then be hydrolyzed to a ketone. The overall transformation corresponds to the reactivity of the dipolar synthon **C**.

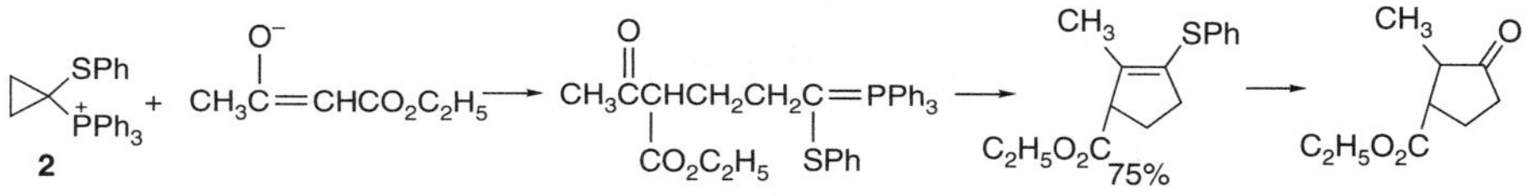

Ref. 16

Many other examples of synthetic equivalent groups have been developed. For example, in Chapter 6 we discussed the use of diene and dienophiles with masked functionality in the Diels-Alder reaction. It should be recognized that there is no absolute difference between what is termed a "reagent" and a "synthetic equivalent group." For example, we think of potassium cyanide as a reagent, but the cyanide ion is a nucleophilic equivalent of a carboxy group. This reactivity is evident in the classical preparation of carboxylic acids from alkyl halides via nitrile intermediates.

$$RX + KCN \longrightarrow RCN \xrightarrow[H^+]{H_2O} RCO_2H$$

The important point is that synthetic analysis and planning should not be restricted to the specific functionalities that appear in the target molecules. These groups can be incorporated as masked equivalents by methods that would not be possible for the functional group itself.

13.1.3. Control of Stereochemistry

The degree of control of stereochemistry that is necessary during synthesis depends on the nature of the molecule and the objective of the synthesis. The issue

15. P. L. Fuchs, *J. Am. Chem. Soc.*, **96**, 1607 (1974).
16. J. P. Marino and R. C. Landick, *Tetrahedron Lett.*, 4531 (1975).

becomes critically important when the target molecule has several stereogenic centers, such as double bonds, ring junctions, and asymmetric carbons. The number of possible stereoisomers is 2^n, where n is the number of stereogenic centers. Failure to control stereochemistry of intermediates in the synthesis of a compound with several centers of stereochemistry leads to a mixture of stereoisomers that will, at best, result in a reduced yield of the desired product and may generate inseparable mixtures. For properties such as biological activity, obtaining the correct stereoisomer is crucial.

We have considered stereoselectivity for many of the reactions that are discussed in the earlier chapters. In ring compounds, for example, stereoselectivity can frequently be predicted on the basis of conformational analysis of the reactant and consideration of the steric and stereoelectronic factors that influence reagent approach. In the *diastereoselective synthesis* of a chiral compound in racemic form, it is necessary to control the *relative configuration* of all stereogenic centers. Thus in planning a synthesis, the stereochemical outcome of all reactions that form new double bonds, ring junctions, or asymmetric carbons must be incorporated into the synthetic plan. In a completely stereoselective synthesis, each successive stereochemical feature is introduced in the proper relationship to existing stereocenters, but this ideal is often difficult to achieve. When a reaction is not completely stereoselective, the product will contain one or more diastereomers of the desired product. This requires either a purification or some manipulation to correct the stereochemistry. Fortunately, diastereomers are usually separable, but the overall efficiency of the synthesis is decreased with each such separation. Thus, high stereoselectivity is an important goal of synthetic planning.

If the compound is to be obtained in enantiomerically pure form, an *enantioselective synthesis* must be developed. As discussed in Section A.2.5, the stereochemical control may be based on chirality in the reactants, auxiliaries, reagents, and/or catalysts. There are several general approaches that are used to obtain enantiomerically pure material by synthesis. One is based on incorporating a *resolution* into the synthetic plan. This approach involves use of racemic or achiral starting materials and resolving some intermediate in the synthesis. In a synthesis based on a resolution, the steps subsequent to the resolution step must meet two criteria: (1) they must not disturb the configuration at existing stereocenters, and (2) new centers of stereochemistry must be introduced with the correct configuration relative to those that already exist. A second general approach is to use an *enantiomerically pure starting material*. Highly enantioselective reactions, such as the Sharpless epoxidation, can be used to prepare enantiomerically pure starting materials. There are a number of naturally occurring materials, or substances derived from them, that are available in enantiomerically pure form.[17]

Enantioselective synthesis can also be based on *chiral reagents*. Examples are hydroboration or reduction using one of the commercial available borane reagents. Again, a completely enantioselective synthesis must be capable of controlling the stereochemistry of all newly introduced stereogenic centers so that they have the proper relationship to the chiral centers that exist in the starting material. When this is not achieved, the desired stereoisomer must be separated and purified. A fourth method for enantioselective synthesis involves the use of a stoichiometric amount of a *chiral auxiliary*. This is an enantiomerically pure material that can control the stereochemistry of one or more reaction steps in such a way as to give product having the desired configuration. When the chiral auxiliary has achieved its purpose, it can be

---

[17.] For a discussion of this approach to enantioselective synthesis, see S. Hanessian, *Total Synthesis of Natural Products: The Chiron Approach*, Pergamon Press, New York, 1983.

eliminated from the molecule. As in syntheses involving resolution or enantiomerically pure starting materials, subsequent steps must give the correct configuration of newly created stereocenters. Another approach to enantioselective synthesis is to use a *chiral catalyst* in a reaction that creates one or more stereocenters. If the catalyst operates with complete efficiency, an enantiomerically pure material will be obtained. Subsequent steps must control the configuration of newly introduced stereocenters.

In practice, any of these approaches might be the most effective for a given synthesis. If they are judged on the basis of absolute efficiency in the use of a chiral material, the ranking is resolution < chiral reactant < chiral reagent < chiral auxiliary < enantioselective catalyst. A resolution process inherently employs only half of the original racemic material. A chiral starting material can, in principle, be used with 100% efficiency, but it is consumed and cannot be reused. A chiral reagent is also consumed, but in principle it can be regenerated, as is done for certain organoboranes (see p. 350). A chiral auxiliary must be used in a stoichiometric amount but it can be recovered. A chiral catalyst, in principle, can produce an unlimited amount of an enantiomerically pure material.

The key issue for synthesis of pure stereoisomers, in either racemic or enantiomerically pure form, is that the configuration at newly created stereocenters be controlled in some way. This can be accomplished by several different methods. Existing functional groups may exert a steric or stereoelectronic influence on the reaction center. For instance, an existing functional group may control the approach of a reagent by coordination, which occurs, for example, in hydroxy-directed cyclopropanation (see p. 919). An existing chiral center may control reactant conformation and, thereby, the direction of approach of a reagent.

Generally, the closer the reaction occurs to an existing stereogenic center, the more likely the reaction is to exhibit high stereoselectivity. For example, the creation of adjacent stereogenic centers in aldol and organometallic addition reactions is generally strongly influenced by adjacent substituents leading to a preference for a *syn* or *anti* disposition of the new substituent. We also encountered some examples of *1,3-asymmetric induction*, as, for example, the role of chelates in reduction of β-hydroxy ketones (p. 412), in chelation control of Mukaiyama addition reactions (p. 94), and in hydroboration (Section p. 342). More remote chiral centers are less likely to influence stereoselectivity and examples of, e.g., 1,4- and 1,5-asymmetric induction, are less common. Whatever the detailed mechanism, the synthetic plan must include the means by which the required stereochemical control is to be achieved. If this cannot be done, the price to be paid is a separation of stereoisomers and the resulting reduction in overall yield.

13.2. Illustrative Syntheses

In this section, we consider several syntheses of six illustrative compounds. We examine the retrosynthetic plans and discuss crucial bond-forming steps and the means of stereochemical control. In this discussion, we have the benefit of hindsight in being able to look at successfully completed syntheses. This retrospective analysis can serve to illustrate the issues that arise in planning a synthesis and provide examples of solutions that have been developed. The individual syntheses also provide many examples of the synthetic transformations presented in the previous chapters and of the use of protective groups in the synthesis of complex molecules. The syntheses shown span a period of several decades and in some cases new reagents and protocols may have been developed since a particular synthesis was completed. Owing to limitations of space, only key steps are discussed although all the steps are shown in the schemes. Usually, only the reagent is shown, although other reaction components such as acids, bases, or solvents may also be of critical importance to the success of the reaction.

13.2.1. Juvabione

Juvabione is a terpene-derived ketoester that has been isolated from various plant sources. There are two stereoisomers, both of which occur naturally with *R*-configuration at C(4) of the cyclohexene ring and are referred to as *erythro*- and *threo*-juvabione. The 7(*S*)-enantiomer is sometimes called epijuvabione. Juvabione exhibits "juvenile hormone" activity in insects; that is, it can modify the process of metamorphosis.[18]

threo-juvabione   *erythro*-juvabione

In considering the retrosynthetic analysis of juvabione, two factors draw special attention to the bond between C(4) and C(7). First, this bond establishes the stereochemistry of the molecule. The C(4) and C(7) carbons are stereogenic centers and their relative configuration determines the diastereomeric structure. In a stereocontrolled synthesis, it is necessary to establish the desired stereochemistry at C(4) and C(7). The C(4)−C(7) bond also connects the side chain to the cyclohexene ring. As a cyclohexane derivative is a logical candidate for one key intermediate, the C(4)−C(7) bond is a potential bond disconnection.

Other bonds that merit attention are those connecting C(7) through C(11). These could be formed by one of the many methods for the synthesis of ketones. Bond disconnections at carbonyl centers can involve the O=C-C(α) (acylation, organometallic addition), the C(α)–C(β) bond (enolate alkylation, aldol addition), or C(β)–C(γ) bond (conjugate addition to enone).

The only other functional group is the conjugated unsaturated ester. This functionality is remote from the stereocenters and the ketone functionality, and does not play a key role in most of the reported syntheses. Most of the syntheses use cyclic starting materials. Those in Schemes 13.4 and 13.5 lead back to a *para*-substituted aromatic ether. The syntheses in Schemes 13.7 and 13.8 begin with an accessible terpene intermediate. The syntheses in Schemes 13.10 and 13.11 start with cyclohexenone. Scheme 13.3 presents a retrosynthetic analysis leading to the key intermediates used for the syntheses in

---

[18.] For a review, see Z. Wimmer and M. Romanuk, *Coll. Czech. Chem. Commun.*, **54**, 2302 (1989).

Scheme 13.3. Retrosynthetic Analysis of Juvabione with Disconnection to 4-Methoxyacetophenone

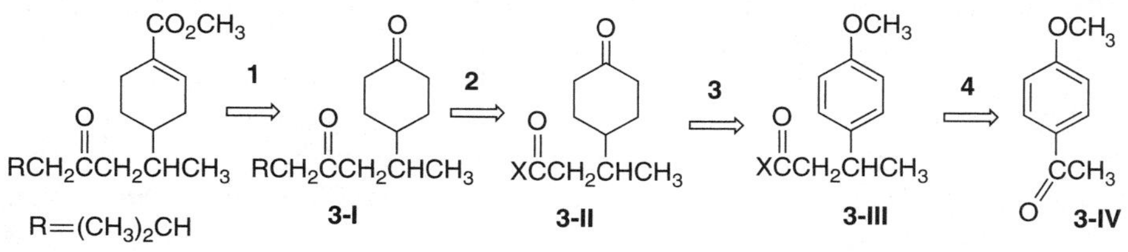

R=(CH₃)₂CH   **3-I**   **3-II**   **3-III**   **3-IV**

Scheme 13.4. Juvabione Synthesis: K. Mori and M. Matsui[a]

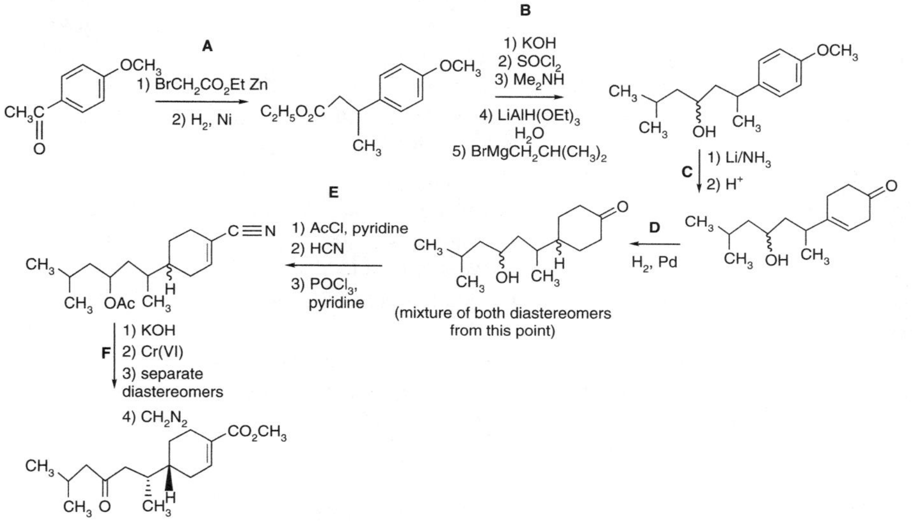

a. K. Mori and M. Matsui, *Tetrahedron*, **24**, 3127 (1968).

Scheme 13.5. Juvabione Synthesis: K. S. Ayyar and G. S. K. Rao[a]

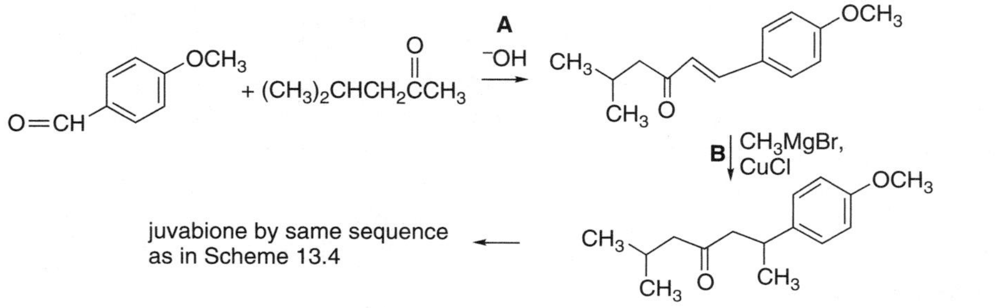

juvabione by same sequence as in Scheme 13.4

a. K. S. Ayyar and G. S. K. Rao, *Can. J. Chem.*, **46**, 1467 (1968).

Schemes 13.4 and 13.5. These syntheses use achiral reactants and provide mixtures of both stereoisomers. The final products are racemic. The first disconnection is that of the ester functionality, which corresponds to a strategic decision that the ester group can be added late in the synthesis. Disconnection **2** identifies the C(9)−C(10) bond as one that can be formed by addition of some nucleophilic group corresponding to C(10)−C(13) to the carbonyl center at C(9). This corresponds to disconnection α shown above. The third retrosynthetic transform recognizes that the cyclohexanone ring could be obtained by a Birch reduction of an appropriately substituted aromatic ether. The methoxy substituent would provide for correct placement of the cyclic carbonyl group. The final disconnection identifies a simple starting material, 4-methoxyacetophenone.

A synthesis corresponding to this pattern that is shown in Scheme 13.4 relies on well-known reaction types. The C(4)−C(7) bond was formed by a Reformatsky reaction. The adduct was dehydrated during work-up and the product was hydrogenated after purification. The ester group was converted to the corresponding aldehyde by Steps **B**-1 through **B**-4. Step **B**-5 introduced the C(10)−C(13) isobutyl group by Grignard addition to an aldehyde. In this synthesis, the relative configuration at C(4) and C(7) was established by the hydrogenation in Step **D**. In principle, this reaction could be diastereoselective if the adjacent chiral center at C(7) strongly influenced the direction of addition of hydrogen. In practice, the reduction was not very selective and a mixture of isomers was obtained. Steps **E** and **F** introduced the C(1) ester group.

The synthesis in Scheme 13.5 also makes use of an aromatic starting material and follows a retrosynthetic plan similar to that in Scheme 13.3. The starting material was 4-methoxybenzaldehyde. This synthesis was somewhat more convergent in that the entire side chain except for C(14) was introduced as a single unit by a mixed aldol condensation in step **A**. The C(14) methyl was introduced by a copper-catalyzed conjugate addition in Step **B**.

Scheme 13.6 is a retrosynthetic outline of the syntheses in Schemes 13.7 to 13.9. The common feature of these syntheses is the use of terpene-derived starting materials. The use of such a starting material is suggested by the terpenoid structure of juvabione, which can be divided into "isoprene units."

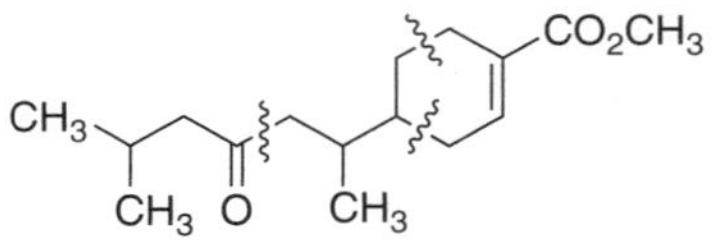

isoprene units in juvabione

The synthesis shown in Scheme 13.7 used limonene as the starting material (R = CH_3 in Scheme 13.6), whereas Schemes 13.8 and 13.9 use the corresponding aldehyde (R = CH=O). The use of these starting materials focuses attention on the means of attaching the C(9)−C(13) side chain. Furthermore, since the starting material is an enantiomerically pure terpene, enantioselectivity controlled by the chiral center at C(4) of the starting material might be feasible. In the synthesis in Scheme 13.7, the C(4)−C(7) stereochemistry was established in the hydroboration that is the first step of the synthesis. This reaction showed only very modest stereoselectivity and a 3:2 mixture of diastereomers was obtained and separated. The subsequent steps do not affect these stereogenic centers. The side chain was elaborated by adding *i*-butyllithium to a nitrile. The synthesis in Scheme 13.7 used a three-step oxidation sequence to oxidize the C(15) methyl group to a carboxy group. The first reaction was oxidation

Scheme 13.6. Retrosynthetic Analysis of Juvabione with Disconnection to the Terpene Limonene

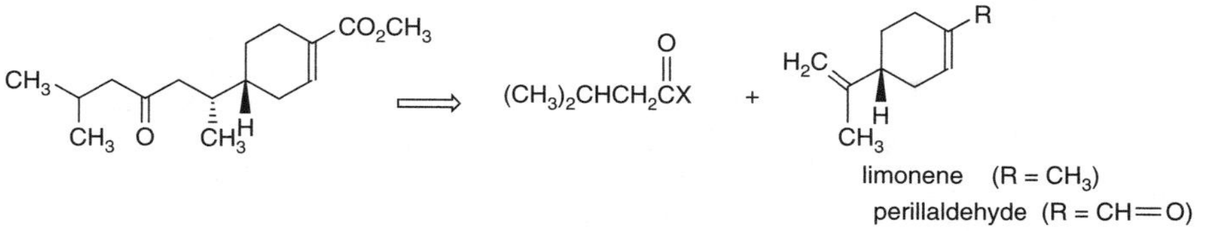

limonene   (R = CH₃)
perillaldehyde (R = CH=O)

Scheme 13.7. Juvabione Synthesis: B. A. Pawson, H.-C. Cheung, S. Gurbaxani, and G. Saucy[a]

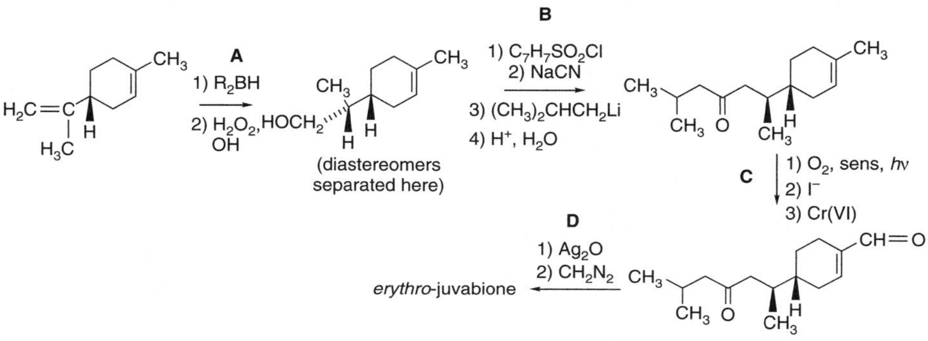

a. B. A. Pawson, H.-C. Cheung, S. Gurbaxani, and G. Saucy, *J. Am. Chem. Soc.*, **92**, 336 (1970).

Scheme 13.8. Juvabione Synthesis: E. Negishi, M. Sabanski, J. J. Katz, and H. C. Brown[a]

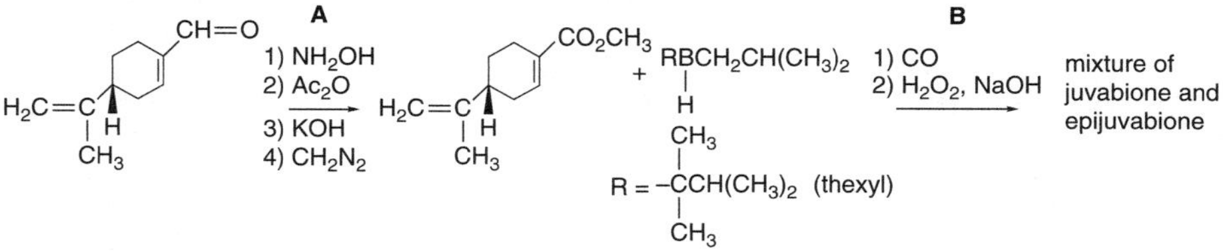

a. E. Negishi, M. Sabanski, J. J. Katz, and H. C. Brown, *Tetrahedron*, **32**, 925 (1976).

Scheme 13.9. Juvabione Synthesis: A. A. Carveiro and I. G. P. Viera[a]

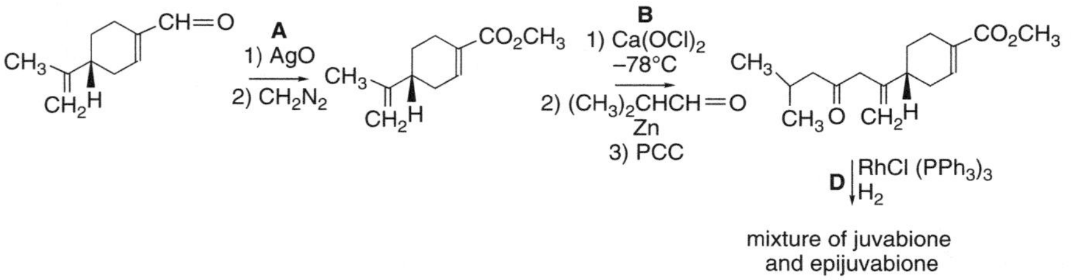

a. A. A. Carveiro and L. G. P. Viera, *J. Braz. Chem. Soc.*, **3**, 124 (1992).

Scheme 13.10. Retrosynthetic Analysis of Juvabione with Alternative Disconnections to Cyclohex-2-enone

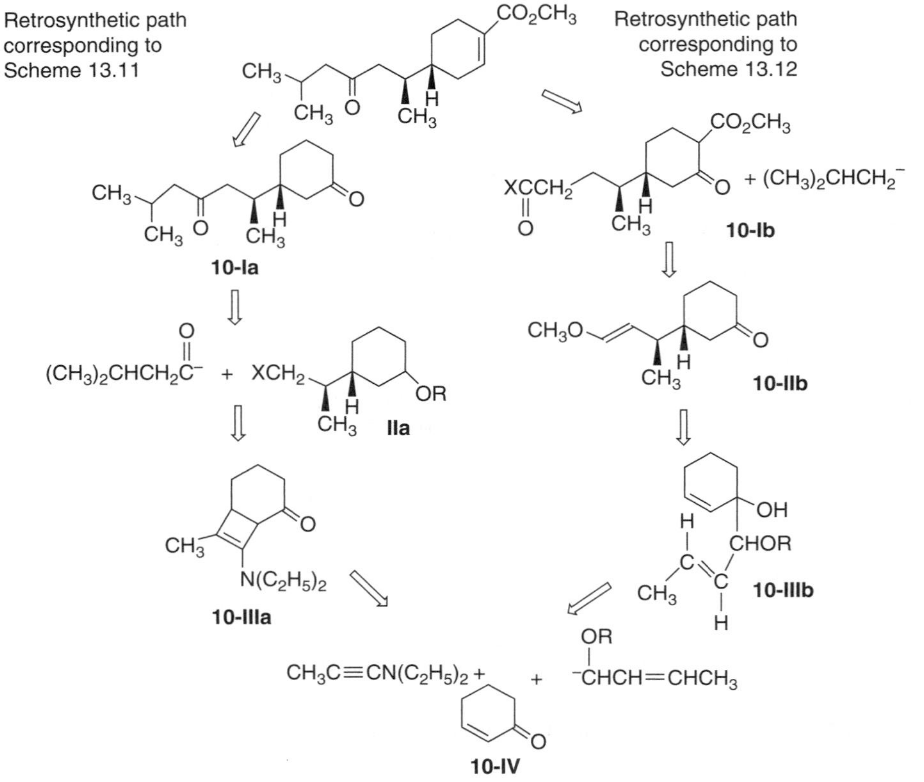

Retrosynthetic path corresponding to Scheme 13.11

Retrosynthetic path corresponding to Scheme 13.12

10-Ia

10-Ib

IIa

10-IIb

10-IIIa

10-IIIb

10-IV

Scheme 13.11. Juvabione Synthesis: J. Ficini, J. D'Angelo, and J. Noire[a]

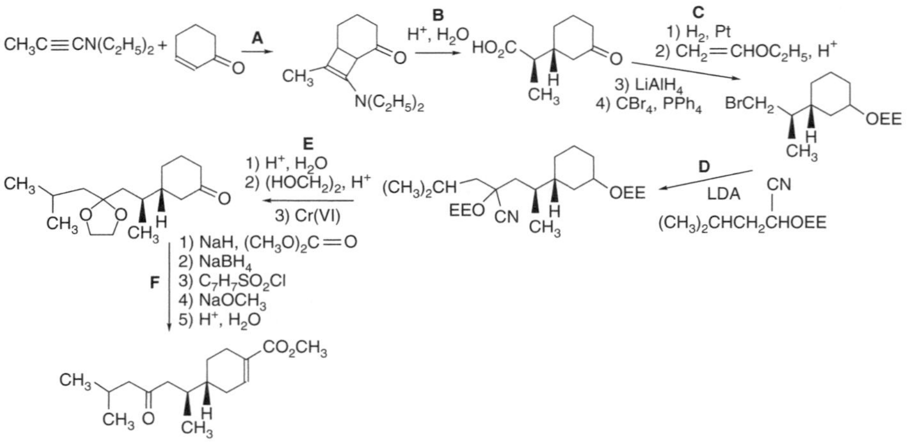

a. J. Ficini, J. D'Angelo, and J. Noire, *J. Am. Chem. Soc.*, **96**, 1213 (1974).

by singlet oxygen to give a mixture of hydroperoxides, with oxygen bound mainly at C(2). The mixture was reduced to the corresponding alcohols, which was then oxidized to the acid via an aldehyde intermediate.

In Scheme 13.8, the side chain was added in one step by a borane carbonylation reaction. This synthesis is very short and the first four steps were used to transform the aldehyde group in the starting material to a methyl ester. The stereochemistry at C(4)–C(7) is established in the hydroboration in Step **B**, in which the C(7)–H bond is formed. A 1:1 mixture of diastereomers resulted, indicating that the configuration at C(4) has little influence on the direction of approach of the borane reagent.

Another synthesis, shown in Scheme 13.9, that starts with the same aldehyde (perillaldehyde) was completed more recently. The C(8)–C(9) bond was established by an allylic chlorination and addition of the corresponding zinc reagent to isobutyraldehyde. In this synthesis, the C(7) stereochemistry was established by a homogeneous hydrogenation of a methylene group, but this reaction also produces both stereoisomers.

The first diastereoselective syntheses of juvabione are described in Schemes 13.11 and 13.12. Scheme 13.10 is a retrosynthetic analysis corresponding to these syntheses, which have certain similarities. Both syntheses started with cyclohexenone, and there is a general similarity in the fragments that were utilized, although the order of construction differs, and both led to (±)-juvabione.

A key step in the synthesis in Scheme 13.11 was a cycloaddition between an electron-rich ynamine and the electron-poor enone. The cyclobutane ring was then opened in a process that corresponds to retrosynthetic step **10-IIa** ⟹ **10-IIIa** in Scheme 13.10. The crucial step for stereochemical control occurs in Step **B**. The stereoselectivity of this step results from preferential protonation of the enamine from the less hindered side of the bicyclic intermediate.

The cyclobutane ring was then cleaved by hydrolysis of the enamine and ring opening of the resulting β-diketone. The relative configuration of the chiral centers is unaffected by subsequent transformations, so the overall sequence is stereoselective. Another key step in this synthesis is Step **D**, which corresponds to the transformation **10-IIa** ⟹ **10-Ia** in the retrosynthesis. A protected cyanohydrin was used as a nucleophilic acyl anion equivalent in this step. The final steps of the synthesis in Scheme 13.11 employed the C(2) carbonyl group to introduce the carboxy group and the C(1)–C(2) double bond.

The stereoselectivity achieved in the synthesis in Scheme 13.12 is the result of a preferred conformation for the base-catalyzed oxy-Cope rearrangement in Step **B**. Although the intermediate used in Step **B** was a mixture of stereoisomers, both gave predominantly the desired relative stereochemistry at C(4) and C(7). The stereoselectivity is based on the preferred chair conformation for the TS of the oxy-Cope rearrangement.

Scheme 13.12. Juvabione Synthesis: D. A. Evans and J. V. Nelson[a]

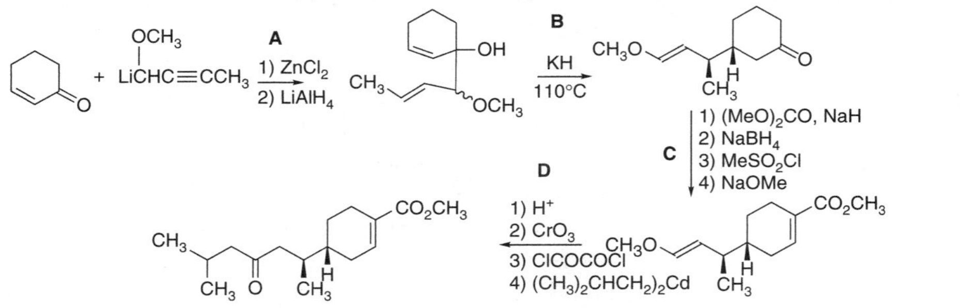

a. D. A. Evans and J. V. Nelson, *J. Am. Chem. Soc.*, **102**, 774 (1980).

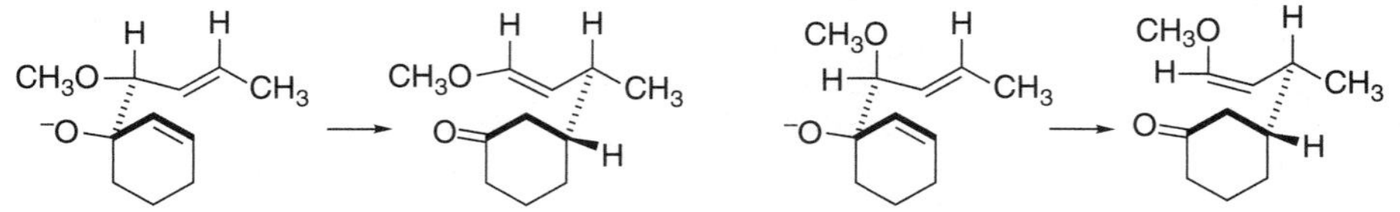

The synthesis in Scheme 13.13 leads diastereospecifically to the *erythro* stereoisomer. An intramolecular enolate alkylation in Step **B** gave a bicyclic intermediate. The relative configuration of C(4) and C(7) was established by the hydrogenation in Step **C**. The hydrogen is added from the less hindered *exo* face of the bicyclic enone. This reaction is an example of the use of geometric constraints of a ring system to control relative stereochemistry.

The *threo* stereoisomer was the major product obtained by the synthesis in Scheme 13.14. This stereochemistry was established by the conjugate addition in Step **A**, where a significant (4–6:1) diastereoselectivity was observed. The C(4)–C(7) stereochemical relationship was retained through the remainder of the synthesis. The other special features of this synthesis are in Steps **B** and **C**. The mercuric acetate–mediated cyclopropane ring opening was facilitated by the alkoxy substituent.[19] The reduction by $NaBH_4$ accomplished both demercuration and reduction of the aldehyde group.

Scheme 13.13. Juvabione Synthesis: A. G. Schultz and J. P. Dittami[a]

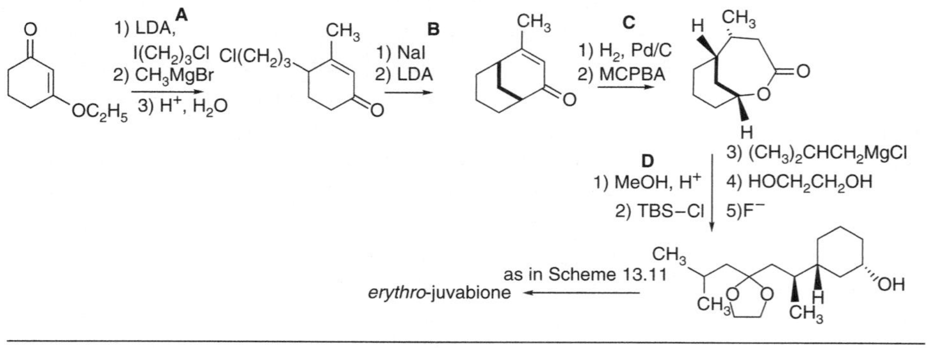

a. A. G. Schultz and J. P. Dittami, *J. Org. Chem.*, **49**, 2615 (1984).

[19.] A. DeBoer and C. H. DePuy, *J. Am. Chem. Soc.*, **92**, 4008 (1970).

Scheme 13.14. Juvabione Synthesis: D. J. Morgans, Jr., and G. B. Feigelson[a]

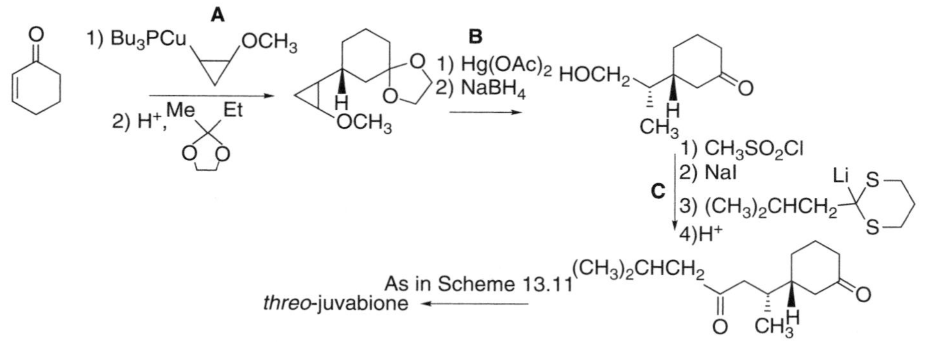

a. D. J. Morgans, Jr., and G. B. Feigelson, *J. Am. Chem. Soc.*, **105**, 5477 (1983).

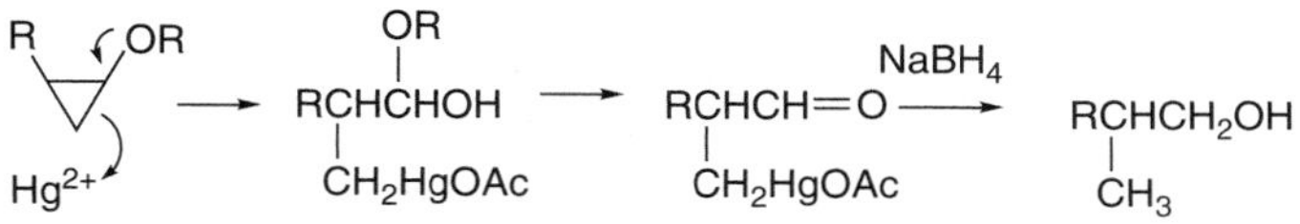

In Step **C** a dithiane anion was used as a nucleophilic acyl anion equivalent to introduce the C(10)–C(13) isobutyl group.

In the synthesis shown in Scheme 13.15, racemates of both *erythro*- and *threo*-juvabione were synthesized by parallel routes. The isomeric intermediates were obtained in greater than 10:1 selectivity by choice of the *E*- or *Z*-silanes used for conjugate addition to cyclohexenone (Michael-Mukaiyama reaction). Further optimization of the stereoselectivity was achieved by the choice of the silyl substituents. The observed stereoselectivity is consistent with synclinal TSs for the addition of the crotyl silane reagents.

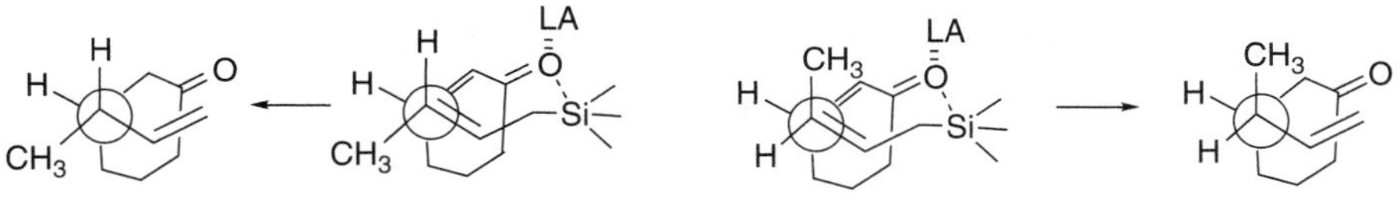

The purified diastereomeric intermediates were then converted to the juvabione stereoisomers.

Except for the syntheses using terpene-derived starting materials (Schemes 13.7, 13.8, and 13.9), the previous juvabione syntheses all gave racemic products. Some of the more recent juvabione syntheses are *enantiospecific*. The synthesis in Scheme 13.16 relied on a chiral sulfoxide that undergoes stereoselective addition to cyclohexenone to establish the correct relative and absolute configuration at C(4) and C(7). The origin of the stereoselectivity is a chelated TS that leads to the observed product.[20]

[20] M. R. Binns, R. K. Haynes, A. G. Katsifis, P. A. Schober, and S. C. Vonwiller, *J. Am. Chem. Soc.*, **110**, 5411 (1988).

Scheme 13.15. Juvabione Synthesis: T. Tokoroyama and L.-R. Pan[a]

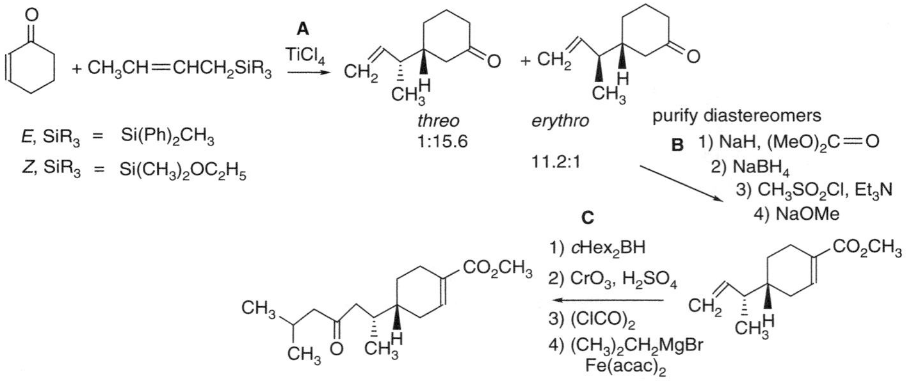

E, SiR_3 = $Si(Ph)_2CH_3$
Z, SiR_3 = $Si(CH_3)_2OC_2H_5$

threo
1:15.6

erythro
11.2:1

purify diastereomers

B 1) NaH, $(MeO)_2C$=O
2) $NaBH_4$
3) CH_3SO_2Cl, Et_3N
4) NaOMe

C
1) $cHex_2BH$
2) CrO_3, H_2SO_4
3) $(ClCO)_2$
4) $(CH_3)_2CH_2MgBr$
$Fe(acac)_2$

a. T. Tokoroyama and L.-R. Pan, *Tetrahedron Lett.*, **30**, 197 (1989).

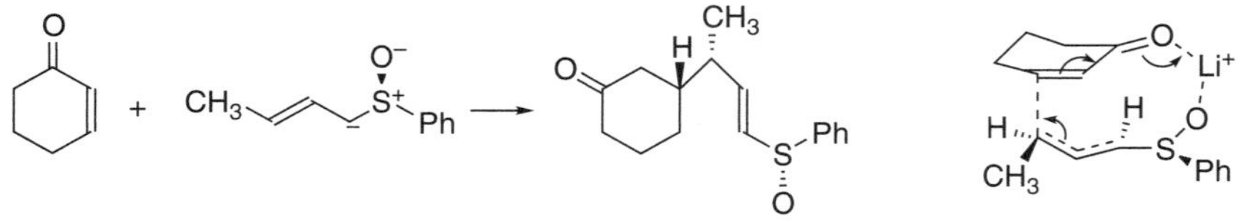

82% of mixture

The sulfoxide substituent was also used to introduce the C(10)–C(13) fragment and was reduced to a vinyl sulfide in Step **B**-1. In Step **C**-1, the vinyl sulfide was hydrolyzed to an aldehyde, which was elaborated by addition of isobutylmagnesium bromide.

Scheme 13.17 depicts a synthesis based on enantioselective reduction of bicyclo[2.2.2]octane-2,6-dione by Baker's yeast.[21] This is an example of desymmetrization (see Part A, Topic 2.2). The unreduced carbonyl group was converted to an alkene by the Shapiro reaction. The alcohol was then reoxidized to a ketone. The enantiomerically pure intermediate was converted to the lactone by Baeyer-Villiger oxidation and an allylic rearrangement. The methyl group was introduced stereoselectively from the *exo* face of the bicyclic lactone by an enolate alkylation in Step **C**-1.

Scheme 13.16. Juvabione Synthesis: H. Watanabe, H. Shimizu, and K. Mori[a]

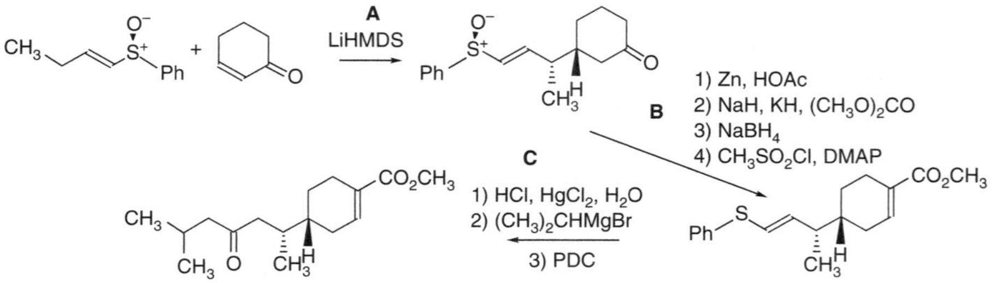

A LiHMDS

B
1) Zn, HOAc
2) NaH, KH, $(CH_3O)_2CO$
3) $NaBH_4$
4) CH_3SO_2Cl, DMAP

C
1) HCl, $HgCl_2$, H_2O
2) $(CH_3)_2CHMgBr$
3) PDC

a. H. Watanabe, H. Shimizu, and K. Mori, *Synthesis*, 1249 (1994).

[21.] K. Mori and F. Nagano, *Biocatalysis*, **3**, 25 (1990).

Scheme 13.17. Juvabione Synthesis: E. Nagano and K. Mori[a]

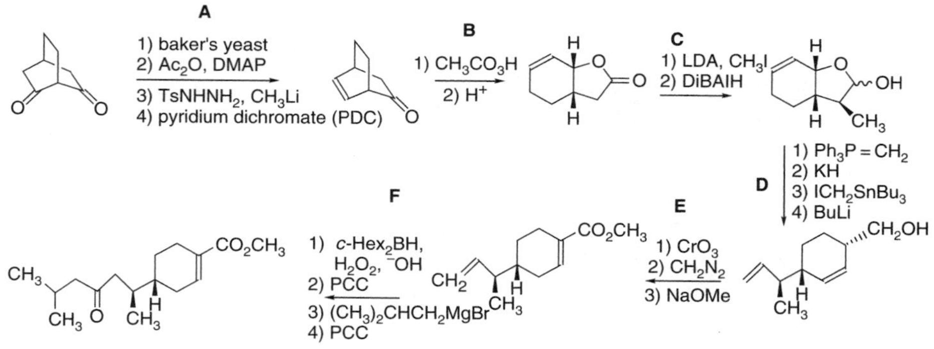

a. E. Nagano and K. Mori, *Biosci. Biotechnol. Biochem.*, **56**, 1589 (1992).

A final crucial step in this synthesis was an anionic [2,3]-sigmatropic rearrangement of an allylic ether in Step **D**-4 to introduce the C(1) carbon.

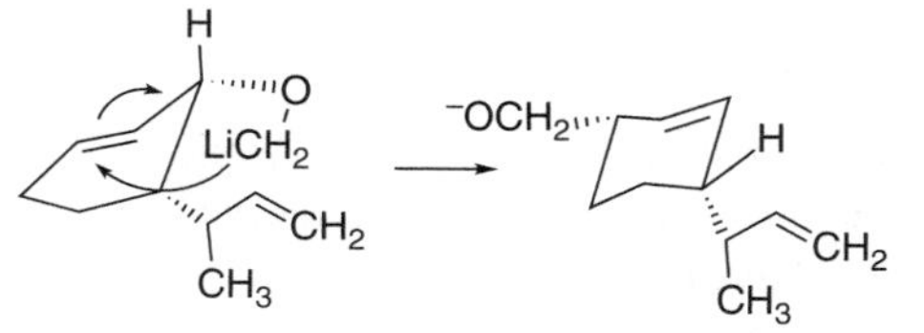

Another enantioselective synthesis, shown in Scheme 13.18, involves a early kinetic resolution of the alcohol intermediate in Step **B**-2 by lipase PS. The stereochemistry at the C(7) methyl group is controlled by the *exo* selectivity in the conjugate addition (Step **D**-1).

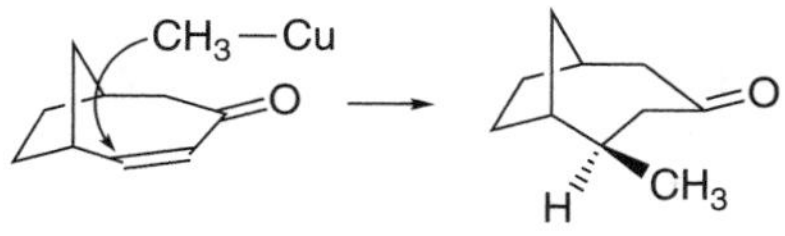

The bicyclic ring is then cleaved by a Baeyer-Villiger reaction in Step **D**-2. Another interesting feature of this synthesis is the ring expansions used in sequences **A** and **F**. Trimethylsilyl enol ethers were treated with Simmons-Smith reagent to form cyclopropyl silyl ethers. These undergo oxidative cleavage and ring expansion when treated with $FeCl_3$ and the α-chloro ketones are then dehydrohalogenated by DBU.[22]

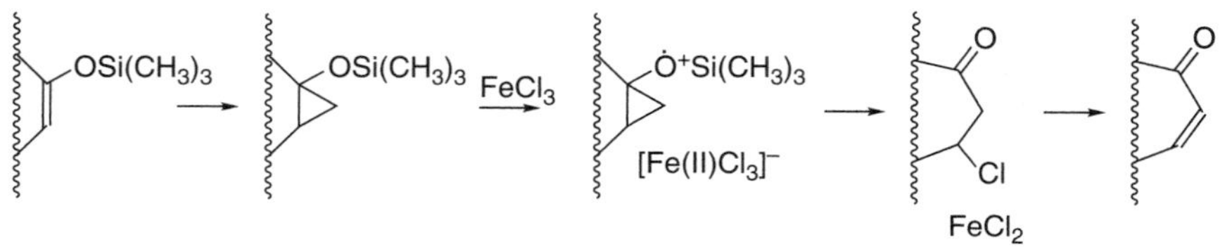

22. V. Ito, S. Fujii, and T. Saegusa, *J. Org. Chem.*, **41**, 2073 (1976).

Scheme 13.18. Juvabione Synthesis: K. Ogasawara and Co-workers[a]

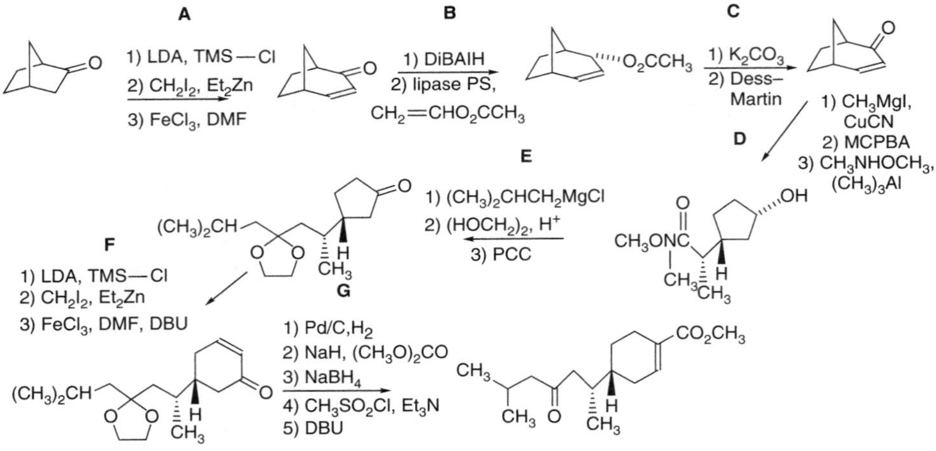

a. H. Nagata, T. Taniguchi, M. Kawamura, and K. Ogasawara, *Tetrahedron Lett.*, **40**, 4207 (1999).

The juvabione synthesis in Scheme 13.19 exploited both the regiochemical and stereochemical features of the starting material, the $Cr(CO)_3$ complex of 4-methoxyphenyltrimethylsilane. The lithium enolate of *t*-butyl propanoate was added, resulting in a 96:4 ratio of *meta:ortho* adducts. The addition was also highly stereoselective, giving a greater than 99:1 preference for the *erythro* stereochemistry. This is consistent with reaction through TS **18-A** in preference to TS **18-B** to avoid a *gauche* interaction between the enolate methyl and the trimethylsilyl substituent.

Scheme 13.19. Juvabione Synthesis: A. J. Pearson, H. Paramahamsan, and J. D. Dudones[a]

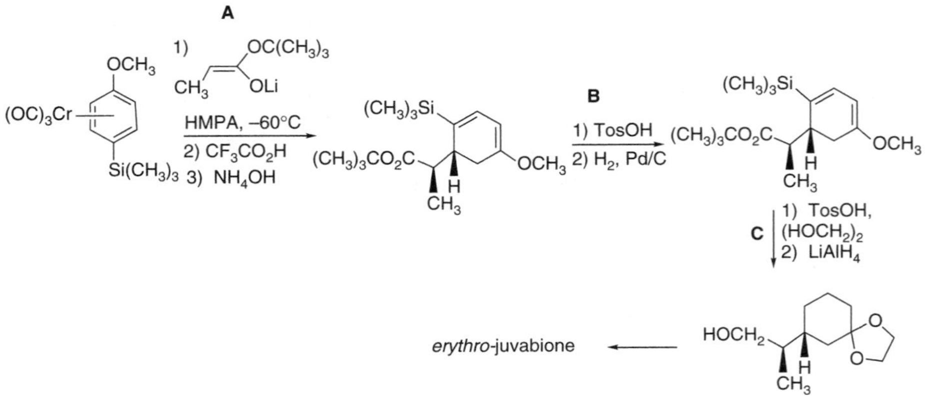

a. A. J. Pearson, H. Paramahamsan, and J. D. Dudones, *Org. Lett.*, **6**, 2121 (2004).

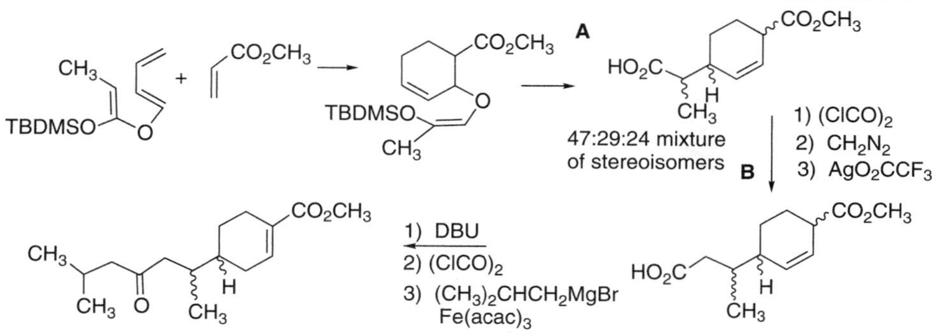

a. N. Soldermann, J. Velker, O. Vallat, H. Stoeckli-Evans, and R. Neier, *Helv. Chim. Acta*, **83**, 2266 (2000).

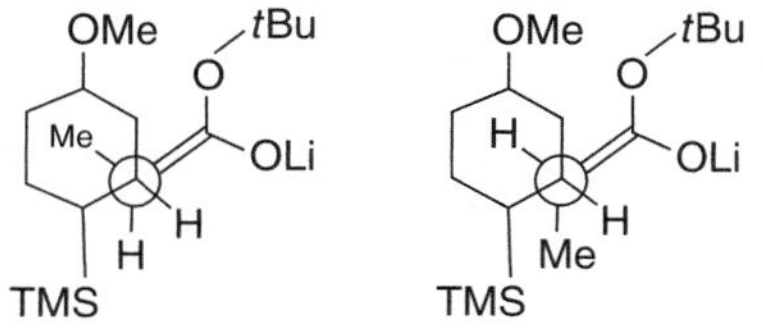

TS **18-A**   TS **18-B**

The reaction product was converted to an intermediate that had previously been converted to *erythro*-juvabione.

The synthesis in Scheme 13.20 features a tandem Diels-Alder reaction and Ireland-Claisen [3,3]-sigmatropic shift as the key steps. Although this strategy was very efficient in constructing the carbon structure, it was not very stereoselective. The major isomer results from an *endo* TS for the Diels-Alder reaction and a [3,3]-sigmatropic rearrangement through a boat TS. Three stereoisomers were obtained in the ratio 47:29:24. These were not separated but were converted to a 4:1 mixture of (±)-juvabione and (±)-epijuvabione by Arndt-Eistert homologation, DBU-based conjugation, and addition of the isobutyl group by a Fe(acac)$_3$-catalyzed Grignard addition.

The synthesis in Scheme 13.21 starts with a lactone that is available in enantiomerically pure form. It was first subjected to an enolate alkylation that was stereocontrolled by the convex shape of the *cis* ring junction (Step **A**). A stereospecific Pd-mediated allylic substitution followed by LiAlH$_4$ reduction generated the first key intermediate (Step **B**). This compound was oxidized with NaIO$_4$, converted to the methyl ester, and subjected to a base-catalyzed conjugation. After oxidation of the primary alcohol to an aldehyde, a Wittig-Horner olefination completed the side chain.

The enantioselective synthesis in Scheme 13.22 is based on stereoselective reduction of an α, β-unsaturated aldehyde generated from (−)-(*S*)-limonene (Step **A**). The reduction was done by Baker's yeast and was completely enantioselective. The diastereoselectivity was not complete, generating an 80:20 mixture, but the diastereomeric alcohols were purified at this stage. After oxidation to the aldehyde, the remainder of the side chain was introduced by a Grignard addition. The ester function

Scheme 13.21. Juvabione Synthesis: E. J Bergner and G. Helmchen[a]

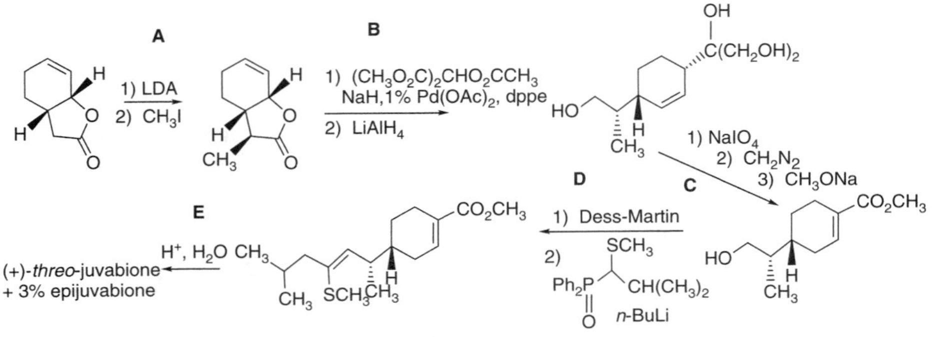

a. E. J. Bergner and G. Helmchen, *J. Org. Chem.*, **65**, 5072 (2000).

was introduced by a base-catalyzed opening of the epoxide to an allylic alcohol (Step **C**-4), which then underwent oxidation with allylic transposition (Step **D**-1).

Several other syntheses of juvabione have also been completed.[23]

13.2.2. Longifolene

Longifolene is a tricyclic sesquiterpene. It is a typical terpene hydrocarbon in terms of the structural complexity. The synthetic challenge lies in construction of the bicyclic ring system. Schemes 13.24 through 13.33 describe nine separate syntheses of longifolene. We wish to particularly emphasize the methods for carbon-carbon bond formation used in these syntheses. There are four stereogenic centers in longifolene,

Scheme 13.22. Juvabione Synthesis. C. Fuganti and S. Serra[a]

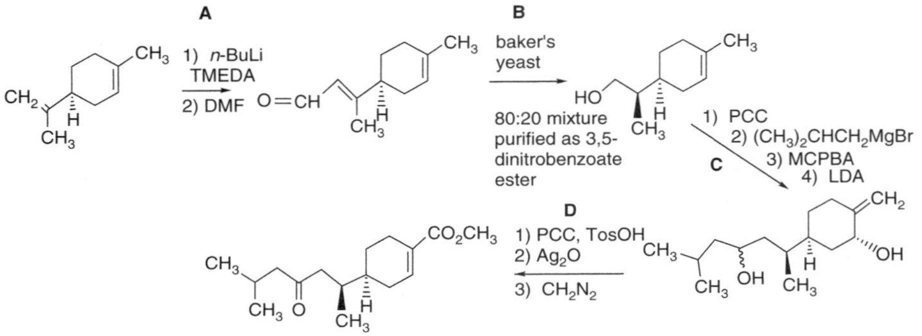

a. C. Fuganti and S. Serra, *J. Chem. Soc., Perkin Trans. 1*, 97 (2000).

23. A. A. Drabkina and Y. S. Tsizin, *J. Gen. Chem. USSR* (English Transl.), **43**, 422, 691 (1973); R. J. Crawford, U. S. Patent, 3,676,506; *Chem. Abstr.*, **77**, 113889e (1972); A. J. Birch, P. L. Macdonald, and V. H. Powell, *J. Chem. Soc. C*, 1469 (1970); B. M. Trost and Y. Tamaru, *Tetrahedron Lett.*, 3797 (1975); M. Fujii, T. Aida, M. Yoshihara, and A. Ohno, *Bull. Chem. Soc. Jpn.*, **63**, 1255 (1990).

but they are not independent of one another because the geometry of the ring system requires that they have a specific relative relationship. That does not mean stereochemistry can be ignored, however, since the formation of the various rings will fail if the reactants do not have the proper stereochemistry.

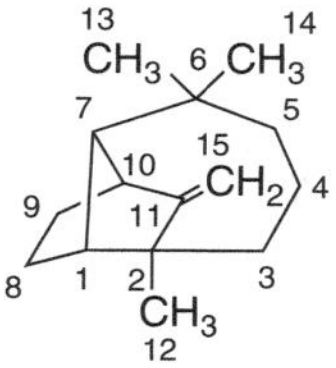

The first successful synthesis of longifolene was described in detail by E. J. Corey and co-workers in 1964. Scheme 13.23 presents a retrosynthetic analysis corresponding to this route. A key disconnection is made on going from **23-I** ⇒ **23-II**. This transformation simplifies the tricyclic to a bicyclic skeleton. For this disconnection to correspond to a reasonable synthetic step, the functionality in the intermediate to be cyclized must engender mutual reactivity between C(7) and C(10). This is achieved in diketone **23-II**, because an enolate generated by deprotonation at C(10) can undergo an intramolecular Michael addition to C(7). The stereochemistry requires that the ring junction be *cis*. Retrosynthetic Step **23-II** ⇒ **23-III** is attractive because it suggests a decalin derivative as a key intermediate. Methods for preparing this type of structure are well developed, since they are useful intermediates in the synthesis of other terpenes as well as steroids. Can a chemical reaction be recognized that would permit **23-III** ⇒ **23-II** to proceed in the synthetic sense? The hydroxy to carbonyl transformation with migration corresponds to the pinacol rearrangement (Section 10.1.2.1). The retrosynthetic transformation **23-II** ⇒ **23-III** corresponds to a workable synthetic step if the group X in **23-III** is a leaving group that could promote the rearrangement. The other transformations in the retrosynthetic plan, **23-III** ⇒ **23-IV** ⇒ **23-V**, are straightforward in concept and lead to identification of **23-V** as a potential starting material.

Scheme 13.23. Retrosynthesis of Longifolene Corresponding to the Synthesis in Scheme 13.24

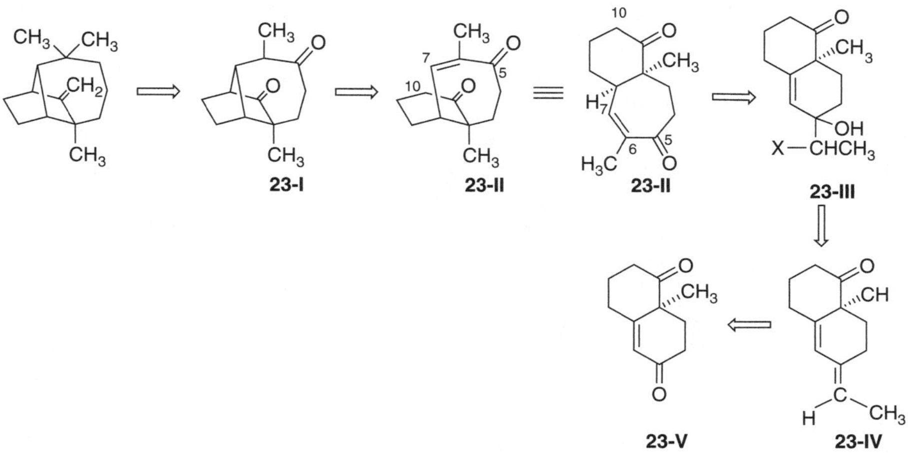

Scheme 13.24. Longifolene Synthesis: E. J. Corey and Co-Workers[a]

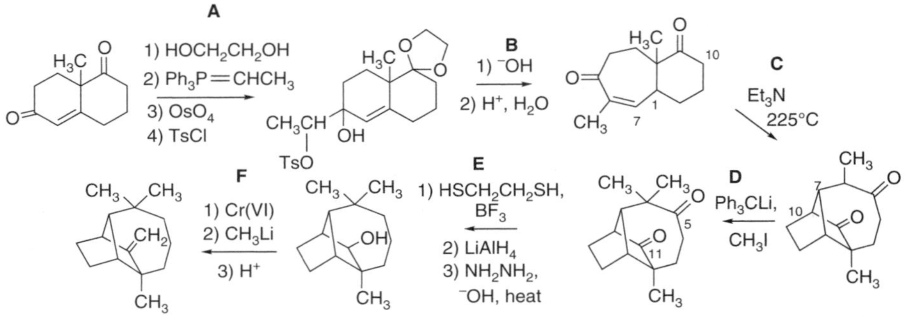

a. E. J. Corey, M. Ohno, R. B. Mitra, and P. A. Vatakencherry, *J. Am. Chem. Soc.*, **86**, 478 (1964).

Compound **23-V** is known as the Wieland-Miescher ketone and can be obtained by Robinson annulation of 2-methylcyclohexane-1,3-dione.

The synthesis was carried out as shown in Scheme 13.24. A diol was formed and selectively tosylated at the secondary hydroxy group (Step **A**-4). Base then promoted the skeletal rearrangement in Step **B**-1 by a pinacol rearrangement corresponding to **23-II** ⇒ **23-III** in the retrosynthesis. The key intramolecular Michael addition was accomplished using triethylamine under high-temperature conditions.

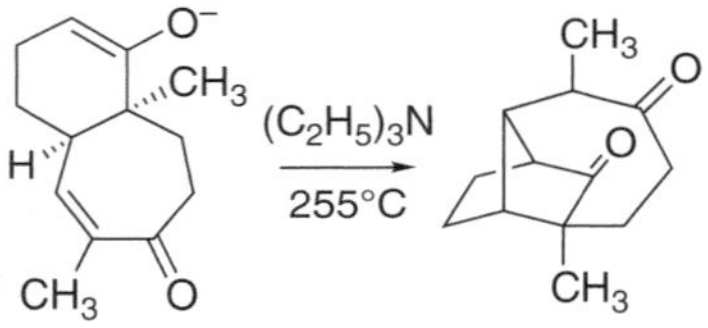

The cyclization requires that the intermediate have a *cis* ring fusion. The stereochemistry of the ring junction was established when the double bond was moved into conjugation in Step **B**-2. The product was not stereochemically characterized, and need not be, because the stereochemically important site at C(1) can be epimerized under the basic cyclization conditions. Thus, the equilibration of the ring junction through a dienol allows the cyclization to proceed to completion from either stereoisomer.

After the crucial cyclization in Step **C**, the subsequent transformations effect the addition of the remaining methyl and methylene groups by well-known methods. Step **E** accomplishes a selective reduction of one of the two carbonyl groups to a methylene by taking advantage of the difference in the steric environment of the two carbonyls. Selective protection of the less hindered C(5) carbonyl was done using a thioketal. The C(11) carbonyl was then reduced to give the alcohol, after which C(5) was reduced to a methylene group under Wolff-Kishner conditions. The hydroxy group at C(11) provided the reactive center necessary to introduce the C(15) methylene group via methyllithium addition and dehydration in Step **F**.

The Wieland-Miescher ketone was also the starting material for the synthesis in Scheme 13.25. The key bond closure was performed on a bicyclo[4.4.0]decane ring system. An enolate was used to open an epoxide ring in Step **B**-2. The ring juncture must be *cis* to permit the intramolecular epoxide ring opening. The required *cis* ring fusion was established during the catalytic hydrogenation in Step **A**.

Scheme 13.25. Longifolene Synthesis: J. E. McMurry and S. J. Isser[a]

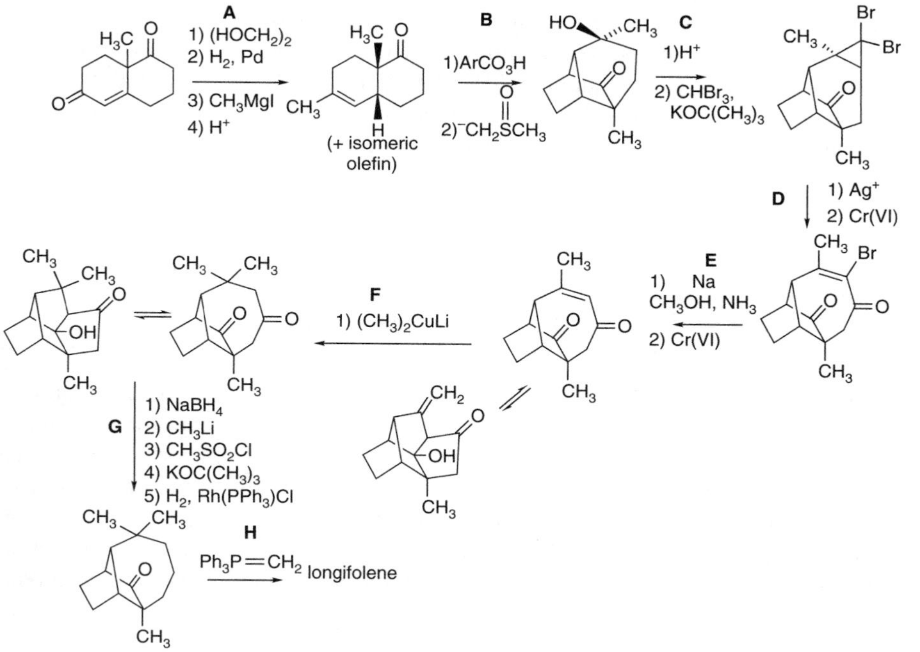

a. J. E. McMurry and S. J. Isser, *J. Am. Chem. Soc.*, **94**, 7132 (1972).

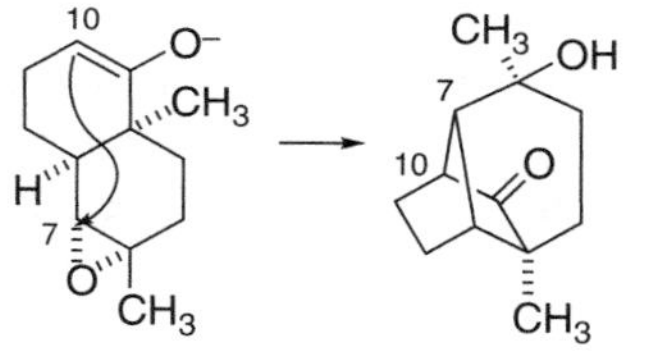

The key cyclization in Step **B**-2 was followed by a sequence of steps that effected a ring expansion via a carbene addition and cyclopropyl halide solvolysis. The products of Steps **E** and **F** are interesting in that the tricyclic structures are largely converted to tetracyclic derivatives by intramolecular aldol reactions. The extraneous bond was broken in Step **G**. First a diol was formed by NaBH$_4$ reduction and this was converted via the lithium alkoxide to a monomesylate. The resulting β-hydroxy mesylate is capable of a concerted fragmentation, which occurred on treatment with potassium *t*-butoxide.

Longifolene has also been synthesized from (±) Wieland-Miescher ketone by a series of reactions that feature an intramolecular enolate alkylation and ring expansion, as shown in Scheme 13.26. The starting material was converted to a dibromo ketone via the *bis*-silyl enol ether in the first sequence of reactions. This intermediate underwent an intramolecular enolate alkylation to form the C(7)–C(10) bond. The ring expansion was then done by conversion of the ketone to a silyl enol ether, cyclopropanation, and treatment of the siloxycyclopropane with $FeCl_3$.

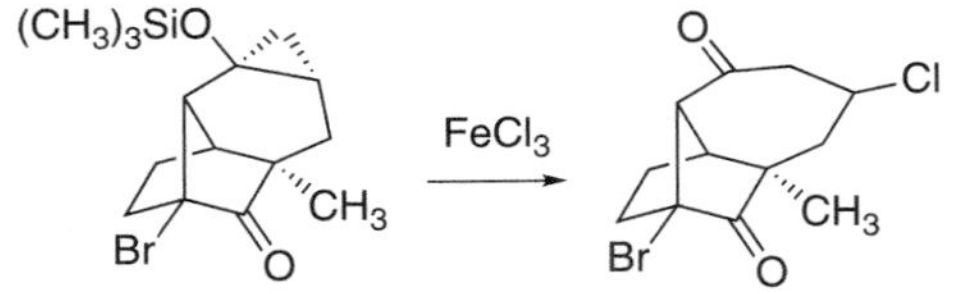

The final stages of the synthesis involved introduction of the final methyl group by Simmons-Smith cyclopropanation and reductive opening of the cyclopropane ring.

A retrosynthetic analysis corresponding to the synthesis in Scheme 13.28 is given in Scheme 13.27. The striking feature of this synthesis is the structural simplicity of the key intermediate **27-IV**. A synthesis according to this scheme generates the tricyclic skeleton in a single step from a monocyclic intermediate. The disconnection **27-III**–**27-IV** corresponds to a cationic cyclization of the highly symmetric allylic cation **27-IVa**.

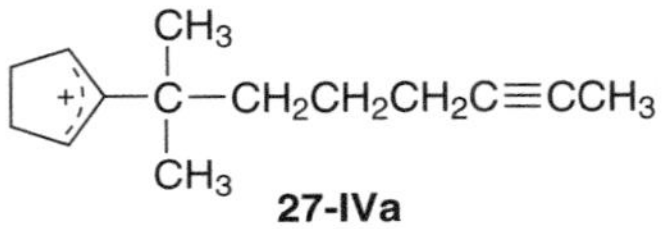

No issues of stereochemistry arise until the carbon skeleton is formed, at which point all of the stereocenters are in the proper relative relationship. The structures of the successive intermediates, assuming a stepwise mechanism for the cationic cyclization, are shown below.

Scheme 13.26. Longifolene Synthesis: S. Karimi and P. Tavares[a]

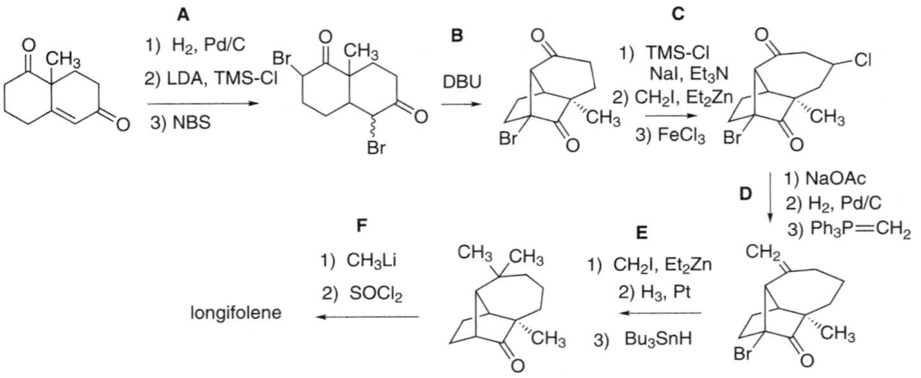

a. S. Karimi, *J. Nat. Prod.*, **64**, 406 (2001); S. Karimi and P. Tavares, *J. Nat. Prod.*, **66**, 520 (2003).

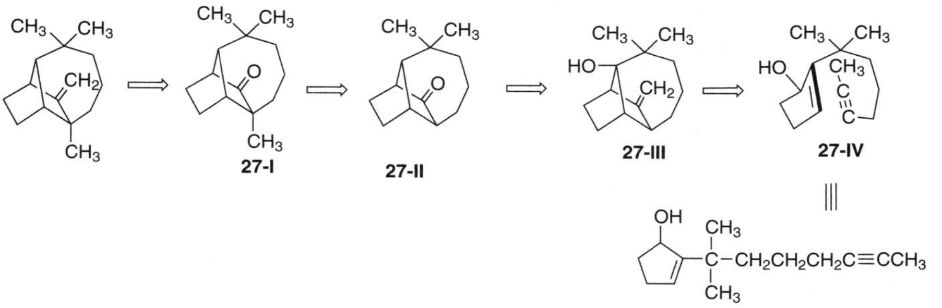

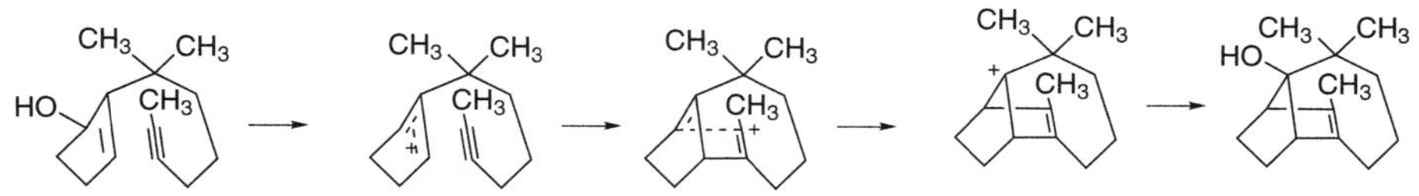

Evidently, these or closely related intermediates are accessible and reactive, since the synthesis was successfully achieved as outlined in Scheme 13.28. In addition to the key cationic cyclization in Step **D**, interesting transformations were carried out in Step **E**, where a bridgehead tertiary alcohol was reductively removed, and in Step **F**, where a methylene group, which was eventually reintroduced, had to be removed. The endocyclic double bond, which is strained because of its bridgehead location, was isomerized to the exocyclic position and then cleaved with RuO_4/IO_4^-. The enolate of the ketone was then used to introduce the C(12) methyl group in Steps **F**-3 and **F**-4.

The synthesis in Scheme 13.29 also uses a remarkably simple starting material to achieve the construction of the tricyclic skeleton. A partial retrosynthesis is outlined below.

Scheme 13.28. Longifolene Synthesis: W. S. Johnson and Co-Workers[a]

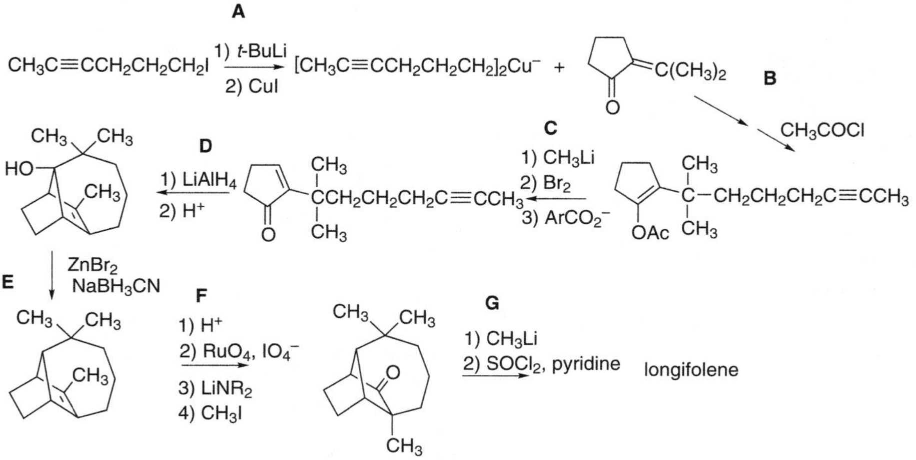

a. R. A. Volkmann, G. C. Anderson, and W. S. Johnson, *J. Am. Chem. Soc.*, **97**, 4777 (1975).

Scheme 13.29. Longifolene Synthesis: W. Oppolzer and T. Godel[a]

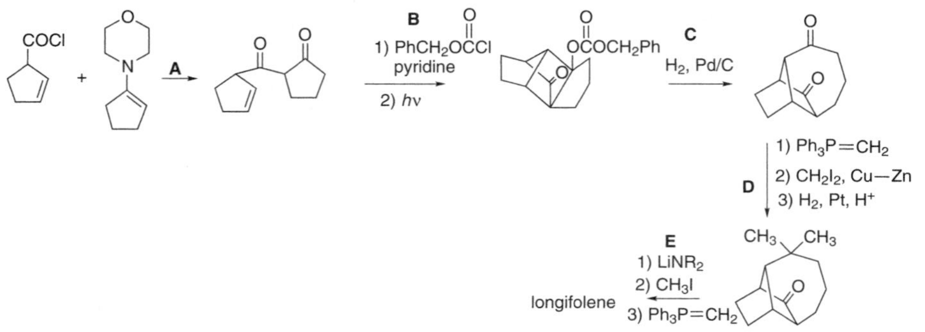

a. W. Oppolzer and T. Godel, *J. Am. Chem. Soc.*, **100**, 2583 (1978); W. Oppolzer and T. Godel, *Helv. Chim. Acta*, **67**, 1154 (1984).

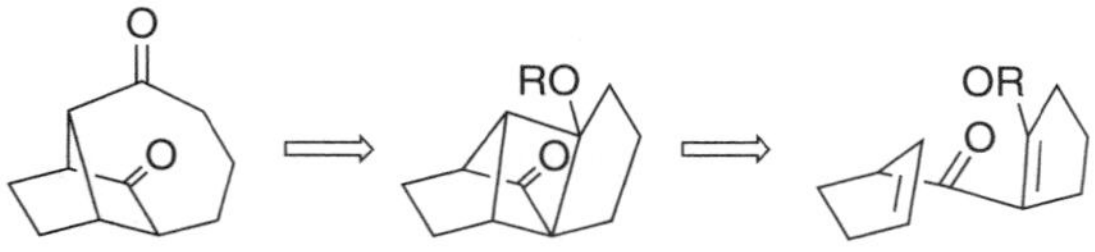

Intermediate **29-I** contains the tricyclic skeleton of longifolene, shorn of its substituents, but containing carbonyl groups suitably placed so that the methyl groups at C(2) and C(6) and the C(11) methylene can be introduced. The retrosynthetic Step **29-I** ⇒ **29-II** corresponds to an intramolecular aldol addition. However, **29-II** is clearly strained relative to **29-I**, and so (with OR = OH) should open to **29-I**.

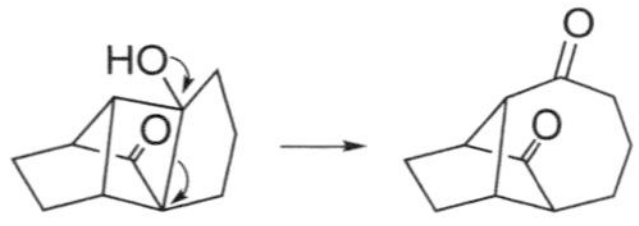

How might **29-II** be obtained? The four-membered ring suggests that a photochemical [2+2] cycloaddition might be useful, and this, in fact, was successful (Scheme 13.29, Step **B**). The cyclopentanone intermediate was converted to an enol carbonate. After photolysis, the carbobenzyloxy group was removed by hydrogenolysis, which led to opening of the strained aldol to the diketo intermediate. After liberation of the hydroxy group, the extra carbon-carbon bond between C(2) and C(6) was broken by a spontaneous retro-aldol reaction. Step **D** in this synthesis is an interesting way of introducing the geminal dimethyl groups. It proceeds through a cyclopropane intermediate that is cleaved by hydrogenolysis. In Step **E**, the C(12) methyl group was introduced by enolate alkylation and the C(15) methylene group was installed by a Wittig reaction.

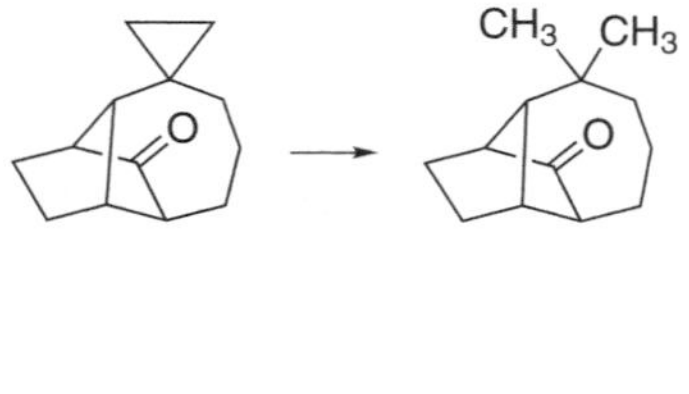

The synthesis of longifolene in Scheme 13.30 commenced with a Birch reduction and tandem alkylation of methyl 2-methoxybenzoate (see Section 5.6.1.2). Step **C** is an intramolecular cycloaddition of a diazoalkane that is generated from an aziridinoimine intermediate.

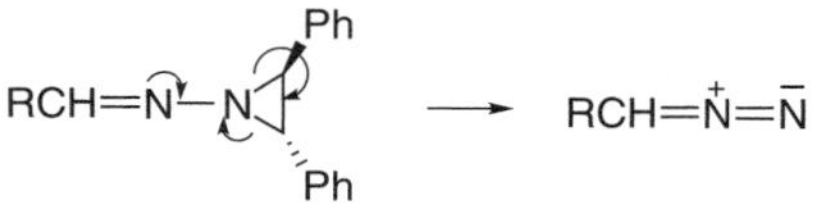

The thermolysis of the adduct generates a diradical (or the corresponding dipolar intermediate), which then closes to the desired carbon skeleton.

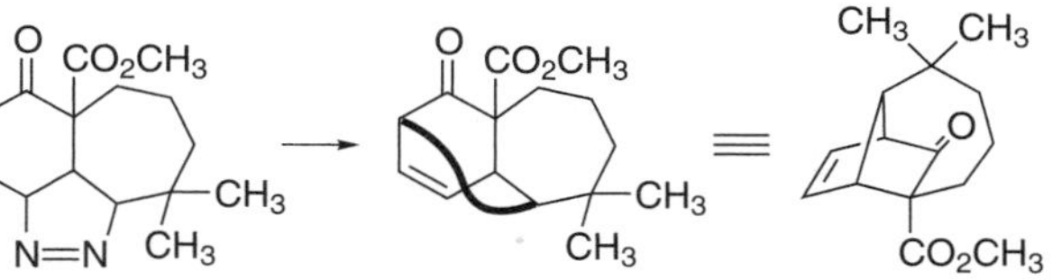

The cyclization product was converted to an intermediate that was used in the longifolene synthesis described in Scheme 12.24.

The synthesis in Scheme 13.30 was also done in such a way as to give enantiomerically pure longifolene. A starting material, whose chirality is derived from the amino acid L-proline, was enantioselectively converted to the product of Step **A** in Scheme 13.30.

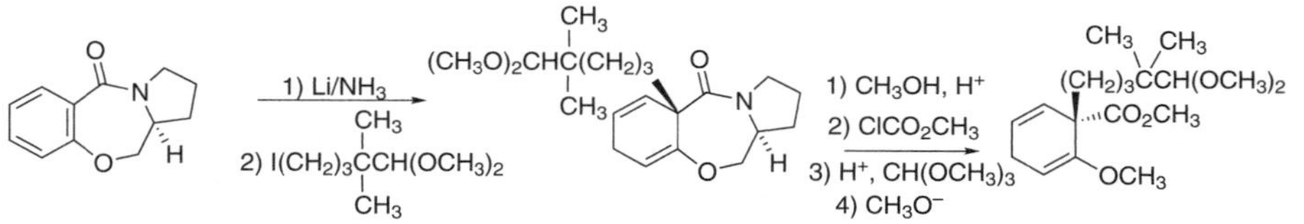

This chiral intermediate, when carried through the reaction sequence in Scheme 13.30, generated the enantiomer of natural longifolene. Thus D-proline would have to be used to generate the natural enantiomer.

Scheme 13.30. Longifolene Synthesis: A. G. Schultz and S. Puig[a]

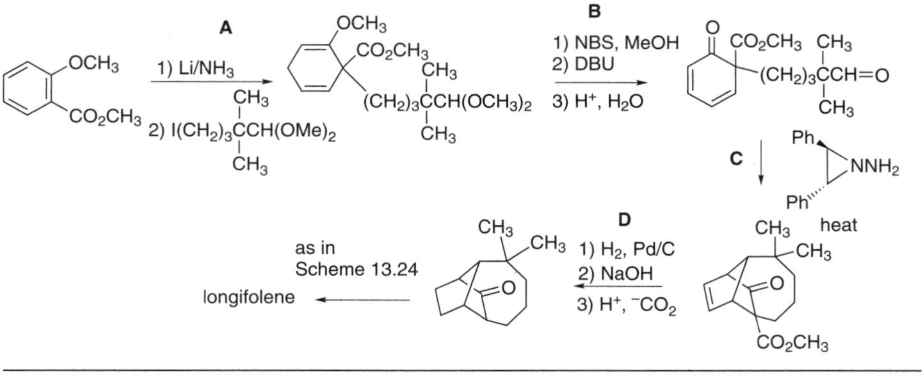

a. A. G. Schultz and S. Puig, *J. Org. Chem.*, **50**, 915 (1985).

An enantiospecific synthesis of longifolene was done starting with camphor, a natural product available in enantiomerically pure form (Scheme 13.31) The tricyclic ring was formed in Step **C** by an intramolecular Mukaiyama reaction. The dimethyl substituents were formed in Step **E-1** by hydrogenolysis of the cyclopropane ring. The final step of the synthesis involved a rearrangement of the tricyclic ring that was induced by solvolysis of the mesylate intermediate.

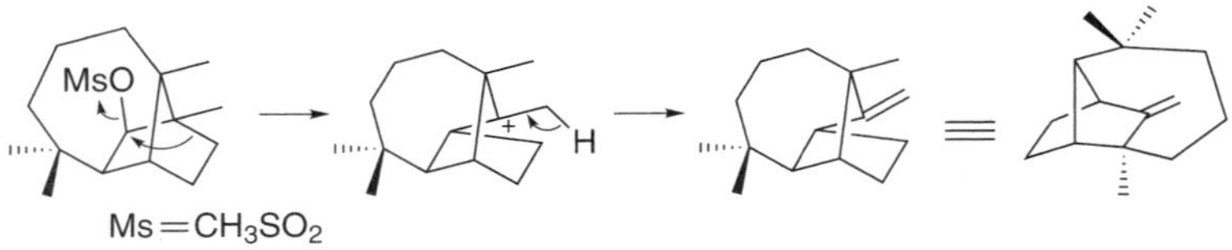

$$Ms = CH_3SO_2$$

Another enantiospecific synthesis of longifolene shown in Scheme 13.32 used an intramolecular Diels-Alder reaction as a key step. An alcohol intermediate was resolved in sequence **B** by formation and separation of a menthyl carbonate. After oxidation, the dihydropyrone ring was introduced by γ-addition of the ester enolate of methyl 3-methylbutenoate, followed by cyclization.

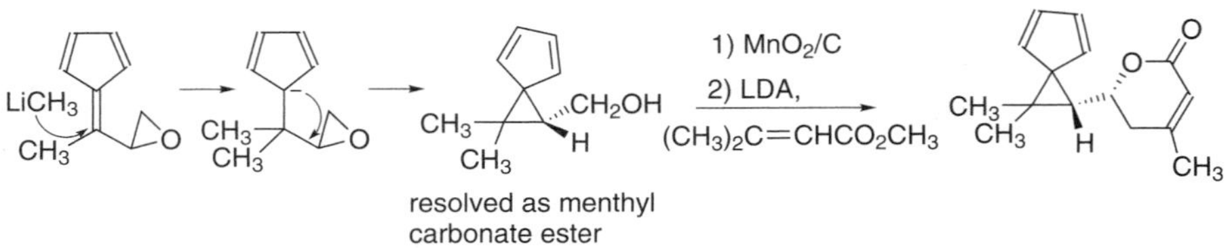

resolved as menthyl
carbonate ester

The dihydropyrone ring then served as the dienophile in the intramolecular Diels-Alder (IMDA) cycloaddition that was conducted in a microwave oven. The cyclopentadiene

Scheme 13.31. Longifolene Synthesis: D. L. Kuo and T. Money[a]

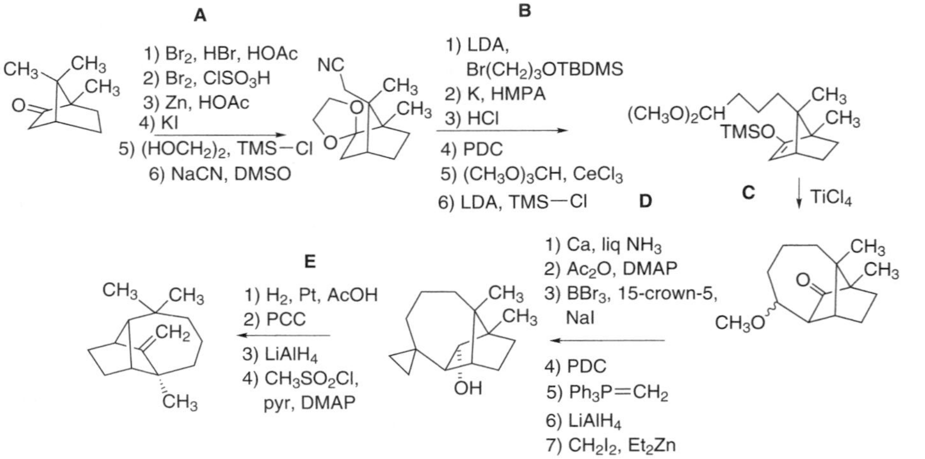

a. D. L. Kuo and T. Money, *Can. J. Chem.*, **66**, 1794 (1988).

Scheme 13.32. Longifolene Synthesis: B. Lei and A. G. Fallis[a]

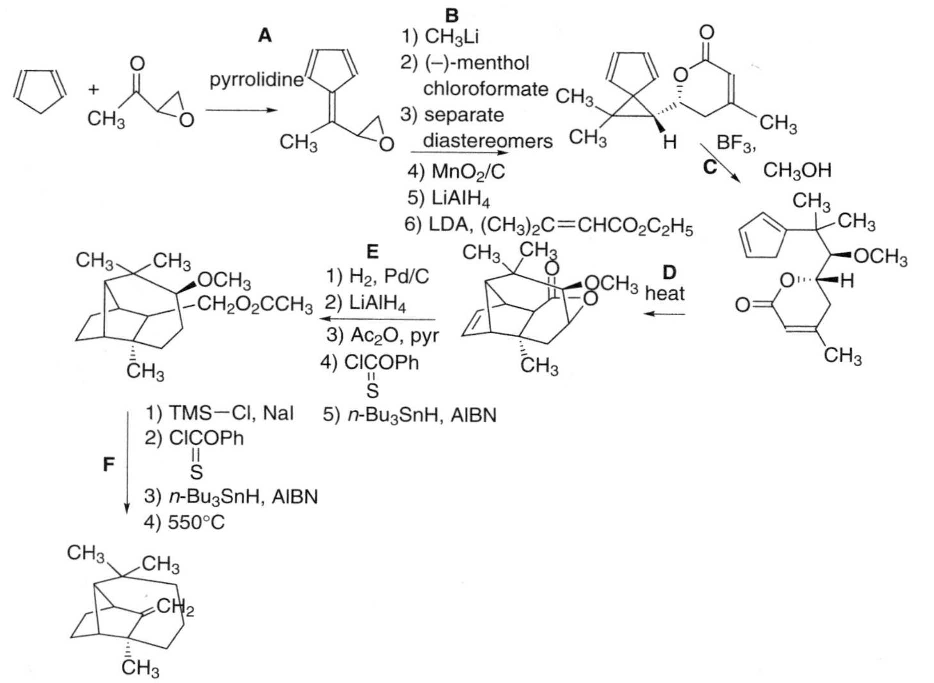

a. B. Lei and A. G. Fallis, *J. Am. Chem. Soc.*, **112**, 4609 (1990); B. Lei and A. G. Fallis, *J. Org. Chem.*, **58**, 2186 (1993).

ring permits rapid equilibration of the diene isomers by 1,5-hydrogen shifts and the most stable IMDA TS leads to the desired product.

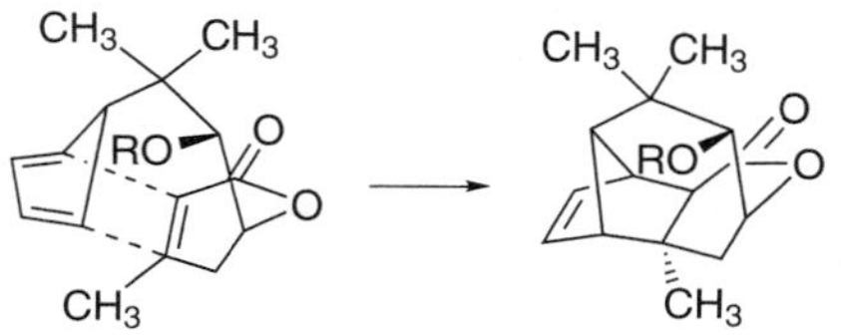

The final step of this synthesis used a high-temperature acetate pyrolysis to introduce the exocyclic double bond of longifolene.

Scheme 13.33 shows broad retrosynthetic formulations of the longifolene syntheses that are discussed in this subsection. Four different patterns of bond formation are represented. In **A**, the C(7)–C(10) bond is formed from a bicyclic intermediate. This pattern corresponds to the syntheses in Schemes 13.24, 13.25, 12.26, and 13.29. In retrosynthesis **B**, there is concurrent formation of the C(1)–C(2) and C(10)–C(11) bonds, as in the synthesis in Scheme 13.28. This is also the pattern found in the synthesis in Scheme 13.32. The synthesis in Scheme 13.29 corresponds to retrosynthesis **C**, in which the C(1)–C(2) and C(6)–C(7) bonds are formed and an extraneous bond between C(2) and C(5) is broken. Finally, retrosynthesis **D**, corresponding to formation of the C(2)–C(3) bond, is represented by the synthesis in Scheme 13.31.

These syntheses of longifolene provide good examples of the approaches that are available for construction of polycyclic ring compounds. In each case, a set of

Scheme 13.33. Summary of Some Retrosynthetic Patterns in Longifolene Syntheses

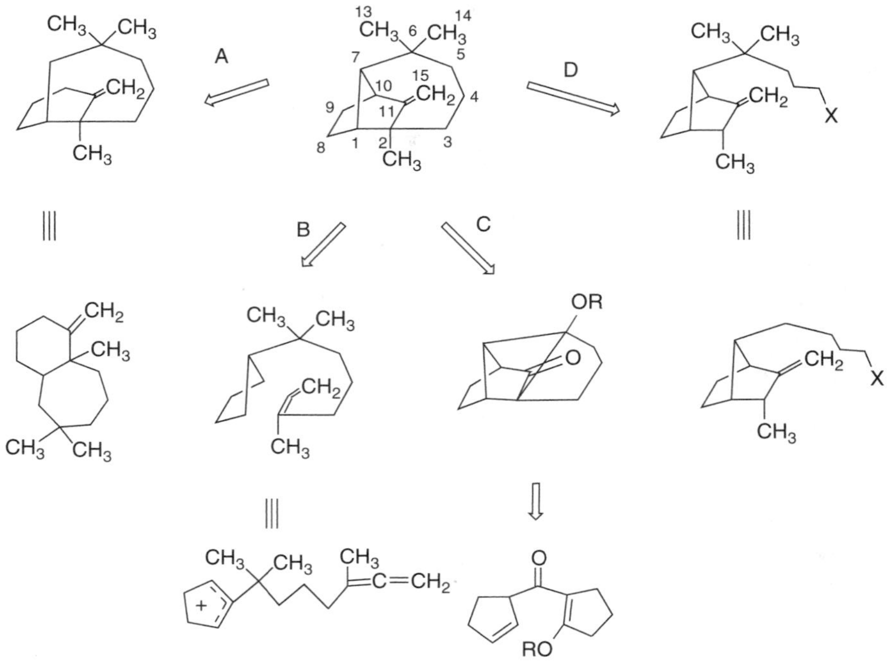

functionalities that have the potential for *intramolecular* reaction was assembled. After assembly of the carbon framework, the final functionality changes were effected. It is the necessity for the formation of the carbon skeleton that determines the functionalities that are present at the ring-closure stage. After the ring structure is established, necessary adjustments of the functionalities are made.

13.2.3. Prelog-Djerassi Lactone

The Prelog-Djerassi lactone (abbreviated here as P-D lactone) was originally isolated as a degradation product during structural investigations of antibiotics. Its open-chain equivalent **3** is typical of the methyl-branched carbon chains that occur frequently in macrolide and polyether antibiotics. The compound serves as a test case for the development of methods of control of stereochemistry in such polymethylated structures. There have been more than 20 different syntheses of P-D lactone.[24] We focus here on some of those that provide enantiomerically pure product, as they illustrate several of the methods for enantioselective synthesis.[25]

24. For references to many of these syntheses, see S. F. Martin and D. G. Guinn, *J. Org. Chem.*, **52**, 5588 (1987); H. F. Chow and I. Fleming, *Tetrahedron Lett.*, **26**, 397 (1985); S. F. Martin and D. E. Guinn, *Synthesis*, 245 (1991).

25. For other syntheses of enantiomerically pure Prelog-Djerassi lactone, see F. E. Ziegler, A. Kneisley, J. K. Thottathil, and R. T. Wester, *J. Am. Chem. Soc.*, **110**, 5434 (1988); A. Nakano, S. Takimoto, J. Inanaga, T. Katsuki, S. Ouchida, K. Inoue, M. Aiga, N. Okukado, and M. Yamaguchi, *Chem. Lett.*, 1019 (1979); K. Suzuki, K. Tomooko, T. Matsumoto, E. Katayama, and G. Tsuchihashi, *Tetrahedron*

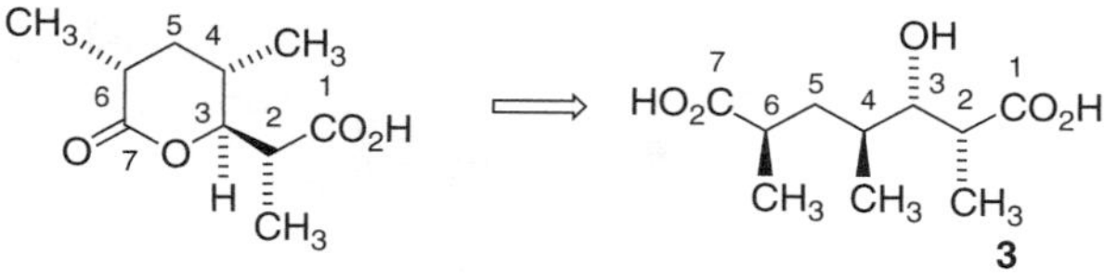

The synthesis in Scheme 13.34 is based on a bicyclic starting material that can be prepared in enantiomerically pure form. In the synthesis, C(7) of the norbornenone starting material becomes C(4) of P-D lactone and the methyl group in the starting material becomes the C(4) methyl substituent. The sequence uses the cyclic starting material to control facial selectivity. The configuration of the C(3) hydroxy and C(2) and C(6) methyl groups must be established relative to the C(4) stereocenter. The *exo*-selective alkylation in Step **A** established the configuration at C(2). The Baeyer-Villiger oxidation in Step **B** was followed by a Lewis acid–mediated allylic rearrangement, which is suprafacial. This stereoselectivity is dictated by the preference for maintaining a *cis* ring juncture at the five-membered rings.

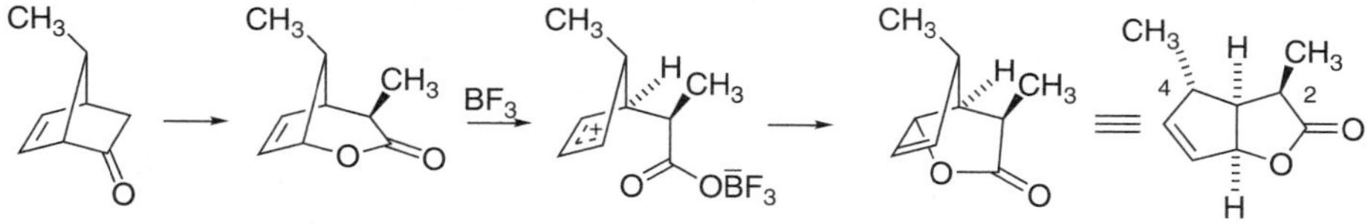

The stereochemistry of the C(3) hydroxy was established in Step **D**. The Baeyer-Villiger oxidation proceeds with retention of configuration of the migrating group (see Section 12.5.2), so the correct stereochemistry is established for the C−O bond. The final stereocenter for which configuration must be established is the methyl group at C(6) that was introduced by an enolate alkylation in Step **E**, but this reaction was not very stereoselective. However, since this center is adjacent to the lactone carbonyl, it can be epimerized through the enolate. The enolate was formed and quenched with acid. The kinetically preferred protonation from the axial direction provides the correct stereochemistry at C(6).

Scheme 13.34. Prelog-Djerassi Lactone Synthesis: P. A. Grieco and Co-Workers[a]

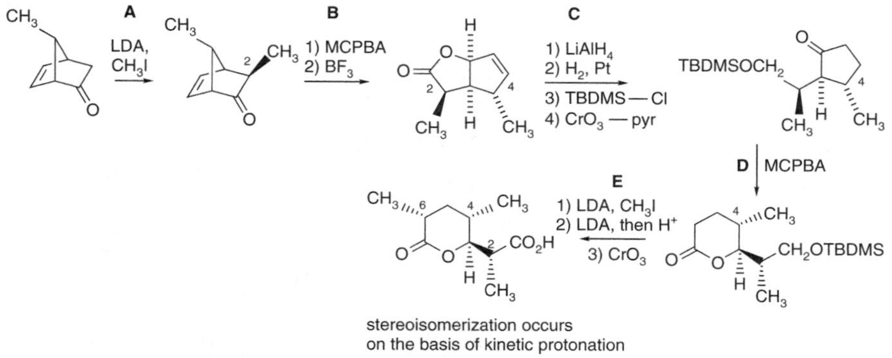

a. P. A. Grieco, Y. Ohfune, Y. Yokoyama, and W. Owens, *J. Am. Chem. Soc.*, **101**, 4749 (1979).

Lett., **26**, 3711 (1985); M. Isobo, Y. Ichikawa, and T. Goto, *Tetrahedron Lett.*, **22**, 4287 (1981); M. Mori, T. Chuman, and K. Kato, *Carbohydrate Res.*, **129**, 73 (1984).

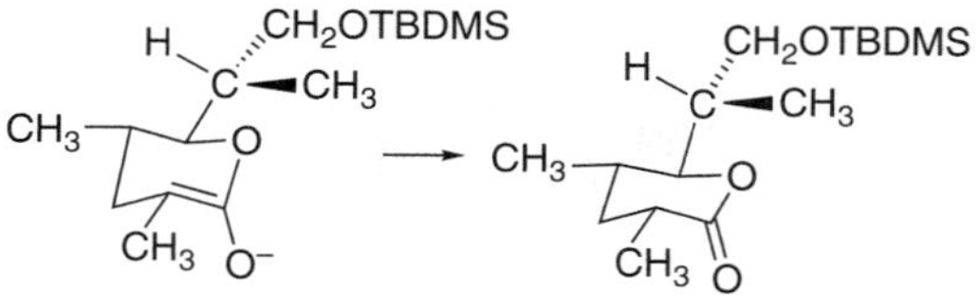

Another synthesis of P-D lactone that is based on an enantiomerically pure starting material is shown in Scheme 13.35. The stereocenter in the starting material is destined to become C(4) in the final product. Steps **A** and **B** served to extend the chain to provide a seven-carbon 1,5-diene. The configuration of two of the three remaining stereo-centers is controlled by the hydroboration step, which is a stereospecific *syn* addition (Section 4.5.1). In 1,5-dienes of this type, an intramolecular hydroboration occurs and establishes the configuration of the two newly formed C—B and C—H bonds.

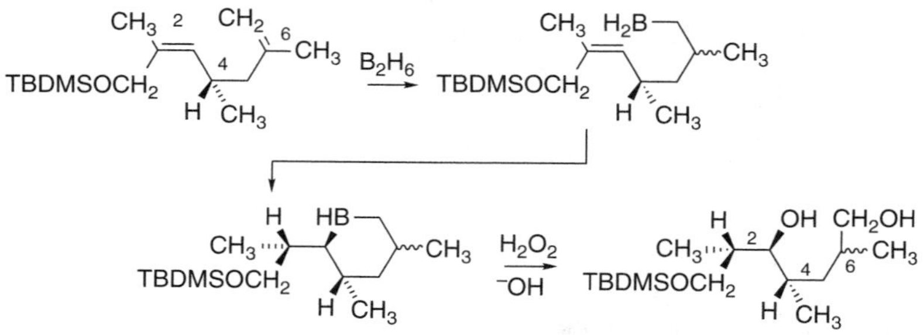

There was, however, no significant selectivity in the initial hydroboration of the terminal double bond. As a result, both configurations are formed at C(6). This problem was overcome using the epimerization process from Scheme 13.34.

The syntheses in Schemes 13.36 to 13.40 are conceptually related. They begin with symmetric achiral derivatives of *meso*-2,4-dimethylglutaric acid and utilize various approaches to the *desymmetrization* of the *meso* starting material. In Scheme 13.36

Scheme 13.35. Prelog-Djerassi Lactone Synthesis: W. C. Still and K. R. Shaw[a]

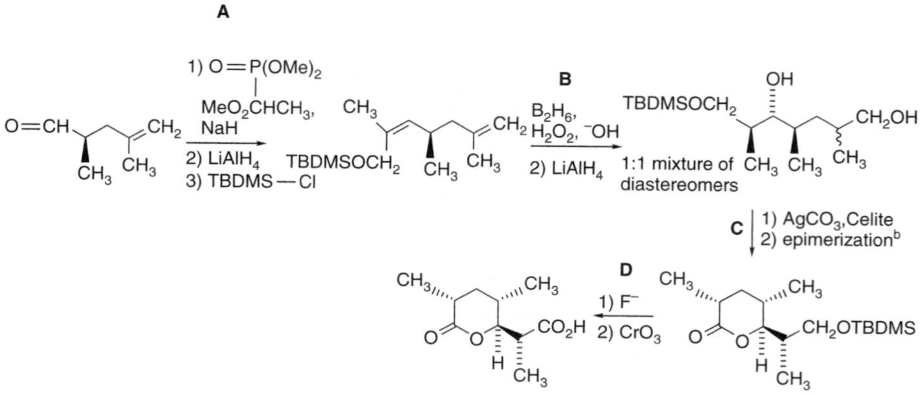

a. W. C. Still and K. R. Shaw, *Tetrahedron Lett.*, **22**, 3725 (1981).
b. Epimerization as in Scheme 13.34.

the starting material was prepared by reduction of the half-ester of *meso*-2,4-dimethylglutaric acid. The use of the *meso*-diacid ensures the correct *relative configuration* of the C(4) and C(6) methyl substituents. The half-acid was resolved and the correct enantiomer was reduced to the aldehyde. The stereochemistry at C(2) and C(3) was established by stereoselective aldol condensation methodology. Both the lithium enolate and the boron enolate methods were employed. The use of bulky enolates enhances the stereoselectivity. The enol derivatives were used in enantiomerically pure form so the condensations are examples of *double stereodifferentiation* (Section 2.1.5.3). The stereoselectivity observed in the reactions is that predicted by a cyclic TS for the aldol condensations.

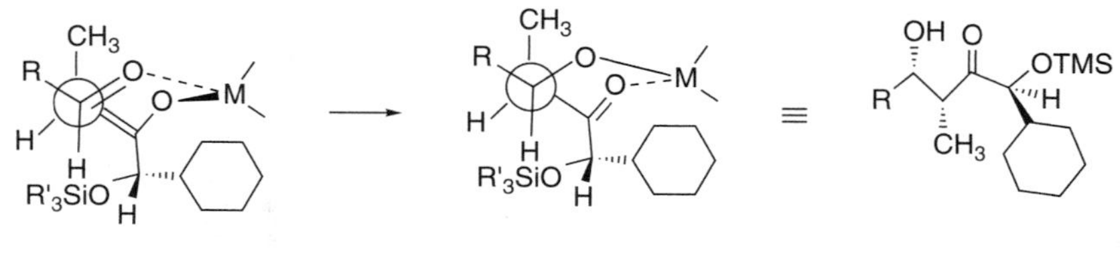

The synthesis in Scheme 13.37 also used a *meso*-3,4-dimethylglutaric acid as the starting material. Both the resolved aldehyde employed in Scheme 13.36 and a resolved half-amide were successfully used as intermediates. The configuration at C(2) and C(3) was controlled by addition of a butenylborane to an aldehyde (see Section 9.1.5). The boronate was used in enantiomerically pure form so that stereoselectivity was enhanced by *double stereodifferentiation*. The allylic additions carried out by the butenylboronates do not appear to have been quite as highly stereoselective as the aldol condensations used in Scheme 13.36, since a minor diastereoisomer was formed in the boronate addition reactions.

The synthesis in Scheme 13.38 is based on an interesting kinetic differentiation in the reactivity of two centers that are structurally identical, but diastereomeric. A *bis*-amide of *meso*-2,4-dimethylglutaric acid and a chiral thiazoline was formed in Step **A**. The thiazoline is derived from the amino acid cysteine. The two amide carbonyls in this *bis*-amide are nonequivalent by virtue of the diastereomeric relationship established

Scheme 13.36. Prelog-Djerassi Lactone Synthesis: S. Masamune and Co-Workers[a]

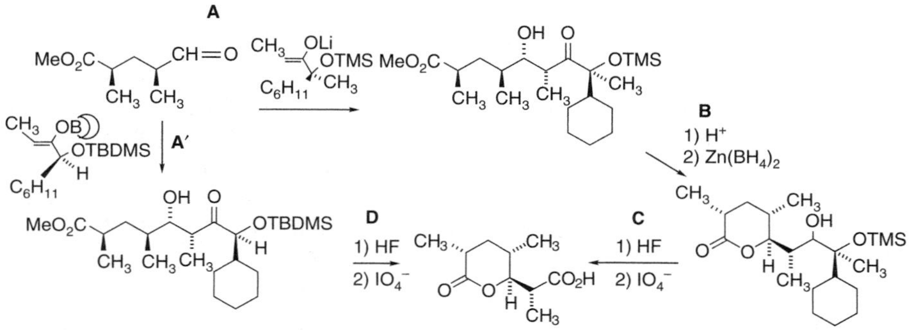

a. S. Masamune, S. A. Ali, D. L. Snitman, and D. S. Garvey, *Angew. Chem. Int. Ed. Engl.*, **19**, 557 (1980); S. Masamune, M. Hirama, S. Mori, S. A. Ali, and D. S. Garvey, *J. Am. Chem. Soc.*, **103**, 1568 (1981).

Scheme 13.37. Prelog-Djerassi Lactone Synthesis: R. W. Hoffmann and Co-Workers[a]

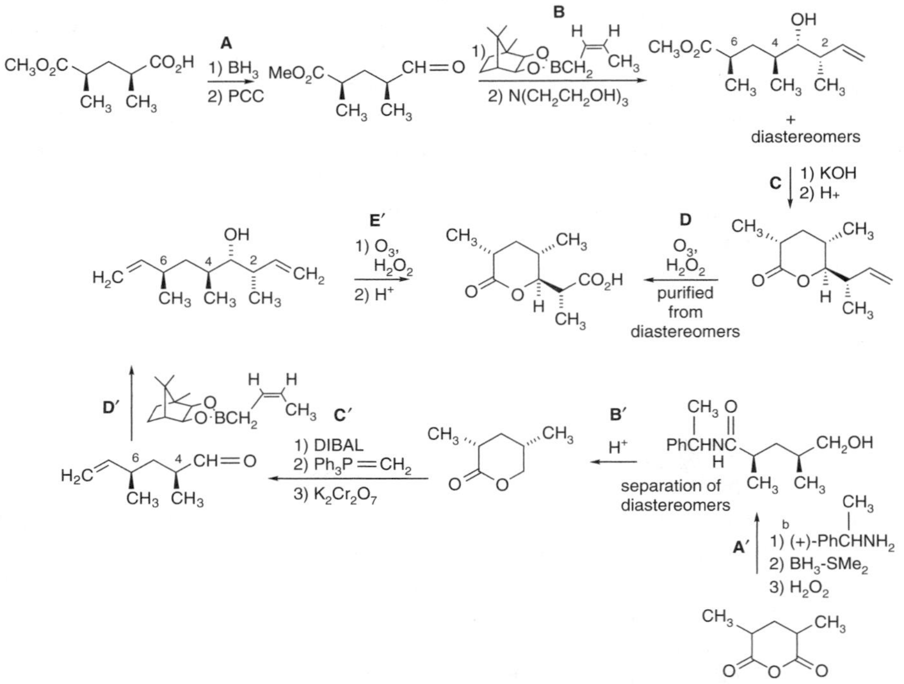

a. R. W. Hoffmann, H.-J. Zeiss, W. Ladner, and S. Tabche, *Chem. Ber.*, **115**, 2357 (1982).
b. Resolved via α-phenylethylamine salt; S. Masamune, S. A. Ali, D. L. Snitman, and D. S. Garvey, *Angew. Chem. Int. Ed. Engl.*, **19**, 557 (1980).

by the stereogenic centers at C(2) and C(4) in the glutaric acid portion of the structure. One of the centers reacted with a 97:3 preference with the achiral amine piperidine.

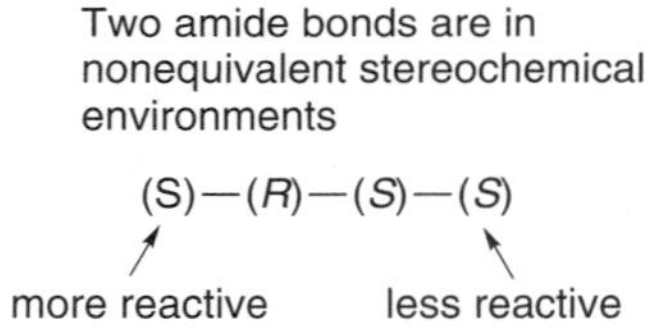

Two amide bonds are in
nonequivalent stereochemical
environments

(S)—(R)—(S)—(S)

more reactive   less reactive

In Step **D** another thiazoline chiral auxiliary, also derived from cysteine, was used to achieve double stereodifferentiation in an aldol addition. A tin enolate was used. The stereoselectivity of this reaction parallels that of aldol reactions carried out with lithium or boron enolates. After the configuration of all the centers was established, the synthesis proceeded to P-D lactone by functional group modifications.

A very short and efficient synthesis based on the desymmetrization principle is shown in Scheme 13.39. *meso*-2,4-Dimethylglutaraldehyde reacted selectively with the diethylboron enolate derived from a bornanesultam chiral auxiliary. This reaction established the stereochemistry at the C(2) and C(3) centers. The dominant aldol product results from an anti-Felkin stereoselectivity with respect to the C(4) center.

Scheme 13.38. Prelog-Djerassi Lactone Synthesis: Y. Nagao and Co-Workers[a]

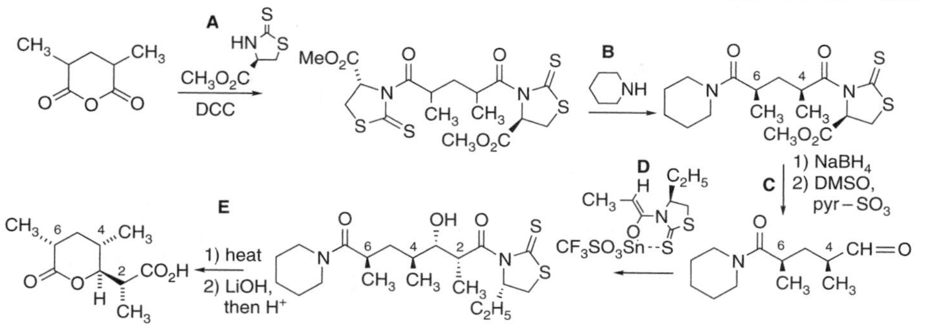

a. Y. Nagao, T. Inoue, K. Hashimoto, Y. Hagiwara, M. Ochai, and E. Fujita, *J. Chem. Soc., Chem. Commun.*, 1419 (1985).

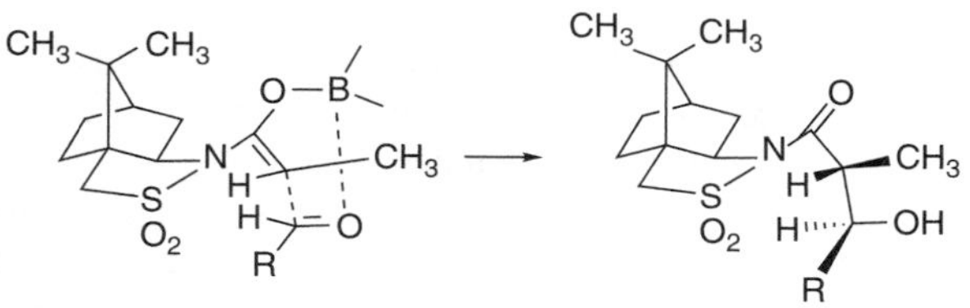

The adduct cyclized to a lactol mixture that was oxidized by TPAP-NMMO to give the corresponding lactones in an 8:1 ratio (86% yield). Hydrolysis in the presence of H_2O_2 gave the P-D lactone and recovered chiral auxiliary.

The synthesis in Scheme 13.40 features a catalytic asymmetric epoxidation (see Section 12.2.1.2). By use of *meso*-2,4-dimethylglutaric anhydride as the starting material, the proper relative configuration at C(4) and C(6) is ensured. The epoxidation directed by the (+)-tartrate catalyst controls the configuration established at C(2) and C(3) by the epoxidation. Although the epoxidation is highly selective in

Scheme 13.39. Prelog-Djerassi Lactone Synthesis: W. Oppolzer and Co-Workers[a]

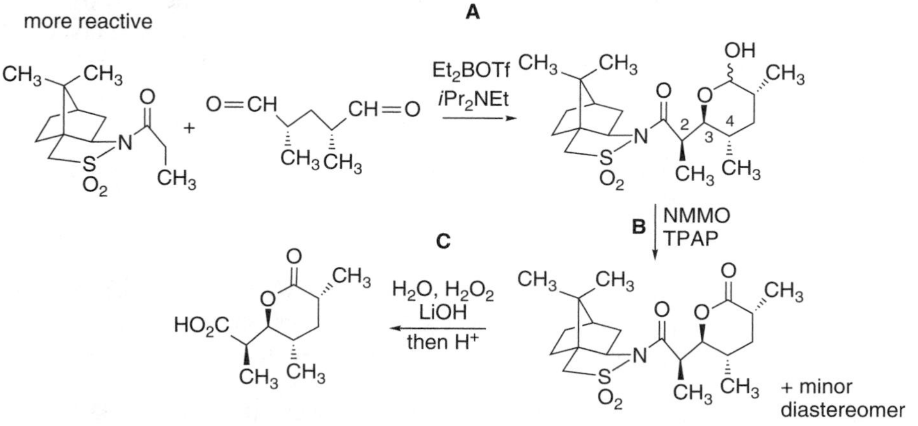

a. W. Oppolzer, E. Walther, C. Perez Balado, and J. De Brabander, *Tetrahedron Lett.*, **38**, 809 (1997).

Scheme 13.40. Prelog-Djerassi Lactone Synthesis: M. Yamaguchi and Co-Workers[a]

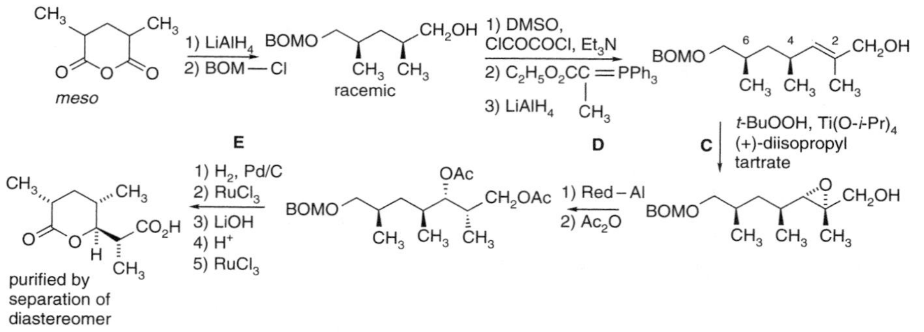

a. M. Honda, T. Katsuki, and M. Yamaguchi, *Tetrahedron Lett.*, **25**, 3857 (1984).

establishing the configuration at C(2) and C(3), the configuration at C(4) and C(6) does not strongly influence the reaction; a mixture of diastereomeric products was formed and then separated at a later stage in the synthesis. The reductive ring opening in Step **D** occurs with dominant inversion to establish the necessary (*R*)-configuration at C(2). The preference for 1,3-diol formation is characteristic of reductive ring opening by Red-Al of epoxides derived from allylic alcohols.[26] Presumably, initial coordination at the hydroxy group and intramolecular delivery of hydride is responsible for this stereoselectivity.

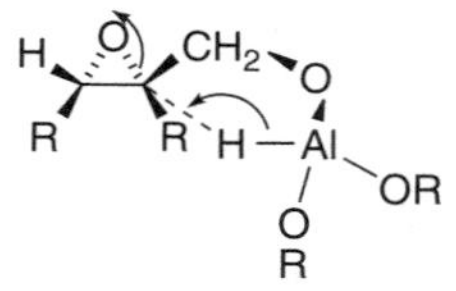

The synthesis in Scheme 13.41 is also built on the desymmetrization concept but uses a very different intermediate. *cis*-5,7-Dimethylcycloheptadiene was acetoxylated with Pd(OAc)$_2$ and the resulting all-*cis*-diacetate intermediate was enantioselectively hydrolyzed with a lipase to give a monoacetate that was protected as the TBDMS ether. An *anti* S$_N$2' displacement by dimethyl cuprate established the correct configuration of the C(2) methyl substituent. Oxidative ring cleavage and lactonization gave the final product.

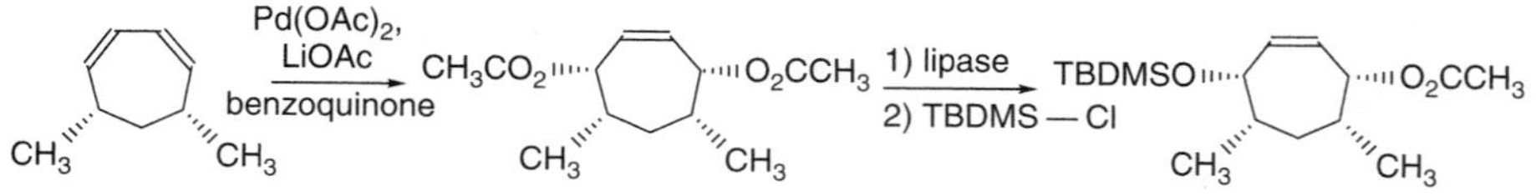

There have been several syntheses of P-D lactone that were based on carbohydrate-derived starting materials. The starting material used in Scheme 13.42 was prepared from a carbohydrate produced in earlier work.[27] The relative stereochemistry at C(4)

26. P. Ma, V. S. Martin, S. Masamune, K. B. Sharpless, and S. M. Viti, *J. Org. Chem.*, **47**, 1378 (1982); S. M. Viti, *Tetrahedron Lett.*, **23**, 4541 (1982); J. M. Finan and Y. Kishi, *Tetrahedron Lett.*, **23**, 2719 (1982).
27. M. B. Yunker, D. E. Plaumann, and B. Fraser-Reid, *Can. J. Chem.*, **55**, 4002 (1977).

Scheme 13.41. Prelog-Djerassi Lactone Synthesis: A. J. Pearson and Y.-S. Lai[a]

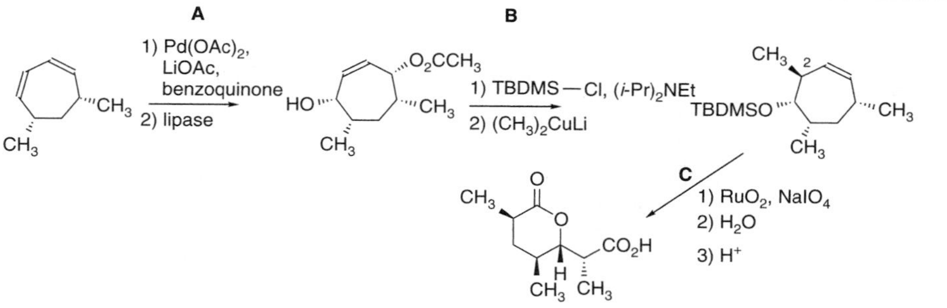

a. A. J. Pearson and Y.-S. Lai, *J. Chem. Soc., Chem. Commun.*, 442 (1988).

and C(6) was established by the hydrogenation in Step **A**-2. This *syn* hydrogenation is not completely stereoselective, but provided a 4:1 mixture favoring the desired stereoisomer. The stereoselectivity is presumably the result of preferential absorption from the less hindered β-face of the molecule. The configuration of C(2) was established by protonation during the hydrolysis of the enol ether in Step **C**-2. This step was not stereoselective, so a separation of diastereomers after the oxidation in Step **C**-3 was required.

The synthesis in Scheme 13.43 also began with carbohydrate-derived starting material and uses catalytic hydrogenation in Step **C**-1 to establish the stereochemical relationship between the C(4) and C(6) methyl groups. As was the case in Scheme 13.42, the configuration at C(2) was not controlled in this synthesis and separation of the diastereomeric products was necessary. This synthesis used an organocopper reagent to introduce both the C(4) and C(2) methyl groups. The former was introduced by S_N2' allylic substitution in Step **B** and the latter by conjugate addition to a nitroalkene in Step **D**.

The synthesis in Scheme 13.44 is also based on a carbohydrate-derived starting material. It controlled the stereochemistry at C(2) by means of the stereoselectivity of the Ireland-Claisen rearrangement in Step **A** (see Section 6.4.2.3). The ester enolate was formed under conditions in which the *E*-enolate is expected to predominate. Heating the resulting silyl enol ether gave a 9:1 preference for the expected stereoisomer. The

Scheme 13.42. Prelog-Djerassi Lactone Synthesis: S. Jarosz and B. Fraser-Reid[a]

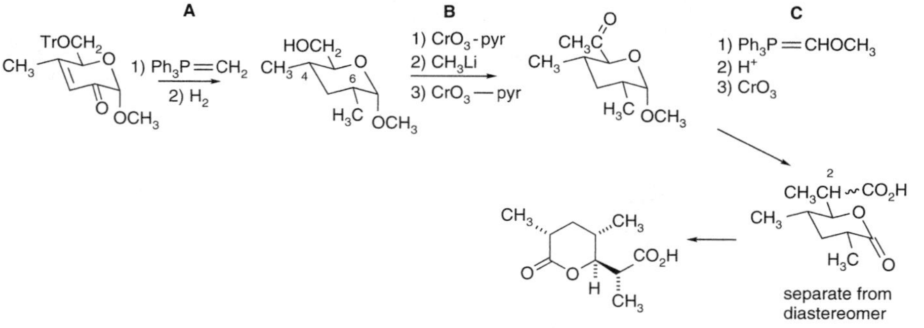

a. S. Jarosz and B. Fraser-Reid, *Tetrahedron Lett.*, **22**, 2533 (1981).

Scheme 13.43. Prelog-Djerassi Lactone Synthesis: N. Kawauchi and H. Hashimoto[a]

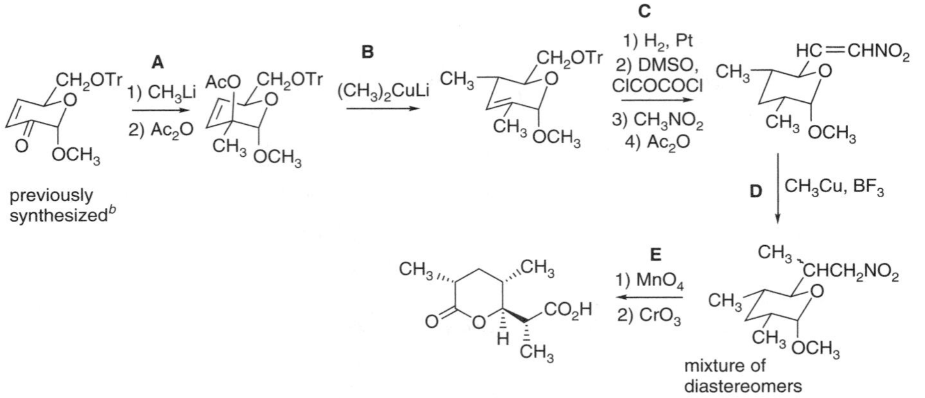

previously
synthesized[b]

mixture of
diastereomers

a. N. Kawauchi and H. Hashimoto, *Bull. Chem. Soc. Jpn.*, **60**, 1441 (1987).
b. N. L. Holder and B. Fraser-Reid, *Can. J. Chem.*, **51**, 3357 (1973).

preferred TS, which is boatlike, minimizes the steric interaction between the bulky silyl substituent and the ring structure.

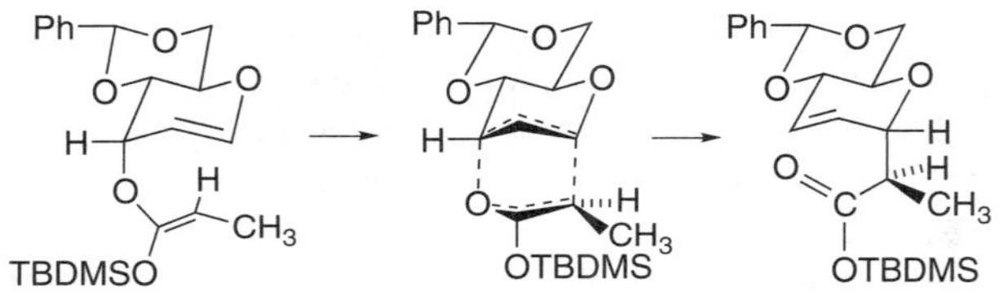

The stereochemistry at C(4) and C(6) was then established. The cuprate addition in Step **C** occurred *anti* to the substituent at C(2) of the pyran ring. After a Wittig

Scheme 13.44. Prelog-Djerassi Lactone Synthesis: R. E. Ireland and J. P. Daub[a]

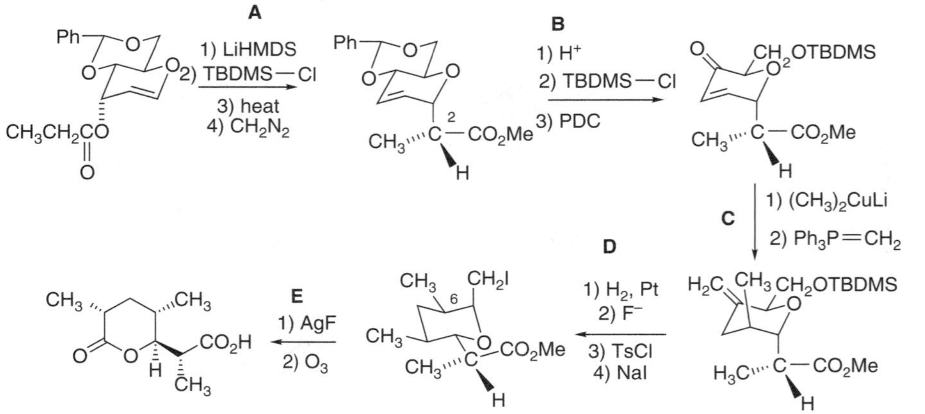

a. R. E. Ireland and J. P. Daub, *J. Org. Chem.*, **46**, 479 (1981).

Scheme 13.45. Prelog-Djerassi Lactone Synthesis: D. A. Evans and J. Bartroli[a]

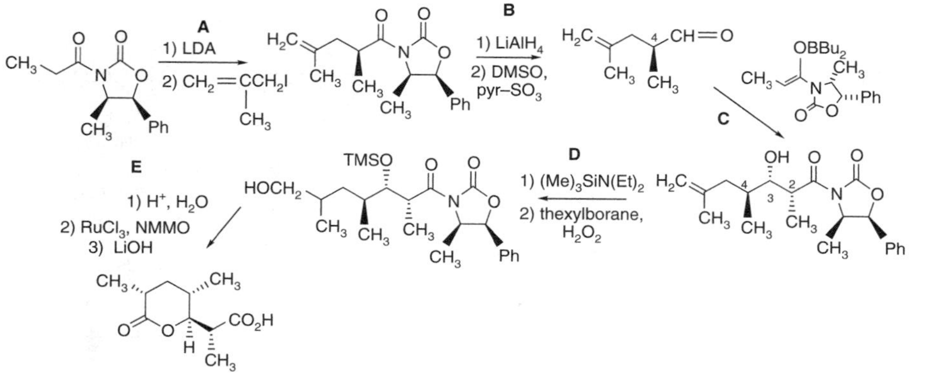

a. D. A. Evans and J. Bartroli, *Tetrahedron Lett.*, **23**, 807 (1982).

methylenation, the catalytic hydrogenation in Step **D** established the stereochemistry at C(6). The lactone carbonyl was introduced by β-elimination and ozonolysis.

The syntheses in Schemes 13.45 and 13.46 illustrate the use of oxazolidinone chiral auxiliaries in enantioselective synthesis. Step **A** in Scheme 13.45 established the configuration at the carbon that becomes C(4) in the product. This is an enolate alkylation in which the steric effect of the oxazolidinone chiral auxiliary directs the approach of the alkylating group. Step **C** also used the oxazolidinone structure. In this case, the enol borinate is formed and condensed with an aldehyde intermediate. This stereoselective aldol addition established the configuration at C(2) and C(3). The configuration at the final stereocenter at C(6) was established by the hydroboration in Step **D**. The selectivity for the desired stereoisomer was 85:15. Stereoselectivity in the same sense has been observed for a number of other 2-methylalkenes in which the remainder of the alkene constitutes a relatively bulky group.[28] A TS such as **45-A** can rationalize this result.

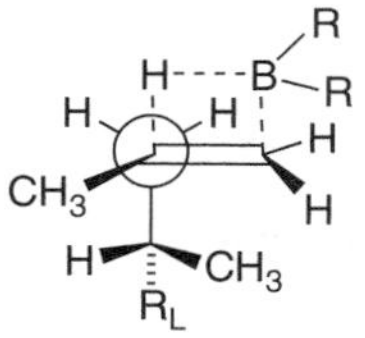

45-A

In the synthesis in Scheme 13.46, a stereoselective aldol addition was used to establish the configuration at C(2) and C(3) in Step **A**. The furan ring was then subjected to an electrophilic addition and solvolytic rearrangement in Step **B**.

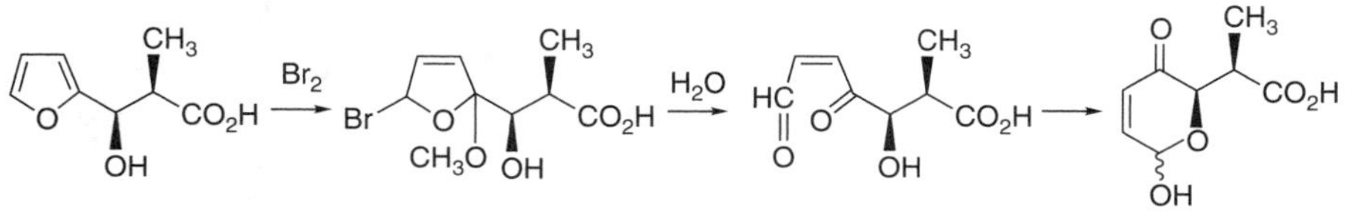

[28] D. A. Evans, J. Bartroli, and T. Godel, *Tetrahedron Lett.*, **23**, 4577 (1982).

Scheme 13.46. Prelog-Djerassi Lactone Synthesis: S. F. Martin and D. E. Guinn[a]

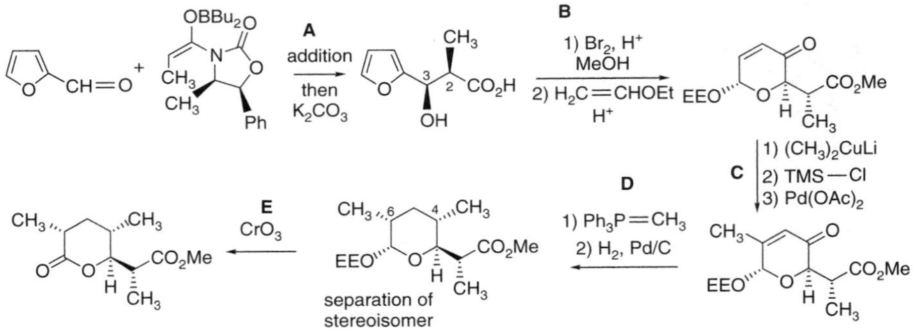

a. S. F. Martin and D. E. Guinn, *J. Org. Chem.*, **52**, 5588 (1987).

The protection of the hemiacetal hydroxyl in Step **B-2** was followed by a purification of the dominant stereoisomer. In Step **C**-1, the addition of the C(6) methyl group gave predominantly the undesired α-stereoisomer. The enolate was trapped as the trimethylsilyl ether and oxidized to the enone by Pd(OAc)$_2$. The enone from sequence **C** was then subjected to a Wittig reaction. As in several of the other syntheses, the hydrogenation in Step **D**-2 was used to establish the configuration at C(4) and C(6).

The synthesis in Scheme 13.47 was also based on use of a chiral auxiliary and provided the TBDMS-protected derivative of P-D lactone in the course of synthesis of the macrolide portion of the antibiotic 10-deoxymethymycin. The relative stereochemistry at C(2)–C(3) was obtained by addition of the dibutylboron enolate of an *N*-propanoyl oxazolidinone. The addition occurs with *syn* anti-Felkin stereochemistry.

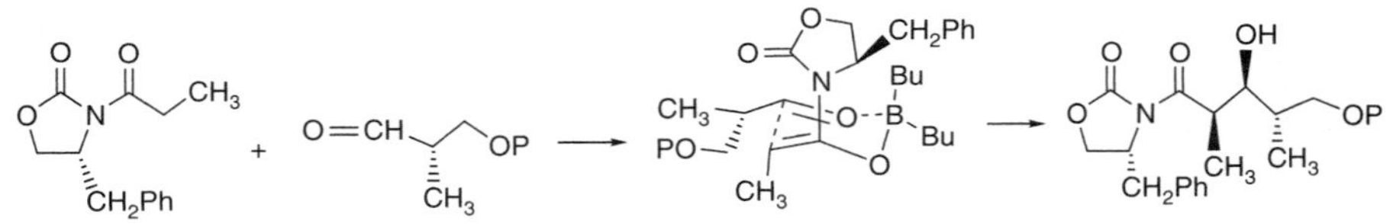

Scheme 13.47. Prelog-Djerassi Lactone Synthesis: R. A. Pilli and Co-Workers[a]

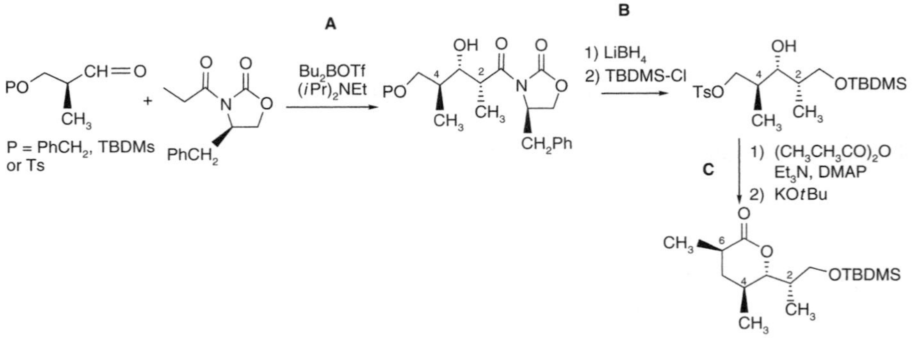

a. R. A. Pilli, C. K. Z. de Andrade, C. R. O. Souto, and A. de Meijere, *J. Org. Chem.*, **63**, 7811 (1998).

Scheme 13.48. Prelog-Djerassi Lactone Synthesis: M. Miyashita and Co-Workers[a]

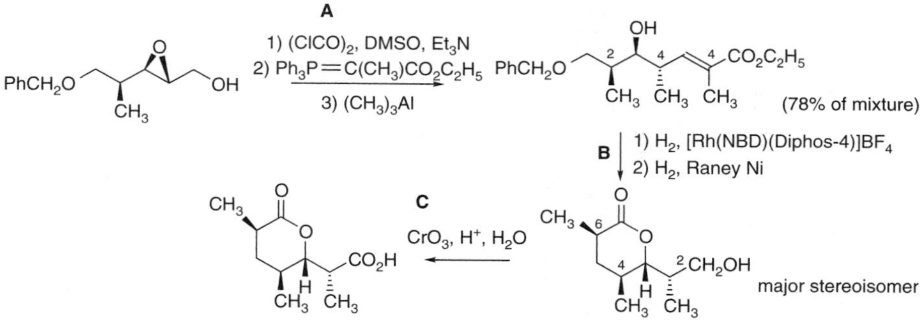

a. M. Miyashita, M. Hoshino, A. Yoshikoshi, K. Kawamine, K. Yoshihara, and H. Irie, *Chem. Lett.*, 1101 (1992).

Removal of the chiral auxiliary and reduction gave an intermediate that had differentiated terminal hydroxy groups. Although the sequence was initially carried out on the benzyl or TBDMS-protected aldehyde, with subsequent removal of the protecting group, it was found that the aldol addition could be carried out directly on the tosylate, providing a shorter route. A propanoyl group was added at Step **C**-1 and provided the remainder of the carbon chain. The lactone ring was closed by an intramolecular enolate alkylation. This step is not highly stereoselective, but equilibration (see Scheme 13.34) gave the desired stereoisomer in a 10:1 ratio.

The synthesis in Scheme 13.48 used stereospecific ring opening of an epoxide by trimethylaluminum to establish the stereochemistry of the C(4) methyl group. The starting material was made by enantiospecific epoxidation of the corresponding allylic alcohol.[29] The hydrogenation in Step **B**-1 achieved about 3:1 stereoselectivity at C(2). Removal of the benzyl protecting group by hydrogenolysis then gave the lactone.

The synthesis in Scheme 13.49 features use of an enantioselective allylic boronate reagent derived from diisopropyl tartrate to establish the C(4) and C(5) stereochemistry. The ring is closed by an olefin metathesis reaction. The C(2) methyl group was introduced by alkylation of the lactone enolate. The alkylation is not stereoselective, but base-catalyzed epimerization favors the desired stereoisomer by 4:1.

Scheme 13.49. Prelog-Djerassi Lactone Synthesis: J. Cossy, D. Bauer, and V. Bellosta[a]

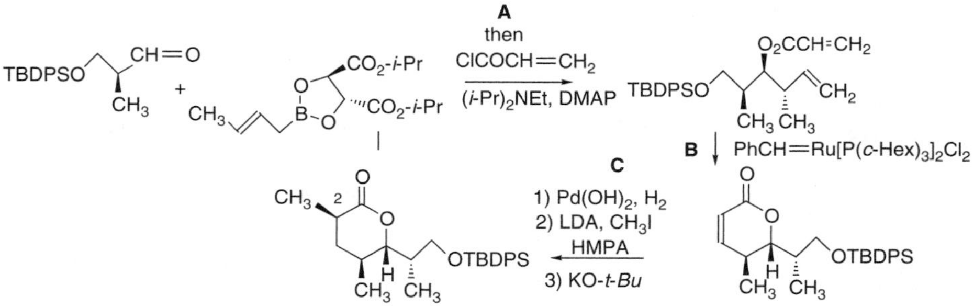

a. J. Cossy, D. Bauer, and V. Bellosta, *Tetrahedron Lett.*, **40**, 4187 (1999).

29. H. Nagaoka and Y. Kishi, *Tetrahedron*, **37**, 3873 (1981).

Scheme 13.50. Prelog-Djerrasi Lactone Synthesis: D. J.-S. Tsai and M. M. Midland[a]

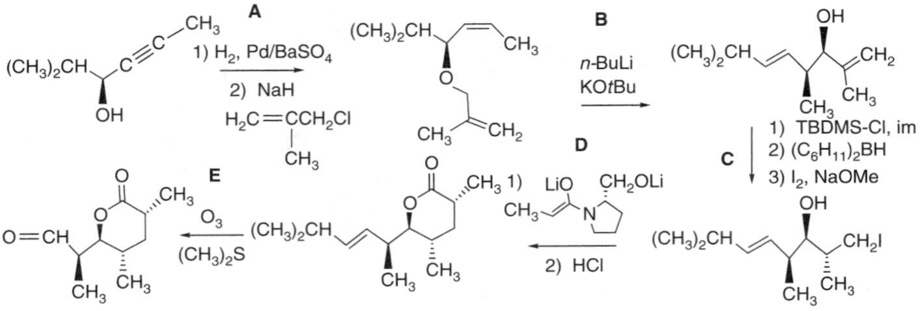

a. D. J.-S. Tsai and M. M. Midland, *J. Am. Chem. Soc.*, **107**, 3915 (1985).

The synthesis in Scheme 13.50 used the stereoselectivity of a [2,3]-sigmatropic rearrangement as the basis of stereochemical control. The starting material was prepared by enantioselective reduction of the corresponding ketone using *S*-Alpine-Borane. The sigmatropic rearrangement of the lithium anion in Step **B** gave 97:3 stereoselectivity for the *syn* isomer (see p. 588). After protection, this intermediate was selectively hydroborated with $(C_6H_{11})_2BH$ and converted to the iodide. The hydroboration in Step **C**-2 establishes the stereochemistry at C(4) with 15:1 stereoselectivity. The iodide was then used in conjunction with a chiral auxiliary to create the C(2)–C(3) bond by alkylation of the amide enolate.

A recent synthesis of P-D lactone (Scheme 13.51) used an enantioselective catalytic approach. A conjugate addition of a silyl ketene acetal derived from an unsaturated ester gave an unsaturated lactone intermediate. The catalyst is CuF-(*S*)-tol-BINAP.[30] The catalytic cycle for the reaction is shown below.

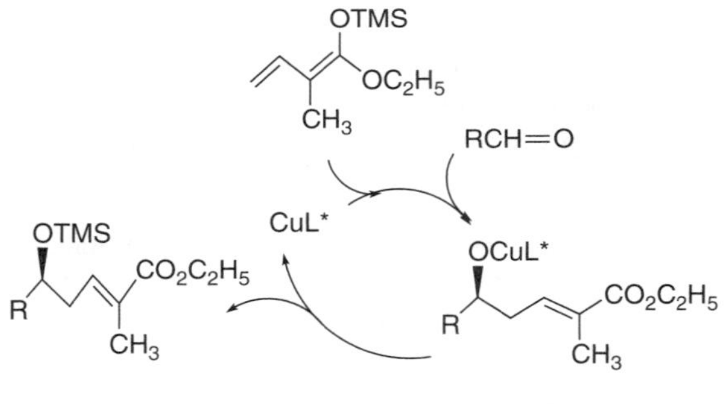

Scheme 13.51. Prelog-Djerassi Lactone Synthesis: J.-M. Campagne and Co-Workers[a]

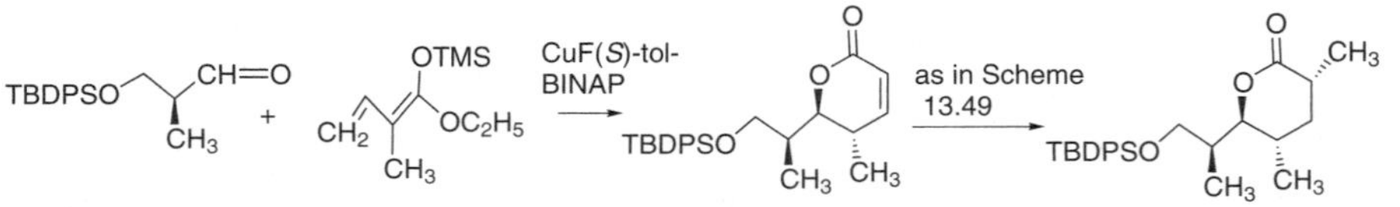

a. G. Bluet, B. Bazan-Tejeda, and J.-M. Campagne, *Org. Lett.*, **3**, 3807 (2001).

[30.] J. Krueger and E. M. Carreira, *J. Am. Chem. Soc.*, **120**, 837 (1998).

The reaction was very stereoselective for the correct P-D lactone configuration. The synthesis, which is outlined in Scheme 13.51, was completed by the sequence shown in Scheme 13.49.

The synthesis shown in Scheme 13.52 started with an enantiomerically pure protected aldehyde. Reaction with a Grignard reagent installed an allylic silane. This reaction gave a mixture of alcohols, but both were converted to the same intermediate by taking advantage of selective formation of *E*- or *Z*-silyl ketene acetal prior to an Ireland-Claisen rearrangement. These stereoconvergent transformations are described on p. 568. Two subsequent steps are noteworthy. In Step **C**-3, a BF₃-mediated opening of the dioxolane ring triggers a desilylation. In Step **E**-1, the diimide reduction occurs with excellent stereoselectivity. This is attributed to a π-stacking interaction with the TBDPS protecting group, since no similar effect was noted with the TBDMS group.

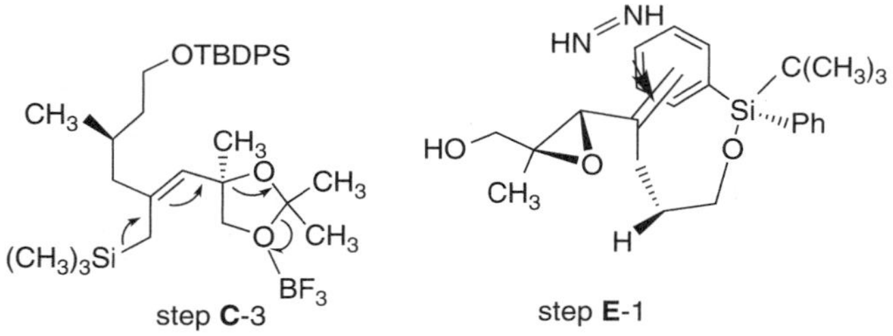

step **C**-3   step **E**-1

The final lactonization and oxidation were done as in Scheme 13.40.

Scheme 13.52. Prelog-Djerassi Lactone Synthesis: P. J. Parsons and Co-Workers[a]

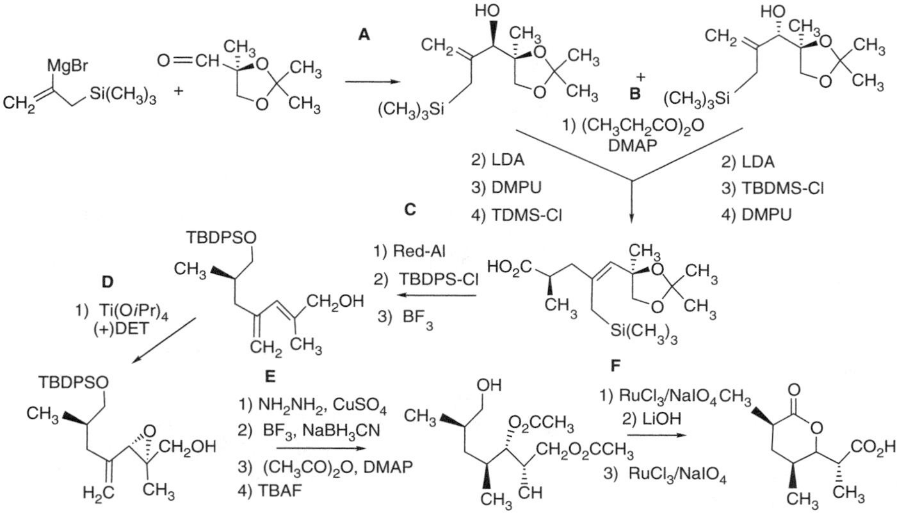

a. S. D. Hiscock, P. B. Hitchcock, and P. J. Parsons, *Tetrahedron*, **54**, 11567 (1998).

13.2.4. Baccatin III and Taxol

Taxol[®31] was first discovered to have anticancer activity during a screening of natural substances,[32] and it is currently an important drug in cancer chemotherapy. Several Taxol analogs differing in the side-chain substitution, such as taxotere, also have good activity.[33] Production of Taxol directly from plant sources presented serious problems because the plants are slow growing and the Taxol content is low. However, the tetracyclic ring system is found in a more available material, Baccatin III, which can be converted to Taxol by introduction of the side chain.[34] The combination of important biological activity, the limited natural sources, and the interesting structure made Taxol a target of synthetic interest during the 1990s. Among the challenging aspects of the structure from a synthetic point of view are the eight-membered ring, the bridgehead double bond, and the large number of oxygen functional groups. Several syntheses of Baccatin III and closely related tetracyclic Taxol precursors have been reported.

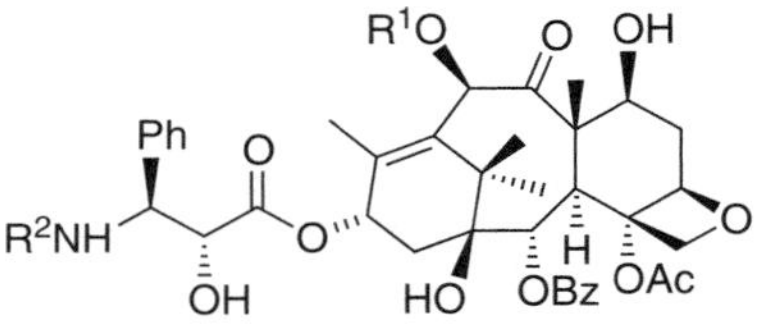

taxol R_1 = Ac, R_2 = PhCO
taxotere R_1 = H, R_2 = $(CH_3)_3CO_2C$

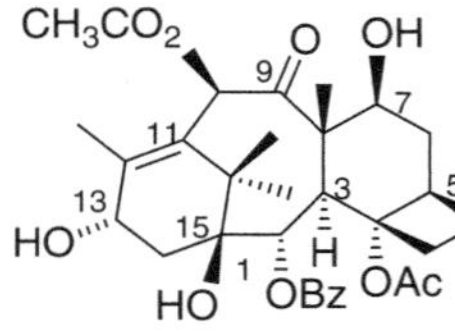

baccatin III

The first synthesis of Taxol was completed by Robert Holton and co-workers and is outlined in Scheme 13.53. One of the key steps occurs early in the synthesis in sequence **A** and effects fragmentation of **4** to **5**. The intermediate epoxide **4** was prepared from a sesquiterpene called "patchino."[35] The epoxide was then converted to **5** by a BF_3-mediated rearrangement.

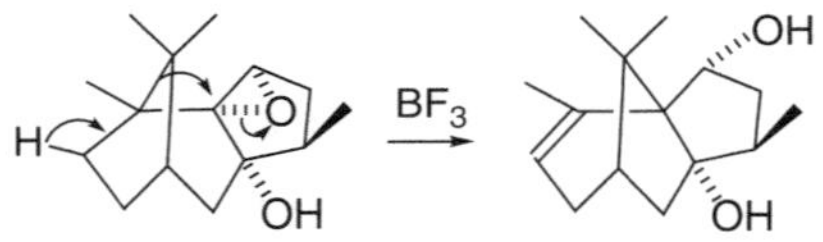

Another epoxidation, followed by fragmentation gave the bicyclic intermediate that contains the eight-membered ring and bridgehead double bond properly positioned for conversion to Taxol (Steps **B**-2 and **B**-3).

[31.] Taxol is a registered trade name of Bristol-Myers Squibb. The generic name is paclitaxel.

[32.] M. C. Wani, H. L. Taylor, M. E. Wall, D. Coggon, and A. McPhail, *J. Am. Chem. Soc.*, **93**, 2325 (1971); M. E. Wall and M. C. Wani, *Alkaloids*, **50**, 509 (1998).

[33.] M. Suffness, ed., *Taxol: Science and Applications*, CRC Press, Boca Raton, FL, 1995.

[34.] J.-N. Denis, A. E. Greene, D. Guenard, F. Gueritte-Vogelein, L. Mangatal, and P. Potier, *J. Am. Chem. Soc.*, **110**, 5917 (1988); R. A. Holton, Z. Zhang, P. A. Clarke, H. Nadizadeh, and D. J. Procter, *Tetrahedron Lett.*, **39**, 2883 (1998).

[35.] R. A. Holton, R. R. Juo, H. B. Kim, A. D. Williams, S. Harusawa, R. E. Lowenthal, and S. Yogai, *J. Am. Chem. Soc.*, **110**, 6558 (1988).

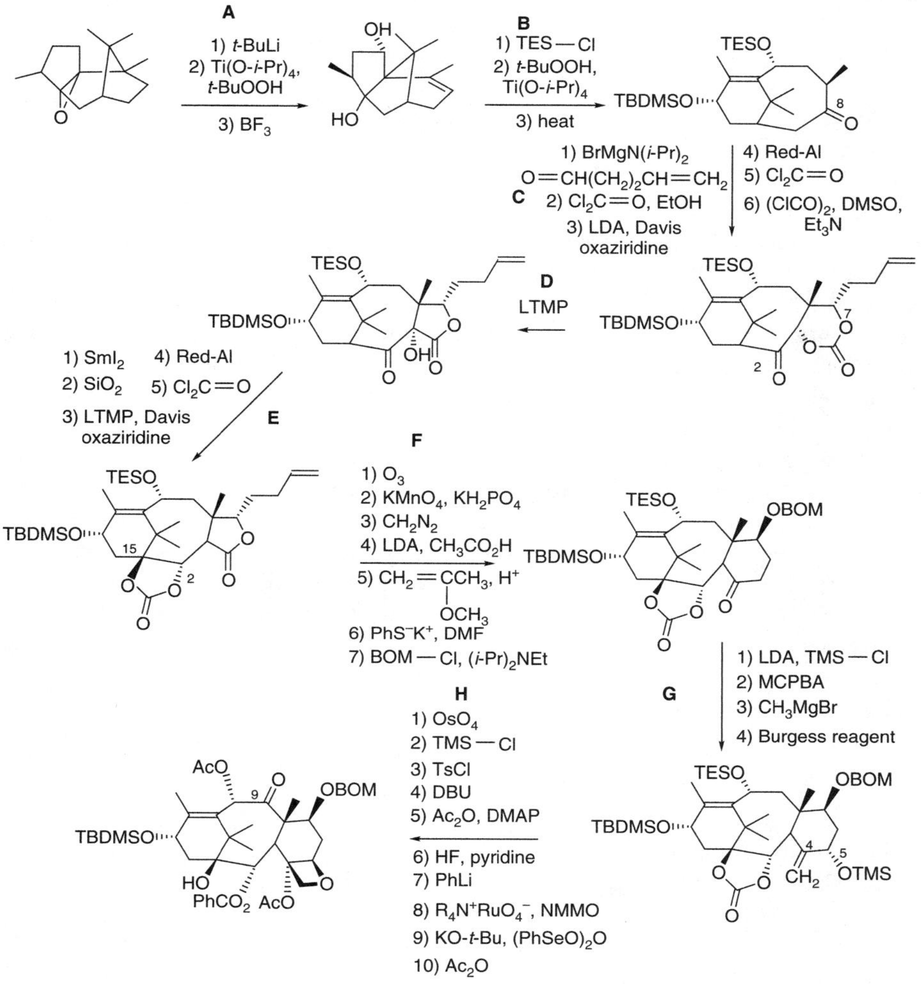

a. R. A. Holton, C. Somoza, H.-B. Kim, F. Liang, R. J. Biediger, P. D. Boatman, M. Shindo, C. C. Smith, S. Kim, H. Nadizadeh, Y. Suzuki, C. Tao, P. Vu, S. Tang, P. Zhang, K. K. Murthi, L. N. Gentile, and J. H. Lin, *J. Am. Chem. Soc.*, **116**, 1597 (1994); R. A. Holton, H.-B. Kim, C. Somoza, F. Liang, R. J. Biediger, P. D. Boatman, M. Shindo, C. C. Smith, S. Kim, H. Nadizadeh, Y. Suzuki, C. Tao, P. Vu, S. Tang, P. Zhang, K. K. Murthi, L. N. Gentile, and J. H. Liu, *J. Am. Chem. Soc.*, **116**, 1599 (1994).

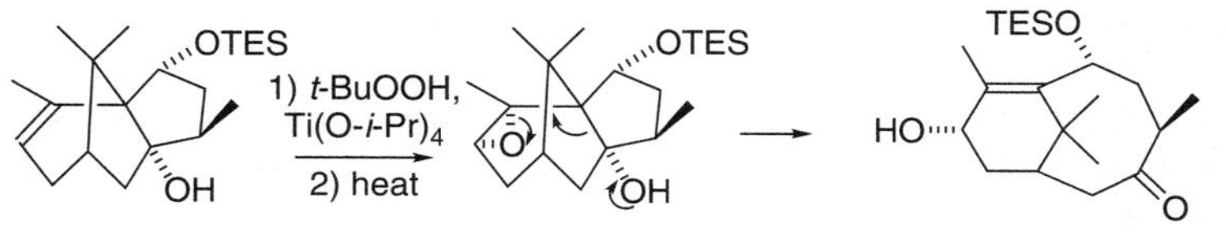

The next phase of the synthesis was construction of the C-ring. An aldol addition was used to introduce a 3-butenyl group at C(8) and the product was trapped as a carbonate ester. The Davis oxaziridine was then used to introduce an oxygen at C(2). After reduction of the C(3) oxygen, a cyclic carbonate was formed, and C(2) was converted to a carbonyl group by Swern oxidation. In Step **D** this carbonate was rearranged to a lactone.

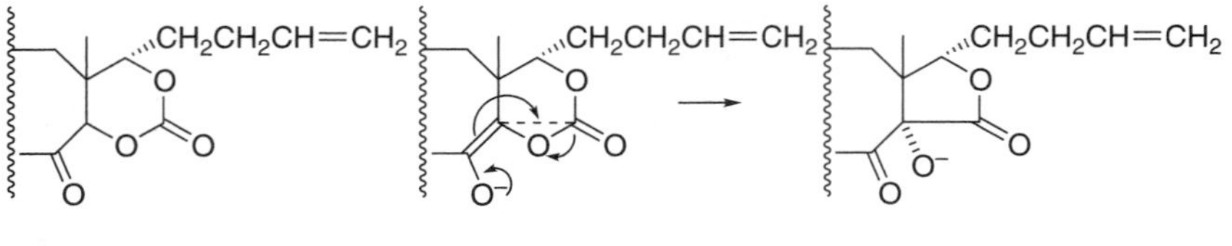

Reaction sequence **E** removed an extraneous oxygen by SmI$_2$ reduction and installed an oxygen at C(15) by enolate oxidation. The C(1) and C(15) hydroxy groups were protected as a carbonate in Step **E**-5. After oxidation of the terminal vinyl group, the C-ring was constructed by a Dieckmann cyclization in Step **F**-4. After temporary protection of the C(7) hydroxy as the MOP derivative, the β-ketoester was subjected to nucleophilic decarboxylation by phenylthiolate and reprotected as the BOM ether (Steps **F**-5, **F**- 6, and **F**-7).

An oxygen substituent was introduced at C(5) by MCPBA oxidation of a silyl enol ether (Steps **G**-1 and **G**-2). An exocyclic methylene group was introduced at C(4) by a methyl Grignard addition followed by dehydration with Burgess reagent (**G**-3). The oxetane ring was constructed in Steps **H**-1 to **H**-4. The double bond was hydroxylated with OsO$_4$ and a sequence of selective transformations of the triol provided the hydroxy tosylate, which undergoes intramolecular nucleophilic substitution to form the oxetane ring.

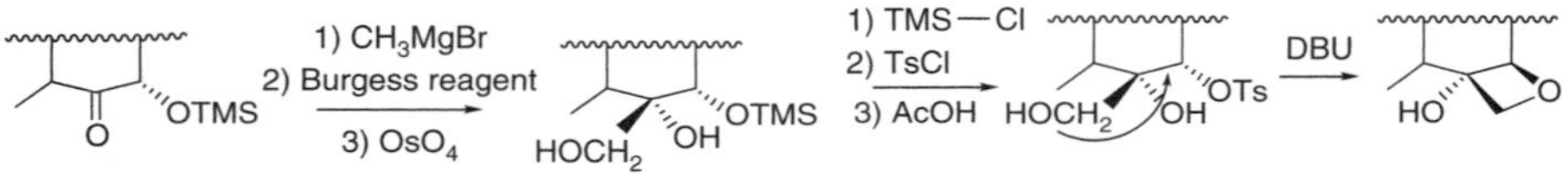

In Step **H**-7 the addition of phenyllithium to the cyclic carbonate group neatly generates the C(2) benzoate group. A similar reaction was used in several other Taxol syntheses.

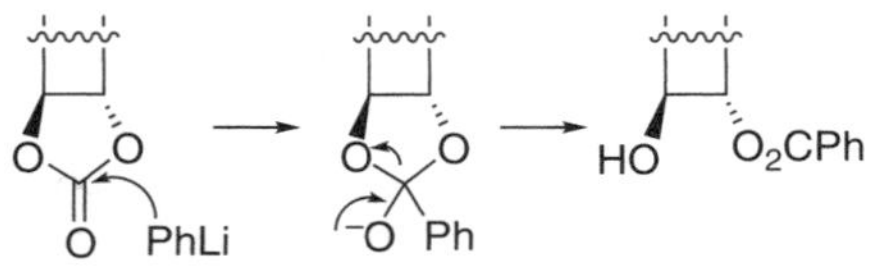

The final phase of the synthesis is introduction of the C(9) oxygen by phenylselenenic anhydride (Step **H**-9) and acetylation.

The Baccatin III synthesis by K. C. Nicolaou and co-workers is summarized in Scheme 13.54. Diels-Alder reactions are prominent in forming the early intermediates. In Step **A** the pyrone ring served as the diene. This reaction was facilitated by phenyl-boronic acid, which brings the diene and dienophile together as a boronate, permitting an intramolecular reaction.

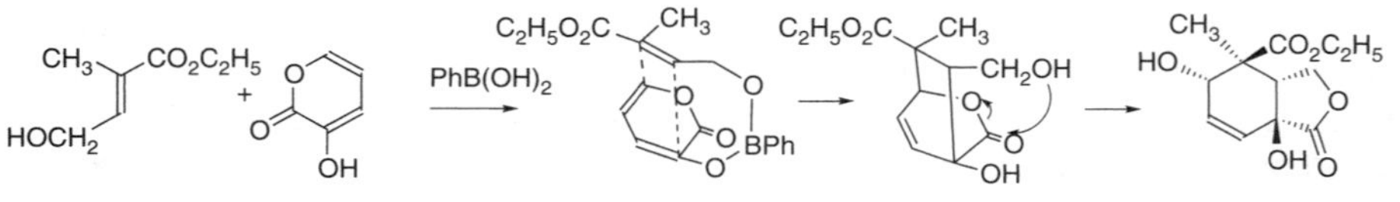

Scheme 13.54. Baccatin III Synthesis: K. C. Nicolaou and Co-Workers[a]

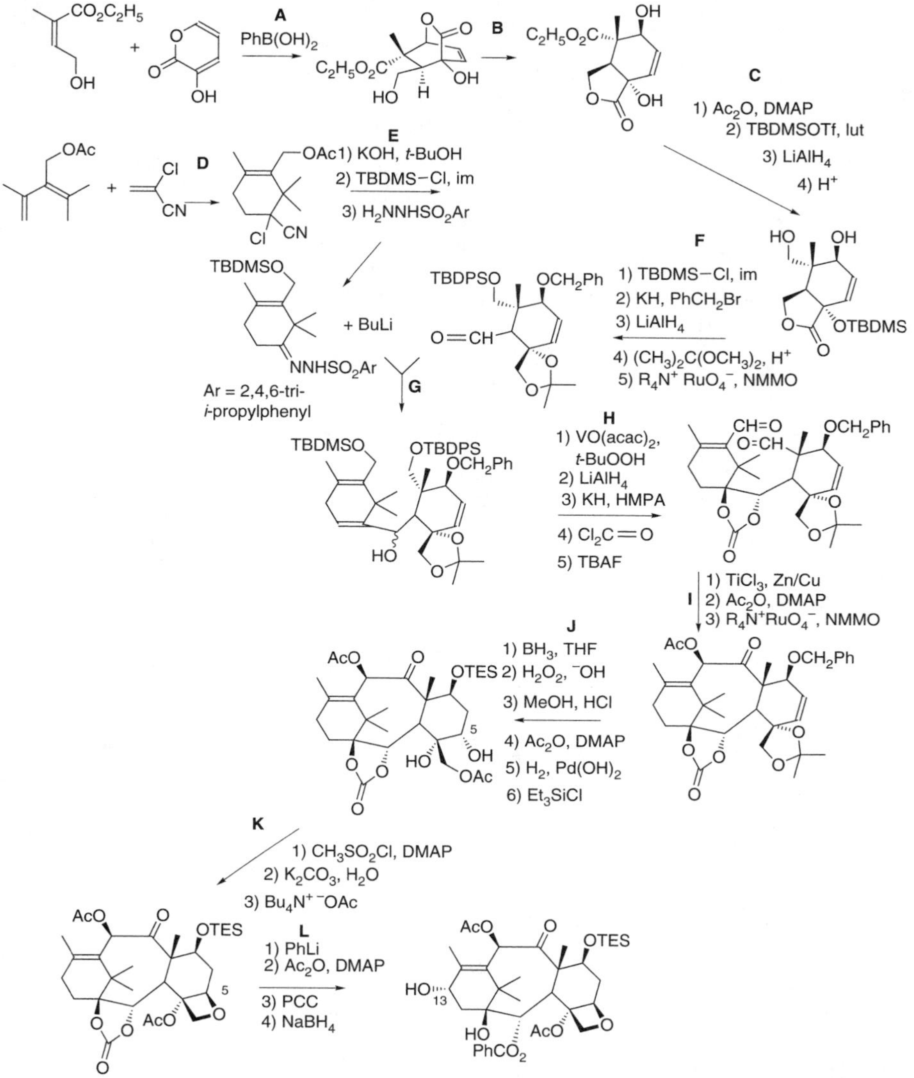

a. K. C. Nicolaou, P. G. Nantermet, H. Ueno, R. K. Guy, E. A. Couladouros, and E. J. Sorenson, *J. Am. Chem. Soc.*, **117**, 624 (1995); K. C. Nicolaou, J.-J. Liu, Z. Yang, H. Ueno, E. J. Sorenson, C. F. Claiborne, R. K. Guy, C.-K. Hwang, M. Nakada, and P. G. Nantermet, *J. Am. Chem. Soc.*, **117**, 634 (1995); K. C. Nicolaou, Z. Zhang, J.-J. Liu, P. G. Nantermet, C. F. Clairborne, J. Renaud, R. K. Guy, and K. S. Shibayama, *J. Am. Chem. Soc.*, **117**, 645 (1995); K. C. Nicolaou, H. Ueno, J.-J. Liu, P. G. Nantermet, Z. Yang, J. Renaud, K. Paulvannan, and R. Chadha, *J. Am. Chem. Soc.*, **117**, 653 (1995).

The formation of the A-ring in Step **D** used α-chloroacrylonitrile as a ketene synthon. The A-ring and C-ring were brought together in Step **G** by an organolithium addition to the aldehyde. The lithium reagent was generated by a Shapiro reaction. An oxygen was introduced at C(1) by hydroxy-directed epoxidation in Step **H**-1 and reductive ring opening of the epoxide in Step **H**-2. The eight-membered B-ring was then closed by a titanium-mediated reductive coupling of a dialdehyde in Step **I**-1. The oxetane

Scheme 13.55. Baccatin III Synthesis: S. J. Danishefsky and Co-Workers[a]

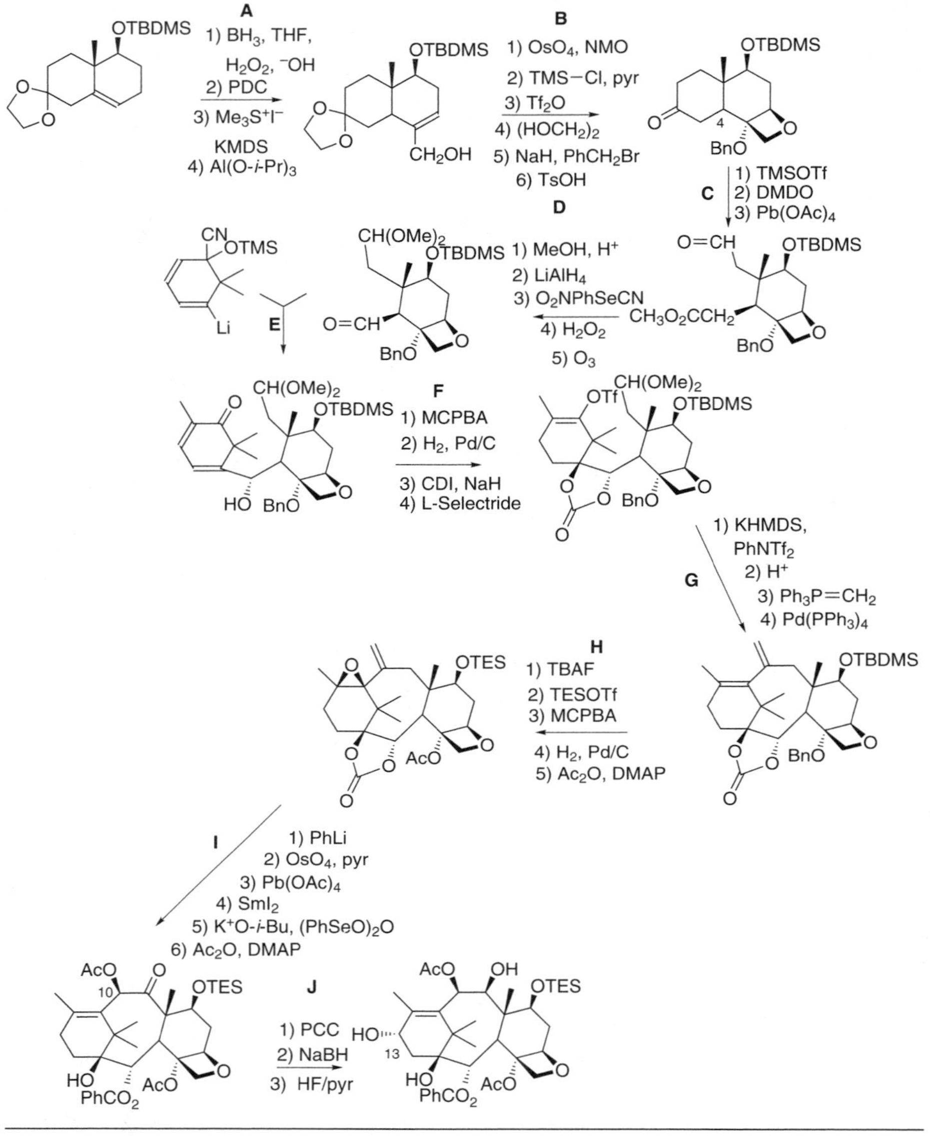

A
1) BH$_3$, THF, H$_2$O$_2$, $^-$OH
2) PDC
3) Me$_3$S$^+$I$^-$ KMDS
4) Al(O-i-Pr)$_3$

B
1) OsO$_4$, NMO
2) TMS−Cl, pyr
3) Tf$_2$O
4) (HOCH$_2$)$_2$
5) NaH, PhCH$_2$Br
6) TsOH

C
1) TMSOTf
2) DMDO
3) Pb(OAc)$_4$

D
1) MeOH, H$^+$
2) LiAlH$_4$
3) O$_2$NPhSeCN
4) H$_2$O$_2$
5) O$_3$

E

F
1) MCPBA
2) H$_2$, Pd/C
3) CDI, NaH
4) L-Selectride

G
1) KHMDS, PhNTf$_2$
2) H$^+$
3) Ph$_3$P=CH$_2$
4) Pd(PPh$_3$)$_4$

H
1) TBAF
2) TESOTf
3) MCPBA
4) H$_2$, Pd/C
5) Ac$_2$O, DMAP

I
1) PhLi
2) OsO$_4$, pyr
3) Pb(OAc)$_4$
4) SmI$_2$
5) K$^+$O-i-Bu, (PhSeO)$_2$O
6) Ac$_2$O, DMAP

J
1) PCC
2) NaBH
3) HF/pyr

a. S. J. Danishefsky, J. J. Masters, W. B. Young, J. T. Link, L. B. Snyder, T. V. Magee, D. K. Jung, R. C. A. Isaacs, W. G. Bornmann, C. A. Alaimo, C. A. Coburn, and M. J. Di Grandi, *J. Am. Chem. Soc.*, **118**, 2843 (1996).

ring was closed in sequence **K** by an intramolecular O-alkylation with inversion at C(5). The C(13) oxygen was introduced late in the synthesis by an allylic oxidation using PCC (Step **L**-3).

The synthesis of S. J. Danishefsky's group is outlined in Scheme 13.55. The starting material is a protected derivative of the Wieland-Miescher ketone. The oxetane ring is formed early in this synthesis. An epoxide is formed using dimethylsulfonium methylide (Step **A**-3) and opened to an allylic alcohol in Step **A**-4. The double bond was dihydroxylated using OsO_4. The cyclization occurs via the C(5) triflate and was done in ethylene glycol. After cyclization, the tertiary hydroxy at C(4) was protected by benzylation and the ketal protecting group was removed. The cyclohexanone ring was then cleaved by oxidation of the silyl enol ether. The A-ring was introduced in Step **E** by use of a functionalized lithium reagent. The closure of the B-ring was done by an intramolecular Heck reaction involving a vinyl triflate at Step **G**-4.

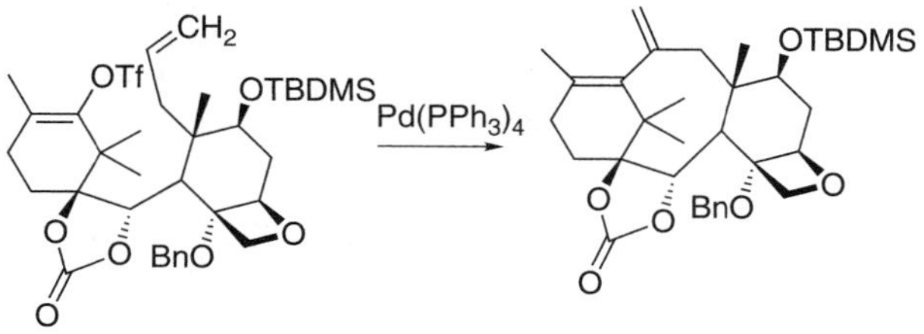

The late functionalization included the introduction of the C(10) and C(13) oxygens, which was done by phenylselenenic anhydride oxidation of the enolate in Step **I**-5 and by allylic oxidation at C(13) in Step **J**-1. These oxidative steps are similar to transformations in the Holton and Nicolaou syntheses.

The synthesis of the Taxol in Scheme 13.56 by P. A. Wender and co-workers at Stanford University began with an oxidation product of the readily available terpene pinene. One of the key early steps was the photochemical rearrangement in Step **B**.

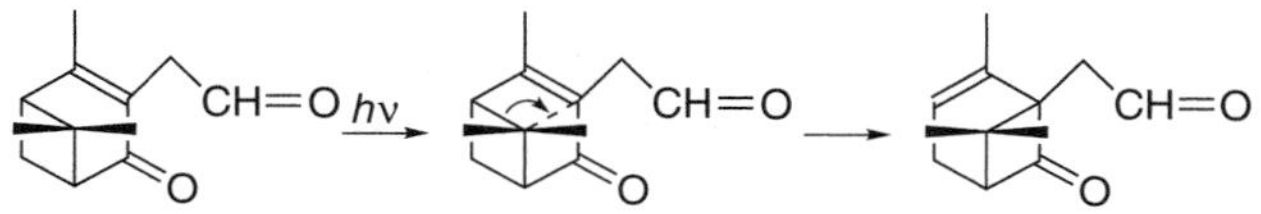

A six-membered ring was then constructed in reaction sequence **C** by addition of lithiated ethyl propynoate and a tandem conjugate addition-cyclization. The C(10) oxygen was introduced by enolate oxidation in Step **D**-2. Another key step is the fragmentation induced by treatment first with MCPBA and then with DABCO (Steps **E**-1 and **E**-2). The four-membered ring is fragmented in the process, forming the eight-membered ring with its bridgehead double bond and providing the C(13) oxygen substituent.

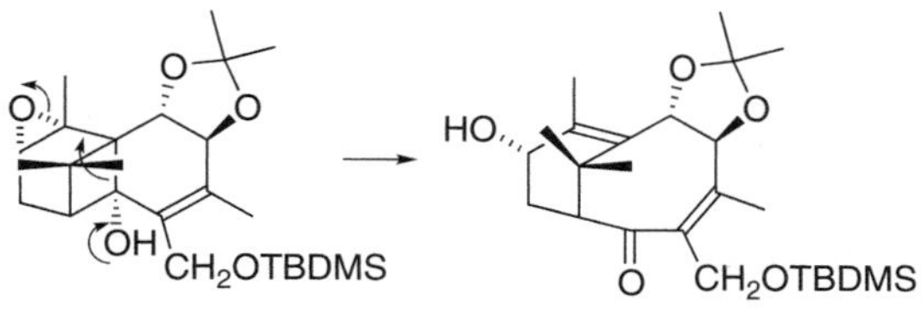

The C(1) oxygen was introduced at Step **F**-1 by enolate oxidation. The **C**-ring was constructed by building up a substituent at C(16) (Steps **G** and **H**). After forming the benzoate at C(2) in Step **H**-4, the C(9) acetoxy ketone undergoes transposition. This is an equilibrium process that goes to about 55% completion. An aldehyde was generated by ozonolysis of the terminal allylic double bond. This group was used to close the C-ring by an aldol cyclization in Step **I**-1. This step completed the construction of the

Scheme 13.56. Baccatin III Synthesis: P. A. Wender and Co-Workers[a]

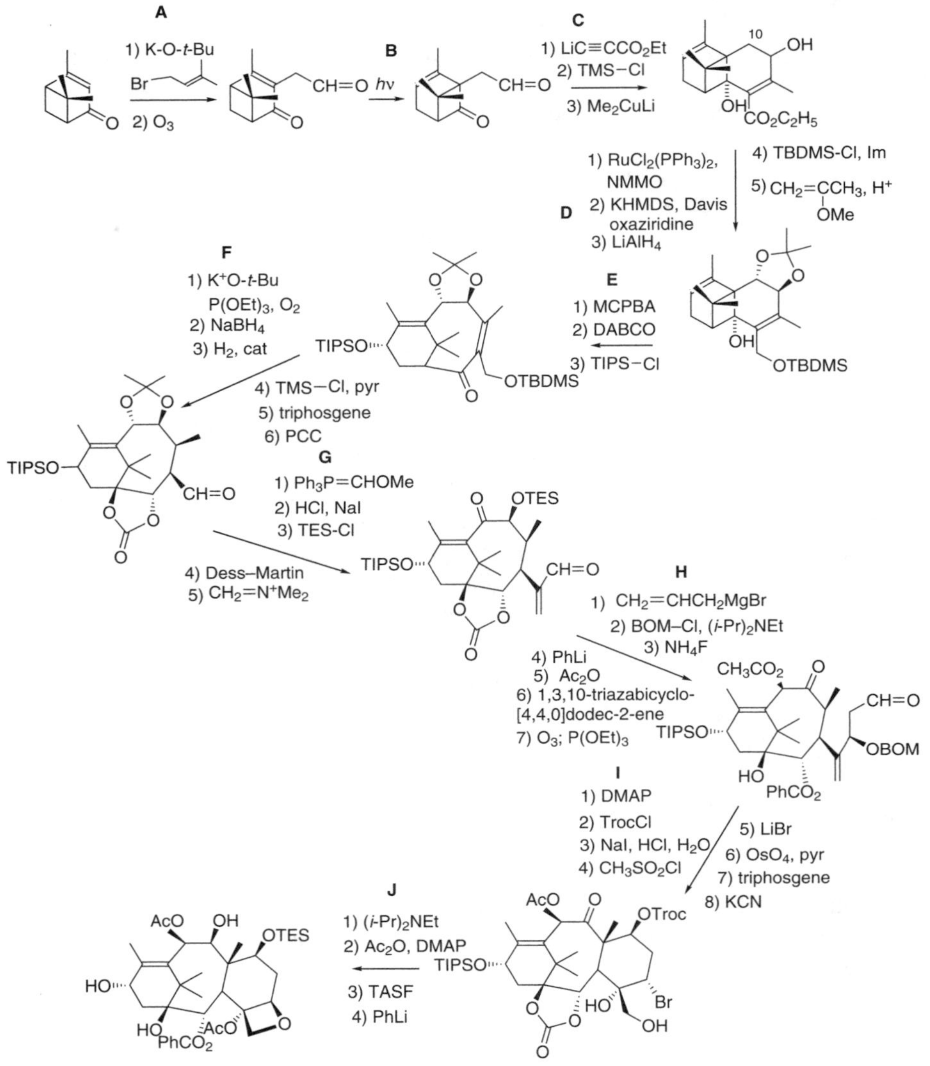

a. P. A. Wender, N. F. Badham, S. P. Conway, P. E. Floreancig, T. E. Glass, C. Granicher, J. B. Houze, J. Janichen, D. Lee, D. G. Marquess, P. L. McGrane, W. Meng, T. P. Mucciaro, M. Muhlebach, M. G. Natchus, H. Paulsen, D. B. Rawlins, J. Satkofsky, A. J. Shuker, J. C. Sutton, R. E. Taylor, and K. Tomooka, *J. Am. Chem. Soc.*, **119**, 2755 (1997); P. A. Wender, N. F. Badham, S. P. Conway, P. E. Floreancig, T. E. Glass, J. B. Houze, N. E. Krauss, D. Lee, D. G. Marquess, P. L. McGrane, W. Meng, M. G. Natchus, A. J. Shuker, J. C. Sutton, and R. E. Taylor, *J. Am. Chem. Soc.*, **119**, 2757 (1997).

carbon framework. The synthesis was completed by formation of the oxetane ring by the sequence **I**-3 to **I**-8, followed by the cyclization in Step **J**-1.

The synthesis of Baccatin III shown in Scheme 13.57, which was completed by a group led by the Japanese chemist Teruaki Mukaiyama, takes a different approach for the previous syntheses. Much of the stereochemistry was built into the B-ring by a series of acyclic aldol additions in Steps **A** through **D**. A silyl ketene acetal derivative

Scheme 13.57. Baccatin III Synthesis: T. Mukaiyama and Co-Workers[a]

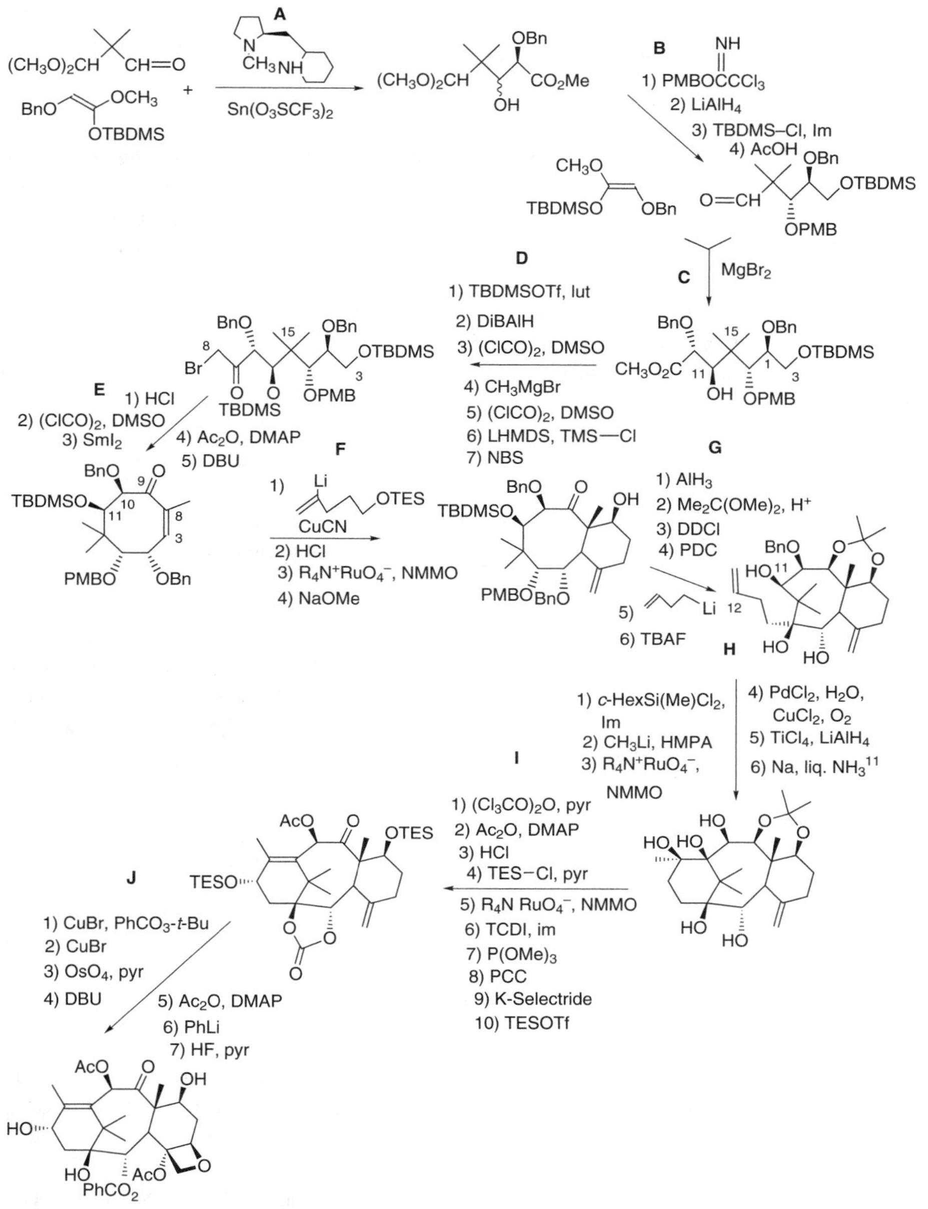

a. T. Mukaiyama, I. Shiina, H. Iwadare, M. Saitoh, T. Nishimura, N. Ohkawa, H. Sakoh, K. Nishimura, Y. Tani, M. Hasegawa, K. Yamada, and K. Saitoh, *Chem. Eur. J.*, **5**, 121 (1999).

of methyl α-benzyloxyacetate served as the nucleophile in Steps **A** and **C**. The C(10)–C(11) bond is formed in Step **C** using $MgBr_2$ to promote the Mukaiyama addition, which forms the correct stereoisomer with 4:1 diastereoselectivity. The B-ring was closed in Step **E**-3 by a samarium-mediated cyclization, forming the C(3)–C(8) bond.

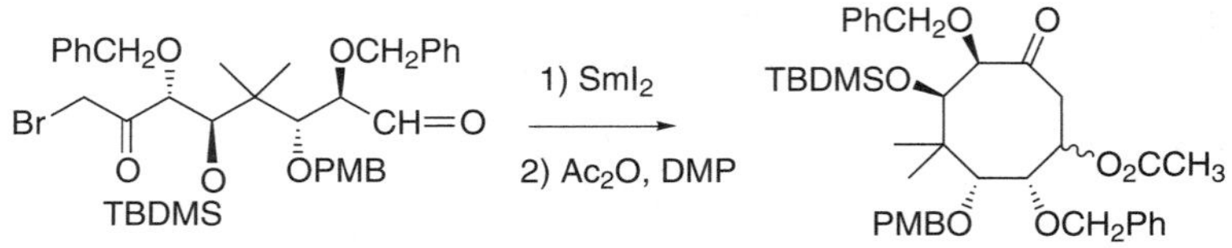

The C(4)–C(7) segment was added by a cuprate conjugate addition in Step **F**-1. The C-ring was then closed using an intramolecular aldol addition in Step **F**-4. The A-ring was closed by a Ti-mediated reductive coupling between carbonyl groups at C(11) and C(12) in Step **H**-5. The C(11)–C(12) double bond was introduced from the diol by deoxygenation of the thiocarbonate (Steps **I**-6 and **I**-7). The final sequence for conversion to Baccatin III, which began with a copper-mediated allylic oxidation at C(5), also involves an allylic rearrangement of the halide that is catalyzed by CuBr. The exocyclic double bond was then used to introduce the final oxygens needed to perform the oxetane ring closure.

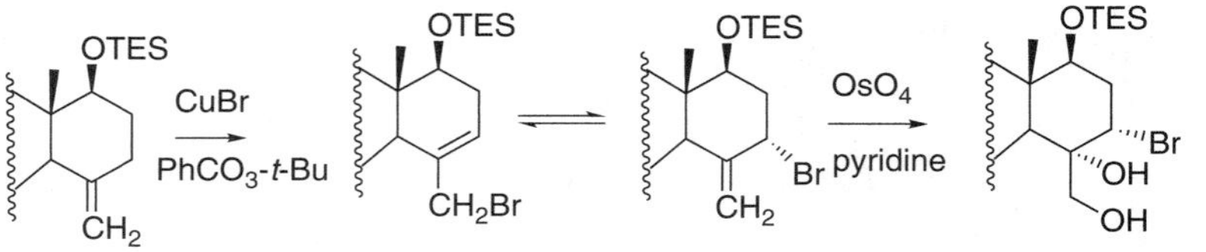

Another Japanese group developed the Baccatin III synthesis shown in Scheme 13.58. The eight-membered B-ring was closed early in the synthesis using a Lewis acid–induced Mukaiyama reaction (Step **B**-1), in which a trimethylsilyl dienol ether served as the nucleophile.

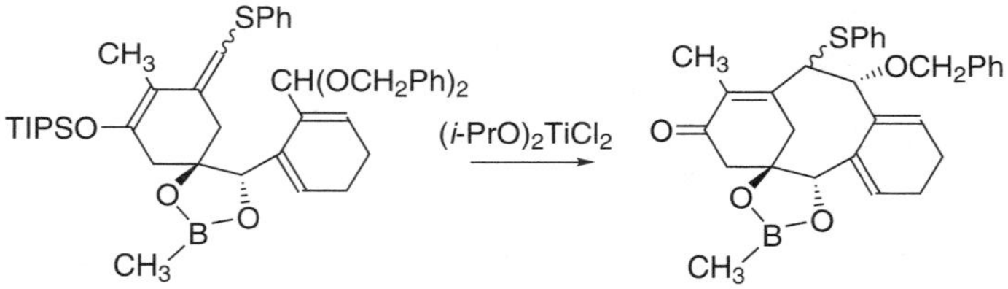

Oxygen was introduced at C(4) and C(7) by a singlet O_2 cycloaddition in Step **C**-1. The peroxide bond was cleaved and the phenylthio group removed by Bu_3SnH in Step **C**-2. The C(19) methyl group was introduced via a cyclopropanation in Step **C**-5, followed by a reduction in Step **D**-1. A Pd-catalyzed cross-coupling reaction was used to introduce a trimethylsilylmethyl group at C(4) via an enol triflate in Step **F**-2. The vinyl silane was then subjected to chlorination in Step **F**-3. The chlorine eventually serves as the leaving group for oxetane ring formation in Step **G**-2.

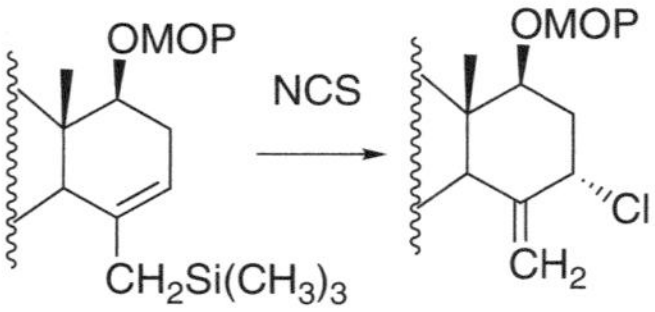

Scheme 13.58. Baccatin III Synthesis: H. Kusama, I. Kuwajima, and Co-Workers[a]

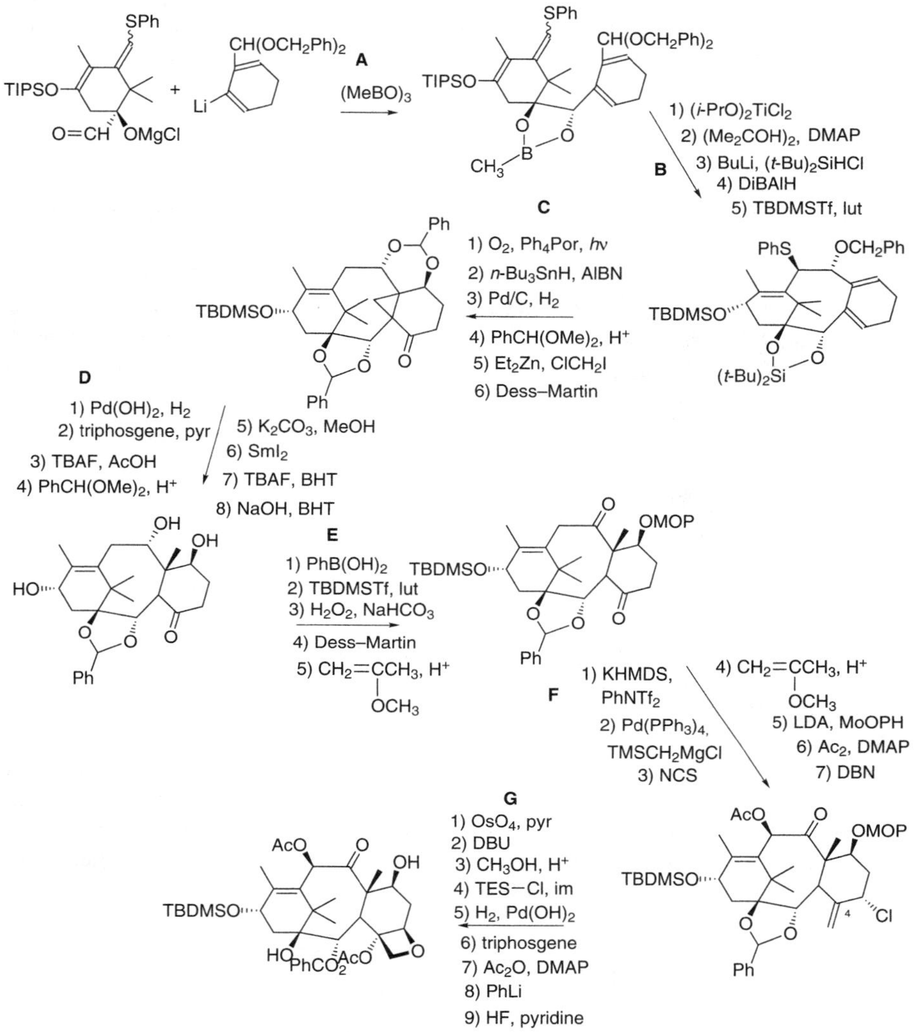

a. K. Morihara, R. Hara, S. Kawahara, T. Nishimori, N. Nakamura, H. Kusama, and I. Kuwajima, *J. Am. Chem. Soc.*, **120**, 12980 (1998); H. Kusama, R. Hara, S. Kawahara, T. Nishimori, H. Kashima, N. Nakamura, K. Morihara, and I. Kuwajima, *J. Am. Chem. Soc.*, **122**, 3811 (2000).

These syntheses of Baccatin III illustrate the versatility of current methodology for ring closure and functional group interconversions. The Holton, Nicolaou, Danishefsky, and Wender syntheses of Baccatin III employ various cyclic intermediates and take advantage of stereochemical features built into these rings to control subsequent reaction stereochemistry. As a reflection of the numerous oxygens in Baccatin III, each of the syntheses makes use of enolate oxidation, alkene hydroxylation, and related oxidation reactions. These syntheses also provide numerous examples of the selective use of protective groups to achieve distinction between the several hydroxy groups that are present in the intermediates. The Mukaiyama synthesis in Scheme 13.57 is somewhat different in approach in that it uses acyclic intermediates to introduce several of the stereocenters. Perhaps because of the structure, none of these syntheses is particularly convergent. The Nicolaou, Danishefsky, and Kusama syntheses achieve some convergence by coupling the A-ring and the C-ring and then forming the B-ring. The Holton and Wender syntheses take advantage of available natural substances as starting materials.

13.2.5. Epothilone A

The epothilones are natural products containing a 16-membered lactone ring that are isolated from mycobacteria. Epothilones A–D differ in the presence of the C(12)–C(13) epoxide and in the C(12) methyl group. Although structurally very different from Taxol, they have a similar mechanism of anticancer action and epothilone A and its analogs are of substantial current interest as chemotherapeutic agents.[36] Schemes 13.59 to 13.66 summarize eight syntheses of epothilone A. Several syntheses of epothilone B have also been completed.[37]

epothilone A
epothilone C(12-13) = CH=CH

epothilone B
epothilone C(12-13) = CH=CH

Two critical objectives for planning the synthesis of epothilone A are the control of the configuration of the stereocenters and the closure of the 16-membered ring. There are eight stereocenters, including the C(16)–C(17) double bond. As the 16-membered lactone ring is quite flexible, it does not impose strong facial stereoselectivity. Instead, the stereoselective synthesis of epothilone A requires building the correct stereochemistry into acyclic precursors that are cyclized later in the synthesis. The stereocenters at C(3), C(6), C(7), and C(8) are adjacent to a potential aldol connection

[36] T. C. Chou, X. G. Zhang, C. R. Harris, S. D. Kuduk, A. Balog, K. A. Savin, J. R. Bertino, and S. J. Danishefsky, *Proc. Natl. Acad. Sci. USA*, **95**, 15978 (1998).

[37] J. Mulzer, A. Mantoulidis, and E. Ohler, *Tetrahedron Lett.*, **39**, 8633 (1998); D. S. Sa. D. F. Meng, P. Bertinato, A. Balog, E. J. Sorensen, S. J. Danishefsky, Y. H. Zheng, T.-C. Chou, L. He, and S. B. Horowitz, *Angew. Chem. Int. Ed. Engl.*, **36**, 757 (1997); A. Balog, C. Harris, K. Savin, S. G. Zhang, T. C. Chou, and S. J. Danishefsky, *Angew. Chem. Int. Ed. Engl.*, **37**, 2675 (1998); D. Shinzer, A. Bauer, and J. Schieber, *Synlett*, 861 (1998); S. A. May and P. A. Grieco, *J. Chem. Soc., Chem. Commun.*, 1597 (1998); K. C. Nicolaou, S. Ninkovic, F. Sarabia, D. Vourloumis, Y. He, H. Vallberg, M. R. V. Finlay, and Z. Yang, *J. Am. Chem. Soc.*, **119**, 7974 (1997); K. C. Nicolaou, D. Hepworth, M. R. V. Finlay, B. Wershkun, and A. Bigot, *J. Chem. Soc., Chem. Commun.*, 519 (1999); D. Schinzer, A. Bauer, and J. Schieber, *Chem. Eur. J.*, **5**, 2492 (1999); J. D. White, R. G. Carter, and K. F. Sundermann, *J. Org. Chem.*, **64**, 684 (1999); J. Mulzer, A. Moantoulidis, and E. Oehler, *J. Org. Chem.*, **65**, 7456 (2000); J. Mulzer, G. Karig, and P. Pojarliev, *Tetrahedron Lett.*, **41**, 7635 (2000); D. Sawada, M. Kanai, and M. Shibasaki, *J. Am. Chem. Soc.*, **122**, 10521 (2000); S. C. Sinha, J. Sun, G. P. Miller, M. Wartmann, and R. A. Lerner, *Chem. Eur. J.*, **7**, 1691 (2001); J. D. White, R. G. Carter, K. F. Sundermann, and M. Wartmann, *J. Am. Chem. Soc.*, **123**, 5407 (2001); H. J. Martin, P. Pojarliev, H. Kahlig, and J. Mulzer, *Chem. Eur. J.*, **7**, 2261 (2001); R. E. Taylor and Y. Chen, *Org. Lett.*, **3**, 2221 (2001); M. Valluri, R. M. Hindupur, P. Bijoy, G. Labadie, J.-C. Jung, and M. A. Avery, *Org. Lett.*, **3**, 3607 (2001); N. Martin and E. J. Thomas, *Tetrahedron Lett.*, **42**, 8373 (2001); M. S. Ermolenko and P. Potier, *Tetrahedron Lett.*, **43**, 2895 (2002); J. Sun and S. C. Sinha, *Angew. Chem. Int. Ed. Engl.*, **41**, 1381 (2002); J.-C. Jung, R. Kache, K. K. Vines, Y.-S. Zheng, P. Bijoy, M. Valluri, and M. A. Avery, *J. Org. Chem.*, **69**, 9269 (2004).

Scheme 13.59. Epothilone A Synthesis by Macrolactonization: K. C. Nicolaou and Co-Workers[a]

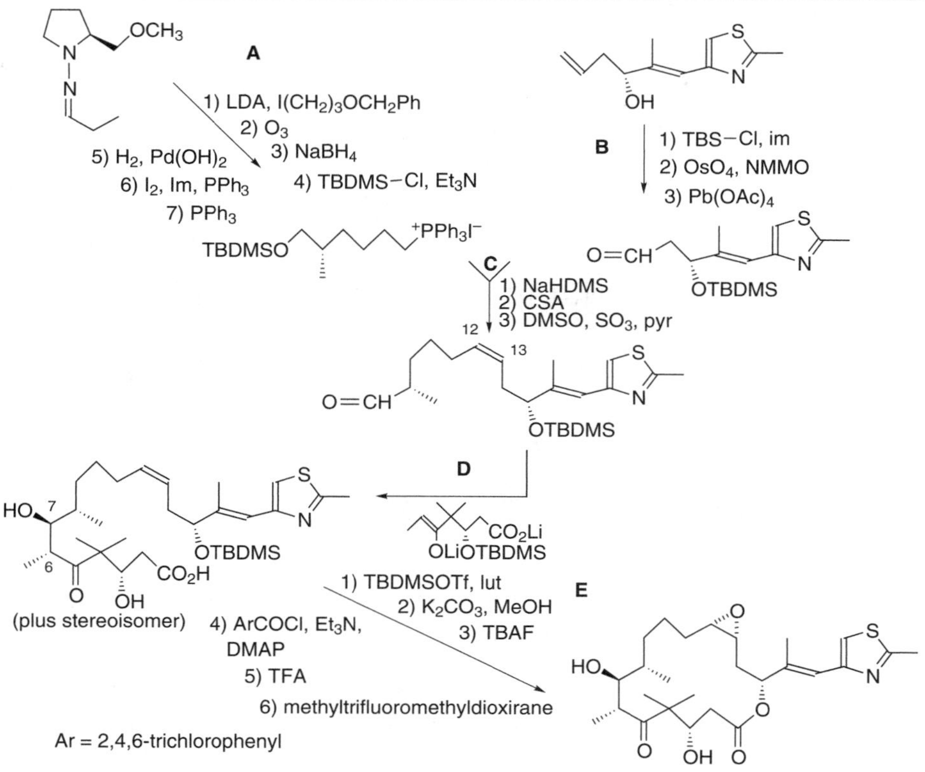
OCH3
A
1) LDA, I(CH2)3OCH2Ph
2) O3
3) NaBH4
4) TBDMS−Cl, Et3N
5) H2, Pd(OH)2
6) I2, Im, PPh3
7) PPh3
TBDMSO
+PPh3I−
OH
B
1) TBS−Cl, im
2) OsO4, NMMO
3) Pb(OAc)4
O=CH
OTBDMS
C
1) NaHDMS
2) CSA
3) DMSO, SO3, pyr
12
13
O=CH
OTBDMS
D
HO
7
6
OTBDMS
CO2H
OH
(plus stereoisomer)
CO2Li
OLiOTBDMS
1) TBDMSOTf, lut
2) K2CO3, MeOH
3) TBAF
4) ArCOCl, Et3N, DMAP
5) TFA
6) methyltrifluoromethyldioxirane
E
HO
OH
Ar = 2,4,6-trichlorophenyl

a. K. C. Nicolaou, F. Sarabia, S. Ninkovic, and Z. Yang, *Angew. Chem. Int. Ed. Engl.*, **36**, 525 (1997); K. C. Nicolaou, S. Ninkovic, F. Sarabia, D. Vourloumis, Y. He, H. Vallberg, M. R. V. Finlay, and Z. Yang, *J. Am. Chem. Soc.*, **119**, 7974 (1997).

between C(6) and C(7) and are amenable to control by aldol methodology. Introduction of the epoxide by epoxidation requires a *Z*-double bond. Several methods for ring closure have been used, but the two most frequently employed are macrolactonization (see Section 3.4) and alkene metathesis (see Section 8.4).

K. C. Nicolaou's group at Scripps Research Institute developed two synthetic routes to epothilone A. One of the syntheses involves closure of the lactone ring as a late step. Three major fragments were synthesized. The bond connection at C(6)–C(7) was made by an aldol reaction. The C(12)–C(13) bond was formed by a Wittig reaction and later epoxidized. The ring was closed by macrolactonization.

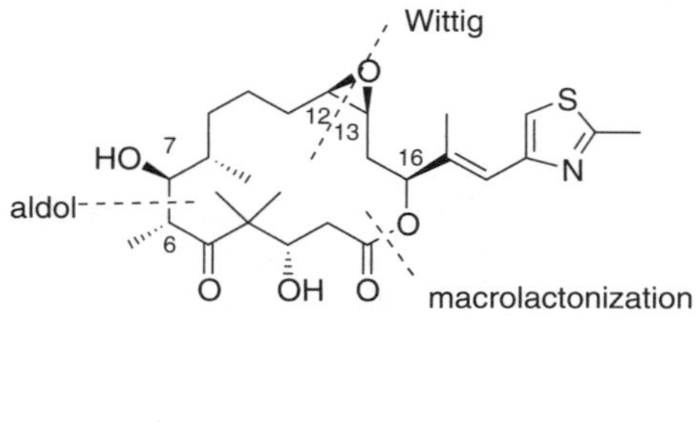
Wittig
12
13
16
HO
7
aldol
6
OH
macrolactonization

Scheme 13.60. Epothilone A Synthesis by Olefin Metathesis: K. C. Nicolaou and Co-Workers[a]

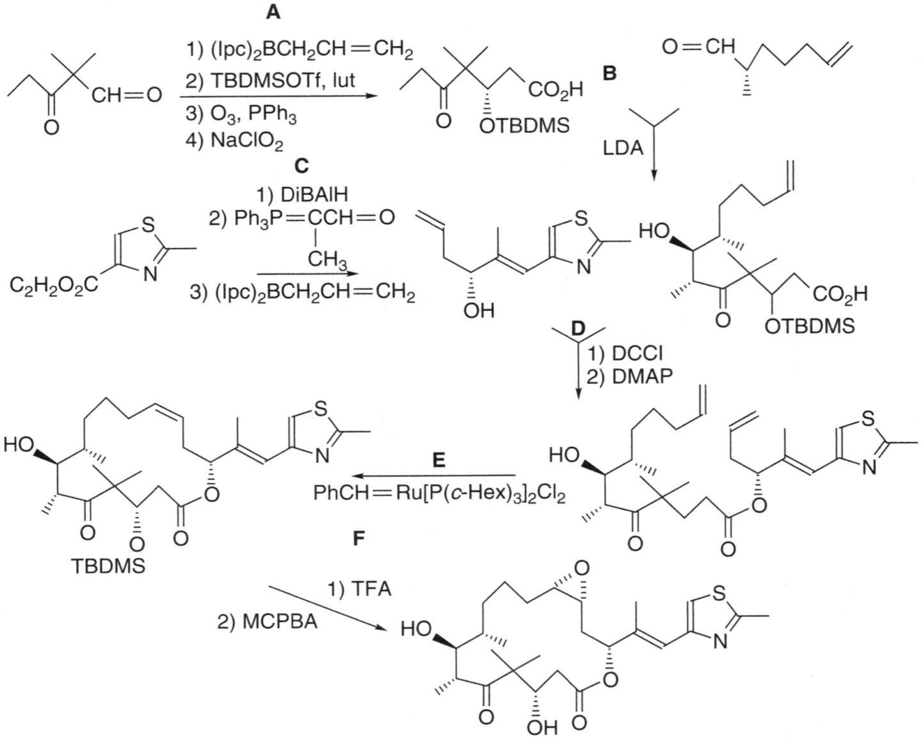

a. Z. Yang, Y. He, D. Vourloumis, H. Vallberg, and K. C. Nicolaou, *Angew. Chem. Int. Ed. Engl.*, **36**, 166 (1997).

This synthesis is shown in Scheme 13.59. Two enantiomerically pure starting materials were brought together by a Wittig reaction in Step **C**. The aldol addition in Step **D** was diastereoselective for the *anti* configuration, but gave a 1:1 mixture with the 6*S*, 7*R*-diastereomer. The stereoisomers were separated after Step **E**-2. The macrolactonization (Step **E**-4) was accomplished by a mixed anhydride (see Section 3.4.1). The final epoxidation was done using 3-methyl-3-trifluoromethyl dioxirane.

The second synthesis from the Nicolaou group is shown in Scheme 13.60. The disconnections were made at the same bonds as in the synthesis in Scheme 13.59. The C(1)–C(6) segment contains a single stereogenic center, which was established in Step **A**-1 by enantioselective allylboration. The C(6)–C(7) configuration was established by the aldol addition in Step **B**. The aldolization was done with the dianion and gave a 2:1 mixture with the 6S, 7R diastereomer. The two fragments were brought together by esterification in Step **D**. The synthesis used an olefin metathesis reaction to construct the 16-membered ring (Step **E**). This reaction gave a 1.4:1 ratio of Z:E product, which was separated by chromatography.

The olefin metathesis reaction was also a key feature of the synthesis of epothilone A completed by a group at the Technical University in Braunschweig, Germany (Scheme 13.61). This synthesis employs a series of stereoselective additions to create the correct substituent stereochemistry. Two enantiomerically pure starting materials

Scheme 13.61. Epothilone A Synthesis: D. Schinzer and Co-Workers[a]

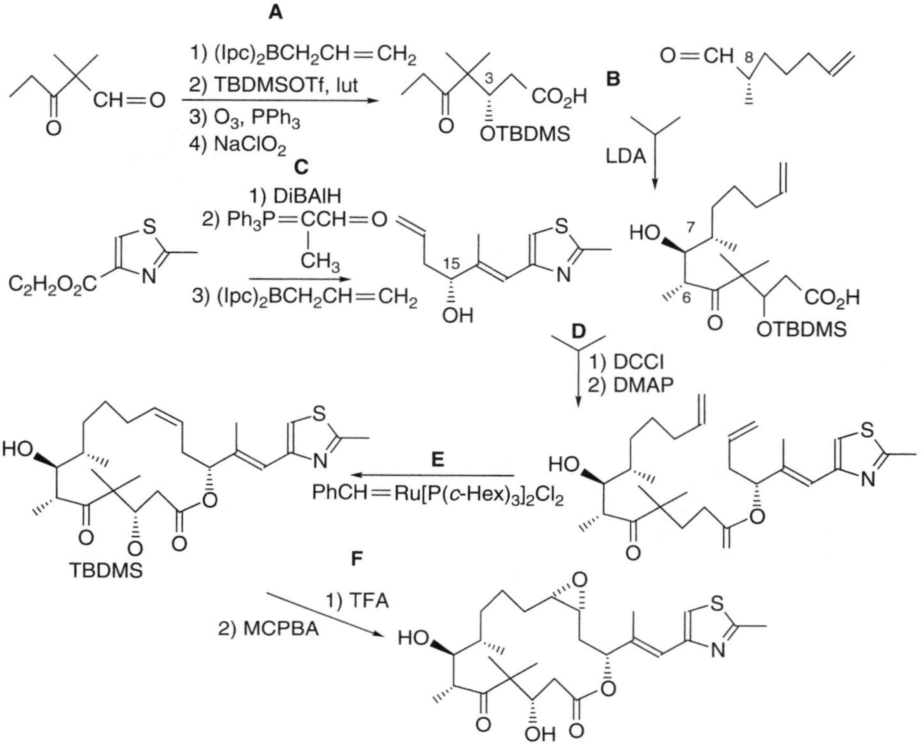

1223

a. D. Schinzer, A. Limberg, A. Bauer, O. M. Bohm, and M. Cordes, *Angew. Chem. Int. Ed. Engl.*, **36**, 523 (1997);
   D. Schinzer, A. Bauer, O. M. Bohm, A. Limberg, and M. Cordes, *Chem. Eur. J.*, **5**, 2483 (1999).

were used, containing the C(3) and C(8) stereocenters. Step **B** used a stereoselective aldol addition to bring these two fragments together and to create the stereocenters at C(6) and C(7). The thiazole ring and the C(13)–C(15) fragment were constructed in sequence **C**. The configuration at C(15) was established by enantioselective allylboration in Step **C**-3. The two segments were coupled by esterification at Step **D**, and the ring was closed by olefin metathesis (Step **E**). The metathesis reaction gave a 1.7:1 ratio favoring the *Z*-isomer. The synthesis was completed by deprotection and epoxidation, after which the stereoisomers were separated by chromatography. This group has also completed a synthesis based on a macrolactonization approach.[38]

Samuel Danishefsky's group at the Sloan Kettering Institute for Cancer Research in New York has also been active in the synthesis of the natural epothilones and biologically active analogs. One of their syntheses also used the olefin metathesis reaction (not shown). The synthesis in Scheme 13.62 used an alternative approach to create the macrocycle, as indicated in the retrosynthetic scheme. The stereochemistry at C(6), C(7), and C(8) was established by a TiCl$_4$-mediated cyclocondensation (Step **A**). The thiazole-containing side chain was created by reaction sequences **F** and **G**. The

[38.] D. Schinzer, A. Bauer, and J. Schieber, *Chem. Eur. J.*, **5**, 2483 (1999).

Scheme 13.62. Epothilone A Synthesis by Macroaldol Cyclization: S. J. Danishefsky and Co-Workers[a]

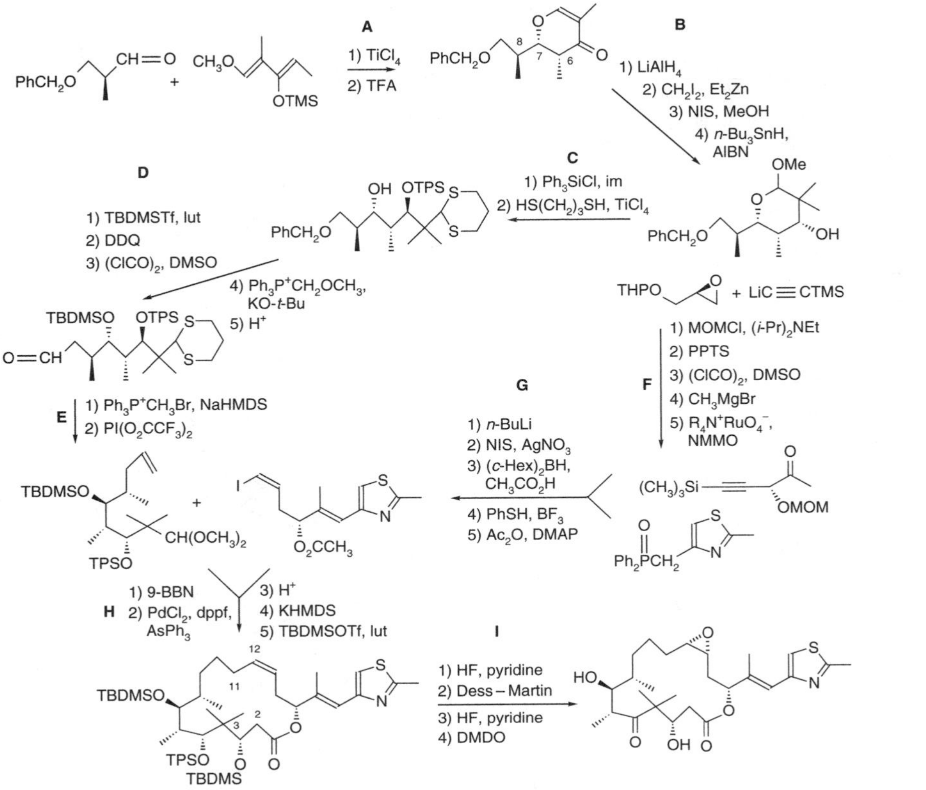

a. A. Balog, D. Meng, T. K. Kamenecka, P. Bertinato, D.-S. Su, E. J. Sorensen, and S. J. Danishefsky, *Angew. Chem. Int. Ed. Engl.*, **35**, 2801 (1996); D. Meng, P. Bertinato, A. Balog, D.-S. Su, T. Kamenecka, E. J. Sorensen, and S. J. Danishefsky, *J. Am. Chem. Soc.*, **119**, 10073 (1997).

Z-vinyl iodide was obtained by hydroboration and protonolysis of an iodoalkyne. The two major fragments were coupled by a Suzuki reaction at Steps **H**-1 and **H**-2 between a vinylborane and vinyl iodide to form the C(11)–C(12) bond. The macrocyclization was done by an aldol addition reaction at Step **H**-4. The enolate of the C(2) acetate adds to the C(3) aldehyde, creating the C(2)–C(3) bond and also establishing the configuration at C(3). The final steps involve selective deprotonation and oxidation at C(5), deprotection at C(3) and C(7), and epoxidation.

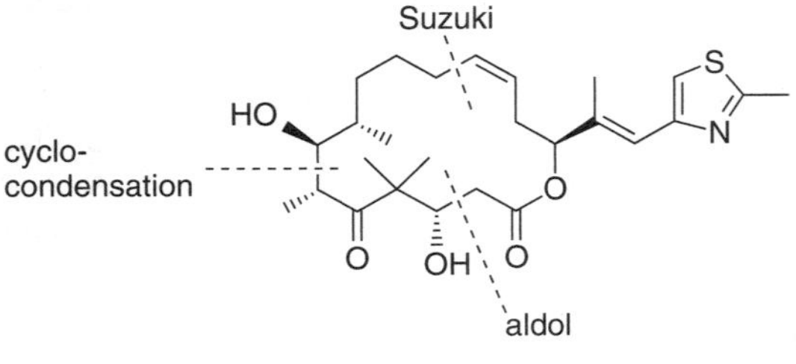

Scheme 13.63. Epothilone A Synthesis: A. Furstner, C. Mathes, and C. W. Lehmann[a]

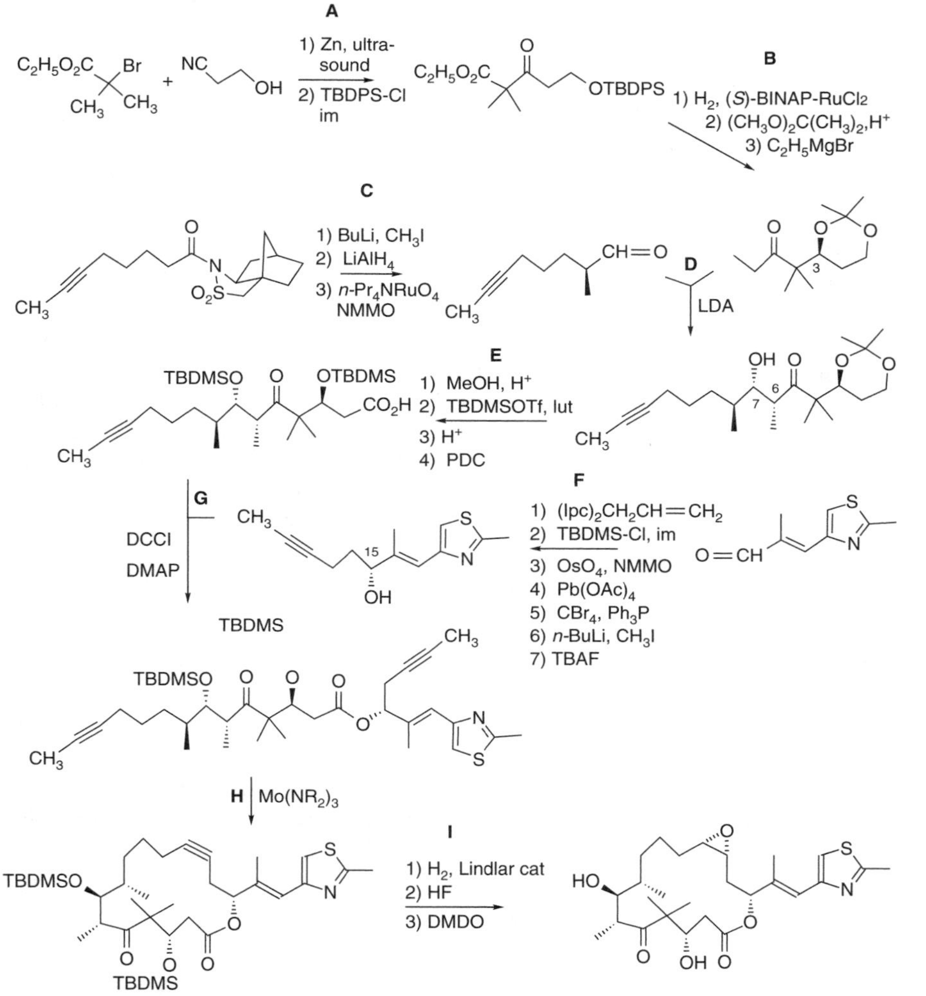

a. A. Furstner, C. Mathes, and C. W. Lehmann, *Chem. Eur. J.*, **7**, 5299 (2001).

The epothilone A synthesis shown in Scheme 13.63 involves an *alkyne metathesis reaction*. The first subunit was constructed using a Reformatsky-type addition to 3-hydroxypropanonitrile. The configuration at C(3) was established by an enantioselective hydrogenation using (*S*)-(BINAP)$RuCl^2$ under acidic conditions. A bornanesultam chiral auxiliary was used to establish the stereochemistry at C(8) by alkylation (Step **C**-1). The stereochemistry at the C(6)–C(7) bond was established by an aldol addition at Step **D**. The thiazole segment was constructed from a conjugated enal, which was subjected to enantioselective allylboration using (+)-$Ipc_2BCH_2CH=CH_2$ in Step **F**-1. This reaction established the configuration at C(5) A terminal alkyne was then installed by the Corey-Fuchs procedure (see p. 835). The lithium acetylide was methylated in situ using CH_3I. A DMAP-DCCI esterification was then used to couple the two major fragments and set the stage for the alkyne metathesis at Step **H**. The catalyst is a molybdenum amide, which is one of a family of catalysts that show good activity in alkyne metathesis. The use of alkyne metathesis avoids the complication of formation of both *Z*- and *E*-isomers, which sometimes occurs in olefin metathesis.

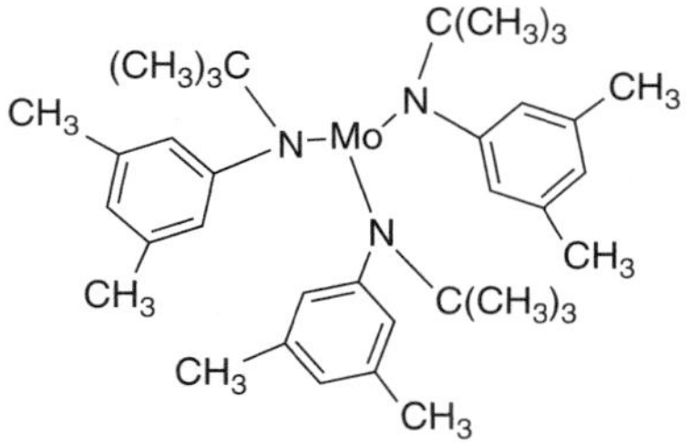

The yield in the metathesis reaction was 80% and was followed by a Lindlar reduction. The synthesis was completed by epoxidation with DMDO.

The synthesis in Scheme 13.64 was carried out by E. Carreira and co-workers at ETH in Zurich, Switzerland. A key step in the synthesis in Scheme 13.64 is a stereoselective cycloaddition using a phosphonyl-substituted nitrile oxide, which was used to form the C(16)–C(17) bond and install the C(15) oxygen.

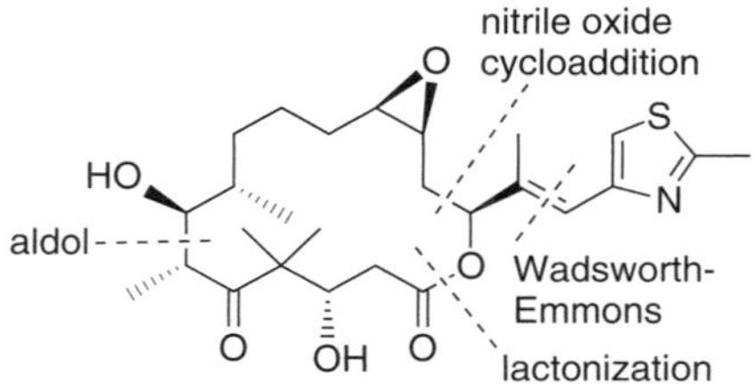

The C(6)–C(15) segment was synthesized by Steps **C**-1 and **C**-2. The stereoselectivity of the cycloaddition reaction between the nitrile oxide and allylic alcohol is the result of a chelated TS involving the Mg alkoxide.[39]

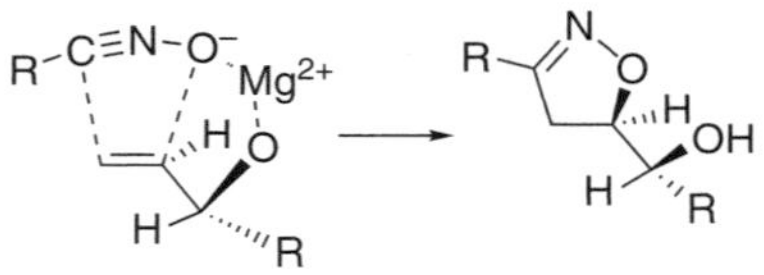

After the cycloaddition, the thiazole ring was introduced via a Wadsworth-Emmons reaction at Step **D**, forming the C(17)–C(18) bond.

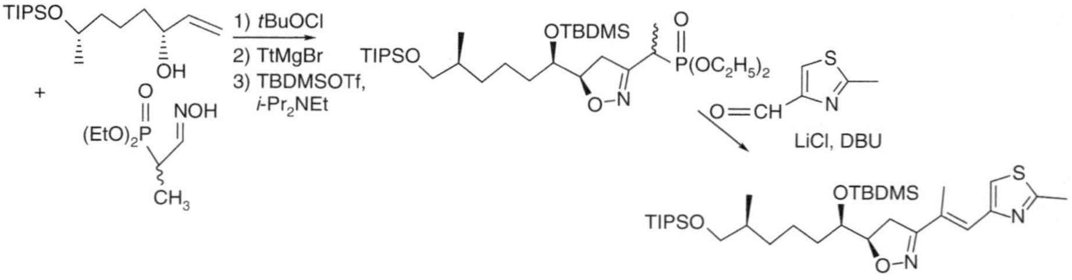

[39] S. Kanemasa, M. Nishiuchi, A. Kamimura, and K. Hori, *J. Am. Chem. Soc.*, **116**, 2324 (1994); S. Fukuda, A. Kanimura, S. Kanemasa, and K. Hori, *Tetrahedron*, **56**, 1637 (2000).

Scheme 13.64. Epolthilone A Synthesis: J. W. Bode and E. M. Carreira[a]

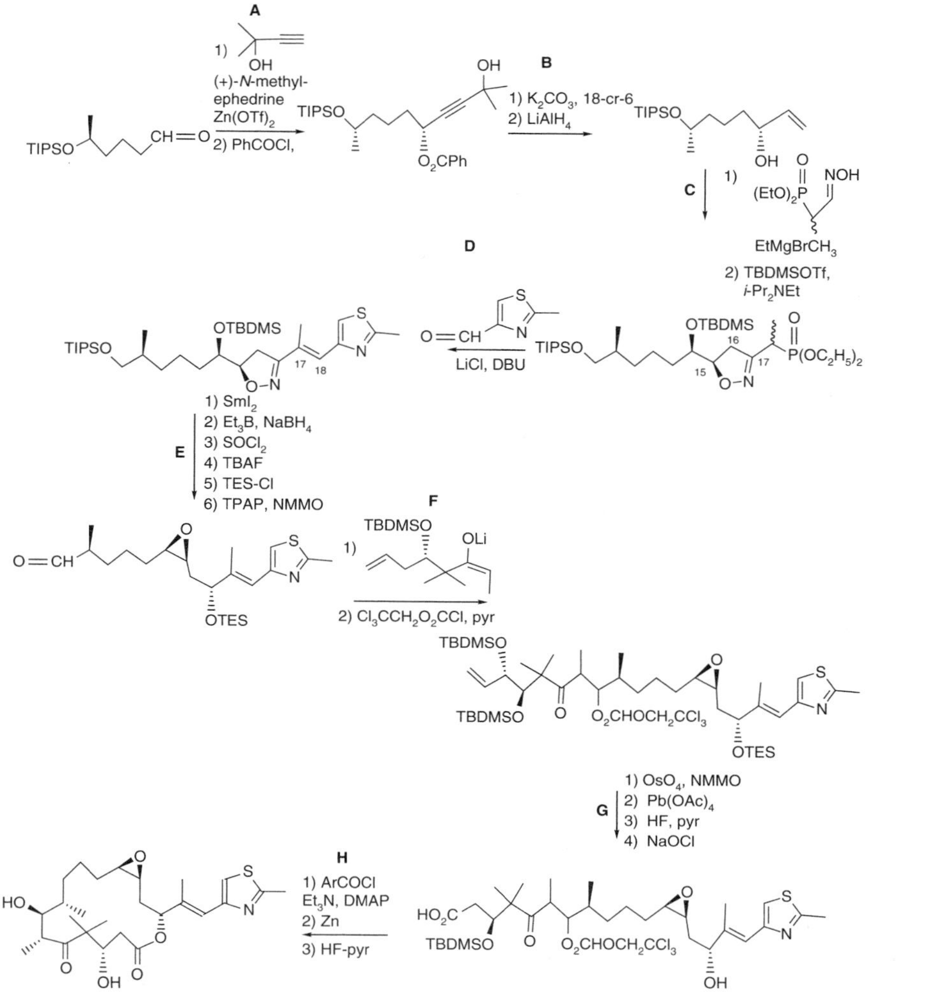

a. J. W. Bode and E. M. Carreira, *J. Am. Chem. Soc.*, **123**, 3611 (2001); *J. Org. Chem.*, **66**, 6410 (2001).

The reduction of the isoxazoline ring after the cycloaddition was not successful with the usual reagents (see p. 532), but SmI$_2$ accomplished the reaction. In contrast to the epoxidation used as the final step in most of the other epothilone A syntheses, the epoxide was introduced through a sulfite intermediate. Deprotection of C(15) leads to intramolecular displacement at the sulfite with formation of the epoxide (Steps **E**-3 and **E**-4).

The C(1)–C(6) and C(7)–C(17) fragments were joined by an aldol addition via a lithium enolate (Step **F**-1), and the ring was closed by a macrolactonization.

The synthesis of epothilone A in Scheme 13.65 features the use of chiral allylic silanes that were obtained by kinetic resolution using *Pseudomonas* AK lipase. The C(5)–C(8) fragment was synthesized by condensing the enantiomerically pure silane with a TBDPS-protected aldehyde in the presence of BF$_3$. The adduct was then subjected to a chelation-controlled aldol addition using TiCl$_4$, adding C(3) and C(4). After protecting group manipulation and oxidation, the chain was extended by two carbons using a Wittig reaction in Step **C**-3. The methyl group at C(8) was added by a stereoselective cuprate conjugate addition in Step **C**-4. The intermediate was then converted to **8** using a DiBAlH reduction under conditions that discriminated between the two ester groups (Step **D**-1). The more hindered group was reduced to the primary alcohol, leaving the less hindered one at the aldehyde level. This selectivity probably arises as a result of the lesser stability of the more hindered partially reduced intermediate. (See p. 401 to review the mechanism of DiBAlH reduction.)

The aldehyde was then converted to the terminal alkene via a Wittig reaction (Step **D**-3).

A kinetic resolution was also used to establish the configuration of the thiazole portion. An allylic aldehyde was subjected to kinetic resolution by ester exchange with vinyl acetate in Step **E**-2 (see Topic 2.2, Part A). The resolved alcohol was protected and subjected to hydroboration, oxidation, and a Wittig reaction to introduce the Z-vinyl iodide. The two fragments were coupled using the Suzuki reaction and the final two carbons were installed by another TiCl$_4$-mediated silyl ketene acetal addition in sequence **H**. The stereochemistry at C(3) presented some problems, but use of the silyl ketene acetal of the isopropyl ester provided an 8:1 mixture favoring the desired diastereomer. The isopropyl ester was used to slow competing lactonization of the intermediate. The macrolactonization was done under the Yamaguchi conditions. The synthesis was completed by epoxidation using the peroxyimidic acid generated in situ from acetonitrile and hydrogen peroxide.

The synthesis shown in Scheme 13.66 starts with the Sharpless asymmetric epoxidation product of geraniol. The epoxide was opened with inversion of configuration by NaBH$_3$CN-BF$_3$. The double bond was cleaved by ozonolysis and converted to the corresponding primary bromide. The terminal alkyne was introduced by alkylation of

Scheme 13.65. Epothilone A Synthesis: B. Zhu and J. S. Panek[a]

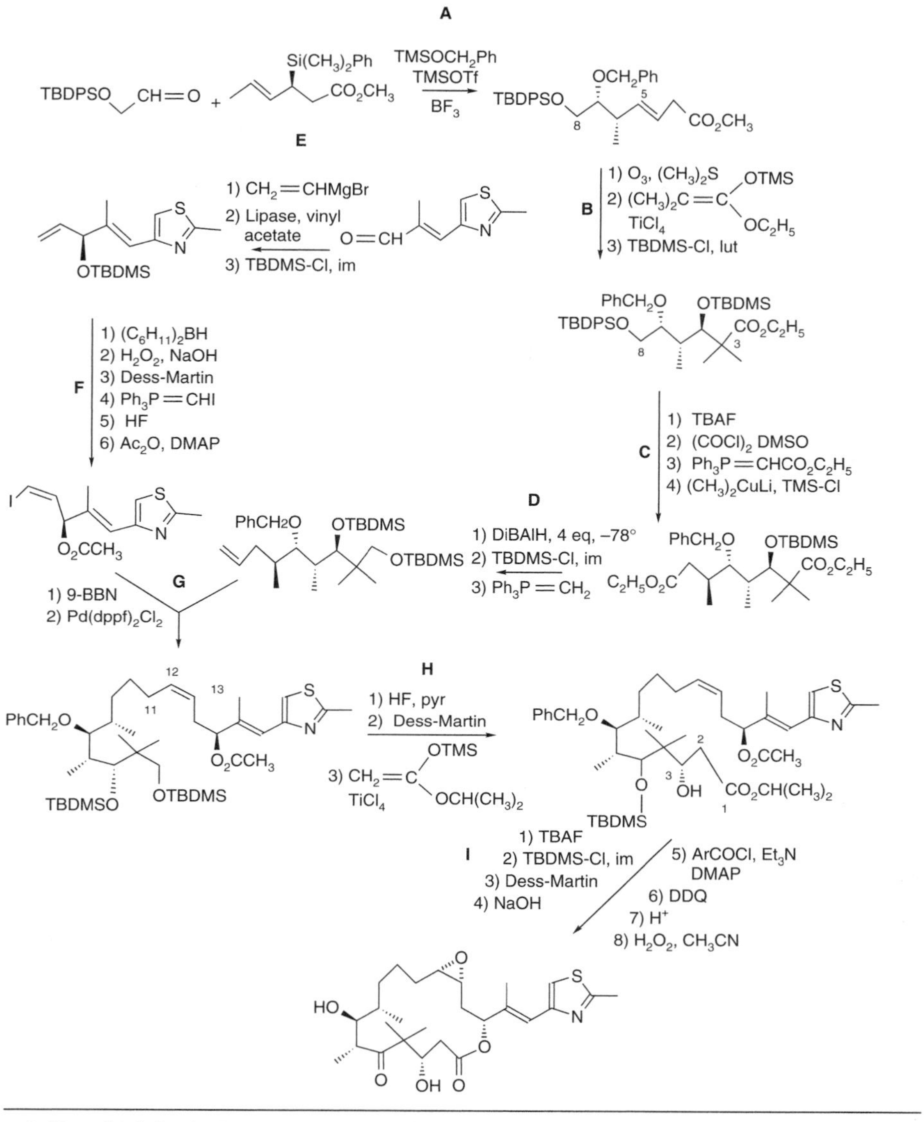

a. B. Zhu and J. S. Panek, *Eur. J. Org. Chem.*, 1701 (2001).

sodium acetylide, completing the synthesis of the C(7)–C(13) segment (Steps **A**-4 to **A**7). The BF$_3$-mediated epoxide ring opening in Step **B**-2 occurred with inversion of configuration, establishing the configuration at C(15). The Z-stereochemistry at the C(12)–C(13) double bond was established by reduction over a Lindlar catalyst. An EE protecting group was used during the Swern oxidation (Step **C**-4) but then replaced by a TBDMS group for the Wittig reaction and beyond. The chirality of the C(1)–C(6) segment was established by a kinetic resolution of an epoxide by selective ring opening

Scheme 13.66. Epothilone A Synthesis: Z.-Y. Liu and Co-Workers[a]

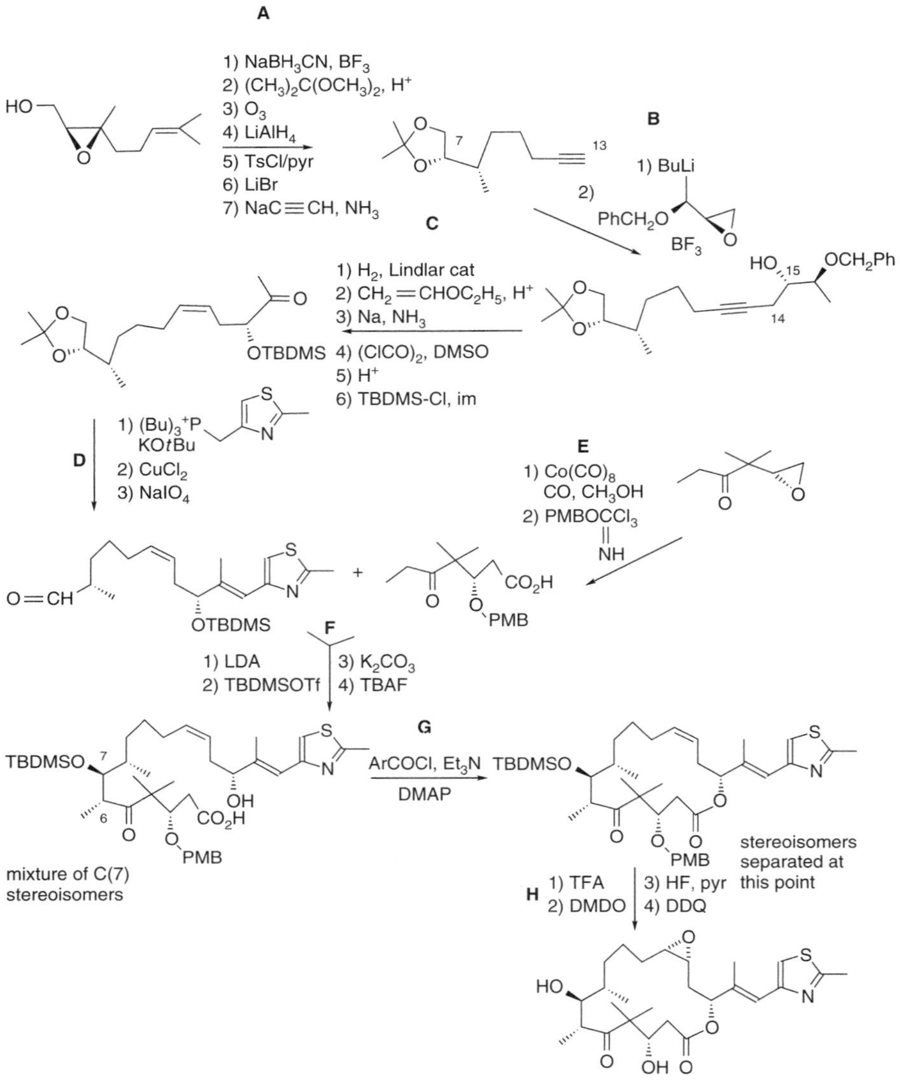

a. Z.-Y. Liu, Z.-C. Chen, C.-Z. Yu, R.-F. Wang, R.-Z. Zhang, C.-S. Huang, Z. Yan, D.-R. Cao, J.-B Sun, and G. Li, *Chem. Eur. J.*, **8**, 3747 (2002).

catalyzed by a chiral salen-Co(III) complex.[40] The resolved epoxide was converted to an ester by a $Co_2(CO)_8$-catalyzed carbonylation in Step **E**-1. The C(6)–C(7) bond was formed by an aldol reaction of a *dianion* of the intermediate. The product was a 1:1 mixture of diastereomers. After protecting group manipulations, this adduct was cyclized by macrolactonization. The two diastereomers were separated prior to completion of the synthesis by deprotection and epoxidation.

40. M. Tokunaga, J. F. Larrow, F. Kakiuchi, and E. N. Jacobsen, *Science*, **277**, 936 (1997).

Although each of the epothilone syntheses has its unique features, there are several recurring themes. Each synthesis uses one or more enantiopure compound as a starting material. All except the Danishefsky synthesis in Scheme 13.62 utilize the ester bond as a major disconnection. Most also use the C(12)–C(13) double bond as a second major disconnection, and several make the synthetic connection by the alkene (or alkyne) metathesis reaction. Others make the C(11)–C(12) disconnection and use a Suzuki coupling reaction in the synthetic sense to form the C(10)–C(11) bond. Wittig reactions figure prominently in the assembly of the thiazole-containing side chain. The configuration of the isolated stereocenter at C(15) is established by use of an enantiopure starting material (Schemes 13–59, 13–62, 13–64, and 13–66), an enantioselective reagent (Schemes 13–60, 13–61, and 13–63), or a kinetic resolution (Scheme 13–65). The stereochemical issues present are in the C(3)–C(8) segment and are addressed mainly by aldol reaction stereoselectivity.

13.2.6. Discodermolide

(+)-Discodermolide is a natural product isolated from a deep-water sponge found in the Caribbean Sea. The compound is probably produced by a symbiotic microorganism and isolation is not currently a practical source of the material. Like Taxol and epothilone A, (+)-discodermolide is a microtubule stabilizing agent with a promising profile of antitumor activity. A significant feature of the discodermolide structure is the three CH_3-OH-CH_3 triads that establish the configuration of nine stereogenic centers. The C(2)–C(4) and C(18)–C(20) triads are *syn, anti*, whereas the C(10)–C(12) triad is *anti, syn*. Seven syntheses are described here. Recently, major elements of two of these syntheses have been combined to provide sufficient material for Phase I clinical trials of (+)-discodermolide.

(+)-Discodermolide

The first (+)-discodermolide synthesis was completed by Stuart Schreiber's group at Harvard University and is outlined in Scheme 13.68. This synthesis was carried through for both enantiomers and established the absolute configuration of the natural material. The retrosynthetic plan outlined in Scheme 13.67 emphasizes the stereochemical triads found at C(2)–C(4), C(10)–C(12) and C(18)–C(20) and was designed to use a common chiral starting material. Each of the segments contains one of the stereochemical triads.

The starting material for the synthesis, methyl (S)-3-hydroxy-2-methylpropanoate, was converted to the corresponding aldehyde by reduction. The aldehyde was then converted to the diastereomeric homoallylic alcohols **9** and **10** using a chiral crotonylboronate (Scheme 13.68). The stereochemistry at C(5) was established by formation of the phenyldioxane ring by conjugate addition of a hemiacetal intermediate in Step **A**-3. After oxidation of C(1) to the aldehyde level the compound was rearranged to **11**, which eventually furnished the lactone terminus. The aldehyde group was introduced

Scheme 13.67. Retrosynthetic Analysis of (+)-Discodermolide to Fragments containing Stereotriads[a]

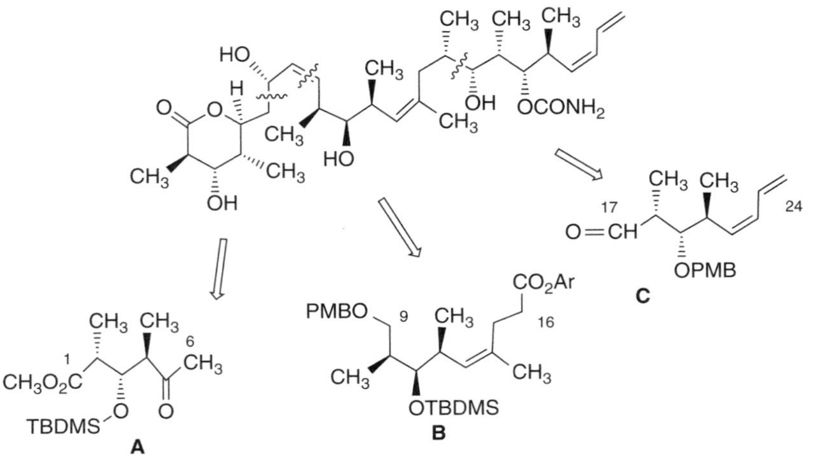

a. D. T. Hung, J. B. Nerenberg, and S. L. Schreiber, *J. Am. Chem. Soc.*, **118**, 11054 (1996).

prior to coupling by reductions of the *N*-methyl-*N*-methyl amide by LiAlH$_4$ (Steps **B**-5 to **B**-7). This fragment was carried through most of the synthesis as the corresponding phenylthio acetal.

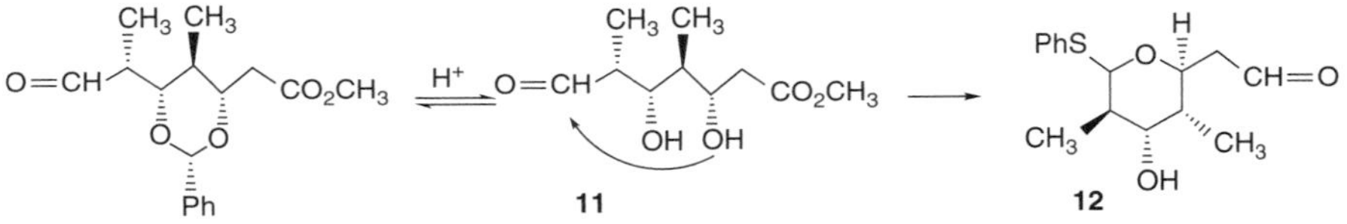

The stereoisomeric alcohol **10** was converted to the C(9)−C(15) fragment by a *Z*-selective Wadsworth-Emmons reaction, followed by reduction of the ester group in Steps **C**-1 to **C**-4. The alcohol was protected as the pivalate ester and then converted to a terminal alkyne using dimethyl diazomethylphosphonate. The C(1)–C(7) and C(8)–C(15) fragments were coupled by a Ni-catalyzed Cr(II) reaction in Step **E**. After reduction to the *Z*-alkene, the allylic alcohol was converted to the bromide via a mesylate. This set the stage for coupling with the C(16)–C(24) segment by enolate alkylation. The C(16) methyl group was installed at this point by a second alkylation (Step **H**-2). When the alkylation was carried out with this methyl group already in place, the C(16) epimer of (+)-discodermolide was obtained. The final conversion to (+)-discodermolide was achieved after carbamoylation of the C(19) hydroxy group. This group promoted stereoselective reduction at C(17) using a bulky hydride reducing agent. Deprotection then gave (+)-discodermolide.

The synthesis of (+)-discodermolide in Scheme 13.69 was completed in James Marshall's laboratory at the University of Virginia and applies allenylmetal methodology at key stages. The starting material was O-protected (*S*)-3-hydroxy-2-methylpropanal. An enantiopure butynyl mesylate was the other starting material. The CH$_3$-OH-CH$_3$ stereochemical triad was established by addition to the aldehyde using Pd-catalyzed reaction with an allenyl zinc reagent generated from a butenyl

Scheme 13.68. Synthesis of Discodermolide: S. L. Schreiber and Co-Workers[a]

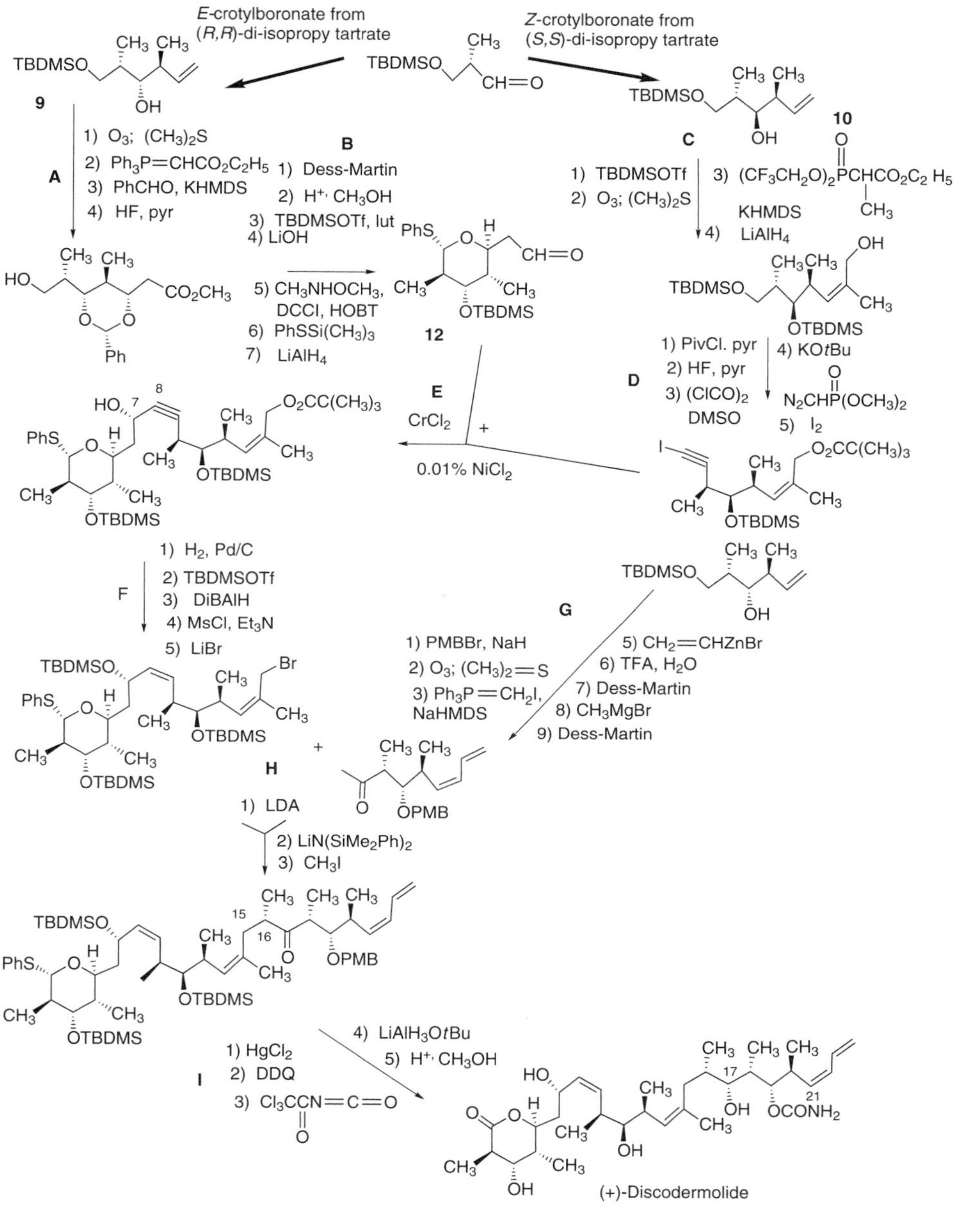

a. D. T. Hung, J. B. Nerenberg, and S. L. Schreiber, *J. Am. Chem. Soc.*, **118**, 11054 (1996).

mesylate (Step **A**). The adduct was cyclized as a 1,3-dioxane and further elaborated to an aldehyde intermediate. Reduction to an allylic alcohol by Red-Al was followed by Sharpless epoxidation. The epoxide was opened by a second Red-Al reduction. After protecting group manipulation, the aldehyde functional group was obtained by Swern oxidation. (Steps **C**-1 to **C**-7). This aldehyde was coupled

Scheme 13.69. Discodermolide Synthesis: J. A. Marshall and Co-Workers[a]

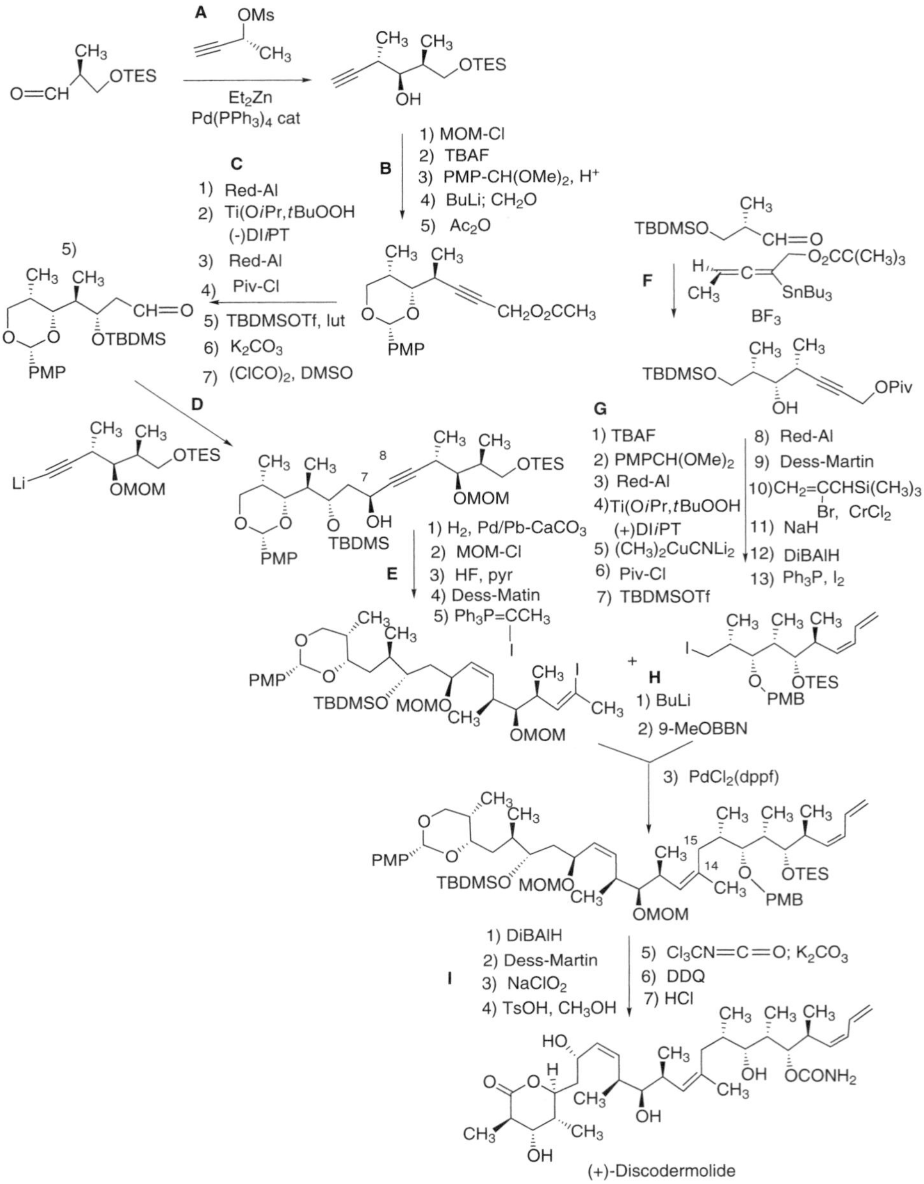

a. J. A. Marshall, Z.-H. Lu, and B. A. Johns, *J. Org. Chem.*, **63**, 817 (1998); J. A. Marshall and B. A. Johns, *J. Org. Chem.*, **63**, 7885 (1998).

with a protected alkyne, forming the C(7)–C(8) bond (Step **D**). Reduction with a Lindlar catalyst gave the *Z*-double bond, which provided C(1)–C(13) of the discodermolide skeleton with the correct stereochemistry. A terminal vinyl iodide including C(14) and its methyl substituent was introduced by a Wittig reaction using 1-iodoethylidenetriphenylphosphorane.

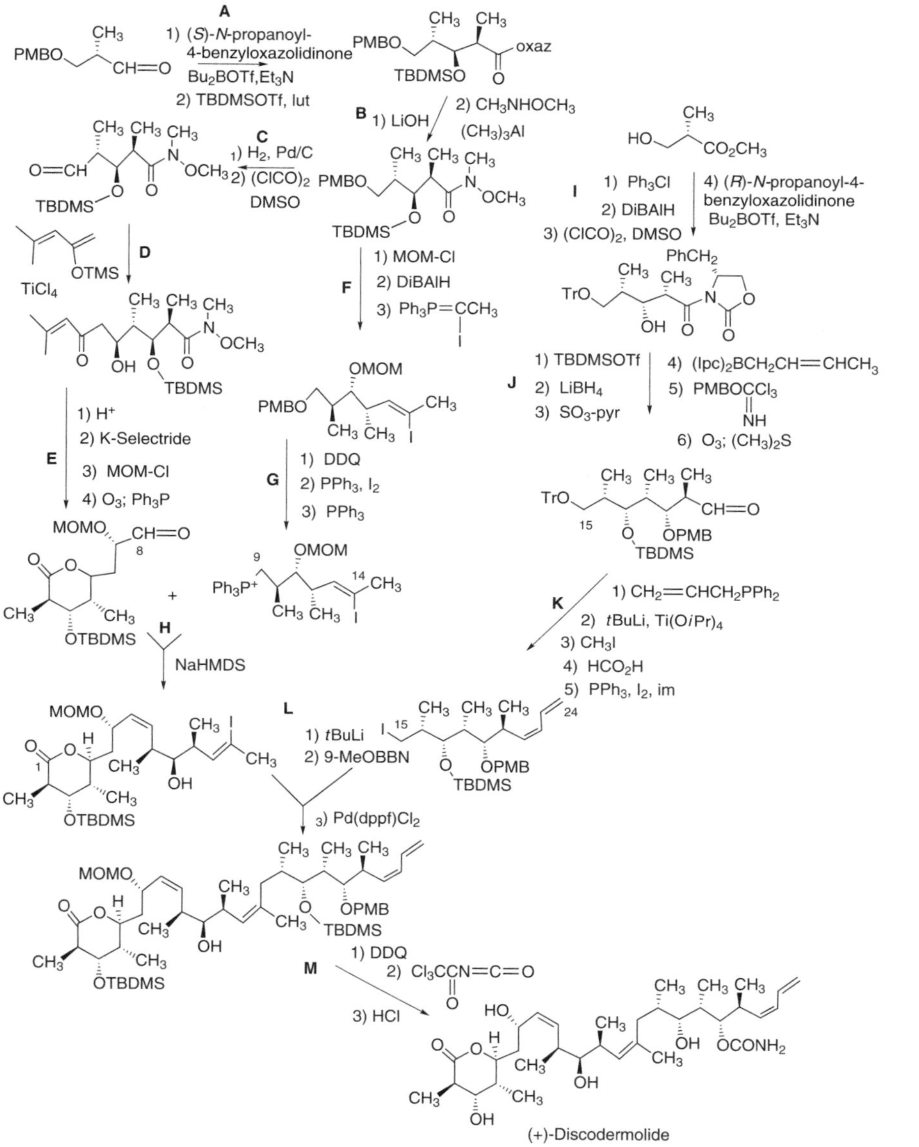

a. A. B. Smith, III, B. S. Freeze, M. Xian, and T. Hirose, *Org. Lett.*, **7**, 1825 (2005).

The C(15)–C(24) segment was constructed by addition of a chiral allenyl-stannane reagent to the starting aldehyde in Step **F**. The propargyl acetate terminus was reduced by DiBAlH, giving an allylic alcohol that was subjected to Sharpless asymmetric epoxidation. The methyl substituent at C(20) was added by nucleophilic opening of the epoxide with dimethylcyanocuprate. This segment was extended to include the terminal diene unit in **G**-9 and **G**-10. The terminal diene unit was introduced by CrCl$_2$-mediated addition in Step **G**-10, followed by base-induced elimination from the β-hydroxysilane.

The two major subunits were coupled by a Suzuki reaction in Step **H**-3. The synthesis was then completed by reductive opening of the 1,3-dioxane ring, oxidation of the terminal alcohol to the carboxylic acid, carbamoylation, deprotection, and lactonization.

The synthesis of discodermolide in Scheme 13.70 was developed by A. B. Smith, III, and co-workers at the University of Pennsylvania. The synthesis shown in the scheme, which is the result of refinement of several previous syntheses from this laboratory, used a common precursor prepared in Steps **A** and **B**. The stereochemistry of the fragments was established by use of oxazolidinone chiral auxiliaries. The boron enolate of *N*-propanoyl-4-benzyloxazolidinone was added to PMP-protected (*S*)-3-hydroxy-2-methylpropanal in Step **A**. The chiral auxiliary was then replaced by an *N*-methoxy-*N*-methylamide in Step **B**. This intermediate was used for the construction of the C(1)–C(8) and C(9)–C(14) segments. The connection between these two fragments was made by a Wittig reaction at Step **H**. The C(15)–C(21) segment was also derived from an oxazolidinone chiral auxiliary, in this case the (*R*)-enantiomer. The configuration at C(20) was established by allylboration (Step **J**-4). The terminal diene was introduced by a Wittig reaction in Step **K**-1. The two major segments were then coupled at the C(14)–C(15) bond by using the Suzuki reaction in Step **L**. The final steps involve deprotection and installation of the carbamoyl group. The overall yield for this version is 9% with a longest linear sequence of 17 steps.

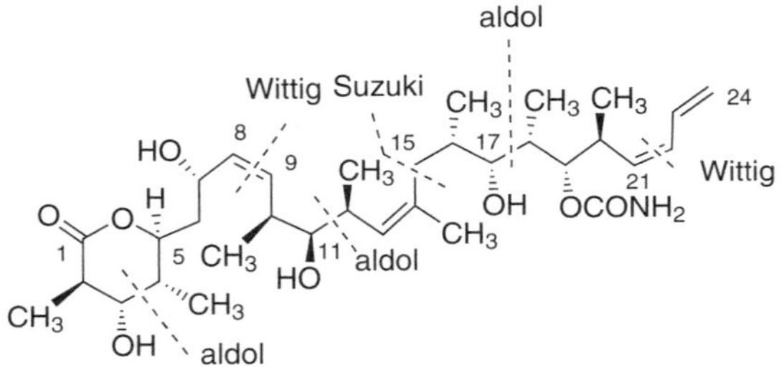

Scheme 13.71 shows the most recent version of a synthesis of (+)-discodermolide developed by Ian Paterson's group at Cambridge University. The synthesis was based on three major subunits and used boron enolate aldol addition reactions to establish the stereochemistry.

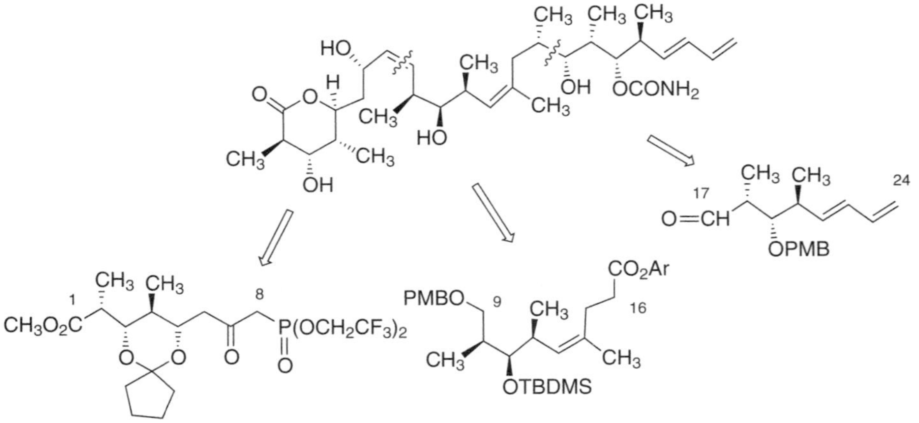

Scheme 13.71. Discodermolide Synthesis: I. Paterson and Co-Workers[a]

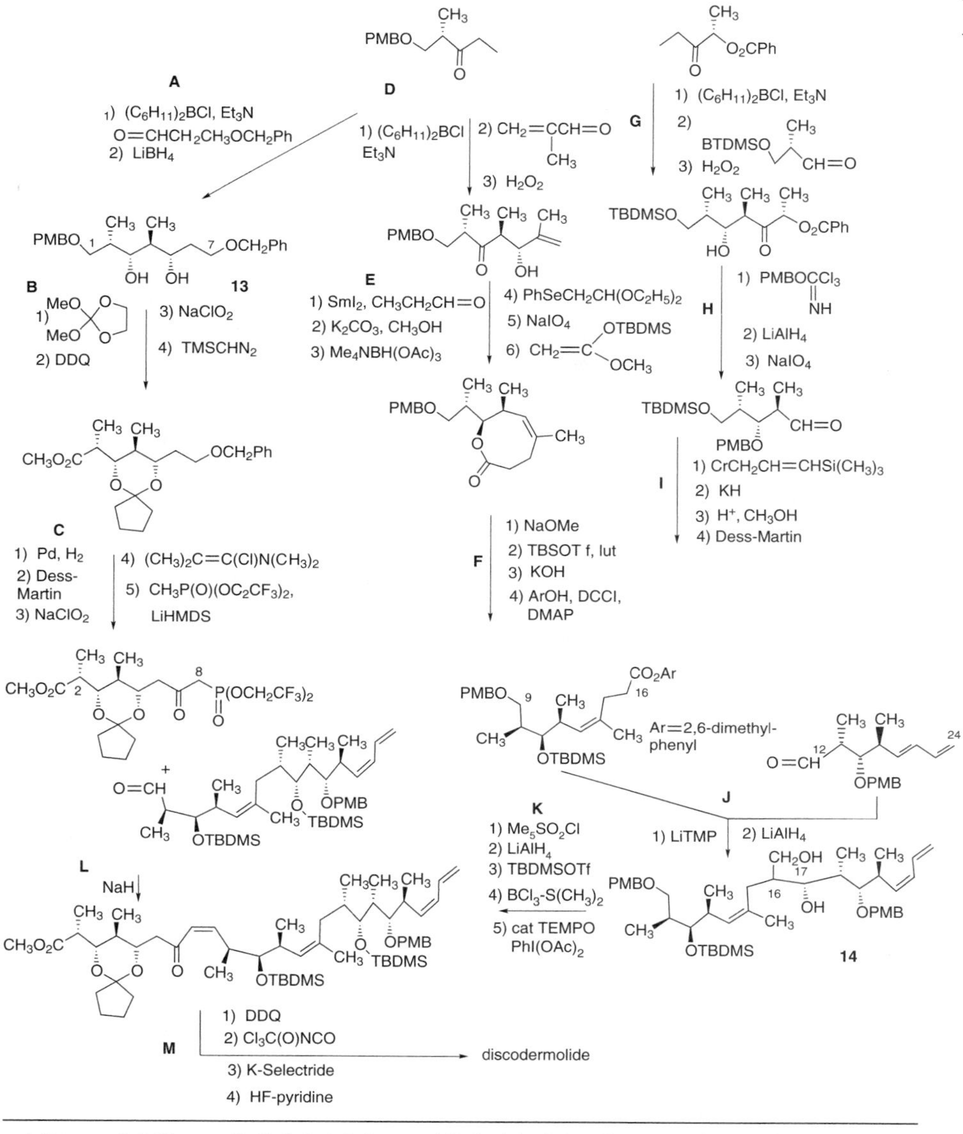

a. I. Paterson, G. J. Florence, K. Gerlach, J. P. Scott, and N. Sereinig, *J. Am. Chem. Soc.*, **123**, 9535 (2001); I. Paterson, O. Delgado, G. J. Florence, I. Lyothier, J. P. Scott, and N. Sereinig, *Org. Lett.*, **5**, 35 (2003); I. Paterson and I. Lyothier, *J. Org. Chem.*, **70**, 5494 (2005).

The synthesis of the C(1)–C(6) subunit was based on addition of an enol boronate to 3-benzyloxypropanal through TS-1. Immediate reduction of the chelate is also stereoselective and provides the intermediate **13**. These steps establish the configuration at C(2)–C(5).

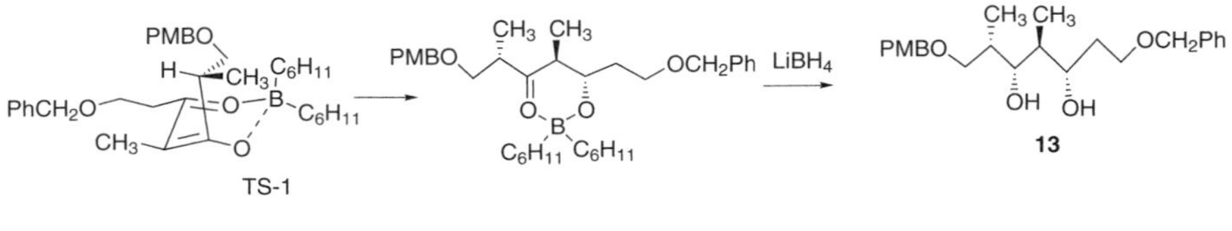

The diol was protected and the C-terminal group converted to a methyl ester in sequence **B**. A phosphonate group was installed at C(7) via an acylation reaction in Step **C**-5. Successive oxidations of the primary and deprotected secondary alcohol gave the C(1)–C(8) intermediate.

The C(9)–C(16) subunit was synthesized from the same starting material. The chain was extended by a boron enolate addition to 2-methylpropenal (Step **D**-2). After introduction of a double bond by selenoxide elimination in Step **E**-4, a Claisen rearrangement was used to generate an eight-membered lactone ring (Step **E**-6).

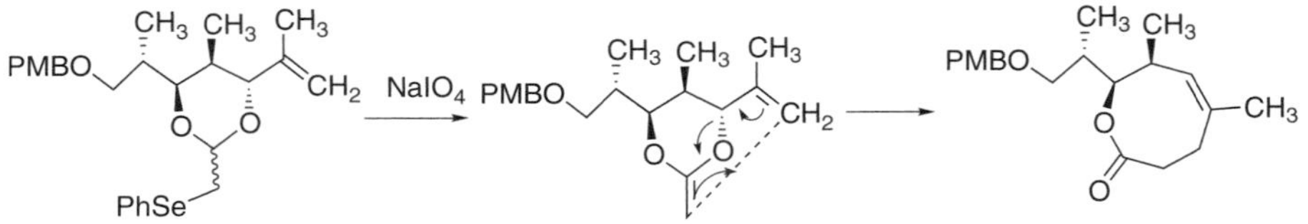

The lactone ring was then opened and the carboxy group converted to a hindered phenolic ester (Step **F**-4), providing the C(9)–C(16) intermediate.

The synthesis of the C(17)–C(24) segment also began with a diastereoselective boron enolate aldol addition. The adduct was protected and converted to an aldehyde in sequence **H**. The terminal diene unit was installed using a γ-silylallyl chromium reagent, which generates a β-hydroxysilane. Peterson elimination using KH then gave the Z-diene.

The three fragments were then coupled. The C(16)–C(17) bond was established by addition of the lithium enolate of the aryl ester in the C(9)–C(16) fragment with the aldehyde group of the C(17)–C(24) fragment. The stereochemistry is consistent with the cyclic aldol addition TS. The adduct was immediately reduced to the diol **14** by LiAlH₄.

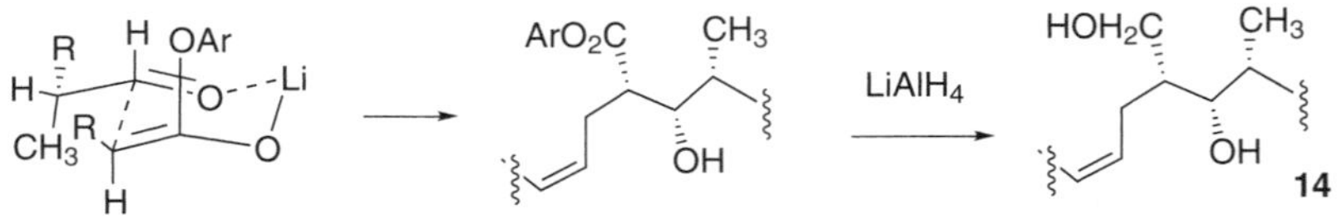

The primary hydroxymethyl group at C(16) was deoxygenated via the mesitylenesulfonate. After removal of the PMP protecting group, a sterically demanding oxidant, TEMPO-PhI(OAc)₂ was used to selectively oxidize the primary alcohol group to an aldehyde. The Still-Gennari version of the Wadsworth-Emmons reaction was used to couple with the C(1)–C(8) fragment in Step **L**. This reaction proceeded with 5:1 *Z* : *E* selectivity and led to isolation of the *Z*-product in 73% yield. The PMB protecting group was then removed and the carbamate group introduced at C(19). The remaining protecting groups were then removed and the lactonization completed the synthesis.

The overall yield was about 12% over a longest linear sequence of 23 steps and about 40 steps total. The major disconnections are illustrated below.

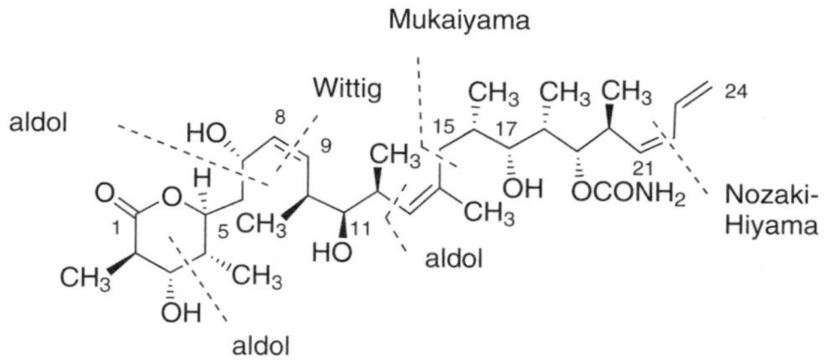

The synthesis outlined in Scheme 13.72 was carried out by James Panek's group at Boston University and is based on three key intermediates that were synthesized from two closely related methyl 3-(dimethylphenylsilyl)hex-4-enoates.

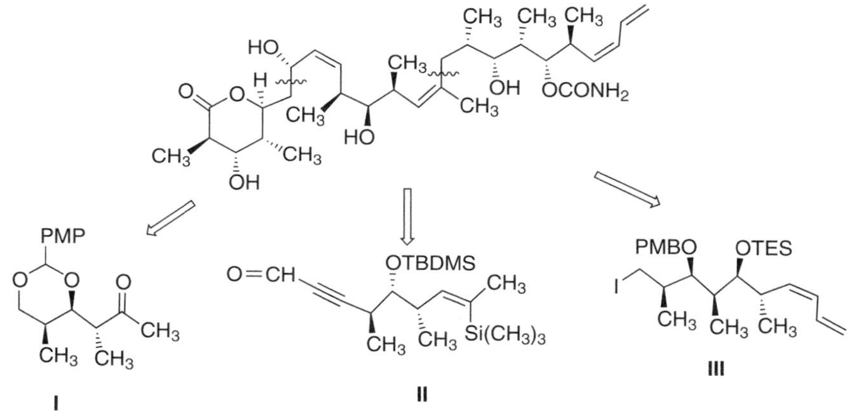

The stereochemistry was controlled by Lewis acid–induced addition of the allylic silanes to aldehydes. The reaction of the silane with O-protected (*S*)-3-hydroxy-2-methylpropanal provides **15**. The silane reacted with the benzyl-protected analog to provide **16**.

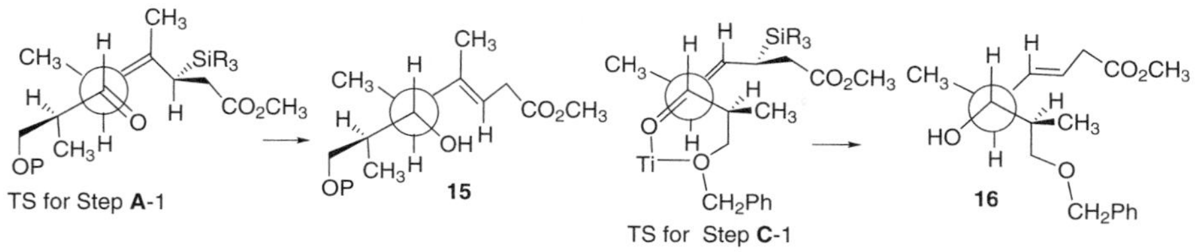

Scheme 13.72. Discodermolide Synthesis: J. S. Panek and A. Arefolov[a]

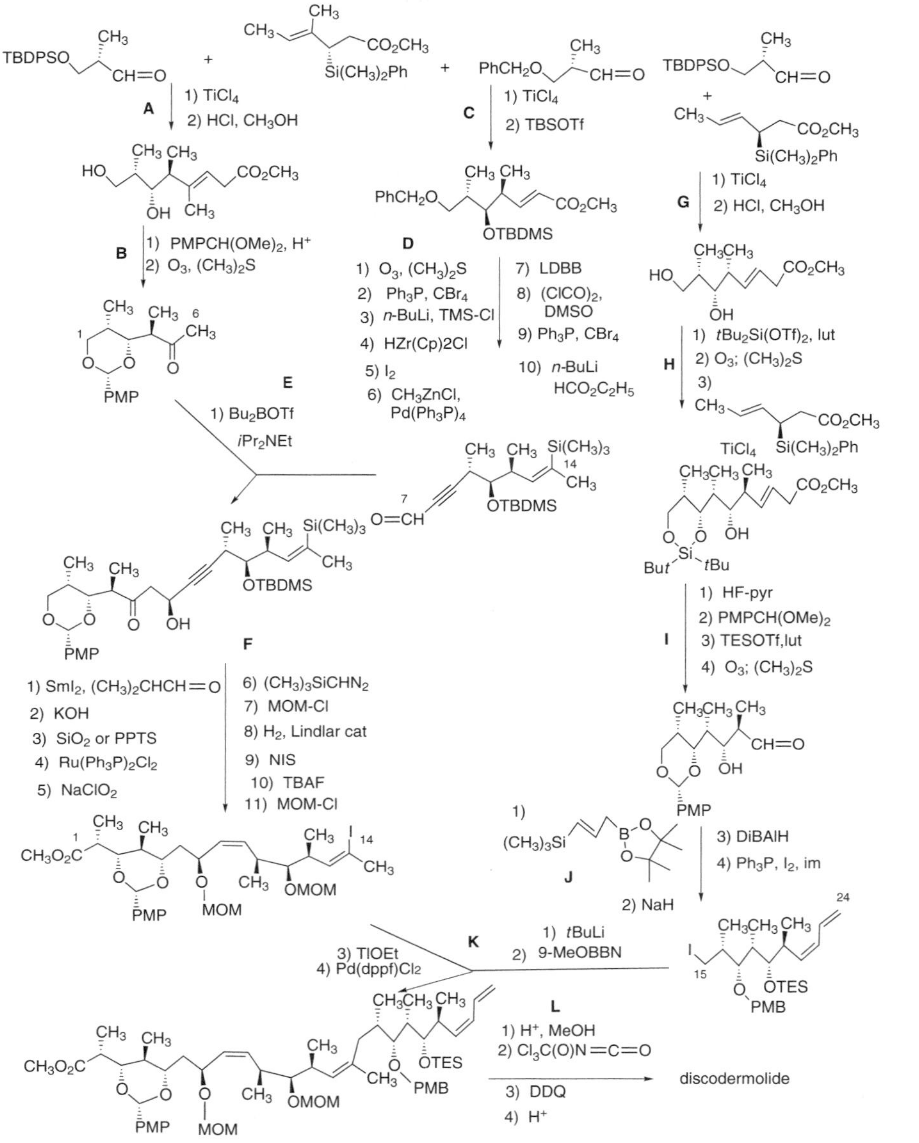

a. A. Arefolov and J. S. Panek, *J. Am. Chem. Soc.*, **127**, 5596 (2005).

These intermediates were then converted to the fragments **I** and **II**, respectively. Intermediate **15** was protected as a cyclic acetal and then ozonized to give segment **I**. In the synthesis of the **II** fragment the adduct was extended by two Corey-Fuchs sequences with in situ functionalization to provide the alkyne intermediate **II** (Steps **D**-2 and **D**-9). Trimethylsilyl and methyl groups were introduced at C(14) and a formyl groups was added at C(8). The fragments **I** and **II** were coupled by boron enolate methodology and a single stereoisomer was obtained in 88% yield (Step **E**).

The coupled fragments were then converted to a vinyl iodide. The key steps were a Z-selective Lindlar reduction and iodinolysis of the vinyl silane, which was done using NIS in acetonitrile (sequence **F**-1 to **F**-11).

The C(15)–C(24) segment **C** was created by two successive additions of the allylic silane synthons (Steps **G**-1 and **H**-3). The unsaturated esters resulting from the additions were subjected to ozonolysis. The terminal diene unit was added using a silyl-substituted allylic boronate and then subjected to base-mediated elimination. The coupling of the **I-II** and **III** segments was done by Suzuki methodology. It was also carried out in somewhat lower yield using a zinc reagent prepared from the vinyl iodide. The synthesis was completed by deprotection and lactonization. There are a total of 42 steps, with the longest linear sequence being 27 steps, in overall 21% yield.

The synthesis of discodermolide in Scheme 13.73 was completed at the University of California, Berkeley by D. C. Myles and co-workers. The synthesis began with a TiCl$_4$-mediated cycloaddition that gave a dihydropyrone intermediate that contains the stereochemistry at C(10)–C(12) and the Z-configuration at the C(13)–C(14) double bond. Reduction and H$^+$-promoted Ferrier rearrangement gave a lactol containing C(9)–C(15) (Steps **B**-1 and **B**-2). This lactol was converted to an allylic iodide, providing one of the key intermediates, **II**.

The stereochemistry at C(18)–C(20) was established using an oxazolidinone chiral auxiliary (Step **D**-1). Carbon-16 and its methyl substituent were added by a Grignard addition in Step **D**-4. The C(9)–C(15) and C(16)–C(21) segments were joined by enolate alkylation (Step **E**). Under optimum conditions, a 6:1 preference for the desired stereoisomer at C(16) was achieved. The stereochemistry at C(17) was established by LiAlH$_4$ reduction in the presence of LiI, with 8:1 stereoselectivity. An iodovinyl group containing C(8) was installed using iodomethylenetriphenylphosphorane, giving a Z : E isomer ratio of 20:1 (Step **G**-2). The terminal diene unit was installed using a γ-silylallylboronate, followed by base-mediated *syn* elimination (Steps **G**-5 and **G**-6). The carbamate group was then installed, completing the synthesis of intermediate **III**.

The synthesis of the C(1)–C(7) fragment began with allylstannylation (Step **H**). The C(1)–C(2) terminus was introduced using the dibutylboron enolate of an oxazolidinone chiral auxiliary. The C(8)–C(24) fragment was added via a NiCl$_2$-CrCl$_2$ coupling. This reaction was improved by inclusion of a chiral *bis*-pyridine ligand. Sequential deprotection and lactonization afforded discodermolide. The overall yield was 1.5% based on a 22-step longest linear sequence.

The synthesis of (+)-discodermolide shown in Scheme 13.74 was developed in the laboratories of the Novartis Pharmaceutical Company and was designed to provide sufficient material for initial clinical trials. The synthesis is largely based on the one

Scheme 13.73. Discodermolide Synthesis: D. C. Myles Co-workers[a]

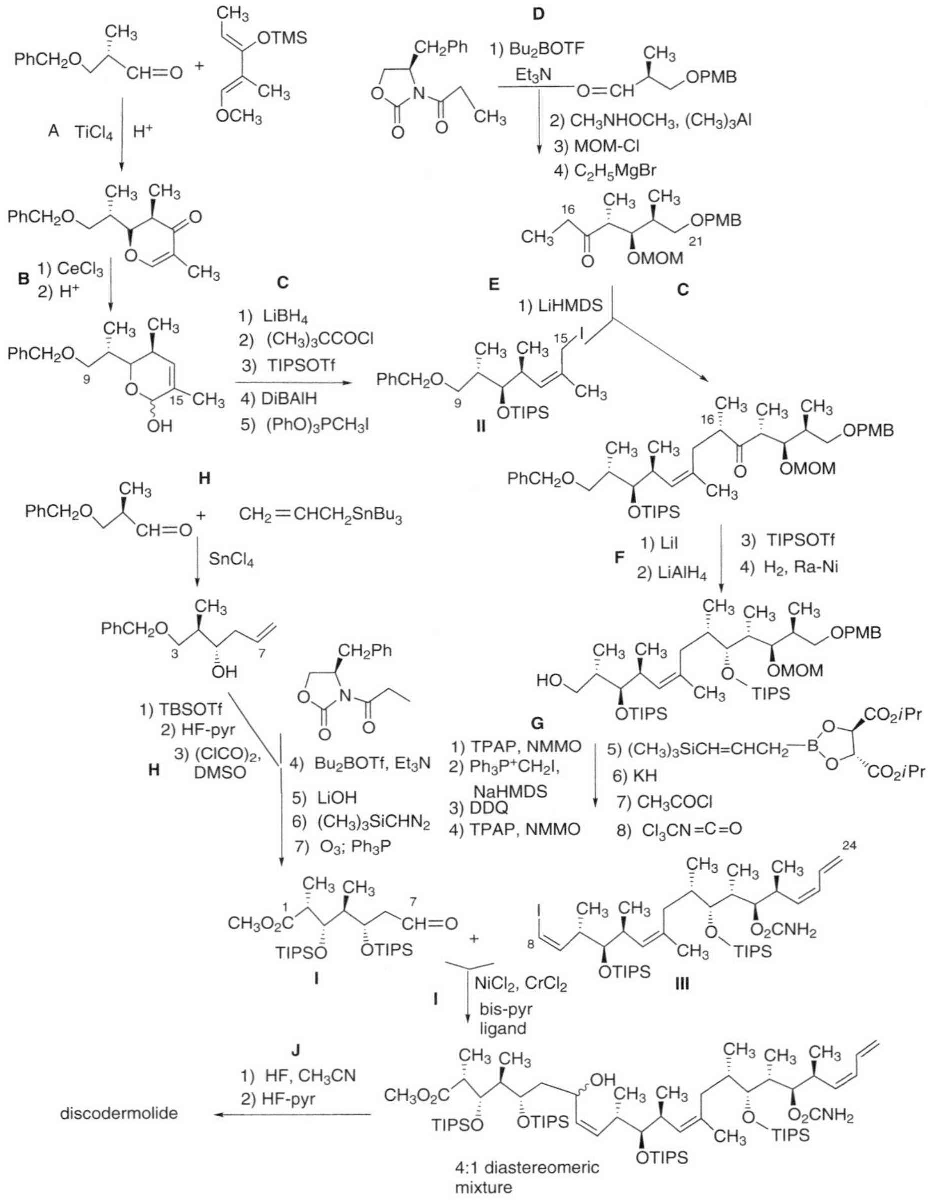

a. S. S. Harried, C. P. Lee, G. Yang, T. I. H. Lee, and D. C. Myles, *J. Org. Chem.*, **68**, 6646 (2003).

by A. B. Smith, III, and co-workers (Scheme 13.70), with the final stages being based on the synthesis in Scheme 13.71. The synthesis begins with a single starting material having one stereogenic center and proceeds through Smith's common intermediate **17** to three segments containing the stereochemical triads.

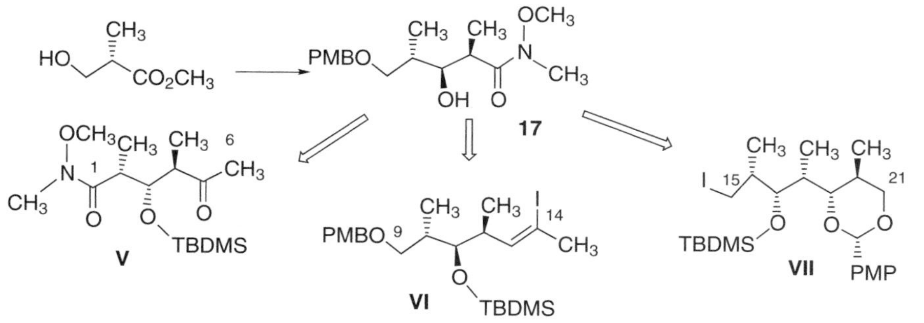

A number of modifications were made to meet scale-up requirements. In the preparation of the common intermediate, LiBH$_4$ was used in place of LiAlH$_4$ in Step **A**-2 and a TEMPO-NaOCl oxidation was used in place of Swern oxidation in Step **A**-3. Some reactions presented difficulty in the scale-up. For example, the boron enolate aldolization in Step **B**-1 gave about 50% yield on the 20- to 25-kg scale as opposed to greater than 75% on a 50-g scale. The amide formation in Step **B**-3 was modified to eliminate the use of trimethylaluminum, and the common intermediate **17** could be prepared on a 30-kg scale using this modified sequence. The synthesis of the C(1)–C(6) segment **V** was done by Steps **C**-1 to **C**-5 in 66% yield on the scale of several kg.

The C(9)–C(14) segment **VI** was prepared by Steps **D**-1 to **D**-3. The formation of the vinyl iodide in Step **D**-3 was difficult and proceeded in only 25–30% yield. The C(15)–C(21) segment **VII** was synthesized from the common intermediate **17** by Steps **E**-1 to **E**-6. A DDQ oxidation led to formation of a 1,3-dioxane ring in Step **E**-1. The *N*-methoxy amide was converted to an aldehyde by LiAlH$_4$ reduction and the chain was extended to include C(14) and C(15) using a boron enolate of an oxazo-lidinone chiral auxiliary. After reductive removal of the chiral auxiliary, the primary alcohol group was converted to a primary iodide. The overall yield for these steps was about 25%.

The C(9)–C(14) and C(15)–C(21) segments were then coupled using Suzuki methodology (Step **F**). The terminal diene unit was then introduced in Steps **G**-1 to **G**-3. The cyclic acetal was reduced with DiBAlH, restoring the PMB protecting group and deprotecting the C(21) hydroxy. This primary alcohol was oxidized to the aldehyde and coupled with an allylic silane using CrCl$_2$, as in Scheme 13.69. The chain was then extended by adding C(7) and C(8) using the *Z*-selective Still-Gennari modification of the Wadsworth-Emmons reaction (Step **H**-3) and the ester was converted to an aldehyde. This permitted the final coupling with the C(1)–C(6) fragment using a boron enolate prepared from (Ipc)$_2$BCl. The optimized procedure gave the product in 50–55% yield with stereoselectivity of about 4:1. A process for converting the minor diastereomer to the desired product was developed. The final reduction was done with [(CH$_3$)$_4$N]$^+$[BH(OAc)$_3$]$^-$. Removal of the final silyl protecting group and lactonization gave (+)-discodermolide. The overall synthesis involved 39 steps.

Scheme 13.74. Discodermolide Synthesis: Novartis Group[a]

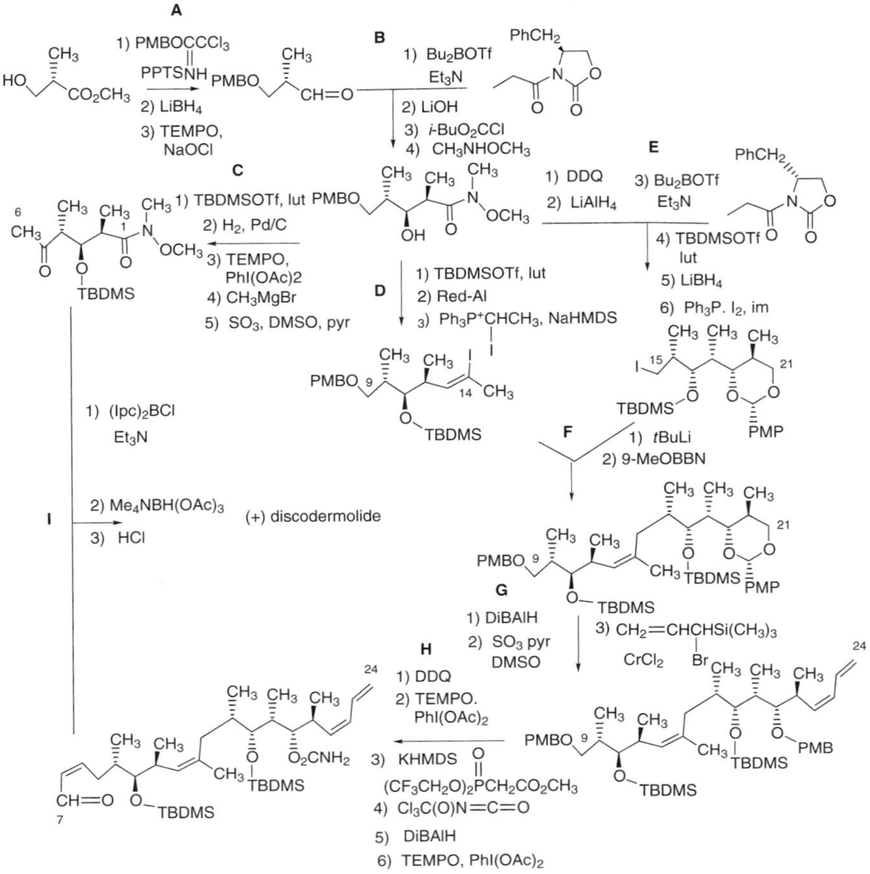

a. S. J. Mickel, G. H. Sedelmeier, D. Niederer, R. Daeffler, A. Osmani, K. Shreiner, M. Seeger-Weibel, B. Berod, K. Schaer, R. Gamboni, S. Chen, W. Chen, C. T. Jagoe, F. R. Kinder, Jr., M. Loo, K. Prasad, O. Repic, W.-C. Shieh, R.-M. Wang, L. Waykole, D. D. Xu, and S. Xue, *Org. Proc. Res. Dev.*, **8**, 92 (2004); S. J. Mickel, G. H. Sedelmeier, D. Niederer, F. Schuerch, D. Grimler, G. Koch, R. Daeffler, A. Osmani, A. Hirni, K. Schaer, R. Gamboni, A. Bach, A. Chaudhary, S. Chen, W. Chen, B. Hu, C. T. Jagoe, H.-Y. Kim, F. R. Kinder, Jr., Y. Liu, Y. Lu, J. McKenna, M. Prasad, T. M. Ramsey, O. Repic, L. Rogersk, W.-C. Shieh, R.-M. Wang, and L. Waykole, *Org. Proc. Res. Dev.*, **8**, 101 (2004); S. J. Mickel, G. H. Sedelmeier, D. Niederer, F. Schuerch, G. Koch, E. Kuesters, R. Daeffler, A. Osmani, M. Seeger-Weibel, E. Schmid, A. Hirni, K. Schaer, R. Gamboni, A. Bach, S. Chen, W. Chen, P. Geng, C. T. Jagoe, F. R. Kinder, Jr., G. T. Lee, J. McKenna, T. M. Ramsey, O. Repic, L. Rogers, W.-C. Shieh, R.-M. Wang, and L. Waykole, *Org. Proc. Res. Dev.*, **8**, 107 (2004); S. J. Mickel, G. H. Sedelmeier, D. Niederer, F. Schuerch, M. Seger, K. Schreiner, R. Daeffler, A. Osmani, D. Bixel, O. Loiseleur, J. Cercus, H. Stettler, K. Schaer, R. Gamboni, A. Bach, G.-P. Chen, W. Chen, P. Geng, G. T. Lee, E. Loesser, J. McKenna, F. R. Kinder, Jr., K. Konigberger, K. Prasad, T. M. Ramsey, N. Reel, O. Repic, L. Rogers, W.-C. Shieh, R.-M. Wang, L. Waykole, S. Xue, G. Florence, and I. Paterson, *Org. Proc. Res. Dev.*, **8**, 113 (2004); S. J. Mickel, D. Niederer, R. Daeffler, A. Osmani, E. Kuesters, E. Schmid, K. Shaer, R. Gamboni, W. Chen, E. Loeser, F. R. Kinder, Jr., K. Koningberger, K. Prasad, T. M. Ramsey, O. Repic, R.-M. Wang, G. Florence, I. Lyothier, and I. Paterson, *Org. Proc. Res. Dev.*, **8**, 122 (2004).

These syntheses of (+)-discodermolide provide examples of the application of several current methods for control of acyclic stereochemistry. They illustrate the use of allylic boronates, allenyl stannanes, oxazolidinone auxiliaries, boron enolates, and allylic silanes to achieve enantioselective formation of key intermediates. Wittig and Suzuki reactions figure prominently in the coupling of key intermediates. Several of the syntheses use β-hydroxy silane elimination to introduce the terminal diene. The discodermolide structure lends itself to a high degree of convergency, and the relationship among the three stereochemical triads permits utilization of common starting materials, which contributes to overall synthetic efficiency. The composite synthesis completed by the Novartis group provides an insight into the logistics of scale-up of a synthesis of this complexity. The synthesis described in Scheme 13.74 produced 60 g of pure (+)-discodermolide. The effort involved about 40 chemists and was carried out over a period of 20 months.

13.3. Solid Phase Synthesis

The syntheses discussed in the previous sections were all carried out in *solution phase* and intermediates were isolated and purified. There is another general approach to multistep synthesis in which the starting material is attached to a solid support. The sequence of synthetic steps is then carried out with the various intermediates remaining attached to the solid support. Called *solid phase synthesis*, this approach has a potential advantage in that excess reagents and by-products can simply be washed away after each step. When the synthesis is complete, the product can be detached from the support. Another potential advantage of solid phase synthesis is that the operations can be automated. A particular sequence for addition of reactants, reagents, and solvents for removal of soluble material can be established. Instruments can then be programmed to carry out these operations.

The most highly developed applications of solid phase methods are in the syntheses of polypeptides and oligonucleotides. These molecules consist of linear sequences of individual amino acids or nucleotides. The connecting bonds are the same for each subunit: amides for polypeptides and phosphate esters for the polynucleotides. The synthesis can be carried out by sequentially adding the amino acids or nucleotides and coupling reagents. The ability to synthesize polypeptides and oligonucleotides of known sequence is of great importance in a number of biological applications. Although these molecules can be synthesized by synthetic manipulations in solution, they are now usually synthesized by solid phase methods, using automated repetitive cycles of deprotection and coupling. Another important application of solid phase synthesis is in combinatorial synthesis, where the goal is to make a large number of related molecules by systematic variation of the individual components.

13.3.1. Solid Phase Polypeptide Synthesis

The techniques for automated solid phase synthesis were first highly developed for polypeptides and the method is abbreviated as *SPPS*. Polypeptide synthesis requires the sequential coupling of the individual amino acids. After each unit is added, it must be deprotected for use in the next coupling step.

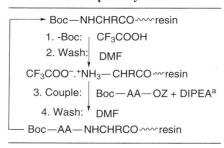

Excellent solution methods involving alternative cycles of deprotection and coupling are available for peptide synthesis,[41] and the techniques have been adapted to solid phase synthesis.[42] The N-protected carboxy terminal amino acid is linked to the solid support, which is usually polystyrene with divinylbenzene cross-linking. The amino group is then deprotected and the second N-protected amino acid is introduced and coupled. The sequence of deprotection and coupling is then continued until the synthesis is complete. Each deprotection and coupling step must go in very high yield. Because of the iterative nature of solid phase synthesis, errors accumulate throughout the process. For the polypeptide to be of high purity, the conversion must be very efficient at each step.

The first version of SPPS to be developed used the *t*-Boc group as the amino-protecting group. *t*-Boc can be cleaved with relatively mild acidic treatment and TFA is usually used. The original coupling reagents utilized for SPPS were carbodiimides. In addition to dicyclohexylcarbodiimide (DCCI), *N, N'*-diisopropylcarbodiimide (DIPCDI) is often used. The mechanism of peptide coupling by carbodiimides was

Scheme 13.75. *t*-Boc Protocol for Solid Phase Peptide Synthesis

Boc—NHCHRCO∼∼∼resin

1. -Boc: | CF$_3$COOH
2. Wash: | DMF

CF$_3$COO$^-$.$^+$NH$_3$—CHRCO∼∼∼resin

3. Couple: | Boc—AA—OZ + DIPEA[a]
4. Wash: | DMF

Boc—AA—NHCHRCO∼∼∼resin

a. OZ = active ester; DIPEA = diisopropylethylamine

41. M. Bodanszky and A. Bodanszky, *The Practice of Peptide Synthesis*, 2nd Edition, Springer Verlag, Berlin, 1994; V. J. Hruby, and J.-P. Mayer, in *Bioorganic Chemistry: Peptides and Proteins*, S. Hecht, ed. Oxford University Press, Oxford, 1998, pp. 27–64.

42. R. B. Merrifield, *Meth. Enzymol.*, **289**, 3 (1997); R. B. Merrifield, in *Peptides: Synthesis, Structure, and Applications*, B. Gutte, ed., Academic Press, San Diego, CA, p. 93; E. Atherton and R. C. Sheppard, *Solid Phase Peptide Synthesis*, IRL Press, Oxford, 1989; P. Lloyd-Williams, F. Albericio, and E. Giralt, *Chemical Synthesis of Peptides and Proteins*, CRC Press, Boca Raton, FL, 1997.

discussed in Section 3.4. Currently, the optimized versions of the *t*-Boc protocol can provide polypeptides of 60–80 residues in high purity.[43] The protocol for using *t*-Boc protection is outlined in Scheme 13.75

A second method that uses the fluorenylmethoxycarboxy (Fmoc) protecting group has been developed.[44] The Fmoc group is stable to mild acid and to hydrogenation, but it is cleaved by basic reagents via fragmentation triggered by deprotonation at the acidic 9-position of the fluorene ring. The protocol for SPPS using the Fmoc group is shown in Scheme 13.76.

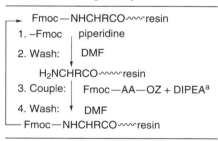

In both the *t*-Boc and Fmoc versions of SPPS, the amino acids with functional groups in the side chain also require protecting groups. These protecting groups are designed to stay in place throughout the synthesis and then are removed when the synthesis is complete. The serine and threonine hydroxyl groups can be protected as benzyl ethers. The ε-amino group of lysine can be protected as the trifluoroacetyl derivative or as a sulfonamide derivative. The imidazole nitrogen of histidine can also be protected as a sulfonamide. The indole nitrogen of tryptophan is frequently protected as a formyl derivative. The exact choice of protecting group depends upon the deprotection-coupling sequence being used.

The original version of SPPS attached the carboxy terminal residue directly to the resin as a benzylic ester using chloromethyl groups attached to the polymer. At the present time the attachment is done using "linking groups." Two of the more common linking groups are shown. These groups have the advantage of permitting

Scheme 13.76. Fmoc Protocol for Solid Phase Peptide Synthesis

Fmoc—NHCHRCO〰resin
1. –Fmoc | piperidine
2. Wash: | DMF

H₂NCHRCO〰resin
3. Couple: | Fmoc—AA—OZ + DIPEA[a]
4. Wash: | DMF

Fmoc—NHCHRCO〰resin

a. OZ = active ester; DIPEA = diisopropylethylamine

43. M. Schnolzer, P. Alewood, A. Jones, D. Alewood, and S. B. H. Kent, *Int. J. Peptide Protein Res.*, **40**, 180 (1992); M. Schnolzer and S. B. H. Kent, *Science*, **256**, 221 (1992).

44. L. A. Carpino and G. Y. Han, *J. Org. Chem.*, **37**, 3404 (1972); G. B. Fields and R. L. Noble, *Int. J. Peptide Protein Res.*, **35**, 161 (1990); D. A. Wellings and E. Atherton, *Meth. Enzymol.*, **289**, 44 (1997); W. C. Chan and P. D. White, ed., *Fmoc Solid Phase Peptide Synthesis: A Practical Approach*, Oxford University Press, Oxford, 2000.

milder conditions for the final removal of the polypeptide from the solid support. The C-terminal amino acid is attached to the hydroxy group of the linker.

Wang linker[45]   Rink linker[46]

In the *t*-Boc protocol, the most common reagent for final removal of the peptide from the solid support is anhydrous hydrogen fluoride. Although this is a hazardous reagent, commercial systems designed for safe handling are available. In the Fmoc protocol milder acidic reagents can be used for cleavage from the resin. The alkoxy-benzyl group at the linker can be cleaved by TFA. Often, a scavenger, such as thioanisole, is used to capture the cations formed by cleavage of *t*-Boc protecting groups from side-chain substituents.

At the present time, the coupling is usually done via an activated ester (see Section 3.4). The coupling reagent and one of several *N*-hydroxy heterocycles are first allowed to react to form the activated ester, followed by coupling with the depro-tected amino group. The most frequently used compounds are *N*-hydroxysuccinimide, 1-hydroxybenzotriazole (HOBt), and 1-hydroxy-7-azabenzotriazole (HOAt).[47]

N-hydroxysuccinimide   HOBt   HOAt

Another family of coupling reagents frequently used with the Fmoc method is related to *N*-hydroxybenzotriazole and *N*-hydroxy 7-azabenzotriazole but also incorpo-rates phosphonium or amidinium groups. The latter can exist in either the O-(uronium) or *N*-(guanidinium) forms.[48] Both can effect coupling. The former are more reactive but isomerize to the latter. Which form is present depends on the protocol of preparation, including the amine used and the time before addition of the carboxylic acid.[49] The

45. S. Wang, *J. Am. Chem. Soc.*, **95**, 1328 (1993).
46. H. Rink, *Tetrahedron Lett.*, **28**, 3787 (1987); M. S. Bernatowicz, S. B. Daniels, and H. Koster, *Tetrahedron Lett.*, **30**, 4645 (1989); R. S. Garigipati, *Tetrahedron Lett.*, **38**, 6807 (1997).
47. F. Albericio and L. A. Carpino, *Meth. Enzymol.*, **289**, 104 (1997).
48. L. A. Carpino, H. Imazumi, A. El-Faham, F. J. Ferrer, C. Zhang, Y. Lee, B. M. Foxman, P. Henklein, C. Hanay, C. Muegge, H. Wenschuh, J. Klose, M. Beyermann, and M. Beinert, *Angew. Chem. Int. Ed. Engl.*, **41**, 441 (2002); T. K. Srivastava, W. Haq, S. Bhanumati, D. Velmurugan, U. Sharma, N. R. Jagannathan, and S. B. Katti, *Protein and Peptide Lett.*, **8**, 39 (2001).
49. L. A. Carpino and A. El-Faham, *Tetrahedron*, **55**, 6813 (1999); L. A. Carpino and F. J. Ferrer, *Org. Lett.*, **3**, 2793 (2001); F. Albericio, J. M. Bofill, A. El-Faham, and S. A. Kates, *J. Org. Chem.*, **63**, 9678 (1998).

phosphonium coupling reagents are believed to form acyloxyphosphonium species that can then be converted to the active ester incorporating the *N*-hydroxy heterocycle.[50]

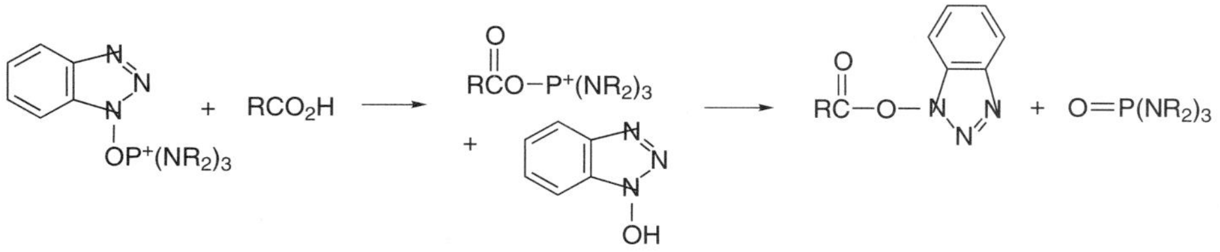

The structures and abbreviations of these reagents are given in Scheme 13.77.

The development of highly efficient protection-deprotection and coupling schemes has made the synthesis of polypeptides derived from the standard amino acids a highly efficient process. Additional challenges can come into play when other amino acids are involved. The HATU reagent, for example, has been applied to *N*-methyl amino acids, as in the case of cyclosporin A, an undecapeptide that is important in preventing transplant rejection. Seven of eleven amino acids are *N*-methylated. The synthesis of cyclosporin analogs has been completed by both solution and solid phase methods. Scheme 13.78 summarizes this synthesis. Fmoc protecting groups were used. Unlike the case of normal amino acids, quantitative coupling was not achieved, even when the coupling cycle was repeated twice for each step. Therefore, after each coupling cycle, a *capping step* using acetic anhydride was done to prevent carrying unextended material to the next phase. The final macrocyclization was done using propylphosphonic anhydride and DMAP, a reaction that presumably proceeds through a mixed phosphonic anhydride.[51]

Scheme 13.77. Phosphonium, Uronium, and Guanidinium Coupling Reagents

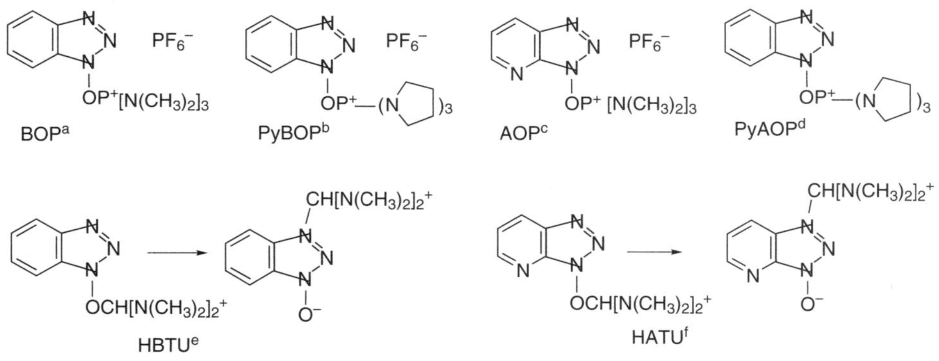

a. B. Castro, J. R. Dormoy, G. Evin, and C. Selve, *Tetrahedron Lett.*, 1219 (1975).
b. J. Coste, D. Le-Nguyen, and B. Castro, *Tetrahedron Lett.*, **31**, 205 (1990).
c. L. A. Carpino, A. El-Faban, C. A. Minor, and F. Albericio, *J. Chem. Soc., Chem. Commun.*, 201 (1994).
d. F. Albericio, M. Cases, J. Alsina, S. A. Triolo, L. A. Carpino, and S. A. Kates, *Tetrahedron Lett.*, **38**, 4853 (1997).
e. R. Knorr, A. Trezciak, W. Barnwarth, and D. Gillessen, *Tetrahedron Lett.*, **30**, 1927 (1989).
f. L. A. Carpino, *J. Am. Chem. Soc.*, **115**, 4397 (1993).

[50.] J. Coste, E. Frerot, and P. Jouin, *J. Org. Chem.*, **59**, 2437 (1994).
[51.] R. M. Wegner, *Helv. Chim. Acta*, **67**, 502 (1984); W. J. Colucci, R. D. Tung, J. A. Petri, and D. H. Rich, *J. Org. Chem.*, **55**, 2895 (1990).

Scheme 13.78. Synthesis of a Cyclosporin Analog by Solid Phase Peptide Synthesis[a.]

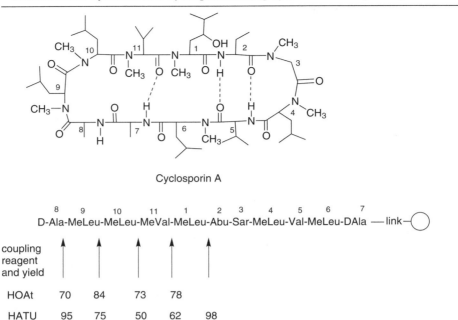

Cyclosporin A

	8	9	10	11	1	2	3	4	5	6	7	
D-Ala-MeLeu-MeLeu-MeVal-MeLeu-Abu-Sar-MeLeu-Val-MeLeu-DAla — link—◯

coupling reagent and yield					
HOAt	70	84	73	78	
HATU	95	75	50	62	98

a. Y. M. Angell, C. Garcia-Echeverria, and D. H. Rich, *Tetrahedron Lett.*, **35**, 5981 (1994); Y. M. Angell, T. L. Thomas, G. R. Flentke, and D. H. Rich, *J. Am. Chem. Soc.*, **117**, 7279 (1995). The analog contains *N*-methylleucine at position 1.

13.3.2. Solid Phase Synthesis of Oligonucleotides

Synthetic oligonucleotides are very important tools in the study and manipulation of DNA, including such techniques as site-directed mutagenesis and DNA amplification by the polymerase chain reaction. The techniques for chemical synthesis of oligonucleotides are highly developed. Very efficient automated methodologies based on solid phase synthesis are used extensively in fields that depend on the availability of defined DNA sequences.[52]

The construction of oligonucleotides proceeds from the four nucleotides by formation of a new phosphorus oxygen bond. The potentially interfering nucleophilic sites on the nucleotide bases are protected. The benzoyl group is usually used for the 6-amino group of adenosine and the 4-amino group of cytidine, whereas the *i*-butyroyl group is used for the 2-amino group of guanosine. These amides are cleaved by ammonia after the synthesis is completed. The nucleotides are protected at the 5′-hydroxy group as ethers, usually with the 4,4′-dimethoxytrityl (DMT) group.

In the early solution phase syntheses of oligonucleotides, coupling of phosphate diesters was used. A mixed 3′-ester with one aryl substituent, usually *o*-chlorophenyl, was coupled with a deprotected 5′-OH nucleotide. The coupling reagents were sulfonyl halides, particularly 2,4,6-tri-*i*-propylbenzenesulfonyl chloride,[53] and the reactions proceeded by formation of reactive sulfonate esters. Coupling conditions

52. S. L. Beaucage and M. H. Caruthers, in *Bioorganic Chemistry: Nucleic Acids*, S. M. Hecht, ed., Oxford University Press, Oxford, 1996, pp. 36–74.
53. C. B. Reese, *Tetrahedron*, **34**, 3143 (1978).

have subsequently been improved and a particularly effective coupling reagent is 1-mesitylenesulfonyl-3-nitrotriazole (MSNT).[54]

Mes = 2,4,6-trimethylphenyl

Current solid phase synthesis of oligonucleotides relies on coupling at the phosphite oxidation level. The individual nucleotides are introduced as phosphoramidites and the technique is called the *phosphoramidite method*.[55] The *N,N*-diisopropyl phosphoramidites are usually used. The third phosphorus substituent is methoxy or 2-cyanoethoxy. The cyanoethyl group is easily removed by mild base (β-elimination) after completion of the synthesis. The coupling is accomplished by tetrazole, which displaces the amine substituent to form a reactive phosphite that undergoes coupling. After coupling, the phosphorus is oxidized to the phosphoryl level by iodine or another oxidant. The most commonly used protecting group for the 5′-OH is the 4,4′-dimethoxytrityl group (DMT), which is removed by mild acid. The typical cycle of deprotection, coupling, and oxidation is outlined in Scheme 13.79. One feature of oligonucleotide synthesis is the use of a *capping step*, an acetylation that follows coupling, the purpose of which is to permanently block any 5′-OH groups that were not successfully coupled. This prevents the addition of a nucleotide at the site in the succeeding cycle, terminates the further growth of this particular oligonucleotide, and avoids the synthesis of oligonucleotides with single-base deletions. The capped oligomers are removed in the final purification.

Silica or porous glass is usually used as the solid phase in oligonucleotide synthesis. The support is functionalized through an amino group attached to the silica surface. There is a secondary linkage through a succinate ester to the terminal 3′-OH group.

Although use of automated oligonucleotide synthesis is widespread, work continues on the optimization of protecting groups, coupling conditions, and deprotection methods, as well as on the automated devices.[56]

54. J. B. Chattapadyaya and C. B. Reese, *Tetrahedron Lett.*, **20**, 5059 (1979).
55. R. L. Letsinger and W. B. Lunsford, *J. Am. Chem. Soc.*, **98**, 3655 (1976); S. L. Beaucage and M. H. Caruthers, *Tetrahedron Lett.*, **22**, 1859 (1981); M. H. Caruthers, *J. Chem. Ed.*, **66**, 577 (1989); S. L. Beaucage and R. P. Iyer, *Tetrahedron*, **48**, 2223 (1992).
56. G. A. Urbina, G. Grubler, A. Weiber, H. Echner, S. Stoeva, J. Schernthaner, W. Gross, and W. Voelter, *Z. Naturforsch.*, **B53**, 1051 (1998); S. Rayner, S. Brignac, R. Bumeiester, Y. Belosludtsev, T. Ward, O. Grant, K. O'Brien, G. A. Evans, and H. R. Garner, *Genome Res.*, **8**, 741 (1998).

Scheme 13.79. Protocol for Automated Solid-Phase Synthesis of Oligonucleotides[a]

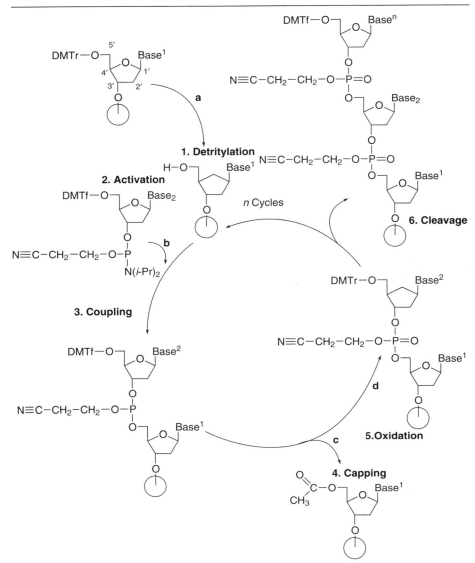

Reagents: **a**: 3% Cl_3CCO_2H in CH_2Cl_2; **b**: 3% tetrazole in CH_3CN; c_1: 10% Ac_2O and 10% 2,6-lutidine in THF; c_2: 7% 1-methylimidazole in THF; **d**: 3% I_2, 2 % H_2O, 2% pyridine in THF.

a. G. A. Urbina, G. Gruebler, A. Weiler, H. Echner, S. Stoeva, J. Schernthaler, W. Grass, and W. Voelter, *Z. Naturforsch.* **B53**, 1051 (1998).

13.4. Combinatorial Synthesis

Over the past decade the techniques of *combinatorial synthesis* have received much attention. Solid phase synthesis of polypeptides and oligonucleotides are especially adaptable to combinatorial synthesis, but the method is not limited to these fields. The goal of combinatorial synthesis is to prepare a large number of related molecules by carrying out a synthetic sequence with several closely related starting materials and reactants. For example, if a linear three-step sequence is done with eight related reactants at each step, a total of 4096 different products are obtained. The product of each step is split into equal portions for the next series of reactions.

$$A\ (8) + B\ (8) \xrightarrow{\text{Step 1}} A\!\!-\!\!B\ (64) \xrightarrow{\text{Step 2}} A\!\!-\!\!B\!\!-\!\!C\ (512) \xrightarrow{\text{Step 3}} A\!\!-\!\!B\!\!-\!\!C\!\!-\!\!D\ (4096)$$
$$\qquad\qquad\qquad\qquad C\ (8) \qquad\qquad\qquad\qquad D\ (8)$$

The objective of traditional multistep synthesis is the preparation of a single pure compound, but combinatorial synthesis is designed to make many related molecules.[57] The purpose is often to have a large collection (library) of compounds for evaluation of biological activity. A goal of combinatorial synthesis is *structural diversity*, that is, systematic variation in subunits and substituents so as to explore the effect of a range of structural entities. In this section, we consider examples of the application of combinatorial methods to several kinds of compounds.

One approach to combinatorial synthesis is to carry out a series of conventional reactions in parallel with one another. For example, a matrix of six starting materials, each treated with eight different reactants will generate 48 reaction products. Splitting each reaction mixture and using a different reactant for each portion can further expand the number of final compounds. However, relatively little savings in effort is achieved by running the reactions in parallel, since each product must be separately isolated and purified. The reaction sequence below was used to create a 48-component library by reacting six amines with each of eight epoxides. Several specific approaches were used to improve the purity of the product and maximize the efficiency of the process. First, the amines were monosilylated to minimize the potential for interference from dialkylation of the amine. The purification process was also chosen to improve efficiency. Since the desired products are basic, they are retained by acidic ion exchange resins. The products were absorbed on the resin and nonbasic impurities were washed out, followed by elution of the products by methanolic ammonia.[58]

A considerable improvement in efficiency can be achieved by solid phase synthesis.[59] The first reactant is attached to a solid support through a linker group, as was described for polypeptide and oligonucleotide synthesis. The individual reaction steps are then conducted on the polymer-bound material. Use of solid phase methodology has several advantages. Excess reagents can be used to drive individual steps to completion and obtain high yields. The purification after each step is also simplified, since excess reagents and by-products are simply rinsed from the solid support. The process can be automated, greatly reducing the manual effort required.

When solid phase synthesis is combined with sample splitting, there is a particularly useful outcome.[60] The solid support can be used in the form of small beads, and

[57]. A. Furka, *Drug Dev. Res.*, **36**, 1 (1995).

[58]. A. J. Shuker, M. G. Siegel, D. P. Matthews, and L. O. Weigel, *Tetrahedron Lett.*, **38**, 6149 (1997).

[59]. A. R. Brown, P. H. H. Hermkens, H. C. J. Ottenheijm, and D. C. Rees, *Synlett*, 817 (1998).

[60]. A. Furka, F. Sebestyen, M. Asgedon, and G. Dibo, *Int. J. Peptide Protein Res.*, **37**, 487 (1991); K. S. Lam, M. Lebl, and V. Krchnak, *Chem. Rev.*, **97**, 411 (1997).

the starting point is a collection of beads, each with one initial starting material. After each reaction step the beads are recombined and divided again. As the collection of beads is split and recombined during the combinatorial synthesis, each bead acquires a particular compound, depending on its history of exposure to the reagents, *but every bead in a particular split has the same compound, since their reaction histories are identical.* Figure 13.1 illustrates this approach for three steps, each using three different reactants. However, in the end all of the beads are together and there must be some means of establishing the identity of the compound attached to any particular bead. In some cases it is possible to detect compounds with the desired property while they are still attached to the bead. This is true for some assays of biological or catalytic activity that can be performed under heterogeneous conditions.

Another approach is to tag the beads with identifying markers that encode the sequence of reactants and thus the structure of the product attached to a particular bead.[61] One method of coding involves attachment of a chemically identifiable tag,

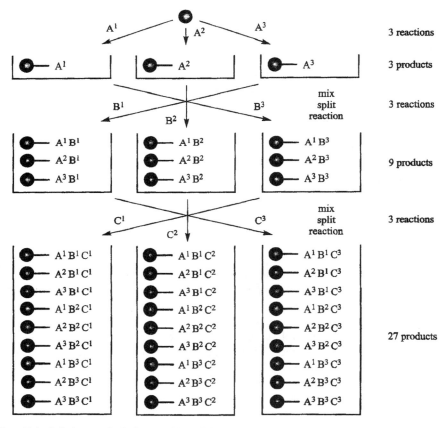

Fig. 13.1. Splitting method for combinatorial synthesis on solid support. Reproduced from F. Balkenhohl, C. von dem Bussche-Huennefeld, A. Lansky, and C. Zechel, *Angew. Chem. Int. Ed. Engl.*, **35**, 2288 (1996), by permission of Wiley-VCH.

61. S. Brenner and R. A. Lerner, *Proc. Natl. Acad. Sci. USA*, **89**, 5381 (1993).

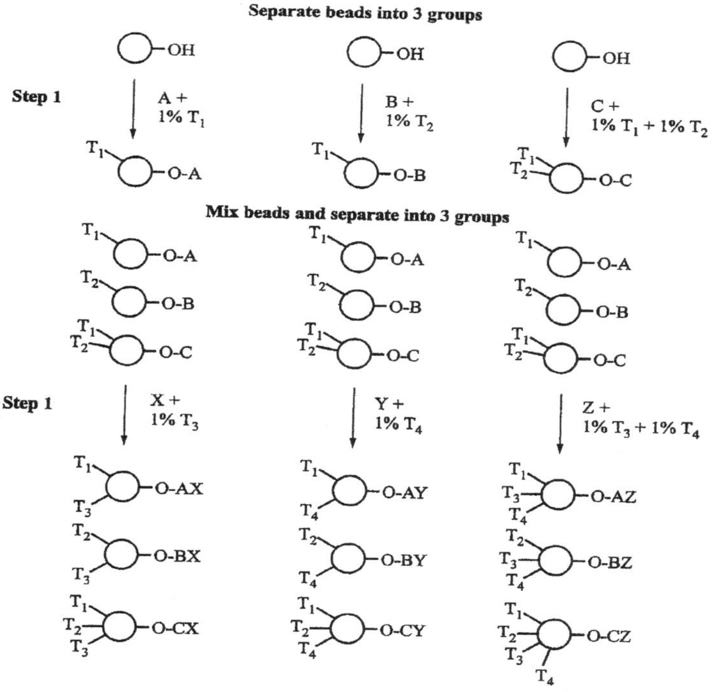

Fig. 13.2. Use of chemical tags to encode the sequence in a combinatorial synthesis on a solid support. Reproduced from W. C. Still, *Acc. Chem. Res.*, **29**, 155 (1996), by permission of the American Chemical Society.

as illustrated in Figure 13.2.[62] After each combinatorial step, a different chemical tag is applied to each of the splits before they are recombined. The tags used for this approach are a series of chlorinated aromatic ethers that can be detected and identified by mass spectrometry. The tags are attached to the polymer support by a Rh-catalyzed carbenoid insertion reaction. Detachment is done by oxidizing the methoxyphenyl linker with CAN. Any bead that shows interesting biological activity can then be identified by analyzing the code provided by the chemical tags for that particular bead.

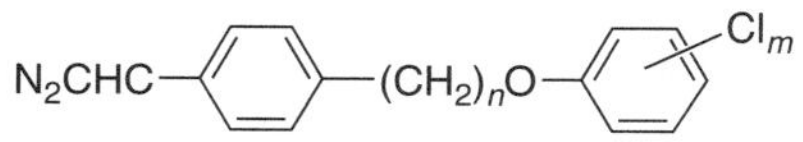

$$n = 2\text{--}11$$
$$m = 2\text{--}5$$

Combinatorial approaches can be applied to the synthesis of any type of molecule that can be built up from a sequence of individual components, for example, in reactions forming heterocyclic rings.[63] The equations below represent an approach to preparing differentially substituted indoles.

[62.] H. P. Nestler, P. A. Bartlett, and W. C. Still, *J. Org. Chem.*, **59**, 4723 (1994); C. Barnes, R. H. Scott, and S. Babasubramanian, *Recent Res. Develop. Org. Chem.*, **2**, 367 (1998).

[63.] A. Netzi, J. M. Ostresh, and R. A. Houghten, *Chem. Rev.*, **97**, 449 (1997).

Ref. 64

There is nothing to prevent incorporation of additional diversity by continuing to build on a side chain at one of the substituent sites.

Another kind of combinatorial synthesis can be applied to reactions that assemble the product from several components in a single step, a *multicomponent reaction*. A particularly interesting four-component reaction is the *Ugi reaction*, which generates dipeptides from an isocyanide, an aldehyde, an amine, and a carboxylic acid.

For example, use of 10 different isocyanides and amines, along with 40 different aldehydes and carboxylic acids has the potential to generate 160,000 different dipeptide analogs.[65] This system was explored by synthesizing arbitrarily chosen sets of 20 compounds that were synthesized in parallel. The biological assay data from these 20 combinations were then used to select the next 20 combinations for synthesis. The synthesis-assay-selection process was repeated 20 times. At the end of this process the average inhibitory concentration of the set of 20 products had been decreased from $1\,mM$ to less than $1\,\mu M$.

A library of over 3000 spirooxindoles was created based on a sequence of four reactions.[66] The synthetic sequence is based on the total synthesis of a natural product called (−)-spirotryprostatin B.[67] A morpholinone chiral auxiliary, aldehyde, and an oxindole condense to give the ring system. Substituents were then added by replacement of the iodine by one of several terminal alkynes. Simultaneous deprotection occurred at the allyl ester. These carboxylic acids were converted to amides using a variety of amines and coupling with PyBOP. The final reaction in the sequence was acylation of the oxindole nitrogen. At each stage in the library creation, certain alkynes or amines reacted poorly and were excluded from the library, which was eventually derived from eight alkynes, twelve amines, and four acylation reagents. As outlined in Scheme 13.80, this synthesis has the potential to prepare 3104 different

[64] H.-C. Zhang and B. E. Maryanoff, *J. Org. Chem.*, **62**, 1804 (1997).

[65] L. Weber, S. Walbaum, C. Broger, and K. Gubernator, *Angew. Chem. Int. Ed. Engl.*, **34**, 2280 (1995).

[66] M. M.-C. Lo, C. S. Neumann, S. Nagayama, E. O. Perlstein, and S. L. Schreiber, *J. Am. Chem. Soc.*, **126**, 16077 (2004).

[67] P. R. Sebahar, H. Osada, T. Usui, and R. M. Williams, *Tetrahedron*, **58**, 6311 (2002).

Scheme 13.80. Creation of a Combinatorial Library of Spirooxindoles[a]

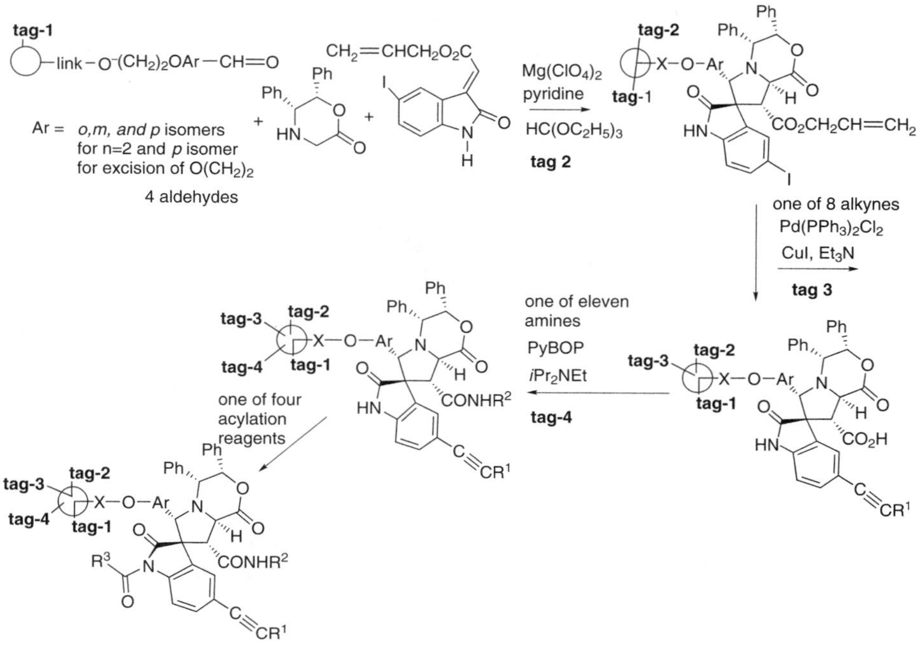

a. M. M.-C. Lo, C. S. Neumann, S. Nagayama, E. O. Perlstein, and S. L. Schreiber, *J. Am. Chem. Soc.*, **126**, 16077 (2004).

Scheme 13.81. Combinatorial Synthesis of Epothilone Analogs Using Microreactors[a]

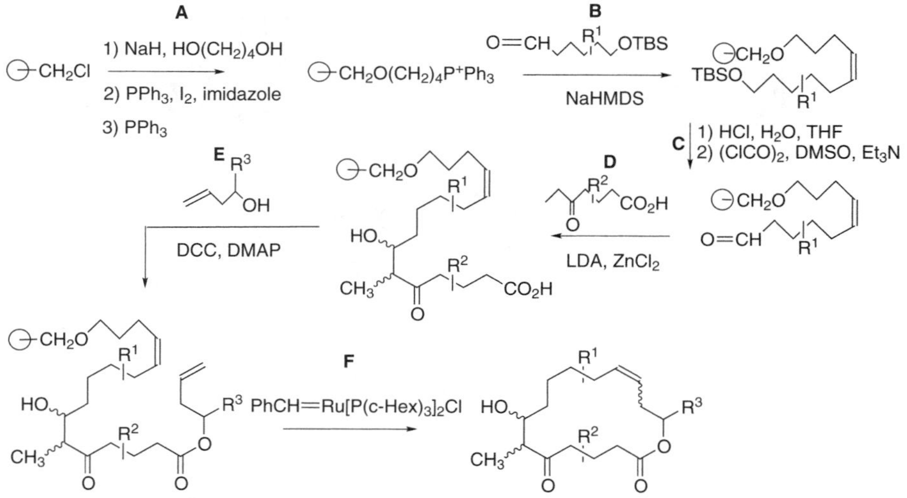

a. K. C. Nicolaou, D. Vorloumis, T. Li, J. Pastor, N. Winssinger, Y. He, S. Ninkovic, F. Sarabia, H. Vallberg, F. Roschanger, N. P. King, M. R. V. Finlay, P. Giannakakou, D. Verdier-Pinard, and E. Hamel, *Angew. Chem. Int. Ed. Engl.*, **36**, 2097 (1997).

compounds, including those lacking a particular substituent (skip) (4 aldehydes $\times 2$ morpholines $\times 1$ oxindole)= 8 core structures $\times (1 + 8 \times 12) \times 4 = 3104$ different compounds. A version of the chemical tagging method was used for coding the beads.[68] Analysis of a sample of the beads indicated that at least 82% of them contained the desired compound in greater than 80% purity.

The epothilone synthesis in Scheme 13.59 (p. 1221) has been used as the basis for a combinatorial approach to epothilone analogs.[69] The acyclic precursors were

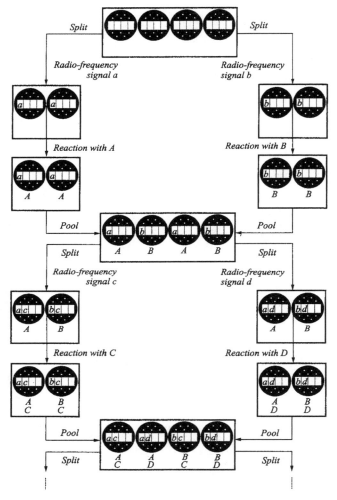

Fig. 13.3. Radio-frequency tagging of microreactors for combinatorial synthesis on a solid support. Reproduced from K. C. Nicolaou, X.-Y. Xiao, Z. Parandoosh, A. Senyei, and M. P. Nova, *Angew. Chem. Int. Ed. Engl.*, **34**, 2289 (1995), by permission of Wiley-VCH.

[68] H. B. Blackwell, L. Perez, R. A. Stavenger, J. A. Tallarico, E. Cope-Etough, M. A. Foley, and S. L. Schreiber, *Chem. Biol.*, **8**, 1167 (2001).

[69] K. C. Nicolaou, N. Wissinger, J. Pastor, S. Ninkovic, F. Sarabia, Y. He, D. Vourloumis, S. Yang, T. Li, P. Giannakakou, and E. Hamel, *Nature*, **387**, 268 (1997); K. C. Nicolaou, D. Vourloumis, T.

synthesized and attached to a solid support resin by Steps **A** and **B** in Scheme 13.81. The cyclization and disconnection from the resin was then done by the olefin metathesis reaction in Step **F**. The aldol condensation in Step **D** was not highly stereoselective. Similarly, olefin metathesis gave a mixture of *E*- and *Z*-stereoisomers, so the product of each combinatorial sequence was a mixture of four isomers. These were separated by thin-layer chromatography prior to bioassay. In this project, reactants **A** (three variations), **B** (three variations), and **C** (five variations) were used, generating 45 possible combinations. The stereoisomeric products increase this to 180 (45 × 4).

In this study a nonchemical means of encoding the identity of each compound was used. The original polymer-bound reagent was placed in a porous microreactor that is equipped with a radiofrequency device that can be used for identification.[70] The porous microreactors permit reagents to diffuse into the polymer-bound reactants, but the polymer cannot diffuse out. At each split, the individual microreactors are coded to identify the reagent that is used. When the synthesis is complete, the sequence of signals recorded in the radiofrequency device identifies the product that has been assembled in that particular reactor. Figure 13.3 illustrates the principle of this coding method.

General References

Protective Groups

T. Greene and P. G. M. Wuts, *Protective Groups in Organic Synthesis*, 3rd Edition, John Wiley & Sons, New York, 1999.
P. J. Kocienski, *Protecting Groups*, G. Thieme, Stuttgart, 1994.
J. F. W. McOmie, ed., Protective Groups in Organic Synthesis, Plenum Press New York, 1973.

Synthetic Equivalents

T. A. Hase, ed., *Umpoled Synthons: A Survey of Sources and Uses in Synthesis*, John Wiley & Sons, New York, 1987.
A. Dondoni, ed., *Advances in the Use of Synthons in Organic Chemistry*, Vols. 1–3, JAI Press, Greenwich, CT, 1993-1995.

Synthetic Analysis and Planning

R. K. Bansal, *Synthetic Approaches to Organic Chemistry*, Jones and Bartlett, Sudbury, MA, 1998.
E. J. Corey and X.-M Chang, *The Logic of Chemical Synthesis*, John Wiley & Sons, New York, 1989.
J.-H. Furhop and G. Penzlin, *Organic Synthesis: Concepts, Methods, and Starting Materials*, Verlag Chemie, Weinheim, 1983.
T.-L. Ho, *Tactics of Organic Synthesis*, John Wiley & Sons, New York, 1994.
T.-L. Ho, *Tandem Organic Reactions*, John Wiley & Sons, New York, 1992.
T. Mukaiyama, *Challenges in Synthetic Organic Chemistry*, Claredon Press, Oxford, 1990.

Li, J. Pastor, N. Wissinger, Y. He, S. Ninkovic, F. Sarabia, H. Vallberg, F. Roschangar, N. P. King, M. R. V. Finlay, P. Giannakakou, D. Verdier-Pinard, and E. Hamel, *Angew. Chem. Int. Ed. Engl.*, **36**, 2097 (1997).
70. K. C. Nicolaou, Y.-Y. Xiao, Z. Parandoosh, A. Senyei, and M. P. Nova, *Angew. Chem. Int. Ed. Engl.*, **34**, 2289 (1995); E. J. Moran, S. Sarshar, J. F. Cargill, M. M. Shahbaz, A. Lio, A. M. M. Mjalli, and R. W. Armstrong, *J. Am. Chem. Soc.*, **117**, 10787 (1995).

F. Serratosa and J. Xicart, *Organic Chemistry in Action: The Design of Organic Synthesis*, Elsevier, New York, 1996.

W. A. Smit, A. F. Bochkov, and R. Caple, *Organic Synthesis: The Science Behind the Art*, Royal Society of Chemistry, Cambridge, 1998.

B. M. Trost, editor-in-chief, *Comprehensive Organic Synthes: Selectivity, Strategy, and Efficiency in Modern Organic Chemistry*, Pergamon Press, New York, 1991.

S. Warren, *Organic Synthesis: The Disconnection Approach*, John Wiley & Sons, New York, 1982.

Stereoselective Synthesis

R. S. Atkinson, *Stereoselective Synthesis*, John Wiley & Sons, New York, 1995.

G. M. Coppola and H. F. Schuster, *Asymmetric Synthesis*, Wiley-Interscience, New York, 1987.

S. Hanessian, *Total Synthesis of Natural Products: The Chiron Approach*, Pergamon Press, New York, 1983.

S. Nogradi, *Stereoselective Syntheses*, Verlag Chemie, Weinheim, 1987.

G. Procter, *Stereoselectivity in Organic Synthesis*, Oxford University Press, Oxford, 1998.

Descriptions of Total Syntheses

N. Anand, J. S. Bindra, and S. Ranganathan, *Art in Organic Synthesis*, 2nd Edition, Wiley-Interscience, New York, 1988.

J. ApSimon, ed., *The Total Synthesis of Natural Products*, Vols. 1–9, Wiley-Interscience, New York, 1973–1992.

S. Danishefsky and S. E. Danishefsky, *Progress in Total Synthesis*, Meredith, NY, 1971.

I. Fleming, *Selected Organic Syntheses*, Wiley-Interscience, New York, 1973.

K. C. Nicolaou and E. J. Sorensen, *Classics in Total Synthesis: Targets, Strategies and Methods*, VCH Publishers, New York, 1996.

Solid Phase Synthesis

K. Burgess, *Solid Phase Organic Synthesis*, John Wiley & Sons, New York, 2000.

Problems

(References for these problems will be found on page 1292.)

13.1. Show how synthetic equivalent groups could be used to carry out each of the following transformations:

(a)

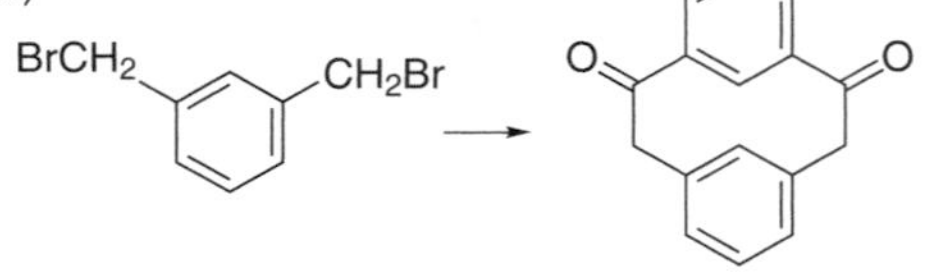

(b)

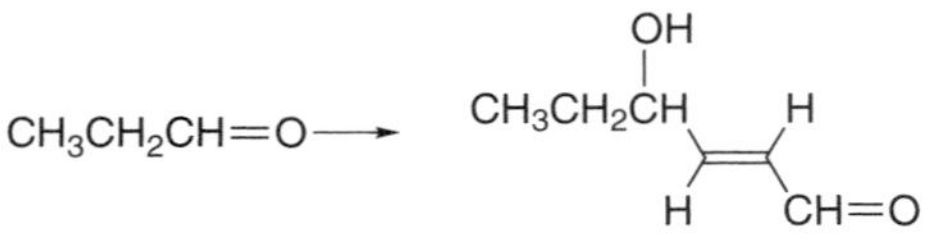

(c)

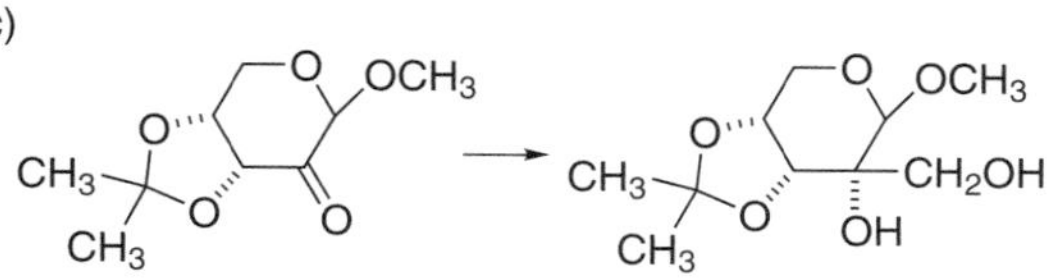

(d)

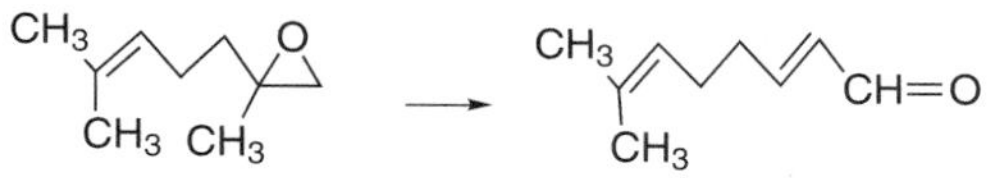

(e)

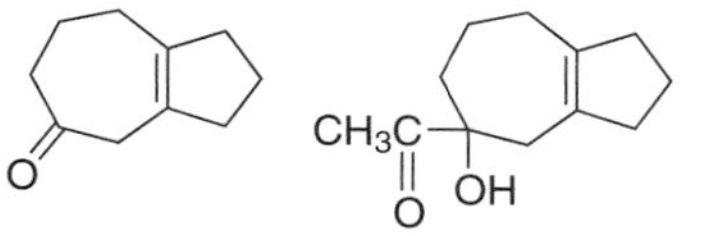

(f)

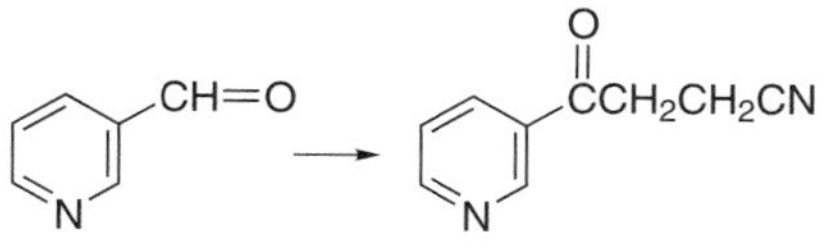

(g)

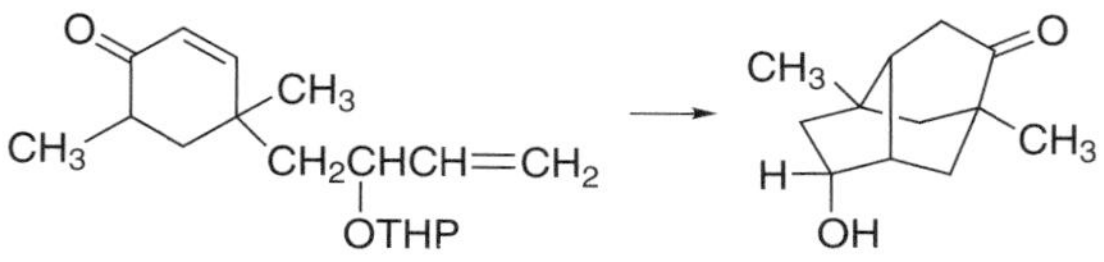

(h)

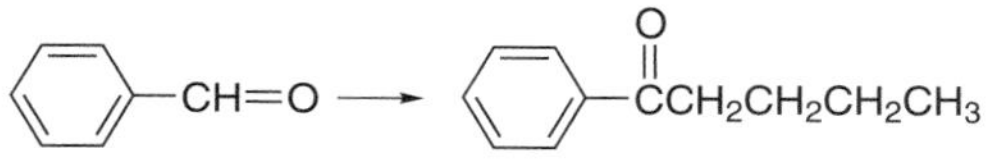

(i)

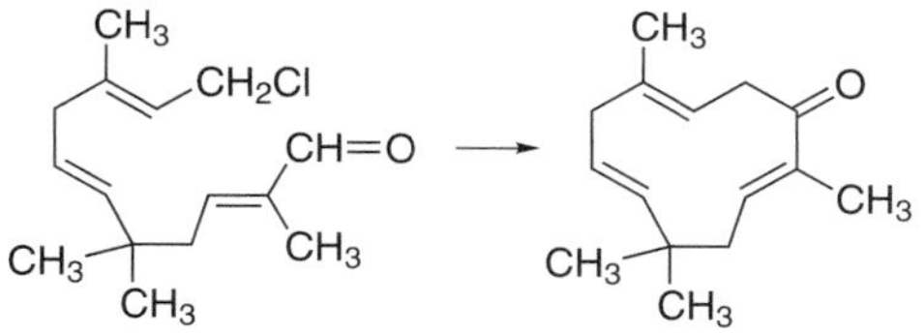

13.2. Indicate a reagent or short synthetic sequence that would accomplish each of the following transformations:

(a)

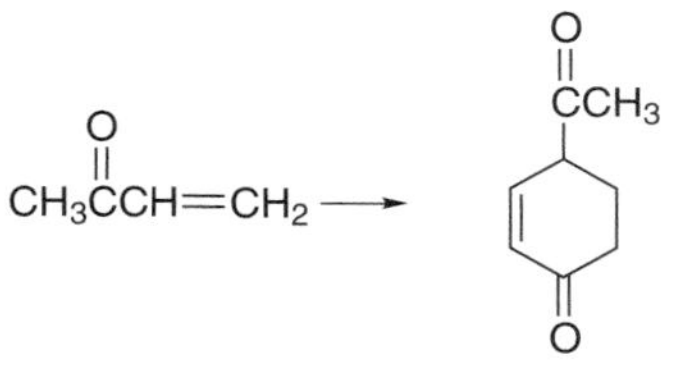

(b)

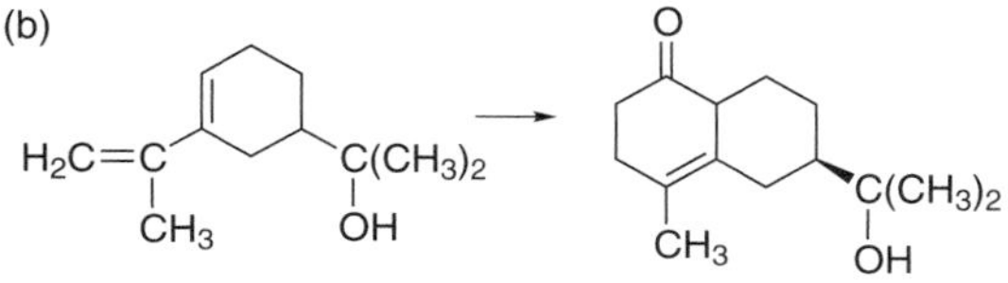

(c)

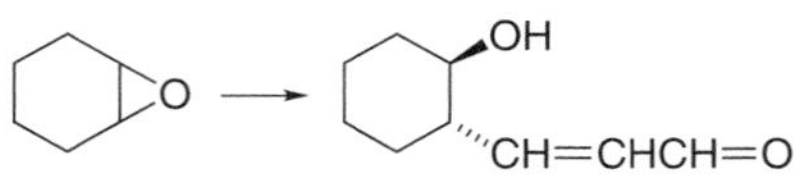

(d)

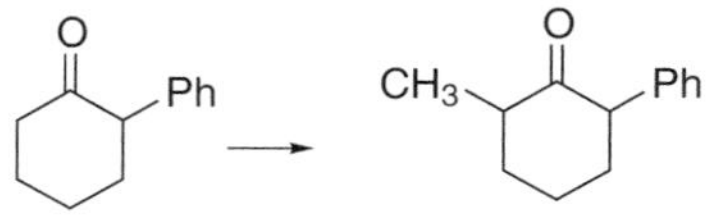

(e

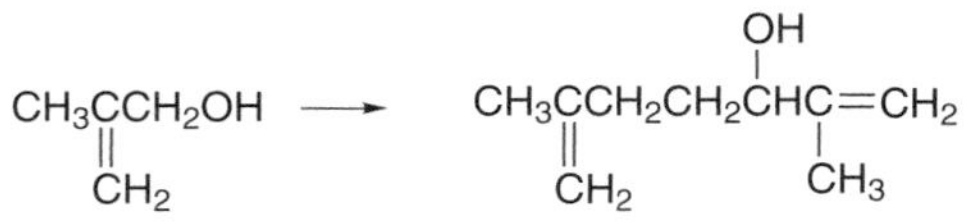

(f)

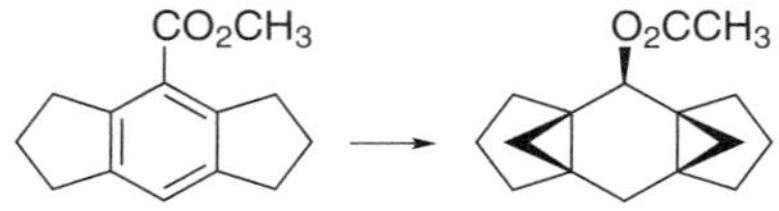

(g)

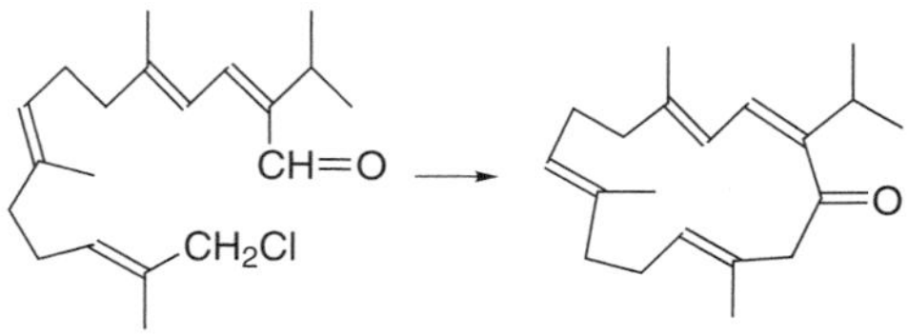

13.3. Indicate reagents or short reaction sequences that could accomplish the synthesis of the target on the left from the starting material on the right.

(a)

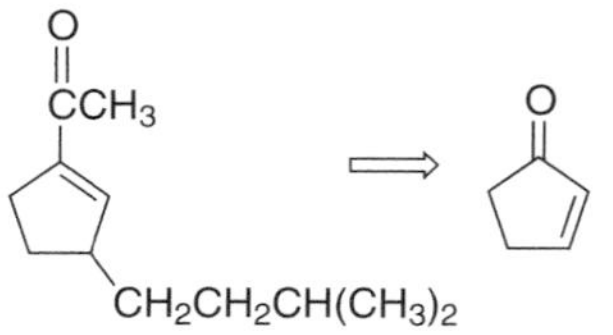

(b)

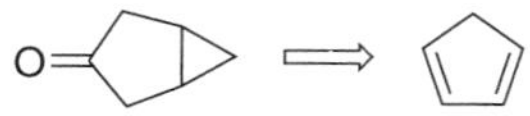

(c)

(d)

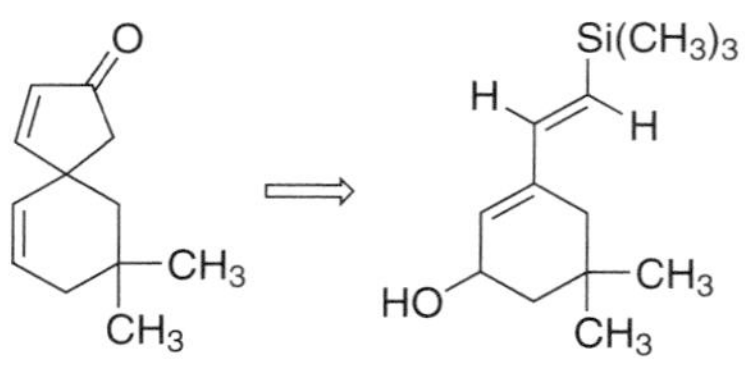

(e)

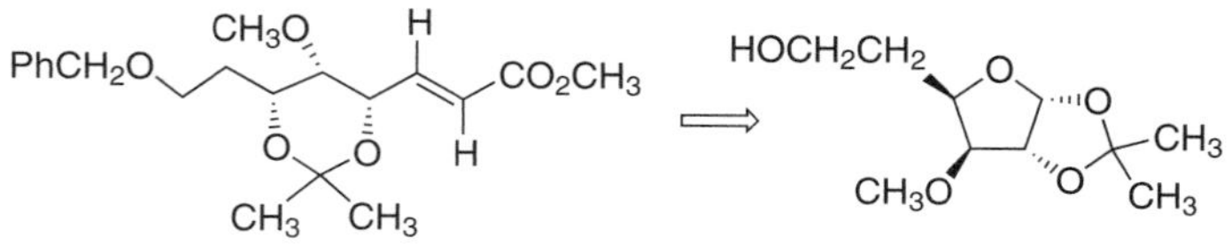

(f)

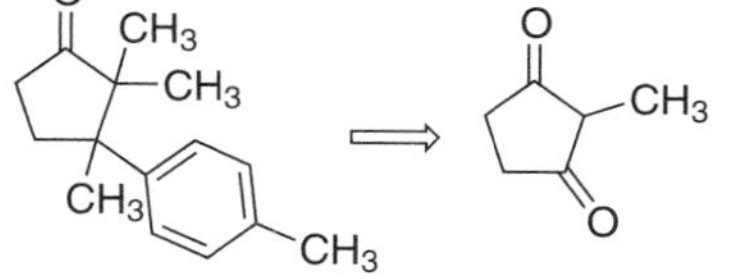

13.4. As they are available from natural sources in enantiomerically pure form, carbohydrates are useful starting materials for syntheses of enantiomerically pure compounds. However, the multiple hydroxy groups require versatile methods for selective protection, reaction, and deprotection. Show how appropriate manipulation of protecting groups and/or selective reagents could be used to effect the following transformations.

(a)

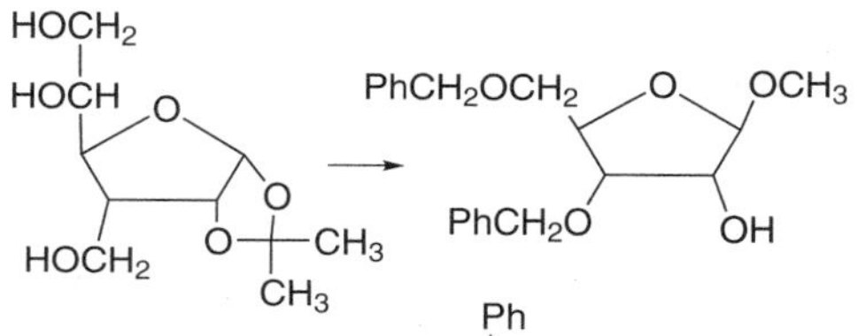

(b)

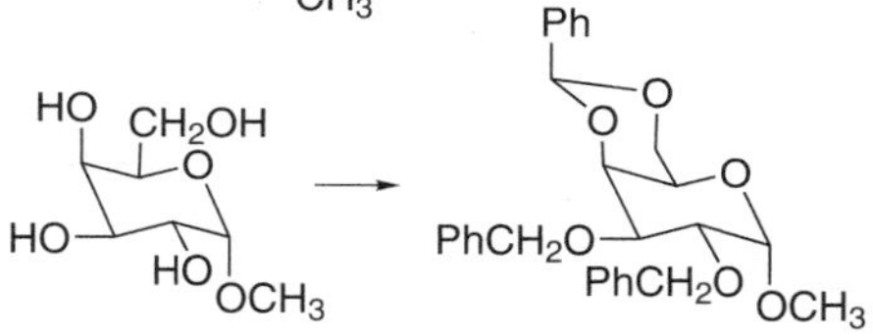

(c)

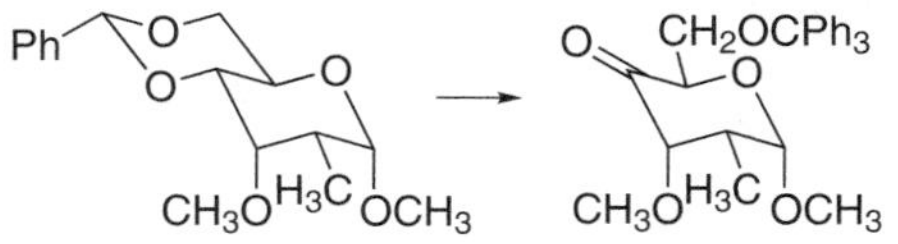

(d)

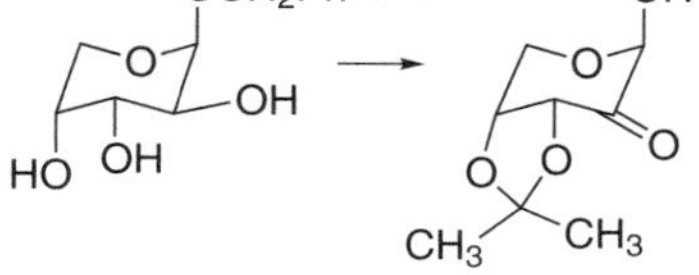

13.5. Several synthetic transformations that are parts of total syntheses of natural products are summarized by retrosynthetic outlines. For each retrosynthetic transform suggest a reagent or short reaction sequence that could accomplish the forward synthetic conversion. The proposed route should be diastereoselective but need not be enantioselective.

(a)

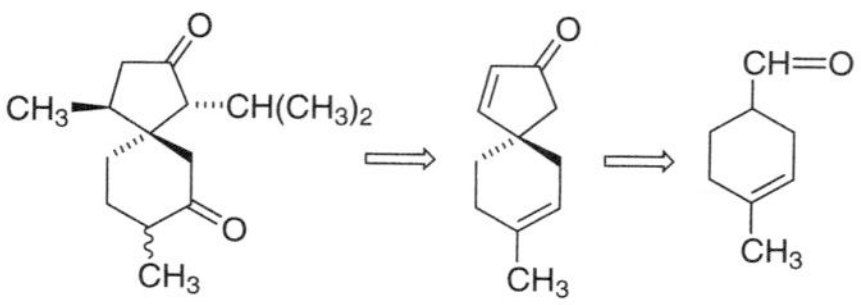

(b)

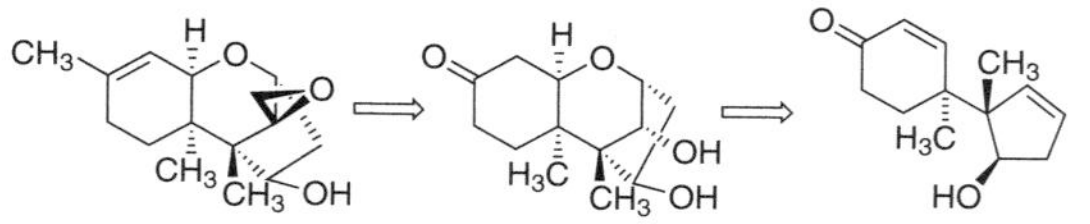

(c)

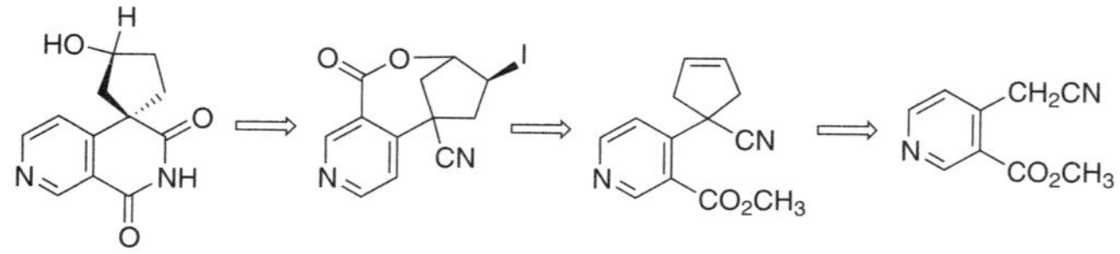

(d)

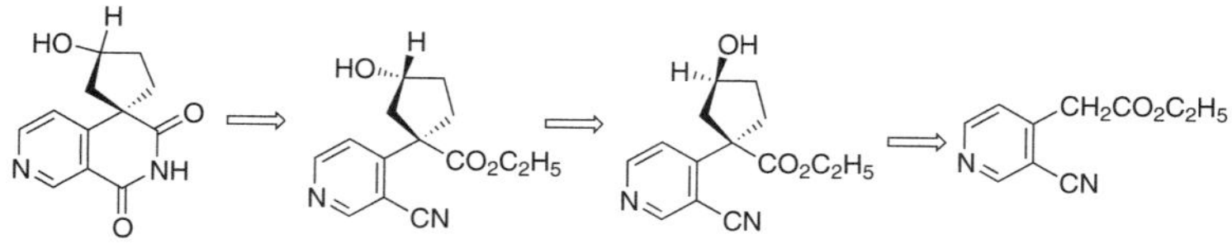

(e)

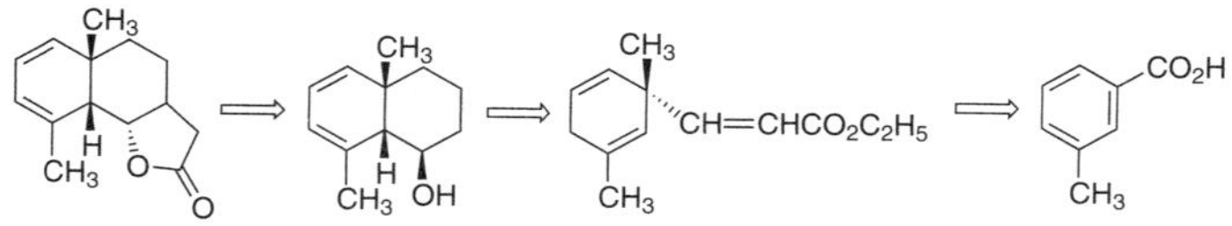

13.6. Diels-Alder reactions are attractive for synthetic application because of the predictable regio- and stereochemistry. There are, however, limitations on the types of compounds that can serve as dienophiles or dienes. As a result, the idea of synthetic equivalence has been exploited by development of dienophiles and dienes that meet the reactivity requirements of the Diels-Alder reaction and can then be converted to the desired structure. For each of the dienophiles and dienes given below, suggest a Diels-Alder reaction and subsequent transformation(s) that would give a product not directly attainable by a Diels-Alder reaction. Give the structure of the diene or dienophile "synthetic equivalent" and indicate why the direct Diels-Alder reaction is not possible.

Dienophiles

(a) $CH_2{=}CHP^+Ph_3$

(b) $CH_2{=}CHSPh$, with $=O$

(c) $CH_2{=}CCO_2C_2H_5$, O_2CCH_3

(d) $CH_2{=}CCCH_3$ with O double bond, O_2CAr Ar = 4-nitrophenyl

Dienes

(e) $CH_2{=}CCH{=}CHOCH_3$, $OSi(CH_3)_3$

(f) $CH_2{=}CC{=}CH_2$, SPh, O_2CCH_3

13.7. One approach to the synthesis of enantiomerically pure compounds is to start with an available enantiomerically pure substance and effect the synthesis by a series of enantiospecific reactions. Devise a sequence of reactions that would be appropriate for the following syntheses based on enantiomerically pure starting materials.

(a)

(b)

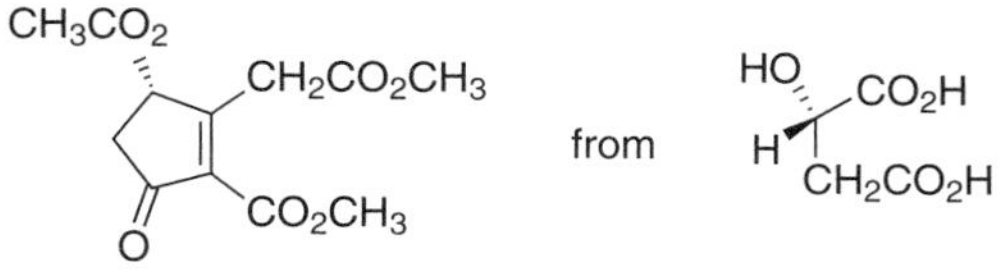

(c)

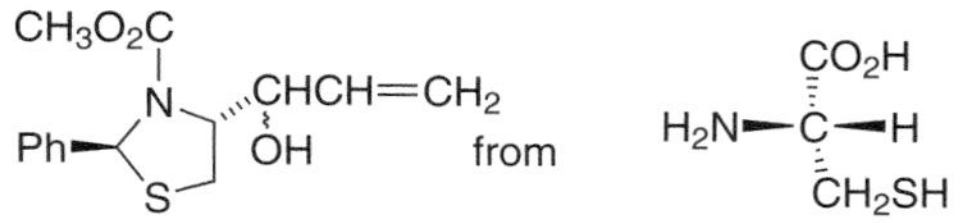

(d)

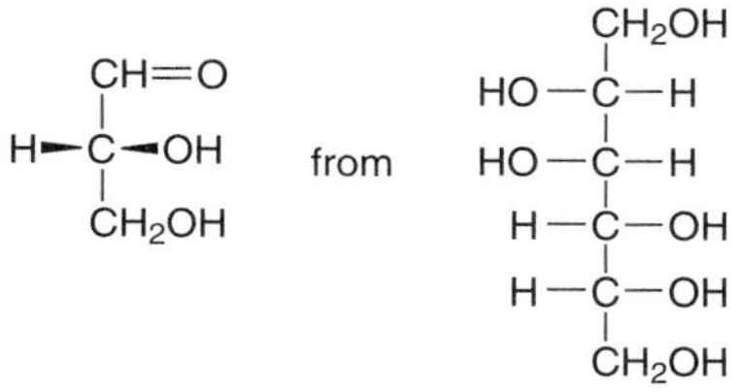

(e)

(f)

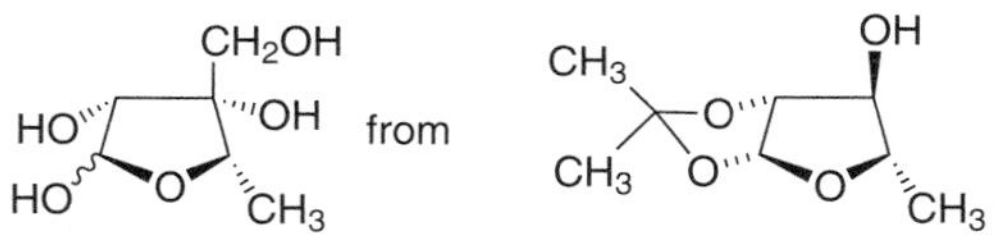

(g)

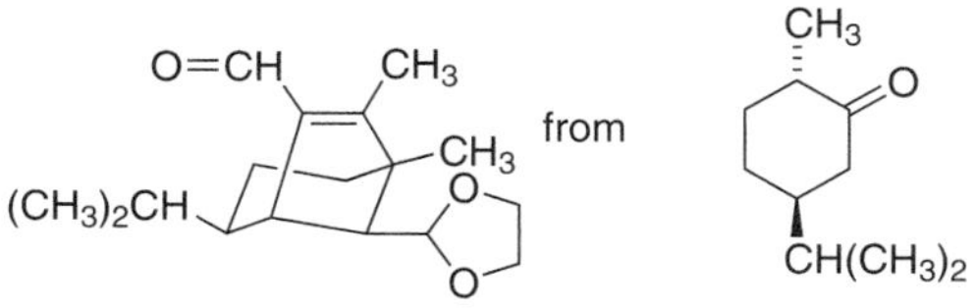

(h)

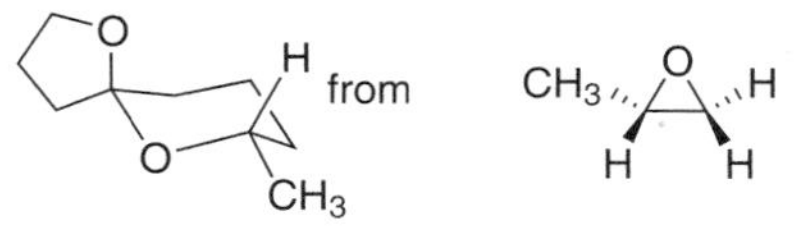

13.8. Several syntheses of terpenoids are outlined in retrosynthetic form. Suggest a reagent or short reaction sequence that could accomplish each lettered transformation in the synthetic direction. The structures refer to racemic material.

(a) Isotelekin

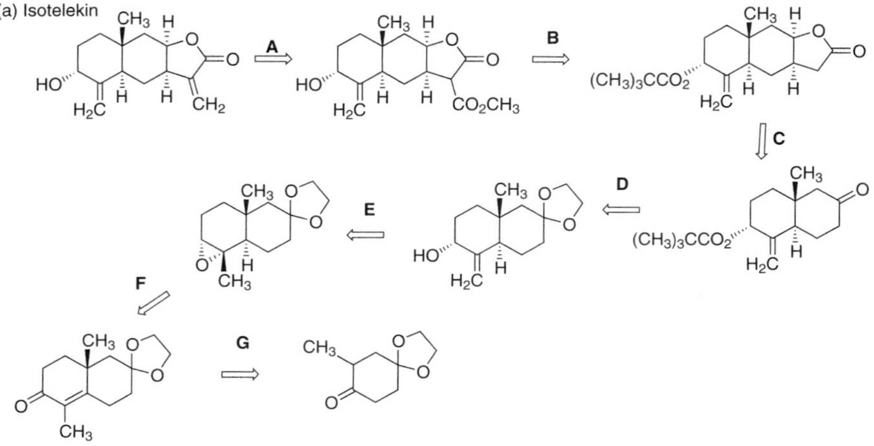

(b) Aromandrene

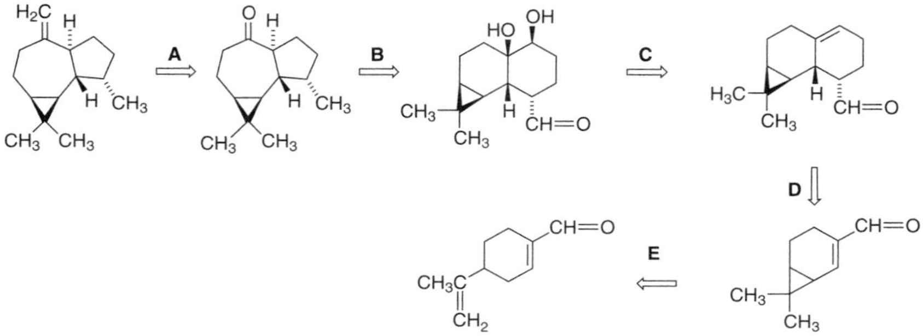

(c) α-Bourbonene

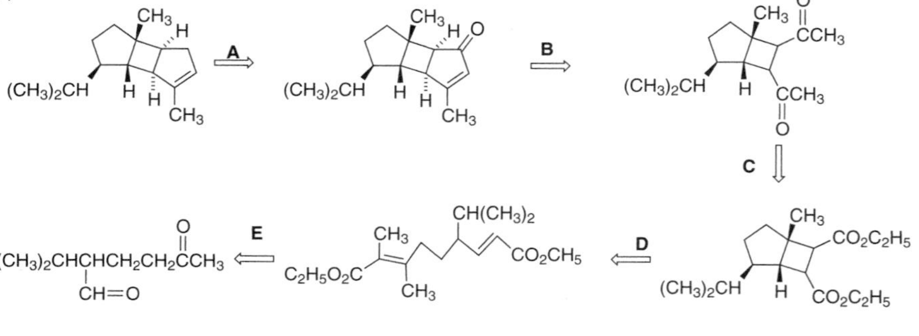

(d) Caryophyllene

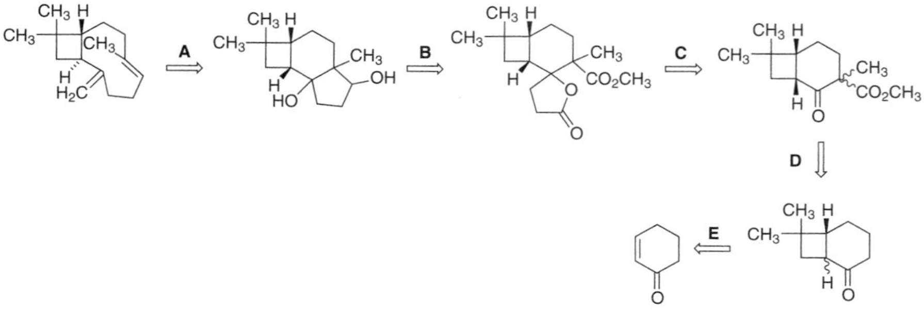

13.9 Use retrosynthetic analysis to suggest syntheses of the following compounds. Develop at least three outline schemes. Discuss the relative merits of the schemes and develop a fully elaborated synthetic plan for the most promising retrosynthetic scheme.

(a)

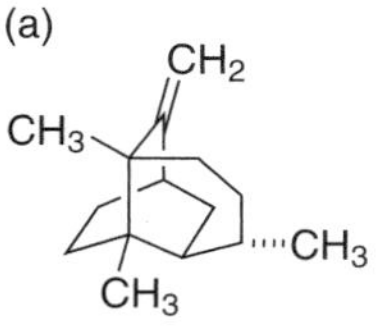

seychellene

(b)

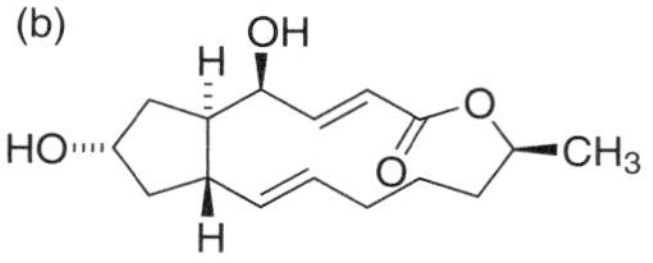

brefeldin A

(c)

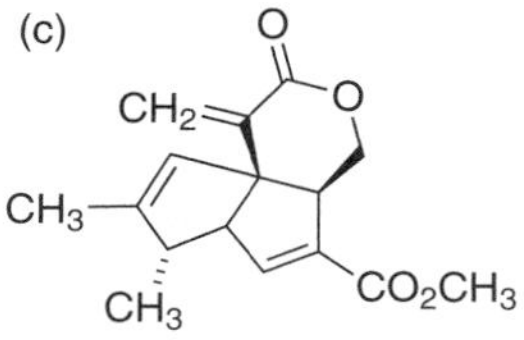

pentalenolactone E

13.10. Suggest a method for diastereoselective synthesis of the following compounds:

(a)

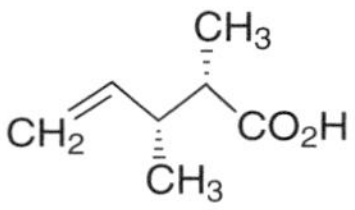

(b)

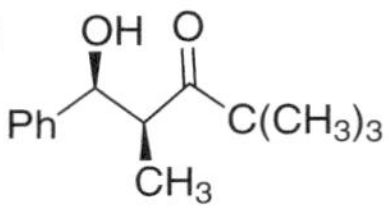

(c)

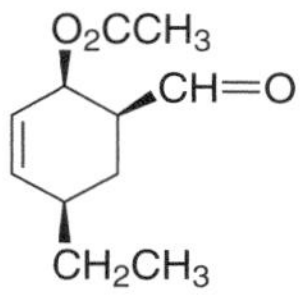

(d)

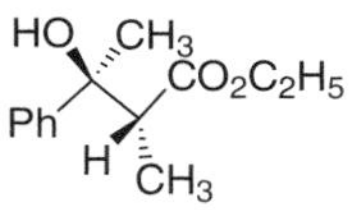

(e)

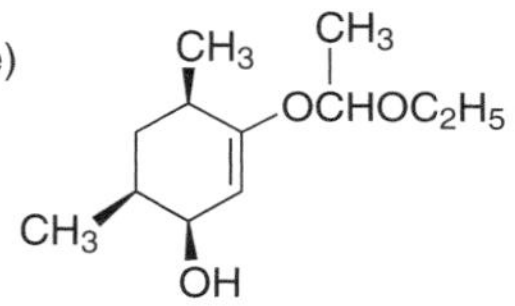

13.11. Devise a route that could be used for synthesis of the desired compound in high enantiomeric purity from the suggested starting material.

(a)

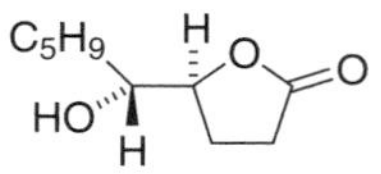

from D-ribose

(b)

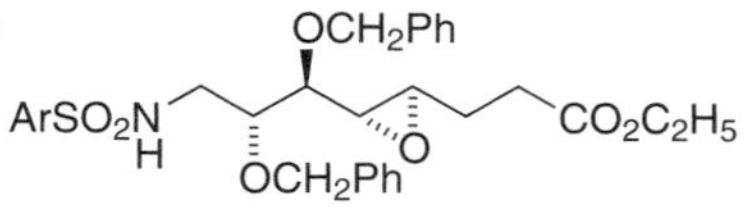

from

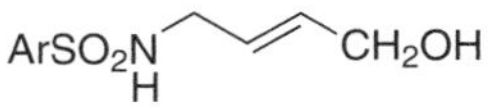

(c)

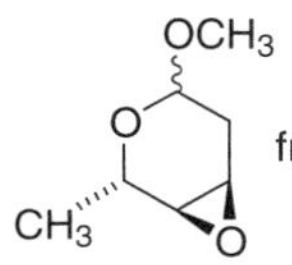

from

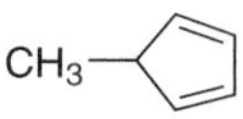

13.12. Select a reagent that will achieve the following syntheses with high enantioselectivity.

(a)

(b)

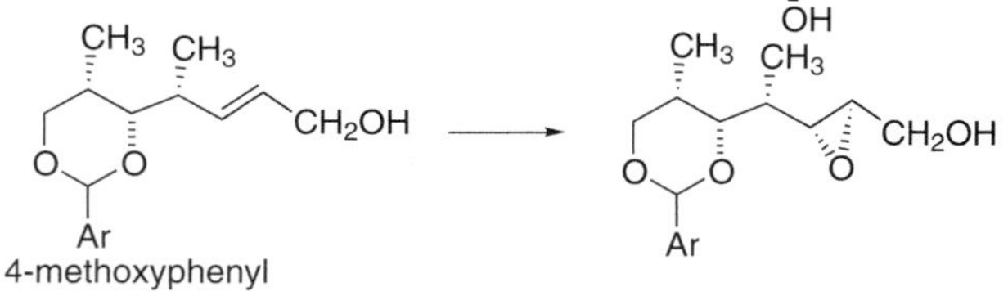

Ar = 4-methoxyphenyl

(c)

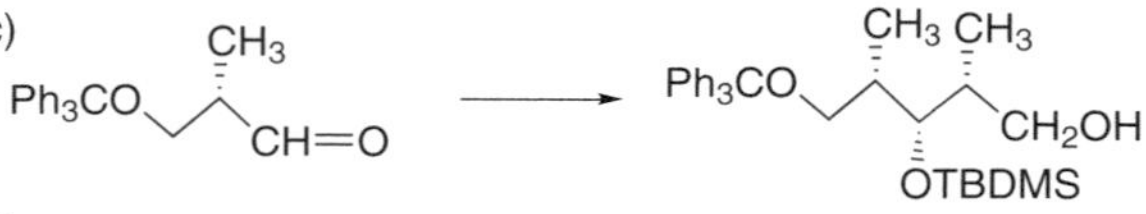

(d)

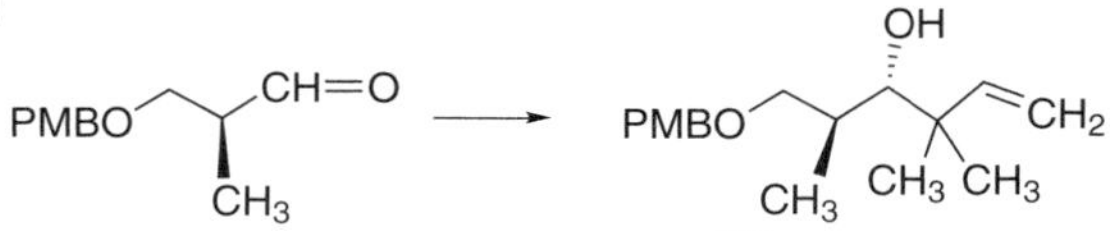

(e)

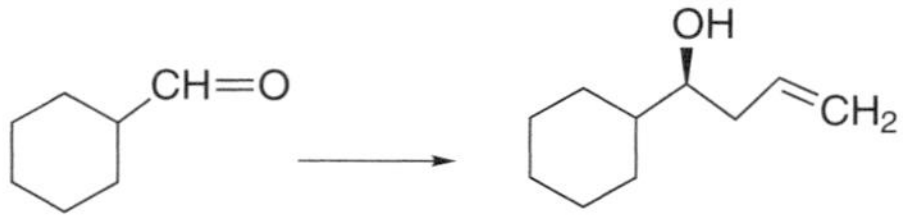

13.13. The following reactions use chiral auxiliaries to achieve enantioselectivity. By
consideration of possible TSs, predict the absolute configuration of the major
product of each reaction.

(a)

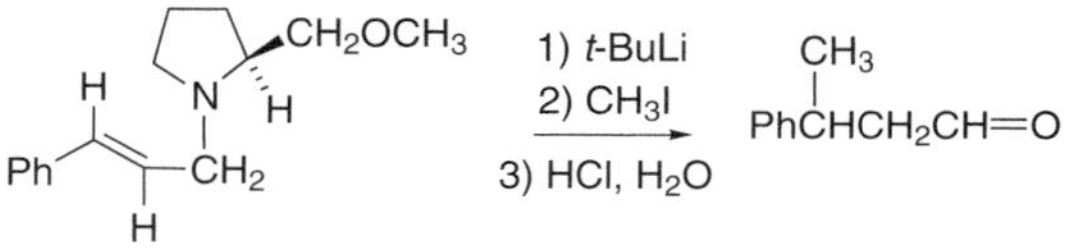

(b)

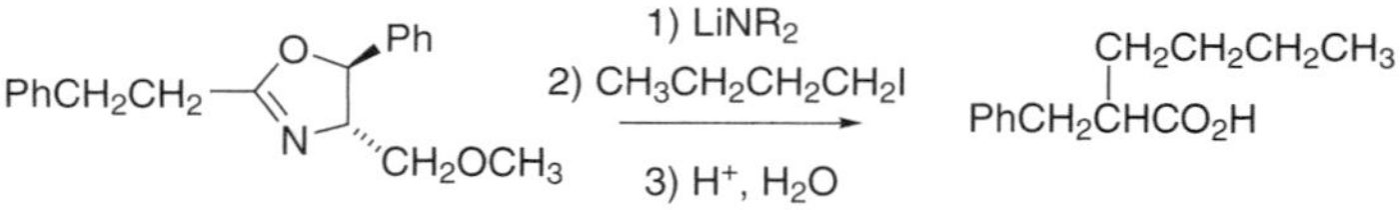

(c)

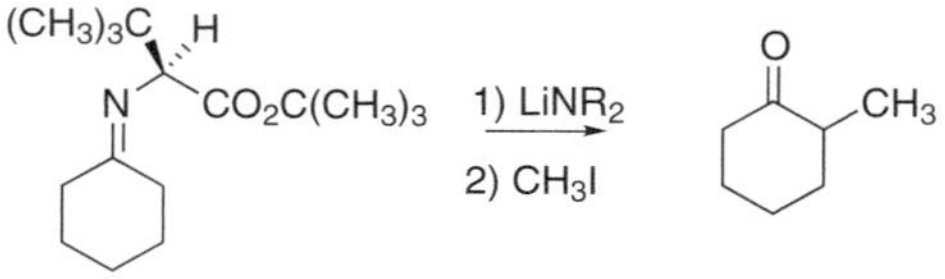

(d)

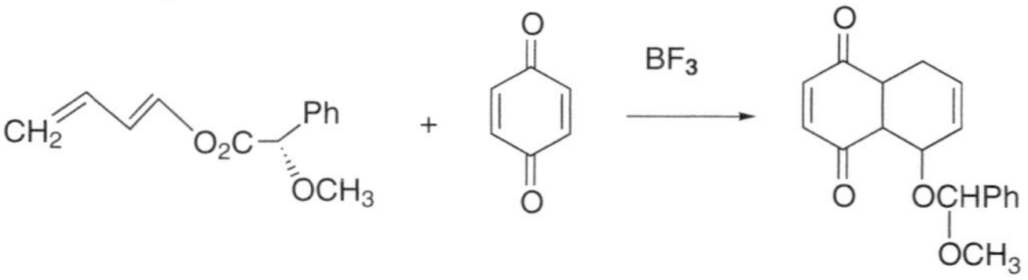

(e)

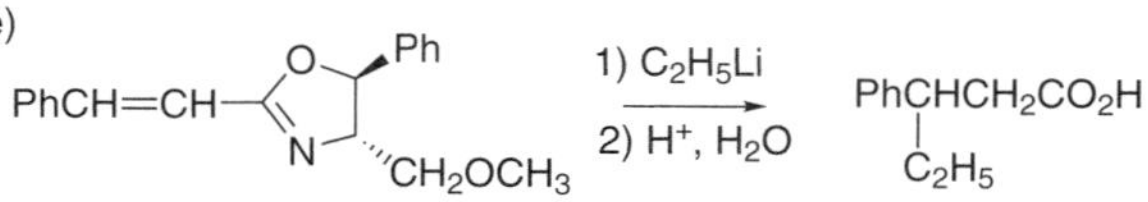

(f)

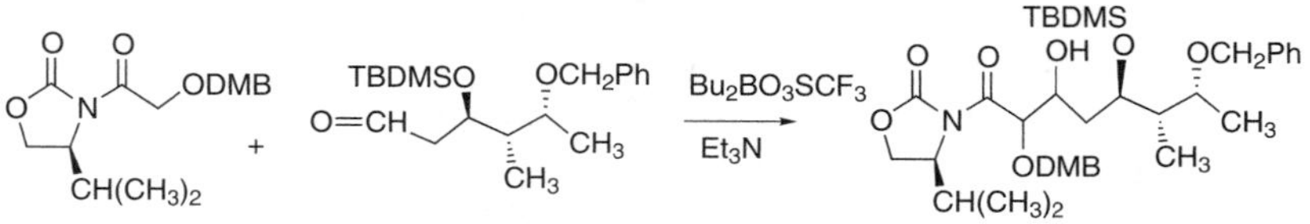

(g)

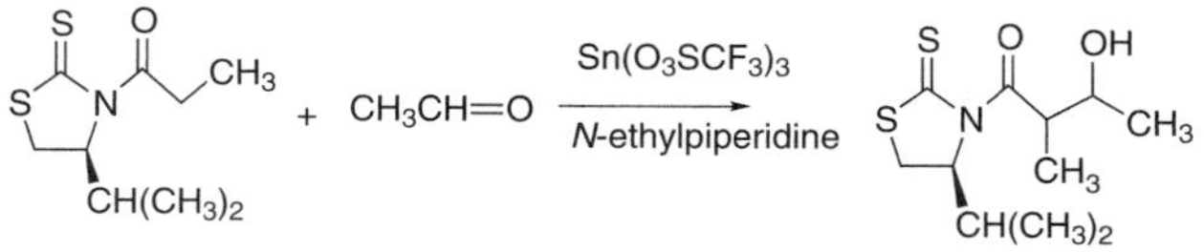

(h)

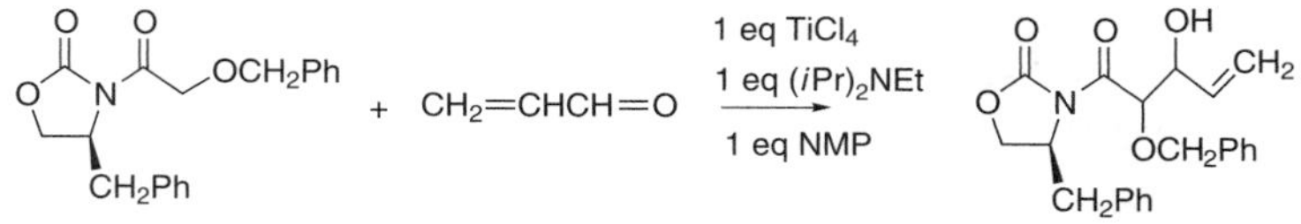

13.14. The macrolide carbonolide B contains six stereogenic centers at sp^3 carbons. Devise a strategy for synthesis of cabonolide B and in particular for establishing the stereochemistry of the C(1)–C(8) segment of the molecule.

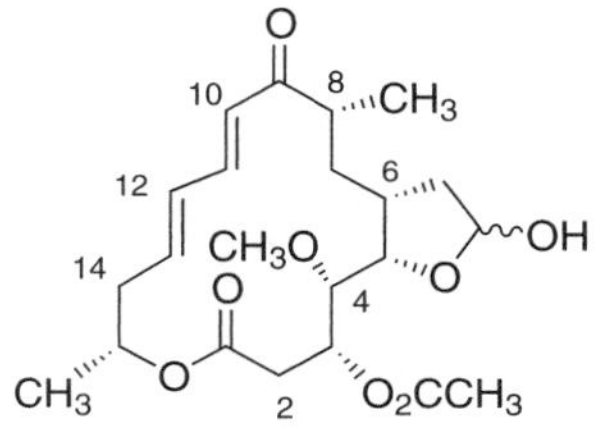

carbonolide B

13.15. 4-(Acylamino)-substituted carboxylate esters and amides can be alkylated with good *anti*-2,4 stereoselectivity using two equivalents of a strong base. The stereoselectivity is independent of the steric bulk of the remainder of the carboxylate structure. Propose a TS that is consistent with these observations.

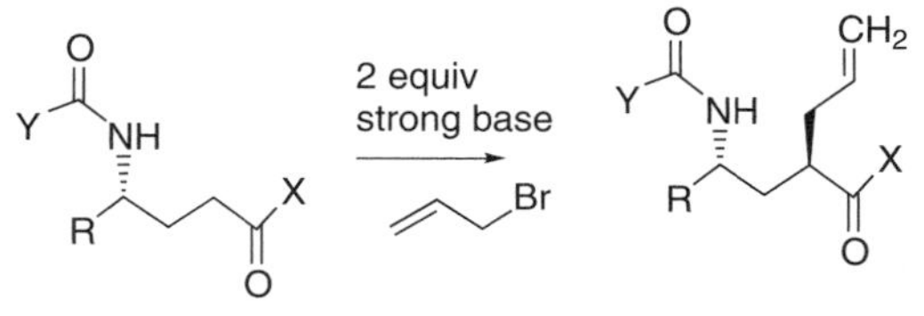

R	Y	X
CH$_3$	CF$_3$	OCH$_3$
(CH$_3$)$_2$CHCH$_2$	CF$_3$	OCH$_3$
PhCH$_2$	CF$_3$	OCH$_3$
PhCH$_2$	CF$_3$	N(CH$_3$)$_2$
PhCH$_2$	CF$_3$	N(CH$_3$)OCH$_3$
CH$_3$	(CH$_3$)$_3$CO	OCH$_3$
(CH$_3$)$_2$CH	(CH$_3$)$_3$CO	OCH$_3$

13.16. Using as a designation of a "step" each numbered reagent or reagent combination in Schemes 13.54 to 13.59 for the synthesis of the Taxol precursors shown there, outline the syntheses in terms of convergence and determine the longest linear sequence (as on p. 1166). In general, these Taxol syntheses are quite linear in character. Is there a structural reason for this tendency toward linearity?

References for Problems

Chapter 1

1a. W. S. Matthews, J. E. Bares, J. E. Bartmess, F. G. Bordwell, F. J. Cornforth, G. E. Drucker, Z. Margolin, R. J. McCallum, G. J. McCollum, and N. E. Vanier, *J. Am. Chem. Soc.*, **97**, 7006 (1975).

b. H. D. Zook, W. L. Kelly, and I. Y. Posey, *J. Org. Chem.*, **33**, 3477 (1968).

2a. H. O. House and M. J. Umen, *J. Org. Chem.*, **38**, 1000 (1973).

b. W. C. Still and M.-Y. Tsai, *J. Am. Chem. Soc.*, **102**, 3654 (1980).

c. H. O. House and B. M. Trost, *J. Org. Chem.*, **30**, 1431 (1965); C. H. Heathcock, C. T. Buse, W. A. Kleschick, M. C. Pirrung, J. E. Sohn, and J. Lampe, *J. Org. Chem.*, **45**, 1066 (1980); L. Xie, K. Vanlandeghem, K. M. Isenberger, and C. Bernier, *J. Org. Chem.*, **68**, 641 (2003).

d. D. Caine and T. L. Smith, Jr., *J. Am. Chem. Soc.*, **102**, 7568 (1980).

e. M. F. Semmelhack, S. Tomoda, and K. M. Hurst, *J. Am. Chem. Soc.*, **102**, 7568 (1980); M. F. Semmelhack, S. Tomoda, H. Nagaoka, S. D. Boettger, and K. M. Hurst, *J. Am. Chem. Soc.*, **104**, 747 (1982).

f. R. A. Lee, C. McAndrews, K. M. Patel, and W. Reusch, *Tetrahedron Lett.*, 965 (1973).

g. R. H. Frazier, Jr., and R. L. Harlow, *J. Org. Chem.*, **45**, 5408 (1980).

h. T. T. Tidwell, *J. Am. Chem. Soc.*, **92**, 1448 (1970); J. M. Jerkunica, S. Borcic, and D. E. Sunko, *Tetrahedron Lett.*, 4465 (1965).

3a. M. Gall and H. O. House, *Org. Synth.*, **52**, 39 (1972).

b. P. S. Wharton and C. E. Sundin, *J. Org. Chem.*, **33**, 4255 (1968).

c. B. W. Rockett and C. R. Hauser, *J. Org. Chem.*, **29**, 1394 (1964).

d. J. Meier, *Bull. Soc. Chim. Fr.*, 290 (1962).

e. M. E. Jung and C. A. McCombs, *Org. Synth.*, **VI**, 445 (1988).

f,g. H. O. House, T. S. B. Sayer, and C. C. Yau, *J. Org. Chem.*, **43**, 2153 (1978).

4a. S. A. Monti and S.-S. Yuan, *Tetrahedron Lett.*, 3627 (1969); J. M. Harless and S. A. Monti, *J. Am. Chem. Soc.*, **96**, 4714 (1974).

b. A. Wissner and J. Meinwald, *J. Org. Chem.*, **38**, 1697 (1973).

c. W. J. Gensler and P. H. Solomon, *J. Org. Chem.*, **38**, 1726 (1973).

d. H. W. Whitlock, Jr., *J. Am. Chem. Soc.*, **84**, 3412 (1962).

e. C. H. Heathcock, R. A. Badger, and J. W. Patterson, Jr., *J. Am. Chem. Soc.*, **89**, 4133 (1967).

f. E. J. Corey and D. S. Watt, *J. Am. Chem. Soc.*, **95**, 2302 (1973).

5. W. G. Kofron and L. G. Wideman, *J. Org. Chem.*, **37**, 555 (1972).

6. C. R. Hauser, T. M. Harris, and T. G. Ledford, *J. Am. Chem. Soc.*, **81**, 4099 (1959).

7a. N. Campbell and E. Ciganek, *J. Chem. Soc.*, 3834 (1956).

b. F. W. Sum and L. Weiler, *J. Am. Chem. Soc.*, **101**, 4401 (1979).

c. K. W. Rosemund, H. Herzberg, and H. Schutt, *Chem. Ber.*, **87**, 1258 (1954).

d. T. Hudlicky, F. J. Koszyk, T. M. Kutchan, and J. P. Sheth, *J. Org. Chem.*, **45**, 5020 (1980).

e. C. R. Hauser and W. R. Dunnavant, *Org. Synth.*, **IV**, 962 (1963).

f. G. Opitz, H. Milderberger, and H. Suhr, *Liebigs Ann. Chem.*, **649**, 47 (1961).

g. K. Wiesner, K. K. Chan, and C. Demerson, *Tetrahedron Lett.*, 2893 (1965).

h. K. Shimo, S. Wakamatsu, and T. Inoue, *J. Org. Chem.*, **26**, 4868 (1961).

i. G. R. Kieczykowski and R. H. Schlessinger, *J. Am. Chem. Soc.*, **100**, 1938 (1978).

8a. E. Wenkert and D. P. Strike, *J. Org. Chem.*, **27**, 1883 (1962).

b. S. J. Etheredge, *J. Org. Chem.*, **31**, 1990 (1966).

c. R. Deghenghi and R. Gaudry, *Tetrahedron Lett.*, 489 (1962).

d. P. A. Grieco and C. C. Pogonowski, *J. Am. Chem. Soc.*, **95**, 3071 (1973).

e. E. M. Kaiser, W. G. Kenyon, and C. R. Hauser, *Org. Synth.*, **V**, 559 (1973).

f. J. Cason, *Org. Synth.*, **IV**, 630 (1963).

g. S. A. Glickman and A. C. Cope, *J. Am. Chem. Soc.*, **67**, 1012 (1945).

h. W. Steglich and L. Zechlin, *Chem. Ber.*, **111**, 3939 (1978).

i. S. F. Brady, M. A. Ilton, and W. S. Johnson, *J. Am. Chem. Soc.*, **99**, 2882 (1968).

9. S. Masamune, *J. Am. Chem. Soc.*, **83**, 1009 (1961).

10a. L. A. Paquette, H.-S. Lin, D. T. Belmont, and J. P. Springer, *J. Org. Chem.*, **51** 4807 (1986); L. A. Paquette, T. T. Belmont, and Y.-L. Hsu, *J. Org. Chem.*, **50**, 4667 (1985).

b. R. K. Boeckman, Jr., D. K. Heckenden, and R. L. Chinn, *Tetrahedron Lett.*, **28**, 3551 (1987).

c. D. Seebach, J. D. Aebi, M. Gander-Coquot, and R. Naef, *Helv. Chim. Acta*, **70**, 1194 (1987).

d. F. E. Ziegler, S. I. Klein, U. K. Pati, and T.-F. Wang, *J. Am. Chem. Soc.*, **107**, 2730 (1985).

e. M. E. Kuehne, *J. Org. Chem.*, **35**, 171 (1970).

f. D. A. Evans, S. L. Bender, and J. Morris, *J. Am. Chem. Soc.*, **110**, 2506 (1988).

g. K. Tomioka, Y.-S. Cho, F. Sato, and K. Koga, *J. Org. Chem.*, **53**, 4094 (1988).

h. K. Tomioka, H. Kawasaki, K. Yasuda, and K. Koga, *J. Am. Chem. Soc.*, **110**, 3597 (1988).

i. M. W. Carson, G. Kim, M. F. Hentemann, D. Trauner, and S. J. Danishefsky, *Angew. Chem. Int. Ed. Engl.*, **40**, 4450 (2001).

j. J. E. Jung, H. Ho, and H.-D. Kim, *Tetrahedron Lett.*, **41**, 1793 (2000).

11a. S. G. Davies and H. J. Sanganee, *Tetrahedron: Asymmetry*, **6**, 671 (1995); A. G. Myers, B. H. Yang, H. Chen, and J. L. Gleason, *J. Am. Chem. Soc.*, **116**, 9361 (1994); K. H. Ahn, A. Lim, and S. Lee, *Tetrahedron: Asymmetry*, **4**, 2435 (1993).

b. D. Enders, H. Eichenauer, U. Baus, H. Schubert, and K. A. M. Kremer, *Tetrahedron*, **40**, 1345 (1984).

12a. T. Kametani, Y. Suzuki, H. Furuyama, and T. Honda, *J. Org. Chem.*, **48**, 31 (1983).

b. R. A. Kjonaas and D. D. Patel, *Tetrahedron Lett.*, **25**, 5467 (1984).

c. D. F. Taber and R. E. Ruckle, Jr., *J. Am. Chem. Soc.*, **108**, 7686 (1986).

d. D. L. Snitman, M.-Y. Tsai, D. S. Watt, C. L. Edwards, and P. L. Stotter, *J. Org. Chem.*, **44**, 2838 (1979).

e. A. G. Schultz and J. P. Dittami, *J. Org. Chem.*, **48**, 2318 (1983).

13a. K. F. McClure and M. Z. Axt, *Bioorg. Med. Chem. Lett.*, **8**, 143 (1998).

b. T. Honda, N. Kimura, and M. Tsubuki, *Tetrahedron: Asymmetry*, **4**, 1475 (1993); T. Honda, F. Ishikawa, K. Kanai, S. Sato, D. Kato, and H. Tominaga, *Heterocycles*, **42**, 109 (1996).

c. I. Vaulot, H.-J. Gais, N. Reuter, E. Schmitz, and R. K. L. Ossenkamp, *Eur. J. Org. Chem.*, 805 (1998); M. Majewski and R. Lazny, *J. Org. Chem.*, **60**, 5825 (1995).

d. H. Pellissier, P.-Y. Michellys, and M. Santelli, *J. Org. Chem.*, **62**, 5588 (1997).

e. K. Narasaka and Y. Ukagi, *Chem. Lett.*, 81 (1986).

14. J. G. Henkel and L. A. Spurlock, *J. Am. Chem. Soc.*, **95**, 8339 (1973).

15. M. S. Newman, V. De Vries, and R. Darlak, *J. Org. Chem.*, **31**, 2171 (1966).

16. P. A. Manis and M. W. Rathke, *J. Org. Chem.*, **45**, 4952 (1980).

Chapter 2

1a. G. Ksander, J. E. McMurry, and N. Johnson, *J. Org. Chem.*, **42**, 1180 (1977).

b. J. Zabicky, *J. Chem. Soc.*, 683 (1961).

c. G. Stork, G. A. Kraus, and G. A. Garcia, *J. Org. Chem.*, **39**, 3459 (1974).

d. H. Midorikawa, *Bull. Chem. Soc. Jpn.*, **27**, 210 (1954).

e. R. A. Auerbach, D. S. Crumrine, D. L. Ellison, and H. O. House, *Org. Synth.*, **54**, 49 (1974).

f. E. C. Du Feu, F. J. McQuillin, and R. Robinson, *J. Chem. Soc.*, 53 (1937).

g. E. Buchta, G. Wolfrum, and H. Ziener, *Chem. Ber.*, **91**, 1552 (1958).

h. L. H. Briggs and E. F. Orgias, *J. Chem. Soc. C*, 1885 (1970).

i. J. A. Proffitt and D. S. Watt, and E. J. Corey, *J. Org. Chem.*, **40**, 127 (1975).

j. U. Hengartner, and V. Chu, *Org. Synth.*, **58**, 83 (1978).

k. E. Giacomini, M. A. Loreto, L. Pellacani, and P. A. Tardella, *J. Org. Chem.*, **45**, 519 (1980).

l. N. Narasimhan and R. Ammanamanchi, *J. Org. Chem.*, **48**, 3945 (1983).

m. M. P. Bosch, F. Camps, J. Coll, A. Guerro, T. Tatsuoka, and J. Meinwald, *J. Org. Chem.*, **51**, 773 (1986).

n. T. A. Spencer, K. K. Schmiegel, and K. L. Williamson, *J. Am. Chem. Soc.*, **85**, 3785 (1963).

2a. M. W. Rathke and D. F. Sullivan, *J. Am. Chem. Soc.*, **95**, 3050 (1973).

b. E. J. Corey, H. Yamamoto, D. K. Herron, and K. Achiwa, *J. Am. Chem. Soc.*, **92**, 6635 (1970).

c. E. J. Corey and D. E. Cane, *J. Org. Chem.*, **36**, 3070 (1971).

d. E. W. Yankee and D. J. Cram, *J. Am. Chem. Soc.*, **92**, 6328 (1970).

e. W. G. Dauben, C. D. Poulter, and C. Suter, *J. Am. Chem. Soc.*, **92**, 7408 (1970).

f. P. A. Grieco and K. Hiroi, *J. Chem. Soc., Chem. Commun.*, 1317 (1972).

g. E. A. Couladouros and A. P. Mihou, *Tetrahedron Lett.*, **40**, 4861 (1999).

h. I. Vlattas, I. T. Harrison, L. Tokes, J. H. Fried, and A. D. Cross, *J. Org. Chem.*, **33**, 4176 (1968).

i. A. T. Nielsen and W. R. Carpenter, *Org. Synth.*, **V**, 288 (1973).

j. M. L. Miles, T. M. Harris, and C. R. Hauser, *Org. Synth.*, **V**, 718 (1973).

k. A. P. Beracierta and D. A. Whiting, *J. Chem. Soc., Perkin Trans. 1*, 1257 (1978).

l. T. Amatayakul, J. R. Cannon, P. Dampawan, T. Dechatiwongse, R. G. F. Giles, D. Huntrakul, K. Kusamran, M. Mokkhasamit, C. L. Raston, V. Reutrakul, and A. H. White, *Aust. J. Chem.*, **32**, 71 (1979).

m. R. M. Coates, S. K. Shah, and R. W. Mason, *J. Am. Chem. Soc.*, **101**, 6765 (1979).

n. K. A. Parker and T. H. Fedynyshyn, *Tetrahedron Lett.*, 1657 (1979).

o. M. Miyashita and A. Yoshikoshi, *J. Am. Chem. Soc.*, **96**, 1917 (1974).

p. E. J. Corey and S. Nozoe, *J. Am. Chem. Soc.*, **85**, 3527 (1963).

q. L. Fitjer and U. Quabeck, *Synth. Commun.*, **15**, 855 (1985).

r. A. Padwa, L. Brodsky, and S. Clough, *J. Am. Chem. Soc.*, **94**, 6767 (1972).

s. W. R. Roush, *J. Am. Chem. Soc.*, **102**, 1390 (1980).

t. C. R. Johnson, K. Mori, and A. Nakanishi, *J. Org. Chem.*, **44**, 2065 (1979).

u. T. Yanami, M. Miyashita, and A. Yoshikoshi, *J. Org. Chem.*, **45**, 607 (1980).

v. D. A. Evans, T. Rovis, M. C. Kozlowski, C. W. Downey, and J. S. Tedrow, *J. Am. Chem. Soc.*, **122**, 9134 (2000).

w. M. Yamaguchi, M. Tsukamoto, S. Tanaka, and I. Hirao, *Tetrahedron Lett.*, **25**, 5661 (1984); M. Yamaguchi, M. Tsukamoto, and I. Hirao, *Tetrahedron Lett.*, **26**, 1723 (1985).

x. D. A. Evans, M. T. Bilodeau, T. C. Somers, J. Clardy, D. Cherry, and Y. Kato, *J. Org. Chem.*, **56**, 5750 (1991).

3a. K. D. Croft, E. L. Ghisalberti, P. R. Jefferies, and A. D. Stuart, *Aust. J. Chem.*, **32**, 2079 (1971).

b. L. H. Briggs and G. W. White, *J. Chem. Soc., C*, 3077 (1971).

c. D. F. Taber and B. P. Gunn, *J. Am. Chem. Soc.*, **101**, 3992 (1979).

d. G. V. Kryshtal, V. V. Kulganek, V. F. Kucherov, and L. A. Yanovskaya, *Synthesis*, 107 (1979).

e. S. F. Brady, M. A. Ilton, and W. S. Johnson, *J. Am. Chem. Soc.*, **90**, 2882 (1968).

f. R. M. Coates and J. E. Shaw, *J. Am. Chem. Soc.*, **92**, 5657 (1970).

g. K. Mitsuhashi and S. Shiotoni, *Chem. Pharm. Bull.*, **18**, 75 (1970).

h. G. Wittig and H.-D. Frommeld, *Chem. Ber.*, **97**, 3548 (1964).

i. R. J. Sundberg, P. A. Bukowick, and F. O. Holcombe, *J. Org. Chem.*, **32**, 2938 (1967).

j. D. R. Howton, *J. Org. Chem.*, **10**, 277 (1945).

k. M. Graff, A. Al Dilaimi, P. Seguineau, M. Rambaud, and J. Villieras, *Tetrahedron Lett.*, **27**, 1577 (1986); T. Yamane and K. Ogasawara, *Synlett*, 925 (1996).

l. Y. Chan and W. W. Epstein, *Org. Synth.*, **53**, 48 (1973).

m. I. Fleming and M. Woolias, *J. Chem. Soc., Perkin Trans. 1*, 827 (1979).

n. F. Johnson, K. G. Paul, D. Favara, R. Ciabatti, and U. Guzzi, *J. Am. Chem. Soc.*, **104**, 2190 (1982).

o. M. Ihara, M. Suzuki, K. Fukumoto, T. Kametani, and C. Kabuto, *J. Am. Chem. Soc.*, **110**, 1963 (1988).

p. H. Hagiwara, T. Okabe, H. Ono, V. R. Kamat, T. Hoshi, T. Suzuki, and M. Ando, *J. Chem. Soc., Perkin Trans. 1*, 895 (2002).

q. S. P. Chavan and M. S. Venkatraman, *Tetrahedron Lett.*, **39**, 6745 (1998).

r. M. Yamaguchi, M. Tsukamoto, S. Tanaka, and I. Hirao, *Tetrahedron Lett.*, **25**, 5661 (1984); M. Yamaguchi, M. Tsukamoto, and I. Hirao, *Tetrahedron Lett.*, **26**, 1723 (1985).

s. D. J. Critcher, S. Connolly, and M. Wills, *J. Org. Chem.*, **62**, 6638 (1997).

4a. W. A. Mosher and R. W. Soeder, *J. Org. Chem.*, **36**, 1561 (1971).

b. M. R. Roberts and R. H. Schlessinger, *J. Am. Chem. Soc.*, **101**, 7626 (1979).

c. J. E. McMurry and T. E. Glass, *Tetrahedron Lett.*, 2575 (1971).

d. D. J. Cram, A. Langemann, and F. Hauck, *J. Am. Chem. Soc.*, **81**, 5750 (1959).

e. W. G. Dauben and J. Ipaktschi, *J. Am. Chem. Soc.*, **95**, 5088 (1973).

f. T. J. Curphy and H. L. Kim, *Tetrahedron Lett.*, 1441 (1968).

g. K. P. Singh and L. Mandell, *Chem. Ber.*, **96**, 2485 (1963).

h. S. D. Lee, T. H. Chan, and K. S. Kwon, *Tetrahedron Lett.*, **25**, 3399 (1984).

i. J. F. Lavallee and P. Deslongchamps, *Tetrahedron Lett.*, **29**, 6033 (1988).

j. K. Aoyagi, H. Nakamura, and Y. Yamamoto, *J. Org. Chem.*, **64**, 4148 (1999).

5. T. T. Howarth, G. P. Murphy, and T. A. Harris, *J. Am. Chem. Soc.*, **91**, 517 (1969).

6a. E. Vedejs, K. A. Snobel, and P. L. Fuchs, *J. Org. Chem.*, **38**, 1178 (1973).

b. P. B. Dervan and M. A. Shippey, *J. Am. Chem. Soc.*, **98**, 1265 (1976).

7a. E. E. Schweizer and G. J. O'Neil, *J. Org. Chem.*, **30**, 2082 (1965); E. E. Schweizer, *J. Am. Chem. Soc.*, **86**, 2744 (1964).

b. G. Buchi and H. Wuest, *Helv. Chim. Acta*, **54**, 1767 (1971).

c. G. H. Posner, S.-B. Lu, and E. Asirvathan, *Tetrahedron Lett.*, **27**, 659 (1986).

d. M. Mikolajczyk, M. Mikina, and A. Jankowiak, *J. Org. Chem.*, **65**, 5127 (2000); M. Mikolajczyk and M. Mikina, *J. Org. Chem.*, **59**, 6760 (1994).

e. W. A. Kleschick and C. H. Heathcock, *J. Org. Chem.*, **43**, 1256 (1978).

f. S. D. Darling, F. N. Muralidharan, and V. B. Muralidharan, *Tetrahedron Lett.*, 2761 (1979).

8. R. B. Woodward, F. Sondheimer, D. Taub, K. Heusler, and W. M. McLamore, *J. Am. Chem. Soc.*, **74**, 4223 (1952).

9. G. Stork, S. D. Darling, I. T. Harrison, and P. S. Wharton, *J. Am. Chem. Soc.*, **84**, 2018 (1962).

10. J. R. Pfister, *Tetrahedron Lett.*, **21**, 1281 (1980).

11. R. M. Jacobson, G. P. Lahm, and J. W. Clader, *J. Org. Chem.*, **45**, 395 (1980).

12a. A. I. Meyers and N. Nazarenko, *J. Org. Chem.*, **38**, 175 (1973).

b. J. A. Marshall and D. J. Schaeffer, *J. Org. Chem.*, **30**, 3642 (1965); W. C. Still and F. L. Van Middlesworth, *J. Org. Chem.*, **42**, 1258 (1977).

c. Y. Fukuda and Y. Okamoto, *Tetrahedron*, **58**, 2513 (2002).

d. E. J. Corey, M. Ohno, R. B. Mitra, and P. A. Vatakencherry, *J. Am. Chem. Soc.*, **86**, 478 (1964).

e. K. Makita, K. Fukumoto, and M. Ihara, *Tetrahedron Lett.*, **38**, 5197 (1997).

13a. R. V. Stevens and A. W. M. Lee, *J. Am. Chem. Soc.*, **101**, 7032 (1979).

b. C. H. Heathcock, E. Kleinman, and E. S. Binkley, *J. Am. Chem. Soc.*, **100**, 3036 (1978).

c. E. J. Corey and R. D. Balanson, *J. Am. Chem. Soc.*, **96**, 6516 (1974).

14a. M. Ertas and D. Seebach, *Helv. Chim. Acta*, **68**, 961 (1985).

b. S. Masamune, W. Choy, F. A. J. Kerdesky, and B. Imperiali, *J. Am. Chem. Soc.*, **103**, 1566 (1981).

c. C. H. Heathcock, C. T. Buse, W. A. Kleschick, M. C. Pirrung, J. E. Sohn, and J. Lampe, *J. Org. Chem.*, **45**, 1066 (1980).

d. R. Noyori, K. Yokoyama, J. Sakata, I. Kuwajima, E. Nakamura, and M. Shimizu, *J. Am. Chem. Soc.*, **99**, 1265 (1977).

e. D. A. Evans, E. Vogel, and J. V. Nelson, *J. Am. Chem. Soc.*, **101**, 6120 (1979); D. A. Evans, J. V. Nelson, E. Vogel, and T. R. Taber, *J. Am. Chem. Soc.*, **103**, 3099 (1981).

f. C. T. Buse and C. H. Heathcock, *J. Am. Chem. Soc.*, **99**, 8109 (1977).

g. R. Mahrwald, B. Costisella, and C. Gundogan, *Synthesis*, 262 (1998).

h. C. Esteve, M. P. Ferrero, P. Romea, F. Urpi, and J. Vilarrasa, *Tetrahedron Lett.*, **40**, 5079, 5083 (1999).

15a. D. Enders, O. F. Prokopenko, G. Raabe, and J. Runsink, *Synthesis*, 1095 (1996).

b. M. T. Reetz and A. Jung, *J. Am. Chem. Soc.*, **105**, 4833 (1983).

16a,b,c,d. E. D. Bergmann, D. Ginsburg, and R. Pappo, *Org. React.*, **10**, 179 (1959).

e. L. Mandell, J. U. Piper, and K. P. Singh, *J. Org. Chem.*, **28**, 3440 (1963).

f. H. O. House, W. A. Kleschick, and E. J. Zaiko, *J. Org. Chem.*, **43**, 3653 (1978).

g. J. E. McMurry and J. Melton, *Org. Synth.*, **56**, 36 (1977).

h. D. F. Taber and B. P. Gunn, *J. Am. Chem. Soc.*, **101**, 3992 (1979).

i. H. Feuer, A. Hirschfield, and E. D. Bergmann, *Tetrhaderon*, **24**, 1187 (1968).

    j. A. M. Baradel, R. Longeray, J. Dreux, and J. Doris, *Bull. Soc. Chim. Fr.*, 255 (1970).

    k. H. H. Baer and K. S. Ong, *Can. J. Chem.*, **46**, 2511 (1968).

    l. A. Wettstein, K. Heusler, H. Ueberwasser, and P. Wieland, *Helv. Chim. Acta*, **40**, 323 (1957).

17. J. D. White and M. Kawasaki, *J. Am. Chem. Soc.*, **112**, 4991 (1990).

18a. C. Somoza, J. Darias, and E. A. Ruveda, *J. Org. Chem.*, **54**, 1539 (1989); (b) E. R. Koft, A. S. Kotnis, and T. A. Broadbent, *Tetrahedron Lett.*, **28**, 2799 (1987); M. Leclaire, R. Levet, and J.-Y. Lallemand, *Synth. Commun.*, **23**, 1923 (1993).

19a. J. Fried in *Heterocyclic Compounds*, Vol. 1., R. C. Elderfield, ed., John Wiley, New York, 1950, p. 358.

    b. R. Charpurlat, J. Heret, and J. Druex, *Bull. Soc. Chim. Fr.*, 2446, 2450 (1967).

    c. A. Miyashita, Y. Matsuoka, A. Numata, and T. Higashino, *Chem. Pharm. Bull.*, **44**, 448 (1996).

    d. M. L. Quesada, R. H. Schlessinger, and W. H. Parsons, *J. Org. Chem.*, **43**, 3968 (1978).

20a. M. T. Reetz and K. Kesseler, *J. Org. Chem.*, **50**, 5434 (1985).

    b. J. Mulzer, A. Mantoulidis, and E. Oehler, *J. Org. Chem.*, **65**, 7456 (2000).

    c. J. G. Solsono, P. Romea, F. Urpi, and J. Vilarrasa, *Org. Lett.*, **5**, 519 (2003).

    d. I. Paterson and R. D. Tillyer, *J. Org. Chem.*, **58**, 4182 (1993).

    e. F. Kuo and P. L. Fuchs, *J. Am. Chem. Soc.*, **109**, 1122 (1987).

    f. M.-H. Filippini, R. Faure, and J. Rodriguez, *J. Org. Chem.*, **60**, 6872 (1995).

21. G. Koch, O. Loiseleur, D. Fuentes, A. Jantsch and K.-H. Altmann, *Org. Lett.*, **4**, 3811 (2002).

22. D. J. Gustin, M. S. Van Nieuwenhze, and W. R. Roush, *Tetrahedron Lett.*, **36**, 3443 (1995).

23a. S. F. Martin and D. E. Guinn, *J. Org. Chem.*, **52**, 5588 (1987).

    b. A. Armstrong, P. A. Barsanti, T. J. Blench, and R. Ogilvie, *Tetrahedron*, **59**, 367 (2003).

    c. A. K. Ghosh and J.-H. Kim, *Tetrahedron Lett.*, **43**, 5621 (2002); A. K. Ghosh and J.-H. Kim, *Tetrahedron Lett.*, **42**, 1227 (2001).

    d. M. Arai, N. Morita, S. Aoyagi, and C. Kibayashi, *Tetrahedron Lett.*, **41**, 1199 (2000).

    e. D. A. Evans, R. L. Dow, T. L. Shih, J. M. Takacs, and R. Zahler, *J. Am. Chem. Soc.*, **112**, 5290 (1990).

    f. N. Murakami, W. Wang, M. Aoki, Y. Tsutsui, M. Sugimoto, and M. Kobayashi, *Tetrahedron Lett.*, **39**, 2349 (1998).

    g. M. T. Crimmins and B. W. King, *J. Am. Chem. Soc.*, **120**, 9084 (1998).

    h. A. B. Smith and B. M. Brandt, *Org. Lett.*, **3**, 1685 (2001).

24. N. Langlois and H. S. Wang, *Synth. Commun.*, **27**, 3133 (1997).

25. A. Bassan, W. Zou, E. Reyes, F. Himo, and A. Cordova, *Angew. Chem. Int. Ed. Engl.*, **44**, 7028 (2005).

Chapter 3

1a. M. E. Kuehne and J. C. Bohnert, *J. Org. Chem.*, **46**, 3443 (1981).

    b. B. C. Barot and H. W. Pinnick, *J. Org. Chem.*, **46**, 2981 (1981).

    c. T. Mukaiyama, S. Shoda, and Y. Watanabe, *Chem. Lett.*, 383 (1977).

    d. H. Loibner and E. Zbiral, *Helv. Chim. Acta*, **59**, 2100 (1976).

    e. E. J. Prisbe, J. Smejkal, J. P. H. Verheyden, and J. G. Moffatt, *J. Org. Chem.*, **41**, 1836 (1976).

    f. B. D. MacKenzie, M. M. Angelo, and J. Wolinsky, *J. Org. Chem.*, **44**, 4042 (1979).

    g. A. I. Meyers, R. K. Smith, and C. E. Whitten, *J. Org. Chem.*, **44**, 2250 (1979).

    h. W. A. Bonner, *J. Org. Chem.*, **32**, 2496 (1967).

    i. B. E. Smith and A. Burger, *J. Am. Chem. Soc.*, **75**, 5891 (1953).

    j. W. D. Klobucar, L. A. Paquette, and J. F. Blount, *J. Org. Chem.*, **46**, 4021 (1981).

    k. G. Grethe, V. Toome, H. L. Lee, M. Uskokovic, and A. Brossi, *J. Org. Chem.*, **33**, 504 (1968).

    l. B. Neiss and W. Steglich, *Org. Synth.*, **63**, 183 (1984).

2. A. W. Friederang and D. S. Tarbell, *J. Org. Chem.*, **33**, 3797 (1968).

3. H. R. Hudson and G. R. de Spinoza, *J. Chem. Soc. Perkin Trans. 1*, 104 (1976).

4a. L. A. Paquette and M. K. Scott, *J. Am. Chem. Soc.*, **94**, 6760 (1972).

    b. P. N. Confalone, G. Pizzolato, E. G. Baggiolini, D. Lollar, and M. R. Uskokovic, *J. Am. Chem. Soc.*, **99**, 7020 (1977).

    c. E. L. Eliel, J. K. Koskimies, and B. Lohri, *J. Am. Chem. Soc.*, **100**, 1614 (1978).

    d. H. Hagiwara, M. Numata, K. Konishi, and Y. Oka, *Chem. Pharm. Bull.*, **13**, 253 (1965).

e. A. S. Kende and T. P. Demuth, *Tetrahedron Lett.*, 715 (1980).

f. P. A. Grieco, D. S. Clark, and G. P. Withers, *J. Org. Chem.*, **44**, 2945 (1979).

g. J. Yu, J. R. Falck, and C. Mioskowski, *J. Org. Chem.*, **57**, 3757 (1992).

h. J. Freedman, M. J. Vaal, and E. W. Huber, *J. Org. Chem.*, **56**, 670 (1991).

5a,b,c. D. Seebach, H.-O. Kalinowski, B. Bastani, G. Crass, H. Daum, H. Dorr, N. P. DuPreez, W. Langer, C. Nussler, H.-A. Oei, and M. Schmidt, *Helv. Chim. Acta*, **60**, 301 (1977).

d. G. L. Baker, S. J. Fritschel, J. R. Stille, and J. K. Stille, *J. Org. Chem.*, **46**, 2954 (1981).

e. M. D. Fryzuk and B. Bosnich, *J. Am. Chem. Soc.*, **99**, 6262 (1977).

f. S. Hanessian and R. Frenette, *Tetrahedron Lett.*, 3391 (1979).

g. K. G. Paul, F. Johnson, and D. Favara, *J. Am. Chem. Soc.*, **98**, 1285 (1976).

h. S. D. Burke, J. Hong, J. R. Lennox, and A. P. Mongin, *J. Org. Chem.*, **63**, 6952 (1998).

i. G. Dujardin, S. Rossignol, and E. Brown, *Synthesis*, 763 (1998).

6a. E. J. Corey, J.-L. Gras, and P. Ulrich, *Tetrahedron Lett.*, 809 (1976).

b. K. C. Nicolaou, S. P. Seitz, and M. R. Pavia, *J. Am. Chem. Soc.*, **103**, 1222 (1981).

c. E. J. Corey and A. Venkateswarlu, *J. Am. Chem. Soc.*, **94**, 6190 (1972).

d.-f. H. H. Meyer, *Liebigs Ann. Chem.*, 732 (1977).

7a. P. Henley-Smith, D. A. Whiting, and A. F. Wood, *J. Chem. Soc. Perkin Trans. 1*, 614 (1980).

b. P. Beak and L. G. Carter, *J. Org. Chem.*, **46**, 2363 (1981).

c. M. E. Jung and T. J. Shaw, *J. Am. Chem. Soc.*, **102**, 6304 (1980).

d. P. N. Swepston, S.-T. Lin, A. Hawkins, S. Humphrey, S. Siegel, and A. W. Cordes, *J. Org. Chem.*, **46**, 3754 (1981).

e. P. J. Maurer and M. J. Miller, *J. Org. Chem.*, **46**, 2835 (1981).

f. N. A. Porter, J. D. Byers, A. E. Ali, and T. E. Eling, *J. Am. Chem. Soc.*, **102**, 1183 (1980).

g. G. A. Olah, B. G. B. Gupta, R. Malhotra, and S. C. Narang, *J. Org. Chem.*, **45**, 1638 (1980).

8a. A. K. Bose, B. Lal, W. Hoffman, III, and M. S. Manhas, *Tetrahedron Lett.*, 1619 (1973).

b. J. B. Hendrickson and S. M. Schwartzman, *Tetrahedron Lett.*, 277 (1975).

c. J. F. King, S. M. Loosmore, J. D. Lock, and M. Aslam, *J. Am. Chem. Soc.*, **100**, 1637 (1978); C. N. Sukenic, and R. G. Bergman, *J. Am. Chem. Soc.*, **98**, 6613 (1976).

d. R. S. Freedlander, T. A. Bryson, R. B. Dunlap, E. M. Schulman, and C. A. Lewis, Jr., *J. Org. Chem.*, **46**, 3519 (1981).

e. A. Trzeciak and W. Bannwarth, *Synthesis*, 1433 (1996).

f. L. M. Beacham, III, *J. Org. Chem.*, **44**, 3100 (1979).

9a. J. Jacobus, M. Raban, and K. Mislow, *J. Org. Chem.*, **33**, 1142 (1968).

b. M. Schmid and R. Barner, *Helv. Chim. Acta*, **62**, 464 (1979).

c. V. Eswarakrishnan and L. Field, *J. Org. Chem.*, **46**, 4182 (1981).

d. R. F. Borch, A. J. Evans, and J. J. Wade, *J. Am. Chem. Soc.*, **99**, 1612 (1977).

e. H. S. Aaron and C. P. Ferguson, *J. Org. Chem.*, **33**, 684 (1968).

10. B. Koppenhoeffer and V. Schuring, *Org. Synth.*, **66**, 151, 160 (1987).

11a. M. Miyashita, A. Yoshikoshi, and P. A. Grieco, *J. Org. Chem.*, **42**, 3772 (1977).

b. E. J. Corey, L. O. Wiegel, D. Floyd, and M. G. Bock, *J. Am. Chem. Soc.*, **100**, 2916 (1978).

c. A. M. Felix, E. P. Heimer, T. J. Lambros, C. Tzougraki, and J. Meienhofer, *J. Org. Chem.*, **43**, 4194 (1978).

d. P. N. Confalone, G. Pizzolato, E. G. Baggionlini, D. Lollar, and M. R. Uskokovic, *J. Am. Chem. Soc.*, **97**, 5936 (1975).

e. A. B. Foster, J. Lehmann, and M. Stacey, *J. Chem. Soc.*, 4649 (1961).

12. B. E. Watkins and H. Rapoport, *J. Org. Chem.*, **47**, 4471 (1982).

13. C. Ahn, R. Correia, and P. DeShong, *J. Org. Chem.*, **67**, 1751 (2002).

14. R. M. Magid, O. S. Fruchey, W. L. Johnson, and T. G. Allen, *J. Org. Chem.*, **44**, 359 (1979).

15a. D. M. Simonovic, A. S. Rao, and S. C. Bhattacharyya, *Tetrahedron*, **19**, 1061 (1963).

b. G. Buchi, W. D. MacLeod, Jr., and J. Padilla, *J. Am. Chem. Soc.*, **86**, 4438 (1964).

c. P. Doyle, I. R. Maclean, W. Parker, and R. A. Raphael, *Proc. Chem. Soc.*, 239 (1963).

d. J. C. Sheehan and K. R. Henery-Logan, *J. Am. Chem. Soc.*, **84**, 2983 (1962).

e. E. J. Corey, M. Ohno, R. B. Mitra, and P. A. Vatakencherry, *J. Am. Chem. Soc.*, **86**, 478 (1964).

16a. R. B. Woodward, R. A. Olofson, and H. Mayer, *Tetrahedron Suppl.*, **8**, 321 (1966); R. B. Woodward and R. A. Olofson, *J. Am. Chem. Soc.*, **83**, 1007 (1961).

b. B. Belleau and G. Malek, *J. Am. Chem. Soc.*, **90**, 1651 (1968).

17a. E. J. Corey, K. C. Nicolaou, and L. S. Melvin, Jr., *J. Am. Chem. Soc.*, **97**, 654 (1975).

b. J. Huang and J. Meinwald, *J. Am. Chem. Soc.*, **103**, 861 (1981).

c. P. Beak and L. G. Carter, *J. Org. Chem.*, **46**, 2363 (1981).

18. T. Mukaiyama, S. Shoda, T. Nakatsuka, and K. Narasaka, _Chem. Lett._, 605 (1978).

19. R. U. Lemieux, K. B. Hendrik, R. V. Stick, and K. James, _J. Am. Chem. Soc._, **97**, 4056 (1975).

20a. T. Mukaiyama, R. Matsueda, and M. Suzuki, _Tetrahedron Lett._ 1901 (1970).

 b. E. J. Corey and D. A. Clark, _Tetrahedron Lett._, 2875 (1979).

21a. J. Y. Lee and B. H. Kim, _Tetrahedron_, **52**, 571 (1996).

 b. J. Y. Lee and B. H. Kim, _Tetrahedron Lett._, **36**, 3361 (1995); I. Fleming and S. K. Ghosh, _J. Chem. Soc., Chem. Commun._, 2287 (1994).

22a. E. M. Acton, R. N. Goerner, H. S. Uh, K. J. Ryan, D. W. Henry, C. E. Cass, and G. A. LePage, _J. Med. Chem._, **22**, 518 (1979).

 b. E. G. Gros, _Carbohydr. Res._, **2**, 56 (1966).

 c. S. Hanessian and G. Rancourt, _Can. J. Chem._, **55**, 1111 (1977).

 d. R. E. Schmidt and A. Gohl, _Chem. Ber._, **112**, 1689 (1979).

Chapter 4

1a. N. Kharasch and C. M. Buess, _J. Am. Chem. Soc._, **71**, 2724 (1949).

 b. A. J. Sisti, _J. Org. Chem._, **33**, 3953 (1968).

 c. H. C. Brown and G. Zweifel, _J. Am. Chem. Soc._, **83**, 1241 (1961).

 d. F. W. Fowler, A. Hassner, and L. A. Levy, _J. Am. Chem. Soc._, **89**, 2077 (1967).

 e. A. Hassner and F. W. Fowler, _J. Org. Chem._, **33**, 2686 (1968).

 f. A. Padwa, T. Blacklock, and A. Tremper, _Org. Synth._, **57**, 83 (1977).

 g. I. Ryu, S. Murai, I. Niwa, and N. Sonoda, _Synthesis_, 874 (1977).

 h. R. A. Amos and J. A. Katzenellenbogen, _J. Org. Chem._, **43**, 560 (1978).

 i. H. C. Brown and G. J. Lynch, _J. Org. Chem._, **46**, 531 (1981).

 j. R. C. Cambie, R. C. Hayward, P. S. Rutledge, T. Smith-Palmer, B. E. Swedlund, and P. D. Woodgate, _J. Chem. Soc. Perkin Trans. 1_, 180 (1979).

 k. A. B. Holmes, K. Russell, E. S. Stern, M. E. Stubbs, and N. K. Wellard, _Tetrahedron Lett._, **25**, 4163 (1984).

 l. N. S. Zefirov, T. N. Velikokhat'ko, and N. K. Sandovaya, _Zh. Org. Khim._ (Engl. Transl.) **19**, 1407 (1983).

 m. F. B. Gonzalez and P. A. Bartlett, _Org. Synth._, **64**, 175 (1985).

2. D. J. Pasto and J. A. Gontarz, _J. Am. Chem. Soc._, **93**, 6902 (1971).

3. D. J. Pasto and J. A. Gontarz, _J. Am. Chem. Soc._, **93**, 6902 (1971).

4. R. Gleiter and G. Mueller, _J. Org. Chem._, **53**, 3912 (1988).

5. G. Stork and R. Borch, _J. Am. Chem. Soc._, **86**, 935 (1964).

6. T. Hori and K. B. Sharpless, _J. Org. Chem._, **43**, 1689 (1978).

7a. E. Kloster-Jensen, E. Kovats, A. Eschenmoser, and E. Heilbronner, _Helv. Chim. Acta_, **39**, 1051 (1956).

 b. P. N. Rao, _J. Org. Chem._, **36**, 2426 (1971).

 c. R. A. Moss and E. Y. Chen, _J. Org. Chem._, **46**, 1466 (1981).

 d. J. M. Jerkunica and T. G. Traylor, _Org. Synth._, **53**, 94 (1973).

 e. G. Zweifel and C. C. Whitney, _J. Am. Chem. Soc._, **89**, 2753 (1967).

 f. W. I. Fanta and W. F. Erman, _J. Org. Chem._, **33**, 1656 (1968).

 g. W. E. Billups, J. H. Cross, and C. V. Smith, _J. Am. Chem. Soc._, **95**, 3438 (1973).

 h. G. W. Kabalka and E. E. Gooch, III, _J. Org. Chem._, **45**, 3578 (1980).

 i. E. J. Corey, G. Weiss, Y. B. Xiang, and A. K. Singh, _J. Am. Chem. Soc._, **109**, 4717 (1987).

 j. W. Oppolzer, H. Hauth, P. Pfaffli, and R. Wenger, _Helv. Chim. Acta_, **60**, 1801 (1977).

 k. G. H. Posner and P. W. Tang, _J. Org. Chem._, **43**, 4131 (1978).

 l. A. V. Bayquen and R. W. Read, _Tetrahedron_, **52**, 13467 (1996).

 m. T. Fukuyama and G. Liu, _J. Am. Chem. Soc._, **118**, 7426 (1996).

8a. E. J. Corey and H. Estreicher, _J. Am. Chem. Soc._, **100**, 6294 (1978).

 b. E. J. Corey and H. Estreicher, _Tetrahedron Lett._, 1113 (1980).

 c. G. A. Olah and M. Nohima, _Synthesis_, 785 (1973).

9. D. J. Pasto and F. M. Klein, _Tetrahedron Lett._, 963 (1967).

10. H. J. Reich, J. M. Renga, and J. L. Reich, _J. Am. Chem. Soc._, **97**, 5434 (1975).

11. H. C. Brown, G. J. Lynch, W. J. Hammar, and L. C. Liu, _J. Org. Chem._, **44**, 1910 (1979).

12a. R. Lavilla, O. Coll, M. Nicolas, and J. Bosch, _Tetrahedron Lett._, **39**, 5089 (1998).

b. A. Garofalo, M. B. Hursthouse, K. M. A. Malik, H. F. Olivio, S. M. Roberts, and V. Sik, _J. Chem. Soc. Perkin Trans. 1_, 1311 (1994).

c. H. Imagawa, T. Shigaraki, T. Suzuki, H. Takao, H. Yamada, T. Sugihara, and M. Nichizawa, _Chem. Pharm. Bull._, **46**, 1341 (1998).

d. A. G. Schultz and S. J. Kirmich, _J. Org. Chem._, **61**, 5626 (1996).

e. I. Fleming and N. J. Lawrence, _J. Chem. Soc. Perkin Trans. 1_, 3309 (1992); 2679 (1998).

f. R. Bittman, H.-S. Byun, K. C. Reddy, P. Samadder, and G. Arthur, _J. Med. Chem._, **40**, 1391 (1997).

g. P. A. Bartlett and J. Myerson, _J. Am. Chem. Soc._, **100**, 3950 (1978).

h. S. Terashima, M. Hayashi, and K. Koga, _Tetrahedron Lett._, 2733 (1980).

i. H. Bernsmann, R. Froehlich, and P. Metz, _Tetrahedron Lett._, **41**, 4347 (2000).

13a. D. A. Evans, J. E. Ellman, and R. L. Dorow, _Tetrahedron Lett._, **28**, 1123 (1987).

b. K. C. Nicolaou, R. L. Magolda, W. J. Sipio, W. E. Barnette, Z. Lysenko, and M. M. Joullie, _J. Am. Chem. Soc._, **102**, 3784 (1980).

c. K. C. Nicolaou, W. E. Barnette, and R. L. Magolda, _J. Am. Chem. Soc._, **103**, 3472 (1981).

d. S. Knapp, A. T. Levorse, and J. A. Potenza, _J. Org. Chem._, **53**, 4773 (1988).

e. T. H. Jones and M. S. Blum, _Tetrahedron Lett._, **22**, 4373 (1981).

14a. W. T. Smith and G. L. McLeod, _Org. Synth._, **IV**, 345 (1963).

b. K. E. Harding, T. H. Marman, and D. Nam, _Tetrahedron Lett._, **29**, 1627 (1988).

15. A. Toshimitsu, K. Terao, and S. Uemura, _J. Org. Chem._, **51**, 1724 (1986).

16a. W. Oppolzer and P. Dudfield, _Tetrahedron Lett._, **26**, 5037 (1985).

b. D. A. Evans, J. A. Ellman, and R. L. Dorow, _Tetrahedron Lett._, **28**, 1123 (1987).

17. T. W. Bell, _J. Am. Chem. Soc._, **103**, 1163 (1981).

18a. H. C. Brown and B. Singaram, _J. Am. Chem. Soc._, **106**, 1797 (1984).

b. S. Masamune, B. M. Kim, J. S. Petersen, T. Sato, S. J. Veenstra, and T. Imai, _J. Am. Chem. Soc._, **107**, 4549 (1985).

19. C. F. Palmer, K. D. Parry, S. M. Roberts, and V. Sik, _J. Chem. Soc. Perkin Trans. 1_, 1021 (1992); C. F. Palmer and R. McCague, _J. Chem. Soc. Perkin Trans. 1_, 2977 (1998); A. Toyota, A. Nishimura, and C. Kaneko, _Heterocycles_, **45**, 2105 (1997).

20a. M. Noguchi, H. Okada, M. Watanabe, K. Okuda, and O. Nakamura, _Tetrahedron_, **52**, 6581 (1996).

b. R. Madsen, C. Roberts, and B. Fraser-Reid, _J. Org. Chem._, **60**, 7920 (1995).

21. M. E. Jung and U. Karama, _Tetrahedron Lett._, **40**, 7907 (1999).

22. B. Fraser and P. Perlmutter, _J. Chem. Soc., Perkin Trans. 1_, 2896 (2002).

23. M. J. Kurth and E. G. Brown, _J. Am. Chem. Soc._, **109**, 6844 (1987); M. J. Kurth, R. L. Beard, M. Olmstead, and J. G. Macmillan, _J. Am. Chem. Soc._, **111**, 3712 (1989); M. J. Kurth, E. G. Brown, E. J. Lewis, and J. C. McKew, _Tetrahedron Lett._, **29**, 1517 (1988).

24. S. Bedford, G. Fenton, D. W. Knight, and D. E. Shaw, _J. Chem. Soc., Perkin Trans. 1_, 1505 (1996).

Chapter 5

1a. W. R. Roush, _J. Am. Chem. Soc._, **102**, 1390 (1980).

b. H. C. Brown, S. C. Kim, and S. Krishnamurthy, _J. Org. Chem._, **45**, 1 (1980).

c. G. W. Kabalka, D. T. C. Yang, and J. D. Baker, Jr., _J. Org. Chem._, **41**, 574 (1976).

d. J. K. Whitesell, R. S. Matthews, M. A. Minton, and A. M. Helbling, _J. Am. Chem. Soc._, **103**, 3468 (1981).

e. K. S. Kim, M. W. Spatz, and F. Johnson, _Tetrahedron Lett._, 331 (1979).

f. M.-H. Rei, _J. Org. Chem._, **44**, 2760 (1979).

g. R. O. Hutchins, D. Kandasamy, F. Dux, III, C. A. Maryanoff, D. Rolstein, B. Goldsmith, W. Burgoyne, F. Cistone, J. Dalessandro, and J. Puglis, _J. Org. Chem._, **43**, 2259 (1978).

h. H. Lindlar, _Helv. Chim. Acta_, **35**, 446 (1952).

i. E. Vedejs, R. A. Buchanan, R. Conrad, G. P. Meier, M. J. Mullins, and Y. Watanabe, _J. Am. Chem. Soc._, **109**, 5878 (1987).

j. C. B. Jackson and G. Pattenden, _Tetrahedron Lett._, **26**, 3393 (1985).

2. D. C. Wigfield and D. J. Phelps, _J. Am. Chem. Soc._, **96**, 543 (1974).

3a. E. J. Corey, T. K. Schaaf, W. Huber, U. Koelliker, and N. M. Weinshenker, _J. Am. Chem. Soc._, **92**, 397 (1970).

b. E. J. Corey and R. Noyori, _Tetrahedron Lett._, 311 (1970).

c. R. F. Borch, _Org. Synth._, **52**, 124 (1972).

d. D. Seyferth and V. A. Mai, *J. Am. Chem. Soc.*, **92**, 7412 (1970).

e. R. V. Stevens and J. T. Lai, *J. Org. Chem.*, **37**, 2138 (1972).

f. M. J. Robins and J. S. Wilson, *J. Am. Chem. Soc.*, **103**, 932 (1981).

g. G. R. Pettit and J. R. Dias, *J. Org. Chem.*, **36**, 3207 (1971).

h. P. A. Grieco, T. Oguri, and S. Gilman, *J. Am. Chem. Soc.*, **102**, 5886 (1980).

i. M. F. Semmelhack, S. Tomoda, and K. M. Hurst, *J. Am. Chem. Soc.*, **102**, 7567 (1980).

j. H. C. Brown and P. Heim, *J. Org. Chem.*, **38**, 912 (1973).

k. R. O. Hutchins and N. R. Natale, *J. Org. Chem.*, **43**, 2299 (1978).

l. M. R. Detty and L. A. Paquette, *J. Am. Chem. Soc.*, **99**, 821 (1977).

m. C. A. Bunnell and P. L. Fuchs, *J. Am. Chem. Soc.*, **99**, 5184 (1977).

n. Y.-J. Wu and D. J. Burnell, *Tetrahedron Lett.*, **29**, 4369 (1988).

o. P. W. Collins, E. Z. Dajani, R. Pappo, A. F. Gasiecki, R. G. Bianchi, and E. M. Woods, *J. Med. Chem.*, **26**, 786 (1983).

4a. F. A. Carey, D. H. Ball, and L. Long, *Carbohydr. Res.*, **3**, 205 (1966).

b. D. J. Cram and R. A. Abd Elhafez, *J. Am. Chem. Soc.*, **74**, 5828 (1952).

c. R. N. Rej, C. Taylor, and G. Eadon, *J. Org. Chem.*, **45**, 126 (1980).

d. M. C. Dart and H. B. Henbest, *J. Chem. Soc.*, 3563 (1960).

e. E. Piers, W. de Waal, and R. W. Britton, *J. Am. Chem. Soc.*, **93**, 5113 (1971).

f. A. L. J. Beckwith and C. Easton, *J. Am. Chem. Soc.*, **100**, 2913 (1978).

g. D. Horton and W. Weckerle, *Carbohydr. Res.*, **44**, 227 (1975).

h. R. A. Holton and R. M. Kennedy, *Tetrahedron Lett.*, **28**, 303 (1987).

i. H. Iida, N. Yamazaki, and C. Kibayashi, *J. Org. Chem.*, **51**, 1069, 3769 (1986).

j. D. A. Evans and M. M. Morrisey, *J. Am. Chem. Soc.*, **106**, 3866 (1984).

k. N. A. Porter, C. B. Ziegler, Jr., F. F. Khouri, and D. H. Roberts, *J. Org. Chem.*, **50**, 2252 (1985).

l. G. Stork and D. E. Kahne, *J. Am. Chem. Soc.*, **105**, 1072 (1983).

m. Y. Yamamoto, K. Matsuoka, and H. Nemoto, *J. Am. Chem. Soc.*, **110**, 4475 (1988).

n. G. Palmisano, B. Danieli, G. Lesma, D. Passerella, and L. Toma, *J. Org. Chem.*, **56**, 2380 (1991).

o. D. A. Evans, S. J. Miller, and M. D. Ennis, *J. Org. Chem.*, **58**, 471 (1993).

p. A. G. Schultz and N. J. Green, *J. Am. Chem. Soc.*, **113**, 4931 (1991).

q. R. Frenette, M. Monette, M. A. Bernstein, R. N. Young, and T. R. Verhoeven, *J. Org. Chem.*, **56**, 3083 (1991).

5a. D. Lenoir, *Synthesis*, 553 (1977).

b. J. A. Marshall and A. E. Greene, *J. Org. Chem.*, **36**, 2035 (1971).

c. B. M. Trost, Y. Nishimura, and K. Yamamoto, *J. Am. Chem. Soc.*, **101**, 1328 (1979); J. E. McMurry, A. Andrus, G. M. Ksander, J. H. Musser, and M. A. Johnson, *J. Am. Chem. Soc.*, **101**, 1330 (1979).

d. R. E. Ireland and C. S. Wilcox, *J. Org. Chem.*, **45**, 197 (1980).

e. P. G. Gassman and T. J. Atkins, *J. Am. Chem. Soc.*, **94**, 7748 (1972).

f. A. Gopalan and P. Magnus, *J. Am. Chem. Soc.*, **102**, 1756 (1980).

g. R. M. Coates, S. K. Shah, and R. W. Mason, *J. Am. Chem. Soc.*, **101**, 6765 (1979); Y.-K. Han and L. A. Paquette, *J. Org. Chem.*, **44**, 3731 (1979).

h. L. P. Kuhn, *J. Am. Chem. Soc.*, **80**, 5950 (1958).

i. R. P. Hatch, J. Shringarpure, and S. M. Weinreb, *J. Org. Chem.*, **43**, 4172 (1978).

j. T. Shono, Y. Matsumura, S. Kashimura, and H. Kyutoko, *Tetrahedron Lett.*, 1205 (1978).

6a,b. K. E. Wiegers and S. G. Smith, *J. Org. Chem.*, **43**, 1126 (1978).

c. D. C. Wigfield and F. W. Gowland, *J. Org. Chem.*, **45**, 653 (1980).

7. D. Caine and T. L. Smith, Jr., *J. Org. Chem.*, **43**, 755 (1978).

8. P. E. A. Dear and F. L. M. Pattison, *J. Am. Chem. Soc.*, **85**, 622 (1963).

9a. S. Danishefsky, M. Hirama, K. Gombatz, T. Harayama, E. Berman, and P. F. Schuda, *J. Am. Chem. Soc.*, **101**, 7020 (1979).

b. A. S. Kende, M. L. King, and D. P. Curran, *J. Org. Chem.*, **46**, 2826 (1981).

c. A. P. Kozikowski and A. Ames, *J. Am. Chem. Soc.*, **103**, 3923 (1981).

d. E. J. Corey, S. G. Pyre, and W. Su, *Tetrahedron Lett.*, **24**, 4883 (1983).

e. T. Rosen and C. Heathcock, *J. Am. Chem. Soc.*, **107**, 3731 (1985).

f. T. Fujisawa and T. Sato, *Org. Synth.*, **66**, 121 (1987).

g. H. J. Liu and M. G. Kulkarni, *Tetrahedron Lett.*, **26**, 4847 (1985).

h. D. A. Evans, S. J. Miller, and M. D. Ennis, *J. Org. Chem.*, **58**, 471 (1993).

i. D. L. J. Clive, K. S. K. Murthy, A. G. H. Wee, J. S. Prasad, G. V. J. da Silva, M. Majewski, P. C. Anderson, C. F. Evans, R. D. Haugen, L. D. Heerze, and J. R. Barrie, *J. Am. Chem. Soc.*, **112**, 3018 (1990).

10a. H. C. Brown and W. C. Dickason, *J. Am. Chem. Soc.*, **92**, 709 (1970).
  b. D. Seyferth, H. Yamazaki, and D. L. Alleston, *J. Org. Chem.*, **28**, 703 (1963).
  c. G. Stork and S. D. Darling, *J. Am. Chem. Soc.*, **82**, 1512 (1960).
11a. R. F. Borch, *Org. Synth.*, **52**, 124 (1972).
  b. R. F. Borch, M. D. Bernstein, and H. D. Durst, *J. Am. Chem. Soc.*, **93**, 2897 (1971).
12. S.-K. Chung and F.-F. Chung, *Tetrahedron Lett.*, 2473 (1979).
13a. D. F. Taber, *J. Org. Chem.*, **41**, 2649 (1976).
  b. D. R. Briggs and W. B. Whalley, *J. Chem. Soc., Perkin Trans. 1*, 1382 (1976).
  c. J. R. Flisak and S. S. Hall, *J. Am. Chem. Soc.*, **112**, 7299 (1990).
14. R. Yoneda, S. Harusawa, and T. Kurihara, *J. Org. Chem.*, **56**, 1827 (1991).
15. S. Kaneko, N. Nakajima, M. Shikano, T. Katoh, and S. Terashima, *Tetrahedron*, **54**, 5485 (1998).
16a. N. J. Leonard and S. Gelfand, *J. Am. Chem. Soc.*, **77**, 3272 (1955).
  b. P. S. Wharton and D. H. Bohlen, *J. Org. Chem.*, **26**, 3615 (1961); W. R. Benn and R. M. Dodson, *J. Org. Chem.*, **29**, 1142 (1964).
  c. G. Lardelli and O. Jeger, *Helv. Chim. Acta*, **32**, 1817 (1949).
  d. R. J. Peterson and P. S. Skell, *Org. Synth.*, **V**, 929 (1973).
17a. N. M. Yoon, C. S. Pak, H. C. Brown, S. Krishnamurthy, and T. P. Stocky, *J. Org. Chem.*, **38**, 2786 (1973).
  b. D. J. Dawson and R. E. Ireland, *Tetrahedron Lett.*, 1899 (1968).
  c. R. O. Hutchins, C. A. Milewski, and B. A. Maryanoff, *J. Am. Chem. Soc.*, **95**, 3662 (1973).
  d. M. J. Kornet, P. A. Thio, and S. I. Tan, *J. Org. Chem.*, **33**, 3637 (1968).
  e. C. T. West, S. J. Donnelly, D. A. Kooistra, and M. P. Doyle, *J. Org. Chem.*, **38**, 2675 (1973).
  f. M. R. Johnson and B. Rickborn, *J. Org. Chem.*, **35**, 1041 (1970).
  g. N. Akubult and M. Balci, *J. Org. Chem.*, **53**, 3338 (1988).
18. H. Iida, N. Yamazaki, and C. Kibayashi, *J. Org. Chem.*, **51**, 3769 (1986).
19a. H. C. Brown, G. G. Pai, and P. K. Jadhav, *J. Am. Chem. Soc.*, **106**, 1531 (1984).
  b. E. J. Corey, R. K. Bakshi, S. Shibata, C.-P. Chen, and V. K. Singh, *J. Am. Chem. Soc.*, **109**, 7925 (1987).
  c. M. Srebnik, P. V. Ramachandran, and H. C. Brown, *J. Org. Chem.*, **53**, 2916 (1988).
20a. L. A. Paquette, T. J. Nitz, R. J. Ross, and J. P. Springer, *J. Am. Chem. Soc.*, **106,** 1446 (1984).
  b. L. N. Mander and M. M. McLachlan, *J. Am. Chem. Soc.*, **125**, 2400 (2003).
  c. S. Bhattacharyya and D. Mukherjee, *Tetrahedron Lett.*, **23**, 4175 (1982).
21. J. A. Marshall, *Acc. Chem. Res.*, **13**, 213 (1980); J. A. Marshall and K. E. Flynn, *J. Am. Chem. Soc.*, **106**, 723 (1984); J. A. Marshall, J. C. Peterson, and L. Lebioda, *J. Am. Chem. Soc.*, **106**, 6006 (1984).
22a. J. P. Guidot, T. Le Gall, and C. Mioskowski, *Tetrahedron Lett.*, **35**, 6671 (1994).
  b. M. Schwaebe and R. D. Little, *J. Org. Chem.*, **61**, 3240 (1996).
  c. T. Kan, S. Hosokawa, S. Naja, M. Oikawa, S. Ito, F. Matsuda, and H. Shirahama, *J. Org. Chem.*, **59**, 5532 (1994).
  d. E. J. Enholm, H. Satici, and A. Trivellas, *J. Org. Chem.*, **54**, 5841 (1989).
  e. E. J. Enholm and A. Trievellas, *Tetrahedron Lett.*, **35**, 1627 (1994).
  f. J. E. Baldwin, S. C. M. Turner, and M. G. Moloney, *Tetrahedron Lett.*, **33**, 1517 (1992).
23a. R. N. Bream, S. V. Ley, B. McDermott, and P. A. Procopiou, *J. Chem. Soc., Perkin Trans. 1*, 2237 (2002).
  b. L. H. P. Teixeira, E. J. Barreiro, and C. A. M. Fraga, *Synth. Commun.*, **27**, 3241 (1997).

Chapter 6

1a. B. M. Trost, S. A. Godleski, and J. P. Genet, *J. Am. Chem. Soc.*, **100**, 3930 (1978).
  b. M. E. Jung and C. A. McCombs, *J. Am. Chem. Soc.*, **100**, 5207 (1978).
  c. L. E. Overman, and P. J. Jessup, *J. Am. Chem. Soc.*, **100**, 5179 (1978).
  d. C. Cupas, W. E. Watts, and P. v. R. Schleyer, *Tetrahedron Lett.*, 2503 (1964).
  e. T. C. Jain, C. M. Banks, and J. E. McCloskey, *Tetrahedron Lett.*, 841 (1970).
  f. G. Buchi and J. E. Powell, Jr., *J. Am. Chem. Soc.*, **89**, 4559 (1967).
  g. M. Raban, F. B. Jones, Jr., E. H. Carlson, E. Banucci, and N. A. LeBel, *J. Org. Chem.*, **35**, 1497 (1970).
  h. H. Yamamoto and H. L. Sham, *J. Am. Chem. Soc.*, **101**, 1609 (1979).
  i. H. O. House, T. S. B. Sayer, and C.-C. Yau, *J. Org. Chem.*, **43**, 2153 (1978).

j. M. C. Pirrung, *J. Am. Chem. Soc.*, **103**, 82 (1981).

k. M. Sevrin and A. Krief, *Tetrahedron Lett.*, 187 (1978).

l. L. A. Paquette, G. D. Crouse, and A. K. Sharma, *J. Am. Chem. Soc.*, **102**, 3972 (1980).

m. N.-K. Chan and G. Saucy, *J. Org. Chem.*, **42**, 3828 (1977).

n,o. J. A. Marshall and J. Lebreton, *J. Org. Chem.*, **53**, 4141 (1988).

p. F. A. J. Kerdesky, R. J. Ardecky, M. V. Lakshmikathan, and M. P. Cava, *J. Am. Chem. Soc.*, **103**, 1992 (1981).

q. B. B. Snider and R. A. H. F. Hui, *J. Org. Chem.*, **50**, 5167 (1985).

r. L. Lambs, N. P. Singh, and J.-F. Biellmann, *J. Org. Chem.*, **57**, 6301 (1992).

s. M. T. Reetz and E. H. Lauterbach, *Tetrahedron Lett.*, **32**, 4481 (1991).

t. B. Coates, D. Montgomery, and P. J. Stevenson, *Tetrahedron Lett.*, **32**, 4199 (1991).

u. K. Honda, S. Inoue, and K. Sato, *J. Org. Chem.*, **57**, 428 (1992).

v. D. Kim, S. K. Ahn, H. Bae, W. J. Choi, and H. S. Kim, *Tetrahedron Lett.*, **38**, 4437 (1997).

w. K. Tanaka, T. Imase, and S. Iwata, *Bull. Chem. Soc. Jpn.* **69**, 2243 (1996).

2a. W. Oppolzer and M. Petrzilka, *J. Am. Chem. Soc.*, **98**, 6722 (1976).

b. A. Padwa and N. Kamigata, *J. Am. Chem. Soc.*, **99**, 1871 (1977).

c. H. W. Gschwend, A. O. Lee, and H.-P. Meier, *J. Org. Chem.*, **38**, 2169 (1973).

d. J. L. Gras and M. Bertrand, *Tetrahedron Lett.*, 4549 (1979).

e. T. Kametani, M. Tsubuki, Y. Shiratori, H. Nemoto, M. Ihara, K. Fukumoto, F. Satoh, and H. Inoue, *J. Org. Chem.*, **42**, 2672 (1977).

3a. K. C. Brannock, A. Bell, R. D. Burpitt, and C. A. Kelly, *J. Org. Chem.*, **29**, 801 (1964).

b. K. Ogura, S. Furukawa, and G. Tsuchihashi, *J. Am. Chem. Soc.*, **102**, 2125 (1980).

c. E. Vedejs, M. J. Arco, D. W. Powell, J. M. Renga, and S. P. Singer, *J. Org. Chem.*, **43**, 4831 (1978).

d. J. J. Tufariello and J. J. Tegeler, *Tetrahedron Lett.*, 4037 (1976).

e. L. A. Paquette, *J. Org. Chem.*, **29**, 2851 (1964).

f. M. E. Monk and Y. K. Kim, *J. Am. Chem. Soc.*, **86**, 2213 (1964).

g. B. Cazes and S. Julia, *Bull. Soc. Chim. Fr.*, 925 (1977).

h. D. L. Boger and D. D. Mullican, *Org. Synth.*, **65**, 98 (1987).

4a. P. E. Eaton and U. R. Chakraborty, *J. Am. Chem. Soc.*, **100**, 3634 (1978).

b. H. Hogeveen and B. J. Nusse, *J. Am. Chem. Soc.*, **100**, 3110 (1978).

c. T. Oida, S. Tanimoto, T. Sugimoto, and M. Okano, *Synthesis*, 131 (1980).

d. J. N. Labovitz, C. A. Henrick, and V. L. Corbin, *Tetrahedron Lett.*, 4209 (1975).

e. W. Steglich and L. Zechlin, *Chem. Ber.*, **111**, 3939 (1978).

f. F. D. Lewis and R. J. DeVoe, *J. Org. Chem.*, **45**, 948 (1980).

g. S. P. Tanis and K. Nakanishi, *J. Am. Chem. Soc.*, **101**, 4398 (1979).

h. S. Danishefsky, M. P. Prisbylla, and S. Hiner, *J. Am. Chem. Soc.*, **100**, 2918 (1978).

i. T. Hudlicky, F. J. Kosyk, T. M. Kutchan, and J. P. Sheth, *J. Org. Chem.*, **45**, 5020 (1980).

j. G. Li, Z. Li, and X. Fang, *Synth. Commun.*, **26**, 2569 (1996).

k. C. Chen and D. J. Hart, *J. Org. Chem.*, **55**, 6236 (1990).

l. S. Chackalamannil, R. J. Davis, Y. Wang, T. Asberom, D. Doller, J. Wong, D. Leone, and A. T. McPhail, *J. Org. Chem.*, **64**, 1932 (1999).

5a. R. Schug and R. Huisgen, *J. Chem. Soc., Chem. Commun.*, 60 (1975).

b. N. Shimizu, M. Ishikawa, K. Ishikura, and S. Nishida, *J. Am. Chem. Soc.*, **96**, 6456 (1974).

c. W. L. Howard and N. B. Loretta, *Org. Synth.*, **V**, 25 (1973).

d. J. Wolinsky and P. B. Login, *J. Org. Chem.*, **35**, 3205 (1970).

6a. S. Danishefsky, M. Hirama, N. Fitsch, and J. Clardy, *J. Am. Chem. Soc.*, **101**, 7013 (1979).

b. D. A. Evans, C. A. Bryan, and C. L. Sims, *J. Am. Chem. Soc.*, **94**, 2891 (1972).

7. J. C. Gilbert and P. D. Selliah, *J. Org. Chem.*, **58**, 6255 (1993).

8. B. Bichan and M. Winnik, *Tetrahedron Lett.*, 3857 (1974).

9a. R. A. Carboni and R. V. Lindsey, Jr., *J. Am. Chem. Soc.*, **81**, 4342 (1969).

b. L. A. Carpino, *J. Am. Chem. Soc.*, **84**, 2196 (1962); **85**, 2144 (1963).

10a. S. Hanessian, P. J. Roy, M. Petrini, P. J. Hodges, R. Di Fabio, and G. Carganico, *J. Org. Chem.*, **55**, 5766 (1990).

b. W. R. Roush and B. B. Brown, *J. Am. Chem. Soc.*, **115**, 2268 (1993).

c. J. W. Coe and W. R. Roush, *J. Org. Chem.*, **54**, 915 (1989).

11a. D. L. J. Clive, G. Chittattu, N. J. Curtis, and S. M. Menchen, *J. Chem. Soc., Chem. Commun.*, 770 (1978).

b. B. W. Metcalf, P. Bey, C. Danzin, M. J. Jung, P. Casara, and J. P. Veveri, *J. Am. Chem. Soc.*, **100**, 2551 (1978).

c. T. Cohen, Z. Kosarych, K. Suzuki, and L.-C. Yu, *J. Org. Chem.*, **50**, 2965 (1985).

d. T. Cohen, M. Bhupathy, and J. R. Matz, *J. Am. Chem. Soc.*, **105**, 520 (1983).

e. R. G. Shea, J. N. Fitzner, J. E. Farkhauser, A. Spaltenstein, P. A. Carpino, R. M. Peevey, D. V. Pratt, B. J. Tenge, and P. B. Hopkins, *J. Org. Chem.*, **51**, 5243 (1986).

f. R. L. Funk, P. M. Novak, and M. M. Abelman, *Tetrahedron Lett.*, **29**, 1493 (1988).

g. C. H. Cummins and R. M. Coates, *J. Org. Chem.*, **51**, 1383 (1986).

h. K. Ogura, S. Furukawa, and G. Tsuchihashi, *J. Am. Chem. Soc.*, **102**, 2125 (1980).

i. H. F. Schmitthenner and S. M. Weinreb, *J. Org. Chem.*, **43**, 3372 (1980).

j. R. A. Gibbs and W. H. Okamura, *J. Am. Chem. Soc.*, **110**, 4062 (1988).

k. E. Vedejs, J. D. Rodgers, and S. J. Wittenberger, *J. Am. Chem. Soc.*, **110**, 4822 (1988).

l. J. Ahman, T. Jarevang, and P. I. Somfai, *J. Org. Chem.*, **61**, 8148 (1996).

m. E. Vedejs and M. Gingras, *J. Am. Chem. Soc.*, **116**, 579 (1994).

12a. N. Ono, A. Kanimura, A. Kaji, *Tetrahedron Lett.*, **27**, 1595 (1986).

b. R. V. C. Carr, R. V. Williams, and L. A. Paquette, *J. Org. Chem.*, **48**, 4976 (1983).

c. C. H. DePuy and P. R. Story, *J. Am. Chem. Soc.*, **82**, 627 (1960).

d. S. Ranganathan, D. Ranganathan, and R. Iyengar, *Tetrahedron*, **32**, 961 (1976).

13a. D. J. Faulkner and M. R. Peterson, *J. Am. Chem. Soc.*, **95**, 553 (1973).

b. N. A. LeBel, N. D. Ojha, J. R. Menke, and R. J. Newland, *J. Org. Chem.*, **37**, 2896 (1972).

c. G. Buchi and H. Wuest, *J. Am. Chem. Soc.*, **96**, 7573 (1974).

d. C. A. Henrick, F. Schaub, and J. B. Siddall, *J. Am. Chem. Soc.*, **94**, 5374 (1972).

e. R. E. Ireland and R. H. Mueller, *J. Am. Chem. Soc.*, **94**, 5897 (1972).

f. E. J. Corey, R. B. Mitra, and H. Uda, *J. Am. Chem. Soc.*, **86**, 485 (1964).

g. R. E. Ireland, P. A. Aristoff, and C. F. Hoyng, *J. Org. Chem.*, **44**, 4318 (1979).

h. W. Sucrow, *Angew. Chem. Int. Ed. Engl.*, **7**, 629 (1968).

i. O. P. Vig, K. L. Matta, and I. Raj, *J. Indian Chem. Soc.*, **41**, 752 (1964).

j. W. Nagata, S. Hirai, T. Okumura, and K. Kawata, *J. Am. Chem. Soc.*, **90**, 1650 (1968).

k. H. O. House, J. Lubinkowski, and J. J. Good, *J. Org. Chem.*, **40**, 86 (1975).

l. L. A. Paquette, G. D. Grouse, and A. K. Sharma, *J. Am. Chem. Soc.*, **102**, 3972 (1980).

m. R. L. Funk and G. L. Bolton, *J. Org. Chem.*, **49**, 5021 (1984).

n. A. P. Kozikowski and C.-S. Li, *J. Org. Chem.*, **52**, 3541 (1987).

o. B. M. Trost and A. C. Lavoie, *J. Am. Chem. Soc.*, **105**, 5075 (1983).

p. A. P. Marchand, S. C. Suri, A. D. Earlywine, D. R. Powell, and D. van der Helm, *J. Org. Chem.*, **49**, 670 (1984).

q. M. Kodoma, Y. Shiobara, H. Sumitomo, K. Fukuzumi, H. Minami, and Y. Miyamoto, *J. Org. Chem.*, **53**, 1437 (1988).

r. K. M. Werner, J. M. de los Santos, and S. M. Weinreb, *J. Org. Chem.*, **64**, 686 (1999).

s. D. Perez, G. Bures, F. Guitian, and L. Castedo, *J. Org. Chem.*, **61**, 1650 (1996).

t. A. K. Mapp and C. H. Heathcock, *J. Org. Chem.*, **64**, 23 (1999).

14a. L. A. Paquette, S. K. Huber, and R. C. Thompson, *J. Org. Chem.*, **58**, 6874 (1993).

b. R. L. Funk, T. Olmstead, M. Parvez, and J. B. Stallmann, *J. Org. Chem.*, **58**, 5873 (1993).

15a. J. J. Turfariello, A. S. Milowsky, M. Al-Nuri, and S. Goldstein, *Tetrahedron Lett.*, **28**, 263 (1987).

b. G. H. Posner, A. Haas, W. Harrison, and C. M. Kinter, *J. Org. Chem.*, **52**, 4836 (1987).

c. F. E. Ziegler, A. Nangia, and G. Schulte, *J. Am. Chem. Soc.*, **109**, 3987 (1987).

d. M. P. Edwards, S. V. Ley, S. G. Lister, B. D. Palmer, and D. J. Williams, *J. Org. Chem.*, **49**, 3503 (1984).

e. R. E. Ireland and M. D. Varney, *J. Org. Chem.*, **48**, 1829 (1983).

f. D. J.-S. Tsai and M. M. Midland, *J. Am. Chem. Soc.*, **107**, 3915 (1985).

g. E. Vedejs, J. M. Dolphin, and H. Mastalerz, *J. Am. Chem. Soc.*, **105**, 127 (1983).

h. T. Zoller, D. Uguen, A. DeCian, J. Fischer, and S. Sable, *Tetrahedron Lett.*, **38**, 3409 (1997).

i. L. Grimaud, J.-P. Ferezou, J. Prunet, and J. Y. Lallemand, *Tetrahedron*, **53**, 9253 (1997).

j. P. M. Wovkulich, K. Shankaran, J. Kliegiel, and M. R. Uskokovic, *J. Org. Chem.*, **58**, 832 (1993).

k. P. A. Grieco and M. D. Kaufman, *Tetrahedron Lett.*, **40**, 1265 (1999); P. Grieco and Y. Dai, *J. Am. Chem. Soc.*, **120**, 5128 (1998).

l. M. E. Jung and B. T. Vu, *J. Org. Chem.*, **61**, 4427 (1996).

16. S. Cossu, S. Battaggia, and O. De Lucchi, *J. Org. Chem.*, **62**, 4162 (1997).

17a. K. Nomura, K. Okazaki, H. Hori, and E. Yoshii, *Chem. Pharm. Bull.*, **34**, 3175 (1996).

b. Y. Tamura, M. Sasho, K. Nakagawa, T. Tsugoshi, and Y. Kita, *J. Org. Chem.*, **49**, 473 (1984).

18. S. D. Burke, D. M. Armistead, and K. Shankaran, *Tetrahedron Lett.*, **27**, 6295 (1986).

19a. C. Siegel and E. R. Thornton, *Tetrahedron Lett.*, **29**, 5225 (1988).

b. D. P. Curran, B. H. Kim, J. Daugherty, and T. A. Heffner, *Tetrahedron Lett.*, **29**, 3555 (1988).

c. H. Waldmann, *J. Org. Chem.*, **53**, 6133 (1988).

20a. T.-Z. Wang, E. Pinard, and L. A. Paquette, *J. Am. Chem. Soc.*, **118**, 1309 (1996).

b. L. Morency and L. Barriault, *Tetrahedron Lett.*, **45**, 6105 (2004).

c. L. Barriault, P. J. A. Ang, and R. M. A. Lavigne, *Org. Lett.*, **6**, 1317 (2004).

d. G. A. Kraus and S. H. Woo, *J. Org. Chem.*, **52**, 4841 (1987).

21. D. A. Evans, D. M. Barnes, J. S. Johnson, T. Lectka, P. von Matt, S. J. Miller, J. A. Murry, R. D. Norcross, E. A. Shaughnessy, and K. R. Campos, *J. Am. Chem. Soc.*, **121**, 7582 (1999).

22. T. B. H. McMurry, A. Work, and B. McKenna, *J. Chem. Soc.*, *Perkin Trans. 1*, 811 (1991).

23. K. Mori, T. Tashiro, and S. Sano, *Tetrahedron Lett.*, **41**, 5243 (2000); T. Tashiro, M. Bando, and K. Mori, *Synthesis*, 1852 (2000).

24. A. S. Raw and E. B. Jang, *Tetrahedron Lett.*, **56**, 3285 (2000).

Chapter 7

1a. H. Neumann and D. Seebach, *Tetrahedron Lett.*, 4839 (1976).

b. P. Canonne, G. Foscolos, and G. Lemay, *Tetrahedron Lett.*, **21**, 155 (1980).

c. T. L. Shih, M. Wyvratt, and H. Mrozik, *J. Org. Chem.*, **52**, 2029 (1987).

d. G. M. Rubottom and C. Kim, *J. Org. Chem.*, **48**, 1550 (1983).

e. S. L. Buchwald, B. T. Watson, R. T. Lum, and W. A. Nugent, *J. Am. Chem. Soc.*, **109**, 7137 (1987); P. D. Brewer, J. Tagat, C. A. Hergueter, and P. Helquist, *Tetrahedron Lett.*, 4573 (1977).

f. T. Okazoe, K. Takai, K. Oshima, and K. Utimoto, *J. Org. Chem.*, **52**, 4410 (1987).

g. J. W. Frankenfeld and J. J. Werner, *J. Org. Chem.*, **34**, 3689 (1969).

h. E. R. Burkhardt and R. D. Rieke, *J. Org. Chem.*, **50**, 416 (1985).

i. G. Veeresa and A. Datta, *Tetrahedron*, **54**, 15673 (1998).

2. R. W. Herr and C. R. Johnson, *J. Am. Chem. Soc.*, **92**, 4979 (1970).

3a. J. S. Sawyer, A. Kucerovy, T. L. Macdonald, and G. J. McGarvey, *J. Am. Chem. Soc.*, **110**, 842 (1988).

b. T. Cohen and J. R. Matz, *J. Am. Chem. Soc.*, **102**, 6900 (1980).

c. C. R. Johnson and J. R. Medlich, *J. Org. Chem.*, **53**, 4131 (1988).

d. B. M. Trost and T. N. Nanninga, *J. Am. Chem. Soc.*, **107**, 1293 (1985).

e. T. Morwick, *Tetrahedron Lett.*, **21**, 3227 (1980).

f. R. F. Cunio and F. J. Clayton, *J. Org. Chem.*, **41**, 1480 (1976).

4a. J. J. Fitt and H. W. Gschwend, *J. Org. Chem.*, **45**, 4258 (1980).

b. S. Akiyama and J. Hooz, *Tetrahedron Lett.*, 4115 (1973).

c. K. P. Klein and C. R. Hauser, *J. Org. Chem.*, **32**, 1479 (1967).

d. B. M. Graybill and D. A. Shirley, *J. Org. Chem.*, **31**, 1221 (1966).

e. M. M. Midland, A. Tramontano, and J. R. Cable, *J. Org. Chem.*, **45**, 28 (1980).

f. W. Fuhrer and H. W. Gschwend, *J. Org. Chem.*, **44**, 1133 (1979).

g. D. F. Taber and R. W. Korsmeyer, *J. Org. Chem.*, **43**, 4925 (1978).

h. B. A. Feit, U. Melamed, R. R. Schmidt, and H. Speer, *Tetrahedron*, **37**, 2143 (1981).

i. R. M. Carlson, *Tetrahedron Lett.*, 111 (1978).

j. R. J. Sundberg, R. Broome, C. P. Walters, and D. Schnur, *J. Heterocycl. Chem.*, **18**, 807 (1981).

k. J. J. Eisch and J. N. Shah, *J. Org. Chem.*, **56**, 2955 (1991).

l. G. P. Crowther, R. J. Sundberg, and A. M. Sarpeshkar, *J. Org. Chem.*, **49**, 4657 (1984).

5a. M. P. Dreyfuss, *J. Org. Chem.*, **28**, 3269 (1963).

b. P. J. Pearce, D. H. Richards, and N. F. Scilly, *Org. Synth.*, **VI**, 240 (1988).

c. U. Schollkopf, H. Kuppers, H.-J. Traencker, and W. Pitteroff, *Liebigs Ann. Chem.*, **704**, 120 (1967); A. Duchene, D. Mouko-Mpegna, J. P. Quintard, *Bull. Soc. Chim. Fr.*, 787 (1985).

d. J. V. Hay and T. M. Harris, *Org. Synth.*, **VI**, 478 (1988).

e. F. Sato, M. Inoue, K. Oguro, and M. Sato, *Tetrahedron Lett.*, 4303 (1979).

f. J. C. H. Hwa and H. Sims, *Org. Synth.*, **V**, 608 (1973).

6a. J. H. Rigby and C. Senanyake, *J. Am. Chem. Soc.*, **109**, 3147 (1987).

b. K. Takai, Y. Kataoka, T. Okazoe, and K. Utimoto, *Tetrahedron Lett.*, **29**, 1065 (1988).

c. E. Nakamura, S. Aoki, K. Sekiya, H. Oshino, and I. Kuwajima, *J. Am. Chem. Soc.*, **109**, 8056 (1987).

d. H. A. Whaley, *J. Am. Chem. Soc.*, **93**, 3767 (1971).

7. J. Barluenga, F. J. Fananas, and M. Yus, *J. Org. Chem.*, **44**, 4798 (1979).

8a,b. W. C. Still and J. H. MacDonald, III, *Tetrahedron Lett.*, **21**, 1031 (1980).

   c. E. Casadevall and Y. Pouet, *Tetrahedron Lett.*, 2841 (1976).

  9. P. Beak, J. E. Hunter, Y. M. Jan, and A. P. Wallin, *J. Am. Chem. Soc.*, **109**, 5403 (1987).

 10. C. J. Kowalski, and M. S. Haque, *J. Org. Chem.*, **50**, 5140 (1985).

 11. C. Fehr, J. Galindo, and R. Perret, *Helv. Chim. Acta*, **70**, 1745 (1987).

12a. M. P. Cooke, Jr., and I. N. Houpis, *Tetrahedron Lett.*, **26**, 4987 (1985).

   b. E. Piers and P. C. Marais, *Tetrahedron Lett.*, **29**, 4053 (1988).

13a. C. Phillips, R. Jacobson, B. Abrahams, H. J. Williams, and C. R. Smith, *J. Org. Chem.*, **45**, 1920 (1980).

   b. T. Cohen and J. R. Matz, *J. Am. Chem. Soc.*, **102**, 6900 (1980).

   c. T. R. Govindachari, P. C. Parthasarathy, H. K. Desai, and K. S. Ramachandran, *Indian J. Chem.*, **13**, 537 (1975).

   d. W. C. Still, *J. Am. Chem. Soc.*, **100**, 1481 (1978).

   e. E. J. Corey and D. R. Williams, *Tetrahedron Lett.*, 3847 (1977).

   f. T. Okazoe, K. Takai, K. Oshiama, and K. Utimoto, *J. Org. Chem.*, **52**, 4410 (1987).

   g. M. A. Adams, A. J. Duggan, J. Smolanoff, and J. Meinwald, *J. Am. Chem. Soc.*, **101**, 5364 (1979).

   h. S. O. de Silva, M. Watanabe, and V. Snieckus, *J. Org. Chem.*, **44**, 4802 (1979).

 14. M. Kitamura, S. Suga, K. Kawai, and R. Noyori, *J. Am. Chem. Soc.*, **108**, 6071 (1986); K. Soai, A. Ookawa, T. Kaba, and K. Ogawa, *J. Am. Chem. Soc.*, **109**, 7111 (1987); M. Kitamura, S. Okada, and R. Noyori, *J. Am. Chem. Soc.*, **111**, 4028 (1989); T. Rasmussen and P.-O. Norrby, *J. Am. Chem. Soc.*, **125**, 5130 (2003).

 15. W.-L. Cheng, Y.-J. Shaw, S.-M. Yeh, P. P. Kanakamma, Y.-H. Chen, C. Chen, J.-C. Shieu, S.-J. Yiin, G.-H. Lee, Y. Wang, and T.-Y Luh, *J. Org. Chem.*, **64**, 532 (1999).

17a. R. F. W. Jackson, I. Rilatt, and P. J. Murry, *Chem. Commun.*, 1242 (2003).

   b. M. I. Calaza, M. R. Paleo, and F. J. Sardina, *J. Am. Chem. Soc.*, **123**, 2095 (2001).

Chapter 8

1a. C. Huynh, F. Derguini-Boumechal, and G. Linstrumelle, *Tetrahedron Lett.*, 1503 (1979).

  b. N. J. LaLima, Jr., and A. B. Levy, *J. Org. Chem.*, **43**, 1279 (1978).

  c. A. Cowell and J. K. Stille, *J. Am. Chem. Soc.*, **102**, 4193 (1980).

  d. T. Sato, M. Kawashima, and T. Fujisawa, *Tetrahedron Lett.*, 2375 (1981).

  e. H. P. Dang and G. Linstrumelle, *Tetrahedron Lett.*, 191 (1978).

  f. B. M. Trost and D. P. Curran, *J. Am. Chem. Soc.*, **102**, 5699 (1980).

  g. D. J. Pasto, S.-K. Chou, E. Fritzen, R. H. Shults, A. Waterhouse, and G. F. Hennion, *J. Org. Chem.*, **43**, 1389 (1978).

  h. B. H. Lipshutz, J. Kozlowski, and R. S. Wilhelm, *J. Am. Chem. Soc.*, **104**, 2305 (1982).

  i. P. A. Grieco and C. V. Srinivasan, *J. Org. Chem.*, **46**, 2591 (1981).

  j. C. Iwata, K. Suzuki, S. Aoki, K. Okamura, M. Yamashita, I. Takahashi, and T. Tanaka, *Chem. Pharm. Bull.*, **34**, 4939 (1988).

  k. A. Alexakis, G. Cahiez, and J. F. Normant, *Org. Synth.*, **VII**, 290 (1990).

  l. J. Tsuji, Y. Kobayashi, H. Kataoka, and T. Takahashi, *Tetrahedron Lett.*, **21**, 1475 (1980).

  m. W. A. Nugent and R. J. McKinney, *J. Org. Chem.*, **50**, 5370 (1985).

  n. R. M. Wilson, K. A. Schnapp, R. K. Merwin, R. Ranganathan, D. L. Moats, and T. T. Conrad, *J. Org. Chem.*, **51**, 4028 (1986).

  o. L. N. Pridgen, *J. Org. Chem.*, **47**, 4319 (1982).

  p. R. Casas, C. Cave, and J. d'Angelo, *Tetrahedron Lett.*, **36**, 1039 (1995).

  q. N. Miyaura, K. Yamada, and A. Suzuki, *Tetrahedron Lett.*, 3437 (1979).

  r. F. K. Steffy, J. P. Godschalx, and J. K. Stille, *J. Am. Chem. Soc.*, **106**, 4833 (1984).

2a. B. H. Lipshutz, M. Koerner, and D. A. Parker, *Tetrahedron Lett.*, **28**, 945 (1987).

  b. B. H. Lipshutz, R. S. Wilhelm, J. A. Kozlowski, and D. Parker, *J. Org. Chem.*, **49**, 3928 (1984).

  c. J. P. Marino, R. Fernandez de la Pradilla, and E. Laborde, *J. Org. Chem.*, **52**, 4898 (1987).

  d. C. R. Johnson and D. S. Dhanoa, *J. Org. Chem.*, **52**, 1885 (1987).

3a. H. Urata, A. Fujita, and T. Fuchikami, *Tetrahedron Lett.*, **29**, 4435 (1988).

  b. Y. Itoh, H. Aoyama, T. Hirao, A. Mochizuki, and T. Saegusa, *J. Am. Chem. Soc.*, **101**, 494 (1979).

  c. P. G. M. Wuts, M. L. Obrzut, and P. A. Thompson, *Tetrahedron Lett.*, **25**, 4051 (1984).

  d. K. Kokubo, K. Matsumasa, M. Miura, and M. Nomura, *J. Org. Chem.*, **61**, 6941 (1996).

e. S. T. Diver and A. J. Giessert, *Synthesis*, 466 (2004).

4a. R. J. Anderson, V. L. Corbin, G. Cotterrel, G. R. Cox, C. A. Henrick, F. Schaub, and J. B. Siddall, *J. Am. Chem. Soc.*, **97**, 1197 (1975).

b. P. de Mayo, L. K. Sydnes, and G. Wenska, *J. Org. Chem.*, **45**, 1549 (1980).

c. Y. Yamamoto, H. Yatagai, and K. Maruyama, *J. Org. Chem.*, **44**, 1744 (1979).

d. H. Shostarez and L. A. Paquette, *J. Am. Chem. Soc.*, **103**, 722 (1981).

e. W. G. Dauben, G. H. Beasley, M. D. Broadhurst, B. Muller, D. J. Peppard, P. Pesnelle, and C. Suter, *J. Am. Chem. Soc.*, **97**, 4973 (1975).

f. L. Watts, J. D. Fitzpatrick, and R. Pettit, *J. Am. Chem. Soc.*, **88**, 623 (1966).

g. J. I. Kim, B. A. Patel, and R. F. Heck, *J. Org. Chem.*, **46**, 1067 (1981).

h. J. A. Marshall, W. F. Huffman, and J. A. Ruth, *J. Am. Chem. Soc.*, **94**, 4691 (1972).

i. H.-A. Hasseberg and H. Gerlach, *Helv. Chim. Acta*, **71**, 957 (1988).

j. R. Alvarez, M. Herrero, S. Lopez, and A. R. de Lera, *Tetrahedron*, **54**, 6793 (1998).

k. T. K. Chakraborty and D. Thippeswamy, *Synlett*, 150 (1999).

l. S. Jinno, T. Okita, and K. Inoyue, *Bioorg. Med. Chem. Lett.*, **9**, 1029 (1999).

m. J. Thibonnet, M. Abarbi, A. Duchene, and J.-L. Parrain, *Synlett*, 141 (1999).

n. G. Giambastiani and G. Poli, *J. Org. Chem.*, **63**, 9608 (1998).

o. N. Miyaura, K. Yamada, H. Suginome, and A. Suzuki, *J. Am. Chem. Soc.*, **107**, 972 (1985).

p. S. Nakamura, Y. Hirata, T. Kurosaki, M. Anada, O. Kataoka, S. Kitagaki, and S. Hashimoto, *Angew. Chem. Int. Ed. Engl.*, **42**, 5351 (2003).

5. B. O'Connor and G. Just, *J. Org. Chem.*, **52**, 1801 (1987); G. Just and B. O'Connor, *Tetrahedron Lett.*, **26**, 1799 (1985).

6. B. H. Lipshutz, R. S. Wilhelm, J. A. Kozlowski, and D. Parker, *J. Org. Chem.*, **49**, 3928 (1984); E. C. Ashby, R. N. DePriest, A. Tuncay, and S. Srivasta, *Tetrahedron Lett.*, **23**, 5251 (1982).

7a. C. G. Chavdarian and C. H. Heathcock, *J. Am. Chem. Soc.*, **97**, 3822 (1975).

b. E. J. Corey and J. G. Smith, *J. Am. Chem. Soc.*, **101**, 1038 (1977).

c. G. Mehta and K. S. Rao, *J. Am. Chem. Soc.*, **108**, 8015 (1986).

d. W. A. Nugent and F. W. Hobbs, Jr., *Org. Synth.*, **66**, 52 (1988).

e. G. F. Cooper, D. L. Wren, D. Y. Jackson, C. C. Beard, E. Galeazzi, A. R. Van Horn, and T. T. Li, *J. Org. Chem.*, **58**, 4280 (1993).

f. R. K. Dieter, J. W. Dieter, C. W. Alexander, and N. S. Bhinderwala, *J. Org. Chem.*, **61**, 2930 (1996).

g. C. R. Johnson and T. D. Penning, *J. Am. Chem. Soc.*, **110**, 4726 (1988).

h. T. Hudlicky and H. F. Olivo, *J. Am. Chem. Soc.*, **114**, 9694 (1992).

8a. C. M. Lentz and G. H. Posner, *Tetrahedron Lett.*, 3769 (1978).

b. A. Marfat, P. R. McGuirk, R. Kramer, and P. Helquist, *J. Am. Chem. Soc.*, **99**, 253 (1977).

c. L. A. Paquette and Y.-K. Han, *J. Am. Chem. Soc.*, **103**, 1831 (1981).

d. A. Alexakis, J. Berlan, and Y. Besace, *Tetrahedron Lett.*, **27**, 1047 (1986).

e. M. Sletzinger, T. R. Verhoeven, R. P. Volante, J. M. McNamara, E. G. Corley, and T. M. H. Liu, *Tetrahedron Lett.*, **26**, 2951 (1985).

9a. E. J. Corey and E. Hamanaka, *J. Am. Chem. Soc.*, **89**, 2758 (1967).

b. Y. Kitagawa, A. Itoh, S. Hashimoto, H. Yamamoto, and H. Nozaki, *J. Am. Chem. Soc.*, **99**, 3864 (1977).

c. B. M. Trost and R. W. Warner, *J. Am. Chem. Soc.*, **105**, 5940 (1983).

d. S. Brandt, A. Marfat, and P. Helquist, *Tetrahedron Lett.*, 2193 (1979).

e. A. Fürstner and H. Weintritt, *J. Am. Chem. Soc.*, **120**, 2817 (1998).

f. Y. Matsuya, T. Kawaguchi, and H. Nemoto, *Org. Lett.*, **5**, 2939 (2003).

10. R. H. Grubbs and R. A. Grey, *J. Am. Chem. Soc.*, **95**, 5765 (1973).

11. H. L. Goering, E. P. Seitz, Jr., and C. C. Tseng, *J. Org. Chem.*, **46**, 5304 (1981).

12. A. Marfat, P. R. McGuirk, and P. Helquist, *J. Org. Chem.*, **44**, 1345 (1979).

13. N. Cohen, W. F. Eichel, R. J. Lopresti, C. Neukom, and G. Saucy, *J. Org. Chem.*, **41**, 3505 (1976).

14a. R. J. Linderman, A. Godfrey, and K. Horne, *Tetrahedron Lett.*, **28**, 3911 (1987).

b. H. Schostarez and L. A. Paquette, *J. Am. Chem. Soc.*, **103**, 722 (1981).

c. Y. Yamamoto, S. Yamamoto, H. Yatagai, Y. Ishihara, and K. Maruyama, *J. Org. Chem.*, **47**, 119 (1982).

d. T. Kawabata, P. A. Grieco, H.-L. Sham, H. Kim, J. Y. Law, and S. Tu, *J. Org. Chem.*, **52**, 3346 (1987).

15a. A. Minato, K. Suzuki, K. Tamao, and M. Kumada, *Tetrahedron Lett.*, **25**, 83 (1984).

b. E. R. Larson and R. A. Raphael, *Tetrahedron Lett.*, 5041 (1979).

c. M. C. Pirrung and S. A. Thompson, *J. Org. Chem.*, **53**, 227 (1988).

d. J. Just and B. O'Connor, *Tetrahedron Lett.*, **29**, 753 (1988).

e. M. F. Semmelhack and A. Yamashita, *J. Am. Chem. Soc.*, **102**, 5924 (1980).

f. A. M. Echavarren and J. K. Stille, *J. Am. Chem. Soc.*, **110**, 4051 (1988).

g. K. Nakamura, H. Okubo, and M. Yamaguchi, *Synlett*, 549 (1999).

h. A. F. Littke, L. Schwarz, and G. C. Fu, *J. Am. Chem. Soc.*, **124**, 6343 (2002).

i. W. A. Moradi and S. Buchwald, *J. Am. Chem. Soc.*, **123**, 7996 (2001).

j. M. Palucki and S. L. Buchwald, *J. Am. Chem. Soc.*, **119**, 11108 (1997).

k. H. Tang, K. Menzel, and G. C. Fu, *Angew. Chem. Int. Ed. Engl.*, **42**, 5079 (2003).

l. Y. Urawa and K. Ogura, *Tetrahedron Lett.*, **44**, 271 (2003).

16a. J. E. Backvall, S. E. Bystrom, and R. E. Nordberg, *J. Org. Chem.* **49**, 4619 (1984).

b. M. F. Semmelhack and C. Bodurow, *J. Am. Chem. Soc.*, **106**, 1496 (1984).

c. D. Valentine, Jr., J. W. Tilley, and R. A. LeMahieu, *J. Org. Chem.*, **46**, 4614 (1981).

d. A. S. Kende, B. Roth, P. J. SanFilippo, and T. J. Blacklock, *J. Am. Chem. Soc.*, **104**, 5808 (1982).

17. E. J. Corey, F. J. Hannon, and N. W. Boaz, *Tetrahedron*, **45**, 545 (1989).

18a. P. A. Bartlett, J. D. Meadows, and E. Ottow, *J. Am. Chem. Soc.*, **106**, 5304 (1984).

b. M. Larcheveque and Y. Petit, *Tetrahedron Lett.*, **28**, 1993 (1987).

c. B. M. Trost and J. D. Oslob, *J. Am. Chem. Soc.*, **121**, 3057 (1999).

19. P. Compain, J. Gore, and J. M. Vatele, *Tetrahedron*, **52**, 10405 (1996); H. D. Doan, J. Gore, and J.-M. Vatele, *Tetrahedron Lett.*, **40**, 6765 (1999).

20. T. Kitamura and M. Mori, *Org. Lett.*, **3**, 1161 (2001).

Chapter 9

1a. A. Suzuki, N. Miyaura, S. Abiko, M. Itoh, H. C. Brown, J. A. Sinclair, and M. M. Midland, *J. Am. Chem. Soc.*, **95**, 3080 (1973).

b. H. Yatagai, Y. Yamamoto, and K. Maruyama, *J. Am. Chem. Soc.*, **102**, 4548 (1980); Y. Yamamoto, H. Yatagai, H. Naruta, and K. Maruyama, *J. Am. Chem. Soc.*, **102**, 7107 (1980).

c. R. Mohan and J. A. Katzenellenbogen, *J. Org. Chem.*, **49**, 1238 (1984).

d. H. C. Brown and T. Imai, *J. Am. Chem. Soc.*, **105**, 6285 (1983).

e. H. C. Brown, N. G. Bhat, and J. B. Campbell, Jr., *J. Org. Chem.*, **51**, 3398 (1986).

2a. H. C. Brown, M. M. Rogic, H. Nambu, and M. W. Rathke, *J. Am. Chem. Soc.*, **91**, 2147 (1969); H. C. Brown, H. Nambu, and M. M. Rogic, *J. Am. Chem. Soc.*, **91**, 6852 (1969).

b. H. C. Brown and R. A. Coleman, *J. Am. Chem. Soc.*, **91**, 4606 (1969).

c. G. Zweifel, R. P. Fisher, J. T. Snow, and C. C. Whitney, *J. Am. Chem. Soc.*, **93**, 6309 (1971).

d. H. C. Brown and M. W. Rathke, *J. Am. Chem. Soc.*, **89**, 2738 (1967).

e. H. C. Brown and M. M. Rogic, *J. Am. Chem. Soc.*, **91**, 2146 (1969); H. C. Brown, H. Nambu, and M. M. Rogic, *J. Am. Chem. Soc.*, **91**, 6852 (1969).

3. See the references to Scheme 9.1.

4a. H. C. Brown, H. D. Lee, and S. U. Kulkarni, *J. Org. Chem.*, **51**, 5282 (1986).

b. J. A. Sikorski, N. G. Bhat, T. E. Cole, K. K. Wang, and H. C. Brown, *J. Org. Chem.*, **51**, 4521 (1986).

c. S. U. Kulkarni, H. D. Lee, and H. C. Brown, *J. Org. Chem.*, **45**, 4542 (1980).

d. M. C. Welch and T. A. Bryson, *Tetrahedron Lett.*, **29**, 521 (1988).

5a. A. Pelter, K. J. Gould, and C. R. Harrison, *Tetrahedron Lett.*, 3327 (1975).

b. A. Pelter and R. A. Drake, *Tetrahedron Lett.*, **29**, 4181 (1988).

c. L. E. Overman and M. J. Sharp, *J. Am. Chem. Soc.*, **110**, 612 (1988).

d. H. C. Brown and S. U. Kulkarni, *J. Org. Chem.*, **44**, 2422 (1979).

6a. W. R. Roush, M. A. Adam, and D. J. Harris, *J. Org. Chem.*, **50**, 2000 (1985).

b. S. J. Danishefsky, S. DeNinno, and P. Lartey, *J. Am. Chem. Soc.*, **109**, 2082 (1987).

c. Y. Nishigaichi, N. Ishida, M. Nishida, and A. Takuwa, *Tetrahedron Lett.*, **37**, 3701 (1996).

d. D. A. Heerding, C. Y. Hong, N. Kado, G. C. Look, and L. E. Overman, *J. Org. Chem.*, **58**, 6947 (1993).

7a,b. H. C. Brown and N. G. Bhat, *J. Org. Chem.*, **53**, 6009 (1988).

c,d. H. C. Brown, D. Basavaiah, S. U. Kulkarni, N. Bhat, and J. V. N. Vara Prasad, *J. Org. Chem.*, **53**, 239 (1988).

8a. J. A. Marshall, S. L. Crooks, and B. S. DeHoff, *J. Org. Chem.*, **53**, 1616 (1988); J. A. Marshall and W. Y. Gung, *Tetrahedron Lett.*, **29**, 1657 (1988).

b. B. M. Trost and T. Sato, *J. Am. Chem. Soc.*, **107**, 719 (1985).

9a. W. E. Fristad, D. S. Dime, T. R. Bailey, and L. A. Paquette, *Tetrahedron Lett.*, 1999 (1979).

b. E. Piers and H. E. Morton, *J. Org. Chem.*, **45**, 4263 (1980).

c. J. C. Bottaro, R. N. Hanson, and D. E. Seitz, *J. Org. Chem.*, **46**, 5221 (1981).

d. Y. Yamamoto and A. Yanagi, *Heterocycles*, **16**, 1161 (1981).

e. M. B. Anderson and P. L. Fuchs, *Synth. Commun.*, **17**, 621 (1987); B. A. Narayanan and W. H. Bunelle, *Tetrahedron Lett.*, **28**, 6261 (1987).

f. A. Hosomi, M. Sato, and H. Sakurai, *Tetrahedron Lett.*, 429 (1979).

10. P. A. Grieco and W. F. Fobare, *Tetrahedron Lett.*, **27**, 5067 (1986).

11. E. J. Corey and W. L. Seibel, *Tetrahedron Lett.*, **27**, 905 (1986).

12a. E. Moret and M. Schlosser, *Tetrahedron Lett.*, **25**, 4491 (1984).

b. L. K. Truesdale, D. Swanson, and R. C. Sun, *Tetrahedron Lett.*, **26**, 5009 (1985).

c. R. L. Funk and G. L. Bolton, *J. Org. Chem.*, **49**, 5021 (1984).

d. L. E. Overman, T. C. Malone, and G. P. Meier, *J. Am. Chem. Soc.*, **105**, 6993 (1983).

13a. H. C. Brown, T. Imai, M. C. Desai, and B. Singaran, *J. Am. Chem. Soc.*, **107**, 4980 (1985).

b,c. H. C. Brown, R. K. Bakshi, and B. Singaran, *J. Am. Chem. Soc.*, **110**, 1529 (1988).

d. H. C. Brown, M. Srebnik, R. R. Bakshi, and T. E. Cole, *J. Am. Chem. Soc.*, **109**, 5420 (1987).

14. K. K. Wang and K.-H. Chu, *J. Org. Chem.*, **49**, 5175 (1984).

15a. W. R. Roush, J. A. Straub, and M. S. Van Nieuwenhze, *J. Org. Chem.*, **56**, 1636 (1991).

b. L. A. Paquette and G. D. Maynard, *J. Am. Chem. Soc.*, **114**, 5018 (1992).

c. C. Y. Hong, N. Kado, and L. E. Overman, *J. Am. Chem. Soc.*, **115**, 11028 (1993).

d. P. V. Ramachandran, G.-M. Chen, and H. C. Brown, *Tetrahedron Lett.* **38**, 2417 (1997).

e. A. B. Charette, C. Mellon, and M. Motamedi, *Tetrahedron Lett.*, **36**, 8561 (1995).

f. C. Masse, M. Yang, J. Solomon, and J. S. Panek, *J. Am. Chem. Soc.*, **120**, 4123 (1998).

16. I. Chataigner, J. Lebreton, F. Zammattio, and J. Villieras, *Tetrahedron Lett.*, **38**, 3719 (1997).

17. J. A. Marshall, B. M. Seletsky, and G. P. Luke, *J. Org. Chem.*, **59**, 3413 (1994); J. A. Marshall, J. A. Jablonowski, and G. P. Luke, *J. Org. Chem.*, **59**, 7825 (1994).

18. K. Maruyama, Y. Ishiara, and Y. Yamamoto, *Tetrahedron Lett.*, **22**, 4235 (1981); Y. Yamamoto, H. Nemoto, R. Kikuchi, H. Komatsu, and I. Suzuki, *J. Am. Chem. Soc.*, **112**, 8598 (1990). For allylic boron additions to this compound see: R. W. Hoffmann, W. Ladner, and K. Ditrich, *Liebigs. Ann. Chem.*, 883 (1989).

19. J. A. Marshall and S. Beaudoin, *J. Org. Chem.*, **59**, 7833 (1994).

Chapter 10

1a. S. Julia and A. Ginebreda, *Synthesis*, 682 (1977).

b. D. S. Breslow, E. I. Edwards, R. Leone, and P. V. R. Schleyer, *J. Am. Chem. Soc.*, **90**, 7097 (1968).

c. D. J. Burton and J. L. Hahnfeld, *J. Org. Chem.*, **42**, 828 (1977).

d. D. Seyferth and S. P. Hopper, *J. Org. Chem.*, **37**, 4070 (1972).

e. G. L. Closs, L. E. Closs, and W. A. Boll, *J. Am. Chem. Soc.*, **85**, 3796 (1963).

f. L. G. Mueller and R. G. Lawton, *J. Org. Chem.*, **44**, 4741 (1979).

g. F. G. Bordwell and M. W. Carlson, *J. Am. Chem. Soc.*, **92**, 3377 (1970).

h. A. Burger and G. H. Harnest, *J. Am. Chem. Soc.*, **65**, 2382 (1943).

i. E. Schmitz, D. Habish, and A. Stark, *Angew. Chem. Int. Ed. Engl.*, **2**, 548 (1963).

j. R. Zurfluh, E. N. Wall, J. B. Sidall, and J. A. Edwards, *J. Am. Chem. Soc.*, **90**, 6224 (1968).

k. M. Nishizawa, H. Takenaka, and Y. Hayashi, *J. Org. Chem.*, **51**, 806 (1986).

l. H. Nishiyama, K. Sakuta, and K. Itoh, *Tetrahedron Lett.*, **25**, 223 (1984).

m. H. Seto, M. Sakaguchi, and Y. Fujimoto, *Chem. Pharm. Bull.*, **33**, 412 (1985).

n. D. F. Taber and E. H. Petty, *J. Org. Chem.*, **47**, 4808 (1982).

o. B. Iddon, D. Price, H. Suschitzsky, and D. J. C. Scopes, *Tetrahedron Lett.*, **24**, 413 (1983).

p. A. Chu and L. N. Mander, *Tetrahedron Lett.*, **29**, 2727 (1988).

q. G. E. Keck and D. F. Kachensky, *J. Org. Chem.*, **51**, 2487 (1986).

r. W. A. Thaler and B. Franzus, *J. Org. Chem.*, **29**, 2226 (1964).

s. B. B. Snider, *J. Org. Chem.*, **41**, 3061 (1976).

t. D. H. R. Barton, J. Guilhem, Y. Herve, P. Potier, and J. Thierry, *Tetrahedron Lett.*, **28**, 1413 (1987).

u. A. M. Gomez, G. O. Danelon, S. Valverde, and J. C. Lopez, *J. Org. Chem.*, **63**, 9626 (1998).

v. S. D. Burke and D. N. Deaton, *Tetrahedron Lett.*, **32**, 4651 (1991).

w. J. A. Wendt and J. Aube, *Tetrahedron Lett.*, **37**, 1531 (1996).

x. G. Mehta, S. Karmarkar, and S. K. Chattopadhyay, *Tetrahedron*, **60**, 5013 (2004).

2a. K. B. Wiberg, B. L. Furtek, and L. K. Olli, *J. Am. Chem. Soc.*, **101**, 7675 (1979).

b. A. E. Greene and J.-P. Depres, *J. Am. Chem. Soc.*, **101**, 4003 (1979).

c. R. A. Moss and E. Y. Chen, *J. Org. Chem.*, **46**, 1466 (1981).

d. B. M. Trost, R. M. Cory, P. H. Scudder, and H. B. Neubold, *J. Am. Chem. Soc.*, **95**, 7813 (1973).

e. T. J. Nitz, E. M. Holt, B. Rubin, and C. H. Stammer, *J. Org. Chem.*, **46**, 2667 (1981).

f. L. N. Mander, J. V. Turner, and B. G. Colmbe, *Aust. J. Chem.*, **27**, 1985 (1974).

g. P. J. Jessup, C. B. Petty, J. Roos, and L. E. Overman, *Org. Synth.*, **59**, 1 (1979).

h. H. Durr, H. Nickels, L. A. Pacala, and M. Jones, Jr., *J. Org. Chem.*, **45**, 973 (1980).

i. G. A Scheisher and J. D. White, *J. Org. Chem.*, **45**, 1864 (1980).

j. M. B. Groen and F. J. Zeelen, *J. Org. Chem.*, **43**, 1961 (1978).

k. R. C. Gadwood, R. M. Lett, and J. E. Wissinger, *J. Am. Chem. Soc.*, **108**, 6343 (1986).

l. V. B. Rao, C. F. George, S. Wolff, and W. C. Agosta, *J. Am. Chem. Soc.*, **107**, 5732 (1985).

m. Y. Araki, T. Endo, M. Tanji, J. Nagasawara, and Y. Ishido, *Tetrahedron Lett.*, **28**, 5853 (1987).

n. G. Stork, P. M. Sher, and H.-L. Chen, *J. Am. Chem. Soc.*, **108**, 6384 (1986).

o. E. J. Corey and M. Kang, *J. Am. Chem. Soc.*, **106**, 5384 (1984).

p. G. E. Keck, D. F. Kachensky, and E. J. Enholm, *J. Org. Chem.*, **50**, 4317 (1985).

q. A. De Mesmaeker, P. Hoffmann, and B. Ernst, *Tetrahedron Lett.*, **30**, 57 (1989).

r. A. K. Singh, R. K. Bakshi, and E. J. Corey, *J. Am. Chem. Soc.*, **109**, 6187 (1987).

s. S. Danishefsky and J. S. Panek, *J. Am. Chem. Soc.*, **109**, 917 (1987).

t. M. Handa, T. Sunazaka, A. Sugawara, Y. Harigara, O. Yoshihiro, K. Otoguro, and S. Omura, *J. Antibiotics*, **56**, 730 (2003).

3a. T. J. Lee, A. Bunge, and H. F. Schaefer, III, *J. Am. Chem. Soc.*, **107**, 137 (1985); M. Rubio, J. Stalring, A. Bernhardsson, R. Lindh, and B. O. Roos, *Theoretical Chem. Acc.*, **105**, 15 (2000); R. Kakkar, R. Garg, and P. Preeti, *Theochem*, **617**, 141 (2000).

b. R. Noyori and M. Yamakawa, *Tetrahedron Lett.*, **21**, 2851 (1980).

c. S. Matzinger, T. Bally, E. V. Patterson, and R. J. McMahon, *J. Am. Chem. Soc.*, **118**, 1535 (1996); P. R. Schreiner, W. L. Karney, P. v. R. Schleyer, W. T. Borden, T. P. Hamilton, and H. F. Schaefer, III, *J. Org. Chem.*, **61**, 7030 (1996).

d. C. Boehme and G. Frenking, *J. Am. Chem. Soc.*, **118**, 2039 (1996).

4. C. D. Poulter, E. C. Friedrich, and S. Winstein, *J. Am. Chem. Soc.*, **91**, 6892 (1969).

5. P. L. Barili, G. Berti, B. Macchia, F. Macchia, and L. Monti, *J. Chem. Soc. C*, 1168 (1970).

6a. R. K. Hill and D. A. Cullison, *J. Am. Chem. Soc.*, **95**, 2923 (1973).

b. A. B. Smith, III, B. H. Toder, S. J. Brancha, and R. K. Dieter, *J. Am. Chem. Soc.*, **103**, 1996 (1981).

c. M. C. Pirrung and J. A. Werner, *J. Am. Chem. Soc.*, **108**, 6060 (1986).

7. E. W. Warnhoff, C. M. Wong, and W. T. Tai, *J. Am. Chem. Soc.*, **90**, 514 (1968).

8a. S. A. Godleski, P. v. R. Schleyer, E. Osawa, Y. Inamoto, and Y. Fujikura, *J. Org. Chem.*, **41**, 2596 (1976).

b. P. E. Eaton, Y. S. Or, and S. J. Branca, *J. Am. Chem. Soc.*, **103**, 2134 (1981).

c. G. H. Posner, K. A. Babiak, G. L. Loomis, W. J. Frazee, R. D. Mittal, and I. L. Karle, *J. Am. Chem. Soc.*, **102**, 7498 (1980).

d. T. Hudlicky, F. J. Koszyk, T. M. Kutchan, and J. P. Sheth, *J. Org. Chem.*, **45**, 5020 (1980).

e. L. A. Paquette and Y.-K. Han, *J. Am. Chem. Soc.*, **103**, 1835 (1981).

f. L. A. Paquette and R. W. Houser, *J. Am. Chem. Soc.*, **91**, 3870 (1969).

g. L. A. Paquette, S. Nakatani, T. M. Zydowsky, S. D. Edmondson, L.-Q. Sun, and R. Skerlj, *J. Org. Chem.*, **64**, 3244 (1999).

9a. Y. Ito, S. Fujii, M. Nakatsuka, F. Kawamoto, and T. Saegusa, *Org. Synth.*, **59**, 113 (1979).

b. P. Nedenskov, H. Heide, and N. Clauson-Kass, *Acta Chem. Scand.*, **16**, 246 (1962).

c. L.-F. Tietze, *J. Am. Chem. Soc.*, **96**, 946 (1974).

d. E. G. Breitholle and A. G. Fallis, *J. Org. Chem.*, **43**, 1964 (1978).

e. E. Y. Chen, *J. Org. Chem.*, **49**, 3245 (1984).

f. G. Mehta and K. S. Rao, *J. Org. Chem.*, **50**, 5537 (1985).

g. T. V. Rajan Babu, *J. Org. Chem.*, **53**, 4522 (1988).

h. W. D. Klobucar, L. A. Paquette, and J. P. Blount, *J. Org. Chem.*, **46**, 4021 (1981).

i. F. E. Ziegler, S. I. Klein, U. K. Pati, and T.-F. Wang, *J. Am. Chem. Soc.*, **107**, 2730 (1985).

j. T. Hudlicky, F. J. Koszyk, D. M. Dochwat, and G. L. Cantrell, *J. Org. Chem.*, **46**, 2911 (1981).

k. R. E. Ireland, W. C. Dow, J. D. Godfrey, and S. Thaisrivongs, *J. Org. Chem.*, **49**, 1001 (1984).

l. C. P. Chuang and D. J. Hart, *J. Org. Chem.*, **48**, 1782 (1983).

m. P. Wender, T. W. von Geldern, and B. H. Levine, *J. Am. Chem. Soc.*, **110**, 4858 (1988).

10a. S. D. Larsen and S. A. Monti, *J. Am. Chem. Soc.*, **99**, 8015 (1977).

b. S. A. Monti and J. M. Harless, *J. Am. Chem. Soc.*, **99**, 2690 (1977).

c. F. T. Bond and C.-Y. Ho, *J. Org. Chem.*, **41**, 1421 (1976).

d. E. Wenkert, R. S. Greenberg, and H.-S. Kim, *Helv. Chim. Acta*, **70**, 2159 (1987).

e. B. B. Snider and M. A. Dombroski, *J. Org. Chem.*, **52**, 5487 (1987).

f. G. A. Kraus and K. Landgrebe, *Tetrahedron Lett.*, **25**, 3939 (1984).

g. S. Kim, S. Lee, and J. S. Koh, *J. Am. Chem. Soc.*, **113**, 5106 (1991).

h. S. Ando, K. P. Minor, and L. E. Overman, *J. Org. Chem.*, **62**, 6379 (1997).

i. A. Johns and J. A. Murphy, *Tetrahedron Lett.*, **29**, 837 (1988).

j. K. S. Feldman and A. K. K. Vong, *Tetrahedron Lett.*, **31**, 823 (1990).

k. D. L. J. Clive and S. Daigneault, *J. Org. Chem.*, **56**, 5285 (1991).

l. M. A. Brodney and A. Padwa, *J. Org. Chem.*, **64**, 556 (1999).

m. S.-H. Chen, S. Huang, and G. P. Roth, *Tetrahedron Lett.*, **36**, 8933 (1995).

n. J. B. Brogan, C. B. Bauer, R. D. Rogers, and C. K. Zerchner, *Tetrahedron Lett.*, **37**, 5053 (1996).

o. G. M. Allan, A. F. Parsons, and J.-F. Pons, *Synlett*, 1431 (2002).

p. J. Barluenga, M. Alvarez-Perez, F. Wuerth, F. Rodriguez, and F. J. Fananas, *Org. Lett.*, **5**, 905 (2003).

q. D. Kim, P. J. Shim, J. Lee, C. W. Park, S. W. Hong, and S. Kim, *J. Org. Chem.*, **65**, 4864 (2000).

r. P. Magnus, L. Diorazio, T. J. Donohoe, M. Giles, P. Rye, J. Tarrant, and S. Thom, *Tetrahedron*, **52**, 14147 (1996).

s. P. Beak, Z. Song, and J. E. Resek, *J. Org. Chem.*, **57**, 944 (1992).

t. T. A. Blumenkopf, G. C. Look, and L. E. Overman, *J. Am. Chem. Soc.*, **112**, 4399 (1990).

u. L. Barriault and I. Denissova, *Org. Lett.*, **4**, 1371 (2002).

11. L. Blanco, N. Slougi, G. Rousseau, and J. M. Conia, *Tetrahedron Lett.*, **22**, 645 (1981).

12. J. A. Marshall and J. A. Ruth, *J. Org. Chem.*, **39**, 1971 (1974).

13. S. D. Burke, M. E. Kort, S. M. S. Strickland, H. M. Organ, and L. A. Silks, III, *Tetrahedron Lett.*, **35**, 1503 (1994).

14. C. A. Grob, H. R. Kiefer, H. J. Lutz, and H. J. Wilkens, *Helv. Chim. Acta*, **50**, 416 (1967).

15. M. P. Doyle, W. E. Buhro, and J. F. Dellaria, Jr., *Tetrahedron Lett.*, 4429 (1979).

16. R. Tsang, J. K. Dickson, Jr., H. Pak, R. Walton, and B. Fraser-Reid, *J. Am. Chem. Soc.*, **109**, 3484 (1987).

17a. K. Takao, H. Ochiai, K. Yoshida, T. Hasizuka, H. Koshimura, K. Tadano, and S. Ogawa, *J. Org. Chem.*, **60**, 8179 (1995).

b. H. Miyabe, M. Ueda, K. Fujii, A. Nishimura, and T. Naito, *J. Org. Chem.*, **68**, 5618 (2003).

c. C.-K. Sha, F.-K. Lee, and C.-J. Chang, *J. Am. Chem. Soc.*, **121**, 9875 (1999).

18. L. N. Mander and M. S. Sherburn, *Terahedron Lett.*, **37**, 4255 (1996).

19. A. Krief, G. L. Lorvelec, and S. Jeanmart, *Tetrahedron Lett.*, **41**, 3871 (2000).

20. R. A. Moss, F. Zheng, J.-M. Fede, Y. Ma, R. R. Sauers, J. P. Toscano, and B. M. Showalter, *J. Am. Chem. Soc.*, **124**, 5258 (2002).

21a. H. Tanada and T. Tsuji, *J. Org. Chem.*, **29**, 849 (1964); P. Story, *Tetrahedron Lett.*, 401 (1962).

b. E. J. Heiba, R. M. Dessau, and W. J. Koehl, Jr., *J. Am. Chem. Soc.*, **90**, 5905 (1968).

Chapter 11

1a. L. Friedman and H. Shechter, *J. Org. Chem.*, **26**, 2522 (1961).

b. E. C. Taylor, F. Kienzle, R. L. Robey, and A. McKillop, *J. Am. Chem. Soc.*, **92**, 2175 (1970).

c. J. Koo, *J. Am. Chem. Soc.*, **75**, 1889 (1953).

d. E. C. Taylor, R. Kienzle, R. L. Robey, A. McKillop, and J. D. Hunt, *J. Am. Chem. Soc.*, **93**, 4845 (1971).

e. G. A. Ropp and E. C. Coyner, *Org. Synth.*, **IV**, 727 (1963).

f. M. Shiratsuchi, K. Kawamura, T. Akashi, M. Fujii, H. Ishihama, and Y. Uchida, *Chem. Pharm. Bull.*, **35**, 632 (1987).

g. D. C. Furlano and K. D. Kirk, *J. Org. Chem.*. **51**, 4073 (1986).

h. C. K. Bradsher, F. C. Brown, and H. K. Porter, *J. Am. Chem. Soc.*, **76**, 2357 (1954).

i. F. A. Macias, D. Marin, D. Chincilla, and J. M. G. Molinillo, *Tetrahedron Lett.*, **43**, 6417 (2002).

2a. E. C. Taylor, E. C. Bingham, and D. K. Johnson, *J. Org. Chem.*, **42**, 362 (1977); G. S. Lal, *J. Org. Chem.*, **58**, 2791 (1993).

b. P. Studt, *Liebigs Ann. Chem.*, 2105 (1978).

c. T. Jojima, H. Takeshiba, and T. Kinoto, *Bull. Chem. Soc. Jpn.*, **52**, 2441 (1979).

d. R. W. Bost and F. Nicholson, *J. Am. Chem. Soc.*, **57**, 2368 (1935).

e. H. Durr, H. Nickels, L. A. Pacala, and M. Jones, Jr., *J. Org. Chem.*, **45**, 973 (1980).

3a. C. L. Perrin and G. A. Skinner, *J. Am. Chem. Soc.*, **93**, 3389 (1971).
  b. R. A. Rossi and J. F. Bunnett, *J. Am. Chem. Soc.*, **94**, 683 (1972).
  c. M. Jones, Jr., and R. H. Levin, *J. Am. Chem. Soc.*, **91**, 6411 (1969).
  d. Y. Naruta, Y. Nishigaichi, and K. Maruyama, *J. Org. Chem.*, **53**, 1192 (1988).
  e. S. P. Khanapure, R. T. Reddy, and E. R. Biehl, *J. Org. Chem.*, **52**, 5685 (1987).
  f. G. Buchi and J. C. Leung, *J. Org. Chem.*, **51**, 4813 (1986).
4a. M. P. Doyle, J. F. Dellaria, Jr., B. Siegfried, and S. W. Bishop, *J. Org. Chem.*, **42**, 3494 (1977).
  b. T. Cohen, A. G. Dietz, Jr., and J. R. Miser, *J. Org. Chem.*, **42**, 2053 (1977).
  c. M. P. Doyle, B. Siegfried, R. C. Elliot, and J. F. Dellaria, Jr., *J. Org. Chem.*, **42**, 2431 (1977).
  d. G. D. Figuly and J. C. Martin, *J. Org. Chem.*, **45**, 3728 (1980).
  e. E. McDonald and R. D. Wylie, *Tetrahedron*, **35**, 1415 (1979).
  f. M. P. Doyle, B. Siegfried, and J. F. Dellaria, Jr., *J. Org. Chem.*, **42**, 2426 (1977).
  g. P. H. Gore and I. M. Khan, *J. Chem. Soc., Perkin Trans. 1*, 2779 (1979).
  h. A. A. Leon, G. Daub, and I. R. Silverman, *J. Org. Chem.*, **49**, 4544 (1984).
  i. S. R. Wilson and L. A. Jacob, *J. Org. Chem.*, **51**, 4833 (1986).
  j. S. A. Khan, M. A. Munawar, and M. Siddiq, *J. Org. Chem.*, **53**, 1799 (1988).
  k. J. R. Beadle, S. H. Korzeniowski, D. E. Rosenberg, B. J. Garcia-Slanga, and G. W. Gokel, *J. Org. Chem.*, **49**, 1594 (1984).
5. B. L. Zenitz and W. H. Hartung, *J. Org. Chem.*, **11**, 444 (1946).
6. T. F. Buckley, III, and H. Rapoport, *J. Am. Chem. Soc.*, **102**, 3056 (1980).
7. G. A. Olah and J. A. Olah, *J. Am. Chem. Soc.*, **98**, 1839 (1976).
8. E. J. Corey, S. Barcza, and G. Klotmann, *J. Am. Chem. Soc.*, **91**, 4782 (1969).
9. E. C. Taylor, F. Kienzle, R. L. Robey, and A. McKillop, *J. Am. Chem. Soc.*, **92**, 2175 (1970).
10. M. Essiz, G. Guillaumet, J.-J. Brunet, and P. Caubere, *J. Org. Chem.*, **45**, 240 (1980).
11. S. P. Khanapure, L. Crenshaw, R. T. Reddy, and E. R. Biehl, *J. Org. Chem.*, **53**, 4915 (1988).
12a. J. H. Boyer and R. S. Burkis, *Org. Synth.*, **V**, 1067 (1973).
  b. H. P. Schultz, *Org. Synth.*, **IV**, 364 (1963); F. D. Gunstone and S. H. Tucker, *Org. Synth.*, **IV**, 160 (1963).
  c. D. H. Hey and M. J. Perkins, *Org. Synth.*, **V**, 51 (1973).
  d. K. Rorig, J. D. Johnston, R. W. Hamilton, and T. J. Telinski, *Org. Synth.*, **IV**, 576 (1963).
  e. K. G. Rutherford and W. Redmond, *Org. Synth.*, **V**, 133 (1973).
  f. M. M. Robinson and B. L. Robinson, *Org. Synth.*, **IV**, 947 (1963).
  g. R. Adams, W. Reifschneider, and A. Ferretti, *Org. Synth.*, **VI**, 21 (1988).
  h. G. H. Cleland, *Org. Synth.*, **VI**, 21 (1988).
13a. R. E. Ireland, C. A. Lipinski, C. J. Kowalski, J. W. Tilley, and D. M. Walba, *J. Am. Chem. Soc.*, **96**, 3333 (1974).
  b. J. J. Korst, J. D. Johnston, K. Butler, E. J. Bianco, L. H. Conover, and R. B. Woodward, *J. Am. Chem. Soc.*, **90**, 439 (1968).
  c. K. A. Parker and J. Kallmerten, *J. Org. Chem.*, **45**, 2614, 2620 (1980).
  d. F. A. Carey and R. M. Giuliano, *J. Org. Chem.*, **46**, 1366 (1981).
  e. R. B. Woodward and T. R. Hoye, *J. Am. Chem. Soc.*, **99**, 8007 (1977).
  f. E. C. Horning, J. Koo, M. S. Fish, and G. N. Walker, *Org. Synth.*, **IV**, 408 (1963); J. Koo, *Org. Synth.*, **V**, 550 (1973).
14a. M. J. Piggott and D. Wege, *Austr. J. Chem.*, **53**, 749 (2000).
  b. D. Perez, E. Guitian, and L. Castedo, *J. Org. Chem.*, **57**, 5911 (1992).
15. H. C. Bell, J. R. Kalman, J. T. Pinhey, and S. Sternhell, *Tetrahedron Lett.*, 3391 (1974).
16. B. Chauncy and E. Gellert, *Austr. J. Chem.*, **22**, 993 (1969); R. I. Duclos, Jr., J. S. Tung, and H. Rapoport, *J. Org. Chem.*, **49**, 5243 (1984).
17. W. G. Miller and C. U. Pittman, Jr., *J. Org. Chem.*, **39**, 1955 (1974).
18. W. Nagata, K. Okada, and T. Aoki, *Synthesis*, 365 (1979).
19. T. J. Doyle, M. Hendrix, D. Van Derveer, S. Javanmard, and J. Haseltine, *Tetrahedron*, **53**, 11153 (1997).
20. T. P. Smyth and B. W. Corby, *Org. Process Res. Dev.* **1**, 264 (1997).

Chapter 12

1a. Y. Butsugan, S. Yoshida, M. Muto, and T. Bito, *Tetrahedron Lett.*, 1129 (1971).
  b. E. J. Corey and H. E. Ensley, *J. Am. Chem. Soc.*, **97**, 6908 (1975).
  c. R. G. Gaughan and C. D. Poulter, *J. Org. Chem.*, **44**, 2441 (1979).

d. E. Vedejs, D. A. Engler, and J. E. Telschow, *J. Org. Chem.*, **43**, 188 (1978).

e. K. Akashi, R. E. Palermo, and K. B. Sharpless, *J. Org. Chem.*, **43**, 2063 (1978).

f. A. Hassner, R. H. Reuss, and H. W. Pinnick, *J. Org. Chem.*, **40**, 3427 (1975).

g. R. N. Mirrington and K. J. Schmalzl, *J. Org. Chem.*, **37**, 2877 (1972).

h. K. B. Sharpless and R. F. Lauer, *J. Org. Chem.*, **39**, 429 (1974).

i. J. A. Marshall and R. C. Andrews, *J. Org. Chem.*, **50**, 1602 (1985).

j. R. H. Schlessinger, J. J. Wood, A. J. Poos, R. A. Nugent, and W. H. Parsons, *J. Org. Chem.*, **48**, 1146 (1983).

k. R. K. Boeckman, Jr., J. E. Starett, Jr., D. G. Nickell, and P.-E. Sum, *J. Am. Chem. Soc.*, **108**, 5549 (1986).

l. E. J. Corey and Y. B. Xiang, *Tetrahedron*, **29**, 995 (1988).

m. D. J. Plata and J. Kallmerten, *J. Am. Chem. Soc.*, **110**, 4041 (1988).

n. B. E. Rossiter, T. Katsuki, and K. B. Sharpless, *J. Am. Chem. Soc.*, **103**, 464 (1981).

o. J. Mulzer, A. Angermann, B. Schubert, and C. Seilz, *J. Org. Chem.*, **51**, 5294 (1986).

p. R. H. Schlessinger and R. A. Nugent, *J. Am. Chem. Soc.*, **104**, 1116 (1982).

q. H. Niwa, T. Mori, T. Hasegawa, and K. Yamada, *J. Org. Chem.*, **51**, 1015 (1986).

2a. J. P. McCormick, W. Tomasik, and M. W. Johnson, *Tetrahedron Lett.*, **22**, 607 (1981).

b. H. C. Brown, J. H. Kawakami, and S. Ikegami, *J. Am. Chem. Soc.*, **92**, 6914 (1970).

c. R. M. Scarborough, Jr., B. H. Toder, and A. B. Smith, III, *J. Am. Chem. Soc.*, **102**, 3904 (1980).

d. B. Rickborn and R. M. Gerkin, *J. Am. Chem. Soc.*, **90**, 4193 (1968).

e. J. A. Marshall and R. A. Ruden, *J. Org. Chem.*, **36**, 594 (1971).

f. G. A. Kraus and B. Roth, *J. Org. Chem.*, **45**, 4825 (1980).

g. T. Sakan and K. Abe, *Tetrahedron Lett.*, 2471 (1968).

h. K. J. Clark, G. I. Fray, R. H. Jaeger, and R. Robinson, *Tetrahedron*, **6**, 217 (1959).

i. T. Kawabata, P. Grieco, H.-L. Sham, H. Kim, J. Y. Jaw, and S. Tu, *J. Org. Chem.*, **52**, 3346 (1987).

j. P. T. Lansbury, J. P. Galbo, and J. P. Springer, *Tetrahedron Lett.*, **29**, 147 (1988).

k. J. P. Marino, R. F. de La Pradilla, and E. Laborde, *J. Org. Chem.*, **52**, 4898 (1987).

l. J. E. Toth, P. R. Hamann, and P. L. Fuchs, *J. Org. Chem.*, **53**, 4694 (1988).

3. E. L. Eliel, S. H. Schroeter, T. J. Brett, F. J. Biros, and J.-C. Richer, *J. Am. Chem. Soc.*, **88**, 3327 (1966).

4. E. E. Royals and J. C. Leffingwell, *J. Org. Chem.*, **31**, 1937 (1966).

5. W. W. Epstein and F. W. Sweat, *Chem. Rev.*, **67**, 247 (1967).

6a. D. P. Higley and R. W. Murray, *J. Am. Chem. Soc.*, **96**, 3330 (1974).

b. R. Criegee and P. Gunther, *Chem. Ber.*, **96**, 1564 (1963).

c. W. P. Keaveny, M. G. Berger, and J. J. Pappas, *J. Org. Chem.*, **32**, 1537 (1967).

7a. S. Isoe, S. B. Hyeon, H. Ichikawa, S. Katsumura, and T. Sakan, *Tetrahedron Lett.*, 5561 (1968).

b. Y. Ogata, Y. Sawaki, and M. Shiroyama, *J. Org. Chem.* **42**, 4061 (1977).

c. F. G. Bordwell and A. C. Knipe, *J. Am. Chem. Soc.*, **93**, 3416 (1971).

d. B. M. Trost, P. R. Bernstein, and P. C. Funfschilling, *J. Am. Chem. Soc.*, **101**, 4378 (1979).

e. C. S. Foote, S. Mazur, P. A. Burns, and D. Lerdal, *J. Am. Chem. Soc.*, **95**, 586 (1973).

f. J. P. Marino, K. E. Pfitzner, and R. A. Olofson, *Tetrahedron*, **27**, 4181 (1971).

g. M. A. Avery, C. Jennings-White, and W. K. M. Chong, *Tetrahedron Lett.*, **28**, 4629 (1987).

h. S. Horvat, P. Karallas, and J. M. White, *J. Chem. Soc., Perkin Trans. 2*, 2151 (1998).

8a. P. N. Confalone, C. Pizzolato, D. L. Confalone, and M. R. Uskokovic, *J. Am. Chem. Soc.*, **102**, 1954 (1980).

b. J. K. Whitesell, R. S. Matthews, M. A. Minton, and A. M. Helbling, *J. Am. Chem. Soc.*, **103**, 3468 (1981).

c. F. A. J. Kerdesky, R. J. Ardecky, M. V. Lakshmikanthan, and M. P. Cava, *J. Am. Chem. Soc.*, **103**, 1992 (1981).

d. R. Fujimoto, Y. Kishi, and J. F. Blount, *J. Am. Chem. Soc.*, **102**, 7154 (1980).

e. S. P. Tanis and K. Nakanishi, *J. Am. Chem. Soc.*, **101**, 4398 (1979).

f. R. B. Miller and R. D. Nash, *J. Org. Chem.*, **38**, 4424 (1973).

g. R. Grewe and I. Hinrichs, *Chem. Ber.*, **97**, 443 (1964).

h. W. Nagata, S. Hirai, K. Kawata, and T. Okumura, *J. Am. Chem. Soc.*, **89**, 5046 (1967).

i. W. G. Dauben, M. Lorber, and D. S. Fullerton, *J. Org. Chem.*, **34**, 3587 (1969).

j. E. E. van Tamelen, M. Shamma, A. W. Burgstahler, J. Wolinsky, R. Tamm, and P. E. Aldrich, *J. Am. Chem. Soc.*, **80**, 5006 (1958).

k. S. D. Burke, C. W. Murtishaw, J. O. Saunders, J. A. Oplinger, and M. S. Dike, *J. Am. Chem. Soc.*, **106**, 4558 (1984).

l. B. M. Trost, P. G. McDougal, and K. J. Haller, *J. Am. Chem. Soc.*, **106**, 383 (1984).

9. B. M. Trost and K. Hiroi, *J. Am. Chem. Soc.*, **97**, 6911 (1975).

10a. F. Delay and G. Ohloff, *Helv. Chim. Acta*, **62**, 2168 (1979).

b. S. Danishefsky, R. Zamboni, M. Kahn, and S. J. Etheridge, *J. Am. Chem. Soc.*, **103**, 3460 (1981).

c. R. Noyori, T. Sato, and Y. Hayakawa, *J. Am. Chem. Soc.*, **100**, 2561 (1978).

d. J. K. Whitesell and R. S. Matthews, *J. Org. Chem.*, **43**, 1650 (1978).

e. R. C. Cambie, M. P. Hay, L. Larsen, C. E. F. Rickard, P. S. Rutledge, and P. D. Woodgate, *Aust. J. Chem.*, **44**, 821 (1991).

f. T. K. M. Shing, C. M. Lee, and H. Y. Lo, *Tetrahedron Lett.*, **42**, 8361 (2001).

g. P. A. Wender and T. P. Mucciaro, *J. Am. Chem. Soc.*, **114**, 5878 (1992).

11a. I. Saito, R. Nagata, K. Yubo, and Y. Matsuura, *Tetrahedron Lett.*, **24**, 4439 (1983).

b. J. R. Wiseman and S. Y. Lee, *J. Org. Chem.*, **51**, 2485 (1986).

c. H. Nishiyama, M. Matsumoto, H. Arai, H. Sakaguchi, and K. Itoh, *Tetrahedron Lett.*, **27**, 1599 (1986).

12a. R. E. Ireland, P. G. M. Wuts, and B. Ernst, *J. Am. Chem. Soc.*, **103**, 3205 (1981).

b. R. M. Scarborough, Jr., B. H. Tober, and A. B. Smith, III, *J. Am. Chem. Soc.*, **102**, 3904 (1980).

c. P. F. Hudrlik, A. M. Hudrlik, G. Nagendrappa, T. Yimenu, E. T. Zellers, and E. Chin, *J. Am. Chem. Soc.*, **102**, 6894 (1980).

d. T. Wakamatsu, K. Akasaka, and Y. Ban, *J. Org. Chem.*, **44**, 2008 (1979).

e. D. A. Evans, C. E. Sacks, R. A. Whitney, and N. G. Mandel, *Tetrahedron Lett.*, 727 (1978).

f. F. Bourelle-Wargnier, M. Vincent, and J. Chuche, *J. Org. Chem.*, **45**, 428 (1980).

g. J. A. Zalikowski, K. E. Gilbert, and W. T. Borden, *J. Org. Chem.*, **45**, 346 (1980).

h. E. Vogel, W. Klug, and A. Breuer, *Org. Synth.*, **55**, 86 (1976).

i. L. D. Spicer, M. W. Bullock, M. Garber, W. Groth, J. J. Hand, D. W. Long, J. L. Sawyer, and R. S. Wayne, *J. Org. Chem.*, **33**, 1350 (1968).

j. B. E. Rossiter, T. Katsuki, and K. B. Sharpless, *J. Am. Chem. Soc.*, **103**, 464 (1981).

k. L. A. Paquette and Y.-K. Han, *J. Am. Chem. Soc.*, **103**, 1831 (1981).

l. M. Muelbacher and C. D. Poulter, *J. Org. Chem.*, **53**, 1026 (1988).

m. P. T. W. Cheng and S. McLean, *Tetrahedron Lett.*, **29**, 3511 (1988).

n. A. B. Smith, III, and R. E. Richmond, Jr., *J. Am. Chem. Soc.*, **105**, 575 (1983).

o. E. J. Corey and Y. B. Xiang, *Tetrahedron Lett.*, **29**, 995 (1988).

p. T. Tanaka, K. Murakami, A. Kanda, D. Patra, S. Yamamoto, N. Satoh, S.-W. Kim, S. M. Abdur Rahman, H. Ohno, and C. Iwata, *J. Org. Chem.* **66**, 7107 (2001).

q. R. K. Boeckman, Jr., J. E. Starrett, Jr., D. G. Nickell, and P.-E. Sum, *J. Am. Chem. Soc.*, **108**, 5549 (1986).

13. Y. Gao and K. B. Sharpless, *J. Org. Chem.*, **53**, 4081 (1988).

14. C. W. Jefford, Y. Wang, and G. Bernardinelli, *Helv. Chim. Acta*, **71**, 2042 (1988).

15. R. W. Murray, M. Singh, B. L. Williams, and H. M. Moncrief, *J. Org. Chem.*, **61**, 1830 (1996).

16a. M. Chini, P. Crotti, L. A. Flippin, and F. Macchia, *J. Org. Chem.*, **55**, 4265 (1990).

b. B. Myrboh, H. Ila, and H. Junjappa, *Synthesis*, 126 (1981); T. Yamauchi, K. Nakao, and K. Fujii, *J. Chem. Soc., Perkin Trans. 1*, 1433 (1987).

c. R. M. Goodman and Y. Kishi, *J. Am. Chem. Soc.*, **120**, 9392 (1998).

17a. J. Moulines, A.-M. Lamidey and V. Desvernes-Breuil, *Synth. Commun.*, **31**, 749 (2001).

b. M. G. Bolster, B. J. M. Jansen, and A. de Groot, *Tetrahedron*, **57**, 5663 (2001).

18. S. Arseniyadis, R. B. Alves, D. V. Yashunsky, Q. Wang, and P. Potier, *Tetrahedron Lett.*, **36**, 1027 (1995); C. Unaleroglu, V. Aviyente, and S. Arseniyadis, *J. Org. Chem.*, **67**, 2447 (2002).

19a. K. D. Eom, J. V. Raman, H. Kim, and J. K. Cha, *J. Am. Chem. Soc.*, **125**, 5415 (2003).

b. Y. Yamano and M. Ito, *Chem. Pharm. Bull.*, **49**, 1662 (2001).

Chapter 13

1a. T. Hylton and V. Boekelheide, *J. Am. Chem. Soc.*, **90**, 6887 (1968).

b. B. W. Erickson, *Org. Synth.*, **54**, 19 (1974); E. J. Corey, B. W. Erickson, and R. Noyori, *J. Am. Chem. Soc.* **93**, 1724 (1971).

c. H. Paulsen, V. Sinnwell, and P. Stadler, *Angew. Chem. Int. Ed. Engl.*, **11**, 149 (1972).

d. S. Torii, K. Uneyama, and M. Isihara, *J. Org. Chem.*, **39**, 3645 (1974).

e. J. A. Marshall and A. E. Greene, *J. Org. Chem.*, **36**, 2035 (1971).

f. E. Leete, M. R. Chedekel, and G. B. Bodem, *J. Org. Chem.*, **37**, 4465 (1972).

g. H. Yamamoto and H. L. Sham, *J. Am. Chem. Soc.*, **101**, 1609 (1979).

h. K. Deuchert, U. Hertenstein, S. Hunig, and G. Weiner, *Chem. Ber.*, **112**, 2045 (1979).

i. T. Takahashi, K. Kitamura, and J. Tsuji, *Tetrahedron Lett.*, **24**, 4695 (1983).

2a. S. Danishefsky and T. Kitahara, *J. Am. Chem. Soc.*, **96**, 7807 (1974).

b. P. S. Wharton, C. E. Sundin, D. W. Johnson, and H. C. Kluender, *J. Org. Chem.*, **37**, 34 (1972).

c. E. J. Corey, B. W. Erickson, and R. Noyori, *J. Am. Chem. Soc.*, **93**, 1724 (1971).

d. R. E. Ireland and J. A. Marshall, *J. Org. Chem.*, **27**, 1615 (1962).

e. W. S. Johnson, T. J. Brocksom, P. Loew, D. H. Rich, L. Werthemann, R. A. Arnold, T. Li, and D. J. Faulkner, *J. Am. Chem. Soc.*, **92**, 4463 (1970).

f. L. Birladeanu, T. Hanafusa, and S. Winstein, *J. Am. Chem. Soc.*, **88**, 2315 (1966); T. Hanafusa, L. Birladeanu, and S. Winstein, *J. Am. Chem. Soc.*, **87**, 3510 (1965).

g. H. Takayanagi, Y. Kitano, and Y. Morinaka, *J. Org. Chem.*, **59**, 2700 (1994).

3a. A. B. Smith, III, and W. C. Agosta, *J. Am. Chem. Soc.*, **96**, 3289 (1974).

b. R. S. Cooke and U. H. Andrews, *J. Am. Chem. Soc.*, **96**, 2974 (1974).

c. L. A. Hulshof and H. Wynberg, *J. Am. Chem. Soc.*, **96**, 2191 (1974).

d. S. D. Burke, C. W. Murtiashaw, M. S. Dike, S. M. S. Strickland, and J. O. Saunders, *J. Org. Chem.*, **46**, 2400 (1981).

e. K. C. Nicolaou, M. R. Pavia, and S. P. Seitz, *J. Am. Chem. Soc.*, **103**, 1224 (1981).

f. J. Cossy, B. Gille, S. BouzBouz, and V. Bellosta, *Tetrahedron Lett.*, **38**, 4069 (1997).

4a. E. M. Acton, R. N. Goerner, H. S. Uh, K. J. Ryan, D. W. Henry, C. E. Cass, and G. A. LePage, *J. Med. Chem.*, **22**, 518 (1979).

b. E. G. Gros, *Carbohdr. Res.*, **2**, 56 (1966).

c. S. Hanessian and G. Rancourt, *Can. J. Chem.*, **55**, 1111 (1977).

d. R. R. Schmidt and A. Gohl, *Chem. Ber.*, **112**, 1689 (1979).

5a. S. F. Martin and T. Chou, *J. Org. Chem.*, **43**, 1027 (1978).

b. W. C. Still and M.-Y. Tsai, *J. Am. Chem. Soc.*, **102**, 3654 (1980).

c. J. C. Bottaro and G. A. Berchtold, *J. Org. Chem.*, **45**, 1176 (1980).

d. A. S. Kende and T. P. Demuth, *Tetrahedron Lett.*, **21**, 715 (1980).

e. J. A. Marshall and P. G. M. Wuts, *J. Org. Chem.*, **43**, 1086 (1978).

6a. R. Bonjouklian and R. A. Ruden, *J. Org. Chem.*, **42**, 4095 (1977).

b. L. A. Paquette, R. E. Moerck, B. Harirchian, and P. D. Magnus, *J. Am. Chem. Soc.*, **100**, 1597 (1978).

c. P. S. Wharton, C. E. Sundin, D. W. Johnson, and H. C. Kluender, *J. Org. Chem.*, **37**, 34 (1972).

d. M. E. Ochoa, M. S. Arias, R. Aguilar, I. Delgado, and J. Tamariz, *Tetrahedron*, **55**, 14535 (1999).

e. S. Danishefsky, T. Kitahara, C. F. Yan, and J. Morris, *J. Am. Chem. Soc.*, **101**, 6996 (1979).

f. B. M. Trost, J. Ippen, and W. C. Vladuchick, *J. Am. Chem. Soc.*, **99**, 8116 (1977).

7a. E. J. Corey, E. J. Trybulski, L. S. Melvin, K. C. Nicolaou, J. A. Secrist, R. Leltt, P. W. Sheldrake, J. R. Falck, D. J. Brunelle, M. F. Haslanger, S. Kim, and S. Yoo, *J. Am. Chem. Soc.*, **100**, 4618 (1978).

b. K. G. Paul, F. Johnson, and D. Favara, *J. Am. Chem. Soc.*, **98**, 1285 (1976).

c. P. N. Confalone, G. Pizzolato, E. G. Baggiolini, D. Lollar, and M. R. Uskokovic, *J. Am. Chem. Soc.*, **97**, 5936 (1975).

d. E. Baer, J. M. Gosheintz, and H. O. L. Fischer, *J. Am. Chem. Soc.*, **61**, 2607 (1939).

e. J. L. Coke and A. B. Richon, *J. Org. Chem.*, **41**, 3516 (1976).

f. J. R. Dyer, W. E. McGonigal, and K. C. Rice, *J. Am. Chem. Soc.*, **87**, 654 (1965).

g. E. J. Corey and S. Nozoe, *J. Am. Chem. Soc.*, **85**, 3527 (1963).

h. R. Jacobson, R. J. Taylor, H. J. Williams, and L. R. Smith, **47**, 1140 (1982).

8a. R. B. Miller and E. S. Behare, *J. Am. Chem. Soc.*, **96**, 8102 (1974).

b. G. Buchi, W. Hofheinz, and J. V. Paukstelis, *J. Am. Chem. Soc.*, **91**, 6473 (1969).

c. M. Brown, *J. Org. Chem.*, **33**, 162 (1968).

d. E. J. Corey, R. B. Mitra, and H. Uda, *J. Am. Chem. Soc.*, **86**, 485 (1964).

9a. E. Piers, R. W. Britton, and W. de Waal, *J. Am. Chem. Soc.*, **93**, 5113 (1971); K. J. Schmalzl and R. N. Mirrington, *Tetrahedron Lett.*, 3219 (1970); N. Fukamiya, M. Kato, and A. Yoshikoshi, *J. Chem. Soc., Chem. Commun.*, 1120 (1971); N. Fukamiya, M. Kato, and A. Yoshikoshi, *J. Chem. Soc. Perkin Trans. 1*, 1843 (1973); G. Frater, *Helv. Chim. Acta*, **57**, 172 (1974); K. Yamada, Y. Kyotani, S. Manabe, and M. Suzuki, *Tetrahedron*, **35**, 293 (1979); M. E. Jung, C. A. McCombs, Y. Takeda, and Y. G. Pan, *J. Am. Chem. Soc.*, **103**, 6677 (1981); S. C. Welch, J. M. Gruber, and P. A. Morrison, *J. Org. Chem.*, **50**, 2676 (1985); S. C. Welch, C. Chou, J. M. Gruber, and J. M. Assercq, *J. Org. Chem.*, **50**, 2668 (1985); H. Hagiwara, A. Okano, and H. Uda, *J. Chem. Soc., Chem. Commun.*, 1047 (1985); G. Stork and N. H. Baird, *Tetrahedron Lett.*, **26**, 5927 (1985); K. V. Bhaskar and G. S. R. S. Rao, *Tetrahedron Lett.*, **30**, 225 (1989); H. Hagaiwara, A. Okano, and H. Uda, *J. Chem. Soc. Perkin Trans. 1*, 2109 (1990); G. S. R. S. Rao and K. V. Bhaskar, *J. Chem. Soc. Perkin Trans. 1*, 2813 (1993).

b. S. Archambaud, K. Aphecetche-Julienne, and A. Guingant, *Synlett*, 139 (2005); Y. Wu, X. Shen, Y.-Q. Yang, Q. Hu, and J.-H. Huang, *J. Org. Chem.*, **69**, 3857 (2004); Y. G. Suh, J.-K. Jung, S.-Y. Seo, K.-H. Min, D.-Y. Shin, Y.-S. Lee, S. H. Kim and H-J. Park, *J. Org. Chem.*, **67**, 4127 (2002); D. Kim, J. Lee, P. J. Shim, J. I. Lim, T. Doi and S. Kim *J. Org. Chem.*, **67**, 772 (2002); D. Kim, J. Lee, P. J. Shim, J. I. Lim, H. Jo, and S. Kim *J. Org. Chem.*, **67**, 764 (2002); B. M. Trost and M. L. Crawley, *J. Am. Chem. Soc.*, **124**, 9328 (2002); Y. Wang, and D. Romo, *Org. Lett.*, **4**, 3231 (2002); R. K. Haynes, W. W.-L Lam, L. L. Yeung, I. D. Williams, A. C. Ridley, S. M. Starling, S. C. Vonwiller, T. W. Hambley, P. Lelandais, *J. Org. Chem.*, **62**, 4552 (1997); P. Ducray, B. Rousseau, and C. Mioskowski, *J. Org. Chem.*, **64**, 3800 (1999); A. B. Argade, R. D. Haugwitz, R. Devraj, J. Kozlowski, P. Fanwick, and M. Cushman, *J. Org. Chem.*, **63**, 273 (1998); K. Tomioka, K. Ishikawa, and T. Nakaik, *Synlett*, 901 (1995); V. Bernades, N. Kann, A. Riera, A. Moyano, M. A. Pericas, and A. E. Greene, *J. Org. Chem.*, **60**, 6670 ((1995); A. J. Carnell, G. Cay, G. Gorins, A. Kompanya-Saeid R. McCague, H. F. Olivo, S. M. Roberts, and A. J. Willets, *J. Chem. Soc. Perkin Trans. 1*, 3431 (1994); G. Solladie and O. Lohse, *J. Org. Chem.*, **58**, 4555 (1993); D. F. Taber, L. J. Silverberg, and E. D. Robinson, *J. Am. Chem. Soc.*, **113**, 6639 (1991); J. Nokami, M. Ohkura, Y. Dan-oh, and Y. Sakamoto, *Tetrahedron Lett.*, **32**, 2409 (1991); S. Hatakeyama, K. Sugawara, M. Kawamura, and S. Takano, *Synlett*, 691 (1990); B. N. Trost, J. Lynch, P. Renaut, and D. H. Steinman, *J. Am. Chem. Soc.*, **108**, 284 (1986); K. Ueno, H. Suemune, S. Saeki, and K. Sakai, *Chem. Pharm. Bull.*, **33**, 4021 (1985); K. Nakatani and S. Isoe, *Tetrahedron Lett.*, **26**, 2209 (1985); H. J. Gais and T. Lied, *Angew. Chem. Int. Ed. Engl.*, **23**, 145 (1984); M. Honda, K. Hirata, H. Sueoka, T. Katsuki, and T. Yamaguchi, *Tetrahedron Lett.*, **22**, 2679 (1981); T. Kitahara and K. Mori, *Tetrahedron*, **40**, 2935 (1984); K. H. Marx, P. Raddatz, and E. Winterfeldt, *Liebigs Ann. Chem.*, 474 (1984); C. Le Drian, and A. E. Greene, *J. Am. Chem. Soc.*, **104**, 5473 (1982); A. E. Greene, C. Le Drian, and P. Crabbe, *J. Am. Chem. Soc.*, **102**, 7583 (1980); P. A. Bartlett and F. R. Green, III, *J. Am. Chem. Soc.*, **100**, 4858 (1978); E. J. Corey and P. Carpino, *Tetrahedron Lett.*, **31**, 7555 (1990); E. J. Corey, R. H. Wollenberg, and D. R. Williams, *Tetrahedron Lett.*, **26**, 2243 (1977); E. J. Corey, and R. H. Wollenberg, *Tetrahedron Lett.*, **25**, 4705 (1976).

c. E. Herrmann, H. J. Gais, B. Rosenstock, G. Raabe, and H. J. Lindner, *Eur. J. Org. Chem.*, 275 (1998); W. Oppolzer, J. Z. Xu, and C. Stone, *Helv. Chim. Acta*, **74**, 465 (1991); K. Mori and M. Tsuji, *Tetrahedron*, **44**, 2835 (1988); D. H. Hua, M. J. Coulter, and I. Badejo, *Tetrahedron Lett.*, **28**, 5465 (1987); J. P. Marino, C. C. Silveira, J. V. Comasseto, and N. Petragnani, *J. Org. Chem.*, **52**, 4140 (1987); J. P. Marino, C. C. Silveira, J. V. Comasseto, and N. Petragnani, *J. Brazil. Chem. Soc.*, **7**, 145 (1996); M. C. Pirrung and S. A. Thomson, *Tetrahedron Lett.*, **27**, 2703 (1986); M. C. Pirrung and S. A. Thomson, *J. Org. Chem.*, **53**, 227 (1988); D. F. Taber and J. L. Schurchardt, *J. Am. Chem. Soc.*, **107**, 5289 (1985); D. F. Taber and J. L. Schuchardt, *Tetrahedron*, **43**, 5677 (1987); D. E. Cane and P. J. Thomas, *J. Am. Chem. Soc.*, **106**, 5295 (1984); T. Ohtsuka, H. Shirahama, and T. Matsumoto, *Tetrahedron Lett.*, **24**, 3851 (1983); L. A. Paquette, G. D. Annis, and H. Schostarez, *J. Am. Chem. Soc.*, **104**, 6646 (1982); W. H. Parsons, R. H. Schlessinger and M. L. Quesada, *J. Am. Chem. Soc.*, **102**, 889 (1979); W. H. Parsons and R. H. Schlessinger, *Bull. Soc. Chim. Fr.*, 327 (1980); S. Danishefsky, M. Hirama, K. Gombatz, T. Harayama, E. Berman and P. F. Schuda, *J. Am. Chem. Soc.*, **100**, 6536 (1978); S. Danishefsky, M. Hirama, K. Gombatz, T. Harayama, E. Berman, and P. F. Schuda, *J. Am. Chem. Soc.*, **101**, 7020 (1979).

10a. R. E. Ireland, R. H. Mueller, and A. K. Willard, *J. Am. Chem. Soc.*, **98**, 2868 (1976).

b. W. A. Kleschick, C. T. Buse, and C. H. Heathcock, *J. Am. Chem. Soc.*, **99**, 247 (1977); P. Fellmann and J. E. Dubois, *Tetrahedron Lett.*, **34**, 1349 (1978).

c. B. M. Trost, S. A. Godleski, and J. P. Genet, *J. Am. Chem. Soc.*, **100**, 3930 (1978).

d. M. Mousseron, M. Mousseron, J. Neyrolles, and Y. Beziat, *Bull. Chim. Soc. Fr.* 1483 (1963); Y. Beziat and M. Mousseron-Canet, *Bull. Chim. Soc. Fr.*, 1187 (1968); N. A. Ross and R. A Bartsch, *J. Org. Chem.*, **68**, 360 (2003); S. A. Babu, M. Yasuda, I. Shibata, and A. Baba, *Org. Lett.*, **6**, 4475 (2004).

e. G. Stork and V. Nair, *J. Am. Chem. Soc.*, **101**, 1315 (1979).

11a. R. D. Cooper, V. B. Jigajimmi, and R. H. Wightman, *Tetrahedron Lett.*, **25**, 5215 (1984).

b. C. E. Adams, F. J. Walker, and K. B. Sharpless, *J. Org. Chem.*, **50**, 420 (1985).

c. G. Grethe, J. Sereno, T. H. Williams, and M. R. Uskokovic, *J. Org. Chem.*, **48**, 5315 (1983).

12a. T. Taniguchi, M. Takeuchi, and K. Ogasawara, *Tetrahedron: Asymmetry*, **9**, 1451 (1998).

b. J. A. Marshall, Z.-H. Lu, and B. A. Johns, *J. Org. Chem.*, **63**, 817 (1998).

c. A. B. Smith, III, B. S. Freeze, M. Xian, and T. Hirose, *Org. Lett.*, **7**, 1825 (2005).

d. P. V. Ramachandran, B. Prabhudas, J. S. Chandra, M. V. R. Reddy, and H. C. Brown, *Tetrahedron Lett.*, **45**, 1011 (2004).

e. W. R. Roush, L. Banfi, J. C. Park, and L. K. Hoong, *Tetrahedron Lett.*, **30**, 6457 (1989).

13a. H. Albrecht, G. Bonnet, D. Enders, and G. Zimmermann, *Tetrahedron Lett.*, 3175 (1980).

  b. A. I. Meyers, G. Knaus, K. Kamata, and M. E. Ford, *J. Am. Chem. Soc.*, **98**, 567 (1976).

  c. S. Hashimoto and K. Koga, *Tetrahedron Lett.*, 573 (1978).

  d. B. M. Trost, D. O'Krongly, and J. L. Balletire, *J. Am. Chem. Soc.*, **102**, 7595 (1980); J. A. Tucker, K. N. Houk, and B. M. Trost, *J. Am. Chem. Soc.*, **112**, 5465 (1990); C. Siegel, and E. R. Thornton, *Tetrahedron: Asymmetry*, **2**, 1413 (1991); J. F. Maddaluno, N. Gresh, and C. Giessner-Prettre, *J. Org. Chem.*, **59**, 793 (1999).

  e. A. I. Meyers, R. K. Smith, and C. E. Whitten, *J. Org. Chem.*, **44**, 2250 (1979).

  f. W. R. Roush, T. G. Marron, and L. A. Pfeifer, *J. Org. Chem.*, **62**, 474 (1997).

  g. D. Zuev and L. A. Paquette, *Org. Lett.*, **2**, 679 (2000).

  h. M. T. Crimmins and J. She, *Synlett*, 1371 (2004).

 14. G. E. Keck, A. Palani, and S. F. McHardy, *J. Org. Chem.*, **59**, 3113 (1994); N. Nakajima, K. Uoto, O. Yonemitsu, and T. Hata, *Chem. Pharm. Bull.*, **39**, 64 (1991); K. C. Nicolaou, M. R. Pavia, and S. P. Seitz, *J. Am. Chem. Soc.*, **103**, 1224 (1981).

 15. S. Hanessian and R. Schaum, *Tetrahedron Lett.*, **38**, 163 (1997).

16a. R. A. Holton, C. Somoza, H.-B. Kim, F. Liang, R. J. Biediger, P. D. Boatman, M. Shindo, C. C. Smith, S. Kim, H. Nadizadeh, Y. Suzuki, C. Tao, P. Va, S. Tang, K. K. Murthi, L. N. Gentile, and J. H. Lin, *J. Am. Chem. Soc.*, **116**, 1597 (1994); R. A. Holton, H.-B. Kim, C. Somoza, F. Liang, R. J. Biediger, P. D. Boatman, M. Shindo, C. C. Smith, S. Kim, H. Nadizadeh, Y. Suzuki, C. Tao, P. Vu, S. Tank, P. Zhang, K. K. Murthi, L. N. Gentile, and J. H. Liu, *J. Am. Chem. Soc.*, **116**, 1599 (1994).

  b. K. C. Nicolaou, P. G. Nanternet, H. Ueno, R. K. Guy, E. A. Couldouros, and E. J. Sorenson, *J. Am. Chem. Soc.*, **117**, 624 (1995); K. C. Nicolaou, J.-J. Liu, Z. Yang, H. Ueno, E. J. Sorenson, C. F. Claiborne, R. K. Guy, C.-K. Hwang, M. Nakada, and P. G. Nanternet, *J. Am. Chem. Soc.*, **117**, 645 (1995); K. C. Nicolaou, H. Ueno, J.-J. Liu, P. G. Nanternet, Z. Yang, J. Renaud, K. Paulvannan, and R. Chadha, *J. Am. Chem. Soc.*, **117**, 653 (1995).

  c. S. J. Danishefsky, J. J. Masters, W. B. Young, J. T. Link, L. B. Snyder, T. V. Magne, D. K. Jung, R. C. A. Isaacs, W. G. Bornmann, C. A. Alaimo, C. A. Coburn, and M. J. Di Grandi, *J. Am. Chem. Soc.*, **118**, 2843 (1996).

  d. P. A. Wender, N. F. Badham, S. P. Conway, P. E. Floreancig, T. E. Glass, C. Granicher, J. B. Houze, J. Janichen, D. Lee, D. G. Marquess, P. L. McGrane, W. Meng, T. P. Mucciaro, M. Mulebach, M. G. Natchus, H. Paulsen, D. B. Rawlins, J. Satkofsky, A. J. Shuker, J. C. Sutton, R. E. Taylor, and K. Tomooka, *J. Am. Chem. Soc.*, **119**, 2755 (1997); P. A. Wender, N. F. Badham, S. P. Conway, P. E. Floeancig, T. E. Glass, J. B. Houze, N. E. Krauss, D. Lee, D. G. Marquessk, P. L. McGrane, W. Meng, M. G. Natchus, A. J. Shuker, J. C. Sutton, and R. E. Taylor, *J. Am. Chem. Soc.*, **119**, 2757 (1997).

  e. T. Mukaiyama, I. Shiina, H. Iwadare, M. Saitoh, T. Nishimura, N. Ohkawa, H. Sakoh, K. Nishimura, Y. Tani, M. Hasegawa, K. Yamada, and K. Saitoh, *Chem. Eur. J.*, **5**, 121 (1999).

  f. K. Morihara, R. Hara, S. Kawahara, T. Nishimori, N. Nakamura, H. Kusama, and I. Kuwajima, *J. Am. Chem. Soc.*, **120**, 12980 (19989); H. Kusama, R. Hara, S. Kawahara, T. Nishimori, H. Kashima, N. Nakamura, K. Morihara, and I. Kuwajima, *J. Am. Chem. Soc.*, **122**, 3811 (2000).

Index

Printed in the United States of America.